INTEGRALS CONTAINING $\sqrt{a + bu}$

14. $\displaystyle\int u\sqrt{a + bu}\, du = \frac{2}{15b^2}(3bu - 2a)(a + bu)^{3/2} + C$

15. $\displaystyle\int u^2\sqrt{a + bu}\, du = \frac{2}{105b^3}(15b^2u^2 - 12abu + 8a^2)(a + bu)^{3/2} + C$

16. $\displaystyle\int u^n\sqrt{a + bu}\, du = \frac{2u^n(a + bu)^{3/2}}{b(2n + 3)} - \frac{2an}{b(2n + 3)}\int u^{n-1}\sqrt{a + bu}\, du$

17. $\displaystyle\int \frac{u\, du}{\sqrt{a + bu}} = \frac{2}{3b^2}(bu - 2a)\sqrt{a + bu} + C$

18. $\displaystyle\int \frac{u^2\, du}{\sqrt{a + bu}} = \frac{2}{15b^3}(3b^2u^2 - 4abu + 8a^2)\sqrt{a + bu} + C$

19. $\displaystyle\int \frac{u^n\, du}{\sqrt{a + bu}} = \frac{2u^n\sqrt{a + bu}}{b(2n + 1)} - \frac{2an}{b(2n + 1)}\int \frac{u^{n-1}\, du}{\sqrt{a + bu}}$

20. $\displaystyle\int \frac{du}{u\sqrt{a + bu}} = \begin{cases} \dfrac{1}{\sqrt{a}}\ln\left|\dfrac{\sqrt{a + bu} - \sqrt{a}}{\sqrt{a + bu} + \sqrt{a}}\right| + C & (a > 0) \\[3mm] \dfrac{2}{\sqrt{-a}}\tan^{-1}\sqrt{\dfrac{a + bu}{-a}} + C & (a < 0) \end{cases}$

21. $\displaystyle\int \frac{du}{u^n\sqrt{a + bu}} = -\frac{\sqrt{a + bu}}{a(n - 1)u^{n-1}} - \frac{b(2n - 3)}{2a(n - 1)}\int \frac{du}{u^{n-1}\sqrt{a + bu}}$

22. $\displaystyle\int \frac{\sqrt{a + bu}\, du}{u} = 2\sqrt{a + bu} + a\int \frac{du}{u\sqrt{a + bu}}$

23. $\displaystyle\int \frac{\sqrt{a + bu}\, du}{u^n} = -\frac{(a + bu)^{3/2}}{a(n - 1)u^{n-1}} - \frac{b(2n - 5)}{2a(n - 1)}\int \frac{\sqrt{a + bu}\, du}{u^{n-1}}$

INTEGRALS CONTAINING $a^2 \pm u^2$ $(a > 0)$

24. $\displaystyle\int \frac{du}{a^2 + u^2} = \frac{1}{a}\tan^{-1}\frac{u}{a} + C$

25. $\displaystyle\int \frac{du}{a^2 - u^2} = \frac{1}{2a}\ln\left|\frac{u + a}{u - a}\right| + C$

26. $\displaystyle\int \frac{du}{u^2 - a^2} = \frac{1}{2a}\ln\left|\frac{u - a}{u + a}\right| + C$

INTEGRALS CONTAINING $\sqrt{u^2 \pm a^2}$ $(a > 0)$

27. $\displaystyle\int \frac{du}{\sqrt{u^2 \pm a^2}} = \ln|u + \sqrt{u^2 \pm a^2}| + C$

28. $\displaystyle\int \sqrt{u^2 \pm a^2}\, du = \frac{u}{2}\sqrt{u^2 \pm a^2} \pm \frac{a^2}{2}\ln|u + \sqrt{u^2 \pm a^2}| + C$

29. $\displaystyle\int u\sqrt{u^2 \pm a^2}\, du = \frac{1}{3}(u^2 \pm a^2)^{3/2} + C$

30. $\displaystyle\int u^2\sqrt{u^2 \pm a^2}\, du = \frac{u}{8}(2u^2 \pm a^2)\sqrt{u^2 \pm a^2} - \frac{a^4}{8}\ln|u + \sqrt{u^2 \pm a^2}| + C$

31. $\displaystyle\int \frac{\sqrt{u^2 + a^2}\, du}{u} = \sqrt{u^2 + a^2} - a\ln\left|\frac{a + \sqrt{u^2 + a^2}}{u}\right| + C$

32. $\displaystyle\int \frac{\sqrt{u^2 - a^2}\, du}{u} = \sqrt{u^2 - a^2} - a\sec^{-1}\frac{u}{a} + C$

33. $\displaystyle\int \frac{\sqrt{u^2 \pm a^2}\, du}{u^2} = -\frac{\sqrt{u^2 \pm a^2}}{u} + \ln|u + \sqrt{u^2 \pm a^2}| + C$

34. $\displaystyle\int \frac{u^2\, du}{\sqrt{u^2 \pm a^2}} = \frac{u}{2}\sqrt{u^2 \pm a^2} \mp \frac{a^2}{2}\ln|u + \sqrt{u^2 \pm a^2}| + C$

35. $\displaystyle\int \frac{du}{u\sqrt{u^2 + a^2}} = -\frac{1}{a}\ln\left|\frac{a + \sqrt{u^2 + a^2}}{u}\right| + C$

36. $\displaystyle\int \frac{du}{u\sqrt{u^2 - a^2}} = \frac{1}{a}\sec^{-1}\frac{u}{a} + C$

37. $\displaystyle\int \frac{du}{u^2\sqrt{u^2 \pm a^2}} = \mp\frac{\sqrt{u^2 \pm a^2}}{a^2 u} + C$

38. $\displaystyle\int (u^2 \pm a^2)^{3/2}\, du = \frac{u}{8}(2u^2 \pm 5a^2)\sqrt{u^2 \pm a^2} + \frac{3a^4}{8}\ln|u + \sqrt{u^2 \pm a^2}| + C$

39. $\displaystyle\int \frac{du}{(u^2 \pm a^2)^{3/2}} = \pm\frac{u}{a^2\sqrt{u^2 \pm a^2}} + C$

INTEGRALS CONTAINING $\sqrt{a^2 - u^2}$ $(a > 0)$

40. $\displaystyle\int \frac{du}{\sqrt{a^2 - u^2}} = \sin^{-1}\frac{u}{a} + C$

41. $\displaystyle\int \sqrt{a^2 - u^2}\, du = \frac{u}{2}\sqrt{a^2 - u^2} + \frac{a^2}{2}\sin^{-1}\frac{u}{a} + C$

42. $\displaystyle\int u^2\sqrt{a^2 - u^2}\, du = \frac{u}{8}(2u^2 - a^2)\sqrt{a^2 - u^2} + \frac{a^4}{8}\sin^{-1}\frac{u}{a} + C$

43. $\displaystyle\int \frac{\sqrt{a^2 - u^2}\, du}{u} = \sqrt{a^2 - u^2} - a\ln\left|\frac{a + \sqrt{a^2 - u^2}}{u}\right| + C$

44. $\displaystyle\int \frac{\sqrt{a^2 - u^2}\, du}{u^2} = -\frac{\sqrt{a^2 - u^2}}{u} - \sin^{-1}\frac{u}{a} + C$

45. $\displaystyle\int \frac{u^2\, du}{\sqrt{a^2 - u^2}} = -\frac{u}{2}\sqrt{a^2 - u^2} + \frac{a^2}{2}\sin^{-1}\frac{u}{a} + C$

46. $\displaystyle\int \frac{du}{u\sqrt{a^2 - u^2}} = -\frac{1}{a}\ln\left|\frac{a + \sqrt{a^2 - u^2}}{u}\right| + C$

47. $\displaystyle\int \frac{du}{u^2\sqrt{a^2 - u^2}} = -\frac{\sqrt{a^2 - u^2}}{a^2 u} + C$

48. $\displaystyle\int (a^2 - u^2)^{3/2}\, du = -\frac{u}{8}(2u^2 - 5a^2)\sqrt{a^2 - u^2} + \frac{3a^4}{8}\sin^{-1}\frac{u}{a} + C$

49. $\displaystyle\int \frac{du}{(a^2 - u^2)^{3/2}} = \frac{u}{a^2\sqrt{a^2 - u^2}} + C$

Calculus

with Analytic Geometry

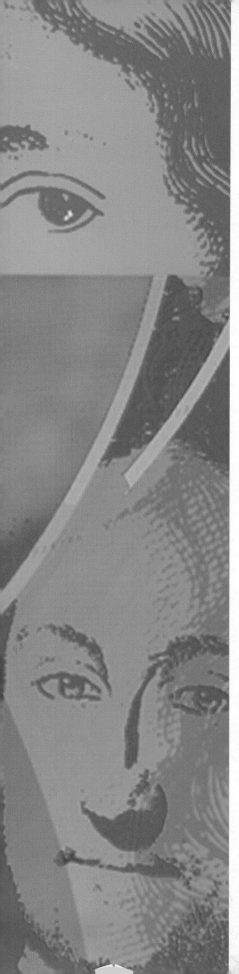

Calculus
with Analytic Geometry

FIFTH EDITION

HOWARD ANTON
Drexel University

in collaboration with
ALBERT HERR, Drexel University

JOHN WILEY & SONS, INC.
New York   Chichester   Brisbane   Toronto   Singapore

Mathematics Editor: Barbara Holland
Developmental Editor: Nancy Perry
Marketing Manager: Susan Elbe
Senior Production Editor: Nancy Prinz
Copy Editor: Lilian Brady
Design Supervisor: Madelyn Lesure
Manufacturing Manager: Susan Stetzer
Photo Editor: Hilary Newman
Illustration Coordinator: Sigmund Malinowski
Production, Cover and Text Design: HRS Electronic Text Management, Inc.
Electronic Illustration: Techsetters, Inc.

This book was set in Times Roman by General Graphic Services and printed and bound by Von Hoffmann Press, Inc. The cover was printed by The Lehigh Press, Inc.

Cover photo credits: Maria Gaetana Agnesi: Culver Pictures, Inc. Archimedes and Carl Friedrich Gauss: The Bettmann Archive. Augustin Louis Cauchy, Gottfried Wilhelm Leibniz and Georg Friedrich Bernhard Riemann: Courtesy of The David Eugene Smith Collection, Columbia University. Leonhard Euler and Sir Isaac Newton: Courtesy of The New York Public Library.

Derive is a registered trademark of Soft Warehouse, Inc.
Maple is a registered trademark of Waterloo Maple Software, Inc.
Mathematica is a registered trademark of Wolfram Research, Inc.

Library of Congress Cataloging in Publication Data:

Anton, Howard.
    Calculus with analytic geometry / Howard Anton.— 5th ed.
      p.   cm.
    Includes index.
    ISBN 0-471-59495-4
     1. Calculus. 2. Geometry, Analytic.   I. Title.
  QA303.A53 1995
  515'.15—dc20   94-35400
                                       CIP

Printed in the United States of America

10 9 8 7 6 5 4 3 2

About Howard Anton

Howard Anton obtained his B.A. from Lehigh University, his M.A. from the University of Illinois, and his Ph.D. from the Polytechnic Institute of Brooklyn, all in mathematics. In the early 1960's he worked for Burroughs Corporation and Avco Corporation at Cape Canaveral, Florida, where he was involved with missile tracking problems for the manned space program. In 1968 he joined the Mathematics Department at Drexel University, where he taught full time until 1983. Since that time he has been an adjunct professor at Drexel and has devoted the majority of his time to textbook writing and activities for mathematical associations. Dr. Anton was President of the EPADEL Section of the Mathematical Association of America (MAA), served on the Board of Governors of that organization, and guided the creation of the Student Chapters of the MAA.

He has published numerous research papers in Functional Analysis, Approximation Theory, and Topology, as well as pedagogical papers on applications of mathematics. He is best known for his textbooks in mathematics, which are among the most widely used in the world. There are currently more than eighty versions of his books including translations into Spanish, Arabic, Portuguese, Italian, Indonesian, French, Japanese, Chinese, and German.

Dr. Anton has an avid interest in computer technology as it relates to mathematical education and publishing. He has developed pedagogical software for teaching linear algebra as well as various software programs for the publishing industry that automate the production of four-color mathematical text and art. For relaxation he enjoys traveling and photography.

About Albert Herr

It is my sad duty to report that Albert J. Herr, my colleague, collaborator, and friend passed away shortly after the first printing of this book appeared. He will be missed greatly by all of those who had the good fortune to work with him.

Al held degrees in Electrical Engineering and Biomedical Engineering from Drexel University. He joined the Department of Mathematics and Computer Science at Drexel in 1964, eventually becoming Assistant Department Head for Undergraduate Programs until his retirement in 1993. Al's career always had a strong focus on the teaching of calculus. In addition to supervising graduate teaching assistants, he coordinated the calculus program at Drexel for many years, wrote the solutions manuals for this text, and assisted in its various revisions. Starting in 1980, he became actively involved in developing software and fostering the use of computers in mathematics education at both the high school and college levels. In 1984 he designed and incorporated the first computer-related materials into the calculus curriculum at Drexel.

In addition to his work in collegiate mathematics, Al actively participated in a variety of programs to stimulate an interest in mathematics and engineering among high school students. He gave numerous invited talks and workshops on the use of computers in mathematics education, and he served as project director of an NSF grant for creating computer graphics mathematics laboratories.

Al received the Lindback Award for excellence in teaching, as well as three outstanding teaching awards from student organizations at Drexel. He also received the Drexel University College of Science Award for dedication and service to students.

To
My Wife Pat
My Children Brian, David, and Lauren
My Mother Shirley

In Memory of
Stephen Girard (1750–1831) Benefactor

PREFACE

ABOUT THIS EDITION

This is a major revision. The goal for this edition is to create a contemporary text that incorporates the best features of calculus reform, yet preserves the main structure of an established and well-tested calculus course. This book is intended for those who want to move forward with calculus reform but do not want to completely dismantle their current course structure with radical or unproved materials. The most salient changes are as follows:

- **Technology** — Each chapter ends with a set of exercises that are designed to be solved using computer algebra systems or graphing calculators. Many of the exercises involve applications, and almost all of them can be solved in a variety of ways that are limited only by the student's imagination.

- **Streamlining** — The text is more than 200 pages shorter than the previous edition. We achieved this by using a less wasteful text design and by rewriting almost every section with the goal of *greater clarity in less space*. No material was omitted or modified for the sake of brevity at the cost of understandability, and the quality of the exposition was ensured by a team of outstanding reviewers that included a Polya award winner (excellence in exposition) and a Lindback award winner (excellence in teaching).

- **Revision of Multivariate Calculus** — The multivariate calculus material was completely rewritten, incorporating the concept of a vector field and focusing more on the major applications of vector analysis to physics and engineering.

- **New Material** — Material not included in previous editions was added: Jacobians, parametric representations of surfaces, Kepler's laws, conics in polar coordinates, integrals with respect to arc length, vector fields, and an appendix with some basic material on complex variables that can serve as a reference for engineers and students who need this material for other courses.

- **Early Transcendental Option** — The chapter on logarithms was completely rewritten. The exposition is greatly improved, and the material is now structured in such a way that much of it can be covered earlier in the text for those who want an earlier treatment of logarithms and exponentials. A free *Early Transcendental Supplement* is available to help implement this option. That supplement breaks the material in Sections 7.1 and 7.2 into self contained units and suggests where those units might be inserted earlier in the text.

- **More Use of Calculator Computations in the Exposition** — We assume in this edition that the student has a numerical calculator available as he or she reads the text, and numerical computations are used more extensively in developing concepts.

OTHER FEATURES

- **Rule of Four** — The term "rule of four" has recently been coined to describe exposition that presents ideas from the symbolic, geometric, computational, and verbal viewpoints. Readers familiar with earlier editions of this text will recognize that this has always been an integral part of my writing style. This style continues in this edition.

- **Early Differential Equations Option** — First-order linear and separable differential equations appear in the chapter on logarithmic and exponential functions (Chapter 7). This allows us to give some nice applications of logarithms and exponentials immediately and also helps meet the needs of those engineering and science students who require this material in courses taken concurrently with calculus. This section can be omitted or deferred with no difficulty.

■ **Early Logarithm and Exponential Option** — Sections 7.1 to 7.3 provide a *preliminary* discussion of logarithms and exponentials that does not rely on the integral definition of the logarithm or on the theory of inverse functions. Thus, these sections can be isolated and presented earlier, reserving the integral definition of the logarithm and the more theoretical material in Sections 7.4 and 7.5 for later coverage.

■ **Trigonometry Review** — Deficiencies in trigonometry plague many students, so I have included a substantial trigonometry review in Appendix B.

■ **Rigor** — The challenge of writing a good calculus book is to strike the right balance between rigor and clarity. My goal is to present precise mathematics to the fullest extent possible for the freshman audience, but where clarity and rigor conflict I choose clarity. However, I believe it to be essential that the student understand the difference between a careful proof and an informal argument, so I try to make it clear to the reader when arguments are informal. Theory involving δ-ϵ arguments appear in separate sections, so they can be bypassed if desired.

■ **Historical Notes** — The biographies and historical notes have been a hallmark of this text from its first edition, and new biographies have been added in this edition. All of the biographical material has been distilled from standard sources with the goal of capturing the personalities of the great mathematicians and bringing them to life for the student.

■ **Section Exercises** — Section exercise sets begin with routine problems and progress gradually toward problems of greater difficulty. Exercises that require a calculator are listed at the beginning of the exercise set and marked with the icon $\boxed{C}$. Many exercise sets contain so-called "spiral" problems, which revisit earlier problem types using concepts from the current section.

ABOUT THE TECHNOLOGY EXERCISES

■ The purpose of the technology exercises is to introduce the student to techniques of problem solving using graphing calculators and/or computer algebra systems such as *Mathematica*™, *Maple*™, or *Derive*™. Many of these exercises involve applications of calculus, and most of them can be solved using *either* a graphing calculator or a computer algebra system. Thus, part of the challenge to the student is to develop a problem-solving strategy that is appropriate for the technology that he or she has available.

■ Many of the problems cannot be solved by a blind, unintelligent use of technology; they may require some preliminary hand calculation to put the problem in an appropriate form or some thoughtful analysis to ensure that solutions are not missed when technology is applied.

■ Many problems will raise issues of accuracy, since some students may be able to avoid decimal approximations using a computer algebra system and other students may obtain different levels of decimal accuracy depending on their strategy and technology. This is the opportunity for an instructor to explore issues of error analysis if so inclined. However, it is not essential.

■ The technology exercises are more open-ended than the exercises at the end of each section, making them more like problems that arise in the real world. Instructors can either leave the students on their own or can provide a level of guidance that fits their own teaching philosophy. Some instructors may want to use these exercises for group projects.

FEATURED IN THIS EDITION

Technology Exercises

Each chapter ends with a set of exercises that are designed to be used with graphing calculators or computer algebra systems. Many of the exercises involve applications and almost all of them can be solved in a variety of ways that are limited only by the student's imagination.

---

654   POLAR COORDINATES AND PARAMETRIC EQUATIONS

16. Use 13.5.1 to show that from $\theta = \alpha$ to $\theta = \beta$ the arc length of $r = 2f(\theta)$ is twice that of $r = f(\theta)$.

17. Suppose that a long thin rod with one end fixed at the pole of a polar coordinate system rotates counter-clockwise at the constant rate of 0.5 rad/sec. At time $t = 0$ a bug on the rod is 10 mm from the pole and is moving outward along the rod at the constant speed of 2 mm/sec.

   (a) Find an equation of the form $r = f(\theta)$ for the path of motion of the bug, assuming that $\theta = 0$ when $t = 0$.

   (b) Find the distance the bug travels along the path in part (a) during the first 5 sec. Round your answer to the nearest tenth of a millimeter.

18. Find all points on the cardioid $r = a(1 + \cos\theta)$ where the tangent is
   (a) horizontal   (b) vertical.

19. Find all points on the limaçon $r = 1 - 2\sin\theta$ where the tangent is horizontal.

◆ TECHNOLOGY EXERCISES   Chapter 13

Most of these exercises require access to a graphing calculator or a computer algebra system (CAS) such as *Mathematica, Maple,* or *Derive*. When you are asked to *find* an answer or to *solve* an equation, you may choose to find an exact result or a numerical approximation, depending on the particular technology you are using and on your own imagination. The form of your answers may differ from those of other students or from those in the answer section of the text, depending on how you solve the problems and the accuracy you use in your numerical approximations. Those exercises that are more appropriate for a CAS than a graphing calculator are labeled with the icon ◆.

1. **Butterfly curve:** The graph of the equation

$$r = \exp(\cos\theta) - 2\cos 4\theta + \sin^3(\theta/4)$$

in polar coordinates is the "butterfly curve" shown in the accompanying figure and on the cover of this text. Generate this curve by letting θ vary over the interval $[0, \alpha)$, where α is chosen so that the curve is traced exactly once.

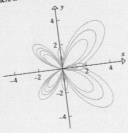

2. **Area in polar coordinates**
   (a) Graph $r = \sin\theta\cos 2\theta$ in polar coordinates. For what value of α will the graph be traced exactly once as θ varies over the interval $[0, \alpha)$?
   (b) Find the area enclosed by the large loop of the graph in part (a).

◆ 3. **Arc length in polar coordinates:** Find the arc length of one petal of the rose $r = 2\sin 3\theta$.

◆ 4. **The orbit of Mars:** If the sun is at the origin of a polar coordinate system, then an equation of the orbit of Mars is

$$r = \frac{2.26 \times 10^8}{1 + 0.0934\cos\theta}$$

where distance is in kilometers.
   (a) Graph the orbit of Mars.
   (b) Find the area swept out by the line from the sun to Mars in one revolution of Mars about the sun.
   (c) Kepler's second law of planetary motion states that

---

by the x-axis. Let V_x be the volume of the solid revolving R about the x-axis and V_y the volume generated by revolving R about the y-axis. Fir $V_y = 2V_x$.

3. **Nuclear cooling tower:** A cooling tower for power plant is to have a height of h feet and the sh solid that is generated by revolving about the y region enclosed by the right branch of the hy $x^2/225 - y^2/1521 = 1$, the y-axis, and the horizonta $y = -h/2$ and $y = h/2$. Assuming that one unit coordinate axes corresponds to one foot, find the heigh the tower is to have a volume of 50,000 ft³.

◆ 4. **Suspension bridges:** Under appropriate conditions, s port cables for suspension bridges are shaped like parabola The main span of the Golden Gate Bridge in San Francisc has a horizontal length of 4200 ft and the central suppor cables sag 470 ft at the middle (see the accompanying fig ure).

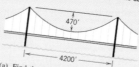

   (a) Find the length of a central support cable.
   (b) Find, to the nearest degree, the acute angle at a support point between the tangent line to the cable and the horizontal.

◆ 5. **Length of the earth's orbit:** The earth moves in an ellipti- cal orbit with the sun at a focus. An equation of the orbit is $x^2/a^2 + y^2/b^2 = 1$ (see the accompanying figure), where $a = 1.49 \times 10^8$ km and $b = 1.48979 \times 10^8$ km.

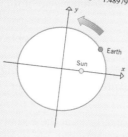

---

level of oil rising
Express your answer in

◆ 8. **Rotated conics:** Consider the conic whose equation is

$$x^2 + xy + 2y^2 - x + 3y + 1 = 0$$

   (a) Use the discriminant to identify the conic.
   (b) Graph the equation by solving for y in terms of x and graphing both solutions.
   (c) Your CAS may be able to graph the equation in the form given. If so, graph the equation in this way.

Clarity

Clarity has been the hallmark of the Anton texts. The fifth edition achieves a rare combination of outstanding exposition and sound mathematics to the fullest extent possible for the freshman audience.

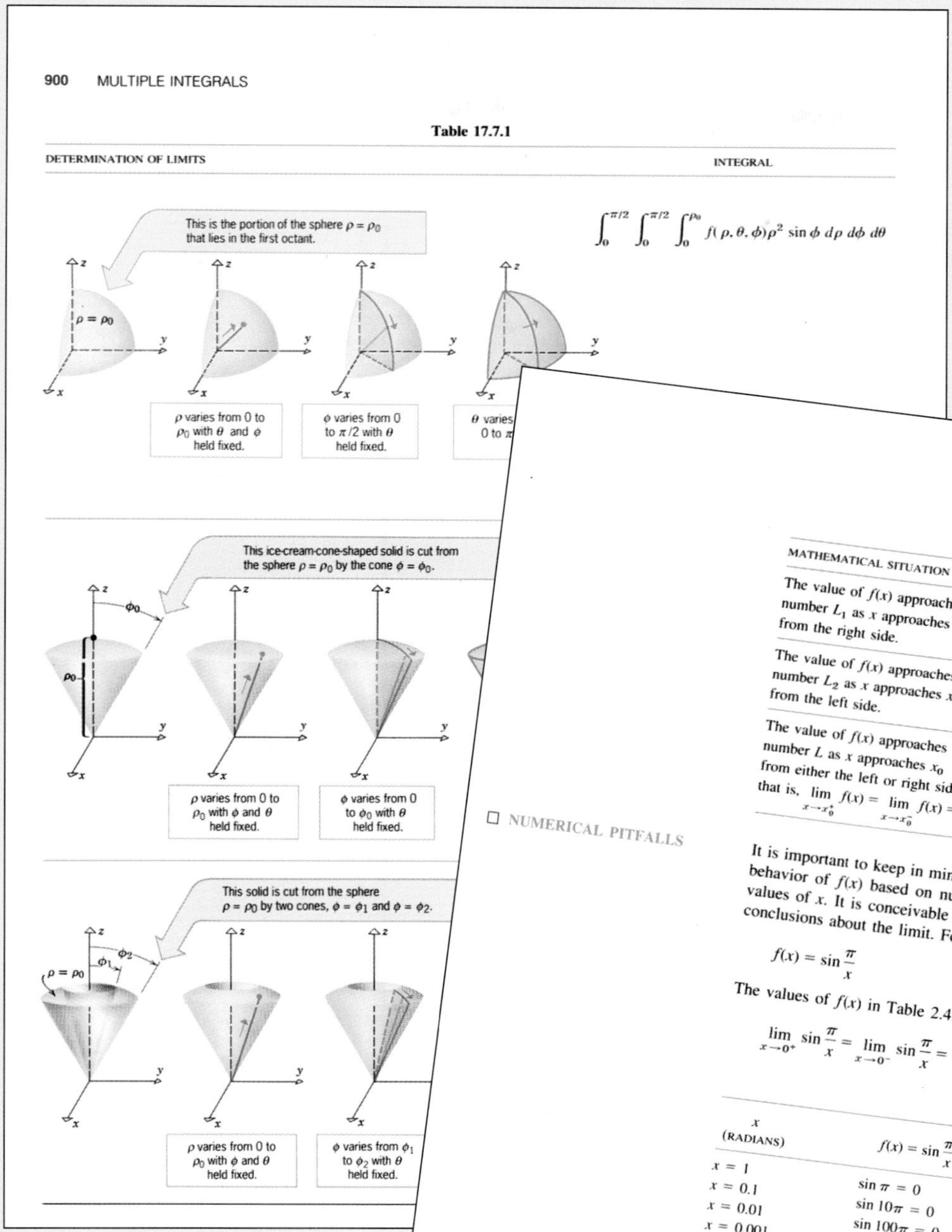

---

900   MULTIPLE INTEGRALS

Table 17.7.1

DETERMINATION OF LIMITS   INTEGRAL

This is the portion of the sphere $\rho = \rho_0$ that lies in the first octant.

$$\int_0^{\pi/2} \int_0^{\pi/2} \int_0^{\rho_0} f(\rho, \theta, \phi)\rho^2 \sin\phi \, d\rho \, d\phi \, d\theta$$

$\rho = \rho_0$

ρ varies from 0 to ρ_0 with θ and ϕ held fixed.

ϕ varies from 0 to $\pi/2$ with θ held fixed.

θ varies 0 to π

This ice-cream-cone-shaped solid is cut from the sphere $\rho = \rho_0$ by the cone $\phi = \phi_0$.

ϕ_0

ρ_0

ρ varies from 0 to ρ_0 with ϕ and θ held fixed.

ϕ varies from 0 to ϕ_0 with θ held fixed.

This solid is cut from the sphere $\rho = \rho_0$ by two cones, $\phi = \phi_1$ and $\phi = \phi_2$.

ϕ_2

ϕ_1

$\rho = \rho_0$

ρ varies from 0 to ρ_0 with ϕ and θ held fixed.

ϕ varies from ϕ_1 to ϕ_2 with θ held fixed.

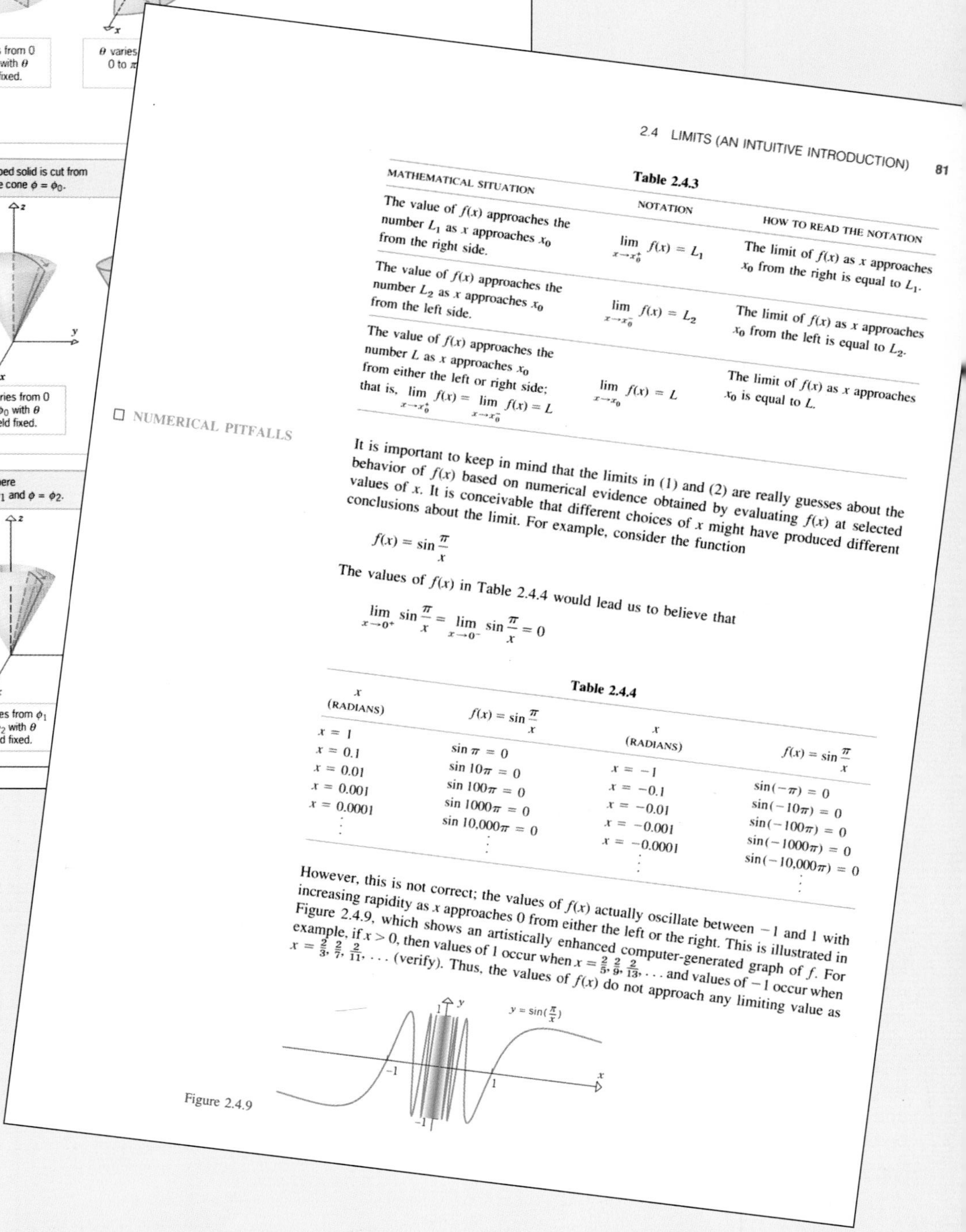

2.4   LIMITS (AN INTUITIVE INTRODUCTION)   81

Table 2.4.3

MATHEMATICAL SITUATION	NOTATION	HOW TO READ THE NOTATION
The value of $f(x)$ approaches the number L_1 as x approaches x_0 from the right side.	$\lim\limits_{x \to x_0^+} f(x) = L_1$	The limit of $f(x)$ as x approaches x_0 from the right is equal to L_1.
The value of $f(x)$ approaches the number L_2 as x approaches x_0 from the left side.	$\lim\limits_{x \to x_0^-} f(x) = L_2$	The limit of $f(x)$ as x approaches x_0 from the left is equal to L_2.
The value of $f(x)$ approaches the number L as x approaches x_0 from either the left or right side; that is, $\lim\limits_{x \to x_0^+} f(x) = \lim\limits_{x \to x_0^-} f(x) = L$	$\lim\limits_{x \to x_0} f(x) = L$	The limit of $f(x)$ as x approaches x_0 is equal to L.

□ NUMERICAL PITFALLS

It is important to keep in mind that the limits in (1) and (2) are really guesses about the behavior of $f(x)$ based on numerical evidence obtained by evaluating $f(x)$ at selected values of x. It is conceivable that different choices of x might have produced different conclusions about the limit. For example, consider the function

$$f(x) = \sin\frac{\pi}{x}$$

The values of $f(x)$ in Table 2.4.4 would lead us to believe that

$$\lim_{x \to 0^+} \sin\frac{\pi}{x} = \lim_{x \to 0^-} \sin\frac{\pi}{x} = 0$$

Table 2.4.4

x (RADIANS)	$f(x) = \sin\dfrac{\pi}{x}$	x (RADIANS)	$f(x) = \sin\dfrac{\pi}{x}$
$x = 1$			
$x = 0.1$	$\sin \pi = 0$		
$x = 0.01$	$\sin 10\pi = 0$	$x = -1$	
$x = 0.001$	$\sin 100\pi = 0$	$x = -0.1$	$\sin(-\pi) = 0$
$x = 0.0001$	$\sin 1000\pi = 0$	$x = -0.01$	$\sin(-10\pi) = 0$
⋮	$\sin 10{,}000\pi = 0$	$x = -0.001$	$\sin(-100\pi) = 0$
		$x = -0.0001$	$\sin(-1000\pi) = 0$
		⋮	$\sin(-10{,}000\pi) = 0$

However, this is not correct; the values of $f(x)$ actually oscillate between -1 and 1 with increasing rapidity as x approaches 0 from either the left or the right. This is illustrated in Figure 2.4.9, which shows an artistically enhanced computer-generated graph of f. For example, if $x > 0$, then values of 1 occur when $x = \frac{2}{3}, \frac{2}{7}, \frac{2}{11}, \ldots$ (verify). Thus, the values of $f(x)$ do not approach any limiting value as $x = \frac{2}{3}, \frac{2}{7}, \frac{2}{11}, \ldots$ (verify). Thus, the values of $f(x)$ do not approach any limiting value as

$y = \sin\left(\frac{\pi}{x}\right)$

Figure 2.4.9

Streamlined Exposition

Almost every section in the fifth edition has been rewritten with the goal of greater clarity in less space. No material was omitted or modified for the sake of brevity at the cost of understandability, and the quality of the exposition was ensured by a team of outstanding, award-winning instructors and expositors.

Revision of Multivariable Calculus

The multivariable calculus material was completely rewritten, incorporating the concept of a vector field and focusing more on the major applications of vector analysis to physics and engineering.

18.1   VECTOR FIELDS

In this section we consider functions that associate vectors with points in 2-space or 3-space. We shall see that such functions play an important role in the study of fluid flow, gravitational force fields, electromagnetic force fields, and a wide range of other applied problems.

☐ VECTOR FIELDS

To motivate the mathematical ideas in this section, consider a unit point mass located at any point in the universe. According to Newton's Universal Law of Gravitation, the earth exerts an attractive force on the mass that is directed toward the earth's center and has a magnitude that is inversely proportional to the square of the distance from the mass to the earth's center (Figure 18.1.1). This association of force vectors with points in space is called the earth's *gravitational field*. A similar idea arises in fluid flow. Imagine a stream in which the water flows horizontally at every level, and consider the layer of water at a

specific depth. At each point of the layer, the water has a certain velocity, which we can represent by a vector at that point (Figure 18.1.2). This association of velocity vectors with points in the two-dimensional layer is called the *flow field* at that layer. These ideas are captured in the following definition.

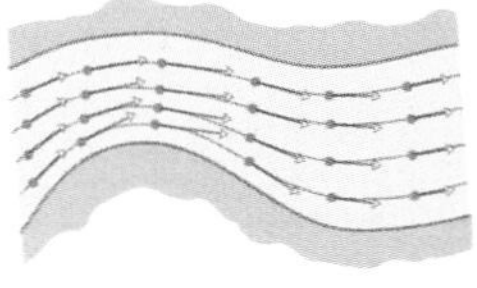

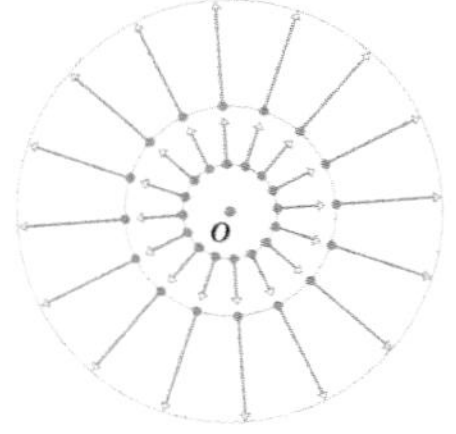

Figure 18.1.1

18.1.1   DEFINITION.   A *vector field* is a function that associates a unique vector $\mathbf{F}(P)$ with each point P in a region of 2-space or 3-space.

Example 1   Let O be a fixed point in 2-space, and for each point P in 2-space define the vector field $\mathbf{F}(P)$ by $\mathbf{F}(P) = \overrightarrow{OP}$. Some typical vectors in this vector field are shown in Figure 18.1.3. In that figure we have followed the standard convention of positioning the vector $\mathbf{F}(P)$ with its initial point at P.   ◄

Observe that the concept of a vector field has been defined without reference to a coordinate system; it is said to be a *coordinate-free* definition. However, for computational purposes it is often desirable to work with vector fields in coordinate systems. If $\mathbf{F}(P)$ is a vector field in 2-space with an xy-coordinate system, then the point P has coordinates (x, y), and the components of the vector $\mathbf{F}(P)$ are functions of x and y. Thus, $\mathbf{F}(P)$ can be expressed as

$$\mathbf{F}(x, y) = f(x, y)\mathbf{i} + g(x, y)\mathbf{j}$$

Similarly, in 3-space with an xyz-coordinate system, a vector field $\mathbf{F}(P)$ can be expressed as

$$\mathbf{F}(x, y, z) = f(x, y, z)\mathbf{i} + g(x, y, z)\mathbf{j} + h(x, y, z)\mathbf{k}$$

Figure 18.1.2

Just as it is impossible to describe a curve completely by plotting finitely many points, so it is impossible to describe a vector field completely by drawing finitely many vectors. Nevertheless, it is often possible to get a useful picture of a vector field by sketching a finite number of vectors that are well chosen.

Example 2   Figure 18.1.4 shows sketches of three vector fields in 2-space. For simplicity, we have omitted the scales and selected vectors that do not overlap; nevertheless, the sketches still provide some useful geometric insight into the behavior of the fields.   ◄

Figure 18.1.3

New Material

Material not included in the previous edition has been added: parametric representation of surfaces, Jacobians, conics in polar coordinates, integrals with respect to arc length, vector fields, Kepler's laws, and an appendix with basic material on complex variables.

☐ PARAMETRIC REPRESENTATION OF SURFACES

We have seen that curves in 3-space can be represented parametrically by three equations involving one parameter. Similarly, surfaces in 3-space can be represented by three equations involving two parameters, say u and v, as

$$x = f(u, v), \quad y = g(u, v), \quad z = h(u, v)$$

or by a single vector-valued function

$$\mathbf{r}(u, v) = x\mathbf{i} + y\mathbf{j} + z\mathbf{k} = f(u, v)\mathbf{i} + g(u, v)\mathbf{j} + h(u, v)\mathbf{k}$$

We can view $\mathbf{r}(u, v) = x\mathbf{i} + y\mathbf{j} + z\mathbf{k}$ as a *radius vector* from the origin to a point (x, y, z) that moves over the surface as u and v vary (Figure 16.1.17).

Example 11   Consider the portion of the paraboloid $z = 4 - x^2 - y^2$ that lies in the first octant (Figure 16.1.18). We can obtain a parametric representation of this surface by letting $x = u$ and $y = v$, from which it follows that $z = 4 - u^2 - v^2$. Thus, the paraboloid can be represented parametrically as

$$x = u, \quad y = v, \quad z = 4 - u^2 - v^2$$

or in vector form as

$$\mathbf{r}(u, v) = u\mathbf{i} + v\mathbf{j} + (4 - u^2 - v^2)\mathbf{k}$$

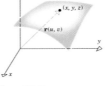

Figure 16.1.17

which implies that the graph of e^x is concave up on $(-\infty, +\infty)$.

Similarly, we can verify that $\ln x$ is increasing and concave down from its first and second derivatives. For all x in $(0, +\infty)$ we have

$$\frac{d}{dx}[\ln x] = \frac{1}{x} > 0$$

which implies that $\ln x$ is increasing on $(0, +\infty)$, and

$$\frac{d^2}{dx^2}[\ln x] = \frac{d}{dx}\left[\frac{1}{x}\right] = -\frac{1}{x^2} < 0$$

which implies that $\ln x$ is concave down on $(0, +\infty)$.

☐ LIMITS INVOLVING $\ln x$ AND e^x

Table 7.3.2

x	e^x	$\ln x$
1	2.718282	0
2	7.389056	0.693147
3	20.08554	1.098612
4	54.59815	1.386294
5	148.4132	1.609438
6	403.4288	1.791759
7	1096.633	1.945910
8	2980.958	2.079442
9	8103.084	2.197225
10	22026.47	2.302585

The following limits, which are consistent with Figure 7.3.1, will be proved later.

$$\lim_{x \to +\infty} e^x = +\infty \qquad \text{and} \qquad \lim_{x \to -\infty} e^x = 0 \qquad (1-2)$$

$$\lim_{x \to +\infty} \ln x = +\infty \qquad \text{and} \qquad \lim_{x \to 0^+} \ln x = -\infty \qquad (3-4)$$

The graph of $y = \ln x$ rises so slowly that Figure 7.3.1 does not adequately convey that

$$\lim_{x \to +\infty} \ln x = +\infty$$

even though we shall prove this to be so later. Moreover, since the graph of e^x is the reflection of the graph of $\ln x$ about the line $y = x$, the slow growth of $\ln x$ corresponds to a rapid growth of e^x. Table 7.3.2, which was generated with a calculator, illustrates the slow growth of $\ln x$ and the rapid growth of e^x.

☐ EXPONENTIAL AND LOGARITHMIC GROWTH

Mathematicians often use powers of x as a "measuring stick" for describing how rapidly a function grows. For example, we shall prove later that if n is any positive integer, then

$$\lim_{x \to +\infty} \frac{e^x}{x^n} = +\infty \qquad \text{and} \qquad \lim_{x \to +\infty} \frac{\ln x}{x^n} = 0$$

Limit (5) tel
n, division b
more rapidly
slowly that d
results in a li
integer powe
The follow

$$\lim_{x \to +\infty} \frac{x}{e}$$

Furthermore, t

$$\lim_{x \to 0^+} x^n$$

☐ OBTAINING GRAPHS USING PROPERTIES OF EXPONENTS AND LOGARITHMS

Graphs of equat
making appropri

Example 1   Sk

Thus, from Formula 90

$$\int e^{\pi x} \sin^{-1}(e^{\pi x})\, dx = \frac{1}{\pi}[u \sin^{-1} u + \sqrt{1 - u^2}] + C$$

$$= \frac{1}{\pi}[e^{\pi x} \sin^{-1}(e^{\pi x}) + \sqrt{1 - e^{2\pi x}}] + C \qquad \blacktriangleleft$$

The form of an answer to an indefinite integration can vary widely between computer algebra systems, depending on the method of integration used by the program and the manner in which it simplifies the result. Indeed, the variation can be so great that it may be difficult to see that results produced by different programs are actually equivalent.

Table 9.1.1 shows how the integrals in Example 1 are evaluated by the computer algebra system's simplification capability.) In the exercises we ask the reader to verify that the results produced by the three computer algebra systems are equivalent to the results obtained from the endpaper Table of Integrals.

☐ USING COMPUTER ALGEBRA SYSTEMS

Table 9.1.1

$$\int x^2\sqrt{7 + 3x}\, dx$$

TABLE OF INTEGRALS	$\frac{2}{2835}(135x^2 - 252x + 392)(7 + 3x)^{3/2}$
Mathematica	$\text{Sqrt}\,[7 + 3x]\left(\frac{784}{405} - \frac{56x}{135} + \frac{2x^2}{15} + \frac{2x^3}{7}\right)$
Maple	$\frac{2}{2835}(7 + 3x)^{3/2}(392 - 252x + 135x^2)$
Derive	$\frac{2(3x + 7)^{3/2}(135x^2 - 252x + 392)}{2835}$

$$\int \sqrt{x - 4x^2}\, dx$$

TABLE OF INTEGRALS	$\frac{8x - 1}{16}\sqrt{x - 4x^2} + \frac{1}{64}\sin^{-1}(8x - 1)$
Mathematica	$\left(-\left(\frac{1}{16}\right) + \frac{x}{2}\right)\text{Sqrt}\,[x - 4x^2] - \frac{\text{ArcSin}\,[1 - 8x]}{64}$
Maple	$-\frac{1}{16}(-8x + 1)\sqrt{x - 4x^2} + \frac{1}{64}\arcsin(8x - 1)$
Derive	$\frac{\text{ASIN}\,(8x - 1)}{64} + \frac{(8x - 1)\sqrt{x(1 - 4x)}}{16}$

$$\int e^{\pi x} \sin^{-1}(e^{\pi x})\, dx$$

TABLE OF INTEGRALS	$\frac{1}{\pi}[e^{\pi x}\sin^{-1}(e^{\pi x}) + \sqrt{1 - e^{2\pi x}}]$
Mathematica	$\frac{\text{Sqrt}\,[1 - E^{2\,\text{Pi}\,x}]}{\text{Pi}} + \frac{E^{\text{Pi}\,x}\,\text{ArcSin}\,[E^{\text{Pi}\,x}]}{\text{Pi}}$
Maple	$\frac{e^{\pi x}\arcsin(e^{\pi x}) + \sqrt{1 - (e^{\pi x})^2}}{\pi}$
Derive	$\frac{e^{\pi x}\,\text{ASIN}\,(e^{\pi x})}{\pi} + \frac{(1 - e^{2\pi x})^{3/2}}{3\pi} + \frac{\sqrt{1 - e^{2\pi x}}(e^{2\pi x} + 2)}{3\pi}$

Calculators and Computers in the Exposition

The student is assumed to have a numerical calculator available as he or she reads the text, and numerical computations are used extensively in developing concepts.

Historical Perspectives

Historical biographies that focus on the personalities of the great mathematicians bring these people to life and give the student a sense of mathematical history.

The relationship between continuity and differentiability was of great historical significance in the development of calculus. In the early nineteenth century mathematicians believed that the graph of a continuous function could not have too many points of nondifferentiability bunched up. They felt that if a continuous function had many points of nondifferentiability, these points, like the tips of a sawblade, would have to be separated from each other and joined by smooth curve segments (Figure 3.2.12). This misconception was shattered by a series of discoveries beginning in 1834. In that year a Bohemian priest, philosopher, and mathematician named Bernhard Bolzano* discovered a procedure for constructing a continuous function that is not differentiable at any point. Later, in 1860, the great German mathematician, Karl Weierstrass** produced the first formula for such a function. The graphs of such functions are impossible to draw; it is as if the corners are so numerous that any segment of the curve, when suitably enlarged, reveals more corners. The discovery of these pathological functions was important in that it made mathematicians distrustful of their geometric intuition and more reliant on precise mathematical proof. However, they remained only mathematical curiosities until the early 1980s, when applications of them began to emerge. During the past 10 years they have started to play a fundamental role in the study of geometric objects called *fractals*. Fractals have revealed an order to natural phenomena that were previously dismissed as random and chaotic.

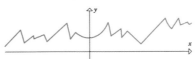

Figure 3.2.12

---

*BERNHARD BOLZANO (1781–1848). Bolzano, the son of an art dealer, was born in Prague, Bohemia (Czechoslovakia). He was educated at the University of Prague, and eventually won enough mathematical fame to be recommended for a mathematics chair there. However, Bolzano became an ordained Roman Catholic priest, and in 1805 he was appointed to a chair of Philosophy at the University of Prague. Bolzano was a man of great human compassion; he spoke out for educational reform, he voiced the right of individual conscience over government demands, and he lectured on the absurdity of war and militarism. His views so disenchanted Emperor Franz I of Austria that the emperor pressed the Archbishop of Prague to have Bolzano recant his statements. Bolzano refused and was then forced to retire in 1824 on a small pension. Bolzano's main contribution to mathematics was philosophical. His work helped convince mathematicians that sound mathematics must ultimately rest on rigorous proof rather than intuition. In addition to his work in mathematics, Bolzano investigated problems concerning space, force, and wave propagation.

**KARL WEIERSTRASS (1815–1897). Weierstrass, the son of a customs officer, was born in Ostenfelde, Germany. As a youth Weierstrass showed outstanding skills in languages and mathematics. However, at the urging of his dominant father, Weierstrass entered the law and commerce program at the University of Bonn. To the chagrin of his family, the rugged and congenial young man concentrated instead on fencing and beer drinking. Four years later he returned home without a degree. In 1839 Weierstrass entered the Academy of Münster to study for a career in secondary education, and he met and studied under an excellent mathematician named Christof Gudermann. Gudermann's ideas greatly influenced the work of Weierstrass. After receiving his teaching certificate, Weierstrass spent the next 15 years in secondary education teaching German, geography, and mathematics. In addition, he taught handwriting to small children. During this period much of Weierstrass's mathematical work was ignored because he was a secondary schoolteacher and not a college professor. Then, in 1854, he published a paper of major importance which created a sensation in the mathematics world and catapulted him to international fame overnight. He was immediately given an honorary Doctorate at the University of Königsberg and began a new career in college teaching at the University of Berlin in 1856. In 1859 the strain of his mathematical research caused a temporary nervous breakdown and led to spells of dizziness that plagued him for the rest of his life. Weierstrass was a brilliant teacher and his classes overflowed with multitudes of auditors. In spite of his fame, he never lost his early beer-drinking congeniality and was always in the company of students, both ordinary and brilliant. Weierstrass was acknowledged as the leading mathematical analyst in the world. He and his students opened the door to the modern school of mathematical analysis.

FUNCTIONS AND LIMITS

2.5.2   THEOREM. *For any polynomial*
$$p(x) = c_0 + c_1 x + \cdots + c_n x^n$$
and any real number a,
$$\lim_{x \to a} p(x) = c_0 + c_1 a + \cdots + c_n a^n = p(a)$$

Proof.
$$\lim_{x \to a} p(x) = \lim_{x \to a} (c_0 + c_1 x + \cdots + c_n x^n)$$
$$= \lim_{x \to a} c_0 + \lim_{x \to a} c_1 x + \cdots + \lim_{x \to a} c_n x^n$$
$$= \lim_{x \to a} c_0 + c_1 \lim_{x \to a} x + \cdots + c_n \lim_{x \to a} x^n$$
$$= c_0 + c_1 a + \cdots + c_n a^n = p(a)$$

Example 2 If we apply Theorem 2.5.2 to the limit pr the intermediate steps and write immediately
$$\lim_{x \to 5} (x^2 - 4x + 3) = 5^2 - 4(5) + 3 = 8$$

The following limits are suggested by the graph of numerical calculations in Table 2.5.2.

LIMITS INVOLVING $1/x$

$$\lim_{x \to 0^+} \frac{1}{x} = +\infty, \quad \lim_{x \to 0^-} \frac{1}{x} = -\infty, \quad \lim_{x \to +\infty} \frac{1}{x} =$$

$$\lim_{x \to 0^+} \frac{1}{x} = +\infty$$

$$\lim_{x \to 0^-} \frac{1}{x} = -\infty$$

$$\lim_{x \to +\infty} \frac{1}{x} = 0$$

$$\lim_{x \to -\infty} \frac{1}{x} = 0$$

Figure 2.5.3

Rule of Four

The term "rule of four" describes the presentations of ideas from symbolic, geometric, computational, and verbal viewpoints. Readers of earlier editions will recognize that this has always been an integral part of the Anton writing style. The style continues in the fifth edition.

Table 2.5.2

	VALUES						CONCLUSION
x	1	10	100	1000	10,000	...	As $x \to +\infty$ the value of $1/x$ decreases toward zero.
$1/x$	1	.1	.01	.001	.0001	...	
x	−1	−10	−100	−1000	−10,000	...	As $x \to -\infty$ the value of $1/x$ increases toward zero.
$1/x$	−1	−.1	−.01	−.001	−.0001	...	
x	1	.1	.01	.001	.0001	...	As $x \to 0^+$ the value of $1/x$ increases without bound.
$1/x$	1	10	100	1000	10,000	...	
x	−1	−.1	−.01	−.001	−.0001	...	As $x \to 0^-$ the value of $1/x$ decreases without bound.
$1/x$	−1	−10	−100	−1000	−10,000	...	

Engineering Applications

The text now includes major applications of vector analysis to engineering and physics.

the value of div $\mathbf{F}$ will not vary much from its value div $\mathbf{F}(P_0)$ at the center, and we can reasonably approximate div $\mathbf{F}$ by the constant div $\mathbf{F}(P_0)$ on G. Thus, the Divergence Theorem implies that the flux $\Phi(G)$ of $\mathbf{F}$ across $\sigma(G)$ can be approximated as

$$\Phi(G) = \iint_{\sigma(G)} \mathbf{F} \cdot \mathbf{n}\, dS = \iiint_G \operatorname{div} \mathbf{F}\, dA \approx \operatorname{div} \mathbf{F}(P_0) \iiint \operatorname{div} \mathbf{F}(P_0)\, \operatorname{vol}(G)$$

from which we obtain the following approximat

$$\operatorname{div} \mathbf{F}(P_0) \approx \frac{\Phi(G)}{\operatorname{vol}(G)}$$

The expression on the right side of (8) is called let the radius of the sphere approach zero [so t plausible that the error in this approximation app the point P_0 is given exactly by

$$\operatorname{div} \mathbf{F}(P_0) = \lim_{\operatorname{vol}(G) \to 0} \frac{\Phi(G)}{\operatorname{vol}(G)}$$

or equivalently,

$$\operatorname{div} \mathbf{F}(P_0) = \lim_{\operatorname{vol}(G) \to 0} \frac{1}{\operatorname{vol}(G)} \iint_{\sigma(G)} \mathbf{F} \cdot \mathbf{n}\, dS$$

☐ INTERPRETATION OF DIVERGENCE IN FLUID FLOW

This limit, called the **flux density of F at the point** of divergence. This results in a definition of diver tion of a coordinate system, unlike the definition

If P_0 is a point in an incompressible fluid at which that $\Phi(G) > 0$ for a sufficiently small sphere G volume of fluid going out through the surface o happen if there is some point *inside* the sphere at net outward flow through the surface would res sphere, contradicting the incompressibility assum would have to be a point *inside* the sphere at whic inward flow through the surface would result in an an incompressible fluid, points at which div $\mathbf{F}(P$ which div $\mathbf{F}(P_0) < 0$ are called **sinks**. Fluid enters sink. In an incompressible fluid without sources

$$\operatorname{div} \mathbf{F}(P) = 0$$

at every point P. In hydrodynamics this is called t **ible fluids** and is sometimes taken as the defining

☐ GAUSS' LAW FOR INVERSE-SQUARE FIELDS

Some of the major principles of physics are conseq shall obtain by applying the Divergence Theorem 18.1.2).

18.7.2 GAUSS' LAW FOR INVERSE-SQUARE FIE

$$\mathbf{F}(\mathbf{r}) = \frac{c}{\|\mathbf{r}\|^3} \mathbf{r}$$

is an inverse-square field in 3-space, and if surrounds the origin and has outward orientatio

$$\Phi = \iint_\sigma \mathbf{F} \cdot \mathbf{n}\, dS = 4\pi c$$

Recall from Formula (3) of Section 18.1 that $\mathbf{F}$ can be expressed

$$\mathbf{F}(x, y, z) = \frac{c}{(x^2 + y^2 + z^2)^{3/2}} (x\mathbf{i} + y\mathbf{j} + z\mathbf{k})$$

Since the components of $\mathbf{F}$ are not continuous at the origin, w Divergence Theorem over the solid enclosed by σ. However, we difficulty by constructing a sphere of radius a centered at the origin sufficiently small that the sphere lies entirely within the region enc 18.7.5). We shall denote the surface of this sphere by σ_a. The solid G e and σ is a three-dimensional *simply connected solid* in which σ is the σ_a is the boundary of a cavity inside the solid. The components of $\mathbf{F}$ sat of the Divergence Theorem on G.

Just as we were able to extend Green's Theorem to multiply connec plane, so it is possible to extend the Divergence Theorem to multiply con space, provided the surface integral in the theorem is taken over the *enti* the outside boundary oriented outward (away from G) and the boundari oriented inward (toward the cavities). Thus, if $\mathbf{F}$ is the inverse-square fie σ_a is oriented inward, then the Divergence Theorem yields

$$\iiint_G \operatorname{div} \mathbf{F}\, dV = \iint_\sigma \mathbf{F} \cdot \mathbf{n}\, dS + \iint_{\sigma_a} \mathbf{F} \cdot \mathbf{n}\, dS$$

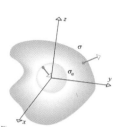

Figure 18.7.5

But we showed in Example 7 of Section 18.1 that div $\mathbf{F} = 0$, so (12) yie

$$\iint_\sigma \mathbf{F} \cdot \mathbf{n}\, dS = -\iint_{\sigma_a} \mathbf{F} \cdot \mathbf{n}\, dS$$

We can evaluate the surface integral over σ_a by expressing the integrand components; however, it is easier to leave it in vector form. At each point on th unit normal $\mathbf{n}$ points inward along a radius from the origin, and hence $\mathbf{n} = -$ (13) yields

$$\iint_\sigma \mathbf{F} \cdot \mathbf{n}\, dS = -\iint_{\sigma_a} \frac{c}{\|\mathbf{r}\|^3} \mathbf{r} \cdot \left(-\frac{\mathbf{r}}{\|\mathbf{r}\|}\right) dS$$

$$= \iint_{\sigma_a} \frac{c}{\|\mathbf{r}\|^4} (\mathbf{r} \cdot \mathbf{r})\, dS$$

$$= \iint_{\sigma_a} \frac{c}{\|\mathbf{r}\|^2}\, dS$$

$$= \frac{c}{a^2} \iint_{\sigma_a} dS \quad \boxed{\|\mathbf{r}\| = a \text{ on } \sigma_a}$$

$$= \frac{c}{a^2} (4\pi a^2) \quad \boxed{\text{The integral is the surface area of the sphere.}}$$

$$= 4\pi c$$

which establishes (10).

It follows from Example 3 of Section 18.1 with $q = 1$ that a single charged particl charge Q located at the origin creates an inverse-square field

$$\mathbf{F}(\mathbf{r}) = \frac{Q}{4\pi \epsilon_0 \|\mathbf{r}\|^3} \mathbf{r}$$

15.7   KEPLER'S LAWS OF PLANETARY MOTION

One of the great advances in the history of astronomy occurred in the early 1600s when Johannes Kepler deduced from empirical data that all planets in our solar system move in elliptical orbits with the sun at a focus. Subsequently, Isaac Newton showed mathematically that such planetary motion is the consequence of an inverse-square law of gravitational attraction. In this section we shall use the concepts developed in the preceding sections of this chapter to derive three basic laws of planetary motion, known as **Kepler's laws**.*

☐ KEPLER'S LAWS

In 1609 Johannes Kepler published a book known as *Astronomia Nova* (or sometimes as *Commentaries on the Motions of Mars*) in which he succeeded in distilling thousands of years of observational astronomy into three beautiful laws of planetary motion.

15.7.1 KEPLER'S LAWS.

- First law (**Law of Orbits**). Each planet moves in an elliptical orbit with the sun at a focus.
- Second law (**Law of Areas**). Equal areas are swept out in equal times by the line from the sun to a planet.
- Third law (**Law of Periods**). The square of a planet's period (the time it takes the planet to complete one orbit about the sun) is proportional to the cube of the length of the semimajor axis of its elliptical orbit.

☐ CENTRAL FORCES

To derive Kepler's laws, we shall assume that the force exerted by the sun on a planet is always directed toward the sun's center. In general, a force that is always directed toward a fixed point is called a **central force**.

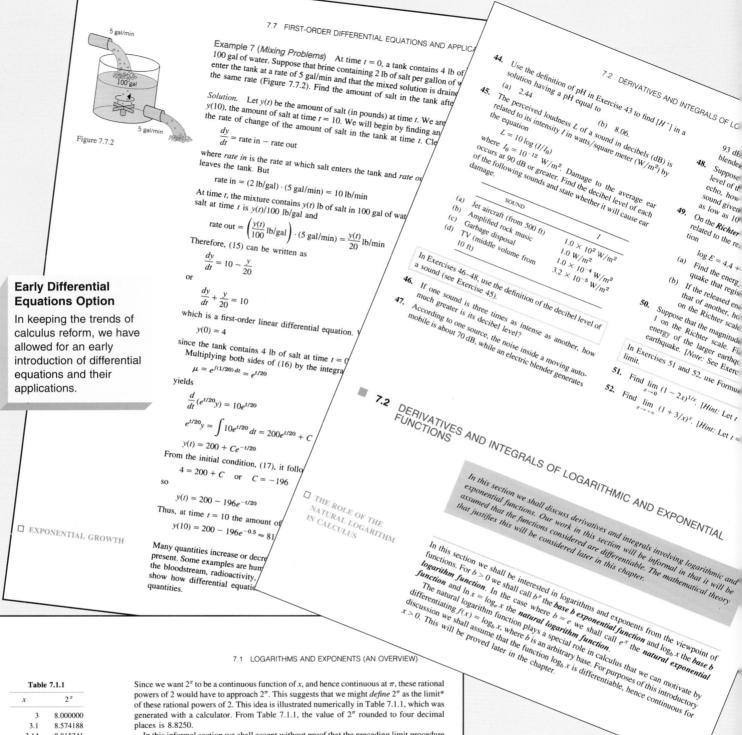

Example 7 (Mixing Problems) At time $t = 0$, a tank contains 4 lb of 100 gal of water. Suppose that brine containing 2 lb of salt per gallon of w... enter the tank at a rate of 5 gal/min and that the mixed solution is drain... the same rate (Figure 7.7.2). Find the amount of salt in the tank afte...

Solution. Let $y(t)$ be the amount of salt (in pounds) at time t. We are... $y(10)$, the amount of salt at time $t = 10$. We will begin by finding an... the rate of change of the amount of salt in the tank at time t. Cle...

$$\frac{dy}{dt} = \text{rate in} - \text{rate out}$$

where *rate in* is the rate at which salt enters the tank and *rate ou*... leaves the tank. But

$$\text{rate in} = (2\ \text{lb/gal}) \cdot (5\ \text{gal/min}) = 10\ \text{lb/min}$$

At time t, the mixture contains $y(t)$ lb of salt in 100 gal of wat... salt at time t is $y(t)/100$ lb/gal and

$$\text{rate out} = \left(\frac{y(t)}{100}\ \text{lb/gal}\right) \cdot (5\ \text{gal/min}) = \frac{y(t)}{20}\ \text{lb/min}$$

Therefore, (15) can be written as

$$\frac{dy}{dt} = 10 - \frac{y}{20}$$

or

$$\frac{dy}{dt} + \frac{y}{20} = 10$$

which is a first-order linear differential equation. Y...

$$y(0) = 4$$

since the tank contains 4 lb of salt at time $t = 0$... Multiplying both sides of (16) by the integra...

$$\mu = e^{\int (1/20)\,dt} = e^{t/20}$$

yields

$$\frac{d}{dt}(e^{t/20}y) = 10e^{t/20}$$

$$e^{t/20}y = \int 10e^{t/20}\,dt = 200e^{t/20} + C$$

$$y(t) = 200 + Ce^{-t/20}$$

From the initial condition, (17), it follo...

$$4 = 200 + C \quad \text{or} \quad C = -196$$

so

$$y(t) = 200 - 196e^{-t/20}$$

Thus, at time $t = 10$ the amount of...

$$y(10) = 200 - 196e^{-0.5} \approx 81...$$

Many quantities increase or decr... present. Some examples are hum... the bloodstream, radioactivity,... show how differential equatio... quantities.

☐ EXPONENTIAL GROWTH

Early Differential Equations Option

In keeping the trends of calculus reform, we have allowed for an early introduction of differential equations and their applications.

Figure 7.7.2

5 gal/min

100 gal

5 gal/min

---

44. Use the definition of pH in Exercise 43 to find $[H^+]$ in a solution having a pH equal to
(a) 2.44   (b) 8.06.

45. The perceived loudness L of a sound in decibels (dB) is related to its intensity I in watts/square meter (W/m²) by the equation
$$L = 10 \log (I/I_0)$$
where $I_0 = 10^{-12}$ W/m². Damage to the average ear occurs at 90 dB or greater. Find the decibel level of each of the following sounds and state whether it will cause ear damage.

SOUND	I
(a) Jet aircraft (from 500 ft)	1.0×10^2 W/m²
(b) Amplified rock music	1.0 W/m²
(c) Garbage disposal	1.0×10^{-4} W/m²
(d) TV (middle volume from 10 ft)	3.2×10^{-5} W/m²

In Exercises 46–48, use the definition of the decibel level of a sound (see Exercise 45).

46. If one sound is three times as intense as another, how much greater is its decibel level?

47. According to one source, the noise inside a moving automobile is about 70 dB, while an electric blender generates...

93 dB... blende...

48. Suppose... level of th... echo, how... sound giver... as low as 10...

49. On the **Richter**... related to the re... tion
(a) $\log E = 4.4 +$... Find the energ... quake that regi...
(b) If the released en... that of another, ho... on the Richter scale...

50. Suppose that the magnitude... 1 on the Richter scale. Fi... energy of the larger earthq... earthquake. [Note: See Exerc... limit.

In Exercises 51 and 52, use Formu...

51. Find $\lim_{x \to 0} (1 - 2x)^{1/x}$. [Hint: Let t...

52. Find $\lim_{x \to +\infty} (1 + 3/x)^x$. [Hint: Let...

7.2 DERIVATIVES AND INTEGRALS OF LOGARITHMIC AND EXPONENTIAL FUNCTIONS

In this section we shall discuss derivatives and integrals involving logarithmic and exponential functions. Our work in this section will be informal in that it will be assumed that the functions considered are differentiable. The mathematical theory that justifies this will be considered later in this chapter.

☐ THE ROLE OF THE NATURAL LOGARITHM IN CALCULUS

In this section we shall be interested in logarithms and exponents from the viewpoint of functions. For $b > 0$ we shall call b^x the **base b exponential function** and $\log_b x$ the **base b logarithm function**. In the case where $b = e$ we shall call e^x the **natural exponential function** and $\ln x = \log_e x$ the **natural logarithm function**. The natural logarithm function plays a special role in calculus that we can motivate by differentiating $f(x) = \log_b x$, where b is an arbitrary base. For purposes of this introductory discussion we shall assume that the function $\log_b x$ is differentiable, hence continuous for $x > 0$. This will be proved later in the chapter.

7.1 LOGARITHMS AND EXPONENTS (AN OVERVIEW)

Table 7.1.1

x	2^x
3	8.000000
3.1	8.574188
3.14	8.815241
3.141	8.821353
3.1415	8.824411
3.14159	8.824962
3.141592	8.824974

Since we want 2^x to be a continuous function of x, and hence continuous at π, these rational powers of 2 would have to approach 2^π. This suggests that we might *define* 2^π as the limit* of these rational powers of 2. This idea is illustrated numerically in Table 7.1.1, which was generated with a calculator. From Table 7.1.1, the value of 2^π rounded to four decimal places is 8.8250.

In this informal section we shall accept without proof that the preceding limit procedure produces a definition of b^x for irrational x such that the following are true:

- b^x is a continuous function for all $b > 0$.
- b^x is a differentiable function for all $b > 0$.
- The standard properties of exponents such as $b^{u+v} = b^u b^v$ continue to hold.

The first two properties are consistent with the graphs shown in Figure 7.1.1b.

☐ REVIEW OF LOGARITHMS

In algebra a logarithm is defined as an exponent. More precisely, if $b > 0$ and $b \neq 1$, then for positive values of x one defines

$$\log_b x$$

(read, "the **logarithm to the base b of x**") to be that power to which b must be raised to produce x. Thus,

$$\log_{10} 100 = 2$$

since 10 must be raised to the second power to produce 100. Similarly,

$$\log_2 8 = 3 \qquad \text{since } 2^3 = 8$$

$$\log \ \frac{1}{...} = -3 \qquad \text{since } 10^{-3} = ...$$

Early Transcendental Option

The logarithm chapter has been rewritten to allow for an early transcendental option. The exposition is now structured so that the basic material can be moved forward for those who want an earlier treatment of logarithms and exponentials.

SUPPLEMENTS

GRAPHING CALCULATOR SUPPLEMENTS

The following supplement contains a collection of problems that are intended to be solved on a graphing calculator. The problems are not specific to a particular brand of calculator. Also provided is an overview of the types of calculators available and general instructions for calculator use.

- *Discovering Calculus with Graphing Calculators, Second Edition*
  ISBN: 0-471-00974-1

The following free supplement provides a brief overview of those aspects of graphing calculators that are relevant to the problems in this text. Topics include: choice of viewing window, roundoff error, techniques for finding roots, and common pitfalls associated with graphing calculators.

- *Graphing Calculator Survival Guide*
  ISBN: 0-471-13172-5

SYMBOLIC ALGEBRA SUPPLEMENTS

The following supplements are collections of problems for the student to solve. Each contains a brief set of instructions for using the software as well as an extensive set of problems utilizing the capabilities of the software. The problems range from very basic to those involving real-world applications.

- *Discovering Calculus with DERIVE™, Second Edition*
  Jerry Johnson, *University of Nevada–Reno*
  Benny Evans, *Oklahoma State University*
  ISBN: 0-471-00972-5

- *Discovering Calculus with MAPLE™*
  Kent Harris, *Western Illinois University*
  Robert J. Lopez, *Rose–Hulman Institute of Technology*
  ISBN: 0-471-55156-2

- *Discovering Calculus with MATHEMATICA™*
  Cecilia A. Knoll, Florida Institute of Technology
  Michael D. Shaw, Florida Institute of Technology
  Jerry Johnson, University of Nevada-Reno
  Benny Evans, Oklahoma State University
  ISBN: 0-471-00976-8

CD-ROM VERSION OF CALCULUS FOR IBM COMPATIBLE COMPUTERS

This supplement is an electronic version of Anton's *Calculus,* the *Student's Solutions Manual,* and the *Calculus Companion* on compact disk for use with IBM compatible computers equipped with a CD-ROM drive. All text material and illustrations are stored on disk with an interconnecting network of hyperlinks that allows the student to access related items that do not appear in proximity in the text. A complete keyword glossary and step-by-step discussions of key concepts are also included.

- *CD-ROM Version of Anton Calculus: An Electronic Study Environment*
  Developed by Smart Books. Inc.
  ISBN: 0-471-55803-6

CD-ROM MULTIMEDIA SUPPLEMENT FOR IBM COMPATIBLE COMPUTERS

This highly interactive multimedia CD provides opportunities for students to ask "what if" questions, change parameters, enter their own functions and see the effects of their mathematical decisions in real time. There are 24 multimedia modules accompanied by a laboratory workbook that covers key concepts and spans the entire calculus sequence.

■ *Calculus Connections: A Multimedia Adventure*
Douglas Quinney, University of Keele
Robert Harding, Cambridge University
IntelliPro, Inc.
ISBN: 0-471-13795-2

EARLY TRANSCENDENTAL SUPPLEMENT

This free supplement is designed for those who want an early treatment of exponentials and logarithms. In this short supplement the material in Section 7.2 is broken into smaller self-contained units for easy placement earlier in the text, and a guide for implementing the early transcendental option is provided.

■ *Early Transcendental Supplement to Accompany Anton Calculus 5/E*
ISBN: 0-471-13173-3

LINEAR ALGEBRA SUPPLEMENT

This free supplement is a brief introduction to those aspects of linear algebra that are of immediate concern to the calculus student. The emphasis is on methods rather than proof.

■ *Linear Algebra Supplement to Accompany Anton Calculus / 5E*
ISBN: 0-471-10677-1

STUDENT STUDY RESOURCES

The following supplement is a tutorial, review, and study aid for the student.

■ *The Calculus Companion to Accompany Anton Calculus / 5E*
William H. Barker and James E. Ward, *Bowdoin College*
ISBN: 0-471-10678-x

The following supplement contains detailed solutions to all odd-numbered exercises.

■ *Student's Solutions Manual to Accompany Anton Calculus / 5E*
Albert Herr, *Drexel University*
ISBN: 0-471-10589-9

RESOURCES FOR THE INSTRUCTOR

There is a resource package for the instructor that includes hard copy and electronic test banks and other materials. These can be obtained by writing on your institutional letterhead to Debra Riegert, Senior Marketing Manager, John Wiley & Sons, Inc., 605 Third Avenue, New York, N.Y., 10158-0012.

ACKNOWLEDGMENTS

It has been my good fortune to have the advice and guidance of many talented people, whose knowledge and skills have enhanced this book in many ways. For their valuable help I thank:

REVIEWERS AND CONTRIBUTORS TO EARLIER EDITIONS

Edith Ainsworth, *University of Alabama*

David Armacost, *Amherst College*

Larry Bates, *University of Calgary*

Irl C. Bivens, *Davidson College*

Harry N. Bixler, *Bernard M. Baruch College, CUNY*

Marilyn Blockus, *San Jose State University*

Ray Boersma, *Front Range Community College*

David Bolen, *Virginia Military Institute*

Daniel Bonar, *Denison University*

George W. Booth, *Brooklyn College*

Mark Bridger, *Northeastern University*

John Brothers, *Indiana University*

Robert C. Bueker, *Western Kentucky University*

Robert Bumcrot, *Hofstra University*

James Caristi, *Valparaiso University*

Chris Christensen, *Northern Kentucky University*

Hannah Clavner, *Drexel University*

David Cohen, *University of California, Los Angeles*

Michael Cohen, *Hofstra University*

Robert Conley, *Precision Visuals*

Terrance Cremeans, *Oakland Community College*

Michael Dagg, *Numerical Solutions, Inc.*

Stephen L. Davis, *Davidson College*

A. L. Deal, *Virginia Military Institute*

Charles Denlinger, *Millersville State College*

Dennis DeTurck, *University of Pennsylvania*

Jacqueline Dewar, *Loyola Marymount University*

Irving Drooyan, *Los Angeles Pierce College*

Tom Drouet, *East Los Angeles College*

Ken Dunn, *Dalhousie University*

Hugh B. Easler, *College of William and Mary*

Joseph M. Egar, *Cleveland State University*

Garret J. Etgen, *University of Houston*

James H. Fife, *University of Richmond*

Barbara Flajnik, *Virginia Military Institute*

Daniel Flath, *University of South Alabama*

Nicholas E. Frangos, *Hofstra University*

Katherine Franklin, *Los Angeles Pierce College*

Michael Frantz, *University of La Verne*

Susan L. Friedman, *Bernard M. Baruch College, CUNY*

William R. Fuller, *Purdue University*

G. S. Gill, *Brigham Young University*

Raymond Greenwell, *Hofstra University*

Gary Grimes, *Mt. Hood Community College*

Jane Grossman, *University of Lowell*

Michael Grossman, *University of Lowell*

Douglas W. Hall, *Michigan State University*

Nancy A. Harrington, *University of Lowell*

Kent Harris, *Western Illinois University*

Albert Herr, *Drexel University*

Peter Herron, *Suffolk County Community College*

Konrad J. Heuvers, *Michigan Technological University*

Robert Higgins, *Quantics Corporation*

Louis F. Hoelzle, *Bucks County Community College*

Herbert Kasube, *Bradley University*

Phil Kavanaugh, *Illinois Wesleyan University*

Maureen Kelly, *Northern Essex Community College*

Harvey B. Keynes, *University of Minnesota*

Paul Kumpel, *SUNY, Stony Brook*

Leo Lampone, *Quantics Corporation*

Bruce Landman, *Hofstra University*

Benjamin Levy, *Lexington H.S., Lexington, Mass.*

Phil Locke, *University of Maine, Orono*

John Lucas, *University of Wisconsin–Oshkosh*

Stanley M. Lukawecki, *Clemson University*

Nicholas Macri, *Temple University*

Melvin J. Maron, *University of Louisville*

Thomas McElligott, *University of Lowell*

Judith McKinney, *California State Polytechnic University, Pomona*

Joseph Meier, *Millersville State College*

Ron Moore, *Ryerson Polytechnical Institute*

Barbara Moses, *Bowling Green State University*

David Nash, *VP Research, Autofacts, Inc.*

Richard Nowakowski, *Dalhousie University*

Robert Phillips, *University of South Carolina at Aiken*

Mark A. Pinsky, *Northeastern University*

David Randall, *Oakland Community College*

William H. Richardson, *Wichita State University*

David Sandell, *U.S. Coast Guard Academy*

George Shapiro, *Brooklyn College*

Donald R. Sherbert, *University of Illinois*

Wolfe Snow, *Brooklyn College*

Ian Spatz, *Brooklyn College*

Jean Springer, *Mount Royal College*

Norton Starr, *Amherst College*

Richard B. Thompson, *The University of Arizona*

William F. Trench, *Trinity University*

Walter W. Turner, *Western Michigan University*

Richard C. Vile, *Eastern Michigan University*

Shirley Wakin, *University of New Haven*

James Warner, *Precision Visuals*

Peter Waterman, *Northern Illinois University*

Evelyn Weinstock, *Glassboro State College*

Candice A. Weston, *University of Lowell*

Yihren Wu, *Hofstra University*

Richard Yuskaitis, *Precision Visuals*

DEVELOPMENT TEAM FOR THE FIFTH EDITION

The following survey respondents critiqued the previous edition and recommended many of the changes that found their way into the new edition.

Robert C. Banash, *St. Ambrose University*
George R. Barnes, *University of Louisville*
John P. Beckwith, *Michigan Technological University*
Joan E. Bell, *Northeastern Oklahoma State University*
Barbara Bohannon, *Hofstra University*
Phyllis Boutilier, *Michigan Technological University*
Stephen L. Brown, *Olivet Nazarene University*
Virginia Buchanan, *Hiram College*
Carlos E. Caballero, *Winthrop University*
Stan R. Chadick, *Northwestern State University*
Hongwei Chen, *Christopher Newport University*
Robert D. Cismowski, *San Bernardino Valley College*
David Clydesdale, *Sauk Valley Community College*
Cecil J. Coone, *State Technical Institute at Memphis*
Norman Cornish, *University of Detroit*
William H. Dent, *Maryville College*
Preston Dinkins, *Southern University*
Scott Eckert, *Cuyamaca College*
Judith Elkins, *Sweet Briar College*
Brett Elliott, *Southeastern Oklahoma State University*
Dorothy M. Fitzgerald, *Golden West College*
Ernesto Franco, *California State University–Fresno*
Daniel B. Gallup, *Pasadena City College*
Mahmood Ghamsary, *Long Beach City College*
Michael Gilpin, *Michigan Technological University*
S. B. Gokhale, *Western Illinois University*
Morton Goldberg, *Broome Community College*
Mordechai Goodman, *Rosary College*
Sid Graham, *Michigan Technological University*
Kent Harris, *Western Illinois University*
Jim Hefferson, *St. Michael College*
Warland R. Hersey, *North Shore Community College*
Konrad J. Heuvers, *Michigan Technological University*

Robert Homolka, *Kansas State University–Salina*
John M. Johnson, *George Fox College*
Wells R. Johnson, *Bowdoin College*
Richard Krikorian, *Westchester Community College*
Fat C. Lam, *Gallaudet University*
James F. Lanahan, *University of Detroit–Mercy*
Kuen Hung Lee, *Los Angeles Trade–Technology College*
Marshall J. Leitman, *Case Western Reserve University*
Darryl A. Linde, *Northeastern Oklahoma State University*
Leland E. Long, *Muscatine Community College*
Mauricio Marroquin, *Los Angeles Valley College*
Larry Matthews, *Concordia College*
Phillip McGill, *Illinois Central College*
Aileen Michaels, *Hofstra University*
Janet S. Milton, *Radford University*
Robert Mitchell, *Rowan College of New Jersey*
Marilyn Molloy, *Our Lady of the Lake University*
Kylene Norman, *Clark State Community College*
Roxie Novak, *Radford University*
Donald Passman, *University of Wisconsin*
Walter M. Patterson, *Lander University*
Edward Peifer, *Ulster County Community College*
Richard Remzowski, *Broome Community College*
Guanshen Ren, *College of Saint Scholastica*
Naomi Rose, *Mercer County Community College*
David Ryeburn, *Simon Fraser University*
Ned W. Schillow, *Lehigh County Community College*
Parashu R. Sharma, *Grambling State University*
Howard Sherwood, *University of Central Florida*
Bhagat Singh, *University of Wisconsin Centers*
Martha Sklar, *Los Angeles City College*
John L. Smith, *Rancho Santiago Community College*
Jean Springer, *Mount Royal College*
David Voss, *Western Illinois University*
Bruce F. White, *Lander University*
Gary L. Wood, *Azusa Pacific University*
Michael L. Zwilling, *Mount Union College*

The following people contributed numerous new and imaginative problems to the text:

Loren Argabright, *Drexel University*
Patricia Clark, *Rochester Institute of Technology*

Lawrence Cusick, *California State University–Fresno*
Benny Evans, *Oklahoma State University*
Rebecca Hill, *Rochester Institute of Technology*
Jerry Johnson, *University of Nevada–Reno*
Michael Zeidler, *Milwaukee Area Technical College*

The following people assisted with the critically important job of preparing the answer section, solutions for the *Student's Solutions Manual*, answers to technology exercises, and preparing the index:

Chris Butler, *Case Western Reserve University*
Stephen L. Davis, *Davidson College*
Michael Dagg, *Numerical Solutions, Inc.*
Blaise DeSesa, *Drexel University*

Clyde Dubbs, *New Mexico Institute of Mining and Technology*
Sheldon Dyck, *Waterloo Maple Software*
Diane Hagglund, *Waterloo Maple Software*
Majid Masso, *Brookdale Community College*
Kylene Norman, *Clark State Community College*
Stanley Ocken, *City College—CUNY*
Sharon Ross, *DeKalb College*
Dennis Schneider, *Knox College*
Dan Seth, *Morehead State University*
Shirley Wakin, *University of New Haven*

The following telesession participants provided crucial feedback on the previous edition and suggestions for the revision:

Chris Butler, *Case Western Reserve University*
Clyde Dubbs, *New Mexico Institute of Mining and Technology*
Della Duncan, *California State University–Fresno*

Kaplana Godbole, *Michigan Technological University*
Ed Hoefer, *Rochester Institute of Technology*
David Patterson, *West Texas A&M*
Father Bernard Portz, *Creighton University*
David Rollins, *University of Central Florida*
Jean Springer, *Mount Royal College*

The following people reviewed various stages of the fifth edition for its content and accuracy:

John Bailey, *Clark State Community College*
Irl Bivens, *Davidson College*
Christopher Butler, *Case Western Reserve University*
Clyde Dubbs, *New Mexico Institute of Mining and Technology*
Della Duncan, *California State University–Fresno*
Edwin Hoefer, *Rochester Institute of Technology*

Susan Friedman, *Bernard M. Baruch College, CUNY*
Marc Frantz, *Indiana University–Purdue University at Indianapolis*
Konrad J. Heuvers, *Michigan Technological University*
Majid Masso, *Brookdale Community College*
Kylene Norman, *Clark State Community College*
Stanley Ocken, *City College–CUNY*
Sharon Ross, *DeKalb College*
Dennis Schneider, *Knox College*

The following people provided additional material for tests and other supplements:

Pasquale Condo, *University of Lowell*
Maureen Kelley, *Northern Essex Community College*
Catherine H. Pirri, *Northern Essex Community College*

THE CONTRIBUTIONS OF ALBERT HERR

This revision was prepared in collaboration with Professor Albert Herr of Drexel University, a gifted, award-winning teacher with years of experience in the calculus classroom. Al participated in every aspect of this revision, and many of the improvements in the exposition and the mathematics are due to him. This edition has benefited greatly from Al's background in engineering mathematics and his wonderful skills as a problem creator—many of the technology problems are his, and many improvements in the text material reflect his ideas. I feel most fortunate to have had this opportunity to learn some new mathematics and new ideas about teaching calculus from such a skilled mathematician and talented teacher.

SPECIAL CONTRIBUTIONS

I am indebted to:

- Barbara Holland, my editor, for her perceptive understanding of contemporary calculus issues, her faith in my work, and doing everything that a fine editor should. She is a joy to work with.

- Ann Berlin, Lucille Buonocore, and Nancy Prinz of the Wiley Production Department for working miracles with a tight production schedule. Thank you all once again.

- Lilian Brady, whose eye for detail and aesthetics of typography created hundreds of purses out of sows' ears.

- Sharon Smith for magically juggling a gaggle of supplements that would stagger a sumo wrestler.

- The group at HRS for biting their tongues at just the right moments.

- The illustration group at Techsetters for tolerating my compulsion for detail and believing that I can see a $\frac{1}{10}$ point difference.

- Irl Bivens of Davidson College whose keen sense of sound pedagogy and good mathematics guided us in structuring this revision.

- Benny Evans of Oklahoma State University and Jerry Johnson of the University of Nevada–Reno for providing a wonderfully imaginative set of technology exercises for us to work with.

- My assistant, Dolores Morgan, for keeping the coffee hot, the project on schedule, and the author sane. I can't thank her enough.

HOWARD ANTON

CONTENTS

■ CHAPTER 17.   MULTIPLE INTEGRALS   851

■ CHAPTER 18.   TOPICS IN VECTOR CALCULUS   919

■ CHAPTER 19.   SECOND-ORDER DIFFERENTIAL EQUATIONS   983

INTRODUCTION

Calculus is the mathematical tool used to analyze changes in physical quantities. It was developed in the seventeenth century to study four major classes of scientific and mathematical problems of the time:

- Find the tangent line to a curve at a point.

- Find the length of a curve, the area of a region, and the volume of a solid.

- Find the maximum or minimum value of a quantity—for example, the maximum and minimum distances of a planet from the sun, or the maximum range attainable for a projectile by varying its angle of fire.

- Given a formula for the distance traveled by a body in any specified amount of time, find the velocity and acceleration of the body at any instant. Conversely, given a formula that specifies the acceleration or velocity at any instant, find the distance traveled by the body in a specified period of time.

These problems were attacked by the greatest minds of the seventeenth century, culminating in the crowning achievement of Gottfried Wilhelm Leibniz and Isaac Newton—the creation of calculus.

Gottfried Wilhelm Leibniz (1646–1716)

This gifted genius was one of the last people to have mastered most major fields of knowledge—an impossible accomplishment in our own era of specialization. He was an expert in law, religion, philosophy, literature, politics, geology, metaphysics, alchemy, history, and mathematics.

Leibniz was born in Leipzig, Germany. His father, a professor of moral philosophy at the University of Leipzig, died when Leibniz was six years old. The precocious boy then gained access to his father's library and began reading voraciously on a wide range of subjects, a habit that he maintained throughout his life. At age 15 he entered the University of Leipzig as a law student and by the age of 20 received a doctorate from the University of Altdorf. Subsequently, Leibniz followed a career in law and international politics, serving as counsel to kings and princes.

During his numerous foreign missions, Leibniz came in contact with outstanding mathematicians and scientists who stimulated his interest in mathematics—most notably, the physicist Christian Huygens. In mathematics Leibniz was self-taught, learning the subject by reading papers and journals. As a result of this fragmented mathematical education, Leibniz often duplicated the results of others, and this ultimately led to a raging conflict as to whether he or Isaac Newton should be regarded as the inventor of calculus. The argument over this question engulfed the scientific circles of England and Europe, with most scientists on the continent supporting Leibniz and those in England supporting Newton. The conflict was unfortunate, and both sides suffered in the end: The continent lost the benefit of Newton's discoveries in astronomy and physics for more than 50 years, and for a long period England became a second-rate country mathematically because its mathematicians were hampered by Newton's inferior calculus notation. It is of interest to note that Newton and Leibniz never went to the lengths of vituperation of their advocates—both were sincere admirers of each other's work. The fact is that both men invented calculus independently; Leibniz invented it 10 years after Newton, in 1685, but he published his results 20 years before Newton.

Leibniz never married. He was moderate in his habits, quick-tempered, but easily appeased, and charitable in his judgment of other people's work. In spite of his great achievements, Leibniz never received the honors showered on Newton, and he spent his final years as a lonely embittered man. At his funeral there was one mourner, his secretary. An eyewitness stated, "He was buried more like a robber than what he really was—an ornament of his country."

Isaac Newton (1642–1727)

Newton was born in the village of Woolsthorpe, England. His father died before he was born and his mother raised him on the family farm. As a youth he showed little evidence of his later brilliance, except for an unusual talent with mechanical devices—he apparently built a working water clock and a toy flour mill powered by a mouse. In 1661 he entered Trinity College in Cambridge with a deficiency in geometry. Fortunately, Newton caught the eye of Isaac Barrow, a gifted mathematician and teacher. Under Barrow's guidance Newton immersed himself in mathematics and science, but he graduated without any special distinction. Because the Plague was spreading rapidly through London, Newton returned to his home in Woolsthorpe and stayed there during the years of 1665 and 1666. In those two momentous years the entire framework of modern science was miraculously created in Newton's mind—he discovered calculus, recognized the underlying principles of planetary motion and gravity, and determined that "white" sunlight was composed of all colors, red to violet. For some reason he kept his discoveries to himself. In 1667 he returned to Cambridge to obtain his Master's degree and upon graduation became a teacher at Trinity. Then in 1669 Newton succeeded his teacher, Isaac Barrow, to the Lucasian chair of mathematics at Trinity, one of the most honored chairs of mathematics in the world. Thereafter, brilliant discoveries flowed from Newton steadily. He formulated the law of gravitation and used it to explain the motion of the moon, the planets, and the tides; he formulated basic theories of light, thermodynamics, and hydrodynamics; and he devised and constructed the first modern reflecting telescope.

Throughout his life Newton was hesitant to publish his major discoveries, revealing them only to a select circle of friends, perhaps because of a fear of criticism or controversy. In 1687, only after intense coaxing by the astronomer, Edmond Halley (Halley's comet), did Newton publish his masterpiece, *Philosophiae Naturalis Principia Mathe-*

Gottfried Leibniz
(Culver Pictures)

Isaac Newton
(Culver Pictures)

matica (The Mathematical Principles of Natural Philosophy). This work is generally considered to be the most important and influential scientific book ever written. In it Newton explained the workings of the solar system and formulated the basic laws of motion which to this day are fundamental in engineering and physics. However, not even the pleas of his friends could convince Newton to publish his discovery of calculus. Only after Leibniz published his results did Newton relent and publish his own work on calculus.

After 35 years as a professor, Newton suffered depression and a nervous breakdown. He gave up research in 1695 to accept a position as warden and later master of the London mint. During the 25 years that he worked at the mint, he did virtually no scientific or mathematical work. He was knighted in 1705 and on his death was buried in Westminster Abbey with all the honors his country could bestow. It is interesting to note that Newton was a learned theologian who viewed the primary value of his work to be its support of the existence of God. Throughout his life he worked passionately to date biblical events by relating them to astronomical phenomena. He was so consumed with this passion that he spent years searching the Book of Daniel for clues to the end of the world and the geography of hell.

Newton described his brilliant accomplishments as follows: "I seem to have been only like a boy playing on the seashore and diverting myself in now and then finding a smoother pebble or prettier shell than ordinary, whilst the great ocean of truth lay all undiscovered before me."

Calculus

with Analytic Geometry

René Descartes (1596–1650)

1 COORDINATES, GRAPHS, LINES

Because numbers are the foundation of all mathematics, it is important to be familiar with the various kinds of numbers and the differences between them. In this section we shall review the basic facts and terminology relating to numbers.

☐ **CLASSIFICATION OF REAL NUMBERS**

The simplest numbers are the ***natural numbers***:

$$1, 2, 3, 4, 5, \ldots$$

The natural numbers form a subset of a larger class of numbers called the ***integers***:

$$\ldots, -4, -3, -2, -1, 0, 1, 2, 3, 4, \ldots$$

The integers in turn are a subset of a still larger class of numbers called the **rational numbers**. With the exception that division by zero is ruled out, the rational numbers are formed by taking ratios of integers. Examples are

$$\frac{2}{3}, \frac{7}{5}, \frac{6}{1}, \frac{0}{9}, -\frac{5}{2} \left(= \frac{-5}{2} = \frac{5}{-2} \right)$$

Observe that every integer is also a rational number because an integer p can be written as the ratio $p = p/1$.

The early Greeks believed that the size of every physical quantity could, in theory, be represented by a rational number. They reasoned that the size of a physical quantity must consist of a certain whole number of units plus some fraction m/n of an additional unit. This idea was shattered in the fifth century B.C. by Hippasus of Metapontum* who demonstrated the existence of **irrational numbers**, that is, numbers that cannot be expressed as the ratio of integers. Using geometric methods, he showed that the hypotenuse of the right triangle in Figure 1.1.1 cannot be expressed as the ratio of integers, thereby proving that $\sqrt{2}$ is an irrational number. Other examples of irrational numbers are

$$\sqrt{3}, \quad \sqrt{5}, \quad 1 + \sqrt{2}, \quad \sqrt[3]{7}, \quad \pi, \quad \cos 19°$$

The rational and irrational numbers together comprise a larger class of numbers, called **real numbers** or sometimes the **real number system**.

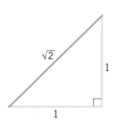

Figure 1.1.1

□ **DIVISION BY ZERO**

In computations with real numbers, division by zero is never allowed because a relationship of the form $y = p/0$ would imply that

$$0 \cdot y = p$$

If p is different from zero, this equation is contradictory; and if p is equal to zero, this equation is satisfied by any number y, so the ratio $0/0$ does not have a unique value—a situation that is mathematically unsatisfactory. For these reasons such symbols as $p/0$ and $0/0$ are not assigned a value; they are said to be **undefined**.

□ **COMPLEX NUMBERS**

Because the square of a real number cannot be negative, the equation

$$x^2 = -1$$

has no solutions in the real number system. In the eighteenth century mathematicians remedied this problem by inventing a new number, which they denoted by

$$i = \sqrt{-1}$$

and which they defined to have the property $i^2 = -1$. This, in turn, led to the development of the **complex numbers**, which are numbers of the form

$$a + bi$$

where a and b are real numbers. Some examples are

$$2 + 3i \qquad 3 - 4i \qquad 6i \qquad \tfrac{2}{3}$$
$$[a = 2, b = 3] \quad [a = 3, b = -4] \quad [a = 0, b = 6] \quad [a = \tfrac{2}{3}, b = 0]$$

Observe that every real number a is also a complex number because it can be written as

$$a = a + 0i$$

Thus, the real numbers are a subset of the complex numbers. Those complex numbers that are not real numbers are called **imaginary numbers**. We shall be concerned primarily with real numbers; however, complex numbers will arise in the course of solving equations. For example, the solutions of the quadratic equation

$$ax^2 + bx + c = 0 \tag{1}$$

---

* HIPPASUS OF METAPONTUM (circa 500 B.C.). A Greek Pythagorean philosopher. According to legend, Hippasus made his discovery at sea and was thrown overboard by fanatic Pythagoreans because his result contradicted their doctrine. The discovery of Hippasus is one of the most fundamental in the entire history of science.

which are given by the **quadratic formula**

$$x = \frac{-b \pm \sqrt{b^2 - 4ac}}{2a}$$

are imaginary if the quantity $b^2 - 4ac$ [called the **discriminant** of (1)] is negative. The hierarchy of numbers is summarized in Figure 1.1.2.

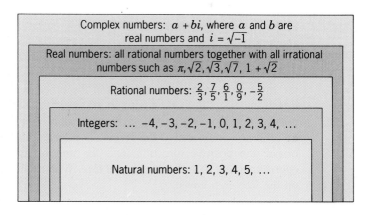

Complex numbers: $a + bi$, where a and b are real numbers and $i = \sqrt{-1}$

Real numbers: all rational numbers together with all irrational numbers such as $\pi, \sqrt{2}, \sqrt{3}, \sqrt{7}, 1 + \sqrt{2}$

Rational numbers: $\frac{2}{3}, \frac{7}{5}, \frac{6}{1}, \frac{0}{9}, -\frac{5}{2}$

Integers: ... $-4, -3, -2, -1, 0, 1, 2, 3, 4, ...$

Natural numbers: 1, 2, 3, 4, 5, ...

Figure 1.1.2

☐ **DECIMAL REPRESENTATION OF REAL NUMBERS**

3.141592653589793238462643383279502884197169
399375105820974944592307816406286208999862803
482534211706798214808651328230664709384460955
050582231725359408128481117450284102701938521
105559644622948954930381964428810975665933441
612847564823378678316527120190914564856692346
034861045432664821339360726024914127372458700
066063155881748815209209628292540917153643678
925903600113305305488204665213841469519415116
094330572703657595919530921861173819326117931
051185480744623799627495673518857527248912279
381830119491298336733624406566430860213949463
952247371907021798609437027705392171762931767
523846748184676694051320005681271452635608277
857713427577896091736371787214684409012249534
301465495853710507922796892589235420199561121
290219608640344181598136297747713099605187072
113499999983729780499510597317328160963185950
244594553469083026425223082533446850352619311
881710100031378387528865875332083814206171776
691473035982534904287554687311595628638823537
875937519577818577805321712268066130019278766
111959092164201989380952572010654858632788659
361533818279682303019520353018529689957736225
994138912497217752834791315155748572424541506
959508295331168617278558907509838175463746493
931925560640092770167113900984882401285836160
356370766010471018194295955962494672782407892
553297746284710404753469620804668425906949129
331367702899894562095073516052569660240580381
501935112533826430035587640247496473263914199
272604269922796782354781636009341721641219924
586315030286182974555076749838505049458858692
699569092721079750930295532116534498720275596
0236480665499119881834797753566369807426542527862655181841757467289097777279380008164706001614524919217321721477235014144197356854816136115735255213347574184946843852332390739143340540724864582518983569485562099921922184272550254256887671790494601653466804988627232791786085784383827967976681454100953883786360950680064225125205117392984896084128488626945604241965285022210661186306744278623209914655049717121137969056363719172874677646575739624138908658326459958133904780275900994657640789512694683983525957098258226205224894077267194782684826014769909026401363944374553050682034962524517493996514314298091906592509372216964615157098583874105978859597729754989301617539284681382686838689427741559918559252459539594310499725246808459872736446958486538367362226260991246080512438843904512441365497627807977156914359977001296160894416948685558484063534220722258284886481584560285060168427394522674676788952521385225499546667278239864565961163548862305774564980355936345681743241125150760694794510965960940252288797108931456691368672287489405601015033086179286809208747609178249385890097149096759852613655497818931297848216829989487226588048575640142704775551323796414515237462343645428584447952658678210511413547357395231134271661021359695362314429524849371871101457654035902799344037420073105785390621983874478084784896833214457138687519435064302184531910484810053706146806749192781911979399520614196634287544406437451237181921799983910159195618146751426912397489409071864942319615679452080951465502252316038819301420937621378559566389377870830390697920773467221825625996615014215030680384477345492026054146659252014974428507325186660021324340881907104863317346496514539057962685610055081066587969981635747363840525714591028970641401109712062804390397595156771577004203378699360072305587631763594218731251471205329281918261861258673215791984148488291644706095752706957220917567116722910981690915280173506712748583222871835209353965725121083579151369882091444210067510334671103141267111369908658516398315019701651511685171437657618351556508849099898599823873455283316355076479185358932261854896321329330898570642046752590709154814165498594616371802709819943099244889575712828905923233260972997120844335732654893823911932597463667305836041428138830320382490375898524374417029132765618093773444030707469211201913020330380197621101100449293215160842444859637669838952286847831235526582131449576857262433441893039686426243410773226978028073189154411010446823252716201052652272111660396665573092547110557853763466820653109896526918620564769312570586356620185581007293606598764861179104533488503461136576867532494416680396265797877185560845529654126654085306143444318586769751456614068007002378776591344017127494704205622305389945613140711270004078547332699390814546646458807972708266830634328587856983052358089330657574067954571637752542021149557615814002501262285941302164715509792592309907965473761255176567513575178296664547791745011299614890304639947132962107340437518957359614589019389713111790429782856475032031986915140287080859904801094121472213179476477726224142548545403321571853061422881375850430633217518297986622371721591607716692547487389866549494501146540628433663937900397692656721463853067360965712091807638327166416274888800786925602902284721040317211860820419000422966171196377921337575114959501566049631862947265473642523081770367515906735023507283540567040386743513622224771589150495309844489333096340878076932599397805419341447377441842631298608099888687413260472156951623965864573021631598193195167353812974167729478672422924654366800980676928238280689964004824354037014163149658979409243237896907069779422362508221688957383798623001593776471651228935786015881617557829735233446042815126272037343146531977774160319906655418763979293344195215413418994854447345673831624993419131814809277771038638773431772075456545322077709212019051660962804909263601975988281613323166636528619326686336062735676303544776280350450777235547105859548702790814356240145171806246436267945612753181340783303362542327839449753824372058353114771199260638133467768796959703098339130771098704085913374641442822772634659470474587847787201927715280731767907707157213444730605700733492436931138350493163128404251219256517980694113528013147013047816437885185290928545201165839341965621349143415956258658655705526904965209858033850722426482939728584783163057777560688876446248246857926039535277348030480290058760758251047470916439613626760449256274204208320856611906254543372131535958450687724602901618766795240616342522577195429162991930645537799140373404328752628889639958794757291746426357455254079091451357111369410911939325191076020825202618798531887705842972591677813149699009019211697173727847684726860849003377024242916513005005168323364350389517029893922334517220138128069650117844087451960121228599371623130171144484640903890644954440061986907548516026327505298349187407866808818338510228334508504860825039302133219715518430635455007668282949304137765527939751754613953984683393638304746119966538581538420568533862186725233402830871123282789212507712629463229563989898935821167456270102183564622013496715188190973038119800497340723961036854066431939509790190699639552453005450580685501956730229219139339185680344903982059551002263535361920419947455385938102343955449597783779023742161727111723643435439478221818528624085140066604433258885698670543154706965747458550332323342107301545940516553790686627333799585115625784322988273723198987571415957811196358330059408730681216028764962867446047746491599505497374256269010490377819868359381465741268049256487985561453723478673303904688383436346553794986419270563872931748723320837601123029911367938627089438799362016295154133714248928307220126901475466847653576164773794675200490757155527819653621323926406160136358155907422020203187277605277219005561484255518792530343513984425322341576233610642506390497500865627109535919465897514131034822769306247435363256916078154781811528436679570611086153315044521274739245449454236828860613408414863776700961207151249140430272538607648236341433462351897576645216413767969031495019108575984423919862916421939949072362346468441173940326591840443780513338945257423995082965912285085558215725031071257012668302402929525220118726767562204154205161841634847565169998116141010029960783869092916030288400269104140792886215078424516709087000699282120660418371806535567252532567532861291042487761825829765157959847035622262934860034158722980534989650226291748788202734209222245339856264766914905562842503912757710284027998066365825488926488025456610172967026640765590429099456815065265305371829412703369313785178609040708667114965583434347693385781711386455873678123014888537181465319095882996170595638726080537678...

Figure 1.1.3

Rational and irrational numbers can be distinguished by their decimal representations. Rational numbers have decimals that are **repeating**, by which we mean that there is some point in the decimal representation at which the digits that follow consist of a fixed block of integers repeated over and over. For example,

$$\frac{4}{3} = 1.333\ldots, \quad \frac{3}{11} = .272727\ldots, \quad \frac{5}{7} = .714285714285714285\ldots$$

      3 repeats   27 repeats   714285 repeats

Repeating decimals that consist of zeros from some point on are called **terminating decimals**. Some examples are

$$\frac{1}{2} = .50000\ldots, \quad \frac{12}{4} = 3.0000\ldots, \quad \frac{8}{25} = .32000\ldots$$

It is usual to omit the repetitive zeros in terminating decimals. For example,

$$\frac{1}{2} = .5, \quad \frac{12}{4} = 3, \quad \frac{8}{25} = .32$$

Moreover, it has become common to denote repeating decimals by writing the repeating digits only once, but with a bar over them to indicate the repetition. For example,

$$\frac{4}{3} = 1.\overline{3}, \quad \frac{3}{11} = .\overline{27}, \quad \frac{5}{7} = .\overline{714285}$$

Rational numbers are represented by repeating decimals, and conversely every repeating decimal represents a rational number. Thus, the *irrational* numbers can be viewed as those real numbers that are represented by *nonrepeating* decimals. For example, the decimal

.101001000100001000001...

does not repeat because the number of zeros between the ones keeps growing. Thus, it represents an irrational number.

Irrational numbers cannot be represented with perfect accuracy in decimal notation. For example, π is only approximated by the decimal 3.14. Moreover, no matter how many decimal places we use, even if we compute π to 2000 places, as in Figure 1.1.3, we still have only an approximation to π.

REMARK. Beginning mathematics students are sometimes taught to approximate π by $\frac{22}{7}$. Note, however, that

$$\frac{22}{7} = 3.\overline{142857}$$

is a rational number whose decimal representation begins to differ from π in the third decimal place.

☐ COORDINATE LINES

In 1637 René Descartes* published a philosophical work called *Discourse on the Method of Rightly Conducting the Reason*. In the back of that book was an appendix that the British philosopher John Stuart Mill described as, "The greatest single step ever made in the progress of the exact sciences." In that appendix René Descartes linked together algebra and geometry, thereby creating a new subject called *analytic geometry*; it gave a way of describing algebraic formulas by geometric curves and, conversely, geometric curves by algebraic formulas.

In analytic geometry, the key step is to establish a correspondence between real numbers and points on a line. This is done by arbitrarily designating one of the two directions along the line as the *positive direction* and the other as the *negative direction*. The positive direction is usually marked with an arrowhead as in Figure 1.1.4; for horizontal lines the positive direction is generally taken to the right. A unit of measurement is then chosen and an arbitrary point, called the *origin*, is selected anywhere along the line. The line, the origin, the positive direction, and the unit of measurement define what is called a *coordinate line* or sometimes a *real line*. With each real number we can associate a point on the line as follows:

Figure 1.1.4

- Associate the origin with the number 0.

- Associate with each positive number r the point that is a distance of r units in the positive direction from the origin.

- Associate with each negative number $-r$ the point that is a distance of r units in the negative direction from the origin.

The real number corresponding to a point on the line is called the *coordinate* of the point.

Example 1   In Figure 1.1.5 we have marked the locations of the points with coordinates -4, -3, -1.75, $-\frac{1}{2}$, $\sqrt{2}$, π, and 4. The locations of π and $\sqrt{2}$, which are approximate, were obtained from their decimal approximations, $\pi \approx 3.14$ and $\sqrt{2} \approx 1.41$. ◀

Figure 1.1.5

It is evident from the way in which real numbers and points on a coordinate line are related that each real number corresponds to a single point and each point corresponds to a single real number. To describe this fact we say that the real numbers and the points on a coordinate line are in *one-to-one correspondence*.

---

*RENÉ DESCARTES (1596–1650). Descartes, a French aristocrat, was the son of a government official. He graduated from the University of Poitiers with a law degree at age 20. After a brief probe into the pleasures of Paris he became a military engineer, first for the Dutch Prince of Nassau and then for the German Duke of Bavaria. It was during his service as a soldier that Descartes began to pursue mathematics seriously and develop his analytic geometry. After the wars, he returned to Paris where he stalked the city as an eccentric, wearing a sword in his belt and a plumed hat. He lived in leisure, seldom arose before 11 A.M., and dabbled in the study of human physiology, philosophy, glaciers, meteors, and rainbows. He eventually moved to Holland, where he published his *Discourse on the Method*, and finally to Sweden where he died while serving as tutor to Queen Christina. Descartes is regarded as a genius of the first magnitude. In addition to major contributions in mathematics and philosophy, he is considered, along with William Harvey, to be a founder of modern physiology.

☐ **ORDER PROPERTIES**

The real numbers can be ordered by size as follows: If $b - a$ is positive, then we say that ***b is greater than a*** or that ***a is less than b*** and write $a < b$. The inequality $a < b$ can also be expressed as $b > a$ when it is more convenient to do so. The inequality $a \leq b$ is defined to mean that either $a < b$ or $a = b$, and the expression $a < b < c$ is defined to mean that $a < b$ and $b < c$.

As one traverses a coordinate line in the positive direction, the real numbers increase in size, so that on a horizontal coordinate line the inequality $a < b$ implies that a is to the left of b, and the inequality $a < b < c$ implies that a is to the left of b and b is to the left of c (Table 1.1.1).

The symbol $a < b \leq c$ means $a < b$ and $b \leq c$. We leave it to the reader to deduce the meanings of such symbols as

$$a \leq b < c, \quad a \leq b \leq c, \quad \text{and} \quad a < b < c < d$$

Table 1.1.1

INEQUALITY	GEOMETRIC INTERPRETATION	ILLUSTRATION
$a < b$ or $b > a$	a is to the left of b.	
$a \leq b$ or $b \geq a$	a is to the left of b or coincides with b.	
$0 < a$ or $a > 0$	a is to the right of the origin.	
$a < 0$ or $0 > a$	a is to the left of the origin.	
$a < b < c$	a is to the left of b and b is to the left of c.	

Example 2   The following inequalities are all correct:

$$3 < 8, \quad -7 < 1.5, \quad -12 \leq -\pi, \quad 5 \leq 5, \quad 0 \leq 2 \leq 4,$$
$$8 \geq 3, \quad 1.5 > -7, \quad -\pi > -12, \quad 5 \geq 5, \quad 3 > 0 > -1 \quad \blacktriangleleft$$

REMARK.   To distinguish verbally between numbers that satisfy $a \geq 0$ and those that satisfy $a > 0$, we shall call a ***nonnegative*** if $a \geq 0$ and ***positive*** if $a > 0$. Thus, a nonnegative number is either positive or zero.

The following properties of inequalities are frequently used in calculus. We omit the proofs.

1.1.1   THEOREM.   *Let a, b, c, and d be real numbers.*

(a)   *If $a < b$ and $b < c$, then $a < c$.*
(b)   *If $a < b$, then $a + c < b + c$ and $a - c < b - c$.*
(c)   *If $a < b$, then $ac < bc$ when c is positive and $ac > bc$ when c is negative.*
(d)   *If $a < b$ and $c < d$, then $a + c < b + d$.*
(e)   *If a and b are both positive or both negative and $a < b$, then $1/a > 1/b$.*

REMARK.   These five properties remain true if $<$ and $>$ are replaced by $\leq$ and $\geq$.

If we call the direction in which an inequality points its *sense*, then parts (b)–(e) of this theorem can be paraphrased informally as follows:

(b) *The sense of an inequality is unchanged if the same number is added to or subtracted from both sides.*

(c) *The sense of an inequality is unchanged if both sides are multiplied by the same positive number, but the sense is reversed if both sides are multiplied by the same negative number.*

(d) *Inequalities with the same sense can be added.*

(e) *If both sides of an inequality have the same sign, then the sense of the inequality is reversed by taking the reciprocal of each side.*

Example 3   The statements in Theorem 1.1.1 are illustrated in Table 1.1.2.

□ **INTERVALS**

We shall assume in this text that you are familiar with the concept of a set and understand the meaning of such symbols as $a \in A$, $a \notin A$, $\varnothing$ (the empty set), $A \cap B$, $A \cup B$, $A = B$, and $A \subset B$, where A and B are sets. If this is not the case, then read the review of sets in Appendix A.

One way to specify a set is to list its members between braces. Thus, the set of all positive integers less than 5 can be written as

$$\{\, 1, 2, 3, 4 \,\}$$

and the set of all positive even integers can be written as

$$\{\, 2, 4, 6, \ldots \,\}$$

where the dots are used to indicate that only some of the members are listed explicitly and the rest can be obtained by continuing the pattern.

When it is inconvenient or impossible to list the members of a set, one can use the set-builder notation

$$\{\, x\!: \underline{\hspace{1cm}} \,\}$$

which is read, "the set of all x such that _____." Where the line is placed, one would state a property that specifies the set. Thus,

$$\{\, x\!:\! x \text{ is a real number and } 2 < x < 3 \,\}$$

is read, "the set of all x such that x is a real number and $2 < x < 3$."

When it is clear that the members of a set are real numbers, we will omit the reference to

Table 1.1.2

STARTING INEQUALITY	OPERATION	RESULTING INEQUALITY
$-2 < 6$	Add 7 to both sides.	$5 < 13$
$-2 < 6$	Subtract 8 from both sides.	$-10 < -2$
$-2 < 6$	Multiply both sides by 3.	$-6 < 18$
$-2 < 6$	Multiply both sides by -3.	$6 > -18$
$3 < 7$	Multiply both sides by 4.	$12 < 28$
$3 < 7$	Multiply both sides by -4.	$-12 > -28$
$3 < 7$	Take reciprocals of both sides.	$\frac{1}{3} > \frac{1}{7}$
$-8 < -6$	Take reciprocals of both sides.	$-\frac{1}{8} > -\frac{1}{6}$
$4 < 5, \; -7 < 8$	Add corresponding sides.	$-3 < 13$

◀

this fact. Thus, the preceding set would usually be written more briefly as

$$\{x : 2 < x < 3\}$$

Of special interest in calculus are sets of real numbers called **intervals**. Geometrically, an interval is a line segment on a coordinate line. If a and b are real numbers such that $a < b$, then the **closed** interval from a to b is denoted by $[a, b]$ and is defined by

$$[a, b] = \{x : a \leq x \leq b\}$$

and the **open** interval from a to b is denoted by (a, b) and is defined by

$$(a, b) = \{x : a < x < b\}$$

The square brackets indicate that the endpoints are included in the interval and the parentheses indicate that they are not.

Table 1.1.3 lists the various types of intervals. In that table, the geometric pictures use solid dots to denote endpoints that are included in the interval and open dots to denote endpoints that are not. As shown in the table, an interval can extend indefinitely in either the positive direction, the negative direction, or both. Intervals of infinite extent are called **infinite intervals** and intervals of finite extent are called **finite intervals**.

A finite interval that includes one endpoint but not the other is called **half-open** (or sometimes **half-closed**). The symbols $-\infty$ (read "negative infinity") and $+\infty$ (read "positive infinity") do not represent numbers: the $+\infty$ indicates that the interval extends indefinitely in the positive direction, and the $-\infty$ indicates that it extends indefinitely in the negative direction.

Infinite intervals of the form $[a, +\infty)$ or $(-\infty, b]$ are considered to be closed because they contain their endpoint and those of the form $(a, +\infty)$ or $(-\infty, b)$ are considered to be open because they do not. The interval $(-\infty, +\infty)$ has no endpoints; it is regarded to be both open and closed.

Table 1.1.3

INTERVAL NOTATION	SET NOTATION	GEOMETRIC PICTURE	CLASSIFICATION
$[a, b]$	$\{x : a \leq x \leq b\}$		Finite; closed
(a, b)	$\{x : a < x < b\}$		Finite; open
$[a, b)$	$\{x : a \leq x < b\}$		Finite; { half-open / half-closed
$(a, b]$	$\{x : a < x \leq b\}$		Finite; { half-open / half-closed
$(-\infty, b]$	$\{x : x \leq b\}$		Infinite; closed
$(-\infty, b)$	$\{x : x < b\}$		Infinite; open
$[a, +\infty)$	$\{x : x \geq a\}$		Infinite; closed
$(a, +\infty)$	$\{x : x > a\}$		Infinite; open
$(-\infty, +\infty)$	$\{x : x \text{ is a real number}\}$		Infinite; open and closed

□ **SOLVING INEQUALITIES**

A **solution** of an inequality in an unknown x is a value for x that makes the inequality a true statement. For example, $x = 1$ is a solution of the inequality $x < 5$, but $x = 7$ is not. The set of all solutions of an inequality is called its **solution set**. It can be shown that if one does not multiply both sides of an inequality by zero or an expression involving an unknown, then the operations in Theorem 1.1.1 will not change the solution set of the inequality. The process of finding the solution set of an inequality is called **solving** the inequality.

Example 4   Solve $3 + 7x \leq 2x - 9$.

Solution.   We shall use the operations of Theorem 1.1.1 to isolate x on one side of the inequality.

$3 + 7x \leq 2x - 9$ Given.

$7x \leq 2x - 12$ We subtracted 3 from both sides.

$5x \leq -12$ We subtracted $2x$ from both sides.

$x \leq -\frac{12}{5}$ We multiplied both sides by $\frac{1}{5}$.

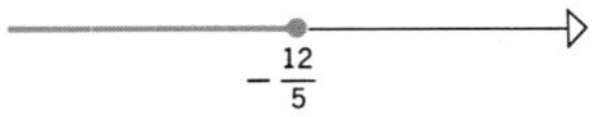

Because we have not multiplied by any expressions involving the unknown x, the last inequality has the same solution set as the first. Thus, the solution set is the interval $(-\infty, -\frac{12}{5}]$ shown in Figure 1.1.6.   ◄

Figure 1.1.6

Example 5   Solve $7 \leq 2 - 5x < 9$.

Solution.   The given inequality is actually a combination of the two inequalities

$7 \leq 2 - 5x$ and $2 - 5x < 9$

We could solve the two inequalities separately, then determine the values of x that satisfy both by taking the intersection of the two solution sets. However, it is possible to work with the combined inequalities in this problem:

$7 \leq 2 - 5x < 9$ Given.

$5 \leq -5x < 7$ We subtracted 2 from each member.

$-1 \geq x > -\dfrac{7}{5}$ We multiplied by $-\frac{1}{5}$ and reversed the sense of the inequalities.

$-\dfrac{7}{5} < x \leq -1$ For clarity, we rewrote the inequalities with the smaller number on the left.

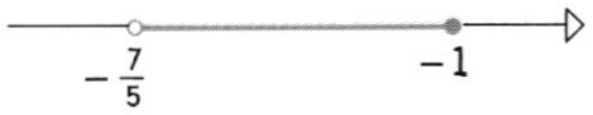

Thus, the solution set is the interval $(-\frac{7}{5}, -1]$ shown in Figure 1.1.7.   ◄

Figure 1.1.7

Example 6   Solve $x^2 - 3x > 10$.

Solution.   By subtracting 10 from both sides, the inequality can be rewritten as

$x^2 - 3x - 10 > 0$

Factoring the left side yields

$(x + 2)(x - 5) > 0$

The values of x for which $x + 2 = 0$ or $x - 5 = 0$ are $x = -2$ and $x = 5$. These points divide the coordinate line into three open intervals,

$(-\infty, -2), \quad (-2, 5), \quad (5, +\infty)$

on each of which the product $(x + 2)(x - 5)$ has constant sign. To determine those signs we shall choose an *arbitrary* point in each interval at which we shall determine the sign; these are called ***test points***. As shown in Figure 1.1.8, we shall use -3, 0, and 6 as our test points. The results can be organized as follows:

INTERVAL	TEST POINT	SIGN OF $(x + 2)(x - 5)$ AT THE TEST POINT
$(-\infty, -2)$	-3	$(-)(-)=+$
$(-2, 5)$	0	$(+)(-)=-$
$(5, +\infty)$	6	$(+)(+)=+$

The pattern of signs in the intervals is shown on the number line in the middle of Figure 1.1.8. We deduce that the solution set is $(-\infty, -2) \cup (5, +\infty)$, which is shown at the bottom of Figure 1.1.8.   ◄

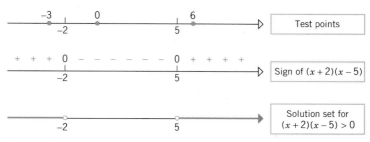

Figure 1.1.8

Example 7   Solve $\dfrac{2x - 5}{x - 2} < 1$.

Solution.   We could start by multiplying both sides by $x - 2$ to eliminate the fraction. However, this would require us to consider the cases $x - 2 > 0$ and $x - 2 < 0$ separately because the sense of the inequality would be reversed in the second case, but not the first. The following approach is simpler:

$$\frac{2x - 5}{x - 2} < 1 \qquad \boxed{\text{Given.}}$$

$$\frac{2x - 5}{x - 2} - 1 < 0 \qquad \boxed{\begin{array}{l}\text{We subtracted 1 from both}\\\text{sides to obtain a 0 on the right.}\end{array}}$$

$$\frac{(2x - 5) - (x - 2)}{x - 2} < 0 \qquad \boxed{\text{We combined terms.}}$$

$$\frac{x - 3}{x - 2} < 0 \qquad \boxed{\text{We simplified.}}$$

The quantity $x - 3$ is zero if $x = 3$, and the quantity $x - 2$ is zero if $x = 2$. These points divide the coordinate line into three open intervals,

$$(-\infty, 2), \quad (2, 3), \quad (3, +\infty)$$

on each of which the quotient $(x - 3)/(x - 2)$ has constant sign. Using 0, 2.5, and 4 as test points (Figure 1.1.9), we obtain the following results:

INTERVAL	TEST POINT	SIGN OF $(x - 3)/(x - 2)$ AT THE TEST POINT
$(-\infty, 2)$	0	$(-)/(-)=+$
$(2, 3)$	2.5	$(-)/(+)=-$
$(3, +\infty)$	4	$(+)/(+)=+$

The signs of the quotient are shown in the middle of Figure 1.1.9. From the figure we see that the solution set consists of all real values of x such that $2 < x < 3$. This is the interval $(2, 3)$ shown at the bottom of Figure 1.1.9. ◀

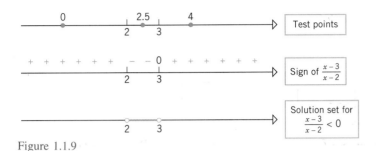

Figure 1.1.9

▶ Exercise Set 1.1   🅲 5, 6

1. Among the terms, *integer*, *rational*, and *irrational*, which ones apply to the given number?

(a) $-\frac{3}{4}$ (b) 0

(c) $\frac{24}{8}$ (d) 0.25

(e) $-\sqrt{16}$ (f) $2^{1/2}$

(g) 0.020202...   (h) 7.000....

2. Which of the terms, *integer*, *rational*, and *irrational*, apply to the given number?

(a) 0.31311311131111...   (b) 0.729999...

(c) 0.376237623762...   (d) $17\frac{4}{5}$.

3. The repeating decimal 0.137137137... can be expressed as a ratio of integers by writing

$$x = 0.137137137...$$
$$1000x = 137.137137137...$$

and subtracting to obtain $999x = 137$ or $x = \frac{137}{999}$. Use this idea, where needed, to express the following decimals as ratios of integers.

(a) 0.123123123...   (b) 12.7777...

(c) 38.07818181...   (d) 0.4296000....

4. Show that the repeating decimal 0.99999... represents the number 1. Since 1.000... is also a decimal representation of 1, this problem shows that a real number can have two different decimal representations. [*Hint:* Use the technique of Exercise 3.]

5. The Rhind Papyrus, which is a fragment of Egyptian mathematical writing from about 1650 B.C., is one of the oldest known examples of written mathematics. It is stated in the papyrus that the area A of a circle is related to its diameter D by

$$A = \left(\tfrac{8}{9}D\right)^2$$

(a) What approximation to π were the Egyptians using?

(b) Use a calculator to determine if this approximation is better or worse than the approximation of $\frac{22}{7}$.

6. The following are all famous approximations to π:

$\dfrac{333}{106}$	Adrian Athoniszoon, c. 1583
$\dfrac{355}{113}$	Tsu Chung-Chi and others
$\dfrac{63}{25}\left(\dfrac{17 + 15\sqrt{5}}{7 + 15\sqrt{5}}\right)$	Ramanujan
$\dfrac{22}{7}$	Archimedes
$\dfrac{223}{71}$	Archimedes

(a) Use a calculator to order these approximations according to size.

(b) Which of these approximations is closest to but larger than π?

(c) Which of these approximations is closest to but smaller than π?

(d) Which of these approximations is most accurate?

7. In each line of the following table, check the blocks, if any, that describe a valid relationship between the real numbers a and b. The first line is already completed as an illustration.

a	b	$a < b$	$a \le b$	$a > b$	$a \ge b$	$a = b$
1	6	✓	✓			
6	1			✓	✓	
−3	5	✓	✓			
5	−3			✓	✓	
−4	−4		✓		✓	✓
0.25	$\frac{1}{3}$	✓	✓			
$-\frac{1}{4}$	$-\frac{3}{4}$			✓	✓	

8. In each line of the following table, check the blocks, if any, that describe a valid relationship between the real numbers $a, b,$ and c.

a	b	c	$a < b < c$	$a \le b \le c$	$a < b \le c$	$a \le b < c$
-1	0	2				
2	4	-3				
$\frac{1}{2}$	$\frac{1}{2}$	$\frac{3}{4}$				
-5	-5	-5				
0.75	1.25	1.25				

9. Which of the following are always correct if $a \le b$?

(a) $a - 3 \le b - 3$ (b) $-a \le -b$

(c) $3 - a \le 3 - b$ (d) $6a \le 6b$

(e) $a^2 \le ab$ (f) $a^3 \le a^2 b$.

10. Which of the following are always correct if $a \le b$ and $c \le d$?

(a) $a + 2c \le b + 2d$ (b) $a - 2c \le b - 2d$

(c) $a - 2c \ge b - 2d$.

11. For what values of a are the following inequalities valid?

(a) $a \le a$ (b) $a < a$.

12. If $a \le b$ and $b \le a$, what can you say about a and b?

13. (a) If $a < b$ is true, does it follow that $a \le b$ must also be true?

(b) If $a \le b$ is true, does it follow that $a < b$ must also be true?

14. In each part, list the elements in the set.

(a) $\{x : x^2 - 5x = 0\}$

(b) $\{x : x$ is an integer satisfying $-2 < x < 3\}$.

15. In each part, express the set in the notation $\{x : \underline{\hspace{1cm}}\}$.

(a) $\{1, 3, 5, 7, 9, \ldots\}$

(b) the set of even integers

(c) the set of irrational numbers

(d) $\{7, 8, 9, 10\}$.

16. Let $A = \{1, 2, 3\}$. Which of the following are equal to A?

(a) $\{0, 1, 2, 3\}$ (b) $\{3, 2, 1\}$

(c) $\{x : (x - 3)(x^2 - 3x + 2) = 0\}$.

17. In Figure 1.1.10, let

$S = $ the set of points inside the square
$T = $ the set of points inside the triangle
$C = $ the set of points inside the circle

and let $a, b,$ and c be the points shown. Answer the following as true or false.

(a) $T \subset C$ (b) $T \subset S$

(c) $a \notin T$ (d) $a \notin S$

(e) $b \in T$ and $b \in C$ (f) $a \in C$ or $a \in T$

(g) $c \in T$ and $c \notin C$.

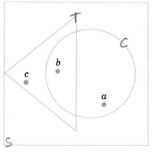

Figure 1.1.10

18. List all subsets of

(a) $\{a_1, a_2, a_3\}$ (b) $\varnothing$.

19. In each part, sketch on a coordinate line all values of x that satisfy the stated condition.

(a) $x \le 4$ (b) $x \ge -3$

(c) $-1 \le x \le 7$ (d) $x^2 = 9$

(e) $x^2 \le 9$ (f) $x^2 \ge 9$.

20. In parts (a)–(d), sketch on a coordinate line all values of x, if any, that satisfy the stated conditions.

(a) $x > 4$ and $x \le 8$
(b) $x \le 2$ or $x \ge 5$
(c) $x > -2$ and $x \ge 3$
(d) $x \le 5$ and $x > 7$.

21. Express in interval notation.

(a) $\{x : x^2 \le 4\}$ (b) $\{x : x^2 > 4\}$.

22. In each part, sketch the set on a coordinate line.

(a) $[-3, 2] \cup [1, 4]$ (b) $[4, 6] \cup [8, 11]$

(c) $(-4, 0) \cup (-5, 1)$ (d) $[2, 4) \cup (4, 7)$

(e) $(-2, 4) \cap (0, 5]$ (f) $[1, 2.3) \cup (1.4, \sqrt{2})$

(g) $(-\infty, -1) \cup (-3, +\infty)$

(h) $(-\infty, 5) \cap [0, +\infty)$.

In Exercises 23–44, solve the inequality and sketch the solution on a coordinate line.

23. $3x - 2 < 8$. **24.** $\frac{1}{5}x + 6 \ge 14$.

25. $4 + 5x \le 3x - 7$. **26.** $2x - 1 > 11x + 9$.

27. $3 \le 4 - 2x < 7$. **28.** $-2 \ge 3 - 8x \ge -11$.

29. $\frac{x}{x - 3} < 4$. **30.** $\frac{x}{8 - x} \ge -2$.

31. $\frac{3x + 1}{x - 2} < 1$. **32.** $\frac{\frac{1}{2}x - 3}{4 + x} > 1$.

33. $\frac{4}{2 - x} \le 1$. **34.** $\frac{3}{x - 5} \le 2$.

35. $x^2 > 9$. **36.** $x^2 \le 5$.

37. $(x - 4)(x + 2) > 0$. **38.** $(x - 3)(x + 4) < 0$.

39. $x^2 - 9x + 20 \le 0$. **40.** $2 - 3x + x^2 \ge 0$.

41. $\frac{2}{x} < \frac{3}{x - 4}$. **42.** $\frac{1}{x + 1} \ge \frac{3}{x - 2}$.

43. $x^3 - x^2 - x - 2 > 0$. **44.** $x^3 - 3x + 2 \le 0$.

In Exercises 45 and 46, find all values of x for which the given expression yields a real number.

45. $\sqrt{x^2 + x - 6}$.

46. $\sqrt{\dfrac{x + 2}{x - 1}}$.

47. Fahrenheit and Celsius temperatures are related by the formula $C = \frac{5}{9}(F - 32)$. If the temperature in degrees Celsius ranges over the interval $25 \leq C \leq 40$ on a certain day, what is the temperature range in degrees Fahrenheit that day?

48. Every integer is either even or odd. The even integers are those that are divisible by 2, so n is even if and only if $n = 2k$ for some integer k. Each odd integer is one unit larger than an even integer, so n is odd if and only if $n = 2k + 1$ for some integer k. Show:

(a) If n is even, then so is n^2.

(b) If n is odd, then so is n^2.

49. Prove the following results about sums of rational and irrational numbers:

(a) rational + rational = rational

(b) rational + irrational = irrational.

50. Prove the following results about products of rational and irrational numbers:

(a) rational · rational = rational

(b) rational · irrational = irrational   (provided the rational factor is nonzero).

51. Show that the sum or product of two irrational numbers can be rational or irrational.

52. Classify the following as rational or irrational and justify your conclusion.

(a) $3 + \pi$

(b) $\frac{3}{4}\sqrt{2}$

(c) $\sqrt{8}\sqrt{2}$

(d) $\sqrt{\pi}$.

(See Exercises 49 and 50.)

53. Prove: The average of two rational numbers is a rational number, but the average of two irrational numbers can be rational or irrational.

54. Can a rational number satisfy $10^x = 3$?

55. Solve: $8x^3 - 4x^2 - 2x + 1 < 0$.

56. Solve: $12x^3 - 20x^2 \geq -11x + 2$.

57. Prove: If a, b, c, and d are positive numbers such that $a < b$ and $c < d$, then $ac < bd$. (This result gives conditions under which inequalities can be "multiplied together.")

■ **1.2   ABSOLUTE VALUE**

In this section we shall review the notion of absolute value. This concept plays an important role in algebraic computations involving radicals and in determining the distance between points on a coordinate line.

☐ ABSOLUTE VALUE

1.2.1   DEFINITION.   The **absolute value** or **magnitude** of a real number a is denoted by $|a|$ and is defined by

$$|a| = \begin{cases} a & \text{if} \quad a \geq 0 \\ -a & \text{if} \quad a < 0 \end{cases}$$

Example 1

$$|5| = 5 \qquad \left|-\tfrac{4}{7}\right| = -\left(-\tfrac{4}{7}\right) = \tfrac{4}{7} \qquad |0| = 0 \qquad \blacktriangleleft$$

Since $5 > 0$ Since $-\tfrac{4}{7} < 0$ Since $0 \geq 0$

REMARK.   Note that the effect of taking the absolute value of a number is to strip away the minus sign if the number is negative and to leave the number unchanged if it is nonnegative. Thus, $|a|$ is a nonnegative number for all values of a and

$$-|a| \leq a \leq |a| \tag{1}$$

REMARK.   Symbols such as $+a$ and $-a$ are deceptive, since it is tempting to conclude that $+a$ is positive and $-a$ is negative. However, this need not be so, since a itself can represent either a positive or negative number. In fact, if a itself is negative, then $-a$ is positive and $+a$ is negative.

Example 2   Solve $|x - 3| = 4$.

Solution.   Depending on whether $x - 3$ is positive or negative, the equation $|x - 3| = 4$ can be written as

$$x - 3 = 4 \quad \text{or} \quad x - 3 = -4$$

Solving these two equations gives $x = 7$ and $x = -1$.   ◄

Example 3   Solve $|3x - 2| = |5x + 4|$.

Solution.   Because two numbers with the same absolute value are either equal or differ only in sign, the given equation will be satisfied if either

$$3x - 2 = 5x + 4 \quad \text{or} \quad 3x - 2 = -(5x + 4)$$

Solving the first equation yields $x = -3$ and solving the second yields $x = -\frac{1}{4}$ (verify); thus, the given equation has the solutions $x = -3$ and $x = -\frac{1}{4}$.   ◄

☐ **RELATIONSHIP BETWEEN SQUARE ROOTS AND ABSOLUTE VALUES**

Recall that a number whose square is a is called a *square root* of a. In algebra it is learned that every positive real number a has two real square roots, one positive and one negative. The positive square root is denoted by $\sqrt{a}$. For example, the number 9 has two square roots, -3 and 3. Since 3 is the positive square root, we have $\sqrt{9} = 3$. In addition, we define $\sqrt{0} = 0$.

REMARK.   Readers who were previously taught to write $\sqrt{9} = \pm 3$ should stop doing so, because it is incorrect.

It is a common error to write $\sqrt{a^2} = a$. Although this equality is correct when a is nonnegative, it is false for negative a. For example, if $a = -4$, then

$$\sqrt{a^2} = \sqrt{(-4)^2} = \sqrt{16} = 4 \neq a$$

A result that is correct for all a is given in the following theorem.

> **1.2.2** THEOREM.   *For any real number a,*
>
> $$\sqrt{a^2} = |a|$$

Proof.   Since $a^2 = (+a)^2 = (-a)^2$, the numbers $+a$ and $-a$ are square roots of a^2. If $a \geq 0$, then $+a$ is the nonnegative square root of a^2, and if $a < 0$, then $-a$ is the nonnegative square root of a^2. Since $\sqrt{a^2}$ denotes the nonnegative square root of a^2, we have

$$\sqrt{a^2} = +a \quad \text{if} \quad a \geq 0$$
$$\sqrt{a^2} = -a \quad \text{if} \quad a < 0$$

That is, $\sqrt{a^2} = |a|$.   ∎

☐ **PROPERTIES OF ABSOLUTE VALUE**

The next theorem, which lists some of the basic properties of absolute value, is based on Theorem 1.2.2 and properties of square roots.

> **1.2.3** THEOREM. *If a and b are real numbers, then*
>
> (a) $|-a| = |a|$ A number and its negative have the same absolute value.
> (b) $|ab| = |a|\,|b|$ The absolute value of a product is the product of the absolute values.
> (c) $|a/b| = |a|/|b|$ The absolute value of a ratio is the ratio of the absolute values.

We shall prove parts (*a*) and (*b*) only.

Proof (a). From Theorem 1.2.2

$$|-a| = \sqrt{(-a)^2} = \sqrt{a^2} = |a|$$

Proof (b). From Theorem 1.2.2 and a basic property of square roots

$$|ab| = \sqrt{(ab)^2} = \sqrt{a^2 b^2} = \sqrt{a^2}\sqrt{b^2} = |a|\,|b| \quad\blacksquare$$

REMARK.   In part (*c*) of Theorem 1.2.3 we did not explicitly state that $b \neq 0$, but this must be so since division by zero is not allowed. Whenever divisions occur in this text, it will be assumed that the denominator is not zero, even if we do not mention it explicitly.

The result in part (*b*) of Theorem 1.2.3 can be extended to three or more factors. More precisely, for any *n* real numbers, $a_1, a_2, \ldots, a_n$, it follows that

$$|a_1 a_2 \cdots a_n| = |a_1|\,|a_2| \cdots |a_n| \tag{2}$$

In the special case where $a_1, a_2, \ldots, a_n$ have the same value, *a*, it follows from (2) that

$$|a^n| = |a|^n \tag{3}$$

☐ **GEOMETRIC INTERPRETATION OF ABSOLUTE VALUE**

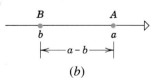

(a)

(b)

Figure 1.2.1

The notion of absolute value arises naturally in distance problems. On a coordinate line, let *A* and *B* be points with coordinates *a* and *b*. Because distance is nonnegative, the distance *d* between *A* and *B* is

$$d = \begin{cases} b - a & \text{if} \quad a < b \\ a - b & \text{if} \quad a > b \\ 0 & \text{if} \quad a = b \end{cases}$$

(Figure 1.2.1). In the first case $b - a$ is positive, so $b - a = |b - a|$; in the second case $b - a$ is negative, so $a - b = -(b - a) = |b - a|$. Thus, in all cases we have the following result:

> **1.2.4** THEOREM (***Distance Formula***). *If A and B are points on a coordinate line with coordinates a and b, respectively, then the distance d between A and B is*
>
> $$d = |b - a| \tag{4}$$

Formula (4) provides a useful geometric interpretation of some common mathematical expressions given in Table 1.2.1.

Table 1.2.1

EXPRESSION	GEOMETRIC INTERPRETATION ON A COORDINATE LINE
$|x - a|$	The distance between x and a
$|x + a|$	The distance between x and $-a$ (since $|x + a| = |x - (-a)|$)
$|x|$	The distance between x and the origin (since $|x| = |x - 0|$)

Inequalities of the forms $|x - a| < k$ and $|x - a| > k$ arise often, so we have summarized the key facts about them in Table 1.2.2.

Table 1.2.2

INEQUALITY ($k > 0$)	GEOMETRIC INTERPRETATION	FIGURE	ALTERNATIVE FORM OF THE INEQUALITY	SOLUTION SET
$|x - a| < k$	x is within k units of a.	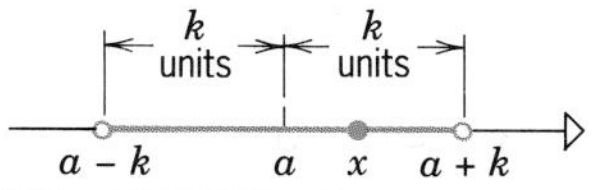 $\xleftarrow{\ k\ \text{units}\ }\xrightarrow{\ k\ \text{units}\ }$ $\quad a - k \quad a \quad x \quad a + k$	$-k < x - a < k$	$(a - k, a + k)$
$|x - a| > k$	x is more than k units away from a.	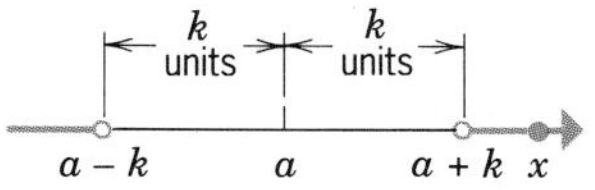 $\xleftarrow{\ k\ \text{units}\ }\xrightarrow{\ k\ \text{units}\ }$ $\quad a - k \quad a \quad a + k \ x$	$\begin{cases} x - a < -k \\ \quad \text{or} \\ x - a > k \end{cases}$	$(-\infty, a - k) \cup (a + k, +\infty)$

REMARK.   In this table we can replace $<$ by $\leq$ and $>$ by $\geq$ if we replace the open dots by closed dots in the illustrations.

Example 4   Solve $|x - 3| < 4$.

Solution.   This inequality can be rewritten as

$$-4 < x - 3 < 4$$

or, on adding 3 throughout,

$$-1 < x < 7$$

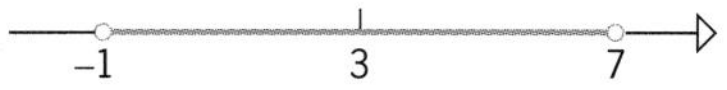

This can be written in interval notation as $(-1, 7)$. (See Figure 1.2.2.)   ◄

Figure 1.2.2

Example 5   Solve $|x + 4| \geq 2$.

Solution.   The given inequality can be rewritten as

$$\begin{cases} x + 4 \leq -2 \\ \quad \text{or} \\ x + 4 \geq 2 \end{cases} \quad \text{or more simply} \quad \begin{cases} x \leq -6 \\ \quad \text{or} \\ x \geq -2 \end{cases}$$

which can be written in set notation as

$$(-\infty, -6] \cup [-2, +\infty)$$

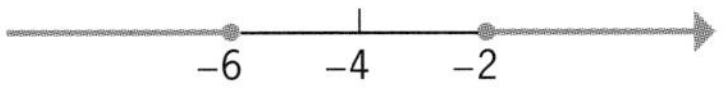

(See Figure 1.2.3.)   ◄

Figure 1.2.3

REMARK.   The problems in the last two examples were solved algebraically. However, we could also have proceeded geometrically. For example, the solution of $|x - 3| < 4$ consists of all x whose distance from 3 is less than 4 units (Figure 1.2.2), and the solution of $|x + 4| \geq 2$ [which can be rewritten as $|x - (-4)| \geq 2$] consists of all x whose distance from -4 is 2 units or more (Figure 1.2.3).

Example 6   Solve $\dfrac{1}{|2x - 3|} > 5$.

Solution.   Observe first that $x = \frac{3}{2}$ is not a solution because this value of x results in a division by zero. Keeping this in mind, we shall begin by applying Theorem 1.1.1(e). Taking reciprocals and reversing the inequality yields

$$|2x - 3| < \tfrac{1}{5}$$

$$|2(x - \tfrac{3}{2})| < \tfrac{1}{5} \quad \boxed{\text{Factor out the coefficient of } x.}$$

$$|2| \, |x - \tfrac{3}{2}| < \tfrac{1}{5} \quad \boxed{\text{Theorem 1.2.3}(b)}$$

$$|x - \tfrac{3}{2}| < \tfrac{1}{10} \quad \boxed{\text{We multiplied both sides by } 1/|2| = 1/2.}$$

$$-\tfrac{1}{10} < x - \tfrac{3}{2} < \tfrac{1}{10} \quad \boxed{\text{Table 1.2.2}}$$

$$\tfrac{7}{5} < x < \tfrac{8}{5} \quad \boxed{\text{We added 3/2 throughout.}}$$

If, as noted above, we eliminate the value $x = \tfrac{3}{2}$ to avoid the division by zero, we see that the solution consists of all x that satisfy

$$\tfrac{7}{5} < x < \tfrac{3}{2} \quad \text{or} \quad \tfrac{3}{2} < x < \tfrac{8}{5}$$

The solution set consists of all x in the set $(\tfrac{7}{5}, \tfrac{3}{2}) \cup (\tfrac{3}{2}, \tfrac{8}{5})$. (See Figure 1.2.4.) ◄

$$\frac{7}{5} \qquad \frac{3}{2} \qquad \frac{8}{5}$$

Figure 1.2.4

☐ **AN INEQUALITY FROM CALCULUS**

The inequality in the following example arises in calculus.

Example 7   Solve

$$0 < |x - a| < \delta \tag{5}$$

where a is any real number and δ (Greek "delta") is a positive real number.

Solution.   The solution set consists of all real values of x that satisfy the two inequalities

$$0 < |x - a| \quad \text{and} \quad |x - a| < \delta$$

The solutions of $|x - a| < \delta$ are those values of x such that

$$a - \delta < x < a + \delta$$

or in interval notation

$$(a - \delta, a + \delta) \tag{6}$$

The inequality $0 < |x - a|$ is satisfied by all real values of x except $x = a$ (why?). Thus, the solution set of $0 < |x - a| < \delta$ is the interval in (6) with the point a removed. This is the set $(a - \delta, a) \cup (a, a + \delta)$ shown in Figure 1.2.5. ◄

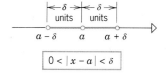

$$\underset{a - \delta}{|} \xleftarrow[\text{units}]{\delta} \underset{a}{|} \xrightarrow[\text{units}]{\delta} \underset{a + \delta}{|}$$

$$\boxed{0 < |x - a| < \delta}$$

Figure 1.2.5

☐ **THE TRIANGLE INEQUALITY**

It is *not* generally true that $|a + b| = |a| + |b|$. For example, if $a = 2$ and $b = -3$, then $a + b = -1$, so that

$$|a + b| = |-1| = 1$$

whereas

$$|a| + |b| = |2| + |-3| = 2 + 3 = 5$$

so $|a + b| \neq |a| + |b|$. It is true, however, that *the absolute value of a sum is always less than or equal to the sum of the absolute values.* This is the content of the following very important theorem, known as the ***triangle inequality***.

1.2.5   THEOREM (***Triangle Inequality***).   *If a and b are any real numbers, then*

$$|a + b| \le |a| + |b| \tag{7}$$

Proof.   From (1)

$$-|a| \le a \le |a| \quad \text{and} \quad -|b| \le b \le |b|$$

Adding these inequalities yields

$$-(|a| + |b|) \le a + b \le (|a| + |b|) \tag{8}$$

We now consider the two cases: $a + b \ge 0$ and $a + b < 0$. In the first of these cases $a + b = |a + b|$, so the right-hand inequality in (8) yields (7). In the second case $a + b = -|a + b|$, so the left-hand inequality in (8) can be written as

$$-(|a| + |b|) \le -|a + b|$$

Multiplying both sides of this inequality by -1 yields (7). ∎

Various applications of the triangle inequality will occur throughout this text.

► Exercise Set 1.2

1. Compute $|x|$ if

(a) $x = 7$ (b) $x = -\sqrt{2}$

(c) $x = k^2$ (d) $x = -k^2$.

2. Rewrite $\sqrt{(x - 6)^2}$ without using a square root or absolute value sign.

In Exercises 3–10, find all values of x for which the given statement is true.

3. $|x - 3| = 3 - x$.   **4.** $|x + 2| = x + 2$.

5. $|x^2 + 9| = x^2 + 9$.   **6.** $|x^2 + 5x| = x^2 + 5x$.

7. $|3x^2 + 2x| = x|3x + 2|$.

8. $|6 - 2x| = 2|x - 3|$.

9. $\sqrt{(x + 5)^2} = x + 5$.

10. $\sqrt{(3x - 2)^2} = 2 - 3x$.

11. Verify $\sqrt{a^2} = |a|$ for $a = 7$ and $a = -7$.

12. Verify the inequalities $-|a| \le a \le |a|$ for $a = 2$ and $a = -5$.

13. Let A and B be points with coordinates a and b. In each part find the distance between A and B.

(a) $a = 9, b = 7$ (b) $a = 2, b = 3$

(c) $a = -8, b = 6$ (d) $a = \sqrt{2}, b = -3$

(e) $a = -11, b = -4$ (f) $a = 0, b = -5$.

14. Is the equality $\sqrt{a^4} = a^2$ valid for all values of a? Explain.

15. Let A and B be points with coordinates a and b. In each part, use the given information to find b.

(a) $a = -3$, B is to the left of A, and $|b - a| = 6$

(b) $a = -2$, B is to the right of A, and $|b - a| = 9$

(c) $a = 5$, $|b - a| = 7$, and $b > 0$.

16. Let E and F be points with coordinates e and f. In each part, determine whether E is to the left or to the right of F on a coordinate line.

(a) $f - e = 4$ (b) $e - f = 4$

(c) $f - e = -6$ (d) $e - f = -7$.

In Exercises 17–24, solve for x.

17. $|6x - 2| = 7$.   **18.** $|3 + 2x| = 11$.

19. $|6x - 7| = |3 + 2x|$.   **20.** $|4x + 5| = |8x - 3|$.

21. $|9x| - 11 = x$.   **22.** $2x - 7 = |x + 1|$.

23. $\left|\dfrac{x + 5}{2 - x}\right| = 6$.   **24.** $\left|\dfrac{x - 3}{x + 4}\right| = 5$.

In Exercises 25–36, solve for x and express the solution in terms of intervals.

25. $|x + 6| < 3$.   **26.** $|7 - x| \le 5$.

27. $|2x - 3| \le 6$.   **28.** $|3x + 1| < 4$.

29. $|x + 2| > 1$.   **30.** $|\frac{1}{2}x - 1| \ge 2$.

31. $|5 - 2x| \ge 4$.   **32.** $|7x + 1| > 3$.

33. $\dfrac{1}{|x - 1|} < 2$.   **34.** $\dfrac{1}{|3x + 1|} \ge 5$.

35. $\dfrac{3}{|2x - 1|} \ge 4$.   **36.** $\dfrac{2}{|x + 3|} < 1$.

37. For which values of x is $\sqrt{(x^2 - 5x + 6)^2} = x^2 - 5x + 6$?

38. Solve $3 \le |x - 2| \le 7$ for x.

39. Solve $|x - 3|^2 - 4|x - 3| = 12$ for x. [*Hint:* First let $u = |x - 3|$.]

40. Verify the triangle inequality $|a + b| \le |a| + |b|$ (Theorem 1.2.5) for

(a) $a = 3, b = 4$ (b) $a = -2, b = 6$

(c) $a = -7, b = -8$ (d) $a = -4, b = 4$.

41. Prove: $|a - b| \le |a| + |b|$.

42. Prove: $|a| - |b| \le |a - b|$.

43. Prove: $||a| - |b|| \le |a - b|$. [*Hint:* Use Exercise 42.]

■ **1.3** COORDINATE PLANES AND GRAPHS

Just as points on a line can be placed in one-to-one correspondence with the real numbers, so points in the plane can be placed in one-to-one correspondence with pairs of real numbers. This fundamental observation will enable us to visualize algebraic equations as geometric curves and, conversely, to represent geometric curves by algebraic equations.

☐ **RECTANGULAR COORDINATE SYSTEMS**

A *rectangular coordinate system* (also called a *Cartesian* coordinate system*) is a pair of perpendicular coordinate lines, called *coordinate axes*, which are placed so that they intersect at their origins. Usually, but not always, one of the lines is horizontal with its positive direction to the right, and the other is vertical with its positive direction up (Figure 1.3.1). As illustrated in this figure, the intersection of the coordinate axes is called the *origin* of the coordinate system. The horizontal axis is usually called the *x-axis* and the vertical axis is called the *y-axis*. A plane in which a rectangular coordinate system has been introduced is called a *coordinate plane*.

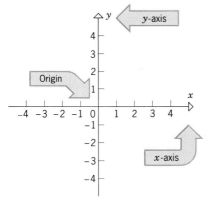

Figure 1.3.1

Labeling the axes with letters x and y is a common convention, but any letters may be used. If the letters x and y are used to label the coordinate axes, then the resulting plane is called an *xy-plane*. In applications it is common to use letters other than x and y to label coordinate axes. Figure 1.3.2 shows a uv-plane and a ts-plane. The first letter in the name of the plane refers to the horizontal axis and the second to the vertical axis.

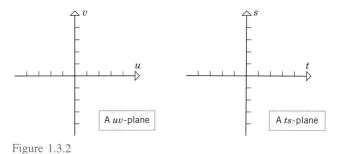

Figure 1.3.2

---

*See p. 4 for footnote on René Descartes.

□ **COORDINATES**

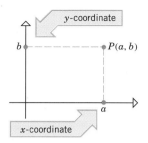

Figure 1.3.3

We shall now show how to establish a one-to-one correspondence between points in a coordinate plane and "ordered pairs" of real numbers. By an ***ordered pair*** of real numbers we mean two real numbers in an assigned order. Thus, there is a "first number" and a "second number." The symbol (a, b) is used to denote the ordered pair of real numbers in which a is the first number and b is the second number. Because order matters, the ordered pairs $(2, -3)$ and $(-3, 2)$ are regarded to be different.

REMARK.   Recall that the symbol (a, b) also denotes the open interval between a and b. The appropriate interpretation will usually be clear from the context.

Every point P in a coordinate plane can be associated with a unique ordered pair of real numbers by drawing two lines through P, one perpendicular to the x-axis and the other perpendicular to the y-axis (Figure 1.3.3). If the first line intersects the x-axis at the point with coordinate a and the second line intersects the y-axis at the point with coordinate b, then we associate the ordered pair of real numbers (a, b) with the point P. The number a is called the ***x-coordinate*** or ***abscissa*** of P and the number b is called the ***y-coordinate*** or ***ordinate*** of P. We shall say that P has ***coordinates*** (a, b) and write $P(a, b)$ when we want to emphasize that the coordinates of P are (a, b).

It is evident that the construction process illustrated in Figure 1.3.3 associates a unique ordered pair of real numbers (a, b) with each point P in the plane. Conversely, if we *start* with an ordered pair of real numbers (a, b) and construct lines perpendicular to the x-axis and y-axis that pass through the points with coordinates a and b, respectively, then these lines intersect at a unique point P in the plane whose coordinates are (a, b). Thus, we have a one-to-one correspondence between ordered pairs of real numbers and points in a coordinate plane.

To ***plot*** a point $P(a, b)$ means to locate the point with coordinates (a, b) in a coordinate plane. For example, in Figure 1.3.4 we have plotted the points

$$P(2, 5), \quad Q(-4, 3), \quad R(-5, -2), \quad \text{and} \quad S(4, -3)$$

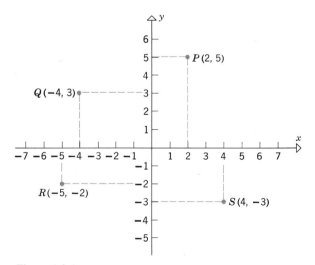

Figure 1.3.4

In a rectangular coordinate system the coordinate axes divide the plane into four regions called ***quadrants***. These are numbered counterclockwise with roman numerals as shown in Figure 1.3.5. As illustrated in Figure 1.3.6, it is easy to determine the quadrant in which a given point lies from the signs of its coordinates: a point with two positive coordinates $(+, +)$ lies in Quadrant I, a point with a negative x-coordinate and a positive y-coordinate $(-, +)$ lies in Quadrant II, and so forth. Points with a zero x-coordinate lie on the y-axis and points with a zero y-coordinate lie on the x-axis.

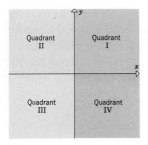

Figure 1.3.5

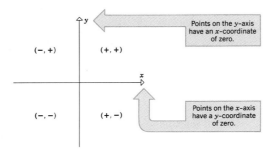

Figure 1.3.6

□ **GRAPHS**

To see how rectangular coordinate systems enable us to describe equations geometrically, assume that we have constructed a rectangular coordinate system and are given an equation involving only two variables, x and y, such as

$$5xy = 2, \quad x^2 + 2y^2 = 7, \quad \text{or} \quad y = x^3 - 7$$

We define a ***solution*** of such an equation to be an ordered pair of real numbers (a, b) such that the equation is satisfied when we substitute $x = a$ and $y = b$. The set of all solutions is called the ***solution set*** of the equation.

Example 1   The pair $(3, 2)$ is a solution of $6x - 4y = 10$ since this equation is satisfied when we substitute $x = 3$ and $y = 2$. However, the pair $(2, 0)$ is not a solution, since the equation is not satisfied when we substitute $x = 2$ and $y = 0$.

We make the following definition.

> **1.3.1**   DEFINITION.   The ***graph*** of an equation in two variables x and y is the set of all points in the xy-plane whose coordinates are members of the solution set of the equation.

x	$y = x^2$	(x, y)
0	0	(0, 0)
1	1	(1, 1)
2	4	(2, 4)
3	9	(3, 9)
−1	1	(−1, 1)
−2	4	(−2, 4)
−3	9	(−3, 9)

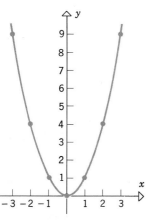

Figure 1.3.7

Example 2   Sketch the graph of $y = x^2$.

Solution.   The solution set of $y = x^2$ has infinitely many members, so that it is impossible to plot them all. However, some sample members of the solution set can be obtained by substituting some arbitrary x values into the right side of $y = x^2$ and solving for the associated values of y.

In Figure 1.3.7, some typical computations are given in the table. The accompanying curve was obtained by plotting the points in the table and connecting them with a smooth curve. This curve approximates the graph of $y = x^2$.   ◄

REMARK.   It should be kept in mind that the curve in Figure 1.3.7 is only an *approximation* to the graph of $y = x^2$. When a graph is obtained by plotting points, whether by hand, calculator, or computer, there is no guarantee that the resulting curve has the correct shape. For example, the two curves in Figure 1.3.8 both pass through the points tabulated in Figure 1.3.7. Moreover, we cannot totally resolve the problem by plotting more points since we will never be sure how the true graph behaves *between* the plotted points. In general, it is only with techniques from calculus that the true shape of a graph can be ascertained.

Example 3   Sketch the graph of $y = x^3$.

Solution.   A table of points and the graph are given in Figure 1.3.9.   ◄

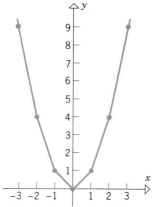

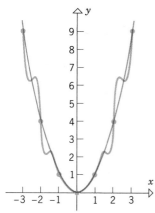

Figure 1.3.8

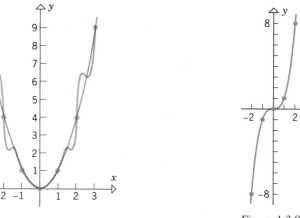

Figure 1.3.9

Example 4 Sketch the graph of $y^2 - 2y - x = 0$.

Solution. To calculate coordinates of points on the graph of an equation in x and y, it is desirable to have y expressed in terms of x or x in terms of y. In this case it is easier to express x in terms of y. We obtain

$$x = y^2 - 2y$$

Some sample members of the solution set can now be obtained by substituting arbitrary y values into the right side of the above equation and computing the associated x values (Figure 1.3.10).   ◄

Example 5 Sketch the graph of $y = 1/x$.

Solution. Because $1/x$ is undefined when $x = 0$, we can plot only points for which $x \neq 0$. This forces a break (commonly called a *discontinuity*) in the graph at $x = 0$ (Figure 1.3.11).   ◄

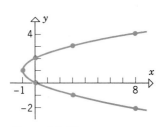

Figure 1.3.10

y	$x = y^2 - 2y$	(x, y)
-2	8	(8, -2)
-1	3	(3, -1)
0	0	(0, 0)
1	-1	(-1, 1)
2	0	(0, 2)
3	3	(3, 3)
4	8	(8, 4)

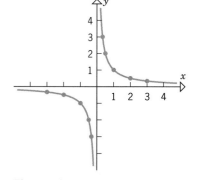

Figure 1.3.11

x	$y = 1/x$	(x, y)
$\frac{1}{3}$	3	$(\frac{1}{3}, 3)$
$\frac{1}{2}$	2	$(\frac{1}{2}, 2)$
1	1	(1, 1)
2	$\frac{1}{2}$	$(2, \frac{1}{2})$
3	$\frac{1}{3}$	$(3, \frac{1}{3})$
$-\frac{1}{3}$	-3	$(-\frac{1}{3}, -3)$
$-\frac{1}{2}$	-2	$(-\frac{1}{2}, -2)$
-1	-1	(-1, -1)
-2	$-\frac{1}{2}$	$(-2, -\frac{1}{2})$
-3	$-\frac{1}{3}$	$(-3, -\frac{1}{3})$

Example 6 Sketch the graph of $y = \sqrt{x}$.

Solution. If $x < 0$, then $\sqrt{x}$ is a complex number. Since the coordinates of points in the xy-plane are real numbers, we can only plot those points for which $x \geq 0$. With the help of a calculator we obtained the table and graph in Figure 1.3.12.   ◄

☐ **INTERCEPTS**

Points where a graph intersects the coordinate axes are of special interest in many problems. As illustrated in Figure 1.3.13, intersections of a graph with the x-axis have the form $(a, 0)$ and intersections with the y-axis have the form $(0, b)$. The number a is called an ***x-intercept*** of the graph and the number b a ***y-intercept***.

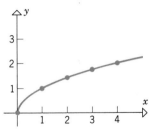

Figure 1.3.12

x	$y = \sqrt{x}$	(x, y)
0	0	$(0, 0)$
1	1	$(1, 1)$
2	$\sqrt{2}$	$(2, \sqrt{2}) \approx (2, 1.4)$
3	$\sqrt{3}$	$(3, \sqrt{3}) \approx (3, 1.7)$
4	2	$(4, 2)$

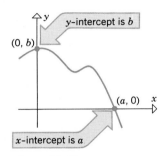

Figure 1.3.13

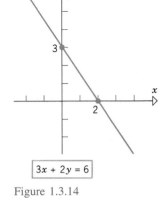

$3x + 2y = 6$

Figure 1.3.14

Example 7   Find all intercepts of

(a) $3x + 2y = 6$ (b) $x = y^2 - 2y$ (c) $y = 1/x$

Solution (a).   To find the x-intercepts we set $y = 0$ and solve for x:

$$3x = 6 \quad \text{or} \quad x = 2$$

To find the y-intercepts we set $x = 0$ and solve for y:

$$2y = 6 \quad \text{or} \quad y = 3$$

As we shall see later, the graph of $3x + 2y = 6$ is the line shown in Figure 1.3.14.

Solution (b).   To find the x-intercepts, set $y = 0$ and solve for x:

$$x = 0$$

Thus, $x = 0$ is the only x-intercept. To find the y-intercepts, set $x = 0$ and solve for y:

$$y^2 - 2y = 0$$

$$y(y - 2) = 0$$

so the y-intercepts are $y = 0$ and $y = 2$. The graph is shown in Figure 1.3.10.

Solution (c).   To find the x-intercepts, set $y = 0$:

$$\frac{1}{x} = 0$$

This equation has no solutions (why?), so there are no x-intercepts. To find y-intercepts we would set $x = 0$ and solve for y. But, substituting $x = 0$ leads to a division by zero, which is not allowed, so there are no y-intercepts either. The graph of the equation is shown in Figure 1.3.11.   ◄

□ **SYMMETRY**

As illustrated in Figure 1.3.15, the points (x, y), $(-x, y)$, $(x, -y)$, and $(-x, -y)$ form the corners of a rectangle. For obvious reasons, the points (x, y) and $(x, -y)$ are said to be *symmetric about the x-axis*, the points (x, y) and $(-x, y)$ *symmetric about the y-axis*, and the points (x, y) and $(-x, -y)$ *symmetric about the origin*.

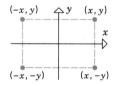

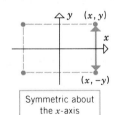

Symmetric about
the x-axis

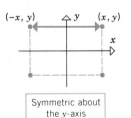

Symmetric about
the y-axis

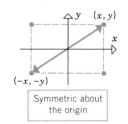

Symmetric about
the origin

Figure 1.3.15

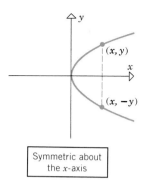

Symmetric about
the x-axis

(a)

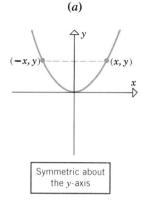

Symmetric about
the y-axis

(b)

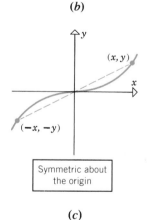

Symmetric about
the origin

(c)

Figure 1.3.16

Analogously, we say that a graph in the xy-plane is

- *symmetric about the x-axis* if for each point (x, y) on the graph the point $(x, -y)$ is also on the graph (Figure 1.3.16a);
- *symmetric about the y-axis* if for each point (x, y) on the graph the point $(-x, y)$ is also on the graph (Figure 1.3.16b);
- *symmetric about the origin* if for each point (x, y) on the graph the point $(-x, -y)$ is also on the graph (Figure 1.3.16c).

Geometrically, a graph is symmetric about the x-axis if folding the plane along the x-axis makes the upper and lower portions of the graph align exactly, and it is symmetric about the y-axis if folding the plane along the y-axis makes the left and right portions of the graph align exactly. A graph is symmetric about the origin if rotating the graph 180° about the origin results in the original graph.

Symmetries can often be detected from the equation of a curve. For example, the graph of

$$y = x^3 \qquad (1)$$

must be symmetric about the origin (see Figure 1.3.9) because for any point (x, y) whose coordinates satisfy (1), the coordinates of the point $(-x, -y)$ also satisfy (1), since substituting these values in (1) yields

$$-y = (-x)^3$$

which simplifies to (1). This discussion suggests the following symmetry tests.

1.3.2 THEOREM (*Symmetry Tests*).

(a) *A plane curve is symmetric about the y-axis if replacing x by −x in its equation produces an equivalent equation.*

(b) *A plane curve is symmetric about the x-axis if replacing y by −y in its equation produces an equivalent equation.*

(c) *A plane curve is symmetric about the origin if replacing both x by −x and y by −y in its equation produces an equivalent equation.*

Example 8 Figure 1.3.7 suggests that the graph of $y = x^2$ is symmetric about the y-axis, but not about the x-axis or the origin. Use Theorem 1.3.2 to verify that this is so.

Solution. To test for symmetry about the y-axis, we substitute $-x$ for x in $y = x^2$, which yields

$$y = (-x)^2$$

This is equivalent to the original equation since squaring the right side yields $y = x^2$. Thus, the graph is symmetric about the y-axis.

To test for symmetry about the x-axis, we substitute $-y$ for y in $y = x^2$, which yields

$$-y = x^2$$

This is not equivalent to $y = x^2$, so the graph is not symmetric about the x-axis.

To test for symmetry about the origin, we substitute $-x$ for x and $-y$ for y in $y = x^2$, which yields

$$-y = (-x)^2 \quad \text{or equivalently} \quad -y = x^2$$

The last equation is not equivalent to $y = x^2$, so the graph is not symmetric about the origin. ◄

□ **SYMMETRY AS A TOOL FOR GRAPHING**

By taking advantage of symmetries when they exist, the work required to obtain a graph can be reduced considerably.

Example 9   Sketch the graph of the equation $y = \frac{1}{8}x^4 - x^2$.

Solution.   The graph is symmetric about the y-axis since substituting $-x$ for x yields

$$y = \frac{1}{8}(-x)^4 - (-x)^2$$

which simplifies to the original equation.

As a consequence of this symmetry, we need only calculate points on the graph that lie in the right half of the xy-plane ($x \geq 0$). The corresponding points in the left half of the xy-plane ($x \leq 0$) can be obtained with no additional computation by using the symmetry. The top part of the table in Figure 1.3.17 was obtained with the help of a calculator. The bottom part was obtained with no computation using the symmetry.   ◄

Example 10   Sketch the graph of $x = y^2$.

Solution.   If we solve $x = y^2$ for y in terms of x, we obtain two solutions, $y = \sqrt{x}$ and $y = -\sqrt{x}$. The graph of $y = \sqrt{x}$ is the portion of the curve $x = y^2$ that lies above or touches the x-axis (since $y = \sqrt{x} \geq 0$), and the graph of $y = -\sqrt{x}$ is the portion that lies below or touches the x-axis (since $y = -\sqrt{x} \leq 0$). However, the curve $x = y^2$ is symmetric about the x-axis because substituting $-y$ for y yields $x = (-y)^2$, which is equivalent to the original equation. Thus, we need only graph $y = \sqrt{x}$ (see Figure 1.3.12) and then reflect it about the x-axis to complete the graph (Figure 1.3.18).   ◄

x	$y = \frac{1}{8}x^4 - x^2$
0	0
0.5	−0.24
1	−0.88
1.5	−1.62
2	−2
2.5	−1.37
3	1.12
3.5	6.51
4	16
−4	16
−3.5	6.51
−3	1.12
−2.5	−1.37
−2	−2
−1.5	−1.62
−1	−0.88
−0.5	−0.24

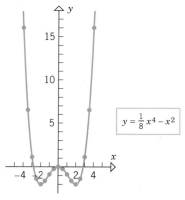

Figure 1.3.17

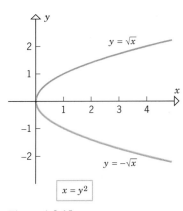

Figure 1.3.18

□ **GRAPHS WITH UNEQUAL SCALES**

In all of our graphs thus far we have used equal scales on the coordinate axes, that is, one unit along the x-axis had the same length as one unit along the y-axis. However, sometimes it is necessary to use unequal scales. For example, to graph $y = x^3$ for values of x between -10 and 10, we would have to allow for y values between $(-10)^3 = -1000$ and $10^3 = 1000$, which is difficult to do within the confines of a standard sheet of paper or printed page; the only solution is to use unequal scales. Although unequal scales distort the graph, the general shape is usually maintained, which is adequate for many purposes. Figure 1.3.19 shows two versions of the graph of $y = x^3$.

In applications where the variables in an equation represent physical quantities, there is usually no reason to use equal scales on the coordinate axes when plotting a graph. However, the units of measure should be indicated on the coordinate axes for clarity. For example, Figure 1.3.20 is a graph of T versus R, where T is the time (in thousands of days) for an object orbiting about the sun to make one revolution and R is the mean distance from the object to the sun (in billions of kilometers).

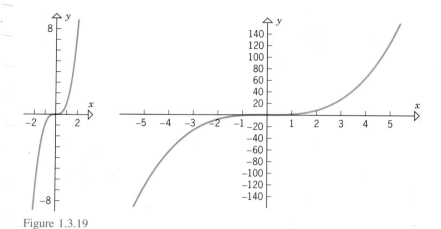

Figure 1.3.19

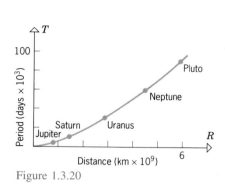

Figure 1.3.20

□ **GRAPHING BY CALCULATOR AND COMPUTER**

Laborious hand-plotting of graphs is rapidly becoming a thing of the past since computers and hand calculators with graphics capabilities can generate complex graphs in a matter of seconds. Although you may have to do a certain amount of hand-plotting as you work through this text, you should keep in mind that in practical applications computers are available to do such work. Thus, in this text the primary focus will be on learning to understand the kind of information that a graph conveys rather than on the numerical process of generating graphs.

□ **A CATALOG OF BASIC GRAPHS**

We conclude this section with a catalog of graphs that arise frequently (Figure 1.3.21). You will find it helpful to be familiar with the general shapes of these graphs without resorting to point-plotting.

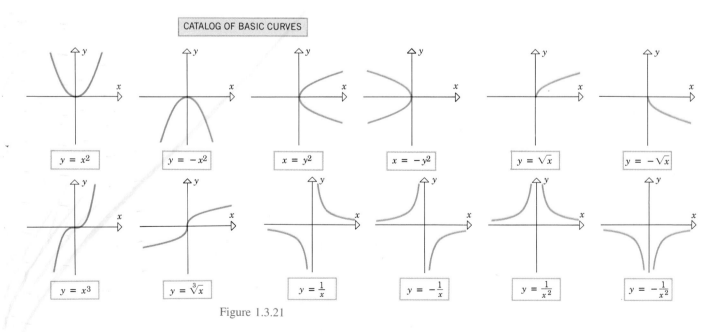

CATALOG OF BASIC CURVES

Figure 1.3.21

▶ Exercise Set 1.3

1. Draw a rectangular coordinate system and locate the points

   (a) $(3, 4)$ (b) $(-2, 5)$ (c) $(-2.5, -3)$ (d) $(1.7, -2)$ (e) $(0, -6)$ (f) $(4, 0)$.

In Exercises 2 and 3, draw a rectangular coordinate system and sketch the set of points whose coordinates (x, y) satisfy the given conditions.

2. (a) $x = 0$ (b) $y = 0$
    (c) $y < 0$ (d) $x \geq 1$ and $y \leq 2$
    (e) $x = 3$ (f) $|x| = 5$.

3. (a) $x = 2$ (b) $y = -3$
    (c) $x \geq 0$ (d) $y = x$
    (e) $y \geq x$ (f) $|x| \geq 1$.

In Exercises 4 and 5, the points lie on a horizontal or vertical line. Determine whether the line is horizontal or vertical.

4. (a) $A(9, 2), B(7, 2)$
    (b) $A(2, -6), B(3, -6)$
    (c) $A(6, 6), B(6, 1)$.

5. (a) $A(-4, \sqrt{2}), B(-4, -3)$
    (b) $A(0, -4), B(3, -4)$
    (c) $A(0, 0), B(0, -5)$.

6. Find the fourth vertex of the rectangle, three of whose vertices are $(-1, 4)$, $(6, 4)$, and $(-1, 9)$.

7. In each part determine if the given ordered pair (x, y) is a solution of $x^2 - 2x + y = 4$.

    (a) $(0, 4)$ (b) $(-3, 7)$
    (c) $\left(\frac{1}{2}, \frac{19}{4}\right)$ (d) $(1 + \sqrt{5 - t}, t)$.

8. Figure 1.3.22 shows only a portion of the graph of an equation in the variables x and y for $x \geq 0$ and $y \geq 0$. Sketch the complete graph if it is known to be

    (a)   symmetric about the x-axis
    (b)   symmetric about the y-axis
    (c)   symmetric about the origin.

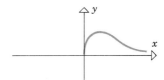

Figure 1.3.22

9. In parts (a)–(d) of Figure 1.3.23, the graphs of equations in the variables x and y are shown. In each case determine whether the graph is symmetric about the x-axis, y-axis, origin, or none of the preceding.

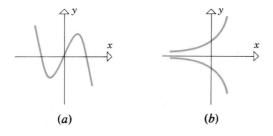

(a)   (b)

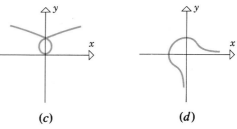

(c)   (d)

Figure 1.3.23

10. Use Theorem 1.3.2 to show that a graph which is symmetric about the x-axis and y-axis must be symmetric about the origin. Give an example to show that the converse is not true.

11. Which of the curves shown in Figure 1.3.21 have symmetry about

    (a)   the x-axis
    (b)   the y-axis
    (c)   the origin?

In Exercises 12 and 13, determine whether the graph is symmetric about the x-axis, the y-axis, or the origin.

12. (a) $x = 5y^2 + 9$
    (b) $x^2 - 2y^2 = 3$
    (c) $xy = 5$.

13. (a) $x^4 = 2y^3 + y$
    (b) $y = \dfrac{x}{3 + x^2}$
    (c) $y^2 = |x| - 5$.

In Exercises 14–23, sketch the graph of the equation. (A calculator will be helpful in some of these problems.)

14. $y = 2x - 3$.   15. $y = 6 - x$.
16. $y = 1 + x^2$.   17. $y = 4 - x^2$.
18. $y = -\sqrt{x + 1}$.   19. $y = \sqrt{x - 4}$.
20. $y = |x|$.   21. $y = |x - 3|$.
22. $xy = -1$.   23. $x^2y = 2$.

In Exercises 24 and 25, sketch the portion of the graph in the first quadrant, and use symmetry to complete the rest of the graph. (A hand calculator will be helpful.)

24. $9x^2 + 4y^2 = 36$.

25. $4x^2 + 16y^2 = 16$.

26. Sketch the graph of $y^2 = 3x$ and explain how this graph is related to the graphs of $y = \sqrt{3x}$ and $y = -\sqrt{3x}$.

27. Sketch the graph of $(x - y)(x + y) = 0$ and explain how it is related to the graphs of $x - y = 0$ and $x + y = 0$.

28. Graph $F = \frac{9}{5}C + 32$ in a CF-coordinate system.

29. Graph $u = 3v^2$ in a uv-coordinate system.

30. Graph $Y = 4X + 5$ in a YX-coordinate system.

1.4 LINES

In this section we shall discuss ways to measure the "steepness" or "slope" of a line in the plane. The ideas we develop here will be important when we discuss equations and graphs of straight lines.

This section assumes a knowledge of the trigonometry material in Section I of Appendix B. Readers who need to review that material are advised to do so before starting this section.

☐ SLOPE

In surveying, the *grade* or *slope* of a hill is defined to be the ratio of its *rise* to its *run* (Figure 1.4.1). We shall now show how the surveyor's notion of slope can be adapted to measure the steepness of a line in the *xy*-plane.

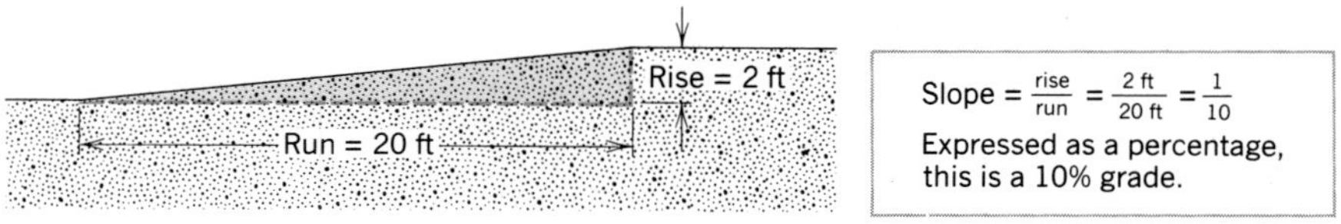

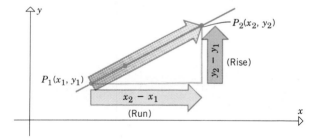

Figure 1.4.1

Consider a particle moving left to right along a *nonvertical* line segment from a point $P_1(x_1, y_1)$ to a point $P_2(x_2, y_2)$. As shown in Figure 1.4.2, the particle moves $y_2 - y_1$ units in the *y*-direction as it travels $x_2 - x_1$ units in the positive *x*-direction. The vertical change $y_2 - y_1$ is called the **rise**, and the horizontal change $x_2 - x_1$ the **run**.

Figure 1.4.2

By analogy with the surveyor's notion of slope we make the following definition.

1.4.1 DEFINITION. If $P_1(x_1, y_1)$ and $P_2(x_2, y_2)$ are points on a nonvertical line, then the **slope** m of the line is defined by

$$m = \frac{\text{rise}}{\text{run}} = \frac{y_2 - y_1}{x_2 - x_1} \tag{1}$$

We make several observations about Definition 1.4.1:

- Definition 1.4.1 does not apply to vertical lines. For such lines we would have $x_2 = x_1$, so (1) would involve a division by zero. The slope of a vertical line is **undefined**. Speaking informally, some people say that a vertical line has **infinite slope**.

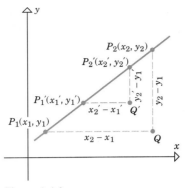

Figure 1.4.3

- When using (1) to calculate the slope of a line through two points, it does not matter which point is called P_1 and which one is called P_2, since reversing the points reverses the sign of both the numerator and denominator of (1), and hence has no effect on the ratio.

- Any two distinct points on a nonvertical line can be used to calculate the slope of the line; that is, the slope m computed from a pair of distinct points P_1 and P_2 on the line will be the same as the slope m' computed from any other pair of distinct points P_1' and P_2' on the line. Although we omit the formal proof, the reader should be able to reach this conclusion from Figure 1.4.3 by using similar triangles to show that

$$m = \frac{y_2 - y_1}{x_2 - x_1} = \frac{y_2' - y_1'}{x_2' - x_1'} = m'$$

Example 1   In each part find the slope of the line through

(a)   the points $(6, 2)$ and $(9, 8)$

(b)   the points $(2, 9)$ and $(4, 3)$

(c)   the points $(-2, 7)$ and $(5, 7)$

Solution.

(a) $m = \dfrac{8 - 2}{9 - 6} = \dfrac{6}{3} = 2$ (b) $m = \dfrac{3 - 9}{4 - 2} = \dfrac{-6}{2} = -3$

(c) $m = \dfrac{7 - 7}{5 - (-2)} = 0$ ◄

☐ **INTERPRETATION OF SLOPE**

Since the slope m of a line is the rise divided by the run, it follows that

rise $= m \cdot$ run

so that as a point travels left to right along the line, there are m units of rise for each unit of run. But the rise is the change in the y value of the point and the run is the change in the x value, so that the slope m is sometimes called the ***rate of change of y with respect to x*** along the line.

Example 2   Figure 1.4.4 shows the three lines determined by the points in Example 1 and interprets the significance of their slopes.   ◄

As illustrated in the last example, the slope of a line can be positive, negative, or zero. A positive slope means that the line is inclined upward to the right, a negative slope means

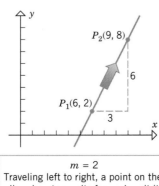

$m = 2$
Traveling left to right, a point on the line rises two units for each unit it moves in the positive x-direction.

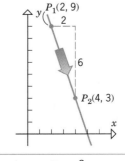

$m = -3$
Traveling left to right, a point on the line falls three units for each unit it moves in the positive x-direction.

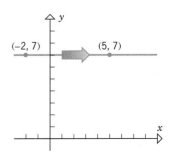

$m = 0$
Traveling left to right, a point on the line neither rises nor falls.

Figure 1.4.4

that it is inclined downward to the right, and a zero slope means that the line is horizontal. An undefined slope means that the line is vertical. In Figure 1.4.5 we have sketched some lines with different slopes.

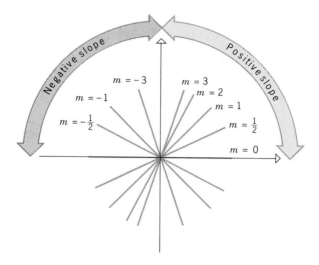

Figure 1.4.5

□ ANGLE OF INCLINATION

If equal scales are used on the coordinate axes, then the slope of a line is related to the angle the line makes with the positive *x*-axis. To establish this relationship we need the following definition.

1.4.2 DEFINITION.   For a line *L* not parallel to the *x*-axis, the ***angle of inclination*** is the smallest angle ϕ measured counterclockwise from the direction of the positive *x*-axis to *L* (Figure 1.4.6). For a line parallel to the *x*-axis, we take $\phi = 0$.

REMARK.   In degree measure the angle of inclination satisfies $0° \leq \phi < 180°$ and in radian measure it satisfies $0 \leq \phi < \pi$.

The following theorem, suggested by Figure 1.4.7, relates the slope of a line to its angle of inclination. We omit the formal proof.

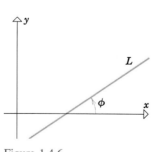

Figure 1.4.6

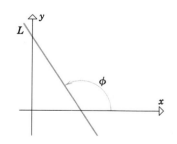

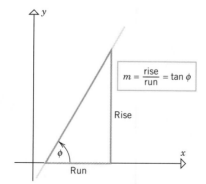

$$m = \frac{\text{rise}}{\text{run}} = \tan \phi$$

Figure 1.4.7

1.4.3 THEOREM.   *For a nonvertical line, the slope m and angle of inclination ϕ are related by*

$$m = \tan \phi \tag{2}$$

REMARK.   If the line *L* is parallel to the *y*-axis, then $\phi = \frac{1}{2}\pi$, so $\tan \phi$ in (2) is undefined. This agrees with the fact that the slope *m* is also undefined in this case.

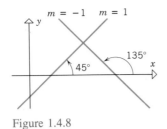

Figure 1.4.8

PARALLEL AND PERPENDICULAR LINES

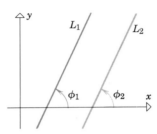

Figure 1.4.9

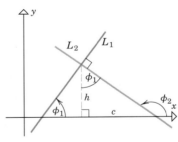

Figure 1.4.10

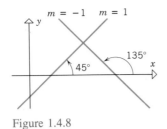

Figure 1.4.11

Example 3 Find the angle of inclination for a line of slope $m = 1$ and also for a line of slope $m = -1$.

Solution. If $m = 1$, then from (2), $\tan \phi = 1$ so that $\phi = \pi/4$ (or in degree measure $\phi = 45°$). If $m = -1$, then from (2), $\tan \phi = -1$. From this equality and the fact that $0 \le \phi < \pi$ we obtain $\phi = 3\pi/4$ or, in degree measure, $\phi = 135°$ (Figure 1.4.8). ◄

As a consequence of Theorem 1.4.3, we obtain the following basic result.

1.4.4 THEOREM. *Let L_1 and L_2 be nonvertical lines with slopes m_1 and m_2, respectively.*

(a) *The lines are parallel if and only if*

$$m_1 = m_2 \tag{3}$$

(b) *The lines are perpendicular if and only if*

$$m_1 m_2 = -1 \tag{4}$$

Proof (a). If L_1 and L_2 are nonvertical parallel lines, then their angles of inclination ϕ_1 and ϕ_2 are equal, since two parallel lines cut by a transversal have equal corresponding angles (Figure 1.4.9). Thus,

$$m_1 = \tan \phi_1 = \tan \phi_2 = m_2$$

Conversely, if L_1 and L_2 have the same slope, say $m_1 = m_2 = m$, then

$$m = \tan \phi_1 = \tan \phi_2 \tag{5}$$

Because $0 \le \phi_1 < \pi$ and $0 \le \phi_2 < \pi$, it follows from (5) that ϕ_1 and ϕ_2 are equal. Thus, L_1 and L_2 are parallel (Figure 1.4.9).

Proof (b). We shall prove first that if L_1 and L_2 are nonvertical perpendicular lines, then (4) holds. Assume that L_1 and L_2 have angles of inclination ϕ_1 and ϕ_2 and that $\phi_1 < \phi_2$ (Figure 1.4.10). Referring to the figure we have

$$m_1 = \tan \phi_1 = c/h \quad \text{and} \quad m_2 = \tan \phi_2 = -h/c$$

so $m_1 m_2 = -1$. The proof of the converse is left as an exercise. ∎

REMARK. Formula (4) can be rewritten in the form $m_2 = -1/m_1$. In words, this tells us that *two nonvertical lines are perpendicular if and only if their slopes are negative reciprocals of one another.*

Example 4 Use slopes to show that the points $A(1, 3)$, $B(3, 7)$, and $C(7, 5)$ are vertices of a right triangle.

Solution. We shall show that the line through A and B is perpendicular to the line through B and C. The slopes of these lines are

$$m_1 = \frac{7 - 3}{3 - 1} = 2 \quad \text{and} \quad m_2 = \frac{5 - 7}{7 - 3} = -\frac{1}{2}$$

| Slope of the line through A and B | Slope of the line through C and D |

Since $m_1 m_2 = -1$, the line through A and B is perpendicular to the line through B and C; thus, ABC is a right triangle (Figure 1.4.11).   ◄

☐ LINES PARALLEL TO THE COORDINATE AXES

We now turn to the problem of finding equations of lines that satisfy specified conditions. The simplest cases are lines parallel to the coordinate axes: A line parallel to the y-axis intersects the x-axis at some point $(a, 0)$. This line consists precisely of those points whose x-coordinate is equal to a (Figure 1.4.12). Similarly, a line parallel to the x-axis intersects the y-axis at some point $(0, b)$. This line consists precisely of those points whose y-coordinate is equal to b (Figure 1.4.12). Thus, we have the following theorem.

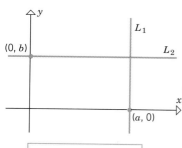

Every point on L_1 has an x-coordinate of a and every point on L_2 has a y-coordinate of b.

Figure 1.4.12

> **1.4.5** THEOREM. *The vertical line through $(a, 0)$ and the horizontal line through $(0, b)$ are represented, respectively, by the equations*
>
> $$x = a \qquad and \qquad y = b$$

Example 5 The graph of $x = -5$ is the vertical line through $(-5, 0)$, and the graph of $y = 7$ is the horizontal line through $(0, 7)$ (Figure 1.4.13).   ◄

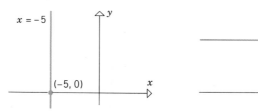

Figure 1.4.13

☐ LINES DETERMINED BY POINT AND SLOPE

There are infinitely many lines that pass through any given point in the plane. However, if we specify the slope of the line in addition to a point on it, then the point and the slope together determine a unique line (Figure 1.4.14).

Let us now consider how to find an equation of a nonvertical line L that passes through a point $P_1(x_1, y_1)$ and has slope m. If $P(x, y)$ is any point on L, different from P_1, then the slope m can be obtained from the points $P(x, y)$ and $P_1(x_1, y_1)$; this gives

$$m = \frac{y - y_1}{x - x_1}$$

which can be rewritten as

$$y - y_1 = m(x - x_1) \tag{6}$$

With the possible exception of (x_1, y_1), we have shown that every point on L satisfies (6). But $x = x_1, y = y_1$ satisfies (6), so that all points on L satisfy (6). We leave it as an exercise to show that every point satisfying (6) lies on L.

In summary, we have the following theorem.

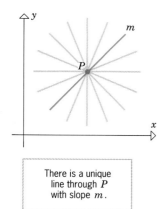

There is a unique line through P with slope m.

Figure 1.4.14

> **1.4.6** THEOREM. *The line passing through $P_1(x_1, y_1)$ and having slope m is given by the equation*
>
> $$y - y_1 = m(x - x_1) \tag{7}$$
>
> *This is called the **point-slope form** of the line.*

Example 6   Find the point-slope form of the line through $(4, -3)$ with slope 5.

Solution.   Substituting the values $x_1 = 4$, $y_1 = -3$, and $m = 5$ in (7) yields the point-slope form $y + 3 = 5(x - 4)$.   ◄

☐ **LINES DETERMINED BY SLOPE AND y-INTERCEPT**

A nonvertical line crosses the y-axis at some point $(0, b)$. If we use this point in the point-slope form of its equation, we obtain

$$y - b = m(x - 0)$$

which we can rewrite as $y = mx + b$. To summarize:

1.4.7   THEOREM.   *The line with y-intercept b and slope m is given by the equation*

$$y = mx + b \qquad (8)$$

*This is called the **slope-intercept form** of the line.*

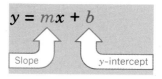

Figure 1.4.15

REMARK.   Note that y is alone on one side of Equation (8). When the equation of a line is written in this way the slope of the line and its y-intercept can be determined by inspection of the equation: the slope is the coefficient of x and the y-intercept is the constant term (Figure 1.4.15).

Example 7

EQUATION	SLOPE	y-INTERCEPT
$y = 3x + 7$	$m = 3$	$b = 7$
$y = -x + \frac{1}{2}$	$m = -1$	$b = \frac{1}{2}$
$y = x$	$m = 1$	$b = 0$
$y = \sqrt{2}x - 8$	$m = \sqrt{2}$	$b = -8$
$y = 2$	$m = 0$	$b = 2$

◄

Example 8   Find the slope-intercept form of the equation of the line that satisfies the stated conditions:

(a)   slope is -9; crosses the y-axis at $(0, -4)$
(b)   slope is 1; passes through the origin
(c)   passes through $(5, -1)$; perpendicular to $y = 3x + 4$
(d)   passes through $(3, 4)$ and $(2, -5)$

Solution (a).   From the given conditions we have $m = -9$ and $b = -4$, so (8) yields $y = -9x - 4$.

Solution (b).   From the given conditions $m = 1$ and the line passes through $(0, 0)$, so $b = 0$. Thus, it follows from (8) that $y = x + 0$ or $y = x$.

Solution (c).   The given line has slope 3, so the line to be determined will have slope $m = -\frac{1}{3}$. Substituting this slope and the given point in the point-slope form (7) and then simplifying yields

$$y - (-1) = -\tfrac{1}{3}(x - 5)$$
$$y = -\tfrac{1}{3}x + \tfrac{2}{3}$$

Solution (d).   We shall first find the point-slope form, then solve for y in terms of x to obtain the stope-intercept form. From the given points the slope of the line is

$$m = \frac{-5 - 4}{2 - 3} = 9$$

We can use either of the given points for (x_1, y_1) in (7). We shall use $(3, 4)$. This yields the point-slope form

$$y - 4 = 9(x - 3)$$

Solving for y in terms of x yields the slope-intercept form

$$y = 9x - 23$$

We leave it for the reader to show that the same equation results if $(2, -5)$ rather than $(3, 4)$ is used for (x_1, y_1) in (7).   ◀

☐ **THE GENERAL EQUATION OF A LINE**

An equation that is expressible in the form

$$Ax + By + C = 0 \tag{9}$$

where A, B, and C are constants and A and B are not both zero, is called a ***first-degree equation*** in x and y. For example,

$$4x + 6y - 5 = 0$$

is a first-degree equation in x and y since it has form (9) with

$$A = 4, \quad B = 6, \quad C = -5$$

In fact, all the equations of lines studied in this section are first-degree equations in x and y.

The following theorem states that the first-degree equations in x and y are precisely the equations whose graphs in the xy-plane are straight lines. (The proof is left as an exercise.)

1.4.8   THEOREM.   *Every first-degree equation in x and y has a straight line as its graph and, conversely, every straight line can be represented by a first-degree equation in x and y.*

Because of this theorem, (9) is sometimes called the ***general equation*** of a line or a ***linear equation*** in x and y.

Example 9   Graph the equation $3x - 4y + 12 = 0$.

Solution.   Since this is a linear equation in x and y, its graph is a straight line. Thus, to sketch the graph we need only plot any two points on the graph and draw the line through them. It is particularly convenient to plot the points where the line crosses the coordinate axes. These points are $(0, 3)$ and $(-4, 0)$ (verify), so the graph is the line in Figure 1.4.16.   ◀

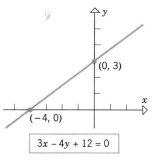

$3x - 4y + 12 = 0$

Figure 1.4.16

Example 10   Find the slope of the line in Example 9.

Solution.   Solving the equation for y yields

$$y = \tfrac{3}{4}x + 3$$

which is the slope-intercept form of the line. Thus, the slope is $m = \tfrac{3}{4}$.   ◀

☐ **SLOPES OF LINES IN APPLIED PROBLEMS**

In applied problems, changing the units of measurement can change the slope of a line, so it is essential to include the units when calculating the slope. The following example illustrates this.

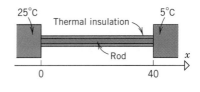

(a)

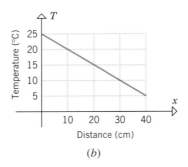

(b)

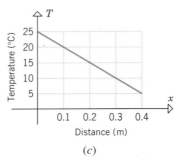

(c)

Figure 1.4.17

Example 11 Suppose that a uniform rod of length 40 cm (= 0.4 m) is thermally insulated around the lateral surface and that the exposed ends of the rod are held at constant temperatures of 25° C and 5° C, respectively (Figure 1.4.17*a*). It is shown in physics that under appropriate conditions the graph of the temperature T versus the distance x from the left-hand end of the rod will be a straight line. Parts (*b*) and (*c*) of Figure 1.4.17 show two such graphs: one in which x is measured in centimeters and one in which it is measured in meters. The slopes in the two cases are

$$m = \frac{5 - 25}{40 - 0} = \frac{-20}{40} = -0.5° \, \text{C/cm} \tag{10}$$

$$m = \frac{5 - 25}{0.4 - 0} = \frac{-20}{0.4} = -50° \, \text{C/m} \tag{11}$$

The slope in (10) implies that the temperature *decreases* at a rate of 0.5° C per centimeter of distance from the left end of the rod. The slope in (11) implies that the temperature decreases at a rate of 50° C per meter of distance from the left end of the rod. The two statements are equivalent physically, even though the slopes differ. ◄

Example 12 Find the slope-intercept form of the equation of the temperature distribution in the preceding example if the temperature T is measured in degrees Celsius (°C) and the distance x is measured in

(a) centimeters   (b) meters

Solution (a). The slope is $m = -0.5°$ C/cm and the intercept on the T-axis is 25°, so

$$T = -0.5x + 25, \quad 0 \le x \le 40$$

where the restriction on x is required because the rod is 40 cm in length. The graph of this equation with the restriction is a line segment rather than a line.

Solution (b). The slope is $m = -50°$ C/m, the intercept on the T-axis is 25°, and the restriction on x is $0 \le x \le 0.4$. Thus, the equation is

$$T = -50x + 25, \quad 0 \le x \le 0.4 \quad ◄$$

Example 13 If an object is thrown straight up near the surface of the earth at time $t = 0$ with an initial velocity of v_0, and if air resistance is ignored, then it is shown in physics that for $t \ge 0$ the object's velocity v is given by the linear equation

$$v = v_0 - gt \tag{12}$$

where g is a constant, called the **_acceleration due to gravity_**, whose value is approximately 9.8 m/sec^2 when distance is in meters and time in seconds. Suppose that $v_0 = 24.5$ m/sec.

(a) Graph velocity versus time for $t \ge 0$, making the t-axis horizontal and the v-axis vertical.

(b) Find the slope of the line and give an interpretation of the slope as a rate of change.

Solution (a). Substituting the given values of g and v_0 in (12) yields

$$v = 24.5 - 9.8t, \quad t \ge 0$$

It follows that $v = 24.5$ when $t = 0$ and that $v = 0$ when $t = 24.5/9.8 = 2.5$, which results in the graph shown in Figure 1.4.18.

Solution (b). The slope is -9.8 m/sec^2, which can be interpreted to mean that the velocity is decreasing at the rate of 9.8 m/sec per second. ◄

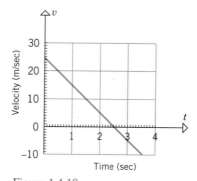

Figure 1.4.18

► **Exercise Set 1.4**   Ⓒ *18, 19, 31, 32, 52–61*

1. Find the slope of the line through
   (a) $(-1, 2)$ and $(3, 4)$
   (b) $(5, 3)$ and $(7, 1)$
   (c) $(4, \sqrt{2})$ and $(-3, \sqrt{2})$
   (d) $(-2, -6)$ and $(-2, 12)$.

2. Find the slopes of the sides of the triangle with vertices $(-1, 2)$, $(6, 5)$, and $(2, 7)$.

3. Use slopes to determine whether the given points lie on the same line.
   (a) $(1, 1)$, $(-2, -5)$, and $(0, -1)$
   (b) $(-2, 4)$, $(0, 2)$, and $(1, 5)$.

4. Draw the line through $(4, 2)$ with slope
   (a) $m = 3$ (b) $m = -2$
   (c) $m = -\frac{3}{4}$.

5. Draw the line through $(-1, -2)$ with slope
   (a) $m = \frac{3}{5}$ (b) $m = -1$ (c) $m = \sqrt{2}$.

6. List the lines in parts (a)–(d) of Figure 1.4.19 in the order of decreasing slope.

(a)

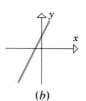
(b)

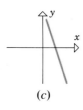
(c)

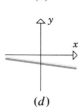
(d)   Figure 1.4.19

7. List the lines in parts (a)–(d) of Figure 1.4.20 in the order of increasing slope.

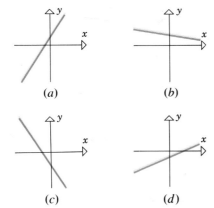
(a)   (b)

(c)   (d)   Figure 1.4.20

8. Find the slope of the lines in parts (a) and (b) of Figure 1.4.21.

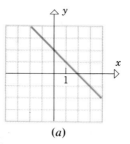
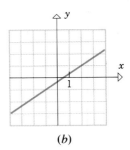
(a)   (b)

Figure 1.4.21

9. Find the slope of the lines in parts (a) and (b) of Figure 1.4.22.

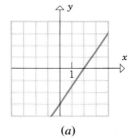
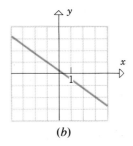
(a)   (b)

Figure 1.4.22

10. A particle, initially at $(7, 5)$, moves along a line of slope $m = -2$ to a new position (x, y).
   (a) Find y if $x = 9$.   (b) Find x if $y = 12$.

11. A particle, initially at $(1, 2)$, moves along a line of slope $m = 3$ to a new position (x, y).
   (a) Find y if $x = 5$.   (b) Find x if $y = -2$.

12. Given that the point $(k, 4)$ is on the line through $(1, 5)$ and $(2, -3)$, find k.

13. Let the point $(3, k)$ lie on the line of slope $m = 5$ through $(-2, 4)$; find k.

14. Find x and y if the line through $(0, 0)$ and (x, y) has slope $\frac{1}{2}$, and the line through (x, y) and $(7, 5)$ has slope 2.

15. Find x if the slope of the line through $(1, 2)$ and $(x, 0)$ is the negative of the slope of the line through $(4, 5)$ and $(x, 0)$.

In Exercises 16 and 17, find the slope of the line whose angle of inclination is given. (You should be able to solve these problems without a calculator.)

16. (a) $45°$ (b) $\dfrac{2\pi}{3}$ (c) $30°$.

17. (a) $\dfrac{\pi}{6}$ (b) $135°$ (c) $60°$.

In Exercises 18 and 19, use a calculator, where necessary, to find (to the nearest degree) the angle of inclination of a line with the given slope.

18. (a) $m = \frac{1}{2}$ (b) $m = -1$
    (c) $m = 2$ (d) $m = -57$.

19. (a) $m = -\frac{1}{2}$ (b) $m = 1$
    (c) $m = -2$ (d) $m = 57$.

20. An equilateral triangle has one vertex at the origin, another on the x-axis, and the third in the first quadrant. Find the slopes of its sides.

21. Find the coordinates of all points P on the x-axis so that the line through $A(1, 2)$ and P is perpendicular to the line through $B(8, 3)$ and P.

22. Use slopes to show that $(3, 1)$, $(6, 3)$, and $(2, 9)$ are vertices of a right triangle.

23. Use slopes to show that $(3, -1)$, $(6, 4)$, $(-3, 2)$, and $(-6, -3)$ are vertices of a parallelogram.

24. Let $P(2, 3)$ be a point on the circle with center $(4, -1)$. Find the slope of the line that is tangent to the circle at P.

25. Graph the equations
    (a) $2x + 5y = 15$ (b) $x = 3$
    (c) $y = -2$ (d) $y = 2x - 7$.

26. Graph the equations
    (a) $\frac{x}{3} - \frac{y}{4} = 1$ (b) $x = -8$
    (c) $y = 0$ (d) $x = 3y + 2$.

27. Graph the equations
    (a) $y = 2x - 1$ (b) $y = 3$
    (c) $y = -2x$.

28. Graph the equations
    (a) $y = 2 - 3x$ (b) $y = \frac{1}{4}x$
    (c) $y = -\sqrt{3}$.

29. Find the slope and y-intercept of
    (a) $y = 3x + 2$ (b) $y = 3 - \frac{1}{4}x$
    (c) $3x + 5y = 8$ (d) $y = 1$
    (e) $\frac{x}{a} + \frac{y}{b} = 1$.

30. Find the slope and y-intercept of
    (a) $y = -4x + 2$ (b) $x = 3y + 2$
    (c) $\frac{x}{2} + \frac{y}{3} = 1$ (d) $y - 3 = 0$
    (e) $a_0 x + a_1 y = 0$ $(a_1 \neq 0)$.

31. To the nearest degree, find the angle of inclination of
    (a) $y = \sqrt{3}\, x + 2$ (b) $y + 2x + 5 = 0$.

32. To the nearest degree, find the angle of inclination of
    (a) $3y = 2 - \sqrt{3}\, x$ (b) $y - 4x + 7 = 0$.

In Exercises 33–46, find the slope-intercept form of the line satisfying the given conditions.

33. Slope $= -2$, y-intercept $= 4$.

34. $m = 5$, $b = -3$.

35. The line is parallel to $y = 4x - 2$ and its y-intercept is 7.

36. The line is parallel to $3x + 2y = 5$ and passes through $(-1, 2)$.

37. The line is perpendicular to $y = 5x + 9$ and has y-intercept 6.

38. The line is perpendicular to $x - 4y = 7$ and passes through $(3, -4)$.

39. The line passes through $(2, 4)$ and $(1, -7)$.

40. The line passes through $(-3, 6)$ and $(-2, 1)$.

41. The line has an angle of inclination of $\phi = \frac{1}{6}\pi$ and a y-intercept of -3.

42. The line has angle of inclination $\phi = \frac{2}{3}\pi$ and passes through the point $(1, 2)$.

43. The y-intercept is 2 and the x-intercept is -4.

44. The y-intercept is b and the x-intercept is a.

45. The line is perpendicular to the y-axis and passes through $(-4, 1)$.

46. The line is parallel to $y = -5$ and passes through $(-1, -8)$.

47. Find an equation for the line that passes through $(5, -2)$ and has angle of inclination $\phi = \frac{1}{2}\pi$.

48. Find an equation for the y-axis.

49. In each part, classify the lines as parallel, perpendicular, or neither.
    (a) $y = 4x - 7$ and $y = 4x + 9$
    (b) $y = 2x - 3$ and $y = 7 - \frac{1}{2}x$
    (c) $5x - 3y + 6 = 0$ and $10x - 6y + 7 = 0$
    (d) $Ax + By + C = 0$ and $Bx - Ay + D = 0$
    (e) $y - 2 = 4(x - 3)$ and $y - 7 = \frac{1}{4}(x - 3)$.

50. In each part, classify the lines as parallel, perpendicular, or neither.
    (a) $y = -5x + 1$ and $y = 3 - 5x$
    (b) $y - 1 = 2(x - 3)$ and $y - 4 = -\frac{1}{2}(x + 7)$
    (c) $4x + 5y + 7 = 0$ and $5x - 4y + 9 = 0$
    (d) $Ax + By + C = 0$ and $Ax + By + D = 0$
    (e) $y = \frac{1}{2}x$ and $x = \frac{1}{2}y$.

51. For what value of k will the line $3x + ky = 4$
    (a) have slope 2
    (b) have y-intercept 5
    (c) pass through the point $(-2, 4)$
    (d) be parallel to the line $2x - 5y = 1$
    (e) be perpendicular to the line $4x + 3y = 2$?

52. Find the slope of the temperature graph in Figure 1.4.17*b* if temperature is in degrees Fahrenheit (°F) and distance is in feet. [*Note:* $T_F = \frac{9}{5}T + 32$, where T_F is the temperature in °F and T is the temperature in °C. Use the approximation 1 cm = 0.03281 ft.]

53. The force F (in newtons) needed to stretch a certain spring x mm beyond its natural length is related linearly to x as shown in Figure 1.4.23. The slope of the line is called the ***spring constant***. Estimate the spring constant in units of newtons/meter (N/m). [*Note:* 1 m = 1000 mm.]

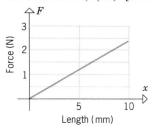

Figure 1.4.23

54. In the interval between 0° C and 100° C, the mass density ρ of mercury (g/cm³) varies linearly with the temperature T (°C), as shown in Figure 1.4.24. Estimate the slope of the line and interpret the slope as a rate of change using the correct units.

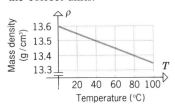

Figure 1.4.24

55. In the interval between 0° C and 40° C, the speed of sound in air (m/sec) varies linearly with the temperature T (°C), as shown in Figure 1.4.25. Estimate the slope of the line and interpret the slope as a rate of change using the correct units.

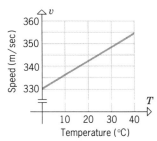

Figure 1.4.25

56. When light of suitably high frequency falls on a metal surface, electrons are emitted (Figure 1.4.26). In his explanation of this phenomenon, Albert Einstein showed in 1905 that the stopping voltage V ($V \geq 0$) needed to prevent emission is related to the frequency f of the light according to the equation

$$V = kf - P$$

where $k = 4.1 \times 10^{-15}$ volts/Hz and $P = 1.77$ volts for sodium metal.

   (a) Graph this equation in an fV-coordinate system for $f \leq 10^{15}$ Hz, and $V \geq 0$.

   (b) What is the frequency below which electrons will not be emitted? [*Hint:* Electrons are not emitted if $V \leq 0$.]

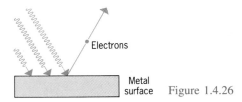

Figure 1.4.26

57. There are two common systems for measuring temperature, Celsius and Fahrenheit. Water freezes at 0° Celsius (0° C) and 32° Fahrenheit (32° F); it boils at 100° C and 212° F.

   (a) Assuming that the Celsius temperature T_C and the Fahrenheit temperature T_F are related by a linear equation, find the equation.

   (b) What is the slope of the line relating T_F and T_C if T_F is plotted on the horizontal axis?

   (c) At what temperature is the Fahrenheit reading equal to the Celsius reading?

   (d) Normal body temperature is 98.6° F. What is it in °C?

58. Thermometers are calibrated using the so-called "triple point" of water, which is 273.16 K on the Kelvin scale and 0.01° C on the Celsius scale. A one-degree difference on the Celsius scale is the same as a one-degree difference on the Kelvin scale, so there is a linear relationship between the temperature T_C in degrees Celsius and the temperature T_K in kelvins.

   (a) Find an equation that relates T_C and T_K.

   (b) Absolute zero (0 K on the Kelvin scale) is the temperature below which a body's temperature cannot be lowered. Express absolute zero in °C.

59. To the extent that water can be assumed to be incompressible, the pressure p in a body of water varies linearly with the distance h below the surface.

   (a) Given that the pressure is 1 atmosphere (1 atm) at the surface and 5.9 atm at a depth of 50 m, find an equation that relates pressure to depth.

   (b) At what depth is the pressure twice that at the surface?

60. A resistance thermometer is a device that determines temperature by measuring the resistance of a fine wire whose resistance varies with temperature. Suppose that the resistance R in ohms (Ω) varies linearly with the temperature T in °C and that $R = 123.4\ \Omega$ when $T = 20° C$ and that $R = 133.9\ \Omega$ when $T = 45° C$.

   (a) Find an equation for R in terms of T.

   (b) If R is measured experimentally as 128.6 Ω, what is the temperature?

61. Suppose that the mass of a spherical mothball decreases with time, due to evaporation, at a rate that is proportional to its surface area. Assuming that it always retains the shape of a sphere, it can be shown that the radius r of the sphere decreases linearly with the time t.

    (a) If, at a certain instant, the radius is 0.80 mm and 4 days later it is 0.75 mm, find an equation for r (in millimeters) in terms of the elapsed time t (in days).

    (b) How long will it take for the mothball to completely evaporate?

62. In physical problems linear equations involving variables other than x and y often arise. In parts (a)–(g) determine whether the equation is linear.

    (a) $3\alpha - 2\beta = 5$

    (b) $A = 2000(1 + 0.06t)$

    (c) $A = \pi r^2$

    (d) $E = mc^2$ (c constant)

    (e) $V = C(1 - rt)$ (r and C constant)

    (f) $V = \frac{1}{3}\pi r^2 h$ (r constant)

    (g) $V = \frac{1}{3}\pi r^2 h$ (h constant).

63. A point moves in the xy-plane in such a way that at any time t its coordinates are given by $x = 5t + 2$ and $y = t - 3$. By expressing y in terms of x, show that the point moves along a straight line.

64. A point moves in the xy-plane in such a way that at any time t its coordinates are given by $x = 1 + 3t^2$ and $y = 2 - t^2$. By expressing y in terms of x, show that the point moves along a straight line path and specify the values of x for which the equation is valid.

65. Find the area of the triangle formed by the coordinate axes and the line through $(1, 4)$ and $(2, 1)$.

66. Draw the graph of $4x^2 - 9y^2 = 0$. [*Hint:* Factor.]

67. Complete the proof of Theorem 1.4.4(b) by showing: If L_1 and L_2 are two nonvertical lines whose slopes satisfy $m_1 m_2 = -1$, then L_1 and L_2 are perpendicular.

68. Prove: If (x, y) satisfies Equation (7), then the point $P(x, y)$ lies on the line with slope m passing through $P_1(x_1, y_1)$.

69. Prove Theorem 1.4.8.

■ **1.5** DISTANCE; CIRCLES; EQUATIONS OF THE FORM $y = ax^2 + bx + c$

> *In this section we shall derive a formula for the distance between two points in a coordinate plane, and we shall use that formula to study equations of circles. We shall also study equations of the form $y = ax^2 + bx + c$. Such equations will be considered in more detail later in the text, but they are introduced here to provide an immediate source of useful examples for subsequent sections.*

☐ **DISTANCE BETWEEN TWO POINTS IN THE PLANE**

Recall from Section 1.2 that if A and B are points on a coordinate line with coordinates a and b, respectively, then the distance between A and B is $|b - a|$. We shall use this result to find the distance d between two arbitrary points $P_1(x_1, y_1)$ and $P_2(x_2, y_2)$ in the plane. If, as in Figure 1.5.1, we form a right triangle with P_1 and P_2 as vertices, then the length of the

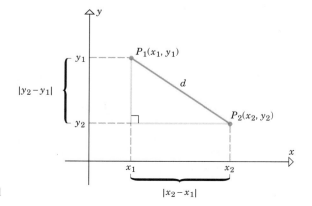

Figure 1.5.1

horizontal side is $|x_2 - x_1|$ and the length of the vertical side is $|y_2 - y_1|$, so it follows from the Theorem of Pythagoras that

$$d = \sqrt{|x_2 - x_1|^2 + |y_2 - y_1|^2}$$

But for every real number a we have $|a|^2 = a^2$ (since a and $|a|$ can differ only in sign). Thus, $|x_2 - x_1|^2 = (x_2 - x_1)^2$ and $|y_2 - y_1|^2 = (y_2 - y_1)^2$, which yields

---

1.5.1   THEOREM.   *The distance d between two points (x_1, y_1) and (x_2, y_2) in a coordinate plane is given by*

$$d = \sqrt{(x_2 - x_1)^2 + (y_2 - y_1)^2} \tag{1}$$

REMARK.   It is assumed in (1) that equal scales are used on the coordinate axes.

Example 1   Find the distance between the points $(-2, 3)$ and $(1, 7)$.

Solution.   If we let (x_1, y_1) be $(-2, 3)$ and let (x_2, y_2) be $(1, 7)$, then (1) yields

$$d = \sqrt{[1 - (-2)]^2 + [7 - 3]^2} = \sqrt{3^2 + 4^2} = \sqrt{25} = 5 \quad \blacktriangleleft$$

REMARK.   When using (1) it does not matter which point is labeled (x_1, y_1) and which is labeled (x_2, y_2). Thus, in Example 1, if we had let (x_1, y_1) be the point $(1, 7)$ and (x_2, y_2) the point $(-2, 3)$, we would have obtained

$$d = \sqrt{[-2 - 1]^2 + [3 - 7]^2} = \sqrt{(-3)^2 + (-4)^2} = \sqrt{25} = 5$$

which is the same result we obtained with the opposite labeling.

REMARK.   The distance between two points P_1 and P_2 in a coordinate plane is commonly denoted by $d(P_1, P_2)$, or $d(P_2, P_1)$.

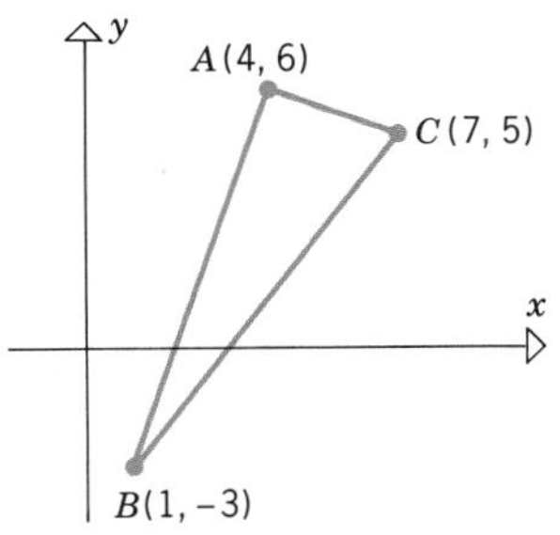

Example 2   It can be shown that the converse of the Theorem of Pythagoras is true; that is, if the sides of a triangle satisfy the relationship $a^2 + b^2 = c^2$, then the triangle must be a right triangle. Use this result to show that the points $A(4, 6)$, $B(1, -3)$, $C(7, 5)$ are vertices of a right triangle.

Solution.   The points and the triangle are shown in Figure 1.5.2. From (1) the lengths of the sides of the triangles are

$$d(A, B) = \sqrt{(1 - 4)^2 + (-3 - 6)^2} = \sqrt{9 + 81} = \sqrt{90}$$

$$d(A, C) = \sqrt{(7 - 4)^2 + (5 - 6)^2} = \sqrt{9 + 1} = \sqrt{10}$$

$$d(B, C) = \sqrt{(7 - 1)^2 + [5 - (-3)]^2} = \sqrt{36 + 64} = \sqrt{100} = 10$$

Since

$$[d(A, B)]^2 + [d(A, C)]^2 = [d(B, C)]^2$$

it follows that $\triangle ABC$ is a right triangle with hypotenuse BC. $\blacktriangleleft$

A(4, 6)

C(7, 5)

B(1, −3)

Figure 1.5.2

☐ THE MIDPOINT FORMULA

It is often necessary to find the coordinates of the midpoint of a line segment joining two points in the plane. To derive the midpoint formula, we shall start with two points on a coordinate line. If we assume that the points have coordinates a and b and that $a \leq b$, then, as shown in Figure 1.5.3a, the distance between a and b is $b - a$, and the coordinate of the midpoint between a and b is

$$a + \tfrac{1}{2}(b - a) = \tfrac{1}{2}a + \tfrac{1}{2}b = \tfrac{1}{2}(a + b)$$

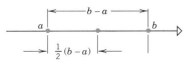

Figure 1.5.3a

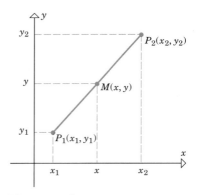

Figure 1.5.3*b*

which is the arithmetic average of a and b. Had the points been labeled with $b \le a$, the same formula would have resulted (verify). Therefore, *the midpoint of two points on a coordinate line is the arithmetic average of their coordinates, regardless of their relative positions.*

If we now let $P_1(x_1, y_1)$ and $P_2(x_2, y_2)$ be any two points in the plane and $M(x, y)$ the midpoint of the line segment joining them (Figure 1.5.3*b*), then it can be shown using similar triangles that x is the midpoint of x_1 and x_2 on the x-axis and y is the midpoint of y_1 and y_2 on the y-axis, so

$$x = \tfrac{1}{2}(x_1 + x_2) \quad \text{and} \quad y = \tfrac{1}{2}(y_1 + y_2)$$

Thus, we have the following result.

1.5.2   THEOREM (***The Midpoint Formula***).   *The midpoint of the line segment joining two points* (x_1, y_1) *and* (x_2, y_2) *in a coordinate plane is*

$$\left(\tfrac{1}{2}(x_1 + x_2), \tfrac{1}{2}(y_1 + y_2)\right) \tag{2}$$

Example 3   Find the midpoint of the line segment joining $(3, -4)$ and $(7, 2)$.

Solution.   From (2) the midpoint is

$$\left(\tfrac{1}{2}(3 + 7), \tfrac{1}{2}(-4 + 2)\right) = (5, -1) \quad \blacktriangleleft$$

☐ CIRCLES

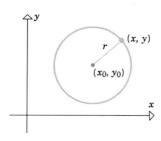

Figure 1.5.4

If (x_0, y_0) is a fixed point in the plane, then the circle of radius r centered at (x_0, y_0) is the set of all points in the plane whose distance from (x_0, y_0) is r (Figure 1.5.4). Thus, a point (x, y) will lie on this circle if and only if

$$\sqrt{(x - x_0)^2 + (y - y_0)^2} = r$$

or equivalently,

$$(x - x_0)^2 + (y - y_0)^2 = r^2 \tag{3}$$

This is called the **standard form of the equation of a circle**.

Example 4   Find an equation for the circle of radius 4 centered at $(-5, 3)$.

Solution.   From (3) with $x_0 = -5$, $y_0 = 3$, and $r = 4$ we obtain

$$(x + 5)^2 + (y - 3)^2 = 16$$

If desired, this equation can be written in an expanded form by squaring the terms, then simplifying:

$$(x^2 + 10x + 25) + (y^2 - 6y + 9) - 16 = 0$$

$$x^2 + y^2 + 10x - 6y + 18 = 0 \quad \blacktriangleleft$$

Example 5   Find an equation for the circle with center $(1, -2)$ that passes through $(4, 2)$.

Solution.   The radius r of the circle is the distance between $(4, 2)$ and $(1, -2)$, so

$$r = \sqrt{(1 - 4)^2 + (-2 - 2)^2} = 5$$

We now know the center and radius, so we can use (3) to obtain the equation

$$(x - 1)^2 + (y + 2)^2 = 25 \quad \text{or} \quad x^2 + y^2 - 2x + 4y - 20 = 0 \quad \blacktriangleleft$$

☐ **FINDING THE CENTER AND RADIUS OF A CIRCLE**

When you encounter an equation of form (3), you will know immediately that its graph is a circle; its center and radius can then be found from the constants that appear in the equation:

$$(x - x_0)^2 \qquad + \qquad (y - y_0)^2 \qquad = \qquad r^2$$

| x-coordinate of the center is x_0 | y-coordinate of the center is y_0 | radius squared |

Example 6

EQUATION OF A CIRCLE	CENTER (x_0, y_0)	RADIUS r
$(x - 2)^2 + (y - 5)^2 = 9$	$(2, 5)$	3
$(x + 7)^2 + (y + 1)^2 = 16$	$(-7, -1)$	4
$x^2 + y^2 = 25$	$(0, 0)$	5
$(x - 4)^2 + y^2 = 5$	$(4, 0)$	$\sqrt{5}$	◀

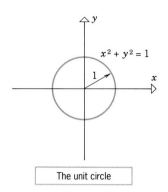

The unit circle

Figure 1.5.5

The circle $x^2 + y^2 = 1$, which is centered at the origin and has radius 1, is of special importance; it is called the **unit circle** (Figure 1.5.5).

☐ **OTHER FORMS FOR THE EQUATION OF A CIRCLE**

An alternative version of Equation (3) can be obtained by squaring the terms and simplifying. This yields an equation of the form

$$x^2 + y^2 + dx + ey + f = 0 \qquad (4)$$

where d, e, and f are constants. (See the final equations in Examples 4 and 5.)

Still another version of the equation of a circle can be obtained by multiplying both sides of (4) by a nonzero constant A. This yields an equation of the form

$$Ax^2 + Ay^2 + Dx + Ey + F = 0 \qquad (5)$$

where A, D, E, and F are constants and $A \neq 0$.

If the equation of a circle is given by (4) or (5), then the center and radius can be found by first rewriting the equation in standard form, then reading off the center and radius from that equation. The following example shows how to do this using the technique of *completing the square*.

Example 7 Find the center and radius of the circle with equation

(a) $x^2 + y^2 - 8x + 2y + 8 = 0$ (b) $2x^2 + 2y^2 + 24x - 81 = 0$

Solution (a). First, group the x-terms, group the y-terms, and take the constant to the right side:

$$(x^2 - 8x) + (y^2 + 2y) = -8$$

Next we want to add the appropriate constant within each set of parentheses to complete the square, and add the same constants to the right side to maintain equality. The appropriate constant is obtained by taking half the coefficient of the first-degree term and squaring it. This yields

$$(x^2 - 8x + 16) + (y^2 + 2y + 1) = -8 + 16 + 1$$

or

$$(x - 4)^2 + (y + 1)^2 = 9$$

Thus from (3), the circle has center $(4, -1)$ and radius 3.

Solution (b). The given equation is of form (5). We shall first divide through by 2 (the coefficient of the squared terms) to reduce the equation to form (4). Then we shall proceed as in part (a) of this example. The computations are as follows:

$$x^2 + y^2 + 12x - \tfrac{81}{2} = 0 \qquad \boxed{\text{We divided through by 2.}}$$

$$(x^2 + 12x) + y^2 = \tfrac{81}{2}$$

$$(x^2 + 12x + 36) + y^2 = \tfrac{81}{2} + 36 \qquad \boxed{\text{We completed the square.}}$$

$$(x + 6)^2 + y^2 = \tfrac{153}{2}$$

From (3), the circle has center $(-6, 0)$ and radius $\sqrt{\tfrac{153}{2}}$.   ◄

□ DEGENERATE CASES OF A CIRCLE

There is no guarantee that an equation of form (5) represents a circle. For example, suppose that we divide both sides of (5) by A, then complete the squares to obtain

$$(x - x_0)^2 + (y - y_0)^2 = k$$

Depending on the value of k, the following situations occur:

- $(k > 0)$ The graph is a circle with center (x_0, y_0) and radius $\sqrt{k}$.
- $(k = 0)$ The only solution of the equation is $x = x_0$, $y = y_0$, so the graph is the single point (x_0, y_0).
- $(k < 0)$ The equation has no real solutions and consequently no graph.

Example 8   Describe the graphs of

(a) $(x - 1)^2 + (y + 4)^2 = -9$ (b) $(x - 1)^2 + (y + 4)^2 = 0$

Solution (a).   There are no real values of x and y that will make the left side of the equation negative. Thus, the solution set of the equation is empty, and the equation has no graph.

Solution (b).   The only values of x and y that will make the left side of the equation 0 are $x = 1$, $y = -4$. Thus, the graph of the equation is the single point $(1, -4)$.   ◄

The following theorem summarizes our observations:

1.5.3   THEOREM.   *An equation of the form*

$$Ax^2 + Ay^2 + Dx + Ey + F = 0 \qquad (6)$$

where $A \neq 0$, represents a circle, or a point, or else has no graph.

REMARK.   The last two cases in Theorem 1.5.3 are called **degenerate cases**. In spite of the fact that these degenerate cases can occur, (6) is often called the **general equation of a circle**.

□ THE GRAPH OF
$y = ax^2 + bx + c$

An equation of the form

$$y = ax^2 + bx + c \quad (a \neq 0) \qquad (7)$$

is called a **quadratic equation in x**. Depending on whether a is positive or negative, the graph, which is called a **parabola**, has one of the two forms shown in Figure 1.5.6. In both cases the parabola is symmetric about a vertical line parallel to the y-axis. This line of symmetry cuts the parabola at a point called the **vertex**. The vertex is the low point on the curve if $a > 0$ and the high point if $a < 0$.

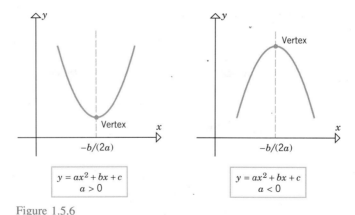

Figure 1.5.6

In the exercises (Exercise 78) we will help the reader show that the x-coordinate of the vertex is given by the formula

$$x = -\frac{b}{2a} \tag{8}$$

With the aid of this formula, a reasonably accurate graph of a quadratic equation in x can be obtained by plotting the vertex and two points on each side of it.

Example 9   Sketch the graph of

    (a) $y = x^2 - 2x - 2$ (b) $y = -x^2 + 4x - 5$

Solution (a).   The equation is of form (7) with $a = 1$, $b = -2$, and $c = -2$, so by (8) the x-coordinate of the vertex is

$$x = -\frac{b}{2a} = 1$$

Using this value and two additional values on each side, we obtain Figure 1.5.7.

Solution (b).   The equation is of form (7) with $a = -1$, $b = 4$, and $c = -5$, so by (8) the x-coordinate of the vertex is

$$x = -\frac{b}{2a} = 2$$

Using this value and two additional values on each side, we obtain the table and graph in Figure 1.5.8.   ◀

Often, the intercepts of a parabola $y = ax^2 + bx + c$ are important to know. The y-intercept, $y = c$, results immediately by setting $x = 0$. However, in order to obtain the x-intercepts, if any, we must set $y = 0$ and then solve the resulting quadratic equation $ax^2 + bx + c = 0$.

Example 10   Solve the inequality

$$x^2 - 2x - 2 > 0$$

Solution.   Because the left side of the inequality does not have readily discernible factors, the test-point method illustrated in Example 6 of Section 1.1 is not convenient to use. Instead, we shall give a graphical solution. The given inequality is satisfied for those values of x where the graph of $y = x^2 - 2x - 2$ is above the x-axis. From Figure 1.5.7 those are

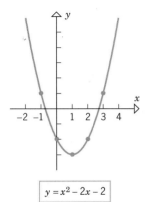

x	$y = x^2 - 2x - 2$
-1	1
0	-2
1	-3
2	-2
3	1

$$y = x^2 - 2x - 2$$

Figure 1.5.7

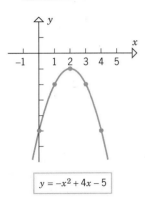

x	$y = -x^2 + 4x - 5$
0	-5
1	-2
2	-1
3	-2
4	-5

$$y = -x^2 + 4x - 5$$

Figure 1.5.8

the values of x to the left of the smaller intercept or to the right of the larger intercept. To find these intercepts we set $y = 0$ to obtain

$$x^2 - 2x - 2 = 0$$

Solving by the quadratic formula gives

$$x = \frac{-b \pm \sqrt{b^2 - 4ac}}{2a} = \frac{2 \pm \sqrt{12}}{2} = 1 \pm \sqrt{3}$$

Thus, the x-intercepts are

$$x = 1 + \sqrt{3} \approx 2.7 \quad \text{and} \quad x = 1 - \sqrt{3} \approx -0.7$$

and the solution set of the inequality is

$$(-\infty, 1 - \sqrt{3}) \cup (1 + \sqrt{3}, +\infty) \quad \blacktriangleleft$$

REMARK.   Note that the decimal approximations of the intercepts calculated in the preceding example agree with the graph in Figure 1.5.7. Observe, however, that we used the exact values of the intercepts to express the solution. The choice of exact versus approximate values is often a matter of judgment that depends on the purpose for which the values are to be used. Numerical approximations often provide a sense of size that exact values do not, but they can introduce severe errors if not used with care.

Example 11   From Figure 1.5.8 we see that the parabola $y = -x^2 + 4x - 5$ has no x-intercepts. This can also be seen algebraically by solving for the x-intercepts. Setting $y = 0$ and solving the resulting equation

$$-x^2 + 4x - 5 = 0$$

by the quadratic formula yields

$$y = \frac{-4 \pm \sqrt{16 - 20}}{-2} = 2 \pm i$$

Because the solutions are complex numbers, there are no (real) x-intercepts. $\blacktriangleleft$

Example 12   A ball is thrown straight up from the surface of the earth at time $t = 0$ sec with an initial velocity of 24.5 m/sec. If air resistance is ignored, it can be shown that the distance s (in meters) of the ball above the ground after t sec is given by

$$s = 24.5t - 4.9t^2 \tag{9}$$

(a)   Graph s versus t, making the t-axis horizontal and the s-axis vertical.

(b)   How high does the ball rise above the ground?

Solution (a).   Equation (9) is of form (7) with $a = -4.9$, $b = 24.5$, and $c = 0$, so by (8) the t-coordinate of the vertex is

$$t = -\frac{b}{2a} = -\frac{24.5}{2(-4.9)} = 2.5 \text{ sec}$$

and consequently the s-coordinate of the vertex is

$$s = 24.5(2.5) - 4.9(2.5)^2 = 30.625 \text{ m}$$

The factored form of (9) is

$$s = 4.9t(5 - t)$$

so the graph has t-intercepts $t = 0$ and $t = 5$. From the vertex and the intercepts we obtain the graph shown in Figure 1.5.9.

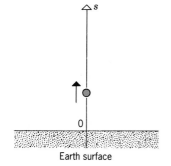

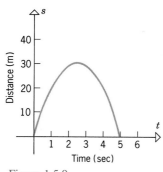

Figure 1.5.9

Solution (b).   From the *s*-coordinate of the vertex we deduce that the ball rises 30.625 m above the ground.   ◄

☐ **THE GRAPH OF**
$x = ay^2 + by + c$

If x and y are interchanged in (7), the resulting equation,

$$x = ay^2 + by + c$$

is called a ***quadratic equation in y***. The graph of such an equation is a parabola with its line of symmetry parallel to the x-axis and its vertex at the point with y-coordinate $y = -b/(2a)$ (Figure 1.5.10). Some problems relating to such equations appear in the exercises.

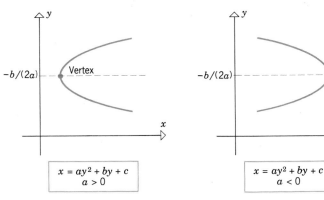

Figure 1.5.10

► Exercise Set 1.5   Ⓒ 81

1. Where in this section did we use the fact that the same unit of measure was used on both coordinate axes?

In Exercises 2–5, find
(a)   the distance between A and B
(b)   the midpoint of the line segment joining A and B.

2. $A(2, 5)$, $B(-1, 1)$.

3. $A(7, 1)$, $B(1, 9)$.

4. $A(2, 0)$, $B(-3, 6)$.

5. $A(-2, -6)$, $B(-7, -4)$.

In Exercises 6–10, use the distance formula to solve the problem.

6. Prove that $(1, 1)$, $(-2, -8)$, and $(4, 10)$ lie on a straight line.

7. Prove that the triangle with vertices $(5, -2)$, $(6, 5)$, $(2, 2)$ is isosceles.

8. Prove that $(1, 3)$, $(4, 2)$, and $(-2, -6)$ are vertices of a right triangle and specify the vertex at which the right angle occurs.

9. Prove that $(0, -2)$, $(-4, 8)$, and $(3, 1)$ lie on a circle with center $(-2, 3)$.

10. Prove that for all values of t the point $(t, 2t - 6)$ is equidistant from $(0, 4)$ and $(8, 0)$.

11. Find k, given that $(2, k)$ is equidistant from $(3, 7)$ and $(9, 1)$.

12. Find x and y if $(4, -5)$ is the midpoint of the line segment joining $(-3, 2)$ and (x, y).

In Exercises 13 and 14, find an equation of the given line.

13. The line is the perpendicular bisector of the line segment joining $(2, 8)$ and $(-4, 6)$.

14. The line is the perpendicular bisector of the line segment joining $(5, -1)$ and $(4, 8)$.

15. Find the point on the line $4x - 2y + 3 = 0$ that is equidistant from $(3, 3)$ and $(7, -3)$. [*Hint:* First find an equation of the line that is the perpendicular bisector of the line segment joining $(3, 3)$ and $(7, -3)$.]

16. Find the distance from the point $(3, -2)$ to the line
(a) $y = 4$ (b) $x = -1$.

17. Find the distance from the point $(2, 1)$ to the line $4x - 3y + 10 = 0$. [*Hint:* Find the foot of the perpendicular dropped from the point to the line.]

18. Find the distance from the point $(8, 4)$ to the line $5x + 12y - 36 = 0$. [*Hint:* See the hint in Exercise 17.]

19. Use the method described in Exercise 17 to prove that the distance d from (x_0, y_0) to the line $Ax + By + C = 0$ is

$$d = \frac{|Ax_0 + By_0 + C|}{\sqrt{A^2 + B^2}}$$

20. Use the formula in Exercise 19 to solve Exercise 17.

21. Use the formula in Exercise 19 to solve Exercise 18.

22. Prove: For any triangle, the perpendicular bisectors of the sides meet at a point. [*Hint:* Position the triangle with one vertex on the y-axis and the opposite side on the x-axis, so that the vertices are $(0, a)$, $(b, 0)$, and $(c, 0)$.]

In Exercises 23 and 24, find the center and radius of each circle.

23. (a) $x^2 + y^2 = 25$

(b) $(x - 1)^2 + (y - 4)^2 = 16$

(c) $(x + 1)^2 + (y + 3)^2 = 5$

(d) $x^2 + (y + 2)^2 = 1$.

24. (a) $x^2 + y^2 = 9$

(b) $(x - 3)^2 + (y - 5)^2 = 36$

(c) $(x + 4)^2 + (y + 1)^2 = 8$

(d) $(x + 1)^2 + y^2 = 1$.

In Exercises 25–32, find the standard equation of the circle satisfying the given conditions.

25. Center $(3, -2)$; radius $= 4$.

26. Center $(1, 0)$; diameter $= \sqrt{8}$.

27. Center $(-4, 8)$; circle is tangent to the x-axis.

28. Center $(5, 8)$; circle is tangent to the y-axis.

29. Center $(-3, -4)$; circle passes through the origin.

30. Center $(4, -5)$; circle passes through $(1, 3)$.

31. A diameter has endpoints $(2, 0)$ and $(0, 2)$.

32. A diameter has endpoints $(6, 1)$ and $(-2, 3)$.

In Exercises 33–44, determine whether the equation represents a circle, a point, or no graph. If the equation represents a circle, find the center and radius.

33. $x^2 + y^2 - 2x - 4y - 11 = 0$.

34. $x^2 + y^2 + 8x + 8 = 0$.

35. $2x^2 + 2y^2 + 4x - 4y = 0$.

36. $6x^2 + 6y^2 - 6x + 6y = 3$.

37. $x^2 + y^2 + 2x + 2y + 2 = 0$.

38. $x^2 + y^2 - 4x - 6y + 13 = 0$.

39. $9x^2 + 9y^2 = 1$.

40. $x^2/4 + y^2/4 = 1$.

41. $x^2 + y^2 + 10y + 26 = 0$.

42. $x^2 + y^2 - 10x - 2y + 29 = 0$.

43. $16x^2 + 16y^2 + 40x + 16y - 7 = 0$.

44. $4x^2 + 4y^2 - 16x - 24y = 9$.

45. Find an equation of

(a) the bottom half of the circle $x^2 + y^2 = 16$

(b) the top half of the circle
$x^2 + y^2 + 2x - 4y + 1 = 0$.

46. Find an equation of

(a) the right half of the circle $x^2 + y^2 = 9$

(b) the left half of the circle $x^2 + y^2 - 4x + 3 = 0$.

47. Graph

(a) $y = \sqrt{25 - x^2}$

(b) $y = \sqrt{5 + 4x - x^2}$.

48. Graph

(a) $x = -\sqrt{4 - y^2}$

(b) $x = 3 + \sqrt{4 - y^2}$.

49. Find an equation of the line that is tangent to the circle $x^2 + y^2 = 25$ at the point $(3, 4)$ on the circle.

50. Find an equation of the line that is tangent to the circle at the point P on the circle

(a) $x^2 + y^2 + 2x = 9$; $P(2, -1)$

(b) $x^2 + y^2 - 6x + 4y = 13$; $P(4, 3)$.

51. For the circle $x^2 + y^2 = 20$ and the point $P(-1, 2)$:

(a) Is P inside, outside, or on the circle?

(b) Find the largest and smallest distances between P and points on the circle.

52. Follow the directions of Exercise 51 for the circle $x^2 + y^2 - 2y - 4 = 0$ and the point $P(3, \frac{5}{2})$.

53. Referring to Figure 1.5.11, find the coordinates of the points T and T', where the lines L and L' are tangent to the circle of radius 1 with center at the origin.

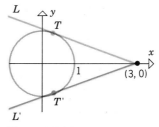

Figure 1.5.11

54. A point (x, y) moves so that its distance to $(2, 0)$ is $\sqrt{2}$ times its distance to $(0, 1)$.

(a) Show that the point moves along a circle.

(b) Find the center and radius.

55. A point (x, y) moves so that the sum of the squares of its distances from $(4, 1)$ and $(2, -5)$ is 45.

(a) Show that the point moves along a circle.

(b) Find the center and radius.

56. Find all values of c for which the system of equations
$$\begin{cases} x^2 - y^2 = 0 \\ (x - c)^2 + y^2 = 1 \end{cases}$$
has 0, 1, 2, 3, or 4 solutions. [*Hint:* Sketch a graph.]

In Exercises 57–70, graph the parabola and label the coordinates of the vertex and the intersections with the coordinate axes.

57. $y = x^2 + 2$.   **58.** $y = x^2 - 3$.

59. $y = x^2 + 2x - 3$.   **60.** $y = x^2 - 3x - 4$.

61. $y = -x^2 + 4x + 5$.   **62.** $y = -x^2 + x$.

63. $y = (x - 2)^2$.   **64.** $y = (3 + x)^2$.

65. $x^2 - 2x + y = 0$.   **66.** $x^2 + 8x + 8y = 0$.

67. $y = 3x^2 - 2x + 1$.   **68.** $y = x^2 + x + 2$.

69. $x = -y^2 + 2y + 2$.   **70.** $x = y^2 - 4y + 5$.

71. Find an equation of
  (a) the right half of the parabola $y = 3 - x^2$
  (b) the left half of the parabola $y = x^2 - 2x$.

72. Find an equation of
  (a) the upper half of the parabola $x = y^2 - 5$
  (b) the lower half of the parabola $x = y^2 - y - 2$.

73. Graph
  (a) $y = \sqrt{x + 5}$ (b) $x = -\sqrt{4 - y}$.

74. Graph
  (a) $y = 1 + \sqrt{4 - x}$ (b) $x = 3 + \sqrt{y}$.

75. If a ball is thrown straight up with an initial velocity of 32 ft/sec, then after t seconds the distance s above its starting height, in feet, is given by $s = 32t - 16t^2$.
  (a) Graph this equation in a ts-coordinate system (t-axis horizontal).
  (b) At what time t will the ball be at its highest point, and how high will it rise?

76. A rectangular field is to be enclosed with 500 ft of fencing along three sides and by a straight stream on the fourth side. Let x be the length of each side perpendicular to the stream, and let y be the length of the side parallel to the stream.

  (a) Express y in terms of x.
  (b) Express the area A of the field in terms of x.
  (c) What is the largest area that can be enclosed?

77. A rectangular plot of land is to be enclosed using two kinds of fencing. Two opposite sides will have heavy-duty fencing costing \$3/ft, while the other two sides will have standard fencing costing \$2/ft. A total of \$600 is available for the fencing. Let x be the length of each side with heavy-duty fencing, and let y be the length of each side with standard fencing.

  (a) Express y in terms of x.
  (b) Find a formula for the area A of the rectangular plot in terms of x.
  (c) What is the largest area that can be enclosed?

78. (a) By completing the square, show that the equation $y = ax^2 + bx + c$ can be rewritten as
$$y = a\left(x + \frac{b}{2a}\right)^2 + \left(c - \frac{b^2}{4a}\right)$$
   if $a \neq 0$.
  (b) Use the result in part (a) to show that the graph of $y = ax^2 + bx + c$ has its high point at $x = -b/(2a)$ if $a < 0$ and its low point there if $a > 0$.

In Exercises 79 and 80, solve the given inequality.

79. (a) $2x^2 + 5x - 1 < 0$ (b) $x^2 - 2x + 3 > 0$.

80. (a) $x^2 + x - 1 > 0$ (b) $x^2 - 4x + 6 < 0$.

81. At time $t = 0$ a ball is thrown straight up from a height of 5 feet above the ground. After t seconds its distance s, in feet, above the ground is given by $s = 5 + 40t - 16t^2$.
  (a) Find the maximum height of the ball above the ground.
  (b) Find, to the nearest tenth of a second, the time when the ball strikes the ground.
  (c) Find, to the nearest tenth of a second, how long the ball will be more than 12 feet above the ground.

82. Find all values of x at which points on the parabola $y = x^2$ lie below the line $y = x + 3$.

◆ TECHNOLOGY EXERCISES   Chapter 1

Most of these exercises require access to a graphing calculator or a computer algebra system (CAS) such as *Mathematica*, *Maple*, or *Derive*. When you are asked to *find* an answer or to *solve* an equation, you may choose to find an exact result or a numerical approximation, depending on the particular technology you are using and on your own imagination. The form of your answers may differ from those of other students or from those in the answer section of the text, depending on how you solve the problems and the accuracy you use in your numerical approximations.

Equations and Inequalities: In Exercises 1–10, solve the given equation or inequality.

1. $x^2 = 2\sqrt{x^3 - 1}$.

2. $x + \sqrt{x} = x^2 - \sqrt{x}$.

3. $x^3 + 1 = 3^x$.

4. $x^2 - 2x = \cos x$.

5. $\dfrac{x + 2}{x^2 + 3} < 0.3$.

6. $\dfrac{\sin x}{x^2 + 1} > 0.1$.

7. $3^x > x^3$.

8. $3^x > x^5$.

9. $|x^3 + 2x - 3| < 5$.

10. $|x^3 - 3x^2 + 1| > 2$.

11. **Distance between a point and a curve:** Write an expression for the distance between the point $P(1, 2)$ and an arbitrary point $(x, \sqrt{x})$ on the curve $y = \sqrt{x}$. Graph this distance versus x, and use the graph to find the x-coordinate of the point on the curve that is closest to the point P.

12. **Distance between a point and a curve:** Write an expression for the distance between the point $P(1, 0)$ and an arbitrary point $(x, 1/x)$ on the curve $y = 1/x$, where $x > 0$. Graph this distance versus x, and use the graph to find the x-coordinate of the point on the curve that is closest to the point P.

In Exercises 13 and 14, use *Archimedes' principle*: *A body wholly or partially immersed in a fluid is buoyed up by a force equal to the weight of the fluid that it displaces.*

13. **Floating sphere:** A hollow metal sphere of diameter 5 feet weighs 108 pounds and floats partially submerged in seawater. Assuming that the weight density of seawater is 63.9 pounds per cubic foot, how far below the surface is the bottom of the sphere? [*Hint:* If a sphere of radius r is submerged to a depth h, then the volume V of the submerged portion is given by the formula $V = \pi h^2(r - h/3)$.]

14. **Floating sphere:** Suppose that a hollow metal sphere of diameter 5 feet and weight w pounds floats in seawater. (See Exercise 13.)

    (a) Graph w versus h for $0 \le h \le 5$.

    (b) Find the weight of the sphere if exactly half of the sphere is submerged.

In Exercises 15 and 16, assume that the following consequence of *Newton's Inverse-Square Law of Gravitational Attraction* holds: *Two homogeneous spherical bodies with masses m_1 and m_2 attract each other with a force of magnitude Gm_1m_2/d^2, where G is a constant and d is the distance between the centers of the bodies.*

15. **Equilibrium point between the earth and the moon:** Suppose that at a certain instant the distance between the center of the earth and the center of the moon is 3.84×10^8 meters. At what distance from the center of the earth will the pull of the earth on a spherical body of mass m be equal in magnitude, but opposite in direction, to the pull of the moon on the body? Use 5.98×10^{24} kilograms as the mass of the earth, and 7.35×10^{22} kilograms as the mass of the moon.

16. **The earth and the sun:** Assume that the centers of the earth and sun are in an xy-coordinate system with the center of the earth at the origin and the center of the sun at the point $(1.49 \times 10^{11}, 0)$ with distance measured in meters. Show that the set of all points of mass m in the xy-plane for which the magnitude of the gravitational pull of the earth is equal to the magnitude of the gravitational pull of the sun forms a circle. Find the center and radius of the circle. [Use 5.98×10^{24} kilograms as the mass of the earth and 1.97×10^{30} kilograms as the mass of the sun.]

17. **Resource management:** A breeding group of 20 bighorn sheep is released in a protected area in Colorado. It is expected that with careful management the number of sheep, N, after t

years will be given by the formula

$$N = \frac{220}{1 + 10(0.83)^t}$$

and that the sheep population will be able to maintain itself without further supervision once the population reaches a size of 80.

(a) How many years must the state of Colorado maintain a program to care for the sheep?

(b) How many bighorn sheep can the environment in the protected area support? [*Hint:* Examine the graph of N versus t for large values of t.]

18. **A changing water table:** Suppose that t days after a heavy rain, the underground water level at a certain location decreases at the rate of r inches per day, where r is given by

$$r = \frac{9}{t(2.7)^{1/t}}$$

(a) Graph r versus t.

(b) For what values of t will the water level be decreasing at a rate of 2 inches per day?

(c) When is the water level decreasing most rapidly, and what is its rate at that time?

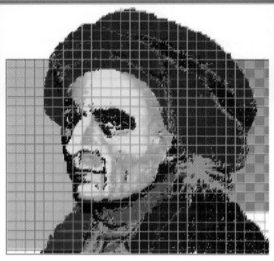

Leonhard Euler (1707–1783)

2 FUNCTIONS AND LIMITS

In this section we shall define one of the most fundamental concepts in mathematics, the notion of a function. We shall discuss the notation used to describe functions and investigate some of their basic properties.

□ THE CONCEPT OF A FUNCTION

Historically, the term "function" was first used by Leibniz in 1673 to denote the dependence of one quantity on another. For example:

- The area of a circle depends on its radius r by the equation $A = \pi r^2$, so we say that "A is a function of r."

- The velocity v of a ball falling freely in the earth's gravitational field increases with time t until it hits the ground, so we say that "v is a function of t."

- In a bacteria culture, the number n of bacteria present after one hour of growth depends on the number N of bacteria present initially, so we say that "n is a function of N."

Table 2.1.1

VALUE OF x	VALUE OF $y = 4x + 1$
$x = 2$	$y = 9$
$x = 1$	$y = 5$
$x = 0$	$y = 1$
$x = -\frac{1}{4}$	$y = 0$
$x = \sqrt{3}$	$y = 4\sqrt{3} + 1$

In general, if a quantity y depends on a quantity x in such a way that each value of x determines exactly one value of y, then we say that "y is a function of x." For example, the equation

$$y = 4x + 1$$

defines y as a function of x because each value assigned to x determines exactly one value of y (Table 2.1.1).

In order to describe functions without stating specific formulas, the Swiss mathematician, Leonhard Euler* (pronounced "oiler") conceived the idea of denoting functions by letters of the alphabet. For example, if we use the letter f to denote a function, then the equation

$$y = f(x) \tag{1}$$

(read, "y equals f of x") indicates that y is a function of x. The quantity x in (1) is called the **independent variable** of f and the quantity y the **dependent variable** of f. This terminology conveys the idea that we are free to assign values to the variable x, but that once a value is assigned to x, a unique value of y is determined. For example, in Table 2.1.1 we arbitrarily assigned values to x, then determined the corresponding y values from the equation $y = 4x + 1$.

REMARK.   It is important to understand that in (1), x and y may represent numerical quantities, but f itself does not represent a numerical quantity; it stands for a "relationship" between y and x.

Functions are commonly specified by formulas that do not explicitly involve a dependent variable. For example, the formula

$$f(x) = x^2 \tag{2}$$

defines a function f that associates the number x^2 with the number x. Thus,

$$f(3) = 3^2 = 9 \qquad \boxed{f \text{ associates 9 with 3.}}$$
$$f(-2) = (-2)^2 = 4 \qquad \boxed{f \text{ associates 4 with } -2.}$$
$$f(0) = 0^2 = 0 \qquad \boxed{f \text{ associates 0 with 0.}}$$
$$f(\sqrt{2}) = (\sqrt{2})^2 = 2 \qquad \boxed{f \text{ associates 2 with } \sqrt{2}.}$$

---

*LEONHARD EULER (1707–1783). Euler was probably the most prolific mathematician who ever lived. It has been said that, "Euler wrote mathematics as effortlessly as most men breathe." He was born in Basel, Switzerland, and was the son of a Protestant minister who had himself studied mathematics. Euler's genius developed early. He attended the University of Basel, where by age 16 he obtained both a Bachelor of Arts degree and a Master's degree in philosophy. While at Basel, Euler had the good fortune to be tutored one day a week in mathematics by a distinguished mathematician, Johann Bernoulli. At the urging of his father, Euler then began to study theology. The lure of mathematics was too great, however, and by age 18 Euler had begun to do mathematical research. Nevertheless, the influence of his father and his theological studies remained, and throughout his life Euler was a deeply religious, unaffected person. At various times Euler taught at St. Petersburg Academy of Sciences (in Russia), the University of Basel, and the Berlin Academy of Sciences. Euler's energy and capacity for work were virtually boundless. His collected works form more than 100 quarto-sized volumes and it is believed that much of his work has been lost. What is particularly astonishing is that Euler was blind for the last 17 years of his life, and this was one of his most productive periods! Euler's flawless memory was phenomenal. Early in his life he memorized the entire *Aeneid* by Virgil and at age 70 could not only recite the entire work, but could also state the first and last sentence on each page of the book from which he memorized the work. His ability to solve problems in his head was beyond belief. He worked out in his head major problems of lunar motion that baffled Isaac Newton and once did a complicated calculation in his head to settle an argument between two students whose computations differed in the fiftieth decimal place.

Following the development of calculus by Leibniz and Newton, results in mathematics developed rapidly in a disorganized way. Euler's genius gave coherence to the mathematical landscape. He was the first mathematician to bring the full power of calculus to bear on problems from physics. He made major contributions to virtually every branch of mathematics as well as to the theory of optics, planetary motion, electricity, magnetism, and general mechanics.

Although Formula (2) does not involve a dependent variable, we can introduce one by letting $y = f(x)$ and rewriting (2) as

$$y = x^2$$

In some situations it is preferable to define a function without using a dependent variable, and in other situations a dependent variable is desirable.

Although f is the symbol most commonly used to denote a function, any symbol can be used. Thus,

$$y = F(x), \quad y = f_1(x), \quad y = g(x), \quad \text{and} \quad y = \phi(x)$$

all express the fact that y is a function of x. It is also not necessary to use the letters x and y for the independent and dependent variables. Any symbols can be used. For example, the equation $s = f(t)$ states that the dependent variable s is a function of the independent variable t.

Example 1 If $f(x) = 3x - 4$, then

$$f(0) = 3 \cdot 0 - 4 = -4$$
$$f(1) = (3 \cdot 1) - 4 = -1$$
$$f(2) = (3 \cdot 2) - 4 = 2$$
$$f(-3) = [3 \cdot (-3)] - 4 = -13$$
$$f(\sqrt{5}) = (3 \cdot \sqrt{5}) - 4 = 3\sqrt{5} - 4 \quad \blacktriangleleft$$

Example 2 If $\phi(x) = \dfrac{1}{x^3 - 1}$, then

$$\phi(\sqrt[3]{7}) = \frac{1}{(\sqrt[3]{7})^3 - 1} = \frac{1}{7 - 1} = \frac{1}{6}$$

$$\phi(5^{1/6}) = \frac{1}{(5^{1/6})^3 - 1} = \frac{1}{5^{3/6} - 1} = \frac{1}{\sqrt{5} - 1}$$

$\phi(1)$ is undefined (division by zero) $\quad \blacktriangleleft$

Example 3 If $F(x) = 2x^2 - 1$, then

$$F(4) = 2(4)^2 - 1 = 31$$
$$F(t) = 2t^2 - 1$$
$$F(k + 1) = 2(k + 1)^2 - 1 = 2k^2 + 4k + 1$$
$$F(k) + 1 = (2k^2 - 1) + 1 = 2k^2 \quad \blacktriangleleft$$

For reasons that will be discussed later, radian measure of angles is preferred over degree measure in calculus. Therefore, in trigonometric expressions such as $\sin x$, $\cos x$, $\tan x$, $\cot x$, $\sec x$, and $\csc x$ it is understood that x is measured in radians. When degree measure is called for, we shall make this clear by writing $\sin x°$, $\cos x°$, and so forth.

Example 4 If $f(x) = \sin x$, then

$$f(0) = \sin(0) = 0, \quad f(\pi/2) = \sin(\pi/2) = 1, \quad f(1) = \sin(1) \approx .841471 \quad \blacktriangleleft$$

Use a calculator set to the radian mode.

Example 5 Consider the following formulas, which are identical except for the symbol used for the independent variable:

$$g(x) = x^2 - 3x + 5 \quad \text{and} \quad g(t) = t^2 - 3t + 5$$

If one substitutes $x = 2$ in the first formula or $t = 2$ in the second formula, the resulting computations are the same, namely,

$$g(2) = 2^2 - 3 \cdot 2 + 5 = 3$$

In general, two formulas that differ only in the symbol used for the independent variable define the *same* function; that is, it is the structure of the formula and not the letter used for the independent variable that determines the function. ◄

☐ **DOMAIN**

In many situations the independent variable of a function is not free to vary arbitrarily. Rather, it must be restricted to lie in some set, called the ***domain*** of the function. If the function is f and $y = f(x)$, then the domain of f can be viewed as the set of allowable values for the independent variable x.

In applications, the domain of a function is often determined by physical considerations; for example, suppose that a square with a side of length x cm is cut from each corner of a piece of cardboard that is 10 cm square, and let y be the area (in cm^2) of the cardboard that remains (Figure 2.1.1).

By subtracting the areas of the four corner squares from the original area, it follows that

$$y = 100 - 4x^2 \tag{3}$$

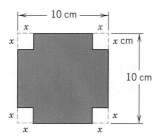

Figure 2.1.1

However, x cannot be negative because it denotes a length, and its value cannot exceed 5 (why?). Thus, x must satisfy the restriction $0 \le x \le 5$. Therefore, even though it is not stated explicitly, the underlying physical meaning of x dictates that the domain of the function defined by Equation (3) is the set

$$\{x : 0 \le x \le 5\} = [0, 5] \tag{4}$$

REMARK.   A practical-minded engineer might argue that $x = 0$ and $x = 5$ should be excluded from the domain because these values do not correspond to "physical" cuts in the cardboard (why?). On the other hand, a mathematician might argue that these values should be regarded as "theoretical" possibilities and thus included. This issue often arises when a physical problem is represented (or "modeled") mathematically. Usually, it will not matter which viewpoint you take as long as you are consistent within a problem and give an appropriate physical interpretation to any answers you obtain.

☐ **RANGE**

If x is a number in the domain of a function f, then the number $f(x)$ that f associates with x is called the ***value of f at x*** or the ***image of x under f***. Thus, if

$$f(x) = x^2$$

then the value of f at $x = 3$ is $f(3) = 3^2 = 9$, and the value of f at $x = -2$ is $f(-2) = (-2)^2 = 4$. Equivalently, if we introduce a dependent variable

$$y = x^2$$

then the value of y at $x = 3$ is $y = 3^2 = 9$, and the value of y at $x = -2$ is $y = (-2)^2 = 4$. The set of all possible values of $f(x)$ as x varies over the domain of f is called the ***range of f***. If $y = f(x)$, then the range of f can be viewed as the set of all possible values for the dependent variable y.

Example 6   We saw above that the function relating the variables x and y in Figure 2.1.1 is given by

$$y = 100 - 4x^2$$

where the domain of the function is the set

$$\{x : 0 \le x \le 5\} = [0, 5]$$

As x varies over this domain, the values of $y = 100 - 4x^2$ vary between $y = 0$ (when $x = 5$) and $y = 100$ (when $x = 0$), so the range of the function is

$$\{y : 0 \le y \le 100\} = [0, 100] \quad \blacktriangleleft$$

☐ **DOMAINS DETERMINED BY PHYSICAL OR GEOMETRIC CONSIDERATIONS**

Restrictions on the independent variable that determine the domain of a function generally come about in one of three ways:

- Physical or geometric considerations.
- Natural restrictions that result from a formula used to define the function.
- Artificial restrictions imposed by a problem solver for one purpose or another.

Example 7 We saw in Example 6 that the function relating the variables x and y in Figure 2.1.1 is given by

$$y = 100 - 4x^2$$

where the domain of this function is the closed interval $[0, 5]$. This domain is a consequence of the physical restrictions on x in the problem. We would generally denote this function and its domain by writing

$$y = 100 - 4x^2, \quad 0 \le x \le 5 \quad \blacktriangleleft$$

Example 8 Assume that it costs 12 cents to manufacture a certain kind of computer chip, and let $f(x)$ be the cost in dollars of manufacturing x such chips. The formula for $f(x)$ is

$$f(x) = 0.12x \tag{5}$$

Physically, x must be a nonnegative integer, so the domain of the function f in (5) is the set $\{x : x = 0, 1, 2, \ldots\}$. To make the domain of f completely clear we would write

$$f(x) = 0.12x, \quad x = 0, 1, 2, \ldots \quad \blacktriangleleft$$

☐ **THE NATURAL DOMAIN**

Not all functions studied in mathematics arise from physical or geometric problems—many are studied purely as mathematical objects. For such functions there are no physical or geometric restrictions on the independent variable. However, restrictions can arise from formulas used to define such functions.

Example 9 If

$$h(x) = \frac{1}{(x - 1)(x - 3)} \tag{6}$$

then the function h is undefined if $x = 1$ or $x = 3$ because division by zero is not defined. However, for all other values of x, the function h is defined and has real values. Thus, the domain of h consists of all real numbers x except $x = 1$ and $x = 3$. In interval notation the domain is

$$(-\infty, 1) \cup (1, 3) \cup (3, +\infty)$$

Figure 2.1.2

(Figure 2.1.2). Although we could explicitly state the domain of h by writing its formula as

$$h(x) = \frac{1}{(x - 1)(x - 3)}, \quad x \ne 1, 3$$

we would not ordinarily do so; we would usually just write (6) and assume that the restrictions $x \ne 1, 3$ are evident from that formula. $\quad \blacktriangleleft$

In general, we make the following convention:

> *If a function is defined by a formula and there is no domain explicitly stated, then it is understood that the domain consists of all real numbers for which the formula makes sense, and the function has a real value. This is called the **natural domain** of the function.*

Example 10   The formula

$$f(x) = x^2$$

makes sense and yields a real value for all real values of x. Thus, the natural domain of f is $(-\infty, +\infty)$.   ◀

Example 11   If

$$g(x) = 2 + \sqrt{x - 1}$$

then g is undefined for $x < 1$, because negative numbers do not have real square roots; the function g is defined and has real values otherwise, and consequently the natural domain of g is $[1, +\infty)$.   ◀

Example 12   If $f(x) = \tan x$, then $f(x)$ is undefined if

$$x = \pm \frac{\pi}{2}, \ \pm \frac{3\pi}{2}, \ \pm \frac{5\pi}{2}, \ldots$$

but has a real value otherwise (Figure 2.1.3). Thus, the natural domain consists of all x except

$$x = \pm \frac{\pi}{2}, \ \pm \frac{3\pi}{2}, \ \pm \frac{5\pi}{2}, \ldots$$ ◀

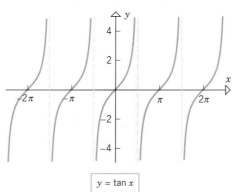

Figure 2.1.3

$$y = \tan x$$

□ **DOMAINS DETERMINED BY SPECIAL RESTRICTIONS**

Sometimes it is desirable to limit the domain of a function to some subset of its natural domain by imposing specific restrictions on the independent variable. Restrictions of this type are often used to focus attention on certain values of the independent variable that may be of special interest in a given problem or situation.

Example 13   The formula

$$f(x) = x^2, \quad 0 \leq x \leq 1$$

defines a function whose domain is the interval $[0, 1]$. Without the stated restriction the domain of f would be the natural domain, $(-\infty, +\infty)$.   ◀

It is a common procedure in algebra to simplify functions by canceling common factors in the numerator and denominator. However, the following example shows that this operation can alter the domain of a function.

Example 14 The function

$$h(x) = \frac{x^2 - 4}{x - 2} \tag{7}$$

has a real value everywhere except at $x = 2$, where we have a division by zero. Thus, the domain of h consists of all x except $x = 2$. However, if we rewrite (7) as

$$h(x) = \frac{(x - 2)(x + 2)}{(x - 2)} \tag{8}$$

then cancel the common factors, we obtain

$$h(x) = x + 2 \tag{9}$$

which *is* defined at $x = 2$ since

$$h(2) = 2 + 2 = 4$$

Thus, our algebraic simplification has altered the domain of the function. In order to cancel the factors in (8) and not alter the domain of h, we must restrict the domain in (9) and write

$$h(x) = x + 2, \quad x \neq 2 \quad \blacktriangleleft$$

REMARK.   In algebra the reader undoubtedly learned to simplify (8) by writing (9). In elementary problems this often causes no difficulty because the domain of the function is irrelevant to the solution of the problem. However, in more advanced problems the domain is often important, so the reader should learn to be precise and place the appropriate restrictions on the domain of the function after canceling factors.

□ **TECHNIQUES FOR FINDING THE RANGE**

Often, the range of a function is evident by inspection.

Example 15 Find the range of

(a) $f(x) = x^2$ (b) $g(x) = 2 + \sqrt{x - 1}$

Solution (a).   Since no domain is stated explicitly, the domain of f is the natural domain $(-\infty, +\infty)$ (see Example 10). To find the range of f it will help to introduce a dependent variable

$$y = x^2$$

As x varies over the domain of f, the corresponding y values (which must be nonnegative) vary over the interval $[0, +\infty)$. This is the range of f.

Solution (b).   Since no domain is stated explicitly, the domain of g is the natural domain $[1, +\infty)$. To determine the range of the function g, let $y = 2 + \sqrt{x - 1}$. As x varies over the interval $[1, +\infty)$, the value of $\sqrt{x - 1}$ varies over the interval $[0, +\infty)$, so the value of $y = 2 + \sqrt{x - 1}$ varies over the interval $[2, +\infty)$. This is the range of g. $\blacktriangleleft$

The next example illustrates a technique that can sometimes be used to find the range of a function when it is not evident by inspection.

Example 16   Find the range of the function $f(x) = \dfrac{x+1}{x-1}$.

Solution.   The natural domain of f consists of all x, except $x = 1$. As in our earlier examples, let us introduce a dependent variable

$$y = \frac{x+1}{x-1}$$

The set of all possible y values is not at all evident from this equation. However, solving this equation for x in terms of y yields

$$(x-1)y = x+1$$
$$xy - y = x+1$$
$$xy - x = y+1$$
$$x(y-1) = y+1$$
$$x = \frac{y+1}{y-1}$$

It is now evident from the right side of this equation that $y = 1$ is not in the range, for otherwise we would have a division by zero. No other values of y are excluded by this equation, so that the range of the function f is $\{y : y \neq 1\} = (-\infty, 1) \cup (1, +\infty)$.   ◄

☐ **FUNCTIONS DEFINED PIECEWISE**

The next example shows that functions must sometimes be defined by formulas that have been "pieced together."

Example 17   The cost of a taxicab ride in a certain metropolitan area is $1.75 for any ride up to and including one mile. After one mile the rider pays an additional amount at the rate of 50 cents per mile (or fraction, thereof). If $f(x)$ is the total cost in dollars for a ride of x miles, then the value of $f(x)$ is

$$f(x) = \begin{cases} 1.75, & 0 < x \leq 1 \\ 1.75 + 0.50(x-1), & 1 < x \end{cases}$$

$1.75 for a ride up to and including one mile

$1.75 for the first mile plus a rate of $0.50 a mile for each mile, or fraction thereof, after the first   ◄

Example 18   From the definition of absolute value, the function $f(x) = |x|$ can be written in an equivalent piecewise form as

$$f(x) = \begin{cases} x, & x \geq 0 \\ -x, & x < 0 \end{cases}$$ ◄

☐ **REVERSING THE ROLES OF x AND y**

Sometimes the form in which an equation is expressed makes it desirable to treat y as an independent variable and x as a dependent variable. For example, the equation

$$x = 4y^5 - 2y^3 + 7y - 5$$

is of the form $x = g(y)$; that is, x is expressed as a function of y. Since it is complicated to solve this equation for y in terms of x, it may be desirable to leave it in this form, treating y as the independent variable and x as the dependent variable.

Sometimes an equation can be solved for y as a function of x or for x as a function of y with equal simplicity. For example, the equation

$$3x + 2y = 6$$

can be written as

$$y = -\tfrac{3}{2}x + 3 \quad \text{or} \quad x = -\tfrac{2}{3}y + 2$$

The choice of forms depends on how the equation will be used.

REMARK.   For all of the functions encountered in this section, the domains and ranges have been sets of real numbers. However, later in this text we shall be interested in functions whose domains and ranges are other mathematical entities.

▶ Exercise Set 2.1   C̲ 57–60

1. Given that $f(x) = 3x^2 + 2$, find
   (a) $f(-2)$ (b) $f(4)$ (c) $f(0)$
   (d) $f(-\sqrt{3})$ (e) $f(a + 1)$ (f) $f(3t)$.

2. Given that $g(x) = \dfrac{x + 1}{x - 1}$, find
   (a) $g(2)$ (b) $g(-2)$ (c) $g(\frac{1}{4})$
   (d) $g(\pi)$ (e) $g(a - 1)$ (f) $g(2t + 1)$.

3. Given that
   $$f(x) = \begin{cases} 1/x, & x > 3 \\ 2x, & x \le 3 \end{cases}$$
   find
   (a) $f(-4)$ (b) $f(4)$ (c) $f(0)$
   (d) $f(3)$ (e) $f(2.9)$ (f) $f(t^2 + 5)$.

4. Given that
   $$g(x) = \begin{cases} \sqrt{x + 1}, & x \ge -1 \\ 3, & x < -1 \end{cases}$$
   find
   (a) $g(0)$ (b) $g(-1.1)$ (c) $g(3)$
   (d) $g(-1)$ (e) $g(-\pi)$ (f) $g(t^2 - 1)$.

In Exercises 5–22, find the natural domain of the given function.

5. $f(x) = \dfrac{1}{x - 3}$.   6. $f(x) = \dfrac{1}{5x + 7}$.

7. $g(x) = \sqrt{x^2 - 3}$.   8. $g(x) = \sqrt{x^2 + 3}$.

9. $h(x) = \sqrt{\dfrac{x - 1}{x + 2}}$.   10. $h(x) = \sqrt{x - 3x^2}$.

11. $\phi(x) = \dfrac{x}{|x| + 1}$.   12. $\phi(x) = \sqrt{3 - \sqrt{x}}$.

13. $F(x) = \sqrt{x - 5} + \sqrt{8 - x}$.

14. $F(x) = 3\sqrt{x} - \sqrt{x^2 - 4}$.

15. $G(x) = \sqrt{x^2 - 2x + 5}$.   16. $G(x) = \sqrt{\dfrac{x^2 - 4}{x - 4}}$.

17. $f(x) = \dfrac{x}{|x|}$.   18. $f(x) = \dfrac{x^2 - 1}{x + 1}$.

19. $g(x) = \sin \sqrt{x}$.   20. $g(x) = \cos \dfrac{1}{x}$.

21. $h(x) = \dfrac{1}{1 - \sin x}$.   22. $h(x) = \dfrac{3}{2 - \cos x}$.

In Exercises 23–36, find the natural domain and the range of the given function.

23. $f(x) = \sqrt{3 - x}$.   24. $f(x) = \sqrt{3x - 2}$.

25. $g(x) = \sqrt{4 - x^2}$.   26. $g(x) = \sqrt{9 - 4x^2}$.

27. $h(x) = 3 + \sqrt{x}$.   28. $h(x) = \dfrac{1}{3 + \sqrt{x}}$.

29. $F(x) = x^2 + 3$.   30. $F(x) = \dfrac{2}{x^2 + 3}$.

31. $G(x) = x^3 + 2$.   32. $G(x) = \dfrac{3}{x}$.

33. $H(x) = 3 \sin x$.   34. $H(x) = \sin^2 \sqrt{x}$.

35. $\phi(x) = 2 + \cos x$.   36. $\phi(x) = \dfrac{5}{3 - \cos 2x}$.

In Exercises 37–40, express the given function in piecewise form without using absolute values.

37. $f(x) = |x| + 3x + 1$.

38. $f(x) = 3 + |2x - 5|$.

39. $g(x) = |x| + |x - 1|$.

40. $g(x) = 3|x - 2| - |x + 1|$.

In Exercises 41–48, find all values of x for which $f(x) = a$.

41. $f(x) = \sqrt{3x - 2}$; $a = 6$.

42. $f(x) = \dfrac{1}{x + 3}$; $a = 5$.

43. $f(x) = x^2 + 5$; $a = 7$.

44. $f(x) = \dfrac{x}{x^2 + 3}$; $a = \frac{1}{4}$.

45. $f(x) = \cos x$; $a = 1$.   46. $f(x) = \sin \dfrac{1}{x}$; $a = 1$.

47. $f(x) = \sin \sqrt{x}$; $a = \frac{1}{2}$.   48. $f(x) = 3 \tan x$; $a = 3$.

49. Express the area A of a circle as a function of its circumference C.

50. Express the area A of an equilateral triangle as a function of
    (a) the length s of each side
    (b) the altitude h.

51. Express the total surface area S of a cube as a function of

(a) the length x of an edge

(b) the volume V of the cube.

52. Express the total surface area S of a right-circular cylinder with given volume V as a function of its radius r.

53. An open box is to be made from an 8 in. × 15 in. piece of sheet metal by cutting out squares with sides of length x from each of the four corners and bending up the sides (Figure 2.1.4). Express the volume V of the box as a function of x, and state the domain of the function.

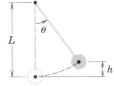

Figure 2.1.4

54. A camera is mounted at a point 3000 ft from the base of a rocket launching pad. The rocket rises vertically when launched. Express the distance x traveled by the rocket as a function of the camera elevation angle θ (Figure 2.1.5).

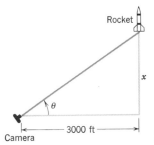

Figure 2.1.5

55. A pendulum of constant length L makes an angle θ with its vertical position (Figure 2.1.6). Express the height h as a function of the angle θ.

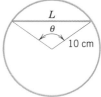

Figure 2.1.6

56. Express the length L of a chord of a circle with radius 10 cm as a function of the central angle θ (Figure 2.1.7).

Figure 2.1.7

57. For a given outside temperature T and wind speed v, the windchill index (WCI) is the equivalent temperature that exposed skin would feel with a wind speed of 4 mi/hr. An empirical formula for the WCI (based on experience and observation) is

$$\text{WCI} = \begin{cases} T, & 0 \le v \le 4 \\ 91.4 + (91.4 - T)(0.0203\,v - 0.304\sqrt{v} - 0.474), & 4 < v < 45 \\ 1.6T - 55, & v \ge 45 \end{cases}$$

where T is the air temperature in °F, v is the wind speed in mi/hr, and WCI is the equivalent temperature in °F. Find the WCI to the nearest degree if the air temperature is 25° F and

(a) $v = 3$ mi/hr

(b) $v = 15$ mi/hr

(c) $v = 46$ mi/hr.

[Adapted from UMAP Module 658, *Windchill*, W. Bosch and L. Cobb, COMAP, Arlington, MA.]

In Exercises 58–60, use the formula for the windchill index described in Exercise 57.

58. Find the air temperature to the nearest degree if the WCI is reported as −60° F with a wind speed of 48 mi/hr.

59. Find the air temperature to the nearest degree if the WCI is reported as −10° F with a wind speed of 8 mi/hr.

60. Find the wind speed to the nearest mile per hour if the WCI is reported as −15° F with an air temperature of 20° F.

61. Criticize the following statement: The function

$$\frac{1 - (1/x)}{1 + (1/x)}$$

can be simplified by multiplying numerator and denominator by x to obtain

$$\frac{1 - (1/x)}{1 + (1/x)} = \frac{x - 1}{x + 1}$$

How would you rewrite the statement to make it accurate?

In Exercises 62–68, simplify the function by canceling factors, and state the natural domain of the simplified function.

62. $f(x) = \dfrac{x^2 - 4}{x + 2}$.

63. $f(x) = \dfrac{(x + 2)(x^2 - 1)}{(x + 2)(x + 1)}$.

64. $f(x) = \dfrac{x^2 + x}{x}$.

65. $f(x) = \dfrac{x + 1 + \sqrt{x + 1}}{\sqrt{x + 1}}$.

66. $f(x) = \dfrac{x^2 - 9}{x - 3}$.

67. $f(x) = \dfrac{x^3 + 2x^2 - 3x}{(x - 1)(x + 3)}$.

68. $f(x) = \dfrac{x + \sqrt{x}}{\sqrt{x}}$.

■ 2.2 OPERATIONS ON FUNCTIONS

> *Just as numbers can be added, subtracted, multiplied, and divided to produce other numbers, so functions can be added, subtracted, multiplied, and divided to produce other functions. In this section we shall discuss these operations and some others that have no analogs in ordinary arithmetic.*

□ **ARITHMETIC OPERATIONS ON FUNCTIONS**

Functions are added, subtracted, multiplied, and divided in a natural way. For example, if

$$f(x) = x \quad \text{and} \quad g(x) = x^2$$

then

$$f(x) + g(x) = x + x^2$$

This formula defines a new function which we call the *sum* of f and g and denote by $f + g$. Thus,

$$(f + g)(x) = f(x) + g(x) = x + x^2$$

In general, we make the following definition.

2.2.1 DEFINITION. Given functions f and g, formulas for the *sum $f + g$*, *difference $f - g$*, *product $f \cdot g$*, and *quotient f/g* are defined by

$$(f + g)(x) = f(x) + g(x)$$
$$(f - g)(x) = f(x) - g(x)$$
$$(f \cdot g)(x)\ \ = f(x) \cdot g(x)$$
$$(f/g)(x)\ \ = f(x)/g(x)$$

For the functions $f + g$, $f - g$, and $f \cdot g$ the domain is defined to be the intersection of the domains of f and g, and for f/g the domain is this intersection with the points where $g(x) = 0$ excluded.

Example 1 Let

$$f(x) = 1 + \sqrt{x - 2} \quad \text{and} \quad g(x) = x - 1$$

Find $(f + g)(x)$, $(f - g)(x)$, $(f \cdot g)(x)$, $(f/g)(x)$, and state the domains of $f + g$, $f - g$, $f \cdot g$, and f/g.

Solution. First, we find formulas for the functions, then we find the domains. The formulas are

$$(f + g)(x) = f(x) + g(x) = (1 + \sqrt{x - 2}) + (x - 1) = x + \sqrt{x - 2} \tag{1}$$
$$(f - g)(x) = f(x) - g(x) = (1 + \sqrt{x - 2}) - (x - 1) = 2 - x + \sqrt{x - 2} \tag{2}$$
$$(f \cdot g)(x) = f(x) \cdot g(x)\ \ = (1 + \sqrt{x - 2})(x - 1) \tag{3}$$
$$(f/g)(x) = f(x)/g(x)\ \ = \frac{1 + \sqrt{x - 2}}{x - 1} \tag{4}$$

The domain of f is the interval $[2, +\infty)$ and the domain of g is the interval $(-\infty, +\infty)$. Thus, the domains of $f + g$, $f - g$, and $f \cdot g$ are the intersections of these two intervals, which is $[2, +\infty)$. But this is precisely the natural domain of (1), (2), and (3), so there is no need to explicitly state the domains. Similarly, the domain of f/g consists of all x in $[2, +\infty)$. Again this is the natural domain in (4). ◄

In the preceding example the formulas for $(f + g)(x)$, $(f - g)(x)$, $(f \cdot g)(x)$, and $(f/g)(x)$ determined the correct domains for the functions $f + g$, $f - g$, $f \cdot g$, and f/g. However, the following example shows that this is not always the case.

Example 2 Find $(f \cdot g)(x)$, where

$$f(x) = 3\sqrt{x} \quad \text{and} \quad g(x) = \sqrt{x}$$

Solution. From Definition 2.2.1,

$$(f \cdot g)(x) = f(x) \cdot g(x) = (3\sqrt{x}) \cdot (\sqrt{x}) = 3x$$

The natural domain of the function $3x$ is $(-\infty, +\infty)$; however, this is not the correct domain of $f \cdot g$. To see why, observe that both f and g have domain $[0, +\infty)$, so that by definition $f \cdot g$ also has domain $[0, +\infty)$, since this is the intersection of the domains of f and g. Thus, the correct formula for $(f \cdot g)(x)$ is

$$(f \cdot g)(x) = 3x, \quad x \geq 0 \quad ◄$$

REMARK. In light of the preceding example, we recommend that the reader develop the habit of indicating the appropriate restrictions on the domain when performing any operations on functions.

Sometimes we shall write f^2 to denote the product $f \cdot f$. For example, if $f(x) = 8x$, then

$$f^2(x) = (f \cdot f)(x) = f(x) \cdot f(x) = (8x) \cdot (8x) = 64x^2$$

In general, if n is a positive integer, then we define

$$f^n(x) = \underbrace{f(x) \cdot f(x) \cdot \cdots \cdot f(x)}_{n \text{ factors}}$$

This notation is especially common with trigonometric functions. For example, $(\sin x)^2$ is generally written as $\sin^2 x$.

☐ **COMPOSITION OF FUNCTIONS**

We shall now consider an operation on functions, called *composition*, which has no direct analog in ordinary arithmetic. Informally stated, the operation of composition is performed by substituting some function for the independent variable of another function. For example, suppose that

$$f(x) = x^2 \quad \text{and} \quad g(x) = x + 1$$

If we substitute $g(x)$ for x in the formula for f, we obtain a new function

$$f(g(x)) = (g(x))^2 = (x + 1)^2$$

which we denote by $f \circ g$. Thus,

$$(f \circ g)(x) = f(g(x)) = (g(x))^2 = (x + 1)^2$$

In general, we make the following definition.

2.2.2 DEFINITION. Given functions f and g, the **composition of f with g**, denoted by $f \circ g$, is the function defined by

$$(f \circ g)(x) = f(g(x))$$

The domain of $f \circ g$ is defined to consist of all x in the domain of g for which $g(x)$ is in the domain of f.

REMARK.   Although the domain of $f \circ g$ may seem complicated at first glance, it is quite natural: To compute $f(g(x))$ one needs x in the domain of g to compute $g(x)$, then one needs $g(x)$ in the domain of f to compute $f(g(x))$.

Example 3   Let $f(x) = x^2 + 3$ and $g(x) = \sqrt{x}$. Find

(a) $(f \circ g)(x)$ (b) $(g \circ f)(x)$

Solution (a).   The formula for $f(g(x))$ is

$$f(g(x)) = [g(x)]^2 + 3 = (\sqrt{x})^2 + 3 = x + 3$$

Since the domain of g is $[0, +\infty)$ and the domain of f is $(-\infty, +\infty)$, the domain of $f \circ g$ consists of all x in $[0, +\infty)$ such that $g(x) = \sqrt{x}$ lies in $(-\infty, +\infty)$; thus, the domain of $f \circ g$ is $[0, +\infty)$. Therefore,

$$(f \circ g)(x) = x + 3, \quad x \geq 0$$

Solution (b).   The formula for $g(f(x))$ is

$$g(f(x)) = \sqrt{f(x)} = \sqrt{x^2 + 3}$$

Since the domain of f is $(-\infty, +\infty)$ and the domain of g is $[0, +\infty)$, the domain of $g \circ f$ consists of all x in $(-\infty, \infty)$ such that $f(x) = x^2 + 3$ lies in $[0, +\infty)$. Thus, the domain of $g \circ f$ is $(-\infty, +\infty)$. Therefore,

$$(g \circ f)(x) = \sqrt{x^2 + 3}$$

There is no need to indicate that the domain is $(-\infty, +\infty)$, since this is the natural domain of $\sqrt{x^2 + 3}$.   ◄

REMARK.   Note that $f \circ g$ and $g \circ f$ are different functions in the preceding example. Thus, the order in which functions are composed can make a difference in the end result. In fact, it is relatively rare that $f \circ g$ and $g \circ f$ are the same.

☐ EXPRESSING A FUNCTION AS A COMPOSITION

Many problems in mathematics are attacked by "decomposing" functions into compositions of simpler functions. For example, consider the function h given by

$$h(x) = (x + 1)^2$$

To evaluate $h(x)$ for a given value of x, we would first compute $x + 1$ and then square the result. These two operations are performed by the functions

$$g(x) = x + 1 \quad \text{and} \quad f(x) = x^2$$

We can express h in terms of f and g by writing

$$h(x) = (x + 1)^2 = [g(x)]^2 = f(g(x))$$

so we have succeeded in expressing h as the composition $h = f \circ g$.

The thought process in this example suggests a general procedure for decomposing a function h into a composition $h = f \circ g$:

- Think about how you would evaluate $h(x)$ for a specific value of x, trying to break the evaluation into two steps performed in succession.

- The first operation in the evaluation will determine a function g and the second a function f.

- The formula for h can then be written as $h(x) = f(g(x))$.

For descriptive purposes, we shall refer to g as the "inside function" and f as the "outside function" in the expression $f(g(x))$. The inside function performs the first operation and the outside function performs the second.

Example 4   Express $h(x) = (x - 4)^5$ as a composition of two functions.

Solution.   To evaluate $h(x)$ for a given value of x we would first compute $x - 4$ and then raise the result to the fifth power. Therefore, the inside function (first operation) is

$$g(x) = x - 4$$

and the outside function (second operation) is

$$f(x) = x^5$$

so $h(x) = f(g(x))$. As a check,

$$f(g(x)) = [g(x)]^5 = (x - 4)^5 = h(x) \quad \blacktriangleleft$$

Example 5   Express $\sin(x^3)$ as a composition of two functions.

Solution.   To evaluate $\sin(x^3)$, we would first compute x^3 and then take the sine, so $g(x) = x^3$ is the inside function and $f(x) = \sin x$ the outside function. Therefore,

$$\sin(x^3) = f(g(x)) \quad \boxed{\text{Where } g(x) = x^3 \text{ and } f(x) = \sin x} \quad \blacktriangleleft$$

Example 6   The following table gives some more examples of decomposing functions into compositions.

Table 2.2.1

FUNCTION	$g(x)$ INSIDE	$f(x)$ OUTSIDE	COMPOSITION
$(x^2 + 1)^{10}$	$x^2 + 1$	x^{10}	$(x^2 + 1)^{10} = f(g(x))$
$\sin^3 x$	$\sin x$	x^3	$\sin^3 x = f(g(x))$
$\tan(x^5)$	x^5	$\tan x$	$\tan(x^5) = f(g(x))$
$\sqrt{4 - 3x}$	$4 - 3x$	$\sqrt{x}$	$\sqrt{4 - 3x} = f(g(x))$
$8 + \sqrt{x}$	$\sqrt{x}$	$8 + x$	$8 + \sqrt{x} = f(g(x))$
$\dfrac{1}{x + 1}$	$x + 1$	$\dfrac{1}{x}$	$\dfrac{1}{x + 1} = f(g(x))$

$\blacktriangleleft$

REMARK.   It should be noted that there is always more than one way to express a function as a composition. For example, here are two ways to express $(x^2 + 1)^{10}$ as a composition that differ from that in Table 2.2.1:

$$(x^2 + 1)^{10} = [(x^2 + 1)^2]^5 = f(g(x)) \quad \boxed{g(x) = (x^2 + 1)^2 \text{ and } f(x) = x^5}$$

$$(x^2 + 1)^{10} = [(x^2 + 1)^3]^{10/3} = f(g(x)) \quad \boxed{g(x) = (x^2 + 1)^3 \text{ and } f(x) = x^{10/3}}$$

☐ **AN EXAMPLE FROM CALCULUS**

Calculations of the type in the next example will arise in a variety of calculus problems later in the text.

Example 7   Let $f(x) = x^2$, and let h be any nonzero real number. Find

$$\frac{f(x + h) - f(x)}{h}$$

and simplify as much as possible.

Solution.

$$\frac{f(x+h) - f(x)}{h} = \frac{(x+h)^2 - x^2}{h} = \frac{x^2 + 2xh + h^2 - x^2}{h}$$

$$= \frac{2xh + h^2}{h} = \frac{h(2x+h)}{h}$$

Canceling the common factor h and being careful to emphasize the restriction on h, we obtain

$$\frac{f(x+h) - f(x)}{h} = 2x + h, \quad h \neq 0 \quad \blacktriangleleft$$

□ **CLASSIFICATION OF FUNCTIONS**

We shall conclude this section by discussing some of the important categories of functions that will occur in this text.

The simplest of all functions are those that assign the same value to every member of the domain. These are called **constant functions**. For example, if f is the constant function defined by $f(x) = 3$, then

$$f(-1) = 3, \quad f(0) = 3, \quad f(\sqrt{2}) = 3, \quad f(9) = 3$$

and so forth.

A function of the form cx^n, where c is a constant and n is a nonnegative integer, is called a **monomial in x**. Examples are

$$2x^3, \quad \pi x^7, \quad 4x^0 (= 4), \quad -6x, \quad x^{17}$$

The functions $4x^{1/2}$ and x^{-3} are *not* monomials because the powers of x are not nonnegative integers. A function that is expressible as the sum of finitely many monomials in x is called a **polynomial in x**. Examples are

$$x^3 + 4x + 7, \quad 3 - 2x^3 + x^{17}, \quad 9, \quad 17 - \tfrac{2}{3}x, \quad x^5$$

The function $(x^2 - 4)^3$ is also a polynomial because it can be expressed as a sum of monomials by performing the cubing operation. Depending on whether one wants the powers written in ascending or descending order, the general formula for a polynomial in x is

$$f(x) = a_0 + a_1 x + a_2 x^2 + \cdots + a_n x^n$$

or

$$f(x) = a_n x^n + a_{n-1} x^{n-1} + \cdots + a_1 x + a_0$$

where n is a nonnegative integer and $a_0, a_1, a_2, \ldots, a_n$ are all constants. The highest power of x that occurs in a nonconstant polynomial is called the **degree** of the polynomial. Thus, $x^3 + 4x + 7$ has degree 3 and $17 - \tfrac{2}{3}x$ has degree 1. A nonzero constant c has degree zero (it can be written as $c = cx^0$). The constant zero is not assigned a degree. First-, second-, and third-degree polynomials are called **linear**, **quadratic**, and **cubic**, respectively. These have the following forms:

DESCRIPTION	GENERAL FORMULA
Linear polynomial	$a_0 + a_1 x \quad (a_1 \neq 0)$
Quadratic polynomial	$a_0 + a_1 x + a_2 x^2 \quad (a_2 \neq 0)$
Cubic polynomial	$a_0 + a_1 x + a_2 x^2 + a_3 x^3 \quad (a_3 \neq 0)$

A function that is expressible as a ratio of two polynomials is called a **rational function**. Examples are

$$\frac{x^5 - 2x^2 + 1}{x^2 - 4}, \quad \frac{x}{x+1}, \quad \frac{1}{x^5}$$

In general, f is a rational function if it is expressible in the form

$$f(x) = \frac{a_0 + a_1 x + \cdots + a_n x^n}{b_0 + b_1 x + \cdots + b_m x^m}$$

The natural domain of f consists of all x where the denominator differs from zero.

The rational functions are part of a broader class of functions called ***explicit algebraic functions***. These are functions that can be evaluated using finitely many additions, subtractions, multiplications, divisions, and root extractions. For example,

$$f(x) = x^{2/3} = (\sqrt[3]{x})^2 \quad \text{and} \quad g(x) = \frac{(x-3)\sqrt[4]{x}}{x^5 + \sqrt{x^2 + 1}}$$

are all explicit algebraic functions.

All remaining functions fall into two categories, ***implicit algebraic functions*** and ***transcendental functions***. We shall not define these terms, but instead refer the interested reader to a classic book in calculus, G. H. Hardy,* *A Course of Pure Mathematics*, Cambridge Press, 1958 (10th edition). Among the transcendental functions are those that involve trigonometric expressions, exponentials, and logarithms.

---

*G. H. HARDY (1877–1947). Hardy was a world-renowned British mathematician. He taught at Cambridge and Oxford universities and was a prolific researcher who produced over 300 research papers. He received numerous medals and honorary degrees for his accomplishments. Hardy's book, *A Course of Pure Mathematics*, which gave the first rigorous English exposition of functions and limits for the college undergraduate, had a great impact on university teaching. Hardy had a rebellious spirit; he once listed among his most ardent wishes: (1) to prove the Riemann hypothesis (a famous unsolved mathematical problem), (2) to make a brilliant play in a crucial cricket match, (3) to prove the nonexistence of God, and (4) to murder Mussolini.

▶ Exercise Set 2.2

1. Let $f(x) = x^2 + 1$. Find
   (a) $f(t)$ (b) $f(t + 2)$ (c) $f(x + 2)$
   (d) $f\left(\dfrac{1}{x}\right)$ (e) $f(x + h)$ (f) $f(-x)$
   (g) $f(\sqrt{x})$ (h) $f(3x)$.

2. Let $g(x) = \sqrt{x}$. Find
   (a) $g(5s + 2)$ (b) $g(\sqrt{x} + 2)$ (c) $3g(5x)$
   (d) $\dfrac{1}{g(x)}$ (e) $g(g(x))$ (f) $g^2(x)$
   (g) $g(1/\sqrt{x})$ (h) $g((x - 1)^2)$.

3. Given that $f(-1) = 4$, $f(2) = 5$, $g(-1) = 3$, and $g(2) = -1$, find
   (a) $(f - g)(-1)$ (b) $(f \cdot g)(-1)$
   (c) $(f/g)(2)$ (d) $(f \circ g)(2)$.

4. Let $f(x) = 3/x$. Find
   (a) $f\left(\dfrac{1}{x}\right) + \dfrac{1}{f(x)}$ (b) $f(x^2) - f^2(x)$.

In Exercises 5–12, find formulas for the following functions, making sure to specify the domain in each case.
   (a) $(f + g)(x)$ (b) $(f - g)(x)$
   (c) $(f \cdot g)(x)$ (d) $(f/g)(x)$
   (e) $(f \circ g)(x)$ (f) $(g \circ f)(x)$

5. $f(x) = 2x$, $g(x) = x^2 + 1$.
6. $f(x) = 3x - 2$, $g(x) = |x|$.
7. $f(x) = \sqrt{x + 1}$, $g(x) = x - 2$.
8. $f(x) = \dfrac{x}{1 + x^2}$, $g(x) = \dfrac{1}{x}$.
9. $f(x) = \sqrt{x - 2}$, $g(x) = \sqrt{x - 3}$.
10. $f(x) = x^3$, $g(x) = \dfrac{1}{\sqrt[3]{x}}$.
11. $f(x) = \sqrt{1 - x^2}$, $g(x) = \sin 3x$.
12. $f(x) = \sin^2 x$, $g(x) = \cos x$.

In Exercises 13–16, find
$$\frac{f(x + h) - f(x)}{h}$$
and simplify as much as possible.

13. $f(x) = 3x^2 - 5$.
14. $f(x) = x^2 + 6x$.
15. $f(x) = 1/x$.
16. $f(x) = 1/x^2$.

17. Let
$$f(x) = \frac{4}{x^2 + 5} \quad \text{and} \quad g(x) = \sqrt{x}$$
Find $(f \circ g)(x)$ and $(g \circ f)(x)$.

18. Let $f(x) = \sqrt{2x - 10}$ and $g(x) = 8x^2 + 5$. Find $(f \circ g)(x)$ and $(g \circ f)(x)$.

19. Let h be defined by $h(x) = 2x - 5$. Find
(a) $(h \circ h)(x)$ (b) $h^2(x)$.

20. Let $g(x) = x^3$ and
$$f(x) = \begin{cases} 5x, & x \le 0 \\ -x, & 0 < x \le 8 \\ \sqrt{x}, & x > 8 \end{cases}$$
Find $(f \circ g)(x)$.

21. Let $f(x) = 1/x$.
(a) If $g(x) = x^2 + 1$, show that $f \circ g$ is defined for all x even though f is not defined when $x = 0$.
(b) Find another function g such that $f \circ g$ is defined for all x.
(c) What property must a function g have in order for $f \circ g$ to be defined for all x?

22. Is it always true that $f \circ g = g \circ f$? Is it ever true that $f \circ g = g \circ f$?

23. Prove or disprove: For any three functions f, g, and h, $f \circ (g \circ h) = (f \circ g) \circ h$.

In Exercises 24–35, express f as a composition of two functions; that is, find functions g and h such that $f = g \circ h$. (Each exercise has more than one correct solution.)

 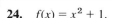

24. $f(x) = x^2 + 1$.   **25.** $f(x) = \sqrt{x + 2}$.

26. $f(x) = \dfrac{1}{x - 3}$.   **27.** $f(x) = (x - 5)^7$.

28. $f(x) = a + bx$.   **29.** $f(x) = |x^2 - 3x + 5|$.

30. $f(x) = 3 \sin(x^2)$.   **31.** $f(x) = \sin^2 x$.

32. $f(x) = \cos^3 2x$.   **33.** $f(x) = \dfrac{3}{5 + \cos x}$.

34. $f(x) = 3 \sin^2 x + 4 \sin x$.

35. $f(x) = \dfrac{\tan x}{3 + \tan x}$.

36. Find functions f, g, and h such that
$$f(g(h(x))) = \sin\sqrt{x^2 + 3x + 7}$$

37. Find functions f, g, and h such that
$$f(g(h(x))) = \sqrt{3 - \sin^2 x}$$

38. Find $(f \circ f)(\pi)$ if $f(x) = \begin{cases} 1 & \text{if } x \text{ is rational} \\ 0 & \text{if } x \text{ is irrational.} \end{cases}$

39. Find $f(x)$ given that $f(x + 1) = x^2 + 3x + 5$. [*Hint:* Let $u = x + 1$ and find $f(u)$.]

40. Find $f(x)$ if $f(3x) = \dfrac{x}{x^2 + 1}$. [*Hint:* Let $u = 3x$ and find $f(u)$.]

41. Given that $f(x) = 0$ only for $x = -1$ and $x = 2$ and that $g(x) = 2x - 1$, find x such that $(f \circ g)(x) = 0$.

42. Find $g(x)$ if $f(x) = 2x - 1$ and $(f \circ g)(x) = x^2$.

43. Find $g(x)$ if $f(x) = \sqrt{x + 5}$ and $(f \circ g)(x) = 3|x|$.

44. Let $f(x) = x^2$, $g(x) = \sin x$, and $h(x) = \cos x$.
(a) Find the exact numerical value of $f(g(0.3)) + f(h(0.3))$
(b) Show that $f(h(x)) - f(g(x)) = h(2x)$.

45. If $f(x + y) = f(x) - f(y)$ for all real x and y, show that $f(x) = 0$ for all real x.

46. If $f(-x) = -f(x)$ for all real x, show that $f(0) = 0$.

For the functions in Exercises 47–50, state which of the following term(s) apply: *monomial*, *polynomial*, *rational function*, *explicit algebraic function*.

47. (a) $3x^7$ (b) $4x^{1/7}$
(c) $\dfrac{1}{5x^6}$ (d) $2x^3 - 1$.

48. (a) $2x^{1/3} + 1$ (b) x^{-2}
(c) $x^{-1/2}$ (d) $(x - 3)^{12}$.

49. (a) $x^2\sqrt{x^3 - 3}$ (b) $\dfrac{x + 1}{x + 2}$
(c) $\pi^{-2} + 1$ (d) $|x|$.

50. (a) $\sqrt{x^2 + \sqrt{x}}$ (b) $\dfrac{x^3 - 2x + 1}{x^2 - 9}$
(c) $\sqrt{\pi} - 3$ (d) $|x - 2|$.

2.3 GRAPHS OF FUNCTIONS

In this section we shall show how to represent functions geometrically by graphs. Such graphs provide a useful way of visualizing the behavior of a function. We shall also develop some basic techniques for using graphs of simple functions to construct graphs of more complicated functions.

☐ **DEFINITION OF THE GRAPH OF A FUNCTION**

We begin with the following definition.

> **2.3.1** DEFINITION. The **graph in the xy-plane of a function f** is defined to be the graph of the equation $y = f(x)$.

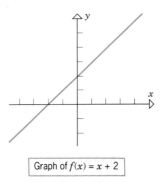

Graph of $f(x) = x + 2$

Figure 2.3.1

Example 1 Sketch the graph of $f(x) = x + 2$.

Solution. By definition, the graph in the xy-plane of f is the graph of the equation $y = x + 2$; this is a line with slope 1 and y-intercept 2 (Figure 2.3.1). ◄

Example 2 Sketch the graph of $f(x) = |x|$.

Solution. By definition, the graph in the xy-plane of f is the graph of $y = |x|$, or equivalently,

$$y = \begin{cases} x, & x \geq 0 \\ -x, & x < 0 \end{cases}$$

The graph coincides with the line $y = x$ for $x \geq 0$ and with the line $y = -x$ for $x < 0$ (Figure 2.3.2). ◄

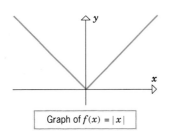

Graph of $f(x) = |x|$

Figure 2.3.2

Example 3 Sketch the graph of

$$\phi(x) = \begin{cases} x + 2, & x \neq 2 \\ 6, & x = 2 \end{cases}$$

Solution. The function ϕ is identical to the function f in Example 1, except at $x = 2$, where we have $\phi(2) = 6$ whereas $f(2) = 4$. Thus, the graph of ϕ is identical to the graph of f shown in Figure 2.3.1, except that the graph of ϕ has a point separated from the line at $x = 2$ (Figure 2.3.3). ◄

Example 4 Sketch the graph of

$$h(x) = \frac{x^2 - 4}{x - 2} \tag{1}$$

Solution. As noted in Example 14 of Section 2.1, this function can be rewritten as

$$h(x) = x + 2, \quad x \neq 2$$

Thus, the function h in (1) is identical to the function f in Example 1, except that h is undefined at $x = 2$. It follows that the graph of h is identical to the graph of f in Figure 2.3.1, except that the graph of h has a hole in it above $x = 2$ (Figure 2.3.4). ◄

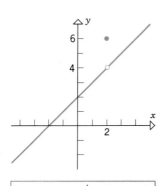

Graph of $\phi(x) = \begin{cases} x + 2, & x \neq 2 \\ 6, & x = 2 \end{cases}$

Figure 2.3.3

Example 5 Sketch the graph of

$$g(x) = \begin{cases} 1, & x \leq 2 \\ x + 2, & x > 2 \end{cases}$$

Solution. By definition, the graph in the xy-plane of g is the graph of

$$y = \begin{cases} 1, & x \leq 2 \\ x + 2, & x > 2 \end{cases}$$

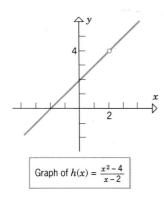

Graph of $h(x) = \frac{x^2-4}{x-2}$

Figure 2.3.4

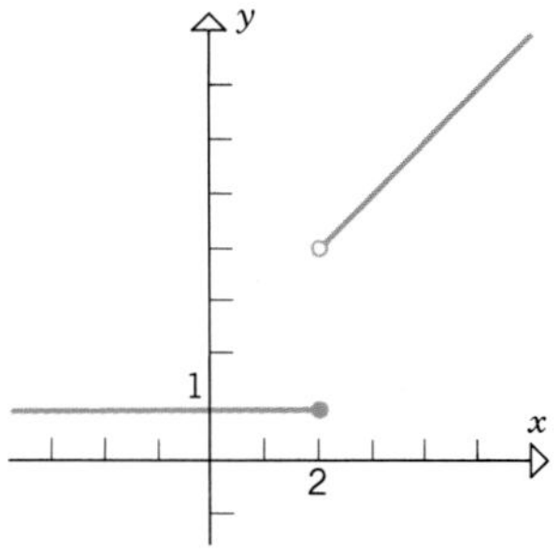

Graph of $g(x) = \begin{cases} 1, & x \le 2 \\ x+2, & x > 2 \end{cases}$

Figure 2.3.5

Thus, for $x \le 2$, we have $y = 1$, and for $x > 2$ we have $y = x + 2$. The graph of $y = 1$ is a horizontal line, and the graph of $y = x + 2$ is the straight line graphed in Example 1. The graph of g is shown in Figure 2.3.5. In that figure we used the heavy dot and open circle above $x = 2$ to emphasize that the value $g(2) = 1$ lies on the horizontal line and not on the inclined line. ◄

☐ GRAPHING FUNCTIONS BY TRANSLATION

Once you have obtained the graph of an equation $y = f(x)$, there are some special techniques that can be used to obtain the graphs of the closely related equations

$$y = f(x) + c, \quad y = f(x) - c, \quad y = f(x + c), \quad y = f(x - c)$$

where c is any positive constant.

If a positive constant is added to or subtracted from $f(x)$, the geometric effect is to translate the graph of the function f parallel to the vertical axis; addition translates the graph in the positive direction and subtraction translates it in the negative direction. This is illustrated in Table 2.3.1. Similarly, if a positive constant is added to or subtracted from the independent variable of a function, the geometric effect is to translate the graph of the function parallel to the horizontal axis; subtraction translates the graph in the positive direction, and addition translates it in the negative direction. This is also illustrated in Table 2.3.1.

Table 2.3.1

OPERATION ON $y = f(x)$	Add a positive constant c to $f(x)$	Subtract a positive constant c from $f(x)$	Add a positive constant c to x	Subtract a positive constant c from x
NEW EQUATION	$y = f(x) + c$	$y = f(x) - c$	$y = f(x + c)$	$y = f(x - c)$
GEOMETRIC EFFECT	Translates the graph of $y = f(x)$ up c units	Translates the graph of $y = f(x)$ down c units	Translates the graph of $y = f(x)$ left c units	Translates the graph of $y = f(x)$ right c units
EXAMPLE	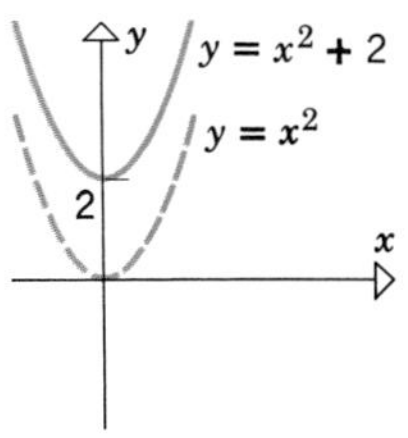 $y = x^2 + 2$, $y = x^2$	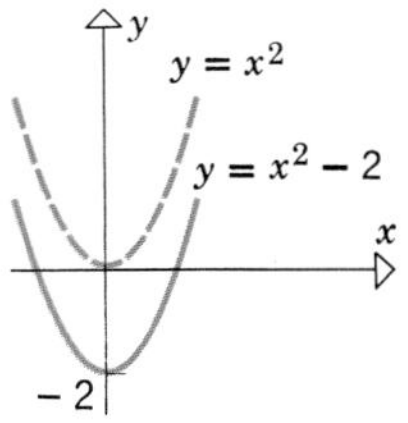 $y = x^2$, $y = x^2 - 2$	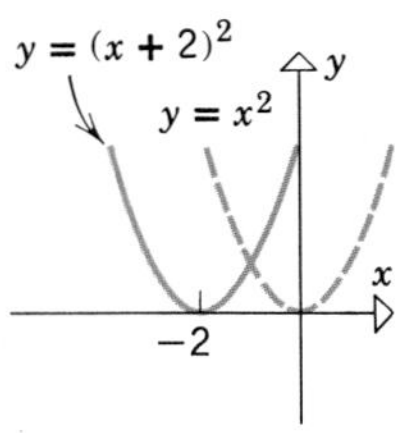 $y = (x+2)^2$, $y = x^2$	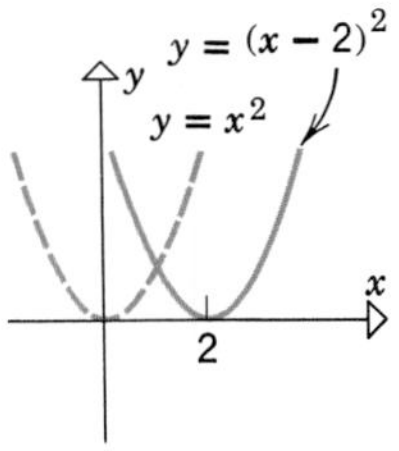 $y = (x-2)^2$, $y = x^2$

Before proceeding to the following examples, it will be helpful to review the graphs in Figure 1.3.21.

Example 6   Sketch the graph of

   (a) $y = \sqrt{x - 3}$ (b) $y = \sqrt{x + 3}$

Solution. The graph of the equation $y = \sqrt{x - 3}$ can be obtained by translating the graph of $y = \sqrt{x}$ right 3 units, and the graph of $y = \sqrt{x + 3}$ by translating the graph of $y = \sqrt{x}$ left 3 units (Figure 2.3.6).   ◄

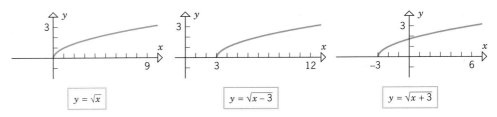

Figure 2.3.6

Example 7   Sketch the graph of $y = |x - 3| + 2$.

Solution. The graph can be obtained by two translations: first translate the graph of $y = |x|$ right 3 units to obtain the graph of $y = |x - 3|$, then translate this graph up 2 units to obtain the graph of $y = |x - 3| + 2$ (Figure 2.3.7).   ◄

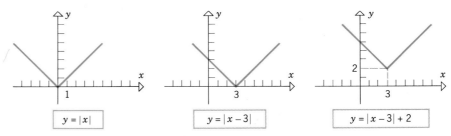

Figure 2.3.7

REMARK.   The graph in the preceding example could also have been obtained by performing the translations in the opposite order: first translating the graph of $y = |x|$ up 2 units to obtain the graph of $y = |x| + 2$, then translating this graph right 3 units to obtain the graph of $y = |x - 3| + 2$.

Example 8   Sketch the graph of $y = x^2 - 4x + 5$.

Solution. Complete the square on the first two terms:

$$y = (x^2 - 4x + 4) - 4 + 5 = (x - 2)^2 + 1$$

In this form we see that the graph can be obtained by translating the graph of $y = x^2$ right 2 units because of the $x - 2$, and up 1 unit because of the $+1$ (Figure 2.3.8).

Alternative Solution. Use the procedure in Example 9 of Section 1.5.   ◄

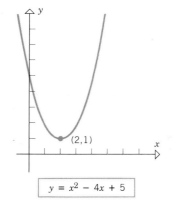

Figure 2.3.8

☐ **REFLECTIONS**

In Section 1.3 (Figure 1.3.16) we observed that $(-x, y)$ is the reflection of (x, y) about the y-axis and $(x, -y)$ is the reflection of (x, y) about the x-axis. Thus, replacing x by $-x$ in an equation reflects its graph about the y-axis and replacing y by $-y$ reflects its graph about the x-axis. In particular,

• the graphs of $y = f(x)$ and $y = f(-x)$ are reflections of one another about the y-axis;
• the graphs of $y = f(x)$ and $-y = f(x)$ [or equivalently, $y = -f(x)$] are reflections of one another about the x-axis.

Examples are given in Table 2.3.2.

Table 2.3.2

OPERATION ON $y = f(x)$	Replace x by $-x$	Multiply $f(x)$ by -1
NEW EQUATION	$y = f(-x)$	$y = -f(x)$
GEOMETRIC EFFECT	Reflects the graph of $y = f(x)$ about the y-axis	Reflects the graph of $y = f(x)$ about the x-axis
EXAMPLE	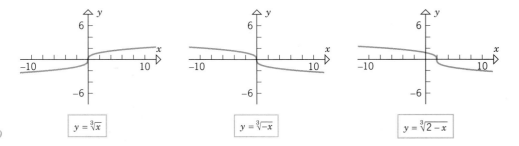	

Example 9 Sketch the graph of $y = \sqrt[3]{2 - x}$.

Solution. The graph can be obtained by a reflection and a translation: first reflect the graph of $y = \sqrt[3]{x}$ about the y-axis to obtain the graph of $y = \sqrt[3]{-x}$, then translate this graph right 2 units to obtain the graph of the equation $y = \sqrt[3]{-(x - 2)} = \sqrt[3]{2 - x}$ (Figure 2.3.9).   ◄

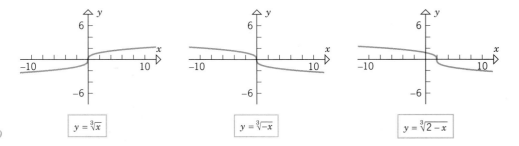

Figure 2.3.9

Example 10 Sketch the graph of $y = 4 - |x - 2|$.

Solution. The graph can be obtained by a reflection and two translations: first reflect the graph of $y = |x|$ about the x-axis to obtain the graph of $y = -|x|$; then translate this graph right 2 units to obtain the graph of $y = -|x - 2|$; and then translate this graph up 4 units to obtain the graph of the equation $y = -|x - 2| + 4 = 4 - |x - 2|$ (Figure 2.3.10).   ◄

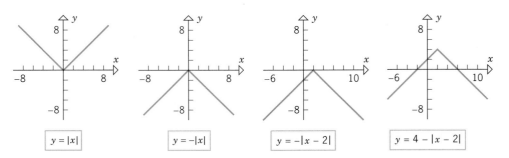

Figure 2.3.10

□ SCALING

Translations and reflections are called **_rigid transformations_** because they do not change the shape of a graph, only its position. We shall now discuss a transformation, called _scaling_, that actually changes the shape of a graph.

If $f(x)$ is multiplied by a *positive* constant c, the geometric effect is to compress or stretch the graph of $y = f(x)$ in the y-direction; a compression occurs if $0 < c < 1$ and a stretching if $c > 1$. This operation is called ***vertical scaling by a factor of c*** (Figure 2.3.11a). Similarly, if x is multiplied by a *positive* constant c, the geometric effect is to compress or stretch the graph of $y = f(x)$ in the x-direction; a compression occurs if $c > 1$ and a stretching if $0 < c < 1$. This operation is called ***horizontal scaling by a factor of c*** (Figure 2.3.11b).

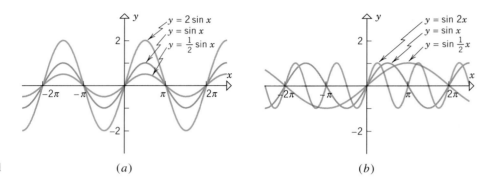

Figure 2.3.11 (a) (b)

□ **VERTICAL LINE TEST**

In all of the examples given in this section so far, we *started* with a function f and by one method or another we *found* the graph of the equation $y = f(x)$. By reversing the situation and starting with a graph, we are led to the following interesting question:

> **2.3.2**   PROBLEM.   *If we draw an arbitrary curve in the xy-plane, must that curve necessarily be the graph of $y = f(x)$ for some function f?*

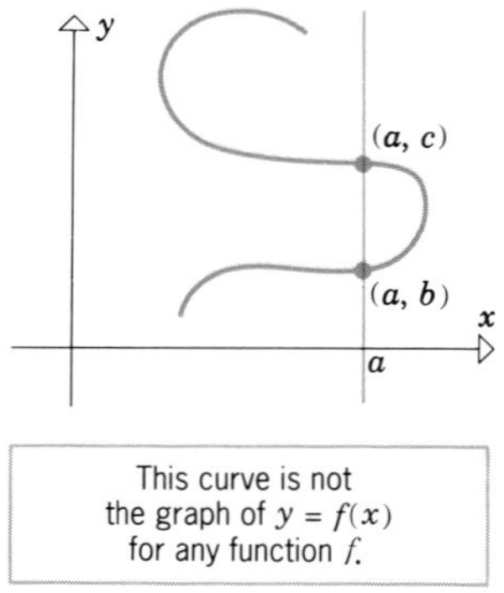

The answer to this question is no! There exist curves in the plane that are not graphs of any functions. To see why this is so, consider the curve in Figure 2.3.12 and the vertical line that intersects it at the two distinct points (a, b) and (a, c). This curve cannot be the graph of

$$y = f(x) \tag{2}$$

for any function f. For if it were, then (a, b) and (a, c), being points on the curve, would both have coordinates satisfying (2). Thus, we would have

$$b = f(a) \quad \text{and} \quad c = f(a)$$

But this is impossible since f cannot assign two different values to a. Therefore, there is no function of x whose graph is the curve in Figure 2.3.12.

This discussion illustrates the following general result, which we shall call the ***vertical line test***.

> **2.3.3**   THE VERTICAL LINE TEST.   *A curve in the xy-plane is the graph of $y = f(x)$ for some function f if and only if no vertical line intersects the curve more than once.*

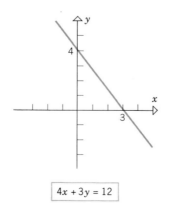

This curve is not
the graph of $y = f(x)$
for any function f.

Figure 2.3.12

Example 11   Is the graph of the equation

$$4x + 3y = 12 \tag{3}$$

the graph of $y = f(x)$ for some function f?

Solution.   We can answer this question by rewriting the equation in the equivalent form

$$y = 4 - \tfrac{4}{3}x$$

$4x + 3y = 12$

Figure 2.3.13

from which it is evident that the graph of (3) is also the graph of the function

$$f(x) = 4 - \tfrac{4}{3}x$$

Observe that the graph of this function, which is shown in Figure 2.3.13, passes the vertical line test.   ◄

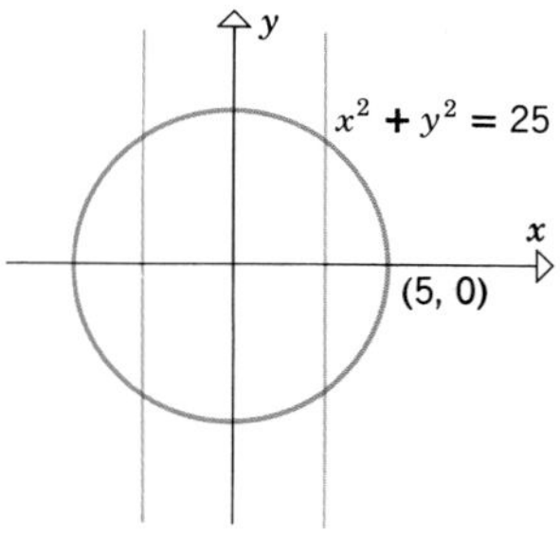

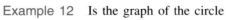

Example 12   Is the graph of the circle

$$x^2 + y^2 = 25 \tag{4}$$

also the graph of $y = f(x)$ for some function f?

Solution.   Since some vertical lines intersect the circle more than once (see Figure 2.3.14), the circle is not the graph of any function of x.

We can also deduce this result algebraically by observing that (4) can be written as

$$y = \pm\sqrt{25 - x^2} \tag{5}$$

But the right side of (5) is not a function of x since it is "multiple-valued," that is, there are values of x that result in more than one value of y. Thus, (4) is not equivalent to an equation of the form $y = f(x)$.   ◄

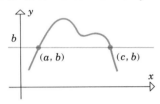

Figure 2.3.14

□ x AS A FUNCTION OF y

Sometimes we will be interested in knowing whether a given curve in the xy-plane is the graph of an equation

$$x = g(y)$$

for some function g. Because g cannot assign two different x values to the same y, the graph of such an equation cannot be cut twice by any horizontal line (Figure 2.3.15). Thus, we have the following analog of the vertical line test.

This curve is not the graph of $x = g(y)$ for any function g.

Figure 2.3.15

2.3.4   THE HORIZONTAL LINE TEST.   *A curve in the xy-plane is the graph of $x = g(y)$ for some function g if and only if no horizontal line intersects the curve more than once.*

Example 13   The graph of the equation

$$y = x^2 \tag{6}$$

(Figure 2.3.16) passes the vertical line test, but not the horizontal line test, so it is the graph of a function of x [namely, $f(x) = x^2$], but it is not the graph of any function of y. This can also be seen algebraically by solving (6) for x in terms of y. This yields

$$x = \pm\sqrt{y} \tag{7}$$

The right side of this equation is "multiple-valued," so it is not the formula for a function of y.   ◄

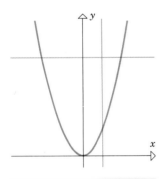

The graph of $y = x^2$ passes the vertical line test but not the horizontal line test.

Figure 2.3.16

REMARK.   When graphing equations of the form $y = f(x)$, we have been keeping the x-axis horizontal and the y-axis vertical. Thus, it would seem natural to graph equations of the form $x = g(y)$ with the y-axis horizontal and the x-axis vertical. However, there are many situations in which one wants to graph $y = f(x)$ and $x = g(y)$ in the same coordinate system, so it is common to keep the x-axis horizontal and the y-axis vertical for equations of both types.

□ IMPLICITLY DEFINED FUNCTIONS

An equation of the form $y = f(x)$ is said to **define y explicitly as a function of x** (the function being f), and an equation of the form $x = g(y)$ is said to **define x explicitly as a function of y** (the function being g). For example,

$$y = 5x^2 \sin x$$

defines y explicitly as a function of x and

$$x = (7y^3 - 2y)^{3/2}$$

defines x explicitly as a function of y.

An equation that is not of the form $y = f(x)$ but whose graph in the xy-plane passes the vertical line test is said to **define y implicitly as a function of x**, and an equation that is not of the form $x = g(y)$ but whose graph in the xy-plane passes the horizontal line test is said to **define x implicitly as a function of y**.

Sometimes one can determine whether an equation defines a function implicitly by solving the equation for y in terms of x or x in terms of y. For example, the equation

$$y^3 - 3y^2 + 3y - x^2 = 1 \tag{8}$$

can be solved for y in terms of x as

$$y^3 - 3y^2 + 3y - 1 = x^2$$
$$(y - 1)^3 = x^2$$
$$y = x^{2/3} + 1 \tag{9}$$

and for x in terms of y as

$$x^2 = y^3 - 3y^2 + 3y - 1$$
$$x = \pm\sqrt{y^3 - 3y^2 + 3y - 1} \tag{10}$$

It follows from (9) that (8) defines y implicitly as a function of x and from (10) that (8) does not define x implicitly as a function of y.

□ **SOME COMPLICATIONS**

Some equations in x and y cannot be solved for x in terms of y or y in terms of x. This can occur either because the algebra required is too complicated or because it is mathematically impossible to do so. For example, one cannot determine by simple algebra whether the equation

$$1 + xy^3 - \sin(x^2 y) = 0$$

implicitly defines y as a function of x, or x as a function of y, or neither. There are theorems studied in advanced courses that can sometimes be used to ascertain whether an equation defines a function implicitly. However, even if it can be shown that the equation does define a function implicitly, there may be no way to produce an explicit formula for the function, so all of its properties would have to be developed indirectly from the given equation. This can be quite complicated in some cases.

□ **FUNCTIONS DEFINED IN A NEIGHBORHOOD OF A POINT**

Sometimes a curve, when considered in its entirety, is not the graph of a function, but some smaller portion of the curve is. As an illustration, we saw in Example 12 that the circle $x^2 + y^2 = 25$ is not the graph of $y = f(x)$ for any function f. However, the upper and lower semicircles pass the vertical line test, so they are each graphs of functions. From (5) the upper semicircle has the equation $y = \sqrt{25 - x^2}$, so it is the graph of the function $f_1(x) = \sqrt{25 - x^2}$, and the lower semicircle has the equation $y = -\sqrt{25 - x^2}$, so it is the graph of the function $f_2(x) = -\sqrt{25 - x^2}$ (Figure 2.3.17a). Similarly, solving $x^2 + y^2 = 25$ for x in terms of y we obtain $x = \pm\sqrt{25 - y^2}$ from which it follows that the right semicircle is the graph of $g_1(y) = \sqrt{25 - y^2}$ and the left semicircle is the graph of $g_2(y) = -\sqrt{25 - y^2}$ (Figure 2.3.17b).

To make the preceding ideas more precise, let C denote the graph of an equation in the xy-plane. We will say that the equation **defines y implicitly as a function of x in a neighborhood of (x_0, y_0)** if there is some rectangle centered at (x_0, y_0), with sides parallel to the coordinate axes, such that the portion of C contained within the rectangle is cut at most once by any vertical line. Similarly, we will say that the equation **defines x implicitly as a function of y in a neighborhood of (x_0, y_0)** if there is some rectangle centered at

(x_0, y_0), with sides parallel to the coordinate axes, such that the portion of C contained within the rectangle is cut at most once by any horizontal line.

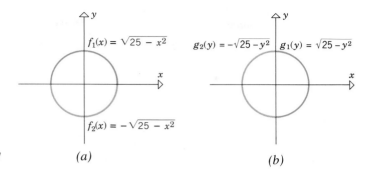

Figure 2.3.17   (a)   (b)

Example 14   The equation of the unit circle $x^2 + y^2 = 1$ does not implicitly define y as a function of x in a neighborhood of the point $P(1, 0)$, because the portion of the circle within any rectangle centered at P is cut twice by some vertical line (Figure 2.3.18a). Similarly, the equation does not implicitly define x as a function of y in a neighborhood of the point $Q(0, 1)$, because the portion of the circle within any rectangle centered at Q is cut twice by some horizontal line (Figure 2.3.18b). However, the equation implicitly defines both y as a function of x and x as a function of y in a neighborhood of the point $(1/\sqrt{2}, 1/\sqrt{2})$ (Figure 2.3.18c).   ◀

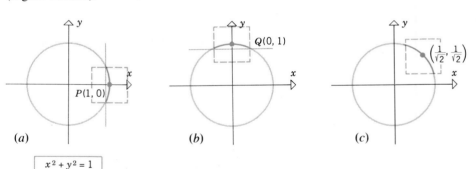

(a)   (b)   (c)

Figure 2.3.18 $x^2 + y^2 = 1$

▶ Exercise Set 2.3

1. Consider the function f graphed in Figure 2.3.19. In each part, find all values of x satisfying the given condition.

    (a) $f(x) = 0$ (b) $f(x) = 3$

    (c) $f(x) \geq 0$ (d) $f(x) \leq 0$.

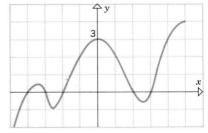

Figure 2.3.19

In Exercises 2–29, sketch the graph of the function.

2. $f(x) = 2x + 1$.

3. $f(x) = 3x - 2$.

4. $G(x) = x, \quad 1 \leq x \leq 2$.

5. $G(x) = x - 2, \quad -1 \leq x \leq 1$.

6. $h(x) = x^2 - 3$.

7. $h(x) = (x - 2)^2$.

8. $F(x) = \sqrt{x + 1}$.

9. $F(x) = \sqrt{3 - x}$.

10. $f(x) = \sqrt{4 - x^2}$.

11. $f(x) = 2 + \sqrt{9 - x^2}$.

12. $g(x) = \sqrt{4x - x^2}$.

13. $g(x) = \sqrt{7 - 6x - x^2}$.

14. $f(x) = 2 \sin x$.

15. $f(x) = 3 \sin 2x$.

16. $f(x) = \cos 2x$.

17. $f(x) = 1 + \cos x$.

18. $f(x) = \dfrac{x^2 - 4}{x + 2}$.

19. $f(x) = \dfrac{x^2 + 2x}{x}$.

20. $g(x) = \dfrac{x^3 - x^2}{x - 1}$.

21. $g(x) = \dfrac{x - x^3}{x}$.

22. $\phi(x) = \dfrac{x}{|x|}$.

23. $\phi(x) = \dfrac{|x-2|}{x-2}$.

24. $g(x) = \begin{cases} x^2, & x \neq 4 \\ 0, & x = 4. \end{cases}$

25. $g(x) = \begin{cases} x-1, & x \neq 1 \\ 3, & x = 1. \end{cases}$

26. $f(x) = \begin{cases} x+2, & x \leq 3 \\ x+4, & x > 3. \end{cases}$

27. $f(x) = \begin{cases} x^2, & x > 1 \\ 2, & x \leq 1. \end{cases}$

28. $h(x) = \begin{cases} 1, & 0 < x \leq 1 \\ 3, & 1 < x \leq 2 \\ -1, & 2 < x \leq 3 \\ 0, & \text{elsewhere.} \end{cases}$

29. $h(x) = \begin{cases} -2, & -2 \leq x < -1 \\ 1, & -1 \leq x < 0 \\ 2, & 0 \leq x < 1 \\ 0, & \text{elsewhere.} \end{cases}$

In Exercises 30–33, express the function in piecewise form without using absolute values and sketch its graph.

30. $f(x) = |x-3| - x$.   **31.** $f(x) = 2x + |2-x|$.

32. $g(x) = |x| + |x-3|$.

33. $g(x) = |x-5| - |x-3|$.

34. Let $f(x) = x^2 - 1$. Sketch the graph of

(a) $y = \frac{1}{2}(f(x) + |f(x)|)$ (b) $y = \frac{1}{2}(f(x) - |f(x)|)$.

[*Hint:* Express the equations in piecewise form.]

35. Given that $f(x) = |x|$ and $g(x) = x$, sketch the graph of the equation.

(a) $y = (f + g)(x)$ (b) $y = (f - g)(x)$

(c) $y = (f \cdot g)(x)$ (d) $y = (f/g)(x)$.

36. A 10-foot ladder leans against a wall. The base of the ladder is x feet from the wall. Express the distance from the top of the ladder to the ground as a function of x, and sketch the graph of the function.

37. Express the area A enclosed by the graph of

$$f(x) = \begin{cases} 2x, & 0 \leq x \leq 1 \\ 2, & x > 1 \end{cases}$$

the x-axis, and the vertical line at $x = c$ $(c \geq 0)$ as a function of c.

38. Find a formula for the function f graphed in Figure 2.3.20.

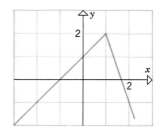

Figure 2.3.20

39. Find a formula for the function g graphed in Figure 2.3.21.

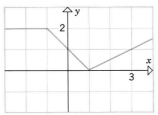

Figure 2.3.21

40. Use Table 2.3.1 and the graph of $y = |x|$ to graph the following:

(a) $y = |x - 4|$ (b) $y = |x| + 4$

(c) $y = |x - 4| + 4$ (d) $y = |x + 5| - 2$.

41. Use Table 2.3.1 and the graph of $y = \sqrt{x}$ to graph the following:

(a) $y = \sqrt{x + 3}$ (b) $y = \sqrt{x} + 3$

(c) $y = \sqrt{x - 3} + 3$ (d) $y = \sqrt{x + 1} - 2$.

42. A function f with domain $[-1, 2]$ has the graph shown in Figure 2.3.22. Use this graph to obtain the graphs of the equations

(a) $y = f(x) - 1$ (b) $y = f(x - 1)$

(c) $y = \frac{1}{2}f(x)$ (d) $y = f(-\frac{1}{2}x)$.

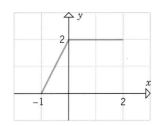

Figure 2.3.22

43. Use the graph of the function f in Exercise 42 to obtain the graphs of the equations

(a) $y = -f(-x)$ (b) $y = f(2 - x)$

(c) $y = 1 - f(2 - x)$ (d) $y = \frac{1}{2}f(2x)$.

44. A function f with domain $[-1, 3]$ has the graph shown in Figure 2.3.23. Use this graph to obtain the graphs of the equations

(a) $y = f(x + 1)$ (b) $y = f(2x)$

(c) $y = f(-x)$ (d) $y = -f(x)$.

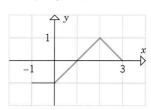

Figure 2.3.23

45. The equation $y = |f(x)|$ can be written as

$$y = \begin{cases} f(x), & f(x) \geq 0 \\ -f(x), & f(x) < 0 \end{cases}$$

which shows that the graph of $y = |f(x)|$ can be obtained from the graph of $y = f(x)$ by retaining the portion that lies on or above the x-axis and reflecting about the x-axis the portion that lies below the x-axis. Use this method to obtain the graph of $y = |2x - 3|$ from the graph of $y = 2x - 3$.

> In Exercises 46 and 47, use the method described in Exercise 45.

46. Sketch the graph of $y = |1 - x^2|$.

47. Sketch the graph of

(a) $f(x) = |\cos x|$

(b) $f(x) = \cos x + |\cos x|$.

48. The *greatest integer function*, $[x]$, is defined to be the greatest integer that is less than or equal to x. For example, $[2.7] = 2$, $[-2.3] = -3$, and $[4] = 4$. Sketch the graph of

(a) $f(x) = [x]$ (b) $f(x) = [x^2]$

(c) $f(x) = [x]^2$ (d) $f(x) = [\sin x]$.

49. A function f is called *even* if $f(-x) = f(x)$ for each x in the domain of f and *odd* if $f(-x) = -f(x)$ for each such x. In each part, classify the function as even, odd, or neither.

(a) $f(x) = x^2$ (b) $f(x) = x^3$

(c) $f(x) = |x|$ (d) $f(x) = x + 1$

(e) $f(x) = \dfrac{x^5 - x}{1 + x^2}$ (f) $f(x) = 2$.

50. In Figure 2.3.24 we have sketched part of the graph of a function f. Complete the graph assuming

(a) f is an even function

(b) f is an odd function.

[See Exercise 49 for terminology.]

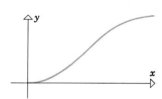

Figure 2.3.24

51. Classify the functions graphed in Figure 2.3.25 as even, odd, or neither. (See Exercise 49 for the definitions of even and odd functions.)

52. Can a function be both even and odd? [See Exercise 49 for terminology.]

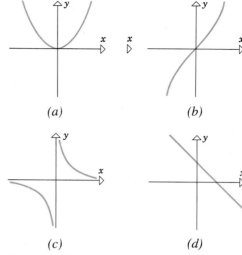

(a)   *(b)*

(c)   *(d)*

Figure 2.3.25

53. Prove that the product of

(a) two even functions is an even function

(b) two odd functions is an even function

(c) an even and an odd function is an odd function.

[See Exercise 49 for terminology.]

54. Let a be a constant and suppose that

$$f(a - x) = f(a + x)$$

for all x. What geometric property must the graph of f have?

55. In each part determine whether the equation defines y as a function of x, or x as a function of y, or both, or neither.

(a) $4x + 2y = -8$ (b) $x^2 y^3 = 1$

(c) $3x^2 + 4y^2 = 12$ (d) $\dfrac{xy}{1 - xy} = 1$.

56. In each part express x explicitly as a function of y.

(a) $xy - x = 1$ (b) $y = \dfrac{x}{1 + x}$

(c) $x^2 + 2xy + y^2 = 0$.

57. In each part express y explicitly as a function of x.

(a) $x^2 y - 1 = 0$ (b) $x = \dfrac{1 - y}{1 + y}$

(c) $y^2 + 2xy + x^2 = 0$.

58. Show that $y^2 + 3xy + x^2 = 0$ does not define y implicitly as a function of x.

59. Show that $y^2 + 4xy + 1 = 0$ does not define y implicitly as a function of x.

60. Find two functions, the union of whose graphs is the graph of the equation in Exercise 59.

61.   In Figure 2.3.26, determine whether the curve is the graph of a function of x, a function of y, both, or neither.

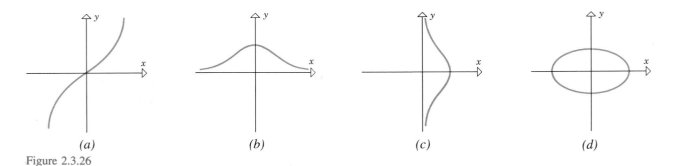

Figure 2.3.26

■ 2.4 LIMITS (AN INTUITIVE INTRODUCTION)

The development of calculus was stimulated in large part by two geometric problems: finding areas of plane regions and finding tangent lines to curves. In this section we shall show that both of these problems are closely related to a fundamental concept of calculus known as a "limit."

□ **THE TANGENT LINE AND AREA PROBLEMS**

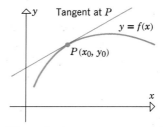

Figure 2.4.1

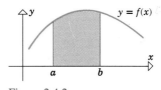

Figure 2.4.2

□ **TANGENT LINES AND LIMITS**

Calculus centers around the following two fundamental problems:

THE TANGENT PROBLEM.   *Given a function f and a point $P(x_0, y_0)$ on its graph, find an equation of the line tangent to the graph at P (Figure 2.4.1).*

THE AREA PROBLEM.   *Given a function f, find the area between the graph of f and an interval [a, b] on the x-axis (Figure 2.4.2).*

Traditionally, that portion of calculus arising from the tangent problem is called *differential calculus* and that arising from the area problem is called *integral calculus*. As we shall see, however, the tangent and area problems are closely related, so that the distinction between differential calculus and integral calculus is often hard to discern.

In order to solve the tangent and area problems it is necessary to have a more precise understanding of the concepts of "tangent line" and "area." As we shall explain in this section, both of these concepts rest on a more fundamental concept, known as a "*limit*."

In plane geometry, a line is called *tangent* to a circle if it meets the circle at precisely one point (Figure 2.4.3a). However, this definition is not satisfactory for other kinds of curves. In Figure 2.4.3b the line meets the curve exactly once, yet is not a tangent, and in Figure 2.4.3c the line is tangent yet meets the curve more than once.

To define the concept of a tangent line so that it applies to curves other than circles, we must view tangent lines another way. Consider a point P on a curve in the xy-plane. If Q is any point on the curve different from P, the line through P and Q is called a *secant line* for

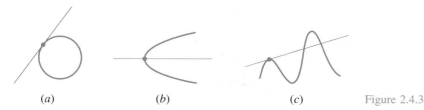

(a)   (b)   (c)   Figure 2.4.3

the curve. Intuition suggests that if we move the point Q along the curve toward P, the secant line will rotate toward a "limiting" position. The line occupying this limiting position we consider to be the ***tangent line*** at P (Figure 2.4.4).

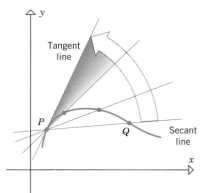

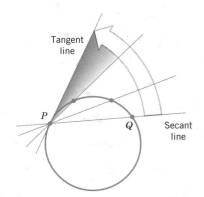

Figure 2.4.4   Figure 2.4.5

As suggested by Figure 2.4.5, this new concept of a tangent line coincides with the traditional concept, when applied to circles.

□ **AREA AS A LIMIT**

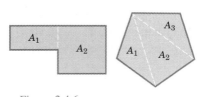

Figure 2.4.6

Just as the general notion of a tangent line leads to the concept of a "limit," so does the general notion of area. Areas of some plane regions can be calculated by subdividing them into a *finite* number of rectangles or triangles, then adding the areas of the constituent parts (Figure 2.4.6). However, for many regions a more general approach is needed. Consider the shaded region in Figure 2.4.7a. We can *approximate* the area of this region by inscribing rectangles of equal width under the curve and adding the areas of these rectangles (Figure 2.4.7b). Moreover, intuition suggests that if we repeat the process using more and more rectangles, then the rectangles will tend to fill in the gaps under the curve and our approximations will "approach" the exact area under the curve as a "limiting value" (Figure 2.4.7c).

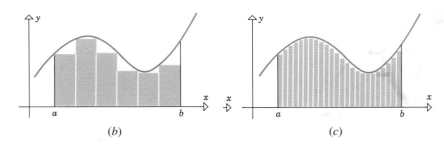

(a)   (b)   (c)

Figure 2.4.7

□ **LIMITS**

In the preceding discussion, we saw that the concepts of a tangent line and the area of a plane region ultimately rest on the notion of a "limit." In this section and the next few we shall investigate the notion of limit in more detail. Our development of limits in this text proceeds in three stages:

Table 2.4.1

x (RADIANS)	$f(x) = \dfrac{\sin x}{x}$
1.0	0.84147
0.9	0.87036
0.8	0.89670
0.7	0.92031
0.6	0.94107
0.5	0.95885
0.4	0.97355
0.3	0.98507
0.2	0.99335
0.1	0.99833
0.01	0.99998

Table 2.4.2

x (RADIANS)	$f(x) = \dfrac{\sin x}{x}$
−1.0	0.84147
−0.9	0.87036
−0.8	0.89670
−0.7	0.92031
−0.6	0.94107
−0.5	0.95885
−0.4	0.97355
−0.3	0.98507
−0.2	0.99335
−0.1	0.99833
−0.01	0.99998

- First we discuss limits intuitively.
- Then we discuss methods for computing limits.
- Finally, we give a precise mathematical discussion of limits.

Limits can be used to describe how a function behaves as the independent variable moves toward a certain value. For example, consider the function

$$f(x) = \frac{\sin x}{x}$$

where x is in radians. Although this function is not defined at $x = 0$, it still makes sense to ask what happens to the values of $f(x)$ as x moves along the x-axis *toward* 0 without actually taking on the value $x = 0$. We will consider two separate cases, the first where x moves toward $x = 0$ from the right side (i.e., along the positive x-axis) and the second where x moves toward $x = 0$ from the left side (i.e., along the negative x-axis).

Table 2.4.1, which was obtained using a calculator set to the radian mode, shows a succession of values of $f(x) = (\sin x)/x$ corresponding to a succession of values of x that move toward 0 from the right side. The results in the table suggest that the values of $f(x)$ approach 1. We call the number 1 the ***limit*** of $f(x) = (\sin x)/x$ as x approaches 0 from the right side and write

$$\lim_{x \to 0^+} \frac{\sin x}{x} = 1 \tag{1}$$

In this expression, "lim" tells us that we are computing a limit; the symbol $x \to 0$ tells us that we are letting x approach 0; and the label "+" on $x \to 0^+$ tells us that x is approaching zero from the *right* side. We can also ask what happens to the values of $f(x) = (\sin x)/x$ as x approaches 0 from the left side. From Table 2.4.2 [or from the fact that $f(-x) = f(x)$] it is evident that the values of $f(x)$ again approach 1. We denote this by writing

$$\lim_{x \to 0^-} \frac{\sin x}{x} = 1 \tag{2}$$

In (2), the label "−" on $x \to 0^-$ tells us that we are computing the limit as x approaches zero from the *left* side.

The limits in (1) and (2) can be visualized geometrically from the computer-generated graph of $f(x) = (\sin x)/x$ shown in Figure 2.4.8. The hole in the graph at $x = 0$ is a consequence of the fact that $(\sin x)/x$ is not defined at that point, even though the limits from the left and right exist there.

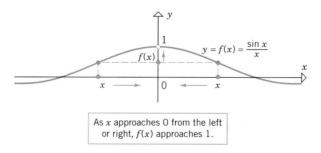

Figure 2.4.8

As x approaches 0 from the left or right, $f(x)$ approaches 1.

☐ **LIMIT NOTATION**

The limits from the right and left in (1) and (2) happened to be the same, which we indicate by writing

$$\lim_{x \to 0} \frac{\sin x}{x} = 1$$

The preceding ideas are summarized in Table 2.4.3.

The expressions $\lim_{x \to x_0^-} f(x)$ and $\lim_{x \to x_0^+} f(x)$ are called the ***one-sided limits of $f(x)$ at x_0*** and the expression $\lim_{x \to x_0} f(x)$ is called the ***two-sided limit of $f(x)$ at x_0***.

Table 2.4.3

MATHEMATICAL SITUATION	NOTATION	HOW TO READ THE NOTATION
The value of $f(x)$ approaches the number L_1 as x approaches x_0 from the right side.	$\displaystyle\lim_{x \to x_0^+} f(x) = L_1$	The limit of $f(x)$ as x approaches x_0 from the right is equal to L_1.
The value of $f(x)$ approaches the number L_2 as x approaches x_0 from the left side.	$\displaystyle\lim_{x \to x_0^-} f(x) = L_2$	The limit of $f(x)$ as x approaches x_0 from the left is equal to L_2.
The value of $f(x)$ approaches the number L as x approaches x_0 from either the left or right side; that is, $\displaystyle\lim_{x \to x_0^+} f(x) = \lim_{x \to x_0^-} f(x) = L$	$\displaystyle\lim_{x \to x_0} f(x) = L$	The limit of $f(x)$ as x approaches x_0 is equal to L.

☐ **NUMERICAL PITFALLS**

It is important to keep in mind that the limits in (1) and (2) are really guesses about the behavior of $f(x)$ based on numerical evidence obtained by evaluating $f(x)$ at selected values of x. It is conceivable that different choices of x might have produced different conclusions about the limit. For example, consider the function

$$f(x) = \sin\frac{\pi}{x}$$

The values of $f(x)$ in Table 2.4.4 would lead us to believe that

$$\lim_{x \to 0^+} \sin\frac{\pi}{x} = \lim_{x \to 0^-} \sin\frac{\pi}{x} = 0$$

Table 2.4.4

x (RADIANS)	$f(x) = \sin\dfrac{\pi}{x}$	x (RADIANS)	$f(x) = \sin\dfrac{\pi}{x}$
$x = 1$	$\sin\pi = 0$	$x = -1$	$\sin(-\pi) = 0$
$x = 0.1$	$\sin 10\pi = 0$	$x = -0.1$	$\sin(-10\pi) = 0$
$x = 0.01$	$\sin 100\pi = 0$	$x = -0.01$	$\sin(-100\pi) = 0$
$x = 0.001$	$\sin 1000\pi = 0$	$x = -0.001$	$\sin(-1000\pi) = 0$
$x = 0.0001$	$\sin 10{,}000\pi = 0$	$x = -0.0001$	$\sin(-10{,}000\pi) = 0$
$\vdots$	$\vdots$	$\vdots$	$\vdots$

However, this is not correct; the values of $f(x)$ actually oscillate between -1 and 1 with increasing rapidity as x approaches 0 from either the left or the right. This is illustrated in Figure 2.4.9, which shows an artistically enhanced computer-generated graph of f. For example, if $x > 0$, then values of 1 occur when $x = \frac{2}{5}, \frac{2}{9}, \frac{2}{13}, \dots$ and values of -1 occur when $x = \frac{2}{3}, \frac{2}{7}, \frac{2}{11}, \dots$ (verify). Thus, the values of $f(x)$ do not approach any limiting value as

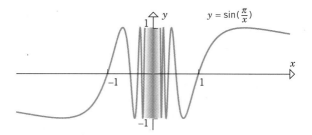

Figure 2.4.9

$x \to 0^+$. Similarly, there is no limiting value as $x \to 0^-$. We were deceived into thinking that the one-sided limits were zero because we accidentally selected x values where the graph of f crosses the x-axis.

In spite of the fact that numerical evidence can sometimes lead to erroneous conclusions about limits, numerical calculations are often a helpful starting point for limit investigations. In subsequent sections we will give a precise mathematical definition of a limit and we will develop a variety of mathematical tools that can be used to determine limits with certainty. The remainder of this section will be devoted to developing a geometric understanding of limits.

Example 1 Let f be the function whose graph is shown in Figure 2.4.10. As x approaches 2 from the left, $f(x)$ approaches 1, so that

$$\lim_{x \to 2^-} f(x) = 1$$

As x approaches 2 from the right, $f(x)$ approaches 3, so that

$$\lim_{x \to 2^+} f(x) = 3$$

Since $f(2) = 1.5$, this example shows that the value of a function at a point and the left- and right-hand limits there can all be different. ◀

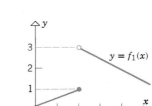

Figure 2.4.10

Example 2 The functions f_1 and f_2 whose graphs are shown in Figure 2.4.11 are identical to the function f in Example 1, except at the point $x = 2$; there we have $f(2) = 1.5$ and $f_1(2) = 1$, while $f_2(2)$ is undefined. However, even though the functions f, f_1, and f_2 behave differently *at* $x = 2$, their limits as x approaches 2 are the same as in Example 1, that is,

$$\lim_{x \to 2^-} f(x) = \lim_{x \to 2^-} f_1(x) = \lim_{x \to 2^-} f_2(x) = 1$$

and

$$\lim_{x \to 2^+} f(x) = \lim_{x \to 2^+} f_1(x) = \lim_{x \to 2^+} f_2(x) = 3 \qquad ◀$$

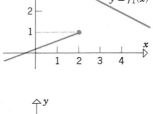

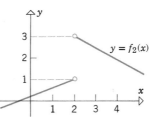

Figure 2.4.11

The preceding example illustrates a general principle about limits that can be stated loosely as follows:

The limit of a function as the independent variable approaches a point does not depend on the value of the function *at* the point.

Thus, if we alter the value of a function f only at $x = x_0$, then we do not affect

$$\lim_{x \to x_0^-} f(x), \quad \lim_{x \to x_0^+} f(x), \quad \text{or} \quad \lim_{x \to x_0} f(x)$$

Function f_2 in Example 2 illustrates another idea: namely, that the left- and right-hand limits can have values at a point where the function itself is undefined.

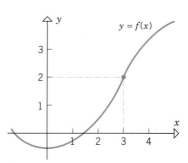

Figure 2.4.12

Example 3 Let f be the function whose graph is shown in Figure 2.4.12. From the graph we see that

$$\lim_{x \to 3^-} f(x) = 2 \quad \text{and} \quad \lim_{x \to 3^+} f(x) = 2$$

so that

$$\lim_{x \to 3} f(x) = 2$$

The graph also shows that $f(3) = 2$. Thus, the value of a function at a point and the one-sided and two-sided limits as x approaches that point may be equal.   ◀

☐ **EXISTENCE OF LIMITS**

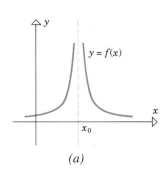

(a)

There is, in general, no guarantee that a function $f(x)$ actually has a limit as $x \to x_0^+$, $x \to x_0^-$, or $x \to x_0$. If there is no limit, then we say that **the limit does not exist**.

Limits can fail to exist for a variety of reasons, but two common causes are

* oscillation;
* unbounded increase or decrease.

For example, we observed earlier that the function $f(x) = \sin(\pi/x)$ does not have a limit as x approaches zero from either the left or the right because the function values oscillate between -1 and 1 without approaching a limiting value (Figure 2.4.9). The following examples illustrate the failure of a limit to exist because of unbounded increase or decrease.

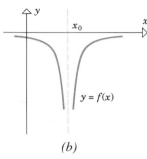

(b)

Figure 2.4.13

Example 4   Let f be the function graphed in Figure 2.4.13a. As x approaches x_0 from the left side or the right side, the value of $f(x)$ gets larger and larger without bound and consequently approaches no fixed finite value. Thus, the limits

$$\lim_{x \to x_0^-} f(x) \quad \text{and} \quad \lim_{x \to x_0^+} f(x)$$

do not exist. To indicate that these limits fail to exist because $f(x)$ is increasing without bound, we write

$$\lim_{x \to x_0^-} f(x) = +\infty \quad \text{and} \quad \lim_{x \to x_0^+} f(x) = +\infty.$$

These two expressions are commonly expressed in the form of a two-sided limit by writing

$$\lim_{x \to x_0} f(x) = +\infty$$

but again, this is just a notation for a type of limit that fails to exist.   ◀

Example 5   Let f be the function graphed in Figure 2.4.13b. As x approaches x_0 from the left side or the right side, the value of $f(x)$ gets smaller and smaller without bound and consequently approaches no fixed finite value. Thus, the limits

$$\lim_{x \to x_0^-} f(x) \quad \text{and} \quad \lim_{x \to x_0^+} f(x)$$

do not exist. To indicate that these limits fail to exist because $f(x)$ is decreasing without bound, we write

$$\lim_{x \to x_0^-} f(x) = -\infty \quad \text{and} \quad \lim_{x \to x_0^+} f(x) = -\infty$$

These two expressions are commonly expressed in the form of a two-sided limit by writing

$$\lim_{x \to x_0} f(x) = -\infty \quad ◀$$

REMARK.   Keep in mind that the symbols $+\infty$ and $-\infty$, as used in the preceding examples, are simply descriptions of limits that fail to exist. These symbols do not represent real numbers and consequently they cannot be manipulated using rules of algebra. For example, it is not correct to write $+\infty - \infty = 0$.

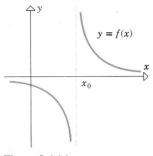

Figure 2.4.14

Example 6 Let f be the function whose graph is shown in Figure 2.4.14. We have

$$\lim_{x \to x_0^-} f(x) = -\infty \quad \text{and} \quad \lim_{x \to x_0^+} f(x) = +\infty$$

However, because these two expressions have opposite signs, there is no special notation for combining them into a two-sided form. ◄

Recall from Table 2.4.3 that in order to write $\lim_{x \to x_0} f(x) = L$ we must have

$$\lim_{x \to x_0^+} f(x) = \lim_{x \to x_0^-} f(x) = L$$

If either of these one-sided limits fails to exist, or if the one-sided limits both exist but have different values, then we shall agree that the two-sided limit does not exist and write

$$\lim_{x \to x_0} f(x) \; does \; not \; exist$$

For example, if f is the function discussed in Example 1 and graphed in Figure 2.4.10, then the two-sided limit

$$\lim_{x \to 2} f(x)$$

does not exist, since the one-sided limits at $x = 2$ have different values.

☐ **LIMITS AT INFINITY**

So far we have used limits to describe how a function behaves as the independent variable approaches a fixed point on the x-axis. However, limits can also be used to describe how a function behaves as the independent variable moves "indefinitely far" from the origin along the x-axis. If x is allowed to increase without bound, then we write $x \to +\infty$ (read, "x approaches plus infinity"), and if x is allowed to decrease without bound, then we write $x \to -\infty$ (read, "x approaches minus infinity").

Example 7 Let f be the function with the graph shown in Figure 2.4.15. As $x \to +\infty$, the graph of f approaches the line $y = 4$ so that the value of $f(x)$ approaches 4. We denote this by writing

$$\lim_{x \to +\infty} f(x) = 4$$

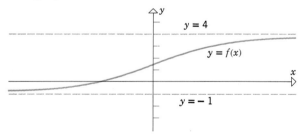

Figure 2.4.15

As $x \to -\infty$, the graph of f tends toward the line $y = -1$, so the value of $f(x)$ approaches -1. We denote this by writing

$$\lim_{x \to -\infty} f(x) = -1 \quad ◄$$

Example 8 Let f be the function with the graph shown in Figure 2.4.16.

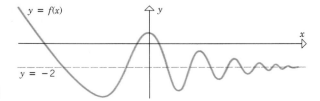

Figure 2.4.16

As x approaches $-\infty$, the value of $f(x)$ increases without bound and consequently approaches no fixed value. Thus,

$$\lim_{x \to -\infty} f(x)$$

does not exist. In this case, we would write

$$\lim_{x \to -\infty} f(x) = +\infty$$

to indicate that the limit fails to exist because $f(x)$ is increasing without bound.

As x approaches $+\infty$, the graph of f oscillates but approaches the line $y = -2$. Thus, the value of $f(x)$ approaches -2 as x approaches $+\infty$ and we write

$$\lim_{x \to +\infty} f(x) = -2 \qquad \blacktriangleleft$$

Example 9 Let f be the function with the graph shown in Figure 2.4.17. As x approaches $-\infty$, the value of $f(x)$ decreases without bound and consequently approaches no fixed finite value. Thus,

$$\lim_{x \to -\infty} f(x)$$

does not exist.

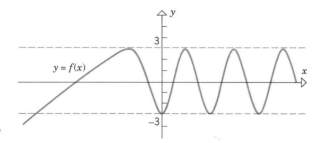

Figure 2.4.17

In this case, we would write

$$\lim_{x \to -\infty} f(x) = -\infty$$

to indicate that the limit fails to exist because $f(x)$ is decreasing without bound.

As x approaches $+\infty$, the value of $f(x)$ oscillates between -3 and $+3$ and consequently approaches no fixed value. Thus,

$$\lim_{x \to +\infty} f(x)$$

does not exist. There is no special notation used to describe limits that fail to exist because of oscillation. $\qquad \blacktriangleleft$

▶ Exercise Set 2.4

1. For the function f graphed to the right, find

(a) $\displaystyle\lim_{x \to 3^-} f(x) = -1$

(b) $\displaystyle\lim_{x \to 3^+} f(x) = 3$

(c) $\displaystyle\lim_{x \to 3} f(x)$ NONE

(d) $f(3) = 1$

(e) $\displaystyle\lim_{x \to -\infty} f(x) = -1$

(f) $\displaystyle\lim_{x \to +\infty} f(x) = 3$

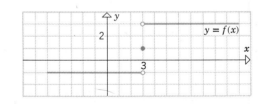

2. For the function f graphed below, find

(a) $\displaystyle\lim_{x \to 2^-} f(x)$ 2

(b) $\displaystyle\lim_{x \to 2^+} f(x)$ 0

(c) $\displaystyle\lim_{x \to 2} f(x)$ NONE

(d) $f(2) = 2$

(e) $\displaystyle\lim_{x \to -\infty} f(x)$ 0

(f) $\displaystyle\lim_{x \to +\infty} f(x)$. 2

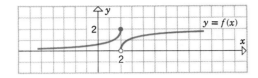

3. For the function g graphed below, find

(a) $\displaystyle\lim_{x \to 4^-} g(x)$

(b) $\displaystyle\lim_{x \to 4^+} g(x)$

(c) $\displaystyle\lim_{x \to 4} g(x)$

(d) $g(4)$

(e) $\displaystyle\lim_{x \to -\infty} g(x)$

(f) $\displaystyle\lim_{x \to +\infty} g(x)$.

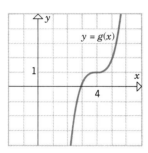

4. For the function g graphed below, find

(a) $\displaystyle\lim_{x \to 0^-} g(x)$ 3

(b) $\displaystyle\lim_{x \to 0^+} g(x)$ 3

(c) $\displaystyle\lim_{x \to 0} g(x)$ 3

(d) $g(0)$ 3

(e) $\displaystyle\lim_{x \to -\infty} g(x)$

(f) $\displaystyle\lim_{x \to +\infty} g(x)$.

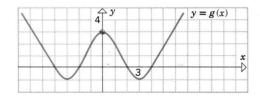

5. For the function F graphed below, find

(a) $\displaystyle\lim_{x \to -2^-} F(x)$

(b) $\displaystyle\lim_{x \to -2^+} F(x)$

(c) $\displaystyle\lim_{x \to -2} F(x)$

(d) $F(-2)$

(e) $\displaystyle\lim_{x \to -\infty} F(x)$

(f) $\displaystyle\lim_{x \to +\infty} F(x)$.

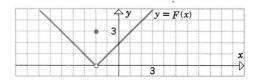

6. For the function F graphed below, find

(a) $\displaystyle\lim_{x \to 3^-} F(x)$ 2

(b) $\displaystyle\lim_{x \to 3^+} F(x)$ 2

(c) $\displaystyle\lim_{x \to 3} F(x)$ 2

(d) $F(3)$ 3

(e) $\displaystyle\lim_{x \to -\infty} F(x)$

(f) $\displaystyle\lim_{x \to +\infty} F(x)$.

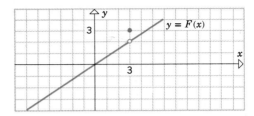

7. For the function ϕ graphed below, find

(a) $\displaystyle\lim_{x \to -2^-} \phi(x)$

(b) $\displaystyle\lim_{x \to -2^+} \phi(x)$

(c) $\displaystyle\lim_{x \to -2} \phi(x)$

(d) $\phi(-2)$

(e) $\displaystyle\lim_{x \to -\infty} \phi(x)$

(f) $\displaystyle\lim_{x \to +\infty} \phi(x)$.

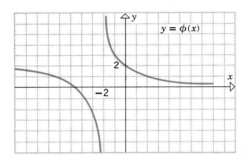

8. For the function ϕ graphed below, find

(a) $\displaystyle\lim_{x \to 4^-} \phi(x)$

(b) $\displaystyle\lim_{x \to 4^+} \phi(x)$

(c) $\displaystyle\lim_{x \to 4} \phi(x)$

(d) $\phi(4)$

(e) $\displaystyle\lim_{x \to -\infty} \phi(x)$ 0

(f) $\displaystyle\lim_{x \to +\infty} \phi(x)$.

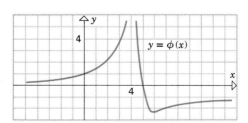

9. For the function f graphed below, find

(a) $\displaystyle\lim_{x \to 3^-} f(x)$

(b) $\displaystyle\lim_{x \to 3^+} f(x)$

(c) $\displaystyle\lim_{x \to 3} f(x)$

(d) $f(3)$

(e) $\displaystyle\lim_{x \to -\infty} f(x)$

(f) $\displaystyle\lim_{x \to +\infty} f(x)$.

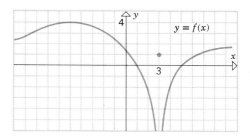

10. For the function f graphed below, find

(a) $\displaystyle\lim_{x\to 0^-} f(x)$ (b) $\displaystyle\lim_{x\to 0^+} f(x)$

(c) $\displaystyle\lim_{x\to 0} f(x)$ (d) $f(0)$

(e) $\displaystyle\lim_{x\to -\infty} f(x)$ (f) $\displaystyle\lim_{x\to +\infty} f(x)$.

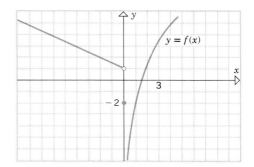

11. For the function G graphed below, find

(a) $\displaystyle\lim_{x\to 0^-} G(x)$ (b) $\displaystyle\lim_{x\to 0^+} G(x)$

(c) $\displaystyle\lim_{x\to 0} G(x)$ (d) $G(0)$

(e) $\displaystyle\lim_{x\to -\infty} G(x)$ (f) $\displaystyle\lim_{x\to +\infty} G(x)$.

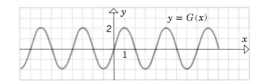

12. For the function G graphed below, find

(a) $\displaystyle\lim_{x\to 0^-} G(x)$ (b) $\displaystyle\lim_{x\to 0^+} G(x)$

(c) $\displaystyle\lim_{x\to 0} G(x)$ (d) $G(0)$

(e) $\displaystyle\lim_{x\to -\infty} G(x)$ (f) $\displaystyle\lim_{x\to +\infty} G(x)$.

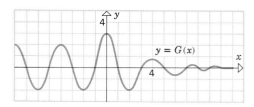

13. Consider the function g graphed below. For what values of x_0 does $\displaystyle\lim_{x\to x_0} g(x)$ exist?

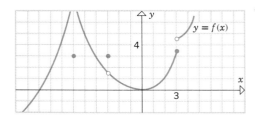

14. Consider the function f graphed below. For what values of x_0 does $\displaystyle\lim_{x\to x_0} f(x)$ exist?

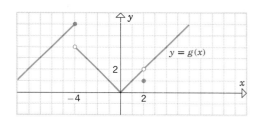

2.5 LIMITS (COMPUTATIONAL TECHNIQUES)

In the last section we concentrated on the graphical interpretation of limits. In this section we shall discuss techniques for finding limits directly from the formula for a function. Again our results will be based on intuition. Mathematical proofs of the results here will have to wait until we have a precise mathematical definition of a limit.

□ **SOME BASIC LIMITS**   Table 2.5.1 lists six basic limits that will form the building blocks for finding more complex limits. The limits in the left side of the table follow from the fact that the constant

Table 2.5.1

LIMIT	EXAMPLE	LIMIT	EXAMPLE
$\lim\limits_{x \to a} k = k$	$\lim\limits_{x \to 2} 3 = 3, \quad \lim\limits_{x \to -2} 3 = 3$	$\lim\limits_{x \to a} x = a$	$\lim\limits_{x \to 5} x = 5, \quad \lim\limits_{x \to 0} x = 0, \quad \lim\limits_{x \to -2} x = -2$
$\lim\limits_{x \to +\infty} k = k$	$\lim\limits_{x \to +\infty} 3 = 3, \quad \lim\limits_{x \to +\infty} 0 = 0$	$\lim\limits_{x \to +\infty} x = +\infty$	(Figure 2.5.2*b*)
$\lim\limits_{x \to -\infty} k = k$	$\lim\limits_{x \to -\infty} 3 = 3, \quad \lim\limits_{x \to -\infty} 0 = 0$	$\lim\limits_{x \to -\infty} x = -\infty$	(Figure 2.5.2*c*)

function $f(x) = k$ has the same value k for all values of x, so its limiting value is also k as x approaches any real number a, or as x approaches $+\infty$ or $-\infty$ (Figure 2.5.1). The limits in the right side of the table should be evident from the graph of $f(x) = x$ (Figure 2.5.2).

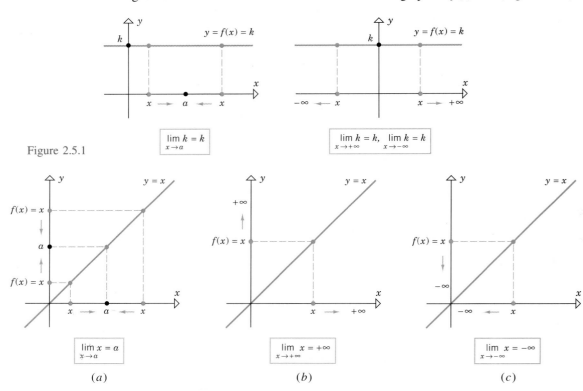

Figure 2.5.1

Figure 2.5.2   (*a*)   (*b*)   (*c*)

The limits studied thus far, together with the fundamental properties of limits given in the following theorem, can be used to help solve more complicated limit problems. Parts of this theorem are proved in Appendix C.

2.5.1   THEOREM.   *Let* lim *stand for one of the limits* $\lim\limits_{x \to a}$, $\lim\limits_{x \to a^-}$, $\lim\limits_{x \to a^+}$, $\lim\limits_{x \to +\infty}$, *or* $\lim\limits_{x \to -\infty}$. *If* $L_1 = \lim f(x)$ *and* $L_2 = \lim g(x)$ *both exist, then*

(*a*) $\lim [f(x) + g(x)] = \lim f(x) + \lim g(x) = L_1 + L_2$

(*b*) $\lim [f(x) - g(x)] = \lim f(x) - \lim g(x) = L_1 - L_2$

(*c*) $\lim [f(x)g(x)] = \lim f(x) \lim g(x) = L_1 L_2$

(*d*) $\lim \dfrac{f(x)}{g(x)} = \dfrac{f(x)}{\lim g(x)} = \dfrac{L_1}{L_2}$ *if* $L_2 \neq 0$

(*e*) $\lim \sqrt[n]{f(x)} = \sqrt[n]{\lim f(x)} = \sqrt[n]{L_1}$ *provided* $L_1 \geq 0$ *if n is even.*

In words, this theorem states:

(a) *The limit of a sum is the sum of the limits.*
(b) *The limit of a difference is the difference of the limits.*
(c) *The limit of a product is the product of the limits.*
(d) *The limit of a quotient is the quotient of the limits provided the limit of the denominator is nonzero.*
(e) *The limit of an nth root is the nth root of the limit.*

REMARK. Although results (a) and (c) are stated for two functions f and g, these results hold as well for any finite number of functions; that is, if the limits $\lim f_1(x), \lim f_2(x), \ldots,$ $\lim f_n(x)$ all exist, then

$$\lim [\, f_1(x) + f_2(x) + \cdots + f_n(x)] = \lim f_1(x) + \lim f_2(x) + \cdots + \lim f_n(x) \qquad (1)$$

$$\lim [\, f_1(x) f_2(x) \cdots f_n(x)] = \lim f_1(x) \lim f_2(x) \cdots \lim f_n(x) \qquad (2)$$

In particular, if $f_1, f_2, \ldots, f_n$ are all the same function f, then (2) reduces to

$$\lim [\, f(x)]^n = [\lim f(x)]^n \qquad (3)$$

From (3) we obtain the useful result

$$\lim_{x \to a} x^n = [\lim_{x \to a} x]^n = a^n \qquad (4)$$

For example,

$$\lim_{x \to 3} x^4 = 3^4 = 81$$

Another useful result follows from part (c) of Theorem 2.5.1 in the special case where one of the factors is a constant k:

$$\lim kf(x) = \lim k \lim f(x) = k \lim f(x) \qquad (5)$$

In words, the first and last expressions in (5) state:

A constant factor can be moved through a limit sign.

☐ **LIMITS OF POLYNOMIALS AS $x \to a$**

Example 1 Find $\lim_{x \to 5} (x^2 - 4x + 3)$ and justify each step.

Solution.

$$\lim_{x \to 5} (x^2 - 4x + 3) = \lim_{x \to 5} x^2 - \lim_{x \to 5} 4x + \lim_{x \to 5} 3 \qquad \boxed{\text{Theorem 2.5.1}(a), (b)}$$

$$= \lim_{x \to 5} x^2 - 4 \lim_{x \to 5} x + \lim_{x \to 5} 3 \qquad \boxed{\text{Equation (5)}}$$

$$= 5^2 - 4(5) + 3 \qquad \boxed{\text{Equation (4)}}$$

$$= 8 \quad \blacktriangleleft$$

The following theorem greatly simplifies the computation of limits of polynomials. It shows that the limit of a polynomial $p(x)$ as x approaches a can be obtained by evaluating the polynomial at a.

2.5.2   THEOREM.   *For any polynomial*

$$p(x) = c_0 + c_1 x + \cdots + c_n x^n$$

and any real number a,

$$\lim_{x \to a} p(x) = c_0 + c_1 a + \cdots + c_n a^n = p(a)$$

Proof.

$$
\begin{aligned}
\lim_{x \to a} p(x) &= \lim_{x \to a} (c_0 + c_1 x + \cdots + c_n x^n) \\
&= \lim_{x \to a} c_0 + \lim_{x \to a} c_1 x + \cdots + \lim_{x \to a} c_n x^n \\
&= \lim_{x \to a} c_0 + c_1 \lim_{x \to a} x + \cdots + c_n \lim_{x \to a} x^n \\
&= c_0 + c_1 a + \cdots + c_n a^n = p(a) \quad \blacksquare
\end{aligned}
$$

Example 2   If we apply Theorem 2.5.2 to the limit problem in Example 1, we can bypass the intermediate steps and write immediately

$$\lim_{x \to 5} (x^2 - 4x + 3) = 5^2 - 4(5) + 3 = 8 \quad \blacktriangleleft$$

☐ **LIMITS INVOLVING** $1/x$

The following limits are suggested by the graph of $f(x) = 1/x$ (Figure 2.5.3) or from the numerical calculations in Table 2.5.2.

$$\lim_{x \to 0^+} \frac{1}{x} = +\infty, \quad \lim_{x \to 0^-} \frac{1}{x} = -\infty, \quad \lim_{x \to +\infty} \frac{1}{x} = 0, \quad \lim_{x \to -\infty} \frac{1}{x} = 0 \qquad (6)$$

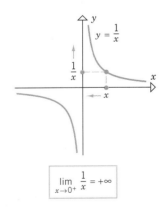

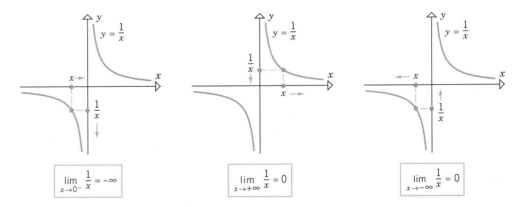

Figure 2.5.3

Table 2.5.2

		VALUES					CONCLUSION
x	1	10	100	1000	10,000	...	As $x \to +\infty$ the value of $1/x$
$1/x$	1	.1	.01	.001	.0001	...	decreases toward zero.
x	-1	-10	-100	-1000	$-10,000$	...	As $x \to -\infty$ the value of $1/x$
$1/x$	-1	$-.1$	$-.01$	$-.001$	$-.0001$	...	increases toward zero.
x	1	.1	.01	.001	.0001	...	As $x \to 0^+$ the value of $1/x$
$1/x$	1	10	100	1000	10,000	...	increases without bound.
x	-1	$-.1$	$-.01$	$-.001$	$-.0001$	...	As $x \to 0^-$ the value of $1/x$
$1/x$	-1	-10	-100	-1000	$-10,000$	...	decreases without bound.

For every real number a the graph of the function $1/(x - a)$ is a translation of the graph of $1/x$ (Figure 2.5.4), so the following limits can be deduced by analyses similar to those that produced the limits in (6):

$$\lim_{x \to a^+} \frac{1}{x - a} = +\infty \qquad \lim_{x \to a^-} \frac{1}{x - a} = -\infty$$

$$\lim_{x \to +\infty} \frac{1}{x - a} = 0 \qquad \lim_{x \to -\infty} \frac{1}{x - a} = 0 \tag{7}$$

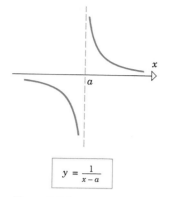

$$y = \frac{1}{x - a}$$

Figure 2.5.4

☐ **LIMITS OF POLYNOMIALS AS $x \to +\infty$ OR $x \to -\infty$**

In Figure 2.5.5 we have graphed the polynomials of the form x^n for $n = 1, 2, 3,$ and 4; and below each figure we have indicated the limits as $x \to +\infty$ and $x \to -\infty$. The results in the figure are special cases of the following general results:

$$\lim_{x \to +\infty} x^n = +\infty, \quad n = 1, 2, 3, \ldots \tag{8}$$

$$\lim_{x \to -\infty} x^n = \begin{cases} +\infty, & n = 2, 4, 6, \ldots \\ -\infty, & n = 1, 3, 5, \ldots \end{cases} \tag{9}$$

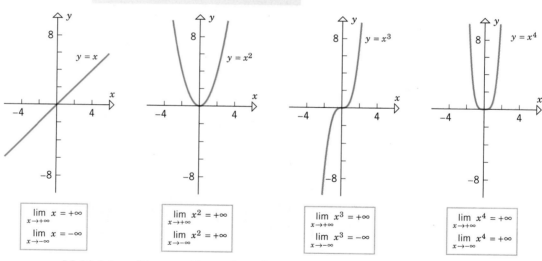

Figure 2.5.5

Multiplying x^n by a positive real number does not affect limits (8) and (9), but multiplying by a negative real number reverses the signs.

Example 3

$$\lim_{x \to +\infty} 2x^5 = +\infty, \qquad \lim_{x \to -\infty} 2x^5 = -\infty$$

$$\lim_{x \to +\infty} -7x^6 = -\infty, \qquad \lim_{x \to -\infty} -7x^6 = -\infty \qquad ◄$$

The following limits are consequences of (6) for positive integer values of n; they are also suggested by the graphs in Figure 2.5.6:

$$\lim_{x \to +\infty} \frac{1}{x^n} = \left(\lim_{x \to +\infty} \frac{1}{x} \right)^n = 0 \qquad \lim_{x \to -\infty} \frac{1}{x^n} = \left(\lim_{x \to -\infty} \frac{1}{x} \right)^n = 0 \tag{10}$$

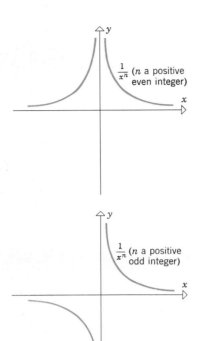

There is an important principle about limits of polynomials which, expressed informally, states that *a polynomial behaves like its term of highest degree as $x \to +\infty$ or $x \to -\infty$.* More precisely, if $c_n \neq 0$, then

$$\lim_{x \to +\infty} (c_0 + c_1 x + \cdots + c_n x^n) = \lim_{x \to +\infty} c_n x^n \qquad (11)$$

$$\lim_{x \to -\infty} (c_0 + c_1 x + \cdots + c_n x^n) = \lim_{x \to -\infty} c_n x^n \qquad (12)$$

We can motivate these results by factoring out the highest power of x from the polynomial and examining the limit of the factored expression. Thus,

$$c_0 + c_1 x + \cdots + c_n x^n = x^n \left(\frac{c_0}{x^n} + \frac{c_1}{x^{n-1}} + \cdots + c_n \right)$$

As $x \to +\infty$ or $x \to -\infty$, it follows from (10) that all of the terms with positive powers of x in the denominator approach 0, so (11) and (12) are certainly plausible.

Example 4

$$\lim_{x \to +\infty} (7x^5 - 4x^3 + 2x - 9) = \lim_{x \to +\infty} 7x^5 = +\infty$$

$$\lim_{x \to -\infty} (-4x^8 + 17x^3 - 5x + 1) = \lim_{x \to -\infty} -4x^8 = -\infty \qquad \blacktriangleleft$$

Figure 2.5.6

☐ **LIMITS OF RATIONAL FUNCTIONS AS $x \to a$**

Recall that a rational function is the ratio of two polynomials. Theorem 2.5.2 and Theorem 2.5.1(d) can often be used in combination to compute limits of rational functions.

Example 5   Find $\lim\limits_{x \to 2} \dfrac{5x^3 + 4}{x - 3}$.

Solution.

$$\lim_{x \to 2} \frac{5x^3 + 4}{x - 3} = \frac{\lim\limits_{x \to 2} (5x^3 + 4)}{\lim\limits_{x \to 2} (x - 3)} = \frac{5 \cdot 2^3 + 4}{2 - 3} = -44 \qquad \blacktriangleleft$$

The method of the preceding example will not work if the limit of the denominator is zero, since Theorem 2.5.1(d) is not applicable in this situation. However, if the numerator and denominator *both* approach zero as x approaches a, then the numerator and denominator will have a common factor of $x - a$ and the limit can often be obtained by first canceling the common factors. The following example illustrates this technique.

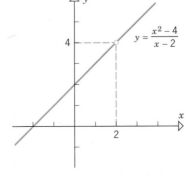

Example 6   Find $\lim\limits_{x \to 2} \dfrac{x^2 - 4}{x - 2}$.

Solution.   The numerator and denominator both have a limit of zero as x approaches 2, so they share a common factor of $x - 2$. The limit can be obtained as follows:

$$\lim_{x \to 2} \frac{x^2 - 4}{x - 2} = \lim_{x \to 2} \frac{(x - 2)(x + 2)}{x - 2} = \lim_{x \to 2} (x + 2) = 4$$

(Figure 2.5.7).   ◀

Figure 2.5.7

REMARK.   Although correct, the second equality in the preceding calculation needs some justification. As noted in Example 14 of Section 2.1, the functions

$$\frac{(x - 2)(x + 2)}{x - 2} \quad \text{and} \quad x + 2$$

are identical, except at the point $x = 2$, where the first function is undefined but the second function is defined. However, this difference has no effect on the limit as x approaches 2, since x remains different from 2 for purposes of the limit.

Example 7 Find

(a) $\displaystyle\lim_{x \to 3} \frac{x^2 - 6x + 9}{x - 3}$ (b) $\displaystyle\lim_{x \to -4} \frac{2x + 8}{x^2 + x - 12}$

Solution (a). The numerator and denominator both have a limit of zero as x approaches 3, so there is a common factor of $x - 3$. We proceed as follows:

$$\lim_{x \to 3} \frac{x^2 - 6x + 9}{x - 3} = \lim_{x \to 3} \frac{(x - 3)^2}{x - 3} = \lim_{x \to 3} (x - 3) = 0$$

Solution (b). The numerator and denominator both have a limit of zero as x approaches -4, so there is a common factor of $x - (-4) = x + 4$. We proceed as follows:

$$\lim_{x \to -4} \frac{2x + 8}{x^2 + x - 12} = \lim_{x \to -4} \frac{2(x + 4)}{(x + 4)(x - 3)} = \lim_{x \to -4} \frac{2}{x - 3} = -\frac{2}{7} \quad \blacktriangleleft$$

If the limit of the denominator is zero, but the limit of the numerator is not, then there are three possibilities for the limit of the rational function as $x \to a$:

- The limit may be $+\infty$.
- The limit may be $-\infty$.
- The limit may be $+\infty$ from one side and $-\infty$ from the other.

Example 8 Find

(a) $\displaystyle\lim_{x \to 4^+} \frac{2 - x}{(x - 4)(x + 2)}$ (b) $\displaystyle\lim_{x \to 4^-} \frac{2 - x}{(x - 4)(x + 2)}$ (c) $\displaystyle\lim_{x \to 4} \frac{2 - x}{(x - 4)(x + 2)}$

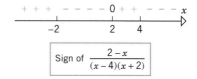

Figure 2.5.8

Solution. In all three parts the limit of the numerator is -2 and the limit of the denominator is 0, so the limit of the ratio does not exist. To be more specific, we need to analyze the sign of the ratio. Using the method of test points discussed in Section 1.1, the reader should be able to obtain the result in Figure 2.5.8. It follows from this figure that as x approaches 4 from the right the ratio is always negative, and as x approaches 4 from the left the sign of the ratio is eventually positive (after x exceeds 2), so

$$\lim_{x \to 4^+} \frac{2 - x}{(x - 4)(x + 2)} = -\infty \quad \text{and} \quad \lim_{x \to 4^-} \frac{2 - x}{(x - 4)(x + 2)} = +\infty$$

Because the one-sided limits have opposite signs, all we can say about the two-sided limit is that it does not exist. $\blacktriangleleft$

☐ LIMITS OF RATIONAL FUNCTIONS AS $x \to +\infty$ OR $x \to -\infty$

If we divide the numerator and denominator of a rational function by the highest power of x that occurs in the denominator, then all the powers of x in the denominator become constants or powers of $1/x$. The following examples show how this observation together with (10), (11), and (12) can be used to find limits of rational functions as $x \to +\infty$ or $x \to -\infty$.

Example 9   Find $\displaystyle\lim_{x \to +\infty} \frac{3x + 5}{6x - 8}$.

Solution.   Divide the numerator and denominator by the highest power of x that occurs in the denominator; this is $x^1 = x$. We obtain

$$\lim_{x \to +\infty} \frac{3x + 5}{6x - 8} = \lim_{x \to +\infty} \frac{3 + 5/x}{6 - 8/x} = \frac{\displaystyle\lim_{x \to +\infty} (3 + 5/x)}{\displaystyle\lim_{x \to +\infty} (6 - 8/x)}$$

$$= \frac{\displaystyle\lim_{x \to +\infty} 3 + \lim_{x \to +\infty} 5/x}{\displaystyle\lim_{x \to +\infty} 6 - \lim_{x \to +\infty} 8/x} = \frac{3 + 5 \displaystyle\lim_{x \to +\infty} 1/x}{6 - 8 \displaystyle\lim_{x \to +\infty} 1/x}$$

$$= \frac{3 + (5 \cdot 0)}{6 - (8 \cdot 0)} = \frac{1}{2} \quad \blacktriangleleft$$

Example 10   Find

(a) $\displaystyle\lim_{x \to -\infty} \frac{4x^2 - x}{2x^3 - 5}$ (b) $\displaystyle\lim_{x \to -\infty} \frac{5x^3 - 2x^2 + 1}{3x + 5}$

Solution (a).   Divide the numerator and denominator by the highest power of x that occurs in the denominator, namely x^3. We obtain

$$\lim_{x \to -\infty} \frac{4x^2 - x}{2x^3 - 5} = \lim_{x \to -\infty} \frac{4/x - 1/x^2}{2 - 5/x^3} = \frac{\displaystyle\lim_{x \to -\infty} (4/x - 1/x^2)}{\displaystyle\lim_{x \to -\infty} (2 - 5/x^3)}$$

$$= \frac{(4 \cdot 0) - 0}{2 - (5 \cdot 0)} = \frac{0}{2} = 0$$

Solution (b).   Divide the numerator and denominator by x to obtain

$$\lim_{x \to -\infty} \frac{5x^3 - 2x^2 + 1}{3x + 5} = \lim_{x \to -\infty} \frac{5x^2 - 2x + 1/x}{3 + 5/x} = +\infty$$

where the final step is justified by the fact that

$$5x^2 - 2x \to +\infty, \quad 1/x \to 0, \quad \text{and} \quad 3 + 5/x \to 3$$

as $x \to -\infty$. $\blacktriangleleft$

☐ **A QUICK METHOD FOR FINDING LIMITS OF RATIONAL FUNCTIONS AS** $x \to +\infty$ **OR** $x \to -\infty$

It can be shown (Exercise 76) that the limit of a *rational function* as $x \to +\infty$ or $x \to -\infty$ is unaffected if all but the highest degree terms in the numerator and denominator are discarded; that is, if $c_n \neq 0$ and $d_m \neq 0$, then

$$\lim_{x \to +\infty} \frac{c_0 + c_1 x + \cdots + c_n x^n}{d_0 + d_1 x + \cdots + d_m x^m} = \lim_{x \to +\infty} \frac{c_n x^n}{d_m x^m} \tag{13}$$

and

$$\lim_{x \to -\infty} \frac{c_0 + c_1 x + \cdots + c_n x^n}{d_0 + d_1 x + \cdots + d_m x^m} = \lim_{x \to -\infty} \frac{c_n x^n}{d_m x^m} \tag{14}$$

Example 11   Use Formulas (13) and (14) to find

(a) $\displaystyle\lim_{x \to +\infty} \frac{3x + 5}{6x - 8}$ (b) $\displaystyle\lim_{x \to -\infty} \frac{4x^2 - x}{2x^3 - 5}$ (c) $\displaystyle\lim_{x \to +\infty} \frac{3 - 2x^4}{x + 1}$

Solution (a).

$$\lim_{x \to +\infty} \frac{3x+5}{6x-8} = \lim_{x \to +\infty} \frac{3x}{6x} = \lim_{x \to +\infty} \frac{1}{2} = \frac{1}{2}$$

which agrees with the result obtained in Example 9.

Solution (b).

$$\lim_{x \to -\infty} \frac{4x^2-x}{2x^3-5} = \lim_{x \to -\infty} \frac{4x^2}{2x^3} = \lim_{x \to -\infty} \frac{2}{x} = 0$$

which agrees with the result obtained in Example 10.

Solution (c).

$$\lim_{x \to +\infty} \frac{3-2x^4}{x+1} = \lim_{x \to +\infty} \frac{-2x^4}{x} = \lim_{x \to +\infty} -2x^3 = -\infty \qquad \blacktriangleleft$$

REMARK.   We emphasize that Formulas (13) and (14) are only applicable if $x \to +\infty$ or $x \to -\infty$; they are not applicable to limits in which x approaches a *finite* number a.

☐ **LIMITS INVOLVING RADICALS**

Example 12   Find $\displaystyle \lim_{x \to +\infty} \sqrt[3]{\frac{3x+5}{6x-8}}$.

Solution.

$$\lim_{x \to +\infty} \sqrt[3]{\frac{3x+5}{6x-8}} \underset{\text{Theorem 2.5.1}(e)}{=} \sqrt[3]{\lim_{x \to +\infty} \frac{3x+5}{6x-8}} \underset{\text{Example 9}}{=} \sqrt[3]{\frac{1}{2}} \qquad \blacktriangleleft$$

 Example 13   Find

(a) $\displaystyle \lim_{x \to +\infty} \frac{\sqrt{x^2+2}}{3x-6}$ (b) $\displaystyle \lim_{x \to -\infty} \frac{\sqrt{x^2+2}}{3x-6}$

In both parts it would be helpful to manipulate the function so that the powers of x become powers of $1/x$. This can be achieved in both cases by dividing the numerator and denominator by $|x|$ and using the fact that $\sqrt{x^2} = |x|$.

Solution (a).   As $x \to +\infty$, the values of x are eventually positive, so we can replace $|x|$ by x where desirable. We obtain

$$\lim_{x \to +\infty} \frac{\sqrt{x^2+2}}{3x-6} = \lim_{x \to +\infty} \frac{\sqrt{x^2+2}/|x|}{(3x-6)/|x|} = \lim_{x \to +\infty} \frac{\sqrt{x^2+2}/\sqrt{x^2}}{(3x-6)/x}$$

$$= \lim_{x \to +\infty} \frac{\sqrt{1+2/x^2}}{3-6/x} = \frac{\displaystyle\lim_{x \to +\infty} \sqrt{1+2/x^2}}{\displaystyle\lim_{x \to +\infty} (3-6/x)}$$

$$= \frac{\sqrt{\displaystyle\lim_{x \to +\infty} (1+2/x^2)}}{\displaystyle\lim_{x \to +\infty} (3-6/x)} = \frac{\sqrt{\displaystyle\lim_{x \to +\infty} 1 + 2 \lim_{x \to +\infty} 1/x^2}}{\displaystyle\lim_{x \to +\infty} 3 - 6 \lim_{x \to +\infty} 1/x}$$

$$= \frac{\sqrt{1+(2 \cdot 0)}}{3-(6 \cdot 0)} = \frac{1}{3}$$

Solution (b).   As $x \to -\infty$, the values of x are eventually negative, so we can replace $|x|$ by $-x$ where desirable. We obtain

$$\lim_{x \to -\infty} \frac{\sqrt{x^2 + 2}}{3x - 6} = \lim_{x \to -\infty} \frac{\sqrt{x^2 + 2}/|x|}{(3x - 6)/|x|} = \lim_{x \to -\infty} \frac{\sqrt{x^2 + 2}/\sqrt{x^2}}{(3x - 6)/(-x)}$$

$$= \lim_{x \to -\infty} \frac{\sqrt{1 + 2/x^2}}{(6/x) - 3} = -\frac{1}{3} \quad \blacktriangleleft$$

☐ **LIMITS OF PIECEWISE DEFINED FUNCTIONS**

For functions that are defined piecewise, a two-sided limit at a point where the formula for the function changes is best obtained by first finding the one-sided limits at the point.

Example 14   Find $\lim_{x \to 3} f(x)$ for $f(x) = \begin{cases} x^2 - 5, & x \leq 3 \\ \sqrt{x + 13}, & x > 3. \end{cases}$

Solution.   As x approaches 3 from the left, the formula for f is

$$f(x) = x^2 - 5$$

so that

$$\lim_{x \to 3^-} f(x) = \lim_{x \to 3^-} (x^2 - 5) = 3^2 - 5 = 4$$

As x approaches 3 from the right, the formula for f is

$$f(x) = \sqrt{x + 13}$$

so that

$$\lim_{x \to 3^+} f(x) = \lim_{x \to 3^+} \sqrt{x + 13} = \sqrt{\lim_{x \to 3^+} (x + 13)} = \sqrt{16} = 4$$

Since the one-sided limits are equal, we have

$$\lim_{x \to 3} f(x) = 4 \quad \blacktriangleleft$$

▶ **Exercise Set 2.5**

Find the limits in Exercises 1–56.

1. $\lim_{x \to 8} 7.$

2. $\lim_{x \to -\infty} (-3).$

3. $\lim_{x \to 0^+} \pi.$

4. $\lim_{x \to -2} 3x.$

5. $\lim_{y \to 3^+} 12y.$

6. $\lim_{h \to +\infty} (-2h).$

7. $\lim_{x \to 5} \sqrt{x^3 - 3x - 1}.$

8. $\lim_{x \to 0^-} (x^4 + 12x^3 - 17x + 2).$

9. $\lim_{y \to -1} (y^6 - 12y + 1).$

10. $\lim_{x \to 3} \frac{x^2 - 2x}{x + 1}.$

11. $\lim_{y \to 2^-} \frac{(y - 1)(y - 2)}{y + 1}.$

12. $\lim_{x \to 0} \frac{6x - 9}{x^3 - 12x + 3}.$

13. $\lim_{x \to 4} \frac{x^2 - 16}{x - 4}.$

14. $\lim_{t \to -2} \frac{t^3 + 8}{t + 2}.$

15. $\lim_{x \to 1^+} \frac{x^4 - 1}{x - 1}.$

16. $\lim_{x \to 2} \frac{x^2 - 4x + 4}{x^2 + x - 6}.$

17. $\lim_{x \to -1} \frac{x^2 + 6x + 5}{x^2 - 3x - 4}.$

18. $\lim_{t \to 1} \frac{t^3 + t^2 - 5t + 3}{t^3 - 3t + 2}.$

19. $\lim_{x \to +\infty} \frac{3x + 1}{2x - 5}.$

20. $\lim_{x \to +\infty} \frac{1}{x - 12}.$

21. $\lim_{y \to -\infty} \frac{3}{y + 4}.$

22. $\lim_{x \to +\infty} \frac{5x^2 + 7}{3x^2 - x}.$

23. $\lim_{x \to -\infty} \frac{x - 2}{x^2 + 2x + 1}.$

24. $\lim_{s \to +\infty} \sqrt[3]{\frac{3s^7 - 4s^5}{2s^7 + 1}}.$

25. $\lim_{x \to -\infty} \frac{\sqrt{5x^2 - 2}}{x + 3}.$

26. $\lim_{x \to +\infty} \frac{\sqrt{5x^2 - 2}}{x + 3}.$

27. $\lim_{y \to -\infty} \frac{2 - y}{\sqrt{7 + 6y^2}}.$

28. $\lim_{y \to +\infty} \frac{2 - y}{\sqrt{7 + 6y^2}}.$

29. $\lim_{x \to -\infty} \frac{\sqrt{3x^4 + x}}{x^2 - 8}.$

30. $\lim_{x \to +\infty} \frac{\sqrt{3x^4 + x}}{x^2 - 8}.$

31. $\lim_{x \to 3^+} \frac{x}{x - 3}.$

32. $\lim_{x \to 3^-} \frac{x}{x - 3}.$

33. $\lim_{x \to 3} \frac{x}{x - 3}.$

34. $\lim_{x \to 2^+} \frac{x}{x^2 - 4}.$

35. $\lim_{x \to 2^-} \frac{x}{x^2 - 4}.$

36. $\lim\limits_{x \to 2} \dfrac{x}{x^2 - 4}$.

37. $\lim\limits_{y \to 6^+} \dfrac{y + 6}{y^2 - 36}$.

38. $\lim\limits_{y \to 6^-} \dfrac{y + 6}{y^2 - 36}$.

39. $\lim\limits_{y \to 6} \dfrac{y + 6}{y^2 - 36}$.

40. $\lim\limits_{x \to 4^+} \dfrac{3 - x}{x^2 - 2x - 8}$.

41. $\lim\limits_{x \to 4^-} \dfrac{3 - x}{x^2 - 2x - 8}$.

42. $\lim\limits_{x \to 4} \dfrac{3 - x}{x^2 - 2x - 8}$.

43. $\lim\limits_{x \to +\infty} \dfrac{7 - 6x^5}{x + 3}$.

44. $\lim\limits_{t \to -\infty} \dfrac{5 - 2t^3}{t^2 + 1}$.

45. $\lim\limits_{t \to +\infty} \dfrac{6 - t^3}{7t^3 + 3}$.

46. $\lim\limits_{x \to 0^+} \dfrac{x}{|x|}$.

47. $\lim\limits_{x \to 0^-} \dfrac{x}{|x|}$.

48. $\lim\limits_{x \to 3^-} \dfrac{1}{|x - 3|}$.

49. $\lim\limits_{x \to 9} \dfrac{x - 9}{\sqrt{x} - 3}$.

50. $\lim\limits_{y \to 4} \dfrac{4 - y}{2 - \sqrt{y}}$.

51. $\lim\limits_{x \to +\infty} \sqrt{x}$.

52. $\lim\limits_{x \to -\infty} \sqrt{5 - x}$.

53. $\lim\limits_{x \to -\infty} (3 - x)$.

54. $\lim\limits_{x \to -\infty} (3 - x^2)$.

55. $\lim\limits_{x \to +\infty} (1 + 2x - 3x^5)$.

56. $\lim\limits_{x \to +\infty} (2x^3 - 100x + 5)$.

57. Find

$$\lim\limits_{x \to a} \dfrac{x}{x + a}$$

where a is an arbitrary constant. [*Hint:* Consider the cases, $a \neq 0$ and $a = 0$ separately.]

58. Let $f(x) = \dfrac{x^3 - 1}{x - 1}$.

(a) Find $\lim\limits_{x \to 1} f(x)$.

(b) Sketch the graph of $y = f(x)$.

59. Let

$$f(x) = \begin{cases} x - 1, & x \leq 3 \\ 3x - 7, & x > 3 \end{cases}$$

Find

(a) $\lim\limits_{x \to 3^-} f(x)$ (b) $\lim\limits_{x \to 3^+} f(x)$

(c) $\lim\limits_{x \to 3} f(x)$.

60. Let

$$g(t) = \begin{cases} t^2, & t \geq 0 \\ t - 2, & t < 0 \end{cases}$$

Find

(a) $\lim\limits_{t \to 0^-} g(t)$ (b) $\lim\limits_{t \to 0^+} g(t)$ (c) $\lim\limits_{t \to 0} g(t)$.

61. Find $\lim\limits_{x \to 3} h(x)$ given that

$$h(x) = \begin{cases} x^2 - 2x + 1, & x \neq 3 \\ 7, & x = 3 \end{cases}$$

62. Let

$$F(x) = \begin{cases} \dfrac{x^2 - 9}{x + 3}, & x \neq -3 \\ k, & x = -3 \end{cases}$$

(a) Find k so that $F(-3) = \lim\limits_{x \to -3} F(x)$.

(b) With k assigned the value $\lim\limits_{x \to -3} F(x)$, show that $F(x)$ can be expressed as a polynomial.

63. (a) Explain why the following calculation is incorrect.

$$\lim\limits_{x \to 0^+} \left(\dfrac{1}{x} - \dfrac{1}{x^2} \right) = \lim\limits_{x \to 0^+} \dfrac{1}{x} - \lim\limits_{x \to 0^+} \dfrac{1}{x^2}$$
$$= +\infty - (+\infty) = 0$$

(b) Show that $\lim\limits_{x \to 0^+} \left(\dfrac{1}{x} - \dfrac{1}{x^2} \right) = -\infty$.

64. Find $\lim\limits_{x \to 0^-} \left(\dfrac{1}{x} + \dfrac{1}{x^2} \right)$.

In Exercises 65–68, first rationalize the numerator, then find the limit.

65. $\lim\limits_{x \to 0} \dfrac{\sqrt{x + 4} - 2}{x}$.

66. $\lim\limits_{x \to 0} \dfrac{\sqrt{x^2 + 4} - 2}{x}$.

67. $\lim\limits_{x \to 0} \dfrac{\sqrt{5x + 9} - 3}{x}$.

68. $\lim\limits_{x \to 3} \dfrac{1 - \sqrt{x - 2}}{x - 3}$.

Find the limits in Exercises 69–74.

69. $\lim\limits_{x \to +\infty} (\sqrt{x^2 + 3} - x)$.

70. $\lim\limits_{x \to +\infty} (\sqrt{2x^2 + 5} - x)$.

71. $\lim\limits_{x \to +\infty} (\sqrt{x^2 + 5x} - x)$.

72. $\lim\limits_{x \to +\infty} (\sqrt{x^2 - 3x} - x)$.

73. $\lim\limits_{x \to +\infty} (\sqrt{x^2 + ax} - x)$.

74. $\lim\limits_{x \to +\infty} (\sqrt{x^2 + ax} - \sqrt{x^2 + bx})$.

75. Let $r(x)$ be a rational function. Under what conditions is it true that $\lim\limits_{x \to a} r(x) = r(a)$?

76. Find

$$\lim\limits_{x \to +\infty} \dfrac{c_0 + c_1 x + \cdots + c_n x^n}{d_0 + d_1 x + \cdots + d_m x^m}$$

where $c_n \neq 0$ and $d_m \neq 0$. [*Hint:* Your answer will depend on whether $m < n$, $m = n$, or $m > n$.]

■ 2.6   LIMITS: A RIGOROUS APPROACH

This section gives a rigorous treatment of two-sided limits. Readers interested in pursuing the theory of one-sided and infinite limits can turn to Appendix C, where we give a rigorous treatment of these topics and also prove some of the basic limit theorems.

☐ **DEFINITION OF A LIMIT**

In the preceding sections we treated limits informally, interpreting

$$\lim_{x \to a} f(x) = L$$

to mean that the values of $f(x)$ approach L as x approaches a from either side (but remains different from a). However, the phrases ''$f(x)$ approaches L'' and ''x approaches a'' are intuitive ideas without precise mathematical definitions. Our goal is to make these ideas precise. However, because the concept of a limit is intricate, we will build up to it in steps by first giving two preliminary definitions of a limit (each of which will capture essential ideas), and then give the final definition as it is most commonly used.

To motivate a precise definition of a limit, consider the function f that is graphed in Figure 2.6.1. We have intentionally placed a hole in the graph at $x = a$ to emphasize that the function f need not be defined at the point a in the ensuing discussion. For the function graphed in the figure, our intuitive understanding of limits suggests that $f(x)$ approaches L as x approaches a. This means that if we pick *any* positive number, say ϵ, and construct an open interval on the y-axis that extends ϵ units above and below L (Figure 2.6.2), then the values of $f(x)$ will ultimately fall inside the interval $(L - \epsilon, L + \epsilon)$ as x approaches a from either side. In Figure 2.6.1 this occurs once x is between x_0 and a on the left side or between a and x_1 on the right side. This suggests the following definition of a two-sided limit.

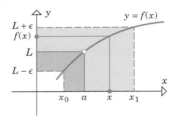

Figure 2.6.1

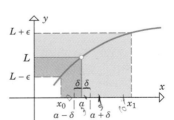

Figure 2.6.2

2.6.1   LIMIT (FIRST PRELIMINARY DEFINITION).   Let $f(x)$ be defined for all x in some open interval containing the number a, with the possible exception that $f(x)$ may not be defined at a. We shall write

$$\lim_{x \to a} f(x) = L$$

if given any number $\epsilon > 0$ we can find an open interval (x_0, x_1) containing the point a such that $f(x)$ satisfies

$$L - \epsilon < f(x) < L + \epsilon$$

for each x in the interval (x_0, x_1), except possibly $x = a$.

In this definition note that the number ϵ is completely arbitrary, except for the restriction that it be positive. In particular, if we choose ϵ to be small, and if we restrict x to the set of values for which $L - \epsilon < f(x) < L + \epsilon$, then the effect of the restriction on x is to force the values of $f(x)$ close to L. Thus, the definition can be paraphrased informally as follows:

The limit of $f(x)$ as x approaches a is L if and only if the values of $f(x)$ can be forced as close as we like to L by restricting x sufficiently close to (but different from) a.

Our next objective is to rephrase the definition of a limit in an alternative form that will be easier to apply. We begin with some observations about Definition 2.6.1.

•   To say that

$$L - \epsilon < f(x) < L + \epsilon \tag{1}$$

holds for all x in the interval (x_0, x_1), except possibly at a, is equivalent to stating that (1) holds for all x in the set

$$(x_0, a) \cup (a, x_1) \tag{2}$$

- If the inequality in (1) holds for all x in the set in (2), then this inequality also holds for all x in any subset of (2). In particular, if we let δ be any positive number that is smaller than both $a - x_0$ and $x_1 - a$, then

$$(a - \delta, a) \cup (a, a + \delta) \tag{3}$$

will be a subset of (2) and (1) will hold for all x in this subset (Figure 2.6.2).

This leads us to the following alternative form of the limit definition.

2.6.2 LIMIT (SECOND PRELIMINARY DEFINITION). Let $f(x)$ be defined for all x in some open interval containing the number a, with the possible exception that $f(x)$ may not be defined at a. We shall write

$$\lim_{x \to a} f(x) = L$$

if given any number $\epsilon > 0$ we can find a number $\delta > 0$ such that $f(x)$ satisfies

$$L - \epsilon < f(x) < L + \epsilon$$

for each x in the set $(a - \delta, a) \cup (a, a + \delta)$.

In preparation for our final form of the limit definition, we make the following observations:

- The inequalities in (1) can be expressed in terms of absolute values as

$$|f(x) - L| < \epsilon$$

- The set in (3) can be expressed in terms of absolute values as the set of all x such that

$$0 < |x - a| < \delta$$

2.6.3 LIMIT DEFINITION (FINAL FORM). Let $f(x)$ be defined for all x in some open interval containing the number a, with the possible exception that $f(x)$ may not be defined at a. We shall write

$$\lim_{x \to a} f(x) = L$$

if given any number $\epsilon > 0$ we can find a number $\delta > 0$ such that

$$|f(x) - L| < \epsilon \quad \text{if } x \text{ satisfies} \quad 0 < |x - a| < \delta$$

In the preceding sections we illustrated various numerical and geometric methods for *guessing* at limits. Now, armed with a precise definition of a limit, we are in a position to decide whether a particular guess is correct. The following example illustrates a method for doing this.

Example 1 Use Definition 2.6.3 to prove that $\lim_{x \to 2} (3x - 5) = 1$.

Solution. We must show that given any positive number ϵ, we can find a positive number δ such that

$$\underbrace{|(3x - 5)}_{f(x)} - \underbrace{1|}_{L} < \epsilon \quad \text{if } x \text{ satisfies} \quad 0 < |x - \underbrace{2}_{a}| < \delta \tag{4}$$

But this "if statement" can be rewritten as

$$|3x - 6| < \epsilon \qquad \text{if} \quad 0 < |x - 2| < \delta$$

$$3|x - 2| < \epsilon \qquad \text{if} \quad 0 < |x - 2| < \delta$$

$$|x - 2| < \epsilon/3 \quad \text{if} \quad 0 < |x - 2| < \delta \qquad (5)$$

One choice of δ that makes the "if statement" in (5) true for any $\epsilon > 0$ is $\delta = \epsilon/3$, since with this choice of δ the right side of the statement becomes

$$0 < |x - 2| < \epsilon/3$$

which implies that $|x - 2| < \epsilon/3$ as required for the left side of the statement. This proves that $\lim\limits_{x \to 2} (3x - 5) = 1$. ◄

REMARK.   The preceding example illustrates the general form of a limit proof: We *assume* that we are given an arbitrary positive number ϵ, and we try to *show* that regardless of the value of ϵ there is a positive number δ such that the "if statement" in Definition 2.6.3 is true. We did this in Example 1 by finding a formula (namely, $\delta = \epsilon/3$) for the unknown number δ in terms of the given number ϵ; we found that formula by manipulating the "if statement" to a form in which the formula became self-evident. We should emphasize that Example 1 is about as easy as a limit proof can get; most limit proofs require a little more algebraic and logical ingenuity. The reader who finds "δ-ϵ" discussions hard going should not become discouraged; the concepts and techniques are intrinsically difficult. In fact, a precise understanding of limits evaded the finest mathematical minds for centuries.

☐ **THE VALUE OF δ IS NOT UNIQUE**

Note that the value of δ in Definition 2.6.3 is not unique. Once one value of δ is found that fulfills the requirements of the definition, any *smaller* positive value for δ will also fulfill these requirements. To see why this is so, assume that we have found a value of δ such that

$$|f(x) - L| < \epsilon \quad \text{if} \quad 0 < |x - a| < \delta \qquad (6)$$

and let δ_1 be any positive number smaller than δ. If x satisfies

$$0 < |x - a| < \delta_1$$

then

$$0 < |x - a| < \delta_1 < \delta$$

Thus $|f(x) - L| < \epsilon$ for this x, since x satisfies the condition $0 < |x - a| < \delta$ in (6). For example, we found in Example 1 that $\delta = \epsilon/3$ fulfills the requirements of Definition 2.6.3. Consequently, any smaller value for δ, such as $\delta = \epsilon/4$, $\delta = \epsilon/5$, or $\delta = \epsilon/6$, does also.

Example 2 Prove that $\lim\limits_{x \to 3} x^2 = 9$.

Solution.   We must show that given any positive number ϵ, we can find a positive number δ such that

$$|x^2 - 9| < \epsilon \quad \text{if} \quad 0 < |x - 3| < \delta \qquad (7)$$

Because $|x - 3|$ occurs on the right side of the "if statement" in (7), it will be helpful to factor the left side to introduce an $|x - 3|$. Since

$$|x^2 - 9| = |(x + 3)(x - 3)| = |x + 3| \, |x - 3|$$

we can rewrite (7) as

$$|x + 3| \, |x - 3| < \epsilon \quad \text{if} \quad 0 < |x - 3| < \delta \qquad (8)$$

If there were some way to find a positive constant k such that

$$|x + 3| < k \qquad (9)$$

then we could write

$$|x + 3| \, |x - 3| < k|x - 3| \quad \text{(if } x \neq 3)$$

This would imply that

$$|x + 3| \, |x - 3| < \epsilon \quad \text{if} \quad k|x - 3| < \epsilon$$

which would enable us to rewrite (9) as

$$k|x - 3| < \epsilon \quad \text{if} \quad 0 < |x - 3| < \delta \tag{10}$$

As in Example 1, we could then in turn rewrite this as

$$|x - 3| < \epsilon/k \quad \text{if} \quad 0 < |x - 3| < \delta$$

from which we see that (10) holds if

$$\delta = \epsilon/k \tag{11}$$

To find a value of k for which (10) holds, we recall our earlier observation that once a value of δ is found, then any *smaller* positive value of δ can be used in its place. This allows us to assume arbitrarily that $\delta \leq 1$, because if $\delta > 1$, then we are free to use the smaller value $\delta = 1$ instead. Thus, the problem of showing that there is a value of δ that satisfies (10) is equivalent to showing that there is a value of δ that does not exceed 1 and such that

$$|x + 3| \, |x - 3| < \epsilon \quad \text{if} \quad 0 < |x - 3| < \delta \leq 1 \tag{12}$$

But it now follows from the restriction on δ in (12) that

$$|x - 3| < 1 \quad \text{or equivalently,} \quad 2 < x < 4$$

This implies that $5 < x + 3 < 7$, which in turn implies that

$$|x + 3| < 7$$

Comparing this inequality to (9) suggests that $k = 7$; and from (11)

$$\delta = \epsilon/k = \epsilon/7$$

In summary, given $\epsilon > 0$ we choose $\delta = \epsilon/7$, provided that $\epsilon/7$ does not exceed 1. If $\epsilon/7$ exceeds 1, we choose $\delta = 1$. More concisely, we take δ to be the minimum of the numbers $\epsilon/7$ and 1. This is sometimes written as

$$\delta = \min(\epsilon/7, 1) \quad \blacktriangleleft$$

REMARK.   In the preceding example the reader may have wondered how we knew to make the restriction $\delta \leq 1$ as opposed to some other restriction such as $\delta \leq \frac{1}{2}$ or $\delta \leq 5$. Actually, our selection was completely arbitrary; any other restriction of the form $\delta \leq c$ would have worked equally well (Exercise 29).

Example 3   Prove that $\displaystyle\lim_{x \to 1/2} \frac{1}{x} = 2$.

Solution.   We must show that given $\epsilon > 0$, there exists a $\delta > 0$ such that

$$|(1/x) - 2| < \epsilon \quad \text{if} \quad 0 < |x - \tfrac{1}{2}| < \delta \tag{13}$$

Because $|x - \frac{1}{2}|$ occurs in (13), it will be helpful to rewrite (13) so that $|x - \frac{1}{2}|$ appears as a factor on the left side. We do this as follows:

$$|(1/x) - 2| = |(2/x)(\tfrac{1}{2} - x)| = |2/x| \, |\tfrac{1}{2} - x| = |2/x| \, |x - \tfrac{1}{2}|$$

Thus, (13) is equivalent to

$$|2/x| \, |x - \tfrac{1}{2}| < \epsilon \quad \text{if} \quad 0 < |x - \tfrac{1}{2}| < \delta \tag{14}$$

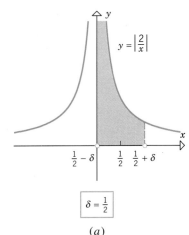

$$\delta = \tfrac{1}{2}$$

(a)

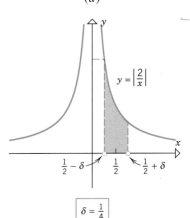

$$\delta = \tfrac{1}{4}$$

(b)

Figure 2.6.3

From here on, the procedure is similar to that used in the last example. By restricting δ we shall try to find a positive constant k such that

$$|2/x| < k \qquad (15)$$

In Figure 2.6.3 we have sketched the graph of $|2/x|$ and marked the set of x values satisfying (13). In part (a) of the figure we took $\delta = \tfrac{1}{2}$ and in part (b) of the figure we took $\delta < \tfrac{1}{2}$. Part (a) of this figure makes it clear that if $\delta = \tfrac{1}{2}$ (or $\delta > \tfrac{1}{2}$), then the values of $|2/x|$ will have no upper bound, making it impossible to satisfy (15). However, part (b) shows that when $\delta < \tfrac{1}{2}$, there is an upper bound k to the values of $|2/x|$. Therefore, the arbitrary restriction $\delta \leq \tfrac{1}{4}$ will be satisfactory to ensure that $|2/x|$ has an upper bound. Thus, showing that there is a value of δ that satisfies (14) is equivalent to showing that there is a value of δ that does not exceed $\tfrac{1}{4}$ and such that

$$|2/x|\,|x - \tfrac{1}{2}| < \epsilon \quad \text{if} \quad 0 < |x - \tfrac{1}{2}| < \delta \leq \tfrac{1}{4} \qquad (16)$$

But it now follows from the restriction on δ in (16) that

$$|x - \tfrac{1}{2}| < \tfrac{1}{4} \quad \text{or equivalently,} \quad \tfrac{1}{4} < x < \tfrac{3}{4}$$

To get a bound on the size of $|2/x|$, we multiply the last set of inequalities through by $\tfrac{1}{2}$ and take reciprocals, which gives

$$8 > 2/x > 8/3$$

This implies that $|2/x| < 8$, so if $x \neq \tfrac{1}{2}$, we have

$$|2/x|\,|x - \tfrac{1}{2}| < \epsilon \quad \text{if} \quad 8|x - \tfrac{1}{2}| < \epsilon$$

or

$$|2/x|\,|x - \tfrac{1}{2}| < \epsilon \quad \text{if} \quad |x - \tfrac{1}{2}| < \epsilon/8$$

Thus,

$$\delta = \epsilon/8$$

In summary, given $\epsilon > 0$ we can choose $\delta = \epsilon/8$, provided that this choice does not violate $\delta \leq \tfrac{1}{4}$. If it does, we choose the value $\delta = \tfrac{1}{4}$ instead. More concisely,

$$\delta = \min\left(\epsilon/8, \tfrac{1}{4}\right) \qquad \blacktriangleleft$$

The next example illustrates a technique for proving that a two-sided limit does not exist in the case where the one-sided limits exist but are not equal.

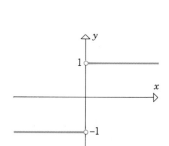

Figure 2.6.4

Example 4 The graph of

$$f(x) = \begin{cases} 1, & x > 0 \\ -1, & x < 0 \end{cases}$$

is shown in Figure 2.6.4. Because the one-sided limits

$$\lim_{x \to 0^+} f(x) = 1 \quad \text{and} \quad \lim_{x \to 0^-} f(x) = -1$$

are not equal, the two-sided limit $\lim_{x \to 0} f(x)$ does not exist. Prove that this is so using Definition 2.6.3.

Solution. We shall assume that there is a limit and obtain a contradiction. Assume that there is a number L such that

$$\lim_{x \to 0} f(x) = L$$

Then given any $\epsilon > 0$, there exists a $\delta > 0$ such that

$$|f(x) - L| < \epsilon \quad \text{if} \quad 0 < |x - 0| < \delta$$

In particular, if we take $\epsilon = 1$, there is a $\delta > 0$ such that

$$|f(x) - L| < 1 \quad \text{if} \quad 0 < |x - 0| < \delta \tag{17}$$

But $x = \delta/2$ and $x = -\delta/2$ both satisfy the requirement of (17), so

$$|f(\delta/2) - L| < 1 \quad \text{and} \quad |f(-\delta/2) - L| < 1 \tag{18}$$

However, $\delta/2$ is positive and $-\delta/2$ is negative, so

$$f(\delta/2) = 1 \quad \text{and} \quad f(-\delta/2) = -1$$

Thus, (18) states that

$$|1 - L| < 1 \quad \text{and} \quad |-1 - L| < 1$$

or equivalently,

$$0 < L < 2 \quad \text{and} \quad -2 < L < 0$$

But this is a contradiction, since no number L can satisfy both of these conditions.   ◄

► Exercise Set 2.6

In Exercises 1–10, we are told that $\lim\limits_{x \to a} f(x) = L$ and we are given a value of ϵ. In each exercise, find a number δ such that $|f(x) - L| < \epsilon$ whenever $0 < |x - a| < \delta$.

1. $\lim\limits_{x \to 4} 2x = 8; \ \epsilon = 0.1$.   **2.** $\lim\limits_{x \to -2} \dfrac{1}{2}x = -1; \ \epsilon = 0.1$.

3. $\lim\limits_{x \to -1} (7x + 5) = -2; \ \epsilon = 0.01$.

4. $\lim\limits_{x \to 3} (5x - 2) = 13; \ \epsilon = 0.01$.

5. $\lim\limits_{x \to 2} \dfrac{x^2 - 4}{x - 2} = 4; \ \epsilon = 0.05$.

6. $\lim\limits_{x \to -1} \dfrac{x^2 - 1}{x + 1} = -2; \ \epsilon = 0.05$.

7. $\lim\limits_{x \to 4} x^2 = 16; \ \epsilon = 0.001$.   **8.** $\lim\limits_{x \to 9} \sqrt{x} = 3; \ \epsilon = 0.001$.

9. $\lim\limits_{x \to 5} \dfrac{1}{x} = \dfrac{1}{5}; \ \epsilon = 0.05$.   **10.** $\lim\limits_{x \to 0} |x| = 0; \ \epsilon = 0.05$.

In Exercises 11–24, use Definition 2.6.3 to prove that the given limit statement is correct.

11. $\lim\limits_{x \to 5} 3x = 15$.   **12.** $\lim\limits_{x \to 3} (4x - 5) = 7$.

13. $\lim\limits_{x \to 2} (2x - 7) = -3$.   **14.** $\lim\limits_{x \to -1} (2 - 3x) = 5$.

15. $\lim\limits_{x \to 0} \dfrac{x^2 + x}{x} = 1$.   **16.** $\lim\limits_{x \to -3} \dfrac{x^2 - 9}{x + 3} = -6$.

17. $\lim\limits_{x \to 1} 2x^2 = 2$.   **18.** $\lim\limits_{x \to 3} (x^2 - 5) = 4$.

19. $\lim\limits_{x \to 1/3} \dfrac{1}{x} = 3$.   **20.** $\lim\limits_{x \to -2} \dfrac{1}{x + 1} = -1$.

21. $\lim\limits_{x \to 4} \sqrt{x} = 2$.   **22.** $\lim\limits_{x \to 6} \sqrt{x + 3} = 3$.

23. $\lim\limits_{x \to 1} f(x) = 3$, where $f(x) = \begin{cases} x + 2, & x \neq 1 \\ 10, & x = 1. \end{cases}$

24. $\lim\limits_{x \to 2} (x^2 + 3x - 1) = 9$.

25. Let $f(x) = \begin{cases} \frac{1}{8}, & x > 0 \\ -\frac{1}{8}, & x < 0. \end{cases}$

Use the method of Example 4 to prove that $\lim\limits_{x \to 0} f(x)$ does not exist.

26. Let $g(x) = \begin{cases} 1 + x, & x > 0 \\ x - 1, & x < 0. \end{cases}$

Prove that $\lim\limits_{x \to 0} g(x)$ does not exist.

27. Prove that $\lim\limits_{x \to 1} \dfrac{1}{x - 1}$ does not exist.

28. (a) In Definition 2.6.3 there is a condition requiring that $f(x)$ be defined for every x in an open interval containing a, except possibly at a itself. What is the purpose of this requirement?

(b) Why is $\lim\limits_{x \to 0} \sqrt{x} = 0$ an incorrect statement?

(c) Is $\lim\limits_{x \to 0.01} \sqrt{x} = 0.1$ a correct statement?

29. Prove the result in Example 2 under the assumption that $\delta \leq 2$ rather than $\delta \leq 1$.

■ 2.7 CONTINUITY

> *A moving physical object cannot vanish at some point and reappear someplace else to continue its motion. Thus, we perceive the path of a moving object as a single, unbroken curve without gaps, jumps, or holes. Such curves can be described as "continuous." In this section we shall express this intuitive idea mathematically and develop some properties of continuous curves.*

□ **DEFINITION OF CONTINUITY**

Before we give any formal definitions, let us consider some of the ways in which curves can be "discontinuous." In Figure 2.7.1 we have graphed some curves which, because of their behavior at the point c, are not continuous at c. The curve in Figure 2.7.1*a* has a hole at the point c because the function f is undefined there. For the curves in Figures 2.7.1*b* and 2.7.1*c*, the function f is defined at c, but

$$\lim_{x \to c} f(x) \tag{1}$$

does not exist, thereby causing a break in the graph. For the curve in Figure 2.7.1*d*, the function f is defined at c, and the limit in (1) exists, yet the graph still has a break at the point c because

$$\lim_{x \to c} f(x) \neq f(c)$$

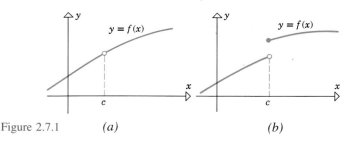
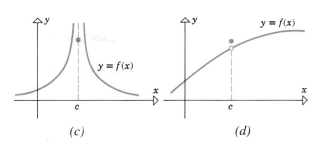

Figure 2.7.1   (a)   (b)   (c)   (d)

Based on this discussion, we see that there is a break or discontinuity in the graph of $y = f(x)$ at a point $x = c$ if any of the following conditions occur:

*   The function f is undefined at c.
*   The limit $\lim_{x \to c} f(x)$ does not exist.
*   The function f is defined at c and the limit $\lim_{x \to c} f(x)$ exists, but the value of the function at c and the value of the limit at c are different.

This suggests the following definition.

2.7.1 DEFINITION.   A function f is said to be ***continuous at a point c*** if the following conditions are satisfied:

  1. $f(c)$ is defined.

  2. $\lim_{x \to c} f(x)$ exists.

  3. $\lim_{x \to c} f(x) = f(c)$.

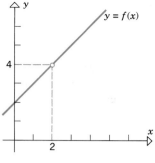

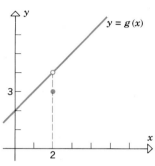

Figure 2.7.2

If one or more of the conditions in this definition fails to hold, then f is called **discontinuous at c** and c is called a **point of discontinuity** of f. If f is continuous at all points of an open interval (a, b), then f is said to be **continuous on (a, b)**. A function that is continuous on $(-\infty, +\infty)$ is said to be **continuous everywhere** or simply **continuous**.

Example 1   Let

$$f(x) = \frac{x^2 - 4}{x - 2} \quad \text{and} \quad g(x) = \begin{cases} \dfrac{x^2 - 4}{x - 2}, & x \neq 2 \\ 3, & x = 2 \end{cases}$$

Both f and g are discontinuous at 2 (Figure 2.7.2), the function f because $f(2)$ is undefined, and the function g because $g(2) = 3$, and

$$\lim_{x \to 2} g(x) = \lim_{x \to 2} \frac{x^2 - 4}{x - 2} = \lim_{x \to 2} (x + 2) = 4$$

so that

$$\lim_{x \to 2} g(x) \neq g(2) \quad \blacktriangleleft$$

REMARK.   Some authors define a function to be continuous at c if just condition 3 in Definition 2.7.1 holds. This is really equivalent to our definition because if condition 3 holds, then 1 and 2 hold automatically. (Why?) We have stated the three conditions for clarity. However, when we want to show that a function is continuous at a point, we shall just show that condition 3 holds.

Example 2   Show that $f(x) = x^2 - 2x + 1$ is a continuous function.

Solution.   We must show that the third condition in Definition 2.7.1 holds for all real numbers c. But, by Theorem 2.5.2

$$\lim_{x \to c} f(x) = \lim_{x \to c} (x^2 - 2x + 1) = c^2 - 2c + 1 = f(c)$$

which shows that the third condition holds. $\blacktriangleleft$

The preceding example is a special case of the following general result.

□ **CONTINUITY OF POLYNOMIALS**

2.7.2   THEOREM.   *Polynomials are continuous functions.*

Proof.   If p is a polynomial and c is any real number, then by Theorem 2.5.2

$$\lim_{x \to c} p(x) = p(c)$$

which proves the continuity of p at c. Since c is an arbitrary real number, p is continuous everywhere.   ∎

Example 3   Show that $f(x) = |x|$ is a continuous function.

Solution.   We can write $f(x)$ as

$$f(x) = |x| = \begin{cases} x & \text{if} \quad x > 0 \\ 0 & \text{if} \quad x = 0 \\ -x & \text{if} \quad x < 0 \end{cases}$$

It follows from Theorem 2.7.2 that $f(x) = |x|$ is continuous if $x > 0$ or $x < 0$ because $|x|$ is identical to the polynomial x in the former case and identical to the polynomial $-x$ in the latter. Thus, $x = 0$ is the only point that remains to be considered. At this point we have $f(0) = |0| = 0$, so it remains to show that

$$\lim_{x \to 0} f(x) = \lim_{x \to 0} |x| = 0 \tag{2}$$

Because the formula for f changes at 0, it will be helpful to consider the one-sided limits at 0 rather than the two-sided limit. We obtain

$$\lim_{x \to 0^+} |x| = \lim_{x \to 0^+} x = 0 \quad \text{and} \quad \lim_{x \to 0^-} |x| = \lim_{x \to 0^-} (-x) = 0$$

Thus, (2) holds and $|x|$ is continuous at $x = 0$. ◀

□ **SOME PROPERTIES OF CONTINUOUS FUNCTIONS**

The following basic result is an immediate consequence of Theorem 2.5.1.

2.7.3 THEOREM. *If the functions f and g are continuous at c, then*

(a) $f + g$ is continuous at c;
(b) $f - g$ is continuous at c;
(c) $f \cdot g$ is continuous at c;
(d) f/g is continuous at c if $g(c) \neq 0$ and is discontinuous at c if $g(c) = 0$.

We shall prove part (d) and leave the remaining proofs as exercises.

Proof (d). If $g(c) = 0$, then f/g is discontinuous at c because $f(c)/g(c)$ is undefined. Assume that $g(c) \neq 0$. We must show that

$$\lim_{x \to c} \frac{f(x)}{g(x)} = \frac{f(c)}{g(c)} \tag{3}$$

Since f and g are continuous at c,

$$\lim_{x \to c} f(x) = f(c) \quad \text{and} \quad \lim_{x \to c} g(x) = g(c)$$

Thus, by Theorem 2.5.1(d)

$$\lim_{x \to c} \frac{f(x)}{g(x)} = \frac{\lim_{x \to c} f(x)}{\lim_{x \to c} g(x)} = \frac{f(c)}{g(c)}$$

which proves (3). ∎

□ **CONTINUITY OF RATIONAL FUNCTIONS**

Example 4 Where is $h(x) = \dfrac{x^2 - 9}{x^2 - 5x + 6}$ continuous?

Solution. The numerator and denominator of h are polynomials and therefore are continuous everywhere by Theorem 2.7.2. Thus, by Theorem 2.7.3(d), the ratio is continuous everywhere except at the points where the denominator is zero. Since the solutions of

$$x^2 - 5x + 6 = 0$$

are $x = 2$ and $x = 3$, $h(x)$ is continuous everywhere except at these two points. ◀

The result in the preceding example is a special case of the following general theorem whose proof is left as an exercise.

> **2.7.4** THEOREM.   *A rational function is continuous everywhere except at the points where the denominator is zero.*

□ **CONTINUITY OF COMPOSITIONS**

The next theorem is useful for calculating limits of compositions of functions.

> **2.7.5** THEOREM.   *Let* lim *stand for one of the limits* $\lim_{x \to c}$, $\lim_{x \to c^-}$, $\lim_{x \to c^+}$, $\lim_{x \to +\infty}$, *or* $\lim_{x \to -\infty}$. *If* $\lim g(x) = L$ *and if the function* f *is continuous at* L, *then* $\lim f(g(x)) = f(L)$. *That is,* $\lim f(g(x)) = f(\lim g(x))$.

For those who have read Section 2.6, the proof of this result appears in Appendix C.

REMARK.   In words, this theorem states that the limit symbol can be moved through a function sign provided the limit of the expression inside the function sign exists and the function is continuous at this limit.

Example 5   We saw in Example 3 that $|x|$ is continuous everywhere, so Theorem 2.7.5 implies that

$$\lim |g(x)| = |\lim g(x)|$$

if $\lim g(x)$ exists. For example,

$$\lim_{x \to 3} |5 - x^2| = |\lim_{x \to 3} (5 - x^2)| = |-4| = 4 \quad \blacktriangleleft$$

The following consequence of Theorem 2.7.5 tells us that compositions of continuous functions are themselves continuous.

> **2.7.6** THEOREM.   *If the function* g *is continuous at the point* c *and the function* f *is continuous at the point* $g(c)$, *then the composition* $f \circ g$ *is continuous at* c.

Proof.   We must show that $f \circ g$ satisfies the third condition of Definition 2.7.1 at c. But this is so since we can write

$$\lim_{x \to c} (f \circ g)(x) = \lim_{x \to c} f(g(x)) = f(\lim_{x \to c} g(x)) = f(g(c)) = (f \circ g)(c) \quad \blacksquare$$

| Theorem 2.7.5 | g is continuous at c. |

Example 6   The function $|5 - x^2|$ is continuous because it is the composition of $|x|$ with $5 - x^2$, both of which are continuous.   ◄

□ **CONTINUITY FROM THE LEFT AND RIGHT**

Figure 2.7.3 shows the graphs of three functions defined only on a closed interval $[a, b]$. Obviously the function shown in Figure 2.7.3a should be regarded as discontinuous at the left-hand endpoint a, the function in Figure 2.7.3b should be regarded as discontinuous at the right-hand endpoint b, and the function in Figure 2.7.3c should be regarded as continuous at both endpoints. However, the definition of continuity (Definition 2.7.1) does not apply at the endpoints, since the two-sided limits appearing in parts 2 and 3 of the definition make no sense. At the left-hand endpoint the only sensible limit is the one-sided limit

$$\lim_{x \to a^+} f(x)$$

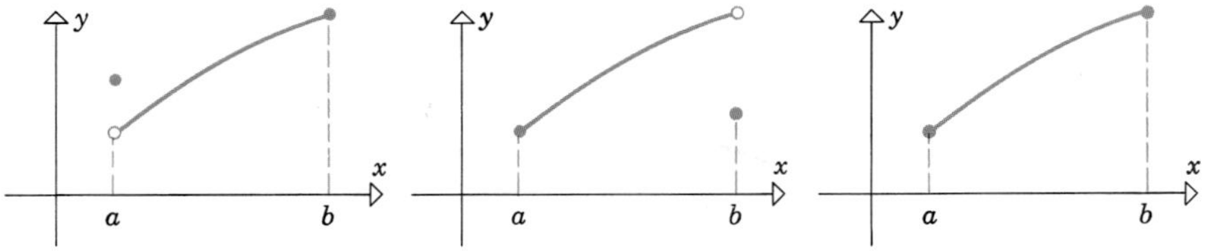

Figure 2.7.3   (a)   (b)   (c)

and at the right-hand endpoint the only sensible limit is the one-sided limit

$$\lim_{x \to b^-} f(x)$$

This motivates the following definitions.

2.7.7 DEFINITION. A function f is called **continuous from the left at the point c** if the conditions in the left column below are satisfied, and is called **continuous from the right at the point c** if the conditions in the right column are satisfied.

1. $f(c)$ is defined.	1′. $f(c)$ is defined.
2. $\lim\limits_{x \to c^-} f(x)$ exists.	2′. $\lim\limits_{x \to c^+} f(x)$ exists.
3. $\lim\limits_{x \to c^-} f(x) = f(c)$.	3′. $\lim\limits_{x \to c^+} f(x) = f(c)$.

2.7.8 DEFINITION. A function f is said to be **continuous on a closed interval $[a, b]$** if the following conditions are satisfied:

1. f is continuous on (a, b).
2. f is continuous from the right at a.
3. f is continuous from the left at b.

The reader should have no trouble deducing the appropriate definitions of continuity on intervals of the form $[a, +\infty)$, $(-\infty, b]$, $[a, b)$, and $(a, b]$.

Example 7 If f denotes the function graphed in Figure 2.7.3a, then

$$f(a) \neq \lim_{x \to a^+} f(x) \quad \text{and} \quad f(b) = \lim_{x \to b^-} f(x)$$

Thus, f is continuous from the left at b, but is not continuous from the right at a.   ◄

Example 8 Show that $f(x) = \sqrt{9 - x^2}$ is continuous on the closed interval $[-3, 3]$.

Solution. Observe that the domain of f is the interval $[-3, 3]$. For c in the interval $(-3, 3)$ we have by Theorem 2.5.1(e)

$$\lim_{x \to c} f(x) = \lim_{x \to c} \sqrt{9 - x^2} = \sqrt{\lim_{x \to c} (9 - x^2)} = \sqrt{9 - c^2} = f(c)$$

so that f is continuous on $(-3, 3)$. Also,

$$\lim_{x \to 3^-} f(x) = \lim_{x \to 3^-} \sqrt{9 - x^2} = \sqrt{\lim_{x \to 3^-} (9 - x^2)} = 0 = f(3)$$

and

$$\lim_{x \to -3^+} f(x) = \lim_{x \to -3^+} \sqrt{9 - x^2} = \sqrt{\lim_{x \to -3^+} (9 - x^2)} = 0 = f(-3)$$

so that f is continuous on $[-3, 3]$.   ◄

□ THE INTERMEDIATE
VALUE THEOREM

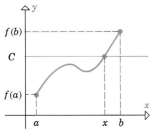

Figure 2.7.4

If f is continuous on a closed interval $[a, b]$, and we draw a horizontal line crossing the y-axis between the numbers $f(a)$ and $f(b)$ (Figure 2.7.4), then it is geometrically obvious that this line will cross the curve $y = f(x)$ at least once over the interval $[a, b]$. Stated another way, a continuous function must assume every possible value between $f(a)$ and $f(b)$ as x varies from a to b. This idea is formalized in the following theorem.

2.7.9 THEOREM (*Intermediate-Value Theorem*). *If f is continuous on a closed interval $[a, b]$ and C is any number between $f(a)$ and $f(b)$, inclusive, then there is at least one number x in the interval $[a, b]$ such that $f(x) = C$.*

This theorem is intuitively obvious, but its proof depends on a mathematically precise development of the real number system, which is beyond the scope of this text. The proof can be found in most advanced calculus texts.

The following useful result is an immediate consequence of the Intermediate-Value Theorem.

2.7.10 THEOREM. *If f is continuous on $[a, b]$, and if $f(a)$ and $f(b)$ have opposite signs, then there is at least one solution of the equation $f(x) = 0$ in the interval (a, b).*

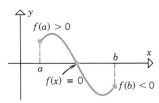

Figure 2.7.5

The proof is left as an exercise, but the result is illustrated in Figure 2.7.5 in the case where $f(a) > 0$ and $f(b) < 0$.

Theorem 2.7.10 is the basis for various numerical techniques for approximating solutions of equations of the form $f(x) = 0$.

Example 9 The equation

$$x^3 - x - 1 = 0$$

cannot be solved readily by factoring because the left side has no simple factors. However, from Figure 2.7.6, which was generated on a computer, we see that there is a solution in the interval $(1, 2)$. This can also be seen from Theorem 2.7.10 by letting $f(x) = x^3 - x - 1$, and noting that $f(1) = -1$ and $f(2) = 5$ have opposite signs.

To pinpoint the solution more exactly, we can divide the interval $(1, 2)$ into 10 equal parts and evaluate $f(x)$ at each point of subdivision using a calculator (Table 2.7.1). From the table we see that $f(1.3)$ and $f(1.4)$ have opposite signs, so there is a solution in the subinterval $(1.3, 1.4)$. If we use the midpoint 1.35 as an approximation to the solution, then our error is at most half the subinterval length or .05. If desired, we could reduce the error to at most .005 by subdividing the subinterval $(1.3, 1.4)$ into 10 parts and repeating the sign analysis process. With sufficiently many subdivisions, the solution can be determined to any degree of accuracy. ◄

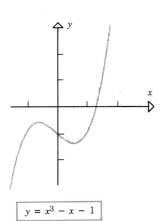

$y = x^3 - x - 1$

Figure 2.7.6

Table 2.7.1

x	1	1.1	1.2	1.3	1.4	1.5	1.6	1.7	1.8	1.9	2
$f(x)$	-1	$-.77$	$-.47$	$-.10$	.34	.88	1.5	2.2	3.0	3.9	5

We conclude this section by stating an alternative form of the continuity definition that will be useful in later sections.

If we let $h = x - c$ in Definition 2.7.1, then the condition that $x \to c$ is equivalent to $h \to 0$. Thus, Definition 2.7.1 can be restated as follows.

☐ **ALTERNATIVE
DEFINITION OF
CONTINUITY**

2.7.11 DEFINITION.   A function f is said to be **continuous at a point c** if the following conditions are satisfied:

1. $f(c)$ is defined.
2. $\lim\limits_{h \to 0} f(c + h)$ exists.
3. $\lim\limits_{h \to 0} f(c + h) = f(c)$.

REMARK.   As with Definition 2.7.1, conditions 1 and 2 hold automatically if condition 3 holds. Thus, to prove that a function is continuous at a point, it is only necessary to show that condition 3 holds.

▶ Exercise Set 2.7   Ⓒ *32–35*

In Exercises 1–4, let f be the function whose graph is shown. On which of the following intervals, if any, is f continuous?

(a) [1, 3]   (b) (1, 3)   (c) [1, 2]
(d) (1, 2)   (e) [2, 3]   (f) (2, 3).

On those intervals where f is discontinuous, state where the discontinuities occur.

1.
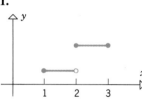

2.
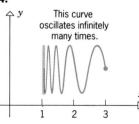

3.
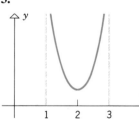

4.
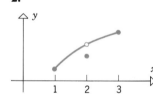

This curve oscillates infinitely many times.

In Exercises 5–16, find the points of discontinuity, if any.

5. $f(x) = x^3 - 2x + 3.$

6. $f(x) = (x - 5)^{17}.$

7. $f(x) = \dfrac{x}{x^2 + 1}.$

8. $f(x) = \dfrac{x}{x^2 - 1}.$

9. $f(x) = \dfrac{x - 4}{x^2 - 16}.$

10. $f(x) = \dfrac{3x + 1}{x^2 + 7x - 2}.$

11. $f(x) = \dfrac{x}{|x| - 3}.$

12. $f(x) = \dfrac{5}{x} + \dfrac{2x}{x + 4}.$

13. $f(x) = |x^3 - 2x^2|.$

14. $f(x) = \dfrac{x + 3}{|x^2 + 3x|}.$

15. $f(x) = \begin{cases} 2x + 3, & x \le 4 \\ 7 + \dfrac{16}{x}, & x > 4. \end{cases}$

16. $f(x) = \begin{cases} \dfrac{3}{x - 1}, & x \ne 1 \\ 3, & x = 1. \end{cases}$

17. Find a value for the constant k, if possible, that will make the function continuous.

(a) $f(x) = \begin{cases} 7x - 2, & x \le 1 \\ kx^2, & x > 1 \end{cases}$

(b) $f(x) = \begin{cases} kx^2, & x \le 2 \\ 2x + k, & x > 2. \end{cases}$

18. On which of the following intervals is

$$f(x) = \frac{1}{\sqrt{x - 2}}$$

continuous?

(a) $[2, +\infty)$ (b) $(-\infty, +\infty)$
(c) $(2, +\infty)$ (d) $[1, 2).$

19. (a) Prove that $f(x) = \sqrt{x}$ is continuous on $[0, +\infty)$.

(b) Prove that if $g(x)$ is continuous and nonnegative, then $\sqrt{g(x)}$ is continuous.

20. Prove that if $g(x)$ is continuous, then $|g(x)|$ is continuous.

21. Find all points of discontinuity of the greatest integer function, $f(x) = [x]$. (See Exercise 48, Section 2.3, for the definition of $[x]$.)

22. A function f is said to have a **removable discontinuity** at $x = c$ if the limit $\lim\limits_{x \to c} f(x)$ exists, but

$$f(c) \ne \lim\limits_{x \to c} f(x)$$

either because $f(c)$ is undefined or the value of $f(c)$ differs from the value of the limit.

(a) Show that the following functions have removable discontinuities at $x = 1$ and sketch their graphs.

$$f(x) = \frac{x^2 - 1}{x - 1} \quad \text{and} \quad g(x) = \begin{cases} 1, & x > 1 \\ 0, & x = 1 \\ 1, & x < 1 \end{cases}$$

(b) The terminology ''removable discontinuity'' is used because a function with a removable discontinuity at $x = c$ can be made continuous at $x = c$ by defining (or redefining) the value of the function at $x = c$ appropriately. Do this for the functions f and g in part (a).

In Exercises 23 and 24, find all points of discontinuity for the functions, and determine whether the discontinuities are removable or not. [See Exercise 22 for terminology.]

---

23. (a) $f(x) = \dfrac{|x|}{x}$ (b) $f(x) = \dfrac{x^2 + 3x}{x + 3}$

    (c) $f(x) = \dfrac{x - 2}{|x| - 2}$.

24. (a) $f(x) = \dfrac{x^2 - 4}{x^3 - 8}$ (b) $f(x) = \begin{cases} 2x - 3, & x \le 2 \\ x^2, & x > 2 \end{cases}$

    (c) $f(x) = \begin{cases} 3x^2 + 5, & x \ne 1 \\ 6, & x = 1. \end{cases}$

25. Prove:

    (a) part (a) of Theorem 2.7.3

    (b) part (b) of Theorem 2.7.3

    (c) part (c) of Theorem 2.7.3.

26. Prove Theorem 2.7.4.

27. Let f and g be discontinuous at c. Give examples to show:

    (a) $f + g$ can be continuous or discontinuous at c

    (b) $f \cdot g$ can be continuous or discontinuous at c.

28. Let f be defined at c. Prove that f is continuous at c if and only if, given $\epsilon > 0$, there exists a $\delta > 0$ such that $|f(x) - f(c)| < \epsilon$ whenever $|x - c| < \delta$.

29. Use Theorem 2.7.9 to prove Theorem 2.7.10.

In Exercises 30 and 31, show that the equation has at least one solution in the given interval.

---

30. $x^3 - 4x + 1 = 0$; $[1, 2]$.

31. $x^3 + x^2 - 2x = 1$; $[-1, 1]$.

32. Figure 2.7.7 shows the graph of $y = x^4 + x - 1$ generated on a computer. Use the method of Example 9 to approximate the real solutions of $x^4 + x - 1 = 0$ with an error of at most .05.

33. Figure 2.7.8 shows the graph of $y = 5 - x - x^4$ generated on a computer. Use the method of Example 9 to approximate the real solutions of $5 - x - x^4 = 0$ with an error of at most .05.

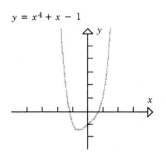
$y = x^4 + x - 1$

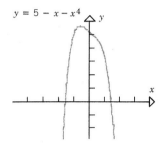
$y = 5 - x - x^4$

Figure 2.7.7   Figure 2.7.8

34. For the equation $x^3 - x - 1 = 0$ discussed in Example 9, show that the real solution is approximately 1.325, with an error of at most .005.

35. Use the fact that $\sqrt{5}$ is a solution of $x^2 - 5 = 0$ to approximate $\sqrt{5}$ with an error of at most

    (a) .05   (b) .005.

36. Prove: If f and g are continuous on $[a, b]$, and $f(a) > g(a)$, $f(b) < g(b)$, then there is at least one solution of the equation $f(x) = g(x)$ in (a, b). [*Hint:* Consider $f(x) - g(x)$.]

37. Construct an example of a function f that is defined at every point in a closed interval, and whose values at the endpoints have opposite signs, but for which the equation $f(x) = 0$ has no solution in the interval.

38. Prove that if a and b are positive, then the equation

    $$\frac{a}{x - 1} + \frac{b}{x - 3} = 0$$

    has at least one solution in the interval $(1, 3)$.

39. Prove: If $p(x)$ is a polynomial of odd degree, then the equation $p(x) = 0$ has at least one real solution.

■ **2.8** LIMITS AND CONTINUITY OF TRIGONOMETRIC FUNCTIONS

In this section we shall derive some basic limits involving trigonometric functions and study the continuity of these functions.

This section requires a knowledge of the trigonometry material in Section II of Appendix B. Readers who need to review that material are advised to do so before starting this section.

☐ **CONTINUITY OF SINE AND COSINE**

In trigonometry the graphs of $y = \sin x$ and $y = \cos x$ are drawn as continuous curves (Figures 2.8.1 and 2.8.2). Our first objective in this section is to show that the functions $\sin x$ and $\cos x$ are indeed continuous. Before we start, however, we recall that in the expressions $\sin x$, $\cos x$, $\tan x$, $\cot x$, $\sec x$, and $\csc x$ the independent variable x can be interpreted as an angle. When x is interpreted in this way it will always be assumed that the angle is measured in radians.

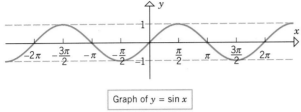

Graph of $y = \sin x$

Figure 2.8.1

Graph of $y = \cos x$

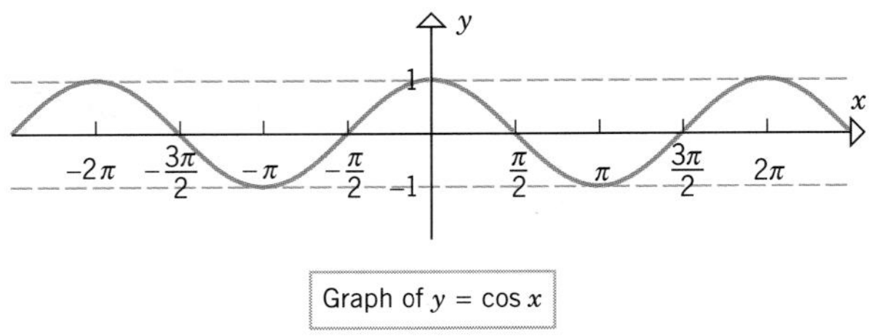

Figure 2.8.2

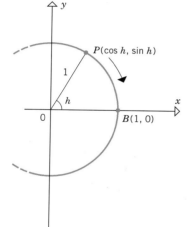

Figure 2.8.3

In Figure 2.8.3 we show a point P moving along the unit circle toward the point $B(1, 0)$. Because the circle has radius 1, the coordinates of the moving point are $P(\cos h, \sin h)$, where h is the angle (in radians) shown in the figure (or possibly a negative angle). It is evident intuitively that as h approaches 0, the coordinates of P approach the corresponding coordinates of B; that is,

$$\lim_{h \to 0} \sin h = 0, \qquad \lim_{h \to 0} \cos h = 1 \qquad\qquad (1a\text{--}b)$$

These limits, whose proofs are discussed in the exercises, tell us that $f(x) = \sin x$ and $g(x) = \cos x$ are continuous at $x = 0$. For example, we have $f(0) = \sin 0 = 0$ and $\lim_{x \to 0} f(x) = \lim_{x \to 0} \sin x = 0$, so $\lim_{x \to 0} f(x) = f(0)$, which is precisely the requirement for $f(x) = \sin x$ to be continuous at $x = 0$. The following theorem shows that they are actually continuous everywhere.

2.8.1   THEOREM.   *The functions* $\sin x$ *and* $\cos x$ *are continuous.*

Proof.   We shall prove the result for $\sin x$, assuming the validity of (1a) and (1b). The proof for $\cos x$ is similar and will be left as an exercise. By the remark following Definition 2.7.11, it suffices to show that the condition

$$\lim_{h \to 0} \sin (c + h) = \sin c \qquad\qquad (2)$$

is satisfied for all values of c. Using the addition formula for the sine function we obtain

$$\lim_{h \to 0} \sin (c + h) = \lim_{h \to 0} [\sin c \cos h + \cos c \sin h]$$

$$= \lim_{h \to 0} [\sin c \cos h] + \lim_{h \to 0} [\cos c \sin h] \qquad\qquad (3)$$

The expressions $\sin c$ and $\cos c$ in (3) do not involve h, so they remain *constant* as $h \to 0$.

This allows us to move these expressions through the limit signs and write

$$\lim_{h \to 0} \sin (c + h) = \sin c \lim_{h \to 0} \cos h + \cos c \lim_{h \to 0} \sin h$$

$$= (\sin c)(1) + (\cos c)(0) = \sin c$$

which proves (2). ∎

☐ **CONTINUITY OF OTHER TRIGONOMETRIC FUNCTIONS**

The continuity properties of $\tan x$, $\cot x$, $\sec x$, and $\csc x$ can be deduced by expressing these functions in terms of $\sin x$ and $\cos x$. For example, $\tan x = \sin x / \cos x$, so by part (d) of Theorem 2.7.3 the function $\tan x$ is continuous everywhere except at the points where $\cos x = 0$. These points of discontinuity are $x = \pm \pi/2, \pm 3\pi/2, \pm 5\pi/2, \ldots$ (see Figure B.28 in Appendix B).

Example 1 Since the functions $\sin x$ and $\cos x$ are continuous everywhere, it follows from Theorem 2.7.5 that we can write

$$\lim [\sin (g(x))] = \sin [\lim g(x)] \quad \text{and} \quad \lim [\cos (g(x))] = \cos [\lim g(x)]$$

if $\lim g(x)$ exists. For example,

$$\lim_{x \to \pi} \left[\sin \left(\frac{x^2}{\pi + x} \right) \right] = \sin \left[\lim_{x \to \pi} \left(\frac{x^2}{\pi + x} \right) \right] = \sin \frac{\pi}{2} = 1$$

$$\lim_{x \to +\infty} \left[\cos \left(\frac{\pi x^2 + 1}{x^2 + 3} \right) \right] = \cos \left[\lim_{x \to +\infty} \left(\frac{\pi x^2 + 1}{x^2 + 3} \right) \right]$$

$$= \cos \left[\lim_{x \to +\infty} \left(\frac{\pi + (1/x^2)}{1 + (3/x^2)} \right) \right] = \cos \pi = -1 \quad ◄$$

☐ **OBTAINING LIMITS BY SQUEEZING**

Our next objective is to establish two fundamental limits:

$$\lim_{h \to 0} \frac{\sin h}{h} = 1 \quad \text{and} \quad \lim_{h \to 0} \frac{1 - \cos h}{h} = 0 \tag{4}$$

Neither of these limits is obvious, though their correctness is strongly suggested by the computer-generated graphs shown in Figures 2.8.4 and 2.8.5. The complication in deducing these limits results from the fact that for each of them the numerator and denominator both approach zero as $h \to 0$. As a result, there are two conflicting influences on the ratio. The numerator approaching 0 drives the magnitude of the ratio toward zero, while the denominator approaching 0 drives the magnitude of the ratio toward $+\infty$. The precise way in which these influences offset one another determines whether the limit exists and what its value is.

In a limit problem where the numerator and denominator both approach zero, it is sometimes possible to circumvent the difficulty by using algebraic manipulations to write

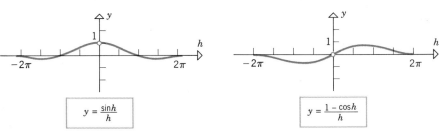

Figure 2.8.4   Figure 2.8.5

the limit in a different form. However, if that is not possible, as here, other methods are required. One such method is to obtain the limit by "squeezing" the function between simpler functions whose limits are known. For example, suppose that we are unable to show that

$$\lim_{x \to a} f(x) = L$$

directly, but we are able to find two functions, g and h, that have the same limit L as $x \to a$ and such that f is "squeezed" between g and h by means of the inequalities

$$g(x) \leq f(x) \leq h(x)$$

It is evident geometrically that $f(x)$ must also approach L as $x \to a$ because the graph of f lies between the graphs of g and h (Figure 2.8.6).

This idea is formalized in the following theorem, which is called the **Squeezing Theorem** or sometimes the **Pinching Theorem**. We omit the proof.

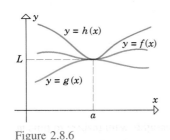

Figure 2.8.6

2.8.2 THEOREM (**The Squeezing Theorem**). *Let f, g, and h be functions satisfying*

$$g(x) \leq f(x) \leq h(x)$$

for all x in some open interval containing the point a, with the possible exception that the inequalities need not hold at a. If g and h have the same limit as x approaches a, say

$$\lim_{x \to a} g(x) = \lim_{x \to a} h(x) = L$$

then f also has this limit as x approaches a, that is,

$$\lim_{x \to a} f(x) = L$$

REMARK. The Squeezing Theorem remains true if $\lim_{x \to a}$ is replaced by $\lim_{x \to a^+}$ or $\lim_{x \to a^-}$. Moreover, for $\lim_{x \to a^+}$ the condition $g(x) \leq f(x) \leq h(x)$ need only hold on an open interval extending to the right from a, and for $\lim_{x \to a^-}$ the condition need only hold on an open interval extending to the left from a. The Squeezing Theorem can also be extended to limits of the form $\lim_{x \to -\infty}$ and $\lim_{x \to +\infty}$.

Example 2   Use the Squeezing Theorem to evaluate the limit

$$\lim_{x \to 0} x^2 \sin^2 \frac{1}{x}$$

Solution.   If $x \neq 0$, we can write

$$0 \leq \sin^2 \frac{1}{x} \leq 1$$

Multiplying through by x^2 yields

$$0 \leq x^2 \sin^2 \frac{1}{x} \leq x^2$$

But $\lim_{x \to 0} 0 = \lim_{x \to 0} x^2 = 0$, so by the Squeezing Theorem

$$\lim_{x \to 0} x^2 \sin^2 \frac{1}{x} = 0 \qquad \blacktriangleleft$$

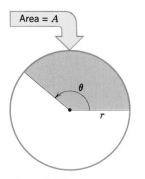

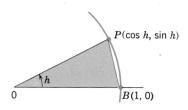

Figure 2.8.7

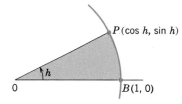

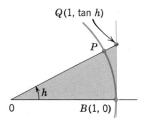

Q(1, tan h)

Figure 2.8.8

Next, we will use the Squeezing Theorem to prove that the limits in (4) are correct. But first it will be helpful to review the formula for the area A of a sector with radius r and a central angle of θ radians (Figure 2.8.7). If $\theta = 2\pi$, the sector is the entire circle and the area is πr^2. For a general sector the area A is proportional to the central angle θ, so we can write

$$\frac{A}{\pi r^2} = \frac{\theta}{2\pi} \qquad \left[\frac{\text{area of the sector}}{\text{area of the circle}} = \frac{\text{central angle of the sector}}{\text{central angle of the circle}}\right]$$

from which we obtain the formula

$$A = \tfrac{1}{2}r^2\theta \tag{5}$$

REMARK.   It is important to keep in mind that the angle θ in the preceding formula is measured in radians. As we progress through this text you will encounter many calculus formulas that build on (5). All such formulas will also require that angles be measured in radians.

2.8.3   THEOREM.

$$\lim_{h \to 0} \frac{\sin h}{h} = 1 \tag{6}$$

Proof.   Assume that h satisfies $0 < h < \pi/2$ and construct the angle h in standard position. The terminal side of h intersects the unit circle at $P(\cos h, \sin h)$ and intersects the vertical line through $B(1, 0)$ at the point $Q(1, \tan h)$ (Figure 2.8.8). From the figure we obtain

$$0 < \text{area of } \triangle OBP < \text{area of sector } OBP < \text{area of } \triangle OBQ$$

But

$$\text{area } \triangle OBP = \tfrac{1}{2}\text{base} \cdot \text{altitude} = \tfrac{1}{2} \cdot 1 \cdot \sin h = \tfrac{1}{2}\sin h$$

$$\text{area sector } OBP = \tfrac{1}{2}(1)^2 \cdot h = \tfrac{1}{2}h$$

$$\text{area } \triangle OBQ = \tfrac{1}{2}\text{base} \cdot \text{altitude} = \tfrac{1}{2} \cdot 1 \cdot \tan h = \tfrac{1}{2}\tan h$$

Therefore,

$$0 < \tfrac{1}{2}\sin h < \tfrac{1}{2}h < \tfrac{1}{2}\tan h \tag{7}$$

Multiplying through by $2/\sin h$ yields

$$1 < \frac{h}{\sin h} < \frac{1}{\cos h}$$

and taking reciprocals yields

$$\cos h < \frac{\sin h}{h} < 1 \tag{8}$$

We have derived (8) under the assumption that $0 < h < \pi/2$. However, the inequalities in (8) are also valid if $-\pi/2 < h < 0$ (Exercise 49), so that (8) holds for all h in the interval $(-\pi/2, \pi/2)$ except $h = 0$. But

$$\lim_{h \to 0} \cos h = 1 \quad \text{and} \quad \lim_{h \to 0} 1 = 1$$

so (6) follows by applying the Squeezing Theorem to (8).   ∎

2.8.4   COROLLARY.

$$\lim_{h \to 0} \frac{1 - \cos h}{h} = 0$$

Proof.   With the help of the trigonometric identity $\sin^2 h = 1 - \cos^2 h$ and the limits in (1a) and (1b) we obtain

$$\lim_{h \to 0} \frac{1 - \cos h}{h} = \lim_{h \to 0} \left[\frac{1 - \cos h}{h} \cdot \frac{1 + \cos h}{1 + \cos h} \right] = \lim_{h \to 0} \frac{\sin^2 h}{h(1 + \cos h)}$$

$$= \left(\lim_{h \to 0} \frac{\sin h}{h} \right) \left(\lim_{h \to 0} \frac{\sin h}{1 + \cos h} \right) = (1) \left(\frac{0}{1 + 1} \right) = 0 \quad \blacksquare$$

The following examples illustrate how the limits obtained in this section can be used to calculate other limits involving trigonometric functions.

Example 3   Find $\lim_{x \to 0} \dfrac{\tan x}{x}$.

Solution.

$$\lim_{x \to 0} \frac{\tan x}{x} = \lim_{x \to 0} \left(\frac{\sin x}{x} \cdot \frac{1}{\cos x} \right) = (1)(1) = 1 \quad \blacktriangleleft$$

Example 4   Find $\lim_{\theta \to 0} \dfrac{\sin 2\theta}{\theta}$.

Solution.   The trick is to multiply and divide by 2, so that the resulting denominator is the same as the expression inside the sine function in the numerator. We obtain

$$\lim_{\theta \to 0} \frac{\sin 2\theta}{\theta} = \lim_{\theta \to 0} 2 \cdot \frac{\sin 2\theta}{2\theta} = 2 \lim_{\theta \to 0} \frac{\sin 2\theta}{2\theta}$$

Now make the substitution $t = 2\theta$ and use the fact that $t \to 0$ as $\theta \to 0$. This yields

$$\lim_{\theta \to 0} \frac{\sin 2\theta}{\theta} = 2 \lim_{\theta \to 0} \frac{\sin 2\theta}{2\theta} = 2 \lim_{t \to 0} \frac{\sin t}{t} = 2(1) = 2 \quad \blacktriangleleft$$

Example 5   Find $\lim_{x \to 0} \dfrac{\sin 3x}{\sin 5x}$.

Solution.

$$\lim_{x \to 0} \frac{\sin 3x}{\sin 5x} = \lim_{x \to 0} \frac{\dfrac{\sin 3x}{x}}{\dfrac{\sin 5x}{x}} = \lim_{x \to 0} \frac{3 \cdot \dfrac{\sin 3x}{3x}}{5 \cdot \dfrac{\sin 5x}{5x}} = \frac{3 \cdot 1}{5 \cdot 1} = \frac{3}{5} \quad \blacktriangleleft$$

□ **LIMITS OF** $\sin x$ **AND** $\cos x$ As $x \to +\infty$ or $x \to -\infty$, the values of $\sin x$ and $\cos x$ oscillate repeatedly between -1 and 1
AS $x \to +\infty$ **OR** $x \to -\infty$ without approaching any fixed real value. Thus, the limits

$$\lim_{x \to +\infty} \sin x, \quad \lim_{x \to -\infty} \sin x, \quad \lim_{x \to +\infty} \cos x, \quad \lim_{x \to -\infty} \cos x$$

do not exist. We shall say that they *fail to exist due to oscillation*.

▶ **Exercise Set 2.8**   [C] 53–56

In Exercises 1–10, find the points of discontinuity, if any.

1. $f(x) = \sin(x^2 - 2)$.   **2.** $f(x) = \cos\left(\dfrac{x}{x - \pi}\right)$.

3. $f(x) = \cot x$.   **4.** $f(x) = \sec x$.

5. $f(x) = \csc x$.   **6.** $f(x) = \dfrac{1}{1 + \sin^2 x}$.

7. $f(x) = |\cos x|$.   **8.** $f(x) = \sqrt{2 + \tan^2 x}$.

9. $f(x) = \dfrac{1}{1 - 2\sin x}$.   **10.** $f(x) = \dfrac{3}{5 + 2\cos x}$.

11. Prove that $\sin(g(x))$ is continuous at every point where $g(x)$ is continuous.

12. Use Theorem 2.7.6 to prove that the following functions are continuous.

   (a) $\sin(x^3 + 7x + 1)$ (b) $|\sin x|$

   (c) $\cos^3(x + 1)$ (d) $\sqrt{3 + \sin 2x}$.

Find the limits in Exercises 13–35.

13. $\lim\limits_{x \to +\infty} \cos\left(\dfrac{1}{x}\right)$.   **14.** $\lim\limits_{x \to +\infty} \sin\left(\dfrac{2}{x}\right)$.

15. $\lim\limits_{x \to +\infty} \sin\left(\dfrac{\pi x}{2 - 3x}\right)$.   **16.** $\lim\limits_{h \to 0} \dfrac{\sin h}{2h}$.

17. $\lim\limits_{\theta \to 0} \dfrac{\sin 3\theta}{\theta}$.   **18.** $\lim\limits_{\theta \to 0^+} \dfrac{\sin \theta}{\theta^2}$.

19. $\lim\limits_{x \to 0^-} \dfrac{\sin x}{|x|}$.   **20.** $\lim\limits_{x \to 0} \dfrac{\sin^2 x}{3x^2}$.

21. $\lim\limits_{x \to 0^+} \dfrac{\sin x}{5\sqrt{x}}$.   **22.** $\lim\limits_{x \to 0} \dfrac{\sin 6x}{\sin 8x}$.

23. $\lim\limits_{x \to 0} \dfrac{\tan 7x}{\sin 3x}$.   **24.** $\lim\limits_{\theta \to 0} \dfrac{\sin^2 \theta}{\theta}$.

25. $\lim\limits_{h \to 0} \dfrac{h}{\tan h}$.   **26.** $\lim\limits_{h \to 0} \dfrac{\sin h}{1 - \cos h}$.

27. $\lim\limits_{\theta \to 0} \dfrac{\theta^2}{1 - \cos \theta}$.   **28.** $\lim\limits_{x \to 0} \dfrac{x}{\cos\left(\frac{1}{2}\pi - x\right)}$.

29. $\lim\limits_{\theta \to 0} \dfrac{\theta}{\cos \theta}$.   **30.** $\lim\limits_{t \to 0} \dfrac{t^2}{1 - \cos^2 t}$.

31. $\lim\limits_{h \to 0} \dfrac{1 - \cos 5h}{\cos 7h - 1}$.   **32.** $\lim\limits_{x \to 0^+} \sin\left(\dfrac{1}{x}\right)$.

33. $\lim\limits_{x \to 0^+} \cos\left(\dfrac{1}{x}\right)$.   **34.** $\lim\limits_{x \to 0} \dfrac{x^2 - 3\sin x}{x}$.

35. $\lim\limits_{x \to 0} \dfrac{2x + \sin x}{x}$.

36. Find a value for the constant k so that

$$f(x) = \begin{cases} \dfrac{\sin 3x}{x}, & x \neq 0 \\ k, & x = 0 \end{cases}$$

will be continuous at $x = 0$.

37. Find a nonzero value for the constant k so that

$$f(x) = \begin{cases} \dfrac{\tan kx}{x}, & x < 0 \\ 3x + 2k^2, & x \geq 0 \end{cases}$$

will be continuous at $x = 0$.

38. Is

$$f(x) = \begin{cases} \dfrac{\sin x}{|x|}, & x \neq 0 \\ 1, & x = 0 \end{cases}$$

continuous at $x = 0$?

39. In each part, find the limit by making the indicated substitution.

   (a) $\lim\limits_{x \to +\infty} x \sin \dfrac{1}{x}$. $\left[\text{Let } t = \dfrac{1}{x}.\right]$

   (b) $\lim\limits_{x \to -\infty} x\left(1 - \cos \dfrac{1}{x}\right)$. $\left[\text{Let } t = \dfrac{1}{x}.\right]$

   (c) $\lim\limits_{x \to \pi} \dfrac{\pi - x}{\sin x}$. [Let $t = \pi - x$.]

40. Find $\lim\limits_{x \to 2} \dfrac{\cos(\pi/x)}{x - 2}$. $\left[\textit{Hint: } \text{Let } t = \dfrac{\pi}{2} - \dfrac{\pi}{x}.\right]$

41. Find $\lim\limits_{x \to 1} \dfrac{\sin(\pi x)}{x - 1}$.

42. Find $\lim\limits_{x \to \pi/4} \dfrac{\tan x - 1}{x - \pi/4}$.

43. Use the Squeezing Theorem to find the following limits.

   (a) $\lim\limits_{x \to +\infty} \dfrac{\sin x}{x}$ (b) $\lim\limits_{x \to +\infty} \dfrac{\cos x}{x}$.

44. Use the Squeezing Theorem to find

$$\lim\limits_{x \to 0^+} x \sin(1/x).$$

45. Let f be a function that satisfies

$$1 - x^2 \leq f(x) \leq \cos x$$

for all x in $(-\pi/2, \pi/2)$. Does $\lim\limits_{x \to 0} f(x)$ exist? If so, find the limit. If not, explain why.

46. Let

$$f(x) = \begin{cases} 1 & \text{if } x \text{ is a rational number} \\ 0 & \text{if } x \text{ is an irrational number} \end{cases}$$

Use the Squeezing Theorem to prove $\lim\limits_{x \to 0} x f(x) = 0$.

47. Prove: If there are constants L and M such that

$$L \leq f(x) \leq M$$

for all x in some open interval containing 0, with the

possible exception that the inequalities may not hold at 0, then

$$\lim_{x \to 0} x f(x) = 0$$

48. Draw pictures analogous to Figure 2.8.6 that illustrate the Squeezing Theorem for limits of the form $\lim_{x \to +\infty} f(x)$ and $\lim_{x \to -\infty} f(x)$.

49. Prove that (8) holds if $-\pi/2 < h < 0$ by substituting $-h$ for h in (8).

50. (a) Prove that $\lim_{h \to 0} \sin h = 0$. [*Hint:* Multiply inequality (7) through by 2 to obtain $0 < \sin h < h$ for $0 < h < \pi/2$ and use the Squeezing Theorem to show that $\lim_{h \to 0^+} \sin h = 0$; then show that $\lim_{h \to 0^-} \sin h = 0$.]

(b) Use the result in part (a) to prove that $\lim_{h \to 0} \cos h = 1$.
[*Hint:* Solve the identity $\sin^2 h + \cos^2 h = 1$ for $\cos h$ under the assumption that $-\pi/2 < h < \pi/2$.]

51. Prove that $\cos x$ is continuous. [*Hint:* Use the identity $\cos(c + h) = \cos c \cos h - \sin c \sin h$.]

52. Prove: If θ is in degrees, then

$$\lim_{\theta \to 0} \frac{\sin \theta}{\theta} = \frac{\pi}{180}$$

53. It follows from Theorem 2.8.3 that if θ is small (near zero) and measured in radians, then one should expect the approximation

$$\sin \theta \approx \theta$$

to be good.

(a) Find $\sin 10°$ using a calculator.

(b) Find $\sin 10°$ using the approximation above.

54. (a) Use the approximation of $\sin \theta$ that is given in Exercise 53 together with the trigonometric identity $\cos 2\alpha = 1 - 2 \sin^2 \alpha$ with $\alpha = \theta/2$ to show that if θ

is small (near zero) and measured in radians, then one should expect the approximation

$$\cos \theta \approx 1 - \tfrac{1}{2} \theta^2$$

to be good.

(b) Find $\cos 10°$ using a calculator.

(c) Find $\cos 10°$ using the approximation above.

55. It follows from Example 3 that if θ is small (near zero) and measured in radians, then one should expect the approximation

$$\tan \theta \approx \theta$$

to be good.

(a) Find $\tan 5°$ using a calculator.

(b) Find $\tan 5°$ using the approximation above.

56. Referring to Figure 2.8.9, suppose that the angle of elevation of the top of a building, as measured from a point L feet from its base, is found to be α degrees.

(a) Use the relationship $h = L \tan \alpha$ to calculate the height of a building for which $L = 500$ ft and $\alpha = 6°$.

(b) Show that if L is large compared to the building height h, then one should expect good results in approximating h by $h \approx \pi L \alpha / 180$.

(c) Use the result in part (b) to approximate the building height h in part (a).

Figure 2.8.9

In Exercises 57 and 58, show that the equation has at least one solution in the given interval.

57. $x = \cos x$; $[0, \pi/2]$.

58. $x + \sin x = 1$; $[0, \pi/6]$.

◆ TECHNOLOGY EXERCISES   Chapter 2

Most of these exercises require access to a graphing calculator or a computer algebra system (CAS) such as *Mathematica*, *Maple*, or *Derive*. When you are asked to *find* an answer or to *solve* an equation, you may choose to find an exact result or a numerical approximation, depending on the particular technology you are using and on your own imagination. The form of your answers may differ from those of other students or from those in the answer section of the text, depending on how you solve the problems and the accuracy you use in your numerical approximations.

Limits: In Exercises 1–6, find the limit of the function by looking at its graph and calculating values for some appropriate choices of x. If you have access to a computer algebra system, compare your answer with the value produced by this system.

1. $\lim\limits_{x \to 0} (1 + x)^{1/x}$.

2. $\lim\limits_{x \to 3} \dfrac{2^x - 8}{x - 3}$.

3. $\lim\limits_{x \to 1} \dfrac{\sin x - \sin 1}{x - 1}$.

4. $\lim\limits_{x \to 0^+} x^{-2}(1.001)^{-1/x}$.

5. $\lim\limits_{x \to +\infty} (\sqrt{x + \sqrt{x}} - \sqrt{x})$.

6. $\lim\limits_{x \to +\infty} (3^x + 5^x)^{1/x}$.

In Exercises 7 and 8, use the following empirical formula for the windchill index (WCI) [see Exercise 57, Section 2.1]:

$$\text{WCI} = \begin{cases} T, & 0 \le v \le 4 \\ 91.4 + (91.4 - T)(0.0203v - 0.304\sqrt{v} - 0.474), & 4 < v < 45 \\ 1.6T - 55, & v \ge 45 \end{cases}$$

where T is the air temperature in °F, v is the wind speed in mi/hr, and WCI is the equivalent temperature in °F.

7. **Windchill index**

(a) Graph WCI versus v over the interval $0 \le v \le 50$ for $T = 20$.

(b) Use your graph to estimate the values of the WCI corresponding to $v = 10, 20, 30, 40$, to the nearest degree.

(c) Use your graph to estimate the values of v corresponding to WCI $= -20, -10, 0, 10$, to the nearest mile per hour.

8. **Windchill index**

(a) Graph T versus v over the interval $4 \le v \le 45$ for WCI $= 0$.

(b) Use your graph to estimate the values of T for WCI $= 0$ corresponding to $v = 10, 20, 30$, to the nearest degree.

9. **Domain and range:** Find the domain and range of the function $f(x) = x^2 - \sqrt{1 + x - x^4}$.

10. **Domain and range:** Find the domain and range of the function

$$f(x) = \frac{\sin x}{x^4 + x^3 + 5}$$

11. **Illustrating the definition of limit:** The limit

$$\lim_{x \to 0} \frac{\sin x}{x} = 1$$

ensures that there is a number δ such that

$$\left| \frac{\sin x}{x} - 1 \right| < 0.001$$

if $0 < |x| < \delta$. Find the largest such δ.

12. **Compound interest:** If \$1000 is invested in an account that pays 7% interest compounded n times each year, then in 10 years there will be $1000(1 + 0.07/n)^{10n}$ dollars in the account. How much money will be in the account in 10 years if the interest is compounded quarterly ($n = 4$)? Monthly ($n = 12$)? Daily ($n = 365$)? How much money will be in the account in 10 years if the interest is compounded *continuously* ($n \to +\infty$)?

13. **Supplies dropped by parachute:** Suppose that a package of medical supplies is dropped from a helicopter straight down by parachute into a remote area. The velocity v (in feet per second) of the package t seconds after it is released is given by $v = 24.61(1 - (0.273^t))$.

(a) Graph v versus t.

(b) What is the terminal velocity of the package? (The *terminal velocity* is defined as the limit of v as $t \to +\infty$.)

(c) How long does it take for the package to reach 98% of its terminal velocity?

14. **Temperature variation:** An oven is preheated and then remains at a constant temperature. A potato is placed in the oven to bake. Suppose that the temperature T (in °F) of the potato t minutes later is given by $T = 400 - 325(0.97)^t$. The potato will be considered done when its temperature is anywhere between 260° F and 280° F.

(a) During what interval of time would the potato be considered done?

(b) How long does it take for the temperature of the potato to reach 95% of the oven temperature?

15. **Solving equations:** There are various numerical methods to obtain approximate solutions of equations of the form $f(x) = 0$. One such method requires that the equation be expressed in the form $x = g(x)$, so that a solution $x = c$ can be interpreted as the value of x where the line $y = x$ intersects the curve $y = g(x)$, as shown in the accompanying figure. If x_1 is an initial estimate of c and the graph of $y = g(x)$ is not too steep in the vicinity of c, then a better approximation can be obtained from $x_2 = g(x_1)$ (see the figure). An even better approximation is obtained from $x_3 = g(x_2)$, and so forth. The formula $x_{n+1} = g(x_n)$ for $n = 1, 2, 3, \ldots$ generates successive approximations $x_2, x_3, x_4, \ldots$ that get closer and closer to c.

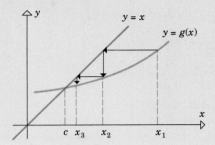

(a) The equation $x^3 - x - 1 = 0$ has only one real solution. Show that this equation can be written as

$$x = g(x) = \sqrt[3]{x + 1}$$

(b) Graph $y = x$ and $y = g(x)$ in the same coordinate system for $-1 \le x \le 3$.

(c) Starting with an arbitrary point x_1, make a sketch illustrating the location of the successive iterates $x_2 = g(x_1)$, $x_3 = g(x_2)$, ...

(d) Use $x_1 = 1$ and calculate $x_2, x_3, \ldots$, continuing until you obtain two consecutive values that differ by less than 10^{-4}. Experiment with other starting values such as $x_1 = 2$ or $x_1 = 1.5$.

16. **Solving equations:** The method described in Exercise 15 will not always work.

(a) The equation $x^3 - x - 1 = 0$ can be expressed as $x = g(x) = x^3 - 1$. Graph $y = x$ and $y = g(x)$ in the same coordinate system. Starting with an arbitrary point x_1, make a sketch illustrating the location of the successive iterates $x_2 = g(x_1)$, $x_3 = g(x_2)$, ...

(b) Use $x_1 = 1$ and calculate the successive iterates x_n for $n = 2, 3, 4, 5, 6$.

Solving Equations: In Exercises 17 and 18, use the method of Exercise 15 to solve the given equation.

17. $x^5 - x - 2 = 0$.

18. $x - \cos x = 0$.

Isaac Newton

Sir Isaac Newton (1642–1727)

3 DIFFERENTIATION

■ 3.1 TANGENT LINES AND RATES OF CHANGE

> *Many physical phenomena involve changing quantities—the speed of a rocket, the inflation of currency, the number of bacteria in a culture, the shock intensity of an earthquake, the voltage of an electrical signal, and so forth. In this section we shall establish a basic relationship between tangent lines and rates of change.*

☐ SLOPE OF A TANGENT LINE

In Section 2.4 we observed informally that if a secant line is drawn between two points P and Q on a curve, and Q is allowed to move along the curve toward P, then we can expect the secant line to rotate toward a "limiting position" which can be regarded as the tangent line to the curve at the point P (Figure 3.1.1). In the next section we will give a precise definition of a tangent line, but for now your intuition will suffice.

For the moment we shall only consider lines tangent to curves of the form $y = f(x)$. If $P(x_0, y_0)$ and $Q(x_1, y_1)$ are distinct points on such a curve, then the secant line connecting P and Q has slope

$$m_{\text{sec}} = \frac{f(x_1) - f(x_0)}{x_1 - x_0} \tag{1}$$

(See Figure 3.1.2.) If we let x_1 approach x_0, as shown in the figure, then Q will approach P along the graph of f, and the secant line through P and Q will approach the tangent line at

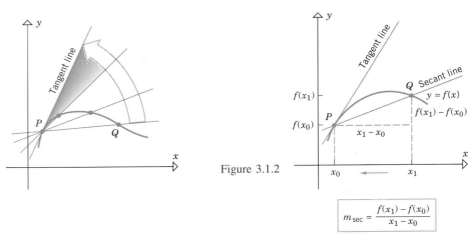

Figure 3.1.1

Figure 3.1.2

$$m_{\sec} = \frac{f(x_1) - f(x_0)}{x_1 - x_0}$$

P. Thus, the slope $m_{\sec}$ of the secant line approaches the slope $m_{\tan}$ of the tangent line as x_1 approaches x_0. Therefore, from (1)

$$m_{\tan} = \lim_{x_1 \to x_0} \frac{f(x_1) - f(x_0)}{x_1 - x_0} \qquad (2)$$

□ AVERAGE AND
INSTANTANEOUS
VELOCITY

While tangent lines are of interest as a matter of pure geometry, much of the impetus for studying them arose in the seventeenth century when scientists recognized that many problems involving objects moving with varying velocity could be reduced to problems involving tangents. To see why this is so, we need to examine critically the meaning of the word "velocity."

If a car travels 75 miles over a straight road in a 3-hour period, then we say that the average velocity of the car is 25 miles per hour (25 mi/hr). More generally, the *average velocity* of an object moving in *one direction* along a line is

$$\text{average velocity} = \frac{\text{distance traveled}}{\text{time elapsed}}$$

Obviously, if an object travels with an average velocity of 25 mi/hr during a 3-hour trip, it need not travel at a fixed velocity of 25 mi/hr; sometimes it may speed up and sometimes it may slow down.

Although average velocity is useful for some purposes, it is not always significant in physical problems. For example, if a moving car strikes a tree, the damage is not determined by the average velocity up to the time of impact, but rather by the *instantaneous velocity* at the precise moment of impact.

A clear understanding of instantaneous velocity evaded scientists until the advent of calculus in the seventeenth century. The subtlety of this concept was nicely described by Morris Kline* who wrote,

> In contrasting average velocity with instantaneous velocity we implicitly utilize a distinction
> between interval and instant. . . . An average velocity is one that concerns what happens over an
> interval of time—3 hours, 5 seconds, one-half second, and so forth. The interval may be small
> or large, but it does represent the passage of a definite amount of time. We use the word instant,

---

* MORRIS KLINE (1908–1992). American mathematician, scholar, and educator. Kline has made numerous contributions to mathematical thought, written extensively on education, especially mathematics education, and has taught, lectured, and served as a consultant throughout his very active career. He is the author of many popular books including *Mathematical Thought from Ancient to Modern Times* and *Why Johnny Can't Add: The Failure of the New Mathematics.* I wish to thank him for permission to use the above quotation, which is taken from *Calculus: An Intuitive and Physical Approach,* Wiley, New York, 1977, p. 17.

however, to state the fact that something happens so fast that no time elapses. The event is momentary. When we say, for example, that it is 3 o'clock, we refer to an instant, a precise moment. If the lapse of time is pictured by length along a line, then an interval (of time) is represented by a line segment, whereas an instant corresponds to a point. The notion of an instant, although it is used in everyday life, is strictly a mathematical idealization.

Our ways of thinking about real events cause us to speak in terms of instants and velocity at an instant, but closer examination shows that the concept of velocity at an instant presents difficulties. Average velocity, which is simply the distance traveled during some interval of time divided by that amount of time, is easily calculated. Suppose, however, that we try to carry over this process to instantaneous velocity. The distance an automobile travels in one instant is 0 and the time that elapses during one instant is also 0. Hence the distance divided by the time is 0/0, which is meaningless. Thus, although instantaneous velocity is a physical reality, there seems to be a difficulty in calculating it, and unless we can calculate it, we cannot work with it mathematically.

Our goal, then, is to define the concept of instantaneous velocity in a way that it can be calculated and worked with mathematically. For this purpose consider a car that moves in a single direction along a straight road on which a coordinate line has been introduced with its positive direction in the direction of motion of the car (Figure 3.1.3). As shown in the figure, if a clock tracks the elapsed time, starting with $t = 0$ initially, then after t seconds have elapsed the car will have some coordinate s that is a function of t, say $s = f(t)$. The graph of this function in a ts-coordinate system is called the ***position versus time curve*** for the motion (Figure 3.1.4).

Elapsed time

Figure 3.1.3

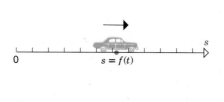

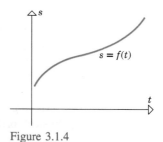

Position versus time curve

Figure 3.1.4

The position versus time curve provides a simple geometric interpretation of the average velocity of the car over a time interval, say from t_0 to t_1. If the car has a coordinate $s_0 = f(t_0)$ at time t_0 and coordinate $s_1 = f(t_1)$ at time t_1, where $t_1 > t_0$, then the distance traveled during the time interval is $s_1 - s_0$ and the time elapsed is $t_1 - t_0$. Thus, the average velocity during the time interval, denoted by v_{ave}, is

$$v_{\text{ave}} = \frac{s_1 - s_0}{t_1 - t_0} = \frac{f(t_1) - f(t_0)}{t_1 - t_0} \tag{3a}$$

which is precisely the slope of the secant line connecting the points (t_0, s_0) and (t_1, s_1) on the position versus time curve (Figure 3.1.5).

Now suppose that we are interested in the instantaneous velocity of the car at time t_0. Intuition suggests that over a small time interval the velocity of the car will not vary much, so if t_1 is close to t_0, then the average velocity of the car over the time interval from t_0 to t_1 will closely approximate the instantaneous velocity of the car at time t_0. Moreover, the smaller the time interval between t_0 and t_1, the better the approximation. This suggests that if we let t_1 get closer and closer to t_0, then the average velocity of the car over the time

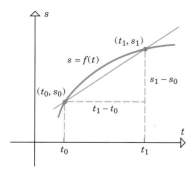

Figure 3.1.5

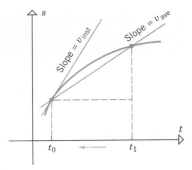

Figure 3.1.6

interval from t_0 to t_1 will get closer and closer to the instantaneous velocity at time t_0. Thus, if we denote the instantaneous velocity of the car at time t_0 by v_{inst}, we have

$$v_{inst} = \lim_{t_1 \to t_0} v_{ave} = \lim_{t_1 \to t_0} \frac{f(t_1) - f(t_0)}{t_1 - t_0} \tag{3b}$$

But the average velocity of the car over the time interval from t_0 to t_1 is the slope of the secant line connecting (t_0, s_0) and (t_1, s_1) on the position versus time curve, and as t_1 approaches t_0 this slope approaches the slope of the tangent line to the position versus time curve at the point (t_0, s_0) (Figure 3.1.6).

Thus, we conclude that the slope of this tangent line is the instantaneous velocity of the car at time t_0. In summary, we make the following observations:

Geometric Interpretation of Average Velocity
For an object moving in the positive direction along a coordinate line, the average velocity of the object between time t_0 and time t_1 is represented geometrically by the slope of the secant line connecting (t_0, s_0) and (t_1, s_1) on the position versus time curve.

Geometric Interpretation of Instantaneous Velocity
For an object moving in the positive direction along a coordinate line, the instantaneous velocity of the object at time t_0 is represented geometrically by the slope of the tangent line at (t_0, s_0) on the position versus time curve.

☐ AVERAGE AND
INSTANTANEOUS
RATES OF CHANGE

Velocity can be viewed as a *rate of change*—the rate of change of position with time, or in algebraic terms, the rate of change of s with t. Rates of change occur in many applications. For example:

- A microbiologist might be interested in the rate at which the number of bacteria in a colony changes with time.

- An engineer might be interested in the rate at which the length of a metal rod changes with temperature.

- An economist might be interested in the rate at which production cost changes with the quantity of a product that is manufactured.

- A medical researcher might be interested in the rate at which the radius of an artery changes with the concentration of alcohol in the bloodstream.

In general, if x and y are any quantities related by an equation $y = f(x)$, we can consider the rate at which y changes with x. As with velocity, we distinguish between an average rate of change represented by the slope of a secant line and an instantaneous rate of change represented by the slope of the tangent line. More precisely, we make the following definitions:

3.1.1   DEFINITION.   If $y = f(x)$, then the *average rate of change of y with respect to x over the interval* $[x_0, x_1]$ is the slope m_{sec} of the secant line joining the points $(x_0, f(x_0))$ and $(x_1, f(x_1))$ on the graph of f (Figure 3.1.7a); that is,

$$m_{sec} = \frac{f(x_1) - f(x_0)}{x_1 - x_0} \tag{4}$$

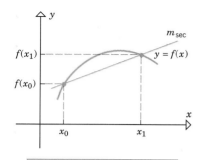

$m_{\sec}$ is the average rate of change of y with respect to x over the interval $[x_0, x_1]$.

(a)

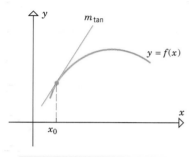

$m_{\tan}$ is the instantaneous rate of change of y with respect to x at the point x_0.

(b)

Figure 3.1.7

3.1.2 DEFINITION. If $y = f(x)$, then the ***instantaneous rate of change of y with respect to x at the point x_0***, is the slope $m_{\tan}$ of the tangent line to the graph of f at the point x_0 (Figure 3.1.7*b*); that is,

$$m_{\tan} = \lim_{x_1 \to x_0} \frac{f(x_1) - f(x_0)}{x_1 - x_0} \tag{5}$$

Example 1 Let $y = x^2 + 1$.

(a) Find the average rate of change of y with respect to x over the interval $[3, 5]$.

(b) Find the instantaneous rate of change of y with respect to x at the point $x = -4$.

(c) Find the instantaneous rate of change of y with respect to x at a general point $x = x_0$.

Solution (a). We will apply Formula (4) with $f(x) = x^2 + 1$, $x_0 = 3$, and $x_1 = 5$. This yields

$$m_{\sec} = \frac{f(x_1) - f(x_0)}{x_1 - x_0} = \frac{f(5) - f(3)}{5 - 3} = \frac{26 - 10}{5 - 3} = 8$$

Thus, on the average, y increases 8 units per unit increase in x over the interval $[3, 5]$.

Solution (b). We will apply Formula (5) with $f(x) = x^2 + 1$ and $x_0 = -4$. This yields

$$m_{\tan} = \lim_{x_1 \to x_0} \frac{f(x_1) - f(x_0)}{x_1 - x_0} = \lim_{x_1 \to -4} \frac{(x_1^2 + 1) - 17}{x_1 + 4}$$

$$= \lim_{x_1 \to -4} \frac{x_1^2 - 16}{x_1 + 4} = \lim_{x_1 \to -4} (x_1 - 4) = -8$$

Because the instantaneous rate of change is negative, y is *decreasing* at the point $x = -4$; it is decreasing at a rate of 8 units per unit increase in x.

Solution (c). We proceed as in part (b).

$$m_{\tan} = \lim_{x_1 \to x_0} \frac{f(x_1) - f(x_0)}{x_1 - x_0} = \lim_{x_1 \to x_0} \frac{(x_1^2 + 1) - (x_0^2 + 1)}{x_1 - x_0}$$

$$= \lim_{x_1 \to x_0} \frac{x_1^2 - x_0^2}{x_1 - x_0} = \lim_{x_1 \to x_0} (x_1 + x_0) = 2x_0$$

Thus, the instantaneous rate of change of y with respect to x at $x = x_0$ is $2x_0$. Observe that the result in part (b) can be obtained from this more general result by letting $x_0 = -4$. ◄

□ **RATES OF CHANGE IN APPLICATIONS**

In applied problems, average and instantaneous rates of change must be accompanied by appropriate units.

Example 2 The limiting factor in athletic endurance is cardiac output. Figure 3.1.8 shows a stress-test graph of cardiac output (measured in liters/min of blood) versus work load (measured in kg · m/min). This graph illustrates the known medical fact that cardiac output increases with the work load, but after reaching a peak value it begins to decrease at very high work loads.

(a) Estimate the average rate of change of cardiac output with respect to work load as the work load increases from 300 to 1200 kg · m/min.

(b) Estimate the instantaneous rate of change of cardiac output with respect to work load at the point where the work load is 300 kg · m/min.

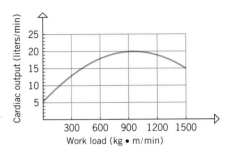

Figure 3.1.8

Solution (a).   Using the estimated points (300, 13) and (1200, 19), the slope of the secant line indicated in Figure 3.1.9a is

$$m_{\text{sec}} \approx \frac{19 - 13}{1200 - 300} \approx 0.0067 \, \frac{\text{liters/min}}{\text{kg} \cdot \text{m/min}}$$

On simplifying the expression for the units we conclude that the average rate of change of cardiac output with respect to work load is approximately 0.0067 liters/kg · m.

Solution (b).   Using the estimated tangent line in Figure 3.1.9b and the estimated points (0, 7) and (900, 25) on this tangent line, we obtain

$$m_{\text{tan}} \approx \frac{25 - 7}{900 - 0} \approx 0.02 \, \frac{\text{liters/min}}{\text{kg} \cdot \text{m/min}}$$

Thus, the instantaneous rate of change of cardiac output with respect to work load is approximately 0.02 liters/kg · m.   ◀

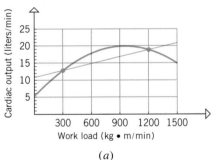

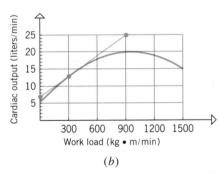

Figure 3.1.9   (*a*)   (*b*)

▶ Exercise Set 3.1   Ⓒ 15, 16, 17

In Exercises 1–4, a function $y = f(x)$ and values of x_0 and x_1 are given.

(a) Find the average rate of change of y with respect to x over the interval $[x_0, x_1]$.
(b) Find the instantaneous rate of change of y with respect to x at the given value of x_0.
(c) Find the instantaneous rate of change of y with respect to x at a general point x_0.
(d) Sketch the graph of $y = f(x)$ together with the secant and tangent lines whose slopes are given by the results in parts (a) and (b).

1. $y = \frac{1}{2}x^2$; $x_0 = 3$, $x_1 = 4$.   **2.** $y = x^3$; $x_0 = 1$, $x_1 = 2$.

3. $y = 1/x$; $x_0 = 2$, $x_1 = 3$.
4. $y = 1/x^2$; $x_0 = 1$, $x_1 = 2$.

In Exercises 5–8, a function f and a value of x_0 are given.

(a) Find the slope of the tangent to the graph of f at a general point x_0.
(b) Use the result in part (a) to find the slope of the tangent at the given value of x_0.

5. $f(x) = x^2 + 1$; $x_0 = 2$.
6. $f(x) = x^2 + 3x + 2$; $x_0 = 2$.
7. $f(x) = \sqrt{x}$; $x_0 = 1$.   **8.** $f(x) = 1/\sqrt{x}$; $x_0 = 4$.

9. Figure 3.1.10 shows the position versus time curve for an elevator that moves upward a distance of 60 meters and then discharges its passengers.

   (a) Estimate the instantaneous velocity of the elevator at $t = 10$ seconds.

   (b) Sketch a velocity versus time curve for the motion of the elevator for $0 \le t \le 20$.

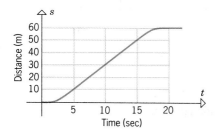

Figure 3.1.10

10. Figure 3.1.11 shows the position versus time curve for a certain particle moving along a straight line. Estimate each of the following from the graph:

   (a) the average velocity over the interval $0 \le t \le 3$

   (b) the values of t at which the instantaneous velocity is zero

   (c) the values of t at which the instantaneous velocity is either a maximum or a minimum

   (d) the instantaneous velocity when $t = 3$ seconds.

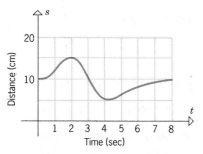

Figure 3.1.11

11. Figure 3.1.12 shows the position versus time curve for a certain particle moving on a straight line.

   (a) Is the particle moving faster at time t_0 or time t_2? Explain.

   (b) At the origin, the tangent is horizontal. What does this tell us about the initial velocity of the particle?

   (c) Is the particle speeding up or slowing down in the interval $[t_0, t_1]$? Explain.

   (d) Is the particle speeding up or slowing down in the interval $[t_1, t_2]$? Explain.

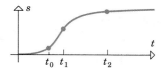

Figure 3.1.12

12. An automobile, initially at rest, begins to move along a straight track. The velocity increases steadily until suddenly the driver sees a concrete barrier in the road and

applies the brakes sharply at time t_0. The car decelerates rapidly, but it is too late—the car crashes into the barrier at time t_1 and instantaneously comes to rest. Sketch a position versus time curve that might represent the motion of the car.

13. If a particle moves at constant velocity, what can you say about its position versus time curve?

14. Figure 3.1.13 shows the position versus time curves of four different particles moving on a straight line. For each particle, determine whether its instantaneous velocity is increasing or decreasing with time.

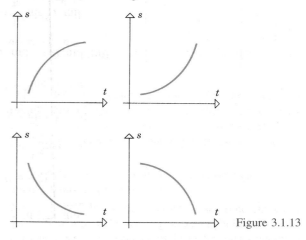

Figure 3.1.13

15. Suppose that the outside temperature versus time curve over a 24-hour period is as shown in Figure 3.1.14.

   (a) Estimate the maximum temperature and the time at which it occurs.

   (b) The temperature rise is fairly linear from 8 A.M. to 2 P.M. Estimate the rate at which the temperature is increasing during this time period.

   (c) Estimate the time at which the temperature is decreasing most rapidly. Estimate the instantaneous rate of change of temperature with respect to time at this instant.

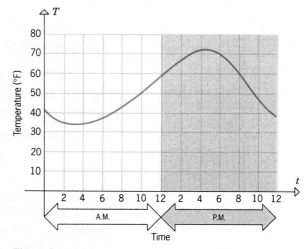

Figure 3.1.14

16. Figure 3.1.15 shows the graph of the pressure p (in atmospheres) versus the volume V (in liters) of 1 mole of an ideal gas at a constant temperature of 300 K (kelvin). Use the tangent lines shown in the figure to estimate the rate of change of pressure with respect to volume at the points where $V = 10$ L and $V = 25$ L.

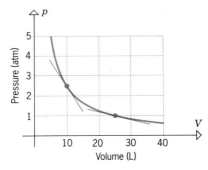

Figure 3.1.15

17. Figure 3.1.16 shows the graph of the height h (in centimeters) versus the age t (in years) of an individual from sometime after birth to age 20.

(a) When is the growth rate greatest?

(b) Estimate the growth rate at age 5.

(c) At approximately what age between 10 and 20 is the growth rate greatest? Estimate the growth rate at this age.

(d) Draw a rough graph of the growth rate versus age.

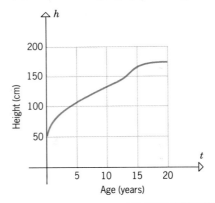

Figure 3.1.16

In Exercises 18–21, use Formulas (3a) and (3b) to find the average and instantaneous velocity.

18. A rock is dropped from a height of 576 ft and falls toward earth in a straight line. In t sec the rock drops a distance of $s = 16t^2$ ft.

(a) How many seconds after release does the rock hit the ground?

(b) What is the average velocity of the rock during the time it is falling?

(c) What is the average velocity of the rock for the first 3 sec?

(d) What is the instantaneous velocity of the rock when it hits the ground?

19. During the first 40 sec of a rocket flight, the rocket is propelled straight up so that in t sec it reaches a height of $s = 5t^3$ ft.

(a) How high does the rocket travel in 40 sec?

(b) What is the average velocity of the rocket during the first 40 sec?

(c) What is the average velocity of the rocket during the first 135 ft of its flight?

(d) What is the instantaneous velocity of the rocket at the end of 40 sec?

20. A particle moves on a line away from its initial position so that after t hr it is $s = 3t^2 + t$ mi from its initial position.

(a) Find the average velocity of the particle over the interval $[1, 3]$.

(b) Find the instantaneous velocity at $t = 1$.

21. A particle moves in one direction along a straight line so that after t min its distance is $s = 6t^4$ ft from the origin.

(a) Find the average velocity of the particle over the interval $[2, 4]$.

(b) Find the instantaneous velocity at $t = 2$.

22. A car is traveling on a straight road that is 120 mi long. For the first 100 mi the car travels at an average velocity of 50 mi/hr. Show that no matter how fast the car travels for the final 20 mi it cannot bring the average velocity up to 60 mi/hr for the entire trip.

23. Use the formula $A = \pi r^2$ for the area of a circle to find

(a) the average rate at which the area of a circle changes with r as the radius increases from $r = 1$ to $r = 2$.

(b) the instantaneous rate at which the area changes with r when $r = 2$.

24. Use the formula $V = l^3$ for the volume of a cube of side l to find

(a) the average rate at which the volume of a cube changes with l as l increases from $l = 2$ to $l = 4$

(b) the instantaneous rate at which the volume of a cube changes with l when $l = 5$.

25. Let $f(x) = x^2$. If we approximate the slope of the tangent line at the point $(x_0, f(x_0))$ by the slope of the secant line between $(x_0, f(x_0))$ and $(x_1, f(x_1))$, show that the error is $|x_1 - x_0|$. (By error we mean $|m_{\tan} - m_{\sec}|$, where $m_{\tan}$ is the slope of the tangent line and $m_{\sec}$ the slope of the secant line.)

■ **3.2** THE DERIVATIVE

> *In this section we shall associate with a given function f a second function f′, called the **derivative of f**, with the property that the value of f′ at a point is the slope of the tangent line to the graph of f at that point. These "slope-producing functions" are of fundamental importance in mathematics.*

□ **DEFINITION OF THE DERIVATIVE**

In the preceding section we showed informally that the slope of the tangent line to the graph of $y = f(x)$ at the point x_0 is given by

$$m_{\tan} = \lim_{x_1 \to x_0} \frac{f(x_1) - f(x_0)}{x_1 - x_0} \tag{1}$$

In keeping with common usage we will rewrite this formula in terms of a variable h defined as

$$h = x_1 - x_0$$

Thus, $x_1 = x_0 + h$ and $h \to 0$ as $x_1 \to x_0$, so (1) can be rewritten as

$$m_{\tan} = \lim_{h \to 0} \frac{f(x_0 + h) - f(x_0)}{h}$$

(Figure 3.2.1). Motivated by the preceding discussion, we make the following definition.

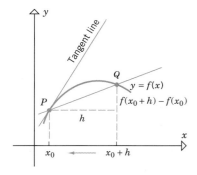

$$m_{\tan} = \lim_{h \to 0} \frac{f(x_0 + h) - f(x_0)}{h}$$

Figure 3.2.1

> **3.2.1**   DEFINITION.   If $P(x_0, y_0)$ is a point on the graph of a function f, then the **tangent line** to the graph of f at P is defined to be the line through P with slope
>
> $$m_{\tan} = \lim_{h \to 0} \frac{f(x_0 + h) - f(x_0)}{h} \tag{2}$$
>
> provided this limit exists.

For brevity, the tangent line at $P(x_0, y_0)$ is often called the **tangent line at x_0**. It follows from Definition 3.2.1 that the point-slope form of the equation of the tangent line at x_0 is

$$y - y_0 = m_{\tan} (x - x_0) \tag{3}$$

REMARK.   It is possible that the limit in (2) may not exist, in which case $m_{\tan}$ will be undefined. We shall consider the geometric significance of this later in this section.

Example 1   Find the slope and an equation of the tangent line to the graph of $f(x) = x^2$ at the point $P(3, 9)$. (See Figure 3.2.2.)

Solution.   We have $x_0 = 3$ and $y_0 = 9$, so from (2)

$$m_{\tan} = \lim_{h \to 0} \frac{f(3 + h) - f(3)}{h} = \lim_{h \to 0} \frac{(3 + h)^2 - 9}{h}$$

$$= \lim_{h \to 0} \frac{(9 + 6h + h^2) - 9}{h} = \lim_{h \to 0} \frac{6h + h^2}{h}$$

$$= \lim_{h \to 0} (6 + h) = 6$$

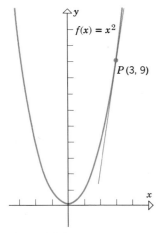

Figure 3.2.2

Thus, from (3) the point-slope form of the tangent line is

$$y - 9 = 6(x - 3)$$

and the slope-intercept form is $y = 6x - 9$.   ◀

Observe that $m_{\tan}$ is a function of x_0, since the slope of the tangent line depends on where along the curve the slope is being computed. For example, in Example 1(c) in Section 3.1 we showed that the slope of the tangent line to the graph of $f(x) = x^2 + 1$ at a point x_0 is

$$m_{\tan} = 2x_0$$

For our present purposes it will be convenient to omit the zero subscript and state more simply that the slope of the tangent line to the graph of $f(x) = x^2 + 1$ at a point x is

$$m_{\tan} = 2x$$

Thus, we have succeeded in associating with the function $f(x) = x^2 + 1$ a second function $f'(x) = 2x$ whose value at the point x is the slope $2x$ of the tangent line to the graph of f at that point.

To generalize the preceding ideas, let us replace x_0 in (2) by x to obtain the following formula for the slope of the tangent line to the graph of f at the point x:

$$m_{\tan} = \lim_{h \to 0} \frac{f(x + h) - f(x)}{h}$$

The function defined by this formula is of such fundamental importance that there is some notation and terminology associated with it.

3.2.2   DEFINITION.   The function f' defined by the formula

$$f'(x) = \lim_{h \to 0} \frac{f(x + h) - f(x)}{h} \tag{4}$$

is called the ***derivative with respect to x*** of the function f. The domain of f' consists of all x for which the limit exists.

Based on our discussion in the preceding section, the derivative of a function f can be interpreted two ways:

Geometric Interpretation of the Derivative
f' is the function whose value at x is the slope of the tangent line to the graph of f at x.

Rate of Change Interpretation of the Derivative
If $y = f(x)$, then f' is the function whose value at x is the instantaneous rate of change of y with respect to x at the point x.

Example 2   Let $f(x) = x^2 + 1$.

(a)   Find $f'(x)$.

(b)   Use the result in part (a) to find the slope of the tangent line to $y = x^2 + 1$ at $x = 2$, $x = 0$, and $x = -2$.

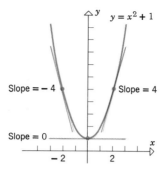

Figure 3.2.3

Solution (a).   From (4)

$$f'(x) = \lim_{h \to 0} \frac{f(x + h) - f(x)}{h} = \lim_{h \to 0} \frac{[(x + h)^2 + 1] - [x^2 + 1]}{h}$$

$$= \lim_{h \to 0} \frac{x^2 + 2xh + h^2 + 1 - x^2 - 1}{h} = \lim_{h \to 0} \frac{2xh + h^2}{h}$$

$$= \lim_{h \to 0} (2x + h) = 2x$$

Solution (b).   From part (a) the slope of the tangent line at any point x is

$$f'(x) = 2x$$

Thus, at $x = 2$, $x = 0$, and $x = -2$ the slopes are

$$f'(2) = 2(2) = 4, \quad f'(0) = 2(0) = 0, \quad \text{and} \quad f'(-2) = 2(-2) = -4$$

respectively (Figure 3.2.3).   ◄

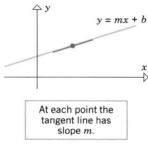

At each point the tangent line has slope m.

Figure 3.2.4

Example 3   It is obvious geometrically that at each point on a line $y = mx + b$, the tangent line coincides with the line itself and thus has slope m (Figure 3.2.4). Therefore, if $f(x) = mx + b$, we should anticipate that $f'(x) = m$ for all x. The following computation shows that this is so.

$$f'(x) = \lim_{h \to 0} \frac{f(x + h) - f(x)}{h} = \lim_{h \to 0} \frac{[m(x + h) + b] - (mx + b)}{h}$$

$$= \lim_{h \to 0} \frac{mx + mh + b - mx - b}{h} = \lim_{h \to 0} \frac{mh}{h} = \lim_{h \to 0} m = m \quad ◄$$

Example 4   Find

(a)   the derivative with respect to x of $f(x) = \sqrt{x}$;

(b)   the slope of the tangent line to the graph of $y = \sqrt{x}$ at $x = 9$;

(c)   the instantaneous rate of change of $y = \sqrt{x}$ with respect to x at $x = 5$.

Solution (a).   From Definition 3.2.2,

$$f'(x) = \lim_{h \to 0} \frac{f(x + h) - f(x)}{h} = \lim_{h \to 0} \frac{\sqrt{x + h} - \sqrt{x}}{h}$$

$$= \lim_{h \to 0} \frac{(\sqrt{x + h} - \sqrt{x})(\sqrt{x + h} + \sqrt{x})}{h(\sqrt{x + h} + \sqrt{x})} = \lim_{h \to 0} \frac{(x + h) - x}{h(\sqrt{x + h} + \sqrt{x})}$$

$$= \lim_{h \to 0} \frac{h}{h(\sqrt{x + h} + \sqrt{x})} = \lim_{h \to 0} \frac{1}{\sqrt{x + h} + \sqrt{x}}$$

$$= \frac{1}{\sqrt{x} + \sqrt{x}} = \frac{1}{2\sqrt{x}}$$

Solution (b).   The slope of the tangent line at $x = 9$ is $f'(9)$, and from part (a) we have $f'(9) = 1/(2\sqrt{9}) = 1/6$.

Solution (c).   The instantaneous rate of change at $x = 5$ is $f'(5)$, and from part (a) we have $f'(5) = 1/(2\sqrt{5})$.   ◄

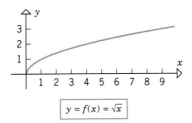

$y = f(x) = \sqrt{x}$

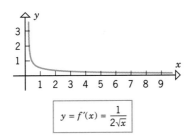

$y = f'(x) = \dfrac{1}{2\sqrt{x}}$

Figure 3.2.5

The graphs of the functions

$$f(x) = \sqrt{x} \quad \text{and} \quad f'(x) = \frac{1}{2\sqrt{x}}$$

from the preceding example are shown in Figure 3.2.5. Observe that

$$\lim_{x \to 0^+} \frac{1}{2\sqrt{x}} = +\infty$$

This implies that the slopes of the tangent lines to the graph of f increase without bound as x approaches zero from the right. (Can you see that this is so from the graph of f?)

Example 5 Find the derivative of $f(x) = \dfrac{x}{x - 9}$.

Solution.

$$
\begin{aligned}
f'(x) &= \lim_{h \to 0} \frac{f(x + h) - f(x)}{h} \\
&= \lim_{h \to 0} \frac{1}{h} \left[\frac{x + h}{(x + h) - 9} - \frac{x}{x - 9} \right] \\
&= \lim_{h \to 0} \frac{1}{h} \left[\frac{(x + h)(x - 9) - x(x + h - 9)}{(x + h - 9)(x - 9)} \right] \\
&= \lim_{h \to 0} \frac{(x^2 - 9x + hx - 9h) - (x^2 + hx - 9x)}{h(x + h - 9)(x - 9)} \\
&= \lim_{h \to 0} \frac{-9h}{h(x + h - 9)(x - 9)} \\
&= \lim_{h \to 0} \frac{-9}{(x + h - 9)(x - 9)} = -\frac{9}{(x - 9)^2} \quad \blacktriangleleft
\end{aligned}
$$

☐ **DERIVATIVE NOTATION**

In the early days of calculus there was vitriolic debate over the proper notation for the derivative. The English mathematicians advocated the notation of Isaac Newton and the Continental mathematicians advocated the notation of Leibniz. Today, there is a variety of different notations in common use, all of which have useful purposes. We now discuss some of these.

The process of finding a derivative is called **differentiation**. It is often useful to think of differentiation as an operation that, when applied to a function f, produces a new function f'. In the case where the independent variable is x, the differentiation operation is often denoted by the symbol

$$\frac{d}{dx}[f(x)]$$

which is read, "**the derivative of f(x) with respect to x**." Thus,

$$\frac{d}{dx}[f(x)] = f'(x) \tag{5}$$

With this notation, the results in Examples 4 and 5 can be written as

$$\frac{d}{dx}[\sqrt{x}] = \frac{1}{2\sqrt{x}} \quad \text{and} \quad \frac{d}{dx}\left[\frac{x}{x - 9} \right] = -\frac{9}{(x - 9)^2} \tag{6}$$

If there is a dependent variable $y = f(x)$, then (5) can be written as

$$\frac{d}{dx}[y] = f'(x) \tag{7}$$

However, it is common to omit the brackets on the left side and write more simply

$$\frac{dy}{dx} = f'(x) \qquad\qquad (8)$$

For example, if $y = \sqrt{x}$, then the first result in (6) can be written as

$$\frac{dy}{dx} = \frac{1}{2\sqrt{x}} \qquad\qquad (9)$$

REMARK.   Later, the symbols dy and dx will be defined separately. However, for the time being, dy/dx should not be regarded as a ratio; rather, it should be considered as a single symbol denoting the derivative.

If the independent variable is not x, then appropriate adjustments in notation have to be made. For example, if $y = f(u)$, then (5) and (8) would be written as

$$\frac{d}{du}[f(u)] = f'(u) \quad \text{and} \quad \frac{dy}{du} = f'(u)$$

In particular, adjusting the notation in (6) yields

$$\frac{d}{du}[\sqrt{u}] = \frac{1}{2\sqrt{u}} \quad \text{and} \quad \frac{d}{du}\left[\frac{u}{u-9}\right] = -\frac{9}{(u-9)^2} \qquad\qquad (10)$$

If $y = f(x)$, then with notations (5) and (8) the value of the derivative of f at a specific point $x = x_0$ can be denoted as

$$\frac{d}{dx}[f(x)]\bigg|_{x=x_0} = f'(x_0) \qquad \text{or} \qquad \frac{dy}{dx}\bigg|_{x=x_0} = f'(x_0)$$

For example, if $y = \sqrt{x}$, then from (6)

$$\frac{d}{dx}[\sqrt{x}]\bigg|_{x=x_0} = \frac{dy}{dx}\bigg|_{x=x_0} = \frac{1}{2\sqrt{x_0}}$$

and from (10)

$$\frac{d}{du}\left[\frac{u}{u-9}\right]\bigg|_{u=10} = -\frac{9}{(10-9)^2} = -9$$

There is yet another useful notation for a derivative: If y is the dependent variable for a function, and if the independent variable is self-evident from the problem, then it is common to denote the derivative simply as y'. Thus, y' would mean dy/dx if the independent variable is x or dy/du if the independent variable is u. Some writers use the notation $D_x[\ \ ]$ to denote the differentiation operation. We shall not use this notation, however.

REMARK.   The variety of notations for derivatives can be confusing at first, but each has its purpose and all are used, so it is important to be familiar with all of them.

☐ EXISTENCE OF
   DERIVATIVES

The derivative of a function f is defined at those points where the limit in (4) exists. If x_0 is such a point, then we say that *f is differentiable at x_0* or *f has a derivative at x_0*. Stated another way, the domain of f' consists of those points where f is differentiable. We say that f is *differentiable on an open interval* (a, b) if it is differentiable at each point in (a, b), and we say that f is a *differentiable function* if it is differentiable on $(-\infty, +\infty)$. At points where f is not differentiable we say that *the derivative of f does not exist.*

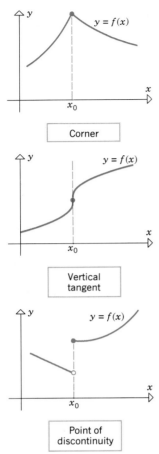

Corner

Vertical tangent

Point of discontinuity

Figure 3.2.6

Informally speaking, the most commonly encountered points of nondifferentiability can be classified as

corners, vertical tangents, or points of discontinuity

(see Figure 3.2.6).

The graph of a function f has a ''corner'' at a point $P(x_0, f(x_0))$ if f is continuous at P and the limiting position for the secant line joining P and Q depends on whether Q approaches P from the left or the right (Figure 3.2.7). At a corner a tangent line does not exist because the slopes of the secant lines do not have a (two-sided) limit.

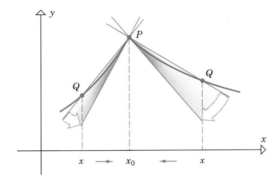

Figure 3.2.7

If the slope of the secant line joining P and Q tends toward $+\infty$ or $-\infty$ as Q approaches P along the graph of f, then f is not differentiable at x_0. Geometrically, such points occur where the secant lines tend toward a vertical limiting position (Figure 3.2.8).

The limiting behavior of secant lines can be very strange at points of discontinuity. For example, in Figure 3.2.9 the slope of the secant line tends toward $+\infty$ as x approaches x_0 from the left and tends toward 0 as x approaches x_0 from the right. Since the slopes of the secant lines do not have a two-sided limit, the function graphed is not differentiable at x_0.

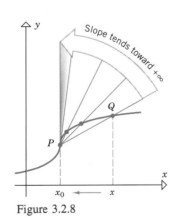

Figure 3.2.8

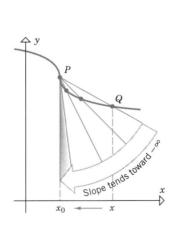

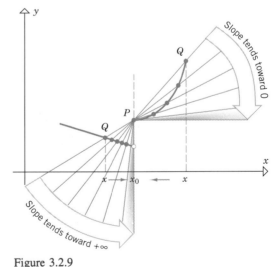

Figure 3.2.9

□ **RELATIONSHIP BETWEEN DIFFERENTIABILITY AND CONTINUITY**

The following major theorem shows that a function must be continuous at each point where it is differentiable.

3.2.3 THEOREM. *If f is differentiable at a point x_0, then f is also continuous at x_0.*

Proof. We shall use Definition 2.7.11 to prove the continuity. We must show that if $f'(x_0)$ exists, then $\lim\limits_{h \to 0} f(x_0 + h) = f(x_0)$ or equivalently, $\lim\limits_{h \to 0} [f(x_0 + h) - f(x_0)] = 0$. But

$$\lim_{h \to 0} [f(x_0 + h) - f(x_0)] = \lim_{h \to 0} \left[\frac{f(x_0 + h) - f(x_0)}{h} \cdot h \right]$$

$$= \lim_{h \to 0} \left[\frac{f(x_0 + h) - f(x_0)}{h} \right] \cdot \lim_{h \to 0} h$$

$$= f'(x_0) \cdot 0 = 0 \quad \blacksquare$$

REMARK.   It follows from the preceding theorem that a function cannot be differentiable at a point of discontinuity.

Theorem 3.2.3 shows that differentiability at a point implies continuity at that point. The converse, however, is false—*a function may be continuous at a point but not differentiable there.* Whenever the graph of a function has a corner at a point, but no break or gap there, we have a point where the function is continuous, but not differentiable.

Example 6   The function $f(x) = |x|$ is continuous for all x and consequently is continuous at $x = 0$ (Example 3, Section 2.7).

(a)   Show that $f(x) = |x|$ is not differentiable at $x = 0$.

(b)   Find $f'(x)$.

Solution (a).   This result is evident geometrically, since the graph of $|x|$ has a corner at $x = 0$ (Figure 3.2.10). However, an analytic argument can be given as follows. From Definition 3.2.2,

$$f'(0) = \lim_{h \to 0} \frac{f(0 + h) - f(0)}{h} = \lim_{h \to 0} \frac{f(h) - f(0)}{h} = \lim_{h \to 0} \frac{|h| - |0|}{h} = \lim_{h \to 0} \frac{|h|}{h}$$

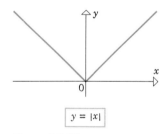

$y = |x|$

Figure 3.2.10

But

$$\frac{|h|}{h} = \begin{cases} 1, & h > 0 \\ -1, & h < 0 \end{cases}$$

so that

$$\lim_{h \to 0^-} \frac{|h|}{h} = -1 \quad \text{and} \quad \lim_{h \to 0^+} \frac{|h|}{h} = 1$$

Thus,

$$f'(0) = \lim_{h \to 0} \frac{|h|}{h}$$

does not exist because the one-sided limits are not equal. Consequently, $f(x) = |x|$ is not differentiable at $x = 0$.

Solution (b).   If $x > 0$, then $f(x) = |x| = x$, so $f'(x) = 1$, and if $x < 0$, then $f(x) = |x| = -x$, so $f'(x) = -1$. Combining these results into a single formula expressed piecewise, we obtain

$$f'(x) = \frac{d}{dx} [|x|] = \begin{cases} 1, & x > 0 \\ -1, & x < 0 \end{cases}$$

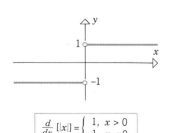

$\dfrac{d}{dx} [|x|] = \begin{cases} 1, & x > 0 \\ -1, & x < 0 \end{cases}$

Figure 3.2.11

The graph of f' is shown in Figure 3.2.11. Observe that f' is not a continuous function, so this example shows that a continuous function can have a derivative that is not continuous.   ◄

The relationship between continuity and differentiability was of great historical significance in the development of calculus. In the early nineteenth century mathematicians believed that the graph of a continuous function could not have too many points of nondifferentiability bunched up. They felt that if a continuous function had many points of nondifferentiability, these points, like the tips of a sawblade, would have to be separated from each other and joined by smooth curve segments (Figure 3.2.12). This misconception was shattered by a series of discoveries beginning in 1834. In that year a Bohemian priest, philosopher, and mathematician named Bernhard Bolzano* discovered a procedure for constructing a continuous function that is not differentiable at any point. Later, in 1860, the great German mathematician, Karl Weierstrass** produced the first formula for such a function. The graphs of such functions are impossible to draw; it is as if the corners are so numerous that any segment of the curve, when suitably enlarged, reveals more corners. The discovery of these pathological functions was important in that it made mathematicians distrustful of their geometric intuition and more reliant on precise mathematical proof. However, they remained only mathematical curiosities until the early 1980s, when applications of them began to emerge. During the past 10 years they have started to play a fundamental role in the study of geometric objects called *fractals*. Fractals have revealed an order to natural phenomena that were previously dismissed as random and chaotic.

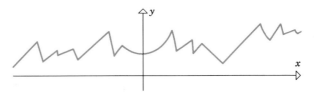

Figure 3.2.12

---

* BERNHARD BOLZANO (1781–1848). Bolzano, the son of an art dealer, was born in Prague, Bohemia (Czechoslovakia). He was educated at the University of Prague, and eventually won enough mathematical fame to be recommended for a mathematics chair there. However, Bolzano became an ordained Roman Catholic priest, and in 1805 he was appointed to a chair of Philosophy at the University of Prague. Bolzano was a man of great human compassion; he spoke out for educational reform, he voiced the right of individual conscience over government demands, and he lectured on the absurdity of war and militarism. His views so disenchanted Emperor Franz I of Austria that the emperor pressed the Archbishop of Prague to have Bolzano recant his statements. Bolzano refused and was then forced to retire in 1824 on a small pension. Bolzano's main contribution to mathematics was philosophical. His work helped convince mathematicians that sound mathematics must ultimately rest on rigorous proof rather than intuition. In addition to his work in mathematics, Bolzano investigated problems concerning space, force, and wave propagation.

** KARL WEIERSTRASS (1815–1897). Weierstrass, the son of a customs officer, was born in Ostenfelde, Germany. As a youth Weierstrass showed outstanding skills in languages and mathematics. However, at the urging of his dominant father, Weierstrass entered the law and commerce program at the University of Bonn. To the chagrin of his family, the rugged and congenial young man concentrated instead on fencing and beer drinking. Four years later he returned home without a degree. In 1839 Weierstrass entered the Academy of Münster to study for a career in secondary education, and he met and studied under an excellent mathematician named Christof Gudermann. Gudermann's ideas greatly influenced the work of Weierstrass. After receiving his teaching certificate, Weierstrass spent the next 15 years in secondary education teaching German, geography, and mathematics. In addition, he taught handwriting to small children. During this period much of Weierstrass's mathematical work was ignored because he was a secondary schoolteacher and not a college professor. Then, in 1854, he published a paper of major importance which created a sensation in the mathematics world and catapulted him to international fame overnight. He was immediately given an honorary Doctorate at the University of Königsberg and began a new career in college teaching at the University of Berlin in 1856. In 1859 the strain of his mathematical research caused a temporary nervous breakdown and led to spells of dizziness that plagued him for the rest of his life. Weierstrass was a brilliant teacher and his classes overflowed with multitudes of auditors. In spite of his fame, he never lost his early beer-drinking congeniality and was always in the company of students, both ordinary and brilliant. Weierstrass was acknowledged as the leading mathematical analyst in the world. He and his students opened the door to the modern school of mathematical analysis.

☐ **DERIVATIVES AT THE ENDPOINTS OF AN INTERVAL**

If a function f is defined on a closed interval $[a, b]$ and is not defined outside of that interval, then the derivative $f'(x)$ is not defined at the endpoints a and b because

$$f'(x) = \lim_{h \to 0} \frac{f(x + h) - f(x)}{h}$$

is a two-sided limit and only one-sided limits make sense at the endpoints. To deal with this situation, we define **derivatives from the left and right**. These are denoted by f'_- and f'_+, respectively, and defined by

$$f'_-(x) = \lim_{h \to 0^-} \frac{f(x + h) - f(x)}{h} \quad \text{and} \quad f'_+(x) = \lim_{h \to 0^+} \frac{f(x + h) - f(x)}{h}$$

At points where $f'_+(x)$ exists we say that the function f is **differentiable from the right** and at points where $f'_-(x)$ exists we say that f is **differentiable from the left**. Geometrically, $f'_+(x)$ is the limit of the slopes of the secant lines approaching x from the right and $f'_-(x)$ is the limit of the slopes of the secant lines approaching x from the left (Figure 3.2.13).

It can be proved that a function f is continuous from the left at those points where it is differentiable from the left and continuous from the right at those points where it is differentiable from the right. (See Definition 2.7.7.)

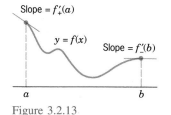

Slope = $f'_+(a)$

$y = f(x)$

Slope = $f'_-(b)$

a b

Figure 3.2.13

3.2.4   DEFINITION.   A function f is said to be **differentiable on a closed interval** $[a, b]$ if the following conditions are satisfied:

1. f is differentiable on (a, b).
2. f is differentiable from the right at a.
3. f is differentiable from the left at b.

The reader should have no trouble formulating the appropriate definitions of differentiability on intervals of the form $[a, +\infty)$, $(-\infty, b]$, $[a, b)$, and $(a, b]$.

▶ Exercise Set 3.2   © *29–32*

In Exercises 1–12, use Definition 3.2.2 to find $f'(x)$.

1. $f(x) = 3x^2$.
2. $f(x) = x^2 - x$.
3. $f(x) = x^3$.
4. $f(x) = 2x^3 + 1$.
5. $f(x) = \sqrt{x + 1}$.
6. $f(x) = x^4$.
7. $f(x) = \dfrac{1}{x}$.
8. $f(x) = \dfrac{1}{x^2}$.
9. $f(x) = ax^2 + b$
      (a, b constants).
10. $f(x) = \dfrac{1}{x + 1}$.
11. $f(x) = \dfrac{1}{\sqrt{x}}$.
12. $f(x) = x^{1/3}$.

In Exercises 13–18, find $f'(a)$ and the equation of the tangent line to the graph of f at the point where $x = a$.

13. f is the function in Exercise 1; $a = 3$.
14. f is the function in Exercise 2; $a = 2$.
15. f is the function in Exercise 3; $a = 0$.
16. f is the function in Exercise 4; $a = -1$.
17. f is the function in Exercise 5; $a = 8$.
18. f is the function in Exercise 6; $a = -2$.

19. Given that $f(3) = -1$ and $f'(3) = 5$, find an equation for the tangent line to the graph of $y = f(x)$ at the point where $x = 3$.

20. Given that $f(-2) = 3$ and $f'(-2) = -4$, find an equation for the tangent line to the graph of $y = f(x)$ at the point where $x = -2$.

21. Let $y = 4x^2 + 2$. Find
   (a) dy/dx
   (b) $dy/dx\big|_{x=1}$.

22. Let $y = (5/x) + 1$. Find
   (a) dy/dx
   (b) $dy/dx\big|_{x=-2}$.

In Exercises 23–26, use Definition 3.2.2 (with the appropriate change in notation) to obtain the derivative requested.

23. Find $f'(t)$ if $f(t) = 4t^2 + t$.
24. Find $g'(u)$ if $g(u) = 5u + 3$.
25. Find $dA/d\lambda$ if $A = 3\lambda^2 - \lambda$.
26. Find dV/dr if $V = \frac{4}{3}\pi r^3$.
27. Match the graphs of the functions shown in (a)–(f) with the graphs of their derivatives in (A)–(F).

(a)
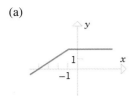

(b)
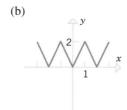

(A)
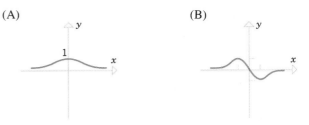

(B)

(c)
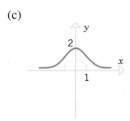

(d)
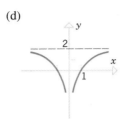

(C)
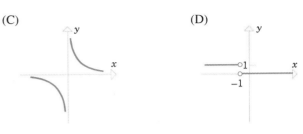

(D)

(e)
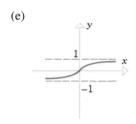

(f)
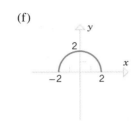

(E)
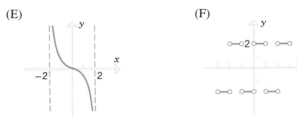

(F)

28. Use the graph of $y = f(x)$ shown in Figure 3.2.14 to estimate the value of $f'(1)$, $f'(3)$, $f'(5)$, and $f'(6)$.

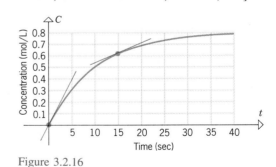

Figure 3.2.14

29. Biological systems in which the exchange of substances occur are sometimes described mathematically by using a *compartment model*. Figure 3.2.15a shows two compartments of equal size separated by a permeable membrane, the left containing a red dye dissolved in water and the right just water. The compartment model states that over time the dye will diffuse from the compartment with higher concentration to the compartment with lower concentration at a rate that is proportional to the difference in concentrations until the concentration of dye is the same in both compartments (Figures 3.2.15b and 3.2.15c). Fig-

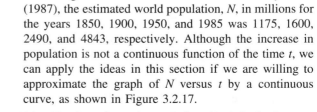

Membrane

(a)   *(b)*   *(c)*

Figure 3.2.15

ure 3.2.16 shows the graph of the concentration C versus the time t for the dye in the right compartment in the case where the left compartment has a concentration of 1.6 mol/L (moles per liter) and the right compartment contains only water at time $t = 0$. Estimate the values of

$$dC/dt\big|_{t=0} \quad \text{and} \quad dC/dt\big|_{t=15}$$

[Adapted from *Compartment Models in Biology*, by Ron Barnes, UMAP Module 676, COMAP, Inc.]

30. According to *The World Almanac and Book of Facts* (1987), the estimated world population, N, in millions for the years 1850, 1900, 1950, and 1985 was 1175, 1600, 2490, and 4843, respectively. Although the increase in population is not a continuous function of the time t, we can apply the ideas in this section if we are willing to approximate the graph of N versus t by a continuous curve, as shown in Figure 3.2.17.

(a) Use the estimated tangent line shown in the figure at the point where $t = 1950$ to approximate the value of dN/dt there. Describe your result as a rate of change.

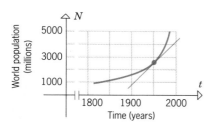

Figure 3.2.17

(b) At any instant, the **growth rate** is defined as

$$\frac{dN/dt}{N}$$

Use your answer to part (a) to approximate the growth rate in 1950. Express the result as a percentage and include the proper units.

31. It is a fact that when a flexible rope is wrapped around a rough cylinder, a small force of magnitude F_0 at one end can resist a large force of magnitude F at the other end (Figure 3.2.18*a*). The size of F depends on the angle θ

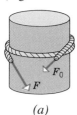
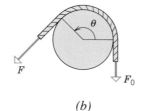

(a)   *(b)*

Figure 3.2.18

through which the rope is wrapped around the cylinder (Figure 3.2.18*b*). For example, Figure 3.2.19 shows the graph of F (in pounds) versus θ (in radians), where F is the magnitude of the force that can be resisted by a force with magnitude $F_0 = 10$ lb for a certain rope and cylinder.

(a) Estimate the values of F and $dF/d\theta$ when $\theta = 10$ radians.

(b) It can be shown that F satisfies the equation $dF/d\theta = \mu F$, where the constant μ is called the **coefficient of friction**. Use the results in part (a) to estimate the value of μ.

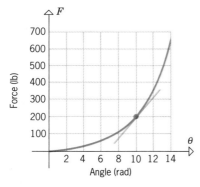

Figure 3.2.19

32. According to **Newton's law of cooling**, the rate of change of an object's temperature is proportional to the difference between the temperature of the object and that of the surrounding medium. Figure 3.2.20 shows the graph of the temperature T (in degrees Fahrenheit) versus time t (in

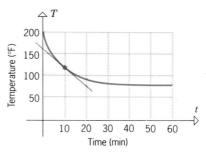

Figure 3.2.20

minutes) for a cup of coffee, initially with a temperature of 200° F, that is allowed to cool in a room with a constant temperature of 75° F.

(a) Estimate T and dT/dt when $t = 10$ min.

(b) Newton's law of cooling can be expressed as

$$\frac{dT}{dt} = k(T - T_0)$$

where k is the constant of proportionality and T_0 is the temperature (assumed constant) of the surrounding medium. Use the results in part (a) to estimate the value of k.

In Exercises 33–40, sketch the graph of the derivative of the function whose graph is shown.

33.
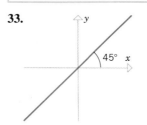

34.
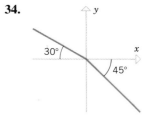

35.
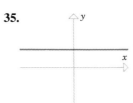

36.
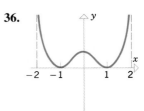

37.
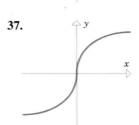

38.
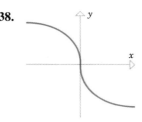

39.
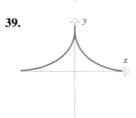

40.
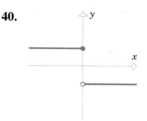

41. Show that $f(x) = \sqrt[3]{x}$ is continuous at $x = 0$ but not differentiable at $x = 0$. Sketch the graph of f.

42. Show that $f(x) = \sqrt[3]{(x-2)^2}$ is continuous at $x = 2$ but not differentiable at $x = 2$. Sketch the graph of f.

43. Show that

$$f(x) = \begin{cases} x^2 + 1, & x \le 1 \\ 2x, & x > 1 \end{cases}$$

is continuous and differentiable at $x = 1$. Sketch the graph of f.

44. Show that

$$f(x) = \begin{cases} x^2 + 2, & x \le 1 \\ x + 2, & x > 1 \end{cases}$$

is continuous but not differentiable at $x = 1$. Sketch the graph of f.

45. Suppose that a function f is differentiable at $x = 1$ and $\lim\limits_{h \to 0} \dfrac{f(1 + h)}{h} = 5$. Find $f(1)$ and $f'(1)$.

46. Suppose that f is a differentiable function with the property that $f(x + y) = f(x) + f(y) + 5xy$ and $\lim\limits_{h \to 0} \dfrac{f(h)}{h} = 3$. Find $f(0)$ and $f'(x)$.

47. Suppose that f has the property $f(x + y) = f(x)f(y)$ for all values of x and y and that $f(0) = f'(0) = 1$. Show that f is differentiable and $f'(x) = f(x)$. [*Hint:* Start by expressing $f'(x)$ as a limit.]

◼ 3.3   TECHNIQUES OF DIFFERENTIATION

Up to now we have obtained derivatives directly from the definition. In this section we shall develop some theorems and formulas that provide more efficient methods.

☐ **DERIVATIVE OF A CONSTANT**

> **3.3.1**   THEOREM.   *If f is a constant function, say $f(x) = c$ for all x, then $f'(x) = 0$; that is,*
>
> $$\frac{d}{dx}[c] = 0$$

Proof.

$$f'(x) = \lim_{h \to 0} \frac{f(x + h) - f(x)}{h} = \lim_{h \to 0} \frac{c - c}{h} = \lim_{h \to 0} 0 = 0 \quad ◼$$

This result is obvious geometrically. If $f(x) = c$ is a constant function, then the graph of f is a line parallel to the x-axis; consequently, the tangent line at each point is horizontal. Therefore, $f'(x) = m_{\text{tan}} = 0$, since a horizontal line has slope 0.

Example 1   If $f(x) = 5$ for all x, then $f'(x) = 0$ for all x; that is,

$$\frac{d}{dx}[5] = 0 \quad ◀$$

☐ **DERIVATIVE OF A POWER**

> **3.3.2**   THEOREM (***The Power Rule***).   *If n is a positive integer, then*
>
> $$\frac{d}{dx}[x^n] = nx^{n-1}$$

? BINOMIAL THEOREM

Proof. Let $f(x) = x^n$. Then

$$f'(x) = \lim_{h \to 0} \frac{f(x + h) - f(x)}{h} = \lim_{h \to 0} \frac{(x + h)^n - x^n}{h}$$

Expanding $(x + h)^n$ by the binomial theorem, we obtain

$$f'(x) = \lim_{h \to 0} \frac{\left[x^n + nx^{n-1}h + \dfrac{n(n-1)}{2!} x^{n-2}h^2 + \cdots + nxh^{n-1} + h^n \right] - x^n}{h}$$

$$= \lim_{h \to 0} \frac{nx^{n-1}h + \dfrac{n(n-1)}{2!} x^{n-2}h^2 + \cdots + nxh^{n-1} + h^n}{h}$$

Canceling a factor of h we obtain

$$f'(x) = \lim_{h \to 0} \left[nx^{n-1} + \frac{n(n-1)}{2!} x^{n-2}h + \cdots + nxh^{n-2} + h^{n-1} \right]$$

Every term but the first has a factor of h, and so every term but the first approaches zero as $h \to 0$. Therefore,

$$f'(x) = nx^{n-1} \quad \blacksquare$$

REMARK. In words, *to differentiate x to a positive integer power, multiply that power by x raised to the next lower integer power.*

Example 2

$$\frac{d}{dx} [x^5] = 5x^4, \quad \frac{d}{dx} [x] = 1 \cdot x^0 = 1, \quad \frac{d}{dx} [x^{12}] = 12x^{11}$$

☐ **DERIVATIVE OF A CONSTANT TIMES A FUNCTION**

3.3.3 THEOREM. *Let c be a constant. If f is differentiable at x, then so is cf, and*

$$\frac{d}{dx} [cf(x)] = c \frac{d}{dx} [f(x)]$$

Proof.

$$\frac{d}{dx} [cf(x)] = \lim_{h \to 0} \frac{cf(x + h) - cf(x)}{h} = \lim_{h \to 0} c \left[\frac{f(x + h) - f(x)}{h} \right]$$

$$= c \lim_{h \to 0} \frac{f(x + h) - f(x)}{h} = c \frac{d}{dx} [f(x)] \quad \blacksquare$$

> A constant factor can be moved through a limit sign.

In function notation, Theorem 3.3.3 states

$$(cf)' = cf'$$

REMARK. In words, *a constant factor can be moved through a derivative sign.*

Example 3

$$\frac{d}{dx}[4x^8] = 4\frac{d}{dx}[x^8] = 4[8x^7] = 32x^7$$

$$\frac{d}{dx}[-x^{12}] = (-1)\frac{d}{dx}[x^{12}] = -12x^{11} \quad \blacktriangleleft$$

☐ **DERIVATIVES OF SUMS AND DIFFERENCES**

> **3.3.4   THEOREM.** *If f and g are differentiable at x, then so is $f + g$, and*
>
> $$\frac{d}{dx}[f(x) + g(x)] = \frac{d}{dx}[f(x)] + \frac{d}{dx}[g(x)]$$

Proof.

$$\frac{d}{dx}[f(x) + g(x)] = \lim_{h \to 0} \frac{[f(x+h) + g(x+h)] - [f(x) + g(x)]}{h}$$

$$= \lim_{h \to 0} \frac{[f(x+h) - f(x)] + [g(x+h) - g(x)]}{h}$$

$$= \lim_{h \to 0} \frac{f(x+h) - f(x)}{h} + \lim_{h \to 0} \frac{g(x+h) - g(x)}{h} \quad \boxed{\text{The limit of a sum is the sum of the limits.}}$$

$$= \frac{d}{dx}[f(x)] + \frac{d}{dx}[g(x)] \quad \blacksquare$$

Theorem 3.3.4 can be written in function notation as

$$(f + g)' = f' + g'$$

By writing $f - g = f + (-1)g$ and then applying Theorems 3.3.3 and 3.3.4 it follows that

$$\frac{d}{dx}[f(x) - g(x)] = \frac{d}{dx}[f(x)] - \frac{d}{dx}[g(x)]$$

or in function notation

$$(f - g)' = f' - g'$$

REMARK.   In words, *the derivative of a sum equals the sum of the derivatives*, and *the derivative of a difference equals the difference of the derivatives.*

Example 4

$$\frac{d}{dx}[x^4 + x^2] = \frac{d}{dx}[x^4] + \frac{d}{dx}[x^2] = 4x^3 + 2x$$

$$\frac{d}{dx}[6x^{11} - 9] = \frac{d}{dx}[6x^{11}] - \frac{d}{dx}[9] = 66x^{10} - 0 = 66x^{10} \quad \blacktriangleleft$$

The result in Theorem 3.3.4 can be extended to any finite number of functions. More precisely, if the functions $f_1, f_2, \ldots, f_n$ are all differentiable at x, then their sum is differentiable at x and

$$\frac{d}{dx}[f_1(x) + f_2(x) + \cdots + f_n(x)] = \frac{d}{dx}[f_1(x)] + \frac{d}{dx}[f_2(x)] + \cdots + \frac{d}{dx}[f_n(x)]$$

Example 5

$$\frac{d}{dx}[3x^8 - 2x^5 + 6x + 1] = \frac{d}{dx}[3x^8] + \frac{d}{dx}[-2x^5] + \frac{d}{dx}[6x] + \frac{d}{dx}[1]$$

$$= 24x^7 - 10x^4 + 6 \quad \blacktriangleleft$$

□ **DERIVATIVE OF A PRODUCT**

3.3.5   THEOREM (*The Product Rule*).   *If f and g are differentiable at x, then so is the product f · g, and*

$$\frac{d}{dx}[f(x)g(x)] = f(x)\frac{d}{dx}[g(x)] + g(x)\frac{d}{dx}[f(x)]$$

Proof.

$$\frac{d}{dx}[f(x)g(x)] = \lim_{h \to 0} \frac{f(x + h) \cdot g(x + h) - f(x) \cdot g(x)}{h}$$

If we add and subtract $f(x + h) \cdot g(x)$ in the numerator, we obtain

$$\frac{d}{dx}[f(x)g(x)] = \lim_{h \to 0} \frac{f(x + h)g(x + h) - f(x + h)g(x) + f(x + h)g(x) - f(x)g(x)}{h}$$

$$= \lim_{h \to 0}\left[f(x + h) \cdot \frac{g(x + h) - g(x)}{h} + g(x) \cdot \frac{f(x + h) - f(x)}{h}\right]$$

$$= \lim_{h \to 0} f(x + h) \cdot \lim_{h \to 0} \frac{g(x + h) - g(x)}{h} + \lim_{h \to 0} g(x) \cdot \lim_{h \to 0} \frac{f(x + h) - f(x)}{h}$$

$$= [\lim_{h \to 0} f(x + h)]\frac{d}{dx}[g(x)] + [\lim_{h \to 0} g(x)]\frac{d}{dx}[f(x)] \tag{1}$$

But

$$\lim_{h \to 0} g(x) = g(x) \tag{2}$$

because $g(x)$ does not involve h and thus remains *constant* as $h \to 0$. Also, it follows from Definition 2.7.11 that

$$\lim_{h \to 0} f(x + h) = f(x) \tag{3}$$

because f is assumed to be differentiable at x and is therefore continuous at x by Theorem 3.2.3. Substituting (2) and (3) into (1) yields

$$\frac{d}{dx}[f(x)g(x)] = f(x)\frac{d}{dx}[g(x)] + g(x)\frac{d}{dx}[f(x)] \quad \blacksquare$$

The product rule can be written in function notation as

$$(f \cdot g)' = f \cdot g' + g \cdot f'$$

REMARK.   In words, *the derivative of a product of two functions is the first function times the derivative of the second plus the second function times the derivative of the first.*

WARNING.   Note that it is *not* true in general that $(f \cdot g)' = f' \cdot g'$; that is, the derivative of a product is *not* generally the product of the derivatives!

Example 6   Find dy/dx if $y = (4x^2 - 1)(7x^3 + x)$.

Solution.   There are two methods that can be used to find dy/dx: We can either use the product rule or we can multiply out the factors in y and then differentiate. We shall give both methods.

Method I.   (*Using the Product Rule.*)

$$\frac{dy}{dx} = \frac{d}{dx}[(4x^2 - 1)(7x^3 + x)]$$

$$= (4x^2 - 1)\frac{d}{dx}[7x^3 + x] + (7x^3 + x)\frac{d}{dx}[4x^2 - 1]$$

$$= (4x^2 - 1)(21x^2 + 1) + (7x^3 + x)(8x) = 140x^4 - 9x^2 - 1$$

Method II.   (*Multiplying First.*)

$$y = (4x^2 - 1)(7x^3 + x) = 28x^5 - 3x^3 - x$$

Thus,

$$\frac{dy}{dx} = \frac{d}{dx}[28x^5 - 3x^3 - x] = 140x^4 - 9x^2 - 1$$

which agrees with the result obtained using the product rule.   ◄

□ **DERIVATIVE OF A QUOTIENT**

3.3.6   THEOREM (*The Quotient Rule*).   *If f and g are differentiable at x and $g(x) \neq 0$, then f/g is differentiable at x and*

$$\frac{d}{dx}\left[\frac{f(x)}{g(x)}\right] = \frac{g(x)\dfrac{d}{dx}[f(x)] - f(x)\dfrac{d}{dx}[g(x)]}{[g(x)]^2}$$

Proof.

$$\frac{d}{dx}\left[\frac{f(x)}{g(x)}\right] = \lim_{h \to 0}\frac{\dfrac{f(x+h)}{g(x+h)} - \dfrac{f(x)}{g(x)}}{h} = \lim_{h \to 0}\frac{f(x+h)\cdot g(x) - f(x)\cdot g(x+h)}{h\cdot g(x)\cdot g(x+h)}$$

Adding and subtracting $f(x)\cdot g(x)$ in the numerator yields

$$\frac{d}{dx}\left[\frac{f(x)}{g(x)}\right] = \lim_{h \to 0}\frac{f(x+h)\cdot g(x) - f(x)\cdot g(x) - f(x)\cdot g(x+h) + f(x)\cdot g(x)}{h\cdot g(x)\cdot g(x+h)}$$

$$= \lim_{h \to 0}\frac{\left[g(x)\cdot\dfrac{f(x+h) - f(x)}{h}\right] - \left[f(x)\cdot\dfrac{g(x+h) - g(x)}{h}\right]}{g(x)\cdot g(x+h)}$$

$$= \frac{\lim\limits_{h \to 0} g(x)\cdot\lim\limits_{h \to 0}\dfrac{f(x+h) - f(x)}{h} - \lim\limits_{h \to 0} f(x)\cdot\lim\limits_{h \to 0}\dfrac{g(x+h) - g(x)}{h}}{\lim\limits_{h \to 0} g(x)\cdot\lim\limits_{h \to 0} g(x+h)}$$

$$= \frac{[\lim\limits_{h \to 0} g(x)]\cdot\dfrac{d}{dx}[f(x)] - [\lim\limits_{h \to 0} f(x)]\cdot\dfrac{d}{dx}[g(x)]}{\lim\limits_{h \to 0} g(x)\cdot\lim\limits_{h \to 0} g(x+h)} \tag{4}$$

Since the expressions $f(x)$ and $g(x)$ do not involve h,

$$\lim_{h \to 0} g(x) = g(x) \quad \text{and} \quad \lim_{h \to 0} f(x) = f(x) \tag{5}$$

Because g is assumed to be differentiable at x, it is continuous at x, and so by Definition 2.7.11

$$\lim_{h \to 0} g(x + h) = g(x) \tag{6}$$

Substituting (5) and (6) into (4) yields

$$\frac{d}{dx}\left[\frac{f(x)}{g(x)}\right] = \frac{g(x)\dfrac{d}{dx}[f(x)] - f(x)\dfrac{d}{dx}[g(x)]}{[g(x)]^2} \qquad \blacksquare$$

The quotient rule can be written in function notation as

$$\left(\frac{f}{g}\right)' = \frac{g \cdot f' - f \cdot g'}{g^2}$$

REMARK.   In words, *the derivative of a quotient of two functions is the denominator times the derivative of the numerator minus the numerator times the derivative of the denominator, all divided by the denominator squared.*

WARNING.   Note that it is *not* generally true that $(f/g)' = f'/g'$; that is, the derivative of a quotient is *not* generally the quotient of the derivatives.

Example 7   Let $y = \dfrac{x^2 - 1}{x^4 + 1}$.

(a)   Find dy/dx.

(b)   At which points does the graph of the equation have a horizontal tangent line?

Solution (a).

$$\frac{dy}{dx} = \frac{d}{dx}\left[\frac{x^2 - 1}{x^4 + 1}\right] = \frac{(x^4 + 1)\dfrac{d}{dx}[x^2 - 1] - (x^2 - 1)\dfrac{d}{dx}[x^4 + 1]}{(x^4 + 1)^2}$$

$$= \frac{(x^4 + 1)(2x) - (x^2 - 1)(4x^3)}{(x^4 + 1)^2} \qquad \boxed{\begin{array}{l}\text{The differentiation is complete.}\\ \text{The rest is simplification.}\end{array}}$$

$$= \frac{-2x^5 + 4x^3 + 2x}{(x^4 + 1)^2} = -\frac{2x(x^4 - 2x^2 - 1)}{(x^4 + 1)^2}$$

Solution (b).   Since dy/dx can be interpreted as the slope of the tangent line to the graph, and since a horizontal tangent line has slope 0, the points where the tangent line is horizontal can be found by solving the equation $dy/dx = 0$, or from part (a)

$$-\frac{2x(x^4 - 2x^2 - 1)}{(x^4 + 1)^2} = 0$$

The solutions of this equation are the values of x for which the numerator is 0:

$$2x(x^4 - 2x^2 - 1) = 0$$

The first factor yields the solution $x = 0$. Other solutions can be found by solving the equation

$$x^4 - 2x^2 - 1 = 0$$

This can be treated as a quadratic equation in x^2 and solved by the quadratic formula. This yields

$$x^2 = \frac{2 \pm \sqrt{8}}{2} = 1 \pm \sqrt{2}$$

The minus sign yields imaginary values for x (why?); these solutions are not relevant to our problem, so we ignore them. The plus sign yields the solutions

$$x = \pm\sqrt{1 + \sqrt{2}}$$

In summary, horizontal tangent lines occur at

$$x = 0, \quad x = \sqrt{1 + \sqrt{2}} \approx 1.55, \quad \text{and} \quad x = -\sqrt{1 + \sqrt{2}} \approx -1.55$$

These results are consistent with the graph of the given equation, which we generated for Figure 3.3.1 using graphing software.   ◀

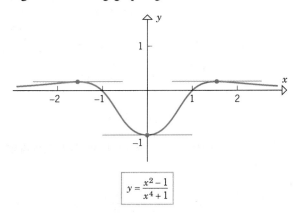

$$y = \frac{x^2 - 1}{x^4 + 1}$$

Figure 3.3.1

☐ **DERIVATIVE OF A RECIPROCAL**

The special case of Theorem 3.3.6 in which f is the constant function 1 is of interest in its own right. We leave it as an exercise for the reader to deduce the following result from Theorem 3.3.6.

3.3.7   THEOREM (*The Reciprocal Rule*).   *If g is differentiable at x and $g(x) \neq 0$, then $1/g$ is differentiable at x and*

$$\frac{d}{dx}\left[\frac{1}{g(x)}\right] = -\frac{\frac{d}{dx}[g(x)]}{[g(x)]^2}$$

The reciprocal rule can be written in function notation as

$$\left(\frac{1}{g}\right)' = -\frac{g'}{g^2}$$

REMARK.   In words, *the derivative of the reciprocal of a function is the negative of the derivative of the function divided by the function squared.*

Example 8

$$\frac{d}{dx}\left[\frac{1}{x}\right] = -\frac{\frac{d}{dx}[x]}{x^2} = -\frac{1}{x^2}$$

$$\frac{d}{dx}\left[\frac{1}{x^3 + 2x - 3}\right] = -\frac{\frac{d}{dx}[x^3 + 2x - 3]}{(x^3 + 2x - 3)^2} = -\frac{3x^2 + 2}{(x^3 + 2x - 3)^2}   ◀$$

REMARK.   The computations in the preceding example could have been done using the quotient rule, but this would have been more work. Where it applies, the reciprocal rule is preferable to the quotient rule.

☐ **THE POWER RULE FOR INTEGER EXPONENTS**

In Theorem 3.3.2 we established the formula

$$\frac{d}{dx}[x^n] = nx^{n-1}$$

for *positive* integer values of n. The following theorem extends this formula so that it applies to *all* integer values of n.

3.3.8   THEOREM.   *If n is any integer, then*

$$\frac{d}{dx}[x^n] = nx^{n-1} \tag{7}$$

Proof.   The result has already been established in the case where $n > 0$. If $n < 0$, then let $m = -n$ so that

$$f(x) = x^{-m} = \frac{1}{x^m}$$

From Theorem 3.3.7,

$$f'(x) = \frac{d}{dx}\left[\frac{1}{x^m}\right] = -\frac{\frac{d}{dx}[x^m]}{(x^m)^2}$$

Since $n < 0$, it follows that $m > 0$, so x^m can be differentiated using Theorem 3.3.2. Thus,

$$f'(x) = -\frac{mx^{m-1}}{x^{2m}} = -mx^{m-1-2m} = -mx^{-m-1} = nx^{n-1}$$

which proves (7). In the case $n = 0$ Formula (7) reduces to

$$\frac{d}{dx}[1] = 0 \cdot x^{-1} = 0$$

which is correct by Theorem 3.3.1.   ∎

Example 9

$$\frac{d}{dx}[x^{-9}] = -9x^{-9-1} = -9x^{-10}$$

$$\frac{d}{dx}\left[\frac{1}{x}\right] = \frac{d}{dx}[x^{-1}] = (-1)x^{-1-1} = -x^{-2} = -\frac{1}{x^2}$$

Note that the last result agrees with that obtained in Example 8.   ◀

In Example 4 of Section 3.2 we showed that

$$\frac{d}{dx}[\sqrt{x}] = \frac{1}{2\sqrt{x}} \tag{8}$$

If we write this result in exponential notation, we obtain

$$\frac{d}{dx}[x^{1/2}] = \frac{1}{2x^{1/2}} = \frac{1}{2}x^{-1/2}$$

which shows that (7) holds for the rational exponent $n = \frac{1}{2}$. Later, we shall show that (7) actually holds for all rational exponents.

☐ HIGHER DERIVATIVES

If the derivative f' of a function f is itself differentiable, then the derivative of f' is denoted by f'' and is called the *second derivative* of f. As long as we have differentiability, we can continue the process of differentiating derivatives to obtain third, fourth, fifth, and even higher derivatives of f. The successive derivatives of f are denoted by

$$f', \quad f'' = (f')', \quad f''' = (f'')', \quad f^{(4)} = (f''')', \quad f^{(5)} = (f^{(4)})', \ldots$$

These are called the first derivative, the second derivative, the third derivative, and so forth. Beyond the third derivative, it is too clumsy to continue using primes, so we switch from primes to integers in parentheses to denote the *order* of the derivative. In this notation it is easy to denote a derivative of arbitrary order by writing

$$f^{(n)} \quad \boxed{\text{The } n\text{th derivative of } f}$$

The significance of the derivatives of order 2 and higher will be discussed later.

Example 10   If $f(x) = 3x^4 - 2x^3 + x^2 - 4x + 2$, then

$$f'(x) \ = 12x^3 - 6x^2 + 2x - 4$$
$$f''(x) \ = 36x^2 - 12x + 2$$
$$f'''(x) \ = 72x - 12$$
$$f^{(4)}(x) = 72$$
$$f^{(5)}(x) = 0$$
$$\vdots$$
$$f^{(n)}(x) = 0 \quad (n \geq 5) \qquad \blacktriangleleft$$

Successive derivatives can also be denoted as follows:

$$f'(x) \ = \frac{d}{dx}[f(x)]$$

$$f''(x) \ = \frac{d}{dx}\left[\frac{d}{dx}[f(x)]\right] = \frac{d^2}{dx^2}[f(x)]$$

$$f'''(x) = \frac{d}{dx}\left[\frac{d^2}{dx^2}[f(x)]\right] = \frac{d^3}{dx^3}[f(x)]$$
$$\vdots \qquad\qquad\qquad \vdots$$

In general, we write

$$f^{(n)}(x) = \frac{d^n}{dx^n}[f(x)]$$

which is read, *the nth derivative of f with respect to x.*

When a dependent variable is involved, say $y = f(x)$, then successive derivatives can be denoted by writing

$$\frac{dy}{dx}, \quad \frac{d^2y}{dx^2}, \quad \frac{d^3y}{dx^3}, \quad \frac{d^4y}{dx^4}, \ldots, \frac{d^ny}{dx^n}, \ldots$$

or more briefly,

$$y', \quad y'', \quad y''', \quad y^{(4)}, \ldots, y^{(n)}, \ldots$$

The following symbols denote values of derivatives at a particular point x_0; their meanings should be self-evident.

$$y''(x_0), \quad f^{(4)}(x_0), \quad \left.\frac{d^3y}{dx^3}\right|_{x=x_0}, \quad \left.\frac{d^2}{dx^2}[x^7 - x]\right|_{x=x_0}$$

Example 11

$$\frac{d^2}{dx^2}[x^5] = \frac{d}{dx}\left[\frac{d}{dx}(x^5)\right] = \frac{d}{dx}[5x^4] = 20x^3$$

Thus,

$$\left.\frac{d^2}{dx^2}[x^5]\right|_{x=2} = 20 \cdot 8 = 160 \quad \blacktriangleleft$$

▶ Exercise Set 3.3

In Exercises 1–28, use the results of this section to find dy/dx.

1. $y = 4x^7$.

2. $y = -3x^{12}$.

3. $y = 3x^8 + 2x + 1$.

4. $y = \frac{1}{2}(x^4 + 7)$.

5. $y = \pi^3$.

6. $y = \sqrt{2}x + (1/\sqrt{2})$.

7. $y = -\frac{1}{3}(x^7 + 2x - 9)$.

8. $y = \frac{x^2 + 1}{5}$.

9. $y = ax^3 + bx^2 + cx + d$ (a, b, c, d constant).

10. $y = \frac{1}{a}\left(x^2 + \frac{1}{b}x + c\right)$ (a, b, c constant).

11. $y = -3x^{-8} + 2\sqrt{x}$.

12. $y = 7x^{-6} - 5\sqrt{x}$.

13. $y = x^{-3} + \frac{1}{x^7}$.

14. $y = \sqrt{x} + \frac{1}{x}$.

15. $y = (3x^2 + 6)(2x - \frac{1}{4})$.

16. $y = (2 - x - 3x^3)(7 + x^5)$.

17. $y = (x^3 + 7x^2 - 8)(2x^{-3} + x^{-4})$.

18. $y = \left(\frac{1}{x} + \frac{1}{x^2}\right)(3x^3 + 27)$.

19. $y = (3x^2 + 1)^2$.

20. $y = (x^5 + 2x)^2$.

21. $y = \frac{1}{5x - 3}$.

22. $y = \frac{3}{\sqrt{x} + 2}$.

23. $y = \frac{3x}{2x + 1}$.

24. $y = \frac{x^2 + 1}{3x}$.

25. $y = \frac{2x - 1}{x + 3}$.

26. $y = \frac{4x + 1}{x^2 - 5}$.

27. $y = \left(\frac{3x + 2}{x}\right)(x^{-5} + 1)$.

28. $y = (2x^7 - x^2)\left(\frac{x - 1}{x + 1}\right)$.

29. If $f(4) = 3$ and $f'(4) = -5$, find $g'(4)$.

   (a) $g(x) = \sqrt{x}\,f(x)$

   (b) $g(x) = \frac{f(x)}{x}$.

30. If $f(3) = -2$ and $f'(3) = 4$, find $g'(3)$.

   (a) $g(x) = 3x^2 - 5f(x)$

   (b) $g(x) = \frac{2x + 1}{f(x)}$.

31. If $f(2) = -3$, $f'(2) = 4$, $g(2) = 1$, and $g'(2) = -5$, find $F'(2)$.

   (a) $F(x) = 5f(x) + 2g(x)$

   (b) $F(x) = f(x) - 3g(x)$

   (c) $F(x) = f(x)g(x)$

   (d) $F(x) = f(x)/g(x)$.

32. If $f(-1) = 2$, $f'(-1) = 3$, $g(-1) = -5$, and $g'(-1) = 4$, find $F'(-1)$.

   (a) $F(x) = 2f(x) - g(x)$

   (b) $F(x) = 4f(x) + 5g(x)$

   (c) $F(x) = f(x)g(x)$

   (d) $F(x) = f(x)/g(x)$.

In Exercises 33–38, the functions involve independent variables other than x. Use the results in this section to find the indicated derivative.

33. Find $\frac{d}{dt}[16t^2]$.

34. $C = 2\pi r$; find $\frac{dC}{dr}$.

35. $V(r) = \pi r^3$; find $V'(r)$.

36. Find $\frac{d}{d\alpha}[2\alpha^{-1} + \alpha]$.

37. $s = \frac{t}{t^3 + 7}$; find $\frac{ds}{dt}$.

38. Find $\frac{d}{d\lambda}\left[\frac{\lambda\lambda_0 + \lambda^6}{2 - \lambda_0}\right]$ (λ_0 is constant).

39. Newton's law of gravitation states that the magnitude F of the force exerted by a point with mass M on a point with mass m is

$$F = \frac{GmM}{r^2}$$

where G is a constant and r is the distance between the bodies. Assuming that the points are moving, find a formula for the instantaneous rate of change of F with respect to r.

40. The volume of a sphere is $V = \frac{4}{3}\pi r^3$. Assuming that the radius is changing, find a formula for the instantaneous rate of change of V with respect to r.

41. Find d^2y/dx^2.

   (a) $y = 7x^3 - 5x^2 + x$ (b) $y = 12x^2 - 2x + 3$

   (c) $y = \dfrac{x+1}{x}$ (d) $y = (5x^2 - 3)(7x^3 + x)$.

42. Find y''.

   (a) $y = 4x^7 - 5x^3 + 2x$ (b) $y = 3x + 2$

   (c) $y = \dfrac{3x-2}{5x}$ (d) $y = (x^3 - 5)(2x + 3)$.

43. Find y'''.

   (a) $y = x^{-5} + x^5$ (b) $y = 1/x$

   (c) $y = ax^3 + bx + c$ $(a, b, c$ constant$)$.

44. Find $\dfrac{d^3y}{dx^3}$.

   (a) $y = 5x^2 - 4x + 7$ (b) $y = 3x^{-2} + 4x^{-1} + x$

   (c) $y = ax^4 + bx^2 + c$ $(a, b, c$ constant$)$.

45. Find

   (a) $f'''(2)$, where $f(x) = 3x^2 - 2$

   (b) $\dfrac{d^2y}{dx^2}\bigg|_{x=1}$, where $y = 6x^5 - 4x^2$

   (c) $\dfrac{d^4}{dx^4}[x^{-3}]\bigg|_{x=1}$

46. Find

   (a) $y'''(0)$, where $y = 4x^4 + 2x^3 + 3$

   (b) $\dfrac{d^4y}{dx^4}\bigg|_{x=1}$, where $y = \dfrac{6}{x^4}$.

47. Show that $y = x^3 + 3x + 1$ satisfies $y''' + xy'' - 2y' = 0$.

48. Show that if $x \neq 0$, then $y = 1/x$ satisfies the equation $x^3y'' + x^2y' - xy = 0$.

49. Find a general formula for $F''(x)$ if $F(x) = xf(x)$ and f and f' are differentiable at x.

50. In the temperature range between $0°\,C$ and $700°\,C$ the resistance R [in ohms (Ω)] of a certain platinum resistance thermometer is given by

$$R = 10 + 0.04124T - 1.779 \times 10^{-5}T^2$$

where T is the temperature in degrees Celsius. Where in the interval from $0°\,C$ to $700°\,C$ is the resistance of the thermometer most sensitive and least sensitive to temperature changes? [*Hint:* Consider the size of dR/dT in the interval $0 \leq T \leq 700$.]

51. At which point(s) does the graph of the equation $y = \frac{1}{3}x^3 - \frac{3}{2}x^2 + 2x$ have a horizontal tangent line?

52. At which point(s) does the graph of $y = \dfrac{x}{x^2 + 9}$ have a horizontal tangent line?

53. Find an equation of the tangent line to the graph of $y = f(x)$ at the point where $x = -3$ if $f(-3) = 2$ and $f'(-3) = 5$.

54. Find an equation for the line that is tangent to $y = (1 - x)/(1 + x)$ at the point where $x = 2$.

55. Find the values of a and b if the tangent to $y = ax^2 + bx$ at $(1, 5)$ has slope $m_{\tan} = 8$.

56. Find the values of a and b if the tangent to

$$y = \frac{a}{x^2} + b$$

at $(2, 4)$ has slope $m_{\tan} = -2$.

57. Find a function $y = ax^2 + bx + c$ whose graph has an x-intercept of 1, a y-intercept of -2, and a tangent line with a slope of -1 at the y-intercept.

58. Find k if the curve $y = x^2 + k$ is tangent to the line $y = 2x$.

59. Find the x-coordinate of the point on the graph of $y = x^2$ where the tangent line is parallel to the secant line that cuts the curve at $x = -1$ and $x = 2$.

60. Find the x-coordinate of the point on the graph of $y = \sqrt{x}$ where the tangent line is parallel to the secant line that cuts the curve at $x = 1$ and $x = 4$.

61. Find the x-coordinate of all points on the graph of $y = 1 - x^2$ at which the tangent line passes through the point $(2, 0)$.

62. Show that any two tangent lines to the parabola $y = ax^2$, $a \neq 0$, intersect at a point that is on the vertical line halfway between the points of tangency.

63. Suppose that L is the tangent line at $x = x_0$ to the graph of the cubic equation $y = ax^3 + bx$. Find the x-coordinate of the point where L intersects the graph a second time.

64. Show that the segment of the tangent line to the graph of $y = 1/x$ that is cut off by the coordinate axes is bisected by the point of tangency.

65. Show that the triangle that is formed by any tangent line to the graph of $y = 1/x$, $x > 0$, and the coordinate axes has an area of 2 square units.

66. Find conditions on a, b, c, and d so that the graph of the polynomial $f(x) = ax^3 + bx^2 + cx + d$ has

   (a) exactly two horizontal tangents

   (b) exactly one horizontal tangent

   (c) no horizontal tangents.

67. Prove: If the graphs of $y = f(x)$ and $y = g(x)$ have parallel tangent lines at $x = c$, then the graph of $y = f(x) - g(x)$ has a horizontal tangent line at $x = c$.

68. (a) Let the functions f, g, and h be differentiable at x. By applying Theorem 3.3.5 twice, show that the product $f \cdot g \cdot h$ is differentiable at x and

$$(f \cdot g \cdot h)'(x)$$
$$= f'(x)g(x)h(x) + f(x)g'(x)h(x) + f(x)g(x)h'(x)$$

(b) State a formula for differentiating a product of n functions.

69. Use the results of Exercise 68 to find

(a) $\dfrac{d}{dx}\left[(2x+1)\left(1+\dfrac{1}{x}\right)(x^{-3}+7)\right]$

(b) $\dfrac{d}{dx}[x^{-5}(x^2+2x)(4-3x)(2x^9+1)]$

(c) $\dfrac{d}{dx}[(x^7+2x-3)^3]$ (d) $\dfrac{d}{dx}[(x^2+1)^{50}]$.

70. Prove: If the function f is differentiable at x, then
$\dfrac{d}{dx}[f^2(x)] = 2f(x)f'(x)$. [*Hint:* Use the product rule.]

71. Use the result in Exercise 70 to find
$$\frac{d}{dx}(2x^3 - 5x^2 + 7x - 2)^2$$

72. Let $f(x) = \sqrt{x}$. Assuming that f is differentiable at x, use the result in Exercise 70 to show that $f'(x) = 1/(2\sqrt{x})$.

In Exercises 73–77, you will have to determine whether a function f is differentiable at a point x_0 where the formula for f changes. Use the following result:

Theorem. *Let f be continuous at x_0 and suppose that*
$$\lim_{x \to x_0^+} f'(x) \quad and \quad \lim_{x \to x_0^-} f'(x)$$
exist. Then f is differentiable at x_0 if and only if these limits are equal. Moreover, in the case of equality
$$f'(x_0) = \lim_{x \to x_0^+} f'(x) = \lim_{x \to x_0^-} f'(x)$$

---

73. Let
$$f(x) = \begin{cases} x^2, & x \le 1 \\ \sqrt{x}, & x > 1 \end{cases}$$
Determine whether f is differentiable at $x = 1$. If so, find the value of the derivative there.

74. Let
$$f(x) = \begin{cases} x^3 + \frac{1}{16}, & x < \frac{1}{2} \\ \frac{3}{4}x^2, & x \ge \frac{1}{2} \end{cases}$$
Determine whether f is differentiable at $x = \frac{1}{2}$. If so, find the value of the derivative there.

75. Let
$$f(x) = \begin{cases} 3x^2, & x \le 1 \\ ax + b, & x > 1 \end{cases}$$
Find the values of a and b so that f will be differentiable at $x = 1$.

76. (a) Let
$$f(x) = \begin{cases} x^2, & x \le 0 \\ x^2 + 1, & x > 0 \end{cases}$$

Show that
$$\lim_{x \to 0^-} f'(x) = \lim_{x \to 0^+} f'(x)$$
but that $f'(0)$ does not exist.

(b) Let
$$f(x) = \begin{cases} x^2, & x \le 0 \\ x^3, & x > 0 \end{cases}$$
Show that $f'(0)$ exists but $f''(0)$ does not.

77. Find all points where f fails to be differentiable. Justify your answer.

(a) $f(x) = |3x - 2|$ (b) $f(x) = |x^2 - 4|$.

78. Prove: If f is differentiable at x and $f(x) \ne 0$, then $1/f(x)$ is differentiable at x and
$$\frac{d}{dx}\left[\frac{1}{f(x)}\right] = -\frac{f'(x)}{[f(x)]^2}$$

79. (a) Find $f^{(n)}(x)$ if $f(x) = x^n$.

(b) Find $f^{(n)}(x)$ if $f(x) = x^k$ and $n > k$, where k is a positive integer.

(c) Find $f^{(n)}(x)$ if
$$f(x) = a_0 + a_1 x + a_2 x^2 + \cdots + a_n x^n$$

80. In each part compute f', f'', f''' and then state the formula for $f^{(n)}$.

(a) $f(x) = 1/x$ (b) $f(x) = 1/x^2$.

[*Hint:* The expression $(-1)^n$ has a value of 1 if n is even and -1 if n is odd. Use this expression in your answer.]

81. (a) Prove:
$$\frac{d^2}{dx^2}[cf(x)] = c\frac{d^2}{dx^2}[f(x)]$$
$$\frac{d^2}{dx^2}[f(x) + g(x)] = \frac{d^2}{dx^2}[f(x)] + \frac{d^2}{dx^2}[g(x)]$$

(b) Do the results in part (a) generalize to nth derivatives? Justify your answer.

82. Prove:
$$(f \cdot g)''(x) = f''(x)g(x) + 2f'(x)g'(x) + f(x)g''(x)$$

83. (a) In our proof of Theorem 3.3.5, we used the fact that $\lim_{h \to 0} f(x+h) = f(x)$ [see Equation (3)]. The argument given used the hypothesis that f is differentiable at x. Find the fallacy in the following proof that makes no assumptions about f: As h approaches 0, the quantity $x + h$ approaches x; consequently,
$$\lim_{h \to 0} f(x+h) = f(x)$$

(b) Let
$$f(x) = \begin{cases} x, & x \ne 1 \\ 3, & x = 1 \end{cases}$$

Show that $\lim\limits_{h \to 0} f(x + h) \neq f(x)$ when $x = 1$.

84. Let $f(x) = x^8 - 2x + 3$ and $x_0 = 2$; find

$$\lim_{h \to 0} \frac{f'(x_0 + h) - f'(x_0)}{h}$$

85. (a) Prove: If $f''(x)$ exists for each x in (a, b), then both f and f' are continuous on (a, b).

(b) What can be said about the continuity of f and its derivatives if $f^{(n)}(x)$ exists for each x in (a, b)?

■ **3.4** DERIVATIVES OF TRIGONOMETRIC FUNCTIONS

The main objective of this section is to obtain formulas for the derivatives of trigonometric functions.

For the purpose of finding derivatives of the trigonometric functions $\sin x$, $\cos x$, $\tan x$, $\cot x$, $\sec x$, and $\csc x$, we shall assume that x is measured in radians. Also, we remind the reader of the limits

$$\lim_{h \to 0} \frac{\sin h}{h} = 1 \quad \text{and} \quad \lim_{h \to 0} \frac{1 - \cos h}{h} = 0$$

derived in Section 2.8.

Let us first consider the problem of differentiating $\sin x$. Let x be any real number. From the definition of a derivative,

$$\frac{d}{dx}[\sin x] = \lim_{h \to 0} \frac{\sin(x + h) - \sin x}{h}$$

$$= \lim_{h \to 0} \frac{\sin x \cos h + \cos x \sin h - \sin x}{h}$$

$$= \lim_{h \to 0} \left[\sin x \left(\frac{\cos h - 1}{h} \right) + \cos x \left(\frac{\sin h}{h} \right) \right]$$

$$= \lim_{h \to 0} \left[\cos x \left(\frac{\sin h}{h} \right) - \sin x \left(\frac{1 - \cos h}{h} \right) \right]$$

Since $\sin x$ and $\cos x$ do not involve h, they remain constant as $h \to 0$; thus,

$$\lim_{h \to 0} (\sin x) = \sin x \quad \text{and} \quad \lim_{h \to 0} (\cos x) = \cos x$$

Consequently,

$$\frac{d}{dx}[\sin x] = \cos x \cdot \lim_{h \to 0} \left(\frac{\sin h}{h} \right) - \sin x \cdot \lim_{h \to 0} \left(\frac{1 - \cos h}{h} \right)$$

$$= \cos x \cdot (1) - \sin x \cdot (0) = \cos x$$

Thus, we have shown that

$$\frac{d}{dx}[\sin x] = \cos x \tag{1}$$

The derivative of $\cos x$ can be obtained similarly, resulting in the formula

$$\frac{d}{dx}[\cos x] = -\sin x \tag{2}$$

The derivatives of the remaining trigonometric functions can be obtained using the relationships

$$\tan x = \frac{\sin x}{\cos x}, \quad \cot x = \frac{\cos x}{\sin x}, \quad \sec x = \frac{1}{\cos x}, \quad \csc x = \frac{1}{\sin x}$$

For example,

$$\frac{d}{dx}[\tan x] = \frac{d}{dx}\left[\frac{\sin x}{\cos x}\right] = \frac{\cos x \cdot \frac{d}{dx}[\sin x] - \sin x \cdot \frac{d}{dx}[\cos x]}{\cos^2 x}$$

$$= \frac{\cos x \cdot \cos x - \sin x \cdot (-\sin x)}{\cos^2 x} = \frac{\cos^2 x + \sin^2 x}{\cos^2 x} = \frac{1}{\cos^2 x} = \sec^2 x$$

Thus,

$$\frac{d}{dx}[\tan x] = \sec^2 x \tag{3}$$

We leave the remaining formulas for the exercises:

$$\frac{d}{dx}[\cot x] = -\csc^2 x \tag{4}$$

$$\frac{d}{dx}[\sec x] = \sec x \tan x \tag{5}$$

$$\frac{d}{dx}[\csc x] = -\csc x \cot x \tag{6}$$

REMARK.   The derivative formulas for the trigonometric functions should be memorized. An easy way of doing this is discussed in Exercise 38. Moreover, we emphasize again that in all of the derivative formulas for the trigonometric functions, x is measured in radians.

Example 1   Find $f'(x)$ if $f(x) = x^2 \tan x$.

Solution.   Using the product rule and Formula (3), we obtain

$$f'(x) = x^2 \cdot \frac{d}{dx}[\tan x] + \tan x \cdot \frac{d}{dx}[x^2] = x^2 \sec^2 x + 2x \tan x \qquad \blacktriangleleft$$

Example 2   Find dy/dx if $y = \frac{\sin x}{1 + \cos x}$.

Solution.   Using the quotient rule together with Formulas (1) and (2) we obtain

$$\frac{dy}{dx} = \frac{(1 + \cos x) \cdot \frac{d}{dx}[\sin x] - \sin x \cdot \frac{d}{dx}[1 + \cos x]}{(1 + \cos x)^2}$$

$$= \frac{(1 + \cos x)(\cos x) - (\sin x)(-\sin x)}{(1 + \cos x)^2}$$

$$= \frac{\cos x + \cos^2 x + \sin^2 x}{(1 + \cos x)^2} = \frac{\cos x + 1}{(1 + \cos x)^2} = \frac{1}{1 + \cos x} \qquad \blacktriangleleft$$

Example 3   Find $y''(\pi/4)$ if $y(x) = \sec x$.

Solution.

$$y'(x) = \sec x \tan x$$

$$y''(x) = \sec x \cdot \frac{d}{dx}[\tan x] + \tan x \cdot \frac{d}{dx}[\sec x]$$

$$= \sec x \cdot \sec^2 x + \tan x \cdot \sec x \tan x$$

$$= \sec^3 x + \sec x \tan^2 x$$

Thus,

$$y''(\pi/4) = \sec^3(\pi/4) + \sec(\pi/4)\tan^2(\pi/4)$$

$$= (\sqrt{2})^3 + (\sqrt{2})(1)^2 = 3\sqrt{2} \quad \blacktriangleleft$$

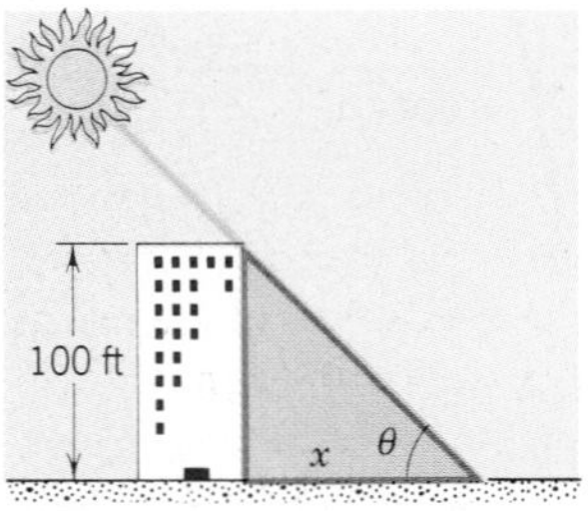

Example 4   Suppose that the rising sun passes directly over a building that is 100 feet high, and let θ be the sun's angle of elevation (Figure 3.4.1). Find the rate at which the length x of the building's shadow is changing with respect to θ when $\theta = 45°$. Express the answer in units of feet/degree.

Figure 3.4.1

Solution.   The variables x and θ are related by $\tan \theta = 100/x$, or equivalently,

$$x = 100 \cot \theta \tag{7}$$

If θ is measured in radians, then Formula (4) is applicable, and (7) yields

$$\frac{dx}{d\theta} = -100 \csc^2 \theta$$

which is the rate of change of shadow length with respect to the elevation angle θ in units of feet/radian. When $\theta = 45°$ (or equivalently, $\theta = \pi/4$ radians), we obtain

$$\left.\frac{dx}{d\theta}\right|_{\theta=\pi/4} = -100 \csc^2(\pi/4) = -200 \text{ feet/radian}$$

Converting radians to degrees yields

$$-200 \frac{\text{feet}}{\text{radian}} \cdot \frac{\pi}{180} \frac{\text{radians}}{\text{degree}} = -\frac{10}{9}\pi \approx -3.49 \text{ feet/degree}$$

Thus, when $\theta = 45°$, the shadow length is decreasing (because of the minus sign) at an approximate rate of 3.49 feet/degree increase in the angle of elevation. $\blacktriangleleft$

▶ Exercise Set 3.4   ⒸC 29–32

In Exercises 1–18, find $f'(x)$.

1. $f(x) = 2\cos x - 3\sin x$.   **2.** $f(x) = \sin x \cos x$.

3. $f(x) = \dfrac{\sin x}{x}$.   **4.** $f(x) = x^2 \cos x$.

5. $f(x) = x^3 \sin x - 5\cos x$.   **6.** $f(x) = \dfrac{\cos x}{x \sin x}$.

7. $f(x) = \sec x - \sqrt{2}\tan x$.   **8.** $f(x) = (x^2 + 1)\sec x$.

9. $f(x) = \sec x \tan x$.   **10.** $f(x) = \dfrac{\sec x}{1 + \tan x}$.

11. $f(x) = x - 4\csc x + 2\cot x$.

12. $f(x) = \csc x \cot x$.

13. $f(x) = \dfrac{\cot x}{1 + \csc x}$.   **14.** $f(x) = \dfrac{\csc x}{\tan x}$.

15. $f(x) = \sin^2 x + \cos^2 x$.   **16.** $f(x) = \dfrac{1}{\cot x}$.

17. $f(x) = \dfrac{\sin x \sec x}{1 + x \tan x}$.   **18.** $f(x) = \dfrac{(x^2 + 1)\cot x}{3 - \cos x \csc x}$.

In Exercises 19–23, find d^2y/dx^2.

19. $y = x\cos x$.   **20.** $y = \csc x$.

21. $y = x\sin x - 3\cos x$.   **22.** $y = x^2 \cos x + 4\sin x$.

23. $y = \sin x \cos x$.

In Exercises 24 and 25, find all points where the graph of f has a horizontal tangent line.

24. (a) $f(x) = \sin x$ (b) $f(x) = \tan x$
    (c) $f(x) = \sec x$.

25. (a) $f(x) = \cos x$ (b) $f(x) = \cot x$
    (c) $f(x) = \csc x$.

26. Find the equation of the line tangent to the graph of $\sin x$ at the point where
    (a) $x = 0$ (b) $x = \pi$ (c) $x = \pi/4$.

27. Find the equation of the line tangent to the graph of $\tan x$ at the point where
    (a) $x = 0$ (b) $x = \pi/4$ (c) $x = -\pi/4$.

28. (a) Show that $y = \cos x$ and $y = \sin x$ are solutions of the equation $y'' + y = 0$.
    (b) Show that $y = A \sin x + B \cos x$ is a solution for all constants A and B.

29. A 10-foot ladder leans against a wall at an angle θ with the horizontal, as shown in Figure 3.4.2. The top of the ladder is x feet above the ground. If the bottom of the ladder is pushed toward the wall, find the rate at which x changes with respect to θ when $\theta = 60°$. Express the answer in units of feet/degree.

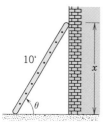

Figure 3.4.2

30. An airplane is flying on a horizontal path at a height of 3800 feet, as shown in Figure 3.4.3. At what rate is the distance s between the airplane and the fixed point P changing with respect to θ when $\theta = 30°$? Express the answer in units of feet/degree.

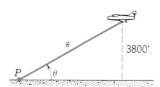

Figure 3.4.3

31. A searchlight is located 50 meters from a straight wall (Figure 3.4.4). Find the rate at which the distance D is changing with θ when $\theta = 45°$. Express the answer in units of meters/degree.

32. An earth-observing satellite can see only a portion of the earth's surface. The satellite has horizon sensors that can

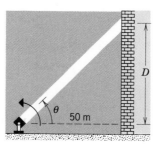

Figure 3.4.4

detect the angle θ shown in Figure 3.4.5. Let r be the radius of the earth (assumed spherical) and h the distance of the satellite from the earth's surface.

(a) Show that $h = r(\csc \theta - 1)$.

(b) Using $r = 6378$ kilometers, and assuming that the satellite is getting closer to the earth, find the rate at which h is changing with respect to θ when $\theta = 30°$. Express the answer in units of kilometers/degree. [Adapted from *Space Mathematics*, NASA, 1985.]

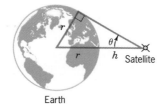

Earth   Figure 3.4.5

33. In each part, determine where f is differentiable.
    (a) $f(x) = \sin x$ (b) $f(x) = \cos x$
    (c) $f(x) = \tan x$ (d) $f(x) = \cot x$
    (e) $f(x) = \sec x$ (f) $f(x) = \csc x$
    (g) $f(x) = \dfrac{1}{1 + \cos x}$ (h) $f(x) = \dfrac{1}{\sin x \cos x}$
    (i) $f(x) = \dfrac{\cos x}{2 - \sin x}$.

34. (a) Derive Formula (2) using the definition of a derivative.
    (b) Use Formulas (1) and (2) to obtain (4).
    (c) Use Formula (2) to obtain (5).
    (d) Use Formula (1) to obtain (6).

35. Let $f(x) = \cos x$. Find all positive integers n for which $f^{(n)}(x) = \sin x$.

36. (a) Show that $\displaystyle\lim_{h \to 0} \frac{\tan h}{h} = 1$.
    (b) Use the result in part (a) to help derive the formula for the derivative of $\tan x$ directly from the definition of a derivative.

37. Without using any trigonometric identities, find
$$\lim_{x \to 0} \frac{\tan (x + y) - \tan y}{x}$$
[*Hint:* Relate the given limit to the definition of the derivative of an appropriate function of y.]

38. Let us agree to call the functions cos x, cot x, and csc x the *cofunctions* of sin x, tan x, and sec x, respectively. Convince yourself that the derivative of any cofunction can be obtained from the derivative of the corresponding function by introducing a minus sign and replacing each function in the derivative by its cofunction. Memorize the derivatives of sin x, tan x, and sec x and then use the above observation to deduce the derivatives of the cofunctions.

39. The derivative formulas for sin x, cos x, tan x, cot x, sec x, and csc x were obtained under the assumption that x is measured in radians. This exercise shows that different (more complicated) formulas result if x is measured in degrees. Prove that if h and x are degree measures, then

(a) $\displaystyle\lim_{h \to 0} \frac{\cos h - 1}{h} = 0$ (b) $\displaystyle\lim_{h \to 0} \frac{\sin h}{h} = \frac{\pi}{180}$

(c) $\displaystyle\frac{d}{dx}[\sin x] = \frac{\pi}{180}\cos x.$

■ 3.5 THE CHAIN RULE

> *In this section we shall derive a formula that expresses the derivative of a composition $f \circ g$ in terms of the derivatives of f and g. This formula will enable us to differentiate complicated functions using known derivatives of simpler functions.*

☐ **DERIVATIVES OF COMPOSITIONS**

3.5.1 PROBLEM. *If we know the derivatives of f and g, how can we use this information to find the derivative of the composition $f \circ g$?*

The key to solving this problem is to introduce dependent variables

$$y = (f \circ g)(x) = f(g(x)) \quad \text{and} \quad u = g(x)$$

so that $y = f(u)$. We are interested in using the known derivatives

$$\frac{dy}{du} = f'(u) \quad \text{and} \quad \frac{du}{dx} = g'(x)$$

to find the unknown derivative

$$\frac{dy}{dx} = \frac{d}{dx}[f(g(x))]$$

Stated another way, we are interested in using the known rates of change dy/du and du/dx to find the unknown rate of change dy/dx. But intuition suggests that rates of change multiply. For example, if y changes at 4 times the rate of change of u and u changes at 2 times the rate of change of x, then y changes at $4 \times 2 = 8$ times the rate of change of x. Thus, Figure 3.5.1 suggests that

$$\frac{dy}{dx} = \frac{dy}{du} \cdot \frac{du}{dx}$$

These ideas are formalized in the following theorem.

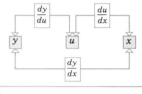

Rates of change multiply:
$$\frac{dy}{dx} = \frac{dy}{du} \cdot \frac{du}{dx}$$

Figure 3.5.1

3.5.2 THEOREM (*The Chain Rule*). *If g is differentiable at the point x and f is differentiable at the point $g(x)$, then the composition $f \circ g$ is differentiable at the point x. Moreover, if*

$$y = f(g(x)) \quad \text{and} \quad u = g(x)$$

then $y = f(u)$ and

$$\frac{dy}{dx} = \frac{dy}{du} \cdot \frac{du}{dx} \tag{1}$$

The proof of this result is given in Section III of Appendix C.

Example 1 Find dy/dx if $y = 4\cos(x^3)$.

Solution. Let $u = x^3$ so that

$$y = 4\cos u$$

By the chain rule,

$$\frac{dy}{dx} = \frac{dy}{du} \cdot \frac{du}{dx} = \frac{d}{du}[4\cos u] \cdot \frac{d}{dx}[x^3]$$

$$= (-4\sin u) \cdot (3x^2) = (-4\sin(x^3)) \cdot (3x^2) = -12x^2\sin(x^3) \quad \blacktriangleleft$$

REMARK. Formula (1) is easy to remember because the left side is exactly what results if we "cancel" the du's on the right side. This "canceling" device provides a good way to remember the chain rule when variables other than x, y, and u are used.

Example 2 Find dw/dt if $w = \tan x$ and $x = 4t^3 + t$.

Solution. In this case, the chain rule takes the form

$$\frac{dw}{dt} = \frac{dw}{dx} \cdot \frac{dx}{dt} = \frac{d}{dx}[\tan x] \cdot \frac{d}{dt}[4t^3 + t]$$

$$= (\sec^2 x)(12t^2 + 1) = (12t^2 + 1)\sec^2(4t^3 + t) \quad \blacktriangleleft$$

☐ **GENERALIZED DERIVATIVE FORMULAS**

Although Formula (1) is useful, it is sometimes unwieldy because it involves so many dependent variables. A simpler version of the chain rule can be obtained by noting that $y = f(u)$ in (1), so

$$\frac{dy}{dx} = \frac{d}{dx}[f(u)] \quad \text{and} \quad \frac{dy}{du} = f'(u)$$

Substituting these expressions in (1) yields the following alternative form of the chain rule.

$$\frac{d}{dx}[f(u)] = f'(u)\frac{du}{dx} \tag{2}$$

This very powerful formula vastly extends our differentiation capabilities. For example, to differentiate the function

$$f(x) = (x^2 - x + 1)^{23} \tag{3}$$

we can let $u = x^2 - x + 1$, so (3) becomes $f(u) = u^{23}$, then apply (2) to obtain

$$\frac{d}{dx}[(x^2 - x + 1)^{23}] = \frac{d}{dx}[u^{23}] = \underbrace{23u^{22}}_{f'(u)}\frac{du}{dx}$$

$$= 23(x^2 - x + 1)^{22}\frac{d}{dx}[x^2 - x + 1]$$

$$= 23(x^2 - x + 1)^{22} \cdot (2x - 1)$$

More generally, if u were any other differentiable function of x, the pattern of computations would be virtually the same. For example, if $u = \cos x$, then

$$\frac{d}{dx}[\cos^{23} x] = \frac{d}{dx}[u^{23}] = 23u^{22}\frac{du}{dx} = 23\cos^{22} x\frac{d}{dx}[\cos x]$$

$$= 23\cos^{22} x \cdot (-\sin x) = -23\sin x\cos^{22} x$$

□ GENERALIZED
DERIVATIVE FORMULAS

In both of the preceding computations, the chain rule took the form

$$\frac{d}{dx}[u^{23}] = 23u^{22}\frac{du}{dx} \tag{4}$$

This formula is a generalization of the more basic formula

$$\frac{d}{dx}[x^{23}] = 23x^{22} \tag{5}$$

In fact, in the special case where $u = x$, Formula (4) reduces to (5) since

$$\frac{d}{dx}[u^{23}] = 23u^{22}\frac{du}{dx} = 23x^{22}\frac{d[x]}{dx} = 23x^{22}$$

Table 3.5.1 contains a list of **generalized derivative formulas** that are consequences of (2).

Table 3.5.1

GENERALIZED DERIVATIVE FORMULAS

$\dfrac{d}{dx}[u^n] = nu^{n-1}\dfrac{du}{dx}$ (n an integer)	$\dfrac{d}{dx}[\sqrt{u}] = \dfrac{1}{2\sqrt{u}}\dfrac{du}{dx}$
$\dfrac{d}{dx}[\sin u] = \cos u\,\dfrac{du}{dx}$	$\dfrac{d}{dx}[\cos u] = -\sin u\,\dfrac{du}{dx}$
$\dfrac{d}{dx}[\tan u] = \sec^2 u\,\dfrac{du}{dx}$	$\dfrac{d}{dx}[\cot u] = -\csc^2 u\,\dfrac{du}{dx}$
$\dfrac{d}{dx}[\sec u] = \sec u \tan u\,\dfrac{du}{dx}$	$\dfrac{d}{dx}[\csc u] = -\csc u \cot u\,\dfrac{du}{dx}$

Example 3   Find

(a) $\dfrac{d}{dx}[\sin(2x)]$ (b) $\dfrac{d}{dx}[\tan(x^2 + 1)]$

Solution (a). Taking $u = 2x$ in the generalized derivative formula for $\sin u$ yields

$$\frac{d}{dx}[\sin(2x)] = \frac{d}{dx}[\sin u] = \cos u\,\frac{du}{dx} = \cos 2x \cdot \frac{d}{dx}[2x]$$

$$= \cos 2x \cdot 2 = 2\cos 2x$$

Solution (b). Taking $u = x^2 + 1$ in the generalized derivative formula for $\tan u$ yields

$$\frac{d}{dx}[\tan(x^2 + 1)] = \frac{d}{dx}[\tan u] = \sec^2 u\,\frac{du}{dx}$$

$$= \sec^2(x^2 + 1)\cdot\frac{d}{dx}[x^2 + 1] = \sec^2(x^2 + 1)\cdot 2x$$

$$= 2x\sec^2(x^2 + 1) \qquad ◀$$

Example 4   Find

(a) $\dfrac{d}{dx}[\sqrt{x^3 + \csc x}]$ (b) $\dfrac{d}{dx}[(1 + x^5\cot x)^{-8}]$

Solution (a).   Taking $u = x^3 + \csc x$ in the generalized derivative formula for $\sqrt{u}$ yields

$$\frac{d}{dx}[\sqrt{x^3 + \csc x}] = \frac{d}{dx}[\sqrt{u}] = \frac{1}{2\sqrt{u}}\frac{du}{dx} = \frac{1}{2\sqrt{x^3 + \csc x}} \cdot \frac{d}{dx}[x^3 + \csc x]$$

$$= \frac{1}{2\sqrt{x^3 + \csc x}} \cdot (3x^2 - \csc x \cot x) = \frac{3x^2 - \csc x \cot x}{2\sqrt{x^3 + \csc x}}$$

Solution (b).   Taking $u = 1 + x^5 \cot x$ in the generalized derivative formula for u^{-8} yields

$$\frac{d}{dx}[(1 + x^5 \cot x)^{-8}] = \frac{d}{dx}[u^{-8}] = -8u^{-9}\frac{du}{dx}$$

$$= -8(1 + x^5 \cot x)^{-9} \cdot \frac{d}{dx}[1 + x^5 \cot x]$$

$$= -8(1 + x^5 \cot x)^{-9} \cdot (x^5(-\csc^2 x) + 5x^4 \cot x)$$

$$= (8x^5 \csc^2 x - 40x^4 \cot x)(1 + x^5 \cot x)^{-9} \qquad \blacktriangleleft$$

As the following example shows, it is sometimes necessary to apply the chain rule more than once to find a derivative.

Example 5   Find

(a) $\dfrac{d}{dx}[\cos^2 \pi x]$ (b) $\dfrac{d}{dx}[\sin(\sqrt{1 + \cos x})]$

Solution (a).   Taking $u = \cos \pi x$ in the generalized derivative formula for u^2 yields

$$\frac{d}{dx}[\cos^2 \pi x] = \frac{d}{dx}[u^2] = 2u\frac{du}{dx}$$

$$= 2\cos \pi x \cdot \frac{d}{dx}[\cos \pi x]$$

$$= 2\cos \pi x \cdot (-\pi \sin \pi x) \qquad \boxed{\text{We used the generalized derivative formula for } \cos u \text{ with } u = \pi x.}$$

$$= -2\pi \sin \pi x \cos \pi x$$

Solution (b).   Taking $u = \sqrt{1 + \cos x}$ in the generalized derivative formula for $\sin u$ yields

$$\frac{d}{dx}[\sin(\sqrt{1 + \cos x})] = \frac{d}{dx}[\sin u] = \cos u\frac{du}{dx}$$

$$= \cos(\sqrt{1 + \cos x}) \cdot \frac{d}{dx}[\sqrt{1 + \cos x}] \qquad \boxed{\text{We use the generalized derivative formula for } \sqrt{u} \text{ with } u = 1 + \cos x.}$$

$$= \cos(\sqrt{1 + \cos x}) \cdot \frac{-\sin x}{2\sqrt{1 + \cos x}}$$

$$= -\frac{\sin x \cos(\sqrt{1 + \cos x})}{2\sqrt{1 + \cos x}} \qquad \blacktriangleleft$$

Example 6   Find $d\mu/dt$ if $\mu = t^2 \sec \sqrt{\omega t}$, where ω is a constant.

Solution.   Because the independent variable is t rather than x, appropriate adjustments in notation have to be made.

$$\frac{d\mu}{dt} = \frac{d}{dt}[t^2 \sec \sqrt{\omega t}] = t^2 \frac{d}{dt}[\sec \sqrt{\omega t}] + (\sec \sqrt{\omega t}) \cdot 2t$$

$$= t^2 \sec \sqrt{\omega t} \tan \sqrt{\omega t} \frac{d}{dt}[\sqrt{\omega t}] + 2t \sec \sqrt{\omega t}$$

> We used the generalized derivative formula for $\sec u$ with $u = \sqrt{\omega t}$.

$$= t^2 \sec \sqrt{\omega t} \tan \sqrt{\omega t} \frac{\omega}{2\sqrt{\omega t}} + 2t \sec \sqrt{\omega t} \quad \blacktriangleleft$$

> We used the generalized derivative formula for $\sqrt{u}$ with $u = \omega t$.

☐ **AN ALTERNATIVE APPROACH TO USING THE CHAIN RULE**

As you become more comfortable with using the chain rule, you may want to dispense with actually writing out the expression for u in your computations. To accomplish this, it is helpful to express Formula (2) in words. If we call u the "inside function" and f the "outside function" in the composition $f(u)$, then (2) states:

The derivative of $f(u)$ is the derivative of the outside function evaluated at the inside function times the derivative of the inside function.

For example,

$$\frac{d}{dx}[\cos (x^2 + 9)] = \underbrace{-\sin (x^2 + 9)}_{\substack{\text{Derivative of} \\ \text{the outside} \\ \text{evaluated at} \\ \text{the inside}}} \cdot \underbrace{2x}_{\substack{\text{Derivative} \\ \text{of the inside}}}$$

$$\frac{d}{dx}[\tan^2 x] = \frac{d}{dx}[(\tan x)^2] = \underbrace{(2 \tan x)}_{\substack{\text{Derivative of} \\ \text{the outside} \\ \text{evaluated at} \\ \text{the inside}}} \cdot \underbrace{(\sec^2 x)}_{\substack{\text{Derivative} \\ \text{of the inside}}} = 2 \tan x \sec^2 x$$

In general, if $f(g(x))$ is a composition of functions in which the inside function g and the outside function f are differentiable, then

$$\frac{d}{dx}[f(g(x))] = \underbrace{f'(g(x))}_{\substack{\text{Derivative of} \\ \text{the outside} \\ \text{evaluated at} \\ \text{the inside}}} \cdot \underbrace{g'(x)}_{\substack{\text{Derivative} \\ \text{of the inside}}} \qquad (6)$$

▶ Exercise Set 3.5   Ⓒ 53, 54

In Exercises 1–39, find $f'(x)$.

1. $f(x) = (x^3 + 2x)^{37}$.

2. $f(x) = (3x^2 + 2x - 1)^6$.

3. $\left(\dfrac{7}{x}\right)^{-2}$.

4. $f(x) = \dfrac{1}{(x^5 - x + 1)^9}$.

5. $\dfrac{4}{- 2x + 1)^3}$.

6. $f(x) = \sqrt{x^3 - 2x + 5}$.

7. $+ 3\sqrt{x}$.

8. $f(x) = \sin^3 x$.

9. $^3)$.

10. $f(x) = \cos^2 (3\sqrt{x})$.

11. $^2)$.

12. $f(x) = 3 \cot^4 x$.

13. $f(x) = 4 \cos^5 x$.

14. $f(x) = \csc (x^3)$.

15. $f(x) = \sin \left(\dfrac{1}{x^2}\right)$.

16. $f(x) = \tan^4 (x^3)$.

17. $f(x) = 2 \sec^2 (x^7)$.

18. $f(x) = \cos^3 \left(\dfrac{x}{x + 1}\right)$.

19. $f(x) = \sqrt{\cos (5x)}$.

20. $f(x) = \sqrt{3x - \sin^2 (4x)}$.

21. $f(x) = [x + \csc (x^3 + 3)]^{-3}$.

22. $f(x) = [x^4 - \sec (4x^2 - 2)]^{-4}$.

23. $f(x) = x^2 \sqrt{5 - x^2}$.

24. $f(x) = \dfrac{x}{\sqrt{1 - x^2}}$.

25. $f(x) = x^3 \sin^2 (5x)$.

26. $f(x) = \sqrt{x} \tan^3 (\sqrt{x})$.

27. $f(x) = x^5 \sec (1/x)$.

28. $f(x) = \dfrac{\sin x}{\sec (3x + 1)}$.

29. $f(x) = \cos (\cos x)$.

30. $f(x) = \sin (\tan 3x)$.

31. $f(x) = \cos^3 (\sin 2x)$.

32. $f(x) = \dfrac{1 + \csc (x^2)}{1 - \cot (x^2)}$.

33. $f(x) = (5x + 8)^{13}(x^3 + 7x)^{12}$.

34. $f(x) = (2x - 5)^2(x^2 + 4)^3$.

35. $f(x) = \left(\dfrac{x - 5}{2x + 1}\right)^3$.

36. $f(x) = \left(\dfrac{1 + x^2}{1 - x^2}\right)^{17}$.

37. $f(x) = \dfrac{(2x + 3)^3}{(4x^2 - 1)^8}$.

38. $f(x) = [1 + \sin^3 (x^5)]^{12}$.

39. $f(x) = [x \sin 2x + \tan^4 (x^7)]^5$.

In Exercises 40–42, find d^2y/dx^2.

40. $y = \sin (3x^2)$.

41. $y = x \cos (5x) - \sin^2 x$.

42. $y = x \tan \left(\dfrac{1}{x}\right)$.

In Exercises 43–46, find an equation for the tangent line to the graph at the specified point.

43. $y = x \cos 3x$, $x = \pi$.

44. $y = \sin (1 + x^3)$, $x = -3$.

45. $y = \sec^3 \left(\dfrac{\pi}{2} - x\right)$, $x = -\dfrac{\pi}{2}$.

46. $y = \left(x - \dfrac{1}{x}\right)^3$, $x = 2$.

In Exercises 47–50, find the indicated derivative.

47. $y = \cot^3 (\pi - \theta)$; find $\dfrac{dy}{d\theta}$.

48. $\lambda = \left(\dfrac{au + b}{cu + d}\right)^6$; find $\dfrac{d\lambda}{du}$ (a, b, c, d constants).

49. $\dfrac{d}{d\omega} [a \cos^2 \pi\omega + b \sin^2 \pi\omega]$ (a, b constants).

50. $x = \csc^2 \left(\dfrac{\pi}{3} - y\right)$; find $\dfrac{dx}{dy}$.

51. If an object suspended from a spring is displaced vertically from its equilibrium position by a small amount and released, and if the air resistance and the mass of the spring are ignored, then the resulting oscillation of the object is called *simple harmonic motion*. For such motion the displacement y from equilibrium in terms of time t is given by

$$y = A \cos \omega t$$

where A is the initial displacement at time $t = 0$, and ω is a constant that depends on the mass of the object and the stiffness of the spring (Figure 3.5.2). The constant $|A|$ is called the *amplitude* of the motion and ω the *angular frequency*.

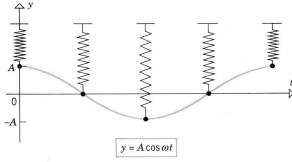

Figure 3.5.2

(a) Show that

$$\dfrac{d^2y}{dt^2} = -\omega^2 y$$

(b) The *period* T is the time required to make one complete oscillation. Show that $T = 2\pi/\omega$.

(c) The *frequency* f of the vibration is the number of oscillations per unit time. Find f in terms of the period T.

(d) Find the amplitude, period, and frequency of an object that is executing simple harmonic motion given by

$$y = 0.6 \cos 15t$$

where t is in seconds and y is in centimeters.

52. Find the value of the constant A so that $y = A \sin 3t$ satisfies the equation

$$\dfrac{d^2y}{dt^2} + 2y = 4 \sin 3t$$

53. Figure 3.5.3 shows the graph of atmospheric pressure p (lb/in²) versus the altitude h (mi) above sea level.

(a) From the graph and the tangent line at $h = 2$ shown on the graph, estimate the values of p and dp/dh at an altitude of 2 mi.

(b) If the altitude of a space vehicle is increasing at the rate of 0.3 mi/sec at the instant when it is 2 mi above sea level, how fast is the pressure changing with time at this instant?

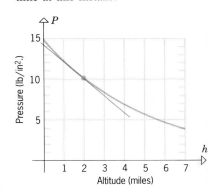

Figure 3.5.3

54. The force F (in pounds) acting at an angle θ with the horizontal that is needed to drag a crate weighing W lb along a horizontal surface at a constant velocity is given by

$$F = \frac{\mu W}{\cos \theta + \mu \sin \theta}$$

where μ is a constant called the **coefficient of sliding friction** between the crate and the surface (Figure 3.5.4). Suppose that the crate weighs 150 lb and that $\mu = 0.3$.

(a)   Find $dF/d\theta$ when $\theta = 30°$. Express the answer in units of pounds/degree.

(b)   Find dF/dt when $\theta = 30°$ if θ is decreasing at the rate of 0.5°/sec at this instant.

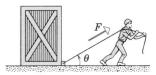

Figure 3.5.4

In Exercises 55–58, find the indicated derivative in terms of x, y, and dy/dx, assuming that y is a differentiable function of x.

55. $\dfrac{d}{dx}(x^2 y^3)$.

56. $\dfrac{d}{dx}\left(\dfrac{x}{y^2}\right)$.

57. $\dfrac{d}{dx}[\sin(xy)]$.

58. $\dfrac{d}{dx}(\sqrt{x^2 + y^2})$.

In Exercises 59–62, find the indicated derivative in terms of x, y, dx/dt, and dy/dt, assuming that x and y are differentiable functions of t.

59. $\dfrac{d}{dt}(x^2 + y^2)$.

60. $\dfrac{d}{dt}[\tan(xy^3)]$.

61. $\dfrac{d}{dt}(x^2 \sqrt{y})$.

62. $\dfrac{d}{dt}\left(\dfrac{y^2}{x}\right)$.

63. Recall that

$$\frac{d}{dx}(|x|) = \begin{cases} 1, & x > 0 \\ -1, & x < 0 \end{cases}$$

Use this result and the chain rule to find

$$\frac{d}{dx}(|\sin x|)$$

for nonzero x in the interval $(-\pi, \pi)$.

64. Use the derivative formula for $\sin x$ and the identity

$$\cos x = \sin\left(\frac{\pi}{2} - x\right)$$

to obtain the derivative formula for $\cos x$.

65. Let

$$f(x) = \begin{cases} x \sin \dfrac{1}{x}, & x \neq 0 \\ 0, & x = 0 \end{cases}$$

(a)   Find $f'(x)$ for $x \neq 0$.

(b)   Show that f is continuous at $x = 0$.

(c)   Use Definition 3.2.2 to show that $f'(0)$ does not exist.

66. Let

$$f(x) = \begin{cases} x^2 \sin \dfrac{1}{x}, & x \neq 0 \\ 0, & x = 0 \end{cases}$$

(a)   Find $f'(x)$ for $x \neq 0$.

(b)   Show that f is continuous at $x = 0$.

(c)   Use Definition 3.2.2 to find $f'(0)$.

(d)   Show that f' is not continuous at $x = 0$.

67. Given the following table of values, find the indicated derivatives in parts (a) and (b).

x	$f(x)$	$f'(x)$
2	1	7
8	5	-3

(a) $g'(2)$, where $g(x) = [f(x)]^3$

(b) $h'(2)$, where $h(x) = f(x^3)$.

68. Given the following table of values, find the indicated derivatives in parts (a) and (b).

x	$f(x)$	$f'(x)$	$g(x)$	$g'(x)$
-1	2	3	2	-3
2	0	4	1	-5

(a) $F'(-1)$, where $F(x) = f(g(x))$

(b) $G'(-1)$, where $G(x) = g(f(x))$.

69. Given that $f'(0) = 2$, $g(0) = 0$, and $g'(0) = 3$, find $(f \circ g)'(0)$.

70. Given that $f'(x) = \sqrt{3x + 4}$ and $g(x) = x^2 - 1$, find $F'(x)$ if $F(x) = f(g(x))$.

71. Given that $f'(x) = \dfrac{x}{x^2 + 1}$ and $g(x) = \sqrt{3x - 1}$, find $F'(x)$ if $F(x) = f(g(x))$.

72. Find $f'(x^2)$ if $\dfrac{d}{dx}[f(x^2)] = x^2$.

73. Find $\dfrac{d}{dx}[f(x)]$ if $\dfrac{d}{dx}[f(3x)] = 6x$.

74. A function f is said to be **even** if $f(-x) = f(x)$ and **odd** if $f(-x) = -f(x)$, for all x in the domain of f. Assuming that f is differentiable, prove:

(a) f' is odd if f is even

(b) f' is even if f is odd.

75. Find a formula for

$$\frac{d}{dx}\left[f\bigl(g(h(x))\bigr)\right]$$

76. Let $y = f_1(u)$, $u = f_2(v)$, $v = f_3(w)$, and $w = f_4(x)$. Express dy/dx in terms of dy/du, dw/dx, du/dv, and dv/dw.

■ 3.6 IMPLICIT DIFFERENTIATION

> *In the preceding sections of this chapter we showed how to differentiate functions defined explicitly by equations of the form $y = f(x)$. In this section we shall learn to find derivatives of functions that are defined implicitly.*

☐ **THE METHOD OF IMPLICIT DIFFERENTIATION**

Consider the equation

$$xy = 1 \tag{1}$$

One way to obtain dy/dx is to rewrite this equation as

$$y = \frac{1}{x} \tag{2}$$

from which it follows that

$$\frac{dy}{dx} = \frac{d}{dx}\left[\frac{1}{x}\right] = -\frac{1}{x^2}$$

However, there is another possibility. We can differentiate both sides of (1) *before* solving for y in terms of x, treating y as a (temporarily unspecified) differentiable function of x. With this approach we obtain

$$\frac{d}{dx}[xy] = \frac{d}{dx}[1]$$

$$x\frac{d}{dx}[y] + y\frac{d}{dx}[x] = 0$$

$$x\frac{dy}{dx} + y = 0$$

$$\frac{dy}{dx} = -\frac{y}{x}$$

If we now substitute (2) into the last expression, we obtain

$$\frac{dy}{dx} = -\frac{1}{x^2}$$

which agrees with the previous computation. This second method of obtaining derivatives is called *implicit differentiation*. It is especially useful when it is inconvenient or impossible to solve explicitly for y in terms of x.

Example 1 By implicit differentiation find dy/dx if $5y^2 + \sin y = x^2$.

Solution.   Differentiating both sides with respect to x and treating y as a differentiable function of x, we obtain

$$\frac{d}{dx}[5y^2 + \sin y] = \frac{d}{dx}[x^2]$$

$$5\frac{d}{dx}[y^2] + \frac{d}{dx}[\sin y] = 2x$$

$$5\left(2y\frac{dy}{dx}\right) + (\cos y)\frac{dy}{dx} = 2x \qquad \boxed{\begin{array}{l}\text{The chain rule was}\\ \text{used here because}\\ y \text{ is a function of } x.\end{array}}$$

$$10y\frac{dy}{dx} + (\cos y)\frac{dy}{dx} = 2x$$

Solving for dy/dx, we obtain

$$\frac{dy}{dx} = \frac{2x}{10y + \cos y} \qquad\qquad (3)$$

Note that this formula for dy/dx involves both the variables x and y. In order to obtain a formula involving x alone we would have to solve the original equation for y in terms of x and substitute in (3). However, it is impossible to do this, so the formula for dy/dx must be left in terms of x and y.   ◄

In problems where dy/dx is used to calculate the slope of a tangent line at a point on a curve, the x and y coordinates of the point are often both known, so that there is no problem using a formula for dy/dx that involves both x and y.

Example 2   Find the slope of the tangent line at the point $(4, 0)$ on the graph of

$$7y^4 + x^3y + x = 4 \qquad\qquad (4)$$

Solution.   It is difficult to solve (4) for y in terms of x, so we shall differentiate implicitly. We obtain

$$\frac{d}{dx}[7y^4 + x^3y + x] = \frac{d}{dx}[4]$$

$$\frac{d}{dx}[7y^4] + \frac{d}{dx}[x^3y] + \frac{d}{dx}[x] = 0$$

$$\frac{d}{dx}[7y^4] + \left(x^3\frac{dy}{dx} + y\frac{d}{dx}[x^3]\right) + \frac{d}{dx}[x] = 0$$

$$28y^3\frac{dy}{dx} + x^3\frac{dy}{dx} + 3yx^2 + 1 = 0 \qquad \boxed{\begin{array}{l}\text{Note the use of}\\ \text{the chain rule in}\\ \text{the first term.}\end{array}}$$

Solving for dy/dx yields

$$\frac{dy}{dx} = -\frac{3yx^2 + 1}{28y^3 + x^3} \qquad\qquad (5)$$

At the point $(4, 0)$ we have $x = 4$ and $y = 0$, so (5) yields

$$m_{\text{tan}} = \frac{dy}{dx}\bigg|_{\substack{x=4\\y=0}} = -\frac{1}{64} \qquad ◄$$

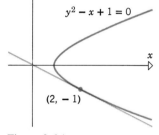

Example 3   Find the slope of the tangent line at $(2, -1)$ to $y^2 - x + 1 = 0$ (Figure 3.6.1).

Figure 3.6.1

Solution.   Differentiating implicitly yields

$$\frac{d}{dx}[y^2 - x + 1] = \frac{d}{dx}[0]$$

$$\frac{d}{dx}[y^2] - \frac{d}{dx}[x] + \frac{d}{dx}[1] = \frac{d}{dx}[0]$$

$$2y\frac{dy}{dx} - 1 = 0$$

$$\frac{dy}{dx} = \frac{1}{2y}$$

At $(2, -1)$ we have $y = -1$, so the slope of the tangent line there is

$$m_{\tan} = \frac{dy}{dx}\bigg|_{\substack{x=2 \\ y=-1}} = -\frac{1}{2}$$

Alternative Solution.   If we solve $y^2 - x + 1 = 0$ for y in terms of x, we obtain

$$y = \sqrt{x - 1} \quad \text{and} \quad y = -\sqrt{x - 1}$$

As shown in Figure 3.6.2, the graph of the first equation is the upper half of the curve $y^2 - x + 1 = 0$ (since $y \geq 0$), and the graph of the second is the lower half (since $y \leq 0$).

Since the point $(2, -1)$ lies on the lower half of the curve, the slope $m_{\tan}$ of the tangent line at this point can be obtained by evaluating the derivative of $y = -\sqrt{x - 1}$ when $x = 2$; thus,

$$\frac{dy}{dx} = \frac{d}{dx}[-\sqrt{x - 1}] = -\frac{1}{2\sqrt{x - 1}} \cdot \frac{d}{dx}[x - 1] = -\frac{1}{2\sqrt{x - 1}}$$

and

$$m_{\tan} = \frac{dy}{dx}\bigg|_{x=2} = -\frac{1}{2}$$

which agrees with the solution obtained by differentiating implicitly.   ◄

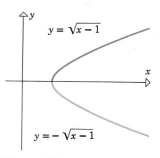

$y = \sqrt{x-1}$

$y = -\sqrt{x-1}$

Figure 3.6.2

Example 4   Use implicit differentiation to find d^2y/dx^2 if $4x^2 - 2y^2 = 9$.

Solution.   Differentiating both sides of $4x^2 - 2y^2 = 9$ implicitly yields

$$8x - 4y\frac{dy}{dx} = 0$$

from which we obtain

$$\frac{dy}{dx} = \frac{2x}{y} \tag{6}$$

Differentiating both sides of (6) implicitly yields

$$\frac{d^2y}{dx^2} = \frac{(y)(2) - (2x)(dy/dx)}{y^2} \tag{7}$$

Substituting (6) into (7) and simplifying, we obtain

$$\frac{d^2y}{dx^2} = \frac{2y - 2x(2x/y)}{y^2} = \frac{2y^2 - 4x^2}{y^3}$$

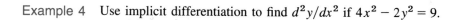

Finally, using the original equation to simplify further, we obtain

$$\frac{d^2y}{dx^2} = \frac{(-9)}{y^3} = -\frac{9}{y^3} \quad \blacktriangleleft$$

☐ **DERIVATIVES OF RATIVE POWERS OF** x

☐ **DERIVATIVES OF RATIONAL POWERS OF** x

In Section 3.3 we showed that the formula

$$\frac{d}{dx}[x^n] = nx^{n-1} \tag{8}$$

holds for integer values of n and for $n = \frac{1}{2}$. By using implicit differentiation we shall show that this formula holds for any rational power of x. More precisely, we shall show that if r is a rational number, then

$$\frac{d}{dx}[x^r] = rx^{r-1} \tag{9}$$

whenever x^r and x^{r-1} are defined. In our computations we shall assume that x^r is differentiable; the justification for this assumption will be considered later.

Let $y = x^r$. Since r is a rational number, it can be expressed as a ratio of integers $r = m/n$. Thus, $y = x^r = x^{m/n}$ can be written as

$$y^n = x^m \quad \text{so that} \quad \frac{d}{dx}[y^n] = \frac{d}{dx}[x^m]$$

By differentiating implicitly with respect to x and using (8), we obtain

$$ny^{n-1}\frac{dy}{dx} = mx^{m-1} \tag{10}$$

But

$$y^{n-1} = [x^{m/n}]^{n-1} = x^{m-(m/n)}$$

Thus, (10) can be written as

$$nx^{m-(m/n)}\frac{dy}{dx} = mx^{m-1}$$

so that

$$\frac{dy}{dx} = \frac{m}{n}x^{(m/n)-1} = rx^{r-1}$$

which establishes (9).

Example 5   From (9)

$$\frac{d}{dx}[x^{4/5}] = \frac{4}{5}x^{(4/5)-1} = \frac{4}{5}x^{-1/5}$$

$$\frac{d}{dx}[x^{-7/8}] = -\frac{7}{8}x^{(-7/8)-1} = -\frac{7}{8}x^{-15/8}$$

$$\frac{d}{dx}[\sqrt[3]{x}] = \frac{d}{dx}[x^{1/3}] = \frac{1}{3}x^{-2/3} = \frac{1}{3\sqrt[3]{x^2}} \quad \blacktriangleleft$$

If u is a differentiable function of x, and r is a rational number, then the chain rule yields the following generalization of (9):

$$\frac{d}{dx}[u^r] = ru^{r-1} \cdot \frac{du}{dx} \tag{11}$$

Example 6

$$\frac{d}{dx}[x^2 - x + 2]^{3/4} = \frac{3}{4}(x^2 - x + 2)^{-1/4} \cdot \frac{d}{dx}[x^2 - x + 2]$$

$$= \frac{3}{4}(x^2 - x + 2)^{-1/4}(2x - 1)$$

$$\frac{d}{dx}[(\sec \pi x)^{-4/5}] = -\frac{4}{5}(\sec \pi x)^{-9/5} \cdot \frac{d}{dx}[\sec \pi x]$$

$$= -\frac{4}{5}(\sec \pi x)^{-9/5} \cdot \sec \pi x \tan \pi x \cdot \pi$$

$$= -\frac{4\pi}{5}(\sec \pi x)^{-4/5} \tan \pi x \qquad \blacktriangleleft$$

☐ **DIFFERENTIABILITY OF IMPLICIT FUNCTIONS**

We conclude this section with some observations about the mathematical assumptions that underlie the method of implicit differentiation.

When differentiating implicitly, it is assumed that y represents a differentiable function of x. If this is not so, the resulting calculations may be nonsense. For example, if we implicitly differentiate the equation

$$x^2 + y^2 + 1 = 0 \tag{12}$$

we obtain

$$2x + 2y\frac{dy}{dx} = 0$$

or

$$\frac{dy}{dx} = -\frac{x}{y} \tag{13}$$

However, the derivative in (13) is meaningless because no function satisfies (12). (The left side of the equation is always greater than zero.) Unfortunately, it can be difficult to determine whether an equation defines y as a function of x and if so, whether the function is differentiable. We leave such questions for a course in advanced calculus. However, the following example should clarify the basic idea.

Example 7   The equation of the unit circle $x^2 + y^2 = 1$ does not define y as a function of x in a neighborhood of the point $P(1, 0)$, because the portion of the circle within any rectangle centered at P is cut twice by some vertical line (Figure 3.6.3a). The equation does, however, define y as a differentiable function of x in a neighborhood of the point $(1/\sqrt{2}, 1/\sqrt{2})$ (Figure 3.6.3b).

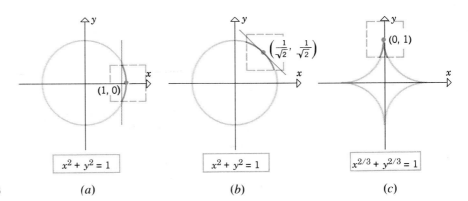

Figure 3.6.3   (a)   (b)   (c)

The graph of the equation $x^{2/3} + y^{2/3} = 1$, called a ***four-cusped hypocycloid*** (Figure 3.6.3c), defines y as a function of x in a neighborhood of the point $(0, 1)$; however, the function is not differentiable at $x = 0$, because of the corner or "cusp" at $(0, 1)$. ◀

▶ Exercise Set 3.6

In Exercises 1–10, find dy/dx.

1. $y = \sqrt[3]{2x - 5}$.

2. $y = \sqrt[3]{2 + \tan (x^2)}$.

3. $y = \left(\dfrac{x - 1}{x + 2}\right)^{3/2}$.

4. $y = \sqrt{\dfrac{x^2 + 1}{x^2 - 5}}$.

5. $y = x^3 (5x^2 + 1)^{-2/3}$.

6. $y = \dfrac{(3 - 2x)^{4/3}}{x^2}$.

7. $y = [\sin (3/x)]^{5/2}$.

8. $y = [\cos (x^3)]^{-1/2}$.

9. $y = \tan [(2x - 1)^{-1/3}]$.

10. $y = [\tan (2x - 1)]^{-1/3}$.

In Exercises 11 and 12, find all rational values of r so that $y = x^r$ satisfies the given equation.

11. $3x^2 y'' + 4xy' - 2y = 0$.   **12.** $16x^2 y'' + 24xy' + y = 0$.

In Exercises 13–29, find dy/dx by implicit differentiation.

13. $x^2 + y^2 = 100$.

14. $x^3 - y^3 = 6xy$.

15. $x^2 y + 3xy^3 - x = 3$.

16. $x^3 y^2 - 5x^2 y + x = 1$.

17. $\dfrac{1}{y} + \dfrac{1}{x} = 1$.

18. $x^2 = \dfrac{x + y}{x - y}$.

19. $\sqrt{x} + \sqrt{y} = 8$.

20. $\sqrt{xy} + 1 = y$.

21. $(x^2 + 3y^2)^{35} = x$.

22. $xy^{2/3} + yx^{2/3} = x^2$.

23. $3xy = (x^3 + y^2)^{3/2}$.

24. $\cos xy = y$.

25. $\sin (x^2 y^2) = x$.

26. $x^2 = \dfrac{\cot y}{1 + \csc y}$.

27. $\tan^3 (xy^2 + y) = x$.

28. $\dfrac{xy^3}{1 + \sec y} = 1 + y^4$.

29. $\sqrt{1 + \sin^3 (xy^2)} = y$.

In Exercises 30–34, use implicit differentiation to find the slope of the tangent line to the given curve at the specified point.

30. $x^2 y - 5xy^2 + 6 = 0$; $(3, 1)$.

31. $x^3 y + y^3 x = 10$; $(1, 2)$.   **32.** $\sin xy = y$; $(\pi/2, 1)$.

33. $x^{2/3} - y^{2/3} - y = 1$; $(1, -1)$.

34. $\sqrt{3 + \tan xy} - 2 = 0$; $(\pi/12, 3)$.

In Exercises 35–39, find the value of dy/dx at the given point in two ways: first by solving for y in terms of x and then by implicit differentiation. (See Example 3.)

35. $xy = 8$; $(2, 4)$.

36. $y^2 - x + 1 = 0$; $(10, 3)$.

37. $x^2 + y^2 = 1$; $(1/\sqrt{2}, -1/\sqrt{2})$.

38. $\dfrac{1 - y}{1 + y} = x$; $(0, 1)$.

39. $y^2 - 3xy + 2x^2 = 4$; $(3, 2)$.

In Exercises 40–45, find d^2y/dx^2 by implicit differentiation.

40. $3x^2 - 4y^2 = 7$.

41. $x^3 + y^3 = 1$.

42. $x^3 y^3 - 4 = 0$.

43. $2xy - y^2 = 3$.

44. $y + \sin y = x$.

45. $x \cos y = y$.

In Exercises 46–49, use implicit differentiation to find the specified derivative.

46. $\sqrt{u} + \sqrt{v} = 5$; du/dv.   **47.** $a^4 - t^4 = 6a^2 t$; da/dt.

48. $y = \sin x$; dx/dy.

49. $a^2 \omega^2 + b^2 \lambda^2 = 1$ (a, b constants); $d\omega/d\lambda$.

50. Use implicit differentiation to show that the equation of the tangent to the curve $y^2 = kx$ at (x_0, y_0) is
$$y_0 y = \tfrac{1}{2}k(x + x_0)$$

51. Find dy/dx if
$$2y^3 t + t^3 y = 1 \quad \text{and} \quad \frac{dt}{dx} = \frac{1}{\cos t}$$

In Exercises 52 and 53, find dy/dt in terms of x, y, and dx/dt, assuming that x and y are differentiable functions of the variable t. [*Hint:* Differentiate both sides of the given equation with respect to t.]

52. $x^3 y^2 + y = 3$.   **53.** $xy^2 = \sin 3x$.

54. At what point(s) is the tangent line to the curve $y^2 = 2x^3$ perpendicular to the line $4x - 3y + 1 = 0$?

55. Find the values of a and b for the curve $x^2 y + ay^2 = b$ if the point $(1, 1)$ is on its graph and the tangent line at $(1, 1)$ has the equation $4x + 3y = 7$.

56. Find the coordinates of the point in the first quadrant at which the tangent line to the curve $x^3 - xy + y^3 = 0$ is parallel to the x-axis.

57. Find equations for two lines through the origin that are tangent to the curve $x^2 - 4x + y^2 + 3 = 0$.

58. (a) Use implicit differentiation to find the slope of the tangent to the four-cusped hypocycloid $x^{2/3} + y^{2/3} = 1$ at $P(-\tfrac{1}{4}\sqrt{2}, \tfrac{1}{4}\sqrt{2})$. (See Figure 3.6.3c.)

   (b) Do part (a) by solving explicitly for y as a function of x and then differentiating.

(c) Trace the curve in Figure 3.6.3c and sketch the tangent line at P.

(d) Where does $x^{2/3} + y^{2/3} = 1$ *not* define y as an implicit function of x?

59. The graph of $8(x^2 + y^2)^2 = 100(x^2 - y^2)$, shown in Figure 3.6.4, is called a **lemniscate**.

(a) Where does this equation *not* define y as an implicit function of x?

(b) Use implicit differentiation to find the equation of the tangent to the curve at the point $(3, 1)$.

(c) Trace the curve in Figure 3.6.4 and sketch the tangent at $(3, 1)$.

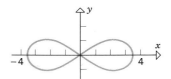

Figure 3.6.4

60. (a) Show that $f(x) = x^{4/3}$ is differentiable at 0, but not twice differentiable at 0.

(b) Show that $f(x) = x^{7/3}$ is twice differentiable at 0, but not three times differentiable at 0.

(c) Find an exponent k such that $f(x) = x^k$ is $(n-1)$ times differentiable at 0, but not n times differentiable at 0.

■ 3.7 Δ-NOTATION; DIFFERENTIALS

> *Often, the key to solving or understanding a problem is to introduce the right notation. In this section, we shall discuss some powerful notational tools.*

☐ **INCREMENTS**

If the value of a variable changes from one number to another, then the final value minus the initial value is called an **increment** in the variable. It is traditional in calculus to denote an increment in a variable x by the symbol Δx ("delta x"). In this notation, "Δx" is not a product of "Δ" and "x." Rather, Δx should be viewed as a single entity representing the *change* in the value of x. Similarly, Δy, Δt, and $\Delta \theta$ denote increments in the variables y, t, and θ.

If $y = f(x)$, and x changes from an initial value x_0 to a final value x_1, then there is a corresponding change in the value of y from $y_0 = f(x_0)$ to $y_1 = f(x_1)$. Stated another way, the increment in x,

$$\Delta x = x_1 - x_0 \tag{1}$$

produces a corresponding increment in y,

$$\Delta y = y_1 - y_0 = f(x_1) - f(x_0) \tag{2}$$

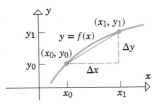

Figure 3.7.1

(Figure 3.7.1). Increments can be either positive, negative, or zero depending on the relative positions of the initial and final points. In Figure 3.7.1, Δx is positive since the final point x_1 is to the right of the initial point x_0. If x_1 were to the left of x_0, then Δx would be negative. Relation (1) can be rewritten as $x_1 = x_0 + \Delta x$, which states that the final value of x is the initial value of x plus the increment in x. With this notation, (2) can be rewritten as

$$\Delta y = f(x_0 + \Delta x) - f(x_0) \tag{3}$$

Sometimes it is convenient to drop the zero subscript on x_0 and use x to denote the initial value as well as the name of the variable. When this is done, $x + \Delta x$ represents the final value of the variable. Similarly, the initial and final values of y may be denoted by y and $y + \Delta y$ rather than y_0 and y_1 (Figure 3.7.2). Moreover, (3) becomes

$$\Delta y = f(x + \Delta x) - f(x) \tag{4}$$

The Δ-notation provides a compact way of rewriting the derivative definition,

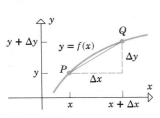

Figure 3.7.2

$$\frac{dy}{dx} = \lim_{h \to 0} \frac{f(x + h) - f(x)}{h} \tag{5}$$

Referring to Figure 3.7.2, the slope of the secant line joining the points P and Q is $\Delta y/\Delta x$. As $\Delta x \to 0$, Q approaches P along the curve $y = f(x)$, so that the slope of the secant line approaches the slope of the tangent line at P; that is,

$$\frac{dy}{dx} = \lim_{\Delta x \to 0} \frac{\Delta y}{\Delta x} \tag{6}$$

This result is expected since $\Delta y = f(x + \Delta x) - f(x)$, so (6) can be written as

$$\frac{dy}{dx} = \lim_{\Delta x \to 0} \frac{f(x + \Delta x) - f(x)}{\Delta x} \tag{7}$$

which is simply a restatement of (5) with Δx used in place of h.

If it is undesirable to introduce a dependent variable y, then we write

$$\Delta f = f(x + \Delta x) - f(x)$$

and (6) as

$$f'(x) = \lim_{\Delta x \to 0} \frac{\Delta f}{\Delta x} \tag{8}$$

Expressions (5), (6), (7), and (8) provide four ways of writing the definition of a derivative. All are commonly used.

☐ DIFFERENTIALS

Up to now we have been viewing the expression dy/dx as a single symbol for the derivative. The symbols "dy" and "dx" in this expression, which are called **differentials**, have had no meaning by themselves. We shall now show how to interpret these symbols so that dy/dx can be viewed as the ratio of dy and dx.

Regard x as fixed and *define dx* to be an independent variable that can be assigned an arbitrary value. If f is differentiable at x, then we *define dy* by the formula

$$dy = f'(x)\,dx \tag{9}$$

If $dx \neq 0$, then we can divide both sides of (9) by dx to obtain

$$\frac{dy}{dx} = f'(x)$$

Thus, we have achieved our goal of defining dy and dx so that their ratio is $f'(x)$.

Because

$$\frac{dy}{dx} = f'(x) = m_{\tan}$$

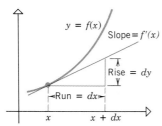

Figure 3.7.3

where $m_{\tan}$ is the slope of the tangent to $y = f(x)$ at x, the differentials dy and dx can be viewed as a corresponding rise and run of this tangent line (Figure 3.7.3).

It is important to understand the distinction between the increment Δy and the differential dy. To see the difference, let us assign the independent variables dx and Δx the same value, so $dx = \Delta x$. Then Δy represents the change in y that occurs when we start at x and travel *along the curve $y = f(x)$* until we have moved $\Delta x \, (= dx)$ units in the x-direction, while dy represents the change in y that occurs if we start at x and travel *along the tangent line* until we have moved $dx \, (= \Delta x)$ units in the x-direction (Figure 3.7.4).

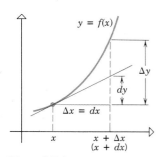

Figure 3.7.4

Example 1 If $y = x^2$, then the relation $dy/dx = 2x$ can be written in the *differential form*

$$dy = 2x\,dx$$

When $x = 3$, this becomes

$$dy = 6\,dx$$

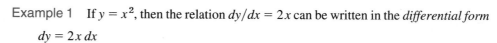

This tells us that if we travel along the tangent to the curve $y = x^2$ at $x = 3$, then a change of dx units in x produces a change of $6\,dx$ units in y. For example, if the change in x is $dx = 4$, then the change in y along the tangent is

$$dy = 6(4) = 24 \quad \text{units} \qquad \blacktriangleleft$$

Example 2   Let $y = \sqrt{x}$. Find dy and Δy if $x = 4$ and $dx = \Delta x = 3$. Then make a sketch of $y = \sqrt{x}$, showing dy and Δy in the picture.

Solution.   From (4) with $f(x) = \sqrt{x}$,

$$f'(x) = \frac{1}{2\sqrt{x}}$$

$$\Delta y = \sqrt{x + \Delta x} - \sqrt{x} = \sqrt{7} - \sqrt{4} \approx .65$$

If $y = \sqrt{x}$, then

$$\frac{dy}{dx} = \frac{1}{2\sqrt{x}}, \quad \text{so} \quad dy = \frac{1}{2\sqrt{x}}\,dx = \frac{1}{2\sqrt{4}}(3) = \frac{3}{4} = .75$$

Figure 3.7.5 shows the curve $y = \sqrt{x}$ together with dy and Δy. $\blacktriangleleft$

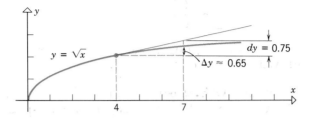

Figure 3.7.5

☐ **TANGENT LINE APPROXIMATIONS**

Figure 3.7.6 suggests that if f is differentiable at x_0, then the tangent line to the curve $y = f(x)$ at x_0 is a reasonably good approximation to the curve $y = f(x)$ for values of x near x_0. Since the tangent line passes through the point $(x_0, f(x_0))$ and has slope $f'(x_0)$, the point-slope form of its equation is

$$y - f(x_0) = f'(x_0)(x - x_0) \quad \text{or} \quad y = f(x_0) + f'(x_0)(x - x_0)$$

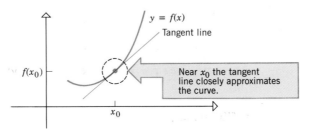

Figure 3.7.6

For values of x close to x_0, the height y of this tangent line will closely approximate the height $f(x)$ of the curve, which yields the approximation

$$f(x) \approx f(x_0) + f'(x_0)(x - x_0) \tag{10}$$

Right side = height of the tangent line

Left side = height of the curve

for x near x_0. If we let $\Delta x = x - x_0$, so that $x = x_0 + \Delta x$, then (10) can be written in the alternative form

$$f(x_0 + \Delta x) \approx f(x_0) + f'(x_0)\,\Delta x \tag{11}$$

which is a good approximation when Δx is near zero. This result is called the ***linear approximation of f near x_0***.

In the event that $f(x_0 + \Delta x)$ is tedious to calculate, but $f(x_0)$ and $f'(x_0)$ are not, then this formula enables us to use the values of $f(x_0)$ and $f'(x_0)$ to approximate $f(x_0 + \Delta x)$.

Example 3   Use (11) to approximate $\sqrt{1.1}$.

Solution.   If we let $f(x) = \sqrt{x}$, then the problem is to approximate $f(1.1)$. But $f'(x) = 1/(2\sqrt{x})$, so $f(1)$ and $f'(1)$ are easy to compute. Thus, we shall apply (11) with

$$x_0 = 1 \quad \text{and} \quad x_0 + \Delta x = 1.1$$

It follows that $\Delta x = .1$, so (11) yields

$$f(1.1) \approx f(1) + f'(1)(.1)$$

or

$$\sqrt{1.1} \approx \sqrt{1} + \frac{1}{2\sqrt{1}}(.1) = 1.05 \quad \blacktriangleleft$$

Example 4   Use (11) to approximate $\cos 62°$.

Solution.   We shall take advantage of the fact that $62°$ is close to $60°$, at which point the trigonometric functions are easy to evaluate. However, to apply (11) we must first convert to radian measure because the derivative formulas for the trigonometric functions are based on the assumption that x is in radians. If we let $f(x) = \cos x$ and note that $62° = 31\pi/90$ radians and $60° = \pi/3$ radians, then the problem is to approximate $f(31\pi/90)$ using the known values of $f(\pi/3)$ and $f'(\pi/3)$. Thus, we will apply (11) with

$$x_0 + \Delta x = \frac{31\pi}{90} \quad \text{and} \quad x_0 = \frac{\pi}{3}$$

It follows that $\Delta x = \pi/90$, so (11) yields

$$f\left(\frac{31\pi}{90}\right) \approx f\left(\frac{\pi}{3}\right) + f'\left(\frac{\pi}{3}\right)\left(\frac{\pi}{90}\right)$$

$$\cos\frac{31\pi}{90} \approx \cos\frac{\pi}{3} - \left(\sin\frac{\pi}{3}\right)\left(\frac{\pi}{90}\right)$$

$$\cos\frac{31\pi}{90} \approx \frac{1}{2} - \frac{\sqrt{3}}{2}\left(\frac{\pi}{90}\right) \approx 0.4697700$$

To seven digits, the value of $\cos 62°$ is 0.4694716, so that the error in the preceding approximation is slightly less than 0.0003. $\blacktriangleleft$

Formula (11) has a useful alternative form, which we shall now derive. Subtract $f(x_0)$ from both sides to obtain

$$f(x_0 + \Delta x) - f(x_0) \approx f'(x_0)\,\Delta x$$

and use (3) to rewrite this as

$$\Delta y \approx f'(x_0)\,\Delta x$$

If we drop the subscript on x_0 and assume that $\Delta x = dx$, then from (9) this can be rewritten as

$$\Delta y \approx dy \tag{12}$$

This formula has applications in the study of ***error propagation***. Suppose that a researcher measures a physical quantity. Because of limitations in the instrumentation and other factors, the researcher will not usually obtain the exact value x of the quantity, but rather will obtain $x + \Delta x$, where Δx is a measurement error. This recorded value may then be used to calculate some other quantity y. In this way the measurement error Δx propagates to produce an error Δy in the calculated value of y.

Example 5   The radius of a sphere is measured to be 50 in. with a possible measurement error of ± 0.02 in. Estimate the possible error in the computed volume of the sphere.

Solution.   The volume of a sphere is

$$V = \tfrac{4}{3}\pi r^3 \tag{13}$$

We are given that the error in the radius is $\Delta r = \pm 0.02$, and we want to find the error ΔV in V. If we consider Δr to be small and we let $dr = \Delta r$, then ΔV can be approximated by dV. Thus, from (13),

$$\Delta V \approx dV = 4\pi r^2\, dr \tag{14}$$

Substituting $r = 50$ and $dr = \pm 0.02$ in (14), we obtain

$$\Delta V \approx 4\pi (2500)(\pm 0.02) \approx \pm 628.32$$

Therefore, the possible error in the calculated volume is approximately ± 628.32 cubic inches (in^3).   ◄

REMARK.   In (14), r represents the exact value of the radius. Since the exact value of r was unknown, we substituted the measured value $r = 50$ to obtain ΔV. This is reasonable since the error Δr was assumed to be small.

If the exact value of a quantity is q and a measurement or calculation results in an error Δq, then $\Delta q/q$ is called the ***relative error*** in the measurement or calculation; when expressed as a percentage, $\Delta q/q$ is called the ***percentage error***. As a practical matter, the exact value q is usually unknown, so that the measured or calculated value of q is used instead; and the relative error is approximated by dq/q.

Example 6   For the sphere in Example 5,

$$\text{relative error in } r \approx \frac{dr}{r} = \frac{\pm 0.02}{50} = \pm 0.0004$$

$$\text{relative error in } V \approx \frac{dV}{V} = \frac{4\pi r^2\, dr}{\tfrac{4}{3}\pi r^3} = 3\frac{dr}{r} = \pm 0.0012$$

Thus, the percentage error in the calculated radius is approximately $\pm 0.04\%$, and the percentage error in the calculated volume is approximately $\pm 0.12\%$.   ◄

Example 7   The side of a square is measured with a possible percentage error of $\pm 5\%$. Use differentials to estimate the possible percentage error in the calculated area of the square.

Solution.   The area of a square with side x is given by

$$A = x^2 \tag{15}$$

and the relative errors in A and x are approximately dA/A and dx/x, respectively. From (15),

$$dA = 2x\, dx$$

so

$$\frac{dA}{A} = \frac{2x\,dx}{A} = \frac{2x\,dx}{x^2} = 2\frac{dx}{x}$$

We are given that $dx/x \approx \pm 0.05$, so that

$$\frac{dA}{A} \approx \pm 2(0.05) = \pm 0.1$$

Thus, the possible percentage error in the calculated area of the square is $\pm 10\%$. ◀

☐ **DIFFERENTIAL FORMULAS**

Just as

$$\frac{d}{dx}[\quad]$$

denotes the derivative of the expression inside the brackets, so

$$d[\quad]$$

denotes the differential of the expression in the brackets. For example,

$$d[x^3] = 3x^2\,dx$$

$$d[f(x)] = f'(x)\,dx$$

The basic rules of differentiation can be expressed in terms of differentials. In the following table, the differential formulas on the right result when the derivative formulas on the left are multiplied through by dx.

DERIVATIVE FORMULA	DIFFERENTIAL FORMULA
$\dfrac{d}{dx}[c] = 0$	$d[c] = 0$
$\dfrac{d}{dx}[cf] = c\dfrac{df}{dx}$	$d[cf] = c\,df$
$\dfrac{d}{dx}[f+g] = \dfrac{df}{dx} + \dfrac{dg}{dx}$	$d[f+g] = df + dg$
$\dfrac{d}{dx}[fg] = f\dfrac{dg}{dx} + g\dfrac{df}{dx}$	$d[fg] = f\,dg + g\,df$
$\dfrac{d}{dx}\left[\dfrac{f}{g}\right] = \dfrac{g\dfrac{df}{dx} - f\dfrac{dg}{dx}}{g^2}$	$d\left[\dfrac{f}{g}\right] = \dfrac{g\,df - f\,dg}{g^2}$

Example 8 Find dy if $y = x \sin x$.

Solution.

$$dy = d[x \sin x] = x\,d[\sin x] + (\sin x)\,d[x]$$

$$= x(\cos x)\,dx + (\sin x)\,dx \quad \boxed{\text{Since } d[x] = 1 \cdot dx = dx}$$

$$= (x \cos x + \sin x)\,dx \quad ◀$$

▶ Exercise Set 3.7   Ⓒ *19, 21, 24–28, 35, 36, 38, 39, 47–50*

1. Let $y = x^2$.
  (a) Find Δy if $\Delta x = 1$ and the initial value of x is $x = 2$.
  (b) Find dy if $dx = 1$ and the initial value of x is $x = 2$.

  (c) Make a sketch of $y = x^2$ and show Δy and dy in the picture.

2. Repeat Exercise 1 with $x = 2$ as the initial value again, but

$\Delta x = -1$ and $dx = -1$.

3. Let $y = 1/x$.

    (a) Find Δy if $\Delta x = 0.5$ and the initial value of x is $x = 1$.

    (b) Find dy if $dx = 0.5$ and the initial value of x is $x = 1$.

    (c) Make a sketch of $y = 1/x$ and show Δy and dy in the picture.

4. Repeat Exercise 3 with $x = 1$ as the initial value again, but $\Delta x = -0.5$ and $dx = -0.5$.

In Exercises 5–8, find general formulas for dy and Δy.

5. $y = x^3$.

6. $y = 8x - 4$.

7. $y = x^2 - 2x + 1$.

8. $y = \sin x$.

In Exercises 9–12, find dy.

9. $y = 4x^3 - 7x^2 + 2x - 1$.

10. $y = \dfrac{1}{x^3 - 1}$.

11. $y = x \cos x$.

12. $y = \dfrac{1 - x^3}{2 - x}$.

In Exercises 13–16, find the limit by expressing it as a derivative.

13. $\lim\limits_{\Delta x \to 0} \dfrac{(x + \Delta x)^2 - x^2}{\Delta x}$.

14. $\lim\limits_{\Delta x \to 0} \dfrac{(3 + \Delta x)^2 - 3^2}{\Delta x}$.

15. $\lim\limits_{\Delta x \to 0} \dfrac{\sin(\pi + \Delta x) - \sin \pi}{\Delta x}$.

16. $\lim\limits_{\Delta x \to 0} \dfrac{5(2 + \Delta x)^4 - 5(2)^4}{\Delta x}$.

In Exercises 17–28, use a linear approximation [Formula (11)] to estimate the value of the given quantity.

17. $(3.02)^4$.

18. $(1.97)^3$.

19. $\sqrt{65}$.

20. $\sqrt{24}$.

21. $\sqrt{80.9}$.

22. $\sqrt{36.03}$.

23. $\sqrt[3]{8.06}$.

24. $\sqrt[3]{63.7}$.

25. $\cos 31°$.

26. $\sin 59°$.

27. $\sin 44°$.

28. $\tan 61°$.

In Exercises 29–32, use dy to approximate Δy when x changes as indicated.

29. $y = \sqrt{3x - 2}$; from $x = 2$ to $x = 2.03$.

30. $y = \sqrt{x^2 + 8}$; from $x = 1$ to $x = 0.97$.

31. $y = \dfrac{x}{x^2 + 1}$; from $x = 2$ to $x = 1.96$.

32. $y = x\sqrt{8x + 1}$; from $x = 3$ to $x = 3.05$.

33. The side of a square is measured to be 10 ft, with a possible error of ± 0.1 ft.

    (a) Use differentials to estimate the error in the calculated area.

    (b) Estimate the percentage errors in the side and the area.

34. The side of a cube is measured to be 25 cm, with a possible error of ± 1 cm.

    (a) Use differentials to estimate the error in the calculated volume.

    (b) Estimate the percentage errors in the side and volume.

35. The hypotenuse of a right triangle is known to be 10 in. exactly, and one of the acute angles is measured to be 30°, with a possible error of $\pm 1°$.

    (a) Use differentials to estimate the errors in the sides opposite and adjacent to the measured angle.

    (b) Estimate the percentage errors in the sides.

36. One side of a right triangle is known to be 25 cm exactly. The angle opposite to this side is measured to be 60°, with a possible error of $\pm 0.5°$.

    (a) Use differentials to estimate the errors in the adjacent side and the hypotenuse.

    (b) Estimate the percentage errors in the adjacent side and hypotenuse.

37. The electrical resistance R of a certain wire is given by $R = k/r^2$, where k is a constant and r is the radius of the wire. Assuming that the radius r has a possible error of $\pm 5\%$, use differentials to estimate the percentage error in R. (Assume k is exact.)

38. A 12-foot ladder leaning against a wall makes an angle θ with the floor. If the top of the ladder is h feet up the wall, express h in terms of θ and then use dh to estimate the change in h if θ changes from 60° to 59°.

39. The area of a right triangle with a hypotenuse of H is calculated using the formula $A = \frac{1}{4}H^2 \sin 2\theta$, where θ is one of the acute angles. Use differentials to approximate the error in calculating A if $H = 4$ cm (exactly) and $\theta = 30° \pm 15'$.

40. The side of a square is measured with a possible percentage error of $\pm 1\%$. Use differentials to estimate the percentage error in the area.

41. The side of a cube is measured with a possible percentage error of $\pm 2\%$. Use differentials to estimate the percentage error in the volume.

42. The volume of a sphere is to be computed from a measured value of its radius. Estimate the maximum permissible percentage error in the measurement if the percentage error in the volume must be kept within $\pm 3\%$. ($V = \frac{4}{3}\pi r^3$ is the volume of a sphere of radius r.)

43. The area of a circle is to be computed from a measured value of its diameter. Estimate the maximum permissible percentage error in the measurement if the percentage error in the area must be kept within $\pm 1\%$.

44. A steel cube with 1-in. sides is coated with 0.01 in. of copper. Use differentials to estimate the volume of copper in the coating. [*Hint:* Let ΔV be the change in the volume of the cube.]

45. A metal rod 15 cm long and 5 cm in diameter is to be covered (except for the ends) with insulation that is 0.001 cm thick. Use differentials to estimate the volume of insulation. [*Hint:* Let ΔV be the change in volume of the rod.]

46. The time required for one complete oscillation of a pendulum is called its **period**. If the length L of the pendulum is measured in feet and the period P in seconds, then the period is given by $P = 2\pi\sqrt{L/g}$, where g is a constant. Use differentials to show that the percentage error in P is approximately half the percentage error in L.

47. If the temperature T of a metal rod of length L is changed by an amount ΔT, then the length will change by the amount $\Delta L = \alpha L \, \Delta T$, where α is called the **coefficient of linear expansion**. For moderate changes in temperature α is taken as constant.

    (a) Suppose that a rod 40 cm long at 20° C is found to be 40.006 cm long when the temperature is raised to 30° C. Find α.

    (b) If an aluminum pole is 180 cm long at 15° C, how long is the pole if the temperature is raised to 40° C? [Take $\alpha = 2.3 \times 10^{-5}$/°C.]

48. If the temperature T of a solid or liquid of volume V is changed by an amount ΔT, then the volume will change by the amount $\Delta V = \beta V \, \Delta T$, where β is called the **coefficient of volume expansion**. For moderate changes in temperature β is taken as constant. Suppose that a tank truck loads 4000 gallons of ethyl alcohol at a temperature of 35° C and delivers its load sometime later at a temperature of 15° C. Using $\beta = 7.5 \times 10^{-4}$/°C for ethyl alcohol, find the number of gallons delivered.

49. Use differentials to show that $(1 + x)^{-1} \approx 1 - x$ when x is near zero. Calculate, to four decimal places, the values of $(1 + x)^{-1}$ and $1 - x$ for $x = 0.1$ and $x = -0.02$.

50. Use differentials to show that $\sqrt{1 + x} \approx 1 + \frac{1}{2}x$ when x is near zero. Calculate, to four decimal places, the values of $\sqrt{1 + x}$ and $1 + \frac{1}{2}x$ for $x = 0.1$ and $x = -0.02$.

51. Let $y = x^k$, where k is a positive rational number. Show that the percentage error in y is approximately k times the percentage error in x.

◆ TECHNOLOGY EXERCISES   Chapter 3

Most of these exercises require access to a graphing calculator or a computer algebra system (CAS) such as *Mathematica*, *Maple*, or *Derive*. When you are asked to *find* an answer or to *solve* an equation, you may choose to find an exact result or a numerical approximation, depending on the particular technology you are using and on your own imagination. The form of your answers may differ from those of other students or from those in the answer section of the text, depending on how you solve the problems and the accuracy you use in your numerical approximations. Those exercises that are more appropriate for a CAS than a graphing calculator are labeled with the icon ◆.

Slope of a tangent line: In Exercises 1–4, zoom in on the graph of f on an interval containing $x = x_0$ until the graph looks like a straight line. Estimate the slope of this line and then check your answer by finding the exact value of $f'(x_0)$.

1. $f(x) = x^2 - 1$, $x_0 = 1.8$.

2. $f(x) = x^3 - x^2 + 1$, $x_0 = 2.3$.

3. $f(x) = \dfrac{x^2}{x - 2}$, $x_0 = 3.5$.

4. $f(x) = \dfrac{x}{x^2 + 1}$, $x_0 = -0.5$.

Derivatives from the definition: In Exercises 5–8, use the difference quotient $[f(2 + h) - f(2)]/h$ with small values of h to approximate $f'(2)$. If you have access to a CAS, see if it can find the exact limit of the difference quotient as $h \to 0$.

5. $f(x) = 2^x$.   **6.** $f(x) = 3^x$.

7. $f(x) = x^x$.   **8.** $f(x) = x^{\sin x}$.

9. Average and instantaneous velocity: At time $t = 0$ a car moves into the passing lane to pass a slow-moving truck.

The average velocity of the car from $t = 1$ to $t = 1 + h$ is

$$v_{\text{ave}} = \frac{3(h + 1)^{2.5} + 580h - 3}{10h}$$

Find the instantaneous velocity of the car at $t = 1$, where time is in seconds and distance is in feet.

10. Sky diving: A sky diver jumps from an airplane. Suppose that the distance she falls during the period before her parachute opens is $s(t) = 986((0.835)^t - 1) + 176t$, where s is in feet and t is in seconds.

(a) Find the diver's average velocity over the intervals $[5, 5.01]$ and $[4.99, 5]$.

(b) Find the diver's instantaneous velocity at $t = 5$.

(c) Graph s versus t for $0 \le t \le 20$. Use your graph to estimate the instantaneous velocity at $t = 15$.

11. Horizontal tangents: Find all values of x where the tangent line to the graph of $y = x^3 - \sin x$ is horizontal.

12. Nondifferentiability: Graph the function

$$f(x) = |x^4 - x - 1| - x$$

and find the values of x where the derivative of this function does not exist.

◆ **13. Derivatives from the definition:** Use a CAS to find the derivative of f from the definition

$$f'(x) = \lim_{h \to 0} \frac{f(x + h) - f(x)}{h}$$

and check the result by finding the derivative by hand.

(a) $f(x) = x^5$

(b) $f(x) = 1/x$

(c) $f(x) = 1/\sqrt{x}$

(d) $f(x) = \dfrac{2x + 1}{x - 1}$

(e) $f(x) = \sqrt{3x^2 + 5}$

(f) $f(x) = \sin 3x.$

Derivatives by computer: In Exercise 13 a CAS was used to find derivatives from the definition of $f'(x)$. A CAS can also find derivatives of many functions directly. In Exercises 14–19,

(a) use a CAS to find $f'(x)$

(b) confirm the result in part (a) by finding $f'(x)$ by hand

(c) use a CAS to find $f''(x)$.

◆ **14.** $f(x) = x^2 \sin x.$

◆ **15.** $f(x) = \sqrt{x} + \cos^2 x.$

◆ **16.** $f(x) = \dfrac{2x^2 - x + 5}{3x + 2}.$

◆ **17.** $f(x) = \dfrac{\tan x}{1 + x^2}.$

◆ **18.** $f(x) = \dfrac{1}{x} \sin \sqrt{x}.$

◆ **19.** $f(x) = \dfrac{\sqrt{x^4 - 3x + 2}}{x(2 - \cos x)}.$

Pierre de Fermat

Pierre de Fermat (1601–1665)

4 APPLICATIONS OF DIFFERENTIATION

◼ **4.1** RELATED RATES

*In this section we shall study **related rates** problems. In such problems one tries to find the rate at which some quantity is changing by relating it to other quantities whose rates of change are known.*

Example 1　Assume that oil spilled from a ruptured tanker spreads in a circular pattern whose radius increases at a constant rate of 2 ft/sec. How fast is the area of the spill increasing when the radius of the spill is 60 ft?

Solution.　Let

　　t = number of seconds elapsed from the time of the spill
　　r = radius of the spill in feet after t seconds
　　A = area of the spill in square feet after t seconds

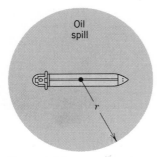

Figure 4.1.1

(See Figure 4.1.1.) At each instant the rate at which the radius is increasing with time is dr/dt, and the rate at which the area is increasing with time is dA/dt. We want to find

$$\left.\frac{dA}{dt}\right|_{r=60} \quad \text{given that} \quad \frac{dr}{dt} = 2 \text{ ft/sec}$$

From the formula for the area of a circle we obtain

$$A = \pi r^2 \tag{1}$$

Because A and r are functions of t, we can differentiate both sides of (1) with respect to t to obtain

$$\frac{dA}{dt} = 2\pi r \frac{dr}{dt}$$

Thus, when $r = 60$ the area of the spill is increasing at the rate of

$$\left.\frac{dA}{dt}\right|_{r=60} = 2\pi(60)(2) = 240\pi \text{ ft}^2/\text{sec}$$

or approximately 754 ft^2/sec.   ◄

With only minor variations, the method used in Example 1 can be used to solve a variety of related rates problems. The method consists of five steps:

Step 1. Draw a figure and label the quantities that vary.

Step 2. Identify the rates of change that are known and the rate of change that is to be found.

Step 3. Find an equation that relates the quantity whose rate of change is to be found to the quantities whose rates of change are known.

Step 4. Differentiate both sides of this equation with respect to time and solve for the derivative that will give the unknown rate of change.

Step 5. Evaluate this derivative at the appropriate point.

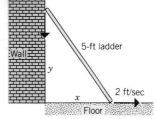

Figure 4.1.2

Example 2 A 5-ft ladder, leaning against a wall (Figure 4.1.2), slips in such a way that its base is moving away from the wall at a rate of 2 ft/sec at the instant when the base is 4 ft from the wall. How fast is the top of the ladder moving down the wall at that instant?

Solution. Let

t = number of seconds after the ladder starts to slip
x = distance in feet from the base of the ladder to the wall
y = distance in feet from the top of the ladder to the floor

(See Figure 4.1.2.) At each instant the rate at which the base moves is dx/dt, and the rate at which the top moves is dy/dt. We want to find

$$\left.\frac{dy}{dt}\right|_{x=4} \quad \text{given that} \quad \left.\frac{dx}{dt}\right|_{x=4} = 2 \text{ ft/sec}$$

From the Theorem of Pythagoras we have

$$x^2 + y^2 = 25 \tag{2}$$

Using the chain rule to differentiate both sides of this equation with respect to t yields

$$2x\frac{dx}{dt} + 2y\frac{dy}{dt} = 0 \quad \text{or} \quad \frac{dy}{dt} = -\frac{x}{y}\frac{dx}{dt} \tag{3}$$

When $x = 4$, it follows from (2) that $y = 3$; moreover, we are given that $dx/dt = 2$, so that (3) yields

$$\frac{dy}{dt}\bigg|_{x=4} = -\frac{4}{3}(2) = -\frac{8}{3} \text{ ft/sec}$$

The negative sign in the answer tells us that y is decreasing, which makes sense physically, since the top of the ladder is moving *down* the wall. ◄

In Figure 4.1.3 we have shown a camera mounted at a point 3000 ft from the base of a rocket launching pad. Let us assume that the rocket rises vertically and the camera is to take a series of photographs of the rocket. Because the rocket will be rising, the elevation angle of the camera will have to vary at just the right rate to keep the rocket in sight. Moreover, because the camera-to-rocket distance will be changing constantly, the camera focusing mechanism will also have to vary at just the right rate to keep the picture sharp. The focusing problem is considered in the exercises, and the elevation problem is addressed in the following example.

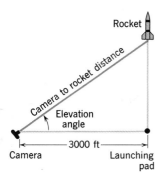

Figure 4.1.3

Example 3   If the rocket shown in Figure 4.1.3 is rising vertically at 880 ft/sec when it is 4000 ft up, how fast must the camera elevation angle change at that instant to keep the rocket in sight?

Solution.   Let

t = number of seconds elapsed from the time of launch
ϕ = camera elevation angle in radians after t seconds
x = height of the rocket in feet after t seconds

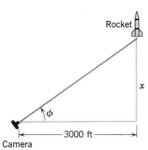

Figure 4.1.4

(See Figure 4.1.4.) At each instant the rate at which the camera elevation angle must change is $d\phi/dt$, and the rate at which the rocket is rising is dx/dt. We want to find

$$\frac{d\phi}{dt}\bigg|_{x=4000} \quad \text{given that} \quad \frac{dx}{dt}\bigg|_{x=4000} = 880 \text{ ft/sec}$$

From Figure 4.1.4 we see that

$$\tan \phi = \frac{x}{3000} \tag{4}$$

Because ϕ and x are functions of t, we can differentiate both sides of (4) with respect to t to obtain

$$(\sec^2 \phi)\frac{d\phi}{dt} = \frac{1}{3000}\frac{dx}{dt} \quad \text{or} \quad \frac{d\phi}{dt} = \frac{1}{3000 \sec^2 \phi}\frac{dx}{dt} \tag{5}$$

When $x = 4000$, it follows that

$$\sec \phi = \frac{5000}{3000} = \frac{5}{3}$$

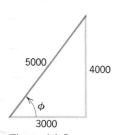

Figure 4.1.5

(Figure 4.1.5), so that from (5)

$$\frac{d\phi}{dt}\bigg|_{x=4000} = \frac{1}{3000(\frac{5}{3})^2} \cdot 880 = \frac{66}{625} \approx 0.11 \text{ radian/sec}$$

If desired, this rate of change can be expressed in degrees per second. The computations are as follows:

$$\frac{d\phi}{dt}\bigg|_{x=4000} = \frac{66}{625}\frac{\text{radian}}{\text{sec}} \cdot \frac{180}{\pi}\frac{\text{degrees}}{\text{radian}} \approx 6.05 \text{ degrees/sec} \quad ◄$$

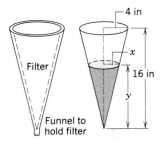

Figure 4.1.6

Example 4   Suppose that a liquid is to be cleared of sediment by pouring it through a cone-shaped filter. Assume that the height of the cone is 16 in. and the radius at the base of the cone is 4 in. (Figure 4.1.6). If the liquid is flowing out of the cone at a rate of 2 in³/min when the level is 8 in. deep, how fast is the depth of the liquid changing at that instant?

Solution.   Let

t = time elapsed from the initial observation (min)
V = volume of liquid in the cone at time t (in³)
y = depth of the liquid in the cone at time t (in.)
x = radius of the liquid surface at time t (in.)

(See Figure 4.1.6.) At each instant the rate at which the volume of liquid is changing is dV/dt, and the rate at which the depth is changing is dy/dt. We want to find

$$\left.\frac{dy}{dt}\right|_{y=8} \quad \text{given that} \quad \left.\frac{dV}{dt}\right|_{y=8} = -2$$

(We must use a minus sign here because V *decreases* as t increases.)

From the formula for the volume of a cone, the volume V, the radius x, and the depth y are related by

$$V = \tfrac{1}{3}\pi x^2 y \tag{6}$$

If we differentiate both sides of (6) with respect to t, the right side will involve the quantity dx/dt. Since we have no direct information about dx/dt, it is desirable to eliminate x from (6) before differentiating. This can be done using similar triangles. From Figure 4.1.6, we see that

$$\frac{x}{y} = \frac{4}{16} \quad \text{or} \quad x = \frac{1}{4}y$$

Substituting this expression in (6) gives

$$V = \frac{\pi}{48} y^3 \tag{7}$$

Differentiating both sides of (7) with respect to t we obtain

$$\frac{dV}{dt} = \frac{\pi}{48}\left(3y^2 \frac{dy}{dt}\right)$$

or

$$\frac{dy}{dt} = \frac{16}{\pi y^2}\frac{dV}{dt}$$

Substituting $y = 8$ and $dV/dt = -2$ into this equation yields

$$\left.\frac{dy}{dt}\right|_{y=8} = \frac{16}{\pi(8)^2}(-2) = -\frac{1}{2\pi} \approx -0.16 \text{ in/min}$$

Thus, when $y = 8$ in., the depth is decreasing at an approximate rate of 0.16 in/min. (The minus sign simply means that the depth y is decreasing as t is increasing.)   ◄

▶ Exercise Set 4.1   ☐ 19, 20

1.   Let A be the area of a square whose sides have length x, and assume that x varies with time.

(a)   How are dA/dt and dx/dt related?

(b)   At a certain instant the sides are 3 ft long and growing at a rate of 2 ft/min. How fast is the area growing at that instant?

2.   Let A be the area of a circle whose radius is r, and assume that r varies with time.

(a) How are dA/dt and dr/dt related?

(b) At a certain instant the radius is 5 in. and is increasing at a rate of 2 in/sec. How fast is the area of the circle increasing at that instant?

3. Let V be the volume of a cylinder having height h and radius r, and assume that h and r vary with time.

(a) How are dV/dt, dh/dt, and dr/dt related?

(b) At a certain instant, the height is 6 in. and increasing at 1 in/sec, while the radius is 10 in. and decreasing at 1 in/sec. How fast is the volume changing at that instant? Is the volume increasing or decreasing at that instant?

4. Let l be the length of a diagonal of a rectangle whose sides have lengths x and y, and assume that x and y vary with time.

(a) How are dl/dt, dx/dt, and dy/dt related?

(b) If x increases at a constant rate of $\frac{1}{2}$ ft/sec and y decreases at a constant rate of $\frac{1}{4}$ ft/sec, how fast is the size of the diagonal changing when $x = 3$ ft and $y = 4$ ft? Is the diagonal increasing or decreasing at that instant?

5. Let θ (in radians) be an acute angle in a right triangle, and let x and y, respectively, be the lengths of the sides adjacent and opposite θ. Suppose also that x and y vary with time.

(a) How are $d\theta/dt$, dx/dt, and dy/dt related?

(b) At a certain instant, $x = 2$ units and is increasing at 1 unit/sec, while $y = 2$ units and is decreasing at $\frac{1}{4}$ unit/sec. How fast is θ changing at that instant? Is θ increasing or decreasing at that instant?

6. Suppose that $z = x^3 y^2$, where both x and y are changing with time. At a certain instant when $x = 1$ and $y = 2$, x is decreasing at the rate of 2 units/sec, and y is increasing at the rate of 3 units/sec. How fast is z changing at this instant? Is z increasing or decreasing?

7. The minute hand of a certain clock is 4 in. long. Starting from the moment when the hand is pointing straight up, how fast is the area of the sector that is swept out by the hand increasing at any instant during the next revolution of the hand?

8. A stone dropped into a still pond sends out a circular ripple whose radius increases at a constant rate of 3 ft/sec. How rapidly is the area enclosed by the ripple increasing at the end of 10 sec?

9. Oil spilled from a ruptured tanker spreads in a circle whose area increases at a constant rate of 6 mi²/hr. How fast is the radius of the spill increasing when the area is 9 mi²?

10. A spherical balloon is inflated so that its volume is increasing at the rate of 3 ft³/min. How fast is the diameter of the balloon increasing when the radius is 1 ft?

11. A spherical balloon is to be deflated so that its radius decreases at a constant rate of 15 cm/min. At what rate

must air be removed when the radius is 9 cm?

12. A 17-ft ladder is leaning against a wall. If the bottom of the ladder is pulled along the ground away from the wall at a constant rate of 5 ft/sec, how fast will the top of the ladder be moving down the wall when it is 8 ft above the ground?

13. A 13-ft ladder is leaning against a wall. If the top of the ladder slips down the wall at a rate of 2 ft/sec, how fast will the foot be moving away from the wall when the top is 5 ft above the ground?

14. A 10-ft plank is leaning against a wall. If at a certain instant the bottom of the plank is 2 ft from the wall and is being pushed toward the wall at the rate of 6 in/sec, how fast is the acute angle that the plank makes with the ground increasing?

15. At a certain instant each edge of a cube is 5 in. long and the volume is increasing at the rate of 2 in³/min. How fast is the surface area of the cube increasing?

16. A rocket, rising vertically, is tracked by a radar station that is on the ground 5 mi from the launchpad. How fast is the rocket rising when it is 4 mi high and its distance from the radar station is increasing at a rate of 2000 mi/hr?

17. For the camera and rocket shown in Figure 4.1.3, at what rate is the camera-to-rocket distance changing when the rocket is 4000 ft up and rising vertically at 880 ft/sec?

18. For the camera and rocket shown in Figure 4.1.3, at what rate is the rocket rising when the elevation angle is $\pi/4$ radians and increasing at a rate of 0.2 radian/sec?

19. A satellite is in an elliptical orbit around Earth. Its distance r (in miles) from the center of Earth is given by

$$r = \frac{4995}{1 + 0.12 \cos \theta}$$

where θ is the angle measured from the point on the orbit nearest the surface of Earth (Figure 4.1.7).

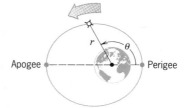

Figure 4.1.7

(a) Find the altitude of the satellite at **perigee** (the point nearest the surface of Earth) and at **apogee** (the point farthest from the surface of Earth). Use 3960 mi as the radius of Earth.

(b) At the instant when θ is 120°, the angle θ is increasing at the rate of 2.7°/min. Find the altitude of the satellite and the rate at which the altitude is changing at this instant. Express the rate in units of mi/min.

20. An aircraft is flying horizontally at a constant height of 4000 ft above a fixed observation point (Figure 4.1.8). At a certain instant the angle of elevation θ is 30° and decreasing, and the speed of the aircraft is 300 mi/hr.

(a) How fast is θ decreasing at this instant? Express the result in units of degrees/sec.

(b) How fast is the distance between the aircraft and the observation point changing at this instant? Express the result in units of ft/sec. Use 1 mi = 5280 ft.

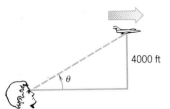

4000 ft

θ

Figure 4.1.8

21. A conical water tank with vertex down has a radius of 10 ft at the top and is 24 ft high. If water flows into the tank at a rate of 20 ft^3/min, how fast is the depth of the water increasing when the water is 16 ft deep?

22. Grain pouring from a chute at the rate of 8 ft^3/min forms a conical pile whose altitude is always twice its radius. How fast is the altitude of the pile increasing at the instant when the pile is 6 ft high?

23. Sand pouring from a chute forms a conical pile whose height is always equal to the diameter. If the height increases at a constant rate of 5 ft/min, at what rate is sand pouring from the chute when the pile is 10 ft high?

24. Wheat is poured through a chute at the rate of 10 ft^3/min, and falls in a conical pile whose bottom radius is always half the altitude. How fast will the circumference of the base be increasing when the pile is 8 ft high?

25. An aircraft is climbing at a 30° angle to the horizontal. How fast is the aircraft gaining altitude if its speed is 500 mi/hr?

26. A boat is pulled into a dock by means of a rope attached to a pulley on the dock (Figure 4.1.9). The rope is attached to the bow of the boat at a point 10 ft below the pulley. If the rope is pulled through the pulley at a rate of 20 ft/min, at what rate will the boat be approaching the dock when 125 ft of rope is out?

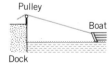

Pulley

Boat

Dock   Figure 4.1.9

27. For the boat in Exercise 26, how fast must the rope be pulled if we want the boat to approach the dock at a rate of 12 ft/min at the instant when 125 ft of rope is out?

28. A man 6 ft tall is walking at the rate of 3 ft/sec toward a streetlight 18 ft high (Figure 4.1.10).

(a) At what rate is his shadow length changing?

(b) How fast is the tip of his shadow moving?

Figure 4.1.10

29. A beacon that makes one revolution every 10 sec is located on a ship anchored 4 kilometers from a straight shoreline. How fast is the beam moving along the shoreline when it makes an angle of 45° with the shore?

30. An aircraft is flying at a constant altitude with a constant speed of 600 mi/hr. An antiaircraft missile is fired on a straight line perpendicular to the flight path of the aircraft so that it will hit the aircraft at a point P (Figure 4.1.11). At the instant the aircraft is 2 mi from the impact point P the missile is 4 mi from P and flying at 1200 mi/hr. At that instant, how rapidly is the distance between missile and aircraft decreasing?

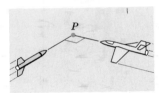

Figure 4.1.11

31. Solve Exercise 30 under the assumption that the angle between the flight paths is 120° instead of the assumption that the paths are perpendicular. [*Hint:* Use the law of cosines.]

32. A police helicopter is flying due north at 100 mi/hr and at a constant altitude of $\frac{1}{2}$ mi. Below, a car is traveling west on a highway at 75 mi/hr. At the moment the helicopter crosses over the highway the car is 2 mi east of the helicopter.

(a) How fast is the distance between the car and helicopter changing at the moment the helicopter crosses the highway?

(b) Is the distance between car and helicopter increasing or decreasing at that moment?

33. A particle is moving along the curve whose equation is

$$\frac{xy^3}{1+y^2} = \frac{8}{5}$$

Assume that the x-coordinate is increasing at the rate of 6 units/sec when the particle is at the point (1, 2).

(a) At what rate is the y-coordinate of the point changing at that instant?

(b) Is the particle rising or falling at that instant?

34. A point P is moving along the curve whose equation is $y = \sqrt{x^3 + 17}$. When P is at (2, 5), y is increasing at the rate of 2 units/sec. How fast is x changing?

35. A point P is moving along the line whose equation is $y = 2x$. How fast is the distance between P and the point $(3, 0)$ changing at the instant when P is at $(3, 6)$ if x is decreasing at the rate of 2 units/sec at that instant?

36. A point P is moving along the curve whose equation is $y = \sqrt{x}$. Suppose that x is increasing at the rate of 4 units/sec when $x = 3$.

   (a) How fast is the distance between P and the point $(2, 0)$ changing at this instant?

   (b) How fast is the angle of inclination of the line segment from P to $(2, 0)$ changing at this instant?

37. A particle is moving along the curve $y = x^2$. Find all values of x at which the rate of change of y with respect to time is three times that of x. [Assume that dx/dt is never zero.]

38. A particle is moving along the curve $16x^2 + 9y^2 = 144$. Find all points (x, y) at which the rates of change of x and y with respect to time are equal. [Assume that dx/dt and dy/dt are never both zero at the same point.]

39. The **thin lens equation** in physics is

$$\frac{1}{s} + \frac{1}{S} = \frac{1}{f}$$

where s is the object distance from the lens, S is the image distance from the lens, and f is the focal length of the lens. Suppose that a certain lens has a focal length of 6 cm and that an object is moving toward the lens at the rate of 2 cm/sec. How fast is the image distance changing at the instant when the object is 10 cm from the lens? Is the image moving away from the lens or toward the lens?

40. Water is stored in a cone-shaped reservoir (vertex down). Assuming the water evaporates at a rate proportional to the surface area exposed to the air, show that the depth of the water will decrease at a constant rate that does not depend on the dimensions of the reservoir.

41. A meteorite enters the earth's atmosphere and burns up at a rate that, at each instant, is proportional to its surface area. Assuming that the meteorite is always spherical, show that the radius decreases at a constant rate.

42. On a certain clock the minute hand is 4 in. long and the hour hand is 3 in. long. How fast is the distance between the tips of the hands changing at 9 o'clock?

43. Coffee is poured at a uniform rate of 20 cm³/sec into a cup whose inside is shaped like a truncated cone (Figure 4.1.12). If the upper and lower radii of the cup are 4 cm and 2 cm and the height of the cup is 6 cm, how fast will the coffee level be rising when the coffee is halfway up? [*Hint:* Extend the cup downward to form a cone.]

Figure 4.1.12

4.2 INTERVALS OF INCREASE AND DECREASE; CONCAVITY

Although point-plotting is useful for determining the general shape of a graph, it only provides an approximation because no matter how many points are plotted, we can only guess at the shape of the graph between those points. In this section we shall show how the derivative can be used to resolve such ambiguities.

☐ **INCREASING AND DECREASING FUNCTIONS**

The terms *increasing*, *decreasing*, and *constant* are used to describe the behavior of a function over an interval as we travel left to right along its graph. For example, the function graphed in Figure 4.2.1 can be described as increasing on the interval $(-\infty, 0]$, decreasing on the interval $[0, 2]$, increasing again on the interval $[2, 4]$, and constant on the interval $[4, +\infty)$.

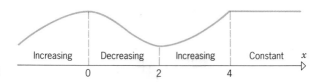

Figure 4.2.1

The following definition, which is illustrated in Figure 4.2.2, expresses these intuitive ideas precisely.

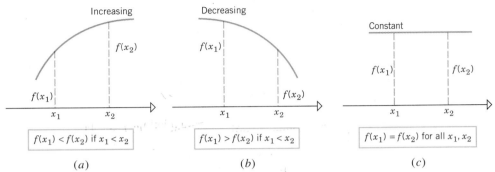

Figure 4.2.2   (a)   (b)   (c)

4.2.1   DEFINITION.   Let f be defined on an interval, and let x_1 and x_2 denote points in that interval.

(a) f is **increasing** on the interval if $f(x_1) < f(x_2)$ whenever $x_1 < x_2$.
(b) f is **decreasing** on the interval if $f(x_1) > f(x_2)$ whenever $x_1 < x_2$.
(c) f is **constant** on the interval if $f(x_1) = f(x_2)$ for all points x_1 and x_2.

Figure 4.2.3 suggests that a differentiable function f is increasing on any interval where its graph has tangent lines with positive slope, is decreasing on any interval where its graph has tangent lines with negative slope, and is constant on any interval where its graph has tangent lines with zero slope. This intuitive observation suggests the following important theorem. (The proof, which requires results not yet discussed, is given in Section 4.9.)

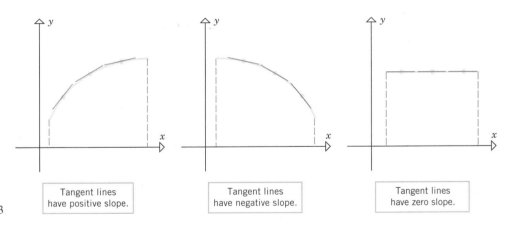

Figure 4.2.3

4.2.2   THEOREM.   *Let f be a function that is continuous on a closed interval $[a, b]$ and differentiable on the open interval (a, b).*

(a)   *If $f'(x) > 0$ for every value of x in (a, b), then f is increasing on $[a, b]$.*
(b)   *If $f'(x) < 0$ for every value of x in (a, b), then f is decreasing on $[a, b]$.*
(c)   *If $f'(x) = 0$ for every value of x in (a, b), then f is constant on $[a, b]$.*

REMARK.   This theorem is stated in a form that is useful for investigating whether a function is increasing, decreasing, or constant on a closed interval $[a, b]$. However, there are a number of variations of the theorem that are applicable to other kinds of intervals. For example, the theorem remains true if the closed interval $[a, b]$ is replaced by a finite or infinite interval (a, b), $[a, b)$, or $(a, b]$, provided f is differentiable on (a, b) and continuous on the entire interval. To take this idea one step further, the theorem also remains true if the closed interval $[a, b]$ is replaced by any interval I on which f is differentiable, because the continuity required in the hypothesis follows from the assumption of differentiability.

Example 1   Find the intervals on which the following functions are increasing and the intervals on which they are decreasing.

(a) $f(x) = x^2 - 4x + 3$ (b) $f(x) = x^3$

Solution (a).   Differentiating f we obtain

$$f'(x) = 2x - 4 = 2(x - 2)$$

It follows that

$$f'(x) < 0 \quad \text{if} \quad -\infty < x < 2$$
$$f'(x) > 0 \quad \text{if} \quad 2 < x < +\infty$$

Since f is continuous at $x = 2$, it follows from Theorem 4.2.2 and the subsequent remark that

f is decreasing on $(-\infty, 2]$

f is increasing on $[2, +\infty)$

Observe that $f'(x) = 0$ at the *point* $x = 2$, but there is no *interval* on which $f'(x) = 0$, so the function is not constant on any interval. For this function the point $x = 2$ marks the place on the x-axis where a transition from decrease to increase occurs. The graph of f is shown in Figure 4.2.4.

Solution (b).   Differentiating f we obtain $f'(x) = 3x^2$. Thus,

$$f'(x) > 0 \quad \text{if} \quad -\infty < x < 0$$
$$f'(x) > 0 \quad \text{if} \quad 0 < x < +\infty$$

Since f is continuous at $x = 0$,

f is increasing on $(-\infty, 0]$

f is increasing on $[0, +\infty)$

Combining these results, it follows that f is increasing on the interval $(-\infty, +\infty)$. (See Exercise 43.) Observe that $f'(x) = 0$ at the point $x = 0$, but unlike the function in part (a), this point does not mark a transition between increase and decrease. The graph of f is shown in Figure 4.2.5.   ◀

Example 2   Find the intervals on which the function

$$f(x) = x^3 - 3x^2 + 1$$

is increasing and the intervals on which it is decreasing.

Solution.   Differentiating f we obtain

$$f'(x) = 3x^2 - 6x = 3x(x - 2)$$

From the sign analysis in Figure 4.2.6 we see that

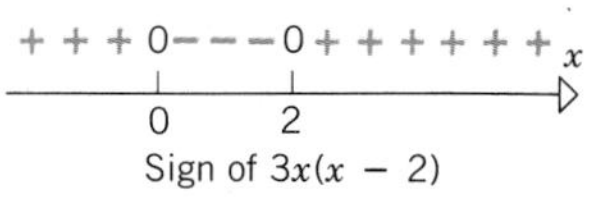

$$f'(x) > 0 \quad \text{if} \quad -\infty < x < 0$$
$$f'(x) < 0 \quad \text{if} \quad 0 < x < 2$$
$$f'(x) > 0 \quad \text{if} \quad 2 < x < +\infty$$

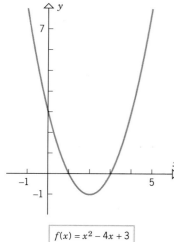

$f(x) = x^2 - 4x + 3$

Figure 4.2.4

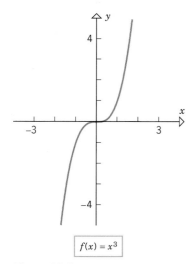

$f(x) = x^3$

Figure 4.2.5

$$+ + + 0 - - - 0 + + + + + +$$
$$ 0 \quad\quad 2$$
Sign of $3x(x - 2)$

Figure 4.2.6

Since f is continuous at $x = 0$ and $x = 2$, it follows that

f is increasing on $(-\infty, 0]$

f is decreasing on $[0, 2]$

f is increasing on $[2, +\infty)$

The graph of f is shown in Figure 4.2.7.   ◄

□ CONCAVITY

$f(x) = x^3 - 3x^2 + 1$

Figure 4.2.7

Although the derivative of a function f can tell us where the graph of f is increasing or decreasing, it does not reveal where the graph has a downward curvature (Figure 4.2.8*a*) or an upward curvature (Figure 4.2.8*b*). To investigate this question, we must study the behavior of the tangent lines shown in the figure.

The curve in part (*a*) lies below its tangent lines and is called *concave down*. As we travel left to right along this curve, the tangent lines rotate clockwise so that their slopes *decrease*. In contrast, the curve in part (*b*) lies above its tangent lines and is called *concave up*. As we travel left to right, its tangent lines rotate counterclockwise so that their slopes *increase*. Since f' is the slope of a tangent line to the graph of f, we are led to the following definition.

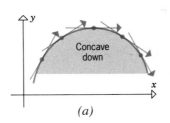

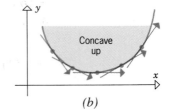

(a)   *(b)*   Figure 4.2.8

4.2.3 DEFINITION.   Let f be differentiable on an interval.

(a) f is called **concave up** on the interval if f' is increasing on the interval.

(b) f is called **concave down** on the interval if f' is decreasing on the interval.

REMARK.   Informally, a curve that is concave up "holds water" and one that is concave down "spills water" (Figure 4.2.9).

REMARK.   Definition 4.2.3 requires the function f to be differentiable. There are more general definitions of concavity that do not require differentiability. However, we shall have no need for that added generality in this text.

Since f'' is the derivative of f', it follows from Theorem 4.2.2 that f' is increasing on an open interval (a, b) if $f''(x) > 0$ for all x in (a, b), and f' is decreasing on (a, b) if $f''(x) < 0$ for all x in (a, b). Thus, we have the following result.

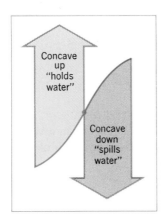

Figure 4.2.9

4.2.4 THEOREM.

(*a*)   *If $f''(x) > 0$ on an open interval (a, b), then f is concave up on (a, b).*

(*b*)   *If $f''(x) < 0$ on an open interval (a, b), then f is concave down on (a, b).*

Example 3   Find open intervals on which the following functions are concave up and open intervals on which they are concave down.

(a) $f(x) = x^2 - 4x + 3$ (b) $f(x) = x^3$ (c) $f(x) = x^3 - 3x^2 + 1$

Solution (a).   Calculating the first two derivatives we obtain

$$f'(x) = 2x - 4 \quad \text{and} \quad f''(x) = 2$$

Since $f''(x) > 0$ for all x, the function f is concave up on $(-\infty, +\infty)$. This is consistent with Figure 4.2.4.

Solution (b).   Calculating the first two derivatives we obtain

$$f'(x) = 3x^2 \quad \text{and} \quad f''(x) = 6x$$

Since $f''(x) < 0$ if $x < 0$ and $f''(x) > 0$ if $x > 0$, the function f is concave down on $(-\infty, 0)$ and concave up on $(0, +\infty)$. This is consistent with Figure 4.2.5.

Solution (c).   Calculating the first two derivatives we obtain

$$f'(x) = 3x^2 - 6x \quad \text{and} \quad f''(x) = 6x - 6$$

Thus, the curve is concave up where

$$6x - 6 > 0 \tag{1}$$

and concave down where

$$6x - 6 < 0 \tag{2}$$

Since (1) holds if $x > 1$ and (2) holds if $x < 1$, we conclude that

f is concave up on $(1, +\infty)$

f is concave down on $(-\infty, 1)$

This is consistent with Figure 4.2.7.   ◀

Points where a graph changes from concave up to concave down or vice versa are of special interest. We make the following definition.

4.2.5   DEFINITION.   If f is continuous on an open interval containing x_0, and if f changes the direction of its concavity at x_0, then the point $(x_0, f(x_0))$ on the graph of f is called an ***inflection point*** of f, and we say that f has an ***inflection point at x_0***. (See Figure 4.2.10.)

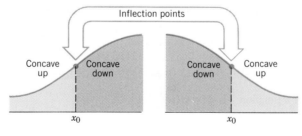

Figure 4.2.10

Referring to Example 3, the function $f(x) = x^2 - 4x + 3$ is concave up on the entire interval $(-\infty, +\infty)$, and hence has no inflection points since there are no changes in concavity. The function $f(x) = x^3$ changes from concave up to concave down at $x = 0$, so an inflection point occurs there. Since $f(0) = 0$, that inflection point is $(0, 0)$ (Figure 4.2.11). Similarly, the function $f(x) = x^3 - 3x^2 + 1$ has an inflection point at $x = 1$ (Figure 4.2.12).

Theorem 4.2.4 is not applicable on intervals in which there are points where $f''(x) = 0$. For such intervals the direction of concavity is best determined by using Definition 4.2.3 directly.

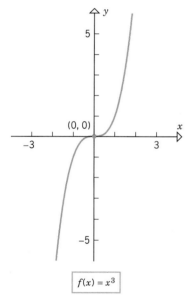

$$f(x) = x^3$$

Figure 4.2.11

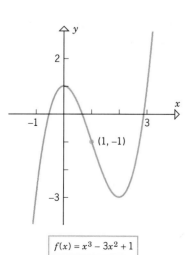

$$f(x) = x^3 - 3x^2 + 1$$

Figure 4.2.12

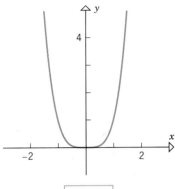

$f(x) = x^4$ Figure 4.2.13

Example 4 Determine where the function $f(x) = x^4$ is concave up and where it is concave down.

Solution. Differentiating yields

$$f'(x) = 4x^3 \quad \text{and} \quad f''(x) = 12x^2$$

Since $f''(x) > 0$ if $x < 0$, and $f''(x) > 0$ if $x > 0$, it follows from Theorem 4.2.4 that f is concave up on the open interval $(-\infty, 0)$ and concave up on the open interval $(0, +\infty)$. However, Theorem 4.2.4 is not applicable on the *entire* interval $(-\infty, +\infty)$ because $f''(0) = 0$. But $f'(x) = 4x^3$ is increasing on $(-\infty, +\infty)$ so it follows from Definition 4.2.3 that f is concave up on $(-\infty, +\infty)$. (See Figure 4.2.13.) ◄

▶ Exercise Set 4.2

Exercises 1 and 2 refer to the function graphed in Figure 4.2.14.

1. Find the largest *open* intervals over which the function is

    (a) increasing   (b) decreasing

    (c) concave up   (d) concave down.

2. Find all values of x where the function has an inflection point.

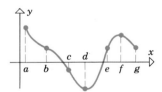

Figure 4.2.14

3. Determine the sign of dy/dx and d^2y/dx^2 at each of the points A, B, and C in Figure 4.2.15.

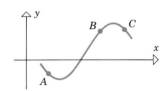

Figure 4.2.15

4. The graph of the *derivative* f' of a function f is shown in Figure 4.2.16. In parts (a)–(f), replace the question mark with <, =, or >.

    (a) $f(0) ? f(1)$ (b) $f(1) ? f(2)$

    (c) $f'(0) ? 0$ (d) $f'(1) ? 0$

    (e) $f''(0) ? 0$ (f) $f''(2) ? 0$.

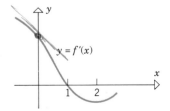

Figure 4.2.16

In Exercises 5–20, find the largest intervals on which f is (a) increasing, (b) decreasing; find the largest open interval on which f is (c) concave up, (d) concave down; and (e) find the x-coordinates of all inflection points.

5. $f(x) = x^2 - 5x + 6$.   6. $f(x) = 4 - 3x - x^2$.

7. $f(x) = (x + 2)^3$.   8. $f(x) = 5 + 12x - x^3$.

9. $f(x) = 3x^3 - 4x + 3$.   10. $f(x) = x^4 - 8x^2 + 16$.

11. $f(x) = 3x^4 - 4x^3$.   12. $f(x) = \dfrac{x}{x^2 + 2}$.

13. $f(x) = \cos x, \ 0 < x < 2\pi$.

14. $f(x) = \sin^2 2x, \ 0 < x < \pi$.

15. $f(x) = \tan x, \ -\pi/2 < x < \pi/2$.

16. $f(x) = x^{2/3}$.   17. $f(x) = \sqrt[3]{x} + 2$.

18. $f(x) = x^{4/3} - x^{1/3}$.   19. $f(x) = x^{1/3}(x + 4)$.

20. $f(x) = \begin{cases} \frac{1}{2}x^2, & x \le 0 \\ -x^2, & x > 0. \end{cases}$

[*Hint:* Refer to the theorem preceding Exercise 73, Section 3.3.]

21. Determine where $f(x) = x + \sin x$ is increasing.

22. Show that

$$\frac{a}{a^2 + 9} > \frac{b}{b^2 + 9}$$

if $3 \le a < b$. [*Hint:* Show that $x/(x^2 + 9)$ is decreasing on $[3, +\infty)$.]

23. Show that $x + 1/x > 2$ if $x > 1$. [*Hint:* Show that $f(x) = x + 1/x$ is increasing on $[1, +\infty)$.]

24. Show that $x < \tan x$ if $0 < x < \pi/2$. [*Hint:* Show that $f(x) = \tan x - x$ is increasing on $[0, \pi/2)$.]

25. Show that $\sqrt[3]{1 + x} < 1 + \frac{1}{3}x$ if $x > 0$. [*Hint:* Show that $f(x) = 1 + \frac{1}{3}x - \sqrt[3]{1 + x}$ is increasing on $[0, +\infty)$.]

26. In each part sketch a continuous curve $y = f(x)$ with the stated properties.

    (a) $f(2) = 4, \ f'(2) = 0, \ f''(x) < 0$ for all x

(b) $f(2) = 4$, $f'(2) = 0$, $f''(x) > 0$ for $x < 2$, $f''(x) < 0$ for $x > 2$

(c) $f(2) = 4$, $f''(x) > 0$ for $x \neq 2$ and
$$\lim_{x \to 2^+} f'(x) = -\infty, \quad \lim_{x \to 2^-} f'(x) = +\infty$$

27. In each part sketch a continuous curve $y = f(x)$ with the stated properties.

(a) $f(2) = 4$, $f'(2) = 0$, $f''(x) > 0$ for all x

(b) $f(2) = 4$, $f'(2) = 0$, $f''(x) < 0$ for $x < 2$, $f''(x) > 0$ for $x > 2$

(c) $f(2) = 4$, $f''(x) < 0$ for $x \neq 2$ and
$$\lim_{x \to 2^+} f'(x) = +\infty, \quad \lim_{x \to 2^-} f'(x) = -\infty$$

28. In parts (a)–(c) sketch a continuous curve $y = f(x)$ with the stated properties.

(a) $f(2) = 4$, $f'(2) = 1$, $f''(x) < 0$ for $x < 2$, $f''(x) > 0$ for $x > 2$

(b) $f(2) = 4$, $f''(x) > 0$ for $x < 2$, $f''(x) < 0$ for $x > 2$, and
$$\lim_{x \to 2^-} f'(x) = +\infty, \quad \lim_{x \to 2^+} f'(x) = +\infty$$

(c) $f(2) = 4$, $f''(x) < 0$ for $x \neq 2$, and
$$\lim_{x \to 2^-} f'(x) = 1, \quad \lim_{x \to 2^+} f'(x) = -1$$

29. In each part sketch a continuous curve $y = f(x)$ with the stated properties.

(a) $f(2) = 4$, $f''(x) < 0$ if $x \neq 2$, and
$$\lim_{x \to 2^-} f'(x) = 0, \quad \lim_{x \to 2^+} f'(x) = +\infty$$

(b) $f(0) = 0$, $f(2) = 4$, $f''(x) = 0$ for all x.

30. Find the inflection points, if any, of $f(x) = (x - a)^3$, where a is constant.

31. Find the inflection points, if any, of $f(x) = (x - a)^4$, where a is constant.

32. Find the inflection points, if any, of $f(x) = ax^2 + bx + c$ $(a \neq 0)$.

In Exercises 33–38, use Definition 4.2.1 to prove the given statement.

---

33. $f(x) = x^2$ is increasing on $[0, +\infty)$.

34. $f(x) = x^2 - 2x$ is decreasing on $(-\infty, 1]$.

35. $f(x) = \sqrt{x}$ is increasing on $[0, +\infty)$.

36. $f(x) = 1/x$ is decreasing on $(0, +\infty)$.

37. If functions f and g are increasing on an interval I, then $f + g$ is increasing on I.

38. If functions f and g are positive and increasing on an interval I, then their product $f \cdot g$ is increasing on I.

39. Give an example of functions f and g that are increasing on an interval I, but $f - g$ is decreasing on I.

40. For the general cubic polynomial
$$f(x) = ax^3 + bx^2 + cx + d \quad (a \neq 0)$$
find conditions on a, b, c, and d to assure that f is always increasing or always decreasing on $(-\infty, +\infty)$.

41. Prove that a third-degree polynomial
$$f(x) = ax^3 + bx^2 + cx + d \quad (a \neq 0)$$
has exactly one inflection point.

42. Prove that an nth-degree polynomial
$$f(x) = a_0 x^n + a_1 x^{n-1} + \cdots + a_n \quad (a_0 \neq 0)$$
has at most $n - 2$ inflection points.

43. Prove:

(a) If f is increasing on the intervals $(a, c]$ and $[c, b)$, then f is increasing on (a, b).

(b) If f is decreasing on the intervals $(a, c]$ and $[c, b)$, then f is decreasing on (a, b).

■ 4.3 RELATIVE EXTREMA; FIRST AND SECOND DERIVATIVE TESTS

In this section we shall develop techniques for finding the highest and lowest points on the graph of a function or, equivalently, the largest and smallest values of the function. The methods that we develop here will have important applications in later sections.

□ RELATIVE MAXIMA
AND MINIMA

The graphs of many functions form hills and valleys. The tops of the hills are called *relative maxima* and the bottoms of the valleys are called *relative minima* (Figure 4.3.1). Just as the top of a hill on the earth's terrain need not be the highest point on earth, so a relative maximum need not be the highest point on the entire graph. However, relative maxima, like tops of hills, are the high points in their *immediate vicinity*, and relative minima, like valley bottoms, are the low points in their *immediate vicinity*. These ideas are captured in the following definitions.

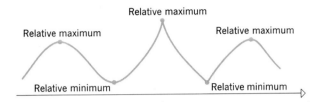

Figure 4.3.1

4.3.1 DEFINITION. A function f is said to have a ***relative maximum*** at x_0 if $f(x_0) \geq f(x)$ for all x in some open interval containing x_0.

4.3.2 DEFINITION. A function f is said to have a ***relative minimum*** at x_0 if $f(x_0) \leq f(x)$ for all x in some open interval containing x_0.

4.3.3 DEFINITION. A function f is said to have a ***relative extremum*** at x_0 if it has either a relative maximum or a relative minimum at x_0.

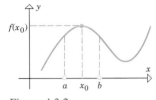

Figure 4.3.2

Example 1 The function f graphed in Figure 4.3.2 has a relative maximum at x_0 because there is an open interval containing x_0 on which $f(x_0) \geq f(x)$ holds. [See interval (a, b) in the figure.] ◄

☐ **CRITICAL POINTS**

Relative extrema can be viewed as the transition points that separate the regions where a graph is increasing from those where it is decreasing. As suggested by Figure 4.3.3, the relative extrema of a function f occur either at points where the graph of f has a horizontal tangent or at points where f is not differentiable.

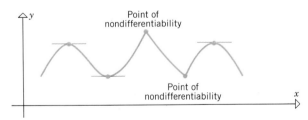

Figure 4.3.3

This is the content of the following theorem (whose proof is given at the end of this section).

4.3.4 THEOREM. *If f has a relative extremum at x_0, then either $f'(x_0) = 0$ or f is not differentiable at x_0.*

Theorem 4.3.4 is of such importance that there is some terminology associated with it.

4.3.5 DEFINITION. A ***critical point*** for a function f is any value of x in the domain of f at which $f'(x) = 0$ or at which f is not differentiable; the critical points where $f'(x) = 0$ are called ***stationary points*** of f.

With the terminology of this definition, Theorem 4.3.4 states:

The relative extrema of a function, if any, occur at critical points.

Example 2   For each of the functions graphed in Figure 4.3.4, x_0 is a critical point. In parts (a), (b), (c), and (d), x_0 is a stationary point because there is a horizontal tangent at x_0, and in the remaining parts x_0 is a critical point, but not a stationary point, because the derivative does not exist there.   ◄

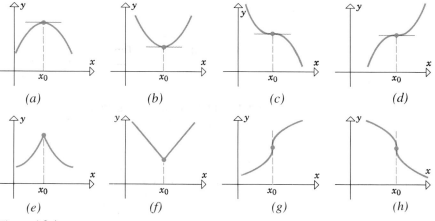

Figure 4.3.4

□ **FIRST AND SECOND**
   DERIVATIVE TESTS

Although the relative extrema of a function must occur at critical points, a relative extremum need not occur at *every* critical point. For example, in parts (c), (d), (g), and (h) of Figure 4.3.4, there is a critical point but no relative extremum at x_0. However, what distinguishes these four cases from the rest is that there is no sign change in the first derivative at x_0. To be specific,

• a relative maximum occurs at those critical points where the sign of the first derivative changes from positive to negative moving in the positive x-direction (Figure 4.3.4a, e);

• a relative minimum occurs at those critical points where the sign of the first derivative changes from negative to positive moving in the positive x-direction (Figure 4.3.4b, f);

• no relative extremum occurs at those critical points where the sign of the first derivative does not change moving in the positive x-direction (Figure 4.3.4c, d, g, h).

These ideas are stated more precisely in the following theorem whose proof is given in Appendix C.

4.3.6   THEOREM (*First Derivative Test*).   *Suppose f is continuous at a critical point x_0.*

(a) *If $f'(x) > 0$ on an open interval extending left from x_0 and $f'(x) < 0$ on an open interval extending right from x_0, then f has a relative maximum at x_0.*

(b) *If $f'(x) < 0$ on an open interval extending left from x_0 and $f'(x) > 0$ on an open interval extending right from x_0, then f has a relative minimum at x_0.*

(c) *If $f'(x)$ has the same sign [either $f'(x) > 0$ or $f'(x) < 0$] on an open interval extending left from x_0 and on an open interval extending right from x_0, then f does not have a relative extremum at x_0.*

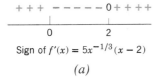

$$+ + + - - - - - 0 + + + +$$

Sign of $f'(x) = 5x^{-1/3}(x - 2)$

(a)

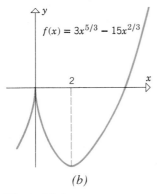

(b)

Figure 4.3.5

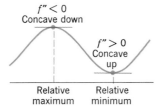

Figure 4.3.6

It follows from this theorem that:

> *The relative extrema, if any, on an open interval where a function f is continuous and not constant occur at those critical points where f' changes sign.*

Example 3   Locate the relative extrema of $f(x) = 3x^{5/3} - 15x^{2/3}$.

Solution.

$$f'(x) = 5x^{2/3} - 10x^{-1/3} = 5x^{-1/3}(x - 2)$$

Since $f'(x)$ does not exist when $x = 0$, and $f'(x) = 0$ when $x = 2$, the critical points of f are $x = 0$ and $x = 2$.

As shown in Figure 4.3.5a, the sign of f' changes from positive to negative at 0 and from negative to positive at 2. Thus, there is a relative maximum at 0 and a relative minimum at 2. Although not needed for the solution, the graph of f is shown in Figure 4.3.5b. ◀

Example 4   Locate the relative extrema of $f(x) = x^3 - 3x^2 + 3x - 1$.

Solution.   Since f is differentiable everywhere, the only possible critical points are stationary points. Differentiating yields

$$f'(x) = 3x^2 - 6x + 3 = 3(x - 1)^2$$

Solving $f'(x) = 0$ yields $x = 1$ as the only critical point. Since $3(x - 1)^2 \geq 0$ for all x, $f'(x)$ does not change sign at $x = 1$, so f does not have a relative extremum at $x = 1$. Thus, f has no relative extrema. ◀

There is another test for relative extrema that is often easier to apply than the first derivative test. It is based on the geometric observation that a function has a relative maximum where its graph has a horizontal tangent line and is concave down (Figure 4.3.6), and has a relative minimum where its graph has a horizontal tangent line and is concave up (Figure 4.3.6).

4.3.7   THEOREM (*Second Derivative Test*).   *Suppose that f is twice differentiable at a stationary point x_0.*

(a)   If $f''(x_0) > 0$, then f has a relative minimum at x_0.
(b)   If $f''(x_0) < 0$, then f has a relative maximum at x_0.

A formal proof of this result is given in Appendix C.

Example 5   Locate and describe the relative extrema of $f(x) = x^4 - 2x^2$.

Solution.

$$f'(x) = 4x^3 - 4x = 4x(x - 1)(x + 1)$$

$$f''(x) = 12x^2 - 4$$

Solving $f'(x) = 0$ yields the stationary points $x = 0$, $x = 1$, and $x = -1$. Since

$$f''(0) = -4 < 0$$

$$f''(1) = 8 > 0$$

$$f''(-1) = 8 > 0$$

there is a relative maximum at $x = 0$ and relative minima at $x = 1$ and $x = -1$. (Figure 4.3.7) ◀

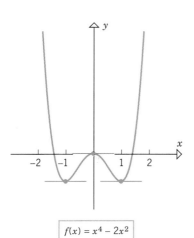

$$f(x) = x^4 - 2x^2$$

Figure 4.3.7

REMARK.   If f is not twice differentiable at the critical point x_0, or if $f''(x_0) = 0$, then we must either rely on the first derivative test or devise more imaginative techniques appropriate to the problem.

REMARK.   In some problems the first derivative test is easier to apply, and in others the second derivative test is easier. Thus, in each problem some thought should be given to determining the test that involves the least amount of computation.

Proof of Theorem 4.3.4.   We shall prove this theorem in the case where there is a relative maximum at x_0. The proof in the case of a relative minimum is similar and is left to the reader.

There are two possibilities—either f is differentiable at x_0 or it is not. If it is not, then x_0 is a critical point for f and we are done. If f is differentiable at x_0, then we must show that $f'(x_0) = 0$. We shall do this by showing that $f'(x_0) \geq 0$ and $f'(x_0) \leq 0$, from which it follows that $f'(x_0) = 0$. From the definition of a derivative we have

$$f'(x_0) = \lim_{h \to 0} \frac{f(x_0 + h) - f(x_0)}{h}$$

so that

$$f'(x_0) = \lim_{h \to 0^+} \frac{f(x_0 + h) - f(x_0)}{h} \tag{1}$$

and

$$f'(x_0) = \lim_{h \to 0^-} \frac{f(x_0 + h) - f(x_0)}{h} \tag{2}$$

Because f has a relative maximum at x_0, there is an open interval (a, b) containing x_0 in which $f(x) \leq f(x_0)$ for all x in (a, b).

Assume that h is sufficiently small so that $x_0 + h$ lies in the interval (a, b). Thus,

$$f(x_0 + h) \leq f(x_0) \quad \text{or equivalently,} \quad f(x_0 + h) - f(x_0) \leq 0$$

Thus, if h is negative,

$$\frac{f(x_0 + h) - f(x_0)}{h} \geq 0 \tag{3}$$

and if h is positive,

$$\frac{f(x_0 + h) - f(x_0)}{h} \leq 0 \tag{4}$$

But an expression that never assumes negative values cannot approach a negative limit and an expression that never assumes positive values cannot approach a positive limit, so that

$$f'(x_0) = \lim_{h \to 0^-} \frac{f(x_0 + h) - f(x_0)}{h} \geq 0 \qquad \boxed{\text{From (2) and (3)}}$$

and

$$f'(x_0) = \lim_{h \to 0^+} \frac{f(x_0 + h) - f(x_0)}{h} \leq 0 \qquad \boxed{\text{From (1) and (4)}}$$

Since $f'(x_0) \geq 0$ and $f'(x_0) \leq 0$, it must be that $f'(x_0) = 0$. ∎

▶ Exercise Set 4.3

In Exercises 1–16, locate the critical points and classify them as stationary points or points of nondifferentiability.

1. $f(x) = x^2 - 5x + 6$.   **2.** $f(x) = 4x^2 + 2x - 5$.

3. $f(x) = x^3 + 3x^2 - 9x + 1$.

4. $f(x) = 2x^3 - 6x + 7$.   **5.** $f(x) = x^4 - 6x^2 - 3$.

6. $f(x) = 3x^4 - 4x^3$.   **7.** $f(x) = \dfrac{x}{x^2 + 2}$.

8. $f(x) = \dfrac{x^2 - 3}{x^2 + 1}$.   **9.** $f(x) = x^{2/3}$.

10. $f(x) = \sqrt[3]{x + 2}$.   **11.** $f(x) = \cos 3x$.

12. $f(x) = x \tan x,\ -\pi/2 < x < \pi/2$.

13. $f(x) = \sin^2 2x,\ 0 < x < 2\pi$.

14. $f(x) = |\sin x|$.   **15.** $f(x) = x^{1/3}(x + 4)$.

16. $f(x) = x^{4/3} - 6x^{1/3}$.

17. Figure 4.3.8 shows the graph of the derivative f' of a function f. Find all values of x, if any, where f has

    (a) relative minima   (b) relative maxima

    (c) inflection points.

18. Repeat Exercise 17 using Figure 4.3.9.

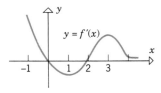

Figure 4.3.8

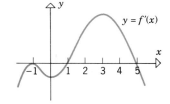

Figure 4.3.9

In Exercises 19–22, the derivative of a continuous function is given. Find all critical points, and determine whether a relative maximum, relative minimum, or neither occurs there.

19. $f'(x) = x^3(x^2 - 5)$.

20. $f'(x) = x^2(2x + 1)(x - 1)$.

21. $f'(x) = \dfrac{9 - 4x^2}{\sqrt[3]{x + 1}}$.

22. $f'(x) = 2 \sin^3 x - \sin^2 x,\ 0 < x < 2\pi$.

In Exercises 23–26, find the relative extrema using (a) the first derivative test; (b) the second derivative test.

23. $f(x) = 1 - 4x - x^2$.

24. $f(x) = 2x^3 - 9x^2 + 12x$.

25. $f(x) = \sin^2 x,\ 0 < x < 2\pi$.

26. $f(x) = \frac{1}{2}x - \sin x,\ 0 < x < 2\pi$.

In Exercises 27–44, use any method to find the relative extrema.

27. $f(x) = x^3 + 5x - 2$.   **28.** $f(x) = x^4 - 2x^2 + 7$.

29. $f(x) = x(x - 1)^2$.   **30.** $f(x) = x^4 + 2x^3$.

31. $f(x) = 2x^2 - x^4$.   **32.** $f(x) = (2x - 1)^5$.

33. $f(x) = x^{4/5}$.   **34.** $f(x) = 2x + x^{2/3}$.

35. $f(x) = \dfrac{x^2}{x^2 + 1}$.   **36.** $f(x) = \dfrac{x}{x + 2}$.

37. $f(x) = |x^2 - 4|$.   **38.** $f(x) = \begin{cases} 9 - x, & x \le 3 \\ x^2 - 3, & x > 3. \end{cases}$

39. $f(x) = \cos^2 x$.

40. $f(x) = \sqrt{3}x + 2 \sin x,\ 0 < x < 2\pi$.

41. $f(x) = \tan(x^2 + 1)$.

42. $f(x) = \dfrac{\sin x}{2 + \cos x},\ 0 < x < 2\pi$.

43. $f(x) = |\sin 2x|,\ 0 < x < 2\pi$.

44. $f(x) = \cos 4x + 2 \sin 2x,\ 0 < x < \pi$.

45. Find a value of k so that $x^2 + \dfrac{k}{x}$ will have a relative extremum at $x = 3$.

46. Find a value of k so that $\dfrac{x}{x^2 + k}$ will have a relative extremum at $x = 2.5$.

47. Recall that the second derivative test (Theorem 4.3.7) does not apply if $f''(x_0) = 0$. Give examples to show that a function f can have a relative maximum at x_0, a relative minimum at x_0, or neither if $f''(x_0) = 0$. [*Hint:* Try functions of the form $f(x) = x^n$.]

48. Let h and g have relative maxima at x_0. Prove or disprove:

    (a) $h + g$ has a relative maximum at x_0

    (b) $h - g$ has a relative maximum at x_0.

49. Sketch some curves that show that the three parts of the first derivative test (Theorem 4.3.6) can be false without the assumption that f is continuous at x_0.

50. Prove: If f is continuous on an open interval I that contains an inflection point at x_0, and if f' exists everywhere in I except perhaps at x_0, then either $f''(x_0) = 0$ or f' is not differentiable at x_0. [*Hint:* Modify the proof of Theorem 4.3.4 using f' and f'' in place of f and f', respectively.]

▪ 4.4 GRAPHS OF POLYNOMIALS AND RATIONAL FUNCTIONS

Graphs of functions are used for many purposes in mathematics and science. In applications where numerical values are read from graphs, extremely accurate point-plotting is required. Such graphs are best generated on a computer. In many mathematical applications, however, accurately plotted points are not needed; what is required is the precise location of the key features such as critical points, inflection points, intervals of increase and decrease, points of discontinuity, and so forth. In this section we shall use the tools developed in the preceding section to investigate the key features of the graphs of polynomials and rational functions.

☐ **GRAPHS OF POLYNOMIALS**

Polynomials are among the simplest functions to graph; they are continuous functions, so their graphs have no breaks or holes, and they are differentiable functions, so their graphs have no sharp corners. The following steps are usually sufficient to obtain the main elements of the graph.

> **A Method for Graphing a Polynomial P(x)**
>
> **Step 1.** Calculate $P'(x)$ and $P''(x)$.
>
> **Step 2.** From $P'(x)$ determine the stationary points and the intervals where P is increasing and decreasing.
>
> **Step 3.** From $P''(x)$ determine the inflection points and the intervals where P is concave up and concave down.
>
> **Step 4.** Plot the intersection with the y-axis, the stationary points, the inflection points, and, if possible, the intersections with the x-axis. Finally, plot any additional points needed to obtain the accuracy desired in the graph.

Example 1 Sketch the graph of $y = x^3 - 3x + 2$.

Solution.

$$\frac{dy}{dx} = 3x^2 - 3 = 3(x - 1)(x + 1) \quad \text{and} \quad \frac{d^2y}{dx^2} = 6x$$

Figure 4.4.1 shows a convenient way of organizing the analysis. The first part of the figure shows the sign pattern of dy/dx, which can be obtained using the method of test points. The

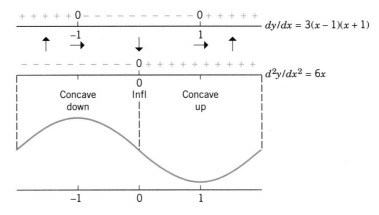

Figure 4.4.1

upward arrows indicate intervals where the graph is increasing, the downward arrows indicate intervals where the graph is decreasing, and the horizontal arrows mark the points where there is a horizontal tangent line. The second part of the figure shows the sign pattern of d^2y/dx^2. The bottom part of the figure combines the information in the previous parts to produce the general graph shape.

All that remains is to refine the sketch by plotting the stationary points, inflection points, and intersections with the coordinate axes. In this case, we will also plot the point where $x = 2$ for some extra precision (Figure 4.4.2). As a final note, the graph in Figure 4.4.2 was drawn as though it increases without bound as $x \to +\infty$ and decreases without bound as $x \to -\infty$. This is consistent with the limits

$$\lim_{x \to +\infty} (x^3 - 3x + 2) = +\infty \quad \text{and} \quad \lim_{x \to -\infty} (x^3 - 3x + 2) = -\infty$$

[See Formulas (8), (9), (11), (12), and Example 3 in Section 2.5.]   ◀

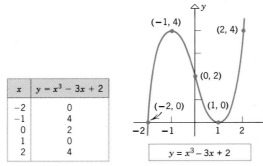

x	$y = x^3 - 3x + 2$
-2	0
-1	4
0	2
1	0
2	4

Figure 4.4.2

□ **GRAPHS OF RATIONAL FUNCTIONS**

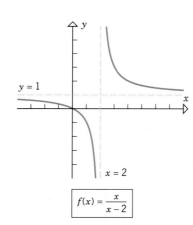

Figure 4.4.3

Recall that if $P(x)$ and $Q(x)$ are polynomials, then their ratio

$$f(x) = \frac{P(x)}{Q(x)}$$

is called a rational function of x. Graphing rational functions is complicated by the fact that discontinuities occur at points where $Q(x) = 0$. In Figure 4.4.3 we have sketched the graph of

$$f(x) = \frac{x}{x - 2}$$

This figure illustrates most of the characteristics that are typical of rational functions. At $x = 2$ the denominator of $f(x)$ is zero, so a discontinuity occurs in the graph at this point. Approaching the point $x = 2$, we have

$$\lim_{x \to 2^+} \frac{x}{x - 2} = +\infty \quad \text{and} \quad \lim_{x \to 2^-} \frac{x}{x - 2} = -\infty$$

The line $x = 2$ is called a *vertical asymptote* for the graph. Also

$$\lim_{x \to +\infty} \frac{x}{x - 2} = 1 \quad \text{and} \quad \lim_{x \to -\infty} \frac{x}{x - 2} = 1$$

so that the graph approaches the line $y = 1$ as $x \to +\infty$ and as $x \to -\infty$. The line $y = 1$ is called a *horizontal asymptote* for the graph. More precisely, we make the following definition.

4.4.1   DEFINITION.   A line $x = x_0$ is called a ***vertical asymptote*** for the graph of a function f if $f(x) \to +\infty$ or $f(x) \to -\infty$ as x approaches x_0 from the right or from the left. A line $y = y_0$ is called a ***horizontal asymptote*** for the graph of f if

$$\lim_{x \to +\infty} f(x) = y_0 \quad \text{or} \quad \lim_{x \to -\infty} f(x) = y_0$$

For rational functions whose numerator and denominator have no common factors, vertical asymptotes occur at points where the denominator is zero.

REMARK.   It is possible for the graph of a rational function to intersect a horizontal asymptote (Exercises 25–28), but it cannot intersect a vertical asymptote.

Example 2   Find all vertical and horizontal asymptotes of

$$\text{(a)} \quad f(x) = \frac{x^2 + 2x}{x^2 - 1} \qquad \text{(b)} \quad f(x) = \frac{x^3 + 1}{x - 2}$$

Solution (a).   The vertical asymptotes occur at the points where $x^2 - 1 = 0$; these are the points $x = -1$ and $x = 1$. Since

$$\lim_{x \to +\infty} f(x) = \lim_{x \to +\infty} \frac{x^2 + 2x}{x^2 - 1} = \lim_{x \to +\infty} \frac{x^2}{x^2} = 1$$

$$\lim_{x \to -\infty} f(x) = \lim_{x \to -\infty} \frac{x^2 + 2x}{x^2 - 1} = \lim_{x \to -\infty} \frac{x^2}{x^2} = 1$$

it follows that $y = 1$ is a horizontal asymptote. (A computer-generated graph of f is shown in Figure 4.4.4a.)

Solution (b).   The denominator is zero at $x = 2$, so a vertical asymptote occurs at this point. Since

$$\lim_{x \to +\infty} f(x) = \lim_{x \to +\infty} \frac{x^3 + 1}{x - 2} = \lim_{x \to +\infty} \frac{x^3}{x} = \lim_{x \to +\infty} x^2 = +\infty$$

$$\lim_{x \to -\infty} f(x) = \lim_{x \to -\infty} \frac{x^3 + 1}{x - 2} = \lim_{x \to -\infty} \frac{x^3}{x} = \lim_{x \to -\infty} x^2 = +\infty$$

there are no horizontal asymptotes. (A computer-generated graph of f is shown in Figure 4.4.4b.)   ◄

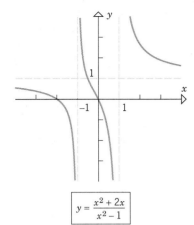

$y = \dfrac{x^2 + 2x}{x^2 - 1}$

(a)

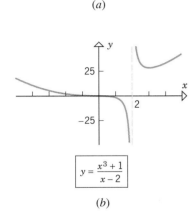

$y = \dfrac{x^3 + 1}{x - 2}$

(b)

Figure 4.4.4

For graphs of rational functions, the properties of interest are:

1. symmetries
2. x-intercepts
3. y-intercept
4. asymptotes
5. intervals of increase and decrease
6. stationary points
7. concavity
8. inflection points

One procedure for organizing the computations is to find the symmetries, the intercepts, and the asymptotes first. Once this is done, a preliminary sketch of the graph can be made by evaluating the function at some well-chosen points. This preliminary sketch can then be refined by calculating the first and second derivatives and using them to incorporate the missing information about stationary points, inflection points, and so forth.

Example 3   Sketch the graph of $y = \dfrac{2x^2 - 8}{x^2 - 16}$.

Solution.

• *Symmetries*: Replacing x by $-x$ does not change the equation, so the graph is symmetric about the y-axis.

• *x-intercepts*: Setting $y = 0$ yields the x-intercepts $x = -2$ and $x = 2$.

- *y-intercept*: Setting $x = 0$ yields the *y*-intercept $y = 1/2$.
- *Vertical asymptotes*: Setting $x^2 - 16 = 0$ yields the vertical asymptotes $x = -4$ and $x = 4$.
- *Horizontal asymptotes*: The limits

$$\lim_{x \to +\infty} \frac{2x^2 - 8}{x^2 - 16} = \lim_{x \to +\infty} \frac{2x^2}{x^2} = 2$$

$$\lim_{x \to -\infty} \frac{2x^2 - 8}{x^2 - 16} = \lim_{x \to -\infty} \frac{2x^2}{x^2} = 2$$

yield the horizontal asymptote $y = 2$.

The set of points where *x*-intercepts or vertical asymptotes occur is $\{-4, -2, 2, 4\}$. These points divide the *x*-axis into the open intervals

$$(-\infty, -4), \quad (-4, -2), \quad (-2, 2), \quad (2, 4), \quad (4, +\infty)$$

Over each of these intervals, *y* cannot change sign (why?). We can find the sign of *y* on each interval by choosing an arbitrary test point in the interval and evaluating $y = f(x)$ at the test points (Table 4.4.1).

Table 4.4.1

INTERVAL	TEST POINT	$y = \dfrac{2x^2 - 8}{x^2 - 16}$	SIGN OF y
$(-\infty, -4)$	$x = -5$	$y = 14/3$	$+$
$(-4, -2)$	$x = -3$	$y = -10/7$	$-$
$(-2, 2)$	$x = 0$	$y = 1/2$	$+$
$(2, 4)$	$x = 3$	$y = -10/7$	$-$
$(4, +\infty)$	$x = 5$	$y = 14/3$	$+$

Using the information in the table, we can draw small curve segments at the intercepts and the test points, using the sign of *y* to position the curve segment properly above or below the *x*-axis; these segments allow us to make an educated guess about the shape of the curve near the vertical and horizontal asymptotes, and we can draw curve segments near the asymptotes as well (Figure 4.4.5*a*). The curve segments suggest the rough shape of the graph. Of course, we may have missed some stationary points and the concavity may be incorrect in places, but this can now be fixed by using information from the first and second derivatives.

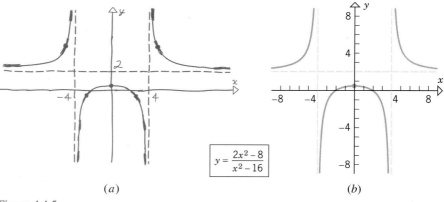

$$y = \frac{2x^2 - 8}{x^2 - 16}$$

(*a*)   (*b*)

Figure 4.4.5

- *Derivatives*:

$$\frac{dy}{dx} = \frac{(x^2 - 16)(4x) - (2x^2 - 8)(2x)}{(x^2 - 16)^2} = -\frac{48x}{(x^2 - 16)^2}$$

$$\frac{d^2y}{dx^2} = \frac{48(16 + 3x^2)}{(x^2 - 16)^3} \quad \text{(verify)}$$

- *Intervals of increase and decrease*: A sign analysis of dy/dx yields

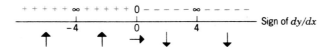

Thus, the graph is increasing on the intervals $(-\infty, -4)$ and $(-4, 0]$; and it is decreasing on the intervals $[0, 4)$ and $(4, +\infty)$. There is a stationary point at $x = 0$.

- *Concavity*: A sign analysis of d^2y/dx^2 yields

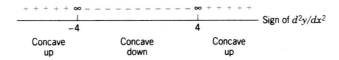

There are changes in concavity at the vertical asymptotes, $x = -4$ and $x = 4$, but there are no inflection points.

Using the information supplied by the derivatives to refine the rough sketch in Figure 4.4.5*a* yields the graph in Figure 4.4.5*b*. ◄

Example 4   Sketch the graph of $y = \dfrac{x^2 - 1}{x^3}$.

Solution.

- *Symmetries*: Replacing x by $-x$ and y by $-y$ yields an equation that simplifies back to the original equation, so the graph is symmetric about the origin.
- *x-intercepts*: Setting $y = 0$ yields the x-intercepts $x = -1$ and $x = 1$.
- *y-intercept*: Setting $x = 0$ leads to a division by zero, so that there is no y-intercept.
- *Vertical asymptotes*: Setting $x^3 = 0$ yields the vertical asymptote $x = 0$.
- *Horizontal asymptotes*: The limits

$$\lim_{x \to +\infty} \frac{x^2 - 1}{x^3} = \lim_{x \to +\infty} \frac{x^2}{x^3} = \lim_{x \to +\infty} \frac{1}{x} = 0$$

$$\lim_{x \to -\infty} \frac{x^2 - 1}{x^3} = \lim_{x \to -\infty} \frac{x^2}{x^3} = \lim_{x \to +\infty} \frac{1}{x} = 0$$

yield the horizontal asymptote $y = 0$.

The set of points where x-intercepts or vertical asymptotes occur, namely $\{-1, 0, 1\}$, divides the x-axis into the open intervals

$$(-\infty, -1), \quad (-1, 0), \quad (0, 1), \quad (1, +\infty)$$

Choosing a test point in each interval and finding the sign of y at the test points yields Table 4.4.2. From the information in the table we obtain the rough sketch of the graph in Figure 4.4.6a.

Table 4.4.2

INTERVAL	TEST POINT	$y = \dfrac{x^2 - 1}{x^3}$	SIGN OF y
$(-\infty, -1)$	$x = -2$	$y = -3/8$	$-$
$(-1, 0)$	$x = -1/2$	$y = 6$	$+$
$(0, 1)$	$x = 1/2$	$y = -6$	$-$
$(1, +\infty)$	$x = 2$	$y = 3/8$	$+$

We now refine the rough sketch.

- *Derivatives*:

$$\frac{dy}{dx} = \frac{x^3(2x) - (x^2 - 1)(3x^2)}{(x^3)^2} = \frac{3 - x^2}{x^4}$$

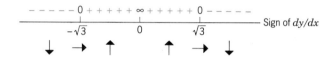

$$\frac{d^2y}{dx^2} = \frac{x^4(-2x) - (3 - x^2)(4x^3)}{(x^4)^2} = \frac{2(x^2 - 6)}{x^5}$$

- *Intervals of increase and decrease*:

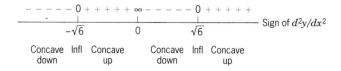

This analysis reveals stationary points at $x = -\sqrt{3}$ and $x = \sqrt{3}$.

- *Concavity*:

This analysis reveals that a change in concavity occurs at the vertical asymptote $x = 0$ and at the points $x = -\sqrt{6}$ and $x = \sqrt{6}$.

The final graph with the stationary points and inflection points plotted is shown in Figure 4.4.6b.   ◀

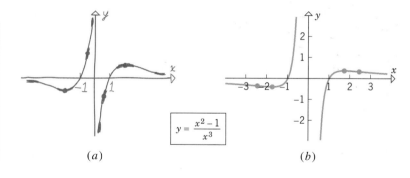

x	$y = \dfrac{x^2 - 1}{x^3}$
$-\sqrt{6} \approx -2.45$	$-\dfrac{5\sqrt{6}}{36} \approx -0.34$
$-\sqrt{3} \approx -1.73$	$-\dfrac{2\sqrt{3}}{9} \approx -0.38$
$\sqrt{3} \approx 1.73$	$\dfrac{2\sqrt{3}}{9} \approx 0.38$
$\sqrt{6} \approx 2.45$	$\dfrac{5\sqrt{6}}{36} \approx 0.34$

Figure 4.4.6

(a) (b)

☐ **COMPUTER-GENERATED
GRAPHS**

Although computers and graphing calculators can generate graphs with great speed and accuracy, the fact that a computer or calculator display can reveal only a finite portion of an entire graph can lead to problems. For example, it is evident that the graph of the polynomial equation $y = (x - 100)^3$ has an inflection point at $x = 100$, since the graph can be obtained by translating the graph of $y = x^3$ in the positive x-direction 100 units (Figure 4.4.7a). However, if one were to program a computer or calculator to graph this equation over the interval $[-40, 40]$, for example, the inflection point, which is the key feature of the graph, would be completely missed. For example, Figure 4.4.7b was generated using such a graphing program. Not only was the inflection point missed but the program was unable to show the y-intercept because of the wide vertical scale. This may create the false impression that the graph has a vertical asymptote at $x = 0$. In short, an intelligent use of computers and calculators for graphing requires a good understanding of the basic principles of graphing that result from the methods of calculus.

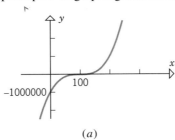

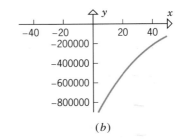

Figure 4.4.7   (a)   (b)

▶ Exercise Set 4.4   Ⓒ 59, 60

In Exercises 1–18, use the techniques illustrated in this section to sketch the graph of the given polynomial. Plot the stationary points and the inflection points.

1. $x^2 - 2x - 3$.
2. $1 + x - x^2$.
3. $x^3 - 3x + 1$.
4. $2x^3 - 6x + 4$.
5. $x^3 + 3x^2 + 5$.
6. $x^2 - x^3$.
7. $2x^3 - 3x^2 + 12x + 9$.
8. $x^3 - 3x^2 + 3$.
9. $(x - 1)^4$.
10. $(x - 1)^5$.
11. $x^4 + 2x^3 - 1$.
12. $x^4 - 2x^2 - 12$.
13. $x^4 - 3x^3 + 3x^2 + 1$.
14. $x^5 - 4x^4 + 4x^3$.
15. $3x^5 - 5x^3$.
16. $3x^4 + 4x^3$.
17. $x(x - 1)^3$.
18. $x^5 + 5x^4$.

In Exercises 19–24, find equations of all vertical and horizontal asymptotes for the graph of the given rational function.

19. $\dfrac{3x}{x - 2}$.
20. $\dfrac{4x + 1}{3x + 2}$.
21. $\dfrac{x^3}{x^2 - 5}$.
22. $\dfrac{3 - x}{x^2 + 1}$.
23. $\dfrac{x^2}{x^2 - 2x - 3}$.
24. $\dfrac{2x^2 + 1}{3x^2 + 6x}$.

In Exercises 25–28, find all values of x where the graph of the given rational function crosses its horizontal asymptote.

25. $\dfrac{x^2}{x^2 + 2x + 5}$.
26. $\dfrac{x^2 - 3x + 2}{x^2}$.
27. $\dfrac{x^2 + 1}{2x^2 - 6x}$.
28. $\dfrac{25 - 9x^2}{x^3}$.

In Exercises 29–47, use the techniques illustrated in this section to sketch the graph of the given rational function. Show any horizontal and vertical asymptotes, and plot the stationary points and, if reasonable, the inflection points.

29. $\dfrac{2x}{x - 3}$.
30. $\dfrac{x}{x^2 - 1}$.
31. $\dfrac{x^2}{x^2 - 1}$.
32. $\dfrac{1}{(x - 1)^2}$.
33. $\dfrac{x}{1 + x^2}$.
34. $1 - \dfrac{1}{x}$.
35. $\dfrac{x - 1}{x - 2}$.
36. $\dfrac{1}{x^2 + 1}$.
37. $x^2 - \dfrac{1}{x}$.
38. $\dfrac{2x^2 - 1}{x^2}$.
39. $\dfrac{1 - x}{x^2}$.
40. $\dfrac{8}{4 - x^2}$.
41. $\dfrac{x - 1}{x^2 - 4}$.
42. $\dfrac{8(x - 2)}{x^2}$.
43. $\dfrac{(x - 1)^2}{x^2}$.
44. $2 + \dfrac{3}{x} - \dfrac{1}{x^3}$.
45. $3 - \dfrac{4}{x} - \dfrac{4}{x^2}$.
46. $\dfrac{x^2 - 1}{x^2 + 1}$.
47. $\dfrac{x^3 - 1}{x^3 + 1}$.

48. **(Oblique Asymptotes)** If a rational function $P(x)/Q(x)$ is

such that the degree of the numerator exceeds the degree of the denominator by *one*, then the graph of $P(x)/Q(x)$ will have an **oblique asymptote**, that is, an asymptote that is neither vertical nor horizontal. To see why, we perform the division of $P(x)$ by $Q(x)$ to obtain

$$\frac{P(x)}{Q(x)} = (ax + b) + \frac{R(x)}{Q(x)}$$

where $ax + b$ is the quotient and $R(x)$ is the remainder. Use the fact that the degree of the remainder $R(x)$ is less than the degree of the divisor $Q(x)$ to help prove

$$\lim_{x \to +\infty} \left[\frac{P(x)}{Q(x)} - (ax + b) \right] = 0$$

$$\lim_{x \to -\infty} \left[\frac{P(x)}{Q(x)} - (ax + b) \right] = 0$$

These results tell us that the graph of the equation $y = P(x)/Q(x)$ "approaches" the line (an oblique asymptote) $y = ax + b$ as $x \to +\infty$ or $x \to -\infty$ (Figure 4.4.8).

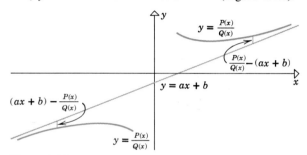

Figure 4.4.8

In Exercises 49–53, sketch the graph of the rational function. Show all vertical, horizontal, and oblique asymptotes (see Exercise 48).

49. $\dfrac{x^2 - 2}{x}$.   **50.** $\dfrac{x^2 - 2x - 3}{x + 2}$.   **51.** $\dfrac{(x - 2)^3}{x^2}$.

52. $\dfrac{4 - x^3}{x^2}$.   **53.** $x + 1 - \dfrac{1}{x} - \dfrac{1}{x^2}$.

54. Find all values of x where the graph of

$$y = \frac{2x^3 - 3x + 4}{x^2}$$

crosses its oblique asymptote. [See Exercise 48.]

55. Let $f(x) = x^2 + 1/x$. Show that the graph of $y = f(x)$ approaches the curve $y = x^2$ "asymptotically" in the sense that

$$\lim_{x \to +\infty} [f(x) - x^2] = 0 \quad \text{and} \quad \lim_{x \to -\infty} [f(x) - x^2] = 0$$

Sketch the graph of $y = f(x)$ showing this asymptotic behavior.

56. Let $f(x) = 3 - x^2 + 2/x$. Show that $y = f(x)$ approaches the curve $y = 3 - x^2$ asymptotically in the sense described in Exercise 55. Sketch the graph of $y = f(x)$ showing this asymptotic behavior.

57. A rectangular plot of land is to be fenced off so that the area enclosed will be 400 ft^2. Let L be the length of fencing needed and x the length of one side of the rectangle. Show that $L = 2x + 800/x$ for $x > 0$, and sketch the graph of L versus x for $x > 0$.

58. A box with a square base and open top is to be made from sheet metal so that its volume is 500 in^3. Let S be the area of the surface of the box and x the length of a side of the square base. Show that $S = x^2 + 2000/x$ for $x > 0$, and sketch the graph of S versus x for $x > 0$.

59. Figure 4.4.9 shows a computer-generated graph of the polynomial $y = 0.1x^5(x - 1)$ using a viewing window of $-2 \le x \le 2.5$ and $-1 \le y \le 5$. Show that the choice of the vertical scale caused the computer to miss important features of the graph. Find the features that were missed and make your own sketch of the graph that shows the missing features.

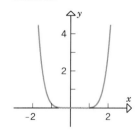

Figure 4.4.9

60. Figure 4.4.10 shows a computer-generated graph of the polynomial $y = 0.1x^5(x + 1)^2$ using a viewing window of $-2 \le x \le 1.5$ and $-0.2 \le y \le 0.2$. Show that the choice of the vertical scale caused the computer to miss important features of the graph. Find the features that were missed and make your own sketch of the graph that shows the missing features.

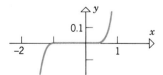

Figure 4.4.10

■ 4.5   OTHER GRAPHING PROBLEMS

In this section we shall examine the graphs of various continuous functions that have characteristics not found in the graphs of polynomials and rational functions.

☐ VERTICAL TANGENT
LINES AND CUSPS

4.5.1   DEFINITION.   The graph of a function f is said to have a **vertical tangent line** at x_0 if f is continuous at x_0 and $|f'(x)|$ approaches $+\infty$ as $x \rightarrow x_0$.

Four common ways in which vertical tangent lines occur are illustrated in Figure 4.5.1. The curve segments in parts (c) and (d) of Figure 4.5.1 are called *cusps*. To be precise, we make the following definition.

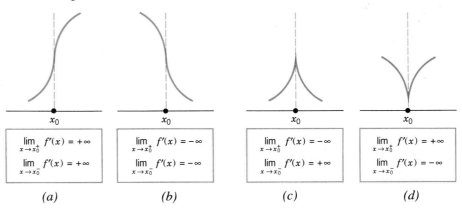

Figure 4.5.1 (a) (b) (c) (d)

4.5.2   DEFINITION.   The graph of a function f is said to have a **cusp** at x_0 if f is continuous at x_0 and $f'(x) \rightarrow +\infty$ as x approaches x_0 from one side, while $f'(x) \rightarrow -\infty$ as x approaches x_0 from the other side.

Example 1   Sketch the graph of $y = (x - 4)^{2/3}$.

Solution.   Let $f(x) = (x - 4)^{2/3}$.

- *Symmetries*: There are no symmetries about the coordinate axes or the origin (verify).
- *x-intercepts*: Setting $y = 0$ yields the x-intercept $x = 4$.
- *y-intercepts*: Setting $x = 0$ yields the y-intercept $y = \sqrt[3]{16}$.
- *Vertical asymptotes*: None, since $f(x) = (x - 4)^{2/3}$ is a continuous function.
- *Horizontal asymptotes*: None, since
$$\lim_{x \to +\infty} (x - 4)^{2/3} = +\infty$$
$$\lim_{x \to -\infty} (x - 4)^{2/3} = +\infty$$
- *Derivatives*:
$$\frac{dy}{dx} = f'(x) = \frac{2}{3}(x - 4)^{-1/3} = \frac{2}{3(x - 4)^{1/3}}$$
$$\frac{d^2y}{dx^2} = f''(x) = -\frac{2}{9}(x - 4)^{-4/3} = -\frac{2}{9(x - 4)^{4/3}}$$

- *Vertical tangent lines*: There is a vertical tangent line and cusp at $x = 4$ of the type in Figure 4.5.1d since $f(x) = (x - 4)^{2/3}$ is continuous at $x = 4$ and

$$\lim_{x \to 4^+} f'(x) = \lim_{x \to 4^+} \frac{2}{3(x - 4)^{1/3}} = +\infty$$

$$\lim_{x \to 4^-} f'(x) = \lim_{x \to 4^-} \frac{2}{3(x - 4)^{1/3}} = -\infty$$

- *Intervals of increase and decrease; concavity*: Combining the preceding information with the following sign analysis of the first and second derivatives yields the graph in Figure 4.5.2. The points in the table accompanying the figure were plotted for some added accuracy. ◀

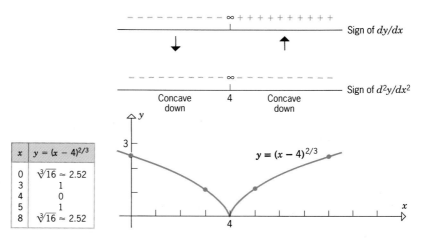

x	$y = (x - 4)^{2/3}$
0	$\sqrt[3]{16} \approx 2.52$
3	1
4	0
5	1
8	$\sqrt[3]{16} \approx 2.52$

Figure 4.5.2

Example 2 Sketch the graph of $y = 6x^{1/3} + 3x^{4/3}$.

Solution. Let $f(x) = 6x^{1/3} + 3x^{4/3} = 3x^{1/3}(2 + x)$.

- *Symmetries*: There are no symmetries about the coordinate axes or the origin (verify).
- *x-intercepts*: Setting $y = 3x^{1/3}(2 + x) = 0$ yields the x-intercepts $x = 0$ and $x = -2$.
- *y-intercept*: Setting $x = 0$ yields the y-intercept $y = 0$.
- *Vertical asymptotes*: None, since $f(x) = 6x^{1/3} + 3x^{4/3}$ is continuous.
- *Horizontal asymptotes*: None, since

$$\lim_{x \to +\infty} (6x^{1/3} + 3x^{4/3}) = \lim_{x \to +\infty} 3x^{1/3}(2 + x) = +\infty$$

$$\lim_{x \to -\infty} (6x^{1/3} + 3x^{4/3}) = \lim_{x \to -\infty} 3x^{1/3}(2 + x) = +\infty$$

- *Derivatives*:

$$\frac{dy}{dx} = f'(x) = 2x^{-2/3} + 4x^{1/3} = 2x^{-2/3}(1 + 2x) = \frac{2(2x + 1)}{x^{2/3}}$$

$$\frac{d^2y}{dx^2} = f''(x) = -\frac{4}{3}x^{-5/3} + \frac{4}{3}x^{-2/3} = \frac{4}{3}x^{-5/3}(-1 + x) = \frac{4(x - 1)}{3x^{5/3}}$$

- *Vertical tangent lines*: There is a vertical tangent line at $x = 0$ of the type in Figure 4.5.1a since

$$\lim_{x \to 0^+} f'(x) = \lim_{x \to 0^+} \frac{2(2x + 1)}{x^{2/3}} = +\infty$$

$$\lim_{x \to 0^-} f'(x) = \lim_{x \to 0^-} \frac{2(2x + 1)}{x^{2/3}} = +\infty$$

• *Intervals of increase and decrease; concavity*: Combining the preceding information with the following sign analysis of the first and second derivatives yields the graph in Figure 4.5.3. In addition to the stationary point at $x = -\frac{1}{2}$ and the inflection points at $x = 0$ and $x = 1$, we plotted two additional points for some extra accuracy.   ◀

REMARK.   The inflection at $x = 1$ is so subtle that it is hardly perceptible on the graph in Figure 4.5.3, yet we know it is there from our analysis of the second derivative.

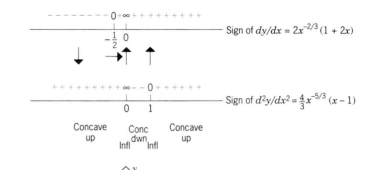

x	$y = 6x^{1/3} + 3x^{4/3}$
-3	≈ 4.3
$-\frac{1}{2}$	≈ -3.6
1	9
2	≈ 15.1

Figure 4.5.3

▶ Exercise Set 4.5 $\boxed{C}$ *14, 17, 18, 24*

In Exercises 1–24, use the techniques illustrated in this section to sketch the graph of the given function.

1. $(x - 2)^{1/3}$.   **2.** $x^{1/4}$.   **3.** $x^{1/5}$.   **4.** $x^{2/5}$.

5. $x^{4/3}$.   **6.** $x^{-1/3}$.   **7.** $1 - x^{2/3}$.   **8.** $\sqrt{x + 2}$.

9. $\sqrt{x^2 - 1}$.   **10.** $\sqrt[3]{x^2 - 4}$.

11. $2x + 3x^{2/3}$.   **12.** $4x - 3x^{4/3}$.

13. $x\sqrt{3 - x}$.   **14.** $4x^{1/3} - x^{4/3}$.

15. $\dfrac{8(\sqrt{x} - 1)}{x}$.   **16.** $\dfrac{1 + \sqrt{x}}{1 - \sqrt{x}}$.   **17.** $\dfrac{\sqrt{x}}{x - 3}$.

18. $x^{2/3}(x - 5)$.   **19.** $x + \sin x$.   **20.** $x - \cos x$.

21. $\sin x + \cos x$.   **22.** $\sqrt{3} \cos x + \sin x$.

23. $\sin^2 x, \ 0 \le x \le 2\pi$.   **24.** $x \tan x, \ -\dfrac{\pi}{2} < x < \dfrac{\pi}{2}$.

■ **4.6**   MAXIMUM AND MINIMUM VALUES OF A FUNCTION

> *Problems concerned with finding the "best" way to perform a task are called* **optimization** *problems. A large class of optimization problems can be reduced to finding the largest or smallest value of a function and determining where this value occurs. In this section we shall develop some mathematical tools for solving such problems.*

□ **ABSOLUTE EXTREMA**

If we imagine the graph of a function f to be a two-dimensional profile of a mountain range (Figure 4.6.1), then the tops of the mountains correspond to the relative maxima and the bottoms of the valleys to the relative minima. Geologically, these are the high and low points of the terrain in their *immediate vicinity*. However, just as a geologist might be interested in finding the highest mountain and deepest valley in an entire mountain range (Figure 4.6.1), so a mathematician might be interested in finding the largest and smallest values of a function over its *entire domain*.

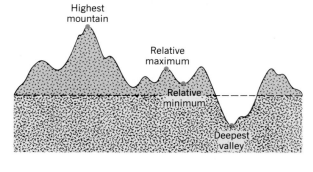

Figure 4.6.1

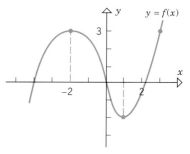

Figure 4.6.2

This leads us to the following definitions.

4.6.1   DEFINITION.   If $f(x_0) \geq f(x)$ for all x in the domain of f, then $f(x_0)$ is called the **absolute maximum value** or simply the **maximum value** of f.

4.6.2   DEFINITION.   If $f(x_0) \leq f(x)$ for all x in the domain of f, then $f(x_0)$ is called the **absolute minimum value** or simply the **minimum value** of f.

4.6.3   DEFINITION.   A number that is either the maximum or the minimum value of a function f is called an **absolute extreme value** or **extreme value** of f. Sometimes the terms **absolute extremum** or **extremum** are also used.

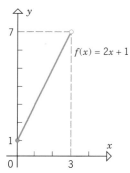

Figure 4.6.3

Often we shall be concerned with the extreme values of f on some specified interval, rather than on the entire domain of f. The meaning of such terms as **maximum value of f on $[a, b]$** or **minimum value of f on (a, b)** should be clear.

Example 1   The function f graphed in Figure 4.6.2 has no maximum value. Its minimum value is 2, and the minimum occurs where $x = 3$.   ◄

Example 2   The function graphed in Figure 4.6.3 has neither a maximum nor a minimum value. However, on the interval $[-2, 2]$ it has both. The maximum value, $f(x) = 3$, occurs at $x = -2$. The minimum value on $[-2, 2]$ is $f(x) = -2$, which occurs at $x = 1$.   ◄

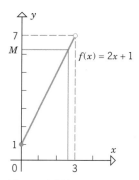

Figure 4.6.4

Example 3   The function $f(x) = 2x + 1$ graphed in Figure 4.6.4 has a minimum, but no maximum value on $[0, 3)$. The minimum value is 1, and this occurs at $x = 0$. The reason why $f(x)$ has no maximum value is subtle, but important to understand. If we had been considering the interval $[0, 3]$ rather than $[0, 3)$, then $f(x)$ would have had a maximum value of 7 occurring at $x = 3$. However, the point $x = 3$ does not lie in the interval $[0, 3)$, so 7 is *not* the maximum value on $[0, 3)$. But no number *less* than 7 can be the maximum either, for if M is any number less than 7, there are values of x in $[0, 3)$ where $f(x) > M$ (see Figure 4.6.5). Thus, $f(x) = 2x + 1$ has no maximum value on $[0, 3)$.   ◄

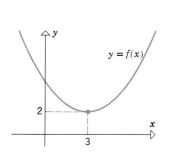

Figure 4.6.5

Given a function f, there are various questions we can ask about its maximum and minimum values; for example:

- Does $f(x)$ have a maximum value?
- If $f(x)$ has a maximum value, what is it?
- If $f(x)$ has a maximum value, where does it occur?

These same questions could, of course, be asked about the minimum value of f. We shall now obtain some results that will help us to answer such questions.

The following theorem gives conditions under which the existence of maximum and minimum values of a function is ensured.

> **4.6.4** THEOREM (*Extreme-Value Theorem*).   *If a function f is continuous on a closed interval $[a, b]$, then f has both a maximum value and a minimum value on $[a, b]$.*

The proof of this theorem is surprisingly difficult, and will be omitted. However, the result is intuitively obvious if we imagine a particle moving along the graph of a continuous function over a closed interval $[a, b]$; during the trip the particle will have to pass through a highest point and a lowest point (Figure 4.6.6).

In the Extreme-Value Theorem the hypotheses that f is continuous and that the interval is closed are essential. If either hypothesis is violated, the existence of maximum or minimum values cannot be guaranteed; this is shown in the next two examples.

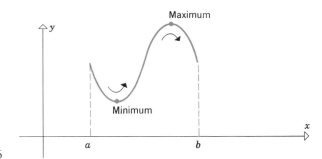

Figure 4.6.6

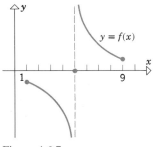

Figure 4.6.7

Example 4   Because $f(x) = 2x + 1$ is a polynomial, it is continuous everywhere. In particular, it is continuous on the interval $[0, 3)$. However, as shown in Example 3, the function f does not have a maximum value on $[0, 3)$. Thus, the closed-interval hypothesis in Theorem 4.6.4 is essential.   ◄

Example 5   The function f graphed in Figure 4.6.7 is defined everywhere on the closed interval $[1, 9]$, yet it has neither a maximum nor minimum value on the interval. Since f has a point of discontinuity on the interval $[1, 9]$, this example shows that the continuity hypothesis in Theorem 4.6.4 is essential.   ◄

The Extreme-Value Theorem is an example of what mathematicians call an ***existence theorem***; it states conditions under which something exists, in this case maximum and minimum values for a function f. However, finding these maximum and minimum values is a separate problem.

Example 6   The polynomial $f(x) = 2x^3 - 15x^2 + 36x$ is continuous everywhere. In particular, it is continuous on the closed interval $[1, 5]$. Thus, the Extreme-Value Theorem tells us that f has both maximum and minimum values on $[1, 5]$. However, this theorem does not tell us what the maximum and minimum values are or where they occur.   ◄

The following theorem is the key tool for finding the extreme values of a function.

4.6.5 THEOREM. *If a function f has an extreme value (either a maximum or a minimum) on an open interval (a, b), then the extreme value occurs at a critical point of f.*

Proof. If f has a maximum value on (a, b) at x_0, then $f(x_0)$ is the largest value of f on (a, b) and therefore the largest value of f in the immediate vicinity of x_0. Thus, f has a relative maximum at x_0 and by Theorem 4.3.4, x_0 is a critical point for f. The proof in the case of a minimum value is similar. ∎

☐ FINDING ABSOLUTE
EXTREMA

This theorem can be used to locate the extreme values of a continuous function f on a finite *closed* interval $[a, b]$. For example, consider the possible locations of the maximum. Either the maximum occurs at an endpoint or it occurs in the open interval (a, b), in which case it occurs at a critical point (Figure 4.6.8). This suggests the following procedure:

How to Find the Extreme Values of a Continuous Function f on a Closed Interval [a, b]

Step 1. Find the critical points of f in (a, b).

Step 2. Evaluate f at all the critical points and at the endpoints a and b.

Step 3. The largest of the values in Step 2 is the maximum value of f on $[a, b]$ and the smallest value is the minimum.

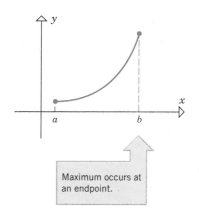

Maximum occurs at an endpoint.

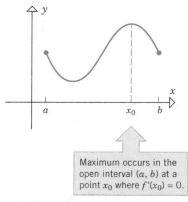

Maximum occurs in the open interval (a, b) at a point x_0 where $f'(x_0) = 0$.

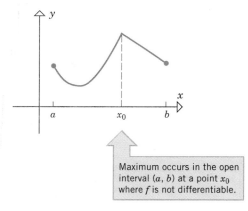

Maximum occurs in the open interval (a, b) at a point x_0 where f is not differentiable.

Figure 4.6.8

Example 7 Find the maximum and minimum values of $f(x) = 2x^3 - 15x^2 + 36x$ on the interval $[1, 5]$, and determine where the maximum and minimum occur.

Solution. Since polynomials are differentiable everywhere, f is differentiable everywhere on the interval $(1, 5)$. Thus, if an extreme value occurs on the open interval $(1, 5)$, it must occur at a point where the derivative is zero. Since

$$f'(x) = 6x^2 - 30x + 36$$

the equation $f'(x) = 0$ becomes $6x^2 - 30x + 36 = 0$, which simplifies to

$$x^2 - 5x + 6 = 0 \quad \text{or} \quad (x - 3)(x - 2) = 0$$

Thus, there are two points in the interval $(1, 5)$ where $f'(x) = 0$, namely $x = 2$ and $x = 3$. Evaluating f at these points and the endpoints, we have

$$f(1) = 2(1)^3 - 15(1)^2 + 36(1) = 23$$

$$f(2) = 2(2)^3 - 15(2)^2 + 36(2) = 28$$

$$f(3) = 2(3)^3 - 15(3)^2 + 36(3) = 27$$

$$f(5) = 2(5)^3 - 15(5)^2 + 36(5) = 55$$

Thus, the minimum value is 23 and the maximum value is 55. The minimum occurs at $x = 1$ and the maximum occurs at $x = 5$. ◀

Example 8 Find the extreme values of $f(x) = 6x^{4/3} - 3x^{1/3}$ on the interval $[-1, 1]$ and determine where these values occur.

Solution. Differentiating, we obtain

$$f'(x) = 8x^{1/3} - x^{-2/3} = x^{-2/3}(8x - 1) = \frac{8x - 1}{x^{2/3}}$$

Table 4.6.1

x	-1	0	$\frac{1}{8}$	1
$f(x)$	9	0	$-\frac{9}{8}$	3

Thus, $f'(x) = 0$ at $x = \frac{1}{8}$ and $f'(x)$ does not exist where $x = 0$. It follows that the critical points of f are $x = 0$, $x = \frac{1}{8}$, both of which lie in the interval $[-1, 1]$. Evaluating f at these critical points and the endpoints we obtain Table 4.6.1. Thus, the minimum value of f on $[-1, 1]$ is $-\frac{9}{8}$, which occurs at $x = \frac{1}{8}$, and the maximum value of f on $[-1, 1]$ is 9, which occurs at $x = -1$. ◀

For a continuous function f on a closed interval $[a, b]$, the Extreme-Value Theorem ensures that f has both a maximum and minimum value; and the procedure for finding these values is mechanical—we simply evaluate f at the critical points and the endpoints. For a continuous function on an open interval, half-open interval, or infinite interval, the problem of finding maximum and minimum values is more complicated for two reasons:

- The function may not have a maximum or minimum value on such an interval.
- If the function does have maximum or minimum values, it often requires some ingenuity to find them.

It is often possible to determine whether a function has extrema by graphing it. Once it is determined that extrema exist, Theorem 4.6.5 can be applied to help locate them.

Example 9 Determine whether the function $f(x) = 3x^4 + 4x^3$ has maximum or minimum values on $(-\infty, +\infty)$ and, if so, find them.

Solution. Using the methods of Section 4.4, the reader should be able to obtain the graph of f shown in Figure 4.6.9. From this graph we see that the function has a minimum value but no maximum. By Theorem 4.6.5 this minimum must occur at a critical point. To locate the critical points, we set $f'(x)$ equal to zero to obtain

$$12x^3 + 12x^2 = 0 \quad \text{or} \quad 12x^2(x + 1) = 0$$

Thus, the critical points are $x = 0$ and $x = -1$. At $x = 0$ we have an inflection point, and at $x = -1$ we have the desired minimum. Substituting $x = -1$ in $f(x) = 3x^4 + 4x^3$ yields $f(-1) = -1$, which is the minimum value of f on $(-\infty, +\infty)$. ◀

If f is a *continuous* function such that

$$\lim_{x \to +\infty} f(x) = \pm\infty \quad \text{and} \quad \lim_{x \to -\infty} f(x) = \pm\infty$$

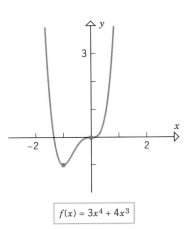

$f(x) = 3x^4 + 4x^3$

Figure 4.6.9

then Table 4.6.2 shows how to ascertain if f has any extrema without actually constructing its graph.

Table 4.6.2

LIMITS	$\lim\limits_{x \to -\infty} f(x) = +\infty$ $\lim\limits_{x \to +\infty} f(x) = +\infty$	$\lim\limits_{x \to -\infty} f(x) = -\infty$ $\lim\limits_{x \to +\infty} f(x) = -\infty$	$\lim\limits_{x \to -\infty} f(x) = -\infty$ $\lim\limits_{x \to +\infty} f(x) = +\infty$	$\lim\limits_{x \to -\infty} f(x) = +\infty$ $\lim\limits_{x \to +\infty} f(x) = -\infty$
CONCLUSION IF f IS CONTINUOUS	f has a minimum but no maximum on $(-\infty, +\infty)$.	f has a maximum but no minimum on $(-\infty, +\infty)$.	f has neither a maximum nor a minimum on $(-\infty, +\infty)$.	f has neither a maximum nor a minimum on $(-\infty, +\infty)$.
GRAPH				

Example 10 Find the maximum and minimum values, if any, of the function
$$f(x) = x^4 + 2x^3 - 1$$
on $(-\infty, +\infty)$.

Solution. Because f is a polynomial it is continuous on $(-\infty, +\infty)$. Moreover,
$$\lim_{x \to +\infty} (x^4 + 2x^3 - 1) = +\infty \quad \text{and} \quad \lim_{x \to -\infty} (x^4 + 2x^3 - 1) = +\infty$$
so f has a minimum but no maximum on $(-\infty, +\infty)$. By Theorem 4.6.5 the minimum occurs at a critical point. Thus, if we evaluate f at each of its critical points, we shall be able to determine where the minimum occurs and find the minimum value. But,
$$f'(x) = 4x^3 + 6x^2 = 2x^2(2x + 3)$$
so $f'(x) = 0$ yields the critical points $x = 0$ and $x = -\frac{3}{2}$. Evaluating f at these points yields
$$f(0) = -1 \quad \text{and} \quad f(-\tfrac{3}{2}) = -\tfrac{43}{16}$$
so the minimum value is $-\frac{43}{16}$ and this occurs at $x = -\frac{3}{2}$. ◄

There is a variation of the results in Table 4.6.2 that is useful for finding extrema on an open interval. If f is continuous on an open interval (a, b) and
$$\lim_{x \to a^+} f(x) = \pm\infty \quad \text{and} \quad \lim_{x \to b^-} f(x) = \pm\infty$$
then Table 4.6.3 shows how to ascertain if f has any extrema on (a, b).

Table 4.6.3

LIMITS	$\lim\limits_{x \to a^+} f(x) = +\infty$ $\lim\limits_{x \to b^-} f(x) = +\infty$	$\lim\limits_{x \to a^+} f(x) = -\infty$ $\lim\limits_{x \to b^-} f(x) = -\infty$	$\lim\limits_{x \to a^+} f(x) = -\infty$ $\lim\limits_{x \to b^-} f(x) = +\infty$	$\lim\limits_{x \to a^+} f(x) = +\infty$ $\lim\limits_{x \to b^-} f(x) = -\infty$
CONCLUSION IF f IS CONTINUOUS ON (a, b)	f has a minimum but no maximum on (a, b).	f has a maximum but no minimum on (a, b).	f has neither a maximum nor a minimum on (a, b).	f has neither a maximum nor a minimum on (a, b).
GRAPH				

We leave it for the reader to give analogs of the results in Tables 4.6.2 and 4.6.3 for intervals of the form $(-\infty, b)$ and $(a, +\infty)$.

Example 11   Find the maximum and minimum values, if any, of

$$f(x) = \frac{1}{x^2 - x}$$

on the open interval $(0, 1)$.

Solution.   We have

$$\lim_{x \to 0^+} f(x) = \lim_{x \to 0^+} \frac{1}{x^2 - x} = \lim_{x \to 0^+} \frac{1}{x(x - 1)} = -\infty$$

$$\lim_{x \to 1^-} f(x) = \lim_{x \to 1^-} \frac{1}{x^2 - x} = \lim_{x \to 1^-} \frac{1}{x(x - 1)} = -\infty$$

so f has a maximum, but no minimum on $(0, 1)$. By Theorem 4.6.5 this maximum occurs at a critical point of f. We have

$$f'(x) = -\frac{2x - 1}{(x^2 - x)^2}$$

so $f'(x) = 0$ if $x = \frac{1}{2}$. Thus, the maximum value of f on $(0, 1)$ occurs at $x = \frac{1}{2}$ and this maximum value is

$$f(\tfrac{1}{2}) = \frac{1}{(\frac{1}{2})^2 - \frac{1}{2}} = -4 \quad \blacktriangleleft$$

Example 12   The function $f(x) = \tan x$ has neither a maximum nor a minimum on $(-\pi/2, \pi/2)$ because $\tan x$ is continuous on this interval and

$$\lim_{x \to -\pi/2^+} \tan x = -\infty, \qquad \lim_{x \to \pi/2^-} \tan x = +\infty$$

(See Figure B.28 in Appendix B.) $\blacktriangleleft$

The following theorem can sometimes be helpful in situations where our previous methods do not apply.

4.6.6   THEOREM.   *Let f be continuous on an interval I and assume that f has exactly one relative extremum on I, say at x_0.*

(a)   *If f has a relative minimum at x_0, then $f(x_0)$ is the minimum value of f on the interval I.*

(b)   *If f has a relative maximum at x_0, then $f(x_0)$ is the maximum value of f on the interval I.*

Although we shall omit the proof, the result is easy to visualize. For example, if f has a relative maximum at x_0 and if $f(x_0)$ is *not* the maximum value of f on I, then the graph of f must make an upward turn somewhere on the interval I, thereby introducing a second relative extremum (Figure 4.6.10). Thus, if there is only one relative extremum on I, an absolute maximum value for f must also occur at x_0.

Where applicable, Theorem 4.6.6 reduces the problem of finding extrema to one of finding relative extrema.

Example 13   Find the maximum and minimum values, if any, of the function

$$f(x) = x^3 - 3x^2 + 4$$

on $(0, +\infty)$.

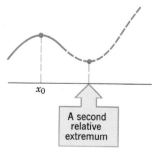

x_0

A second relative extremum

Figure 4.6.10

Solution. We have

$$\lim_{x \to 0^+} (x^3 - 3x^2 + 4) = 4$$

$$\lim_{x \to +\infty} (x^3 - 3x^2 + 4) = +\infty$$

Because the first limit is finite, Tables 4.6.2 and 4.6.3 do not apply. Moreover, the second limit tells us that f definitely has no maximum on $(0, +\infty)$. Thus, there are two possibilities: either f has no extreme values on $(0, +\infty)$ or f has a minimum on $(0, +\infty)$. To determine which is the case, we shall look for critical points. We have

$$f'(x) = 3x^2 - 6x = 3x(x - 2)$$

so $f'(x) = 0$ yields the critical points $x = 0$ and $x = 2$. However, $x = 2$ is the only critical point in the interval $(0, +\infty)$, so we shall try to apply Theorem 4.6.6. Since

$$f''(x) = 6x - 6$$

we have $f''(2) = 6 > 0$. Thus, a relative minimum occurs at $x = 2$ by the second derivative test and consequently a minimum value for f occurs at $x = 2$ by Theorem 4.6.6. This minimum value is $f(2) = 0$. ◀

▶ Exercise Set 4.6

In Exercises 1–14, find the maximum and minimum values of f on the given closed interval and state where these values occur.

1. $f(x) = 4x^2 - 4x + 1$; $[0, 1]$.

2. $f(x) = 8x - x^2$; $[0, 6]$.

3. $f(x) = (x - 1)^3$; $[0, 4]$.

4. $f(x) = 2x^3 - 3x^2 - 12x$; $[-2, 3]$.

5. $f(x) = \dfrac{3x}{\sqrt{4x^2 + 1}}$; $[-1, 1]$.

6. $f(x) = \dfrac{x}{x^2 + 2}$; $[-1, 4]$.

7. $f(x) = x^{2/3}(20 - x)$; $[-1, 20]$.

8. $f(x) = (x^2 + x)^{2/3}$; $[-2, 3]$.

9. $f(x) = x - \tan x$; $[-\pi/4, \pi/4]$.

10. $f(x) = \sin x - \cos x$; $[0, \pi]$.

11. $f(x) = 2 \sec x - \tan x$; $[0, \pi/4]$.

12. $f(x) = \sin^2 x + \cos x$; $[-\pi, \pi]$.

13. $f(x) = 1 + |9 - x^2|$; $[-5, 1]$.

14. $f(x) = |6 - 4x|$; $[-3, 3]$.

In Exercises 15–26, find the maximum and minimum values of f on the given interval, if they exist.

15. $f(x) = x^2 - 3x - 1$; $(-\infty, +\infty)$.

16. $f(x) = 3 - 4x - 2x^2$; $(-\infty, +\infty)$.

17. $f(x) = 4x^3 - 3x^4$; $(-\infty, +\infty)$.

18. $f(x) = x^4 + 4x$; $(-\infty, +\infty)$.

19. $f(x) = (x^2 - 1)^2$; $(-\infty, +\infty)$.

20. $f(x) = (x - 1)^2(x + 2)^2$; $(-\infty, +\infty)$.

21. $f(x) = x^3 - 3x - 2$; $(-\infty, +\infty)$.

22. $f(x) = x^3 - 9x + 1$; $(-\infty, +\infty)$.

23. $f(x) = 1 + \dfrac{1}{x}$; $(0, +\infty)$.

24. $f(x) = \dfrac{x}{x^2 + 1}$; $[0, +\infty)$.

25. $f(x) = \dfrac{x^2}{x + 1}$; $(-5, -1)$.

26. $f(x) = \dfrac{x + 3}{x - 3}$; $[-5, 5]$.

27. Find the maximum and minimum values of

$$f(x) = 2 \sin 2x + \sin 4x$$

and state where the maximum and minimum values occur. [*Hint:* Since f is periodic, this problem can be solved by first finding the maximum and minimum values on an appropriate closed interval.]

28. Find the maximum and minimum values of

$$f(x) = 3 \cos \frac{x}{3} + 2 \cos \frac{x}{2}$$

[See the hint in the preceding exercise.]

29. Find the maximum and minimum values of the function $f(x) = \sin (\cos x)$ on $[0, 2\pi]$.

30. Find the maximum and minimum values of the function $f(x) = \cos (\sin x)$ on $[0, \pi]$.

31. Find the maximum and minimum values of

$$f(x) = \begin{cases} 4x - 2, & x < 1 \\ (x - 2)(x - 3), & x \geq 1 \end{cases}$$

on $[\frac{1}{2}, \frac{7}{2}]$.

32. Let $f(x) = x^2 + px + q$. Find values of p and q such that

$f(1) = 3$ is an extreme value of f on $[0, 2]$. Is this value a maximum or minimum?

33. Let $f(x) = (x - a)^p$, where p is an integer greater than 1 and a is any real number. Find the relative extrema of f if

(a) p is even   (b) p is odd.

34. What is the smallest possible slope for a tangent to $y = x^3 - 3x^2 + 5x$?

35. (a) Show that

$$f(x) = \frac{64}{\sin x} + \frac{27}{\cos x}$$

has a minimum value, but no maximum value on the interval $(0, \pi/2)$.

(b) Find the minimum value.

36. Prove that the minimum value of

$$f(x) = x^2 + \frac{16x^2}{(8 - x)^2}, \quad x > 8$$

occurs at $x = 4(2 + \sqrt[3]{2})$.

37. Find the maximum value of the function $\sin^2 \theta \cos \theta$ for $0 \le \theta \le \pi/2$.

38. Prove: For every positive value of t the function $f(x) = x + t/x$ has a minimum value but no maximum value on $(0, +\infty)$.

39. Find values of a_0, a_1, and a_2 such that the graph of $f(x) = a_0 + a_1 x + a_2 x^2$ passes through the point $(0, 9)$ and f has a minimum value of 1 at $x = 2$.

40. Prove that $\sin x \le x$ for all x in the interval $[0, 2\pi]$. [*Hint:* Find the minimum value of $x - \sin x$ on the given interval.]

41. Prove that

$$1 - \tfrac{1}{2}x^2 \le \cos x$$

for all x in the interval $[0, 2\pi]$. [See the hint in the preceding exercise.]

42. (a) Sketch the graph of a function f that is continuous on the open interval $(-1, 1)$ and has both maximum and minimum values on the interval.

(b) Sketch the graph of a function f that is defined everywhere on the open interval $(-1, 1)$ and is not continuous everywhere on $(-1, 1)$, yet has both maximum and minimum values on $(-1, 1)$.

(c) Do the functions in parts (a) and (b) violate the Extreme-Value Theorem? Explain.

43. Prove: If $ax^2 + bx + c = 0$ has two distinct real roots, then the midpoint between these roots is a stationary point for $f(x) = ax^2 + bx + c$.

44. Prove Theorem 4.6.5 in the case where the extreme value is a minimum.

45. Let $f(x) = ax^2 + bx + c$, where $a > 0$. Prove that $f(x) \ge 0$ for all x if and only if $b^2 - 4ac \le 0$. [*Hint:* Find the minimum of $f(x)$.]

46. Assume that the function f is differentiable on an open interval (a, b) and L is a nonvertical line that does not intersect the curve $y = f(x)$ over the interval (a, b). Show that if there is a point on this curve over the interval (a, b) that is closest to or farthest from L, then the tangent line at this point is parallel to L. [*Hint:* Let $y = mx + b$ be the equation of L, and assume without proof that the perpendicular distance between a point on $y = f(x)$ and L is largest (smallest) at the same place where the vertical distance between the curve and the line is largest (smallest). (See Figure 4.6.11.)]

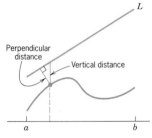

Figure 4.6.11

47. Use the result in Exercise 46 to find the coordinates of all points on the graph of $y = -x^2$ closest to and farthest from the line $y = 2 - x$ for $-1 \le x \le \tfrac{3}{2}$.

48. Use the result in Exercise 46 to find the coordinates of all points on the graph of $y = x^3$ closest to and farthest from the line $y = \tfrac{4}{3}x - 1$ for $-1 \le x \le 1$.

■ **4.7 APPLIED MAXIMUM AND MINIMUM PROBLEMS**

> *In this section we shall show how the methods developed in the preceding section can be used to solve some applied optimization problems.*

☐ **CLASSIFICATION OF OPTIMIZATION PROBLEMS**

The applied optimization problems considered in this section fall into two categories:

• Problems that reduce to maximizing or minimizing a continuous function over a finite closed interval.

• Problems that reduce to maximizing or minimizing a continuous function over an infinite interval or over a finite interval that is not closed (i.e., open or half-closed).

For problems of the first type the Extreme-Value Theorem (4.6.4) guarantees the existence of a solution, and the solution can be obtained by examining the values of the function at the critical points and the interval endpoints. However, for problems of the second type, there is no general guarantee that a solution exists. Thus, part of the problem is to determine whether there actually is a solution. If there is a solution, some ingenuity may be required to find it. We begin with some problems of the first type.

□ **PROBLEMS INVOLVING FINITE CLOSED INTERVALS**

Example 1 Find the dimensions of a rectangle with perimeter 100 ft whose area is as large as possible.

Solution. Let

x = length of the rectangle (ft)
y = width of the rectangle (ft)
A = area of the rectangle (ft^2)

Then

$$A = xy \tag{1}$$

Since the perimeter of the rectangle is 100 ft, the variables x and y are related by the equation

$$2x + 2y = 100 \quad \text{or} \quad y = 50 - x \tag{2}$$

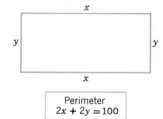

Figure 4.7.1

(See Figure 4.7.1.) Substituting (2) in (1) yields

$$A = x(50 - x) = 50x - x^2 \tag{3}$$

Because x represents a length it cannot be negative, and because the two sides of length x cannot have a combined length exceeding the total perimeter of 100 ft, the variable x must satisfy

$$0 \leq x \leq 50 \tag{4}$$

Thus, we have reduced the problem to that of finding the value (or values) of x in [0, 50], for which A is maximum. Since A is a polynomial in x, it is continuous on [0, 50], and so the maximum must occur at an endpoint of this interval or at a critical point.

From (3) we obtain

$$\frac{dA}{dx} = 50 - 2x$$

Setting $dA/dx = 0$ we obtain

$$50 - 2x = 0$$

or $x = 25$. Thus, the maximum occurs at one of the points

$$x = 0, \quad x = 25, \quad x = 50$$

Substituting these values in (3) yields Table 4.7.1, which tells us that the maximum area of 625 ft^2 occurs at $x = 25$. From (2) the corresponding value of y is $y = 25$, so the rectangle of perimeter 100 ft with greatest area is a square with sides of length 25 ft. ◀

Table 4.7.1

x	0	25	50
A	0	625	0

REMARK. In (4) we included $x = 0$ and $x = 50$ as possible values for x. Because $x = 50$ implies $y = 0$ from (2), these x values correspond to rectangles with two sides of length zero. One can argue that these x values should not be allowed because a "true" rectangle cannot have sides of length zero.

Actually, it is just a matter of the assumptions we choose to make. If we view Example 1 as a purely mathematical problem, then there is nothing wrong with allowing sides of length zero. However, if we view it as a practical problem in which the rectangle is to be constructed from physical material, we would not want to allow $x = 0$ or $x = 50$ and (4) should be replaced by $0 < x < 50$, in which case the interval is not closed and the problem would have to be handled by methods discussed later in this section.

Example 1 illustrates the following five-step procedure that can be used for solving many applied maximum and minimum problems.

Step 1. Draw an appropriate figure and label the quantities relevant to the problem.

Step 2. Find a formula for the quantity to be maximized or minimized.

Step 3. Using the conditions stated in the problem to eliminate variables, express the quantity to be maximized or minimized as a function of one variable.

Step 4. Find the interval of possible values for this variable from the physical restrictions in the problem.

Step 5. If applicable, use the techniques of the preceding section to obtain the maximum or minimum.

Example 2 An open box is to be made from a 16-in. by 30-in. piece of cardboard by cutting out squares of equal size from the four corners and bending up the sides (Figure 4.7.2). What size should the squares be to obtain a box with largest possible volume?

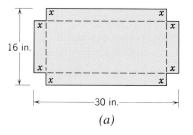

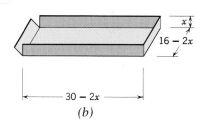

Figure 4.7.2   *(a)*   *(b)*

Solution. Let

x = length (in inches) of the sides of the squares to be cut out
V = volume (in cubic inches) of the resulting box

Because we are removing a square of side x from each corner, the resulting box will have dimensions $16 - 2x$ by $30 - 2x$ by x (Figure 4.7.2b). Since the volume of a box is the product of its dimensions, we have

$$V = (16 - 2x)(30 - 2x)x = 480x - 92x^2 + 4x^3 \qquad (5)$$

The variable x in this expression is subject to certain restrictions. Because x represents a length it cannot be negative, and because the width of the cardboard is 16 in. we cannot cut out squares whose sides are more than 8 in. long. Thus, the variable x in (5) must satisfy

$$0 \le x \le 8$$

We have thus reduced our problem to finding the value (or values) of x in the interval $[0, 8]$ for which (5) is maximum. Since the right side of (5) is a polynomial in x, it is continuous on the closed interval $[0, 8]$ and consequently we can use the methods developed in the preceding section to find the maximum.

From (5) we obtain

$$\frac{dV}{dx} = 480 - 184x + 12x^2 = 4(120 - 46x + 3x^2)$$

Setting $dV/dx = 0$ yields

$$120 - 46x + 3x^2 = 0$$

which can be solved by the quadratic formula to obtain the critical points

$$x = \frac{10}{3} \quad \text{and} \quad x = 12$$

Since $x = 12$ falls outside the interval $[0, 8]$, the maximum value of V occurs either at the critical point $x = \frac{10}{3}$ or at one of the endpoints $x = 0$, $x = 8$. Substituting these values in (5) yields Table 4.7.2, which tells us that the greatest possible volume $V = \frac{19600}{27} \text{in}^3 \approx 726 \text{ in}^3$ occurs when we cut out squares whose sides have length $\frac{10}{3}$ in. ◀

Table 4.7.2

x	0	$\frac{10}{3}$	8
V	0	$\frac{19600}{27}$	0

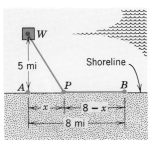

Figure 4.7.3

Example 3 An offshore oil well is located in the ocean at a point W, which is 5 mi from the closest shorepoint A on a straight shoreline (Figure 4.7.3). The oil is to be piped to a shorepoint B that is 8 mi from A by piping it on a straight line under water from W to some shorepoint P between A and B and then on to B via a pipe along the shoreline. If the cost of laying pipe is \$100,000 per mile under water and \$75,000 per mile over land, where should the point P be located to minimize the cost of laying the pipe?

REMARK. Since the shortest distance between two points is a straight line, a pipeline directly from W to B uses the least amount of pipe. However, the pipe, being completely under water, would be expensive to lay. At the other extreme, a pipeline from W to A to B uses the least amount of expensive underwater pipe, but uses the greatest total amount of pipe. Thus, it seems plausible that by piping to some point P between A and B, one might incur less total cost than by piping to either of the extreme locations.

Solution. Let

$x =$ distance (in miles) between A and P
$c =$ cost (in thousands of dollars) for the entire pipeline

From Figure 4.7.3 the length of pipe under water is the distance between W and P. By the Theorem of Pythagoras, that length is

$$\sqrt{x^2 + 25} \tag{6}$$

Also from Figure 4.7.3, the length of pipe over land is the distance between P and B, which is

$$8 - x \tag{7}$$

From (6) and (7) it follows that the total cost c (in thousands of dollars) for the pipeline is

$$c = 100\sqrt{x^2 + 25} + 75(8 - x) \tag{8}$$

Because the distance between A and B is 8 mi, the distance x between A and P must satisfy

$$0 \leq x \leq 8$$

We have thus reduced our problem to finding the value (or values) of x in the interval $[0, 8]$ for which (8) is a minimum. Since c is a continuous function of x on the closed interval $[0, 8]$, we can use the methods developed in the preceding section to find the minimum.
From (8) we obtain

$$\frac{dc}{dx} = \frac{100x}{\sqrt{x^2 + 25}} - 75 = 25\left(\frac{4x}{\sqrt{x^2 + 25}} - 3\right)$$

Setting $dc/dx = 0$ yields

$$\frac{4x}{\sqrt{x^2 + 25}} - 3 = 0 \tag{9}$$

or

$$4x = 3\sqrt{x^2 + 25}$$
$$16x^2 = 9(x^2 + 25)$$
$$7x^2 = 225$$
$$x = \pm\, 15/\sqrt{7}$$

The number $-15/\sqrt{7}$ is not a solution of (9) and must be discarded, leaving $x = 15/\sqrt{7}$ as the only critical point. Since this point lies in the interval [0, 8], the minimum must occur at one of the points

$$x = 0, \quad x = 15/\sqrt{7}, \quad x = 8$$

Substituting these values in (8) yields Table 4.7.3, which tells us that the least possible cost for the pipeline, approximately $c = 930.719 = \$930,719$, occurs if the point P is located at a distance of $15/\sqrt{7} \approx 5.67$ mi from A.   ◄

Table 4.7.3

x	0	$15/\sqrt{7}$	8
c	1100	$100\sqrt{\dfrac{225}{7} + 25} + 75\left(8 - \dfrac{15}{\sqrt{7}}\right)$ $= 600 + 125\sqrt{7} \approx 930.719$	$100\sqrt{89} \approx 943.398$

Example 4 Find the radius and height of the right-circular cylinder of largest volume that can be inscribed in a right-circular cone with radius 6 in. and height 10 in. (Figure 4.7.4a).

Solution. Let

r = radius (in inches) of the cylinder
h = height (in inches) of the cylinder
V = volume (in cubic inches) of the cylinder

The formula for the volume of the inscribed cylinder is

$$V = \pi r^2 h \tag{10}$$

To eliminate one of the variables in (10) we need a relationship between r and h. Using similar triangles (Figure 4.7.4b) we obtain

$$\frac{10 - h}{r} = \frac{10}{6} \quad \text{or} \quad h = 10 - \tfrac{5}{3}r \tag{11}$$

Substituting (11) into (10) we obtain

$$V = \pi r^2(10 - \tfrac{5}{3}r) = 10\pi r^2 - \tfrac{5}{3}\pi r^3$$

which expresses V in terms of r alone. Because r represents a radius it cannot be negative, and because the radius of the inscribed cylinder cannot exceed the radius of the cone, the variable r must satisfy

$$0 \le r \le 6$$

Thus, we have reduced the problem to that of finding the value (or values) of r in [0, 6] for which (12) is a maximum. Since V is a continuous function of r on [0, 6], the methods developed in the preceding section apply.

From (12) we obtain

$$\frac{dV}{dr} = 20\pi r - 5\pi r^2 = 5\pi r(4 - r)$$

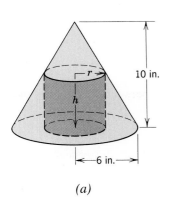

(a)

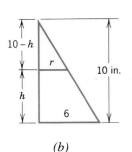

(b)

Figure 4.7.4

Setting $dV/dr = 0$ gives

$$5\pi r(4 - r) = 0$$

so $r = 0$ and $r = 4$ are critical points. Since these lie in the interval $[0, 6]$, the maximum must occur at one of the points

$$r = 0, \quad r = 4, \quad r = 6$$

Substituting these values in (12) yields Table 4.7.4, which tells us that the maximum volume $V = \frac{160}{3}\pi \approx 168$ in^3 occurs when the inscribed cylinder has radius 4 in. When $r = 4$ it follows from (11) that $h = \frac{10}{3}$. Thus, the inscribed cylinder of largest volume has radius $r = 4$ in. and height $h = \frac{10}{3}$ in. ◄

Table 4.7.4

r	0	4	6
V	0	$\frac{160}{3}\pi$	0

☐ **PROBLEMS INVOLVING INTERVALS THAT ARE NOT FINITE AND CLOSED**

Example 5 A closed cylindrical can is to hold 1 liter (1000 cm^3) of liquid. How should we choose the height and radius to minimize the amount of material needed to manufacture the can?

Solution. Let

$h =$ height (in cm) of the can
$r =$ radius (in cm) of the can
$S =$ surface area (in cm^2) of the can

Assuming there is no waste or overlap, the amount of material needed for manufacture will be the same as the surface area of the can. Since the can consists of two circular disks of radius r and a rectangular sheet with dimensions h by $2\pi r$ (Figure 4.7.5), the surface area will be

$$S = 2\pi r^2 + 2\pi rh \tag{13}$$

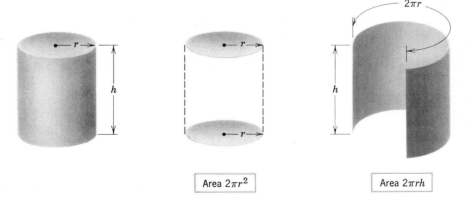

Figure 4.7.5

Area $2\pi r^2$ Area $2\pi rh$

We shall now eliminate one of the variables in (13) so that S will be expressed as a function of one variable. Since the volume of the can is 1000 cm^3, it follows from the formula $V = \pi r^2 h$ for the volume of a cylinder that

$$1000 = \pi r^2 h \quad \text{or} \quad h = \frac{1000}{\pi r^2} \tag{14–15}$$

Substituting (15) in (13) yields

$$S = 2\pi r^2 + \frac{2000}{r} \tag{16}$$

Because r represents a radius it must be positive;* thus, we have reduced the problem to

---

*The value $r = 0$ must be excluded in this case, otherwise (16) cannot be satisfied.

finding a value of r in $(0, +\infty)$ for which (16) is minimum (provided such a value exists). But S is a continuous function of r on $(0, +\infty)$ with

$$\lim_{r \to 0^+} \left(2\pi r^2 + \frac{2000}{r}\right) = +\infty \quad \text{and} \quad \lim_{r \to +\infty} \left(2\pi r^2 + \frac{2000}{r}\right) = +\infty$$

so (16) has a minimum, but no maximum on $(0, +\infty)$. This minimum must occur at a critical point, so we calculate

$$\frac{dS}{dr} = 4\pi r - \frac{2000}{r^2} \tag{17}$$

Setting $dS/dr = 0$ gives

$$4\pi r - \frac{2000}{r^2} = 0 \quad \text{or} \quad r = \frac{10}{\sqrt[3]{2\pi}} \tag{18}$$

Since (18) is the only critical point in the interval $(0, +\infty)$, this value of r yields the minimum value of S. From (15) the value of h corresponding to this r is

$$h = \frac{1000}{\pi(10/\sqrt[3]{2\pi})^2} = \frac{20}{\sqrt[3]{2\pi}} = 2r$$

It is not accidental here that the minimum occurs when the height of the can is equal to the diameter of its base (Exercise 27).

Second Solution. The conclusion that a minimum occurs at the value of r in (18) can be deduced from Theorem 4.6.6 and the second derivative test by noting that

$$\frac{d^2S}{dr^2} = 4\pi + \frac{4000}{r^3}$$

is positive if $r > 0$ and hence is positive if $r = 10/\sqrt[3]{2\pi}$. This implies that a relative minimum, and therefore a minimum, occurs at the critical point $r = 10/\sqrt[3]{2\pi} \approx 5.4$.

Third Solution. Using the methods of Section 4.4, the reader should be able to obtain the graph in Figure 4.7.6. From this graph we see that S has a minimum, and this minimum occurs at the critical point $r = 10/\sqrt[3]{2\pi}$ calculated in our first solution of this problem [see (18)]. ◄

REMARK. Note that S has no maximum on $(0, +\infty)$. Thus, had we asked for the dimensions of the can requiring the maximum amount of material for its manufacture, there would have been no solution to the problem. Optimization problems with no solution are sometimes called ***ill posed***.

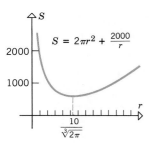

Figure 4.7.6

Example 6 Find a point on the curve $y = x^2$ that is closest to the point $(18, 0)$.

Solution. The distance L between $(18, 0)$ and an arbitrary point (x, y) on the curve $y = x^2$ (Figure 4.7.7) is given by

$$L = \sqrt{(x - 18)^2 + (y - 0)^2}$$

Since (x, y) lies on the curve, x and y satisfy $y = x^2$; thus,

$$L = \sqrt{(x - 18)^2 + x^4} \tag{19}$$

Because there are no restrictions on x, the problem reduces to finding a value of x in $(-\infty, +\infty)$ for which (19) is minimum, provided such a value exists.

In problems of minimizing or maximizing a distance, there is a trick that is helpful for simplifying the computations. It is based on the observation that the distance and the square of the distance have their maximum or minimum at the same point (see Exercise 62). Thus, the minimum value of L in (19) and the minimum value of

$$S = L^2 = (x - 18)^2 + x^4 \tag{20}$$

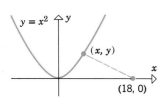

Figure 4.7.7

occur at the same x value.

From (20),

$$\frac{dS}{dx} = 2(x - 18) + 4x^3 = 4x^3 + 2x - 36 \tag{21}$$

so that the critical points satisfy $4x^3 + 2x - 36 = 0$ or equivalently,

$$2x^3 + x - 18 = 0 \tag{22}$$

To solve for x we shall begin by searching for integer solutions. This task can be simplified by using the fact that all integer solutions, if there are any, to a polynomial equation with integer coefficients

$$a_n x^n + \cdots + a_1 x + a_0 = 0$$

must be divisors of the constant term a_0. This result is usually proved in algebra courses. Thus, the only possible integer solutions of (22) are the divisors of -18: ± 1, ± 2, ± 3, ± 6, ± 9, ± 18. Successively substituting these values in (22) we find that $x = 2$ is a solution; therefore, $x - 2$ is a factor of the left side of (22). After dividing by the factor $x - 2$ we can rewrite (22) as

$$(x - 2)(2x^2 + 4x + 9) = 0$$

Thus, the remaining solutions of (22) satisfy the quadratic equation

$$2x^2 + 4x + 9 = 0$$

But these solutions are imaginary numbers (use the quadratic formula), so that $x = 2$ is the only real solution of (22) and consequently the only critical point of S. To determine the nature of this critical point we shall use the second derivative test. From (21),

$$\frac{d^2S}{dx^2} = 12x^2 + 2, \quad \text{so} \quad \left.\frac{d^2S}{dx^2}\right|_{x=2} = 50 > 0$$

which shows that a relative minimum occurs at $x = 2$. Since $x = 2$ is the only relative extremum for L, it follows from Theorem 4.6.6 that an absolute minimum value of L also occurs at $x = 2$. Thus, the point on the curve $y = x^2$ closest to $(18, 0)$ is

$$(x, y) = (x, x^2) = (2, 4) \quad \blacktriangleleft$$

□ AN APPLICATION TO ECONOMICS

Three functions of importance to an economist or a manufacturer are

$C(x)$ = total cost of producing x units of a product during some time period

$R(x)$ = total revenue received from selling x units of the product during the time period

$P(x)$ = total profit obtained by selling x units of the product during the time period

These are called, respectively, the **cost function**, **revenue function**, and **profit function**. If all units produced are sold, then these are related by

$$P(x) = R(x) - C(x) \tag{23}$$

[profit] = [revenue] − [cost]

The total cost $C(x)$ of producing x units can be expressed as a sum

$$C(x) = a + M(x) \tag{24}$$

where a is a constant, called **overhead**, and $M(x)$ is a function representing **manufacturing cost**. The overhead, which includes such fixed costs as rent and insurance, does not depend on x; it must be paid even if nothing is produced. On the other hand, the manufacturing cost $M(x)$, which includes such items as cost of materials and labor, depends on the number of items manufactured. It is shown in economics that with suitable simplifying assumptions, $M(x)$ can be expressed in the form

$$M(x) = bx + cx^2$$

where b and c are constants. Substituting this in (24) yields

$$C(x) = a + bx + cx^2 \tag{25}$$

If a manufacturing firm can sell all the items it produces for p dollars apiece, then its total revenue $R(x)$ (in dollars) will be

$$R(x) = px \tag{26}$$

and its total profit $P(x)$ (in dollars) will be

$$P(x) = [\text{total revenue}] - [\text{total cost}] = R(x) - C(x) = px - C(x)$$

Thus, if the cost function is given by (25),

$$P(x) = px - (a + bx + cx^2) \tag{27}$$

Depending on such factors as number of employees, amount of machinery available, economic conditions, and competition, there will be some upper limit l on the number of items a manufacturer is capable of producing and selling. Thus, during a fixed time period the variable x in (27) will satisfy

$$0 \le x \le l$$

By determining the value or values of x in $[0, l]$ that maximize (27), the firm can determine how many units of its product must be manufactured and sold to yield the greatest profit. This is illustrated in the following numerical example.

Example 7 A liquid form of penicillin manufactured by a pharmaceutical firm is sold in bulk at a price of \$200 per unit. If the total production cost (in dollars) for x units is

$$C(x) = 500,000 + 80x + 0.003x^2$$

and if the production capacity of the firm is at most 30,000 units in a specified time, how many units of penicillin must be manufactured and sold in that time to maximize the profit?

Solution. Since the total revenue for selling x units is $R(x) = 200x$, the profit $P(x)$ on x units will be

$$P(x) = R(x) - C(x) = 200x - (500,000 + 80x + 0.003x^2) \tag{28}$$

Since the production capacity is at most 30,000 units, x must lie in the interval $[0, 30,000]$. From (28)

$$\frac{dP}{dx} = 200 - (80 + 0.006x) = 120 - 0.006x$$

Setting $dP/dx = 0$ gives

$$120 - 0.006x = 0 \quad \text{or} \quad x = 20,000$$

Since this critical point lies in the interval $[0, 30,000]$, the maximum profit must occur at one of the points

$$x = 0, \quad x = 20,000, \quad \text{or} \quad x = 30,000$$

Substituting these values in (28) yields Table 4.7.5, which tells us that the maximum profit $P = \$700,000$ occurs when $x = 20,000$ units are manufactured and sold in the specified time. ◄

Table 4.7.5

x	0	20,000	30,000
$P(x)$	$-500,000$	700,000	400,000

▶ Exercise Set 4.7

1. Express the number 10 as a sum of two nonnegative numbers whose product is as large as possible.

2. How should two nonnegative numbers be chosen so that their sum is 1 and the sum of their squares is
   (a) as large as possible
   (b) as small as possible?

3. Find a number in the closed interval $\left[\frac{1}{2}, \frac{3}{2}\right]$ such that the sum of the number and its reciprocal is
   (a) as small as possible
   (b) as large as possible.

4. A rectangular field is to be bounded by a fence on three sides and by a straight stream on the fourth side. Find the dimensions of the field with maximum area that can be enclosed with 1000 feet of fence.

5. A rectangular plot of land is to be fenced in using two kinds of fencing. Two opposite sides will use heavy-duty fencing selling for $3 a foot, while the remaining two sides will use standard fencing selling for $2 a foot. What are the dimensions of the rectangular plot of greatest area that can be fenced in at a cost of $6000?

6. A rectangle is to be inscribed in a right triangle having sides of length 6 in., 8 in., and 10 in. Find the dimensions of the rectangle with greatest area assuming the rectangle is positioned as in Figure 4.7.8a.

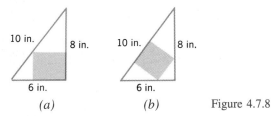

10 in.   8 in.   10 in.   8 in.

6 in.   6 in.

(a)   (b)   Figure 4.7.8

7. Solve the problem in Exercise 6 assuming the rectangle is positioned as in Figure 4.7.8b.

8. A rectangle has its two lower corners on the x-axis and its two upper corners on the curve $y = 16 - x^2$. For all such rectangles, what are the dimensions of the one with largest area?

9. Find the dimensions of the rectangle with maximum area that can be inscribed in a circle of radius 10.

10. Find the dimensions of the rectangle of greatest area that can be inscribed in a semicircle of radius R (Figure 4.7.9).

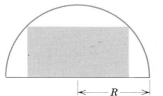

←— R —→   Figure 4.7.9

11. A rectangular area of 3200 ft^2 is to be fenced off. Two opposite sides will use fencing costing $1 per foot and the remaining sides will use fencing costing $2 per foot. Find the dimensions of the rectangle of least cost.

12. Show that among all rectangles with perimeter p, the square has the maximum area.

13. Show that among all rectangles with area A, the square has the minimum perimeter.

14. A wire of length 12 in. can be bent into a circle, bent into a square, or cut into two pieces to make both a circle and a square. How much wire should be used for the circle if the total area enclosed by the figure(s) is to be
    (a) a maximum   (b) a minimum?

15. A church window consisting of a rectangle topped by a semicircle is to have a perimeter p. Find the radius of the semicircle if the area of the window is to be maximum.

16. A church window consists of a blue semicircular section surmounting a clear rectangular section as shown in Figure 4.7.10. The blue glass lets through half as much light per unit area as the clear glass. Find the radius r of the window that admits the most light if the perimeter of the entire window is to be p feet.

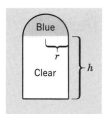

Blue

r

Clear   h

Figure 4.7.10

17. A sheet of cardboard 12 in. square is used to make an open box by cutting squares of equal size from the four corners and folding up the sides. What size squares should be cut to obtain a box with largest possible volume?

18. A square sheet of cardboard of side k is used to make an open box by cutting squares of equal size from the four corners and folding up the sides. What size squares should be cut from the corners to obtain a box with largest possible volume?

19. An open box is to be made from a 3-ft by 8-ft rectangular piece of sheet metal by cutting out squares of equal size from the four corners and bending up the sides. Find the maximum volume that the box can have.

20. A closed rectangular container with a square base is to have a volume of 2250 in^3. The material for the top and bottom of the container will cost $2 per in^2, and the material for the sides will cost $3 per in^2. Find the dimensions of the container of least cost.

21. A closed rectangular container with a square base is to have a volume of 2000 cm^3. It costs twice as much per square centimeter for the top and bottom as it does for the sides. Find the dimensions of the container of least cost.

22. A container with square base, vertical sides, and open top

is to be made from 1000 ft^2 of material. Find the dimensions of the container with greatest volume.

23. A rectangular container with two square sides and an open top is to have a volume of V cubic units. Find the dimensions of the container of minimum surface area.

24. Find the dimensions of the right-circular cylinder of largest volume that can be inscribed in a sphere of radius R.

25. Find the dimensions of the right-circular cylinder of greatest surface area that can be inscribed within a sphere of radius R.

26. Show that the right-circular cylinder of greatest volume that can be inscribed in a right-circular cone has volume that is $\frac{4}{9}$ the volume of the cone (Figure 4.7.11).

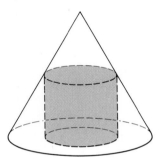

Figure 4.7.11

27. A closed, cylindrical can is to have a volume of V cubic units. Show that the can of minimum surface area is achieved when the height is equal to the diameter of the base.

28. A closed cylindrical can is to have a surface area of S square units. Show that the can of maximum volume is achieved when the height is equal to the diameter of the base.

29. A cylindrical can, open at the top, is to hold 500 cm^3 of liquid. Find the height and radius that minimize the amount of material needed to manufacture the can.

30. A soup can in the shape of a right-circular cylinder of radius r and height h is to have a prescribed volume V. The top and bottom are cut from squares as shown in Figure 4.7.12. If the shaded corners are wasted, but there is no other waste, find the ratio r/h for the can requiring the least material (including waste).

Figure 4.7.12

31. A box-shaped wire frame consists of two identical wire squares whose vertices are connected by four straight wires of equal length (Figure 4.7.13). If the frame is to be made from a wire of length L, what should the dimensions be to obtain a box of greatest volume?

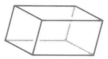

Figure 4.7.13

32. Suppose that the sum of the surface areas of a sphere and a cube is a constant.

(a) Show that the sum of their volumes is smallest when the diameter of the sphere is equal to the length of an edge of the cube.

(b) When will the sum of their volumes be greatest?

33. Find the height and radius of the cone of slant height L whose volume is as large as possible.

34. A cone is made from a circular sheet of radius R by cutting out a sector and gluing the cut edges of the remaining piece together (Figure 4.7.14). What is the maximum volume attainable for the cone?

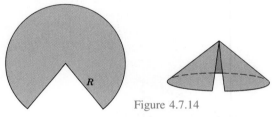

Figure 4.7.14

35. A cone-shaped paper drinking cup is to hold 10 cm^3 of water. Find the height and radius of the cup that will require the least amount of paper.

36. Find the dimensions of the isosceles triangle of least area that can be circumscribed about a circle of radius R.

37. Find the height and radius of the right-circular cone with least volume that can be circumscribed about a sphere of radius R.

38. A trapezoid is inscribed in a semicircle of radius 2 so that one side is along the diameter (Figure 4.7.15). Find the maximum possible area for the trapezoid. [*Hint:* Express the area of the trapezoid in terms of θ.]

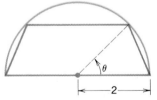

Figure 4.7.15

39. A drainage channel is to be made so that its cross section is a trapezoid with equally sloping sides (Figure 4.7.16). If the sides and bottom all have a length of 5 ft, how should the angle θ ($0 \leq \theta \leq \pi/2$) be chosen to yield the greatest cross-sectional area?

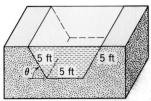

Figure 4.7.16

40. A lamp is suspended above the center of a round table of radius r. How high above the table should the lamp be placed to achieve maximum illumination at the edge of the table? [Assume that the illumination I is directly proportional to the cosine of the angle of incidence ϕ of the light rays and inversely proportional to the square of the distance l from the light source (Figure 4.7.17).]

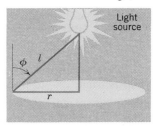

Figure 4.7.17

41. A plank is used to reach over a fence 8 ft high to support a wall that is 1 ft behind the fence (Figure 4.7.18). What is the length of the shortest plank that can be used? [*Hint:* Express the length of the plank in terms of the angle θ shown in the figure.]

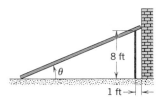

Figure 4.7.18

42. A commercial cattle ranch currently allows 20 steers per acre of grazing land; on the average its steers weigh 2000 lb at market. Estimates by the Agriculture Department indicate that the average market weight per steer will be reduced by 50 lb for each additional steer added per acre of grazing land. How many steers per acre should be allowed in order for the ranch to get the largest possible total market weight for its cattle?

43. (a) A chemical manufacturer sells sulfuric acid in bulk at a price of \$100 per unit. If the daily total production cost in dollars for x units is

$$C(x) = 100{,}000 + 50x + 0.0025x^2$$

and if the daily production capacity is at most 7000 units, how many units of sulfuric acid must be manufactured and sold daily to maximize the profit?

(b) Would it benefit the manufacturer to expand the daily production capacity?

44. A firm determines that x units of its product can be sold daily at p dollars per unit, where

$$x = 1000 - p$$

The cost of producing x units per day is

$$C(x) = 3000 + 20x$$

(a) Find the revenue function $R(x)$.

(b) Find the profit function $P(x)$.

(c) Assuming that the production capacity is at most 500 units per day, determine how many units the company must produce and sell each day to maximize the profit.

(d) Find the maximum profit.

(e) What price per unit must be charged to obtain the maximum profit?

45. In a certain chemical manufacturing process, the daily weight y of defective chemical output depends on the total weight x of all output according to the empirical formula

$$y = 0.01x + 0.00003x^2$$

where x and y are in pounds. If the profit is \$100 per pound of nondefective chemical produced and the loss is \$20 per pound of defective chemical produced, how many pounds of chemical should be produced daily to maximize the profit?

46. The cost c (in dollars per hour) to run an ocean liner at a constant speed v (in miles per hour) is given by $c = a + bv^n$, where a, b, and n are positive constants with $n > 1$. Find the speed needed to make the cheapest 3000-mi run.

47. Two particles, A and B, are in motion in the xy-plane. Their coordinates at each instant of time t ($t \geq 0$) are given by $x_A = t$, $y_A = 2t$, $x_B = 1 - t$, and $y_B = t$. Find the minimum distance between A and B.

48. Follow the directions of Exercise 47, with $x_A = t$, $y_A = t^2$, $x_B = 2t$, and $y_B = 2$.

49. Prove that $(1, 0)$ is the closest point on the curve $x^2 + y^2 = 1$ to $(2, 0)$.

50. Find all points on the curve $y = \sqrt{x}$ for $0 \leq x \leq 3$ that are closest to, and at the greatest distance from, the point $(2, 0)$.

51. Find all points on the curve $x^2 - y^2 = 1$ closest to $(0, 2)$.

52. Find a point on the curve $x = 2y^2$ closest to $(0, 9)$.

53. Find the coordinates of the point P on the curve $y = 1/x^2$ ($x > 0$) where the segment of the tangent line at P that is cut off by the coordinate axes will have its shortest length.

54. Find the x-coordinate of the point P on the parabola $y = 1 - x^2$ for $0 < x \leq 1$ where the triangle that is enclosed by the tangent line at P and the coordinate axes has the smallest area.

55. Suppose that a line L of variable slope passes through $(1, 3)$ and intersects the coordinate axes at the points $(a, 0)$ and $(0, b)$ where a and b are positive. Find the slope of L for which the area of the triangle with vertices $(a, 0)$, $(0, b)$, and $(0, 0)$ is

(a) maximum   (b) minimum.

56. A triangle is inscribed in a segment of the parabola $y = kx^2$ ($k > 0$), as shown in Figure 4.7.19. Show that the area of the triangle is greatest when $x = (a + b)/2$, and find the maximum area. [*Hint:* First show that the area of the triangle is $(b - a) \cdot D/2$, where D is the vertical distance between the line and the point (x, kx^2).]

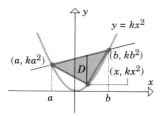

Figure 4.7.19

57. Where on the curve $y = (1 + x^2)^{-1}$ does the tangent line have the greatest slope?

58. A man is on the bank of a river that is 1 mi wide. He wants to travel to a town on the opposite bank, but 1 mi upstream. He intends to row on a straight line to some point P on the opposite bank and then walk the remaining distance along the bank (Figure 4.7.20). To what point should he row in order to reach his destination in the least time if

(a) he can walk 5 mi/hr and row 3 mi/hr

(b) he can walk 5 mi/hr and row 4 mi/hr?

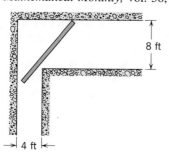

Figure 4.7.20

59. A pipe of negligible diameter is to be carried horizontally around a corner from a hallway 8 ft wide into a hallway 4 ft wide (Figure 4.7.21). What is the maximum length that the pipe can have? [An interesting discussion of this problem in the case where the diameter of the pipe is not neglected is given by Norman Miller in the *American Mathematical Monthly*, vol. 56, 1949, pp. 177–179.]

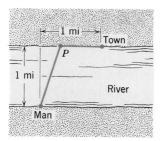

Figure 4.7.21

60. If an unknown physical quantity x is measured n times, the measurements $x_1, x_2, \ldots, x_n$ often vary because of uncontrollable factors such as temperature, atmospheric pressure, and so forth. Thus, a scientist is often faced with the problem of using n different observed measurements to obtain an estimate $\bar{x}$ of an unknown quantity x. One method for making such an estimate is based on the **least squares principle**, which states that the estimate $\bar{x}$ should be chosen to minimize

$$s = (x_1 - \bar{x})^2 + (x_2 - \bar{x})^2 + \cdots + (x_n - \bar{x})^2$$

which is the sum of the squares of the deviations between the estimate $\bar{x}$ and the measured values. Show that the estimate resulting from the least squares principle is

$$\bar{x} = \frac{1}{n}(x_1 + x_2 + \cdots + x_n)$$

that is, $\bar{x}$ is the arithmetic average of the observed values.

61. Suppose that the intensity of a point light source is directly proportional to the strength of the source and inversely proportional to the square of the distance from the source. Two point light sources with strengths of S and $8S$ are separated by a distance of 90 cm. Where on the line segment between the two sources is the intensity a minimum?

62. Prove: If $f(x) \geq 0$ on an interval I and if $f(x)$ has a maximum value on I at x_0, then $\sqrt{f(x)}$ also has a maximum value at x_0. Similarly, for minimum values. [*Hint:* Use the fact that $\sqrt{x}$ is an increasing function on the interval $[0, +\infty)$.]

63. Fermat's* (biography on p. 228) principle in optics states that light traveling from one point to another follows that path for which the total travel time is minimum. In a uniform medium, the paths of "minimum time" and "shortest distance" turn out to be the same, so that light, if unobstructed, travels along a straight line. Assume that we have a light source, a flat mirror, and an observer in a uniform medium. If a light ray leaves the source, bounces off the mirror, and travels on to the observer, then its path will consist of two line segments, as shown in Figure 4.7.22. According to Fermat's principle, the path will be such that the total travel time t is minimum or, since the medium is uniform, the path will be such that the total distance traveled from A to P to B is as small as possible. Assuming the minimum occurs when $dt/dx = 0$, show that the light ray will strike the mirror at the point P where the "angle of incidence" θ_1 equals the "angle of reflection" θ_2.

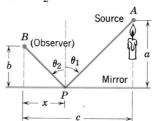

Figure 4.7.22

64. Fermat's principle (Exercise 63) also explains why light rays traveling between air and water undergo bending (refraction). Imagine that we have two uniform media (such as air and water) and a light ray traveling from a source A in one medium to an observer B in the other medium (Figure 4.7.23). It is known that light travels at a constant speed in a uniform medium, but more slowly in a dense medium (such as water) than in a thin medium (such as air). Consequently, the path of shortest time from A to B is not necessarily a straight line, but rather some broken line path A to P to B allowing the light to take greatest advantage of its higher speed through the thin medium. Snell's† law of refraction states that the path of the light ray will be such that

$$\frac{\sin \theta_1}{v_1} = \frac{\sin \theta_2}{v_2}$$

where v_1 is the speed of light in the first medium, v_2 is the speed of light in the second medium, and θ_1 and θ_2 are the angles shown in Figure 4.7.23. Show that this follows from the assumption that the path of minimum time occurs when $dt/dx = 0$.

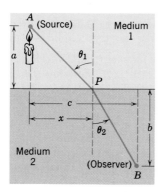

Figure 4.7.23

65. A farmer wants to walk at a constant rate from her barn to a straight river, fill her pail, and carry it to her house in the least time.

(a) Explain how this problem relates to Fermat's principle and the light-reflection problem in Exercise 63.

(b) Use the result of Exercise 63 to describe geometrically the best path for the farmer to take.

(c) Use part (b) to determine where the farmer should fill her pail if her house and barn are located as in Figure 4.7.24.

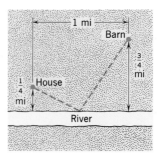

Figure 4.7.24

---

*PIERRE DE FERMAT (1601–1665). Fermat, the son of a successful French leather merchant, was a lawyer who practiced mathematics as a hobby. He received a Bachelor of Civil Laws degree from the University of Orleans in 1631 and subsequently held various government positions, including a post as councillor to the Toulouse parliament. Although he was apparently financially successful, confidential documents of that time suggest that his performance in office and as a lawyer was poor, perhaps because he devoted so much time to mathematics. Throughout his life, Fermat fought all efforts to have his mathematical results published. He had the unfortunate habit of scribbling his work in the margins of books and often sent his results to friends without keeping copies for himself. As a result, he never received credit for many major achievements until his name was raised from obscurity in the mid-nineteenth century. It is now known that Fermat, simultaneously and independently of Descartes, developed analytic geometry. Unfortunately, Descartes and Fermat argued bitterly over various problems so that there was never any real cooperation between these two great geniuses.

Fermat solved many fundamental calculus problems. He obtained the first procedure for differentiating polynomials, and solved many important maximization, minimization, area, and tangent problems. His work served to inspire Isaac Newton. Fermat is best known for his work in number theory, the study of properties and relationships between whole numbers. He was the first mathematician to make substantial contributions to this field after the ancient Greek mathematician Diophantus. Unfortunately, none of Fermat's contemporaries appreciated his work in this area, a fact that eventually pushed Fermat into isolation and obscurity in later life.

In addition to his work in calculus and number theory, Fermat was one of the founders of probability theory and made major contributions to the theory of optics. Outside mathematics, Fermat was a classical scholar of some note, was fluent in French, Italian, Spanish, Latin, and Greek, and he composed a considerable amount of Latin poetry.

One of the great mysteries of mathematics is shrouded in Fermat's work in number theory. In the margin of a book by Diophantus, Fermat scribbled that for integer values of n greater than 2, the equation $x^n + y^n = z^n$ has no nonzero integer solutions for x, y, and z. He stated, "I have discovered a truly marvelous proof of this, which however the margin is not large enough to contain." This result, which became known as "Fermat's last theorem," appeared to be true, but its proof evaded the greatest mathematical geniuses for 300 years until Professor Andrew Wiles of Princeton University presented a proof in June 1993 in a dramatic series of three lectures that drew international media attention (see *New York Times*, June 27, 1993). Mathematicians are currently investigating a possible gap in that proof. A prize of 100,000 German marks was offered in 1908 for the solution, but it is worthless today because of inflation.

---

†WILLEBRORD VAN ROIJEN SNELL (1591–1626). Dutch mathematician. Snell, who succeeded his father to the post of Professor of Mathematics at the University of Leiden in 1613, is most famous for the result on light refraction that bears his name. Although this phenomenon was studied as far back as the ancient Greek astronomer Ptolemy, until Snell's work the relationship was incorrectly thought to be $\theta_1/v_1 = \theta_2/v_2$. Snell's law was published by Descartes in 1638 without giving proper credit to Snell. Snell also discovered a method for determining distances by triangulation that founded the modern technique of mapmaking.

■ **4.8** NEWTON'S METHOD

> *In Section 2.7 we showed how to approximate a solution of an equation $f(x) = 0$ to any degree of accuracy using the Intermediate-Value Theorem. In this section we shall study a technique, called Newton's Method, that is generally more efficient.*

In beginning algebra one learns that the solution of a first-degree equation $ax + b = 0$ is given by the formula $x = -b/a$, and the solutions of a second-degree equation $ax^2 + bx + c = 0$ are given by the quadratic formula. Formulas also exist for the solutions of all third- and fourth-degree equations, although they are too complicated to be of practical use. In 1826 it was shown by the Norwegian mathematician Niels Henrik Abel* that it is impossible to construct a similar formula for the solutions of a *general* fifth-degree equation or higher. Thus, for a *specific* fifth-degree polynomial equation such as

$$x^5 - 9x^4 + 2x^3 - 5x^2 + 17x - 8 = 0$$

it may be difficult or impossible to find exact values for all of the solutions. Similar difficulties occur for trigonometric equations such as

$$x - \cos x = 0$$

as well as equations of other types. For such equations the solutions are generally approximated in some way, often by the method we shall now discuss.

☐ **NEWTON'S METHOD**

We note first that the solutions of $f(x) = 0$ are the values of x where the graph of f crosses the x-axis. Suppose that $x = r$ is the solution we are seeking. Even if we cannot find the value of r exactly, it is usually possible to approximate it by graphing f and applying

---

*NIELS HENRIK ABEL (1802–1829). Norwegian mathematician. Abel was the son of a poor Lutheran minister and a remarkably beautiful mother from whom he inherited strikingly good looks. In his brief life of 26 years Abel lived in virtual poverty and suffered a succession of adversities; yet he managed to prove major results that altered the mathematical landscape forever. At the age of thirteen he was sent away from home to a school whose better days had long passed. By a stroke of luck the school had just hired a teacher named Bernt Michael Holmboe, who quickly discovered that Abel had extraordinary mathematical ability. Together, they studied the calculus texts of Euler and works of Newton and the later French mathematicians. By the time he graduated, Abel was familiar with most of the great mathematical literature. In 1820 his father died, leaving the family in dire financial straits. Abel was able to enter the University of Christiania in Oslo only because he was granted a free room and several professors supported him directly from their salaries. The University had no advanced courses in mathematics, so Abel took a preliminary degree in 1822 and then continued to study mathematics on his own. In 1824 he published at his own expense the proof that it is impossible to solve the general fifth-degree polynomial equation algebraically. With the hope that this landmark paper would lead to his recognition and acceptance by the European mathematical community, Abel sent the paper to the great German mathematician Gauss, who casually declared it to be a ''monstrosity'' and tossed it aside. However, in 1826 Abel's paper on the fifth-degree equation and other work was published in the first issue of a new journal, founded by his friend, Leopold Crelle. In the summer of 1826 he completed a landmark work on transcendental functions, which he submitted to the French Academy of Sciences in the hope of establishing himself as a major mathematician, for many young mathematicians had gained quick distinction by having their work accepted by the Academy. However, Abel waited in vain because the paper was either ignored or misplaced by one of the referees, and it did not surface again until two years after his death. That paper was later described by one major mathematician as ''. . . the most important mathematical discovery that has been made in our century. . . .'' After submitting his paper, Abel returned to Norway, ill with tuberculosis and in heavy debt. While eking out a meager living as a tutor, he continued to produce great work and his fame spread. Soon great efforts were being made to secure a suitable mathematical position for him. Fearing that his great work had been lost by the Academy, he mailed a proof of the main result to Crelle in January of 1829. In April he suffered a violent hemorrhage and died. Two days later Crelle wrote to inform him that an appointment had been secured for him in Berlin and his days of poverty were over! Abel's great paper was finally published by the Academy twelve years after his death.

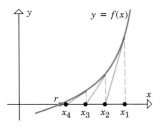

Figure 4.8.1

Theorem 2.7.10 to estimate where the graph crosses the x-axis. If we let x_1 denote our initial approximation to r, then we can generally improve on this approximation by moving along the tangent line to $y = f(x)$ at x_1 until we meet the x-axis at a point x_2 (Figure 4.8.1). Usually, x_2 will be closer to r than x_1. To improve the approximation further, we can repeat the process by moving along the tangent line to $y = f(x)$ at x_2 until we meet the x-axis at a point x_3. Continuing in this way we can generate a succession of values $x_1, x_2, x_3, x_4, \ldots$ that will usually get closer and closer to r. This procedure for approximating r is called **Newton's Method**.

To implement Newton's Method analytically, we must derive a formula that will tell us how to calculate each improved approximation from the preceding approximation. For this purpose, we note that the point-slope form of the tangent line to $y = f(x)$ at the initial approximation x_1 is

$$y - f(x_1) = f'(x_1)(x - x_1) \tag{1}$$

If $f'(x_1) \neq 0$, then this line is not parallel to the x-axis and consequently it crosses the x-axis at some point $(x_2, 0)$. Substituting the coordinates of this point in (1) yields

$$-f(x_1) = f'(x_1)(x_2 - x_1)$$

Solving for x_2 we obtain

$$x_2 = x_1 - \frac{f(x_1)}{f'(x_1)} \tag{2}$$

The next approximation can be obtained more easily. If we view x_2 as the starting approximation and x_3 the new approximation, we can simply apply (2) with x_2 in place of x_1 and x_3 in place of x_2. This yields

$$x_3 = x_2 - \frac{f(x_2)}{f'(x_2)} \tag{3}$$

provided $f'(x_2) \neq 0$. In general, if x_n is the nth approximation, then it is evident from the pattern in (2) and (3) that the improved approximation x_{n+1} is given by

> **Newton's Method**
>
> $$x_{n+1} = x_n - \frac{f(x_n)}{f'(x_n)}, \quad n = 1, 2, 3, \ldots \tag{4}$$

Example 1   Use Newton's Method to approximate the real solutions of
$$x^3 - x - 1 = 0$$

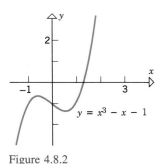

Figure 4.8.2

Solution.   Let $f(x) = x^3 - x - 1$, so $f'(x) = 3x^2 - 1$ and (4) becomes

$$x_{n+1} = x_n - \frac{x_n^3 - x_n - 1}{3x_n^2 - 1} \tag{5}$$

From the graph of f in Figure 4.8.2, we see that the given equation has only one real solution. This solution lies between 1 and 2 because $f(1) = -1 < 0$ and $f(2) = 5 > 0$. We shall use $x_1 = 1.5$ as our first approximation ($x_1 = 1$ or $x_1 = 2$ would also be reasonable choices).

Letting $n = 1$ in (5) and substituting $x_1 = 1.5$ yields

$$x_2 = 1.5 - \frac{(1.5)^3 - 1.5 - 1}{3(1.5)^2 - 1} = 1.34782609$$

(We used a calculator that displays nine digits.) Next, we let $n = 2$ in (5) and substitute $x_2 = 1.34782609$ to obtain

$$x_3 = 1.34782609 - \frac{(1.34782609)^3 - (1.34782609) - 1}{3(1.34782609)^2 - 1} = 1.32520040$$

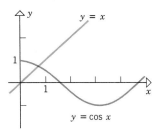

Figure 4.8.3

If we continue this process until two identical approximations are generated in succession, we obtain

$$x_1 = 1.5$$
$$x_2 = 1.34782609$$
$$x_3 = 1.32520040$$
$$x_4 = 1.32471817$$
$$x_5 = 1.32471796$$
$$x_6 = 1.32471796$$

At this stage there is no need to continue further because we have reached the accuracy limit of our calculator, and all subsequent approximations that the calculator generates will be the same. Thus, the solution is approximately $x \approx 1.32471796$. ◀

Example 2 It is evident from Figure 4.8.3 that if x is in radians, then the equation

$$\cos x = x$$

has a solution between 0 and 1. Use Newton's Method to approximate it.

Solution. Rewrite the equation as

$$x - \cos x = 0$$

and apply (4) with $f(x) = x - \cos x$. Since $f'(x) = 1 + \sin x$, (4) becomes

$$x_{n+1} = x_n - \frac{x_n - \cos x_n}{1 + \sin x_n} \tag{6}$$

From Figure 4.8.3, the solution seems closer to $x = 1$ than $x = 0$, so we shall use $x_1 = 1$ (radian) as our initial approximation. Letting $n = 1$ in (6) and substituting $x_1 = 1$ yields

$$x_2 = 1 - \frac{1 - \cos 1}{1 + \sin 1} = .750363868$$

Next, letting $n = 2$ in (6) and substituting this value of x_2 yields

$$x_3 = .750363868 - \frac{.750363868 - \cos (.750363868)}{1 + \sin (.750363868)} = .739112891$$

If we continue this process until two identical approximations are generated in succession, we obtain

$$x_1 = 1$$
$$x_2 = .750363868$$
$$x_3 = .739112891$$
$$x_4 = .739085133$$
$$x_5 = .739085133$$

Thus, to the accuracy limit of our calculator, the solution of the equation $\cos x = x$ is $x \approx .739085133$. ◀

□ **SOME DIFFICULTIES WITH NEWTON'S METHOD**

Newton's Method does not always work. For example, if $f'(x_n) = 0$ for some n, then (4) involves a division by zero, making it impossible to generate x_{n+1}. However, this is to be expected because the tangent line to $y = f(x)$ is parallel to the x-axis where $f'(x_n) = 0$, and so this tangent line does not cross the x-axis to generate the next approximation (Figure 4.8.4).

Sometimes the values of x_n produced by Newton's Method do not converge to a solution. For example, consider the equation

$$x^{1/3} = 0$$

which has $x = 0$ as its only solution, and try to approximate this solution by Newton's

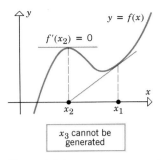

Figure 4.8.4

Method with a starting value of $x_0 = 1$. Letting $f(x) = x^{1/3}$, Formula (4) becomes

$$x_{n+1} = x_n - \frac{(x_n)^{1/3}}{\frac{1}{3}(x_n)^{-2/3}} = x_n - 3x_n = -2x_n$$

Beginning with $x_1 = 1$, the successive values generated by this formula are

$$x_1 = 1, \quad x_2 = -2, \quad x_3 = 4, \quad x_4 = -8, \ldots$$

which obviously do not converge to $x = 0$. Figure 4.8.5 illustrates what is happening geometrically.

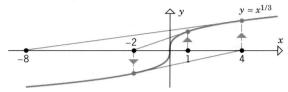

Figure 4.8.5

For a statement of conditions under which Newton's Method works and a discussion of error questions, the reader should consult a book on numerical analysis. In situations where Newton's method fails, the method discussed in Section 2.7 can be used; but when Newton's Method works, it is the method of choice because it converges more rapidly.

▶ Exercise Set 4.8   C̲ *All*

In this exercise set, use a calculator, and keep as many decimal places as it can display.

1. Approximate $\sqrt{2}$ by applying Newton's Method to the equation $x^2 - 2 = 0$.

2. Approximate $\sqrt{7}$ by applying Newton's Method to the equation $x^2 - 7 = 0$.

3. Approximate $\sqrt[3]{6}$ by applying Newton's Method to the equation $x^3 - 6 = 0$.

In Exercises 4–7, the equation has one real solution. Approximate it by Newton's Method.

4. $x^3 + x - 1 = 0$.
5. $x^3 - x + 3 = 0$.
6. $x^5 - x + 1 = 0$.
7. $x^5 + x^4 - 5 = 0$.

In Exercises 8–15, the equation has one solution satisfying the given conditions. Approximate it by Newton's Method.

8. $2x^2 + 4x - 3 = 0;\ x < 0$.
9. $2x^2 + 4x - 3 = 0;\ x > 0$.
10. $x^4 + x - 3 = 0;\ x > 0$.
11. $x^4 + x - 3 = 0;\ x < 0$.
12. $x^5 - 5x^3 - 2 = 0;\ x > 0$.
13. $2 \sin x = x;\ x > 0$.
14. $\sin x = x^2;\ x > 0$.
15. $x - \tan x = 0;\ \pi/2 < x < 3\pi/2$.

16. Many calculators compute reciprocals using the approximation $1/a \approx x_{n+1}$, where

$$x_{n+1} = x_n(2 - ax_n), \quad n = 1, 2, 3, \ldots$$

and x_1 is an initial approximation to $1/a$. This formula makes it possible to use multiplications and subtractions (which can be done quickly) to perform divisions that would be slow to obtain directly.

(a) Apply Newton's Method to

$$f(x) = (1/x) - a$$

to derive this approximation.

(b) Use the formula to approximate $\frac{1}{17}$.

17. The *mechanic's rule* for approximating square roots states that $\sqrt{a} \approx x_{n+1}$, where

$$x_{n+1} = \frac{1}{2}\left(x_n + \frac{a}{x_n}\right), \quad n = 1, 2, 3, \ldots$$

and x_1 is any positive approximation to $\sqrt{a}$.

(a) Apply Newton's Method to

$$f(x) = x^2 - a$$

to derive the mechanic's rule.

(b) Use the mechanic's rule to approximate $\sqrt{10}$.

In Exercises 18–22, find the x-coordinates of all points of intersection of the given curves. Use Newton's Method, if needed.

18. $y = x^2 + 1$ and $y = x^3$.
19. $y = x^3$ and $y = \frac{1}{2}x - 1$.

20. $y = \frac{1}{4}x^2$ and $y = \frac{2x}{x^2 + 1}$.

21. $y = x^2$ and $y = \sqrt{2x + 1}$.

22. $y = \frac{1}{8}x^3 + 1$ and $y = \cos 2x$.

In Exercises 23 and 24, use Newton's Method to approximate all real values of y satisfying the given equation for the indicated value of x.

23. $xy^4 + x^3y = 1$; $x = 1$.

24. $xy - \cos\left(\frac{1}{2}xy\right) = 0$; $x = 2$.

In Exercises 25–30, use Newton's Method.

25. Find the minimum value of
$$f(x) = \frac{1}{4}x^4 + x^2 + 5x$$

26. Find the maximum value of $f(x) = x \sin x$ on the interval $[0, \pi]$.

27. Find the coordinates of the point on the parabola $y = x^2$ that is nearest the point $(1, 0)$.

28. Find the dimensions of the rectangle of largest area that can be inscribed under the curve $y = \cos x$ for $0 \le x \le \pi/2$, as shown in Figure 4.8.6.

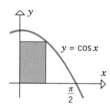

Figure 4.8.6

29. Find the central angle θ of a circle that subtends an arc whose length is 1.5 times the length of its chord (Figure 4.8.7). Give θ to the nearest degree. [*Hint:* Recall that $r\theta$ is the length of arc subtended by θ, where θ is in radians and r is the radius of the circle.]

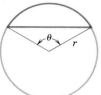

Figure 4.8.7

30. A *segment* of a circle is the region enclosed by an arc and its chord (Figure 4.8.8). If r is the radius of the circle and θ the angle subtended at the center of the circle, then it can be shown that the area A of the segment is $A = \frac{1}{2}r^2(\theta - \sin\theta)$, where θ is in radians. Find the value of θ for which the area of the segment is one-fourth the area of the circle. Give θ to the nearest degree.

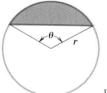

Figure 4.8.8

4.9 ROLLE'S THEOREM; MEAN-VALUE THEOREM

In this section we shall discuss a result called the Mean-Value Theorem. There are so many major consequences of this theorem that it is regarded as one of the most fundamental results in calculus.

☐ **ROLLE'S THEOREM**

We shall begin with a special case of the Mean-Value Theorem, called **Rolle's*** (biography on p. 234) **Theorem**. Geometrically, Rolle's Theorem states that between any two points, a and b, where a "well-behaved" curve $y = f(x)$ crosses the x-axis, there must be at least one place where the tangent line to the curve is horizontal (Figure 4.9.1).

The precise statement of this result is as follows.

4.9.1   THEOREM (*Rolle's Theorem*).   *Let f be differentiable on (a, b) and continuous on $[a, b]$. If $f(a) = f(b) = 0$, then there is at least one point c in (a, b) where $f'(c) = 0$.*

Proof.   Either $f(x)$ is equal to zero for all x in $[a, b]$ or it is not. If it is, then $f'(x) = 0$ for all x in (a, b), since f is constant on (a, b). Thus, for any c in (a, b)
$$f'(c) = 0$$

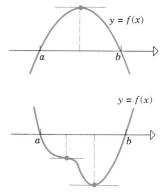

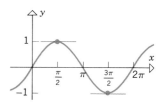

Figure 4.9.1

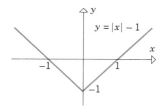

Figure 4.9.2

If $f(x)$ is not equal to zero for all x in $[a, b]$, then there must be a point x in (a, b) where $f(x) > 0$ or $f(x) < 0$. We shall consider the first case and leave the second as an exercise.

Since f is continuous on $[a, b]$, it follows from the Extreme-Value Theorem that f has a maximum value at some point c in $[a, b]$. Since $f(a) = f(b) = 0$ and $f(x) > 0$ at some point in (a, b), the point c cannot be an endpoint; it must lie in (a, b). By hypothesis, f is differentiable everywhere on (a, b). In particular, it is differentiable at c so that $f'(c) = 0$ by Theorem 4.6.5.   ∎

Example 1   The function

$$f(x) = \sin x$$

is both continuous and differentiable everywhere, hence is continuous on $[0, 2\pi]$ and differentiable on $(0, 2\pi)$. Moreover,

$$f(0) = \sin 0 = 0 \quad \text{and} \quad f(2\pi) = \sin 2\pi = 0$$

so that f satisfies the hypotheses of Rolle's Theorem on the interval $[0, 2\pi]$. Since $f'(c) = \cos c$, Rolle's Theorem guarantees that there is at least one point c in $(0, 2\pi)$ such that

$$\cos c = 0 \tag{1}$$

Solving (1) yields two values for c, namely $c_1 = \pi/2$ and $c_2 = 3\pi/2$ (Figure 4.9.2).   ◄

REMARK.   In the preceding example we were able to find c because (1) was easy to solve. However, frequently the equation $f'(c) = 0$ is sufficiently complicated that an exact value for c cannot be obtained. This rarely causes problems, since one is usually more interested in knowing that c exists than in finding its value.

Example 2   The function

$$f(x) = |x| - 1$$

has the property that $f(-1) = 0$ and $f(1) = 0$, yet there is no point in the interval $(-1, 1)$ where the graph of f has a horizontal tangent line (Figure 4.9.3). This does *not* contradict Rolle's Theorem because the function f is not differentiable at every point of the interval $(-1, 1)$.   ◄

Figure 4.9.3

☐ **THE MEAN-VALUE THEOREM**

Rolle's Theorem is a special case of the ***Mean-Value Theorem***, which states that between any two points A and B on a "well-behaved" curve $y = f(x)$, there must be at least one

---

*MICHEL ROLLE (1652–1719). French mathematician. Rolle, the son of a shopkeeper, received only an elementary education. He married early and as a young man struggled hard to support his family on the meager wages of a transcriber for notaries and attorneys. In spite of his financial problems and minimal education, Rolle studied algebra and Diophantine analysis (a branch of number theory) on his own. Rolle's fortune changed dramatically in 1682 when he published an elegant solution of a difficult, unsolved problem in Diophantine analysis. The public recognition of his achievement led to a patronage under minister Louvois, a job as an elementary mathematics teacher, and eventually to a short-term administrative post in the Ministry of War. In 1685 he joined the Académie des Sciences in a low-level position for which he received no regular salary until 1699. He stayed there until he died of apoplexy in 1719.

While Rolle's forté was always Diophantine analysis, his most important work was a book on the algebra of equations, called *Traité d'algèbre,* published in 1690. In that book Rolle firmly established the notation $\sqrt[n]{a}$ [earlier written as $\sqrt{\overline{n}\ a}$] for the nth root of a, and proved a polynomial version of the theorem that today bears his name. (Rolle's Theorem was named by Giusto Bellavitis in 1846.) Ironically, Rolle was one of the most vocal early antagonists of calculus. He strove intently to demonstrate that it gave erroneous results and was based on unsound reasoning. He quarreled so vigorously on the subject that the Académie des Sciences was forced to intervene on several occasions. Among his several achievements, Rolle helped advance the currently accepted size order for negative numbers. Descartes, for example, viewed -2 as smaller than -5. Rolle preceded most of his contemporaries by adopting the current convention in 1691.

place where the tangent line to the curve is parallel to the secant line joining A and B (Figure 4.9.4).

Noting that the slope of the secant line joining $A(a, f(a))$ and $B(b, f(b))$ is

$$\frac{f(b) - f(a)}{b - a}$$

and the slope of the tangent at c is $f'(c)$, the Mean-Value Theorem can be stated precisely as follows.

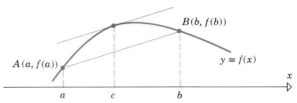

Figure 4.9.4

4.9.2   THEOREM (*Mean-Value Theorem*).   *Let f be differentiable on (a, b) and continuous on $[a, b]$. Then there is at least one point c in (a, b) where*

$$f'(c) = \frac{f(b) - f(a)}{b - a} \tag{2}$$

Motivation for the Proof of Theorem 4.9.2.   Figure 4.9.4 suggests that (2) will hold (i.e., the tangent line will be parallel to the secant line) at a point c where the vertical distance between the curve and the secant line is maximum. Thus, to prove the Mean-Value Theorem it is natural to begin by looking for a formula for the vertical distance $v(x)$ between the curve $y = f(x)$ and the secant line joining $(a, f(a))$ and $(b, f(b))$.

Proof of Theorem 4.9.2.   Since the two-point form of the equation of the secant line joining $(a, f(a))$ and $(b, f(b))$ is

$$y - f(a) = \frac{f(b) - f(a)}{b - a}(x - a)$$

or equivalently,

$$y = \frac{f(b) - f(a)}{b - a}(x - a) + f(a)$$

the difference $v(x)$ between the height of the graph of f and the height of the secant line is

$$v(x) = f(x) - \left[\frac{f(b) - f(a)}{b - a}(x - a) + f(a)\right] \tag{3}$$

Since $f(x)$ is continuous on $[a, b]$ and differentiable on (a, b), so is $v(x)$. Moreover,

$$v(a) = 0 \quad \text{and} \quad v(b) = 0$$

so that $v(x)$ satisfies the hypotheses of Rolle's Theorem on the interval $[a, b]$. Thus, there is a point c in (a, b) such that $v'(c) = 0$. But from Equation (3)

$$v'(x) = f'(x) - \frac{f(b) - f(a)}{b - a}$$

so

$$v'(c) = f'(c) - \frac{f(b) - f(a)}{b - a}$$

Thus, at the point c in (a, b), where $v'(c) = 0$, we have

$$f'(c) = \frac{f(b) - f(a)}{b - a} \qquad ■$$

Example 3   Let $f(x) = x^3 + 1$. Show that f satisfies the hypotheses of the Mean-Value Theorem on the interval $[1, 2]$ and find all values of c in this interval whose existence is guaranteed by the theorem.

Solution.   Because f is a polynomial, f is continuous and differentiable everywhere, hence is continuous on $[1, 2]$ and differentiable on $(1, 2)$. Thus, the hypotheses of the Mean-Value Theorem are satisfied with $a = 1$ and $b = 2$. But

$$f(a) = f(1) = 2, \quad f(b) = f(2) = 9$$
$$f'(x) = 3x^2, \quad f'(c) = 3c^2$$

so that the equation

$$f'(c) = \frac{f(b) - f(a)}{b - a}$$

becomes $3c^2 = 7$, which has two solutions:

$$c = \sqrt{7/3} \quad \text{and} \quad c = -\sqrt{7/3}$$

Only the first is in the interval $(1, 2)$, so $c = \sqrt{7/3}$ is the number whose existence is guaranteed by the Mean-Value Theorem.   ◄

□ **CONSEQUENCES OF THE MEAN-VALUE THEOREM**

We stated earlier that the Mean-Value Theorem is the starting point for many important results in calculus. We shall conclude this section by using it to prove Theorem 4.2.2, which was one of our main graphing tools.

4.2.2   THEOREM (*Revisited*).   *Let f be a function that is continuous on a closed interval $[a, b]$ and differentiable on the open interval (a, b).*

(a)   *If $f'(x) > 0$ for every value of x in (a, b), then f is increasing on $[a, b]$.*
(b)   *If $f'(x) < 0$ for every value of x in (a, b), then f is decreasing on $[a, b]$.*
(c)   *If $f'(x) = 0$ for every value of x in (a, b), then f is constant on $[a, b]$.*

Proof (a).   Let x_1 and x_2 be points in $[a, b]$ such that $x_1 < x_2$. We must show that $f(x_1) < f(x_2)$. Because the hypotheses of the Mean-Value Theorem are satisfied on the entire interval $[a, b]$, they are satisfied on the subinterval $[x_1, x_2]$. Thus, there is some point c in the open interval (x_1, x_2) such that

$$f'(c) = \frac{f(x_2) - f(x_1)}{x_2 - x_1}$$

or equivalently,

$$f(x_2) - f(x_1) = f'(c)(x_2 - x_1) \tag{4}$$

Since c is in the open interval (x_1, x_2), it follows that $a < c < b$; thus, $f'(c) > 0$. But $x_2 - x_1 > 0$ since we assumed that $x_1 < x_2$. Thus, it follows from (4) that $f(x_2) - f(x_1) > 0$ or, equivalently, $f(x_1) < f(x_2)$, which is what we were to prove. The proofs of parts (b) and (c) are similar and are left as exercises.   ▌

We know from earlier work that the derivative of a constant is zero. Part (c) of Theorem 4.2.2 is a converse; it states that a function whose derivative is zero at every point of an interval must be constant on that interval. The following useful theorem is a consequence of this result.

4.9.3   THEOREM.   *If f and g are continuous on a closed interval $[a, b]$, and if $f'(x) = g'(x)$ for all x in the open interval (a, b), then f and g differ by a constant on $[a, b]$; that is, there is a constant k such that $f(x) - g(x) = k$ for all x in $[a, b]$.*

Proof. Let $h(x) = f(x) - g(x)$. Then for every x in (a, b)

$$h'(x) = f'(x) - g'(x) = 0$$

Thus, $h(x) = f(x) - g(x)$ is constant on $[a, b]$ by Theorem 4.2.2(c). ∎

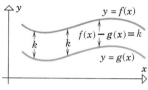

Figure 4.9.5

REMARK. This theorem remains true if the closed interval $[a, b]$ is replaced by a finite or infinite interval (a, b), $[a, b)$, or $(a, b]$, provided f and g are differentiable on (a, b) and continuous on the entire interval.

Theorem 4.9.3 has a useful geometric interpretation. It implies that two functions with the same derivative at each point of an interval have graphs that are vertical translations of one another over the interval (Figure 4.9.5).

▶ Exercise Set 4.9

In Exercises 1–6, verify that the hypotheses of Rolle's Theorem are satisfied on the given interval and find all values of c that satisfy the conclusion of the theorem.

1. $f(x) = x^2 - 6x + 8$; $[2, 4]$.

2. $f(x) = x^3 - 3x^2 + 2x$; $[0, 2]$.

3. $f(x) = \cos x$; $[\pi/2, 3\pi/2]$.

4. $f(x) = \dfrac{x^2 - 1}{x - 2}$; $[-1, 1]$.

5. $f(x) = \frac{1}{2}x - \sqrt{x}$; $[0, 4]$.

6. $f(x) = \dfrac{1}{x^2} - \dfrac{4}{3x} + \dfrac{1}{3}$; $[1, 3]$.

In Exercises 7–12, verify that the hypotheses of the Mean-Value Theorem are satisfied on the given interval and find all values of c that satisfy the conclusion of the theorem.

7. $f(x) = x^2 + x$; $[-4, 6]$.

8. $f(x) = x^3 + x - 4$; $[-1, 2]$.

9. $f(x) = \sqrt{x + 1}$; $[0, 3]$.

10. $f(x) = x + \dfrac{1}{x}$; $[3, 4]$.

11. $f(x) = \sqrt{25 - x^2}$; $[-5, 3]$.

12. $f(x) = \dfrac{1}{x - 1}$; $[2, 5]$.

13. Let $f(x) = \tan x$.

(a) Show that there is no point c in $(0, \pi)$ such that $f'(c) = 0$, even though $f(0) = f(\pi) = 0$.

(b) Explain why the result in part (a) does not violate Rolle's Theorem.

14. Let $f(x) = x^{2/3}$, $a = -1$, and $b = 8$.

(a) Show that there is no point c in (a, b) such that

$$f'(c) = \frac{f(b) - f(a)}{b - a}$$

(b) Explain why the result in part (a) does not violate the Mean-Value Theorem.

In Exercises 15–18, use the Mean-Value Theorem to prove the given statement.

15. $|\sin x - \sin y| \le |x - y|$ for all real values of x and y.

16. $|\tan x + \tan y| \ge |x + y|$ for all real values of x and y in the interval $(-\pi/2, \pi/2)$.

17. If $0 < x < y$, then $\sqrt{xy} < \frac{1}{2}(x + y)$.

$$\left[\text{Hint: First show that } \sqrt{y} - \sqrt{x} < \frac{y - x}{2\sqrt{x}}. \right]$$

18. If $f(1) = 0$ and $f'(x) = 1/x$ for all x in $(0, +\infty)$, then $f(x) \le x - 1$ for all x in $(0, +\infty)$. [*Hint:* Consider the cases $x > 1$, $0 < x < 1$, and $x = 1$ separately.]

19. Prove: If $f(x) = a_2 x^2 + a_1 x + a_0$ $(a_2 \ne 0)$, then the number c whose existence is guaranteed by the Mean-Value Theorem occurs at the midpoint of the interval $[a, b]$.

In Exercises 20–26, use Rolle's Theorem to prove the given statement.

20. The equation $6x^5 - 4x + 1 = 0$ has at least one solution in the interval $(0, 1)$.

$$\left[\text{Hint: } \frac{d}{dx}(x^6 - 2x^2 + x) = 6x^5 - 4x + 1. \right]$$

21. The equation $3ax^2 + 2bx = a + b$ has at least one solution in the interval $(0, 1)$. [*Hint:* Consider the function $f(x) = ax^3 + bx^2 - (a + b)x$.]

22. If $f'(x)$ exists and $f'(x) \ne 0$ for all x in some open interval I, then $f(x) = 0$ can have at most one solution in I.

23. $x^3 + 4x - 1 = 0$ must have fewer than two distinct real solutions.

24. If $b^2 - 3ac < 0$, then $ax^3 + bx^2 + cx + d = 0$ $(a \ne 0)$ can have at most one real solution.

25. If $f(x) = x^5 + ax^4 + bx^3 + cx^2 + dx + e$ and $2a^2 < 5b$, then the equation $f(x) = 0$ cannot have more than three distinct real solutions. [*Hint:* Consider $f'''(x)$.]

26. If $f(x) = x^4 - 7x + 2$, then the equation $f(x) = 0$ has exactly two distinct real solutions. [*Hint:* First find $f(0)$,

$f(1)$, and $f(2)$ to show that there are at least two distinct real solutions.]

In Exercises 27 and 28, use part (c) of Theorem 4.2.2 to prove the given statement.

27. If f and g are functions for which $f'(x) = g(x)$ and $g'(x) = -f(x)$ for all x, then $f^2(x) + g^2(x)$ is a constant.

28. If f and g are functions for which $f'(x) = g(x)$ and $g'(x) = f(x)$ for all x, then $f^2(x) - g^2(x)$ is a constant.

In Exercises 29–32, use Theorem 4.9.3.

29. Let $f(x) = (x - 1)^3$ and $g(x) = (x^2 + 3)(x - 3)$. Show that $f(x) - g(x)$ is a constant. Find the constant.

30. Let $f(x) = \dfrac{x + 2}{3 - x}$ and $g(x) = -\dfrac{5}{x - 3}$ for $x > 3$. Show that $f(x) - g(x)$ is a constant. Find the constant.

31. Let $g(x) = x^3 - 4x + 6$. Find $f(x)$ so that $f'(x) = g'(x)$ and $f(1) = 2$.

32. Let $g(x) = \sqrt{x^2 + 7}$. Find $f(x)$ so that $f'(x) = g'(x)$ and $f(-3) = 1$.

33. Let f and g be continuous on $[a, b]$ and differentiable on (a, b). Prove: If $f(a) = g(a)$ and $f(b) = g(b)$, then there is a point c in (a, b) where $f'(c) = g'(c)$.

34. Prove part (b) of Theorem 4.2.2.

35. Prove part (c) of Theorem 4.2.2.

36. (a) Prove: If $f''(x) > 0$ for all x in (a, b), then $f'(x) = 0$ at most once in (a, b).

(b) Give a geometric interpretation of the result in (a).

37. Prove: If f is continuous on $[a, b]$ and differentiable on (a, b), and if $f(a) = f(b)$, then there is a point c in (a, b) where $f'(c) = 0$.

38. Let $s = f(t)$ be the position versus time curve for a particle moving in the positive direction along a coordinate line. Prove: If f satisfies the hypotheses of the Mean-Value Theorem on a time interval $[a, b]$, then there will be an instant t_0 in (a, b) where the instantaneous velocity at time t_0 equals the average velocity over the interval $[a, b]$.

39. Complete the proof of Rolle's Theorem by considering the case where $f(x) < 0$ at some point x in (a, b).

40. Use the Mean-Value Theorem to prove

$$1.71 < \sqrt{3} < 1.75$$

[*Hint:* Let $f(x) = \sqrt{x}$, $a = 3$, and $b = 4$ in the Mean-Value Theorem.]

41. An automobile starts from rest and travels 4 miles along a straight road in 5 minutes. Use the Mean-Value Theorem to show that at some instant during the trip its velocity is exactly 48 mi/hr.

42. At 11 A.M. on a certain morning the outside temperature was 76° F. At 11 P.M. that evening it had dropped to 52° F.

(a) Use the Mean-Value Theorem to show that at some instant during this period the temperature was decreasing at the rate of 2° F/hr.

(b) Suppose you know that the temperature reached a high of 88° F sometime between 11 A.M. and 11 P.M. Show that at some instant during this period the temperature was decreasing at a rate greater than 3° F/hr.

■ 4.10 MOTION ALONG A LINE (RECTILINEAR MOTION)

In this section we shall study the motion of a particle along a line, sometimes called **rectilinear motion**. *Examples are a piston moving up and down in a cylinder, a rock tossed straight up and returning to earth straight down, and the back and forth vibrations of a spring. In later sections we shall study motion along curves in two- and three-dimensional space.*

☐ VELOCITY AND
ACCELERATION

To study the motion of a particle along a line, it is usually desirable to coordinatize the line. We do this by selecting an arbitrary origin, positive direction, and unit of length. We also choose a unit for measuring time and let t be the amount of time elapsed from some arbitrary initial observation. Thus, at the initial observation we have $t = 0$.

As the particle moves along the coordinate line its coordinate s will vary as a function of time t. This function, denoted by $s(t)$, is called the **position function** of the particle. The rate at which the particle's coordinate changes with time is called the *velocity* of the particle, and the rate at which the velocity changes with time is called the *acceleration* of the particle. More precisely, we make the following definition.

4.10.1 DEFINITION. If $s(t)$ is the position function of a particle moving on a coordinate line, then the ***instantaneous velocity*** at time t is defined by

$$v(t) = s'(t) = \frac{ds}{dt} \tag{1}$$

and the ***instantaneous acceleration*** at time t is defined by

$$a(t) = v'(t) = \frac{dv}{dt} \tag{2}$$

Since $v(t) = s'(t) = ds/dt$, the instantaneous acceleration can also be written as

$$a(t) = s''(t) = \frac{d^2s}{dt^2} \tag{3}$$

In everyday language the terms ''speed'' and ''velocity'' are often used interchangeably; however, in mathematics and science there is a distinction: The ***instantaneous speed*** of a particle moving on a coordinate line is defined to be the absolute value of the velocity. Thus,

$$\begin{bmatrix} \text{instantaneous} \\ \text{speed} \end{bmatrix} = |v(t)| = |s'(t)| = \left| \frac{ds}{dt} \right| \tag{4}$$

If $v(t) > 0$ at a given time t, then the coordinate s of the particle is increasing at that instant, which means that the particle is moving in the positive direction along the line; similarly, if $v(t) < 0$, then s is decreasing at time t and the particle is moving in the negative direction (Figures 4.10.1a and 4.10.1b). The speed of a particle is always nonnegative; it tells how fast the particle is moving, but provides no information about the direction of motion.

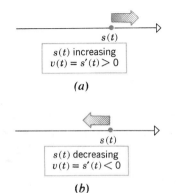

(a)

(b)

Figure 4.10.1

□ **INTERPRETING THE SIGN OF ACCELERATION**

We will say that a particle is ***speeding up*** when its instantaneous speed is increasing and ***slowing down*** when its instantaneous speed is decreasing. It is important to understand that a positive acceleration does not necessarily mean that the particle is speeding up and a negative acceleration does not necessarily mean that it is slowing down; the sign of the velocity also plays a role. We leave it as an exercise to show that a particle is speeding up when the velocity and acceleration have the same sign and is slowing down when they have opposite signs. Thus, a particle with negative velocity is speeding up when its acceleration is negative and slowing down when its acceleration is positive. A particle with positive velocity is speeding up when its acceleration is positive and is slowing down when it is negative.

□ **THE POSITION VERSUS TIME CURVE**

The graph of $s(t)$ in a ts-coordinate system is called the ***position versus time curve***; most of the significant information about the motion of a particle on a coordinate line can be obtained from that curve:

• Where $s(t) > 0$, the particle is on the positive side of the origin on the line of motion.

• Where $s(t) < 0$, the particle is on the negative side of the origin on the line of motion.

• Where a tangent line has positive slope we have $v(t) = s'(t) > 0$, so the particle is moving in the positive direction along its line of motion.

• Where a tangent line has negative slope we have $v(t) = s'(t) < 0$, so the particle is moving in the negative direction along its line of motion.

- Where a tangent line is horizontal we have $v(t) = s'(t) = 0$, so the particle is momentarily stopped.
- Where the graph is concave up we have $a(t) = v'(t) = s''(t) > 0$, so the velocity is increasing.
- Where the graph is concave down we have $a(t) = v'(t) = s''(t) < 0$, so the velocity is decreasing.

Example 1 The following table illustrates the above observations.

POSITION VERSUS TIME CURVE	CHARACTERISTICS OF THE CURVE AT $t = t_0$	BEHAVIOR OF PARTICLE AT TIME $t = t_0$
	• $s(t_0) > 0$ • Tangent line has positive slope. • Curve is concave down.	• Particle is on the positive side of the origin. • Particle is moving in the positive direction. • Velocity is decreasing. • Particle is slowing down.
	• $s(t_0) > 0$ • Tangent line has negative slope. • Curve is concave down.	• Particle is on the positive side of the origin. • Particle is moving in the negative direction. • Velocity is decreasing. • Particle is speeding up.
	• $s(t_0) < 0$ • Tangent line has negative slope. • Curve is concave up.	• Particle is on the negative side of the origin. • Particle is moving in the negative direction. • Velocity is increasing. • Particle is slowing down.
	• $s(t_0) > 0$ • Tangent line has zero slope. • Curve is concave down.	• Particle is on the positive side of the origin. • Particle is momentarily stopped. • Velocity is decreasing.

◀

Since acceleration is the rate at which velocity changes with time, it is expressed in units of velocity per unit of time. For example, if t is in seconds and s is in meters, then velocity units are meters per second (m/sec), and acceleration units are meters per second per second, which would be written as

$$(\text{m/sec})/\text{sec} \quad \text{or} \quad \text{m/sec}^2$$

Example 2 Let $s = s(t) = t^3 - 6t^2$ be the position function of a particle moving on a coordinate line, where t is in seconds and s is in meters. Then the instantaneous velocity, speed, and acceleration are given by

$$v(t) = \frac{ds}{dt} = 3t^2 - 12t, \quad |v(t)| = |3t^2 - 12t|, \quad a(t) = \frac{dv}{dt} = 6t - 12$$

At time $t = 1$, the instantaneous position, velocity, speed, and acceleration of the particle are

$$s(1) = -5, \quad v(1) = -9, \quad |v(1)| = 9, \quad a(1) = -6$$

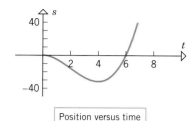

Position versus time

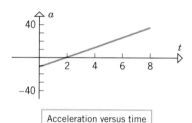

Velocity versus time

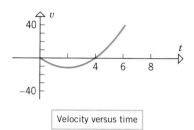

Acceleration versus time

Figure 4.10.2

Thus, the particle is 5 meters to the left of the origin and is moving with a velocity of -9 m/sec (i.e., moving in the negative direction with a speed of 9 m/sec), and has an acceleration of -6 m/sec^2. Since the acceleration and velocity have the same sign at $t = 1$, the particle is speeding up at that instant. At time $t = 4$, the instantaneous position, velocity, speed, and acceleration of the particle are

$$s(4) = -32, \quad v(4) = 0, \quad |v(4)| = 0, \quad a(4) = 12$$

Thus, the particle is 32 meters to the left of the origin, and is momentarily stopped, but is accelerating at the rate of 12 m/sec^2. The position versus time curve and the graphs of velocity and acceleration versus time are shown in Figure 4.10.2.   ◀

The next example gives a schematic way of describing the motion of a particle moving on a coordinate line.

Example 3   The position function of a particle moving on a coordinate line is given by $s(t) = 2t^3 - 21t^2 + 60t + 3$, where s is in feet and t is in seconds. Describe the motion of the particle for $t \geq 0$.

Solution.   The velocity and acceleration at time t are

$$v(t) = s'(t) = 6t^2 - 42t + 60 = 6(t - 2)(t - 5)$$
$$a(t) = v'(t) = 12t - 42 = 12(t - \tfrac{7}{2})$$

At each instant we can determine the direction of motion from the sign of $v(t)$ and whether the particle is speeding up or slowing down from the signs of $v(t)$ and $a(t)$ together (Figures 4.10.3a and 4.10.3b). The motion of the particle is described schematically by the curved

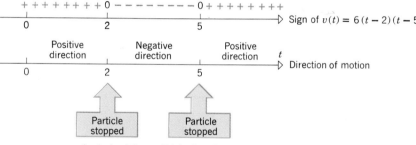

Figure 4.10.3a

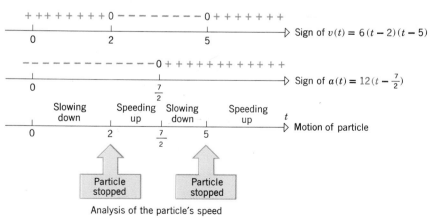

Figure 4.10.3b

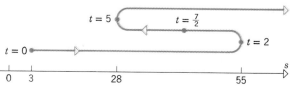

Figure 4.10.3c

line in Figure 4.10.3c. At time $t = 0$ the particle is at the point $s(0) = 3$ moving right with velocity $v(0) = 60$ ft/sec, but slowing down with acceleration $a(0) = -42$ ft/sec^2. The particle continues moving right until time $t = 2$, when it stops at the point $s(2) = 55$, reverses direction, and begins to speed up with an acceleration of $a(2) = -18$ ft/sec^2. At time $t = \frac{7}{2}$ the particle begins to slow down, but continues moving left until time $t = 5$, when it stops at the point $s(5) = 28$, reverses direction again, and begins to speed up with acceleration $a(5) = 18$ ft/sec^2. The particle then continues moving right thereafter with increasing speed.   ◄

REMARK.   The curved line in Figure 4.10.3c is descriptive only. The actual path of the particle is back and forth on the coordinate line.

► Exercise Set 4.10

1. The graphs of three position functions are shown in Figure 4.10.4. In each case determine the signs of the velocity and acceleration, then determine whether the particle is speeding up or slowing down.

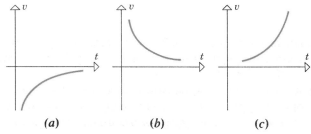

(a)   *(b)*   *(c)*

Figure 4.10.4

2. The graphs of three velocity functions are shown in Figure 4.10.5. In each case determine the sign of the acceleration, then determine whether the particle is speeding up or slowing down.

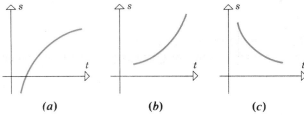

(a)   *(b)*   *(c)*

Figure 4.10.5

3. The position function of a particle moving on a coordinate line is shown in Figure 4.10.6.

(a) Is the particle moving left or right at time t_0?

(b) Is the acceleration positive or negative at time t_0?

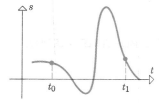

Figure 4.10.6

(c) Is the particle speeding up or is it slowing down at time t_0?

(d) Is the particle speeding up or is it slowing down at time t_1?

4. Match the graphs of the position functions shown in Figure 4.10.7 with their velocity functions shown in Figure 4.10.8.

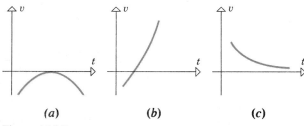

(a)   *(b)*   *(c)*

Figure 4.10.7

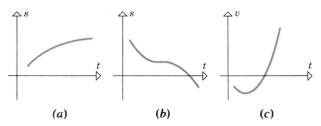

(a)   *(b)*   *(c)*

Figure 4.10.8

5. Figure 4.10.9 shows the velocity versus time graph for a

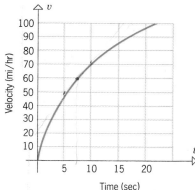

Figure 4.10.9

test run of the Grand Prix GTP. Using this graph, estimate

(a) the acceleration at 60 mi/hr (in units of ft/sec^2)

(b) the time at which the maximum acceleration occurs.

[Data from *Car and Driver*, October 1990.]

6. Figure 4.10.10 shows the position versus time graph for an elevator that ascends 40 m from one stop to the next.

(a) Estimate the velocity when the elevator is halfway up.

(b) Sketch rough graphs of the velocity versus time curve and the acceleration versus time curve.

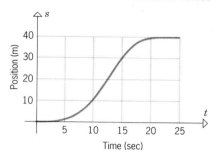

Figure 4.10.10

In Exercises 7–22, s is the position in feet, and t is the time in seconds, for a particle moving on a coordinate line.

7. (a) Let $s(t) = t^3 - 6t^2$. Make a table showing the position, velocity, speed, and acceleration at times $t = 1$, $t = 2$, $t = 3$, $t = 4$, and $t = 5$.

(b) At each of these times specify the direction of motion, if any, and whether the particle is speeding up, slowing down, or neither.

8. Let $s = \dfrac{100}{t^2 + 12}$ for $t \geq 0$. Find the maximum speed of the particle and the direction of motion of the particle when it has this speed.

9. Let $s = 5t^2 - 22t$.

(a) Find the maximum speed of the particle during the time interval $1 \leq t \leq 3$.

(b) When, during the time interval $1 \leq t \leq 3$, is the particle farthest from the origin? What is its position at that instant?

In Exercises 10–15, describe the motion of the particle for $t \geq 0$ (as in Example 3) and make a sketch as in Figure 4.10.3c.

10. $s = -3t + 2$.

11. $s = 1 + 6t - t^2$.

12. $s = t^3 - 6t^2 + 9t + 1$.

13. $s = t^3 - 9t^2 + 24t$.

14. $s = t + \dfrac{9}{t + 1}$.

15. $s = \begin{cases} \cos t, & 0 \leq t \leq 2\pi \\ 1, & t > 2\pi. \end{cases}$

16. Let $s = t^3 - 6t^2 + 1$.

(a) Find s and v when $a = 0$.

(b) Find s and a when $v = 0$.

17. Let $s = 4t^{3/2} - 3t^2$ for $t > 0$.

(a) Find s and v when $a = 0$.

(b) Find s and a when $v = 0$.

18. Let $s = \sqrt{2t^2 + 1}$. Find $\lim\limits_{t \to +\infty} v$.

19. (a) Use the chain rule to show that for a particle in rectilinear motion $a = v(dv/ds)$.

(b) Let $s = \sqrt{3t + 7}$, $t \geq 0$. Find a formula for v in terms of s and use the equation in part (a) to find the acceleration when $s = 5$.

20. If $s = t/(t^2 + 5)$ is the position function of a moving particle for $t \geq 0$, at what instant of time will the particle start to reverse its direction of motion, and where is it at that instant?

21. Suppose that the position functions of two particles, P_1 and P_2, in motion along the same line are

$$s_1 = \tfrac{1}{2}t^2 - t + 3 \quad \text{and} \quad s_2 = -\tfrac{1}{4}t^2 + t + 1$$

respectively, for $t \geq 0$.

(a) Prove that P_1 and P_2 do not collide.

(b) How close can P_1 and P_2 get to one another?

(c) During what intervals of time are they moving in opposite directions?

22. Let $s_A = 15t^2 + 10t + 20$ and $s_B = 5t^2 + 40t$, $t \geq 0$, be the position functions of cars A and B that are moving along parallel straight lanes of a highway.

(a) How far is car A ahead of car B when $t = 0$?

(b) At what instants of time are the cars next to one another?

(c) At what instant of time do they have the same velocity? Which car is ahead at this instant?

23. Figure 4.10.11 shows the velocity versus distance graph for a 222 Remington Magnum 55 grain pointed soft point bullet.

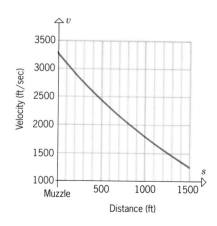

Figure 4.10.11

(a) Use the graph to estimate the value of dv/ds when the velocity is 2000 ft/sec.

(b) Use the result in part (a) and the chain rule to approximate the acceleration when the velocity is 2000 ft/sec. [*Hint:* See Exercise 19.]

[Data from the *Shooter's Bible,* No. 82, Stoeger Publishing Co., 1991.]

24. For a particle moving on a coordinate line, the ***average velocity*** v_{ave} and ***average acceleration*** a_{ave} over a time interval $[t_0, t_1]$ are defined by

$$v_{ave} = \frac{\text{change in position}}{\text{time elapsed}} = \frac{s(t_1) - s(t_0)}{t_1 - t_0}$$

$$a_{ave} = \frac{\text{change in velocity}}{\text{time elapsed}} = \frac{v(t_1) - v(t_0)}{t_1 - t_0}$$

(a) Interpret v_{ave} geometrically on the graph of $s(t)$.

(b) Interpret a_{ave} geometrically on the graph of $v(t)$.

(c) Find the average velocity and acceleration over the time interval $2 \le t \le 4$ for a particle with position function $s(t) = t^3 - 3t^2$.

25. (a) Show that the average velocity v_{ave} over $[t_0, t_1]$ approaches the instantaneous velocity $v(t_0)$ as t_1 approaches t_0. [See Exercise 24 for definitions of v_{ave} and a_{ave}.]

(b) Show that the average acceleration a_{ave} over $[t_0, t_1]$ approaches the instantaneous acceleration $a(t_0)$ as t_1 approaches t_0.

26. Prove that a particle is speeding up if the velocity and acceleration have the same sign, and slowing down if they have opposite signs. [*Hint:* Let $r(t) = |v(t)| = \sqrt{v^2(t)}$, and find $r'(t)$.]

◆ TECHNOLOGY EXERCISES   Chapter 4

Most of these exercises require access to a graphing calculator or a computer algebra system (CAS) such as *Mathematica, Maple,* or *Derive*. When you are asked to *find* an answer or to *solve* an equation, you may choose to find an exact result or a numerical approximation, depending on the particular technology you are using and on your own imagination. The form of your answers may differ from those of other students or from those in the answer section of the text, depending on how you solve the problems and the accuracy you use in your numerical approximations. Those exercises that are more appropriate for a CAS than a graphing calculator are labeled with the icon ◆.

1. **Absolute maximum and minimum:** Let

$$f(x) = \frac{x^3 + 2}{x^4 + 1}$$

(a) Graph $y = f(x)$.

(b) Zoom in on the graph of f to approximate the coordinates of the absolute maximum and minimum points.

(c) Find $f'(x)$ and solve the equation $f'(x) = 0$ to check the results in part (b).

2. **Concavity:** Let $f(x) = x^4 + 5\cos x$.

(a) Graph f and find the intervals on which it is concave up.

(b) Graph f' and find the intervals on which it is increasing.

(c) Graph f'' and find the intervals on which it is positive.

(d) What is the relationship among the results in parts (a), (b), and (c)? Explain.

Relative maxima and minima: In Exercises 3–5, find the critical points and the relative extrema of f, and graph $y = f(x)$.

3. $f(x) = x^2 - \sin x$.

4. $f(x) = \sqrt{x^4 + 1} - \sqrt{x^2 + 1}$.

5. $f(x) = \frac{x}{x^2 + \sin x + 1}$.

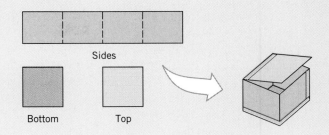

Sides

Bottom   Top

A matter of scale: Unless the viewing window is chosen carefully, the important features of a graph may not be evident. In Exercises 6 and 7, it is important to choose the horizontal scale carefully to reveal features that are close together.

6. (a) Graph $f(x) = \frac{1}{3}x^3 - \frac{1}{400}x$ over the interval $[-5, 5]$.

   (b) Find the critical points of f, and at each such point determine whether a relative maximum, a relative minimum, or neither occurs.

   (c) Confirm the results in part (b) by graphing f in an appropriate viewing window.

7. (a) Graph $f(x) = \frac{1}{5}x^5 - \frac{7}{8}x^4 + \frac{1}{3}x^3 + \frac{7}{2}x^2 - 6x$ over the interval $[-5, 5]$.

   (b) Use the first and second derivatives of f to find the values of x where the relative extrema and inflection points occur.

   (c) Using the results of part (b), zoom in on various portions of the graph to reveal the finer detail that is present. Note that it will not be possible to display all of the important features in a single window.

8. **Vertical and horizontal asymptotes:** Let

$$f(x) = \frac{2x^3 + x^2 - 15x + 7}{(2x - 1)(3x^2 + x - 1)}$$

   Graph $y = f(x)$ and find the equations of all horizontal and vertical asymptotes. Explain why there is no vertical asymptote at $x = \frac{1}{2}$, even though the denominator of f is zero at that point.

Oblique asymptotes: In Exercises 9 and 10, the graph of the function has an oblique asymptote as described in Exercise 48, Section 4.4. Find an equation of the asymptote, and graph the function and the asymptote.

9. $f(x) = \frac{x^3 - 8}{x^2 + 1}$.   10. $f(x) = \frac{2x^3 - 9x + 4}{x^2}$.

11. **Optimization:** A contractor wants to bid on an order to make 75,000 boxes from cardboard that costs 8 cents per square foot. Each box must have a square base and a volume of 4 cubic feet. The sides of the box will be made from a single piece of cardboard, folded three times and taped at the seam. The top and bottom will be made from separate pieces of cardboard with the bottom taped on all four edges and the top taped on just one edge to create a lid that opens (see the figure). Taping costs 3 cents per foot.

   (a) Find the dimensions of the box of minimum cost.

   (b) How much should the contractor bid (in dollars) to make a profit of 17%?

   (c) How much should the bid in part (b) be increased if the customer insists on cubical boxes?

12. **Distance from a point to a curve:** Find the coordinates of the point on the curve $y = \sin x$ that is closest to the point $(1, 0)$.

13. **Motion along a line:** A particle moves on a coordinate line so that its position at time t is given by

$$s(t) = \frac{t^2 + 1}{t^4 + 1}, \quad t \geq 0$$

   where t is in seconds and s is in meters.

   (a) Find the time interval during which the particle is moving in the positive direction.

   (b) Find the maximum speed of the particle and the time at which the maximum is attained.

14. **Motion in a plane:** As shown in the figure on the following page, suppose that a boat enters the river at the point $(1, 0)$ and maintains a heading toward the origin. As a result of the strong current, the boat follows the path

$$y = \frac{x^{10/3} - 1}{2x^{2/3}}$$

   where x and y are in miles.

   (a) Graph the path taken by the boat.

   (b) Can the boat reach the origin? If not, discuss its fate and find how close it comes to the origin.

   (c) What is the velocity of the boat in the x-direction at the instant when it is closest to the origin if the velocity in the y-direction is -4 miles per hour at this instant?

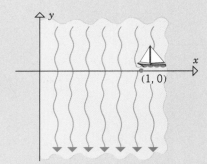

15. Population growth: Suppose that the number of individuals at time t in a certain wildlife population is given by

$$N(t) = \frac{340}{1 + 9(0.77)^t}, \quad t \geq 0$$

where t is in years. At what instant of time is the size of the population increasing most rapidly?

16. Distance from Earth to the sun: According to *Kepler's laws*, the planets in our solar system move in elliptical orbits around the sun. If a planet's closest approach to the sun occurs at time $t = 0$, then the distance r from the center of the planet to the center of the sun at some later time t can be determined from the equation

$$r = a(1 - e \cos \phi)$$

where a is the average distance between centers, e is a positive constant that measures the "flatness" of the elliptical orbit, and ϕ is the solution of *Kepler's equation*

$$\frac{2\pi t}{T} = \phi - e \sin \phi$$

in which T is the time its takes for one complete orbit of the planet. Find the distance from Earth to the sun when $t = 90$ days. [First find ϕ from Kepler's equation, and then use this value of ϕ to find the distance. Use $a = 150 \times 10^6$ km, $e = 0.0167$, and $T = 365$ days.]

17. Distance from Mars to the sun: Using the formulas in Exercise 16, find the distance from Mars to the sun when $t = 1$ year. For Mars use $a = 228 \times 10^6$ km, $e = 0.0934$, and $T = 1.88$ years.

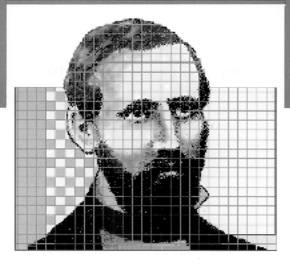

Georg Friedrich Bernhard Riemann
(1826–1866)

5 INTEGRATION

■ **5.1** INTRODUCTION

In this chapter we shall study the second major problem of calculus:

> THE AREA PROBLEM. *Given a function f that is continuous and nonnegative on an interval [a, b], find the area between the graph of f and the interval [a, b] on the x-axis* (Figure 5.1.1).

Area formulas for basic geometric figures such as rectangles, polygons, and circles date back to the earliest written records of mathematics. The first real advance beyond the elementary level of area computation was made by the Greek mathematician, Archimedes,* (see p. 248) who devised an ingenious but cumbersome method for obtaining areas, called the *method of exhaustion.* Using this technique Archimedes was able to obtain areas bounded by parabolas and spirals. By the early seventeenth century several mathematicians had learned to obtain such areas more simply by calculating appropriate limits. However, both the method of exhaustion and its successor lacked generality. For each different problem one had to devise special procedures or formulas that worked, and more often than

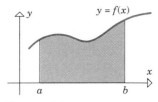

Figure 5.1.1

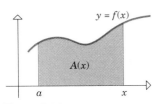

Figure 5.1.2

not these were difficult or impossible to obtain. The major breakthrough in solving the general area problem was made independently by Newton and Leibniz when they discovered that areas could be obtained by reversing the process of differentiation. This major discovery, which marked the real beginning of calculus, was circulated by Newton in 1669 and then published in 1711 in a paper entitled, *De Analysi per Aequationes Numero Terminorum Infinitas* (*On Analysis by Means of Equations with Infinitely Many Terms*). Independently, Leibniz discovered the same result around 1673 and stated it in an unpublished manuscript dated November 11, 1675.

In the remainder of this introductory section, we will discover the fascinating solution of the area problem for ourselves. As with most mathematical discoveries, our initial work will be intuitive; in later sections of this chapter these ideas will be made more precise.

Assume that we are interested in finding the area of the region in Figure 5.1.1. As it turns out, it will actually be easier for us to solve a whole class of area problems simultaneously. Instead of limiting the discussion to the case where the right-hand endpoint of the interval is b, we will allow this endpoint to be any real number x greater than or equal to a (Figure 5.1.2). As indicated in the figure, we will denote the desired area by $A(x)$. We have used the

---

*ARCHIMEDES (287 B.C.–212 B.C.). Greek mathematician and scientist. Born in Syracuse, Sicily, Archimedes was the son of the astronomer Pheidias and possibly related to Heiron II, king of Syracuse. Most of the facts about his life come from the Roman biographer, Plutarch, who inserted a few tantalizing pages about him in the massive biography of the Roman soldier, Marcellus. In the words of one writer, ''the account of Archimedes is slipped like a tissue-thin shaving of ham in a bull-choking sandwich.''

Archimedes ranks with Newton and Gauss as one of the three greatest mathematicians who ever lived, and he is certainly the greatest mathematician of antiquity. His mathematical work is so modern in spirit and technique that it is barely distinguishable from that of a seventeenth-century mathematician, yet it was all done without benefit of algebra or a convenient number system. Among his mathematical achievements, Archimedes developed a general method (exhaustion) for finding areas and volumes, and he used the method to find areas bounded by parabolas and spirals and to find volumes of cylinders, paraboloids, and segments of spheres. He gave a procedure for approximating π and bounded its value between $3\frac{10}{71}$ and $3\frac{1}{7}$. In spite of the limitations of the Greek numbering system, he devised methods for finding square roots and invented a method based on the Greek myriad (10,000) for representing numbers as large as 1 followed by 80 million billion zeros.

Of all his mathematical work, Archimedes was most proud of his discovery of the method for finding the volume of a sphere—he showed that the volume of a sphere is two-thirds the volume of the smallest cylinder that can contain it. At his request, the figure of a sphere and cylinder was engraved on his tombstone.

In addition to mathematics, Archimedes worked extensively in mechanics and hydrostatics. Nearly every schoolchild knows Archimedes as the absent-minded scientist who, on realizing that a floating object displaces its weight of liquid, leaped from his bath and ran naked through the streets of Syracuse shouting, ''Eureka, Eureka!''—(meaning, ''I have found it!''). Archimedes actually created the discipline of hydrostatics and used it to find equilibrium positions for various floating bodies. He laid down the fundamental postulates of mechanics, discovered the laws of levers, and calculated centers of gravity for various flat surfaces and solids. In the excitement of discovering the mathematical laws of the lever, he is said to have declared, ''Give me a place to stand and I will move the earth.''

Although Archimedes was apparently more interested in pure mathematics than its applications, he was an engineering genius. During the second Punic war, when Syracuse was attacked by the Roman fleet under the command of Marcellus, it was reported by Plutarch that Archimedes' military inventions held the fleet at bay for three years. He invented super catapults that showered the Romans with rocks weighing a quarter ton or more, and fearsome mechanical devices with iron ''beaks and claws'' that reached over the city walls, grasped the ships, and spun them against the rocks. After the first repulse, Marcellus called Archimedes a ''geometrical Briareus (a hundred-armed mythological monster) who uses our ships like cups to ladle water from the sea.''

Eventually the Roman army was victorious and contrary to Marcellus' specific orders the 75-year-old Archimedes was killed by a Roman soldier. According to one report of the incident, the soldier cast a shadow across the sand in which Archimedes was working on a mathematical problem. When the annoyed Archimedes yelled, ''Don't disturb my circles,'' the soldier flew into a rage and cut the old man down.

With his death the Greek gift of mathematics passed into oblivion, not to be fully resurrected again until the sixteenth century. Unfortunately, there is no known accurate likeness or statue of this great man.

function notation here to emphasize that the area of the region depends on the value of the right-hand endpoint x.

Isaac Newton once said,

If I have seen farther than others, it is because I have stood on the shoulders of giants.

We are fortunate here in being able to stand on the shoulders of two giants, Newton and Leibniz, for it was their idea to find $A(x)$ by first finding the *derivative* of $A(x)$, then using the known derivative $A'(x)$ to determine $A(x)$ itself. Thus, we consider the problem of finding

$$A'(x) = \lim_{h \to 0} \frac{A(x + h) - A(x)}{h} \tag{1}$$

For simplicity, consider the case where $h > 0$. The numerator on the right side of (1) is the difference of two areas: the area between a and $x + h$ minus the area between a and x (Figure 5.1.3a). If we let c be the midpoint between x and $x + h$, then this difference of areas can be approximated by the area of a rectangle with base h and height $f(c)$ (Figure 5.1.3b). Thus,

$$\frac{A(x + h) - A(x)}{h} \approx \frac{f(c) \cdot h}{h} = f(c) \tag{2}$$

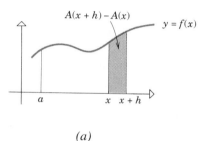

Figure 5.1.3   (a)   (b)

It seems plausible from Figure 5.1.3b that the error in approximation (2) will approach zero as $h \to 0$. If we accept this to be so, then it follows from (1) and (2) that

$$A'(x) = \lim_{h \to 0} \frac{A(x + h) - A(x)}{h} = \lim_{h \to 0} f(c) \tag{3}$$

Since c is the midpoint between x and $x + h$, it follows that $c \to x$ as $h \to 0$. But we have assumed f to be a continuous function, so $f(c) \to f(x)$ as $c \to x$. Therefore,

$$\lim_{h \to 0} f(c) = f(x)$$

Thus, it follows from (3) that

$$A'(x) = f(x) \tag{4}$$

This is the result we were looking for; it tells us that the derivative of the area function $A(x)$ is the function whose graph forms the upper boundary of the region in Figure 5.1.2. The following example illustrates how this formula can be used to find the area under a parabola.

Example 1   We will use Formula (4) to find the area of the region under the graph of the parabola $y = x^2$ over the interval $[0, 1]$. (See Figure 5.1.4.)

As in the derivation of (4), we will start by considering the more general problem of finding the area $A(x)$ under $y = x^2$ over the interval $[0, x]$. (See Figure 5.1.5.) Since the function whose graph forms the upper boundary of the region is $f(x) = x^2$, it follows from (4) that

$$A'(x) = x^2 \tag{5}$$

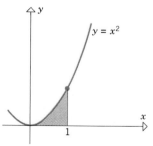

Figure 5.1.4

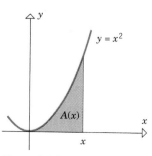

Figure 5.1.5

Thus, to find $A(x)$ we must look for a function whose derivative is x^2. This is called an *antidifferentiation* problem because we are trying to find $A(x)$ by "undoing" a differentiation. By simply guessing we see that

$$A(x) = \tfrac{1}{3}x^3$$

is one solution to (5). But this is not the only solution, since it follows from Theorem 4.9.3 that

$$A(x) = \tfrac{1}{3}x^3 + C \tag{6}$$

also satisfies (5) for any real value of C. We still have some work to do since this formula involves an unknown constant C that must be determined. This is where our decision to solve the area problem for a general right-hand endpoint helps. If we consider the case where $x = 0$, then the interval $[0, x]$ reduces to a single point. If we agree that the area above a single point should be taken as zero, then it follows on substituting $x = 0$ in (6) that

$$A(0) = 0 + C = 0 \quad \text{or} \quad C = 0$$

so (6) simplifies to

$$A(x) = \tfrac{1}{3}x^3 \tag{7}$$

which is the formula for the area under $y = x^2$ over the interval $[0, x]$. For the area over the interval $[0, 1]$ we set $x = 1$ in (7); this yields

$$A(1) = \tfrac{1}{3} \text{ (square units)} \qquad \blacktriangleleft$$

We note that our success in finding the area in the preceding example hinged on our ability to *guess* at a function $A(x)$ satisfying $A'(x) = x^2$. Had we not been able to find such a function, we could not have found the area. One of the goals in subsequent sections is to develop a systematic approach to finding functions from their derivatives, where possible.

We conclude this introductory section by noting that there are numerous mathematical gaps in our work here. For example, we used the term "area" freely, assuming its meaning to be understood and its properties known. Although we calculated the area under a parabola, we never stated precisely what is meant by the area under a parabola or, more generally, by the area under a curve $y = f(x)$. Until the latter part of the nineteenth century, mathematicians found their intuitive concept of area to be perfectly satisfactory. However, in the late 1800s they began to encounter problems that could only be resolved by defining the concept of area precisely and developing its properties in a mathematically rigorous manner. Research in this field has continued to the present day. One of the goals in this chapter is to give more precision to the concept of area.

■ 5.2   ANTIDERIVATIVES; THE INDEFINITE INTEGRAL

In this section we shall develop some basic results that will ultimately help us to obtain systematic procedures for finding a function from its derivative.

☐ ANTIDERIVATIVES

5.2.1   DEFINITION.   A function F is called an *antiderivative* of a function f on a given interval if $F'(x) = f(x)$ for all x in that interval.

Example 1   The functions

$$\tfrac{1}{3}x^3, \quad \tfrac{1}{3}x^3 + 2, \quad \tfrac{1}{3}x^3 - \pi, \quad \tfrac{1}{3}x^3 + C \quad (C \text{ any constant})$$

are antiderivatives of $f(x) = x^2$ on the interval $(-\infty, +\infty)$ since the derivative of each is x^2.   ◄

This example shows that a function can have many antiderivatives. In fact, if $F(x)$ is any antiderivative of $f(x)$ and C is any constant, then

$$F(x) + C$$

is also an antiderivative of $f(x)$ since

$$\frac{d}{dx}[F(x) + C] = \frac{d}{dx}[F(x)] + \frac{d}{dx}[C] = f(x) + 0 = f(x)$$

It is reasonable to ask if there are antiderivatives of f that cannot be obtained by adding a constant to F. The answer is *no*. To see this, let $G(x)$ be any other antiderivative of $f(x)$; then

$$\frac{d}{dx}[F(x)] = \frac{d}{dx}[G(x)] = f(x)$$

so that by Theorem 4.9.3, F and G differ only by a constant C on I, that is,

$$G(x) = F(x) + C$$

for x in I. The following theorem summarizes these observations.

5.2.2   THEOREM.   *If $F(x)$ is any antiderivative of $f(x)$ on a given interval, then for any value of C the function $F(x) + C$ is also an antiderivative of $f(x)$ on that interval; moreover every antiderivative of $f(x)$ on the interval is expressible in the form $F(x) + C$, where C is a constant.*

REMARK.   Do not confuse derivatives and antiderivatives. For example, the *derivative* of $f(x) = x^2$ is $f'(x) = 2x$, but the *antiderivatives* of $f(x) = x^2$ are the functions of the form $F(x) = \tfrac{1}{3}x^3 + C$.

☐ **THE INDEFINITE INTEGRAL**

The process of finding antiderivatives is called ***antidifferentiation*** or ***integration***. If there is some function F such that

$$\frac{d}{dx}[F(x)] = f(x)$$

then the functions of the form $F(x) + C$ are the antiderivatives of $f(x)$. We denote this by writing

$$\int f(x)\, dx = F(x) + C \tag{1}$$

The symbol $\int$ is called an ***integral sign***, and $f(x)$ is called the ***integrand***. Statement (1) is read, "the ***indefinite integral*** of $f(x)$ equals $F(x)$ plus C." The adjective "indefinite" is used because the right side of (1) is not a definite function, but rather a whole set of possible functions; the constant C is called the ***constant of integration***.

We observed in Example 1 and in the preceding remark that the antiderivatives of $f(x) = x^2$ are the functions of the form $F(x) = \tfrac{1}{3}x^3 + C$. Thus,

$$\int x^2\, dx = \tfrac{1}{3}x^3 + C$$

☐ **SOME MATTERS OF NOTATION**

The dx symbols in the differentiation and antidifferentiation operations

$$\frac{d}{dx}[\quad] \quad \text{and} \quad \int [\quad]\, dx$$

serve to identify the independent variable. If an independent variable other than x is used, say t, the notation must be adjusted appropriately. Thus,

$$\frac{d}{dt}[F(t)] = f(t) \quad \text{and} \quad \int f(t)\, dt = F(t) + C$$

are equivalent statements* (see p. 253).

Example 2

DERIVATIVE FORMULA	EQUIVALENT INTEGRATION FORMULA
$\dfrac{d}{dx}[x^3] = 3x^2$	$\displaystyle\int 3x^2\, dx = x^3 + C$
$\dfrac{d}{dx}[\sin x] = \cos x$	$\displaystyle\int \cos x\, dx = \sin x + C$
$\dfrac{d}{dt}[\tan t] = \sec^2 t$	$\displaystyle\int \sec^2 t\, dt = \tan t + C$
$\dfrac{d}{du}[u^{3/2}] = \frac{3}{2}u^{1/2}$	$\displaystyle\int \frac{3}{2}u^{1/2}\, du = u^{3/2} + C$

◀

For simplicity, the dx is sometimes absorbed into the integrand. For example,

$$\int 1\, dx \quad \text{can be written as} \quad \int dx$$

$$\int \frac{1}{x^2}\, dx \quad \text{can be written as} \quad \int \frac{dx}{x^2}$$

☐ **INTEGRATION FORMULAS**

Antidifferentiation is often educated guesswork. By looking only at the derivative of a function we try to guess the function itself. It simplifies this guessing process if we keep in mind that every differentiation formula produces a companion integration formula. Some basic integration formulas are given in Table 5.2.1.

Table 5.2.1

DIFFERENTIATION FORMULA	INTEGRATION FORMULA
1. $\dfrac{d}{dx}[x] = 1$	$\displaystyle\int dx = x + C$
2. $\dfrac{d}{dx}\left[\dfrac{x^{r+1}}{r+1}\right] = x^r \quad (r \neq -1)$	$\displaystyle\int x^r\, dx = \dfrac{x^{r+1}}{r+1} + C \quad (r \neq -1)$
3. $\dfrac{d}{dx}[\sin x] = \cos x$	$\displaystyle\int \cos x\, dx = \sin x + C$
4. $\dfrac{d}{dx}[-\cos x] = \sin x$	$\displaystyle\int \sin x\, dx = -\cos x + C$
5. $\dfrac{d}{dx}[\tan x] = \sec^2 x$	$\displaystyle\int \sec^2 x\, dx = \tan x + C$
6. $\dfrac{d}{dx}[-\cot x] = \csc^2 x$	$\displaystyle\int \csc^2 x\, dx = -\cot x + C$
7. $\dfrac{d}{dx}[\sec x] = \sec x \tan x$	$\displaystyle\int \sec x \tan x\, dx = \sec x + C$
8. $\dfrac{d}{dx}[-\csc x] = \csc x \cot x$	$\displaystyle\int \csc x \cot x\, dx = -\csc x + C$

Example 3 From the second integration formula in Table 5.2.1 we obtain

$$\int x^2 \, dx = \frac{x^3}{3} + C \quad \boxed{r = 2}$$

$$\int x^3 \, dx = \frac{x^4}{4} + C \quad \boxed{r = 3}$$

$$\int \frac{1}{x^5} \, dx = \int x^{-5} \, dx = \frac{x^{-5+1}}{-5+1} + C = -\frac{1}{4x^4} + C \quad \boxed{r = -5}$$

$$\int \sqrt{x} \, dx = \int x^{\frac{1}{2}} \, dx = \frac{x^{\frac{1}{2}+1}}{\frac{1}{2}+1} + C = \frac{2}{3}x^{\frac{3}{2}} + C = \frac{2}{3}(\sqrt{x})^3 + C \quad \boxed{r = \frac{1}{2}} \quad \blacktriangleleft$$

☐ **PROPERTIES OF THE INDEFINITE INTEGRAL**

If we differentiate an antiderivative of $f(x)$, we obtain $f(x)$ back again. Thus,

$$\frac{d}{dx}\left[\int f(x) \, dx\right] = f(x) \tag{2}$$

This result is helpful for proving the following basic properties of antiderivatives.

5.2.3 THEOREM.

(a) *A constant factor can be moved through an integral sign; that is,*

$$\int cf(x) \, dx = c \int f(x) \, dx$$

(b) *An antiderivative of a sum is the sum of the antiderivatives; that is,*

$$\int [f(x) + g(x)] \, dx = \int f(x) \, dx + \int g(x) \, dx$$

(c) *An antiderivative of a difference is the difference of the antiderivatives; that is,*

$$\int [f(x) - g(x)] \, dx = \int f(x) \, dx - \int g(x) \, dx$$

Proof. In each part we must show that the expression on the right side of the equation is an antiderivative of the integrand on the left side of the equation. This can be done using (2) as follows:

$$\frac{d}{dx}\left[c\int f(x) \, dx\right] = c\frac{d}{dx}\left[\int f(x) \, dx\right] = cf(x)$$

$$\frac{d}{dx}\left[\int f(x) \, dx + \int g(x) \, dx\right] = \frac{d}{dx}\left[\int f(x) \, dx\right] + \frac{d}{dx}\left[\int g(x) \, dx\right] = f(x) + g(x)$$

$$\frac{d}{dx}\left[\int f(x) \, dx - \int g(x) \, dx\right] = \frac{d}{dx}\left[\int f(x) \, dx\right] - \frac{d}{dx}\left[\int g(x) \, dx\right]$$

$$= f(x) - g(x) \qquad \blacksquare$$

---

*This notation was devised by Leibniz. In his early papers Leibniz used the notation "omn." (an abbreviation for the Latin word "omnes") to denote integration. Then on October 29, 1675 he wrote, "It will be useful to write ∫ for omn., thus ∫ ℓ for omn. ℓ . . . ." Two or three weeks later he refined the notation further and wrote ∫[ ] dx rather than ∫ alone. This notation is so useful and so powerful that its development by Leibniz must be regarded as a major milestone in the history of mathematics and science.

When applying Theorem 5.2.3, it is best to put in the constant of integration at the *very end* of the computations to obtain the simplest form of the answer. This is illustrated in the following example.

Example 4   Evaluate

(a) $\displaystyle\int 4 \cos x \, dx$ (b) $\displaystyle\int (x + x^2) \, dx$

Solution (a).

$$\underset{\underset{\text{Theorem 5.2.3}(a)}{\uparrow}}{\int 4 \cos x \, dx = 4 \int} \cos x \, dx \underset{\underset{\text{Table 5.2.1}}{\uparrow}}{= 4\, (\sin x + C)} = 4 \sin x + 4C$$

Since C is an arbitrary constant, so is $4C$. However, this form of the arbitrary constant is unnecessarily complicated. A simpler form of the answer results by deferring the insertion of the constant until the end. This procedure yields

$$\int 4 \cos x \, dx = 4 \int \cos x \, dx = 4 \sin x + C$$

Solution (b).

$$\underset{\underset{\text{Theorem 5.2.3}(b)}{\uparrow}}{\int (x + x^2) \, dx = \int x \, dx +} \int x^2 \, dx \underset{\underset{\text{Table 5.2.1}}{\uparrow}}{= \frac{x^2}{2} + \frac{x^3}{3}} + C \quad \blacktriangleleft$$

Parts (*b*) and (*c*) of Theorem 5.2.3 can be extended to more than two functions, which in combination with part (*a*) results in the following general formula:

$$\int [c_1 f_1(x) + c_2 f_2(x) + \cdots + c_n f_n(x)] \, dx$$
$$= c_1 \int f_1(x) \, dx + c_2 \int f_2(x) \, dx + \cdots + c_n \int f_n(x) \, dx$$

Example 5

$$\int (3x^6 - 2x^2 + 7x + 1) \, dx = 3 \int x^6 \, dx - 2 \int x^2 \, dx + 7 \int x \, dx + \int 1 \, dx$$
$$= \frac{3x^7}{7} - \frac{2x^3}{3} + \frac{7x^2}{2} + x + C \quad \blacktriangleleft$$

Sometimes it is useful to rewrite an integrand in a different form before performing the integration.

Example 6   Evaluate

(a) $\displaystyle\int \frac{\cos x}{\sin^2 x} \, dx$ (b) $\displaystyle\int \frac{t^2 - 2t^4}{t^4} \, dt$

Solution (a).

$$\int \frac{\cos x}{\sin^2 x} \, dx = \int \frac{1}{\sin x} \frac{\cos x}{\sin x} \, dx = \int \csc x \cot x \, dx \underset{\underset{\text{Formula 8 in Table 5.2.1}}{\uparrow}}{=} -\csc x + C$$

Solution (b).

$$\int \frac{t^2 - 2t^4}{t^4} \, dt = \int \left(\frac{1}{t^2} - 2 \right) dt = \int (t^{-2} - 2) \, dt$$

$$= \frac{t^{-1}}{-1} - 2t + C = -\frac{1}{t} - 2t + C \quad \blacktriangleleft$$

☐ **INTEGRAL CURVES**

From Theorem 5.2.2 the antiderivatives of a function f differ by a constant on any interval. Thus, on any interval the graphs of the antiderivatives of f form a family of curves that are vertical translations of one another. These are called *integral curves* of f. Some integral curves of $f(x) = x^2$ are shown in Figure 5.2.1.

In many problems one is interested in finding a function whose derivative satisfies specified conditions. The following example illustrates a geometric problem of this type.

Example 7 Suppose that a point moves along some unknown curve $y = f(x)$ in the xy-plane in such a way that at each point (x, y) on the curve, the tangent line has slope x^2. Find an equation for the curve given that it passes through the point $(2, 1)$.

Solution. We know that $dy/dx = x^2$, so

$$y = \int x^2 \, dx = \tfrac{1}{3} x^3 + C \tag{3}$$

Since the curve passes through $(2, 1)$, a specific value for C can be found by using the fact that $y = 1$ if $x = 2$. Substituting these values in (3) yields

$$1 = \tfrac{1}{3} (2^3) + C \quad \text{or} \quad C = -\tfrac{5}{3}$$

so the curve is $y = \tfrac{1}{3} x^3 - \tfrac{5}{3}$.   ◀

Observe that in the preceding example the requirement that the unknown curve pass through the point $(2, 1)$ enabled us to determine a specific value for the constant of integration, thereby isolating the single integral curve $y = \tfrac{1}{3} x^3 - \tfrac{5}{3}$ from the family $y = \tfrac{1}{3} x^3 + C$ (Figure 5.2.2).

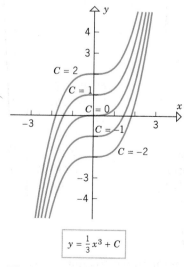

$$y = \tfrac{1}{3} x^3 + C$$

Figure 5.2.1

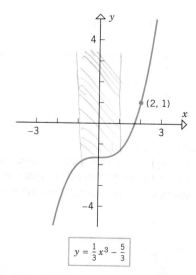

$$y = \tfrac{1}{3} x^3 - \tfrac{5}{3}$$

Figure 5.2.2

► Exercise Set 5.2

In Exercises 1–30, evaluate the integrals and check your results by differentiating the answers.

1. $\int x^8 \, dx.$

2. $\int \frac{1}{x^6} \, dx.$

3. $\int x^{5/7} \, dx.$

4. $\int \sqrt[3]{x^2} \, dx.$

5. $\int \frac{4}{\sqrt{t}} \, dt.$

6. $\int \frac{1}{2x^3} \, dx.$

7. $\int x^3 \sqrt{x} \, dx.$

8. $\int (u^3 - 2u + 7) \, du.$

9. $\int (x^{-3} + \sqrt{x} - 3x^{1/4} + x^2) \, dx.$

10. $\int (x^{2/3} - 4x^{-1/5} + 4) \, dx.$

11. $\int \left(\frac{7}{y^{3/4}} - \sqrt[3]{y} + 4\sqrt{y} \right) dy.$

12. $\int (2 + y^2)^2 \, dy.$

13. $\int x(1 + x^3) \, dx.$

14. $\int (1 + x^2)(2 - x) \, dx.$

15. $\int x^{1/3}(2 - x)^2 \, dx.$

16. $\int \frac{1 - 2t^3}{t^3} \, dt.$

17. $\int \frac{x^5 + 2x^2 - 1}{x^4} \, dx.$

18. $\int \left[\frac{1}{t^2} - \cos t \right] dt.$

19. $\int [4 \sin x + 2 \cos x] \, dx.$

20. $\int [4 \sec^2 x + \csc x \cot x] \, dx.$

21. $\int \sec x(\sec x + \tan x) \, dx.$

22. $\int [\sqrt{\theta} - \csc^2 \theta] \, d\theta.$

23. $\int \sec x(\tan x + \cos x) \, dx.$

24. $\int \frac{dy}{\csc y}.$

25. $\int \frac{\sin x}{\cos^2 x} \, dx.$

26. $\int \frac{\sin 2x}{\cos x} \, dx.$

27. $\int [1 + \sin^2 \theta \csc \theta] \, d\theta.$

28. $\int \left[\phi + \frac{2}{\sin^2 \phi} \right] d\phi.$

29. $\int \frac{\cos^3 \theta - 5}{\cos^2 \theta} \, d\theta.$

30. $\int \frac{1}{1 + \sin x} \, dx.$ [*Hint:* Multiply numerator and denominator by $1 - \sin x$.]

31. Find the antiderivative $F(x)$ of $f(x) = \sqrt[3]{x}$ that satisfies $F(1) = 2$.

32. Find a function f such that $f'(x) + \sin x = 0$ and $f(0) = 2$.

33. Find the general form of a function whose second derivative is $\sqrt{x}$. [*Hint:* Solve the equation $f''(x) = \sqrt{x}$ for $f(x)$ by integrating both sides twice.]

34. Find a function f such that $f''(x) = x + \cos x$ and such that

$f(0) = 1$ and $f'(0) = 2$. [*Hint:* Integrate both sides of the equation twice.]

In Exercises 35–37, find an equation of the curve that satisfies the given conditions.

35. At each point (x, y) on the curve the slope is $2x + 1$; the curve passes through the point $(-3, 0)$.

36. At each point (x, y) on the curve the slope equals the square of the distance between the point and the y-axis; the point $(-1, 2)$ is on the curve.

37. At each point (x, y) on the curve $d^2y/dx^2 = 6x$; the line $y = 5 - 3x$ is tangent to the curve at the point where $x = 1$.

38. Suppose that a uniform metal rod 50 cm long is insulated laterally, and the temperatures at the exposed ends are maintained at $25° \text{C}$ and $85° \text{C}$, respectively. Assume that an x-axis is chosen as in Figure 5.2.3 and that the temperature $T(x)$ at each point x satisfies the equation

$$\frac{d^2T}{dx^2} = 0$$

Find $T(x)$ for $0 \leq x \leq 50$.

Figure 5.2.3

In Exercises 39–42, find the derivative and state a corresponding integration formula.

39. $\frac{d}{dx}[\sqrt{x^3 + 5}].$

40. $\frac{d}{dx}\left[\frac{x}{x^2 + 3} \right].$

41. $\frac{d}{dx}[\sin(2\sqrt{x})].$

42. $\frac{d}{dx}[\sin x - x \cos x].$

43. Show that
$$F(x) = \tfrac{1}{6}(3x + 4)^2 \quad \text{and} \quad G(x) = \tfrac{3}{2}x^2 + 4x$$
are both antiderivatives of the same function by
(a) differentiation
(b) showing that $F(x)$ and $G(x)$ differ by a constant (Theorem 5.2.2).

44. Work Exercise 43 using
$$F(x) = \frac{x^2}{x^2 + 5} \quad \text{and} \quad G(x) = -\frac{5}{x^2 + 5}$$

45. Find $f(x)$ if
$$\int f(x) \, dx = 5x^3 - 3x + C$$

46. Find $g(t)$ if
$$\int g(t) \, dt = \frac{1}{\sqrt{4 - t^2}} + C$$

In Exercises 47–49, use a trigonometric identity to help evaluate the integral.

47. $\int \tan^2 x \, dx.$ **48.** $\int \cot^2 x \, dx.$

49. $\int \sin^2 (x/2) \, dx.$ [*Hint:* $\sin^2 \theta = \frac{1}{2}(1 - \cos 2\theta).$]

50. (a)   Let

$$F(x) = \begin{cases} x, & x > 0 \\ -x, & x < 0 \end{cases}$$

and

$$F_1(x) = \begin{cases} x + 2, & x > 0 \\ -x + 3, & x < 0 \end{cases}$$

Show that $F'(x) = F_1'(x) = f(x)$, where

$$f(x) = \begin{cases} 1, & x > 0 \\ -1, & x < 0 \end{cases}$$

but that $F_1(x) - F(x)$ is not a constant.

(b)   Does this violate Theorem 5.2.2? Explain.

■ **5.3** INTEGRATION BY SUBSTITUTION

In this section we shall discuss a technique, called **substitution**, *that can often be used to transform complicated integration problems into simpler ones.*

☐ *u*-SUBSTITUTION

The method of substitution hinges on the following formula in which u stands for a differentiable function of x.

$$\int \left[f(u) \frac{du}{dx} \right] dx = \int f(u) \, du \qquad (1)$$

To justify this formula, let F be an antiderivative of f, so that

$$\frac{d}{du} [F(u)] = f(u)$$

or, equivalently,

$$\int f(u) \, du = F(u) + C \qquad (2)$$

If u is a differentiable function of x, the chain rule implies that

$$\frac{d}{dx} [F(u)] = \frac{d}{du} [F(u)] \cdot \frac{du}{dx} = f(u) \frac{du}{dx}$$

or, equivalently,

$$\int \left[f(u) \frac{du}{dx} \right] dx = F(u) + C \qquad (3)$$

Formula (1) follows from (2) and (3).   ■

The following example illustrates how Formula (1) is used.

Example 1   Evaluate $\int (x^2 + 1)^{50} \cdot 2x \, dx.$

Solution. If we let $u = x^2 + 1$, then $du/dx = 2x$, so the given integral can be written as

$$\int (x^2 + 1)^{50} \cdot 2x\, dx = \int \left[u^{50} \frac{du}{dx} \right] dx = \int u^{50}\, du$$

$$\int \left[f(u) \frac{du}{dx} \right] dx \qquad \int f(u)\, du$$

$$= \frac{u^{51}}{51} + C = \frac{(x^2 + 1)^{51}}{51} + C \qquad \blacktriangleleft$$

The method illustrated in this example can be summarized as follows:

Integration by Substitution

Step 1. Make a choice for u, say $u = g(x)$.

Step 2. Compute $du/dx = g'(x)$.

Step 3. Make the substitution $u = g(x)$, $du = g'(x)\, dx$.

At this stage, the *entire* integral must be in terms of u; no x's should remain. If this is not the case, try a different choice of u.

Step 4. Evaluate the resulting integral.

Step 5. Replace u by $g(x)$, so the final answer is in terms of x.

REMARK.   There are no hard and fast rules for choosing u—making an appropriate choice will come with experience. However, we note that there are many integration problems in which no choice of u will work. In such cases other methods need to be used, some of which will be discussed later.

Example 2   The easiest substitutions occur when the integrand is the derivative of a known function, except for a constant added to or subtracted from the independent variable. For example,

$$\int \sin (x + 9)\, dx = \int \sin u\, du = -\cos u + C = -\cos (x + 9) + C$$

$$u = x + 9$$
$$du = 1 \cdot dx = dx$$

$$\int (x - 8)^{23}\, dx = \int u^{23}\, du = \frac{u^{24}}{24} + C = \frac{(x - 8)^{24}}{24} + C \qquad \blacktriangleleft$$

$$u = x - 8$$
$$du = 1 \cdot dx = dx$$

Another easy u-substitution occurs when the integrand is the derivative of a known function, except for a constant that multiplies or divides the independent variable. The following example illustrates two ways to evaluate such integrals.

Example 3   Evaluate $\int \cos 5x\, dx$.

Solution.

$$\int \cos 5x \, dx = \int (\cos u) \cdot \frac{1}{5} \, du = \frac{1}{5} \int \cos u \, du = \frac{1}{5} \sin u + C = \frac{1}{5} \sin 5x + C$$

$u = 5x$
$du = 5 \, dx \text{ or } dx = \frac{1}{5} \, du$

Alternative Solution.   There is a variation of the preceding method that some people prefer. The substitution $u = 5x$ requires $du = 5 \, dx$. If there were a factor of 5 in the integrand, then we could group the 5 and dx together to form the du required by the substitution. Since there is no factor of 5, we shall insert one and compensate by putting a factor of $\frac{1}{5}$ in front of the integral. The computations are as follows:

$$\int \cos 5x \, dx = \frac{1}{5} \int \cos 5x \cdot 5 \, dx = \frac{1}{5} \int \cos u \, du = \frac{1}{5} \sin u + C = \frac{1}{5} \sin 5x + C \qquad \blacktriangleleft$$

$u = 5x$
$du = 5 \, dx$

Example 4   Evaluate $\int \sin^2 x \cos x \, dx$.

Solution.   If we let $u = \sin x$, then

$$du/dx = \cos x, \quad \text{so} \quad du = \cos x \, dx$$

Thus,

$$\int \sin^2 x \cos x \, dx = \int u^2 \, du = \frac{u^3}{3} + C = \frac{\sin^3 x}{3} + C \qquad \blacktriangleleft$$

Example 5   Evaluate $\int \frac{\cos \sqrt{x}}{\sqrt{x}} \, dx$.

Solution.   If we let $u = \sqrt{x}$, then

$$\frac{du}{dx} = \frac{1}{2\sqrt{x}}, \quad \text{so} \quad du = \frac{1}{2\sqrt{x}} \, dx$$

Thus,

$$\int \frac{\cos \sqrt{x}}{\sqrt{x}} \, dx = \int 2 \cos u \, du = 2 \int \cos u \, du = 2 \sin u + C = 2 \sin \sqrt{x} + C \qquad \blacktriangleleft$$

Example 6

$$\int \frac{dx}{(\frac{1}{3}x - 8)^5} = \int \frac{3 \, du}{u^5} = 3 \int u^{-5} \, du = -\frac{3}{4} u^{-4} + C = -\frac{3}{4} \left(\frac{1}{3}x - 8 \right)^{-4} + C \qquad \blacktriangleleft$$

$u = \frac{1}{3}x - 8$
$du = \frac{1}{3} \, dx \text{ or } dx = 3 \, du$

Example 7   With the help of Theorem 5.2.3, a complicated integral can sometimes be computed by expressing it as a sum of simpler integrals. For example,

$$\int (x + \sec^2 \pi x)\, dx = \int x\, dx + \int \sec^2 \pi x\, dx = \frac{x^2}{2} + \int \sec^2 \pi x\, dx$$

$$= \frac{x^2}{2} + \frac{1}{\pi} \int \sec^2 u\, du$$

$$\boxed{\begin{array}{c} u = \pi x \\ du = \pi\, dx \text{ or } dx = \frac{1}{\pi}\, du \end{array}}$$

$$= \frac{x^2}{2} + \frac{1}{\pi} \tan u + C = \frac{x^2}{2} + \frac{1}{\pi} \tan \pi x + C \qquad \blacktriangleleft$$

Example 8   Evaluate $\int t^4 \sqrt[3]{3 - 5t^5}\, dt$.

Solution.   After some possible false starts most readers would eventually hit on the following substitution:

$$\int t^4 \sqrt[3]{3 - 5t^5}\, dt = -\frac{1}{25} \int \sqrt[3]{u}\, du = -\frac{1}{25} \int u^{1/3}\, du$$

$$\boxed{\begin{array}{c} u = 3 - 5t^5 \\ du = -25t^4\, dt \text{ or } -\frac{1}{25}\, du = t^4\, dt \end{array}}$$

$$= -\frac{1}{25} \frac{u^{4/3}}{4/3} + C = -\frac{3}{100} (3 - 5t^5)^{4/3} + C \qquad \blacktriangleleft$$

Example 9   Evaluate $\int x^2 \sqrt{x - 1}\, dx$.

Solution.   Let

$$u = x - 1 \quad \text{so that} \quad du = dx \tag{4}$$

From the first equality in (4)

$$x^2 = (u + 1)^2 = u^2 + 2u + 1$$

so that

$$\int x^2 \sqrt{x - 1}\, dx = \int (u^2 + 2u + 1) \sqrt{u}\, du = \int (u^{5/2} + 2u^{3/2} + u^{1/2})\, du$$

$$= \tfrac{2}{7} u^{7/2} + \tfrac{4}{5} u^{5/2} + \tfrac{2}{3} u^{3/2} + C$$

$$= \tfrac{2}{7} (x - 1)^{7/2} + \tfrac{4}{5} (x - 1)^{5/2} + \tfrac{2}{3} (x - 1)^{3/2} + C \qquad \blacktriangleleft$$

REMARK.   Not every function can be integrated in terms of familiar functions using u-substitution. For example, you will not find any u-substitutions that will integrate the following in terms of functions encountered thus far in this text (try):

$$\int \frac{1}{x}\, dx, \quad \int \frac{dx}{\sqrt{1 - x^2}}, \quad \int \sin (x^2)\, dx$$

☐ **INTEGRATION USING COMPUTER ALGEBRA SYSTEMS**

Recent advances in programs called *computer algebra systems* now make it possible to evaluate many indefinite integrals on computers and calculators. For example, the following integral was evaluated in a matter of seconds using a symbolic manipulator.

$$\int \frac{x}{\sqrt{1 + x}}\, dx = -2\sqrt{1 + x} + \tfrac{2}{3} \sqrt{(1 + x)^3} + C$$

In spite of the existence of computer algebra systems, a basic level of competence in the evaluation of indefinite integrals is necessary. Just as one would not want to rely on a calculator to compute $2 + 2$, so one would not want to rely on a computer algebra system to integrate $f(x) = x^2$. Moreover, many of the techniques that we will develop for evaluating integrals are applicable to other kinds of mathematical problems.

▶ Exercise Set 5.3

1. Evaluate the integrals by making the indicated substitutions.

(a) $\int 2x(x^2 + 1)^{23}\, dx; \quad u = x^2 + 1$

(b) $\int \cos^3 x \sin x\, dx; \quad u = \cos x$

(c) $\int \frac{1}{\sqrt{x}} \sin \sqrt{x}\, dx; \quad u = \sqrt{x}$

(d) $\int \frac{3x\, dx}{\sqrt{4x^2 + 5}}; \quad u = 4x^2 + 5.$

2. Evaluate the integrals by making the indicated substitutions.

(a) $\int \sec^2 (4x + 1)\, dx; \quad u = 4x + 1$

(b) $\int y\sqrt{1 + 2y^2}\, dy; \quad u = 1 + 2y^2$

(c) $\int \sqrt{\sin \pi\theta} \cos \pi\theta\, d\theta; \quad u = \sin \pi\theta$

(d) $\int (2x + 7)(x^2 + 7x + 3)^{4/5}\, dx; \quad u = x^2 + 7x + 3.$

3. Evaluate the integrals by making the indicated substitutions.

(a) $\int \cot x \csc^2 x\, dx; \quad u = \cot x$

(b) $\int (1 + \sin t)^9 \cos t\, dt; \quad u = 1 + \sin t$

(c) $\int x^2\sqrt{1 + x}\, dx; \quad u = 1 + x$

(d) $\int [\csc (\sin x)]^2 \cos x\, dx; \quad u = \sin x.$

In Exercises 4–29, evaluate the integrals.

4. $\int (3x - 1)^5\, dx.$

5. $\int x(2 - x^2)^3\, dx.$

6. $\int \sin 3x\, dx.$

7. $\int \cos 8x\, dx.$

8. $\int \sec^2 5x\, dx.$

9. $\int \sec 4x \tan 4x\, dx.$

10. $\int \sqrt{3t + 1}\, dt.$

11. $\int t\sqrt{7t^2 + 12}\, dt.$

12. $\int \frac{x}{\sqrt{4 - 5x^2}}\, dx.$

13. $\int \frac{x^2}{\sqrt{x^3 + 1}}\, dx.$

14. $\int \frac{1}{(1 - 3x)^2}\, dx.$

15. $\int \frac{x}{(4x^2 + 1)^3}\, dx.$

16. $\int x \cos (3x^2)\, dx.$

17. $\int \frac{\sin (5/x)}{x^2}\, dx.$

18. $\int \frac{\sec^2 (\sqrt{x})}{\sqrt{x}}\, dx.$

19. $\int x^2 \sec^2 (x^3)\, dx.$

20. $\int \cos^3 2t \sin 2t\, dt.$

21. $\int \sin^5 3t \cos 3t\, dt.$

22. $\int \frac{\sin 2\theta}{(5 + \cos 2\theta)^3}\, d\theta.$

23. $\int \cos 4\theta \sqrt{2 - \sin 4\theta}\, d\theta.$

24. $\int \tan^3 5x \sec^2 5x\, dx.$

25. $\int \sec^3 2x \tan 2x\, dx.$

26. $\int [\sin (\sin \theta)] \cos \theta\, d\theta.$

27. $\int [\sec^2 (\cos 3\theta)] \sin 3\theta\, d\theta.$

28. $\int \sqrt[n]{a + bx}\, dx \quad (b \neq 0).$

29. $\int \sin^n (a + bx) \cos (a + bx)\, dx \quad (n > 0, b \neq 0).$

In Exercises 30–36, evaluate the integrals. These are a little trickier than those in the preceding exercises.

30. $\int (4x^2 - 12x + 9)^{2/3}\, dx.$

31. $\int x\sqrt{x - 3}\, dx.$

32. $\int x^2\sqrt{2 - x}\, dx.$

33. $\int \frac{y\, dy}{\sqrt{y + 1}}.$

34. $\int \sin^3 2\theta\, d\theta.$

[Hint: Use the identity $\sin^2 x + \cos^2 x = 1$.]

35. $\int \tan^2 3\theta\, d\theta.$

[Hint: Use a trigonometric identity.]

36. $\int \sqrt{1 + x^{-2/3}}\, dx \quad (x > 0).$

37. Evaluate the integral in Example 9 by making the substitution $u = \sqrt{x} - 1$. Check your answer against the one obtained in the example.

38. (a) Evaluate the integral $\int \sin x \cos x\, dx$ by two methods: first by letting $u = \sin x$, then by letting $u = \cos x$.

(b) Explain why the two apparently different answers obtained in part (a) are really equivalent.

39. (a) Evaluate $\int (5x - 1)^2\, dx$ by two methods: first square and integrate, then let $u = 5x - 1$.

(b) Explain why the two apparently different answers obtained in part (a) are really equivalent.

40. Find a function f such that $f'(x) = \sqrt{3x+1}$ and $f(1) = 5$.

41. Find a function f such that $f'(x) = 6 - 5 \sin 2x$ and $f(0) = 3$.

In Exercises 42–45, express the given integral in terms of the function f. [*Hint:* $\int f'(u)\,du = f(u) + C$.]

42. $\displaystyle\int f'(5x)\,dx.$

43. $\displaystyle\int f'(3x+2)\,dx.$

44. $\displaystyle\int xf'(3x^2)\,dx.$

45. $\displaystyle\int \frac{1}{x^2} f'(2/x)\,dx.$

■ 5.4 SIGMA NOTATION

*In this section we shall digress from the main theme of this chapter to introduce a notation that can be used to write lengthy sums in a compact form. This material will be helpful when we discuss area in the next section. The notation uses the uppercase Greek letter Σ (sigma) and is called **sigma notation** or **summation notation**.*

☐ SIGMA NOTATION

To illustrate how sigma notation works, consider the sum

$$1^2 + 2^2 + 3^2 + 4^2 + 5^2$$

in which each term is of the form k^2, where k is one of the integers from 1 to 5. In sigma notation this sum can be written as

$$\sum_{k=1}^{5} k^2$$

which is read, "the summation of k^2, where k runs from 1 to 5." The notation tells us to form the sum of the terms that result when we substitute successive integers for k in the expression k^2, starting with $k = 1$ and ending with $k = 5$.

More generally, if $f(k)$ is a function of k, and a and b are integers such that $a \le b$, then

$$\sum_{k=a}^{b} f(k) \tag{1}$$

denotes the sum of the terms that result when we substitute successive integers for k, starting with $k = a$ and ending with $k = b$.

Example 1

$$\sum_{k=4}^{8} k^3 = 4^3 + 5^3 + 6^3 + 7^3 + 8^3$$

$$\sum_{k=1}^{5} 2k = 2 \cdot 1 + 2 \cdot 2 + 2 \cdot 3 + 2 \cdot 4 + 2 \cdot 5 = 2 + 4 + 6 + 8 + 10$$

$$\sum_{k=0}^{5} (2k + 1) = 1 + 3 + 5 + 7 + 9 + 11$$

$$\sum_{k=0}^{5} (-1)^k (2k + 1) = 1 - 3 + 5 - 7 + 9 - 11$$

$$\sum_{k=-3}^{1} k^3 = (-3)^3 + (-2)^3 + (-1)^3 + 0^3 + 1^3 = -27 - 8 - 1 + 0 + 1$$

$$\sum_{k=1}^{3} k \sin\left(\frac{k\pi}{5}\right) = \sin\frac{\pi}{5} + 2 \sin\frac{2\pi}{5} + 3 \sin\frac{3\pi}{5} \qquad \blacktriangleleft$$

The numbers a and b in (1) are called, respectively, the **lower** and **upper limits of summation**; and the letter k is called the **index of summation**. It is not essential to use k as the index of summation; any letter not reserved for another purpose will do. For example,

$$\sum_{i=1}^{6} \frac{1}{i}, \quad \sum_{j=1}^{6} \frac{1}{j}, \quad \text{and} \quad \sum_{n=1}^{6} \frac{1}{n}$$

all denote the sum

$$1 + \frac{1}{2} + \frac{1}{3} + \frac{1}{4} + \frac{1}{5} + \frac{1}{6}$$

If the upper and lower limits of summation are the same, then the "sum" in (1) reduces to one term. For example,

$$\sum_{k=2}^{2} k^3 = 2^3 \quad \text{and} \quad \sum_{i=1}^{1} \frac{1}{i+2} = \frac{1}{1+2} = \frac{1}{3}$$

In the sums

$$\sum_{i=1}^{5} 2, \quad \sum_{k=3}^{6} 7, \quad \text{and} \quad \sum_{j=0}^{2} x^3$$

the expression to the right of the Σ sign does not involve the index of summation. In such cases, we take all the terms in the sum to be the same, with one term for each allowable value of the summation index. Thus,

$$\sum_{i=1}^{5} 2 = 2 + 2 + 2 + 2 + 2$$

$$\sum_{k=3}^{6} 7 = 7 + 7 + 7 + 7$$

$$\sum_{j=0}^{2} x^3 = x^3 + x^3 + x^3$$

A sum can be written in more than one way with sigma notation by changing the limits of summation. For example, the sum of the first five positive even integers can be written in the following ways:

$$\sum_{k=1}^{5} 2k = 2 + 4 + 6 + 8 + 10$$

$$\sum_{k=0}^{4} (2k + 2) = 2 + 4 + 6 + 8 + 10$$

$$\sum_{k=2}^{6} (2k - 2) = 2 + 4 + 6 + 8 + 10$$

☐ **CHANGING THE INDEX OF SUMMATION**

On occasion we shall want to change the sigma notation for a given sum to a sigma notation with different limits of summation. The following example illustrates a method for doing this.

Example 2 Express

$$\sum_{k=3}^{7} 5^{k-2}$$

in sigma notation so that the lower limit of summation is 0 rather than 3.

Solution. If we define a new summation index j by means of the formula

$$j = k - 3 \tag{2}$$

then j runs from 0 up to 4 as k runs from 3 up to 7. From (2), $k = j + 3$, so

$$\sum_{k=3}^{7} 5^{k-2} = \sum_{j=0}^{4} 5^{(j+3)-2} = \sum_{j=0}^{4} 5^{j+1}$$

As a check, the reader can verify that

$$\sum_{j=0}^{4} 5^{j+1} \quad \text{and} \quad \sum_{k=3}^{7} 5^{k-2}$$

both denote the sum $5 + 5^2 + 5^3 + 5^4 + 5^5$. ◄

REMARK.   In the solution of Example 2 the summation index was changed from k to j. If it is desirable to keep the same symbol for the summation index, we can change the j back to k *at the very end* and express the final result as

$$\sum_{k=0}^{4} 5^{k+1} \quad \text{instead of} \quad \sum_{j=0}^{4} 5^{j+1}$$

When we want to represent a general sum we shall use letters with subscripts. For example, a general sum with five terms might be written as

$$a_1 + a_2 + a_3 + a_4 + a_5$$

or in sigma notation as

$$\sum_{k=1}^{5} a_k, \quad \sum_{j=1}^{5} a_j, \quad \text{or} \quad \sum_{m=1}^{5} a_m$$

A general sum with n terms might be written as

$$b_1 + b_2 + \cdots + b_n$$

or in sigma notation as

$$\sum_{k=1}^{n} b_k, \quad \sum_{j=1}^{n} b_j, \quad \text{or} \quad \sum_{m=1}^{n} b_m$$

□ **PROPERTIES OF SIGMA NOTATION**

The following properties of sigma notation will help to manipulate sums.

5.4.1   THEOREM.

(a) $\displaystyle\sum_{k=1}^{n} ca_k = c \sum_{k=1}^{n} a_k$

(b) $\displaystyle\sum_{k=1}^{n} (a_k + b_k) = \sum_{k=1}^{n} a_k + \sum_{k=1}^{n} b_k$

(c) $\displaystyle\sum_{k=1}^{n} (a_k - b_k) = \sum_{k=1}^{n} a_k - \sum_{k=1}^{n} b_k$

We shall prove parts (a) and (b) and leave (c) as an exercise.

Proof (a).

$$\sum_{k=1}^{n} ca_k = ca_1 + ca_2 + \cdots + ca_n = c(a_1 + a_2 + \cdots + a_n) = c \sum_{k=1}^{n} a_k$$

Proof (b).

$$\sum_{k=1}^{n} (a_k + b_k) = (a_1 + b_1) + (a_2 + b_2) + \cdots + (a_n + b_n)$$

$$= (a_1 + a_2 + \cdots + a_n) + (b_1 + b_2 + \cdots + b_n)$$

$$= \sum_{k=1}^{n} a_k + \sum_{k=1}^{n} b_k \quad \blacksquare$$

REMARK.   Loosely phrased, this theorem states: *A constant factor can be moved through a sigma sign; sigma of a sum equals the sum of the sigmas; and sigma of a difference equals the difference of the sigmas.*

☐ **SUMMATION FORMULAS**

The following formulas will be used in our later work.

5.4.2   THEOREM.

(a) $\displaystyle\sum_{k=1}^{n} k = 1 + 2 + 3 + \cdots + n = \frac{n(n + 1)}{2}$

(b) $\displaystyle\sum_{k=1}^{n} k^2 = 1^2 + 2^2 + 3^2 + \cdots + n^2 = \frac{n(n + 1)(2n + 1)}{6}$

(c) $\displaystyle\sum_{k=1}^{n} k^3 = 1^3 + 2^3 + 3^3 + \cdots + n^3 = \left[\frac{n(n + 1)}{2} \right]^2$

We shall prove parts (*a*) and (*b*) and leave part (*c*) as an exercise.

Proof (a).   If we write the terms of

$$\sum_{k=1}^{n} k = 1 + 2 + 3 + \cdots + (n - 2) + (n - 1) + n \tag{3}$$

in the opposite order, we obtain

$$\sum_{k=1}^{n} k = n + (n - 1) + (n - 2) + \cdots + 3 + 2 + 1 \tag{4}$$

Adding (3) and (4) term by term yields

$$2\sum_{k=1}^{n} k = \underbrace{(n + 1) + (n + 1) + (n + 1) + \cdots + (n + 1)}_{n \text{ terms}} = n(n + 1)$$

Thus,

$$\sum_{k=1}^{n} k = \frac{n(n + 1)}{2}$$

Proof (b).   This proof begins with a trick. Since

$$(k + 1)^3 - k^3 = k^3 + 3k^2 + 3k + 1 - k^3 = 3k^2 + 3k + 1$$

we obtain

$$\sum_{k=1}^{n} [(k + 1)^3 - k^3] = \sum_{k=1}^{n} (3k^2 + 3k + 1) \tag{5}$$

Writing out the left side of (5) yields

$$[2^3 - 1^3] + [3^3 - 2^3] + [4^3 - 3^3] + \cdots + [(n + 1)^3 - n^3] \tag{6}$$

Observe that each term in (6) cancels part of the next term, so that the entire sum collapses like a folding telescope (hence, is called a ***telescoping sum***), leaving only $-1^3 + (n+1)^3$. Thus, (5) can be rewritten as

$$-1 + (n+1)^3 = \sum_{k=1}^{n} (3k^2 + 3k + 1) \tag{7}$$

or, from Theorem 5.4.1,

$$-1 + (n+1)^3 = 3\sum_{k=1}^{n} k^2 + 3\sum_{k=1}^{n} k + \sum_{k=1}^{n} 1 \tag{8}$$

But

$$\sum_{k=1}^{n} 1 = \underbrace{1 + 1 + \cdots + 1}_{n \text{ terms}} = n$$

and by part (*a*) of this theorem

$$\sum_{k=1}^{n} k = \frac{n(n+1)}{2}$$

Thus, (8) can be written as

$$-1 + (n+1)^3 = 3\sum_{k=1}^{n} k^2 + 3\frac{n(n+1)}{2} + n$$

Therefore,

$$\sum_{k=1}^{n} k^2 = \frac{1}{3}\left[(n+1)^3 - 3\frac{n(n+1)}{2} - (n+1) \right]$$

$$= \frac{n+1}{6}[2(n+1)^2 - 3n - 2]$$

$$= \frac{n+1}{6}(2n^2 + n) = \frac{n(n+1)(2n+1)}{6} \quad \blacksquare$$

Example 3   Evaluate $\sum_{k=1}^{30} k(k+1)$.

Solution.

$$\sum_{k=1}^{30} k(k+1) = \sum_{k=1}^{30} (k^2 + k) = \sum_{k=1}^{30} k^2 + \sum_{k=1}^{30} k$$

$$= \frac{30(31)(61)}{6} + \frac{30(31)}{2} = 9920 \quad \boxed{\text{Theorem } 5.4.2(a),\,(b)} \quad \blacktriangleleft$$

REMARK.   In formulas such as

$$\sum_{k=1}^{n} k^2 = \frac{n(n+1)(2n+1)}{6}$$

or

$$1^2 + 2^2 + \cdots + n^2 = \frac{n(n+1)(2n+1)}{6}$$

the left side of the equality is said to express the sum in ***open form*** and the right side is said to express it in ***closed form***; the open form just indicates the terms to be added, while the closed form is an explicit formula for their sum.

Example 4 Express $\displaystyle\sum_{k=1}^{n} (3 + k)^2$ in closed form.

Solution.

$$\sum_{k=1}^{n} (3 + k)^2 = \sum_{k=1}^{n} (9 + 6k + k^2) = \sum_{k=1}^{n} 9 + 6 \sum_{k=1}^{n} k + \sum_{k=1}^{n} k^2$$

$$= 9n + 6\frac{n(n + 1)}{2} + \frac{n(n + 1)(2n + 1)}{6}$$

$$= \frac{1}{3}n^3 + \frac{7}{2}n^2 + \frac{73}{6}n \quad \blacktriangleleft$$

▶ Exercise Set 5.4

1. Evaluate

(a) $\displaystyle\sum_{k=1}^{3} k^3$

(b) $\displaystyle\sum_{j=2}^{6} (3j - 1)$

(c) $\displaystyle\sum_{i=-4}^{1} (i^2 - i)$

(d) $\displaystyle\sum_{n=0}^{5} 1.$

2. Evaluate

(a) $\displaystyle\sum_{k=1}^{4} k \sin\frac{k\pi}{2}$

(b) $\displaystyle\sum_{j=0}^{5} (-1)^j$

(c) $\displaystyle\sum_{i=7}^{20} \pi$

(d) $\displaystyle\sum_{m=3}^{5} 2^{m+1}.$

In Exercises 3–16, express in sigma notation, but do not evaluate.

3. $1 + 2 + 3 + \cdots + 10.$

4. $3 \cdot 1 + 3 \cdot 2 + 3 \cdot 3 + \cdots + 3 \cdot 20.$

5. $1 \cdot 2 + 2 \cdot 3 + 3 \cdot 4 + \cdots + 49 \cdot 50.$

6. $1 + 2 + 2^2 + 2^3 + 2^4.$

7. $2 + 4 + 6 + 8 + \cdots + 20.$

8. $1 + 3 + 5 + 7 + \cdots + 15.$

9. $1 - 3 + 5 - 7 + 9 - 11.$

10. $1 - \dfrac{1}{2} + \dfrac{1}{3} - \dfrac{1}{4} + \dfrac{1}{5}.$ **11.** $-1 + \dfrac{1}{2} - \dfrac{1}{3} + \dfrac{1}{4} - \dfrac{1}{5}.$

12. $1 + \cos\dfrac{\pi}{7} + \cos\dfrac{2\pi}{7} + \cos\dfrac{3\pi}{7}.$

13. $\sin\dfrac{\pi}{8} + \sin\dfrac{3\pi}{8} + \sin\dfrac{5\pi}{8} + \sin\dfrac{7\pi}{8}.$

14. $2 + 4 + 8 + 16 + 32.$ **15.** $\dfrac{1}{2} + \dfrac{2}{3} + \dfrac{3}{4} + \dfrac{4}{5} + \dfrac{5}{6}.$

16. $15 + 24 + 35 + \cdots + (n^2 - 1).$

17. Express in sigma notation.

(a) $a_1 - a_2 + a_3 - a_4 + a_5$

(b) $-b_0 + b_1 - b_2 + b_3 - b_4 + b_5$

(c) $a_0 + a_1 x + a_2 x^2 + \cdots + a_n x^n$

(d) $a^5 + a^4 b + a^3 b^2 + a^2 b^3 + ab^4 + b^5.$

In Exercises 18–25, use Theorem 5.4.2 to evaluate the sums.

18. $\displaystyle\sum_{k=1}^{100} k.$

19. $\displaystyle\sum_{k=3}^{100} k.$

20. $\displaystyle\sum_{k=1}^{100} (7k + 1).$

21. $\displaystyle\sum_{k=1}^{20} k^2.$

22. $\displaystyle\sum_{k=4}^{20} k^2.$

23. $\displaystyle\sum_{k=1}^{6} (4k^3 - 2k + 1).$

24. $\displaystyle\sum_{k=1}^{6} (k - k^3).$

25. $\displaystyle\sum_{k=1}^{30} k(k - 2)(k + 2).$

In Exercises 26–30, express the sum in closed form.

26. (a) $\displaystyle\sum_{k=1}^{n} (4k - 3)$

(b) $\displaystyle\sum_{k=1}^{n-1} k^2.$

27. $\displaystyle\sum_{k=1}^{n} \frac{3k}{n}.$

28. $\displaystyle\sum_{k=1}^{n-1} \frac{k^2}{n}.$

29. $\displaystyle\sum_{k=1}^{n-1} \frac{k^3}{n^2}.$

30. $\displaystyle\sum_{k=1}^{n} \left(\frac{5}{n} - \frac{2k}{n}\right).$

In Exercises 31–35, the limit of a function of n is given. Express the function of n in closed form, then find the limit. [*Note:* Although n assumes only integer values, the limits can be calculated using the same techniques that we have been using for functions of a real-valued variable x. Functions of integer-valued variables will be studied in more detail later.]

31. $\displaystyle\lim_{n \to +\infty} \frac{1 + 2 + 3 + \cdots + n}{n^2}.$

32. $\lim\limits_{n \to +\infty} \dfrac{1^2 + 2^2 + 3^2 + \cdots + n^2}{n^3}$.

33. $\lim\limits_{n \to +\infty} \sum\limits_{k=1}^{n} \dfrac{5k}{n^2}$.   **34.** $\lim\limits_{n \to +\infty} \sum\limits_{k=1}^{n-1} \dfrac{2k^2}{n^3}$.

35. $\lim\limits_{n \to +\infty} \sum\limits_{k=1}^{n-1} \left(\dfrac{9}{n} - \dfrac{k}{n^2} \right)$.

36. Show that
$$1 \cdot 2 + 2 \cdot 3 + \cdots + n(n+1) = \tfrac{1}{3}n(n+1)(n+2)$$

37. Show that the sum of the first n consecutive positive odd integers is n^2.

> When each term of a sum cancels part of the next term, leaving only portions of the first and last terms at the end, the sum is said to **telescope**. In Exercises 38–43, evaluate the telescoping sum.

38. $\sum\limits_{k=1}^{50} \left(\dfrac{1}{k} - \dfrac{1}{k+1} \right)$.   **39.** $\sum\limits_{k=5}^{17} (3^k - 3^{k-1})$.

40. $\sum\limits_{k=1}^{100} (2^{k+1} - 2^k)$.   **41.** $\sum\limits_{k=2}^{20} \left(\dfrac{1}{k^2} - \dfrac{1}{(k-1)^2} \right)$.

42. $\sum\limits_{k=1}^{n} (a_k - a_{k+1})$.   **43.** $\sum\limits_{k=1}^{n} (a_k - a_{k-1})$.

44. (a) Show that
$$\dfrac{1}{1 \cdot 2} + \dfrac{1}{2 \cdot 3} + \dfrac{1}{3 \cdot 4} + \cdots + \dfrac{1}{n(n+1)} = \dfrac{n}{n+1}$$
$$\left[\text{Hint: } \dfrac{1}{n(n+1)} = \dfrac{1}{n} - \dfrac{1}{n+1}. \right]$$
(b) Use the result in part (a) to find
$$\lim\limits_{n \to +\infty} \sum\limits_{k=1}^{n} \dfrac{1}{k(k+1)}$$

45. (a) Show that
$$\dfrac{1}{1 \cdot 3} + \dfrac{1}{3 \cdot 5} + \cdots + \dfrac{1}{(2n-1)(2n+1)} = \dfrac{n}{2n+1}$$
$$\left[\text{Hint: } \dfrac{1}{(2n-1)(2n+1)} = \dfrac{1}{2}\left(\dfrac{1}{2n-1} - \dfrac{1}{2n+1} \right). \right]$$
(b) Use the result in part (a) to find
$$\lim\limits_{n \to +\infty} \sum\limits_{k=1}^{n} \dfrac{1}{(2k-1)(2k+1)}$$

46. Evaluate

(a) $\sum\limits_{j=0}^{m} m$ (b) $\sum\limits_{n=4}^{4} 5$

(c) $\sum\limits_{k=1}^{n} x$ (d) $\sum\limits_{i=1}^{n} i^2 c$.

47. Evaluate

(a) $\sum\limits_{k=1}^{n} n$ (b) $\sum\limits_{i=0}^{0} (-3)$

(c) $\sum\limits_{k=1}^{n} kx$ **(d)** $\sum\limits_{k=m}^{n} c \quad (n \ge m)$.

48. Express $1 + 2 + 2^2 + 2^3 + 2^4 + 2^5$ in sigma notation with

(a) $j = 0$ as the lower limit of summation

(b) $j = 1$ as the lower limit of summation

(c) $j = 2$ as the lower limit of summation.

49. Express
$$\sum\limits_{k=4}^{18} k(k-3)$$
in sigma notation with

(a) $k = 0$ as the lower limit of summation

(b) $k = 5$ as the lower limit of summation.

50. Express
$$\sum\limits_{k=5}^{9} k2^{k+4}$$
in sigma notation with

(a) $k = 1$ as the lower limit of summation

(b) $k = 13$ as the upper limit of summation.

51. Simplify
$$\sum\limits_{k=11}^{28} (k-10) \sin\left(\dfrac{\pi}{k-10} \right)$$
by changing the limits of summation.

52. Which of the following are valid identities?

(a) $\sum\limits_{i=1}^{n} a_i b_i = \sum\limits_{i=1}^{n} a_i \sum\limits_{i=1}^{n} b_i$

(b) $\sum\limits_{i=1}^{n} \dfrac{a_i}{b_i} = \sum\limits_{i=1}^{n} a_i \bigg/ \sum\limits_{i=1}^{n} b_i$

(c) $\sum\limits_{i=1}^{n} a_i^2 = \left(\sum\limits_{i=1}^{n} a_i \right)^2$.

53. By writing out the sums, determine whether the following are valid identities.

(a) $\displaystyle\int \left[\sum\limits_{i=1}^{n} f_i(x) \right] dx = \sum\limits_{i=1}^{n} \left[\int f_i(x)\, dx \right]$

(b) $\dfrac{d}{dx} \left[\sum\limits_{i=1}^{n} f_i(x) \right] = \sum\limits_{i=1}^{n} \left[\dfrac{d}{dx} [f_i(x)] \right]$.

54. Let
$$S = \sum\limits_{k=0}^{n} ar^k$$
Show that $S - rS = a - ar^{n+1}$ and hence that
$$\sum\limits_{k=0}^{n} ar^k = \dfrac{a - ar^{n+1}}{1 - r} \quad (r \ne 1)$$
(A sum of this form is called a **geometric sum**.)

55. Use Exercise 54 to evaluate

(a) $\displaystyle\sum_{k=1}^{20} 3^k$ (b) $\displaystyle\sum_{k=5}^{30} 2^k$ (c) $\displaystyle\sum_{k=0}^{100} (-1)^{k+1}\frac{1}{2^k}$.

56. Use Exercise 54 to express $\displaystyle\sum_{k=1}^{n} \sin^k \theta$ in closed form.

57. Evaluate

$$\sum_{i=1}^{4}\left(\sum_{j=1}^{5}(i+j)\right)$$

58. Let $\bar{x}$ denote the arithmetic average of the n numbers $x_1, x_2, \ldots, x_n$. Use Theorem 5.4.1 to prove that

$$\sum_{i=1}^{n}(x_i - \bar{x}) = 0$$

59. Prove part (c) of Theorem 5.4.1.

60. Prove part (c) of Theorem 5.4.2. [*Hint:* Begin with the difference $(k+1)^4 - k^4$ and follow the steps used to prove part (b) of the theorem.]

■ 5.5 AREAS AS LIMITS

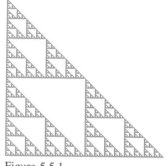

Figure 5.5.1

So far we have used the term "area" freely, assuming its meaning is to be self-evident. However, to work with areas mathematically we must ultimately define the concept of area precisely. It is natural to hope that we might be able to define the area of an arbitrary region in the plane. Surprisingly, there is no way to do this. It became clear in the late nineteenth century that there are regions in the plane of such complexity that any attempt to assign them areas in the traditional sense would ultimately lead to mathematical inconsistencies (the blue region in Figure 5.5.1, for example). Although we shall not be concerned with such regions, it is important to know that they exist. In this section we shall show how to compute areas using limits and shall develop a precise definition of area.

☐ DEFINITION OF AREA

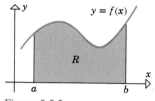

Figure 5.5.2

Figure 5.5.2 shows a region R that is bounded below by the x-axis, on the sides by the lines $x = a$ and $x = b$, and above by a curve $y = f(x)$, where f is continuous and nonnegative on $[a, b]$. In Section 5.1 we showed how to compute the area of this kind of region using antiderivatives, but that development was based on an imprecise intuitive concept of area. We shall now remedy that problem by developing a precise definition of the area of R.

Recall that we defined the slope of a tangent line as a limit of slopes of secant lines. Analogously, we shall define the area of R as a limit of areas of simple regions whose areas are known. We take it as a *definition* that the area of a rectangle is the product of its length and width and the area of a region composed of rectangles is the sum of the areas of those rectangles. To define the area of region R of the form shown in Figure 5.5.2, we build on these definitions as follows:

- Choose an arbitrary positive integer n, and divide the interval $[a, b]$ into n subintervals of width $(b-a)/n$ by inserting $n-1$ equally spaced points between a and b, say

  $$x_1, x_2, \ldots, x_{n-1}$$

  (Figure 5.5.3a). These points of subdivision form what is called a **regular partition** of the interval $[a, b]$.

- Next, draw vertical lines through the points $a, x_1, x_2, \ldots, x_{n-1}, b$ to divide the region R into n strips of uniform width (Figure 5.5.3b).

- Now we want to approximate the area of each strip by the area of a rectangle. For this purpose choose an *arbitrary* point in each subinterval, say

  $$x_1^*, x_2^*, \ldots, x_n^*$$

  and over each subinterval construct a rectangle whose height is the value of the function f at the selected arbitrary point (Figure 5.5.3c). The union of these rectangles forms a region R_n that we can regard as a reasonable approximation to the entire region R. The area of this approximating region can be obtained by adding the areas of the component rectangles.

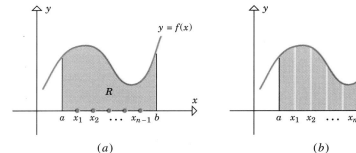

Figure 5.5.3

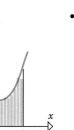

Figure 5.5.4

- If we now allow n to increase, the widths of the rectangles will get smaller, and the approximation of the area of R by the area of R_n will get better as the thinner rectangles fill in the gaps under the curve and the portions that overlap the curve diminish (Figure 5.5.4). Thus, we can *define* the area of R to be the limit of the area of the approximating region R_n as $n \to +\infty$; that is,

$$A = \text{area}(R) = \lim_{n \to +\infty} [\text{area}(R_n)] \tag{1}$$

REMARK. There is a difference between writing $\lim_{n \to +\infty}$ and writing $\lim_{x \to +\infty}$, where n represents a positive integer, and x has no such restriction. Later we will study limits of the type $\lim_{n \to +\infty}$ in detail, but for now let it suffice to say that the computational techniques we have used for limits of the type $\lim_{x \to +\infty}$ will also work for $\lim_{n \to +\infty}$.

For computational purposes, the definition in (1) can be written in a more useful form as follows: For n subdivisions of the interval $[a, b]$, each of the approximating rectangles has width $(b - a)/n$, which it is customary to denote by

$$\Delta x = \frac{b - a}{n}$$

The heights of the approximating rectangles are the values of f at the points

$$x_1^*, x_2^*, \ldots, x_n^*$$

so the approximating rectangles that make up the region R_n have areas

$$f(x_1^*)\,\Delta x, f(x_2^*)\,\Delta x, \ldots, f(x_n^*)\,\Delta x$$

(Figure 5.5.5), and the total area of the region R_n is given by

$$\text{area}(R_n) = f(x_1^*)\,\Delta x + f(x_2^*)\,\Delta x + \cdots + f(x_n^*)\,\Delta x$$

or in sigma notation,

$$\text{area}(R_n) = \sum_{k=1}^{n} f(x_k^*)\,\Delta x$$

With this notation (1) can be expressed as

$$A = \lim_{n \to +\infty} \sum_{k=1}^{n} f(x_k^*)\,\Delta x \tag{2}$$

which we can take as a precise definition of the area of the region R.

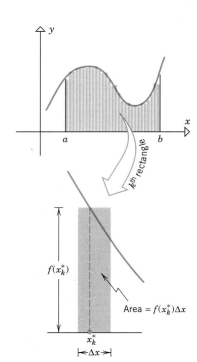

Figure 5.5.5

SOME TECHNICAL CONSIDERATIONS

In (2) the points $x_1^*, x_2^*, \ldots, x_n^*$ can be chosen arbitrarily, so it is conceivable that different choices of these points might produce different values of A, in which case this would not be an acceptable definition of area. Fortunately, this does not happen; it is proved in advanced courses that when f is continuous (as we have assumed), the same value of A results no matter how the points are chosen. Typically, the points are chosen in some systematic fashion, some common choices being:

- The left endpoint of each subinterval
- The right endpoint of each subinterval
- The midpoint of each subinterval

If, as shown in Figure 5.5.6, the subinterval $[a, b]$ is divided by points $x_1, x_2, x_3, \ldots, x_{n-1}$ into n equal parts each of length Δx, and if we let $x_0 = a$ and $x_n = b$, then

$$x_k = a + k\,\Delta x \quad \text{for } k = 0, 1, 2, \ldots, n$$

so

- $x_k^* = x_{k-1} = a + (k-1)\,\Delta x$ Left endpoint   (3)
- $x_k^* = x_k = a + k\,\Delta x$ Right endpoint   (4)
- $x_k^* = \frac{1}{2}(x_{k-1} + x_k) = a + (k - \frac{1}{2})\,\Delta x$ Midpoint   (5)

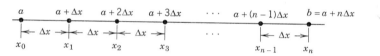

Figure 5.5.6

Example 1 Use (2) with x_k^* as the right endpoint of each subinterval to find the area under the line $y = x$ over the interval $[1, 2]$.

Solution. If we subdivide the interval $[1, 2]$ into n equal parts, then each part will have length

$$\Delta x = \frac{b - a}{n} = \frac{2 - 1}{n} = \frac{1}{n}$$

and from (4)

$$x_k^* = a + k\,\Delta x = 1 + \frac{k}{n}$$

(Figure 5.5.7). Thus, the kth rectangle has area

$$f(x_k^*)\,\Delta x = x_k^*\,\Delta x = \left(1 + \frac{k}{n}\right)\Delta x = \left(1 + \frac{k}{n}\right)\frac{1}{n}$$

and the sum of the areas of the n rectangles is

$$\sum_{k=1}^{n} f(x_k^*)\,\Delta x = \sum_{k=1}^{n}\left[\left(1 + \frac{k}{n}\right)\frac{1}{n}\right] = \frac{1}{n}\sum_{k=1}^{n} 1 + \frac{1}{n^2}\sum_{k=1}^{n} k$$

$$= \frac{1}{n}\cdot n + \frac{1}{n^2}\left[\tfrac{1}{2}n(n+1)\right] = \frac{3}{2} + \frac{1}{2n}$$

Theorem 5.4.2(*a*)

$\frac{n}{n} + \frac{1}{2}n^2 + \frac{1}{2}n$

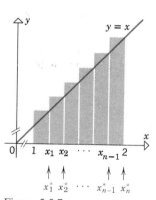

Figure 5.5.7

Thus, from (2) the area is

$$A = \lim_{n \to \infty} \sum_{k=1}^{n} f(x_k^*)\,\Delta x = \lim_{n \to \infty}\left(\frac{3}{2} + \frac{1}{2n}\right) = \frac{3}{2} + 0 = \frac{3}{2}$$

The region whose area we have computed is a trapezoid with height $h = 1$ and bases $b_1 = 1$ and $b_2 = 2$. From plane geometry the area of this trapezoid is

$$A = \tfrac{1}{2}h(b_1 + b_2) = \tfrac{1}{2}(1)(1 + 2) = \tfrac{3}{2}$$

which agrees with the result that we obtained here.   ◄

REMARK.   In the preceding example, the height of each rectangle is the maximum value of the function in each subinterval. These are called *circumscribed* rectangles and their areas overestimate the actual areas of the strips.

Example 2   Use (2) with x_k^* as the left endpoint of each subinterval to find the area under the line $y = x$ over the interval $[1, 2]$.

Solution.   As in Example 1, $\Delta x = 1/n$. From (3)

$$x_k^* = a + (k - 1)\,\Delta x = 1 + \frac{k - 1}{n}$$

(Figure 5.5.8), so rectangle k has area

$$f(x_k^*)\,\Delta x = x_k^*\,\Delta x = \left(1 + \frac{k - 1}{n}\right)\Delta x = \left(1 + \frac{k - 1}{n}\right)\frac{1}{n}$$

and the sum of the areas of the n rectangles is

$$\sum_{k=1}^{n} f(x_k^*)\,\Delta x = \sum_{k=1}^{n}\left[\left(1 + \frac{k - 1}{n}\right)\frac{1}{n}\right] = \frac{1}{n}\sum_{k=1}^{n} 1 + \frac{1}{n^2}\sum_{k=1}^{n}(k - 1)$$

But

$$\sum_{k=1}^{n}(k - 1) = 0 + 1 + 2 + \cdots + (n - 1) = \sum_{k=1}^{n-1} k = \tfrac{1}{2}(n - 1)n$$

Theorem 5.4.2(a) with $n - 1$ substituted for n

so

$$\sum_{k=1}^{n} f(x_k^*)\,\Delta x = \frac{1}{n}(n) + \frac{1}{2}\frac{(n - 1)n}{n^2} = 1 + \frac{n - 1}{2n} = \frac{3}{2} - \frac{1}{2n}$$

From (2), the area under the curve is

$$A = \lim_{n \to \infty}\sum_{k=1}^{n} f(x_k^*)\,\Delta x = \lim_{n \to \infty}\left(\frac{3}{2} - \frac{1}{2n}\right) = \frac{3}{2}$$

This agrees with the result of Example 1, where we used the right endpoints of the subintervals.   ◄

REMARK.   In the preceding example, the height of each rectangle is the minimum value of the function in each subinterval. These are called *inscribed* rectangles and their areas underestimate the actual areas of the strips.

Example 3   Use (2) with x_k^* as the right endpoint of each subinterval to find the area under the parabola $y = 9 - x^2$ over the interval $[0, 3]$.

Solution.   Each subinterval will have length

$$\Delta x = \frac{b - a}{n} = \frac{3 - 0}{n} = \frac{3}{n}$$

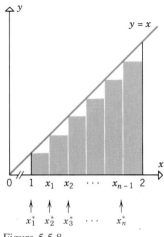

Figure 5.5.8

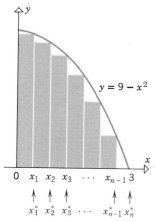

Figure 5.5.9

and from (4)

$$x_k^* = a + k\,\Delta x = 0 + \frac{3k}{n} = \frac{3k}{n}$$

(Figure 5.5.9). Thus, rectangle k has area

$$f(x_k^*)\,\Delta x = [9 - (x_k^*)^2]\,\Delta x = \left(9 - \frac{9k^2}{n^2}\right)\frac{3}{n} = \frac{27}{n} - \frac{27k^2}{n^3}$$

and the sum of these areas is

$$\sum_{k=1}^{n} f(x_k^*) \cdot \Delta x = \sum_{k=1}^{n}\left(\frac{27}{n} - \frac{27k^2}{n^3}\right) = \sum_{k=1}^{n}\frac{27}{n} - \sum_{k=1}^{n}\frac{27k^2}{n^3}$$

$$= n \cdot \frac{27}{n} - \frac{27}{n^3}\sum_{k=1}^{n} k^2 = 27 - \frac{27}{n^3} \cdot \frac{n(n+1)(2n+1)}{6}$$

Theorem 5.4.2(b)

Thus, from (2)

$$A = \lim_{n \to +\infty} \sum_{k=1}^{n} f(x_k^*) \cdot \Delta x = \lim_{n \to +\infty}\left[27 - \frac{27}{n^3} \cdot \frac{n(n+1)(2n+1)}{6}\right]$$

$$= 27 - \lim_{n \to +\infty} \frac{27}{6}\left(1 + \frac{1}{n}\right)\left(2 + \frac{1}{n}\right) = 27 - \frac{27}{6}(1)(2) = 18 \qquad \blacktriangleleft$$

☐ **NUMERICAL APPROXIMATIONS OF AREA**

As evidenced by the preceding examples, the computations required to find an exact area from (2) are tedious, even for simple functions. In many situations they are impractical or impossible to carry out, and one must settle for a numerical approximation of the area using the sum

$$\sum_{k=1}^{n} f(x_k^*)\,\Delta x$$

with a value of n that is sufficiently large to produce the required accuracy. For this purpose it is convenient to rewrite this sum in the form

$$\sum_{k=1}^{n} f(x_k^*)\,\Delta x = \Delta x \sum_{k=1}^{n} f(x_k^*) = \Delta x\,[f(x_1^*) + f(x_2^*) + \cdots + f(x_n^*)] \qquad (6)$$

where $\Delta x = (b - a)/n$. The calculation requires only the sum of the values of the function at n points, followed by a multiplication by Δx. A programmable calculator or a computer is useful for such computations, especially when n is large.

Depending on how each x_k^* is chosen, Formula (6) is called the **left endpoint approximation**, the **right endpoint approximation**, or the **midpoint approximation** of the exact area (Figure 5.5.10). As we shall see later, the midpoint approximation is usually more accurate than the endpoint approximations.

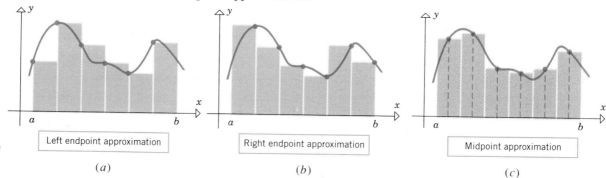

Figure 5.5.10

| Left endpoint approximation | Right endpoint approximation | Midpoint approximation |

(a)   (b)   (c)

Example 4 Using (6) with $n = 10$, $n = 20$, and $n = 50$, find for each of these values of n the left endpoint, right endpoint, and midpoint approximations of the area under the curve $y = 9 - x^2$ over the interval $[0, 3]$.

Solution. Details of the computations for the case $n = 10$ are shown to six decimal places in Table 5.5.1 and the results of all computations are given in Table 5.5.2.

Table 5.5.1

$n = 10$, $\Delta x = (b - a)/n = (3 - 0)/10 = 0.3$

	LEFT ENDPOINT APPROXIMATION		RIGHT ENDPOINT APPROXIMATION		MIDPOINT APPROXIMATION	
k	x_k^*	$9 - (x_k^*)^2$	x_k^*	$9 - (x_k^*)^2$	x_k^*	$9 - (x_k^*)^2$
1	0.0	9.000000	0.3	8.910000	0.15	8.977500
2	0.3	8.910000	0.6	8.640000	0.45	8.797500
3	0.6	8.640000	0.9	8.190000	0.75	8.437500
4	0.9	8.190000	1.2	7.560000	1.05	7.897500
5	1.2	7.560000	1.5	6.750000	1.35	7.177500
6	1.5	6.750000	1.8	5.760000	1.65	6.277500
7	1.8	5.760000	2.1	4.590000	1.95	5.197500
8	2.1	4.590000	2.4	3.240000	2.25	3.937500
9	2.4	3.240000	2.7	1.710000	2.55	2.497500
10	2.7	1.710000	3.0	0.000000	2.85	0.877500
		64.350000		55.350000		60.075000
$\displaystyle \Delta x \sum_{k=1}^{n} f(x_k^*)$		(.3)(64.350000) = 19.305000		(.3)(55.350000) = 16.605000		(.3)(60.075000) = 18.022500

Table 5.5.2

n	LEFT ENDPOINT APPROXIMATION	RIGHT ENDPOINT APPROXIMATION	MIDPOINT APPROXIMATION
10	19.305000	16.605000	18.022500
20	18.663750	17.313750	18.005625
50	18.268200	17.728200	18.000900

From Example 3, the exact area is 18. Notice that the midpoint approximation is better than either the left or right endpoint approximations. ◀

▶ Exercise Set 5.5   Ⓒ *18–21*

In Exercises 1–4, divide the interval $[a, b]$ into $n = 4$ subintervals of equal length, and then compute

$$\sum_{k=1}^{4} f(x_k^*) \, \Delta x$$

with x_k^* as (a) the left endpoint of each subinterval and (b) the right endpoint of each subinterval.

1. $y = 3x + 1$; $a = 2$, $b = 6$.   **2.** $y = 1/x$; $a = 1$, $b = 9$.

3. $y = \cos x$; $a = -\pi/2$, $b = \pi/2$.

4. $y = 2x - x^2$; $a = 1$, $b = 2$.

In Exercises 5–10, use (2) with x_k^* as the *right* endpoint of each subinterval to find the area under the curve $y = f(x)$ over the interval $[a, b]$.

5. $y = \frac{1}{2}x$; $a = 1$, $b = 4$.

6. $y = 5 - x$; $a = 0$, $b = 5$.

7. $y = x^2$; $a = 0$, $b = 1$.

8. $y = 4 - \frac{1}{4}x^2$; $a = 0$, $b = 3$.

9. $y = x^3$; $a = 2$, $b = 6$.

10. $y = 1 - x^3$; $a = -3$, $b = -1$.

In Exercises 11–13, use (2) with x_k^* as the *left* endpoint of each subinterval to find the area under the curve $y = f(x)$ over the interval $[a, b]$.

11. Exercise 5.

12. Exercise 6.

13. Exercise 7.

14. Use (2) with x_k^* as the left endpoint of each subinterval to find the area under $y = mx$ over the interval $[a, b]$, where $m > 0$ and $a \geq 0$.

15. (a) Show that the area under $y = x^3$ over the interval $[0, b]$ is $b^4/4$.

(b) Find a formula for the area under $y = x^3$ over the interval $[a, b]$, where $a \geq 0$.

16. Find the area between the curve $y = \sqrt{x}$ and the interval $0 \leq y \leq 1$ on the y-axis. [*Hint:* Subdivide the interval on the y-axis.]

17. Assuming that $a > 0$, find the area under $y = x$ over the interval $[a, b]$ by

(a) using (2) with x_k^* as the left endpoint of each subinterval

(b) using (2) with x_k^* as the right endpoint of each subinterval

(c) using an area formula from plane geometry.

In Exercises 18–21, use a computer or calculator to obtain an approximate value for the area under the curve with $n = 10$, 20, and 50 subintervals by using the (a) left endpoint, (b) right endpoint, and (c) midpoint approximations. [*Note:* A programmable calculator is best for this problem. If your calculator is not programmable, then just do the case for $n = 10$.]

18. $y = 1/x$; $[1, 2]$.

19. $y = 1/x^2$; $[1, 3]$.

20. $y = \sqrt{x}$; $[0, 4]$.

21. $y = \sin x$; $[0, \pi/2]$.

■ **5.6** THE DEFINITE INTEGRAL

In mathematics and science there are a variety of concepts, such as length, volume, density, probability, work, and others, whose properties are remarkably similar to properties of area. In this section we shall introduce the notion of a "definite integral," which is the unifying thread that relates these concepts. Our work in this section will focus on ideas rather than computations. In the next section we shall develop the tools that are necessary for efficient calculations.

In the preceding section we developed a definition of area using approximating rectangles with equal widths. Although rectangles with equal widths are convenient for computations, they are not essential; we can just as well express the area of the region in Figure 5.5.2 as a limit of areas of regions composed of rectangles with different widths (Figure 5.6.1), provided we ensure that the width of *every* rectangle decreases to zero as the number of rectangles approaches infinity. For a subdivision of $[a, b]$ into n rectangles of equal width, the width of each rectangle is $\Delta x = (b - a)/n$, so we are guaranteed that $\Delta x \to 0$ as $n \to +\infty$. However, for rectangles in which the widths can vary, letting $n \to +\infty$ does not of itself guarantee that the width of each rectangle approaches zero. For example, suppose that we were to continually divide only the left half of the interval $[a, b]$ into more and more subintervals, but were always to leave the right half of the interval alone (Figure 5.6.2). With this construction the number of rectangles increases to infinity, but the sum of the areas of the rectangles does not approach the area under the curve because the "error" on the right half of the interval never decreases. To remedy this problem, we need to ensure that the widths of *all* rectangles decrease to zero as the number of rectangles increases to infinity.

Let us suppose that the interval $[a, b]$ is divided into n subintervals whose widths are

$$\Delta x_1, \Delta x_2, \ldots, \Delta x_n$$

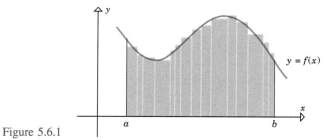

Figure 5.6.1

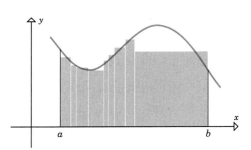

Figure 5.6.2

The subintervals are said to form a **partition** of the interval $[a, b]$, and the largest of the subinterval widths is called the **mesh size** of the partition. The mesh size is denoted by the symbol

$$\max \Delta x_k$$

which is read, "the maximum of the Δx_k's." For example, Figure 5.6.3 shows a partition of the interval $[0, 6]$ into four subintervals with a mesh size of 2.

If an interval $[a, b]$ is partitioned into n subintervals, and if x_k^* is an arbitrary point in the kth subinterval, then

$$f(x_k^*) \, \Delta x_k$$

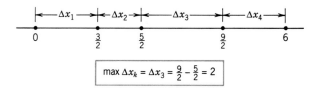

Figure 5.6.3

is the area of a rectangle of height $f(x_k^*)$ and width Δx_k, so

$$\tag{1}$$

is the sum of the shaded rectangular areas in Figure 5.6.4.

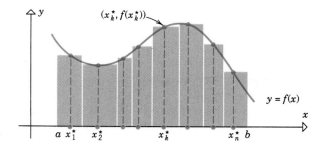

Figure 5.6.4

If we now increase n in such a way that

$$\max \Delta x_k \to 0$$

then the width of *every* rectangle tends to zero because no width exceeds the maximum; thus, (1) approaches the exact area under the curve. We denote this by writing

$$A = \lim_{\max \Delta x_k \to 0} \sum_{k=1}^{n} f(x_k^*) \, \Delta x_k$$

Thus, we are led to the following more general definition of area that allows for approximating areas using rectangles of variable width.

> **5.6.1** DEFINITION (*Area Under a Curve*).   If the function f is continuous on $[a, b]$ and if $f(x) \geq 0$ for all x in $[a, b]$, then the **area** under the curve $y = f(x)$ over the interval $[a, b]$ is defined by
>
> $$A = \lim_{\max \Delta x_k \to 0} \sum_{k=1}^{n} f(x_k^*) \, \Delta x_k$$

□ **DEFINITE INTEGRALS OF CONTINUOUS FUNCTIONS WITH NONNEGATIVE VALUES**

The limit in Definition 5.6.1 is so important that there is some special notation for it. We write

$$\int_a^b f(x) \, dx = \lim_{\max \Delta x_k \to 0} \sum_{k=1}^{n} f(x_k^*) \, \Delta x_k \tag{2}$$

The expression on the left side of this equation is called the **definite integral of f from a to b**, and the numbers b and a are called the **upper and lower limits of integration**, respectively. The reason for the use of the integral sign will become clear in the next section where we shall establish a relationship between the definite integral and the indefinite integral studied earlier.

It follows from Definition 5.6.1 and (2) that the area A under the curve $y = f(x)$ over the interval $[a, b]$ can be expressed in definite integral notation as

$$A = \begin{bmatrix} \text{area under} \\ y = f(x) \\ \text{over } [a, b] \end{bmatrix} = \int_a^b f(x) \, dx \tag{3}$$

One of our goals is to develop efficient methods for evaluating definite integrals; however, the definite integrals in the following example can be evaluated using area formulas from plane geometry.

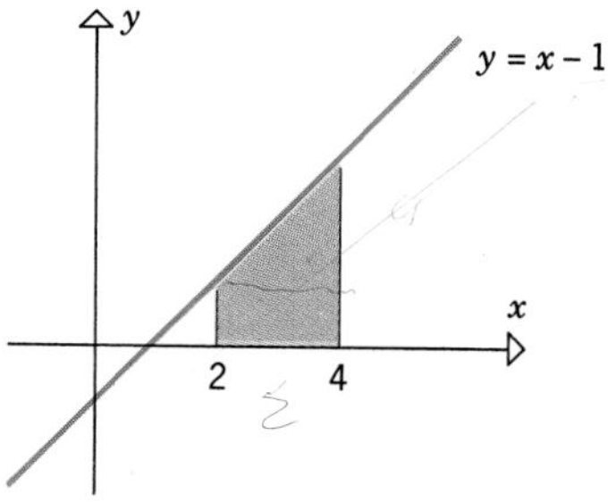

Figure 5.6.5

Example 1   Use appropriate area formulas from plane geometry to evaluate the following definite integrals.

$$\text{(a)} \quad \int_2^4 (x - 1) \, dx \quad \text{(b)} \quad \int_0^1 \sqrt{1 - x^2} \, dx$$

Solution (a).   From (3), the integral represents the area under the graph of $y = x - 1$ over the interval $[2, 4]$. (See Figure 5.6.5.) This region is a trapezoid whose parallel sides have lengths 1 and 3 and whose height (the distance between the parallel sides) is 2. Thus,

$$\int_2^4 (x - 1) \, dx = \frac{1}{2}(1 + 3) \cdot 2 = 4$$

(See the inside text cover for a review of area formulas from plane geometry.)

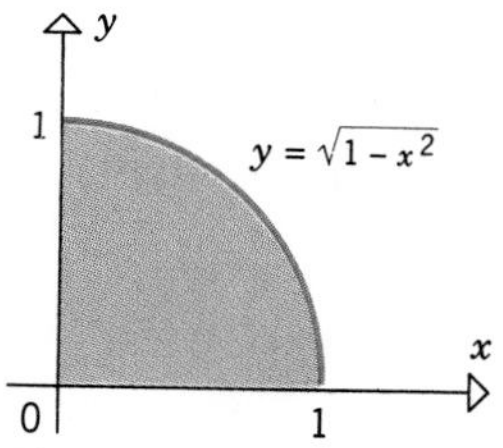

Figure 5.6.6

Solution (b).   From (3), the integral represents the area under the graph of $y = \sqrt{1 - x^2}$ over the interval $[0, 1]$. This is the region in the first quadrant bounded by the coordinate axes and the circle $x^2 + y^2 = 1$. (See Figure 5.6.6.) The area of this region is one-fourth of the area enclosed by the entire circle, so

$$\int_0^1 \sqrt{1 - x^2} \, dx = \frac{1}{4}(\pi \cdot 1^2) = \frac{\pi}{4} \quad \blacktriangleleft$$

The sum

$$\sum_{k=1}^{n} f(x_k^*)\,\Delta x_k$$

which appears in (2), is called a ***Riemann* sum*** in honor of the German mathematician Bernhard Riemann, who first formulated many of the concepts about definite integrals. Thus, (2) defines the definite integral as a limit of Riemann sums. It is proved in advanced courses that for nonnegative continuous functions this limit must exist. As a result, we need not worry about the existence of the limit that defines the area A in Definition 5.6.1.

Although we defined the definite integral as a notation for area, limits of Riemann sums occur in a wide variety of applications that have no immediate connection with area problems. In many of these applications the functions involved assume both positive and negative values and have discontinuities, so we will want to extend the concept of a definite integral to allow for such functions.

☐ **DEFINITE INTEGRALS OF CONTINUOUS FUNCTIONS WITH POSITIVE AND NEGATIVE VALUES**

Thus far we have defined the definite integral for nonnegative continuous functions on an interval $[a, b]$. We will now extend the definition of the definite integral to include functions that are continuous and have both positive and negative values on $[a, b]$. It can be proved that for such functions the limit

$$\lim_{\max \Delta x_k \to 0} \sum_{k=1}^{n} f(x_k^*)\,\Delta x_k$$

always exists, and thus we define

$$\int_a^b f(x)\,dx = \lim_{\max \Delta x_k \to 0} \sum_{k=1}^{n} f(x_k^*)\,\Delta x_k \tag{4}$$

To motivate a geometric interpretation of this integral, consider the Riemann sum shown in Figure 5.6.7.

---

*GEORG FRIEDRICH BERNHARD RIEMANN (1826–1866). German mathematician. Bernhard Riemann, as he is commonly known, was the son of a Protestant minister. He received his elementary education from his father and showed brilliance in arithmetic at an early age. In 1846 he enrolled at Göttingen University to study theology and philology, but he soon transferred to mathematics. He studied physics under W. E. Weber and mathematics under Karl Friedrich Gauss, whom some people consider to be the greatest mathematician who ever lived. In 1851 Riemann received his Ph.D. under Gauss, after which he remained at Göttingen to teach. In 1862, one month after his marriage, Riemann suffered an attack of pleuritis, and for the remainder of his life was an extremely sick man. He finally succumbed to tuberculosis in 1866 at age 39.

An interesting story surrounds Riemann's work in geometry. For his introductory lecture prior to becoming an associate professor, Riemann submitted three possible topics to Gauss. Gauss surprised Riemann by choosing the topic Riemann liked the least, the foundations of geometry. The lecture was like a scene from a movie. The old and failing Gauss, a giant in his day, watching intently as his brilliant and youthful protégé skillfully pieced together portions of the old man's own work into a complete and beautiful system. Gauss is said to have gasped with delight as the lecture neared its end, and on the way home he marveled at his student's brilliance. Gauss died shortly thereafter. The results presented by Riemann that day eventually evolved into a fundamental tool that Einstein used some 50 years later to develop relativity theory.

In addition to his work in geometry, Riemann made major contributions to the theory of complex functions and mathematical physics. The notion of the definite integral, as it is presented in most basic calculus courses, is due to him. Riemann's early death was a great loss to mathematics, for his mathematical work was brilliant and of fundamental importance.

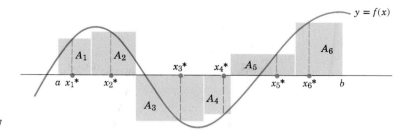

Figure 5.6.7

On those intervals where $f(x_k^*)$ is positive, the product $f(x_k^*)\,\Delta x_k$ is the area A_k of the rectangle with height $f(x_k^*)$ and base Δx_k, and on those intervals where $f(x_k^*)$ is negative, the product $f(x_k^*)\,\Delta x_k$ is the negative, $-A_k$, of such an area.

For the particular Riemann sum in Figure 5.6.7 we have

$$\sum_{k=1}^{6} f(x_k^*)\,\Delta x_k = f(x_1^*)\,\Delta x_1 + f(x_2^*)\,\Delta x_2 + \cdots + f(x_6^*)\,\Delta x_6$$

$$= A_1 + A_2 - A_3 - A_4 + A_5 + A_6$$

$$= (A_1 + A_2 + A_5 + A_6) - (A_3 + A_4)$$

Geometrically, this Riemann sum is the difference of two areas—the total area of the rectangles above the x-axis minus the total area of the rectangles below the x-axis. If we were to allow the number of subdivisions to increase so that $\max \Delta x_k \to 0$, then the rectangles would begin to fill out the regions between the curve $y = f(x)$ and the interval $[a, b]$ with less and less error (Figure 5.6.8). This suggests that for the function in the figure the limit of the Riemann sums would also be the difference of two areas, namely

$$\int_a^b f(x)\,dx = (A_I + A_{III}) - A_{II} = \begin{bmatrix} \text{area above} \\ [a, b] \end{bmatrix} - \begin{bmatrix} \text{area below} \\ [a, b] \end{bmatrix}$$

The area above an interval $[a, b]$ but below $y = f(x)$ minus the area below $[a, b]$ but above $y = f(x)$ will be called the *net signed area* between $y = f(x)$ and the interval $[a, b]$. More precisely, we make the following definition.

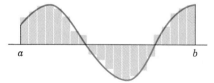

Figure 5.6.8

5.6.2 DEFINITION. If the function f is continuous on $[a, b]$, and can assume both positive and negative values, then the ***net signed area*** A between $y = f(x)$ and the interval $[a, b]$ is defined by

$$\int_a^b f(x)\,dx = \lim_{\max \Delta x_k \to 0} \sum_{k=1}^{n} f(x_k^*)\,\Delta x_k$$

The net signed area between $y = f(x)$ and $[a, b]$ can be positive, negative, or zero; it is positive when there is more area above the interval than below, negative when there is more area below than above, and zero when the area above and below are equal.

Example 2 Evaluate $\displaystyle\int_2^4 (1 - x)\,dx$.

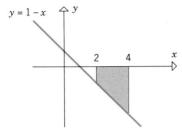

Figure 5.6.9

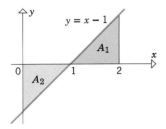

Figure 5.6.10

☐ **DEFINITE INTEGRALS OF FUNCTIONS WITH DISCONTINUITIES**

Solution. The integrand is negative over the integral [2, 4] (Figure 5.6.9), so the integral is the negative of the area of the trapezoid shaded in the figure. The area of that trapezoid is 4, so

$$\int_2^4 (x-1)\,dx = -4 \qquad \blacktriangleleft$$

Example 3 Use an appropriate area formula from plane geometry to evaluate the definite integral

$$\int_0^2 (x-1)\,dx$$

Solution. From Definition 5.6.2, the integral represents the net signed area between the line $y = x - 1$ and the interval [0, 2]. (See Figure 5.6.10.) Calculating the areas of the two triangular regions in the figure and subtracting yields

$$\int_0^2 (x-1)\,dx = A_1 - A_2 = \tfrac{1}{2} - \tfrac{1}{2} = 0 \qquad \blacktriangleleft$$

We noted earlier without proof that if f is continuous on $[a, b]$, then the limit of the Riemann sums must exist. Thus, in Definitions 5.6.1 and 5.6.2, the existence of a numerical value for the area and the net signed area was not an issue because we assumed continuity of f. However, for functions with discontinuities, the limit of the Riemann sums may or may not exist depending on the number and the nature of the discontinuities. We make the following definition.

5.6.3 DEFINITION. If the function f is defined on the closed interval $[a, b]$, then f is called **Riemann integrable** on $[a, b]$ or more simply **integrable** on $[a, b]$ if the limit

$$\lim_{\max \Delta x_k \to 0} \sum_{k=1}^n f(x_k^*)\,\Delta x_k$$

exists. If f is integrable on $[a, b]$, then we define the **definite integral** of f from a to b by

$$\int_a^b f(x)\,dx = \lim_{\max \Delta x_k \to 0} \sum_{k=1}^n f(x_k^*)\,\Delta x_k$$

In our discussion up to now, and in Definition 5.6.3, we assumed the lower limit of integration to be less than the upper limit of integration. The following definition extends the concept of the definite integral to allow for equal limits of integration and for a lower limit of integration that is greater than the upper limit of integration.

5.6.4 DEFINITION.

(a) If a is in the domain of f, we define

$$\int_a^a f(x)\,dx = 0$$

(b) If f is integrable on $[a, b]$, then we define

$$\int_b^a f(x)\,dx = -\int_a^b f(x)\,dx$$

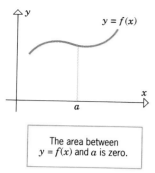

The area between
$y = f(x)$ and a is zero.

Figure 5.6.11

REMARK.   Part (a) of this definition is consistent with the intuitive idea that the area between a point on the x-axis and a curve $y = f(x)$ should be zero (Figure 5.6.11). Part (b) of the definition is simply a useful convention; it states that interchanging the limits of integration reverses the sign of the integral.

Example 4

(a) $\displaystyle \int_1^1 x^2\, dx = 0$

(b) $\displaystyle \int_1^0 \sqrt{1 - x^2}\, dx = -\int_0^1 \sqrt{1 - x^2}\, dx = -\frac{\pi}{4}$ ◀

Example 1(b)

☐ **PROPERTIES OF THE DEFINITE INTEGRAL**

The following properties of definite integrals follow from the definition of a definite integral. We shall omit the proofs.

> **5.6.5** THEOREM.   *If f and g are integrable on $[a, b]$ and if c is a constant, then cf, $f + g$, and $f - g$ are integrable on $[a, b]$ and*
>
> (a) $\displaystyle \int_a^b cf(x)\, dx = c \int_a^b f(x)\, dx$
>
> (b) $\displaystyle \int_a^b [f(x) + g(x)]\, dx = \int_a^b f(x)\, dx + \int_a^b g(x)\, dx$
>
> (c) $\displaystyle \int_a^b [f(x) - g(x)]\, dx = \int_a^b f(x)\, dx - \int_a^b g(x)\, dx$

Part (b) of this theorem can be extended to more than two functions. More precisely,

$$\int_a^b [f_1(x) + f_2(x) + \cdots + f_n(x)]\, dx$$
$$= \int_a^b f_1(x)\, dx + \int_a^b f_2(x)\, dx + \cdots + \int_a^b f_n(x)\, dx \tag{5}$$

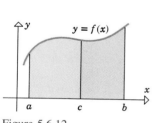

Figure 5.6.12

If f is continuous and nonnegative on $[a, b]$ and if c is a point between a and b, then it is evident that the area under $y = f(x)$ over the interval $[a, b]$ can be split into two parts, the area under the curve from a to c plus the area under the curve from c to b (Figure 5.6.12), that is,

$$\int_a^b f(x)\, dx = \int_a^c f(x)\, dx + \int_c^b f(x)\, dx$$

This is a special case of the following theorem about definite integrals, which we state without proof.

> **5.6.6** THEOREM.   *If f is integrable on a closed interval containing the three points a, b, and c, then*
>
> $$\int_a^b f(x)\, dx = \int_a^c f(x)\, dx + \int_c^b f(x)\, dx \tag{6}$$
>
> *no matter how the points are ordered.*

Example 5   Suppose that

$$\int_1^5 f(x)\, dx = -1, \quad \int_3^5 f(x)\, dx = 3, \quad \text{and} \quad \int_3^5 g(x)\, dx = 4$$

Find

(a) $\int_3^5 [2f(x) - g(x)]\, dx$ (b) $\int_1^3 f(x)\, dx$

Solution (a).   From Theorem 5.6.5

$$\int_3^5 [2f(x) - g(x)]\, dx = 2\int_3^5 f(x)\, dx - \int_3^5 g(x)\, dx = 2(3) - 4 = 2$$

Solution (b).   From Theorem 5.6.6 with $a = 1$, $b = 5$, and $c = 3$,

$$\int_1^5 f(x)\, dx = \int_1^3 f(x)\, dx + \int_3^5 f(x)\, dx$$

so

$$\int_1^3 f(x)\, dx = \int_1^5 f(x)\, dx - \int_3^5 f(x)\, dx = -1 - 3 = -4 \quad \blacktriangleleft$$

□ **INEQUALITIES INVOLVING DEFINITE INTEGRALS**

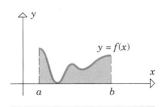

Net signed area ≥ 0

Figure 5.6.13

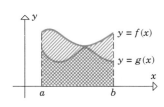

Area under f ≥ area under g

Figure 5.6.14

If f is a nonnegative continuous function on $[a, b]$, then the graph of $y = f(x)$ does not cross over the x-axis, so it is evident intuitively that the net signed area between $y = f(x)$ and the x-axis must be greater than or equal to zero (Figure 5.6.13). Similarly, if f and g are nonnegative continuous functions on $[a, b]$, and if $f(x) \geq g(x)$ for all x in $[a, b]$, then the area under $y = f(x)$ must be greater than or equal to the area under $y = g(x)$ over $[a, b]$, since the graph of $y = f(x)$ does not cross over the graph of $y = g(x)$ (Figure 5.6.14). The following theorem, which we state without proof, generalizes these observations to any integrable functions.

5.6.7   THEOREM.

(*a*)   *If f is integrable on $[a, b]$ and $f(x) \geq 0$ for all x in $[a, b]$, then*

$$\int_a^b f(x)\, dx \geq 0$$

(*b*)   *If f and g are integrable on $[a, b]$ and $f(x) \geq g(x)$ for all x in $[a, b]$, then*

$$\int_a^b f(x)\, dx \geq \int_a^b g(x)\, dx$$

REMARK.   If $b > a$, then this theorem remains true if $\geq$ is replaced by $>$, $\leq$, or $<$ throughout.

Part (*b*) of Theorem 5.6.7 states that one can integrate both sides of an inequality relating integrable functions without altering the sense of the inequality.

Example 6   Show that $\int_0^1 \dfrac{\cos x}{2x^3 - 5}\, dx$ is negative.

Solution.   Since $0 \leq x \leq 1$, it follows that the integrand is negative because $\cos x > 0$ and $2x^3 - 5 < 0$. Thus from part (*a*) of Theorem 5.6.7, with $\geq$ replaced by $<$, the integral is negative.   ◄

☐ CONDITIONS FOR
INTEGRABILITY

The problem of determining precisely which functions are integrable is quite complex and beyond the scope of this text. However, we will discuss a few results about integrability that are useful to know. We start with a definition.

> **5.6.8** DEFINITION.   A function f is said to be **bounded** on an interval $[a, b]$ if there is a positive number M such that
>
> $$-M \le f(x) \le M$$
>
> for all x in $[a, b]$. Geometrically, this means that the graph of f on the interval $[a, b]$ lies between the lines $y = -M$ and $y = M$.

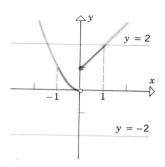

Figure 5.6.15

Example 7   The function

$$f(x) = \begin{cases} x^2, & x < 0 \\ x + 1, & x \ge 0 \end{cases}$$

is bounded on the interval $[-1, 1]$, since its graph on this interval lies between the lines $y = -2$ and $y = 2$ on this interval (Figure 5.6.15). However, the function

$$g(x) = \begin{cases} 1/x, & x \ne 0 \\ 0, & x = 0 \end{cases}$$

is not bounded on the interval $[-1, 1]$ (or on any closed interval containing the origin), since its graph on this interval cannot be enclosed between two horizontal lines (Figure 5.6.16). In general, a function that approaches $+\infty$ or $-\infty$ somewhere in an interval is not bounded on that interval.   ◄

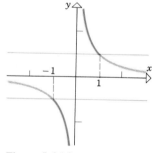

Figure 5.6.16

The following theorem, which we state without proof, lists three of the most important facts about integrability.

> **5.6.9** THEOREM.   *Let f be a function that is defined at all points in the interval $[a, b]$.*
>
> (a)   *If f is continuous on $[a, b]$, then f is integrable on $[a, b]$.*
> (b)   *If f is bounded on $[a, b]$ and has only finitely many points of discontinuity on $[a, b]$, then f is integrable on $[a, b]$.*
> (c)   *If f is not bounded on $[a, b]$, then f is not integrable on $[a, b]$.*

Example 8   The function f in Example 7 is integrable on the interval $[-1, 1]$ because it is bounded and has only finitely many points of discontinuity on this interval (one discontinuity). However, the function g in Example 7 is not integrable on $[-1, 1]$ because it is not bounded on this interval.   ◄

■ MORE ON LIMITS OF RIEMANN SUMS

Limits of Riemann sums are somewhat different from the kinds of limits discussed in Chapter 2. Loosely phrased, the expression

$$\lim_{\max \Delta x_k \to 0} \sum_{k=1}^{n} f(x_k^*) \, \Delta x_k = L$$

is intended to convey the idea that we can force the Riemann sums to be as close as we please to L, regardless of how $x_1^*, x_2^*, \dots, x_n^*$ are chosen, by making the mesh size of the partition sufficiently small. This idea is captured by the following definition.

5.6.10 DEFINITION.   We shall write

$$\lim_{\max \Delta x_k \to 0} \sum_{k=1}^{n} f(x_k^*)\, \Delta x_k = L \tag{7}$$

if given any number $\epsilon > 0$, there is a number $\delta > 0$ such that

$$\left| \sum_{k=1}^{n} f(x_k^*)\, \Delta x_k - L \right| < \epsilon$$

whenever $\max \Delta x_k < \delta$, regardless of how $x_1^*, x_2^*, \ldots, x_n^*$ are chosen.

It can be shown that when a number L satisfying (7) exists, it is unique (Exercise 41). We denote this number by

$$\lim_{\max \Delta x_k \to 0} \sum_{k=1}^{n} f(x_k^*)\, \Delta x_k = \int_{a}^{b} f(x)\, dx \tag{8}$$

REMARK.   Some writers use the symbol $\|\Delta\|$ rather than $\max \Delta x_k$ for the mesh size of the partition, in which case (8) would be written as

$$\lim_{\|\Delta\| \to 0} \sum_{k=1}^{n} f(x_k^*)\, \Delta x_k = \int_{a}^{b} f(x)\, dx$$

▶ Exercise Set 5.6

In Exercises 1–4, find the value of

(a) $\displaystyle\sum_{k=1}^{n} f(x_k^*)\, \Delta x_k$ (b) $\max \Delta x_k$.

1. $f(x) = x + 1$; $a = 0, b = 4$; $n = 3$;
    $\Delta x_1 = 1, \Delta x_2 = 1$; $\Delta x_3 = 2$;
    $x_1^* = \frac{1}{3}, x_2^* = \frac{3}{2}, x_3^* = 3$.

2. $f(x) = \cos x$; $a = 0, b = 2\pi$; $n = 4$;
    $\Delta x_1 = \pi/2, \Delta x_2 = 3\pi/4, \Delta x_3 = \pi/2, \Delta x_4 = \pi/4$;
    $x_1^* = \pi/4, x_2^* = \pi, x_3^* = 3\pi/2, x_4^* = 7\pi/4$.

3. $f(x) = 4 - x^2$; $a = -3, b = 4$; $n = 4$;
    $\Delta x_1 = 1, \Delta x_2 = 2, \Delta x_3 = 1, \Delta x_4 = 3$;
    $x_1^* = -\frac{5}{2}, x_2^* = -1, x_3^* = \frac{1}{4}, x_4^* = 3$.

4. $f(x) = x^3$; $a = -3, b = 3$; $n = 4$;
    $\Delta x_1 = 2, \Delta x_2 = 1, \Delta x_3 = 1, \Delta x_4 = 2$;
    $x_1^* = -2, x_2^* = 0, x_3^* = 0, x_4^* = 2$.

5. Find the smallest and largest values that the Riemann sum

    $$\sum_{k=1}^{3} f(x_k^*)\, \Delta x_k$$

    can have on the interval $[0, 4]$ if $f(x) = x^2 - 3x + 4$ and $\Delta x_1 = 1, \Delta x_2 = 2, \Delta x_3 = 1$.

In Exercises 6–8, use the given values of a and b to express the following limits as definite integrals. (Do not evaluate the integrals.)

6. $\displaystyle\lim_{\max \Delta x_k \to 0} \sum_{k=1}^{n} (x_k^*)^3\, \Delta x_k$; $a = 1, b = 2$.

7. $\displaystyle\lim_{\max \Delta x_k \to 0} \sum_{k=1}^{n} 4x_k^*(1 - 3x_k^*)\, \Delta x_k$; $a = -3, b = 3$.

8. $\displaystyle\lim_{\max \Delta x_k \to 0} \sum_{k=1}^{n} (\sin^2 x_k^*)\, \Delta x_k$; $a = 0, b = \pi/2$.

In Exercises 9–11, express the definite integrals as limits. (Do not evaluate.)

9. $\displaystyle\int_{1}^{2} 2x\, dx$.   10. $\displaystyle\int_{-\pi/2}^{\pi/2} (1 + \cos x)\, dx$.

11. $\displaystyle\int_{0}^{1} \frac{x}{x + 1}\, dx$.

12. Show on a sketch the area that is represented by

    (a) $\displaystyle\int_{1}^{4} \sqrt{x}\, dx$ (b) $\displaystyle\int_{1}^{3} \frac{1}{x}\, dx$

    (c) $\displaystyle\int_{-1}^{2} \sqrt{9 - x^2}\, dx$ (d) $\displaystyle\int_{0}^{\pi/2} \sin x\, dx$.

13. Given the areas shown in Figure 5.6.17, find

    (a) $\displaystyle\int_{a}^{b} f(x)\, dx$ (b) $\displaystyle\int_{b}^{c} f(x)\, dx$

    (c) $\displaystyle\int_{a}^{c} f(x)\, dx$ (d) $\displaystyle\int_{a}^{d} f(x)\, dx$.

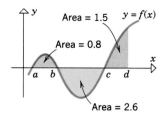

Figure 5.6.17

In Exercises 14–27, evaluate the definite integrals by using area formulas from plane geometry, where needed.

14. (a) $\int_0^3 x\,dx$
    (b) $\int_{-2}^{-1} x\,dx$
    (c) $\int_{-1}^4 x\,dx$
    (d) $\int_{-5}^5 x\,dx.$

15. Let $f(x) = 1 - \frac{1}{2}x.$
    (a) $\int_0^2 f(x)\,dx$
    (b) $\int_{-1}^1 f(x)\,dx$
    (c) $\int_2^3 f(x)\,dx$
    (d) $\int_0^3 f(x)\,dx.$

16. $\int_{-10}^{-5} 6\,dx.$

17. $\int_0^2 (2 - 4x)\,dx.$

18. $\int_0^3 |x - 2|\,dx.$

19. $\int_{-1}^2 |2x - 3|\,dx.$

20. $\int_0^2 \sqrt{4 - x^2}\,dx.$

21. $\int_{-1}^1 \sqrt{1 - x^2}\,dx.$

22. $\int_{-3}^0 (2 + \sqrt{9 - x^2})\,dx.$

23. $\int_0^{10} \sqrt{10x - x^2}\,dx.$ [*Hint:* Complete the square.]

24. $\int_{-\pi/3}^{\pi/3} \sin x\,dx.$

25. $\int_0^\pi \cos x\,dx.$

26. $\int_0^3 f(x)\,dx,$ where
$$f(x) = \begin{cases} 2x, & x \le 1 \\ 2, & x > 1 \end{cases}$$

27. $\int_{-2}^2 f(x)\,dx,$ where
$$f(x) = \begin{cases} 3, & x \le 0 \\ x + 3, & x > 0 \end{cases}$$

28. The function $f(x) = \sqrt{x}$ is continuous on $[0, 4]$, so by Theorem 5.6.9 the integral $\int_0^4 \sqrt{x}\,dx$ exists. Find its value by using Definition 5.6.3. Use subintervals of unequal length by taking the points
$$4k^2/n^2 \quad (k = 1, 2, \ldots, n - 1)$$
as the points of subdivision of $[0, 4]$, and let x_k^* be the right-hand endpoint of the kth subinterval.

29. Find $\int_{-1}^2 [f(x) + 2g(x)]\,dx$ if
$$\int_{-1}^2 f(x)\,dx = 5 \quad \text{and} \quad \int_{-1}^2 g(x)\,dx = -3$$

30. Find $\int_1^4 [3f(x) - g(x)]\,dx$ if
$$\int_1^4 f(x)\,dx = 2 \quad \text{and} \quad \int_1^4 g(x)\,dx = 10$$

31. Find $\int_1^5 f(x)\,dx$ if
$$\int_0^1 f(x)\,dx = -2 \quad \text{and} \quad \int_0^5 f(x)\,dx = 1$$

32. Find $\int_3^{-2} f(x)\,dx$ if
$$\int_{-2}^1 f(x)\,dx = 2 \quad \text{and} \quad \int_1^3 f(x)\,dx = -6$$

In Exercises 33–38, determine whether the value of the integral is positive or negative. [*Hint:* See Theorem 5.6.7.]

33. $\int_2^3 \frac{\sqrt{x}}{1 - x}\,dx.$

34. $\int_{-3}^{-1} \frac{x^4}{\sqrt{3 - x}}\,dx.$

35. $\int_0^4 \frac{x^2}{3 - \cos x}\,dx.$

36. $\int_{-2}^2 \frac{x^3 - 9}{|x| + 1}\,dx.$

37. $\int_2^0 x^2 \sin \sqrt{x}\,dx.$

38. $\int_0^{-1} \sqrt[3]{x^2 - 2}\,dx.$

In Exercises 39 and 40, use Theorem 5.6.9 to determine whether the function is integrable on the given interval.

39. $f(x) = \begin{cases} 1/x^2, & x > 0 \\ 0, & x = 0 \end{cases}$ on $[0, 1].$

40. $f(x) = \begin{cases} \sin 1/x, & x \ne 0 \\ 0, & x = 0 \end{cases}$ on $[-1, 1].$

41. Prove: If f is integrable on $[a, b]$, then the value of $\int_a^b f(x)\,dx$ is unique. [*Hint:* Suppose there are two different values, L_1 and L_2, and obtain a contradiction.]

42. Prove: If f is integrable on $[a, b]$ and c is any constant, then cf is integrable on $[a, b]$ and
$$\int_a^b cf(x)\,dx = c \int_a^b f(x)\,dx$$

43. Prove: If f and g are integrable on $[a, b]$, then $f + g$ is integrable on $[a, b]$ and
$$\int_a^b [f(x) + g(x)]\,dx = \int_a^b f(x)\,dx + \int_a^b g(x)\,dx$$

44. Prove that the function
$$f(x) = \begin{cases} 1 & \text{if } x \text{ is rational} \\ 0 & \text{if } x \text{ is irrational} \end{cases}$$
is not integrable on any closed interval $[a, b]$. [*Hint:* Every interval contains rational and irrational numbers.]

45. Assuming the functions involved are all integrable on $[a, b]$, and $c_1, c_2, \ldots, c_k$ are constants, which of the following are always valid?

(a) $\displaystyle\int_a^b f(x)g(x)\,dx = \int_a^b f(x)\,dx \int_a^b g(x)\,dx$

(b) $\displaystyle\int_a^b [c_1 f(x) + c_2 g(x)]\,dx$
$$= c_1 \int_a^b f(x)\,dx + c_2 \int_a^b g(x)\,dx$$

(c) $\displaystyle\int_a^b \left(\sum_{k=1}^n c_k f_k(x) \right) dx = \sum_{k=1}^n \left[c_k \int_a^b f_k(x)\,dx \right]$

(d) $\displaystyle\int_a^b [f(x)]^n\,dx = \left[\int_a^b f(x)\,dx \right]^n$

(e) $\displaystyle\int_a^b \sqrt{f(x)}\,dx = \sqrt{\int_a^b f(x)\,dx}.$

46. Express Equation (5) in sigma notation.

5.7 THE FIRST FUNDAMENTAL THEOREM OF CALCULUS

In the preceding section we defined the concept of a definite integral but did not give any general methods for evaluating them. In this section we shall give a method for using antiderivatives to evaluate definite integrals.

☐ **THE FIRST FUNDAMENTAL THEOREM OF CALCULUS**

The following theorem is the basic tool for evaluating definite integrals.

> **5.7.1** THEOREM (*The First Fundamental Theorem of Calculus*). *If f is continuous on $[a, b]$ and if F is an antiderivative of f on $[a, b]$, then*
> $$\int_a^b f(x)\,dx = F(b) - F(a) \tag{1}$$

Proof. Let $x_1, x_2, \ldots, x_{n-1}$ be any points in $[a, b]$ such that
$$a < x_1 < x_2 < \cdots < x_{n-1} < b$$

These points divide $[a, b]$ into n subintervals
$$[a, x_1], [x_1, x_2], \ldots, [x_{n-1}, b] \tag{2}$$

whose lengths, as usual, we denote by
$$\Delta x_1, \Delta x_2, \ldots, \Delta x_n \tag{3}$$

By hypothesis, $F'(x) = f(x)$ for all x in $[a, b]$, so F satisfies the hypotheses of the Mean-Value Theorem (4.9.2) on each subinterval in (2). Hence, by the Mean-Value Theorem we can find points $x_1^*, x_1^*, \ldots, x_n^*$ in the respective subintervals in (2) such that

$$F(x_1) - F(a) = F'(x_1^*)(x_1 - a) = f(x_1^*)\,\Delta x_1$$

$$F(x_2) - F(x_1) = F'(x_2^*)(x_2 - x_1) = f(x_2^*)\,\Delta x_2$$

$$F(x_3) - F(x_2) = F'(x_3^*)(x_3 - x_2) = f(x_3^*)\,\Delta x_3$$

$$\vdots \qquad\qquad \vdots \qquad\qquad \vdots$$

$$F(b) - F(x_n) = F'(x_n^*)(b - x_{n-1}) = f(x_n^*)\,\Delta x_n$$

Adding the preceding equations yields

$$F(b) - F(a) = \sum_{k=1}^n f(x_k^*)\,\Delta x_k \tag{4}$$

Let us now increase n in such a way that $\max \Delta x_k \to 0$. Since f is assumed to be continuous, the right side of (4) approaches $\int_a^b f(x)\,dx$, by Theorem 5.6.9(a). However, the left side of (4) is a constant that is independent of n; thus,

$$F(b) - F(a) = \lim_{\max \Delta x_k \to 0} \sum_{k=1}^{n} f(x_k^*)\,\Delta x_k = \int_a^b f(x)\,dx \quad \blacksquare$$

The difference $F(b) - F(a)$ is commonly denoted by $F(x)\big]_a^b$ so that (1) can be written as

$$\int_a^b f(x)\,dx = F(x)\bigg]_a^b \tag{5}$$

Some other common notations are

$$\int_a^b f(x)\,dx = \left[F(x)\right]_a^b \quad \text{and} \quad \int_a^b f(x)\,dx = F(x)\bigg]_{x=a}^b$$

The latter notation emphasizes that the limits of integration refer to the variable x. This can be important in problems where more than one variable occurs.

Example 1 Evaluate $\displaystyle\int_1^2 x\,dx$.

Solution. The function $F(x) = \frac{1}{2}x^2$ is an antiderivative of $f(x) = x$; thus, from (5)

$$\int_1^2 x\,dx = \frac{1}{2}x^2\bigg]_1^2 = \frac{1}{2}(2)^2 - \frac{1}{2}(1)^2 = 2 - \frac{1}{2} = \frac{3}{2} \quad \blacktriangleleft$$

REMARK.   The bracket notation $[F(x)]_a^b$ has some properties in common with the definite integral. We leave it as an exercise to show that

$$[cF(x)]_a^b = c[F(x)]_a^b$$

$$[F(x) + G(x)]_a^b = F(x)]_a^b + G(x)]_a^b$$

$$[F(x) - G(x)]_a^b = F(x)]_a^b - G(x)]_a^b$$

□ **THE RELATIONSHIP BETWEEN DEFINITE AND INDEFINITE INTEGRALS**

When applying the First Fundamental Theorem of Calculus, it does not matter which antiderivative of f is used, for if F is any antiderivative of f on $[a, b]$, then all others have the form

$$F(x) + C \quad \boxed{\text{Theorem 5.2.2}}$$

Thus,

$$[F(x) + C]_a^b = [F(b) + C] - [F(a) + C]$$

$$= F(b) - F(a) = F(x)]_a^b$$

$$= \int_a^b f(x)\,dx \tag{6}$$

which shows that all antiderivatives of f on $[a, b]$ yield the same value for $\int_a^b f(x)\,dx$. Since

$$\int f(x)\,dx = F(x) + C$$

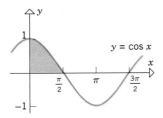

Figure 5.7.1

it follows from (6) that

$$\int_a^b f(x)\,dx = \left[\int f(x)\,dx\right]_a^b \qquad (7)$$

which relates the definite and indefinite integrals of f.

Example 2   Use the First Fundamental Theorem of Calculus to find the area under the curve $y = \cos x$ over the interval $[0, \pi/2]$. (See Figure 5.7.1.)

Solution.   Since $\cos x \ge 0$ for $0 \le x \le \pi/2$, the area is

$$A = \int_0^{\pi/2} \cos x\,dx = \left[\int \cos x\,dx\right]_0^{\pi/2}$$

From (7)

$$= \sin x \Big]_0^{\pi/2} = \sin \frac{\pi}{2} - \sin 0 = 1 \qquad \blacktriangleleft$$

REMARK.   In the third equality of this example we took the constant of integration for the indefinite integral to be $C = 0$. This is justified because we can select any antiderivative of f on $[a, b]$, in particular the one for which $C = 0$.

Example 3   Evaluate $\displaystyle\int_0^3 (x^3 - 4x + 1)\,dx$.

Solution.

$$\int_0^3 (x^3 - 4x + 1)\,dx = \left[\frac{x^4}{4} - 4 \cdot \frac{x^2}{2} + x\right]_0^3$$
$$= \left(\tfrac{81}{4} - 18 + 3\right) - (0) = \tfrac{21}{4} \qquad \blacktriangleleft$$

Formula (1) in Theorem 5.7.1 is applicable in the cases where $a = b$ or $b < a$. The following example illustrates this.

Example 4

(a) $\displaystyle\int_1^1 x^2\,dx = \frac{x^3}{3}\bigg]_1^1 = \frac{1}{3} - \frac{1}{3} = 0$

(b) $\displaystyle\int_4^0 x\,dx = \frac{x^2}{2}\bigg]_4^0 = \left[\frac{0}{2} - \frac{16}{2}\right] = -8$

This is consistent with the result that would be obtained by first reversing the limits of integration in accordance with Definition 5.6.4(b):

$$\int_4^0 x\,dx = -\int_0^4 x\,dx = -\frac{x^2}{2}\bigg]_0^4 = -\left[\frac{16}{2} - \frac{0}{2}\right] = -8 \qquad \blacktriangleleft$$

Example 5   Evaluate $\displaystyle\int_0^6 f(x)\,dx$ if

$$f(x) = \begin{cases} x^2, & x < 2 \\ 3x - 2, & x \ge 2 \end{cases}$$

Solution.   From Theorem 5.6.6

$$\int_0^6 f(x)\,dx = \int_0^2 f(x)\,dx + \int_2^6 f(x)\,dx = \int_0^2 x^2\,dx + \int_2^6 (3x-2)\,dx$$

$$= \frac{x^3}{3}\Big]_0^2 + \left[\frac{3x^2}{2} - 2x\right]_2^6 = \left(\frac{8}{3} - 0\right) + (42-2) = \frac{128}{3} \quad \blacktriangleleft$$

Example 6   Evaluate $\displaystyle\int_{-1}^2 |x|\,dx$.

Solution.   Since $|x| = x$ when $x \geq 0$ and $|x| = -x$ when $x \leq 0$,

$$\int_{-1}^2 |x|\,dx = \int_{-1}^0 |x|\,dx + \int_0^2 |x|\,dx$$

$$= \int_{-1}^0 (-x)\,dx + \int_0^2 x\,dx$$

$$= -\frac{x^2}{2}\Big]_{-1}^0 + \frac{x^2}{2}\Big]_0^2 = \frac{1}{2} + 2 = \frac{5}{2} \quad \blacktriangleleft$$

REMARK.   Theorem 5.7.1 is only applicable to those continuous functions that have antiderivatives. We have not yet addressed the question of which functions actually have antiderivatives. This matter will be taken up in Section 5.9, at which time we will show that *all* continuous functions have antiderivatives.

☐ **THE MEAN-VALUE THEOREM FOR INTEGRALS**

Our next objective is to obtain a result called the *Mean-Value Theorem for Integrals*. In this section we will use this result to define the notion of "average value" for continuous functions, and in later sections we will use it to derive some very fundamental theorems.

Let f be a continuous nonnegative function on $[a, b]$, and let m and M be the minimum and maximum values of $f(x)$ on this interval. Consider the rectangles of heights m and M over the interval $[a, b]$ (Figure 5.7.2). It is clear geometrically from this figure that the area

$$A = \int_a^b f(x)\,dx$$

under $y = f(x)$ is at least as large as the area of the rectangle of height m and no larger than the area of the rectangle of height M. It seems reasonable, therefore, that there is a rectangle over the interval $[a, b]$ of some appropriate height $f(x^*)$ between m and M whose area is precisely A; that is,

$$\int_a^b f(x)\,dx = f(x^*)(b-a)$$

(Figure 5.7.3). This is a special case of the following result.

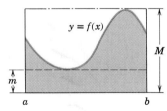

Figure 5.7.2

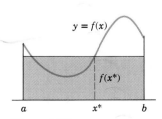

Figure 5.7.3

5.7.2   THEOREM (*The Mean-Value Theorem for Integrals*).   *If f is continuous on a closed interval $[a, b]$, then there is at least one number x^* in $[a, b]$ such that*

$$\int_a^b f(x)\,dx = f(x^*)(b-a) \tag{8}$$

Proof.   By the Extreme-Value Theorem (4.6.4), f assumes a maximum value M and a minimum value m on $[a, b]$. Thus, for all x in $[a, b]$,

$$m \leq f(x) \leq M$$

and from Theorem 5.6.7(*b*)

$$\int_a^b m \, dx \le \int_a^b f(x) \, dx \le \int_a^b M \, dx$$

or

$$m(b - a) \le \int_a^b f(x) \, dx \le M(b - a)$$

or

$$m \le \frac{1}{b - a} \int_a^b f(x) \, dx \le M \tag{9}$$

Since (9) states that

$$\frac{1}{b - a} \int_a^b f(x) \, dx \tag{10}$$

is a number between *m* and *M*, and since $f(x)$ assumes the values *m* and *M* on $[a, b]$, it follows from the Intermediate-Value Theorem (2.7.9) that $f(x)$ must assume the value (10) at some point x^* in $[a, b]$; that is,

$$\frac{1}{b - a} \int_a^b f(x) \, dx = f(x^*) \quad \text{or} \quad \int_a^b f(x) \, dx = f(x^*)(b - a) \quad \blacksquare$$

Example 7   Since $f(x) = x^2$ is continuous on the interval $[1, 4]$, the Mean-Value Theorem for Integrals guarantees that there is a number x^* in $[1, 4]$ such that

$$\int_1^4 x^2 \, dx = f(x^*)(4 - 1) = (x^*)^2(4 - 1) = 3(x^*)^2$$

But

$$\int_1^4 x^2 \, dx = \left.\frac{x^3}{3}\right]_1^4 = 21$$

so that

$$3(x^*)^2 = 21 \quad \text{or} \quad (x^*)^2 = 7 \quad \text{or} \quad x^* = \pm\sqrt{7}$$

Thus, $x^* = \sqrt{7} \approx 2.65$ is the number in the interval $[1, 4]$ whose existence is guaranteed by the Mean-Value Theorem for Integrals.   ◀

☐ **AVERAGE VALUE**

The number $f(x^*)$ in Theorem 5.7.2 is closely related to the familiar notion of an *arithmetic average*. To see this, divide the interval $[a, b]$ into *n* subintervals of equal length

$$\Delta x = \frac{b - a}{n} \tag{11}$$

and choose arbitrary points $x_1^*, x_2^*, \ldots, x_n^*$ in successive subintervals. Then the arithmetic average of the numbers $f(x_1^*), f(x_2^*), \ldots, f(x_n^*)$ is

$$\text{ave} = \frac{1}{n}[f(x_1^*) + f(x_2^*) + \cdots + f(x_n^*)]$$

or from (11)

$$\text{ave} = \frac{1}{b - a}[f(x_1^*) \Delta x + f(x_2^*) \Delta x + \cdots + f(x_n^*) \Delta x]$$

$$= \frac{1}{b - a} \sum_{k=1}^{n} f(x_k^*) \Delta x$$

Taking the limit as $n \to +\infty$ yields

$$\lim_{n \to +\infty} \frac{1}{b-a} \sum_{k=1}^{n} f(x_k^*) \, \Delta x = \frac{1}{b-a} \int_a^b f(x) \, dx$$

Since this equation describes what happens when we compute the average of "more and more" values of $f(x)$, we are led to the following definition.

> **5.7.3** DEFINITION.   If f is integrable on $[a, b]$, then the **average value** (or **mean value**) of f on $[a, b]$ is defined to be
>
> $$f_{\text{ave}} = \frac{1}{b-a} \int_a^b f(x) \, dx$$
>
> If $y = f(x)$, then f_{ave} is also called the **average value of y with respect to x** over $[a, b]$.

Example 8   Find f_{ave} over the interval $[1, 4]$ for $f(x) = x^2$.

Solution.

$$f_{\text{ave}} = \frac{1}{b-a} \int_a^b f(x) \, dx = \frac{1}{4-1} \int_1^4 x^2 \, dx = \tfrac{1}{3}(21) = 7 \qquad \blacktriangleleft$$

REMARK.   In light of Definition 5.7.3, the quantity $f(x^*)$ in the Mean-Value Theorem for Integrals (5.7.2) is just the average value of $f(x)$ over $[a, b]$. Thus, the Mean-Value Theorem for Integrals can be stated in the form

$$\int_a^b f(x) \, dx = (b-a) f_{\text{ave}}$$

▶ Exercise Set 5.7   Ⓒ 29, 30, 41, 42

In Exercises 1–22, evaluate the definite integrals using the First Fundamental Theorem of Calculus.

1. $\displaystyle\int_2^3 x^3 \, dx.$

2. $\displaystyle\int_{-1}^1 x^4 \, dx.$

3. $\displaystyle\int_{-1}^2 x(1 + x^3) \, dx.$

4. $\displaystyle\int_{-3}^0 (x^2 - 4x + 7) \, dx.$

5. $\displaystyle\int_1^2 (t^2 - 2t + 8) \, dt.$

6. $\displaystyle\int_0^1 (x^5 - x^3 + 2x) \, dx.$

7. $\displaystyle\int_1^3 \frac{1}{x^2} \, dx.$

8. $\displaystyle\int_1^2 \frac{1}{x^6} \, dx.$

9. $\displaystyle\int_1^2 \left(\frac{1}{x^3} - \frac{2}{x^2} + x^{-4} \right) dx.$

10. $\displaystyle\int_{-2}^{-1} \left(u^{-4} + 3u^{-2} - \frac{1}{u^5} \right) du.$

11. $\displaystyle\int_1^9 \sqrt{x} \, dx.$

12. $\displaystyle\int_1^4 x^{-3/5} \, dx.$

13. $\displaystyle\int_4^9 2y\sqrt{y} \, dy.$

14. $\displaystyle\int_1^8 (5x^{2/3} - 4x^{-2}) \, dx.$

15. $\displaystyle\int_1^4 \left(\frac{3}{\sqrt{x}} - 5\sqrt{x} - x^{-3/2} \right) dx.$

16. $\displaystyle\int_4^9 (4y^{-1/2} + 2y^{1/2} + y^{-5/2}) \, dy.$

17. $\displaystyle\int_{-\pi/2}^{\pi/2} \sin \theta \, d\theta.$

18. $\displaystyle\int_0^{\pi/4} \sec^2 \theta \, d\theta.$

19. $\displaystyle\int_{-\pi/4}^{\pi/4} \cos x \, dx.$

20. $\displaystyle\int_0^1 (x - \sec x \tan x) \, dx.$

21. $\displaystyle\int_{\pi/6}^{\pi/2} \left(x + \frac{2}{\sin^2 x} \right) dx.$

22. $\displaystyle\int_a^{4a} (a^{1/2} - x^{1/2})\, dx$ (*a* a positive constant).

In Exercises 23–28, use Theorem 5.6.6 to evaluate the integrals.

23. $\displaystyle\int_0^2 |2x - 3|\, dx.$ **24.** $\displaystyle\int_1^5 |x - 2|\, dx.$

25. $\displaystyle\int_0^{3\pi/4} |\cos x|\, dx.$ **26.** $\displaystyle\int_{-1}^2 \sqrt{2 + |x|}\, dx.$

27. $\displaystyle\int_{-2}^3 f(x)\, dx$, where $f(x) = \begin{cases} -x, & x \geq 0 \\ x^2, & x < 0. \end{cases}$

28. $\displaystyle\int_0^4 f(x)\, dx$, where $f(x) = \begin{cases} \sqrt{x}, & 0 \leq x < 1 \\ 1/x^2, & x \geq 1. \end{cases}$

In Exercises 29 and 30, find the midpoint approximation of the integral using $n = 20$ subintervals, then find the exact value of the integral using the First Fundamental Theorem of Calculus.

29. $\displaystyle\int_1^3 \frac{1}{x^2}\, dx.$ **30.** $\displaystyle\int_0^{\pi/2} \sin x\, dx.$

In Exercises 31 and 32, find the limit by evaluating an appropriate definite integral. [*Hint:* Interpret the expression as a limit of Riemann sums in which the interval [0, 1] is divided into subintervals of equal width.]

31. $\displaystyle\lim_{n \to +\infty} \frac{\sqrt{1} + \sqrt{2} + \sqrt{3} + \cdots + \sqrt{n}}{n^{3/2}}.$

32. $\displaystyle\lim_{n \to +\infty} \frac{1^4 + 2^4 + 3^4 + \cdots + n^4}{n^5}.$

33. Find the area under the curve $y = x^2 + 1$ over the interval [0, 3]. Make a sketch of the region.

34. Find the area that is above the *x*-axis, but below the curve $y = (1 - x)(x - 2)$. Make a sketch of the region.

35. Find the area under the curve $y = 3 \sin x$ over the interval $[0, 2\pi/3]$. Sketch the region.

36. Find the area below the interval $[-2, -1]$, but above the curve $y = x^3$. Make a sketch of the region.

37. Find the total area between the curve $y = x^2 - 3x - 10$ and the interval $[-3, 8]$. Make a sketch of the region. [*Hint:* Find the portion of area above the interval and the portion of area below the interval separately.]

38. Prove:
(a) $[cF(x)]_a^b = c[F(x)]_a^b.$
(b) $[F(x) + G(x)]_a^b = F(x)]_a^b + G(x)]_a^b$
(c) $[F(x) - G(x)]_a^b = F(x)]_a^b - G(x)]_a^b$

39. Use the fact that
$$\frac{x}{x^4 + 1} < \frac{1}{x^3} \quad \text{for} \quad 1 \leq x \leq 2$$

to show that
$$\int_1^2 \frac{x}{x^4 + 1}\, dx < \frac{3}{8}$$

40. Use the fact that
$$\frac{1}{\sqrt{x^3 + 1}} < \frac{1}{x^{3/2}} \quad \text{for} \quad 4 \leq x \leq 9$$
to show that
$$\int_4^9 \frac{1}{\sqrt{x^3 + 1}}\, dx < \frac{1}{3}$$

41. Use the inequality
$$\frac{1}{x^{1.01}} \leq \frac{1}{x} \leq \frac{1}{x^{0.99}}$$
which holds for $x \geq 1$, to find numbers *m* and *M* such that $m < M$ and
$$m \leq \int_1^3 \frac{1}{x}\, dx \leq M$$
[*Note:* A calculator will be helpful.]

42. Repeat Exercise 41 using the inequality
$$\frac{1}{x^{1.001}} \leq \frac{1}{x} \leq \frac{1}{x^{0.999}}$$
which holds for $x \geq 1$.

In Exercises 43–49, find the average value of the function over the given interval.

43. $f(x) = 3x$; [1, 3].   **44.** $f(x) = x^2$; [−1, 2].
45. $f(x) = \sin x$; $[0, \pi]$.   **46.** $f(x) = \cos x$; $[0, \pi]$.
47. $f(x) = \sqrt{x}$; [0, 4].   **48.** $f(x) = 1/\sqrt{x}$; [1, 4].
49. $f(x) = \sqrt{4 - x^2}$; [−2, 2]. [*Hint:* Evaluate the integral by using a formula from geometry.]

50. (a) Find f_{ave} of $f(x) = 2x$ over [0, 4].
(b) Find a point x^* in [0, 4] such that $f(x^*) = f_{\text{ave}}$.
(c) Sketch the graph of $f(x) = 2x$ over [0, 4] and construct a rectangle over the interval whose area is the same as the area under the graph of *f* over the interval.

51. (a) Find f_{ave} of $f(x) = x^2$ over [0, 2].
(b) Find a point x^* in [0, 2] such that $f(x^*) = f_{\text{ave}}$.
(c) Sketch the graph of $f(x) = x^2$ over [0, 2] and construct a rectangle over the interval whose area is the same as the area under the graph of *f* over the interval.

In Exercises 52–55, find the average value of $f(x)$ over the interval and find all values of x^* described in the Mean-Value Theorem for Integrals.

52. $f(x) = 1/x^2$; [1, 3].   **53.** $f(x) = \sqrt{x}$; [0, 9].
54. $f(x) = \sin x$; $[-\pi, \pi]$.
55. $f(x) = \alpha x + \beta$ $(\alpha \neq 0)$; $[x_0, x_1]$.

56. Let $s(t)$ be the position function of a particle moving on a coordinate line. In Section 4.10 (Exercise 24) we defined the average velocity v_{ave} over a time interval $[t_0, t_1]$ by

$$v_{ave} = \frac{\text{change in position}}{\text{time elapsed}} = \frac{s(t_1) - s(t_0)}{t_1 - t_0}$$

(a) Show that

$$v_{ave} = \frac{1}{t_1 - t_0} \int_{t_0}^{t_1} v(t)\, dt$$

so that v_{ave} can also be interpreted as the average value of the velocity function over the time interval $[t_0, t_1]$.

(b) Use the result in part (a) to compute the average velocity over the time interval $[0, 5]$ for a particle moving on a coordinate line with velocity function $v(t) = 32t$.

57. For a particle moving on a coordinate line, show that the average acceleration a_{ave} over a time interval $[t_0, t_1]$ (as defined in Exercise 24 of Section 4.10) can be written as

$$a_{ave} = \frac{1}{t_1 - t_0} \int_{t_0}^{t_1} a(t)\, dt$$

which is the average of the acceleration function over the time interval $[t_0, t_1]$.

58. Consider a particle moving on a coordinate line. Use Exercises 56 and 57 to find

(a) v_{ave} for $1 \le t \le 4$ if $v(t) = 3t^3 + 2$

(b) a_{ave} for $2 \le t \le 9$ if $a(t) = t^{1/2}$

(c) v_{ave} for $t_0 \le t \le t_1$ if $v(t) = 32t + v_0$
   (v_0 constant).

59. Water is run at a constant rate of $1\ \text{ft}^3/\text{min}$ to fill a cylindrical tank of radius 3 ft and height 5 ft. Assuming the tank is initially empty, find the average force on the bottom over the time period required to fill the tank (density of water $= 62.4\ \text{lb/ft}^3$).

60. Prove: If $f(x) = k$ is constant on $[a, b]$, then $f_{ave} = k$ on $[a, b]$.

61. (a) Prove: If f is continuous on $[a, b]$, then

$$\int_a^b [f(x) - f_{ave}]\, dx = 0$$

(b) Does there exist a constant $c \neq f_{ave}$ such that

$$\int_a^b [f(x) - c]\, dx = 0?$$

5.8 EVALUATING DEFINITE INTEGRALS BY SUBSTITUTION; APPROXIMATION BY RIEMANN SUMS

In this section we shall discuss methods of evaluating definite integrals for which a substitution is needed, and we shall discuss methods for approximating the value of a definite integral using Riemann sums.

We shall consider two methods for evaluating a definite integral

$$\int_a^b h(x)\, dx$$

by substitution.

Method 1

First evaluate the indefinite integral

$$\int h(x)\, dx$$

by substitution, as discussed in Section 5.3, and then use the relationship

$$\int_a^b h(x)\, dx = \left[\int h(x)\, dx \right]_a^b$$

to evaluate the definite integral.

Method 2

Avoid the indefinite integral altogether by first expressing the definite integral in the form

$$\int_a^b h(x)\,dx = \int_a^b f(g(x))g'(x)\,dx \tag{1}$$

and then making the substitution

$$u = g(x) \quad \text{and} \quad du = g'(x)\,dx$$

directly into the definite integral (1). However, to do this we must change the x-limits of integration to corresponding u-limits of integration: Since $u = g(x)$, it follows that

$$u = g(a) \quad \text{if} \quad x = a$$

$$u = g(b) \quad \text{if} \quad x = b$$

Thus, if (1) is expressed in terms of u, we obtain

$$\int_a^b h(x)\,dx = \int_{g(a)}^{g(b)} f(u)\,du$$

With a good choice for the substitution, the new definite integral involving u may be easier to evaluate than the original.

Example 1 Use the two methods above to evaluate $\displaystyle\int_0^2 x(x^2 + 1)^3\,dx$.

Method 1

If we let

$$u = x^2 + 1 \quad \text{so that} \quad du = 2x\,dx \tag{2}$$

then we obtain

$$\int x(x^2 + 1)^3\,dx = \frac{1}{2}\int u^3\,du = \frac{u^4}{8} + C = \frac{(x^2 + 1)^4}{8} + C$$

Thus,

$$\int_0^2 x(x^2 + 1)^3\,dx = \left[\int x(x^2 + 1)^3\,dx\right]_{x=0}^2 = \frac{(x^2 + 1)^4}{8}\Bigg]_{x=0}^2$$

$$= \frac{625}{8} - \frac{1}{8} = 78$$

Method 2

For the substitution in (2) we have

$$u = 1 \quad \text{if} \quad x = 0$$

$$u = 5 \quad \text{if} \quad x = 2$$

Thus,

$$\int_0^2 x(x^2 + 1)^3\,dx = \frac{1}{2}\int_1^5 u^3\,du = \frac{u^4}{8}\Bigg]_{u=1}^5 = \frac{625}{8} - \frac{1}{8} = 78$$

which agrees with the result obtained by Method 1.   ◄

The choice of methods for evaluating a definite integral by substitution is purely a matter of taste. However, since Method 2 requires some thought to obtain the limits of integration, we shall give a few more examples using that method.

Example 2   Evaluate $\displaystyle\int_0^{\pi/4} \cos(\pi - x)\, dx$.

Solution.   Let

$$u = \pi - x \quad \text{so that} \quad du = -dx$$

With this substitution we have

$$u = \pi \qquad \text{if} \quad x = 0$$

$$u = 3\pi/4 \quad \text{if} \quad x = \pi/4$$

so

$$\int_0^{\pi/4} \cos(\pi - x)\, dx = \int_{\pi}^{3\pi/4} \cos u\, (-du)$$

$$= -\int_{\pi}^{3\pi/4} \cos u\, du = -\sin u \Big]_{\pi}^{3\pi/4}$$

$$= -[\sin(3\pi/4) - \sin(\pi)]$$

$$= -[1/\sqrt{2} - 0] = -1/\sqrt{2} \qquad \blacktriangleleft$$

Example 3   Evaluate $\displaystyle\int_0^{\pi/8} \sin^5 2x \cos 2x\, dx$.

Solution.   Let $u = \sin 2x$ so that

$$du = 2 \cos 2x\, dx \quad \text{or} \quad \tfrac{1}{2} du = \cos 2x\, dx$$

With this substitution we have

$$u = \sin(0) = 0 \quad \text{if} \quad x = 0$$

$$u = \sin(\pi/4) = 1/\sqrt{2} \quad \text{if} \quad x = \pi/8$$

so

$$\int_0^{\pi/8} \sin^5 2x \cos 2x\, dx = \frac{1}{2} \int_0^{1/\sqrt{2}} u^5\, du = \frac{1}{2} \cdot \frac{u^6}{6} \Big]_0^{1/\sqrt{2}}$$

$$= \frac{1}{2}\left[\frac{1}{6(\sqrt{2})^6} - 0 \right] = \frac{1}{96} \qquad \blacktriangleleft$$

☐ **NUMERICAL APPROXIMATION OF DEFINITE INTEGRALS**

In cases where a definite integral cannot be evaluated exactly, one must settle for a numerical approximation. We shall study methods for obtaining such approximations in detail later, but for now we note that the methods used in Section 5.5 to approximate areas can also be used to approximate definite integrals by Riemann sums. To see this, observe from Definition 5.6.3 that when the mesh size of the partition is small, one can expect the approximation

$$\int_a^b f(x)\, dx \approx \sum_{k=1}^{n} f(x_k^*)\, \Delta x_k \tag{3}$$

to be good. In practice, it is common to use a regular partition of the interval $[a, b]$, so that the subintervals all have the same width Δx. In this case (3) can be written as

$$\int_a^b f(x)\, dx \approx \sum_{k=1}^{n} f(x_k^*)\, \Delta x = \Delta x\, [\, f(x_1^*) + f(x_2^*) + \cdots + f(x_n^*)] \tag{4}$$

As in Section 5.5, this formula produces a *left endpoint approximation*, a *right endpoint approximation*, or a *midpoint approximation*, depending on the choice of the points x_k^*.

Example 4 Approximate the value of $\int_0^1 \sqrt{1 - x^2}\, dx$ using the left endpoint, right endpoint, and midpoint approximations, each with $n = 10$, $n = 20$, $n = 50$, and $n = 100$ subintervals.

Solution. The results of using Formula (4) are shown to nine decimal places in the following table:

n	LEFT ENDPOINT APPROXIMATION	RIGHT ENDPOINT APPROXIMATION	MIDPOINT APPROXIMATION
10	0.826129582	0.726129582	0.788102858
20	0.807116220	0.757116220	0.786357647
50	0.794567128	0.774567128	0.785641388
100	0.790104258	0.780104258	0.785484214

As shown in Example 1 of Section 5.6, the exact value of the integral is $\pi/4$. This is consistent with the preceding computations, since $\pi/4 \approx 0.785398163$. ◄

We conclude this section with the theorem that justifies the substitution method used in the preceding examples.

5.8.1   THEOREM.   *If g' is continuous on $[a, b]$ and f is continuous and has an antiderivative on an interval containing the values of $g(x)$ for $a \le x \le b$, then*

$$\int_a^b f(g(x))g'(x)\, dx = \int_{g(a)}^{g(b)} f(u)\, du$$

Proof.   Let $u = g(x)$ and let F be an antiderivative of f on an interval containing the values of $g(x)$ for $a \le x \le b$. Then by the chain rule

$$\frac{d}{dx} F(g(x)) = \frac{d}{dx} F(u) = \frac{dF}{du}\frac{du}{dx} = f(u)\frac{du}{dx} = f(g(x))g'(x)$$

for each x in $[a, b]$. Thus, $F(g(x))$ is an antiderivative of $f(g(x))g'(x)$ on $[a, b]$. Therefore, by the First Fundamental Theorem of Calculus (Theorem 5.7.1)

$$\int_a^b f(g(x))g'(x)\, dx = F(g(x))\Big]_a^b = F(g(b)) - F(g(a)) = \int_{g(a)}^{g(b)} f(u)\, du \quad ■$$

► **Exercise Set 5.8**   Ⓒ 34–37

---

1.   In each part express the integral in terms of the variable u, but do not evaluate.

(a) $\displaystyle\int_0^2 (x + 1)^7\, dx; \quad u = x + 1$

(b) $\displaystyle\int_{-1}^2 x\sqrt{8 - x^2}\, dx; \quad u = 8 - x^2$

(c) $\displaystyle\int_{-1}^1 \sin(\pi\theta)\, d\theta; \quad u = \pi\theta$

(d) $\displaystyle\int_0^{\pi/4} \tan^2 x \sec^2 x\, dx; \quad u = \tan x$

(e) $\displaystyle\int_0^1 x^3\sqrt{x^2 + 3}\, dx; \quad u = x^2 + 3$

(f) $\displaystyle\int_0^3 (x + 2)(x - 3)^{20}\, dx; \quad u = x - 3.$

In Exercises 2–11, evaluate the integrals two ways: first by a *u*-substitution in the definite integral, and then by a *u*-substitution in the corresponding indefinite integral.

2. $\displaystyle\int_{1}^{2} (4x - 2)^3 \, dx.$ **3.** $\displaystyle\int_{0}^{1} (2x + 1)^4 \, dx.$

4. $\displaystyle\int_{1}^{2} (4 - 3x)^8 \, dx.$ **5.** $\displaystyle\int_{-1}^{0} (1 - 2x)^3 \, dx.$

6. $\displaystyle\int_{-5}^{0} x\sqrt{4 - x} \, dx.$ **7.** $\displaystyle\int_{0}^{8} x\sqrt{1 + x} \, dx.$

8. $\displaystyle\int_{0}^{\pi/6} 2 \cos 3x \, dx.$ **9.** $\displaystyle\int_{0}^{\pi/2} 4 \sin (x/2) \, dx.$

10. $\displaystyle\int_{1-\pi}^{1+\pi} \sec^2 \left(\tfrac{1}{4}x - \tfrac{1}{4}\right) dx.$ **11.** $\displaystyle\int_{-2}^{-1} \frac{x}{(x^2 + 2)^3} \, dx.$

In Exercises 12–29, evaluate the integrals by any method.

12. $\displaystyle\int_{0}^{1} \sqrt[3]{a + bx} \, dx \quad (b \neq 0).$

13. $\displaystyle\int_{0}^{1} \frac{du}{\sqrt{3u + 1}}.$ **14.** $\displaystyle\int_{1}^{2} \sqrt{5x - 1} \, dx.$

15. $\displaystyle\int_{-1}^{1} \frac{x^2 \, dx}{\sqrt{x^3 + 9}}.$ **16.** $\displaystyle\int_{-1}^{0} 6t^2(t^3 + 1)^{19} \, dt.$

17. $\displaystyle\int_{1}^{3} \frac{x + 2}{\sqrt{x^2 + 4x + 7}} \, dx.$ **18.** $\displaystyle\int_{1}^{2} \frac{dx}{x^2 - 6x + 9}.$

19. $\displaystyle\int_{-3\pi/4}^{-\pi/4} \sin x \cos x \, dx.$ **20.** $\displaystyle\int_{0}^{\pi/4} \sqrt{\tan x} \sec^2 x \, dx.$

21. $\displaystyle\int_{0}^{\sqrt{\pi}} 5x \cos (x^2) \, dx.$

22. $\displaystyle\int_{0}^{2\pi/t} t^2 \sin tx \, dx \quad (t \text{ a positive constant}).$

23. $\displaystyle\int_{\pi^2}^{4\pi^2} \frac{1}{\sqrt{x}} \sin \sqrt{x} \, dx.$ **24.** $\displaystyle\int_{-\pi/4}^{\pi} \sin \theta \cos \theta \, d\theta.$

25. $\displaystyle\int_{0}^{\pi/2} \sin^2 3x \cos 3x \, dx.$

26. $\displaystyle\int_{0}^{\pi/4} \frac{\cos 2x}{\sqrt{7 - 3 \sin 2x}} \, dx.$ **27.** $\displaystyle\int_{\pi/12}^{\pi/9} \sec^2 3\theta \, d\theta.$

28. $\displaystyle\int_{-1}^{4} \frac{x \, dx}{\sqrt{5 + x}}.$ **29.** $\displaystyle\int_{0}^{1} \frac{y^2 \, dy}{\sqrt{4 - 3y}}.$

30. Find the area under the curve $y = 1/(3x + 1)^2$ over the interval $[0, 1]$.

31. Find the area under the curve $y = 3 \cos 2x$ over the interval $[0, \pi/8]$.

32. Electricity is supplied to homes in the form of *alternating current*, which means that the voltage has a sinusoidal waveform described by
$$V = V_p \sin (2\pi f t)$$

where V_p is the amplitude (or peak voltage) measured in volts, t is time in seconds, and f is the frequency in cycles per second (Figure 5.8.1). (A cycle is the electrical term for one period of the waveform.) Alternating current voltmeters read what is called the *rms* (*root-mean-square*) value of V. This is the square root of the average value of V^2 over one cycle. Show that
$$V_{\text{rms}} = \frac{V_p}{\sqrt{2}}$$

[*Hint:* One cycle corresponds to $0 \leq t \leq 1/f$. Use the identity $\sin^2 \theta = \tfrac{1}{2}(1 - \cos 2\theta)$ to help evaluate the integral.]

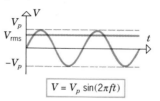

$\boxed{V = V_p \sin(2\pi f t)}$ Figure 5.8.1

33. The standard electrical outlets in some countries supply alternating current with an rms voltage of 120 volts at a frequency of 60 cycles/second. Use the result in Exercise 32 to find the peak voltage at such an outlet.

34. Verify the results shown in the table in Example 4 for the case where

    (a) $n = 10$ (b) $n = 20$

    (c) $n = 50$ (d) $n = 100$.

In Exercises 35–37, find the midpoint approximation of the integral using $n = 10, 20,$ and 50 subintervals. [*Note:* A programmable calculator is best for this problem. If your calculator is not programmable, then just do the case for $n = 10$.]

35. $\displaystyle\int_{1}^{2} \frac{1}{x} \, dx.$ **36.** $\displaystyle\int_{0}^{2} \sin \sqrt{x} \, dx.$

37. $\displaystyle\int_{0}^{1} \frac{4}{x^2 + 1} \, dx.$ [*Note:* The exact value is π.]

In Exercises 38–41, make the indicated substitution and evaluate the resulting definite integral by using an appropriate area formula from plane geometry.

38. $\displaystyle\int_{0}^{5/3} \sqrt{25 - 9x^2} \, dx;$ let $u = 3x.$

39. $\displaystyle\int_{-3}^{1} \sqrt{3 - 2x - x^2} \, dx;$ let $u = x + 1$ after completing the square.

40. $\displaystyle\int_{0}^{2} x\sqrt{16 - x^4} \, dx;$ let $u = x^2.$

41. $\displaystyle\int_{\pi/3}^{\pi/2} \sin \theta \sqrt{1 - 4 \cos^2 \theta} \, d\theta;$ let $u = 2 \cos \theta.$

42. Find $\displaystyle\int_{0}^{3} f(3x) \, dx$ if $\displaystyle\int_{0}^{9} f(x) \, dx = 5.$

43. Find $\displaystyle\int_{1/2}^{1} \frac{1}{x^2} f(1/x)\,dx$ if $\displaystyle\int_{1}^{2} f(x)\,dx = 3$.

44. Find $\displaystyle\int_{-2}^{0} xf(x^2)\,dx$ if $\displaystyle\int_{0}^{4} f(x)\,dx = 1$.

45. Find $\displaystyle\int_{0}^{1} f(3x+1)\,dx$ if $\displaystyle\int_{1}^{4} f(x)\,dx = 5$.

46. Prove: If m and n are positive integers, then
$$\int_{0}^{1} x^m (1-x)^n\,dx = \int_{0}^{1} x^n (1-x)^m\,dx$$
[*Hint:* Do not evaluate; use a substitution.]

47. Prove: If n is a positive integer, then
$$\int_{0}^{\pi/2} \sin^n x\,dx = \int_{0}^{\pi/2} \cos^n x\,dx$$
[*Hint:* Do not evaluate; use a trigonometric identity and a substitution.]

48. Let f be a function that is integrable on $[-a, a]$.

(a) Prove: If $f(-x) = -f(x)$ for all x in $[-a, a]$, then
$$\int_{-a}^{a} f(x)\,dx = 0$$
Give a geometric explanation of this result.

(b) Prove: If $f(-x) = f(x)$ for all x in $[-a, a]$, then
$$\int_{-a}^{a} f(x)\,dx = 2\int_{0}^{a} f(x)\,dx$$
Give a geometric explanation of this result.

In Exercises 49–52, use the results in Exercise 48 to express the value of the integral in terms of k.

49. $\displaystyle\int_{-2}^{2} \frac{x^4}{\sqrt{x^2+3}}\,dx$ if $\displaystyle\int_{0}^{2} \frac{x^4}{\sqrt{x^2+3}}\,dx = k$.

50. $\displaystyle\int_{-3}^{3} \frac{x^3}{x^2+5}\,dx$ if $\displaystyle\int_{0}^{3} \frac{x^3}{x^2+5}\,dx = k$.

51. $\displaystyle\int_{-1}^{1} \tan^3 x\,dx$ if $\displaystyle\int_{0}^{1} \tan^3 x\,dx = k$.

52. $\displaystyle\int_{0}^{1} \frac{\cos x}{x^2+1}\,dx$ if $\displaystyle\int_{-1}^{1} \frac{\cos x}{x^2+1}\,dx = k$.

53. Find the limit
$$\lim_{n \to +\infty} \sum_{k=1}^{n} \frac{\sin(k\pi/n)}{n}$$
by evaluating an appropriate definite integral over the interval $[0, 1]$.

54. Show that if f and g are continuous functions, then
$$\int_{0}^{t} f(t-x)g(x)\,dx = \int_{0}^{t} f(x)g(t-x)\,dx$$

55. (a) Let $I = \displaystyle\int_{0}^{a} \frac{f(x)}{f(x) + f(a-x)}\,dx$. Show that $I = \dfrac{a}{2}$.
[*Hint:* Let $u = a - x$, and then express the integrand as the sum of two fractions.]

(b) Use the result of part (a) to find
$$\int_{0}^{3} \frac{\sqrt{x}}{\sqrt{x} + \sqrt{3-x}}\,dx.$$

(c) Use the result of part (a) to find
$$\int_{0}^{\pi/2} \frac{\sin x}{\sin x + \cos x}\,dx.$$

56. Let $I = \displaystyle\int_{-1}^{1} \frac{1}{1+x^2}\,dx$. Show that the substitution $x = 1/u$ results in
$$I = -\int_{-1}^{1} \frac{1}{1+u^2}\,du = -I$$
so $2I = 0$, or $I = 0$, which is false (why?). Explain.

5.9   THE SECOND FUNDAMENTAL THEOREM OF CALCULUS

The First Fundamental Theorem of Calculus tells us how to evaluate the definite integral of a continuous function if we can find an antiderivative for that function, but it does not address the question of which functions actually have antiderivatives; that is the purpose of the Second Fundamental Theorem of Calculus, which we shall discuss in this section.

☐ DUMMY VARIABLES

Before turning to the question of "existence" of antiderivatives, it will be helpful to consider some notational matters. Sometimes it is convenient to use a letter other than x for the variable of integration in a definite integral. For example,

$$\int_a^b f(t)\, dt, \quad \int_a^b f(u)\, du, \quad \int_a^b f(y)\, dy$$

These integrals all have the same value. As an example, the following integrals have the same limits of integration but different variables of integration:

$$\int_1^3 x^2\, dx = \left.\frac{x^3}{3}\right]_{x=1}^3 = \frac{27}{3} - \frac{1}{3} = \frac{26}{3}$$

$$\int_1^3 t^2\, dt = \left.\frac{t^3}{3}\right]_{t=1}^3 = \frac{27}{3} - \frac{1}{3} = \frac{26}{3}$$

$$\int_1^3 u^2\, du = \left.\frac{u^3}{3}\right]_{u=1}^3 = \frac{27}{3} - \frac{1}{3} = \frac{26}{3}$$

In general, we have the following result:

> *The value of a definite integral is unaffected if we change the letter used for the variable of integration, but do not change the limits of integration.*

Because the letter used for the variable of integration has no effect on the final value of the definite integral, it is sometimes called a **dummy variable**.

□ **DEFINITE INTEGRALS WITH A VARIABLE UPPER LIMIT OF INTEGRATION**

In the work to follow we shall consider definite integrals of the form

$$\int_a^x \underline{\qquad}$$

where the upper limit x is allowed to vary. For such integrals we shall use a letter different from x (often t) for the variable of integration; thus, we would write

$$\int_a^x f(t)\, dt \quad \text{rather than} \quad \int_a^x f(x)\, dx$$

This avoids using x in two different ways (as a variable of integration and as a limit of integration), which can cause errors.

Example 1 Evaluate $\displaystyle\int_2^x t^2\, dt$.

Solution.

$$\int_2^x t^2\, dt = \left.\frac{t^3}{3}\right]_{t=2}^x = \frac{x^3}{3} - \frac{8}{3} \qquad \blacktriangleleft$$

Note that the final expression in the preceding example is a function of x alone. In general, an expression of the form

$$\int_a^x f(t)\, dt$$

represents a function of x—the variable t does not enter into the final result.

□ **THE SECOND FUNDAMENTAL THEOREM OF CALCULUS**

In Section 5.1 we showed informally that if f is a nonnegative continuous function, and if $A(x)$ is the area under the curve $y = f(x)$ over the interval $[a, x]$, then

$$A'(x) = f(x) \tag{1}$$

[See Figure 5.1.2 and Formula (4) in Section 5.1.] If we express $A(x)$ as the definite integral

$$A(x) = \int_a^x f(t)\, dt$$

then it follows from (1) that

$$\frac{d}{dx}\left[\int_a^x f(t)\, dt\right] = f(x)$$

The following theorem shows that this result holds for all continuous functions.

5.9.1   THEOREM (*The Second Fundamental Theorem of Calculus*).   *Let f be a continuous function on an interval I, and let a be any point in I. If F is defined by*

$$F(x) = \int_a^x f(t)\, dt \tag{2}$$

then $F'(x) = f(x)$ at each point x in the interval I.

REMARK.   In this theorem, if x is an endpoint of the interval I, then $F'(x)$ must be interpreted as the derivative from the left or right, as appropriate.

Proof.   If x is not an endpoint of the interval I, then it follows from the definition of a derivative that

$$F'(x) = \lim_{h \to 0} \frac{F(x + h) - F(x)}{h}$$

$$= \lim_{h \to 0} \frac{1}{h}\left[\int_a^{x+h} f(t)\, dt - \int_a^x f(t)\, dt\right]$$

$$= \lim_{h \to 0} \frac{1}{h}\left[\int_a^{x+h} f(t)\, dt + \int_x^a f(t)\, dt\right]$$

$$= \lim_{h \to 0} \frac{1}{h}\int_x^{x+h} f(t)\, dt \qquad \boxed{\text{Theorem 5.6.6}}$$

Applying the Mean-Value Theorem for Integrals (5.7.2) to the last expression, we obtain

$$F'(x) = \lim_{h \to 0} \frac{1}{h}[f(t^*) \cdot h] = \lim_{h \to 0} f(t^*) \tag{3}$$

where t^* is some number between x and $x + h$. Because t^* is between x and $x + h$, it follows that $t^* \to x$ as $h \to 0$. Thus, $f(t^*) \to f(x)$ as $h \to 0$, since f is assumed continuous at x. Therefore, it follows from (3) that $F'(x) = f(x)$. (If x is an endpoint of the interval I, then the two-sided limits in the proof must be replaced by the appropriate one-sided limits, but otherwise the arguments are identical.)   ■

The result in Theorem 5.9.1 can be expressed by the formula

$$\frac{d}{dx}\left[\int_a^x f(t)\, dt\right] = f(x) \tag{4}$$

In words, this formula states:

> *Where the integrand is continuous, the derivative of a definite integral with respect to its upper limit is equal to the integrand evaluated at the upper limit.*

Example 2   Since $f(x) = x^3$ is a continuous function, it follows from Formula (4) that

$$\frac{d}{dx}\left[\int_1^x t^3\, dt\right] = x^3$$

As a check, let us evaluate the integral, then differentiate:

$$\int_1^x t^3 \, dt = \frac{t^4}{4}\Bigg]_{t=1}^{x} = \frac{x^4}{4} - \frac{1}{4}$$

Differentiating this function yields x^3 as above.

Example 3   Since

$$f(x) = \frac{\sin x}{x}$$

is continuous on any interval that does not contain the origin, it follows from (4) that on the interval $(0, +\infty)$ we have

$$\frac{d}{dx}\left[\int_1^x \frac{\sin t}{t} \, dt \right] = \frac{\sin x}{x}$$

Unlike the preceding example, there is no way to evaluate the integral in terms of familiar functions, so Formula (4) provides the only simple method for finding the derivative.   ◄

□ **EXISTENCE OF ANTIDERIVATIVES FOR CONTINUOUS FUNCTIONS**

If f is a continuous function on an interval I and a is any point in I, then it follows from (4) that the function

$$F(x) = \int_a^x f(t) \, dt \qquad (5)$$

is an antiderivative of f on I. Thus, thanks to the Second Fundamental Theorem of Calculus, we now know that every function that is continuous on an interval has an antiderivative on that interval.

It should be noted that unlike the indefinite integral

$$\int f(x) \, dx$$

which represents an arbitrary antiderivative of f (because of the constant of integration), the function defined by (5) is not an arbitrary antiderivative of f; it is the specific antiderivative whose value at $x = a$ is zero because

$$F(a) = \int_a^a f(t) \, dt = 0$$

Example 4   The function defined by

$$F(x) = \int_1^x t^3 \, dt$$

is that antiderivative of $f(x) = x^3$ whose value at $x = 1$ is zero. As a check, in Example 2 we found that

$$F(x) = \frac{x^4}{4} - \frac{1}{4}$$

so $F(1) = 0$, as expected.   ◄

□ **FUNCTIONS DEFINED BY INTEGRALS**

Although we know that continuous functions have antiderivatives, there are two possible impediments that can prevent us from applying the formula

$$\int_a^b f(x) \, dx = F(b) - F(a) \qquad (6)$$

even if f is continuous:

- We may not be clever enough to find a usable formula for an antiderivative F.
- An antiderivative F may not be expressible in terms of familiar functions.

To illustrate the second situation, suppose that we are interested in evaluating the integral

$$\int_1^3 \frac{1}{x}\,dx \tag{7}$$

Because $f(x) = 1/x$ is continuous on any interval that does not contain the origin, the function f is integrable on the interval $[1, 3]$, and we are guaranteed by the Second Fundamental Theorem of Calculus that there is some antiderivative for f on this interval. One such antiderivative is

$$F(x) = \int_1^x \frac{1}{t}\,dt \tag{8}$$

However, this formula is of little direct help in evaluating (7) because applying (6) to (7) with this antiderivative yields

$$\int_1^3 \frac{1}{x}\,dx = F(3) - F(1) = \int_1^3 \frac{1}{t}\,dt - \int_1^1 \frac{1}{t}\,dt = \int_1^3 \frac{1}{t}\,dt$$

which is a true but useless equation for the purpose of obtaining a numerical value for (7).

Although we might try looking for an antiderivative formula for $1/x$ that does not involve an integral, we would not have any success since it can be proved that no antiderivative of $1/x$ can be expressed in terms of finitely many polynomials, rational functions, trigonometric functions, or any other kinds of functions encountered thus far in this text. In short, our repertoire of basic functions is too limited, at present, to produce an antiderivative of $1/x$ that is simpler than (8). However, this problem does not totally preclude the possibility of obtaining a numerical value for (7), since this integral can be approximated to any degree of accuracy using Riemann sums. In Chapter 9 we shall discuss a variety of methods for approximating the value of a definite integral.

Since (8) cannot be expressed in terms of familiar functions, this formula actually defines a new function for us. This function is of special importance and will be studied in detail in Chapter 7. Many of the most important functions in scientific applications arise as integrals from the Second Fundamental Theorem of Calculus.

▶ Exercise Set 5.9

1. Given that $\int_{-1}^5 f(x)\,dx = 3$, find

   (a) $\int_{-1}^5 f(t)\,dt$ (b) $\int_{-1}^5 f(u)\,du.$

2. Define $F(x)$ by

   $$F(x) = \int_{\pi/4}^x \cos 2t\,dt$$

   (a) Use the Second Fundamental Theorem of Calculus to find $F'(x)$.

   (b) Check the result in part (a) by first integrating and then differentiating.

3. Define $F(x)$ by

   $$F(x) = \int_1^x (t^3 + 1)\,dt$$

   (a) Use the Second Fundamental Theorem of Calculus to find $F'(x)$.

   (b) Check the result in part (a) by first integrating and then differentiating.

   In Exercises 4–7, use the Second Fundamental Theorem of Calculus to find the derivative.

4. $\dfrac{d}{dx} \displaystyle\int_0^x \dfrac{dt}{1 + \sqrt{t}}.$ 5. $\dfrac{d}{dx} \displaystyle\int_1^x \sin\left(\sqrt{t}\right) dt.$

6. $\dfrac{d}{dx} \displaystyle\int_0^x \dfrac{t}{\cos t}\,dt.$ 7. $\dfrac{d}{dx} \displaystyle\int_0^x |t|\,dt.$

   In Exercises 8–11, express the antiderivatives as integrals.

8. The antiderivative of $1/(1 + x^2)$ on the interval $(-\infty, +\infty)$ whose value at $x = 1$ is 0.

9. The antiderivative of $1/(x-1)$ on the interval $(1, +\infty)$ whose value at $x = 2$ is 0.

10. The antiderivative of $1/(x-1)$ on the interval $(-\infty, 1)$ whose value at $x = -3$ is 0.

11. The antiderivative of $1/(x-1)$ on the interval $(-\infty, 1)$ whose value at $x = 0$ is 0.

12. (a) Over what open interval does the formula

$$F(x) = \int_1^x \frac{1}{t^2 - 9} \, dt$$

represent an antiderivative of

$$f(x) = \frac{1}{x^2 - 9} \, ?$$

(b) Find a point where the graph of F crosses the x-axis.

13. (a) Over what open interval does the formula

$$F(x) = \int_1^x \frac{dt}{t}$$

represent an antiderivative of $f(x) = 1/x$?

(b) Find a point where the graph of F crosses the x-axis.

14. Let $F(x) = \int_2^x \sqrt{3t^2 + 1} \, dt$. Find

(a) $F(2)$ (b) $F'(2)$ (c) $F''(2)$.

15. Let $F(x) = \int_0^x \frac{\cos t}{t^2 + 3} \, dt$. Find

(a) $F(0)$ (b) $F'(0)$ (c) $F''(0)$.

16. Let $F(x) = \int_0^x \frac{t-3}{t^2 + 7} \, dt$ for $-\infty < x < +\infty$.

(a) Find the value of x where F attains its minimum value.

(b) Find intervals over which F is only increasing or only decreasing.

(c) Find open intervals over which F is only concave up or only concave down.

In Exercises 17–20, determine the domain of $F(x)$ and specify all values of x for which $F(x)$ is positive, negative, or zero. (Do not perform the integration.)

17. $F(x) = \int_1^x \frac{t^4}{t^2 + 3} \, dt.$ **18.** $F(x) = \int_2^x \frac{2-t}{t^2 + 1} \, dt.$

19. $F(x) = \int_{-1}^x \sqrt{4 - t^2} \, dt.$

20. $F(x) = \int_{-3}^x t^2 \sqrt{2 - t} \, dt.$

In Exercises 21–23, express $F(x)$ in a piecewise form that does not use integrals.

21. $F(x) = \int_{-1}^x |t| \, dt.$

22. $F(x) = \int_0^x f(t) \, dt,$ where $f(x) = \begin{cases} x, & 0 \le x \le 2 \\ 2, & x > 2. \end{cases}$

23. $F(x) = \int_{-1}^x f(t) \, dt,$ where $f(x) = \begin{cases} x^2, & x \le 0 \\ 2x, & x > 0. \end{cases}$

24. Use the Second Fundamental Theorem of Calculus and the chain rule to show that

$$\frac{d}{dx} \int_a^{g(x)} f(t) \, dt = f(g(x))g'(x)$$

In Exercises 25 and 26, use the result in Exercise 24 to perform the differentiation.

25. $\dfrac{d}{dx} \displaystyle\int_1^{x^3} \frac{1}{t} \, dt.$ **26.** $\dfrac{d}{dx} \displaystyle\int_3^{\sin x} \frac{1}{1 + t^2} \, dt.$

27. Prove that the function

$$F(x) = \int_0^x \frac{1}{1 + t^2} \, dt + \int_0^{1/x} \frac{1}{1 + t^2} \, dt$$

is constant on the interval $(0, +\infty)$.

28. Use Exercise 24 and Theorem 5.6.6 to show that

$$\frac{d}{dx} \int_{h(x)}^{g(x)} f(t) \, dt = f(g(x))g'(x) - f(h(x))h'(x)$$

29. Use the result in Exercise 28 to perform the following differentiations:

(a) $\dfrac{d}{dx} \displaystyle\int_{x^2}^{x^3} \sin^2 t \, dt$ (b) $\dfrac{d}{dx} \displaystyle\int_{-x}^x \frac{1}{1 + t} \, dt.$

30. Prove that the function

$$F(x) = \int_x^{3x} \frac{1}{t} \, dt$$

is constant on the interval $(0, +\infty)$ by using Exercise 28 to find $F'(x)$.

31. Prove: If f is continuous on an open interval I and b is any point in I, then at each point in I

$$\frac{d}{dx} \int_x^b f(t) \, dt = -f(x)$$

32. Prove: If f is continuous on an open interval I and a is any point in I, then

$$F(x) = \int_a^x f(t) \, dt$$

is continuous on I.

◆ **TECHNOLOGY EXERCISES** Chapter 5

Most of these exercises require access to a graphing calculator or a computer algebra system (CAS) such as *Mathematica*, *Maple*, or *Derive*. When you are asked to *find* an answer or to *solve* an equation, you may choose to find an exact result or a numerical approximation, depending on the particular technology you are using and on your own imagination. The form of your answers may differ from those of other students or from those in the answer section of the text, depending on how you solve the problems and the accuracy you use in your numerical approximations. Those exercises that are more appropriate for a CAS than a graphing calculator are labeled with the icon ◆.

Indefinite integrals: In Exercises 1–4, perform the integration by hand and then use a CAS to check your result.

◆ **1.** $\int \dfrac{\cos 3x}{\sqrt{5 + 2 \sin 3x}}\, dx.$ ◆ **2.** $\int \dfrac{\sqrt{3 + \sqrt{x}}}{\sqrt{x}}\, dx.$

◆ **3.** $\int \dfrac{x^2}{(ax^3 + b)^2}\, dx,\ a \neq 0.$

◆ **4.** $\int x \sec^2(ax^2)\, dx,\ a \neq 0.$

Antiderivatives: In Exercises 5–8, use a CAS to find the function f that satisfies the given conditions.

◆ **5.** $f'(x) = \dfrac{1}{x^2\sqrt{9 + x^2}},\ f(4) = 1.$

◆ **6.** $f'(x) = x \sin^2 x,\ f(1) = 2.$

◆ **7.** $f'(x) = x^2 \cos 3x,\ f(\pi/2) = -1.$

◆ **8.** $f'(x) = \dfrac{x^3}{(4 + x^2)^{3/2}},\ f(0) = -2.$

Definite integrals: In Exercises 9–12, find the value of the definite integral.

◆ **9.** $\int_0^2 \dfrac{x}{\sqrt{1 + x^3}}\, dx.$ ◆ **10.** $\int_{-1}^3 \dfrac{1}{\sqrt{1 + x^4}}\, dx.$

◆ **11.** $\int_{-\pi/2}^{\pi/2} \sqrt{2 - \sin x}\, dx.$ ◆ **12.** $\int_0^{\sqrt{2\pi}} \cos(x^2)\, dx.$

13. Area: Find the area of the region in the first quadrant that lies below the curve $y = x + x^2 - x^3$ and above the x-axis.

14. Area: Find the area of the region in the first quadrant that lies below the curve $y = 3 \sin x - x - 0.5$ and above the x-axis.

Equations containing definite integrals: In Exercises 15–18, find all values of k that satisfy the equation.

15. $\int_1^k (x^3 - 2x - 1)\, dx = 0,\ k > 1.$

16. $\int_0^k (x^2 + \sin 2x)\, dx = 3,\ k > 0.$

17. $\int_1^4 \dfrac{x}{\sqrt{1 + kx^2}}\, dx = 2.5,\ k > 0.$

18. $\int_0^2 \sin(kx)\, dx = 0.8,\ k > 0.$

◆ **19. The Second Fundamental Theorem of Calculus:** Let

$$f(x) = \dfrac{x}{\sqrt{2 + x^3}} \quad \text{and} \quad F(x) = \int_{-1}^x f(t)\, dt$$

(a) Graph $y = f(x)$ and $y = F(x)$ over the interval $[-1, 3]$.

(b) Find the value of x for which $x > -1$ and $F(x) = 0$.

(c) Find the x-coordinates of the relative extrema and inflection points of the graph of F.

◆ **20. The Second Fundamental Theorem of Calculus:** Let

$$f(x) = \sqrt{x^3 + 2} \times \sqrt{3x^2 + 1}$$

and

$$F(x) = \int_0^x f(t)\, dt$$

(a) Graph $y = f(x)$ and $y = F(x)$ over the interval $[-\sqrt[3]{2}, 5]$.

(b) Find the x-coordinates of the relative extrema and inflection points of the graph of F.

21. The Fresnel sine integral

$$S(x) = \int_0^x \sin\left(\frac{\pi}{2} t^2\right) dt$$

Graph $y = S(x)$ over the interval $[0, 3]$ and find the maximum value of $S(x)$ for x in the interval $[0, 3]$.

22. The sine integral

$$\text{Si}(x) = \int_0^x \frac{\sin t}{t} dt$$

Graph $y = \text{Si}(x)$ over the interval $[0, 3\pi]$, and find the maximum value of $\text{Si}(x)$ for x in the interval $[0, 3\pi]$.

23. The complete elliptic integral of the second kind

$$E(x) = \int_0^{\pi/2} \sqrt{1 - x \sin^2 t}\, dt, \quad 0 \le x \le 1$$

Graph $y = E(x)$ over the interval $[0, 1]$, and find the value of x for which $E(x) = 1.5$.

24. The Bessel function of order zero

$$J_0(x) = \frac{1}{\pi} \int_0^{\pi} \cos(x \sin t)\, dt$$

Graph $y = J_0(x)$ over the interval $[0, 8]$, and find the smallest positive value of x for which $J_0(x) = 0$.

Area as a limit: In Exercises 25 and 26, use a CAS to find the Riemann sum

$$\sum_{k=1}^{n} f(x_k^*)\, \Delta x$$

as a function of n, then find the area under the graph by taking the limit as $n \to +\infty$.

25. Find the area under the graph of $f(x) = x^3$ over the interval $[-1, 3]$ as a limit of Riemann sums in which x_k^* is the *right* endpoint of each subinterval.

26. Find the area under the graph of $f(x) = 5x - x^2$ over the interval $[0, 5]$ as a limit of Riemann sums in which x_k^* is the *left* endpoint of each subinterval.

Gottfried Wilhelm Leibniz (1646–1716)

6 APPLICATIONS OF THE DEFINITE INTEGRAL

> *In this section we shall discuss methods for calculating the area between curves in the plane.*

☐ **AREA BETWEEN** $y = f(x)$
AND $y = g(x)$

FIRST AREA PROBLEM.   *Suppose that f and g are continuous functions on an interval* [a, b] *and*

$$f(x) \geq g(x) \quad for \quad a \leq x \leq b$$

[*This means that the curve y = f(x) lies above the curve y = g(x) and that the two can touch but not cross.*] *Find the area A of the region bounded above by y = f(x), below by y = g(x), and on the sides by the lines x = a and x = b (Figure 6.1.1a).*

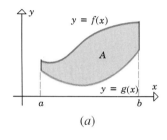
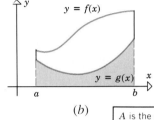
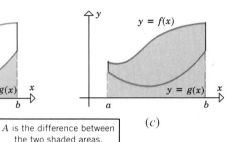

(a)   (b)   (c)

A is the difference between
the two shaded areas.

Figure 6.1.1

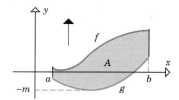

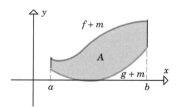

Figure 6.1.2

If f and g are nonnegative on $[a, b]$, then as illustrated in Figure 6.1.1b, we have

$$A = [\text{area under } f] - [\text{area under } g]$$

or equivalently,

$$A = \int_a^b f(x)\, dx - \int_a^b g(x)\, dx = \int_a^b [f(x) - g(x)]\, dx \tag{1}$$

This formula can be extended to the case where g has negative values by translating the graphs of f and g upward some suitable distance m until the graph of g is no longer below the x-axis (Figure 6.1.2). This translation does not affect the area A between the curves, so we have

$$A = \int_a^b [f(x) + m]\, dx - \int_a^b [g(x) + m]\, dx = \int_a^b [f(x) - g(x)]\, dx$$

which is the same as (1). Thus, in all cases we have the following result.

6.1.1   AREA FORMULA.   If f and g are continuous functions on the interval $[a, b]$, and if $f(x) \geq g(x)$ for all x in $[a, b]$, then the area of the region bounded above by $y = f(x)$, below by $y = g(x)$, on the left by the line $x = a$, and on the right by the line $x = b$ is

$$A = \int_a^b [f(x) - g(x)]\, dx \tag{2}$$

When the region is complicated, it may require some careful thought to determine the integrand and limits of integration in (2). Here is a systematic procedure that you can follow to set up this formula.

> **Step 1.**   Sketch the region and then draw a vertical line segment through the region at an arbitrary point x, connecting the top and bottom boundaries (Figure 6.1.3a).
>
> **Step 2.**   The top endpoint of the line segment sketched in Step 1 will be $f(x)$, the bottom one $g(x)$, and the length of the line segment will be $f(x) - g(x)$. This is the integrand in (2).
>
> **Step 3.**   To determine the limits of integration, imagine moving the line segment left and then right. The leftmost position at which the line segment intersects the region is $x = a$ and the rightmost is $x = b$ (Figures 6.1.3b and 6.1.3c).

REMARK.   In Step 1 above, it is not necessary to make an extremely accurate sketch of the region; sufficient accuracy to determine which curve is the upper boundary and which curve is the lower boundary is all that is required.

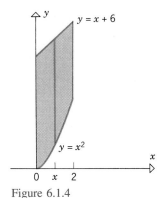

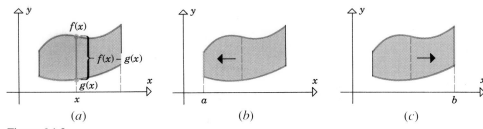

Figure 6.1.3

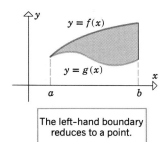

The left-hand boundary reduces to a point.

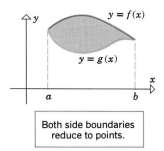

Both side boundaries reduce to points.

Figure 6.1.5

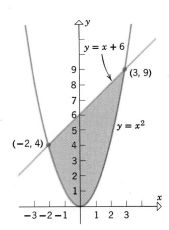

Figure 6.1.4

Example 1 Find the area of the region bounded above by $y = x + 6$, bounded below by $y = x^2$, and bounded on the sides by the lines $x = 0$ and $x = 2$.

Solution. The region and a vertical line segment through it are shown in Figure 6.1.4. The line segment extends from $f(x) = x + 6$ on the top to $g(x) = x^2$ on the bottom. If the line segment is moved through the region, its leftmost position will be $x = 0$ and its rightmost will be $x = 2$. Thus, from (2)

$$A = \int_0^2 [(x + 6) - x^2]\, dx = \left[\frac{x^2}{2} + 6x - \frac{x^3}{3} \right]_0^2 = \frac{34}{3} - 0 = \frac{34}{3} \quad \blacktriangleleft$$

Sometimes the upper curve $y = f(x)$ intersects the lower curve $y = g(x)$ at either the left-hand boundary $x = a$, the right-hand boundary $x = b$, or both. When this happens the side of the region where the upper and lower curves intersect reduces to a point, rather than a vertical line segment (Figure 6.1.5).

Example 2 Find the area of the region that is enclosed between the curves $y = x^2$ and $y = x + 6$.

Solution. A sketch of the region (Figure 6.1.6) shows that the lower boundary is $y = x^2$ and the upper boundary is $y = x + 6$. At the endpoints of the region, the upper and lower boundaries have the same y-coordinates; thus, to find the endpoints we equate

$$y = x^2 \quad \text{and} \quad y = x + 6 \tag{3}$$

This yields

$$x^2 = x + 6 \quad \text{or} \quad x^2 - x - 6 = 0 \quad \text{or} \quad (x + 2)(x - 3) = 0$$

from which we obtain

$$x = -2 \quad \text{and} \quad x = 3$$

Although the y-coordinates of the endpoints are not essential to our solution, they may be obtained from (3) by substituting $x = -2$ and $x = 3$ in either equation. This yields $y = 4$ and $y = 9$, so the upper and lower boundaries intersect at $(-2, 4)$ and $(3, 9)$.

From (2) with $f(x) = x + 6$, $g(x) = x^2$, $a = -2$, and $b = 3$, we obtain the area

$$A = \int_{-2}^3 [(x + 6) - x^2]\, dx = \left[\frac{x^2}{2} + 6x - \frac{x^3}{3} \right]_{-2}^3$$

$$= \frac{27}{2} - \left(-\frac{22}{3} \right) = \frac{125}{6} \quad \blacktriangleleft$$

Example 3 Find the area of the region enclosed by $x = y^2$ and $y = x - 2$.

Solution. To make an accurate sketch of the region we need to know where the curves $x = y^2$ and $y = x - 2$ intersect. In Example 2 we found intersections by equating the expressions for y. Here it is easier to rewrite the latter equation as $x = y + 2$ and equate the expressions for x, namely

$$x = y^2 \quad \text{and} \quad x = y + 2 \tag{4}$$

This yields

$$y^2 = y + 2 \quad \text{or} \quad y^2 - y - 2 = 0 \quad \text{or} \quad (y + 1)(y - 2) = 0$$

from which we obtain $y = -1$, $y = 2$. Substituting these values in either equation in (4) we see that the corresponding x values are $x = 1$ and $x = 4$, respectively, so the points of intersection are $(1, -1)$ and $(4, 2)$ (Figure 6.1.7a).

To apply Formula (2), the equations of the boundaries must be written so that y is expressed explicitly as a function of x. The upper boundary can be written as $y = \sqrt{x}$ (rewrite $x = y^2$ as $y = \pm\sqrt{x}$ and choose the $+$ for the upper portion of the curve). The lower portion of the boundary consists of two parts: $y = -\sqrt{x}$ for $0 \le x \le 1$ and $y = x - 2$ for $1 \le x \le 4$ (Figure 6.1.7b). Because of this change in the formula for the lower boundary, it is necessary to divide the region into two parts and find the area of each part separately.

From (2) with $f(x) = \sqrt{x}$, $g(x) = -\sqrt{x}$, $a = 0$, and $b = 1$, we obtain

$$A_1 = \int_0^1 [\sqrt{x} - (-\sqrt{x})]\, dx = 2 \int_0^1 \sqrt{x}\, dx$$

$$= 2 \left[\frac{2}{3} x^{3/2} \right]_0^1 = \frac{4}{3} - 0 = \frac{4}{3}$$

From (2) with $f(x) = \sqrt{x}$, $g(x) = x - 2$, $a = 1$, and $b = 4$, we obtain

$$A_2 = \int_1^4 [\sqrt{x} - (x - 2)]\, dx = \int_1^4 (\sqrt{x} - x + 2)\, dx$$

$$= \left[\frac{2}{3} x^{3/2} - \frac{1}{2} x^2 + 2x \right]_1^4 = \left(\frac{16}{3} - 8 + 8 \right) - \left(\frac{2}{3} - \frac{1}{2} + 2 \right) = \frac{19}{6}$$

Thus, the area of the entire region is

$$A = A_1 + A_2 = \tfrac{4}{3} + \tfrac{19}{6} = \tfrac{9}{2} \quad \blacktriangleleft$$

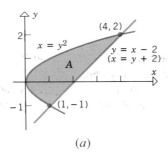
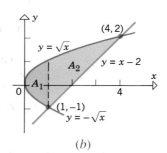

Figure 6.1.7   (a)   (b)

☐ AREA BETWEEN $x = v(y)$ AND $x = w(y)$

Sometimes it is possible to avoid splitting a region into parts by integrating with respect to y rather than x. We shall now show how this can be done.

SECOND AREA PROBLEM. *Suppose that w and v are continuous functions of y on an interval $[c, d]$ and that*

$$w(y) \ge v(y) \quad \text{for} \quad c \le y \le d$$

[*This means that the curve $x = w(y)$ lies to the right of the curve $x = v(y)$ and that the two can touch but not cross.*] *Find the area A of the region bounded on the left by $x = v(y)$, on the right by $x = w(y)$, and above and below by the lines $y = d$ and $y = c$* (Figure 6.1.8).

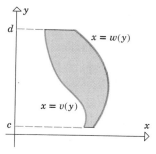

Figure 6.1.8

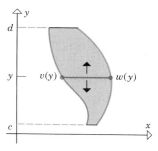

Figure 6.1.9

Proceeding as in the derivation of (2), but with the roles of x and y reversed, yields the following result.

6.1.2 AREA FORMULA.   If w and v are continuous functions and if $w(y) \geq v(y)$ for all y in $[c, d]$, then the area of the region bounded on the left by $x = v(y)$, on the right by $x = w(y)$, below by $y = c$, and above by $y = d$ is

$$A = \int_c^d [w(y) - v(y)] \, dy \qquad (5)$$

The procedure for finding the integrand and limits of integration in (5) is similar to that used for (2): Draw a horizontal line segment through the region from boundary to boundary at an arbitrary point y (Figure 6.1.9). The right-hand endpoint of the line segment will be $w(y)$, the left-hand endpoint $v(y)$, and the length of the line segment will be $w(y) - v(y)$, which is the integrand in (5). To determine the limits of integration, imagine moving the line segment down and then up (Figure 6.1.9). The lowest position at which the line segment intersects the region is $y = c$ and the highest is $y = d$.

In Example 3, where we integrated with respect to x to find the area of the region enclosed by $x = y^2$ and $y = x - 2$, we had to split the region into parts and evaluate two integrals. In the next example we shall see that by integrating with respect to y no splitting of the region is necessary.

Example 4   Find the area of the region enclosed by $x = y^2$ and $y = x - 2$, integrating with respect to y.

Solution.   From Figure 6.1.7 the left boundary is $x = y^2$, the right boundary is $y = x - 2$, and the region extends over the interval $-1 \leq y \leq 2$. However, to apply (5) the equations for the boundaries must be written so that x is expressed explicitly as a function of y. Thus, we rewrite $y = x - 2$ as $x = y + 2$. It now follows from (5) that

$$A = \int_{-1}^2 [(y + 2) - y^2] \, dy = \left[\frac{y^2}{2} + 2y - \frac{y^3}{3} \right]_{-1}^2 = \frac{9}{2}$$

which agrees with the result obtained in Example 3.   ◀

REMARK.   The choice between Formulas (2) and (5) is generally dictated by the shape of the region, and one would usually choose the formula that requires the least amount of splitting. However, if the integral(s) resulting by one method are difficult to evaluate, then the other method might be preferable, even if it requires more splitting.

▶ **Exercise Set 6.1**   [C] *39, 40*

In Exercises 1–4, find the area of the shaded region.

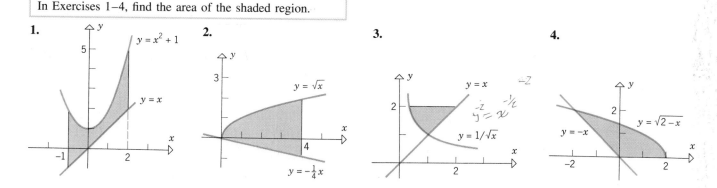

5. Find the area of the region enclosed by the curves $y = x^2$ and $y = 4x$ by integrating

(a)   with respect to x (b)   with respect to y.

6. Find the area of the region enclosed by the curves $y^2 = 4x$ and $y = 2x - 4$ by integrating

(a)   with respect to x (b)   with respect to y.

7. Find the area of the region enclosed by the curves $y^2 = 2x$ and $y = 2x - 2$ by integrating

(a)   with respect to x (b)   with respect to y.

In Exercises 8–27, sketch the region enclosed by the curves and find its area by any method.

---

8. $y = x^3, y = x, x = 0, x = 1/2$.

9. $y = x^2, y = \sqrt{x}, x = 1/4, x = 1$.

10. $y = x^3 - 4x, y = 0, x = 0, x = 2$.

11. $y = \cos 2x, y = 0, x = \pi/4, x = \pi/2$.

12. $y = x^3 - 4x^2 + 3x, y = 0, x = 0, x = 3$.

13. $x = y^2 - 4y, x = 0, y = 0, y = 4$.

14. $x = \sin y, x = 0, y = \pi/4, y = 3\pi/4$.

15. $y = \sec^2 x, y = 2, x = -\pi/4, x = \pi/4$.

16. $x^2 = y, x = y - 2$.

17. $y = x^2 + 4, x + y = 6$.

18. $y = x^3, y = -x, y = 8$.

19. $y^2 = -x, y = x - 6, y = -1, y = 4$.

20. $y = x, y = 4x, y = -x + 2$.

21. $y = 2 + |x - 1|, y = -\frac{1}{5}x + 7$.

22. $y = x^3 - 4x, y = 0, x = -2, x = 2$.

23. $x = y^3 - y, x = 0$.

24. $y = x^3 - 2x^2, y = 2x^2 - 3x, x = 0, x = 3$.

25. $y = \sin x, y = \cos x, x = 0, x = 2\pi$.

26. $y = \sqrt{x + 2}, y = x, y = 0$.

27. $y = 1/x^2, y = x, y = 4$.

28. Find the area between the curve $y = \sin x$ and the line segment joining the points $(0, 0)$ and $(5\pi/6, 1/2)$ on the curve.

29. Find the area enclosed by the curve $y = \sqrt{x}$, the tangent to the curve at $x = 4$, and the y-axis.

30. Let A be the area enclosed by $y = 1/x^2, y = 0, x = 1$, and $x = b$ $(b > 1)$.

(a)   Find A.   (b)   Find $\lim\limits_{b \to +\infty} A$.

31. Let A be the area enclosed by $y = 1/\sqrt{x}, y = 0, x = 1$, and $x = b$ $(b > 1)$.

(a)   Find A.   (b)   Find $\lim\limits_{b \to +\infty} A$.

32. Find a vertical line $x = k$ that divides the area enclosed by $x = \sqrt{y}, x = 2$, and $y = 0$ into two equal parts.

33. Find a horizontal line $y = k$ that divides the area between $y = x^2$ and $y = 9$ into two equal parts.

34. (a)   Find the area of the region enclosed by the parabola $y = 2x - x^2$ and the x-axis.

(b)   Find the value of m so that the line $y = mx$ divides the region in part (a) into two regions of equal area.

35. Show that the area of the ellipse in Figure 6.1.10 is πab. [*Hint:* Use a formula from geometry to help evaluate the definite integral.]

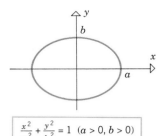

$$\boxed{\dfrac{x^2}{a^2} + \dfrac{y^2}{b^2} = 1 \ \ (a > 0, b > 0)}$$

Figure 6.1.10

36. Find the area of the region enclosed between the curve $x^{1/2} + y^{1/2} = a^{1/2}$ and the coordinate axes.

37. Suppose that f and g are integrable on $[a, b]$, but neither $f(x) \geq g(x)$ nor $g(x) \geq f(x)$ holds for all x in $[a, b]$ [i.e., the curves $y = f(x)$ and $y = g(x)$ are intertwined].

(a)   What is the geometric significance of the integral

$$\int_a^b [f(x) - g(x)]\, dx?$$

(b)   What is the geometric significance of the integral

$$\int_a^b |f(x) - g(x)|\, dx?$$

38. A rectangle with edges parallel to the coordinate axes has one vertex at the origin and the diagonally opposite vertex on the curve $y = kx^m$ at the point where $x = b$ $(b > 0, k > 0$, and $m \geq 0)$. Show that the fraction of the area of the rectangle that lies between the curve and the x-axis depends on m but not on k or b.

In Exercises 39 and 40, find the area to at least four decimal places. Use Newton's Method (Section 4.8) to approximate the x-coordinates of the points of intersection.

---

39. Approximate the area of the region that lies below the curve $y = \sin x$ and above the line $y = 0.2x$ where $x \geq 0$.

40. Approximate the area of the region enclosed by the graphs of $y = x^2$ and $y = \cos x$.

6.2 VOLUMES BY SLICING; DISKS AND WASHERS

In this section we shall use definite integrals to find volumes of three-dimensional solids.

□ **CYLINDERS**

A right-circular cylinder (Figure 6.2.1, center) can be generated by translating a plane circular disk along a line perpendicular to the disk. In general, we define a ***right cylinder*** to be any solid that can be generated by translating a plane region along a line or ***axis*** perpendicular to the region. All of the solids in Figure 6.2.1 are right cylinders, the center one being a ***right-circular cylinder***. Observe that all cross sections of a right cylinder taken perpendicular to the axis are identical in size and shape.

Cross sections perpendicular to the axis are identical in size and shape.

Figure 6.2.1

□ **THE METHOD OF SLICING**

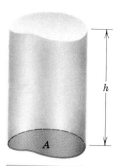

Volume = $A \cdot h$

Figure 6.2.2

If a right cylinder is generated by moving a plane region of area A through a distance h (Figure 6.2.2), then the volume V of the cylinder is *defined* to be

$$V = A \cdot h$$

that is, *the volume of a right cylinder is the cross-sectional area times the height*.

Volumes of solids that are neither right cylinders nor composed of finitely many right cylinders can be obtained by a technique called "slicing." To illustrate the idea, suppose that a solid S extends along the x-axis and is bounded on the left and right by planes perpendicular to the x-axis at $x = a$ and $x = b$ (Figure 6.2.3). Because the solid S is not assumed to be a right cylinder, its cross sections perpendicular to the x-axis can vary from point to point; we will denote by $A(x)$ the area of the cross section at x (Figure 6.2.3).

Let us divide the interval $[a, b]$ into n subintervals with widths

$$\Delta x_1, \Delta x_2, \ldots, \Delta x_n$$

by inserting points

$$x_1, x_2, \ldots, x_{n-1}$$

between a and b, and let us pass a plane perpendicular to the x-axis through each of these points. As illustrated in Figure 6.2.4, these planes cut the solid S into n slices

$$S_1, S_2, \ldots, S_n$$

Consider a typical slice S_k. In general, this slice may not be a right cylinder because its cross section can vary. However, if the slice is very thin, the cross section will not vary

Figure 6.2.3

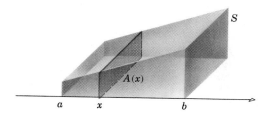

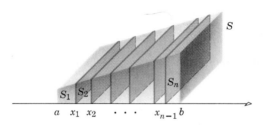

Figure 6.2.4

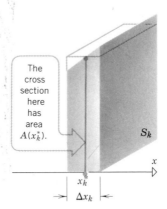

Figure 6.2.5

much. Therefore, if we choose an arbitrary point x_k^* in the kth subinterval, each cross section of slice S_k will be approximately the same as the cross section at x_k^*, and we can approximate slice S_k by a right cylinder of thickness Δx_k and cross-sectional area $A(x_k^*)$ (Figure 6.2.5).

Thus, the volume V_k of slice S_k is approximately the volume of this cylinder, namely,

$$V_k \approx A(x_k^*)\,\Delta x_k$$

and the volume V of the entire solid is approximately

$$V = V_1 + V_2 + \cdots + V_n \approx \sum_{k=1}^{n} A(x_k^*)\,\Delta x_k \tag{1}$$

If we now increase the number of slices in such a way that $\max \Delta x_k \to 0$, then the slices will become thinner and thinner and our approximations will get better and better. Thus, intuition suggests that approximation (1) will approach the exact value of the volume V as $\max \Delta x_k \to 0$, that is,

$$V = \lim_{\max \Delta x_k \to 0} \sum_{k=1}^{n} A(x_k^*)\,\Delta x_k \tag{2}$$

Since the right side of (2) is just the definite integral

$$\int_a^b A(x)\,dx$$

we are led to the following result.

VOLUMES BY CROSS SECTIONS PERPENDICULAR TO THE x-AXIS

6.2.1   VOLUME FORMULA.   Let S be a solid bounded by two parallel planes perpendicular to the x-axis at $x = a$ and $x = b$. If, for each x in $[a, b]$, the cross-sectional area of S perpendicular to the x-axis is $A(x)$, then the volume of the solid is

$$V = \int_a^b A(x)\,dx \tag{3}$$

provided $A(x)$ is integrable.

There is a similar result for cross sections perpendicular to the y-axis.

VOLUMES BY CROSS SECTIONS PERPENDICULAR TO THE y-AXIS

6.2.2   VOLUME FORMULA.   Let S be a solid bounded by two parallel planes perpendicular to the y-axis at $y = c$ and $y = d$. If, for each y in $[c, d]$, the cross-sectional area of S perpendicular to the y-axis is $A(y)$, then the volume of the solid is

$$V = \int_c^d A(y)\,dy \tag{4}$$

provided $A(y)$ is integrable.

Example 1   Derive the formula for the volume of a right pyramid whose altitude is h and whose base is a square with sides of length a.

Solution.   As illustrated in Figure 6.2.6a, we introduce a rectangular coordinate system so that the y-axis passes through the apex and is perpendicular to the base, and the x-axis passes through the base and is parallel to a side of the base.

At any point y in the interval $[0, h]$ on the y-axis, the cross section perpendicular to the y-axis is a square. If s denotes the length of a side of this square, then by similar triangles (Figure 6.2.6b)

$$\frac{\frac{1}{2}s}{\frac{1}{2}a} = \frac{h-y}{h} \quad \text{or} \quad s = \frac{a}{h}(h-y)$$

Thus, the area $A(y)$ of the cross section at y is

$$A(y) = s^2 = \frac{a^2}{h^2}(h-y)^2$$

and by (4) the volume is

$$V = \int_0^h A(y)\,dy = \int_0^h \frac{a^2}{h^2}(h-y)^2\,dy = \frac{a^2}{h^2}\int_0^h (h-y)^2\,dy$$

$$= \frac{a^2}{h^2}\left[-\frac{1}{3}(h-y)^3\right]_{y=0}^h = \frac{a^2}{h^2}\left[0 + \frac{1}{3}h^3\right] = \frac{1}{3}a^2h$$

That is, the volume is $\frac{1}{3}$ of the area of the base times the altitude.   ◀

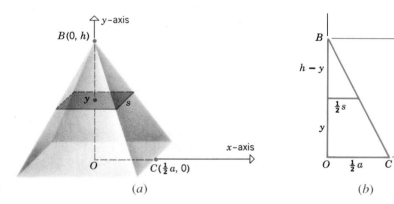

Figure 6.2.6   (a)   (b)

☐ **VOLUMES OF SOLIDS OF REVOLUTION**

Let f be nonnegative and continuous on $[a, b]$, and let R be the region bounded above by the graph of f, below by the x-axis, and on the sides by the lines $x = a$ and $x = b$ (Figure 6.2.7a). When this region is revolved about the x-axis, it generates a solid having circular cross sections (Figure 6.2.7b). Since the cross section at x has radius $f(x)$, the cross-sectional area is

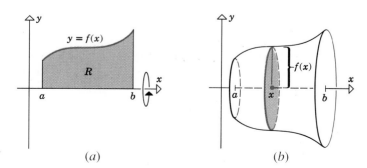

Figure 6.2.7   (a)   (b)

$$A(x) = \pi [f(x)]^2$$

Therefore, from (3), the volume of the solid is

$$V = \int_a^b \pi [f(x)]^2 \, dx \tag{5}$$

Because the cross sections are circular or disk shaped, the application of this formula is called the ***method of disks***.

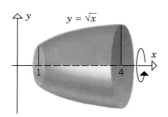

Figure 6.2.8

Example 2   Find the volume of the solid that is obtained when the region under the curve $y = \sqrt{x}$ over the interval $[1, 4]$ is revolved about the x-axis (Figure 6.2.8).

Solution.   From (5), the volume is

$$V = \int_a^b \pi [f(x)]^2 \, dx = \int_1^4 \pi x \, dx = \frac{\pi x^2}{2}\bigg]_1^4 = 8\pi - \frac{\pi}{2} = \frac{15\pi}{2} \quad \blacktriangleleft$$

Example 3   Derive the formula for the volume of a sphere of radius r.

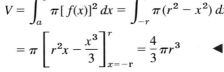

Solution.   As indicated in Figure 6.2.9, a sphere of radius r can be generated by revolving the upper half of the circle

$$x^2 + y^2 = r^2$$

about the x-axis. Since the upper half of this circle is the graph of $y = f(x) = \sqrt{r^2 - x^2}$, it follows from (5) that the volume of the sphere is

$$V = \int_a^b \pi [f(x)]^2 \, dx = \int_{-r}^r \pi (r^2 - x^2) \, dx$$

$$= \pi \left[r^2 x - \frac{x^3}{3} \right]_{x=-r}^r = \frac{4}{3} \pi r^3 \quad \blacktriangleleft$$

Figure 6.2.9

We shall now consider more general solids of revolution. Suppose that f and g are nonnegative continuous functions such that

$$g(x) \leq f(x) \quad \text{for} \quad a \leq x \leq b$$

and let R be the region enclosed between the graphs of these functions and the lines $x = a$ and $x = b$ (Figure 6.2.10a). When this region is revolved about the x-axis, it generates a solid having annular or washer-shaped cross sections (Figure 6.2.10b). Since the cross section at x has inner radius $g(x)$ and outer radius $f(x)$, its area is

$$A(x) = \pi [f(x)]^2 - \pi [g(x)]^2 = \pi ([f(x)]^2 - [g(x)]^2)$$

Therefore, from (3), the volume of the solid is

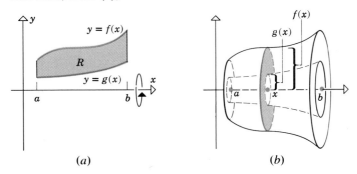

Figure 6.2.10 (a) (b)

$$V = \int_a^b \pi([f(x)]^2 - [g(x)]^2)\, dx \tag{6}$$

The application of this formula is called the **method of washers**.

Example 4 Find the volume of the solid generated when the region between the graphs of $f(x) = \frac{1}{2} + x^2$ and $g(x) = x$ over the interval $[0, 2]$ is revolved about the *x*-axis (Figure 6.2.11).

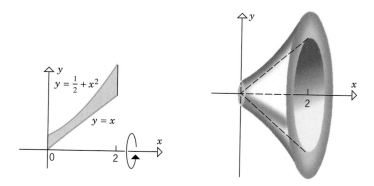

Figure 6.2.11

Solution. From (6) the volume is

$$V = \int_a^b \pi([f(x)]^2 - [g(x)]^2)\, dx = \int_0^2 \pi([\tfrac{1}{2} + x^2]^2 - x^2)\, dx$$

$$= \int_0^2 \pi\left(\frac{1}{4} + x^4\right) dx = \pi\left[\frac{x}{4} + \frac{x^5}{5}\right]_0^2 = \frac{69\pi}{10} \quad \blacktriangleleft$$

The methods of disks and washers have analogs for regions revolved about the *y*-axis. If the region of Figure 6.2.12a is revolved about the *y*-axis, the cross sections of the resulting solid taken perpendicular to the *y*-axis are disks, and it follows from Formula 6.2.2 that the volume of the solid in Figure 6.2.12b is

$$V = \int_c^d \pi[u(y)]^2\, dy \tag{7}$$

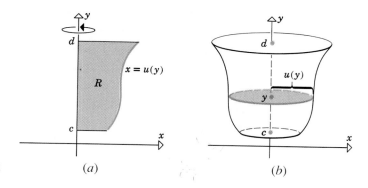

Figure 6.2.12   (a)   (b)

Also, if the region of Figure 6.2.13a is revolved about the *y*-axis, the cross sections taken perpendicular to the *y*-axis are washers, and it follows from Formula 6.2.2 that the volume of the solid in Figure 6.2.13b is

$$V = \int_c^d \pi \left([u(y)]^2 - [v(y)]^2\right) dy \qquad (8)$$

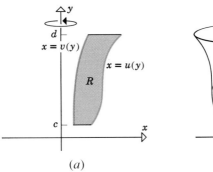
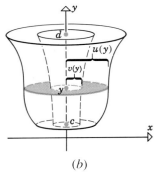

Figure 6.2.13   (a)   (b)

Example 5 Find the volume of the solid generated when the region enclosed by $y = \sqrt{x}$, $y = 2$, and $x = 0$ is revolved about the y-axis (Figure 6.2.14).

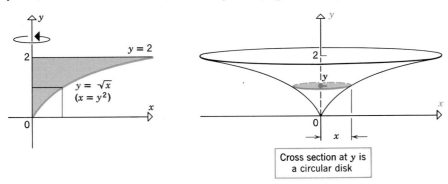

Cross section at y is a circular disk

Figure 6.2.14

Solution. The cross sections taken perpendicular to the y-axis are disks, so we shall apply (7). But first we must rewrite $y = \sqrt{x}$ as $x = y^2$. Thus, from (7) with $u(y) = y^2$, the volume is

$$V = \int_c^d \pi [u(y)]^2 \, dy = \int_0^2 \pi y^4 \, dy = \left. \frac{\pi y^5}{5} \right]_0^2 = \frac{32\pi}{5} \qquad \blacktriangleleft$$

▶ Exercise Set 6.2

In Exercises 1–4, find the volume of the solid that results when the shaded region is revolved about the indicated axis.

In Exercises 5–16, find the volume of the solid that results when the region enclosed by the given curves is revolved about the x-axis.

1.

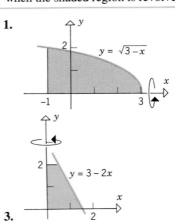

2.

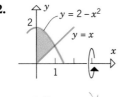

3.

4.

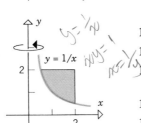

5. $y = x^2, x = 0, x = 2, y = 0.$

6. $y = \sec x, x = \pi/4, x = \pi/3, y = 0.$

7. $y = 1 + x^3, x = 1, x = 2, y = 0.$

8. $y = 1/x, x = 1, x = 4, y = 0.$ **9.** $y = 9 - x^2, y = 0.$

10. $y = \sqrt{\cos x}, x = \pi/4, x = \pi/2, y = 0.$

11. $y = x^2, y = 4x.$ **12.** $y = x^2, y = 9.$

13. $y = \sin x, y = \cos x, x = 0, x = \pi/4.$ [*Hint:* Use the identity $\cos 2x = \cos^2 x - \sin^2 x.$]

14. $y = x^2 + 1, y = x + 3.$

15. $y = \sqrt{x}, y = x.$ **16.** $y = x^2, y = x^3.$

In Exercises 17–28, find the volume of the solid that results when the region enclosed by the given curves is revolved about the y-axis.

17. $y = x^3, x = 0, y = 1$.

18. $x = 1 - y^2, x = 0$.

19. $x = \sqrt{1 + y}, x = 0, y = 3$.

20. $x = \sqrt{\cos y}, y = 0, y = \pi/2, x = 0$.

21. $x = \csc y, y = \pi/4, y = 3\pi/4, x = 0$.

22. $y = 2/x, y = 1, y = 3, x = 0$.

23. $x = \sqrt{9 - y^2}, y = 1, y = 3, x = 0$.

24. $y = x^2 - 1, x = 2, y = 0$.

25. $y = 1 + x^3, x = 1, y = 9$.

26. $y = x^2, x = y^2$.

27. $x = y^2, x = y + 2$.

28. $x = 1 - y^2, x = 2 + y^2, y = -1, y = 1$.

29. Find the volume of the solid that results when the region enclosed by the semicircle $y = \sqrt{25 - x^2}$ and the line $y = 3$ is revolved about the x-axis.

30. Find the volume of the torus that results when the region enclosed by the circle of radius r with center at $(h, 0)$, $h > r$, is revolved about the y-axis. [*Hint:* Use an appropriate formula from plane geometry to help evaluate the definite integral.]

31. Let V be the volume of the solid that results when the region enclosed by $y = 1/x$, $y = 0$, $x = 1$, and $x = b$ ($b > 1$) is revolved about the x-axis.
  (a) Find V.   (b) Find $\lim\limits_{b \to +\infty} V$.

32. Let h, k, and m be constants such that $h > 0$, $k > 0$, and $m \geq 0$, and let S be the solid that results when the region between the curve $y = kx^m$ and the interval $[0, h]$ on the x-axis is revolved about the x-axis. Let V be the volume of the solid, and let B be the area of the cross section of the solid at $x = h$.
  (a) Show that $V = hB$ when $m = 0$.
  (b) Show that $V = \frac{1}{2}hB$ when $m = \frac{1}{2}$.
  (c) Show that $V = \frac{1}{3}hB$ when $m = 1$.
  (d) Find m so that $V = \dfrac{1}{n} hB$, where n is a positive integer.

33. Let V be the volume of the solid that results when the region enclosed by $y = \sqrt{x}$, $y = 0$, and $x = b$ ($b > 0$) is revolved about the x-axis. Find the value of b for which $V = 2$.

34. Let V be the volume of the solid that results when the region enclosed by $y = 1/x$, $y = 0$, $x = 2$, and $x = b$ ($0 < b < 2$) is revolved about the x-axis. Find the value of b for which $V = 3$.

35. Find the volume of the solid that results when the region enclosed by $y = \sqrt{x}$, $y = 0$, and $x = 9$ is revolved about the line $x = 9$.

36. Find the volume of the solid that results when the region in Exercise 35 is revolved about the line $y = 3$.

37. Find the volume of the solid that results when the region enclosed by $x = y^2$ and $x = y$ is revolved about the line $y = -1$.

38. Find the volume of the solid that results when the region in Exercise 37 is revolved about the line $x = -1$.

39. Find the volume of the solid that results when the region above the x-axis and below the ellipse
$$\frac{x^2}{a^2} + \frac{y^2}{b^2} = 1 \quad (a > 0, b > 0)$$
is revolved about the x-axis.

40. Find the volume of the solid generated when the region enclosed by $y = \sqrt{x}$, $y = 6 - x$, and $y = 0$ is revolved about the x-axis. [*Hint:* Split the solid into two parts.]

41. Find the volume of the solid generated when the region enclosed by $y = \sqrt{x + 1}$, $y = \sqrt{2x}$, and $y = 0$ is revolved about the x-axis. [*Hint:* Split the solid into two parts.]

42. Let R_1, R_2, R_3, and R_4 be the regions indicated in Figure 6.2.15. Express the following as definite integrals:
  (a) The volume of the solid generated when R_1 is revolved about the x-axis.
  (b) The volume of the solid generated when R_2 is revolved about the x-axis.
  (c) The volume of the solid generated when R_3 is revolved about the y-axis.
  (d) The volume of the solid generated when R_4 is revolved about the y-axis.

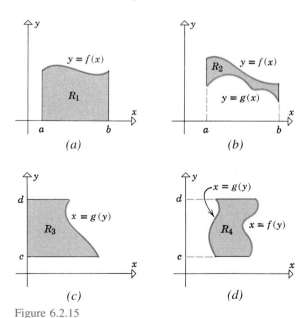

Figure 6.2.15

43. Derive the formula for the volume of a right-circular cone with radius r and height h.

44. A hole of radius $r/2$ is drilled through the center of a sphere of radius r. Find the volume of the remaining solid.

45. A cylindrical hole is drilled all the way through the center of a sphere (Figure 6.2.16). Show that the volume of the remaining solid depends only on the length L of the hole, and not on the size of the sphere.

Figure 6.2.16

46. (a) A vat, shaped like a hemisphere of radius r ft, is filled with a fluid to a depth of h ft. Find the volume of the fluid.

(b) If fluid enters a hemispherical vat with a radius of 10 ft at a rate of $\frac{1}{2}$ ft^3/min, how fast will the fluid be rising when the depth is 5 ft?

47. A cocktail glass with a bowl shaped like a hemisphere of diameter 8 cm contains a cherry with a diameter of 2 cm (Figure 6.2.17). If the glass is filled to a depth of h cm, what is the volume of liquid it contains? [*Hint:* First consider the case where the cherry is partially submerged, then the case where it is totally submerged.]

Figure 6.2.17

48. A *general cylinder* is any solid that can be generated by translating a plane region along a line passing through but not contained in the region. Derive a formula for the volume of a general cylinder with base area A and height h (Figure 6.2.18).

Figure 6.2.18

49. A nose cone for a space reentry vehicle is designed so that a cross section, taken x ft from the tip and perpendicular to the axis of symmetry, is a circle of radius $\frac{1}{4}x^2$ ft. Find the volume of the nose cone given that its length is 20 ft.

50. A certain solid is 1 ft high, and a horizontal cross section taken x ft above the bottom of the solid is an annulus of inner radius x^2 and outer radius $\sqrt{x}$. Find the volume of the solid.

51. Find the volume of the solid whose base is the region bounded between the curves $y = x$ and $y = x^2$, and whose cross sections perpendicular to the x-axis are squares.

52. Find the volume of the solid whose base is the triangular region with vertices $(0, 0)$, $(a, 0)$, and $(0, b)$, where $a > 0$

and $b > 0$, and whose cross sections perpendicular to the y-axis are semicircles.

53. The base of a certain solid is the region enclosed by the circle $x^2 + y^2 = 9$, and each cross section perpendicular to the x-axis is an equilateral triangle with one side across the base. Find the volume of the solid.

54. The base of a certain solid is the region enclosed by $y = \sqrt{x}$, $y = 0$, and $x = 4$. Every cross section perpendicular to the x-axis is a semicircle with its diameter across the base. Find the volume of the solid.

55. The base of a certain solid is the region enclosed by $y = \sin x$, $y = 0$, $x = \pi/4$, and $x = 3\pi/4$; and every cross section perpendicular to the x-axis is a square with one side across the base. Find the volume of the solid. [*Hint:* To help with the integration, use the identity $\sin^2 x = \frac{1}{2}(1 - \cos 2x)$.]

56. The base of a certain solid is the region enclosed by $y = 1/x$, $y = 0$, $x = 1$, and $x = 3$. Every cross section perpendicular to the x-axis is an isosceles right triangle with its hypotenuse across the base. Find the volume of the solid.

57. A wedge is cut from a right-circular cylinder of radius r by two planes, one perpendicular to the axis of the cylinder and the other making an angle θ with the first. Find the volume of the wedge by slicing perpendicular to the y-axis as shown in Figure 6.2.19.

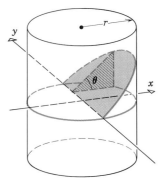

Figure 6.2.19

58. Find the volume of the wedge described in Exercise 57 by slicing perpendicular to the x-axis.

59. Two right-circular cylinders of radius r have axes that intersect at right angles. Find the volume of the solid common to the two cylinders. [*Hint:* One-eighth of the solid is sketched in Figure 6.2.20.]

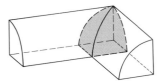

Figure 6.2.20

60. In 1635 Bonaventura Cavalieri, a student of Galileo, stated the following result, called *Cavalieri's principle*: *If two solids have the same height, and if the areas of their cross sections taken parallel to and at equal distances* *from their bases are always equal, then the solids have the same volume.* Prove this result.

61. Suppose that an open container of unspecified shape contains water that is evaporating into the air (Figure 6.2.21). Let $V(t)$ and $h(t)$ be the volume and depth of the water, respectively, at time t, and let $A(h)$ be the area of the water surface when the depth is h. Assume that the volume of water decreases at a rate that is proportional to the area of the water surface, that is,

$$\frac{dV}{dt} = -kA(h)$$

where k is a positive constant.

(a) Show that

$$\frac{dh}{dt} = -k$$

that is, the depth decreases at a constant rate regardless of the shape of the container. [*Hint:* At any depth h, $V = \int_0^h A(x)\ dx$.]

(b) If at time $t = 0$ the depth is h_0, how long will it take for all of the water to evaporate? [*Hint:* Solve the equation in part (a) for h in terms of t.]

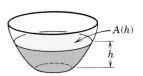

Figure 6.2.21

■ **6.3** VOLUMES BY CYLINDRICAL SHELLS

The methods discussed so far for computing volumes of solids depend on our ability to compute cross-sectional areas of the solid. In this section we shall develop an alternative technique for computing volumes that can sometimes be applied when it is inconvenient or difficult to determine the cross-sectional areas or if the integration is too difficult.

☐ **CYLINDRICAL SHELLS**

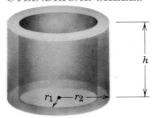

Figure 6.3.1

☐ **VOLUME OF A CYLINDRICAL SHELL**

A *cylindrical shell* is a solid enclosed by two concentric right-circular cylinders (Figure 6.3.1). The volume V of a cylindrical shell having inner radius r_1, outer radius r_2, and height h can be written as

$$V = [\text{area of cross section}] \cdot [\text{height}] = (\pi r_2^2 - \pi r_1^2)h$$

$$= \pi(r_2 + r_1)(r_2 - r_1)h = 2\pi \cdot [\tfrac{1}{2}(r_1 + r_2)] \cdot h \cdot (r_2 - r_1)$$

But $\tfrac{1}{2}(r_1 + r_2)$ is the average radius of the shell and $r_2 - r_1$ is its thickness, so

$$V = 2\pi \cdot [\text{average radius}] \cdot [\text{height}] \cdot [\text{thickness}] \tag{1}$$

We shall now show how this formula can be used to find the volume of a solid of revolution.

Let R be a plane region bounded above by a continuous curve $y = f(x)$, bounded below by the x-axis, and bounded on the left and right, respectively, by the lines $x = a$ and $x = b$. Let S be the solid generated by revolving the region R about the y-axis (Figure 6.3.2).

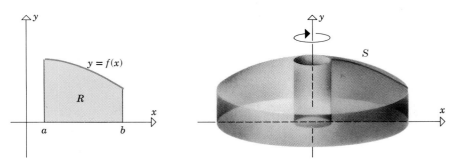

Figure 6.3.2

To find the volume of S, let us divide the interval $[a, b]$ into n subintervals with widths $\Delta x_1, \Delta x_2, \ldots, \Delta x_n$ by inserting points $x_1, x_2, \ldots, x_{n-1}$ between a and b, and let us draw a vertical line through each of these points to divide the region R into n strips $R_1, R_2, \ldots, R_n$ (Figure 6.3.3a).

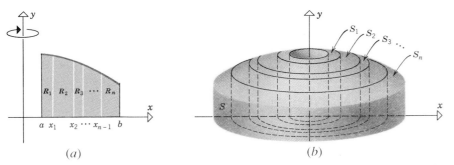

Figure 6.3.3 (a) (b)

These strips, when revolved about the y-axis, generate solids $S_1, S_2, \ldots, S_n$. As illustrated in Figure 6.3.3b, these solids are nested one inside the other and together form the entire solid S. Thus, the volume of the solid S can be obtained by adding together the volumes of solids $S_1, S_2, \ldots, S_n$:

$$V(S) = V(S_1) + V(S_2) + \cdots + V(S_n) \tag{2}$$

Consider a typical strip R_k and the solid S_k that it generates (Figure 6.3.4). Although solid S_k resembles a cylindrical shell, it will not, in general, be a cylindrical shell because it can have a curved upper surface. However, if the interval width

$$\Delta x_k = x_k - x_{k-1}$$

is small, we can obtain a good approximation to the region R_k by a rectangle of width Δx_k and height $f(x_k^*)$, where

$$x_k^* = \frac{x_k + x_{k-1}}{2}$$

is the midpoint of the interval $[x_{k-1}, x_k]$ (Figure 6.3.5a). This rectangle, when revolved about the y-axis, generates a cylindrical shell, which is a good approximation to the solid S_k (Figure 6.3.5b).

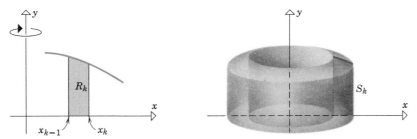

Figure 6.3.4

This cylindrical shell has thickness Δx_k, height $f(x_k^*)$, and average radius x_k^*, so by (1)

$$V(S_k) \approx [\text{volume of approximating cylindrical shell}] = 2\pi x_k^* f(x_k^*) \, \Delta x_k$$

Thus, from (2), the volume, $V(S)$, of the entire solid S is approximately

$$\sum_{k=1}^{n} 2\pi x_k^* f(x_k^*) \, \Delta x_k \tag{3}$$

If we now divide $[a, b]$ into more and more subintervals in such a way that $\max \Delta x_k \to 0$, then intuition suggests that our approximations will tend to get better and (3) will approach the exact value of the volume, that is,

$$V(S) = \lim_{\max \Delta x_k \to 0} \sum_{k=1}^{n} 2\pi x_k^* f(x_k^*) \, \Delta x_k \tag{4}$$

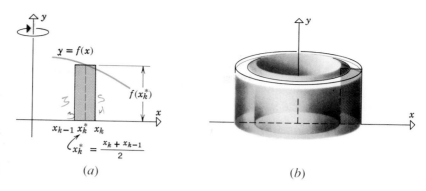

Figure 6.3.5 (a) (b)

Because the right side of (4) is just the definite integral

$$\int_a^b 2\pi x f(x)\, dx$$

we are led to the following result.

☐ **CYLINDRICAL SHELLS CENTERED ON THE y-AXIS**

> **6.3.1** VOLUME FORMULA. Let R be a plane region bounded above by a continuous curve $y = f(x)$, below by the x-axis, and on the left and right, respectively, by the lines $x = a$ and $x = b$. Then the volume of the solid generated by revolving R about the y-axis is given by
>
> $$V = \int_a^b 2\pi x f(x)\, dx \tag{5}$$

Example 1 Use cylindrical shells to find the volume of the solid generated when the region enclosed between $y = \sqrt{x}$, $x = 1$, $x = 4$, and the x-axis is revolved about the y-axis (Figure 6.3.6).

Solution. Since $f(x) = \sqrt{x}$, $a = 1$, and $b = 4$, Formula (5) yields

$$V = \int_1^4 2\pi x \sqrt{x}\, dx = 2\pi \int_1^4 x^{3/2}\, dx$$

$$= \left[2\pi \cdot \frac{2}{5} x^{5/2} \right]_1^4 = \frac{4\pi}{5}[32 - 1] = \frac{124\pi}{5} \quad \blacktriangleleft$$

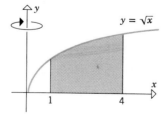

Figure 6.3.6

Cutaway view of the solid

☐ **VARIATIONS OF THE METHOD OF CYLINDRICAL SHELLS**

The method of cylindrical shells is applicable in a variety of situations that do not fit the conditions required by Formula (5). For example, the solid may be generated by a revolution about the y-axis or some other line, or the region may be enclosed between two curves rather than a curve and a coordinate axis. Rather than develop separate formulas for each of these situations, we shall discuss a general way of thinking about the method of cylindrical shells that will enable the reader to adapt the method to a variety of situations. Consider Formula (5). At each point x in $[a, b]$ the vertical line through x cuts the region R in a line segment that we can view as the vertical "cross section" of R at x (Figure 6.3.7a). When the region R is revolved about the y-axis, the vertical cross section at x generates the *surface* of a right-circular cylinder having height $f(x)$ and radius x (Figure 6.3.7b). The area of this surface is

$$2\pi x f(x)$$

(see Figure 6.3.7c), which is precisely the integrand in (5). Thus, 6.3.1 can be paraphrased as follows:

> *The volume V by cylindrical shells is the integral of the surface area generated by an arbitrary cross section of R taken parallel to the axis about which R is revolved.*

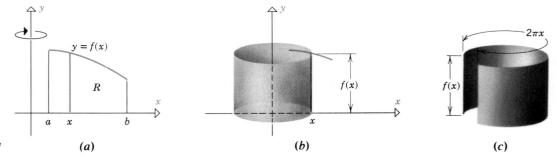

Figure 6.3.7   *(a)*   *(b)*   *(c)*

The following examples illustrate how this result can be used to compute volumes by cylindrical shells in situations where Formula (5) is not applicable.

Example 2   Use cylindrical shells to find the volume of the solid generated when the region R in the first quadrant enclosed between $y = x$ and $y = x^2$ is revolved about the y-axis (Figure 6.3.8).

Solution.   At each x in $[0, 1]$ the cross section of R parallel to the y-axis generates a cylindrical surface of height $x - x^2$ and radius x. Since the area of this surface is

$$2\pi x(x - x^2)$$

the volume of the solid is

$$V = \int_0^1 2\pi x(x - x^2)\, dx = 2\pi \int_0^1 (x^2 - x^3)\, dx$$

$$= 2\pi \left[\frac{x^3}{3} - \frac{x^4}{4} \right]_0^1 = 2\pi \left[\frac{1}{3} - \frac{1}{4} \right] = \frac{\pi}{6} \quad \blacktriangleleft$$

This solid looks like a bowl with a cone-shaped interior.

Figure 6.3.8

Example 3   Use cylindrical shells to find the volume of the solid generated when the region R under $y = x^2$ over the interval $[0, 2]$ is revolved about the x-axis (Figure 6.3.9).

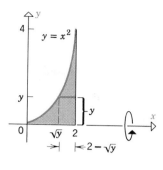

Figure 6.3.9

Solution.   At each y in the interval $0 \leq y \leq 4$, the cross section of R parallel to the x-axis generates a cylindrical surface of height $2 - \sqrt{y}$ and radius y. Since the area of this surface is

$$2\pi y(2 - \sqrt{y})$$

the volume of the solid is

$$V = \int_0^4 2\pi y(2 - \sqrt{y})\,dy = 2\pi \int_0^4 (2y - y^{3/2})\,dy$$

$$= 2\pi \left[y^2 - \frac{2}{5} y^{5/2} \right]_0^4 = \frac{32\pi}{5} \qquad \blacktriangleleft$$

The volume in the preceding example can also be obtained by the method of disks [Formula (5) of Section 6.2]. The computations are

$$V = \int_0^2 \pi (x^2)^2\,dx = \int_0^2 \pi x^4\,dx = \frac{\pi x^5}{5} \Big]_0^2 = \frac{32}{5}\pi$$

▶ Exercise Set 6.3

In Exercises 1–4, use cylindrical shells to find the volume of the solid generated when the shaded region is revolved about the indicated axis.

In Exercises 5–12, use cylindrical shells to find the volume of the solid generated when the region enclosed by the given curves is revolved about the y-axis.

1.

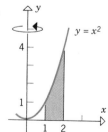

2.

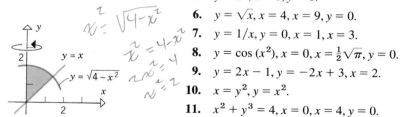

5. $y = x^3, x = 1, y = 0$.

6. $y = \sqrt{x}, x = 4, x = 9, y = 0$.

7. $y = 1/x, y = 0, x = 1, x = 3$.

8. $y = \cos(x^2), x = 0, x = \frac{1}{2}\sqrt{\pi}, y = 0$.

9. $y = 2x - 1, y = -2x + 3, x = 2$.

10. $x = y^2, y = x^2$.

11. $x^2 + y^3 = 4, x = 0, x = 4, y = 0$.

12. $y = 2x - x^2, y = 0$.

In Exercises 13–16, use cylindrical shells to find the volume of the solid generated when the region enclosed by the given curves is revolved about the x-axis.

13. $y^2 = x, y = 1, x = 0$.

14. $x = 2y, y = 2, y = 3, x = 0$.

3.

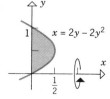

4.

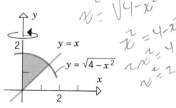

15. $y = x^2, x = 1, y = 0.$

16. $xy = 4, x + y = 5.$

17. (a) Show by differentiation that

$$\int x \sin x \, dx = \sin x - x \cos x + C$$

(b) Find the volume of the solid generated when the region enclosed by $y = \sin x$ and $y = 0$ for $0 \le x \le \pi$ is revolved about the y-axis.

18. (a) Show by differentiation that

$$\int x \cos x \, dx = \cos x + x \sin x + C$$

(b) Find the volume of the solid generated when the region enclosed by $y = \cos x$, $y = 0$, and $x = 0$ for $0 \le x \le \pi/2$ is revolved about the y-axis.

19. (a) Use cylindrical shells to find the volume of the solid that is generated when the region under the curve $y = x^3 - 3x^2 + 2x$ over $[0, 1]$ is revolved about the y-axis.

(b) For this problem, is the method of cylindrical shells easier or harder than the method of slicing discussed in the last section? Explain.

20. Use cylindrical shells to find the volume of the solid that is generated when the region that is enclosed by $y = 1/x^3$, $x = 1$, $x = 2$, $y = 0$ is revolved about the line $x = -1$.

21. Use cylindrical shells to find the volume of the solid that is generated when the region that is enclosed by $y = x^3$, $y = 1$, $x = 0$ is revolved about the line $y = 1$.

22. Let R_1 and R_2 be regions of the form shown in Figure 6.3.10. Use cylindrical shells to find a formula for the volume of the solid that results when

(a) region R_1 is revolved about the y-axis

(b) region R_2 is revolved about the x-axis.

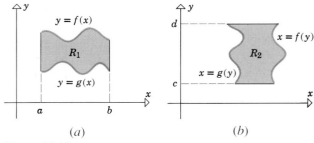

Figure 6.3.10

23. Use cylindrical shells to find the volume of the cone generated when the triangle with vertices $(0, 0)$, $(0, r)$, $(h, 0)$, where $r > 0$ and $h > 0$, is revolved about the x-axis.

24. The region enclosed between the curve $y^2 = kx$ and the line $x = \frac{1}{4}k$ is revolved about the line $x = \frac{1}{2}k$. Use cylindrical shells to find the volume of the resulting solid. (Assume $k > 0$.)

25. A round hole of radius a is drilled through the center of a solid sphere of radius r. Use cylindrical shells to find the volume of the portion removed. (Assume $r > a$.)

26. Use cylindrical shells to find the volume of the torus obtained by revolving the circle

$$x^2 + y^2 = a^2$$

about the line $x = b$, where $b > a$. [*Hint:* It may help in the integration to think of an integral as an area.]

27. Let V_x and V_y be the volumes of the solids that result when the region enclosed by $y = 1/x$, $y = 0$, $x = \frac{1}{2}$, and $x = b$ $(b > \frac{1}{2})$ is revolved about the x-axis and y-axis, respectively. Is there a value of b for which $V_x = V_y$?

■ **6.4** LENGTH OF A PLANE CURVE

> *In this section we shall use definite integrals to find arc lengths of plane curves. To start, we shall consider only curves that are graphs of functions. In a later section, we shall extend our results to more general curves.*

☐ **ARC LENGTH**

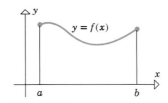

Figure 6.4.1

If f' is continuous on an interval, we shall say that $y = f(x)$ is a ***smooth curve*** (or f is a ***smooth function***) on that interval. We shall restrict our discussion of arc length to smooth curves in order to eliminate some complications that would otherwise occur.

> **6.4.1** ARC LENGTH PROBLEM. *Suppose that f is a smooth function on the interval $[a, b]$. Find the arc length L of the curve $y = f(x)$ over the interval $[a, b]$ (Figure 6.4.1).*

In order to solve Problem 6.4.1 we must first define the term "arc length" precisely. For motivation, consider the graph of a smooth curve $y = f(x)$ over an interval $[a, b]$, and as shown in Figure 6.4.2, divide the interval $[a, b]$ into n subintervals with widths $\Delta x_1, \Delta x_2, \ldots, \Delta x_n$ by inserting points $x_1, x_2, \ldots, x_{n-1}$ between a and b. Let $P_0, P_1, \ldots, P_n$ be the points on the curve whose x-coordinates are $a, x_1, x_2, \ldots, x_{n-1}, b$ and join these points with straight line segments. These segments form a ***polygonal path*** that we can regard as an approximation to the curve $y = f(x)$. Intuition suggests that the length of the approximating polygonal path will approach the length of the curve if we increase the number of points in such a way that the lengths of the line segments in the polygonal path approach zero.

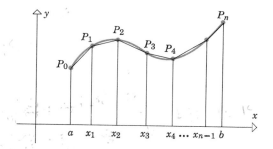

Figure 6.4.2

To examine this idea more closely, let us isolate a typical subinterval, say the kth (Figure 6.4.3). As suggested by this figure, the length L_k of the kth line segment in the polygonal path is given by

$$L_k = \sqrt{(\Delta x_k)^2 + (\Delta y_k)^2} = \sqrt{(\Delta x_k)^2 + [f(x_k) - f(x_{k-1})]^2} \tag{1}$$

By the Mean-Value Theorem (4.9.2) there is a point x_k^* between x_{k-1} and x_k such that

$$\frac{f(x_k) - f(x_{k-1})}{x_k - x_{k-1}} = f'(x_k^*) \quad \text{or} \quad f(x_k) - f(x_{k-1}) = f'(x_k^*)\,\Delta x_k$$

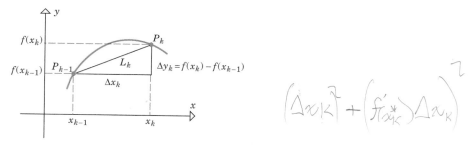

Figure 6.4.3

Thus, (1) can be rewritten as

$$L_k = \sqrt{1 + [f'(x_k^*)]^2}\,\Delta x_k$$

(verify), which means that the length of the *entire* polygonal path is

$$\sum_{k=1}^{n} L_k = \sum_{k=1}^{n} \sqrt{1 + [f'(x_k^*)]^2}\,\Delta x_k$$

If we now increase the number of subintervals in such a way that max $\Delta x_k \to 0$, then the length of the polygonal path will approach the arc length L of the curve $y = f(x)$ over $[a, b]$. Thus,

$$L = \lim_{\text{max}\,\Delta x_k \to 0} \sum_{k=1}^{n} \sqrt{1 + [f'(x_k^*)]^2}\,\Delta x_k \tag{2}$$

Since the right side of (2) is just the definite integral

$$\int_a^b \sqrt{1 + [f'(x)]^2}\,dx$$

we are led to the following result.

6.4.2 ARC LENGTH FORMULAS. If f is a smooth function on $[a, b]$, then the **arc length** L of the curve $y = f(x)$ from $x = a$ to $x = b$ is defined by

$$L = \int_a^b \sqrt{1 + [f'(x)]^2}\, dx = \int_a^b \sqrt{1 + \left(\frac{dy}{dx}\right)^2}\, dx \tag{3}$$

Similarly, for a curve expressed in the form $x = g(y)$, where g' is continuous on $[c, d]$, the arc length L from $y = c$ to $y = d$ is defined by

$$L = \int_c^d \sqrt{1 + [g'(y)]^2}\, dy = \int_c^d \sqrt{1 + \left(\frac{dx}{dy}\right)^2}\, dy \tag{4}$$

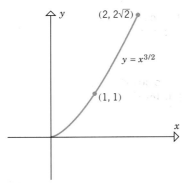

Figure 6.4.4

Example 1 Find the arc length of the curve $y = x^{3/2}$ from $(1, 1)$ to $(2, 2\sqrt{2})$ (Figure 6.4.4) using

(a) Formula (3)   (b) Formula (4)

Solution (a).   Since $f(x) = x^{3/2}$,

$$f'(x) = \tfrac{3}{2} x^{1/2}$$

Thus, from (3), the arc length from $x = 1$ to $x = 2$ is

$$L = \int_1^2 \sqrt{1 + \frac{9}{4} x}\, dx$$

To evaluate this integral we make the u-substitution

$$u = 1 + \tfrac{9}{4} x, \quad du = \tfrac{9}{4}\, dx$$

and then change the x-limits of integration ($x = 1, x = 2$) to the corresponding u-limits ($u = \frac{13}{4}, u = \frac{22}{4}$):

$$L = \frac{4}{9} \int_{13/4}^{22/4} u^{1/2}\, du = \frac{8}{27} u^{3/2} \Big]_{13/4}^{22/4} = \frac{8}{27} \left[\left(\frac{22}{4}\right)^{3/2} - \left(\frac{13}{4}\right)^{3/2} \right]$$

$$= \frac{22\sqrt{22} - 13\sqrt{13}}{27}$$

Solution (b).   Solving $y = x^{3/2}$ for x in terms of y, we obtain $x = y^{2/3}$. Hence $g(y) = y^{2/3}$ and

$$g'(y) = \tfrac{2}{3} y^{-1/3}$$

Thus, from (4), the arc length from $y = 1$ to $y = 2\sqrt{2}$ is

$$L = \int_1^{2\sqrt{2}} \sqrt{1 + \frac{4}{9} y^{-2/3}}\, dy = \frac{1}{3} \int_1^{2\sqrt{2}} y^{-1/3} \sqrt{9 y^{2/3} + 4}\, dy$$

To evaluate this integral we make the u-substitution

$$u = 9 y^{2/3} + 4, \quad du = 6 y^{-1/3}\, dy$$

and change the y-limits of integration ($y = 1, y = 2\sqrt{2}$) to the corresponding u-limits ($u = 13, u = 22$). This gives

$$L = \frac{1}{18} \int_{13}^{22} u^{1/2} \, du = \frac{1}{27} u^{3/2} \Big]_{13}^{22}$$

$$= \frac{1}{27} \left[(22)^{3/2} - (13)^{3/2} \right] = \frac{22\sqrt{22} - 13\sqrt{13}}{27}$$

This result agrees with that in part (a); however, the integration here is more tedious. In problems where there is a choice between using (3) or (4), it is often the case that one of the formulas leads to a simpler integral than the other.   ◄

▶ Exercise Set 6.4 $\boxed{C}$ 13, 14

1. Find the arc length of the curve $y = 2x$ from $(1, 2)$ to $(2, 4)$ using

   (a)   Formula (3)   (b)   Formula (4)

   (c)   the Theorem of Pythagoras.

2. Find the arc length of the curve $y = mx + b$ from $x = k_1$ to $x = k_2$ $(m \neq 0, k_2 > k_1)$ using

   (a)   Formula (3)   (b)   Formula (4)

   (c)   the Theorem of Pythagoras.

3. Find the arc length of the curve $y = 3x^{3/2} - 1$ from $x = 0$ to $x = 1$.

4. Find the arc length of the curve $x = \frac{1}{3}(y^2 + 2)^{3/2}$ from $y = 0$ to $y = 1$.

5. Find the arc length of the curve $y = x^{2/3}$ from $x = 1$ to $x = 8$.

6. Find the arc length of the curve $y = \dfrac{x^4}{16} + \dfrac{1}{2x^2}$ from $x = 2$ to $x = 3$.

7. Find the arc length of the curve $24xy = y^4 + 48$ from $y = 2$ to $y = 4$.

8. Find the arc length of the curve $x = \frac{1}{8}y^4 + \frac{1}{4}y^{-2}$ from $y = 1$ to $y = 4$.

9. Consider the curve $y = x^{2/3}$.

   (a)   Sketch the portion of the curve between $x = -1$ and $x = 8$.

   (b)   Explain why Formula (3) cannot be used to find the arc length of the curve sketched in part (a).

   (c)   Find the arc length of the curve sketched in part (a).

10. Find the arc length in the second quadrant of the curve $x^{2/3} + y^{2/3} = a^{2/3}$ from the point $x = -a$ to $x = -\frac{1}{8}a$ $(a > 0)$.

11. Let $y = f(x)$ be a smooth curve on the closed interval $[a, b]$. Prove that if there are nonnegative numbers m and M such that $m \leq f'(x) \leq M$ for all x in $[a, b]$, then the arc length L of $y = f(x)$ over the interval $[a, b]$ satisfies the inequalities

$$(b - a)\sqrt{1 + m^2} \leq L \leq (b - a)\sqrt{1 + M^2}$$

12. Use the result of Exercise 11 to show that the arc length L of $y = \sin x$ over the interval $0 \leq x \leq \pi/4$ satisfies

$$\frac{\pi}{4}\sqrt{\frac{3}{2}} \leq L \leq \frac{\pi}{4}\sqrt{2}$$

In Exercises 13 and 14, use the midpoint approximation with $n = 20$ subintervals to approximate the arc length of the curve over the given interval.

13. $y = x^2$; $[0, 2]$.   14. $y = \sin x$; $[0, \pi]$.

■ **6.5**   AREA OF A SURFACE OF REVOLUTION

In this section we shall apply the definite integral to the problem of finding the area of a surface of revolution.

☐ **DEFINITION OF SURFACE AREA**

6.5.1   SURFACE AREA PROBLEM.   *Let f be a smooth, nonnegative function on $[a, b]$. Find the area of the surface generated by revolving the portion of the curve $y = f(x)$ between $x = a$ and $x = b$ about the x-axis (Figure 6.5.1).*

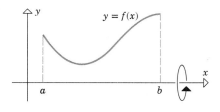

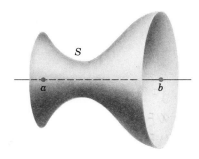

Figure 6.5.1

In order to solve this problem we must first define the term ''surface area'' precisely. For motivation, let us approximate the curve $y = f(x)$ by a polygonal path of straight line segments connecting the points on the curve that have x-coordinates $a, x_1, x_2, \ldots, x_{n-1}, b$ (Figure 6.5.2a). As usual, let

$$\Delta x_1, \Delta x_2, \ldots, \Delta x_n$$

be the widths of the subintervals determined by these x-coordinates. If these widths are small, then the surface generated by revolving the polygonal path about the x-axis will have approximately the same area as the surface generated by revolving the curve $y = f(x)$ about the x-axis (Figure 6.5.2b).

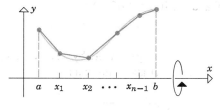

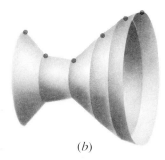

Figure 6.5.2   (a)   (b)

Observe that the surface generated by the polygonal path is made of parts, each of which is a frustum of a cone. Thus, the lateral area of each of these parts can be obtained from the formula

$$S = \pi(r_1 + r_2)l \tag{1}$$

Figure 6.5.3

for the lateral area S of a frustum of slant height l and base radii r_1 and r_2 (Figure 6.5.3). A derivation of this formula is discussed in Exercise 17. Intuition suggests that the area of the approximating surface will approach the desired area of the surface if we increase the number of subdivisions in such a way that the lengths of the line segments in the polygonal path approach zero.

To examine this idea more closely, let us isolate a typical section of the approximating surface, say the kth (Figure 6.5.4). Applying Formula (1) above and Formula (1) of the preceding seciton, we obtain as the lateral area S_k of this kth section

$$S_k = \pi[f(x_{k-1}) + f(x_k)]\sqrt{(\Delta x_k)^2 + [f(x_k) - f(x_{k-1})]^2} \tag{2}$$

By the Mean-Value Theorem (4.9.2) there is a point x_k^* between x_{k-1} and x_k such that

$$\frac{f(x_k) - f(x_{k-1})}{x_k - x_{k-1}} = f'(x_k^*) \quad \text{or} \quad f(x_k) - f(x_{k-1}) = f'(x_k^*)\,\Delta x_k$$

Thus, (2) can be rewritten as

$$S_k = \pi[f(x_{k-1}) + f(x_k)]\sqrt{1 + [f'(x_k^*)]^2}\,\Delta x_k \tag{3}$$

Since the arithmetic average of two numbers lies between those numbers, the quantity $\frac{1}{2}[f(x_{k-1}) + f(x_k)]$ is between $f(x_{k-1})$ and $f(x_k)$. Thus, because f is continuous on the

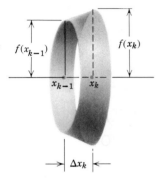

Figure 6.5.4

interval $[x_{k-1}, x_k]$, the Intermediate-Value Theorem (2.7.9) implies that there exists a point x_k^{**} in this interval such that

$$\tfrac{1}{2}[f(x_{k-1}) + f(x_k)] = f(x_k^{**})$$

Thus, (3) can be rewritten as

$$S_k = 2\pi f(x_k^{**})\sqrt{1 + [f'(x_k^{*})]^2}\,\Delta x_k$$

so that the area of the *entire* polygonal surface is

$$\sum_{k=1}^{n} S_k = \sum_{k=1}^{n} 2\pi f(x_k^{**})\sqrt{1 + [f'(x_k^{*})]^2}\,\Delta x_k$$

If we now increase the number of subintervals in such a way that max $\Delta x_k \to 0$, then the area of the approximating polygonal surface will approach the exact surface area S. Thus,

$$S = \lim_{\max \Delta x_k \to 0} \sum_{k=1}^{n} 2\pi f(x_k^{**})\sqrt{1 + [f'(x_k^{*})]^2}\,\Delta x_k \tag{4}$$

If it were the case that $x_k^{**} = x_k^{*}$, then the right side of (4) would be the definite integral

$$\int_a^b 2\pi f(x)\sqrt{1 + [f'(x)]^2}\,dx \tag{5}$$

However, it is proved in advanced calculus that this is true even if $x_k^{*} \neq x_k^{**}$ because of the continuity of f and f'. In light of (4) and (5) we are led to the following results.

6.5.2 SURFACE AREA FORMULAS. Let f be a smooth, nonnegative function on $[a, b]$. Then the **surface area S** generated by revolving the portion of the curve $y = f(x)$ between $x = a$ and $x = b$ about the x-axis is

$$S = \int_a^b 2\pi f(x)\sqrt{1 + [f'(x)]^2}\,dx \tag{6}$$

For a curve expressed in the form $x = g(y)$, where g' is continuous on $[c, d]$, and $g(y) \geq 0$ for $c \leq y \leq d$, the surface area S generated by revolving the portion of the curve from $y = c$ to $y = d$ about the y-axis is given by

$$S = \int_c^d 2\pi g(y)\sqrt{1 + [g'(y)]^2}\,dy \tag{7}$$

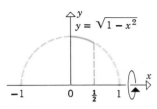

Figure 6.5.5

Example 1 Find the surface area of the portion of the sphere generated by revolving the curve

$$y = \sqrt{1 - x^2}, \quad 0 \leq x \leq \tfrac{1}{2}$$

about the x-axis (Figure 6.5.5).

Solution. Since $f(x) = \sqrt{1 - x^2}$,

$$f'(x) = -\frac{x}{\sqrt{1 - x^2}}$$

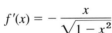

Thus, (6) yields

$$S = \int_0^{1/2} 2\pi\sqrt{1 - x^2}\,\sqrt{1 + \frac{x^2}{1 - x^2}}\,dx = \int_0^{1/2} 2\pi\,dx = 2\pi x\bigg]_0^{1/2} = \pi \quad \blacktriangleleft$$

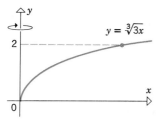

Figure 6.5.6

Example 2 Find the surface area generated by revolving the curve

$$y = \sqrt[3]{3x}, \quad 0 \le y \le 2$$

about the y-axis (Figure 6.5.6).

Solution. We shall apply (7) after rewriting $y = \sqrt[3]{3x}$ as $x = g(y) = \frac{1}{3}y^3$. Thus,

$$g'(y) = y^2$$

so that from (7) we obtain

$$S = \int_0^2 2\pi \left(\frac{1}{3}y^3\right)\sqrt{1 + y^4}\, dy = \frac{2\pi}{3}\int_0^2 y^3\sqrt{1 + y^4}\, dy \qquad (8)$$

The u-substitution

$$u = 1 + y^4, \quad du = 4y^3\, dy$$

yields

$$\int y^3\sqrt{1 + y^4}\, dy = \frac{1}{4}\int \sqrt{u}\, du = \frac{1}{4}\cdot\frac{2}{3}u^{3/2} + C = \frac{1}{6}(1 + y^4)^{3/2} + C$$

Thus, from (8),

$$S = \frac{2\pi}{3}\left[\frac{1}{6}(1 + y^4)^{3/2}\right]_0^2 = \frac{\pi}{9}(17^{3/2} - 1) \qquad \blacktriangleleft$$

▶ Exercise Set 6.5

In Exercises 1–6, find the area of the surface generated by revolving the given curve about the x-axis.

1. $y = 7x, 0 \le x \le 1$.

2. $y = \sqrt{x}, 1 \le x \le 4$.

3. $y = \sqrt{4 - x^2}, -1 \le x \le 1$.

4. $x = \sqrt[3]{y}, 1 \le y \le 8$.

5. $y = \sqrt{x} - \frac{1}{3}x^{3/2}, 1 \le x \le 3$.

6. $y = \frac{1}{3}x^3 + \frac{1}{4}x^{-1}, 1 \le x \le 2$.

In Exercises 7–12, find the area of the surface generated by revolving the given curve about the y-axis.

7. $x = 9y + 1, 0 \le y \le 2$.

8. $x = y^3, 0 \le y \le 1$.

9. $x = \sqrt{9 - y^2}, -2 \le y \le 2$.

10. $x = 2\sqrt{1 - y}, -1 \le y \le 0$.

11. $8xy^2 = 2y^6 + 1, 1 \le y \le 2$.

12. $x = |y - 11|, 0 \le y \le 2$.

13. The lateral area S of a right-circular cone with height h and base radius r is $S = \pi r\sqrt{r^2 + h^2}$. Obtain this result using (6).

14. Show that the area of the surface of a sphere of radius r is $4\pi r^2$. [*Hint:* Revolve the semicircle $y = \sqrt{r^2 - x^2}$ about the x-axis.]

15. The portion of the surface of a sphere between two parallel planes that cut the sphere and are h units apart is called a **zone** of altitude h. Show that the area of the surface of a zone depends only on the radius r of the sphere and the altitude h of the zone, and not on the location of the zone. [*Hint:* Let $-r \le a \le r - h$. Revolve the portion of the curve $y = \sqrt{r^2 - x^2}$ for $a \le x \le a + h$ about the x-axis.]

16. Assume that f is smooth and $f(x) \ge 0$ for $a \le x \le b$. Derive a formula for the surface area generated when the curve $y = f(x)$, $a \le x \le b$ is revolved about the line $y = -k$ $(k > 0)$.

17. **(Formula for Surface Area of a Frustum)**

    (a) If a cone of slant height l and base radius r is cut along a lateral edge and laid flat, it becomes a sector of a circle of radius l (Figure 6.5.7). Use the formula $\frac{1}{2}l^2\theta$ for the area of a sector with radius l and central angle θ (in radians) to show that the lateral surface area of the cone is $\pi r l$.

    (b) Use the result in part (a) to obtain Formula (1) for the lateral surface area of a frustum.

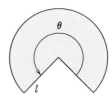

Figure 6.5.7

18. Let $y = f(x)$ be a smooth curve on the interval $[a, b]$ and assume that $f(x) \geq 0$ for $a \leq x \leq b$. By the Extreme-Value Theorem (4.6.4), the function f has a maximum value K and a minimum value k on $[a, b]$. Prove: If L is the arc length of the curve $y = f(x)$ between $x = a$ and $x = b$ and if S is the area of the surface that is generated by revolving this curve about the x-axis, then

$$2\pi k L \leq S \leq 2\pi K L$$

19. Let $y = f(x)$ be a smooth curve on $[a, b]$ and assume that $f(x) \geq 0$ for $a \leq x \leq b$. Let A be the area under the curve $y = f(x)$ between $x = a$ and $x = b$ and let S be the area of the surface obtained when this section of curve is revolved about the x-axis.

(a) Prove that $2\pi A \leq S$.

(b) For what functions f is $2\pi A = S$?

■ **6.6** APPLICATION OF INTEGRATION TO RECTILINEAR MOTION

In Section 4.10 we used the derivative to define the notions of instantaneous velocity and acceleration for a particle moving along a line. In this section we shall resume the study of such motion, using the integration tools developed in the preceding chapter.

☐ **FINDING POSITION AND VELOCITY BY INTEGRATION**

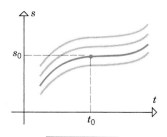

There is a unique position function such that $s(t_0) = s_0$.

Figure 6.6.1

Recall from Section 4.10 that if the position function of a particle moving on a coordinate line is $s(t)$, then its instantaneous velocity and instantaneous acceleration are given by the formulas

$$v(t) = s'(t) = \frac{ds}{dt} \quad \text{and} \quad a(t) = v'(t) = \frac{dv}{dt} = \frac{d^2s}{dt^2}$$

respectively. It follows from these formulas that $s(t)$ is an antiderivative of $v(t)$ and $v(t)$ is an antiderivative of $a(t)$, that is,

$$s(t) = \int v(t)\, dt \quad \text{and} \quad v(t) = \int a(t)\, dt \qquad (1\text{--}2)$$

Thus, if the velocity function of a particle is known, then its position function can be obtained from (1) by integration provided there is sufficient additional information to determine the constant of integration. In particular, we can determine the constant of integration if we know the position s_0 of the particle at some time t_0, since this additional information determines a unique antiderivative of $v(t)$. (See Figure 6.6.1.) Similarly, if the acceleration function of a particle is known, then its velocity function can be obtained from (2) by integration if we know the velocity v_0 of the particle at some time t_0. (See Figure 6.6.2.)

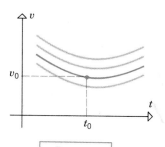

There is a unique velocity function such that $v(t_0) = v_0$.

Figure 6.6.2

Example 1 Find the position function of a particle moving with velocity $v(t) = \cos \pi t$ along a straight line, assuming the particle is at $s = 4$ when $t = 0$.

Solution. The position function is

$$s(t) = \int v(t)\, dt = \int \cos \pi t\, dt = \frac{1}{\pi} \sin \pi t + C$$

Since $s = 4$ when $t = 0$, it follows that

$$4 = s(0) = \frac{1}{\pi} \sin 0 + C = C$$

Thus,

$$s(t) = \frac{1}{\pi} \sin \pi t + 4 \qquad \blacktriangleleft$$

◻ MOTION NEAR THE EARTH'S SURFACE

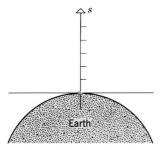

Figure 6.6.3

It is a fact of physics that an object moving on a vertical line near the earth's surface and subject only to the force of gravity moves with constant acceleration.* This constant, denoted by the letter g, is approximately 32 ft/sec² or 9.8 m/sec², depending on whether distance is measured in feet or meters.

From this simple principle it is possible to describe completely the vertical motion of a particle moving freely near the earth's surface, provided the initial position and velocity of the particle are known. To see how, assume that a coordinate line has been superimposed on the line of motion with the origin at the surface of the earth and the positive direction upward (Figure 6.6.3). Moreover, let us assume that time is measured in seconds, distance is measured in feet or meters, and that

the position at time $t = 0$ is s_0

the velocity at time $t = 0$ is v_0

where the initial displacement s_0 and initial velocity v_0 are known. Recall that a particle speeds up when its velocity and acceleration have the same sign and slows down when they have opposite signs. Because we have chosen the positive direction to be up, and because a particle moving up (positive velocity) slows down under the force of gravity and a particle moving down (negative velocity) speeds up under the force of gravity, it follows that the acceleration a of the particle must always be negative, that is,

$$a(t) = -g$$

Since the acceleration of the particle is constant, we obtain

$$v(t) = \int a(t) \, dt = \int -g \, dt = -gt + C_1 \tag{3}$$

To determine the constant of integration we use the fact that the velocity is v_0 when $t = 0$. Thus,

$$v_0 = v(0) = -g \cdot 0 + C_1 = C_1$$

Substituting this in (3) yields

$$v(t) = -gt + v_0 \tag{4}$$

Since v_0 is a constant, it follows that

$$s(t) = \int v(t) \, dt = \int (-gt + v_0) \, dt = -\tfrac{1}{2} gt^2 + v_0 t + C_2 \tag{5}$$

To determine the constant C_2 we use the fact that the position is s_0 when $t = 0$. Thus,

$$s_0 = s(0) = -\tfrac{1}{2} g \cdot 0 + v_0 \cdot 0 + C_2 = C_2$$

Substituting this in (5) yields

$$s(t) = -\tfrac{1}{2} gt^2 + v_0 t + s_0 \tag{6}$$

Example 2   A rock, initially at rest, is dropped from a height of 400 ft. Assuming gravity is the only force acting, how long does it take for the rock to hit the ground and what is its speed at the time of impact?

Solution.   Since distance is in feet, we take $g = 32$ ft/sec². Initially, we have $s_0 = 400$ and $v_0 = 0$, so from (6)

$$s(t) = -16t^2 + 400$$

---

*Strictly speaking, the acceleration changes with the distance from the earth's center. However, near the surface of the earth the acceleration is approximately constant.

Impact occurs when $s(t) = 0$. Solving this equation for t, we obtain

$$-16t^2 + 400 = 0$$

$$t^2 = 25$$

$$t = \pm 5$$

where t is in seconds. Since the rock is released at $t = 0$, the time of impact is positive, so we can discard the negative solution and conclude that it takes 5 sec for the rock to hit the ground. Substituting $t = 5$ and $v_0 = 0$ in (4), we obtain the velocity at time of impact,

$$v(5) = -32(5) + 0 = -160 \text{ ft/sec}$$

Thus, the speed at impact is $|v(5)| = 160$ ft/sec.   ◄

Example 3   A ball is thrown directly upward from a point 8 m above the ground with an initial velocity of 49 m/sec. Assuming gravity is the only force acting on the ball after its release, how high will the ball travel?

Solution.   Since distance is in meters, we take $g = 9.8$ m/sec^2. Initially, we have $s_0 = 8$ and $v_0 = 49$, so from (4) and (6)

$$v(t) = -9.8t + 49$$

$$s(t) = -4.9t^2 + 49t + 8$$

The ball will rise until $v(t) = 0$, that is, until $-9.8t + 49 = 0$ or $t = 5$. At this instant the height above the ground will be

$$s(5) = -4.9(5)^2 + 49(5) + 8 = 130.5 \text{ m}$$ ◄

☐ **DISPLACEMENT**

We conclude this section with an application of the First Fundamental Theorem of Calculus (Theorem 5.7.1) to rectilinear motion. It follows from the First Fundamental Theorem of Calculus and (1) that

$$\int_{t_1}^{t_2} v(t)\, dt = s(t) \Big]_{t_1}^{t_2} = s(t_2) - s(t_1) \tag{7}$$

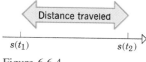

$s(t_1)$ $s(t_2)$

Figure 6.6.4

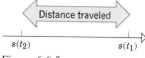

$s(t_2)$ $s(t_1)$

Figure 6.6.5

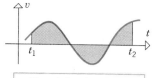

The net signed area is the displacement of the particle during the interval $[t_1, t_2]$.

Figure 6.6.6

Since $s(t_1)$ is the position of the particle at time t_1 and $s(t_2)$ is the position at time t_2, the difference, $s(t_2) - s(t_1)$, is the change in position or ***displacement*** of the particle during the time interval $[t_1, t_2]$. Assuming a horizontal line of motion, a positive displacement means that the particle is farther to the right at time t_2 than at time t_1; a negative displacement means that the particle is farther to the left. In the case where $v(t) \geq 0$ throughout the time interval $[t_1, t_2]$, the particle moves in the positive direction only; thus, the displacement $s(t_2) - s(t_1)$ is the same as the distance traveled by the particle (Figure 6.6.4).

In the case where $v(t) \leq 0$ throughout the time interval $[t_1, t_2]$, the particle moves in the negative direction only; thus, the displacement $s(t_2) - s(t_1)$ is the negative of the distance traveled by the particle (Figure 6.6.5). In the case where $v(t)$ assumes both positive and negative values during the time interval $[t_1, t_2]$, the particle moves back and forth and the displacement is the distance traveled in the positive direction minus the distance traveled in the negative direction.

Geometrically, (7) can be interpreted as the net signed area between the velocity versus time curve and the interval $[t_1, t_2]$, so that this net signed area represents the displacement of the particle (Figure 6.6.6).

☐ **DISTANCE TRAVELED**

If we want to find the total distance traveled in this case (distance traveled in the positive direction *plus* the distance traveled in the negative direction), we must integrate the absolute value of the velocity function (which is the speed of the particle), that is,

$$\begin{bmatrix} \text{total distance} \\ \text{traveled during} \\ \text{time interval} \\ [t_1, t_2] \end{bmatrix} = \int_{t_1}^{t_2} |v(t)|\, dt \qquad\qquad (8)$$

Example 4 A particle moves on a coordinate line so that its velocity at time t is $v(t) = t^2 - 2t$ m/sec. Find

(a) the displacement of the particle during the time interval $0 \le t \le 3$;

(b) the distance traveled by the particle during the time interval $0 \le t \le 3$.

Solution (a). From (7) the displacement is

$$\int_0^3 v(t)\, dt = \int_0^3 (t^2 - 2t)\, dt = \left[\frac{t^3}{3} - t^2 \right]_0^3 = 0$$

Thus, the particle is at the same position at time $t = 3$ as at $t = 0$.

Solution (b). The velocity can be written as $v(t) = t^2 - 2t = t(t - 2)$, from which we see that $v(t) \le 0$ for $0 \le t \le 2$ and $v(t) \ge 0$ for $2 \le t \le 3$. Thus, it follows from (8) that the distance traveled is

$$\int_0^3 |v(t)|\, dt = \int_0^2 -v(t)\, dt + \int_2^3 v(t)\, dt$$

$$= \int_0^2 -(t^2 - 2t)\, dt + \int_2^3 (t^2 - 2t)\, dt$$

$$= -\left[\frac{t^3}{3} - t^2 \right]_0^2 + \left[\frac{t^3}{3} - t^2 \right]_2^3 = \frac{4}{3} + \frac{4}{3} = \frac{8}{3}\ \text{m} \qquad \blacktriangleleft$$

▶ Exercise Set 6.6

In Exercises 1–8, use the given information to find the position function of the particle.

1. $v(t) = 2t - 3;\ s(1) = 5.$

2. $v(t) = 3t^2;\ s(0) = 0.$

3. $v(t) = t^3 - 2t^2 + 1;\ s(0) = 1.$

4. $v(t) = 1 + \sin t;\ s(0) = -3.$

5. $a(t) = 4;\ v(0) = 1;\ s(0) = 0.$

6. $a(t) = t^2 - 3t + 1;\ v(0) = 0;\ s(0) = 0.$

7. $a(t) = 4\cos 2t;\ v(0) = -1;\ s(0) = -3.$

8. $a(t) = 1/\sqrt{2t + 3},\ t \ge 0;\ v(3) = 1;\ s(3) = 0.$

9. In each part use the given information to find the position, velocity, speed, and acceleration at time $t = 1$.

   (a) $v = \sin \frac{1}{2}\pi t;\ s = 0$ when $t = 0$.

   (b) $a = -3t;\ s = 1$ and $v = 0$ when $t = 0$.

10. A car traveling 60 mi/hr along a straight road decelerates at a constant rate of 10 ft/sec^2.

    (a) How long will it take until the speed is 45 mi/hr? [*Note:* 60 mi/hr = 88 ft/sec.]

    (b) How far will the car travel before coming to a stop?

11. A car traveling 60 mi/hr skids 180 ft after its brakes are applied. Find the acceleration of the car, assuming that it is constant.

12. A particle moving along a straight line is accelerating at a constant rate of 3 m/sec^2. Find the initial velocity if the particle moves 40 m in the first 4 sec.

In Exercises 13–24, assume that the only force acting is the earth's gravity, and apply Formulas (4) and (6) to solve the problems.

13. A projectile is launched vertically upward from ground level with an initial velocity of 112 ft/sec.

   (a) Find the velocity at $t = 3$ sec and $t = 5$ sec.

   (b) How high will the projectile rise?

   (c) Find the speed of the projectile when it hits the ground.

14. A projectile fired downward from a height of 112 ft reaches the ground in 2 sec. What is its initial velocity?

15. A projectile is fired vertically upward from ground level with an initial velocity of 16 ft/sec.

   (a) How long will it take for the projectile to hit the ground?

   (b) How long will the projectile be moving upward?

16. A rock is dropped from the top of the Washington Monument, which is 555 ft high.

   (a) How long will it take for the rock to hit the ground?

   (b) What is the speed of the rock at impact?

17. A helicopter pilot drops a package when the helicopter is 200 ft above the ground and rising at a speed of 20 ft/sec.

   (a) How long will it take for the package to hit the ground?

   (b) What will be its speed at impact?

18. A stone is thrown downward with an initial speed of 96 ft/sec from a height of 112 ft.

   (a) How long will it take for the stone to hit the ground?

   (b) What will be its speed at impact?

19. A projectile is fired vertically upward with an initial velocity of 49 m/sec from a tower 150 m high.

   (a) How long will it take for the projectile to reach its maximum height?

   (b) What is the maximum height?

   (c) How long will it take for the projectile to pass its starting point on the way down?

   (d) What is the velocity when it passes the starting point on the way down?

   (e) How long will it take for the projectile to hit the ground?

   (f) What will be its speed at impact?

20. A man drops a stone from a bridge. What is the height of the bridge if

   (a) the stone hits the water 4 sec later

   (b) the sound of the splash reaches the man 4 sec later? (Take 1080 ft/sec as the speed of sound.)

21. A projectile fired upward from ground level is to reach a height of 1000 ft. What must its initial velocity be?

22. A stone is released from rest from a point 40 ft above the ground. Find a formula for v in terms of s.

23. A projectile is fired upward from ground level. Find the initial velocity if it hits the ground 8 sec later. How high does it go?

24. Show that $v^2 = v_0^2 - 2g(s - s_0)$.

25. The graph of a velocity function over the interval $[t_1, t_2]$ is shown in Figure 6.6.7.

   (a) Is the acceleration positive or is it negative?

   (b) Is the acceleration increasing or is it decreasing?

   (c) Is the displacement positive or is it negative?

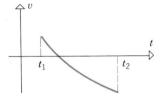

Figure 6.6.7

In Exercises 26–31, $v(t)$ is the velocity (meters/second) of a particle moving on a coordinate line. Find the displacement and the distance traveled by the particle during the given time interval.

26. $v(t) = 2t - 4$; $0 \le t \le 4$.

27. $v(t) = t^2 + t - 2$; $0 \le t \le 2$.

28. $v(t) = |t - 3|$; $0 \le t \le 5$.

29. $v(t) = \cos t$; $0 \le t \le \pi$.

30. $v(t) = 3 \sin t$; $\pi/4 \le t \le \pi$.

31. $v(t) = t^3 - 3t^2 + 2t$; $0 \le t \le 3$.

In Exercises 32–35, $a(t)$ is the acceleration in m/sec^2 of a particle moving on a coordinate line, and v_0 is its velocity at time $t = 0$. Find the displacement and the distance traveled by the particle during the given time interval.

32. $a(t) = -2$; $v_0 = 3$; $1 \le t \le 4$.

33. $a(t) = t - 2$; $v_0 = 0$; $1 \le t \le 5$.

34. $a(t) = \sin t$; $v_0 = 1$; $\dfrac{\pi}{4} \le t \le \dfrac{\pi}{2}$.

35. $a(t) = \dfrac{1}{\sqrt{5t + 1}}$; $v_0 = 2$; $0 \le t \le 3$.

36. Suppose that the velocity of a car moving along a straight road increases from 5 to 50 ft/sec in 20 sec. If you know that the acceleration did not increase, but could have decreased, what is the minimum distance traveled by the car during this time period?

■ 6.7   WORK

In this section we shall use the integration tools developed in the preceding chapter to study the concept of work that arises in physics and engineering.

☐ **WORK DONE BY A CONSTANT FORCE**

If an object moves a distance d along a line while subjected to a *constant* force F applied in the direction of motion, then physicists and engineers define the **work** W done on the object by the force F to be

$$W = F \cdot d$$
$$[\text{work}] = [\text{force}] \cdot [\text{distance}] \tag{1}$$

Common units for measuring force are **pounds** (lb) in the British engineering system and **dynes** or **newtons** (N) in the metric system. (One dyne is the force required to give a mass of 1 g an acceleration of 1 cm/sec², and one newton is the force required to give a mass of 1 kg an acceleration of 1 m/sec².) The most common units of work are foot-pounds (ft·lb), dyne-centimeters (dyne·cm), and newton-meters (N·m). One newton-meter is also called a **joule** (J). One foot-pound is approximately 1.36 J.

Example 1   An object moves 5 ft along a line while subjected to a constant force of 100 lb in its direction of motion. The work done is

$$W = F \cdot d = 100 \cdot 5 = 500 \text{ ft·lb}$$

An object moves 25 m along a line while subjected to a constant force of 4 N in its direction of motion. The work done is

$$W = F \cdot d = 4 \cdot 25 = 100 \text{ N·m} = 100 \text{ J} \qquad ◀$$

☐ **WORK DONE BY A VARIABLE FORCE**

Calculus is needed when it is required to calculate the work done by a *variable* force. We shall consider the following problem.

> **6.7.1**   PROBLEM.   *Suppose that an object moves in the positive direction along a coordinate line while subject to a force $F(x)$, in the direction of motion, whose magnitude depends on the coordinate x. Find the work done by the force when the object moves over an interval $[a, b]$.*

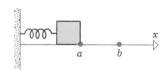

Figure 6.7.1

For example, Figure 6.7.1 shows a block subjected to the force of a compressed spring. As the block moves from a to b the spring expands and the force it applies diminishes. Thus, the force $F(x)$ applied by the spring varies with x.

Before we can solve Problem 6.7.1 we must define precisely what is meant by the work done by a variable force. To motivate this definition, let us divide the interval $[a, b]$ into n subintervals with widths $\Delta x_1, \Delta x_2, \ldots, \Delta x_n$ by inserting points $x_1, x_2, \ldots, x_{n-1}$ between a and b. If we denote by W_k the work done when the object moves across the kth subinterval, then total work W done when the object moves across the entire interval $[a, b]$ will be

$$W = W_1 + W_2 + \cdots + W_n$$

Let us estimate W_k. If the kth subinterval is short and if $F(x)$ is continuous, then the force will not vary much over this subinterval; it will be almost constant. We can approximate this nearly constant force by $F(x_k^*)$, where x_k^* is any point in the kth subinterval. Thus, from (1), the work done over the kth subinterval is approximately

$$W_k \approx F(x_k^*) \, \Delta x_k$$

and the work W done over the entire interval $[a, b]$ is approximately

$$\sum_{k=1}^{n} F(x_k^*) \, \Delta x_k$$

If we now increase the number of subintervals in such a way that $\max \Delta x_k \to 0$, then intuition suggests that our approximations will tend to get better; hence,

$$W = \lim_{\max \Delta x_k \to 0} \sum_{k=1}^{n} F(x_k^*) \, \Delta x_k$$

Since the limit on the right is just the definite integral

$$\int_a^b F(x) \, dx$$

we are led to the following definition.

6.7.2 DEFINITION. If an object moves in the positive direction over the interval $[a, b]$ while subjected to a variable force $F(x)$ in the direction of motion, then the **work** done by the force is

$$W = \int_a^b F(x) \, dx \tag{2}$$

Hooke's law [Robert Hooke (1635–1703), English physicist] states that under appropriate conditions a spring stretched x units beyond its equilibrium position pulls back with a force

$$F(x) = kx$$

where k is a constant (called the *spring constant*). The value of k depends on such factors as the thickness of the spring, the material used in its composition, and the units of force and distance.

Example 2 A spring whose natural length is 2.4 m exerts a force of 5 N when stretched 1 m beyond its natural length.

(a) Find the spring constant k.

(b) How much work is required to stretch the spring from its natural length to a length of 4.2 m?

Solution (a). From Hooke's law,

$$F(x) = kx$$

From the data, $F(x) = 5$ N when $x = 1$ m, so $5 = k \cdot 1$. Thus, the spring constant is $k = 5$ newtons per meter (N/m). This means that the force $F(x)$ required to stretch the spring x meters is

$$F(x) = 5x \tag{3}$$

Solution (b). Place the spring along a coordinate line as shown in Figure 6.7.2. We want to find the work W required to stretch the spring over the interval from $x = 0$ to $x = 1.8$. From (2) and (3) the work W required is

$$W = \int_a^b F(x) \, dx = \int_0^{1.8} 5x \, dx = \frac{5x^2}{2} \Big]_0^{1.8} = 8.1 \text{ J} \quad \blacktriangleleft$$

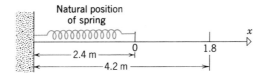

Figure 6.7.2

Example 3   A cylindrical water tank of radius 10 ft and height 30 ft is half filled with water. How much work is required to pump all the water over the upper rim of the tank?

Solution.   Introduce a coordinate line as shown in Figure 6.7.3. Imagine the water to be divided into n thin layers with thicknesses

$$\Delta x_1, \Delta x_2, \ldots, \Delta x_n$$

The force required to move the kth layer equals the weight of the layer, which can be found by multiplying its volume by the weight density of water ($62.4 \, \text{lb/ft}^3$ or $9810 \, \text{N/m}^3$). Since the kth layer is a cylinder of radius $r = 10$ ft and height Δx_k, the force required to move it is

$$\begin{bmatrix} \text{force to move} \\ \text{the } k\text{th layer} \end{bmatrix} = (\pi r^2 \, \Delta x_k) \cdot [\text{weight density of water}]$$

$$= (\pi (10)^2 \, \Delta x_k)(62.4) = 6240\pi \, \Delta x_k$$

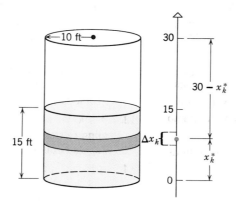

Figure 6.7.3

Because the kth layer has a finite thickness, the upper and lower surfaces are at different distances from the origin. However, if the layer is thin, the difference in these distances is small, and we can reasonably assume that the entire layer is concentrated at a single distance x_k^* from the origin (Figure 6.7.3). With this assumption, the work W_k required to pump the kth layer over the rim will be approximately

$$W_k \approx \underbrace{(30 - x_k^*)}_{\text{Distance}} \cdot \underbrace{6240\pi \, \Delta x_k}_{\text{Force}}$$

and the work W required to pump all n layers will be approximately

$$W = \sum_{k=1}^{n} W_k \approx \sum_{k=1}^{n} (30 - x_k^*)(6240\pi) \, \Delta x_k$$

To find the *exact* value of the work we take the limit as $\max \Delta x_k \to 0$. This yields

$$W = \lim_{\max \Delta x_k \to 0} \sum_{k=1}^{n} (30 - x_k^*)(6240\pi) \, \Delta x_k = \int_0^{15} (30 - x)(6240\pi) \, dx$$

$$= 6240\pi \left(30x - \frac{x^2}{2} \right) \Bigg]_0^{15} = 2{,}106{,}000\pi \; \text{ft} \cdot \text{lb} \approx 6{,}616{,}194 \; \text{ft} \cdot \text{lb} \qquad \blacktriangleleft$$

▶ Exercise Set 6.7   Ⓒ *9*

1. Find the work done when

   (a) a constant force of 30 lb along the *x*-axis moves an object from $x = -2$ to $x = 5$ ft

   (b) a variable force of $F(x) = 1/x^2$ lb along the *x*-axis moves an object from $x = 1$ to $x = 6$ ft.

2. A spring whose natural length is 15 cm exerts a force of 45 N when stretched to a length of 20 cm.

   (a) Find the spring constant (in newtons/meter).

   (b) Find the work that is done in stretching the spring 3 cm beyond its natural length.

   (c) Find the work done in stretching the spring from a length of 20 cm to a length of 25 cm.

3. A spring exerts a force of 100 N when it is stretched 0.2 m beyond its natural length. How much work is required to stretch the spring 0.8 m beyond its natural length?

4. Assume that a force of 6 N is required to compress a spring from a natural length of 4 m to a length of $3\frac{1}{2}$ m. Find the work required to compress the spring from its natural length to a length of 2 m. (Hooke's law applies to compression as well as extension.)

5. Assume that 10 ft·lb of work is required to stretch a spring 1 ft beyond its natural length. What is the spring constant?

6. A cylindrical tank of radius 5 ft and height 9 ft is two-thirds filled with water. Find the work required to pump all the water over the upper rim.

7. Solve Exercise 6 assuming that the tank is two-thirds filled with a liquid that weighs ρ lb/ft³.

8. A cone-shaped water reservoir is 20 ft in diameter across the top and 15 ft deep. If the reservoir is filled to a depth of 10 ft, how much work is required to pump all the water to the top of the reservoir?

9. The vat shown in Figure 6.7.4 contains water to a depth of 2 m. Find the work required to pump all the water to the top of the vat. [Use 9810 N/m³ as the weight density of water.]

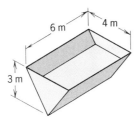

Figure 6.7.4

10. A cylindrical tank lying on its side (Figure 6.7.5) is filled with a liquid weighing 50 lb/ft³. Find the work required to pump all the liquid to a level 1 ft above the top of the tank.

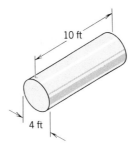

Figure 6.7.5

11. A swimming pool is built in the shape of a rectangular parallelepiped 10 ft deep, 15 ft wide, and 20 ft long.

    (a) If the pool is filled to 1 ft below the top, how much work is required to pump all the water into a drain at the top edge of the pool?

    (b) A one-horsepower motor can do 550 ft·lb of work per second. What size motor is required to empty the pool in one hour?

12. A water tower in the shape of a hemisphere of radius 10 ft is to be filled by pumping water over the top edge from a lake 200 ft below the top. How much work is required to fill the tank? [*Hint:* Solve without calculus.]

13. A 100-ft length of steel chain weighing 15 lb/ft is dangling from a pulley. How much work is required to wind the chain onto the pulley?

14. A rocket weighing 3 tons is filled with 40 tons of liquid fuel. In the initial part of the flight, fuel is burned off at a constant rate of 2 tons per 1000 ft of vertical height. How much work is done in lifting the rocket to 3000 ft?

15. (**Satellite Problem**) The weight of an object is the force exerted on it by the earth's gravity. Thus, if a person weighs 100 lb on the surface of the earth, the earth's gravity is pulling on that person with a force of 100 lb. It is a fundamental law of physics that the force the earth exerts on an object varies inversely as the square of its distance from the earth's center. Thus an object's weight $F(x)$ is related to its distance *x* from the earth's center by a formula of the form

    $$F(x) = k/x^2$$

    where *k* is a constant of proportionality depending on the mass of the object and the units of force and distance.

    (a) Assuming that the earth is a sphere of radius 4000 mi, find the constant *k* in the formula above for a satellite that weighs 6000 lb on the earth's surface.

    (b) How much work must be performed to lift this satellite to an orbital position 1000 mi above the earth's surface?

16. **(Coulomb's Law)** It follows from Coulomb's law in physics that two like electrostatic charges repel each other with a force inversely proportional to the square of the distance between them. Suppose that two charges A and B repel with a force of k newtons when they are positioned at points $A(-a, 0)$ and $B(a, 0)$, where a is measured in meters. Find the work W required to move charge A along the x-axis to the origin if charge B remains stationary.

■ **6.8** FLUID PRESSURE AND FORCE

> *In this section we shall use the definite integral to calculate the force exerted by a liquid on a submerged surface.*

☐ **FLUID PRESSURE**

If a flat surface of area A is submerged horizontally at a depth h in a container of fluid, then the fluid exerts a force F normal to the surface and given by

$$F = \rho h A \tag{1}$$

where ρ is the **weight density** of the fluid (weight per unit volume). The weight density of water is approximately 62.4 lb/ft^3, for example. It is a physical fact that the force F in (1) does not depend on the shape of the container. Thus, if the three containers in Figure 6.8.1 have bases of the same area and are filled with a fluid to the same depth, h, then each container will have the same fluid force on its base.

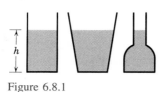

Figure 6.8.1

Pressure is defined to be force per unit area. Thus, from (1), the pressure p exerted at each point of a flat surface of area A submerged horizontally at a depth h is given by

$$p = \frac{F}{A} = \rho h \tag{2}$$

In the British engineering system weight and force are commonly measured in pounds (lb) and area in square feet (ft^2), in which case weight density would be measured in units of pounds per cubic foot (lb/ft^3) and pressure in units of pounds per square foot (lb/ft^2).

In the metric system weight and force are commonly measured in newtons (N) and area in square meters (m^2), in which case weight density can be measured in units of newtons per cubic meter (N/m^3) and pressure in units of newtons per square meter (N/m^2). One newton per square meter is also called a **pascal** (Pa). One Pa is approximately 0.02089 lb/ft^2. The weight density of water is approximately 9810 N/m^3.

Example 1 If a flat circular plate of radius $r = 2$ m is submerged horizontally in water so that the top surface is at a depth of 3 m, then the force on the top surface of the plate is

$$F = \rho h A = \rho h (\pi r^2) = (9810)(3)(4\pi) = 117,720\pi \text{ N} \approx 369,828 \text{ N}$$

and the pressure at each point on the plate surface is

$$p = \rho h = (9810)(3) = 29,430 \text{ N/m}^2 = 29,430 \text{ Pa} \qquad ◀$$

☐ **PASCAL'S PRINCIPLE**

Pascal's (see p. 343) *principle* in physics states that *fluid pressure is the same in all directions*. Thus, if a flat surface is submerged vertically (or at any angle at all), the pressure at a point of depth h is exactly the same as the pressure at a point of depth h on a horizontal surface (Figure 6.8.2). However, (1) cannot be applied to obtain the total force acting on a flat surface that is not submerged horizontally, because the depth h can vary from point to point on the surface. It is in such problems that calculus comes into play.

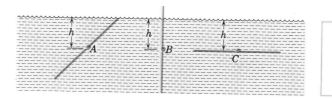

Figure 6.8.2

By Pascal's Principle
the fluid pressure
at points A, B,
and C is the same.

Suppose that we want to compute the total fluid force against a submerged vertical surface. As shown in Figure 6.8.3, let us introduce a vertical x-axis whose positive direction is downward and whose origin is at any convenient point.

As indicated in Figure 6.8.3a, let us assume that the submerged portion of the surface extends from $x = a$ to $x = b$ on the x-axis. Also, suppose that a point x on the axis lies $h(x)$ units below the surface and that the cross section of the plate at x has width $w(x)$.

Next, divide the interval $[a, b]$ into n subintervals with lengths

$$\Delta x_1, \Delta x_2, \ldots, \Delta x_n$$

and in each subinterval choose an arbitrary point x_k^*. Following a familiar pattern, we approximate the section of plate along the kth subinterval by a rectangle of length $w(x_k^*)$ and width Δx_k (Figure 6.8.3b). Because the upper and lower edges of this rectangle are at different depths, (1) cannot be used to calculate the force on this rectangle. However, if Δx_k is small, the difference in depth between the upper and lower edges is small, and we can reasonably assume that the entire rectangle is concentrated at a single depth $h(x_k^*)$ below the surface. With this assumption, (1) can be used to *approximate* the force F_k on the kth rectangle. We obtain

$$F_k \approx \rho \underbrace{h(x_k^*)}_{\text{Depth}} \cdot \underbrace{w(x_k^*) \, \Delta x_k}_{\text{Area of rectangle}}$$

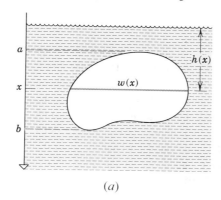

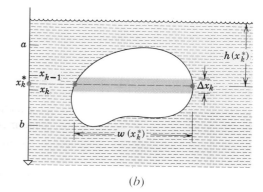

Figure 6.8.3 (a) (b)

---

*BLAISE PASCAL (1623–1662). French mathematician and scientist. Pascal's mother died when he was three years old and his father, a highly educated magistrate, personally provided the boy's early education. Although Pascal showed an inclination for science and mathematics, his father refused to tutor him in those subjects until he mastered Latin and Greek. Pascal's sister and primary biographer claimed that he independently discovered the first thirty-two propositions of Euclid without ever reading a book on geometry. (However, it is generally agreed that the story is apocryphal.) Nevertheless, the precocious Pascal published a highly respected essay on conic sections by the time he was sixteen years old. Descartes, who read the essay, thought it so brilliant that he could not believe that it was written by such a young man. By age 18 his health began to fail and until his death he was in frequent pain. However, his creativity was unimpaired.

Pascal's contributions to physics include the discovery that air pressure decreases with altitude and the principle of fluid pressure that bears his name. However, the originality of his work is questioned by some historians. Pascal made major contributions to a branch of mathematics called "projective geometry," and he helped to develop probability theory through a series of letters with Fermat.

In 1646, Pascal's health problems resulted in a deep emotional crisis that led him to become increasingly concerned with religious matters. Although born a Catholic, he converted to a religious doctrine called Jansenism and spent most of his final years writing on religion and philosophy.

Thus, the total force F on the plate is approximately

$$F = \sum_{k=1}^{n} F_k \approx \sum_{k=1}^{n} \rho h(x_k^*) w(x_k^*) \, \Delta x_k$$

To find the *exact* value of the force we take the limit as $\max \Delta x_k \to 0$. This yields

$$F = \lim_{\max \Delta x_k \to 0} \sum_{k=1}^{n} \rho h(x_k^*) w(x_k^*) \, \Delta x_k = \int_a^b \rho h(x) w(x) \, dx$$

which suggests the following result.

6.8.1   FORMULA FOR FLUID FORCE.   Assume that a flat surface is immersed vertically in a liquid of weight density ρ and that the submerged portion extends from $x = a$ to $x = b$ on a vertical x-axis. For $a \le x \le b$, let $w(x)$ be the width of the surface at x and let $h(x)$ be the depth of the point x. Then the total **fluid force** on the surface is

$$F = \int_a^b \rho h(x) w(x) \, dx \tag{3}$$

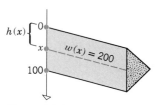

Figure 6.8.4

Example 2   The face of a dam is a vertical rectangle of height 100 ft and width 200 ft (Figure 6.8.4). Find the total fluid force exerted on the face when the water surface is level with the top of the dam.

Solution.   Introduce an x-axis with its origin at the water surface as shown in Figure 6.8.4. At a point x on this axis, the width of the dam in feet is $w(x) = 200$ and the depth in feet is $h(x) = x$. Thus, from (3) with $\rho = 62.4 \, \text{lb/ft}^3$ (the weight density of water) we obtain as the total force on the face

$$F = \int_0^{100} (62.4)(x)(200) \, dx = 12{,}480 \int_0^{100} x \, dx$$

$$= 12{,}480 \left. \frac{x^2}{2} \right]_0^{100} = 62{,}400{,}000 \, \text{lb} \quad \blacktriangleleft$$

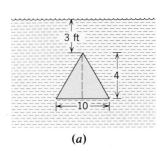

(a)

Example 3   A plate in the form of an isosceles triangle with base 10 ft and altitude 4 ft is submerged vertically in oil as shown in Figure 6.8.5a. Find the fluid force F against the plate surface if the oil has weight density $\rho = 30 \, \text{lb/ft}^3$.

Solution.   Introduce an x-axis as shown in Figure 6.8.5b. By similar triangles, the width of the plate, in feet, at a depth of $h(x) = (3 + x)$ ft satisfies

$$\frac{w(x)}{10} = \frac{x}{4}, \quad \text{so} \quad w(x) = \frac{5}{2} x$$

Thus, it follows from (3) that the force on the plate is

$$F = \int_a^b \rho h(x) w(x) \, dx = \int_0^4 (30)(3 + x) \left(\frac{5}{2} x \right) dx$$

$$= 75 \int_0^4 (3x + x^2) \, dx = 75 \left[\frac{3x^2}{2} + \frac{x^3}{3} \right]_0^4 = 3400 \, \text{lb} \quad \blacktriangleleft$$

(b)

Figure 6.8.5

▶ Exercise Set 6.8

1. A flat square plate with 3-ft sides and negligible thickness is submerged horizontally in a liquid. Find the force and pressure on a surface of the plate if

   (a) the liquid is water and the plate is at a depth of 5 ft

   (b) the liquid has weight density 40 lb/ft³ and the plate is at a depth of 10 ft.

> In Exercises 2–7, the flat surfaces shown are submerged vertically in water. Find the fluid force against the surface.

2.
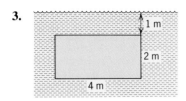

3.
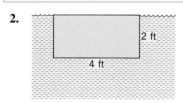

4.
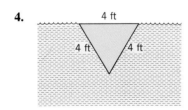

5.
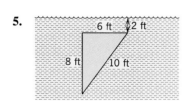

6.
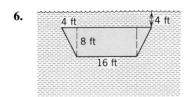

7.
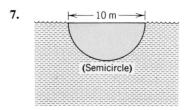

8. An oil tank is shaped like a right-circular cylinder of diameter 4 ft. Find the total fluid force against one end when the axis is horizontal and the tank is half filled with oil of weight density 50 lb/ft³.

9. A square plate of side a feet is dipped in a liquid of weight density ρ lb/ft³. Find the fluid force on the plate if a vertex is at the surface and a diagonal is perpendicular to the surface.

10. Figure 6.8.6 shows a dam whose face is an inclined rectangle. Find the fluid force on the face when the water is level with the top of this dam.

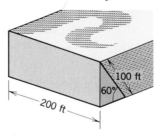

Figure 6.8.6

11. Figure 6.8.7 shows a rectangular swimming pool whose bottom is an inclined plane. Find the fluid force on the bottom when the pool is filled to the top.

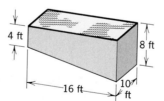

Figure 6.8.7

12. An observation window on a submarine is a square with 2-ft sides. Using ρ_0 for the weight density of seawater, find the fluid force on the window when the submarine has descended so that the window is vertical and its top is at a depth of h feet.

13. (a) Show: If the submarine in Exercise 12 descends vertically at a constant rate, then the fluid force on the window increases at a constant rate.

   (b) At what rate is the force on the window increasing if the submarine is descending vertically at 20 ft/min?

◆ **TECHNOLOGY EXERCISES**   **Chapter 6**

Most of these exercises require access to a graphing calculator or a computer algebra system (CAS) such as *Mathematica*, *Maple*, or *Derive*. When you are asked to *find* an answer or to *solve* an equation, you may choose to find an exact result or a numerical approximation, depending on the particular technology you are using and on your own imagination. The form of your answers may differ from those of other students or from those in the answer section of the text, depending on how you solve the problems and the accuracy you use in your numerical approximations. Those exercises that are more appropriate for a CAS than a graphing calculator are labeled with the icon ◆.

1. **Area:** Find the area of the region enclosed by the curves $y = x^2 - 1$ and $y = 2 \sin x$.

2. **Area:** Find a positive value of k such that the region in the first quadrant enclosed by $y = x^2$, $y = x^2/4$, and the horizontal lines $y = k$ and $y = k + 1$ has an area of 3 square units.

3. **Area:** Find a positive value of k such that the region enclosed by $y = \sqrt{x}$, $y = x/2 + 1$, $x = 0$, and $x = k$ has an area of 2 square units.

4. **Area:** Referring to the following figure, find the value of k so that the areas of the shaded regions are equal. [*Note:* This exercise is based on Problem A1 of the Fifty-Fourth Annual William Lowell Putnam Mathematical Competition.]

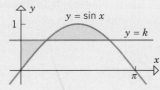

5. **Volume:** Find the volume of the solid that results when the region in the first quadrant enclosed by the coordinate axes and the curve $y = 1 + x^2 - x^3$ is revolved about the y-axis.

◆ 6. **Volume:** Consider the region to the left of the vertical line $x = k$ $(0 < k < \pi)$ and between the curve $y = \sin x$ and the x-axis. Find the value of k so that the solid generated by revolving the region about the y-axis has a volume of 8 cubic units.

7. **Volume:** Suppose that the region described in Exercise 3 is revolved about the x-axis to generate a solid. Find the value of k so that the resulting solid has a volume of 10 cubic units.

8. **Volume:** Consider the region in the first quadrant enclosed by the coordinate axes, the line $x = 2$, and the curve $y = 1/(1 + kx)^{3/2}$ $(k > 0)$. Find the value of k so that the solid generated by revolving the region about the x-axis has a volume of 4 cubic units.

◆ 9. **Arc length:** Find the value of k $(k > 1)$ so that the curve $y = 1/x$ has a length of 2 units over the interval $[1, k]$.

◆ 10. **Arc length:** Find the value of k $(k > 0)$ such that the curve $y = k \sin x$ has a length of 5 units over the interval $[0, \pi]$.

◆ 11. **Distance traveled by a basketball:** A basketball player makes a successful shot from the free throw line. Suppose that the path of the ball from the moment of release to the moment it enters the hoop is described by

$$y = 2.15 + 2.09x - 0.41x^2, \quad 0 \le x \le 4.6$$

where x is the horizontal distance (in meters) from the point of release, and y is the vertical distance (in meters) above the floor. Find the distance the ball travels from the moment it is released to the moment it enters the hoop.

◆ 12. **Deflection of a beam:** As shown in the following figure, a horizontal beam with dimensions 2 in × 6 in × 16 ft is fixed at both ends and is subjected to a uniformly distributed load of 120 lb/ft. As a result of the load, the centerline of the beam undergoes a deflection that is described by

$$y = -1.67 \times 10^{-8}(x^4 - 2Lx^3 + L^2x^2)$$

$(0 \le x \le 192)$, where L is the length of the unloaded beam, x is the horizontal distance along the beam measured from the left end, and y is the deflection of the centerline (L, x, and y in inches).

(a) Graph y versus x for $0 \le x \le 192$.

(b) Find the maximum deflection of the centerline.

(c) Find the length of the centerline of the loaded beam.

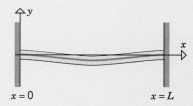

where t is in seconds and v is in cm/sec. Find the total distance that the object travels during the first 6 seconds of vibration.

◆ **15. Work:** Suppose that an object moves in the positive direction on an x-axis while subject to the force

$$F(x) = \frac{x}{\sqrt{1 + kx}}, \quad k > 0$$

where x is in meters and F is in newtons. Find the value of k so that the work done by F over the interval $[0, 2]$ is 0.8 J (J = joules).

13. Rectilinear motion: Suppose that a particle moves along a horizontal line so that its velocity v at time t is given by

$$v(t) = 5t - 10 \sin t, \quad t \geq 0$$

where t is in seconds and v is in cm/sec. Assume that the particle is moving to the right if $v > 0$, and to the left if $v < 0$.

(a) What is the maximum distance attained to the left of its starting position?

(b) When does the particle arrive back at the point where it started?

◆ **14. Vibrating spring:** Suppose that an object suspended from a spring vibrates so that the velocity of the object in the vertical direction is given by

$$v(t) = (0.8)^t(3 \cos 5t - 5 \sin 5t), \quad t \geq 0$$

16. Work: Suppose that an object moves in the positive direction on an x-axis while subject to the force

$$F(x) = \frac{x}{\sqrt{1 + x^3}}, \quad x \geq 0$$

where x is in meters and F is in newtons. The object moves 2 m from an unspecified starting point $x = a$ ($a \geq 0$).

(a) Find a definite integral that gives the work done by F as a function of a.

(b) Find the value of a for which the work done by F is maximum. What is that maximum work? [*Hint:* See Exercise 28, Section 5.9.]

Maria Gaetana Agnesi (1718–1799)

7 LOGARITHMIC AND EXPONENTIAL FUNCTIONS

▉ 7.1 LOGARITHMS AND EXPONENTS (AN OVERVIEW)

Although logarithms were first introduced as a computational tool, today calculus provides wide-ranging applications of logarithms in both mathematics and science. Our goal in this chapter is to discuss logarithms and exponents from the calculus viewpoint. We shall proceed in three steps:

- *In Sections 7.1 to 7.3 we shall develop the basic properties of logarithms and exponents informally, with the primary emphasis on algebraic computations, differentiation, integration, and graphing. Full justification for many of the details will be deferred to Section 7.5.*

- *In Section 7.4 we shall introduce the concept of inverse functions, which is the basic idea on which the precise definition of logarithms and exponents rests.*

- *In Section 7.5 we shall define logarithms and exponents precisely, and fill in most of the technical details that were omitted in Sections 7.1 to 7.3.*

> Many of the examples and exercises in this section require the use of a scientific calculator that has the capability of evaluating logarithms and powers of real numbers. Henceforth, we shall assume that you have access to such a calculator and know how to use it to compute powers and logarithms. Consult your calculator manual, if necessary.

☐ **IRRATIONAL EXPONENTS**

In algebra, integer exponents and rational exponents of a positive number b are defined by

$$b^n = \underbrace{b \times b \times \cdots \times b}_{n \text{ factors}}, \quad b^{-n} = \frac{1}{b^n}, \quad b^0 = 1,$$

$$b^{p/q} = \sqrt[q]{b^p} = (\sqrt[q]{b})^p, \quad b^{-p/q} = \frac{1}{b^{p/q}}$$

Observe that these definitions do not provide a meaning to b^x for irrational values of x, so that such expressions as

$$2^\pi, \quad 3^{\sqrt{2}}, \quad \text{and} \quad \pi^{-\sqrt{7}}$$

are as yet undefined.

Depending on whether

$$0 < b < 1, \quad b = 1, \quad \text{or} \quad b > 1$$

the graph of $y = b^x$ will have one of the shapes shown in Figure 7.1.1a. Because b^x is not yet defined for irrational values of x, each of the graphs consists of densely packed dots separated by holes at the irrational values of x. One of our objectives in this chapter is to define b^x for irrational values of x in such a way that b^x becomes a continuous function (Figure 7.1.1b).

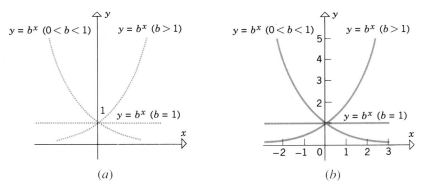

Figure 7.1.1   (a)   (b)

REMARK. We have limited the discussion of b^x to the case where $b > 0$, since negative values of b can lead to imaginary numbers; for example, if $b = -2$ and $x = \frac{1}{2}$, then $b^x = (-2)^{1/2} = \sqrt{-2}$.

There are various methods for defining irrational exponents. One approach is to define irrational powers of b as limits of rational powers of b. For example, to define 2^π we might start with the decimal representation of π, namely

$$3.1415926\ldots$$

From this decimal we can obtain a sequence of rational numbers that get closer and closer to π, namely

$$3, \quad 3.1, \quad 3.14, \quad 3.141, \quad 3.1415, \quad 3.14159, \ldots$$

and from these we obtain the following sequence of rational powers of 2:

$$2^3, \quad 2^{3.1}, \quad 2^{3.14}, \quad 2^{3.141}, \quad 2^{3.1415}, \quad 2^{3.14159}, \ldots$$

Table 7.1.1

x	2^x
3	8.000000
3.1	8.574188
3.14	8.815241
3.141	8.821353
3.1415	8.824411
3.14159	8.824962
3.141592	8.824974

Since we want 2^x to be a continuous function of x, and hence continuous at π, these rational powers of 2 would have to approach 2^π. This suggests that we might *define* 2^π as the limit* of these rational powers of 2. This idea is illustrated numerically in Table 7.1.1, which was generated with a calculator. From Table 7.1.1, the value of 2^π rounded to four decimal places is 8.8250.

In this informal section we shall accept without proof that the preceding limit procedure produces a definition of b^x for irrational x such that the following are true:

- b^x is a continuous function for all $b > 0$.
- b^x is a differentiable function for all $b > 0$.
- The standard properties of exponents such as $b^{u+v} = b^u b^v$ continue to hold.

The first two properties are consistent with the graphs shown in Figure 7.1.1b.

☐ **REVIEW OF LOGARITHMS**

In algebra a logarithm is defined as an exponent. More precisely, if $b > 0$ and $b \neq 1$, then for positive values of x one defines

$$\log_b x$$

(read, "the **logarithm to the base b of x**") to be that power to which b must be raised to produce x. Thus,

$$\log_{10} 100 = 2$$

since 10 must be raised to the second power to produce 100. Similarly,

$$\log_2 8 = 3 \qquad \boxed{\text{since } 2^3 = 8}$$

$$\log_{10} \tfrac{1}{1000} = -3 \qquad \boxed{\text{since } 10^{-3} = \tfrac{1}{1000}}$$

$$\log_{10} 1 = 0 \qquad \boxed{\text{since } 10^0 = 1}$$

$$\log_3 81 = 4 \qquad \boxed{\text{since } 3^4 = 81}$$

In general, if $y = \log_b x$, then y is that power to which b must be raised to produce x, so $x = b^y$. Conversely, if $x = b^y$, then $y = \log_b x$, so the statements

$$y = \log_b x \quad \text{and} \quad x = b^y$$

are equivalent. By substituting each equation in the other we obtain

$$\log_b b^y = y \quad \text{and} \quad b^{\log_b x} = x$$

We could have reversed the roles of x and y in the preceding discussion, so the first of these equations is also correct if y is replaced by x. Thus, we have the following important relationships:

$$\boxed{\log_b b^x = x} \qquad \text{and} \qquad \boxed{b^{\log_b x} = x} \tag{1}$$

The earliest logarithms studied were the logarithms with base 10, which are called **common logarithms**. For such logarithms it is usual to suppress explicit reference to the base and write "log" rather than "$\log_{10}$." Thus, (1) implies that

$$\log 10^x = x \quad \text{and} \quad 10^{\log x} = x$$

The reader should already be familiar with the properties of logarithms listed in the following theorem.

---

*The limit described here is different from those studied earlier in the text, since here we are taking the limit of a *sequence* of numerical values, whereas earlier we studied limits of $f(x)$ in which x varied over an interval. Later in the text we will define limits of sequences precisely, but for now your intuition about such limits will suffice.

7.1.1 THEOREM.

(a) $\log_b 1 = 0$

(b) $\log_b b = 1$

(c) $\log_b ac = \log_b a + \log_b c$

(d) $\log_b \dfrac{a}{c} = \log_b a - \log_b c$

(e) $\log_b a^r = r \log_b a$

(f) $\log_b \dfrac{1}{c} = -\log_b c$

We shall prove (a) and (c) and leave the remaining proofs as exercises.

Proof (a). Since $b^0 = 1$, it follows that $\log_b 1 = 0$.

Proof (c). Let

$$x = \log_b a \quad \text{and} \quad y = \log_b c \tag{2}$$

so

$$b^x = a \quad \text{and} \quad b^y = c$$

Therefore,

$$ac = b^x b^y = b^{x+y} \quad \text{or equivalently,} \quad \log_b ac = x + y$$

Thus, from (2)

$$\log_b ac = \log_b a + \log_b c \quad \blacksquare$$

□ **THE NUMBER** *e*
NATURAL LOGARITHMS

The most important logarithms in applications are called ***natural logarithms***; these have a certain irrational base that is denoted by the letter *e*. Most scientific calculators have a key for evaluating powers of *e*. We leave it for the reader to verify with a calculator that the value of *e* to six decimal places is

$$e \approx 2.718282 \tag{3}$$

This number was discovered by the Swiss mathematician Leonard Euler (see p. 52), who showed that $y = e$ is a horizontal asymptote of the graph of

$$y = \left(1 + \frac{1}{x}\right)^x \tag{4}$$

(Figure 7.1.2). Euler suggested the applicability of *e* to logarithms in an unpublished paper he wrote in 1728.

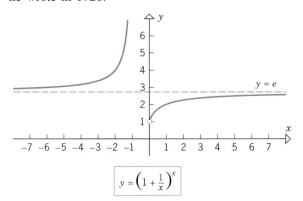

$$y = \left(1 + \tfrac{1}{x}\right)^x$$

Figure 7.1.2

The fact that $y = e$ is a horizontal asymptote of (4) as $x \to +\infty$ and as $x \to -\infty$ is expressed by the limits

$$e = \lim_{x \to +\infty} \left(1 + \frac{1}{x}\right)^x \qquad \text{and} \qquad e = \lim_{x \to -\infty} \left(1 + \frac{1}{x}\right)^x \tag{5–6}$$

Later, we will show that these limits can be derived from the following useful limit,

$$e = \lim_{x \to 0} (1 + x)^{1/x} \tag{7}$$

which we state here for reference.

Limit (5) is illustrated numerically in Table 7.1.2, which was generated using a calculator. Compare the results in the table to (3).

It is fairly standard to denote the natural logarithm of x by $\ln x$ (read, "ell en of x") rather than $\log_e x$. Thus, $\ln x$ is that power to which e must be raised to produce x. For example,

$$\ln 1 = 0, \qquad \ln e = 1, \qquad \ln 1/e = -1, \qquad \ln (e^2) = 2$$

$$\boxed{\text{Since } e^0 = 1} \qquad \boxed{\text{Since } e^1 = e} \qquad \boxed{\text{Since } e^{-1} = 1/e} \qquad \boxed{\text{Since } e^2 = e^2}$$

In general, the statements

$$y = \ln x \quad \text{and} \quad x = e^y$$

are equivalent. Moreover, for the case where the base is e the formulas in (1) become

$$\ln e^x = x \quad \text{and} \quad e^{\ln x} = x \tag{8}$$

Table 7.1.2

x	$1 + \dfrac{1}{x}$	$\left(1 + \dfrac{1}{x}\right)^x$
1	2	2.000000
10	1.1	2.593742
100	1.01	2.704814
1,000	1.001	2.716924
10,000	1.0001	2.718146
100,000	1.00001	2.718268
1,000,000	1.000001	2.718280

REMARK.   It should be noted that some calculators, computer programs, and mathematical literature use "log" rather than "ln" to denote the natural logarithm, so it is a good idea to check the convention being followed.

☐ **GRAPHS OF** $y = \ln x$ **AND** $y = e^x$

Figure 7.1.3 shows tables of values for $\ln x$ and e^x together with the graphs of $y = \ln x$ and $y = e^x$ that result by plotting the values in the tables. The table of values for $\ln x$ was obtained with a calculator, but the table of values for e^x was obtained by interchanging the columns of the table for $\ln x$. This is justified because the relationship $y = \ln x$ implies that $x = e^y$, so that the values on the left side of the table for $\ln x$ are also the values that result when the values on the right side are used as exponents of e.

Figure 7.1.3 suggests that the graphs of $y = \ln x$ and $y = e^x$ are symmetric about the line $y = x$. Later, we will prove this to be true, and we will show that in general the graphs of $y = \log_b x$ and $y = b^x$ are symmetric about the line $y = x$. (See Figure 7.1.4 for the case $b > 1$.)

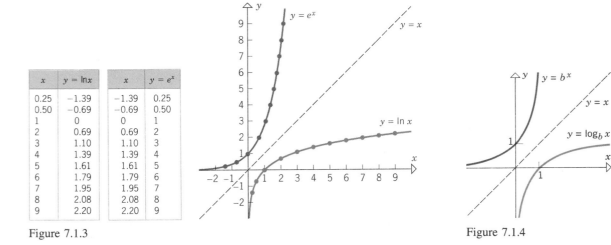

x	$y = \ln x$		x	$y = e^x$
0.25	−1.39		−1.39	0.25
0.50	−0.69		−0.69	0.50
1	0		0	1
2	0.69		0.69	2
3	1.10		1.10	3
4	1.39		1.39	4
5	1.61		1.61	5
6	1.79		1.79	6
7	1.95		1.95	7
8	2.08		2.08	8
9	2.20		2.20	9

Figure 7.1.3

Figure 7.1.4

☐ **CHANGE OF BASE FORMULA FOR LOGARITHMS**

The following theorem provides a formula for expressing a logarithm with a given base in terms of logarithms with a different base. This formula can be used, for example, to express common logarithms in terms of natural logarithms or conversely.

7.1.2 THEOREM.

$$\log_b x = \frac{\log_a x}{\log_a b} \tag{9}$$

Proof. Let $y = \log_b x$. Thus,

$$b^y = x$$
$$\log_a b^y = \log_a x$$
$$y \log_a b = \log_a x \quad \boxed{\text{By Theorem 7.1.1}(e)}$$
$$y = \frac{\log_a x}{\log_a b}$$
$$\log_b x = \frac{\log_a x}{\log_a b} \quad \blacksquare$$

Example 1 Scientific calculators generally provide keys for evaluating common logarithms and natural logarithms, but have no keys for evaluating logarithms with other bases. Use a calculator to evaluate $\log_2 5$ by expressing this logarithm in terms of natural logarithms.

Solution. From (9) with $b = 2$ and $a = e$ we obtain

$$\log_2 5 = \frac{\ln 5}{\ln 2} \approx 2.321928 \quad \blacktriangleleft$$

☐ **EXPANDING AND CONDENSING LOGARITHMIC EXPRESSIONS**

The results in Theorem 7.1.1 are commonly used in two ways:

• To expand a single logarithm into sums, differences, and multiples of logarithms.
• To condense sums, differences, and multiples of logarithms into a single logarithm.

Example 2   Express

$$\log \frac{xy^5}{\sqrt{z}}$$

in terms of sums, differences, and multiples of $\log x$, $\log y$, and $\log z$.

Solution.

$$\log \frac{xy^5}{\sqrt{z}} = \log xy^5 - \log \sqrt{z}$$

$$= \log x + \log y^5 - \log z^{1/2}$$

$$= \log x + 5 \log y - \tfrac{1}{2} \log z \quad \blacktriangleleft$$

Example 3   In each part rewrite the expression as a single logarithm.

(a) $5 \log 2 + \log 3 - \log 8$

(b) $\tfrac{1}{3} \ln x - \ln (x^2 - 1) + 2 \ln (x + 3)$

Solution (a).

$$5 \log 2 + \log 3 - \log 8 = \log 2^5 + \log 3 - \log 8$$

$$= \log 32 + \log 3 - \log 8$$

$$= \log (32 \cdot 3) - \log 8$$

$$= \log \frac{32 \cdot 3}{8} = \log 12$$

Solution (b).

$$\tfrac{1}{3} \ln x - \ln (x^2 - 1) + 2 \ln (x + 3) = \ln x^{1/3} - \ln (x^2 - 1) + \ln (x + 3)^2$$

$$= \ln \frac{\sqrt[3]{x}(x + 3)^2}{x^2 - 1} \quad \blacktriangleleft$$

REMARK.   Expressions of the form $\log_b (u + v)$ and $\log_b (u - v)$ have no useful simplifications in terms of $\log_b u$ and $\log_b v$. In particular,

$$\log_b (u + v) \neq \log_b u + \log_b v$$

$$\log_b (u - v) \neq \log_b u - \log_b v$$

☐ **SOLVING EQUATIONS OF THE FORM** $\log_b f(x) = k$

Equations of the form $\log_b f(x) = k$ can be solved by converting them to exponential form.

Example 4   Solve for x.

(a) $\log x = 2$ (b) $\ln (x + 1) = 5$

Solution (a).   Rewriting the equation in exponential form yields

$$x = 10^2 = 100$$

Solution (b).   Rewriting the equation in exponential form yields

$$x + 1 = e^5 \quad \text{or} \quad x = e^5 - 1$$

If desired, a calculator can be used to obtain a decimal approximation of this exact solution; to six decimal places this approximation is $x \approx 147.413159$. $\quad \blacktriangleleft$

☐ SOLVING EQUATIONS OF
THE FORM $b^{f(x)} = k$

Equations of the form $b^{f(x)} = k$ can be solved by taking the logarithm of both sides. Usually common logarithms or natural logarithms are used.

Example 5 Solve for x.

(a) $5^x = 7$ (b) $e^x = 9$

Solution (a). Taking the common logarithm of both sides yields

$$\log 5^x = \log 7$$

$$x \log 5 = \log 7$$

$$x = \frac{\log 7}{\log 5} \approx 1.209062$$

Alternatively, taking the natural logarithm of both sides yields

$$\ln 5^x = \ln 7$$

$$x \ln 5 = \ln 7$$

$$x = \frac{\ln 7}{\ln 5} \approx 1.209062$$

Solution (b). Because the equation involves a power of e, natural logarithms will result in simpler computations than common logarithms; taking the natural logarithm of both sides yields

$$\ln e^x = \ln 9$$

$$x = \ln 9 \approx 2.197225 \qquad ◄$$

Example 6 A satellite has a radioisotope power supply for which the power output in watts is given by the equation

$$P = 75e^{-t/125} \tag{10}$$

where t is the time in days that the supply is used.

(a)   What will the power output be after one year of use?
(b)   How long will it take until the satellite is operating at half power?

Solution (a). Substituting $t = 365$ yields

$$P = 75e^{-365/125} = 75e^{-2.92} \approx 4.05$$

Thus, the power output will be approximately 4.05 watts after one year.

Solution (b). To find the initial power output, we substitute $t = 0$ into (10); this yields

$$P = 75e^0 = 75 \text{ watts}$$

Thus, it follows from (10) that the satellite will be operating at half power when t satisfies

$$37.5 = 75e^{-t/125}$$

Dividing through by 75, then taking the natural logarithm of both sides yields

$$\tfrac{1}{2} = e^{-t/125}$$

$$\ln\left(\tfrac{1}{2}\right) = \ln\left(e^{-t/125}\right)$$

$$-\ln 2 = -t/125 \qquad \boxed{\text{See Theorem 7.1.1}(f).}$$

$$t = 125\ln 2 \approx 86.64$$

Thus, the satellite will be operating at half power on the 86th day.   ◄

☐ OTHER TYPES OF
LOGARITHMIC AND
EXPONENTIAL
EQUATIONS

The methods used in the preceding examples can be adapted to solve other types of logarithmic and exponential equations.

Example 7   Solve $\dfrac{e^x - e^{-x}}{2} = 1$ for x.

Solution.   Multiplying both sides of the given equation by 2 yields

$$e^x - e^{-x} = 2$$

or equivalently,

$$e^x - \frac{1}{e^x} = 2$$

Multiplying through by e^x yields

$$e^{2x} - 1 = 2e^x \quad \text{or} \quad e^{2x} - 2e^x - 1 = 0$$

This is really a quadratic equation in disguise; this can be seen by rewriting the equation as

$$(e^x)^2 - 2e^x - 1 = 0$$

and letting

$$u = e^x \tag{11}$$

to obtain

$$u^2 - 2u - 1 = 0$$

Solving for u by the quadratic formula yields

$$u = \frac{2 \pm \sqrt{4+4}}{2} = \frac{2 \pm \sqrt{8}}{2} = 1 \pm \sqrt{2}$$

Thus, from (11)

$$e^x = 1 \pm \sqrt{2}$$

But e^x cannot be negative, so we discard the negative value $1 - \sqrt{2}$; thus,

$$e^x = 1 + \sqrt{2}$$

$$\ln e^x = \ln(1 + \sqrt{2})$$

$$x = \ln(1 + \sqrt{2}) \approx 0.881 \qquad ◄$$

▶ Exercise Set 7.1   © 3, 4, 7, 8, 39, 41–50

In Exercises 1 and 2, simplify the expression without using a calculator.

In Exercises 3 and 4, use a calculator to obtain an approximate value of the expression. Round the answer to four decimal places.

1. (a) $-8^{2/3}$ (b) $(-8)^{2/3}$ (c) $8^{-2/3}$.

2. (a) 2^{-4} (b) $4^{1.5}$ (c) $9^{-0.5}$.

3. (a) $2^{1.57}$ (b) $5^{-2.1}$.

4. (a) $\sqrt[5]{24}$ (b) $\sqrt[8]{0.6}$.

In Exercises 5 and 6, find the exact value of the logarithm without using a calculator.

5. (a) $\log_2 16$ (b) $\log_2 \left(\frac{1}{32}\right)$

     (c) $\log_4 4$ (d) $\log_9 3$.

6. (a) $\log_{10} (0.001)$ (b) $\log_{10} (10^4)$

     (c) $\ln (e^3)$ (d) $\ln (\sqrt{e})$.

In Exercises 7 and 8, use a calculator to obtain an approximate value of the logarithm. Round the answer to four decimal places.

7. (a) $\log 23.2$ (b) $\ln 0.74$.

8. (a) $\log 0.3$ (b) $\ln \pi$.

In Exercises 9 and 10, use properties of logarithms (Theorem 7.1.1) to help expand the logarithm in terms of r, s, and t, where $r = \ln a$, $s = \ln b$, and $t = \ln c$.

9. (a) $\ln a^2 \sqrt{bc}$ (b) $\ln \dfrac{b}{a^3 c}$.

10. (a) $\ln \dfrac{\sqrt[3]{c}}{ab}$ (b) $\ln \sqrt{\dfrac{ab^3}{c^2}}$.

In Exercises 11 and 12, expand the logarithm in terms of sums, differences, and multiples of simpler logarithms.

11. (a) $\log (10x\sqrt{x-3})$ (b) $\ln \dfrac{x^2 \sin^3 x}{\sqrt{x^2+1}}$.

12. (a) $\log \dfrac{\sqrt[3]{x+2}}{\cos 5x}$ (b) $\ln \sqrt{\dfrac{x^2+1}{x^3+5}}$.

In Exercises 13–15, rewrite the expression as a single logarithm.

13. $4 \log 2 - \log 3 + \log 16$.

14. $\frac{1}{2} \log x - 3 \log (\sin 2x) + 2$.

15. $2 \ln (x+1) + \frac{1}{3} \ln x - \ln (\cos x)$.

In Exercises 16–25, solve for x without using a calculator.

16. $\log_{10} (1 + x) = 3$.   **17.** $\log_{10} (\sqrt{x}) = -1$.

18. $\ln (x^2) = 4$.   **19.** $\ln (1/x) = -2$.

20. $\log_3 (3^x) = 7$.   **21.** $\log_5 (5^{2x}) = 8$.

22. $\log_{10} x^2 + \log_{10} x = 30$.

23. $\log_{10} x^{3/2} - \log_{10} \sqrt{x} = 5$.

24. $\ln 4x - 3 \ln (x^2) = \ln 2$.

25. $\ln (1/x) + \ln (2x^3) = \ln 3$.

In Exercises 26–35, solve for x without using a calculator. Use the natural logarithm when logarithms are needed.

26. $3^x = 2$.   **27.** $5^{-2x} = 3$.

28. $3e^{-2x} = 5$.   **29.** $2e^{3x} = 7$.

30. $e^x - 2xe^x = 0$.   **31.** $xe^{-x} + 2e^{-x} = 0$.

32. $e^{2x} - e^x = 6$.   **33.** $e^{-2x} - 3e^{-x} = -2$.

34. $\dfrac{e^x}{1 - e^x} = 2$.   **35.** $\dfrac{1}{1 - 3e^{2x}} = 4$.

36. In each part, sketch the graph of $f(x)$.

(a) $f(x) = 2^x$ (b) $f(x) = (1/2)^x$

(c) $f(x) = 3^x$ (d) $f(x) = (1/3)^x$.

37. Find the fallacy in the following "proof" that $\frac{1}{8} > \frac{1}{4}$: Multiply both sides of the inequality $3 > 2$ by $\log \frac{1}{2}$ to get

$$3 \log \tfrac{1}{2} > 2 \log \tfrac{1}{2}$$
$$\log \left(\tfrac{1}{2}\right)^3 > \log \left(\tfrac{1}{2}\right)^2$$
$$\log \tfrac{1}{8} > \log \tfrac{1}{4}$$
$$\tfrac{1}{8} > \tfrac{1}{4}$$

38. (a) Prove part (b) of Theorem 7.1.1.

(b) Prove part (d) of Theorem 7.1.1.

(c) Prove part (e) of Theorem 7.1.1.

(d) Prove part (f) of Theorem 7.1.1.

39. Use a calculator and Formula (9) to find the values of $\log_2 7.35$ and $\log_5 0.6$ rounded to four decimal-place accuracy (use either common or natural logarithms).

40. Find the exact value of $(\log_2 81)(\log_3 32)$ without using a calculator. [*Hint:* Let $x = a$ in Formula (9).]

41. If equipment in the satellite of Example 6 requires 15 watts to operate correctly, what is the operational lifetime of the power supply?

42. The equation $Q = 12e^{-0.055t}$ gives the mass Q in grams of radioactive Potassium-42 that will remain from some initial quantity after t hours of radioactive decay.

(a) How many grams were there initially?

(b) How many grams remain after 4 hours?

(c) How long will it take to reduce the amount of radioactive Potassium-42 to half of the initial amount?

43. The acidity of a substance is measured by its pH value, which is defined by the formula

$$pH = -\log [H^+]$$

where the symbol $[H^+]$ denotes the concentration of hydrogen ions measured in moles per liter. Distilled water has a pH of 7; a substance is called *acidic* if it has pH < 7 and *basic* if it has pH > 7. Find the pH of each of the following substances and state whether it is acidic or basic.

SUBSTANCE	$[H^+]$
(a) Arterial blood	3.9×10^{-8} mol/L
(b) Tomatoes	6.3×10^{-5} mol/L
(c) Milk	4.0×10^{-7} mol/L
(d) Coffee	1.2×10^{-6} mol/L

44. Use the definition of pH in Exercise 43 to find $[H^+]$ in a solution having a pH equal to

(a) 2.44   (b) 8.06.

45. The perceived loudness L of a sound in decibels (dB) is related to its intensity I in watts/square meter (W/m^2) by the equation

$$L = 10 \log (I/I_0)$$

where $I_0 = 10^{-12}$ W/m^2. Damage to the average ear occurs at 90 dB or greater. Find the decibel level of each of the following sounds and state whether it will cause ear damage.

SOUND		I
(a)	Jet aircraft (from 500 ft)	1.0×10^2 W/m^2
(b)	Amplified rock music	1.0 W/m^2
(c)	Garbage disposal	1.0×10^{-4} W/m^2
(d)	TV (middle volume from 10 ft)	3.2×10^{-5} W/m^2

In Exercises 46–48, use the definition of the decibel level of a sound (see Exercise 45).

46. If one sound is three times as intense as another, how much greater is its decibel level?

47. According to one source, the noise inside a moving automobile is about 70 dB, while an electric blender generates 93 dB. Find the ratio of the intensity of the noise of the blender to that of the automobile.

48. Suppose that the decibel level of an echo is $\frac{2}{3}$ the decibel level of the original sound. If each echo results in another echo, how many echoes will be heard from a 120-dB sound given that the average human ear can hear a sound as low as 10 dB?

49. On the **Richter scale**, the magnitude M of an earthquake is related to the released energy E in joules (J) by the equation

$$\log E = 4.4 + 1.5M$$

(a) Find the energy E of the 1906 San Francisco earthquake that registered $M = 8.2$ on the Richter scale.

(b) If the released energy of one earthquake is 10 times that of another, how much greater is its magnitude on the Richter scale?

50. Suppose that the magnitudes of two earthquakes differ by 1 on the Richter scale. Find the ratio of the released energy of the larger earthquake to that of the smaller earthquake. [*Note:* See Exercise 49 for terminology.]

In Exercises 51 and 52, use Formula (5) or (7) to find the limit.

51. Find $\lim_{x \to 0} (1 - 2x)^{1/x}$. [*Hint:* Let $t = -2x$.]

52. Find $\lim_{x \to +\infty} (1 + 3/x)^x$. [*Hint:* Let $t = 3/x$.]

▪ 7.2 DERIVATIVES AND INTEGRALS OF LOGARITHMIC AND EXPONENTIAL FUNCTIONS

In this section we shall discuss derivatives and integrals involving logarithmic and exponential functions. Our work in this section will be informal in that it will be assumed that the functions considered are differentiable. The mathematical theory that justifies this will be considered later in this chapter.

☐ **THE ROLE OF THE NATURAL LOGARITHM IN CALCULUS**

In this section we shall be interested in logarithms and exponents from the viewpoint of functions. For $b > 0$ we shall call b^x the **base b exponential function** and $\log_b x$ the **base b logarithm function**. In the case where $b = e$ we shall call e^x the **natural exponential function** and $\ln x = \log_e x$ the **natural logarithm function**.

The natural logarithm function plays a special role in calculus that we can motivate by differentiating $f(x) = \log_b x$, where b is an arbitrary base. For purposes of this introductory discussion we shall assume that the function $\log_b x$ is differentiable, hence continuous for $x > 0$. This will be proved later in the chapter.

Using the definition of a derivative we obtain

$$\frac{d}{dx}[\log_b x] = \lim_{h \to 0} \frac{\log_b (x + h) - \log_b x}{h}$$

$$= \lim_{h \to 0} \frac{1}{h} \log_b \left(\frac{x + h}{x} \right) \qquad \boxed{\text{Theorem 7.1.1}(d)}$$

$$= \lim_{h \to 0} \frac{1}{h} \log_b \left(1 + \frac{h}{x} \right)$$

$$= \lim_{v \to 0} \frac{1}{vx} \log_b (1 + v) \qquad \boxed{\begin{array}{l} \text{Let } v = h/x \text{ and note} \\ \text{that } v \to 0 \text{ as } h \to 0. \end{array}}$$

$$= \frac{1}{x} \lim_{v \to 0} \frac{1}{v} \log_b (1 + v) \qquad \boxed{\begin{array}{l} 1/x \text{ does not vary with} \\ v, \text{ so it can be moved} \\ \text{through the limit sign.} \end{array}}$$

$$= \frac{1}{x} \lim_{v \to 0} \log_b (1 + v)^{1/v} \qquad \boxed{\text{Theorem 7.1.1}(e)}$$

$$= \frac{1}{x} \log_b \left[\lim_{v \to 0} (1 + v)^{1/v} \right] \qquad \boxed{\begin{array}{l} \log_b x \text{ is continuous,} \\ \text{so the remark following} \\ \text{Theorem 2.7.5 applies.} \end{array}}$$

$$= \frac{1}{x} \log_b e \qquad \boxed{\begin{array}{l} \text{Formula (7) of} \\ \text{Section 7.1} \end{array}}$$

Thus,

$$\frac{d}{dx}[\log_b x] = \frac{1}{x} \log_b e, \quad x > 0$$

which from Theorem 7.1.2 can be rewritten as (verify)

$$\frac{d}{dx}[\log_b x] = \frac{1}{x \ln b}, \quad x > 0 \tag{1}$$

In the special case where $b = e$ we have $\ln b = \ln e = 1$, so this formula becomes

$$\frac{d}{dx}[\ln x] = \frac{1}{x}, \quad x > 0 \tag{2}$$

Thus, among all possible choices for the base b, the one that yields the simplest derivative formula for $\log_b x$ is $b = e$. This is one of the main reasons why the natural logarithm plays such a special role in calculus.

☐ **DERIVATIVES AND INTEGRALS INVOLVING** $\ln x$

If $u(x) > 0$, and if the function u is differentiable at x, then the following generalizations of (1) and (2) are consequences of the chain rule [Formula (2) of Section 3.5]:

$$\frac{d}{dx}[\log_b u] = \frac{1}{u \ln b} \cdot \frac{du}{dx} \qquad \text{and} \qquad \frac{d}{dx}[\ln u] = \frac{1}{u} \cdot \frac{du}{dx} \tag{3-4}$$

Example 1   Find $\dfrac{d}{dx}[\ln (x^2 + 1)]$.

Solution.   From (4) with $u = x^2 + 1$,

$$\frac{d}{dx}[\ln (x^2 + 1)] = \frac{1}{x^2 + 1} \cdot \frac{d}{dx}[x^2 + 1] = \frac{1}{x^2 + 1} \cdot 2x = \frac{2x}{x^2 + 1} \qquad \blacktriangleleft$$

When possible, the properties of logarithms in Theorem 7.1.1 should be used to convert products, quotients, and exponents into sums, differences, and constant multiples *before* differentiating a function involving natural logarithms.

Example 2

$$\frac{d}{dx}\left[\ln\left(\frac{x^2\sin x}{\sqrt{1+x}}\right)\right] = \frac{d}{dx}\left[2\ln x + \ln(\sin x) - \frac{1}{2}\ln(1+x)\right]$$

$$= \frac{2}{x} + \frac{\cos x}{\sin x} - \frac{1}{2(1+x)}$$

$$= \frac{2}{x} + \cot x - \frac{1}{2+2x} \quad \blacktriangleleft$$

Example 3   Find $\dfrac{d}{dx}\left[\ln|x|\right]$.

Solution.   We shall consider the cases where $x > 0$ and $x < 0$ separately. (The case where $x = 0$ is excluded because $\ln 0$ is undefined.) If $x > 0$, then $|x| = x$, so that

$$\frac{d}{dx}\left[\ln|x|\right] = \frac{d}{dx}\left[\ln x\right] = \frac{1}{x}$$

If $x < 0$, then $|x| = -x$, so that from (4) with $u = -x$,

$$\frac{d}{dx}\left[\ln|x|\right] = \frac{d}{dx}\left[\ln(-x)\right] = \frac{1}{(-x)}\cdot\frac{d}{dx}\left[-x\right] = \frac{1}{x}$$

Thus,

$$\frac{d}{dx}\left[\ln|x|\right] = \frac{1}{x} \quad \text{if } x \neq 0 \quad \blacktriangleleft \tag{5}$$

Example 4   From (5) and the chain rule,

$$\frac{d}{dx}\left[\ln|\sin x|\right] = \frac{1}{\sin x}\cdot\frac{d}{dx}\left[\sin x\right] = \frac{\cos x}{\sin x} = \cot x \quad \blacktriangleleft$$

Formula (5) states that the function $\ln|x|$ is an antiderivative of $1/x$ everywhere except at $x = 0$. In contrast, the function $\ln x$ is an antiderivative of $1/x$ only for $x > 0$. The companion integration formula for (5) is

$$\int\frac{1}{u}\,du = \ln|u| + C \tag{6}$$

Example 5   Evaluate $\displaystyle\int\frac{3x^2}{x^3+5}\,dx$.

Solution.   Make the substitution

$$u = x^3 + 5, \quad du = 3x^2\,dx$$

so that

$$\int\frac{3x^2}{x^3+5}\,dx = \int\frac{1}{u}\,du = \ln|u| + C = \ln|x^3+5| + C \quad \blacktriangleleft$$

Formula (6)

Example 6   Evaluate $\int \tan x \, dx$.

Solution.

$$\int \tan x \, dx = \int \frac{\sin x}{\cos x} \, dx = -\int \frac{1}{u} \, du = -\ln |u| + C = -\ln |\cos x| + C \qquad \blacktriangleleft$$

$$u = \cos x$$
$$du = -\sin x \, dx$$

REMARK.   The last two examples illustrate an important point: Any integral of the form

$$\int \frac{g'(x)}{g(x)} \, dx$$

(where the numerator of the integrand is the derivative of the denominator) can be evaluated by the *u*-substitution $u = g(x)$, $du = g'(x) \, dx$, since this substitution yields

$$\int \frac{g'(x)}{g(x)} \, dx = \int \frac{du}{u} = \ln |u| + C = \ln |g(x)| + C$$

□ **LOGARITHMIC DIFFERENTIATION**

We now consider an important technique, called *logarithmic differentiation*, that is useful for differentiating functions that are composed of products, quotients, and powers.

Example 7   The derivative of

$$y = \frac{x^2 \sqrt[3]{7x - 14}}{(1 + x^2)^4} \tag{7}$$

is messy to calculate directly. However, if we first take the natural logarithm of both sides and then use its properties, we can write

$$\ln y = 2 \ln x + \tfrac{1}{3} \ln (7x - 14) - 4 \ln (1 + x^2)$$

Differentiating both sides with respect to x yields

$$\frac{1}{y} \frac{dy}{dx} = \frac{2}{x} + \frac{7/3}{7x - 14} - \frac{8x}{1 + x^2} \tag{8}$$

Thus, on solving for dy/dx and using (7) we obtain

$$\frac{dy}{dx} = \frac{x^2 \sqrt[3]{7x - 14}}{(1 + x^2)^4} \left[\frac{2}{x} + \frac{1}{3x - 6} - \frac{8x}{1 + x^2} \right] \qquad \blacktriangleleft \tag{9}$$

Since $\ln y$ is defined only for $y > 0$, logarithmic differentiation of $y = f(x)$ is valid only on intervals where $f(x)$ is positive. Thus, the derivative obtained in the preceding example is valid on the interval $(2, +\infty)$, since the given function is positive for $x > 2$. However, the formula is actually valid on the interval $(-\infty, 2)$ as well. This can be seen by taking absolute values before proceeding with the logarithmic differentiation and noting that $\ln |y|$ is defined for all y except $y = 0$. If we do this and simplify using properties of logarithms and absolute values, we obtain

$$\ln |y| = 2 \ln |x| + \tfrac{1}{3} \ln |7x - 14| - 4 \ln |1 + x^2|$$

Differentiating both sides with respect to x yields (8), and hence results in (9).

In general, if the derivative of $y = f(x)$ is to be obtained by logarithmic differentiation, then the same formula for dy/dx will result regardless of whether one first takes absolute values or not. Thus, a derivative formula obtained by logarithmic differentiation is valid except perhaps at points where $f(x)$ is zero. The formula may, in fact, be valid at those points as well, but it is not guaranteed.

☐ **DERIVATIVES OF IRRATIONAL POWERS OF x**

In Section 3.6 we showed that the formula

$$\frac{d}{dx}[x^r] = rx^{r-1} \tag{10}$$

holds for rational values of r. By using logarithmic differentiation we shall show that this formula holds if r is any *real* number (rational or irrational). In our computations we shall assume that x^r is differentiable and that the standard rules of exponents apply to all real exponents; this will be justified later.

Let $y = x^r$, where r is a real number. Proceeding by logarithmic differentiation we obtain

$$\ln y = \ln x^r = r \ln x$$

$$\frac{d}{dx}[\ln y] = \frac{d}{dx}[r \ln x]$$

$$\frac{1}{y}\frac{dy}{dx} = \frac{r}{x}$$

$$\frac{dy}{dx} = \frac{r}{x}y = \frac{r}{x}x^r = rx^{r-1}$$

which establishes (10) for real values of r.

Example 8

$$\frac{d}{dx}[x^\pi] = \pi x^{\pi-1} \qquad \blacktriangleleft$$

☐ **DERIVATIVES INVOLVING b^x**

Our next objective is to obtain a derivative formula for b^x (and hence for e^x). For this purpose we shall assume that b^x is differentiable for all x. This will be proved later in the chapter.

To obtain the derivative of b^x, we shall let $y = b^x$ and use logarithmic differentiation:

$$\ln y = \ln b^x = x \ln b$$

$$\frac{1}{y}\frac{dy}{dx} = \ln b$$

$$\frac{dy}{dx} = y \ln b = b^x \ln b$$

Thus,

$$\frac{d}{dx}[b^x] = b^x \ln b \tag{11}$$

In the special case where $b = e$ we have $\ln e = 1$, so that (11) becomes

$$\frac{d}{dx}[e^x] = e^x \tag{12}$$

If u is a differentiable function of x, then it follows from (11) and (12) that

$$\frac{d}{dx}[b^u] = b^u \ln b \cdot \frac{du}{dx} \qquad \text{and} \qquad \frac{d}{dx}[e^u] = e^u \cdot \frac{du}{dx} \tag{13-14}$$

Example 9 It is important to distinguish between differentiating b^x (variable exponent) and x^b (variable base). Compare, for example, the derivative of x^π obtained in Example 8 to the following derivative, which is obtained from Formula (11):

$$\frac{d}{dx}[\pi^x] = \pi^x \ln \pi \qquad \blacktriangleleft$$

Example 10 The following computations use (13) and (14).

$$\frac{d}{dx}[2^{\sin x}] = (2^{\sin x})(\ln 2) \cdot \frac{d}{dx}[\sin x] = (2^{\sin x})(\ln 2)(\cos x)$$

$$\frac{d}{dx}[e^{-2x}] = e^{-2x} \cdot \frac{d}{dx}[-2x] = -2e^{-2x}$$

$$\frac{d}{dx}[e^{x^3}] = e^{x^3} \cdot \frac{d}{dx}[x^3] = 3x^2 e^{x^3}$$

$$\frac{d}{dx}[e^{\cos x}] = e^{\cos x} \cdot \frac{d}{dx}[\cos x] = -(\sin x)e^{\cos x} \qquad \blacktriangleleft$$

☐ **INTEGRALS INVOLVING** b^x

Associated with derivatives (13) and (14) are the companion integration formulas

$$\int b^u \, du = \frac{b^u}{\ln b} + C \qquad \text{and} \qquad \int e^u \, du = e^u + C \qquad (15\text{--}16)$$

Example 11

$$\int 2^x \, dx = \frac{2^x}{\ln 2} + C \qquad \blacktriangleleft$$

Example 12 Evaluate $\int e^{5x} \, dx$.

Solution. Let $u = 5x$ so that $du = 5 \, dx$ or $dx = \frac{1}{5} \, du$, which yields

$$\int e^{5x} \, dx = \frac{1}{5} \int e^u \, du = \frac{1}{5} e^u + C = \frac{1}{5} e^{5x} + C \qquad \blacktriangleleft$$

Example 13

$$\int e^{-x} \, dx = -\int e^u \, du = -e^u + C = -e^{-x} + C$$

$$u = -x$$
$$du = -dx$$

$$\int x^2 e^{x^3} \, dx = \frac{1}{3} \int e^u \, du = \frac{1}{3} e^u + C = \frac{1}{3} e^{x^3} + C \qquad \blacktriangleleft$$

$$u = x^3$$
$$du = 3x^2 \, dx$$

Example 14 Evaluate $\displaystyle\int_0^{\ln 3} e^x (1 + e^x)^{1/2} \, dx$.

Solution. Make the u-substitution

$$u = 1 + e^x, \quad du = e^x \, dx$$

and change the x-limits of integration ($x = 0, x = \ln 3$) to the u-limits ($u = 1 + e^0 = 2$, $u = 1 + e^{\ln 3} = 1 + 3 = 4$).

$$\int_0^{\ln 3} e^x(1 + e^x)^{1/2}\, dx = \int_2^4 u^{1/2}\, du = \frac{2}{3} u^{3/2}\Big]_2^4 = \frac{2}{3}[4^{3/2} - 2^{3/2}]$$

$$= \frac{16}{3} - \frac{4\sqrt{2}}{3} \qquad \blacktriangleleft$$

REMARK.   For ease of printing, the exponential function is often denoted by $\exp x$ rather than e^x. For example, in this notation the following statements are equivalent:

$$e^{x_1 + x_2} = e^{x_1} e^{x_2} \qquad \text{and} \qquad \exp(x_1 + x_2) = \exp x_1 \cdot \exp x_2$$

$$e^{\ln x} = x \qquad \text{and} \qquad \exp(\ln x) = x$$

$$\ln(e^x) = x \qquad \text{and} \qquad \ln(\exp x) = x$$

▶ Exercise Set 7.2   Ⓒ 62, 63, 145, 146, 147

In Exercises 1–20, find dy/dx.

1. $y = \ln 2x$.
2. $y = \ln(x^3)$.
3. $y = (\ln x)^2$.
4. $y = \ln(\sin x)$.
5. $y = \ln|\tan x|$.
6. $y = \ln(2 + \sqrt{x})$.
7. $y = \ln\left(\dfrac{x}{1 + x^2}\right)$.
8. $y = \ln(\ln x)$.
9. $y = \ln|x^3 - 7x^2 - 3|$.
10. $y = x^3 \ln x$.
11. $y = \sqrt{\ln x}$.
12. $y = \cos(\ln x)$.
13. $y = \sin\left(\dfrac{5}{\ln x}\right)$.
14. $y = \sqrt{1 + \ln^2 x}$.
15. $y = x^3 \ln(3 - 2x)$.
16. $y = x[\ln(x^2 - 2x)]^3$.
17. $y = (x^2 + 1)[\ln(x^2 + 1)]^2$.
18. $y = \dfrac{\ln x}{1 + \ln x}$.
19. $y = \dfrac{x^2}{1 + \ln x}$.
20. $y = \ln\left|\dfrac{1 - \cos \pi x}{1 + \cos \pi x}\right|$.

21. Find dy/dx by implicit differentiation if $y + \ln xy = 1$.
22. Find dy/dx by implicit differentiation if $y = \ln(x \tan y)$.

In Exercises 23–34, evaluate the indefinite integrals.

23. $\displaystyle\int \frac{dx}{2x}$.
24. $\displaystyle\int \frac{5x^4}{x^5 + 1}\, dx$.
25. $\displaystyle\int \frac{x^2}{x^3 - 4}\, dx$.
26. $\displaystyle\int \frac{t + 1}{t}\, dt$.
27. $\displaystyle\int \frac{\sec^2 x}{\tan x}\, dx$.
28. $\displaystyle\int \cot x\, dx$.
29. $\displaystyle\int \frac{\sin 3\theta}{1 + \cos 3\theta}\, d\theta$.
30. $\displaystyle\int \frac{dx}{x \ln x}$.

31. $\displaystyle\int \frac{x^3}{x^2 + 1}\, dx$.
32. $\displaystyle\int \frac{1}{x} \cos(\ln x)\, dx$.
33. $\displaystyle\int \frac{1}{y}(\ln y)^3\, dy$.
34. $\displaystyle\int \frac{dx}{\sqrt{x}(1 - 2\sqrt{x})}$.

In Exercises 35–38, evaluate the definite integral.

35. $\displaystyle\int_0^1 \frac{1}{3x + 2}\, dx$.
36. $\displaystyle\int_1^4 \frac{3}{1 - 2x}\, dx$.
37. $\displaystyle\int_{-1}^0 \frac{x}{x^2 + 5}\, dx$.
38. $\displaystyle\int_1^4 \frac{1}{\sqrt{x}(1 + \sqrt{x})}\, dx$.

In Exercises 39–42, use the method shown in Example 2 to help perform the indicated differentiation.

39. $\dfrac{d}{dx}\left[\ln \dfrac{\cos x}{\sqrt{4 - 3x^2}}\right]$.
40. $\dfrac{d}{dx}\left[\ln \sqrt{\dfrac{x - 1}{x + 1}}\right]$.
41. $\dfrac{d}{dx}[\ln(\sqrt{x}\sqrt[3]{x + 3}\sqrt[5]{3x - 2})]$.
42. $\dfrac{d}{dx}\left[\ln\left(\dfrac{\sqrt{x}\sqrt[3]{x + 1}}{\sin x \sec x}\right)\right]$.

In Exercises 43–46, obtain dy/dx by logarithmic differentiation.

43. $y = x\sqrt[3]{1 + x^2}$.
44. $y = \sqrt[5]{\dfrac{x - 1}{x + 1}}$.
45. $y = \dfrac{(x^2 - 8)^{1/3}\sqrt{x^3 + 1}}{x^6 - 7x + 5}$.
46. $y = \dfrac{\sin x \cos x \tan^3 x}{\sqrt{x}}$.

47. Find

(a) $\dfrac{d}{dx}[\log_x e]$

(b) $\dfrac{d}{dx}[\log_x 2]$.

In Exercises 48 and 49, interpret the expression as a limit of Riemann sums in which the interval [0, 1] is divided into n subintervals of equal width, then find the limit by evaluating an appropriate definite integral.

48. $\displaystyle\lim_{n \to +\infty} \sum_{k=1}^{n} \frac{1}{n + k}$.

49. $\displaystyle\lim_{n \to +\infty} \sum_{k=1}^{n} \frac{k}{n^2 + k^2}$.

50. Find $\displaystyle\lim_{h \to 0} \frac{\ln(1 + h)}{h}$. [*Hint:* Let $f(x) = \ln x$ and consider $f'(1)$ directly from the definition of the derivative.]

51. Find the minimum value of $x^2 - \ln x$.

52. Find the minimum value of $\sqrt{x} \ln x$.

53. Find the relative extrema of $(\ln x)^2 / x$.

54. Prove: $\ln x < \sqrt{x}$. [*Hint:* Show that $\sqrt{x} - \ln x$ is always positive.]

55. Prove: $\ln x \le x - 1$.
[*Hint:* Show that $x - 1 - \ln x \ge 0$.]

56. Prove: $\ln x \ge 1 - 1/x$.
[*Hint:* Show that $1 - 1/x - \ln x \le 0$.]

57. *Boyle's law* in physics states that under appropriate conditions the pressure p exerted by a gas is related to its volume v by $pv = c$, where c is a constant depending on the units and various physical factors. Prove: When such a gas increases in volume from v_0 to v_1 the average pressure it exerts with respect to volume is

$$p_{\text{ave}} = \frac{c}{v_1 - v_0} \ln \frac{v_1}{v_0}$$

58. Consider the region enclosed by the curve $y = 1/x$, the x-axis, and the lines $x = 1$ and $x = b$, where $b > 1$.

(a) Let A be the area of the region. Find A and $\displaystyle\lim_{b \to +\infty} A$.

(b) Let V be the volume of the solid generated when the region is revolved about the x-axis. Find V and $\displaystyle\lim_{b \to +\infty} V$.

59. Find the area of the region enclosed by $y = \tan x$, $y = 0$, and $x = \pi/3$. [*Hint:* See Example 6.]

60. Find the volume of the solid that is generated when the region enclosed by $y = 1/x^2$, $y = 4$, and $x = 3$ is revolved about the y-axis.

61. Find the volume of the solid that is generated when the region enclosed by $y = 1/\sqrt{x}$, $y = 0$, $x = 1$, and $x = 4$ is revolved about the x-axis.

In Exercises 62 and 63, use Newton's Method (Section 4.8) to approximate the solutions of the equation. Show as many decimal places as your calculator displays.

62. $\sin x = \ln x$.

63. $\ln x = x - 2$.

In Exercises 64–79, find dy/dx.

64. $y = e^{7x}$.

65. $y = e^{-5x^2}$.

66. $y = e^{1/x}$.

67. $y = x^3 e^x$.

68. $y = \sin(e^x)$.

69. $y = \dfrac{e^x - e^{-x}}{e^x + e^{-x}}$.

70. $y = \dfrac{e^x}{\ln x}$.

71. $y = e^{x \tan x}$.

72. $y = \exp(\sqrt{1 + 5x^3})$.

73. $y = e^{(x - e^{3x})}$.

74. $y = \ln(\cos e^x)$.

75. $y = \ln(1 - xe^{-x})$.

76. $y = \sqrt{1 + e^x}$.

77. $y = e^{ax} \cos bx$ (a, b constant).

78. $y = \dfrac{a}{1 + be^{-x}}$ (a, b constant).

79. $y = e^{\ln(x^3 + 1)}$.

In Exercises 80–83, find $f'(x)$ by Formula (13) and then by logarithmic differentiation.

80. $f(x) = 2^x$.

81. $f(x) = 3^{-x}$.

82. $f(x) = \pi^{\sin x}$.

83. $f(x) = \pi^{x \tan x}$.

84. (a) Explain why Formula (11) cannot be used to find $(d/dx)[x^x]$.

(b) Find this derivative by logarithmic differentiation.

85. Assuming that u and v are differentiable functions of x, use logarithmic differentiation to derive a formula for $d(u^v)/dx$ in terms of u, v, du/dx, and dv/dx. What formula results when the base u is constant? When the exponent v is constant?

In Exercises 86–91, find dy/dx by logarithmic differentiation.

86. $y = x^{\sin x}$.

87. $y = (x^3 - 2x)^{\ln x}$.

88. $y = (x^2 + 3)^{\ln x}$.

89. $y = (\ln x)^{\tan x}$.

90. $y = (1 + x)^{1/x}$.

91. $y = x^{(e^x)}$.

92. Show that $y = e^{3x}$ and $y = e^{-3x}$ both satisfy the equation $y'' - 9y = 0$.

93. Show that for any constants A and B, the function $y = Ae^{2x} + Be^{-4x}$ satisfies the equation
$$y'' + 2y' - 8y = 0$$

94. Show that for any constants A and k, the function $y = Ae^{kt}$ satisfies the equation $dy/dt = ky$.

95. Let $f(x) = e^{kx}$ and $g(x) = e^{-kx}$. Find

(a) $f^{(n)}(x)$

(b) $g^{(n)}(x)$.

96. Find dy/dt if $y = e^{-\lambda t}(A \sin \omega t + B \cos \omega t)$, where A, B, λ, and ω are constants.

97. Find $f'(x)$ if
$$f(x) = \frac{1}{\sqrt{2\pi}\sigma} \exp\left[-\frac{1}{2}\left(\frac{x - \mu}{\sigma}\right)^2\right]$$
where μ and σ are constants and $\sigma \ne 0$.

In Exercises 98–125, evaluate the integrals.

98. $\displaystyle\int \frac{dx}{e^x}.$

99. $\displaystyle\int e^{-5x}\, dx.$

100. $\displaystyle\int e^{\tan x} \sec^2 x\, dx.$

101. $\displaystyle\int e^{\sin x} \cos x\, dx.$

102. $\displaystyle\int x^3 e^{x^4}\, dx.$

103. $\displaystyle\int x^2 e^{-2x^3}\, dx.$

104. $\displaystyle\int \frac{e^x + e^{-x}}{e^x - e^{-x}}\, dx.$

105. $\displaystyle\int \frac{e^x}{1 + e^x}\, dx.$

106. $\displaystyle\int \sqrt{e^x}\, dx.$

107. $\displaystyle\int e^{2t} \sqrt{1 + e^{2t}}\, dt.$

108. $\displaystyle\int (x + 3) \exp(x^2 + 6x)\, dx.$

109. $\displaystyle\int \sin x \exp(\cos x)\, dx.$

110. $\displaystyle\int e^x \sin(1 + e^x)\, dx.$

111. $\displaystyle\int e^{-x} \sec^2(2 - e^{-x})\, dx.$

112. $\displaystyle\int 2^{5x}\, dx.$

113. $\displaystyle\int \pi^{\sin x} \cos x\, dx.$

114. $\displaystyle\int [ex^2 + (\tfrac{1}{2} \ln 2) \sin x]\, dx.$

115. $\displaystyle\int (x \ln 3 - 4\pi e^2 \cos x)\, dx.$

116. $\displaystyle\int e^{2 \ln x}\, dx.$

117. $\displaystyle\int [\ln(e^x) + \ln(e^{-x})]\, dx.$

118. $\displaystyle\int \frac{dy}{\sqrt{y}e^{\sqrt{y}}}.$

119. $\displaystyle\int \frac{e^{\sqrt{y}}}{\sqrt{y}}\, dy.$

120. $\displaystyle\int_0^{\ln 2} e^{-3x}\, dx.$

121. $\displaystyle\int_0^{\ln 5} e^x(3 - 4e^x)\, dx.$

122. $\displaystyle\int_1^{\sqrt{2}} x 4^{-x^2}\, dx.$

123. $\displaystyle\int_1^2 (3 - e^x)\, dx.$

124. $\displaystyle\int_0^e \frac{dx}{x + e}.$

125. $\displaystyle\int_{-\ln 3}^{\ln 3} \frac{e^x}{e^x + 4}\, dx.$

126. Show that the equation $e^{1/x} - e^{-1/x} = 0$ has no solution.

127. Prove: If $f'(x) = f(x)$ for all x in $(-\infty, +\infty)$, then $f(x)$ has the form $f(x) = ke^x$ for some constant k. [*Hint:* Let $g(x) = e^{-x} f(x)$ and find $g'(x)$.]

128. A particle moves along the x-axis so that its x-coordinate at time t is given by $x = ae^{kt} + be^{-kt}$. Show that its acceleration is proportional to x.

129. Find the intersections of the curves $y = 2^x$ and $y = 3^{x+1}$.

130. Find a point on the graph of $y = e^{3x}$ at which the tangent line passes through the origin.

131. Find $f'(x)$ if $f(x) = x^e$.

132. Show that

(a) $y = xe^{-x}$ satisfies the equation
$xy' = (1 - x)y$

(b) $y = xe^{-x^2/2}$ satisfies the equation
$xy' = (1 - x^2)y.$

133. The equilibrium constant k of a balanced chemical reaction changes with the absolute temperature T according to the law

$$k = k_0 \exp\left(-\frac{q(T - T_0)}{2T_0 T}\right)$$

where k_0, q, and T_0 are constants. Find the rate of change of k with respect to T.

134. Find a function $y = f(x)$ such that $e^y - e^{-y} = x$.

135. Evaluate

$$\int \frac{e^{2x}}{e^x + 3}\, dx$$

[*Hint:* Divide $e^x + 3$ into e^{2x}.]

136. Find

$$\lim_{n \to +\infty} \frac{e^{1/n} + e^{2/n} + e^{3/n} + \cdots + e^{n/n}}{n}$$

[*Hint:* Interpret this as a limit of Riemann sums in which the interval [0, 1] is divided into n subintervals of equal width.]

137. Find the maximum value of $x^3 e^{-2x}$.

138. Prove: $e^x \geq 1 + x$. [*Hint:* Show that $1 + x - e^x \leq 0$.]

139. Find the area of the region enclosed by $y = e^x$, $y = 3$, and $x = 0$.

140. Find the positive value of k such that the area under the graph of $y = e^{2x}$ over the interval [0, k] is 3 square units.

141. Find the volume of the solid that is generated when the region enclosed by $y = e^x$, $y = 0$, $x = 0$, and $x = \ln 3$ is revolved about the x-axis.

142. (a) Show by differentiation that $xe^x - e^x$ is an antiderivative of xe^x.

(b) Find the volume of the solid that is generated when the region enclosed by $y = e^x$, $y = 0$, $x = 0$, and $x = 2$ is revolved about the y-axis.

143. Evaluate $\int_1^5 \ln x\, dx$. [*Hint:* Interpret the integral as an area and integrate with respect to y.]

144. Evaluate $\int_1^3 x \ln x\, dx$. [*Hint:* Relate the integral to the volume of a solid by the method of shells and use the method of washers to find the volume.]

145. Find all values of x for which $e^x < 2 - x^2$. State your answer to at least five decimal places. [*Hint:* Use Newton's Method (Section 4.8).]

146. Find the three solutions of $x^2 = 2^x$. [*Hint:* You should be able to find two of them by inspection. Use Newton's Method (Section 4.8) to approximate the third one to at least five decimal places.]

147. In physics, the **spectral radiancy** S of an ideal radiator at a constant temperature is given by **Planck's radiation law**

$$S = \frac{af^3}{e^{bf} - 1}$$

where a and b are positive constants and f is the frequency of the radiant photons.

(a) Show that the frequency at which S attains its maximum value satisfies the equation
$$3e^{-bf} + bf - 3 = 0$$

(b) Assuming that $f > 0$, use Newton's Method (Section 4.8) to approximate the value of bf that satisfies the equation in part (a). Express your answer to at least five decimal places. [*Hint:* Let $x = bf$.]

(c) For a radiator with
$$b = \frac{4.8043 \times 10^{-11}}{T}$$

where T is the temperature in kelvins, express the frequency at which S attains its maximum value as a function of the temperature T.

148. Sketch the graph of $y = x^{1/\ln x}$. [*Hint:* First find a simpler form for $x^{1/\ln x}$.]

149. Find the minimum value of x^x, $x > 0$.

150. Find the maximum value of $x^{1/x}$, $x > 0$.

151. Let A denote the area under the curve $y = e^{-2x}$ over the interval $[0, b]$. Express A as a function of b, and then find the value of b for which $A = \frac{1}{4}$. What is the limiting value of A as $b \to +\infty$?

152. In each part find the indicated limit by interpreting the expression as a derivative and evaluating the derivative.

(a) $\displaystyle\lim_{h \to 0} \frac{10^h - 1}{h}$

(b) $\displaystyle\lim_{h \to 0} \frac{e^{(3+h)^2} - e^9}{h}$

(c) $\displaystyle\lim_{h \to 0} \frac{\ln(e^2 + h) - 2}{h}$

(d) $\displaystyle\lim_{x \to 1} \frac{2^x - 2}{x - 1}$.

■ 7.3 GRAPHS INVOLVING EXPONENTIALS AND LOGARITHMS

In this section we shall discuss techniques for graphing equations involving exponential and logarithmic functions. In the course of our work we will use various limits that are plausible, but which will not actually be derived precisely until Section 7.5. Thus, our work here is informal, as in the preceding sections.

☐ **SOME PROPERTIES OF e^x AND $\ln x$**

In Section 7.1 we obtained graphs of $y = e^x$ and $y = \ln x$ by plotting points (Figure 7.1.3). For reference, those curves are shown in Figure 7.3.1. We noted in Section 7.2 that the two curves are reflections of one another about the line $y = x$. These graphs suggest that e^x and $\ln x$ have the properties listed in Table 7.3.1.

We can verify that e^x is increasing and that its graph is concave up from its first and second derivatives. For all x in $(-\infty, +\infty)$ we have
$$\frac{d}{dx}[e^x] = e^x > 0$$

which implies that e^x is increasing on $(-\infty, +\infty)$, and
$$\frac{d^2}{dx^2}[e^x] = \frac{d}{dx}[e^x] > 0$$

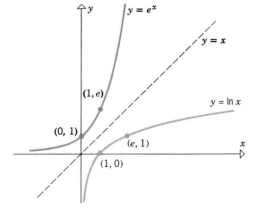

Figure 7.3.1

Table 7.3.1

PROPERTIES OF e^x	PROPERTIES OF $\ln x$
$e^x > 0$ for all x	$\ln x > 0$ if $x > 1$ $\ln x < 0$ if $0 < x < 1$ $\ln x = 0$ if $x = 1$
e^x is increasing on $(-\infty, +\infty)$	$\ln x$ is increasing on $(0, +\infty)$
The graph of e^x is concave up on $(-\infty, +\infty)$	The graph of $\ln x$ is concave down on $(0, +\infty)$

which implies that the graph of e^x is concave up on $(-\infty, +\infty)$.

Similarly, we can verify that $\ln x$ is increasing and concave down from its first and second derivatives. For all x in $(0, +\infty)$ we have

$$\frac{d}{dx}[\ln x] = \frac{1}{x} > 0$$

which implies that $\ln x$ is increasing on $(0, +\infty)$, and

$$\frac{d^2}{dx^2}[\ln x] = \frac{d}{dx}\left[\frac{1}{x}\right] = -\frac{1}{x^2} < 0$$

which implies that $\ln x$ is concave down on $(0, +\infty)$.

☐ **LIMITS INVOLVING**
$\ln x$ AND e^x

Table 7.3.2

x	e^x	$\ln x$
1	2.718282	0
2	7.389056	0.693147
3	20.08554	1.098612
4	54.59815	1.386294
5	148.4132	1.609438
6	403.4288	1.791759
7	1096.633	1.945910
8	2980.958	2.079442
9	8103.084	2.197225
10	22026.47	2.302585

The following limits, which are consistent with Figure 7.3.1, will be proved later.

$$\lim_{x \to +\infty} e^x = +\infty \qquad \text{and} \qquad \lim_{x \to -\infty} e^x = 0 \qquad (1\text{-}2)$$

$$\lim_{x \to +\infty} \ln x = +\infty \qquad \text{and} \qquad \lim_{x \to 0^+} \ln x = -\infty \qquad (3\text{-}4)$$

The graph of $y = \ln x$ rises so slowly that Figure 7.3.1 does not adequately convey that

$$\lim_{x \to +\infty} \ln x = +\infty$$

even though we shall prove this to be so later. Moreover, since the graph of e^x is the reflection of the graph of $\ln x$ about the line $y = x$, the slow growth of $\ln x$ corresponds to a rapid growth of e^x. Table 7.3.2, which was generated with a calculator, illustrates the slow growth of $\ln x$ and the rapid growth of e^x.

☐ **EXPONENTIAL AND**
LOGARITHMIC GROWTH

Mathematicians often use powers of x as a "measuring stick" for describing how rapidly a function grows. For example, we shall prove later that if n is any positive integer, then

$$\lim_{x \to +\infty} \frac{e^x}{x^n} = +\infty \qquad \text{and} \qquad \lim_{x \to +\infty} \frac{\ln x}{x^n} = 0 \qquad (5\text{-}6)$$

Limit (5) tells us that e^x increases so rapidly that no matter how large we choose the integer n, division by x^n does not prevent the growth toward $+\infty$. Thus, we say that e^x *increases more rapidly than any positive integer power of x*. Limit (6) tells us that $\ln x$ increases so slowly that dividing by any positive integer power of x prevents its growth toward $+\infty$ and results in a limit of zero. Thus, we say that $\ln x$ *increases more slowly than any positive integer power of x*.

The following alternative forms of (5) and (6) can be obtained by taking reciprocals:

$$\lim_{x \to +\infty} \frac{x^n}{e^x} = 0 \qquad \text{and} \qquad \lim_{x \to +\infty} \frac{x^n}{\ln x} = +\infty \qquad (7\text{-}8)$$

Furthermore, the following limit can be obtained from (6) by (Exercise 39):

$$\lim_{x \to 0^+} x^n \ln x = 0 \qquad (9)$$

☐ **OBTAINING GRAPHS**
USING PROPERTIES OF
EXPONENTS AND
LOGARITHMS

Graphs of equations involving exponents and logarithms can sometimes be obtained by making appropriate reflections and translations of known graphs.

Example 1 Sketch the graph of $y = 1/e^x$.

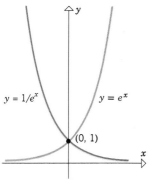

$y = 1/e^x$ $y = e^x$

(0, 1)

Figure 7.3.2

Solution.   The given equation can be rewritten as

$$y = e^{-x}$$

which is the equation that results on replacing x by $-x$ in $y = e^x$. Thus, the graphs of $y = 1/e^x$ and $y = e^x$ are reflections of one another about the y-axis (Figure 7.3.2).   ◀

   Example 1 is a special case of a general result about exponential functions: Replacing x by $-x$ in the equation $y = b^x$ yields

$$y = b^{-x} = \frac{1}{b^x} = \left(\frac{1}{b}\right)^x$$

from which it follows that *exponential functions with reciprocal bases have graphs that are reflections of one another about the y-axis.*
   The following limits, which are consequences of (1) and (2), are consistent with Figure 7.3.2.

$$\lim_{x \to +\infty} e^{-x} = \lim_{x \to +\infty} \frac{1}{e^x} = 0 \quad \text{and} \quad \lim_{x \to -\infty} e^{-x} = \lim_{x \to -\infty} \frac{1}{e^x} = +\infty \qquad (10\text{--}11)$$

Example 2   Sketch the graph of

$$y = \ln\left(\frac{1}{x-1}\right)$$

Solution.   The equation can be rewritten as

$$y = -\ln(x - 1) \quad \text{or equivalently,} \quad -y = \ln(x - 1)$$

Since the last equation results by first replacing y by $-y$ in the equation $y = \ln x$, then replacing x by $x - 1$, the graph of the given equation can be obtained by first reflecting the graph of $y = \ln x$ about the x-axis, then translating the resulting graph 1 unit to the right (Figure 7.3.3).   ◀

Figure 7.3.3

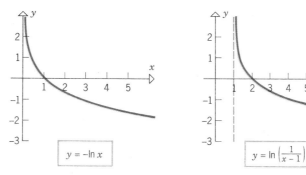

$y = -\ln x$

$y = \ln\left(\frac{1}{x-1}\right)$

Example 3   Sketch the graph of $y = e^{-x^2/2}$.

Solution.

- *Symmetries:*   Replacing *x* **by** $-x$ does not change the equation, so the graph is symmetric about the y-axis.
- *x-intercepts:*   Setting $y = 0$ yields the equation $e^{-x^2/2} = 0$, which has no solutions since all powers of e have positive values; thus, there are no x-intercepts.
- *y-intercepts:*   Setting $x = 0$ yields the y-intercept $y = 1$.
- *Vertical asymptotes:*   None, since $e^{-x^2/2}$ is a continuous function.

- *Horizontal asymptotes:*   Since $x^2/2 \to +\infty$ as $x \to +\infty$ or $x \to -\infty$, it follows from (10) that

$$\lim_{x \to +\infty} e^{-x^2/2} = \lim_{x \to -\infty} e^{-x^2/2} = 0$$

   Thus, $y = 0$ is a horizontal asymptote.

- *Derivatives:*

$$\frac{dy}{dx} = e^{-x^2/2} \frac{d}{dx}\left[-\frac{x^2}{2} \right] = -xe^{-x^2/2}$$

$$\frac{d^2y}{dx^2} = -x\frac{d}{dx}[e^{-x^2/2}] + e^{-x^2/2}\frac{d}{dx}[-x]$$

$$= x^2 e^{-x^2/2} - e^{-x^2/2}$$

$$= (x^2 - 1)e^{-x^2/2}$$

- *Intervals of increase and decrease:*   Since $e^{-x^2/2} > 0$ for all x, the sign of dy/dx is the same as that of $-x$.

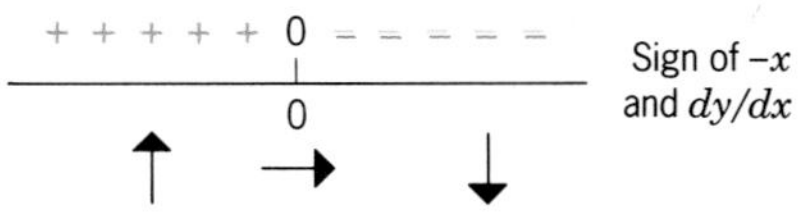

$$\underbrace{+\ +\ +\ +\ +\ 0}_{0}\ -\ -\ -\ -\ -\qquad \text{Sign of } -x \text{ and } dy/dx$$

$$\uparrow \qquad \rightarrow \qquad \downarrow$$

   This analysis reveals a relative maximum at $x = 0$.

- *Concavity:*   Since $e^{-x^2/2} > 0$ for all x, the sign of d^2y/dx^2 is the same as that of $x^2 - 1$.

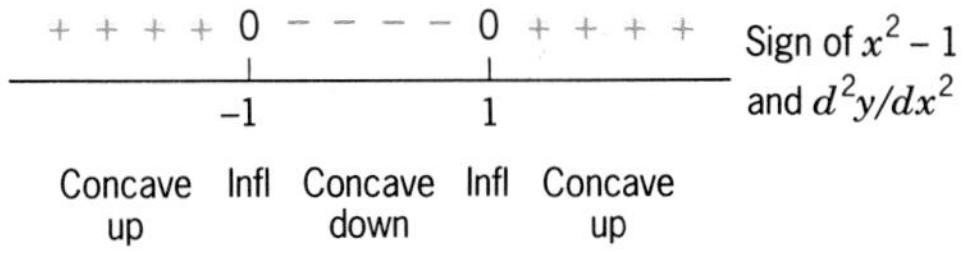

$$+\ +\ +\ +\ \underset{-1}{0}\ -\ -\ -\ -\ \underset{1}{0}\ +\ +\ +\ + \qquad \text{Sign of } x^2 - 1 \text{ and } d^2y/dx^2$$

   Concave   Infl   Concave   Infl   Concave
     up   down   up

   This analysis reveals inflection points at $x = 1$ and $x = -1$.

   Combining the preceding information yields the graph in Figure 7.3.4.   ◀

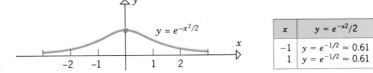

x	$y = e^{-x^2/2}$
-1	$y = e^{-1/2} \approx 0.61$
1	$y = e^{-1/2} \approx 0.61$

Figure 7.3.4

Example 4   Sketch the graph of $y = \dfrac{\ln x}{x}$.

Solution.

- *Domain:*   Since $\ln x$ is not defined for $x \le 0$, the graph lies entirely to the right of the y-axis.

- *Symmetries:*   None.

- *x-intercepts:*   Setting $y = 0$ yields the equation

$$\frac{\ln x}{x} = 0$$

   This yields a single x-intercept of $x = 1$, since this is the only value of x for which $\ln x = 0$.

- *y-intercepts:*   There are no y-intercepts since $\ln x$ is undefined at $x = 0$.

- *Vertical asymptotes:*   It follows from (4) that

$$\lim_{x \to 0^+} \frac{\ln x}{x} = -\infty$$

so the graph has a vertical asymptote at $x = 0$.

- *Horizontal asymptotes:*   It follows from (6) with $n = 1$ that

$$\lim_{x \to +\infty} \frac{\ln x}{x} = 0$$

so $y = 0$ is a horizontal asymptote.

- *Derivatives:*

$$\frac{dy}{dx} = \frac{x\left(\dfrac{1}{x}\right) - (\ln x)(1)}{x^2} = \frac{1 - \ln x}{x^2}$$

$$\frac{d^2y}{dx^2} = \frac{x^2\left(-\dfrac{1}{x}\right) - (1 - \ln x)(2x)}{x^4} = \frac{2x \ln x - 3x}{x^4} = \frac{2 \ln x - 3}{x^3}$$

- *Intervals of increase and decrease:*   Since $x^2 > 0$ for $x > 0$, the sign of dy/dx is the same as that of $1 - \ln x$. But

$$\ln e = 1$$

$$\ln x > 1 \quad \text{if} \quad x > e$$

$$\ln x < 1 \quad \text{if} \quad x < e$$

so the signs for $1 - \ln x$ are as shown in the following diagram.

This analysis reveals a relative maximum at $x = e$.

- *Concavity:*   Since we are only concerned with positive values of x, the sign of d^2y/dx^2 will be determined by the sign of the numerator, $2 \ln x - 3$. Thus, $d^2y/dx^2 > 0$ if

$$2 \ln x - 3 > 0$$

$$\ln x > \tfrac{3}{2}$$

$$e^{\ln x} > e^{3/2} \qquad \boxed{\text{Because } e^x \text{ is an increasing function}}$$

$$x > e^{3/2}$$

Similarly, $2 \ln x - 3 < 0$ if $x < e^{3/2}$, and $2 \ln x - 3 = 0$ if $x = e^{3/2}$. Thus, the signs of $2 \ln x - 3$ are as shown in the following diagram.

This analysis reveals an inflection point at $x = e^{3/2}$.

Combining the preceding information yields the graph in Figure 7.3.5.   ◀

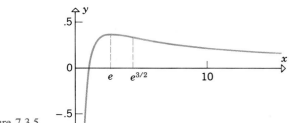

Figure 7.3.5

x	$y = \ln x/x$
e	$1/e \approx 0.37$
$e^{3/2} \approx 4.48$	$3/2e^{3/2} \approx 0.33$

► Exercise Set 7.3   Ⓒ 26–30, 44, 45, 46

In Exercises 1–8, find the limits.

1. (a) $\lim\limits_{x \to +\infty} 4e^{3x}$ (b) $\lim\limits_{x \to -\infty} 4e^{3x}$.

2. (a) $\lim\limits_{x \to +\infty} 3e^{-2x}$ (b) $\lim\limits_{x \to -\infty} 3e^{-2x}$.

3. (a) $\lim\limits_{x \to +\infty} (e^x + e^{-x})$ (b) $\lim\limits_{x \to -\infty} (e^x + e^{-x})$.

4. (a) $\lim\limits_{x \to +\infty} (e^x - e^{-x})$ (b) $\lim\limits_{x \to -\infty} (e^x - e^{-x})$.

5. (a) $\lim\limits_{x \to +\infty} (1 - e^{-x^2})$ (b) $\lim\limits_{x \to -\infty} (1 - e^{-x^2})$.

6. (a) $\lim\limits_{x \to +\infty} e^{1/x}$ (b) $\lim\limits_{x \to -\infty} e^{1/x}$.

7. (a) $\lim\limits_{x \to 0^+} e^{1/x}$ (b) $\lim\limits_{x \to 0^-} e^{1/x}$.

8. (a) $\lim\limits_{x \to 0^+} e^{-1/x^2}$ (b) $\lim\limits_{x \to 0^-} e^{-1/x^2}$.

In Exercises 9–12, sketch the graph. Avoid plotting points.

9. $y = e^x - 1$.   10. $y = e^{x-1}$.

11. $y = 1 - e^{-x}$.   12. $y = e^{1-x}$.

13. Let $f(x) = e^{|x|}$.
    (a) Is f continuous at $x = 0$?
    (b) Is f differentiable at $x = 0$?
    (c) Sketch the graph of f.

14. In each part determine whether the limit exists. If so, find it.
    (a) $\lim\limits_{x \to +\infty} e^x \cos x$ (b) $\lim\limits_{x \to -\infty} e^x \cos x$.

15. Sketch the graphs of e^x, $-e^x$, and $e^x \cos x$, all in the same figure, and label any points of intersection.

In Exercises 16–19, find the limits.

16. $\lim\limits_{x \to +\infty} \dfrac{e^x + e^{-x}}{e^x - e^{-x}}$.

17. $\lim\limits_{x \to -\infty} \dfrac{e^x + e^{-x}}{e^x - e^{-x}}$.

18. $\lim\limits_{x \to +\infty} \dfrac{2 + e^x}{1 + 3e^x}$.

19. $\lim\limits_{x \to +\infty} e^{(-e^x)}$.

20. One of the fundamental functions of mathematical statistics is
$$f(x) = \frac{1}{\sqrt{2\pi}\sigma} \exp\left[-\frac{1}{2}\left(\frac{x - \mu}{\sigma}\right)^2\right]$$
where μ and σ are constants such that $\sigma > 0$ and $-\infty < \mu < +\infty$.
    (a) Locate the inflection points and relative extreme points.
    (b) Find $\lim\limits_{x \to +\infty} f(x)$ and $\lim\limits_{x \to -\infty} f(x)$.
    (c) Sketch the graph of f.

21. Prove that
$$\lim_{h \to 0} \frac{e^h - 1}{h} = 1$$
by expressing the derivative of e^x at $x = 0$ as a limit.

In Exercises 22–25, use the method of Exercise 21 to find the limits.

22. $\lim\limits_{x \to 0} \dfrac{e^{2x} - e^x}{x}$.

23. $\lim\limits_{x \to 0} \dfrac{1 - e^{-x}}{x}$.

24. $\lim\limits_{x \to a} \dfrac{e^x - e^a}{x - a}$.

25. $\lim\limits_{x \to +\infty} x(e^{1/x} - 1)$.

In Exercises 26–30, Formulas (5) and (7) will be helpful.

26. (a) Find $\lim\limits_{x \to +\infty} xe^{-2x}$ and $\lim\limits_{x \to -\infty} xe^{-2x}$.
    (b) Sketch the graph of $y = xe^{-2x}$ and label all relative extrema and inflection points.

27. (a) Find $\lim\limits_{x \to +\infty} xe^x$ and $\lim\limits_{x \to -\infty} xe^x$.
    (b) Sketch the graph of $y = xe^x$ and label all relative extrema and inflection points.

28. (a) Find $\lim\limits_{x \to +\infty} x^2 e^{2x}$ and $\lim\limits_{x \to -\infty} x^2 e^{2x}$.
    (b) Sketch the graph of $y = x^2 e^{2x}$ and label all relative extrema and inflection points.

29. (a) Find $\lim\limits_{x \to +\infty} x^2/e^{2x}$ and $\lim\limits_{x \to -\infty} x^2/e^{2x}$.

  (b) Sketch the graph of $y = x^2/e^{2x}$ and label all relative extrema and inflection points.

30. (a) Find $\lim\limits_{x \to +\infty} e^x/x$, $\lim\limits_{x \to -\infty} e^x/x$, $\lim\limits_{x \to 0^+} e^x/x$, and $\lim\limits_{x \to 0^-} e^x/x$.

  (b) Sketch the graph of $y = e^x/x$ and label all relative extrema and inflection points.

31. Let c_1, c_2, k_1, and k_2 be positive constants. In each part make a sketch that shows the relative positions of the curves $y = c_1 e^{k_1 x}$ and $y = c_2 e^{k_2 x}$ when

  (a) $c_1 = c_2$ and $k_1 < k_2$

  (b) $c_1 < c_2$ and $k_1 = k_2$

  (c) $c_1 < c_2$ and $k_2 < k_1$.

32. Repeat Exercise 31 for the curves $y = c_1 e^{-k_1 x}$ and $y = c_2 e^{-k_2 x}$.

33. Find the area of the region enclosed by the curve $y = e^{-x}$ and the line through the points $(0, 1)$ and $(1, 1/e)$.

34. A rectangle has one side along the x-axis, one vertex at the origin, and another vertex on the curve $y = e^{-x}$, $x > 0$. Find the maximum area that the rectangle can have.

35. (a) Use the inequality $\ln x < \sqrt{x}$ (see Exercise 54, Section 7.2) to show that

$$1/x > e^{-\sqrt{x}} \quad (x > 0)$$

  (b) If n is a positive integer, then from part (a)

$$1/x^n > e^{-n\sqrt{x}}, \quad \text{so} \quad e^x/x^n > e^{x-n\sqrt{x}}$$

Use this result to show that $\lim\limits_{x \to +\infty} e^x/x^n = +\infty$.

36. Use Formula (5) to deduce that $\lim\limits_{x \to +\infty} x^n/e^x = 0$.

37. Given that $\ln x < \sqrt{x}$ (see Exercise 54, Section 7.2), use the Squeezing Theorem (Theorem 2.8.2) to prove that if n is any positive integer, then

$$\lim_{x \to +\infty} \frac{\ln x}{x^n} = 0$$

[*Hint:* If $x > 1$, then $0 < \ln x < \sqrt{x}$.]

38. Use Formula (6) to deduce that $\lim\limits_{x \to +\infty} x^n/\ln x = +\infty$.

39. Use Formula (6) to deduce that $\lim\limits_{x \to 0^+} x^n \ln x = 0$, where n is a positive integer. [*Hint:* Let $x = 1/t$.]

In Exercises 40–43, sketch the graph. Avoid plotting points.

40. $y = \ln |x|$.

41. $y = \ln \dfrac{1}{x}$.

42. $y = \ln \sqrt{x}$.

43. $y = \ln(x - 1)$.

In Exercises 44–46, sketch the graph. Formulas (6) and (9) will be helpful.

44. $y = x \ln x$.   **45.** $y = (\ln x)/x^2$.   **46.** $y = x^2 \ln x$.

47. Show that $x^e \leq e^x$ for all $x > 0$, with equality only for $x = e$. [*Hint:* From Example 4, $(\ln x)/x \leq 1/e$.]

▪ 7.4 INVERSE FUNCTIONS

In the preceding sections we developed the basic properties of logarithmic and exponential functions informally, assuming differentiability and continuity, where needed. One purpose of this section is to lay the foundation for a precise definition of the logarithm that will enable us to prove those properties in the next section. However, our work here will lead us to investigate the very concept of solving equations, a topic with consequences that reach far beyond the study of logarithms.

□ **INVERSE FUNCTIONS**

The concept of solving an equation $y = f(x)$ for x as a function of y, say $x = g(y)$, is one of the most important ideas in mathematics. Sometimes solving such equations is a simple process. For example, by basic algebra we can solve

$$y = 5x + 1 \quad (y = f(x))$$

for x as a function of y:

$$x = \tfrac{1}{5}(y - 1) \quad (x = g(y))$$

Sometimes solving an equation $y = f(x)$ for x in terms of y does not result in a single expression of x as a function of y. For example, by basic algebra we can solve

$$y = x^2$$

for x in terms of y as

$$x = \pm \sqrt{y}$$

which is not a single function of y, but can be viewed as two functions of y:

$$x = \sqrt{y} \quad \text{and} \quad x = -\sqrt{y}$$

There are times when the function f is sufficiently complicated that it is impossible to find any finite sequence of algebraic operations that yields a solution of $y = f(x)$ for x in terms of y. For example, there is no finite sequence of algebraic operations that will yield a solution of

$$y = \frac{\sin x + x^3}{e^x}$$

for x in terms of y. It is important to understand, however, that one's inability to obtain a formula for a function by algebraic methods does not negate the existence of that function. Thus, even though we may not be able to produce algebraic steps to rewrite $y = f(x)$ as $x = g(y)$ does not mean that there is no such g. To pursue this matter in more detail requires a better understanding for what it means for an equation $y = f(x)$ to have a *solution* for x as a function of y. This is one of the main goals of this section.

To obtain a better understanding of what it means for an equation $y = f(x)$ to have a solution for x as a function of y, consider the equations

$$y = x^3 + 1 \qquad \boxed{y = f(x)}$$
$$x = \sqrt[3]{y - 1} \qquad \boxed{x = g(y)}$$

each of which can be obtained from the other by basic algebra (verify). The functions f and g in this example have the property that when they are composed in either order they can "cancel out" the effect of one another in the sense that

$$f(g(y)) = [g(y)]^3 + 1 = (\sqrt[3]{y - 1})^3 + 1 = y$$
$$g(f(x)) = \sqrt[3]{f(x) - 1} = \sqrt[3]{(x^3 + 1) - 1} = x$$

Pairs of functions that cancel the effect of one another are of such importance that there is some terminology associated with them.

7.4.1 DEFINITION. If the functions f and g satisfy the two conditions

$$f(g(x)) = x \text{ for every } x \text{ in the domain of } g$$

$$g(f(x)) = x \text{ for every } x \text{ in the domain of } f$$

then we say that ***f* is an inverse of *g*** and ***g* is an inverse of *f*** or, alternatively, that ***f* and *g* are inverse functions**.

Example 1   The functions $f(x) = 2x$ and $g(x) = \frac{1}{2}x$ are inverse functions since

$$f(g(x)) = f(\tfrac{1}{2}x) = 2(\tfrac{1}{2}x) = x$$

$$g(f(x)) = g(2x) = \tfrac{1}{2}(2x) = x$$

Similarly, $f(x) = x^{1/3}$ and $g(x) = x^3$ are inverse functions since

$$f(g(x)) = f(x^3) = (x^3)^{1/3} = x$$

$$g(f(x)) = g(x^{1/3}) = (x^{1/3})^3 = x \qquad \blacktriangleleft$$

It can be shown that a function cannot have two different inverses; that is, if f has an inverse, then it is unique and we are entitled to talk about *the* inverse of f. The inverse of f is commonly denoted by f^{-1} (read, "f inverse"). Thus, it follows from Definition 7.1.1 that

$$
\begin{aligned}
f(f^{-1}(x)) &= x \quad \text{for every } x \text{ in the domain of } f^{-1} \\
f^{-1}(f(x)) &= x \quad \text{for every } x \text{ in the domain of } f
\end{aligned}
\tag{1}
$$

WARNING.   The symbol f^{-1} does not mean $1/f$.

Example 2   It follows from Example 1 that if $f(x) = \frac{1}{2}x$ and $g(x) = 2x$, then

$$f^{-1}(x) = 2x \quad \text{and} \quad g^{-1}(x) = \tfrac{1}{2}x$$

Similarly, if $f(x) = x^{1/3}$ and $g(x) = x^3$, then

$$f^{-1}(x) = x^3 \quad \text{and} \quad g^{-1}(x) = x^{1/3} \quad \blacktriangleleft$$

☐ **DOMAIN AND RANGE OF INVERSE FUNCTIONS**

The domains and ranges of inverse functions have the following simple relationships:

$$
\begin{aligned}
\text{range of } f^{-1} &= \text{domain of } f \\
\text{domain of } f^{-1} &= \text{range of } f
\end{aligned}
\tag{2}
$$

These results follow immediately from (1). For example, it follows from the first equation in (1) that the range of f^{-1} is a subset of the domain of f since $f(f^{-1}(x))$ is defined for all x in the domain of f^{-1}. On the other hand, the second equation in (1) shows that the domain of f is a subset of the range of f^{-1}. This proves the first relationship in (2). The second relationship is proved similarly.

☐ **SOLVING** $y = f(x)$ **FOR** x **AS A FUNCTION OF** y

The problem of finding the inverse of a function f is closely linked to the problem of solving the equation $y = f(x)$ for x as a function of y. To see this, suppose that f has an inverse and apply f^{-1} to both sides of the equation $y = f(x)$. This yields

$$f^{-1}(y) = f^{-1}(f(x)) = x$$

which expresses x as a function of y. This suggests the following definition.

> **7.4.2**   DEFINITION.   If the function f has an inverse, then we say that $y = f(x)$ can be *solved for x as a function of y* and we call $x = f^{-1}(y)$ the *solution* of $y = f(x)$ for x as a function of y.

This definition is consistent with results learned in algebra. For example, from algebra the solution of $y = x^3$ for x in terms of y is $x = y^{1/3}$. But if we let $f(x) = x^3$, then from Example 2 we have $f^{-1}(y) = y^{1/3}$, so the solution for x as a function of y according to the preceding definition is $x = y^{1/3}$, which agrees with the result from algebra.

☐ **EXISTENCE OF INVERSE FUNCTIONS**

We now turn to the problem of determining which functions have inverses, or equivalently, which equations of the form $y = f(x)$ can be solved for x as a function of y. The following theorem answers this question.

> **7.4.3**   THEOREM (*Horizontal Line Test*).   *The following are equivalent statements (i.e., all are true or all are false):*
>
> (a)   *The function f has an inverse.*
> (b)   *The graph of f is cut at most once by any horizontal line.*
> (c)   *The function f does not have the same value at two distinct points in its domain; that is, if x_1 and x_2 are points in the domain such that $x_1 \neq x_2$, then $f(x_1) \neq f(x_2)$.*

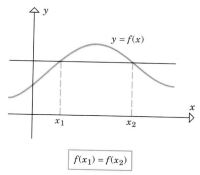

Figure 7.4.1

The equivalence of parts (*b*) and (*c*) is self-evident, since multiple intersections of the graph of *f* with a horizontal line correspond to points on the *x*-axis at which *f* has the same value (Figure 7.4.1). Thus, we need only prove that parts (*a*) and (*c*) are equivalent.

Proof [(*a*) *implies* (*c*)].   Assume that *f* has an inverse. If x_1 and x_2 are points in the domain of *f* such that $f(x_1) = f(x_2)$, then applying f^{-1} to both sides yields

$$f^{-1}(f(x_1)) = f^{-1}(f(x_2)) \quad \text{or} \quad x_1 = x_2$$

Thus, if $x_1 \neq x_2$, it must be that $f(x_1) \neq f(x_2)$.

Proof [(*c*) *implies* (*a*)].   Let $y = f(x)$, and assume that *f* does not have the same value at two distinct points in its domain; that is, each *y* in the range of *f* comes from a unique *x* in the domain. Thus, we can define a function *g* whose domain is the range of *f* by the equation $g(y) = x$; that is,

$$g(f(x)) = x \tag{3}$$

for all *x* in the domain of *f*. If we now substitute $x = g(y)$ into $y = f(x)$, we obtain $y = f(g(y))$ or equivalently,

$$f(g(y)) = y \tag{4}$$

for all *y* in the domain of *g*. Equations (3) and (4) establish that *g* is the inverse of *f*. ∎

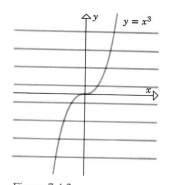

Figure 7.4.2

A function that satisfies any (and hence all) of the three properties in Theorem 7.4.3 is said to be ***one-to-one***. Thus, to say that *f* is one-to-one is the same as saying that *f* has an inverse or that the equation $y = f(x)$ can be solved for *x* as a function of *y*.

Example 3   The function $f(x) = x^3$ is one-to-one because two different numbers cannot have the same cube; that is, if $x_1 \neq x_2$, then $(x_1)^3 \neq (x_2)^3$. Geometrically, no horizontal line cuts the graph of $f(x) = x^3$ more than once (Figure 7.4.2). This implies that $f(x) = x^3$ has an inverse and that the equation of $y = x^3$ can be solved for *x* as a function of *y*. The solution for *x* as a function of *y* is $x = y^{1/3}$, so that the inverse function of *f* (expressed with *y* as the independent variable) is $f^{-1}(y) = y^{1/3}$. ◀

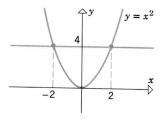

The graph of $y = x^2$ is cut more than once by some horizontal line.

Figure 7.4.3

Example 4   The function $f(x) = x^2$ is not one-to-one because there are distinct points in the domain that have the same square; for example, $f(-2) = 4$ and $f(2) = 4$. Geometrically, there are horizontal lines that cut the graph of $y = x^2$ more than once (Figure 7.4.3). This implies that *f* does not have an inverse and that the equation $y = x^2$ cannot be solved for *x* as a (*single*) function of *y*. Indeed, we noted earlier that solving this equation for *x* in terms of *y* produces *more than one* function of *y*, for example, $x = \sqrt{y}$ and $x = -\sqrt{y}$. ◀

☐ **FINDING A FORMULA FOR** f^{-1}

If *f* is one-to-one, and if *f* is sufficiently simple that the equation $y = f(x)$ can be solved by algebra for *x* as a function of *y*, then the resulting equation $x = f^{-1}(y)$ will provide a formula for f^{-1} in which *y* is the independent variable. If one were interested in a formula for f^{-1} in which *x* is the independent variable, one could reverse the roles of *x* and *y* at the start. Thus, if the necessary algebra can be performed, the following procedure will produce a formula for $f^{-1}(x)$.

Step 1.   Interchange x and y in the equation $y = f(x)$ to produce the equation $x = f(y)$.

Step 2.   Solve the equation $x = f(y)$ for y as a function of x.

Step 3.   The resulting equation in Step 2 will be $y = f^{-1}(x)$, the right side of which is the formula for $f^{-1}(x)$.

Example 5   Find the inverse of $f(x) = \sqrt{3x - 2}$.

Solution.   We first introduce a dependent variable y; this yields

$$y = \sqrt{3x - 2}$$

Interchanging x and y in this equation and then solving for y we obtain

$$x = \sqrt{3y - 2}$$

$$x^2 = 3y - 2$$

$$y = \tfrac{1}{3}(x^2 + 2)$$

from which it follows that

$$f^{-1}(x) = \tfrac{1}{3}(x^2 + 2)$$

At this point we have successfully produced a formula for $f^{-1}(x)$; however, we are not quite done, since there is no guarantee that the natural domain associated with this formula is the correct domain for f^{-1}. This requires a separate analysis as follows: The range of $f(x) = \sqrt{3x - 2}$ is $[0, +\infty)$, so this interval is also the domain of f^{-1}. Thus, the inverse of f is

$$f^{-1}(x) = \tfrac{1}{3}(x^2 + 2), \quad x \geq 0 \qquad \blacktriangleleft$$

□ **GRAPHS OF INVERSE FUNCTIONS**

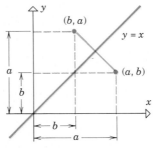

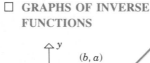

The line $y = x$ is the perpendicular bisector of the line segment joining (a, b) and (b, a).

Figure 7.4.4

We shall now examine the relationship between the graph of a one-to-one function and the graph of its inverse. For this purpose we note that the points (a, b) and (b, a) are symmetric about the line $y = x$ in the sense that this line is the perpendicular bisector of the line segment joining the points (Figure 7.4.4). This being the case, it follows that interchanging the x and y coordinates of a point reflects that point about the line $y = x$, and interchanging the x and y variables in an equation reflects the graph of that equation about the line $y = x$. It follows from this that the graph of $y = f^{-1}(x)$, which is the same as the graph of $x = f(y)$, is the reflection about $y = x$ of the graph of $y = f(x)$, as illustrated in Figure 7.4.5. In summary, we have the following theorem.

7.4.4   THEOREM.   *If f is a one-to-one function, then the graphs of $y = f(x)$ and $y = f^{-1}(x)$ are reflections of one another about the line $y = x$; that is, each is the mirror image of the other.*

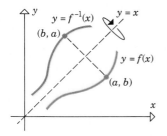

Figure 7.4.5

Example 6 Figure 7.4.6 shows the graphs of the inverse functions discussed in Examples 2 and 5. ◀

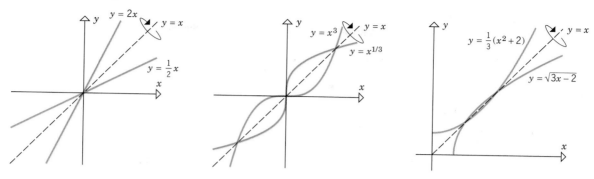

Figure 7.4.6

REMARK. In Section 7.1 we noted that the graphs of $\log_b x$ and b^x are reflections of one another about the line $y = x$ (Figure 7.1.4). In the next section we shall prove that $\log_b x$ and b^x are, in fact, inverse functions.

☐ **INCREASING AND DECREASING FUNCTIONS HAVE INVERSES**

The graph of an increasing function or a decreasing function is cut at most once by any horizontal line (Figure 7.4.7), so that all such functions are one-to-one. This is the content of the following theorem.

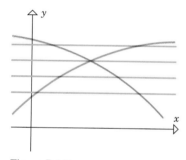

Figure 7.4.7

> **7.4.5** THEOREM. *If the domain of f is an interval, and if f is either an increasing function or a decreasing function on that interval, then f is one-to-one and hence has an inverse.*

Proof. If f is increasing and $x_1 < x_2$ are distinct points in its domain, then $f(x_1) < f(x_2)$ (Definition 4.2.1), so that $f(x_1) \neq f(x_2)$. Thus, f is one-to-one and therefore has an inverse. The proof for decreasing functions is similar. ∎

Recall from Theorem 4.2.2 and the subsequent discussion that f is increasing on an interval if $f'(x) > 0$ on the interval and is decreasing if $f'(x) < 0$. This result makes it easy to apply Theorem 7.4.5 to differentiable functions.

Example 7 The function of $f(x) = x^5 + 7x^3 + 4x + 1$ is increasing on $(-\infty, +\infty)$ since

$$f'(x) = 5x^4 + 21x^2 + 4 > 0$$

for all x. Thus, f has an inverse and the equation

$$y = x^5 + 7x^3 + 4x + 1$$

has a solution for y as a function of x. However, this equation is sufficiently complicated that algebraic steps for obtaining this solution are not evident. Thus, we cannot produce a formula for f^{-1} even though we know that the inverse exists. ◀

☐ **CONTINUITY OF INVERSE FUNCTIONS**

Because the graphs of f and f^{-1} are reflections of one another about the line $y = x$, it is intuitively obvious that if the graph of f has no breaks, then neither will the graph of f^{-1}. This suggests the following result, which we state without proof.

> **7.4.6** THEOREM. *If a function f is continuous and has an inverse, then f^{-1} is also continuous.*

Thus, even though we were unable to produce a formula for f^{-1} in Example 7, the continuity of $f(x) = x^5 + 7x^3 + 4x + 1$ guarantees that f^{-1} is a continuous function.

☐ **DIFFERENTIABILITY OF INVERSE FUNCTIONS**

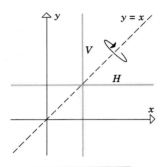

The vertical line V reflects into the horizontal line H and conversely.

Figure 7.4.8

Because the graphs of f and f^{-1} are reflections of one another about the line $y = x$, it is intuitively obvious that if the graph of f has no corners, then neither will the graph of f^{-1}. However, since reflection about $y = x$ converts horizontal lines into vertical lines (Figure 7.4.8), a point where the graph of f has a horizontal tangent line will be reflected into a point where f^{-1} has a vertical tangent line, and this will be a point of nondifferentiability for f^{-1}.

A relationship between the derivatives of f and f^{-1} can be obtained as follows. Let (x_0, y_0) be a point on the graph of f^{-1}, and suppose that f is differentiable at y_0 and $f'(y_0)$ is nonzero (Figure 7.4.9). Thus, the tangent line to the graph of f at the point (y_0, x_0) has an equation of the form

$$y = mx + b \tag{5}$$

where $m \neq 0$. The reflection of this tangent line about the line $y = x$ is the tangent line at (x_0, y_0) to the graph of f^{-1}; this line has the equation

$$x = my + b$$

which we obtained by interchanging x and y in (5). Solving this equation for y yields

$$y = \frac{1}{m}x - \frac{b}{m}$$

which tells us that the slope of the tangent line to $y = f^{-1}(x)$ at (x_0, y_0) and the slope of the tangent line to $y = f(x)$ at (y_0, x_0) are reciprocals (Figure 7.4.9). Stating this relationship in terms of derivatives, we obtain

$$(f^{-1})'(x_0) = \frac{1}{f'(y_0)} \tag{6}$$

But (x_0, y_0) lies on the graph of f^{-1}, so $y_0 = f^{-1}(x_0)$; thus, (6) can be rewritten as

$$(f^{-1})'(x_0) = \frac{1}{f'(f^{-1}(x_0))}$$

which is the relationship between the derivative of f and the derivative of f^{-1}. In summary, we have the following result.

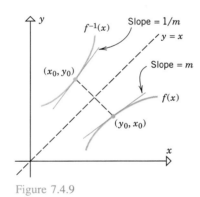

Figure 7.4.9

7.4.7 THEOREM. *Suppose that the function f has an inverse and that the value of $f^{-1}(x)$ varies over an interval on which f has a nonzero derivative as x varies over an interval I. Then f^{-1} is differentiable on I and the derivative of f^{-1} is given by the formula*

$$(f^{-1})'(x) = \frac{1}{f'(f^{-1}(x))} \tag{7}$$

Formula (7) can be expressed in a less forbidding form by letting

$$y = f^{-1}(x) \quad \text{so that} \quad x = f(y)$$

Thus,

$$\frac{dy}{dx} = (f^{-1})'(x) \quad \text{and} \quad \frac{dx}{dy} = f'(y) = f'(f^{-1}(x))$$

Substituting these expressions in (7) yields the following alternative version of that formula:

$$\frac{dy}{dx} = \frac{1}{dx/dy} \tag{8}$$

If an explicit formula can be obtained for the inverse of a function, then the differentiability and the derivative of the inverse can generally be deduced from that formula. However, if no explicit formula for the inverse can be obtained, then Theorem 7.4.7 is the primary mathematical tool for establishing differentiability of the inverse. Once the differentiability has been established, a derivative formula for the inverse can be obtained either by implicit differentiation or by using Formulas (7) or (8). The following example illustrates this.

Example 8   In Example 7 we saw that the function $f(x) = x^5 + 7x^3 + 4x + 1$ has an inverse.

(a)   Show that f^{-1} is differentiable on the interval $(-\infty, +\infty)$.
(b)   Find the derivative of f^{-1} by using Formula (8).
(c)   Find the derivative of f^{-1} by implicit differentiation.

Solution (a).   Let I denote the interval $(-\infty, +\infty)$. We must show that as x varies over I, the value of $f^{-1}(x)$ varies over an interval on which f has a nonzero derivative. But this is so because the derivative of f is

$$f'(x) = 5x^4 + 21x^2 + 4$$

which is positive (hence, nonzero) for all x.

Solution (b).   If we let $y = f^{-1}(x)$, then

$$x = f(y) = y^5 + 7y^3 + 4y + 1 \tag{9}$$

from which it follows that

$$\frac{dx}{dy} = 5y^4 + 21y^2 + 4$$

$$\frac{dy}{dx} = \frac{1}{dx/dy} = \frac{1}{5y^4 + 21y^2 + 4} \tag{10}$$

Because (9) is too complicated to solve for y in terms of x, we must leave (10) in terms of y.

Solution (c).   Differentiating (9) implicitly with respect to x yields

$$\frac{d}{dx}[x] = \frac{d}{dx}[y^5 + 7y^3 + 4y + 1]$$

$$1 = 5y^4\frac{dy}{dx} + 21y^2\frac{dy}{dx} + 4\frac{dy}{dx}$$

$$1 = (5y^4 + 21y^2 + 4)\frac{dy}{dx}$$

$$\frac{dy}{dx} = \frac{1}{5y^4 + 21y^2 + 4}$$

which agrees with (10).   ◄

▶ Exercise Set 7.4

1. In (a)–(d), determine whether f and g are inverse functions.

   (a) $f(x) = 4x$, $g(x) = \frac{1}{4}x$

   (b) $f(x) = 3x + 1$, $g(x) = 3x - 1$

   (c) $f(x) = \sqrt[3]{x - 2}$, $g(x) = x^3 + 2$

   (d) $f(x) = x^4$, $g(x) = \sqrt[4]{x}$.

In Exercises 2–14, determine whether the function has an inverse.

---

2. $f(x) = 1 - x$.   3. $f(x) = 3x + 2$.

4. $f(x) = x^2 - 2x + 1$.   5. $f(x) = 2 - x - x^2$.

6. $f(x) = x^3 - 3x + 2$.   7. $f(x) = x^3 - x - 1$.

8. $f(x) = x^3 - 3x^2 + 3x - 1$.

9. $f(x) = x^3 + 3x^2 + 3x + 1$.

10. $f(x) = x^5 + 8x^3 + 2x - 1$.

11. $f(x) = 2x^5 + x^3 + 3x + 2$.

12. $f(x) = x + \dfrac{1}{x}$, $x > 0$.

13. $f(x) = \sin x$, $-\pi/2 < x < \pi/2$.

14. $f(x) = \tan x$, $-\pi/2 < x < \pi/2$.

In Exercises 15–29, find $f^{-1}(x)$.

---

15. $f(x) = x^5$.   16. $f(x) = 6x$.

17. $f(x) = 7x - 6$.   18. $f(x) = \dfrac{x + 1}{x - 1}$.

19. $f(x) = 3x^3 - 5$.   20. $f(x) = \sqrt[5]{4x + 2}$.

21. $f(x) = \sqrt[3]{2x - 1}$.

22. $f(x) = 5/(x^2 + 1)$, $x \geq 0$.

23. $f(x) = 3/x^2$, $x < 0$.

24. $f(x) = e^{2x+1}$.   25. $f(x) = e^{1/x}$.

26. $f(x) = 4\ln(x + 1)$.   27. $f(x) = 1 - \ln(3x)$.

28. $f(x) = \begin{cases} 2x, & x \leq 0 \\ x^2, & x > 0. \end{cases}$ 29. $f(x) = \begin{cases} 5/2 - x, & x < 2 \\ 1/x, & x \geq 2. \end{cases}$

In Exercises 30–36, use Formula (8) to find the derivative of f^{-1}, and check your work by differentiating implicitly.

---

30. $f(x) = 2x^3 + 5x + 3$.   31. $f(x) = 5x^3 + x - 7$.

32. $f(x) = 1/x^2$, $x > 0$.

33. $f(x) = \tan 2x$, $-\pi/4 < x < \pi/4$.

34. $f(x) = 5x - \sin 2x$.   35. $f(x) = 2x^5 + x^3 + 1$.

36. $f(x) = x^7 + 2x^5 + x^3$.

37. Let $f(x) = x^2$, $x > 1$, and $g(x) = \sqrt{x}$.

   (a) Show that $f(g(x)) = x$, $x > 1$, and $g(f(x)) = x$, $x > 1$.

   (b) Show that f and g are *not* inverses of one another by showing that the graphs of $y = f(x)$ and $y = g(x)$ are not reflections of one another about $y = x$.

   (c) Do parts (a) and (b) contradict Definition 7.4.1? Explain.

38. Let $f(x) = ax^2 + bx + c$, $a > 0$. Find f^{-1} if the domain of f is restricted to

   (a) $x \geq -b/(2a)$ (b) $x \leq -b/(2a)$.

In Exercises 39–43, find $f^{-1}(x)$ and its domain.

---

39. $f(x) = (x + 2)^4$, $x \geq 0$.

40. $f(x) = \sqrt{x + 3}$.   41. $f(x) = -\sqrt{3 - 2x}$.

42. $f(x) = 3x^2 + 5x - 2$, $x \geq 0$.

43. $f(x) = x - 5x^2$, $x \geq 1$.

44. Prove that if $a^2 + bc \neq 0$, then the graph of

   $$f(x) = \frac{ax + b}{cx - a}$$

   is symmetric about the line $y = x$.

45. (a) Show that $f(x) = (3 - x)/(1 - x)$ is its own inverse.

   (b) What does the result in part (a) tell you about the graph of f?

46. Suppose that a line of nonzero slope m intersects the x-axis at $(x_0, 0)$. Find an equation for the reflection of this line about $y = x$.

47. (a) Show that $f(x) = x^3 - 3x^2 + 2x$ is not one-to-one on $(-\infty, +\infty)$.

   (b) Find the largest value of k such that f is one-to-one on the interval $(-k, k)$.

48. (a) Show that $f(x) = x^4 - 2x^3$ is not one-to-one on $(-\infty, +\infty)$.

   (b) Find the smallest value of k such that f is one-to-one on the interval $[k, +\infty)$.

49. Let $f(x) = 2x^3 + 5x + 3$. Find x if $f^{-1}(x) = 1$.

50. Let $f(x) = \dfrac{x^3}{x^2 + 1}$. Find x if $f^{-1}(x) = 2$.

In Exercises 51–54, a function f and the coordinates of a point on the graph of $y = f^{-1}(x)$ are given. Find the slope of the tangent line to the graph of $y = f^{-1}(x)$ at the given point.

---

51. $f(x) = x^3 + x$; $(10, 2)$.

52. $f(x) = x^5 + 2x^3 + x + 4$; $(0, -1)$.

53. $f(x) = \sin 2x$, $0 \leq x \leq \pi/4$; $(\frac{1}{2}, \pi/12)$.

54. $f(x) = x^3 - \dfrac{2}{x}$; $(-1, 1)$.

55. Let $f(x) = \displaystyle\int_1^x \sqrt[3]{1 + t^2}\, dt$.

(a) Without integrating, show that f is one-to-one on the interval $(-\infty, +\infty)$.

(b) Find $\dfrac{d}{dx}\,[f^{-1}(x)]\,\Big|_{x=0}$

56. (a) Prove: If f and g are one-to-one, then so is the composition $f \circ g$.

(b) Prove: If f and g are one-to-one, then
$$(f \circ g)^{-1} = g^{-1} \circ f^{-1}$$

57. Sketch the graph of a function that is one-to-one on $(-\infty, +\infty)$, yet not increasing on $(-\infty, +\infty)$ and not decreasing on $(-\infty, +\infty)$.

58. Prove: A one-to-one function f cannot have two different inverses.

59. Let $F(x) = f(2g(x))$ where $f(x) = x^4 + x^3 + 1$ for $0 \le x \le 2$, and $g(x) = f^{-1}(x)$. Find $F'(3)$.

■ 7.5 LOGARITHMIC AND EXPONENTIAL FUNCTIONS (RIGOROUS APPROACH)

In this section we shall give precise mathematical definitions of the functions $\ln x$, $\log_b x$, e^x, *and* b^x, *and we shall use these definitions to obtain the properties of these functions that were stated without proof in preceding sections.*

□ **DEFINITION OF THE NATURAL LOGARITHM**

Our first objective is to define the function $\ln x$ in a manner that will be consistent with the properties

$$\frac{d}{dx}\,[\ln x] = \frac{1}{x} \quad \text{and} \quad \ln 1 = 0$$

that were obtained informally in Sections 7.1 and 7.2. In words, we want $\ln x$ to be that antiderivative of $1/x$ with a value of 0 when $x = 1$. To obtain such a function, we turn to the Second Fundamental Theorem of Calculus (Theorem 5.9.1). It follows from that theorem that the function

$$F(x) = \int_1^x \frac{1}{t}\,dt$$

has these properties, since

$$\frac{d}{dx}\,[F(x)] = \frac{d}{dx}\left[\int_1^x \frac{1}{t}\,dt\right] = \frac{1}{x} \quad \text{and} \quad F(1) = \int_1^1 \frac{1}{t}\,dt = 0$$

This suggests the following definition of $\ln x$.

7.5.1 DEFINITION. The ***natural logarithm*** function is defined by the formula

$$\ln x = \int_1^x \frac{1}{t}\,dt, \quad x > 0 \tag{1}$$

□ **GEOMETRIC INTERPRETATION OF $\ln x$**

Some basic properties of $\ln x$ can be obtained by interpreting the integral in the preceding definition as an area. For example, the familiar properties

$$\ln x > 0 \quad \text{if} \quad x > 1$$
$$\ln x < 0 \quad \text{if} \quad 0 < x < 1$$
$$\ln x = 0 \quad \text{if} \quad x = 1$$

can be deduced as follows: If $x > 1$, then the integral represents the area under the curve $y = 1/t$ over the interval $[1, x]$ on the t-axis (Figure 7.5.1a) and hence is positive; if $0 < x < 1$, then the integral represents the negative of the area under the curve $y = 1/t$ over the interval $[x, 1]$ (Figure 7.5.1b) and hence is negative. If $x = 1$, then the interval $[1, x]$ reduces to a single point, and hence the area under the curve $y = 1/t$ over the interval $[1, x]$ is zero.

The next theorem follows from the fact that

$$\frac{d}{dx}[\ln x] = \frac{1}{x}$$

is positive for $x > 0$.

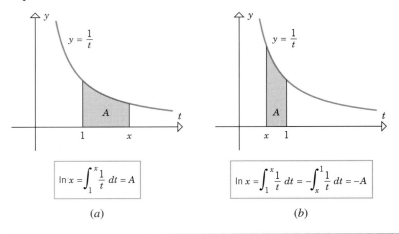

Figure 7.5.1   (a)   (b)

7.5.2   THEOREM.   *The function* $\ln x$ *is increasing and hence one-to-one on the interval* $(0, +\infty)$.

This result is consistent with the graph of $\ln x$ shown in Figure 7.3.1.

□ NUMERICAL
   APPROXIMATION
   OF $\ln x$

In Section 5.8 we discussed three methods for approximating definite integrals numerically: left endpoint approximation, right endpoint approximation, and midpoint approximation. These methods can be applied to (1) to obtain numerical approximations of natural logarithms. The following example illustrates this.

Example 1   Approximate $\ln 2$ using a midpoint approximation with $n = 10$.

Solution.   From (1) the exact value of $\ln 2$ is

$$\ln 2 = \int_1^2 \frac{1}{t}\, dt$$

We shall apply Formula (4) of Section 5.8 (with t in place of x), choosing $t_1^*, t_2^*, \ldots, t_n^*$ as the midpoints of the intervals. The computations to six decimal places are shown in Table 7.5.1. To get a sense of the accuracy of these computations, a calculator set to display six digits gives $\ln 2 \approx 0.693147$, so the magnitude of the error in the midpoint approximation with $n = 10$ is about 0.000311. Greater accuracy can be obtained by increasing n. For example, the midpoint approximation with $n = 100$ yields $\ln 2 \approx 0.693144$, which is correct to five decimal places.   ◄

□ PROPERTIES OF THE
   NATURAL LOGARITHM

The following theorem shows that $\ln x$, as defined in Definition 7.5.1, has the basic properties of logarithms that were noted without proof in Section 7.1.

Table 7.5.1

	$n = 10$	
	$\Delta t = (b-a)/n = (2-1)/10 = 0.1$	
k	t_k^*	$1/t_k^*$
1	1.05	0.952381
2	1.15	0.869565
3	1.25	0.800000
4	1.35	0.740741
5	1.45	0.689655
6	1.55	0.645161
7	1.65	0.606061
8	1.75	0.571429
9	1.85	0.540541
10	1.95	0.512821
		6.928355

$$\Delta t \sum_{k=1}^{n} f(t_k^*) = (0.1)(6.928355)$$
$$= 0.692836$$

7.5.3 THEOREM. *For any positive numbers a and c and any rational number r,*

$(a)\quad \ln ac = \ln a + \ln c \qquad (b)\quad \ln \dfrac{1}{c} = -\ln c$

$(c)\quad \ln \dfrac{a}{c} = \ln a - \ln c \qquad (d)\quad \ln a^r = r \ln a$

Proof (a). Consider the function $f(x) = \ln ax$. Treating a as constant and differentiating with respect to x, we obtain

$$f'(x) = \frac{d}{dx}[\ln ax] = \frac{1}{ax} \cdot \frac{d}{dx}[ax] = \frac{1}{ax} \cdot a = \frac{1}{x}$$

which shows that $\ln ax$ and $\ln x$ have the same derivative on $(0, +\infty)$. Thus, these functions differ by a constant on this interval; that is, there is a constant k such that

$$\ln ax - \ln x = k \tag{2}$$

on $(0, +\infty)$. If we let $x = 1$ in this equation and use the fact that $\ln 1 = 0$, we obtain $\ln a = k$, so that (2) can be written as

$$\ln ax - \ln x = \ln a$$

In particular, if $x = c$, we obtain

$$\ln ac - \ln c = \ln a \quad \text{or} \quad \ln ac = \ln a + \ln c$$

Proofs (b) and (c). Using part (a) we can write

$$0 = \ln 1 = \ln\left(c \cdot \frac{1}{c}\right) = \ln c + \ln \frac{1}{c}$$

so that

$$\ln \frac{1}{c} = -\ln c$$

which proves (b). To prove (c) we write

$$\ln \frac{a}{c} = \ln\left(a \cdot \frac{1}{c}\right) = \ln a + \ln \frac{1}{c} = \ln a - \ln c$$

Proof (d). The functions $\ln x^r$ and $r \ln x$ have the same derivative on $(0, +\infty)$, since

$$\frac{d}{dx}[\ln x^r] = \frac{1}{x^r} \cdot \frac{d}{dx}[x^r] = \frac{1}{x^r} \cdot rx^{r-1} = \frac{r}{x}$$

and

$$\frac{d}{dx}[r \ln x] = r\frac{d}{dx}[\ln x] = \frac{r}{x}$$

Thus, there is a constant k such that

$$\ln x^r - r \ln x = k$$

If we let $x = 1$ in this equation and use the fact that $\ln 1 = 0$, we obtain $k = 0$ (verify), so that this equation can be written as

$$\ln x^r - r \ln x = 0 \quad \text{or} \quad \ln x^r = r \ln x$$

Letting $x = a$ completes the proof. ∎

Example 2 Property (d) in Theorem 7.5.3 yields the following useful result:

$$\ln \sqrt[n]{a} = \ln a^{1/n} = \frac{1}{n} \ln a \qquad \blacktriangleleft$$

The limits in the following theorem were observed without proof in Section 7.3 [see (3) and (4) in that section]. We shall now prove their validity.

7.5.4   THEOREM.

(a) $\displaystyle\lim_{x \to +\infty} \ln x = +\infty$ (b) $\displaystyle\lim_{x \to 0^+} \ln x = -\infty$

Proof (a).   We shall show that for any positive integer N (no matter how large) the values of $\ln x$ eventually exceed N as $x \to +\infty$. It will then follow that $\displaystyle\lim_{x \to +\infty} \ln x = +\infty$. To do this we shall need the inequality

$$\ln 2 > \tfrac{1}{2} \tag{3}$$

which can be obtained from Figure 7.5.2 by comparing the shaded area (which represents $\ln 2$) to the area of rectangle of height $\frac{1}{2}$ over the interval $[1, 2]$.

Since $\ln x$ is an increasing function, it follows from (3) that for $x > 2^N$ we must have

$$\ln x > \ln 2^N = N \ln 2 > N \left(\frac{1}{2} \right) = \frac{N}{2}$$

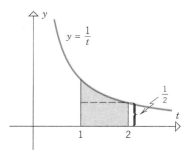

Figure 7.5.2

Therefore, $\displaystyle\lim_{x \to +\infty} \ln x = +\infty$, since $N/2$ can be made arbitrarily large by choosing N appropriately.

Proof (b).   If we let $v = 1/x$, then $v \to +\infty$ as $x \to 0^+$, so that from Theorem 7.5.3(b) and part (a) of this theorem we obtain

$$\lim_{x \to 0^+} \ln x = \lim_{v \to +\infty} \left(\ln \frac{1}{v} \right) = \lim_{v \to +\infty} (-\ln v) = - \lim_{v \to +\infty} \ln v = -\infty \quad \blacksquare$$

The function $\ln x$ is defined for x in the interval $(0, +\infty)$, and it follows from Theorem 7.5.4 that as x varies over this interval the values of $\ln x$ vary from $-\infty$ to $+\infty$. Thus, we have the following theorem.

7.5.5   THEOREM.

(a)   *The domain of* $\ln x$ *is* $(0, +\infty)$.
(b)   *The range of* $\ln x$ *is* $(-\infty, +\infty)$.

☐ **DEFINITION OF** e^x

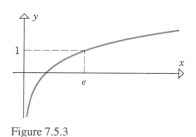

Figure 7.5.3

In Section 7.1 we introduced the number e informally and stated a limit that can be used to approximate its value [see (5) in Section 7.1]. We shall now give a precise definition of the number e.

Because $\ln x$ is an increasing, continuous function on the interval $(0, +\infty)$, and because its values vary from $-\infty$ to $+\infty$ as x varies over this interval, we are assured that the curve $y = \ln x$ crosses every horizontal line exactly once and that the intersection occurs over the interval $(0, +\infty)$. This implies that there is a unique positive real number x such that $\ln x = 1$. We *define* this number to be e (Figure 7.5.3); that is,

$$\ln e = 1 \tag{4}$$

(This agrees with the value of $\ln e$ obtained informally in Section 7.1.)

In Section 7.1 we noted informally that $\ln e^x = x$ and $e^{\ln x} = x$ [see (8) in that section]. We know now from our work in Section 7.4 that equations of this type are characteristic of inverse functions. This suggests the following definition, which we can make because $\ln x$ is a one-to-one function and hence has an inverse.

> **7.5.6** DEFINITION.   The inverse of the natural logarithm function $\ln x$ is denoted by e^x and is called the ***natural exponential function***.

The results in the next theorem follow from Theorem 7.5.5 and Formulas (1) and (2) of Section 7.4.

> **7.5.7** THEOREM.
>
> (a) *domain of e^x = range of $\ln x = (-\infty, +\infty)$*
> (b) *range of e^x = domain of $\ln x = (0, +\infty)$*
> (c) $\ln e^x = x$ *for* $-\infty < x < +\infty$
> (d) $e^{\ln x} = x$ *for* $x > 0$

The results in this theorem are consistent with the graphs of $\ln x$ and e^x shown in Figure 7.3.1. Moreover, it follows from part (d) that $\ln x$ is that power to which e must be raised to produce x, which is consistent with the interpretation of $\ln x$ given in Section 7.1.

☐ **DEFINITION OF b^x**

One of the goals stated in Section 7.1 was to define irrational exponents in a way that would make b^x a continuous function. Definition 7.5.6 does this for the case where $b = e$, since the continuity of $\ln x$ implies the continuity of e^x by Theorem 7.4.6. We shall now show how to exploit Definition 7.5.6 to achieve the same result for an arbitrary positive base.

It follows from Theorem 7.5.3(d) that if b is a positive real number and r is a rational number, then

$$\ln b^r = r \ln b$$

from which it follows that

$$e^{\ln b^r} = e^{r \ln b}$$

From Theorem 7.5.7(d) this can be rewritten as

$$b^r = e^{r \ln b} \tag{5}$$

Motivated by this formula, which is valid for rational r, we make the following definition.

> **7.5.8** DEFINITION.   If x is a real number and b is a positive real number, then the function b^x is defined by the formula
>
> $$b^x = e^{x \ln b}$$
>
> and is called the ***base b exponential function***.

Later, we will show that b^x is differentiable and hence continuous, but the continuity can also be seen by noting that b^x is the composition of e^x and $x \ln b$, both of which are continuous functions; hence b^x itself is continuous.

Example 3   We leave it for the reader to check the following computations with a calculator:

$$2^\pi = e^{\pi \ln 2} \approx 8.824978$$

$$7^{-\sqrt{2}} = e^{-\sqrt{2} \ln 7} \approx 0.063804$$

$$\sqrt{5}^{\sqrt{3}} = e^{\sqrt{3} \ln \sqrt{5}} \approx 4.030192 \qquad \blacktriangleleft$$

☐ **LAWS OF REAL**
EXPONENTS

The following theorem shows that part (d) of Theorem 7.5.3 is valid for all real exponents.

7.5.9   THEOREM.   *For any positive number a and any real number x,*

$$\ln a^x = x \ln a$$

Proof.   From Definition 7.5.8 we have

$$a^x = e^{x \ln a}$$

so from Theorem 7.5.7(c)

$$\ln a^x = \ln e^{x \ln a} = x \ln a \quad \blacksquare$$

The following theorem shows that the laws of exponents that were developed in algebra for rational exponents hold for all real exponents.

7.5.10   THEOREM.   *If a and b are positive real numbers, then for all real numbers k and l the following laws of exponents hold.*

(a) $a^0 = 1$ (b) $a^1 = a$

(c) $a^k \cdot a^l = a^{k+l}$ (d) $a^{-l} = \dfrac{1}{a^l}$

(e) $\dfrac{a^k}{a^l} = a^{k-l}$ (f) $a^k \cdot b^k = (ab)^k$

(g) $\dfrac{a^k}{b^k} = \left(\dfrac{a}{b}\right)^k$ (h) $(a^k)^l = a^{kl}$

Proof.   We shall prove (c). The proofs of the rest are similar and will be omitted. We can write

$$\ln (a^k \cdot a^l) = \ln a^k + \ln a^l = k \ln a + l \ln a$$

$$= (k + l) \ln a = \ln (a^{k+l})$$

Thus, we have $a^k \cdot a^l = a^{k+l}$ since the two sides have the same natural logarithm. $\blacksquare$

☐ **DIFFERENTIABILITY OF**
e^x, b^x, AND x^r

The following theorem establishes the differentiability of e^x and b^x, which was assumed without proof in Section 7.2.

7.5.11   THEOREM.

(a)   *The function e^x is differentiable on $(-\infty, +\infty)$, and its derivative is*

$$\frac{d}{dx}[e^x] = e^x$$

(b)   *The function b^x is differentiable on $(-\infty, +\infty)$, and its derivative is*

$$\frac{d}{dx}[b^x] = b^x \ln b$$

Proof (a).   Because $\ln x$ is differentiable, and its derivative

$$\frac{d}{dx}[\ln x] = \frac{1}{x}$$

is nonzero on $(0, +\infty)$, it follows from Theorem 7.4.7 with $f(x) = \ln x$ and $f^{-1}(x) = e^x$ that e^x is differentiable on $(-\infty, +\infty)$, and its derivative is

$$\frac{d}{dx}\underbrace{[e^x]}_{f^{-1}(x)} = \underbrace{\frac{1}{1/e^x}}_{f'(f^{-1}(x))} = e^x$$

Proof (b). The function $b^x = e^{x \ln b}$ is the composition of e^x and $x \ln b$, both of which are differentiable functions on $(-\infty, +\infty)$. Thus, by the chain rule the function b^x is differentiable on $(-\infty, +\infty)$, and its derivative is

$$\frac{d}{dx}[b^x] = \frac{d}{dx}[e^{x \ln b}] = e^{x \ln b} \ln b = b^x \ln b \quad \blacksquare$$

In Section 3.6 we obtained the derivative formula

$$\frac{d}{dx}[x^r] = rx^{r-1}$$

for *rational* r by assuming differentiability and using implicit differentiation. The following theorem proves the differentiability of x^r for positive x and shows that the formula is valid for any *real* r in that case.

7.5.12 THEOREM. *For any real number r the function x^r is differentiable on $(0, +\infty)$, and its derivative is*

$$\frac{d}{dx}[x^r] = rx^{r-1}$$

Proof. The function $x^r = e^{r \ln x}$ is the composition of e^x and $r \ln x$. The function $r \ln x$ is differentiable on $(0, +\infty)$, and the function e^x is differentiable on $(-\infty, +\infty)$, so by the chain rule the composition x^r is differentiable on $(0, +\infty)$, and its derivative is

$$\frac{d}{dx}[x^r] = \frac{d}{dx}[e^{r \ln x}] = e^{r \ln x}\left(\frac{r}{x}\right) = x^r\left(\frac{r}{x}\right) = rx^{r-1} \quad \blacksquare$$

☐ **GENERAL LOGARITHMS**

Motivated by Theorem 7.1.2, we make the following definition.

7.5.13 DEFINITION. For $b > 0$ and $b \neq 1$, the **base b logarithm function** is defined by the formula

$$\log_b x = \frac{\ln x}{\ln b}, \quad x > 0 \tag{6}$$

In the special case where $b = e$, it follows from this definition that

$$\log_e x = \ln x$$

Thus, the natural logarithm function is the same as the base e logarithm function, which is consistent with the results obtained informally in Section 7.2

In Section 7.1 we showed informally that $\log_b b^x = x$ and $b^{\log_b x} = x$ [see (1) in that section]. We know from our work in Section 7.4 that equations of this type are characteristic of inverse functions, so the following theorem should not be surprising.

7.5.14 THEOREM. *For $b > 0$ and $b \neq 1$, the functions b^x and $\log_b x$ are inverses.*

Proof. We leave it for the exercises to show that

$$\text{domain } b^x = \text{range } \log_b x = (-\infty, +\infty)$$

$$\text{range } b^x = \text{domain } \log_b x = (0, +\infty)$$

so that b^x and $\log_b x$ meet the domain and range requirements for inverse functions. To complete the proof, we must establish the formulas in (1) of Section 7.4. This can be done as follows:

$$\log_b b^x = \frac{\ln b^x}{\ln b} = \frac{x \ln b}{\ln b} = x$$

$$b^{\log_b x} = e^{(\log_b x)\ln b} = e^{(\ln x/\ln b)\ln b} = e^{\ln x} = x \qquad \blacksquare$$

Because b^x and $\log_b x$ are inverse functions, their graphs are reflections of one another about the line $y = x$. We leave it for the exercises to show that the graphs are of the forms shown in Figure 7.5.4.

Figure 7.5.4

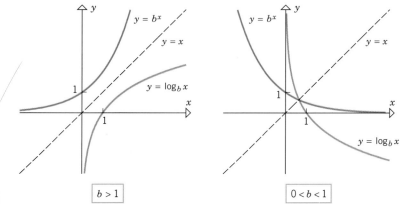

$$\boxed{b > 1} \qquad\qquad \boxed{0 < b < 1}$$

☐ **SOME FUNDAMENTAL LIMITS**

The following theorem establishes three fundamental limits that were discussed informally in Section 7.1.

7.5.15   THEOREM.

(a) $\displaystyle\lim_{x \to 0} (1 + x)^{1/x} = e$

(b) $\displaystyle\lim_{x \to +\infty} \left(1 + \frac{1}{x}\right)^x = e$

(c) $\displaystyle\lim_{x \to -\infty} \left(1 + \frac{1}{x}\right)^x = e$

Parts (b) and (c) can be proved by making an appropriate substitution in the limit in (a); we leave those parts for the exercises.

Proof (a).   Our proof builds on the differentiability of $\ln x$, and more specifically on the derivative of $\ln x$ at the point $x = 1$, namely

$$\frac{d}{dx}[\ln x]\bigg|_{x=1} = \frac{1}{x}\bigg|_{x=1} = 1$$

If we express this relationship using the definition of a derivative, we obtain

$$1 = \lim_{h \to 0} \frac{\ln(1+h) - \ln 1}{h} = \lim_{h \to 0} \frac{\ln(1+h)}{h} = \lim_{h \to 0} \ln(1+h)^{1/h}$$

Thus, it follows that

$$e = e^{\lim_{h \to 0} \ln(1+h)^{1/h}}$$

which from the continuity of e^x can be written as

$$e = \lim_{h \to 0} e^{\ln(1+h)^{1/h}} = \lim_{h \to 0} (1+h)^{1/h} \quad \blacksquare$$

▶ **Exercise Set 7.5** 　C　8

In Exercises 1–6, find the domain of $f(x)$.

1. $f(x) = \ln(3x + 2)$.   **2.** $f(x) = \ln(1 - 2x)$.

3. $f(x) = \ln(4 - x^2)$.   **4.** $f(x) = \ln|5 - 3x|$.

5. $f(x) = \sqrt{1 + \ln x}$.   **6.** $f(x) = \ln(\ln x)$.

7. (a) Find the domain of $\ln(x^2)$.

    (b) For what values of x does $\ln(x^2) = 2\ln x$?

8. Approximate $\ln 3$ using the midpoint approximation with $n = 20$. Find the magnitude of the error in this approximation.

9. Simplify the expression and state the values of x for which your simplification is valid.

    (a) $e^{-\ln x}$ (b) $e^{\ln x^2}$

    (c) $\ln(e^{-x^2})$ (d) $\ln(1/e^x)$

    (e) $\exp(3\ln x)$ (f) $\ln(xe^x)$

    (g) $\ln(e^{x - \sqrt[3]{x}})$ (h) $e^{x - \ln x}$.

10. Let $f(x) = e^{-2x}$. Find the simplest exact value of $f(\ln 3)$.

11. Let $f(x) = e^x + 3e^{-x}$. Find the simplest exact value of $f(\ln 2)$.

In Exercises 12 and 13, use Definition 7.5.8 to express the given quantity as a power of e.

12. (a) 3^π (b) $2^{\sqrt{2}}$.

13. (a) π^{-x} (b) x^{2x}, $x > 0$.

14. Prove that $\lim_{x \to +\infty} e^x = +\infty$ by showing that for any $N > 0$, there is a point x_0 such that $e^x > N$ whenever $x > x_0$.

15. Prove that $\lim_{x \to -\infty} e^x = 0$ by showing that for any $\epsilon > 0$, there is a point x_0 such that $0 < e^x < \epsilon$ whenever $x < x_0$.

16. Prove: For $b > 0$ and $b \neq 1$

    domain b^x = range $\log_b x = (-\infty, +\infty)$

    range b^x = domain $\log_b x = (0, +\infty)$

    [*Hint:* Use Definitions 7.5.8 and 7.5.13.]

17. Show that the graphs of

$$y = b^x \quad \text{and} \quad y = \log_b x$$

for $b > 1$ and $0 < b < 1$ are of the forms shown in Figure 7.5.4. [*Hint:* See Definition 7.5.13.]

18. Prove parts (b) and (c) of Theorem 7.5.15 by making an appropriate substitution and using part (a).

19. (a) Give a geometric argument to show that

$$\frac{1}{x+1} < \int_x^{x+1} \frac{1}{t} \, dt < \frac{1}{x}, \quad x > 0$$

    (b) Use the result in part (a) to prove that

$$\frac{1}{x+1} < \ln\left(1 + \frac{1}{x}\right) < \frac{1}{x}, \quad x > 0$$

    (c) Use the result in part (b) to prove that

$$e^{\frac{x}{x+1}} < \left(1 + \frac{1}{x}\right)^x < e, \quad x > 0$$

    and hence that

$$\lim_{x \to +\infty} \left(1 + \frac{1}{x}\right)^x = e$$

    (d) Use the inequality in part (c) to prove that

$$\left(1 + \frac{1}{x}\right)^x < e < \left(1 + \frac{1}{x}\right)^{x+1}, \quad x > 0$$

■ 7.6 THE HYPERBOLIC FUNCTIONS

*In this section we shall study certain combinations of e^x and e^{-x}, called **hyperbolic functions**. These functions have numerous engineering applications and arise naturally in many mathematical problems. It will become evident as we progress that the hyperbolic functions have many properties in common with the trigonometric functions. This similarity is reflected in the names of the hyperbolic functions.*

□ **DEFINITIONS OF HYPERBOLIC FUNCTIONS**

7.6.1 DEFINITION. The **hyperbolic sine** and **hyperbolic cosine** functions, denoted by **sinh** and **cosh**, respectively, are defined by

$$\sinh x = \frac{e^x - e^{-x}}{2} \quad \text{and} \quad \cosh x = \frac{e^x + e^{-x}}{2}$$

REMARK. We note that sinh rhymes with "cinch" and cosh rhymes with "gosh."

The graph of $\cosh x = \frac{1}{2}e^x + \frac{1}{2}e^{-x}$ can be obtained by separately graphing $\frac{1}{2}e^x$ and $\frac{1}{2}e^{-x}$ and then adding the y-coordinates together at each point (Figure 7.6.1a). This graphing technique, called **addition of ordinates**, can also be used to graph $\sinh x = \frac{1}{2}e^x - \frac{1}{2}e^{-x}$ (Figure 7.6.1b).

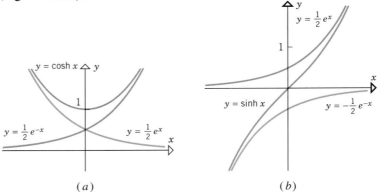

Figure 7.6.1   (a)   (b)

As an illustration of how hyperbolic functions occur in physical problems, consider a homogeneous flexible cable hanging suspended between two points (e.g., an electrical transmission line suspended between two poles). The cable forms a curve called a **catenary** (from the Latin "catena" meaning chain). If, as in Figure 7.6.2, a coordinate system is introduced in such a way that the low point of the cable lies on the y-axis, then it can be shown using principles of physics that the equation of the curve formed by the cable is

$$y = a \cosh (x/a)$$

where a depends on the tension and physical properties of the cable.

The remaining hyperbolic functions, **hyperbolic tangent, hyperbolic cotangent, hyperbolic secant**, and **hyperbolic cosecant** are defined in terms of sinh and cosh as follows.

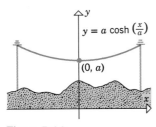

Figure 7.6.2

$$\tanh x = \frac{\sinh x}{\cosh x} = \frac{e^x - e^{-x}}{e^x + e^{-x}}, \qquad \operatorname{sech} x = \frac{1}{\cosh x} = \frac{2}{e^x + e^{-x}},$$

$$\coth x = \frac{\cosh x}{\sinh x} = \frac{e^x + e^{-x}}{e^x - e^{-x}}, \qquad \operatorname{csch} x = \frac{1}{\sinh x} = \frac{2}{e^x - e^{-x}}$$

The graphs of these functions are shown in Figure 7.6.3 (see Exercises 39–42).

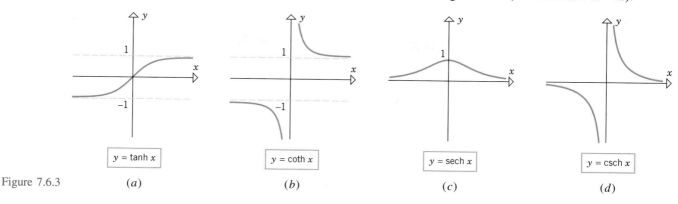

Figure 7.6.3   (a)   (b)   (c)   (d)

☐ **HYPERBOLIC IDENTITIES**

The hyperbolic functions satisfy various identities similar to the identities for the trigonometric functions. The most fundamental of these is

$$\cosh^2 x - \sinh^2 x = 1 \tag{1}$$

which can be proved by writing

$$\cosh^2 x - \sinh^2 x = \left(\frac{e^x + e^{-x}}{2}\right)^2 - \left(\frac{e^x - e^{-x}}{2}\right)^2$$

$$= \frac{1}{4}(e^{2x} + 2e^0 + e^{-2x}) - \frac{1}{4}(e^{2x} - 2e^0 + e^{-2x})$$

$$= 1$$

If we divide (1) by $\cosh^2 x$, we obtain

$$1 - \tanh^2 x = \operatorname{sech}^2 x \tag{2}$$

and if we divide (1) by $\sinh^2 x$, we obtain

$$\coth^2 x - 1 = \operatorname{csch}^2 x \tag{3}$$

The addition formulas for sinh and cosh are

$$\sinh(x + y) = \sinh x \cosh y + \cosh x \sinh y \tag{4a}$$
$$\cosh(x + y) = \cosh x \cosh y + \sinh x \sinh y \tag{4b}$$

These are easier to prove than the corresponding identities for the trigonometric functions. For example, to prove (4a) we use the relations

$$\cosh x + \sinh x = e^x \tag{5a}$$
$$\cosh x - \sinh x = e^{-x} \tag{5b}$$

which follow immediately from Definition 7.6.1. From (5a) and (5b) it follows that

$$\sinh(x + y) = \frac{e^{(x+y)} - e^{-(x+y)}}{2} = \frac{e^x e^y - e^{-x} e^{-y}}{2}$$

$$= \tfrac{1}{2}[(\cosh x + \sinh x)(\cosh y + \sinh y) - (\cosh x - \sinh x)(\cosh y - \sinh y)]$$

$$= \sinh x \cosh y + \cosh x \sinh y$$

which proves (4a). The proof of (4b) is similar.

If we let $x = y$ in (4a) and (4b), we obtain the following analogs of the trigonometric double-angle formulas:

$$\sinh 2x = 2 \sinh x \cosh x \tag{6a}$$

$$\cosh 2x = \cosh^2 x + \sinh^2 x \tag{6b}$$

By using (1) and (6b) the reader should be able to show that

$$\cosh 2x = 2 \sinh^2 x + 1 \tag{7a}$$

$$\cosh 2x = 2 \cosh^2 x - 1 \tag{7b}$$

To obtain subtraction formulas for sinh and cosh we shall use the addition formulas and the relations

$$\cosh (-x) = \cosh x \tag{8a}$$

$$\sinh (-x) = -\sinh x \tag{8b}$$

which follow immediately from Definition 7.6.1. If we replace y by $-y$ in (4a) and (4b) and then apply (8a) and (8b), we obtain

$$\sinh (x - y) = \sinh x \cosh y - \cosh x \sinh y \tag{9a}$$

$$\cosh (x - y) = \cosh x \cosh y - \sinh x \sinh y \tag{9b}$$

☐ **WHY HYPERBOLIC FUNCTIONS ARE SO NAMED**

If t is any real number, then the point $(\cos t, \sin t)$ lies on the circle $x^2 + y^2 = 1$ because

$$\cos^2 t + \sin^2 t = 1$$

(Figure 7.6.4). For this reason sine and cosine are called *circular functions*. Analogously, for any real number t the point $(\cosh t, \sinh t)$ lies on the curve $x^2 - y^2 = 1$ because

$$\cosh^2 t - \sinh^2 t = 1$$

(Figure 7.6.5). As we shall see later, this curve is called a *hyperbola* and accordingly sinh and cosh are called *hyperbolic functions*.

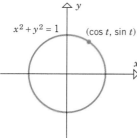

Figure 7.6.4

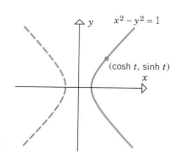

Figure 7.6.5

☐ **DERIVATIVE AND INTEGRAL FORMULAS**

Derivative formulas for sinh x and cosh x follow easily from Definition 7.6.1. For example,

$$\frac{d}{dx} [\cosh x] = \frac{d}{dx} \left[\frac{e^x + e^{-x}}{2} \right] = \frac{1}{2} \left(\frac{d}{dx} [e^x] + \frac{d}{dx} [e^{-x}] \right)$$

$$= \frac{e^x - e^{-x}}{2} = \sinh x$$

Similarly,

$$\frac{d}{dx} [\sinh x] = \cosh x$$

The derivatives of the remaining hyperbolic functions can be obtained by first expressing them in terms of sinh and cosh. For example,

$$\frac{d}{dx}[\tanh x] = \frac{d}{dx}\left[\frac{\sinh x}{\cosh x}\right] = \frac{\cosh x \dfrac{d}{dx}[\sinh x] - \sinh x \dfrac{d}{dx}[\cosh x]}{\cosh^2 x}$$

$$= \frac{\cosh^2 x - \sinh^2 x}{\cosh^2 x} = \frac{1}{\cosh^2 x} = \text{sech}^2 x$$

Derivations of differentiation formulas for the remaining hyperbolic functions are exercises. For reference we summarize the generalized derivative formulas.

$$\frac{d}{dx}[\sinh u] = \cosh u \frac{du}{dx}, \qquad \frac{d}{dx}[\coth u] = -\text{csch}^2 u \frac{du}{dx}$$

$$\frac{d}{dx}[\cosh u] = \sinh u \frac{du}{dx}, \qquad \frac{d}{dx}[\text{sech } u] = -\text{sech } u \tanh u \frac{du}{dx}$$

$$\frac{d}{dx}[\tanh u] = \text{sech}^2 u \frac{du}{dx}, \qquad \frac{d}{dx}[\text{csch } u] = -\text{csch } u \coth u \frac{du}{dx}$$

REMARK.   Except for a difference in the pattern of signs, these formulas are analogous to the formulas for the derivatives of the trigonometric functions.

These differentiation formulas produce the following companion integration formulas.

$$\int \sinh u \, du = \cosh u + C, \qquad \int \text{csch}^2 u \, du = -\coth u + C$$

$$\int \cosh u \, du = \sinh u + C, \qquad \int \text{sech } u \tanh u \, du = -\text{sech } u + C$$

$$\int \text{sech}^2 u \, du = \tanh u + C, \qquad \int \text{csch } u \coth u \, du = -\text{csch } u + C$$

Example 1

$$\frac{d}{dx}[\cosh (x^3)] = \sinh (x^3) \cdot \frac{d}{dx}[x^3] = 3x^2 \sinh (x^3)$$

$$\frac{d}{dx}[\ln (\tanh x)] = \frac{1}{\tanh x} \cdot \frac{d}{dx}[\tanh x] = \frac{\text{sech}^2 x}{\tanh x} \qquad \blacktriangleleft$$

Example 2

$$\int \sinh^5 x \cosh x \, dx = \frac{1}{6}\sinh^6 x + C \qquad \boxed{\begin{array}{l} u = \sinh x \\ du = \cosh x \, dx \end{array}}$$

$$\int \tanh x \, dx = \int \frac{\sinh x}{\cosh x} \, dx$$

$$= \ln |\cosh x| + C \qquad \boxed{\begin{array}{l} u = \cosh x \\ du = \sinh x \, dx \end{array}}$$

$$= \ln (\cosh x) + C$$

We are justified in writing $|\cosh x| = \cosh x$ because $\cosh x$ is positive for all x. In fact, $\cosh x \geq 1$ for all x. This is geometrically clear from Figure 7.6.1a and can be proved using (1) (Exercise 41). $\blacktriangleleft$

▶ **Exercise Set 7.6** $\boxed{C}$ 48, 54, 55

---

1. In each part, a value for one of the hyperbolic functions is given at x_0. Find the values of the remaining five hyperbolic functions at x_0.

(a) $\sinh x_0 = -2$

(b) $\cosh x_0 = \frac{5}{4}, x_0 < 0$

(c) $\tanh x_0 = -\frac{4}{5}$

(d) $\coth x_0 = 2$

(e) $\operatorname{sech} x_0 = \frac{15}{17}, x_0 > 0$

(f) $\operatorname{csch} x_0 = -1$.

In Exercises 2–14, establish the identities.

2. $\cosh(x + y) = \cosh x \cosh y + \sinh x \sinh y$.

3. $\cosh 2x = 2 \sinh^2 x + 1$.   **4.** $\cosh 2x = 2 \cosh^2 x - 1$.

5. $\cosh(-x) = \cosh x$.   **6.** $\sinh(-x) = -\sinh x$.

7. $\tanh(x + y) = \dfrac{\tanh x + \tanh y}{1 + \tanh x \tanh y}$.

8. $\tanh(x - y) = \dfrac{\tanh x - \tanh y}{1 - \tanh x \tanh y}$.

9. $\tanh 2x = \dfrac{2 \tanh x}{1 + \tanh^2 x}$.

10. $\cosh \frac{1}{2} x = \sqrt{\frac{1}{2}(\cosh x + 1)}$.

11. $\sinh \frac{1}{2} x = \pm \sqrt{\frac{1}{2}(\cosh x - 1)}$.

12. $\sinh x + \sinh y = 2 \sinh\left(\dfrac{x + y}{2}\right) \cosh\left(\dfrac{x - y}{2}\right)$.

13. $\cosh x + \cosh y = 2 \cosh\left(\dfrac{x + y}{2}\right) \cosh\left(\dfrac{x - y}{2}\right)$.

14. $\cosh 3x = 4 \cosh^3 x - 3 \cosh x$.

15. Derive the differentiation formula for

(a) $\sinh x$

(b) $\coth x$

(c) $\operatorname{sech} x$

(d) $\operatorname{csch} x$.

In Exercises 16–25, find dy/dx.

16. $y = \cosh(x^4)$.

17. $y = \sinh(4x - 8)$.

18. $y = \ln(\tanh 2x)$.

19. $y = \coth(\ln x)$.

20. $y = \operatorname{sech}(e^{2x})$.

21. $y = \operatorname{csch}(1/x)$.

22. $y = \sinh^3(2x)$.

23. $y = \sqrt{4x + \cosh^2(5x)}$.

24. $y = \sinh(\cos 3x)$.

25. $y = x^3 \tanh^2(\sqrt{x})$.

In Exercises 26–35, evaluate the integral.

26. $\displaystyle\int \cosh(2x - 3)\, dx$.

27. $\displaystyle\int \sinh^6 x \cosh x\, dx$.

28. $\displaystyle\int \operatorname{csch}^2(3x)\, dx$.

29. $\displaystyle\int \sqrt{\tanh x}\, \operatorname{sech}^2 x\, dx$.

30. $\displaystyle\int \coth^2 x \operatorname{csch}^2 x\, dx$.

31. $\displaystyle\int \tanh x\, dx$.

32. $\displaystyle\int \dfrac{e^x - e^{-x}}{e^x + e^{-x}}\, dx$.

33. $\displaystyle\int \tanh x \operatorname{sech}^3 x\, dx$.

34. $\displaystyle\int \dfrac{\sinh 2x}{3 + 5 \cosh 2x}\, dx$.

35. $\displaystyle\int \dfrac{\cosh(\sqrt{x})}{\sqrt{x}}\, dx$.

36. In each part, find the exact numerical value:

(a) $\sinh(\ln 3)$

(b) $\cosh(-\ln 2)$

(c) $\tanh(2 \ln 5)$

(d) $\sinh(-3 \ln 2)$.

37. In each part, express as a rational function:

(a) $\cosh(\ln x)$

(b) $\sinh(\ln x)$

(c) $\tanh(2 \ln x)$

(d) $\cosh(-\ln x)$.

38. Prove the following facts about $\tanh x$:

(a) $-1 < \tanh x < 1$ for all x

(b) $\displaystyle\lim_{x \to +\infty} \tanh x = 1$

(c) $\displaystyle\lim_{x \to -\infty} \tanh x = -1$.

39. Determine the intervals on which $y = \tanh x$ is positive, negative, increasing, decreasing, concave up, and concave down. Then use these results and Exercise 38 to obtain the graph of $y = \tanh x$.

40. Use the results found in Exercises 38 and 39 to help obtain the graph of $y = \coth x$.

41. (a) Prove: $\cosh x \geq 1$ for all x.

(b) Prove: $0 < \operatorname{sech} x \leq 1$ for all x.

(c) Obtain the graph of $y = \operatorname{sech} x$.

42. (a) Show that $\displaystyle\lim_{x \to +\infty} \operatorname{csch} x = \lim_{x \to -\infty} \operatorname{csch} x = 0$.

(b) Show that
$$\lim_{x \to 0^+} \operatorname{csch} x = +\infty \quad \text{and} \quad \lim_{x \to 0^-} \operatorname{csch} x = -\infty$$

(c) Obtain the graph of $y = \operatorname{csch} x$.

43. Prove: $(\sinh x + \cosh x)^n = \sinh nx + \cosh nx$.

44. (a) It is sometimes said that $\sinh x$ and $\cosh x$ *behave* *like* $\frac{1}{2} e^x$ for x large and positive. Give an informal justification of this statement.

(b) Make a sketch showing the graphs of $y = \sinh x$, $y = \cosh x$, and $y = \frac{1}{2} e^x$, all in the same figure.

45. Find

(a) $\displaystyle\lim_{x \to +\infty} \dfrac{\cosh x}{e^x}$

(b) $\displaystyle\lim_{x \to +\infty} \dfrac{\sinh ax}{e^x}$ $(a > 0, \text{ constant})$.

[*Hint:* In part (b) consider the cases $a > 1$, $a = 1$, and $0 < a < 1$ separately.]

46. Show that
$$\int_{-a}^{a} e^{tx}\, dx = \dfrac{2 \sinh at}{t}$$
where t is a nonzero constant.

47. Find the area enclosed by $y = \sinh 2x$, $y = 0$, and $x = \ln 3$.

48. Use Newton's Method to approximate the positive value of the constant a such that the area enclosed by $y = \cosh ax$, $y = 0$, $x = 0$, and $x = 1$ is 2 square units. Express your answer to at least five decimal places.

49. Find the volume of the solid that is generated when the region enclosed by $y = \cosh 2x$, $y = \sinh 2x$, $x = 0$, and $x = 5$ is revolved about the x-axis.

50. Find the volume of the solid that is generated when the region enclosed by $y = \operatorname{sech} x$, $y = 0$, $x = 0$, and $x = \ln 2$ is revolved about the x-axis.

51. Find the arc length of $y = \cosh x$ between $x = 0$ and $x = \ln 2$.

52. Find the arc length of the catenary $y = a \cosh (x/a)$ between $x = 0$ and $x = x_1$ $(x_1 > 0)$.

53. A cable is suspended between two poles as shown in Figure 7.6.2. The equation of the curve formed by the cable is $y = a \cosh (x/a)$, where a is a positive constant. Suppose that the x-coordinates of the points of support are $x = -b$ and $x = b$, where $b > 0$.

(a) Show that the length L of the cable is given by

$$L = 2a \sinh \frac{b}{a}$$

(b) Show that the sag S (the vertical distance between the highest and lowest points on the cable) is given by

$$S = a \cosh \frac{b}{a} - a$$

Exercises 54 and 55 refer to the hanging cable described in Exercise 53.

54. Assuming that the cable is 120 ft long and the poles are 100 ft apart, approximate the sag in the cable by using Newton's Method to approximate a. Express your final answer to the nearest tenth of a foot. [*Hint:* First let $u = 50/a$.]

55. Assuming that the poles are 400 ft apart and the sag in the cable is 30 ft, approximate the length of the cable by using Newton's Method to approximate a. Express your final answer to the nearest tenth of a foot. [*Hint:* First let $u = 200/a$.]

◼ 7.7 FIRST-ORDER DIFFERENTIAL EQUATIONS AND APPLICATIONS

Many of the principles in science and engineering concern relationships between changing quantities. Since rates of change are represented mathematically by derivatives, such principles often lead to equations involving unknown functions and their derivatives. In this section we shall study some basic equations of this type, and in Chapter 19 we go a little further into this topic.

☐ TERMINOLOGY

A *differential equation* is an equation involving one or more derivatives of an unknown function. The *order* of a differential equation is defined to be the order of the highest derivative it contains. Here are some examples:

DIFFERENTIAL EQUATION	ORDER
$\dfrac{dy}{dx} = 3y$	1
$\dfrac{d^2y}{dx^2} - 6\dfrac{dy}{dx} + 8y = 0$	2
$y' - y = e^{2x}$	1
$\dfrac{d^3y}{dt^3} - t\dfrac{dy}{dt} + (t^2 - 1)y = e^t$	3

In the first three equations, $y = y(x)$ is an unknown function of x, and in the last equation $y = y(t)$ is an unknown function of t. The variable t is generally used in applied problems involving time. Except in such problems, we shall use x as the independent variable.

A function $y = y(x)$ is a **solution** of a differential equation if the equation is satisfied for all x in some interval when y and its derivatives are substituted. For example, $y = e^{2x}$ is a solution of the differential equation

$$\frac{dy}{dx} - y = e^{2x} \tag{1}$$

since substituting y and its derivative on the left side of this equation yields

$$\frac{dy}{dx} - y = 2e^{2x} - e^{2x} = e^{2x}$$

This is not the only solution; for any choice of the constant C the function

$$y = Ce^x + e^{2x} \tag{2}$$

is also a solution, since

$$\frac{dy}{dx} - y = (Ce^x + 2e^{2x}) - (Ce^x + e^{2x}) = e^{2x}$$

One can prove that *all* solutions of (1) can be obtained by substituting values for the constant C. A solution of a differential equation from which all solutions can be derived by substituting values for arbitrary constants is called the **general solution** of the equation. Usually, the general solution of an nth-order differential equation contains n arbitrary constants. For example, (2), which is the general solution of the first-order equation (1), contains one arbitrary constant.

☐ **INITIAL-VALUE PROBLEMS**

When a physical problem leads to a differential equation, there are usually conditions in the problem that determine specific values for the arbitrary constants in the general solution of the equation. For a first-order equation, a condition that specifies the value of the unknown function $y(x)$ at some point $x = x_0$ is called an **initial condition**. A first-order differential equation together with one initial condition constitutes a **first-order initial-value problem**.

Example 1 The solution of the initial-value problem

$$\frac{dy}{dx} - y = e^{2x}, \quad y(0) = 3$$

can be obtained by substituting the initial condition $x = 0$, $y = 3$ in the general solution (2) to find C. We obtain

$$3 = Ce^0 + e^0 = C + 1$$

Thus, $C = 2$, and the solution of the initial-value problem, which is obtained by substituting this value of C in (2), is

$$y = 2e^x + e^{2x} \quad \blacktriangleleft$$

☐ **FIRST-ORDER EQUATIONS**

The simplest first-order differential equations are of the form

$$\frac{dy}{dx} = f(x) \tag{3}$$

Such equations can sometimes be solved by integration. For example, if

$$\frac{dy}{dx} = x^3 \tag{4}$$

then

$$y = \int x^3 \, dx = \frac{x^4}{4} + C$$

is the general solution of (4).

☐ **FIRST-ORDER SEPARABLE EQUATIONS**

First-order differential equations in which the right side of (3) involves both x and y rather than x alone can be complicated to solve exactly, and one must often settle for numerical approximations of solutions. However, if the equation can be expressed in the form

$$\frac{dy}{dx} = \frac{g(x)}{h(y)} \tag{5}$$

then the general solution can sometimes be found by rewriting the equation in the differential form

$$h(y) \, dy = g(x) \, dx \tag{6}$$

and integrating both sides to obtain

$$\int h(y) \, dy = \int g(x) \, dx + C$$

A first-order differential equation of form (5) is said to be **separable**, and the process of rewriting (5) in form (6) is called **separating the variables**.

Example 2   Solve the initial-value problem:

$$\frac{dy}{dx} = -4xy^2, \quad y(0) = 1$$

Solution.   We first solve the differential equation by separating variables:

$$\frac{dy}{y^2} = -4x \, dx$$

$$\int \frac{dy}{y^2} = \int -4x \, dx$$

$$-\frac{1}{y} = -2x^2 + C$$

or on multiplying by -1, taking reciprocals, and writing K in place of $-C$,

$$y = \frac{1}{2x^2 + K} \tag{7}$$

The initial condition, $y(0) = 1$, requires that $y = 1$ when $x = 0$. Substituting these values in (7) yields $K = 1$. Thus, the solution of the initial-value problem is

$$y = \frac{1}{2x^2 + 1} \quad \blacktriangleleft$$

The graph of a solution of a differential equation is sometimes called an **integral curve**. Because the general solution of a first-order linear equation involves an arbitrary constant, it produces a family of integral curves. Some integral curves for the differential equation in the preceding example are shown in Figure 7.7.1. As illustrated in that figure, the initial condition ($y = 1$ if $x = 0$) isolates the unique integral curve in the family that passes through the point (0, 1).

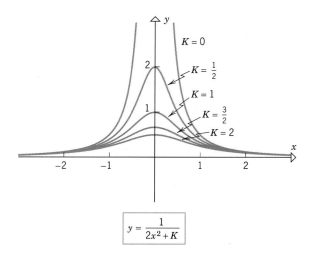

$$y = \frac{1}{2x^2 + K}$$

Figure 7.7.1

☐ IMPLICIT SOLUTIONS

Sometimes solutions of differential equations are expressed as implicitly defined functions. For example,

$$\ln y = xy + C \tag{8}$$

defines a solution of

$$\frac{dy}{dx} = \frac{y^2}{1 - xy} \tag{9}$$

for any value of the constant C, since implicit differentiation of (8) yields

$$\frac{1}{y}\frac{dy}{dx} = x\frac{dy}{dx} + y \quad \text{or} \quad \frac{dy}{dx} - xy\frac{dy}{dx} = y^2$$

from which (9) follows.

Example 3   Solve the equation $x(y - 1)\dfrac{dy}{dx} = y.$

Solution.   Separating the variables and integrating yields

$$x(y - 1)\, dy = y\, dx$$

$$\frac{y - 1}{y}\, dy = \frac{dx}{x}$$

$$\int \frac{y - 1}{y}\, dy = \int \frac{dx}{x}$$

$$\int \left(1 - \frac{1}{y}\right) dy = \int \frac{dx}{x}$$

$$y - \ln |y| = \ln |x| + C$$

or, using properties of logarithms,

$$y = \ln |xy| + C$$

This gives a solution implicitly as a function of x; in this case there is no simple formula for y explicitly as a function of x.   ◀

☐ FIRST-ORDER LINEAR EQUATIONS

Not every first-order differential equation is separable. For example, it is impossible to separate the variables in the equation

$$\frac{dy}{dx} + 2xy = xe^{-x^2}$$

However, this equation can be solved by a different method that we shall now consider. A first-order differential equation is called *linear* if it is expressible in the form

$$\frac{dy}{dx} + p(x)y = q(x) \tag{10}$$

where the functions $p(x)$ and $q(x)$ may or may not be constant. Some examples are

$$\frac{dy}{dx} + x^2 y = e^x, \qquad \frac{dy}{dx} + (\sin x)y + x^3 = 0, \qquad \frac{dy}{dx} + 5y = 2$$

$$\boxed{p(x) = x^2, \, q(x) = e^x} \qquad \boxed{p(x) = \sin x, \, q(x) = -x^3} \qquad \boxed{p(x) = 5, \, q(x) = 2}$$

One procedure for solving (10) is based on the observation that if we define $\mu = \mu(x)$ by

$$\mu = e^{\int p(x)\, dx}$$

then

$$\frac{d\mu}{dx} = e^{\int p(x)\, dx} \cdot \frac{d}{dx} \int p(x)\, dx = \mu p(x)$$

Thus,

$$\frac{d}{dx}(\mu y) = \mu \frac{dy}{dx} + \frac{d\mu}{dx} y = \mu \frac{dy}{dx} + \mu p(x)y \tag{11}$$

If (10) is multiplied through by μ, and then simplified using (11), it becomes

$$\mu \frac{dy}{dx} + \mu p(x)y = \mu q(x)$$

$$\frac{d}{dx}(\mu y) = \mu q(x)$$

This equation can be solved by integrating both sides to obtain

$$\mu y = \int \mu q(x)\, dx + C \quad \text{or} \quad y = \frac{1}{\mu}\left[\int \mu q(x)\, dx + C \right]$$

To summarize, (10) can be solved in three steps:

Step 1. Calculate

$$\mu = e^{\int p(x)\, dx}$$

This is called the *integrating factor*. Since any μ will suffice, we can take the constant of integration to be zero in this step.

Step 2. Multiply both sides of (10) by μ and express the result as

$$\frac{d}{dx}(\mu y) = \mu q(x)$$

Step 3. Integrate both sides of the equation obtained in Step 2 and then solve for y. Be sure to include a constant of integration in this step.

Example 4 Solve the equation $\dfrac{dy}{dx} - 4xy = x$.

Solution. Since $p(x) = -4x$, the integrating factor is

$$\mu = e^{\int(-4x)\, dx} = e^{-2x^2}$$

If we multiply the given equation by μ, we obtain

$$\frac{d}{dx}(e^{-2x^2}y) = xe^{-2x^2}$$

Integrating both sides of this equation yields

$$e^{-2x^2}y = \int xe^{-2x^2}\,dx = -\tfrac{1}{4}e^{-2x^2} + C$$

and then multiplying both sides by e^{2x^2} yields

$$y = -\tfrac{1}{4} + Ce^{2x^2} \qquad \blacktriangleleft$$

REMARK. The equation in Example 4 can also be solved by separating variables.

Example 5 Solve the equation $x\dfrac{dy}{dx} - y = x$, where $x > 0$.

Solution. To put the equation in form (10), we divide through by x to obtain

$$\frac{dy}{dx} - \frac{1}{x}y = 1 \tag{12}$$

Since $p(x) = -1/x$, the integrating factor is

$$\mu = e^{\int -(1/x)\,dx} = e^{-\ln|x|} = \frac{1}{|x|} = \frac{1}{x}$$

(The absolute value was dropped because of the assumption that $x > 0$.) If we multiply (12) by μ, we obtain

$$\frac{d}{dx}\left(\frac{1}{x}y\right) = \frac{1}{x}$$

Integrating both sides of this equation yields

$$\frac{1}{x}y = \int \frac{1}{x}\,dx = \ln x + C \quad \text{or} \quad y = x\ln x + Cx \qquad \blacktriangleleft$$

□ **APPLICATIONS**

We conclude this section with some applications of first-order differential equations.

Example 6 (*Geometry*) Find a curve in the xy-plane that passes through $(0, 3)$ and whose tangent line at a point (x, y) has slope $2x/y^2$.

Solution. Since the slope of the tangent line is dy/dx, we have

$$\frac{dy}{dx} = \frac{2x}{y^2} \tag{13}$$

and, since the curve passes through $(0, 3)$, we have the initial condition

$$y(0) = 3 \tag{14}$$

Equation (13) is separable and can be written as

$$y^2\,dy = 2x\,dx$$

so

$$\int y^2\,dy = \int 2x\,dx \quad \text{or} \quad \tfrac{1}{3}y^3 = x^2 + C$$

From the initial condition, (14), it follows that $C = 9$, and so the curve has the equation

$$\tfrac{1}{3}y^3 = x^2 + 9 \quad \text{or} \quad y = (3x^2 + 27)^{1/3} \qquad \blacktriangleleft$$

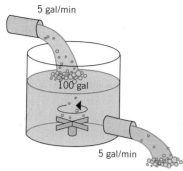

5 gal/min

100 gal

5 gal/min

Figure 7.7.2

Example 7 (*Mixing Problems*)   At time $t = 0$, a tank contains 4 lb of salt dissolved in 100 gal of water. Suppose that brine containing 2 lb of salt per gallon of water is allowed to enter the tank at a rate of 5 gal/min and that the mixed solution is drained from the tank at the same rate (Figure 7.7.2). Find the amount of salt in the tank after 10 min ($t = 10$).

Solution.   Let $y(t)$ be the amount of salt (in pounds) at time t. We are interested in finding $y(10)$, the amount of salt at time $t = 10$. We will begin by finding an expression for dy/dt, the rate of change of the amount of salt in the tank at time t. Clearly,

$$\frac{dy}{dt} = \text{rate in} - \text{rate out} \qquad (15)$$

where *rate in* is the rate at which salt enters the tank and *rate out* is the rate at which salt leaves the tank. But

$$\text{rate in} = (2 \text{ lb/gal}) \cdot (5 \text{ gal/min}) = 10 \text{ lb/min}$$

At time t, the mixture contains $y(t)$ lb of salt in 100 gal of water; thus, the concentration of salt at time t is $y(t)/100$ lb/gal and

$$\text{rate out} = \left(\frac{y(t)}{100} \text{ lb/gal} \right) \cdot (5 \text{ gal/min}) = \frac{y(t)}{20} \text{ lb/min}$$

Therefore, (15) can be written as

$$\frac{dy}{dt} = 10 - \frac{y}{20}$$

or

$$\frac{dy}{dt} + \frac{y}{20} = 10 \qquad (16)$$

which is a first-order linear differential equation. We also have the initial condition

$$y(0) = 4 \qquad (17)$$

since the tank contains 4 lb of salt at time $t = 0$.

Multiplying both sides of (16) by the integrating factor

$$\mu = e^{\int (1/20) \, dt} = e^{t/20}$$

yields

$$\frac{d}{dt} (e^{t/20} y) = 10 e^{t/20}$$

$$e^{t/20} y = \int 10 e^{t/20} \, dt = 200 e^{t/20} + C$$

$$y(t) = 200 + C e^{-t/20}$$

From the initial condition, (17), it follows that

$$4 = 200 + C \quad \text{or} \quad C = -196$$

so

$$y(t) = 200 - 196 e^{-t/20}$$

Thus, at time $t = 10$ the amount of salt in the tank is

$$y(10) = 200 - 196 e^{-0.5} \approx 81.1 \text{ lb} \qquad \blacktriangleleft$$

□ **EXPONENTIAL GROWTH**

Many quantities increase or decrease with time in proportion to the amount of the quantity present. Some examples are human population, bacteria in a culture, drug concentration in the bloodstream, radioactivity, and the values of certain kinds of investments. We shall show how differential equations can be used to study the growth and decay of such quantities.

> **7.7.1** DEFINITION.   A quantity is said to have an ***exponential growth (decay) model*** if at each instant of time its rate of increase (decrease) is proportional to the amount of the quantity present.

Consider a quantity with an exponential growth or decay model and denote by $y(t)$ the amount of the quantity present at time t. We assume that $y(t) > 0$ for all t. Since the rate of change of $y(t)$ is proportional to the amount present, it follows that $y(t)$ satisfies

$$\frac{dy}{dt} = ky \tag{18}$$

where k is a constant of proportionality.

The constant k in (18) is called the ***growth constant*** if $k > 0$ and the ***decay constant*** if $k < 0$. If $k > 0$, then $dy/dt > 0$, so that y is increasing with time (a growth model). If $k < 0$, then $dy/dt < 0$, so that y is decreasing with time (a decay model).

REMARK.   The constant k in (18) is sometimes called the ***growth rate*** (a negative growth rate meaning that y is decreasing). Strictly speaking, this is not correct since the growth rate of y is not k, but rather dy/dt ($= ky$). Although it is rarely done, it would be more accurate to call k the ***relative growth rate***, since it follows from (18) that

$$k = \frac{dy/dt}{y}$$

However, it is so common to call k the growth rate that we shall use this standard terminology in this section. The growth rate k is usually expressed as a percentage; thus, a growth rate of 3% means $k = 0.03$ and a growth rate of -500% means $k = -5$.

If we are given a quantity y with an exponential growth or decay model, and if we know the amount y_0 present at some initial time $t = 0$, then we can find the amount present at any time t by solving the initial-value problem

$$\frac{dy}{dt} = ky, \quad y(0) = y_0 \tag{19}$$

This differential equation is both separable and linear and thus can be solved by either of the methods we have studied in this section. We shall treat it as a linear equation. Rewriting the differential equation as

$$\frac{dy}{dt} - ky = 0$$

and multiplying through by the integrating factor

$$\mu = e^{\int -k\,dt} = e^{-kt}$$

yields

$$\frac{d}{dt}(e^{-kt}y) = 0$$

After integrating,

$$e^{-kt}y = C \quad \text{or} \quad y = Ce^{kt}$$

From the initial condition, $y(0) = y_0$, it follows that $C = y_0$; thus, the solution of (19) is

$$y(t) = y_0 e^{kt} \tag{20}$$

Exponential models have proved useful in studies of population growth. Although populations (e.g., people, bacteria, and flowers) grow in discrete steps, we can apply the results of this section if we are willing to approximate the population graph by a continuous curve $y = y(t)$ (Figure 7.7.3).

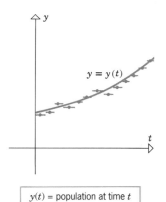

$y(t) =$ population at time t

Figure 7.7.3

Example 8 (*Population Growth*)   According to United Nations data, the world population at the beginning of 1990 was approximately 5.3 billion and growing at a rate of about 2% per year. Assuming an exponential growth model, estimate the world population at the beginning of the year 2015.

Solution.   Let

t = time elapsed from the beginning of 1990 (in years)

y = world population (in billions)

Since the beginning of 1990 corresponds to $t = 0$, it follows from the given data that

$$y_0 = y(0) = 5.3 \text{ (billion)}$$

Since the growth rate is 2% ($k = 0.02$), it follows from (20) that the world population at time t will be

$$y(t) = y_0 e^{kt} = 5.3 e^{0.02t} \qquad (21)$$

Since the beginning of the year 2015 corresponds to an elapsed time of $t = 25$ years ($2015 - 1990 = 25$), it follows from (21) that the world population by the year 2015 will be

$$y(25) = 5.3 e^{0.02(25)} = 5.3 e^{0.5} \approx 5.3(1.6487) = 8.7381 \text{ (billion)}$$

which is a population of approximately 8.7 billion.   ◀

☐ **DOUBLING AND HALVING TIME**

If a quantity y has an exponential growth model, then the time required for it to double in size is called the ***doubling time***. Similarly, if y has an exponential decay model, then the time required for it to reduce in value by half is called the ***halving time***. As it turns out, doubling and halving times depend only on the growth rate and not on the amount present initially. To see why, suppose y has an exponential growth model so that

$$y = y_0 e^{kt} \quad (k > 0)$$

At any fixed time t_1 let

$$y_1 = y_0 e^{kt_1} \qquad (22)$$

be the value of y, and let T denote the amount of time required for y to double in size. Thus, at time $t_1 + T$ the value of y will be $2y_1$, so

$$2y_1 = y_0 e^{k(t_1 + T)} = y_0 e^{kt_1} e^{kT}$$

or, from (22),

$$2y_1 = y_1 e^{kT}$$

Thus,

$$2 = e^{kT} \quad \text{and} \quad \ln 2 = kT$$

Therefore, the doubling time T is

$$T = \frac{1}{k} \ln 2 \qquad (23)$$

which does not depend on y_0 or t_1. We leave it as an exercise to show that the halving time for a quantity with an exponential decay model ($k < 0$) is

$$T = -\frac{1}{k} \ln 2 \qquad (24)$$

Example 9   It follows from (23) that at the current 2% annual growth rate, the doubling time for the world population is

$$T = \frac{1}{0.02} \ln 2 \approx \frac{1}{0.02}(0.6931) = 34.655$$

or approximately 35 years. Thus, with a continued 2% annual growth rate the population of 5.3 billion in 1990 will double to 10.6 billion by the year 2025 and will double again to 21.2 billion by 2060.   ◄

Radioactive elements continually undergo a process of disintegration called **radioactive decay**. It is a physical fact that at each instant of time the rate of decay is proportional to the amount of the element present. Consequently, the amount of any radioactive element has an exponential decay model. For radioactive elements, halving time is called **half-life**.

Example 10 (*Radioactive Decay*)   The radioactive element carbon-14 has a half-life of 5750 years. If 100 grams of this element are present initially, how much will be left after 1000 years?

Solution.   From (24) the decay constant is

$$k = -\frac{1}{T} \ln 2 \approx -\frac{1}{5750}(0.6931) \approx -0.00012$$

Thus, if we take $t = 0$ to be the present time, then $y_0 = y(0) = 100$, so that (20) implies that the amount of carbon-14 after 1000 years will be

$$y(1000) = 100e^{-0.00012(1000)} = 100e^{-0.12} \approx 100(0.88692) = 88.692$$

Thus, about 89 grams of carbon-14 will remain.   ◄

▶ Exercise Set 7.7   ⓒ 27, 29–32, 34–41, 46

In Exercises 1–6, solve the given separable differential equation. Where convenient, express the solution explicitly as a function of x.

1. $\dfrac{dy}{dx} = \dfrac{y}{x}$.

2. $\dfrac{dy}{dx} = \dfrac{x^3}{(1 + x^4)y}$.

3. $\sqrt{1 + x^2}\, y' + x(1 + y) = 0$.

4. $3 \tan y - \dfrac{dy}{dx} \sec x = 0$.

5. $e^{-y} \sin x - y' \cos^2 x = 0$.

6. $\dfrac{dy}{dx} = 1 - y + x^2 - yx^2$.

In Exercises 7–12, solve the given first-order linear differential equation. Where convenient, express the solution explicitly as a function of x.

7. $\dfrac{dy}{dx} + 3y = e^{-2x}$.

8. $\dfrac{dy}{dx} - \dfrac{5}{x} y = x$ $(x > 0)$.

9. $y' + y = \cos(e^x)$.

10. $2\dfrac{dy}{dx} + 4y = 1$.

11. $x^2 y' + 3xy + 2x^5 = 0$ $(x > 0)$.

12. $\dfrac{dy}{dx} + y - \dfrac{1}{1 + e^x} = 0$.

In Exercises 13–18, solve the initial-value problems.

13. $\dfrac{dy}{dx} - xy = x$, $y(0) = 3$.

14. $2y\dfrac{dy}{dx} = 3x^2(x^3 + 1)^{-1/2}$, $y(2) = 1$.

15. $\dfrac{dy}{dt} + y = 2$, $y(0) = 1$.

16. $y' - xe^y = 2e^y$, $y(0) = 0$.

17. $y^2 t \dfrac{dy}{dt} - t + 1 = 0$, $y(1) = 3$ $(t > 0)$.

18. $y' \cosh x + y \sinh x = \cosh^2 x$, $y(0) = \frac{1}{4}$.

In Exercises 19–22, find an equation of the curve in the xy-plane that passes through the given point and whose tangent at (x, y) has the given slope.

19. $(1, -1)$; slope $= \dfrac{y^2}{3\sqrt{x}}$.

20. $(1, 1)$; slope $= \dfrac{3x^2}{2y}$.

21. $(2, 0)$; slope $= xe^y$.

22. $(1, -1)$; slope $= 2y + 3$, $y > -\frac{3}{2}$.

The following discussion is needed for Exercises 23 and 24. Suppose that a tank containing a liquid is vented to the air at the top and has an outlet at the bottom through which the liquid can drain. It follows from **Torricelli's law** in physics that if the outlet is opened at time $t = 0$, then at each instant the depth of the liquid $h(t)$ and the area $A(h)$ of the liquid's surface are related by

$$A(h)\frac{dh}{dt} = -k\sqrt{h}$$

where k is a positive constant that depends on such factors as the viscosity of the liquid and the cross-sectional area of the outlet. Use this result in Exercises 23 and 24, assuming that h is in feet, $A(h)$ is in square feet, and t is in seconds. A calculator will be useful.

23. Suppose that the cylindrical tank in Figure 7.7.4 is filled to a depth of 4 ft at time $t = 0$ and that the constant in Torricelli's law is $k = 0.025$.

    (a)   Find $h(t)$.

    (b)   How many minutes will it take for the tank to drain completely?

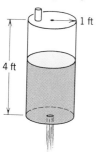

Figure 7.7.4

24. Follow the directions of Exercise 23 for the cylindrical tank in Figure 7.7.5, assuming that the tank is filled to a depth of 4 ft at time $t = 0$ and that the constant in Torricelli's law is $k = 0.025$.

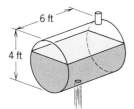

Figure 7.7.5

25. Suppose that a particle moving along the x-axis encounters a resisting force that results in an acceleration of $a = dv/dt = -0.04v^2$. Given that $x = 0$ cm and $v = 50$ cm/sec at time $t = 0$, find the velocity v and position x as a function of t for $t \geq 0$.

26. Suppose that a particle moving along the x-axis encounters a resisting force that results in an acceleration of $a = dv/dt = -0.02\sqrt{v}$. Given that $x = 0$ cm and $v = 9$ cm/sec at time $t = 0$, find the velocity v and position x as a function of t for $t \geq 0$.

27. A rocket, fired upward from rest at time $t = 0$, has an initial mass of m_0 (including its fuel). Assuming that the fuel is consumed at a constant rate k, the mass m of the rocket, while fuel is being burned, will be given by $m = m_0 - kt$. It can be shown that if air resistance is neglected and the fuel gases are expelled at a constant speed c relative to the rocket, then the velocity v of the rocket will satisfy the equation

$$m\frac{dv}{dt} = ck - mg$$

where g is the acceleration due to gravity.

    (a)   Find $v(t)$ keeping in mind that the mass m is a function of t.

    (b)   Suppose that the fuel accounts for 80% of the initial mass of the rocket and that all of the fuel is consumed in 100 sec. Find the velocity of the rocket in meters/second at the instant the fuel is exhausted. (Use $g = 9.8$ m/sec^2 and $c = 2500$ m/sec.)

28. An object of mass m is dropped from rest at time $t = 0$. If we assume that the only forces acting on the object are the constant force due to gravity and a retarding force of air resistance, which is proportional to the velocity $v(t)$ of the object, then the velocity will satisfy the equation

$$m\frac{dv}{dt} = mg - kv$$

where k is a positive constant and g is the acceleration due to gravity.

    (a)   Find $v(t)$.

    (b)   Use the result in part (a) to find $\lim_{t \to +\infty} v(t)$.

    (c)   Find the distance $x(t)$ that the object has fallen at time t given that $x = 0$ when $t = 0$.

29. An object of constant mass is projected upward from the surface of the earth. Neglecting air resistance, the gravitational force exerted by the earth on the object results in an acceleration of

$$a = dv/dt = -gR^2/x^2$$

where x is the distance of the object from the center of the earth, R is the radius of the earth, and g is the acceleration due to gravity at the surface of the earth. From the chain rule $dv/dt = v\,dv/dx$, so v and x must satisfy the equation

$$v\,dv/dx = -gR^2/x^2$$

    (a)   Find v^2 in terms of x given that $v = v_0$ when $x = R$.

    (b)   Use the result in part (a) to show that the velocity cannot reach zero if $v_0 \geq \sqrt{2gR}$.

    (c)   The minimum value of v_0 that is required for the object not to fall back to the earth is called the **escape velocity**. Assuming that $g = 32$ ft/sec^2 and $R = 3960$ mi, show that the escape velocity is approximately 6.9 mi/sec.

30. A bullet of mass m, fired straight up with an initial velocity of v_0, is slowed by the force of gravity and a drag force of air resistance kv^2, where g is the constant acceleration due to gravity and k is a positive constant. As the bullet moves upward, its velocity v satisfies the equation

$$m \frac{dv}{dt} = -(kv^2 + mg)$$

(a) Show that if x is the position of the bullet at time t (Figure 7.7.6), then

$$mv \frac{dv}{dx} = -(kv^2 + mg)$$

Figure 7.7.6

(b) Express x in terms of v given that $x = 0$ when $v = v_0$.

(c) Assuming that $v_0 = 988$ m/sec, $g = 9.8$ m/sec^2, $m = 3.56 \times 10^{-3}$ kg, and $k = 7.3 \times 10^{-6}$, use the result in part (b) to find out how high the bullet rises. [*Hint:* Find the velocity of the bullet at its highest point.]

31. At time $t = 0$, a tank contains 25 lb of salt dissolved in 50 gal of water. Then brine containing 4 lb of salt per gallon of water is allowed to enter the tank at a rate of 2 gal/min and the mixed solution is drained from the tank at the same rate.

(a) How much salt is in the tank at an arbitrary time t?

(b) How much salt is in the tank after 25 min?

32. A tank initially contains 200 gal of pure water. Then at time $t = 0$ brine containing 5 lb of salt per gallon of water is allowed to enter the tank at a rate of 10 gal/min and the mixed solution is drained from the tank at the same rate.

(a) How much salt is in the tank at an arbitrary time t?

(b) How much salt is in the tank after 30 min?

33. A tank with a 1000-gal capacity initially contains 500 gal of brine containing 50 lb of salt. At time $t = 0$, pure water is added at a rate of 20 gal/min and the mixed solution is drained off at a rate of 10 gal/min. How much salt is in the tank when it reaches the point of overflowing?

34. The number of bacteria in a certain culture grows exponentially at a rate of 1% per hour. Assuming that 10,000 bacteria are present initially, find

(a) the number of bacteria present at any time t

(b) the number of bacteria present after 5 hr

(c) the time required for the number of bacteria to reach 45,000.

35. Polonium-210 is a radioactive element with a half-life of 140 days. Assume that a sample has a mass of 10 mg initially.

(a) Find a formula for the amount that will remain after t days.

(b) How much will remain after 10 weeks?

36. In a certain chemical reaction a substance decomposes at a rate proportional to the amount present. Tests show that under appropriate conditions 15,000 grams will reduce to 5000 grams in 10 hours.

(a) Find a formula for the amount that will remain from a 15,000-gram sample after t hours.

(b) How long will it take for 50% of an initial sample of y_0 grams to decompose?

37. One hundred fruit flies are placed in a breeding container that can support a population of at most 5000 flies. If the population grows exponentially at a rate of 2% per day, how long will it take for the container to reach capacity?

38. In 1960 the American scientist W. F. Libby won the Nobel prize for his discovery of carbon dating, a method for determining the age of certain fossils. Carbon dating is based on the fact that nitrogen is converted to radioactive carbon-14 by cosmic radiation in the upper atmosphere. This radioactive carbon is absorbed by plant and animal tissue through the life processes while the plant or animal lives. However, when the plant or animal dies the absorption process stops and the amount of carbon-14 decreases through radioactive decay. Suppose that tests on a fossil show that 70% of its carbon-14 has decayed. Estimate the age of the fossil, assuming a half-life of 5750 years for carbon-14.

39. Forty percent of a radioactive substance decays in 5 years. Find the half-life of the substance.

40. Assume that if the temperature is constant, then the atmospheric pressure p varies with the altitude h (above sea level) in such a way that $dp/dh = kp$, where k is a constant.

(a) Find a formula for p in terms of k, h, and the atmospheric pressure p_0 at sea level.

(b) Given that p measures 15 lb/in^2 at sea level and 12 lb/in^2 at 5000 ft above sea level, find the pressure at 10,000 ft (assuming temperature is constant).

41. The town of Grayrock had a population of 10,000 in 1980 and 12,000 in 1990.

(a) Assuming an exponential growth model, estimate the population in 2000.

(b) What is the doubling time for the town's population?

42. Prove: If a quantity A has an exponential growth or decay model and A has values A_1 and A_2 at times t_1 and t_2, respectively, then the growth rate k is

$$k = \frac{1}{t_1 - t_2} \ln\left(\frac{A_1}{A_2}\right)$$

43. Newton's law of cooling states that the rate at which an object cools is proportional to the difference in temperature between the object and the surrounding medium. Show that if C is the constant temperature of a surrounding medium, then the temperature $T(t)$ at time t of a cooling object is given by

$$T(t) = (T_0 - C)e^{kt} + C$$

where T_0 is the temperature of the object at $t = 0$ and k is a negative constant.

44. A liquid with an initial temperature of $200°$ is enclosed in a metal container that is held at a constant temperature of $80°$. If the liquid cools to $120°$ in 30 min, what will the temperature be after 1 hr? (Use the result of Exercise 43.)

45. Suppose that P dollars is invested at an annual interest rate of $r \times 100\%$. If the accumulated interest is credited to the account at the end of the year, then the interest is said to be *compounded annually*; if it is credited at the end of each six-month period, then it is said to be *compounded semiannually*; and if it is credited at the end of each three-month period, then it is said to be *compounded quarterly*. The more frequently the interest is compounded, the better it is for the investor since more of the interest is itself earning interest.

   (a) Show that if interest is compounded n times a year at equally spaced intervals, then the value A of the investment after t years is

$$A = P\left(1 + \frac{r}{n}\right)^{nt}$$

   (b) One can imagine interest to be compounded each day, each hour, each minute, and so forth. Carried to the limit one can conceive of interest compounded at each instant of time; this is called *continuous compounding*. Thus, from part (a), the value A of P dollars after t years when invested at an annual rate of $r \times 100\%$, compounded continuously, is

$$A = \lim_{n \to +\infty} P\left(1 + \frac{r}{n}\right)^{nt}$$

   Use the fact that $\lim_{x \to 0} (1 + x)^{1/x} = e$ to prove that $A = Pe^{rt}$.

   (c) Use the result in part (b) to show that money invested at continuous compound interest increases at a rate proportional to the amount present.

46. (a) If $1000 is invested at 8% per year compounded continuously (Exercise 45), what will the investment be worth after 5 years?

   (b) If it is desired that an investment at 8% per year compounded continuously should have a value of $10,000 after 10 years, how much should be invested now?

   (c) How long does it take for an investment at 8% per year compounded continuously to double in value?

47. Derive Formula (24) for halving time.

48. Let a quantity have an exponential growth model with growth rate k. How long does it take for the quantity to triple in size?

◆ TECHNOLOGY EXERCISES   Chapter 7

Most of these exercises require access to a graphing calculator or a computer algebra system (CAS) such as *Mathematica, Maple,* or *Derive*. When you are asked to *find* an answer or to *solve* an equation, you may choose to find an exact result or a numerical approximation, depending on the particular technology you are using and on your own imagination. The form of your answers may differ from those of other students or from those in the answer section of the text, depending on how you solve the problems and the accuracy you use in your numerical approximations. Those exercises that are more appropriate for a CAS than a graphing calculator are labeled with the icon ◆.

◆ **1. Area:** Find the value of k $(k > 1)$ such that the region enclosed by $y = \ln x$, $y = 0$, and $x = k$ has an area of 3 square units.

2. Area: Find the value of k $(k > 0)$ such that the area of the region between $y = xe^{-\sqrt{x}}$, $y = 0$, and the vertical lines $x = k$ and $x = k + 1$ is maximum. Find this maximum area. [*Hint:* See Exercise 28, Section 5.9.]

3. Volume: Consider the region enclosed by $y = e^{kx}$, $y = 0$, $x = 0$, and $x = 2$. Find the value of k so that the solid generated by revolving the region about the y-axis has a volume of 1.5 cubic units.

4. Arc length

(a) Find the value of k ($k > 1$) so that the curve $y = \ln x$ has a length of 5 units for $1 \le x \le k$.

(b) For the value of k obtained in part (a), what is the length of the curve $y = e^x$ for $0 \le x \le \ln k$? Solve this part by inspection and justify your conclusion.

5. Rectilinear motion: Suppose that a particle moves along a line so that its velocity v at time t is given by

$$v(t) = 1 - 4e^{-t}, \quad t \ge 0$$

where t is in seconds and v is in centimeters per second (cm/sec). Find the value of t such that the total distance traveled by the particle over the interval $[0, t]$ is

(a) 1 cm   (b) 4 cm.

6. Solving equations

(a) Show that for $x > 0$ and $k \ne 0$ the equations

$$x^k = e^x \quad \text{and} \quad \frac{\ln x}{x} = \frac{1}{k}$$

have the same solutions.

(b) Use the graph of $y = (\ln x)/x$ (see Example 4, Section 7.3) to determine the values of k for which the equation $x^k = e^x$ has two distinct positive solutions.

(c) Find the positive solutions of $x^8 = e^x$.

> **Points of tangency:** In Exercises 7–9, find the value of k so that the graphs of $y = f(x)$ and $y = \ln x$ share a common tangent line at their point of intersection. Confirm your result by graphing both $y = f(x)$ and $y = \ln x$ in the same coordinate system.

7. $f(x) = \sqrt{x} + k$.   **8.** $f(x) = k\sqrt{x}$.

9. $f(x) = x^k$.

10. Points of tangency: Find the value of b so that the line $y = x$ is tangent to the graph of $y = \log_b x$. Confirm your result by graphing both $y = x$ and $y = \log_b x$ in the same coordinate system.

11. Relative extrema: Find the relative extrema of $f(x) = x^{\sin x}$ for $0 < x < 6$.

12. Relative extrema: Find the relative extrema of $f(x) = (\sin x)^x$ for $0 < x < 2$.

13. Prime number theory

(a) In number theory, $\pi(n)$ denotes the number of prime numbers less than or equal to the positive integer n. It can be shown that $\pi(n) \approx n/(\ln n)$ for large n. Use this result to approximate the number of primes that are less than or equal to 100,000. [*Note:* For comparison, the exact number of such primes is 9592.]

(b) For large values of n the integral

$$\int_2^n \frac{1}{\ln t} \, dt$$

gives a better approximation of $\pi(n)$ than does $n/(\ln n)$. Use this integral to approximate the number of primes that are less than or equal to 100,000.

14. The error function

$$\text{erf}(x) = \frac{2}{\sqrt{\pi}} \int_0^x e^{-t^2} \, dt$$

Graph $y = \text{erf}(x)$ over the interval $[-3, 3]$, and find the value of x for which $\text{erf}(x) = 0.5$. Make a conjecture about $\lim\limits_{x \to +\infty} \text{erf}(x)$ and $\lim\limits_{x \to -\infty} \text{erf}(x)$.

15. Rotation of a hollow tube: Suppose that a hollow tube rotates with a constant angular velocity of ω radians/sec about a horizontal axis at one end of the tube, as shown in the accompanying figure. Assume that an object is free to slide without friction in the tube while the tube is rotating. Let r be the distance from the object to the pivot point at time $t \ge 0$, and assume that the object is at rest and $r = 0$ when $t = 0$. It can be shown that if the tube is horizontal at time $t = 0$, then

$$r = \frac{g}{2\omega^2} [\sinh(\omega t) - \sin(\omega t)]$$

during the period that the object is in the tube. Assume that t is in seconds and r is in meters, and use $g = 9.8$ m/sec^2 and $\omega = 2$ rad/sec.

(a) Graph r versus t for $0 \le t \le 1$.

(b) Assuming that the tube has a length of 1 m, how long does it take for the object to reach the end of the tube?

(c) Use the result of part (b) to find dr/dt at the instant that the object reaches the end of the tube.

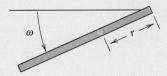

16. Hanging cable: A cable is suspended between two poles as shown in Figure 7.6.2. The equation of the curve formed by the cable is $y = a \cosh(x/a)$, where a is a positive constant. If the points of support have the same elevation and the x-coordinates of these points are $x = -b$ and $x = b$ $(b > 0)$, then the length L of the cable and the sag S (the vertical distance between the highest and lowest points on the cable) are given by

$$L = 2a \sinh(b/a) \quad \text{and} \quad S = a \cosh(b/a) - a$$

Assume that the poles are 100 m apart and the sag in the cable is 20 m.

(a) Find the length of the cable.

(b) Find to the nearest degree the angle at the supports between the tangent line to the cable and a horizontal line.

17. Inverse of a function

(a) Find the smallest value of k so that the function $f(x) = x^2 + 3\sin x$ is one-to-one for $x \geq k$.

(b) For the value of k found in part (a), find the domain and range of f^{-1} and the value of $f^{-1}(5)$ for the function $f(x) = x^2 + 3\sin x$, $x \geq k$.

18. Inverse of a function

(a) Find the largest value of k so that the function $f(x) = 1 + 2x + x^3 - x^4$ is one-to-one for $x \leq k$.

(b) For the value of k found in part (a), find the domain and range of f^{-1} and the value of $f^{-1}(-1)$ for the function $f(x) = 1 + 2x + x^3 - x^4$, $x \leq k$.

Euler's Method: The differential equation

$$\frac{dy}{dx} = e^{-xy}$$

is neither separable nor linear, and hence cannot be solved by the methods discussed in this section. The right side of this equation is an example of a function of two variables, x and y. A general function of this type is commonly denoted by $f(x, y)$. Such functions will be discussed in detail later, but for now we only need the notation. There are various numerical methods for approximating the solution of an initial-value problem of the form

$$\frac{dy}{dx} = f(x, y), \quad y(x_0) = y_0$$

many of which are based on the following idea: From the initial conditions, the point (x_0, y_0) is known to lie on the solution curve $y = y(x)$. If we denote the derivative dy/dx by y', and if we let Δx be a small increment in x, then it follows from Formula (11) of Section 3.7 that the y-coordinate of the point on the solution curve corresponding to $x = x_0 + \Delta x$ can be approximated by

$$y(x_0 + \Delta x) \approx y_0 + y'(x_0)\,\Delta x$$

or, since $y'(x) = f(x, y)$,

$$y(x_0 + \Delta x) \approx y_0 + f(x_0, y_0)\,\Delta x$$

This suggests the following algorithm, known as **Euler's Method**, for approximating points on the solution curve of the initial-value problem:

Step 1. Choose a positive number h to serve as an increment or step size along the x-axis, and let

$$x_1 = x_0 + h, \quad x_2 = x_1 + h, \quad x_3 = x_2 + h, \ldots$$

Step 2. Compute successively

$$y_1 = y_0 + f(x_0, y_0)h$$
$$y_2 = y_1 + f(x_1, y_1)h$$
$$y_3 = y_2 + f(x_2, y_2)h$$
$$\vdots$$
$$y_{n+1} = y_n + f(x_n, y_n)h$$

For $n = 1, 2, 3, \ldots$ the point (x_n, y_n) approximates the point on the solution curve with x-coordinate x_n.

19. (a) Use Euler's Method with $h = 0.05$ to approximate $y(0.3)$ for the initial-value problem $dy/dx = y \cos x$, $y(0) = 1$.

(b) Find the exact solution of the initial-value problem $dy/dx = y \cos x$, $y(0) = 1$, and use it to find the percentage error in the approximation of $y(0.3)$ obtained in part (a). [*Note:* The percentage error is given by $100(Q - Q^*)/Q$, where Q^* is the approximate value of a quantity and Q is its true value.]

20. Use Euler's Method with $h = 0.01$ to approximate $y(0.12)$ for the initial-value problem $dy/dx = e^{-xy}$, $y(0) = 1$.

Isaac Barrow (1630–1677)

8 INVERSE TRIGONOMETRIC AND HYPERBOLIC FUNCTIONS

■ 8.1 INVERSE TRIGONOMETRIC FUNCTIONS

Because the six basic trigonometric functions are periodic, each of their values is repeated infinitely many times. As a result, none of these functions are one-to-one and consequently none of them have inverses. In this section we shall impose restrictions on the domains of the trigonometric functions that yield one-to-one functions. We shall develop the algebraic properties of these functions in this section and in the next section we shall obtain their derivatives and discuss some of their applications to calculus.

☐ **INVERSE SINE**

Figure 8.1.1 shows the graph of $y = \sin x$ with the segment of the graph over the interval $[-\pi/2, \pi/2]$ emphasized in solid color. This emphasized segment is the graph of a function f that has the same values as $\sin x$, but whose domain is the interval $[-\pi/2, \pi/2]$. Since no horizontal line cuts the graph of f more than once, this restricted form of the sine function has an inverse. This leads us to the following definition.

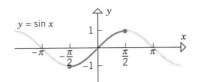

Figure 8.1.1

8.1.1 DEFINITION. The *inverse sine* function, denoted by $\sin^{-1}$, is defined to be the inverse of the function $\sin x$, $-\pi/2 \le x \le \pi/2$.

REMARK.   The symbol $\sin^{-1} x$ is *never* used to denote $1/\sin x$. If desired, $1/\sin x$ can be rewritten as $(\sin x)^{-1}$ or $\csc x$. The function $\sin^{-1} x$ is also called the *arcsine* function, in which case it is denoted by $\arcsin x$. Although we shall not use this terminology or notation, it is commonly used on calculators and in mathematical computer software.

Since $\sin x$, $-\pi/2 \le x \le \pi/2$, has domain $[-\pi/2, \pi/2]$ and range $[-1, 1]$, it follows that

$$\text{domain of } \sin^{-1} x = [-1, 1]$$

$$\text{range of } \sin^{-1} x = [-\pi/2, \pi/2]$$

and

$$\sin^{-1}(\sin x) = x \quad \text{if} \quad -\frac{\pi}{2} \le x \le \frac{\pi}{2} \tag{1a}$$

$$\sin(\sin^{-1} x) = x \quad \text{if} \quad -1 \le x \le 1 \tag{1b}$$

Moreover, the graph of $y = \sin^{-1} x$ can be obtained by reflecting the graph of $y = \sin x$, $-\pi/2 \le x \le \pi/2$ (which is the solid color curve in Figure 8.1.1), about the line $y = x$ (Figure 8.1.2). Observe that $\sin^{-1} x$ is a continuous function since it is the inverse of a continuous function (Theorem 7.4.6).

Let us consider the equation

$$\sin y = x$$

where $-1 \le x \le 1$ and $-\pi/2 \le y \le \pi/2$. If we take $\sin^{-1}$ of both sides of this equation and use Formula (1) with y replacing x, we obtain

$$y = \sin^{-1} x$$

Thus, we have the following result.

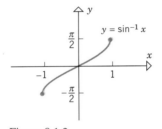

Figure 8.1.2

8.1.2 THEOREM.

$$y = \sin^{-1} x \quad \text{is equivalent to} \quad \sin y = x \quad \text{if} \quad \begin{cases} -1 \le x \le 1 \\ -\pi/2 \le y \le \pi/2 \end{cases}$$

☐ **GEOMETRIC INTERPRETATION OF $\sin^{-1} x$**

If we think of y as an angle in radian measure, then the algebraic requirement in Theorem 8.1.2 that y lie in the interval $[-\pi/2, \pi/2]$ imposes the geometric requirement that $\sin^{-1} x$ be an angle that terminates in the first or fourth quadrant or on an axis adjacent to those quadrants. Thus, from a geometric viewpoint, $\sin^{-1} x$ is an angle whose sine is x and which terminates in the first or fourth quadrant (or on an adjacent axis).

Example 1   Find

(a) $\sin^{-1}\left(\frac{1}{2}\right)$ (b) $\sin^{-1}(-1/\sqrt{2})$ (c) $\sin^{-1}(-1)$

Solution (a).   Let $y = \sin^{-1}\left(\frac{1}{2}\right)$. From Theorem 8.1.2 this equation is equivalent to

$$\sin y = \tfrac{1}{2}, \quad -\pi/2 \le y \le \pi/2$$

Viewing y as an angle in radian measure, it follows that y terminates in the first quadrant since $\sin y > 0$. Thus, we are looking for an angle terminating in the first quadrant whose sine is $1/2$. The desired angle is $y = \pi/6$, so $\sin^{-1}\left(\frac{1}{2}\right) = \pi/6$.

Solution (*b*). Let $y = \sin^{-1}(-1/\sqrt{2})$. This is equivalent to

$$\sin y = -1/\sqrt{2}, \quad -\pi/2 \le y \le \pi/2$$

Viewing y as an angle in radian measure, it follows that y terminates in the fourth quadrant since $\sin y < 0$. Thus, we are looking for an angle terminating in the fourth quadrant whose sine is $-1/\sqrt{2}$. The desired angle is $y = -\pi/4$, so $\sin^{-1}(-1/\sqrt{2}) = -\pi/4$.

Solution (*c*). Let $y = \sin^{-1}(-1)$. This is equivalent to

$$\sin y = -1$$

from which it follows that $y = -\pi/2$. Thus, $\sin^{-1}(-1) = -\pi/2$. ◄

☐ **INVERSE COSINE, TANGENT, AND SECANT**

As with the definition of $\sin^{-1} x$, inverses of the remaining trigonometric functions can be defined by placing appropriate restrictions on the domains of those functions. Since the inverse cotangent and cosecant are of lesser importance, we shall focus for the moment on the definitions of the inverse cosine, tangent, and secant. The inverses of these functions are defined to be the inverses of the functions whose graphs are shown in solid blue in Figure 8.1.3.

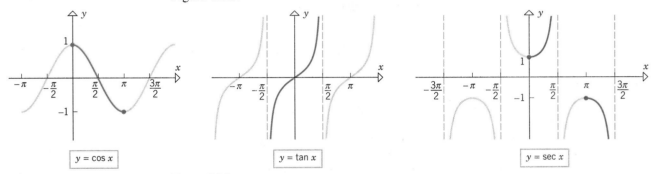

Figure 8.1.3

Observe that in all three cases the restricted functions pass the horizontal line test, and hence have inverses.

8.1.3 DEFINITION.

(a) The *inverse cosine* function, denoted by $\cos^{-1}$, is defined to be the inverse of the function $\cos x$, $0 \le x \le \pi$.

(b) The *inverse tangent* function, denoted by $\tan^{-1}$, is defined to be the inverse of the function $\tan x$, $-\pi/2 < x < \pi/2$.

(c) The *inverse secant* function, denoted by $\sec^{-1}$, is defined to be the inverse of the function $\sec x$, $0 \le x < \pi/2$ or $\pi \le x < 3\pi/2$.

The graphs of $\cos^{-1} x$, $\tan^{-1} x$, and $\sec^{-1} x$, which are the reflections about the line $y = x$ of the graphs in solid blue in Figure 8.1.3, are shown in Figure 8.1.4.

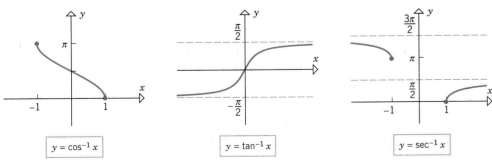

Figure 8.1.4

REMARK.   There is no universal agreement among mathematicians about the range of $\sec^{-1} x$. For example, some writers define $\sec^{-1} x$ by restricting the domain of $\sec x$ so that $0 \le x < \pi/2$ or $\pi/2 < x \le \pi$. Although there are some advantages to this definition, our definition will produce simpler derivative and integration formulas in the next section.

The results in Table 8.1.1 are immediate consequences of Definition 8.1.3. The reader should confirm that domains and ranges are in agreement with Figure 8.1.4.

Table 8.1.1

FUNCTION	DOMAIN	RANGE	BASIC RELATIONSHIPS
$\cos^{-1} x$	$[-1, 1]$	$[0, \pi]$	$\cos^{-1}(\cos x) = x$ if $0 \le x \le \pi$ $\cos(\cos^{-1} x) = x$ if $-1 \le x \le 1$
$\tan^{-1} x$	$(-\infty, +\infty)$	$(-\pi/2, \pi/2)$	$\tan^{-1}(\tan x) = x$ if $-\pi/2 < x < \pi/2$ $\tan(\tan^{-1} x) = x$ if $-\infty < x < +\infty$
$\sec^{-1} x$	$(-\infty, -1] \cup [1, +\infty)$	$[0, \pi/2) \cup [\pi, 3\pi/2)$	$\sec^{-1}(\sec x) = x$ if $0 \le x < \pi/2$ or $\pi \le x < 3\pi/2$ $\sec(\sec^{-1} x) = x$ if $x \le -1$ or $x \ge 1$

The following theorem is analogous to Theorem 8.1.2.

8.1.4   THEOREM.

$y = \cos^{-1} x$ *is equivalent to* $\cos y = x$ *if* $\begin{cases} -1 \le x \le 1 \\ 0 \le y \le \pi \end{cases}$

$y = \tan^{-1} x$ *is equivalent to* $\tan y = x$ *if* $\begin{cases} -\infty < x < +\infty \\ -\pi/2 < y < \pi/2 \end{cases}$

$y = \sec^{-1} x$ *is equivalent to* $\sec y = x$ *if* $\begin{cases} x \ge 1 \\ 0 \le y < \pi/2 \end{cases}$ or $\begin{cases} x \le -1 \\ \pi \le y < 3\pi/2 \end{cases}$

If we think of y as an angle in radian measure, then the algebraic requirement in the first part of Theorem 8.1.4 that y lie in the interval $[0, \pi]$ imposes the geometric requirement that y be an angle that terminates in the first or second quadrant or on an axis adjacent to those quadrants. Thus, from a geometric viewpoint, $\cos^{-1} x$ is an angle whose cosine is x and which terminates in the first or second quadrant or on an adjacent axis. Similarly, $\tan^{-1} x$ is an angle whose tangent is x and which terminates in the first or fourth quadrant or on the positive x-axis, and $\sec^{-1} x$ is an angle whose secant is x and which terminates in the first or third quadrant or on the x-axis (verify).

Example 2   Find $\cos^{-1}(-\sqrt{3}/2)$.

Solution.   Let $y = \cos^{-1}(-\sqrt{3}/2)$. This is equivalent to

$$\cos y = -\sqrt{3}/2$$

Viewing y as an angle in radian measure, it follows that y terminates in the second quadrant since $\cos y < 0$. Thus, we are looking for an angle terminating in the second quadrant whose cosine is $-\sqrt{3}/2$. The desired angle is $y = 5\pi/6$, so

$$\cos^{-1}(-\sqrt{3}/2) = 5\pi/6 \qquad \blacktriangleleft$$

As evidenced by Figure 8.1.4, the graph of $y = \tan^{-1} x$ has two horizontal asymptotes: $y = \pi/2$ as x approaches $+\infty$ and $y = -\pi/2$ as x approaches $-\infty$. Expressing these observations as limits we obtain

$$\lim_{x \to -\infty} \tan^{-1} x = -\pi/2 \qquad \text{and} \qquad \lim_{x \to +\infty} \tan^{-1} x = \pi/2 \qquad (2\text{–}3)$$

☐ **INVERSE COTANGENT AND COSECANT**

The *inverse cotangent* and *inverse cosecant* functions are of lesser importance, so we shall just summarize their properties in Table 8.1.2.

Table 8.1.2

FUNCTION	DOMAIN (x VALUE)	RANGE (y VALUE)	EQUIVALENT STATEMENTS
$\cot^{-1} x$	$(-\infty, +\infty)$	$(0, \pi)$	$y = \cot^{-1} x$ and $x = \cot y$
$\csc^{-1} x$	$(-\infty, -1] \cup [1, +\infty)$	$(-\pi, -\pi/2] \cup (0, \pi/2]$	$y = \csc^{-1} x$ and $x = \csc y$

We leave it for the reader to show that the graphs of $y = \cot^{-1} x$ and $y = \csc^{-1} x$ are as shown in Figures 8.1.5 and 8.1.6.

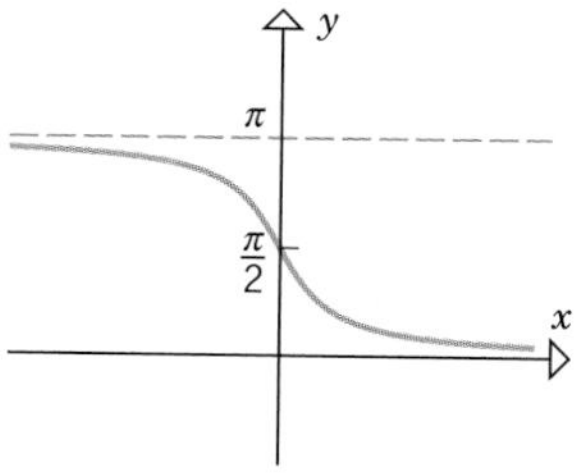

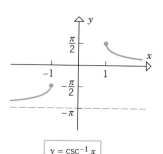

$y = \cot^{-1} x$

Figure 8.1.5

$y = \csc^{-1} x$

Figure 8.1.6

☐ **IDENTITIES INVOLVING INVERSE TRIGONOMETRIC FUNCTIONS**

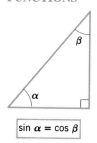

$\sin \alpha = \cos \beta$

Figure 8.1.7

In the definitions of the six inverse trigonometric functions, the restrictions on the domains are designed not only to produce one-to-one functions, but also to ensure that certain "natural" identities hold for the inverse trigonometric functions. For example, if α and β are acute complementary angles, then from basic trigonometry, $\sin \alpha$ and $\cos \beta$ are equal (Figure 8.1.7). Let us write $x = \sin \alpha = \cos \beta$ so that

$$\alpha = \sin^{-1} x \quad \text{and} \quad \beta = \cos^{-1} x$$

Since $\alpha + \beta = \pi/2$, we obtain the identity

$$\sin^{-1} x + \cos^{-1} x = \frac{\pi}{2} \qquad (4)$$

[Because α and β were assumed to be nonnegative acute angles, this derivation is only valid for $0 \le x \le 1$; for a derivation that is valid for all x in $[-1, 0]$ see Exercise 25(a).] Similarly, we can obtain the identities

$$\tan^{-1} x + \cot^{-1} x = \frac{\pi}{2} \qquad \text{and} \qquad \sec^{-1} x + \csc^{-1} x = \frac{\pi}{2} \qquad (5\text{–}6)$$

☐ **SIMPLIFYING EXPRESSIONS INVOLVING INVERSE TRIGONOMETRIC FUNCTIONS**

In the next section we will encounter functions that are compositions of trigonometric and inverse trigonometric functions. The following examples illustrate techniques for simplifying such functions.

Example 3 Simplify the function $\cos (\sin^{-1} x)$.

Solution.   The idea is to express cosine in terms of sine in order to take advantage of the simplification $\sin(\sin^{-1} x) = x$. Thus, we start with the identity

$$\cos^2 y = 1 - \sin^2 y$$

and substitute $y = \sin^{-1} x$ to obtain

$$\cos^2(\sin^{-1} x) = 1 - \sin^2(\sin^{-1} x) = 1 - x^2$$

and on taking square roots

$$\left|\cos(\sin^{-1} x)\right| = \sqrt{1 - x^2}$$

Since $-\pi/2 \le \sin^{-1} x \le \pi/2$, it follows that $\cos(\sin^{-1} x)$ is nonnegative; thus, we can drop the absolute value sign and write

$$\cos(\sin^{-1} x) = \sqrt{1 - x^2} \qquad (7)$$

Alternative Solution.   In the case where $0 \le x \le 1$, an alternative geometric solution is possible. We let

$$y = \sin^{-1} x$$

and construct a right triangle with an angle of y as shown in Figure 8.1.8. It follows from the definition of y that $\sin y = x$, so we may assume that the side opposite the angle y has length x and the hypotenuse has length 1 since these values yield

$$\sin y = \frac{\text{opposite side}}{\text{hypotenuse}} = \frac{x}{1} = x$$

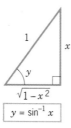

$y = \sin^{-1} x$

Figure 8.1.8

By the Theorem of Pythagoras, the side that is adjacent to the angle y has length $\sqrt{1 - x^2}$, so

$$\cos y = \frac{\text{adjacent side}}{\text{hypotenuse}} = \frac{\sqrt{1 - x^2}}{1} = \sqrt{1 - x^2}$$

or

$$\cos(\sin^{-1} x) = \sqrt{1 - x^2}$$

which agrees with (7).   ◄

Example 4   Use the triangle method to simplify the function $\sec(\tan^{-1} x)$.

Solution.   In the case where $x \ge 0$, we can let

$$y = \tan^{-1} x$$

and construct a right triangle with an angle of y as shown in Figure 8.1.9. It follows from the definition of y that $\tan y = x$, so we can assume that the side opposite the angle y has length x and the side adjacent to y has length 1 since these values yield

$$\tan y = \frac{\text{opposite side}}{\text{adjacent side}} = \frac{x}{1} = x$$

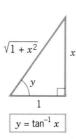

$y = \tan^{-1} x$

Figure 8.1.9

By the Theorem of Pythagoras the hypotenuse has length $\sqrt{1 + x^2}$, so it follows from the triangle that

$$\sec(\tan^{-1} x) = \sec y = \frac{\text{hypotenuse}}{\text{adjacent side}} = \frac{\sqrt{1 + x^2}}{1} = \sqrt{1 + x^2}$$

We note that although this relationship was derived subject to the restriction $x \ge 0$, it is also valid for $x < 0$ [Exercise 25(b)].   ◄

▶ Exercise Set 8.1   Ⓒ *27, 31–41*

---

1. Find the exact value of
  (a) $\sin^{-1}(-1)$
  (b) $\cos^{-1}(-1)$
  (c) $\tan^{-1}(-1)$
  (d) $\cot^{-1}(1)$
  (e) $\sec^{-1}(1)$
  (f) $\csc^{-1}(1)$.

2. Find the exact value of
  (a) $\sin^{-1}(\tfrac{1}{2}\sqrt{3})$
  (b) $\cos^{-1}(\tfrac{1}{2})$
  (c) $\tan^{-1}(1)$
  (d) $\cot^{-1}(-1)$
  (e) $\sec^{-1}(-2)$
  (f) $\csc^{-1}(-2)$.

3. Given that $\theta = \sin^{-1}(-\tfrac{1}{2}\sqrt{3})$, find the exact values of $\cos\theta$, $\tan\theta$, $\cot\theta$, $\sec\theta$, and $\csc\theta$.

4. Given that $\theta = \cos^{-1}(\tfrac{1}{2})$, find the exact values of $\sin\theta$, $\tan\theta$, $\cot\theta$, $\sec\theta$, and $\csc\theta$.

5. Given that $\theta = \tan^{-1}(\tfrac{4}{3})$, find the exact values of $\sin\theta$, $\cos\theta$, $\cot\theta$, $\sec\theta$, and $\csc\theta$.

6. Make a table that lists the six inverse trigonometric functions together with their domains and ranges.

7. Find the exact value of
  (a) $\sin^{-1}(\sin \pi/7)$
  (b) $\sin^{-1}(\sin \pi)$
  (c) $\sin^{-1}(\sin 5\pi/7)$
  (d) $\sin^{-1}(\sin 630)$.

8. Find the exact value of
  (a) $\cos^{-1}(\cos \pi/7)$
  (b) $\cos^{-1}(\cos \pi)$
  (c) $\cos^{-1}(\cos 12\pi/7)$
  (d) $\cos^{-1}(\cos 200)$.

9. For which values of x is it true that
  (a) $\cos^{-1}(\cos x) = x$
  (b) $\cos(\cos^{-1} x) = x$
  (c) $\tan^{-1}(\tan x) = x$
  (d) $\tan(\tan^{-1} x) = x$
  (e) $\csc^{-1}(\csc x) = x$
  (f) $\csc(\csc^{-1} x) = x$?

In Exercises 10–15, find the exact value of the given quantity.

---

10. $\sec[\sin^{-1}(-\tfrac{3}{4})]$.

11. $\sin[2\cos^{-1}(\tfrac{3}{5})]$.

12. $\tan^{-1}[\sin(-\pi/2)]$.

13. $\sin^{-1}[\cot(\pi/4)]$.

14. $\sin[\sin^{-1}(\tfrac{2}{3}) + \cos^{-1}(\tfrac{1}{3})]$.

15. $\tan[2\sec^{-1}(\tfrac{3}{2})]$.

16. Prove:
$$\tan^{-1} x + \tan^{-1} y = \tan^{-1}\left(\frac{x+y}{1-xy}\right)$$
provided $-\pi/2 < \tan^{-1} x + \tan^{-1} y < \pi/2$.
[*Hint:* Use an identity for $\tan(\alpha + \beta)$.]

17. Use the result in Exercise 16 to show that
  (a) $\tan^{-1}\tfrac{1}{2} + \tan^{-1}\tfrac{1}{3} = \pi/4$
  (b) $2\tan^{-1}\tfrac{1}{3} + \tan^{-1}\tfrac{1}{7} = \pi/4$.

In Exercises 18 and 19, complete the identity by using the triangle method that was illustrated in Examples 3 and 4 of this section (see Figures 8.1.8 and 8.1.9).

---

18. (a) $\sin(\cos^{-1} x) = ?$
  (b) $\tan(\cos^{-1} x) = ?$
  (c) $\csc(\tan^{-1} x) = ?$
  (d) $\sin(\tan^{-1} x) = ?$

19. (a) $\cos(\tan^{-1} x) = ?$
  (b) $\tan(\cot^{-1} x) = ?$
  (c) $\sin(\sec^{-1} x) = ?$
  (d) $\cot(\csc^{-1} x) = ?$

20. Use Figure 8.1.10 to deduce the results in parts (a) and (b).
  (a) $\tan^{-1} 1 + \tan^{-1} 2 + \tan^{-1} 3 = \pi$.
    [*Hint:* Consider angles A, B, and C.]
  (b) $\tan^{-1} 1 + \tan^{-1}\tfrac{1}{2} + \tan^{-1}\tfrac{1}{3} = \pi/2$.
    [*Hint:* Consider angles α, β, and γ.]

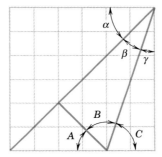

Figure 8.1.10

21. Sketch the graphs of
  (a) $y = \sin^{-1} 2x$
  (b) $y = \tan^{-1}\tfrac{1}{2}x$.

22. Sketch the graphs of
  (a) $y = \cos^{-1}\tfrac{1}{3}x$
  (b) $y = 2\cot^{-1} 2x$.

23. Prove:
  (a) $\sin^{-1}(-x) = -\sin^{-1} x$
  (b) $\tan^{-1}(-x) = -\tan^{-1} x$.

24. Prove:
  (a) $\cos^{-1}(-x) = \pi - \cos^{-1} x$
  (b) $\sec^{-1}(-x) = \pi + \sec^{-1} x$, if $x \geq 1$.

25. (a) In the text we proved that
$$\sin^{-1} x + \cos^{-1} x = \frac{\pi}{2}$$
  for $0 \leq x \leq 1$. Use this result together with Exercises 23(a) and 24(a) to prove this identity for $-1 \leq x < 0$.
  (b) In Example 4 we proved that
$$\sec(\tan^{-1} x) = \sqrt{1 + x^2}$$
  for $x \geq 0$. Use this result together with Exercise 23(b) to prove this identity for $x < 0$.

26. Prove:
  (a) $\cot^{-1} x = \tan^{-1}\dfrac{1}{x}$, if $x > 0$

(b) $\sec^{-1} x = \cos^{-1}\dfrac{1}{x}, \quad$ if $x \geq 1$

(c) $\csc^{-1} x = \sin^{-1}\dfrac{1}{x}, \quad$ if $x \geq 1.$

27. Most scientific calculators have keys for the values of only $\sin^{-1} x$, $\cos^{-1} x$, and $\tan^{-1} x$. The formulas in Exercise 26 show how a calculator can be used to obtain values of $\cot^{-1} x$, $\sec^{-1} x$, and $\csc^{-1} x$ for positive values of x. Use these formulas and a calculator to find numerical values for each of the following inverse trigonometric functions. Express your answers in degrees, rounded to the nearest tenth of a degree.

(a) $\cot^{-1} 0.7$ (b) $\sec^{-1} 1.2$ (c) $\csc^{-1} 2.3.$

In Exercises 28–30, solve the given equation for x in terms of k using the corresponding inverse trigonometric function. Note that x may not be in the range of the inverse function.

28. $\cos x = k, \quad$ if $0 < k < 1$ and $3\pi/2 < x < 2\pi.$

29. $\tan x = k, \quad$ if $k < 0$ and $\pi/2 < x < \pi.$

30. $\sin 2x = k, \quad$ if $0 < k < 1$ and $0 < x < \pi/2.$
[Hint: Consider the two cases: $0 < 2x < \pi/2$ and $\pi/2 < 2x < \pi.$]

In Exercises 31–36, use a calculator to approximate the solution of the equation. Where radians are used, express your answer to four decimal places, and where degrees are used, express it to the nearest tenth of a degree. [Note: In each part, the solution is not in the range of the corresponding inverse trigonometric function.]

31. $\sin x = 0.37, \quad \pi/2 < x < \pi.$

32. $\cos x = -0.85, \quad \pi < x < 3\pi/2.$

33. $\tan x = 3.16, \quad -\pi < x < -\pi/2.$

34. $\sin \theta = -0.61, \quad 180° < \theta < 270°.$

35. $\cos \theta = 0.23, \quad -90° < \theta < 0°.$

36. $\tan \theta = -0.45, \quad 90° < \theta < 180°.$

37. An Earth-observing satellite has horizon sensors that can measure the angle θ shown in Figure 8.1.11. Let R be the radius of Earth (assumed spherical) and h the distance between the satellite and Earth's surface.

(a) Show that $\sin \theta = \dfrac{R}{R + h}.$

(b) Find θ, to the nearest degree, for a satellite that is 10,000 km from Earth's surface (use $R = 6378$ km).

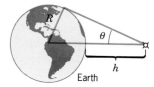

Figure 8.1.11

38. The number of hours of daylight on a given day at a given point on Earth's surface depends on the latitude λ of the point, the angle γ through which Earth has moved in its orbital plane during the time period from the vernal equinox (March 21), and the angle of inclination ϕ of Earth's axis of rotation measured from ecliptic north ($\phi \approx 23.55°$). The number of hours of daylight h can be approximated by the formula

$$h = \begin{cases} 24, & D \geq 1 \\ 12 + \frac{2}{15}\sin^{-1} D, & |D| < 1 \\ 0, & D \leq -1 \end{cases}$$

where

$$D = \frac{\sin\phi \, \sin\gamma \, \tan\lambda}{\sqrt{1 - \sin^2\phi \, \sin^2\gamma}}$$

and $\sin^{-1} D$ is in degree measure. Given that Fairbanks, Alaska, is located at a latitude of $\lambda = 65°$ N and that $\gamma = 90°$ on June 20 and $\gamma = 270°$ on December 20, approximate

(a) the maximum number of daylight hours at Fairbanks to one decimal place;

(b) the minimum number of daylight hours at Fairbanks to one decimal place.

[Note: This problem was adapted from TEAM, A Path to Applied Mathematics, The Mathematical Association of America, Washington, D.C., 1985.]

39. A soccer player kicks a ball with an initial speed of 14 m/sec at an angle θ with the horizontal (Figure 8.1.12). The ball lands 18 m down the field. If air resistance is neglected, then the ball will have a parabolic trajectory and the horizontal range R will be given by

$$R = \frac{v^2}{g}\sin 2\theta$$

where v is the initial speed of the ball and g is the acceleration due to gravity. Using $g = 9.8$ m/sec^2, approximate two values of θ, to the nearest degree, at which the ball could have been kicked. Which angle results in the shorter time of flight? Why?

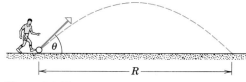

Figure 8.1.12

40. The law of cosines states that
$$c^2 = a^2 + b^2 - 2ab\cos\theta$$
where a, b, and c are the lengths of the sides of a triangle and θ is the angle formed by sides a and b. Find θ, to the nearest degree, for the triangle with $a = 2$, $b = 3$, and $c = 4.$

41. An airplane is flying at a constant height of 3000 ft above water at a speed of 400 ft/sec. The pilot is to release a survival package so that it lands in the water at a sighted point P. If air resistance is neglected, then the package will follow a parabolic trajectory whose equation relative

to the coordinate system in Figure 8.1.13 is

$$y = 3000 - \frac{g}{2v^2}x^2$$

where g is the acceleration due to gravity and v is the speed of the airplane. Using $g = 32$ ft/sec^2, find the "line of sight" angle θ, to the nearest degree, that will result in the package hitting the target point.

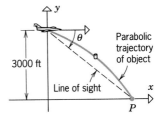

Figure 8.1.13

42. A camera is positioned x feet from the base of a missile launching pad (Figure 8.1.14). If a missile of length a feet is launched vertically, show that when the base of the

missile is b feet above the camera lens, the angle θ subtended at the lens by the missile is

$$\theta = \cot^{-1}\frac{x}{a+b} - \cot^{-1}\frac{x}{b}$$

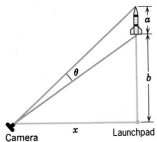

Figure 8.1.14

43. Prove:

(a) $\quad \sin^{-1} x = \tan^{-1}\dfrac{x}{\sqrt{1-x^2}}$

(b) $\quad \cos^{-1} x = \dfrac{\pi}{2} - \tan^{-1}\dfrac{x}{\sqrt{1-x^2}}.$

8.2 DERIVATIVES AND INTEGRALS INVOLVING INVERSE TRIGONOMETRIC FUNCTIONS

In this section we shall obtain the derivatives of the inverse trigonometric functions and the related integration formulas.

☐ **DERIVATIVE FORMULAS**

8.2.1 THEOREM.

$$\frac{d}{dx}[\sin^{-1} x] = \frac{1}{\sqrt{1-x^2}}, \qquad \frac{d}{dx}[\cos^{-1} x] = -\frac{1}{\sqrt{1-x^2}} \qquad (1\text{–}2)$$

$$\frac{d}{dx}[\tan^{-1} x] = \frac{1}{1+x^2}, \qquad \frac{d}{dx}[\cot^{-1} x] = -\frac{1}{1+x^2} \qquad (3\text{–}4)$$

$$\frac{d}{dx}[\sec^{-1} x] = \frac{1}{x\sqrt{x^2-1}}, \qquad \frac{d}{dx}[\csc^{-1} x] = -\frac{1}{x\sqrt{x^2-1}} \qquad (5\text{–}6)$$

Proof (1). First, we will show that $g(x) = \sin^{-1} x$ is differentiable on the interval $(-1, 1)$, then we will derive Formula (1). The function $g(x) = \sin^{-1} x$ is the inverse of the function

$$f(x) = \sin x, \quad -\pi/2 \le x \le \pi/2$$

As x varies over the interval $(-1, 1)$, the value of $g(x) = \sin^{-1} x$ varies over the interval $(-\pi/2, \pi/2)$ on which f is differentiable and has a *nonzero* derivative (verify). Thus, it follows from Theorem 7.4.7 that g is differentiable on the interval $(-1, 1)$.

To obtain (1) let

$$y = \sin^{-1} x \quad \text{or equivalently,} \quad x = \sin y$$

Differentiating the second equation implicitly with respect to x yields

$$\frac{d}{dx}[x] = \frac{d}{dx}[\sin y]$$

$$1 = \cos y \cdot \frac{dy}{dx}$$

$$\frac{dy}{dx} = \frac{1}{\cos y} = \frac{1}{\cos(\sin^{-1} x)} \qquad (7)$$

As illustrated in Example 3 of Section 8.1,

$$\cos(\sin^{-1} x) = \sqrt{1 - x^2}$$

Thus, (7) simplifies to

$$\frac{dy}{dx} = \frac{1}{\sqrt{1 - x^2}}$$

which establishes Formula (1).

Proof (3). As in the proof of (1), the differentiability of $g(x) = \tan^{-1} x$ follows from Theorem 7.4.7. (We omit the details.) To obtain (3), let

$$y = \tan^{-1} x \quad \text{or equivalently,} \quad x = \tan y$$

Differentiating the second equation implicitly with respect to x yields

$$\frac{d}{dx}[x] = \frac{d}{dx}[\tan y]$$

$$1 = \sec^2 y \cdot \frac{dy}{dx}$$

$$\frac{dy}{dx} = \frac{1}{\sec^2 y} = \frac{1}{\sec^2(\tan^{-1} x)} \qquad (8)$$

It follows from Example 4 of Section 8.1 that

$$\sec^2(\tan^{-1} x) = 1 + x^2$$

Thus, (8) simplifies to

$$\frac{dy}{dx} = \frac{1}{1 + x^2}$$

which establishes Formula (3).

Proof (5). As in the proof of (1), the differentiability of $g(x) = \sec^{-1} x$ follows from Theorem 7.4.7. (We omit the details.) To obtain (5), let

$$y = \sec^{-1} x \quad \text{or equivalently,} \quad x = \sec y$$

Differentiating the second equation implicitly with respect to x yields

$$1 = \sec y \tan y \cdot \frac{dy}{dx}$$

$$\frac{dy}{dx} = \frac{1}{\sec y \tan y} = \frac{1}{\sec(\sec^{-1} x) \tan(\sec^{-1} x)} \qquad (9)$$

We can simplify (9) using the methods of Examples 3 and 4 of Section 8.1. We do this now for $x > 1$, leaving the case $x < -1$ for the reader. As illustrated in Figure 8.2.1, construct a right triangle with an acute angle of $y = \sec^{-1} x$. Since $\sec y = x$, we can assume that the side adjacent to y has length 1 and that the hypotenuse has length x $(x > 1)$. It then

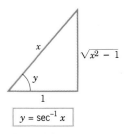

$y = \sec^{-1} x$

Figure 8.2.1

follows from the Theorem of Pythagoras that the side opposite to y has length $\sqrt{x^2 - 1}$. From the triangle we obtain

$$\sec(\sec^{-1} x) = \sec y = x$$

$$\tan(\sec^{-1} x) = \tan y = \sqrt{x^2 - 1}$$

Thus, (9) simplifies to

$$\frac{dy}{dx} = \frac{1}{x\sqrt{x^2 - 1}}$$

which establishes Formula (5).

Proofs (2), (4), *and* (6).   Formulas (2), (4), and (6) follow from Formulas (1), (3), and (5) above and from identities (4), (5), and (6) in Section 8.1. For example,

$$\frac{d}{dx}[\cos^{-1} x] = \frac{d}{dx}\left[\frac{\pi}{2} - \sin^{-1} x\right] = -\frac{d}{dx}[\sin^{-1} x] = -\frac{1}{\sqrt{1 - x^2}}$$

We leave the rest of the details for the reader.   ∎

☐ **GENERALIZED DERIVATIVE FORMULAS**

If u is a differentiable function of x, then applying the chain rule to (1)–(6) yields the following generalized derivative formulas:

$$\frac{d}{dx}[\sin^{-1} u] = \frac{1}{\sqrt{1 - u^2}}\frac{du}{dx}, \qquad \frac{d}{dx}[\cos^{-1} u] = -\frac{1}{\sqrt{1 - u^2}}\frac{du}{dx} \qquad (10\text{--}11)$$

$$\frac{d}{dx}[\tan^{-1} u] = \frac{1}{1 + u^2}\frac{du}{dx}, \qquad \frac{d}{dx}[\cot^{-1} u] = -\frac{1}{1 + u^2}\frac{du}{dx} \qquad (12\text{--}13)$$

$$\frac{d}{dx}[\sec^{-1} u] = \frac{1}{u\sqrt{u^2 - 1}}\frac{du}{dx}, \qquad \frac{d}{dx}[\csc^{-1} u] = -\frac{1}{u\sqrt{u^2 - 1}}\frac{du}{dx} \qquad (14\text{--}15)$$

Example 1   Find dy/dx if

(a) $y = \sin^{-1}(x^3)$ (b) $y = \sec^{-1}(e^x)$

Solution (a).   From (10)

$$\frac{dy}{dx} = \frac{1}{\sqrt{1 - (x^3)^2}}(3x^2) = \frac{3x^2}{\sqrt{1 - x^6}}$$

Solution (b).   From (14)

$$\frac{dy}{dx} = \frac{1}{e^x\sqrt{(e^x)^2 - 1}}(e^x) = \frac{1}{\sqrt{e^{2x} - 1}} \qquad \blacktriangleleft$$

Example 2   If the rocket shown in Figure 8.2.2 is rising vertically at 880 ft/sec when it is 4000 ft up, how fast must the camera elevation angle change at that instant to keep the rocket in sight?

Solution.   We solved this related rates problem in Example 3 of Section 4.1 by differentiating both sides of the equation

$$\tan \phi = \frac{x}{3000} \qquad (16)$$

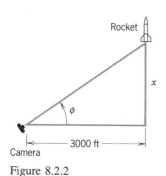

Figure 8.2.2

with respect to the time t, after which we solved for $d\phi/dt$.

Now that we can differentiate $\tan^{-1}$, an alternative solution is possible. Solving (16) for ϕ yields

$$\phi = \tan^{-1}\left(\frac{x}{3000}\right)$$

so

$$\frac{d\phi}{dt} = \frac{1}{1 + \left(\dfrac{x}{3000}\right)^2} \cdot \frac{1}{3000}\frac{dx}{dt} \tag{17}$$

We are given that $dx/dt = 880$ when $x = 4000$, so it follows from (17) that

$$\left.\frac{d\phi}{dt}\right|_{x=4000} = \frac{1}{1 + \left(\dfrac{4000}{3000}\right)^2} \cdot \frac{880}{3000} = \frac{66}{625} \approx 0.11 \text{ radian/second}$$

which agrees with the result obtained in Example 3 of Section 4.1.   ◀

☐ **INTEGRATION FORMULAS**

Differentiation formulas (1)–(6) yield useful integration formulas. Those most commonly needed are

$$\int \frac{du}{\sqrt{1 - u^2}} = \sin^{-1} u + C \tag{18}$$

$$\int \frac{du}{1 + u^2} = \tan^{-1} u + C \tag{19}$$

$$\int \frac{du}{u\sqrt{u^2 - 1}} = \sec^{-1} u + C \tag{20}$$

Example 3   Evaluate $\displaystyle\int \frac{dx}{1 + 3x^2}$.

Solution.   Substituting

$$u = \sqrt{3}\,x, \quad du = \sqrt{3}\,dx$$

yields

$$\int \frac{dx}{1 + 3x^2} = \frac{1}{\sqrt{3}} \int \frac{du}{1 + u^2} = \frac{1}{\sqrt{3}} \tan^{-1} u + C = \frac{1}{\sqrt{3}} \tan^{-1}(\sqrt{3}\,x) + C \quad ◀$$

Example 4   Evaluate $\displaystyle\int \frac{e^x}{\sqrt{1 - e^{2x}}}\,dx$.

Solution.   Substituting

$$u = e^x, \quad du = e^x\,dx$$

yields

$$\int \frac{e^x}{\sqrt{1 - e^{2x}}}\,dx = \int \frac{du}{\sqrt{1 - u^2}} = \sin^{-1} u + C = \sin^{-1}(e^x) + C \quad ◀$$

Example 5 Evaluate $\displaystyle\int \frac{dx}{a^2 + x^2}$, where $a \neq 0$ is a constant.

Solution. If we had a 1 in place of a^2, we could use (19). Thus, we look for a u-substitution that will enable us to replace the a^2 with a 1. If we let

$$x = au, \quad dx = a\,du$$

then

$$\int \frac{dx}{a^2 + x^2} = \int \frac{a\,du}{a^2 + a^2 u^2} = \frac{1}{a}\int \frac{du}{1 + u^2} = \frac{1}{a}\tan^{-1} u + C = \frac{1}{a}\tan^{-1}\frac{x}{a} + C \quad \blacktriangleleft$$

The method of Example 5 leads to the following generalizations of (18), (19), and (20) for $a > 0$:

$$\int \frac{du}{\sqrt{a^2 - u^2}} = \sin^{-1}\frac{u}{a} + C \tag{21}$$

$$\int \frac{du}{a^2 + u^2} = \frac{1}{a}\tan^{-1}\frac{u}{a} + C \tag{22}$$

$$\int \frac{du}{u\sqrt{u^2 - a^2}} = \frac{1}{a}\sec^{-1}\frac{u}{a} + C \tag{23}$$

Example 6 Evaluate $\displaystyle\int \frac{dx}{\sqrt{2 - x^2}}$.

Solution. Applying (21) with $u = x$ and $a = \sqrt{2}$ yields

$$\int \frac{dx}{\sqrt{2 - x^2}} = \sin^{-1}\frac{x}{\sqrt{2}} + C \quad \blacktriangleleft$$

▶ Exercise Set 8.2   C 49, 50

In Exercises 1–11, find dy/dx.

1. (a) $y = \sin^{-1}\left(\frac{1}{3}x\right)$ (b) $y = \cos^{-1}(2x + 1)$.

2. (a) $y = \tan^{-1}(x^2)$ (b) $y = \cot^{-1}(\sqrt{x})$.

3. (a) $y = \sec^{-1}(x^7)$ (b) $y = \csc^{-1}(e^x)$.

4. (a) $y = (\tan x)^{-1}$ (b) $y = \dfrac{1}{\tan^{-1} x}$.

5. (a) $y = \sin^{-1}\left(\dfrac{1}{x}\right)$ (b) $y = \cos^{-1}(\cos x)$.

6. (a) $y = \ln(\cos^{-1} x)$ (b) $y = \sqrt{\cot^{-1} x}$.

7. (a) $y = e^x \sec^{-1} x$ (b) $y = x^2(\sin^{-1} x)^3$.

8. (a) $y = \sin^{-1} x + \cos^{-1} x$ (b) $y = \sec^{-1} x + \csc^{-1} x$.

9. (a) $y = \tan^{-1}\left(\dfrac{1 - x}{1 + x}\right)$ (b) $y = (1 + x\csc^{-1} x)^{10}$.

10. (a) $y = \sin^{-1}(e^{-3x})$ (b) $y = \tan^{-1}(xe^{2x})$.

11. (a) $y = \tan^{-1}\sqrt{\dfrac{1 - x}{1 + x}}$ (b) $y = \sin^{-1}(x^2 \ln x)$.

In Exercises 12 and 13, find dy/dx by implicit differentiation.

12. $x^3 + x\tan^{-1} y = e^y$.

13. $\sin^{-1}(xy) = \cos^{-1}(x - y)$.

In Exercises 14–27, evaluate the integral.

14. $\displaystyle\int_0^{1/\sqrt{2}} \frac{dx}{\sqrt{1 - x^2}}$.

15. $\displaystyle\int_{-1}^{1} \frac{dx}{1 + x^2}$.

16. $\displaystyle\int_{\sqrt{2}}^{2} \frac{dx}{x\sqrt{x^2 - 1}}$.

17. $\displaystyle\int_{-\sqrt{2}}^{-2/\sqrt{2}} \frac{dx}{x\sqrt{x^2 - 1}}$.

18. $\displaystyle\int \frac{dx}{\sqrt{1-4x^2}}.$

19. $\displaystyle\int \frac{dx}{1+16x^2}.$

20. $\displaystyle\int \frac{dx}{x\sqrt{9x^2-1}}.$

21. $\displaystyle\int \frac{e^x}{1+e^{2x}}\,dx.$

22. $\displaystyle\int_{\ln 2}^{\ln(2/\sqrt{3})} \frac{e^{-x}\,dx}{\sqrt{1-e^{-2x}}}$

23. $\displaystyle\int_1^3 \frac{dx}{\sqrt{x}(x+1)}.$

24. $\displaystyle\int \frac{t}{t^4+1}\,dt.$

25. $\displaystyle\int \frac{\sec^2 x\,dx}{\sqrt{1-\tan^2 x}}.$

26. $\displaystyle\int \frac{\sin\theta}{\cos^2\theta+1}\,d\theta.$

27. $\displaystyle\int \frac{dx}{x\sqrt{1-(\ln x)^2}}.$

28. Derive integration formulas (21) and (23).

In Exercises 29–34, use (21), (22), and (23) to evaluate the integrals.

---

29. (a) $\displaystyle\int \frac{dx}{\sqrt{9-x^2}}$ (b) $\displaystyle\int \frac{dx}{5+x^2}$

(c) $\displaystyle\int \frac{dx}{x\sqrt{x^2-\pi}}.$

30. (a) $\displaystyle\int \frac{e^x}{4+e^{2x}}\,dx$ (b) $\displaystyle\int \frac{dx}{\sqrt{9-4x^2}}$

(c) $\displaystyle\int \frac{dy}{y\sqrt{5y^2-3}}.$

31. $\displaystyle\int_0^1 \frac{x}{\sqrt{4-3x^4}}\,dx.$

32. $\displaystyle\int_1^2 \frac{1}{\sqrt{x}\sqrt{4-x}}\,dx.$

33. $\displaystyle\int_0^{2/\sqrt{2}} \frac{1}{4+9x^2}\,dx.$

34. $\displaystyle\int_1^{\sqrt{2}} \frac{x}{3+x^4}\,dx.$

35. Find the area of the region enclosed by the graphs of $y=1/\sqrt{1-9x^2}$, $y=0$, $x=0$, and $x=1/6$.

36. Find the area of the region enclosed by the graphs of $y=\sin^{-1}x$, $x=0$, and $y=\pi/2$.

37. Find the volume of the solid generated when the region bounded by $x=2$, $x=-2$, $y=0$, and $y=1/\sqrt{4+x^2}$ is revolved about the x-axis.

38. (a) Find the volume V of the solid generated when the region bounded by $y=1/(1+x^4)$, $y=0$, $x=1$, and $x=b\,(b>1)$ is revolved about the y-axis.

(b) Find $\displaystyle\lim_{b\to+\infty} V$.

39. Evaluate $\int_0^1 \sin^{-1}x\,dx$. [*Hint:* Interpret the integral as the area of a region in the xy-plane, and integrate with respect to y.]

40. In Exercise 42 of Section 8.1, how far from the launching pad should the camera be positioned to maximize the angle θ subtended at the lens by the missile?

41. Given points $A(2,1)$ and $B(5,4)$, find the point P in the interval $[2,5]$ on the x-axis that maximizes angle APB.

42. An aircraft is flying at an altitude of 4 mi at a speed of 800 mi/hr in a direction away from a tracking station on the ground. How fast is the angle of elevation changing when the aircraft is over a point 10 mi from the station?

43. A lighthouse is located 3 mi off a straight shore. If the light revolves at 2 revolutions per minute, how fast is the beam moving along the coastline at a point 2 mi down the coast?

44. A 25-ft ladder is leaning against a vertical wall. If the bottom of the ladder slides away from the base of the wall at the rate of 4 ft/sec, how fast is the angle between the ladder and the wall changing when the top of the ladder is 20 ft above the ground?

45. The lower edge of a painting, 10 ft in height, is 2 ft above an observer's eye level. Assuming that the best view is obtained when the angle subtended at the observer's eye by the painting is maximum, how far from the wall should the observer stand?

46. Use Theorem 4.9.3 to prove that

(a) $2\sin^{-1}\sqrt{x} = \sin^{-1}(2x-1) + \pi/2$ for $0\le x\le 1$

(b) $\sin^{-1}(\tanh x) = \tan^{-1}(\sinh x)$.

47. Use Theorem 4.9.2 (the Mean-Value Theorem) to prove that

$$\frac{x}{1+x^2} < \tan^{-1}x < x \quad (x>0)$$

48. Find $\displaystyle\lim_{n\to+\infty}\sum_{k=1}^{n}\frac{n}{n^2+k^2}$. [*Hint:* Interpret this as the limit of a Riemann sum in which the interval $[0,1]$ is divided into n subintervals of equal width.]

49. A student wants to find the area enclosed by the graphs of $y=1/\sqrt{1-x^2}$, $y=0$, $x=0$, and $x=0.8$.

(a) Show that the exact area is $\sin^{-1}0.8$.

(b) The student uses a calculator to approximate the result in part (a) to two decimal places and obtains an incorrect answer of 53.13. What was the student's error? Find the correct approximation.

Exercise 50 requires separation of variables covered in Section 7.7.

---

50. Refer to Exercise 30, Section 7.7.

(a) Solve the differential equation

$$m\frac{dv}{dt} = -(kv^2+mg)$$

for v in terms of t given that $v=v_0$ when $t=0$.

(b) Find, to the nearest tenth of a second, how long it takes the bullet to reach its highest point. Use the numerical values of v_0, g, m, and k given in part (c) of Exercise 30, Section 7.7.

■ **8.3** INVERSE HYPERBOLIC FUNCTIONS

In this section we shall discuss inverses of hyperbolic functions. We shall be interested primarily in the integration formulas that these functions produce.

□ **INVERSES OF HYPERBOLIC FUNCTIONS**

Referring to Figures 7.6.1 and 7.6.3, it it evident that $\sinh x$, $\tanh x$, $\coth x$, and $\operatorname{csch} x$ all pass the horizontal line test and hence are one-to-one. We denote the inverses of these functions by $\sinh^{-1} x$, $\tanh^{-1} x$, $\coth^{-1} x$, and $\operatorname{csch}^{-1} x$. The functions $\cosh x$ and $\operatorname{sech} x$ are not one-to-one, but can be made so by restricting their domains to the interval $[0, +\infty)$, as indicated by the curves in solid color in Figure 8.3.1. We define the inverses of these restricted functions to be $\cosh^{-1} x$ and $\operatorname{sech}^{-1} x$.

The graphs of the inverse hyperbolic functions are shown in Figure 8.3.2; they are the reflections about the line $y = x$ of the graphs in Figures 7.6.1, 7.6.3, and 8.3.1.

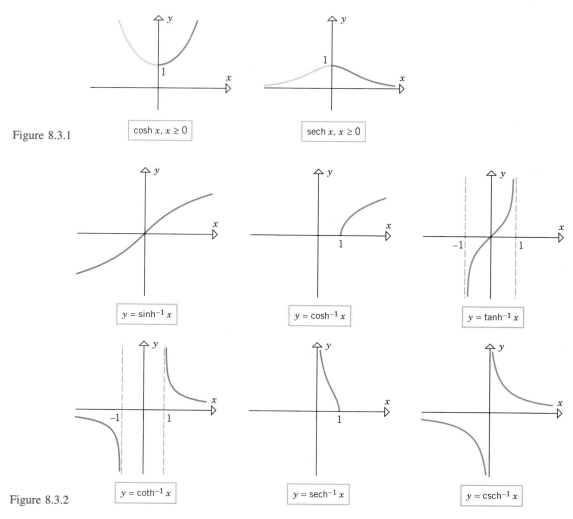

Figure 8.3.1

$\cosh x$, $x \geq 0$

$\operatorname{sech} x$, $x \geq 0$

$y = \sinh^{-1} x$

$y = \cosh^{-1} x$

$y = \tanh^{-1} x$

$y = \coth^{-1} x$

$y = \operatorname{sech}^{-1} x$

$y = \operatorname{csch}^{-1} x$

Figure 8.3.2

Table 8.3.1 and Theorem 8.3.1 summarize some key properties of the inverse hyperbolic functions.

The reader should confirm that the domains and ranges listed in Table 8.3.1 are consistent with Figure 8.3.2.

Table 8.3.1

FUNCTION	DOMAIN	RANGE	BASIC RELATIONSHIPS
$\sinh^{-1} x$	$(-\infty, +\infty)$	$(-\infty, +\infty)$	$\sinh^{-1}(\sinh x) = x$ if $-\infty < x < +\infty$ $\sinh(\sinh^{-1} x) = x$ if $-\infty < x < +\infty$
$\cosh^{-1} x$	$[1, +\infty)$	$[0, +\infty)$	$\cosh^{-1}(\cosh x) = x$ if $x \geq 0$ $\cosh(\cosh^{-1} x) = x$ if $x \geq 1$
$\tanh^{-1} x$	$(-1, 1)$	$(-\infty, +\infty)$	$\tanh^{-1}(\tanh x) = x$ if $-\infty < x < +\infty$ $\tanh(\tanh^{-1} x) = x$ if $-1 < x < 1$
$\coth^{-1} x$	$(-\infty, -1) \cup (1, +\infty)$	$(-\infty, 0) \cup (0, +\infty)$	$\coth^{-1}(\coth x) = x$ if $x < 0$ or $x > 0$ $\coth(\coth^{-1} x) = x$ if $x < -1$ or $x > 1$
$\text{sech}^{-1} x$	$(0, 1]$	$[0, +\infty)$	$\text{sech}^{-1}(\text{sech } x) = x$ if $x \geq 0$ $\text{sech}(\text{sech}^{-1} x) = x$ if $0 < x \leq 1$
$\text{csch}^{-1} x$	$(-\infty, 0) \cup (0, +\infty)$	$(-\infty, 0) \cup (0, +\infty)$	$\text{csch}^{-1}(\text{csch } x) = x$ if $x < 0$ or $x > 0$ $\text{csch}(\text{csch}^{-1} x) = x$ if $x < 0$ or $x > 0$

8.3.1 THEOREM.

$y = \sinh^{-1} x$ *is equivalent to* $\sinh y = x$ *for all x, y*

$y = \cosh^{-1} x$ *is equivalent to* $\cosh y = x$ *if* $\begin{cases} x \geq 1 \\ y \geq 0 \end{cases}$

$y = \tanh^{-1} x$ *is equivalent to* $\tanh y = x$ *if* $\begin{cases} -1 < x < 1 \\ -\infty < y < +\infty \end{cases}$

$y = \coth^{-1} x$ *is equivalent to* $\coth y = x$ *if* $\begin{cases} |x| > 1 \\ y \neq 0 \end{cases}$

$y = \text{sech}^{-1} x$ *is equivalent to* $\text{sech } y = x$ *if* $\begin{cases} 0 < x \leq 1 \\ y \geq 0 \end{cases}$

$y = \text{csch}^{-1} x$ *is equivalent to* $\text{csch } y = x$ *if* $\begin{cases} x \neq 0 \\ y \neq 0 \end{cases}$

□ **LOGARITHMIC FORMS OF INVERSE HYPERBOLIC FUNCTIONS**

Because the hyperbolic functions are expressible in terms of e^x, it should not be surprising that the inverse hyperbolic functions are expressible in terms of natural logarithms; the following theorem shows that this is so.

8.3.2 THEOREM. *The following relationships hold, subject to the restrictions on x in Theorem 8.3.1.*

$$\sinh^{-1} x = \ln(x + \sqrt{x^2 + 1}), \qquad \cosh^{-1} x = \ln(x + \sqrt{x^2 - 1})$$

$$\tanh^{-1} x = \frac{1}{2}\ln\frac{1 + x}{1 - x}, \qquad \coth^{-1} x = \frac{1}{2}\ln\frac{x + 1}{x - 1}$$

$$\text{sech}^{-1} x = \ln\left(\frac{1 + \sqrt{1 - x^2}}{x}\right), \qquad \text{csch}^{-1} x = \ln\left(\frac{1}{x} + \frac{\sqrt{1 + x^2}}{|x|}\right)$$

Proof. We shall prove the formula for $\sinh^{-1} x$ and leave the others as exercises. We begin by writing $y = \sinh^{-1} x$ in the alternative form

$$x = \sinh y = \frac{e^y - e^{-y}}{2} \quad \text{or} \quad e^y - 2x - e^{-y} = 0$$

Multiplying the last equation through by e^y we obtain

$$e^{2y} - 2xe^y - 1 = 0$$

and applying the quadratic formula yields

$$e^y = \frac{2x \pm \sqrt{4x^2 + 4}}{2} = x \pm \sqrt{x^2 + 1}$$

Since $e^y > 0$, the solution involving the minus sign is extraneous and must be discarded. Thus,

$$e^y = x + \sqrt{x^2 + 1}$$

Taking natural logarithms yields

$$y = \ln(x + \sqrt{x^2 + 1}) \quad \text{or} \quad \sinh^{-1} x = \ln(x + \sqrt{x^2 + 1}) \qquad \blacksquare$$

☐ **DERIVATIVES OF INVERSE HYPERBOLIC FUNCTIONS**

The differentiability and the formulas for the derivatives of the inverse hyperbolic functions can be established using Theorem 7.4.7. However, the derivative formulas can also be obtained by differentiating the logarithmic expressions for the inverse hyperbolic functions. For example,

$$\frac{d}{dx}[\sinh^{-1} x] = \frac{d}{dx}[\ln(x + \sqrt{x^2 + 1})] = \frac{1}{x + \sqrt{x^2 + 1}}\left(1 + \frac{x}{\sqrt{x^2 + 1}}\right)$$

$$= \frac{\sqrt{x^2 + 1} + x}{(x + \sqrt{x^2 + 1})(\sqrt{x^2 + 1})} = \frac{1}{\sqrt{x^2 + 1}}$$

We leave it for the exercises to derive the remaining derivative formulas, which are given in the following theorem.

8.3.3 THEOREM.

$$\frac{d}{dx}[\sinh^{-1} x] = \frac{1}{\sqrt{1 + x^2}}$$

$$\frac{d}{dx}[\cosh^{-1} x] = \frac{1}{\sqrt{x^2 - 1}} \qquad (x > 1)$$

$$\frac{d}{dx}[\tanh^{-1} x] = \frac{1}{1 - x^2} \qquad (|x| < 1)$$

$$\frac{d}{dx}[\coth^{-1} x] = \frac{1}{1 - x^2} \qquad (|x| > 1)$$

$$\frac{d}{dx}[\text{sech}^{-1} x] = -\frac{1}{x\sqrt{1 - x^2}} \qquad (0 < x < 1)$$

$$\frac{d}{dx}[\text{csch}^{-1} x] = -\frac{1}{|x|\sqrt{1 + x^2}} \qquad (x \neq 0)$$

REMARK.   Observe that $\tanh^{-1} x$ and $\coth^{-1} x$ have the same derivative formula, but the formulas are valid on different intervals. This difference becomes important when we integrate $1/(1 - x^2)$, since we must be careful to write

$$\int \frac{1}{1 - x^2} \, dx = \begin{cases} \tanh^{-1} x + C & \text{if} \quad |x| < 1 \\ \coth^{-1} x + C & \text{if} \quad |x| > 1 \end{cases}$$

The following integration formulas follow from Theorem 8.3.3.

8.3.4   THEOREM.

$$\int \frac{du}{\sqrt{1 + u^2}} = \sinh^{-1} u + C$$

$$\int \frac{du}{\sqrt{u^2 - 1}} = \cosh^{-1} u + C \qquad (u > 1)$$

$$\int \frac{du}{1 - u^2} = \begin{cases} \tanh^{-1} u + C & \text{if } |u| < 1 \\ \coth^{-1} u + C & \text{if } |u| > 1 \end{cases}$$

$$\int \frac{du}{u\sqrt{1 - u^2}} = -\operatorname{sech}^{-1} |u| + C$$

$$\int \frac{du}{u\sqrt{1 + u^2}} = -\operatorname{csch}^{-1} |u| + C$$

The first three integration formulas follow immediately from the corresponding differentiation formulas. The last two require additional work (see Exercises 25 and 26). By using Theorem 8.3.2 these formulas can also be expressed in terms of the natural logarithm. In particular, we leave it as an exercise to show that the third formula can be written as

$$\int \frac{du}{1 - u^2} = \frac{1}{2} \ln \left| \frac{1 + u}{1 - u} \right| + C \tag{1}$$

▶ Exercise Set 8.3   Ⓒ *22, 23, 24, 33*

1. (a) Prove: $\cosh^{-1} x = \ln(x + \sqrt{x^2 - 1})$, $x \geq 1$.

   (b) Use part (a) to obtain the derivative of $\cosh^{-1} x$.

2. (a) Prove: $\tanh^{-1} x = \frac{1}{2} \ln \frac{1 + x}{1 - x}$, $-1 < x < 1$.

   (b) Use part (a) to obtain the derivative of $\tanh^{-1} x$.

3. (a) Prove:

   $$\operatorname{sech}^{-1} x = \cosh^{-1}(1/x), \quad 0 < x \leq 1$$
   $$\coth^{-1} x = \tanh^{-1}(1/x), \quad |x| > 1$$
   $$\operatorname{csch}^{-1} x = \sinh^{-1}(1/x), \quad x \neq 0$$

   (b) Use part (a) and the derivatives of $\sinh^{-1} x$, $\cosh^{-1} x$, and $\tanh^{-1} x$ to find the derivatives of $\operatorname{sech}^{-1} x$, $\coth^{-1} x$, and $\operatorname{csch}^{-1} x$.

   (c) Use part (a) and the logarithmic expressions for $\sinh^{-1} x$, $\cosh^{-1} x$, and $\tanh^{-1} x$ to derive the logarithmic expressions for $\operatorname{sech}^{-1} x$, $\coth^{-1} x$, and $\operatorname{csch}^{-1} x$.

4. Without referring to the text, state the domains of the six inverse hyperbolic functions.

In Exercises 5 and 6, find the logarithmic equivalent of each of the given expressions.

5. (a) $\cosh^{-1}(3)$ (b) $\sinh^{-1}(-2)$.

6. (a) $\tanh^{-1}(3/4)$ (b) $\coth^{-1}(-5/4)$.

In Exercises 7–15, find dy/dx.

7. (a) $y = \sinh^{-1}\left(\frac{1}{3}x\right)$ (b) $y = \cosh^{-1}(2x + 1)$.

8. (a) $y = \tanh^{-1}(x^2)$ (b) $y = \coth^{-1}(\sqrt{x})$.

9. (a) $y = \operatorname{sech}^{-1}(x^7)$ (b) $y = \operatorname{csch}^{-1}(e^x)$.

10. (a) $y = (\tanh^{-1} x)^2$ (b) $y = \dfrac{1}{\tanh^{-1} x}$.

11. (a) $y = \sinh^{-1}(1/x)$ (b) $y = \cosh^{-1}(\cosh x)$.

12. (a) $y = \ln(\cosh^{-1} x)$ (b) $y = \sqrt{\coth^{-1} x}$.

13. (a) $y = e^x \operatorname{sech}^{-1} x$ (b) $y = x^2 (\sinh^{-1} x)^3$.

14. (a) $y = \sinh^{-1}(\tanh x)$

    (b) $y = \cosh^{-1}(\sinh^{-1} x)$.

15. (a) $y = \tanh^{-1}\left(\dfrac{1 - x}{1 + x}\right)$

    (b) $y = (1 + x \operatorname{csch}^{-1} x)^{10}$.

In Exercises 16–21, evaluate the integral.

16. $\displaystyle\int \frac{dx}{\sqrt{1 + 9x^2}}$.

17. $\displaystyle\int \frac{dx}{\sqrt{x^2 - 2}} \quad (x > \sqrt{2})$.

18. $\displaystyle\int \frac{dx}{\sqrt{9x^2 - 25}} \quad (x > 5/3)$.

19. $\displaystyle\int \frac{dx}{\sqrt{1-e^{2x}}}.$

20. $\displaystyle\int \frac{\sin\theta\, d\theta}{\sqrt{1+\cos^2\theta}}.$

21. $\displaystyle\int \frac{dx}{x\sqrt{1+x^6}}.$

In Exercises 22–24, use a calculator to obtain a numerical value for the integral.

22. $\displaystyle\int_0^{1/2} \frac{dx}{1-x^2}.$

23. $\displaystyle\int_2^3 \frac{dx}{1-x^2}.$

24. $\displaystyle\int_0^{\sqrt{3}} \frac{dt}{\sqrt{t^2+1}}.$

25. Show that
$$\frac{d}{dx}[\text{sech}^{-1}|x|] = -\frac{1}{x\sqrt{1-x^2}}$$

26. Show that
$$\frac{d}{dx}[\text{csch}^{-1}|x|] = -\frac{1}{x\sqrt{1+x^2}}$$

27. Derive integration formula (1).

28. Let a be a positive constant. Derive an integration formula for

(a) $\displaystyle\int \frac{du}{\sqrt{a^2+u^2}}$

(b) $\displaystyle\int \frac{du}{\sqrt{u^2-a^2}}$

(c) $\displaystyle\int \frac{du}{a^2-u^2}.$

[*Hint*: See Example 5 of Section 8.2.]

29. Our derivation of the differentiation formula for $\sinh^{-1}x$ used the logarithmic expression for this function. Show that the derivative can also be obtained by the method we used in Section 8.2 to obtain the derivative formula for $\sin^{-1}x$.

30. Find

(a) $\displaystyle\lim_{x\to+\infty} \sinh^{-1}x$

(b) $\displaystyle\lim_{x\to+\infty} \coth^{-1}x$

(c) $\displaystyle\lim_{x\to 0^+} \text{csch}^{-1}x$

(d) $\displaystyle\lim_{x\to+\infty} (\cosh^{-1}x - \ln x).$

31. Show that
$$\int \frac{1}{\sqrt{x^2-1}}\,dx = -\cosh^{-1}(-x) + C$$
if $x < -1$.

32. Use the result in Exercise 31 to show that
$$\int \frac{1}{\sqrt{x^2-1}}\,dx = \ln|x + \sqrt{x^2-1}| + C$$
if $x < -1$.

Exercises 33 and 34 require separation of variables covered in Section 7.7.

33. A thin flexible chain of length 2 ft is held stretched out on the top of a table with $\frac{1}{4}$ ft of the chain hanging over the edge (Figure 8.3.3). The chain is released and allowed to slide off the table. Assuming that the surface of the table is frictionless, the velocity v of the chain will satisfy the differential equation
$$v\frac{dv}{dx} = \frac{g}{L}x$$
where L is the total length of the chain, g is the acceleration due to gravity (32 ft/sec^2), and x is the distance from the top of the table to the hanging end of the chain.

(a) Solve the differential equation for v in terms of x.

(b) Use the result in part (a) and the fact that $v = dx/dt$ to find x in terms of the time t. Approximate the time it will take for the chain to slip completely off the table. Express your answer to the nearest tenth of a second.

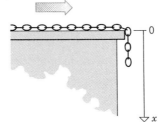

Figure 8.3.3

34. An object of mass m, released from rest at time $t = 0$, from a point above the surface of the earth is acted on by the force of gravity and a drag force that is proportional to the square of the velocity $v(t)$. The differential equation of motion is
$$m\frac{dv}{dt} = mg - kv^2$$
where g is the constant acceleration due to gravity and k is a positive constant.

(a) Solve the differential equation for $v(t)$.

(b) Use the result in part (a) to find $\displaystyle\lim_{t\to+\infty} v(t)$.

(c) Find the distance $x(t)$ of the object from its starting point.

◆ TECHNOLOGY EXERCISES   Chapter 8

Most of these exercises require access to a graphing calculator or a computer algebra system (CAS) such as *Mathematica, Maple,* or *Derive.* When you are asked to *find* an answer or to *solve* an equation, you may choose to find an exact result or a numerical approximation, depending on the particular technology you are using and on your own imagination. The form of your answers may differ from those of other students or from those in the answer section of the text, depending on how you solve the problems and the accuracy you use in your numerical approximations. Those exercises that are more appropriate for a CAS than a graphing calculator are labeled with the icon ◆.

1. **Maximum value:** Consider the curves $y = \sin^{-1} x$ and $y = 2 \tan^{-1} x$ for $0 \leq x \leq 1$.

   (a) Find the maximum vertical separation between the curves.

   (b) Find the maximum horizontal separation between the curves.

2. **Maximum value:** Consider the curves $y = \sin^{-1} x$ and $y = \tan x$ for $0 \leq x \leq 1$.

   (a) Find the two points of intersection of the curves.

   (b) Find the maximum vertical separation between the curves.

3. **Maximum value:** Find the value of $k \, (k > 0)$ such that $f(x) = e^{-x} \tan^{-1}(kx)$ attains its maximum value at $x = 0.8$. Using this value of k, confirm your result by graphing $f(x)$.

4. **Rectilinear motion:** Suppose that a particle moves along a line so that its velocity v at time t is given by

   $$v(t) = \frac{3}{t^2 + 1} - 0.5t, \quad t \geq 0$$

   where t is in seconds and v is in centimeters per second (cm/sec). Find the times at which the particle is 2 cm from its starting position.

5. **Area:** Find the area of the region in the first quadrant enclosed by $y = \sin 2x$ and $y = \sin^{-1} x$.

6. **Area:** Find the value of $k \, (0 < k < 1)$ so that the region enclosed by $y = 1/\sqrt{1 - x^2}$, $y = x$, $x = 0$, and $x = k$ has an area of 1 square unit.

7. **Area:** Find the value of $k \, (k > 0)$ so that the area of the region enclosed by $y = 1/(1 + 3x^2)$, $y = 0$, $x = 0$, and $x = k$ is equal to the area of the region enclosed by $y = 1/(1 + 3x^2)$, $y = 1$, and $x = k$.

8. **Area:** Find the value of $k \, (k > 0)$ so that the region enclosed by $y = 1/(1 + kx^2)$, $y = 0$, $x = 0$, and $x = 2$ has an area of 0.6 square unit.

◆ 9. **Volume:** Consider the region enclosed by $y = \sin^{-1} x$, $y = 0$, and $x = 1$. Find the volume of the solid generated by revolving the region about the x-axis using

   (a) disks   (b) cylindrical shells.

10. **Volume:** Consider the region that is enclosed by $y = 1/\sqrt{4 + x^2}$, $y = 0$, $x = 0$, and $x = k \, (k > 0)$. Find the value of k so that the volume of the solid generated by revolving the region about the x-axis is equal to the volume of the solid generated by revolving the region about the y-axis.

◆ 11. **Cylindrical storage tank:** Suppose that a tank for storing liquid is a right-circular cylinder oriented horizontally as shown in the accompanying figure. The tank has a length of 6 ft and a diameter of 3 ft. It can be shown that if the depth of the liquid in the tank is h, then the volume V of the liquid is given by

   $$V = \frac{V_0}{2\pi}(\theta - \sin \theta)$$

   where $\theta = 2 \cos^{-1}[(r - h)/r]$ for $0 \leq h \leq 2r$. In this formula, V_0 is the total volume of the tank and r is the radius.

   (a) Find the total volume V_0 in gallons (assume that 1 gallon = 231 in^3).

   (b) Assuming that r and h are in inches and V is in gallons, use the value of V_0 from part (a) to obtain the graph of V versus h for $0 \leq h \leq 36$.

   (c) Use the graph obtained in part (b) to estimate dV/dh if $h = 18$ inches. Check your estimate by calculating dV/dh at $h = 18$ directly.

(d) Find the depth of the liquid if there are 70 gallons of liquid in the tank.

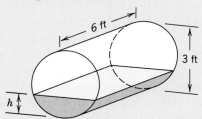

12. **Motion of a projectile:** Suppose that an object of mass m is projected straight up at time $t = 0$ with an initial velocity of v_0 and that the only forces acting on the object are the force of gravity (assumed constant) and a drag force kv^2 due to air resistance, where k is a constant and v is the velocity of the object. It can be shown that the time t_{up} that it takes for the object to reach its highest point is given by

$$t_{up} = \sqrt{\frac{m}{kg}} \tan^{-1}\left(\sqrt{\frac{k}{mg}}\, v_0\right)$$

where g is the acceleration due to gravity. The time t_{down} that it takes for the object to fall from its highest point to its starting point is

$$t_{down} = \sqrt{\frac{m}{kg}} \sinh^{-1}\left(\sqrt{\frac{k}{mg}}\, v_0\right)$$

Assume that t is measured in seconds and v_0 in meters per second (m/sec), and use

$$m = 0.003 \text{ kg}, \quad g = 9.8 \text{ m/sec}^2, \quad k = 10^{-5} \text{ kg/m}.$$

(a) Graph t_{up} and t_{down} versus v_0 in the same coordinate system for $0 \le v_0 \le 1000$ m/sec.

(b) Find $\lim\limits_{v_0 \to +\infty} t_{up}$ and $\lim\limits_{v_0 \to +\infty} t_{down}$.

(c) Find t_{up} and t_{down} if $v_0 = 900$ m/sec.

(d) Find the value of v_0 for which $t_{down} = 2t_{up}$.

Archimedes (287 B.C.–212 B.C.)

9 TECHNIQUES OF INTEGRATION

9.1 REVIEW; INTEGRATION USING TABLES AND COMPUTER ALGEBRA SYSTEMS

Although we can now integrate a wide variety of functions, there remain many important kinds of indefinite integrals that we cannot yet evaluate. The purpose of this chapter is to develop some additional techniques of integration and also to systematize the procedure of integration.

□ **A REVIEW OF FAMILIAR INTEGRATION FORMULAS**

In Section 5.3 we showed how to evaluate indefinite integrals by the method of substitution. The underlying idea was to make an appropriate *u*-substitution that would transform the given integral into a more basic integral that could be evaluated easily. The following is a list of basic integrals that we have encountered thus far.

CONSTANTS, POWERS, EXPONENTIALS   See Section 7.2.

1. $\displaystyle\int du = u + C$

2. $\displaystyle\int a\,du = a\int du = au + C$

3. $\displaystyle\int u^r\,du = \frac{u^{r+1}}{r+1} + C, r \neq -1$

4. $\displaystyle\int \frac{du}{u} = \ln|u| + C$

5. $\displaystyle\int e^u\,du = e^u + C$

6. $\displaystyle\int a^u\,du = \frac{a^u}{\ln a} + C, a > 0$

TRIGONOMETRIC FUNCTIONS   See Sections 5.2 and 7.2.

7. $\displaystyle\int \sin u\,du = -\cos u + C$

8. $\displaystyle\int \cos u\,du = \sin u + C$

9. $\displaystyle\int \sec^2 u\,du = \tan u + C$

10. $\displaystyle\int \csc^2 u\,du = -\cot u + C$

11. $\displaystyle\int \sec u \tan u\,du = \sec u + C$

12. $\displaystyle\int \csc u \cot u\,du = -\csc u + C$

13. $\displaystyle\int \tan u\,du = -\ln|\cos u| + C$

14. $\displaystyle\int \cot u\,du = \ln|\sin u| + C$

HYPERBOLIC FUNCTIONS   See Section 7.6.

15. $\displaystyle\int \sinh u\,du = \cosh u + C$

16. $\displaystyle\int \cosh u\,du = \sinh u + C$

17. $\displaystyle\int \text{sech}^2 u\,du = \tanh u + C$

18. $\displaystyle\int \text{csch}^2 u\,du = -\coth u + C$

19. $\displaystyle\int \text{sech}\, u \tanh u\,du = -\text{sech}\, u + C$

20. $\displaystyle\int \text{csch}\, u \coth u\,du = -\text{csch}\, u + C$

ALGEBRAIC FUNCTIONS   See Sections 8.2 and 8.3.

21. $\displaystyle\int \frac{du}{\sqrt{a^2 - u^2}} = \sin^{-1}\frac{u}{a} + C$

22. $\displaystyle\int \frac{du}{a^2 + u^2} = \frac{1}{a}\tan^{-1}\frac{u}{a} + C$

23. $\displaystyle\int \frac{du}{u\sqrt{u^2 - a^2}} = \frac{1}{a}\sec^{-1}\frac{u}{a} + C$

24. $\displaystyle\int \frac{du}{\sqrt{a^2 + u^2}} = \ln(u + \sqrt{u^2 + a^2}) + C$

25. $\displaystyle\int \frac{du}{\sqrt{u^2 - a^2}} = \ln|u + \sqrt{u^2 - a^2}| + C$

26. $\displaystyle\int \frac{du}{a^2 - u^2} = \frac{1}{2a}\ln\left|\frac{a+u}{a-u}\right| + C$

27. $\displaystyle\int \frac{du}{u\sqrt{a^2 - u^2}} = -\frac{1}{a}\ln\left|\frac{a + \sqrt{a^2 - u^2}}{u}\right| + C$

28. $\displaystyle\int \frac{du}{u\sqrt{a^2 + u^2}} = -\frac{1}{a}\ln\left|\frac{a + \sqrt{a^2 + u^2}}{u}\right| + C$

REMARK.   Formulas 24–28 are generalizations of those in Section 8.3. Readers who did not cover that section can ignore those formulas for now; we shall develop alternative methods for evaluating them in this chapter.

☐ **USING TABLES OF INTEGRALS**

Mathematicians have formulated extensive tables of integrals, an example of which appears in the endpapers of this text. We shall refer to this as *the endpaper Table of Integrals*. More comprehensive tables are published in such reference books as the *CRC Standard Mathematical Tables and Formulae*, CRC Press, Inc., 1991. Until recently, engineers relied heavily on such tables to evaluate difficult integrals arising in applications. However, such tables are rapidly giving way to computer algebra systems such as *Mathematica*, *Maple*, *Derive*, and others, which are capable of evaluating indefinite integrals. Some handheld calculators now also have such capabilities. In the examples that follow we shall illustrate some techniques for using integral tables, and we shall compare the results obtained to those produced by *Mathematica*, *Maple*, and *Derive*.

Example 1 Use the endpaper Table of Integrals to evaluate

(a) $\displaystyle\int x^2\sqrt{7 + 3x}\, dx$ (b) $\displaystyle\int \sqrt{x - 4x^2}\, dx$ (c) $\displaystyle\int e^{\pi x}\sin^{-1}(e^{\pi x})\, dx$

Solution (a). The integral involves an expression of the form

$$\sqrt{a + bu}$$

where $a = 7$, $b = 3$, and $u = x$, so we are led to the section of the table with the heading *Integrals Containing $\sqrt{a + bu}$*. Under that heading we find that the integral in Formula 15 matches the given integral with the above values of a, b, and u. Since $u = x$, it follows that $du = dx$, so Formula 15 yields

$$\int x^2\sqrt{7 + 3x}\, dx = \frac{2}{2835}(135x^2 - 252x + 392)(7 + 3x)^{3/2} + C$$

Solution (b). The integral does not involve any expressions that match the table headings precisely, but if the factor multiplying x^2 had been 1 rather than 4, we would have had a match with

$$2au - u^2$$

by taking $a = \frac{1}{2}$ and $u = x$. This suggests that we consider making the u-substitution

$$u = 2x$$

since this substitution has the effect of converting $4x^2$ to u^2. With this substitution we have

$$x = \tfrac{1}{2}u \quad \text{and} \quad dx = \tfrac{1}{2}\, du$$

so the given integral becomes

$$\int \sqrt{x - 4x^2}\, dx = \frac{1}{2}\int \sqrt{\tfrac{1}{2}u - u^2}\, du$$

Thus, from Formula 50 with $a = \frac{1}{4}$ we obtain

$$\int \sqrt{x - 4x^2}\, dx = \frac{1}{2}\left[\frac{u - \frac{1}{4}}{2}\sqrt{\tfrac{1}{2}u - u^2} + \frac{1}{32}\sin^{-1}\left(\frac{u - \frac{1}{4}}{\frac{1}{4}}\right)\right] + C$$

$$= \frac{1}{2}\left[\frac{2x - \frac{1}{4}}{2}\sqrt{x - 4x^2} + \frac{1}{32}\sin^{-1}\left(\frac{2x - \frac{1}{4}}{\frac{1}{4}}\right)\right] + C$$

$$= \frac{8x - 1}{16}\sqrt{x - 4x^2} + \frac{1}{64}\sin^{-1}(8x - 1) + C$$

Solution (c). The integral does not involve any expressions that match the table headings; however, if we make the substitution

$$u = e^{\pi x} \quad \text{so that} \quad du = \pi e^{\pi x}\, dx$$

we obtain

$$\int e^{\pi x}\sin^{-1}(e^{\pi x})\, dx = \frac{1}{\pi}\int \sin^{-1} u\, du$$

Thus, from Formula 90

$$\int e^{\pi x} \sin^{-1}(e^{\pi x})\,dx = \frac{1}{\pi}\left[u\sin^{-1}u + \sqrt{1 - u^2}\right] + C$$

$$= \frac{1}{\pi}\left[e^{\pi x}\sin^{-1}(e^{\pi x}) + \sqrt{1 - e^{2\pi x}}\right] + C \quad \blacktriangleleft$$

☐ **USING COMPUTER ALGEBRA SYSTEMS**

The form of an answer to an indefinite integration can vary widely between computer algebra systems, depending on the method of integration used by the program and the manner in which it simplifies the result. Indeed, the variation can be so great that it may be difficult to see that results produced by different programs are actually equivalent.

Table 9.1.1 shows how the integrals in Example 1 are evaluated by *Mathematica*, *Maple*, and *Derive*. (Some of these were simplified using the computer algebra system's simplification capability.) In the exercises we ask the reader to verify that the results produced by the three computer algebra systems are equivalent to the results obtained from the endpaper Table of Integrals.

Table 9.1.1

$$\int x^2\sqrt{7 + 3x}\,dx$$

TABLE OF INTEGRALS	$\dfrac{2}{2835}(135x^2 - 252x + 392)(7 + 3x)^{3/2}$
Mathematica	$\text{Sqrt}\,[7 + 3x]\left(\dfrac{784}{405} - \dfrac{56x}{135} + \dfrac{2x^2}{15} + \dfrac{2x^3}{7}\right)$
Maple	$\dfrac{2}{2835}(7 + 3x)^{3/2}(392 - 252x + 135x^2)$
Derive	$\dfrac{2(3x + 7)^{3/2}(135x^2 - 252x + 392)}{2835}$

$$\int \sqrt{x - 4x^2}\,dx$$

TABLE OF INTEGRALS	$\dfrac{8x - 1}{16}\sqrt{x - 4x^2} + \dfrac{1}{64}\sin^{-1}(8x - 1)$
Mathematica	$\left(-\left(\dfrac{1}{16}\right) + \dfrac{x}{2}\right)\text{Sqrt}\,[x - 4x^2] - \dfrac{\text{ArcSin}\,[1 - 8x]}{64}$
Maple	$-\dfrac{1}{16}(-8x + 1)\sqrt{x - 4x^2} + \dfrac{1}{64}\arcsin(8x - 1)$
Derive	$\dfrac{\text{ASIN}(8x - 1)}{64} + \dfrac{(8x - 1)\sqrt{x(1 - 4x)}}{16}$

$$\int e^{\pi x}\sin^{-1}(e^{\pi x})\,dx$$

TABLE OF INTEGRALS	$\dfrac{1}{\pi}[e^{\pi x}\sin^{-1}(e^{\pi x}) + \sqrt{1 - e^{2\pi x}}]$
Mathematica	$\dfrac{\text{Sqrt}\,[1 - E^{2\,\text{Pi}\,x}]}{\text{Pi}} + \dfrac{E^{\text{Pi}\,x}\,\text{ArcSin}\,[E^{\text{Pi}\,x}]}{\text{Pi}}$
Maple	$\dfrac{e^{\pi x}\arcsin(e^{\pi x}) + \sqrt{1 - (e^{\pi x})^2}}{\pi}$
Derive	$\dfrac{e^{\pi x}\,\text{ASIN}(e^{\pi x})}{\pi} + \dfrac{(1 - e^{2\pi x})^{3/2}}{3\pi} + \dfrac{\sqrt{1 - e^{2\pi x}}(e^{2\pi x} + 2)}{3\pi}$

▶ **Exercise Set 9.1** Ⓒ *53–66*

Review: Without looking at the text, complete the following integration formulas and then check your results by referring to the list of formulas at the beginning of this section.

Constants, Powers, Exponentials

$$\int du =$$ $$\int a\,du =$$

$$\int u^r\,du =$$ $$\int \frac{du}{u} =$$

$$\int e^u\,du =$$ $$\int a^u\,du =$$

Trigonometric Functions

$$\int \sin u\,du =$$ $$\int \cos u\,du =$$

$$\int \sec^2 u\,du =$$ $$\int \csc^2 u\,du =$$

$$\int \sec u \tan u\,du =$$ $$\int \csc u \cot u\,du =$$

$$\int \tan u\,du =$$ $$\int \cot u\,du =$$

Algebraic Functions

$$\int \frac{du}{\sqrt{1-u^2}} =$$ $$\int \frac{du}{1+u^2} =$$

$$\int \frac{du}{u\sqrt{u^2-1}} =$$ $$\int \frac{du}{\sqrt{1+u^2}} =$$

$$\int \frac{du}{\sqrt{u^2-1}} =$$ $$\int \frac{du}{1-u^2} =$$

$$\int \frac{du}{u\sqrt{1-u^2}} =$$ $$\int \frac{du}{u\sqrt{1+u^2}} =$$

Hyperbolic Functions

$$\int \sinh u\,du =$$ $$\int \cosh u\,du =$$

$$\int \operatorname{sech}^2 u\,du =$$ $$\int \operatorname{csch}^2 u\,du =$$

$$\int \operatorname{sech} u \tanh u\,du =$$

$$\int \operatorname{csch} u \coth u\,du =$$

In Exercises 1–24:
(a) Use the endpaper Table of Integrals to evaluate the integral.
(b) If you have access to a computer algebra system such as *Mathematica*, *Maple*, or *Derive*, use it to evaluate the integral.
(c) Confirm that the results obtained in parts (a) and (b) are equivalent.

1. $\displaystyle\int \frac{3x}{4x-1}\,dx.$

2. $\displaystyle\int \frac{x}{(2-3x)^2}\,dx.$

3. $\displaystyle\int \frac{1}{x(2x+5)}\,dx.$

4. $\displaystyle\int \frac{1}{x^2(1-5x)}\,dx.$

5. $\displaystyle\int x\sqrt{2x-3}\,dx.$

6. $\displaystyle\int \frac{x}{\sqrt{2-x}}\,dx.$

7. $\displaystyle\int \frac{1}{x\sqrt{4-3x}}\,dx.$

8. $\displaystyle\int \frac{1}{x\sqrt{3x-4}}\,dx.$

9. $\displaystyle\int \frac{1}{5-x^2}\,dx.$

10. $\displaystyle\int \frac{1}{x^2-9}\,dx.$

11. $\displaystyle\int \sqrt{x^2-3}\,dx.$

12. $\displaystyle\int \frac{\sqrt{x^2+5}}{x^2}\,dx.$

13. $\displaystyle\int \frac{x^2}{\sqrt{x^2+4}}\,dx.$

14. $\displaystyle\int \frac{1}{x^2\sqrt{x^2-2}}\,dx.$

15. $\displaystyle\int \sqrt{9-x^2}\,dx.$

16. $\displaystyle\int \frac{\sqrt{4-x^2}}{x^2}\,dx.$

17. $\displaystyle\int \frac{\sqrt{3-x^2}}{x}\,dx.$

18. $\displaystyle\int \frac{1}{x\sqrt{6x-x^2}}\,dx.$

19. $\displaystyle\int \sin 3x \sin 2x\,dx.$

20. $\displaystyle\int \sin 2x \cos 5x\,dx.$

21. $\displaystyle\int x^3 \ln x\,dx.$

22. $\displaystyle\int \frac{\ln x}{\sqrt{x}}\,dx.$

23. $\displaystyle\int e^{-2x} \sin 3x\,dx.$

24. $\displaystyle\int e^x \cos 2x\,dx.$

In Exercises 25–36:
(a) Make the indicated *u*-substitution and then use the endpaper Table of Integrals to evaluate the resulting integral.
(b) If you have access to a computer algebra system such as *Mathematica*, *Maple*, or *Derive*, use it to evaluate the original integral (no substitution).
(c) Confirm that the results obtained in parts (a) and (b) are equivalent.

25. $\displaystyle\int \frac{e^{4x}}{(4-3e^{2x})^2}\,dx, \quad u = e^{2x}.$

26. $\int \dfrac{\cos 2x}{(\sin 2x)(3 - \sin 2x)}\, dx,\ u = \sin 2x.$

27. $\int \dfrac{1}{\sqrt{x}(9x + 4)}\, dx,\ u = 3\sqrt{x}.$

28. $\int \dfrac{\cos 4x}{9 + \sin^2 4x}\, dx,\ u = \sin 4x.$

29. $\int \dfrac{1}{\sqrt{9x^2 - 4}}\, dx,\ u = 3x.$

30. $\int x\sqrt{2x^4 + 3}\, dx,\ u = \sqrt{2}x^2.$

31. $\int \dfrac{x^5}{\sqrt{5 - 9x^4}}\, dx,\ u = 3x^2.$

32. $\int \dfrac{1}{x^2\sqrt{3 - 4x^2}}\, dx,\ u = 2x.$

33. $\int \dfrac{\sin^2 (\ln x)}{x}\, dx,\ u = \ln x.$

34. $\int e^{-2x}\cos^2 (e^{-2x})\, dx,\ u = e^{-2x}.$

35. $\int x e^{-2x}\, dx,\ u = -2x.$

36. $\int \ln (5x - 1)\, dx,\ u = 5x - 1.$

In Exercises 37–48:
(a) Make an appropriate u-substitution and then use the endpaper Table of Integrals to evaluate the resulting integral.
(b) If you have access to a computer algebra system such as *Mathematica*, *Maple*, or *Derive*, use it to evaluate the original integral (no substitution).
(c) Confirm that the results obtained in parts (a) and (b) are equivalent.

37. $\int \dfrac{\sin 3x}{(\cos 3x)(\cos 3x + 1)^2}\, dx.$

38. $\int \dfrac{\ln x}{x\sqrt{4\ln x - 1}}\, dx.$

39. $\int \dfrac{x}{16x^4 - 1}\, dx.$

40. $\int \dfrac{e^x}{3 - 4e^{2x}}\, dx.$

41. $\int e^x\sqrt{3 - 4e^{2x}}\, dx.$

42. $\int \dfrac{\sqrt{4 - 9x^2}}{x^2}\, dx.$

43. $\int \sqrt{5x - 9x^2}\, dx.$

44. $\int \dfrac{1}{x\sqrt{x - 5x^2}}\, dx.$

45. $\int x \sin 3x\, dx.$

46. $\int \cos \sqrt{x}\, dx.$

47. $\int e^{-\sqrt{x}}\, dx.$

48. $\int x \ln (2 - 3x^2)\, dx.$

In Exercises 49–52:
(a) Complete the square, make an appropriate u-substitution, and then use the endpaper Table of Integrals to evaluate the resulting integral.
(b) If you have access to a computer algebra system such as *Mathematica*, *Maple*, or *Derive*, use it to evaluate the original integral (no completion of square or substitution).
(c) Confirm that the results obtained in parts (a) and (b) are equivalent.

49. $\int \dfrac{1}{x^2 + 4x - 5}\, dx.$

50. $\int \sqrt{3 - 2x - x^2}\, dx.$

51. $\int \dfrac{x}{\sqrt{5 + 4x - x^2}}\, dx.$

52. $\int \dfrac{x}{x^2 + 6x + 13}\, dx.$

In Exercises 53 and 54, use any method to solve for x.

53. $\displaystyle\int_2^x \dfrac{1}{t(4 - t)}\, dt = 0.5,\ 2 < x < 4.$

54. $\displaystyle\int_1^x \dfrac{1}{t\sqrt{2t - 1}}\, dt = 1,\ x > \tfrac{1}{2}.$

In Exercises 55–58, use any method to find the area of the region enclosed by the curves.

55. $y = \sqrt{25 - x^2},\ y = 0,\ x = 0,\ x = 4.$

56. $y = \sqrt{9x^2 - 4},\ y = 0,\ x = 2.$

57. $y = \dfrac{1}{25 - 16x^2},\ y = 0,\ x = 0,\ x = 1.$

58. $y = \sqrt{x}\ \ln x,\ y = 0,\ x = 4.$

In Exercises 59–62, use any method to find the volume of the solid generated when the region enclosed by the curves is revolved about the y-axis.

59. $y = \cos x$ and $y = 0$ for $0 \le x \le \pi/2$.

60. $y = \sqrt{x - 4},\ y = 0,\ x = 8.$

61. $y = e^{-x},\ y = 0,\ x = 0,\ x = 3.$

62. $y = \ln x,\ y = 0,\ x = 5.$

In Exercises 63 and 64, use any method to find the arc length of the curve.

63. $y = 2x^2$ for $0 \le x \le 2$.

64. $y = 3\ln x$ for $1 \le x \le 3$.

In Exercises 65 and 66, use any method to find the area of the surface generated by revolving the curve about the x-axis.

65. $y = \sin x$ for $0 \leq x \leq \pi$.

66. $y = 1/x$ for $1 \leq x \leq 4$.

■ **9.2** INTEGRATION BY PARTS

In this section, we shall develop a technique that will help us to evaluate a wide variety of integrals that do not fit any of the basic integration formulas.

☐ **DERIVATION OF THE FORMULA FOR INTEGRATION BY PARTS**

If f and g are differentiable functions, then by the rule for differentiating products

$$\frac{d}{dx}[f(x)g(x)] = f(x)g'(x) + g(x)f'(x)$$

Integrating both sides we obtain

$$\int \frac{d}{dx}[f(x)g(x)]\,dx = \int f(x)g'(x)\,dx + \int g(x)f'(x)\,dx$$

or

$$f(x)g(x) + C = \int f(x)g'(x)\,dx + \int g(x)f'(x)\,dx$$

or

$$\int f(x)g'(x)\,dx = f(x)g(x) - \int g(x)f'(x)\,dx + C$$

Since the integral on the right will produce another constant of integration, there is no need to keep the C in this last equation; thus, we obtain

$$\int f(x)g'(x)\,dx = f(x)g(x) - \int f'(x)g(x)\,dx \tag{1}$$

which is called the formula for **integration by parts**. By using this formula we can sometimes reduce a hard integration problem to an easier one.

In practice, it is usual to rewrite (1) by letting

$$u = f(x), \quad du = f'(x)\,dx$$

$$v = g(x), \quad dv = g'(x)\,dx$$

This yields the following alternative form for (1).

$$\int u\,dv = uv - \int v\,du \tag{2}$$

Example 1   Evaluate $\int xe^x\,dx$.

Solution.   To apply (2) we must write the integral in the form

$$\int u\,dv$$

One way to do this is to let

$$u = x \quad \text{and} \quad dv = e^x \, dx$$

so that

$$du = dx \quad \text{and} \quad v = \int e^x \, dx = e^x$$

Thus, from (2)

$$\int xe^x \, dx = \int \underbrace{x}_{u} \underbrace{e^x \, dx}_{dv} = \underbrace{x}_{u} \underbrace{e^x}_{v} - \int \underbrace{e^x}_{v} \underbrace{dx}_{du} = xe^x - e^x + C \quad \blacktriangleleft$$

REMARK.   In the calculation of v from dv above, we omitted the constant of integration and wrote $v = \int e^x \, dx = e^x$. Had we included a constant of integration and written $v = \int e^x \, dx = e^x + C_1$, the constant C_1 would have eventually canceled out [Exercise 56(a)]. This is always the case in integration by parts [Exercise 56(b)], so we shall usually omit the constant when calculating v from dv.

To use integration by parts successfully, the choice of u and dv must be made so that the new integral is easier than the original. For example, had we decided above to let

$$u = e^x, \quad dv = x \, dx, \quad du = e^x \, dx, \quad v = \int x \, dx = \frac{x^2}{2}$$

then we would have obtained

$$\int xe^x \, dx = \int u \, dv = uv - \int v \, du = \frac{x^2}{2} e^x - \frac{1}{2} \int x^2 e^x \, dx$$

For this choice of u and dv the new integral is actually more complicated than the original. It is difficult to give hard and fast rules for choosing u and dv. It is a matter of experience that comes with lots of practice.

The next example shows that it is sometimes necessary to use integration by parts more than once in the same problem.

Example 2   Evaluate $\int x^2 e^{-x} \, dx$.

Solution.   Let

$$u = x^2, \quad dv = e^{-x} \, dx, \quad du = 2x \, dx, \quad v = \int e^{-x} \, dx = -e^{-x}$$

so that

$$\int x^2 e^{-x} \, dx = \int u \, dv = uv - \int v \, du = -x^2 e^{-x} + 2 \int xe^{-x} \, dx \tag{3}$$

The last integral is similar to the original except that we have replaced x^2 by x. Another integration by parts applied to $\int xe^{-x} \, dx$ will complete the problem. We let

$$u = x, \quad dv = e^{-x} \, dx, \quad du = dx, \quad v = \int e^{-x} \, dx = -e^{-x}$$

so that

$$\int xe^{-x} \, dx = \int u \, dv = uv - \int v \, du$$

$$= -xe^{-x} + \int e^{-x} \, dx$$

$$= -xe^{-x} - e^{-x} + C_1$$

Substituting in (3) we obtain

$$\int x^2 e^{-x}\, dx = -x^2 e^{-x} + 2(-xe^{-x} - e^{-x} + C_1)$$
$$= -x^2 e^{-x} - 2xe^{-x} - 2e^{-x} + 2C_1$$
$$= -(x^2 + 2x + 2)e^{-x} + C$$

where $C = 2C_1$.   ◄

Example 3   Evaluate $\int \ln x\, dx$.

Solution.   Let

$$u = \ln x, \quad dv = dx, \quad du = \frac{1}{x}\, dx, \quad v = \int dx = x$$

so that

$$\int \ln x\, dx = \int u\, dv = uv - \int v\, du = x \ln x - \int x\left(\frac{1}{x}\right) dx$$
$$= x \ln x - \int dx = x \ln x - x + C \quad ◄$$

Example 4   Evaluate $\int e^x \cos x\, dx$.

Solution.   Let

$$u = e^x, \quad dv = \cos x\, dx, \quad du = e^x\, dx, \quad v = \int \cos x\, dx = \sin x$$

Thus,

$$\int e^x \cos x\, dx = \int u\, dv = uv - \int v\, du$$
$$= e^x \sin x - \int e^x \sin x\, dx \tag{4}$$

Since the integral $\int e^x \sin x\, dx$ is similar in form to the original integral $\int e^x \cos x\, dx$, it seems that nothing has been accomplished. However, let us integrate this new integral by parts; we let

$$u = e^x, \quad dv = \sin x\, dx, \quad du = e^x\, dx, \quad v = \int \sin x\, dx = -\cos x$$

Thus,

$$\int e^x \sin x\, dx = \int u\, dv = uv - \int v\, du$$
$$= -e^x \cos x + \int e^x \cos x\, dx$$

Substituting in (4) yields

$$\int e^x \cos x\, dx = e^x \sin x - \left[-e^x \cos x + \int e^x \cos x\, dx\right]$$

or

$$\int e^x \cos x\, dx = e^x \sin x + e^x \cos x - \int e^x \cos x\, dx$$

which is an equation we can solve for the unknown integral. We obtain

$$2 \int e^x \cos x \, dx = e^x \sin x + e^x \cos x$$

and hence

$$\int e^x \cos x \, dx = \tfrac{1}{2} e^x \sin x + \tfrac{1}{2} e^x \cos x + C \quad \blacktriangleleft$$

☐ INTEGRATION BY PARTS FOR DEFINITE INTEGRALS

For definite integrals the formula corresponding to (2) is

$$\int_a^b u \, dv = uv \Big]_a^b - \int_a^b v \, du \tag{5}$$

REMARK.   It is important to keep in mind that the variables u and v in this formula are functions of x, and that the limits of integration in (5) are limits on the variable x. Sometimes it is helpful to emphasize this by writing (5) as

$$\int_{x=a}^{x=b} u \, dv = uv \Big]_{x=a}^{x=b} - \int_{x=a}^{x=b} v \, du \tag{6}$$

The next example illustrates how integration by parts can be used to integrate the inverse trigonometric functions.

Example 5   Evaluate $\displaystyle\int_0^1 \tan^{-1} x \, dx$.

Solution.   Let

$$u = \tan^{-1} x, \quad dv = dx, \quad du = \frac{1}{1+x^2} dx, \quad v = \int dx = x$$

Thus,

$$\int_0^1 \tan^{-1} x \, dx = \int_0^1 u \, dv = uv \Big]_0^1 - \int_0^1 v \, du$$

$$= x \tan^{-1} x \Big]_0^1 - \int_0^1 \frac{x}{1+x^2} dx$$

> The limits of integration refer to x; that is, $x = 0$ and $x = 1$.

But

$$\int_0^1 \frac{x}{1+x^2} dx = \frac{1}{2} \int_0^1 \frac{2x}{1+x^2} dx = \frac{1}{2} \ln(1+x^2) \Big]_0^1 = \frac{1}{2} \ln 2$$

so

$$\int_0^1 \tan^{-1} x \, dx = x \tan^{-1} x \Big]_0^1 - \frac{1}{2} \ln 2 = \left(\frac{\pi}{4} - 0\right) - \frac{1}{2} \ln 2$$

$$= \frac{\pi}{4} - \ln \sqrt{2} \quad \blacktriangleleft$$

☐ REDUCTION FORMULAS

Integration by parts can be used to derive *reduction formulas* for integrals. These are formulas that express an integral involving a power of a function in terms of an integral that involves a *lower* power of that function. For example, if n is a positive integer and $n \geq 2$, then integration by parts can be used to obtain the reduction formulas

$$\int \sin^n x \, dx = -\frac{1}{n} \sin^{n-1} x \cos x + \frac{n-1}{n} \int \sin^{n-2} x \, dx \tag{7}$$

$$\int \cos^n x \, dx = \frac{1}{n} \cos^{n-1} x \sin x + \frac{n-1}{n} \int \cos^{n-2} x \, dx \tag{8}$$

To illustrate how such formulas can be obtained, let us derive (8). We begin by writing $\cos^n x$ as $\cos^{n-1} x \cdot \cos x$ and letting

$$u = \cos^{n-1} x \qquad\qquad dv = \cos x \, dx$$

$$du = (n-1) \cos^{n-2} x (-\sin x) \, dx \qquad v = \int \cos x \, dx = \sin x$$

$$= -(n-1) \cos^{n-2} x \sin x \, dx$$

so that

$$\int \cos^n x \, dx = \int \cos^{n-1} x \cos x \, dx = \int u \, dv = uv - \int v \, du$$

$$= \cos^{n-1} x \sin x + (n-1) \int \sin^2 x \cos^{n-2} x \, dx$$

$$= \cos^{n-1} x \sin x + (n-1) \int (1 - \cos^2 x) \cos^{n-2} x \, dx$$

$$= \cos^{n-1} x \sin x + (n-1) \int \cos^{n-2} x \, dx - (n-1) \int \cos^n x \, dx$$

Transposing the last term on the right to the left side yields

$$n \int \cos^n x \, dx = \cos^{n-1} x \sin x + (n-1) \int \cos^{n-2} x \, dx$$

from which (8) follows.

Reduction formulas (7) and (8) reduce the exponent of sin (or cos) by 2. Thus, if the formulas are applied repeatedly, the exponent can eventually be reduced to 0 if n is even or 1 if n is odd, at which point the integration can be completed. This is illustrated in the following example.

Example 6 Evaluate

(a) $\displaystyle\int \cos^3 x \, dx$ (b) $\displaystyle\int \cos^4 x \, dx$

Solution (a). From (8) with $n = 3$

$$\int \cos^3 x \, dx = \frac{1}{3} \cos^2 x \sin x + \frac{2}{3} \int \cos x \, dx = \frac{1}{3} \cos^2 x \sin x + \frac{2}{3} \sin x + C$$

Solution (b). From (8) with $n = 4$

$$\int \cos^4 x \, dx = \frac{1}{4} \cos^3 x \sin x + \frac{3}{4} \int \cos^2 x \, dx$$

and from (8) with $n = 2$

$$\int \cos^2 x \, dx = \frac{1}{2} \cos x \sin x + \frac{1}{2} \int dx = \frac{1}{2} \cos x \sin x + \frac{1}{2} x + C_1$$

so that

$$\int \cos^4 x \, dx = \frac{1}{4} \cos^3 x \sin x + \frac{3}{4} \left(\frac{1}{2} \cos x \sin x + \frac{1}{2} x + C_1 \right)$$

$$= \frac{1}{4} \cos^3 x \sin x + \frac{3}{8} \cos x \sin x + \frac{3}{8} x + C$$

where $C = \frac{3}{4} C_1$. ◀

▶ **Exercise Set 9.2**

In Exercises 1–28, evaluate the integral.

1. $\int xe^{-x}\, dx.$

2. $\int xe^{3x}\, dx.$

3. $\int x^2 e^x\, dx.$

4. $\int x^2 e^{-2x}\, dx.$

5. $\int x \sin 2x\, dx.$

6. $\int x \cos 3x\, dx.$

7. $\int x^2 \cos x\, dx.$

8. $\int x^2 \sin x\, dx.$

9. $\int \sqrt{x} \ln x\, dx.$

10. $\int x \ln x\, dx.$

11. $\int (\ln x)^2\, dx.$

12. $\int \frac{\ln x}{\sqrt{x}}\, dx.$

13. $\int \ln (2x + 3)\, dx.$

14. $\int \ln (x^2 + 4)\, dx.$

15. $\int \sin^{-1} x\, dx.$

16. $\int \cos^{-1} (2x)\, dx.$

17. $\int \tan^{-1} (2x)\, dx.$

18. $\int x \tan^{-1} x\, dx.$

19. $\int e^x \sin x\, dx.$

20. $\int e^{-3\theta} \sin 3\theta\, d\theta.$

21. $\int e^{ax} \sin bx\, dx.$

22. $\int e^{2x} \cos 3x\, dx.$

23. $\int \sin (\ln x)\, dx.$

24. $\int \cos (\ln x)\, dx.$

25. $\int x \sec^2 x\, dx.$

26. $\int x \tan^2 x\, dx.$

27. $\int x^3 e^{x^2}\, dx.$

28. $\int \frac{xe^x}{(x + 1)^2}\, dx.$

In Exercises 29–41, evaluate the definite integral.

29. $\int_0^1 xe^{-5x}\, dx.$

30. $\int_0^2 xe^{2x}\, dx.$

31. $\int_1^e x^2 \ln x\, dx.$

32. $\int_{\sqrt{e}}^e \frac{\ln x}{x^2}\, dx.$

33. $\int_{-2}^2 \ln (x + 3)\, dx.$

34. $\int_0^{1/2} \sin^{-1} x\, dx.$

35. $\int_2^4 \sec^{-1} \sqrt{\theta}\, d\theta.$

36. $\int_1^2 x \sec^{-1} x\, dx.$

37. $\int_0^{\pi/2} x \sin 4x\, dx.$

38. $\int_0^{\pi} (x + x \cos x)\, dx.$

39. $\int_1^3 \sqrt{x} \tan^{-1} \sqrt{x}\, dx.$

40. $\int_0^2 \ln (x^2 + 1)\, dx.$

41. $\int_0^1 \frac{x^3}{\sqrt{x^2 + 1}}\, dx.$

42. Solve Exercise 41 without using integration by parts. [*Hint:* Let $u = \sqrt{x^2 + 1}$.]

43. (a) Find the area of the region enclosed by $y = \ln x$, the line $x = e$, and the x-axis.

   (b) Find the volume of the solid generated when the region in part (a) is revolved about the x-axis.

44. Find the area of the region between $y = x \sin x$ and $y = x$ for $0 \le x \le \pi/2$.

45. Find the volume of the solid generated when the region between $y = \sin x$ and $y = 0$ for $0 \le x \le \pi$ is revolved about the y-axis.

46. Find the volume of the solid generated when the region enclosed between $y = \cos x$ and $y = 0$ for $0 \le x \le \pi/2$ is revolved about the y-axis.

47. Use reduction formula (7) to evaluate

   (a) $\int \sin^3 x\, dx$ (b) $\int_0^{\pi/4} \sin^4 x\, dx.$

48. Use reduction formula (8) to evaluate

   (a) $\int \cos^5 x\, dx$ (b) $\int_0^{\pi/2} \cos^6 x\, dx.$

49. Use reduction formula (8) to help evaluate

   (a) $\int \cos^3 5x\, dx$ (b) $\int x \cos^4 (x^2)\, dx.$

   [*Hint:* First make a substitution.]

50. Use reduction formula (7) to help evaluate

   (a) $\int \sin^4 2x\, dx$ (b) $\int \frac{\sin^3 \sqrt{x}}{\sqrt{x}}\, dx.$

   [*Hint:* First make a substitution.]

51. Derive reduction formula (7).

In Exercises 52 and 53, derive the reduction formula in part (a) and use it to evaluate the integral in (b).

52. (a) $\int \sec^n x\, dx = \frac{\sec^{n-2} x \tan x}{n - 1} + \frac{n - 2}{n - 1} \int \sec^{n-2} x\, dx$

   (b) $\int \sec^4 x\, dx.$

53. (a) $\int x^n e^x\, dx = x^n e^x - n \int x^{n-1} e^x\, dx$

   (b) $\int x^3 e^x\, dx.$

54. Use the reduction formula in part (a) of Exercise 53 to help evaluate

   (a) $\int x^2 e^{3x}\, dx$ (b) $\int_0^1 xe^{-\sqrt{x}}\, dx.$

   [*Hint:* First make a substitution.]

55. Let f be a function whose second derivative is continuous on $[-1, 1]$. Show that

$$\int_{-1}^{1} xf''(x)\,dx = f'(1) + f'(-1) + f(-1) - f(1)$$

56. (a) In Example 1, let

$$u = x, \quad dv = e^x\,dx,$$

$$du = dx, \quad v = \int e^x\,dx = e^x + C_1$$

and show that the constant C_1 cancels out, thus giving the same solution obtained by omitting C_1.

(b) Show that in general

$$uv - \int v\,du = u(v + C_1) - \int (v + C_1)\,du$$

thereby justifying the omission of the constant of integration when calculating v in integration by parts.

57. Use integration by parts on $\int \dfrac{1}{x}\,dx$ with $u = 1/x$ and $dv = dx$. Explain.

9.3 INTEGRATING POWERS OF SINE AND COSINE

> In this section we shall study methods for evaluating integrals of the form
>
> $$\int \sin^m x \cos^n x\,dx$$
>
> where m and n are nonnegative integers.

Integrals of the form

$$\int \sin^m x\,dx \quad \text{and} \quad \int \cos^n x\,dx$$

can be evaluated using reduction formulas (7) and (8) in the preceding section. We shall now give alternative methods for evaluating such integrals that sometimes result in simpler forms of the antiderivatives. We shall need the trigonometric identities

$$\sin^2 x = \tfrac{1}{2}(1 - \cos 2x) \quad \text{and} \quad \cos^2 x = \tfrac{1}{2}(1 + \cos 2x) \tag{1a–b}$$

which follow from the double-angle formulas

$$\cos 2x = 1 - 2\sin^2 x \quad \text{and} \quad \cos 2x = 2\cos^2 x - 1$$

Example 1 Evaluate

(a) $\displaystyle\int \sin^2 x\,dx$ (b) $\displaystyle\int \cos^2 x\,dx$

Solution. It follows from (1a) and (1b) that

$$\int \sin^2 x\,dx = \frac{1}{2}\int (1 - \cos 2x)\,dx = \frac{1}{2}x - \frac{1}{4}\sin 2x + C$$

$$\int \cos^2 x\,dx = \frac{1}{2}\int (1 + \cos 2x)\,dx = \frac{1}{2}x + \frac{1}{4}\sin 2x + C \quad ◀$$

If m and n are both positive integers, then the integral

$$\int \sin^m x \cos^n x\,dx$$

can be evaluated by one of three procedures, depending on whether m and n are odd or even. The procedures are outlined in Table 9.3.1.

Table 9.3.1

$\int \sin^m x \cos^n x \, dx$	PROCEDURE	RELEVANT IDENTITIES
n odd	• Split off a factor of $\cos x$. • Apply the relevant identity. • Make the substitution $u = \sin x$.	$\cos^2 x = 1 - \sin^2 x$
m odd	• Split off a factor of $\sin x$. • Apply the relevant identity. • Make the substitution $u = \cos x$.	$\sin^2 x = 1 - \cos^2 x$
$\begin{cases} m \text{ even} \\ n \text{ even} \end{cases}$	• Use the relevant identities to reduce the powers on $\sin x$ and $\cos x$.	$\begin{cases} \sin^2 x = \frac{1}{2}(1 - \cos 2x) \\ \cos^2 x = \frac{1}{2}(1 + \cos 2x) \end{cases}$

Example 2 Evaluate

(a) $\displaystyle\int \cos^3 x \, dx$　　(b) $\displaystyle\int \sin^4 x \, dx$

Solution (a).

$$\int \cos^3 x \, dx = \int \cos^2 x \cos x \, dx$$

$$= \int (1 - \sin^2 x) \cos x \, dx$$

$$= \int (1 - u^2) \, du \qquad u = \sin x, \, du = \cos x \, dx$$

$$= u - \frac{u^3}{3} + C$$

$$= \sin x - \frac{1}{3} \sin^3 x + C$$

Solution (b).

$$\int \sin^4 x \, dx = \int (\sin^2 x)^2 \, dx = \int [\tfrac{1}{2}(1 - \cos 2x)]^2 \, dx$$

$$= \frac{1}{4} \int (1 - 2\cos 2x + \cos^2 2x) \, dx$$

To finish, we apply (1b) to $\cos^2 2x$ to obtain

$$\cos^2 2x = \frac{1}{2}(1 + \cos 4x) = \frac{1}{2} + \frac{1}{2}\cos 4x$$

which gives

$$\int \sin^4 x \, dx = \frac{1}{4} \int \left(\frac{3}{2} - 2\cos 2x + \frac{1}{2}\cos 4x \right) dx$$

$$= \frac{3}{8} x - \frac{1}{4} \sin 2x + \frac{1}{32} \sin 4x + C \qquad \blacktriangleleft$$

Example 3 Evaluate $\displaystyle\int \sin^4 x \cos^5 x \, dx$.

Solution. Since $n = 5$ is odd, we shall follow the first procedure in Table 9.3.1:

$$\int \sin^4 x \cos^5 x \, dx = \int \sin^4 x \cos^4 x \cos x \, dx$$

$$= \int \sin^4 x (1 - \sin^2 x)^2 \cos x \, dx$$

$$= \int u^4 (1 - u^2)^2 \, du$$

$$= \int (u^4 - 2u^6 + u^8) \, du$$

$$= \frac{1}{5} u^5 - \frac{2}{7} u^7 + \frac{1}{9} u^9 + C$$

$$= \frac{1}{5} \sin^5 x - \frac{2}{7} \sin^7 x + \frac{1}{9} \sin^9 x + C \quad \blacktriangleleft$$

Example 4 Evaluate $\int \sin^4 x \cos^4 x \, dx$.

Solution. Since $m = n = 4$, we shall follow the third procedure in Table 9.3.1:

$$\int \sin^4 x \cos^4 x \, dx = \int (\sin^2 x)^2 (\cos^2 x)^2 \, dx$$

$$= \int (\tfrac{1}{2}[1 - \cos 2x])^2 (\tfrac{1}{2}[1 + \cos 2x])^2 \, dx$$

$$= \frac{1}{16} \int (1 - \cos^2 2x)^2 \, dx$$

$$= \frac{1}{16} \int \sin^4 2x \, dx$$

> Note that this can be obtained more directly from the original integral using identity $\sin x \cos x = \frac{1}{2} \sin 2x$.

To finish, we shall let $u = 2x$, $du = 2 \, dx$ and then use the result in Example 2(b).

$$\int \sin^4 x \cos^4 x \, dx = \frac{1}{32} \int \sin^4 u \, du$$

$$= \frac{1}{32} \left(\frac{3}{8} u - \frac{1}{4} \sin 2u + \frac{1}{32} \sin 4u \right) + C$$

$$= \frac{3}{128} x - \frac{1}{128} \sin 4x + \frac{1}{1024} \sin 8x + C \quad \blacktriangleleft$$

Integrals of the form

$$\int \sin mx \cos nx \, dx, \quad \int \sin mx \sin nx \, dx, \quad \int \cos mx \cos nx \, dx$$

can be found using the product to sum formulas from trigonometry (see Formulas (47–49) of Appendix B.I.).

Example 5 Evaluate $\int \sin 7x \cos 3x \, dx$.

Solution. Since

$$\sin 7x \cos 3x = \frac{1}{2}(\sin 4x + \sin 10x)$$

we can write

$$\int \sin 7x \cos 3x\, dx = \frac{1}{2}\int (\sin 4x + \sin 10x)\, dx$$

$$= -\frac{1}{8}\cos 4x - \frac{1}{20}\cos 10x + C \quad \blacktriangleleft$$

▶ Exercise Set 9.3

In Exercises 1–30, perform the indicated integration.

1. $\displaystyle\int \cos^5 x \sin x\, dx.$

2. $\displaystyle\int \sin^4 3x \cos 3x\, dx.$

3. $\displaystyle\int \sin ax \cos ax\, dx \quad (a \neq 0).$

4. $\displaystyle\int \cos^2 3x\, dx.$

5. $\displaystyle\int \sin^2 5\theta\, d\theta.$

6. $\displaystyle\int \cos^3 at\, dt \quad (a \neq 0).$

7. $\displaystyle\int \cos^4 (x/4)\, dx.$

8. $\displaystyle\int \sin^5 x\, dx.$

9. $\displaystyle\int \cos^5 \theta\, d\theta.$

10. $\displaystyle\int \sin^3 x \cos^3 x\, dx.$

11. $\displaystyle\int \sin^2 2t \cos^3 2t\, dt.$

12. $\displaystyle\int \sin^4 x \cos^5 x\, dx.$

13. $\displaystyle\int \cos^4 x \sin^3 x\, dx.$

14. $\displaystyle\int \sin^3 2x \cos^2 2x\, dx.$

15. $\displaystyle\int \sin^5 \theta \cos^4 \theta\, d\theta.$

16. $\displaystyle\int \cos^{1/5} x \sin x\, dx.$

17. $\displaystyle\int \sin^2 x \cos^2 x\, dx.$

18. $\displaystyle\int \sin^2 x \cos^4 x\, dx.$

19. $\displaystyle\int \sin x \cos 2x\, dx.$

20. $\displaystyle\int \sin 3\theta \cos 2\theta\, d\theta.$

21. $\displaystyle\int \sin x \cos (x/2)\, dx.$

22. $\displaystyle\int \sin ax \cos bx\, dx \quad (a > 0, b > 0, a \neq b).$

23. $\displaystyle\int \frac{\sin x}{\cos^8 x}\, dx.$

24. $\displaystyle\int \sqrt{\cos \theta} \sin \theta\, d\theta.$

25. $\displaystyle\int_0^{\pi/4} \cos^3 x\, dx.$

26. $\displaystyle\int_{-\pi}^{\pi} \cos^2 5\theta\, d\theta.$

27. $\displaystyle\int_0^{\pi/3} \sin^4 3x \cos^3 3x\, dx.$

28. $\displaystyle\int_0^{\pi/2} \sin^2 \frac{x}{2} \cos^2 \frac{x}{2}\, dx.$

29. $\displaystyle\int_0^{\pi/6} \sin 2x \cos 4x\, dx.$

30. $\displaystyle\int_0^{2\pi} \sin^2 kx\, dx \quad (k \neq 0).$

31. Let m, n be distinct nonnegative integers. Prove:

(a) $\displaystyle\int_0^{2\pi} \sin mx \cos nx\, dx = 0$

(b) $\displaystyle\int_0^{2\pi} \cos mx \cos nx\, dx = 0$

(c) $\displaystyle\int_0^{2\pi} \sin mx \sin nx\, dx = 0.$

32. The region bounded below by the x-axis and above by the portion of $y = \sin x$ from $x = 0$ to $x = \pi$ is revolved about the x-axis. Find the volume of the resulting solid.

33. Find the volume of the solid that results when the region enclosed by $y = \cos x$, $y = \sin x$, $x = 0$, and $x = \pi/4$ is revolved about the x-axis.

34. (a) Use Formula (7) in Section 9.2 to show that

$$\int_0^{\pi/2} \sin^n x\, dx = \frac{n-1}{n}\int_0^{\pi/2} \sin^{n-2} x\, dx$$

(b) Use this result to derive the **Wallis sine formulas**:

$$\int_0^{\pi/2} \sin^n x\, dx = \frac{\pi}{2}\cdot \frac{1\cdot 3\cdot 5\cdots(n-1)}{2\cdot 4\cdot 6\cdots n} \quad \binom{n\ \text{even}}{\text{and}\ \geq 2}$$

$$\int_0^{\pi/2} \sin^n x\, dx = \frac{2\cdot 4\cdot 6\cdots(n-1)}{3\cdot 5\cdot 7\cdots n} \quad \binom{n\ \text{odd}}{\text{and}\ \geq 3}$$

35. Use the Wallis formulas in Exercise 34 to evaluate

(a) $\displaystyle\int_0^{\pi/2} \sin^3 x\, dx$

(b) $\displaystyle\int_0^{\pi/2} \sin^4 x\, dx$

(c) $\displaystyle\int_0^{\pi/2} \sin^5 x\, dx$

(d) $\displaystyle\int_0^{\pi/2} \sin^6 x\, dx.$

36. Use Formula (8) in Section 9.2 and the method of Exercise 34 to derive the **Wallis cosine formulas**:

$$\int_0^{\pi/2} \cos^n x\, dx = \frac{2\cdot 4\cdot 6\cdots(n-1)}{3\cdot 5\cdot 7\cdots n} \quad \binom{n\ \text{odd}}{\text{and}\ \geq 3}$$

$$\int_0^{\pi/2} \cos^n x\, dx = \frac{\pi}{2}\cdot \frac{1\cdot 3\cdot 5\cdots(n-1)}{2\cdot 4\cdot 6\cdots n} \quad \binom{n\ \text{even}}{\text{and}\ \geq 2}$$

◼ **9.4** INTEGRATING POWERS OF SECANT AND TANGENT

> *In this section we shall discuss methods for evaluating integrals of the form*
>
> $$\int \tan^m x \sec^n x \, dx$$
>
> *where m and n are nonnegative integers.*

We begin with the two basic integral formulas

$$\int \tan x \, dx = \ln |\sec x| + C \tag{1}$$

$$\int \sec x \, dx = \ln |\sec x + \tan x| + C \tag{2}$$

the first of which can be obtained by writing

$$\int \tan x \, dx = \int \frac{\sin x}{\cos x} \, dx$$

$$= -\ln |\cos x| + C$$

$$\boxed{\begin{aligned} u &= \cos x \\ du &= -\sin x \, dx \end{aligned}}$$

$$= \ln |\sec x| + C$$

$$\boxed{\ln |\cos x| = -\ln \frac{1}{|\cos x|}}$$

Integral (2) requires a trick. We write

$$\int \sec x \, dx = \int \sec x \left(\frac{\sec x + \tan x}{\sec x + \tan x} \right) dx$$

$$= \int \frac{\sec^2 x + \sec x \tan x}{\sec x + \tan x} \, dx$$

$$= \ln |\sec x + \tan x| + C$$

$$\boxed{\begin{aligned} u &= \sec x + \tan x \\ du &= (\sec^2 x + \sec x \tan x) \, dx \end{aligned}}$$

Higher powers of secant and tangent can be evaluated using the reduction formulas

$$\int \sec^n x \, dx = \frac{\sec^{n-2} x \tan x}{n-1} + \frac{n-2}{n-1} \int \sec^{n-2} x \, dx \tag{3}$$

$$\int \tan^m x \, dx = \frac{\tan^{m-1} x}{m-1} - \int \tan^{m-2} x \, dx \tag{4}$$

[See Exercise 52(a) of Section 9.2 and Exercise 41 of this section.]

Example 1   Evaluate

(a) $\displaystyle\int \sec^3 x \, dx$ (b) $\displaystyle\int \tan^5 x \, dx$

Solution (a).   From (3) with $n = 3$,

$$\int \sec^3 x \, dx = \frac{\sec x \tan x}{2} + \frac{1}{2} \int \sec x \, dx$$

$$= \tfrac{1}{2} \sec x \tan x + \tfrac{1}{2} \ln |\sec x + \tan x| + C$$

Solution (b).   We shall use Formula (4) twice.

$$\int \tan^5 x \, dx = \frac{\tan^4 x}{4} - \int \tan^3 x \, dx$$

$$= \frac{\tan^4 x}{4} - \left[\frac{\tan^2 x}{2} - \int \tan x \, dx \right]$$

$$= \tfrac{1}{4} \tan^4 x - \tfrac{1}{2} \tan^2 x + \ln |\sec x| + C \qquad \blacktriangleleft$$

If m and n are positive integers, then the integral

$$\int \tan^m x \sec^n x \, dx$$

can be evaluated by one of the three procedures in Table 9.4.1.

Table 9.4.1

$\int \tan^m x \sec^n x \, dx$	PROCEDURE	RELEVANT IDENTITIES
n even	• Split off a factor of $\sec^2 x$. • Apply the relevant identity. • Make the substitution $u = \tan x$.	$\sec^2 x = \tan^2 x + 1$
m odd	• Split off a factor of $\sec x \tan x$. • Apply the relevant identity. • Make the substitution $u = \sec x$.	$\tan^2 x = \sec^2 x - 1$
$\begin{cases} m \text{ even} \\ n \text{ odd} \end{cases}$	• Use the relevant identity to reduce the integrand to powers of $\sec x$ alone. • Then use the reduction formula for powers of $\sec x$.	$\tan^2 x = \sec^2 x - 1$

Example 2   Evaluate $\displaystyle\int \tan^2 x \sec^4 x \, dx$.

Solution.   Since $n = 4$ is even, we shall follow the first procedure in Table 9.4.1:

$$\int \tan^2 x \sec^4 x \, dx = \int \tan^2 x \sec^2 x \sec^2 x \, dx$$

$$= \int \tan^2 x \, (\tan^2 x + 1) \sec^2 x \, dx$$

$$= \int u^2 (u^2 + 1) \, du$$

$$= \tfrac{1}{5} u^5 + \tfrac{1}{3} u^3 + C = \tfrac{1}{5} \tan^5 x + \tfrac{1}{3} \tan^3 x + C \qquad \blacktriangleleft$$

Example 3   Evaluate $\displaystyle\int \tan^3 x \sec^3 x \, dx$.

Solution.   Since $m = 3$ is odd, we shall follow the second procedure in Table 9.4.1:

$$\int \tan^3 x \sec^3 x \, dx = \int \tan^2 x \sec^2 x \, (\sec x \tan x) \, dx$$

$$= \int (\sec^2 x - 1) \sec^2 x \, (\sec x \tan x) \, dx$$

$$= \int (u^2 - 1) u^2 \, du$$

$$= \tfrac{1}{5} u^5 - \tfrac{1}{3} u^3 + C = \tfrac{1}{5} \sec^5 x - \tfrac{1}{3} \sec^3 x + C \qquad \blacktriangleleft$$

Example 4   Evaluate $\int \tan^2 x \sec x \, dx$.

Solution.   Since $m = 2$ and $n = 1$, we shall follow the third procedure in Table 9.4.1:

$$\int \tan^2 x \sec x \, dx = \int (\sec^2 x - 1) \sec x \, dx = \int \sec^3 x \, dx - \int \sec x \, dx$$

See Example 1.

$$= \tfrac{1}{2} \sec x \tan x + \tfrac{1}{2} \ln \left| \sec x + \tan x \right| - \ln \left| \sec x + \tan x \right| + C$$

$$= \tfrac{1}{2} \sec x \tan x - \tfrac{1}{2} \ln \left| \sec x + \tan x \right| + C \qquad \blacktriangleleft$$

Example 5   Instead of using reduction formula (3), the first procedure of Table 9.4.1 can be used to integrate an even power of sec x. For example,

$$\int \sec^6 x \, dx = \int \sec^4 x \sec^2 x \, dx = \int (\sec^2 x)^2 \sec^2 x \, dx$$

$$= \int (\tan^2 x + 1)^2 \sec^2 x \, dx = \int (u^2 + 1)^2 \, du \qquad \boxed{\text{Let } u = \tan x, \\ du = \sec^2 x \, dx.}$$

$$= \int (u^4 + 2u^2 + 1) \, du = \tfrac{1}{5} u^5 + \tfrac{2}{3} u^3 + u + C$$

$$= \tfrac{1}{5} \tan^5 x + \tfrac{2}{3} \tan^3 x + \tan x + C \qquad \blacktriangleleft$$

REMARK.   With the aid of the identity $1 + \cot^2 x = \csc^2 x$ the techniques in Table 9.4.1 can be adapted to treat integrals of the form

$$\int \cot^m x \csc^n x \, dx$$

Also, there are reduction formulas for powers of csc and cot that are analagous to Formulas (3) and (4).

▶ **Exercise Set 9.4**

In Exercises 1–34, perform the indicated integration.

1. $\int \sec^2 (3x + 1) \, dx.$

2. $\int \tan 5x \, dx.$

3. $\int e^{-2x} \tan (e^{-2x}) \, dx.$

4. $\int \cot 3x \, dx.$

5. $\int \sec 2x \, dx.$

6. $\int \dfrac{\sec (\sqrt{x})}{\sqrt{x}} \, dx.$

7. $\int \tan^2 x \sec^2 x \, dx.$

8. $\int \tan^5 x \sec^4 x \, dx.$

9. $\int \tan^3 4x \sec^4 4x \, dx.$

10. $\int \tan^4 \theta \sec^4 \theta \, d\theta.$

11. $\int \sec^5 x \tan^3 x \, dx.$

12. $\int \tan^5 \theta \sec \theta \, d\theta.$

13. $\int \tan^4 x \sec x \, dx.$

14. $\int \tan^2 \frac{x}{2} \sec^3 \frac{x}{2} \, dx.$

15. $\int \tan 2t \sec^3 2t \, dt.$

16. $\int \tan x \sec^5 x \, dx.$

17. $\int \sec^4 x \, dx.$

18. $\int \sec^5 x \, dx.$

19. $\int \sec^6 (\pi x) \, dx.$

20. $\int \tan^3 4x \, dx.$

21. $\int \tan^4 x \, dx.$

22. $\int \tan^7 \theta \, d\theta.$

23. $\int x \tan^2 (x^2) \sec^2 (x^2) \, dx.$

24. $\int \tan^2 (1 - 2x) \sec (1 - 2x) \, dx.$

25. $\int \cot^3 x \csc^3 x \, dx.$

26. $\int \cot^2 3t \sec 3t \, dt.$

27. $\int \cot^3 x \, dx.$

28. $\int \csc^4 x \, dx.$

29. $\int \sqrt{\tan x} \sec^4 x \, dx.$

30. $\int \tan x \sec^{3/2} x \, dx.$

31. $\int_0^{\pi/6} \tan^2 2x \, dx.$

32. $\int_0^{\pi/6} \sec^3 \theta \tan \theta \, d\theta.$

33. $\int_0^{\pi/2} \tan^5 \frac{x}{2} \, dx.$

34. $\int_{\pi/4}^{\pi/2} \csc^3 x \cot x \, dx.$

35. Find the arc length of the curve $y = \ln (\cos x)$ over the interval $[0, \pi/4]$.

36. Find the volume of the solid generated when the region enclosed by $y = \tan x$, $y = 1$, and $x = 0$ is revolved about the x-axis.

37. (a) Show that
$$\int \csc x \, dx = -\ln |\csc x + \cot x| + C$$
(b) Show that the result in part (a) can also be written as
$$\int \csc x \, dx = \ln |\csc x - \cot x| + C$$
and
$$\int \csc x \, dx = \ln |\tan \tfrac{1}{2} x| + C$$

38. Rewrite $\sin x + \cos x$ in the form
$$A \sin (x + \phi)$$
and use your result together with Exercise 37 to evaluate
$$\int \frac{dx}{\sin x + \cos x}$$

39. Use the method of Exercise 38 to evaluate
$$\int \frac{dx}{a \sin x + b \cos x} \quad (a, b \text{ not both zero})$$

40. Use integration by parts and Formula (2) to evaluate $\int \sec^3 x \, dx$.

41. Derive reduction formula (4).

■ 9.5 TRIGONOMETRIC SUBSTITUTIONS

In this section we shall show how to evaluate integrals that contain expressions of the form
$$\sqrt{a^2 - x^2}, \quad \sqrt{x^2 + a^2}, \quad \text{and} \quad \sqrt{x^2 - a^2}$$
$(a > 0)$ by making substitutions involving trigonometric functions. We shall also show how the method of completing the square can sometimes be used to help evaluate integrals involving expressions of the form $ax^2 + bx + c$.

The basic idea for evaluating an integral that involves one of the radicals described above is to make a substitution that will eliminate the radical. For example, to eliminate the radical in the expression $\sqrt{a^2 - x^2}$ we can make the substitution

$$x = a \sin \theta, \quad -\pi/2 \le \theta \le \pi/2 \tag{1}$$

which yields

$$\sqrt{a^2 - x^2} = \sqrt{a^2 - a^2 \sin^2 \theta} = \sqrt{a^2(1 - \sin^2 \theta)}$$
$$= a\sqrt{\cos^2 \theta} = a |\cos \theta| = a \cos \theta$$

$\underset{\boxed{\cos \theta \ge 0 \text{ since } -\pi/2 \le \theta \le \pi/2}}{\uparrow}$

The purpose of the restriction on θ in (1) is twofold. First, it enables us to replace $|\cos \theta|$ by $\cos \theta$, thereby simplifying the resulting calculations, and second, the interval to which θ is restricted is the range of the function $\sin^{-1}$ (see Section 8.1), so (1) can be rewritten as

$$\theta = \sin^{-1}(x/a)$$

where needed. Table 9.5.1 lists the trigonometric substitutions that we will be using in this section.

Table 9.5.1

EXPRESSION IN THE INTEGRAND	SUBSTITUTION	RESTRICTION ON θ	TRIGONOMETRIC IDENTITY NEEDED FOR SIMPLIFICATION
$\sqrt{a^2 - x^2}$	$x = a \sin \theta$	$-\pi/2 \leq \theta \leq \pi/2$	$a^2 - a^2 \sin^2 \theta = a^2 \cos^2 \theta$
$\sqrt{a^2 + x^2}$	$x = a \tan \theta$	$-\pi/2 < \theta < \pi/2$	$a^2 + a^2 \tan^2 \theta = a^2 \sec^2 \theta$
$\sqrt{x^2 - a^2}$	$x = a \sec \theta$	$\begin{cases} 0 \leq \theta < \pi/2 & \text{(if } x \geq a) \\ \pi \leq \theta < 3\pi/2 & \text{(if } x \leq -a) \end{cases}$	$a^2 \sec^2 \theta - a^2 = a^2 \tan^2 \theta$

Example 1 Evaluate $\displaystyle\int \frac{dx}{x^2 \sqrt{4 - x^2}}$.

Solution.   To eliminate the radical we make the substitution

$$x = 2 \sin \theta, \quad -\pi/2 \leq \theta \leq \pi/2$$

$$\frac{dx}{d\theta} = 2 \cos \theta \quad \text{or} \quad dx = 2 \cos \theta \, d\theta$$

This yields

$$\int \frac{dx}{x^2 \sqrt{4 - x^2}} = \int \frac{2 \cos \theta \, d\theta}{(2 \sin \theta)^2 \sqrt{4 - 4 \sin^2 \theta}}$$

$$= \int \frac{2 \cos \theta \, d\theta}{(2 \sin \theta)^2 (2 \cos \theta)} = \frac{1}{4} \int \frac{d\theta}{\sin^2 \theta}$$

$$= \frac{1}{4} \int \csc^2 \theta \, d\theta = -\frac{1}{4} \cot \theta + C$$

To complete the solution we must express $\cot \theta$ in terms of x. This can be done using trigonometric identities or by representing the substitution $x = 2 \sin \theta$ ($\sin \theta = x/2$) as in Figure 9.5.1. From the figure we obtain

$$\cot \theta = \frac{\sqrt{4 - x^2}}{x}$$

so that

$$\int \frac{dx}{x^2 \sqrt{4 - x^2}} = -\frac{1}{4} \cot \theta + C = -\frac{1}{4} \frac{\sqrt{4 - x^2}}{x} + C \quad \blacktriangleleft$$

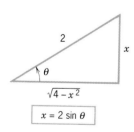

$x = 2 \sin \theta$

Figure 9.5.1

Example 2 Evaluate $\displaystyle\int \frac{dx}{\sqrt{x^2 + a^2}}$.

Solution.   To eliminate the radical we make the substitution

$$x = a \tan \theta, \quad -\pi/2 < \theta < \pi/2 \tag{2}$$

$$\frac{dx}{d\theta} = a \sec^2 \theta \quad \text{or} \quad dx = a \sec^2 \theta \, d\theta$$

This yields

$$\int \frac{dx}{\sqrt{x^2 + a^2}} = \int \frac{a \sec^2 d\theta}{\sqrt{a^2 \tan^2 \theta + a^2}} = \int \frac{a \sec^2 \theta \, d\theta}{a \, |\sec \theta|} \uparrow \int \frac{a \sec^2 \theta \, d\theta}{a \sec \theta}$$

$$\boxed{\sec \theta > 0 \text{ since } -\pi/2 < \theta < \pi/2}$$

$$= \int \sec \theta \, d\theta = \ln |\sec \theta + \tan \theta| + C$$

To express the solution in terms of x, we can represent (2) by the triangle in Figure 9.5.2, from which we obtain

$$\sec \theta = \frac{\sqrt{x^2 + a^2}}{a} \quad \text{and} \quad \tan \theta = \frac{x}{a}$$

so

$$\int \frac{dx}{\sqrt{x^2 + a^2}} = \ln \left| \frac{\sqrt{x^2 + a^2}}{a} + \frac{x}{a} \right| + C$$

or if preferred we can rewrite the expression on the right as

$$\ln |\sqrt{x^2 + a^2} + x| - \ln a + C$$

and combine the constant $\ln a$ with the constant of integration to obtain

$$\int \frac{dx}{\sqrt{x^2 + a^2}} = \ln |\sqrt{x^2 + a^2} + x| + C'$$

Moreover, $\sqrt{x^2 + a^2} + x > 0$ for all x, so we can drop the absolute value sign and write

$$\int \frac{dx}{\sqrt{x^2 + a^2}} = \ln (\sqrt{x^2 + a^2} + x) + C' \tag{3}$$

◀

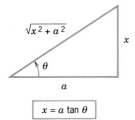

$$x = a \tan \theta$$

Figure 9.5.2

Example 3 Evaluate $\displaystyle\int \frac{\sqrt{x^2 - 25}}{x} \, dx$.

Solution. To eliminate the radical, we make the substitution

$$x = 5 \sec \theta, \quad 0 \le \theta < \pi/2 \quad \text{or} \quad \pi \le \theta < 3\pi/2 \tag{4}$$

$$\frac{dx}{d\theta} = 5 \sec \theta \tan \theta \quad \text{or} \quad dx = 5 \sec \theta \tan \theta \, d\theta$$

Thus,

$$\int \frac{\sqrt{x^2 - 25}}{x} \, dx = \int \frac{\sqrt{25 \sec^2 \theta - 25}}{5 \sec \theta} (5 \sec \theta \tan \theta) \, d\theta$$

$$= \int \frac{5 \, |\tan \theta|}{5 \sec \theta} (5 \sec \theta \tan \theta) \, d\theta$$

$$= 5 \int \tan^2 \theta \, d\theta$$

$$\boxed{\begin{array}{l} \tan \theta \ge 0 \text{ since} \\ 0 \le \theta < \pi/2 \quad \text{or} \quad \pi \le \theta < 3\pi/2 \end{array}}$$

$$= 5 \int (\sec^2 \theta - 1) \, d\theta$$

$$= 5 \tan \theta - 5\theta + C$$

To express the solution in terms of x, we can represent (4) by the triangle in Figure 9.5.3, from which we obtain

$$\tan \theta = \frac{\sqrt{x^2 - 25}}{5}$$

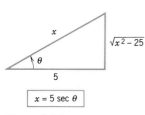

$$x = 5 \sec \theta$$

Figure 9.5.3

so that

$$\int \frac{\sqrt{x^2 - 25}}{x} dx = \sqrt{x^2 - 25} - 5 \sec^{-1}\left(\frac{x}{5}\right) + C \quad \blacktriangleleft$$

The integral in the next example will arise frequently in later sections.

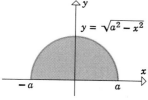

Figure 9.5.4

Example 4 Evaluate $\int_{-a}^{a} \sqrt{a^2 - x^2}\, dx \ (a > 0)$.

Solution. This integral can be evaluated by the substitution $x = a \sin \theta$ (verify); however, a better approach is to observe that the integral represents the area of a semicircle of radius a (Figure 9.5.4). Thus,

$$\int_{-a}^{a} \sqrt{a^2 - x^2}\, dx = \frac{1}{2}\pi a^2 \quad \blacktriangleleft$$

☐ **INTEGRALS INVOLVING** $ax^2 + bx + c$

Integrals that involve a quadratic expression $ax^2 + bx + c$, where $a \neq 0$ and $b \neq 0$, can often be evaluated by first completing the square, then making an appropriate substitution. The following examples illustrate this idea.

Example 5 Evaluate $\int \frac{x}{x^2 - 4x + 8} dx$.

Solution. Completing the square yields

$$x^2 - 4x + 8 = (x^2 - 4x + 4) + 8 - 4 = (x - 2)^2 + 4$$

Thus, the substitution

$$u = x - 2, \quad du = dx$$

yields

$$\int \frac{x}{x^2 - 4x + 8} dx = \int \frac{x}{(x - 2)^2 + 4} dx = \int \frac{u + 2}{u^2 + 4} du$$

$$= \int \frac{u}{u^2 + 4} du + 2 \int \frac{du}{u^2 + 4}$$

$$= \frac{1}{2} \int \frac{2u}{u^2 + 4} du + 2 \int \frac{du}{u^2 + 4}$$

$$= \frac{1}{2} \ln (u^2 + 4) + 2\left(\frac{1}{2}\right) \tan^{-1}\frac{u}{2} + C$$

$$= \frac{1}{2} \ln [(x - 2)^2 + 4] + \tan^{-1}\left(\frac{x - 2}{2}\right) + C \quad \blacktriangleleft$$

Example 6 Evaluate $\int \frac{dx}{\sqrt{5 - 4x - 2x^2}}$.

Solution. Completing the square yields

$$5 - 4x - 2x^2 = 5 - 2(x^2 + 2x) = 5 - 2(x^2 + 2x + 1) + 2$$

$$= 5 - 2(x + 1)^2 + 2 = 7 - 2(x + 1)^2$$

Thus,

$$\int \frac{dx}{\sqrt{5 - 4x - 2x^2}} = \int \frac{dx}{\sqrt{7 - 2(x + 1)^2}}$$

$$= \int \frac{du}{\sqrt{7 - 2u^2}} \qquad \boxed{\begin{array}{l} u = x + 1 \\ du = dx \end{array}}$$

$$= \frac{1}{\sqrt{2}} \int \frac{du}{\sqrt{(7/2) - u^2}}$$

$$= \frac{1}{\sqrt{2}} \sin^{-1}\left(\frac{u}{\sqrt{7/2}}\right) + C \qquad \boxed{\begin{array}{l} \text{Formula (21),} \\ \text{Section 9.1} \\ \text{with } a = \sqrt{7/2} \end{array}}$$

$$= \frac{1}{\sqrt{2}} \sin^{-1}\left(\sqrt{2/7}(x + 1)\right) + C \qquad \blacktriangleleft$$

▶ Exercise Set 9.5

In Exercises 1–30, perform the indicated integration.

1. $\int \sqrt{4 - x^2}\, dx.$

2. $\int \sqrt{1 - 4x^2}\, dx.$

3. $\int \frac{x^2}{\sqrt{9 - x^2}}\, dx.$

4. $\int \frac{dx}{x^2\sqrt{16 - x^2}}.$

5. $\int \frac{dx}{(4 + x^2)^2}.$

6. $\int \frac{dx}{(x^2 + 1)^{3/2}}.$

7. $\int \frac{\sqrt{x^2 - 9}}{x}\, dx.$

8. $\int \frac{dx}{x^2\sqrt{x^2 - 16}}.$

9. $\int \frac{x^3}{\sqrt{2 - x^2}}\, dx.$

10. $\int x^3\sqrt{5 - x^2}\, dx.$

11. $\int \frac{dx}{(3 + x^2)^{3/2}}.$

12. $\int \frac{x^2}{\sqrt{5 + x^2}}\, dx.$

13. $\int \frac{dx}{x^2\sqrt{4x^2 - 9}}.$

14. $\int \frac{\sqrt{1 + t^2}}{t}\, dt.$

15. $\int \frac{dx}{(1 - x^2)^{3/2}}.$

16. $\int \frac{dx}{x^2\sqrt{x^2 + 25}}.$

17. $\int \frac{dx}{\sqrt{x^2 - 1}}.$

18. $\int \frac{dx}{1 + 2x^2 + x^4}.$

19. $\int \frac{dx}{x^2\sqrt{9 - 4x^2}}.$

20. $\int \frac{x^2}{\sqrt{x^2 - 25}}\, dx.$

21. $\int \frac{dx}{(9x^2 - 1)^{3/2}}.$

22. $\int \frac{\cos\theta}{\sqrt{2 - \sin^2\theta}}\, d\theta.$

23. $\int e^x\sqrt{1 - e^{2x}}\, dx.$

24. $\int_0^{1/3} \frac{dx}{(4 - 9x^2)^2}.$

25. $\int_0^4 x^3\sqrt{16 - x^2}\, dx.$

26. $\int_{\sqrt{2}}^2 \frac{\sqrt{2x^2 - 4}}{x}\, dx.$

27. $\int_{\sqrt{2}}^2 \frac{dx}{x^2\sqrt{x^2 - 1}}.$

28. $\int_{-1/\sqrt{2}}^{1/\sqrt{2}} (1 - 2x^2)^{3/2}\, dx.$

29. $\int_1^3 \frac{dx}{x^4\sqrt{x^2 + 3}}.$

30. $\int_0^3 \frac{x^3}{(3 + x^2)^{5/2}}\, dx.$

31. The integral

$$\int \frac{x}{x^2 + 4}\, dx$$

can be evaluated either by a trigonometric substitution or by the substitution $u = x^2 + 4$. Do it both ways and show that the results are equivalent.

32. By integrating, show that the area of a circle of radius r is πr^2. [*Hint:* $x^2 + y^2 = r^2$ is the equation of such a circle.]

33. Find the arc length of the curve $y = \ln x$ from $x = 1$ to $x = 2$.

34. Find the arc length of the curve $y = x^2$ from $x = 0$ to $x = 1$.

35. Find the area of the surface generated when the curve in Exercise 34 is revolved about the x-axis.

36. Find the volume of the solid generated when the region enclosed by $x = y(1 - y^2)^{1/4}$, $y = 0$, $y = 1$, and $x = 0$ is revolved about the y-axis.

In Exercises 37 and 38 the trigonometric substitutions $x = a \sec\theta$ and $x = a \tan\theta$ lead to difficult integrals; for such integrals it is possible to use the **hyperbolic substitutions**

$x = a \sinh u$ for integrals involving $\sqrt{x^2 + a^2}$

$x = a \cosh u$ for integrals involving $\sqrt{x^2 - a^2}$

These substitutions are useful because in each case the hyperbolic identity

$$a^2 \cosh^2 u - a^2 \sinh^2 u = a^2$$

removes the radical.

---

37. (a) Evaluate

$$\int \frac{dx}{\sqrt{x^2 + 9}}$$

using the hyperbolic substitution that is suggested above.

(b) Evaluate the integral in part (a) by a trigonometric substitution and show that the results in parts (a) and (b) agree.

38. Follow the directions of Exercise 37 for the integral

$$\int \sqrt{x^2 - 1}\, dx, \quad x \geq 1$$

In Exercises 39–53, perform the integrations.

39. $\displaystyle\int \frac{dx}{x^2 - 4x + 13}.$

40. $\displaystyle\int \frac{dx}{\sqrt{2x - x^2}}.$

41. $\displaystyle\int \frac{dx}{\sqrt{8 + 2x - x^2}}.$

42. $\displaystyle\int \frac{dx}{16x^2 + 16x + 5}.$

43. $\displaystyle\int \frac{dx}{\sqrt{x^2 - 6x + 10}}.$

44. $\displaystyle\int \frac{x}{x^2 + 6x + 10}\, dx.$

45. $\displaystyle\int \sqrt{3 - 2x - x^2}\, dx.$

46. $\displaystyle\int \frac{e^x}{\sqrt{1 + e^x + e^{2x}}}\, dx.$

47. $\displaystyle\int \frac{dx}{2x^2 + 4x + 7}.$

48. $\displaystyle\int \frac{\cos\theta}{\sin^2\theta - 6\sin\theta + 12}\, d\theta.$

49. $\displaystyle\int \frac{2x + 5}{x^2 + 2x + 5}\, dx.$

50. $\displaystyle\int \frac{2x + 3}{4x^2 + 4x + 5}\, dx.$

51. $\displaystyle\int \frac{x + 3}{\sqrt{x^2 + 2x + 2}}\, dx.$

52. $\displaystyle\int_1^2 \frac{dx}{\sqrt{4x - x^2}}.$

53. $\displaystyle\int_0^1 \sqrt{x(4 - x)}\, dx.$

■ 9.6 INTEGRATING RATIONAL FUNCTIONS; PARTIAL FRACTIONS

Recall that a rational function is the quotient of two polynomials. In general, rational functions can be quite difficult to integrate. However, in this section we shall give a method that, in theory, can be used to express any rational function as a sum of simple rational functions that can be integrated by the methods studied in earlier sections.

☐ **PARTIAL FRACTIONS**

In algebra one learns to combine two or more fractions into a single fraction by finding a common denominator; for example,

$$\frac{2}{x - 4} + \frac{3}{x + 1} = \frac{2(x + 1) + 3(x - 4)}{(x - 4)(x + 1)} = \frac{5x - 10}{x^2 - 3x - 4} \tag{1}$$

However, the left side of (1) is easier to integrate than the right side:

$$\int \frac{5x - 10}{x^2 - 3x - 4}\, dx = \int \frac{2}{x - 4}\, dx + \int \frac{3}{x + 1}\, dx$$

$$= 2\ln|x - 4| + 3\ln|x + 1| + C$$

Thus, for purposes of integration it would be desirable to have a method for obtaining the left side of (1) starting with the right side. To see how this can be done, let us factor the denominator on the right side of (1) and *assume* that there exist unknown constants A and B such that

$$\frac{5x - 10}{(x - 4)(x + 1)} = \frac{A}{x - 4} + \frac{B}{x + 1} \tag{2}$$

The terms on the right side of (2) are called ***partial fractions*** (because they each constitute *part* of the expression on the left side), and the entire right side of (2) is called the ***partial fraction decomposition*** of the left side. Observe that the denominators of the partial fractions are the factors of the denominator on the left side.

To find the constants A and B, we first multiply (2) through by $(x - 4)(x + 1)$ to clear fractions:

$$5x - 10 = A(x + 1) + B(x - 4) \tag{3}$$

This equation is actually an identity that holds for all real values of x; thus, it holds for any values of x that we choose to substitute. In particular, we can substitute $x = 4$ (which makes the second term on the right zero) and $x = -1$ (which makes the first term on the right zero). These substitutions yield

$$A = 2 \quad \text{and} \quad B = 3$$

which agrees with (1).

As an alternative method for finding A and B, we can multiply out on the right side of (3) and collect like powers of x:

$$5x - 10 = (A + B)x + (A - 4B) \tag{4}$$

Equating coefficients of like powers of x on the two sides yields

$$A + B = \quad 5$$

$$A - 4B = -10$$

which is a system of two linear equations in the unknowns A and B. Solving this system yields $A = 2$ and $B = 3$ as before. This method is justified because (4) is an identity holding for all x, and two polynomials are equal for all x if and only if their corresponding coefficients are equal (Exercise 62).

The critical step in the preceding work was the (correct) guess that the partial fraction decomposition had form (2). Once we knew the form of the decomposition, it was a straightforward matter to find A and B. One of the goals of this section is to explain how to obtain the form of a partial fraction decomposition.

There is a theorem in advanced algebra which states that every rational function $P(x)/Q(x)$ in which the *degree of the numerator is less than the degree of the denominator* can be expressed as a sum

$$\frac{P(x)}{Q(x)} = F_1(x) + F_2(x) + \cdots + F_n(x) \tag{5}$$

where $F_1(x), F_2(x), \ldots, F_n(x)$ are rational functions of the form

$$\frac{A}{(ax + b)^k} \quad \text{or} \quad \frac{Ax + B}{(ax^2 + bx + c)^k}$$

The terms $F_1(x), F_2(x), \ldots, F_n(x)$ on the right side of (5) are called **partial fractions**, and the entire right side is called the **partial fraction decomposition** of the left side. We will see below that the number of terms of each type depends on the factors of $Q(x)$; however, first let us review some results about factoring polynomials.

In theory, every polynomial $Q(x)$ with real coefficients can be factored into a product of linear and quadratic factors with real coefficients. For example, the polynomial

$$Q(x) = x^3 - 3x^2 + x - 3$$

factors into

$$Q(x) = (x - 3)(x^2 + 1)$$

The quadratic factor $x^2 + 1$ cannot be further decomposed into linear factors without using imaginary numbers $[x^2 + 1 = (x - i)(x + i)]$. Such quadratic factors are said to be **irreducible**. Although it is theoretically possible to factor every polynomial into linear and irreducible quadratic factors with real coefficients, it can be difficult to do so in many cases.

☐ **FINDING THE FORM OF A PARTIAL FRACTION DECOMPOSITION**

The first step in finding the partial fraction decomposition of a rational function $P(x)/Q(x)$ *whose numerator has smaller degree than the denominator* is to factor $Q(x)$ completely into linear and irreducible quadratic factors, and then collect all repeated factors so that $Q(x)$ is expressed as a product of distinct factors of the form

$$(ax + b)^m \quad \text{and} \quad (ax^2 + bx + c)^m$$

From these factors we can determine the form of the partial fraction decomposition using two rules that we shall now discuss.

☐ **LINEAR FACTORS**

If all of the factors of $Q(x)$ are linear, then the partial fraction decomposition of $P(x)/Q(x)$ can be determined by using the following rule:

> **LINEAR FACTOR RULE.** For each factor of the form $(ax + b)^m$, the partial fraction decomposition contains the following sum of m partial fractions:
>
> $$\frac{A_1}{ax + b} + \frac{A_2}{(ax + b)^2} + \cdots + \frac{A_m}{(ax + b)^m}$$
>
> where $A_1, A_2, \ldots, A_m$ are constants to be determined.

Example 1 Evaluate $\displaystyle\int \frac{dx}{x^2 + x - 2}$.

Solution. The integrand can be written as

$$\frac{1}{x^2 + x - 2} = \frac{1}{(x - 1)(x + 2)}$$

which is a rational function whose numerator has smaller degree than the denominator. According to the linear factor rule, the factor $x - 1$ introduces one term (since $m = 1$) of the form

$$\frac{A}{x - 1}$$

and the factor $x + 2$ introduces one term of the form

$$\frac{B}{x + 2}$$

so that the partial fraction decomposition is

$$\frac{1}{(x - 1)(x + 2)} = \frac{A}{x - 1} + \frac{B}{x + 2} \tag{6}$$

where A and B are constants to be determined so that (6) becomes an *identity*. To find these constants we can multiply both sides of (6) by $(x - 1)(x + 2)$ to obtain

$$1 = A(x + 2) + B(x - 1) \tag{7}$$

As previously discussed, we can find A and B by substituting the values of x that make the terms on the right side of (7) zero, namely $x = -2$ and $x = 1$. Substituting $x = -2$ yields

$$1 = -3B \quad \text{or} \quad B = -\frac{1}{3}$$

and substituting $x = 1$ yields

$$1 = 3A \quad \text{or} \quad A = \frac{1}{3}$$

Thus, (6) becomes

$$\frac{1}{(x - 1)(x + 2)} = \frac{1/3}{x - 1} + \frac{-1/3}{x + 2}$$

and

$$\int \frac{dx}{(x-1)(x+2)} = \frac{1}{3} \int \frac{dx}{x-1} - \frac{1}{3} \int \frac{dx}{x+2}$$

$$= \frac{1}{3} \ln |x-1| - \frac{1}{3} \ln |x+2| + C$$

$$= \frac{1}{3} \ln \left| \frac{x-1}{x+2} \right| + C$$

Alternative Solution.   The constants A and B in (7) can also be determined by collecting like powers of x,

$$1 = (A + B)x + (2A - B)$$

and then equating the coefficients of the like powers of x on both sides to obtain

$$A + B = 0$$

$$2A - B = 1$$

The solution of this system of linear equations is $A = 1/3$ and $B = -1/3$, which agrees with the results obtained above.   ◀

When $Q(x)$ does not have repeated linear factors, as in the preceding example, then the unknown constants in the partial fraction decomposition can all be found by substituting appropriate values of x. However, if there are repeated linear factors, then it is not as easy to obtain all of the constants by substituting values of x. For such problems one can either find all of the constants by equating like powers of x and solving the resulting system of linear equations or one can use a combination of the two methods, first finding as many constants as possible by substituting values, then finding the rest by equating appropriate like powers of x. The following example illustrates the latter procedure.

Example 2   Evaluate $\int \frac{2x+4}{x^3 - 2x^2} \, dx$.

Solution.   The integrand can be rewritten as

$$\frac{2x+4}{x^3 - 2x^2} = \frac{2x+4}{x^2(x-2)}$$

Although x^2 is a quadratic factor, it is *not* irreducible since $x^2 = xx$. Thus, by the linear factor rule, x^2 introduces two terms (since $m = 2$) of the form

$$\frac{A}{x} + \frac{B}{x^2}$$

and the factor $x - 2$ introduces one term (since $m = 1$) of the form

$$\frac{C}{x-2}$$

so the partial fraction decomposition is

$$\frac{2x+4}{x^2(x-2)} = \frac{A}{x} + \frac{B}{x^2} + \frac{C}{x-2} \tag{8}$$

Multiplying by $x^2(x-2)$ yields

$$2x + 4 = Ax(x-2) + B(x-2) + Cx^2 \tag{9}$$

which, after multiplying out and collecting like powers of x, becomes

$$2x + 4 = (A+C)x^2 + (-2A + B)x - 2B \tag{10}$$

To determine A, B, and C we shall use a combination of the methods illustrated in Example 1. Let $x = 0$ and $x = 2$ in (9) to obtain

$$B = -2 \quad \text{and} \quad C = 2$$

Equating corresponding coefficients of x^2 in (10) gives

$$A + C = 0, \quad \text{so} \quad A = -C = -2$$

and (8) becomes

$$\frac{2x + 4}{x^2(x - 2)} = \frac{-2}{x} + \frac{-2}{x^2} + \frac{2}{x - 2}$$

Thus,

$$\int \frac{2x + 4}{x^2(x - 2)}\, dx = -2 \int \frac{dx}{x} - 2 \int \frac{dx}{x^2} + 2 \int \frac{dx}{x - 2}$$

$$= -2 \ln |x| + \frac{2}{x} + 2 \ln |x - 2| + C$$

$$= 2 \ln \left| \frac{x - 2}{x} \right| + \frac{2}{x} + C \quad \blacktriangleleft$$

☐ **QUADRATIC FACTORS**

If some of the factors of $Q(x)$ are irreducible quadratics, then the contribution of those factors to the partial fraction decomposition of $P(x)/Q(x)$ can be determined from the following rule:

QUADRATIC FACTOR RULE.   For each factor of the form $(ax^2 + bx + c)^m$, the partial fraction decomposition contains the following sum of m partial fractions:

$$\frac{A_1 x + B_1}{ax^2 + bx + c} + \frac{A_2 x + B_2}{(ax^2 + bx + c)^2} + \cdots + \frac{A_m x + B_m}{(ax^2 + bx + c)^m}$$

where $A_1, A_2, \ldots, A_m, B_1, B_2, \ldots, B_m$ are constants to be determined.

Example 3   Evaluate $\displaystyle \int \frac{x^2 + x - 2}{3x^3 - x^2 + 3x - 1}\, dx$.

Solution.   The denominator in the integrand can be factored by grouping:

$$\frac{x^2 + x - 2}{3x^3 - x^2 + 3x - 1} = \frac{x^2 + x - 2}{x^2(3x - 1) + (3x - 1)} = \frac{x^2 + x - 2}{(3x - 1)(x^2 + 1)}$$

By the linear factor rule, the factor $3x - 1$ introduces one term:

$$\frac{A}{3x - 1}$$

and by the quadratic factor rule, the factor $x^2 + 1$ introduces one term:

$$\frac{Bx + C}{x^2 + 1}$$

Thus, the partial fraction decomposition is

$$\frac{x^2 + x - 2}{(3x - 1)(x^2 + 1)} = \frac{A}{3x - 1} + \frac{Bx + C}{x^2 + 1} \tag{11}$$

Multiplying by $(3x - 1)(x^2 + 1)$ yields

$$x^2 + x - 2 = A(x^2 + 1) + (Bx + C)(3x - 1) \tag{12}$$

To determine A, B, and C, we multiply out and collect like terms:

$$x^2 + x - 2 = (A + 3B)x^2 + (-B + 3C)x + (A - C) \tag{13}$$

Equating corresponding coefficients gives

$$
\begin{aligned}
A + 3B \quad &= \quad 1 \\
-\,B + 3C &= \quad 1 \\
A \quad\;\; -\,C &= -2
\end{aligned}
$$

To solve this system, subtract the third equation from the first to eliminate A. Then use the resulting equation together with the second equation to solve for B and C. Finally, determine A from the first or third equation. This yields (verify)

$$A = -\frac{7}{5}, \quad B = \frac{4}{5}, \quad C = \frac{3}{5}$$

Thus, (11) becomes

$$\frac{x^2 + x - 2}{(3x - 1)(x^2 + 1)} = \frac{-\frac{7}{5}}{3x - 1} + \frac{\frac{4}{5}x + \frac{3}{5}}{x^2 + 1}$$

and

$$\int \frac{x^2 + x - 2}{(3x - 1)(x^2 + 1)}\, dx = -\frac{7}{5} \int \frac{dx}{3x - 1} + \frac{4}{5} \int \frac{x}{x^2 + 1}\, dx + \frac{3}{5} \int \frac{dx}{x^2 + 1}$$

$$= -\frac{7}{15} \ln |3x - 1| + \frac{2}{5} \ln (x^2 + 1) + \frac{3}{5} \tan^{-1} x + C \quad \blacktriangleleft$$

REMARK.   The partial fraction decomposition in the preceding example can also be obtained by a combination of the two methods we have discussed. Substituting $x = \frac{1}{3}$ in (12) yields $A = -\frac{7}{5}$, then equating the corresponding constant terms and the corresponding coefficients of x^2 in (13) yields equations that express B and C in terms of the known value of A. (Verify that this procedure yields the same results as those obtained in the example.)

Example 4   Evaluate $\displaystyle\int \frac{3x^4 + 4x^3 + 16x^2 + 20x + 9}{(x + 2)(x^2 + 3)^2}\, dx$.

Solution.   Observe first that the degree of the numerator (which is 4) is less than the degree of the denominator (which is 5), so the method of partial fractions is applicable. By the linear factor rule, the factor $x + 2$ introduces one term:

$$\frac{A}{x + 2}$$

and by the quadratic factor rule, the factor $(x^2 + 3)^2$ introduces two terms (since $m = 2$):

$$\frac{Bx + C}{x^2 + 3} + \frac{Dx + E}{(x^2 + 3)^2}$$

Thus, the partial fraction decomposition of the integrand is

$$\frac{3x^4 + 4x^3 + 16x^2 + 20x + 9}{(x + 2)(x^2 + 3)^2} = \frac{A}{x + 2} + \frac{Bx + C}{x^2 + 3} + \frac{Dx + E}{(x^2 + 3)^2} \tag{14}$$

Multiplying by $(x + 2)(x^2 + 3)^2$ yields

$$3x^4 + 4x^3 + 16x^2 + 20x + 9$$

$$= A(x^2 + 3)^2 + (Bx + C)(x^2 + 3)(x + 2) + (Dx + E)(x + 2) \tag{15}$$

which, after multiplying out and collecting like powers of x, becomes

$$3x^4 + 4x^3 + 16x^2 + 20x + 9$$
$$= (A + B)x^4 + (2B + C)x^3 + (6A + 3B + 2C + D)x^2$$
$$+ (6B + 3C + 2D + E)x + (9A + 6C + 2E) \quad (16)$$

Equating corresponding coefficients in (16) yields the following system of five linear equations in five unknowns:

$$\begin{aligned} A + B &= 3 \\ 2B + C &= 4 \\ 6A + 3B + 2C + D &= 16 \\ 6B + 3C + 2D + E &= 20 \\ 9A + 6C + 2E &= 9 \end{aligned} \qquad (17)$$

This system is tedious to solve, but the work can be reduced by first substituting $x = -2$ in (15), which yields $A = 1$. Substituting this known value of A in (17) yields the simpler system

$$\begin{aligned} B &= 2 \\ 2B + C &= 4 \\ 3B + 2C + D &= 10 \\ 6B + 3C + 2D + E &= 20 \\ 6C + 2E &= 0 \end{aligned} \qquad (18)$$

This system can be solved by starting at the top and working down, first substituting $B = 2$ in the second equation to get $C = 0$, then substituting the known values of B and C in the third equation to get $D = 4$, and so forth. This yields

$$A = 1, \quad B = 2, \quad C = 0, \quad D = 4, \quad E = 0$$

Thus, (14) becomes

$$\frac{3x^4 + 4x^3 + 16x^2 + 20x + 9}{(x + 2)(x^2 + 3)^2} = \frac{1}{x + 2} + \frac{2x}{x^2 + 3} + \frac{4x}{(x^2 + 3)^2}$$

and so

$$\int \frac{3x^4 + 4x^3 + 16x^2 + 20x + 9}{(x + 2)(x^2 + 3)^2}\, dx$$

$$= \int \frac{dx}{x + 2} + \int \frac{2x}{x^2 + 3}\, dx + 4\int \frac{x}{(x^2 + 3)^2}\, dx$$

$$= \ln|x + 2| + \ln(x^2 + 3) - \frac{2}{x^2 + 3} + C \quad \blacktriangleleft$$

☐ **INTEGRATING IMPROPER RATIONAL FUNCTIONS**

As noted earlier, partial fraction decomposition applies only to rational functions in which the degree of the numerator is less than the degree of the denominator; these are called **proper rational functions**. Rational functions in which the degree of the numerator is greater than or equal to the degree of the denominator are called **improper rational functions**. The following example shows that improper rational functions can be integrated by first performing a long division, and then working with the remainder term.

Example 5 Evaluate $\displaystyle\int \frac{3x^4 + 3x^3 - 5x^2 + x - 1}{x^2 + x - 2}\, dx.$

Solution.   Since the integrand is an improper rational function, we cannot use a partial fraction decomposition directly. However, if we perform the long division

$$
\begin{array}{r}
3x^2 + 1 \\
x^2 + x - 2 \overline{\smash{\big)}\ 3x^4 + 3x^3 - 5x^2 + x - 1} \\
\underline{3x^4 + 3x^3 - 6x^2} \\
x^2 + x - 1 \\
\underline{x^2 + x - 2} \\
1
\end{array}
$$

we can write the integrand as the quotient plus the remainder over the divisor, that is,

$$
\frac{3x^4 + 3x^3 - 5x^2 + x - 1}{x^2 + x - 2} = (3x^2 + 1) + \frac{1}{x^2 + x - 2}
$$

Thus,

$$
\int \frac{3x^4 + 3x^3 - 5x^2 + x - 1}{x^2 + x - 2}\, dx = \int (3x^2 + 1)\, dx + \int \frac{dx}{x^2 + x - 2}
$$

The second integral on the right now involves a proper rational function and can thus be evaluated by a partial fraction decomposition. Using the result of Example 1 we obtain

$$
\int \frac{3x^4 + 3x^3 - 5x^2 + x - 1}{x^2 + x - 2}\, dx = x^3 + x + \frac{1}{3} \ln \left| \frac{x-1}{x+2} \right| + C \qquad \blacktriangleleft
$$

■ FACTORING POLYNOMIALS

The method of partial fractions depends on our ability to carry out the necessary factorization. This is not always easy to do. The following results, usually proved in algebra courses, are helpful to know.

9.6.1   THEOREM (*Factor Theorem*).   *If $p(x)$ is a polynomial and r is a solution of the equation $p(x) = 0$, then $x - r$ is a factor of $p(x)$.*

9.6.2   THEOREM.   *Let*

$$
p(x) = a_0 x^n + a_1 x^{n-1} + \cdots + a_{n-1}x + a_n
$$

be a polynomial of degree n with integer coefficients.

(a)   *If r is an integer solution of $p(x) = 0$, then the constant term a_n is an integer multiple of r.*

(b)   *If c/d is a rational solution of $p(x) = 0$, and if c/d is expressed in lowest terms, then the constant term a_n is an integer multiple of c; and the leading coefficient a_0 is an integer multiple of d.*

Example 6   For the equation

$$
x^3 + x^2 - 10x + 8 = 0
$$

it follows from part (a) of the preceding theorem that the only possible integer solutions are $\pm 1,\ \pm 2,\ \pm 4,\ \pm 8$. By substitution, or by using synthetic division, the reader can show that $1, 2, -4$ are solutions and the rest are not. It follows that

$$
x^3 + x^2 - 10x + 8 = (x - 1)(x - 2)(x + 4) \qquad \blacktriangleleft
$$

Example 7   For the equation

$$2x^3 + x^2 - 6x - 3 = 0$$

the only possible numerators for rational solutions are ± 1, ± 3, and the only possible denominators are ± 1, ± 2; thus, the only possible rational solutions are

$$\pm 1, \ \pm 3, \ \pm \tfrac{1}{2}, \ \pm \tfrac{3}{2}$$

By substitution, or by using synthetic division, the reader can show that $-\tfrac{1}{2}$ is a solution, but the rest are not. It follows by division that

$$2x^3 + x^2 - 6x - 3 = (x + \tfrac{1}{2})(2x^2 - 6) = 2(x + \tfrac{1}{2})(x + \sqrt{3})(x - \sqrt{3})$$

Therefore, the solutions of the equation are $x = -\tfrac{1}{2}$, $x = -\sqrt{3}$, and $x = \sqrt{3}$.   ◄

▶ Exercise Set 9.6 $\boxed{C}$ 54

In Exercises 1–8, write out the form of the partial fraction decomposition. (Do not find the numerical values of the coefficients.)

1. $\dfrac{3x - 1}{(x - 2)(x + 5)}$.

2. $\dfrac{5}{x(x^2 - 9)}$.

3. $\dfrac{2x - 3}{x^3 - x^2}$.

4. $\dfrac{x^2}{(x + 2)^3}$.

5. $\dfrac{1 - 5x^2}{x^3(x^2 + 1)}$.

6. $\dfrac{2x}{(x - 1)(x^2 + 5)}$.

7. $\dfrac{4x^3 - x}{(x^2 + 5)^2}$.

8. $\dfrac{1 - 3x^4}{(x - 2)(x^2 + 1)^2}$.

In Exercises 9–46, perform the integrations.

9. $\displaystyle\int \dfrac{dx}{x^2 + 3x - 4}$.

10. $\displaystyle\int \dfrac{dx}{x^2 + 8x + 7}$.

11. $\displaystyle\int \dfrac{x}{x^2 - 5x + 6}\,dx$.

12. $\displaystyle\int \dfrac{5x - 4}{x^2 - 4x}\,dx$.

13. $\displaystyle\int \dfrac{11x + 17}{2x^2 + 7x - 4}\,dx$.

14. $\displaystyle\int \dfrac{5x - 5}{3x^2 - 8x - 3}\,dx$.

15. $\displaystyle\int \dfrac{dx}{(x - 1)(x + 2)(x - 3)}$.

16. $\displaystyle\int \dfrac{dx}{x(x^2 - 1)}$.

17. $\displaystyle\int \dfrac{2x^2 - 9x - 9}{x^3 - 9x}\,dx$.

18. $\displaystyle\int \dfrac{2x^2 + 4x - 8}{x^3 - 4x}\,dx$.

19. $\displaystyle\int \dfrac{x^2 + 2}{x + 2}\,dx$.

20. $\displaystyle\int \dfrac{x^2 - 4}{x - 1}\,dx$.

21. $\displaystyle\int \dfrac{3x^2 - 10}{x^2 - 4x + 4}\,dx$.

22. $\displaystyle\int \dfrac{x^2}{x^2 - 3x + 2}\,dx$.

23. $\displaystyle\int \dfrac{x^3}{x^2 - 3x + 2}\,dx$.

24. $\displaystyle\int \dfrac{x^3}{x^2 - x - 6}\,dx$.

25. $\displaystyle\int \dfrac{x^5 + 2x^2 + 1}{x^3 - x}\,dx$.

26. $\displaystyle\int \dfrac{2x^5 - x^3 - 1}{x^3 - 4x}\,dx$.

27. $\displaystyle\int \dfrac{2x^2 + 3}{x(x - 1)^2}\,dx$.

28. $\displaystyle\int \dfrac{3x^2 - x + 1}{x^3 - x^2}\,dx$.

29. $\displaystyle\int \dfrac{x^2 + x - 16}{(x + 1)(x - 3)^2}\,dx$.

30. $\displaystyle\int \dfrac{2x^2 - 2x - 1}{x^3 - x^2}\,dx$.

31. $\displaystyle\int \dfrac{x^2}{(x + 2)^3}\,dx$.

32. $\displaystyle\int \dfrac{2x^2 + 3x + 3}{(x + 1)^3}\,dx$.

33. $\displaystyle\int \dfrac{2x^2 - 1}{(4x - 1)(x^2 + 1)}\,dx$.

34. $\displaystyle\int \dfrac{dx}{x(x^2 + x + 1)}$.

35. $\displaystyle\int \dfrac{dx}{x^4 - 16}$.

36. $\displaystyle\int \dfrac{dx}{x^3 + x}$.

37. $\displaystyle\int \dfrac{x^3 + 3x^2 + x + 9}{(x^2 + 1)(x^2 + 3)}\,dx$.

38. $\displaystyle\int \dfrac{x^3 + x^2 + x + 2}{(x^2 + 1)(x^2 + 2)}\,dx$.

39. $\displaystyle\int \dfrac{x^3 - 3x^2 + 2x - 3}{x^2 + 1}\,dx$.

40. $\displaystyle\int \dfrac{x^4 + 6x^3 + 10x^2 + x}{x^2 + 6x + 10}\,dx$.

41. $\displaystyle\int \dfrac{x^2 + 1}{(x^2 + 2x + 3)^2}\,dx$.

42. $\displaystyle\int \dfrac{x^5 + x^4 + 4x^3 + 4x^2 + 4x + 4}{(x^2 + 2)^3}\,dx$.

43. $\displaystyle\int \dfrac{\cos\theta}{\sin^2\theta + 4\sin\theta - 5}\,d\theta$.

44. $\displaystyle\int \dfrac{e^t}{e^{2t} - 4}\,dt$.

45. $\displaystyle\int \dfrac{dx}{1 + e^x}$.

46. $\displaystyle\int \dfrac{\sec^2\theta}{\tan^3\theta - \tan^2\theta}\,d\theta$.

47. (a) Find constants a and b such that

$$x^4 + 1 = (x^2 + ax + 1)(x^2 + bx + 1)$$

(b) Use the result in part (a) to show that

$$\int_0^1 \dfrac{x}{x^4 + 1}\,dx = \dfrac{\pi}{8}$$

48. Find the area of the region that is enclosed by $y = (x - 3)/(x^3 + x^2)$, $y = 0$, $x = 1$, and $x = 2$.

49. Find the volume of the solid generated when the region enclosed by $y = x^2/(9 - x^2)$, $y = 0$, $x = 0$, and $x = 2$ is revolved about the x-axis.

In Exercises 50–53, solve the differential equations.

50. $\dfrac{dy}{dx} = y^2 + y$.

51. $\dfrac{dy}{dx} = y^2 - 5y + 6$.

52. $\dfrac{dy}{dt} = t^2 y^2 - 4t^2 y$.

53. $t(t - 1)\dfrac{dy}{dt} - (y^2 + y) = 0$.

54. In some chemical reactions in which two reacting substances combine to form a resultant substance, the concentration of the resultant changes with time at a rate that is proportional to the product of the concentrations of the reacting substances. If x is the concentration of the resultant at any time $t \geq 0$, and a and b are the concentrations of the reactants at time $t = 0$, then x satisfies the differential equation

$$\frac{dx}{dt} = k(a - x)(b - x) \quad (a \neq b)$$

(a) Assuming that $x = 0$ when $t = 0$, solve the equation for t in terms of x, a, b, and k.

(b) Assuming that the concentrations are measured in moles/liter, time is measured in minutes, and the values of the constants are $a = 0.10$, $b = 0.06$, and $k = 0.3$, find the length of time that it will take for the concentration of the resultant to reach 0.02 mol/L. Express your answer to the nearest tenth of a minute.

55. The differential equation

$$\frac{dy}{dt} = ay - by^2 \quad (a > 0, b > 0)$$

which is called the **logistic equation**, arises in the study of human population growth.

(a) By solving the equation, show that its general solution is

$$y = \frac{a}{b + Ce^{-at}}$$

where C is an arbitrary constant.

(b) Find $\lim\limits_{t \to +\infty} y(t)$.

56. (a) Use the result in part (a) of Exercise 55 to find the solution of the logistic equation that satisfies the initial condition $y(0) = \frac{1}{2}$ in the case where $a = 2$ and $b = 1$.

(b) Sketch the graph of the solution in part (a).

In Exercises 57–60, use Theorems 9.6.1 and 9.6.2.

57. Find all rational solutions, if any, and use your results to factor the polynomial into a product of linear and irreducible quadratic factors.

(a) $x^3 - 6x^2 + 11x - 6 = 0$

(b) $x^3 - 3x^2 + x - 20 = 0$

(c) $x^4 - 5x^3 + 7x^2 - 5x + 6 = 0$.

58. Find all rational solutions, if any, and use your results to factor the polynomial into a product of linear and irreducible quadratic factors.

(a) $8x^3 + 4x^2 - 2x - 1 = 0$

(b) $6x^4 - 7x^3 + 6x^2 - 1 = 0$

(c) $9x^4 - 56x^3 + 57x^2 + 98x - 24 = 0$.

59. Evaluate

$$\int \frac{dx}{x^4 - 3x^3 - 7x^2 + 27x - 18}$$

60. Evaluate

$$\int \frac{dx}{16x^3 - 4x^2 + 4x - 1}$$

61. Use Theorem 9.6.2 to prove:

(a) $\sqrt{2}$ is irrational.

(b) If a is a positive integer, then $\sqrt{a}$ is either an integer or is irrational.

62. (a) Prove: $a_0 x^n + a_1 x^{n-1} + \cdots + a_n = 0$ for all x if and only if $a_0 = a_1 = \cdots = a_n = 0$.

(b) Use the result in part (a) to prove:

$$a_0 x^n + a_1 x^{n-1} + \cdots + a_n$$
$$= b_0 x^n + b_1 x^{n-1} + \cdots + b_n$$

for all x if and only if $a_0 = b_0$, $a_1 = b_1, \ldots, a_n = b_n$.

■ 9.7 MISCELLANEOUS SUBSTITUTIONS

In this section we shall consider some integrals that do not fit into any of the categories previously studied.

□ **INTEGRALS INVOLVING RATIONAL EXPONENTS**

Integrals involving rational powers of x can often be simplified by substituting

$$u = x^{1/n}$$

where n is the least common multiple of the denominators of the exponents. The effect of this substitution is to replace fractional exponents with integer exponents, which are easier to work with.

Example 1 Evaluate $\displaystyle\int \frac{\sqrt{x}}{1 + \sqrt[3]{x}}\, dx$.

Solution. The integrand involves $x^{1/2}$ and $x^{1/3}$, so we make the substitution

$$u = x^{1/6}$$

or

$$x = u^6 \quad \text{and} \quad dx = 6u^5\, du$$

Thus,

$$\int \frac{\sqrt{x}}{1 + \sqrt[3]{x}}\, dx = \int \frac{(u^6)^{1/2}}{1 + (u^6)^{1/3}}\, (6u^5)\, du = 6 \int \frac{u^8}{1 + u^2}\, du$$

By long division

$$\frac{u^8}{1 + u^2} = u^6 - u^4 + u^2 - 1 + \frac{1}{1 + u^2}$$

Thus,

$$\int \frac{\sqrt{x}}{1 + \sqrt[3]{x}}\, dx = 6 \int \left(u^6 - u^4 + u^2 - 1 + \frac{1}{1 + u^2} \right) du$$

$$= \tfrac{6}{7} u^7 - \tfrac{6}{5} u^5 + 2u^3 - 6u + 6 \tan^{-1} u + C$$

$$= \tfrac{6}{7} x^{7/6} - \tfrac{6}{5} x^{5/6} + 2x^{1/2} - 6x^{1/6} + 6 \tan^{-1} (x^{1/6}) + C \quad \blacktriangleleft$$

Example 2 Evaluate $\displaystyle\int \frac{dx}{2 + 2\sqrt{x}}$.

Solution. The integrand contains $\sqrt{x} = x^{1/2}$, so we make the substitution

$$u = x^{1/2}$$

or

$$x = u^2 \quad \text{and} \quad dx = 2u\, du$$

This yields

$$\int \frac{dx}{2 + 2\sqrt{x}} = \int \frac{2u}{2 + 2u}\, du$$

$$= \int \left(1 - \frac{1}{1 + u} \right) du \quad \boxed{\text{Long division}}$$

$$= u - \ln |1 + u| + C = \sqrt{x} - \ln |1 + \sqrt{x}| + C \quad \blacktriangleleft$$

The following example illustrates a variation on the above idea.

Example 3 Evaluate $\displaystyle\int \sqrt{1 + e^x}\, dx$.

Solution. Let

$$u = \sqrt{1 + e^x}$$

To express dx in terms of du, it is helpful to solve this equation for x and then differentiate. We obtain

$$e^x = u^2 - 1$$

$$x = \ln(u^2 - 1)$$

$$\frac{dx}{du} = \frac{2u}{u^2 - 1}, \quad dx = \frac{2u}{u^2 - 1}\, du$$

Thus,

$$\int \sqrt{1 + e^x}\, dx = \int u \left(\frac{2u}{u^2 - 1} \right) du$$

$$= \int \frac{2u^2}{u^2 - 1}\, du$$

$$= \int \left(2 + \frac{2}{u^2 - 1} \right) du \quad \boxed{\text{Long division}}$$

$$= 2u + \int \left(\frac{1}{u - 1} - \frac{1}{u + 1} \right) du \quad \boxed{\text{Partial fractions}}$$

$$= 2u + \ln|u - 1| - \ln|u + 1| + C$$

$$= 2u + \ln \left| \frac{u - 1}{u + 1} \right| + C$$

$$= 2\sqrt{1 + e^x} + \ln \left[\frac{\sqrt{1 + e^x} - 1}{\sqrt{1 + e^x} + 1} \right] + C \quad \blacktriangleleft$$

◻ **INTEGRALS CONTAINING RATIONAL FUNCTIONS OF $\sin x$ AND $\cos x$**

Functions that consist of finitely many sums, differences, products, and quotients of $\sin x$ and $\cos x$ are called *rational functions of $\sin x$ and $\cos x$*. Some examples are

$$\frac{\sin x + 3\cos^2 x}{\cos x + 4 \sin x}, \quad \frac{\sin x}{1 + \cos x - \cos^2 x}, \quad \frac{3 \sin^5 x}{1 + 4 \sin x}$$

A method for integrating such functions can be based on the trigonometric identities

$$\sin x = 2 \sin(x/2) \cos(x/2) \tag{1}$$

$$\cos x = \cos^2(x/2) - \sin^2(x/2) \tag{2}$$

If we let

$$u = \tan(x/2), \quad -\pi < x < \pi \tag{3}$$

then as suggested by Figure 9.7.1,

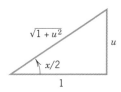

Figure 9.7.1

$$\sin(x/2) = \frac{u}{\sqrt{1 + u^2}} \quad \text{and} \quad \cos(x/2) = \frac{1}{\sqrt{1 + u^2}}$$

Substituting these expressions in (1) and (2) yields

$$\sin x = 2 \left(\frac{u}{\sqrt{1 + u^2}} \right) \left(\frac{1}{\sqrt{1 + u^2}} \right) = \frac{2u}{1 + u^2} \tag{4}$$

$$\cos x = \left(\frac{1}{\sqrt{1 + u^2}} \right)^2 - \left(\frac{u}{\sqrt{1 + u^2}} \right)^2 = \frac{1 - u^2}{1 + u^2} \tag{5}$$

The combination of (3), (4), and (5) yields the following substitution formulas, which are often effective for integrating rational functions of $\sin x$ and $\cos x$.

$$u = \tan (x/2) \quad (-\pi < x < \pi) \qquad x = 2 \tan^{-1} u$$

$$\sin x = \frac{2u}{1 + u^2} \qquad \cos x = \frac{1 - u^2}{1 + u^2} \qquad dx = \frac{2}{1 + u^2} \, du$$

The last formula follows from differentiating $x = 2 \tan^{-1} u$ with respect to u.

Example 4 Evaluate $\displaystyle\int \frac{dx}{1 + \sin x}$.

Solution. The preceding substitution yields

$$\int \frac{dx}{1 + \sin x} = \int \frac{1}{1 + \left(\dfrac{2u}{1 + u^2}\right)} \left(\frac{2}{1 + u^2}\right) du = \int \frac{2}{1 + 2u + u^2} \, du$$

$$= \int \frac{2}{(1 + u)^2} \, du = -\frac{2}{1 + u} + C = -\frac{2}{1 + \tan (x/2)} + C \qquad \blacktriangleleft$$

REMARK. The method of the preceding example can lead to cumbersome partial fraction decompositions and consequently should be used only in absence of finding a simpler method.

▶ Exercise Set 9.7

In Exercises 1–26, perform the integrations.

1. $\displaystyle\int x\sqrt{x - 2} \, dx.$

2. $\displaystyle\int_0^8 \frac{x}{\sqrt{x + 1}} \, dx.$

3. $\displaystyle\int_4^8 \frac{\sqrt{x - 4}}{x} \, dx.$

4. $\displaystyle\int_0^9 \frac{\sqrt{x}}{x + 9} \, dx.$

5. $\displaystyle\int_0^4 \frac{1}{3 + \sqrt{x}} \, dx.$

6. $\displaystyle\int \frac{x^5}{\sqrt{x^3 + 1}} \, dx.$

7. $\displaystyle\int x^5\sqrt{x^3 + 1} \, dx.$

8. $\displaystyle\int \frac{1}{x\sqrt{x^3 - 1}} \, dx.$

9. $\displaystyle\int \frac{dx}{\sqrt{x} + \sqrt[3]{x}}.$

10. $\displaystyle\int \frac{dx}{x - x^{3/5}}.$

11. $\displaystyle\int \frac{dv}{v(1 - v^{1/4})}.$

12. $\displaystyle\int \frac{x^{2/3}}{x + 1} \, dx.$

13. $\displaystyle\int \frac{dt}{t^{1/2} - t^{1/3}}.$

14. $\displaystyle\int \frac{1 + \sqrt{x}}{1 - \sqrt{x}} \, dx.$

15. $\displaystyle\int \frac{x^3}{\sqrt{1 + x^2}} \, dx.$

16. $\displaystyle\int \frac{x}{(x + 3)^{1/5}} \, dx.$

17. $\displaystyle\int \sin \sqrt{x} \, dx.$

18. $\displaystyle\int e^{\sqrt{x}} \, dx.$

19. $\displaystyle\int \frac{1}{\sqrt{e^x + 1}} \, dx.$

20. $\displaystyle\int_0^{\ln 2} \sqrt{e^x - 1} \, dx.$

21. $\displaystyle\int \frac{dx}{1 + \sin x + \cos x}.$

22. $\displaystyle\int \frac{dx}{2 + \sin x}.$

23. $\displaystyle\int_{\pi/2}^\pi \frac{d\theta}{1 - \cos \theta}.$

24. $\displaystyle\int \frac{dx}{4 \sin x - 3 \cos x}.$

25. $\displaystyle\int \frac{\cos x}{2 - \cos x} \, dx.$

26. $\displaystyle\int \frac{dx}{\sin x + \tan x}.$

27. (a) Use the substitution $u = \tan (x/2)$ to show that

$$\int \sec x \, dx = \ln \left| \frac{1 + \tan (x/2)}{1 - \tan (x/2)} \right| + C$$

(b) Use the result in part (a) to show that

$$\int \sec x \, dx = \ln \left| \tan \left(\frac{\pi}{4} + \frac{x}{2} \right) \right| + C$$

(c) Show that this agrees with (2) of Section 9.4.

28. (a) Use the substitution $u = \tan (x/2)$ to show that

$$\int \csc x \, dx = \frac{1}{2} \ln \left[\frac{1 - \cos x}{1 + \cos x} \right] + C$$

(b) Show that this agrees with the result in Exercise 37(a) of Section 9.4.

29. Find a substitution that can be used to integrate rational functions of sinh x and cosh x and use your substitution to evaluate

$$\int \frac{dx}{2\cosh x + \sinh x}$$

without expressing the integrand in terms of e^x and e^{-x}.

In Exercises 30–33, use the substitution $x = 1/u$ to help evaluate the integral. (Assume that $x > 0$.)

30. $\displaystyle\int \frac{\sqrt{4 - x^2}}{x^4}\,dx.$

31. $\displaystyle\int \frac{1}{x^2\sqrt{3 - x^2}}\,dx.$

32. $\displaystyle\int \frac{1}{x^2\sqrt{x^2 + 1}}\,dx.$

33. $\displaystyle\int \frac{\sqrt{x^2 - 5}}{x^4}\,dx.$

■ **9.8 NUMERICAL INTEGRATION; SIMPSON'S RULE**

The most direct way to evaluate a definite integral is to find an antiderivative of the integrand and apply the First Fundamental Theorem of Calculus. If an antiderivative cannot be found, then the value of the integral can be approximated using methods that we shall study in this section.

☐ **RIEMANN SUM APPROXIMATIONS**

The simplest approximations of definite integrals use Riemann sums. In Formula (4) of Section 5.8 we gave the following formula for approximating a definite integral by a Riemann sum in which the interval $[a, b]$ is divided into subintervals of width $\Delta x = (b - a)/n$, and x_k^* denotes a point in the kth subinterval:

$$\int_a^b f(x)\,dx \approx \Delta x\,[f(x_1^*) + f(x_2^*) + \cdots + f(x_n^*)]$$

In this section we will denote the values of f at the endpoints of the subintervals by

$$y_0 = f(a), \quad y_1 = f(x_1), \quad y_2 = f(x_2), \ldots, \quad y_{n-1} = f(x_{n-1}), \quad y_n = f(b)$$

and we will denote the values of f at the midpoints of the subintervals by

$$y_{m_1}, y_{m_2}, \ldots, y_{m_n}$$

(Figure 9.8.1).

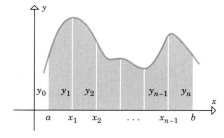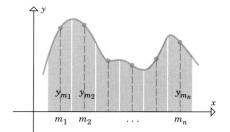

Figure 9.8.1

With this notation the left endpoint, right endpoint, and midpoint approximations discussed in Sections 5.5 and 5.8 can be expressed as shown in Table 9.8.1.

Table 9.8.1

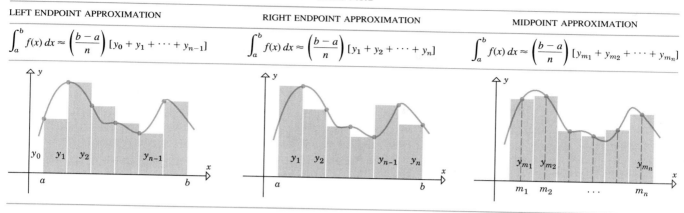

LEFT ENDPOINT APPROXIMATION	RIGHT ENDPOINT APPROXIMATION	MIDPOINT APPROXIMATION
$\int_a^b f(x)\,dx \approx \left(\dfrac{b-a}{n}\right)[y_0 + y_1 + \cdots + y_{n-1}]$	$\int_a^b f(x)\,dx \approx \left(\dfrac{b-a}{n}\right)[y_1 + y_2 + \cdots + y_n]$	$\int_a^b f(x)\,dx \approx \left(\dfrac{b-a}{n}\right)[y_{m_1} + y_{m_2} + \cdots + y_{m_n}]$

☐ **TRAPEZOIDAL APPROXIMATION**

The left-hand and right-hand endpoint approximations are rarely used in applications; however, if we take the average of the left-hand and right-hand endpoint approximations, we obtain a result, called the ***trapezoidal approximation***, which is commonly used:

> **Trapezoidal Approximation**
> $$\int_a^b f(x)\,dx \approx \left(\frac{b-a}{2n}\right)[y_0 + 2y_1 + \cdots + 2y_{n-1} + y_n] \tag{1}$$

The name, *trapezoidal approximation*, can be explained by considering the case where $f(x) \geq 0$ on $[a, b]$, so that $\int_a^b f(x)\,dx$ represents the area under $f(x)$ over $[a, b]$. Geometrically, the trapezoidal approximation formula results if we approximate this area by the sum of the trapezoidal areas shown in Figure 9.8.2 (Exercise 41).

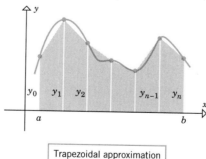

Figure 9.8.2

Trapezoidal approximation

Example 1 In Table 9.8.2 we have approximated

$$\ln 2 = \int_1^2 \frac{1}{x}\,dx$$

using the midpoint approximation and the trapezoidal approximation. In each case we used $n = 10$ subdivisions of the interval $[1, 2]$, so that

$$\underbrace{\frac{b-a}{n} = \frac{2-1}{10} = 0.1}_{\text{Midpoint}} \quad \text{and} \quad \underbrace{\frac{b-a}{2n} = \frac{2-1}{20} = 0.05}_{\text{Trapezoidal}} \quad \blacktriangleleft$$

Table 9.8.2

	Midpoint Approximation			Trapezoidal Approximation			
i	MIDPOINT m_i	$y_{m_i} = f(m_i) = 1/m_i$	i	ENDPOINT x_i	$y_i = f(x_i) = 1/x_i$	MULTIPLIER w_i	$w_i y_i$
1	1.05	0.952380952	0	1.0	1.000000000	1	1.000000000
2	1.15	0.869565217	1	1.1	0.909090909	2	1.818181818
3	1.25	0.800000000	2	1.2	0.833333333	2	1.666666667
4	1.35	0.740740741	3	1.3	0.769230769	2	1.538461538
5	1.45	0.689655172	4	1.4	0.714285714	2	1.428571429
6	1.55	0.645161290	5	1.5	0.666666667	2	1.333333333
7	1.65	0.606060606	6	1.6	0.625000000	2	1.250000000
8	1.75	0.571428571	7	1.7	0.588235294	2	1.176470588
9	1.85	0.540540541	8	1.8	0.555555556	2	1.111111111
10	1.95	0.512820513	9	1.9	0.526315789	2	1.052631579
		6.928353603	10	2.0	0.500000000	1	0.500000000
							13.875428063

$$\int_1^2 \frac{1}{x}\,dx \approx (0.1)(6.928353603) = 0.692835360$$

$$\int_1^2 \frac{1}{x}\,dx \approx (0.05)(13.875428063) = 0.693771403$$

REMARK.   In Example 1 we rounded the numerical values to nine places to the right of the decimal point; we shall follow this procedure throughout this section. If your calculator cannot produce this many places, then you will have to make the appropriate adjustments. What is important here is that you understand the principles involved.

□ COMPARISON OF THE MIDPOINT AND TRAPEZOIDAL APPROXIMATIONS

The value of ln 2 rounded to nine decimal places is

$$\ln 2 = \int_1^2 \frac{1}{x}\,dx \approx 0.693147181 \qquad (2)$$

so that the midpoint approximation in Example 1 produced a more accurate result than the trapezoidal approximation (verify). To see why this should be so, we need to look at the midpoint approximation from another viewpoint. [For simplicity in the explanations, we shall assume that $f(x) \geq 0$, but the conclusions will be true without this assumption.] For differentiable functions, the midpoint approximation is sometimes called the *tangent line approximation* because over each subinterval the area of the rectangle used in the midpoint approximation is equal to the area of the trapezoid whose upper boundary is the tangent line to $y = f(x)$ at the midpoint of the interval (Figure 9.8.3). The equality of these areas follows from the fact that the shaded triangles in Figure 9.8.3 are congruent.

Let us denote the midpoint and trapezoidal approximations of

$$\int_a^b f(x)\,dx$$

with n subintervals by M_n and T_n, respectively. We define the **errors** in these approximations to be

$$E_M = \int_a^b f(x)\,dx - M_n \quad \text{and} \quad E_T = \int_a^b f(x)\,dx - T_n$$

The quantities $|E_M|$ and $|E_T|$, which describe the sizes of the errors without regard to sign, are also important; these are called the **magnitudes** of the errors.

In Figure 9.8.4a we have isolated a subinterval of $[a, b]$ on which the graph of a function f is concave down, and we have shaded the areas that represent the magnitudes of the errors in the midpoint and trapezoidal approximations over the subinterval. In Figure 9.8.4b we show a succession of four illustrations which make it evident that the magnitude

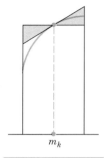

m_k

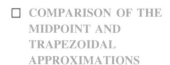

The shaded triangles have equal areas

Figure 9.8.3

of the error from the midpoint approximation is less than that from the trapezoidal approximation. If the graph of f were concave up, analogous figures would lead to the same conclusion. (This argument, due to Frank Buck, appeared in *The College Mathematics Journal*, Vol. 16, No. 1, 1985.)

Figure 9.8.4*a* also suggests that on a subinterval where the graph is concave down, the midpoint approximation is larger than the value of the integral and the trapezoidal approximation is smaller. On an interval where the graph is concave up it is the other way around. In summary, we have the following result, which we state without formal proof.

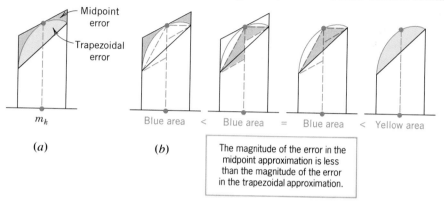

Figure 9.8.4

9.8.1   THEOREM.   *Let f be continuous on $[a, b]$, and let $|E_M|$ and $|E_T|$ be the magnitudes of the errors that result from the midpoint and trapezoidal approximations of $\int_a^b f(x)\,dx$ using n subintervals.*

(a)   *If the graph of f is either concave up or concave down on (a, b), then $|E_M| < |E_T|$, that is, the magnitude of the error from the midpoint approximation is less than that from the trapezoidal approximation.*

(b)   *If the graph of f is concave down on (a, b), then*

$$T_n < \int_a^b f(x)\,dx < M_n$$

(c)   *If the graph of f is concave up on (a, b), then*

$$M_n < \int_a^b f(x)\,dx < T_n$$

Example 2   We observed earlier that the midpoint approximation of $\ln 2$ obtained in Example 1 was more accurate than the trapezoidal approximation. This is consistent with part (*a*) of Theorem 9.8.1, since $f(x) = 1/x$ is continuous on $[1, 2]$ and concave up on $(1, 2)$. Moreover, a comparison of the two approximations to (2) shows that the midpoint approximation is smaller than $\ln 2$ and the trapezoidal approximation is larger. This is consistent with part (*c*) of Theorem 9.8.1.   ◀

REMARK.   Do not erroneously conclude that the midpoint approximation is always better than the trapezoidal approximation; for functions with inflection points, the trapezoidal approximation can be more accurate.

□ **SIMPSON'S RULE**

Intuition suggests that we might improve on the midpoint and trapezoidal approximations by replacing the linear upper boundaries of the approximating strips in Figure 9.8.2 by curved upper boundaries chosen to fit the shape of the curve $y = f(x)$ more closely. This is the idea behind **Simpson's*** (see biography on page 476) **rule**, which uses parabolic curves of the form

$$y = ax^2 + bx + c \tag{3}$$

to approximate sections of the curve $y = f(x)$. [Recall from Section 1.5 that (3) is the equation of a parabola with axis of symmetry parallel to the y-axis.]

To simplify the description of Simpson's rule we shall assume that $f(x) \geq 0$ on $[a, b]$ so that we can interpret $\int_a^b f(x)\, dx$ as an area. However, the method is valid without this assumption. The heart of Simpson's rule is the formula

$$A = \frac{h}{3}[Y_0 + 4Y_1 + Y_2] \tag{4}$$

which gives the area under the curve

$$y = ax^2 + bx + c$$

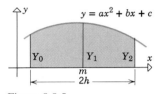

Figure 9.8.5

over an arbitrary interval of width $2h$. In this formula Y_0, Y_1, and Y_2 represent the y values at the left-hand endpoint, the midpoint m, and the right-hand endpoint of the interval (Figure 9.8.5).

To derive (4), observe that the left-hand endpoint of the interval is $m - h$ and the right-hand endpoint is $m + h$, so the area A under $y = ax^2 + bx + c$ over this interval is

$$A = \int_{m-h}^{m+h} (ax^2 + bx + c)\, dx = \left. \frac{a}{3}x^3 + \frac{b}{2}x^2 + cx \right]_{m-h}^{m+h}$$

$$= \frac{a}{3}[(m+h)^3 - (m-h)^3] + \frac{b}{2}[(m+h)^2 - (m-h)^2] + c[(m+h) - (m-h)]$$

or on simplifying,

$$A = \frac{h}{3}[a(6m^2 + 2h^2) + b(6m) + 6c] \tag{5}$$

But the values of $y = ax^2 + bx + c$ at the left-hand endpoint, the midpoint, and the right-hand endpoint are, respectively,

$$Y_0 = a(m - h)^2 + b(m - h) + c$$

$$Y_1 = am^2 + bm + c$$

$$Y_2 = a(m + h)^2 + b(m + h) + c$$

from which it follows that

$$Y_0 + 4Y_1 + Y_2 = a(6m^2 + 2h^2) + b(6m) + 6c \tag{6}$$

Thus, (4) follows from (5) and (6).

Simpson's rule is obtained by dividing the interval $[a, b]$ into an *even* number of subintervals of equal width h and applying Formula (4) to approximate the area under $y = f(x)$ over successive pairs of subintervals. The sum of these approximations then serves as an estimate of $\int_a^b f(x)\, dx$. More precisely, let $[a, b]$ be divided into n subintervals of width $h = (b - a)/n$ (n even) and let

$$y_0, y_1, \ldots, y_n$$

---

*THOMAS SIMPSON (1710–1761). English mathematician. Simpson was the son of a weaver. He was trained to follow in his father's footsteps and had little formal education in his early life. His interest in science and mathematics was aroused in 1724, when he witnessed an eclipse of the sun and received two books from a peddler, one on astrology and the other on arithmetic. Simpson quickly absorbed their contents and soon became a successful local fortune teller. His improved financial situation enabled him to give up weaving and marry his landlady, an older woman. Then in 1733 some mysterious ''unfortunate incident'' forced him to move. He settled in Derby, where he taught in an evening school and worked at weaving during the day. In 1736 he moved to London and published his first mathematical work in a periodical called the *Ladies' Diary* (of which he later became the editor). In 1737 he published a successful calculus textbook that enabled him to give up weaving completely and concentrate on textbook writing and teaching. His fortunes improved further in 1740 when one Robert Heath accused him of plagiarism. The publicity was marvelous, and Simpson proceeded to dash off a succession of best-selling textbooks: *Algebra* (ten editions plus translations), *Geometry* (twelve editions plus translations), *Trigonometry* (five editions plus translations), and numerous others.

It is interesting to note that Simpson did not discover the rule that bears his name. It was a well-known result by Simpson's time.

be the values of $y = f(x)$ at the subinterval endpoints

$$a = x_0, x_1, \ldots, x_n = b$$

By (4) the area under $y = f(x)$ over the first two subintervals is approximately

$$\frac{h}{3} [y_0 + 4y_1 + y_2]$$

and the area over the second pair of subintervals is approximately

$$\frac{h}{3} [y_2 + 4y_3 + y_4]$$

and the area over the last pair of subintervals is approximately

$$\frac{h}{3} [y_{n-2} + 4y_{n-1} + y_n]$$

Adding all the approximations, collecting terms, and replacing h by $(b - a)/n$ yields

Simpson's Rule

$$\int_a^b f(x) \, dx \approx \left(\frac{b - a}{3n} \right) [y_0 + 4y_1 + 2y_2 + 4y_3 + 2y_4 + \cdots$$
$$+ 2y_{n-2} + 4y_{n-1} + y_n]$$

We shall denote the Simpson's rule approximation with n subintervals by S_n and define the error in this approximation to be

$$E_S = \int_a^b f(x) \, dx - S_n$$

Example 3 In Table 9.8.3 we have approximated

$$\ln 2 = \int_1^2 \frac{1}{x} \, dx$$

by Simpson's rule using $n = 10$ subdivisions so that

$$\frac{b - a}{3n} = \frac{2 - 1}{3(10)} = \frac{1}{30}$$

Observe, by comparing this result to (2), that Simpson's rule produced a more accurate approximation of $\ln 2$ than either of the methods in Example 1.   ◄

Table 9.8.3 Simpson's Rule

i	ENDPOINT x_i	$y_i = f(x_i) = 1/x_i$	MULTIPLIER w_i	$w_i y_i$
0	1.0	1.000000000	1	1.000000000
1	1.1	0.909090909	4	3.636363636
2	1.2	0.833333333	2	1.666666666
3	1.3	0.769230769	4	3.076923076
4	1.4	0.714285714	2	1.428571428
5	1.5	0.666666667	4	2.666666668
6	1.6	0.625000000	2	1.250000000
7	1.7	0.588235294	4	2.352941176
8	1.8	0.555555556	2	1.111111112
9	1.9	0.526315789	4	2.105263156
10	2.0	0.500000000	1	0.500000000
				20.794506918

$$\int_1^2 \frac{1}{x} \, dx \approx (\tfrac{1}{30})(20.794506918) = 0.693150231$$

□ **ERROR ESTIMATES**

With all the methods studied in this section, there are two sources of error: the *intrinsic* or *truncation error* due to the approximation formula and the *roundoff* error introduced in the calculations. In general, increasing n reduces the truncation error but increases the roundoff error, since more computations are required for larger n. In practical applications, it is important to know how large n must be taken to ensure that a specified degree of accuracy is obtained. The analysis of roundoff error is complicated and will not be considered here. However, the following theorems, which are proved in books on **numerical analysis**, provide estimates of the truncation errors in the midpoint, trapezoidal, and Simpson's rule approximations.

9.8.2 THEOREM (*Midpoint and Trapezoidal Error Estimates*). *If f'' is continuous on $[a, b]$ and if K_2 is the maximum value of $|f''(x)|$ on $[a, b]$, then for n subdivisions of $[a, b]$*

$$(a) \quad |E_M| \leq \frac{(b-a)^3 K_2}{24n^2} \qquad (b) \quad |E_T| \leq \frac{(b-a)^3 K_2}{12n^2} \qquad (7\text{–}8)$$

9.8.3 THEOREM (*Simpson Error Estimate*). *If $f^{(4)}$ is continuous on $[a, b]$ and if K_4 is the maximum value of $|f^{(4)}(x)|$ on $[a, b]$, then for n subdivisions of $[a, b]$*

$$|E_S| \leq \frac{(b-a)^5 K_4}{180n^4} \qquad (9)$$

The quantities on the right sides of (7), (8), and (9) (and any numbers larger than these) are called **upper bounds** on the magnitudes of the errors.

Example 4 Find an upper bound on the magnitude of the error from approximating

$$\ln 2 = \int_1^2 \frac{1}{x} \, dx$$

using $n = 10$ subintervals by (a) the trapezoidal approximation, (b) the midpoint approximation, and (c) Simpson's rule.

Solution. We shall apply Formulas (7), (8), and (9) with

$$f(x) = \frac{1}{x}, \quad a = 1, \quad b = 2, \quad \text{and} \quad n = 10$$

We have

$$f'(x) = -\frac{1}{x^2}, \quad f''(x) = \frac{2}{x^3}, \quad f'''(x) = -\frac{6}{x^4}, \quad f^{(4)}(x) = \frac{24}{x^5}$$

Thus,

$$|f''(x)| = \left|\frac{2}{x^3}\right| = \frac{2}{x^3}, \qquad |f^{(4)}(x)| = \left|\frac{24}{x^5}\right| = \frac{24}{x^5} \qquad (10\text{–}11)$$

where we have dropped the absolute values because $f''(x)$ and $f^{(4)}(x)$ have positive values for $1 \leq x \leq 2$. Since (10) and (11) are continuous and decreasing on $[1, 2]$, both functions have their maximum values at $x = 1$; for (10) this maximum value is 2 and for (11) it is 24, so we can take $K_2 = 2$ in (7) and (8) and $K_4 = 24$ in (9). This yields

$$|E_T| \leq \frac{(b-a)^3 K_2}{12n^2} = \frac{1^3 \cdot 2}{12 \cdot 10^2} \approx 0.001666667$$

$$|E_M| \leq \frac{(b-a)^3 K_2}{24n^2} = \frac{1^3 \cdot 2}{24 \cdot 10^2} \approx 0.000833333$$

$$|E_S| \leq \frac{(b-a)^5 K_4}{180n^4} = \frac{1^5 \cdot 24}{180 \cdot 10^4} \approx 0.000013333 \qquad \blacktriangleleft$$

Table 9.8.4 shows that the estimates in the preceding example are consistent with the computations in Examples 1 and 3. In the table we have obtained approximate values of $|E_T|$, $|E_M|$, and $|E_S|$ by computing the absolute value of the difference between the value of ln 2 (accurate to nine decimal places) and the approximations obtained in Examples 1 and 3. Observe that these values of $|E_T|$, $|E_M|$, and $|E_S|$ satisfy the upper bounds obtained in Example 4 and, in fact, are considerably smaller than the upper bounds. It is quite common that the actual errors in the approximations are substantially smaller than the upper bounds.

Table 9.8.4

ln 2 (NINE DECIMAL PLACES)	APPROXIMATION	ABSOLUTE VALUE OF THE DIFFERENCE
0.693147181	$T_{10} = 0.693771403$	$|E_T| \approx 0.000624222$
0.693147181	$M_{10} = 0.692835360$	$|E_M| \approx 0.000311821$
0.693147181	$S_{10} = 0.693150231$	$|E_S| \approx 0.000003050$

Example 5 How many subintervals might be used in approximating

$$\ln 2 = \int_1^2 \frac{1}{x}\,dx$$

by Simpson's rule to ensure that the magnitude of the error is less than 10^{-6}?

Solution. We want to find an even value of n (the number of subintervals) such that $|E_S| < 10^{-6}$, which, from (9), can be achieved by choosing n so that

$$\frac{(b-a)^5 K_4}{180 n^4} < 10^{-6}$$

Using the values of a, b, and K_4 obtained in Example 4, this inequality can be written as

$$\frac{24}{180 n^4} < 10^{-6} \quad \text{or} \quad \frac{2}{15 n^4} < 10^{-6}$$

which can be rewritten as

$$n^4 > \frac{2 \times 10^6}{15} \quad \text{or} \quad n > \sqrt[4]{\frac{2 \times 10^6}{15}}$$

The right side of this inequality is slightly larger than 19.1 (verify), so the smallest even integer satisfying the inequality is $n = 20$. ◄

In cases where it is difficult to find the values of K_2 and K_4 required in Formulas (7), (8), and (9), these constants may be replaced by any larger constants if such constants are easier to find. For example, if $K_2 < K$, then

$$|E_T| \leq \frac{(b-a)^3 K_2}{12 n^2} < \frac{(b-a)^3 K}{12 n^2} \tag{12}$$

so the right side of (12) is also an upper bound on the value of $|E_T|$ (although it is larger and therefore less desirable than the upper bound using K_2). The following example illustrates this idea.

Example 6 How many subintervals might be used in approximating

$$\int_0^1 \cos(x^2)\,dx$$

by the midpoint approximation to ensure that the magnitude of the error is less than 5×10^{-4}?

Solution. We want to find a value of n such that $|E_M| < 5 \times 10^{-4}$. From (7) using $a = 0$ and $b = 1$ we obtain

$$|E_M| \leq \frac{K_2}{24n^2} \tag{13}$$

We have

$$f(x) = \cos(x^2)$$

so

$$f'(x) = -2x \sin(x^2)$$

$$f''(x) = -4x^2 \cos(x^2) - 2\sin(x^2) = -(4x^2 \cos(x^2) + 2\sin(x^2))$$

so

$$|f''(x)| = |4x^2 \cos(x^2) + 2\sin(x^2)| \tag{14}$$

It is tedious to find the maximum value of (14) on $[0, 1]$, so we will use a larger value instead. Such a value can be obtained by observing that

$$|4x^2 \cos(x^2) + 2\sin(x^2)| \leq |4x^2 \cos(x^2)| + |2\sin(x^2)| \qquad \boxed{\begin{array}{c} \text{Theorem 1.2.5} \\ \text{(the triangle inequality)} \end{array}}$$

$$= 4x^2 |\cos(x^2)| + 2|\sin(x^2)| \qquad \boxed{\begin{array}{l} |\cos(x^2)| \leq 1, \\ |\sin(x^2)| \leq 1, \\ x^2 \leq 1 \text{ (on } [0, 1]) \end{array}}$$

$$\leq 4 \cdot 1 + 2 \cdot 1 = 6$$

Thus, $K_2 \leq 6$, so from (13)

$$|E_M| \leq \frac{K_2}{24n^2} \leq \frac{6}{24n^2} = \frac{1}{4n^2}$$

It follows that $|E_M| < 5 \times 10^{-4}$ if

$$\frac{1}{4n^2} < 5 \times 10^{-4} \quad \text{or equivalently,} \quad n > \sqrt{500} > 22.3$$

Thus, $n = 23$ subintervals will ensure the desired accuracy (as will any larger number of subintervals). ◀

☐ **A COMPARISON OF THE THREE METHODS**

Of the three methods studied in this section, Simpson's rule generally produces more accurate results than the midpoint or trapezoidal approximations for the same amount of work. To make this plausible, let us express (7), (8), and (9) in terms of the subinterval width

$$\Delta x = \frac{b - a}{n}$$

We obtain

$$|E_M| \leq \frac{1}{24} K_2 (b - a)(\Delta x)^2 \tag{15}$$

$$|E_T| \leq \frac{1}{12} K_2 (b - a)(\Delta x)^2 \tag{16}$$

$$|E_S| \leq \frac{1}{180} K_4 (b - a)(\Delta x)^4 \tag{17}$$

(verify). Thus, for Simpson's rule the upper bound on the magnitude of the error is proportional to $(\Delta x)^4$, whereas it is proportional to $(\Delta x)^2$ for the midpoint and trapezoidal approximations. Thus, reducing the interval width by a factor of 10, for example, reduces the error bound by a factor of $1/100$ for the midpoint and trapezoidal approximations but reduces it by a factor of $1/10,000$ for Simpson's rule. This suggests that the accuracy of

Simpson's rule improves much more rapidly than that of the other approximations as n increases.

As a final note, observe that if $f(x)$ is a polynomial of degree 3 or less, then $f^{(4)}(x) = 0$ for all x, so $K_4 = 0$ in (9) and consequently $|E_S| = 0$. Thus, Simpson's rule gives exact results for polynomials of degree 3 or less. Similarly, the midpoint and trapezoidal approximations give exact results for polynomials of degree 1 or less. (You should also be able to see that this is so geometrically.)

▶ **Exercise Set 9.8** ☐ 1–27, 29, 30, 33–40

In Exercises 1–6, use $n = 10$ subdivisions to approximate the value of the integral by (a) the midpoint rule, (b) the trapezoidal rule, and (c) Simpson's rule. In each case find the exact value of the integral and approximate the magnitude of the error. Express your answers to at least four decimal places.

1. $\displaystyle\int_0^3 \sqrt{x + 1}\, dx.$ **2.** $\displaystyle\int_1^4 \frac{1}{\sqrt{x}}\, dx.$

3. $\displaystyle\int_0^\pi \sin x\, dx.$ **4.** $\displaystyle\int_0^1 \cos x\, dx.$

5. $\displaystyle\int_1^3 e^{-x}\, dx.$ **6.** $\displaystyle\int_{-1}^1 \frac{1}{2x + 3}\, dx.$

In Exercises 7–12, use inequalities (7), (8), and (9) to find upper bounds on the magnitudes of the errors in parts (a), (b), and (c) of the indicated exercise.

7. Exercise 1. **8.** Exercise 2.

9. Exercise 3. **10.** Exercise 4.

11. Exercise 5. **12.** Exercise 6.

In Exercises 13–18, use inequalities (7), (8), and (9) to find a value for n to ensure that the magnitude of the error will be less than the given value if n subdivisions are used to approximate the integral by (a) the midpoint rule, (b) the trapezoidal rule, and (c) Simpson's rule.

13. Exercise 1; 5×10^{-4}. **14.** Exercise 2; 5×10^{-4}.

15. Exercise 3; 10^{-3}. **16.** Exercise 4; 10^{-3}.

17. Exercise 5; 10^{-6}. **18.** Exercise 6; 10^{-6}.

In Exercises 19–24, use $n = 10$ subdivisions to approximate the value of the integral by (a) the midpoint rule, (b) the trapezoidal rule, and (c) Simpson's rule. Express your answers to at least four decimal places.

19. $\displaystyle\int_0^1 e^{-x^2}\, dx.$ **20.** $\displaystyle\int_0^2 \frac{x}{\sqrt{1 + x^3}}\, dx.$

21. $\displaystyle\int_1^2 \sqrt{1 + x^3}\, dx.$ **22.** $\displaystyle\int_0^\pi \frac{1}{2 - \sin x}\, dx.$

23. $\displaystyle\int_0^2 \sin(x^2)\, dx.$ **24.** $\displaystyle\int_1^3 \sqrt{\ln x}\, dx.$

In Exercises 25 and 26, the exact value of the integral is π (verify). Use $n = 10$ subdivisions to approximate the integral by (a) the midpoint rule, (b) the trapezoidal rule, and (c) Simpson's rule. Approximate the magnitude of the error and express your answers to at least four decimal places.

25. $\displaystyle\int_0^1 \frac{4}{1 + x^2}\, dx.$ **26.** $\displaystyle\int_0^2 \sqrt{4 - x^2}\, dx.$

27. In Example 5 we showed that taking $n = 20$ subdivisions ensures that the magnitude of the error in approximating

$$\ln 2 = \int_1^2 \frac{1}{x}\, dx$$

by Simpson's rule is less than 10^{-6}. Use Simpson's rule with $n = 14$ subdivisions to approximate this integral, and show that the magnitude of the resulting error is less than 10^{-6} by comparing your result to (2), which is accurate to nine decimal places. Does this contradict the fact that we obtained $n = 20$ in Example 5? Explain.

28. In parts (a) and (b), determine whether an approximation of the integral by the trapezoidal rule would be less than or would be greater than the exact value of the integral.

(a) $\displaystyle\int_1^2 e^{-x^2}\, dx$ (b) $\displaystyle\int_0^{0.5} e^{-x^2}\, dx$

In Exercises 29 and 30, find a value for n to ensure that the magnitude of the error in approximating the integral by the midpoint rule will be less than 10^{-4}.

29. $\displaystyle\int_0^2 x \sin x\, dx.$ **30.** $\displaystyle\int_0^1 e^{\cos x}\, dx.$

In Exercises 31 and 32, show that inequalities (7) and (8) are of no value in finding an upper bound on the magnitude of the error that results from approximating the integral by either the midpoint rule or the trapezoidal rule.

31. $\displaystyle\int_0^1 \sqrt{x}\, dx.$ **32.** $\displaystyle\int_0^1 \sin \sqrt{x}\, dx.$

In Exercises 33 and 34, use Simpson's rule with $n = 10$ subdivisions to approximate the length of the curve. Round your answer to three decimal places.

33. $y = \sin x,\ 0 \le x \le \pi.$ **34.** $y = 1/x,\ 1 \le x \le 3.$

Numerical integration methods can be used in problems where only measured or experimentally determined values of the integrand are available. In Exercises 35–40, use Simpson's rule to estimate the value of the integral.

35. A graph of the speed v versus time t for a test run of an Infiniti G20 automobile is shown in Figure 9.8.6. Estimate the speeds at $t = 0, 5, 10, 15$, and 20 sec from the graph, convert to feet/second using 1 mi/hr $= 22/15$ ft/sec, and use these speeds to approximate the number of feet traveled during the first 20 sec. Round your answer to the nearest foot. [*Hint:* Distance traveled $= \int_0^{20} v(t) \, dt$.] [Data from *Road and Track*, October 1990.]

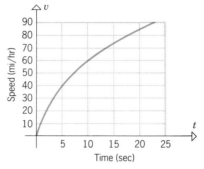

Figure 9.8.6

36. A graph of the acceleration a versus time t for an object moving on a straight line is shown in Figure 9.8.7. Estimate the accelerations at $t = 0, 1, 2, \ldots, 8$ sec from the graph and use them to approximate the change in velocity from $t = 0$ to $t = 8$ sec. Round your answer to the nearest tenth centimeter/second.
[*Hint:* Change in velocity $= \int_0^8 a(t) \, dt$.]

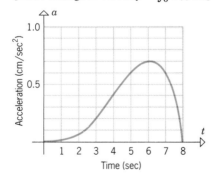

Figure 9.8.7

37. Table 9.8.5 gives the speeds, in miles/second, at various times for a test rocket that was fired upward from the surface of Earth. Use these values to approximate the number of miles traveled during the first 180 sec. Round your answer to the nearest tenth of a mile.
[*Hint:* Distance traveled $= \int_0^{180} v(t) \, dt$.]

38. Table 9.8.6 gives the speeds of a bullet at various distances from the muzzle of a rifle. Use these values to approximate the number of seconds for the bullet to travel 1800 ft. Express your answer to the nearest hundredth of a second. [*Hint:* If v is the speed of the bullet and x is the distance traveled, then $v = dx/dt$ so that $dt/dx = 1/v$ and $t = \int_0^{1800} (1/v) \, dx$.]

Table 9.8.5

TIME t (sec)	SPEED v (mi/sec)
0	0.00
30	0.03
60	0.08
90	0.16
120	0.27
150	0.42
180	0.65

Table 9.8.6

DISTANCE x (ft)	SPEED v (ft/sec)
0	3100
300	2908
600	2725
900	2549
1200	2379
1500	2216
1800	2059

39. Measurements of a pottery shard recovered from an archaeological dig reveal that the shard came from a pot with a flat bottom and circular cross sections (Figure 9.8.8). The figure shows interior radius measurements of the shard made every 4 cm from the bottom of the pot to the top. Use those values to approximate the interior volume of the pot to the nearest tenth of a liter (1 L = 1000 cm^3). [*Hint:* Use 6.2.2 (volume by cross sections) to set up an appropriate integral for the volume.]

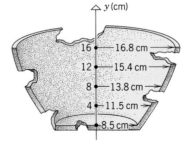

Figure 9.8.8

40. Engineers want to construct a straight and level road 600 ft long and 75 ft wide by cutting through an intervening hill (Figure 9.8.9). Heights of the hill above the centerline of the proposed road, as obtained at various points from a contour map of the region, are shown in Table 9.8.7. To estimate the construction costs, the engineers need to know the volume of earth that must be removed. Approximate this volume, rounded to the nearest cubic foot. [*Hint:* First, set up an integral for the cross-sectional area of the cut along the centerline of the road, then assume that the height of the hill does not vary between the centerline and edges of the road.]

Table 9.8.7

HORIZONTAL DISTANCE x (ft)	HEIGHT h (ft)
0	0
100	7
200	16
300	24
400	25
500	16
600	0

41. Derive the trapezoidal rule by summing the areas of the trapezoids in Figure 9.8.2.

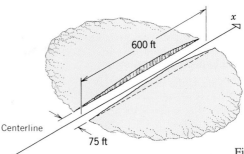

Figure 9.8.9

42. Let f be a function that is positive, continuous, decreasing, and concave down on the interval $[a, b]$. Assuming that $[a, b]$ is subdivided into n equal subintervals, arrange the following approximations of $\int_a^b f(x)\, dx$ in order of increasing value: left endpoint, right endpoint, midpoint, and trapezoidal.

◆ **TECHNOLOGY EXERCISES**   Chapter 9

Most of these exercises require access to a graphing calculator or a computer algebra system (CAS) such as *Mathematica*, *Maple*, or *Derive*. When you are asked to *find* an answer or to *solve* an equation, you may choose to find an exact result or a numerical approximation, depending on the particular technology you are using and on your own imagination. The form of your answers may differ from those of other students or from those in the answer section of the text, depending on how you solve the problems and the accuracy you use in your numerical approximations. Those exercises that are more appropriate for a CAS than a graphing calculator are labeled with the icon ◆.

Techniques of integration: Some integrals that can be evaluated by hand cannot be evaluated by all computer algebra systems. In Exercises 1–10, evaluate the integral by hand, and determine if it can be evaluated on your CAS.

◆ **1.** $\int \dfrac{x^3}{\sqrt{1 - x^8}}\, dx.$

◆ **2.** $\int \dfrac{(x - 2)^3}{\sqrt{4x - x^2}}\, dx.$

◆ **3.** $\int \dfrac{\sqrt{1 + x} + \sqrt{1 - x}}{\sqrt{1 + x} - \sqrt{1 - x}}\, dx.$

◆ **4.** $\int (\cos^{32} x \sin^{30} x - \cos^{30} x \sin^{32} x)\, dx.$

◆ **5.** $\int \dfrac{\sqrt{x + 1}}{(x - 1)^{5/2}}\, dx.$ [*Hint:* Let $x - 1 = 1/u$.]

◆ **6.** $\int \dfrac{1}{x^{10} + x}\, dx.$ [*Hint:* Rewrite the denominator as $x^{10}(1 + x^{-9})$ and let $u = 1 + x^{-9}$.]

◆ **7.** $\int \dfrac{1}{x(3x^5 + 2)}\, dx.$ [*Hint:* See Exercise 6.]

◆ **8.** $\int \dfrac{3x^6 - 2}{x(2x^6 + 5)}\, dx.$ [*Hint:* Rewrite as the sum of two integrals and see Exercise 6.]

◆ **9.** $\int \sqrt{x - \sqrt{x^2 - 4}}\, dx.$
[*Hint:* $\frac{1}{2}(\sqrt{x + 2} - \sqrt{x - 2})^2 = ?$]

◆ **10.** $\int \sqrt{1 + \sqrt{1 - x^2}}\, dx.$
[*Hint:* $\frac{1}{2}(\sqrt{1 + x} + \sqrt{1 - x})^2 = ?$]

◆ **11. Techniques of integration**

(a) Evaluate the integral

$$\int \frac{1}{\sqrt{2x - x^2}} \, dx$$

four ways: using the substitution $u = \sqrt{x}$; using the substitution $u = \sqrt{2 - x}$; completing the square under the radical; and using a CAS.

(b) Show that the results in part (a) are equivalent.

◆ **12. Techniques of integration:** Consider

$$\int \frac{1}{x^3 - x} \, dx$$

(a) Evaluate the integral using the substitution $x = \sec \theta$. For what values of x is your result valid?

(b) Evaluate the integral using the substitution $x = \sin \theta$. For what values of x is your result valid?

(c) Evaluate the integral using the method of partial fractions. For what values of x is your result valid?

(d) Evaluate the integral using a CAS. For what values of x is your result valid?

◆ **13. Error bound for the midpoint approximation:** Let $f(x) = \cos(x^2)$.

(a) Find the maximum value of $|f''(x)|$ on the interval $[0, 1]$.

(b) How large must n be in the midpoint approximation of $\int_0^1 f(x) \, dx$ to ensure that the magnitude of the error is less than 5×10^{-4}? Compare your result with that obtained in Example 6, Section 9.8.

(c) Evaluate the integral using the midpoint approximation with the value of n obtained in part (b).

◆ **14. Error bound for the trapezoidal approximation:** Let $f(x) = \sqrt{1 + x^3}$.

(a) Find the maximum value of $|f''(x)|$ on the interval $[0, 1]$.

(b) How large must n be in the trapezoidal approximation of $\int_0^1 f(x) \, dx$ to ensure that the magnitude of the error is less than 10^{-3}?

(c) Evaluate the integral using the trapezoidal approximation with the value of n obtained in part (b).

◆ **15. Error bound for Simpson's rule:** Let $f(x) = \cos(x^2)$.

(a) Find the maximum value of $|f^{(4)}(x)|$ on the interval $[0, 1]$.

(b) How large must the value of n be in the approximation of $\int_0^1 f(x) \, dx$ by Simpson's rule to ensure that the magnitude of the error is less than 10^{-4}?

(c) Evaluate the integral using Simpson's rule with the value of n obtained in part (b).

◆ **16. Error bound for Simpson's rule:** Let $f(x) = \sqrt{1 + x^3}$.

(a) Find the maximum value of $|f^{(4)}(x)|$ on the interval $[0, 1]$.

(b) How large must the value of n be in the approximation of $\int_0^1 f(x) \, dx$ by Simpson's rule to ensure that the magnitude of the error is less than 10^{-5}?

(c) Evaluate the integral using Simpson's rule with the value of n obtained in part (b).

Partial fractions: In Exercises 17 and 18, find the partial fraction decomposition.

◆ **17.** $\dfrac{5x^3 - 18x^2 + 23x - 21}{x^4 - 5x^3 + 13x^2 - 24x + 20}$.

◆ **18.** $\dfrac{-2x^5 + 26x^4 + 15x^3 + 6x^2 + 20x + 43}{x^6 - x^5 - 18x^4 - 2x^3 - 39x^2 - x - 20}$.

◆ **19. Partial fractions:** Find conditions that a and b must satisfy for the partial fraction decomposition

$$\frac{x^2 + ax + b}{x^3(x^2 + 1)^2} = \frac{A}{x} + \frac{B}{x^2} + \frac{C}{x^3} + \frac{Dx + E}{x^2 + 1} + \frac{Fx + G}{(x^2 + 1)^2}$$

to have the property

(a) $A = 0$ (b) $B = 0$ (c) $F = 0$.

◆ **20. Partial fractions:** Find conditions that a and b must satisfy for the partial fraction decomposition

$$\frac{x^2 + ax + b}{(x + 2)^3(x^2 + x + 3)} = \frac{A}{x + 2} + \frac{B}{(x + 2)^2} + \frac{C}{(x + 2)^3}$$
$$+ \frac{Dx + E}{x^2 + x + 3}$$

to have the property

(a) $A = 0$ (b) $B = 0$ (c) $E = 0$.

Augustin Louis Cauchy (1789–1857)

10 IMPROPER INTEGRALS; L'HÔPITAL'S RULE

10.1 IMPROPER INTEGRALS

In this section we shall extend the concept of the definite integral to include

- *integrals over infinite intervals;*
- *integrals in which the integrand becomes infinite within the interval of integration.*

*These are called **improper integrals**.*

□ **INTEGRALS OVER INFINITE INTERVALS**

In the definition of $\int_a^b f(x)\,dx$, it is assumed that the interval $[a, b]$ is finite. If f is continuous on the interval $[a, +\infty)$, then we define the improper integral $\int_a^{+\infty} f(x)\,dx$ as a limit in the following way:

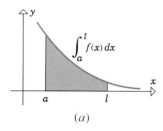

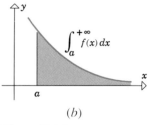

Figure 10.1.1

$$\int_a^{+\infty} f(x)\,dx = \lim_{l \to +\infty} \int_a^l f(x)\,dx \qquad (1)$$

If this limit exists, the improper integral is said to **converge**, and the value of the limit is the value assigned to the integral. If the limit does not exist, then the improper integral is said to **diverge**, in which case it is not assigned a value.

If f is nonnegative and continuous on $[a, +\infty)$, then (1) has an important geometric interpretation. For each value of $l > a$, the definite integral $\int_a^l f(x)\,dx$ represents the area under the curve $y = f(x)$ over the finite interval $[a, l]$ (Figure 10.1.1a). As we let l approach $+\infty$, this area tends toward the area under $y = f(x)$ over the entire interval $[a, +\infty)$ (Figure 10.1.1b). Thus, $\int_a^{+\infty} f(x)\,dx$ can be regarded as the area under $y = f(x)$ over the interval $[a, +\infty)$.

Example 1   Evaluate $\displaystyle\int_1^{+\infty} \frac{dx}{x^2}$.

Solution.   We begin by replacing the infinite upper limit with a finite upper limit l:

$$\int_1^l \frac{dx}{x^2} = -\frac{1}{x}\Bigg]_1^l = -\frac{1}{l} - (-1) = 1 - \frac{1}{l}$$

Thus,

$$\int_1^{+\infty} \frac{dx}{x^2} = \lim_{l \to +\infty} \int_1^l \frac{dx}{x^2} = \lim_{l \to +\infty} \left(1 - \frac{1}{l}\right) = 1$$

so the given integral converges to 1.   ◀

Example 2   Evaluate $\displaystyle\int_1^{+\infty} \frac{dx}{x}$.

Solution.

$$\int_1^{+\infty} \frac{dx}{x} = \lim_{l \to +\infty} \int_1^l \frac{dx}{x} = \lim_{l \to +\infty} \left[\ln|x|\right]_1^l = \lim_{l \to +\infty} \ln|l| = +\infty$$

Thus, the integral diverges.   ◀

It is worthwhile to reflect on the results of Examples 1 and 2. Why is it that $\int_1^{+\infty} dx/x^2$ converges while $\int_1^{+\infty} dx/x$ diverges when, in fact, the graphs of $1/x$ and $1/x^2$ look similar over the interval $[1, +\infty)$ (Figure 10.1.2)? The explanation is that $1/x^2$ approaches zero more rapidly than $1/x$ as $x \to +\infty$, so that area accumulates under the curve $y = 1/x^2$ less rapidly than under the curve $y = 1/x$. When we calculate

$$\lim_{l \to +\infty} \int_1^l \frac{dx}{x} \quad \text{and} \quad \lim_{l \to +\infty} \int_1^l \frac{dx}{x^2}$$

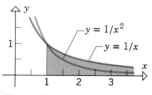

Figure 10.1.2

the difference is enough that the first limit is infinite while the second is finite.

If f is continuous on the interval $(-\infty, b]$, then we define the improper integral $\int_{-\infty}^b f(x)\,dx$ as a limit in the following way:

$$\int_{-\infty}^b f(x)\,dx = \lim_{l \to -\infty} \int_l^b f(x)\,dx$$

As before, the improper integral is said to **converge** if the limit exists and **diverge** if it does not. If f is nonnegative, then the integral represents the area under $y = f(x)$ over the interval $(-\infty, b]$.

Example 3 Evaluate $\displaystyle\int_{-\infty}^{1} e^x\,dx$.

Solution. We begin by replacing the infinite lower limit with a finite lower limit l:

$$\int_{l}^{1} e^x\,dx = e^x\bigg]_{l}^{1} = e^1 - e^l = e - e^l$$

Thus,

$$\int_{-\infty}^{1} e^x\,dx = \lim_{l\to-\infty}\int_{l}^{1} e^x\,dx = \lim_{l\to-\infty}(e - e^l) = e \qquad \blacktriangleleft$$

If the two improper integrals $\int_{-\infty}^{0} f(x)\,dx$ and $\int_{0}^{+\infty} f(x)\,dx$ both converge, then we say that $\int_{-\infty}^{+\infty} f(x)\,dx$ **converges** and we define

$$\int_{-\infty}^{+\infty} f(x)\,dx = \int_{-\infty}^{0} f(x)\,dx + \int_{0}^{+\infty} f(x)\,dx \qquad (2)$$

If either integral on the right side of (2) diverges, then we say that $\int_{-\infty}^{+\infty} f(x)\,dx$ **diverges**. If f is nonnegative, then the integral $\int_{-\infty}^{+\infty} f(x)\,dx$ represents the area under $y = f(x)$ over the interval $(-\infty, +\infty)$.

Example 4 Evaluate $\displaystyle\int_{-\infty}^{+\infty} \frac{dx}{1 + x^2}$.

Solution.

$$\int_{0}^{+\infty} \frac{dx}{1 + x^2} = \lim_{l\to+\infty}\int_{0}^{l} \frac{dx}{1 + x^2} = \lim_{l\to+\infty}\left[\tan^{-1} x\right]_{0}^{l} = \lim_{l\to+\infty}(\tan^{-1} l) = \frac{\pi}{2}$$

$$\int_{-\infty}^{0} \frac{dx}{1 + x^2} = \lim_{l\to-\infty}\int_{l}^{0} \frac{dx}{1 + x^2} = \lim_{l\to-\infty}\left[\tan^{-1} x\right]_{l}^{0} = \lim_{l\to-\infty}(-\tan^{-1} l) = \frac{\pi}{2}$$

Thus,

$$\int_{-\infty}^{+\infty} \frac{dx}{1 + x^2} = \int_{-\infty}^{0} \frac{dx}{1 + x^2} + \int_{0}^{+\infty} \frac{dx}{1 + x^2} = \frac{\pi}{2} + \frac{\pi}{2} = \pi \qquad \blacktriangleleft$$

REMARK. In (2), the decision to split the integral $\int_{-\infty}^{+\infty} f(x)\,dx$ at $x = 0$ is arbitrary. The split can be made at any other point $x = c$ without affecting the convergence, the divergence, or the value of the integral; that is,

$$\int_{-\infty}^{+\infty} f(x)\,dx = \int_{-\infty}^{c} f(x)\,dx + \int_{c}^{+\infty} f(x)\,dx$$

☐ **INTEGRALS WHOSE INTEGRANDS BECOME INFINITE**

Recall from Theorem 5.6.9(c) that if a function f is not bounded on an interval $[a, b]$, then f is not integrable on $[a, b]$. For example, the integral

$$\int_{0}^{3} \frac{1}{(x - 2)^{2/3}}\,dx$$

does not exist since the integrand approaches $+\infty$ at the point $x = 2$ in the interval of integration. It is possible, however, to extend the concept of an integral to include functions that are not bounded by defining the integral of such a function as a limit of integrals of bounded functions in an appropriate way. These limits are also called **improper integrals**. We shall first consider the case where the integrand of f approaches $+\infty$ or $-\infty$ at one of the endpoints of integration.

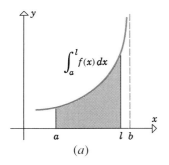

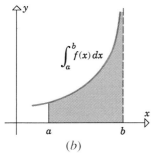

Figure 10.1.3

If f is continuous on $[a, b)$ but fails to have a limit as x approaches b from the left [for example, if $f(x) \to +\infty$ or $f(x) \to -\infty$], then we define the improper integral $\int_a^b f(x)\,dx$ as a limit in the following way:

$$\int_a^b f(x)\,dx = \lim_{l \to b^-} \int_a^l f(x)\,dx \tag{3}$$

To visualize this definition geometrically, consider the case where f is nonnegative on the interval $[a, b)$. For each number l satisfying $a \le l < b$, the integral $\int_a^l f(x)\,dx$ represents the area under $y = f(x)$ over the interval $[a, l]$ (Figure 10.1.3a). As we let l approach b from the left, these areas tend to fill out the entire area under $y = f(x)$ over the interval $[a, b)$ (Figure 10.1.3b). As before, the improper integral $\int_a^b f(x)\,dx$ is said to **converge** or **diverge** depending on whether the limit in (3) does or does not exist.

Example 5   Evaluate $\displaystyle\int_0^1 \frac{dx}{\sqrt{1 - x}}$.

Solution.   The integral is improper because the integrand approaches $+\infty$ as x approaches the upper limit 1 from the left. From (3),

$$\int_0^1 \frac{dx}{\sqrt{1 - x}} = \lim_{l \to 1^-} \int_0^l \frac{dx}{\sqrt{1 - x}} = \lim_{l \to 1^-} \left[-2\sqrt{1 - x} \right]_0^l$$
$$= \lim_{l \to 1^-} \left[-2\sqrt{1 - l} + 2 \right] = 2 \quad \blacktriangleleft$$

If f is continuous on $(a, b]$, but fails to have a limit as x approaches a from the right [say $f(x) \to +\infty$ or $f(x) \to -\infty$], then we define the improper integral $\int_a^b f(x)\,dx$ as

$$\int_a^b f(x)\,dx = \lim_{l \to a^+} \int_l^b f(x)\,dx$$

The improper integral **converges** or **diverges** depending on whether the limit does or does not exist.

Example 6   The integral

$$\int_1^2 \frac{dx}{1 - x}$$

is improper because the integrand approaches $-\infty$ as x approaches the lower limit 1 from the right. The integral diverges because

$$\int_1^2 \frac{dx}{1 - x} = \lim_{l \to 1^+} \int_l^2 \frac{dx}{1 - x} = \lim_{l \to 1^+} \left[-\ln|1 - x| \right]_l^2$$
$$= \lim_{l \to 1^+} \left[-\ln|-1| + \ln|1 - l| \right]$$
$$= \lim_{l \to 1^+} \ln|1 - l| = -\infty \quad \blacktriangleleft$$

Let f be continuous on the interval $[a, b]$ with the exception that at some point c satisfying $a < c < b$, $f(x)$ becomes infinite (tends to $+\infty$ or $-\infty$) as x approaches c from the left or the right. If the two improper integrals $\int_a^c f(x)\,dx$ and $\int_c^b f(x)\,dx$ both converge, then we say that the improper integral $\int_a^b f(x)\,dx$ **converges**, and we define

$$\int_a^b f(x)\,dx = \int_a^c f(x)\,dx + \int_c^b f(x)\,dx \tag{4}$$

If either integral on the right side of (4) diverges, then we say that $\int_a^b f(x)\,dx$ **diverges**.

Example 7 Evaluate $\displaystyle\int_1^4 \frac{dx}{(x-2)^{2/3}}$.

Solution. The integrand approaches $+\infty$ as $x \to 2$, so we use (4) to write

$$\int_1^4 \frac{dx}{(x-2)^{2/3}} = \int_1^2 \frac{dx}{(x-2)^{2/3}} + \int_2^4 \frac{dx}{(x-2)^{2/3}} \tag{5}$$

But

$$\int_1^2 \frac{dx}{(x-2)^{2/3}} = \lim_{l \to 2^-} \int_1^l \frac{dx}{(x-2)^{2/3}} = \lim_{l \to 2^-}[3(l-2)^{1/3} - 3(1-2)^{1/3}] = 3$$

$$\int_2^4 \frac{dx}{(x-2)^{2/3}} = \lim_{l \to 2^+} \int_l^4 \frac{dx}{(x-2)^{2/3}} = \lim_{l \to 2^+}[3(4-2)^{1/3} - 3(l-2)^{1/3}] = 3\sqrt[3]{2}$$

Thus, from (5)

$$\int_1^4 \frac{dx}{(x-2)^{2/3}} = 3 + 3\sqrt[3]{2} \qquad \blacktriangleleft$$

WARNING. It is sometimes tempting to apply the First Fundamental Theorem of Calculus directly to an improper integral without taking the appropriate limits. To illustrate what can go wrong with this procedure, suppose we ignore the fact that the integral

$$\int_0^2 \frac{dx}{(x-1)^2} \tag{6}$$

is improper and write

$$\int_0^2 \frac{dx}{(x-1)^2} = -\frac{1}{x-1}\Bigg]_0^2 = -1 - (1) = -2$$

This result is clearly nonsense because the integrand is never negative and consequently the integral cannot be negative! To evaluate (6) correctly we should write

$$\int_0^2 \frac{dx}{(x-1)^2} = \int_0^1 \frac{dx}{(x-1)^2} + \int_1^2 \frac{dx}{(x-1)^2} \tag{7}$$

But

$$\int_0^1 \frac{dx}{(x-1)^2} = \lim_{l \to 1^-}\int_0^l \frac{dx}{(x-1)^2} = \lim_{l \to 1^-}\left[-\frac{1}{l-1} - 1\right] = +\infty$$

so that (6) diverges.

▶ Exercise Set 10.1   ⓒ *63–66*

In Exercises 1–32, evaluate the integrals that converge.

1. $\displaystyle\int_0^{+\infty} e^{-x}\,dx$.

2. $\displaystyle\int_1^{+\infty} \frac{dx}{x^3}$.

3. $\displaystyle\int_1^{+\infty} \frac{dx}{\sqrt{x}}$.

4. $\displaystyle\int_{-1}^{+\infty} \frac{x}{1+x^2}\,dx$.

5. $\displaystyle\int_4^{+\infty} \frac{2}{x^2-1}\,dx$.

6. $\displaystyle\int_0^{+\infty} xe^{-x^2}\,dx$.

7. $\displaystyle\int_e^{+\infty} \frac{1}{x\ln^3 x}\,dx$.

8. $\displaystyle\int_2^{+\infty} \frac{1}{x\sqrt{\ln x}}\,dx$.

9. $\displaystyle\int_a^{+\infty} \frac{x\,dx}{(x^2+1)^2}$.

10. $\displaystyle\int_0^{+\infty} \frac{dx}{a^2+b^2x^2}$, $a,b>0$.

11. $\displaystyle\int_{-\infty}^{0} \frac{dx}{(2x-1)^3}.$

12. $\displaystyle\int_{-\infty}^{2} \frac{dx}{x^2+4}.$

13. $\displaystyle\int_{-\infty}^{0} e^{3x}\,dx.$

14. $\displaystyle\int_{-\infty}^{0} \frac{e^x\,dx}{3-2e^x}.$

15. $\displaystyle\int_{-\infty}^{+\infty} x^3\,dx.$

16. $\displaystyle\int_{-\infty}^{+\infty} \frac{x}{\sqrt{x^2+2}}\,dx.$

17. $\displaystyle\int_{-\infty}^{+\infty} \frac{x}{(x^2+3)^2}\,dx.$

18. $\displaystyle\int_{-\infty}^{+\infty} \frac{e^{-t}}{1+e^{-2t}}\,dt.$

19. $\displaystyle\int_{3}^{4} \frac{dx}{(x-3)^2}.$

20. $\displaystyle\int_{0}^{8} \frac{dx}{\sqrt[3]{x}}.$

21. $\displaystyle\int_{0}^{\pi/2} \tan x\,dx.$

22. $\displaystyle\int_{0}^{9} \frac{dx}{\sqrt{9-x}}.$

23. $\displaystyle\int_{0}^{1} \frac{dx}{\sqrt{1-x^2}}.$

24. $\displaystyle\int_{-3}^{1} \frac{x\,dx}{\sqrt{9-x^2}}.$

25. $\displaystyle\int_{0}^{\pi/6} \frac{\cos x}{\sqrt{1-2\sin x}}\,dx.$

26. $\displaystyle\int_{0}^{\pi/4} \frac{\sec^2 x}{1-\tan x}\,dx.$

27. $\displaystyle\int_{0}^{3} \frac{dx}{x-2}.$

28. $\displaystyle\int_{-2}^{2} \frac{dx}{x^2}.$

29. $\displaystyle\int_{-1}^{8} x^{-1/3}\,dx.$

30. $\displaystyle\int_{0}^{4} \frac{dx}{(x-2)^{2/3}}.$

31. $\displaystyle\int_{0}^{+\infty} \frac{1}{x^2}\,dx.$

32. $\displaystyle\int_{1}^{+\infty} \frac{dx}{x\sqrt{x^2-1}}.$

In Exercises 33–36, make the given substitution and evaluate the resulting integral.

33. $\displaystyle\int_{0}^{+\infty} \frac{e^{-\sqrt{x}}}{\sqrt{x}}\,dx; \quad u=\sqrt{x}.$

34. $\displaystyle\int_{0}^{+\infty} \frac{dx}{\sqrt{x}(x+4)}; \quad u=\sqrt{x}.$

35. $\displaystyle\int_{0}^{+\infty} \frac{e^{-x}}{\sqrt{1-e^{-x}}}\,dx; \quad u=1-e^{-x}.$

36. $\displaystyle\int_{0}^{+\infty} \frac{e^{-x}}{\sqrt{1-e^{-2x}}}\,dx; \quad u=e^{-x}.$

In Exercises 37 and 38, find the value of a.

37. $\displaystyle\int_{0}^{+\infty} e^{-ax}\,dx = 5.$

38. $\displaystyle\int_{0}^{+\infty} \frac{1}{x^2+a^2}\,dx = 1, \quad a>0.$

In Exercises 39 and 40, evaluate the integral given that $\int_{0}^{+\infty} e^{-x^2}\,dx = \frac{1}{2}\sqrt{\pi}.$

39. $\displaystyle\int_{0}^{+\infty} \frac{e^{-x}}{\sqrt{x}}\,dx.$

40. $\displaystyle\int_{0}^{+\infty} e^{-a^2x^2}\,dx, \ a>0.$

41. Evaluate $\displaystyle\int_{0}^{+\infty} \frac{\sin x}{\sqrt{x}}\,dx$ given that

$$\int_{0}^{+\infty} \sin(x^2)\,dx = \frac{1}{2}\sqrt{\frac{\pi}{2}}$$

42. The *average speed*, $\bar{v}$, of the molecules of an ideal gas is given by

$$\bar{v} = \frac{4}{\sqrt{\pi}}\left(\frac{M}{2RT}\right)^{3/2} \int_{0}^{+\infty} v^3 e^{-Mv^2/(2RT)}\,dv$$

and the *root-mean-square speed*, v_{rms}, by

$$v_{\text{rms}}^2 = \frac{4}{\sqrt{\pi}}\left(\frac{M}{2RT}\right)^{3/2} \int_{0}^{+\infty} v^4 e^{-Mv^2/(2RT)}\,dv$$

where v is the molecular speed, T is the gas temperature, M is the molecular weight of the gas, and R is the gas constant.

(a) Show that $\bar{v} = \sqrt{\dfrac{8RT}{\pi M}}$ given that

$$\int_{0}^{+\infty} x^3 e^{-a^2x^2}\,dx = \frac{1}{2a^4}$$

(b) Show that $v_{\text{rms}} = \sqrt{3RT/M}$ given that

$$\int_{0}^{+\infty} x^4 e^{-a^2x^2}\,dx = \frac{3\sqrt{\pi}}{8a^5}, \quad a>0$$

43. (a) It is possible for an improper integral to diverge without becoming infinite. Show that $\int_{0}^{+\infty} \cos x\,dx$ does this.

(b) Show that $\displaystyle\int_{0}^{+\infty} \frac{\cos\sqrt{x}}{\sqrt{x}}\,dx$ diverges.

(c) Show that $\displaystyle\int_{0}^{1} \frac{\cos(1/x)}{x^2}\,dx$ diverges.

44. Given that $I = \int_{0}^{\pi/2} \ln(\tan x)\,dx$ converges, show that the value of the integral is zero. [*Hint:* Use the substitution $x = \pi/2 - u$ to show that $I = -I$.]

45. Evaluate $\displaystyle\int_{0}^{+\infty} e^{-x}\cos x\,dx.$

46. For what values of p does $\displaystyle\int_{0}^{+\infty} e^{px}\,dx$ converge?

47. Show that $\displaystyle\int_{0}^{1} \frac{dx}{x^p}$ converges if $p<1$ and diverges if $p \geq 1$.

48. Show that $\displaystyle\int_{1}^{+\infty} \frac{dx}{x^p}$ converges if $p>1$ and diverges if $p \leq 1$.

49. It can be shown that if f and g are continuous and $0 \leq f(x) \leq g(x)$ for all $x \geq a$, then $\int_{a}^{+\infty} f(x)\,dx$ converges if $\int_{a}^{+\infty} g(x)\,dx$ converges, and

$$\int_{a}^{+\infty} f(x)\,dx \leq \int_{a}^{+\infty} g(x)\,dx$$

Use this result to obtain an upper bound for each of the following improper integrals.

(a) $\displaystyle\int_{2}^{+\infty} \frac{x}{x^5+1}\,dx$

(b) $\displaystyle\int_{1}^{+\infty} e^{-x^2}\,dx.$

50. It can be shown that if f and g are continuous and $0 \le f(x) \le g(x)$ for all $x \ge a$, then $\int_a^{+\infty} g(x)\, dx$ diverges if $\int_a^{+\infty} f(x)\, dx$ diverges. Use this result to show that each of the following improper integrals diverges.

(a) $\displaystyle\int_2^{+\infty} \frac{\sqrt{x^3 + 1}}{x}\, dx$ (b) $\displaystyle\int_0^{+\infty} \frac{e^x}{2x + 1}\, dx$.

51. Find the area of the region between the x-axis and the curve $y = e^{-3x}$ for $x \ge 0$.

52. Find the area of the region between the x-axis and the curve $y = 8/(x^2 - 4)$ for $x \ge 3$.

53. Let R be the region to the right of $x = 1$ that is bounded by the x-axis and the curve $y = 1/x$.

(a) Show that the solid obtained by revolving R about the x-axis has a finite volume.

(b) Show that the solid of part (a) has an infinite surface area. [*Hint:* See Exercise 50.]
[*Note:* It has been suggested that by filling this solid with paint and letting it seep through to the surface one could paint an infinite surface area with a finite amount of paint!]

54. Suppose that the region between the x-axis and the curve $y = e^{-x}$ for $x \ge 0$ is revolved about the x-axis.

(a) Find the volume of the solid that is generated.

(b) Find the surface area of the solid.

55. In electromagnetic theory, the magnetic potential at a point on the axis of a circular coil is given by

$$u = \frac{2\pi NIr}{k} \int_a^{+\infty} \frac{dx}{(r^2 + x^2)^{3/2}}$$

where N, I, r, k, and a are constants. Find u.

56. Sketch the region whose area is $\displaystyle\int_0^{+\infty} \frac{dx}{1 + x^2}$, and use your sketch to show that

$$\int_0^{+\infty} \frac{dx}{1 + x^2} = \int_0^1 \sqrt{\frac{1 - y}{y}}\, dy$$

57. (**Satellite Problem**) In Exercise 15 of Section 6.7, we determined the work required to lift a 6000-lb satellite to a specified orbital position. The result in part (a) of that problem will be needed here.

(a) Find a definite integral that represents the work required to lift a 6000-lb satellite to a position l miles above the earth's surface.

(b) Find a definite integral that represents the work required to lift a 6000-lb satellite an "infinite distance" above the earth's surface. Evaluate the integral. [*Note:* The result obtained here is sometimes called the work required to "escape" the earth's gravity.]

58. The **Laplace transform** of a function $f(t)$, denoted by $\mathcal{L}\{f(t)\}$, is defined by

$$\mathcal{L}\{f(t)\} = \int_0^{+\infty} e^{-st} f(t)\, dt$$

where s is regarded as a constant for the integration. The Laplace transform has the effect of "transforming" the function $f(t)$ into a function of s. Show that

$$\mathcal{L}\{e^{2t}\} = \frac{1}{s - 2} \quad \text{if } s > 2$$

In Exercises 59–62, find $\mathcal{L}\{f(t)\}$ as defined in Exercise 58.

59. $f(t) = 1, \ s > 0$.

60. $f(t) = \begin{cases} 0, & t < 3 \\ 1, & t \ge 3 \end{cases}, \ s > 0$.

61. $f(t) = \sin t, \ s > 0$.

62. $f(t) = \cos t, \ s > 0$.

In Exercises 63 and 64, transform the given improper integral into a proper integral by making the stated u-substitution, then approximate the proper integral by Simpson's rule (Section 9.8) with $n = 10$ subdivisions. Round your answer to three decimal places.

63. $\displaystyle\int_0^1 \frac{\cos x}{\sqrt{x}}\, dx; \ u = \sqrt{x}$.

64. $\displaystyle\int_0^1 \frac{\sin x}{\sqrt{1 - x}}\, dx; \ u = \sqrt{1 - x}$.

An improper integral over an infinite interval can be approximated by first replacing the infinite limit(s) of integration by finite limit(s), then using a numerical integration technique, such as Simpson's rule, to approximate the integral with finite limit(s). This is illustrated in Exercises 65 and 66.

65. (a) It can be shown that

$$\int_0^{+\infty} \frac{1}{x^6 + 1}\, dx = \frac{\pi}{3}$$

Approximate this integral by applying Simpson's rule with $n = 20$ subdivisions to the integral

$$\int_0^K \frac{1}{x^6 + 1}\, dx$$

with $K = 4$. Round your answer to three decimal places and compare it to $\pi/3$ rounded to three decimal places.

(b) Use the result in Exercise 49 and the fact that $1/(x^6 + 1) < 1/x^6$ for $x \ge 4$ to show that

$$\int_0^{+\infty} \frac{1}{x^6 + 1}\, dx = \int_0^4 \frac{1}{x^6 + 1}\, dx + E$$

where $0 < E < 2 \times 10^{-4}$. [*Note:* E is the error in part (a) that results from replacing $+\infty$ by $K = 4$. The other two sources of error in the approximation are the error from Simpson's rule and the roundoff error.]

66. (a)   As noted above Exercise 39,

$$\int_0^{+\infty} e^{-x^2}\,dx = \tfrac{1}{2}\sqrt{\pi}$$

Approximate this integral by applying Simpson's rule with $n = 10$ subdivisions to the integral

$$\int_0^K e^{-x^2}\,dx$$

with $K = 3$. Round your answer to four decimal places and compare it to $\tfrac{1}{2}\sqrt{\pi}$ rounded to four decimal places.

(b)   Use the result in Exercise 49 and the fact that $e^{-x^2} < xe^{-x^2}$ for $x \geq 3$ to show that

$$\int_0^{+\infty} e^{-x^2}\,dx = \int_0^3 e^{-x^2}\,dx + E$$

where $0 < E < 7 \times 10^{-5}$. [See note in Exercise 65.]

■ 10.2   L'HÔPITAL'S RULE (INDETERMINATE FORMS OF TYPE 0/0)

In this section we shall develop an important new technique for finding limits of functions.

☐ L'HÔPITAL'S RULE

In each of the limits

$$\lim_{x \to 2} \frac{x^2 - 4}{x - 2} \quad \text{and} \quad \lim_{x \to 0} \frac{\sin x}{x} \qquad (1)$$

the numerator and denominator both approach zero. It is customary to describe such limits as **indeterminate forms of type 0/0**. As we shall see, a limit of this type can have any real number whatsoever as its value or can diverge. The value of such a limit, if it converges, is not generally evident by inspection, so the term "indeterminate" is used to convey the idea that the limit cannot be determined without some additional work.

In Example 6 of Section 2.5 we evaluated the first limit in (1) by canceling the common factor $x - 2$ from the numerator and denominator, and in Theorem 2.8.3 we resorted to a rather intricate geometric argument to obtain the second limit in (1). Because geometric arguments and the technique of canceling factors apply only to a limited range of problems, it is desirable to have a general method for handling indeterminate forms. This is provided by **L'Hôpital's* rule**, which we now discuss.

> **10.2.1**   THEOREM (**L'Hôpital's Rule for Form 0/0**).   *Let* $\lim$ *stand for one of the limits* $\lim_{x \to a}$, $\lim_{x \to a^+}$, $\lim_{x \to a^-}$, $\lim_{x \to +\infty}$, *or* $\lim_{x \to -\infty}$, *and suppose that* $\lim f(x) = 0$ *and* $\lim g(x) = 0$. *If* $\lim [f'(x)/g'(x)]$ *has a finite value L, or if this limit is* $+\infty$ *or* $-\infty$, *then*
>
> $$\lim \frac{f(x)}{g(x)} = \lim \frac{f'(x)}{g'(x)}$$

---

* GUILLAUME FRANCOIS ANTOINE DE L'HÔPITAL (1661–1704). French mathematician. L'Hôpital, born to parents of the French high nobility, held the title of Marquis de Sainte-Mesme Comte d'Autrement. He showed mathematical talent quite early and at age 15 solved a difficult problem about cycloids posed by Pascal. As a young man he served briefly as a cavalry officer, but resigned because of nearsightedness. In his own time he gained fame as the author of the first textbook ever published on differential calculus, *L'Analyse des Infiniment Petits pour l'Intelligence des Lignes Courbes* (1696). L'Hôpital's rule appeared for the first time in that book. Actually, L'Hôpital's rule and most of the material in the calculus text were due to John Bernoulli, who was L'Hôpital's teacher. L'Hôpital dropped his plans for a book on integral calculus when Leibniz informed him that he intended to write such a text. L'Hôpital was apparently generous and personable, and his many contacts with major mathematicians provided the vehicle for disseminating major discoveries in calculus throughout Europe.

REMARK.   There are some hypotheses implicit in this theorem. For example, in the case where $x \rightarrow a$, the statement

$$\lim_{x \rightarrow a} \frac{f'(x)}{g'(x)} = L$$

requires that f'/g' be defined in some open interval I containing a (except possibly at a). This implies that f and g are differentiable and $g'(x) \neq 0$ in I (except possibly at a). Similar hypotheses are implicit in the other cases.

In essence, L'Hôpital's rule enables us to replace one limit problem with another that may be simpler. In each of the following examples we shall employ the following three-step process:

Step 1.   Check that $\lim f(x)/g(x)$ is an indeterminate form. If it is not, then L'Hôpital's rule cannot be used.

Step 2.   Differentiate f and g separately.

Step 3.   Find $\lim f'(x)/g'(x)$. If this limit is finite, $+\infty$, or $-\infty$, then it is equal to $\lim f(x)/g(x)$.

Example 1   Use L'Hôpital's rule to evaluate

(a) $\displaystyle \lim_{x \rightarrow 2} \frac{x^2 - 4}{x - 2}$ (b) $\displaystyle \lim_{x \rightarrow 0} \frac{\sin 2x}{x}$

Solution (a).   Since

$$\lim_{x \rightarrow 2} (x^2 - 4) = 0 \quad \text{and} \quad \lim_{x \rightarrow 2} (x - 2) = 0$$

the given limit is an indeterminate form of type $0/0$. Thus, L'Hôpital's rule applies and we can write

$$\lim_{x \rightarrow 2} \frac{x^2 - 4}{x - 2} = \lim_{x \rightarrow 2} \frac{\dfrac{d}{dx}[x^2 - 4]}{\dfrac{d}{dx}[x - 2]} = \lim_{x \rightarrow 2} \frac{2x}{1} = 4$$

Observe that this agrees with the result obtained in Example 6 of Section 2.5 by factoring.

Solution (b).   Since

$$\lim_{x \rightarrow 0} \sin 2x = 0 \quad \text{and} \quad \lim_{x \rightarrow 0} x = 0$$

the given limit is an indeterminate form of type $0/0$. Thus, L'Hôpital's rule applies and we can write

$$\lim_{x \rightarrow 0} \frac{\sin 2x}{x} = \lim_{x \rightarrow 0} \frac{\dfrac{d}{dx}[\sin 2x]}{\dfrac{d}{dx}[x]} = \lim_{x \rightarrow 0} \frac{2 \cos 2x}{1} = 2$$

Observe that this agrees with the result that we obtained in Example 4 of Section 2.8 by substitution.   ◄

REMARK.   To be rigorous, in each of the preceding examples the first equality is not justified until the limit on the right is shown to exist. However, for simplicity we shall usually arrange the computations as shown when applying L'Hôpital's rule.

Example 2   Evaluate $\lim\limits_{x \to \pi/2} \dfrac{1 - \sin x}{\cos x}$.

Solution.   Since

$$\lim_{x \to \pi/2} (1 - \sin x) = \lim_{x \to \pi/2} \cos x = 0$$

the given limit is an indeterminate form of type $0/0$. Thus, by L'Hôpital's rule

$$\lim_{x \to \pi/2} \frac{1 - \sin x}{\cos x} = \lim_{x \to \pi/2} \frac{\dfrac{d}{dx}[1 - \sin x]}{\dfrac{d}{dx}[\cos x]} = \lim_{x \to \pi/2} \frac{-\cos x}{-\sin x} = \frac{0}{-1} = 0 \quad \blacktriangleleft$$

Example 3   Evaluate $\lim\limits_{x \to 0} \dfrac{e^x}{x^2}$.

Solution.   We have

$$\lim_{x \to 0} e^x = 1 \quad \text{and} \quad \lim_{x \to 0} x^2 = 0$$

so the given problem is not an indeterminate form of type $0/0$ and consequently we cannot apply L'Hôpital's rule. By inspection

$$\lim_{x \to 0} \frac{e^x}{x^2} = +\infty \quad \blacktriangleleft$$

WARNING.   Applying L'Hôpital's rule to limits that are not indeterminate forms can lead to erroneous results. As an illustration, two applications of L'Hôpital's rule in the preceding example would have led to the *incorrect* conclusion that the limit is $1/2$.

Example 4   Evaluate $\lim\limits_{x \to 0} \dfrac{e^x - 1}{x^3}$.

Solution.   Since

$$\lim_{x \to 0} (e^x - 1) = \lim_{x \to 0} x^3 = 0$$

the given limit is an indeterminate form of type $0/0$. Thus, by L'Hôpital's rule

$$\lim_{x \to 0} \frac{e^x - 1}{x^3} = \lim_{x \to 0} \frac{\dfrac{d}{dx}[e^x - 1]}{\dfrac{d}{dx}[x^3]} = \lim_{x \to 0} \frac{e^x}{3x^2} = +\infty \quad \blacktriangleleft$$

As the following example shows, it is sometimes necessary to apply L'Hôpital's rule more than once in the same problem.

Example 5   Evaluate $\lim\limits_{x \to 0} \dfrac{1 - \cos x}{x^2}$.

Solution.   Since

$$\lim_{x \to 0} (1 - \cos x) = \lim_{x \to 0} x^2 = 0$$

the given limit is an indeterminate form of type 0/0. Thus, by L'Hôpital's rule

$$\lim_{x \to 0} \frac{1 - \cos x}{x^2} = \lim_{x \to 0} \frac{\sin x}{2x}$$

However, the new limit is also an indeterminate form of type 0/0, so we apply L'Hôpital's rule again. This yields

$$\lim_{x \to 0} \frac{1 - \cos x}{x^2} = \lim_{x \to 0} \frac{\sin x}{2x} = \lim_{x \to 0} \frac{\cos x}{2} = \frac{1}{2} \quad \blacktriangleleft$$

Example 6 Evaluate $\displaystyle \lim_{x \to +\infty} \frac{x^{-4/3}}{\sin (1/x)}$.

Solution. Since

$$\lim_{x \to +\infty} x^{-4/3} = \lim_{x \to +\infty} \sin (1/x) = 0$$

the given limit is an indeterminate form of type 0/0. Thus, by L'Hôpital's rule

$$\lim_{x \to +\infty} \frac{x^{-4/3}}{\sin (1/x)} = \lim_{x \to +\infty} \frac{-\frac{4}{3}x^{-7/3}}{(-1/x^2) \cos (1/x)} = \lim_{x \to +\infty} \frac{\frac{4}{3}x^{-1/3}}{\cos (1/x)} = \frac{0}{1} = 0 \quad \blacktriangleleft$$

Example 7 Evaluate $\displaystyle \lim_{x \to 0^-} \frac{\tan x}{x^2}$.

Solution. Since

$$\lim_{x \to 0^-} \tan x = \lim_{x \to 0^-} x^2 = 0$$

the given limit is an indeterminate form of type 0/0. Thus, by L'Hôpital's rule

$$\lim_{x \to 0^-} \frac{\tan x}{x^2} = \lim_{x \to 0^-} \frac{\sec^2 x}{2x} = -\infty \quad \blacktriangleleft$$

WARNING. When applying L'Hôpital's rule to $\lim f(x)/g(x)$, the derivatives of $f(x)$ and $g(x)$ are taken separately to yield the new limit, $\lim [f'(x)/g'(x)]$. Do not make the mistake of differentiating $f(x)/g(x)$ according to the quotient rule.

■ PROOFS

□ **THEORY BEHIND L'HÔPITAL'S RULE**

The proof of L'Hôpital's rule depends on the following result, called the ***Extended Mean-Value Theorem*** or sometimes the ***Cauchy*** (see biography p. 496) **Mean-Value Theorem***.

10.2.2 THEOREM (*Extended Mean-Value Theorem*). *Let the functions f and g be differentiable on (a, b) and continuous on $[a, b]$. If $g'(x) \neq 0$ for all x in (a, b), then there is at least one point c in (a, b) such that*

$$\frac{f'(c)}{g'(c)} = \frac{f(b) - f(a)}{g(b) - g(a)} \tag{2}$$

Proof. Observe first that $g(b) - g(a) \neq 0$, since otherwise it would follow from the Mean-Value Theorem (4.9.2) that $g'(x) = 0$ at some point x in (a, b), contradicting our hypothesis. For convenience, introduce a new function F defined by

$$F(x) = [f(b) - f(a)]g(x) - [g(b) - g(a)]f(x) \tag{3}$$

It follows from our assumptions about f and g that F is continuous on $[a, b]$ and differentiable on (a, b). Moreover, $F(b) = F(a)$ (verify), so that the Mean-Value Theorem (4.9.2) implies that there is at least one point c in (a, b) where $F'(c) = 0$. Thus, from (3)

$$[f(b) - f(a)]g'(c) - [g(b) - g(a)]f'(c) = 0$$

or

$$\frac{f'(c)}{g'(c)} = \frac{f(b) - f(a)}{g(b) - g(a)} \qquad \blacksquare$$

Note that in the special case where $g(x) = x$, Formula (2) reduces to

$$f'(c) = \frac{f(b) - f(a)}{b - a}$$

which is precisely the conclusion of the Mean-Value Theorem (4.9.2). Thus, the Extended Mean-Value Theorem is, in fact, an extension of the Mean-Value Theorem.

☐ **PROOF OF L'HÔPITAL'S RULE**

Since there are numerous cases of L'Hôpital's rule, we shall prove just one of them. The techniques used in the proof of this one case are typical of those used in the other cases.

Proof of Theorem 10.2.1. We shall consider only the case where L is finite and lim is the two-sided limit $\lim_{x \to a}$, with a finite.

By hypothesis,

$$\lim_{x \to a} \frac{f'(x)}{g'(x)} = L \tag{4}$$

As noted in the remark following the statement of Theorem 10.2.1, (4) implies that there is an interval $(a, r]$ extending to the right of a and an interval $[l, a)$ extending to the left of a on

---

which $f'(x)$ and $g'(x)$ are defined and $g'(x) \neq 0$. For convenience, define two new functions F and G by

$$F(x) = \begin{cases} f(x), & x \neq a \\ 0, & x = a \end{cases} \quad \text{and} \quad G(x) = \begin{cases} g(x), & x \neq a \\ 0, & x = a \end{cases}$$

Both F and G are continuous on $[l, r]$; they are continuous on the intervals $[l, a)$ and $(a, r]$ because on these intervals $F(x) = f(x)$, $G(x) = g(x)$, and f and g are continuous (since they are differentiable). The continuity of F and G at a follows from the fact that $\lim_{x \to a} f(x)/g(x)$ is an indeterminate form of type 0/0, so

$$\lim_{x \to a} F(x) = \lim_{x \to a} f(x) = 0 \quad \text{and} \quad \lim_{x \to a} G(x) = \lim_{x \to a} g(x) = 0$$

Thus, F and G satisfy the hypotheses of the Extended Mean-Value Theorem (10.2.2) on $[l, a]$ and $[a, r]$. Moreover, the definitions of F and G imply that

$$F'(c) = f'(c) \quad \text{and} \quad G'(c) = g'(c) \tag{5}$$

at any point c (different from a) in one of these intervals. If we choose a point $x \neq a$ in one of the intervals $[l, a]$ or $[a, r]$ and apply the Extended Mean-Value Theorem to $[x, a]$ (or $[a, x]$), we conclude that there is a number c between a and x such that

$$\frac{F(x) - F(a)}{G(x) - G(a)} = \frac{F'(c)}{G'(c)} \tag{6}$$

From (5) together with the fact that $F(a) = G(a) = 0$ and $F(x) = f(x)$, $G(x) = g(x)$, we can rewrite (6) as

$$\frac{f(x)}{g(x)} = \frac{f'(c)}{g'(c)}$$

Thus,

$$\lim_{x \to a} \frac{f(x)}{g(x)} = \lim_{x \to a} \frac{f'(c)}{g'(c)} \tag{7}$$

Since c is between a and x, it follows that $c \to a$ as $x \to a$. This fact together with (4) yields

$$\lim_{x \to a} \frac{f'(c)}{g'(c)} = \lim_{c \to a} \frac{f'(c)}{g'(c)} = L$$

Thus, from (7)

$$\lim_{x \to a} \frac{f(x)}{g(x)} = L \qquad \blacksquare$$

▶ Exercise Set 10.2

In Exercises 1–32, find the limits.

1. $\displaystyle \lim_{x \to 1} \frac{\ln x}{x - 1}$.

2. $\displaystyle \lim_{x \to 0} \frac{\sin 2x}{\sin 5x}$.

3. $\displaystyle \lim_{x \to 0} \frac{e^x - 1}{\sin x}$.

4. $\displaystyle \lim_{x \to 3} \frac{x - 3}{3x^2 - 13x + 12}$.

5. $\displaystyle \lim_{\theta \to 0} \frac{\tan \theta}{\theta}$.

6. $\displaystyle \lim_{t \to 0} \frac{te^t}{1 - e^t}$.

7. $\displaystyle \lim_{x \to 1} \frac{\ln x}{\tan \pi x}$.

8. $\displaystyle \lim_{x \to c} \frac{x^{1/3} - c^{1/3}}{x - c}$.

9. $\displaystyle \lim_{x \to \pi^+} \frac{\sin x}{x - \pi}$.

10. $\displaystyle \lim_{x \to 0^+} \frac{\sin x}{x^2}$.

11. $\displaystyle \lim_{x \to \frac{1}{2}\pi^-} \frac{\cos x}{\sqrt{\frac{1}{2}\pi - x}}$.

12. $\displaystyle \lim_{x \to 0^+} \frac{1 - \cos x}{x^3}$.

13. $\displaystyle \lim_{x \to 0} \frac{e^x + e^{-x} - 2}{1 - \cos 2x}$.

14. $\displaystyle \lim_{x \to 0} \frac{2 \cosh x - 2}{1 - \cos 2x}$.

15. $\displaystyle \lim_{x \to 0} \frac{\sin 3x}{\sinh 2x}$.

16. $\displaystyle \lim_{x \to 0} \frac{1 - e^{-2x}}{x^2 + 3x}$.

17. $\displaystyle \lim_{x \to \pi/4} \frac{\sin 2x}{4x^2 - \pi^2}$.

18. $\displaystyle \lim_{x \to 2} \frac{\ln(5x - 9)}{x^3 - 8}$.

19. $\displaystyle \lim_{x \to 0} \frac{x - \ln(x + 1)}{1 - \cos 2x}$.

20. $\displaystyle \lim_{x \to 0} \frac{x - \tan^{-1} x}{x^3}$.

21. $\displaystyle \lim_{x \to 0} \frac{2 - x^2 - 2 \cos x}{x^4}$.

22. $\displaystyle \lim_{x \to +\infty} \frac{\ln(1 + 3/x)}{\sin(2/x)}$.

23. $\lim\limits_{x \to 0} \dfrac{x - \sin x}{x^3}$.

24. $\lim\limits_{x \to -1} \dfrac{x^2 - 1}{\ln(3x + 4)}$.

25. $\lim\limits_{x \to 0} \dfrac{e^{ax} - e^{bx}}{x}$.

26. $\lim\limits_{x \to 0} \dfrac{a^x - 1}{x}$, $a > 0$.

27. $\lim\limits_{x \to 0} \dfrac{x - \tan x}{\sin x - x}$.

28. $\lim\limits_{x \to \pi} \dfrac{\sin^2 x}{1 + \cos 3x}$.

29. $\lim\limits_{x \to +\infty} \dfrac{\frac{1}{2}\pi - \tan^{-1} x}{\ln\left(1 + \dfrac{1}{x^2}\right)}$.

30. $\lim\limits_{x \to 0^+} \dfrac{\tan x}{\tan 2x}$.

31. $\lim\limits_{x \to 0^+} \dfrac{\ln(\cos x)}{\ln(\cos 3x)}$.

32. $\lim\limits_{\theta \to 0} \dfrac{\sin^2 \theta - \sin(\theta^2)}{\theta^4}$.

33. (a) Find the error in the following calculation:

$$\lim_{x \to 1} \frac{x^3 - x^2 + x - 1}{x^3 - x^2} = \lim_{x \to 1} \frac{3x^2 - 2x + 1}{3x^2 - 2x}$$

$$= \lim_{x \to 1} \frac{6x - 2}{6x - 2} = 1$$

  (b) Find the correct answer.

34. Find $\lim\limits_{x \to 1} \dfrac{x^4 - 4x^3 + 6x^2 - 4x + 1}{x^4 - 3x^3 + 3x^2 - x}$.

35. Find all values of k and l such that

$$\lim_{x \to 0} \frac{k + \cos lx}{x^2} = -4$$

36. Find $\lim\limits_{x \to 0} \dfrac{(2 + x)\ln(1 - x)}{(1 - e^x)\cos x}$.

37. Find $\lim\limits_{x \to 1} \sqrt{\dfrac{\ln x}{x^4 - 1}}$. $\left[\text{\textit{Hint:} First find } \lim\limits_{x \to 1} \dfrac{\ln x}{x^4 - 1}.\right]$

38. Find

$$\lim_{x \to 0^+} \frac{x}{\sqrt{1 - e^{-3x^2}}} \quad \text{and} \quad \lim_{x \to 0^-} \frac{x}{\sqrt{1 - e^{-3x^2}}}$$

$$\left[\text{\textit{Hint:} First find } \lim_{x \to 0} \frac{x^2}{1 - e^{-3x^2}}.\right]$$

39. Find $\lim\limits_{x \to +\infty} x \ln\left(\dfrac{x + 1}{x - 1}\right)$.

  [*Hint:* Rewrite the problem as an indeterminate form of type 0/0.]

40. (a) Explain why L'Hôpital's rule does not apply to the problem

$$\lim_{x \to 0} \frac{x^2 \sin(1/x)}{\sin x}$$

  (b) Find the limit.

41. Find $\lim\limits_{x \to 0^+} \dfrac{x \sin(1/x)}{\sin x}$ if it exists.

42. (a) Given that $k \neq 0$ and $x > 0$, show that

$$\int_1^x \frac{1}{t^{1-k}} \, dt = \frac{x^k - 1}{k}$$

  (b) Use the result in part (a) to make a guess at the value of the limit

$$\lim_{k \to 0} \frac{x^k - 1}{k}$$

  (c) Use L'Hôpital's rule to substantiate your guess.

  [This exercise was motivated by an article by Henry C. Finlayson, which appeared in the *American Math. Monthly*, Vol 94, No. 5, May 1987, p. 450.]

43. In Figure 10.2.1, let $T(\theta)$ be the area of the right triangle ABC, and $S(\theta)$ the area of the segment of the circle formed by chord AB and the arc of the unit circle subtended by the central angle θ. Find $\lim\limits_{\theta \to 0^+} T(\theta)/S(\theta)$.

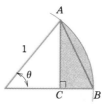

Figure 10.2.1

■ **10.3** OTHER INDETERMINATE FORMS
$$\left(\frac{\infty}{\infty},\ 0 \cdot \infty,\ 0^0,\ \infty^0,\ 1^\infty,\ \infty - \infty\right)$$

In this section we shall continue our study of indeterminate forms and introduce another version of L'Hôpital's rule.

☐ **SOME NOTATION**

In the four parts of Figure 10.3.1, the one-sided limits at a are either $+\infty$ or $-\infty$. When we want to indicate that one of these four situations occurs without specifying which one, we shall write

$$\lim_{x\to a} f(x) = \infty$$

Similarly,

$$\lim_{x\to +\infty} f(x) = \infty \quad \text{means} \quad \lim_{x\to +\infty} f(x) = +\infty \quad \text{or} \quad \lim_{x\to +\infty} f(x) = -\infty$$

$$\lim_{x\to -\infty} f(x) = \infty \quad \text{means} \quad \lim_{x\to -\infty} f(x) = +\infty \quad \text{or} \quad \lim_{x\to -\infty} f(x) = -\infty$$

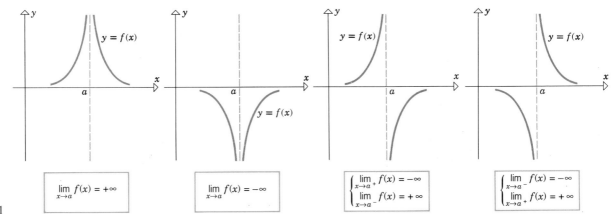

Figure 10.3.1

☐ **INDETERMINATE FORMS OF TYPE ∞/∞**

An ***indeterminate form of type*** ∞/∞ is a limit $\lim f(x)/g(x)$ in which $\lim f(x) = \infty$ and $\lim g(x) = \infty$. Some examples are

$$\lim_{x\to 0^+} \frac{\ln x}{\csc x} \qquad \boxed{\text{Numerator} \to -\infty,\ \text{denominator} \to +\infty}$$

$$\lim_{x\to +\infty} \frac{x}{e^x} \qquad \boxed{\text{Numerator} \to +\infty,\ \text{denominator} \to +\infty}$$

$$\lim_{x\to 0} \frac{1 + 1/x}{\cot x} \qquad \boxed{\text{Numerator} \to \infty,\ \text{denominator} \to \infty}$$

The following version of L'Hôpital's rule, usually proved in advanced courses, is applicable to problems like these.

> **10.3.1** THEOREM (***L'Hôpital's Rule for Form*** ∞/∞).   *Let* $\lim$ *stand for one of the limits* $\lim\limits_{x\to a}$, $\lim\limits_{x\to a^+}$, $\lim\limits_{x\to a^-}$, $\lim\limits_{x\to +\infty}$, *or* $\lim\limits_{x\to -\infty}$, *and suppose that* $\lim f(x) = \infty$ *and* $\lim g(x) = \infty$. *If* $\lim [f'(x)/g'(x)]$ *has a finite value* L, *or if this limit is* $+\infty$ *or* $-\infty$, *then*
>
> $$\lim \frac{f(x)}{g(x)} = \lim \frac{f'(x)}{g'(x)}$$

REMARK.   The remark about hypotheses that follows Theorem 10.2.1 is also applicable here.

Example 1 Evaluate $\lim\limits_{x \to +\infty} \dfrac{x}{e^x}$.

Solution.

$$\lim_{x \to +\infty} x = \lim_{x \to +\infty} e^x = +\infty$$

so that the given limit is an indeterminate form of type ∞/∞. Thus, by L'Hôpital's rule

$$\lim_{x \to +\infty} \frac{x}{e^x} = \lim_{x \to +\infty} \frac{\dfrac{d}{dx}[x]}{\dfrac{d}{dx}[e^x]} = \lim_{x \to +\infty} \frac{1}{e^x} = 0 \qquad \blacktriangleleft$$

Example 2 Evaluate $\lim\limits_{x \to 0^+} \dfrac{\ln x}{\csc x}$.

Solution.

$$\lim_{x \to 0^+} \ln x = -\infty \quad \text{and} \quad \lim_{x \to 0^+} \csc x = +\infty$$

so that the given limit is an indeterminate form of type ∞/∞. Thus, by L'Hôpital's rule

$$\lim_{x \to 0^+} \frac{\ln x}{\csc x} = \lim_{x \to 0^+} \frac{1/x}{-\csc x \cot x} \qquad (1)$$

This last limit is again an indeterminate form of type ∞/∞. Moreover, any additional applications of L'Hôpital's rule will yield powers of $1/x$ in the numerator and expressions involving $\csc x$ and $\cot x$ in the denominator; thus, repeated application of L'Hôpital's rule simply produces new indeterminate forms. We must try something else. The last limit in (1) can be rewritten as

$$\lim_{x \to 0^+} \left(-\frac{\sin x}{x} \tan x \right) = -\lim_{x \to 0^+} \frac{\sin x}{x} \cdot \lim_{x \to 0^+} \tan x = -(1)(0) = 0$$

Thus,

$$\lim_{x \to 0^+} \frac{\ln x}{\csc x} = 0 \qquad \blacktriangleleft$$

☐ **INDETERMINATE FORMS OF TYPE $0 \cdot \infty$**

A limit of a product, $\lim f(x)g(x)$, is called an ***indeterminate form of type $0 \cdot \infty$*** if $\lim f(x) = 0$ and $\lim g(x) = \infty$. Limit problems of this type can be converted to the form $0/0$ by writing

$$f(x)g(x) = \frac{f(x)}{1/g(x)}$$

or to the form ∞/∞ **by writing**

$$f(x)g(x) = \frac{g(x)}{1/f(x)}$$

The limit can then possibly be evaluated by L'Hôpital's rule.

Example 3 Evaluate $\lim\limits_{x \to 0^+} x \ln x$.

Solution. Since

$$\lim_{x \to 0^+} x = 0 \quad \text{and} \quad \lim_{x \to 0^+} \ln x = -\infty$$

the given problem is an indeterminate form of type $0 \cdot \infty$. We shall convert the problem to the form ∞/∞ and apply L'Hôpital's rule as follows:

$$\lim_{x \to 0^+} x \ln x = \lim_{x \to 0^+} \frac{\ln x}{1/x} = \lim_{x \to 0^+} \frac{1/x}{-1/x^2} = \lim_{x \to 0^+} (-x) = 0 \quad \blacktriangleleft$$

In Example 3 we could have converted the problem to form $0/0$ by writing

$$x \ln x = \frac{x}{1/\ln x}$$

However, this is less desirable than the first choice because of the relatively complicated derivative of $1/\ln x$.

WARNING.   It is tempting to argue that an indeterminate form of type $0 \cdot \infty$ has value 0 since "zero times anything is zero." However, this is fallacious since $0 \cdot \infty$ is not a product of numbers, but rather a statement about limits. For example, the following limits are of the form $0 \cdot \infty$:

$$\lim_{x \to 0^+} x \cdot \frac{1}{x} = 1, \quad \lim_{x \to 0^+} x^2 \cdot \frac{1}{x} = 0, \quad \lim_{x \to 0^+} \sqrt{x} \cdot \frac{1}{x} = +\infty$$

Example 4   Evaluate $\displaystyle\lim_{x \to \pi/4} (1 - \tan x) \sec 2x$.

Solution.   The given problem is an indeterminate form of type $0 \cdot \infty$. We convert it to type $0/0$ and apply L'Hôpital's rule as follows:

$$\lim_{x \to \pi/4} (1 - \tan x) \sec 2x = \lim_{x \to \pi/4} \frac{1 - \tan x}{1/\sec 2x} = \lim_{x \to \pi/4} \frac{1 - \tan x}{\cos 2x}$$

$$= \lim_{x \to \pi/4} \frac{-\sec^2 x}{-2 \sin 2x} = \frac{-2}{-2} = 1 \quad \blacktriangleleft$$

□ **INDETERMINATE FORMS OF TYPE** $0^0, \; \infty^0, \; 1^\infty$

Limits of the form

$$\lim_{x \to a} f(x)^{g(x)}$$

give rise to *indeterminate forms of the types* **0^0, ∞^0, and 1^∞**. (The meaning of these symbols should be clear.) All three types are treated by first introducing a dependent variable

$$y = f(x)^{g(x)}$$

and then calculating

$$\lim_{x \to a} \ln y = \lim_{x \to a} [\ln (f(x)^{g(x)})] = \lim_{x \to a} [g(x) \ln f(x)]$$

Once the value of $\displaystyle\lim_{x \to a} \ln y$ is known, it is a simple matter to determine

$$\lim_{x \to a} y = \lim_{x \to a} f(x)^{g(x)}$$

as our examples will show.

In the next example we shall use L'Hôpital's rule to derive the limit formula for e stated in Theorem 7.5.15a.

Example 5   Show that $\lim\limits_{x \to 0} (1 + x)^{1/x} = e$.

Solution.   Since

$$\lim_{x \to 0} (1 + x) = 1 \quad \text{and} \quad \lim_{x \to 0} 1/x = \infty$$

the given limit is an indeterminate form of type 1^{∞}. As discussed above, we introduce a dependent variable

$$y = (1 + x)^{1/x}$$

and take the natural logarithm of both sides:

$$\ln y = \ln (1 + x)^{1/x} = \frac{1}{x} \ln (1 + x) = \frac{\ln (1 + x)}{x}$$

The limit

$$\lim_{x \to 0} \ln y = \lim_{x \to 0} \frac{\ln (1 + x)}{x}$$

is an indeterminate form of type $0/0$, so by L'Hôpital's rule,

$$\lim_{x \to 0} \ln y = \lim_{x \to 0} \frac{\ln (1 + x)}{x} = \lim_{x \to 0} \frac{1/(1 + x)}{1} = 1$$

Since $\ln y \to 1$ as $x \to 0$, it follows from the continuity of the natural exponential function that $e^{\ln y} \to e^1$ or equivalently, $y \to e$ as $x \to 0$. Therefore,

$$\lim_{x \to 0} (1 + x)^{1/x} = e \qquad \blacktriangleleft$$

☐ **INDETERMINATE FORMS OF TYPE** $\infty - \infty$

If a limit problem such as

$$\lim [\, f(x) - g(x)] \quad \text{or} \quad \lim [\, f(x) + g(x)]$$

leads to one of the expressions

$$(+\infty) - (+\infty), \quad (-\infty) - (-\infty),$$

$$(+\infty) + (-\infty), \quad (-\infty) + (+\infty)$$

then one term tends to push the expression in the positive direction and the other tends to push it in the negative direction, resulting in an indeterminate form. These are called *indeterminate forms of type* $\infty - \infty$. Such forms can sometimes be treated by combining the two terms into one and manipulating the result into one of the previous forms.

Example 6   Evaluate $\lim\limits_{x \to 0^+} \left(\dfrac{1}{x} - \dfrac{1}{\sin x} \right)$.

Solution.   Since

$$\lim_{x \to 0^+} \frac{1}{x} = +\infty \quad \text{and} \quad \lim_{x \to 0^+} \frac{1}{\sin x} = +\infty$$

the given limit is an indeterminate form of type $\infty - \infty$. Combining terms yields

$$\lim_{x \to 0^+} \left(\frac{1}{x} - \frac{1}{\sin x} \right) = \lim_{x \to 0^+} \left(\frac{\sin x - x}{x \sin x} \right)$$

which is an indeterminate form of type $0/0$. Applying L'Hôpital's rule twice yields

$$\lim_{x \to 0^+} \left(\frac{\sin x - x}{x \sin x} \right) = \lim_{x \to 0^+} \frac{\cos x - 1}{\sin x + x \cos x}$$

$$= \lim_{x \to 0^+} \frac{-\sin x}{\cos x + \cos x - x \sin x} = \frac{0}{2} = 0 \qquad \blacktriangleleft$$

Example 7   Evaluate $\lim\limits_{x\to0^+} (\cot x - \ln x)$.

Solution.   Since

$$\lim_{x\to0^+}\cot x = +\infty \quad \text{and} \quad \lim_{x\to0^+}\ln x = -\infty$$

we are dealing with a problem of the type $(+\infty) - (-\infty)$. This is not an indeterminate form. The first term tends to make the limit large and because of the subtraction, the second term also tends to make the limit large. Thus,

$$\lim_{x\to0^+}(\cot x - \ln x) = +\infty \quad \blacktriangleleft$$

REMARK.   In general, limits of the type $0/\infty$, $\infty/0$, 0^∞, $\infty\cdot\infty$, $+\infty + (+\infty)$, $+\infty - (-\infty)$, $-\infty + (-\infty)$, $-\infty - (+\infty)$ are not indeterminate forms (Exercise 45).

▶ Exercise Set 10.3 $\boxed{C}$ *47, 61, 62*

In Exercises 1–42, find the limits.

1. $\lim\limits_{x\to+\infty}\dfrac{\ln x}{x}$.

2. $\lim\limits_{x\to+\infty}\dfrac{e^{3x}}{x^2}$.

3. $\lim\limits_{x\to0^+}\dfrac{\cot x}{\ln x}$.

4. $\lim\limits_{x\to0^+}\dfrac{1-\ln x}{e^{1/x}}$.

5. $\lim\limits_{x\to+\infty}\dfrac{x\ln x}{x+\ln x}$.

6. $\lim\limits_{x\to+\infty}\dfrac{x^3-2x+1}{4x^3+2}$.

7. $\lim\limits_{x\to+\infty}\dfrac{x^{100}}{e^x}$.

8. $\lim\limits_{x\to0^+}\dfrac{\ln(\sin x)}{\ln(\tan x)}$.

9. $\lim\limits_{x\to+\infty} xe^{-x}$.

10. $\lim\limits_{x\to\pi^-} (x-\pi)\tan\frac12 x$.

11. $\lim\limits_{x\to+\infty} x\sin\dfrac{\pi}{x}$.

12. $\lim\limits_{x\to0^+}\tan x\ln x$.

13. $\lim\limits_{x\to+\infty} x(e^{\sin(2/x)}-1)$.

14. $\lim\limits_{x\to1} x^{1/(1-x)}$.

15. $\lim\limits_{x\to+\infty}(1-3/x)^x$.

16. $\lim\limits_{x\to0}(1+2x)^{-3/x}$.

17. $\lim\limits_{x\to0}(e^x+x)^{1/x}$.

18. $\lim\limits_{x\to+\infty}(1+a/x)^{bx}$.

19. $\lim\limits_{x\to+\infty}(1+1/x^2)^x$.

20. $\lim\limits_{x\to+\infty}\left(\dfrac{x+1}{x+2}\right)^x$.

21. $\lim\limits_{x\to+\infty}(1+1/x)^{x^2}$.

22. $\lim\limits_{x\to0}(1+\sin 2x)^{1/x}$.

23. $\lim\limits_{x\to1}(2-x)^{\tan(\pi/2)x}$.

24. $\lim\limits_{x\to+\infty}[\cos(2/x)]^{x^2}$.

25. $\lim\limits_{x\to0^+} x^{\sin x}$.

26. $\lim\limits_{x\to0^+} x^x$.

27. $\lim\limits_{x\to0^+}(\sin x)^{3/\ln x}$.

28. $\lim\limits_{x\to0^+}(e^{2x}-1)^{1/\ln x}$.

29. $\lim\limits_{x\to(1/2)\pi^-}(\tan x)^{\cos x}$.

30. $\lim\limits_{x\to+\infty}(\ln x)^{1/x}$.

31. $\lim\limits_{x\to+\infty}(1+x^2)^{1/\ln x}$.

32. $\lim\limits_{x\to+\infty}(3^x+5^x)^{1/x}$.

33. $\lim\limits_{\theta\to0}\left(\dfrac{1}{1-\cos\theta}-\dfrac{2}{\sin^2\theta}\right)$.

34. $\lim\limits_{x\to0}\left(\dfrac{1}{x^2}-\dfrac{\cos 3x}{x^2}\right)$.

35. $\lim\limits_{x\to0}(\csc x - 1/x)$.

36. $\lim\limits_{x\to0}\left(\dfrac{1}{x}-\dfrac{1}{e^x-1}\right)$.

37. $\lim\limits_{x\to+\infty}(\sqrt{x^2+x}-x)$.

38. $\lim\limits_{x\to+\infty}[\ln x - \ln(1+x)]$.

39. $\lim\limits_{x\to+\infty}[x-\ln(x^2+1)]$.

40. $\lim\limits_{x\to+\infty}[x-\ln(1+2e^x)]$.

41. $\lim\limits_{x\to0^+}\dfrac{\cot x}{\cot 2x}$.

42. $\lim\limits_{x\to(1/2)\pi^-}\dfrac{4\tan x}{1+\sec x}$.

43. Show that for any positive integer n

   (a) $\lim\limits_{x\to+\infty}\dfrac{x^n}{e^x}=0$

   (b) $\lim\limits_{x\to+\infty}\dfrac{e^x}{x^n}=+\infty$.

44. Show that for any positive integer n

   (a) $\lim\limits_{x\to+\infty}\dfrac{\ln x}{x^n}=0$

   (b) $\lim\limits_{x\to+\infty}\dfrac{x^n}{\ln x}=+\infty$.

45. Limits of the type

$$0/\infty,\ \infty/0,\ 0^\infty,\ \infty\cdot\infty,\ +\infty+(+\infty),$$
$$+\infty-(-\infty),\ -\infty+(-\infty),\ -\infty-(+\infty)$$

are *not* indeterminate forms. Find the following limits by inspection.

   (a) $\lim\limits_{x\to0^+}\dfrac{x}{\ln x}$

   (b) $\lim\limits_{x\to+\infty}\dfrac{x^3}{e^{-x}}$

   (c) $\lim\limits_{x\to(1/2)\pi^-}(\cos x)^{\tan x}$

   (d) $\lim\limits_{x\to0^+}(\ln x)\cot x$

   (e) $\lim\limits_{x\to(1/2)\pi^-}\left(\dfrac{1}{\frac12\pi-x}+\tan x\right)$

   (f) $\lim\limits_{x\to0^+}\left(\dfrac{1}{x}-\ln x\right)$

   (g) $\lim\limits_{x\to-\infty}(x+x^3)$

   (h) $\lim\limits_{x\to+\infty}\left(\ln\left(\dfrac{1}{x}\right)-e^x\right)$.

46. Find $\lim\limits_{x \to +\infty} (e^x - x^2)$.

47. Sketch the graph of x^x, $x > 0$.

48. Sketch the graph of $(1/x) \tan x$, $-\pi/2 < x < \pi/2$.

In Exercises 49–51, evaluate the improper integral.

49. $\displaystyle\int_0^1 \ln x \, dx$.   **50.** $\displaystyle\int_1^{+\infty} \frac{\ln x}{x^2} \, dx$.

51. $\displaystyle\int_0^{+\infty} xe^{-3x} \, dx$.

52. Find the area between the x-axis and the curve $y = (\ln x - 1)/x^2$ for $x \geq e$.

53. Find the volume of the solid that is generated when the region between the x-axis and the curve $y = e^{-x}$ for $x \geq 0$ is revolved about the y-axis.

54. (a) Show that $\int_1^{+\infty} e^{t^2} \, dt = +\infty$. [*Hint:* Don't try to carry out the integration. Instead, compare the sizes of $\int_1^l e^t \, dt$ and $\int_1^l e^{t^2} \, dt$.]

   (b) Use L'Hôpital's rule to evaluate
$$\lim_{x \to +\infty} \frac{1}{x} \int_1^x e^{t^2} \, dt$$

55. (a) Show that
$$\int_0^{+\infty} \sqrt{1 + t^3} \, dt = +\infty$$
   [*Hint:* $\sqrt{1 + t^3} \geq t^{3/2}$ for $t \geq 0$.]

   (b) Use L'Hôpital's rule to evaluate
$$\lim_{x \to +\infty} \frac{\displaystyle\int_0^{2x} \sqrt{1 + t^3} \, dt}{x^{5/2}}$$

56. (a) Show that
$$\int_0^{+\infty} \sqrt{3 + e^{-2x}} \, dx = +\infty$$
   [*Hint:* $\sqrt{3 + e^{-2x}} > \sqrt{3}$ for $x \geq 0$.]

   (b) Let f_{ave} denote the average value of the function $f(x) = \sqrt{3 + e^{-2x}}$ on the interval $[0, a]$, where $a > 0$. Find $\lim\limits_{a \to +\infty} f_{\text{ave}}$.

In Exercises 57–60, verify that L'Hôpital's rule is of no help in finding the limit, then find the limit by some other method.

57. $\lim\limits_{x \to +\infty} \dfrac{x + \sin 2x}{x}$.   **58.** $\lim\limits_{x \to +\infty} \dfrac{2x - \sin x}{3x + \sin x}$.

59. $\lim\limits_{x \to +\infty} \dfrac{x(2 + \sin 2x)}{x + 1}$.   **60.** $\lim\limits_{x \to +\infty} \dfrac{x(2 + \sin x)}{x^2 + 1}$.

61. (a) Show that if k is a positive constant, then
$$\lim_{x \to +\infty} x(k^{1/x} - 1) = \ln k$$
   [*Hint:* Let $t = 1/x$.]

(b) If n is a positive integer, then it follows from part (a) with $x = n$ that the approximation
$$n(\sqrt[n]{k} - 1) \approx \ln k$$
should be good when n is large. Use this result and the square root key on a calculator to approximate the values of $\ln 0.3$ and $\ln 2$ with $n = 1024$, then compare the values obtained with values of the logarithms generated directly from the calculator. [*Hint:* The nth roots for which n is a power of 2 can be obtained as successive square roots.]

62. (a) Show that $\lim\limits_{x \to \pi/2} (\pi/2 - x) \tan x = 1$.

   (b) Show that
$$\lim_{x \to \pi/2} \left(\frac{1}{\pi/2 - x} - \tan x \right) = 0$$

   (c) It follows from part (b) that the approximation
$$\tan x \approx \frac{1}{\pi/2 - x}$$
should be good for values of x near $\pi/2$. Use a calculator to find $\tan x$ and $1/(\pi/2 - x)$ for $x = 1.57$; compare the results.

63. Find $\lim\limits_{x \to +\infty} xf(t/\sqrt{x})$ if $\lim\limits_{x \to 0^+} f(x) = \lim\limits_{x \to 0^+} f'(x) = 0$ and $\lim\limits_{x \to 0^+} f''(x) = a$. [*Hint:* Let $u = t/\sqrt{x}$.]

64. Let f and g be differentiable functions for which $\lim\limits_{x \to +\infty} f(x) = +\infty$, $\lim\limits_{x \to +\infty} g(x) = +\infty$, $\lim\limits_{x \to +\infty} f'(x) = a$, and $\lim\limits_{x \to +\infty} g'(x) = b$, where $a \neq 0$ and $b \neq 0$. Find
$$\lim_{x \to +\infty} \frac{\ln f(x)}{\ln g(x)}$$

In Exercises 65 and 66, find the Laplace transform of $f(t)$. (See Exercise 58 of Section 10.1 for the definition of the Laplace transform.)

65. $f(t) = t$, $s > 0$.   **66.** $f(t) = t^2$, $s > 0$.

67. There is a myth that circulates among beginning calculus students which states that all indeterminate forms of types 0^0, ∞^0, and 1^∞ have value 1 because "anything to the zero power is 1" and "1 to any power is 1." The fallacy is that 0^0, ∞^0, and 1^∞ are not powers of numbers, but rather descriptions of limits. The following examples, which were transmitted to me by the late Professor Jack Staib of Drexel University, show that such indeterminate forms can have any positive real value:

   (a) $\lim\limits_{x \to 0^+} [x^{(\ln a)/(1 + \ln x)}] = 0^0 = a$

   (b) $\lim\limits_{x \to +\infty} [x^{(\ln a)/(1 + \ln x)}] = \infty^0 = a$

   (c) $\lim\limits_{x \to 0} [(x + 1)^{(\ln a)/x}] = 1^\infty = a$.

Prove these results.

68. The *Gamma function*, $\Gamma(x)$, is defined as

$$\Gamma(x) = \int_0^{+\infty} t^{x-1} e^{-t} \, dt$$

It can be shown that this improper integral converges if and only if $x > 0$.

(a) Find $\Gamma(1)$.

(b) Prove: $\Gamma(x + 1) = x\Gamma(x)$ for all $x > 0$. [*Hint:* Use integration by parts.]

(c) Use the results in parts (a) and (b) to find $\Gamma(2)$, $\Gamma(3)$, and $\Gamma(4)$.

(d) Show that $\Gamma(\frac{1}{2}) = \sqrt{\pi}$ given that

$$\int_0^{+\infty} e^{-x^2} \, dx = \frac{1}{2}\sqrt{\pi}$$

(e) Use the results in parts (b) and (d) to show that $\Gamma(\frac{3}{2}) = \frac{1}{2}\sqrt{\pi}$ and $\Gamma(\frac{5}{2}) = \frac{3}{4}\sqrt{\pi}$.

69. Refer to the Gamma function defined in Exercise 68 to show that

(a) $\int_0^1 (\ln x)^n \, dx = (-1)^n \Gamma(n + 1), \quad n > 0$.
[*Hint:* Let $t = -\ln x$.]

(b) $\int_0^{+\infty} e^{-x^n} \, dx = \Gamma(1/n + 1), \quad n > 0$.
[*Hint:* Let $t = x^n$. Use the result in part (b) of Exercise 68.]

◆ TECHNOLOGY EXERCISES   Chapter 10

Most of these exercises require access to a graphing calculator or a computer algebra system (CAS) such as *Mathematica*, *Maple*, or *Derive*. When you are asked to *find* an answer or to *solve* an equation, you may choose to find an exact result or a numerical approximation, depending on the particular technology you are using and on your own imagination. The form of your answers may differ from those of other students or from those in the answer section of the text, depending on how you solve the problems and the accuracy you use in your numerical approximations. Those exercises that are more appropriate for a CAS than a graphing calculator are labeled with the icon ◆.

Improper integrals: In Exercises 1–4, use a CAS to evaluate the improper integral.

◆ **1.** $\int_1^{+\infty} \frac{\sin x}{x^2} \, dx$.

◆ **2.** $\int_0^{+\infty} \frac{1}{x^8 + x + 1} \, dx$.

◆ **3.** $\int_0^{+\infty} \frac{1}{\sqrt{1 + x^3}} \, dx$.

◆ **4.** $\int_1^{+\infty} \frac{\ln x}{e^x} \, dx$.

Improper integrals: In Exercises 5–8, use a CAS to verify the following results.

◆ **5.** $\int_0^{+\infty} \frac{1}{1 + x^6} \, dx = \frac{\pi}{3}$.

◆ **6.** $\int_0^{+\infty} \frac{\sin x}{\sqrt{x}} \, dx = \sqrt{\frac{\pi}{2}}$.

◆ **7.** $\int_0^{+\infty} \frac{x^3}{e^x - 1} \, dx = \frac{\pi^4}{15}$.

◆ **8.** $\int_0^1 \frac{\ln x}{1 + x} \, dx = -\frac{\pi^2}{12}$.

Change of variable in improper integrals: It is sometimes possible to convert an improper integral into a "proper" integral having the same value by making an appropriate substitution. In Exercises 9 and 10, evaluate the integral by making the indicated substitution. Investigate what happens if you evaluate the integral directly using a CAS.

◆ **9.** $\int_0^1 \sqrt{\frac{1 + x}{1 - x}} \, dx; \ u = \sqrt{1 - x}$.

◆ **10.** $\int_1^{+\infty} \frac{e^{1/x}}{x^{5/2}} \, dx; \ x = 1/u$.

◆ **11. Arc length:** Find the length of the curve $y = \sqrt{9 - 4x^2}$ over the interval $[0, \frac{3}{2}]$.

◆ **12. Volume:** The volume of a solid of infinite extent can be treated as an improper integral. Consider the region in the

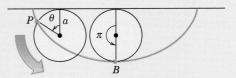

fourth quadrant between $y = e^x \ln x$ and the x-axis. Find the volume of the solid generated if the region is revolved about

(a) the x-axis   (b) the y-axis.

◆ **13. The tautochrone:** If a circle of radius a rolls along a straight line without slipping, then a fixed point P on the circle traces a curve called a **cycloid**. Assume that P is initially on the line and let θ be the angle shown in the accompanying figure, so that the position of P is determined by θ. Suppose that a wire has the shape of the cycloid. If a bead is released from rest at the point P for which $\theta = \theta_0$ $(0 \leq \theta_0 < \pi)$, and the bead slides without friction from P to the bottom B, then it can be shown that the time that it takes for the bead to reach the bottom is given by

$$\sqrt{\frac{a}{g}} \int_{\theta_0}^{\pi} \sqrt{\frac{1 - \cos \theta}{\cos \theta_0 - \cos \theta}} \, d\theta$$

where g is the acceleration due to gravity. You will show below that the value of the integral is π for all values of θ_0 $(0 \leq \theta_0 < \pi)$. This implies that the time it takes for the bead to reach the bottom is $\pi \sqrt{a/g}$, regardless of where it started. The path is called a **tautochrone** (from Greek words meaning "equal time").

(a) Determine if your CAS can evaluate the integral above.

(b) There is a good possibility that your CAS could not evaluate the integral; however, it may be able to evaluate the integral that results from multiplying the numerator and denominator inside the radical by $1 + \cos \theta$ and then making the change of variable $u = -\cos \theta$. Show that the resulting integral is

$$\int_{-\cos \theta_0}^{1} \frac{1}{\sqrt{(\cos \theta_0 + u)(1 - u)}} \, du$$

and determine if your CAS can evaluate it.

(c) In this particular case it is actually possible to evaluate the integral by hand by substituting

$$1 - \cos \theta = 2 \sin^2 (\theta/2)$$
$$\cos \theta = 2 \cos^2 (\theta/2) - 1$$
$$\cos \theta_0 = 2 \cos^2 (\theta_0/2) - 1$$

and then making the change of variable

$$u = \cos (\theta/2)/\cos (\theta_0/2)$$

Evaluate the integral using this procedure.

◆ **14. The simple pendulum:** A *simple pendulum* consists of a mass that swings in a vertical plane at the end of a massless rod of length L, as shown in the accompanying figure. Suppose that a simple pendulum is displaced through an angle θ_0 and released from rest. It can be shown that in the absence of friction, the time T required for the pendulum to make one complete swing (so that it returns to its initial position) is given by

$$T = \sqrt{\frac{8L}{g}} \int_0^{\theta_0} \frac{1}{\sqrt{\cos \theta - \cos \theta_0}} \, d\theta \tag{1}$$

where θ is the angle the pendulum makes with the vertical at any time during its motion. The improper integral in (1) is difficult to evaluate numerically. By a substitution outlined below it can be shown that the period can be expressed as

$$T = 4\sqrt{\frac{L}{g}} \int_0^{\pi/2} \frac{1}{\sqrt{1 - k^2 \sin^2 \phi}} \, d\phi \tag{2}$$

where $k = \sin (\theta_0/2)$. The integral in (2) is called a **complete elliptic integral of the first kind** and is more easily evaluated by standard numerical techniques.

(a) Use (2) to find the period of a simple pendulum for which $L = 18$ in., $\theta_0 = 20°$, and $g = 32$ ft/sec².

(b) Find, to the nearest degree, the value of θ_0 for the pendulum in part (a) so that the period is 1.5 sec.

(c) Obtain (2) from (1) by substituting
$$\cos \theta = 1 - 2 \sin^2 (\theta/2)$$
$$\cos \theta_0 = 1 - 2 \sin^2 (\theta_0/2)$$
$$k = \sin (\theta_0/2)$$

and then making the change of variable

$$\sin \phi = \sin (\theta/2)/\sin (\theta_0/2) = \sin (\theta/2)/k$$

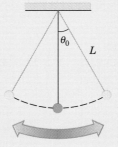

15. When technology needs help: Let

$$f(x) = \frac{1}{x^{10}(1.0001)^{1/x}}, \quad x > 0$$

(a) Find the maximum value of $f(x)$ by hand.

(b) Find $\lim\limits_{x \to 0^+} f(x)$ and $\lim\limits_{x \to 0^+} f'(x)$.

(c) Use your available technology to try to generate a graph of f that clearly shows the characteristics obtained in parts (a) and (b).

16. When technology is not enough: Let

$$f(x) = \frac{1}{x^{0.01} e^{\sqrt{-\ln x}}}$$

(a) Find the minimum value of $f(x)$ by hand.

(b) Find $\lim\limits_{x \to 0^+} f(x)$.

(c) Use your available technology to graph f on the interval $(0, 1]$.

(d) Does the graph in part (c) show the characteristics obtained in parts (a) and (b)? If not, try to show those characteristics by graphing f over a smaller interval.

Brook Taylor (signature)

Brook Taylor (1685–1731)

11 INFINITE SERIES

■ 11.1 SEQUENCES

This chapter is concerned with the study of "infinite series," which, loosely speaking, are sums with infinitely many terms. The material in this chapter has far-reaching applications in engineering and science and is the cornerstone for many branches of mathematics. In this initial section we shall develop some preliminary results that are important in their own right.

□ DEFINITION OF A
SEQUENCE

In everyday language, we use the term "sequence" to suggest a succession of objects or events given in a specified order. *Informally* speaking, the term "sequence" in mathematics is used to describe an unending succession of numbers. Some possibilities are

$$1, 2, 3, 4, \ldots, \qquad 1, \tfrac{1}{2}, \tfrac{1}{3}, \tfrac{1}{4}, \ldots,$$

$$2, 4, 6, 8, \ldots, \qquad 1, -1, 1, -1, \ldots$$

In each case, the three dots are used to suggest that the sequence continues indefinitely, following the obvious pattern. The numbers in a sequence are called the ***terms*** of the sequence. The terms may be described according to the positions they occupy. Thus, a sequence has a *first term* a_1, a *second term* a_2, a *third term* a_3, and so forth. Because a sequence continues indefinitely, there is no last term.

The most common way to specify a sequence is to give a formula that relates the terms in the sequence to their term numbers. For example, in the sequence

$$2, 4, 6, 8, \ldots$$

each term is twice the term number; that is, the nth term in the sequence is given by the formula $2n$. We denote this by writing the sequence as

$$2, 4, 6, 8, \ldots, 2n, \ldots$$

or more compactly in ***bracket notation*** as

$$\{2n\}_{n=1}^{+\infty}$$

which conveys that the sequence can be generated by successively substituting the integer values $n = 1, 2, 3, \ldots$ into the formula $2n$.

Example 1 List the first five terms of the sequence $\{2^n\}_{n=1}^{+\infty}$.

Solution. Substituting $n = 1, 2, 3, 4, 5$ into the formula 2^n yields

$$2^1, 2^2, 2^3, 2^4, 2^5, \ldots$$

or, equivalently,

$$2, 4, 8, 16, 32, \ldots \quad \blacktriangleleft$$

Example 2 Express the following sequences in bracket notation.

(a) $\dfrac{1}{2}, \dfrac{2}{3}, \dfrac{3}{4}, \dfrac{4}{5}, \ldots$ (b) $\dfrac{1}{2}, \dfrac{1}{4}, \dfrac{1}{8}, \dfrac{1}{16}, \ldots$

(c) $\dfrac{1}{2}, -\dfrac{2}{3}, \dfrac{3}{4}, -\dfrac{4}{5}, \ldots$ (d) $1, 3, 5, 7, \ldots$

Solution (a). The nth term in the following table is obtained by observing that for each term in the sequence the numerator is the same as the term number, and the denominator is one greater than the term number.

TERM NUMBER	1	2	3	4	$\cdots$	n	$\cdots$
TERM	$\dfrac{1}{2}$	$\dfrac{2}{3}$	$\dfrac{3}{4}$	$\dfrac{4}{5}$	$\cdots$	$\dfrac{n}{n+1}$	$\cdots$

Thus, the sequence can be written as $\left\{\dfrac{n}{n+1}\right\}_{n=1}^{+\infty}$.

Solution (b). The nth term in the following table is obtained by rewriting the denominators in the sequence as powers of 2 and observing that for each term the exponent of the denominator is the same as the term number.

TERM NUMBER	1	2	3	4	$\cdots$	n	$\cdots$
TERM	$\dfrac{1}{2}$	$\dfrac{1}{2^2}$	$\dfrac{1}{2^3}$	$\dfrac{1}{2^4}$	$\cdots$	$\dfrac{1}{2^n}$	$\cdots$

Thus, the sequence can be written as $\left\{\dfrac{1}{2^n}\right\}_{n=1}^{+\infty}$.

Solution (c).   This sequence is identical to that in part (a), except for the alternating signs. Thus, the *n*th term in the sequence can be obtained by multiplying the *n*th term in part (a) by $(-1)^{n+1}$. This factor produces the correct alternating signs, since its successive values, starting with $n = 1$, are $1,\ -1,\ 1,\ -1,\ldots$. Thus, the sequence can be written as

$$\left\{(-1)^{n+1}\frac{n}{n+1}\right\}_{n=1}^{+\infty}$$

Solution (d).   The *n*th term in the following table is obtained by observing that each term is one less than twice the term number.

TERM NUMBER	1	2	3	4	$\cdots$	n	$\cdots$
TERM	1	3	5	7	$\ldots$	$2n-1$	$\ldots$

Thus, the sequence can be written as $\{2n-1\}_{n=1}^{+\infty}$.

Frequently we shall want to write down a sequence without specifying the numerical values of the terms. We do this by writing

$$a_1, a_2, \ldots, a_n, \ldots$$

or in bracket notation

$$\{a_n\}_{n=1}^{+\infty}$$

REMARK.   Sometimes it will be convenient to omit the limits in the bracket notation; thus, the preceding sequence can also be written as $\{a_n\}$. Moreover, there is nothing special about the letters a and n; any other letters may be used.

At the start of this section we described a sequence as an unending succession of numbers. However, this is not a satisfactory mathematical definition, since the word "succession" is itself an undefined term. To motivate an appropriate definition, consider the sequence of even integers

$$2, 4, 6, 8, \ldots, 2n, \ldots$$

We can think of the *n*th term as a formula for the function

$$f(n) = 2n, \quad n = 1, 2, 3, \ldots$$

whose domain is the set of positive integers, and we can think of the sequence as a listing of the function values

$$f(1), f(2), f(3), \ldots, f(n), \ldots$$

This suggests the following definition.

11.1.1   DEFINITION.   A *sequence* or *infinite sequence* is a function whose domain is the set of positive integers; that is, $\{a_n\}_{n=1}^{+\infty}$ is an alternative notation for the function $f(n) = a_n$, $n = 1, 2, 3, \ldots$.

□ GRAPHS OF SEQUENCES

Because sequences are functions, we may inquire about the graph of a sequence. For example, the graph of the sequence $\{1/n\}_{n=1}^{+\infty}$ is the graph of the equation

$$y = \frac{1}{n}, \quad n = 1, 2, 3, \ldots$$

Because the right side of this equation is defined only for positive integer values of n, the graph consists of a succession of isolated points (Figure 11.1.1a). This is in marked distinction to the graph of

$$y = \frac{1}{x}, \quad x \geq 1$$

which is a continuous curve (Figure 11.1.1b).

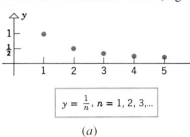

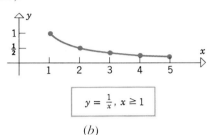

$$y = \frac{1}{n}, n = 1, 2, 3, \ldots$$

$$y = \frac{1}{x}, x \geq 1$$

Figure 11.1.1   (a)   (b)

☐ **LIMIT OF A SEQUENCE**

In Figure 11.1.2 we have sketched the graphs of four sequences, each of which behaves differently as n increases:

- The terms in the sequence $\{n + 1\}$ increase without bound.
- The terms in the sequence $\{(-1)^{n+1}\}$ oscillate between -1 and 1.
- The terms in the sequence $\{n/(n + 1)\}$ increase toward a "limiting value" of 1.
- The terms in the sequence $\{1 + (-\frac{1}{2})^n\}$ also tend toward a "limiting value" of 1, but do so in an oscillatory fashion.

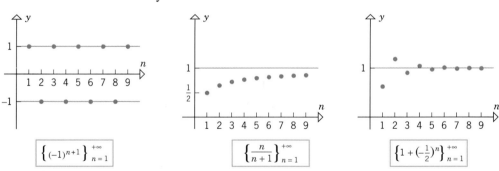

$$\left\{n + 1\right\}_{n=1}^{+\infty} \qquad \left\{(-1)^{n+1}\right\}_{n=1}^{+\infty} \qquad \left\{\frac{n}{n+1}\right\}_{n=1}^{+\infty} \qquad \left\{1 + (-\tfrac{1}{2})^n\right\}_{n=1}^{+\infty}$$

Figure 11.1.2

Our next goal is to make the concept of the "limit" of a sequence precise. Figure 11.1.3 conveys the basic idea: A sequence $\{a_n\}$ converges to a limit L if for any positive number ϵ there is a point in the sequence after which all terms lie between the lines $y = L + \epsilon$ and $y = L - \epsilon$; that is, eventually the terms in the sequence lie within ϵ units of L. Phrased another way, *a sequence $\{a_n\}$ converges to a limit L if the terms in the sequence eventually become arbitrarily close to L.* The following definition makes these ideas precise.

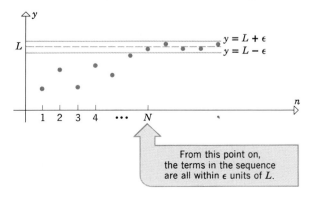

$y = L + \epsilon$

$y = L - \epsilon$

From this point on, the terms in the sequence are all within ϵ units of L.

Figure 11.1.3

11.1.2 DEFINITION.   A sequence $\{a_n\}$ is said to **converge** to the **limit** L if given any $\epsilon > 0$, there is a positive integer N such that $|a_n - L| < \epsilon$ for $n \geq N$. In this case we write

$$\lim_{n \to +\infty} a_n = L$$

A sequence that does not converge to some finite limit is said to **diverge**.

Example 3   The first two sequences in Figure 11.1.2 diverge, while the second two sequences converge to 1; that is,

$$\lim_{n \to +\infty} \frac{n}{n+1} = 1 \quad \text{and} \quad \lim_{n \to +\infty} \left(1 + (-\tfrac{1}{2})^n\right) = 1 \quad \blacktriangleleft$$

The following theorem, which we state without proof, shows that the familiar properties of limits apply to sequences.

11.1.3   THEOREM.   *Suppose that the sequences $\{a_n\}$ and $\{b_n\}$ converge to limits L_1 and L_2, respectively, and c is a constant. Then*

(a) $\displaystyle \lim_{n \to +\infty} c = c$

(b) $\displaystyle \lim_{n \to +\infty} ca_n = c \lim_{n \to +\infty} a_n = cL_1$

(c) $\displaystyle \lim_{n \to +\infty} (a_n + b_n) = \lim_{n \to +\infty} a_n + \lim_{n \to +\infty} b_n = L_1 + L_2$

(d) $\displaystyle \lim_{n \to +\infty} (a_n - b_n) = \lim_{n \to +\infty} a_n - \lim_{n \to +\infty} b_n = L_1 - L_2$

(e) $\displaystyle \lim_{n \to +\infty} (a_n b_n) = \lim_{n \to +\infty} a_n \cdot \lim_{n \to +\infty} b_n = L_1 L_2$

(f) $\displaystyle \lim_{n \to +\infty} \left(\frac{a_n}{b_n}\right) = \frac{\displaystyle \lim_{n \to +\infty} a_n}{\displaystyle \lim_{n \to +\infty} b_n} = \frac{L_1}{L_2}$ $\quad$ (*if* $L_2 \neq 0$)

REMARK.   It follows from this theorem that the algebraic techniques used to find limits of the form $\displaystyle \lim_{x \to +\infty}$ can also be used for limits of the form $\displaystyle \lim_{n \to +\infty}$.

Example 4   In each part, determine whether the sequence converges or diverges. If it converges, find the limit.

(a) $\left\{\dfrac{n}{2n+1}\right\}_{n=1}^{+\infty}$ $\qquad$ (b) $\left\{(-1)^{n+1} \dfrac{n}{2n+1}\right\}_{n=1}^{+\infty}$

(c) $\left\{(-1)^{n+1} \dfrac{1}{n}\right\}_{n=1}^{+\infty}$ $\qquad$ (d) $\{8 - 2n\}_{n=1}^{+\infty}$ $\qquad$ (e) $\left\{\dfrac{n}{e^n}\right\}_{n=1}^{+\infty}$

Solution (a).   Dividing numerator and denominator by n yields

$$\lim_{n \to +\infty} \frac{n}{2n+1} = \lim_{n \to +\infty} \frac{1}{2 + 1/n} = \frac{\displaystyle \lim_{n \to +\infty} 1}{\displaystyle \lim_{n \to +\infty} (2 + 1/n)} = \frac{\displaystyle \lim_{n \to +\infty} 1}{\displaystyle \lim_{n \to +\infty} 2 + \lim_{n \to +\infty} 1/n}$$

$$= \frac{1}{2+0} = \frac{1}{2}$$

Thus, the sequence converges to $\tfrac{1}{2}$.

Solution (b).   This sequence is the same as that in part (a), except for the factor of $(-1)^{n+1}$, which oscillates between $+1$ and -1. Thus, the terms in this sequence oscillate between positive and negative values, with the odd-numbered terms being identical to those in part (a) and the even-numbered terms being the negatives of those in part (a). Since the sequence in part (a) has a limit of $\frac{1}{2}$, it follows that the odd-numbered terms in this sequence approach $\frac{1}{2}$, while the even-numbered terms approach $-\frac{1}{2}$. Therefore, this sequence has no limit—it diverges.

Solution (c). Since $\lim\limits_{n \to +\infty} 1/n = 0$, the product $(-1)^{n+1}(1/n)$ oscillates between positive and negative values, with the odd-numbered terms approaching 0 through positive values and the even-numbered terms approaching 0 through negative values. Thus,

$$\lim_{n \to +\infty} (-1)^{n+1} \frac{1}{n} = 0$$

so the sequence converges to 0.

Solution (d). $\lim\limits_{n \to +\infty} (8 - 2n) = -\infty$, so the sequence $\{8 - 2n\}_{n=1}^{+\infty}$ diverges.

Solution (e). We want to find $\lim\limits_{n \to +\infty} n/e^n$, which is an indeterminate form of type ∞/∞. Unfortunately, we cannot apply L'Hôpital's rule directly since e^n and n are not differentiable functions (n assumes only integer values). However, we can apply L'Hôpital's rule to the related problem $\lim\limits_{x \to +\infty} x/e^x$ to obtain

$$\lim_{x \to +\infty} \frac{x}{e^x} = \lim_{x \to +\infty} \frac{1}{e^x} = 0$$

We conclude from this that $\lim\limits_{n \to +\infty} n/e^n = 0$ since the values of n/e^n and x/e^x are the same when x is a positive integer. ◄

Example 5 Show that $\lim\limits_{n \to +\infty} \sqrt[n]{n} = 1$.

Solution.

$$\lim_{n \to +\infty} \sqrt[n]{n} = \lim_{n \to +\infty} n^{1/n} = \lim_{n \to +\infty} e^{(1/n)\ln n} = e^0 = 1 \quad ◄$$

> By L'Hôpital's rule $\lim\limits_{n \to +\infty} (1/n)\ln n = 0$

Sometimes the even-numbered and odd-numbered terms of a sequence behave sufficiently differently that it is desirable to investigate their convergence separately. The following theorem, whose proof is omitted, is helpful for that purpose.

11.1.4 THEOREM. *A sequence converges to a limit L if and only if the sequences of even-numbered terms and odd-numbered terms both converge to L.*

Example 6 The sequence

$$\frac{1}{2}, \frac{1}{3}, \frac{1}{2^2}, \frac{1}{3^2}, \frac{1}{2^3}, \frac{1}{3^3}, \dots$$

converges to 0, since the even-numbered terms and the odd-numbered terms both converge to 0, and the sequence

$$1, \tfrac{1}{2}, 1, \tfrac{1}{3}, 1, \tfrac{1}{4}, \dots$$

diverges, since the odd-numbered terms converge to 1 and the even-numbered terms converge to 0. ◄

☐ **SEQUENCES DEFINED RECURSIVELY**

Sometimes sequences are defined by specifying one or more initial terms and giving a formula that relates each subsequent term to terms that precede it. Such sequences are said to be defined *recursively*, and the defining formula is called a *recursion formula*.

Example 7 Let $\{a_n\}$ be the sequence defined by $a_1 = 1$ and the recursion formula

$$a_{n+1} = \tfrac{1}{2}(a_n + 3/a_n) \tag{1}$$

for $n \geq 1$. The values of $a_2, a_3, a_4, \ldots$ can be obtained by successively substituting $n = 1, 2, 3, \ldots$ in (1):

$$a_2 = \tfrac{1}{2}(a_1 + 3/a_1) = \tfrac{1}{2}(1 + 3) = 2 \qquad \boxed{n=1}$$

$$a_3 = \tfrac{1}{2}(a_2 + 3/a_2) = \tfrac{1}{2}(2 + \tfrac{3}{2}) = \tfrac{7}{4} \qquad \boxed{n=2}$$

$$a_4 = \tfrac{1}{2}(a_3 + 3/a_3) = \tfrac{1}{2}(\tfrac{7}{4} + \tfrac{12}{7}) = \tfrac{97}{56} \qquad \boxed{n=3}$$

$$\vdots \qquad \blacktriangleleft$$

REMARK. The recursion formula in the preceding example results from applying Newton's Method to the equation $x^2 - 3 = 0$ [verify using Formula (4) of Section 4.8]. It will be shown in the exercises of the next section that the resulting sequence converges to $\sqrt{3}$.

▶ Exercise Set 11.1

In Exercises 1–18, list the first five terms of the sequence, determine whether the sequence converges, and if so find the limit. (When writing out the terms of the sequence, you need not find numerical values; leave the terms in the first form you obtain.)

1. $\left\{\dfrac{n}{n+2}\right\}_{n=1}^{+\infty}$.

2. $\left\{\dfrac{n^2}{2n+1}\right\}_{n=1}^{+\infty}$.

3. $\{2\}_{n=1}^{+\infty}$.

4. $\left\{\ln\left(\dfrac{1}{n}\right)\right\}_{n=1}^{+\infty}$.

5. $\left\{\dfrac{\ln n}{n}\right\}_{n=1}^{+\infty}$.

6. $\left\{n \sin\dfrac{\pi}{n}\right\}_{n=1}^{+\infty}$.

7. $\{1 + (-1)^n\}_{n=1}^{+\infty}$.

8. $\left\{\dfrac{(-1)^{n+1}}{n^2}\right\}_{n=1}^{+\infty}$.

9. $\left\{(-1)^n \dfrac{2n^3}{n^3+1}\right\}_{n=1}^{+\infty}$.

10. $\left\{\dfrac{n}{2^n}\right\}_{n=1}^{+\infty}$.

11. $\left\{\dfrac{(n+1)(n+2)}{2n^2}\right\}_{n=1}^{+\infty}$.

12. $\left\{\dfrac{\pi^n}{4^n}\right\}_{n=1}^{+\infty}$.

13. $\{\cos(3/n)\}_{n=1}^{+\infty}$.

14. $\left\{\cos\dfrac{\pi n}{2}\right\}_{n=1}^{+\infty}$.

15. $\{n^2 e^{-n}\}_{n=1}^{+\infty}$.

16. $\{\sqrt{n^2 + 3n} - n\}_{n=1}^{+\infty}$.

17. $\left\{\left(\dfrac{n+3}{n+1}\right)^n\right\}_{n=1}^{+\infty}$.

18. $\left\{\left(1 - \dfrac{2}{n}\right)^n\right\}_{n=1}^{+\infty}$.

In Exercises 19–26, express the sequence in the notation $\{a_n\}_{n=1}^{+\infty}$, determine whether the sequence converges, and if so find its limit.

19. $\dfrac{1}{2}, \dfrac{3}{4}, \dfrac{5}{6}, \dfrac{7}{8}, \ldots$

20. $0, \dfrac{1}{2^2}, \dfrac{2}{3^2}, \dfrac{3}{4^2}, \ldots$

21. $\dfrac{1}{3}, \dfrac{1}{9}, \dfrac{1}{27}, \dfrac{1}{81}, \ldots$

22. $-1, 2, -3, 4, -5, \ldots$

23. $\left(1 - \dfrac{1}{2}\right), \left(\dfrac{1}{2} - \dfrac{1}{3}\right), \left(\dfrac{1}{3} - \dfrac{1}{4}\right), \left(\dfrac{1}{4} - \dfrac{1}{5}\right), \ldots$

24. $3, \dfrac{3}{2}, \dfrac{3}{2^2}, \dfrac{3}{2^3}, \ldots$

25. $(\sqrt{2} - \sqrt{3}), (\sqrt{3} - \sqrt{4}), (\sqrt{4} - \sqrt{5}), \ldots$

26. $\dfrac{1}{3^5}, -\dfrac{1}{3^6}, \dfrac{1}{3^7}, -\dfrac{1}{3^8}, \ldots$

27. Let $\{a_n\}$ be the sequence defined recursively by $a_1 = \sqrt{6}$ and $a_{n+1} = \sqrt{6 + a_n}$ for $n \geq 1$.

    (a) List the first three terms of the sequence.

    (b) It can be shown that the sequence $\{a_n\}$ converges. Assuming this to be so, find its limit L. [*Hint:* $\lim\limits_{n \to +\infty} a_n = \lim\limits_{n \to +\infty} a_{n+1} = L$.]

28. Let a_1 and k be any positive real numbers and let $\{a_n\}$ be the sequence defined recursively by $a_{n+1} = \tfrac{1}{2}(a_n + k/a_n)$ for $n \geq 1$. Assuming that this sequence converges, find its limit using the hint in Exercise 27(b).

29. The *Fibonacci sequence* is defined by $a_{n+2} = a_n + a_{n+1}$ for $n \geq 1$, where $a_1 = a_2 = 1$.

(a) List the first eight terms of the sequence.

(b) Find $\lim\limits_{n \to +\infty} (a_{n+1}/a_n)$ assuming that it exists.

[*Hint:* $\lim\limits_{n \to +\infty} (a_{n+1}/a_n) = \lim\limits_{n \to +\infty} (a_{n+2}/a_{n+1})$.]

30. The nth term a_n of the sequence $1, 2, 1, 4, 1, 6, \ldots$ is best written in the form

$$a_n = \begin{cases} 1, & \text{if } n \text{ is odd} \\ n, & \text{if } n \text{ is even} \end{cases}$$

since a single formula applicable to all terms would be too complicated to be useful. By considering even and odd terms separately, find a similar expression for the nth term of the sequence

(a) $1, \dfrac{1}{2^2}, 3, \dfrac{1}{2^4}, 5, \dfrac{1}{2^6}, \ldots$

(b) $1, \dfrac{1}{3}, \dfrac{1}{3}, \dfrac{1}{5}, \dfrac{1}{5}, \dfrac{1}{7}, \dfrac{1}{7}, \dfrac{1}{9}, \dfrac{1}{9}, \ldots$

31. Consider the sequence $\{a_n\}_{n=1}^{+\infty}$, where

$$a_n = \frac{1}{n^2} + \frac{2}{n^2} + \cdots + \frac{n}{n^2}$$

(a) Write out the first four terms of the sequence.

(b) Find the limit of the sequence. [*Hint:* Sum up the terms in the formula for a_n.]

32. Follow the directions in Exercise 31 with

$$a_n = \frac{1^2}{n^3} + \frac{2^2}{n^3} + \cdots + \frac{n^2}{n^3}$$

33. If we accept the fact that the sequence $\{1/n\}_{n=1}^{+\infty}$ converges to the limit $L = 0$, then according to Definition 11.1.2, for every $\epsilon > 0$, there exists an integer N such that $|a_n - L| = |(1/n) - 0| < \epsilon$ when $n \geq N$. In each part, find

the smallest possible value of N for the given value of ϵ.

(a) $\epsilon = 0.5$ (b) $\epsilon = 0.1$ (c) $\epsilon = 0.001$.

34. If we accept the fact that the sequence

$$\left\{ \frac{n}{n+1} \right\}_{n=1}^{+\infty}$$

converges to the limit $L = 1$, then according to Definition 11.1.2, for every $\epsilon > 0$ there exists an integer N such that

$$|a_n - L| = \left| \frac{n}{n+1} - 1 \right| < \epsilon$$

when $n \geq N$. In each part, find the smallest value of N for the given value of ϵ.

(a) $\epsilon = 0.25$ (b) $\epsilon = 0.1$ (c) $\epsilon = 0.001$.

35. Use Definition 11.1.2 to prove that

(a) the sequence $\{1/n\}_{n=1}^{+\infty}$ converges to 0

(b) the sequence $\left\{ \dfrac{n}{n+1} \right\}_{n=1}^{+\infty}$ converges to 1.

36. Consider the sequence $\{a_n\}_{n=1}^{+\infty}$ whose nth term is

$$a_n = \sum_{k=1}^{n} \frac{1}{1 + k/n} \cdot \frac{1}{n}$$

Show that $\lim\limits_{n \to +\infty} a_n = \ln 2$. [*Hint:* Interpret $\lim\limits_{n \to +\infty} a_n$ as a definite integral.]

37. (a) Show that a polygon with n equal sides inscribed in a circle of radius r has perimeter $p_n = 2rn \sin(\pi/n)$.

(b) By finding the limit of the sequence $\{p_n\}_{n=1}^{+\infty}$, derive the formula for the circumference of the circle.

38. Find $\lim\limits_{n \to +\infty} r^n$, where r is a real number. [*Hint:* Consider the cases $|r| < 1$, $|r| > 1$, $r = 1$, and $r = -1$ separately.]

39. Find the limit of the sequence $\{(2^n + 3^n)^{1/n}\}_{n=1}^{+\infty}$.

■ **11.2** MONOTONE SEQUENCES

Sometimes the critical information about a sequence is whether it converges or not, with the limit being of lesser interest. In this section we shall discuss results that are used to study convergence of sequences.

□ TERMINOLOGY

We begin with some terminology.

11.2.1 DEFINITION. A sequence $\{a_n\}$ is called

increasing if $\quad a_1 < a_2 < a_3 < \cdots < a_n < \cdots$

nondecreasing if $\quad a_1 \leq a_2 \leq a_3 \leq \cdots \leq a_n \leq \cdots$

decreasing if $\quad a_1 > a_2 > a_3 > \cdots > a_n > \cdots$

nonincreasing if $\quad a_1 \geq a_2 \geq a_3 \geq \cdots \geq a_n \geq \cdots$

A sequence that is either nondecreasing or nonincreasing is called **monotone**, and a sequence that is increasing or decreasing is called **strictly monotone**. Observe that a strictly monotone sequence is monotone, but not conversely.

Example 1

$$\frac{1}{2}, \frac{2}{3}, \frac{3}{4}, \ldots, \frac{n}{n+1}, \ldots \qquad \text{is increasing}$$

$$1, \frac{1}{2}, \frac{1}{3}, \ldots, \frac{1}{n}, \ldots \qquad \text{is decreasing}$$

$$1, 1, 2, 2, 3, 3, \ldots \qquad \text{is nondecreasing}$$

$$1, 1, \frac{1}{2}, \frac{1}{2}, \frac{1}{3}, \frac{1}{3}, \ldots \qquad \text{is nonincreasing}$$

All four of these sequences are monotone, but the sequence

$$1, -\frac{1}{2}, \frac{1}{3}, -\frac{1}{4}, \ldots, (-1)^{n+1}\frac{1}{n}, \ldots$$

is not. The first and second sequences are strictly monotone. ◀

☐ **TESTING FOR MONOTONICITY**

In order for a sequence to be increasing, *all* pairs of successive terms, a_n and a_{n+1}, must satisfy $a_n < a_{n+1}$, or equivalently, $a_{n+1} - a_n > 0$. More generally, monotone sequences can be classified as follows:

DIFFERENCE BETWEEN SUCCESSIVE TERMS	CLASSIFICATION
$a_{n+1} - a_n > 0$	Increasing
$a_{n+1} - a_n < 0$	Decreasing
$a_{n+1} - a_n \geq 0$	Nondecreasing
$a_{n+1} - a_n \leq 0$	Nonincreasing

Frequently, one can *guess* whether a sequence is increasing, decreasing, nondecreasing, or nonincreasing after writing out some of the initial terms. However, to be certain that the guess is correct, a precise mathematical proof is needed. The following example illustrates a method for doing this.

Example 2 Show that

$$\frac{1}{2}, \frac{2}{3}, \frac{3}{4}, \ldots, \frac{n}{n+1}, \ldots$$

is an increasing sequence.

Solution. It is intuitively clear that the sequence is increasing. To prove that this is so, let

$$a_n = \frac{n}{n+1}$$

We can obtain a_{n+1} by replacing n by $n+1$ in this formula. This yields

$$a_{n+1} = \frac{n+1}{(n+1)+1} = \frac{n+1}{n+2}$$

Thus, for $n \geq 1$

$$a_{n+1} - a_n = \frac{n+1}{n+2} - \frac{n}{n+1} = \frac{n^2 + 2n + 1 - n^2 - 2n}{(n+1)(n+2)}$$

$$= \frac{1}{(n+1)(n+2)} > 0$$

which proves that the sequence is increasing. ◀

If a_n and a_{n+1} are any successive terms in an increasing sequence, then $a_n < a_{n+1}$. If the terms in the sequence are all positive, then we can divide both sides of this inequality by a_n to obtain $1 < a_{n+1}/a_n$ or equivalently, $a_{n+1}/a_n > 1$. More generally, monotone sequences with *positive* terms can be classified as follows:

RATIO OF SUCCESSIVE TERMS	CLASSIFICATION
$a_{n+1}/a_n > 1$	Increasing
$a_{n+1}/a_n < 1$	Decreasing
$a_{n+1}/a_n \geq 1$	Nondecreasing
$a_{n+1}/a_n \leq 1$	Nonincreasing

Example 3   Show that the sequence in Example 2 is increasing by examining the ratio of successive terms.

Solution.   As shown in the solution of Example 2,

$$a_n = \frac{n}{n+1} \quad \text{and} \quad a_{n+1} = \frac{n+1}{n+2}$$

Thus,

$$\frac{a_{n+1}}{a_n} = \frac{(n+1)/(n+2)}{n/(n+1)} = \frac{n+1}{n+2} \cdot \frac{n+1}{n} = \frac{n^2 + 2n + 1}{n^2 + 2n} \tag{1}$$

Since the numerator in (1) exceeds the denominator, the ratio exceeds 1, that is, $a_{n+1}/a_n > 1$ for $n \geq 1$. This proves that the sequence is increasing.   ◄

The following example illustrates still a third technique for determining whether a sequence is increasing or decreasing.

Example 4   In Examples 2 and 3 we proved that the sequence

$$\frac{1}{2}, \frac{2}{3}, \frac{3}{4}, \ldots, \frac{n}{n+1}, \ldots$$

is increasing by considering the difference and ratio of successive terms. Alternatively, we can proceed as follows. Let

$$f(x) = \frac{x}{x+1}$$

so the *n*th term in the given sequence is $a_n = f(n)$. The function f is increasing for $x \geq 1$ since

$$f'(x) = \frac{(x+1)(1) - x(1)}{(x+1)^2} = \frac{1}{(x+1)^2} > 0$$

Thus,

$$a_n = f(n) < f(n+1) = a_{n+1}$$

which proves that the given sequence is increasing.   ◄

In general, if $f(n) = a_n$ is the *n*th term of a sequence, and if f is differentiable for $x \geq 1$, then we have the following results:

DERIVATIVE OF f FOR $x \geq 1$	CLASSIFICATION OF THE SEQUENCE WITH $a_n = f(n)$
$f'(x) > 0$	Increasing
$f'(x) < 0$	Decreasing
$f'(x) \geq 0$	Nondecreasing
$f'(x) \leq 0$	Nonincreasing

☐ EVENTUALLY MONOTONE SEQUENCES

If the terms of a sequence fail to have a certain property from the start, but the terms have that property from some point on, then we say that the sequence has the property *eventually*. For example, a sequence $\{a_n\}$ would be called *eventually monotone* if there is some integer N such that the sequence is *monotone* for $n \geq N$.

We shall illustrate this idea with a sequence involving **factorials**. For this purpose recall that if n is a positive integer, then $n!$ (read "n factorial") denotes the product of the first n positive integers, that is,

$$n! = 1 \cdot 2 \cdot 3 \cdots n \quad \text{or equivalently,} \quad n! = n(n-1)(n-2) \cdots 1$$

Furthermore, it is agreed by convention that $0! = 1$.

Example 5 Show that the sequence $\left\{ \dfrac{10^n}{n!} \right\}_{n=1}^{+\infty}$ is eventually decreasing.

Solution. We have

$$a_n = \frac{10^n}{n!} \quad \text{and} \quad a_{n+1} = \frac{10^{n+1}}{(n+1)!}$$

so

$$\frac{a_{n+1}}{a_n} = \frac{10^{n+1}/(n+1)!}{10^n/n!} = \frac{10^{n+1}\, n!}{10^n\,(n+1)!} = 10\,\frac{n!}{(n+1)n!} = \frac{10}{n+1} \tag{2}$$

From (2), $a_{n+1}/a_n < 1$ for all $n \geq 10$, so the sequence is eventually decreasing. ◄

☐ CONVERGENCE OF MONOTONE SEQUENCES

The following two theorems, whose proofs are discussed at the end of this section, show that a monotone sequence either converges or becomes infinite—divergence by oscillation cannot occur.

11.2.2 THEOREM. *If $a_1 \leq a_2 \leq a_3 \leq \cdots \leq a_n \leq \cdots$ is a nondecreasing sequence, then there are two possibilities:*

(a) *There is a constant M, called an **upper bound** for the sequence, such that $a_n \leq M$ for all n, in which case the sequence converges to a limit L satisfying $L \leq M$.*

(b) *No upper bound exists, in which case $\lim\limits_{n \to +\infty} a_n = +\infty$.*

> **11.2.3   THEOREM.** *If $a_1 \geq a_2 \geq a_3 \geq \cdots \geq a_n \geq \cdots$ is a nonincreasing sequence, then there are two possibilities*:
>
> (a)   *There is a constant M, called a **lower bound** for the sequence, such that $a_n \geq M$ for all n, in which case the sequence converges to a limit L satisfying $L \geq M$.*
> (b)   *No lower bound exists, in which case $\lim\limits_{n \to +\infty} a_n = -\infty$.*

It should be noted that these results do not give a method for obtaining limits; they tell us only whether a limit exists.

Example 6   Use Theorems 11.2.2 and 11.2.3 to show that the sequence $\left\{ \dfrac{n}{n+1} \right\}_{n=1}^{+\infty}$ converges.

Solution.   We showed in Examples 2 and 3 that the given sequence is increasing (hence nondecreasing). It is evident that the number $M = 1$ is an upper bound for the sequence since

$$a_n = \frac{n}{n+1} < 1, \quad n = 1, 2, \ldots$$

Thus, by Theorem 11.2.2 the sequence converges to some limit L such that $L \leq M = 1$. This is indeed the case since

$$\lim_{n \to +\infty} \frac{n}{n+1} = \lim_{n \to +\infty} \frac{1}{1 + 1/n} = 1 \qquad \blacktriangleleft$$

☐ **AN INTUITIVE VIEW OF CONVERGENCE**

Informally stated, the convergence or divergence of a sequence does not depend on the behavior of the "initial terms" of the sequence, but rather on the behavior of the "tail end." Thus, for a sequence $\{a_n\}$ to converge to a limit L, it does not matter if the initial terms are far from L, just so the terms in the sequence are eventually arbitrarily close to L. This being the case, one can add, delete, or alter *finitely* many terms without affecting the convergence, divergence, or the limit (if it exists).

As one would expect from the preceding discussion, it can be proved that Theorems 11.2.2 and 11.2.3 hold for sequences that are eventually nondecreasing and eventually nonincreasing, respectively.

Example 7   Show that the sequence $\left\{ \dfrac{10^n}{n!} \right\}_{n=1}^{+\infty}$ converges and find its limit.

Solution.   We showed in Example 5 that the sequence is eventually decreasing. Since all terms in the sequence are positive, it is bounded below by $M = 0$, and hence Theorem 11.2.3 guarantees that it converges to a limit L such that $L \geq 0$. However, the limit is not evident directly from the formula $10^n/n!$ for the nth term, so we will need some ingenuity to obtain it.

Recall from Formula (2) of Example 5 that successive terms in the given sequence are related by the recursion formula

$$a_{n+1} = \frac{10}{n+1} a_n \tag{3}$$

where $a_n = 10^n/n!$. We shall take the limit as $n \to +\infty$ of both sides of (3) and use the fact that

$$\lim_{n \to +\infty} a_{n+1} = \lim_{n \to +\infty} a_n = L$$

We obtain

$$L = \lim_{n \to +\infty} a_{n+1} = \lim_{n \to +\infty} \left(\frac{10}{n+1} a_n \right) = \lim_{n \to +\infty} \frac{10}{n+1} \lim_{n \to +\infty} a_n = 0 \cdot L = 0$$

so that

$$L = \lim_{n \to +\infty} \frac{10^n}{n!} = 0 \quad \blacktriangleleft$$

REMARK.   In the exercises we will show that the technique illustrated in the preceding example can be adapted to obtain the following limit

$$\lim_{n \to +\infty} \frac{x^n}{n!} = 0 \tag{4}$$

for any real value of x (Exercise 25). This result, which shows that $n!$ eventually increases more rapidly than any positive integer power of x, will be useful in our later work.

■ SOME PROOFS

In this text we have not been concerned with a rigorous development of the real number system; we have simply accepted the familiar properties of real numbers without proof, and indeed, we have not even attempted to define the term "real number." Although this is sufficient for many purposes, it was recognized by the late nineteenth century that the study of limits and functions in calculus requires a precise axiomatic formulation of the real numbers analogous to the axiomatic development of Euclidean geometry. Although we will not attempt to pursue this development, we will need to discuss one of the axioms about real numbers in order to prove Theorems 11.2.2 and 11.2.3. But first we shall introduce some terminology.

If S is a nonempty set of real numbers, then we call u an **upper bound** for S if u is greater than or equal to every number in S, and we call l a **lower bound** for S if l is smaller than or equal to every number in S. For example, if S is the set of numbers in the interval $(1, 3)$, then $u = 4$, 10, and 100 are upper bounds for S and $l = -10$, 0, and $\frac{1}{2}$ are lower bounds for S. Observe also that $u = 3$ is the smallest of all upper bounds and $l = 1$ is the largest of all lower bounds. The existence of a smallest upper bound and a greatest lower bound for S is not accidental; it is a consequence of the following axiom.

> **11.2.4**   AXIOM (*The Completeness Axiom*).   *If a nonempty set S of real numbers has an upper bound, then it has a smallest upper bound (called the **least upper bound**), and if a nonempty set S of real numbers has a lower bound, then it has a largest lower bound (called the **greatest lower bound**).*

Proof of Theorem 11.2.2.

(*a*)   Assume there exists a number M such that $a_n \le M$ for $n = 1, 2, \ldots$. Then M is an upper bound for the set of terms in the sequence. By the Completeness Axiom there is a least upper bound for the terms, call it L. Now let ϵ be any positive number. Since L is the least upper bound for the terms, $L - \epsilon$ is not an upper bound for the terms, which means that there is at least one term a_N such that

$$a_N > L - \epsilon$$

Moreover, since $\{a_n\}$ is a nondecreasing sequence, we must have

$$a_n \ge a_N > L - \epsilon \tag{5}$$

when $n \geq N$. But a_n cannot exceed L since L is an upper bound for the terms. This observation together with (5) tells us that $L \geq a_n > L - \epsilon$ for $n \geq N$, so all terms from the Nth on are within ϵ units of L. This is exactly the requirement to have

$$\lim_{n \to +\infty} a_n = L$$

Finally, $L \leq M$ since M is an upper bound for the terms and L is the least upper bound. This proves part (a).

(b) If there is no number M such that $a_n \leq M$ for $n = 1, 2, \ldots$, then no matter how large we choose M, there is a term a_N such that

$$a_N > M$$

and, since the sequence is nondecreasing,

$$a_n \geq a_N > M$$

when $n \geq N$. Thus, the terms in the sequence become arbitrarily large as n increases. That is,

$$\lim_{n \to +\infty} a_n = +\infty \quad \blacksquare$$

The proof of Theorem 11.2.3 will be omitted since it is similar to the proof of 11.2.2.

▶ Exercise Set 11.2

In Exercises 1–6, use $a_{n+1} - a_n$ to show that the given sequence $\{a_n\}$ is strictly monotone and classify it as increasing or decreasing.

1. $\left\{\dfrac{1}{n}\right\}_{n=1}^{+\infty}$.

2. $\left\{1 - \dfrac{1}{n}\right\}_{n=1}^{+\infty}$.

3. $\left\{\dfrac{n}{2n+1}\right\}_{n=1}^{+\infty}$.

4. $\left\{\dfrac{n}{4n-1}\right\}_{n=1}^{+\infty}$.

5. $\{n - 2^n\}_{n=1}^{+\infty}$.

6. $\{n - n^2\}_{n=1}^{+\infty}$.

In Exercises 7–12, use a_{n+1}/a_n to show that the given sequence $\{a_n\}$ is strictly monotone and classify it as increasing or decreasing.

7. $\left\{\dfrac{n}{2n+1}\right\}_{n=1}^{+\infty}$.

8. $\left\{\dfrac{2^n}{1+2^n}\right\}_{n=1}^{+\infty}$.

9. $\{ne^{-n}\}_{n=1}^{+\infty}$.

10. $\left\{\dfrac{10^n}{(2n)!}\right\}_{n=1}^{+\infty}$.

11. $\left\{\dfrac{n^n}{n!}\right\}_{n=1}^{+\infty}$.

12. $\left\{\dfrac{5^n}{2^{(n^2)}}\right\}_{n=1}^{+\infty}$.

In Exercises 13–18, use differentiation to show that the sequence is strictly monotone and classify it as increasing or decreasing.

13. $\left\{\dfrac{n}{2n+1}\right\}_{n=1}^{+\infty}$.

14. $\left\{3 - \dfrac{1}{n}\right\}_{n=1}^{+\infty}$.

15. $\left\{\dfrac{1}{n + \ln n}\right\}_{n=1}^{+\infty}$.

16. $\{ne^{-2n}\}_{n=1}^{+\infty}$.

17. $\left\{\dfrac{\ln(n+2)}{n+2}\right\}_{n=1}^{+\infty}$.

18. $\{\tan^{-1} n\}_{n=1}^{+\infty}$.

In Exercises 19–24, use any method to show that the sequence is eventually increasing or eventually decreasing.

19. $\{2n^2 - 7n\}_{n=1}^{+\infty}$.

20. $\{n^3 - 4n^2\}_{n=1}^{+\infty}$.

21. $\left\{\dfrac{n}{n^2 + 10}\right\}_{n=1}^{+\infty}$.

22. $\left\{n + \dfrac{17}{n}\right\}_{n=1}^{+\infty}$.

23. $\left\{\dfrac{n!}{3^n}\right\}_{n=1}^{+\infty}$.

24. $\{n^5 e^{-n}\}_{n=1}^{+\infty}$.

25. Prove:

$$\lim_{n \to +\infty} \frac{x^n}{n!} = 0$$

for any real value of x. [*Hint:* Consider the cases $x = 0$ and $x \neq 0$. For $x \neq 0$, show that $\lim_{n \to +\infty} |x|^n/n! = 0$ and use the Squeezing Theorem (Theorem 2.8.2).]

26. Show that

$$\lim_{n \to +\infty} \frac{n!}{n^n} = 0$$

27. Let $\{a_n\}$ be the sequence defined recursively by $a_1 = \sqrt{2}$ and $a_{n+1} = \sqrt{2 + a_n}$ for $n \geq 1$.

   (a) List the first three terms of the sequence.

   (b) Show that $a_n < 2$ for $n \geq 1$.

   (c) Show that $a_{n+1}^2 - a_n^2 = (2 - a_n)(1 + a_n)$ for $n \geq 1$.

   (d) Use the results in parts (b) and (c) to show that $\{a_n\}$ is an increasing sequence. [*Hint:* If x and y are positive real numbers such that $x^2 - y^2 > 0$, then it follows by factoring that $x - y > 0$.]

(e)   Show that $\{a_n\}$ converges and find its limit L.

28.   Let $\{a_n\}$ be the sequence defined recursively by $a_1 = 1$ and $a_{n+1} = \frac{1}{2}(a_n + 3/a_n)$ for $n \geq 1$ (see Example 7, Section 11.1).

(a)   Show that $a_n \geq \sqrt{3}$ for $n \geq 2$. [*Hint:* What is the minimum value of $\frac{1}{2}(x + 3/x)$ for $x > 0$?]

(b)   Show that $\{a_n\}$ is eventually nonincreasing. [*Hint:* Examine $a_{n+1} - a_n$ or a_{n+1}/a_n and use the result in part (a).]

(c)   Show that $\{a_n\}$ converges and find its limit L.

29.   (a)   Show that if $\{a_n\}_{n=1}^{+\infty}$ is a nonincreasing sequence, then $\{-a_n\}_{n=1}^{+\infty}$ is a nondecreasing sequence.

(b)   Use part (a) and Theorem 11.2.2 to help prove Theorem 11.2.3.

30.   (a)   Deduce the inequalities

$$\int_1^n \ln x \, dx < \ln n! < \int_1^{n+1} \ln x \, dx$$

from Figure 11.2.1 for $n \geq 2$ by comparing appropriate areas.

(b)   Use the result in part (a) to show that

$$\frac{n^n}{e^{n-1}} < n! < \frac{(n+1)^{n+1}}{e^n}, \quad n > 1$$

31.   Use the Squeezing Theorem (Theorem 2.8.2) and the result in Exercise 30(b) to show that

$$\lim_{n \to +\infty} \frac{\sqrt[n]{n!}}{n} = \frac{1}{e}$$

32.   Use the left inequality in Exercise 30(b) to show that $\lim_{n \to +\infty} \sqrt[n]{n!} = +\infty$.

33.   (a)   Show that $\left\{ \dfrac{n^n}{n!e^n} \right\}_{n=1}^{+\infty}$ is a decreasing sequence.

[*Hint:* From Exercise 20(d) of Section 7.5 we have $(1 + 1/x)^x < e$ for $x > 0$.]

(b)   Does the sequence converge? Explain.

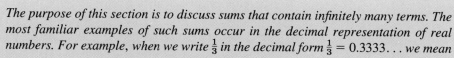

Figure 11.2.1

11.3   INFINITE SERIES

The purpose of this section is to discuss sums that contain infinitely many terms. The most familiar examples of such sums occur in the decimal representation of real numbers. For example, when we write $\frac{1}{3}$ in the decimal form $\frac{1}{3} = 0.3333\ldots$ we mean

$$\frac{1}{3} = 0.3 + 0.03 + 0.003 + 0.0003 + \cdots$$

which suggests that the decimal representation of $\frac{1}{3}$ can be viewed as a sum of infinitely many real numbers.

SUMS OF INFINITE SERIES

Our first objective is to define what is meant by the "sum" of infinitely many real numbers. We begin with some terminology.

11.3.1   DEFINITION.   An *infinite series* is an expression that can be written in the form

$$\sum_{k=1}^{\infty} u_k = u_1 + u_2 + u_3 + \cdots + u_k + \cdots$$

The numbers $u_1, u_2, u_3, \ldots$ are called the ***terms*** of the series.

Since it is physically impossible to add infinitely many numbers together, sums of infinite series are defined and computed by an indirect limiting process. To motivate the basic idea, consider the infinite decimal

$$0.3333\ldots \tag{1}$$

which can be viewed as the infinite series

$$0.3 + 0.03 + 0.003 + 0.0003 + \cdots$$

or equivalently,

$$\frac{3}{10} + \frac{3}{10^2} + \frac{3}{10^3} + \frac{3}{10^4} + \cdots \tag{2}$$

Since (1) is the decimal expansion of $\frac{1}{3}$, any reasonable definition for the sum of an infinite series should yield $\frac{1}{3}$ for the sum of (2). To obtain such a definition, consider the following sequence of (finite) sums:

$$s_1 = \frac{3}{10} = 0.3$$

$$s_2 = \frac{3}{10} + \frac{3}{10^2} = 0.33$$

$$s_3 = \frac{3}{10} + \frac{3}{10^2} + \frac{3}{10^3} = 0.333$$

$$s_4 = \frac{3}{10} + \frac{3}{10^2} + \frac{3}{10^3} + \frac{3}{10^4} = 0.3333$$

$$\vdots$$

The sequence of numbers $s_1, s_2, s_3, s_4, \ldots$ can be viewed as a succession of approximations to the "sum" of the infinite series, which we want to be $\frac{1}{3}$. As we progress through the sequence, more and more terms of the infinite series are used, and the approximations get better and better, suggesting that the desired sum of $\frac{1}{3}$ might be the *limit* of this sequence of approximations. To see that this is so, we must calculate the limit of the general term in the sequence of approximations, namely

$$s_n = \frac{3}{10} + \frac{3}{10^2} + \cdots + \frac{3}{10^n} \tag{3}$$

The problem of calculating

$$\lim_{n \to +\infty} s_n = \lim_{n \to +\infty} \left(\frac{3}{10} + \frac{3}{10^2} + \cdots + \frac{3}{10^n} \right)$$

is complicated by the fact that both the last term and the number of terms in the sum change with n. It is best to rewrite such limits in a closed form in which the number of terms does not vary, if possible. (See the remark following Example 3 in Section 5.4.) To do this, we multiply both sides of (3) by $\frac{1}{10}$ to obtain

$$\frac{1}{10} s_n = \frac{3}{10^2} + \frac{3}{10^3} + \cdots + \frac{3}{10^n} + \frac{3}{10^{n+1}} \tag{4}$$

and then subtract (4) from (3) to obtain

$$s_n - \frac{1}{10} s_n = \frac{3}{10} - \frac{3}{10^{n+1}}$$

$$\frac{9}{10} s_n = \frac{3}{10} \left(1 - \frac{1}{10^n} \right)$$

$$s_n = \frac{1}{3} \left(1 - \frac{1}{10^n} \right)$$

Since $1/10^n \to 0$ as $n \to +\infty$, it follows that

$$\lim_{n \to +\infty} s_n = \lim_{n \to +\infty} \frac{1}{3}\left(1 - \frac{1}{10^n}\right) = \frac{1}{3}$$

which we denote by writing

$$\frac{1}{3} = \frac{3}{10} + \frac{3}{10^2} + \frac{3}{10^3} + \cdots + \frac{3}{10^n} + \cdots$$

Motivated by the preceding example, we are now ready to define the general concept of the "sum" of an infinite series

$$u_1 + u_2 + u_3 + \cdots + u_k + \cdots$$

We begin with some terminology: Let s_n denote the sum of the first n terms of the series. Thus,

$$s_1 = u_1$$

$$s_2 = u_1 + u_2$$

$$s_3 = u_1 + u_2 + u_3$$

$$\vdots$$

$$s_n = u_1 + u_2 + u_3 + \cdots + u_n = \sum_{k=1}^{n} u_k$$

The number s_n is called the **nth partial sum** of the series and the sequence $\{s_n\}_{n=1}^{+\infty}$ is called the **sequence of partial sums**.

WARNING.   In everyday language the words "sequence" and "series" are often used interchangeably. However, this is not so in mathematics—mathematically, a sequence is a *succession* and a series is a *sum*. It is essential that you keep this distinction in mind.

As n increases, the partial sum $s_n = u_1 + u_2 + \cdots + u_n$ includes more and more terms of the series. Thus, if s_n tends toward a limit as $n \to +\infty$, it is reasonable to view this limit as the sum of *all* the terms in the series. This suggests the following definition.

11.3.2   DEFINITION.   Let $\{s_n\}$ be the sequence of partial sums of the series $u_1 + u_2 + u_3 + \cdots + u_k + \cdots$. If the sequence $\{s_n\}$ converges to a limit S, then the series is said to **converge**, and S is called the **sum** of the series. We denote this by writing

$$S = \sum_{k=1}^{\infty} u_k$$

If the sequence of partial sums diverges, then the series is said to **diverge**. A divergent series has no sum.

Example 1   Determine whether the series

$$1 - 1 + 1 - 1 + 1 - 1 + \cdots$$

converges or diverges. If it converges, find the sum.

Solution.   It is tempting to conclude that the sum of the series is zero by arguing that the positive and negative terms cancel one another. However, this is *not correct*; the problem is that algebraic operations that hold for finite sums do not carry over to infinite series in all cases. Later, we shall discuss conditions under which familiar algebraic operations can be applied to infinite series, but for this example we turn directly to Definition 11.3.2. The partial sums are

$$s_1 = 1$$

$$s_2 = 1 - 1 = 0$$

$$s_3 = 1 - 1 + 1 = 1$$

$$s_4 = 1 - 1 + 1 - 1 = 0$$

and so forth. Thus, the sequence of partial sums is

$$1, 0, 1, 0, 1, 0, \ldots$$

Since this is a divergent sequence, the given series diverges and consequently has no sum. ◀

☐ GEOMETRIC SERIES

The series encountered thus far are examples of **geometric series**. A geometric series is one of the form

$$a + ar + ar^2 + ar^3 + \cdots + ar^{k-1} + \cdots \quad (a \neq 0)$$

where each term is obtained by multiplying the preceding one by a constant r. The multiplier r is called the **ratio** for the series. Some examples of geometric series are

$$1 + 2 + 4 + 8 + \cdots + 2^{k-1} + \cdots \qquad \boxed{a = 1, \, r = 2}$$

$$\frac{3}{10} + \frac{3}{10^2} + \frac{3}{10^3} + \cdots + \frac{3}{10^k} + \cdots \qquad \boxed{a = \frac{3}{10}, \, r = \frac{1}{10}}$$

$$\frac{1}{2} - \frac{1}{4} + \frac{1}{8} - \frac{1}{16} + \cdots + (-1)^{k+1} \frac{1}{2^k} + \cdots \qquad \boxed{a = \frac{1}{2}, \, r = -\frac{1}{2}}$$

$$1 + 1 + 1 + \cdots + 1 + \cdots \qquad \boxed{a = 1, \, r = 1}$$

$$1 - 1 + 1 - 1 + \cdots + (-1)^{k+1} + \cdots \qquad \boxed{a = 1, \, r = -1}$$

The following theorem is the fundamental result on convergence of geometric series.

11.3.3   THEOREM.   *A geometric series*

$$a + ar + ar^2 + \cdots + ar^{k-1} + \cdots \quad (a \neq 0)$$

converges if $|r| < 1$ and diverges if $|r| \geq 1$. If the series converges, then the sum is

$$\frac{a}{1 - r} = a + ar + ar^2 + \cdots + ar^{k-1} + \cdots$$

Proof.   Let us treat the case $|r| = 1$ first. If $r = 1$, then the series is

$$a + a + a + \cdots + a + \cdots$$

so the nth partial sum is $s_n = na$ and $\lim_{n \to +\infty} s_n = \lim_{n \to +\infty} na = \pm\infty$ (the sign depending on whether a is positive or negative). This proves divergence. If $r = -1$, the series is

$$a - a + a - a + \cdots$$

so the sequence of partial sums is

$$a, 0, a, 0, a, 0, \ldots$$

which diverges.

Now let us consider the case where $|r| \neq 1$. The nth partial sum of the series is

$$s_n = a + ar + ar^2 + \cdots + ar^{n-1} \tag{5}$$

Multiplying both sides of (5) by r yields

$$rs_n = ar + ar^2 + \cdots + ar^{n-1} + ar^n \tag{6}$$

and subtracting (6) from (5) gives

$$s_n - rs_n = a - ar^n$$

or

$$(1 - r)s_n = a - ar^n \tag{7}$$

Since $r \neq 1$ in the case we are considering, this can be rewritten as

$$s_n = \frac{a - ar^n}{1 - r} = \frac{a}{1 - r} - \frac{ar^n}{1 - r} \tag{8}$$

If $|r| < 1$, then $\lim\limits_{n \to +\infty} r^n = 0$ (can you see why?), so $\{s_n\}$ converges. From (8)

$$\lim_{n \to +\infty} s_n = \frac{a}{1 - r}$$

If $|r| > 1$, then either $r > 1$ or $r < -1$. In the case $r > 1$, $\lim\limits_{n \to +\infty} r^n = +\infty$, and in the case $r < -1$, r^n oscillates between positive and negative values that grow in magnitude, so $\{s_n\}$ diverges in both cases. ∎

Example 2   The series

$$5 + \frac{5}{4} + \frac{5}{4^2} + \cdots + \frac{5}{4^{k-1}} + \cdots$$

is a geometric series with $a = 5$ and $r = \frac{1}{4}$. Since $|r| = \frac{1}{4} < 1$, the series converges and the sum is

$$\frac{a}{1 - r} = \frac{5}{1 - \frac{1}{4}} = \frac{20}{3} \quad \blacktriangleleft$$

Example 3   Find the rational number represented by the repeating decimal

$$0.784784784\ldots$$

Solution.   We can write

$$0.784784784\ldots = 0.784 + 0.000784 + 0.000000784 + \cdots$$

so the given decimal is the sum of a geometric series with $a = 0.784$ and $r = 0.001$. Thus,

$$0.784784784\ldots = \frac{a}{1 - r} = \frac{0.784}{1 - 0.001} = \frac{0.784}{0.999} = \frac{784}{999} \quad \blacktriangleleft$$

Example 4   Determine whether the series

$$\sum_{k=1}^{\infty} \frac{1}{k(k + 1)} = \frac{1}{1 \cdot 2} + \frac{1}{2 \cdot 3} + \frac{1}{3 \cdot 4} + \frac{1}{4 \cdot 5} + \cdots$$

converges or diverges. If it converges, find the sum.

Solution.   The nth partial sum of the series is

$$s_n = \sum_{k=1}^{n} \frac{1}{k(k + 1)} = \frac{1}{1 \cdot 2} + \frac{1}{2 \cdot 3} + \frac{1}{3 \cdot 4} + \cdots + \frac{1}{n(n + 1)}$$

To calculate $\lim\limits_{n \to +\infty} s_n$ we shall rewrite s_n in closed form. This can be accomplished by using the method of partial fractions to obtain (verify)

$$\frac{1}{k(k+1)} = \frac{1}{k} - \frac{1}{k+1}$$

from which we obtain the telescoping sum

$$s_n = \sum_{k=1}^{n} \left(\frac{1}{k} - \frac{1}{k+1} \right)$$

$$= \left(1 - \frac{1}{2} \right) + \left(\frac{1}{2} - \frac{1}{3} \right) + \left(\frac{1}{3} - \frac{1}{4} \right) + \cdots + \left(\frac{1}{n} - \frac{1}{n+1} \right)$$

$$= 1 + \left(-\frac{1}{2} + \frac{1}{2} \right) + \left(-\frac{1}{3} + \frac{1}{3} \right) + \cdots + \left(-\frac{1}{n} + \frac{1}{n} \right) - \frac{1}{n+1}$$

$$= 1 - \frac{1}{n+1}$$

so

$$\sum_{k=1}^{\infty} \frac{1}{k(k+1)} = \lim_{n \to +\infty} s_n = \lim_{n \to +\infty} \left(1 - \frac{1}{n+1} \right) = 1 \qquad \blacktriangleleft$$

☐ HARMONIC SERIES

One of the most important of all diverging series is the **harmonic series**,

$$\sum_{k=1}^{\infty} \frac{1}{k} = 1 + \frac{1}{2} + \frac{1}{3} + \frac{1}{4} + \frac{1}{5} + \cdots$$

which arises in connection with the overtones produced by a vibrating musical string. It is not immediately evident that this series diverges. However, the divergence will become apparent when we examine the partial sums in detail. Because the terms in the series are all positive, the partial sums

$$s_1 = 1, \ s_2 = 1 + \tfrac{1}{2}, \ s_3 = 1 + \tfrac{1}{2} + \tfrac{1}{3}, \ s_4 = 1 + \tfrac{1}{2} + \tfrac{1}{3} + \tfrac{1}{4}, \ldots$$

form an increasing sequence

$$s_1 < s_2 < s_3 < \cdots < s_n < \cdots$$

Thus, by Theorem 11.2.2 we can prove divergence by demonstrating that there is no constant M that is greater than or equal to *every* partial sum. To this end, we shall consider some selected partial sums, namely $s_2, s_4, s_8, s_{16}, s_{32}, \ldots$. Note that the subscripts are successive powers of 2, so that these are the partial sums of the form s_{2^n}. These partial sums satisfy the inequalities

$$s_2 = 1 + \tfrac{1}{2} > \tfrac{1}{2} + \tfrac{1}{2} = \tfrac{2}{2}$$

$$s_4 = s_2 + \tfrac{1}{3} + \tfrac{1}{4} > s_2 + \left(\tfrac{1}{4} + \tfrac{1}{4} \right) = s_2 + \tfrac{1}{2} > \tfrac{3}{2}$$

$$s_8 = s_4 + \tfrac{1}{5} + \tfrac{1}{6} + \tfrac{1}{7} + \tfrac{1}{8} > s_4 + \left(\tfrac{1}{8} + \tfrac{1}{8} + \tfrac{1}{8} + \tfrac{1}{8} \right) = s_4 + \tfrac{1}{2} > \tfrac{4}{2}$$

$$s_{16} = s_8 + \tfrac{1}{9} + \tfrac{1}{10} + \tfrac{1}{11} + \tfrac{1}{12} + \tfrac{1}{13} + \tfrac{1}{14} + \tfrac{1}{15} + \tfrac{1}{16}$$

$$> s_8 + \left(\tfrac{1}{16} + \tfrac{1}{16} + \tfrac{1}{16} + \tfrac{1}{16} + \tfrac{1}{16} + \tfrac{1}{16} + \tfrac{1}{16} + \tfrac{1}{16} \right) = s_8 + \tfrac{1}{2} > \tfrac{5}{2}$$

$$\vdots$$

$$s_{2^n} > \frac{n+1}{2}$$

If M is any constant, we can find a positive integer n such that $(n+1)/2 > M$. But for this n

$$s_{2^n} > \frac{n+1}{2} > M$$

so that no constant M is greater than or equal to *every* partial sum of the harmonic series. This proves divergence.

▶ Exercise Set 11.3

1. In each part, find the first four partial sums; find a closed form for the nth partial sum; and determine whether the series converges (if so, give the sum).

(a) $\displaystyle\sum_{k=1}^{\infty} \frac{2}{5^{k-1}}$

(b) $\displaystyle\sum_{k=1}^{\infty} \frac{1}{(k+1)(k+2)}$

(c) $\displaystyle\sum_{k=1}^{\infty} \frac{2^{k-1}}{4}$.

In Exercises 2–16, determine whether the series converges or diverges. If it converges, find the sum.

2. $\displaystyle\sum_{k=1}^{\infty} \frac{1}{5^k}$.

3. $\displaystyle\sum_{k=1}^{\infty} \left(-\frac{3}{4}\right)^{k-1}$.

4. $\displaystyle\sum_{k=1}^{\infty} \left(\frac{2}{3}\right)^{k+2}$.

5. $\displaystyle\sum_{k=1}^{\infty} (-1)^{k-1}\frac{7}{6^{k-1}}$.

6. $\displaystyle\sum_{k=1}^{\infty} 4^{k-1}$.

7. $\displaystyle\sum_{k=1}^{\infty} \left(-\frac{3}{2}\right)^{k+1}$.

8. $\displaystyle\sum_{k=1}^{\infty} \left(\frac{1}{k+3} - \frac{1}{k+4}\right)$.

9. $\displaystyle\sum_{k=1}^{\infty} \frac{1}{(k+2)(k+3)}$.

10. $\displaystyle\sum_{k=1}^{\infty} \left(\frac{1}{2^k} - \frac{1}{2^{k+1}}\right)$.

11. $\displaystyle\sum_{k=1}^{\infty} \frac{1}{9k^2 + 3k - 2}$.

12. $\displaystyle\sum_{k=2}^{\infty} \frac{1}{k^2 - 1}$.

13. $\displaystyle\sum_{k=1}^{\infty} \frac{4^{k+2}}{7^{k-1}}$.

14. $\displaystyle\sum_{k=1}^{\infty} (e/\pi)^{k-1}$.

15. $\displaystyle\sum_{k=1}^{\infty} (-1/2)^k$.

16. $\displaystyle\sum_{k=3}^{\infty} \frac{5}{k-2}$.

In Exercises 17–22, express the repeating decimal as a fraction.

17. $0.4444\ldots$

18. $0.9999\ldots$

19. $5.373737\ldots$

20. $0.159159159\ldots$

21. $0.782178217821\ldots$

22. $0.451141414\ldots$

23. Find a closed form for the nth partial sum of the series

$$\ln\frac{1}{2} + \ln\frac{2}{3} + \ln\frac{3}{4} + \cdots + \ln\frac{n}{n+1} + \cdots$$

and determine whether the series converges.

24. A ball is dropped from a height of 10 m. Each time it strikes the ground it bounces vertically to a height that is $\frac{3}{4}$ of the preceding height. Find the total distance the ball will travel if it is allowed to bounce indefinitely.

25. Show: $\displaystyle\sum_{k=2}^{\infty} \ln(1 - 1/k^2) = -\ln 2$.

26. Show: $\displaystyle\sum_{k=1}^{\infty} \frac{\sqrt{k+1} - \sqrt{k}}{\sqrt{k^2 + k}} = 1$.

27. Show: $\displaystyle\sum_{k=1}^{\infty} \left(\frac{1}{k} - \frac{1}{k+2}\right) = \frac{3}{2}$.

28. (a) Find A and B such that

$$\frac{6^k}{(3^{k+1} - 2^{k+1})(3^k - 2^k)} = \frac{2^k A}{3^k - 2^k} + \frac{2^k B}{3^{k+1} - 2^{k+1}}$$

(b) Use the result in part (a) to help show that

$$\sum_{k=1}^{\infty} \frac{6^k}{(3^{k+1} - 2^{k+1})(3^k - 2^k)} = 2$$

[This problem appeared in the Forty-Fifth Annual William Lowell Putnam Mathematical Competition.]

29. Show: $\dfrac{1}{1\cdot 3} + \dfrac{1}{3\cdot 5} + \dfrac{1}{5\cdot 7} + \cdots = \dfrac{1}{2}$.

30. Show: $\dfrac{1}{1\cdot 3} + \dfrac{1}{2\cdot 4} + \dfrac{1}{3\cdot 5} + \cdots = \dfrac{3}{4}$.

31. Use geometric series to show that

(a) $\displaystyle\sum_{k=0}^{\infty} (-1)^k x^k = \frac{1}{1+x}$ if $-1 < x < 1$

(b) $\displaystyle\sum_{k=0}^{\infty} (x-3)^k = \frac{1}{4-x}$ if $2 < x < 4$

(c) $\displaystyle\sum_{k=0}^{\infty} (-1)^k x^{2k} = \frac{1}{1+x^2}$ if $-1 < x < 1$.

In Exercises 32–35, find all values of x for which the series converges, and for these values find its sum.

32. $x - x^3 + x^5 - x^7 + x^9 - \cdots$.

33. $\dfrac{1}{x^2} + \dfrac{2}{x^3} + \dfrac{4}{x^4} + \dfrac{8}{x^5} + \dfrac{16}{x^6} + \cdots$.

34. $e^{-x} + e^{-2x} + e^{-3x} + e^{-4x} + e^{-5x} + \cdots$.

35. $\sin x - \frac{1}{2}\sin^2 x + \frac{1}{4}\sin^3 x - \frac{1}{8}\sin^4 x + \cdots$.

36. Prove the following decimal equality assuming that $a_n \neq 9$:

$$0.a_1 a_2 \ldots a_n 9999\ldots = 0.a_1 a_2 \ldots (a_n + 1)0000\ldots$$

37. Let a_1 be any real number and define

$$a_{n+1} = \frac{1}{2}(a_n + 1) \text{ for } n = 1, 2, 3, \ldots$$

Show that the sequence $\{a_n\}_{n=1}^{+\infty}$ converges and find its limit. [*Hint:* Express a_n in terms of a_1.]

38. Lines L_1 and L_2 form an angle θ, $0 < \theta < \pi/2$, at their point of intersection P (Figure 11.3.1). A point P_0 is chosen that is on L_1 and a units from P. Starting from P_0 a zig-zag path is constructed by successively going back and forth between L_1 and L_2 along a perpendicular from one line to the other. Find the following sums in terms of θ:

(a) $P_0 P_1 + P_1 P_2 + P_2 P_3 + \cdots$

(b) $P_0 P_1 + P_2 P_3 + P_4 P_5 + \cdots$

(c) $P_1 P_2 + P_3 P_4 + P_5 P_6 + \cdots$.

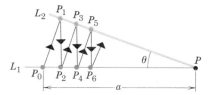

Figure 11.3.1

39. As shown in Figure 11.3.2, suppose that an angle θ is bisected using a straightedge and compass to produce ray R_1, then the angle between R_1 and the initial side is bisected to produce ray R_2. Thereafter, rays $R_3, R_4, R_5, \ldots$ are constructed in succession by bisecting the angle between the preceding two rays. Show that the sequence of angles that these rays make with the initial side has a limit of $\theta/3$. [This problem is based on *Trisection of an Angle in an Infinite Number of Steps* by Eric Kincannon, which appeared in *The College Mathematics Journal*, Vol. 21, No. 5, November 1990.]

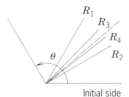

Figure 11.3.2

40. In his *Treatise on the Configurations of Qualities and Motions* (written in the 1350s), the French Bishop of Lisieux, Nicole Oresme, used a geometric method to find the sum of the series

$$\sum_{k=1}^{\infty} \frac{k}{2^k} = \frac{1}{2} + \frac{2}{4} + \frac{3}{8} + \frac{4}{16} + \cdots$$

In Figure 11.3.3a each term in the series is represented by the area of a rectangle. In Figure 11.3.3b the configuration in part (a) has been divided into rectangles with areas $A_1, A_2, A_3, \ldots$. Find the sum $A_1 + A_2 + A_3 + \cdots$. [For convenience, the horizontal and vertical scales in Figure 11.3.3 are different.]

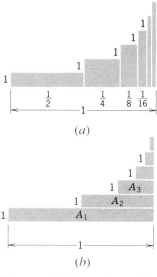

Figure 11.3.3

41. The great Swiss mathematician Leonhard Euler (biography on p. 52) made occasional errors in his pioneering work on infinite series. For example, Euler deduced that

$$\frac{1}{2} = 1 - 1 + 1 - 1 + \cdots$$

and

$$-1 = 1 + 2 + 4 + 8 + \cdots$$

by substituting $x = -1$ and $x = 2$ in the formula

$$\frac{1}{1-x} = 1 + x + x^2 + x^3 + \cdots$$

What was the error in his reasoning?

■ **11.4** CONVERGENCE TESTS

> *In the preceding section we found sums of series and investigated convergence by first writing the nth partial sum s_n in closed form and then examining its limit. However, it is relatively rare that the nth partial sum of a series can be written in closed form; for most series, convergence or divergence is determined by using convergence tests, some of which we shall introduce in this section. Once it is established that a series converges, the sum of the series can generally be approximated to any degree of accuracy by a partial sum with sufficiently many terms.*

□ THE DIVERGENCE TEST

The kth term in an infinite series Σu_k is sometimes called the *general term* of the series. The following theorem establishes a relationship between the limit of the general term and the convergence properties of a series.

11.4.1 THEOREM (*The Divergence Test*).

(a) *If* $\lim\limits_{k \to +\infty} u_k \neq 0$, *then the series* Σu_k *diverges.*

(b) *If* $\lim\limits_{k \to +\infty} u_k = 0$, *then the series* Σu_k *may either converge or diverge.*

Proof (a). To prove this result, it suffices to show that if the series converges, then $\lim\limits_{k \to +\infty} u_k = 0$ (why?). We shall prove this alternative form of (*a*).

Let us assume that the series converges. The general term u_k can be written as

$$u_k = s_k - s_{k-1} \tag{1}$$

where s_k is the sum of the first k terms and s_{k-1} is the sum of the first $k - 1$ terms. If S denotes the sum of the series, then $\lim\limits_{k \to +\infty} s_k = S$, and since $(k - 1) \to +\infty$ as $k \to +\infty$, we also have $\lim\limits_{k \to +\infty} s_{k-1} = S$. Thus, from (1)

$$\lim_{k \to +\infty} u_k = \lim_{k \to +\infty} (s_k - s_{k-1}) = S - S = 0$$

Proof (b). To prove this result, it suffices to produce both a convergent series and a divergent series for which $\lim\limits_{k \to +\infty} u_k = 0$. The following series both have this property:

$$1/2 + 1/2^2 + \cdots + 1/2^k + \cdots \quad \text{and} \quad 1 + 1/2 + 1/3 + \cdots + 1/k + \cdots$$

The first is a convergent geometric series and the second is the divergent harmonic series. ■

The alternative form of part (*a*) given in the preceding proof is sufficiently important that we state it separately for future reference.

11.4.2 THEOREM. *If the series* Σu_k *converges, then* $\lim\limits_{k \to +\infty} u_k = 0$.

Example 1   The series

$$\sum_{k=1}^{\infty} \frac{k}{k+1} = \frac{1}{2} + \frac{2}{3} + \frac{3}{4} + \cdots + \frac{k}{k+1} + \cdots$$

diverges since

$$\lim_{k \to +\infty} \frac{k}{k+1} = \lim_{k \to +\infty} \frac{1}{1 + 1/k} = 1 \neq 0 \quad \blacktriangleleft$$

WARNING.   The converse of Theorem 11.4.2 is false. To prove that a series converges it does not suffice to show that $\lim_{k \to +\infty} u_k = 0$, since this property may hold for divergent as well as convergent series. For example, the kth term of the divergent harmonic series $1 + 1/2 + 1/3 + \cdots + 1/k + \cdots$ approaches zero as $k \to +\infty$, and the kth term of the convergent geometric series $1/2 + 1/2^2 + \cdots + 1/2^k + \cdots$ tends to zero as $k \to +\infty$.

☐ **ALGEBRAIC PROPERTIES OF INFINITE SERIES**

For brevity, the proof of the following result is omitted.

11.4.3   THEOREM.

(a)   *If Σu_k and Σv_k are convergent series, then $\Sigma(u_k + v_k)$ and $\Sigma(u_k - v_k)$ are convergent series and the sums of these series are related by*

$$\sum_{k=1}^{\infty} (u_k + v_k) = \sum_{k=1}^{\infty} u_k + \sum_{k=1}^{\infty} v_k$$

$$\sum_{k=1}^{\infty} (u_k - v_k) = \sum_{k=1}^{\infty} u_k - \sum_{k=1}^{\infty} v_k$$

(b)   *If c is a nonzero constant, then the series Σu_k and $\Sigma c u_k$ both converge or both diverge. In the case of convergence, the sums are related by*

$$\sum_{k=1}^{\infty} c u_k = c \sum_{k=1}^{\infty} u_k$$

(c)   *Convergence or divergence is unaffected by deleting a finite number of terms from a series; in particular, for any positive integer K, the series*

$$\sum_{k=1}^{\infty} u_k = u_1 + u_2 + u_3 + \cdots$$

$$\sum_{k=K}^{\infty} u_k = u_K + u_{K+1} + u_{K+2} + \cdots$$

both converge or both diverge.

REMARK.   Do not read too much into part (c) of this theorem. Although the convergence is not affected when a finite number of terms is deleted from the beginning of a convergent series, the *sum* of a convergent series is changed by the removal of these terms.

Example 2   Find the sum of the series

$$\sum_{k=1}^{\infty} \left(\frac{3}{4^k} - \frac{2}{5^{k-1}} \right)$$

Solution.   The series

$$\sum_{k=1}^{\infty} \frac{3}{4^k} = \frac{3}{4} + \frac{3}{4^2} + \frac{3}{4^3} + \cdots$$

is a convergent geometric series $(a = \frac{3}{4}, r = \frac{1}{4})$, and the series

$$\sum_{k=1}^{\infty} \frac{2}{5^{k-1}} = 2 + \frac{2}{5} + \frac{2}{5^2} + \frac{2}{5^3} + \cdots$$

is also a convergent geometric series $(a = 2, r = \frac{1}{5})$. Thus, from Theorems 11.4.3(a) and 11.3.3 the given series converges and

$$\sum_{k=1}^{\infty} \left(\frac{3}{4^k} - \frac{2}{5^{k-1}} \right) = \sum_{k=1}^{\infty} \frac{3}{4^k} - \sum_{k=1}^{\infty} \frac{2}{5^{k-1}} = \frac{\frac{3}{4}}{1 - \frac{1}{4}} - \frac{2}{1 - \frac{1}{5}} = -\frac{3}{2} \quad \blacktriangleleft$$

Example 3 Determine whether the following series converge or diverge.

(a) $\displaystyle\sum_{k=1}^{\infty} \frac{5}{k} = 5 + \frac{5}{2} + \frac{5}{3} + \cdots + \frac{5}{k} + \cdots$

(b) $\displaystyle\sum_{k=10}^{\infty} \frac{1}{k} = \frac{1}{10} + \frac{1}{11} + \frac{1}{12} + \cdots$

Solution. The first series is a constant times the divergent harmonic series, and hence diverges by part (b) of Theorem 11.4.3. The second series results by deleting the first nine terms from the divergent harmonic series, and hence diverges by part (c) of Theorem 11.4.3. $\quad \blacktriangleleft$

☐ **THE INTEGRAL TEST**

The expressions

$$\sum_{k=1}^{\infty} \frac{1}{k^2} \quad \text{and} \quad \int_{1}^{+\infty} \frac{1}{x^2}\, dx$$

are related in that the integrand in the improper integral results when the index k in the general term of the series is replaced by x and the limits of summation in the series are replaced by the corresponding limits of integration. The following theorem, which is proved at the end of this section, shows that there is a relationship between the convergence of the series and the integral.

11.4.4 THEOREM (*The Integral Test*). *Let* Σu_k *be a series with positive terms, and let* $f(x)$ *be the function that results when* k *is replaced by* x *in the formula for* u_k. *If* f *is decreasing and continuous on the interval* $[a, +\infty)$, *then*

$$\sum_{k=1}^{\infty} u_k \quad \text{and} \quad \int_{a}^{+\infty} f(x)\, dx$$

both converge or both diverge.

Example 4 Use the integral test to determine whether the following series converge or diverge.

(a) $\displaystyle\sum_{k=1}^{\infty} \frac{1}{k}$ \quad (b) $\displaystyle\sum_{k=1}^{\infty} \frac{1}{k^2}$

Solution (a). We already know that this is the divergent harmonic series, so the integral test will simply provide another way of establishing the divergence. If we replace k by x in the general term $1/k$, we obtain the function $f(x) = 1/x$, which is decreasing and continuous for $x \geq 1$ (as required to apply the integral test with $a = 1$). Since

$$\int_{1}^{+\infty} \frac{1}{x}\, dx = \lim_{l \to +\infty} \int_{1}^{l} \frac{1}{x}\, dx = \lim_{l \to +\infty} [\ln l - \ln 1] = +\infty$$

the integral diverges and consequently so does the series.

Solution (b).   If we replace k by x in the general term $1/k^2$, we obtain the function $f(x) = 1/x^2$, which is decreasing and continuous for $x \geq 1$. Since

$$\int_1^{+\infty} \frac{1}{x^2}\, dx = \lim_{l \to +\infty} \int_1^l \frac{dx}{x^2} = \lim_{l \to +\infty} \left[-\frac{1}{x} \right]_1^l = \lim_{l \to +\infty} \left[1 - \frac{1}{l} \right] = 1$$

the integral converges and consequently the series converges by the integral test with $a = 1$.   ◀

REMARK.   In part (b) of the preceding example, do *not* erroneously conclude that the sum of the series is 1 because the value of the corresponding integral is 1. It can be proved that the sum of the series is actually $\pi^2/6$ and, indeed, the sum of the first two terms alone exceeds 1.

□ *p*-SERIES

The series in Example 4 are special cases of a class of series called ***p*-series** or **hyperharmonic series**. A *p*-series is an infinite series of the form

$$\sum_{k=1}^{\infty} \frac{1}{k^p} = 1 + \frac{1}{2^p} + \frac{1}{3^p} + \cdots + \frac{1}{k^p} + \cdots$$

where $p > 0$. Examples of *p*-series are

$$\sum_{k=1}^{\infty} \frac{1}{k} = 1 + \frac{1}{2} + \frac{1}{3} + \cdots + \frac{1}{k} + \cdots \qquad \boxed{p = 1}$$

$$\sum_{k=1}^{\infty} \frac{1}{k^2} = 1 + \frac{1}{2^2} + \frac{1}{3^2} + \cdots + \frac{1}{k^2} + \cdots \qquad \boxed{p = 2}$$

$$\sum_{k=1}^{\infty} \frac{1}{\sqrt{k}} = 1 + \frac{1}{\sqrt{2}} + \frac{1}{\sqrt{3}} + \cdots + \frac{1}{\sqrt{k}} + \cdots \qquad \boxed{p = \tfrac{1}{2}}$$

The following theorem tells when a *p*-series converges.

11.4.5   THEOREM   (*Convergence of p-Series*).

$$\sum_{k=1}^{\infty} \frac{1}{k^p} = 1 + \frac{1}{2^p} + \frac{1}{3^p} + \cdots + \frac{1}{k^p} + \cdots$$

converges if $p > 1$ and diverges if $0 < p \leq 1$.

Proof.   To establish this result when $p \neq 1$, we shall use the integral test.

$$\int_1^{+\infty} \frac{1}{x^p}\, dx = \lim_{l \to +\infty} \int_1^l x^{-p}\, dx = \lim_{l \to +\infty} \frac{x^{1-p}}{1-p} \bigg]_1^l = \lim_{l \to +\infty} \left[\frac{l^{1-p}}{1-p} - \frac{1}{1-p} \right]$$

If $p > 1$, then $1 - p < 0$, so $l^{1-p} \to 0$ as $l \to +\infty$. Thus, the integral converges [its value is $-1/(1-p)$] and consequently the series also converges. For $0 < p < 1$, it follows that $1 - p > 0$ and $l^{1-p} \to +\infty$ as $l \to +\infty$, so the integral and the series diverge. The case $p = 1$ is the harmonic series, which was previously shown to diverge.   ∎

Example 5

$$1 + \frac{1}{\sqrt[3]{2}} + \frac{1}{\sqrt[3]{3}} + \cdots + \frac{1}{\sqrt[3]{k}} + \cdots$$

diverges since it is a *p*-series with $p = \frac{1}{3} < 1$.   ◀

□ PROOF OF THE
   INTEGRAL TEST

Before we can prove the integral test, we need a basic result about convergence of series with *nonnegative* terms. If $u_1 + u_2 + u_3 + \cdots + u_k + \cdots$ is such a series, then its sequence of partial sums is nondecreasing, that is,

$$s_1 \leq s_2 \leq s_3 \leq \cdots \leq s_n \leq \cdots$$

Thus, from Theorem 11.2.2 the sequence of partial sums converges to a limit S if and only if it has some upper bound M, in which case $S \leq M$. If no upper bound exists, then the sequence of partial sums diverges. Since convergence of the sequence of partial sums corresponds to convergence of the series, we have the following theorem.

11.4.6 THEOREM. *If Σu_k is a series with nonnegative terms, and if there is a constant M such that*

$$s_n = u_1 + u_2 + \cdots + u_n \leq M$$

for every n, then the series converges and the sum S satisfies $S \leq M$. If no such M exists, then the series diverges.

In words, this theorem implies that *a series with nonnegative terms converges if and only if its sequence of partial sums is bounded above.*

Proof of Theorem 11.4.4. We need only show that the series converges when the integral converges and that the series diverges when the integral diverges. The remaining cases are logical implications of these. For simplicity, we will limit the proof to the case where $a = 1$. Assume that $f(x)$ satisfies the hypotheses of the theorem for $x \geq 1$. Since

$$f(1) = u_1, f(2) = u_2, \ldots, f(n) = u_n, \ldots$$

the values of $u_1, u_2, \ldots, u_n, \ldots$ can be interpreted as the areas of the rectangles shown in Figure 11.4.1.

The following inequalities (for $n > 1$) result by comparing the areas under the curve $y = f(x)$ to the areas of the rectangles in Figure 11.4.1:

$$\int_1^{n+1} f(x)\,dx < u_1 + u_2 + \cdots + u_n = s_n \qquad \boxed{\text{Figure 11.4.1}a}$$

$$s_n - u_1 = u_2 + u_3 + \cdots + u_n < \int_1^n f(x)\,dx \qquad \boxed{\text{Figure 11.4.1}b}$$

These inequalities can be combined as

$$\int_1^{n+1} f(x)\,dx < s_n < u_1 + \int_1^n f(x)\,dx \qquad (2)$$

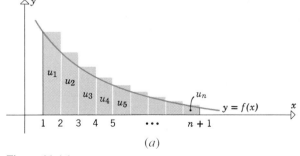

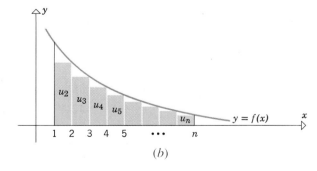

(a)   (b)

Figure 11.4.1

If the integral $\int_1^{+\infty} f(x)\,dx$ converges to a finite value L, then from the right-hand inequality in (2)

$$s_n < u_1 + \int_1^n f(x)\,dx < u_1 + \int_1^{+\infty} f(x)\,dx = u_1 + L$$

Thus, each partial sum is less than the finite constant $u_1 + L$, and the series converges by Theorem 11.4.6. On the other hand, if the integral $\int_1^{+\infty} f(x)\, dx$ diverges, then

$$\lim_{n \to +\infty} \int_1^{n+1} f(x)\, dx = +\infty$$

so that from the left-hand inequality in (2), $\lim_{n \to +\infty} s_n = +\infty$. This implies that the series also diverges. ∎

▶ **Exercise Set 11.4** Ⓒ 37–40

In Exercises 1–4, use Theorem 11.4.3 to find the sum of the series.

1. $\displaystyle\sum_{k=1}^{\infty} \left[\frac{1}{2^k} + \frac{1}{4^k} \right].$

2. $\displaystyle\sum_{k=1}^{\infty} \left[\frac{1}{5^k} - \frac{1}{k(k+1)} \right].$

3. $\displaystyle\sum_{k=2}^{\infty} \left[\frac{1}{k^2 - 1} - \frac{7}{10^{k-1}} \right].$

4. $\displaystyle\sum_{k=1}^{\infty} \left[\frac{7}{3^k} + \frac{6}{(k+3)(k+4)} \right].$

5. In each part, determine whether the given p-series converges or diverges.

(a) $\displaystyle\sum_{k=1}^{\infty} \frac{1}{k^3}$

(b) $\displaystyle\sum_{k=1}^{\infty} \frac{1}{\sqrt{k}}$

(c) $\displaystyle\sum_{k=1}^{\infty} k^{-1}$

(d) $\displaystyle\sum_{k=1}^{\infty} k^{-2/3}$

(e) $\displaystyle\sum_{k=1}^{\infty} k^{-4/3}$

(f) $\displaystyle\sum_{k=1}^{\infty} \frac{1}{\sqrt[4]{k}}$

(g) $\displaystyle\sum_{k=1}^{\infty} \frac{1}{\sqrt[3]{k^5}}$

(h) $\displaystyle\sum_{k=1}^{\infty} \frac{1}{k^\pi}.$

In Exercises 6–8, use the divergence test to show that the series diverges.

6. (a) $\displaystyle\sum_{k=1}^{\infty} \frac{k+1}{k+2}$

(b) $\displaystyle\sum_{k=1}^{\infty} \ln k.$

7. (a) $\displaystyle\sum_{k=1}^{\infty} \frac{k^2 + k + 3}{2k^2 + 1}$

(b) $\displaystyle\sum_{k=1}^{\infty} \left(1 + \frac{1}{k} \right)^k.$

8. (a) $\displaystyle\sum_{k=1}^{\infty} \cos k\pi$

(b) $\displaystyle\sum_{k=1}^{\infty} \frac{e^k}{k}.$

In Exercises 9–30, determine whether the series converges or diverges.

9. $\displaystyle\sum_{k=1}^{\infty} \frac{1}{k+6}.$

10. $\displaystyle\sum_{k=1}^{\infty} \frac{3}{5k}.$

11. $\displaystyle\sum_{k=1}^{\infty} \frac{1}{5k+2}.$

12. $\displaystyle\sum_{k=1}^{\infty} \frac{k}{1+k^2}.$

13. $\displaystyle\sum_{k=1}^{\infty} \frac{1}{1+9k^2}.$

14. $\displaystyle\sum_{k=1}^{\infty} \frac{1}{(4+2k)^{3/2}}.$

15. $\displaystyle\sum_{k=1}^{\infty} \frac{1}{\sqrt{k+5}}.$

16. $\displaystyle\sum_{k=1}^{\infty} \frac{1}{\sqrt[k]{e}}.$

17. $\displaystyle\sum_{k=1}^{\infty} \frac{1}{\sqrt[3]{2k-1}}.$

18. $\displaystyle\sum_{k=3}^{\infty} \frac{\ln k}{k}.$

19. $\displaystyle\sum_{k=1}^{\infty} \frac{k}{\ln(k+1)}.$

20. $\displaystyle\sum_{k=1}^{\infty} ke^{-k^2}.$

21. $\displaystyle\sum_{k=1}^{\infty} \frac{1}{(k+1)[\ln(k+1)]^2}.$

22. $\displaystyle\sum_{k=1}^{\infty} \frac{k^2 + 1}{k^2 + 3}.$

23. $\displaystyle\sum_{k=1}^{\infty} \left(1 + \frac{1}{k} \right)^{-k}.$

24. $\displaystyle\sum_{k=1}^{\infty} \frac{1}{\sqrt{k^2 + 1}}.$

25. $\displaystyle\sum_{k=1}^{\infty} \frac{\tan^{-1} k}{1+k^2}.$

26. $\displaystyle\sum_{k=1}^{\infty} \operatorname{sech}^2 k.$

27. $\displaystyle\sum_{k=5}^{\infty} 7k^{-p} \quad (p > 1).$

28. $\displaystyle\sum_{k=1}^{\infty} 7(k+5)^{-p} \quad (p \le 1).$

29. $\displaystyle\sum_{k=1}^{\infty} k^2 \sin^2 \left(\frac{1}{k} \right).$

30. $\displaystyle\sum_{k=1}^{\infty} k^2 e^{-k^3}.$

31. Prove: $\displaystyle\sum_{k=2}^{\infty} \frac{1}{k(\ln k)^p}$ converges if $p > 1$ and diverges if $p \le 1$.

32. Prove: $\displaystyle\sum_{k=3}^{\infty} \frac{1}{k(\ln k)[\ln(\ln k)]^p}$ converges if $p > 1$ and diverges if $p \le 1$.

33. Prove: If Σu_k converges and Σv_k diverges, then $\Sigma(u_k + v_k)$ diverges and $\Sigma(u_k - v_k)$ diverges. [*Hint:* Assume that $\Sigma(u_k + v_k)$ converges and use Theorem 11.4.3 to obtain a contradiction. Similarly, for $\Sigma(u_k - v_k)$.]

34. Find examples to show that $\Sigma(u_k + v_k)$ and $\Sigma(u_k - v_k)$ may converge or may diverge if Σu_k and Σv_k both diverge.

35. With the help of Exercise 33, determine whether the given series in parts (a)–(d) converge or diverge.

(a) $\displaystyle\sum_{k=1}^{\infty} \left[\left(\frac{2}{3} \right)^{k-1} + \frac{1}{k} \right]$

(b) $\sum_{k=1}^{\infty} \left[\dfrac{k^2}{1+k^2} + \dfrac{1}{k(k+1)} \right]$

(c) $\sum_{k=1}^{\infty} \left[\dfrac{1}{3k+2} + \dfrac{1}{k^{3/2}} \right]$

(d) $\sum_{k=2}^{\infty} \left[\dfrac{1}{k(\ln k)^2} - \dfrac{1}{k^2} \right]$.

36. If the sum S of a convergent series $\sum_{k=1}^{\infty} u_k$ of positive terms is approximated by its nth partial sum s_n, then the error in the approximation is $S - s_n$. Let $f(x)$ be the function that results when k is replaced by x in the formula for u_k. Show that if f is decreasing for $x \geq n$, then

$$\int_{n+1}^{+\infty} f(x)\,dx < S - s_n < \int_{n}^{+\infty} f(x)\,dx$$

The result in Exercise 36 can be written as

$$s_n + \int_{n+1}^{+\infty} f(x)\,dx < S < s_n + \int_{n}^{+\infty} f(x)\,dx$$

which provides an upper and lower bound on the sum S of the series. Use this result in part (a) of Exercises 37 and 38.

37. (a) Find the partial sum s_{10} of the series $\sum_{k=1}^{\infty} 1/k^3$, and use it to obtain an upper and lower bound on the sum S of the series. Express your results to four decimal places.

(b) Use the right-hand inequality in Exercise 36 to find a value of n to ensure that the error in approximating S by s_n is less than 10^{-3}.

38. (a) Find the partial sum s_6 of the series $\sum_{k=1}^{\infty} 1/k^4$, and use it to obtain an upper and lower bound on the sum S

of the series. Express your results to five decimal places. [*Note:* Compare your bounds with the exact value of S, which is $\pi^4/90$.]

(b) Use the right-hand inequality in Exercise 36 to find a value of n to ensure that the error in approximating S by s_n is less than 10^{-5}.

39. Let s_n be the nth partial sum of the divergent series $\sum_{k=1}^{\infty} 1/k$.

(a) Use inequality (2) to show that for $n \geq 2$

$$\ln(n+1) < s_n < 1 + \ln n$$

Then find integer upper and lower bounds for $s_{1,000,000}$.

(b) Find a value of n to ensure that $s_n > 100$.

40. Let s_n be the nth partial sum of the divergent series $\sum_{k=1}^{\infty} 1/\sqrt{k}$.

(a) Use inequality (2) to show that for $n \geq 2$

$$2\sqrt{n+1} - 2 < s_n < 2\sqrt{n} - 1$$

Then find integer upper and lower bounds for $s_{10,000}$.

(b) Find a value of n to ensure that $s_n > 100$.

41. Let $\sum u_k$ be a series with positive terms and $f(x)$ the function that results when k is replaced by x in the formula for u_k. Suppose that f is decreasing and continuous on the interval $[1, +\infty)$. Let $\{a_n\}_{n=1}^{+\infty}$ be the sequence defined by

$$a_n = \sum_{k=1}^{n} u_k - \int_{1}^{n} f(x)\,dx$$

(a) Use inequality (2) to show that $0 < a_n < u_1$.

(b) Show that $\{a_n\}$ is a decreasing sequence by considering $a_{n+1} - a_n$.

(c) Prove that the sequence $\{a_n\}$ converges.

11.5 ADDITIONAL CONVERGENCE TESTS

In this section we shall develop some additional convergence tests for series with positive terms.

☐ **THE COMPARISON TEST**

The following result is a convergence test in its own right and will also serve as the foundation for other convergence tests that we will develop.

11.5.1 THEOREM (*The Comparison Test*). *Let $\sum a_k$ and $\sum b_k$ be series with nonnegative terms and suppose that*

$$a_1 \leq b_1, a_2 \leq b_2, a_3 \leq b_3, \ldots, a_k \leq b_k, \ldots$$

(a) *If the "bigger series" $\sum b_k$ converges, then the "smaller series" $\sum a_k$ also converges.*

(b) *If the "smaller series" $\sum a_k$ diverges, then the "bigger series" $\sum b_k$ also diverges.*

Proof (a). Suppose that the series Σb_k converges and its sum is B. Then for all n

$$b_1 + b_2 + \cdots + b_n \leq \sum_{k=1}^{\infty} b_k = B$$

From our hypothesis it follows that

$$a_1 + a_2 + \cdots + a_n \leq b_1 + b_2 + \cdots + b_n \leq B$$

Thus, each partial sum of the series Σa_k is less than or equal to B, which implies that Σa_k converges by Theorem 11.4.6.

Proof (b). This part is really just an alternative phrasing of part (a). If Σa_k diverges, then Σb_k must diverge since convergence of Σb_k would imply convergence of Σa_k, contrary to the hypothesis. ∎

REMARK. As one would expect, it is not essential in Theorem 11.5.1 that the condition $a_k \leq b_k$ hold for all k, as stated; the conclusions of the theorem remain true if this condition is eventually true.

☐ **THE RATIO TEST**

Since the comparison test requires a little ingenuity to use, we shall wait until the next section before applying it. For now, we shall use the comparison test to develop some other tests that are easier to apply.

11.5.2 THEOREM (*The Ratio Test*). *Let Σu_k be a series with positive terms and suppose that*

$$\rho = \lim_{k \to +\infty} \frac{u_{k+1}}{u_k}$$

(a) *If $\rho < 1$, the series converges.*
(b) *If $\rho > 1$ or $\rho = +\infty$, the series diverges.*
(c) *If $\rho = 1$, the series may converge or diverge, so that another test must be tried.*

Proof (a). The number ρ must be nonnegative since it is the limit of u_{k+1}/u_k, which is positive for all k. In this part of the proof we assume that $\rho < 1$, so that $0 \leq \rho < 1$.

We will prove convergence by showing that the terms of the given series are eventually less than the terms of a convergent geometric series. For this purpose, choose any real number r such that $0 < \rho < r < 1$. Since the limit of u_{k+1}/u_k is ρ, and $\rho < r$, the terms of the sequence $\{u_{k+1}/u_k\}$ must eventually be less than r. Thus, there is a positive integer K such that for $k \geq K$ we have

$$\frac{u_{k+1}}{u_k} < r \quad \text{or} \quad u_{k+1} < ru_k$$

This yields the inequalities

$$u_{K+1} < ru_K$$
$$u_{K+2} < ru_{K+1} < r^2 u_K$$
$$u_{K+3} < ru_{K+2} < r^3 u_K \tag{1}$$
$$u_{K+4} < ru_{K+3} < r^4 u_K$$
$$\vdots$$

But $0 < r < 1$, so

$$ru_K + r^2 u_K + r^3 u_K + \cdots$$

is a convergent geometric series. From the inequalities in (1) and the comparison test it follows that

$$u_{K+1} + u_{K+2} + u_{K+3} + \cdots$$

must also be a convergent series. Thus, $u_1 + u_2 + u_3 + \cdots + u_k + \cdots$ converges by Theorem 11.4.3(c).

Proof (b).   In this part we will prove divergence by showing that the limit of the general term is not zero. Since the limit of u_{k+1}/u_k is ρ and $\rho > 1$, the terms in the sequence $\{u_{k+1}/u_k\}$ must eventually be greater than 1. Thus, there is a positive integer K such that for $k \geq K$ we have

$$\frac{u_{k+1}}{u_k} > 1 \quad \text{or} \quad u_{k+1} > u_k$$

This yields the inequalities

$$u_{K+1} > u_K$$

$$u_{K+2} > u_{K+1} > u_K$$

$$u_{K+3} > u_{K+2} > u_K \tag{2}$$

$$u_{K+4} > u_{K+3} > u_K$$

$$\vdots$$

Since $u_K > 0$, it follows from the inequalities in (2) that $\lim\limits_{k \to +\infty} u_k \neq 0$, and thus $u_1 + u_2 + \cdots + u_k + \cdots$ diverges by part (a) of Theorem 11.4.1. The proof in the case where $\rho = +\infty$ is omitted.

Proof (c).   The divergent harmonic series and the convergent *p*-series with $p = 2$ both have $\rho = 1$ (verify), so the ratio test does not distinguish between convergence and divergence when $\rho = 1$.   ∎

Example 1   Use the ratio test to determine whether the following series converge or diverge.

(a) $\displaystyle\sum_{k=1}^{\infty} \frac{1}{k!}$ (b) $\displaystyle\sum_{k=1}^{\infty} \frac{k}{2^k}$ (c) $\displaystyle\sum_{k=1}^{\infty} \frac{k^k}{k!}$ (d) $\displaystyle\sum_{k=1}^{\infty} \frac{(2k)!}{4^k}$

Solution (a).   The series converges, since

$$\rho = \lim_{k \to +\infty} \frac{u_{k+1}}{u_k} = \lim_{k \to +\infty} \frac{1/(k+1)!}{1/k!} = \lim_{k \to +\infty} \frac{k!}{(k+1)!} = \lim_{k \to +\infty} \frac{1}{k+1} = 0 < 1$$

Solution (b).   The series converges, since

$$\rho = \lim_{k \to +\infty} \frac{u_{k+1}}{u_k} = \lim_{k \to +\infty} \frac{k+1}{2^{k+1}} \cdot \frac{2^k}{k} = \frac{1}{2} \lim_{k \to +\infty} \frac{k+1}{k} = \frac{1}{2} < 1$$

Solution (c).   The series diverges, since

$$\rho = \lim_{k \to +\infty} \frac{u_{k+1}}{u_k} = \lim_{k \to +\infty} \frac{(k+1)^{k+1}}{(k+1)!} \cdot \frac{k!}{k^k} = \lim_{k \to +\infty} \frac{(k+1)^k}{k^k}$$

$$\boxed{\text{See Theorem } 7.5.15(b)}$$

$$= \lim_{k \to +\infty} \left(1 + \frac{1}{k}\right)^k = e > 1$$

Solution (d).   The series diverges, since

$$\rho = \lim_{k \to +\infty} \frac{u_{k+1}}{u_k} = \lim_{k \to +\infty} \frac{[2(k+1)]!}{4^{k+1}} \cdot \frac{4^k}{(2k)!} = \lim_{k \to +\infty} \left(\frac{(2k+2)!}{(2k)!} \cdot \frac{1}{4} \right)$$

$$= \frac{1}{4} \lim_{k \to +\infty} (2k+2)(2k+1) = +\infty \qquad \blacktriangleleft$$

Example 2   Determine whether the series

$$1 + \frac{1}{3} + \frac{1}{5} + \frac{1}{7} + \cdots + \frac{1}{2k-1} + \cdots$$

converges or diverges.

Solution.   The ratio test is of no help since

$$\rho = \lim_{k \to +\infty} \frac{u_{k+1}}{u_k} = \lim_{k \to +\infty} \frac{1}{2(k+1)-1} \cdot \frac{2k-1}{1} = \lim_{k \to +\infty} \frac{2k-1}{2k+1} = 1$$

However, the integral test proves that the series diverges since

$$\int_1^{+\infty} \frac{dx}{2x-1} = \lim_{l \to +\infty} \int_1^l \frac{dx}{2x-1} = \lim_{l \to +\infty} \frac{1}{2} \ln(2x-1) \Big]_1^l = +\infty \qquad \blacktriangleleft$$

☐ THE ROOT TEST

Sometimes the following result is easier to apply than the ratio test.

> **11.5.3**   THEOREM (*The Root Test*).   *Let Σu_k be a series with positive terms and suppose that*
>
> $$\rho = \lim_{k \to +\infty} \sqrt[k]{u_k} = \lim_{k \to +\infty} (u_k)^{1/k}$$
>
> (a)   *If $\rho < 1$, the series converges.*
> (b)   *If $\rho > 1$, or $\rho = +\infty$, the series diverges.*
> (c)   *If $\rho = 1$, the series may converge or diverge, so that another test must be tried.*

Since the proof of the root test is similar to the proof of the ratio test, we shall omit it.

Example 3   Use the root test to determine whether the following series converge or diverge.

(a) $\displaystyle\sum_{k=2}^{\infty} \left(\frac{4k-5}{2k+1} \right)^k$ (b) $\displaystyle\sum_{k=1}^{\infty} \frac{1}{(\ln(k+1))^k}$

Solution (a).   The series diverges, since

$$\rho = \lim_{k \to +\infty} (u_k)^{1/k} = \lim_{k \to +\infty} \frac{4k-5}{2k+1} = 2 > 1$$

Solution (b).   The series converges, since

$$\rho = \lim_{k \to +\infty} (u_k)^{1/k} = \lim_{k \to +\infty} \frac{1}{\ln(k+1)} = 0 < 1 \qquad \blacktriangleleft$$

☐ COMMENTS ON
NOTATION

Until now we have written most of our infinite series with the summation index beginning at 1. If the summation index begins at some other integer, it is always possible to rewrite the series so that the summation index starts at 1. For example, the series

$$\sum_{k=0}^{\infty} \frac{2^k}{k!} = 1 + 2 + \frac{2^2}{2!} + \frac{2^3}{3!} + \cdots \tag{3}$$

can be written as

$$\sum_{k=1}^{\infty} \frac{2^{k-1}}{(k-1)!} = 1 + 2 + \frac{2^2}{2!} + \frac{2^3}{3!} + \cdots \tag{4}$$

However, the index of summation need not start at 1 to apply the convergence tests. For example, we can apply the ratio test to (3) without converting to the more complicated form (4). Doing so yields

$$\rho = \lim_{k \to +\infty} \frac{u_{k+1}}{u_k} = \lim_{k \to +\infty} \frac{2^{k+1}}{(k+1)!} \cdot \frac{k!}{2^k} = \lim_{k \to +\infty} \frac{2}{k+1} = 0$$

which shows that the series converges since $\rho < 1$.

▶ Exercise Set 11.5 $\boxed{C}$ 39–42

In Exercises 1–6, apply the ratio test. According to the test, does the series converge, does the series diverge, or are the results inconclusive?

1. $\displaystyle\sum_{k=1}^{\infty} \frac{3^k}{k!}$.

2. $\displaystyle\sum_{k=1}^{\infty} \frac{4^k}{k^2}$.

3. $\displaystyle\sum_{k=2}^{\infty} \frac{1}{5k}$.

4. $\displaystyle\sum_{k=1}^{\infty} k \left(\frac{1}{2}\right)^k$.

5. $\displaystyle\sum_{k=1}^{\infty} \frac{k!}{k^3}$.

6. $\displaystyle\sum_{k=1}^{\infty} \frac{k}{k^2 + 1}$.

In Exercises 7–10, apply the root test. According to the test, does the series converge, does the series diverge, or are the results inconclusive?

7. $\displaystyle\sum_{k=1}^{\infty} \left(\frac{3k+2}{2k-1}\right)^k$.

8. $\displaystyle\sum_{k=1}^{\infty} \left(\frac{k}{100}\right)^k$.

9. $\displaystyle\sum_{k=1}^{\infty} \frac{k}{5^k}$.

10. $\displaystyle\sum_{k=1}^{\infty} (1 - e^{-k})^k$.

In Exercises 11–32, use any appropriate test to determine whether the series converges.

11. $\displaystyle\sum_{k=1}^{\infty} \frac{2^k}{k^3}$.

12. $\displaystyle\sum_{k=1}^{\infty} \frac{1}{k^2}$.

13. $\displaystyle\sum_{k=0}^{\infty} \frac{7^k}{k!}$.

14. $\displaystyle\sum_{k=1}^{\infty} \frac{1}{2k+1}$.

15. $\displaystyle\sum_{k=1}^{\infty} \frac{k^2}{5^k}$.

16. $\displaystyle\sum_{k=1}^{\infty} \frac{k! \, 10^k}{3^k}$.

17. $\displaystyle\sum_{k=1}^{\infty} k^{50} e^{-k}$.

18. $\displaystyle\sum_{k=1}^{\infty} \frac{k^2}{k^3 + 1}$.

19. $\displaystyle\sum_{k=1}^{\infty} k \left(\frac{2}{3}\right)^k$.

20. $\displaystyle\sum_{k=1}^{\infty} k^k$.

21. $\displaystyle\sum_{k=2}^{\infty} \frac{1}{k \ln k}$.

22. $\displaystyle\sum_{k=1}^{\infty} \frac{2^k}{k^3 + 1}$.

23. $\displaystyle\sum_{k=1}^{\infty} \left(\frac{4}{7k-1}\right)^k$.

24. $\displaystyle\sum_{k=1}^{\infty} \frac{(k!)^2 2^k}{(2k+2)!}$.

25. $\displaystyle\sum_{k=0}^{\infty} \frac{(k!)^2}{(2k)!}$.

26. $\displaystyle\sum_{k=1}^{\infty} \frac{1}{k^2 + 25}$.

27. $\displaystyle\sum_{k=1}^{\infty} \frac{1}{1 + \sqrt{k}}$.

28. $\displaystyle\sum_{k=1}^{\infty} \frac{k!}{k^k}$.

29. $\displaystyle\sum_{k=1}^{\infty} \frac{\ln k}{e^k}$.

30. $\displaystyle\sum_{k=1}^{\infty} \frac{k!}{e^{k^2}}$.

31. $\displaystyle\sum_{k=0}^{\infty} \frac{(k+4)!}{4! \, k! \, 4^k}$.

32. $\displaystyle\sum_{k=1}^{\infty} \left(\frac{k}{k+1}\right)^{k^2}$.

In Exercises 33–35, show that the series converges.

33. $1 + \dfrac{1 \cdot 2}{1 \cdot 3} + \dfrac{1 \cdot 2 \cdot 3}{1 \cdot 3 \cdot 5} + \dfrac{1 \cdot 2 \cdot 3 \cdot 4}{1 \cdot 3 \cdot 5 \cdot 7} + \cdots$.

34. $1 + \dfrac{1 \cdot 3}{3!} + \dfrac{1 \cdot 3 \cdot 5}{5!} + \dfrac{1 \cdot 3 \cdot 5 \cdot 7}{7!} + \cdots$.

35. $\dfrac{2!}{1} + \dfrac{3!}{1 \cdot 4} + \dfrac{4!}{1 \cdot 4 \cdot 7} + \dfrac{5!}{1 \cdot 4 \cdot 7 \cdot 10} + \cdots$.

36. For which positive values of α does $\sum_{k=1}^{\infty} \alpha^k / k^\alpha$ converge?

37. (a) Show: $\lim_{k \to +\infty} (\ln k)^{1/k} = 1$. [*Hint:* Let $y = (\ln x)^{1/x}$ and find $\lim_{x \to +\infty} \ln y$.]

    (b) Use the result in part (a) and the root test to show that $\sum_{k=1}^{\infty} (\ln k)/3^k$ converges.

    (c) Show that the series converges using the ratio test.

38. If the sum S of a convergent series $\sum_{k=1}^{\infty} u_k$ of positive terms is approximated by the nth partial sum s_n, then the error in the approximation is defined as

$$S - s_n = \sum_{k=n+1}^{\infty} u_k = u_{n+1} + u_{n+2} + u_{n+3} + \cdots$$

Let $r_k = u_{k+1}/u_k$.

(a) Prove: If r_k is *decreasing* for all $k \geq n+1$ and if $r_{n+1} < 1$, then

$$S - s_n < \frac{u_{n+1}}{1 - r_{n+1}}$$

[*Hint:* Show that the sum of the series $\sum_{k=n+1}^{\infty} u_k$ is less than the sum of the convergent geometric series

$$u_{n+1} + r_{n+1}u_{n+1} + r_{n+1}^2 u_{n+1} + \cdots]$$

(b) Prove: If r_k is *increasing* for all $k \geq n+1$ and if $\lim_{k \to +\infty} r_k = \rho < 1$, then

$$S - s_n < \frac{u_{n+1}}{1 - \rho}$$

[*Hint:* Show that the sum of the series $\sum_{k=n+1}^{\infty} u_k$ is less than the sum of the convergent geometric series

$$u_{n+1} + \rho u_{n+1} + \rho^2 u_{n+1} + \cdots]$$

In Exercises 39–42, use the results in Exercise 38. For each exercise do the following:

(a) Compute the stated partial sum and find an upper bound on the error in approximating S by the partial sum. Express your results to five decimal places.

(b) Find a value of n to ensure that s_n will approximate S with an error that is less than 10^{-5}.

39. $S = \sum_{k=1}^{\infty} \dfrac{1}{k!}$; s_5.

40. $S = \sum_{k=1}^{\infty} \dfrac{k}{3^k}$; s_8.

41. $S = \sum_{k=1}^{\infty} \dfrac{1}{k2^k}$; s_7.

42. $S = \sum_{k=1}^{\infty} \dfrac{1}{\sqrt{k3^k}}$; s_4.

▪ 11.6 THE LIMIT COMPARISON TEST

In this section we shall discuss procedures for applying the comparison test and we shall state an alternative version of this test that is easier to work with. Before starting, we remind the reader that the comparison test applies only to series with positive terms.

☐ SOME USEFUL INFORMAL PRINCIPLES

The convergence tests that we will discuss in this section require a little more ingenuity than those in the preceding sections in that they will require us to make an initial *guess* at whether the series in question is likely to converge or diverge. The subsequent steps in the test will depend on that initial guess.

To help with the guessing process in the first step we have formulated some principles that sometimes *suggest* whether a series is likely to converge or diverge. We have called these "informal principles" because they are not intended as formal theorems. In fact, we shall not guarantee that they *always* work. However, they work often enough to be useful as a starting point for the comparison test.

11.6.1 INFORMAL PRINCIPLE.   *Constant terms in the denominator of u_k can usually be deleted without affecting the convergence or divergence of the series.*

Example 1   Use the above principle to help guess whether the following series converge or diverge.

(a) $\displaystyle\sum_{k=1}^{\infty} \frac{1}{2^k + 1}$ (b) $\displaystyle\sum_{k=5}^{\infty} \frac{1}{\sqrt{k} - 2}$ (c) $\displaystyle\sum_{k=1}^{\infty} \frac{1}{(k + \frac{1}{2})^3}$

Solution (*a*).   Deleting the constant 1 suggests that

$$\sum_{k=1}^{\infty} \frac{1}{2^k + 1} \quad \text{behaves like} \quad \sum_{k=1}^{\infty} \frac{1}{2^k}$$

The modified series is a convergent geometric series, so the given series is likely to converge.

Solution (b). Deleting the -2 suggests that

$$\sum_{k=5}^{\infty} \frac{1}{\sqrt{k} - 2} \quad \text{behaves like} \quad \sum_{k=5}^{\infty} \frac{1}{\sqrt{k}}$$

The modified series is a portion of a divergent p-series $(p = \frac{1}{2})$, so the given series is likely to diverge.

Solution (c). Deleting the $\frac{1}{2}$ suggests that

$$\sum_{k=1}^{\infty} \frac{1}{(k + \frac{1}{2})^3} \quad \text{behaves like} \quad \sum_{k=1}^{\infty} \frac{1}{k^3}$$

The modified series is a convergent p-series $(p = 3)$, so the given series is likely to converge. ◀

11.6.2 INFORMAL PRINCIPLE. *If a polynomial in k appears as a factor in the numerator or denominator of u_k, all but the highest power of k in the polynomial may usually be deleted without affecting the convergence or divergence of the series.*

Example 2 Use the above principle to help guess whether the following series converge or diverge.

$$\text{(a)} \quad \sum_{k=1}^{\infty} \frac{1}{\sqrt{k^3 + 2k}} \qquad \text{(b)} \quad \sum_{k=1}^{\infty} \frac{6k^4 - 2k^3 + 1}{k^5 + k^2 - 2k}$$

Solution (a). Deleting the term $2k$ suggests that

$$\sum_{k=1}^{\infty} \frac{1}{\sqrt{k^3 + 2k}} \quad \text{behaves like} \quad \sum_{k=1}^{\infty} \frac{1}{\sqrt{k^3}} = \sum_{k=1}^{\infty} \frac{1}{k^{3/2}}$$

Since the modified series is a convergent p-series $(p = \frac{3}{2})$, the given series is likely to converge.

Solution (b). Deleting all but the highest powers of k in the numerator and also in the denominator suggests that

$$\sum_{k=1}^{\infty} \frac{6k^4 - 2k^3 + 1}{k^5 + k^2 - 2k} \quad \text{behaves like} \quad \sum_{k=1}^{\infty} \frac{6k^4}{k^5} = \sum_{k=1}^{\infty} 6 \left(\frac{1}{k} \right)$$

Since each term in the modified series is a constant times the corresponding term in the divergent harmonic series, the given series is likely to diverge. ◀

☐ **THE LIMIT COMPARISON TEST**

The following result, which is proved at the end of the section, can be used to establish convergence or divergence by examining the limit of the ratio of the general term of the series in question with the general term of a series whose convergence properties are known.

11.6.3 THEOREM (*The Limit Comparison Test*). *Let Σa_k and Σb_k be series with positive terms and suppose that*

$$\rho = \lim_{k \to +\infty} \frac{a_k}{b_k}$$

If ρ is finite and $\rho > 0$, then the series both converge or both diverge.

The cases where $\rho = 0$ or $\rho = +\infty$ are discussed in the exercises (Exercise 44).

Example 3 Use the limit comparison test to determine whether the following series converge or diverge.

(a) $\displaystyle\sum_{k=1}^{\infty} \frac{1}{2k^2 - k}$ (b) $\displaystyle\sum_{k=1}^{\infty} \frac{1}{k - \frac{1}{4}}$ (c) $\displaystyle\sum_{k=1}^{\infty} \frac{3k^3 - 2k^2 + 4}{k^5 - k^3 + 2}$

Solution (*a*).   Using Principle 11.6.2, the given series behaves like the series

$$\sum_{k=1}^{\infty} \frac{1}{2k^2} = \frac{1}{2}\sum_{k=1}^{\infty} \frac{1}{k^2} \tag{1}$$

which is a constant times a convergent *p*-series. Thus, the given series is likely to converge. To prove this, we apply Theorem 11.6.3 with

$$a_k = \frac{1}{2k^2 - k} \quad \text{and} \quad b_k = \frac{1}{2k^2}$$

We obtain

$$\rho = \lim_{k \to +\infty} \frac{a_k}{b_k} = \lim_{k \to +\infty} \frac{2k^2}{2k^2 - k} = \lim_{k \to +\infty} \frac{2}{2 - 1/k} = 1$$

Since ρ is finite and positive, it follows from Theorem 11.6.3 that the given series converges, since (1) converges.

Solution (*b*).   Using Principle 11.6.1, the series behaves like the divergent harmonic series

$$\sum_{k=1}^{\infty} \frac{1}{k} \tag{2}$$

Thus, the given series is likely to diverge. To prove this, we apply Theorem 11.6.3 with

$$a_k = \frac{1}{k - \frac{1}{4}} \quad \text{and} \quad b_k = \frac{1}{k}$$

We obtain

$$\rho = \lim_{k \to +\infty} \frac{a_k}{b_k} = \lim_{k \to +\infty} \frac{k}{k - \frac{1}{4}} = \lim_{k \to +\infty} \frac{1}{1 - \frac{1}{4k}} = 1$$

Since ρ is finite and positive, it follows from Theorem 11.6.3 that the given series diverges, since (2) diverges.

Solution (*c*).   From Principle 11.6.2, the series behaves like

$$\sum_{k=1}^{\infty} \frac{3k^3}{k^5} = \sum_{k=1}^{\infty} \frac{3}{k^2} \tag{3}$$

which converges since it is a constant times a convergent *p*-series. Thus, the given series is likely to converge. To prove this, we apply the limit comparison test to series (3) and the given series. We obtain

$$\rho = \lim_{k \to +\infty} \frac{\dfrac{3k^3 - 2k^2 + 4}{k^5 - k^3 + 2}}{\dfrac{3}{k^2}} = \lim_{k \to +\infty} \frac{3k^5 - 2k^4 + 4k^2}{3k^5 - 3k^3 + 6} = 1$$

Since ρ is finite and nonzero, it follows from Theorem 11.6.3 that the given series converges, since (3) converges. ◄

☐ **THE COMPARISON TEST**

We shall now discuss some techniques for applying the comparison test (Theorem 11.5.1). However, as a practical matter, you should try the limit comparison test before attempting the comparison test, since it is usually easier to apply; the comparison test should be viewed as a last resort.

There are two basic steps required to apply the comparison test to a series Σu_k of positive terms:

- Guess at whether the series Σu_k converges or diverges.
- Find a series that proves the guess to be correct. Thus, if the guess is divergence, we must find a divergent series whose terms are "smaller" than the corresponding terms of Σu_k, and if the guess is convergence, we must find a convergent series whose terms are "bigger" than the corresponding terms of Σu_k.

Example 4   Use the comparison test to determine whether the following series converge or diverge.

(a) $\displaystyle\sum_{k=1}^{\infty} \frac{1}{k - \frac{1}{4}}$ (b) $\displaystyle\sum_{k=1}^{\infty} \frac{1}{\sqrt[3]{k} + 5}$ (c) $\displaystyle\sum_{k=1}^{\infty} \frac{1}{2k^2 + k}$

Solution (a).   In part (b) of Example 3 we used the limit comparison test to show that this series diverges. To reach the same conclusion by the comparison test, we note (as in Example 3) that the series behaves like the divergent harmonic series, and hence is likely to diverge. Thus, our goal is to find a divergent series that is "smaller" than the given series. We can do this by dropping the constant $-\frac{1}{4}$ in the denominator, thereby *decreasing* the size of the general term:

$$\frac{1}{k - \frac{1}{4}} > \frac{1}{k} \quad \text{for } k = 1, 2, \ldots$$

Thus, the given series diverges by the comparison test, since $\displaystyle\sum_{k=1}^{\infty} \frac{1}{k}$ diverges.

Solution (b).   Using Principle 11.6.1, the series behaves like the divergent p-series

$$\sum_{k=1}^{\infty} \frac{1}{\sqrt[3]{k}} \tag{4}$$

$\left(p = \frac{1}{3}\right)$, and hence is likely to diverge. As in the preceding example, our goal is to find a divergent series that is "smaller" than the given series; but here we cannot achieve this by dropping the constant in the denominator, since that would decrease the denominator, thereby increasing the size of the general term, which is contrary to our objective of finding a smaller series. Instead, we will increase the denominator by replacing the constant 5 with a quantity that is eventually greater than 5. A convenient choice for this quantity is $\sqrt[3]{k}$, which is greater than 5 for $k > 125$. Thus,

$$\frac{1}{\sqrt[3]{k} + 5} \geq \frac{1}{\sqrt[3]{k} + \sqrt[3]{k}} = \frac{1}{2\sqrt[3]{k}} \quad \text{for } k = 125, 126, \ldots$$

But the series $\displaystyle\sum_{k=125}^{\infty} \frac{1}{2\sqrt[3]{k}}$ diverges, since (4) diverges (why?), and hence the given series diverges by the comparison test.

Solution (c).   Using Principle 11.6.2, the series behaves like the convergent series

$$\sum_{k=1}^{\infty} \frac{1}{2k^2} = \frac{1}{2} \sum_{k=1}^{\infty} \frac{1}{k^2} \tag{5}$$

and hence is likely to converge. Thus, our goal is to find a "bigger" convergent series. We can do this by dropping the k from the denominator, since that decreases the denominator and *increases* the size of the general term:

$$\frac{1}{2k^2 + k} < \frac{1}{2k^2} \quad \text{for } k = 1, 2, \dots$$

Thus, the given series converges, since (5) converges.   ◀

■ PROOF OF THE LIMIT COMPARISON TEST

We conclude this section with a proof of the limit comparison test.

Proof of Theorem 11.6.3.   We need only show that Σb_k converges when Σa_k converges and that Σb_k diverges when Σa_k diverges, since the remaining cases are logical implications of these (why?). The idea of the proof is to apply the comparison test to Σa_k and suitable multiples of Σb_k. For this purpose let ϵ be any positive number. Since

$$\rho = \lim_{k \to +\infty} \frac{a_k}{b_k}$$

it follows that eventually the terms in the sequence $\{a_k/b_k\}$ must be within ϵ units of ρ; that is, there is a positive integer K such that for $k \geq K$ we have

$$\rho - \epsilon < \frac{a_k}{b_k} < \rho + \epsilon$$

In particular, if we take $\epsilon = \rho/2$, then for $k \geq K$ we have

$$\frac{1}{2}\rho < \frac{a_k}{b_k} < \frac{3}{2}\rho \quad \text{or} \quad \frac{1}{2}\rho b_k < a_k < \frac{3}{2}\rho b_k$$

Thus, by the comparison test we can conclude that

$$\sum_{k=K}^{\infty} \frac{1}{2}\rho b_k \quad \text{converges if} \quad \sum_{k=K}^{\infty} a_k \quad \text{converges} \tag{6}$$

$$\sum_{k=K}^{\infty} \frac{3}{2}\rho b_k \quad \text{diverges if} \quad \sum_{k=K}^{\infty} a_k \quad \text{diverges} \tag{7}$$

But the convergence or divergence of a series is not affected by deleting finitely many terms or by multiplying the general term by a nonzero constant, so (6) and (7) imply that

$$\sum_{k=1}^{\infty} b_k \quad \text{converges if} \quad \sum_{k=1}^{\infty} a_k \quad \text{converges}$$

$$\sum_{k=1}^{\infty} b_k \quad \text{diverges if} \quad \sum_{k=1}^{\infty} a_k \quad \text{diverges} \quad ■$$

▶ Exercise Set 11.6

In Exercises 1–6, use the limit comparison test to determine whether the series converges or diverges.

1. $\displaystyle\sum_{k=1}^{\infty} \frac{4k^2 - 2k + 6}{8k^7 + k - 8}$.

2. $\displaystyle\sum_{k=1}^{\infty} \frac{1}{9k + 6}$.

3. $\displaystyle\sum_{k=1}^{\infty} \frac{5}{3^k + 1}$.

4. $\displaystyle\sum_{k=1}^{\infty} \frac{k(k + 3)}{(k + 1)(k + 2)(k + 5)}$.

5. $\displaystyle\sum_{k=1}^{\infty} \frac{1}{\sqrt[3]{8k^2 - 3k}}$.

6. $\displaystyle\sum_{k=1}^{\infty} \frac{1}{(2k + 3)^{17}}$.

In Exercises 7–12, prove that the series converges by the comparison test.

7. $\displaystyle\sum_{k=1}^{\infty} \frac{1}{3^k + 5}$.

8. $\displaystyle\sum_{k=1}^{\infty} \frac{2}{k^4 + k}$.

9. $\displaystyle\sum_{k=1}^{\infty} \frac{1}{5k^2 - k}$.

10. $\displaystyle\sum_{k=1}^{\infty} \frac{k}{8k^3 + 2k^2 - 1}$.

11. $\displaystyle\sum_{k=1}^{\infty} \frac{2^k - 1}{3^k + 2k}$.

12. $\displaystyle\sum_{k=1}^{\infty} \frac{5 \sin^2 k}{k!}$.

In Exercises 13–18, prove that the series diverges by the comparison test.

13. $\displaystyle\sum_{k=1}^{\infty} \frac{3}{k - \frac{1}{4}}$.

14. $\displaystyle\sum_{k=1}^{\infty} \frac{1}{\sqrt{k + 8}}$.

15. $\displaystyle\sum_{k=1}^{\infty} \frac{9}{\sqrt{k + 1}}$.

16. $\displaystyle\sum_{k=2}^{\infty} \frac{k + 1}{k^2 - k}$.

17. $\displaystyle\sum_{k=1}^{\infty} \frac{k^{4/3}}{8k^2 + 5k + 1}$.

18. $\displaystyle\sum_{k=1}^{\infty} \frac{k^{-1/2}}{2 + \sin^2 k}$.

In Exercises 19–34, use any method to determine whether the series converges or diverges. In some cases, you may have to use tests from earlier sections.

19. $\displaystyle\sum_{k=1}^{\infty} \frac{1}{k^3 + 2k + 1}$.

20. $\displaystyle\sum_{k=1}^{\infty} \frac{1}{(3 + k)^{2/5}}$.

21. $\displaystyle\sum_{k=1}^{\infty} \frac{1}{9k - 2}$.

22. $\displaystyle\sum_{k=1}^{\infty} \frac{\ln k}{k}$.

23. $\displaystyle\sum_{k=1}^{\infty} \frac{\sqrt{k}}{k^3 + 1}$.

24. $\displaystyle\sum_{k=1}^{\infty} \frac{4}{2 + 3^k k}$.

25. $\displaystyle\sum_{k=1}^{\infty} \frac{1}{\sqrt{k(k + 1)}}$.

26. $\displaystyle\sum_{k=1}^{\infty} \frac{2 + (-1)^k}{5^k}$.

27. $\displaystyle\sum_{k=1}^{\infty} \frac{2 + \sqrt{k}}{(k + 1)^3 - 1}$.

28. $\displaystyle\sum_{k=1}^{\infty} \frac{4 + |\cos k|}{k^3}$.

29. $\displaystyle\sum_{k=1}^{\infty} \frac{1}{4 + 2^{-k}}$.

30. $\displaystyle\sum_{k=1}^{\infty} \frac{\sqrt{k} \ln k}{k^3 + 1}$.

31. $\displaystyle\sum_{k=1}^{\infty} \frac{\tan^{-1} k}{k^2}$.

32. $\displaystyle\sum_{k=1}^{\infty} \frac{5^k + k}{k! + 3}$.

33. $\displaystyle\sum_{k=1}^{\infty} \frac{\ln k}{k\sqrt{k}}$.

34. $\displaystyle\sum_{k=1}^{\infty} \frac{\cos (1/k)}{k^2}$.

35. Use the limit comparison test to show that the series
$$\sum_{k=1}^{\infty} (1 - \cos (1/k)) \text{ converges.}$$
$$\left[\text{\textit{Hint:} Compare with the series } \sum_{k=1}^{\infty} 1/k^2. \right]$$

36. Use the limit comparison test to show that the series
$$\sum_{k=1}^{\infty} \sin (\pi/k) \text{ diverges.}$$
$$\left[\text{\textit{Hint:} Compare with the series } \sum_{k=1}^{\infty} \pi/k. \right]$$

37. Use the comparison test to determine whether $\displaystyle\sum_{k=1}^{\infty} \frac{\ln k}{k^2}$ converges or diverges. [*Hint:* $\ln x < \sqrt{x}$ by Exercise 54 of Section 7.2.]

38. Determine whether $\displaystyle\sum_{k=2}^{\infty} \frac{1}{(\ln k)^2}$ converges or diverges. [*Hint:* See the hint in Exercise 37.]

39. Let a, b, and p be positive constants. For which values of p does the series $\displaystyle\sum_{k=1}^{\infty} \frac{1}{(a + bk)^p}$ converge?

40. (a) Show that $k^k \geq k!$ and use this result to prove that the series $\displaystyle\sum_{k=1}^{\infty} k^{-k}$ converges by the comparison test.

(b) Prove convergence using the root test.

41. Use the limit comparison test to investigate convergence of $\displaystyle\sum_{k=1}^{\infty} \frac{(k + 1)^2}{(k + 2)!}$.

42. Use the limit comparison test to investigate convergence of the series $1 + \frac{1}{3} + \frac{1}{5} + \frac{1}{7} + \cdots$.

43. Prove that $\sum_{k=1}^{\infty} 1/k!$ converges by comparison with a suitable geometric series.

44. Let Σa_k and Σb_k be series with positive terms. Prove:

(a) If $\displaystyle\lim_{k \to +\infty} (a_k/b_k) = 0$ and Σb_k converges, then Σa_k converges.

(b) If $\displaystyle\lim_{k \to +\infty} (a_k/b_k) = +\infty$ and Σb_k diverges, then Σa_k diverges.

11.7 ALTERNATING SERIES; CONDITIONAL CONVERGENCE

So far our emphasis has been on series with positive terms. In this section we shall discuss series containing both positive and negative terms.

□ **ALTERNATING SERIES**

Of special importance are series whose terms are alternately positive and negative. These are called **alternating series**. Some examples are

$$1 - 1 + 1 - 1 + \cdots + (-1)^{k+1} + \cdots$$

$$1 - \frac{1}{2} + \frac{1}{3} - \frac{1}{4} + \cdots + (-1)^{k+1} \frac{1}{k} + \cdots$$

$$-1 + \frac{1}{2!} - \frac{1}{3!} + \frac{1}{4!} - \cdots + (-1)^k \frac{1}{k!} + \cdots$$

In general, an alternating series has one of the following two forms:

$$\sum_{k=1}^{\infty} (-1)^{k+1} a_k = a_1 - a_2 + a_3 - a_4 + \cdots \tag{1}$$

$$\sum_{k=1}^{\infty} (-1)^{k} a_k = -a_1 + a_2 - a_3 + a_4 - \cdots \tag{2}$$

where the a_k's are assumed to be positive in both cases.

The following theorem is the key result on convergence of alternating series.

11.7.1 THEOREM (*Alternating Series Test*). *An alternating series of either form (1) or form (2) converges if the following two conditions are satisfied:*

(*a*) $a_1 > a_2 > a_3 > \cdots > a_k > \cdots$

(*b*) $\lim\limits_{k \to +\infty} a_k = 0$

Proof. We will consider only alternating series of form (1). The idea of the proof is to show that if conditions (*a*) and (*b*) hold, then the sequences of even-numbered and odd-numbered partial sums converge to a common limit S. It will then follow from Theorem 11.1.4 that the entire sequence of partial sums converges to S.

Figure 11.7.1 shows how successive partial sums satisfying conditions (*a*) and (*b*) appear when plotted on a horizontal axis. The even-numbered partial sums

$$s_2, s_4, s_6, s_8, \ldots, s_{2n}, \ldots$$

form an increasing sequence bounded above by a_1, and the odd-numbered partial sums

$$s_1, s_3, s_5, \ldots, s_{2n-1}, \ldots$$

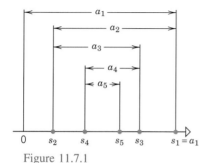

Figure 11.7.1

form a decreasing sequence bounded below by 0. Thus, by Theorems 11.2.2 and 11.2.3, the even-numbered partial sums converge to some limit S_E and the odd-numbered partial sums converge to some limit S_O. To complete the proof we must show that $S_E = S_O$. But the $(2n)$-th term in the series is $-a_{2n}$, so that $s_{2n} - s_{2n-1} = -a_{2n}$, which can be written as

$$s_{2n-1} = s_{2n} + a_{2n}$$

However, $2n \to +\infty$ and $2n - 1 \to +\infty$ as $n \to +\infty$, so that

$$S_O = \lim_{n \to +\infty} s_{2n-1} = \lim_{n \to +\infty} (s_{2n} + a_{2n}) = S_E + 0 = S_E$$

which completes the proof. ∎

REMARK.   As might be expected, it is not essential for condition (*a*) in the alternating series test to hold for all terms; an alternating series will converge if condition (*b*) is true and condition (*a*) holds eventually.

Example 1   Use the alternating series test to show that the following series converge.

(a) $\displaystyle\sum_{k=1}^{\infty} (-1)^{k+1} \frac{1}{k}$ (b) $\displaystyle\sum_{k=1}^{\infty} (-1)^{k+1} \frac{k+3}{k(k+1)}$

Solution (*a*).   The two conditions in the alternating series test are satisfied since

$$a_k = \frac{1}{k} > \frac{1}{k+1} = a_{k+1} \quad \text{and} \quad \lim_{k \to +\infty} a_k = \lim_{k \to +\infty} \frac{1}{k} = 0$$

Solution (*b*).   The two conditions in the alternating series test are satisfied, since

$$\frac{a_{k+1}}{a_k} = \frac{k+4}{(k+1)(k+2)} \cdot \frac{k(k+1)}{k+3} = \frac{k^2 + 4k}{k^2 + 5k + 6} = \frac{k^2 + 4k}{(k^2 + 4k) + (k + 6)} < 1$$

so

$$a_k > a_{k+1}$$

and

$$\lim_{k \to +\infty} a_k = \lim_{k \to +\infty} \frac{k+3}{k(k+1)} = \lim_{k \to +\infty} \frac{\dfrac{1}{k} + \dfrac{3}{k^2}}{1 + \dfrac{1}{k}} = 0 \quad \blacktriangleleft$$

REMARK.   The series in part (a) of the preceding example is called the ***alternating harmonic series***. It is important to keep in mind that this series converges, whereas the harmonic series diverges.

REMARK.   If an alternating series violates condition (*b*) of the alternating series test, then the series must diverge by the divergence test (11.4.1). However, if condition (*b*) is satisfied, but condition (*a*) is not, the series can either converge or diverge.*

☐ **APPROXIMATING SUMS OF ALTERNATING SERIES**

The following theorem is concerned with the error that results when the sum of an alternating series is approximated by a partial sum.

11.7.2   THEOREM.   *If an alternating series satisfies the conditions of the alternating series test, and if the sum S of the series is approximated by the nth partial sum s_n, thereby resulting in an error of $S - s_n$, then*

$$|S - s_n| < a_{n+1}$$

Moreover, the sign of the error is the same as that of the coefficient of a_{n+1} in the series.

---

*The interested reader will find some nice examples in an article by R. Lariviere, "On a Convergence Test for Alternating Series," *Mathematics Magazine*, Vol. 29, 1956, p. 88.

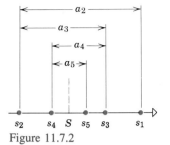

Figure 11.7.2

Proof. We shall prove the theorem for series of form (1). Referring to Figure 11.7.2 and keeping in mind our observation in the proof of Theorem 11.7.1 that the odd-numbered partial sums form a decreasing sequence converging to S and the even-numbered partial sums form an increasing sequence converging to S, we see that successive partial sums oscillate from one side of S to the other in smaller and smaller steps with the odd-numbered partial sums being greater than S and the even-numbered partial sums being less than S. Thus, the sum S falls between any two successive partial sums; that is, for any positive integer n

$$s_n < S < s_{n+1} \quad \text{or} \quad s_{n+1} < S < s_n$$

depending on whether n is even or odd. In either case,

$$|S - s_n| < |s_{n+1} - s_n| \tag{3}$$

But $s_{n+1} - s_n = \pm a_{n+1}$ (the sign depending on whether n is even or odd), so (3) yields $|S - s_n| < a_{n+1}$, which completes the proof. ∎

REMARK. In words, the preceding theorem states that for series satisfying the conditions of the alternating series test, the magnitude of the error that results from approximating the sum of the series by the nth partial sum is less than the magnitude of the first term of the series *after* the partial sum.

Example 2 As shown in Example 1, the alternating harmonic series

$$1 - \frac{1}{2} + \frac{1}{3} - \frac{1}{4} + \cdots + (-1)^{k+1} \frac{1}{k} + \cdots$$

satisfies the conditions of the alternating series test; hence, the series has a sum S, which we know must lie between any two successive partial sums. In particular, it must lie between

$$s_7 = 1 - \frac{1}{2} + \frac{1}{3} - \frac{1}{4} + \frac{1}{5} - \frac{1}{6} + \frac{1}{7} = \frac{319}{420}$$

and

$$s_8 = 1 - \frac{1}{2} + \frac{1}{3} - \frac{1}{4} + \frac{1}{5} - \frac{1}{6} + \frac{1}{7} - \frac{1}{8} = \frac{533}{840}$$

so

$$\frac{533}{840} < S < \frac{319}{420} \tag{4}$$

Later in this chapter we shall show that the sum S of the alternating harmonic series is $\ln 2$. If we accept this to be so for now, it follows from (4) that

$$\frac{533}{840} < \ln 2 < \frac{319}{420}$$

or with the help of a calculator,

$$0.6345 < \ln 2 < 0.7596$$

The value of $\ln 2$, rounded to four decimal places, is 0.6931, which is consistent with these inequalities. It follows from Theorem 11.7.2 that

$$\left| \ln 2 - s_7 \right| = \left| \ln 2 - \frac{319}{420} \right| < a_8 = \frac{1}{8}$$

and

$$\left| \ln 2 - s_8 \right| = \left| \ln 2 - \frac{533}{840} \right| < a_9 = \frac{1}{9} \quad \blacktriangleleft$$

ABSOLUTE AND
CONDITIONAL
CONVERGENCE

The series

$$1 - \frac{1}{2} - \frac{1}{2^2} + \frac{1}{2^3} + \frac{1}{2^4} - \frac{1}{2^5} - \frac{1}{2^6} + \cdots$$

does not fit in any of the categories studied so far—it has mixed signs, but is not alternating. We shall now develop some convergence tests that can be applied to such series.

11.7.3 DEFINITION.   A series

$$\sum_{k=1}^{\infty} u_k = u_1 + u_2 + \cdots + u_k + \cdots$$

is said to **converge absolutely** if the series of absolute values

$$\sum_{k=1}^{\infty} |u_k| = |u_1| + |u_2| + \cdots + |u_k| + \cdots$$

converges.

Example 3   The series

$$1 - \frac{1}{2} - \frac{1}{2^2} + \frac{1}{2^3} + \frac{1}{2^4} - \frac{1}{2^5} - \frac{1}{2^6} + \cdots$$

converges absolutely since the series of absolute values

$$1 + \frac{1}{2} + \frac{1}{2^2} + \frac{1}{2^3} + \frac{1}{2^4} + \frac{1}{2^5} + \frac{1}{2^6} + \cdots$$

is a convergent geometric series. On the other hand, the alternating harmonic series

$$1 - \frac{1}{2} + \frac{1}{3} - \frac{1}{4} + \frac{1}{5} - \cdots$$

does not converge absolutely since the series of absolute values

$$1 + \frac{1}{2} + \frac{1}{3} + \frac{1}{4} + \frac{1}{5} + \cdots$$

diverges.

Absolute convergence is of importance because of the following theorem.

11.7.4 THEOREM.   *If the series*

$$\sum_{k=1}^{\infty} |u_k| = |u_1| + |u_2| + \cdots + |u_k| + \cdots$$

converges, then so does the series

$$\sum_{k=1}^{\infty} u_k = u_1 + u_2 + \cdots + u_k + \cdots$$

In other words, if a series converges absolutely, then it converges.

Proof.   Our proof is based on a trick. We shall show that the series

$$\sum_{k=1}^{\infty} (u_k + |u_k|) \tag{5}$$

converges. Since $\Sigma |u_k|$ is assumed to converge, it will then follow from Theorem 11.4.3(a) that Σu_k converges, since

$$\sum_{k=1}^{\infty} u_k = \sum_{k=1}^{\infty} [(u_k + |u_k|) - |u_k|]$$

For all k, the value of $u_k + |u_k|$ is either 0 or $2|u_k|$, depending on whether u_k is negative or not. Thus, for all values of k

$$0 \leq u_k + |u_k| \leq 2|u_k| \tag{6}$$

But $\Sigma 2|u_k|$ is a convergent series since it is a constant times the convergent series $\Sigma |u_k|$. Thus, from (6), series (5) converges by the comparison test. ∎

Example 4   In Example 3 we showed that

$$1 - \frac{1}{2} - \frac{1}{2^2} + \frac{1}{2^3} + \frac{1}{2^4} - \frac{1}{2^5} - \frac{1}{2^6} + \cdots$$

converges absolutely. It follows from Theorem 11.7.4 that the series converges. ◀

Example 5   Show that the series $\displaystyle\sum_{k=1}^{\infty} \frac{\cos k}{k^2}$ converges.

Solution.   Since $|\cos k| \leq 1$ for all k,

$$\left| \frac{\cos k}{k^2} \right| \leq \frac{1}{k^2}$$

Thus,

$$\sum_{k=1}^{\infty} \left| \frac{\cos k}{k^2} \right|$$

converges by the comparison test, and consequently

$$\sum_{k=1}^{\infty} \frac{\cos k}{k^2}$$

converges.   ◀

If $\Sigma |u_k|$ *diverges*, no conclusion can be drawn about the convergence or divergence of Σu_k. For example, consider the two series

$$1 - \frac{1}{2} + \frac{1}{3} - \frac{1}{4} + \cdots + (-1)^{k+1} \frac{1}{k} + \cdots \tag{7}$$

$$-1 - \frac{1}{2} - \frac{1}{3} - \frac{1}{4} - \cdots - \frac{1}{k} - \cdots \tag{8}$$

Series (7), the alternating harmonic series, converges, whereas series (8), being a constant times the harmonic series, diverges. Yet in each case the series of absolute values is

$$1 + \frac{1}{2} + \frac{1}{3} + \cdots + \frac{1}{k} + \cdots$$

which diverges. A series such as (7), which is convergent, but not absolutely convergent, is called ***conditionally convergent***.

☐ **THE RATIO TEST FOR ABSOLUTE CONVERGENCE**

The following version of the ratio test is useful for investigating absolute convergence.

11.7.5 THEOREM (*Ratio Test for Absolute Convergence*). *Let* Σu_k *be a series with nonzero terms and suppose that*

$$\rho = \lim_{k \to +\infty} \frac{|u_{k+1}|}{|u_k|}$$

(a) *If* $\rho < 1$, *the series* Σu_k *converges absolutely and therefore converges.*

(b) *If* $\rho > 1$ *or if* $\rho = +\infty$, *then the series* Σu_k *diverges.*

(c) *If* $\rho = 1$, *no conclusion about convergence or absolute convergence can be drawn from this test.*

The proof is discussed in the exercises.

Example 6   The series

$$\sum_{k=1}^{\infty} (-1)^k \frac{2^k}{k!}$$

converges absolutely since

$$\rho = \lim_{k \to +\infty} \frac{|u_{k+1}|}{|u_k|} = \lim_{k \to +\infty} \frac{2^{k+1}}{(k+1)!} \cdot \frac{k!}{2^k} = \lim_{k \to +\infty} \frac{2}{k+1} = 0 < 1 \qquad \blacktriangleleft$$

The following review is included as a ready reference to convergence tests.

Review of Convergence Tests

NAME	STATEMENT	COMMENTS
Divergence Test (11.4.1)	If $\lim_{k \to +\infty} u_k \neq 0$, then Σu_k diverges.	If $\lim_{k \to +\infty} u_k = 0$, Σu_k may or may not converge.
Integral Test (11.4.4)	Let Σu_k be a series with positive terms, and let $f(x)$ be the function that results when k is replaced by x in the formula for u_k. If f is decreasing and continuous for $x \geq 1$, then $$\sum_{k=1}^{\infty} u_k \quad \text{and} \quad \int_{1}^{+\infty} f(x)\, dx$$ both converge or both diverge.	Use this test when $f(x)$ is easy to integrate. This test only applies to series that have positive terms.
Comparison Test (11.5.1)	Let Σa_k and Σb_k be series with nonnegative terms such that $$a_1 \leq b_1,\ a_2 \leq b_2, \ldots, a_k \leq b_k, \ldots$$ If Σb_k converges, then Σa_k converges, and if Σa_k diverges, then Σb_k diverges.	Use this test as a last resort; other tests are often easier to apply. This test only applies to series with nonnegative terms.

(continued on p. 554)

Review of Convergence Tests (Continued)

NAME	STATEMENT	COMMENTS
Ratio Test (11.5.2)	Let Σu_k be a series with positive terms and suppose that $$\rho = \lim_{k \to +\infty} \frac{u_{k+1}}{u_k}$$ (a) Series converges if $\rho < 1$. (b) Series diverges if $\rho > 1$ or $\rho = +\infty$. (c) No conclusion if $\rho = 1$.	Try this test when u_k involves factorials or kth powers.
Root Test (11.5.3)	Let Σu_k be a series with positive terms such that $$\rho = \lim_{k \to +\infty} \sqrt[k]{u_k}$$ (a) Series converges if $\rho < 1$. (b) Series diverges if $\rho > 1$ or $\rho = +\infty$. (c) No conclusion if $\rho = 1$.	Try this test when u_k involves kth powers.
Limit Comparison Test (11.6.3)	Let Σa_k and Σb_k be series with positive terms such that $$\rho = \lim_{k \to +\infty} \frac{a_k}{b_k}$$ If $0 < \rho < +\infty$, then both series converge or both diverge.	This is easier to apply than the comparison test, but still requires some skill in choosing the series Σb_k for comparison.
Alternating Series Test (11.7.1)	The series $$a_1 - a_2 + a_3 - a_4 + \cdots$$ $$-a_1 + a_2 - a_3 + a_4 - \cdots$$ converge if (a) $a_1 > a_2 > a_3 > \cdots$ (b) $\lim_{k \to +\infty} a_k = 0$	This test applies only to alternating series. It is assumed that $a_k > 0$ for all k.
Ratio Test for Absolute Convergence (11.7.5)	Let Σu_k be a series with nonzero terms such that $$\rho = \lim_{k \to +\infty} \frac{|u_{k+1}|}{|u_k|}$$ (a) Series converges absolutely if $\rho < 1$. (b) Series diverges if $\rho > 1$ or $\rho = +\infty$. (c) No conclusion if $\rho = 1$.	The series need not have positive terms and need not be alternating to use this test.

▶ Exercise Set 11.7   [C] 38–46, 56

In Exercises 1–6, determine whether the given alternating series converges or diverges.

1. $\displaystyle\sum_{k=1}^{\infty} \frac{(-1)^{k+1}}{2k+1}$.

2. $\displaystyle\sum_{k=1}^{\infty} (-1)^{k+1} \frac{k}{3^k}$.

3. $\displaystyle\sum_{k=1}^{\infty} (-1)^{k+1} \frac{k+1}{3k+1}$.

4. $\displaystyle\sum_{k=1}^{\infty} (-1)^{k+1} \frac{k+4}{k^2+k}$.

5. $\displaystyle\sum_{k=1}^{\infty} (-1)^{k+1} e^{-k}$.

6. $\displaystyle\sum_{k=3}^{\infty} (-1)^k \frac{\ln k}{k}$.

In Exercises 7–12, use the ratio test for absolute convergence to determine whether the series converges absolutely or diverges.

7. $\displaystyle\sum_{k=1}^{\infty} \left(-\frac{3}{5}\right)^k$.

8. $\displaystyle\sum_{k=1}^{\infty} (-1)^{k+1} \frac{2^k}{k!}$.

9. $\displaystyle\sum_{k=1}^{\infty} (-1)^{k+1} \frac{3^k}{k^2}$.

10. $\displaystyle\sum_{k=1}^{\infty} (-1)^k \left(\frac{k}{5^k}\right)$.

11. $\displaystyle\sum_{k=1}^{\infty} (-1)^k \left(\frac{k^3}{e^k}\right)$.

12. $\displaystyle\sum_{k=1}^{\infty} (-1)^{k+1} \frac{k^k}{k!}$.

In Exercises 13–30, classify the series as absolutely convergent, conditionally convergent, or divergent.

13. $\sum_{k=1}^{\infty} \frac{(-1)^{k+1}}{3k}$.

14. $\sum_{k=1}^{\infty} \frac{(-1)^{k+1}}{k^{4/3}}$.

15. $\sum_{k=1}^{\infty} \frac{(-4)^k}{k^2}$.

16. $\sum_{k=1}^{\infty} \frac{(-1)^{k+1}}{k!}$.

17. $\sum_{k=1}^{\infty} \frac{\cos k\pi}{k}$.

18. $\sum_{k=3}^{\infty} \frac{(-1)^k \ln k}{k}$.

19. $\sum_{k=1}^{\infty} (-1)^{k+1} \left(\frac{k+2}{3k-1} \right)^k$.

20. $\sum_{k=1}^{\infty} \frac{(-1)^{k+1}}{k^2+1}$.

21. $\sum_{k=1}^{\infty} (-1)^{k+1} \frac{k+2}{k(k+3)}$.

22. $\sum_{k=1}^{\infty} \frac{(-1)^{k+1}k^2}{k^3+1}$.

23. $\sum_{k=1}^{\infty} \sin \frac{k\pi}{2}$.

24. $\sum_{k=1}^{\infty} \frac{\sin k}{k^3}$.

25. $\sum_{k=2}^{\infty} \frac{(-1)^k}{k \ln k}$.

26. $\sum_{k=1}^{\infty} \frac{(-1)^k}{\sqrt{k(k+1)}}$.

27. $\sum_{k=2}^{\infty} \left(-\frac{1}{\ln k} \right)^k$.

28. $\sum_{k=1}^{\infty} \frac{(-1)^{k+1}}{\sqrt{k+1}+\sqrt{k}}$.

29. $\sum_{k=2}^{\infty} \frac{(-1)^k(k^2+1)}{k^3+2}$.

30. $\sum_{k=1}^{\infty} \frac{k \cos k\pi}{k^2+1}$.

In Exercises 31–34, the series satisfies the conditions of the alternating series test. For the stated value of n, use Theorem 11.7.2 to find an upper bound on the magnitude of the error that results if the sum of the series is approximated by the nth partial sum.

31. $\sum_{k=1}^{\infty} \frac{(-1)^{k+1}}{k}$; $n = 7$.

32. $\sum_{k=1}^{\infty} \frac{(-1)^{k+1}}{k!}$; $n = 5$.

33. $\sum_{k=1}^{\infty} \frac{(-1)^{k+1}}{\sqrt{k}}$; $n = 99$.

34. $\sum_{k=1}^{\infty} \frac{(-1)^{k+1}}{(k+1)\ln(k+1)}$; $n = 3$.

In Exercises 35–38, the series satisfies the conditions of the alternating series test. Use Theorem 11.7.2 to find a value of n for which the nth partial sum is ensured to approximate the sum of the series to the stated accuracy.

35. $\sum_{k=1}^{\infty} \frac{(-1)^{k+1}}{k}$; $|\text{error}| < 0.0001$.

36. $\sum_{k=1}^{\infty} \frac{(-1)^{k+1}}{k!}$; $|\text{error}| < 0.00001$.

37. $\sum_{k=1}^{\infty} \frac{(-1)^{k+1}}{\sqrt{k}}$; $|\text{error}| < 0.005$.

38. $\sum_{k=1}^{\infty} \frac{(-1)^{k+1}}{(k+1)\ln(k+1)}$; $|\text{error}| < 0.1$.

In Exercises 39 and 40, use Theorem 11.7.2 to find an upper bound on the magnitude of the error that results if s_{10} is used to approximate the sum of the given *geometric* series. Compute s_{10} rounded to four decimal places and compare this value with the exact sum of the series.

39. $\frac{3}{4} - \frac{3}{8} + \frac{3}{16} - \frac{3}{32} + \cdots$.

40. $1 - \frac{2}{3} + \frac{4}{9} - \frac{8}{27} + \cdots$.

In Exercises 41–44, the series satisfies the conditions of the alternating series test. Use Theorem 11.7.2 to find a value of n for which s_n is ensured to approximate the sum of the series with an error that is less than 10^{-4} in magnitude. Compute s_n rounded to five decimal places and compare this value with the sum of the series.

41. $\sin 1 = 1 - \frac{1}{3!} + \frac{1}{5!} - \frac{1}{7!} + \cdots$.

42. $\cos 1 = 1 - \frac{1}{2!} + \frac{1}{4!} - \frac{1}{6!} + \cdots$.

43. $\ln \frac{3}{2} = \frac{1}{1 \cdot 2} - \frac{1}{2 \cdot 2^2} + \frac{1}{3 \cdot 2^3} - \frac{1}{4 \cdot 2^4} + \cdots$.

44. $\frac{\pi}{16} = \frac{1}{1^5 + 4 \cdot 1} - \frac{1}{3^5 + 4 \cdot 3}$
$$+ \frac{1}{5^5 + 4 \cdot 5} - \frac{1}{7^5 + 4 \cdot 7} + \cdots .$$

45. For the series $\frac{\pi^2}{12} = 1 - \frac{1}{2^2} + \frac{1}{3^2} - \frac{1}{4^2} + \cdots$

   (a) use Theorem 11.7.2 to find a value of n for which s_n is ensured to approximate the sum of the series with an error that is less than 5×10^{-3} in magnitude

   (b) compute s_{10} and show that the magnitude of the error is less than 5×10^{-3}, thus showing that the value of n obtained from Theorem 11.7.2 is a conservative estimate.

46. For the series $\frac{\pi}{4} = 1 - \frac{1}{3} + \frac{1}{5} - \frac{1}{7} + \cdots$

   (a) use Theorem 11.7.2 to find a value of n for which s_n is ensured to approximate the sum of the series with an error that is less than 10^{-2} in magnitude

   (b) compute s_{26} and show that the magnitude of the error is less than 10^{-2}, thus showing that the value of n obtained from Theorem 11.7.2 is a conservative estimate.

47. Prove: If Σa_k converges absolutely, then Σa_k^2 converges.

48. Show that the converse of the result in Exercise 47 is false by finding a series for which Σa_k^2 converges, but $\Sigma |a_k|$ diverges.

49. Prove Theorem 11.7.1 for series of the form
$$-a_1 + a_2 - a_3 + a_4 - \cdots + (-1)^k a_k + \cdots$$

50. Prove Theorem 11.7.5. [*Hint:* Theorem 11.7.4 will help in part (*a*). For part (*b*), it may help to review the proof of Theorem 11.5.2.]

51. The sum of an absolutely convergent series is independent of the order in which the terms are added, but the terms of a conditionally convergent series can be rearranged to converge to any given value, or even diverge. For example, let S be the sum of the conditionally convergent alternating harmonic series,
$$S = 1 - \frac{1}{2} + \frac{1}{3} - \frac{1}{4} + \frac{1}{5} - \frac{1}{6} + \cdots$$

Rearrange the terms in this series to get
$$\left(1 - \frac{1}{2} - \frac{1}{4}\right) + \left(\frac{1}{3} - \frac{1}{6} - \frac{1}{8}\right) + \left(\frac{1}{5} - \frac{1}{10} - \frac{1}{12}\right) + \cdots$$

Show that this rearrangement results in a series that converges to $S/2$. [*Hint:* Add the first two terms within each pair of parentheses.]

52. Based on the discussion in Exercise 51, rearrange the terms in the convergent series
$$1 - \frac{1}{\sqrt{2}} + \frac{1}{\sqrt{3}} - \frac{1}{\sqrt{4}} + \frac{1}{\sqrt{5}} - \frac{1}{\sqrt{6}} + \cdots$$
as
$$\left(1 + \frac{1}{\sqrt{3}} - \frac{1}{\sqrt{2}}\right) + \left(\frac{1}{\sqrt{5}} + \frac{1}{\sqrt{7}} - \frac{1}{\sqrt{4}}\right) + \cdots$$
$$= \sum_{k=1}^{\infty} \left(\frac{1}{\sqrt{4k-3}} + \frac{1}{\sqrt{4k-1}} - \frac{1}{\sqrt{2k}}\right)$$

Show that this rearrangement results in a series that diverges to $+\infty$. [*Hint:* Note that

$$1/\sqrt{4k-3} > 1/\sqrt{4k} \quad \text{and} \quad 1/\sqrt{4k-1} > 1/\sqrt{4k}.$$

Show that the kth term in the rearranged series is greater than $(1 - 1/\sqrt{2})/\sqrt{k}$.]

In Exercises 53–55, use parts (*a*) and (*b*) of Theorem 11.4.3 and the fact that the sum of an absolutely convergent series is independent of the order in which the terms are added.

53. Given: $\dfrac{\pi^2}{6} = 1 + \dfrac{1}{2^2} + \dfrac{1}{3^2} + \dfrac{1}{4^2} + \cdots$.

Show: $\dfrac{\pi^2}{8} = 1 + \dfrac{1}{3^2} + \dfrac{1}{5^2} + \dfrac{1}{7^2} + \cdots$.

54. Given: $\dfrac{\pi^4}{90} = 1 + \dfrac{1}{2^4} + \dfrac{1}{3^4} + \dfrac{1}{4^4} + \cdots$.

Show: $\dfrac{\pi^4}{96} = 1 + \dfrac{1}{3^4} + \dfrac{1}{5^4} + \dfrac{1}{7^4} + \cdots$.

55. Given: $\dfrac{\pi^2}{6} = 1 + \dfrac{1}{2^2} + \dfrac{1}{3^2} + \dfrac{1}{4^2} + \cdots$.

Show: $\dfrac{\pi^2}{12} = 1 - \dfrac{1}{2^2} + \dfrac{1}{3^2} - \dfrac{1}{4^2} + \cdots$.

56. A small bug moves back and forth along a straight line as follows: it walks D units, stops and reverses direction, walks $D/2$ units, stops and reverses direction, walks $D/3$ units, stops and reverses direction, walks $D/4$ units, stops and reverses direction, and so forth, until it stops for the 1000th time. Given that $D = 180$ cm, find upper and lower bounds for

(a) the final distance between the bug and its starting point; [*Hint:* Use Theorem 11.7.2 and the fact that $1 - \frac{1}{2} + \frac{1}{3} - \frac{1}{4} + \cdots = \ln 2$.]

(b) the total distance traveled by the bug. [*Hint:* Use inequality (2) in Section 11.4.]

■ **11.8** POWER SERIES

> *In previous sections we studied series with constant terms. In this section we shall consider series whose terms involve variables. Such series are of fundamental importance in many branches of mathematics and the physical sciences.*

☐ **POWER SERIES IN** *x*

If $c_0, c_1, c_2, \ldots$ are constants and x is a variable, then a series of the form
$$\sum_{k=0}^{\infty} c_k x^k = c_0 + c_1 x + c_2 x^2 + \cdots + c_k x^k + \cdots \tag{1}$$

is called a *power series in x*. Some examples are

$$\sum_{k=0}^{\infty} x^k = 1 + x + x^2 + x^3 + \cdots$$

$$\sum_{k=0}^{\infty} \frac{x^k}{k!} = 1 + x + \frac{x^2}{2!} + \frac{x^3}{3!} + \cdots$$

$$\sum_{k=0}^{\infty} (-1)^k \frac{x^{2k}}{(2k)!} = 1 - \frac{x^2}{2!} + \frac{x^4}{4!} - \frac{x^6}{6!} + \cdots$$

If a numerical value is substituted for x in a power series $\Sigma c_k x^k$, then the resulting series of constants may either converge or diverge. The problem of determining those values of x for which a given power series converges is addressed by the following theorem whose proof is omitted.

11.8.1 THEOREM. *For any power series in x, exactly one of the following is true*:

(a) *The series converges only for $x = 0$.*

(b) *The series converges absolutely (and hence converges) for all real values of x.*

(c) *The series converges absolutely (and hence converges) for all x in some finite open interval $(-R, R)$, and diverges if $x < -R$ or $x > R$. At either of the points $x = R$ or $x = -R$ the series may converge absolutely, converge conditionally, or diverge, depending on the particular series.*

☐ RADIUS AND INTERVAL OF CONVERGENCE

Theorem 11.8.1 states that the set of values for which a power series in x converges is always an interval centered at 0; we call this the ***interval of convergence*** (Figure 11.8.1). In part (a) of Theorem 11.8.1 the interval of convergence reduces to a single point (the origin), in which case we say that the series has ***radius of convergence $R = 0$***; in part (b) the interval of convergence is infinite (the entire real line), in which case we say that the series has ***radius of convergence $R = +\infty$***; and in part (c) the interval extends between $-R$ and R, in which case we say that the series has ***radius of convergence R***.

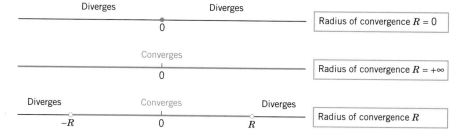

Figure 11.8.1

As illustrated in the following example, the main tool for finding the interval of convergence of a power series is the ratio test for absolute convergence (Theorem 11.7.5).

Example 1 Find the interval of convergence and radius of convergence of the following power series.

(a) $\displaystyle\sum_{k=0}^{\infty} x^k$ (b) $\displaystyle\sum_{k=0}^{\infty} \frac{x^k}{k!}$ (c) $\displaystyle\sum_{k=0}^{\infty} k! x^k$ (d) $\displaystyle\sum_{k=0}^{\infty} \frac{(-1)^k x^k}{3^k (k+1)}$

Solution (a). We shall apply the ratio test for absolute convergence. We have

$$\rho = \lim_{k \to +\infty} \left| \frac{u_{k+1}}{u_k} \right| = \lim_{k \to +\infty} \left| \frac{x^{k+1}}{x^k} \right| = \lim_{k \to +\infty} |x| = |x|$$

so the ratio test for absolute convergence implies that the series converges absolutely if $\rho = |x| < 1$ and diverges if $\rho = |x| > 1$. The test is inconclusive if $|x| = 1$ (i.e., if $x = 1$ or $x = -1$), so convergence at these points must be investigated separately. At these points the series becomes

$$\sum_{k=0}^{\infty} 1^k = 1 + 1 + 1 + 1 + \cdots \qquad \boxed{x = 1}$$

$$\sum_{k=0}^{\infty} (-1)^k = 1 - 1 + 1 - 1 + \cdots \qquad \boxed{x = -1}$$

both of which diverge; thus, the interval of convergence for the given power series is $(-1, 1)$, and the radius of convergence is $R = 1$.

Solution (b). Applying the ratio test for absolute convergence, we obtain

$$\rho = \lim_{k \to +\infty} \left| \frac{u_{k+1}}{u_k} \right| = \lim_{k \to +\infty} \left| \frac{x^{k+1}}{(k+1)!} \cdot \frac{k!}{x^k} \right| = \lim_{k \to +\infty} \left| \frac{x}{k+1} \right| = 0$$

Since $\rho < 1$ for all x, the series converges absolutely for all x. Thus, the interval of convergence is $(-\infty, +\infty)$ and the radius of convergence is $R = +\infty$.

Solution (c). If $x \neq 0$, then the ratio test for absolute convergence yields

$$\rho = \lim_{k \to +\infty} \left| \frac{u_{k+1}}{u_k} \right| = \lim_{k \to +\infty} \left| \frac{(k+1)! x^{k+1}}{k! x^k} \right| = \lim_{k \to +\infty} |(k+1)x| = +\infty$$

Therefore, the series diverges for all nonzero values of x. Consequently, the interval of convergence is the single point $x = 0$ and the radius of convergence is $R = 0$.

Solution (d). Since $|(-1)^k| = |(-1)^{k+1}| = 1$, we obtain

$$\rho = \lim_{k \to +\infty} \left| \frac{u_{k+1}}{u_k} \right| = \lim_{k \to +\infty} \left| \frac{x^{k+1}}{3^{k+1}(k+2)} \cdot \frac{3^k(k+1)}{x^k} \right|$$

$$= \lim_{k \to +\infty} \left[\frac{|x|}{3} \cdot \left(\frac{k+1}{k+2} \right) \right]$$

$$= \frac{|x|}{3} \lim_{k \to +\infty} \left(\frac{1 + 1/k}{1 + 2/k} \right) = \frac{|x|}{3}$$

The ratio test for absolute convergence implies that the series converges absolutely if $|x| < 3$ and diverges if $|x| > 3$. The ratio test fails to provide any information when $|x| = 3$, so the cases $x = -3$ and $x = 3$ need separate analyses. Substituting $x = -3$ in the given series yields

$$\sum_{k=0}^{\infty} \frac{(-1)^k(-3)^k}{3^k(k+1)} = \sum_{k=0}^{\infty} \frac{(-1)^k(-1)^k 3^k}{3^k(k+1)} = \sum_{k=0}^{\infty} \frac{1}{k+1}$$

which is the divergent harmonic series $1 + \frac{1}{2} + \frac{1}{3} + \frac{1}{4} + \cdots$. Substituting $x = 3$ in the given series yields

$$\sum_{k=0}^{\infty} \frac{(-1)^k 3^k}{3^k(k+1)} = \sum_{k=0}^{\infty} \frac{(-1)^k}{k+1} = 1 - \frac{1}{2} + \frac{1}{3} - \frac{1}{4} + \cdots$$

which is the conditionally convergent alternating harmonic series. Thus, the interval of convergence for the given series is $(-3, 3]$ and the radius of convergence is $R = 3$. ◀

☐ **POWER SERIES IN** $x - a$

If a is a constant and $x - a$ is substituted for x in (1), then the resulting series,

$$\sum_{k=0}^{\infty} c_k(x - a)^k = c_0 + c_1(x - a) + c_2(x - a)^2 + \cdots + c_k(x - a)^k + \cdots$$

is called a **power series in x − a**. Some examples are

$$\sum_{k=0}^{\infty} \frac{(x-1)^k}{k+1} = 1 + \frac{(x-1)}{2} + \frac{(x-1)^2}{3} + \frac{(x-1)^3}{4} + \cdots \qquad \boxed{a = 1}$$

$$\sum_{k=0}^{\infty} \frac{(-1)^k (x+3)^k}{k!} = 1 - (x+3) + \frac{(x+3)^2}{2!} - \frac{(x+3)^3}{3!} + \cdots \qquad \boxed{a = -3}$$

The first of these is a power series in $x - 1$ and the second is a power series in $x + 3$. Note that a power series in x is the special case of a power series in $x - a$ in which $a = 0$.

The conditions for convergence of a power series in $x - a$ can be obtained by substituting $x - a$ for x in Theorem 11.8.1. This leads to the following theorem.

11.8.2   THEOREM.   *For a power series $\Sigma c_k (x - a)^k$, exactly one of the following is true:*

(a)   *The series converges only for $x = a$.*

(b)   *The series converges absolutely (and hence converges) for all real values of x.*

(c)   *The series converges absolutely (and hence converges) for all x in some finite open interval $(a - R, a + R)$ and diverges if $x < a - R$ or $x > a + R$. At either of the points $x = a - R$ or $x = a + R$, the series may converge absolutely, converge conditionally, or diverge, depending on the particular series.*

It follows from the preceding theorem that the set of values for which a power series in $x - a$ converges is always an interval centered at $x = a$; we call this the **interval of convergence** (Figure 11.8.2). In part (a) of Theorem 11.8.2 the interval of convergence reduces to the single point $x = a$, in which case we say that the series has **radius of convergence $R = 0$**; and in part (b) the interval of convergence is infinite (the entire real line), in which case we say that the series has **radius of convergence $R = +\infty$**; and in part (c) the interval extends between $a - R$ and $a + R$, in which case we say that the series has **radius of convergence R**.

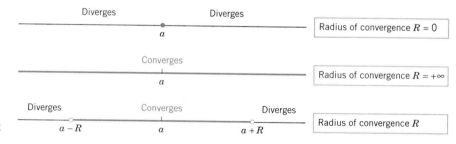

Figure 11.8.2

Example 2   Find the interval of convergence and radius of convergence of the series

$$\sum_{k=1}^{\infty} \frac{(x-5)^k}{k^2}$$

Solution.   We apply the ratio test for absolute convergence.

$$\rho = \lim_{k \to +\infty} \left| \frac{u_{k+1}}{u_k} \right| = \lim_{k \to +\infty} \left| \frac{(x-5)^{k+1}}{(k+1)^2} \cdot \frac{k^2}{(x-5)^k} \right|$$

$$= \lim_{k \to +\infty} \left[|x - 5| \left(\frac{k}{k+1} \right)^2 \right]$$

$$= |x - 5| \lim_{k \to +\infty} \left(\frac{1}{1 + 1/k} \right)^2 = |x - 5|$$

Thus, the series converges absolutely if $|x - 5| < 1$, or $-1 < x - 5 < 1$, or $4 < x < 6$. The series diverges if $x < 4$ or $x > 6$.

To determine the convergence behavior at the endpoints $x = 4$ and $x = 6$, we substitute these values in the given series. If $x = 6$, the series becomes

$$\sum_{k=1}^{\infty} \frac{1^k}{k^2} = \sum_{k=1}^{\infty} \frac{1}{k^2} = 1 + \frac{1}{2^2} + \frac{1}{3^2} + \frac{1}{4^2} + \cdots$$

which is a convergent p-series $(p = 2)$. If $x = 4$, the series becomes

$$\sum_{k=1}^{\infty} \frac{(-1)^k}{k^2} = -1 + \frac{1}{2^2} - \frac{1}{3^2} + \frac{1}{4^2} - \cdots$$

Since this series converges absolutely, the interval of convergence for the given series is $[4, 6]$. The radius of convergence is $R = 1$ (Figure 11.8.3). ◀

REMARK.   The ratio test should never be used for testing for convergence at the endpoints of the interval of convergence, since the ratio ρ is always 1 at those points (why?).

Figure 11.8.3
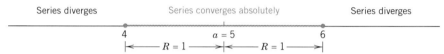

▶ Exercise Set 11.8

In Exercises 1–24, find the radius of convergence and the interval of convergence.

1. $\displaystyle\sum_{k=0}^{\infty} \frac{x^k}{k+1}$.

2. $\displaystyle\sum_{k=0}^{\infty} 3^k x^k$.

3. $\displaystyle\sum_{k=0}^{\infty} \frac{(-1)^k x^k}{k!}$.

4. $\displaystyle\sum_{k=0}^{\infty} \frac{k!}{2^k} x^k$.

5. $\displaystyle\sum_{k=1}^{\infty} \frac{5^k}{k^2} x^k$.

6. $\displaystyle\sum_{k=2}^{\infty} \frac{x^k}{\ln k}$.

7. $\displaystyle\sum_{k=1}^{\infty} \frac{x^k}{k(k+1)}$.

8. $\displaystyle\sum_{k=0}^{\infty} \frac{(-2)^k x^{k+1}}{k+1}$.

9. $\displaystyle\sum_{k=1}^{\infty} (-1)^{k-1} \frac{x^k}{\sqrt{k}}$.

10. $\displaystyle\sum_{k=0}^{\infty} \frac{(-1)^k x^{2k}}{(2k)!}$.

11. $\displaystyle\sum_{k=0}^{\infty} (-1)^k \frac{x^{2k+1}}{(2k+1)!}$.

12. $\displaystyle\sum_{k=1}^{\infty} (-1)^k \frac{x^{3k}}{k^{3/2}}$.

13. $\displaystyle\sum_{k=0}^{\infty} \frac{3^k}{k!} x^k$.

14. $\displaystyle\sum_{k=2}^{\infty} (-1)^{k+1} \frac{x^k}{k(\ln k)^2}$.

15. $\displaystyle\sum_{k=0}^{\infty} \frac{x^k}{1 + k^2}$.

16. $\displaystyle\sum_{k=0}^{\infty} \frac{(x-3)^k}{2^k}$.

17. $\displaystyle\sum_{k=1}^{\infty} (-1)^{k+1} \frac{(x+1)^k}{k}$.

18. $\displaystyle\sum_{k=0}^{\infty} (-1)^k \frac{(x-4)^k}{(k+1)^2}$.

19. $\displaystyle\sum_{k=0}^{\infty} \left(\frac{3}{4}\right)^k (x+5)^k$.

20. $\displaystyle\sum_{k=1}^{\infty} \frac{(2k+1)!}{k^3} (x-2)^k$.

21. $\displaystyle\sum_{k=1}^{\infty} (-1)^k \frac{(x+1)^{2k+1}}{k^2+4}$.

22. $\displaystyle\sum_{k=1}^{\infty} \frac{(\ln k)(x-3)^k}{k}$.

23. $\displaystyle\sum_{k=0}^{\infty} \frac{\pi^k (x-1)^{2k}}{(2k+1)!}$.

24. $\displaystyle\sum_{k=0}^{\infty} \frac{(2x-3)^k}{4^{2k}}$.

In Exercises 25–27, show the first four terms of the series, and find the radius of convergence.

25. $\displaystyle\sum_{k=1}^{\infty} \frac{1 \cdot 2 \cdot 3 \cdots k}{1 \cdot 4 \cdot 7 \cdots (3k-2)} x^k$.

26. $\displaystyle\sum_{k=1}^{\infty} (-1)^k \frac{1 \cdot 2 \cdot 3 \cdots k}{1 \cdot 3 \cdot 5 \cdots (2k-1)} x^{2k+1}$.

27. $\displaystyle\sum_{k=1}^{\infty} \frac{1 \cdot 3 \cdot 5 \cdots (2k-1)}{(2k-2)!} x^k$.

28. Use the root test to find the interval of convergence of
$$\sum_{k=2}^{\infty} \frac{x^k}{(\ln k)^k}.$$

29. Find the interval of convergence of $\displaystyle\sum_{k=0}^{\infty} \frac{(x-a)^k}{b^k}$, where $b > 0$.

30. Find the radius of convergence of the power series
$$\sum_{k=0}^{\infty} \frac{(pk)!}{(k!)^p} x^k, \text{ where } p \text{ is a positive integer.}$$

31. Find the radius of convergence of the power series
$$\sum_{k=0}^{\infty} \frac{(k+p)!}{k!(k+q)!} x^k, \text{ where } p \text{ and } q \text{ are positive integers.}$$

32. Prove: If $\displaystyle\lim_{k \to +\infty} |c_k|^{1/k} = L$, where $L \neq 0$, then $1/L$ is the radius of convergence of the power series $\sum_{k=0}^{\infty} c_k x^k$.

33. Prove: If the power series $\sum_{k=0}^{\infty} c_k x^k$ has radius of convergence R, then the series $\sum_{k=0}^{\infty} c_k x^{2k}$ has radius of convergence $\sqrt{R}$.

34. Prove: If the interval of convergence of the series $\sum_{k=0}^{\infty} c_k(x - a)^k$ is $(a - R, a + R]$, then the series converges conditionally at $a + R$.

■ **11.9** TAYLOR AND MACLAURIN SERIES

One of the early applications of calculus was the computation of approximate numerical values for functions such as $\sin x$, $\ln x$, and e^x. One common method for obtaining such values is to approximate the function by a polynomial, then use that polynomial to compute the desired numerical values. In this section we shall discuss procedures for approximating functions by polynomials, and in the next section we shall investigate the errors in these approximations. In this section we shall also see how polynomial approximations lead naturally to the important problem of finding a power series that converges to a specified function.

□ **APPROXIMATING FUNCTIONS BY POLYNOMIALS**

The problem of primary interest in this section can be phrased informally as follows:

PROBLEM.   *Given a function f and a point a on the x-axis, find a polynomial of specified degree that best approximates the function f in the "vicinity" of the point a.*

As stated, the problem is somewhat vague in that we have not specified any requirements on f such as continuity, differentiability, and so forth, and it is not evident what we mean by the "best approximation in the vicinity of a point." However, the problem is suggestive enough to get us started, and we shall resolve the ambiguities as we progress.

Suppose that we are interested in approximating a function f in the vicinity of the point $a = 0$ by a polynomial

$$p(x) = c_0 + c_1 x + \cdots + c_n x^n \tag{1}$$

Because $p(x)$ has $n + 1$ coefficients, it seems reasonable that we should be able to impose $n + 1$ conditions on this polynomial to achieve a good approximation to $f(x)$. Because the point $a = 0$ is the center of interest, our strategy will be to choose the coefficients of $p(x)$ so that the value of p and its first n derivatives are the same as the value of f and its first n derivatives at $a = 0$. By forcing this high degree of "match" at $a = 0$, it is reasonable to hope that $f(x)$ and $p(x)$ will remain close over some interval (possibly quite small) centered at $a = 0$. Thus, we shall assume that f can be differentiated n times at 0, and we shall try to find the coefficients in (1) such that

$$f(0) = p(0), \quad f'(0) = p'(0), \quad f''(0) = p''(0), \quad \ldots, \quad f^{(n)}(0) = p^{(n)}(0) \tag{2}$$

We have

$$p(x) \quad = c_0 + c_1 x + c_2 x^2 + c_3 x^3 + \cdots + c_n x^n$$

$$p'(x) \quad = c_1 + 2c_2 x + 3c_3 x^2 + \cdots + nc_n x^{n-1}$$

$$p''(x) \quad = 2c_2 + 3 \cdot 2c_3 x + \cdots + n(n-1)c_n x^{n-2}$$

$$p'''(x) \quad = 3 \cdot 2c_3 + \cdots + n(n-1)(n-2)c_n x^{n-3}$$

$$\vdots$$

$$p^{(n)}(x) = n(n-1)(n-2) \cdots (1)c_n$$

Thus, to satisfy (2) we must have

$$f(0) \quad = p(0) \quad = c_0$$

$$f'(0) \quad = p'(0) \quad = c_1$$

$$f''(0) \quad = p''(0) \quad = 2c_2 = 2!c_2$$

$$f'''(0) \quad = p'''(0) \quad = 3 \cdot 2c_3 = 3!c_3$$

$$\vdots$$

$$f^{(n)}(0) = p^{(n)}(0) = n(n-1)(n-2)\cdots(1)c_n = n!c_n$$

which yields the following values for the coefficients of $p(x)$:

$$c_0 = f(0), \quad c_1 = f'(0), \quad c_2 = \frac{f''(0)}{2!}, \quad c_3 = \frac{f'''(0)}{3!}, \quad \ldots, \quad c_n = \frac{f^{(n)}(0)}{n!}$$

☐ **MACLAURIN**
POLYNOMIALS

The polynomial that results by using these coefficients in (1) is called the *nth Maclaurin**
polynomial for f. In summary, we have the following definition.

11.9.1   DEFINITION.   If f can be differentiated n times at 0, then we define the **nth**
Maclaurin polynomial for f to be

$$p_n(x) = f(0) + f'(0)x + \frac{f''(0)}{2!}x^2 + \frac{f'''(0)}{3!}x^3 + \cdots + \frac{f^{(n)}(0)}{n!}x^n \tag{3}$$

This polynomial has the property that its value and the values of its first n derivatives
match the value of $f(x)$ and its first n derivatives when $x = 0$.

Example 1   Find the Maclaurin polynomials p_0, p_1, p_2, p_3, and p_n for e^x.

Solution.   Let $f(x) = e^x$. Thus,

$$f'(x) = f''(x) = f'''(x) = \cdots = f^{(n)}(x) = e^x$$

and

$$f(0) = f'(0) = f''(0) = f'''(0) = \cdots = f^{(n)}(0) = e^0 = 1$$

---

*COLIN MACLAURIN (1698–1746). Scottish mathematician. Maclaurin's father, a minister, died when the boy was
only six months old, and his mother when he was nine years old. He was then raised by an uncle who was also a
minister. Maclaurin entered Glasgow University as a divinity student, but transferred to mathematics after one
year. He received his Master's degree at age 17 and, in spite of his youth, began teaching at Marischal College in
Aberdeen, Scotland. He met Isaac Newton during a visit to London in 1719 and from that time on became
Newton's disciple. During that era, some of Newton's analytic methods were bitterly attacked by major
mathematicians and much of Maclaurin's important mathematical work resulted from his efforts to defend
Newton's ideas geometrically. Maclaurin's work, *A Treatise of Fluxions* (1742), was the first systematic
formulation of Newton's methods. The treatise was so carefully done that it was a standard of mathematical rigor
in calculus until the work of Cauchy in 1821.

Maclaurin was an outstanding experimentalist. He devised numerous ingenious mechanical devices, made
important astronomical observations, performed actuarial computations for insurance societies, and helped to
improve maps of the islands around Scotland.

Therefore,

$$p_0(x) = f(0) = 1$$

$$p_1(x) = f(0) + f'(0)x = 1 + x$$

$$p_2(x) = f(0) + f'(0)x + \frac{f''(0)}{2!}x^2 = 1 + x + \frac{x^2}{2!} = 1 + x + \frac{1}{2}x^2$$

$$p_3(x) = f(0) + f'(0)x + \frac{f''(0)}{2!}x^2 + \frac{f'''(0)}{3!}x^3$$

$$= 1 + x + \frac{x^2}{2!} + \frac{x^3}{3!} = 1 + x + \frac{1}{2}x^2 + \frac{1}{6}x^3$$

$$p_n(x) = f(0) + f'(0)x + \frac{f''(0)}{2!}x^2 + \cdots + \frac{f^{(n)}(0)}{n!}x^n$$

$$= 1 + x + \frac{x^2}{2!} + \cdots + \frac{x^n}{n!} \quad \blacktriangleleft$$

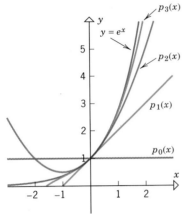

Figure 11.9.1

Figure 11.9.1 shows the graphs of e^x (in blue) and the graphs of the first four Maclaurin polynomials. Note that the graphs of $p_1(x)$, $p_2(x)$, and $p_3(x)$ are virtually indistinguishable from the graph of e^x near the origin, so that these polynomials are good approximations of e^x for x near 0. However, the farther x is from 0, the poorer these approximations become. This is typical of the Maclaurin polynomials for a function $f(x)$; they provide good approximations of $f(x)$ near 0, but the accuracy diminishes as x progresses away from 0. (In the next section we shall investigate the accuracy of such approximations.)

To obtain polynomial approximations of $f(x)$ that have their best accuracy near a general point $x = a$, it will be convenient to express the polynomials in powers of $x - a$, so that they have the form

$$p(x) = c_0 + c_1(x - a) + c_2(x - a)^2 + \cdots + c_n(x - a)^n \tag{4}$$

Since the point $x = a$ is the center of interest, we want to choose the coefficients so that the values of $p(x)$ and its first n derivatives match the values of $f(x)$ and its first n derivatives at a. We leave it as an exercise for the reader to show that if f can be differentiated n times at a, then the desired coefficients are

$$c_0 = f(a), \quad c_1 = f'(a), \quad c_2 = \frac{f''(a)}{2!}, \quad c_3 = \frac{f'''(a)}{3!}, \quad \ldots, \quad c_n = \frac{f^{(n)}(a)}{n!}$$

Substituting these values in (4) we obtain a polynomial called the *nth Taylor* polynomial about $x = a$ for f.*

---

*BROOK TAYLOR (1685–1731). English mathematician. Taylor was born of well-to-do parents. Musicians and artists were entertained frequently in the Taylor home, which undoubtedly had a lasting influence on young Brook. In later years, Taylor published a definitive work on the mathematical theory of perspective and obtained major mathematical results about the vibrations of strings. There also exists an unpublished work, On Musick, that was intended to be part of a joint paper with Isaac Newton. Taylor's life was scarred with unhappiness, illness, and tragedy. Because his first wife was not rich enough to suit his father, the two men argued bitterly and parted ways. Subsequently, his wife died in childbirth. Then, after he remarried, his second wife also died in childbirth, though his daughter survived. Taylor's most productive period was from 1714 to 1719, during which time he wrote on a wide range of subjects—magnetism, capillary action, thermometers, perspective, and calculus. In his final years, Taylor devoted his writing efforts to religion and philosophy. According to Taylor, the results that bear his name were motivated by coffeehouse conversations about works of Newton on planetary motion and works of Halley (''Halley's comet'') on roots of polynomials.

Taylor's writing style was so terse and hard to understand that he never received credit for many of his innovations.

11.9.2   DEFINITION.   If f can be differentiated n times at a, then we define the **nth Taylor polynomial for f about $x = a$** to be

$$p_n(x) = f(a) + f'(a)(x - a) + \frac{f''(a)}{2!}(x - a)^2$$

$$+ \frac{f'''(a)}{3!}(x - a)^3 + \cdots + \frac{f^{(n)}(a)}{n!}(x - a)^n \quad (5)$$

REMARK.   Observe that with $a = 0$, the nth Taylor polynomial for f is the nth Maclaurin polynomial for f; that is, the Maclaurin polynomials are special cases of the Taylor polynomials.

Example 2   Find the first four Taylor polynomials for $\ln x$ about $x = 2$.

Solution.   Let $f(x) = \ln x$. Thus,

$$\begin{aligned}
f(x) &= \ln x & f(2) &= \ln 2 \\
f'(x) &= 1/x & f'(2) &= 1/2 \\
f''(x) &= -1/x^2 & f''(2) &= -1/4 \\
f'''(x) &= 2/x^3 & f'''(2) &= 1/4
\end{aligned}$$

Substituting in (5) with $a = 2$ yields

$$p_0(x) = f(2) = \ln 2$$

$$p_1(x) = f(2) + f'(2)(x - 2) = \ln 2 + \tfrac{1}{2}(x - 2)$$

$$p_2(x) = f(2) + f'(2)(x - 2) + \frac{f''(2)}{2!}(x - 2)^2 = \ln 2 + \tfrac{1}{2}(x - 2) - \tfrac{1}{8}(x - 2)^2$$

$$p_3(x) = f(2) + f'(2)(x - 2) + \frac{f''(2)}{2!}(x - 2)^2 + \frac{f'''(2)}{3!}(x - 2)^3$$

$$= \ln 2 + \tfrac{1}{2}(x - 2) - \tfrac{1}{8}(x - 2)^2 + \tfrac{1}{24}(x - 2)^3$$

The graph of $\ln x$ (in blue) and its first four Taylor polynomials about $x = 2$ are shown in Figure 11.9.2. As expected, these polynomials produce their best approximations of $\ln x$ near 2.   ◀

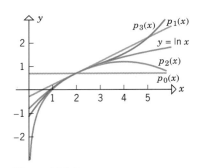

Figure 11.9.2

□ **SIGMA NOTATION FOR TAYLOR AND MACLAURIN POLYNOMIALS**

Frequently, it is convenient to express the defining formula for the Taylor polynomial in sigma notation. To do this, we use the notation $f^{(k)}(a)$ to denote the kth derivative of f at $x = a$, and we make the added convention that $f^{(0)}(a)$ denotes $f(a)$. This enables us to write

$$\sum_{k=0}^{n} \frac{f^{(k)}(a)}{k!}(x - a)^k = f(a) + f'(a)(x - a) + \frac{f''(a)}{2!}(x - a)^2 + \cdots + \frac{f^{(n)}(a)}{n!}(x - a)^n$$

In particular, the nth Maclaurin polynomial for $f(x)$ can be written as

$$\sum_{k=0}^{n} \frac{f^{(k)}(0)}{k!}x^k = f(0) + f'(0)x + \frac{f''(0)}{2!}x^2 + \cdots + \frac{f^{(n)}(0)}{n!}x^n \quad (6)$$

□ **TAYLOR AND MACLAURIN SERIES**

For a fixed value of x near a, one would expect that the approximation of $f(x)$ by its Taylor polynomial $p_n(x)$ about $x = a$ should improve as n increases, since increasing n has the effect of matching higher and higher derivatives of $f(x)$ with those of $p_n(x)$ at $x = a$. Indeed, its seems plausible that one might be able to achieve any desired degree of accuracy by choosing n sufficiently large; that is, the values of $p_n(x)$ might actually converge to $f(x)$ as $n \to +\infty$. Should this happen, we would have

$$f(x) = \lim_{n \to +\infty} \sum_{k=0}^{n} \frac{f^{(k)}(a)}{k!} (x-a)^k = \sum_{k=0}^{\infty} \frac{f^{(k)}(a)}{k!} (x-a)^k$$

In the next section we shall study conditions under which the series on the right actually converges to $f(x)$. For the remainder of this section, we will focus on the computational aspects of finding these series. We make the following definition.

11.9.3 DEFINITION.   If f has derivatives of all orders at a, then we define the **Taylor series for f about x = a** to be

$$\sum_{k=0}^{\infty} \frac{f^{(k)}(a)}{k!} (x-a)^k = f(a) + f'(a)(x-a)$$

$$+ \frac{f''(a)}{2!} (x-a)^2 + \cdots + \frac{f^{(k)}(a)}{k!} (x-a)^k + \cdots \quad (7)$$

In the special case where $a = 0$, the Taylor series for f is called the **Maclaurin series for f**. In this case the series has the form

$$\sum_{k=0}^{\infty} \frac{f^{(k)}(0)}{k!} x^k = f(0) + f'(0)x + \frac{f''(0)}{2!} x^2 + \cdots + \frac{f^{(k)}(0)}{k!} x^k + \cdots \quad (8)$$

REMARK.   Because the summation index in the Taylor series for f about $x = a$ starts at 0, it is convenient to think of the initial term in the series as the zeroth term. With this convention the nth term in the series is the term involving $(x - a)^n$. It then follows that the nth partial sum of the Taylor series is the nth Taylor polynomial.

Example 3   Find the Maclaurin series for

(a) e^x (b) $\sin x$ (c) $\cos x$

Solution (*a*).   In Example 1 we found the nth Maclaurin polynomial for the function e^x to be

$$\sum_{k=0}^{n} \frac{x^k}{k!} = 1 + x + \frac{x^2}{2!} + \cdots + \frac{x^n}{n!}$$

Thus, the Maclaurin series for e^x is

$$\sum_{k=0}^{\infty} \frac{x^k}{k!} = 1 + x + \frac{x^2}{2!} + \frac{x^3}{3!} + \cdots + \frac{x^k}{k!} + \cdots$$

Solution (*b*).   In the Maclaurin polynomials for $\sin x$, only the odd powers of x appear explicitly. To see this, let $f(x) = \sin x$; thus,

$$f(x) \ = \sin x \qquad f(0) \ = 0$$

$$f'(x) = \cos x \qquad f'(0) = 1$$

$$f''(x) = -\sin x \qquad f''(0) = 0$$

$$f'''(x) = -\cos x \qquad f'''(0) = -1$$

Since $f^{(4)}(x) = \sin x = f(x)$, the pattern $0, 1, 0, -1$ will repeat over and over as we evaluate successive derivatives at 0. Therefore, the successive Maclaurin polynomials for $\sin x$ are

$$p_0(x) = 0$$

$$p_1(x) = 0 + x$$

$$p_2(x) = 0 + x + 0$$

$$p_3(x) = 0 + x + 0 - \frac{x^3}{3!}$$

$$p_4(x) = 0 + x + 0 - \frac{x^3}{3!} + 0$$

$$p_5(x) = 0 + x + 0 - \frac{x^3}{3!} + 0 + \frac{x^5}{5!}$$

$$p_6(x) = 0 + x + 0 - \frac{x^3}{3!} + 0 + \frac{x^5}{5!} + 0$$

$$p_7(x) = 0 + x + 0 - \frac{x^3}{3!} + 0 + \frac{x^5}{5!} + 0 - \frac{x^7}{7!}$$

$$\vdots$$

Because of the zero terms, each even-numbered Maclaurin polynomial [after $p_0(x)$] is the same as the preceding odd-numbered Maclaurin polynomial; that is,

$$p_{2n+1}(x) = p_{2n+2}(x) = x - \frac{x^3}{3!} + \frac{x^5}{5!} - \frac{x^7}{7!} + \cdots + (-1)^n \frac{x^{2n+1}}{(2n+1)!} \quad (n = 0, 1, 2, \ldots)$$

Thus, the Maclaurin series for $\sin x$ is

$$\sum_{k=0}^{\infty} (-1)^k \frac{x^{2k+1}}{(2k+1)!} = x - \frac{x^3}{3!} + \frac{x^5}{5!} - \frac{x^7}{7!} + \cdots + (-1)^k \frac{x^{2k+1}}{(2k+1)!} + \cdots$$

The graphs of $\sin x$, $p_1(x)$, $p_3(x)$, $p_5(x)$, and $p_7(x)$ are shown in Figure 11.9.3.

Solution (c).   In the Maclaurin polynomials for $\cos x$, only the even powers of x appear explicitly; the computations are similar to those in part (b). The reader should be able to show that

$$p_0(x) = p_1(x) = 1$$

$$p_2(x) = p_3(x) = 1 - \frac{x^2}{2!}$$

$$p_4(x) = p_5(x) = 1 - \frac{x^2}{2!} + \frac{x^4}{4!}$$

$$p_6(x) = p_7(x) = 1 - \frac{x^2}{2!} + \frac{x^4}{4!} - \frac{x^6}{6!}$$

In general, the Maclaurin polynomials for $\cos x$ are

$$p_{2n}(x) = p_{2n+1}(x) = 1 - \frac{x^2}{2!} + \frac{x^4}{4!} - \cdots + (-1)^n \frac{x^{2n}}{(2n)!} \quad (n = 0, 1, 2, \ldots)$$

from which it follows that the Maclaurin series for $\cos x$ is

$$\sum_{k=0}^{\infty} (-1)^k \frac{x^{2k}}{(2k)!} = 1 - \frac{x^2}{2!} + \frac{x^4}{4!} - \frac{x^6}{6!} + \cdots + (-1)^k \frac{x^{2k}}{(2k)!} + \cdots$$

The graphs of $\cos x$, $p_0(x)$, $p_2(x)$, $p_4(x)$, and $p_6(x)$ are shown in Figure 11.9.4.   ◄

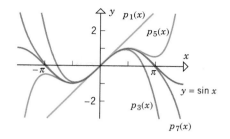

Figure 11.9.3

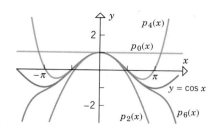

Figure 11.9.4

Example 4   Find the Taylor series about $x = 1$ for $1/x$.

Solution.   Let $f(x) = 1/x$ so that

$$f(x) \quad = \frac{1}{x} \qquad\qquad f(1) \quad = 1$$

$$f'(x) \quad = -\frac{1}{x^2} \qquad\qquad f'(1) \quad = -1$$

$$f''(x) \quad = \frac{2}{x^3} \qquad\qquad f''(1) \quad = 2!$$

$$f'''(x) \quad = -\frac{3 \cdot 2}{x^4} \qquad\qquad f'''(1) \quad = -3!$$

$$f^{(4)}(x) = \frac{4 \cdot 3 \cdot 2}{x^5} \qquad\qquad f^{(4)}(1) = 4!$$

$$\vdots \qquad\qquad\qquad\qquad \vdots$$

$$f^{(k)}(x) = (-1)^k \frac{k!}{x^{k+1}} \qquad f^{(k)}(1) = (-1)^k k!$$

$$\vdots \qquad\qquad\qquad\qquad \vdots$$

Thus, substituting in (7) with $a = 1$ yields

$$\sum_{k=0}^{\infty} \frac{(-1)^k k!}{k!} (x - 1)^k = \sum_{k=0}^{\infty} (-1)^k (x - 1)^k$$

$$= 1 - (x - 1) + (x - 1)^2 - (x - 1)^3 + \cdots \qquad \blacktriangleleft$$

We conclude this section by emphasizing again that we have no guarantee that the Maclaurin and Taylor series of a function f converge, or if they do converge, that the sum is $f(x)$. Convergence questions will be addressed in the next section.

▶ Exercise Set 11.9

In Exercises 1–12, find the fourth Maclaurin polynomial ($n = 4$) for the given function.

1. e^{-2x}.

2. $\dfrac{1}{1 + x}$.

3. $\sin 2x$.

4. $e^x \cos x$.

5. $\tan x$.

6. $x^3 - x^2 + 2x + 1$.

7. xe^x.

8. $\tan^{-1} x$.

9. $\sec x$.

10. $\sqrt{1 + x}$.

11. $\ln(3 + 2x)$.

12. $\sinh x$.

In Exercises 13–22, find the third Taylor polynomial ($n = 3$) about $x = a$ for the given function.

13. e^x; $a = 1$.

14. $\ln x$; $a = 1$.

15. $\sqrt{x}$; $a = 4$.

16. $x^4 + x - 3$; $a = -2$.

17. $\cos x$; $a = \dfrac{\pi}{4}$.

18. $\tan x$; $a = \dfrac{\pi}{3}$.

19. $\sin \pi x$; $a = -\dfrac{1}{3}$.

20. $\csc x$; $a = \dfrac{\pi}{2}$.

21. $\tan^{-1} x$; $a = 1$.

22. $\cosh x$; $a = \ln 2$.

In Exercises 23–31, find the Maclaurin series for the given function. Express your answer in sigma notation.

23. e^{-x}.

24. e^{ax}.

25. $\dfrac{1}{1+x}$.

26. xe^x.

27. $\ln(1+x)$.

28. $\sin \pi x$.

29. $\cos\left(\dfrac{x}{2}\right)$.

30. $\sinh x$.

31. $\cosh x$.

In Exercises 32–39, find the Taylor series about $x = a$ for the given function. Express your answer in sigma notation.

32. $\dfrac{1}{x}$; $a = 3$.

33. $\dfrac{1}{x}$; $a = -1$.

34. e^x; $a = 2$.

35. $\ln x$; $a = 1$.

36. $\cos x$; $a = \pi/2$.

37. $\sin \pi x$; $a = 1/2$.

38. $\dfrac{1}{x+2}$; $a = 3$.

39. $\sinh x$; $a = \ln 4$.

40. Prove: The value of $p_n(x) = \displaystyle\sum_{k=0}^{n} \dfrac{f^{(k)}(a)}{k!}(x-a)^k$ and its first n derivatives match the value of $f(x)$ and its first n derivatives at $x = a$.

■ 11.10 TAYLOR FORMULA WITH REMAINDER; CONVERGENCE OF TAYLOR SERIES

In the preceding section we anticipated the possibility that under appropriate conditions the Taylor series about $x = a$ for a function f might converge to $f(x)$ for values of x near a. In this section we shall establish conditions under which convergence occurs.

□ **TAYLOR'S THEOREM**

If p_n denotes the nth Taylor polynomial about $x = a$ for a function f, and if we approximate $f(x)$ by $p_n(x)$ at a point x, then the difference $f(x) - p_n(x)$ is denoted by

$$R_n(x) = f(x) - p_n(x) \tag{1}$$

and is called the **nth remainder for f** about $x = a$. This equation can be rewritten as

$$f(x) = p_n(x) + R_n(x)$$

or more explicitly as

$$f(x) = \sum_{k=0}^{n} \frac{f^{(k)}(a)}{k!}(x-a)^k + R_n(x) \tag{2}$$

It is evident from (1) that $p_n(x) \to f(x)$ as $n \to +\infty$ if and only if $R_n(x) \to 0$ as $n \to +\infty$. But $p_n(x)$ is the nth partial sum of the Taylor series for f about $x = a$, so we have the following theorem.

11.10.1 THEOREM. *The equality*

$$f(x) = \sum_{k=0}^{\infty} \frac{f^{(k)}(a)}{k!}(x-a)^k$$

holds if and only if $\displaystyle\lim_{n \to +\infty} R_n(x) = 0$.

Thus, the problem of determining whether the Taylor series for f about $x = a$ converges to $f(x)$ at a point x reduces to determining whether the nth remainder for f has a limit of zero at the point. The following theorem, which is proved at the end of this section, provides a formula for the nth remainder that will be useful for investigating this convergence question.

11.10.2   THEOREM (*Taylor's Theorem*).   *Suppose that a function f can be differentiated $n + 1$ times at each point in an interval containing the point a, and let $R_n(x)$ be the nth remainder for f about $x = a$. Then for each x in that interval there is at least one point c between a and x such that*

$$R_n(x) = \frac{f^{(n+1)}(c)}{(n+1)!}(x-a)^{n+1} \tag{3}$$

REMARK.   In this theorem the statement that c is "between" a and x should be interpreted to mean that c is in the interval (a, x) if $a < x$, or in the interval (x, a) if $x < a$, or $c = a$ if $x = a$.

□ **LAGRANGE'S FORM OF THE REMAINDER**

Historically, (3) was not discovered by Taylor but rather by Joseph Louis Lagrange.* For this reason this formula is commonly called **Lagrange's form of the remainder**. In this formula values for c are unspecified; all that can be said about them is that they depend on a, x, and n and that they lie between a and x. Other formulas for the remainder can be found in advanced calculus texts.

Substituting (3) in (2) yields

$$f(x) = \sum_{k=0}^{n} \frac{f^{(k)}(a)}{k!}(x-a)^k + \frac{f^{(n+1)}(c)}{(n+1)!}(x-a)^{n+1}$$

---

*JOSEPH LOUIS LAGRANGE (1736–1813). French–Italian mathematician and astronomer. Lagrange, the son of a public official, was born in Turin, Italy. (Baptismal records list his name as Giuseppe Lodovico Lagrangia.) Although his father wanted him to be a lawyer, Lagrange was attracted to mathematics and astronomy after reading a memoir by the astronomer Halley. At age 16 he began to study mathematics on his own and by age 19 was appointed to a professorship at the Royal Artillery School in Turin. The following year Lagrange sent Euler solutions to some famous problems using new methods that eventually blossomed into a branch of mathematics called calculus of variations. These methods and Lagrange's applications of them to problems in celestial mechanics were so monumental that by age 25 he was regarded by many of his contemporaries as the greatest living mathematician.

In 1776, on the recommendation of Euler, he was chosen to succeed Euler as the director of the Berlin Academy. During his stay in Berlin, Lagrange distinguished himself not only in celestial mechanics, but also in algebraic equations and the theory of numbers. After twenty years in Berlin, he moved to Paris at the invitation of Louis XVI. He was given apartments in the Louvre and treated with great honor, even during the revolution.

Napoleon was a great admirer of Lagrange and showered him with honors—count, senator, and Legion of Honor. The years Lagrange spent in Paris were devoted primarily to didactic treatises summarizing his mathematical conceptions. One of Lagrange's most famous works is a memoir, *Mécanique Analytique*, in which he reduced the theory of mechanics to a few general formulas from which all other necessary equations could be derived.

It is an interesting historical fact that Lagrange's father speculated unsuccessfully in several financial ventures, so his family was forced to live quite modestly. Lagrange himself stated that if his family had money, he would not have made mathematics his vocation. In spite of his fame, Lagrange was always a shy and modest man. On his death, he was buried with honor in the Pantheon.

or equivalently,

$$f(x) = f(a) + f'(a)(x - a) + \frac{f''(a)}{2!}(x - a)^2 + \cdots$$

$$+ \frac{f^{(n)}(a)}{n!}(x - a)^n + \frac{f^{(n+1)}(c)}{(n + 1)!}(x - a)^{n+1} \quad (4)$$

which is called **Taylor's formula with remainder**. In the special case where $a = 0$ this formula becomes

$$f(x) = f(0) + f'(0)x + \frac{f''(0)}{2!}x^2 + \cdots + \frac{f^{(n)}(0)}{n!}x^n + \frac{f^{(n+1)}(c)}{(n + 1)!}x^{n+1} \quad (5)$$

In (4) the number c is between a and x and in (5) it is between 0 and x.

Example 1   From Example 1 of the preceding section, the nth Maclaurin polynomial for $f(x) = e^x$ is

$$1 + x + \frac{x^2}{2!} + \cdots + \frac{x^n}{n!}$$

Since $f^{(n+1)}(x) = e^x$, it follows that

$$f^{(n+1)}(c) = e^c$$

Thus, from (5)

$$e^x = 1 + x + \frac{x^2}{2!} + \cdots + \frac{x^n}{n!} + \frac{e^c}{(n + 1)!}x^{n+1} \quad (6)$$

where c is between 0 and x. Note that (6) is valid for all real values of x since the hypotheses of Taylor's Theorem are satisfied on the interval $(-\infty, +\infty)$ (verify).   ◀

Example 2   Show that the Maclaurin series for e^x converges to e^x for all x; that is,

$$e^x = \sum_{k=0}^{\infty} \frac{x^k}{k!} = 1 + x + \frac{x^2}{2!} + \frac{x^3}{3!} + \frac{x^4}{4!} + \cdots \quad -\infty < x < +\infty$$

Solution.   We must show that $R_n(x) \to 0$ for all x as $n \to +\infty$. Thus, using Lagrange's form of the remainder given in (6) we must show that for all x

$$\lim_{n \to +\infty} R_n(x) = \lim_{n \to +\infty} \frac{e^c}{(n + 1)!}x^{n+1} = 0 \quad (7)$$

In Lagrange's remainder formula the value of x does not depend on n, so for purposes of proving the preceding limit, we can treat x as a constant. However, c does depend on n, so it must be treated as a function of n. This creates a complication, since we have no explicit formula for this function, and hence we cannot obtain the limit directly. To circumvent this difficulty, we will resort to an indirect approach based on the Squeezing Theorem (2.8.2). For this purpose, let r be the right endpoint of the closed interval whose endpoints are 0 and x. Thus, $r = 0$ or $r = x$, depending on whether $x \leq 0$ or $x > 0$. Since the exponential function is an increasing function, and since c lies between 0 and x, we are guaranteed that $e^c \leq e^r$. Therefore,

$$0 \leq |R_n(x)| = \left| \frac{e^c}{(n + 1)!}x^{n+1} \right| = e^c \frac{|x|^{n+1}}{(n + 1)!} \leq e^r \frac{|x|^{n+1}}{(n + 1)!}$$

Since r is constant, it follows from Formula (4) of Section 11.2, with $n + 1$ in place of n and $|x|$ in place of x, that

$$\lim_{n \to +\infty} e^r \frac{|x|^{n+1}}{(n+1)!} = e^r \lim_{n \to +\infty} \frac{|x|^{n+1}}{(n+1)!} = e^r \cdot 0 = 0$$

From this limit and the Squeezing Theorem, the preceding inequalities imply that for all x

$$\lim_{n \to +\infty} |R_n(x)| = 0 \quad \text{and consequently} \quad \lim_{n \to +\infty} R_n(x) = 0 \quad \blacktriangleleft$$

Example 3   Show that the Maclaurin series for $\cos x$ converges to $\cos x$ for all x; that is,

$$\cos x = \sum_{k=0}^{\infty} (-1)^k \frac{x^{2k}}{(2k)!} = 1 - \frac{x^2}{2!} + \frac{x^4}{4!} - \frac{x^6}{6!} + \cdots \qquad -\infty < x < +\infty$$

Solution.   As in the preceding example, we must show that $R_n(x) \to 0$ for all x as $n \to +\infty$. For this purpose let $f(x) = \cos x$, so that for all x we have

$$f^{(n+1)}(x) = \pm \cos x \quad \text{or} \quad f^{(n+1)}(x) = \pm \sin x$$

In all of these cases $|f^{(n+1)}(x)| \leq 1$, so that for all possible values of c

$$|f^{(n+1)}(c)| \leq 1$$

Hence, using Lagrange's form of the remainder we have

$$0 \leq |R_n(x)| = \left| \frac{f^{(n+1)}(c)}{(n+1)!} x^{n+1} \right| \leq \frac{|x|^{n+1}}{(n+1)!}$$

But

$$\lim_{n \to +\infty} \frac{|x|^{n+1}}{(n+1)!} = 0$$

so the Squeezing Theorem implies that $\lim_{n \to +\infty} R_n(x) = 0$ for all x. $\blacktriangleleft$

Example 4   Find the Taylor series for $\sin x$ about $x = \pi/2$ and show that the series converges to $\sin x$ for all x.

Solution.   Let $f(x) = \sin x$. Thus,

$$f(x) = \sin x \qquad f\left(\frac{\pi}{2}\right) = \sin \frac{\pi}{2} = 1$$

$$f'(x) = \cos x \qquad f'\left(\frac{\pi}{2}\right) = \cos \frac{\pi}{2} = 0$$

$$f''(x) = -\sin x \qquad f''\left(\frac{\pi}{2}\right) = -\sin \frac{\pi}{2} = -1$$

$$f'''(x) = -\cos x \qquad f'''\left(\frac{\pi}{2}\right) = -\cos \frac{\pi}{2} = 0$$

Since $f^{(4)}(x) = \sin x$, the pattern 1, 0, -1, 0 will repeat over and over as we evaluate successive derivatives at $\pi/2$. Thus, the Taylor series representation about $x = \pi/2$ for $\sin x$ is

$$\sin x = 1 - \frac{1}{2!}\left(x - \frac{\pi}{2}\right)^2 + \frac{1}{4!}\left(x - \frac{\pi}{2}\right)^4 - \frac{1}{6!}\left(x - \frac{\pi}{2}\right)^6 + \cdots \qquad (8)$$

which we are trying to show is valid for all x. As shown in Example 3, $|f^{(n+1)}(c)| \leq 1$, so

$$0 \leq |R_n(x)| = \left| \frac{f^{(n+1)}(c)}{(n+1)!} \left(x - \frac{\pi}{2} \right)^{n+1} \right| \leq \frac{\left| x - \frac{\pi}{2} \right|^{n+1}}{(n+1)!}$$

From Formula (4) of Section 11.2, with $|x - \pi/2|$ replacing x and $(n+1)!$ replacing $n!$, it follows that

$$\lim_{n \to +\infty} \frac{\left| x - \frac{\pi}{2} \right|^{n+1}}{(n+1)!} = 0$$

so that by the same argument given in Example 3 we have $\lim_{n \to +\infty} R_n(x) = 0$ for all x. This shows that (8) is valid for all x. ◀

REMARK. It can be shown that the Taylor series for e^x, $\sin x$, and $\cos x$ about any point $x = a$ converges to these functions for all x.

□ **CONSTRUCTING MACLAURIN SERIES BY SUBSTITUTION**

Sometimes Maclaurin series can be obtained by substituting in other Maclaurin series.

Example 5 Using the Maclaurin series

$$e^x = 1 + x + \frac{x^2}{2!} + \frac{x^3}{3!} + \frac{x^4}{4!} + \cdots \qquad -\infty < x < +\infty$$

we can derive the Maclaurin series for e^{-x} by substituting $-x$ for x to obtain

$$e^{-x} = 1 + (-x) + \frac{(-x)^2}{2!} + \frac{(-x)^3}{3!} + \frac{(-x)^4}{4!} + \cdots \qquad -\infty < -x < +\infty$$

or

$$e^{-x} = 1 - x + \frac{x^2}{2!} - \frac{x^3}{3!} + \frac{x^4}{4!} - \cdots \qquad -\infty < x < +\infty$$

From the Maclaurin series for e^x and e^{-x} we can obtain the Maclaurin series for $\cosh x$ by writing

$$\cosh x = \frac{1}{2}(e^x + e^{-x}) = \frac{1}{2}\left(\left[1 + x + \frac{x^2}{2!} + \frac{x^3}{3!} + \frac{x^4}{4!} + \cdots \right] \right.$$
$$\left. + \left[1 - x + \frac{x^2}{2!} - \frac{x^3}{3!} + \frac{x^4}{4!} + \cdots \right] \right)$$

or

$$\cosh x = 1 + \frac{x^2}{2!} + \frac{x^4}{4!} + \cdots \qquad -\infty < x < +\infty \qquad ◀$$

REMARK. Although we could have derived the preceding Maclaurin series directly, indirect methods are useful when it is messy to calculate the higher derivatives required for a Maclaurin series. There is, however, a loose thread in the logic of Example 5. We have produced a power series in x that converges to $\cosh x$ for all x. But isn't it conceivable that we have produced a power series in x *different* from the Maclaurin series? In Section 11.12 we shall show that if a power series in $x - a$ converges to $f(x)$ on some interval containing a, then the series must be the Taylor series for f about $x = a$. Thus, we are assured the series obtained for $\cosh x$ is, in fact, the Maclaurin series.

Example 6 Using the Maclaurin series

$$\frac{1}{1-x} = 1 + x + x^2 + x^3 + \cdots \qquad -1 < x < 1$$

we can derive the Maclaurin series for $1/(1 - 2x^2)$ by substituting $2x^2$ for x to obtain

$$\frac{1}{1 - 2x^2} = 1 + (2x^2) + (2x^2)^2 + (2x^2)^3 + \cdots \quad -1 < 2x^2 < 1$$

or

$$\frac{1}{1 - 2x^2} = 1 + 2x^2 + 4x^4 + 8x^6 + \cdots = \sum_{k=0}^{\infty} 2^k x^{2k} \quad -1 < 2x^2 < 1$$

Since $2x^2 \geq 0$ for all x, the convergence condition $-1 < 2x^2 < 1$ can be written in the equivalent form $0 \leq 2x^2 < 1$ or $0 \leq x^2 < 1/2$ or $-1/\sqrt{2} < x < 1/\sqrt{2}$. ◄

□ **BINOMIAL SERIES**

If m is a real number, then the Maclaurin series for $(1 + x)^m$ is called the **binomial series**; it is given by (verify)

$$1 + mx + \frac{m(m - 1)}{2!}x^2 + \frac{m(m - 1)(m - 2)}{3!}x^3 + \cdots$$

REMARK.   If m is a nonnegative integer, then $f(x) = (1 + x)^m$ is a polynomial of degree m, so

$$f^{(m+1)}(0) = f^{(m+2)}(0) = f^{(m+3)}(0) = \cdots = 0$$

and the binomial series reduces to the familiar binomial expansion

$$(1 + x)^m = 1 + mx + \frac{m(m - 1)}{2!}x^2 + \frac{m(m - 1)(m - 2)}{3!}x^3 + \cdots + x^m$$

which is valid for $-\infty < x < +\infty$.

It can be proved that if m is not a nonnegative integer, then the binomial series converges to $(1 + x)^m$ if $|x| < 1$. Thus, for such values of x

$$(1 + x)^m = 1 + mx + \frac{m(m - 1)}{2!}x^2 + \cdots$$

$$+ \frac{m(m - 1)(m - 2) \cdots (m - k + 1)}{k!}x^k + \cdots \quad (9)$$

or in sigma notation

$$(1 + x)^m = 1 + \sum_{k=1}^{\infty} \frac{m(m - 1) \cdots (m - k + 1)}{k!}x^k \quad \text{if } |x| < 1 \quad (10)$$

Example 7   Express $1/\sqrt{1 + x}$ as a binomial series.

Solution.   Substituting $m = -\frac{1}{2}$ in (9) yields

$$\frac{1}{\sqrt{1 + x}} = 1 - \frac{1}{2}x + \frac{(-\frac{1}{2})(-\frac{1}{2} - 1)}{2!}x^2 + \frac{(-\frac{1}{2})(-\frac{1}{2} - 1)(-\frac{1}{2} - 2)}{3!}x^3$$

$$+ \cdots + \frac{(-\frac{1}{2})(-\frac{3}{2})(-\frac{5}{2}) \cdots (-\frac{1}{2} - k + 1)}{k!}x^k + \cdots$$

$$= 1 - \frac{1}{2}x + \frac{1 \cdot 3}{2^2 \cdot 2!}x^2 - \frac{1 \cdot 3 \cdot 5}{2^3 \cdot 3!}x^3 + \cdots$$

$$+ (-1)^k \frac{1 \cdot 3 \cdot 5 \cdots (2k - 1)}{2^k k!}x^k + \cdots \quad ◄$$

For reference, we have listed in Table 11.10.1 the Maclaurin series for a number of important functions, and we have indicated the interval over which the series converges to the function. It should be noted that the intervals of convergence stated for $\ln(1 + x)$ and $\tan^{-1} x$ are somewhat difficult to obtain directly. However, these intervals can be obtained by indirect methods that we shall study in the last section of this chapter.

Table 11.10.1

MACLAURIN SERIES	INTERVAL OF CONVERGENCE
$\dfrac{1}{1 - x} = \displaystyle\sum_{k=0}^{\infty} x^k = 1 + x + x^2 + x^3 + \cdots$	$-1 < x < 1$
$e^x = \displaystyle\sum_{k=0}^{\infty} \dfrac{x^k}{k!} = 1 + x + \dfrac{x^2}{2!} + \dfrac{x^3}{3!} + \dfrac{x^4}{4!} + \cdots$	$-\infty < x < +\infty$
$\sin x = \displaystyle\sum_{k=0}^{\infty} (-1)^k \dfrac{x^{2k+1}}{(2k+1)!} = x - \dfrac{x^3}{3!} + \dfrac{x^5}{5!} - \dfrac{x^7}{7!} + \cdots$	$-\infty < x < +\infty$
$\cos x = \displaystyle\sum_{k=0}^{\infty} (-1)^k \dfrac{x^{2k}}{(2k)!} = 1 - \dfrac{x^2}{2!} + \dfrac{x^4}{4!} - \dfrac{x^6}{6!} + \cdots$	$-\infty < x < +\infty$
$\ln(1 + x) = \displaystyle\sum_{k=0}^{\infty} (-1)^k \dfrac{x^{k+1}}{k + 1} = x - \dfrac{x^2}{2} + \dfrac{x^3}{3} - \dfrac{x^4}{4} + \cdots$	$-1 < x \le 1$
$\tan^{-1} x = \displaystyle\sum_{k=0}^{\infty} (-1)^k \dfrac{x^{2k+1}}{2k + 1} = x - \dfrac{x^3}{3} + \dfrac{x^5}{5} - \dfrac{x^7}{7} + \cdots$	$-1 \le x \le 1$
$\sinh x = \displaystyle\sum_{k=0}^{\infty} \dfrac{x^{2k+1}}{(2k+1)!} = x + \dfrac{x^3}{3!} + \dfrac{x^5}{5!} + \dfrac{x^7}{7!} + \cdots$	$-\infty < x < +\infty$
$\cosh x = \displaystyle\sum_{k=0}^{\infty} \dfrac{x^{2k}}{(2k)!} = 1 + \dfrac{x^2}{2!} + \dfrac{x^4}{4!} + \dfrac{x^6}{6!} + \cdots$	$-\infty < x < +\infty$
$(1 + x)^m = 1 + \displaystyle\sum_{k=1}^{\infty} \dfrac{m(m - 1) \cdots (m - k + 1)}{k!} x^k$	$-1 < x < 1*$ $(m \ne 0, 1, 2, \ldots)$

*The behavior at the endpoints depends on m: For $m \ge 0$ the series converges absolutely at both endpoints; for $m \le -1$ the series diverges at both endpoints; and for $-1 < m < 0$ the series converges conditionally at $x = 1$ and diverges at $x = -1$.

■ PROOF OF TAYLOR'S THEOREM

Proof of Theorem 11.10.2. By hypothesis, f can be differentiated $n + 1$ times at each point in an interval containing the point a. Choose any point b in this interval. To be specific, we shall assume that $b > a$. (The cases $b < a$ and $b = a$ are left to the reader.) Let $p_n(x)$ be the nth Taylor polynomial for $f(x)$ about $x = a$ and define

$$h(x) = f(x) - p_n(x) \tag{11}$$

$$g(x) = (x - a)^{n+1} \tag{12}$$

Because $f(x)$ and $p_n(x)$ have the same value and the same first n derivatives at $x = a$, it follows that

$$h(a) = h'(a) = h''(a) = \cdots = h^{(n)}(a) = 0 \tag{13}$$

Also, we leave it for the reader to show that

$$g(a) = g'(a) = g''(a) = \cdots = g^{(n)}(a) = 0 \tag{14}$$

and that $g(x)$ and its first n derivatives are nonzero when $x \neq a$.

It is straightforward to check that h and g satisfy the hypotheses of the Extended Mean-Value Theorem (10.2.2) on the interval $[a, b]$, so that there is a point c_1 with $a < c_1 < b$ such that

$$\frac{h(b) - h(a)}{g(b) - g(a)} = \frac{h'(c_1)}{g'(c_1)} \tag{15}$$

or, from (13) and (14),

$$\frac{h(b)}{g(b)} = \frac{h'(c_1)}{g'(c_1)} \tag{16}$$

If we now apply the Extended Mean-Value Theorem to h' and g' over the interval $[a, c_1]$, we may deduce that there is a point c_2 with $a < c_2 < c_1 < b$ such that

$$\frac{h'(c_1) - h'(a)}{g'(c_1) - g'(a)} = \frac{h''(c_2)}{g''(c_2)}$$

or, from (13) and (14),

$$\frac{h'(c_1)}{g'(c_1)} = \frac{h''(c_2)}{g''(c_2)}$$

which, when combined with (16), yields

$$\frac{h(b)}{g(b)} = \frac{h''(c_2)}{g''(c_2)}$$

It should now be clear that if we continue in this way, applying the Extended Mean-Value Theorem to the successive derivatives of h and g, we shall eventually obtain a relationship of the form

$$\frac{h(b)}{g(b)} = \frac{h^{(n+1)}(c_{n+1})}{g^{(n+1)}(c_{n+1})} \tag{17}$$

where $a < c_{n+1} < b$. However, $p_n(x)$ is a polynomial of degree n, so that its $(n + 1)$-st derivative is zero. Thus, from (11)

$$h^{(n+1)}(c_{n+1}) = f^{(n+1)}(c_{n+1}) \tag{18}$$

Also, from (12), the $(n + 1)$-st derivative of $g(x)$ is the constant $(n + 1)!$, so that

$$g^{(n+1)}(c_{n+1}) = (n + 1)! \tag{19}$$

Substituting (18) and (19) into (17) yields

$$\frac{h(b)}{g(b)} = \frac{f^{(n+1)}(c_{n+1})}{(n + 1)!}$$

Letting $c = c_{n+1}$, and using (11) and (12), it follows that

$$f(b) - p_n(b) = \frac{f^{(n+1)}(c)}{(n + 1)!}(b - a)^{n+1}$$

But this is precisely Formula (3) in Taylor's Theorem (Theorem 11.10.2), with the exception that the variable here is b rather than x. Thus, to finish we need only replace b by x. ∎

▶ Exercise Set 11.10

For the functions in Exercises 1–14, find Lagrange's form of the remainder for the given values of a and n.

1. e^{2x}; $a = 0$; $n = 5$.

2. $\cos x$; $a = 0$; $n = 8$.

3. $\dfrac{1}{x + 1}$; $a = 0$; $n = 4$.

4. $\tan x$; $a = 0$; $n = 2$.

5. xe^x; $a = 0$; $n = 3$.

6. $\ln(1 + x)$; $a = 0$; $n = 5$.

7. $\tan^{-1} x$; $a = 0$; $n = 2$.

8. $\sinh x$; $a = 0$; $n = 6$.

9. $\sqrt{x}$; $a = 4$; $n = 3$.

10. $\dfrac{1}{x}$; $a = 1$; $n = 5$.

11. $\sin x$; $a = \dfrac{\pi}{6}$; $n = 4$.

12. $\cos \pi x$; $a = \dfrac{1}{2}$; $n = 2$.

13. $\dfrac{1}{(1+x)^2}$; $a = -2$; $n = 5$.

14. $\csc x$; $a = \dfrac{\pi}{2}$; $n = 1$.

In Exercises 15–18, find Lagrange's form of the remainder $R_n(x)$ when $a = 0$.

15. $f(x) = \dfrac{1}{1-x}$.

16. $f(x) = e^{-x}$.

17. $f(x) = e^{2x}$.

18. $f(x) = \ln(1+x)$.

19. Prove: The Maclaurin series for $\sin x$ converges to $\sin x$ for all x.

20. Prove: The Taylor series for $\sin x$ about $x = \pi/4$ converges to $\sin x$ for all x.

21. Prove: The Taylor series for e^x about $x = 1$ converges to e^x for all x.

22. (a) Prove: The Maclaurin series for $\ln(1+x)$ converges to $\ln(1+x)$ if $0 \le x \le 1$. [*Remark:* The Maclaurin series actually converges to $\ln(1+x)$ on the interval $(-1, 1]$, but the Lagrange form of the remainder is not strong enough to establish this fact.]

    (b) Use $x = 1$ in the Maclaurin series for $\ln(1+x)$, Table 11.10.1, to show that
$$\ln 2 = 1 - \frac{1}{2} + \frac{1}{3} - \frac{1}{4} + \cdots$$

23. Prove: The Taylor series for e^x about any point $x = a$ converges to e^x for all x.

24. Prove: The Taylor series for $\sin x$ about any point $x = a$ converges to $\sin x$ for all x.

25. Prove: The Taylor series for $\cos x$ about any point $x = a$ converges to $\cos x$ for all x.

26. Derive (9). (Do not try to prove convergence.)

In Exercises 27–50, use the Maclaurin series in Table 11.10.1 to obtain the Maclaurin series for the given function. In each case give the first four terms of the series and specify the interval on which the series converges to the function.

27. e^{-2x}.

28. $x^2 e^x$.

29. xe^{-x}.

30. e^{x^2}.

31. $\sin 2x$.

32. $\cos 2x$.

33. $x^2 \cos x$.

34. $\sin(x^2)$.

35. $\sin^2 x$. [*Hint:* $\sin^2 x = \frac{1}{2}(1 - \cos 2x)$.]

36. $\cos^2 x$. [*Hint:* $\cos^2 x = \frac{1}{2}(1 + \cos 2x)$.]

37. $\ln(1 - x^2)$.

38. $\ln(1 + 2x)$.

39. $\dfrac{1}{1 - 4x^2}$.

40. $\dfrac{x}{1-x}$.

41. $\dfrac{x^2}{1 + 3x}$.

42. $\dfrac{x}{1 + x^2}$.

43. $x \sinh 2x$.

44. $\cosh(x^2)$.

45. $\sqrt{1 + 3x}$.

46. $\sqrt{1 + x^2}$.

47. $\dfrac{1}{(1 - 2x)^2}$.

48. $\dfrac{x}{(1 + 2x)^3}$.

49. $\dfrac{x}{\sqrt{1 - x^2}}$.

50. $x(1 - x^2)^{3/2}$.

51. Use the Maclaurin series for $1/(1 - x)$ to express $1/x$ in powers of $x - 1$. Find the interval of convergence.
$$\left[\textit{Hint:}\ \frac{1}{x} = \frac{1}{1 + (x - 1)}. \right]$$

52. Show that the series
$$1 - \frac{x}{2!} + \frac{x^2}{4!} - \frac{x^3}{6!} + \cdots$$

converges to the function
$$f(x) = \begin{cases} \cos \sqrt{x}, & x \ge 0 \\ \cosh \sqrt{-x}, & x < 0 \end{cases}$$

[*Hint:* Use the Maclaurin series for $\cos x$ and $\cosh x$ to obtain series for $\cos \sqrt{x}$ where $x \ge 0$, and $\cosh \sqrt{-x}$ where $x \le 0$.]

53. If m is any real number, and k is a nonnegative integer, then we define the **binomial coefficients** $\dbinom{m}{k}$ by the formulas $\dbinom{m}{0} = 1$ and
$$\binom{m}{k} = \frac{m(m - 1)(m - 2) \cdots (m - k + 1)}{k!}$$

for $k \ge 1$.

Express Formula (10) in terms of binomial coefficients.

In Exercises 54–57, use the known Maclaurin series for e^x, $\sin x$, and $\cos x$ to help find the sum of the given series.

54. $2 + \dfrac{4}{2!} + \dfrac{8}{3!} + \dfrac{16}{4!} + \cdots$.

55. $\pi - \dfrac{\pi^3}{3!} + \dfrac{\pi^5}{5!} - \dfrac{\pi^7}{7!} + \cdots$.

56. $1 - \dfrac{e^2}{2!} + \dfrac{e^4}{4!} - \dfrac{e^6}{6!} + \cdots$.

57. $1 - \ln 3 + \dfrac{(\ln 3)^2}{2!} - \dfrac{(\ln 3)^3}{3!} + \cdots$.

58. (a) Use the Maclaurin series for $\dfrac{1}{1 - x}$ to help find the Maclaurin series for the function
$$f(x) = \frac{x}{1 - x^2}$$

    (b) Use the result in part (a) to help find $f^{(5)}(0)$ and $f^{(6)}(0)$.

59. (a) Use the Maclaurin series for $\cos x$ to find the Maclaurin series for $f(x) = x^2 \cos 2x$.

(b) Use the result in part (a) to help find $f^{(5)}(0)$.

60. The purpose of this exercise is to show that the Taylor series of a function f may possibly converge to a value different from $f(x)$ for certain x. Let

$$f(x) = \begin{cases} e^{-1/x^2}, & x \neq 0 \\ 0, & x = 0 \end{cases}$$

(a) Use the definition of a derivative to show that $f'(0) = 0$.

(b) With some difficulty it can be shown that $f^{(n)}(0) = 0$ for $n \geq 2$. Accepting this fact, show that the Maclaurin series of f converges for all x, but converges to $f(x)$ only at the point $x = 0$.

■ 11.11 COMPUTATIONS USING TAYLOR SERIES

In this section we shall show how Taylor and Maclaurin series can be used to obtain approximate values for trigonometric functions and logarithms.

☐ **ACCURACY OF APPROXIMATIONS**

In practical applications where a function f is approximated by a Taylor polynomial, it is important to have some way of estimating the error in the approximation. For this purpose we introduce the following terminology: An approximation is said to be *accurate to n decimal places* if the magnitude of the error is less than 0.5×10^{-n}. For example,

DESCRIPTION	MAGNITUDE OF THE ERROR IS LESS THAN	
1 decimal-place accuracy	0.05	$= 0.5 \times 10^{-1}$
2 decimal-place accuracy	0.005	$= 0.5 \times 10^{-2}$
3 decimal-place accuracy	0.0005	$= 0.5 \times 10^{-3}$
4 decimal-place accuracy	0.00005	$= 0.5 \times 10^{-4}$

☐ **APPROXIMATING e**

In the preceding section we showed that for all x

$$e^x = 1 + x + \frac{x^2}{2!} + \frac{x^3}{3!} + \frac{x^4}{4!} + \cdots + \frac{x^k}{k!} + \cdots$$

In particular, if we let $x = 1$, we obtain the following expression for e as the sum of an infinite series

$$e = 1 + 1 + \frac{1}{2!} + \frac{1}{3!} + \frac{1}{4!} + \cdots + \frac{1}{k!} + \cdots \tag{1}$$

Thus, we can approximate e to any degree of accuracy using an appropriate partial sum

$$e \approx 1 + 1 + \frac{1}{2!} + \frac{1}{3!} + \cdots + \frac{1}{n!} \tag{2}$$

The following example shows how Lagrange's remainder formula can be used to investigate the accuracy of such an approximation.

Example 1 Use (2) to approximate e to four decimal-place accuracy.

Solution. It follows from Taylor's formula with remainder that

$$e^x = 1 + x + \frac{x^2}{2!} + \cdots + \frac{x^n}{n!} + \frac{e^c}{(n + 1)!} x^{n+1}$$

where c is between 0 and x (see Example 1, Section 11.10). Thus, in the case $x = 1$ we obtain

$$e = 1 + 1 + \frac{1}{2!} + \cdots + \frac{1}{n!} + \frac{e^c}{(n+1)!}$$

where c is between 0 and 1. This tells us that the magnitude of the error in approximation (2) is

$$|R_n(1)| = \left| \frac{e^c}{(n+1)!} \right| = \frac{e^c}{(n+1)!} \tag{3}$$

where $0 < c < 1$. Since $c < 1$, it follows that

$$e^c < e^1 = e$$

so that from (3)

$$|R_n(1)| < \frac{e}{(n+1)!} \tag{4}$$

This inequality provides an upper bound on the magnitude of the error R_n. Unfortunately, this inequality is not very useful since the right side involves the quantity e, which we are trying to estimate. However, if we use the fact that $e < 3$, then we can replace (4) with the following less precise but more useful result:

$$|R_n(1)| < \frac{3}{(n+1)!} \tag{4a}$$

It follows that if we choose n so that

$$|R_n(1)| < \frac{3}{(n+1)!} < 0.5 \times 10^{-4} = 0.00005 \tag{5}$$

then approximation (2) will be accurate to four decimal places. An appropriate value for n may be found by trial and error. For example, using a calculator one can evaluate $3/(n+1)!$ for $n = 0, 1, 2, \ldots$ until a value of n satisfying (5) is obtained. We leave it for the reader to show that $n = 8$ is the first positive integer satisfying (5). Thus, to four decimal-place accuracy

$$e \approx 1 + 1 + \frac{1}{2!} + \frac{1}{3!} + \frac{1}{4!} + \frac{1}{5!} + \frac{1}{6!} + \frac{1}{7!} + \frac{1}{8!} \approx 2.7183 \tag{6}$$

◄

REMARK. It should be noted that there are two types of errors that result when computing with series. The first, called *truncation error*, is the error that results when the entire series is approximated by a partial sum. The second kind of error, called *roundoff error*, results when decimal approximations are used. For example, (6) involves a truncation error of at most 0.5×10^{-4} (four decimal-place accuracy). However, to obtain the numerical value in (6) we used a calculator to evaluate the left side, resulting in roundoff error due to the limitations of the calculator. The problem of controlling roundoff error is surprisingly difficult and is studied in courses in a branch of mathematics called *numerical analysis*. As a rule of thumb, to achieve n decimal-place accuracy in a final result, one should use more than $n + 1$ decimal places in each intermediate computation, then round off to n decimal places at the end. However, even this procedure may occasionally not produce n decimal-place accuracy. For purposes of this text, we recommend that you perform all intermediate calculations with the maximum number of decimal places that your calculator will allow, then round off the final result.

□ **AN UPPER BOUND ON THE REMAINDER**

The technique illustrated in Example 1 can be applied to a variety of problems involving approximations by Taylor polynomials. To see how, recall from Taylor's Theorem that the absolute value of the error that results when $f(x)$ is approximated by its nth Taylor polynomial $p_n(x)$ about $x = a$ is

$$|R_n(x)| = |f(x) - p_n(x)| = \left| \frac{f^{(n+1)}(c)}{(n+1)!}(x-a)^{n+1} \right| \tag{7}$$

In this formula, c is an unknown number between a and x, so that the value of $f^{(n+1)}(c)$ usually cannot be determined. However, it is frequently possible to determine an upper bound on the size of $|f^{(n+1)}(c)|$; that is, one can often find a constant M such that $|f^{(n+1)}(c)| \le M$. For such an M, it follows from (7) that

$$|R_n(x)| = |f(x) - p_n(x)| \le \frac{M}{(n+1)!} |x-a|^{n+1} \tag{8}$$

which gives an upper bound on the magnitude of the error $R_n(x)$. The example that follows uses this result.

☐ **APPROXIMATING TRIGONOMETRIC FUNCTIONS**

To approximate the value of a function f at a point x_0 using a Taylor series, two factors enter into the selection of the point $x = a$ for the series. First, the point $x = a$ must be selected so that f and its derivatives can be evaluated at a, since those values are needed to find Taylor series. Second, it is desirable to choose a as close as possible to x_0 since the "rate of convergence" of a Taylor series is usually most rapid close to a; that is, fewer terms are required in a partial sum to achieve a given level of accuracy. For example, to approximate $\sin 3°$ $(= \pi/60$ radians), it would be reasonable to take $a = 0$, since $\pi/60$ is close to 0, and the successive derivatives of $\sin x$ are easy to evaluate at 0.

Example 2 Use the Maclaurin series for $\sin x$ to approximate $\sin 3°$ to five decimal-place accuracy.

Solution. In the Maclaurin series

$$\sin x = x - \frac{x^3}{3!} + \frac{x^5}{5!} - \frac{x^7}{7!} + \cdots \tag{9}$$

the angle x is assumed to be in radians (because the differentiation formulas for the trigonometric functions were derived with this assumption). Since $3° = \pi/60$ radians, it follows from (9) that

$$\sin 3° = \sin \frac{\pi}{60} = \left(\frac{\pi}{60} \right) - \frac{\left(\frac{\pi}{60} \right)^3}{3!} + \frac{\left(\frac{\pi}{60} \right)^5}{5!} - \frac{\left(\frac{\pi}{60} \right)^7}{7!} + \cdots \tag{10}$$

We must now decide how many terms in this series must be kept in order to obtain five decimal-place accuracy. We shall consider two possible approaches, one using Lagrange's remainder formula, and the other exploiting the fact that (10) satisfies the conditions of the alternating series test.

If we let $f(x) = \sin x$, then the magnitude of the error that results when $\sin x$ is approximated by its nth Maclaurin polynomial is

$$|R_n(x)| = \left| \frac{f^{(n+1)}(c)}{(n+1)!} x^{n+1} \right|$$

where c is between 0 and x. Since $f^{(n+1)}(c)$ is either $\pm \sin c$ or $\pm \cos c$, it follows that $|f^{(n+1)}(c)| \le 1$, so from (8) with $a = 0$ and $M = 1$ we have

$$|R_n(x)| \le \frac{|x|^{n+1}}{(n+1)!}$$

In particular, if $x = \pi/60$, then

$$|R_n(\pi/60)| \leq \frac{\left(\dfrac{\pi}{60}\right)^{n+1}}{(n+1)!}$$

Thus, for five decimal-place accuracy, we must choose n so that

$$\frac{\left(\dfrac{\pi}{60}\right)^{n+1}}{(n+1)!} < 0.5 \times 10^{-5} = 0.000005$$

By trial and error with the help of a calculator or computer, the reader can check that $n = 3$ is the smallest n that works. Thus, in (10) we need only keep terms up to the third power for five decimal-place accuracy, that is,

$$\sin 3° \approx \left(\frac{\pi}{60}\right) - \frac{\left(\dfrac{\pi}{60}\right)^3}{3!} \approx 0.05234 \tag{11}$$

An alternative approach to determining n uses the fact that (10) satisfies the conditions of the alternating series test, Theorem 11.7.1. (Verify.) Thus, by Theorem 11.7.2, if we use only those terms up to and including

$$\pm \frac{\left(\dfrac{\pi}{60}\right)^m}{m!} \quad \boxed{m \text{ is an odd positive integer.}}$$

then the magnitude of the error will be at most

$$\frac{\left(\dfrac{\pi}{60}\right)^{m+2}}{(m+2)!}$$

Thus, for five decimal-place accuracy, we look for the first positive odd integer m such that

$$\frac{\left(\dfrac{\pi}{60}\right)^{m+2}}{(m+2)!} < 0.5 \times 10^{-5} = 0.000005$$

By trial and error, $m = 3$ is the first such integer. Thus, to five decimal-place accuracy

$$\sin 3° \approx \left(\frac{\pi}{60}\right) - \frac{\left(\dfrac{\pi}{60}\right)^3}{3!} \approx 0.05234 \tag{12}$$

which is the same as our earlier result.   ◄

☐ **APPROXIMATING LOGARITHMS**

The Maclaurin series

$$\ln(1 + x) = x - \frac{x^2}{2} + \frac{x^3}{3} - \frac{x^4}{4} + \cdots \quad -1 < x \leq 1 \tag{13}$$

is the starting point for the approximation of natural logarithms. Unfortunately, the usefulness of this series is limited because of its slow convergence and the restriction $-1 < x \leq 1$. However, if we replace x by $-x$ in this series, we obtain

$$\ln(1 - x) = -x - \frac{x^2}{2} - \frac{x^3}{3} - \frac{x^4}{4} - \cdots \quad -1 \leq x < 1 \tag{14}$$

and on subtracting (14) from (13) we obtain

$$\ln\left(\frac{1+x}{1-x}\right) = 2\left(x + \frac{x^3}{3} + \frac{x^5}{5} + \frac{x^7}{7} + \cdots\right) \quad -1 < x < 1 \tag{15}$$

Series (15), first obtained by James Gregory* in 1668, can be used to compute the natural logarithm of any positive number y by letting

$$y = \frac{1 + x}{1 - x}$$

or equivalently,

$$x = \frac{y - 1}{y + 1} \tag{16}$$

and noting that $-1 < x < 1$. For example, to compute $\ln 2$ we let $y = 2$ in (16), which yields $x = 1/3$. Substituting this value in (15) gives

$$\ln 2 = 2 \left[(1/3) + \frac{(1/3)^3}{3} + \frac{(1/3)^5}{5} + \frac{(1/3)^7}{7} + \cdots \right]$$

Adding the four terms shown and then rounding to four decimal places at the end yields

$$\ln 2 \approx 0.6931$$

It is of interest to note that in the case $x = 1$, series (13) yields

$$\ln 2 = 1 - \frac{1}{2} + \frac{1}{3} - \frac{1}{4} + \frac{1}{5} - \cdots$$

Although this result gives the sum of the alternating harmonic series, the series converges too slowly to be of computational value.

☐ APPROXIMATING π

If we let $x = 1$ in the Maclaurin series

$$\tan^{-1} x = x - \frac{x^3}{3} + \frac{x^5}{5} - \frac{x^7}{7} + \cdots \qquad -1 \le x \le 1 \tag{17}$$

we obtain

$$\frac{\pi}{4} = \tan^{-1} 1 = 1 - \frac{1}{3} + \frac{1}{5} - \frac{1}{7} + \cdots$$

or

$$\pi = 4 \left[1 - \frac{1}{3} + \frac{1}{5} - \frac{1}{7} + \cdots \right]$$

This famous series, obtained by Leibniz in 1674, converges too slowly to be of computational importance. A more practical procedure for approximating π uses the identity

$$\frac{\pi}{4} = \tan^{-1} \frac{1}{2} + \tan^{-1} \frac{1}{3} \tag{18}$$

By using this identity and series (17) to approximate $\tan^{-1} \frac{1}{2}$ and $\tan^{-1} \frac{1}{3}$, the value of π can be effectively approximated to any degree of accuracy.

---

*JAMES GREGORY (1638–1675). Scottish mathematician and astronomer. Gregory, the son of a minister, was famous in his time as the inventor of the Gregorian reflecting telescope, so named in his honor. Although he is not generally ranked with the great mathematicians, much of his work relating to calculus was studied by Leibniz and Newton and undoubtedly influenced some of their discoveries. There is a manuscript, discovered posthumously, which shows that Gregory had anticipated Taylor series well before Taylor.

▶ Exercise Set 11.11   ⒸⒾ 1–17, 19

1. Use inequality (4a) to find a value of n to ensure that (2) will approximate e to
   (a) five decimal-place accuracy
   (b) ten decimal-place accuracy.

In Exercises 2–8, apply Lagrange's form of the remainder.

2. Use $x = -1$ in the Maclaurin series for e^x to approximate $1/e$ to three decimal-place accuracy.

3. Use $x = \frac{1}{2}$ in the Maclaurin series for e^x to approximate $\sqrt{e}$ to four decimal-place accuracy.

4. Use the Maclaurin series for $\sin x$ to approximate $\sin 4°$ to five decimal-place accuracy.

5. Use the Maclaurin series for $\cos x$ to approximate $\cos(\pi/20)$ to four decimal-place accuracy.

6. Use an appropriate Taylor series for $\sin x$ to approximate $\sin 85°$ to four decimal-place accuracy.

7. Use the first two terms of series (15) to approximate $\ln 1.25$. Round your answer to three decimal places.

8. Use the first six terms of series (15) to approximate $\ln 3$. Round your answer to four decimal places.

9. Use the Maclaurin series for $\tan^{-1} x$ to approximate $\tan^{-1} 0.1$ to three decimal-place accuracy. [*Hint:* Use the fact that (17) is an alternating series.]

10. Use the Maclaurin series for $\sinh x$ to approximate $\sinh 0.5$ to three decimal-place accuracy.

11. Use the Maclaurin series for $\cosh x$ to approximate $\cosh 0.1$ to four decimal-place accuracy.

12. Use an appropriate Taylor series for $\sqrt[3]{x}$ to approximate $\sqrt[3]{28}$ to three decimal-place accuracy.

13. Find an interval of values for x containing $x = 0$ over which $\sin x$ can be approximated by $x - x^3/3!$ with three decimal-place accuracy ensured.

14. Find an interval of values for x over which e^x can be approximated by $1 + x + x^2/2!$ with three decimal-place accuracy ensured.

15. Find an upper bound on the magnitude of the error in the approximation
$$\cos x \approx 1 - x^2/2! + x^4/4!$$
if $-0.2 \le x \le 0.2$.

16. Find an upper bound on the magnitude of the error in the approximation $\ln(1 + x) \approx x$ if $|x| < 0.01$.

17. In each part find the number of terms of the stated series that are required to ensure that the sum of the terms approximates $\ln 2$ to six decimal-place accuracy.

    (a) The alternating harmonic series

    (b) Series (15).

18. Prove identity (18).

19. Approximate $\tan^{-1} \frac{1}{2}$ and $\tan^{-1} \frac{1}{3}$ to three decimal-place accuracy, and then use identity (18) to approximate π.

■ **11.12 DIFFERENTIATION AND INTEGRATION OF POWER SERIES**

In this section we shall show that if a power series in $x - a$ converges in some interval and has a sum of $f(x)$ in that interval, then the power series must be the Taylor series about $x = a$ for the function f. This result is important for many reasons, but it is important for computational purposes because it will enable us to find Taylor series without using the defining formula for such series. This is often necessary in cases where successive derivatives of f are prohibitively complicated to compute. In this section we shall also consider conditions under which a Taylor series can be differentiated or integrated term by term.

☐ **POWER SERIES**
REPRESENTATIONS OF
FUNCTIONS

If a function f is expressed as the sum of a power series for all x in some interval, then we shall say that the power series *represents* f on the interval or that the series is a *power series representation of f* on the interval. For example, we know from our study of geometric series that

$$\frac{1}{1 - x} = 1 + x + x^2 + x^3 + \cdots \quad -1 < x < 1$$

Thus, the series $1 + x + x^2 + x^3 + \cdots$ represents the function

$$f(x) = \frac{1}{1 - x}$$

on the interval $(-1, 1)$.

The following theorems show that if a function f is represented by a power series on an interval, then differentiating the series term by term produces a power series representation of f' on the interval, and integrating the series term by term produces a power series representation of the integral of f on the interval.

11.12.1 THEOREM (*Differentiation of Power Series*). *If a function f is represented by a power series, say*

$$f(x) = \sum_{k=0}^{\infty} c_k(x - a)^k$$

where the series has a nonzero radius of convergence R, then:

(a) *The series of differentiated terms*

$$\sum_{k=0}^{\infty} \frac{d}{dx}[c_k(x - a)^k] = \sum_{k=1}^{\infty} kc_k(x - a)^{k-1}$$

has radius of convergence R.

(b) *The function f is differentiable on the interval $(a - R, a + R)$, and for every x in this interval*

$$f'(x) = \sum_{k=0}^{\infty} \frac{d}{dx}[c_k(x - a)^k]$$

To paraphrase this theorem informally, *a power series representation of a function can be differentiated term by term on any open interval within the interval of convergence.*

Example 1   To illustrate this theorem, we shall use the Maclaurin series

$$\sin x = x - \frac{x^3}{3!} + \frac{x^5}{5!} - \frac{x^7}{7!} + \cdots \qquad -\infty < x < +\infty$$

$$\cos x = 1 - \frac{x^2}{2!} + \frac{x^4}{4!} - \frac{x^6}{6!} + \cdots \qquad -\infty < x < +\infty$$

$$e^x = 1 + x + \frac{x^2}{2!} + \frac{x^3}{3!} + \frac{x^4}{4!} + \cdots \qquad -\infty < x < +\infty$$

to obtain the familiar derivative formulas

$$\frac{d}{dx}[\sin x] = \cos x \quad \text{and} \quad \frac{d}{dx}[e^x] = e^x$$

Differentiating the Maclaurin series for $\sin x$ and e^x term by term yields

$$\frac{d}{dx}[\sin x] = \frac{d}{dx}\left[x - \frac{x^3}{3!} + \frac{x^5}{5!} - \frac{x^7}{7!} + \cdots\right]$$

$$= 1 - 3\frac{x^2}{3!} + 5\frac{x^4}{5!} - 7\frac{x^6}{7!} + \cdots$$

$$= 1 - \frac{x^2}{2!} + \frac{x^4}{4!} - \frac{x^6}{6!} + \cdots = \cos x$$

$$\frac{d}{dx}[e^x] = \frac{d}{dx}\left[1 + x + \frac{x^2}{2!} + \frac{x^3}{3!} + \cdots\right]$$

$$= 1 + 2\frac{x}{2!} + 3\frac{x^2}{3!} + 4\frac{x^3}{4!} + \cdots$$

$$= 1 + x + \frac{x^2}{2!} + \frac{x^3}{3!} + \cdots = e^x \qquad \blacktriangleleft$$

11.12.2   THEOREM (*Integration of Power Series*).   *If a function f is represented by a power series, say*

$$f(x) = \sum_{k=0}^{\infty} c_k(x - a)^k$$

where the series has a nonzero radius of convergence R, then:

(a)   *The series of integrated terms*

$$\sum_{k=0}^{\infty}\left[\int c_k(x-a)^k\, dx\right] = \sum_{k=0}^{\infty}\frac{c_k}{k+1}(x-a)^{k+1}$$

has radius of convergence R.

(b)   *The function f is continuous on the interval $(a - R, a + R)$ and for all x in this interval*

$$\int f(x)\, dx = \sum_{k=0}^{\infty}\left[\int c_k(x-a)^k\, dx\right] + C$$

(c)   *For all α and β in the interval $(a - R, a + R)$, the series*

$$\sum_{k=0}^{\infty}\left[\int_\alpha^\beta c_k(x-a)^k\, dx\right]$$

converges absolutely and

$$\int_\alpha^\beta f(x)\, dx = \sum_{k=0}^{\infty}\left[\int_\alpha^\beta c_k(x-a)^k\, dx\right]$$

To paraphrase this theorem informally, *a power series representation of a function can be integrated term by term on any interval within the interval of convergence.*

REMARK.   Note that in part (b) a separate constant of integration is not introduced for each term in the series; rather, a single constant C is added to the entire series.

Example 2   To illustrate part (b) of the preceding theorem, we shall use the Maclaurin series for $\sin x$ and $\cos x$ to obtain the familiar integration formula

$$\int \cos x\, dx = \sin x + C$$

Integrating the Maclaurin series for $\cos x$ term by term yields

$$\int \cos x\, dx = \int\left[1 - \frac{x^2}{2!} + \frac{x^4}{4!} - \frac{x^6}{6!} + \cdots\right] dx$$

$$= \left[x - \frac{x^3}{3(2!)} + \frac{x^5}{5(4!)} - \frac{x^7}{7(6!)} + \cdots\right] + C$$

$$= \left[x - \frac{x^3}{3!} + \frac{x^5}{5!} - \frac{x^7}{7!} + \cdots\right] + C = \sin x + C \quad \blacktriangleleft$$

Example 3   The integral

$$\int_0^1 e^{-x^2}\, dx$$

cannot be evaluated directly because there is no elementary antiderivative of e^{-x^2}. However, it is possible to approximate the integral by some numerical technique such as

Simpson's rule. Still another possibility is to represent e^{-x^2} by its Maclaurin series and then integrate term by term in accordance with part (c) of Theorem 11.12.2. This produces a series that converges to the integral.

The simplest way to obtain the Maclaurin series for e^{-x^2} is to replace x by $-x^2$ in the Maclaurin series

$$e^x = 1 + x + \frac{x^2}{2!} + \frac{x^3}{3!} + \frac{x^4}{4!} + \cdots$$

to obtain

$$e^{-x^2} = 1 - x^2 + \frac{x^4}{2!} - \frac{x^6}{3!} + \frac{x^8}{4!} - \cdots$$

Therefore,

$$\int_0^1 e^{-x^2}\, dx = \int_0^1 \left[1 - x^2 + \frac{x^4}{2!} - \frac{x^6}{3!} + \frac{x^8}{4!} - \cdots \right] dx$$

$$= \left[x - \frac{x^3}{3} + \frac{x^5}{5(2!)} - \frac{x^7}{7(3!)} + \frac{x^9}{9(4!)} - \cdots \right]_0^1$$

$$= 1 - \frac{1}{3} + \frac{1}{5 \cdot 2!} - \frac{1}{7 \cdot 3!} + \frac{1}{9 \cdot 4!} - \cdots \qquad (1)$$

Thus, we have found a series that converges to the value of the integral $\int_0^1 e^{-x^2}\, dx$. Using the first three terms in this series we obtain the approximation

$$\int_0^1 e^{-x^2}\, dx \approx 1 - \frac{1}{3} + \frac{1}{10} = \frac{23}{30} \approx 0.767$$

Since series (1) satisfies the conditions of the alternating series test, it follows from Theorem 11.7.2 that the magnitude of the error in this approximation is at most $1/(7 \cdot 3!) = 1/42 \approx 0.0238$. Greater accuracy can be obtained by using more terms in the series.   ◄

□ **POWER SERIES REPRESENTATIONS MUST BE TAYLOR SERIES**

The following theorem shows that if a function f is represented by a power series in $x - a$ on an interval, then that series must be the Taylor series for f; thus, Taylor series are the only power series that can represent functions on an interval.

11.12.3   THEOREM.   *If*

$$f(x) = c_0 + c_1(x - a) + c_2(x - a)^2 + \cdots + c_n(x - a)^n + \cdots$$

for all x in some open interval containing a, then the series is the Taylor series for f about a.

Proof.   By repeated application of Theorem 11.12.1(b) we obtain

$$f(x)\ = c_0 + c_1(x - a) + c_2(x - a)^2 + c_3(x - a)^3 + c_4(x - a)^4 + \cdots$$

$$f'(x) = c_1 + 2c_2(x - a) + 3c_3(x - a)^2 + 4c_4(x - a)^3 + \cdots$$

$$f''(x) = 2!c_2 + (3 \cdot 2)c_3(x - a) + (4 \cdot 3)c_4(x - a)^2 + \cdots$$

$$f'''(x) = 3!c_3 + (4 \cdot 3 \cdot 2)c_4(x - a) + \cdots$$

$$\vdots$$

On substituting $x = a$, all the powers of $x - a$ drop out leaving

$$f(a) = c_0, \quad f'(a) = c_1, \quad f''(a) = 2!c_2, \quad f'''(a) = 3!c_3, \quad \ldots$$

from which we obtain

$$c_0 = f(a), \quad c_1 = f'(a), \quad c_2 = \frac{f''(a)}{2!}, \quad c_3 = \frac{f'''(a)}{3!}, \quad \ldots$$

which shows that the coefficients $c_0, c_1, c_2, c_3, \ldots$ are precisely the coefficients in the Taylor series about a for $f(x)$. ∎

REMARK.   The preceding theorem tells us that no matter how we arrive at a power series in $x - a$ converging to $f(x)$, be it by substitution, by integration, by differentiation, or by algebraic manipulation, the resulting series will be the Taylor series about a for $f(x)$.

Example 4   Find the Maclaurin series for $\tan^{-1} x$.

Solution.   We could calculate this Maclaurin series directly. However, we can also exploit Theorem 11.12.2 by first observing that

$$\int \frac{1}{1 + x^2}\, dx = \tan^{-1} x + C$$

and then integrating the Maclaurin series for $1/(1 + x^2)$ term by term. Since

$$\frac{1}{1 - x} = 1 + x + x^2 + x^3 + x^4 + \cdots \qquad -1 < x < 1$$

it follows on replacing x by $-x^2$ that

$$\frac{1}{1 + x^2} = 1 - x^2 + x^4 - x^6 + x^8 - \cdots \qquad -1 < x < 1 \tag{2}$$

Thus,

$$\tan^{-1} x + C = \int \frac{1}{1 + x^2}\, dx = \int [1 - x^2 + x^4 - x^6 + x^8 - \cdots]\, dx$$

or

$$\tan^{-1} x = \left[x - \frac{x^3}{3} + \frac{x^5}{5} - \frac{x^7}{7} + \frac{x^9}{9} - \cdots \right] - C$$

The constant of integration may be evaluated by substituting $x = 0$ and using the condition $\tan^{-1} 0 = 0$. This gives $C = 0$, so that

$$\tan^{-1} x = x - \frac{x^3}{3} + \frac{x^5}{5} - \frac{x^7}{7} + \frac{x^9}{9} - \cdots \tag{3}$$

We are guaranteed by Theorem 11.12.3 that the series we have produced is the Maclaurin series for $\tan^{-1} x$ and that it actually converges to $\tan^{-1} x$ for $-1 < x < 1$.   ◄

REMARK.   Theorems 11.12.1 and 11.12.2 say nothing about the behavior of the differentiated and integrated series at the endpoints $a - R$ and $a + R$. Indeed, by differentiating termwise, convergence may be lost at one or both endpoints; and by integrating termwise, convergence may be gained at one or both endpoints. As an illustration, we derived series (3) for $\tan^{-1} x$ by integrating series (2) for $1/(1 + x^2)$. We omit the proof that the series for $\tan^{-1} x$ converges to $\tan^{-1} x$ on the interval $[-1, 1]$, while the series for $1/(1 + x^2)$ converges to $1/(1 + x^2)$ only on $(-1, 1)$. Thus, convergence was gained at both endpoints by integrating.

☐ MISCELLANEOUS
TECHNIQUES FOR
OBTAINING TAYLOR
SERIES

$$1 - x^2 + \frac{x^4}{2} - \cdots$$
$$\times \quad x - \frac{x^3}{3} + \frac{x^5}{5} - \cdots$$
$$\overline{\quad x - x^3 + \frac{x^5}{2} - \cdots}$$
$$-\frac{x^3}{3} + \frac{x^5}{3} - \frac{x^7}{6} + \cdots$$
$$\frac{x^5}{5} - \frac{x^7}{5} + \cdots$$
$$\overline{x - \frac{4}{3}x^3 + \frac{31}{30}x^5 - \cdots}$$

We conclude this section with some techniques for obtaining Taylor series that would be messy to obtain directly.

Example 5 Find the first three terms that occur in the Maclaurin series for $e^{-x^2} \tan^{-1} x$.

Solution. Using the series for e^{-x^2} and $\tan^{-1} x$ obtained in Examples 3 and 4 gives

$$e^{-x^2} \tan^{-1} x = \left(1 - x^2 + \frac{x^4}{2} - \cdots\right)\left(x - \frac{x^3}{3} + \frac{x^5}{5} - \cdots\right)$$

Multiplying, as shown in the margin, we obtain

$$e^{-x^2} \tan^{-1} x = x - \frac{4}{3}x^3 + \frac{31}{30}x^5 - \cdots \qquad -1 < x < 1$$

More terms in the series can be obtained by including more terms in the factors. ◄

$$x + \frac{x^3}{3} + \frac{2x^5}{15} + \cdots$$
$$1 - \frac{x^2}{2} + \frac{x^4}{24} - \cdots \overline{\Big)\; x - \frac{x^3}{6} + \frac{x^5}{120} - \cdots}$$
$$\underline{x - \frac{x^3}{2} + \frac{x^5}{24} - \cdots}$$
$$\frac{x^3}{3} - \frac{x^5}{30} + \cdots$$
$$\underline{\frac{x^3}{3} - \frac{x^5}{6} + \cdots}$$
$$\frac{2x^5}{15} + \cdots$$

Example 6 Find the first three nonzero terms in the Maclaurin series for $\tan x$.

Solution. Instead of computing the series directly, we write

$$\tan x = \frac{\sin x}{\cos x} = \frac{x - \dfrac{x^3}{3!} + \dfrac{x^5}{5!} - \cdots}{1 - \dfrac{x^2}{2!} + \dfrac{x^4}{4!} - \cdots}, \qquad -\frac{\pi}{2} < x < \frac{\pi}{2}$$

Dividing, as shown in the margin, we obtain

$$\tan x = x + \frac{x^3}{3} + \frac{2x^5}{15} + \cdots \qquad -\frac{\pi}{2} < x < \frac{\pi}{2} \qquad ◄$$

▶ **Exercise Set 11.12** ⃞C 14–21

In Exercises 1–4, obtain the stated results by differentiating or integrating Maclaurin series term by term.

1. (a) $\dfrac{d}{dx}[e^x] = e^x$ (b) $\displaystyle\int e^x \, dx = e^x + C.$

2. (a) $\dfrac{d}{dx}[\cos x] = -\sin x$

(b) $\displaystyle\int \sin x \, dx = -\cos x + C.$

3. (a) $\dfrac{d}{dx}[\sinh x] = \cosh x$

(b) $\displaystyle\int \sinh x \, dx = \cosh x + C.$

4. (a) $\dfrac{d}{dx}[\ln(1+x)] = \dfrac{1}{1+x}$

(b) $\displaystyle\int \frac{1}{1+x} \, dx = \ln(1+x) + C.$

5. Derive the Maclaurin series for $1/(1+x)^2$ by differentiating an appropriate Maclaurin series term by term.

6. By differentiating an appropriate series, show that

$$\sum_{k=1}^{\infty} kx^k = \frac{x}{(1-x)^2} \qquad \text{for } -1 < x < 1$$

$$\left[\text{Hint: Consider } x\frac{d}{dx}\left[\frac{1}{1-x}\right].\right]$$

7. By integrating an appropriate series, show that

$$\sum_{k=1}^{\infty} \frac{x^k}{k} = \ln\left(\frac{1}{1-x}\right) \qquad \text{for } -1 < x < 1$$

$$\left[\text{Hint: } \ln\left(\frac{1}{1-x}\right) = -\ln(1-x).\right]$$

8. Use the result of Exercise 6 to find the sum of the series

$$\frac{1}{3} + \frac{2}{3^2} + \frac{3}{3^3} + \frac{4}{3^4} + \cdots$$

9. Use the result of Exercise 7 to find the sum of the series

$$\frac{1}{4} + \frac{1}{2(4^2)} + \frac{1}{3(4^3)} + \frac{1}{4(4^4)} + \cdots$$

10. Find the sum

$$\sum_{k=0}^{\infty} \frac{k+1}{k!} = 1 + 2 + \frac{3}{2!} + \frac{4}{3!} + \frac{5}{4!} + \cdots$$

[*Hint:* Differentiate the Maclaurin series for xe^x.]

11. Find the sum of the series

$$2 + 6x + 12x^2 + 20x^3 + \cdots$$

[*Hint:* Find the second derivative of the Maclaurin series for $1/(1-x)$.]

12. Find the sum $\displaystyle\sum_{k=1}^{\infty} \frac{k^2}{4^k}$. [*Hint:* Differentiate the Maclaurin series for $1/(1-x)$, multiply by x, differentiate, and multiply by x again.]

13. Let $\displaystyle f(x) = \sum_{k=0}^{\infty} (-1)^k \frac{x^{k+1}}{k+1}$

$$= x - \frac{x^2}{2} + \frac{x^3}{3} - \frac{x^4}{4} + \cdots$$

(a) Use the ratio test to show that the series converges for all x in the interval $(-1, 1)$.

(b) Use part (a) of Theorem 11.12.1 to find a power series for $f'(x)$. What is its interval of convergence?

(c) From the series obtained in part (b), deduce that $f'(x) = 1/(1+x)$ and hence that

$$f(x) = \ln(1+x) \quad \text{for } -1 < x < 1$$

[*Remark:* The Lagrange form of the remainder can be used to show that the Maclaurin series for $\ln(1+x)$ converges to $\ln(1+x)$ for $x = 1$ as well.]

In Exercises 14–21, use series to approximate the value of the integral to three decimal-place accuracy.

14. $\displaystyle\int_0^1 \sin x^2 \, dx.$

15. $\displaystyle\int_0^1 \cos\sqrt{x} \, dx.$

16. $\displaystyle\int_0^{0.1} \frac{\sin x}{x} \, dx.$

17. $\displaystyle\int_0^{1/2} \frac{dx}{1+x^4}.$

18. $\displaystyle\int_0^{1/2} \tan^{-1} 2x^2 \, dx.$

19. $\displaystyle\int_0^{0.1} e^{-x^3} \, dx.$

20. $\displaystyle\int_0^{0.2} \sqrt[3]{1+x^4} \, dx.$

21. $\displaystyle\int_0^{1/2} \frac{dx}{\sqrt[4]{x^2+1}}.$

In Exercises 22–29, use any method to find the first four nonzero terms in the Maclaurin series of the given function.

22. $x^4 e^x.$

23. $e^{-x^2} \cos x.$

24. $\dfrac{x^2}{1+x^4}.$

25. $\dfrac{\sin x}{e^x}.$

26. $\tanh x.$

27. $x \ln(1-x^2).$

28. $\dfrac{\ln(1+x)}{1-x}.$

29. $x^2 e^{4x}\sqrt{1+x}.$

30. Obtain the familiar result, $\lim_{x \to 0}(\sin x)/x = 1$, by finding a power series for $(\sin x)/x$ and taking the limit term by term.

31. Use the method of Exercise 30 to find the limits.

(a) $\displaystyle\lim_{x \to 0} \frac{1 - \cos x}{\sin x}$

(b) $\displaystyle\lim_{x \to 0} \frac{\ln\sqrt{1+x} - \sin 2x}{x}.$

32. (a) Use the relationship

$$\int \frac{1}{\sqrt{1-x^2}} \, dx = \sin^{-1} x + C$$

to find the first four nonzero terms in the Maclaurin series for $\sin^{-1} x$.

(b) Express the series in sigma notation.

(c) What is the radius of convergence?

33. (a) Use the relationship

$$\int \frac{1}{\sqrt{1+x^2}} \, dx = \sinh^{-1} x + C$$

to find the first four nonzero terms in the Maclaurin series for $\sinh^{-1} x$.

(b) Express the series in sigma notation.

(c) What is the radius of convergence?

34. Prove: If the power series $\sum_{k=0}^{\infty} a_k x^k$ and $\sum_{k=0}^{\infty} b_k x^k$ have the same sum on an interval $(-r, r)$, then $a_k = b_k$ for all values of k.

◆ **TECHNOLOGY EXERCISES**   Chapter 11

Most of these exercises require access to a graphing calculator or a computer algebra system (CAS) such as *Mathematica*, *Maple*, or *Derive*. When you are asked to *find* an answer or to *solve* an equation, you may choose to find an exact result or a numerical approximation, depending on the particular technology you are using and on your own imagination. The form of your answers may differ from those of other students or from those in the answer section of the text, depending on how you solve the problems and the accuracy you use in your numerical approximations. Those exercises that are more appropriate for a CAS than a graphing calculator are labeled with the icon ◆.

◆ **1. An approximation for π:** The Maclaurin series for $\tan^{-1} x$ converges on the interval $[-1, 1]$ and is given by

$$\tan^{-1} x = x - \frac{x^3}{3} + \frac{x^5}{5} - \frac{x^7}{7} + \cdots \qquad (1)$$

If $x = 1$, then we obtain

$$\frac{\pi}{4} = 1 - \frac{1}{3} + \frac{1}{5} - \frac{1}{7} + \cdots$$

(a) Use Theorem 11.7.2 to find a value for n to ensure that the nth partial sum s_n for the series for $\pi/4$ will approximate $\pi/4$ with an error that is less than 10^{-3} in magnitude.

(b) Using the value of n obtained in part (a), find an approximation for π and a bound on the magnitude of the error in this approximation.

(c) Verify that the bound obtained in part (b) is correct by comparing your approximation with the value for π produced by a calculator or CAS.

2. A better approximation for π: A formula for $\pi/4$ discovered by the British astronomer and mathematician John Machin in 1706 is

$$\frac{\pi}{4} = 4 \tan^{-1} \frac{1}{5} - \tan^{-1} \frac{1}{239}$$

(a) Let s_m be the mth partial sum for (1) in Exercise 1 with $x = 1/5$, and let s'_n be the nth partial sum for (1) with $x = 1/239$. Use Theorem 11.7.2 to find values of m and n to ensure that s_m and s'_n will approximate $\tan^{-1}(1/5)$ and $\tan^{-1}(1/239)$, respectively, with errors less than 10^{-7}.

(b) Using the values of m and n obtained in part (a), find an approximation for π from Machin's formula.

(c) Show that the magnitude of the error in the approximation to π obtained in part (b) is less than 2×10^{-6},

and verify that this bound is correct by comparing your approximation with the value for π produced by a calculator or CAS. [*Hint:* Use the triangle inequality (Theorem 1.2.5) to find the bound.]

◆ **3. An even better approximation for π:** In 1914, the brilliant Indian mathematician Ramanujan showed that

$$\frac{1}{\pi} = \frac{\sqrt{8}}{9801} \sum_{k=0}^{\infty} \frac{(4k)! \, (1103 + 26{,}390k)}{(k!)^4 \, 396^{4k}}$$

(a) Use the first two terms in Ramanujan's formula to obtain an approximation for π. [*Note:* Because of the high degree of accuracy in this formula, we suggest that you display 20 digits.]

(b) Compare your approximation for π in part (a) with the value for π in Figure 1.1.3.

◆ **4. Approximating tan (1/2):** Let $f(x) = \tan x$, and let $p_5(x)$ be the Maclaurin polynomial of degree 5 for $f(x)$.

(a) Use Lagrange's form of the remainder to find a bound on the magnitude of the error in using $p_5(1/2)$ to approximate $\tan (1/2)$.

(b) Find $p_5(1/2)$ and verify that the bound found in part (a) is correct by comparing $p_5(1/2)$ with the value of $\tan (1/2)$ obtained from a calculator or computer.

◆ **5. Maclaurin polynomials for $\tan x$:** Graph $f(x) = \tan x$ together with the Maclaurin polynomials $p_1(x)$, $p_3(x)$, $p_5(x)$, and $p_7(x)$ in the same coordinate system over the interval $[-1.5, 1.5]$.

◆ **6. Maclaurin series for rational functions:** Let

$$f(x) = \frac{2x^3 + 5x - 9}{x^4 - x^3 - 3x^2 - 3x - 18}$$

(a) Find the partial fraction decomposition of $f(x)$.

(b) Use the result of part (a) to find the radius of convergence of the Maclaurin series for $f(x)$.

◆ **7. The Riemann zeta function:** For real values of $x > 1$, the function defined by the convergent p-series

$$\zeta(x) = \sum_{k=1}^{\infty} \frac{1}{k^x}$$

is called the **Riemann zeta function**.

(a) Let s_n be the nth partial sum for the series for $\zeta(3.7)$. Find a value for n to ensure that s_n will approximate $\zeta(3.7)$ with an error that is less than 10^{-5}. [*Hint:* From Exercise 36 in Section 11.4, if S is the sum of the series, then

$$S - s_n < \int_n^{+\infty} f(x)\, dx$$

where $f(x) = 1/x^{3.7}$.]

(b) Approximate $\zeta(3.7)$ by s_n using the value of n obtained in part (a).

◆ **8. Euler's constant:** The number γ defined by

$$\gamma = \lim_{n \to +\infty} \left(\sum_{k=1}^{n} \frac{1}{k} - \ln n \right)$$

$$= \lim_{n \to +\infty} \left(1 + \frac{1}{2} + \frac{1}{3} + \cdots + \frac{1}{n} - \ln n \right)$$

is called **Euler's constant**. Approximate Euler's constant by finding the value of

$$\sum_{k=1}^{n} \frac{1}{k} - \ln n$$

for $n = 10{,}000$. [*Note:* For comparison, the value of γ to ten decimal places is 0.5772156649.]

◆ **9. The Bernoulli numbers:** Let $f(x) = x/(e^x - 1)$. Note that f and its derivatives are defined everywhere except at $x = 0$; however, if we define $f(0) = 1$, then it can be shown that $f^{(k)}(x)$ is continuous at $x = 0$ for $k = 0, 1, 2, 3, \ldots$, and thus f has a Maclaurin series of the form

$$\frac{x}{e^x - 1} = \sum_{k=0}^{\infty} \frac{B_k}{k!} x^k$$

where $B_k = f^{(k)}(0)$, $k = 0, 1, 2, 3, \ldots$. The number B_k is called the kth **Bernoulli number**. Use the continuity of $f^{(k)}(x)$ at $x = 0$ to find B_0 through B_6 from the formula

$$B_k = \lim_{x \to 0} f^{(k)}(x)$$

Approximation of functions: The nth Taylor polynomial for a function $f(x)$ about $x = a$ is generally a good approximation for $f(x)$ when x is near a, but the accuracy usually deteriorates as the distance between x and a increases. In Exercises 10 and 11 we will consider alternative approximations that may produce better accuracy than a Taylor polynomial.

10. Polynomial approximations

(a) Let $E_1(x) = \cos x - p_4(x)$, where

$$p_4(x) = 1 - \frac{x^2}{2!} + \frac{x^4}{4!}$$

is the fourth-degree Maclaurin polynomial approximation for $\cos x$. Graph $E_1(x)$ for $-1 \le x \le 1$. Find the maximum value of $|E_1(x)|$ on the interval $[-1, 1]$.

(b) Using methods from numerical analysis, it can be shown that the polynomial

$$p(x) = 0.99995795 - 0.49924045 x^2 + 0.03962674 x^4$$

is a good approximation for $\cos x$ on the interval $[-1, 1]$. Let $E_2(x) = \cos x - p(x)$. Graph $E_2(x)$ for $-1 \le x \le 1$. Find the maximum value of $|E_2(x)|$ on the interval $[-1, 1]$.

(c) Discuss the advantages and disadvantages of the approximations $p_4(x)$ and $p(x)$ for $\cos x$ in parts (a) and (b).

11. Rational approximations

(a) Let $E_1(x) = e^x - p_5(x)$, where

$$p_5(x) = 1 + x + \frac{x^2}{2!} + \frac{x^3}{3!} + \frac{x^4}{4!} + \frac{x^5}{5!}$$

is the fifth-degree Maclaurin polynomial approximation for e^x. Graph $E_1(x)$ for $-1 \le x \le 1$. Find the maximum value of $|E_1(x)|$ on the interval $[-1, 1]$.

(b) Using methods from numerical analysis, it can be shown that the rational function

$$r(x) = \frac{1 + \frac{3}{5}x + \frac{3}{20}x^2 + \frac{1}{60}x^3}{1 - \frac{2}{5}x + \frac{1}{20}x^2}$$

is a good approximation for e^x on the interval $[-1, 1]$. Let $E_2(x) = e^x - r(x)$. Graph $E_2(x)$ for $-1 \le x \le 1$. Find the maximum value of $|E_2(x)|$ on the interval $[-1, 1]$.

(c) Discuss the advantages and disadvantages of the approximations $p_5(x)$ and $r(x)$ for e^x in parts (a) and (b).

C. Mac Laurin

Colin Maclaurin (1698–1746)

12 TOPICS IN ANALYTIC GEOMETRY

■ **12.1** INTRODUCTION TO THE CONIC SECTIONS

The surface shown at the left in Figure 12.1.1 is called a **double right-circular cone** or sometimes simply a **cone**. It is the surface in three-dimensional space generated by a line that revolves about a fixed **axis** in such a way that the line passes through a fixed point on the axis and always makes the same angle with the axis. The fixed point is called the **vertex** of the cone. The cone consists of two parts, called **nappes**, that intersect at the vertex.

The curves that can be obtained as intersections of a cone and a plane are called **conics** or **conic sections**, the most important of which are circles, ellipses, parabolas, and hyperbolas (Figure 12.1.1). A **circle** is obtained by cutting a cone with a plane that is perpendicular to the axis and does not contain the vertex. If the cutting plane is tilted slightly and intersects only one nappe, the resulting intersection is an **ellipse**. If the cutting plane is tilted still further so that it is parallel to a line on the surface of the cone, but intersects only one nappe, the resulting intersection is a **parabola**. Finally, if the plane intersects both nappes, but does not contain the vertex, the resulting intersection is a **hyperbola**.

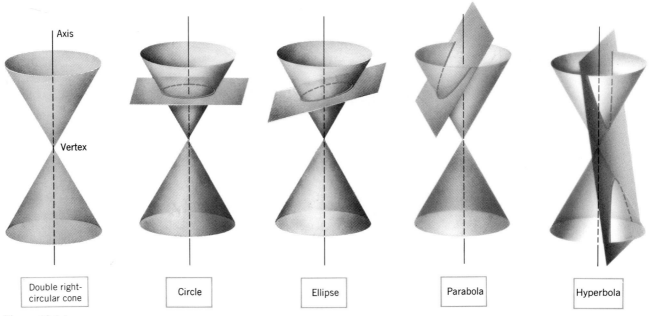

Figure 12.1.1

By choosing the cutting plane to pass through the vertex, it is possible to obtain a point, a line, or a pair of lines for the intersection (Figure 12.1.2). These are called **_degenerate conic sections_**.

According to the Alexandrian geographer and astronomer Eratosthenes, the conic sections were first discovered by Menaechmus, a geometer and astronomer in Plato's academy. Although it is not known for certain what motivated the discovery of the conic sections, it is commonly believed that they resulted from the study of construction problems. Menaechmus, for example, used them to solve the problem of "doubling the cube," that is, constructing a cube whose volume is twice that of a given cube. Another theory suggests that the conic sections may have originated as a result of work on sundials. With the advent of analytic geometry and calculus, conic sections gained importance in the physical sciences. In 1609 Johannes Kepler published a book known as _Astronomia Nova_ in which he presented his landmark discovery that the path of each planet about the sun is an ellipse. Galileo and Newton showed that objects subject to gravitational forces can also move along paths that are parabolas or hyperbolas.

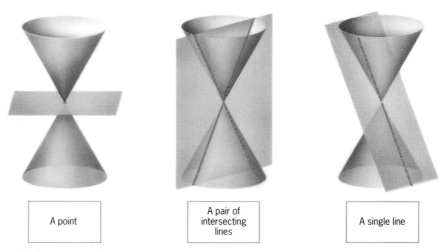

Figure 12.1.2

Today, properties of conic sections are used in the construction of telescopes, radar antennas, medical equipment, navigational systems, and in the determination of satellite orbits. In this chapter we shall develop the basic geometric properties and equations of conic sections. For simplicity, our working definitions of the conic sections will be based on their geometric properties rather than their interpretation as intersections of a plane with a cone. This will enable us to keep our work in a two-dimensional setting.

■ 12.2 THE PARABOLA; TRANSLATION OF COORDINATE AXES

In this section we shall discuss properties of parabolas.

□ **DEFINITION OF A PARABOLA**

12.2.1 DEFINITION. A *parabola* is the set of all points in the plane that are equidistant from a given line and a given point not on the line.

The given line is called the *directrix* of the parabola, and the given point the *focus* (Figure 12.2.1). A parabola is symmetric about the line that passes through the focus at right angles to the directrix. This line, called the *axis* or the *axis of symmetry* of the parabola, meets the parabola at a point called the *vertex*.

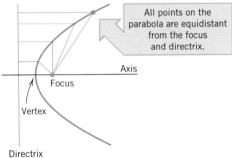

All points on the parabola are equidistant from the focus and directrix.

Axis

Focus

Vertex

Figure 12.2.1   Directrix

□ **GEOMETRIC PROPERTIES OF A PARABOLA**

It is traditional in the study of parabolas to denote the distance between the focus and the vertex by p. The vertex is equidistant from the focus and the directrix, so the distance between the vertex and the directrix is also p; consequently, the distance between the focus and the directrix is $2p$ (Figure 12.2.2). As illustrated in that figure, the parabola passes through two of the corners of a box that extends from the vertex to the focus along the axis of symmetry and extends $2p$ units above and $2p$ units below the axis of symmetry.

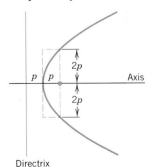

p p $2p$ Axis $2p$

Figure 12.2.2   Directrix

□ **STANDARD EQUATIONS OF A PARABOLA**

The equation of a parabola is simplest if the coordinate axes are positioned so that the vertex is at the origin and the axis of symmetry is along the x-axis or y-axis. The four possible such orientations are shown in Figure 12.2.3. These are called the *standard positions* of a parabola, and the resulting equations are called the ***standard equations*** of a parabola.

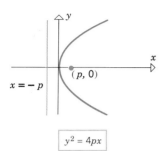

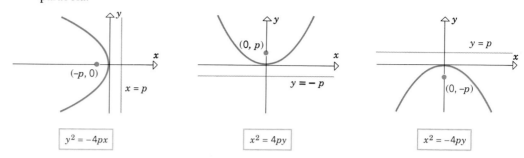

Figure 12.2.3

To illustrate how the equations in Figure 12.2.3 are obtained, we shall derive the equation for the parabola on the left. Let $P(x, y)$ be any point on the parabola. Since P is equidistant from the focus and directrix, the distances PF and PD in Figure 12.2.4 are equal; that is,

$$PF = PD \tag{1}$$

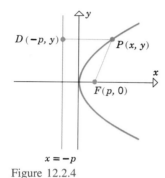

Figure 12.2.4

where $D(-p, y)$ is the foot of the perpendicular from P to the directrix. From the distance formula, the distances PF and PD are

$$PF = \sqrt{(x - p)^2 + y^2} \quad \text{and} \quad PD = \sqrt{(x + p)^2} \tag{2}$$

Substituting in (1) and squaring yields

$$(x - p)^2 + y^2 = (x + p)^2 \tag{3}$$

and after simplifying

$$y^2 = 4px \tag{4}$$

Conversely, any point $P(x, y)$ satisfying (4) also satisfies (3) (reverse the steps in the simplification); thus, from (2), $PF = PD$, which shows that P is equidistant from the focus and directrix. Therefore, each point satisfying (4) lies on the parabola.

The remaining equations in Figure 12.2.3 have similar derivations; they are left as exercises.

☐ **A TECHNIQUE FOR GRAPHING PARABOLAS**

Parabolas can be graphed from their *standard equations* using four basic steps:

- Determine whether the axis of symmetry is along the x-axis or the y-axis. This can be ascertained from the quadratic term in the equation; referring to Figure 12.2.3, the axis of symmetry is along the x-axis if the equation has a y^2-term, and it is along the y-axis if it has an x^2-term.

- Determine which way the parabola opens. If the axis of symmetry is along the x-axis, then the parabola opens to the right if the coefficient of x is positive, and it opens to the left if the coefficient is negative. If the axis of symmetry is along the y-axis, then the parabola opens up if the coefficient of y is positive, and it opens down if the coefficient is negative.

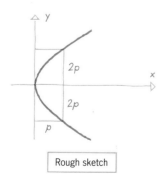

Rough sketch

Figure 12.2.5

- Determine the value of p and draw a box extending p units from the origin along the axis of symmetry in the direction in which the parabola opens and extending $2p$ units on each side of the axis of symmetry.

- Using the box as a guide, sketch the parabola so that its vertex is at the origin and it passes through the corners of the box (Figure 12.2.5).

Example 1   Sketch the graphs of the parabolas

(a) $x^2 = 12y$ (b) $y^2 + 8x = 0$

showing the focus and directrix of each.

Solution (a).   This equation involves x^2, so the axis of symmetry is along the y-axis, and the coefficient of y is positive, so the parabola opens upward. From the coefficient of y, we obtain $4p = 12$ or $p = 3$. Drawing a box extending $p = 3$ units up from the origin and $2p = 6$ units to the left and $2p = 6$ units to the right of the y-axis, then using corners of the box as a guide, yields the graph in Figure 12.2.6.

The focus lies $p = 3$ units from the vertex along the axis of symmetry in the direction in which the parabola opens, so its coordinates are $(0, 3)$. The directrix is perpendicular to the axis of symmetry at a distance of $p = 3$ units from the vertex on the opposite side from the focus, so its equation is $y = -3$.

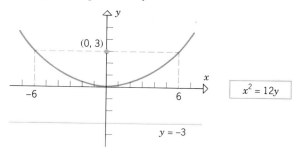

Figure 12.2.6

Solution (b).   We first rewrite the equation in the standard form

$$y^2 = -8x$$

This equation involves y^2, so the axis of symmetry is along the x-axis, and the coefficient of x is negative, so the parabola opens to the left. From the coefficient of x we obtain $4p = 8$, so $p = 2$. Drawing a box extending $p = 2$ units left from the origin and $2p = 4$ units above and $2p = 4$ units below the x-axis, then using corners of the box as a guide, yields the graph in Figure 12.2.7.   ◄

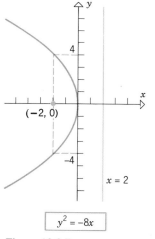

Figure 12.2.7

Example 2   Find an equation for the parabola that is symmetric about the y-axis, has its vertex at the origin, and passes through the point $(5, 2)$.

Solution.   Since the parabola is symmetric about the y-axis and has its vertex at the origin, the equation is of the form

$$x^2 = 4py   \quad \text{or} \quad   x^2 = -4py$$

where the sign depends on whether the parabola opens up or down. But the parabola must open up, since it passes through the point $(5, 2)$, which lies in the first quadrant. Thus, the equation is of the form

$$x^2 = 4py \tag{5}$$

Since the parabola passes through $(5, 2)$, we must have $5^2 = 4p \cdot 2$ or $4p = \frac{25}{2}$. Therefore, (5) becomes

$$x^2 = \frac{25}{2}y   \quad ◄$$

☐ **TRANSLATION OF AXES**

If a parabola has its axis of symmetry parallel to one of the coordinate axes, and its vertex is not at the origin, then the location of the focus and directrix can be found by introducing an auxiliary coordinate system with its origin at the vertex and its axes parallel to the original axes. Before we can discuss the details, we need some preliminary results about translating coordinate axes.

In Figure 12.2.8*a* we have translated the axes of an *xy*-coordinate system to obtain a new *x'y'*-coordinate system whose origin O' is at the point $(x, y) = (h, k)$. As a result, a point P in the plane will have both (x, y)-coordinates and (x', y')-coordinates. As suggested by Figure 12.2.8*b*, these coordinates are related by

$$x' = x - h, \quad y' = y - k \tag{6}$$

or, equivalently,

$$x = x' + h, \quad y = y' + k \tag{7}$$

These are called the **translation equations**.

As an illustration, if the new origin is at $(h, k) = (4, -1)$ and the *xy*-coordinates of a point P are $(2, 5)$, then the *x'y'*-coordinates of P are

$$x' = x - h = 2 - 4 = -2 \quad \text{and} \quad y' = y - k = 5 - (-1) = 6$$

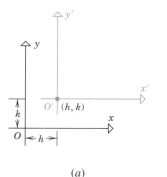   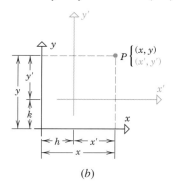

Figure 12.2.8   (*a*)   (*b*)

□ **TRANSLATED PARABOLAS**

Let us now consider a parabola with vertex V at (h, k) in an *xy*-coordinate system and axis of symmetry parallel to the *y*-axis. If we translate the axes so that the vertex V is at the origin of an *x'y'*-coordinate system, then in *x'y'*-coordinates the equation of the parabola will be

$$(x')^2 = 4py' \quad \text{or} \quad (x')^2 = -4py'$$

(see Figure 12.2.9, for example) and from the translation equations (6), the corresponding equation in *xy*-coordinates will be

> **Parabola with Vertex (h, k) and Axis Parallel to y-axis**
>
> $$(x - h)^2 = \pm 4p(y - k) \tag{8}$$
>
> where the $+$ sign occurs if the parabola opens in the positive *y*-direction and the $-$ sign if it opens in the negative *y*-direction.

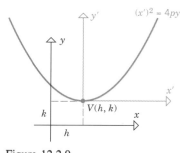

Figure 12.2.9

Similarly,

> **Parabola with Vertex (h, k) and Axis Parallel to x-axis**
>
> $$(y - k)^2 = \pm 4p(x - h) \tag{9}$$
>
> where the $+$ sign occurs if the parabola opens in the positive *x*-direction and the $-$ sign if it opens in the negative *x*-direction.

Sometimes (8) and (9) appear in expanded form, in which case some algebraic manipulations are required to identify the parabola. The following example illustrates this.

Example 3 Show that the curve

$$y^2 - 8x - 6y - 23 = 0$$

is a parabola. Sketch its graph, and show the focus and directrix.

Solution. Because the equation involves y to the second power and x to the first power, and because the expanded form of (9) has the same property, we shall try to rewrite the equation in form (9). To do this, we first collect all the y-terms on one side:

$$y^2 - 6y = 8x + 23$$

Next, we complete the square on the y-terms by adding 9 to both sides:

$$(y - 3)^2 = 8x + 32$$

Finally, we factor out the coefficient of the x-term to obtain

$$(y - 3)^2 = 8(x + 4)$$

From (9) with the plus sign, the equation represents a parabola with vertex at $(h, k) = (-4, 3)$ and axis of symmetry parallel to the x-axis. Moreover, $4p = 8$ or $p = 2$. Since the coefficient of p is positive, the parabola opens in the positive x-direction. This places the focus 2 units to the right of the vertex or at the point $(-2, 3)$. The directrix is 2 units to the left of the vertex (and parallel to the y-axis), so its equation is $x = -6$. The parabola is sketched in Figure 12.2.10. ◄

The procedure used in Example 3 can be used to prove the following general result.

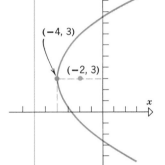

Directrix
$x = -6$

Figure 12.2.10

12.2.2 THEOREM. *The graph of*

$$y = Ax^2 + Bx + C \quad (A \neq 0)$$

is a parabola with axis of symmetry parallel to the y-axis; the parabola opens in the positive y-direction if $A > 0$ and in the negative y-direction if $A < 0$. The graph of

$$x = Ay^2 + By + C \quad (A \neq 0)$$

is a parabola with axis of symmetry parallel to the x-axis; the parabola opens in the positive x-direction if $A > 0$ and in the negative x-direction if $A < 0$.

Example 4 Find an equation for the parabola that has its vertex at $(1, 2)$ and its focus at $(4, 2)$.

Solution. Since the focus and vertex are on a horizontal line, and since the focus is to the right of the vertex, the parabola opens to the right and its equation has the form

$$(y - k)^2 = 4p(x - h)$$

Since the vertex and focus are three units apart we have $p = 3$, and since the vertex is at $(h, k) = (1, 2)$ we obtain

$$(y - 2)^2 = 12(x - 1) \quad ◄$$

☐ **REFLECTION PROPERTIES OF PARABOLAS**

Parabolas have important applications in the design of telescopes, radar antennas, and lighting systems. This is due to the following property of parabolas (Exercise 44).

12.2.3 THEOREM (*A Geometric Property of Parabolas*). *The tangent line at a point P on a parabola makes equal angles with the line through P parallel to the axis of symmetry and the line through P and the focus (Figure 12.2.11).*

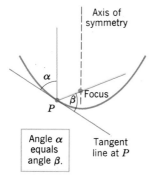

Figure 12.2.11

It is a principle of physics that when light is reflected from a point P on a surface, the angle of incidence equals the angle of reflection; that is, the angle between the incoming ray and the tangent line at P equals the angle between the outgoing ray and the tangent line at P. Therefore, if the reflecting surface has parabolic cross sections with a common focus, then it follows from Theorem 12.2.3 that all light rays entering parallel to the axis will be reflected through the focus (Figure 12.2.12a). In reflecting telescopes, this principle is used to reflect the (approximately) parallel rays of light from the stars or planets off a parabolic mirror to an eyepiece at the focus of the parabola. Conversely, if a light source is located at the focus of a parabolic reflector, it follows from Theorem 12.2.3 that the reflected rays will form a beam parallel to the axis (Figure 12.2.12b). The parabolic reflectors in flashlights and automobile headlights utilize this principle. The optical principles just discussed also apply to radar signals, which explains the parabolic shape of many radar antennas.

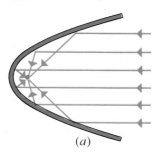

Figure 12.2.12 (a) (b)

▶ Exercise Set 12.2

In Exercises 1–16, sketch the parabola. Show the focus, vertex, and directrix.

1. $y^2 = 6x$.

2. $y^2 = -10x$.

3. $x^2 = -9y$.

4. $x^2 = 4y$.

5. $5y^2 = 12x$.

6. $x^2 - 40y = 0$.

7. $(y - 3)^2 = 6(x - 2)$.

8. $(y + 1)^2 = -7(x - 4)$.

9. $(x + 2)^2 = -(y + 2)$.

10. $(x - \frac{1}{2})^2 = 2(y - 1)$.

11. $x^2 - 4x + 2y = 1$.

12. $y^2 - 6y - 2x + 1 = 0$.

13. $x = y^2 - 4y + 2$.

14. $y = 4x^2 + 8x + 5$.

15. $-y^2 + 2y + x = 0$.

16. $y = 1 - 4x - x^2$.

In Exercises 17–31, find an equation for the parabola satisfying the given conditions.

17. Vertex $(0, 0)$;
       focus $(3, 0)$.

18. Vertex $(0, 0)$;
       focus $(0, -4)$.

19. Vertex $(0, 0)$;
       directrix $x = 7$.

20. Vertex $(0, 0)$;
       directrix $y = \frac{1}{2}$.

21. Vertex $(0, 0)$; symmetric about the x-axis; passes through $(2, 2)$.

22. Vertex $(0, 0)$; symmetric about the y-axis; passes through $(-1, 3)$.

23. Focus $(0, -3)$;
       directrix $y = 3$.

24. Focus $(6, 0)$;
       directrix $x = -6$.

25. Axis $y = 0$; passes through $(3, 2)$ and $(2, -3)$.

26. Axis $x = 0$; passes through $(2, -1)$ and $(-4, 5)$.

27. Focus $(3, 0)$;
       directrix $x = 0$.

28. Vertex $(4, -5)$;
       focus $(1, -5)$.

29. Vertex $(1, 1)$;
       directrix $y = -2$.

30. Focus $(-1, 4)$;
       directrix $x = 5$.

31. Vertex $(5, -3)$; axis parallel to the y-axis; passes through $(9, 5)$.

32. Use the definition of a parabola (12.2.1) to find the equation of the parabola with focus $(2, 1)$ and directrix $x + y + 1 = 0$. [*Hint:* Use the result of Exercise 19, Section 1.5.]

33. (a) Find an equation for the parabola with axis parallel to the y-axis and passing through $(0, 3)$, $(2, 0)$, and $(3, 2)$.

   (b) Find an equation for the parabola with axis parallel to the x-axis and passing through the points in part (a).

34. Prove: The line tangent to the parabola $x^2 = 4py$ at the point (x_0, y_0) is

$$y = \frac{x_0}{2p}x - y_0$$

35. Find the vertex, focus, and directrix of the parabola $y = Ax^2 + Bx + C$ $(A \neq 0)$.

36. (a) Find an equation for the parabolic arch with base b and height h, shown in the following figure.

   (b) Find the area under the arch.

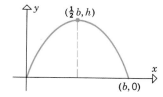

37. A parabolic arch spans a road 40 feet wide. How high is the arch if a center section of the road 20 feet wide has a minimum clearance of 12 feet?

38. Let C be a circle of radius r and L a line that does not intersect C but is in the same plane as C. Show that the centers of all circles that do not enclose C and are tangent to both C and L lie on a parabola. Specify the location of the focus and directrix of the parabola.

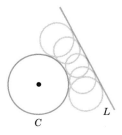

39. Prove: The vertex is the closest point on a parabola to the focus.

40. A comet moves in a parabolic orbit with the sun at its focus. When the comet is 40 million miles from the center of the sun, the line from the sun to the comet makes an angle of 60° with the axis of the parabola as shown in the

following figure. How close will the comet come to the center of the sun? [*Hint:* See Exercise 39.]

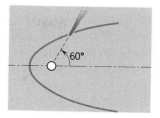

41. How far from the vertex should a light source be placed on the axis of the following parabolic reflector to produce a beam of parallel rays?

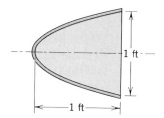

42. Derive the equation $x^2 = 4py$ in Figure 12.2.3.

43. Derive the equation $y^2 = -4px$ in Figure 12.2.3.

44. Prove Theorem 12.2.3. [*Hint:* Choose coordinate axes so that the parabola has the equation $x^2 = 4py$. Show that the tangent line at $P(x_0, y_0)$ intersects the y-axis at $Q(0, -y_0)$ and that the triangle whose three vertices are at P, Q, and the focus is isosceles.]

■ 12.3 THE ELLIPSE

In this section we shall discuss properties of ellipses.

☐ **DEFINITION OF AN ELLIPSE**

12.3.1   DEFINITION.   An *ellipse* is the set of all points in the plane, the sum of whose distances from two fixed points is a given positive constant that is greater than the distance between the fixed points.

The two fixed points are called the *foci* (plural of "focus"), and the midpoint of the line segment joining the foci is called the *center* of the ellipse (Figure 12.3.1). To help visualize Definition 12.3.1, imagine that two ends of a string are tacked to the foci and a pencil traces a curve as it is held tight against the string (Figure 12.3.2). The resulting curve will be an ellipse since the sum of the distances to the foci is a constant, namely the total length of the string. Note that if the foci coincide, the ellipse reduces to a circle.

For ellipses other than circles, the line segment through the foci and across the ellipse is called the *major axis* (Figure 12.3.3), and the line segment across the ellipse, through the center, and perpendicular to the major axis is called the *minor axis*. The endpoints of the

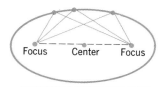

For all points on the ellipse, the sum of the distances to the foci is the same.

Figure 12.3.1

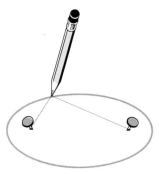

Figure 12.3.2

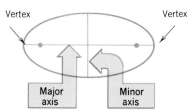

Figure 12.3.3

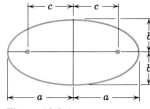

Figure 12.3.4

major axis are called **vertices**. It is also traditional in the study of ellipses to denote the length of the major axis by $2a$, the length of the minor axis by $2b$, and the distance between the foci by $2c$ (Figure 12.3.4). The numbers a and b are called the **semiaxes** of the ellipse.

☐ **GEOMETRIC PROPERTIES OF AN ELLIPSE**

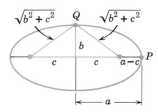

Figure 12.3.5

There is a basic relationship between the numbers a, b, and c that can be obtained by considering a point P at the end of the major axis and a point Q at the end of the minor axis. Because P and Q both lie on the ellipse, the sum of the distances from each of them to the foci will be the same. If we express this fact as an equation, we obtain (see Figure 12.3.5)

$$2\sqrt{b^2 + c^2} = (a - c) + (a + c)$$

from which it follows that

$$a = \sqrt{b^2 + c^2} \tag{1}$$

or equivalently,

$$c = \sqrt{a^2 - b^2} \tag{2}$$

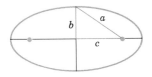

Figure 12.3.6

From (1), the distance from a focus to an end of the minor axis is a (Figure 12.3.6), which implies that for *all* points on the ellipse the sum of the distances to the foci is $2a$.

It also follows from (1) that $a \geq b$ with the equality holding only when $c = 0$. Geometrically, this means that the major axis of an ellipse is at least as large as the minor axis, and that the two axes have equal length only when the foci coincide, in which case the ellipse is a circle.

☐ **STANDARD EQUATIONS OF AN ELLIPSE**

The equation of an ellipse is simplest if the coordinate axes are positioned so that the center of the ellipse is at the origin and the foci are on the x-axis or y-axis. The two possible such orientations are shown in Figure 12.3.7. These are called the **standard positions** of an ellipse, and the resulting equations are called the **standard equations** of an ellipse.

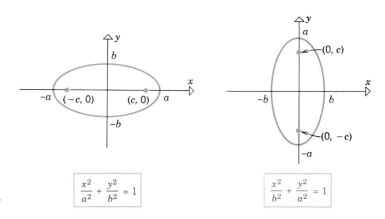

Figure 12.3.7

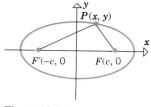

Figure 12.3.8

To illustrate how the equations in Figure 12.3.7 are obtained, we shall derive the equation for the ellipse on the left. Let $P(x, y)$ be any point on that ellipse. Since the sum of the distances from P to the foci is $2a$, it follows (Figure 12.3.8) that

$$PF' + PF = 2a$$

so

$$\sqrt{(x + c)^2 + y^2} + \sqrt{(x - c)^2 + y^2} = 2a$$

Transposing the second radical to the right side of the equation and squaring yields

$$(x + c)^2 + y^2 = 4a^2 - 4a\sqrt{(x - c)^2 + y^2} + (x - c)^2 + y^2$$

and, on simplifying,

$$\sqrt{(x - c)^2 + y^2} = a - \frac{c}{a}x \tag{3}$$

Squaring again and simplifying yields

$$\frac{x^2}{a^2} + \frac{y^2}{a^2 - c^2} = 1$$

which, by virtue of (1), can be written as

$$\frac{x^2}{a^2} + \frac{y^2}{b^2} = 1 \tag{4}$$

Conversely, it can be shown that any point whose coordinates satisfy (4) has $2a$ as the sum of its distances from the foci, so that such a point is on the ellipse.

□ **A TECHNIQUE FOR GRAPHING ELLIPSES**

Ellipses can be graphed from their *standard equations* using three basic steps:

- Determine whether the major axis is on the x-axis or the y-axis. This can be ascertained from the sizes of the denominators in the equation. Referring to Figure 12.3.7, and keeping in mind that $a^2 > b^2$ (since $a > b$), the major axis is along the x-axis if x^2 has the larger denominator, and it is along the y-axis if y^2 has the larger denominator. If the denominators are equal, the ellipse is a circle.

- Determine the values of a and b and draw a box extending a units on each side of the center along the major axis and b units on each side of the center along the minor axis.

- Using the box as a guide, sketch the ellipse so that its center is at the origin and it touches the sides of the box where the sides intersect the coordinate axes (Figure 12.3.9).

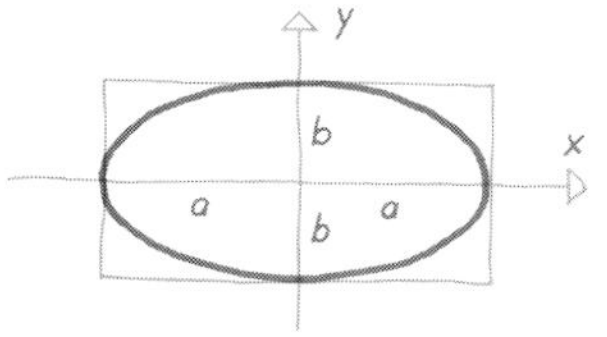

Figure 12.3.9

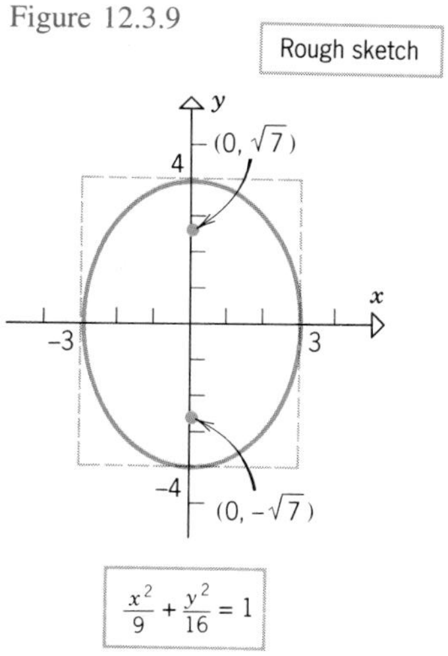

Rough sketch

Example 1 Sketch the graphs of the ellipses

(a) $\dfrac{x^2}{9} + \dfrac{y^2}{16} = 1$ (b) $x^2 + 2y^2 = 4$

showing the foci of each.

Solution (a). Since y^2 has the larger denominator, the major axis is along the y-axis. Moreover, since $a^2 > b^2$, we must have $a^2 = 16$ and $b^2 = 9$, so

$$a = 4 \quad \text{and} \quad b = 3$$

Drawing a box extending 4 units on each side of the origin along the y-axis and 3 units on each side of the origin along the x-axis as a guide yields the graph in Figure 12.3.10.

The foci lie c units on each side of the center along the major axis, where c is given by (2). From the values of a^2 and b^2 above, we obtain

$$c = \sqrt{a^2 - b^2} = \sqrt{16 - 9} = \sqrt{7} \approx 2.6$$

Thus, the coordinates of the foci are $(0, \sqrt{7})$ and $(0, -\sqrt{7})$, since they lie on the y-axis.

$\dfrac{x^2}{9} + \dfrac{y^2}{16} = 1$

Figure 12.3.10

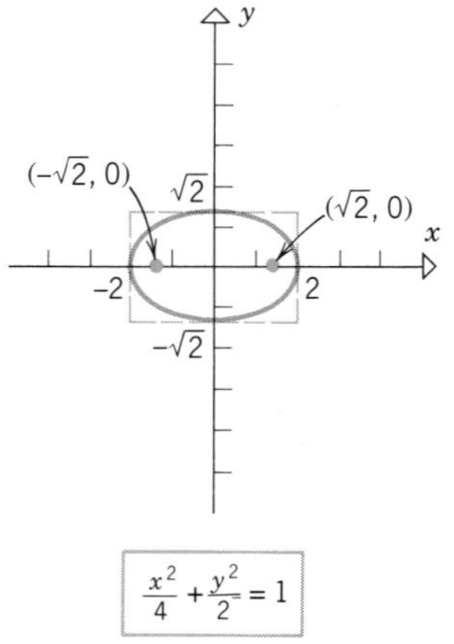

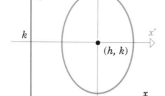

Figure 12.3.11

Solution (b).   We first rewrite the equation in the standard form

$$\frac{x^2}{4} + \frac{y^2}{2} = 1$$

Since x^2 has the larger denominator, the major axis lies along the x-axis, and we have $a^2 = 4$ and $b^2 = 2$. Drawing a box extending $a = 2$ on each side of the origin along the x-axis and extending $b = \sqrt{2} \approx 1.4$ units on each side of the origin along the y-axis as a guide yields the graph in Figure 12.3.11.

From (2), we obtain

$$c = \sqrt{a^2 - b^2} = \sqrt{2} \approx 1.4$$

Thus, the coordinates of the foci are $(\sqrt{2}, 0)$ and $(-\sqrt{2}, 0)$, since they lie on the x-axis.   ◄

Example 2   Find an equation for the ellipse with foci $(0, \pm 2)$ and major axis with endpoints $(0, \pm 4)$.

Solution.   From Figure 12.3.7, the equation has the form

$$\frac{x^2}{b^2} + \frac{y^2}{a^2} = 1$$

and from the given information, $a = 4$ and $c = 2$. It follows from (1) that

$$b^2 = a^2 - c^2 = 16 - 4 = 12$$

so the equation of the ellipse is

$$\frac{x^2}{12} + \frac{y^2}{16} = 1$$ ◄

□ **TRANSLATED ELLIPSES**

If the axes of an ellipse are parallel to the coordinate axes but the center is not at the origin, then the equation of the ellipse may be determined by translation of axes. Specifically, if the center of the ellipse is at the point with xy-coordinates (h, k), and if we translate the xy-axes so that the origin of the $x'y'$-system is at the center (h, k), then the equation of the ellipse in the $x'y'$-system will be

$$\frac{(x')^2}{a^2} + \frac{(y')^2}{b^2} = 1 \quad \text{or} \quad \frac{(x')^2}{b^2} + \frac{(y')^2}{a^2} = 1$$

depending on the orientation of the major and minor axes (Figure 12.3.12). Thus, from the translation equations $x' = x - h$, $y' = y - k$, the equation of the ellipse in the xy-system will be one of the following:

Figure 12.3.12

Ellipse with Center (h, k) and Major Axis Parallel to x-axis

$$\frac{(x - h)^2}{a^2} + \frac{(y - k)^2}{b^2} = 1 \qquad (a \geq b) \tag{5}$$

Ellipse with Center (h, k) and Major Axis Parallel to y-axis

$$\frac{(x - h)^2}{b^2} + \frac{(y - k)^2}{a^2} = 1 \qquad (a \geq b) \tag{6}$$

Sometimes (5) and (6) appear in expanded form, in which case it is necessary to complete the squares before the ellipse can be analyzed.

Example 3   Show that the curve

$$16x^2 + 9y^2 - 64x - 54y + 1 = 0$$

is an ellipse. Sketch the ellipse and show the location of the foci.

Solution. Our objective is to rewrite the equation in one of the forms, (5) or (6). To do this, group the x-terms and y-terms and take the constant to the right side:

$$(16x^2 - 64x) + (9y^2 - 54y) = -1$$

Next, factor out the coefficients of x^2 and y^2 and complete the squares:

$$16(x^2 - 4x + 4) + 9(y^2 - 6y + 9) = -1 + 64 + 81$$

or

$$16(x - 2)^2 + 9(y - 3)^2 = 144$$

Finally, divide through by 144 to introduce a 1 on the right side:

$$\frac{(x-2)^2}{9} + \frac{(y-3)^2}{16} = 1$$

This is an equation of form (6), with $h = 2$, $k = 3$, $a^2 = 16$, and $b^2 = 9$. Thus, the given equation is an ellipse with center $(2, 3)$ and major axis parallel to the y-axis. Since $a = 4$, the major axis extends 4 units above and 4 units below the center, so its endpoints are $(2, 7)$ and $(2, -1)$ (Figure 12.3.13). Since $b = 3$, the minor axis extends 3 units to the left and 3 units to the right of the center, so its endpoints are $(-1, 3)$ and $(5, 3)$. Since

$$c = \sqrt{a^2 - b^2} = \sqrt{16 - 9} = \sqrt{7}$$

the foci lie $\sqrt{7}$ units above and below the center, placing them at the points $(2, 3 + \sqrt{7})$ and $(2, 3 - \sqrt{7})$. ◄

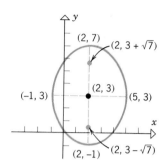

Figure 12.3.13

□ **FOCUS–DIRECTRIX PROPERTY OF ELLIPSES**

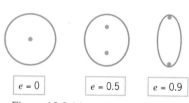

Figure 12.3.14

The ratio c/a is called the **eccentricity** of an ellipse. The eccentricity is commonly denoted by the letter e (not to be confused with the base e of the natural logarithm), so that $e = c/a$ and $c = ae$. It follows that $0 \le e < 1$, since $0 \le c < a$. As illustrated in Figure 12.3.14, the eccentricity can be viewed as a measure of the "flatness" of an ellipse; circles have an eccentricity of 0, and with the major axis kept constant, the closer the eccentricity is to 1, the flatter the ellipse. For ellipses other than circles, Equation (3) can be expressed in terms of e as

$$\sqrt{(x-c)^2 + y^2} = e\left(\frac{a}{e} - x\right) \tag{7}$$

which provides an alternative geometric viewpoint about how a noncircular ellipse can be generated: The quantity on the left side of (7) is the distance PF between any point $P(x, y)$ on the ellipse and the focus $F(c, 0)$, whereas the quantity $a/e - x$ is the perpendicular distance PD between P and the vertical line $x = a/e$ (Figure 12.3.14). It follows that

$$PF = ePD$$

which tells us that *a noncircular ellipse can be viewed as the set of all points in the plane whose distance from a fixed point F (a **focus**) is proportional to the distance from a fixed line D that does not contain the focus (e being the constant of proportionality).* The line D is called the **directrix** of the ellipse corresponding to F. By symmetry, an ellipse has two foci and two directrices; for example, the ellipse in Figure 12.3.15 can also be generated from the focus $(-c, 0)$ and the corresponding directrix $x = -a/e$.

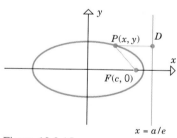

Figure 12.3.15

□ **REFLECTION PROPERTIES OF ELLIPSES**

In the exercises the reader is asked to prove the following result.

12.3.2 THEOREM (*A Geometric Property of Ellipses*). *A line tangent to an ellipse at a point P makes equal angles with the lines joining P to the foci (Figure 12.3.16).*

It follows from Theorem 12.3.2 that a ray of light emanating from one focus of an ellipse will be reflected through the other focus. This property of ellipses is used in "whispering galleries.'' Such rooms have ceilings whose cross sections are elliptical in shape with common foci. As a result, if a person standing at one focus whispers, the sound waves are reflected by the ceiling to the other focus, making it possible for a person at that focus to hear the whispered sound.

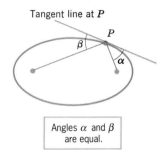

Figure 12.3.16

Angles α and β are equal.

▶ Exercise Set 12.3 $\boxed{C}$ 52

In Exercises 1–14, sketch the ellipse. Label the foci and the ends of the major and minor axes.

1. $\dfrac{x^2}{16} + \dfrac{y^2}{9} = 1.$ **2.** $\dfrac{x^2}{4} + \dfrac{y^2}{25} = 1.$

3. $9x^2 + y^2 = 9.$ **4.** $4x^2 + 9y^2 = 36.$

5. $x^2 + 3y^2 = 2.$ **6.** $16x^2 + 4y^2 = 1.$

7. $9(x - 1)^2 + 16(y - 3)^2 = 144.$

8. $(x + 3)^2 + 4(y - 5)^2 = 16.$

9. $3(x + 2)^2 + 4(y + 1)^2 = 12.$

10. $\frac{1}{4}x^2 + \frac{1}{9}(y + 2)^2 - 1 = 0.$

11. $x^2 + 9y^2 + 2x - 18y + 1 = 0.$

12. $9x^2 + 4y^2 + 18x - 24y + 9 = 0.$

13. $4x^2 + y^2 + 8x - 10y = -13.$

14. $5x^2 + 9y^2 - 20x + 54y = -56.$

In Exercises 15–26, find an equation for the ellipse satisfying the given conditions.

15. Ends of major axis $(\pm 3, 0)$; ends of minor axis $(0, \pm 2)$.

16. Ends of major axis $(0, \pm \sqrt{5})$; ends of minor axis $(\pm 1, 0)$.

17. Length of major axis 26; foci $(\pm 5, 0)$.

18. Length of minor axis 16; foci $(0, \pm 6)$.

19. Foci $(\pm 1, 0)$; $b = \sqrt{2}$.

20. Foci $(\pm 3, 0)$; $a = 4$.

21. $c = 2\sqrt{3}$; $a = 4$; center at the origin; foci on a coordinate axis (two answers).

22. $b = 3$; $c = 4$; center at the origin; foci on a coordinate axis (two answers).

23. Ends of major axis $(\pm 6, 0)$; passes through $(2, 3)$.

24. Center at $(0, 0)$; major and minor axes along the coordinate axes; passes through $(3, 2)$ and $(1, 6)$.

25. Foci $(1, 2)$ and $(1, 4)$; minor axis of length 2.

26. Foci $(2, 1)$ and $(2, -3)$; major axis of length 6.

27. Find an equation of the ellipse traced by a point which moves so that the sum of its distances to $(4, 1)$ and $(4, 5)$ is 12.

28. Find an equation of the ellipse traced by a point which moves so that the sum of its distances to $(0, 0)$ and $(1, 1)$ is 4.

In Exercises 29 and 30, find the eccentricity of the ellipse.

29. $x^2 + 4y^2 = 16.$ **30.** $25x^2 + 9y^2 = 25.$

In Exercises 31 and 32, find an equation of the ellipse that has the given eccentricity and foci.

31. $e = 2/3$, foci $(\pm 4, 0)$.   **32.** $e = 3/5$, foci $(0, \pm 3)$.

In Exercises 33 and 34, use the focus–directrix definition of an ellipse to find an equation of the ellipse.

33. $e = 4/5$; focus at $(0, 0)$; corresponding directrix $x = 2$.

34. $e = 1/2$; focus at $(0, 0)$; corresponding directrix $y = -1$.

In Exercises 35–38, find all intersections of the given curves, and make a sketch of the curves that shows the points of intersection.

35. $x^2 + 4y^2 = 40$ and $x + 2y = 8.$

36. $y^2 = 2x$ and $x^2 + 2y^2 = 12.$

37. $x^2 + 9y^2 = 36$ and $x^2 + y^2 = 20.$

38. $16x^2 + 9y^2 = 36$ and $5x^2 + 18y^2 = 45.$

39. Find two values of k such that the line $x + 2y = k$ is tangent to the ellipse $x^2 + 4y^2 = 8$. Find the points of tangency.

40. Prove: The line tangent to the ellipse $x^2/a^2 + y^2/b^2 = 1$ at the point (x_0, y_0) has the equation $xx_0/a^2 + yy_0/b^2 = 1.$

Exercises 41–44 refer to an ellipse with major and minor axes of lengths $2a$ and $2b$, respectively (Figure 12.3.17).

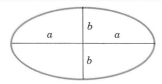

Figure 12.3.17

41. Find the area enclosed by the ellipse in Figure 12.3.17.

42. Find the volume of the solid that results when the region enclosed by the ellipse in Figure 12.3.17 is revolved about

(a) the major axis   (b) the minor axis.

43. Show that if $2c$ is the distance between the foci of the ellipse in Figure 12.3.17, then the area of the surface generated by revolving the ellipse about the major axis is

$$2\pi ab \left(\frac{b}{a} + \frac{a}{c} \sin^{-1} \frac{c}{a} \right)$$

44. Show that if $2c$ is the distance between the foci of the ellipse in Figure 12.3.17, then the area of the surface generated by revolving the ellipse about the minor axis is

$$2\pi ab \left(\frac{a}{b} + \frac{b}{c} \ln \frac{a+c}{b} \right)$$

45. The tank of an oil truck is 18 feet long and has elliptical cross sections that are 6 feet wide and 4 feet high (Figure 12.3.18). Find a formula for the volume of oil in the tank in terms of the depth h of the oil.

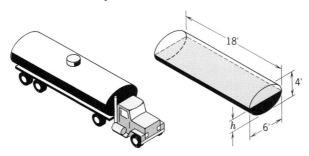

Figure 12.3.18

46. A semielliptic arch spans a highway 50 feet wide. How high is the arch if a center section of the highway 30 feet wide has a minimum clearance of 14 feet?

47. Find the area of the square that can be inscribed in the ellipse $x^2/a^2 + y^2/b^2 = 1$.

48. Suppose that the base of a solid is elliptical with a major axis of length 9 and a minor axis of length 4. Find the volume of the solid if

(a) the cross sections perpendicular to the major axis are squares

(b) the cross sections perpendicular to the minor axis are equilateral triangles.

49. Suppose that you want to draw an ellipse that has given values for the lengths of the major and minor axes by using the method shown in Figure 12.3.2. Assuming that the axes are drawn, explain how a compass can be used to locate the positions for the tacks.

50. Show that the curve of intersection of a plane and a right-circular cylinder is an ellipse, assuming that the plane is neither perpendicular to nor parallel to the axis of the cylinder. [*Hint:* Let θ be the angle that the plane makes with a circular cross section of the cylinder. Introduce xy- and $x'y'$-axes as shown in Figure 12.3.19, and find x' and y' in terms of x and y.]

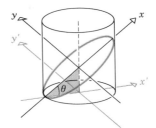

Figure 12.3.19

51. A carpenter needs to cut an elliptical hole in a sloped roof through which a circular vent pipe of diameter D is to be inserted vertically (Figure 12.3.20). The carpenter wants to draw the outline of the hole on the roof using a pencil, a piece of string, and two tacks, as illustrated in Figure 12.3.2. The center point of the ellipse is known, and common sense suggests that its major axis must be perpendicular to the drip line of the roof. The carpenter needs to determine the length L of the string and the distance T between a tack and the center point. The architect's plans show that the pitch of the roof is p (pitch = rise over run, as in Figure 12.3.20). Find T and L in terms of D and p. [*Note:* This exercise is based on an article by William H. Enos, which appeared in the *Mathematics Teacher*, Feb. 1991, p. 148.]

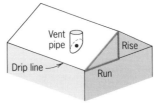

Figure 12.3.20

52. The planets and comets in our solar system have elliptical orbits with the sun at a focus. The point closest to the sun in an orbit is called the *perihelion*, and the point farthest from the sun is called the *aphelion*; these points are at the ends of the major axis of the orbit (Figure 12.3.21).

(a) Find the eccentricity of the earth's orbit given that the ratio of the distance from the sun at the perihelion to the distance from the sun at the aphelion is 59/61.

(b) Find the distance between the earth and the sun at the perihelion given that the semimajor axis of the orbit has a length of 93 million miles.

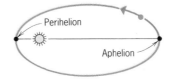

Figure 12.3.21

53. Let C_1 and C_2 be circles, in the same plane, with radii r_1 and r_2, respectively. Assume that $r_1 < r_2$ and that C_1 lies entirely inside C_2, but that C_1 and C_2 do not have the same center (Figure 12.3.22). Show that the centers of all circles that are outside C_1 and inside C_2, and are tangent to both C_1 and C_2, lie on an ellipse whose foci are the centers of C_1 and C_2. Find the length of the major axis and the location of the center of the ellipse.

Figure 12.3.22

54. Given two intersecting lines, let L_2 be the line with the larger angle of inclination ϕ_2, and let L_1 be the line with the smaller angle of inclination ϕ_1. We define the **angle θ between L_1 and L_2** by $\theta = \phi_2 - \phi_1$ (Figure 12.3.23).

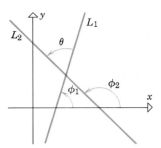

Figure 12.3.23

(a) Prove: If L_1 and L_2 are not perpendicular, then

$$\tan \theta = \frac{m_2 - m_1}{1 + m_1 m_2}$$

where L_1 and L_2 have slopes m_1 and m_2.

(b) Prove Theorem 12.3.2. [*Hint:* Introduce a coordinate system so that the ellipse has the equation $x^2/a^2 + y^2/b^2 = 1$, and use part (a).]

55. Derive the equation $x^2/b^2 + y^2/a^2 = 1$ in Figure 12.3.7.

◼ 12.4 THE HYPERBOLA

In this section we shall discuss properties of hyperbolas.

☐ **DEFINITION OF A HYPERBOLA**

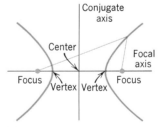

Figure 12.4.1

12.4.1 DEFINITION. A *hyperbola* is the set of all points in the plane, the difference of whose distances from two fixed distinct points is a given positive constant that is less than the distance between the fixed points.

In this definition the "difference" of the distances is understood to mean the distance to the farther point minus the distance to the closer point. The two fixed points are called the *foci*, and the midpoint of the line segment joining the foci is called the *center* of the hyperbola (Figure 12.4.1). The line through the foci is called the *focal axis* (or *transverse axis*), and the line through the center and perpendicular to the focal axis is called the *conjugate axis*. The hyperbola intersects the focal axis at two points, called *vertices*. The two separate parts of a hyperbola are called the *branches*.

☐ **ASYMPTOTES OF A HYPERBOLA**

Associated with every hyperbola is a pair of lines, called the *asymptotes* of the hyperbola. These lines intersect at the center of the hyperbola and have the property that as a point P moves along the hyperbola away from the center, the distance between P and one of the asymptotes approaches zero (Figure 12.4.2).

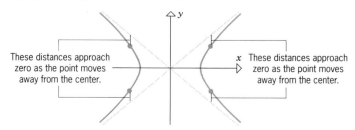

Figure 12.4.2

GEOMETRIC PROPERTIES OF A HYPERBOLA

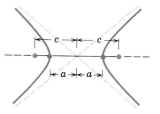

Figure 12.4.3

It is traditional in the study of hyperbolas to denote the distance between the vertices by $2a$, the distance between the foci by $2c$ (Figure 12.4.3), and to define the quantity b as

$$b = \sqrt{c^2 - a^2} \qquad (1)$$

This relationship, which can also be expressed as

$$c = \sqrt{a^2 + b^2} \qquad (2)$$

is pictured geometrically in Figure 12.4.4. As illustrated in that figure, the asymptotes pass through the center of the hyperbola and the corners of a box extending b units on each side of the center along the conjugate axis and a units on each side of the center along the focal axis. We shall verify this later in this section.

If V is one vertex of a hyperbola, then, as illustrated in Figure 12.4.5, the distance from V to the farther focus minus the distance from V to the closer focus is

$$[(c - a) + 2a] - (c - a) = 2a$$

Thus, for *all* points on a hyperbola, the distance to the farther focus minus the distance to the closer focus is $2a$. (Why?)

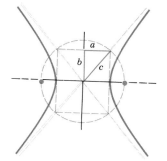

Figure 12.4.4

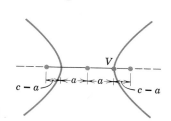

Figure 12.4.5

STANDARD EQUATIONS OF A HYPERBOLA

The equation of a hyperbola is simplest if the coordinate axes are positioned so that the center of the hyperbola is at the origin and the foci are on the x-axis or y-axis. The two possible such orientations are shown in Figure 12.4.6. These are called the ***standard positions*** of a hyperbola, and the resulting equations are called the ***standard equations*** of a hyperbola.

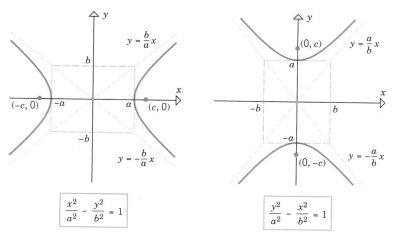

Figure 12.4.6

To illustrate how the equations in Figure 12.4.6 are obtained, we shall derive the equation for the hyperbola on the left. If $P(x, y)$ is any point on the hyperbola, then the distance from P to the farther focus minus the distance from P to the closer focus is $2a$.

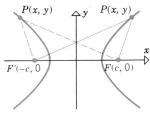

Figure 12.4.7

Depending on which focus is farther from P (Figure 12.4.7), this condition leads either to the equation

$$PF' - PF = 2a$$

from which we obtain

$$\sqrt{(x+c)^2 + y^2} - \sqrt{(x-c)^2 + y^2} = 2a \tag{3}$$

or to the equation

$$PF - PF' = 2a$$

from which we obtain

$$\sqrt{(x-c)^2 + y^2} - \sqrt{(x+c)^2 + y^2} = 2a \tag{4}$$

Rewriting (3) yields

$$\sqrt{(x+c)^2 + y^2} = 2a + \sqrt{(x-c)^2 + y^2}$$

then squaring both sides and simplifying yields

$$\sqrt{(x-c)^2 + y^2} = \frac{c}{a}x - a \tag{5}$$

Squaring and simplifying again then yields

$$\frac{x^2}{a^2} - \frac{y^2}{c^2 - a^2} = 1$$

which, by (1), can be written as

$$\frac{x^2}{a^2} - \frac{y^2}{b^2} = 1 \tag{6}$$

We leave it as an exercise to show that (4) also simplifies to (6). Thus, each point $P(x, y)$ on the hyperbola satisfies (6), regardless of the branch on which it lies. Conversely, it can be shown that for any point P whose coordinates satisfy (6), the distance from P to the farther focus minus the distance from P to the closer focus is $2a$, so that such a point must lie on the hyperbola.

To prove that the asymptotes of (6) have the equations shown in Figure 12.4.6, we rewrite (6) in the form

$$y^2 = \frac{b^2}{a^2}(x^2 - a^2)$$

which is equivalent to the pair of equations

$$y = \frac{b}{a}\sqrt{x^2 - a^2} \quad \text{and} \quad y = -\frac{b}{a}\sqrt{x^2 - a^2}$$

Thus, in the first quadrant, the vertical distance between the line $y = (b/a)x$ and the hyperbola can be written (Figure 12.4.8) as

$$\frac{b}{a}x - \frac{b}{a}\sqrt{x^2 - a^2}$$

But this distance tends to zero as $x \to +\infty$ since

$$\lim_{x \to +\infty} \left(\frac{b}{a}x - \frac{b}{a}\sqrt{x^2 - a^2} \right) = \lim_{x \to +\infty} \frac{b}{a}(x - \sqrt{x^2 - a^2})$$

$$= \lim_{x \to +\infty} \frac{b}{a} \frac{(x - \sqrt{x^2 - a^2})(x + \sqrt{x^2 - a^2})}{x + \sqrt{x^2 - a^2}}$$

$$= \lim_{x \to +\infty} \frac{ab}{x + \sqrt{x^2 - a^2}} = 0$$

The analysis in the remaining quadrants is similar.

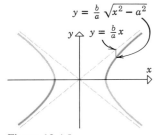

Figure 12.4.8

REMARK.   There is a trick that can be used to avoid memorizing the equations of the asymptotes of a hyperbola. They can be obtained, when needed, by substituting 0 for the 1 on the right side of the hyperbola equation, and then solving for y in terms of x. For example, for the hyperbola

$$\frac{x^2}{a^2} - \frac{y^2}{b^2} = 1$$

we would write

$$\frac{x^2}{a^2} - \frac{y^2}{b^2} = 0 \quad \text{or} \quad y^2 = \frac{b^2}{a^2}x^2 \quad \text{or} \quad y = \pm\frac{b}{a}x$$

which are the equations for the asymptotes.

☐ **A TECHNIQUE FOR GRAPHING HYPERBOLAS**

Hyperbolas can be graphed from their *standard equations* using four basic steps:

* Determine whether the focal axis is on the x-axis or the y-axis. This can be ascertained from the location of the minus sign in the equation. Referring to Figure 12.4.6, the focal axis is along the x-axis when the minus sign precedes the y^2-term, and it is along the y-axis when the minus sign precedes the x^2-term.
* Determine the values of a and b and draw a box extending a units on either side of the center along the focal axis and b units on either side of the center along the conjugate axis. (The squares of a and b can be read directly from the equation.)
* Draw the asymptotes along the diagonals of the box.
* Using the box and the asymptotes as a guide, sketch the graph of the hyperbola (Figure 12.4.9).

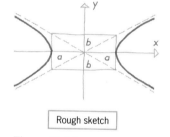

Rough sketch

Figure 12.4.9

Example 1   Sketch the graphs of the hyperbolas

(a) $\dfrac{x^2}{4} - \dfrac{y^2}{9} = 1$ (b) $\dfrac{y^2}{7} - \dfrac{x^2}{2} = 1$

showing their vertices, foci, and asymptotes.

Solution (a).   The minus sign precedes the y^2-term, so the focal axis is along the x-axis. From the denominators in the equation we obtain

$$a^2 = 4 \quad \text{and} \quad b^2 = 9$$

Since a and b are positive, we must have $a = 2$ and $b = 3$. Recalling that the vertices lie a units on each side of the center on the focal axis, it follows that their coordinates in this case are $(2, 0)$ and $(-2, 0)$. Drawing a box extending $a = 2$ units along the x-axis on each side of the origin and $b = 3$ units on each side of the origin along the y-axis, then drawing the asymptotes along the diagonals of the box as a guide, yields the graph in Figure 12.4.10.

To obtain equations for the asymptotes, we substitute 0 for 1 in the given equation; this yields

$$\frac{x^2}{4} - \frac{y^2}{9} = 0 \quad \text{or} \quad y = \pm\frac{3}{2}x$$

The foci lie c units on each side of the center along the focal axis, where c is given by (2). From the values of a^2 and b^2 above we obtain

$$c = \sqrt{a^2 + b^2} = \sqrt{4 + 9} = \sqrt{13} \approx 3.6$$

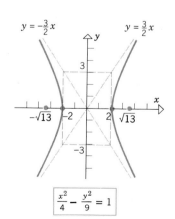

Figure 12.4.10

Since the foci lie on the x-axis in this case, their coordinates are $(\sqrt{13}, 0)$ and $(-\sqrt{13}, 0)$.

Solution (b).   The minus sign precedes the x^2-term, so the focal axis is along the y-axis. From the denominators in the equation we obtain $a^2 = 7$ and $b^2 = 2$, from which it follows that

$$a = \sqrt{7} \approx 2.6 \quad \text{and} \quad b = \sqrt{2} \approx 1.4$$

Thus, the vertices are at $(0, \sqrt{7})$ and $(0, -\sqrt{7})$. Drawing a box extending $a = \sqrt{7}$ units on either side of the origin along the y-axis and $b = \sqrt{2}$ units on either side of the origin along the x-axis, then drawing the asymptotes, yields the graph in Figure 12.4.11. Substituting 0 for 1 in the given equation yields

$$\frac{y^2}{7} - \frac{x^2}{2} = 0 \quad \text{or} \quad y = \pm \sqrt{\frac{7}{2}} x$$

which are the equations of the asymptotes. Also,

$$c = \sqrt{a^2 + b^2} = \sqrt{7 + 2} = 3$$

so the foci, which lie on the y-axis in this case, are at $(0, 3)$ and $(0, -3)$.   ◄

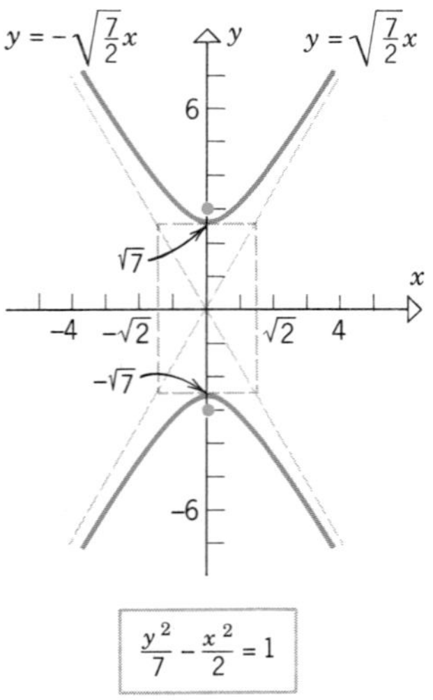

Figure 12.4.11

Example 2   Find the equation of the hyperbola with vertices $(0, \pm 8)$ and asymptotes $y = \pm \frac{4}{3}x$.

Solution.   Since the vertices are on the y-axis, the equation of the hyperbola has the form $y^2/a^2 - x^2/b^2 = 1$ and the asymptotes are

$$y = \pm \frac{a}{b} x$$

From the location of the vertices we have $a = 8$, so the given equations of the asymptotes yield

$$y = \pm \frac{a}{b} x = \pm \frac{8}{b} x = \pm \frac{4}{3} x$$

from which it follows that $b = 6$. Thus, the hyperbola has the equation

$$\frac{y^2}{64} - \frac{x^2}{36} = 1 \quad ◄$$

☐ **TRANSLATED HYPERBOLAS**

If the center of a hyperbola is at the point (h, k) and its focal and conjugate axes are parallel to the coordinate axes, then its equation has one of the following forms:

> ***Hyperbola with Center (h, k) and Focal Axis Parallel to x-axis***
>
> $$\frac{(x - h)^2}{a^2} - \frac{(y - k)^2}{b^2} = 1 \tag{7}$$

> ***Hyperbola with Center (h, k) and Focal Axis Parallel to y-axis***
>
> $$\frac{(y - k)^2}{a^2} - \frac{(x - h)^2}{b^2} = 1 \tag{8}$$

The derivations are similar to those for the parabola and ellipse and will be omitted. As before, the equations of the asymptotes can be obtained by replacing the 1 by a 0 on the right side of (7) or (8), and then solving for y in terms of x.

Example 3   Show that the curve

$$x^2 - y^2 - 4x + 8y - 21 = 0$$

is a hyperbola. Sketch the hyperbola and show the foci, vertices, and asymptotes in the figure.

Solution.   Our objective is to rewrite the equation in form (7) or (8) by completing the squares. To do this, we first group the x-terms and y-terms and take the constant to the right side:

$$(x^2 - 4x) - (y^2 - 8y) = 21$$

Next, complete the squares:

$$(x^2 - 4x + 4) - (y^2 - 8y + 16) = 21 + 4 - 16$$

or

$$(x - 2)^2 - (y - 4)^2 = 9$$

Finally, divide by 9 to introduce a 1 on the right side:

$$\frac{(x - 2)^2}{9} - \frac{(y - 4)^2}{9} = 1 \tag{9}$$

This is an equation of form (7) with $h = 2$, $k = 4$, $a^2 = 9$, and $b^2 = 9$. Thus, the equation represents a hyperbola with center $(2, 4)$ and focal axis parallel to the x-axis. Since $a = 3$, the vertices are located 3 units to the left and 3 units to the right of the center, or at the points $(-1, 4)$ and $(5, 4)$. From (2), $c = \sqrt{a^2 + b^2} = \sqrt{9 + 9} = 3\sqrt{2}$, so the foci are located $3\sqrt{2}$ units to the left and right of the center, or at the points $(2 - 3\sqrt{2}, 4)$ and $(2 + 3\sqrt{2}, 4)$.

The equations of the asymptotes may be found by substituting 0 for 1 in (9) to obtain

$$\frac{(x - 2)^2}{9} - \frac{(y - 4)^2}{9} = 0 \quad \text{or} \quad y - 4 = \pm(x - 2)$$

which yields the asymptotes

$$y = x + 2 \quad \text{and} \quad y = -x + 6$$

With the aid of a box extending $a = 3$ units left and right of the center and $b = 3$ units above and below the center, we obtain the sketch in Figure 12.4.12.   ◀

A hyperbola (such as that in the last example) is called **equilateral** if $a = b$. The asymptotes of an equilateral hyperbola are perpendicular. (Verify.)

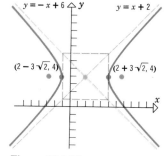

Figure 12.4.12

☐ **FOCUS–DIRECTRIX PROPERTY OF HYPERBOLAS**

The ratio c/a is called the **eccentricity** of a hyperbola. The eccentricity is commonly denoted by the letter e, so that $e = c/a$ and $c = ae$. It follows that $e > 1$ since, $c > a > 0$. Equation (5) can be expressed in terms of e as

$$\sqrt{(x - c)^2 + y^2} = e\left(x - \frac{a}{e}\right) \tag{10}$$

which provides an alternative geometric viewpoint about how a hyperbola can be generated: If $x - a/e > 0$, then this quantity is the perpendicular distance PD between any point $P(x, y)$ on the right branch of the hyperbola and the line $x = a/e$, whereas the quantity on

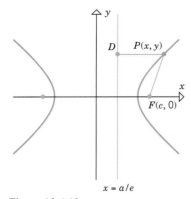

Figure 12.4.13

the left side of (10) is the distance of PF between P and the focus $F(c, 0)$ (Figure 12.4.13). It follows that

$$PF = ePD \tag{11}$$

Similarly, the equation for the left branch of the hyperbola can be written as

$$\sqrt{(x - c)^2 + y^2} = e \left(\frac{a}{e} - x \right)$$

from which it follows that (11) also holds for any point on the left branch of the hyperbola. Thus, *a hyperbola can be viewed as the set of all points in the plane whose distance from a fixed point F (a **focus**) is proportional to the distance from a fixed line D that does not contain the focus (e being the constant of proportionality).* The line D is called the ***directrix*** of the hyperbola corresponding to F. By symmetry, a hyperbola has two foci and two directrices; for example, the hyperbola in Figure 12.4.13 can also be generated from the focus $(-c, 0)$ and the corresponding directrix $x = -a/e$.

☐ **HYPERBOLIC NAVIGATION**

By measuring the difference in reception times of synchronized radio signals from two widely spaced transmitters, a ship's electronic equipment can determine the difference $2a$ in its distances from the two transmitters. This information places the ship somewhere on the hyperbola whose foci are at the transmitters and whose points have $2a$ as the difference in their distances from the foci. By using two pairs of transmitters, the position of a ship can be determined as the intersection of two hyperbolas (Figure 12.4.14).

Figure 12.4.14

☐ **REFLECTION PROPERTIES OF HYPERBOLAS**

In the exercises, the reader is asked to prove the following result.

12.4.2 THEOREM (*A Geometric Property of Hyperbolas*). *A line tangent to a hyperbola at a point P makes equal angles with the lines joining P to the foci* (Figure 12.4.15).

It follows from Theorem 12.4.2 that a ray of light emanating from one focus of a hyperbola will be reflected back along the line from the opposite focus (Figure 12.4.16). Light reflection properties of hyperbolas are used to advantage in the design of high-quality telescopes.

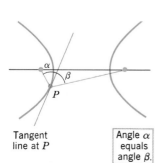

Tangent line at P | Angle α equals angle β.

Figure 12.4.15

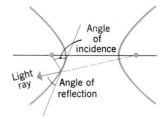

Figure 12.4.16

▶ Exercise Set 12.4

In Exercises 1–16, sketch the hyperbola. Find the coordinates of the vertices and foci, and find equations for the asymptotes.

1. $\dfrac{x^2}{16} - \dfrac{y^2}{4} = 1.$　　　　**2.** $\dfrac{y^2}{9} - \dfrac{x^2}{25} = 1.$

3. $9y^2 - 4x^2 = 36.$　　　**4.** $16x^2 - 25y^2 = 400.$

5. $8x^2 - y^2 = 8.$　　　　**6.** $3x^2 - y^2 = -9.$

7. $x^2 - y^2 = 1.$　　　　　**8.** $4y^2 - 4x^2 = 1.$

9. $\dfrac{(x-2)^2}{9} - \dfrac{(y-4)^2}{4} = 1.$

10. $\dfrac{(y+4)^2}{3} - \dfrac{(x-2)^2}{5} = 1.$

11. $(y+3)^2 - 9(x+2)^2 = 36.$

12. $16(x+1)^2 - 8(y-3)^2 = 16.$

13. $x^2 - 4y^2 + 2x + 8y - 7 = 0.$

14. $4x^2 - 9y^2 + 16x + 54y - 29 = 0.$

15. $16x^2 - y^2 - 32x - 6y = 57.$

16. $4y^2 - x^2 + 40y - 4x = -60.$

In Exercises 17–30, find an equation for the hyperbola satisfying the given conditions.

17. Vertices $(\pm 2, 0)$; foci $(\pm 3, 0)$.

18. Vertices $(0, \pm 3)$; foci $(0, \pm 5)$.

19. Vertices $(\pm 1, 0)$; asymptotes $y = \pm 2x.$

20. Vertices $(0, \pm 3)$; asymptotes $y = \pm x.$

21. Asymptotes $y = \pm \frac{3}{2}x$; $b = 4.$

22. Foci $(0, \pm 5)$; asymptotes $y = \pm 2x.$

23. Passes through $(5, 9)$; asymptotes $y = \pm x.$

24. Asymptotes $y = \pm \frac{3}{4}x$; $c = 5.$

25. Vertices $(\pm 2, 0)$; passes through $(4, 2).$

26. Foci $(\pm 3, 0)$; asymptotes $y = \pm 2x.$

27. Vertices $(4, -3)$ and $(0, -3)$; foci 6 units apart.

28. Foci $(1, 8)$ and $(1, -12)$; vertices 4 units apart.

29. Vertices $(2, 4)$ and $(10, 4)$; foci 10 units apart.

30. Asymptotes $y = 2x + 1$ and $y = -2x + 3$; passes through the origin.

31. Find the equation of the hyperbola traced by a point which moves so that the difference between its distances to $(0, 0)$ and $(1, 1)$ is 1.

32. Find the equation of the hyperbola traced by a point which moves so that the difference between its distances to $(4, -3)$ and $(-2, 5)$ is 6.

In Exercises 33 and 34, find the eccentricity of the hyperbola.

33. $16x^2 - 9y^2 = 144.$　　　**34.** $x^2 - 4y^2 + 4 = 0.$

In Exercises 35 and 36, find an equation of the hyperbola that has the given eccentricity and foci.

35. $e = 5/3$, foci $(\pm 5, 0)$.　　**36.** $e = 2$, foci $(0, \pm 1)$.

In Exercises 37 and 38, use the focus–directrix definition of a hyperbola to find an equation of the hyperbola.

37. $e = 3$, a focus at $(0, 0)$, corresponding directrix $x = 1.$

38. $e = 4$, a focus at $(0, 0)$, corresponding directrix $y = 2.$

In Exercises 39–42, find all intersections of the given curves, and make a sketch of the curves that shows the points of intersection.

39. $x^2 - 4y^2 = 36$ and $x - 2y - 20 = 0.$

40. $y^2 - 8x^2 = 5$ and $y - 2x^2 = 0.$

41. $3x^2 - 7y^2 = 5$ and $9y^2 - 2x^2 = 1.$

42. $x^2 - y^2 = 1$ and $x^2 + y^2 = 7.$

43. Find the coordinates of all points on the hyperbola $4x^2 - y^2 = 4$ where the two lines that pass through the point and the foci are perpendicular.

44. Prove that the line that is tangent to the hyperbola $x^2/a^2 - y^2/b^2 = 1$ at the point (x_0, y_0) has the equation $xx_0/a^2 - yy_0/b^2 = 1.$

45. A line tangent to the hyperbola $4x^2 - y^2 = 36$ intersects the y-axis at the point $(0, 4)$. Find the point(s) of tangency.

46. Let R be the region enclosed between the hyperbola $b^2x^2 - a^2y^2 = a^2b^2$ and the line $x = c$ through the focus $(c, 0)$. Find the volume of the solid generated by revolving R about

(a) the x-axis　　　　　(b) the y-axis.

47. The "flatness" of a hyperbola is related to its eccentricity. What happens to the shape of a hyperbola if the distance between the vertices is kept constant, but the eccentricity approaches 1? Approaches $+\infty$?

48. Let C_1 and C_2 be circles in the same plane with unequal radii r_1 and r_2, respectively. Assume that C_1 and C_2 do not intersect and that neither circle is inside the other (Figure 12.4.17). Show that the centers of all circles that are outside C_1 and C_2, and are tangent to both C_1 and C_2, lie on a branch of a hyperbola whose foci are the centers of C_1 and C_2. Where is the center of the hyperbola?

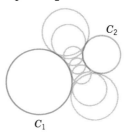

C_1　　　　　Figure 12.4.17

49. Suppose that the report of a gun is heard first by one observer and then t seconds later by a second observer in another location. Assuming that the gun and the ear levels of the observers are all in the same horizontal plane and that the speed of sound is constant, show that the gun was fired somewhere on a branch of a hyperbola whose foci coincide with the observers. [*Hint:* Let v be the speed of sound and d_1 and d_2 the horizontal distances between the gun and the observers (Figure 12.4.18).]

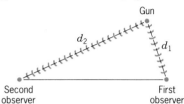

Gun

d_2 d_1

Second   First
observer   observer   Figure 12.4.18

50. Show that Equation (4) simplifies to (6).

51. Derive the equation $y^2/a^2 - x^2/b^2 = 1$ in Figure 12.4.6.

52. Prove Theorem 12.4.2. [*Hint:* Introduce coordinate axes so the hyperbola has the equation $x^2/a^2 - y^2/b^2 = 1$, and use the result of Exercise 54(a) in Section 12.3.]

53. (a) Show that if an ellipse and a hyperbola have the same foci, then at each point of intersection their tangent lines are perpendicular. [*Hint:* Use the results in Exercise 44 above and Exercise 40 of Section 12.3.]

(b) Give a geometric proof of the result in part (a) using the reflection properties in Theorems 12.3.2 and 12.4.2.

■ 12.5 ROTATION OF AXES; SECOND-DEGREE EQUATIONS

In previous sections we obtained equations of conic sections with axes parallel to the coordinate axes. In this section we shall study the equations of conics that are "tilted" relative to the coordinate axes. This will lead us to investigate rotations of coordinate axes.

□ QUADRATIC EQUATIONS
IN x AND y

We have seen in the preceding sections of this chapter that equations of the form

$$Ax^2 + Cy^2 + Dx + Ey + F = 0 \qquad (1)$$

represent conic sections. Equation (1) is a special case of the more general equation

$$Ax^2 + Bxy + Cy^2 + Dx + Ey + F = 0 \qquad (2)$$

which, if A, B, and C are not all zero, is called a *second-degree equation* or *quadratic equation* in x and y. We shall show later that the graph of any second-degree equation is a conic section (possibly a degenerate conic section). If $B = 0$, then (2) reduces to (1) and the conic section has its axis or axes parallel to the coordinate axes. However, if $B \neq 0$, then (2) contains a "cross-product" term Bxy, and the graph of the conic section represented by the equation has its axis or axes "tilted" relative to the coordinate axes. As an illustration, consider the ellipse with foci $F_1(1, 2)$ and $F_2(-1, -2)$ and such that the sum of the distances from each point $P(x, y)$ on the ellipse to the foci is 6 units. Expressing this condition as an equation, we obtain (Figure 12.5.1)

$$\sqrt{(x - 1)^2 + (y - 2)^2} + \sqrt{(x + 1)^2 + (y + 2)^2} = 6$$

Squaring both sides, then isolating the remaining radical, then squaring again ultimately yields

$$8x^2 - 4xy + 5y^2 = 36$$

as the equation of the ellipse. This is an equation of form (2) with $A = 8$, $B = -4$, $C = 5$, $D = 0$, $E = 0$, $F = -36$.

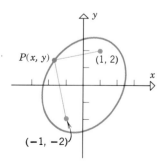

$P(x, y)$ $(1, 2)$

$(-1, -2)$

Figure 12.5.1

☐ **ROTATION OF AXES**

To study conics that are tilted relative to the coordinate axes, it is frequently helpful to rotate the coordinate axes, so that the rotated coordinate axes are parallel to the axes of the conic. Before we can discuss the details, we need to develop some ideas about rotation of coordinate axes.

In Figure 12.5.2a the axes of an xy-coordinate system have been rotated about the origin through an angle θ to produce a new $x'y'$-coordinate system. As shown in the figure, each point P in the plane has coordinates (x', y') as well as coordinates (x, y). To see how the two are related, let r be the distance from the common origin to the point P, and let α be the angle shown in Figure 12.5.2b. It follows that

$$x = r \cos(\theta + \alpha), \quad y = r \sin(\theta + \alpha) \tag{3}$$

and

$$x' = r \cos \alpha, \quad y' = r \sin \alpha \tag{4}$$

Using familiar trigonometric identities, the relationships in (3) can be written as

$$x = r \cos \theta \cos \alpha - r \sin \theta \sin \alpha$$

$$y = r \sin \theta \cos \alpha + r \cos \theta \sin \alpha$$

and on substituting (4) in these equations we obtain the following relationships called the ***rotation equations***:

$$\begin{aligned} x &= x' \cos \theta - y' \sin \theta \\ y &= x' \sin \theta + y' \cos \theta \end{aligned} \tag{5}$$

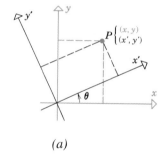
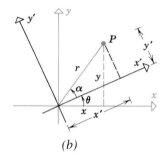

Figure 12.5.2   *(a)*   *(b)*

Example 1   Suppose that the axes of an xy-coordinate system are rotated through an angle of $\theta = 45°$ to obtain an $x'y'$-coordinate system. Find the equation of the curve

$$x^2 - xy + y^2 - 6 = 0$$

in $x'y'$-coordinates.

Solution.   Substituting $\sin \theta = \sin 45° = 1/\sqrt{2}$ and $\cos \theta = \cos 45° = 1/\sqrt{2}$ in (5) yields the rotation equations

$$x = \frac{x'}{\sqrt{2}} - \frac{y'}{\sqrt{2}} \quad \text{and} \quad y = \frac{x'}{\sqrt{2}} + \frac{y'}{\sqrt{2}}$$

Substituting these into the given equation yields

$$\left(\frac{x'}{\sqrt{2}} - \frac{y'}{\sqrt{2}} \right)^2 - \left(\frac{x'}{\sqrt{2}} - \frac{y'}{\sqrt{2}} \right)\left(\frac{x'}{\sqrt{2}} + \frac{y'}{\sqrt{2}} \right) + \left(\frac{x'}{\sqrt{2}} + \frac{y'}{\sqrt{2}} \right)^2 - 6 = 0$$

or

$$\frac{x'^2 - 2x'y' + y'^2 - x'^2 + y'^2 + x'^2 + 2x'y' + y'^2}{2} = 6$$

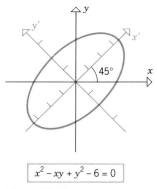

$$x^2 - xy + y^2 - 6 = 0$$

Figure 12.5.3

or

$$\frac{x'^2}{12} + \frac{y'^2}{4} = 1$$

which is the equation of an ellipse (Figure 12.5.3). ◄

If the rotation equations (5) are solved for x' and y' in terms of x and y, one obtains (Exercise 14):

$$x' = x \cos \theta + y \sin \theta$$
$$y' = -x \sin \theta + y \cos \theta$$

(6)

Example 2 Find the new coordinates of the point $(2, 4)$ if the coordinate axes are rotated through an angle of $\theta = 30°$.

Solution.   Using the rotation equations in (6) with $x = 2$, $y = 4$, $\cos \theta = \cos 30° = \sqrt{3}/2$, and $\sin \theta = \sin 30° = 1/2$, we obtain

$$x' = 2(\sqrt{3}/2) + 4(1/2) = \sqrt{3} + 2$$
$$y' = -2(1/2) + 4(\sqrt{3}/2) = -1 + 2\sqrt{3}$$

Thus, the new coordinates are $(\sqrt{3} + 2, -1 + 2\sqrt{3})$. ◄

☐ **ELIMINATING THE CROSS-PRODUCT TERM**

In Example 1, we were able to identify the curve $x^2 - xy + y^2 - 6 = 0$ as an ellipse because the rotation of axes eliminated the xy-term, thereby reducing the equation to a familiar form. This occurred because the new $x'y'$-axes were aligned with the axes of the ellipse. The following theorem tells how to determine an appropriate rotation of axes to eliminate the cross-product term of a second-degree equation in x and y.

12.5.1   THEOREM.   *If the equation*

$$Ax^2 + Bxy + Cy^2 + Dx + Ey + F = 0$$

(7)

is such that $B \neq 0$, and if an $x'y'$-coordinate system is obtained by rotating the xy-axes through an angle θ satisfying

$$\cot 2\theta = \frac{A - C}{B}$$

(8)

then, in $x'y'$-coordinates, Equation (7) will have the form

$$A'x'^2 + C'y'^2 + D'x' + E'y' + F' = 0$$

Proof.   Substituting (5) into (7) and simplifying yields

$$A'x'^2 + B'x'y' + C'y'^2 + D'x' + E'y' + F' = 0$$

where

$$A' = A \cos^2 \theta + B \cos \theta \sin \theta + C \sin^2 \theta$$
$$B' = B (\cos^2 \theta - \sin^2 \theta) + 2(C - A) \sin \theta \cos \theta$$
$$C' = A \sin^2 \theta - B \sin \theta \cos \theta + C \cos^2 \theta$$

(9)

$$D' = D \cos \theta + E \sin \theta$$
$$E' = -D \sin \theta + E \cos \theta$$
$$F' = F$$

(Verify.) To complete the proof we must show that $B' = 0$ if

$$\cot 2\theta = \frac{A - C}{B}$$

or equivalently,

$$\frac{\cos 2\theta}{\sin 2\theta} = \frac{A - C}{B} \qquad (10)$$

However, by using the trigonometric double-angle formulas, we can rewrite B' in the form

$$B' = B \cos 2\theta - (A - C) \sin 2\theta$$

Thus, $B' = 0$ if θ satisfies (10).   ∎

REMARK.   It is always possible to satisfy (8) with an angle θ in the range $0 < \theta < \pi/2$. We shall always use such a value of θ.

Example 3   Identify and sketch the curve $xy = 1$.

Solution.   As a first step, we shall rotate the coordinate axes to eliminate the cross-product term. Comparing the given equation to (7), we have

$$A = 0, \quad B = 1, \quad C = 0$$

Thus, the desired angle of rotation must satisfy

$$\cot 2\theta = \frac{A - C}{B} = \frac{0 - 0}{1} = 0$$

This condition can be met by taking $2\theta = \pi/2$ or $\theta = \pi/4 = 45°$. Substituting $\cos \theta = \cos 45° = 1/\sqrt{2}$ and $\sin \theta = \sin 45° = 1/\sqrt{2}$ in (5) yields

$$x = \frac{x'}{\sqrt{2}} - \frac{y'}{\sqrt{2}} \quad \text{and} \quad y = \frac{x'}{\sqrt{2}} + \frac{y'}{\sqrt{2}}$$

Substituting these in the equation $xy = 1$ yields

$$\left(\frac{x'}{\sqrt{2}} - \frac{y'}{\sqrt{2}} \right) \left(\frac{x'}{\sqrt{2}} + \frac{y'}{\sqrt{2}} \right) = 1 \quad \text{or} \quad \frac{x'^2}{2} - \frac{y'^2}{2} = 1$$

which is the equation in the $x'y'$-coordinate system of an equilateral hyperbola with vertices at $(\sqrt{2}, 0)$ and $(-\sqrt{2}, 0)$ in that coordinate system (Figure 12.5.4).   ◀

In problems where it is inconvenient to solve

$$\cot 2\theta = \frac{A - C}{B}$$

for θ, the values of $\sin \theta$ and $\cos \theta$ needed for the rotation equations may be obtained by first calculating $\cos 2\theta$ and then computing $\sin \theta$ and $\cos \theta$ from the identities

$$\sin \theta = \sqrt{\frac{1 - \cos 2\theta}{2}} \quad \text{and} \quad \cos \theta = \sqrt{\frac{1 + \cos 2\theta}{2}}$$

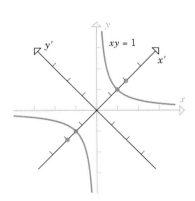

Figure 12.5.4

Example 4   Identify and sketch the curve

$$153x^2 - 192xy + 97y^2 - 30x - 40y - 200 = 0$$

Solution.   We have $A = 153$, $B = -192$, and $C = 97$, so

$$\cot 2\theta = \frac{A - C}{B} = -\frac{56}{192} = -\frac{7}{24}$$

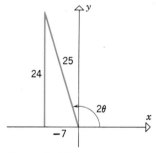

Figure 12.5.5

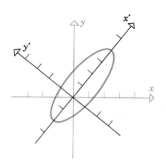

Figure 12.5.6

Since θ is to be chosen in the range $0 < \theta < \pi/2$, this relationship is represented by the triangle in Figure 12.5.5. From that triangle we obtain $\cos 2\theta = -7/25$, which implies that

$$\cos \theta = \sqrt{\frac{1 + \cos 2\theta}{2}} = \sqrt{\frac{1 - 7/25}{2}} = \frac{3}{5}$$

$$\sin \theta = \sqrt{\frac{1 - \cos 2\theta}{2}} = \sqrt{\frac{1 + 7/25}{2}} = \frac{4}{5}$$

Substituting these values in (5) yields the rotation equations

$$x = \tfrac{3}{5}x' - \tfrac{4}{5}y' \quad \text{and} \quad y = \tfrac{4}{5}x' + \tfrac{3}{5}y'$$

and substituting these in turn in the given equation yields

$$\tfrac{153}{25}(3x' - 4y')^2 - \tfrac{192}{25}(3x' - 4y')(4x' + 3y') + \tfrac{97}{25}(4x' + 3y')^2$$
$$- \tfrac{30}{5}(3x' - 4y') - \tfrac{40}{5}(4x' + 3y') - 200 = 0$$

which simplifies to

$$25x'^2 + 225y'^2 - 50x' - 200 = 0$$

or

$$x'^2 + 9y'^2 - 2x' - 8 = 0$$

Completing the square yields

$$\frac{(x' - 1)^2}{9} + y'^2 = 1$$

which is the equation in the $x'y'$-coordinate system of an ellipse with center $(1, 0)$ in that coordinate system and semiaxes $a = 3$ and $b = 1$ (Figure 12.5.6).   ◄

■ THE DISCRIMINANT

It is possible to describe the graph of a second-degree equation without rotating coordinate axes.

> **12.5.2 THEOREM.** *Consider a second-degree equation*
>
> $$Ax^2 + Bxy + Cy^2 + Dx + Ey + F = 0 \qquad (11)$$
>
> (a) *If $B^2 - 4AC < 0$, the equation represents an ellipse, a circle, a point, or else has no graph.*
>
> (b) *If $B^2 - 4AC > 0$, the equation represents a hyperbola or a pair of intersecting lines.*
>
> (c) *If $B^2 - 4AC = 0$, the equation represents a parabola, a line, a pair of parallel lines, or else has no graph.*

The quantity $B^2 - 4AC$ in this theorem is called the **discriminant** of the quadratic equation. To see why Theorem 12.5.2 is true, we need a fact about the discriminant. It can be shown (Exercise 19) that if the coordinate axes are rotated through any angle θ, and if

$$A'x'^2 + B'x'y' + C'y'^2 + D'x' + E'y' + F' = 0 \qquad (12)$$

is the equation resulting from (11) after rotation, then

$$B^2 - 4AC = B'^2 - 4A'C' \qquad (13)$$

In other words, the discriminant of a quadratic equation is not altered by rotating the coordinate axes. For this reason the discriminant is said to be **invariant** under a rotation of coordinate axes. In particular, if we choose the angle of rotation to eliminate the cross-product term, then (12) becomes

$$A'x'^2 + C'y'^2 + D'x' + E'y' + F' = 0 \tag{14}$$

and since $B' = 0$, (13) tells us that

$$B^2 - 4AC = -4A'C' \tag{15}$$

Part (a) of Theorem 12.5.2 can now be proved as follows. If $B^2 - 4AC < 0$, then from (15), $A'C' > 0$, so (14) may be divided through by $A'C'$ and written in the form

$$\frac{1}{C'}\left(x'^2 + \frac{D'}{A'}x'\right) + \frac{1}{A'}\left(y'^2 + \frac{E'}{C'}y'\right) = -\frac{F'}{A'C'}$$

Since $A'C' > 0$, the numbers A' and C' have the same sign. We may assume this sign to be positive, since Equation (14) can be multiplied through by -1 to achieve this, if necessary. By completing the squares, we can rewrite the last equation in the form

$$\frac{(x'-h)^2}{(\sqrt{C'})^2} + \frac{(y'-k)^2}{(\sqrt{A'})^2} = K$$

There are three possibilities: $K > 0$, in which case the graph is either a circle or an ellipse, depending on whether or not the denominators are equal; $K < 0$, in which case there is no graph, since the left side is nonnegative for all x' and y'; or $K = 0$, in which case the graph is the single point (h, k), since the equation is satisfied only by $x' = h$ and $y' = k$. The proofs of parts (b) and (c) require a similar kind of analysis. ■

Example 5 Use the discriminant to identify the graph of

$$8x^2 - 3xy + 5y^2 - 7x + 6 = 0$$

Solution. We have

$$B^2 - 4AC = (-3)^2 - 4(8)(5) = -151$$

Since the discriminant is negative, the equation represents an ellipse, a point, or else has no graph. (Why can't the graph be a circle?) ◄

In cases where a quadratic equation represents a point, a line, a pair of parallel lines, a pair of intersecting lines, or has no graph, we say that equation represents a ***degenerate conic section***. Thus, if we allow for possible degeneracy, it follows from Theorem 12.5.2 that *every quadratic equation has a conic section as its graph*.

▶ **Exercise Set 12.5**

---

1. Let an $x'y'$-coordinate system be obtained by rotating an xy-coordinate system through an angle of $\theta = 60°$.

   (a) Find the $x'y'$-coordinates of the point whose xy-coordinates are $(-2, 6)$.

   (b) Find an equation of the curve $\sqrt{3}xy + y^2 = 6$ in $x'y'$-coordinates.

   (c) Sketch the curve in part (b), showing both xy-axes and $x'y'$-axes.

2. Let an $x'y'$-coordinate system be obtained by rotating an xy-coordinate system through an angle of $\theta = 30°$.

   (a) Find the $x'y'$-coordinates of the point whose xy-coordinates are $(1, -\sqrt{3})$.

   (b) Find an equation of the curve $2x^2 + 2\sqrt{3}xy = 3$ in $x'y'$-coordinates.

   (c) Sketch the curve in part (b), showing both xy-axes and $x'y'$-axes.

In Exercises 3–12, rotate the coordinate axes to remove the xy-term. Then name the conic and sketch its graph.

3. $xy = -9$.

4. $x^2 - xy + y^2 - 2 = 0$.

5. $x^2 + 4xy - 2y^2 - 6 = 0$.

6. $31x^2 + 10\sqrt{3}xy + 21y^2 - 144 = 0$.

7. $x^2 + 2\sqrt{3}xy + 3y^2 + 2\sqrt{3}x - 2y = 0.$

8. $34x^2 - 24xy + 41y^2 - 25 = 0.$

9. $9x^2 - 24xy + 16y^2 - 80x - 60y + 100 = 0.$

10. $5x^2 - 6xy + 5y^2 - 8\sqrt{2}x + 8\sqrt{2}y = 8.$

11. $52x^2 - 72xy + 73y^2 + 40x + 30y - 75 = 0.$

12. $6x^2 + 24xy - y^2 - 12x + 26y + 11 = 0.$

13. Let an $x'y'$-coordinate system be obtained by rotating an xy-coordinate system through an angle θ. Prove: For every choice of θ, the equation $x^2 + y^2 = r^2$ becomes $x'^2 + y'^2 = r^2$. Give a geometric explanation.

14. Derive (6) by solving the rotation equations in (5) for x' and y' in terms of x and y.

15. Let an $x'y'$-coordinate system be obtained by rotating an xy-coordinate system through an angle of 45°. Use (6) to find an equation of the curve $3x'^2 + y'^2 = 6$ in xy-coordinates.

16. Let an $x'y'$-coordinate system be obtained by rotating an xy-coordinate system through an angle of 30°. Use (5) to find an equation in $x'y'$-coordinates of the curve $y = x^2$.

17. Show that the graph of the equation

$$\sqrt{x} + \sqrt{y} = 1$$

is a portion of a parabola. [*Hint:* First rationalize the equation and then perform a rotation of axes.]

18. Derive the expression for B' in (9).

19. Use (9) to prove that $B^2 - 4AC = B'^2 - 4A'C'$ for all values of θ.

20. Use (9) to prove that $A + C = A' + C'$ for all values of θ.

21. Prove: If $A = C$ in (7), then the cross-product term can be eliminated by rotating through 45°.

22. Prove: If $B \neq 0$, then the graph of $x^2 + Bxy + F = 0$ is a hyperbola if $F \neq 0$ and two intersecting lines if $F = 0$.

In Exercises 23–27, use the discriminant to identify the graph of the given equation.

23. $x^2 - xy + y^2 - 2 = 0.$

24. $x^2 + 4xy - 2y^2 - 6 = 0.$

25. $x^2 + 2\sqrt{3}xy + 3y^2 + 2\sqrt{3}x - 2y = 0.$

26. $6x^2 + 24xy - y^2 - 12x + 26y + 11 = 0.$

27. $34x^2 - 24xy + 41y^2 - 25 = 0.$

28. Each of the following represents a degenerate conic section. Where possible, sketch the graph.

    (a) $x^2 - y^2 = 0$

    (b) $x^2 + 3y^2 + 7 = 0$

    (c) $8x^2 + 7y^2 = 0$

    (d) $x^2 - 2xy + y^2 = 0$

    (e) $9x^2 + 12xy + 4y^2 - 36 = 0$

    (f) $x^2 + y^2 - 2x - 4y = -5.$

29. Prove parts (b) and (c) of Theorem 12.5.2.

◆ **TECHNOLOGY EXERCISES**   Chapter 12

Most of these exercises require access to a graphing calculator or a computer algebra system (CAS) such as *Mathematica*, *Maple*, or *Derive*. When you are asked to *find* an answer or to *solve* an equation, you may choose to find an exact result or a numerical approximation, depending on the particular technology you are using and on your own imagination. The form of your answers may differ from those of other students or from those in the answer section of the text, depending on how you solve the problems and the accuracy you use in your numerical approximations. Those exercises that are more appropriate for a CAS than a graphing calculator are labeled with the icon ◆.

1. **Distance between a point and a curve:** Find the maximum and minimum distance between the point $(1, -1)$ and a point on the ellipse $x^2/9 + y^2/5 = 1$.

2. **Volume:** Let R be the region in the first quadrant that lies to the right of the vertical line $x = k$ $(0 < k < 4)$ and is bounded above by the ellipse $x^2/16 + y^2/9 = 1$ and below

by the x-axis. Let V_x be the volume of the solid generated by revolving R about the x-axis and V_y the volume of the solid generated by revolving R about the y-axis. Find k so that $V_y = 2V_x$.

3. **Nuclear cooling tower:** A cooling tower for a nuclear power plant is to have a height of h feet and the shape of the solid that is generated by revolving about the y-axis the region enclosed by the right branch of the hyperbola $x^2/225 - y^2/1521 = 1$, the y-axis, and the horizontal lines $y = -h/2$ and $y = h/2$. Assuming that one unit on the coordinate axes corresponds to one foot, find the height h if the tower is to have a volume of 50,000 ft^3.

◆ 4. **Suspension bridges:** Under appropriate conditions, support cables for suspension bridges are shaped like parabolas. The main span of the Golden Gate Bridge in San Francisco has a horizontal length of 4200 ft and the central support cables sag 470 ft at the middle (see the accompanying figure).

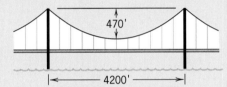

(a) Find the length of a central support cable.

(b) Find, to the nearest degree, the acute angle at a support point between the tangent line to a central cable and the horizontal.

◆ 5. **Length of the earth's orbit:** The earth moves in an elliptical orbit with the sun at a focus. An equation of the orbit is $x^2/a^2 + y^2/b^2 = 1$ (see the accompanying figure), where $a = 1.49 \times 10^8$ km and $b = 1.48979 \times 10^8$ km.

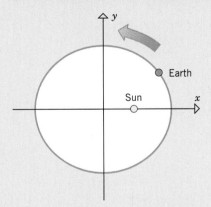

(a) Find the total distance traveled by the earth in one orbit about the sun.

(b) Find the average speed of the earth in kilometers per hour over one orbit (use 365 days as the time for one orbit).

◆ 6. **Location of a ship at sea:** A ship at sea starts at the point $(0, 3)$ in an xy-coordinate system and travels into the first quadrant in such a way that its distance from $(0, 4)$ minus its distance to $(0, -4)$ remains constant. Assuming that all coordinates are in miles, find, to the nearest tenth of a mile, the coordinates of the ship when it has traveled 100 miles.

◆ 7. **Calibrating a dipstick:** The tank of an oil truck is 18 ft long and has elliptical cross sections that are 6 ft wide and 4 ft high (see the accompanying figure). From Exercise 45, Section 12.3, the volume V of oil in the tank in terms of the depth h of the oil is given by

$$V = 27 \left[4 \sin^{-1}\frac{h-2}{2} + (h-2)\sqrt{4h - h^2} + 2\pi \right]$$

where V is in cubic feet and h is in feet.

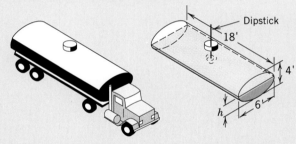

(a) Determine how many inches from the bottom of a dipstick the calibration marks should be placed to indicate when the tank is $\frac{1}{4}$, $\frac{1}{2}$, and $\frac{3}{4}$ full.

(b) Suppose that the tank is filled with oil at the constant rate of 0.5 ft^3/sec. How fast is the level of oil rising when the tank is $\frac{1}{4}$ full? Express your answer in inches/second.

◆ 8. **Rotated conics:** Consider the conic whose equation is

$$x^2 + xy + 2y^2 - x + 3y + 1 = 0$$

(a) Use the discriminant to identify the conic.

(b) Graph the equation by solving for y in terms of x and graphing both solutions.

(c) Your CAS may be able to graph the equation in the form given. If so, graph the equation in this way.

◆ **9. Rotated conics:** Consider the conic whose equation is

$$2x^2 + 9xy + y^2 - 6x + y - 4 = 0$$

(a) Use the discriminant to identify the conic.

(b) Graph the equation by solving for y in terms of x and graphing both solutions.

(c) Your CAS may be able to graph the equation in the form given. If so, graph the equation in this way.

Johann Bernoulli (1667–1748)

13 POLAR COORDINATES AND PARAMETRIC EQUATIONS

13.1 POLAR COORDINATES

*Up to now, we have specified the location of points in the plane by means of coordinates relative to a rectangular coordinate system consisting of two perpendicular coordinate axes. In this section, we shall introduce a new kind of coordinate system, called a **polar coordinate system**, that will be easier to work with for many problems.*

POLAR COORDINATE SYSTEMS

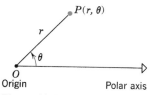

Figure 13.1.1

A **polar coordinate system** in a plane consists of a fixed point O, called the **origin** (or **pole**) and a ray emanating from the origin, called the **polar axis**. In such a coordinate system we can associate with each point P in the plane a pair of **polar coordinates** (r, θ), where r is the distance from P to the origin and θ is an angle from the polar axis to the ray OP (Figure 13.1.1). The number r is called the **radial coordinate** of P and the number θ is called the **angular coordinate** (or **polar angle**) of P. In Figure 13.1.2, the points $(6, 45°)$, $(5, 120°)$, $(3, 225°)$, and $(4, 330°)$ are plotted in polar coordinate systems. If P is the origin, then $r = 0$, but there is no clearly defined polar angle. We shall agree that any arbitrary angle can be used in this case, so $(0, \theta)$ are polar coordinates of the origin for all choices of θ.

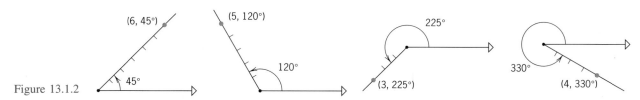

Figure 13.1.2

The polar coordinates of a point are not unique. For example, the polar coordinates

$$(1, 315°), \quad (1, -45°), \quad \text{and} \quad (1, 675°)$$

all represent the same point (Figure 13.1.3). In general, if a point P has polar coordinates (r, θ), then for any integer $n = 0, 1, 2, \ldots$

$$(r, \theta + n \cdot 360°) \quad \text{and} \quad (r, \theta - n \cdot 360°)$$

are also polar coordinates of P.

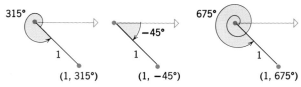

Figure 13.1.3

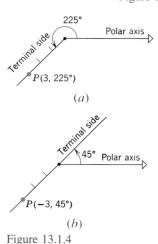

(a)

(b)

Figure 13.1.4

As defined above, the radial coordinate r of a point P is nonnegative, since it represents the distance from P to the origin. It will be convenient to allow for negative values of r as well. To motivate an appropriate definition, consider the point P with polar coordinates $(3, 225°)$. We can reach this point by rotating the polar axis through an angle of $225°$ and then moving 3 units from the origin along the terminal side of the angle (Figure 13.1.4a), or we can reach the point P by rotating the polar axis through an angle of $45°$ and then moving 3 units from the origin along the *extension* of the terminal side (Figure 13.1.4b). This suggests that the point $(3, 225°)$ might also be denoted by $(-3, 45°)$, with the minus sign serving to indicate that the point is on the extension of the angle's terminal side rather than on the terminal side itself. In general, the terminal side of the angle $\theta + 180°$ is the extension of the terminal side of θ, so we define negative radial coordinates by agreeing that

$$(-r, \theta) \quad \text{and} \quad (r, \theta + 180°)$$

are polar coordinates of the same point.

☐ **RELATIONSHIP BETWEEN POLAR AND RECTANGULAR COORDINATES**

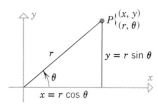

Figure 13.1.5

Frequently, it is helpful to use both polar and rectangular coordinates in the same problem. To do this, we let the positive x-axis of the rectangular coordinate system serve as the polar axis for the polar coordinate system. When this is done, each point P has both polar coordinates (r, θ) and rectangular coordinates (x, y). As suggested by Figure 13.1.5, these coordinates are related by the equations

$$\begin{aligned} x &= r \cos \theta \\ y &= r \sin \theta \end{aligned} \quad \text{and} \quad \begin{aligned} r^2 &= x^2 + y^2 \\ \tan \theta &= \frac{y}{x} \end{aligned} \qquad (1a\text{–}b)$$

Example 1   Find the rectangular coordinates of the point P whose polar coordinates are $(6, 135°)$.

Solution.   Substituting the polar coordinates $r = 6$ and $\theta = 135°$ in (1a) yields

$$x = 6 \cos 135° = 6 \left(-\sqrt{2}/2\right) = -3\sqrt{2}$$

$$y = 6 \sin 135° = 6 \left(\sqrt{2}/2\right) = 3\sqrt{2}$$

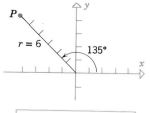

The point P has polar coordinates (6, 135°) and rectangular coordinates $(-3\sqrt{2}, 3\sqrt{2})$

Figure 13.1.6

Thus, the rectangular coordinates of P are $(-3\sqrt{2}, 3\sqrt{2})$ (Figure 13.1.6).   ◀

REMARK.   For many purposes it does not matter whether polar angles are measured in degrees or radians. However, in problems involving derivatives or integrals they must be measured in radians, since the derivative formulas for the trigonometric functions were derived under this assumption. From here on, we shall generally use radian measure for polar angles.

Example 2   Find polar coordinates of the point P whose rectangular coordinates are $(-2, 2\sqrt{3})$.

Solution.   We will find polar coordinates (r, θ) of P such that $r > 0$ and $0 \le \theta < 2\pi$. From the first equation in (1b),

$$r^2 = x^2 + y^2 = (-2)^2 + (2\sqrt{3})^2 = 4 + 12 = 16$$

so $r = 4$. From the second equation in (1b),

$$\tan\theta = \frac{y}{x} = \frac{2\sqrt{3}}{-2} = -\sqrt{3}$$

From this and the fact that $(-2, 2\sqrt{3})$ lies in the second quadrant, it follows that $\theta = 2\pi/3$. Thus, $(4, 2\pi/3)$ are polar coordinates of P. All other polar coordinates of P have the form

$$\left(4, \frac{2\pi}{3} + 2n\pi\right) \text{or} \left(-4, \frac{5\pi}{3} + 2n\pi\right)$$

where n is an integer.   ◀

Polar coordinates provide a new way of graphing certain equations. As an example, consider the equation

$$r = \sin\theta \tag{2}$$

By substituting values for θ at increments of $\pi/6$ ($= 30°$) and calculating r, we can construct the following table:

Table 13.1.1

θ (RADIANS)	0	$\dfrac{\pi}{6}$	$\dfrac{\pi}{3}$	$\dfrac{\pi}{2}$	$\dfrac{2\pi}{3}$	$\dfrac{5\pi}{6}$	π	$\dfrac{7\pi}{6}$	$\dfrac{4\pi}{3}$	$\dfrac{3\pi}{2}$	$\dfrac{5\pi}{3}$	$\dfrac{11\pi}{6}$	2π
$r = \sin\theta$	0	$\dfrac{1}{2}$	$\dfrac{\sqrt{3}}{2}$	1	$\dfrac{\sqrt{3}}{2}$	$\dfrac{1}{2}$	0	$-\dfrac{1}{2}$	$-\dfrac{\sqrt{3}}{2}$	-1	$-\dfrac{\sqrt{3}}{2}$	$-\dfrac{1}{2}$	0

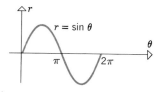

Figure 13.1.7

The pairs of values listed in this table can be plotted in one of two ways. One possibility is to view θ and r as *rectangular* coordinates of a point and plot each point in the θr-plane. This yields points on the familiar sine curve (Figure 13.1.7). Alternatively, we can view each pair (r, θ) as *polar* coordinates of a point and plot each point in a polar coordinate system (Figure 13.1.8). Note that there are 13 pairs listed in Table 13.1.1, but only 6 points plotted in Figure 13.1.8. This is because the pairs from $\theta = \pi$ on yield duplicates of the preceding points. For example, $(-1/2, 7\pi/6)$ and $(1/2, \pi/6)$ represent the same point.

The points in Figure 13.1.8 appear to lie on a circle. That this is indeed the case may be seen by expressing (2) in terms of x and y. To do this we first multiply (2) through by r to obtain

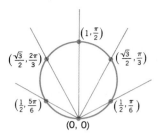

Figure 13.1.8

$$r^2 = r \sin \theta$$

which can be rewritten using (1a) and (1b) as

$$x^2 + y^2 = y \quad \text{or} \quad x^2 + y^2 - y = 0$$

or on completing the square,

$$x^2 + (y - \tfrac{1}{2})^2 = \tfrac{1}{4}$$

This is a circle of radius $\frac{1}{2}$ centered at the point $(0, \frac{1}{2})$ in the xy-plane.

Example 3   Sketch the curve $r = \sin 2\theta$ in polar coordinates.

Solution.   A rough sketch of the graph may be obtained without plotting points by observing how the value of r changes with θ (Figure 13.1.9).   ◄

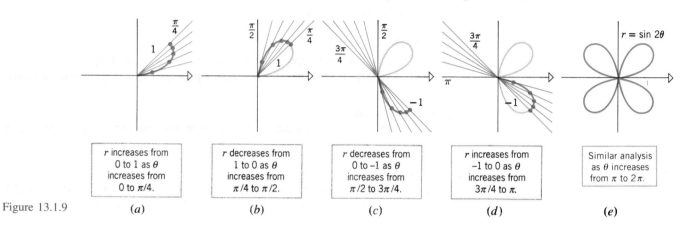

Figure 13.1.9   *(a)*   *(b)*   *(c)*   *(d)*   *(e)*

▶ Exercise Set 13.1

1. Plot the following points in polar coordinates.
   (a) $(3, \pi/4)$
   (b) $(5, 2\pi/3)$
   (c) $(1, \pi/2)$
   (d) $(4, 7\pi/6)$
   (e) $(2, 4\pi/3)$
   (f) $(0, \pi)$.

2. Plot the following points in polar coordinates.
   (a) $(2, -\pi/3)$
   (b) $(3/2, -7\pi/4)$
   (c) $(-3, 3\pi/2)$
   (d) $(-5, -\pi/6)$
   (e) $(-6, -\pi)$
   (f) $(-1, 9\pi/4)$.

3. Find rectangular coordinates of the points whose polar coordinates are given.
   (a) $(6, \pi/6)$
   (b) $(7, 2\pi/3)$
   (c) $(8, 9\pi/4)$
   (d) $(5, 0)$
   (e) $(7, 17\pi/6)$
   (f) $(0, \pi)$.

4. Find rectangular coordinates of the points whose polar coordinates are given.
   (a) $(-8, \pi/4)$
   (b) $(7, -\pi/4)$
   (c) $(-6, -5\pi/6)$
   (d) $(0, -\pi)$
   (e) $(-2, -3\pi/2)$
   (f) $(-5, 0)$.

5. The following points are given in rectangular coordinates. Express the points in polar coordinates with $r \geq 0$ and $0 \leq \theta < 2\pi$.
   (a) $(-5, 0)$
   (b) $(2\sqrt{3}, -2)$
   (c) $(0, -2)$
   (d) $(-8, -8)$
   (e) $(-3, 3\sqrt{3})$
   (f) $(1, 1)$.

6. Express the points in Exercise 5 in polar coordinates with $r \geq 0$ and $-\pi < \theta \leq \pi$.

7. Express the points in Exercise 5 in polar coordinates with $r \leq 0$ and $0 \leq \theta < 2\pi$.

8. In each part find polar coordinates satisfying the stated conditions for the point whose rectangular coordinates are $(-\sqrt{3}, 1)$.
   (a) $r \geq 0$ and $0 \leq \theta < 2\pi$
   (b) $r \leq 0$ and $0 \leq \theta < 2\pi$
   (c) $r \geq 0$ and $-2\pi < \theta \leq 0$
   (d) $r \leq 0$ and $-\pi < \theta \leq \pi$.

In Exercises 9–20, identify the curve by transforming to rectangular coordinates.

9. $r = 2$.
10. $r = 3$.
11. $r \sin \theta = 4$.
12. $r = 5 \sec \theta$.
13. $r = 3 \cos \theta$.
14. $r = 2 \sin \theta$.

15. $r^2 \sin 2\theta = 8$.

16. $r^2 \cos 2\theta = 9$.

17. $r = \dfrac{2}{1 + \sin \theta}$.

18. $r = \dfrac{6}{2 - \cos \theta}$.

19. $r = \dfrac{6}{3 \cos \theta + 2 \sin \theta}$.

20. $r = \sec \theta \tan \theta$.

In Exercises 21–32, express the equations in polar coordinates.

21. $x = 7$.

22. $y = -3$.

23. $x^2 + y^2 = 9$.

24. $x^2 + y^2 = 5$.

25. $x^2 + y^2 - 6y = 0$.

26. $x^2 + y^2 + 4x = 0$.

27. $x^2 = 9y$.

28. $x^2 - y^2 = 4$.

29. $4xy = 9$.

30. $(x^2 + y^2)^2 = 16(x^2 - y^2)$.

31. $(x^2 + y^2)^2 = 2xy$.

32. $x^2(x^2 + y^2) = y^2$.

33. Sketch the graph of $r^2 - 3r + 2 = 0$ in polar coordinates. [*Hint:* Factor.]

In Exercises 34–37, use the method of Example 3 to sketch the curve in polar coordinates.

34. $r = \cos 2\theta$.

35. $r = 2(1 + \sin \theta)$.

36. $r = 4 \cos 3\theta$.

37. $r = 1 - \cos \theta$.

38. Prove that the distance between the points with polar coordinates (r_1, θ_1) and (r_2, θ_2) is

$$d = \sqrt{r_1^2 + r_2^2 - 2r_1 r_2 \cos (\theta_1 - \theta_2)}$$

39. Prove: In polar coordinates, the equation $r = a \sin \theta + b \cos \theta$ represents a circle.

40. Prove: The area of the triangle whose vertices have polar coordinates $(0, 0)$, (r_1, θ_1), and (r_2, θ_2) is

$$A = \tfrac{1}{2} r_1 r_2 \sin (\theta_2 - \theta_1)$$

(Assume that $0 \le \theta_1 < \theta_2 \le \pi$ and r_1 and r_2 are positive.)

41. Prove: Equations (1a) hold if r is negative.

42. Prove that

$$rr_1 \sin (\theta - \theta_1) + rr_2 \sin (\theta_2 - \theta)$$
$$+ r_1 r_2 \sin (\theta_1 - \theta_2) = 0$$

is the polar equation of the line through the points (r_1, θ_1) and (r_2, θ_2).

■ 13.2 GRAPHS IN POLAR COORDINATES

In this section we shall graph various curves in polar coordinates. Since we shall be working with both rectangular xy-coordinates and polar coordinates, we shall assume throughout that the positive x-axis coincides with the polar axis.

☐ **LINES IN POLAR COORDINATES**

Polar equations of vertical and horizontal lines can be obtained by substituting $x = r \cos \theta$ and $y = r \sin \theta$ in the equations $x = a$ and $y = b$. This yields the following polar equations for the vertical line through $(a, 0)$ and the horizontal line through $(0, b)$:

$$r \cos \theta = a \qquad \text{and} \qquad r \sin \theta = b \qquad (1–2)$$

As shown by the dashed red lines in Figures 13.2.1 and 13.2.2, these equations make sense geometrically, since each point P on the vertical line produces a value of a for $r \cos \theta$, and each point P on the horizontal line produces a value of b for $r \sin \theta$.

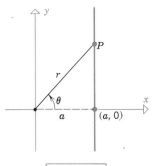

Figure 13.2.1

$r \cos \theta = a$

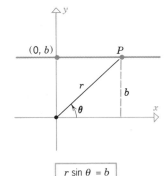

Figure 13.2.2

$r \sin \theta = b$

A line through the origin that makes an angle of $\theta = \theta_0$ (radians) with the polar axis consists of all points of the form $P(r, \theta_0)$. It follows that the line has the equation

$$\theta = \theta_0 \tag{3}$$

since this equation is satisfied by the coordinates of P, regardless of the value of r (Figure 13.2.3). A polar equation for an arbitrary line can be obtained by substituting $x = r \cos \theta$ and $y = r \sin \theta$ in the equation $Ax + By + C = 0$; this yields the **general polar form of a line**,

$$r(A \cos \theta + B \sin \theta) + C = 0 \tag{4}$$

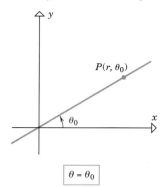

Figure 13.2.3

Example 1 Sketch the graphs of the following equations in polar coordinates.

(a) $r \cos \theta = 3$ (b) $r \sin \theta = -2$ (c) $\theta = \dfrac{3\pi}{4}$

Solution. From Equations (1), (2), and (3), each graph is a line; they are shown in Figure 13.2.4. ◀

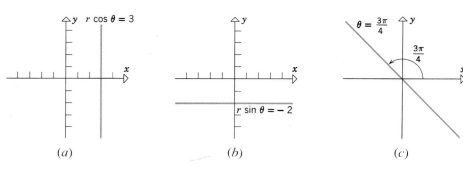

Figure 13.2.4   *(a)*   *(b)*   *(c)*

☐ **CIRCLES IN POLAR COORDINATES**

A circle of radius a that is centered at the origin consists of all points of the form $P(a, \theta)$. It follows that the circle has the equation

$$r = a \tag{5}$$

since this equation is satisfied by the coordinates of P, regardless of the value of θ (Figure 13.2.5).

We shall also be interested in polar equations of circles that are centered on the x-axis or y-axis and pass through the origin. To obtain such equations, recall from plane geometry that a triangle that is inscribed in a circle with a diameter for a side must be a right triangle. Thus, from Figure 13.2.6, the equation of a circle that is centered on the x-axis and passes through the origin has an equation of the form

$$r = 2a \cos \theta \qquad \text{or} \qquad r = -2a \cos \theta \tag{6a–6b}$$

where (6a) applies to circles centered on the positive *x*-axis and (6b) applies to circles centered on the negative *x*-axis. Similarly, from Figure 13.2.7 a circle that is centered on the *y*-axis and passes through the origin has an equation of the form

$$r = 2a \sin \theta \qquad \text{or} \qquad r = -2a \sin \theta \qquad\qquad (7a\text{--}7b)$$

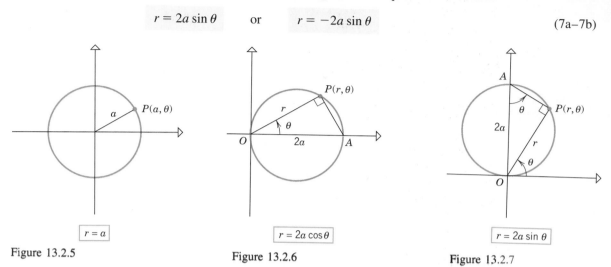

Figure 13.2.5 $r = a$

Figure 13.2.6 $r = 2a \cos \theta$

Figure 13.2.7 $r = 2a \sin \theta$

Example 2 Sketch the graphs of the following equations in polar coordinates.

(a) $r = 3$ (b) $r = 4 \cos \theta$ (c) $r = -5 \sin \theta$

Solution. These equations are of forms (5), (6a), and (7b), respectively, and therefore represent circles. The graphs are shown in Figure 13.2.8.   ◀

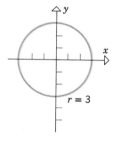 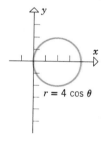 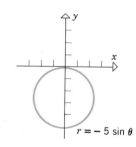

Figure 13.2.8 $r = 3$ $r = 4 \cos \theta$ $r = -5 \sin \theta$

□ **SYMMETRY TESTS**

The symmetry tests in the following theorem will be helpful for graphing equations in polar coordinates. Although we will omit the proof, the results should be evident from Figure 13.2.9.

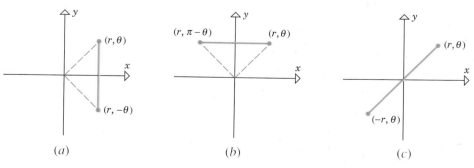

Figure 13.2.9   (*a*)   (*b*)   (*c*)

13.2.1   THEOREM (Symmetry Tests).

(a)   *A curve in polar coordinates is symmetric about the x-axis if replacing θ by −θ in its equation produces an equivalent equation (Figure 13.2.9a).*

(b)   *A curve in polar coordinates is symmetric about the y-axis if replacing θ by π − θ in its equation produces an equivalent equation (Figure 13.2.9b).*

(c)   *A curve in polar coordinates is symmetric about the origin if replacing r by −r in its equation produces an equivalent equation (Figure 13.2.9c).*

□ **CARDIOIDS AND LIMAÇONS**

Equations of the form

$$r = a + b \sin \theta, \qquad r = a - b \sin \theta \tag{8a–8b}$$

$$r = a + b \cos \theta, \qquad r = a - b \cos \theta \tag{8c–8d}$$

produce polar curves called **limaçons** (from the Latin word "limax," for a slug, a snail-like creature). There are four possible shapes for a limaçon* that can be determined from the ratio a/b when $a > 0$ and $b > 0$ (Figure 13.2.10).

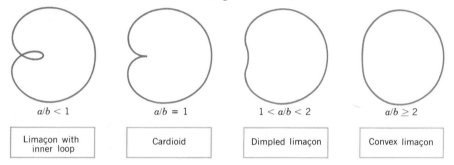

| $a/b < 1$ | $a/b = 1$ | $1 < a/b < 2$ | $a/b \geq 2$ |

| Limaçon with inner loop | Cardioid | Dimpled limaçon | Convex limaçon |

Figure 13.2.10

The position of the limaçon relative to the polar axis depends on whether $\sin \theta$ or $\cos \theta$ appears in the equation and whether the + or − occurs. Because of the heart-shaped appearance of the curve in the case $a = b$, limaçons of this type are called **cardioids** (from the Greek word "kardia," for heart).

Example 3   Assuming a to be a positive constant, sketch the graph of the equation $r = a(1 - \cos \theta)$ in polar coordinates.

Solution.   This is an equation of form (8d) with $a = b$; therefore, it represents a cardioid. Since $\cos(-\theta) = \cos \theta$, the equation is unaltered when θ is replaced by $-\theta$; thus, the cardioid is symmetric about the x-axis. This being the case, we can obtain the entire curve by first sketching the portion of the cardioid above the x-axis and then reflecting this portion about the x-axis.

As θ varies from 0 to π, $\cos \theta$ decreases steadily from 1 to -1, and $1 - \cos \theta$ increases steadily from 0 to 2. Thus, as θ varies from 0 to π, the value of $r = a(1 - \cos \theta)$ will increase steadily from an initial value of $r = 0$ to a final value of $r = 2a$. Using this

---

*For a detailed discussion of limaçons and their shapes, the reader is referred to the article by Jane T. Grossman and Michael P. Grossman, "Dimple or No Dimple," *Two-Year College Mathematics Journal*, Vol. 13, No. 1, 1982, pp. 52–55.

information, and by plotting the points in Table 13.2.1, we obtain the graph in Figure 13.2.11a.

Table 13.2.1

θ	0	$\pi/3$	$\pi/2$	$2\pi/3$	π
$r = a(1 - \cos\theta)$	0	$a/2$	a	$3a/2$	$2a$

On reflecting the curve in Figure 13.2.11a about the x-axis, we obtain the entire cardioid (Figure 13.2.11b). ◀

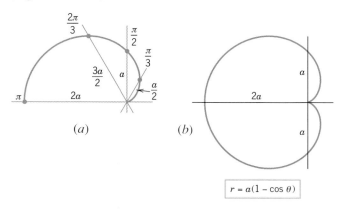

(a) (b)

$r = a(1 - \cos\theta)$

Figure 13.2.11

REMARK.   We have sketched the cardioid in Figure 13.2.11a with the line $\theta = 0$ tangent to the curve at the origin. Although it is not evident from our discussion that this is the case, we will prove in Section 13.5 that if a polar curve passes through the origin when $\theta = \theta_0$, then the line $\theta = \theta_0$ is tangent to the curve at the origin.

☐ **LEMNISCATES**

If $a > 0$, then equations of the form

$$r^2 = a^2 \cos 2\theta, \qquad r^2 = -a^2 \cos 2\theta \tag{9a–9b}$$
$$r^2 = a^2 \sin 2\theta, \qquad r^2 = -a^2 \sin 2\theta \tag{9c–9d}$$

represent propeller-shaped curves, called **lemniscates** (from the Greek word "lemniscos" for a looped ribbon resembling the number 8) (Figure 13.2.12). The lemniscates are centered at the origin, but the position relative to the polar axis depends on the sign preceding the a^2 and whether $\sin 2\theta$ or $\cos 2\theta$ appears in the equation.

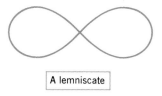

A lemniscate

Figure 13.2.12

Example 4   Sketch the curve

$$r^2 = 4 \cos 2\theta \tag{10}$$

in polar coordinates.

Solution.   The equation is of type (9a) with $a = 2$, and thus represents a lemniscate. Solving (10) for r in terms of θ yields two functions of θ:

$$r = 2\sqrt{\cos 2\theta} \quad \text{and} \quad r = -2\sqrt{\cos 2\theta} \tag{11a–b}$$

We leave it for the reader to apply the symmetry tests in Theorem 13.2.1 and show that the graph of each of these equations is symmetric about the x-axis and the y-axis.

Therefore, we can obtain each graph by first sketching the portion of the graph in the range $0 \leq \theta \leq \pi/2$ and then reflecting that portion about the *x*- and *y*-axes.

We shall consider the graph of (11a) first. As θ varies from 0 to $\pi/4$, the value of $\cos 2\theta$ decreases steadily from 1 to 0, so that $r = 2\sqrt{\cos 2\theta}$ decreases steadily from 2 to 0. With this information and the points plotted from Table 13.2.2, we obtain the sketch in Figure 13.2.13*a*. (Note that the curve passes through the origin when $\theta = \pi/4$, so the line $\theta = \pi/4$ is tangent to the curve at the origin by the remark following Example 3.)

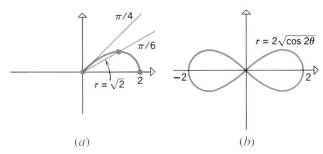

Figure 13.2.13   (*a*)   (*b*)

For θ in the range $\pi/4 < \theta < \pi/2$, the quantity $\cos 2\theta$ is negative, so there are no real values of *r* satisfying (11a). Thus, there are no points on the graph for such θ. The entire graph is obtained by reflecting the curve in Figure 13.2.13*a* about the *x*-axis and then reflecting the resulting curve about the *y*-axis (Figure 13.2.13*b*).

Table 13.2.2

θ	0	$\pi/6$	$\pi/4$
$r = 2\sqrt{\cos 2\theta}$	2	$\sqrt{2}$	0

To graph $r = -2\sqrt{\cos 2\theta}$, we simply note that (11a) and (11b) differ only in the sign of *r*, so the graph of (11b) is identical to that in Figure 13.2.13*b*, but traced in a ''diagonally opposite'' manner. For example, as θ varies from 0 to $\pi/4$, the portion of the graph generated is in the third quadrant because *r* is negative (Figure 13.2.14). To summarize, the graph of (10) actually consists of two superimposed lemniscates, one being the graph of (11a) and the other the graph of (11b). ◄

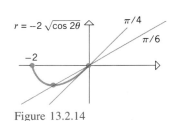

Figure 13.2.14

□ SPIRALS

A curve that ''winds around the origin'' infinitely many times in such a way that *r* increases (or decreases) steadily as θ increases is called a ***spiral***. The most common example is the ***spiral of Archimedes***, which has an equation of the form

$$r = a\theta \quad (\theta \geq 0) \qquad \text{or} \qquad r = a\theta \quad (\theta \leq 0) \qquad (12a\text{--}b)$$

In these equations, θ is in radians and *a* is positive.

Example 5   Sketch the curve

$$r = \theta \quad (\theta \geq 0)$$

in polar coordinates.

Solution.   This is an equation of form (12a) with $a = 1$; thus, it represents an Archimedean spiral. Since $r = 0$ when $\theta = 0$, the origin is on the curve and the polar axis is tangent to the spiral.

A reasonably accurate sketch may be obtained by plotting the intersections of the spiral with the *x*- and *y*-axes and noting that *r* increases steadily as θ increases. These intersections occur when

$$\theta = 0, \pi/2, \pi, 3\pi/2, 2\pi, 5\pi/2, \ldots$$

at which points *r* has these same values (Figure 13.2.15). ◄

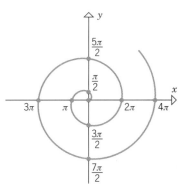

Figure 13.2.15

Starting from the origin, the spiral in Figure 13.2.15 loops counterclockwise around the origin. This is true of all Archimedean spirals of form (12a); Archimedean spirals of form (12b) loop clockwise around the origin.

☐ **ROSE CURVES**

Equations of the form

$$r = a \sin n\theta \qquad \text{and} \qquad r = a \cos n\theta \qquad (13\text{a–b})$$

represent flower-shaped curves called **_roses_**. The rose has n equally spaced petals or loops if n is odd and $2n$ equally spaced petals if n is even (Figure 13.2.16). The orientation of the rose relative to the polar axis depends on the sign of the constant a and whether $\sin \theta$ or $\cos \theta$ appears in the equation. (A rose curve occurred in Example 3 of the preceding section.)

Figure 13.2.16

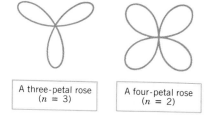

A three-petal rose
($n = 3$)

A four-petal rose
($n = 2$)

If $n = 1$ in (13a) or (13b), then we obtain the equation of a circle, which can be regarded as a one-petal rose.

REMARK.   It can be shown that a rose with an even number of petals is traced out exactly once as θ varies over the interval $0 \leq \theta < 2\pi$ (Example 3 in the preceding section), and a rose with an odd number of petals is traced out exactly once as θ varies over the interval $0 \leq \theta < \pi$ (Exercise 60).

☐ **CONICS IN POLAR COORDINATES**

Kepler's* first law of planetary motion states that each planet moves in an elliptical orbit about the sun with the sun at a focus of the ellipse. As we will show in Section 15.7, this is a consequence of a more general law, which states that a body in motion under an inverse square law force (such as gravity) directed toward a fixed point (such as the center of the sun) will move along a path that is a conic section with that point as a focus. Since the study of such motion is usually done in polar coordinates using the focus–directrix representation of conics, we summarize the results on this topic obtained in Sections 12.2, 12.3, and 12.4 as follows:

---

*JOHANNES KEPLER (1571–1630). German astronomer and physicist. Kepler, whose work provided our contemporary view of planetary motion, led a fascinating but ill-starred life. His alcoholic father made him work in a family-owned tavern as a child, later withdrawing him from elementary school and hiring him out as a field laborer, where the boy contracted smallpox, permanently crippling his hands and impairing his eyesight. In later years, Kepler's first wife and several children died, his mother was accused of witchcraft, and being a Protestant he was often subjected to persecution by Catholic authorities. He was often impoverished, eking out a living as an astrologer and prognosticator. Looking back on his unhappy childhood, Kepler described his father as "criminally inclined" and "quarrelsome" and his mother as "garrulous" and "bad-tempered." However, it was his mother who left an indelible mark on the six-year-old Kepler by showing him the comet of 1577; and in later life he personally prepared her defense against the witchcraft charges. Kepler became acquainted with the work of Copernicus as a student at the University of Tübingen, where he received his master's degree in 1591. He continued on as a theological student, but at the urging of the university officials he abandoned his clerical studies and accepted a position as a mathematician and teacher in Graz, Austria. However, he was expelled from the city

(continued on next page)

Let a fixed point (the focus) and a fixed line (the directrix) be given so that the directrix does not contain the focus. If a point P moves in the plane determined by the focus and the directrix so that the distance between P and the focus is proportional to the distance between P and the directrix, then the curve traced out by P is a conic.

If we let F be the focus, PF the distance between P and the focus, PD the distance between P and the directrix, and e (the eccentricity) the constant of proportionality, then

$$PF = ePD \tag{14}$$

where the curve is an ellipse if $0 < e < 1$, a parabola if $e = 1$, and a hyperbola if $e > 1$.

If, as in Figure 13.2.17, the focus F is at the pole and the directrix is perpendicular to the polar axis and k units to the right of the pole, then $PF = r$ and $PD = k - r \cos \theta$. Thus, from (14)

$$r = e(k - r \cos \theta)$$

which can be solved for r to obtain

$$r = \frac{ek}{1 + e \cos \theta}$$

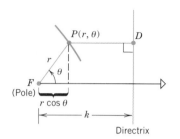

Figure 13.2.17

For the case where the directrix is perpendicular to the polar axis and k units to the *left* of the pole, the equation becomes

$$r = \frac{ek}{1 - e \cos \theta}$$

In summary, an equation of the form

$$r = \frac{ek}{1 \pm e \cos \theta} \tag{15}$$

represents a conic section with eccentricity e, a focus at the pole, and a directrix perpendicular to the polar axis and k units to the right of the pole in the $+$ case and k units to the left of the pole in the $-$ case. Similarly, an equation of the form

$$r = \frac{ek}{1 \pm e \sin \theta} \tag{16}$$

represents a conic section with eccentricity e, a focus at the pole, and a directrix *parallel* to the polar axis and k units above the pole in the $+$ case and k units below the pole in the $-$ case.

(continued)

Example 6 Sketch the graph of $r = \dfrac{6}{2 + \cos\theta}$ in polar coordinates.

Solution. Divide numerator and denominator by 2 to get

$$r = \frac{3}{1 + \frac{1}{2}\cos\theta}$$

The graph is an ellipse with $e = 1/2$, one focus at the pole, and a vertical directrix. The points on the graph that correspond to $\theta = 0$, $\pi/2$, π, and $3\pi/2$ are $(2, 0)$, $(3, \pi/2)$, $(6, \pi)$, and $(3, 3\pi/2)$. The graph is shown in Figure 13.2.18. ◀

Example 7 Sketch the graph of $r = \dfrac{2}{1 + 2\sin\theta}$ in polar coordinates.

Solution. The graph is a hyperbola with $e = 2$, one focus at the pole, and a horizontal directrix. The points $(2, 0)$, $(2/3, \pi/2)$, $(2, \pi)$, and $(-2, 3\pi/2)$ are on the graph. The vertices are at $(2/3, \pi/2)$ and $(-2, 3\pi/2)$ with the center midway between them. Because r is undefined when $\sin\theta = -1/2$, the asymptotes (which pass through the center) must be parallel to the lines $\theta = 7\pi/6$ and $\theta = 11\pi/6$. The graph is shown in Figure 13.2.19. ◀

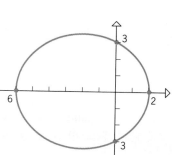

Figure 13.2.18

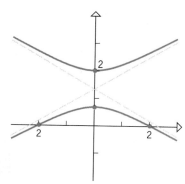

Figure 13.2.19

▶ Exercise Set 13.2   C 38, 48

In Exercises 1–36, sketch the curve in polar coordinates and give its name.

1. $\theta = \pi/6$.

2. $\theta = -3\pi/4$.

3. $r = 5$.

4. $r = 4\sin\theta$.

5. $r = -6\cos\theta$.

6. $r = -3\sin\theta$.

7. $2r = \cos\theta$.

8. $r = 1 + \sin\theta$.

9. $r = 3(1 - \sin\theta)$.

10. $r - 2 = 2\cos\theta$.

11. $r = 4 - 4\cos\theta$.

12. $r = -5 + 5\sin\theta$.

13. $r = -1 - \cos\theta$.

14. $r = 1 + 2\sin\theta$.

15. $r = 1 - 2\cos\theta$.

16. $r = 4 + 3\cos\theta$.

17. $r = 3 + 2\sin\theta$.

18. $r = 3 - \cos\theta$.

19. $r = 2 + \sin\theta$.

20. $r - 5 = 3\sin\theta$.

21. $r = 3 + 4\cos\theta$.

22. $r = -3 - 4\sin\theta$.

23. $r = 5 - 2\cos\theta$.

24. $r^2 = 9\cos 2\theta$.

25. $r^2 = -9\cos 2\theta$.

26. $r^2 = \sin 2\theta$.

27. $r^2 = -16\sin 2\theta$.

28. $r = 4\theta$ $(\theta \geq 0)$.

29. $r = 4\theta$ $(\theta \leq 0)$.

30. $r = 4\theta$.

31. $r = \cos 2\theta$.

32. $r = 3\sin 2\theta$.

33. $r = \sin 3\theta$.

34. $r = 2\cos 3\theta$.

35. $r = 9\sin 4\theta$.

36. $r = \cos 5\theta$.

In Exercises 37–40, sketch the curve in polar coordinates.

37. $r = 4\cos\theta + 4\sin\theta$.
   [*Hint:* Convert to rectangular coordinates.]

38. $r = e^\theta$ (logarithmic spiral).

39. $r = \sin(\theta/2)$.

40. $r^2 = \theta$ (parabolic spiral).

In Exercises 41–46, find all vertical and horizontal asymptotes, if any, by expressing the rectangular coordinates x and y as functions of θ, and then sketch the curve.

41. $r = 4 \tan \theta$ (kappa curve). [*Hint:* Show that $x \to 4$ or $x \to -4$ and $y \to +\infty$ or $y \to -\infty$ as θ approaches an odd multiple of $\pi/2$.]

42. $r = 2 + 2 \sec \theta$ (conchoid of Nicomedes). [*Hint:* Show that $x \to 2$ and $y \to +\infty$ or $y \to -\infty$ as θ approaches an odd multiple of $\pi/2$.]

43. $r = 2 \sin \theta \tan \theta$ (cissoid). [*Hint:* See the hint in Exercise 42.]

44. $r = 1/\theta$ $(\theta > 0)$ (hyperbolic spiral). [*Hint:* Find $\lim\limits_{\theta \to 0^+} y$ to show that there is a horizontal asymptote.]

45. $r = 1/\sqrt{\theta}$ $(\theta > 0)$. [*Hint:* Find $\lim\limits_{\theta \to 0^+} y$ to show that there is a horizontal asymptote.]

46. $r = 1/\theta^2$ $(\theta > 0)$. [*Hint:* Find $\lim\limits_{\theta \to 0^+} y$ to show that there is no horizontal asymptote.]

47. Find the maximum value of the y-coordinate of points on the cardioid $r = 1 + \cos \theta$ for θ in the interval $[0, \pi]$.

48. (a) Show that the maximum value of the y-coordinate of points on the curve $r = 1/\sqrt{\theta}$ for θ in the interval $(0, \pi]$ occurs when $\tan \theta = 2\theta$.

(b) Use Newton's Method to solve the equation in part (a) for θ to at least four decimal-place accuracy.

(c) Use the result of part (b) to approximate the maximum value of y for $0 < \theta \le \pi$.

In Exercises 49 and 50, take the *width* of a petal of a rose curve to be the dimension shown in Figure 13.2.20.

Petal width

Figure 13.2.20

49. Show that the width of a petal of the four-petal rose $r = \cos 2\theta$ is $2\sqrt{6}/9$. [*Hint:* Express y in terms of θ and find the maximum value of y for $0 \le \theta \le \pi/4$.]

50. The width of a petal of the rose $r = \cos n\theta$ is twice the maximum value of y for $0 \le \theta \le \pi/(2n)$. Show that the value of θ for which y attains this maximum value satisfies the equation $n \tan n\theta \tan \theta = 1$.

51. Find the minimum value of the x-coordinate of points on the cardioid $r = 1 + \cos \theta$ for θ in the interval $[0, \pi]$.

52. Show that the minimum value of the x-coordinate of points on the curve $r = a + b \cos \theta$, $a > 0$, $b > 0$, for θ in the interval $[0, \pi]$ is $-a^2/(4b)$ if $a < 2b$ and is $b - a$ if $a \ge 2b$.

In Exercises 53–58, find the eccentricity e and sketch a graph of the conic.

53. $r = \dfrac{3}{2 - 2 \cos \theta}$.

54. $r = \dfrac{5}{3 + 3 \sin \theta}$.

55. $r = \dfrac{3}{2 + \sin \theta}$.

56. $r = \dfrac{4}{3 - 2 \cos \theta}$.

57. $r = \dfrac{4}{2 + 3 \cos \theta}$.

58. $r = \dfrac{3}{3 - 4 \sin \theta}$.

59. What is the geometric relationship between the graphs of $r = f(\theta)$ and $r = f(\theta + \alpha)$ in polar coordinates if α is a constant?

60. Prove that a rose with an even number of petals is traced out exactly once as θ varies over the interval $0 \le \theta < 2\pi$ and a rose with an odd number of petals is traced out exactly once as θ varies over the interval $0 \le \theta < \pi$.

13.3 AREA IN POLAR COORDINATES

In this section we shall show how to find areas of regions that are bounded by curves whose equations are given in polar coordinates.

☐ **THE AREA PROBLEM IN POLAR COORDINATES**

13.3.1 PROBLEM. *Find the area of the region R that is enclosed by the curve $r = f(\theta)$ and the lines $\theta = \alpha$ and $\theta = \beta$ in polar coordinates, where $0 \le \beta - \alpha \le 2\pi$ and $f(\theta)$ is continuous and nonnegative for $\alpha \le \theta \le \beta$ (Figure 13.3.1).*

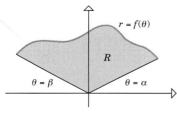

Figure 13.3.1

In rectangular coordinates we attacked the area problem stated at the beginning of Section 5.1 by dividing the region into strips. Here we shall divide the region into "wedges" by constructing radial lines

$$\theta = \theta_1, \theta = \theta_2, \ldots, \theta = \theta_{n-1}$$

such that

$$\alpha < \theta_1 < \theta_2 < \cdots < \theta_{n-1} < \beta$$

As shown in Figure 13.3.2, these lines divide the region R into n wedge-like subregions with areas $A_1, A_2, \ldots, A_n$ and central angles $\Delta\theta_1, \Delta\theta_2, \ldots, \Delta\theta_n$. The area of the entire region can be written as

$$A = A_1 + A_2 + \cdots + A_n = \sum_{k=1}^{n} A_k \tag{1}$$

Let us now concentrate on the areas $A_1, A_2, \ldots, A_n$ of the subregions. If $\Delta\theta_k$ is not too large, we can approximate the area A_k by the area of a *sector* having central angle $\Delta\theta_k$ and radius $f(\theta_k^*)$, where θ_k^* is an arbitrary angle terminating in the kth subregion (Figure 13.3.3). Thus, from (1) and the area formula for a sector [Formula (5), Section 2.8],

$$A = \sum_{k=1}^{n} A_k \approx \sum_{k=1}^{n} \frac{1}{2}[f(\theta_k^*)]^2 \, \Delta\theta_k \tag{2}$$

If we now increase n in such a way that max $\Delta\theta_k \to 0$, then the sectors will become better and better approximations to the subregions and (2) will approach the exact value of the area A; that is,

$$A = \lim_{\max \Delta\theta_k \to 0} \sum_{k=1}^{n} \frac{1}{2}[f(\theta_k^*)]^2 \, \Delta\theta_k = \int_{\alpha}^{\beta} \frac{1}{2}[f(\theta)]^2 \, d\theta$$

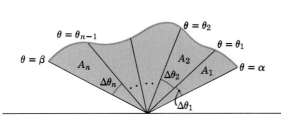

Figure 13.3.2

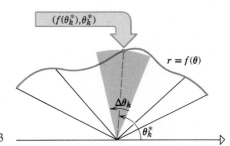

Figure 13.3.3

Thus, we have the following solution of Problem 13.3.1.

13.3.2 AREA IN POLAR COORDINATES.   If $f(\theta)$ is continuous and nonnegative for $\alpha \le \theta \le \beta$, and if $0 \le \beta - \alpha \le 2\pi$, then the area A enclosed by the polar curve $r = f(\theta)$ and the lines $\theta = \alpha$ and $\theta = \beta$ is

$$A = \int_{\alpha}^{\beta} \frac{1}{2}[f(\theta)]^2 \, d\theta = \int_{\alpha}^{\beta} \frac{1}{2} r^2 \, d\theta \tag{3}$$

The hardest part of applying (3) is determining the limits of integration. This can be done as follows:

Step 1. Sketch the region R whose area is to be determined.

Step 2. Draw an arbitrary "radial line" from the origin to the boundary curve $r = f(\theta)$.

Step 3. Ask, "Over what interval of values must θ vary in order for the radial line to sweep out the region R?"

Step 4. Your answer in Step 3 will determine the lower and upper limits of integration.

Example 1 Find the area of the region in the first quadrant within the cardioid $r = 1 - \cos\theta$.

Solution. The region and a typical radial line are shown in Figure 13.3.4. For the radial line to sweep out the region, θ must vary from 0 to $\pi/2$. Thus, from (3) with $\alpha = 0$ and $\beta = \pi/2$,

$$A = \int_0^{\pi/2} \frac{1}{2} r^2 \, d\theta = \frac{1}{2} \int_0^{\pi/2} (1 - \cos\theta)^2 \, d\theta$$

$$= \frac{1}{2} \int_0^{\pi/2} (1 - 2\cos\theta + \cos^2\theta) \, d\theta$$

With the help of the identity $\cos^2\theta = \frac{1}{2}(1 + \cos 2\theta)$, this can be rewritten as

$$A = \frac{1}{2} \int_0^{\pi/2} \left(\frac{3}{2} - 2\cos\theta + \frac{1}{2}\cos 2\theta \right) d\theta$$

$$= \frac{1}{2} \left[\frac{3}{2}\theta - 2\sin\theta + \frac{1}{4}\sin 2\theta \right]_0^{\pi/2} = \frac{3}{8}\pi - 1 \quad \blacktriangleleft$$

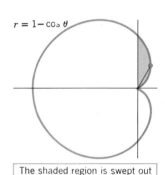

$r = 1 - \cos\theta$

The shaded region is swept out by the radial line as θ varies from 0 to $\frac{\pi}{2}$.

Figure 13.3.4

Example 2 Find the entire area within the cardioid of Example 1.

Solution. For the radial line to sweep out the entire cardioid, θ must vary from 0 to 2π. Thus, from (3) with $\alpha = 0$ and $\beta = 2\pi$,

$$A = \int_0^{2\pi} \frac{1}{2} r^2 \, d\theta = \frac{1}{2} \int_0^{2\pi} (1 - \cos\theta)^2 \, d\theta$$

If we proceed as in Example 1, this reduces to

$$A = \frac{1}{2} \int_0^{2\pi} \left(\frac{3}{2} - 2\cos\theta + \frac{1}{2}\cos 2\theta \right) d\theta = \frac{3\pi}{2}$$

Alternative Solution. Since the cardioid is symmetric about the x-axis, we can calculate the portion of area above the x-axis and double the result. In the portion of the cardioid above the x-axis, θ ranges from 0 to π, so that

$$A = 2 \int_0^{\pi} \frac{1}{2} r^2 \, d\theta = \int_0^{\pi} (1 - \cos\theta)^2 \, d\theta = \frac{3\pi}{2} \quad \blacktriangleleft$$

□ **USING SYMMETRY TO AVOID NEGATIVE VALUES OF r**

It is important to keep in mind that Formula (3) requires $r = f(\theta)$ to be nonnegative. For example, suppose that we were interested in finding the area of the region enclosed by the rose $r = \sin 2\theta$ shown in Figure 13.1.9. Although this region is swept out by a radial line as θ varies from 0 to 2π, the value of

$$r = \sin 2\theta$$

is negative over portions of this interval (if $\pi/2 < \theta < \pi$ or $3\pi/2 < \theta < 2\pi$). Thus, Formula (3) is not applicable over the entire interval $0 \le \theta \le 2\pi$. The following example shows how to avoid this difficulty by calculating a portion of the area and using symmetry.

Example 3 Find the area of the region enclosed by the rose curve $r = \sin 2\theta$.

Solution. The petal in the first quadrant is swept out as θ varies over the interval $0 \le \theta \le \pi/2$ (Figure 13.1.9a and b), and for these values of θ, the value of $r = \sin 2\theta$ is nonnegative, so Formula (3) can be used to find the area of this petal. By symmetry, the area enclosed by one petal is one-fourth of the entire area A, so

$$A = 4 \int_0^{\pi/2} \frac{1}{2} r^2 \, d\theta = 2 \int_0^{\pi/2} \sin^2 2\theta \, d\theta$$

$$= 2 \int_0^{\pi/2} \frac{1}{2} (1 - \cos 4\theta) \, d\theta = \int_0^{\pi/2} (1 - \cos 4\theta) \, d\theta$$

$$= \left[\theta - \frac{1}{4} \sin 4\theta \right]_0^{\pi/2} = \pi/2 \quad \blacktriangleleft$$

☐ **USING NEGATIVE ANGLES AS LIMITS OF INTEGRATION**

It was assumed in our derivation of Formula (3) that $\alpha < \beta$. Sometimes, this is best achieved by using negative angles for one or both limits of integration. The following example illustrates this.

Example 4 Find the area of the region that is inside the cardioid $r = 4 + 4 \cos \theta$ and outside the circle $r = 6$.

Solution. To sketch the region, we need to know where the circle and cardioid intersect. To find these points, we equate the given expressions for r. This yields

$$4 + 4 \cos \theta = 6$$

from which we obtain $\cos \theta = \frac{1}{2}$ or

$$\theta = \frac{\pi}{3} \quad \text{and} \quad \theta = -\frac{\pi}{3}$$

The region is sketched in Figure 13.3.5. The desired area can be obtained by subtracting the shaded areas in parts (b) and (c) of Figure 13.3.5. Thus,

$$A = \int_{-\pi/3}^{\pi/3} \frac{1}{2} (4 + 4 \cos \theta)^2 \, d\theta - \int_{-\pi/3}^{\pi/3} \frac{1}{2} (6)^2 \, d\theta$$

$$= \int_{-\pi/3}^{\pi/3} \frac{1}{2} [(4 + 4 \cos \theta)^2 - 36] \, d\theta = \int_{-\pi/3}^{\pi/3} (16 \cos \theta + 8 \cos^2 \theta - 10) \, d\theta$$

$$= \left[16 \sin \theta + (4\theta + 2 \sin 2\theta) - 10\theta \right]_{-\pi/3}^{\pi/3} = 18\sqrt{3} - 4\pi$$

Alternative Solution. Using symmetry, we can calculate the portion of the area above the x-axis and double it to obtain the entire area A. This yields (verify)

$$A = 2 \int_0^{\pi/3} \frac{1}{2} [(4 + 4 \cos \theta)^2 - 36] \, d\theta = 2(9\sqrt{3} - 2\pi) = 18\sqrt{3} - 4\pi$$

which agrees with the preceding result. $\quad \blacktriangleleft$

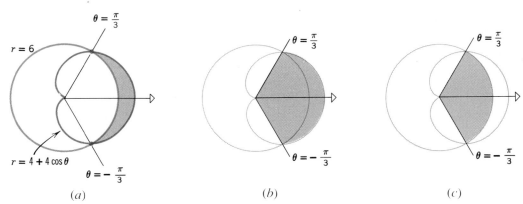

Figure 13.3.5   (*a*)   (*b*)   (*c*)

WARNING. In the last example we determined the intersections of the curves $r = 4 + 4 \cos \theta$ and $r = 6$ by equating the right-hand sides and solving for θ. However, it may not be possible to find *all* the intersections of two polar curves, $r = f_1(\theta)$ and $r = f_2(\theta)$, by solving the equation $f_1(\theta) = f_2(\theta)$ for θ. This is because every point has infinitely many pairs of polar coordinates. It is possible, therefore, that a point of intersection may have no single pair of polar coordinates that satisfies both equations. For example, the cardioids

$$r = 1 - \cos \theta \quad \text{and} \quad r = 1 + \cos \theta \tag{4}$$

intersect at three points, the origin, the point $(1, \pi/2)$, and the point $(1, 3\pi/2)$ (Figure 13.3.6). Equating the right-hand sides of the equations in (4) yields $1 - \cos \theta = 1 + \cos \theta$ or $\cos \theta = 0$, so

$$\theta = \frac{\pi}{2} + k\pi, \quad k = 0, \pm 1, \pm 2, \ldots$$

Substituting any of these values in (4) yields $r = 1$, so that we have found only two distinct points of intersection, $(1, \pi/2)$ and $(1, 3\pi/2)$; the origin has been missed.

When calculating intersections in polar coordinates, it is a good idea to make a sketch of the curves to determine how many intersections there should be.

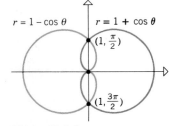

Figure 13.3.6

▶ Exercise Set 13.3

In Exercises 1–21, find the area of the region described.

1. The region in the first quadrant enclosed by the first loop of the spiral $r = \theta$ $(\theta \geq 0)$ and the lines $\theta = \pi/6$ and $\theta = \pi/3$.

2. The region in the first quadrant within the cardioid $r = 1 + \sin \theta$.

3. The region that is enclosed by the cardioid $r = 2 + 2 \cos \theta$.

4. The region outside the cardioid $r = 2 - 2 \cos \theta$ and inside the circle $r = 4$.

5. The region inside the circle $r = 5 \sin \theta$ and outside the limaçon $r = 2 + \sin \theta$.

6. The region enclosed by the inner loop of the limaçon $r = 1 + 2 \cos \theta$.

7. The region inside the cardioid $r = 2 + 2 \cos \theta$ and outside the circle $r = 3$.

8. The region inside the circle $r = 2a \sin \theta$.

9. The region enclosed by the curve $r^2 = \sin 2\theta$.

10. The region that is common to the circles $r = 4 \cos \theta$ and $r = 4 \sin \theta$.

11. The region enclosed by the rose $r = 4 \cos 3\theta$.

12. The region inside the rose $r = 2a \cos 2\theta$ and outside the circle $r = a\sqrt{2}$.

13. The region common to the circle $r = 3 \cos \theta$ and the cardioid $r = 1 + \cos \theta$.

14. The region between the loops of the limaçon $r = \frac{1}{2} + \cos \theta$.

15. The region inside the circle $r = 10$ and to the right of the line $r \cos \theta = 6$.

16. The region inside the cardioid $r = a(1 + \sin \theta)$ and outside the circle $r = a \sin \theta$.

17. The region enclosed by $r = a \sec^2 \frac{1}{2}\theta$ and the rays whose polar angles are $\theta = 0$ and $\theta = \frac{1}{2}\pi$.

18. The region inside the cardioid $r = 2 + 2\cos\theta$ and to the right of the line $r\cos\theta = 3/2$.

19. The region swept out by a radial line from the origin to the curve $r = 3e^{-2\theta}$ as θ varies over the interval $0 \leq \theta \leq \pi$.

20. The region swept out by a radial line from the origin to the curve $r = 2/\theta$ as θ varies over the interval $1 \leq \theta \leq 3$.

21. The region swept out by a radial line from the origin to the curve $r = 1/\sqrt{\theta}$ as θ varies over the interval $1/9 \leq \theta \leq 4$.

22. (a) Find the error: The area inside the lemniscate $r^2 = a^2\cos 2\theta$ is

$$A = \int_0^{2\pi} \frac{1}{2} r^2\, d\theta = \int_0^{2\pi} \frac{1}{2} a^2\cos 2\theta\, d\theta$$

$$= \frac{1}{4} a^2 \sin 2\theta \Big]_0^{2\pi} = 0$$

(b) Find the correct area.

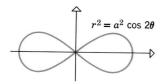

$r^2 = a^2 \cos 2\theta$

23. Find the area inside the lemniscate $r^2 = 4\cos 2\theta$ and outside the circle $r = \sqrt{2}$.

24. A radial line is drawn from the origin to the spiral $r = a\theta$ ($a > 0$ and $\theta \geq 0$). Find the area swept out during the

second revolution of the radial line that was not swept out during the first revolution.

25. The graph of $x^3 - xy + y^3 = 0$, called a **folium of Descartes**, is shown in the figure below.

(a) Show that in polar coordinates its equation is

$$r = \frac{\sin\theta\cos\theta}{\cos^3\theta + \sin^3\theta}$$

(b) Find the area of the region that is enclosed by the loop.
   [*Hint:* Show that $r = \sec\theta\tan\theta/(1 + \tan^3\theta)$.]

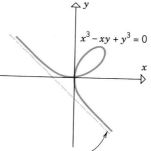

$x^3 - xy + y^3 = 0$

Asymptote: $x + y + 1/3 = 0$

26. Show that if $f(\theta)$ is continuous and nonnegative for $\alpha \leq \theta \leq \beta$ and $0 \leq \beta - \alpha \leq 2\pi$, then the area enclosed by $r = 2f(\theta)$ and the lines $\theta = \alpha$ and $\theta = \beta$ is four times the area enclosed by $r = f(\theta)$ and the lines $\theta = \alpha$ and $\theta = \beta$.

13.4 PARAMETRIC EQUATIONS

Because the graph in the xy-plane of a function f can be cut at most once by any vertical line, there are many important curves in the xy-plane that are not graphs of equations of the form y = f(x) (circles, for example). In this section we shall show that such curves can be obtained from pairs of functions, one used to specify the x-coordinate of a point on the graph and the other used to specify the y-coordinate.

□ **PARAMETRIC EQUATIONS**

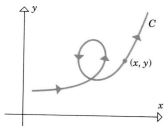

Figure 13.4.1

Imagine a particle moving along a curve C in the xy-plane (Figure 13.4.1). The x- and y-coordinates of the particle are functions of time, say

$$x = f(t), \quad y = g(t)$$

Physicists and engineers call these the **equations of motion** of the particle and the curve C the **trajectory** of the particle.

Example 1   Sketch the trajectory over the time interval $0 \leq t \leq 4$ of the particle that moves in the xy-plane with the equations of motion

$$x = \tfrac{1}{2}t^3 - 6t, \quad y = \tfrac{1}{2}t^2$$

Solution. The trajectory, shown in Figure 13.4.2, was found by calculating the *x*- and *y*-coordinates of the particle for $t = 0, 1, 2, 3, 4$, plotting the points (x, y), and connecting successive points with a smooth curve. The arrows on the curve indicate the direction of motion. ◄

t	$x = \frac{1}{2}t^3 - 6t$	$y = \frac{1}{2}t^2$	(x, y)
0	0	0	$(0, 0)$
1	$-\frac{11}{2}$	$\frac{1}{2}$	$(-\frac{11}{2}, \frac{1}{2})$
2	-8	2	$(-8, 2)$
3	$-\frac{9}{2}$	$\frac{9}{2}$	$(-\frac{9}{2}, \frac{9}{2})$
4	8	8	$(8, 8)$

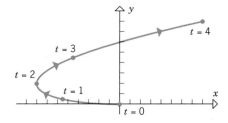

Figure 13.4.2

The physicists' concept of equations of motion has been adapted by mathematicians to describe curves in the *xy*-plane by pairs of equations

$$x = x(t), \quad y = y(t)$$

that express the coordinates (x, y) of a point on a curve as functions of an auxiliary variable *t*. The variable *t* is called a ***parameter***, and the equations are called ***parametric equations*** for the curve. The parameter *t* need not represent time; it should be viewed simply as an independent variable that varies over some interval of real numbers. If no restrictions on *t* are stated explicitly or implied by the equations, then it is understood that *t* varies between $-\infty$ and $+\infty$. To indicate that *t* is restricted to an interval $[a, b]$, we will write

$$x = x(t), \quad y = y(t) \qquad (a \le t \le b)$$

In these parametric equations we have followed a common practice of writing the equations as $x = x(t)$ and $y = y(t)$ rather than $x = f(t)$ and $y = g(t)$. This rarely causes any confusion and often simplifies problems by reducing the number of different letters involved. When this notation is used the derivatives of *x* and *y* with respect to *t* may be denoted by $x'(t)$ and $y'(t)$. It should also be noted that it is not essential to use *t* for the parameter; any letter not reserved for another purpose can be used.

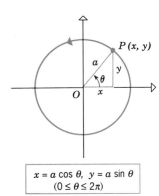

$x = a \cos \theta, \ y = a \sin \theta$
$(0 \le \theta \le 2\pi)$

Figure 13.4.3

Example 2 A circle of radius *a* centered at the origin can be represented by the parametric equations

$$x = a \cos \theta, \quad y = a \sin \theta \qquad (0 \le \theta \le 2\pi) \tag{1}$$

in which the parameter θ can be interpreted as the angle measured counterclockwise from the positive *x*-axis to the radius *OP* from the origin to the point on the circle with coordinates $P(x, y)$ (Figure 13.4.3). As θ varies from 0 to 2π, the point *P* makes one complete counterclockwise revolution. By changing the interval over which the parameter θ varies, one can obtain different portions of the circle. For example,

$$x = a \cos \theta, \quad y = a \sin \theta \qquad (0 \le \theta \le \pi) \tag{2}$$

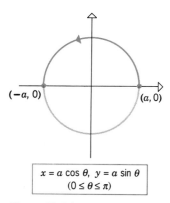

$x = a \cos \theta, \ y = a \sin \theta$
$(0 \le \theta \le \pi)$

Figure 13.4.4

represents just the upper half of the circle (Figure 13.4.4). ◄

☐ **ORIENTATION OF A PARAMETRIC CURVE**

In the preceding example, the circle is traced in a counterclockwise direction as the parameter θ increases, so we say that the parametric equations (1) impose a *counterclockwise orientation* on the circle. In general, if a curve is represented by parametric equations, and if suitable conditions are imposed on the equations to ensure that the curve is traced continuously and in a specific direction as the parameter increases, then we call this the ***direction of increasing parameter*** or sometimes the ***orientation*** of the curve imposed by the equations. As in Figures 13.3.3 and 13.3.4, the orientation can be shown by arrows on the curve.

It should be noted that the orientation of a curve is determined by the parametric equations and that different parametric equations can produce different orientations for the same curve. For example, the parametric equations

$$x = a \cos \theta, \quad y = -a \sin \theta \qquad (0 \le \theta \le 2\pi) \tag{3}$$

represent the circle of radius a, centered at the origin, with *clockwise* orientation, as can be seen from Figure 13.4.5 by rewriting these equations as

$$x = a \cos (-\theta), \quad y = a \sin (-\theta) \qquad (0 \le \theta \le 2\pi)$$

It is important to keep in mind that different parametric equations can trace the same curve; for example, (1) and (3) trace the same circle, but they do so with opposite motions. Thus, we make the distinction between a **curve**, which is a set of points, and a ***parametric curve***, which is a set of points traced in a particular way by parametric equations.

REMARK. Not all parametric equations produce curves with orientations. In absence of restricting conditions, the curve might be traced discontinuously, with the point $(x(t), y(t))$ leaping around sporadically or perhaps with the point moving back and forth along the curve and failing to determine a definite direction. In the next chapter we will discuss restrictions that eliminate such erratic behavior and produce curves with orientations.

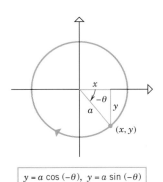

$y = a \cos (-\theta), \ y = a \sin (-\theta)$
$(0 \le \theta \le 2\pi)$

Figure 13.4.5

Example 3 Let m denote the slope of the tangent line to the parabola $y = x^2$ at an arbitrary point (x, y). Find parametric equations for the parabola in terms of the parameter m, and indicate the orientation imposed on the parabola by the parametric equations.

Solution. Our objective is to express both x and y in terms of m. But

$$m = \frac{dy}{dx} = 2x$$

so

$$x = \frac{1}{2} \frac{dy}{dx} = \frac{1}{2} m \quad \text{and} \quad y = x^2 = \left(\frac{1}{2} m \right)^2 = \frac{1}{4} m^2$$

which leads to the parametric equations

$$x = \frac{1}{2} m, \quad y = \frac{1}{4} m^2$$

The direction of increasing parameter, indicated in Figure 13.4.6, is obtained by noting that x increases as m increases, since $x = \frac{1}{2}m$. ◄

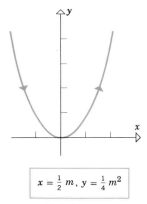

$x = \frac{1}{2} m, \ y = \frac{1}{4} m^2$

Figure 13.4.6

☐ **ELIMINATING THE PARAMETER**

Sometimes a curve given parametrically can be recognized by eliminating the parameter. The following example illustrates this.

Example 4 Sketch the curve

$$x = 2t - 3, \quad y = 6t - 7$$

and indicate its orientation.

Solution. Solving the first equation for t, we obtain

$$t = \frac{x + 3}{2}$$

Substituting this into the second equation yields

$$y = 6 \left(\frac{x + 3}{2} \right) - 7 \quad \text{or} \quad y = 3x + 2$$

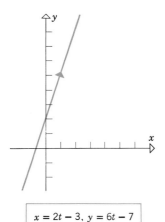

$x = 2t - 3, \ y = 6t - 7$

Figure 13.4.7

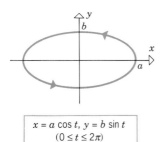

$x = a \cos t, \ y = b \sin t$
$(0 \le t \le 2\pi)$

Figure 13.4.8

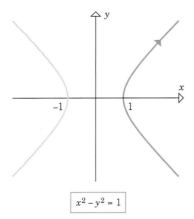

$x^2 - y^2 = 1$

Figure 13.4.9

which is the line shown in Figure 13.4.7. The process of eliminating the parameter loses the orientation information, so we must look to the original parametric equations for that information. The direction of increasing t, shown in Figure 13.4.7, can be obtained from the original equations by noting that x increases as t increases (or that y increases as t increases).   ◀

Example 5   Sketch the curve

$$x = a \cos t, \quad y = b \sin t \qquad (0 \le t \le 2\pi)$$

where $0 < b < a$, and indicate its orientation.

Solution.   We can eliminate t by writing

$$\frac{x}{a} = \cos t \quad \text{and} \quad \frac{y}{b} = \sin t$$

from which it follows that

$$\frac{x^2}{a^2} + \frac{y^2}{b^2} = 1$$

This is the equation of the ellipse shown in Figure 13.4.8. The counterclockwise orientation can be deduced from the parametric equations by noting that $x = a \cos t$ decreases from a to $-a$ for $0 \le t \le \pi$, and then increases from $-a$ to a for $\pi \le t \le 2\pi$, while $y = b \sin t$ increases for $0 \le t \le \pi/2$, decreases for $\pi/2 \le t \le 3\pi/2$, and increases for $3\pi/2 \le t \le 2\pi$.   ◀

Example 6   Sketch the curve

$$x = \cosh t, \quad y = \sinh t$$

Solution.   We can eliminate the parameter by using the identity

$$\cosh^2 t - \sinh^2 t = 1$$

[Formula (1) of Section 7.6] to obtain

$$x^2 - y^2 = 1$$

This is the equation of the hyperbola shown in Figure 13.4.9. However, the original parametric equations represent only the right branch of this hyperbola, since $x = \cosh t \ge 1$ for all t. The orientation shown in the figure can be deduced by noting that $y = \sinh t$ increases as t varies from $-\infty$ to $+\infty$.   ◀

REMARK.   The preceding example shows that the process of eliminating the parameter from parametric equations can alter the extent of the graph, so that it is important to be alert to this possibility.

☐ **REPRESENTING GRAPHS OF FUNCTIONS PARAMETRICALLY**

Although parametric equations are of special importance for curves that are not graphs of functions, curves of the form $y = f(x)$ or $x = g(y)$ can also be represented parametrically by introducing a parameter t that is equal to the independent variable.

Example 7   The curves $y = 4x^2 - 1$ and $x = 5y^3 - y$ can be represented parametrically as

$$x = t, \quad y = 4t^2 - 1 \qquad \text{and} \qquad x = 5t^3 - t, \quad y = t$$

respectively.   ◀

REMARK. A given curve can always be represented parametrically in infinitely many different ways. For example, a parametrization of $y = 4x^2 - 1$ different from that in Example 7 can be obtained by making the substitution $x = s + 1$, which yields the following new parametric equations with s as a parameter:

$$x = s + 1, \quad y = 4(s + 1)^2 - 1 = 4s^2 + 8s + 3$$

Different substitutions would yield different parametric equations for the curve.

Example 8 If $r = f(\theta)$ is a polar curve, then we can obtain parametric equations for this curve with θ as the parameter by substituting $r = f(\theta)$ in the formulas

$$x = r \cos \theta, \quad y = r \sin \theta$$

which relate rectangular and polar coordinates. For example, the circle $r = 4 \cos \theta$ can be represented by the parametric equations

$$x = 4 \cos^2 \theta, \quad y = 4 \cos \theta \sin \theta = 2 \sin 2\theta$$

If we impose the added restriction, $0 \le \theta \le \pi$, then the point (x, y) will traverse the circle exactly once as θ varies over $[0, \pi]$ (Figure 13.4.10). ◄

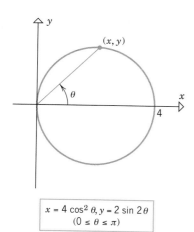

$x = 4 \cos^2 \theta, y = 2 \sin 2\theta$
$(0 \le \theta \le \pi)$

Figure 13.4.10

□ **TANGENT LINES TO CURVES DEFINED PARAMETRICALLY**

It can be proved that if $x(t)$ and $y(t)$ have continuous first derivatives with respect to t, and if $dx/dt \ne 0$, then y is a differentiable function of x. In this case it follows from the chain rule that

$$\frac{dy}{dx} = \frac{dy/dt}{dx/dt} \qquad (4)$$

This formula can be used to find dy/dx directly from the parametric equations of a curve without eliminating the parameter.

REMARK. Formula (4) does not apply if $dx/dt = 0$; however, we will agree that a curve has a **vertical tangent line** at a point (x, y) where $dx/dt = 0$ and $dy/dt \ne 0$. Points (x, y) where $dx/dt = 0$ and $dy/dt = 0$ are called **singular points** for the parametric equations. It is hard to make general statements about the behavior of curves at singular points; case-by-case analyses are required.

Example 9 The curve represented by the parametric equations

$$x = t^2, \quad y = t^3$$

is called a **semicubical parabola**. The parameter t can be eliminated by cubing x and squaring y, from which it follows that $y^2 = x^3$. The graph of this equation, shown in Figure 13.4.11, consists of two branches: an upper branch obtained by graphing $y = x^{3/2}$ and a lower branch obtained by graphing $y = -x^{3/2}$. The two branches meet at the origin; the origin is a singular point for the parametric equations, since it corresponds to $t = 0$ and

$$dx/dt = 2t \quad \text{and} \quad dy/dt = 3t^2$$

are both zero for $t = 0$. However, there is a horizontal tangent line at the origin since

$$\frac{dy}{dx} = \frac{dy/dt}{dx/dt} = \frac{3t^2}{2t} = \frac{3}{2}t \quad \text{if} \quad t \ne 0$$

and hence $\lim_{t \to 0^+} dy/dx = \lim_{t \to 0^-} dy/dx = 0$. ◄

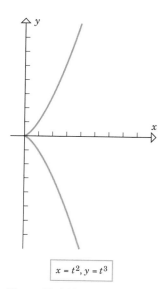

$x = t^2, y = t^3$

Figure 13.4.11

Example 10 Without eliminating the parameter, find dy/dx and d^2y/dx^2 at the point $(1, 1)$ on the semicubical parabola given by the parametric equations in Example 9.

Solution.   From Example 9 we have

$$\frac{dy}{dx} = \frac{3}{2} t \tag{5}$$

and from (4) applied to $y' = dy/dx$ we have

$$\frac{d^2y}{dx^2} = \frac{dy'}{dx} = \frac{dy'/dt}{dx/dt} = \frac{3/2}{2t} = \frac{3}{4t} \tag{6}$$

Since the point $(1, 1)$ on the curve corresponds to $t = 1$ in the parametric equations, it follows from (5) and (6) that

$$\left.\frac{dy}{dx}\right|_{t=1} = \frac{3}{2}(1) = \frac{3}{2} \quad \text{and} \quad \left.\frac{d^2y}{dx^2}\right|_{t=1} = \frac{3}{4}$$

The reader may wish to check the preceding calculations by finding dy/dx and d^2y/dx^2 at $x = 1$ from the formula $y = x^{3/2}$ for the upper branch of the semicubical parabola. The tangent line at $(1, 1)$ with its slope of $m = 3/2$ is shown in Figure 13.4.12. Note that the semicubical parabola is concave up at $(1, 1)$ in keeping with the fact that d^2y/dx^2 is positive at that point.   ◄

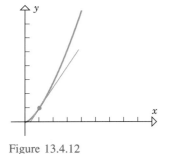

Figure 13.4.12

□ ARC LENGTH OF
CURVES DEFINED
PARAMETRICALLY

The following result provides a method for finding the arc length of a curve from parametric equations for the curve. Its derivation is similar to that of 6.4.2 and will be omitted.

13.4.1   ARC-LENGTH FORMULA FOR PARAMETRIC CURVES.   If no segment of the curve represented by the parametric equations

$$x = x(t), \quad y = y(t) \quad (a \le t \le b)$$

is traced more than once as t increases from a to b, and if dx/dt and dy/dt are continuous functions for $a \le t \le b$, then the arc length L of the curve is given by

$$L = \int_a^b \sqrt{\left(\frac{dx}{dt}\right)^2 + \left(\frac{dy}{dt}\right)^2} \, dt \tag{7}$$

In the exercises we ask the reader to show that Formulas (3) and (4) of Section 6.4 can be derived as special cases of (7).

Example 11   Use (7) to find the circumference of a circle of radius a from the parametric equations

$$x = a \cos t, \quad y = a \sin t \quad (0 \le t \le 2\pi)$$

Solution.

$$L = \int_0^{2\pi} \sqrt{\left(\frac{dx}{dt}\right)^2 + \left(\frac{dy}{dt}\right)^2} \, dt = \int_0^{2\pi} \sqrt{(-a \sin t)^2 + (a \cos t)^2} \, dt$$

$$= \int_0^{2\pi} a \, dt = at \Big]_0^{2\pi} = 2\pi a \quad ◄$$

■ THE CYCLOID

If a wheel rolls along a straight line without slipping, then a point on the rim of the wheel traces a curve called a *cycloid* (Figure 13.4.13). The cycloid is of special interest because it provides the solution of two famous mathematical problems: the ***brachistochrone problem*** (from Greek words meaning "shortest time") and the ***tautochrone problem*** (from Greek words meaning "equal time"). In June of 1696, Johann Bernoulli* posed the brachistochrone problem in the form of a challenge to other mathematicians. The problem was to determine the shape of a wire along which a bead might slide from a point P to another point Q, not directly below, in the *shortest time*.

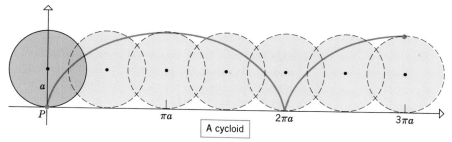

Figure 13.4.13

A cycloid

---

* BERNOULLI. An amazing Swiss family that included several generations of outstanding mathematicians and scientists. Nikolaus Bernoulli (1623–1708), a druggist, fled from Antwerp to escape religious persecution and ultimately settled in Basel, Switzerland. There he had three sons, Jakob I (also called Jacques or James), Nikolaus, and Johann I (also called Jean or John). The Roman numerals are used to distinguish family members with identical names.

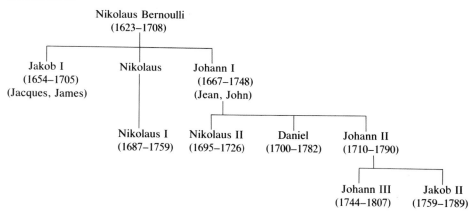

Following Newton and Leibniz, the Bernoulli brothers, Jakob I and Johann I, are considered by some to be the two most important founders of calculus. Jakob I was self-taught in mathematics. His father wanted him to study for the ministry, but he turned to mathematics and in 1686 became a professor at the University of Basel. When he started working in mathematics, he knew nothing of Newton's and Leibniz' work. He eventually became familiar with Newton's results, but because so little of Leibniz' work was published, Jakob duplicated many of Leibniz' results.

  Jakob's younger brother Johann I was urged to enter into business by his father. Instead, he turned to medicine and studied mathematics under the guidance of his older brother. He eventually became a mathematics professor at Gröningen in Holland, and then, when Jakob died in 1705, Johann succeeded him as mathematics professor at Basel. Throughout their lives, Jakob I and Johann I had a mutual passion for criticizing each other's work, which frequently erupted into ugly confrontations. Leibniz tried to mediate the disputes, but Jakob, who resented Leibniz' superior intellect, accused him of siding with Johann, and thus Leibniz became entangled in the arguments. The brothers often worked on common problems that they posed as challenges to one another. Johann, interested in gaining fame, often used unscrupulous means to make himself appear the originator of his brother's results; Jakob occasionally retaliated. Thus, it is often difficult to determine who deserves credit for many results. However, both men made major contributions to the development of calculus. In addition to his work on calculus, Jakob helped establish fundamental principles in probability, including the Law of Large Numbers, which is a cornerstone of modern probability theory. Johann was Euler's teacher at Basel and L'Hôpital's tutor in France.

(continued on next page)

Figure 13.4.14

At first, one might conjecture that the wire should form a straight line, since that shape yields the shortest distance from P to Q. However, the correct answer is half of one arch of an inverted cycloid (Figure 13.4.14). In essence, this shape allows the bead to fall rapidly at first, building up sufficient initial speed to reach Q in the shortest time, even though the path does not provide the shortest distance from P to Q. This problem was solved by Newton, Leibniz, L'Hôpital, Johann Bernoulli, and his older brother Jakob Bernoulli. It was formulated and incorrectly solved years earlier by Galileo, who gave the arc of a circle as the answer.

In the tautochrone problem the object is to find the shape of a wire from P to Q such that two beads started at *any* points on the wire between P and Q reach Q in the same amount of time. Again, the answer is half of one arch of a cycloid.

We conclude this section by deriving parametric equations for a cycloid.

Example 12   Let P be a fixed point on the rim of a wheel of radius a. Suppose that the wheel rolls along the positive x-axis of a rectangular coordinate system and that P is initially at the origin. When the wheel rolls a given distance, the radial line from the center C to the point P rotates through an angle ϕ (Figure 13.4.15), which we shall use as our parameter.

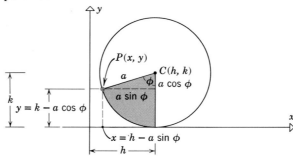

Figure 13.4.15

(It is customary to regard ϕ as a positive angle, even though it is generated by a clockwise rotation.) Our objective is to express the coordinates of $P(x, y)$ in terms of the angle ϕ. Figure 13.4.15 suggests that the coordinates of $P(x, y)$ and the coordinates of the wheel's center $C(h, k)$ are related by

$$x = h - a \sin \phi, \quad y = k - a \cos \phi \tag{8}$$

Since the height of the wheel's center is the radius of the wheel, we have $k = a$; and since the distance h moved by the center is the same as the circular arc length subtended by ϕ (why?), we have $h = a\phi$. Therefore, (8) yields

$$x = a\phi - a \sin \phi, \quad y = a - a \cos \phi \quad (0 \leq \phi < +\infty)$$

These are the parametric equations of the cycloid.   ◀

———————————

(continued)

Among the other members of the Bernoulli family, Daniel, son of Johann I, is the most famous. He was a professor of mathematics at St. Petersburg Academy in Russia and subsequently a professor of anatomy and then physics at Basel. He did work in calculus and probability, but is best known for his work in physics. A basic law of fluid flow, called Bernoulli's principle, is named in his honor. He won the annual prize of the French Academy 10 times for work on vibrating strings, tides of the sea, and kinetic theory of gases.

Johann II succeeded his father as professor of mathematics at Basel. His research was on the theory of heat and sound. Nikolaus I was a mathematician and law scholar who worked on probability and series. On the recommendation of Leibniz, he was appointed professor of mathematics at Padua and then went to Basel as a professor of logic and then law. Nikolaus II was professor of jurisprudence in Switzerland and then professor of mathematics at St. Petersburg Academy. Johann III was a professor of mathematics and astronomy in Berlin and Jakob II succeeded his uncle Daniel as professor of mathematics at St. Petersburg Academy in Russia. Truly an incredible family!

▶ Exercise Set 13.4 $\boxed{C}$ 49

1. (a) Sketch the curve $x = t$, $y = t^2$ by plotting the points corresponding to $t = -3, -2, -1, 0, 1, 2, 3$.

   (b) Show that the curve is a parabola by eliminating the parameter t.

2. Sketch the curve $x = t^3 - 4t$, $y = 2t$, $-3 \le t \le 3$ by plotting some appropriate points.

> In Exercises 3–15, sketch the curve by eliminating the parameter t, and indicate the direction of increasing t.

3. $x = \cos t$, $y = \sin t$, $0 \le t \le 2\pi$.

4. $x = 1 + \cos t$, $y = 3 - \sin t$, $0 \le t \le 2\pi$.

5. $x = 3t - 4$, $y = 6t + 2$.

6. $x = t - 3$, $y = 3t - 7$, $0 \le t \le 3$.

7. $x = 2 \cos t$, $y = 5 \sin t$, $0 \le t \le 2\pi$.

8. $x = \sqrt{t}$, $y = 2t + 4$.

9. $x = 3 + 2 \cos t$, $y = 2 + 4 \sin t$, $0 \le t \le 2\pi$.

10. $x = 2 \cosh t$, $y = 4 \sinh t$.

11. $x = 4 \sin 2\pi t$, $y = 4 \cos 2\pi t$, $0 \le t \le 1$.

12. $x = \sec t$, $y = \tan t$, $\pi \le t < 3\pi/2$.

13. $x = \cos 2t$, $y = \sin t$, $-\pi/2 \le t \le \pi/2$.

14. $x = 4t + 3$, $y = 16t^2 - 9$.

15. $x = t^2$, $y = 2 \ln t$, $t \ge 1$.

> In Exercises 16–25, sketch the curve.

16. $x = 3e^{-t} - 2$, $y = 4e^{-t} - 1$, $t \ge 0$.

17. $x = 3e^{-2t} - 1$, $y = e^{-t}$, $t \ge 0$.

18. $x = \sec^2 t$, $y = \tan^2 t$.

19. $x = 2 \sin^2 t$, $y = 3 \cos^2 t$.

20. $x = \sec t$, $y = \tan t$, $-\pi/2 < t < \pi/2$.

21. $x = \cos t$, $y = \sin^2 t$.

22. $x = \cos 2t$, $y = 2 \cos t$.

23. $x = \sin^2 t$, $y = 1 + \cos t$.

24. $x = \cos 3t$, $y = 2 \sin^2 3t - 1$.

25. $x = \cos(e^{-t})$, $y = \sin(e^{-t})$, $t \ge 0$.

26. Find parametric equations of the upper semicircle

$$y = \sqrt{a^2 - x^2} \ (a > 0)$$

in terms of the parameter t, where t is the distance between $(-a, 0)$ and any point on the semicircle.

27. Find parametric equations of the curve $y = \sqrt{x}$ in terms of the parameter t, using for t the x-coordinate of the point where the tangent line to the curve crosses the x-axis.

> In Exercises 28–34, find dy/dx at the point corresponding to the given value of the parameter without eliminating the parameter.

28. $x = t^2 + 4$, $y = 8t$; $t = 2$.

29. $x = \cos t$, $y = \sin t$; $t = 3\pi/4$.

30. $x = t + 5$, $y = 5t - 7$; $t = 1$.

31. $x = \sqrt{t}$, $y = 2t + 4$; $t = 9$.

32. $x = \sec \theta$, $y = \tan \theta$; $\theta = \pi/3$.

33. $x = 4 \cos 2\pi s$, $y = 3 \sin 2\pi s$; $s = -1/4$.

34. $x = \sinh t$, $y = \cosh t$; $t = 0$.

> In Exercises 35–38, find d^2y/dx^2 at the point corresponding to the given value of the parameter without eliminating the parameter.

35. $x = \frac{1}{2}t^2$, $y = \frac{1}{3}t^3$; $t = 2$.

36. $x = \cos \phi$, $y = \sin \phi$; $\phi = \pi/4$.

37. $x = \sqrt{t}$, $y = 2t + 4$; $t = 1$.

38. $x = \sec t$, $y = \tan t$; $t = \pi/3$.

> In Exercises 39–47, find the arc length of the curve.

39. $x = 4t + 3$, $y = 3t - 2$, $0 \le t \le 2$.

40. $x = \cos^3 t$, $y = \sin^3 t$, $0 \le t \le \pi/2$.

41. $x = \frac{1}{3}t^3$, $y = \frac{1}{2}t^2$, $0 \le t \le 1$.

42. $x = \frac{1}{3}t^3$, $y = \frac{1}{2}t^2$, $-1 \le t \le 0$.

43. $x = \cos 2t$, $y = \sin 2t$, $0 \le t \le \pi/2$.

44. $x = e^t(\sin t + \cos t)$, $y = e^t(\cos t - \sin t)$, $1 \le t \le 4$.

45. $x = (1 + t)^2$, $y = (1 + t)^3$, $0 \le t \le 1$.

46. $x = e^t \cos t$, $y = e^t \sin t$, $0 \le t \le \pi/2$.

47. One arch of the cycloid
$$x = a(t - \sin t), \quad y = a(1 - \cos t) \qquad (a > 0).$$

48. Show that the total arc length of the ellipse $x = a \cos t$, $y = b \sin t$, $0 \le t \le 2\pi$ for $a > b > 0$ is given by

$$4a \int_0^{\pi/2} \sqrt{1 - e^2 \cos^2 t} \, dt$$

where $e = \sqrt{a^2 - b^2}/a$.

49. (a) Show that the total arc length of the ellipse
$$x = 2 \cos t, y = \sin t, 0 \le t \le 2\pi$$
is given by
$$4 \int_0^{\pi/2} \sqrt{1 + 3 \sin^2 t} \, dt$$

   (b) Use Simpson's rule with $n = 10$ subdivisions to approximate the total arc length of the ellipse in part (a). Round your answer to two decimal places.

   (c) Suppose that the parametric equations in part (a) describe the path of a particle moving in the xy-plane, where t is time in seconds and x and y are in centimeters. Use Simpson's rule with $n = 10$ subdivisions to approximate the distance traveled from $t = 1.5$ sec to $t = 4.8$ sec. Round your answer to two decimal places.

50. Find parametric equations for the rose $r = 2 \cos 2\theta$ using θ as the parameter.

51. Find parametric equations for the limaçon $r = 2 + 3 \sin \theta$ using θ as the parameter.

52. Find the equation of the tangent line to the curve $x = 2t + 4$, $y = 8t^2 - 2t + 4$ at the point where $t = 1$.

53. Find the equation of the tangent line to the curve $x = e^t$, $y = e^{-t}$ at the point where $t = 2$.

54. Find all values of t at which the curve $x = 2 \cos t$, $y = 4 \sin t$ has a tangent that is

    (a) horizontal   (b) vertical.

55. Find all values of the parameter t at which the curve $x = 2t^3 - 15t^2 + 24t + 7$, $y = t^2 + t + 1$ has a tangent line that is

    (a) horizontal   (b) vertical.

56. Show that the curve $x = t^3 - 4t$, $y = t^2$ intersects itself at the point $(0, 4)$, and find equations for two tangent lines to the curve at the point of intersection.

57. Show that the curve with parametric equations

$$x = t^2 - 3t + 5, \quad y = t^3 + t^2 - 10t + 9$$

intersects itself at the point $(3, 1)$, and find equations for two tangent lines to the curve at the point of intersection.

58. A point traces the circle $x^2 + y^2 = 25$ so that $dx/dt = 8$ when the point reaches $(4, 3)$. Find dy/dt there.

59. Describe the curve whose parametric equations are $x = a \cos t + h$, $y = b \sin t + k$, $0 \le t \le 2\pi$.

60. A *hypocycloid* is a curve traced by a point P on the circumference of a circle that rolls inside a larger fixed circle. Suppose that the fixed circle has radius a, the rolling circle has radius b, and the fixed circle is centered at the origin. Let ϕ be the angle shown in the following figure, and assume that the point P is at $(a, 0)$ when $\phi = 0$. Show that the hypocycloid generated is given by the parametric equations

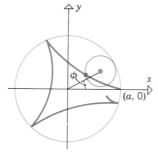

$$x = (a - b) \cos \phi + b \cos\left(\frac{a - b}{b}\phi\right)$$

$$y = (a - b) \sin \phi - b \sin\left(\frac{a - b}{b}\phi\right)$$

61. If $b = \frac{1}{4}a$ in Exercise 60, then the resulting curve is called a four-cusped hypocycloid.

    (a) Sketch this curve.

    (b) Show that the curve is given by the parametric equations $x = a \cos^3 \phi$, $y = a \sin^3 \phi$.

    (c) Show that the curve is given by the equation $x^{2/3} + y^{2/3} = a^{2/3}$ in rectangular coordinates.

Exercises 62–67 require the formulas developed in the following discussion: If $x'(t)$ and $y'(t)$ are continuous functions and if no segment of the curve

$$x = x(t), \quad y = y(t) \quad (a \le t \le b)$$

is traced more than once, then it can be shown that the area of the surface generated by revolving this curve about the x-axis is

$$S = \int_a^b 2\pi y(t)\sqrt{[x'(t)]^2 + [y'(t)]^2}\, dt$$

and the area of the surface generated by revolving the curve about the y-axis is

$$S = \int_a^b 2\pi x(t)\sqrt{[x'(t)]^2 + [y'(t)]^2}\, dt$$

[The derivations are similar to those used to obtain Formulas (6) and (7) in Section 6.5.]

---

62. Find the area of the surface generated by revolving $x = t^2$, $y = 2t$, $0 \le t \le 4$ about the x-axis.

63. Find the area of the surface generated by revolving $x = e^t \cos t$, $y = e^t \sin t$, $0 \le t \le \pi/2$ about the x-axis.

64. Find the area of the surface generated by revolving $x = \cos^2 t$, $y = \sin^2 t$, $0 \le t \le \pi/2$ about the y-axis.

65. Find the area of the surface generated by revolving $x = t$, $y = 2t^2$, $0 \le t \le 1$ about the y-axis.

66. By revolving the semicircle $x = r \cos t$, $y = r \sin t$, $0 \le t \le \pi$ about the x-axis, show that the surface area of a sphere of radius r is $4\pi r^2$.

67. The equations

$$x = a\phi - a \sin \phi, \quad y = a - a \cos \phi \quad (0 \le \phi \le 2\pi)$$

represent one arch of a cycloid. Show that the surface area generated by revolving this curve about the x-axis is $S = 64\pi a^2/3$.

68. Show that Formulas (3) and (4) of Section 6.4 can be derived from Formula (7) by choosing appropriate parametrizations.

■ 13.5 TANGENT LINES AND ARC LENGTH IN POLAR COORDINATES

> *In this section we shall use our results on parametric equations to derive formulas for arc length and slopes of tangent lines to polar curves.*

□ PARAMETRIC EQUATIONS FOR POLAR CURVES

Recall from Example 8 of the preceding section that if $r = f(\theta)$ is a curve in polar coordinates, then we can obtain parametric equations for this curve with θ as the parameter by substituting $r = f(\theta)$ into the equations

$$x = r \cos \theta, \quad y = r \sin \theta$$

which relate polar and rectangular coordinates. This yields the parametric equations

$$x = f(\theta) \cos \theta, \quad y = f(\theta) \sin \theta \tag{1}$$

Example 1 The cardioid $r = 1 - \cos \theta$ can be represented by the parametric equations

$$x = (1 - \cos \theta) \cos \theta, \quad y = (1 - \cos \theta) \sin \theta \quad (0 \le \theta \le 2\pi)$$

where the restriction on θ ensures that the cardioid is traced exactly once as θ varies from 0 to 2π. ◄

□ TANGENT LINES TO POLAR CURVES

If $r = f(\theta)$, and if f is a differentiable function of θ, then it follows from (1) that x and y are differentiable functions of θ and

$$\frac{dx}{d\theta} = -f(\theta) \sin \theta + f'(\theta) \cos \theta = -r \sin \theta + \frac{dr}{d\theta} \cos \theta$$

$$\frac{dy}{d\theta} = f(\theta) \cos \theta + f'(\theta) \sin \theta = r \cos \theta + \frac{dr}{d\theta} \sin \theta \tag{2}$$

Thus, if $dx/d\theta$ and $dy/d\theta$ are continuous and if $dx/d\theta \ne 0$, then y is a differentiable function of x, and from Formula (4) of Section 13.4 with θ in place of t, the derivative of y with respect to x is

$$\frac{dy}{dx} = \frac{dy/d\theta}{dx/d\theta} = \frac{r \cos \theta + \sin \theta \dfrac{dr}{d\theta}}{-r \sin \theta + \cos \theta \dfrac{dr}{d\theta}} \tag{3}$$

Formula (3) provides the slope of the tangent line to the curve $r = f(\theta)$ at the point with polar coordinates (r, θ). This formula was derived assuming that $dx/d\theta \ne 0$. If $dx/d\theta = 0$ and $dy/d\theta \ne 0$, then the curve has a vertical tangent line at (r, θ), and if both $dx/d\theta = 0$ and $dy/d\theta = 0$, then there is a singular point at (r, θ). The behavior of polar curves at singular points must be determined on a case-by-case basis.

Example 2 Find the slope of the tangent line to the circle $r = 4 \cos \theta$ at the point where $\theta = \pi/4$.

Solution. From (3) with $r = 4 \cos \theta$ we obtain (verify)

$$\frac{dy}{dx} = \frac{-4 \sin^2 \theta + 4 \cos^2 \theta}{-8 \sin \theta \cos \theta} = \frac{4 \cos 2\theta}{-4 \sin 2\theta} = -\cot 2\theta$$

Thus, at the point where $\theta = \pi/4$ the slope of the tangent line is

$$m = \frac{dy}{dx}\bigg|_{\theta=\pi/4} = -\cot(\pi/2) = 0$$

which implies that the circle has a horizontal tangent line at the point where $\theta = \pi/4$ (Figure 13.5.1).   ◀

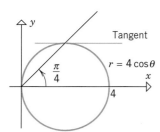

Figure 13.5.1

Example 3   Find the points on the cardioid $r = 1 - \cos\theta$ at which there is a vertical tangent line or a singular point.

Solution.   As shown in Example 1, the cardioid can be represented parametrically by the equations

$$x = (1 - \cos\theta)\cos\theta, \quad y = (1 - \cos\theta)\sin\theta \quad (0 \le \theta \le 2\pi)$$

Differentiating these equations with respect to θ and then simplifying yields (verify)

$$\frac{dx}{d\theta} = \sin\theta(2\cos\theta - 1) \quad \text{and} \quad \frac{dy}{d\theta} = (1 - \cos\theta)(1 + 2\cos\theta)$$

It follows from these equations that $dx/d\theta = 0$ if $\sin\theta = 0$ or $\cos\theta = 1/2$, that is, if

$$\theta = 0, \quad \pi/3, \quad \pi, \quad 5\pi/3, \quad \text{or} \quad 2\pi$$

Vertical tangent lines occur at those points where $dx/d\theta = 0$ and $dy/d\theta \ne 0$, that is, where $\theta = \pi/3$, π, and $5\pi/3$ (Figure 13.5.2). Singular points occur where $dx/d\theta = 0$ and $dy/d\theta = 0$, that is, where $\theta = 0$ or $\theta = 2\pi$. Since $r = 0$ for these values of θ, the origin is the only singular point for the cardioid. We leave it for the reader to prove that there is a horizontal tangent line at the origin by showing that

$$\lim_{\theta \to 0^+} dy/dx = \lim_{\theta \to 2\pi^-} dy/dx = 0 \quad ◀$$

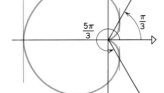

Figure 13.5.2

REMARK.   If a polar curve $r = f(\theta)$ passes through the origin when $\theta = \theta_0$, and if $dr/d\theta \ne 0$ when $\theta = \theta_0$, then it follows from (3) on substituting $r = 0$ and $\theta = \theta_0$ that the slope of the tangent line at $\theta = \theta_0$ is

$$m = dy/dx = \tan\theta_0$$

(verify). But we know from Theorem 1.4.3 that $m = dy/dx = \tan\phi$, where ϕ is the angle of inclination of the tangent line. Thus, $\tan\phi = \tan\theta_0$, which implies that the line $\theta = \theta_0$ is tangent to the curve at the origin. In summary, *if a polar curve $r = f(\theta)$ passes through the origin when $\theta = \theta_0$, and if $dr/d\theta \ne 0$ at θ_0, then $\theta = \theta_0$ is tangent to the curve at the origin.*

☐ ARC LENGTH OF A
POLAR CURVE

The arc-length formula for a parametric curve can be used to obtain a formula for the arc length of a polar curve.

13.5.1   ARC-LENGTH FORMULA FOR POLAR CURVES.   If no segment of the polar curve $r = f(\theta)$ is traced more than once as θ increases from α to β, and if $dr/d\theta$ is continuous for $\alpha \le \theta \le \beta$, then the arc length L from $\theta = \alpha$ to $\theta = \beta$ is

$$L = \int_\alpha^\beta \sqrt{r^2 + \left(\frac{dr}{d\theta}\right)^2}\, d\theta \tag{4}$$

Proof.   It follows from (2) that (verify)

$$\left(\frac{dx}{d\theta}\right)^2 + \left(\frac{dy}{d\theta}\right)^2 = r^2 + \left(\frac{dr}{d\theta}\right)^2$$

Thus, (4) follows from Formula (7) of Section 13.4 with θ in place of t and α and β in place of a and b.   ■

Example 4 Find the arc length of the spiral $r = e^{\theta}$ between $\theta = 0$ and $\theta = 1$.

Solution.

$$L = \int_{\alpha}^{\beta} \sqrt{r^2 + \left(\frac{dr}{d\theta}\right)^2} \, d\theta = \int_{0}^{1} \sqrt{(e^{\theta})^2 + (e^{\theta})^2} \, d\theta$$

$$= \int_{0}^{1} \sqrt{2} \, e^{\theta} \, d\theta = \sqrt{2} e^{\theta} \Big]_{0}^{1} = \sqrt{2}(e - 1) \quad \blacktriangleleft$$

Example 5 Find the total arc length of the cardioid $r = 1 + \cos\theta$.

Solution. The cardioid is traced out once as θ varies from $\theta = 0$ to $\theta = 2\pi$. Thus,

$$L = \int_{\alpha}^{\beta} \sqrt{r^2 + \left(\frac{dr}{d\theta}\right)^2} \, d\theta = \int_{0}^{2\pi} \sqrt{(1 + \cos\theta)^2 + (-\sin\theta)^2} \, d\theta$$

$$= \sqrt{2} \int_{0}^{2\pi} \sqrt{1 + \cos\theta} \, d\theta$$

$$= 2 \int_{0}^{2\pi} \sqrt{\cos^2 \tfrac{1}{2}\theta} \, d\theta \quad \boxed{\text{Identity (27a) of Appendix B}}$$

$$= 2 \int_{0}^{2\pi} \left| \cos \tfrac{1}{2}\theta \right| \, d\theta$$

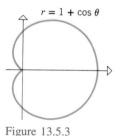

$r = 1 + \cos\theta$

Figure 13.5.3

Since $\cos \tfrac{1}{2}\theta$ changes sign at π, we must split the last integral into the sum of two integrals: the integral from 0 to π plus the integral from π to 2π. However, the integral from π to 2π is equal to the integral from 0 to π, since the cardioid is symmetric about the x-axis (Figure 13.5.3). Thus,

$$L = 2 \int_{0}^{2\pi} \left| \cos \tfrac{1}{2}\theta \right| \, d\theta = 4 \int_{0}^{\pi} \cos \tfrac{1}{2}\theta \, d\theta$$

$$= 8 \sin \tfrac{1}{2}\theta \Big]_{0}^{\pi} = 8 \quad \blacktriangleleft$$

▶ **Exercise Set 13.5**   ⓒ 15, 17

In Exercises 1–6, find the slope of the tangent to the curve at the point with the given value of θ.

1. $r = 2\cos\theta$; $\theta = \pi/3$.   **2.** $r = 1 + \sin\theta$; $\theta = \pi/4$.

3. $r = 1/\theta$; $\theta = 2$.   **4.** $r = a\sec 2\theta$; $\theta = \pi/6$.

5. $r = \cos 3\theta$; $\theta = 3\pi/4$.   **6.** $r = 4 - 3\sin\theta$; $\theta = \pi$.

In Exercises 7–13, find the arc length of the curve.

7. $r = e^{3\theta}$ from $\theta = 0$ to $\theta = 2$.

8. The entire circle $r = a$.

9. The entire circle $r = 2a\cos\theta$.

10. $r = \sin^2(\theta/2)$ from $\theta = 0$ to $\theta = \pi$.

11. $r = a\theta^2$ from $\theta = 0$ to $\theta = \pi$.

12. $r = \sin^3(\theta/3)$ from $\theta = 0$ to $\theta = \pi/2$.

13. The entire cardioid $r = a(1 - \cos\theta)$.

14. (a) Find the arc length L of the curve $r = e^{-a\theta}$, $a > 0$, for $0 \le \theta \le \theta_0$.

   (b) Find $\displaystyle\lim_{\theta_0 \to +\infty} L$.

15. (a) Show that the arc length of one petal of the rose $r = \cos n\theta$ is given by

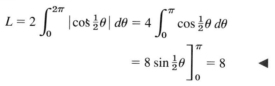

$$2 \int_{0}^{\pi/(2n)} \sqrt{1 + (n^2 - 1)\sin^2 n\theta} \, d\theta$$

   (b) Use the result in part (a) and Simpson's rule with ten subdivisions to approximate the arc length of one petal of the four-petal rose $r = \cos 2\theta$. Round your answer to two decimal places.

16. Use 13.5.1 to show that from $\theta = \alpha$ to $\theta = \beta$ the arc length of $r = 2f(\theta)$ is twice that of $r = f(\theta)$.

17. Suppose that a long thin rod with one end fixed at the pole of a polar coordinate system rotates counter-clockwise at the constant rate of 0.5 rad/sec. At time $t = 0$ a bug on the rod is 10 mm from the pole and is moving outward along the rod at the constant speed of 2 mm/sec.

(a) Find an equation of the form $r = f(\theta)$ for the path of motion of the bug, assuming that $\theta = 0$ when $t = 0$.

(b) Find the distance the bug travels along the path in part (a) during the first 5 sec. Round your answer to the nearest tenth of a millimeter.

18. Find all points on the cardioid $r = a\,(1 + \cos\theta)$ where the tangent is

(a) horizontal   (b) vertical.

19. Find all points on the limaçon $r = 1 - 2\sin\theta$ where the tangent is horizontal.

◆ TECHNOLOGY EXERCISES   Chapter 13

Most of these exercises require access to a graphing calculator or a computer algebra system (CAS) such as *Mathematica*, *Maple*, or *Derive*. When you are asked to *find* an answer or to *solve* an equation, you may choose to find an exact result or a numerical approximation, depending on the particular technology you are using and on your own imagination. The form of your answers may differ from those of other students or from those in the answer section of the text, depending on how you solve the problems and the accuracy you use in your numerical approximations. Those exercises that are more appropriate for a CAS than a graphing calculator are labeled with the icon ◆.

1. Butterfly curve: The graph of the equation

$$r = \exp\,(\cos\theta) - 2\cos 4\theta + \sin^3\,(\theta/4)$$

in polar coordinates is the "butterfly curve" shown in the accompanying figure and on the cover of this text. Generate this curve by letting θ vary over the interval $[0, \alpha)$, where α is chosen so that the curve is traced exactly once.

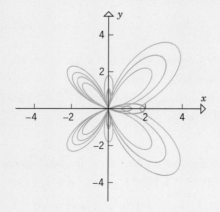

2. Area in polar coordinates

(a) Graph $r = \sin\theta \cos 2\theta$ in polar coordinates. For what value of α will the graph be traced exactly once as θ varies over the interval $[0, \alpha)$?

(b) Find the area enclosed by the large loop of the graph in part (a).

◆ **3. Arc length in polar coordinates:** Find the arc length of one petal of the rose $r = 2\sin 3\theta$.

◆ **4. The orbit of Mars:** If the sun is at the origin of a polar coordinate system, then an equation of the orbit of Mars is

$$r = \frac{2.26 \times 10^8}{1 + 0.0934 \cos\theta}$$

where distance is in kilometers.

(a) Graph the orbit of Mars.

(b) Find the area swept out by the line from the sun to Mars in one revolution of Mars about the sun.

(c) Kepler's second law of planetary motion states that

equal areas are swept out in equal times by the line from the sun to a planet. Use the result in part (b) to find the rate in km^2/yr at which area is swept out by the line from the sun to Mars, given that Mars makes one orbit in 1.88 years.

(d) Suppose that Mars moves from the point where $\theta = 0$ to the point where $\theta = \theta_0$ in $\frac{1}{2}$ year (see the accompanying figure). Find the distance from the sun to Mars when $\theta = \theta_0$. [*Hint:* First use part (c) to find the area swept out by the radial line from the sun to Mars during the time interval, then set up an appropriate integral to find θ_0.]

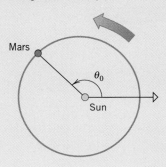

5. **The average speed of Mars:** Use the equation of the orbit of Mars in Exercise 4 to find the average speed, in kilometers per hour, of Mars in one orbit about the sun, given that Mars makes one orbit in 1.88 years (take 365 days in one year).

6. **The highest point on a polar curve:** Find the maximum value of the y-coordinate of a point on the curve $r = 2 + 3 \cos 2\theta$.

7. **Graphing equations:** One method for graphing an equation of the form $x = f(y)$ $(a \le y \le b)$ in an xy-coordinate system is to solve the equation for y as a function (or functions) of x and graph the resulting functions. However, this method is not applicable if the equation cannot be solved explicitly for y in terms of x. In this case the graph can be obtained by expressing $x = f(y)$ $(a \le y \le b)$ parametrically as $x = f(t)$, $y = t$ $(a \le t \le b)$ and graphing these parametric equations.

(a) Use the parametric method to graph
$$x = y^2 - 1.8y - 1.2 \quad (-1 \le y \le 3)$$
in an xy-coordinate system.

(b) From Technology Exercise 14 in Chapter 1, if a hollow metal sphere of diameter 5 feet and weight w pounds floats submerged to a depth h in seawater, then
$$w = 63.9\pi h^2 (2.5 - h/3), \quad 0 \le h \le 5$$
Use the parametric method to graph this equation in a wh-coordinate system.

8. **Distance between a point and a curve:** Let $P(2, \pi/6)$ be a point in polar coordinates. Find the maximum and minimum distance between P and a point on the limaçon $r = 1 + 3 \cos \theta$. [*Suggestion:* Convert the coordinates of P to rectangular coordinates and represent the limaçon in parametric form with θ as the parameter.]

9. **Motion in a plane:** Suppose that two particles moving in the xy-plane have equations of motion $x = t - 1$, $y = e^{-t}$ $(t \ge 0)$ and $x = \cos t$, $y = \sin t$ $(t \ge 0)$. Given that x and y are in meters and t is in seconds, how close do the particles get to one another?

10. **Distance traveled:** A particle starts at the point $(3, 0)$ and moves along the ellipse $x = 3 \cos t$, $y = 2 \sin t$ in the direction of increasing t. Given that x and y are in centimeters, find the coordinates of the particle when it has traveled a distance of 10 cm along the curve.

11. **Rotation of an object:** A child whirls an object attached to a string, letting out string as the object moves. Suppose that the object moves in a vertical plane with a horizontal x-axis and that the path of the object is given by the parametric equations $x = (1 + t/2) \cos 6.2t$, $y = (1 + t/2) \sin 6.2t$ $(t \ge 0)$, where x and y are in feet and t is in seconds.

(a) Graph the path of the object for $0 \le t \le 4$.

(b) Find the distance that the object travels during the time interval $0 \le t \le 4$.

(c) If the string is released, then the object will move tangentially to the path of motion. Assuming that the string is released when $t = 4$, find, to the nearest degree, the acute angle between the tangent line to the path and the horizontal at this instant.

12. **Motion of a projectile:** Suppose that a projectile of mass m is fired from ground level at the origin of an xy-coordinate system with an initial speed of v meters per second at an angle θ with the horizontal (see the accompanying

figure). If the projectile encounters air resistance that is proportional to the velocity, and if k ($k > 0$) is the constant of proportionality, then the coordinates of the projectile t seconds later are given by

$$x = \left(\frac{mv}{k} \cos \theta\right)(1 - e^{-kt/m})$$

$$y = -\frac{mg}{k}t + \left(\frac{m^2g}{k^2} + \frac{mv}{k}\sin \theta\right)(1 - e^{-kt/m})$$

where g is the acceleration due to gravity. Assume that $v = 100$ m/sec, $\theta = 60°$, $g = 9.8$ m/sec^2, $m = 0.1$ kg, and $k = 0.001$ kg/sec.

(a) Find the maximum height of the projectile.

(b) Where does the projectile strike the ground?

(c) Find, to the nearest degree, the acute angle formed with the horizontal at the point where the projectile strikes the ground.

(d) Find the distance the projectile travels before it strikes the ground.

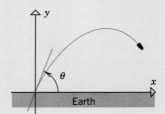

Jakob Bernoulli (1654–1705)

14 THREE-DIMENSIONAL SPACE; VECTORS

■ **14.1** RECTANGULAR COORDINATES IN 3-SPACE; SPHERES; CYLINDRICAL SURFACES

In this section we shall discuss coordinate systems in three-dimensional space and some basic facts about surfaces in three dimensions.

□ **RECTANGULAR COORDINATE SYSTEMS**

In the remainder of this text we shall call three-dimensional space *3-space*, two-dimensional space (a plane) *2-space*, and one-dimensional space (a line) *1-space*. Just as points in 2-space can be placed in one-to-one correspondence with pairs of real numbers using two perpendicular coordinate lines, so points in 3-space can be placed in one-to-one correspondence with triples of real numbers by using three mutually perpendicular coordinate lines, called the *x-axis*, the *y-axis*, and the *z-axis*, positioned so that their origins coincide (Figure 14.1.1). The three coordinate axes form a three-dimensional *rectangular coordinate system* (or *Cartesian coordinate system*). The point of intersection of the coordinate axes is called the *origin* of the coordinate system.

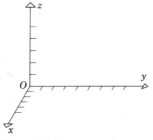

Figure 14.1.1

Rectangular coordinate systems in 3-space fall into two categories: ***left-handed*** and ***right-handed***. A right-handed system has the property that when the fingers of the right hand are cupped so that they curve from the positive x-axis toward the positive y-axis, the thumb points (roughly) in the direction of the positive z-axis (Figure 14.1.2a). A system that is not right-handed is called left-handed (Figure 14.1.2b). We shall use only right-handed coordinate systems.

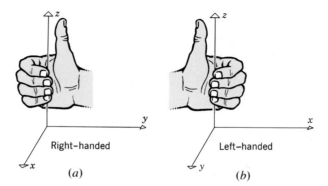

Figure 14.1.2   (*a*)   (*b*)

The coordinate axes, taken in pairs, determine three ***coordinate planes***: the ***xy-plane***, the ***xz-plane***, and the ***yz-plane***. To each point P in 3-space we can assign a triple of real numbers by passing three planes through P parallel to the coordinate planes and letting a, b, and c be the coordinates of the intersections of those planes with the x-axis, y-axis, and z-axis, respectively (Figure 14.1.3). We call a, b, and c the ***x-coordinate***, ***y-coordinate***, and ***z-coordinate*** of P, respectively, and we denote the point P by (a, b, c) or by $P(a, b, c)$. Figure 14.1.4 shows the points $(4, 5, 6)$ and $(-3, 2, -4)$.

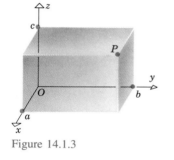

Figure 14.1.3

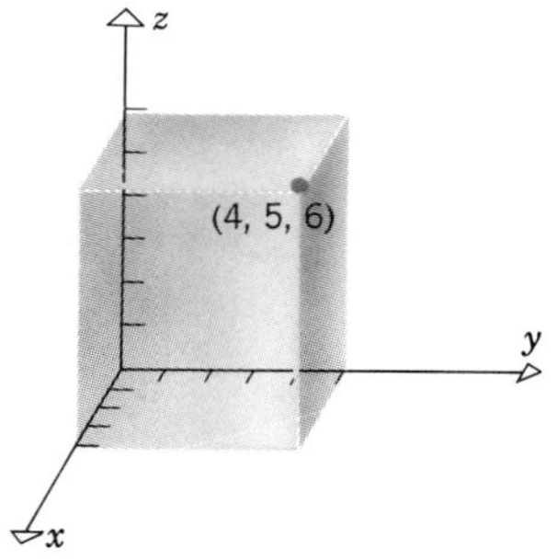

Figure 14.1.4   (*a*)   (*b*)

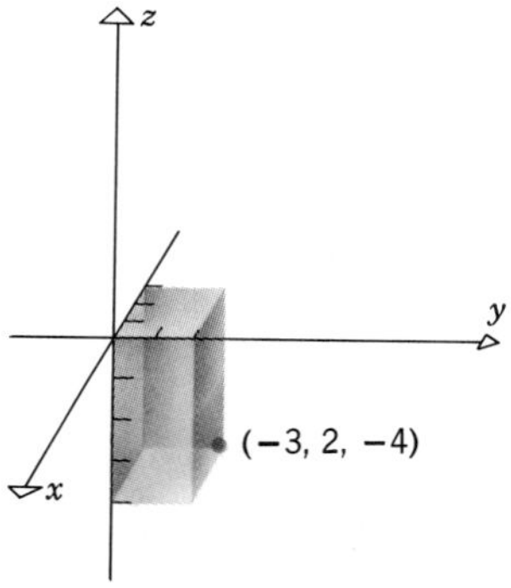

Just as the coordinate axes in a two-dimensional coordinate system divide 2-space into four quadrants, so the coordinate planes of a three-dimensional coordinate system divide 3-space into eight parts, called ***octants*** (count them). Those points having three positive coordinates form the ***first octant***; the remaining octants have no standard numbering.

The reader should be able to visualize the following results about three-dimensional rectangular coordinate systems:

REGION	DESCRIPTION
xy-plane	Consists of all points of the form $(x, y, 0)$
xz-plane	Consists of all points of the form $(x, 0, z)$
yz-plane	Consists of all points of the form $(0, y, z)$
x-axis	Consists of all points of the form $(x, 0, 0)$
y-axis	Consists of all points of the form $(0, y, 0)$
z-axis	Consists of all points of the form $(0, 0, z)$

☐ **DISTANCE, MIDPOINT, AND SPHERES**

A formula for the distance between two points in 3-space can be obtained by starting with a box with dimensions a, b, and c and considering the right triangle in Figure 14.1.5 that has one side along the diagonal of the base, one side along a vertical edge, and its hypotenuse along a diagonal of the box. By the Theorem of Pythagoras, the side in the base of the box has length $\sqrt{a^2 + b^2}$, and by a second application of the Theorem of Pythagoras the hypotenuse has length

$$d = \sqrt{(\sqrt{a^2 + b^2})^2 + c^2} = \sqrt{a^2 + b^2 + c^2} \tag{1}$$

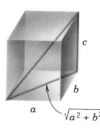

Figure 14.1.5

If we now apply (1) to the box in Figure 14.1.6, which has $P_1(x_1, y_1, z_1)$ and $P_2(x_2, y_2, z_2)$ as diagonal corners and hence sides of length

$$|x_2 - x_1|, \quad |y_2 - y_1|, \quad \text{and} \quad |z_2 - z_1|$$

then we find that the distance d between P_1 and P_2 is

$$d = \sqrt{(x_2 - x_1)^2 + (y_2 - y_1)^2 + (z_2 - z_1)^2} \tag{2}$$

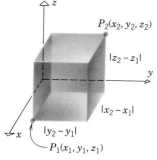

Figure 14.1.6

(where we have dropped the absolute value signs, since $|x|^2 = x^2$ for all x.)

Observe that Formula (2) has the same form as the distance formula in 2-space, except for an additional term to account for the z-coordinate. Similarly, we shall prove later that the formula for the midpoint of a line segment in 3-space has the same form as the formula in 2-space, except for an additional coordinate:

$$\begin{bmatrix} \text{The midpoint of the} \\ \text{line segment joining} \\ P_1(x_1, y_1, z_1) \text{ and } P_2(x_2, y_2, z_2) \end{bmatrix} = \left(\tfrac{1}{2}(x_1 + x_2), \tfrac{1}{2}(y_1 + y_2), \tfrac{1}{2}(z_1 + z_2)\right) \tag{3}$$

Example 1 Find the distance d between the points $(4, -1, 3)$ and $(2, 3, -1)$, and find the midpoint M of the line segment that joins them.

Solution. From Formulas (2) and (3)

$$d = \sqrt{(4 - 2)^2 + (-1 - 3)^2 + (3 + 1)^2} = \sqrt{36} = 6$$

$$M = \left(\tfrac{1}{2}(4 + 2), \tfrac{1}{2}(-1 + 3), \tfrac{1}{2}(3 - 1)\right) = (3, 1, 1) \quad \blacktriangleleft$$

In 2-space, the set of points satisfying an equation in x and y is usually a *curve* in the xy-plane. Similarly, in 3-space the set of points satisfying an equation in x, y, and z is usually a *surface* in an xyz-coordinate system. The surface is called the **graph** of the equation. For example, Figure 14.1.7 shows a sphere in 3-space with center (x_0, y_0, z_0) and radius r. The sphere consists of all points whose distance from (x_0, y_0, z_0) is r, so from (2) it consists of those points (x, y, z) whose coordinates satisfy

$$\sqrt{(x - x_0)^2 + (y - y_0)^2 + (z - z_0)^2} = r$$

or equivalently,

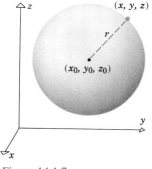

Figure 14.1.7

$$(x - x_0)^2 + (y - y_0)^2 + (z - z_0)^2 = r^2 \tag{4}$$

This is called the ***standard form of the equation of the sphere*** with center (x_0, y_0, z_0) and radius r. Observe that this equation has the same form as the standard form of the equation of a circle in 2-space, except for the additional term to account for the z-coordinate. Some examples are given in the following table.

EQUATION	GRAPH
$(x - 3)^2 + (y - 2)^2 + (z - 1)^2 = 9$	Sphere with center $(3, 2, 1)$ and radius 3
$(x + 1)^2 + y^2 + (z + 4)^2 = 5$	Sphere with center $(-1, 0, -4)$ and radius $\sqrt{5}$
$x^2 + y^2 + z^2 = 1$	Sphere with center $(0, 0, 0)$ and radius 1

If the terms in (4) are squared out and like terms are then collected, then the resulting equation has the form

$$x^2 + y^2 + z^2 + Gx + Hy + Iz + J = 0 \tag{5}$$

As the following example shows, the center and radius of a sphere expressed in this form can be obtained by completing the squares.

Example 2   Find the center and radius of the sphere

$$x^2 + y^2 + z^2 - 2x - 4y + 8z + 17 = 0$$

Solution.   We can put the equation in the form of (4) by completing the squares:

$$(x^2 - 2x) + (y^2 - 4y) + (z^2 + 8z) = -17$$

$$(x^2 - 2x + 1) + (y^2 - 4y + 4) + (z^2 + 8z + 16) = -17 + 21$$

$$(x - 1)^2 + (y - 2)^2 + (z + 4)^2 = 4$$

which is the equation of the sphere with center $(1, 2, -4)$ and radius 2.   ◄

In general, completing the squares in (5) produces an equation of the form

$$(x - x_0)^2 + (y - y_0)^2 + (z - z_0)^2 = k$$

which represents a sphere of radius $\sqrt{k}$ and center (x_0, y_0, z_0) if $k > 0$, the point (x_0, y_0, z_0) if $k = 0$, or has no graph if $k < 0$ (why?). The following theorem summarizes these observations.

14.1.1   THEOREM.   *An equation of the form*

$$x^2 + y^2 + z^2 + Gx + Hy + Iz + J = 0$$

represents either a sphere or a point, or else has no graph.

☐ **CYLINDRICAL SURFACES**

Many important surfaces in 3-space can be generated by translating a plane curve along a line. For example, the surface in Figure 14.1.8 is obtained by translating the curve $y = x^2$ in the xy-plane along a line parallel to the z-axis. The process of translating a curve along a line to generate a surface is called ***extrusion***, and surfaces that are generated by extrusion are called ***cylindrical surfaces***.

A surface in an xyz-system that is obtained by extrusion along a line parallel to a coordinate axis has an equation with only two of the three variables x, y, and z. For

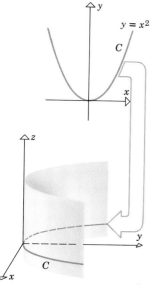

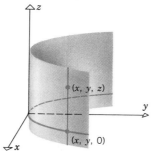

example, the surface in Figure 14.1.8 is represented by the equation $y = x^2$ (with no z variable). This can be seen by observing that if (x, y) satisfies the equation $y = x^2$, then for *arbitrary* values of z the points of the form (x, y, z) lie on the surface; they are directly above or below the point $(x, y, 0)$ on the curve $y = x^2$ in the xy-plane (Figure 14.1.9).

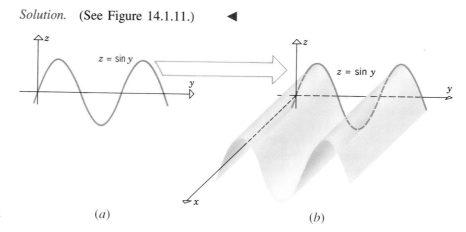

Figure 14.1.9

Figure 14.1.8

In summary, we have the following theorem.

14.1.2   THEOREM.   *An equation containing only two of the three variables x, y, and z represents a cylindrical surface in 3-space. The surface is obtained by extrusion parallel to the axis corresponding to the missing variable.*

REMARK.   Just as $x = a$ can represent a point on the x-axis or a line in the xy-plane parallel to the y-axis, so an equation with two variables, such as $y = x^2$, can represent either a curve in the xy-plane in 2-space or a cylindrical surface in an xyz-coordinate system in 3-space. The appropriate interpretation will usually be clear from the context in which the equation appears.

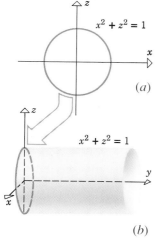

Figure 14.1.10

Example 3   Sketch the graph of $x^2 + z^2 = 1$ in 3-space.

Solution.   Since y does not appear in this equation, the graph is a cylindrical surface generated by extrusion parallel to the y-axis. It is helpful to begin with a sketch of the equation in 2-space. In the xz-plane the curve $x^2 + z^2 = 1$ is a circle (Figure 14.1.10a). Thus, in 3-space this equation represents a right-circular cylinder parallel to the y-axis (Figure 14.1.10b).   ◄

Example 4   Sketch the graph of $z = \sin y$ in 3-space.

Solution.   (See Figure 14.1.11.)   ◄

Figure 14.1.11   (a)   (b)

▶ Exercise Set 14.1

1. Plot the points P and Q in a right-handed coordinate system. Then, find the distance between them and the midpoint of the line segment joining them.

   (a) $P(0, 0, 0)$; $Q(2, 1, 3)$

   (b) $P(5, 2, 3)$; $Q(4, 1, 6)$

   (c) $P(-2, -1, 3)$; $Q(3, 0, 5)$

   (d) $P(-1, -1, -3)$; $Q(4, 3, -2)$.

2. A cube of side 4 has its geometric center at the origin and its faces parallel to the coordinate planes. Sketch the cube and give the coordinates of the vertices.

3. A rectangular parallelepiped has its faces parallel to the coordinate planes and has $(4, 2, -2)$ and $(-6, 1, 1)$ as endpoints of a diagonal. Sketch the parallelepiped and give the coordinates of the vertices.

4. Show that $(4, 5, 2)$, $(1, 7, 3)$, and $(2, 4, 5)$ are vertices of an equilateral triangle.

5. (a) Show that $(2, 1, 6)$, $(4, 7, 9)$, and $(8, 5, -6)$ are the vertices of a right triangle.

   (b) Which vertex is at the 90° angle?

   (c) Find the area of the triangle.

6. Find the distance from the point $(-5, 2, -3)$ to the

   (a) xy-plane   (b) xz-plane   (c) yz-plane

   (d) x-axis   (e) y-axis   (f) z-axis.

7. Show that the distance from a point (x_0, y_0, z_0) to the z-axis is $\sqrt{x_0^2 + y_0^2}$, and find the distances from the point to the x- and y-axes.

In Exercises 8–11, find an equation for the sphere with center C and radius r.

8. $C(0, 0, 0)$; $r = 8$.

9. $C(-2, 4, -1)$; $r = 6$.

10. $C(5, -2, 4)$; $r = \sqrt{7}$.

11. $C(0, 1, 0)$; $r = 3$.

12. In each part find an equation for the sphere with center $(-3, 5, -4)$ and satisfying the given condition.

   (a) Tangent to the xy-plane

   (b) Tangent to the xz-plane

   (c) Tangent to the yz-plane.

13. In each part, find an equation for the sphere with center $(2, -1, -3)$ and satisfying the given condition.

   (a) Tangent to the xy-plane

   (b) Tangent to the xz-plane

   (c) Tangent to the yz-plane.

In Exercises 14–19, find the standard equation of the sphere satisfying the given conditions.

14. Center $(1, 0, -1)$; diameter $= 8$.

15. A diameter has endpoints $(-1, 2, 1)$ and $(0, 2, 3)$.

16. Center $(-1, 3, 2)$ and passing through the origin.

17. Center $(3, -2, 4)$ and passing through $(7, 2, 1)$.

18. Center $(-3, 5, -4)$; tangent to the sphere of radius 1 centered at the origin (two answers).

19. Center $(0, 0, 0)$; tangent to the sphere of radius 1 centered at $(3, -2, 4)$ (two answers).

In Exercises 20–25, describe the surface whose equation is given.

20. $x^2 + y^2 + z^2 - 2x - 6y - 8z + 1 = 0$.

21. $x^2 + y^2 + z^2 + 10x + 4y + 2z - 19 = 0$.

22. $x^2 + y^2 + z^2 - y = 0$.

23. $2x^2 + 2y^2 + 2z^2 - 2x - 3y + 5z - 2 = 0$.

24. $x^2 + y^2 + z^2 + 2x - 2y + 2z + 3 = 0$.

25. $x^2 + y^2 + z^2 - 3x + 4y - 8z + 25 = 0$.

26. Find the largest and smallest distances between the point $P(1, 1, 1)$ and the sphere $x^2 + y^2 + z^2 - 2y + 6z - 6 = 0$.

27. Find the largest and smallest distances between the origin and the sphere $x^2 + y^2 + z^2 + 2x - 2y - 4z - 3 = 0$.

28. Describe the set of all points in 3-space whose coordinates satisfy the inequality $x^2 + y^2 + z^2 - 2x + 8z \leq 8$.

29. Describe the set of all points in 3-space whose coordinates satisfy the inequality $y^2 + z^2 + 6y - 4z > 3$.

30. The distance between a point $P(x, y, z)$ and the point $A(1, -2, 0)$ is twice the distance between P and the point $B(0, 1, 1)$. Show that the set of all such points is a sphere, and find the center and radius of the sphere.

31. A bowling ball of radius R is placed inside a box just large enough to hold it, and it is secured for shipping by packing a Styrofoam sphere into each corner of the box. Find the radius of the largest Styrofoam sphere that can be used. [*Hint:* Take the origin of a Cartesian coordinate system at a corner of the box with the coordinate axes along the edges.] (See Figure 14.1.12.)

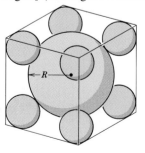

Figure 14.1.12

In Exercises 32–39, sketch the surface whose equation is given.

32. (a) $y = x$ (b) $y = z$ (c) $x = z$.

33. (a) $x^2 + y^2 = 25$ (b) $y^2 + z^2 = 25$

   (c) $x^2 + z^2 = 25$.

34. (a) $y = x^2$ (b) $z = x^2$ (c) $y = z^2$.

35. (a) $y = e^x$ (b) $x = \ln z$ (c) $yz = 1$.

36. (a) $2x + 3y = 6$ (b) $2x + z = 3$.

37. (a) $y = \sin x$ (b) $z = \cos x$.

38. (a) $z = 1 - y^2$ (b) $z = \sqrt{3 - x}$.

39. (a) $4x^2 + 9z^2 = 36$ (b) $y^2 - 4z^2 = 4$.

40. In each part of Figure 14.1.13 find an equation for the right-circular cylinder of radius a shown.

41. Show that for all values of θ and ϕ, the point

$(a \sin \phi \cos \theta, \ a \sin \phi \sin \theta, \ a \cos \phi)$ lies on the sphere $x^2 + y^2 + z^2 = a^2$.

42. Consider the equation

$$x^2 + y^2 + z^2 + Gx + Hy + Iz + J = 0$$

and let $K = G^2 + H^2 + I^2 - 4J$.

(a) Prove that the equation represents a sphere, a point, or no graph, according to whether $K > 0$, $K = 0$, or $K < 0$.

(b) In the case where $K > 0$, find the center and radius of the sphere.

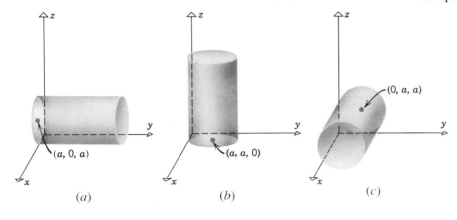

(a)   (b)   (c)   Figure 14.1.13

■ 14.2 VECTORS

> *Many physical quantities such as area, length, mass, and temperature are completely described once the magnitude of the quantity is given. Such quantities are called scalars. Other physical quantities, called **vectors**, are not completely determined until both a magnitude and a direction are specified. For example, wind movement is usually described by giving the speed and the direction, say 20 mi/hr northeast. The wind speed and wind direction together form a vector quantity called the wind velocity. Other examples of vectors are force and displacement. In this section we shall develop the basic mathematical properties of vectors.*

☐ **VECTORS VIEWED GEOMETRICALLY**

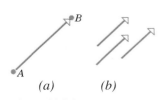

(a)   (b)

Figure 14.2.1

Vectors can be represented geometrically as directed line segments or arrows in two- or three-dimensional space; the direction of the arrow specifies the direction of the vector and the length of the arrow describes its magnitude. The tail of the arrow is called the *initial point* of the vector, and the tip of the arrow the *terminal point*. We shall denote vectors by lowercase boldface type such as **a**, **k**, **v**, **w**, and **x**. When discussing vectors, we shall refer to real numbers as *scalars*. Scalars will be denoted by lowercase italic type such as a, k, v, w, and x.

If, as in Figure 14.2.1a, the initial point of a vector **v** is A and the terminal point is B, we write $\mathbf{v} = \overrightarrow{AB}$. Vectors having the same length and same direction, such as those in Figure 14.2.1b, are called *equivalent*. Since we want a vector to be determined solely by its length and direction, equivalent vectors are regarded as *equal* even though they may be located in different positions. If **v** and **w** are equivalent, we write $\mathbf{v} = \mathbf{w}$.

The vector whose initial and terminal points coincide has length zero. We call this the *zero vector* and denote it by **0**. The zero vector has no natural direction, so we shall agree that it can be assigned any direction that is convenient for a problem at hand.

There are various algebraic operations that are performed on vectors, all of whose definitions originated in physics. We begin with vector addition.

> **14.2.1** DEFINITION. If **v** and **w** are any two nonzero vectors, then the *sum* **v** + **w** is the vector determined as follows: Position the vector **w** so that its initial point coincides with the terminal point of **v**. The vector **v** + **w** is represented by the arrow from the initial point of **v** to the terminal point of **w** (Figure 14.2.2*a*).

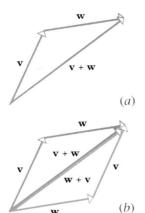

In Figure 14.2.2*b* we have constructed two sums, **v** + **w** (purple arrows) and **w** + **v** (green arrows). It is evident that

$$\mathbf{v} + \mathbf{w} = \mathbf{w} + \mathbf{v}$$

and that the sum coincides with the diagonal of the parallelogram determined by **v** and **w** when these vectors are located so that they have the same initial point.

Definition 14.2.1 does not apply if either of the vectors is zero, but we shall agree that for all vectors **v**

$$\mathbf{0} + \mathbf{v} = \mathbf{v} + \mathbf{0} = \mathbf{v}$$

We now consider another basic operation on vectors.

Figure 14.2.2

> **14.2.2** DEFINITION. If **v** is a nonzero vector and k is a nonzero real number (scalar), then the *scalar multiple* $k\mathbf{v}$ is defined to be the vector whose length is $|k|$ times the length of **v** and whose direction is the same as that of **v** if $k > 0$ and opposite to that of **v** if $k < 0$. We define $k\mathbf{v} = \mathbf{0}$ if $k = 0$ or $\mathbf{v} = \mathbf{0}$.

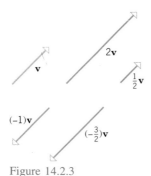

Figure 14.2.3 shows the geometric relationship between a vector **v** and various scalar multiples of it. Observe that the vector $(-1)\mathbf{v}$ has the same length as **v** but is oppositely directed. We call this vector the *negative of* **v** and denote it by

$$(-1)\mathbf{v} = -\mathbf{v}$$

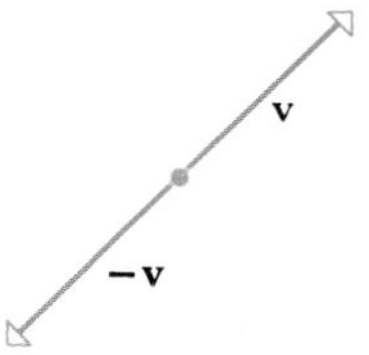

Figure 14.2.3

(Figure 14.2.4). In addition, we define $-\mathbf{0} = (-1)\mathbf{0} = \mathbf{0}$. Subtraction of vectors is defined as follows.

> **14.2.3** DEFINITION. If **v** and **w** are any two vectors, then the *difference* **v** − **w** is defined by
>
> $$\mathbf{v} - \mathbf{w} = \mathbf{v} + (-\mathbf{w})$$

Figure 14.2.4

The difference **v** − **w** can be obtained geometrically by the parallelogram method shown in Figure 14.2.5*a*, or more directly, as in Figure 14.2.5*b*, by drawing the vector from the terminal point of **w** to the terminal point of **v**. We leave it for the reader to deduce that

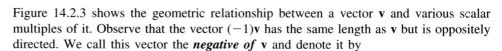

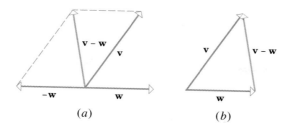

Figure 14.2.5   (a)   (b)

$$\mathbf{v} + (-\mathbf{v}) = \mathbf{v} - \mathbf{v} = \mathbf{0}$$

☐ **VECTORS IN COORDINATE SYSTEMS**

Problems involving vectors can often be simplified by introducing a rectangular coordinate system, in which case one must distinguish between vectors in 2-space and vectors in 3-space.

If **v** is a vector in 2-space or 3-space with its initial point at the origin of a rectangular coordinate system (Figure 14.2.6), then the coordinates (v_1, v_2) or (v_1, v_2, v_3) of the terminal point are called the *components* of **v** and we write

$$\mathbf{v} = \langle v_1, v_2 \rangle \quad \text{or} \quad \mathbf{v} = \langle v_1, v_2, v_3 \rangle$$

depending on whether the vector is in 2-space or 3-space.

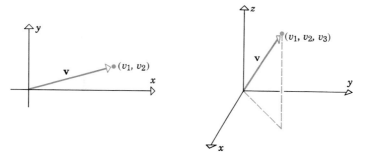

Figure 14.2.6

REMARK.   We emphasize that a vector must be positioned with its initial point at the origin in order for the components to be the coordinates of the terminal point. Later we shall show how to obtain the components of a vector that does not have its initial point at the origin.

If the zero vector is positioned with its initial point at the origin, then the terminal point is also at the origin, so

$$\mathbf{0} = \langle 0, 0 \rangle \quad \text{and} \quad \mathbf{0} = \langle 0, 0, 0 \rangle$$
$$\text{2-space} \qquad\qquad \text{3-space}$$

Since equivalent vectors have the same length and direction, such vectors must have the same terminal point when they are positioned with their initial points at the origin. Hence, equivalent vectors have the same components. Conversely, vectors with the same components have the same length and direction, and hence must be equivalent. Thus, for vectors **v** and **w** in either 2-space or 3-space we have **v** = **w** if and only if corresponding components of **v** and **w** are the same. For example,

$$\langle a, b, c \rangle = \langle 1, -4, 2 \rangle$$

if and only if $a = 1$, $b = -4$, and $c = 2$.

☐ **ARITHMETIC OPERATIONS ON VECTORS**

The following theorem shows how to perform arithmetic operations on vectors using components.

14.2.4   THEOREM.   *If $\mathbf{v} = \langle v_1, v_2 \rangle$ and $\mathbf{w} = \langle w_1, w_2 \rangle$ are vectors in 2-space and k is any scalar, then*

$$\mathbf{v} + \mathbf{w} = \langle v_1 + w_1, v_2 + w_2 \rangle \tag{1a}$$

$$\mathbf{v} - \mathbf{w} = \langle v_1 - w_1, v_2 - w_2 \rangle \tag{1b}$$

$$k\mathbf{v} = \langle kv_1, kv_2 \rangle \tag{1c}$$

Similarly, if $\mathbf{v} = \langle v_1, v_2, v_3 \rangle$ and $\mathbf{w} = \langle w_1, w_2, w_3 \rangle$ are vectors in 3-space and k is any scalar, then

$$\mathbf{v} + \mathbf{w} = \langle v_1 + w_1, v_2 + w_2, v_3 + w_3 \rangle \tag{2a}$$

$$\mathbf{v} - \mathbf{w} = \langle v_1 - w_1, v_2 - w_2, v_3 - w_3 \rangle \tag{2b}$$

$$k\mathbf{v} = \langle kv_1, kv_2, kv_3 \rangle \tag{2c}$$

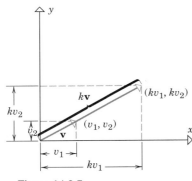

Figure 14.2.7

We shall not prove this theorem. However, results (1a) and (1c) should be evident from Figure 14.2.7. Similar figures in 3-space can be used to motivate (2a) and (2c). Formulas (1b) and (2b) can be obtained from parts (a) and (c) by writing $\mathbf{v} - \mathbf{w} = \mathbf{v} + (-1)\mathbf{w}$.

Example 1   If $\mathbf{v} = \langle -2, 0, 1 \rangle$ and $\mathbf{w} = \langle 3, 5, -4 \rangle$, then

$$\mathbf{v} + \mathbf{w} = \langle -2, 0, 1 \rangle + \langle 3, 5, -4 \rangle = \langle 1, 5, -3 \rangle$$

$$3\mathbf{v} = \langle -6, 0, 3 \rangle$$

$$-\mathbf{w} = \langle -3, -5, 4 \rangle$$

$$\mathbf{w} - 2\mathbf{v} = \langle 3, 5, -4 \rangle - \langle -4, 0, 2 \rangle = \langle 7, 5, -6 \rangle \quad ◀$$

REMARK.   Except for the number of components, there is no difference between arithmetic computations on vectors in 2-space and 3-space.

☐ **VECTORS WITH INITIAL POINT NOT AT THE ORIGIN**

If a vector in 2-space or 3-space is positioned with its initial point at the origin, then the coordinates of the terminal point are the components of the vector. If the vector does not have its initial point at the origin (Figure 14.2.8), then the following theorem shows that the components can be obtained by subtracting the coordinates of the initial point from the coordinates of the terminal point.

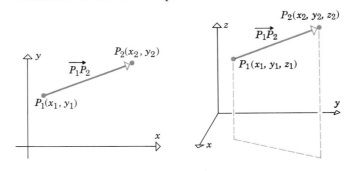

Figure 14.2.8

14.2.5 THEOREM. *If $\overrightarrow{P_1P_2}$ is a vector in 2-space with initial point $P_1(x_1, y_1)$ and terminal point $P_2(x_2, y_2)$, then*

$$\overrightarrow{P_1P_2} = \langle x_2 - x_1, y_2 - y_1 \rangle \tag{3a}$$

Similarly, if $\overrightarrow{P_1P_2}$ is a vector in 3-space with initial point $P_1(x_1, y_1, z_1)$ and terminal point $P_2(x_2, y_2, z_2)$, then

$$\overrightarrow{P_1P_2} = \langle x_2 - x_1, y_2 - y_1, z_2 - z_1 \rangle \tag{3b}$$

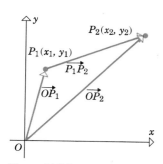

Figure 14.2.9

Proof. We shall give the proof in 2-space. The proof in 3-space is similar. The vector $\overrightarrow{P_1P_2}$ is the difference of vectors $\overrightarrow{OP_2}$ and $\overrightarrow{OP_1}$ (Figure 14.2.9). Thus,

$$\overrightarrow{P_1P_2} = \overrightarrow{OP_2} - \overrightarrow{OP_1} = \langle x_2, y_2 \rangle - \langle x_1, y_1 \rangle = \langle x_2 - x_1, y_2 - y_1 \rangle \quad \blacksquare$$

Example 2 In 2-space the vector from $P_1(1, 3)$ to $P_2(4, -2)$ is

$$\overrightarrow{P_1P_2} = \langle 4 - 1, -2 - 3 \rangle = \langle 3, -5 \rangle$$

and in 3-space the vector from $A(0, -2, 5)$ to $B(3, 4, -1)$ is

$$\overrightarrow{AB} = \langle 3 - 0, 4 - (-2), -1 - 5 \rangle = \langle 3, 6, -6 \rangle \quad \blacktriangleleft$$

In the preceding section we stated without proof that the midpoint $M(x, y, z)$ of the line segment joining the points $P_1(x_1, y_1, z_1)$ and $P_2(x_2, y_2, z_2)$ has coordinates

$$x = \tfrac{1}{2}(x_1 + x_2), \quad y = \tfrac{1}{2}(y_1 + y_2), \quad z = \tfrac{1}{2}(z_1 + z_2) \tag{4}$$

To see that this is so we need only observe that

$$\overrightarrow{P_1M} = \tfrac{1}{2}\overrightarrow{P_1P_2}$$

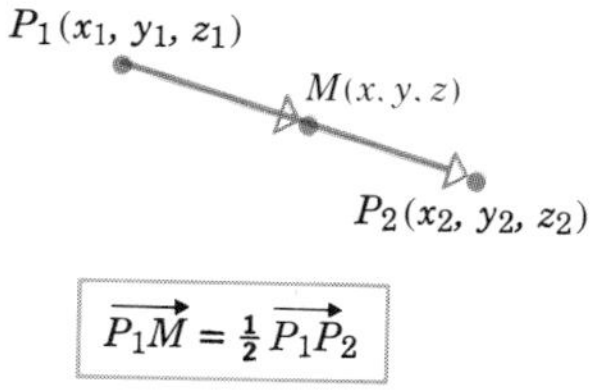

$P_1(x_1, y_1, z_1)$

$M(x, y, z)$

$P_2(x_2, y_2, z_2)$

$$\overrightarrow{P_1M} = \tfrac{1}{2}\overrightarrow{P_1P_2}$$

Figure 14.2.10

(Figure 14.2.10), so that

$$\langle x - x_1, y - y_1, z - z_1 \rangle = \tfrac{1}{2}\langle x_2 - x_1, y_2 - y_1, z_2 - z_1 \rangle$$

or, on equating components,

$$x - x_1 = \tfrac{1}{2}(x_2 - x_1), \quad y - y_1 = \tfrac{1}{2}(y_2 - y_1), \quad z - z_1 = \tfrac{1}{2}(z_2 - z_1)$$

from which (4) follows.

☐ **RULES OF VECTOR ARITHMETIC**

The following theorem shows that many of the familiar rules of ordinary arithmetic also hold for vector arithmetic.

14.2.6 THEOREM. *For any vectors $\mathbf{u}$, $\mathbf{v}$, and $\mathbf{w}$ and any scalars k and l, the following relationships hold*:

(a) $\mathbf{u} + \mathbf{v} = \mathbf{v} + \mathbf{u}$
(b) $(\mathbf{u} + \mathbf{v}) + \mathbf{w} = \mathbf{u} + (\mathbf{v} + \mathbf{w})$
(c) $\mathbf{u} + \mathbf{0} = \mathbf{0} + \mathbf{u} = \mathbf{u}$
(d) $\mathbf{u} + (-\mathbf{u}) = \mathbf{0}$
(e) $k(l\mathbf{u}) = (kl)\mathbf{u}$
(f) $k(\mathbf{u} + \mathbf{v}) = k\mathbf{u} + k\mathbf{v}$
(g) $(k + l)\mathbf{u} = k\mathbf{u} + l\mathbf{u}$
(h) $1\mathbf{u} = \mathbf{u}$

Before discussing the proof, we note that we have developed two approaches to vectors: *geometric*, in which vectors are represented by arrows or directed line segments, and *analytic*, in which vectors are represented by pairs or triples of numbers called components. As a consequence, the results in this theorem can be established either geometrically or analytically. As an illustration, we shall prove part (*b*) both ways. The remaining proofs are left as exercises.

Proof (b) (Analytic in 2-space).   Let $\mathbf{u} = \langle u_1, u_2 \rangle$, $\mathbf{v} = \langle v_1, v_2 \rangle$, and $\mathbf{w} = \langle w_1, w_2 \rangle$. Then

$$
\begin{aligned}
(\mathbf{u} + \mathbf{v}) + \mathbf{w} &= (\langle u_1, u_2 \rangle + \langle v_1, v_2 \rangle) + \langle w_1, w_2 \rangle \\
&= \langle u_1 + v_1, u_2 + v_2 \rangle + \langle w_1, w_2 \rangle \\
&= \langle (u_1 + v_1) + w_1, (u_2 + v_2) + w_2 \rangle \\
&= \langle u_1 + (v_1 + w_1), u_2 + (v_2 + w_2) \rangle \\
&= \langle u_1, u_2 \rangle + \langle v_1 + w_1, v_2 + w_2 \rangle \\
&= \mathbf{u} + (\mathbf{v} + \mathbf{w})
\end{aligned}
$$

Proof (b) (Geometric).   Let $\mathbf{u}$, $\mathbf{v}$, and $\mathbf{w}$ be represented by $\overrightarrow{PQ}$, $\overrightarrow{QR}$, and $\overrightarrow{RS}$ as shown in Figure 14.2.11. Then

$$\mathbf{v} + \mathbf{w} = \overrightarrow{QS} \quad \text{and} \quad \mathbf{u} + (\mathbf{v} + \mathbf{w}) = \overrightarrow{PS}$$

$$\mathbf{u} + \mathbf{v} = \overrightarrow{PR} \quad \text{and} \quad (\mathbf{u} + \mathbf{v}) + \mathbf{w} = \overrightarrow{PS}$$

Therefore,

$$(\mathbf{u} + \mathbf{v}) + \mathbf{w} = \mathbf{u} + (\mathbf{v} + \mathbf{w}) \qquad \blacksquare$$

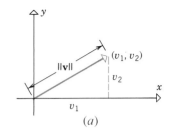

Figure 14.2.11

REMARK.   In light of part (*b*) of this theorem, the symbol $\mathbf{u} + \mathbf{v} + \mathbf{w}$ is unambiguous since the same result is obtained no matter where parentheses are inserted. Moreover, if the vectors $\mathbf{u}$, $\mathbf{v}$, and $\mathbf{w}$ are placed "tip to tail," then the sum $\mathbf{u} + \mathbf{v} + \mathbf{w}$ is the vector from the initial point of $\mathbf{u}$ to the terminal point of $\mathbf{w}$ (Figure 14.2.11).

☐ **LENGTH OF A VECTOR**

Geometrically, the **length** of a vector $\mathbf{v}$, also called the **norm** of $\mathbf{v}$, is the distance between its initial and terminal points. The length (or norm) of $\mathbf{v}$ is denoted by $\|\mathbf{v}\|$. It follows from the distance formulas in 2-space and 3-space that the norm of a vector $\mathbf{v} = \langle v_1, v_2 \rangle$ in 2-space is given by

$$\|\mathbf{v}\| = \sqrt{v_1^2 + v_2^2} \tag{5a}$$

and the norm of a vector $\mathbf{v} = \langle v_1, v_2, v_3 \rangle$ in 3-space is given by

$$\|\mathbf{v}\| = \sqrt{v_1^2 + v_2^2 + v_3^2} \tag{5b}$$

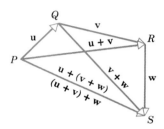

(Figures 14.2.12*a* and 14.2.12*b*).

Example 3   Find the norm of $\mathbf{v} = \langle -2, 3 \rangle$ and $\mathbf{w} = \langle 2, 3, 6 \rangle$.

Solution.   From (5a) and (5b)

$$\|\mathbf{v}\| = \sqrt{(-2)^2 + 3^2} = \sqrt{13}$$

$$\|\mathbf{w}\| = \sqrt{2^2 + 3^2 + 6^2} = \sqrt{49} = 7 \qquad \blacktriangleleft$$

Recall from Definition 14.2.2 that the length of $k\mathbf{v}$ is $|k|$ times the length of $\mathbf{v}$. Expressed as an equation, this statement says that

$$\|k\mathbf{v}\| = |k|\, \|\mathbf{v}\| \tag{6}$$

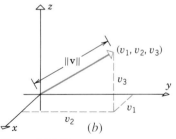

Figure 14.2.12

This formula applies to both vectors in 2-space and 3-space.

□ **UNIT VECTORS**

Vectors of length 1 are called **_unit vectors_**. The unit vectors that run along the positive coordinate axes of a Cartesian coordinate system are especially important: In 2-space or 3-space the unit vectors along the positive x- and y-axes are denoted by **i** and **j**, respectively, and in 3-space the unit vector along the positive z-axis is denoted by **k**. Thus, we have (Figure 14.2.13)

$$\mathbf{i} = \langle 1, 0 \rangle, \qquad \mathbf{j} = \langle 0, 1 \rangle \qquad \boxed{\text{In 2-space}}$$

$$\mathbf{i} = \langle 1, 0, 0 \rangle, \quad \mathbf{j} = \langle 0, 1, 0 \rangle, \quad \mathbf{k} = \langle 0, 0, 1 \rangle \qquad \boxed{\text{In 3-space}}$$

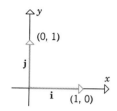

Every vector in 2-space is expressible uniquely in terms of **i** and **j**, and every vector in 3-space is expressible uniquely in terms of **i**, **j**, and **k** as follows:

$$\mathbf{v} = \langle v_1, v_2 \rangle = \langle v_1, 0 \rangle + \langle 0, v_2 \rangle = v_1\langle 1, 0 \rangle + v_2\langle 0, 1 \rangle = v_1\mathbf{i} + v_2\mathbf{j}$$

$$\mathbf{v} = \langle v_1, v_2, v_3 \rangle = v_1\langle 1, 0, 0 \rangle + v_2\langle 0, 1, 0 \rangle + v_3\langle 0, 0, 1 \rangle = v_1\mathbf{i} + v_2\mathbf{j} + v_3\mathbf{k}$$

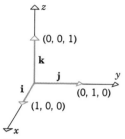

REMARK. The notations $\langle v_1, v_2, v_3 \rangle$ and $v_1\mathbf{i} + v_2\mathbf{j} + v_3\mathbf{k}$ are interchangeable, as are $\langle v_1, v_2 \rangle$ and $v_1\mathbf{i} + v_2\mathbf{j}$. For example, (5b) can be written as

$$\|v_1\mathbf{i} + v_2\mathbf{j} + v_3\mathbf{k}\| = \sqrt{v_1^2 + v_2^2 + v_3^2}$$

Figure 14.2.13

Example 4

2-SPACE	3-SPACE
$\langle 2, 3 \rangle = 2\mathbf{i} + 3\mathbf{j}$	$\langle 2, -3, 4 \rangle = 2\mathbf{i} - 3\mathbf{j} + 4\mathbf{k}$
$\langle -4, 0 \rangle = -4\mathbf{i} + 0\mathbf{j} = -4\mathbf{i}$	$\langle 0, 3, 0 \rangle = 3\mathbf{j}$
$\langle 0, 0 \rangle = 0\mathbf{i} + 0\mathbf{j} = \mathbf{0}$	$\langle 0, 0, 0 \rangle = 0\mathbf{i} + 0\mathbf{j} + 0\mathbf{k} = \mathbf{0}$
$(3\mathbf{i} + 2\mathbf{j}) + (4\mathbf{i} + \mathbf{j}) = 7\mathbf{i} + 3\mathbf{j}$	$(3\mathbf{i} + 2\mathbf{j} - \mathbf{k}) - (4\mathbf{i} - \mathbf{j} + 2\mathbf{k}) = -\mathbf{i} + 3\mathbf{j} - 3\mathbf{k}$
$5(6\mathbf{i} - 2\mathbf{j}) = 30\mathbf{i} - 10\mathbf{j}$	$2(\mathbf{i} + \mathbf{j} - \mathbf{k}) + 4(\mathbf{i} - \mathbf{j}) = 6\mathbf{i} - 2\mathbf{j} - 2\mathbf{k}$
$\|2\mathbf{i} - 3\mathbf{j}\| = \sqrt{2^2 + (-3)^2} = \sqrt{13}$	$\|\mathbf{i} + 2\mathbf{j} - 3\mathbf{k}\| = \sqrt{1^2 + 2^2 + (-3)^2} = \sqrt{14}$

◄

If **v** is a nonzero vector, then it follows from (6) with $k = 1/\|\mathbf{v}\|$ that

$$\left\| \frac{1}{\|\mathbf{v}\|}\mathbf{v} \right\| = \left| \frac{1}{\|\mathbf{v}\|} \right| \|\mathbf{v}\| = \frac{1}{\|\mathbf{v}\|}\|\mathbf{v}\| = 1$$

which tells us that multiplying a nonzero vector by the reciprocal of its length produces a unit vector. We call the process of multiplying **v** by $1/\|\mathbf{v}\|$ **_normalizing_** **v**. Thus, by normalizing, a nonzero vector **v** in 2-space or 3-space can be expressed as

$$\mathbf{v} = \|\mathbf{v}\| \left(\frac{1}{\|\mathbf{v}\|}\mathbf{v} \right) \tag{7}$$

which is the length of **v** times a unit vector in the same direction as **v**.

Example 5   The vector $\mathbf{v} = \langle 3, 4 \rangle$ has length $\|\mathbf{v}\| = \sqrt{3^2 + 4^2} = 5$, so that

$$\frac{1}{\|\mathbf{v}\|}\mathbf{v} = \tfrac{1}{5}\langle 3, 4 \rangle = \langle \tfrac{3}{5}, \tfrac{4}{5} \rangle$$

is a unit vector in the same direction as **v**, and from (7) we can express **v** as

$$\langle 3, 4 \rangle = 5\langle \tfrac{3}{5}, \tfrac{4}{5} \rangle$$

which is the length of **v** times a unit vector in the same direction as **v**.   ◄

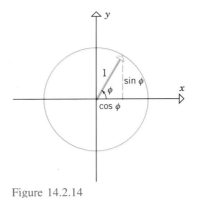

Figure 14.2.14

The components of a unit vector **u** in 2-space have a useful geometric interpretation: If ϕ denotes the angle from the positive x-axis to **u**, then, as suggested by Figure 14.2.14, the x- and y-components of **u** are $\cos \phi$ and $\sin \phi$, respectively; that is,

$$\mathbf{u} = \langle \cos \phi, \sin \phi \rangle = (\cos \phi)\mathbf{i} + (\sin \phi)\mathbf{j}$$

Thus, if **v** is any nonzero vector in 2-space, then it follows from (7) that

$$\mathbf{v} = \|\mathbf{v}\| \langle \cos \phi, \sin \phi \rangle \qquad \text{or equivalently,} \qquad \mathbf{v} = \|\mathbf{v}\| \cos \phi \, \mathbf{i} + \|\mathbf{v}\| \sin \phi \, \mathbf{j} \qquad (8)$$

where ϕ is the angle from the positive x-axis to **v**.

Example 6   Find the vector of length 2 that makes an angle of $\pi/4$ with the positive x-axis.

Solution.   From (8)

$$\mathbf{v} = 2 \cos \frac{\pi}{4} \mathbf{i} + 2 \sin \frac{\pi}{4} \mathbf{j} = \sqrt{2}\mathbf{i} + \sqrt{2}\mathbf{j} \qquad \blacktriangleleft$$

▶ Exercise Set 14.2

In Exercises 1–4, sketch the vectors with the initial point at the origin.

1. (a) $\langle 2, 5 \rangle$
(b) $\langle -5, -4 \rangle$
(c) $\langle 2, 0 \rangle$.

2. (a) $\langle -3, 7 \rangle$
(b) $\langle 6, -2 \rangle$
(c) $\langle 0, -8 \rangle$.

3. (a) $\langle 1, -2, 2 \rangle$
(b) $\langle 2, 2, -1 \rangle$.

4. (a) $\langle -1, 3, 2 \rangle$
(b) $\langle 3, 4, 2 \rangle$.

In Exercises 5–8, sketch the vectors with the initial point at the origin.

5. (a) $-5\mathbf{i} + 3\mathbf{j}$
(b) $3\mathbf{i} - 2\mathbf{j}$
(c) $-6\mathbf{j}$.

6. (a) $4\mathbf{i} + 2\mathbf{j}$
(b) $-2\mathbf{i} - \mathbf{j}$
(c) $4\mathbf{i}$.

7. (a) $-\mathbf{i} + 2\mathbf{j} + 3\mathbf{k}$
(b) $2\mathbf{i} + 3\mathbf{j} - \mathbf{k}$.

8. (a) $2\mathbf{j} - \mathbf{k}$
(b) $\mathbf{i} - \mathbf{j} + 2\mathbf{k}$.

In Exercises 9–12, find the components of the vector $\overrightarrow{P_1 P_2}$.

9. (a) $P_1(3, 5)$, $P_2(2, 8)$
(b) $P_1(7, -2)$, $P_2(0, 0)$
(c) $P_1(-6, -2)$, $P_2(-4, -1)$
(d) $P_1(0, 0)$, $P_2(-8, 7)$.

10. (a) $P_1(1, 3)$, $P_2(4, 1)$
(b) $P_1(6, -4)$, $P_2(0, 0)$
(c) $P_1(-8, -1)$, $P_2(-3, -2)$
(d) $P_1(0, 0)$, $P_2(-3, -5)$.

11. (a) $P_1(5, -2, 1)$, $P_2(2, 4, 2)$
(b) $P_1(-1, 3, 5)$, $P_2(0, 0, 0)$.

12. (a) $P_1(0, 0, 0)$, $P_2(-1, 6, 1)$
(b) $P_1(4, 1, -3)$, $P_2(9, 1, -3)$.

13. Find the terminal point of $\mathbf{v} = 3\mathbf{i} - 2\mathbf{j}$ if the initial point is $(1, -2)$.

14. Find the terminal point of $\mathbf{v} = \langle 7, 6 \rangle$ if the initial point is $(2, -1)$.

15. Find the initial point of $\mathbf{v} = \langle -2, 4 \rangle$ if the terminal point is $(2, 0)$.

16. Find the terminal point of $\mathbf{v} = \mathbf{i} + 2\mathbf{j} - 3\mathbf{k}$ if the initial point is $(-2, 1, 4)$.

17. Find the initial point of $\mathbf{v} = \langle -3, 1, 2 \rangle$ if the terminal point is $(5, 0, -1)$.

18. Let $\mathbf{u} = \langle 1, 3 \rangle$, $\mathbf{v} = \langle 2, 1 \rangle$, and $\mathbf{w} = \langle 4, -1 \rangle$. Find
(a) $\mathbf{u} - \mathbf{w}$
(b) $7\mathbf{v} + 3\mathbf{w}$
(c) $-\mathbf{w} + \mathbf{v}$
(d) $3(\mathbf{u} - 7\mathbf{v})$
(e) $-3\mathbf{v} - 8\mathbf{w}$
(f) $2\mathbf{v} - (\mathbf{u} + \mathbf{w})$.

19. Let $\mathbf{u} = 2\mathbf{i} + 3\mathbf{j}$, $\mathbf{v} = 4\mathbf{i}$, $\mathbf{w} = -\mathbf{i} - 2\mathbf{j}$. Find
(a) $\mathbf{w} - \mathbf{v}$
(b) $6\mathbf{u} + 4\mathbf{w}$
(c) $-\mathbf{v} - 2\mathbf{w}$
(d) $4(3\mathbf{u} + \mathbf{v})$
(e) $-8(\mathbf{v} + \mathbf{w}) + 2\mathbf{u}$
(f) $3\mathbf{w} - (\mathbf{v} - \mathbf{w})$.

20. Let $\mathbf{u} = \langle 2, -1, 3 \rangle$, $\mathbf{v} = \langle 4, 0, -2 \rangle$, $\mathbf{w} = \langle 1, 1, 3 \rangle$. Find
(a) $\mathbf{u} - \mathbf{w}$
(b) $7\mathbf{v} + 3\mathbf{w}$
(c) $-\mathbf{w} + \mathbf{v}$
(d) $3(\mathbf{u} - 7\mathbf{v})$
(e) $-3\mathbf{v} - 8\mathbf{w}$
(f) $2\mathbf{v} - (\mathbf{u} + \mathbf{w})$.

21. Let $\mathbf{u} = 3\mathbf{i} - \mathbf{k}$, $\mathbf{v} = \mathbf{i} - \mathbf{j} + 2\mathbf{k}$, $\mathbf{w} = 3\mathbf{j}$. Find
(a) $\mathbf{w} - \mathbf{v}$
(b) $6\mathbf{u} + 4\mathbf{w}$
(c) $-\mathbf{v} - 2\mathbf{w}$
(d) $4(3\mathbf{u} + \mathbf{v})$
(e) $-8(\mathbf{v} + \mathbf{w}) + 2\mathbf{u}$
(f) $3\mathbf{w} - (\mathbf{v} - \mathbf{w})$.

In Exercises 22–25, compute the norm of **v**.

22. (a) $\mathbf{v} = \langle 3, 4 \rangle$ (b) $\mathbf{v} = -\mathbf{i} + 7\mathbf{j}$
    (c) $\mathbf{v} = -3\mathbf{j}$.

23. (a) $\mathbf{v} = \langle 1, -1 \rangle$ (b) $\mathbf{v} = \langle 2, 0 \rangle$
    (c) $\mathbf{v} = \sqrt{2}\mathbf{i} - \sqrt{7}\mathbf{j}$.

24. (a) $\mathbf{v} = \mathbf{i} + \mathbf{j} + \mathbf{k}$ (b) $\mathbf{v} = \langle -1, 2, 4 \rangle$.

25. (a) $\mathbf{v} = -3\mathbf{i} + 2\mathbf{j} + \mathbf{k}$ (b) $\mathbf{v} = \langle 0, -3, 0 \rangle$.

26. Let $\mathbf{u} = \langle 1, -3 \rangle$, $\mathbf{v} = \langle 1, 1 \rangle$, and $\mathbf{w} = \langle 2, -4 \rangle$. Find
    (a) $\|\mathbf{u} + \mathbf{v}\|$ (b) $\|\mathbf{u}\| + \|\mathbf{v}\|$
    (c) $\|-2\mathbf{u}\| + 2\|\mathbf{v}\|$ (d) $\|3\mathbf{u} - 5\mathbf{v} + \mathbf{w}\|$.

27. Let $\mathbf{u} = 2\mathbf{i} - 5\mathbf{j}$, $\mathbf{v} = 2\mathbf{i}$, and $\mathbf{w} = 3\mathbf{i} + 4\mathbf{j}$. Find
    (a) $\|\mathbf{v} + \mathbf{w}\|$ (b) $\|\mathbf{v}\| + \|\mathbf{w}\|$
    (c) $\|-3\mathbf{u}\| + 4\|\mathbf{v}\|$ (d) $\|\mathbf{u} - \mathbf{v} - \mathbf{w}\|$
    (e) $\dfrac{1}{\|\mathbf{w}\|}\mathbf{w}$ (f) $\left\|\dfrac{1}{\|\mathbf{w}\|}\mathbf{w}\right\|$.

28. Let $\mathbf{u} = \langle 2, -1, 0 \rangle$ and $\mathbf{v} = \langle 0, 1, -1 \rangle$. Find
    (a) $\|\mathbf{u} + \mathbf{v}\|$ (b) $\|\mathbf{u}\| + \|\mathbf{v}\|$
    (c) $\|3\mathbf{u}\|$ (d) $\|2\mathbf{u} - 3\mathbf{v}\|$.

29. Let $\mathbf{u} = \mathbf{i} - 3\mathbf{j} + 2\mathbf{k}$, $\mathbf{v} = \mathbf{i} + \mathbf{j}$, and $\mathbf{w} = 2\mathbf{i} + 2\mathbf{j} - 4\mathbf{k}$. Find
    (a) $\|\mathbf{u} + \mathbf{v}\|$ (b) $\|\mathbf{u}\| + \|\mathbf{v}\|$
    (c) $\|-2\mathbf{u}\| + 2\|\mathbf{v}\|$ (d) $\|3\mathbf{u} - 5\mathbf{v} + \mathbf{w}\|$
    (e) $\dfrac{1}{\|\mathbf{w}\|}\mathbf{w}$ (f) $\left\|\dfrac{1}{\|\mathbf{w}\|}\mathbf{w}\right\|$.

30. Let $\mathbf{u} = \langle -1, 1 \rangle$, $\mathbf{v} = \langle 0, 1 \rangle$, and $\mathbf{w} = \langle 3, 4 \rangle$. Find the vector $\mathbf{x}$ that satisfies $\mathbf{u} - 2\mathbf{x} = \mathbf{x} - \mathbf{w} + 3\mathbf{v}$.

31. Let $\mathbf{u} = \langle 1, 3 \rangle$, $\mathbf{v} = \langle 2, 1 \rangle$, $\mathbf{w} = \langle 4, -1 \rangle$. Find the vector $\mathbf{x}$ that satisfies $2\mathbf{u} - \mathbf{v} + \mathbf{x} = 7\mathbf{x} + \mathbf{w}$.

32. Find $\mathbf{u}$ and $\mathbf{v}$ if $\mathbf{u} + \mathbf{v} = \langle 2, -3 \rangle$ and $3\mathbf{u} + 2\mathbf{v} = \langle -1, 2 \rangle$.

33. Find $\mathbf{u}$ and $\mathbf{v}$ if $\mathbf{u} + 2\mathbf{v} = 3\mathbf{i} - \mathbf{k}$ and $3\mathbf{u} - \mathbf{v} = \mathbf{i} + \mathbf{j} + \mathbf{k}$.

34. Give a geometric argument to show that if $\mathbf{u}$ and $\mathbf{v}$ are nonzero and $\mathbf{u}$ is not parallel to $\mathbf{v}$, then any vector $\mathbf{w}$ in the plane of $\mathbf{u}$ and $\mathbf{v}$ can be written as $\mathbf{w} = c_1\mathbf{u} + c_2\mathbf{v}$ for a suitable choice of c_1 and c_2.

35. Give a geometric argument to show that if $\mathbf{u}$, $\mathbf{v}$, and $\mathbf{w}$ are not coplanar, then any vector $\mathbf{z}$ can be written as $\mathbf{z} = c_1\mathbf{u} + c_2\mathbf{v} + c_3\mathbf{w}$ for a suitable choice of scalars c_1, c_2, and c_3.

36. Let $\mathbf{u} = 2\mathbf{i} - \mathbf{j}$ and $\mathbf{v} = 4\mathbf{i} + 2\mathbf{j}$. Find scalars c_1 and c_2 such that $c_1\mathbf{u} + c_2\mathbf{v} = -4\mathbf{j}$.

37. Let $\mathbf{u} = \langle 1, -3 \rangle$ and $\mathbf{v} = \langle -2, 6 \rangle$. Show that there do not exist scalars c_1 and c_2 such that $c_1\mathbf{u} + c_2\mathbf{v} = \langle 3, 5 \rangle$.

38. Let $\mathbf{u} = \langle 1, 0, 1 \rangle$, $\mathbf{v} = \langle 3, 2, 0 \rangle$, and $\mathbf{w} = \langle 0, 1, 1 \rangle$. Find scalars c_1, c_2, and c_3 such that
$$c_1\mathbf{u} + c_2\mathbf{v} + c_3\mathbf{w} = \langle -1, 1, 5 \rangle$$

39. Let $\mathbf{u} = \mathbf{i} - \mathbf{j}$, $\mathbf{v} = 3\mathbf{i} + \mathbf{k}$, and $\mathbf{w} = 4\mathbf{i} - \mathbf{j} + \mathbf{k}$. Show that

there do not exist scalars c_1, c_2, and c_3 such that
$$c_1\mathbf{u} + c_2\mathbf{v} + c_3\mathbf{w} = 2\mathbf{i} + \mathbf{j} - \mathbf{k}$$

40. Let $\mathbf{v} = 4\mathbf{i} - 3\mathbf{j}$. Find all scalars k such that $\|k\mathbf{v}\| = 3$.

41. Verify parts (b), (e), (f), and (g) of Theorem 14.2.6 for $\mathbf{u} = \langle 1, -3 \rangle$, $\mathbf{v} = \langle 6, 6 \rangle$, $\mathbf{w} = \langle -8, 1 \rangle$, $k = 3$, and $l = 6$.

42. Find a unit vector having the same direction as $-\mathbf{i} + 4\mathbf{j}$.

43. Find a unit vector oppositely directed to $3\mathbf{i} - 4\mathbf{j}$.

44. Find a unit vector having the same direction as $2\mathbf{i} - \mathbf{j} - 2\mathbf{k}$.

45. Find a unit vector oppositely directed to $6\mathbf{i} - 4\mathbf{j} + 2\mathbf{k}$.

46. Find a unit vector having the same direction as the vector from the point $A(-3, 2)$ to the point $B(1, -1)$.

47. Find a unit vector having the same direction as the vector from the point $A(-1, 0, 2)$ to the point $B(3, 1, 1)$.

48. Find a vector having the same direction as the vector $\mathbf{v} = -2\mathbf{i} + 3\mathbf{j}$ but with three times the length of $\mathbf{v}$.

49. Find a vector oppositely directed to $\mathbf{v} = \langle 3, -4 \rangle$ but with half the length of $\mathbf{v}$.

50. Find a vector with the same direction as $\mathbf{v} = \langle 7, 0, -6 \rangle$ but with twice the length of $\mathbf{v}$.

51. Find a vector oppositely directed to $\mathbf{v} = -3\mathbf{i} + 4\mathbf{j} + \mathbf{k}$ but with twice the length of $\mathbf{v}$.

52. Let $\mathbf{r} = \langle x, y \rangle$. Describe the set of points (x, y) for which $\|\mathbf{r}\| = 1$.

53. Let $\mathbf{r}_0 = \langle x_0, y_0 \rangle$ and $\mathbf{r} = \langle x, y \rangle$. Describe the set of all points (x, y) for which $\|\mathbf{r} - \mathbf{r}_0\| = 1$.

54. Let $\mathbf{r}_1 = \langle x_1, y_1 \rangle$, $\mathbf{r}_2 = \langle x_2, y_2 \rangle$, and $\mathbf{r} = \langle x, y \rangle$. Describe the set of all points (x, y) for which $\|\mathbf{r} - \mathbf{r}_1\| + \|\mathbf{r} - \mathbf{r}_2\| = k$, where $k > \|\mathbf{r}_2 - \mathbf{r}_1\|$.

55. Let $\mathbf{r}_0 = \langle x_0, y_0, z_0 \rangle$ and $\mathbf{r} = \langle x, y, z \rangle$. Describe the set of all points (x, y, z) for which
    (a) $\|\mathbf{r}\| = 2$ (b) $\|\mathbf{r} - \mathbf{r}_0\| = 3$
    (c) $\|\mathbf{r} - \mathbf{r}_0\| \le 1$.

56. Find two unit vectors in 2-space parallel to the line $y = 3x + 2$.

57. (a) Find two unit vectors in 2-space parallel to the line $x + y = 4$.
    (b) Find two unit vectors in 2-space perpendicular to the line in part (a).

58. Let P be the point $(2, 3)$ and Q the point $(7, -4)$. Use vectors to find the point on the line segment joining P and Q that is $\frac{3}{4}$ of the way from P to Q.

59. For the points P and Q in Exercise 58, use vectors to find the point on the line segment joining P and Q that is $\frac{3}{4}$ of the way from Q to P.

60. Find a unit vector in 2-space making an angle of $135°$ with the x-axis.

61. (a) Find a unit vector in 2-space making an angle of $\pi/3$ with the positive x-axis.

(b) Find a vector of length 4 in 2-space making an angle of $3\pi/4$ with the positive x-axis.

62. Use vectors to find the length of the diagonal of the parallelogram determined by $\mathbf{i} + \mathbf{j}$ and $\mathbf{i} - 2\mathbf{j}$.

63. Use vectors to find the fourth vertex of a parallelogram, three of whose vertices are $(0, 0)$, $(1, 3)$, and $(2, 4)$. [*Note:* There is more than one answer.]

64. Prove: $\|\mathbf{u} + \mathbf{v}\| \leq \|\mathbf{u}\| + \|\mathbf{v}\|$ geometrically.

65. Prove parts (a), (c), and (e) of Theorem 14.2.6 analytically in 2-space.

66. Prove parts (d), (g), and (h) of Theorem 14.2.6 analytically in 2-space.

67. Prove part (f) of Theorem 14.2.6 geometrically.

68. Use vectors to prove that the line segment joining the midpoints of two sides of a triangle is parallel to the third side and half as long.

69. Use vectors to prove that the midpoints of the sides of a quadrilateral are the vertices of a parallelogram.

70. Find $\overrightarrow{AB} + \overrightarrow{BC} + \overrightarrow{CA}$, where A, B, and C are any three distinct points.

In Exercises 71–73, let A, B, C, and D be any four points in 3-space.

---

71. If M is the midpoint of BC, show that $\overrightarrow{AB} + \overrightarrow{AC} = 2\overrightarrow{AM}$.

72. If M and N are the midpoints of AC and BD, show that $\overrightarrow{AB} + \overrightarrow{CD} = 2\overrightarrow{MN}$.

73. If M and N are the midpoints of AC and BD, show that $\overrightarrow{AB} + \overrightarrow{AD} + \overrightarrow{CB} + \overrightarrow{CD} = 4\overrightarrow{MN}$.

74. Let A and B be distinct points on a straight line L. If a point P different from B is on L, then $\overrightarrow{AP} = t\overrightarrow{PB}$ for some value of the scalar t. Let $\mathbf{a}$, $\mathbf{b}$, and $\mathbf{r}$ be vectors from the origin to the points A, B, and P, respectively. Show that $\mathbf{r} = (\mathbf{a} + t\mathbf{b})/(1 + t)$.

75. Let $\mathbf{r}_1, \mathbf{r}_2, \ldots, \mathbf{r}_n$ be vectors from the origin to points $P_1, P_2, \ldots, P_n$, respectively. The **centroid** of points $P_1, P_2, \ldots, P_n$ is defined as the point P for which

$$\sum_{k=1}^{n} \overrightarrow{PP_k} = \mathbf{0}.$$ Let $\mathbf{r}$ be the vector from the origin to P.

Show that $\mathbf{r} = \dfrac{1}{n} \displaystyle\sum_{k=1}^{n} \mathbf{r}_k$. [*Hint:* Write $\overrightarrow{PP_k}$ as a difference of vectors $\mathbf{r}$ and $\mathbf{r}_k$.]

76. Let $P_1, P_2, \ldots, P_n$ be consecutive vertices of a polygon in 2-space, all of whose interior angles are less than π. Let $\mathbf{r}_k$ be the vector from the origin to the point P_k for $k = 1, 2, \ldots, n$. From Exercise 75, the endpoint of the vector

$$\mathbf{r} = \frac{1}{n} \sum_{k=1}^{n} \mathbf{r}_k$$

drawn from the origin is the centroid of the points P_1, $P_2, \ldots, P_n$. It can be shown that the centroid is the balance point of the polygon.

(a) Find the centroid of the triangle with vertices $P_1(1, 1)$, $P_2(3, 3)$, and $P_3(5, 0)$.

(b) Cut the polygon described in part (a) out of cardboard and show that it balances when the tip of a pencil is placed at the centroid.

77. Repeat Exercise 76 for the quadrilateral with vertices $P_1(-1, 2)$, $P_2(2, 3)$, $P_3(5, -2)$, and $P_4(0, -1)$.

■ **14.3 DOT PRODUCT; PROJECTIONS**

> *In this section we shall introduce a type of multiplication of vectors in 2-space and 3-space. We shall also discuss the arithmetic properties of this multiplication and give some of its applications.*

☐ ANGLE BETWEEN
VECTORS

Let $\mathbf{u}$ and $\mathbf{v}$ be two nonzero vectors in 2-space or 3-space, and assume these vectors have been positioned so that their initial points coincide. By the **angle between** $\mathbf{u}$ *and* $\mathbf{v}$, we shall mean the angle θ determined by $\mathbf{u}$ and $\mathbf{v}$ that satisfies $0 \leq \theta \leq \pi$ (Figure 14.3.1).

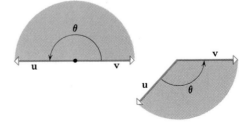

Figure 14.3.1

14.3.1 DEFINITION. If **u** and **v** are vectors in 2-space or 3-space and θ is the angle between **u** and **v**, then the ***dot product*** or ***Euclidean inner product*** $\mathbf{u} \cdot \mathbf{v}$ is defined by

$$\mathbf{u} \cdot \mathbf{v} = \begin{cases} \|\mathbf{u}\| \|\mathbf{v}\| \cos \theta, & \text{if } \mathbf{u} \neq \mathbf{0} \text{ and } \mathbf{v} \neq \mathbf{0} \\ 0, & \text{if } \mathbf{u} = \mathbf{0} \text{ or } \mathbf{v} = \mathbf{0} \end{cases}$$

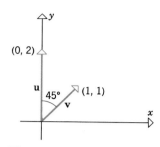

Figure 14.3.2

Example 1 As shown in Figure 14.3.2, the angle between vectors $\mathbf{u} = \langle 0, 2 \rangle$ and $\mathbf{v} = \langle 1, 1 \rangle$ is 45°. Thus,

$$\mathbf{u} \cdot \mathbf{v} = \|\mathbf{u}\| \|\mathbf{v}\| \cos \theta = \sqrt{0^2 + 2^2} \sqrt{1^2 + 1^2} \cos 45° = (2)(\sqrt{2}) \frac{1}{\sqrt{2}} = 2 \quad \blacktriangleleft$$

☐ **FORMULA FOR THE DOT PRODUCT**

For purposes of computation, it is desirable to have a formula that expresses the dot product of two vectors in terms of the components of the vectors. We shall derive such a formula for vectors in 3-space and just state the corresponding formula for vectors in 2-space.

Let $\mathbf{u} = \langle u_1, u_2, u_3 \rangle$ and $\mathbf{v} = \langle v_1, v_2, v_3 \rangle$ be two nonzero vectors. If, as in Figure 14.3.3, θ is the angle between **u** and **v**, then the law of cosines yields

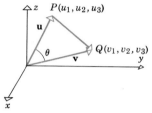

Figure 14.3.3

$$\|\overrightarrow{PQ}\|^2 = \|\mathbf{u}\|^2 + \|\mathbf{v}\|^2 - 2\|\mathbf{u}\| \|\mathbf{v}\| \cos \theta \tag{1}$$

Since $\overrightarrow{PQ} = \mathbf{v} - \mathbf{u}$, we can rewrite (1) as

$$\|\mathbf{u}\| \|\mathbf{v}\| \cos \theta = \tfrac{1}{2}(\|\mathbf{u}\|^2 + \|\mathbf{v}\|^2 - \|\mathbf{v} - \mathbf{u}\|^2)$$

or

$$\mathbf{u} \cdot \mathbf{v} = \tfrac{1}{2}(\|\mathbf{u}\|^2 + \|\mathbf{v}\|^2 - \|\mathbf{v} - \mathbf{u}\|^2)$$

Substituting

$$\|\mathbf{u}\|^2 = u_1^2 + u_2^2 + u_3^2, \quad \|\mathbf{v}\|^2 = v_1^2 + v_2^2 + v_3^2$$

and

$$\|\mathbf{v} - \mathbf{u}\|^2 = (v_1 - u_1)^2 + (v_2 - u_2)^2 + (v_3 - u_3)^2$$

we obtain, after simplifying,

$$\mathbf{u} \cdot \mathbf{v} = u_1 v_1 + u_2 v_2 + u_3 v_3 \tag{2a}$$

This formula also holds if $\mathbf{u} = \mathbf{0}$ or $\mathbf{v} = \mathbf{0}$. If $\mathbf{u} = \langle u_1, u_2 \rangle$ and $\mathbf{v} = \langle v_1, v_2 \rangle$ are two vectors in 2-space, then the formula corresponding to (2a) is

$$\mathbf{u} \cdot \mathbf{v} = u_1 v_1 + u_2 v_2 \tag{2b}$$

If **u** and **v** are nonzero vectors, the formula in Definition 14.3.1 can be written as

$$\cos \theta = \frac{\mathbf{u} \cdot \mathbf{v}}{\|\mathbf{u}\| \|\mathbf{v}\|} \tag{3}$$

Example 2 Consider the vectors

$$\mathbf{u} = 2\mathbf{i} - \mathbf{j} + \mathbf{k} \quad \text{and} \quad \mathbf{v} = \mathbf{i} + \mathbf{j} + 2\mathbf{k}$$

Find $\mathbf{u} \cdot \mathbf{v}$ and determine the angle θ between **u** and **v**.

Solution.

$$\mathbf{u} \cdot \mathbf{v} = u_1 v_1 + u_2 v_2 + u_3 v_3 = (2)(1) + (-1)(1) + (1)(2) = 3$$

For the given vectors, $\|\mathbf{u}\| = \|\mathbf{v}\| = \sqrt{6}$, so that

$$\cos\theta = \frac{3}{\sqrt{6}\sqrt{6}} = \frac{1}{2}$$

Thus, $\theta = 60°$ ◄

Example 3 Find the angle between a diagonal of a cube and one of its edges.

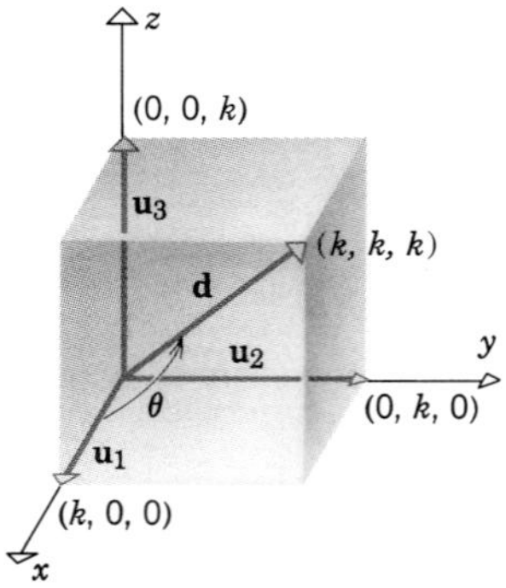

Solution. Let k be the length of an edge and let us introduce a coordinate system as shown in Figure 14.3.4.

If we let $\mathbf{u}_1 = \langle k, 0, 0\rangle$, $\mathbf{u}_2 = \langle 0, k, 0\rangle$, and $\mathbf{u}_3 = \langle 0, 0, k\rangle$, then the vector

$$\mathbf{d} = \langle k, k, k\rangle = \mathbf{u}_1 + \mathbf{u}_2 + \mathbf{u}_3$$

is a diagonal of the cube. The angle θ between $\mathbf{d}$ and the edge $\mathbf{u}_1$ satisfies

$$\cos\theta = \frac{\mathbf{u}_1 \cdot \mathbf{d}}{\|\mathbf{u}_1\|\|\mathbf{d}\|} = \frac{k^2}{(k)(\sqrt{3k^2})} = \frac{1}{\sqrt{3}}$$

The same results hold for $\mathbf{u}_2$ and $\mathbf{u}_3$. Thus, with the help of a calculator

$$\theta = \cos^{-1} 1/\sqrt{3} \approx 54°44' \quad ◄$$

The sign of the dot product provides useful information about the angle between two vectors.

14.3.2 THEOREM. *If $\mathbf{u}$ and $\mathbf{v}$ are nonzero vectors in 2-space or 3-space, and if θ is the angle between them, then*

θ is acute	if and only if	$\mathbf{u} \cdot \mathbf{v} > 0$
θ is obtuse	if and only if	$\mathbf{u} \cdot \mathbf{v} < 0$
$\theta = \pi/2$	if and only if	$\mathbf{u} \cdot \mathbf{v} = 0$

Proof. Since $\mathbf{u}$ and $\mathbf{v}$ are nonzero vectors, $\|\mathbf{u}\| > 0$ and $\|\mathbf{v}\| > 0$. Thus,

$$\mathbf{u} \cdot \mathbf{v} = \|\mathbf{u}\|\|\mathbf{v}\|\cos\theta$$

is positive, negative, or zero according to whether $\cos\theta$ is positive, negative, or zero. Since $0 \leq \theta \leq \pi$, it follows that θ is acute if and only if $\cos\theta > 0$; θ is obtuse if and only if $\cos\theta < 0$; and $\theta = \pi/2$ if and only if $\cos\theta = 0$. ∎

Example 4 If $\mathbf{u} = \mathbf{i} - 2\mathbf{j} + 3\mathbf{k}$, $\mathbf{v} = -3\mathbf{i} + 4\mathbf{j} + 2\mathbf{k}$, and $\mathbf{w} = 3\mathbf{i} + 6\mathbf{j} + 3\mathbf{k}$, then

$$\mathbf{u} \cdot \mathbf{v} = (1)(-3) + (-2)(4) + (3)(2) = -5$$

$$\mathbf{v} \cdot \mathbf{w} = (-3)(3) + (4)(6) + (2)(3) = 21$$

$$\mathbf{u} \cdot \mathbf{w} = (1)(3) + (-2)(6) + (3)(3) = 0$$

Therefore $\mathbf{u}$ and $\mathbf{v}$ make an obtuse angle, $\mathbf{v}$ and $\mathbf{w}$ make an acute angle, and $\mathbf{u}$ and $\mathbf{w}$ are perpendicular. ◄

□ **ORTHOGONAL VECTORS**

Perpendicular vectors are also called ***orthogonal*** vectors. In light of Theorem 14.3.2, two nonzero vectors are orthogonal if and only if their dot product is zero. If we agree to consider $\mathbf{u}$ and $\mathbf{v}$ to be perpendicular when either or both of these vectors is $\mathbf{0}$, then we can state without exception that two vectors $\mathbf{u}$ and $\mathbf{v}$ are orthogonal (perpendicular) if and only if $\mathbf{u} \cdot \mathbf{v} = 0$.

Figure 14.3.4

□ DIRECTION COSINES

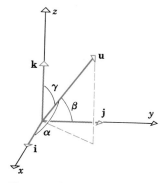

Figure 14.3.5

Of special interest are the angles α, β, and γ that a vector **u** in 3-space makes with the vectors **i**, **j**, and **k** (Figure 14.3.5). These are called the *direction angles* of **u**. The numbers $\cos \alpha$, $\cos \beta$, and $\cos \gamma$ are called the *direction cosines* of **u**. Formulas for the direction cosines follow easily from (3).

14.3.3   THEOREM.   *The three direction cosines of a nonzero vector* $\mathbf{u} = u_1\mathbf{i} + u_2\mathbf{j} + u_3\mathbf{k}$ *in 3-space are*

$$\cos \alpha = \frac{u_1}{\|\mathbf{u}\|}, \quad \cos \beta = \frac{u_2}{\|\mathbf{u}\|}, \quad \cos \gamma = \frac{u_3}{\|\mathbf{u}\|}$$

Proof.   Since $\mathbf{u} \cdot \mathbf{i} = (u_1)(1) + (u_2)(0) + (u_3)(0) = u_1$, it follows that

$$\cos \alpha = \frac{\mathbf{u} \cdot \mathbf{i}}{\|\mathbf{u}\|\,\|\mathbf{i}\|} = \frac{u_1}{\|\mathbf{u}\|}$$

Similarly for $\cos \beta$ and $\cos \gamma$.   ∎

The direction cosines of a vector $\mathbf{u} = u_1\mathbf{i} + u_2\mathbf{j} + u_3\mathbf{k}$ can be obtained by simply reading off the components of the unit vector $\mathbf{u}/\|\mathbf{u}\|$ since

$$\frac{\mathbf{u}}{\|\mathbf{u}\|} = \frac{u_1}{\|\mathbf{u}\|}\mathbf{i} + \frac{u_2}{\|\mathbf{u}\|}\mathbf{j} + \frac{u_3}{\|\mathbf{u}\|}\mathbf{k} = (\cos \alpha)\mathbf{i} + (\cos \beta)\mathbf{j} + (\cos \gamma)\mathbf{k}$$

Example 5   Find the direction cosines of the vector $\mathbf{u} = 2\mathbf{i} - 4\mathbf{j} + 4\mathbf{k}$, and approximate the direction angles to the nearest degree.

Solution. $\|\mathbf{u}\| = \sqrt{4 + 16 + 16} = 6$, so that $\mathbf{u}/\|\mathbf{u}\| = \frac{1}{3}\mathbf{i} - \frac{2}{3}\mathbf{j} + \frac{2}{3}\mathbf{k}$. Thus,

$$\cos \alpha = \tfrac{1}{3}, \quad \cos \beta = -\tfrac{2}{3}, \quad \cos \gamma = \tfrac{2}{3}$$

With the help of a calculator that can compute inverse trigonometric functions, one obtains

$$\alpha = \cos^{-1}\left(\tfrac{1}{3}\right) \approx 71°, \quad \beta = \cos^{-1}\left(-\tfrac{2}{3}\right) \approx 132°, \quad \gamma = \cos^{-1}\left(\tfrac{2}{3}\right) \approx 48° \quad \blacktriangleleft$$

□ PROPERTIES OF THE DOT PRODUCT

The following arithmetic properties of the dot product are useful in calculations involving vectors.

14.3.4   THEOREM.   *If* **u**, **v**, *and* **w** *are vectors in 2- or 3-space and* k *is a scalar, then*

(*a*) $\mathbf{u} \cdot \mathbf{v} = \mathbf{v} \cdot \mathbf{u}$
(*b*) $\mathbf{u} \cdot (\mathbf{v} + \mathbf{w}) = \mathbf{u} \cdot \mathbf{v} + \mathbf{u} \cdot \mathbf{w}$
(*c*) $k(\mathbf{u} \cdot \mathbf{v}) = (k\mathbf{u}) \cdot \mathbf{v} = \mathbf{u} \cdot (k\mathbf{v})$
(*d*) $\mathbf{v} \cdot \mathbf{v} = \|\mathbf{v}\|^2$

We shall prove parts (*c*) and (*d*) for vectors in 3-space and omit the remaining proofs.

Proof (c).   Let $\mathbf{u} = \langle u_1, u_2, u_3 \rangle$ and $\mathbf{v} = \langle v_1, v_2, v_3 \rangle$; then

$$k(\mathbf{u} \cdot \mathbf{v}) = k(u_1v_1 + u_2v_2 + u_3v_3) = (ku_1)v_1 + (ku_2)v_2 + (ku_3)v_3 = (k\mathbf{u}) \cdot \mathbf{v}$$

Similarly, $k(\mathbf{u} \cdot \mathbf{v}) = \mathbf{u} \cdot (k\mathbf{v})$.

Proof (d). $\mathbf{v} \cdot \mathbf{v} = v_1v_1 + v_2v_2 + v_3v_3 = v_1{}^2 + v_2{}^2 + v_3{}^2 = \|\mathbf{v}\|^2$.   ∎

Part (*d*) of the last theorem is sometimes expressed in the following alternative form:

$$\|\mathbf{v}\| = \sqrt{\mathbf{v} \cdot \mathbf{v}} \tag{4}$$

□ **ORTHOGONAL PROJECTIONS OF VECTORS**

In many applications it is of interest to "decompose" a vector $\mathbf{u}$ into a sum of two vectors, one parallel to a specified nonzero vector $\mathbf{b}$ and the other perpendicular to $\mathbf{b}$. If $\mathbf{u}$ and $\mathbf{b}$ are positioned so that their initial points coincide at a point Q, we may decompose the vector $\mathbf{u}$ as follows (Figure 14.3.6): Drop a perpendicular from the tip of $\mathbf{u}$ to the line through $\mathbf{b}$ and construct the vector $\mathbf{w}_1$ from Q to the foot of this perpendicular; next, form the difference

$$\mathbf{w}_2 = \mathbf{u} - \mathbf{w}_1$$

As indicated in Figure 14.3.6, the vector $\mathbf{w}_1$ is parallel to $\mathbf{b}$, the vector $\mathbf{w}_2$ is perpendicular to $\mathbf{b}$, and

$$\mathbf{u} = \mathbf{w}_1 + \mathbf{w}_2 = \mathbf{w}_1 + (\mathbf{u} - \mathbf{w}_1) \tag{5}$$

Figure 14.3.6

The vector $\mathbf{w}_1$ is called the ***vector component of $\mathbf{u}$ along*** $\mathbf{b}$, and the vector $\mathbf{w}_2$ is called the ***vector component of $\mathbf{u}$ orthogonal to*** $\mathbf{b}$. The vector component of $\mathbf{u}$ along $\mathbf{b}$ is also called the ***orthogonal projection of $\mathbf{u}$ on*** $\mathbf{b}$ and is denoted by $\text{proj}_{\mathbf{b}}\,\mathbf{u}$. With this notation the decomposition in (5) can be expressed as

$$\mathbf{u} = \text{proj}_{\mathbf{b}}\,\mathbf{u} + (\mathbf{u} - \text{proj}_{\mathbf{b}}\,\mathbf{u})$$

Vector component + along $\mathbf{b}$ Vector component orthogonal to $\mathbf{b}$

The following theorem gives formulas for calculating $\text{proj}_{\mathbf{b}}\,\mathbf{u}$ and $\mathbf{u} - \text{proj}_{\mathbf{b}}\,\mathbf{u}$.

14.3.5 THEOREM.   *If $\mathbf{u}$ and $\mathbf{b}$ are vectors in 2-space or 3-space and if $\mathbf{b} \neq \mathbf{0}$, then*

$$\text{proj}_{\mathbf{b}}\,\mathbf{u} = \frac{\mathbf{u} \cdot \mathbf{b}}{\|\mathbf{b}\|^2}\mathbf{b} \qquad (\textit{vector component of }\mathbf{u}\textit{ along }\mathbf{b}) \tag{6}$$

$$\mathbf{u} - \text{proj}_{\mathbf{b}}\,\mathbf{u} = \mathbf{u} - \frac{\mathbf{u} \cdot \mathbf{b}}{\|\mathbf{b}\|^2}\mathbf{b} \qquad \begin{array}{l}(\textit{ vector component of }\mathbf{u}\\ \textit{orthogonal to }\mathbf{b})\end{array} \tag{7}$$

Proof.   We shall begin by expressing the vector $\text{proj}_{\mathbf{b}}\,\mathbf{u}$ as its length times a unit vector in the same direction. If θ is the angle between $\mathbf{u}$ and $\mathbf{b}$, then a unit vector in the same direction as $\text{proj}_{\mathbf{b}}\,\mathbf{u}$ is $\mathbf{b}/\|\mathbf{b}\|$ in the case where $0 \leq \theta \leq \pi/2$ or $-\mathbf{b}/\|\mathbf{b}\|$ in the case where $\pi/2 < \theta \leq \pi$ (Figure 14.3.7). From this and Theorem 14.3.2 it follows that

$$\text{proj}_{\mathbf{b}}\,\mathbf{u} = \begin{cases} \|\text{proj}_{\mathbf{b}}\,\mathbf{u}\|\,(\mathbf{b}/\|\mathbf{b}\|) & \text{if} \quad \mathbf{u} \cdot \mathbf{b} \geq 0 \\ \|\text{proj}_{\mathbf{b}}\,\mathbf{u}\|\,(-\mathbf{b}/\|\mathbf{b}\|) & \text{if} \quad \mathbf{u} \cdot \mathbf{b} < 0 \end{cases} \tag{8}$$

Again referring to Figure 14.3.7, it follows that

$$\|\text{proj}_{\mathbf{b}}\,\mathbf{u}\| = \|\mathbf{u}\|\,|\cos\theta| \tag{9}$$

or on multiplying by $\|\mathbf{b}\|/\|\mathbf{b}\|$

$$\|\text{proj}_{\mathbf{b}}\,\mathbf{u}\| = \frac{\|\mathbf{u}\|\,\|\mathbf{b}\|\,|\cos\theta|}{\|\mathbf{b}\|} = \frac{|\mathbf{u} \cdot \mathbf{b}|}{\|\mathbf{b}\|} \tag{10}$$

Substituting this formula in (8) yields (6) in all cases (verify). Formula (7) follows directly from (6). ∎

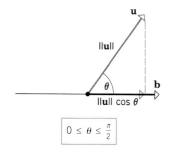

$$0 \leq \theta \leq \frac{\pi}{2}$$

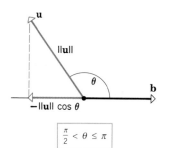

$$\frac{\pi}{2} < \theta \leq \pi$$

Figure 14.3.7

Example 6   Let $\mathbf{u} = 2\mathbf{i} - \mathbf{j} + 3\mathbf{k}$ and $\mathbf{b} = 4\mathbf{i} - \mathbf{j} + 2\mathbf{k}$. Find the vector component of $\mathbf{u}$ along $\mathbf{b}$ and the vector component of $\mathbf{u}$ orthogonal to $\mathbf{b}$.

Solution.

$$\mathbf{u} \cdot \mathbf{b} = (2)(4) + (-1)(-1) + (3)(2) = 15$$

$$\|\mathbf{b}\|^2 = 4^2 + (-1)^2 + 2^2 = 21$$

Thus, the vector component of **u** along **b** is

$$\text{proj}_{\mathbf{b}} \mathbf{u} = \frac{\mathbf{u} \cdot \mathbf{b}}{\|\mathbf{b}\|^2} \mathbf{b} = \frac{15}{21} (4\mathbf{i} - \mathbf{j} + 2\mathbf{k}) = \frac{20}{7}\mathbf{i} - \frac{5}{7}\mathbf{j} + \frac{10}{7}\mathbf{k}$$

and the vector component of **u** orthogonal to **b** is

$$\mathbf{u} - \text{proj}_{\mathbf{b}} \mathbf{u} = (2\mathbf{i} - \mathbf{j} + 3\mathbf{k}) - \left(\frac{20}{7}\mathbf{i} - \frac{5}{7}\mathbf{j} + \frac{10}{7}\mathbf{k} \right) = -\frac{6}{7}\mathbf{i} - \frac{2}{7}\mathbf{j} + \frac{11}{7}\mathbf{k}$$

As a check, the reader may wish to verify that the vectors $\mathbf{u} - \text{proj}_{\mathbf{b}} \mathbf{u}$ and **b** are perpendicular by showing that their dot product is zero. ◀

☐ **WORK**

It follows from Section 6.7 that the work W done by a constant force **F** of magnitude $\|\mathbf{F}\|$ acting in the direction of motion on a particle moving from P to Q on a line is

$$W = (\text{force}) \times (\text{distance}) = \|\mathbf{F}\| \|\overrightarrow{PQ}\|$$

If the force **F** is constant, but makes an angle θ with the direction of motion (Figure 14.3.8), then we *define* the work done by **F** to be

$$W = (\|\mathbf{F}\| \cos \theta) \|\overrightarrow{PQ}\| = \mathbf{F} \cdot \overrightarrow{PQ} \tag{11}$$

The quantity $\|\mathbf{F}\| \cos \theta$ is the "component" of force in the direction of motion and $\|\overrightarrow{PQ}\|$ is the distance traveled by the particle.

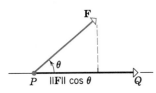

Figure 14.3.8

Example 7 A wagon is pulled horizontally by exerting a continual force of 10 lb on the handle at an angle of 60° with the horizontal. How much work is done in moving the wagon 50 ft?

Solution. Introduce an xy-coordinate system so that the wagon moves from $P(0, 0)$ to $Q(50, 0)$ along the x-axis (Figure 14.3.9). In this coordinate system

$$\overrightarrow{PQ} = 50\mathbf{i}$$

and

$$\mathbf{F} = (10 \cos 60°)\mathbf{i} + (10 \sin 60°)\mathbf{j} = 5\mathbf{i} + 5\sqrt{3}\mathbf{j}$$

so that the work done is

$$W = \mathbf{F} \cdot \overrightarrow{PQ} = (5\mathbf{i} + 5\sqrt{3}\mathbf{j}) \cdot (50\mathbf{i}) = 250 \text{ (foot-pounds)} ◀$$

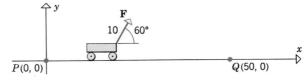

Figure 14.3.9

▶ **Exercise Set 14.3**

1. Find $\mathbf{u} \cdot \mathbf{v}$.
    (a) $\mathbf{u} = \mathbf{i} + 2\mathbf{j}$, $\mathbf{v} = 6\mathbf{i} - 8\mathbf{j}$
    (b) $\mathbf{u} = \langle -7, -3 \rangle$, $\mathbf{v} = \langle 0, 1 \rangle$
    (c) $\mathbf{u} = \mathbf{i} - 3\mathbf{j} + 7\mathbf{k}$, $\mathbf{v} = 8\mathbf{i} - 2\mathbf{j} - 2\mathbf{k}$
    (d) $\mathbf{u} = \langle -3, 1, 2 \rangle$, $\mathbf{v} = \langle 4, 2, -5 \rangle$.

2. In each part of Exercise 1, find the cosine of the angle θ between **u** and **v**.

3. Determine whether **u** and **v** make an acute angle, an obtuse angle, or are orthogonal.
    (a) $\mathbf{u} = 7\mathbf{i} + 3\mathbf{j} + 5\mathbf{k}$, $\mathbf{v} = -8\mathbf{i} + 4\mathbf{j} + 2\mathbf{k}$

(b) $\mathbf{u} = 6\mathbf{i} + \mathbf{j} + 3\mathbf{k}$, $\mathbf{v} = 4\mathbf{i} - 6\mathbf{k}$

(c) $\mathbf{u} = \langle 1, 1, 1 \rangle$, $\mathbf{v} = \langle -1, 0, 0 \rangle$

(d) $\mathbf{u} = \langle 4, 1, 6 \rangle$, $\mathbf{v} = \langle -3, 0, 2 \rangle$.

4.   Find the orthogonal projection of $\mathbf{u}$ on $\mathbf{a}$.

(a) $\mathbf{u} = 2\mathbf{i} + \mathbf{j}$, $\mathbf{a} = -3\mathbf{i} + 2\mathbf{j}$

(b) $\mathbf{u} = \langle 2, 6 \rangle$, $\mathbf{a} = \langle -9, 3 \rangle$

(c) $\mathbf{u} = -7\mathbf{i} + \mathbf{j} + 3\mathbf{k}$, $\mathbf{a} = 5\mathbf{i} + \mathbf{k}$

(d) $\mathbf{u} = \langle 0, 0, 1 \rangle$, $\mathbf{a} = \langle 8, 3, 4 \rangle$.

5.   In each part of Exercise 4, find the vector component of $\mathbf{u}$ orthogonal to $\mathbf{a}$.

6.   Find $\|\text{proj}_{\mathbf{a}}\, \mathbf{u}\|$.

(a) $\mathbf{u} = 2\mathbf{i} - \mathbf{j}$, $\mathbf{a} = 3\mathbf{i} + 4\mathbf{j}$

(b) $\mathbf{u} = \langle 4, 5 \rangle$, $\mathbf{a} = \langle 1, -2 \rangle$

(c) $\mathbf{u} = 2\mathbf{i} - \mathbf{j} + 3\mathbf{k}$, $\mathbf{a} = \mathbf{i} + 2\mathbf{j} + 2\mathbf{k}$

(d) $\mathbf{u} = \langle 4, -1, 7 \rangle$, $\mathbf{a} = \langle 2, 3, -6 \rangle$.

7.   Verify part (c) of Theorem 14.3.4 for $\mathbf{u} = 6\mathbf{i} - \mathbf{j} + 2\mathbf{k}$, $\mathbf{v} = 2\mathbf{i} + 7\mathbf{j} + 4\mathbf{k}$, and $k = -5$.

8.   Find two vectors in 2-space of norm 1 that are orthogonal to $3\mathbf{i} - 2\mathbf{j}$.

9.   Let $\mathbf{u} = \langle 1, 2 \rangle$, $\mathbf{v} = \langle 4, -2 \rangle$, and $\mathbf{w} = \langle 6, 0 \rangle$. Find

(a) $\mathbf{u} \cdot (7\mathbf{v} + \mathbf{w})$ (b) $\|(\mathbf{u} \cdot \mathbf{w})\mathbf{w}\|$

(c) $\|\mathbf{u}\|(\mathbf{v} \cdot \mathbf{w})$ (d) $(\|\mathbf{u}\|\mathbf{v}) \cdot \mathbf{w}$.

10.   Explain why each of the following expressions makes no sense.

(a) $\mathbf{u} \cdot (\mathbf{v} \cdot \mathbf{w})$ (b) $(\mathbf{u} \cdot \mathbf{v}) + \mathbf{w}$

(c) $\|\mathbf{u} \cdot \mathbf{v}\|$ (d) $k \cdot (\mathbf{u} + \mathbf{v})$.

11.   Use vectors to find the cosines of the interior angles of the triangle with vertices $(-1, 0)$, $(2, -1)$, and $(1, 4)$.

12.   Find two unit vectors in 2-space that make an angle of $45°$ with $4\mathbf{i} + 3\mathbf{j}$.

13.   Show that $A(2, -1, 1)$, $B(3, 2, -1)$, and $C(7, 0, -2)$ are vertices of a right triangle. At which vertex is the right angle?

14.   Find k so that the vector from the point $A(1, -1, 3)$ to the point $B(3, 0, 5)$ is perpendicular to the vector from A to the point $P(k, k, k)$.

15.   Let $\mathbf{r_0} = \langle x_0, y_0 \rangle$ and $\mathbf{r} = \langle x, y \rangle$. Describe the set of all points (x, y) for which

(a) $\mathbf{r} \cdot \mathbf{r_0} = 0$ (b) $(\mathbf{r} - \mathbf{r_0}) \cdot \mathbf{r_0} = 0$

(c) $\mathbf{r} \cdot (\mathbf{r} - \mathbf{r_0}) = 0$.

16.   Suppose that $\mathbf{a} \cdot \mathbf{b} = \mathbf{a} \cdot \mathbf{c}$ and $\mathbf{a} \neq \mathbf{0}$. Does it follow that $\mathbf{b} = \mathbf{c}$? Explain.

17.   Let $\mathbf{a} = k\mathbf{i} + \mathbf{j}$ and $\mathbf{b} = 4\mathbf{i} + 3\mathbf{j}$. Find k so that

(a) $\mathbf{a}$ and $\mathbf{b}$ are orthogonal

(b)   the angle between $\mathbf{a}$ and $\mathbf{b}$ is $\pi/4$

(c)   the angle between $\mathbf{a}$ and $\mathbf{b}$ is $\pi/6$

(d) $\mathbf{a}$ and $\mathbf{b}$ are parallel.

18.   Find the direction cosines of $\mathbf{u}$ and estimate the direction angles to the nearest degree.

(a) $\mathbf{u} = \mathbf{i} + \mathbf{j} - \mathbf{k}$ (b) $\mathbf{u} = 2\mathbf{i} - 2\mathbf{j} + \mathbf{k}$

(c) $\mathbf{u} = 3\mathbf{i} - 2\mathbf{j} - 6\mathbf{k}$ (d) $\mathbf{u} = 3\mathbf{i} - 4\mathbf{k}$.

19.   Prove: The direction cosines of a vector satisfy the equation $\cos^2 \alpha + \cos^2 \beta + \cos^2 \gamma = 1$.

20.   Prove: Two nonzero vectors $\mathbf{u_1}$ and $\mathbf{u_2}$ are perpendicular if and only if their direction cosines satisfy

$$\cos \alpha_1 \cos \alpha_2 + \cos \beta_1 \cos \beta_2 + \cos \gamma_1 \cos \gamma_2 = 0$$

21.   Given the points $A(2, -3)$, $B(5, 1)$, and $P(1, 0)$,

(a)   find $\|\text{proj}_{\overrightarrow{AB}}\, \overrightarrow{AP}\|$

(b)   use the Pythagorean Theorem and the result of part (a) to find the distance from P to the line through A and B.

22.   Follow the directions of Exercise 21 for the points $A(1, 1, 0)$, $B(-2, 3, -4)$, and $P(-3, 1, 2)$.

23.   Follow the directions of Exercise 21 for the points $A(2, 1, -3)$, $B(0, 2, -1)$, and $P(4, 3, 0)$.

24.   A boat travels 100 meters due north while the wind exerts a force of 50 newtons toward the northeast. How much work does the wind do?

25.   Find the work done by a force $\mathbf{F} = -3\mathbf{j}$ (pounds) applied to a point that moves on a line from $(1, 3)$ to $(4, 7)$. Assume that distance is measured in feet.

26.   Let $\mathbf{u}$ and $\mathbf{v}$ determine a parallelogram. Use vectors to prove that the diagonals of the parallelogram are perpendicular if and only if the sides are equal in length.

27.   Let $\mathbf{u}$ and $\mathbf{v}$ determine a parallelogram. Use vectors to prove that the parallelogram is a rectangle if and only if the diagonals are equal in length.

28.   Prove: $\|\mathbf{u} + \mathbf{v}\|^2 + \|\mathbf{u} - \mathbf{v}\|^2 = 2\|\mathbf{u}\|^2 + 2\|\mathbf{v}\|^2$.

29.   Prove: $\mathbf{u} \cdot \mathbf{v} = \frac{1}{4}\|\mathbf{u} + \mathbf{v}\|^2 - \frac{1}{4}\|\mathbf{u} - \mathbf{v}\|^2$.

30.   Find, to the nearest degree, the angle between the diagonal of a cube and a diagonal of one of its faces.

31.   Find, to the nearest degree, the acute angle formed by two diagonals of a cube.

32.   Find, to the nearest degree, the angles that a diagonal of a box with dimensions 10 in. by 15 in. by 25 in. makes with the edges of the box.

33.   Prove: If vectors $\mathbf{v_1}$, $\mathbf{v_2}$, and $\mathbf{v_3}$ are nonzero and mutually perpendicular, then any vector $\mathbf{v}$ can be written as

$$\mathbf{v} = c_1\mathbf{v_1} + c_2\mathbf{v_2} + c_3\mathbf{v_3}$$

where $c_i = (\mathbf{v} \cdot \mathbf{v_i})/\|\mathbf{v_i}\|^2$, $i = 1, 2, 3$.

34.   Show that the three vectors

$$\mathbf{v_1} = 3\mathbf{i} - \mathbf{j} + 2\mathbf{k}, \quad \mathbf{v_2} = \mathbf{i} + \mathbf{j} - \mathbf{k}, \quad \mathbf{v_3} = \mathbf{i} - 5\mathbf{j} - 4\mathbf{k}$$

are mutually perpendicular. Use the result of Exercise 33 to find scalars c_1, c_2, and c_3 so that

$$c_1\mathbf{v_1} + c_2\mathbf{v_2} + c_3\mathbf{v_3} = \mathbf{i} - \mathbf{j} + \mathbf{k}$$

35. Prove: If $\mathbf{v}$ is orthogonal to $\mathbf{w}_1$ and $\mathbf{w}_2$, then $\mathbf{v}$ is orthogonal to $k_1\mathbf{w}_1 + k_2\mathbf{w}_2$ for all scalars k_1 and k_2.

36. Let $\mathbf{u}$ and $\mathbf{v}$ be nonzero vectors, and let $k = \|\mathbf{u}\|$ and $l = \|\mathbf{v}\|$. Prove that

$$\mathbf{w} = l\mathbf{u} + k\mathbf{v}$$

bisects the angle between $\mathbf{u}$ and $\mathbf{v}$.

■ 14.4 CROSS PRODUCT

In many applications of vectors to problems in geometry, physics, and engineering, it is of interest to construct a vector in 3-space that is perpendicular to two given vectors. In this section we shall introduce a type of vector multiplication that facilitates this construction.

□ **DETERMINANTS**

Determinants are functions that assign numerical values to square arrays of numbers. For example, if a_1, a_2, b_1, and b_2 are real numbers, then a **2 × 2 determinant** is defined by

$$\begin{vmatrix} a_1 & a_2 \\ b_1 & b_2 \end{vmatrix} = a_1 b_2 - a_2 b_1$$

For example,

$$\begin{vmatrix} 3 & -2 \\ 4 & 5 \end{vmatrix} = (3)(5) - (-2)(4) = 15 + 8 = 23$$

A **3 × 3 determinant** is defined in terms of 2 × 2 determinants by

$$\begin{vmatrix} a_1 & a_2 & a_3 \\ b_1 & b_2 & b_3 \\ c_1 & c_2 & c_3 \end{vmatrix} = a_1 \begin{vmatrix} b_2 & b_3 \\ c_2 & c_3 \end{vmatrix} - a_2 \begin{vmatrix} b_1 & b_3 \\ c_1 & c_3 \end{vmatrix} + a_3 \begin{vmatrix} b_1 & b_2 \\ c_1 & c_2 \end{vmatrix}$$

The right side of this formula is easily remembered by noting that a_1, a_2, and a_3 are the entries in the first "row" of the left side, and the 2 × 2 determinants on the right side arise by deleting the first row and an appropriate column from the left side. The pattern is as follows:

$$\begin{vmatrix} a_1 & a_2 & a_3 \\ b_1 & b_2 & b_3 \\ c_1 & c_2 & c_3 \end{vmatrix} = a_1 \begin{vmatrix} a_1 & a_2 & a_3 \\ b_1 & b_2 & b_3 \\ c_1 & c_2 & c_3 \end{vmatrix} - a_2 \begin{vmatrix} a_1 & a_2 & a_3 \\ b_1 & b_2 & b_3 \\ c_1 & c_2 & c_3 \end{vmatrix} + a_3 \begin{vmatrix} a_1 & a_2 & a_3 \\ b_1 & b_2 & b_3 \\ c_1 & c_2 & c_3 \end{vmatrix}$$

For example,

$$\begin{vmatrix} 3 & -2 & -5 \\ 1 & 4 & -4 \\ 0 & 3 & 2 \end{vmatrix} = 3 \begin{vmatrix} 4 & -4 \\ 3 & 2 \end{vmatrix} - (-2) \begin{vmatrix} 1 & -4 \\ 0 & 2 \end{vmatrix} + (-5) \begin{vmatrix} 1 & 4 \\ 0 & 3 \end{vmatrix}$$

$$= 3(20) + 2(2) - 5(3) = 49$$

Higher-order determinants can be defined as well, but we will not need them in this text.

□ **CROSS PRODUCT**

14.4.1 DEFINITION. If $\mathbf{u} = \langle u_1, u_2, u_3 \rangle$ and $\mathbf{v} = \langle v_1, v_2, v_3 \rangle$ are vectors in 3-space, then the *cross product* $\mathbf{u} \times \mathbf{v}$ is the vector defined by

$$\mathbf{u} \times \mathbf{v} = \begin{vmatrix} u_2 & u_3 \\ v_2 & v_3 \end{vmatrix} \mathbf{i} - \begin{vmatrix} u_1 & u_3 \\ v_1 & v_3 \end{vmatrix} \mathbf{j} + \begin{vmatrix} u_1 & u_2 \\ v_1 & v_2 \end{vmatrix} \mathbf{k} \qquad (1)$$

Formula (1) can be remembered by writing it in the form

$$\mathbf{u} \times \mathbf{v} = \begin{vmatrix} \mathbf{i} & \mathbf{j} & \mathbf{k} \\ u_1 & u_2 & u_3 \\ v_1 & v_2 & v_3 \end{vmatrix} \tag{2}$$

However, this is just a mnemonic device since the entries in a determinant must be numbers, not vectors.

Example 1   Find $\mathbf{u} \times \mathbf{v}$, where $\mathbf{u} = \langle 1, 2, -2 \rangle$ and $\mathbf{v} = \langle 3, 0, 1 \rangle$.

Solution.

$$\mathbf{u} \times \mathbf{v} = \begin{vmatrix} \mathbf{i} & \mathbf{j} & \mathbf{k} \\ 1 & 2 & -2 \\ 3 & 0 & 1 \end{vmatrix}$$

$$= \begin{vmatrix} 2 & -2 \\ 0 & 1 \end{vmatrix} \mathbf{i} - \begin{vmatrix} 1 & -2 \\ 3 & 1 \end{vmatrix} \mathbf{j} + \begin{vmatrix} 1 & 2 \\ 3 & 0 \end{vmatrix} \mathbf{k} = 2\mathbf{i} - 7\mathbf{j} - 6\mathbf{k} \quad \blacktriangleleft$$

Observe that the cross product of two vectors is another vector, whereas the dot product of two vectors is a scalar. Moreover, the cross product is defined only for vectors in 3-space, whereas the dot product is defined for vectors in 2-space and 3-space.

The following theorem gives an important relationship between dot product and cross product and also shows that $\mathbf{u} \times \mathbf{v}$ is orthogonal to both $\mathbf{u}$ and $\mathbf{v}$.

14.4.2   THEOREM.   *If $\mathbf{u}$ and $\mathbf{v}$ are vectors in 3-space, then*

(a) $\mathbf{u} \cdot (\mathbf{u} \times \mathbf{v}) = 0$ ($\mathbf{u} \times \mathbf{v}$ *is orthogonal to* $\mathbf{u}$)
(b) $\mathbf{v} \cdot (\mathbf{u} \times \mathbf{v}) = 0$ ($\mathbf{u} \times \mathbf{v}$ *is orthogonal to* $\mathbf{v}$)
(c) $\|\mathbf{u} \times \mathbf{v}\|^2 = \|\mathbf{u}\|^2 \|\mathbf{v}\|^2 - (\mathbf{u} \cdot \mathbf{v})^2$ (*Lagrange's identity*)

Proof.   Let $\mathbf{u} = \langle u_1, u_2, u_3 \rangle$ and $\mathbf{v} = \langle v_1, v_2, v_3 \rangle$.

(a)   By definition

$$\mathbf{u} \times \mathbf{v} = \begin{vmatrix} u_2 & u_3 \\ v_2 & v_3 \end{vmatrix} \mathbf{i} - \begin{vmatrix} u_1 & u_3 \\ v_1 & v_3 \end{vmatrix} \mathbf{j} + \begin{vmatrix} u_1 & u_2 \\ v_1 & v_2 \end{vmatrix} \mathbf{k}$$

which may be rewritten as

$$\mathbf{u} \times \mathbf{v} = \langle u_2 v_3 - u_3 v_2, \; u_3 v_1 - u_1 v_3, \; u_1 v_2 - u_2 v_1 \rangle \tag{3}$$

so that

$$\mathbf{u} \cdot (\mathbf{u} \times \mathbf{v}) = u_1(u_2 v_3 - u_3 v_2) + u_2(u_3 v_1 - u_1 v_3) + u_3(u_1 v_2 - u_2 v_1) = 0$$

(b)   Similar to (a).

(c)   From (3) we obtain

$$\|\mathbf{u} \times \mathbf{v}\|^2 = (u_2 v_3 - u_3 v_2)^2 + (u_3 v_1 - u_1 v_3)^2 + (u_1 v_2 - u_2 v_1)^2 \tag{4}$$

Moreover,

$$\|\mathbf{u}\|^2 \|\mathbf{v}\|^2 - (\mathbf{u} \cdot \mathbf{v})^2 = (u_1{}^2 + u_2{}^2 + u_3{}^2)(v_1{}^2 + v_2{}^2 + v_3{}^2)$$
$$- (u_1 v_1 + u_2 v_2 + u_3 v_3)^2 \tag{5}$$

Lagrange's identity can be established by ''multiplying out'' the right sides of (4) and (5) and verifying their equality.   ∎

Example 2 Let $\mathbf{u} = \langle 1, 2, -2 \rangle$ and $\mathbf{v} = \langle 3, 0, 1 \rangle$. In Example 1 we showed that

$$\mathbf{u} \times \mathbf{v} = \langle 2, -7, -6 \rangle$$

Thus,

$$\mathbf{u} \cdot (\mathbf{u} \times \mathbf{v}) = (1)(2) + (2)(-7) + (-2)(-6) = 0$$

and

$$\mathbf{v} \cdot (\mathbf{u} \times \mathbf{v}) = (3)(2) + (0)(-7) + (1)(-6) = 0$$

so that the vector $\mathbf{u} \times \mathbf{v}$ is orthogonal to both $\mathbf{u}$ and $\mathbf{v}$ as guaranteed by Theorem 14.4.2. ◄

If $\mathbf{u}$ and $\mathbf{v}$ are nonzero vectors in 3-space, then the length of $\mathbf{u} \times \mathbf{v}$ has a useful geometric interpretation. Lagrange's identity, given in Theorem 14.4.2, states that

$$\|\mathbf{u} \times \mathbf{v}\|^2 = \|\mathbf{u}\|^2 \|\mathbf{v}\|^2 - (\mathbf{u} \cdot \mathbf{v})^2$$

If θ denotes the angle between $\mathbf{u}$ and $\mathbf{v}$, then $\mathbf{u} \cdot \mathbf{v} = \|\mathbf{u}\| \|\mathbf{v}\| \cos \theta$, so the preceding equation can be rewritten as

$$\begin{aligned} \|\mathbf{u} \times \mathbf{v}\|^2 &= \|\mathbf{u}\|^2 \|\mathbf{v}\|^2 - \|\mathbf{u}\|^2 \|\mathbf{v}\|^2 \cos^2 \theta \\ &= \|\mathbf{u}\|^2 \|\mathbf{v}\|^2 (1 - \cos^2 \theta) \\ &= \|\mathbf{u}\|^2 \|\mathbf{v}\|^2 \sin^2 \theta \end{aligned}$$

Since $0 \leq \theta \leq \pi$, it follows that $\sin \theta \geq 0$, so

$$\|\mathbf{u} \times \mathbf{v}\| = \|\mathbf{u}\| \|\mathbf{v}\| \sin \theta \tag{6}$$

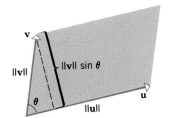

But $\|\mathbf{v}\| \sin \theta$ is the altitude of the parallelogram determined by $\mathbf{u}$ and $\mathbf{v}$ (Figure 14.4.1). Thus, from (7), the area A of this parallelogram is given by

$$A = (\text{base})(\text{altitude}) = \|\mathbf{u}\| \|\mathbf{v}\| \sin \theta = \|\mathbf{u} \times \mathbf{v}\| \tag{7}$$

In other words, the length of $\mathbf{u} \times \mathbf{v}$ is numerically equal to the area of the parallelogram determined by $\mathbf{u}$ and $\mathbf{v}$.

Figure 14.4.1

Example 3 Find the area of the triangle that is determined by the points $P_1(2, 2, 0)$, $P_2(-1, 0, 2)$, and $P_3(0, 4, 3)$.

Solution. The area A of the triangle is half the area of the parallelogram determined by the vectors $\overrightarrow{P_1 P_2}$ and $\overrightarrow{P_1 P_3}$ (Figure 14.4.2). But $\overrightarrow{P_1 P_2} = \langle -3, -2, 2 \rangle$ and $\overrightarrow{P_1 P_3} = \langle -2, 2, 3 \rangle$, so

$$\overrightarrow{P_1 P_2} \times \overrightarrow{P_1 P_3} = \langle -10, 5, -10 \rangle$$

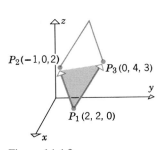

Figure 14.4.2

(verify), and consequently

$$A = \tfrac{1}{2} \| \overrightarrow{P_1 P_2} \times \overrightarrow{P_1 P_3} \| = \tfrac{15}{2} \qquad ◄$$

It follows from (7) that $\mathbf{u} \times \mathbf{v} = \mathbf{0}$ if and only if

$$\mathbf{u} = \mathbf{0}, \quad \text{or} \quad \mathbf{v} = \mathbf{0}, \quad \text{or} \quad \sin \theta = 0$$

In all three cases the vectors $\mathbf{u}$ and $\mathbf{v}$ are parallel. For the first two cases this is true because $\mathbf{0}$ is parallel to every vector, and in the third case, $\sin \theta = 0$ implies that the angle θ between $\mathbf{u}$ and $\mathbf{v}$ is $\theta = 0$ or $\theta = \pi$. In summary, we have the following result.

14.4.3 THEOREM. *If $\mathbf{u}$ and $\mathbf{v}$ are vectors in 3-space, then $\mathbf{u} \times \mathbf{v} = \mathbf{0}$ if and only if $\mathbf{u}$ and $\mathbf{v}$ are parallel vectors.*

The main arithmetic properties of the cross product are listed in the next theorem.

14.4.4   THEOREM.   *If* **u**, **v**, *and* **w** *are any vectors in 3-space and k is any scalar, then*

(a) $\mathbf{u} \times \mathbf{v} = -(\mathbf{v} \times \mathbf{u})$
(b) $\mathbf{u} \times (\mathbf{v} + \mathbf{w}) = (\mathbf{u} \times \mathbf{v}) + (\mathbf{u} \times \mathbf{w})$
(c) $(\mathbf{u} + \mathbf{v}) \times \mathbf{w} = (\mathbf{u} \times \mathbf{w}) + (\mathbf{v} \times \mathbf{w})$
(d) $k(\mathbf{u} \times \mathbf{v}) = (k\mathbf{u}) \times \mathbf{v} = \mathbf{u} \times (k\mathbf{v})$
(e) $\mathbf{u} \times \mathbf{0} = \mathbf{0} \times \mathbf{u} = \mathbf{0}$
(f) $\mathbf{u} \times \mathbf{u} = \mathbf{0}$

We shall prove (a) and leave the remaining proofs as exercises.

Proof (a).   First note that interchanging the rows of a 2 × 2 determinant changes the sign of the determinant, since

$$\begin{vmatrix} c & d \\ a & b \end{vmatrix} = bc - ad = -(ad - bc) = -\begin{vmatrix} a & b \\ c & d \end{vmatrix}$$

It follows that interchanging **u** and **v** in (1) interchanges the rows of the three determinants on the right side of (1) and thereby changes the sign of each component in the cross product. Thus, $\mathbf{u} \times \mathbf{v} = -(\mathbf{v} \times \mathbf{u})$. This means that (unlike the dot product) the cross product is not commutative. ∎

REMARK.   The fact that interchanging the rows of a 2 × 2 determinant reverses the sign of the determinant is important in its own right. We leave it for the reader to prove the analogous result for 3 × 3 determinants; that is, interchanging any two rows reverses the sign of the determinant.

Cross products of the unit vectors **i**, **j**, and **k** are of special interest. We obtain, for example,

$$\mathbf{i} \times \mathbf{j} = \begin{vmatrix} \mathbf{i} & \mathbf{j} & \mathbf{k} \\ 1 & 0 & 0 \\ 0 & 1 & 0 \end{vmatrix} = \begin{vmatrix} 0 & 0 \\ 1 & 0 \end{vmatrix}\mathbf{i} - \begin{vmatrix} 1 & 0 \\ 0 & 0 \end{vmatrix}\mathbf{j} + \begin{vmatrix} 1 & 0 \\ 0 & 1 \end{vmatrix}\mathbf{k} = \mathbf{k}$$

The reader should have no trouble obtaining the following list of cross products:

$$\mathbf{i} \times \mathbf{j} = \mathbf{k} \qquad \mathbf{j} \times \mathbf{k} = \mathbf{i} \qquad \mathbf{k} \times \mathbf{i} = \mathbf{j}$$
$$\mathbf{j} \times \mathbf{i} = -\mathbf{k} \qquad \mathbf{k} \times \mathbf{j} = -\mathbf{i} \qquad \mathbf{i} \times \mathbf{k} = -\mathbf{j}$$
$$\mathbf{i} \times \mathbf{i} = \mathbf{0} \qquad \mathbf{j} \times \mathbf{j} = \mathbf{0} \qquad \mathbf{k} \times \mathbf{k} = \mathbf{0}$$

The diagram in Figure 14.4.3 is helpful for remembering these results. In this diagram, the cross product of two consecutive vectors going clockwise is the next vector around, and the cross product of two consecutive vectors going counterclockwise is the negative of the next vector around.

WARNING.   It is *not* true in general that $\mathbf{u} \times (\mathbf{v} \times \mathbf{w}) = (\mathbf{u} \times \mathbf{v}) \times \mathbf{w}$. For example, we have

$$\mathbf{i} \times (\mathbf{j} \times \mathbf{j}) = \mathbf{i} \times \mathbf{0} = \mathbf{0} \quad \text{and} \quad (\mathbf{i} \times \mathbf{j}) \times \mathbf{j} = \mathbf{k} \times \mathbf{j} = -\mathbf{i}$$

so that

$$\mathbf{i} \times (\mathbf{j} \times \mathbf{j}) \neq (\mathbf{i} \times \mathbf{j}) \times \mathbf{j}$$

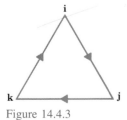

Figure 14.4.3

We know from Theorem 14.4.2 that $\mathbf{u} \times \mathbf{v}$ is orthogonal to both $\mathbf{u}$ and $\mathbf{v}$. It can be shown that if $\mathbf{u}$ and $\mathbf{v}$ are nonzero vectors, then the direction of $\mathbf{u} \times \mathbf{v}$ can be determined using the following "right-hand rule"* (Figure 14.4.4). Let θ be the angle between $\mathbf{u}$ and $\mathbf{v}$, and suppose that $\mathbf{u}$ is rotated through the angle θ until it coincides with $\mathbf{v}$. If the fingers of the right hand are cupped so that they point in the direction of rotation, then the thumb indicates (roughly) the direction of $\mathbf{u} \times \mathbf{v}$. The reader may find it instructive to practice this rule with the products

$$\mathbf{i} \times \mathbf{j} = \mathbf{k}, \qquad \mathbf{j} \times \mathbf{k} = \mathbf{i}, \qquad \mathbf{k} \times \mathbf{i} = \mathbf{j}$$

Equation (7) and the right-hand rule show that both the magnitude and direction of $\mathbf{u} \times \mathbf{v}$ can be determined purely geometrically; that is, $\mathbf{u} \times \mathbf{v}$ does not depend on the coordinate system being used. Thus, we say that the definition of $\mathbf{u} \times \mathbf{v}$ is ***coordinate free***. This result is important to physicists and engineers, since it allows them to choose any convenient coordinate system with assurance that the choice will not affect the end result of any vector computations.

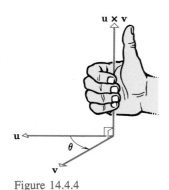

Figure 14.4.4

☐ **TRIPLE SCALAR PRODUCTS**

If $\mathbf{a} = \langle a_1, a_2, a_3 \rangle$, $\mathbf{b} = \langle b_1, b_2, b_3 \rangle$, and $\mathbf{c} = \langle c_1, c_2, c_3 \rangle$ are vectors in 3-space, then the number

$$\mathbf{a} \cdot (\mathbf{b} \times \mathbf{c})$$

is called the ***triple scalar product*** of $\mathbf{a}$, $\mathbf{b}$, and $\mathbf{c}$. The triple scalar product may be conveniently calculated from the formula

$$\mathbf{a} \cdot (\mathbf{b} \times \mathbf{c}) = \begin{vmatrix} a_1 & a_2 & a_3 \\ b_1 & b_2 & b_3 \\ c_1 & c_2 & c_3 \end{vmatrix} \tag{8}$$

The validity of this formula may be seen by writing

$$\mathbf{a} \cdot (\mathbf{b} \times \mathbf{c}) = \mathbf{a} \cdot \left(\begin{vmatrix} b_2 & b_3 \\ c_2 & c_3 \end{vmatrix} \mathbf{i} - \begin{vmatrix} b_1 & b_3 \\ c_1 & c_3 \end{vmatrix} \mathbf{j} + \begin{vmatrix} b_1 & b_2 \\ c_1 & c_2 \end{vmatrix} \mathbf{k} \right)$$

$$= \begin{vmatrix} b_2 & b_3 \\ c_2 & c_3 \end{vmatrix} a_1 - \begin{vmatrix} b_1 & b_3 \\ c_1 & c_3 \end{vmatrix} a_2 + \begin{vmatrix} b_1 & b_2 \\ c_1 & c_2 \end{vmatrix} a_3$$

$$= \begin{vmatrix} a_1 & a_2 & a_3 \\ b_1 & b_2 & b_3 \\ c_1 & c_2 & c_3 \end{vmatrix}$$

Example 4 Calculate the triple scalar product $\mathbf{a} \cdot (\mathbf{b} \times \mathbf{c})$ of the vectors

$$\mathbf{a} = 3\mathbf{i} - 2\mathbf{j} - 5\mathbf{k}, \quad \mathbf{b} = \mathbf{i} + 4\mathbf{j} - 4\mathbf{k}, \quad \mathbf{c} = 3\mathbf{j} + 2\mathbf{k}$$

Solution.

$$\mathbf{a} \cdot (\mathbf{b} \times \mathbf{c}) = \begin{vmatrix} 3 & -2 & -5 \\ 1 & 4 & -4 \\ 0 & 3 & 2 \end{vmatrix}$$

$$= 3 \begin{vmatrix} 4 & -4 \\ 3 & 2 \end{vmatrix} - (-2) \begin{vmatrix} 1 & -4 \\ 0 & 2 \end{vmatrix} + (-5) \begin{vmatrix} 1 & 4 \\ 0 & 3 \end{vmatrix}$$

$$= 60 + 4 - 15 = 49 \quad \blacktriangleleft$$

---

*Recall that we agreed to consider only right-handed coordinate systems in this text. Had we used left-handed systems instead, a "left-hand rule" would apply here.

REMARK.   The symbol $(\mathbf{a} \cdot \mathbf{b}) \times \mathbf{c}$ makes no sense since we cannot form the cross product of a scalar and a vector. Thus, no ambiguity arises if we write $\mathbf{a} \cdot \mathbf{b} \times \mathbf{c}$ rather than $\mathbf{a} \cdot (\mathbf{b} \times \mathbf{c})$.

Recall from the remark following Theorem 14.4.4 that interchanging two rows of a 3×3 determinant reverses the sign of the determinant. It follows from this that

$$\mathbf{a} \cdot (\mathbf{b} \times \mathbf{c}) = \mathbf{c} \cdot (\mathbf{a} \times \mathbf{b}) = \mathbf{b} \cdot (\mathbf{c} \times \mathbf{a}) \tag{9}$$

since the 3×3 determinants that represent these products can be obtained from one another by *two* row interchanges. (Verify.) These relationships may be remembered by moving the vectors $\mathbf{a}$, $\mathbf{b}$, and $\mathbf{c}$ clockwise around the vertices of the triangle in Figure 14.4.5. We also note that from part (*a*) of Theorem 14.3.4, the first equality in (9) can be written as $\mathbf{a} \cdot (\mathbf{b} \times \mathbf{c}) = (\mathbf{a} \times \mathbf{b}) \cdot \mathbf{c}$, which, on dropping parentheses, yields

$$\mathbf{a} \cdot \mathbf{b} \times \mathbf{c} = \mathbf{a} \times \mathbf{b} \cdot \mathbf{c} \tag{10}$$

This shows that interchanging the dot and cross does not change the value of a triple scalar product.

The triple scalar product $\mathbf{a} \cdot (\mathbf{b} \times \mathbf{c})$ has a useful geometric interpretation. If we assume, for the moment, that the vectors $\mathbf{a}$, $\mathbf{b}$, $\mathbf{c}$ do not all lie in the same plane when they are positioned with a common initial point, then the three vectors form adjacent sides of a parallelepiped (Figure 14.4.6). If the parallelogram determined by $\mathbf{b}$ and $\mathbf{c}$ is regarded as the base of the parallelepiped, then the area of the base is $\|\mathbf{b} \times \mathbf{c}\|$, and the height h is the length of the orthogonal projection of $\mathbf{a}$ on $\mathbf{b} \times \mathbf{c}$ (Figure 14.4.6). Therefore, by Formula (10) of Section 14.3 we have

$$h = \|\text{proj}_{\mathbf{b} \times \mathbf{c}} \, \mathbf{a}\| = \frac{|\mathbf{a} \cdot (\mathbf{b} \times \mathbf{c})|}{\|\mathbf{b} \times \mathbf{c}\|}$$

It follows that the volume V of the parallelepiped is

$$V = (\text{area of base}) \cdot \text{height} = \|\mathbf{b} \times \mathbf{c}\| \frac{|\mathbf{a} \cdot (\mathbf{b} \times \mathbf{c})|}{\|\mathbf{b} \times \mathbf{c}\|}$$

or, on simplifying,

$$V = \begin{bmatrix} \text{volume of parallelepiped with} \\ \text{adjacent sides } \mathbf{a}, \mathbf{b}, \text{ and } \mathbf{c} \end{bmatrix} = |\mathbf{a} \cdot (\mathbf{b} \times \mathbf{c})| \tag{11}$$

REMARK.   It follows from this formula that

$$\mathbf{a} \cdot (\mathbf{b} \times \mathbf{c}) = \pm V$$

where the $+$ or $-$ results depending on whether $\mathbf{a}$ makes an acute or obtuse angle with $\mathbf{b} \times \mathbf{c}$ (Theorem 14.3.2).

If the vectors $\mathbf{a}$, $\mathbf{b}$, and $\mathbf{c}$ do not lie in a plane when they are positioned with a common initial point, then they determine a parallelepiped of positive volume. Thus, from (11), $\mathbf{a} \cdot (\mathbf{b} \times \mathbf{c}) \neq 0$ if $\mathbf{a}$, $\mathbf{b}$, and $\mathbf{c}$ do not lie in a plane. It follows, therefore, that if $\mathbf{a} \cdot (\mathbf{b} \times \mathbf{c}) = 0$, then $\mathbf{a}$, $\mathbf{b}$, and $\mathbf{c}$ lie in a plane. Conversely, it can be shown that if $\mathbf{a}$, $\mathbf{b}$, and $\mathbf{c}$ lie in a plane, then $\mathbf{a} \cdot (\mathbf{b} \times \mathbf{c}) = 0$. In summary, we have the following result.

14.4.5   THEOREM.   *If the vectors* $\mathbf{a} = \langle a_1, a_2, a_3 \rangle$, $\mathbf{b} = \langle b_1, b_2, b_3 \rangle$, *and* $\mathbf{c} = \langle c_1, c_2, c_3 \rangle$ *have the same initial point, then they lie in a plane if and only if*

$$\mathbf{a} \cdot (\mathbf{b} \times \mathbf{c}) = \begin{vmatrix} a_1 & a_2 & a_3 \\ b_1 & b_2 & b_3 \\ c_1 & c_2 & c_3 \end{vmatrix} = 0$$

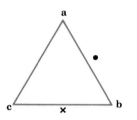

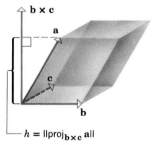

$h = \|\text{proj}_{\mathbf{b} \times \mathbf{c}} \, \mathbf{a}\|$

Figure 14.4.6

▶ Exercise Set 14.4

In Exercises 1–4, find $\mathbf{a} \times \mathbf{b}$.

1. $\mathbf{a} = \langle 1, 2, -3 \rangle$, $\mathbf{b} = \langle -4, 1, 2 \rangle$.

2. $\mathbf{a} = 3\mathbf{i} + 2\mathbf{j} - \mathbf{k}$, $\mathbf{b} = -\mathbf{i} - 3\mathbf{j} + \mathbf{k}$.

3. $\mathbf{a} = \langle 0, 1, -2 \rangle$, $\mathbf{b} = \langle 3, 0, -4 \rangle$.

4. $\mathbf{a} = 4\mathbf{i} + \mathbf{k}$, $\mathbf{b} = 2\mathbf{i} - \mathbf{j}$.

5. Let $\mathbf{u} = \langle 2, -1, 3 \rangle$, $\mathbf{v} = \langle 0, 1, 7 \rangle$, and $\mathbf{w} = \langle 1, 4, 5 \rangle$. Find

   (a) $\mathbf{u} \times (\mathbf{v} \times \mathbf{w})$ (b) $(\mathbf{u} \times \mathbf{v}) \times \mathbf{w}$

   (c) $\mathbf{u} \times (\mathbf{v} - 2\mathbf{w})$ (d) $(\mathbf{u} \times \mathbf{v}) - 2\mathbf{w}$

   (e) $(\mathbf{u} \times \mathbf{v}) \times (\mathbf{v} \times \mathbf{w})$ (f) $(\mathbf{v} \times \mathbf{w}) \times (\mathbf{u} \times \mathbf{v})$.

6. Find a vector orthogonal to both $\mathbf{u}$ and $\mathbf{v}$.

   (a) $\mathbf{u} = -7\mathbf{i} + 3\mathbf{j} + \mathbf{k}$, $\mathbf{v} = 2\mathbf{i} + 4\mathbf{k}$

   (b) $\mathbf{u} = \langle -1, -1, -1 \rangle$, $\mathbf{v} = \langle 2, 0, 2 \rangle$.

7. Verify Theorem 14.4.2 for the vectors $\mathbf{u} = \mathbf{i} - 5\mathbf{j} + 6\mathbf{k}$ and $\mathbf{v} = 2\mathbf{i} + \mathbf{j} + 2\mathbf{k}$.

8. Verify Theorem 14.4.4 for $\mathbf{u} = \langle 2, 0, -1 \rangle$, $\mathbf{v} = \langle 6, 7, 4 \rangle$, $\mathbf{w} = \langle 1, 1, 1 \rangle$, and $k = -3$.

9. Find all unit vectors parallel to the yz-plane that are perpendicular to the vector $3\mathbf{i} - \mathbf{j} + 2\mathbf{k}$.

10. Prove: If θ is the angle between $\mathbf{u}$ and $\mathbf{v}$ and $\mathbf{u} \cdot \mathbf{v} \neq 0$, then $\tan \theta = \|\mathbf{u} \times \mathbf{v}\|/(\mathbf{u} \cdot \mathbf{v})$.

11. Simplify $(\mathbf{u} + \mathbf{v}) \times (\mathbf{u} - \mathbf{v})$.

12. Find the area of the parallelogram determined by the vectors $\mathbf{u}$ and $\mathbf{v}$.

   (a) $\mathbf{u} = \mathbf{i} - \mathbf{j} + 2\mathbf{k}$, $\mathbf{v} = 3\mathbf{j} + \mathbf{k}$

   (b) $\mathbf{u} = 2\mathbf{i} + 3\mathbf{j}$, $\mathbf{v} = -\mathbf{i} + 2\mathbf{j} - 2\mathbf{k}$.

13. Find the area of the triangle having vertices P, Q, and R.

   (a) $P(1, 5, -2)$, $Q(0, 0, 0)$, $R(3, 5, 1)$

   (b) $P(2, 0, -3)$, $Q(1, 4, 5)$, $R(7, 2, 9)$.

14. Use the cross product to find the sine of the angle between the vectors $\mathbf{a} = 2\mathbf{i} + 3\mathbf{j} - 6\mathbf{k}$ and $\mathbf{b} = 2\mathbf{i} + 3\mathbf{j} + 6\mathbf{k}$.

15. (a) Find the area of the triangle having vertices $A(1, 0, 1)$, $B(0, 2, 3)$, and $C(2, 1, 0)$.

   (b) Use the result of part (a) to find the length of the altitude from vertex C to side AB.

16. Show that if $\mathbf{u}$ is a vector from any point on a line to a point P not on the line, and $\mathbf{v}$ is a vector parallel to the line, then the distance between P and the line is given by $\|\mathbf{u} \times \mathbf{v}\|/\|\mathbf{v}\|$.

17. Use the result of Exercise 16 to find the distance between the point P and the line through the points A and B.

   (a) $P(-3, 1, 2)$, $A(1, 1, 0)$, $B(-2, 3, -4)$

   (b) $P(4, 3, 0)$, $A(2, 1, -3)$, $B(0, 2, -1)$.

18. Find all unit vectors in the plane of $\mathbf{u} = 3\mathbf{i} + \mathbf{k}$ and $\mathbf{v} = \mathbf{i} - \mathbf{j} - \mathbf{k}$ that are perpendicular to the vector $\mathbf{w} = \mathbf{i} + 2\mathbf{j}$.

19. What is wrong with the expression $\mathbf{u} \times \mathbf{v} \times \mathbf{w}$?

In Exercises 20–23, find $\mathbf{a} \cdot (\mathbf{b} \times \mathbf{c})$.

20. $\mathbf{a} = \langle 1, -2, 2 \rangle$, $\mathbf{b} = \langle 0, 3, 2 \rangle$, $\mathbf{c} = \langle -4, 1, -3 \rangle$.

21. $\mathbf{a} = 2\mathbf{i} - 3\mathbf{j} + \mathbf{k}$, $\mathbf{b} = 4\mathbf{i} + \mathbf{j} - 3\mathbf{k}$, $\mathbf{c} = \mathbf{j} + 5\mathbf{k}$.

22. $\mathbf{a} = \langle 2, 1, 0 \rangle$, $\mathbf{b} = \langle 1, -3, 1 \rangle$, $\mathbf{c} = \langle 4, 0, 1 \rangle$.

23. $\mathbf{a} = \mathbf{i}$, $\mathbf{b} = \mathbf{i} + \mathbf{j}$, $\mathbf{c} = \mathbf{i} + \mathbf{j} + \mathbf{k}$.

24. Suppose that $\mathbf{u} \cdot (\mathbf{v} \times \mathbf{w}) = 3$. Find

   (a) $\mathbf{u} \cdot (\mathbf{w} \times \mathbf{v})$ (b) $(\mathbf{v} \times \mathbf{w}) \cdot \mathbf{u}$

   (c) $\mathbf{w} \cdot (\mathbf{u} \times \mathbf{v})$ (d) $\mathbf{v} \cdot (\mathbf{u} \times \mathbf{w})$

   (e) $(\mathbf{u} \times \mathbf{w}) \cdot \mathbf{v}$ (f) $\mathbf{v} \cdot (\mathbf{w} \times \mathbf{w})$.

25. Find the volume of the parallelepiped with sides $\mathbf{a}$, $\mathbf{b}$, and $\mathbf{c}$.

   (a) $\mathbf{a} = \langle 2, -6, 2 \rangle$, $\mathbf{b} = \langle 0, 4, -2 \rangle$, $\mathbf{c} = \langle 2, 2, -4 \rangle$

   (b) $\mathbf{a} = 3\mathbf{i} + \mathbf{j} + 2\mathbf{k}$, $\mathbf{b} = 4\mathbf{i} + 5\mathbf{j} + \mathbf{k}$, $\mathbf{c} = \mathbf{i} + 2\mathbf{j} + 4\mathbf{k}$.

26. Determine whether $\mathbf{u}$, $\mathbf{v}$, and $\mathbf{w}$ lie in the same plane.

   (a) $\mathbf{u} = \langle 1, -2, 1 \rangle$, $\mathbf{v} = \langle 3, 0, -2 \rangle$, $\mathbf{w} = \langle 5, -4, 0 \rangle$

   (b) $\mathbf{u} = 5\mathbf{i} - 2\mathbf{j} + \mathbf{k}$, $\mathbf{v} = 4\mathbf{i} - \mathbf{j} + \mathbf{k}$, $\mathbf{w} = \mathbf{i} - \mathbf{j}$

   (c) $\mathbf{u} = \langle 4, -8, 1 \rangle$, $\mathbf{v} = \langle 2, 1, -2 \rangle$, $\mathbf{w} = \langle 3, -4, 12 \rangle$.

27. Consider the parallelepiped with sides

$$\mathbf{a} = 3\mathbf{i} + 2\mathbf{j} + \mathbf{k}$$
$$\mathbf{b} = \mathbf{i} + \mathbf{j} + 2\mathbf{k}$$
$$\mathbf{c} = \mathbf{i} + 3\mathbf{j} + 3\mathbf{k}$$

   (a) Find the volume.

   (b) Find the area of the face determined by $\mathbf{a}$ and $\mathbf{c}$.

   (c) Find the angle between $\mathbf{a}$ and the plane containing the face determined by $\mathbf{b}$ and $\mathbf{c}$.

28. Find a vector $\mathbf{n}$ perpendicular to the plane determined by the points $A(0, -2, 1)$, $B(1, -1, -2)$, and $C(-1, 1, 0)$.

29. Simplify $(\mathbf{u} + k\mathbf{v}) \times \mathbf{v}$.

30. Let $\mathbf{u}$, $\mathbf{v}$, and $\mathbf{w}$ be vectors in 3-space with the same initial point and such that no two are collinear. Prove:

   (a) $\mathbf{u} \times (\mathbf{v} \times \mathbf{w})$ lies in the plane determined by $\mathbf{v}$ and $\mathbf{w}$

   (b) $(\mathbf{u} \times \mathbf{v}) \times \mathbf{w}$ lies in the plane determined by $\mathbf{u}$ and $\mathbf{v}$.

31. Prove parts (b) and (c) of Theorem 14.4.4.

32. Prove parts (d), (e), and (f) of Theorem 14.4.4.

33. Prove that $\mathbf{x} \times (\mathbf{y} \times \mathbf{z}) = (\mathbf{x} \cdot \mathbf{z})\mathbf{y} - (\mathbf{x} \cdot \mathbf{y})\mathbf{z}$. [*Hint:* First prove the result for the case $\mathbf{z} = \mathbf{i}$, then for $\mathbf{z} = \mathbf{j}$, and then for $\mathbf{z} = \mathbf{k}$. Finally, prove it for an arbitrary vector $\mathbf{z} = z_1\mathbf{i} + z_2\mathbf{j} + z_3\mathbf{k}$.]

34. For the vectors $\mathbf{a} = \mathbf{i} + 3\mathbf{j} - \mathbf{k}$, $\mathbf{b} = \mathbf{i} + \mathbf{j} + 2\mathbf{k}$, and $\mathbf{c} = 3\mathbf{i} - \mathbf{j} + 2\mathbf{k}$, calculate $\mathbf{a} \times (\mathbf{b} \times \mathbf{c})$ using Exercise 33, then check your result by calculating directly.

35. Prove: If $\mathbf{a}$, $\mathbf{b}$, $\mathbf{c}$, and $\mathbf{d}$ lie in the same plane when positioned with a common initial point, then

$$(\mathbf{a} \times \mathbf{b}) \times (\mathbf{c} \times \mathbf{d}) = \mathbf{0}$$

36. It is a theorem of solid geometry that the volume of a tetrahedron is $\frac{1}{3}$(area of base) · (height). Use this result to prove that the volume of a tetrahedron whose sides are the vectors **a**, **b**, and **c** is $\frac{1}{6}|\mathbf{a} \cdot (\mathbf{b} \times \mathbf{c})|$.

37. Use the result of Exercise 36 to find the volume of the tetrahedron with vertices P, Q, R, and S.

(a) $P(-1, 2, 0)$, $Q(2, 1, -3)$, $R(1, 0, 1)$, $S(3, -2, 3)$

(b) $P(0, 0, 0)$, $Q(1, 2, -1)$, $R(3, 4, 0)$, $S(-1, -3, 4)$.

■ **14.5   PARAMETRIC EQUATIONS OF LINES**

> *In this section we shall discuss parametric equations of lines in 2-space and 3-space. In 3-space, parametric equations of lines are especially important because they are generally the most convenient form for representing such lines algebraically.*

☐ **LINES DETERMINED BY A POINT AND A VECTOR**

A line in 2-space or 3-space can be determined uniquely by specifying a point on the line and a nonzero vector parallel to the line (Figure 14.5.1). The following theorem gives parametric equations of the line through a point P_0 and parallel to a nonzero vector **v**.

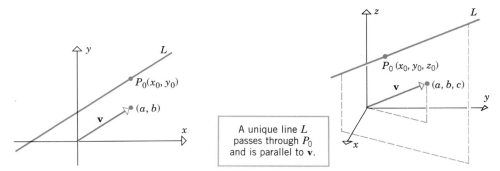

A unique line L passes through P_0 and is parallel to **v**.

Figure 14.5.1

14.5.1   THEOREM.

(a)   *The line in 2-space that passes through the point $P_0(x_0, y_0)$ and is parallel to the nonzero vector* $\mathbf{v} = \langle a, b \rangle = a\mathbf{i} + b\mathbf{j}$ *has parametric equations*

$$x = x_0 + at, \quad y = y_0 + bt \tag{1a}$$

(b)   *The line in 3-space that passes through the point $P_0(x_0, y_0, z_0)$ and is parallel to the nonzero vector* $\mathbf{v} = \langle a, b, c \rangle = a\mathbf{i} + b\mathbf{j} + c\mathbf{k}$ *has parametric equations*

$$x = x_0 + at, \quad y = y_0 + bt, \quad z = z_0 + ct \tag{1b}$$

We shall prove part (b). The proof of (a) is similar.

Proof (b).   If L is the line in 3-space passing through the point $P_0(x_0, y_0, z_0)$ and parallel to the nonzero vector $\mathbf{v} = \langle a, b, c \rangle$, it is clear (Figure 14.5.2) that L consists precisely of those points $P(x, y, z)$ for which the vector $\overrightarrow{P_0P}$ is parallel to **v**. In other words, the point $P(x, y, z)$ is on L if and only if $\overrightarrow{P_0P}$ is a scalar multiple of **v**, say

$$\overrightarrow{P_0P} = t\mathbf{v}$$

This equation can be written as

$$\langle x - x_0, y - y_0, z - z_0 \rangle = \langle ta, tb, tc \rangle$$

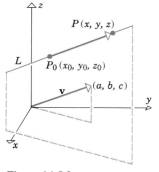

Figure 14.5.2

Equating components yields

$$x - x_0 = ta, \quad y - y_0 = tb, \quad z - z_0 = tc$$

from which (1b) follows. ∎

REMARK.   Although it is not stated explicitly, it is understood in (1a) and (1b) that $-\infty < t < +\infty$, which reflects the fact that lines extend indefinitely.

Example 1   Find parametric equations of the line

(a)   passing through $(4, 2)$ and parallel to $\mathbf{v} = \langle -1, 5 \rangle$;

(b)   passing through $(1, 2, -3)$ and parallel to $\mathbf{v} = 4\mathbf{i} + 5\mathbf{j} - 7\mathbf{k}$;

(c)   passing through the origin in 3-space and parallel to $\mathbf{v} = \langle 1, 1, 1 \rangle$.

Solution (a).   From (1a) with $x_0 = 4$, $y_0 = 2$, $a = -1$, and $b = 5$ we obtain

$$x = 4 - t, \quad y = 2 + 5t$$

Solution (b).   From (1b) we obtain

$$x = 1 + 4t, \quad y = 2 + 5t, \quad z = -3 - 7t$$

Solution (c).   From (1b) with $x_0 = 0$, $y_0 = 0$, $z_0 = 0$, $a = 1$, $b = 1$, and $c = 1$ we obtain

$$x = t, \quad y = t, \quad z = t \quad \blacktriangleleft$$

Example 2

(a)   Find parametric equations of the line L passing through the points $P_1(2, 4, -1)$ and $P_2(5, 0, 7)$.

(b)   Where does the line intersect the xy-plane?

Solution (a).   Since the vector $\overrightarrow{P_1P_2} = \langle 3, -4, 8 \rangle$ is parallel to L and $P_1(2, 4, -1)$ lies on L, the line L is given by

$$x = 2 + 3t, \quad y = 4 - 4t, \quad z = -1 + 8t$$

Had we used P_2 as the point on L rather than P_1, we would have obtained the equations

$$x = 5 + 3t, \quad y = -4t, \quad z = 7 + 8t$$

Although these equations look different from those obtained using P_1, the two sets of equations are actually equivalent in that both generate L as t varies from $-\infty$ to $+\infty$.

Solution (b).   From the first parametrization in part (a) the line intersects the xy-plane at the point where $z = -1 + 8t = 0$, that is, when $t = \frac{1}{8}$. Substituting this value of t in the parametric equations for L yields the point of intersection

$$(x, y, z) = \left(\tfrac{19}{8}, \tfrac{7}{2}, 0 \right) \quad \blacktriangleleft$$

Example 3   Let L_1 and L_2 be the lines

$$L_1: x = 1 + 4t, \quad y = 5 - 4t, \quad z = -1 + 5t$$

$$L_2: x = 2 + 8t, \quad y = 4 - 3t, \quad z = 5 + t$$

(a)   Are the lines parallel?

(b)   Do the lines intersect?

Solution (a).   The line L_1 is parallel to the vector $4\mathbf{i} - 4\mathbf{j} + 5\mathbf{k}$, and the line L_2 is parallel to the vector $8\mathbf{i} - 3\mathbf{j} + \mathbf{k}$. These vectors are not parallel since neither is a scalar multiple of the other. Thus, the lines are not parallel.

Solution (b).   In order for the lines to intersect at some point (x_0, y_0, z_0) these coordinates would have to satisfy the equations of both L_1 and L_2. In other words, there would have to exist values t_1 and t_2 for the parameters such that

$$x_0 = 1 + 4t_1, \quad y_0 = 5 - 4t_1, \quad z_0 = -1 + 5t_1$$

and

$$x_0 = 2 + 8t_2, \quad y_0 = 4 - 3t_2, \quad z_0 = 5 + t_2$$

This leads to three conditions on t_1 and t_2,

$$1 + 4t_1 = 2 + 8t_2$$
$$5 - 4t_1 = 4 - 3t_2 \qquad\qquad\qquad (2)$$
$$-1 + 5t_1 = 5 + t_2$$

We shall try to solve these equations for t_1 and t_2. If we obtain a solution, then the lines intersect; if we find that there is no solution, then the lines do not intersect, since the three conditions cannot be satisfied.

The first two equations in (2) may be solved by adding them together to obtain

$$6 = 6 + 5t_2$$

or $t_2 = 0$. Substituting $t_2 = 0$ in the first equation yields

$$1 + 4t_1 = 2$$

or $t_1 = \frac{1}{4}$. However, the values $t_1 = \frac{1}{4}, t_2 = 0$ do not satisfy the third equation in (2), so there is no simultaneous solution to the three equations. Thus, the lines do not intersect.   ◀

Two lines in 3-space that are not parallel and do not intersect (such as those in Example 3) are called *skew* lines. As illustrated in Figure 14.5.3, any two skew lines lie in parallel planes.

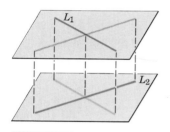

Parallel planes containing skew lines L_1 and L_2 may be determined by translating each line until it intersects the other.

Figure 14.5.3

☐ **LINE SEGMENTS**

Sometimes one is not interested in an entire line, but rather some *segment* of a line. Parametric equations of a line segment can be obtained by finding parametric equations for the entire line, then restricting the parameter appropriately so that only the desired segment is generated as the parameter varies.

Example 4   Find parametric equations for the line segment that joins the points $P_1(2, 4, -1)$ and $P_2(5, 0, 7)$.

Solution.   From Example 2, the line through P_1 and P_2 has parametric equations $x = 2 + 3t$, $y = 4 - 4t$, $z = -1 + 8t$. With these equations, the point P_1 corresponds to $t = 0$ and P_2 to $t = 1$. Thus, the line segment from P_1 to P_2 is given by

$$x = 2 + 3t, \quad y = 4 - 4t, \quad z = -1 + 8t \qquad (0 \le t \le 1) \quad ◀$$

☐ **VECTOR EQUATIONS OF LINES**

We shall now show how vector notation can be used to express the parametric equations of a line in a more compact form. Because two vectors are equal if and only if their components are equal, (1a) and (1b) can be written as

$$\langle x, y \rangle = \langle x_0 + at, y_0 + bt \rangle$$

and

$$\langle x, y, z \rangle = \langle x_0 + at, y_0 + bt, z_0 + ct \rangle$$

or, equivalently, as

$$\langle x, y \rangle = \langle x_0, y_0 \rangle + t \langle a, b \rangle \qquad (3a)$$

and

$$\langle x, y, z \rangle = \langle x_0, y_0, z_0 \rangle + t \langle a, b, c \rangle \qquad (3b)$$

In 2-space let us introduce the vectors $\mathbf{r}$, $\mathbf{r}_0$, and $\mathbf{v}$ given by

$$\mathbf{r} = \langle x, y \rangle, \quad \mathbf{r}_0 = \langle x_0, y_0 \rangle, \quad \mathbf{v} = \langle a, b \rangle \qquad (4a)$$

and in 3-space

$$\mathbf{r} = \langle x, y, z \rangle, \quad \mathbf{r}_0 = \langle x_0, y_0, z_0 \rangle, \quad \mathbf{v} = \langle a, b, c \rangle \qquad (4b)$$

Substituting (4a) and (4b) in (3a) and (3b), respectively, yields the equation

$$\mathbf{r} = \mathbf{r}_0 + t\mathbf{v} \qquad (5)$$

in both cases. We call this the ***vector equation of a line*** in 2-space or 3-space. In this equation $\mathbf{v}$ is a nonzero vector parallel to the line and $\mathbf{r}_0$ is a vector whose components are the coordinates of a point on the line.

Figure 14.5.4 illustrates how Equation (5) can be interpreted geometrically: $\mathbf{r}_0$ can be viewed as a vector from the origin to a point P_0 on L, $t\mathbf{v}$ is a scalar multiple of a nonzero vector $\mathbf{v}$ that is parallel to L, and $\mathbf{r} = \mathbf{r}_0 + t\mathbf{v}$ can be interpreted as a vector from the origin to a point on L. As the parameter t varies from $-\infty$ to $+\infty$, the terminal point of $\mathbf{r}$ traces out the line L.

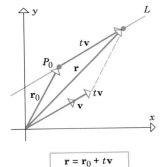

$$\boxed{\mathbf{r} = \mathbf{r}_0 + t\mathbf{v}}$$

Figure 14.5.4

Example 5   The equation

$$\langle x, y, z \rangle = \langle -1, 0, 2 \rangle + t \langle 1, 5, -4 \rangle$$

is of form (5) with

$$\mathbf{r}_0 = \langle -1, 0, 2 \rangle \quad \text{and} \quad \mathbf{v} = \langle 1, 5, -4 \rangle$$

Thus, the equation represents the line in 3-space that passes through the point $(-1, 0, 2)$ and is parallel to the vector $\langle 1, 5, -4 \rangle$.   ◄

Example 6   Find a vector equation of the line in 3-space that passes through the points $P_1(2, 4, -1)$ and $P_2(5, 0, 7)$.

Solution.   The vector

$$\overrightarrow{P_1 P_2} = \langle 3, -4, 8 \rangle$$

is parallel to the line, so it can be used as $\mathbf{v}$ in (5). For $\mathbf{r}_0$ we can use either the vector from the origin to P_1 or the vector from the origin to P_2. Using the former yields

$$\mathbf{r}_0 = \langle 2, 4, -1 \rangle$$

Thus, a vector equation of the line through P_1 and P_2 is

$$\langle x, y, z \rangle = \langle 2, 4, -1 \rangle + t \langle 3, -4, 8 \rangle$$

(Note that equating corresponding components on the two sides of this vector equation yields the parametric equations for the line that were obtained in Example 2.)   ◄

▶ Exercise Set 14.5   Ⓒ 53, 54, 57

In Exercises 1–8, find parametric equations for the line through P_1 and P_2.

1. $P_1(3, -2)$, $P_2(5, 1)$.   2. $P_1(0, 1)$, $P_2(-3, -4)$.
3. $P_1(4, 1)$, $P_2(4, 3)$.   4. $P_1(5, 2)$, $P_2(3, 7)$.
5. $P_1(5, -2, 1)$, $P_2(2, 4, 2)$.
6. $P_1(-1, 3, 5)$, $P_2(-1, 3, 2)$.
7. $P_1(0, 0, 0)$, $P_2(-1, 6, 1)$.
8. $P_1(4, 0, 7)$, $P_2(-1, -1, 2)$.

In Exercises 9–16, find parametric equations for the line segment joining P_1 and P_2.

9. $P_1(3, -2)$, $P_2(5, 1)$.   10. $P_1(0, 1)$, $P_2(-3, -4)$.
11. $P_1(4, 1)$, $P_2(4, 3)$.   12. $P_1(5, 2)$, $P_2(3, 7)$.
13. $P_1(5, -2, 1)$, $P_2(2, 4, 2)$.
14. $P_1(-1, 3, 5)$, $P_2(-1, 3, 2)$.
15. $P_1(0, 0, 0)$, $P_2(-1, 6, 1)$.
16. $P_1(4, 0, 7)$, $P_2(-1, -1, 2)$.

In Exercises 17–25, find parametric equations for the line.

17. The line through $(-5, 2)$ and parallel to $2\mathbf{i} - 3\mathbf{j}$.
18. The line through $(0, 3)$ and parallel to the line $x = -5 + t$, $y = 1 - 2t$.
19. The line tangent to the circle $x^2 + y^2 = 25$ at the point $(3, -4)$.
20. The line tangent to the parabola $y = x^2$ at the point $(-2, 4)$.
21. The line through $(-1, 2, 4)$ and parallel to $3\mathbf{i} - 4\mathbf{j} + \mathbf{k}$.
22. The line through $(2, -1, 5)$ and parallel to $\langle -1, 2, 7 \rangle$.
23. The line through $(-2, 0, 5)$ and parallel to the line $x = 1 + 2t$, $y = 4 - t$, $z = 6 + 2t$.
24. The line through the origin and parallel to the line $x = t$, $y = -1 + t$, $z = 2$.
25. The line through $(3, 7, 0)$ and parallel to the x-axis.
26. Where does the line $x = 1 + 3t$, $y = 2 - t$ intersect
    (a) the x-axis   (b) the y-axis?
27. Where does the line $x = 2t$, $y = 3 + 4t$ intersect the parabola $y = x^2$?
28. Where does the line $x = -1 + 2t$, $y = 3 + t$, $z = 4 - t$ intersect
    (a) the xy-plane   (b) the xz-plane
    (c) the yz-plane?
29. Where does the line $x = -2$, $y = 4 + 2t$, $z = -3 + t$ intersect
    (a) the xy-plane   (b) the xz-plane
    (c) the yz-plane?

30. Where does the line $x = 2 - t$, $y = 3t$, $z = -1 + 2t$ intersect the plane $2y + 3z = 6$?
31. Where does the line $x = 1 + t$, $y = 3 - t$, $z = 2t$ intersect the cylinder $x^2 + y^2 = 16$?
32. Find parametric equations for the line through (x_0, y_0, z_0) and (x_1, y_1, z_1).
33. Find parametric equations for the line through (x_1, y_1, z_1) and parallel to the line
    $$x = x_0 + at, \quad y = y_0 + bt, \quad z = z_0 + ct$$
34. Prove: If a, b, and c are nonzero, then each point on the line $x = x_0 + at$, $y = y_0 + bt$, $z = z_0 + ct$ satisfies
    $$\frac{x - x_0}{a} = \frac{y - y_0}{b} = \frac{z - z_0}{c}$$
    and conversely, each point (x, y, z) satisfying these equations lies on the line. (These are called the **symmetric equations** of the line.)
35. Show that the lines
    $$x = 2 + t, \quad y = 2 + 3t, \quad z = 3 + t$$
    and
    $$x = 2 + t, \quad y = 3 + 4t, \quad z = 4 + 2t$$
    intersect and find the point of intersection.
36. Show that the lines
    $$x + 1 = 4t, \quad y - 3 = t, \quad z - 1 = 0$$
    and
    $$x + 13 = 12t, \quad y - 1 = 6t, \quad z - 2 = 3t$$
    intersect and find the point of intersection.
37. Show that the lines
    $$x = 1 + 7t, \quad y = 3 + t, \quad z = 5 - 3t$$
    and
    $$x = 4 - t, \quad y = 6, \quad z = 7 + 2t$$
    are skew.
38. Show that the lines
    $$x = 2 + 8t, \quad y = 6 - 8t, \quad z = 10t$$
    and
    $$x = 3 + 8t, \quad y = 5 - 3t, \quad z = 6 + t$$
    are skew.
39. Determine whether P_1, P_2, and P_3 lie on the same line.
    (a) $P_1(6, 9, 7)$, $P_2(9, 2, 0)$, $P_3(0, -5, -3)$
    (b) $P_1(1, 0, 1)$, $P_2(3, -4, -3)$, $P_3(4, -6, -5)$.
40. Find k_1 and k_2 so that the point $(k_1, 1, k_2)$ lies on the line passing through $(0, 2, 3)$ and $(2, 7, 5)$.
41. Find the point on the line segment joining $P_1(1, 4, -3)$ and $P_2(1, 5, -1)$ that is $\frac{2}{3}$ of the way from P_1 to P_2.
42. In each part, determine whether the lines are parallel.
    (a) $x = 3 - 2t$, $y = 4 + t$, $z = 6 - t$
        and
        $x = 5 - 4t$, $y = -2 + 2t$, $z = 7 - 2t$

(b) $x = 5 + 3t,$ $y = 4 - 2t,$ $z = -2 + 3t$
and
$x = -1 + 9t,$ $y = 5 - 6t,$ $z = 3 + 8t.$

43. Show that the equations
$$x = 3 - t,   y = 1 + 2t$$
and
$$x = -1 + 3t,   y = 9 - 6t$$
represent the same line.

44. Show that the equations
$$x = 1 + 3t,   y = -2 + t,   z = 2t$$
and
$$x = 4 - 6t,   y = -1 - 2t,   z = 2 - 4t$$
represent the same line.

45. Find a vector equation of the line in 2-space that passes through the points P_1 and P_2.
(a) $P_1(2, -1),$ $P_2(-5, 3)$
(b) $P_1(0, 3),$ $P_2(4, 3).$

46. Find a vector equation of the line in 3-space that passes through the points P_1 and P_2.
(a) $P_1(3, -1, 2),$ $P_2(0, 1, 1)$
(b) $P_1(2, 4, 1),$ $P_2(2, -1, 1).$

47. Describe the line segment represented by the vector equation $\langle x, y \rangle = \langle 1, 0 \rangle + t\langle -2, 3 \rangle,$ $0 \le t \le 2.$

48. Describe the line segment represented by the vector equation $\langle x, y, z \rangle = \langle -2, 1, 4 \rangle + t\langle 3, 0, -1 \rangle,$ $0 \le t \le 3.$

In Exercises 49–52, use the result in Exercise 16, Section 14.4.

49. Find the distance between the point $(-2, 1, 1)$ and the line $x = 3 - t, y = t, z = 1 + 2t.$

50. Find the distance between the point $(1, 4, -3)$ and the line $x = 2 + t, y = -1 - t, z = 3t.$

51. Verify that the lines $x = 2 - t,$ $y = 2t,$ $z = 1 + t$ and $x = 1 + 2t, y = 3 - 4t, z = 5 - 2t$ are parallel, and find the distance between them.

52. Verify that the lines $x = 2t, y = 3 + 4t, z = 2 - 6t$ and $x = 1 + 3t, y = 6t, z = -9t$ are parallel, and find the distance between them.

53. Let L_1 and L_2 be the lines whose parametric equations are
$$x = 1 + 2t,   y = 2 - t,   z = 4 - 2t$$
and
$$x = 9 + t,   y = 5 + 3t,   z = -4 - t$$
(a) Show that L_1 and L_2 intersect at the point $(7, -1, -2).$
(b) Find, to the nearest degree, the acute angle between L_1 and L_2 at their intersection.
(c) Find parametric equations for the line that is perpendicular to L_1 and L_2 and passes through their point of intersection.

54. Let L_1 and L_2 be the lines whose parametric equations are
$$x = 4t,   y = 1 - 2t,   z = 2 + 2t$$
and
$$x = 1 + t,   y = 1 - t,   z = -1 + 4t$$
(a) Show that L_1 and L_2 intersect at the point $(2, 0, 3).$
(b) Find, to the nearest degree, the acute angle between L_1 and L_2 at their intersection.
(c) Find parametric equations for the line that is perpendicular to L_1 and L_2 and passes through their point of intersection.

55. Find parametric equations for the line that contains the point $(0, 2, 1)$ and intersects the line $x = 2t, y = 1 - t, z = 2 + t$ at a right angle.

56. Find parametric equations for the line that contains the point $(3, 1, -2)$ and intersects the line $x = -2 + 2t, y = 4 + 2t, z = 2 + t$ at a right angle.

57. Let $x = 4 - t, y = 1 + 2t, z = 2 + t$ and $x = t, y = 1 + t, z = 1 + 2t$ be the straight-line paths of motion of two particles. Suppose that t is the time in seconds and that $x,$ $y,$ and z are measured in centimeters.
(a) How far apart are the particles when $t = 0$?
(b) How close can the particles get?

■ **14.6** PLANES IN 3-SPACE

In this section we shall use vectors to derive equations of planes in 3-space, and we shall use these equations to solve some basic geometric problems.

☐ **PLANES PARALLEL TO THE COORDINATE PLANES**

The plane parallel to the xy-plane and passing through the point $(0, 0, k)$ on the z-axis consists of all points (x, y, z) for which $z = k$ (Figure 14.6.1); thus, the equation of this plane is $z = k.$ Similarly, $x = k$ represents a plane parallel to the yz-plane and passing through $(k, 0, 0),$ while $y = k$ represents a plane parallel to the xz-plane and passing through $(0, k, 0).$

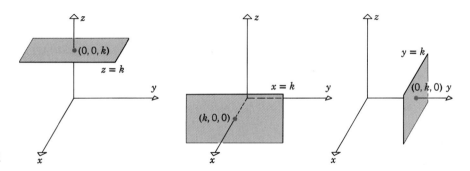

Figure 14.6.1

☐ **PLANES DETERMINED BY A POINT AND A NORMAL VECTOR**

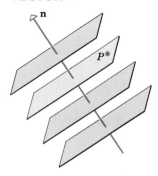

The colored plane is uniquely determined by the point P and the vector $\mathbf{n}$ perpendicular to the plane.

Figure 14.6.2

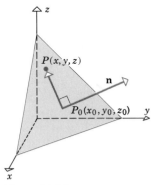

Figure 14.6.3

A plane in 3-space is uniquely determined by specifying a point in the plane and a vector perpendicular to the plane (Figure 14.6.2). A vector perpendicular to a plane is called a *normal* to the plane.

Suppose that we want the equation of the plane passing through the point $P_0(x_0, y_0, z_0)$ and perpendicular to the nonzero vector $\mathbf{n} = \langle a, b, c \rangle$. It is evident from Figure 14.6.3 that the plane consists precisely of those points $P(x, y, z)$ for which the vector $\overrightarrow{P_0P}$ is perpendicular to $\mathbf{n}$; or, phrased as an equation,

$$\mathbf{n} \cdot \overrightarrow{P_0P} = 0 \tag{1}$$

Since $\overrightarrow{P_0P} = \langle x - x_0, y - y_0, z - z_0 \rangle$, (1) can be rewritten as

$$a(x - x_0) + b(y - y_0) + c(z - z_0) = 0 \tag{2}$$

We shall call this the *point-normal form* of the equation of a plane.

Example 1   Find an equation of the plane passing through the point $(3, -1, 7)$ and perpendicular to the vector $\mathbf{n} = \langle 4, 2, -5 \rangle$.

Solution.   From (2), a point-normal form of the equation is

$$4(x - 3) + 2(y + 1) - 5(z - 7) = 0 \quad \blacktriangleleft$$

By multiplying out and collecting terms, (2) can be rewritten in the form

$$ax + by + cz + d = 0 \tag{3}$$

where a, b, c, and d are constants, and a, b, and c are not all zero. To illustrate, the equation in Example 1 can be rewritten as

$$4x + 2y - 5z + 25 = 0$$

As our next theorem shows, every equation of form (3) represents a plane in 3-space.

14.6.1   THEOREM.   *If a, b, c, and d are constants, and a, b, and c are not all zero, then the graph of the equation*

$$ax + by + cz + d = 0$$

is a plane having the vector $\mathbf{n} = \langle a, b, c \rangle$ *as a normal.*

Proof.   By hypothesis, a, b, and c are not all zero. Assume, for the moment, that $a \neq 0$, and note that the equation $ax + by + cz + d = 0$ can be rewritten in the form $a[x + (d/a)] + by + cz = 0$. But this is a point-normal form of the plane passing through the point $(-d/a, 0, 0)$ and having $\mathbf{n} = \langle a, b, c \rangle$ as a normal.

If $a = 0$, then either $b \neq 0$ or $c \neq 0$. A straightforward modification of the above argument will handle these other cases. ∎

The equation in Theorem 14.6.1 is called the *general form* of the equation of a plane.

Example 2 Determine whether the planes

$$3x - 4y + 5z = 0 \quad \text{and} \quad -6x + 8y - 10z - 4 = 0$$

are parallel.

Solution. It is clear geometrically that two planes are parallel if and only if their normals are parallel vectors. A normal to the first plane is

$$\mathbf{n_1} = \langle 3, -4, 5 \rangle$$

and a normal to the second plane is

$$\mathbf{n_2} = \langle -6, 8, -10 \rangle$$

Since $\mathbf{n_2}$ is a scalar multiple of $\mathbf{n_1}$, the normals are parallel. Thus, the planes are also parallel. ◀

A plane cannot be specified by giving a point on it and *one* vector parallel to it, since there are infinitely many such planes (Figure 14.6.4*a*). However, as shown in Figure 14.6.4*b*, a plane is uniquely determined by giving a point in the plane and *two* nonparallel vectors that are parallel to the plane. A plane is also uniquely determined by specifying three noncollinear points in the plane (Figure 14.6.4*c*).

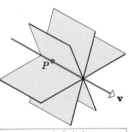

There are infinitely many planes containing P and parallel to $\mathbf{v}$.

(*a*)

Example 3 Find an equation of the plane through the points $P_1(1, 2, -1)$, $P_2(2, 3, 1)$, and $P_3(3, -1, 2)$.

Solution. Since the points P_1, P_2, and P_3 lie in the plane, the vectors $\overrightarrow{P_1P_2} = \langle 1, 1, 2 \rangle$ and $\overrightarrow{P_1P_3} = \langle 2, -3, 3 \rangle$ are parallel to the plane. Therefore,

$$\overrightarrow{P_1P_2} \times \overrightarrow{P_1P_3} = \begin{vmatrix} \mathbf{i} & \mathbf{j} & \mathbf{k} \\ 1 & 1 & 2 \\ 2 & -3 & 3 \end{vmatrix} = 9\mathbf{i} + \mathbf{j} - 5\mathbf{k}$$

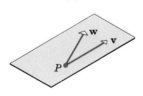

There is a unique plane through P that is parallel to both $\mathbf{v}$ and $\mathbf{w}$.

(*b*)

is normal to the plane, since it is perpendicular to both $\overrightarrow{P_1P_2}$ and $\overrightarrow{P_1P_3}$. By using this normal and the point $P_1(1, 2, -1)$ in the plane, we obtain the point-normal form

$$9(x - 1) + (y - 2) - 5(z + 1) = 0$$

which may be rewritten as

$$9x + y - 5z - 16 = 0 \quad ◀$$

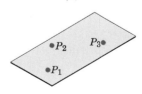

There is a unique plane through three noncollinear points.

(*c*)

Figure 14.6.4

Example 4 Determine whether the line

$$x = 3 + 8t, \quad y = 4 + 5t, \quad z = -3 - t$$

is parallel to the plane $x - 3y + 5z = 12$.

Solution. The vector $\mathbf{v} = \langle 8, 5, -1 \rangle$ is parallel to the line and the vector $\mathbf{n} = \langle 1, -3, 5 \rangle$ is normal to the plane. In order for the line and plane to be parallel, the vectors $\mathbf{v}$ and $\mathbf{n}$ must be perpendicular. But this is not so, since the dot product

$$\mathbf{v} \cdot \mathbf{n} = (8)(1) + (5)(-3) + (-1)(5) = -12$$

is nonzero. Thus, the line and plane are not parallel. ◀

Example 5   Find the intersection of the line and plane in Example 4.

Solution.   If we let (x_0, y_0, z_0) be the point of intersection, then the coordinates of this point satisfy both the equation of the plane and the parametric equations of the line. Thus,

$$x_0 - 3y_0 + 5z_0 = 12 \tag{4}$$

and for some value of t, say $t = t_0$,

$$x_0 = 3 + 8t_0, \quad y_0 = 4 + 5t_0, \quad z_0 = -3 - t_0 \tag{5}$$

Substituting (5) in (4) yields

$$(3 + 8t_0) - 3(4 + 5t_0) + 5(-3 - t_0) = 12$$

Solving for t_0 yields $t_0 = -3$ and on substituting this value in (5), we obtain

$$(x_0, y_0, z_0) = (-21, -11, 0) \quad \blacktriangleleft$$

Two intersecting planes determine two angles of intersection, an acute angle θ ($0 \leq \theta \leq 90°$) and its supplement $180° - \theta$ (Figure 14.6.5a). If $\mathbf{n}_1$ and $\mathbf{n}_2$ are normals to the planes, then the angle between $\mathbf{n}_1$ and $\mathbf{n}_2$ is θ or $180° - \theta$, depending on the directions of the normals (Figure 14.6.5b). Thus, the angles of intersection of two planes may be determined from the normals.

Example 6   Find the acute angle of intersection between the two planes

$$2x - 4y + 4z = 7 \quad \text{and} \quad 6x + 2y - 3z = 2$$

Solution.   From the given equations, we obtain the normals $\mathbf{n}_1 = \langle 2, -4, 4 \rangle$ and $\mathbf{n}_2 = \langle 6, 2, -3 \rangle$. Since the dot product $\mathbf{n}_1 \cdot \mathbf{n}_2 = -8$ is negative, the normals make an obtuse angle. To obtain normals making an acute angle, we can reverse the direction of either $\mathbf{n}_1$ or $\mathbf{n}_2$. To be specific, let us reverse the direction of $\mathbf{n}_1$. Thus, the acute angle θ between the planes will be the angle between $-\mathbf{n}_1 = \langle -2, 4, -4 \rangle$ and $\mathbf{n}_2 = \langle 6, 2, -3 \rangle$. From (3) of Section 14.3 we obtain

$$\cos \theta = \frac{(-\mathbf{n}_1) \cdot \mathbf{n}_2}{\|-\mathbf{n}_1\| \|\mathbf{n}_2\|} = \frac{8}{\sqrt{36}\sqrt{49}} = \frac{4}{21}$$

With the aid of a calculator one obtains

$$\theta = \cos^{-1}\left(\frac{4}{21}\right) \approx 79° \quad \blacktriangleleft$$

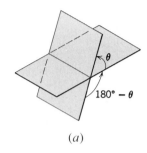

(a)

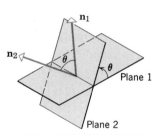

In this figure the angle between $\mathbf{n}_1$ and $\mathbf{n}_2$ is θ. If the direction of one of the normals is reversed, then the angle between the normals would be $180° - \theta$.

(b)

Figure 14.6.5

☐ **DISTANCE PROBLEMS INVOLVING PLANES**

We conclude this section by discussing three basic "distance problems" in 3-space:

- Find the distance between a point and a plane.
- Find the distance between two parallel planes.
- Find the distance between two skew lines.

The three problems are related. If we can find the distance between a point and a plane, then we can find the distance between parallel planes by computing the distance between one of the planes and an arbitrary point P_0 in the other plane (Figure 14.6.6a). Moreover, we can find the distance between two skew lines by computing the distance between parallel planes containing them (Figure 14.6.6b).

14.6.2   THEOREM.   *The distance D between a point $P_0(x_0, y_0, z_0)$ and the plane $ax + by + cz + d = 0$ is*

$$D = \frac{|ax_0 + by_0 + cz_0 + d|}{\sqrt{a^2 + b^2 + c^2}} \tag{6}$$

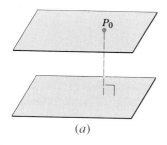

(a)

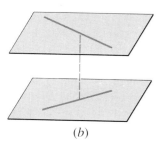

(b)

Figure 14.6.6

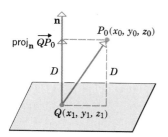

Figure 14.6.7

Proof. Let $Q(x_1, y_1, z_1)$ be any point in the plane, and position the normal $\mathbf{n} = \langle a, b, c \rangle$ so that its initial point is at Q. As illustrated in Figure 14.6.7, the distance D is equal to the length of the orthogonal projection of $\overrightarrow{QP_0}$ on $\mathbf{n}$. Thus, from (10) of Section 14.3,

$$D = \|\text{proj}_{\mathbf{n}} \overrightarrow{QP_0}\| = \frac{|\overrightarrow{QP_0} \cdot \mathbf{n}|}{\|\mathbf{n}\|}$$

But

$$\overrightarrow{QP_0} = \langle x_0 - x_1, y_0 - y_1, z_0 - z_1 \rangle$$
$$\overrightarrow{QP_0} \cdot \mathbf{n} = a(x_0 - x_1) + b(y_0 - y_1) + c(z_0 - z_1)$$
$$\|\mathbf{n}\| = \sqrt{a^2 + b^2 + c^2}$$

Thus,

$$D = \frac{|a(x_0 - x_1) + b(y_0 - y_1) + c(z_0 - z_1)|}{\sqrt{a^2 + b^2 + c^2}} \tag{7}$$

Since the point $Q(x_1, y_1, z_1)$ lies in the plane, its coordinates satisfy the equation of the plane, so that

$$ax_1 + by_1 + cz_1 + d = 0$$

or

$$d = -ax_1 - by_1 - cz_1$$

Substituting this expression in (7) yields (6). ∎

REMARK. See Exercise 31 for an analog of Formula (6) in 2-space.

Example 7 Find the distance D between the point $(1, -4, -3)$ and the plane

$$2x - 3y + 6z = -1$$

Solution. To apply (6), we first rewrite the equation of the plane in the form

$$2x - 3y + 6z + 1 = 0$$

Then

$$D = \frac{|(2)(1) + (-3)(-4) + 6(-3) + 1|}{\sqrt{2^2 + (-3)^2 + 6^2}} = \frac{|-3|}{7} = \frac{3}{7} \quad \blacktriangleleft$$

Example 8 The planes

$$x + 2y - 2z = 3 \quad \text{and} \quad 2x + 4y - 4z = 7$$

are parallel since their normals, $\langle 1, 2, -2 \rangle$ and $\langle 2, 4, -4 \rangle$, are parallel vectors. Find the distance between these planes.

Solution. To find the distance D between the planes, we may select an arbitrary point in one of the planes and compute its distance to the other plane. By setting $y = z = 0$ in the equation $x + 2y - 2z = 3$, we obtain the point $P_0(3, 0, 0)$ in this plane. From (6), the distance from P_0 to the plane $2x + 4y - 4z = 7$ is

$$D = \frac{|(2)(3) + 4(0) + (-4)(0) - 7|}{\sqrt{2^2 + 4^2 + (-4)^2}} = \frac{1}{6} \quad \blacktriangleleft$$

Example 9 It was shown in Example 3 of Section 14.5 that the lines

$$L_1: x = 1 + 4t, \quad y = 5 - 4t, \quad z = -1 + 5t$$

and

$$L_2: \ x = 2 + 8t, \quad y = 4 - 3t, \quad z = 5 + t$$

are skew. Find the distance between them.

Solution.   Let P_1 and P_2 denote parallel planes containing L_1 and L_2, respectively (Figure 14.6.8). To find the distance D between L_1 and L_2, we shall calculate the distance from a point in P_1 to the plane P_2. Setting $t = 0$ in the equations of L_1 yields the point $Q_1(1, 5, -1)$ in plane P_1. (The value $t = 0$ was chosen for simplicity. Any value of t will suffice.) Next we shall obtain an equation for plane P_2.

The vector $\mathbf{u}_1 = \langle 4, -4, 5 \rangle$ is parallel to line L_1, and therefore also parallel to planes P_1 and P_2. Similarly, $\mathbf{u}_2 = \langle 8, -3, 1 \rangle$ is parallel to L_2 and hence parallel to P_1 and P_2. Therefore, the cross product

$$\mathbf{n} = \mathbf{u}_1 \times \mathbf{u}_2 = \begin{vmatrix} \mathbf{i} & \mathbf{j} & \mathbf{k} \\ 4 & -4 & 5 \\ 8 & -3 & 1 \end{vmatrix} = 11\mathbf{i} + 36\mathbf{j} + 20\mathbf{k}$$

is normal to both P_1 and P_2. Using this normal and the point $Q_2(2, 4, 5)$ found by setting $t = 0$ in the equations of L_2, we obtain an equation for P_2:

$$11(x - 2) + 36(y - 4) + 20(z - 5) = 0$$

or

$$11x + 36y + 20z - 266 = 0$$

The distance between $Q_1(1, 5, -1)$ and this plane is

$$D = \frac{|(11)(1) + (36)(5) + (20)(-1) - 266|}{\sqrt{11^2 + 36^2 + 20^2}} = \frac{95}{\sqrt{1817}}$$

This is also the distance between L_1 and L_2.   ◄

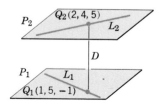

Figure 14.6.8

► Exercise Set 14.6   [C] 11

In Exercises 1–4, find an equation of the plane passing through P and having $\mathbf{n}$ as a normal.

1. $P(2, 6, 1)$; $\mathbf{n} = \langle 1, 4, 2 \rangle$.

2. $P(-1, -1, 2)$; $\mathbf{n} = \langle -1, 7, 6 \rangle$.

3. $P(1, 0, 0)$; $\mathbf{n} = \langle 0, 0, 1 \rangle$.

4. $P(0, 0, 0)$; $\mathbf{n} = \langle 2, -3, -4 \rangle$.

5. Find an equation of the plane passing through the given points.

    (a) $(-2, 1, 1)$, $(0, 2, 3)$, and $(1, 0, -1)$

    (b) $(3, 2, 1)$, $(2, 1, -1)$, and $(-1, 3, 2)$.

6. Determine whether the planes are parallel.

    (a) $3x - 2y + z = 4$ and $6x - 4y + 3z = 7$

    (b) $2x - 8y - 6z - 2 = 0$ and $-x + 4y + 3z - 5 = 0$

    (c) $y = 4x - 2z + 3$ and $x = \frac{1}{4}y + \frac{1}{2}z$.

7. Determine whether the line and plane are parallel.

    (a) $x = 4 + 2t, \ y = -t, \ z = -1 - 4t$;
    $3x + 2y + z - 7 = 0$

    (b) $x = t, \ y = 2t, \ z = 3t; \ x - y + 2z = 5$.

8. Determine whether the planes are perpendicular.

    (a) $x - y + 3z - 2 = 0, \ 2x + z = 1$

    (b) $3x - 2y + z = 1, \ 4x + 5y - 2z = 4$.

9. Determine whether the line and plane are perpendicular.

    (a) $x = -1 + 2t, \ y = 4 + t, \ z = 1 - t$;
    $4x + 2y - 2z = 7$

    (b) $x = 3 - t, \ y = 2 + t, \ z = 1 - 3t; \ 2x + 2y - 5 = 0$.

10. Find the point of intersection of the line and plane.

    (a) $x = t, \ y = t, \ z = t; \ 3x - 2y + z - 5 = 0$

    (b) $x = 1 + t, \ y = -1 + 3t, \ z = 2 + 4t; \ x - y + 4z = 7$

    (c) $x = 2 - t, \ y = 3 + t, \ z = t; \ 2x + y + z = 1$.

11. Find the acute angle of intersection of the planes (to the nearest degree).

    (a) $x = 0$ and $2x - y + z - 4 = 0$

    (b) $x + 2y - 2z = 5$ and $6x - 3y + 2z = 8$.

12. Find an equation of the plane through $(-1, 4, -3)$ and perpendicular to the line $x - 2 = t, \ y + 3 = 2t, \ z = -t$.

13. Find an equation of

    (a) the xy-plane   (b) the xz-plane

    (c) the yz-plane.

14. Find an equation of the plane that contains the point (x_0, y_0, z_0) and is

    (a) parallel to the xy-plane

    (b) parallel to the yz-plane

    (c) parallel to the xz-plane.

15. Find an equation of the plane through the origin that is parallel to the plane $4x - 2y + 7z + 12 = 0$.

16. Find an equation of the plane containing the line $x = -2 + 3t$, $y = 4 + 2t$, $z = 3 - t$ and perpendicular to the plane $x - 2y + z = 5$.

17. Find an equation of the plane through $(-1, 4, 2)$ and containing the line of intersection of the planes $4x - y + z - 2 = 0$ and $2x + y - 2z - 3 = 0$.

18. Show that the points $(1, 0, -1)$, $(0, 2, 3)$, $(-2, 1, 1)$, and $(4, 2, 3)$ lie in the same plane.

19. Find parametric equations of the line through $(5, 0, -2)$ that is parallel to the planes $x - 4y + 2z = 0$ and $2x + 3y - z + 1 = 0$.

20. Find an equation of the plane through $(-1, 2, -5)$ and perpendicular to the planes $2x - y + z = 1$ and $x + y - 2z = 3$.

21. Find an equation of the plane through $(1, 2, -1)$ and perpendicular to the line of intersection of the planes $2x + y + z = 2$ and $x + 2y + z = 3$.

22. Find a plane through the points $P_1(-2, 1, 4)$, $P_2(1, 0, 3)$ and perpendicular to the plane $4x - y + 3z = 2$.

23. Show that the lines

$$x = -2 + t, \quad y = 3 + 2t, \quad z = 4 - t$$

and

$$x = 3 - t, \quad y = 4 - 2t, \quad z = t$$

are parallel and find an equation of the plane they determine.

24. Find an equation of the plane that contains the point $(2, 0, 3)$ and the line $x = -1 + t$, $y = t$, $z = -4 + 2t$.

25. Find an equation of the plane, each of whose points is equidistant from $(2, -1, 1)$ and $(3, 1, 5)$.

26. Find an equation of the plane containing the line $x = 3t$, $y = 1 + t$, $z = 2t$ and parallel to the intersection of the planes $2x - y + z = 0$ and $y + z + 1 = 0$.

27. Show that the line $x = 0$, $y = t$, $z = t$

    (a) lies in the plane $6x + 4y - 4z = 0$

    (b) is parallel to and below the plane $5x - 3y + 3z = 1$

    (c) is parallel to and above the plane $6x + 2y - 2z = 3$.

28. Show that the lines

$$x + 1 = 4t, \quad y - 3 = t, \quad z - 1 = 0$$

and

$$x + 13 = 12t, \quad y - 1 = 6t, \quad z - 2 = 3t$$

intersect and find an equation of the plane they determine.

29. Find parametric equations of the line of intersection of the planes

    (a) $-2x + 3y + 7z + 2 = 0$ and
          $x + 2y - 3z + 5 = 0$

    (b) $3x - 5y + 2z = 0$ and $z = 0$.

30. Show that the plane whose intercepts with the coordinate axes are $x = a$, $y = b$, and $z = c$ has equation

$$\frac{x}{a} + \frac{y}{b} + \frac{z}{c} = 1$$

provided a, b, and c are nonzero.

31. (a) Prove that the distance D between a point $P_0(x_0, y_0)$ and the line $ax + by + c = 0$ is given by

$$D = \frac{|ax_0 + by_0 + c|}{\sqrt{a^2 + b^2}}$$

        [*Hint:* First establish that the vector $\mathbf{n} = a\mathbf{i} + b\mathbf{j}$ is normal to the line by showing that for any distinct points $P_1(x_1, y_1)$ and $P_2(x_2, y_2)$ on the line, the vector $\mathbf{n}$ is perpendicular to $\overrightarrow{P_1 P_2}$.]

    (b) Use the formula in part (a) to find the distance between the point $P(1, -2)$ and the line $3x + 4y + 7 = 0$.

32. Use the formula in part (a) of Exercise 31 to find the distance between the point $P(-3, 5)$ and the line $y = -2x + 1$.

> In Exercises 33–35, find the distance between the point and the plane.

33. $(1, -2, 3)$; $2x - 2y + z = 4$.

34. $(0, 1, 5)$; $3x + 6y - 2z - 5 = 0$.

35. $(7, 2, -1)$; $20x - 4y - 5z = 0$.

> In Exercises 36–38, find the distance between the given parallel planes.

36. $2x - 3y + 4z = 7$, $4x - 6y + 8z = 3$.

37. $-2x + y + z = 0$, $6x - 3y - 3z - 5 = 0$.

38. $x + y + z = 1$, $x + y + z = -1$.

> In Exercises 39–41, find the distance between the given skew lines.

39. $x = 1 + 7t$, $y = 3 + t$, $z = 5 - 3t$; $x = 4 - t$, $y = 6$, $z = 7 + 2t$.

40. $x = 3 - t$, $y = 4 + 4t$, $z = 1 + 2t$; $x = t$, $y = 3$, $z = 2t$.

41. $x = 2 + 4t$, $y = 6 - 4t$, $z = 5t$; $x = 3 + 8t$, $y = 5 - 3t$, $z = 6 + t$.

42. Show that the line $x = -1 + t$, $y = 3 + 2t$, $z = -t$ and the plane $2x - 2y - 2z + 3 = 0$ are parallel, and find the distance between them.

43. Prove: The planes

$$a_1x + b_1y + c_1z = d_1$$

and

$$a_2x + b_2y + c_2z = d_2$$

are perpendicular if and only if $a_1a_2 + b_1b_2 + c_1c_2 = 0$.

44. Let $\mathbf{r}_0 = \langle x_0, y_0, z_0 \rangle$ and $\mathbf{r} = \langle x, y, z \rangle$. Describe the set of all points (x, y, z) for which

(a) $\mathbf{r} \cdot \mathbf{r}_0 = 0$ (b) $(\mathbf{r} - \mathbf{r}_0) \cdot \mathbf{r}_0 = 0$.

45. Find an equation of the sphere with its center at $(2, 1, -3)$ and tangent to the plane $x - 3y + 2z = 4$.

46. Find where the line that passes through the point $(3, 1, 0)$ and is normal to the plane $2x + y - z = 0$ intersects the plane.

■ 14.7 QUADRIC SURFACES

In this section we shall study an important class of surfaces that are the three-dimensional analogs of the conic sections.

☐ **MESH PLOTS OF SURFACES**

In 2-space the general shape of a curve can be obtained by plotting points. However, for surfaces in 3-space, point-plotting is not generally helpful since too many points are needed to obtain even a crude picture of the surface. It is better to build up the shape of the surface using curves obtained by cutting the surface with planes parallel to the coordinate planes. This is called a ***mesh plot*** of the surface. For example, Figure 14.7.1 shows a mesh plot of the surface $z = x^3 - 3xy^2$. (This surface is called a "monkey saddle" because a monkey sitting astride the x-axis has a place for its two legs and tail.)

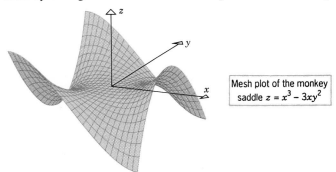

Mesh plot of the monkey saddle $z = x^3 - 3xy^2$

Figure 14.7.1

☐ **TRACES OF SURFACES**

Figure 14.7.2

The curve of intersection of a surface with a plane is called the ***trace*** of the surface in the plane (Figure 14.7.2). Mesh plots are usually built up from traces in planes parallel to the coordinate planes. Equations for such traces can be obtained by substituting the equation of the plane into the equation of the surface. For example, the trace of the monkey saddle of Figure 14.7.1 in the plane $x = 3$ is obtained by substituting $x = 3$ into

$$z = x^3 - 3xy^2$$

which yields

$$z - 27 = -9y^2 \quad (x = 3) \tag{1}$$

This is a parabola with vertex at the point $(x, y, z) = (3, 0, 27)$ opening in the negative z-direction (why?). Observe that this result is consistent with Figure 14.7.1.

REMARK.   In (1) we explicitly noted the restriction $x = 3$ in parentheses. This is necessary because the equation by itself does not convey the information that the trace lies in the plane $x = 3$.

☐ **THE QUADRIC SURFACES**

Earlier in the text, we saw that the graph in two dimensions of a second-degree equation in x and y,

$$Ax^2 + Bxy + Cy^2 + Dx + Ey + F = 0$$

is a conic section (possibly degenerate). In three dimensions, the graph of a second-degree equation in x, y, and z,

$$Ax^2 + By^2 + Cz^2 + Dxy + Exz + Fyz + Gx + Hy + Iz + J = 0$$

is called a *quadric surface* or a *quadric*. The most important quadric surfaces, shown in Table 14.7.1, are the *ellipsoids*, *hyperboloids of one and two sheets*, *elliptic cones*, *elliptic paraboloids*, and *hyperbolic paraboloids*.

The simplest equations for the quadric surfaces result when the surfaces are positioned in certain "standard positions" relative to the coordinate axes. Table 14.7.1 illustrates

Table 14.7.1

SURFACE	EQUATION	SURFACE	EQUATION
ELLIPSOID	$\dfrac{x^2}{a^2} + \dfrac{y^2}{b^2} + \dfrac{z^2}{c^2} = 1$ The traces in the coordinate planes are ellipses, as are the traces in planes parallel to the coordinate planes.	ELLIPTIC CONE	$z^2 = \dfrac{x^2}{a^2} + \dfrac{y^2}{b^2}$ The trace in the xy-plane is a point (the origin), and the traces in planes parallel to the xy-plane are ellipses. The traces in the yz- and xz-planes are pairs of lines intersecting at the origin. The traces in planes parallel to these are hyperbolas.
HYPERBOLOID OF ONE SHEET	$\dfrac{x^2}{a^2} + \dfrac{y^2}{b^2} - \dfrac{z^2}{c^2} = 1$ The trace in the xy-plane is an ellipse, as are the traces in planes parallel to the xy-plane. The traces in the yz-plane and xz-plane are hyperbolas, as are the traces in planes parallel to these.	ELLIPTIC PARABOLOID	$z = \dfrac{x^2}{a^2} + \dfrac{y^2}{b^2}$ The trace in the xy-plane is a point (the origin), and the traces in planes parallel to and above the xy-plane are ellipses. The traces in the yz- and xz-planes are parabolas, as are the traces in planes parallel to these.
HYPERBOLOID OF TWO SHEETS	$\dfrac{x^2}{a^2} + \dfrac{y^2}{b^2} - \dfrac{z^2}{c^2} = -1$ There is no trace in the xy-plane. In planes parallel to the xy-plane, which intersect the surface, the traces are ellipses. In the yz- and xz-planes, the traces are hyperbolas, as are the traces in planes parallel to these.	HYPERBOLIC PARABOLOID	$z = \dfrac{y^2}{b^2} - \dfrac{x^2}{a^2}$ The trace in the xy-plane is a pair of lines intersecting at the origin. The traces in planes parallel to the xy-plane are hyperbolas. The hyperbolas above the xy-plane open in the y-direction, and those below in the x-direction. The traces in the yz- and xz-planes are parabolas, as are the traces in planes parallel to these.

some typical standard positions and the equations that result. The table also describes the traces of the quadric surfaces in planes parallel to the coordinate planes. As we shall see, such traces are important in studying properties of quadric surfaces. The constants a, b, and c that appear in the equations in the table are all assumed to be positive.

Example 1   To illustrate how the traces described in Table 14.7.1 were obtained, we shall consider the case of the elliptic cone

$$z^2 = \frac{x^2}{a^2} + \frac{y^2}{b^2} \tag{2}$$

The other cases are similar. The trace of (2) in the plane $z = 0$ (the xy-plane) is

$$\frac{x^2}{a^2} + \frac{y^2}{b^2} = 0 \qquad (z = 0)$$

which implies that $x = 0$, $y = 0$, $z = 0$. Thus, the trace is the single point $(0, 0, 0)$.

For $k \neq 0$, the trace of (2) in the plane $z = k$ is (after simplification)

$$\frac{x^2}{(ak)^2} + \frac{y^2}{(bk)^2} = 1 \qquad (z = k)$$

which is an ellipse. As $|k|$ increases, so do $(ak)^2$ and $(bk)^2$. Thus, the dimensions of these ellipses increase as the planes containing them recede from the xy-plane. This is consistent with the picture of the elliptic cone in Table 14.7.1.

The trace of (2) in the xz-plane is

$$z^2 = \frac{x^2}{a^2} \qquad (y = 0)$$

which is equivalent to the pair of equations

$$z = \frac{x}{a} \quad \text{and} \quad z = -\frac{x}{a} \qquad (y = 0)$$

These are the equations of two lines in the xz-plane that intersect at the origin. Similarly, the trace of (2) in the yz-plane is the pair of intersecting lines

$$z = \frac{y}{b} \quad \text{and} \quad z = -\frac{y}{b} \qquad (x = 0)$$

For $k \neq 0$, the trace of (2) in the plane $y = k$, which is parallel to the xz-plane, is (after simplification)

$$\frac{z^2}{k^2/b^2} - \frac{x^2}{a^2 k^2/b^2} = 1 \qquad (y = k)$$

This is a hyperbola in the plane $y = k$ opening along a line parallel to the z-axis. Similarly, for $k \neq 0$ the trace in the plane $x = k$, which is parallel to the yz-plane, is (after simplification)

$$\frac{z^2}{k^2/a^2} - \frac{y^2}{b^2 k^2/a^2} = 1 \qquad (x = k)$$

This is a hyperbola in the plane $x = k$ opening along a line parallel to the z-axis. ◀

REMARK.   An elliptic cone in which the elliptical cross sections are circles is called a *circular cone*; similarly, elliptic paraboloids with circular cross sections are called *circular paraboloids*.

☐ **TECHNIQUES FOR GRAPHING QUADRIC SURFACES**

Accurate graphs of quadric surfaces are best left to computers. However, the techniques that follow can be used to obtain rough sketches of these surfaces that are useful for many purposes.

ELLIPSOIDS   A sketch of the ellipsoid

$$\frac{x^2}{a^2} + \frac{y^2}{b^2} + \frac{z^2}{c^2} = 1 \quad (a > 0, b > 0, c > 0) \tag{3}$$

can be obtained by first plotting the intersections with the coordinate axes, then sketching the elliptical traces in the coordinate planes, and then sketching the surface itself using the traces as a guide.

Example 2   Sketch the graph of the ellipsoid

$$\frac{x^2}{4} + \frac{y^2}{16} + \frac{z^2}{9} = 1 \tag{4}$$

Solution.   The x-intercepts, obtained by setting $y = 0$ and $z = 0$ in (4), are $x = \pm 2$. Similarly, the y-intercepts are $y = \pm 4$, and the z-intercepts are $z = \pm 3$. From these intercepts we obtain the elliptical traces and the ellipsoid sketched in Figure 14.7.3.   ◀

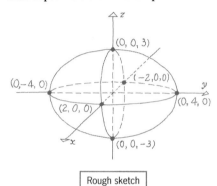

Rough sketch

Figure 14.7.3

HYPERBOLOIDS OF ONE SHEET   A sketch of the hyperboloid of one sheet

$$\frac{x^2}{a^2} + \frac{y^2}{b^2} - \frac{z^2}{c^2} = 1 \quad (a > 0, b > 0, c > 0) \tag{5}$$

can be obtained by first sketching the elliptical trace in the xy-plane, then the elliptical traces in the planes $z = c$ and $z = -c$, and then the hyperbolic curves that join the endpoints of the axes of these ellipses.

Example 3   Sketch the graph of the hyperboloid of one sheet

$$x^2 + y^2 - \frac{z^2}{4} = 1 \tag{6}$$

Solution.   The trace in the xy-plane, obtained by setting $z = 0$ in (6), is

$$x^2 + y^2 = 1 \quad (z = 0)$$

which is a circle of radius 1 centered on the z-axis. The traces in the planes $z = 2$ and $z = -2$, obtained by setting $z = \pm 2$ in (6), are given by

$$x^2 + y^2 = 2 \quad (z = \pm 2)$$

which are circles of radius $\sqrt{2}$ centered on the z-axis. Joining these circles by the hyperbolic traces in the vertical coordinate planes yields Figure 14.7.4.   ◀

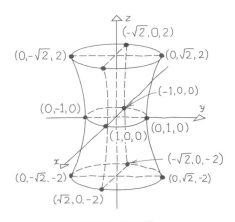

Figure 14.7.4

Rough sketch

HYPERBOLOIDS OF TWO SHEETS   A sketch of the hyperboloid of two sheets

$$\frac{x^2}{a^2} + \frac{y^2}{b^2} - \frac{z^2}{c^2} = -1 \quad (a > 0, b > 0, c > 0) \tag{7}$$

can be obtained by first plotting the intersections with the z-axis, then sketching the elliptical traces in the planes $z = 2c$ and $z = -2c$, and then sketching the hyperbolic traces that connect the z-axis intersections and the endpoints of the axes of the ellipses. (It is not essential to use the planes $z = \pm 2c$; any pair of horizontal planes above $z = c$ and below $z = -c$ would do just as well.)

Example 4   Sketch the graph of the hyperboloid of two sheets

$$x^2 + \frac{y^2}{4} - z^2 = -1 \tag{8}$$

Solution.   The z-intercepts, obtained by setting $x = 0$ and $y = 0$ in (8), are $z = \pm 1$. The traces in the planes $z = 2$ and $z = -2$, obtained by setting $z = \pm 2$ in (8), are given by

$$\frac{x^2}{3} + \frac{y^2}{12} = 1 \qquad (z = \pm 2)$$

Sketching these ellipses and the hyperbolic traces in the vertical coordinate planes yields Figure 14.7.5.   ◄

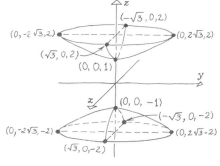

Figure 14.7.5

Rough sketch

ELLIPTIC CONES   A sketch of the elliptic cone

$$z^2 = \frac{x^2}{a^2} + \frac{y^2}{b^2} \quad (a > 0, b > 0) \tag{9}$$

can be obtained by first sketching the elliptical traces in the planes $z = \pm 1$, then sketching the linear traces that connect the endpoints of the axes of the ellipses.

Example 5   Sketch the graph of the elliptic cone

$$z^2 = x^2 + \frac{y^2}{4} \tag{10}$$

Solution.   The traces of (10) in the planes $z = \pm 1$ are given by

$$x^2 + \frac{y^2}{4} = 1 \qquad (z = \pm 1)$$

Sketching these ellipses and the linear traces in the vertical coordinate planes yields the graph in Figure 14.7.6.   ◄

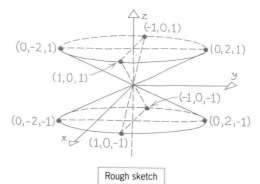

Figure 14.7.6

Rough sketch

ELLIPTIC PARABOLOIDS   A sketch of the elliptic paraboloid

$$z = \frac{x^2}{a^2} + \frac{y^2}{b^2} \quad (a > 0, b > 0) \tag{11}$$

can be obtained by first sketching the elliptical trace in the plane $z = 1$, then sketching the parabolic traces (in the vertical coordinate planes) whose vertices are at the origin and pass through the endpoints of the axes of the ellipse.

Example 6   Sketch the graph of the elliptic paraboloid

$$z = \frac{x^2}{4} + \frac{y^2}{9} \tag{12}$$

Solution.   The trace of (12) in the plane $z = 1$ is

$$\frac{x^2}{4} + \frac{y^2}{9} = 1 \qquad (z = 1)$$

Sketching this ellipse and the parabolic traces in the vertical coordinate planes yields the graph in Figure 14.7.7.   ◄

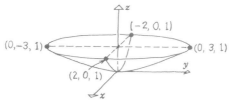

Figure 14.7.7

Rough sketch

HYPERBOLIC PARABOLOIDS   The graph of a hyperbolic paraboloid is difficult to draw, but fortunately a rough sketch showing the orientation of the surface relative to the coordinate axes suffices for most purposes. The orientation of the hyperbolic paraboloid

$$z = \frac{y^2}{b^2} - \frac{x^2}{a^2} \quad (a > 0, b > 0) \tag{13}$$

can be obtained by first sketching the two parabolic traces that pass through the origin; one of these results from setting $x = 0$ in (13) and the other from setting $y = 0$. After the parabolic traces are drawn, sketch the hyperbolic traces in the planes $z = \pm 1$ with their proper orientation. Finally, sketch in any missing edges.

Example 7   Sketch the graph of the hyperbolic paraboloid

$$z = \frac{y^2}{4} - \frac{x^2}{9} \tag{14}$$

Solution.   Setting $x = 0$ in (14) yields

$$z = \frac{y^2}{4} \quad (x = 0)$$

which is a parabola in the yz-plane with vertex at the origin and opening in the positive z-direction (why?), and setting $y = 0$ yields

$$z = -\frac{x^2}{9} \quad (y = 0)$$

which is a parabola in the xz-plane with vertex at the origin and opening in the negative z-direction.

The trace in the plane $z = 1$ is

$$\frac{y^2}{4} - \frac{x^2}{9} = 1 \quad (z = 1)$$

which is a hyperbola that opens along a line parallel to the y-axis (verify), and the trace in the plane $z = -1$ is

$$\frac{x^2}{9} - \frac{y^2}{4} = 1 \quad (z = -1)$$

which is a hyperbola that opens along a line parallel to the x-axis. Combining all of the above information leads to the sketch in Figure 14.7.8.   ◀

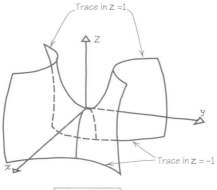

Figure 14.7.8

Rough sketch

REMARK.   The hyperbolic paraboloid (13) has an interesting behavior near the origin. The trace in the xz-plane has a relative maximum at the origin, and the trace in the yz-plane has a relative minimum there (see Figure 14.7.8, for example). On this surface the origin is commonly described as a **saddle point** or a **minimax point**.

□ REFLECTIONS OF QUADRIC SURFACES

In our earlier study of curves in the xy-plane, we observed that interchanging the variables x and y in an equation has the effect of reflecting the graph of the equation about the line $y = x$. Analogous results occur in xyz-coordinate systems. For example, interchanging the variables x and z in the equation of a surface has the geometric effect of reflecting that surface symmetrically about the plane $x = z$. Thus, the equation

$$x^2 = \frac{z^2}{a^2} + \frac{y^2}{b^2}$$

represents an elliptic cone opening along the x-axis rather than the z-axis as shown in Table 14.7.1.

We also note that replacing z by $-z$ in the equation of the elliptic paraboloid in Table 14.7.1 has the geometric effect of reflecting the surface symmetrically about the xy-plane, thereby producing an elliptic paraboloid that opens down. The equation for this surface can be written as

$$z = -\left(\frac{x^2}{a^2} + \frac{y^2}{b^2}\right) \tag{15}$$

□ TRANSLATION OF AXES IN 3-SPACE

Let an $x'y'z'$-coordinate system be obtained by translating an xyz-coordinate system so that the $x'y'z'$-origin is at the point whose xyz-coordinates are $(x, y, z) = (h, k, l)$ (Figure 14.7.9). It can be shown that the $x'y'z'$-coordinates and xyz-coordinates of a point P are related by

$$x' = x - h, \quad y' = y - k, \quad z' = z - l \tag{16}$$

[Compare this to (6) in Section 12.2.]

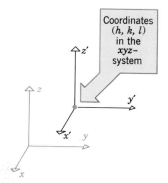

Coordinates (h, k, l) in the xyz-system

Figure 14.7.9

Example 8   Sketch the surface $z = 1 - x^2 - y^2$.

Solution.   Rewrite the equation in the form

$$z - 1 = -(x^2 + y^2) \tag{17}$$

If we translate the coordinate axes so that the new origin is at the point $(h, k, l) = (0, 0, 1)$, then the translation equations (16) are

$$x' = x, \quad y' = y, \quad z' = z - 1$$

so that in $x'y'z'$-coordinates (17) becomes

$$z' = -(x'^2 + y'^2)$$

which is of form (15) with $a = b = 1$. Thus, the surface is a circular paraboloid, opening down, with vertex $(0, 0, 1)$ in xyz-coordinates (Figure 14.7.10).   ◀

Example 9   Sketch the surface

$$4x^2 + 4y^2 + z^2 + 8y - 4z = -4$$

Solution.   Completing the squares yields

$$4x^2 + 4(y + 1)^2 + (z - 2)^2 = -4 + 4 + 4$$

or

$$x^2 + (y + 1)^2 + \frac{(z - 2)^2}{4} = 1 \qquad (18)$$

If we translate the coordinate axes so that the new origin is $(h, k, l) = (0, -1, 2)$, then the translation equations (16) are

$$x' = x, \quad y' = y + 1, \quad z' = z - 2$$

so that in $x'y'z'$-coordinates (18) becomes

$$x'^2 + y'^2 + \frac{z'^2}{4} = 1$$

which is of form (3) with $a = 1$, $b = 1$, and $c = 2$. Thus, the surface is the ellipsoid shown in Figure 14.7.11.   ◀

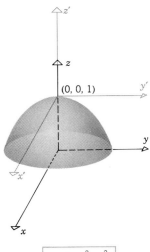

Figure 14.7.10 $z = 1 - x^2 - y^2$

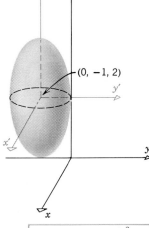

Figure 14.7.11 $x^2 + (y + 1)^2 + \frac{(z-2)^2}{4} = 1$

▶ **Exercise Set 14.7**

In Exercises 1–6, find an equation for the trace of the surface in the given plane and identify the trace.

1. $4x^2 + y^2 + z^2 = 4$:

(a) $z = 0$ (b) $x = 1/2$ (c) $y = 1$.

2. $9x^2 - y^2 + 4z^2 = 9$:

(a) $z = 0$ (b) $x = 2$ (c) $y = 4$.

3. $9x^2 - y^2 - z^2 = 16$:

(a) $y = 0$ (b) $x = 2$ (c) $z = 2$.

4. $x^2 + 4y^2 - 9z^2 = 0$:

(a) $x = 0$ (b) $y = 1$ (c) $z = 1$.

5. $z = 9x^2 + 4y^2$:

    (a) $x = 0$ (b) $y = 2$ (c) $z = 4$.

6. $z = x^2 - 4y^2$:

    (a) $y = 0$ (b) $x = 1$ (c) $z = 4$.

In Exercises 7–28, name and sketch the quadric surface.

7. $x^2 + y^2/4 + z^2/9 = 1$.   **8.** $x^2 + 4y^2 + 9z^2 = 36$.

9. $4x^2 + y^2 + 4z^2 = 16$.   **10.** $4x^2 + 4y^2 + z^2 = 9$.

11. $x^2/4 + y^2/9 - z^2/16 = 1$.

12. $4x^2 - y^2 + 4z^2 = 16$.

13. $2y^2 - x^2 + 2z^2 = 8$.   **14.** $x^2 + y^2 - z^2 = 9$.

15. $y^2 - 2x^2 - 2z^2 = 1$.   **16.** $x^2 - 3y^2 - 3z^2 = 9$.

17. $9z^2 - 4y^2 - 9x^2 = 36$.   **18.** $y^2 - 4x^2 - z^2 = 4$.

19. $4z^2 = x^2 + 4y^2$.   **20.** $y^2 = x^2 + z^2$.

21. $x^2 - 3y^2 - 3z^2 = 0$.   **22.** $9x^2 + 4y^2 - 36z^2 = 0$.

23. $y = x^2 + z^2$.   **24.** $z - 3x^2 - 3y^2 = 0$.

25. $4z = x^2 + 2y^2$.   **26.** $x - y^2 - 4z^2 = 0$.

27. $z = x^2/4 - y^2/9$.   **28.** $z = y^2 - x^2$.

29. The following equations represent quadric surfaces with orientations different from those in Table 14.7.1. Name and sketch the surface.

    (a) $\dfrac{z^2}{c^2} - \dfrac{y^2}{b^2} + \dfrac{x^2}{a^2} = 1$ (b) $\dfrac{x^2}{a^2} - \dfrac{y^2}{b^2} - \dfrac{z^2}{c^2} = 1$

    (c) $x = \dfrac{y^2}{b^2} + \dfrac{z^2}{c^2}$ (d) $x^2 = \dfrac{y^2}{b^2} + \dfrac{z^2}{c^2}$

    (e) $y = \dfrac{z^2}{c^2} - \dfrac{x^2}{a^2}$ (f) $y = -\left(\dfrac{x^2}{a^2} + \dfrac{z^2}{c^2}\right)$.

In Exercises 30–33, sketch the graph of the equation.

30. $z = \sqrt{1 - x^2 - y^2}$.   **31.** $z = \sqrt{x^2 + y^2}$.

32. $z = \sqrt{1 + x^2 + y^2}$.   **33.** $z = \sqrt{x^2 + y^2 - 1}$.

In Exercises 34–39, name and sketch the surface.

34. $\dfrac{(x-1)^2}{4} + \dfrac{(y-2)^2}{9} + \dfrac{(z-4)^2}{16} = 1$.

35. $z = (x+2)^2 + (y-3)^2 - 9$.

36. $4x^2 - y^2 + 16(z-2)^2 = 100$.

37. $9x^2 + y^2 + 4z^2 - 18x + 2y + 16z = 10$.

38. $z^2 = 4x^2 + y^2 + 8x - 2y + 4z$.

39. $z = 4 - x^2 - y^2 - 2y$.

40. Obtain the results in Table 14.7.1 for the ellipsoid $x^2/a^2 + y^2/b^2 + z^2/c^2 = 1$.

41. Obtain the results in Table 14.7.1 for the hyperboloid of one sheet $x^2/a^2 + y^2/b^2 - z^2/c^2 = 1$.

42. Obtain the results in Table 14.7.1 for the hyperboloid of two sheets $x^2/a^2 + y^2/b^2 - z^2/c^2 = -1$.

43. Obtain the results in Table 14.7.1 for the elliptic paraboloid $z = x^2/a^2 + y^2/b^2$.

44. Obtain the results in Table 14.7.1 for the hyperbolic paraboloid $z = y^2/b^2 - x^2/a^2$.

In Exercises 45–52, find an equation of the projection onto the xy-plane of the curve of intersection of the surfaces. Identify the curve. [*Hint:* The values of z on both surfaces are the same along the curve of intersection.]

45. The paraboloids $z = x^2 + y^2$ and $z = 4 - x^2 - y^2$.

46. The paraboloids $z = x^2 + y^2$ and $z = 1 - 4x^2 - y^2$.

47. The paraboloid $z = x^2 + y^2$ and the plane $z = 2x$.

48. The paraboloid $z = 4 - x^2 - y^2$ and the parabolic cylinder $z = y^2$.

49. The cone $z^2 = x^2 + y^2$ and the plane $z = y + 1$.

50. The cone $z^2 = x^2 + y^2$ and the portion of the parabolic cylinder given by $z = 2\sqrt{y}$.

51. The ellipsoid $x^2 + y^2 + 4z^2 = 5$ and the portion of the parabolic cylinder given by $z = \sqrt{x}$.

52. The ellipsoid $x^2 + 2y^2 + z^2 = 2$ and the plane $z = x$.

53. For the elliptic paraboloid

$$z = \frac{x^2}{9} + \frac{y^2}{4}$$

    (a) find the focus and vertex of the (parabolic) trace in the plane $x = k$

    (b) find the foci and the endpoints of the major and minor axes of the (elliptic) trace in the plane $z = k$.

54. Use the method of slicing to find the volume of the ellipsoid

$$\frac{x^2}{a^2} + \frac{y^2}{b^2} + \frac{z^2}{c^2} = 1$$

[*Hint:* The area of the ellipse $x^2/a^2 + y^2/b^2 = 1$ is πab.]

55. Find an equation of the surface consisting of the points $P(x, y, z)$ for which the distance between P and the plane $z = -1$ is equal to the distance between P and the point $(0, 0, 1)$. Identify the surface.

56. Find an equation of the surface consisting of the points $P(x, y, z)$ for which the distance between P and the plane $z = -1$ is twice the distance between P and the point $(0, 0, 1)$. Identify the surface.

57. (a) Show that the lines

$$x = 3 + t, \quad y = 2 + t, \quad z = 5 + 2t$$

    and

$$x = 3 + t, \quad y = 2 - t, \quad z = 5 + 10t$$

    both lie completely on the hyperbolic paraboloid $z = x^2 - y^2$.

(b) Let $P_0(x_0, y_0, z_0)$ be any point on the surface $z = x^2 - y^2$. Show that it is always possible to find two lines with equations of the form $x = x_0 + t$, $y = y_0 + at$, $z = z_0 + bt$ that pass through P_0 and lie completely on the surface.

58. (a) Show that the lines $x = 2 + \frac{3}{5}t$, $y = 1 + \frac{4}{5}t$, $z = 2 + t$ and $x = 2 + t$, $y = 1$, $z = 2 + t$ both lie completely on the hyperboloid $x^2 + y^2 - z^2 = 1$.

(b) Let $P_0(x_0, y_0, z_0)$ be any point on the surface $x^2 + y^2 - z^2 = 1$. Show that it is always possible to find two lines with equations of the form $x = x_0 + at$, $y = y_0 + bt$, $z = z_0 + t$ that pass through P_0 and lie completely on the surface.

■ 14.8 CYLINDRICAL AND SPHERICAL COORDINATES

> *In this section we shall discuss two new types of coordinate systems in three dimensions that are extremely important in applied problems. These coordinate systems produce simpler equations than rectangular coordinate systems for surfaces with various kinds of symmetries.*

☐ **CYLINDRICAL AND SPHERICAL COORDINATES**

To have a useful coordinate system in three dimensions, each point in space must be associated with a triple of real numbers (the coordinates of the point), and each triple of real numbers must determine a unique point. Figure 14.8.1 illustrates three possible ways for doing this. The rectangular coordinates (x, y, z) of a point P are shown in part (a) of the figure, the *cylindrical coordinates* (r, θ, z) of P are shown in part (b), and the *spherical coordinates* (ρ, θ, ϕ) of P are shown in part (c).

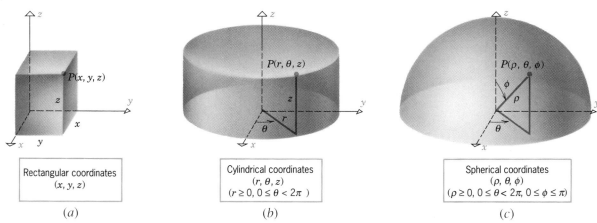

| Rectangular coordinates (x, y, z) | Cylindrical coordinates (r, θ, z) $(r \geq 0, 0 \leq \theta < 2\pi)$ | Spherical coordinates (ρ, θ, ϕ) $(\rho \geq 0, 0 \leq \theta < 2\pi, 0 \leq \phi \leq \pi)$ |

Figure 14.8.1 (a) (b) (c)

The restrictions on the values of the cylindrical and spherical coordinates noted in Figure 14.8.1 are fairly standard; they ensure that each point not on the z-axis has a unique set of coordinates.

☐ **CONSTANT SURFACES**

In rectangular coordinates the surfaces represented by equations of the form

$$x = x_0, \quad y = y_0, \quad \text{and} \quad z = z_0$$

where x_0, y_0, and z_0 are constants, are planes parallel to the yz-plane, xz-plane, and xy-plane, respectively (Figure 14.8.2a). In cylindrical coordinates the surfaces represented by equations of the form

$$r = r_0, \quad \theta = \theta_0, \quad \text{and} \quad z = z_0$$

where r_0, θ_0, and z_0 are constants, are shown in Figure 14.8.2b.

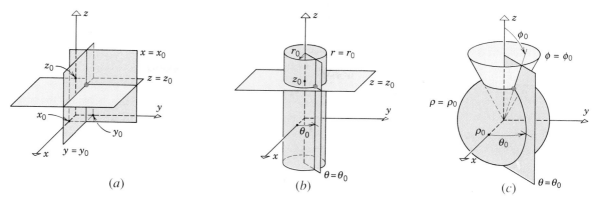

Figure 14.8.2

(a) (b) (c)

- The surface $r = r_0$ is a right-circular cylinder of radius r_0 centered on the z-axis. At each point (r, θ, z) on this cylinder, r has the value r_0, but θ and z are unrestricted except for our blanket assumption that $0 \leq \theta < 2\pi$.

- The surface $\theta = \theta_0$ is a half-plane attached along the z-axis and making an angle θ_0 with the positive x-axis. At each point (r, θ, z) on this surface, θ has the value θ_0, but r and z are unrestricted except for our blanket assumption that $r \geq 0$.

- The surface $z = z_0$ is a horizontal plane. At each point (r, θ, z) on this plane, z has the value z_0, but r and θ are unrestricted except for our blanket assumptions.

In spherical coordinates the surfaces represented by equations of the form

$$\rho = \rho_0, \quad \theta = \theta_0, \quad \text{and} \quad \phi = \phi_0$$

where ρ_0, θ_0, and ϕ_0 are constants, are shown in Figure 14.8.2c.

- The surface $\rho = \rho_0$ consists of all points whose distance ρ from the origin is ρ_0. Assuming ρ_0 to be nonnegative, this is a sphere of radius ρ_0 centered at the origin.

- As in cylindrical coordinates, the surface $\theta = \theta_0$ is a half-plane attached along the z-axis, making an angle of θ_0 with the positive x-axis.

- The surface $\phi = \phi_0$ consists of all points from which a line segment to the origin makes an angle of ϕ_0 with the positive z-axis. Depending on whether $0 < \phi_0 < \pi/2$ or $\pi/2 < \phi_0 < \pi$, this will be a cone opening up or opening down. (If $\phi_0 = \pi/2$, then the cone is flat and the surface is the xy-plane.)

☐ **CONVERTING COORDINATES**

Frequently, the coordinates of a point are known in one type of coordinate system and it is of interest to find the coordinates in one of the other types. Table 14.8.1 lists the formulas for making such coordinate conversions.

Table 14.8.1

CONVERSION	FORMULAS
$(r, \theta, z) \rightarrow (x, y, z)$	$x = r\cos\theta, \quad y = r\sin\theta, \quad z = z$
$(x, y, z) \rightarrow (r, \theta, z)$	$r = \sqrt{x^2 + y^2}, \quad \tan\theta = y/x, \quad z = z$
$(\rho, \theta, \phi) \rightarrow (r, \theta, z)$	$r = \rho\sin\phi, \quad \theta = \theta, \quad z = \rho\cos\phi$
$(r, \theta, z) \rightarrow (\rho, \theta, \phi)$	$\rho = \sqrt{r^2 + z^2}, \quad \theta = \theta, \quad \tan\phi = r/z$
$(\rho, \theta, \phi) \rightarrow (x, y, z)$	$x = \rho\sin\phi\cos\theta, \quad y = \rho\sin\phi\sin\theta, \quad z = \rho\cos\phi$
$(x, y, z) \rightarrow (\rho, \theta, \phi)$	$\rho = \sqrt{x^2 + y^2 + z^2}, \quad \tan\theta = y/x, \quad \cos\phi = z/\sqrt{x^2 + y^2 + z^2}$

The formulas in this table can be derived by considering the diagrams in Figure 14.8.3. Part (a) of the figure illustrates the relationship between the rectangular coordinates (x, y, z) and the cylindrical coordinates (r, θ, z) of a point P; it shows that the value of z is the same in

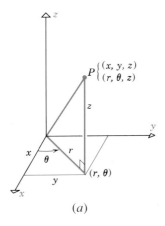

(a)

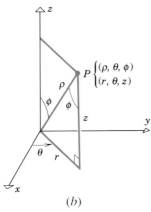

(b)

Figure 14.8.3

both coordinate systems and that (r, θ) is a pair of polar coordinates for the point (x, y) in the xy-plane. Thus, it follows from the relationship between polar and rectangular coordinates [Formulas (1a–b) of Section 13.1] that

$$x = r \cos \theta, \quad y = r \sin \theta, \quad z = z \tag{1}$$

Part (b) of Figure 14.8.3 illustrates the relationship between the spherical coordinates (ρ, θ, ϕ) and the cylindrical coordinates (r, θ, z) of a point P:

$$r = \rho \sin \phi, \quad \theta = \theta, \quad z = \rho \cos \phi \tag{2}$$

Substituting (2) into (1) yields the following relationships between the rectangular coordinates (x, y, z) and the spherical coordinates (ρ, θ, ϕ) of a point P:

$$x = \rho \sin \phi \cos \theta, \quad y = \rho \sin \phi \sin \theta, \quad z = \rho \cos \phi \tag{3}$$

We leave it as an exercise for the reader to deduce the remaining three conversion formulas in Table 14.8.1 from (1), (2), and (3).

Example 1   Find the rectangular coordinates of the point whose cylindrical coordinates are $(r, \theta, z) = (4, \pi/3, -3)$.

Solution.   From (1)

$$x = 4 \cos \frac{\pi}{3} = 2, \quad y = 4 \sin \frac{\pi}{3} = 2\sqrt{3}, \quad z = -3 \quad \blacktriangleleft$$

Example 2   Find an equation in cylindrical coordinates of the surface whose equation in rectangular coordinates is $z = x^2 + y^2 - 2x + y$.

Solution.   From (1)

$$z = r^2 - 2r \cos \theta + r \sin \theta \quad \blacktriangleleft$$

Example 3   Find an equation in rectangular coordinates of the surface whose equation in cylindrical coordinates is $r = 4 \cos \theta$.

Solution.   Multiplying both sides of the given equation by r yields the equation $r^2 = 4r \cos \theta$, then using the relationships $x^2 + y^2 = r^2$ and $x = r \cos \theta$, which follow from (1), yields

$$x^2 + y^2 = 4x \quad \text{or equivalently,} \quad (x - 2)^2 + y^2 = 4$$

This is a right-circular cylinder parallel to the z-axis. $\blacktriangleleft$

Example 4   Find the rectangular coordinates of the point whose spherical coordinates (ρ, θ, ϕ) are $(4, \pi/3, \pi/4)$.

Solution.   From (3)

$$x = \rho \sin \phi \cos \theta = 4 \sin \frac{\pi}{4} \cos \frac{\pi}{3} = \sqrt{2}$$

$$y = \rho \sin \phi \sin \theta = 4 \sin \frac{\pi}{4} \sin \frac{\pi}{3} = \sqrt{6}$$

$$z = \rho \cos \phi = 4 \cos \frac{\pi}{4} = 2\sqrt{2} \quad \blacktriangleleft$$

Example 5 Find an equation of the paraboloid $z = x^2 + y^2$ in spherical coordinates.

Solution. Substituting (3) in this equation yields

$$\rho \cos \phi = \rho^2 \sin^2 \phi \cos^2 \theta + \rho^2 \sin^2 \phi \sin^2 \theta$$

$$= \rho^2 \sin^2 \phi (\cos^2 \theta + \sin^2 \theta)$$

$$= \rho^2 \sin^2 \phi$$

which simplifies to $\rho \sin^2 \phi = \cos \phi$. ◀

☐ **SPHERICAL COORDINATES IN NAVIGATION**

Spherical coordinates are related to longitude and latitude coordinates used in navigation. Let us construct a right-hand rectangular coordinate system with origin at the earth's center, positive z-axis passing through the north pole, and positive x-axis passing through the prime meridian (Figure 14.8.4). If we assume the earth to be a perfect sphere of radius $\rho = 4000$ miles, then each point on the earth has spherical coordinates of the form $(4000, \theta, \phi)$, where ϕ and θ determine the latitude and longitude of the point. It is useful to specify longitudes in degrees east or west of the prime meridian and latitudes in degrees north or south of the equator. However, it is a simple matter to determine ϕ and θ from such data.

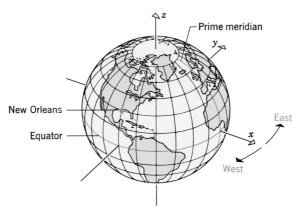

Figure 14.8.4

Example 6 The city of New Orleans is located at 90° West longitude and 30° North latitude. Find its spherical and rectangular coordinates relative to the coordinate axes of Figure 14.8.4. (Take miles as the unit of distance.)

Solution. A longitude of 90° West corresponds to $\theta = 360° - 90° = 270°$ or $\theta = 3\pi/2$ radians; and a latitude of 30° North corresponds to $\phi = 90° - 30° = 60°$ or $\phi = \pi/3$ radians. Thus, the spherical coordinates (ρ, θ, ϕ) of New Orleans are $(4000, 3\pi/2, \pi/3)$.
From (3), the rectangular coordinates of New Orleans are

$$x = 4000 \sin \frac{\pi}{3} \cos \frac{3\pi}{2} = 4000 \frac{\sqrt{3}}{2} (0) = 0 \text{ miles}$$

$$y = 4000 \sin \frac{\pi}{3} \sin \frac{3\pi}{2} = 4000 \frac{\sqrt{3}}{2} (-1) = -2000\sqrt{3} \text{ miles}$$

$$z = 4000 \cos \frac{\pi}{3} = 4000 \left(\frac{1}{2}\right) = 2000 \text{ miles}$$ ◀

▶ Exercise Set 14.8

1. Convert from rectangular to cylindrical coordinates.
 (a) $(4\sqrt{3}, 4, -4)$ (b) $(-5, 5, 6)$
 (c) $(0, 2, 0)$ (d) $(4, -4\sqrt{3}, 6)$
 (e) $(\sqrt{2}, -\sqrt{2}, 1)$ (f) $(0, 0, 1)$.

2. Convert from cylindrical to rectangular coordinates.
 (a) $(4, \pi/6, 3)$ (b) $(8, 3\pi/4, -2)$
 (c) $(5, 0, 4)$ (d) $(7, \pi, -9)$
 (e) $(6, 5\pi/3, 7)$ (f) $(1, \pi/2, 0)$.

3. Convert from rectangular to spherical coordinates.
 (a) $(1, \sqrt{3}, -2)$ (b) $(1, -1, \sqrt{2})$
 (c) $(0, 3\sqrt{3}, 3)$ (d) $(-5\sqrt{3}, 5, 0)$
 (e) $(4, 4, 4\sqrt{6})$ (f) $(1, -\sqrt{3}, -2)$.

4. Convert from spherical to rectangular coordinates.
 (a) $(5, \pi/6, \pi/4)$ (b) $(7, 0, \pi/2)$
 (c) $(1, \pi, 0)$ (d) $(2, 3\pi/2, \pi/2)$
 (e) $(1, 2\pi/3, 3\pi/4)$ (f) $(3, 7\pi/4, 5\pi/6)$.

5. Convert from cylindrical to spherical coordinates.
 (a) $(\sqrt{3}, \pi/6, 3)$ (b) $(1, \pi/4, -1)$
 (c) $(2, 3\pi/4, 0)$ (d) $(6, 1, -2\sqrt{3})$
 (e) $(4, 5\pi/6, 4)$ (f) $(2, 0, -2)$.

6. Convert from spherical to cylindrical coordinates.
 (a) $(5, \pi/4, 2\pi/3)$ (b) $(1, 7\pi/6, \pi)$
 (c) $(3, 0, 0)$ (d) $(4, \pi/6, \pi/2)$
 (e) $(5, \pi/2, 0)$ (f) $(6, 0, 3\pi/4)$.

In Exercises 7–14, an equation is given in cylindrical coordinates. Express the equation in rectangular coordinates and sketch the graph.

7. $r = 3$.   **8.** $\theta = \pi/4$.
9. $z = r^2$.   **10.** $z = r\cos\theta$.
11. $r = 4\sin\theta$.   **12.** $r = 2\sec\theta$.
13. $r^2 + z^2 = 1$.   **14.** $r^2\cos 2\theta = z$.

In Exercises 15–22, an equation is given in spherical coordinates. Express the equation in rectangular coordinates and sketch the graph.

15. $\rho = 3$.   **16.** $\theta = \pi/3$.
17. $\phi = \pi/4$.   **18.** $\rho = 2\sec\phi$.
19. $\rho = 4\cos\phi$.   **20.** $\rho\sin\phi = 1$.
21. $\rho\sin\phi = 2\cos\theta$.   **22.** $\rho - 2\sin\phi\cos\theta = 0$.

In Exercises 23–34, an equation of a surface is given in rectangular coordinates. Find an equation of the surface in (a) cylindrical coordinates and (b) spherical coordinates.

23. $z = 3$.   **24.** $y = 2$.
25. $z = 3x^2 + 3y^2$.   **26.** $z = \sqrt{3x^2 + 3y^2}$.
27. $x^2 + y^2 = 4$.   **28.** $x^2 + y^2 - 6y = 0$.
29. $x^2 + y^2 + z^2 = 9$.   **30.** $z^2 = x^2 - y^2$.
31. $2x + 3y + 4z = 1$.   **32.** $x^2 + y^2 - z^2 = 1$.
33. $x^2 = 16 - z^2$.   **34.** $x^2 + y^2 + z^2 = 2z$.

In Exercises 35–38, describe the three-dimensional region that satisfies the given inequalities.

35. $r^2 \le z \le 4$.
36. $0 \le r \le 2\sin\theta$, $0 \le z \le 3$.
37. $1 \le \rho \le 3$.
38. $0 \le \phi \le \pi/6$, $0 \le \rho \le 2$.

39. St. Petersburg (Leringrad), Russia, is located at 30° East longitude and 60° North latitude. Find its spherical and rectangular coordinates relative to the coordinate axes of Figure 14.8.4. Take miles as the unit of distance and assume the earth to be a sphere of radius 4000 miles.

40. (a) Show that the curve of intersection of the surfaces $z = \sin\theta$ and $r = a$ (cylindrical coordinates) is an ellipse.
 (b) Sketch a portion of the surface $z = \sin\theta$ for $0 \le \theta \le \pi/2$.

41. Sketch the surface whose equation in spherical coordinates is $\rho = a(1 - \cos\phi)$. [*Hint:* The surface is shaped like a familiar fruit.]

◆ TECHNOLOGY EXERCISES   Chapter 14

Most of these exercises require access to a graphing calculator or a computer algebra system (CAS) such as *Mathematica*, *Maple*, or *Derive*. When you are asked to *find* an answer or to *solve* an equation, you may choose to find an exact result or a numerical approximation, depending on the particular technology you are using and on your own imagination. The form of your answers may differ from those of other students or from those in the answer section of the text, depending on how you solve the problems and the accuracy you use in your numerical approximations. Those exercises that are more appropriate for a CAS than a graphing calculator are labeled with the icon ◆.

◆ 1. **Minimum angle between two vectors:** For each x in $(-\infty, +\infty)$, let $\mathbf{u}(x)$ be the vector from the origin to the point $P(x, y)$ on the curve $y = x^2 + 1$, and $\mathbf{v}(x)$ the vector from the origin to the point $Q(x, y)$ on the line $y = -x - 1$.

   (a) Find, to the nearest degree, the minimum angle between $\mathbf{u}(x)$ and $\mathbf{v}(x)$ for x in $(-\infty, +\infty)$.

   (b) Find all values of x such that the vectors $\mathbf{u}(x)$ and $\mathbf{v}(x)$ are orthogonal.

◆ 2. **Cones and vectors**

   (a) Find all unit vectors in 3-space that make an angle of 1 radian with both of the vectors $\mathbf{i} + 2\mathbf{j} + \mathbf{k}$ and $2\mathbf{i} - 2\mathbf{j} + \mathbf{k}$.

   (b) Let $\mathbf{v_0}$ be a vector with its initial point at the origin. All vectors with initial points at the origin that make an angle ϕ_0 with $\mathbf{v_0}$ lie on a cone whose axis is along $\mathbf{v_0}$ (see the accompanying figure). What does this imply about the vectors obtained in part (a)?

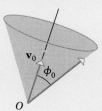

◆ 3. **The angle between two vectors:** Let $\mathbf{u}$ and $\mathbf{v}$ be unit vectors. Assume that $\mathbf{u}$ is in the xy-plane and makes an angle θ_1 with the positive x-axis, and that $\mathbf{v}$ is in the yz-plane and makes an angle θ_2 with the positive z-axis, as shown in the accompanying figure.

   (a) Show that if θ is the angle between the vectors $\mathbf{u}$ and $\mathbf{v}$, then $\cos \theta = \sin \theta_1 \sin \theta_2$.

   (b) Use the formula in part (a) to find the angle θ if $\theta_1 = \theta_2 = 45°$.

   (c) Use the formula in part (a) to find, to the nearest degree, the angle θ_1 ($0° \le \theta_1 < 360°$) if $\theta_2 = 60°$ and $\theta = 45°$.

   (d) Suppose that $\theta_1 = t$ and $\theta_2 = 2t$, where t is time ($t \ge 0$). Use the formula in part (a) to find, to the nearest degree, the maximum and minimum values of θ.

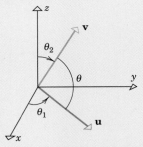

4. **Distance between two moving points:** Suppose that two particles moving in 3-space have equations of motion $x = 3 \cos 2t$, $y = 2 \sin 2t$, $z = 0$ ($t \ge 0$) and $x = 4 - 3t$, $y = -1 + t$, $z = -3 + t$ ($t \ge 0$). Given that x, y, and z are in meters and t is in seconds, how close do the particles get to one another?

◆ 5. **Distance between a point and a line:** From Exercise 16, Section 14.4, if $\mathbf{u}$ is a vector from any point on a line to a point P not on the line, and $\mathbf{v}$ is a vector parallel to the line, then the distance between P and the line is given by $\|\mathbf{u} \times \mathbf{v}\|/\|\mathbf{v}\|$. Use this result to help find the minimum distance between a point on the curve $x = 3 \cos 2t$, $y = 2 \sin 2t$, $z = 0$ and the line $x = 4 - 3t$, $y = -1 + t$, $z = -3 + t$.

◆ 6. **Extreme distances between a point and a curve:** Find the maximum and minimum distances between the point $(1, 1, 2)$ and a point on the curve of intersection of the plane $z = y + 2$ and the cylinder $x^2 + y^2 = 4$.

7. **Intersection of a line and a surface:** Find the coordinates of all points where the line $x = 2 + 3t$, $y = t$, $z = -1 + t$ intersects the cylindrical surface $2z = \sin x$.

8. **Area of a triangle:** Find the minimum area of a triangle if two of its vertices are $(2, -1, 0)$ and $(3, 2, 2)$ and its third vertex is on the curve $y = \ln x$ in the xy-plane.

9. **Maximum temperature on a line:** Suppose that the temperature T at a point (x, y, z) is given by $T = 25x^2yz$. Find the maximum value of T along the portion of the line $x = t$, $y = 1 + t$, $z = 3 - 2t$ that extends from the xz-plane to the xy-plane.

10. **Distance on the surface of the earth:** A ship at sea is at point A that is 60° West longitude and 40° North latitude. The ship travels to point B that is 40° West longitude and 20° North latitude. Assuming that the earth is a sphere with radius 6370 kilometers (km), find the shortest distance the ship can travel in going from A to B, given that the shortest distance between two points on a sphere is along the arc of the great circle joining the points. [*Hint:* See Figure 14.8.4 and Example 6, and consider the angle between the vectors from the center of the earth to the points A and B. (If you are not familiar with the term "great circle," consult a dictionary.)]

Karl Weierstrass (1815–1897)

15 VECTOR-VALUED FUNCTIONS

15.1 INTRODUCTION TO VECTOR-VALUED FUNCTIONS

In this section we shall show how vectors can be used to express parametric equations in a more compact form. As part of our work we shall discuss functions that associate vectors with real numbers. This new category of functions has important applications in science and engineering.

□ **PARAMETRIC CURVES IN 3-SPACE**

Recall from Section 13.4 that a parametric curve in 2-space is represented by a pair of parametric equations $x = x(t)$, $y = y(t)$, where the parameter t varies over some finite or infinite interval of real values. Similarly, we can represent a *parametric curve* in 3-space by three equations $x = x(t)$, $y = y(t)$, $z = z(t)$. Under appropriate conditions, the parametric curve will be traced in a specific direction as t increases; as in 2-space, we call this the *direction of increasing parameter* or the *orientation* of the parametric curve. If no restrictions on t are stated explicitly or implied by the equations, then it is understood that t varies over the interval $(-\infty, +\infty)$. In Section 14.5 we discussed parametric equations of lines in 3-space. Here we will be concerned with other kinds of parametric curves as well.

Example 1   Sketch the graph of the parametric equations

$$x = a \cos t, \quad y = a \sin t, \quad z = ct$$

where a and c are positive constants.

Solution.   As the parameter t increases, the value of $z = ct$ also increases, so the point (x, y, z) moves upward. However, as t increases, the point (x, y, z) also moves in a path directly over the circle

$$x = a \cos t, \quad y = a \sin t$$

in the xy-plane. The combination of these upward and circular motions produces a corkscrew-shaped curve that wraps around a right-circular cylinder of radius a centered on the z-axis (Figure 15.1.1). This curve is called a **circular helix**.   ◀

Example 2   Show that the graph of the parametric equations

$$x = t, \quad y = t^2, \quad z = t^3$$

is the intersection of the parabolic cylinder $y = x^2$ and the cubic cylinder $z = x^3$. Sketch the portion of the graph for $t \geq 0$. The curve is called a **twisted cubic**.

Solution.   Eliminating the parameter t in the equations for x and y yields $y = x^2$, so the curve lies on the parabolic cylinder with this equation. Similarly, eliminating t in the equations for x and z yields $z = x^3$, so the curve also lies on the cubic cylinder with this equation (Figure 15.1.2). The curve is traced by a point that starts at the origin when $t = 0$, and then moves upward as t increases, since x, y, and z increase with t.   ◀

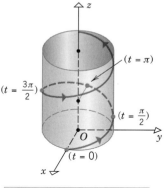

Figure 15.1.1
$$x = a \cos t, \ y = a \sin t, \ z = ct$$

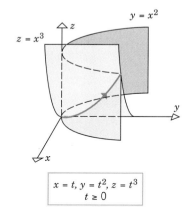

Figure 15.1.2
$$x = t, y = t^2, z = t^3$$
$$t \geq 0$$

☐ **PARAMETRIC EQUATIONS IN VECTOR FORM**

Parametric equations in 2-space and 3-space can be expressed in a useful vector form. For example, in 2-space the two equations

$$x = x(t), \quad y = y(t)$$

can be written as the single vector equation

$$x\mathbf{i} + y\mathbf{j} = x(t)\mathbf{i} + y(t)\mathbf{j} \tag{1}$$

and in 3-space the three equations

$$x = x(t), \quad y = y(t), \quad z = z(t)$$

can be written as the single vector equation

$$x\mathbf{i} + y\mathbf{j} + z\mathbf{k} = x(t)\mathbf{i} + y(t)\mathbf{j} + z(t)\mathbf{k} \tag{2}$$

If we let $\mathbf{r} = x\mathbf{i} + y\mathbf{j}$ and $\mathbf{r}(t) = x(t)\mathbf{i} + y(t)\mathbf{j}$ in 2-space and let $\mathbf{r} = x\mathbf{i} + y\mathbf{j} + z\mathbf{k}$ and

$\mathbf{r}(t) = x(t)\mathbf{i} + y(t)\mathbf{j} + z(t)\mathbf{k}$ in 3-space, then both (1) and (2) can be expressed as

$$\mathbf{r} = \mathbf{r}(t) \tag{3}$$

For example, the vector form of the twisted cubic in Example 2 is

$$\mathbf{r} = t\mathbf{i} + t^2\mathbf{j} + t^3\mathbf{k}$$

and the parametric equations corresponding to the vector equation

$$\mathbf{r} = (t^3 + 1)\mathbf{i} + 3\mathbf{j} + e^t\mathbf{k}$$

are

$$x = t^3 + 1, \quad y = 3, \quad z = e^t$$

☐ VECTOR-VALUED FUNCTIONS

Recall that a function is a rule that assigns to each element in its domain one and only one element in its range. Thus far, we have considered primarily functions for which the domain and range are sets of real numbers; such functions are called ***real-valued functions of a real variable*** or sometimes simply ***real-valued functions***. In contrast, the function $\mathbf{r}(t)$ in (3) associates a vector in 2-space or 3-space with a real value of t; such functions are called ***vector-valued functions of a real variable*** or more simply ***vector-valued functions***. For example, if

$$\mathbf{r}(t) = t\mathbf{i} + t^2\mathbf{j} + t^3\mathbf{k}$$

then the vectors associated with $t = 1, -2$, and 0 are

$$\mathbf{r}(1) = \mathbf{i} + \mathbf{j} + \mathbf{k}, \quad \mathbf{r}(-2) = -2\mathbf{i} + 4\mathbf{j} - 8\mathbf{k}, \quad \mathbf{r}(0) = 0\mathbf{i} + 0\mathbf{j} + 0\mathbf{k} = \mathbf{0}$$

If $\mathbf{r}(t) = x(t)\mathbf{i} + y(t)\mathbf{j}$ in 2-space or if $\mathbf{r}(t) = x(t)\mathbf{i} + y(t)\mathbf{j} + z(t)\mathbf{k}$ in 3-space, then the real-valued functions $x(t)$, $y(t)$, and $z(t)$ are called the ***component functions*** or the ***components*** of $\mathbf{r}(t)$. The ***domain*** of $\mathbf{r}(t)$ is the set of allowable values for t. If the domain is not specified explicitly, then it is understood to consist of all values of t for which every component is defined and yields a real value; this is called the ***natural domain*** of $\mathbf{r}(t)$. Thus, the natural domain of $\mathbf{r}(t)$ is the intersection of the natural domains of its components. For example, the components of

$$\mathbf{r}(t) = \ln(t - 1)\mathbf{i} + e^t\mathbf{j} + \sqrt{t}\mathbf{k}$$

are

$$x(t) = \ln(t - 1), \quad y(t) = e^t, \quad z(t) = \sqrt{t}$$

and the natural domain of $\mathbf{r}(t)$ is the set of t values such that $t > 1$.

☐ GRAPHS OF VECTOR-VALUED FUNCTIONS

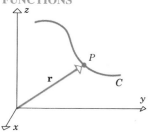

If $\mathbf{r}(t)$ is a vector-valued function in 2-space or 3-space, then we define the ***graph*** of $\mathbf{r}(t)$ [or of $\mathbf{r} = \mathbf{r}(t)$] to be the parametric curve represented by the parametric equations corresponding to $\mathbf{r}(t)$. For example, the graph of

$$\mathbf{r}(t) = t\mathbf{i} + t^2\mathbf{j} + t^3\mathbf{k}$$

is the parametric curve represented by the equations

$$x = t, \quad y = t^2, \quad z = t^3$$

which is the twisted cubic shown in Figure 15.1.2 (for $t \geq 0$).

Up to now, we have imagined a parametric curve C to be traced by moving point P. However, if the curve is viewed as the graph of a vector-valued function, then we can also imagine the curve to be traced by the tip of the vector $\mathbf{r}$ whose initial point is at the origin and whose terminal point is at P. We call $\mathbf{r}$ the ***radius vector*** or ***position vector*** for C (Figure 15.1.3).

As t varies, the tip of the radius vector $\mathbf{r}$ traces out the curve C.

Figure 15.1.3

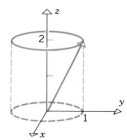

$$\mathbf{r} = (\cos t)\,\mathbf{i} + (\sin t)\,\mathbf{j}$$

Figure 15.1.4

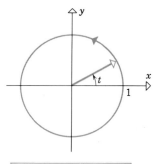

$$\mathbf{r} = (\cos t)\,\mathbf{i} + (\sin t)\,\mathbf{j} + 2\mathbf{k}$$

Figure 15.1.5

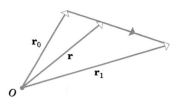

$$\mathbf{r} = (1 - t)\mathbf{r}_0 + t\mathbf{r}_1 \quad (0 \le t \le 1)$$

Figure 15.1.6

Example 3   Sketch the graph and a radius vector of

(a) $\mathbf{r}(t) = (\cos t)\mathbf{i} + (\sin t)\mathbf{j}, \quad 0 \le t \le 2\pi$

(b) $\mathbf{r}(t) = (\cos t)\mathbf{i} + (\sin t)\mathbf{j} + 2\mathbf{k}, \quad 0 \le t \le 2\pi$

Solution (a).   The corresponding parametric equations are

$$x = \cos t, \quad y = \sin t \qquad (0 \le t \le 2\pi)$$

so from Example 2 of Section 13.4 the graph is a circle of radius 1, centered at the origin, and oriented counterclockwise. The graph in 2-space and a radius vector are shown in Figure 15.1.4.

Solution (b).   The corresponding parametric equations are

$$x = \cos t, \quad y = \sin t, \quad z = 2 \qquad (0 \le t \le 2\pi)$$

From the last equation, the tip of the radius vector traces a curve in the plane $z = 2$, and from the first two equations and part (a), the curve is a circle of radius 1 centered on the z-axis and traced counterclockwise looking down the z-axis. The graph and a radius vector are shown in Figure 15.1.5.   ◄

Example 4   Let $\mathbf{r}_0$ and $\mathbf{r}_1$ be distinct nonzero vectors in 2-space or 3-space with their initial points at the origin. Show that the graph of the equation

$$\mathbf{r} = (1 - t)\mathbf{r}_0 + t\mathbf{r}_1 \quad (0 \le t \le 1) \tag{4}$$

is the line segment joining the tips of $\mathbf{r}_0$ and $\mathbf{r}_1$ and oriented from $\mathbf{r}_0$ to $\mathbf{r}_1$ (Figure 15.1.6).

Solution.   We shall give the solution in 2-space; the solution in 3-space is similar. Suppose that

$$\mathbf{r} = \langle x, y \rangle, \quad \mathbf{r}_0 = \langle x_0, y_0 \rangle, \quad \mathbf{r}_1 = \langle x_1, y_1 \rangle$$

Then (4) can be expressed as

$$\langle x, y \rangle = (1 - t)\langle x_0, y_0 \rangle + t\langle x_1, y_1 \rangle \quad (0 \le t \le 1)$$

or in parametric form as

$$x = x_0 + (x_1 - x_0)t, \quad y = y_0 + (y_1 - y_0)t \quad (0 \le t \le 1)$$

From part (a) of Theorem 14.5.1 these are parametric equations of some portion of the line that is parallel to $\mathbf{r}_1 - \mathbf{r}_0 = \langle x_1 - x_0, y_1 - y_0 \rangle$ and passes through the tip of $\mathbf{r}_0 = \langle x_0, y_0 \rangle$, which is the line through the tips of $\mathbf{r}_0$ and $\mathbf{r}_1$. As t varies from 0 to 1, the vector $\mathbf{r} = (1 - t)\mathbf{r}_0 + t\mathbf{r}_1$ varies from $\mathbf{r}_0$ to $\mathbf{r}_1$, and its tip traces out the line segment from the tip of $\mathbf{r}_0$ to the tip of $\mathbf{r}_1$.   ◄

□ **NORM OF A VECTOR-VALUED FUNCTION**

If $\mathbf{r}(t)$ is a vector-valued function, then $\|\mathbf{r}(t)\|$ is a real-valued function. Moreover, $\|\mathbf{r}(t)\|$ is continuous if $\mathbf{r}(t)$ is continuous (why?). For example, the vector-valued function

$$\mathbf{r}(t) = t\mathbf{i} + (t - 1)\mathbf{j}$$

is continuous, since its components are continuous, and the real-valued function

$$\|\mathbf{r}(t)\| = \sqrt{t^2 + (t-1)^2} = \sqrt{2t^2 - 2t + 1}$$

is continuous, since the function inside the radical is continuous and positive for all t.

▶ Exercise Set 15.1

In Exercises 1–4, find the domain of $\mathbf{r}(t)$ and the value of $\mathbf{r}(t_0)$.

1. $\mathbf{r}(t) = (\cos t)\mathbf{i} - 3t\mathbf{j}; \; t_0 = \pi.$

2. $\mathbf{r}(t) = \langle \sqrt{3t+1}, t^2 \rangle; \; t_0 = 1.$

3. $\mathbf{r}(t) = (\cos \pi t)\mathbf{i} - (\ln t)\mathbf{j} + \sqrt{t-2}\,\mathbf{k}; \; t_0 = 3.$

4. $\mathbf{r}(t) = \langle 2e^{-t}, \sin^{-1} t, \ln(1-t) \rangle; \; t_0 = 0.$

In Exercises 5–8, express the parametric equations as a single vector equation of the form $\mathbf{r} = x(t)\mathbf{i} + y(t)\mathbf{j}$ or $\mathbf{r} = x(t)\mathbf{i} + y(t)\mathbf{j} + z(t)\mathbf{k}$.

5. $x = 3 \cos t, \; y = t + \sin t.$

6. $x = t^2 + 1, \; y = e^{-2t}.$

7. $x = 2t, \; y = 2 \sin 3t, \; z = 5 \cos 3t.$

8. $x = t \sin t, \; y = \ln t, \; z = \cos^2 t.$

In Exercises 9–12, find the parametric equations that correspond to the given vector equation.

9. $\mathbf{r} = 3t^2\mathbf{i} - 2\mathbf{j}.$

10. $\mathbf{r} = (\sin^2 t)\mathbf{i} + (1 - \cos 2t)\mathbf{j}.$

11. $\mathbf{r} = (2t - 1)\mathbf{i} - 3\sqrt{t}\mathbf{j} + (\sin 3t)\mathbf{k}.$

12. $\mathbf{r} = te^{-t}\mathbf{i} - 5t^2\mathbf{k}.$

In Exercises 13–18, describe the graph of the equation.

13. $\mathbf{r} = (2 - 3t)\mathbf{i} - 4t\mathbf{j}.$

14. $\mathbf{r} = (3 \sin 2t)\mathbf{i} + (3 \cos 2t)\mathbf{j}.$

15. $\mathbf{r} = 2t\mathbf{i} - 3\mathbf{j} + (1 + 3t)\mathbf{k}.$

16. $\mathbf{r} = 3\mathbf{i} + (2 \cos t)\mathbf{j} + (2 \sin t)\mathbf{k}.$

17. $\mathbf{r} = (3 \cos t)\mathbf{i} + (2 \sin t)\mathbf{j} - \mathbf{k}.$

18. $\mathbf{r} = -2\mathbf{i} + t\mathbf{j} + (t^2 - 1)\mathbf{k}.$

19. Find the slope of the line in 2-space that is represented by the vector equation $\mathbf{r} = (1 - 2t)\mathbf{i} - (2 - 3t)\mathbf{j}$.

20. Find the y-intercept of the line in 2-space that is represented by the vector equation $\mathbf{r} = (3 + 2t)\mathbf{i} + 5t\mathbf{j}$.

21. Find the coordinates of the point where the line $\mathbf{r} = (2 + t)\mathbf{i} + (1 - 2t)\mathbf{j} + 3t\mathbf{k}$ intersects the xz-plane.

22. Find the coordinates of the point where the line

$$\mathbf{r} = t\mathbf{i} + (1 + 2t)\mathbf{j} - 3t\mathbf{k}$$

intersects the plane $3x - y - z = 2$.

In Exercises 23–34, sketch the graph of $\mathbf{r}(t)$ and show the direction of increasing t.

23. $\mathbf{r}(t) = 2\mathbf{i} + t\mathbf{j}.$

24. $\mathbf{r}(t) = \langle 3t - 4, 6t + 2 \rangle.$

25. $\mathbf{r}(t) = (1 + \cos t)\mathbf{i} + (3 - \sin t)\mathbf{j}, \; 0 \le t \le 2\pi.$

26. $\mathbf{r}(t) = \langle 2 \cos t, 5 \sin t \rangle, \; 0 \le t \le 2\pi.$

27. $\mathbf{r}(t) = (\cosh t)\mathbf{i} + (\sinh t)\mathbf{j}.$

28. $\mathbf{r}(t) = \sqrt{t}\mathbf{i} + (2t + 4)\mathbf{j}.$

29. $\mathbf{r}(t) = t\mathbf{i} + t\mathbf{j} + t\mathbf{k}.$

30. $\mathbf{r}(t) = (1 + 3t)\mathbf{i} + (-1 + t)\mathbf{j} + 2t\mathbf{k}.$

31. $\mathbf{r}(t) = (2 \cos t)\mathbf{i} + (2 \sin t)\mathbf{j} + t\mathbf{k}.$

32. $\mathbf{r}(t) = (9 \cos t)\mathbf{i} + (4 \sin t)\mathbf{j} + t\mathbf{k}.$

33. $\mathbf{r}(t) = t\mathbf{i} + t^2\mathbf{j} + 2\mathbf{k}.$

34. $\mathbf{r}(t) = t\mathbf{i} + t\mathbf{j} + (\sin t)\mathbf{k}, \; 0 \le t \le 2\pi.$

35. Show that the graph of

$$\mathbf{r} = (t \sin t)\mathbf{i} + (t \cos t)\mathbf{j} + t^2\mathbf{k}$$

lies on the paraboloid $z = x^2 + y^2$.

36. Show that the graph of

$$\mathbf{r} = t\mathbf{i} + \frac{1+t}{t}\mathbf{j} + \frac{1 - t^2}{t}\mathbf{k}, \quad t > 0$$

lies in the plane $x - y + z + 1 = 0$.

37. Show that the graph of

$$\mathbf{r} = (\sin t)\mathbf{i} + (2 \cos t)\mathbf{j} + (\sqrt{3} \sin t)\mathbf{k}$$

is a circle, and find its center and radius. [*Hint:* Show that the curve lies on both a sphere and a plane.]

38. Show that the graph of

$$\mathbf{r} = (3 \cos t)\mathbf{i} + (3 \sin t)\mathbf{j} + (3 \sin t)\mathbf{k}$$

is an ellipse, and find the lengths of the major and minor axes. [*Hint:* Show that the graph lies on both a circular cylinder and a plane and use the result in Exercise 50 of Section 12.3.]

39. For the helix $\mathbf{r} = (a \cos t)\mathbf{i} + (a \sin t)\mathbf{j} + ct\mathbf{k}$, find c ($c > 0$) so that the helix will make one complete turn in a distance of 3 units measured along the z-axis.

40. How many revolutions will the circular helix

$$\mathbf{r} = (a \cos t)\mathbf{i} + (a \sin t)\mathbf{j} + 0.2t\mathbf{k}$$

make in a distance of 10 units measured along the z-axis?

41. Show that the curve $\mathbf{r} = (t \cos t)\mathbf{i} + (t \sin t)\mathbf{j} + t\mathbf{k}$, $t \geq 0$, lies on the cone $z = \sqrt{x^2 + y^2}$. Describe the curve.

42. Describe the curve $\mathbf{r} = (a \cos t)\mathbf{i} + (b \sin t)\mathbf{j} + ct\mathbf{k}$, where a, b, and c are positive constants such that $a \neq b$.

In Exercises 43–48, find parametric equations of the curve of intersection of the surfaces. [*Note:* The answer is not unique.]

43. The cone $z = \sqrt{x^2 + y^2}$ and the plane $z = y + 2$. Identify the curve.

44. The paraboloid $z = x^2 + y^2$ and the plane $x = -2$. Identify the curve.

45. The circular cylinder $x^2 + y^2 = 9$ and the parabolic cylinder $z = x^2$.

46. The paraboloid $z = 4 - x^2 - y^2$ and the circular cylinder $x^2 + y^2 = 1$. Identify the curve.

47. The elliptic paraboloid $z = x^2 + 4y^2$ and the plane $z = 2x$. [*Hint:* Find the orthogonal projection of the curve onto the xy-plane.]

48. The cone $z = \sqrt{x^2 + y^2}$ and the parabolic cylinder $z = 2\sqrt{y}$. [*Hint:* See the hint in Exercise 47.]

■ **15.2** CALCULUS OF VECTOR-VALUED FUNCTIONS

In this section we shall define limits, derivatives, and integrals of vector-valued functions and discuss their properties.

☐ **LIMITS, DERIVATIVES, AND INTEGRALS**

As shown in Table 15.2.1, limits, derivatives, and integrals of vector-valued functions can be defined by taking the limits, derivatives, and integrals of the components.

Table 15.2.1

2-SPACE $\mathbf{r}(t) = x(t)\mathbf{i} + y(t)\mathbf{j}$	3-SPACE $\mathbf{r}(t) = x(t)\mathbf{i} + y(t)\mathbf{j} + z(t)\mathbf{k}$
$\lim\limits_{t \to a} \mathbf{r}(t) = \left(\lim\limits_{t \to a} x(t)\right)\mathbf{i} + \left(\lim\limits_{t \to a} y(t)\right)\mathbf{j}$	$\lim\limits_{t \to a} \mathbf{r}(t) = \left(\lim\limits_{t \to a} x(t)\right)\mathbf{i} + \left(\lim\limits_{t \to a} y(t)\right)\mathbf{j} + \left(\lim\limits_{t \to a} z(t)\right)\mathbf{k}$
$\mathbf{r}'(t) = x'(t)\mathbf{i} + y'(t)\mathbf{j}$	$\mathbf{r}'(t) = x'(t)\mathbf{i} + y'(t)\mathbf{j} + z'(t)\mathbf{k}$
$\int \mathbf{r}(t)\,dt = \left(\int x(t)\,dt\right)\mathbf{i} + \left(\int y(t)\,dt\right)\mathbf{j}$	$\int \mathbf{r}(t)\,dt = \left(\int x(t)\,dt\right)\mathbf{i} + \left(\int y(t)\,dt\right)\mathbf{j} + \left(\int z(t)\,dt\right)\mathbf{k}$
$\int_a^b \mathbf{r}(t)\,dt = \left(\int_a^b x(t)\,dt\right)\mathbf{i} + \left(\int_a^b y(t)\,dt\right)\mathbf{j}$	$\int_a^b \mathbf{r}(t)\,dt = \left(\int_a^b x(t)\,dt\right)\mathbf{i} + \left(\int_a^b y(t)\,dt\right)\mathbf{j} + \left(\int_a^b z(t)\,dt\right)\mathbf{k}$

The definitions in Table 15.2.1 assume that the operations on the components can be performed. If, for example, the limit of any component of $\mathbf{r}(t)$ does not exist, then we shall agree that the limit of $\mathbf{r}(t)$ ***does not exist***. The limit definition is also applicable to one-sided and infinite limits. In addition, the standard terminology and notation relating to derivatives and integrals continues to apply. For example, a vector-valued function is said to be ***differentiable*** (***integrable***) if and only if each component is differentiable (integrable). Moreover, the derivative of $\mathbf{r}(t)$ can also be expressed in any of the following notations:

$$\frac{d}{dt}[\mathbf{r}(t)], \quad \frac{d\mathbf{r}}{dt}, \quad \mathbf{r}'(t), \quad \text{and} \quad \mathbf{r}'$$

Example 1 Let

$$\mathbf{r}(t) = t^2\mathbf{i} + e^t\mathbf{j} - 2 \cos \pi t\mathbf{k}$$

Then

$$\lim_{t \to 0} \mathbf{r}(t) = \left(\lim_{t \to 0} t^2\right)\mathbf{i} + \left(\lim_{t \to 0} e^t\right)\mathbf{j} - \left(\lim_{t \to 0} 2 \cos \pi t\right)\mathbf{k}$$

$$= \mathbf{j} - 2\mathbf{k}$$

$$\mathbf{r}'(t) = 2t\mathbf{i} + e^t\mathbf{j} + 2\pi \sin \pi t \mathbf{k}$$

$$\mathbf{r}'(1) = 2\mathbf{i} + e\mathbf{j}$$

$$\int_0^1 \mathbf{r}(t)\, dt = \frac{t^3}{3}\bigg]_0^1 \mathbf{i} + e^t\bigg]_0^1 \mathbf{j} - \frac{2}{\pi}\sin \pi t\bigg]_0^1 \mathbf{k}$$

$$= \tfrac{1}{3}\mathbf{i} + (e - 1)\mathbf{j} \quad \blacktriangleleft$$

Recall that indefinite integration of a real-valued function produces a constant of integration C that is an arbitrary real number. Analogously, indefinite integration of a vector-valued function produces a constant of integration $\mathbf{C}$ that is an arbitrary vector. This is illustrated in the following example.

Example 2

$$\int (2t\mathbf{i} + 3t^2\mathbf{j})\, dt = \left(\int 2t\, dt\right)\mathbf{i} + \left(\int 3t^2\, dt\right)\mathbf{j}$$

$$= (t^2 + C_1)\mathbf{i} + (t^3 + C_2)\mathbf{j}$$

$$= (t^2\mathbf{i} + t^3\mathbf{j}) + C_1\mathbf{i} + C_2\mathbf{j} = t^2\mathbf{i} + t^3\mathbf{j} + \mathbf{C}$$

where $\mathbf{C} = C_1\mathbf{i} + C_2\mathbf{j}$ is an arbitrary vector constant of integration. $\quad \blacktriangleleft$

☐ **PROPERTIES OF DERIVATIVES AND INTEGRALS**

Because limits, derivatives, and integrals of vector-valued functions are defined in terms of the same operations on components, most of the standard theorems on limits, derivatives, and integrals carry over to vector-valued functions. The following two theorems, whose proofs are left as exercises, list the standard properties of differentiation and integration of vector-valued functions.

15.2.1 THEOREM (*Rules of Differentiation*). *In either 2-space or 3-space let* $\mathbf{r}(t)$, $\mathbf{r}_1(t)$, *and* $\mathbf{r}_2(t)$ *be vector-valued functions,* $f(t)$ *a real-valued function,* k *a scalar, and* $\mathbf{c}$ *a fixed (constant) vector. Then the following rules of differentiation hold:*

(a) $\dfrac{d}{dt}[\mathbf{c}] = \mathbf{0}$

(b) $\dfrac{d}{dt}[k\mathbf{r}(t)] = k\dfrac{d}{dt}[\mathbf{r}(t)]$

(c) $\dfrac{d}{dt}[\mathbf{r}_1(t) + \mathbf{r}_2(t)] = \dfrac{d}{dt}[\mathbf{r}_1(t)] + \dfrac{d}{dt}[\mathbf{r}_2(t)]$

(d) $\dfrac{d}{dt}[\mathbf{r}_1(t) - \mathbf{r}_2(t)] = \dfrac{d}{dt}[\mathbf{r}_1(t)] - \dfrac{d}{dt}[\mathbf{r}_2(t)]$

(e) $\dfrac{d}{dt}[f(t)\mathbf{r}(t)] = f(t)\dfrac{d}{dt}[\mathbf{r}(t)] + \dfrac{d}{dt}[f(t)]\mathbf{r}(t)$

15.2.2   THEOREM (*Rules of Integration*).   *In either 2-space or 3-space let* $\mathbf{r}(t)$, $\mathbf{r}_1(t)$, *and* $\mathbf{r}_2(t)$ *be vector-valued functions, and let k be a scalar. Then the following rules of integration hold:*

(*a*) $\displaystyle\int k\mathbf{r}(t)\,dt = k\int \mathbf{r}(t)\,dt$

(*b*) $\displaystyle\int [\mathbf{r}_1(t) + \mathbf{r}_2(t)]\,dt = \int \mathbf{r}_1(t)\,dt + \int \mathbf{r}_2(t)\,dt$

(*c*) $\displaystyle\int [\mathbf{r}_1(t) - \mathbf{r}_2(t)]\,dt = \int \mathbf{r}_1(t)\,dt - \int \mathbf{r}_2(t)\,dt$

REMARK.   The results in the preceding theorem are also valid for definite integrals of vector-valued functions.

☐ **FURTHER PROPERTIES OF DERIVATIVES**

The derivative of a vector-valued function was defined in terms of the derivatives of its components. The following theorem provides a formula for the derivative of $\mathbf{r}(t)$ that does not require breaking up the function into components. [Compare the formula in this theorem to Formula (4) of Section 3.2.]

15.2.3   THEOREM.   *If* $\mathbf{r}(t)$ *is a vector-valued function in 2-space or 3-space, then the derivative of* $\mathbf{r}(t)$ *can be expressed as*

$$\mathbf{r}'(t) = \lim_{h \to 0} \frac{\mathbf{r}(t + h) - \mathbf{r}(t)}{h} \tag{1}$$

provided this limit exists.

Proof.   For simplicity, we give the proof in 2-space; the proof in 3-space is identical, except for the additional component. Assume that $\mathbf{r}(t) = x(t)\mathbf{i} + y(t)\mathbf{j}$, so

$$\mathbf{r}'(t) = x'(t)\mathbf{i} + y'(t)\mathbf{j}$$

$$= \lim_{h \to 0} \frac{[x(t + h) - x(t)]}{h}\mathbf{i} + \lim_{h \to 0} \frac{[y(t + h) - y(t)]}{h}\mathbf{j}$$

$$= \lim_{h \to 0} \frac{[x(t + h)\mathbf{i} + y(t + h)\mathbf{j}] - [x(t)\mathbf{i} + y(t)\mathbf{j}]}{h}$$

$$= \lim_{h \to 0} \frac{\mathbf{r}(t + h) - \mathbf{r}(t)}{h} \qquad \blacksquare$$

☐ **FURTHER PROPERTIES OF INTEGRALS**

We leave it for the reader to show that the following analog of Formula (2) in Section 5.2 holds for vector-valued functions in 2-space or 3-space.

$$\frac{d}{dt}\left[\int \mathbf{r}(t)\,dt\right] = \mathbf{r}(t) \tag{2}$$

This shows that an indefinite integral of $\mathbf{r}(t)$ is, in fact, the set of antiderivatives of $\mathbf{r}(t)$, just as for real-valued functions. We also leave it as an exercise to show that if $\mathbf{R}(t)$ is any antiderivative of $\mathbf{r}(t)$ in the sense that $\mathbf{R}'(t) = \mathbf{r}(t)$, then

$$\int \mathbf{r}(t)\,dt = \mathbf{R}(t) + \mathbf{C} \tag{3}$$

where $\mathbf{C}$ is an arbitrary vector constant of integration. Moreover,

$$\int_a^b \mathbf{r}(t)\,dt = \mathbf{R}(t)\Big]_a^b = \mathbf{R}(b) - \mathbf{R}(a) \tag{4}$$

which is the extension of the First Fundamental Theorem of Calculus (Theorem 5.7.1) to vector-valued functions.

Example 3 In Example 2 we showed that $\mathbf{R}(t) = t^2\mathbf{i} + t^3\mathbf{j}$ is an antiderivative of $\mathbf{r}(t) = 2t\mathbf{i} + 3t^2\mathbf{j}$. (It is the antiderivative that results when $\mathbf{C} = \mathbf{0}$.) Thus, from (4)

$$\int_0^2 \mathbf{r}(t)\,dt = \int_0^2 (2t\mathbf{i} + 3t^2\mathbf{j})\,dt = \Big[\mathbf{R}(t)\Big]_0^2$$

$$= \Big[t^2\mathbf{i} + t^3\mathbf{j}\Big]_0^2 = (4\mathbf{i} + 8\mathbf{j}) - (0\mathbf{i} + 0\mathbf{j}) = 4\mathbf{i} + 8\mathbf{j}$$

The reader may want to check that the same result can be obtained by integrating the components of $\mathbf{r}(t)$ term by term. ◀

Example 4 Find $\mathbf{r}(t)$ given that $\mathbf{r}'(t) = 3\mathbf{i} + 2t\mathbf{j}$ and $\mathbf{r}(1) = 2\mathbf{i} + 5\mathbf{j}$.

Solution. Integrating $\mathbf{r}'(t)$ to obtain $\mathbf{r}(t)$ yields

$$\mathbf{r}(t) = \int \mathbf{r}'(t)\,dt = \int (3\mathbf{i} + 2t\mathbf{j})\,dt = 3t\mathbf{i} + t^2\mathbf{j} + \mathbf{C}$$

where $\mathbf{C}$ is a vector constant of integration. To find $\mathbf{C}$ we substitute $t = 1$ in this equation and use the given value of $\mathbf{r}(1)$ to obtain

$$\mathbf{r}(1) = 3\mathbf{i} + \mathbf{j} + \mathbf{C} = 2\mathbf{i} + 5\mathbf{j}$$

so that $\mathbf{C} = -\mathbf{i} + 4\mathbf{j}$. Thus,

$$\mathbf{r}(t) = 3t\mathbf{i} + t^2\mathbf{j} - \mathbf{i} + 4\mathbf{j} = (3t - 1)\mathbf{i} + (t^2 + 4)\mathbf{j} \quad ◀$$

□ **GEOMETRIC INTERPRETATION OF LIMITS AND DERIVATIVES**

Although we have defined limits and derivatives of vector-valued functions in terms of the corresponding operations on components, it is desirable to have vector interpretations of limits and derivatives that do not require us to break the vectors into components. Limits of vector-valued functions can be interpreted geometrically as follows:

> **15.2.4** GEOMETRIC INTERPRETATION OF LIMITS.   *If $\mathbf{r}(t)$ is a vector-valued function in 2-space or 3-space, then*
>
> $$\lim_{t \to a} \mathbf{r}(t) = \mathbf{L}$$
>
> *if and only if the radius vector $\mathbf{r} = \mathbf{r}(t)$ approaches $\mathbf{L}$ in both length and direction as $t \to a$ (Figure 15.2.1).*

This result should be evident from the fact that the components of $\mathbf{r}(t)$ approach the components of $\mathbf{L}$ as $t \to a$.

The definition of continuity for vector-valued functions in 2-space and 3-space is similar to that for real-valued functions: We shall say that $\mathbf{r}(t)$ is *continuous at t_0* if $\mathbf{r}(t_0)$ is defined and

$$\lim_{t \to t_0} \mathbf{r}(t) = \mathbf{r}(t_0)$$

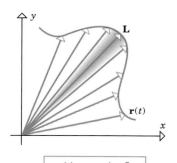

$\mathbf{r}(t)$ approaches $\mathbf{L}$ in length and direction if $\lim_{t \to a} \mathbf{r}(t) = \mathbf{L}$.

Figure 15.2.1

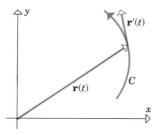

Figure 15.2.2

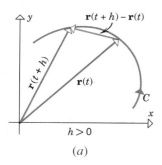

(a)

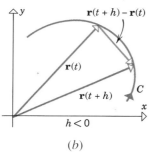

(b)

Figure 15.2.3

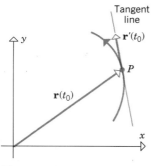

Figure 15.2.4

It can be shown that $\mathbf{r}(t)$ is continuous at t_0 if and only if each component of $\mathbf{r}(t)$ is continuous at t_0 (Exercise 67). As with real-valued functions, we shall call $\mathbf{r}(t)$ *continuous everywhere* or simply *continuous* if $\mathbf{r}(t)$ is continuous at all real values of t. Geometrically, the graph of a continuous vector-valued function is an unbroken curve.

The derivative of a vector-valued function can be interpreted geometrically as follows:

> **15.2.5**   GEOMETRIC INTERPRETATION OF THE DERIVATIVE. *Suppose that C is the graph of a vector-valued function $\mathbf{r}(t)$ in 2-space or 3-space and that $\mathbf{r}'(t)$ exists and is nonzero for a given value of t. If the vector $\mathbf{r}'(t)$ is positioned with its initial point at the terminal point of the radius vector $\mathbf{r}(t)$ (Figure 15.2.2), then $\mathbf{r}'(t)$ is tangent to C and points in the direction of increasing parameter.*

To make this result plausible, let C be the graph of $\mathbf{r}(t)$, and for a fixed value of the parameter t construct the position vectors $\mathbf{r}(t)$ and $\mathbf{r}(t + h)$. If $h > 0$, then the tip of the vector $\mathbf{r}(t + h)$ is in the direction of increasing parameter from the tip of $\mathbf{r}(t)$, and if $h < 0$, it is the other way around (Figure 15.2.3). In either case, the difference $\mathbf{r}(t + h) - \mathbf{r}(t)$ coincides with the secant line through the tips of the vectors $\mathbf{r}(t)$ and $\mathbf{r}(t + h)$. Moreover, since h is a scalar, the vector

$$\frac{1}{h}[\mathbf{r}(t + h) - \mathbf{r}(t)] = \frac{\mathbf{r}(t + h) - \mathbf{r}(t)}{h} \tag{5}$$

also coincides with this secant line. If $h < 0$, then vector (5) is oppositely directed to $\mathbf{r}(t + h) - \mathbf{r}(t)$, and if $h > 0$, these vectors have the same direction. In both cases, vector (5) points in the direction of increasing parameter. If the limit of (5) exists as $h \rightarrow 0$, then the secant lines through the tips of $\mathbf{r}(t + h)$ and $\mathbf{r}(t)$ tend toward the tangent line to C at the tip of $\mathbf{r}(t)$. Thus, the vector

$$\mathbf{r}'(t) = \lim_{h \to 0} \frac{\mathbf{r}(t + h) - \mathbf{r}(t)}{h} \tag{6}$$

is tangent to the curve C at the tip of $\mathbf{r}(t)$ and points in the direction of increasing parameter.

Motivated by the preceding discussion, we make the following definition.

> **15.2.6**   DEFINITION.   Let P be a point on the graph of a vector-valued function $\mathbf{r}(t)$, and let $\mathbf{r}(t_0)$ be the radius vector from the origin to P (Figure 15.2.4). If $\mathbf{r}'(t_0)$ exists and $\mathbf{r}'(t_0) \neq \mathbf{0}$, then we call $\mathbf{r}'(t_0)$ the *tangent vector* to the graph of $\mathbf{r}(t)$ at $\mathbf{r}(t_0)$.

REMARK.   Observe that the tangent vector can fail to exist at a point either because the derivative in (6) does not exist or because the derivative is the zero vector at the point.

If a vector-valued function $\mathbf{r}(t)$ has a tangent vector $\mathbf{r}'(t_0)$ at a point on its graph, then the line that is parallel to $\mathbf{r}'(t_0)$ and passes through the tip of the radius vector $\mathbf{r}(t_0)$ is called the *tangent line* to the graph of $\mathbf{r}(t)$ at $\mathbf{r}(t_0)$ (Figure 15.2.4). It follows from Formula (5) of Section 14.5 that a vector equation of the tangent line is

$$\mathbf{r} = \mathbf{r}(t_0) + t\mathbf{r}'(t_0) \tag{7}$$

Example 5   Find parametric equations of the tangent line to the circular helix

$$x = \cos t, \quad y = \sin t, \quad z = t$$

at the point where $t = \pi/6$.

Solution. We shall first use Formula (7) to find a vector equation of the tangent line, then we shall equate components to obtain the parametric equations. A vector equation $\mathbf{r} = \mathbf{r}(t)$ of the helix is

$$x\mathbf{i} + y\mathbf{j} + z\mathbf{k} = (\cos t)\mathbf{i} + (\sin t)\mathbf{j} + t\mathbf{k}$$

Thus,

$$\mathbf{r}(t) = (\cos t)\mathbf{i} + (\sin t)\mathbf{j} + t\mathbf{k}$$

$$\mathbf{r}'(t) = (-\sin t)\mathbf{i} + (\cos t)\mathbf{j} + \mathbf{k}$$

At the point where $t = \pi/6$, these vectors are

$$\mathbf{r}\left(\frac{\pi}{6}\right) = \frac{\sqrt{3}}{2}\mathbf{i} + \frac{1}{2}\mathbf{j} + \frac{\pi}{6}\mathbf{k} \quad \text{and} \quad \mathbf{r}'\left(\frac{\pi}{6}\right) = -\frac{1}{2}\mathbf{i} + \frac{\sqrt{3}}{2}\mathbf{j} + \mathbf{k}$$

so from (7) with $t_0 = \pi/6$ a vector equation of the tangent line is

$$\mathbf{r} = \mathbf{r}\left(\frac{\pi}{6}\right) + t\mathbf{r}'\left(\frac{\pi}{6}\right) = \left(\frac{\sqrt{3}}{2}\mathbf{i} + \frac{1}{2}\mathbf{j} + \frac{\pi}{6}\mathbf{k}\right) + t\left(-\frac{1}{2}\mathbf{i} + \frac{\sqrt{3}}{2}\mathbf{j} + \mathbf{k}\right)$$

Simplifying, then equating the resulting components with the corresponding components of $\mathbf{r} = x\mathbf{i} + y\mathbf{j} + z\mathbf{k}$ yields the parametric equations

$$x = \frac{\sqrt{3}}{2} - \frac{1}{2}t, \quad y = \frac{1}{2} + \frac{\sqrt{3}}{2}t, \quad z = \frac{\pi}{6} + t \quad \blacktriangleleft$$

◻ **DERIVATIVES OF DOT AND CROSS PRODUCTS**

The following rules, which are derived in the exercises, provide a method for differentiating dot products in 2-space and 3-space and cross products in 3-space.

$$\frac{d}{dt}[\mathbf{r}_1(t) \cdot \mathbf{r}_2(t)] = \mathbf{r}_1(t) \cdot \frac{d\mathbf{r}_2}{dt} + \frac{d\mathbf{r}_1}{dt} \cdot \mathbf{r}_2(t) \tag{8}$$

$$\frac{d}{dt}[\mathbf{r}_1(t) \times \mathbf{r}_2(t)] = \mathbf{r}_1(t) \times \frac{d\mathbf{r}_2}{dt} + \frac{d\mathbf{r}_1}{dt} \times \mathbf{r}_2(t) \tag{9}$$

REMARK. In (8) the order of the factors in each term on the right does not matter, but in (9) it does.

In plane geometry one learns that a tangent line to a circle is perpendicular to the radius at the point of tangency. Consequently, if a point moves along a circular arc in 2-space, one would expect the radius vector and the tangent vector at any point on the arc to be perpendicular. This is the motivation for the following useful theorem, which is applicable in both 2-space and 3-space.

15.2.7 THEOREM. *If $\mathbf{r}(t)$ is a vector-valued function in 2-space or 3-space and $\|\mathbf{r}(t)\|$ is constant for all t, then*

$$\mathbf{r}(t) \cdot \mathbf{r}'(t) = 0 \tag{10}$$

that is, $\mathbf{r}(t)$ and $\mathbf{r}'(t)$ are orthogonal vectors for all t.

Proof. It follows from (8) with $\mathbf{r}_1(t) = \mathbf{r}_2(t) = \mathbf{r}(t)$ that

$$\frac{d}{dt}[\mathbf{r}(t) \cdot \mathbf{r}(t)] = \mathbf{r}(t) \cdot \frac{d\mathbf{r}}{dt} + \frac{d\mathbf{r}}{dt} \cdot \mathbf{r}(t)$$

or, equivalently,

$$\frac{d}{dt}[\|\mathbf{r}(t)\|^2] = 2\mathbf{r}(t) \cdot \frac{d\mathbf{r}}{dt} \tag{11}$$

But $\|\mathbf{r}(t)\|^2$ is constant, so its derivative is zero. Thus

$$2\mathbf{r}(t) \cdot \frac{d\mathbf{r}}{dt} = 0$$

from which (10) follows. ∎

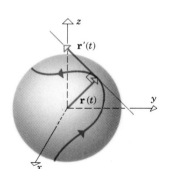

Figure 15.2.5

Example 6 Just as a tangent line to a circular arc in 2-space is perpendicular to the radius at the point of tangency, so a tangent vector to a curve on the surface of a sphere in 3-space is perpendicular to the radius vector at the point of tangency (Figure 15.2.5). To see that this is so, suppose that the graph of $\mathbf{r}(t)$ lies on the surface of the sphere of radius $k > 0$ centered at the origin. For each value of t we have $\|\mathbf{r}(t)\| = k$, so by Theorem 15.2.7

$$\mathbf{r}(t) \cdot \mathbf{r}'(t) = 0$$

and hence the radius vector $\mathbf{r}(t)$ and the tangent vector $\mathbf{r}'(t)$ are perpendicular. ◀

▶ **Exercise Set 15.2** ⬛ 33, 35, 36

In Exercises 1–8, find the limit.

1. $\lim\limits_{t \to 3} (t^2\mathbf{i} + 2t\mathbf{j})$.

2. $\lim\limits_{t \to \pi/4} \langle \cos t, \sin t \rangle$.

3. $\lim\limits_{t \to 0^+} \left(\sqrt{t}\,\mathbf{i} + \dfrac{\sin t}{t}\mathbf{j} \right)$.

4. $\lim\limits_{t \to +\infty} \left\langle \dfrac{t^2+1}{3t^2+2}, \dfrac{1}{t} \right\rangle$.

5. $\lim\limits_{t \to 2} (t\mathbf{i} - 3\mathbf{j} + t^2\mathbf{k})$.

6. $\lim\limits_{t \to \pi} \langle \cos 3t, e^{-t}, \sqrt{t} \rangle$.

7. $\lim\limits_{t \to +\infty} \left(\tan^{-1} t\,\mathbf{i} + \dfrac{t}{t^2+3}\mathbf{j} + \cos\dfrac{2}{t}\mathbf{k} \right)$.

8. $\lim\limits_{t \to 1} \left\langle \dfrac{3}{t^2}, \dfrac{\ln t}{t^2-1}, \sin 2t \right\rangle$.

In Exercises 9 and 10, prove that $\mathbf{r}$ is continuous at t_0.

9. $\mathbf{r}(t) = (3 \sin t)\mathbf{i} - 2t\mathbf{j}$; $t_0 = \pi/2$.

10. $\mathbf{r}(t) = 5\mathbf{i} - \sqrt{3t+1}\,\mathbf{j} + e^{2t}\mathbf{k}$; $t_0 = 1$.

In Exercises 11–14, find $\mathbf{r}'(t)$.

11. $\mathbf{r}(t) = (4 + 5t)\mathbf{i} + (t - t^2)\mathbf{j}$.

12. $\mathbf{r}(t) = 4\mathbf{i} - (\cos t)\mathbf{j}$.

13. $\mathbf{r}(t) = \dfrac{1}{t}\mathbf{i} + (\tan t)\mathbf{j} + e^{2t}\mathbf{k}$.

14. $\mathbf{r}(t) = (\tan^{-1} t)\mathbf{i} + (t \cos t)\mathbf{j} - \sqrt{t}\,\mathbf{k}$.

In Exercises 15–22, find $\mathbf{r}'(t_0)$; then sketch the graph of $\mathbf{r}(t)$ and the tangent vector $\mathbf{r}'(t_0)$.

15. $\mathbf{r}(t) = \langle t, t^2 \rangle$; $t_0 = 2$.

16. $\mathbf{r}(t) = (\cos t)\mathbf{i} + (\sin t)\mathbf{j}$; $t_0 = 3\pi/4$.

17. $\mathbf{r}(t) = \langle e^{-t}, e^{2t} \rangle$; $t_0 = \ln 2$.

18. $\mathbf{r}(t) = (\cos 2t)\mathbf{i} - (4 \sin t)\mathbf{j}$; $t_0 = \pi$.

19. $\mathbf{r}(t) = (2 \sin t)\mathbf{i} + \mathbf{j} + (2 \cos t)\mathbf{k}$; $t_0 = \pi/2$.

20. $\mathbf{r}(t) = (\cos t)\mathbf{i} + (\sin t)\mathbf{j} + t\mathbf{k}$; $t_0 = \pi/4$.

21. $\mathbf{r}(t) = 3\mathbf{i} + t\mathbf{j} + (2 - t^2)\mathbf{k}$; $t_0 = 1$.

22. $\mathbf{r}(t) = t\mathbf{i} + 2t\mathbf{j} + t^2\mathbf{k}$; $t_0 = 2$.

In Exercises 23–26, find parametric equations of the line tangent to the graph of $\mathbf{r}(t)$ at the point where $t = t_0$.

23. $\mathbf{r}(t) = t^2\mathbf{i} + (2 - \ln t)\mathbf{j}$; $t_0 = 1$.

24. $\mathbf{r}(t) = e^{2t}\mathbf{i} - (2 \cos 3t)\mathbf{j}$; $t_0 = 0$.

25. $\mathbf{r}(t) = (2 \cos \pi t)\mathbf{i} + (2 \sin \pi t)\mathbf{j} + 3t\mathbf{k}$; $t_0 = \frac{1}{3}$.

26. $\mathbf{r}(t) = (\ln t)\mathbf{i} + e^{-t}\mathbf{j} + t^3\mathbf{k}$; $t_0 = 2$.

In Exercises 27–30, find a vector equation of the line tangent to the graph of $\mathbf{r}(t)$ at the point P_0 on the curve.

27. $\mathbf{r}(t) = (2t - 1)\mathbf{i} + \sqrt{3t + 4}\,\mathbf{j}$; $P_0(-1, 2)$.

28. $\mathbf{r}(t) = (4 \cos t)\mathbf{i} - 3t\mathbf{j}$; $P_0(2, -\pi)$.

29. $\mathbf{r}(t) = t^2\mathbf{i} - \dfrac{1}{t + 1}\mathbf{j} + (4 - t^2)\mathbf{k}$; $P_0(4, 1, 0)$.

30. $\mathbf{r}(t) = (\sin t)\mathbf{i} + (\cosh t)\mathbf{j} + (\tan^{-1} t)\mathbf{k}$; $P_0(0, 1, 0)$.

31. Find an equation of the plane that is perpendicular to the curve $\mathbf{r} = (3 \sin t)\mathbf{i} - (2 \cos t)\mathbf{j} + t\mathbf{k}$ at the point where $t = \pi/2$. [*Note:* A plane is considered to be perpendicular to a curve at a point if it is perpendicular to the tangent line at that point.]

32. Find an equation of the plane that is perpendicular to the curve $\mathbf{r} = 3t^2\mathbf{i} + \sqrt{t + 5}\,\mathbf{j} - 2t\mathbf{k}$ at the point $P(3, 2, 2)$ on the curve. [See note in Exercise 31.]

33. (a) Find the points where the curve
$$\mathbf{r} = t\mathbf{i} + t^2\mathbf{j} - 3t\mathbf{k}$$
intersects the plane $2x - y + z = -2$.

(b) For the curve and plane in part (a), find, to the nearest degree, the acute angle that the tangent line to the curve makes with a line normal to the plane at each point of intersection.

34. Find where the tangent line to the curve
$$\mathbf{r} = e^{-2t}\mathbf{i} + (\cos t)\mathbf{j} + (3 \sin t)\mathbf{k}$$
at the point $(1, 1, 0)$ intersects the yz-plane.

In Exercises 35 and 36, show that the graphs of $\mathbf{r}_1(t)$ and $\mathbf{r}_2(t)$ intersect at the point P. Find, to the nearest degree, the acute angle between the tangent lines to the graphs of $\mathbf{r}_1(t)$ and $\mathbf{r}_2(t)$ at the point P.

35. $\mathbf{r}_1(t) = t^2\mathbf{i} + t\mathbf{j} + 3t^3\mathbf{k}$,
$\mathbf{r}_2(t) = (t - 1)\mathbf{i} + \frac{1}{4}t^2\mathbf{j} + (5 - t)\mathbf{k};\ P(1, 1, 3)$.

36. $\mathbf{r}_1(t) = 2e^{-t}\mathbf{i} + (\cos t)\mathbf{j} + (t^2 + 3)\mathbf{k}$,
$\mathbf{r}_2(t) = (1 - t)\mathbf{i} + t^2\mathbf{j} + (t^3 + 4)\mathbf{k};\ P(2, 1, 3)$.

In Exercises 37–48, evaluate the integral.

37. $\displaystyle\int (3\mathbf{i} + 4t\mathbf{j})\,dt$. **38.** $\displaystyle\int [(\cos t)\mathbf{i} + (\sin t)\mathbf{j}]\,dt$.

39. $\displaystyle\int_0^{\pi/3} \langle \cos 3t, -\sin 3t \rangle\,dt$.

40. $\displaystyle\int_0^1 (t^2\mathbf{i} + t^3\mathbf{j})\,dt$. **41.** $\displaystyle\int_1^9 (t^{1/2}\mathbf{i} + t^{-1/2}\mathbf{j})\,dt$.

42. $\displaystyle\int [(t \sin t)\mathbf{i} + \mathbf{j}]\,dt$. **43.** $\displaystyle\int \langle te^t, \ln t \rangle\,dt$.

44. $\displaystyle\int_0^2 \|t\mathbf{i} + t^2\mathbf{j}\|\,dt$. **45.** $\displaystyle\int \left[t^2\mathbf{i} - 2t\mathbf{j} + \frac{1}{t}\mathbf{k} \right]\,dt$.

46. $\displaystyle\int \langle e^{-t}, e^t, 3t^2 \rangle\,dt$. **47.** $\displaystyle\int_0^1 (e^{2t}\mathbf{i} + e^{-t}\mathbf{j} + t\mathbf{k})\,dt$.

48. $\displaystyle\int_{-3}^3 \langle (3 - t)^{3/2}, (3 + t)^{3/2}, 1 \rangle\,dt$.

49. Suppose that a particle moves through 3-space along the curve $\mathbf{r} = t\mathbf{i} - 3t^2\mathbf{j} + \mathbf{k}$ and that it is subjected to a force of $\mathbf{F} = 3x\mathbf{i} - 2\mathbf{j} + yz\mathbf{k}$ when it is at the point (x, y, z).

(a) Find $\mathbf{F}$ in terms of t for points on the path.

(b) Find $\displaystyle\int_0^2 \mathbf{F} \cdot \frac{d\mathbf{r}}{dt}\,dt$. [*Note:* Later, we shall see that this is the work done by the force as the particle moves along the curve from the point where $t = 0$ to the point where $t = 2$.]

50. Find $\mathbf{r}(t)$ given that $\mathbf{r}'(t) = t^2\mathbf{i} + 2t\mathbf{j}$ and $\mathbf{r}(0) = \mathbf{i} + \mathbf{j}$.

51. Find $\mathbf{r}(t)$ given that $\mathbf{r}'(t) = (\cos t)\mathbf{i} + (\sin t)\mathbf{j}$ and $\mathbf{r}(0) = \mathbf{i} - \mathbf{j}$.

52. Find $\mathbf{r}(t)$ given that $\mathbf{r}''(t) = \mathbf{i} + e^t\mathbf{j}$, $\mathbf{r}(0) = 2\mathbf{i}$, and $\mathbf{r}'(0) = \mathbf{j}$.

53. Find $\mathbf{r}(t)$ given that $\mathbf{r}''(t) = 12t^2\mathbf{i} - 2\mathbf{j}$, $\mathbf{r}'(0) = \mathbf{0}$, and $\mathbf{r}(0) = 2\mathbf{i} - 4\mathbf{j}$.

54. Find $\mathbf{r}(t)$ given that $\mathbf{r}'(t) = e^{-2t}\mathbf{i} + (\cos t)\mathbf{j} - \mathbf{k}$ and $\mathbf{r}(0) = 3\mathbf{j} + 2\mathbf{k}$.

55. Find $\mathbf{r}(t)$ given that $\mathbf{r}'(t) = 2\mathbf{i} + \dfrac{t}{t^2 + 1}\mathbf{j} + t\mathbf{k}$ and $\mathbf{r}(1) = \mathbf{0}$.

56. Find $\mathbf{r}(t)$ given that $\mathbf{r}''(t) = (4 \sin 2t)\mathbf{i} + 6t\mathbf{j} + e^{-t}\mathbf{k}$, $\mathbf{r}(0) = 2\mathbf{i}$, and $\mathbf{r}'(0) = \mathbf{k}$.

57. Calculate $(d/dt)[\mathbf{r}_1(t) \cdot \mathbf{r}_2(t)]$ two ways: first using Formula (8), and then by differentiating $\mathbf{r}_1(t) \cdot \mathbf{r}_2(t)$ directly.

(a) $\mathbf{r}_1(t) = 2t\mathbf{i} + 3t^2\mathbf{j} + t^3\mathbf{k}$, $\mathbf{r}_2(t) = t^4\mathbf{k}$

(b) $\mathbf{r}_1(t) = 3 \sec t\mathbf{i} - t\mathbf{j} + \ln t\mathbf{k}$, $\mathbf{r}_2(t) = 4t\mathbf{i} - \sin t\mathbf{k}$.

58. Calculate $(d/dt)[\mathbf{r}_1(t) \times \mathbf{r}_2(t)]$ two ways: first using Formula (9), and then by differentiating $\mathbf{r}_1(t) \times \mathbf{r}_2(t)$ directly.

(a) $\mathbf{r}_1(t) = 2t\mathbf{i} + 3t^2\mathbf{j} + t^3\mathbf{k}$, $\mathbf{r}_2(t) = t^4\mathbf{k}$

(b) $\mathbf{r}_1(t) = 3 \sec t\mathbf{i} - t\mathbf{j} + \ln t\mathbf{k}$, $\mathbf{r}_2(t) = 4t\mathbf{i} - \sin t\mathbf{k}$.

59. Prove: $(d/dt)[\mathbf{r}(t) \times \mathbf{r}'(t)] = \mathbf{r}(t) \times \mathbf{r}''(t)$. [*Hint:* Use Formula (9).]

60. Let $\mathbf{r} = \mathbf{r}(t)$. Show that
$$\frac{d}{dt}[\|\mathbf{r}\|] = \frac{1}{\|\mathbf{r}\|} \mathbf{r} \cdot \mathbf{r}'$$
[*Hint:* Consider $\mathbf{r} \cdot \mathbf{r}$.]

61. Use part (e) of Theorem 15.2.1 and Exercise 60 to derive the formula
$$\frac{d}{dt}\left[\frac{\mathbf{r}}{\|\mathbf{r}\|} \right] = \frac{1}{\|\mathbf{r}\|}\mathbf{r}' - \frac{\mathbf{r} \cdot \mathbf{r}'}{\|\mathbf{r}\|^3}\mathbf{r}$$

62. Let $\mathbf{u} = \mathbf{u}(t)$, $\mathbf{v} = \mathbf{v}(t)$, and $\mathbf{w} = \mathbf{w}(t)$ be differentiable vector-valued functions. Use Formulas (8) and (9) to show that
$$\frac{d}{dt}[\mathbf{u} \cdot (\mathbf{v} \times \mathbf{w})]$$
$$= \frac{d\mathbf{u}}{dt} \cdot [\mathbf{v} \times \mathbf{w}] + \mathbf{u} \cdot \left[\frac{d\mathbf{v}}{dt} \times \mathbf{w} \right] + \mathbf{u} \cdot \left[\mathbf{v} \times \frac{d\mathbf{w}}{dt} \right]$$

63. Let $u_1, u_2, u_3, v_1, v_2, v_3, w_1, w_2$, and w_3 be differentiable functions of t. Use Exercise 62 to show that
$$\frac{d}{dt}\begin{vmatrix} u_1 & u_2 & u_3 \\ v_1 & v_2 & v_3 \\ w_1 & w_2 & w_3 \end{vmatrix}$$
$$= \begin{vmatrix} u_1' & u_2' & u_3' \\ v_1 & v_2 & v_3 \\ w_1 & w_2 & w_3 \end{vmatrix} + \begin{vmatrix} u_1 & u_2 & u_3 \\ v_1' & v_2' & v_3' \\ w_1 & w_2 & w_3 \end{vmatrix} + \begin{vmatrix} u_1 & u_2 & u_3 \\ v_1 & v_2 & v_3 \\ w_1' & w_2' & w_3' \end{vmatrix}$$

64. Prove Theorem 15.2.1 for 2-space.

65. Derive Formulas (8) and (9) for 3-space.

66. Let lim stand for any one of the limit symbols $\lim_{t \to a}$, $\lim_{t \to a^+}$, $\lim_{t \to a^-}$, $\lim_{t \to +\infty}$, or $\lim_{t \to -\infty}$. Prove for 2-space:

(a) If k is a scalar and $\lim \mathbf{r}(t)$ exists, then $\lim k\mathbf{r}(t) = k \lim \mathbf{r}(t)$.

(b) If $\lim \mathbf{r}_1(t)$ and $\lim \mathbf{r}_2(t)$ exist, then
$\lim [\mathbf{r}_1(t) + \mathbf{r}_2(t)] = \lim \mathbf{r}_1(t) + \lim \mathbf{r}_2(t)$,
$\lim [\mathbf{r}_1(t) - \mathbf{r}_2(t)] = \lim \mathbf{r}_1(t) - \lim \mathbf{r}_2(t)$.

67. Prove for 2-space: $\mathbf{r}$ is continuous at t_0 if and only if each component of $\mathbf{r}$ is continuous at t_0.

68. Prove for 2-space:

(a) $\int k\mathbf{r}(t)\,dt = k\int \mathbf{r}(t)\,dt$, where k is a scalar constant

(b) $\int [\mathbf{r}_1(t) + \mathbf{r}_2(t)]\,dt = \int \mathbf{r}_1(t)\,dt + \int \mathbf{r}_2(t)\,dt.$

69. Prove for 2-space: If $\mathbf{R}'(t) = \mathbf{r}(t)$ on an interval $[a, b]$, then

(a) $\int \mathbf{r}(t)\,dt = \mathbf{R}(t) + \mathbf{C}$, where $\mathbf{C}$ is an arbitrary vector constant

(b) $\int_a^b \mathbf{r}(t)\,dt = \mathbf{R}(b) - \mathbf{R}(a).$

■ 15.3 CHANGE OF PARAMETER; ARC LENGTH

We observed in earlier sections that a given curve in 2-space or 3-space can be represented parametrically in more than one way. For example, in Section 13.4 we gave two parametric representations of a circle—one in which the circle was traced clockwise and the other in which it was traced counterclockwise. Sometimes it is desirable to change a given parametric representation of a curve to an alternative representation that is better suited for the problem at hand. In this section we shall investigate issues associated with changes of parameter, and we shall show that arc length plays a special role in parametric representations of curves.

☐ **SMOOTH PARAMETRIZATIONS**

In classical applications, there is relatively little interest in parametric representations of curves in 2-space or 3-space in which the curve is traced in a discontinuous or erratic fashion. Thus, we impose restrictions on the kind of parametric representations that will be considered. We shall say that $\mathbf{r}(t)$ is *smoothly parametrized* or that $\mathbf{r}$ is a *smooth function of t* if $\mathbf{r}'(t)$ is continuous and $\mathbf{r}'(t) \neq \mathbf{0}$ for any value of t. Stated another way, the components of $\mathbf{r}(t)$ have continuous derivatives with respect to t and are not all zero for any value of t. Thus, in 3-space

$$\mathbf{r}(t) = x(t)\mathbf{i} + y(t)\mathbf{j} + z(t)\mathbf{k}$$

is a smooth function of t if $x'(t)$, $y'(t)$, and $z'(t)$ are continuous and there is no value of t at which all three derivatives are zero.

It can be shown that if $\mathbf{r}(t)$ is a smoothly parametrized function, then the angles between the tangent vector $\mathbf{r}'(t)$ and the unit vectors $\mathbf{i}$, $\mathbf{j}$, and $\mathbf{k}$ are continuous functions of t (Exercise 38). Thus, a smoothly parametrized function $\mathbf{r}(t)$ is said to have a *continuously turning tangent vector*.

Example 1   Determine whether the following vector-valued functions have continuously turning tangent vectors.

(a) $\mathbf{r}(t) = a\cos t\,\mathbf{i} + a\sin t\,\mathbf{j} + ct\,\mathbf{k}$ $(a > 0, c > 0)$

(b) $\mathbf{r}(t) = t^2\mathbf{i} + t^3\mathbf{j}$

Solution (a).   We have

$$\mathbf{r}'(t) = -a\sin t\,\mathbf{i} + a\cos t\,\mathbf{j} + c\mathbf{k}$$

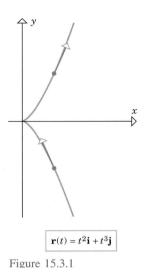

$$\mathbf{r}(t) = t^2\mathbf{i} + t^3\mathbf{j}$$

Figure 15.3.1

□ **CHANGE OF PARAMETER**

All three components are continuous functions, and there is no value of t for which all three components are zero (verify), so $\mathbf{r}(t)$ has a continuously turning tangent vector. The graph of $\mathbf{r}(t)$ is the circular helix in Figure 15.1.1.

Solution (b).   We have

$$\mathbf{r}'(t) = 2t\mathbf{i} + 3t^2\mathbf{j}$$

Although both components are continuous functions, they are both equal to zero if $t = 0$, so $\mathbf{r}(t)$ does not have a continuously turning tangent vector. The graph of $\mathbf{r}(t)$, which is shown in Figure 15.3.1, is a semicubical parabola traced in the upward direction (see Example 9 of Section 13.4). Observe that for values of t slightly less than zero the angle between $\mathbf{r}'(t)$ and $\mathbf{i}$ is near π, and for values of t slightly larger than zero the angle is near 0; hence there is a sudden reversal in the direction of the tangent vector as t increases through $t = 0$.   ◄

A *change of parameter* in a vector-valued function $\mathbf{r}(t)$ is a substitution $t = g(\tau)$ which produces a new vector-valued function $\mathbf{r}(g(\tau))$ having the same graph as $\mathbf{r}(t)$, but possibly traced differently as the parameter τ increases.

Example 2   Find a change of parameter $t = g(\tau)$ for the circle

$$\mathbf{r}(t) = \cos t\,\mathbf{i} + \sin t\,\mathbf{j}   (0 \le t \le 2\pi)$$

such that

(a)   the circle is traced counterclockwise as τ increases over the interval $[0, 1]$;

(b)   the circle is traced clockwise as τ increases over the interval $[0, 1]$.

Solution (a).   The circle is traced counterclockwise if t increases (Example 2 of Section 13.4). Thus, if we choose g to be an increasing function, then it will follow from the relationship $t = g(\tau)$ that t increases when τ increases, thereby ensuring that the circle is traced counterclockwise as τ increases. We also want to choose g so that t increases from 0 to 2π as τ increases from 0 to 1. A simple choice of g that satisfies all of the required criteria is the linear function graphed in Figure 15.3.2a. The equation of this line is

$$t = g(\tau) = 2\pi\tau \tag{1}$$

which is the desired change of parameter. The resulting representation of the circle in terms of the parameter τ is

$$\mathbf{r}(g(\tau)) = \cos 2\pi\tau\,\mathbf{i} + \sin 2\pi\tau\,\mathbf{j}   (0 \le \tau \le 1)$$

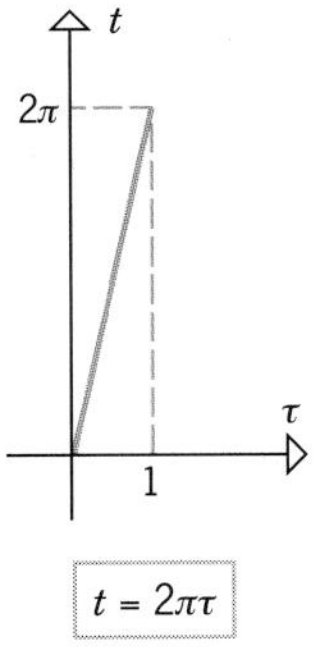

Solution (b).   To ensure that the circle is traced clockwise, we shall choose g to be a decreasing function such that τ decreases from 2π to 0 as τ increases from 0 to 1. A simple choice of g that achieves this is the linear function

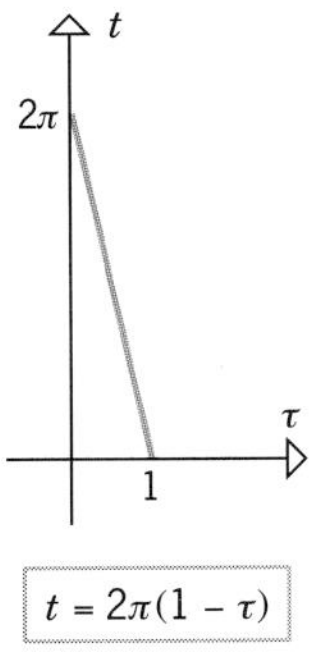

$$t = g(\tau) = 2\pi(1 - \tau) \tag{2}$$

graphed in Figure 15.3.2b. The resulting representation of the circle in terms of the parameter τ is

$$\mathbf{r}(g(\tau)) = \cos(2\pi(1 - \tau))\mathbf{i} + \sin(2\pi(1 - \tau))\mathbf{j}   (0 \le \tau \le 1)$$

which simplifies to (verify)

$$\mathbf{r}(g(\tau)) = \cos 2\pi\tau\,\mathbf{i} - \sin 2\pi\tau\,\mathbf{j}   (0 \le \tau \le 1)   ◄$$

$t = 2\pi\tau$

(a)

$t = 2\pi(1 - \tau)$

(b)

Figure 15.3.2

The following version of the chain rule for vector-valued functions relates the derivatives of $\mathbf{r}(t)$ and $\mathbf{r}(g(\tau))$ under a change of parameter $t = g(\tau)$. The proof is left as an exercise.

15.3.1 THEOREM (*Chain Rule*). *Let $\mathbf{r}(t)$ be a vector-valued function in 2-space or 3-space that is differentiable with respect to t. If $t = g(\tau)$ is a change of parameter in which g is differentiable with respect to τ, then $\mathbf{r}(g(\tau))$ is differentiable with respect to τ and*

$$\frac{d\mathbf{r}}{d\tau} = \frac{d\mathbf{r}}{dt}\frac{dt}{d\tau} \tag{3}$$

When making a change of parameter $t = g(\tau)$ in a vector-valued function $\mathbf{r}(t)$, it will be important to ensure that $\mathbf{r}(g(\tau))$ is smooth if $\mathbf{r}(t)$ is smooth. A change of parameter for which this is true is called a ***smooth change of parameter***. It follows from (3) that $t = g(\tau)$ is a smooth change of parameter if $dt/d\tau$ is continuous and $dt/d\tau \neq 0$ for any value of τ, since these conditions imply that $d\mathbf{r}/d\tau$ is continuous and nonzero if $d\mathbf{r}/dt$ is continuous and nonzero. Smooth changes of parameter fall into two categories—those for which $dt/d\tau > 0$ for all τ and those for which $dt/d\tau < 0$ for all τ.

A smooth change of parameter for which $dt/d\tau > 0$ for all τ will be called a ***positive change of parameter***, and a smooth change of parameter for which $dt/d\tau < 0$ for all τ will be called a ***negative change of parameter***. A positive change of parameter preserves the orientation of a parametric curve and a negative change of parameter reverses the orientation.

Example 3 In Example 2 the change of parameter given by (1) is positive since $dt/d\tau = 2\pi > 0$, and the change of parameter given by (2) is negative since $dt/d\tau = -2\pi < 0$. The positive change of parameter preserved the orientation of the circle, and the negative change of parameter reversed it.   ◀

□ **ARC LENGTH IN 3-SPACE**

In Theorem 13.4.1 we showed that in 2-space the arc length L of a parametric curve

$$x = x(t), \quad y = y(t) \qquad (a \leq t \leq b)$$

is given by

$$L = \int_a^b \sqrt{\left(\frac{dx}{dt}\right)^2 + \left(\frac{dy}{dt}\right)^2}\,dt \tag{4}$$

This result generalizes to curves in 3-space exactly as one would expect: If no segment of the curve

$$x = x(t), \quad y = y(t), \quad z = z(t) \qquad (a \leq t \leq b)$$

is traced more than once as t increases from a to b, and if dx/dt, dy/dt, and dz/dt are continuous for $a \leq t \leq b$, then the arc length L of the curve is given by

$$L = \int_a^b \sqrt{\left(\frac{dx}{dt}\right)^2 + \left(\frac{dy}{dt}\right)^2 + \left(\frac{dz}{dt}\right)^2}\,dt \tag{5}$$

Example 4 Find the arc length of that portion of the circular helix

$$x = \cos t, \quad y = \sin t, \quad z = t$$

from $t = 0$ to $t = \pi$.

Solution.   From (5) the arc length is

$$L = \int_0^\pi \sqrt{\left(\frac{dx}{dt}\right)^2 + \left(\frac{dy}{dt}\right)^2 + \left(\frac{dz}{dt}\right)^2}\, dt$$

$$= \int_0^\pi \sqrt{(-\sin t)^2 + (\cos t)^2 + 1}\, dt = \int_0^\pi \sqrt{2}\, dt = \sqrt{2}\,\pi \quad \blacktriangleleft$$

If the parametric equations corresponding to Formulas (4) and (5) are expressed in vector form, say

$$\mathbf{r}(t) = x(t)\mathbf{i} + y(t)\mathbf{j} \quad \text{or} \quad \mathbf{r}(t) = x(t)\mathbf{i} + y(t)\mathbf{j} + z(t)\mathbf{k}$$

then

$$\frac{d\mathbf{r}}{dt} = \frac{dx}{dt}\mathbf{i} + \frac{dy}{dt}\mathbf{j} \quad \text{or} \quad \frac{d\mathbf{r}}{dt} = \frac{dx}{dt}\mathbf{i} + \frac{dy}{dt}\mathbf{j} + \frac{dz}{dt}\mathbf{k}$$

Thus, Formulas (4) and (5) can both be expressed in vector form as

$$L = \int_a^b \left\| \frac{d\mathbf{r}}{dt} \right\| dt \tag{6}$$

☐ **ARC LENGTH AS A PARAMETER**

For many purposes the best parameter to use for representing a curve in 2-space or 3-space parametrically is the length of arc measured along the curve from some fixed reference point. This can be done as follows:

Step 1. Select an arbitrary point on the curve C to serve as a ***reference point***.

Step 2. Starting from the reference point, choose one direction along the curve to be the ***positive direction*** and the other to be the ***negative direction***.

Step 3. If P is a point on the curve, let s be the "signed" arc length along C from the reference point to P, where s is positive if P is in the positive direction from the reference point, and s is negative if P is in the negative direction. Figure 15.3.3 illustrates this idea.

By this procedure, a unique point P on the curve is determined when a value for s is given. For example, $s = 2$ determines the point that is 2 units along the curve in the positive direction from the reference point, and $s = -\frac{3}{2}$ determines the point that is $\frac{3}{2}$ units along the curve in the negative direction from the reference point.

Let us now treat s as a variable. As the value of s changes, the corresponding point P moves along C and the coordinates of P become functions of s. Thus, in 2-space the coordinates of P are $(x(s), y(s))$, and in 3-space they are $(x(s), y(s), z(s))$. Therefore, in 2-space or 3-space the curve C is given by the parametric equations

$$x = x(s), \quad y = y(s) \quad \text{or} \quad x = x(s), \quad y = y(s), \quad z = z(s)$$

A parametric representation of a curve with arc length as the parameter is called an ***arc-length parametrization*** of the curve. Note that a given curve will generally have infinitely many different arc-length parametrizations, since the reference point and orientation can be chosen arbitrarily.

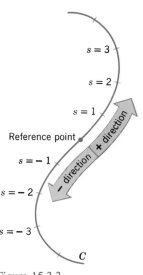

Figure 15.3.3

Example 5   Find the arc-length parametrization of the circle $x^2 + y^2 = a^2$ with counterclockwise orientation and $(a, 0)$ as the reference point.

Solution. The circle with counterclockwise orientation can be represented by the parametric equations

$$x = a \cos t, \quad y = a \sin t \qquad (0 \le t \le 2\pi) \tag{7}$$

in which t can be interpreted as the angle in radian measure from the positive x-axis to the radius from the origin to the point $P(x, y)$ (Figure 15.3.4). If we take the positive direction for measuring the arc length to be counterclockwise, and we take $(a, 0)$ to be the reference point, then s and t are related by

$$s = at \quad \text{or} \quad t = s/a$$

Making this change of variable in (7) and noting that s increases from 0 to $2\pi a$ as t increases from 0 to 2π yields the following arc-length parametrization of the circle

$$x = a \cos (s/a), \quad y = a \sin (s/a) \qquad (0 \le s \le 2\pi a) \quad \blacktriangleleft$$

In the preceding example we used basic trigonometry to find a formula for changing the parameter from t to arc length s. However, it is only in the simplest cases that geometric methods can be used to find arc-length parametrizations. The following theorem will provide a more general method for doing this.

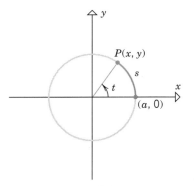

Figure 15.3.4

15.3.2 THEOREM. *Let C be the graph of a smooth vector-valued function* $\mathbf{r}(t)$ *in 2-space or 3-space, and let* $\mathbf{r}(t_0)$ *be any point on C. Then the following formula defines a positive change of parameter from t to s, where s is an arc-length parameter having* $\mathbf{r}(t_0)$ *as its reference point:*

$$s = \int_{t_0}^{t} \left\| \frac{d\mathbf{r}}{du} \right\| du \tag{8}$$

(Figure 15.3.5).

Proof. From (6) with u as the variable of integration instead of t, the integral represents the arc length of that portion of C between $\mathbf{r}(t_0)$ and $\mathbf{r}(t)$ if $t > t_0$ and the negative of that arc length if $t < t_0$. Thus, s is the arc-length parameter with $\mathbf{r}(t_0)$ as its reference point and its positive direction in the direction of increasing t. $\blacksquare$

For reference, we note that in 2-space and 3-space, respectively, Formula (8) can be expressed in component form as

$$s = \int_{t_0}^{t} \sqrt{\left(\frac{dx}{du} \right)^2 + \left(\frac{dy}{du} \right)^2} \, du \tag{9}$$

$$s = \int_{t_0}^{t} \sqrt{\left(\frac{dx}{du} \right)^2 + \left(\frac{dy}{du} \right)^2 + \left(\frac{dz}{du} \right)^2} \, du \tag{10}$$

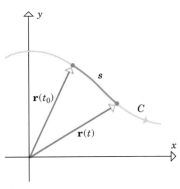

Figure 15.3.5

Example 6 Find the arc-length parametrization of the line

$$x = 2t + 1, \quad y = 3t - 2 \tag{11}$$

that has reference point $(1, -2)$ and the same orientation as the original line.

Solution. In Formula (9) we used u as the variable of integration because t was needed as a limit of integration. To apply (9), we first rewrite the given parametric equations with u in place of t; this gives

$$x = 2u + 1, \quad y = 3u - 2$$

from which we obtain

$$\frac{dx}{du} = 2, \quad \frac{dy}{du} = 3$$

From (11) we see that the reference point $(1, -2)$ corresponds to $t = t_0 = 0$, so (9) yields

$$s = \int_{t_0}^{t} \sqrt{\left(\frac{dx}{du}\right)^2 + \left(\frac{dy}{du}\right)^2}\, du = \int_{0}^{t} \sqrt{13}\, du = \sqrt{13}\, u \,\Big]_{u=0}^{u=t} = \sqrt{13}\, t$$

Therefore, $t = s/\sqrt{13}$. Substituting this expression in (11) and simplifying yields the parametric equations

$$x = \frac{2}{\sqrt{13}} s + 1, \quad y = \frac{3}{\sqrt{13}} s - 2 \quad \blacktriangleleft$$

Example 7 Let $\mathbf{u} = \langle u_1, u_2, u_3 \rangle$ be a vector of length 1, and let l be the line that is parallel to $\mathbf{u}$ and passes through the point $P_0(x_0, y_0, z_0)$. Find the arc-length parametrization of l with the reference point at P_0 and the positive direction in the direction of $\mathbf{u}$.

Solution. If $P(x, y, z)$ is any point on the line, then the vector $\overrightarrow{P_0 P}$ is parallel to $\mathbf{u}$ and hence is some scalar multiple of $\mathbf{u}$, say

$$\overrightarrow{P_0 P} = s\mathbf{u} \tag{12}$$

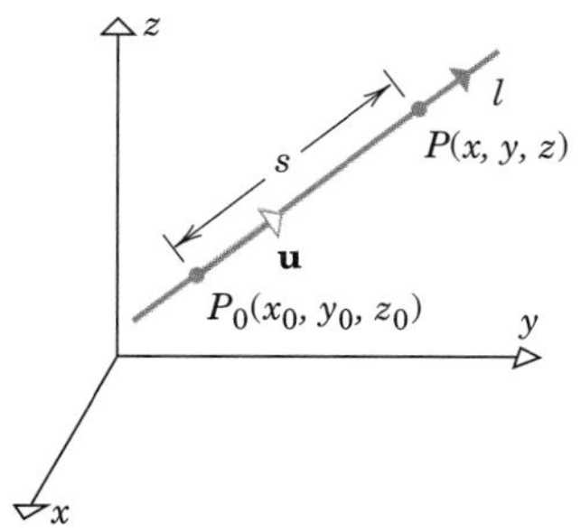

(Figure 15.3.6). Because $\mathbf{u}$ has length 1, it follows that s can be interpreted as the signed arc length from P_0 to P, where $s = 0$ if P_0 and P coincide, $s > 0$ if P is distinct from P_0 and $\overrightarrow{P_0 P}$ has the same direction as $\mathbf{u}$, and $s < 0$ if P is distinct from P_0 and $\overrightarrow{P_0 P}$ is oppositely directed to $\mathbf{u}$. It follows from (12) that

$$\langle x - x_0, y - y_0, z - z_0 \rangle = s\langle u_1, u_2, u_3 \rangle$$

from which we obtain the parametric equations

$$x = x_0 + su_1, \quad y = y_0 + su_2, \quad z = z_0 + su_3$$

(We leave it for the reader to show that the same result holds in 2-space with two components rather than three.) $\blacktriangleleft$

Figure 15.3.6

REMARK. Formula (8) tends to produce integrals that cannot be evaluated easily, so it is only in relatively simple cases, such as those in the preceding examples, that explicit formulas for arc-length parametrizations can be obtained.

Because arc-length parameters for a curve C are intimately related to the geometric characteristics of C, arc-length parametrizations have properties that are not enjoyed by other parametrizations. For example, the following theorem shows that if a smooth curve is represented parametrically using an arc-length parameter, then the tangent vectors all have length 1.

15.3.3 THEOREM. *If $\mathbf{r}(t)$ is a smooth vector-valued function in 2-space or 3-space, and if s is an arc-length parameter, then $\|d\mathbf{r}/ds\| = 1$ for all s.*

Proof.   It follows from the chain rule (3) with s in place of τ that

$$\frac{d\mathbf{r}}{ds} = \frac{d\mathbf{r}}{dt}\frac{dt}{ds} = \frac{1}{ds/dt}\frac{d\mathbf{r}}{dt} \tag{13}$$

But from (8) and the Second Fundamental Theorem of Calculus (5.9.1) we have

$$\frac{ds}{dt} = \left\|\frac{d\mathbf{r}}{dt}\right\| \tag{14}$$

so (13) can be written as

$$\frac{d\mathbf{r}}{ds} = \frac{1}{\|d\mathbf{r}/dt\|}\frac{d\mathbf{r}}{dt} \tag{15}$$

Recalling that multiplying a nonzero vector by the reciprocal of its length (normalizing) produces a vector of length 1, we conclude from (15) that $\|d\mathbf{r}/ds\| = 1$.   ■

Example 8   In Example 5 we showed that the circle $x^2 + y^2 = a^2$ can be represented parametrically in terms of arc length as

$$x = a\cos(s/a), \quad y = a\sin(s/a) \qquad (0 \leq s \leq 2\pi a)$$

Writing these equations in vector form, and then differentiating with respect to s to obtain the tangent vector, yields

$$\mathbf{r}(s) = a\cos(s/a)\mathbf{i} + a\sin(s/a)\mathbf{j}$$

$$\frac{d\mathbf{r}}{ds} = -\sin(s/a)\mathbf{i} + \cos(s/a)\mathbf{j}$$

As guaranteed by Theorem 15.3.3, every tangent vector has length 1, since

$$\|d\mathbf{r}/ds\| = \sqrt{[-\sin(s/a)]^2 + [\cos(s/a)]^2} = \sqrt{1} = 1 \qquad \blacktriangleleft$$

The component forms of Formula (14) in 2-space and 3-space will be of sufficient interest in later sections that we provide them here for reference:

$$\frac{ds}{dt} = \left\|\frac{d\mathbf{r}}{dt}\right\| = \sqrt{\left(\frac{dx}{dt}\right)^2 + \left(\frac{dy}{dt}\right)^2} \tag{16}$$

$$\frac{ds}{dt} = \left\|\frac{d\mathbf{r}}{dt}\right\| = \sqrt{\left(\frac{dx}{dt}\right)^2 + \left(\frac{dy}{dt}\right)^2 + \left(\frac{dz}{dt}\right)^2} \tag{17}$$

REMARK.   It is of interest to note that (16) and (17) do not involve t_0, and hence do not depend on where the reference point for s is chosen. This is to be expected, since changing the position of the reference point shifts s by a constant (the arc length between the reference points), and this constant drops out on differentiating.

▶ Exercise Set 15.3   ⓒ *18*

1. Show that $\mathbf{r}_1(t) = t\mathbf{i} + t^2\mathbf{j}$ and $\mathbf{r}_2(t) = t^3\mathbf{i} + t^6\mathbf{j}$ have the same graphs, but that $\mathbf{r}_1'(t)$ is never zero, whereas $\mathbf{r}_2'(t) = \mathbf{0}$ for some value of t.

2. Show that the graphs of

$$\mathbf{r}_1(t) = (\cos t)\mathbf{i} + (\sin t)\mathbf{j} + t\mathbf{k}$$

and

$$\mathbf{r}_2(t) = \cos(t^3)\mathbf{i} + \sin(t^3)\mathbf{j} + t^3\mathbf{k}$$

are the same, but that $\mathbf{r}_1'(t)$ is never zero, whereas $\mathbf{r}_2'(t) = \mathbf{0}$ for some value of t.

In Exercises 3–6, determine whether $\mathbf{r}$ is a smooth function of the parameter t.

3. $\mathbf{r} = t^3\mathbf{i} + (3t^2 - 2t)\mathbf{j} + t^2\mathbf{k}$.

4. $\mathbf{r} = \cos(t^2)\mathbf{i} + \sin(t^2)\mathbf{j} + e^{-t}\mathbf{k}$.

5. $\mathbf{r} = te^{-t}\mathbf{i} + (t^2 - 2t)\mathbf{j} + \cos(\pi t)\mathbf{k}$.

6. $\mathbf{r} = \sin(\pi t)\mathbf{i} + (2t - \ln t)\mathbf{j} + (t^2 - t)\mathbf{k}$.

> In Exercises 7–9, calculate $d\mathbf{r}/d\tau$ by the chain rule, and then check your result by expressing $\mathbf{r}$ in terms of τ and differentiating.

7. $\mathbf{r} = t\mathbf{i} + t^2\mathbf{j}$; $t = 4\tau + 1$.

8. $\mathbf{r} = \langle 3\cos t, 3\sin t\rangle$; $t = \pi\tau$.

9. $\mathbf{r} = e^t\mathbf{i} + 4e^{-t}\mathbf{j}$; $t = \tau^2$.

10. Prove Theorem 15.3.1 for 2-space.

> In Exercises 11–16, find the arc length of the curve.

11. $\mathbf{r}(t) = (4 + 3t)\mathbf{i} + (2 - 2t)\mathbf{j} + (5 + t)\mathbf{k}$; $3 \le t \le 4$.

12. $\mathbf{r}(t) = 3\cos t\mathbf{i} + 3\sin t\mathbf{j} + t\mathbf{k}$; $0 \le t \le 2\pi$.

13. $\mathbf{r}(t) = t^3\mathbf{i} + t\mathbf{j} + \frac{1}{2}\sqrt{6}t^2\mathbf{k}$; $1 \le t \le 3$.

14. $x = \cos^3 t$, $y = \sin^3 t$, $z = 2$; $0 \le t \le \pi/2$.

15. $\mathbf{r}(t) = \langle e^t, e^{-t}, \sqrt{2}t\rangle$; $0 \le t \le 1$.

16. $x = \frac{1}{2}t$, $y = \frac{1}{3}(1 - t)^{3/2}$, $z = \frac{1}{3}(1 + t)^{3/2}$; $-1 \le t \le 1$.

17. Find the arc length of the circular helix $x = a\cos t$, $y = a\sin t$, $z = ct$ for $0 \le t \le t_0$.

18. Copper tubing with an outside diameter of $\frac{1}{2}$ in. is to be wrapped in a circular helix around a cylindrical core that has a 12-in. diameter. What length of tubing will make one complete turn around the cylinder in a distance of 20 in. measured along the axis of the cylinder?

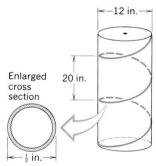

> In Exercises 19–27, find parametric equations for the curve using arc length s as a parameter. Use the point on the curve where $t = 0$ as the reference point.

19. $\mathbf{r}(t) = (3t - 2)\mathbf{i} + (4t + 3)\mathbf{j}$.

20. $\mathbf{r}(t) = (3\cos 2t)\mathbf{i} + (3\sin 2t)\mathbf{j}$; $0 \le t \le \pi$.

21. $\mathbf{r}(t) = (3 + \cos t)\mathbf{i} + (2 + \sin t)\mathbf{j}$; $0 \le t \le 2\pi$.

22. $\mathbf{r}(t) = (\cos^3 t)\mathbf{i} + (\sin^3 t)\mathbf{j}$; $0 \le t \le \pi/2$.

23. $\mathbf{r}(t) = \frac{1}{3}t^3\mathbf{i} + \frac{1}{2}t^2\mathbf{j}$; $t \ge 0$.

24. $\mathbf{r}(t) = (1 + t)^2\mathbf{i} + (1 + t)^3\mathbf{j}$; $0 \le t \le 1$.

25. $\mathbf{r}(t) = (e^t \cos t)\mathbf{i} + (e^t \sin t)\mathbf{j}$; $0 \le t \le \pi/2$.

26. $\mathbf{r}(t) = \sin(e^t)\mathbf{i} + \cos(e^t)\mathbf{j} + \sqrt{3}e^t\mathbf{k}$; $t \ge 0$.

27. $\mathbf{r}(t) = (t\cos t)\mathbf{i} + (t\sin t)\mathbf{j} + \frac{2}{3}\sqrt{2}t^{3/2}\mathbf{k}$; $t \ge 0$.

28. Find parametric equations for the cycloid

$$x = at - a\sin t$$
$$y = a - a\cos t \qquad (0 \le t \le 2\pi)$$

using arc length as the parameter. Take $(0, 0)$ as the reference point.

29. Use the result in Exercise 17 to show that the circular helix

$$\mathbf{r} = (a\cos t)\mathbf{i} + (a\sin t)\mathbf{j} + ct\mathbf{k}$$

can be expressed as

$$\mathbf{r} = \left(a\cos\frac{s}{w}\right)\mathbf{i} + \left(a\sin\frac{s}{w}\right)\mathbf{j} + \frac{cs}{w}\mathbf{k}$$

where $w = \sqrt{a^2 + c^2}$ and s is an arc-length parameter with reference point at $(a, 0, 0)$.

30. Recall from Formula (5) of Section 14.5 that $\mathbf{r} = \mathbf{r}_0 + t\mathbf{v}$ is the vector equation of a line in 2-space or 3-space if $\mathbf{v} \ne \mathbf{0}$. Show that $\mathbf{r} = \mathbf{r}_0 + s\mathbf{v}/\|\mathbf{v}\|$ is a vector equation of the line in terms of an arc-length parameter s whose reference point is at the tip of $\mathbf{r}_0$.

31. A thread with negligible thickness is unwound from a spool of radius a, with the unwound piece always held straight.

(a) Show that for $\theta \ge 0$ the curve traced out by the free end of the thread has parametric equations
$$x = a(\cos\theta + \theta\sin\theta)$$
$$y = a(\sin\theta - \theta\cos\theta)$$
(see accompanying figure).

(b) Find parametric equations for the curve of part (a) using arc length as the parameter, where $\theta = 0$ is the reference point.

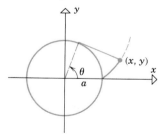

32. Show that in cylindrical coordinates a curve given by the parametric equations $r = r(t)$, $\theta = \theta(t)$, $z = z(t)$, for $a \le t \le b$, has arc length

$$L = \int_a^b \sqrt{\left(\frac{dr}{dt}\right)^2 + r^2\left(\frac{d\theta}{dt}\right)^2 + \left(\frac{dz}{dt}\right)^2}\, dt$$

[*Hint:* Use the relationships $x = r\cos\theta$, $y = r\sin\theta$.]

33. Use the formula in Exercise 32 to find the arc length of the following curves.

(a) $r = e^{2t}$, $\theta = t$, $z = e^{2t}$; $0 \le t \le \ln 2$

(b) $r = t^2$, $\theta = \ln t$, $z = \frac{1}{3}t^3$; $1 \le t \le 2$.

34. Show that in spherical coordinates a curve given by the parametric equations $\rho = \rho(t)$, $\theta = \theta(t)$, $\phi = \phi(t)$, for $a \le t \le b$, has arc length

$$L = \int_a^b \sqrt{\left(\frac{d\rho}{dt}\right)^2 + \rho^2 \sin^2 \phi \left(\frac{d\theta}{dt}\right)^2 + \rho^2 \left(\frac{d\phi}{dt}\right)^2}\; dt$$

[*Hint:* $x = \rho \sin \phi \cos \theta$, $y = \rho \sin \phi \sin \theta$, $z = \rho \cos \phi$.]

35. Use the formula in Exercise 34 to find the arc length of the following curves.

(a) $\rho = e^{-t}, \theta = 2t, \phi = \pi/4;\ 0 \le t \le 2$

(b) $\rho = 2t, \theta = \ln t, \phi = \pi/6;\ 1 \le t \le 5$.

36. Let $x = \cos t$, $y = \sin t$, $z = t^{3/2}$. Find

(a) $\|\mathbf{r}'(t)\|$ (b) $\dfrac{ds}{dt}$ (c) $\displaystyle\int_0^2 \|\mathbf{r}'(t)\|\, dt$.

37. Let $\mathbf{r}(t) = (\ln t)\mathbf{i} + 2t\mathbf{j} + t^2\mathbf{k}$. Find

(a) $\|\mathbf{r}'(t)\|$ (b) $\dfrac{ds}{dt}$ (c) $\displaystyle\int_1^3 \|\mathbf{r}'(t)\|\, dt$.

38. Prove: If $\mathbf{r}(t)$ is a smoothly parametrized function, then the angles between $\mathbf{r}'(t)$ and the vectors $\mathbf{i}$, $\mathbf{j}$, and $\mathbf{k}$ are continuous functions of t.

■ 15.4   UNIT TANGENT AND NORMAL VECTORS

> *In this section we shall discuss some geometric properties of vector-valued functions. Our work here will have important applications to the study of motion along a curved path in 2-space or 3-space.*

□ **UNIT TANGENT VECTORS**

Recall from Section 15.2 that if $\mathbf{r}(t)$ is a smooth vector-valued function in 2-space or 3-space with graph C, then the vector $\mathbf{r}'(t)$ is tangent to the graph of C if this vector is positioned so that its initial point is at the terminal point of the radius vector $\mathbf{r}(t)$. Moreover, $\mathbf{r}'(t)$ points in the direction of increasing t. Thus, if $\mathbf{r}'(t) \ne \mathbf{0}$, then the vector $\mathbf{T}(t)$ defined by

$$\mathbf{T}(t) = \frac{\mathbf{r}'(t)}{\|\mathbf{r}'(t)\|} \tag{1}$$

is tangent to C, points in the direction of increasing t, and has length 1 (Figure 15.4.1). We call $\mathbf{T}(t)$ the **unit tangent vector** to C at t. In the special case where C is parametrized by an arc-length parameter s, the tangent vectors have length 1 (Theorem 15.3.3), so (1) simplifies to

$$\mathbf{T}(s) = \mathbf{r}'(s) \tag{2}$$

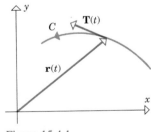

REMARK.   Unless stated otherwise, we shall always assume that the unit tangent vector $\mathbf{T}(t)$ is positioned with its initial point at the terminal point of $\mathbf{r}(t)$ as in Figure 15.4.1. This will ensure that $\mathbf{T}(t)$ is actually tangent to the graph of $\mathbf{r}(t)$ and not simply parallel to a tangent vector.

Figure 15.4.1

Example 1   Find the unit tangent vector to the graph of $\mathbf{r}(t) = t^2\mathbf{i} + t^3\mathbf{j}$ at the point where $t = 2$.

Solution.   Since

$$\mathbf{r}'(t) = 2t\mathbf{i} + 3t^2\mathbf{j}$$

we obtain

$$\mathbf{T}(2) = \frac{\mathbf{r}'(2)}{\|\mathbf{r}'(2)\|} = \frac{4\mathbf{i} + 12\mathbf{j}}{\sqrt{160}} = \frac{4\mathbf{i} + 12\mathbf{j}}{4\sqrt{10}} = \frac{1}{\sqrt{10}}\mathbf{i} + \frac{3}{\sqrt{10}}\mathbf{j}$$

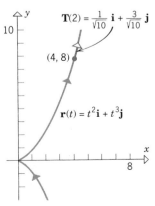

Figure 15.4.2

The graph of $\mathbf{r}(t)$ and the vector $\mathbf{T}(2)$ are shown in Figure 15.4.2.   ◄

□ UNIT NORMAL VECTORS

Let C be the graph of a smooth vector-valued function $\mathbf{r}(t)$ in 2-space or 3-space. In 2-space there are two unit vectors that are perpendicular to the unit tangent vector $\mathbf{T}(t)$, and in 3-space there are infinitely many such vectors (Figure 15.4.3). Since $\|\mathbf{T}(t)\| = 1$ is constant, it follows from Theorem 15.2.7 with $\mathbf{T}$ in place of $\mathbf{r}$ that $\mathbf{T}'(t)$ is perpendicular to $\mathbf{T}(t)$. If $\mathbf{T}'(t) \neq \mathbf{0}$, then we define the ***principal unit normal vector*** to C at t by

$$\mathbf{N}(t) = \frac{\mathbf{T}'(t)}{\|\mathbf{T}'(t)\|} \tag{3}$$

The vector $\mathbf{N}(t)$ has length 1, is perpendicular to $\mathbf{T}(t)$, and has the same direction as $\mathbf{T}'(t)$. For simplicity we shall refer to $\mathbf{N}(t)$ as the ***unit normal vector***.

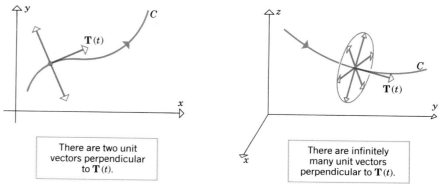

Figure 15.4.3

There are two unit vectors perpendicular to $\mathbf{T}(t)$.

There are infinitely many unit vectors perpendicular to $\mathbf{T}(t)$.

REMARK.   Unless stated otherwise, we shall always assume that $\mathbf{N}(t)$ is positioned with its initial point at the terminal point of $\mathbf{r}(t)$ (see Figure 15.4.4).

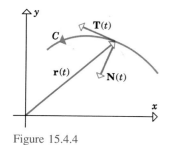

Figure 15.4.4

Example 2   Find $\mathbf{T}(t)$ and $\mathbf{N}(t)$ for the circular helix

$$x = a \cos t, \quad y = a \sin t, \quad z = ct$$

where $a > 0$.

Solution.   The radius vector for the helix is

$$\mathbf{r}(t) = (a \cos t)\mathbf{i} + (a \sin t)\mathbf{j} + (ct)\mathbf{k}$$

Thus,

$$\mathbf{r}'(t) = (-a \sin t)\mathbf{i} + (a \cos t)\mathbf{j} + c\mathbf{k}$$

$$\|\mathbf{r}'(t)\| = \sqrt{(-a \sin t)^2 + (a \cos t)^2 + c^2} = \sqrt{a^2 + c^2}$$

$$\mathbf{T}(t) = \frac{\mathbf{r}'(t)}{\|\mathbf{r}'(t)\|} = -\frac{a \sin t}{\sqrt{a^2 + c^2}}\mathbf{i} + \frac{a \cos t}{\sqrt{a^2 + c^2}}\mathbf{j} + \frac{c}{\sqrt{a^2 + c^2}}\mathbf{k}$$

$$\mathbf{T}'(t) = -\frac{a \cos t}{\sqrt{a^2 + c^2}}\mathbf{i} - \frac{a \sin t}{\sqrt{a^2 + c^2}}\mathbf{j}$$

$$\|\mathbf{T}'(t)\| = \sqrt{\left(-\frac{a \cos t}{\sqrt{a^2 + c^2}}\right)^2 + \left(-\frac{a \sin t}{\sqrt{a^2 + c^2}}\right)^2} = \sqrt{\frac{a^2}{a^2 + c^2}} = \frac{a}{\sqrt{a^2 + c^2}}$$

$$\mathbf{N}(t) = \frac{\mathbf{T}'(t)}{\|\mathbf{T}'(t)\|} = (-\cos t)\mathbf{i} - (\sin t)\mathbf{j}$$

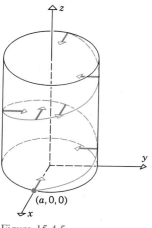

Figure 15.4.5

Because the $\mathbf{k}$ component of $\mathbf{N}(t)$ is zero, this vector lies in a horizontal plane for every value of t. Moreover, in the exercises we ask the reader to show that $\mathbf{N}(t)$ points directly toward the z-axis for all t (Figure 15.4.5).   ◄

In the special case where the curve C is the graph of $\mathbf{r}(s)$, and s is an arc-length parameter, it follows from (2) that (3) simplifies to

$$\mathbf{N}(s) = \frac{\mathbf{r}''(s)}{\|\mathbf{r}''(s)\|} \tag{4}$$

Example 3 In Example 5 of the preceding section we showed that the circle $x^2 + y^2 = a^2$ can be parametrized in terms of arc length as

$$\mathbf{r}(s) = a\cos(s/a)\mathbf{i} + a\sin(s/a)\mathbf{j} \quad (0 \le s \le 2\pi a)$$

Find $\mathbf{N}(s)$.

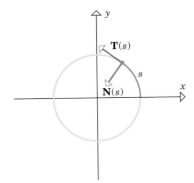

Figure 15.4.6

Solution. The first and second derivatives of $\mathbf{r}$ with respect to s are

$$\mathbf{r}'(s) = -\sin(s/a)\mathbf{i} + \cos(s/a)\mathbf{j}$$

$$\mathbf{r}''(s) = -(1/a)\cos(s/a)\mathbf{i} - (1/a)\sin(s/a)\mathbf{j}$$

so

$$\|\mathbf{r}''(s)\| = \sqrt{(-1/a)^2 \cos^2(s/a) + (-1/a)^2 \sin^2(s/a)} = 1/a$$

Thus, from (4)

$$\mathbf{N}(s) = \mathbf{r}''(s)/\|\mathbf{r}''(s)\| = -\cos(s/a)\mathbf{i} - \sin(s/a)\mathbf{j}$$

Observe that $\mathbf{N}(s)$ is a negative scalar multiple of the radius vector $\mathbf{r}(s)$, so that $\mathbf{N}(s)$ points toward the center of the circle for all s (Figure 15.4.6).   ◄

☐ INWARD UNIT NORMAL VECTORS IN 2-SPACE

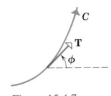

Figure 15.4.7

If $\mathbf{T}$ is a unit tangent vector to a curve C in 2-space, then as shown in Figure 15.4.3, there are two unit vectors perpendicular to $\mathbf{T}$. To determine which of these is the principal unit normal $\mathbf{N}$, we shall let $\phi = \phi(t)$ be the counterclockwise angle from the direction of the positive x-axis to $\mathbf{T}$ (Figure 15.4.7). Since $\mathbf{T}$ has length 1, it follows from Formula (8) of Section 14.2 that $\mathbf{T}$ can be expressed in terms of ϕ as

$$\mathbf{T} = (\cos\phi)\mathbf{i} + (\sin\phi)\mathbf{j} \tag{5}$$

From this formula and the chain rule we obtain

$$\mathbf{T}'(t) = \frac{d\mathbf{T}}{dt} = \frac{d\mathbf{T}}{d\phi}\frac{d\phi}{dt} = [(-\sin\phi)\mathbf{i} + (\cos\phi)\mathbf{j}]\frac{d\phi}{dt} = \mathbf{n}(t)\phi'(t) \tag{6}$$

where

$$\mathbf{n} = \mathbf{n}(t) = (-\sin\phi)\mathbf{i} + (\cos\phi)\mathbf{j} \tag{7}$$

Formula (6) implies that $\mathbf{n}(t)$ and $\mathbf{T}'(t)$ have the same direction if $\phi'(t) > 0$ (i.e., ϕ increasing) and opposite directions if $\phi'(t) < 0$ (i.e., ϕ decreasing). But $\mathbf{N}(t)$ has the same direction as $\mathbf{T}'(t)$ [see (3)], so $\mathbf{n}(t)$ and $\mathbf{N}(t)$ have the same direction if $\phi'(t) > 0$ and opposite directions if $\phi'(t) < 0$. However, $\mathbf{n}(t)$ lies 90° counterclockwise from $\mathbf{T}(t)$, since (7) can be written as (verify)

$$\mathbf{n} = \mathbf{n}(t) = \cos(\phi + \pi/2)\mathbf{i} + \sin(\phi + \pi/2)\mathbf{j}$$

and hence $\mathbf{N}(t)$ always points "inward" toward the concave side of the curve (Figure 15.4.8). For this reason $\mathbf{N}$ is sometimes referred to as the ***inward unit normal***. Observe that this is consistent with Example 3, where we found that the principal unit normal for the circle always pointed inward toward the center.

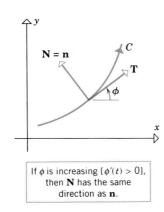

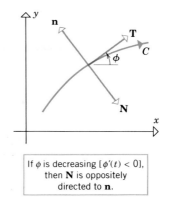

Figure 15.4.8

If ϕ is increasing $[\phi'(t) > 0]$, then **N** has the same direction as **n**.

If ϕ is decreasing $[\phi'(t) < 0]$, then **N** is oppositely directed to **n**.

☐ **BINORMAL VECTORS IN 3-SPACE**

If $\mathbf{r}(t)$ is a vector-valued function in 3-space with graph C, then for each value of t at which $\mathbf{T}(t)$ and $\mathbf{N}(t)$ exist, we define

$$\mathbf{B}(t) = \mathbf{T}(t) \times \mathbf{N}(t) \tag{8}$$

to be the **binormal vector** to C at t. For each value of t the binormal vector **B** has length 1 and is perpendicular to both **T** and **N**. At each point P on the curve C the vectors **T**, **N**, and **B** determine three mutually perpendicular planes through P whose names are shown in Figure 15.4.9. Moreover, as illustrated in Figure 14.4.4, the vectors are oriented by a "right-hand rule" in the sense that if the fingers of the right hand are cupped to point in the direction of rotation from **T** to **N**, then the thumb indicates the direction of **B**.

Three mutually perpendicular unit vectors in 3-space are sometimes called a **triad**. Whereas the **ijk**-triad is constant relative to the xyz-coordinate axes, the **TNB**-triad varies from point to point along the graph of $\mathbf{r}(t)$, since $\mathbf{T} = \mathbf{T}(t)$, $\mathbf{N} = \mathbf{N}(t)$, and $\mathbf{B} = \mathbf{B}(t)$ are functions of t (Figure 15.4.10). The **TNB**-triad is sometimes described as a *moving* triad.

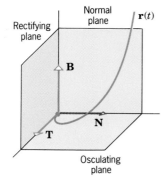

Figure 15.4.9

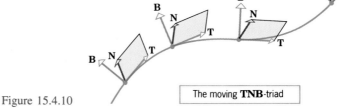

The moving **TNB**-triad

Figure 15.4.10

► Exercise Set 15.4

In Exercises 1–8, find the unit tangent vector **T** and the unit normal vector **N** for the given value of t. Then sketch the vectors and a portion of the curve containing the point of tangency.

1. $\mathbf{r}(t) = (5 \cos t)\mathbf{i} + (5 \sin t)\mathbf{j}$; $t = \pi/3$.

2. $\mathbf{r}(t) = 2t\mathbf{i} + 4t^2\mathbf{j}$; $t = 1$.

3. $\mathbf{r}(t) = (t^2 - 1)\mathbf{i} + t\mathbf{j}$; $t = 1$.

4. $\mathbf{r}(t) = e^t\mathbf{i} + e^{-t}\mathbf{j}$; $t = 0$.

5. $\mathbf{r}(t) = \frac{1}{2}t^2\mathbf{i} + \frac{1}{3}t^3\mathbf{j}$; $t = 1$.

6. $\mathbf{r}(t) = (\ln t)\mathbf{i} + t\mathbf{j}$; $t = e$.

7. $\mathbf{r}(t) = (4 \cos t)\mathbf{i} + (9 \sin t)\mathbf{j}$; $t = \pi/4$.

8. $\mathbf{r}(t) = (\ln \sin t)\mathbf{i} + (\ln \cos t)\mathbf{j}$; $t = \pi/6$.

In Exercises 9–16, find the unit tangent vector **T** and the unit normal vector **N** for the given value of t.

9. $\mathbf{r}(t) = 4\cos t\mathbf{i} + 4\sin t\mathbf{j} + t\mathbf{k}$; $t = \pi/2$.

10. $x = e^t, y = e^{-t}, z = t$; $t = 0$.

11. $\mathbf{r}(t) = t\mathbf{i} + \frac{1}{2}t^2\mathbf{j} + \frac{1}{3}t^3\mathbf{k}$; $t = 0$.

12. $x = \sin t, y = \cos t, z = \frac{1}{2}t^2$; $t = 0$.

13. $\mathbf{r}(t) = 3\cos t\mathbf{i} + 4\sin t\mathbf{j} + t\mathbf{k}$; $t = \pi/2$.

14. $x = e^t\cos t, y = e^t\sin t, z = e^t$; $t = 0$.

15. $\mathbf{r}(t) = \mathbf{i} + t\mathbf{j} + t^2\mathbf{k}$; $t = 1$.

16. $x = \cosh t, y = \sinh t, z = t$; $t = \ln 2$.

17. Let C be the curve given parametrically in terms of arc length by
$$x = \tfrac{3}{5}s + 1, \quad y = \tfrac{4}{5}s - 2 \qquad (0 \le s \le 10)$$
   (a) Find $\mathbf{T} = \mathbf{T}(s)$.
   (b) Sketch the curve and the vector $\mathbf{T}(5)$.

18. Let C be the curve given parametrically in terms of arc length by
$$x = 2\sin(s/2), \quad y = 2\cos(s/2) \qquad (0 \le s \le 4\pi)$$
   (a) Find $\mathbf{T} = \mathbf{T}(s)$ and $\mathbf{N} = \mathbf{N}(s)$.
   (b) Sketch the curve and the vectors $\mathbf{T}(4\pi/3)$ and $\mathbf{N}(4\pi/3)$.

19. Show that the vector $\mathbf{N}(t) = (-\cos t)\mathbf{i} - (\sin t)\mathbf{j}$ found in Example 2 points directly toward the z-axis.

20. From Formula (3), the vector **N** points in the direction of
$$\mathbf{T}' = \frac{d}{dt}[\mathbf{r}'/\|\mathbf{r}'\|]$$
Replace **r** by **r'** in the formula of Exercise 61, Section

15.2, to show that **N** points in the direction of the vector $\mathbf{u} = \|\mathbf{r}'\|^2\mathbf{r}'' - (\mathbf{r}' \cdot \mathbf{r}'')\mathbf{r}'$ and hence that **N** can also be obtained from the formula $\mathbf{N} = \mathbf{u}/\|\mathbf{u}\|$.

In Exercises 21–28, use the formula in Exercise 20 to find **N** for the given value of t.

21. Exercise 1.   22. Exercise 2.

23. Exercise 3.   24. Exercise 4.

25. Exercise 9.   26. Exercise 10.

27. Exercise 11.   28. Exercise 12.

In Exercises 29 and 30, find $\mathbf{B} = \mathbf{T} \times \mathbf{N}$.

29. $\mathbf{r}(t) = (3\sin t)\mathbf{i} + (3\cos t)\mathbf{j} + 4t\mathbf{k}$.

30. $\mathbf{r}(t) = (a\cos t)\mathbf{i} + (a\sin t)\mathbf{j} + bt\mathbf{k}$ $(a \ne 0, b \ne 0)$.

31. Let S be the plane with the equation
$$a(x - x_0) + b(y - y_0) + c(z - z_0) = 0$$
and let $\mathbf{n} = a\mathbf{i} + b\mathbf{j} + c\mathbf{k}$ and $\mathbf{r}_0 = x_0\mathbf{i} + y_0\mathbf{j} + z_0\mathbf{k}$.
   (a) Show that a curve with the vector equation $\mathbf{r} = \mathbf{r}(t)$ lies in the plane S if and only if $\mathbf{n} \cdot (\mathbf{r}(t) - \mathbf{r}_0) = 0$ for all t.
   (b) Show that if the curve with the vector equation $\mathbf{r} = \mathbf{r}(t)$ lies in the plane S, then at each point where **T** and **N** are defined, these vectors also lie in S. [*Hint:* Use the result in part (a) to show that **r'** and **r''** are perpendicular to **n**, then use Formula (1) to show that $\mathbf{T} \cdot \mathbf{n} = 0$ and Exercise 20 to show that $\mathbf{N} \cdot \mathbf{n} = 0$.]

■ 15.5 CURVATURE

In this section we shall consider the problem of obtaining a numerical measure of how sharply a curve in 2-space or 3-space bends. Our results will have applications in geometry and in the study of motion along a curved path.

□ CURVATURE

Suppose that s is an arc-length parameter for a smooth curve C in 2-space or 3-space. Figure 15.5.1 suggests that for a curve in 2-space the "sharpness" of the bend in C is closely related to $d\mathbf{T}/ds$, which is the rate of change of the unit tangent vector **T** with respect to s. (Keep in mind that **T** has constant length, so only its direction changes.) If C is a straight line (no bend), then the direction of **T** remains constant (Figure 15.5.1*a*); if C bends slightly, then **T** undergoes a gradual change of direction (Figure 15.5.1*b*); and if C bends sharply, then **T** undergoes a rapid change of direction (Figure 15.5.1*c*).

In 3-space, bends in a curve are not limited to a single plane—they can occur in all possible directions. In 3-space, **T** is perpendicular to the normal plane (Figure 15.4.9), so $d\mathbf{T}/ds$ relates to the rate at which the normal plane turns as s increases. Similarly, $d\mathbf{N}/ds$

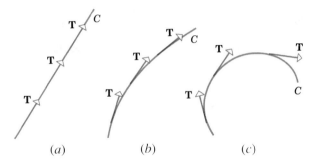

Figure 15.5.1   (a)   (b)   (c)

and $d\mathbf{B}/ds$ relate to the rates at which the rectifying and osculating planes turn as s increases. In this text we shall focus primarily on $d\mathbf{T}/ds$, which is the most important of these derivatives in applications. We make the following definition.

15.5.1 DEFINITION.   If C is a smooth curve in 2-space or 3-space, and if s is an arc-length parameter for C, then the **curvature** of C, denoted by $\kappa = \kappa(s)$ (κ = Greek ''kappa''), is defined by

$$\kappa = \left\| \frac{d\mathbf{T}}{ds} \right\| \tag{1}$$

Observe that in (1) it is the *length* of the vector $d\mathbf{T}/ds$ that is the measure of curvature; thus, κ is a nonnegative real number. In general, the curvature varies from point to point along a curve; however, the following example shows that for circles in 2-space curvature is constant, as one might expect.

Example 1   In Example 5 of Section 15.3 we showed that the circle of radius a, centered at the origin, can be parametrized in terms of arc length as

$$\mathbf{r}(s) = a \cos(s/a)\mathbf{i} + a \sin(s/a)\mathbf{j} \quad (0 \leq s \leq 2\pi a)$$

Thus, from (1)

$$\frac{d\mathbf{T}}{ds} = \mathbf{r}''(s) = -\frac{1}{a}\cos(s/a)\mathbf{i} - \frac{1}{a}\sin(s/a)\mathbf{j}$$

$$\kappa = \left\| \frac{d\mathbf{T}}{ds} \right\| = \sqrt{\left[-\frac{1}{a}\cos(s/a) \right]^2 + \left[-\frac{1}{a}\sin(s/a) \right]^2} = \frac{1}{a}$$

so the circle has constant curvature $1/a$.   ◀

☐ **FORMULAS FOR CURVATURE**

Formula (1) is not useful for computational purposes unless the curve happens to be expressed in terms of an arc-length parameter. The following theorem provides two formulas for curvature in terms of a general parameter t.

15.5.2 THEOREM.   *If $\mathbf{r}(t)$ is a smooth vector-valued function in 2-space or 3-space, then for each value of t at which $\mathbf{T}'(t)$ and $\mathbf{r}''(t)$ exist, the curvature κ can be expressed as*

(a) $$\kappa = \kappa(t) = \frac{\|\mathbf{T}'(t)\|}{\|\mathbf{r}'(t)\|} \tag{2}$$

(b) $$\kappa = \kappa(t) = \frac{\|\mathbf{r}'(t) \times \mathbf{r}''(t)\|}{\|\mathbf{r}'(t)\|^3} \tag{3}$$

Proof (a). It follows from (1), Formula (13) of Section 15.3, and the chain rule that

$$\kappa = \left\| \frac{d\mathbf{T}}{ds} \right\| = \left\| \frac{d\mathbf{T}/dt}{ds/dt} \right\| = \left\| \frac{d\mathbf{T}/dt}{\|d\mathbf{r}/dt\|} \right\| = \frac{\|\mathbf{T}'(t)\|}{\|\mathbf{r}'(t)\|}$$

Proof (b). It follows from Formula (1) of Section 15.4 that

$$\mathbf{r}'(t) = \|\mathbf{r}'(t)\| \, \mathbf{T}(t) \tag{4}$$

so

$$\mathbf{r}''(t) = \|\mathbf{r}'(t)\|' \, \mathbf{T}(t) + \|\mathbf{r}'(t)\| \, \mathbf{T}'(t) \tag{5}$$

But from Formula (3) of Section 15.4 and part (a) of this theorem we have

$$\mathbf{T}'(t) = \|\mathbf{T}'(t)\| \, \mathbf{N}(t) \quad \text{and} \quad \|\mathbf{T}'(t)\| = \kappa \|\mathbf{r}'(t)\|$$

so

$$\mathbf{T}'(t) = \kappa \|\mathbf{r}'(t)\| \, \mathbf{N}(t)$$

Substituting this into (5) yields

$$\mathbf{r}''(t) = \|\mathbf{r}'(t)\|' \, \mathbf{T}(t) + \kappa \|\mathbf{r}'(t)\|^2 \, \mathbf{N}(t) \tag{6}$$

Thus, from (4) and (6)

$$\mathbf{r}'(t) \times \mathbf{r}''(t) = \|\mathbf{r}'(t)\| \, \|\mathbf{r}'(t)\|' \, (\mathbf{T}(t) \times \mathbf{T}(t)) + \kappa \|\mathbf{r}'(t)\|^3 \, (\mathbf{T}(t) \times \mathbf{N}(t))$$

But the cross product of a vector with itself is zero [Theorem 14.4.4(*f*)], so this equation simplifies to

$$\mathbf{r}'(t) \times \mathbf{r}''(t) = \kappa \|\mathbf{r}'(t)\|^3 \, (\mathbf{T}(t) \times \mathbf{N}(t))$$

It follows from this equation and the fact that $\mathbf{T}(t) \times \mathbf{N}(t)$ is a unit vector (why?) that

$$\|\mathbf{r}'(t) \times \mathbf{r}''(t)\| = \kappa \|\mathbf{r}'(t)\|^3$$

Formula (3) now follows. ∎

REMARKS. Formula (2) is useful if $\mathbf{T}(t)$ happens to be known or is easy to obtain; however, Formula (3) will usually be easier to apply, since it involves only $\mathbf{r}(t)$ and its derivatives. We also note that cross products were defined only for vectors in 3-space, so to use Formula (4) in 2-space one must first write the 2-space function $\mathbf{r}(t) = x(t)\mathbf{i} + y(t)\mathbf{j}$ as the 3-space function $\mathbf{r}(t) = x(t)\mathbf{i} + y(t)\mathbf{j} + 0\mathbf{k}$ with a zero $\mathbf{k}$ component.

Example 2 Find $\kappa(t)$ for the circular helix

$$x = a \cos t, \quad y = a \sin t, \quad z = ct$$

where $a > 0$.

Solution. The radius vector for the helix is

$$\mathbf{r}(t) = (a \cos t)\mathbf{i} + (a \sin t)\mathbf{j} + (ct)\mathbf{k}$$

Thus,

$$\mathbf{r}'(t) = (-a \sin t)\mathbf{i} + (a \cos t)\mathbf{j} + c\mathbf{k}$$

$$\mathbf{r}''(t) = (-a \cos t)\mathbf{i} + (-a \sin t)\mathbf{j}$$

so

$$\mathbf{r}'(t) \times \mathbf{r}''(t) = \begin{vmatrix} \mathbf{i} & \mathbf{j} & \mathbf{k} \\ -a \sin t & a \cos t & c \\ -a \cos t & -a \sin t & 0 \end{vmatrix}$$

$$= (ac \sin t)\mathbf{i} - (ac \cos t)\mathbf{j} + a^2\mathbf{k}$$

Therefore,

$$\|\mathbf{r}'(t)\| = \sqrt{(-a \sin t)^2 + (a \cos t)^2 + c^2} = \sqrt{a^2 + c^2}$$

and

$$\|\mathbf{r}'(t) \times \mathbf{r}''(t)\| = \sqrt{(ac \sin t)^2 + (-ac \cos t)^2 + a^4}$$
$$= \sqrt{a^2 c^2 + a^4} = a\sqrt{a^2 + c^2}$$

so

$$\kappa(t) = \frac{\|\mathbf{r}'(t) \times \mathbf{r}''(t)\|}{\|\mathbf{r}'(t)\|^3} = \frac{a\sqrt{a^2 + c^2}}{(\sqrt{a^2 + c^2})^3} = \frac{a}{a^2 + c^2}$$

Note that κ does not depend on t, and hence the helix has constant curvature. ◀

Example 3 The graph of the vector equation

$$\mathbf{r} = (2 \cos t)\mathbf{i} + (3 \sin t)\mathbf{j} \quad (0 \leq t \leq 2\pi)$$

is the ellipse in Figure 15.5.2 (see Example 5 of Section 13.4). Find the curvature of the ellipse at the endpoints of the major and minor axes.

Solution. We have

$$\mathbf{r}'(t) = (-2 \sin t)\mathbf{i} + (3 \cos t)\mathbf{j} \quad \text{and} \quad \mathbf{r}''(t) = (-2 \cos t)\mathbf{i} + (-3 \sin t)\mathbf{j}$$

so

$$\mathbf{r}'(t) \times \mathbf{r}''(t) = \begin{vmatrix} \mathbf{i} & \mathbf{j} & \mathbf{k} \\ -2 \sin t & 3 \cos t & 0 \\ -2 \cos t & -3 \sin t & 0 \end{vmatrix} = [(6 \sin^2 t) + (6 \cos^2 t)]\mathbf{k} = 6\mathbf{k}$$

Therefore,

$$\|\mathbf{r}'(t)\| = \sqrt{(-2 \sin t)^2 + (3 \cos t)^2} = \sqrt{4 \sin^2 t + 9 \cos^2 t}$$

and

$$\|\mathbf{r}'(t) \times \mathbf{r}''(t)\| = 6$$

so

$$\kappa = \kappa(t) = \frac{\|\mathbf{r}'(t) \times \mathbf{r}''(t)\|}{\|\mathbf{r}'(t)\|^3} = \frac{6}{[4 \sin^2 t + 9 \cos^2 t]^{3/2}} \tag{7}$$

The endpoints of the minor axis are $(2, 0)$ and $(-2, 0)$, which correspond to $t = 0$ and $t = \pi$, respectively. Substituting these values in (7) yields the same curvature at both points, namely

$$\kappa = \kappa(0) = \kappa(\pi) = \frac{6}{9^{3/2}} = \frac{6}{27} = \frac{2}{9}$$

The endpoints of the major axis are $(0, 3)$ and $(0, -3)$, which correspond to $t = \pi/2$ and $t = 3\pi/2$, respectively; from (7) the curvature at these points is

$$\kappa = \kappa\left(\frac{\pi}{2}\right) = \kappa\left(\frac{3\pi}{2}\right) = \frac{6}{4^{3/2}} = \frac{3}{4}$$

Thus, the curvature is greater at the ends of the major axis than at the ends of the minor axis, as one would expect. ◀

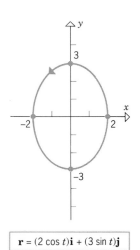

$\mathbf{r} = (2 \cos t)\mathbf{i} + (3 \sin t)\mathbf{j}$

Figure 15.5.2

☐ RADIUS OF CURVATURE

In the preceding example, we found the curvature at the ends of the minor axis to be $2/9$ and the curvature at the ends of the major axis to be $3/4$. To obtain a better understanding of the meaning of these numbers, recall from Example 1 that a circle of radius a has a constant curvature of $1/a$; thus, the curvature of the ellipse at the ends of the minor axis is the same as that of a circle of radius 9/2, and the curvature at the ends of the major axis is the same as that of a circle of radius 4/3 (Figure 15.5.3).

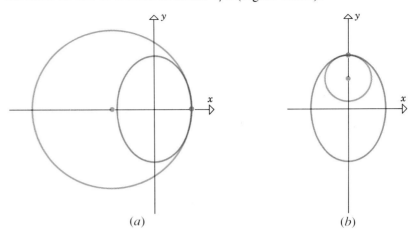

(a) (b) Figure 15.5.3

Figure 15.5.4

In general, if a curve C in 2-space has nonzero curvature κ at a point P, then the circle of radius $\rho = 1/\kappa$ sharing a common tangent with C at P, and centered on the concave side of the curve at P, is called the ***circle of curvature*** or ***osculating circle*** at P (Figure 15.5.4). The osculating circle and the curve C not only touch at P but they have equal curvatures at that point. In this sense, the osculating circle is the circle that best approximates the curve C near P. The radius ρ of the osculating circle at P is called the ***radius of curvature*** at P, and the center of the circle is called the ***center of curvature*** at P (Figure 15.5.4).

□ **A BASIC RELATIONSHIP BETWEEN T, N, AND κ**

The following formula, which will be used in the next section, provides a fundamental relationship between **T**, **N**, and κ.

$$\frac{d\mathbf{T}}{ds} = \kappa \mathbf{N} \tag{8}$$

To derive this relationship, we use Formula (3) of Section 15.4 with s in place of t to obtain

$$\mathbf{N}(s) = \frac{d\mathbf{T}/ds}{\|d\mathbf{T}/ds\|} = \frac{1}{\kappa}\frac{d\mathbf{T}}{ds}$$

from which (8) follows.

□ **INTERPRETATION OF CURVATURE IN 2-SPACE**

A simple geometric interpretation of curvature in 2-space can be obtained by considering the angle ϕ measured counterclockwise from the direction of the positive x-axis to **T** (Figure 15.4.7). By Formula (5) of Section 15.4, we can write $\mathbf{T} = (\cos\phi)\mathbf{i} + (\sin\phi)\mathbf{j}$ from which we obtain

$$\frac{d\mathbf{T}}{d\phi} = (-\sin\phi)\mathbf{i} + (\cos\phi)\mathbf{j}$$

By the chain rule

$$\frac{d\mathbf{T}}{ds} = \frac{d\mathbf{T}}{d\phi}\frac{d\phi}{ds}$$

so

$$\left\|\frac{d\mathbf{T}}{ds}\right\| = \left|\frac{d\phi}{ds}\right|\left\|\frac{d\mathbf{T}}{d\phi}\right\| = \left|\frac{d\phi}{ds}\right|\sqrt{(-\sin\phi)^2 + \cos^2\phi} = \left|\frac{d\phi}{ds}\right|$$

It now follows from (1) that

$$\kappa = \left|\frac{d\phi}{ds}\right| \qquad (9)$$

Thus, curvature in 2-space can be interpreted as the magnitude of the rate of change of ϕ with respect to s—the greater the curvature, the more rapidly ϕ changes with s (Figure 15.5.5). In the case of a straight line, the angle ϕ is constant (Figure 5.5.6) and consequently $\kappa = |d\phi/ds| = 0$, which means that a straight line has zero curvature at every point.

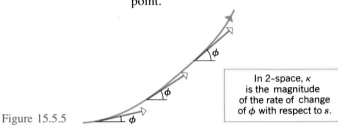

Figure 15.5.5

In 2-space, κ is the magnitude of the rate of change of ϕ with respect to s.

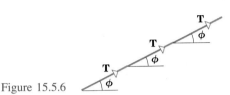

Figure 15.5.6

▶ Exercise Set 15.5   Ⓒ 57, 58

In Exercises 1–14, use Formula (3) to find the curvature at the indicated point.

1. $\mathbf{r}(t) = t^2\mathbf{i} + t^3\mathbf{j}$; $t = \frac{1}{2}$.

2. $\mathbf{r}(t) = (4\cos t)\mathbf{i} + (\sin t)\mathbf{j}$; $t = \pi/2$.

3. $\mathbf{r}(t) = e^{3t}\mathbf{i} + e^{-t}\mathbf{j}$; $t = 0$.

4. $x = 1 - t^3$, $y = t - t^2$; $t = 1$.

5. $x = t\cos t$, $y = t\sin t$; $t = 1$.

6. $x = 2abt$, $y = b^2t^2$; $t = 1$ $(a > 0, b > 0)$.

7. $\mathbf{r}(t) = 4\cos t\,\mathbf{i} + 4\sin t\,\mathbf{j} + t\mathbf{k}$; $t = \pi/2$.

8. $x = e^t$, $y = e^{-t}$, $z = t$; $t = 0$.

9. $\mathbf{r}(t) = t\mathbf{i} + \frac{1}{2}t^2\mathbf{j} + \frac{1}{3}t^3\mathbf{k}$; $t = 0$.

10. $x = \sin t$, $y = \cos t$, $z = \frac{1}{2}t^2$; $t = 0$.

11. $\mathbf{r}(t) = 3\cos t\,\mathbf{i} + 4\sin t\,\mathbf{j} + t\mathbf{k}$; $t = \pi/2$.

12. $x = e^t\cos t$, $y = e^t\sin t$, $z = e^t$; $t = 0$.

13. $\mathbf{r}(t) = \mathbf{i} + t\mathbf{j} + t^2\mathbf{k}$; $t = 1$.

14. $x = \cosh t$, $y = \sinh t$, $z = t$; $t = \ln 2$.

15. Use Formula (3) to show that for a plane curve described by $y = f(x)$ the curvature $\kappa(x)$ is
$$\kappa(x) = \frac{|d^2y/dx^2|}{[1 + (dy/dx)^2]^{3/2}}$$
[*Hint:* Let x be the parameter so that $\mathbf{r}(x) = x\mathbf{i} + y\mathbf{j} = x\mathbf{i} + f(x)\mathbf{j}$.]

16. Show that the formula for κ in Exercise 15 can also be written as $\kappa = |y''\cos^3\phi|$, where ϕ is the angle of inclination of the tangent line to the graph of $y = f(x)$.

In Exercises 17–22, use the formula of Exercise 15 to find the curvature at the indicated point.

17. $y = \sin x$; $x = \pi/2$.

18. $y = x^3/3$; $x = 0$.

19. $y = 1/x$; $x = 1$.

20. $y = e^{-x}$; $x = 1$.

21. $y = \tan x$; $x = \pi/4$.

22. $y^2 - 4x^2 = 9$; $(2, 5)$.

23. Use the formula of Exercise 15 to find the curvature of $y = \ln\cos x$ for $-\pi/2 < x < \pi/2$. For what value of x is the curvature maximum?

24. Use Formula (3) to show that for a plane curve described by $\mathbf{r} = x(t)\mathbf{i} + y(t)\mathbf{j}$ the curvature is
$$\kappa = \frac{|x'y'' - y'x''|}{(x'^2 + y'^2)^{3/2}}$$
where a prime denotes differentiation with respect to t.

In Exercises 25–30, use the formula of Exercise 24 to find the curvature in the indicated problem.

25. Exercise 1.   26. Exercise 2.

27. Exercise 3.   28. Exercise 4.

29. Exercise 5.   30. Exercise 6.

31. Use the formula of Exercise 24 to find the curvature at $t = 0$ and $t = \pi/2$ of the curve $x = a\cos t$, $y = b\sin t$ $(a > 0, b > 0)$.

32. Use Formula (3) to show that for a curve in polar coordinates described by $r = f(\theta)$ the curvature is
$$\kappa = \frac{\left|r^2 + 2\left(\dfrac{dr}{d\theta}\right)^2 - r\dfrac{d^2r}{d\theta^2}\right|}{\left[r^2 + \left(\dfrac{dr}{d\theta}\right)^2\right]^{3/2}}$$
[*Hint:* Let θ be the parameter and use the relationships $x = r\cos\theta$, $y = r\sin\theta$.]

In Exercises 33–36, use the formula of Exercise 32 to find the curvature at the indicated point.

33. $r = 2 \sin \theta$; $\theta = \pi/6$.   **34.** $r = \theta$; $\theta = 1$.

35. $r = a(1 + \cos \theta)$; $\theta = \pi/2$.

36. $r = e^{2\theta}$; $\theta = 1$.

37. Find $\kappa(t)$ for the cycloid

$$x = a(t - \sin t), \quad y = a(1 - \cos t)$$

for $0 < t < \pi$ and sketch the graph of $\kappa(t)$.

38. Find $\kappa(t)$ for the curve $x = e^{-t} \cos t$, $y = e^{-t} \sin t$ and sketch the graph of $\kappa(t)$.

In Exercises 39–45, sketch the curve, calculate the radius of curvature at the indicated point, and sketch the osculating circle.

39. $y = \sin x$; $x = \pi/2$.   **40.** $x = t, y = t^2$; $t = 1$.

41. $y = \frac{1}{2}x^2$; $x = -1$.   **42.** $x = t, y = \sqrt{t}$; $t = 2$.

43. $y = \ln x$; $x = 1$.   **44.** $x = 2t, y = 4/t$; $t = 1$.

45. $x = t - \sin t, y = 1 - \cos t$; $t = \pi$.

46. Find the radius of curvature of $y = \cos x$ at $x = 0$ and $x = \pi$. Sketch the osculating circles at those points.

47. Find the radius of curvature of the ellipse $x = 2 \cos t$, $y = \sin t$, $0 \le t \le 2\pi$ at $t = 0$ and $t = \pi/2$. Sketch the osculating circles at those points.

48. Consider the curve $y = x^4 - 2x^2$.

(a) Find the radius of curvature at each relative extremum.

(b) Sketch the curve and show the osculating circles at the relative extrema.

49. Find the radius of curvature of the parabola $y^2 = 4px$ at $(0, 0)$.

50. At what point(s) does $y = e^x$ have maximum curvature?

51. At what point(s) does $4x^2 + 9y^2 = 36$ have minimum radius of curvature?

52. Find the value of x, $x > 0$, where $y = x^3$ has maximum curvature.

53. Find the maximum and minimum values of the radius of curvature for the curve $x = \cos t$, $y = \sin t$, $z = \cos t$.

54. Find the minimum value of the radius of curvature for the curve $x = e^t$, $y = e^{-t}$, $z = \sqrt{2}t$.

55. Show that the curvature of the polar curve $r = e^{a\theta}$ is inversely proportional to r.

56. Show that the curvature of the polar curve $r^2 = a^2 \cos 2\theta$ is directly proportional to r ($r > 0$).

57. Given that $d\mathbf{T}/ds = 0.03\mathbf{i} - 0.04\mathbf{j}$ at a certain point P on a curve, find the value of $|d\phi/ds|$, where ϕ is the angle of inclination of the tangent line to the curve at P. Assuming that s is measured in centimeters, express your answer in units of $^\circ$/cm.

58. Assuming that arc length is measured in inches along the curve $y = x^3$, find the magnitude of the rate of change with respect to arc length of the angle of inclination of the tangent line to the curve in units of $^\circ$/in. at the point where $x = 1$.

In Exercises 59–62, we shall say that a **smooth transition** occurs at a point P on a curve if the curvature κ is continuous at P. Use the formula for κ given in Exercise 15 to help solve the problem.

59. Show that the transition at $x = 0$ from the horizontal line $y = 0$ for $x \le 0$ to the parabola $y = x^2$ for $x > 0$ is not smooth, whereas the transition to $y = x^3$ for $x > 0$ is smooth.

60. (a) Sketch the graph of the curve defined piecewise by $y = x^2$ for $x < 0$, $y = x^4$ for $x \ge 0$.

(b) Show that for the curve in part (a) the transition at $x = 0$ is not smooth.

61. The figure below shows the arc of a circle of radius r with center at $(0, r)$. Find the value of a so that there is a smooth transition from the circle to the parabola $y = ax^2$ at the point where $x = 0$.

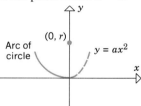

62. Find a, b, and c so that there is a smooth transition at $x = 0$ from the curve $y = e^x$ for $x \le 0$ to the parabola $y = ax^2 + bx + c$ for $x > 0$. [*Hint:* The curvature is continuous at those points where y'' is continuous.]

63. Let C be a smooth curve in the xy-plane that is tangent to the x-axis at the origin, and let s be an arc-length parameter for C with its reference point at the origin ($x = 0$ and $y = 0$ if $s = 0$). Suppose that the curvature κ is proportional to s and that s is chosen so that the constant of proportionality is positive ($\kappa = as$ where $a > 0$).

(a) Assuming that $d\phi/ds \ge 0$, use Formula (9) to show that $\phi = \frac{1}{2}as^2$.

(b) Use the result in part (a) and the fact that $\mathbf{T} = (\cos \phi)\mathbf{i} + (\sin \phi)\mathbf{j}$ to show that

$$x = \int_0^s \cos\left(\tfrac{1}{2}au^2\right) du, \quad y = \int_0^s \sin\left(\tfrac{1}{2}au^2\right) du$$

are parametric equations of the curve.

In Exercises 64–66, we assume that s is an arc-length parameter for a smooth curve C in 3-space and that $d\mathbf{T}/ds$ and $d\mathbf{N}/ds$ exist at each point on the curve. (This implies that $d\mathbf{B}/ds$ exists as well, since $\mathbf{B} = \mathbf{T} \times \mathbf{N}$.)

64. (a) Show that $d\mathbf{B}/ds$ is perpendicular to $\mathbf{B}$. [*Hint:* See Theorem 15.2.7.]

(b) Show that $d\mathbf{B}/ds$ is perpendicular to $\mathbf{T}$. [*Hint:* Use the fact that $\mathbf{B}$ is perpendicular to both $\mathbf{T}$ and $\mathbf{N}$, and differentiate $\mathbf{B} \cdot \mathbf{T}$ with respect to s.]

(c) Use the results in parts (a) and (b) to show that $d\mathbf{B}/ds$ is a scalar multiple of $\mathbf{N}$.

(d) It follows from part (c) that there is a real-valued function $\tau = \tau(s)$ (τ = Greek "tau"), called the *torsion,* such that $d\mathbf{B}/ds = -\tau\mathbf{N}$. Use the result in Exercise 31(b), Section 15.4, to show that if C lies entirely in a plane, then $\tau(s) = 0$ for all values of s. [*Note:* $d\mathbf{B}/ds$ is related to the rate at which the osculating plane turns (Figure 15.4.9), so the torsion is sometimes viewed as a measure of the tendency for C to twist out of the osculating plane.]

65. Let κ be the curvature of C and τ the torsion (defined in Exercise 64). By differentiating $\mathbf{N} = \mathbf{B} \times \mathbf{T}$ with respect to s, show that $d\mathbf{N}/ds = -\kappa\mathbf{T} + \tau\mathbf{B}$.

66. The following derivatives, known as the *Frenet–Serret formulas,* are fundamental in the theory of curves in 3-space:

$$d\mathbf{T}/ds = \kappa\mathbf{N} \qquad \text{[Formula (8)]}$$
$$d\mathbf{N}/ds = -\kappa\mathbf{T} + \tau\mathbf{B} \qquad \text{[Exercise 65]}$$
$$d\mathbf{B}/ds = -\tau\mathbf{N} \qquad \text{[Exercise 64(d)]}$$

Use the first two Frenet–Serret formulas and the fact that $\mathbf{r}'(s) = \mathbf{T}$ if $\mathbf{r} = \mathbf{r}(s)$ to show that

$$\tau = \frac{[\mathbf{r}'(s) \times \mathbf{r}''(s)] \cdot \mathbf{r}'''(s)}{\|\mathbf{r}''(s)\|^2} \quad \text{and} \quad \mathbf{B} = \frac{\mathbf{r}'(s) \times \mathbf{r}''(s)}{\|\mathbf{r}''(s)\|}$$

67. Use the results in Exercise 66 and the results in Exercise 29 of Section 15.3 to show that for the circular helix

$$\mathbf{r} = (a \cos t)\mathbf{i} + (a \sin t)\mathbf{j} + ct\mathbf{k}$$

with $a > 0$ the torsion and the binormal vector are

$$\tau = \frac{c}{w^2}$$

and

$$\mathbf{B} = \left(\frac{c}{w} \sin \frac{s}{w}\right)\mathbf{i} - \left(\frac{c}{w} \cos \frac{s}{w}\right)\mathbf{j} + \left(\frac{a}{w}\right)\mathbf{k}$$

where $w = \sqrt{a^2 + c^2}$ and s has its reference point at $(a, 0, 0)$.

68. Suppose that the arc-length parameter in the Frenet–Serret formulas of Exercise 66 is a function $s = s(t)$ of a general parameter t and that primes are used to denote derivatives with respect to t.

(a) Use the first two Frenet–Serret formulas to show that $\mathbf{T}' = \kappa s'\mathbf{N}$ and $\mathbf{N}' = -\kappa s'\mathbf{T} + \tau s'\mathbf{B}$. [*Hint:* Use the chain rule.]

(b) Show that Formulas (4) and (6) can be written in the form

$$\mathbf{r}'(t) = s'\mathbf{T} \quad \text{and} \quad \mathbf{r}''(t) = s''\mathbf{T} + \kappa(s')^2\mathbf{N}$$

(c) Use the results in parts (a) and (b) to show that

$$\mathbf{r}'''(t) = [s''' - \kappa^2(s')^3]\mathbf{T}$$
$$+ [3\kappa s's'' + \kappa'(s')^2]\mathbf{N} + \kappa\tau(s')^3\mathbf{B}$$

(d) Use the results in parts (b) and (c) to show that

$$\tau = \frac{[\mathbf{r}'(t) \times \mathbf{r}''(t)] \cdot \mathbf{r}'''(t)}{\|\mathbf{r}'(t) \times \mathbf{r}''(t)\|^2}$$

In Exercises 69–72, use the formula in Exercise 68(d) to find the torsion $\tau = \tau(t)$.

69. The twisted cubic $\mathbf{r}(t) = 2t\mathbf{i} + t^2\mathbf{j} + \frac{1}{3}t^3\mathbf{k}$.

70. The circular helix $\mathbf{r}(t) = (a \cos t)\mathbf{i} + (a \sin t)\mathbf{j} + bt\mathbf{k}$.

71. $\mathbf{r}(t) = e^t\mathbf{i} + e^{-t}\mathbf{j} + \sqrt{2}t\mathbf{k}$.

72. $\mathbf{r}(t) = (t - \sin t)\mathbf{i} + (1 - \cos t)\mathbf{j} + t\mathbf{k}$.

■ **15.6 MOTION ALONG A CURVE**

In Section 4.10 we considered the motion of a particle moving along a coordinate line. For such particles, there are only two possible directions of motion—the positive direction or the negative direction. For particles moving in 2-space or 3-space the situation is much more complicated, since there are infinitely many directions of motion possible. In this section we shall show how vectors can be used to study the motion of a particle moving along a curve in 2-space or 3-space.

□ **VELOCITY, ACCELERATION, AND SPEED**

Suppose that a particle is moving through 2-space or 3-space along a smooth curve C. At each instant of time we will regard the direction of motion to be the direction of the unit tangent vector, and we will regard the speed of the particle to be the instantaneous rate of change with respect to time of the arc length traveled by the particle, as measured from some arbitrary reference point (Figure 15.6.1*a*). The speed and the direction of motion together determine what is called the "velocity vector" of the particle. Mathematically, this vector can be written in the form

$$\mathbf{v}(t) = \frac{ds}{dt}\mathbf{T}(t) \tag{1}$$

which is a vector of length ds/dt with the same direction as $\mathbf{T}(t)$ (Figure 15.6.1*b*).

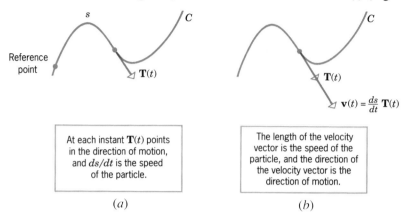

At each instant $\mathbf{T}(t)$ points in the direction of motion, and ds/dt is the speed of the particle.

The length of the velocity vector is the speed of the particle, and the direction of the velocity vector is the direction of motion.

Figure 15.6.1   (*a*)   (*b*)

Recall that for a particle moving on a coordinate line, the velocity function is the derivative of the position function. The same is true for a particle moving along a smooth curve in 2-space or 3-space. To see that this is so, suppose that the position function of the particle is $\mathbf{r}(t)$. From the definition of the unit tangent vector [Formula (1) of Section 15.4] and Formula (17) of Section 15.3 we have

$$\mathbf{T}(t) = \frac{\mathbf{r}'(t)}{\|\mathbf{r}'(t)\|} \quad \text{and} \quad \frac{ds}{dt} = \|\mathbf{r}'(t)\|$$

Substituting these expressions in (1) and simplifying yields

$$\mathbf{v}(t) = \mathbf{r}'(t)$$

which shows that the velocity function is the derivative of the position function. As was the case for motion along a coordinate line, the *acceleration function* of a particle moving along a curve in 2-space or 3-space is defined to be the derivative of the velocity function. Thus, we have the following definitions.

15.6.1   DEFINITION.   If a particle moves along a curve C in 2-space or 3-space so that its position vector at time t is $\mathbf{r}(t)$, then the *instantaneous velocity*, *instantaneous acceleration*, and *instantaneous speed* of the particle at time t are defined by

$$\text{velocity} = \mathbf{v}(t) = \frac{d\mathbf{r}}{dt} \tag{2}$$

$$\text{acceleration} = \mathbf{a}(t) = \frac{d\mathbf{v}}{dt} = \frac{d^2\mathbf{r}}{dt^2} \tag{3}$$

$$\text{speed} = \|\mathbf{v}(t)\| = \frac{ds}{dt} \tag{4}$$

---

where s is the total arc length traveled by the particle at time t, measured from some arbitrary reference point on C.

Example 1 A particle moves along a circular path in such a way that its x- and y-coordinates at time t are

$$x = 2 \cos t, \quad y = 2 \sin t$$

(a) Find the instantaneous velocity and speed of the particle at time t.

(b) Sketch the path of the particle, and show the position and velocity vectors at time $t = \pi/4$ with the velocity vector drawn so that its initial point is at the tip of the position vector.

(c) Show that at each instant the acceleration vector is perpendicular to the velocity vector.

Solution (a). At time t, the position vector is

$$\mathbf{r}(t) = 2 \cos t\mathbf{i} + 2 \sin t\mathbf{j}$$

so the instantaneous velocity and speed are

$$\mathbf{v}(t) = \frac{d\mathbf{r}}{dt} = -2 \sin t\mathbf{i} + 2 \cos t\mathbf{j}$$

$$\|\mathbf{v}(t)\| = \sqrt{(-2 \sin t)^2 + (2 \cos t)^2} = 2$$

Solution (b). The graph of the parametric equations is a circle of radius 2 centered at the origin. At time $t = \pi/4$ the position and velocity vectors of the particles are

$$\mathbf{r}(\pi/4) = 2 \cos (\pi/4)\mathbf{i} + 2 \sin (\pi/4)\mathbf{j} = \sqrt{2}\mathbf{i} + \sqrt{2}\mathbf{j}$$

$$\mathbf{v}(\pi/4) = -2 \sin (\pi/4)\mathbf{i} + 2 \cos (\pi/4)\mathbf{j} = -\sqrt{2}\mathbf{i} + \sqrt{2}\mathbf{j}$$

These vectors and the circle are shown in Figure 15.6.2.

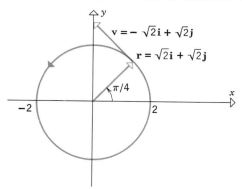

Figure 15.6.2

Solution (c). At time t, the acceleration vector is

$$\mathbf{a}(t) = \frac{d\mathbf{v}}{dt} = -2 \cos t\mathbf{i} - 2 \sin t\mathbf{j}$$

One way of showing that $\mathbf{v}(t)$ and $\mathbf{a}(t)$ are perpendicular is to show that their dot product is zero (try it). However, it is easier to observe that $\mathbf{a}(t)$ is the negative of $\mathbf{r}(t)$, which implies that $\mathbf{v}(t)$ and $\mathbf{a}(t)$ are perpendicular, since at each point on a circle the radius and tangent line are perpendicular. ◄

REMARKS. In Example 1 we omitted all units for simplicity. We shall follow this practice throughout this section, except in some applied problems that appear later. Moreover, as in Example 1, we shall follow the standard convention of drawing velocity and acceleration vectors with their initial points at the tip of the position vector, rather than at the origin.

Since the velocity function $\mathbf{v}(t)$ of a particle moving in 2-space or 3-space can be obtained by differentiating the position function $\mathbf{r}(t)$, it follows that $\mathbf{r}(t)$ can be obtained by integrating $\mathbf{v}(t)$. Similarly, $\mathbf{v}(t)$ can be obtained by integrating $\mathbf{a}(t)$. However, integrating $\mathbf{v}(t)$ does not produce a unique position function, and integrating $\mathbf{a}(t)$ does not produce a unique velocity function, since constants of integration occur. To determine these constants, some additional information is needed. For example, if the position vector of the particle is known at some specific point in time, then the constant resulting from integrating $\mathbf{v}(t)$ can be determined from this information, and if the velocity vector of the particle is known at some point in time, then the constant resulting from integrating $\mathbf{a}(t)$ can be determined.

Example 2 A particle moves through 3-space in such a way that its velocity at time t is

$$\mathbf{v}(t) = \mathbf{i} + t\mathbf{j} + t^2\mathbf{k}$$

Find the coordinates of the particle at time $t = 1$ given that the particle is at the point $(-1, 2, 4)$ at time $t = 0$.

Solution. Integrating the velocity function to obtain the position function yields

$$\mathbf{r}(t) = \int \mathbf{v}(t)\, dt = \int (\mathbf{i} + t\mathbf{j} + t^2\mathbf{k})\, dt = t\mathbf{i} + \frac{t^2}{2}\mathbf{j} + \frac{t^3}{3}\mathbf{k} + \mathbf{C} \tag{5}$$

where $\mathbf{C}$ is a vector constant of integration. Since the coordinates of the particle at time $t = 0$ are $(-1, 2, 4)$, the position vector at time $t = 0$ is

$$\mathbf{r}(0) = -\mathbf{i} + 2\mathbf{j} + 4\mathbf{k} \tag{6}$$

It follows on substituting $t = 0$ in (5) and equating the result with (6) that $\mathbf{C} = -\mathbf{i} + 2\mathbf{j} + 4\mathbf{k}$. Substituting this value of $\mathbf{C}$ in (5) and simplifying yields

$$\mathbf{r}(t) = (t - 1)\mathbf{i} + \left(\frac{t^2}{2} + 2\right)\mathbf{j} + \left(\frac{t^3}{3} + 4\right)\mathbf{k}$$

Thus, at time $t = 1$ the position vector of the particle is

$$\mathbf{r}(1) = 0\mathbf{i} + \frac{5}{2}\mathbf{j} + \frac{13}{3}\mathbf{k}$$

so its coordinates at that instant are $(0, \frac{5}{2}, \frac{13}{3})$. ◀

☐ **DISPLACEMENT AND DISTANCE TRAVELED**

If a particle travels along a curve C in 2-space or 3-space, the *displacement* of the particle over the time interval $t_1 \le t \le t_2$ is commonly denoted by $\Delta\mathbf{r}$, and is defined as

$$\Delta\mathbf{r} = \mathbf{r}(t_2) - \mathbf{r}(t_1) \tag{7}$$

(Figure 15.6.3). The displacement vector, which describes the change in position of the particle during the time interval, can be obtained by integrating the velocity function from t_1 to t_2:

$$\Delta\mathbf{r} = \int_{t_1}^{t_2} \mathbf{v}(t)\, dt = \int_{t_1}^{t_2} \frac{d\mathbf{r}}{dt}\, dt = \mathbf{r}(t)\Big]_{t_1}^{t_2} = \mathbf{r}(t_2) - \mathbf{r}(t_1) \tag{8}$$

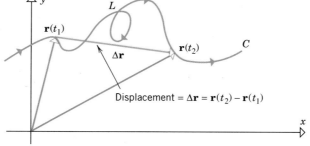

Figure 15.6.3

The distance L traveled by the particle along C during the time interval $t_1 \leq t \leq t_2$ can be obtained by integrating the magnitude of the velocity function from t_1 to t_2 [Formula (6) of Section 15.3]:

$$L = \int_{t_1}^{t_2} \|\mathbf{v}(t)\| \, dt = \int_{t_1}^{t_2} \left\| \frac{d\mathbf{r}}{dt} \right\| \, dt \tag{9}$$

Example 3 Suppose that a particle travels along a circular helix in 3-space so that its position vector at time t is

$$\mathbf{r}(t) = (4 \cos \pi t)\mathbf{i} + (4 \sin \pi t)\mathbf{j} + t\mathbf{k}$$

Find the displacement and distance traveled by the particle during the time interval $1 \leq t \leq 5$.

Solution.   We have

$$\mathbf{v}(t) = \frac{d\mathbf{r}}{dt} = (-4\pi \sin \pi t)\mathbf{i} + (4\pi \cos \pi t)\mathbf{j} + \mathbf{k}$$

$$\|\mathbf{v}(t)\| = \sqrt{(-4\pi \sin \pi t)^2 + (4\pi \cos \pi t)^2 + 1} = \sqrt{16\pi^2 + 1}$$

Since $\mathbf{r}(t)$ is known, the displacement can be found directly from (7); it is

$$\Delta\mathbf{r} = \mathbf{r}(5) - \mathbf{r}(1)$$
$$= [(4 \cos 5\pi)\mathbf{i} + (4 \sin 5\pi)\mathbf{j} + 5\mathbf{k}] - [(4 \cos \pi)\mathbf{i} + (4 \sin \pi)\mathbf{j} + \mathbf{k}]$$
$$= (-4\mathbf{i} + 5\mathbf{k}) - (-4\mathbf{i} + \mathbf{k}) = 4\mathbf{k}$$

which tells us that the change in the position of the particle over the time interval was 4 units straight up.

From (9), the distance L traveled by the particle during the time interval is

$$L = \int_1^5 \sqrt{16\pi^2 + 1} \, dt = 4\sqrt{16\pi^2 + 1} \text{ units} \quad \blacktriangleleft$$

☐ **NORMAL AND TANGENTIAL COMPONENTS OF ACCELERATION**

In applications, it is often desirable to resolve the velocity and acceleration vectors into vector components that are parallel to the unit tangent and unit normal vectors. The following theorem explains how this can be done.

15.6.2   THEOREM.   *For a particle moving along a curve C in 2-space or 3-space, the velocity and acceleration vectors can be written as*

$$\mathbf{v} = \frac{ds}{dt} \mathbf{T} \tag{10}$$

$$\mathbf{a} = \frac{d^2s}{dt^2} \mathbf{T} + \kappa \left(\frac{ds}{dt} \right)^2 \mathbf{N} \tag{11}$$

where s is an arc-length parameter for the curve, and $\mathbf{T} = \mathbf{T}(t)$, $\mathbf{N} = \mathbf{N}(t)$, *and* $\kappa = \kappa(t)$ *denote the unit tangent vector, unit normal vector, and curvature (Figure 15.6.4).*

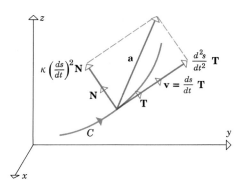

Figure 15.6.4

Proof. Formula (10) is just a restatement of (1). To obtain (11), we differentiate both sides of (10) with respect to t; this yields

$$\mathbf{a} = \frac{d}{dt}\left(\frac{ds}{dt}\mathbf{T}\right) = \frac{d^2s}{dt^2}\mathbf{T} + \frac{ds}{dt}\frac{d\mathbf{T}}{dt}$$

$$= \frac{d^2s}{dt^2}\mathbf{T} + \frac{ds}{dt}\frac{d\mathbf{T}}{ds}\frac{ds}{dt}$$

$$= \frac{d^2s}{dt^2}\mathbf{T} + \left(\frac{ds}{dt}\right)^2\frac{d\mathbf{T}}{ds}$$

$$= \frac{d^2s}{dt^2}\mathbf{T} + \left(\frac{ds}{dt}\right)^2 \kappa\mathbf{N} \qquad \boxed{\begin{array}{l}\text{Formula (8)}\\\text{of Section 15.5}\end{array}}$$

from which (11) follows. ∎

Formula (11) should make sense from your experience as an automobile passenger. If a car increases its speed rapidly, the effect is to press the passenger back against the seat. This occurs because the rapid increase in speed results in a large value for

$$\frac{d}{dt}\left(\frac{ds}{dt}\right) = \frac{d^2s}{dt^2}$$

which produces a large tangential component of acceleration in Formula (11). On the other hand, if the car rounds a turn in the road, the passenger is thrown to the side: the greater the curvature in the road or the greater the speed of the car, the greater the force with which the passenger is thrown. This occurs because ds/dt (speed) and κ (curvature) both enter as factors in the normal component of acceleration in (11).

The coefficients of $\mathbf{T}$ and $\mathbf{N}$ in (11) are commonly denoted by

$$a_T = \frac{d^2s}{dt^2} \qquad a_N = \kappa\left(\frac{ds}{dt}\right)^2 \qquad\qquad\qquad (12\text{a–b})$$

These numbers are called the *scalar tangential component of acceleration* and the *scalar normal component of acceleration*, respectively. Using this notation, (11) can be expressed as

$$\mathbf{a} = a_T\mathbf{T} + a_N\mathbf{N} \qquad\qquad\qquad (13)$$

The vectors $a_T\mathbf{T}$ and $a_N\mathbf{N}$ are called the *vector tangential component of acceleration* and *vector normal component of acceleration*, respectively.

It should be noted that Formula (13) applies to motion in both 2-space and 3-space. What is interesting is that the 3-space formula does not involve the binormal vector $\mathbf{B}$, so the acceleration vector always lies in the plane of $\mathbf{T}$ and $\mathbf{N}$ (the osculating plane), even for the most twisting paths of motion (Figure 15.6.5).

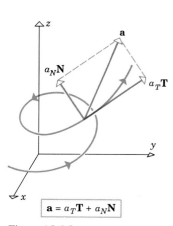

Figure 15.6.5

Although Formulas (12a) and (12b) provide useful insight into the physical properties of the tangential and normal components of acceleration, they are not always the best formulas to use for computations. We shall now derive some more useful formulas that express the scalar normal and tangential components of acceleration directly in terms of the velocity and acceleration of the particle. If, as shown in Figure 15.6.6, we let θ be the angle between the vector $\mathbf{a}$ and the vector $a_T\mathbf{T}$, then we obtain

$$a_T = \|\mathbf{a}\|\cos\theta \quad \text{and} \quad a_N = \|\mathbf{a}\|\sin\theta$$

With the help of Definition 14.3.1 and Formula (7) of Section 14.4, these equations can be rewritten as

$$a_T = \frac{\|\mathbf{v}\|\,\|\mathbf{a}\|\cos\theta}{\|\mathbf{v}\|} = \frac{\mathbf{v}\cdot\mathbf{a}}{\|\mathbf{v}\|} \quad \text{and} \quad a_N = \frac{\|\mathbf{v}\|\,\|\mathbf{a}\|\sin\theta}{\|\mathbf{v}\|} = \frac{\|\mathbf{v}\times\mathbf{a}\|}{\|\mathbf{v}\|}$$

Summarizing, we have

$$a_T = \frac{\mathbf{v}\cdot\mathbf{a}}{\|\mathbf{v}\|} \quad \text{and} \quad a_N = \frac{\|\mathbf{v}\times\mathbf{a}\|}{\|\mathbf{v}\|} \tag{14a–b}$$

These formulas are applicable in both 2-space and 3-space, but to compute the cross product in 2-space, the vectors $\mathbf{v}$ and $\mathbf{a}$ must be treated as vectors in 3-space with a zero $\mathbf{k}$ component. Note also that a_N is nonnegative. Thus, for motion in 2-space the vector $a_N\mathbf{N}$ points "inward" toward the concave side of the curve because $\mathbf{N}$ does. (See the subsection of 15.4 on Inward Unit Normal Vectors in 2-Space.) For this reason, the vector $a_N\mathbf{N}$ is also called the ***inward vector component of acceleration*** in 2-space.

Before proceeding to a numerical example, it is worth noting that by rewriting Formula (12b) as

$$\kappa = \frac{a_N}{(ds/dt)^2} = \frac{a_N}{\|\mathbf{v}\|^2} \tag{15}$$

and combining it with (14b) we can write

$$\kappa = \frac{\|\mathbf{v}\times\mathbf{a}\|}{\|\mathbf{v}\|^3} \tag{16}$$

which expresses curvature in terms of velocity and acceleration. [Compare this result with Formula (3) of Section 15.5.]

Example 4 Suppose that a particle moves through 3-space so that its position vector at time t is

$$\mathbf{r}(t) = t\mathbf{i} + t^2\mathbf{j} + t^3\mathbf{k}$$

(The path is the twisted cubic shown in Figure 15.1.2.)

(a) Find the scalar tangential and normal components of acceleration at time t.

(b) Find the scalar tangential and normal components of acceleration at time $t = 1$.

(c) Find the vector tangential and normal components of acceleration at time $t = 1$.

(d) Find the curvature of the path at the point where the particle is located at time $t = 1$.

Solution (a). We have

$$\mathbf{v}(t) = \mathbf{r}'(t) = \mathbf{i} + 2t\mathbf{j} + 3t^2\mathbf{k}$$

$$\mathbf{a}(t) = \mathbf{v}'(t) = 2\mathbf{j} + 6t\mathbf{k}$$

$$\|\mathbf{v}(t)\| = \sqrt{1 + 4t^2 + 9t^4}$$

$$\mathbf{v}(t)\cdot\mathbf{a}(t) = 4t + 18t^3$$

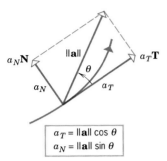

$$a_T = \|\mathbf{a}\|\cos\theta$$
$$a_N = \|\mathbf{a}\|\sin\theta$$

Figure 15.6.6

$$\mathbf{v}(t) \times \mathbf{a}(t) = \begin{vmatrix} \mathbf{i} & \mathbf{j} & \mathbf{k} \\ 1 & 2t & 3t^2 \\ 0 & 2 & 6t \end{vmatrix} = 6t^2\mathbf{i} - 6t\mathbf{j} + 2\mathbf{k}$$

Thus, from (14a) and (14b)

$$a_T = \frac{\mathbf{v} \cdot \mathbf{a}}{\|\mathbf{v}\|} = \frac{4t + 18t^3}{\sqrt{1 + 4t^2 + 9t^4}}$$

$$a_N = \frac{\|\mathbf{v} \times \mathbf{a}\|}{\|\mathbf{v}\|} = \frac{\sqrt{36t^4 + 36t^2 + 4}}{\sqrt{1 + 4t^2 + 9t^4}} = 2\sqrt{\frac{9t^4 + 9t^2 + 1}{9t^4 + 4t^2 + 1}}$$

Solution (b).   At time $t = 1$, the components a_T and a_N in part (a) are

$$a_T = \frac{22}{\sqrt{14}} \approx 5.88 \quad \text{and} \quad a_N = 2\sqrt{\frac{19}{14}} \approx 2.33$$

Solution (c).   Since $\mathbf{T}$ and $\mathbf{v}$ have the same direction, $\mathbf{T}$ can be obtained by normalizing $\mathbf{v}$, that is,

$$\mathbf{T}(t) = \frac{\mathbf{v}(t)}{\|\mathbf{v}(t)\|}$$

At time $t = 1$ we have

$$\mathbf{T}(1) = \frac{\mathbf{v}(1)}{\|\mathbf{v}(1)\|} = \frac{\mathbf{i} + 2\mathbf{j} + 3\mathbf{k}}{\|\mathbf{i} + 2\mathbf{j} + 3\mathbf{k}\|} = \frac{1}{\sqrt{14}}(\mathbf{i} + 2\mathbf{j} + 3\mathbf{k})$$

From this and part (b) we obtain the vector tangential component of acceleration:

$$a_T\mathbf{T}(1) = \frac{22}{\sqrt{14}}\mathbf{T}(1) = \tfrac{11}{7}(\mathbf{i} + 2\mathbf{j} + 3\mathbf{k}) = \tfrac{11}{7}\mathbf{i} + \tfrac{22}{7}\mathbf{j} + \tfrac{33}{7}\mathbf{k}$$

To find the normal vector component of acceleration, we rewrite $\mathbf{a} = a_T\mathbf{T} + a_N\mathbf{N}$ as

$$a_N\mathbf{N} = \mathbf{a} - a_T\mathbf{T}$$

Thus, at time $t = 1$ the normal vector component of acceleration is

$$a_N\mathbf{N}(1) = \mathbf{a}(1) - a_T\mathbf{T}(1)$$
$$= (2\mathbf{j} + 6\mathbf{k}) - (\tfrac{11}{7}\mathbf{i} + \tfrac{22}{7}\mathbf{j} + \tfrac{33}{7}\mathbf{k})$$
$$= -\tfrac{11}{7}\mathbf{i} - \tfrac{8}{7}\mathbf{j} + \tfrac{9}{7}\mathbf{k}$$

Solution (d).   We shall apply Formula (16) with $t = 1$. From part (a)

$$\|\mathbf{v}(1)\| = \sqrt{14} \quad \text{and} \quad \mathbf{v}(1) \times \mathbf{a}(1) = 6\mathbf{i} - 6\mathbf{j} + 2\mathbf{k}$$

Thus, at time $t = 1$

$$\kappa = \frac{\|\mathbf{v} \times \mathbf{a}\|}{\|\mathbf{v}\|^3} = \frac{\sqrt{76}}{(\sqrt{14})^3} = \frac{1}{14}\sqrt{\frac{38}{7}} \approx 0.17 \quad \blacktriangleleft$$

☐ **MOTION OF A PROJECTILE**

Our work in this section provides many of the mathematical tools required to study such fundamental physical phenomena as planetary motion and the motion of objects in the earth's gravitational field. As an illustration, we shall investigate the following problem.

15.6.3 PROBLEM.   *An object of mass m is fired, thrown, or released at some known point near the surface of the earth with a known initial velocity vector. Find the trajectory of the object.*

To solve this problem we need a principle from physics known as *Newton's Second Law of Motion*; it states that when an object of mass m is subjected to a force $\mathbf{F}$, it undergoes an acceleration $\mathbf{a}$ satisfying

$$\mathbf{F} = m\mathbf{a} \tag{17}$$

We shall make three simplifying assumptions:

1. The mass of the object is constant.
2. The only force acting on the object is the force of the earth's gravity. (Thus, air resistance and the gravitational effect of other planets and celestial objects are ignored.)
3. The object remains sufficiently close to the earth that the earth's gravity can be assumed constant.

We shall assume that the mass of the object is m and that t (measured in seconds) is the time elapsed from the initial firing or release, so that this firing or release occurs at time $t = 0$. We shall denote the known initial velocity vector by $\mathbf{v_0}$ and the known initial position vector by $\mathbf{r_0}$.

As shown in Figure 15.6.7, we shall introduce an xy-coordinate system whose origin is on the surface of the earth and whose positive y-axis points up and passes through the initial position of the object. Thus, at time $t = 0$ the object has coordinates $(0, s_0)$, which are assumed known, and the initial position vector is $\mathbf{r_0} = s_0\mathbf{j}$.

It is shown in physics that the downward force of the earth's gravity on an object of mass m is

$$\mathbf{F} = -mg\mathbf{j} \tag{18}$$

where g is a constant approximately equal to 32 ft/sec^2 if distance is measured in feet and 9.8 m/sec^2 if distance is measured in meters. Substituting (18) into (17) yields

$$m\mathbf{a} = -mg\mathbf{j}$$

or on canceling m from both sides

$$\mathbf{a} = -g\mathbf{j} \tag{19}$$

Observe that the acceleration vector $\mathbf{a}$ is constant because (19) does not involve t. Moreover, since the value of g is known, the acceleration vector $\mathbf{a}$ is also known; thus, we can find the position function of the object by integrating the known acceleration twice, once to obtain the velocity $\mathbf{v}(t)$, and again to obtain $\mathbf{r}(t)$. Thus, Problem 15.6.3 has been reduced to solving the vector differential equation

$$\frac{d^2\mathbf{r}}{dt^2} = -g\mathbf{j} \tag{20}$$

subject to the initial conditions

$$\mathbf{r}(0) = \mathbf{r_0} = s_0\mathbf{j} \tag{21}$$

$$\mathbf{v}(0) = \mathbf{v_0} \tag{22}$$

Integrating (20) with respect to t (keeping in mind that $-g\mathbf{j}$ is constant) yields

$$\mathbf{v}(t) = -gt\mathbf{j} + \mathbf{c_1} \tag{23}$$

where $\mathbf{c_1}$ is a vector constant of integration. Substituting $t = 0$ in (23) and using initial condition (22) yields

$$\mathbf{v_0} = \mathbf{c_1}$$

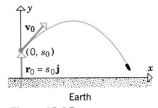

Figure 15.6.7

so that (23) may be written as

$$\mathbf{v}(t) = -gt\mathbf{j} + \mathbf{v}_0 \tag{24}$$

This formula specifies the velocity of the object at any time t. Integrating $\mathbf{v}(t)$ with respect to t (keeping in mind that $\mathbf{v}_0$ is constant) yields

$$\mathbf{r}(t) = -\tfrac{1}{2}gt^2\mathbf{j} + \mathbf{v}_0 t + \mathbf{c}_2 \tag{25}$$

where $\mathbf{c}_2$ is another vector constant of integration. Substituting $t = 0$ in (25) and using initial condition (21) yields

$$s_0\mathbf{j} = \mathbf{c}_2$$

so that (25) can be written as

$$\mathbf{r}(t) = (-\tfrac{1}{2}gt^2 + s_0)\mathbf{j} + \mathbf{v}_0 t \tag{26}$$

This is the result we were seeking—the position function of the object, expressed in terms of the initial velocity and position.

REMARK. Observe that the mass of the object does not enter into the final formulas for velocity and position. Physically, this means that the mass has no influence on the trajectory or the velocity of the object—these are completely determined by the initial position and velocity. This explains the famous observation of Galileo that two objects of different mass, released from the same height, will reach the ground at the same time if air resistance is neglected.

There is a useful alternative form of (26) that expresses the position function of the object in terms of its initial speed and the angle that the initial velocity vector makes with the x-axis. As shown in Figure 15.6.8, suppose that the initial speed $\|\mathbf{v}_0\|$ is denoted by v_0 and the angle that $\mathbf{v}_0$ makes with the x-axis is denoted by α. Thus, the vector $\mathbf{v}_0$ can be expressed as

$$\mathbf{v}_0 = (v_0 \cos \alpha)\mathbf{i} + (v_0 \sin \alpha)\mathbf{j}$$

Substituting this expression in (26) and combining like components yields

$$\mathbf{r}(t) = (v_0 \cos \alpha)t\mathbf{i} + (s_0 + (v_0 \sin \alpha)t - \tfrac{1}{2}gt^2)\mathbf{j} \tag{27}$$

which is equivalent to the parametric equations

$$x = (v_0 \cos \alpha)t, \quad y = s_0 + (v_0 \sin \alpha)t - \tfrac{1}{2}gt^2 \tag{28}$$

The parametric equations reveal that the trajectory of the object is a parabolic arc. To see that this is so, we can solve the first equation for t, then substitute in the second to eliminate the parameter. This yields (verify)

$$y = s_0 + (\tan \alpha)x - \left(\frac{g}{2v_0^2 \cos^2 \alpha} \right)x^2$$

which is an equation of a parabola since the right side is a quadratic polynomial in x.

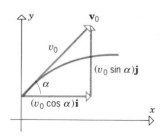

Figure 15.6.8

Example 5 A shell, fired from a cannon, has a muzzle speed (the speed as it leaves the barrel) of 800 ft/sec. The barrel makes an angle of 45° with the horizontal and, for simplicity, the barrel opening is assumed to be at ground level.

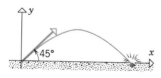

Figure 15.6.9

(a) Find parametric equations for the shell's trajectory relative to the coordinate system in Figure 15.6.9.

(b) How high does the shell rise?

(c) How far does the shell travel horizontally?

(d) What is the speed of the shell at its point of impact with the ground?

Solution (a). From (28) with $v_0 = 800$, $\alpha = 45°$, $s_0 = 0$ (since the shell starts at ground level), and $g = 32$ (since distance is measured in feet), we obtain the parametric equations

$$x = (800 \cos 45°)t, \quad y = (800 \sin 45°)t - 16t^2 \qquad (t \geq 0)$$

which simplify to

$$x = 400\sqrt{2}t, \quad y = 400\sqrt{2}t - 16t^2 \qquad (t \geq 0) \tag{29}$$

Solution (b). The maximum height of the shell is the maximum value of y in (29), which occurs when $dy/dt = 0$, that is, when

$$400\sqrt{2} - 32t = 0 \quad \text{or} \quad t = \frac{25\sqrt{2}}{2}$$

Substituting this value of t in (29) yields

$$y = 5000 \text{ ft}$$

as the maximum height of the shell.

Solution (c). The shell will hit the ground when $y = 0$. From (29), this occurs when

$$400\sqrt{2}t - 16t^2 = 0 \quad \text{or} \quad t(400\sqrt{2} - 16t) = 0$$

The solution $t = 0$ corresponds to the initial position of the shell and the solution $t = 25\sqrt{2}$ to the time of impact. Substituting the latter value in the equation for x in (29) yields

$$x = 20{,}000 \text{ ft}$$

as the horizontal distance traveled by the shell.

Solution (d). From (29), the position function of the shell is

$$\mathbf{r}(t) = (400\sqrt{2}t)\mathbf{i} + (400\sqrt{2}t - 16t^2)\mathbf{j}$$

so that the velocity function is

$$\mathbf{v}(t) = \mathbf{r}'(t) = 400\sqrt{2}\mathbf{i} + (400\sqrt{2} - 32t)\mathbf{j}$$

From part (c), impact occurs when $t = 25\sqrt{2}$, so that the velocity vector at this point is

$$\mathbf{v}(25\sqrt{2}) = 400\sqrt{2}\mathbf{i} + [400\sqrt{2} - 32(25\sqrt{2})]\mathbf{j} = 400\sqrt{2}\mathbf{i} - 400\sqrt{2}\mathbf{j}$$

Thus, the speed at impact is

$$\|\mathbf{v}(25\sqrt{2})\| = \sqrt{(400\sqrt{2})^2 + (-400\sqrt{2})^2} = 800 \text{ ft/sec} \qquad \blacktriangleleft$$

▶ Exercise Set 15.6   ©️ *17, 61, 65–68, 72, 74, 75*

In Exercises 1–6, $\mathbf{r}(t)$ is the position vector of a particle moving in the plane. Find the velocity, acceleration, and speed at an arbitrary time t; then sketch the path of the particle together with the velocity and acceleration vectors at the indicated time t.

1. $\mathbf{r}(t) = 3 \cos t\mathbf{i} + 3 \sin t\mathbf{j}$; $t = \pi/3$.

2. $\mathbf{r}(t) = t\mathbf{i} + t^2\mathbf{j}$; $t = 2$.

3. $\mathbf{r}(t) = e^t\mathbf{i} + e^{-t}\mathbf{j}$; $t = 0$.

4. $\mathbf{r}(t) = (2 + 4t)\mathbf{i} + (1 - t)\mathbf{j}$; $t = 1$.

5. $\mathbf{r}(t) = \cosh t\mathbf{i} + \sinh t\mathbf{j}$; $t = \ln 2$.

6. $\mathbf{r}(t) = 4 \cos t\mathbf{i} + 9 \sin t\mathbf{j}$; $t = \pi/2$.

In Exercises 7–12, find the velocity, speed, and acceleration at the given time t of a particle moving along the given curve.

7. $\mathbf{r}(t) = t\mathbf{i} + \frac{1}{2}t^2\mathbf{j} + \frac{1}{3}t^3\mathbf{k}$; $t = 1$.

8. $x = 1 + 3t, y = 2 - 4t, z = 7 + t$; $t = 2$.

9. $\mathbf{r}(t) = 2 \cos t\mathbf{i} + 2 \sin t\mathbf{j} + t\mathbf{k}$; $t = \pi/4$.

10. $\mathbf{r}(t) = 3t\mathbf{i} + 2t^2\mathbf{j} + \ln t\mathbf{k};\ t = 1.$

11. $\mathbf{r}(t) = e^t \sin t\mathbf{i} + e^t \cos t\mathbf{j} + t\mathbf{k};\ t = \pi/2.$

12. $\mathbf{r}(t) = 2t\mathbf{i} + t^2\mathbf{j} + \ln t\mathbf{k};\ t = 2.$

13. Suppose that the position vector of a particle moving in the plane is $\mathbf{r} = 12\sqrt{t}\mathbf{i} + t^{3/2}\mathbf{j},\ t > 0$. Find the minimum speed of the particle and its location when it has this speed.

14. Suppose that the motion of a particle is described by $\mathbf{r} = (t - t^2)\mathbf{i} - t^2\mathbf{j}$. Find the minimum speed of the particle and its location when it has this speed.

15. Find the maximum and minimum speeds of a particle whose motion is described by the position vector $\mathbf{r} = \sin 3t\mathbf{i} - 2\cos 3t\mathbf{j}$.

16. Find the maximum and minimum speeds of a particle whose motion is described by the position vector $\mathbf{r} = 3\cos 2t\mathbf{i} + \sin 2t\mathbf{j} + 4t\mathbf{k}$.

17. Find, to the nearest degree, the angle between $\mathbf{v}$ and $\mathbf{a}$ for $\mathbf{r} = t^3\mathbf{i} + t^2\mathbf{j}$ when $t = 1$.

18. Show that the angle between $\mathbf{v}$ and $\mathbf{a}$ is constant for $\mathbf{r} = e^t \cos t\mathbf{i} + e^t \sin t\mathbf{j}$. Find the angle.

19. Where on the path

$$\mathbf{r} = (t^2 - 5t)\mathbf{i} + (2t + 1)\mathbf{j} + 3t^2\mathbf{k}$$

are the velocity and acceleration vectors orthogonal?

20. Prove: If the speed of a particle is constant, then the acceleration and velocity vectors are orthogonal. [*Hint:* Consider $\mathbf{v} \cdot \mathbf{v}$.]

In Exercises 21 and 22, the position vectors of two particles are given. Show that the particles move along the same path but the speed of the first is constant and the speed of the second is not.

21. $\mathbf{r}_1 = 2\cos 3t\mathbf{i} + 2\sin 3t\mathbf{j},$
$\mathbf{r}_2 = 2\cos (t^2)\mathbf{i} + 2\sin (t^2)\mathbf{j}\quad (t \geq 0).$

22. $\mathbf{r}_1 = (3 + 2t)\mathbf{i} + t\mathbf{j} + (1 - t)\mathbf{k},$
$\mathbf{r}_2 = (5 - 2t^3)\mathbf{i} + (1 - t^3)\mathbf{j} + t^3\mathbf{k}.$

In Exercises 23–30, use the given information to find the position and velocity vectors of the particle.

23. $\mathbf{a}(t) = -32\mathbf{j};\ \mathbf{v}(0) = \mathbf{0};\ \mathbf{r}(0) = \mathbf{0}.$

24. $\mathbf{a}(t) = t\mathbf{j};\ \mathbf{v}(0) = \mathbf{i} + \mathbf{j};\ \mathbf{r}(0) = \mathbf{0}.$

25. $\mathbf{a}(t) = -\cos t\mathbf{i} - \sin t\mathbf{j};\ \mathbf{v}(0) = \mathbf{i};\ \mathbf{r}(0) = \mathbf{j}.$

26. $\mathbf{a}(t) = \mathbf{i} + e^{-t}\mathbf{j};\ \mathbf{v}(0) = 2\mathbf{i} + \mathbf{j};\ \mathbf{r}(0) = \mathbf{i} - \mathbf{j}.$

27. $\mathbf{a}(t) = \mathbf{i} + t\mathbf{k};\ \mathbf{v}(0) = \mathbf{0};\ \mathbf{r}(0) = \mathbf{j}.$

28. $\mathbf{a}(t) = (\sin 2t)\mathbf{j};\ \mathbf{v}(0) = \mathbf{i} - \mathbf{k};\ \mathbf{r}(0) = \mathbf{0}.$

29. $\mathbf{a}(t) = \sin t\mathbf{i} + \cos t\mathbf{j} + e^t\mathbf{k};\ \mathbf{v}(0) = \mathbf{k};\ \mathbf{r}(0) = -\mathbf{i} + \mathbf{k}.$

30. $\mathbf{a}(t) = (t + 1)^{-2}\mathbf{j} - e^{-2t}\mathbf{k}, t > -1;\ \mathbf{v}(0) = 3\mathbf{i} - \mathbf{j};$
$\mathbf{r}(0) = 2\mathbf{k}.$

31. Prove: If the acceleration of a moving particle is zero for all t, then the particle moves on a straight line.

32. Show that for a particle in motion over the time interval $[t_1, t_2]$ the change in velocity $\mathbf{v}(t_2) - \mathbf{v}(t_1)$ is given by

$$\mathbf{v}(t_2) - \mathbf{v}(t_1) = \int_{t_1}^{t_2} \mathbf{a}(t)\, dt$$

In Exercises 33–38, find the displacement and the distance traveled over the indicated time interval.

33. $\mathbf{r}(t) = t^2\mathbf{i} + \frac{1}{3}t^3\mathbf{j};\ 1 \leq t \leq 3.$

34. $\mathbf{r}(t) = (1 - 3\sin t)\mathbf{i} + 3\cos t\mathbf{j};\ 0 \leq t \leq 3\pi/2.$

35. $\mathbf{r}(t) = 2\sin 3t\mathbf{i} + 2\cos 3t\mathbf{j};\ 0 \leq t \leq 2\pi.$

36. $\mathbf{r}(t) = 3\sin 2t\mathbf{i} - 3\cos 2t\mathbf{j} + 8t\mathbf{k};\ 0 \leq t \leq 3\pi/4.$

37. $\mathbf{r}(t) = e^t\mathbf{i} + e^{-t}\mathbf{j} + \sqrt{2}t\mathbf{k};\ 0 \leq t \leq \ln 3.$

38. $\mathbf{r}(t) = \cos 2t\mathbf{i} + (1 - \cos 2t)\mathbf{j} + (3 + \frac{1}{2}\cos 2t)\mathbf{k};$
$0 \leq t \leq \pi.$

In Exercises 39–48, find the scalar tangential and normal components of acceleration at the indicated time t.

39. $\mathbf{r}(t) = 2\cos t\mathbf{i} + 2\sin t\mathbf{j};\ t = \pi/3.$

40. $\mathbf{r}(t) = t\mathbf{i} + t^2\mathbf{j};\ t = 1.$

41. $\mathbf{r}(t) = e^{-t}\mathbf{i} + e^t\mathbf{j};\ t = 0.$

42. $\mathbf{r}(t) = \cos (t^2)\mathbf{i} + \sin (t^2)\mathbf{j};\ t = \sqrt{\pi}/2.$

43. $\mathbf{r}(t) = (t^3 - 2t)\mathbf{i} + (t^2 - 4)\mathbf{j};\ t = 1.$

44. $\mathbf{r}(t) = e^t \cos t\mathbf{i} + e^t \sin t\mathbf{j};\ t = \pi/4.$

45. $\mathbf{r}(t) = t\mathbf{i} + t^2\mathbf{j} + t^3\mathbf{k};\ t = 1.$

46. $\mathbf{r}(t) = e^t\mathbf{i} + e^{-2t}\mathbf{j} + t\mathbf{k};\ t = 0.$

47. $\mathbf{r}(t) = 3\sin t\mathbf{i} + 2\cos t\mathbf{j} - \sin 2t\mathbf{k};\ t = \pi/2.$

48. $\mathbf{r}(t) = 2\mathbf{i} + t^3\mathbf{j} - 16\ln t\mathbf{k};\ t = 1.$

In Exercises 49–52, $\mathbf{v}$ and $\mathbf{a}$ are given at a certain instant of time. Find a_T, a_N, $\mathbf{T}$, and $\mathbf{N}$ at this instant.

49. $\mathbf{v} = -4\mathbf{j},\ \mathbf{a} = 2\mathbf{i} + 3\mathbf{j}.$

50. $\mathbf{v} = \mathbf{i} + 2\mathbf{j},\ \mathbf{a} = 3\mathbf{i}.$

51. $\mathbf{v} = 2\mathbf{i} + 2\mathbf{j} + \mathbf{k},\ \mathbf{a} = \mathbf{i} + 2\mathbf{k}.$

52. $\mathbf{v} = 3\mathbf{i} - 4\mathbf{k},\ \mathbf{a} = \mathbf{i} - \mathbf{j} + 2\mathbf{k}.$

In Exercises 53–56, the speed $\|\mathbf{v}\|$ of a particle at an arbitrary time t is given. Find the scalar tangential component of acceleration at the indicated time.

53. $\|\mathbf{v}\| = \sqrt{3t^2 + 4};\ t = 2.$

54. $\|\mathbf{v}\| = \sqrt{t^2 + e^{-3t}};\ t = 0.$

55. $\|\mathbf{v}\| = \sqrt{(4t - 1)^2 + \cos^2 \pi t};\ t = \frac{1}{4}.$

56. $\|\mathbf{v}\| = \sqrt{t^4 + 5t^2 + 3};\ t = 1.$

In Exercises 57–60, find the curvature of the path of motion of the particle at the instant at which the velocity and acceleration vectors are given.

57. Exercise 49.   **58.** Exercise 50.

59. Exercise 51.   **60.** Exercise 52.

61. The nuclear accelerator at the Enrico Fermi Laboratory is circular with a radius of 1 km. Find the scalar normal component of acceleration of a proton moving around the accelerator with a constant speed of 2.9×10^5 km/sec.

In Exercises 62–64, use the formula for $\kappa(x)$ in Exercise 15 of Section 15.5 to help solve the problem.

62. Suppose that a particle moves with nonzero acceleration along the curve $y = f(x)$. Show that the acceleration vector is tangent to the curve at each point where $f''(x) = 0$.

63. A particle moves along the parabola $y = x^2$ with a constant speed of 3 units/sec. Find the normal component of acceleration as a function of x.

64. A particle moves along the curve $y = e^x$ with a constant speed of 2 units/sec. Find the normal component of acceleration as a function of x.

65. A shell is fired from ground level with a muzzle speed of 320 ft/sec and elevation angle of 60°. Find

   (a) parametric equations for the shell's trajectory

   (b) the maximum height reached by the shell

   (c) the horizontal distance traveled by the shell

   (d) the speed of the shell at impact.

66. Solve Exercise 65 assuming that the muzzle speed is 980 m/sec and the elevation angle is 45°.

67. A rock is thrown downward from the top of a building, 168 ft high, at an angle of 60° with the horizontal. How far from the base of the building will the rock land if its initial speed is 80 ft/sec?

68. Solve Exercise 67 assuming that the rock is thrown horizontally at a speed of 80 ft/sec.

69. A shell is to be fired from ground level at an elevation angle of 30°. What should the muzzle speed be in order for the maximum height of the shell to be 2500 ft?

70. A shell, fired from ground level at an elevation angle of 45°, hits the ground 24,500 m away. Calculate the muzzle speed of the shell.

71. Find two elevation angles that will enable a shell, fired from ground level with a muzzle speed of 800 ft/sec, to hit a ground-level target 10,000 ft away.

72. A ball rolls off a table 4 ft high while moving at a constant speed of 5 ft/sec.

   (a) How long does it take for the ball to hit the floor after it leaves the table?

   (b) At what speed does the ball hit the floor?

   (c) If a ball were dropped from table height (initial speed 0) at the same time the rolling ball leaves the table, which ball would hit the ground first?

73. A shell is fired from ground level at an elevation angle of α and a muzzle speed of v_0.

   (a) Show that the maximum height reached by the shell is

   $$\text{maximum height} = \frac{(v_0 \sin \alpha)^2}{2g}$$

   (b) The *horizontal range R* of the shell is the horizontal distance traveled when the shell returns to ground level. Show that $R = (v_0^2 \sin 2\alpha)/g$. For what elevation angle will the range be maximum? What is the maximum range?

74. A shell is fired from ground level with an elevation angle α and a muzzle speed of v_0. Find the angle that should be used to hit a target at ground level that is at a distance of $3/4$ the maximum range of the shell. Express your answer to the nearest tenth of a degree. [*Hint:* See Exercise 73(b).]

75. At time $t = 0$ a baseball that is 5 ft above the ground is hit with a bat. The ball leaves the bat with a speed of 80 ft/sec at an angle of 30° above the horizontal.

   (a) How long will it take for the baseball to hit the ground? Express your answer to the nearest hundredth of a second.

   (b) Use the result in part (a) to find the horizontal distance traveled by the ball. Express your answer to the nearest tenth of a foot.

76. At time $t = 0$ a projectile is fired from a height h above level ground at an elevation angle of α with a speed v. Let R be the horizontal distance to the point where the projectile hits the ground.

   (a) Show that α and R must satisfy the equation

   $$g(\sec^2 \alpha)R^2 - 2v^2(\tan \alpha)R - 2v^2h = 0$$

   (b) If g, h, and v are constant, then the equation in part (a) defines R implicitly as a function of α. Let R_0 be the maximum value of R and α_0 the value of α when $R = R_0$. Use implicit differentiation to find $dR/d\alpha$ and show that

   $$\tan \alpha_0 = \frac{v^2}{gR_0}$$

   [*Hint:* Assume that $dR/d\alpha = 0$ when R is maximum.]

   (c) Use the results in parts (a) and (b) to show that

   $$R_0 = \frac{v}{g}\sqrt{v^2 + 2gh}$$

   and

   $$\alpha_0 = \tan^{-1} \frac{v}{\sqrt{v^2 + 2gh}}$$

■ **15.7** KEPLER'S LAWS OF PLANETARY MOTION

> *One of the great advances in the history of astronomy occurred in the early 1600s when Johannes Kepler* deduced from empirical data that all planets in our solar system move in elliptical orbits with the sun at a focus. Subsequently, Isaac Newton showed mathematically that such planetary motion is the consequence of an inverse-square law of gravitational attraction. In this section we shall use the concepts developed in the preceding sections of this chapter to derive three basic laws of planetary motion, known as **Kepler's laws**.*

□ **KEPLER'S LAWS**

In 1609 Johannes Kepler published a book known as *Astronomia Nova* (or sometimes as *Commentaries on the Motions of Mars*) in which he succeeded in distilling thousands of years of observational astronomy into three beautiful laws of planetary motion.

15.7.1   KEPLER'S LAWS.

- First law (***Law of Orbits***).   Each planet moves in an elliptical orbit with the sun at a focus.

- Second law (***Law of Areas***).   Equal areas are swept out in equal times by the line from the sun to a planet.

- Third law (***Law of Periods***).   The square of a planet's period (the time it takes the planet to complete one orbit about the sun) is proportional to the cube of the length of the semimajor axis of its elliptical orbit.

□ **CENTRAL FORCES**

To derive Kepler's laws, we shall assume that the force exerted by the sun on a planet is always directed toward the sun's center. In general, a force that is always directed toward a fixed point is called a ***central force***.

Suppose that a particle P of mass m moves with acceleration $\mathbf{a}$ under the influence of a central force $\mathbf{F}$ that is directed toward a fixed point O, and let $\mathbf{r} = \mathbf{r}(t)$ be the position vector for the particle from O to P (Figure 15.7.1). We will show that the particle moves in a *plane* containing O.

It follows from Newton's Second Law of Motion ($\mathbf{F} = m\mathbf{a}$) that $\mathbf{a}$ is in the same direction as $\mathbf{F}$, and consequently $\mathbf{a}$ and $\mathbf{r}$ are oppositely directed vectors. Thus, it follows from Theorem 14.4.3 that

$$\mathbf{r} \times \mathbf{a} = \mathbf{0}$$

Since the velocity and acceleration of the particle are given by $\mathbf{v} = d\mathbf{r}/dt$ and $\mathbf{a} = d\mathbf{v}/dt$, respectively, we have

$$\frac{d}{dt}(\mathbf{r} \times \mathbf{v}) = \mathbf{r} \times \frac{d\mathbf{v}}{dt} + \frac{d\mathbf{r}}{dt} \times \mathbf{v} = (\mathbf{r} \times \mathbf{a}) + (\mathbf{v} \times \mathbf{v}) = \mathbf{0} + \mathbf{0} = \mathbf{0}$$

Integrating the left and right sides of this equation with respect to t yields

$$\mathbf{r} \times \mathbf{v} = \mathbf{b} \tag{1}$$

where $\mathbf{b}$ is a constant vector. This implies that $\mathbf{r}$ is perpendicular to $\mathbf{b}$, and hence the particle moves along a path that lies in the plane that is perpendicular to $\mathbf{b}$ and contains the fixed point O. This shows that each planet moves in a plane through the center of the sun. Astronomers call this plane the ***ecliptic*** of the planet.

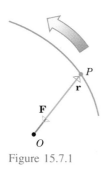

Figure 15.7.1

---

*See biography on p. 633.

We shall now derive Kepler's laws: first the Law of Areas, then the Law of Orbits, and finally the Law of Periods. In all of these derivations we shall assume that the sun and planets are homogeneous spheres, so that they can be treated as if their masses were concentrated at their centers. When we speak of the distance between the sun and a planet, we will mean the distance between their centers. Moreover, we will ignore the gravitational effects of all other celestial bodies on the sun and the planet.

☐ **KEPLER'S SECOND LAW**

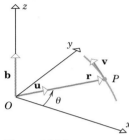

Figure 15.7.2

To establish Kepler's second law we introduce a Cartesian xyz-coordinate system with its origin O at the sun's center and its positive z-axis in the direction of $\mathbf{b} = \mathbf{r} \times \mathbf{v}$. Because $\mathbf{b}$, $\mathbf{r}$, and $\mathbf{v}$ are related by the right-hand rule, the polar angle θ from the positive x-axis to P increases with time t (Figure 15.7.2). It will be convenient to express $\mathbf{r}$ as $\mathbf{r} = r\mathbf{u}$, where $r = \|\mathbf{r}\|$ and

$$\mathbf{u} = \cos\theta\mathbf{i} + \sin\theta\mathbf{j} \tag{2}$$

is a unit vector in the direction of $\mathbf{r}$. It then follows that

$$\mathbf{v} = \frac{d\mathbf{r}}{dt} = \frac{d}{dt}(r\mathbf{u}) = r\frac{d\mathbf{u}}{dt} + \frac{dr}{dt}\mathbf{u}$$

and hence from (1)

$$\mathbf{b} = \mathbf{r} \times \mathbf{v} = (r\mathbf{u}) \times \left(r\frac{d\mathbf{u}}{dt} + \frac{dr}{dt}\mathbf{u}\right) = r^2\mathbf{u} \times \frac{d\mathbf{u}}{dt} + r\frac{dr}{dt}\mathbf{u} \times \mathbf{u} = r^2\mathbf{u} \times \frac{d\mathbf{u}}{dt} \tag{3}$$

But (2) implies that

$$\frac{d\mathbf{u}}{dt} = \frac{d\mathbf{u}}{d\theta}\frac{d\theta}{dt} = (-\sin\theta\mathbf{i} + \cos\theta\mathbf{j})\frac{d\theta}{dt}$$

so

$$\mathbf{u} \times \frac{d\mathbf{u}}{dt} = \frac{d\theta}{dt}\mathbf{k} \tag{4}$$

Substituting (4) in (3) yields

$$\mathbf{b} = r^2\frac{d\theta}{dt}\mathbf{k} \tag{5}$$

which shows that $r^2(d\theta/dt)$ must be constant because $\mathbf{b}$ is constant.

If the path of P is described in polar coordinates by an equation $r = f(\theta)$, then from Formula (2) of Section 13.3 the area A swept out by the radius vector as it varies from some initial angle θ_0 to an arbitrary angle θ can be expressed as a function of θ by

$$A = \int_{\theta_0}^{\theta} \frac{1}{2}[f(\phi)]^2 \, d\phi$$

where we have used ϕ as the variable of integration rather than θ, since θ is already being used as an independent variable in the upper limit of integration. Applying the Second Fundamental Theorem of Calculus (Theorem 5.9.1) and the chain rule we obtain

$$\frac{dA}{dt} = \frac{dA}{d\theta}\frac{d\theta}{dt} = \frac{1}{2}[f(\theta)]^2\frac{d\theta}{dt} = \frac{1}{2}r^2\frac{d\theta}{dt} \tag{6}$$

Thus dA/dt must be constant because $r^2(d\theta/dt)$ is constant. This shows that area is swept out at a constant rate and hence that equal areas are swept out in equal times by the line from the sun to the planet (Figure 15.7.3). It also follows from this that a planet moves more rapidly when it is closer to the sun.

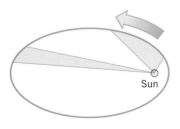

Sun

Equal areas are swept out in equal lengths of time.

Figure 15.7.3

☐ **KEPLER'S FIRST LAW**

Kepler's second law made no assumptions about the magnitude of the force $\mathbf{F}$. However, to obtain Kepler's first law we need *Newton's inverse-square law of gravitational attraction*, namely, that *every particle of matter in the universe attracts every other particle of matter with a force that is proportional to the product of their masses and inversely proportional to the square of the distance between them.* Thus, the central force **F** exerted by the sun on a planet can be expressed as

$$\mathbf{F} = -\frac{GmM}{r^2}\mathbf{u} \tag{7}$$

where m is the mass of the planet, M is the mass of the sun, G is a constant (called the **universal gravitational constant**), and $\mathbf{r} = r\mathbf{u}$ is the position vector from the center of the sun to the center of the planet. The negative sign in (7) indicates that the force **F** acts opposite to **r** (toward the sun's center).

It follows from (7) and Newton's Second Law of Motion ($\mathbf{F} = m\mathbf{a}$) that the acceleration of the planet is given by

$$\mathbf{a} = -\frac{GM}{r^2}\mathbf{u} \tag{8}$$

Observe that the acceleration does not depend on the mass of the planet.

It will be convenient to orient the coordinate axes so that at time $t = 0$ the planet is at its closest to the sun and on the positive x-axis (i.e., $\theta = 0$ if $t = 0$). Under this assumption, the velocity vector at time $t = 0$ is perpendicular to the radius vector and consequently the radius vector and the velocity vector at this time are of the form

$$\mathbf{r} = r_0\mathbf{i} \quad \text{and} \quad \mathbf{v} = v_0\mathbf{j} \tag{9}$$

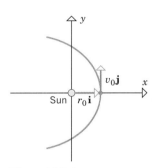

Figure 15.7.4

(Figure 15.7.4). It follows from this assumption that at time $t = 0$ we have

$$\mathbf{b} = \mathbf{r} \times \mathbf{v} = (r_0\mathbf{i}) \times (v_0\mathbf{j}) = r_0 v_0 \mathbf{k} \tag{10}$$

and hence from (5) that

$$r^2\frac{d\theta}{dt} = r_0 v_0 \tag{11}$$

It follows from (2), (5), and (8) that

$$\mathbf{a} \times \mathbf{b} = -\frac{GM}{r^2}(\cos\theta\mathbf{i} + \sin\theta\mathbf{j}) \times \left(r^2\frac{d\theta}{dt}\mathbf{k}\right)$$

$$= GM(-\sin\theta\mathbf{i} + \cos\theta\mathbf{j})\frac{d\theta}{dt}$$

$$= GM\frac{d\mathbf{u}}{dt}$$

From this formula and the fact that $d\mathbf{b}/dt = \mathbf{0}$ (since **b** is constant) and $d\mathbf{v}/dt = \mathbf{a}$, we obtain

$$\frac{d}{dt}(\mathbf{v} \times \mathbf{b}) = \mathbf{v} \times \frac{d\mathbf{b}}{dt} + \frac{d\mathbf{v}}{dt} \times \mathbf{b} = \mathbf{a} \times \mathbf{b} = GM\frac{d\mathbf{u}}{dt}$$

and integrating the left and right sides of this equation with respect to t yields

$$\mathbf{v} \times \mathbf{b} = GM\mathbf{u} + \mathbf{C} \tag{12}$$

where **C** is a vector constant of integration. We leave it for the reader to evaluate $\mathbf{v} \times \mathbf{b}$ at time $t = 0$ to show that

$$\mathbf{C} = (r_0 v_0^2 - GM)\mathbf{i} \tag{13}$$

Next, we take the dot product of both sides of (12) with **r** to obtain

$$\mathbf{r} \cdot (\mathbf{v} \times \mathbf{b}) = GM\mathbf{r} \cdot \mathbf{u} + \mathbf{r} \cdot \mathbf{C} \tag{14}$$

But

$$\mathbf{r} \cdot (\mathbf{v} \times \mathbf{b}) = (\mathbf{r} \times \mathbf{v}) \cdot \mathbf{b} = \mathbf{b} \cdot \mathbf{b} = r_0^2 v_0^2$$

$$\mathbf{r} \cdot \mathbf{u} = (r\mathbf{u}) \cdot \mathbf{u} = r(\mathbf{u} \cdot \mathbf{u}) = r$$

$$\mathbf{r} \cdot \mathbf{C} = r(\cos\theta\mathbf{i} + \sin\theta\mathbf{j}) \cdot (r_0 v_0^2 - GM)\mathbf{i} = r(r_0 v_0^2 - GM)\cos\theta$$

so (14) can be written as

$$r_0^2 v_0^2 = GMr + r(r_0 v_0^2 - GM)\cos\theta$$

which when solved for r gives

$$r = \frac{r_0^2 v_0^2}{GM + (r_0 v_0^2 - GM)\cos\theta} = \frac{\dfrac{r_0^2 v_0^2}{GM}}{1 + \left(\dfrac{r_0 v_0^2}{GM} - 1\right)\cos\theta} \tag{15}$$

or simply

$$r = \frac{d}{1 + e\cos\theta} \tag{16}$$

where

$$d = \frac{r_0^2 v_0^2}{GM} \quad\text{and}\quad e = \frac{r_0 v_0^2}{GM} - 1 \tag{17–18}$$

From (16) and the fact that r is smallest when $\theta = 0$, it follows that $e \geq 0$ (why?), so Equation (16) is the polar form of a conic with eccentricity e and a focus at the origin (Section 13.2). This implies that the path of an object subjected only to the sun's gravitational force is a circle if $e = 0$, an ellipse if $0 < e < 1$, a parabola if $e = 1$, or a hyperbola if $e > 1$ (Figure 15.7.5). As indicated by (18), the value of e depends on r_0 and v_0, so at the time the planets in our solar system were created those position and velocity vectors were such that elliptical orbits resulted.

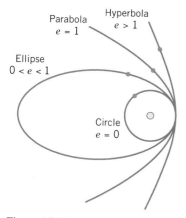

Figure 15.7.5

□ **KEPLER'S THIRD LAW**

To derive Kepler's third law, we let a and b be the semimajor and semiminor axes of the planet's elliptical orbit, and we recall that the area of such an ellipse is πab. It follows from (6) and (11) that the radius vector from the sun to a planet sweeps out area at the constant rate of $\frac{1}{2} r_0 v_0$ (units of area per unit of time). Thus, if one revolution around the sun is completed in T units of time (the period), then the area swept out during that revolution is

$$\pi ab = \tfrac{1}{2} r_0 v_0 T \tag{19}$$

The vertices of a planet's elliptical orbit are reached when $\theta = 0$ and when $\theta = \pi$. When $\theta = 0$ the planet is at its closest point to the sun, and when $\theta = \pi$ it is at its farthest point. If we denote the distances from the planet to the center of the sun at those points by $r_{\min}$ and $r_{\max}$, respectively, then substituting $\theta = 0$ and $\theta = \pi$ in (16) yields

$$r_{\min} = \frac{d}{1 + e} \quad\text{and}\quad r_{\max} = \frac{d}{1 - e}$$

so the length of the major axis is

$$2a = r_{\min} + r_{\max} = \frac{d}{1 + e} + \frac{d}{1 - e} = \frac{2d}{1 - e^2}$$

and hence

$$a = \frac{d}{1 - e^2} \tag{20}$$

Also, because $b^2 = a^2 - c^2$ and $e = c/a$ for an ellipse (Section 12.3) and $d = a(1 - e^2)$ from (20), it follows that $b^2 = a^2(1 - e^2) = ad$; thus, from (17) and (19)

$$T^2 = \frac{4\pi^2 a^2 (ad)}{r_0^2 v_0^2} = \frac{4\pi^2 a^3}{r_0^2 v_0^2} d = \frac{4\pi^2 a^3}{r_0^2 v_0^2} \frac{r_0^2 v_0^2}{GM} = \frac{4\pi^2}{GM} a^3 \tag{21}$$

which establishes Kepler's third law in which $4\pi^2/GM$ is the constant of proportionality.

□ **ARTIFICIAL SATELLITES**

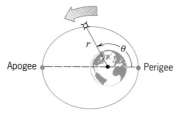

Figure 15.7.6

Kepler's laws apply to all celestial bodies subjected to a central gravitational force; we need only interpret the value of M in (7) to be the mass of the body exerting the central force and m to be the mass of the body subjected to the force. In particular, Kepler's laws apply to artificial satellites under the central force of the earth's gravity. For elliptical orbits about the sun, the point where the distance from the planet to the sun's center is minimum is called the *perihelion*, and the point where the distance is maximum is called the *aphelion*. We will refer to the actual distances as the *perihelion and aphelion distances*. For elliptical orbits about the earth, the minimum and maximum distances to the earth's center occur at points called the *perigee* and *apogee*, respectively (Figure 15.7.6), and we will refer to the actual distances as the *perigee and apogee distances*. We will refer to the distances to the *surface* of the earth as the *perigee and apogee altitudes*.

For orbits about the sun it is often convenient to measure distances in *astronomical units*, where one astronomical unit (AU) is the length of the semimajor axis a of the earth's orbit (approximately 150×10^6 km or 92.9×10^6 miles). From Kepler's third law the ratio T^2/a^3 is constant for all elliptical orbits about a central point with a given mass M. Since the period T of rotation of the earth about the sun is 1 year and $a = 1$ AU for the earth's orbit, we find that $T^2/a^3 = 1$, so that

$$T^2 = a^3 \tag{22}$$

for all elliptical orbits about the sun, provided that T is measured in years and a in astronomical units.

Example 1   The value of a for the elliptical orbit of Halley's comet (last seen in 1986) about the sun is approximately 18.1 AU. Find the period of its orbit in years.

Solution.   We are given that $a = 18.1$, so from (22)

$$T = a^{3/2} = (18.1)^{3/2} \approx 77 \text{ years} \qquad \blacktriangleleft$$

Example 2   The eccentricity of the orbit of Halley's comet is approximately 0.97. Find the perihelion and the aphelion distances from the sun in miles.

Solution.   The perihelion and aphelion distances, $r_{\min}$ and $r_{\max}$, can be obtained from the formulas (see Exercise 1)

$$r_{\min} = a(1 - e) \quad \text{and} \quad r_{\max} = a(1 + e)$$

From Example 1 $a = 18.1$, so the perihelion distance is

$$r_{\min} = a(1 - e) = 18.1(1 - 0.97) \approx 0.54 \text{ AU} \approx 5 \times 10^7 \text{ miles}$$

and the aphelion distance is

$$r_{\max} = a(1 + e) = 18.1(1 + 0.97) \approx 35.7 \text{ AU} \approx 3.3 \times 10^9 \text{ miles} \qquad \blacktriangleleft$$

▶ **Exercise Set 15.7**   ⃞C 3, 6–15

In exercises that require numerical calculations, use the following values:

>   radius of the earth $= 6370$ km $= 3960$ mi
>   1 AU $= 150 \times 10^6$ km $= 92.9 \times 10^6$ mi
> GM (for the earth) $= 3.99 \times 10^5$ km^3/sec^2
> $= 1.238 \times 10^{12}$ mi^3/hr^2

1. Using the notation in Example 2, show that $r_{\min} = a(1 - e)$ and $r_{\max} = a(1 + e)$.

2. (a)   Use the results in Exercise 1 to show that
$$e = \frac{r_{\max} - r_{\min}}{r_{\max} + r_{\min}}$$
   (b)   Show that

$$r_{max} = r_{min} \frac{1 + e}{1 - e}$$

3. (a) Use Formula (18) to show that

$$v_0 = \sqrt{\frac{2GM}{r_0}}$$

is the minimum speed needed when $\theta = 0$ so that an object will escape from the pull of a central force due to mass M.

(b) Use the result in part (a) to find the minimum speed needed when $\theta = 0$ for a space probe at an altitude of 300 km above the surface of the earth to escape from the gravitational pull of the earth.

4. If **v** is the velocity of an object at any point in its orbit, then $v = \|\mathbf{v}\|$ is its speed.

(a) Show that

$$v = \frac{v_0}{1 + e} \sqrt{e^2 + 2e \cos \theta + 1}$$

[*Hint:* Use (12) along with (2), (10), (13), and (18). Note that $\|\mathbf{v} \times \mathbf{b}\| = \|\mathbf{v}\| \|\mathbf{b}\|$ because **v** and **b** are perpendicular.]

(b) Use the result in part (a) to find the speed of an object when it reaches an end of the minor axis of an elliptical orbit.

5. Use the result in part (a) of Exercise 4 to show that for an object moving in an elliptical orbit

$$v_{max} = v_{min} \frac{1 + e}{1 - e}$$

where v_{min} and v_{max} are the minimum and maximum speeds of the object respectively.

6. (a) Use the result in part (a) of Exercise 4, and (18), to show that for an object moving in a circular orbit of radius r_0 the speed v is constant and

$$v = v_0 = \sqrt{\frac{GM}{r_0}}$$

(b) Use the result in part (a) to find the speed of a spacecraft in a circular orbit about the earth at an altitude of 200 km above the surface of the earth.

7. (a) A *geosynchronous orbit* is a circular orbit about the equator of the earth in which an object appears to remain stationary over some point on the equator. Find the altitude in miles of a communications satellite that is in geosynchronous orbit about the earth. [*Hint:* The earth makes one revolution about its axis of rotation in 24 hours.]

(b) Use the result in part (a) of Exercise 6 to find the speed in miles per hour of a satellite that is in geosynchronous orbit about the earth.

8. (a) Assume that T is measured in days and a in kilometers (km). Find the value of T^2/a^3 for planets that orbit the sun. [*Hint:* The period of the earth's orbit is about 365 days.]

(b) Use the result in part (a) to find the period in days for the orbit of the planet Mercury, given that the length of its semimajor axis is 57.95×10^6 km.

9. Find the period in years for the orbit of the planet Pluto, given that $a = 39.5$ AU.

10. (a) The eccentricity of the moon's orbit about the earth is 0.055, and the length of the semimajor axis is about 238,900 miles. Find the perigee and apogee distances of the moon from the earth.

(b) Find the period in days of the moon's orbit.

11. (a) Vanguard 1 was launched in March 1958 with perigee and apogee altitudes above the earth of 649 km and 4340 km, respectively. Find the length of the semimajor axis of its orbit.

(b) Use the result in part (a) of Exercise 2 to find the eccentricity of its orbit.

(c) Find its period of rotation in minutes.

12. (a) Intelsat 5 was launched in December 1980 with perigee and apogee altitudes above the earth of 35,143 km and 35,707 km, respectively. Find the length of the semimajor axis of its orbit.

(b) Use the result in part (a) of Exercise 2 to find the eccentricity of its orbit.

(c) Find its period of rotation in hours.

13. (a) Suppose that a space probe is in a circular orbit at an altitude of 180 miles above the surface of the earth. Use the result in part (a) of Exercise 6 to find its speed.

(b) During a very short period of time, a thruster rocket on the space probe is fired to increase the speed of the probe by 600 mi/hr in its direction of motion. Find the eccentricity of the resulting elliptical orbit, and use the result in part (b) of Exercise 2 to find the apogee altitude.

14. Suppose that the perigee altitude of an earth satellite is 600 km and that the period is 100 min. Find the eccentricity of the orbit and the apogee altitude.

15. (a) Use Formula (20) to show that (16) can be expressed as

$$r = \frac{a(1 - e^2)}{1 + e \cos \theta}$$

(b) Use the result in part (a) to find an equation of the orbit of Mercury given that $a = 57.9 \times 10^6$ km, $e = 0.206$.

◆ TECHNOLOGY EXERCISES   Chapter 15

Most of these exercises require access to a graphing calculator or a computer algebra system (CAS) such as *Mathematica*, *Maple*, or *Derive*. When you are asked to *find* an answer or to *solve* an equation, you may choose to find an exact result or a numerical approximation, depending on the particular technology you are using and on your own imagination. The form of your answers may differ from those of other students or from those in the answer section of the text, depending on how you solve the problems and the accuracy you use in your numerical approximations. Those exercises that are more appropriate for a CAS than a graphing calculator are labeled with the icon ◆.

1. **Distance from a point to a curve:** Find the shortest distance between the point $(1, 2, 1)$ and a point on the curve

$$\mathbf{r}(t) = \frac{1}{t}\mathbf{i} + \ln t\mathbf{j} + \sqrt{t}\mathbf{k}$$

2. **Distance from a point to a curve:** Find the maximum and minimum distances from the point $(1, 2, -1)$ to a point on the curve of intersection of the plane $z = y/2$ and the ellipsoid $x^2/4 + y^2/9 + z^2/4 = 1$. [*Suggestion:* Find parametric equations for the curve of intersection.]

3. **Distance between moving particles:** Given that two particles moving in 3-space have equations of motion $x = 2\cos t$, $y = 3\sin t$, $z = t$ $(t \geq 0)$ and $x = t$, $y = t^2$, $z = t^3$ $(t \geq 0)$, how close do the particles get to one another and when are they closest? (Assume that x, y, and z are in feet and t is in minutes.)

4. **Speed of a particle:** Suppose that the equation of motion of a moving particle is given by

$$\mathbf{r}(t) = (\sin t + \cos t)\mathbf{i} + (\sin t - t)\mathbf{j}$$

where $\mathbf{r}$ is in meters and t is in seconds.

(a) Graph $\mathbf{r}(t)$ for $0 \leq t \leq 2\pi$.

(b) Find the maximum speed of the particle over the interval $0 \leq t \leq 2\pi$.

5. **Maximum temperature along a curve:** Suppose that the temperature T at a point (x, y, z) is given by $T = 30ze^{-x^2-2y^2}$. Find the maximum value of T along the curve $x = t\cos t$, $y = t$, $z = e^{-t}$.

◆ 6. **Trajectory of a ski jump:** At time $t = 0$ a skier leaves the end of a ski jump with a speed of v_0 ft/sec at an angle α with the horizontal (see the accompanying figure). The skier lands 259 ft down the incline 2.9 sec later. Use (28) of Section 15.6 to find

(a) v_0 and α

(b) the distance traveled by the skier.

(Use $g = 32$ ft/sec^2 as the acceleration due to gravity.)

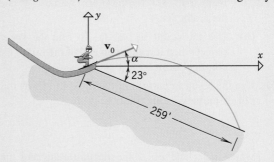

◆ 7. **Tangential and normal components of acceleration:** Suppose that a particle moves through 3-space so that its position vector at time t is

$$\mathbf{r}(t) = t\cos t\mathbf{i} + t\sin t\mathbf{j} + 3t\mathbf{k}, \quad t \geq 0$$

where $\mathbf{r}$ is in meters and t is in seconds.

(a) Show that the particle moves on a circular cone.

(b) Find, to the nearest degree, the angle between the velocity and acceleration vectors when $t = 1.5$.

(c) Find the scalar tangential and normal components of acceleration when $t = 1.5$.

◆ 8. **Curvature:** Consider the curve

$$\mathbf{r}(t) = 3\cos t\mathbf{i} + 4\sin t\mathbf{j} + \sin 2t\mathbf{k}$$

(a) Show that the curve lies on an elliptic cylinder.

(b) Find the maximum value and the minimum value of the curvature.

◆   9. **Path along a hill:** Suppose that a hill has the shape of the circular paraboloid $z = 1 - x^2 - y^2$, where x, y, and z are in miles. At time $t = 0$ a hiker at the point $(1, 0, 0)$ starts to walk up the hill with equations of motion given by

$$x = e^{-0.2t} \cos t, \ y = e^{-0.2t} \sin t, \ z = 1 - e^{-0.4t}$$

where t is in hours.

(a) Verify that the path of motion lies on the hill.

(b) Find, to the nearest tenth of a degree, the acute angle that the tangent line to the path makes with the xy-plane at time $t = 1$.

(c) How long does it take the hiker to travel 2 miles along the path?

◆   10. **An amusement park ride:** An amusement park ride has cars that move along a track as shown in the accompanying figure. The cars are attached to arms that are 20 feet long and are connected to a common point on the axis of rotation (the z-axis). The arms swing up and down as the cars move along a track whose elevation is given by $z = 4 \sin^2 \theta$. Assuming that all distances are measured in feet, find the length of the track. [*Suggestion:* Convert to spherical coordinates and use the formula for arc length in Exercise 34, Section 15.3, with θ as the parameter.]

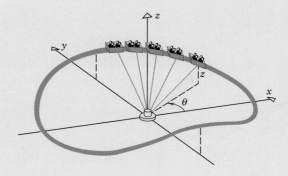

$\mathcal{O}.\,\mathcal{L}.\,\text{Lagrange}$

Joseph Louis Lagrange (1736–1813)

16 PARTIAL DERIVATIVES

16.1 FUNCTIONS OF TWO OR MORE VARIABLES

In previous sections we studied real-valued functions of a real variable and vector-valued functions of a real variable. In this section we shall consider real-valued functions of two or more real variables.

☐ **NOTATION AND TERMINOLOGY**

There are many familiar formulas in which a given variable depends on two or more other variables. For example, the area A of a triangle depends on the base length b and height h by the formula $A = \frac{1}{2}bh$; the volume V of a rectangular box depends on the length l, the width w, and the height h by the formula $V = lwh$; and the arithmetic average $\bar{x}$ of n real numbers, $x_1, x_2, \ldots, x_n$, depends on those numbers by the formula $\bar{x} = (x_1 + x_2 + \cdots + x_n)/n$. Thus, we say that

A is a function of the two variables b and h;

V is a function of the three variables l, w, and h;

$\bar{x}$ is a function of the n variables $x_1, x_2, \ldots, x_n$.

The terminology and notation for functions of two or more variables is similar to that used for functions of one variable. For example, the expression

$$z = f(x, y)$$

means that z is a function of x and y in the sense that a unique value of the *dependent variable* z is determined by specifying values for the *independent variables* x and y. Similarly,

$$w = f(x, y, z)$$

expresses w as a function of x, y, and z, and

$$u = f(x_1, x_2, \ldots, x_n)$$

expresses u as a function of $x_1, x_2, \ldots, x_n$.

The functional relationship $z = f(x, y)$ has a useful geometric interpretation. When values of the independent variables x and y are specified, a point (x, y) in the xy-plane is determined. Thus, the dependent variable z may be viewed as a numerical value associated with the point (x, y). Similarly, the functional relationship $w = f(x, y, z)$ associates the numerical value w with the point (x, y, z) in 3-space.

The following definitions summarize this discussion.

16.1.1 DEFINITION.   A *function f of two real variables*, x and y, is a rule that assigns a unique real number $f(x, y)$ to each point (x, y) in some set D of the xy-plane.

16.1.2 DEFINITION.   A *function f of three variables*, x, y, and z, is a rule that assigns a unique real number $f(x, y, z)$ to each point (x, y, z) in some set D of three-dimensional space.

The set D in these definitions is the *domain* of the function; it is the set of points at which the function is defined. If a function f is specified by a formula and the domain of f is not stated explicitly, then it is understood that the domain consists of all points at which the formula has no divisions by zero and produces only real numbers; this is called the *natural domain* of the function.

REMARK.   In more advanced courses the notion of "n-dimensional space" for $n > 3$ is defined, and a *function f of n real variables*, $x_1, x_2, \ldots, x_n$, is regarded as a rule that assigns a unique real number $f(x_1, x_2, \ldots, x_n)$ to each "point" $(x_1, x_2, \ldots, x_n)$ in some set of n-dimensional space. However, we shall not pursue that idea in this text.

Example 1   Let

$$f(x, y) = 3x^2\sqrt{y} - 1$$

Find $f(1, 4)$, $f(0, 9)$, $f(t^2, t)$, $f(ab, 9b)$, and the natural domain of f.

Solution.   By substitution

$$f(1, 4) = 3(1)^2\sqrt{4} - 1 = 5$$
$$f(0, 9) = 3(0)^2\sqrt{9} - 1 = -1$$
$$f(t^2, t) = 3(t^2)^2\sqrt{t} - 1 = 3t^4\sqrt{t} - 1$$
$$f(ab, 9b) = 3(ab)^2\sqrt{9b} - 1 = 9a^2b^2\sqrt{b} - 1$$

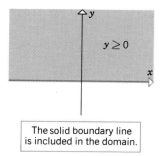

The solid boundary line is included in the domain.

Figure 16.1.1

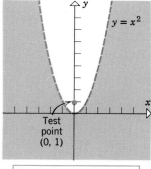

$y = x^2$

Test point (0, 1)

The dashed boundary does not belong to the domain.

Figure 16.1.2

Because of the $\sqrt{y}$, we must have $y \geq 0$ to avoid imaginary values for $f(x, y)$. Thus, the natural domain of f consists of all points in the xy-plane that are on or above the x-axis. (See Figure 16.1.1.) ◄

Example 2 Sketch the natural domain of the function $f(x, y) = \ln(x^2 - y)$.

Solution. $\ln(x^2 - y)$ is defined only when $0 < x^2 - y$ or $y < x^2$. To sketch this region, we use the fact that the curve $y = x^2$ separates the region where $y < x^2$ from the region where $y > x^2$. To determine the region where $y < x^2$ holds, we can select an arbitrary "test point" off the boundary $y = x^2$ and determine whether $y < x^2$ or $y > x^2$ at the test point. For example, if we choose the test point $(x, y) = (0, 1)$, then $x^2 = 0$, $y = 1$, so that this point lies in the region where $y > x^2$. Thus, the region where $y < x^2$ is the one that does *not* contain the test point (Figure 16.1.2). ◄

Example 3 Let
$$f(x, y, z) = \sqrt{1 - x^2 - y^2 - z^2}$$
Find $f(0, \frac{1}{2}, -\frac{1}{2})$ and the natural domain of f.

Solution. By substitution,
$$f(0, \tfrac{1}{2}, -\tfrac{1}{2}) = \sqrt{1 - (0)^2 - (\tfrac{1}{2})^2 - (-\tfrac{1}{2})^2} = \sqrt{\tfrac{1}{2}}$$

Because of the square root sign, we must have $0 \leq 1 - x^2 - y^2 - z^2$ in order to have a real value for $f(x, y, z)$. Rewriting this inequality in the form
$$x^2 + y^2 + z^2 \leq 1$$

we see that the natural domain of f consists of all points on or within the sphere $x^2 + y^2 + z^2 = 1$. ◄

☐ **GRAPHS OF FUNCTIONS OF TWO VARIABLES**

Recall that for a function f of one variable, the graph of $f(x)$ in the xy-plane was defined to be the graph of the equation $y = f(x)$. Similarly, if f is a function of two variables, we define the **graph** of $f(x, y)$ in xyz-space to be the graph of the equation $z = f(x, y)$. In general, such a graph will be a surface in 3-space.

Example 4 Describe the graph of the function $f(x, y) = 1 - x - \frac{1}{2}y$ in xyz-space.

Solution. By definition, the graph of the given function is the graph of the equation
$$z = 1 - x - \tfrac{1}{2}y \quad \text{or equivalently,} \quad x + \tfrac{1}{2}y + z = 1$$

which is a plane. A triangular portion of the plane can be sketched by plotting the intersections with the coordinate axes and joining them with line segments (Figure 16.1.3). ◄

Example 5 Sketch the graphs of the following functions in xyz-space.

(a) $f(x, y) = \sqrt{1 - x^2 - y^2}$ (b) $f(x, y) = -\sqrt{x^2 + y^2}$

Solution (a). By definition, the graph of the given function is the graph of the equation
$$z = \sqrt{1 - x^2 - y^2} \tag{1}$$

After squaring both sides, this can be rewritten as
$$x^2 + y^2 + z^2 = 1$$

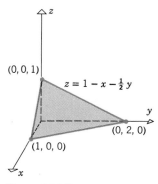

$z = 1 - x - \frac{1}{2}y$

$(0, 0, 1)$

$(0, 2, 0)$

$(1, 0, 0)$

Figure 16.1.3

which represents a sphere of radius 1, centered at the origin. Since (1) imposes the added condition that $z \geq 0$, the graph is just the upper hemisphere (Figure 16.1.4).

Solution (b).   The graph of the given function is the graph of the equation

$$z = -\sqrt{x^2 + y^2} \tag{2}$$

After squaring, we obtain

$$z^2 = x^2 + y^2$$

which is the equation of a right-circular cone [(9) of Section 14.7]. Since (2) imposes the condition that $z \leq 0$, the graph is just the lower nappe of the cone (Figure 16.1.5).   ◄

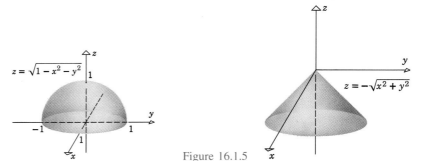

Figure 16.1.4   Figure 16.1.5

☐ **LEVEL CURVES**

When it is necessary to introduce a dependent variable for a function $f(x, y, z)$ of three variables, we shall usually use the letter w and write $w = f(x, y, z)$. Because this equation involves four variables, it cannot be graphed in three dimensions—''four dimensions'' are needed. Thus, there is no *direct* way to represent a function of three or more variables geometrically. However, we shall now discuss some methods for representing functions geometrically that can be applied to functions of three variables.

We are all familiar with topographic (or contour) maps in which a three-dimensional landscape, such as a mountain range, is represented by two-dimensional contour lines or curves of constant elevation. Consider, for example, the model hill and its contour map shown in Figure 16.1.6. The contour map is constructed by passing planes of constant elevation through the hill, projecting the resulting contours onto a flat surface, and labeling the contours with their elevations. In Figure 16.1.6, note how the two gullies appear as indentations in the contour lines and how the curves are close together on the contour map where the hill has a steep slope and become more widely spaced where the slope is gradual.

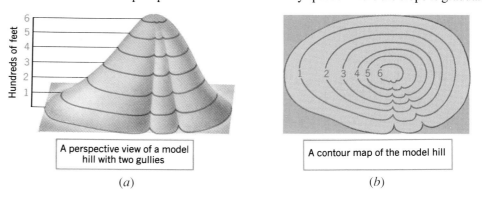

A perspective view of a model hill with two gullies   A contour map of the model hill

Figure 16.1.6   (a)   (b)

Contour maps are useful for studying functions of two variables. If the surface $z = f(x, y)$ is cut by the horizontal plane $z = k$, then for points on the intersection we have $f(x, y) = k$. The projection of this intersection onto the xy-plane is called the ***level curve of height k*** or the ***level curve with constant k*** (Figure 16.1.7).

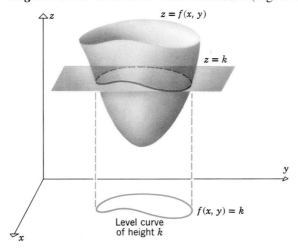

Figure 16.1.7

Example 6   The graph of the function $f(x, y) = x^2 + \frac{1}{4}y^2$ in xyz-space is the elliptic paraboloid shown in Figure 16.1.8a. The level curves have equations of the form

$$x^2 + \tfrac{1}{4}y^2 = k \tag{3}$$

For $k > 0$ these are ellipses; for $k = 0$ it is the single point $(0, 0)$; and for $k < 0$ there are no level curves, since (3) is not satisfied by any real values of x and y. Some sample level curves are shown in Figure 16.1.8b.   ◀

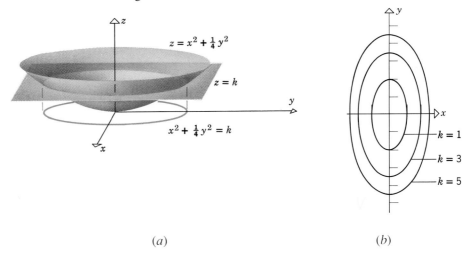

Figure 16.1.8   (a)   (b)

Example 7   The graph of the function $f(x, y) = 2 - x - y$ in xyz-space is the plane shown in Figure 16.1.9a. The level curves have equations of the form $2 - x - y = k$, or $y = -x + (2 - k)$. These form a family of parallel lines each of which has a slope of -1 (Figure 16.1.9b).   ◀

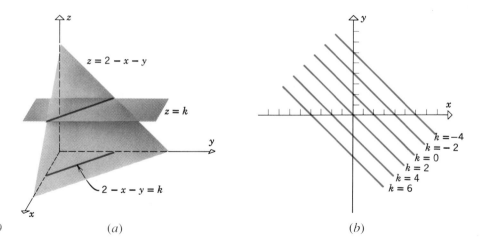

Figure 16.1.9   (a)   (b)

Example 8   The graph of the function $f(x, y) = y^2 - x^2$ in *xyz*-space is the hyperbolic paraboloid (saddle curve) shown in Figure 16.1.10*a*. The level curves have equations of the form $y^2 - x^2 = k$. For $k > 0$ these curves are hyperbolas opening along lines parallel to the *y*-axis; for $k < 0$ they are hyperbolas opening along lines parallel to the *x*-axis; and for $k = 0$ the level curve consists of the intersecting lines $y + x = 0$ and $y - x = 0$ (Figure 16.1.10*b*).   ◄

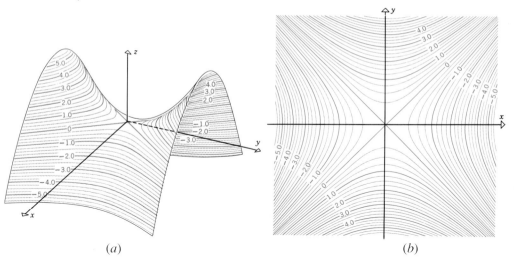

Figure 16.1.10   (a)   (b)

☐ **LEVEL SURFACES**

The concept of a level curve for a function of two variables can be extended to functions of three variables. If *k* is a constant, then an equation of the form $f(x, y, z) = k$ will, in general, represent a surface in three-dimensional space (e.g., $x^2 + y^2 + z^2 = 1$ represents a sphere). The graph of this surface is called the ***level surface with constant k*** for the function *f*.

REMARK.   The term "level surface" can be confusing. A level surface need *not* be level in the sense of being horizontal. It is simply a surface on which all values of *f* are the same.

Example 9   Describe the level surfaces of

(a) $f(x, y, z) = x^2 + y^2 + z^2$ (b) $f(x, y, z) = z^2 - x^2 - y^2$

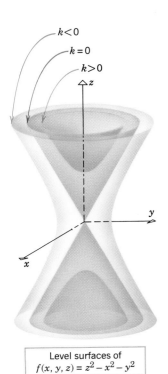

Level surfaces of
$f(x, y, z) = z^2 - x^2 - y^2$

Figure 16.1.12

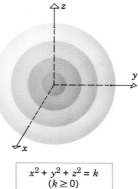

$x^2 + y^2 + z^2 = k$
$(k \geq 0)$

Figure 16.1.11

Solution (*a*). The level surfaces have equations of the form

$$x^2 + y^2 + z^2 = k$$

For $k > 0$ the graph of this equation is a sphere of radius $\sqrt{k}$, centered at the origin; for $k = 0$ the graph is the single point $(0, 0, 0)$; and for $k < 0$ there is no level surface (Figure 16.1.11).

Solution (*b*). The level surfaces have equations of the form

$$z^2 - x^2 - y^2 = k$$

As discussed in Section 14.7, this equation represents a cone if $k = 0$, a hyperboloid of two sheets if $k > 0$, and a hyperboloid of one sheet if $k < 0$ (Figure 16.1.12). ◄

☐ **OPEN AND CLOSED SETS**

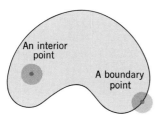

Figure 16.1.13

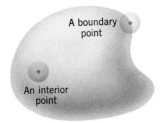

Figure 16.1.14

In our study of functions of one variable, the domains of the functions we encountered were generally intervals. For functions of two or three variables the situation is more complicated, so we shall need to discuss some terminology about sets in 2-space and 3-space that will be helpful when we want to accurately describe the domain of a function of two or three variables.

If D is a set of points in 2-space, then a point (x_0, y_0) is called an ***interior point*** of D if there is *some* circular disk with positive radius, centered at (x_0, y_0), and containing only points in D (Figure 16.1.13). A point (x_0, y_0) is called a ***boundary point*** of D if *every* circular disk with positive radius and centered at (x_0, y_0) contains both points in D and points not in D (Figure 16.1.13). Similarly, if D is a set of points in 3-space, then a point (x_0, y_0, z_0) is called an ***interior point*** of D if there is some spherical ball with positive radius, centered at (x_0, y_0, z_0), and containing only points in D (Figure 16.1.14). A point (x_0, y_0, z_0) is called a ***boundary point*** of D if *every* spherical ball with positive radius and centered at (x_0, y_0, z_0) contains both points in D and points not in D (Figure 16.1.14).

For a set D in either 2-space or 3-space, the set of all boundary points of D is called the ***boundary*** of D and the set of all interior points of D is called the ***interior*** of D.

Recall that an open interval (a, b) on a coordinate line contains *neither* of its endpoints and a closed interval $[a, b]$ contains *both* of its endpoints. Analogously, a set D in 2-space or 3-space is called ***open*** if it contains *none* of its boundary points and ***closed*** if it contains *all* of its boundary points. The set D of all points in 2-space has no boundary; it is regarded as both open and closed. Similarly, the set D of all points in 3-space is both open and closed.

Example 10   Let D be the set of points in the xy-plane that are inside or on the circle of radius 1 centered at the origin. The set D, its interior I, and its boundary B can be expressed in set notation as

$$D = \{(x, y) : x^2 + y^2 \leq 1\}, \quad I = \{(x, y) : x^2 + y^2 < 1\}, \quad B = \{(x, y) : x^2 + y^2 = 1\}$$

respectively (Figure 16.1.15). The set D is closed and the set I is open.   ◄

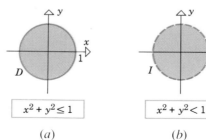
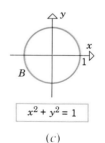

Figure 16.1.15

$$\boxed{x^2 + y^2 \leq 1}$$
$$\boxed{x^2 + y^2 < 1}$$
$$\boxed{x^2 + y^2 = 1}$$

(a)   (b)   (c)

☐ **BOUNDED SETS**

Just as we distinguished between finite intervals and infinite intervals on the real line, so we shall want to distinguish between regions of "finite extent" and regions of "infinite extent" in 2-space and 3-space. A set of points in 2-space is called **bounded** if the entire set can be contained within some rectangle, and is called **unbounded** if there is no rectangle that contains all the points of the set. Similarly, a set of points in 3-space is **bounded** if the entire set can be contained within some box, and is unbounded otherwise (Figure 16.1.16).

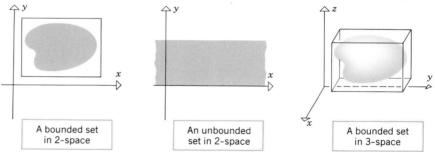

Figure 16.1.16

A bounded set in 2-space   An unbounded set in 2-space   A bounded set in 3-space

☐ **PARAMETRIC REPRESENTATION OF SURFACES**

We have seen that curves in 3-space can be represented parametrically by three equations involving one parameter. Similarly, surfaces in 3-space can be represented by three equations involving two parameters, say u and v, as

$$x = f(u, v), \quad y = g(u, v), \quad z = h(u, v)$$

or by a single vector-valued function

$$\mathbf{r}(u, v) = x\mathbf{i} + y\mathbf{j} + z\mathbf{k} = f(u, v)\mathbf{i} + g(u, v)\mathbf{j} + h(u, v)\mathbf{k}$$

We can view $\mathbf{r}(u, v) = x\mathbf{i} + y\mathbf{j} + z\mathbf{k}$ as a **radius vector** from the origin to a point (x, y, z) that moves over the surface as u and v vary (Figure 16.1.17).

Example 11   Consider the portion of the paraboloid $z = 4 - x^2 - y^2$ that lies in the first octant (Figure 16.1.18). We can obtain a parametric representation of this surface by letting $x = u$ and $y = v$, from which it follows that $z = 4 - u^2 - v^2$. Thus, the paraboloid can be represented parametrically as

$$x = u, \quad y = v, \quad z = 4 - u^2 - v^2$$

or in vector form as

$$\mathbf{r}(u, v) = u\mathbf{i} + v\mathbf{j} + (4 - u^2 - v^2)\mathbf{k}$$

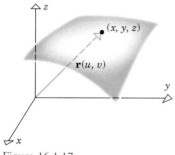

Figure 16.1.17

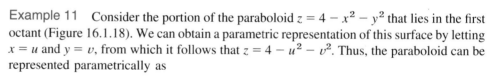

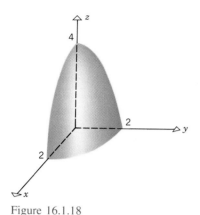

Figure 16.1.18

where u and v satisfy

$$u^2 + v^2 \le 4, \quad u \ge 0, \quad \text{and} \quad v \ge 0 \tag{4}$$

(These conditions ensure that $x^2 + y^2 \le 4$, $x \ge 0$, and $y \ge 0$, which gives the portion of the paraboloid in the first octant, rather than the entire paraboloid.) ◄

If we think of (u, v) as a point that varies over some region in a uv-plane, then a vector function $\mathbf{r}(u, v)$ associates points on a surface with points in that region, just as $\mathbf{r}(t)$ associates points on a curve with points in an interval on the t-axis. If u is allowed to change, while v is kept constant, then the radius vector traces out a curve on the surface called a ***u-curve***; and if v is allowed to change, while u is kept constant, the radius vector traces out a ***v-curve***. The u-curves (v constant) are associated with horizontal lines in the uv-plane, and the v-curves (u constant) are associated with vertical lines (Figure 16.1.19). The u-curves and v-curves usually cover the surface like a net and are useful for visualizing the shape of the surface.

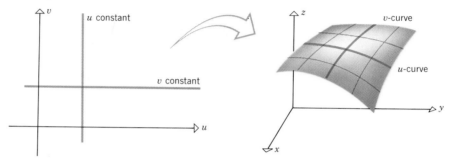

Figure 16.1.19

Example 12   Describe the u-curves and v-curves for the surface in Example 11.

Solution.   From (4), the allowable values of u and v correspond to points in the quarter-circle of the uv-plane shown in Figure 16.1.20. If v is constant, say $v = v_0$, where $0 \le v_0 \le 2$, then u varies from 0 to $\sqrt{4 - v_0^2}$, resulting in the u-curve

$$x = u, \quad y = v_0, \quad z = 4 - u^2 - v_0^2$$

or equivalently,

$$z - (4 - v_0^2) = -x^2 \quad (y = v_0, 0 \le x \le \sqrt{4 - v_0^2})$$

These u-curves are portions of parabolas that are parallel to the xz-plane, open down, and have their vertices over the y-axis at the points $(0, v_0, 4 - v_0^2)$. They are the green curves in Figure 16.1.20. Similarly, if u is constant, say $u = u_0$, where $0 \le u_0 \le 2$, then v varies from 0 to $\sqrt{4 - u_0^2}$, resulting in the v-curve

$$z - (4 - u_0^2) = -y^2 \quad (x = u_0, 0 \le y \le \sqrt{4 - u_0^2})$$

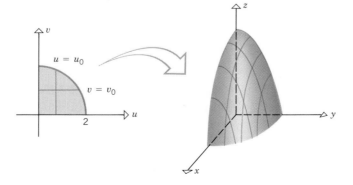

Figure 16.1.20

These v-curves are portions of parabolas that are parallel to the yz-plane, open down, and have their vertices over the x-axis at the points $(u_0, 0, 4 - u_0^2)$. They are the blue curves in Figure 16.1.20. ◀

Example 13 The surface in Example 11 can be parametrized in another way by expressing its equation in cylindrical coordinates (r, θ, z). To convert $z = 4 - x^2 - y^2$ to cylindrical coordinates, we substitute $x = r \cos \theta$ and $y = r \sin \theta$, which yields $z = 4 - (r \cos \theta)^2 - (r \sin \theta)^2 = 4 - r^2$. Thus, the surface can be represented parametrically in terms of the parameters r and θ as

$$x = r \cos \theta, \quad y = r \sin \theta, \quad z = 4 - r^2$$

or in vector form as

$$\mathbf{r}(r, \theta) = r \cos \theta \mathbf{i} + r \sin \theta \mathbf{j} + (4 - r^2)\mathbf{k}$$

where r and θ satisfy $0 \le r \le 2$ and $0 \le \theta \le \pi/2$, since we have only the portion of the paraboloid in the first octant. These parametric equations associate points on the surface with points (r, θ) in an $r\theta$-plane (Figure 16.1.21). The r-curves (θ constant, say $\theta = \theta_0$), which are shown in green, are portions of parabolas that extend from the top of the paraboloid down to the xy-plane and lie in the plane making an angle θ_0 with the x-axis. The θ-curves (r constant, say $r = r_0$), which are shown in blue in Figure 16.1.21, are quarter-circles of radius r_0 centered on the z-axis and lying in planes parallel to the xy-plane. ◀

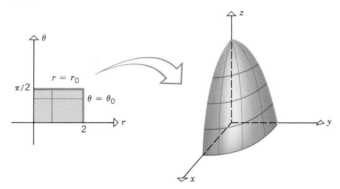

Figure 16.1.21

Example 14 Find a parametric representation of the portion of the cylindrical surface $x^2 + z^2 = 9$ for which $0 \le y \le 5$ in terms of the parameters u and v shown in Figure 16.1.22. The parameter u is the distance from a point $P(x, y, z)$ on the surface to the xz-plane, and v is the angle shown in the figure.

Solution. The radius of the cylinder is 3, so it is evident from the figure that $y = u$, $x = 3 \cos v$, and $z = 3 \sin v$. Thus, the surface can be represented parametrically as

$$x = 3 \cos v, \quad y = u, \quad z = 3 \sin v$$

or in vector form as

$$\mathbf{r}(u, v) = 3 \cos v \mathbf{i} + u \mathbf{j} + 3 \sin v \mathbf{k}$$

where $0 \le u \le 5$ and $0 \le v < 2\pi$ (Figure 16.1.23). The u-curves (v constant, say $v = v_0$) are line segments parallel to the y-axis that extend from the xz-plane to the plane $y = 5$, and the v-curves (u constant, say $u = u_0$) are circles of radius 3 centered on the y-axis and lying in planes parallel to the xz-plane (Figure 16.1.23). ◀

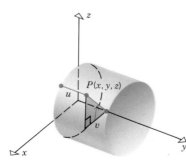

Figure 16.1.22

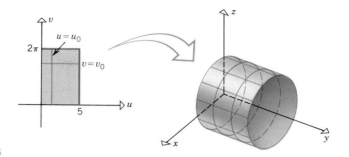

Figure 16.1.23

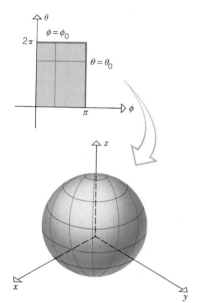

Figure 16.1.24

□ **COMPUTER-GENERATED SURFACES**

Example 15 By eliminating the parameters, show that if $a > 0$, then

$$\mathbf{r}(\phi, \theta) = a \sin \phi \cos \theta \mathbf{i} + a \sin \phi \sin \theta \mathbf{j} + a \cos \phi \mathbf{k}$$

for $0 \leq \phi \leq \pi$ and $0 \leq \theta \leq 2\pi$ represents a sphere of radius a centered at the origin, and describe the ϕ-curves and θ-curves.

Solution. It follows from the formula for $\mathbf{r}(\phi, \theta)$ that

$$x = a \sin \phi \cos \theta, \quad y = a \sin \phi \sin \theta, \quad z = a \cos \phi \tag{5}$$

so

$$x^2 + y^2 + z^2 = a^2 \sin^2 \phi \cos^2 \theta + a^2 \sin^2 \phi \sin^2 \theta + a^2 \cos^2 \phi$$
$$= a^2 \sin^2 \phi + a^2 \cos^2 \phi = a^2$$

which, in Cartesian coordinates, is the equation of a sphere of radius a centered at the origin. By comparing the formulas in (5) to the third set of formulas in Table 14.8.1, we observe that ϕ and θ can be interpreted geometrically as spherical coordinates of a point on the surface. It follows that a ϕ-curve (θ constant, say $\theta = \theta_0$) is a semicircle of radius a, centered at the origin and lying in a plane that makes an angle of θ_0 with the xz-plane; and a θ-curve (ϕ constant) is a circle centered on the z-axis and lying in a plane parallel to the xy-plane (Figure 16.1.24). In the language of cartographers, the ϕ-curves are *lines of longitude* and the θ-curves are *lines of latitude*. ◄

In recent years computers have extended our ability to visualize complicated three-dimensional surfaces. Many computer programs draw surfaces within a box whose edges are parallel to the coordinate axes. As illustrated with the elliptic paraboloid in Figure 16.1.25, this sometimes produces artificial-looking cuts in the surface; however, this can be

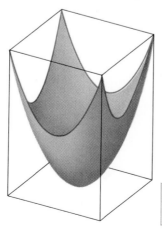

The box cuts the elliptic paraboloid, producing parabolic traces in the sides of the box.

Figure 16.1.25

useful for visualizing the surface since the cuts are traces of the surface parallel to the coordinate planes. Some examples of other computer-generated surfaces in three-dimensional space are shown in Figure 16.1.26.

SOME COMPUTER GENERATED SURFACES IN 3-SPACE

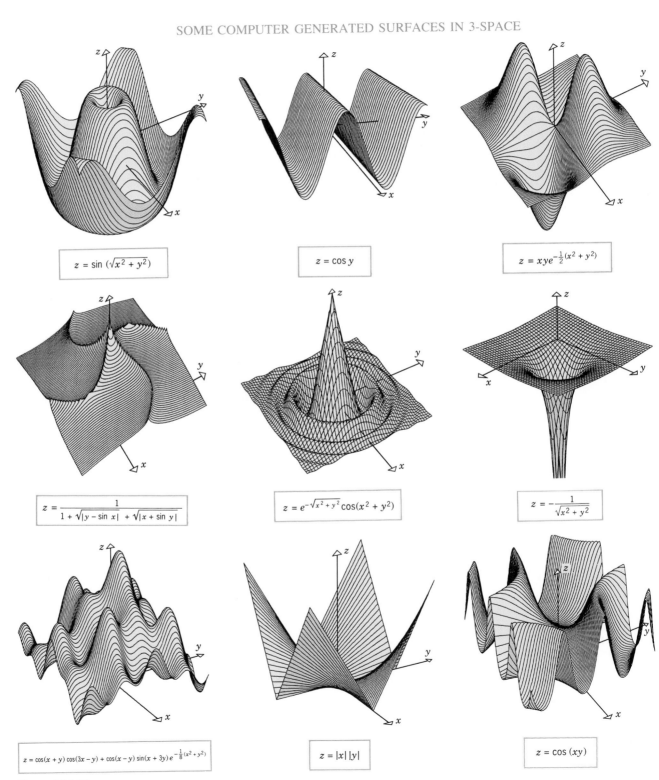

$z = \sin(\sqrt{x^2 + y^2})$

$z = \cos y$

$z = xye^{-\frac{1}{2}(x^2 + y^2)}$

$z = \dfrac{1}{1 + \sqrt{|y - \sin x|} + \sqrt{|x + \sin y|}}$

$z = e^{-\sqrt{x^2 + y^2}}\cos(x^2 + y^2)$

$z = -\dfrac{1}{\sqrt{x^2 + y^2}}$

$z = \cos(x + y)\cos(3x - y) + \cos(x - y)\sin(x + 3y)e^{-\frac{1}{8}(x^2 + y^2)}$

$z = |x||y|$

$z = \cos(xy)$

Figure 16.1.26

► Exercise Set 16.1

1. Let $f(x, y) = x^2 y + 1$. Find
  (a) $f(2, 1)$ (b) $f(1, 2)$ (c) $f(0, 0)$
  (d) $f(1, -3)$ (e) $f(3a, a)$ (f) $f(ab, a - b)$.

2. Let $f(x, y) = x + \sqrt[3]{xy}$. Find
  (a) $f(t, t^2)$ (b) $f(x, x^2)$ (c) $f(2y^2, 4y)$.

3. Let $f(x, y) = xy + 3$. Find
  (a) $f(x + y, x - y)$ (b) $f(xy, 3x^2 y^3)$.

4. Let $g(x) = x \sin x$. Find
  (a) $g(x/y)$ (b) $g(xy)$ (c) $g(x - y)$.

5. Find $F(g(x), h(y))$ if $F(x, y) = xe^{xy}$, $g(x) = x^3$, and $h(y) = 3y + 1$.

6. Find $g(u(x, y), v(x, y))$ if $g(x, y) = y \sin(x^2 y)$, $u(x, y) = x^2 y^3$, and $v(x, y) = \pi xy$.

7. Let $f(x, y) = x + 3x^2 y^2$, $x(t) = t^2$, and $y(t) = t^3$. Find
  (a) $f(x(t), y(t))$ (b) $f(x(0), y(0))$
  (c) $f(x(2), y(2))$.

8. Let $g(x, y) = ye^{-3x}$, $x(t) = \ln(t^2 + 1)$, and $y(t) = \sqrt{t}$. Find $g(x(t), y(t))$.

9. Let $f(x, y, z) = xy^2 z^3 + 3$. Find
  (a) $f(2, 1, 2)$ (b) $f(-3, 2, 1)$
  (c) $f(0, 0, 0)$ (d) $f(a, a, a)$
  (e) $f(t, t^2, -t)$ (f) $f(a + b, a - b, b)$.

10. Let $f(x, y, z) = zxy + x$. Find
  (a) $f(x + y, x - y, x^2)$ (b) $f(xy, y/x, xz)$.

11. Find $F(f(x), g(y), h(z))$ if $F(x, y, z) = ye^{xyz}$, $f(x) = x^2$, $g(y) = y + 1$, and $h(z) = z^2$.

12. Find $g(u(x, y, z), v(x, y, z), w(x, y, z))$ if $g(x, y, z) = z \sin xy$, $u(x, y, z) = x^2 z^3$, $v(x, y, z) = \pi xyz$, and $w(x, y, z) = xy/z$.

13. Let $f(x, y, z) = x^2 y^2 z^4$, $x(t) = t^3$, $y(t) = t^2$, and $z(t) = t$. Find
  (a) $f(x(t), y(t), z(t))$ (b) $f(x(0), y(0), z(0))$
  (c) $f(x(2), y(2), z(2))$.

In Exercises 14–19, sketch the domain of f. Use solid lines for portions of the boundary included in the domain and dashed lines for portions not included.

14. $f(x, y) = xy\sqrt{y - 1}$.

15. $f(x, y) = \ln(1 - x^2 - y^2)$.

16. $f(x, y) = \sqrt{x^2 + y^2 - 4}$.

17. $f(x, y) = \dfrac{1}{x - y^2}$.

18. $f(x, y) = \ln xy$.   **19.** $f(x, y) = \sqrt{\dfrac{x^2 + y^2}{x^2 - y^2}}$.

In Exercises 20–27, describe the domain of f.

20. $f(x, y) = \sin^{-1}(x + y)$.   **21.** $f(x, y) = xe^{-\sqrt{y+2}}$.

22. $f(x, y) = \dfrac{\sqrt{4 - x^2}}{y^2 + 3}$.   **23.** $f(x, y) = \ln(y - 2x)$.

24. $f(x, y, z) = \sqrt{25 - x^2 - y^2 - z^2}$.

25. $f(x, y, z) = \dfrac{xyz}{x + y + z}$.   **26.** $f(x, y, z) = e^{xyz}$.

27. $f(x, y, z) = z + \ln(1 - x^2 - y^2)$.

In Exercises 28–39, sketch the graph of f.

28. $f(x, y) = 3$.   **29.** $f(x, y) = 4 - 2x - 4y$.

30. $f(x, y) = \sqrt{9 - x^2 - y^2}$.

31. $f(x, y) = \sqrt{x^2 + y^2}$.   **32.** $f(x, y) = x^2 + y^2$.

33. $f(x, y) = x^2 - y^2$.   **34.** $f(x, y) = 4 - x^2 - y^2$.

35. $f(x, y) = -\sqrt{1 - x^2/4 - y^2/9}$.

36. $f(x, y) = \sqrt{x^2 + y^2 - 1}$.

37. $f(x, y) = \sqrt{x^2 + y^2 + 1}$.

38. $f(x, y) = x^2$.   **39.** $f(x, y) = y + 1$.

In Exercises 40–47, sketch the level curve $z = k$ for the specified values of k.

40. $z = 3x + y$; $k = -2, -1, 0, 1, 2$.
41. $z = x^2 + y^2$; $k = 0, 1, 2, 3, 4$.
42. $z = y/x$; $k = -2, -1, 0, 1, 2$.
43. $z = x^2 + y$; $k = -2, -1, 0, 1, 2$.
44. $z = x^2 + 9y^2$; $k = 0, 1, 2, 3, 4$.
45. $z = x^2 - y^2$; $k = -2, -1, 0, 1, 2$.
46. $z = y \csc x$; $k = -2, -1, 0, 1, 2$.
47. $z = \sqrt{\dfrac{x + y}{x - y}}$; $k = 0, 1, 2, 3, 4$.

In Exercises 48–51, sketch the level surface $f(x, y, z) = k$.

48. $f(x, y, z) = 4x - 2y + z$; $k = 1$.
49. $f(x, y, z) = 4x^2 + y^2 + 4z^2$; $k = 16$.
50. $f(x, y, z) = x^2 + y^2 - z^2$; $k = 0$.
51. $f(x, y, z) = z - x^2 - y^2 + 4$; $k = 7$.

In Exercises 52–55, describe the level surfaces.

52. $f(x, y, z) = 3x - y + 2z$.
53. $f(x, y, z) = (x - 2)^2 + y^2 + z^2$.
54. $f(x, y, z) = z - x^2 - y^2$.
55. $f(x, y, z) = x^2 + z^2$.

56. Let $f(x, y) = yx^2 + 1$. Find an equation of the level curve that passes through the point
  (a) $(1, 2)$ (b) $(-2, 4)$ (c) $(0, 0)$.

57. Let $f(x, y) = x^2 - 2x^3 + 3xy$. Find an equation of the level curve that passes through the point

(a) $(-1, 1)$ (b) $(0, 0)$ (c) $(2, -1)$.

58. Let $f(x, y) = ye^x$. Find an equation of the level curve that passes through the point

(a) $(\ln 2, 1)$ (b) $(0, 3)$ (c) $(1, -2)$.

59. Let $f(x, y, z) = x^2 + y^2 - z$. Find an equation of the level surface that passes through the point

(a) $(1, -2, 0)$ (b) $(1, 0, 3)$ (c) $(0, 0, 0)$.

60. Let $f(x, y, z) = xyz + 3$. Find an equation of the level surface that passes through the point

(a) $(1, 0, 2)$ (b) $(-2, 4, 1)$ (c) $(0, 0, 0)$.

61. If $V(x, y)$ is the voltage or potential at a point (x, y) in the xy-plane, then the level curves of V are called **equipotential curves**. Along such a curve, the voltage remains constant. Given that

$$V(x, y) = \frac{8}{\sqrt{16 + x^2 + y^2}}$$

sketch the equipotential curves at which $V = 2.0$, $V = 1.0$, and $V = 0.5$.

62. If $T(x, y)$ is the temperature at a point (x, y) on a thin metal plate in the xy-plane, then the level curves of T are called **isothermal curves**. All points on such a curve are at the same temperature. Suppose that a plate occupies the first quadrant and $T(x, y) = xy$.

(a) Sketch the isothermal curves on which $T = 1, T = 2$, and $T = 3$.

(b) An ant, initially at $(1, 4)$, wants to walk on the plate so that the temperature along its path remains constant. What path should the ant take?

In Exercises 63–65, use the contour map shown in Figure 16.1.27 (all elevations in hundreds of feet).

63. For points A and B,

(a) which one is higher?

(b) which one is on the steeper slope?

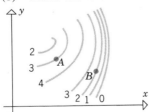

Figure 16.1.27

64. Starting at point A, will the elevation begin to increase or decrease if we travel so that

(a) y remains constant and x increases

(b) y remains constant and x decreases

(c) x remains constant and y increases

(d) x remains constant and y decreases?

65. Starting at point B, will the elevation begin to increase or decrease if we travel so that

(a) y remains constant and x increases

(b) y remains constant and x decreases

(c) x remains constant and y increases

(d) x remains constant and y decreases?

In Exercises 66 and 67, classify the given set of points as open, closed, or neither.

66. The set of points (x, y) in the xy-plane that satisfy the inequality

(a) $1 \le x^2 + y^2 \le 3$

(b) $0 \le y < 2x + 1$ and $0 \le x \le 1$

(c) $\sqrt{y} < x < 1$

(d) $0 < y < e^x$ and $0 < x < 2$.

67. The set of points (x, y, z) in 3-space that satisfy the inequality

(a) $x^2 + y^2 + z^2 < 3$

(b) $0 \le z \le x^2 + y^2$ for $x^2 + y^2 < 1$

(c) $0 \le z \le 5 - 3x - 2y$ for $x \ge 0$ and $y \ge 0$

(d) $y^2 + z^2 \le 5$ for $1 \le x \le 2$.

In Exercises 68 and 69, classify the given set as bounded or unbounded.

68. The set of points (x, y) in the xy-plane that satisfy the inequality

(a) $x^2 + y^2 < 100$ (b) $x^2 + y^2 \ge 100$

(c) $-1 < y < 2$

(d) $y \le 3 - x$ and $-1 < x < 1$.

69. The set of points in 3-space that satisfy the inequality

(a) $x^2 + y^2 + z^2 \le 3$

(b) $0 < z < 2 - y$ and $0 < x < 1$

(c) $x^2 + y^2 < 4$ (d) $z \ge 2 - x - 3y$.

In Exercises 70–74, find a parametric representation of the surface in terms of the parameters u and v, where $u = x$ and $v = y$.

70. $2z - 3x + 4y = 5$.   **71.** $z = x^2$.   **72.** $y^2 - 3z = 5$.

73. The portion of the cylinder $x^2 + z^2 = 4$ on or above the xy-plane and on or between the planes $y = 1$ and $y = 3$.

74. The portion of the sphere $x^2 + y^2 + z^2 = 25$ on or above the plane $z = 3$.

In Exercises 75–82, find a parametric representation of the surface in terms of the parameters r and θ, where (r, θ, z) are cylindrical coordinates of a point on the surface.

75. $z = \dfrac{1}{1 + x^2 + y^2}$.   **76.** $z = e^{-(x^2 + y^2)}$.

77. $z = 2xy$.   **78.** $z = x^2 - y^2$.

79. The portion of the sphere $x^2 + y^2 + z^2 = 9$ on or above the plane $z = 2$.

80. The portion of the cone $z = \sqrt{x^2 + y^2}$ on or below the plane $z = 3$.

81. The portion of the plane $z + 2y = 3$ on or inside the cylinder $x^2 + y^2 = 4$.

82. The surface of revolution that is generated by revolving about the z-axis the curve $z = 1/x^2$ (in the plane $y = 0$) for $\frac{1}{2} \leq x \leq 2$.

83. Find a parametric representation of the cone
$$z = \sqrt{3x^2 + 3y^2}$$
in terms of parameters ρ and θ, where (ρ, θ, ϕ) are spherical coordinates of a point on the surface.

84. Find a parametric representation of the cylinder $x^2 + y^2 = 9$ in terms of parameters θ and ϕ, where (ρ, θ, ϕ) are spherical coordinates.

85. Find a parametric representation of the portion of the elliptic cylinder $x^2 + 4y^2 = 9$ for $0 \leq z \leq 5$.

86. Find a parametric representation of the *torus* that is generated by revolving the circle $(x - a)^2 + z^2 = b^2$ (in the plane $y = 0$), where $0 < b < a$, about the z-axis. Use the angles u and v shown in Figure 16.1.28 as parameters.

In Exercises 87–92, describe the surface by eliminating the parameters to obtain an equation for the surface in rectangular coordinates.

87. $x = 2u + v$, $y = u - v$, $z = 3v$ for $-\infty < u < +\infty$ and $-\infty < v < +\infty$.

88. $x = u \cos v, y = u^2, z = u \sin v$ for $0 \leq u \leq 2$ and $0 \leq v < 2\pi$.

89. $x = 3 \sin u, y = 2 \cos u, z = 2v$ for $0 \leq u < 2\pi$ and $1 \leq v \leq 2$.

90. $x = \sqrt{u} \cos v, y = \sqrt{u} \sin v, z = u$ for $0 \leq u \leq 4$ and $0 \leq v < 2\pi$.

91. $\mathbf{r}(u, v) = 3u \cos v\mathbf{i} + 4u \sin v\mathbf{j} + u\mathbf{k}$ for $0 \leq u \leq 1$ and $0 \leq v < 2\pi$.

92. $\mathbf{r}(u, v) = \sin u \cos v\mathbf{i} + 2 \sin u \sin v\mathbf{j} + 3 \cos u\mathbf{k}$ for $0 \leq u \leq \pi$ and $0 \leq v < 2\pi$.

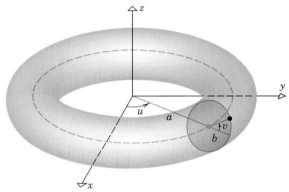

Figure 16.1.28

16.2 LIMITS AND CONTINUITY

In this section we shall introduce the notions of limit and continuity for functions of two or more variables. We shall not go into great detail; our objective is to develop the basic ideas accurately, and to obtain results needed in later sections. A more extensive study of these topics is usually given in advanced calculus.

□ **LIMITS ALONG CURVES**

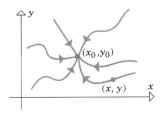

Figure 16.2.1

For a function of one variable there are two one-sided limits at a point x_0, namely
$$\lim_{x \to x_0^+} f(x) \quad \text{and} \quad \lim_{x \to x_0^-} f(x)$$
reflecting the fact that there are only two directions from which x can approach x_0, the right or the left. For functions of two or three variables the situation is more complicated because there are infinitely many different curves along which one point can approach another (Figure 16.2.1). Our first objective in this section is to define the limit of $f(x, y)$ as (x, y) approaches a point (x_0, y_0) along a curve C (and similarly for functions of three variables). Limits along curves are the two-variable and three-variable analogs of one-sided limits for functions of one variable. Later in this section we shall define analogs of two-sided limits.

If C is a smooth parametric curve in 2-space or 3-space that is represented by the equations

$$x = x(t), \quad y = y(t) \qquad \text{or} \qquad x = x(t), \quad y = y(t), \quad z = z(t)$$

and if $x_0 = x(t_0)$, $y_0 = y(t_0)$, and $z_0 = z(t_0)$, then the limits

$$\lim_{\substack{(x,y)\to(x_0,y_0)\\ \text{(along } C)}} f(x, y) \quad \text{and} \quad \lim_{\substack{(x,y,z)\to(x_0,y_0,z_0)\\ \text{(along } C)}} f(x, y, z)$$

are defined by

$$\lim_{\substack{(x,y)\to(x_0,y_0)\\ \text{(along } C)}} f(x, y) = \lim_{t\to t_0} f(x(t), y(t)) \tag{1}$$

$$\lim_{\substack{(x,y,z)\to(x_0,y_0,z_0)\\ \text{(along } C)}} f(x, y, z) = \lim_{t\to t_0} f(x(t), y(t), z(t)) \tag{2}$$

Simply stated, limits along parametric curves are obtained by substituting the parametric equations into the formula for the function f and computing the appropriate limit of the resulting function of one variable. A geometric interpretation of the limit along a curve for a function of two variables is shown in Figure 16.2.2: As the point $(x(t), y(t))$ moves along the curve C in the xy-plane toward (x_0, y_0), the point $(x(t), y(t), f(x(t), y(t)))$ moves directly above it along the graph of $z = f(x, y)$ with $f(x(t), y(t))$ approaching the limiting value L. In that figure we followed a common practice of omitting the zero z-coordinate for points in the xy-plane.

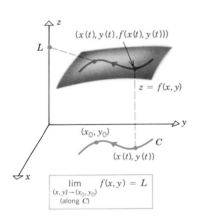

$$\lim_{\substack{(x,y)\to(x_0,y_0)\\ \text{(along } C)}} f(x, y) = L$$

Figure 16.2.2

REMARK.   In both (1) and (2), the limit of the function of t has to be treated as a one-sided limit if (x_0, y_0) or (x_0, y_0, z_0) is an endpoint of C.

Example 1   Let

$$f(x, y) = \frac{xy}{x^2 + y^2}$$

Find the limit of $f(x, y)$ as $(x, y) \to (0, 0)$ along

(a)   the x-axis   (b)   the y-axis
(c)   the parabola $y = x^2$ (d)   the line $y = x$

Solution (a).   The x-axis has parametric equations $x = t$, $y = 0$, with $(0, 0)$ corresponding to $t = 0$, so

$$\lim_{\substack{(x,y)\to(0,0)\\ \text{(along } y=0)}} f(x, y) = \lim_{t\to 0} f(t, 0) = \lim_{t\to 0} \frac{0}{t^2} = \lim_{t\to 0} 0 = 0$$

Solution (b).   The y-axis has parametric equations $x = 0$, $y = t$, with $(0, 0)$ corresponding to $t = 0$, so

$$\lim_{\substack{(x,y)\to(0,0)\\ \text{(along } x=0)}} f(x, y) = \lim_{t\to 0} f(0, t) = \lim_{t\to 0} \frac{0}{t^2} = \lim_{t\to 0} 0 = 0$$

Solution (c).   The parabola $y = x^2$ has parametric equations $x = t$, $y = t^2$, with $(0, 0)$ corresponding to $t = 0$, so

$$\lim_{\substack{(x,y)\to(0,0)\\ \text{(along } y=x^2)}} f(x, y) = \lim_{t\to 0} f(t, t^2) = \lim_{t\to 0} \frac{t^3}{t^2 + t^4} = \lim_{t\to 0} \frac{t}{1 + t^2} = 0$$

Solution (d). The line $y = x$ has parametric equations $x = t$, $y = t$, with $(0, 0)$ corresponding to $t = 0$, so

$$\lim_{\substack{(x,y)\to(0,0)\\ (\text{along } y=x)}} f(x, y) = \lim_{t\to 0} f(t, t) = \lim_{t\to 0} \frac{t^2}{2t^2} = \lim_{t\to 0} \frac{1}{2} = \frac{1}{2} \qquad \blacktriangleleft$$

REMARK. The graph of the function f in the preceding example is shown in Figure 16.2.3 with the axes drawn in red for clarity. The limits in parts (a), (b), and (d) can be visualized from this graph if you keep in mind that, except at the origin, where f is undefined, the x- and y-axes lie on the surface (why?). The limit in part (c) can also be visualized from the graph, but it may take some effort.

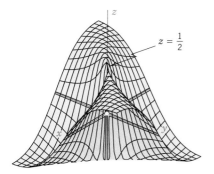

$$z = \frac{xy}{x^2 + y^2}$$

Figure 16.2.3

Example 2 Find

$$\lim_{\substack{(x,y,z)\to(-1,0,\pi)\\ (\text{along } C)}} \frac{x^2 + y^2 + x}{z - \pi}$$

where C is the circular helix with parametric equations $x = \cos t$, $y = \sin t$, $z = t$.

Solution. The point $(-1, 0, \pi)$ corresponds to $t = \pi$, so

$$\lim_{\substack{(x,y,z)\to(-1,0,\pi)\\ (\text{along } C)}} \frac{x^2 + y^2 + x}{z - \pi} = \lim_{t\to \pi} \frac{\cos^2 t + \sin^2 t + \cos t}{t - \pi}$$

$$= \lim_{t\to \pi} \frac{1 + \cos t}{t - \pi} = 0 \qquad \blacktriangleleft$$

where L'Hôpital's rule was applied to evaluate the last limit.

☐ **GENERAL LIMITS OF FUNCTIONS OF TWO AND THREE VARIABLES**

Although limits along specific curves are useful for many purposes, they do not always tell the complete story about the limiting behavior of a function; what is required is a limit concept that accounts for the behavior of the function in an *entire vicinity* of a point, not just along smooth curves passing through the point. As illustrated in Figure 16.2.4, we will want the statement

$$\lim_{(x, y)\to(x_0, y_0)} f(x, y) = L$$

to mean that the value of $f(x, y)$ can be made as close as we like to L (say within ϵ units of L) by restricting (x, y) to lie within (but not at the center of) some sufficiently small circle

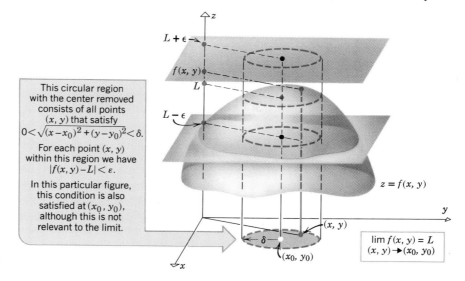

Figure 16.2.4

centered at (x_0, y_0) (say a circle of radius δ). This idea is conveyed by Definition 16.2.1 and for functions of three variables by Definition 16.2.2.

16.2.1   DEFINITION.   Let f be a function of two variables. We shall write

$$\lim_{(x,y)\to(x_0,y_0)} f(x, y) = L \tag{3}$$

if given any number $\epsilon > 0$, we can find a number $\delta > 0$ such that $f(x, y)$ satisfies

$$|f(x, y) - L| < \epsilon$$

whenever (x, y) lies in the domain of f and the distance between (x, y) and (x_0, y_0) satisfies

$$0 < \sqrt{(x - x_0)^2 + (y - y_0)^2} < \delta$$

16.2.2   DEFINITION.   Let f be a function of three variables. We shall write

$$\lim_{(x,y,z)\to(x_0,y_0,z_0)} f(x, y, z) = L \tag{4}$$

if given any number $\epsilon > 0$, we can find a number $\delta > 0$ such that $f(x, y, z)$ satisfies

$$|f(x, y, z) - L| < \epsilon$$

whenever (x, y, z) lies in the domain of f and the distance between (x, y, z) and (x_0, y_0, z_0) satisfies

$$0 < \sqrt{(x - x_0)^2 + (y - y_0)^2 + (z - z_0)^2} < \delta$$

When convenient, (3) and (4) can also be written in the alternative notations

$$f(x, y) \to L \quad \text{as} \quad (x, y) \to (x_0, y_0)$$

and

$$f(x, y, z) \to L \quad \text{as} \quad (x, y, z) \to (x_0, y_0, z_0)$$

☐ **PROPERTIES OF LIMITS**

We note without proof that the basic properties of limits given in Theorem 2.5.1 hold for limits along curves and for the limits in the preceding definitions, so that computations involving such limits can be performed in the usual way.

Example 3

$$\lim_{(x,y)\to(1,4)} [5x^3y^2 - 9] = \lim_{(x,y)\to(1,4)} [5x^3y^2] - \lim_{(x,y)\to(1,4)} 9$$

$$= 5\left[\lim_{(x,y)\to(1,4)} x\right]^3 \left[\lim_{(x,y)\to(1,4)} y\right]^2 - 9$$

$$= 5(1)^3(4)^2 - 9 = 71 \quad \blacktriangleleft$$

☐ **RELATIONSHIPS BETWEEN GENERAL LIMITS AND LIMITS ALONG SMOOTH CURVES**

The following theorem, which we state without proof, establishes an important relationship between general limits and limits along smooth curves.

16.2.3   THEOREM.   *If a function $f(x, y)$ has a limit L as (x, y) approaches a point (x_0, y_0), then $f(x, y)$ approaches the same limit L as (x, y) approaches (x_0, y_0) along any smooth curve that lies in the domain of f. Similarly, for functions of three variables.*

It follows from this theorem that if one can find two different smooth curves containing (x_0, y_0) along which $f(x, y)$ has different limits as (x, y) approaches (x_0, y_0), or if one can find any single smooth curve containing (x_0, y_0) such that the limit of $f(x, y)$ does not exist as (x, y) approaches (x_0, y_0), then

$$\lim_{(x,y) \to (x_0, y_0)} f(x, y)$$

does not exist. Similarly for functions of three variables.

Example 4   The limit

$$\lim_{(x,y) \to (0,0)} \frac{xy}{x^2 + y^2}$$

does not exist because in Example 1 we found two different smooth curves along which this limit had different values. For example, we saw that

$$\lim_{\substack{(x,y) \to (0,0) \\ (\text{along } x=0)}} \frac{xy}{x^2 + y^2} = 0 \quad \text{and} \quad \lim_{\substack{(x,y) \to (0,0) \\ (\text{along } y=x)}} \frac{xy}{x^2 + y^2} = \frac{1}{2} \quad \blacktriangleleft$$

□ **CONTINUITY**

Informally stated, a function of one variable is continuous if its graph is an unbroken curve without jumps or holes. To extend this idea to functions of two variables, imagine that the graph of $z = f(x, y)$ is molded from a thin sheet of clay that has been hollowed or pinched into peaks and valleys. We shall consider f to be continuous if the clay surface has no tears or holes. The functions graphed in Figure 16.2.5 fail to be continuous because of their behavior at $(0, 0)$.

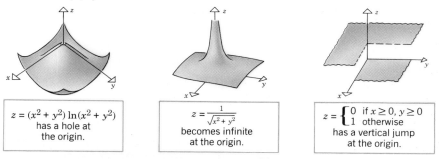

$z = (x^2 + y^2)\ln(x^2 + y^2)$ has a hole at the origin.

$z = \dfrac{1}{\sqrt{x^2 + y^2}}$ becomes infinite at the origin.

$z = \begin{cases} 0 & \text{if } x \geq 0, y \geq 0 \\ 1 & \text{otherwise} \end{cases}$ has a vertical jump at the origin.

Figure 16.2.5

The precise definitions of continuity for functions of two and three variables are similar to the definition for a function of one variable. Recall that $f(x)$ is continuous at a point x_0 if

$$\lim_{x \to x_0} f(x) = f(x_0)$$

Analogously, we define $f(x, y)$ to be **continuous at** (x_0, y_0) and $f(x, y, z)$ to be **continuous at** (x_0, y_0, z_0) if

$$\lim_{(x, y) \to (x_0, y_0)} f(x, y) = f(x_0, y_0) \quad \text{and} \quad \lim_{(x, y, z) \to (x_0, y_0, z_0)} f(x, y, z) = f(x_0, y_0, z_0)$$

respectively. Moreover, a function f of two or three variables that is continuous at each point of a region R in 2-space or 3-space is said to be **continuous on** R, and if f is continuous at every point in 2-space or 3-space, then f is said to be **continuous everywhere** or simply **continuous**.

□ **COMPOSITIONS OF CONTINUOUS FUNCTIONS**

The following theorem, which we state without proof, will help us to identify continuous functions of two variables.

16.2.4   THEOREM.

(a)   *If g and h are continuous functions of one variable, then $f(x, y) = g(x)h(y)$ is a continuous function of x and y.*

(b)   *If g is a continuous function of one variable and h is a continuous function of two variables, then their composition $f(x, y) = g(h(x, y))$ is a continuous function of x and y.*

Example 5   The function $f(x, y) = 3x^2y^5$ is continuous because it is the product of the continuous functions $g(x) = 3x^2$ and $h(y) = y^5$.

In general, any function of the form $f(x, y) = Ax^m y^n$ (*m* and *n* nonnegative integers) is continuous because it is the product of the continuous functions Ax^m and y^n.   ◄

Example 6   Since $g(x) = \sin x$ is a continuous function of one variable and $h(x, y) = xy^2$ is a continuous function of two variables, it follows that $g(h(x, y)) = g(xy^2) = \sin(xy^2)$ is a continuous function of *x* and *y*. By a similar argument, each of the following is continuous:

$$(x^4 y^5)^{1/3}, \quad e^{xy}, \quad \cosh(x^3 y) \quad \blacktriangleleft$$

Example 7   By Example 6, e^{xy} is continuous. Thus, $\cos(e^{xy})$ is continuous by part (*b*) of Theorem 16.2.4.   ◄

Theorem 16.2.4 is one of a whole class of theorems about continuity of functions in any number of variables. The content of these theorems can be summarized informally with three basic principles:

- A composition of continuous functions is continuous.

- A sum, difference, or product of continuous functions is continuous.

- A quotient of continuous functions is continuous, except where the denominator is zero.

Example 8   The following functions are continuous since they are sums, differences, products, and compositions of continuous functions:

$$3 - 2x^2 yz + 9x^4 y^8 z^3, \quad e^{xy} \cos(xy^2 + 1), \quad (3z + ye^x)^{17} \quad \blacktriangleleft$$

Example 9   Since the function

$$f(x, y) = \frac{x^3 y^2}{1 - xy}$$

is a quotient of continuous functions, it is continuous except where $1 - xy = 0$. Thus, $f(x, y)$ is continuous everywhere except on the hyperbola $xy = 1$.   ◄

Example 10   Evaluate

$$\lim_{(x,y) \to (-1,2)} \frac{xy}{x^2 + y^2}$$

Solution.   Since $f(x, y) = xy/(x^2 + y^2)$ is continuous at $(-1, 2)$ (why?), it follows from the definition of continuity for functions of two variables that

$$\lim_{(x,y) \to (-1,2)} \frac{xy}{x^2 + y^2} = \frac{(-1)(2)}{(-1)^2 + (2)^2} = -\frac{2}{5} \quad \blacktriangleleft$$

☐ **LIMITS AT POINTS OF DISCONTINUITY**

At a point of discontinuity the method of Example 10 cannot be used. However, sometimes such limits can be obtained by converting the given function to polar coordinates.

Example 11 Find

$$\lim_{(x,y)\to(0,0)} (x^2 + y^2) \ln (x^2 + y^2)$$

Solution. Let (r, θ) be the coordinates of the point (x, y) with $r \geq 0$. Then we have

$$x = r \cos \theta, \quad y = r \sin \theta, \quad r^2 = x^2 + y^2$$

Moreover, since $r \geq 0$ we have $r = \sqrt{x^2 + y^2}$, so that $r \to 0^+$ if and only if $(x, y) \to (0, 0)$. Thus, we can rewrite the given limit as

$$\lim_{(x,y)\to(0,0)} (x^2 + y^2) \ln (x^2 + y^2) = \lim_{r\to 0^+} r^2 \ln r^2$$

$$= \lim_{r\to 0^+} \frac{2 \ln r}{1/r^2} \qquad \boxed{\text{This converts the limit to an indeterminate form of type } \infty/\infty.}$$

$$= \lim_{r\to 0^+} \frac{2/r}{-2/r^3} \qquad \boxed{\text{L'Hôpital's rule}}$$

$$= \lim_{r\to 0^+} (-r^2) = 0$$

This result is consistent with the graph of f shown in Figure 16.2.5. ◄

▶ **Exercise Set 16.2**

In Exercises 1–8, sketch the region where the function f is continuous.

1. $f(x, y) = y \ln (1 + x)$.   **2.** $f(x, y) = \sqrt{x - y}$.

3. $f(x, y) = \dfrac{x^2 y}{\sqrt{25 - x^2 - y^2}}$.

4. $f(x, y) = \ln (2x - y + 1)$.

5. $f(x, y) = \cos \left(\dfrac{xy}{1 + x^2 + y^2} \right)$.

6. $f(x, y) = e^{(1-xy)}$.   **7.** $f(x, y) = \sin^{-1}(xy)$.

8. $f(x, y) = \tan^{-1}(y - x)$.

In Exercises 9–12, describe the region on which the function f is continuous.

9. $f(x, y, z) = 3x^2 e^{yz} \cos (xyz)$.

10. $f(x, y, z) = \ln (4 - x^2 - y^2 - z^2)$.

11. $f(x, y, z) = \dfrac{y + 1}{x^2 + z^2 - 1}$.

12. $f(x, y, z) = \sin \sqrt{x^2 + y^2 + 3z^2}$.

In Exercises 13–35, find the limit, if it exists.

13. $\displaystyle\lim_{(x,y)\to(1,3)} (4xy^2 - x)$.   **14.** $\displaystyle\lim_{(x,y)\to(1/2,\pi)} (xy^2 \sin xy)$.

15. $\displaystyle\lim_{(x,y)\to(-1,2)} \frac{xy^3}{x + y}$.   **16.** $\displaystyle\lim_{(x,y)\to(1,-3)} e^{2x-y^2}$.

17. $\displaystyle\lim_{(x,y)\to(0,0)} \ln (1 + x^2 y^3)$.   **18.** $\displaystyle\lim_{(x,y)\to(4,-2)} x\sqrt[3]{y^3 + 2x}$.

19. $\displaystyle\lim_{(x,y)\to(0,0)} \frac{x - y}{x^2 + y^2}$. [*Hint:* Let $(x, y) \to (0, 0)$ along the line $y = 0$.]

20. $\displaystyle\lim_{(x,y)\to(0,0)} \frac{3}{x^2 + 2y^2}$.   **21.** $\displaystyle\lim_{(x,y)\to(0,0)} \frac{\sin (x^2 + y^2)}{x^2 + y^2}$.

22. $\displaystyle\lim_{(x,y)\to(0,0)} \frac{1 - \cos (x^2 + y^2)}{x^2 + y^2}$.

23. $\displaystyle\lim_{(x,y)\to(0,0)} \frac{x^4 - y^4}{x^2 + y^2}$.   **24.** $\displaystyle\lim_{(x,y)\to(0,0)} \frac{x^4 - 16y^4}{x^2 + 4y^2}$.

25. $\displaystyle\lim_{(x,y)\to(0,0)} \frac{xy}{3x^2 + 2y^2}$.   **26.** $\displaystyle\lim_{(x,y)\to(0,0)} \frac{1 - x^2 - y^2}{x^2 + y^2}$.

27. $\displaystyle\lim_{(x,y)\to(0,0)} e^{-1/(x^2+y^2)}$.   **28.** $\displaystyle\lim_{(x,y)\to(0,0)} \frac{e^{-1/\sqrt{x^2+y^2}}}{\sqrt{x^2 + y^2}}$.

29. $\displaystyle\lim_{(x,y)\to(0,0)} y \ln (x^2 + y^2)$.

30. $\displaystyle\lim_{(x,y)\to(0,0)} x \ln (|x| + |y|)$.

31. $\displaystyle\lim_{(x,y,z)\to(2,-1,2)} \frac{xz^2}{\sqrt{x^2 + y^2 + z^2}}$.

32. $\lim\limits_{(x,y,z)\to(2,0,-1)} \ln(2x + y - z)$.

33. $\lim\limits_{(x,y,z)\to(0,0,0)} \dfrac{\sin(x^2 + y^2 + z^2)}{\sqrt{x^2 + y^2 + z^2}}$.

34. $\lim\limits_{(x,y,z)\to(0,0,0)} \dfrac{\sin\sqrt{x^2 + y^2 + z^2}}{x^2 + y^2 + z^2}$.

35. $\lim\limits_{(x,y,z)\to(0,0,0)} \dfrac{yz}{x^2 + y^2 + z^2}$.

[*Hint:* First let $(x, y, z) \to (0, 0, 0)$ along the z-axis and then along the line $x = t$, $y = t$, $z = t$.]

36. Show that $\dfrac{xy}{x^2 + y^2}$ can be made to approach any value in the interval $[-\frac{1}{2}, \frac{1}{2}]$ by letting $(x, y) \to (0, 0)$ along some line $y = mx$.

37. (a) Show that the value of $\dfrac{x^2 y}{x^4 + y^2}$ approaches zero as $(x, y) \to (0, 0)$ along any straight line $y = mx$.

(b) Show that $\lim\limits_{(x,y)\to(0,0)} \dfrac{x^2 y}{x^4 + y^2}$ does not exist by letting $(x, y) \to (0, 0)$ along the parabola $y = x^2$.

38. (a) Show that the value of $\dfrac{x^3 y}{2x^6 + y^2}$ approaches 0 as $(x, y) \to (0, 0)$ along any straight line $y = mx$, or along any parabola $y = kx^2$.

(b) Show that $\lim\limits_{(x,y)\to(0,0)} \dfrac{x^3 y}{2x^6 + y^2}$ does not exist by letting $(x, y) \to (0, 0)$ along the curve $y = x^3$.

39. (a) Show that the value of $\dfrac{xyz}{x^2 + y^4 + z^4}$ approaches 0 as $(x, y, z) \to (0, 0, 0)$ along any line $x = at$, $y = bt$, $z = ct$.

(b) Show that the limit $\lim\limits_{(x,y,z)\to(0,0,0)} \dfrac{xyz}{x^2 + y^4 + z^4}$ does

not exist by letting $(x, y, z) \to (0, 0, 0)$ along the curve $x = t^2$, $y = t$, $z = t$.

40. Find

$$\lim_{(x,y)\to(0,1)} \tan^{-1}\left[\frac{x^2 + 1}{x^2 + (y - 1)^2}\right]$$

41. Find

$$\lim_{(x,y)\to(0,1)} \tan^{-1}\left[\frac{x^2 - 1}{x^2 + (y - 1)^2}\right]$$

42. Let $f(x, y) = \begin{cases} \dfrac{\sin(x^2 + y^2)}{x^2 + y^2}, & (x, y) \neq (0, 0) \\ 1, & (x, y) = (0, 0). \end{cases}$

Show that f is continuous at $(0, 0)$.

43. Let $f(x, y) = \dfrac{x^2}{x^2 + y^2}$. Is it possible to define $f(0, 0)$ so that f will be continuous at $(0, 0)$?

44. Let $f(x, y) = xy \ln(x^2 + y^2)$. Is it possible to define $f(0, 0)$ so that f will be continuous at $(0, 0)$?

In Exercises 45 and 46, use Definition 16.2.1 to prove the given statement. [*Hint:* Let $r = \sqrt{x^2 + y^2}$.]

45. $\lim\limits_{(x,y)\to(0,0)} (x^2 + y^2) = 0$.

46. $\lim\limits_{(x,y)\to(0,0)} \dfrac{x^2 y^2}{\sqrt{x^2 + y^2}} = 0$.

In Exercises 47 and 48, use Definition 16.2.2 to prove the given statement. [*Hint:* Let $\rho = \sqrt{x^2 + y^2 + z^2}$.]

47. $\lim\limits_{(x,y,z)\to(0,0,0)} (x^2 + y^2 + z^2) = 0$.

48. $\lim\limits_{(x,y,z)\to(0,0,0)} e^{\sqrt{x^2+y^2+z^2}} = 1$.

■ **16.3** PARTIAL DERIVATIVES

*If f is a function of two or more independent variables and all but one of those variables are held fixed, then the derivative of f with respect to that one remaining independent variable is called a **partial derivative** of f. In this section we shall show how to compute partial derivatives and discuss their geometric significance.*

□ **PARTIAL DERIVATIVES OF FUNCTIONS OF TWO VARIABLES**

Let f be a function of x and y. If we hold y constant, say $y = y_0$, and view x as a variable, then $f(x, y_0)$ is a function of x alone. If this function is differentiable at $x = x_0$, then the value of this derivative is denoted by

$$f_x(x_0, y_0) \tag{1}$$

and is called the ***partial derivative of f with respect to x*** at the point (x_0, y_0). Similarly, if we hold x constant, say $x = x_0$, then $f(x_0, y)$ is a function of y alone. If this function is differentiable at $y = y_0$, then the value of this derivative is denoted by

$$f_y(x_0, y_0) \tag{2}$$

and is called the ***partial derivative of f with respect to y*** at (x_0, y_0).

The partial derivatives of $f(x, y)$ have a simple geometric interpretation. Let P be a point on the intersection of the surface $z = f(x, y)$ and the plane $y = y_0$. If y is held constant at $y = y_0$ and x is allowed to vary, then the point P moves along the curve C_1 that is the intersection of the surface with the vertical plane $y = y_0$ (Figure 16.3.1a). Thus, the partial derivative $f_x(x_0, y_0)$ can be interpreted as the slope (change in z per unit increase in x) of the tangent line to the curve C_1 at the point (x_0, y_0). Similarly, if x is held constant, say $x = x_0$, and y is allowed to vary, then the point P moves along the curve C_2 that is the intersection of the surface with the vertical plane $x = x_0$. Thus, the partial derivative $f_y(x_0, y_0)$ can be interpreted as the slope of the tangent line (change in z per unit increase in y) to the curve C_2 at the point (x_0, y_0) (Figure 16.3.1b).

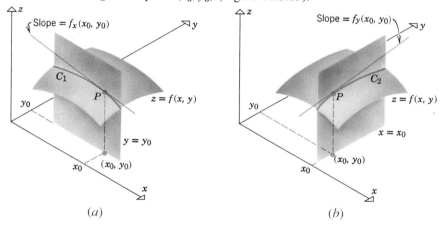

Figure 16.3.1 (a) (b)

REMARK.   In this text we shall only consider partial derivatives at *interior points* of the domain of f. Partial derivatives at boundary points lead to complications that are best left for more advanced courses.

The values of $f_x(x_0, y_0)$ and $f_y(x_0, y_0)$ are usually obtained by finding expressions for $f_x(x, y)$ and $f_y(x, y)$ at a general point (x, y) and then substituting $x = x_0$ and $y = y_0$ in these expressions. To obtain $f_x(x, y)$ we differentiate $f(x, y)$ with respect to x, *treating y as a constant*; and to obtain $f_y(x, y)$ we differentiate $f(x, y)$ with respect to y, *treating x as a constant*.

Example 1   Find $f_x(1, 2)$ and $f_y(1, 2)$ if $f(x, y) = 2x^3y^2 + 2y + 4x$.

Solution.   Treating y as a constant and differentiating with respect to x, we obtain

$$f_x(x, y) = 6x^2y^2 + 4$$

Treating x as a constant and differentiating with respect to y, we obtain

$$f_y(x, y) = 4x^3y + 2$$

Substituting $x = 1$ and $y = 2$ in these partial-derivative formulas yields

$$f_x(1, 2) = 6(1)^2(2)^2 + 4 = 28$$
$$f_y(1, 2) = 4(1)^3(2) + 2 = 10 \quad \blacktriangleleft$$

☐ **PARTIAL-DERIVATIVE NOTATION**

If $z = f(x, y)$, then the partial derivatives f_x and f_y can also be denoted by the symbols*

$$\frac{\partial f}{\partial x}, \quad \frac{\partial z}{\partial x} \quad \text{and} \quad \frac{\partial f}{\partial y}, \quad \frac{\partial z}{\partial y}$$

Some typical notations for the partial derivatives at a point (x_0, y_0) are

$$\frac{\partial f}{\partial x}\bigg|_{x=x_0,\, y=y_0}, \quad \frac{\partial z}{\partial y}\bigg|_{(x_0,\, y_0)}, \quad \frac{\partial f}{\partial x}\bigg|_{(x_0,\, y_0)}, \quad \frac{\partial f}{\partial x}(x_0, y_0)$$

Example 2   Find $\partial z/\partial x$ and $\partial z/\partial y$ if $z = x^4 \sin (xy^3)$.

Solution.

$$\frac{\partial z}{\partial x} = \frac{\partial}{\partial x}[x^4 \sin (xy^3)] = x^4 \frac{\partial}{\partial x}[\sin (xy^3)] + \sin (xy^3) \cdot \frac{\partial}{\partial x}(x^4)$$

$$= x^4 \cos (xy^3) \cdot y^3 + \sin (xy^3) \cdot 4x^3 = x^4 y^3 \cos (xy^3) + 4x^3 \sin (xy^3)$$

$$\frac{\partial z}{\partial y} = \frac{\partial}{\partial y}[x^4 \sin (xy^3)] = x^4 \frac{\partial}{\partial y}[\sin (xy^3)] + \sin (xy^3) \cdot \frac{\partial}{\partial y}(x^4)$$

$$= x^4 \cos (xy^3) \cdot 3xy^2 + \sin (xy^3) \cdot 0 = 3x^5 y^2 \cos (xy^3) \quad \blacktriangleleft$$

Example 3   Suppose that a point Q moves along the intersection of the sphere $x^2 + y^2 + z^2 = 1$ with the plane $x = \frac{2}{3}$. At what rate is z changing with respect to y when the point is at $P(\frac{2}{3}, \frac{1}{3}, \frac{2}{3})$?

Solution.   Since the z-coordinate of the point $P(\frac{2}{3}, \frac{1}{3}, \frac{2}{3})$ is positive, this point lies on the upper hemisphere

$$z = \sqrt{1 - x^2 - y^2} \tag{3}$$

and hence for each fixed value of x, the rate of change of z with respect to y on the upper hemisphere is

$$\frac{\partial z}{\partial y} = \frac{\partial}{\partial y}[(1 - x^2 - y^2)^{1/2}] = \frac{1}{2}(1 - x^2 - y^2)^{-1/2}(-2y) = -\frac{y}{\sqrt{1 - x^2 - y^2}}$$

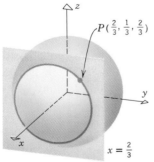

$P(\frac{2}{3}, \frac{1}{3}, \frac{2}{3})$

$x = \frac{2}{3}$

Figure 16.3.2

In particular, if $x = \frac{2}{3}$ (Figure 16.3.2), then it follows from this equation that the rate of change of z with respect to y at the point P is

$$\frac{\partial z}{\partial y}\bigg|_{x=\frac{2}{3},\, y=\frac{1}{3}} = -\frac{\frac{1}{3}}{\sqrt{1 - (\frac{2}{3})^2 - (\frac{1}{3})^2}} = -\frac{1}{2}$$

Alternative Solution.   Instead of solving $x^2 + y^2 + z^2 = 1$ explicitly for z as a function of x and y, we can obtain $\partial z/\partial y$ by implicit differentiation. Differentiating both sides of $x^2 + y^2 + z^2 = 1$ with respect to y and treating z as a function of x and y yields

$$\frac{\partial}{\partial y}[x^2 + y^2 + z^2] = \frac{\partial}{\partial y}[1]$$

$$2y + 2z\frac{\partial z}{\partial y} = 0$$

$$\frac{\partial z}{\partial y} = -\frac{y}{z}$$

Substituting the y- and z-coordinates of the point $(\frac{2}{3}, \frac{1}{3}, \frac{2}{3})$ yields $-\frac{1}{2}$ as before. $\blacktriangleleft$

---

*The symbol ∂ is called a partial derivative sign. It is derived from the Cyrillic alphabet.

Example 4 Suppose that $D = \sqrt{x^2 + y^2}$ is the length of the diagonal of a rectangle whose sides have lengths x and y that are allowed to vary. Find a formula for the rate of change of D with respect to x if x varies with y held constant, and use this formula to find the rate of change of D with respect to x at the instant when $x = 3$ and $y = 4$.

Solution. The instantaneous rate of change of D with respect to x with y held constant is

$$\frac{\partial D}{\partial x} = \frac{1}{2}(x^2 + y^2)^{-1/2}(2x) = \frac{x}{\sqrt{x^2 + y^2}}$$

from which it follows that

$$\left.\frac{\partial D}{\partial x}\right|_{x=3,\, y=4} = \frac{3}{\sqrt{3^2 + 4^2}} = \frac{3}{5}$$

Thus, D is increasing at a rate of $\frac{3}{5}$ unit per unit increase in x at the point $(3, 4)$. ◀

□ **HIGHER-ORDER PARTIAL DERIVATIVES**

Since the partial derivatives $\partial f/\partial x$ and $\partial f/\partial y$ are functions of x and y, each can have partial derivatives. This gives rise to four possible *second-order* partial derivatives of f, which are defined by

$$\frac{\partial^2 f}{\partial x^2} = \frac{\partial}{\partial x}\left(\frac{\partial f}{\partial x}\right), \qquad \frac{\partial^2 f}{\partial y^2} = \frac{\partial}{\partial y}\left(\frac{\partial f}{\partial y}\right)$$

$$\frac{\partial^2 f}{\partial x\, \partial y} = \frac{\partial}{\partial x}\left(\frac{\partial f}{\partial y}\right), \qquad \frac{\partial^2 f}{\partial y\, \partial x} = \frac{\partial}{\partial y}\left(\frac{\partial f}{\partial x}\right)$$

Henceforth, we shall call $\partial f/\partial x$ and $\partial f/\partial y$ the *first-order* partial derivatives of f.

Example 5 Find the second-order partial derivatives of $f(x, y) = x^2 y^3 + x^4 y$.

Solution. We have

$$\frac{\partial f}{\partial x} = 2xy^3 + 4x^3 y \quad \text{and} \quad \frac{\partial f}{\partial y} = 3x^2 y^2 + x^4$$

so that

$$\frac{\partial^2 f}{\partial x^2} = \frac{\partial}{\partial x}\left(\frac{\partial f}{\partial x}\right) = \frac{\partial}{\partial x}(2xy^3 + 4x^3 y) = 2y^3 + 12x^2 y$$

$$\frac{\partial^2 f}{\partial y^2} = \frac{\partial}{\partial y}\left(\frac{\partial f}{\partial y}\right) = \frac{\partial}{\partial y}(3x^2 y^2 + x^4) = 6x^2 y$$

$$\frac{\partial^2 f}{\partial x\, \partial y} = \frac{\partial}{\partial x}\left(\frac{\partial f}{\partial y}\right) = \frac{\partial}{\partial x}(3x^2 y^2 + x^4) = 6xy^2 + 4x^3$$

$$\frac{\partial^2 f}{\partial y\, \partial x} = \frac{\partial}{\partial y}\left(\frac{\partial f}{\partial x}\right) = \frac{\partial}{\partial y}(2xy^3 + 4x^3 y) = 6xy^2 + 4x^3 \quad ◀$$

REMARK. The derivatives

$$\frac{\partial^2 f}{\partial y\, \partial x} \quad \text{and} \quad \frac{\partial^2 f}{\partial x\, \partial y}$$

are called the *mixed second-order partial derivatives* or *mixed second-partials*. For most functions that arise in applications, these mixed partial derivatives are equal (as in the last example). In the next section we shall state precise conditions under which equality holds.

By successively differentiating, we can obtain third-order partial derivatives and higher. Some possibilities are

$$\frac{\partial^3 f}{\partial x^3} = \frac{\partial}{\partial x}\left(\frac{\partial^2 f}{\partial x^2}\right), \qquad \frac{\partial^3 f}{\partial y^2\,\partial x} = \frac{\partial}{\partial y}\left(\frac{\partial^2 f}{\partial y\,\partial x}\right)$$

$$\frac{\partial^3 f}{\partial y\,\partial x^2} = \frac{\partial}{\partial y}\left(\frac{\partial^2 f}{\partial x^2}\right), \qquad \frac{\partial^4 f}{\partial y^2\,\partial x^2} = \frac{\partial}{\partial y}\left(\frac{\partial^3 f}{\partial y\,\partial x^2}\right)$$

Higher-order partial derivatives can be denoted more compactly with subscript notation. For example,

$$\frac{\partial^2 f}{\partial y\,\partial x} = \frac{\partial}{\partial y}\left(\frac{\partial f}{\partial x}\right) = \frac{\partial}{\partial y}(f_x) = (f_x)_y$$

It is usual to drop the parentheses and write simply

$$\frac{\partial^2 f}{\partial y\,\partial x} = f_{xy}$$

Note that in "∂" notation the sequence of differentiations is obtained by reading from right to left, but in the subscript notation it is left to right. Some other examples are

$$f_{xx} = \frac{\partial^2 f}{\partial x^2}, \quad f_{yyx} = \frac{\partial^3 f}{\partial x\,\partial y^2}, \quad f_{xxyy} = \frac{\partial^4 f}{\partial y^2\,\partial x^2}$$

Example 6   Let $f(x, y) = y^2 e^x + y$. Find f_{xyy}.

Solution.

$$f_{xyy} = \frac{\partial^3 f}{\partial y^2\,\partial x} = \frac{\partial^2}{\partial y^2}\left(\frac{\partial f}{\partial x}\right) = \frac{\partial^2}{\partial y^2}(y^2 e^x) = \frac{\partial}{\partial y}(2ye^x) = 2e^x \qquad \blacktriangleleft$$

☐ **PARTIAL DERIVATIVES OF FUNCTIONS WITH MORE THAN TWO VARIABLES**

For a function $f(x, y, z)$ of three variables, there are three *partial derivatives*:

$$f_x(x, y, z), \quad f_y(x, y, z), \quad f_z(x, y, z)$$

The partial derivative f_x is calculated by holding y and z constant and differentiating with respect to x. For f_y the variables x and z are held constant, and for f_z the variables x and y are held constant. If a dependent variable

$$w = f(x, y, z)$$

is used, then the three partial derivatives of f may be denoted by

$$\frac{\partial w}{\partial x}, \quad \frac{\partial w}{\partial y}, \quad \text{and} \quad \frac{\partial w}{\partial z}$$

Example 7   If $f(x, y, z) = x^3 y^2 z^4 + 2xy + z$, then

$$f_x(x, y, z) = 3x^2 y^2 z^4 + 2y$$

$$f_y(x, y, z) = 2x^3 yz^4 + 2x$$

$$f_z(x, y, z) = 4x^3 y^2 z^3 + 1$$

$$f_z(-1, 1, 2) = 4(-1)^3(1)^2(2)^3 + 1 = -31 \qquad \blacktriangleleft$$

Example 8   If $f(\rho, \theta, \phi) = \rho^2 \cos\phi \sin\theta$, then

$$f_\rho(\rho, \theta, \phi) = 2\rho \cos\phi \sin\theta$$

$$f_{\rho\phi}(\rho, \theta, \phi) = -2\rho \sin\phi \sin\theta$$

$$f_{\rho\phi\theta}(\rho, \theta, \phi) = -2\rho \sin\phi \cos\theta \qquad \blacktriangleleft$$

In general, if $f(v_1, v_2, \ldots, v_n)$ is a function of n variables, there are n partial derivatives of f, each of which is obtained by holding $n - 1$ of the variables fixed and differentiating the function f with respect to the remaining variable. If $w = f(v_1, v_2, \ldots, v_n)$, then these partial derivatives are denoted by

$$\frac{\partial w}{\partial v_1}, \frac{\partial w}{\partial v_2}, \ldots, \frac{\partial w}{\partial v_n}$$

where $\partial w / \partial v_i$ is obtained by holding all variables except v_i fixed and differentiating with respect to v_i.

Example 9 Find

$$\frac{\partial}{\partial x_i}[\sqrt{x_1^2 + x_2^2 + \cdots + x_n^2}]$$

for $i = 1, 2, \ldots, n$.

Solution. For each $i = 1, 2, \ldots, n$ we obtain

$$\frac{\partial}{\partial x_i}[\sqrt{x_1^2 + x_2^2 + \cdots + x_n^2}] = \frac{1}{2\sqrt{x_1^2 + x_2^2 + \cdots + x_n^2}} \cdot \frac{\partial}{\partial x_i}[x_1^2 + x_2^2 + \cdots + x_n^2]$$

$$= \frac{1}{2\sqrt{x_1^2 + x_2^2 + \cdots + x_n^2}}[2x_i] \quad \boxed{\text{All terms except } x_i^2 \text{ are constant.}}$$

$$= \frac{x_i}{\sqrt{x_1^2 + x_2^2 + \cdots + x_n^2}} \quad \blacktriangleleft$$

☐ **PARTIAL DERIVATIVES OF VECTOR-VALUED FUNCTIONS**

Partial derivatives of vector-valued functions of two or more variables are defined by taking partial derivatives of the components. For example, if

$$\mathbf{r}(u, v) = x(u, v)\mathbf{i} + y(u, v)\mathbf{j} + z(u, v)\mathbf{k}$$

then

$$\frac{\partial \mathbf{r}}{\partial u} = \frac{\partial x}{\partial u}\mathbf{i} + \frac{\partial y}{\partial u}\mathbf{j} + \frac{\partial z}{\partial u}\mathbf{k}$$

$$\frac{\partial \mathbf{r}}{\partial v} = \frac{\partial x}{\partial v}\mathbf{i} + \frac{\partial y}{\partial v}\mathbf{j} + \frac{\partial z}{\partial v}\mathbf{k}$$

These derivatives can also be expressed as limits:

$$\frac{\partial \mathbf{r}}{\partial u} = \lim_{h \to 0} \frac{\mathbf{r}(u + h, v) - \mathbf{r}(u, v)}{h} \tag{4}$$

$$\frac{\partial \mathbf{r}}{\partial v} = \lim_{k \to 0} \frac{\mathbf{r}(u, v + k) - \mathbf{r}(u, v)}{k} \tag{5}$$

Suppose now that σ denotes the surface represented parametrically by $\mathbf{r}(u, v)$ for (u, v) in a region R of the uv-plane and that $\partial \mathbf{r}/\partial u$ and $\partial \mathbf{r}/\partial v$ are continuous on R. It follows from 15.2.5 that if $\partial \mathbf{r}/\partial u \neq \mathbf{0}$ at a point (u_0, v_0), then $\partial \mathbf{r}/\partial u$ is tangent to the u-curve at the point $\mathbf{r}(u_0, v_0)$ and points in the direction of increasing u; and similarly, if $\partial \mathbf{r}/\partial v \neq \mathbf{0}$ at (u_0, v_0), then $\partial \mathbf{r}/\partial v$ is tangent to the v-curve at $\mathbf{r}(u_0, v_0)$ and points in the direction of increasing v. Thus, if $\partial \mathbf{r}/\partial u \times \partial \mathbf{r}/\partial v \neq \mathbf{0}$, then the vector

$$\mathbf{n} = \frac{\dfrac{\partial \mathbf{r}}{\partial u} \times \dfrac{\partial \mathbf{r}}{\partial v}}{\left\| \dfrac{\partial \mathbf{r}}{\partial u} \times \dfrac{\partial \mathbf{r}}{\partial v} \right\|} \tag{6}$$

is a unit vector that is perpendicular to both $\partial \mathbf{r}/\partial u$ and $\partial \mathbf{r}/\partial v$ at the point $\mathbf{r}(u_0, v_0)$. We call $\mathbf{n}$ the ***unit normal*** to σ at the point $\mathbf{r}(u_0, v_0)$, and we define the ***tangent plane*** to σ at $\mathbf{r}(u_0, v_0)$ to be the plane through $\mathbf{r}(u_0, v_0)$ that has $\mathbf{n}$ as a normal (Figure 16.3.3).

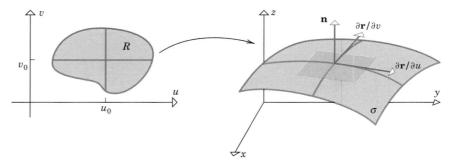

Figure 16.3.3

Example 10   Show that the tangent plane is perpendicular to the radius vector at each point of the sphere of radius a

$$\mathbf{r}(\phi, \theta) = a \sin \phi \cos \theta \mathbf{i} + a \sin \phi \sin \theta \mathbf{j} + a \cos \phi \mathbf{k}$$

where $0 \le \phi \le \pi$ and $0 \le \theta \le 2\pi$.

Solution.   It is sufficient to show that at each point of the sphere the unit normal vector $\mathbf{n}$ is a scalar multiple of $\mathbf{r}$ (and hence parallel to $\mathbf{r}$). But

$$\frac{\partial \mathbf{r}}{\partial \phi} \times \frac{\partial \mathbf{r}}{\partial \theta} = \begin{vmatrix} \mathbf{i} & \mathbf{j} & \mathbf{k} \\ \dfrac{\partial x}{\partial \phi} & \dfrac{\partial y}{\partial \phi} & \dfrac{\partial z}{\partial \phi} \\ \dfrac{\partial x}{\partial \theta} & \dfrac{\partial y}{\partial \theta} & \dfrac{\partial z}{\partial \theta} \end{vmatrix} = \begin{vmatrix} \mathbf{i} & \mathbf{j} & \mathbf{k} \\ a \cos \phi \cos \theta & a \cos \phi \sin \theta & -a \sin \phi \\ -a \sin \phi \sin \theta & a \sin \phi \cos \theta & 0 \end{vmatrix}$$

$$= a^2 \sin^2 \phi \cos \theta \mathbf{i} + a^2 \sin^2 \phi \sin \theta \mathbf{j} + a^2 \sin \phi \cos \phi \mathbf{k}$$

and hence

$$\left\| \frac{\partial \mathbf{r}}{\partial \phi} \times \frac{\partial \mathbf{r}}{\partial \theta} \right\| = \sqrt{a^4 \sin^4 \phi \cos^2 \theta + a^4 \sin^4 \phi \sin^2 \theta + a^4 \sin^2 \phi \cos^2 \phi}$$

$$= \sqrt{a^4 \sin^4 \phi + a^4 \sin^2 \phi \cos^2 \phi}$$

$$= a^2 \sqrt{\sin^2 \phi} = a^2 |\sin \phi| = a^2 \sin \phi$$

Thus, it follows from (6) that

$$\mathbf{n} = \sin \phi \cos \theta \mathbf{i} + \sin \phi \sin \theta \mathbf{j} + \cos \phi \mathbf{k} = \frac{1}{a} \mathbf{r}$$

which is also evident geometrically.   ◀

▶ Exercise Set 16.3

In Exercises 1–6, find $\partial z/\partial x$ and $\partial z/\partial y$.

1. $z = 3x^3 y^2$.

2. $z = 4x^2 - 2y + 7x^4 y^5$.

3. $z = 4e^{x^2 y^3}$.   **4.** $z = \cos(x^5 y^4)$.

5. $z = x^3 \ln(1 + xy^{-3/5})$.   **6.** $z = e^{xy} \sin 4y^2$.

In Exercises 7–12, find $f_x(x, y)$ and $f_y(x, y)$.

7. $f(x, y) = \sqrt{3x^5 y - 7x^3 y}$.

8. $f(x, y) = \dfrac{x + y}{x - y}$.

9. $f(x, y) = y^{-3/2} \tan^{-1}(x/y)$.

10. $f(x, y) = x^3 e^{-y} + y^3 \sec \sqrt{x}$.

11. $f(x, y) = (y^2 \tan x)^{-4/3}$.

12. $f(x, y) = \cosh(\sqrt{x}) \sinh^2(xy^2)$.

13. Given $f(x, y) = 9 - x^2 - 7y^3$, find

   (a) $f_x(3, 1)$ (b) $f_y(3, 1)$.

14. Given $f(x, y) = x^2 y e^{xy}$, find

   (a) $\left. \dfrac{\partial f}{\partial x} \right|_{(1, 1)}$ (b) $\left. \dfrac{\partial f}{\partial y} \right|_{(1, 1)}$.

15. Given $z = \sqrt{x^2 + 4y^2}$, find

   (a) $\left. \dfrac{\partial z}{\partial x} \right|_{(1, 2)}$ (b) $\left. \dfrac{\partial z}{\partial y} \right|_{(1, 2)}$.

16. Given $w = x^2 \cos xy$, find

   (a) $\dfrac{\partial w}{\partial x}\left(\tfrac{1}{2}, \pi\right)$ (b) $\dfrac{\partial w}{\partial y}\left(\tfrac{1}{2}, \pi\right)$.

In Exercises 17–20, calculate $\partial z / \partial x$ and $\partial z / \partial y$ using implicit differentiation. Leave your answers in terms of x, y, and z.

---

17. $(x^2 + y^2 + z^2)^{3/2} = 1$.   18. $\ln(2x^2 + y - z^3) = x$.

19. $x^2 + z \sin xyz = 0$.   20. $e^{xy} \sinh z - z^2 x + 1 = 0$.

In Exercises 21–26, find f_{xx}, f_{yy}, f_{xy}, and f_{yx}.

---

21. $f(x, y) = 4x^2 - 8xy^4 + 7y^5 - 3$.

22. $f(x, y) = \sqrt{x^2 + y^2}$.   23. $f(x, y) = e^x \cos y$.

24. $f(x, y) = e^{x - y^2}$.   25. $f(x, y) = \ln(4x - 5y)$.

26. $f(x, y) = (x^2 - y^2)/(x^2 + y^2)$.

27. Given $f(x, y) = x^3 y^5 - 2x^2 y + x$, find

   (a) f_{xxy} (b) f_{yxy} (c) f_{yyy}.

28. Given $z = (2x - y)^5$, find

   (a) $\dfrac{\partial^3 z}{\partial y \, \partial x \, \partial y}$ (b) $\dfrac{\partial^3 z}{\partial x^2 \, \partial y}$ (c) $\dfrac{\partial^4 z}{\partial x^2 \, \partial y^2}$.

29. Given $f(x, y) = y^3 e^{-5x}$, find

   (a) $f_{xyy}(0, 1)$ (b) $f_{xxx}(0, 1)$ (c) $f_{yyxx}(0, 1)$.

30. Given $w = e^y \cos x$, find

   (a) $\left. \dfrac{\partial^3 w}{\partial y^2 \, \partial x} \right|_{(\pi/4, 0)}$ (b) $\left. \dfrac{\partial^3 w}{\partial x^2 \, \partial y} \right|_{(\pi/4, 0)}$.

31. Express the following derivatives in "∂" notation.

   (a) f_{xxx} (b) f_{xyy} (c) f_{yyxx} (d) f_{xyyy}.

32. Express the derivatives in "subscript" notation.

   (a) $\dfrac{\partial^3 f}{\partial y^2 \, \partial x}$ (b) $\dfrac{\partial^4 f}{\partial x^4}$

   (c) $\dfrac{\partial^4 f}{\partial y^2 \, \partial x^2}$ (d) $\dfrac{\partial^5 f}{\partial x^2 \, \partial y^3}$.

In Exercises 33–37, find $\partial w / \partial x$, $\partial w / \partial y$, and $\partial w / \partial z$.

---

33. $w = x^2 y^4 z^3 + xy + z^2 + 1$.

34. $w = ye^z \sin x$.

35. $w = \dfrac{x^2 - y^2}{y^2 + z^2}$.

36. $w = y^3 e^{2x + 3z}$.

37. $w = \sqrt{x^2 + y^2 + z^2}$.

In Exercises 38–42, find f_x, f_y, and f_z.

---

38. $f(x, y, z) = z \ln(x^2 y \cos z)$.

39. $f(x, y, z) = \tan^{-1}\left(\dfrac{1}{xy^2 z^3}\right)$.

40. $f(x, y, z) = y^{-3/2} \sec\left(\dfrac{xz}{y}\right)$.

41. $f(x, y, z) = \cosh(\sqrt{z}) \sinh^2(x^2 yz)$.

42. $f(x, y, z) = \left(\dfrac{xz}{1 - z^2 - y^2}\right)^{-3/4}$.

43. Let $f(x, y, z) = 5x^2 yz^3$. Find

   (a) $f_x(1, -1, 2)$ (b) $f_y(1, -1, 2)$

   (c) $f_z(1, -1, 2)$.

44. Let $f(x, y, z) = y^2 e^{xz}$. Find

   (a) $\partial f / \partial x|_{(1, 1, 1)}$ (b) $\partial f / \partial y|_{(1, 1, 1)}$

   (c) $\partial f / \partial z|_{(1, 1, 1)}$.

45. Let $w = \sqrt{x^2 + 4y^2 - z^2}$. Find

   (a) $\partial w / \partial x|_{(2, 1, -1)}$ (b) $\partial w / \partial y|_{(2, 1, -1)}$

   (c) $\partial w / \partial z|_{(2, 1, -1)}$.

46. Let $w = x \sin xyz$. Find

   (a) $\dfrac{\partial w}{\partial x}\left(1, \tfrac{1}{2}, \pi\right)$ (b) $\dfrac{\partial w}{\partial y}\left(1, \tfrac{1}{2}, \pi\right)$

   (c) $\dfrac{\partial w}{\partial z}\left(1, \tfrac{1}{2}, \pi\right)$.

In Exercises 47–50, find $\partial w / \partial x$, $\partial w / \partial y$, and $\partial w / \partial z$ using implicit differentiation. Leave your answers in terms of x, y, z, and w.

---

47. $(x^2 + y^2 + z^2 + w^2)^{3/2} = 4$.

48. $\ln(2x^2 + y - z^3 + 3w) = z$.

49. $w^2 + w \sin xyz = 1$.

50. $e^{xy} \sinh w - z^2 w + 1 = 0$.

51. Let $f(x, y, z) = x^3 y^5 z^7 + xy^2 + y^3 z$. Find

   (a) f_{xy} (b) f_{yz} (c) f_{xz} (d) f_{zz}

   (e) f_{zyy} (f) f_{xxy} (g) f_{zyx} (h) f_{xxyz}.

52. Let $w = (4x - 3y + 2z)^5$. Find

   (a) $\dfrac{\partial^2 w}{\partial x \, \partial z}$ (b) $\dfrac{\partial^3 w}{\partial x \, \partial y \, \partial z}$ (c) $\dfrac{\partial^4 w}{\partial z^2 \, \partial y \, \partial x}$.

53. Show that the functions in parts (a)–(c) satisfy **Laplace's equation**

$$\frac{\partial^2 f}{\partial x^2} + \frac{\partial^2 f}{\partial y^2} = 0$$

(a) $f(x, y) = e^x \sin y + e^y \cos x$

(b) $f(x, y) = \ln (x^2 + y^2)$

(c) $f(x, y) = \tan^{-1} \dfrac{2xy}{x^2 - y^2}$.

54. Show that $u(x, y)$ and $v(x, y)$ satisfy

$$\frac{\partial u}{\partial x} = \frac{\partial v}{\partial y} \quad \text{and} \quad \frac{\partial u}{\partial y} = -\frac{\partial v}{\partial x}$$

(These are called the **Cauchy–Riemann equations**.)

(a) $u = x^2 - y^2$, $v = 2xy$

(b) $u = e^x \cos y$, $v = e^x \sin y$

(c) $u = \ln (x^2 + y^2)$, $v = 2 \tan^{-1} (y/x)$.

55. A point moves along the intersection of the elliptic paraboloid $z = x^2 + 3y^2$ and the plane $x = 2$. At what rate is z changing with y when the point is at $(2, 1, 7)$?

56. A point moves along the intersection of the surface $z = \sqrt{29 - x^2 - y^2}$ and the plane $y = 3$. At what rate is z changing with x when the point is at $(4, 3, 2)$?

57. Find the slope of the tangent line at $(-1, 1, 5)$ to the curve of intersection of the surface $z = x^2 + 4y^2$ and

(a) the plane $x = -1$ (b) the plane $y = 1$.

58. Find the slope of the tangent line at $(2, 1, 2)$ to the curve of intersection of the surface $x^2 + y^2 + z^2 = 9$ and

(a) the plane $x = 2$ (b) the plane $y = 1$.

59. The volume V of a right-circular cylinder is given by $V = \pi r^2 h$, where r is the radius and h is the height.

(a) Find a formula for the instantaneous rate of change of V with respect to r if h remains constant.

(b) Find a formula for the instantaneous rate of change of V with respect to h if r remains constant.

(c) Suppose that h has a constant value of 4 in., but r varies. Find the rate of change of V with respect to r at the instant when $r = 6$ in.

(d) Suppose that r has a constant value of 8 in., but h varies. Find the instantaneous rate of change of V with respect to h at the instant when $h = 10$ in.

60. The volume V of a right-circular cone is given by

$$V = \frac{\pi}{24} d^2 \sqrt{4s^2 - d^2}$$

where s is the slant height and d is the diameter of the base.

(a) Find a formula for the instantaneous rate of change of V with respect to s if d remains constant.

(b) Find a formula for the instantaneous rate of change of V with respect to d if s remains constant.

(c) Suppose that d has a constant value of 16 cm, but s varies. Find the rate of change of V with respect to s at the instant when $s = 10$ cm.

(d) Suppose that s has a constant value of 10 cm, but d varies. Find the rate of change of V with respect to d at the instant when $d = 16$ cm.

61. According to the ideal gas law, the pressure, temperature, and volume of a gas are related by $P = kT/V$. Suppose that for a certain gas, $k = 10$.

(a) Find the instantaneous rate of change of pressure (lb/in^2) with respect to temperature if the temperature is 80 K and the volume remains fixed at 50 in^3.

(b) Find the instantaneous rate of change of volume with respect to pressure if the volume is 50 in^3 and the temperature remains fixed at 80 K.

62. Find parametric equations for the tangent line at $(1, 3, 3)$ to the curve of intersection of the surface $z = x^2 y$ and

(a) the plane $x = 1$ (b) the plane $y = 3$.

63. Suppose that $\sin (x + z) + \sin (x - y) = 1$. Use implicit differentiation to find $\partial z/\partial x$, $\partial z/\partial y$, and $\partial^2 z/\partial x \, \partial y$ in terms of x, y, and z.

64. The volume of a right-circular cone of radius r and height h is $V = \frac{1}{3}\pi r^2 h$. Show that if the height remains constant while the radius changes, then the volume satisfies

$$\frac{\partial V}{\partial r} = \frac{2V}{r}$$

65. The temperature at a point (x, y) on a metal plate in the xy-plane is $T(x, y) = x^3 + 2y^2 + x$ degrees. Assume that distance is measured in centimeters and find the rate at which temperature changes with distance if we start at the point $(1, 2)$ and move

(a) to the right and parallel to the x-axis

(b) upward and parallel to the y-axis.

66. When two resistors having resistances R_1 ohms and R_2 ohms are connected in parallel, their combined resistance R in ohms is $R = R_1 R_2/(R_1 + R_2)$. Show:

$$\frac{\partial^2 R}{\partial R_1{}^2} \frac{\partial^2 R}{\partial R_2{}^2} = \frac{4R^2}{(R_1 + R_2)^4}$$

67. Prove: If $u(x, y)$ and $v(x, y)$ each have equal mixed second partials, and if u and v satisfy the Cauchy–Riemann equations (Exercise 54), then u and v both satisfy Laplace's equation (Exercise 53).

68. Recall that for a function of one variable the derivative $f'(x)$ can be expressed as the limit

$$f'(x) = \lim_{h \to 0} \frac{f(x + h) - f(x)}{h}$$

Express $f_x(x, y)$ and $f_y(x, y)$ as limits.

In Exercises 69 and 70, the figures show some of the isotherms (curves of constant temperature) of a temperature function $T(x, y)$ for a thin metal plate in the xy-plane. Based on the figures, determine at the point P

(a) the signs of $\partial T/\partial x$ and $\partial T/\partial y$, and which is largest in absolute value

(b) the signs of $\partial^2 T/\partial x^2$, $\partial^2 T/\partial y^2$, $\partial^2 T/\partial y \, \partial x$, and $\partial^2 T/\partial x \, \partial y$.

69.

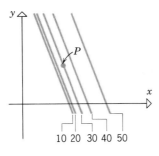

70.

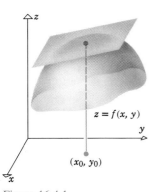

71. Let $f(x, y) = (x^2 + y^2)^{2/3}$. Show that

$$f_x(x, y) = \begin{cases} \dfrac{4x}{3(x^2 + y^2)^{1/3}}, & (x, y) \neq (0, 0) \\ 0, & (x, y) = (0, 0) \end{cases}$$

[This problem, due to Don Cohen, appeared in *Mathematics and Computer Education*, Vol. 25, No. 2, 1991, p. 179.]

72. Let $f(x, y) = (x^3 + y^3)^{1/3}$.

(a) Show that $f_y(0, 0) = 1$.

(b) At what points, if any, does $f_y(x, y)$ fail to exist?

■ **16.4 DIFFERENTIABILITY AND CHAIN RULES FOR FUNCTIONS OF TWO VARIABLES**

In this section we shall extend the notion of differentiability to functions of two variables and derive versions of the chain rule for these functions. We have restricted the discussion in this section to functions of two variables because some of the results we shall discuss have geometric interpretations that only apply to such functions. In a later section we shall extend the concepts developed here to functions of three or more variables.

☐ **DIFFERENTIABILITY OF FUNCTIONS OF TWO VARIABLES**

Recall that a function f of one variable is called differentiable at x_0 if it has a derivative at x_0, or, in other words, if the limit

$$f'(x_0) = \lim_{\Delta x \to 0} \frac{f(x_0 + \Delta x) - f(x_0)}{\Delta x} \tag{1}$$

exists. A function f that is differentiable at a point x_0 enjoys two important properties:

- $f(x)$ is continuous at x_0.
- The curve $y = f(x)$ has a nonvertical tangent line at x_0.

Our primary objective in this section is to extend the notion of differentiability to functions of two variables in such a way that the natural analogs of these two properties hold. More precisely, when $f(x, y)$ is differentiable at (x_0, y_0), we shall want it to be the case that

- $f(x, y)$ is continuous at (x_0, y_0).
- The surface $z = f(x, y)$ has a nonvertical tangent plane at (x_0, y_0) (Figure 16.4.1). (A precise definition of a tangent plane will be given later.)

It would not be unreasonable to guess that a function f of two variables should be called differentiable at (x_0, y_0) if the two partial derivatives $f_x(x_0, y_0)$ and $f_y(x_0, y_0)$ exist at

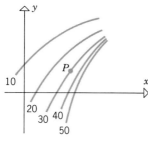

Figure 16.4.1

(x_0, y_0). Unfortunately, this condition is not strong enough to meet our objectives, since there are functions that have partial derivatives at a point, but are not continuous at that point. For example, the function

$$f(x, y) = \begin{cases} -1 & \text{if } x > 0 \text{ and } y > 0 \\ 0 & \text{otherwise} \end{cases}$$

is discontinuous at $(0, 0)$, but has partial derivatives at $(0, 0)$; these derivatives are $f_x(0, 0) = 0$ and $f_y(0, 0) = 0$. The discontinuity and the values of the partial derivatives should be evident from Figure 16.4.2.

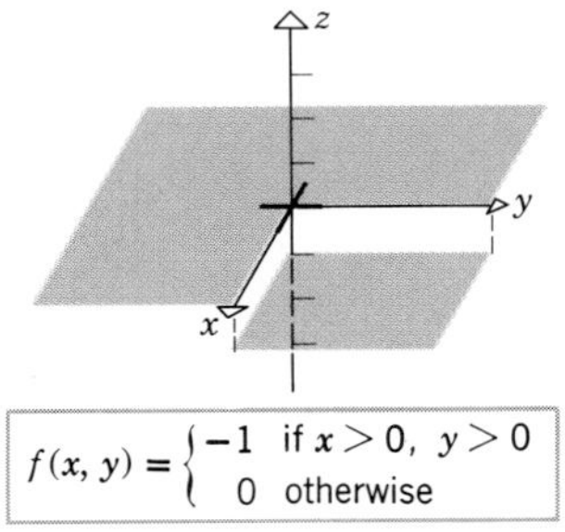

To motivate an appropriate definition of differentiability for functions of two variables, it will be helpful to reexamine the concept of differentiability for functions of one variable. Assuming, for the moment, that f is a function of one variable that is differentiable at $x = x_0$, it follows from Formula (8) of Section 3.7 that (1) can be rewritten as

$$f'(x_0) = \lim_{\Delta x \to 0} \frac{\Delta f}{\Delta x} \tag{2}$$

or, equivalently, as

$$\lim_{\Delta x \to 0} \left[\frac{\Delta f}{\Delta x} - f'(x_0) \right] = 0 \tag{3}$$

where

$$\Delta f = f(x_0 + \Delta x) - f(x_0)$$

If we now define ϵ by

$$\epsilon = \frac{\Delta f}{\Delta x} - f'(x_0) \tag{4}$$

then it follows from this formula that

$$\Delta f = f'(x_0) \Delta x + \epsilon \Delta x \tag{5}$$

where ϵ is a function of Δx. Using (4), the limit in (3) can be rewritten as

$$\lim_{\Delta x \to 0} \epsilon = 0 \tag{6}$$

Formulas (5) and (6) suggest the following alternative definition of differentiability for functions of one variable.

16.4.1   DEFINITION.   A function f of one variable is said to be ***differentiable*** at x_0 if there exists a number $f'(x_0)$ such that Δf can be written in the form

$$\Delta f = f'(x_0) \Delta x + \epsilon \Delta x \tag{7}$$

where ϵ is a function of Δx such that $\epsilon \to 0$ as $\Delta x \to 0$.

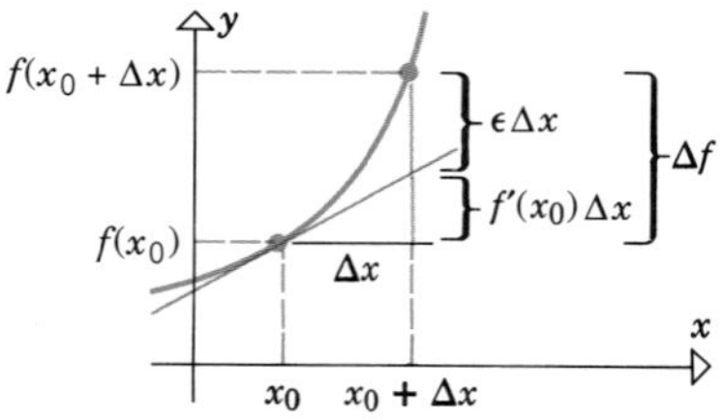

Although this definition of differentiability is more complicated than that given earlier in the text, it provides the basis for extending the notion of differentiability to functions of two or more variables. A geometric interpretation of the terms appearing in (7) is shown in Figure 16.4.3. The term Δf represents the change in height that results when a point moves along the graph of f as the x-coordinate changes from x_0 to $x_0 + \Delta x$; the term $f'(x_0) \Delta x$ represents the change in height that results when a point moves along the tangent line at $(x_0, f(x_0))$ as the x-coordinate changes from x_0 to $x_0 + \Delta x$; finally, the term $\epsilon \Delta x$ represents the difference between Δf and $f'(x_0) \Delta x$. It is evident from Figure 16.4.3 that $\epsilon \Delta x \to 0$ as $\Delta x \to 0$. However, (7) actually makes the stronger statement that $\epsilon \to 0$ as $\Delta x \to 0$. This is not at all evident from Figure 16.4.3, but it follows from (6).

$$f(x, y) = \begin{cases} -1 & \text{if } x > 0, \ y > 0 \\ 0 & \text{otherwise} \end{cases}$$

Figure 16.4.2

Figure 16.4.3

If f is a function of x and y, then the symbol Δf, called the **increment** of f, denotes the change in the value of $f(x, y)$ that results when (x, y) varies from some initial position (x_0, y_0) to some new position $(x_0 + \Delta x, y_0 + \Delta y)$; thus,

$$\Delta f = f(x_0 + \Delta x, y_0 + \Delta y) - f(x_0, y_0) \tag{8}$$

(See Figure 16.4.4.) If a dependent variable $z = f(x, y)$ is used, then we shall sometimes write Δz rather than Δf.

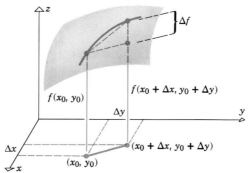

Figure 16.4.4

Motivated by Definition 16.4.1, we now make the following definition of differentiability for functions of two variables.

16.4.2 DEFINITION.   A function f of two variables is said to be **differentiable** at (x_0, y_0) if $f_x(x_0, y_0)$ and $f_y(x_0, y_0)$ exist and Δf can be written in the form

$$\Delta f = f_x(x_0, y_0)\,\Delta x + f_y(x_0, y_0)\,\Delta y + \epsilon_1\,\Delta x + \epsilon_2\,\Delta y \tag{9}$$

where ϵ_1 and ϵ_2 are functions of Δx and Δy such that $\epsilon_1 \to 0$ and $\epsilon_2 \to 0$ as $(\Delta x, \Delta y) \to (0, 0)$.

A function is called **differentiable on a region R** of the xy-plane if it is differentiable at each point of R. A function that is differentiable on the entire xy-plane is called **everywhere differentiable** or simply **differentiable**.

REMARK.   Before proceeding further, we note that for functions of one variable, the terms ''differentiable'' and ''has a derivative'' are synonymous. However, for functions of two variables, differentiability requires more than the existence of partial derivatives.

☐ **RELATIONSHIP BETWEEN DIFFERENTIABILITY AND CONTINUITY**

Earlier, we set two goals for our definition of differentiability: We wanted a function that is differentiable at (x_0, y_0) to be continuous at (x_0, y_0), and we wanted its graph to have a nonvertical tangent plane at (x_0, y_0). The next theorem shows that the continuity criterion is met; the existence of a nonvertical tangent plane will be demonstrated in the next section.

16.4.3 THEOREM.   *If f is differentiable at (x_0, y_0), then f is continuous at (x_0, y_0).*

Proof.   We must prove that

$$\lim_{(x, y) \to (x_0, y_0)} f(x, y) = f(x_0, y_0)$$

which, on letting $x = x_0 + \Delta x$ and $y = y_0 + \Delta y$, is equivalent to

$$\lim_{(\Delta x, \Delta y) \to (0, 0)} f(x_0 + \Delta x, y_0 + \Delta y) = f(x_0, y_0)$$

which from (8) is equivalent to

$$\lim_{(\Delta x, \Delta y) \to (0,0)} \Delta f = 0$$

But f is assumed to be differentiable at (x_0, y_0), so it follows from (9) that

$$\Delta f = f_x(x_0, y_0) \, \Delta x + f_y(x_0, y_0) \, \Delta y + \epsilon_1 \, \Delta x + \epsilon_2 \, \Delta y$$

where $\epsilon_1 \to 0$, $\epsilon_2 \to 0$ as $(\Delta x, \Delta y) \to (0, 0)$. Thus,

$$\lim_{(\Delta x, \Delta y) \to (0,0)} \Delta f = \lim_{(\Delta x, \Delta y) \to (0,0)} [f_x(x_0, y_0) \, \Delta x + f_y(x_0, y_0) \, \Delta y + \epsilon_1 \, \Delta x + \epsilon_2 \, \Delta y] = 0$$

which completes the proof. ∎

The next theorem, whose proof is usually studied in advanced calculus courses, provides simple conditions for a function of two variables to be differentiable at a point.

> **16.4.4**   THEOREM.   *If f has first-order partial derivatives at each point in some circular region centered at (x_0, y_0), and if these partial derivatives are continuous at (x_0, y_0), then f is differentiable at (x_0, y_0).*

Example 1   Show that $f(x, y) = x^3 y^4$ is a differentiable function.

Solution.   The partial derivatives $f_x = 3x^2 y^4$ and $f_y = 4x^3 y^3$ are defined and continuous everywhere in the xy-plane. Thus, the hypotheses of Theorem 16.4.4 are satisfied at each point (x_0, y_0) in the xy-plane, so $f(x, y) = x^3 y^4$ is everywhere differentiable.   ◄

The following important corollary follows from Theorems 16.4.3 and 16.4.4.

> **16.4.5**   COROLLARY.   *If f has first-order partial derivatives at each point of some circular region centered at (x_0, y_0), and if these partial derivatives are continuous at (x_0, y_0), then f is continuous at (x_0, y_0).*

☐ **EQUALITY OF MIXED PARTIALS**

The following theorem, which we state without proof, shows that with appropriate continuity restrictions the mixed second-order partial derivatives of a function of two variables are equal.

> **16.4.6**   THEOREM.   *Let f be a function of two variables. If f_x, f_y, f_{xy}, and f_{yx} are continuous on an open set, then $f_{xy} = f_{yx}$ at each point of the set.*

Example 2   Let $f(x, y) = 2e^{xy} \sin y$. It should be evident from the form of this function and from Theorem 16.2.4 that f and all its partial derivatives are continuous everywhere. Thus, Theorem 16.4.6 guarantees that $f_{xy} = f_{yx}$ everywhere. This is, in fact, the case since

$$f_x(x, y) = 2ye^{xy} \sin y = (2y \sin y)e^{xy}$$

$$f_{xy}(x, y) = (2y \sin y)(xe^{xy}) + e^{xy}(2y \cos y + 2 \sin y)$$

$$= 2e^{xy}(xy \sin y + y \cos y + \sin y)$$

$$f_y(x, y) = 2e^{xy} \cos y + 2xe^{xy} \sin y$$

$$f_{yx}(x, y) = 2ye^{xy} \cos y + 2xye^{xy} \sin y + 2e^{xy} \sin y$$

$$= 2e^{xy}(xy \sin y + y \cos y + \sin y)$$

so $f_{xy}(x, y) = f_{yx}(x, y)$ for all (x, y).   ◄

In general, the order of differentiation in an nth-order partial derivative can be changed without affecting the final result whenever the function and all its partial derivatives of order n or less are continuous. For example, if f and its partial derivatives of the first, second, and third orders are continuous on an open set, then at each point of that set,

$$f_{xyy} = f_{yxy} = f_{yyx}$$

or in another notation,

$$\frac{\partial^3 f}{\partial y^2 \, \partial x} = \frac{\partial^3 f}{\partial y \, \partial x \, \partial y} = \frac{\partial^3 f}{\partial x \, \partial y^2}$$

☐ **CHAIN RULES**

If y is a differentiable function of x and x is a differentiable function of t, then the chain rule for functions of one variable states that

$$\frac{dy}{dt} = \frac{dy}{dx}\frac{dx}{dt}$$

We shall now derive versions of the chain rule for functions of two variables.

Assume that z is a function of x and y, say

$$z = f(x, y) \tag{10}$$

and suppose that x and y, in turn, are functions of a single variable t, say

$$x = x(t), \quad y = y(t)$$

On substituting these functions of t in (10), we obtain the relationship

$$z = f\big(x(t), y(t)\big)$$

which expresses z as a function of the single variable t. Thus, we may ask for the derivative dz/dt, and we may inquire about its relationship to the derivatives $\partial z/\partial x$, $\partial z/\partial y$, dx/dt, and dy/dt.

16.4.7   THEOREM (*Chain Rule*). *If $x = x(t)$ and $y = y(t)$ are differentiable at t, and if $z = f(x, y)$ is differentiable at the point $(x(t), y(t))$, then $z = f(x(t), y(t))$ is differentiable at t, and*

$$\frac{dz}{dt} = \frac{\partial z}{\partial x}\frac{dx}{dt} + \frac{\partial z}{\partial y}\frac{dy}{dt} \tag{11}$$

Proof. From the derivative definition for functions of one variable

$$\frac{dz}{dt} = \lim_{\Delta t \to 0} \frac{\Delta z}{\Delta t} \tag{12}$$

Since $z = f(x, y)$ is differentiable at the point $(x, y) = (x(t), y(t))$, we can express Δz in the form

$$\Delta z = \frac{\partial z}{\partial x}\Delta x + \frac{\partial z}{\partial y}\Delta y + \epsilon_1 \Delta x + \epsilon_2 \Delta y \tag{13}$$

where the partial derivatives are evaluated at the point $(x(t), y(t))$ and $\epsilon_1 \to 0$, $\epsilon_2 \to 0$ as $(\Delta x, \Delta y) \to (0, 0)$. Thus, from (12) and (13),

$$\frac{dz}{dt} = \lim_{\Delta t \to 0} \frac{\Delta z}{\Delta t} = \lim_{\Delta t \to 0}\left[\frac{\partial z}{\partial x}\frac{\Delta x}{\Delta t} + \frac{\partial z}{\partial y}\frac{\Delta y}{\Delta t} + \epsilon_1 \frac{\Delta x}{\Delta t} + \epsilon_2 \frac{\Delta y}{\Delta t}\right] \tag{14}$$

But

$$\lim_{\Delta t \to 0} \frac{\Delta x}{\Delta t} = \frac{dx}{dt} \quad \text{and} \quad \lim_{\Delta t \to 0} \frac{\Delta y}{\Delta t} = \frac{dy}{dt}$$

Therefore, if we can show that $\epsilon_1 \to 0$, $\epsilon_2 \to 0$ as $\Delta t \to 0$, then the proof will be complete, since (14) will reduce to (11). But $\Delta x \to 0$ and $\Delta y \to 0$ as $\Delta t \to 0$, since

$$\lim_{\Delta t \to 0} \Delta x = \lim_{\Delta t \to 0} \frac{\Delta x}{\Delta t} \Delta t = \frac{dx}{dt} \cdot 0 = 0$$

and similarly for Δy. Thus, as Δt tends to zero, $(\Delta x, \Delta y) \to (0, 0)$, which implies that $\epsilon_1 \to 0$, $\epsilon_2 \to 0$. ∎

Example 3   Suppose that

$$z = x^2 y, \quad x = t^2, \quad y = t^3$$

Use the chain rule to find dz/dt, and check the result by expressing z as a function of t and differentiating directly.

Solution.   By the chain rule

$$\frac{dz}{dt} = \frac{\partial z}{\partial x} \frac{dx}{dt} + \frac{\partial z}{\partial y} \frac{dy}{dt} = (2xy)(2t) + (x^2)(3t^2)$$

$$= (2t^5)(2t) + (t^4)(3t^2) = 7t^6$$

Alternatively, we may express z directly as a function of t,

$$z = x^2 y = (t^2)^2 (t^3) = t^7$$

and then differentiate to obtain $dz/dt = 7t^6$. However, this procedure is not always convenient. ◄

Example 4   Suppose that

$$z = \sqrt{xy + y}, \quad x = \cos \theta, \quad y = \sin \theta$$

Use the chain rule to find $dz/d\theta$ when $\theta = \pi/2$.

Solution.   From the chain rule with θ in place of t,

$$\frac{dz}{d\theta} = \frac{\partial z}{\partial x} \frac{dx}{d\theta} + \frac{\partial z}{\partial y} \frac{dy}{d\theta}$$

we obtain

$$\frac{dz}{d\theta} = \frac{1}{2} (xy + y)^{-1/2} (y)(-\sin \theta) + \frac{1}{2} (xy + y)^{-1/2} (x + 1)(\cos \theta)$$

When $\theta = \pi/2$, we have

$$x = \cos \frac{\pi}{2} = 0, \quad y = \sin \frac{\pi}{2} = 1$$

Substituting $x = 0$, $y = 1$, $\theta = \pi/2$ in the formula for $dz/d\theta$ yields

$$\left. \frac{dz}{d\theta} \right|_{\theta = \pi/2} = \frac{1}{2} (1)(1)(-1) + \frac{1}{2} (1)(1)(0) = -\frac{1}{2}$$ ◄

REMARK.   There are many variations in derivative notations, each of which gives the chain rule a different look. If $z = f(x, y)$, where x and y are functions of t, then some possibilities are

$$\frac{dz}{dt} = f_x \frac{dx}{dt} + f_y \frac{dy}{dt}$$

$$\frac{df}{dt} = \frac{\partial f}{\partial x} \frac{dx}{dt} + \frac{\partial f}{\partial y} \frac{dy}{dt}$$

$$\frac{df}{dt} = f_x x'(t) + f_y y'(t)$$

The reader may be able to construct other variations as well.

In the special case where $z = F(x, y)$ and y is a differentiable function of x, chain-rule formula (11) yields

$$\frac{dz}{dx} = \frac{\partial F}{\partial x}\frac{dx}{dx} + \frac{\partial F}{\partial y}\frac{dy}{dx} = \frac{\partial F}{\partial x} + \frac{\partial F}{\partial y}\frac{dy}{dx} \tag{15}$$

This result can be used to find derivatives of functions that are defined implicitly. Suppose that the equation

$$F(x, y) = 0 \tag{16}$$

defines y implicitly as a differentiable function of x, and we are interested in finding dy/dx. Differentiating both sides of (16) with respect to x and applying (15) yields

$$\frac{\partial F}{\partial x} + \frac{\partial F}{\partial y}\frac{dy}{dx} = 0$$

or

$$\frac{dy}{dx} = -\frac{\partial F/\partial x}{\partial F/\partial y} \tag{17}$$

provided $\partial F/\partial y \neq 0$.

Example 5 Given that

$$x^3 + y^2 x - 3 = 0$$

find dy/dx using (17) and check the result using implicit differentiation.

Solution. By (17) with $F(x, y) = x^3 + y^2 x - 3$

$$\frac{dy}{dx} = -\frac{\partial F/\partial x}{\partial F/\partial y} = -\frac{3x^2 + y^2}{2yx}$$

On the other hand, differentiating the given equation implicitly yields

$$3x^2 + y^2 + x\left(2y\frac{dy}{dx}\right) - 0 = 0 \quad \text{or} \quad \frac{dy}{dx} = -\frac{3x^2 + y^2}{2yx}$$

which agrees with the result obtained by (17). ◀

In Theorem 16.4.7 the variables x and y are each functions of a single variable t. We now consider the case where x and y are each functions of two variables. Let

$$z = f(x, y) \tag{18}$$

and suppose that x and y are functions of u and v, say

$$x = x(u, v), \quad y = y(u, v)$$

On substituting these functions of u and v into (18), we obtain the relationship

$$z = f(x(u, v), y(u, v))$$

which expresses z as a function of the two variables u and v. Thus, we may ask for the partial derivatives $\partial z/\partial u$ and $\partial z/\partial v$; and we may inquire about the relationship between these derivatives and the derivatives $\partial z/\partial x$, $\partial z/\partial y$, $\partial x/\partial u$, $\partial x/\partial v$, $\partial y/\partial u$, and $\partial y/\partial v$.

> **16.4.8** THEOREM (**Chain Rule**). *If $x = x(u, v)$ and $y = y(u, v)$ have first-order partial derivatives at the point (u, v), and if $z = f(x, y)$ is differentiable at the point $(x(u, v), y(u, v))$, then $z = f(x(u, v), y(u, v))$ has first-order partial derivatives at (u, v) given by*
>
> $$\frac{\partial z}{\partial u} = \frac{\partial z}{\partial x}\frac{\partial x}{\partial u} + \frac{\partial z}{\partial y}\frac{\partial y}{\partial u} \qquad \text{and} \qquad \frac{\partial z}{\partial v} = \frac{\partial z}{\partial x}\frac{\partial x}{\partial v} + \frac{\partial z}{\partial y}\frac{\partial y}{\partial v}$$

Proof. If v is held fixed, then $x = x(u, v)$ and $y = y(u, v)$ become functions of u alone. Thus, we are back to the case of Theorem 16.4.7. If we apply that theorem with u in place of t, and if we use ∂ rather than d to indicate that the variable v is fixed, we obtain

$$\frac{\partial z}{\partial u} = \frac{\partial z}{\partial x}\frac{\partial x}{\partial u} + \frac{\partial z}{\partial y}\frac{\partial y}{\partial u}$$

The formula for $\partial z/\partial v$ is derived similarly. ∎

Example 6 Given that

$$z = e^{xy}, \quad x = 2u + v, \quad y = u/v$$

find $\partial z/\partial u$ and $\partial z/\partial v$ using the chain rule.

Solution.

$$\frac{\partial z}{\partial u} = \frac{\partial z}{\partial x}\frac{\partial x}{\partial u} + \frac{\partial z}{\partial y}\frac{\partial y}{\partial u} = (ye^{xy})(2) + (xe^{xy})(1/v) = \left[2y + \frac{x}{v}\right]e^{xy}$$

$$= \left[\frac{2u}{v} + \frac{2u + v}{v}\right]e^{(2u+v)(u/v)} = \left[\frac{4u}{v} + 1\right]e^{(2u+v)(u/v)}$$

$$\frac{\partial z}{\partial v} = \frac{\partial z}{\partial x}\frac{\partial x}{\partial v} + \frac{\partial z}{\partial y}\frac{\partial y}{\partial v} = (ye^{xy})(1) + (xe^{xy})\left(-\frac{u}{v^2}\right)$$

$$= \left[y - x\left(\frac{u}{v^2}\right)\right]e^{xy} = \left[\frac{u}{v} - (2u + v)\left(\frac{u}{v^2}\right)\right]e^{(2u+v)(u/v)}$$

$$= -\frac{2u^2}{v^2}e^{(2u+v)(u/v)} \qquad \blacktriangleleft$$

The chain rules are useful in related rates problems.

Example 7 At what rate is the area of a rectangle changing if its length is 15 ft and increasing at 3 ft/sec while its width is 6 ft and increasing at 2 ft/sec?

Solution. Let

$x = $ length of the rectangle in feet

$y = $ width of the rectangle in feet

$A = $ area of the rectangle in square feet

$t = $ time in seconds

We are given that

$$\frac{dx}{dt} = 3 \quad \text{and} \quad \frac{dy}{dt} = 2 \tag{19}$$

at the instant when

$$x = 15, \quad y = 6 \tag{20}$$

We want to find dA/dt at that instant.

From the area formula $A = xy$, we obtain

$$\frac{dA}{dt} = \frac{\partial A}{\partial x}\frac{dx}{dt} + \frac{\partial A}{\partial y}\frac{dy}{dt} = y\frac{dx}{dt} + x\frac{dy}{dt}$$

Substituting (19) and (20) in this equation yields

$$\frac{dA}{dt} = 6(3) + 15(2) = 48$$

Thus, the area is increasing at a rate of 48 ft^2/sec at the given instant.   ◀

▶ Exercise Set 16.4

1. Find Δf given that $f(x, y) = x^2 y$, $(x_0, y_0) = (1, 3)$, $\Delta x = 0.1$, and $\Delta y = 0.2$.

2. Find Δz given that $z = 3x^2 - 2y$, $(x_0, y_0) = (-2, 4)$, $\Delta x = 0.02$, and $\Delta y = -0.03$.

3. Find the increment of $f(x, y) = x/y$ as (x, y) varies from $(-1, 2)$ to $(3, 1)$.

4. Find the increment of $g(u, v) = 2uv - v^3$ as (u, v) varies from $(0, 1)$ to $(4, -2)$.

In Exercises 5–8, use Definition 16.4.2 to establish the differentiability of the given function. [*Remark:* ϵ_1 and ϵ_2 are not unique.]

5. $f(x, y) = xy$.

6. $f(x, y) = x^2 + y^2$.

7. $f(x, y) = x^2 y$.

8. $f(x, y) = 3x + y^2$.

9. Let $f(x, y) = \sqrt{x^2 + y^2}$.

   (a) Show that f is continuous at $(0, 0)$.

   (b) Show that $f_x(0, 0)$ does not exist, and hence that f is not differentiable at $(0, 0)$. [*Hint:* Express $f_x(0, 0)$ as a limit (see Exercise 68, Section 16.3).]

10. Let

$$f(x, y) = \begin{cases} 5 - 3x - 2y, & x \geq 0 \text{ or } y \geq 0 \\ 0, & x < 0 \text{ and } y < 0 \end{cases}$$

Show that $f_x(0, 0)$ and $f_y(0, 0)$ exist, but f is not continuous at $(0, 0)$.

11. Let

$$f(x, y) = \begin{cases} \dfrac{xy}{x^2 + y^2}, & (x, y) \neq (0, 0) \\ 0, & (x, y) = (0, 0) \end{cases}$$

Prove: $f_x(0, 0)$ and $f_y(0, 0)$ exist, but f is not continuous at $(0, 0)$. [*Hint:* Show that $f_x(0, 0) = 0$ and $f_y(0, 0) = 0$ by expressing these derivatives as limits (see Exercise 68, Section 16.3). To prove that f is not continuous at $(0, 0)$, show that

$$\lim_{(x,y)\to(0,0)} f(x, y)$$

does not exist by letting $(x, y) \to (0, 0)$ along $y = 0$ and along $y = x$.]

In Exercises 12–16, find f_{xy} and f_{yx} and verify their equality.

12. $f(x, y) = 2xy - 3y^2$.

13. $f(x, y) = 4x^3 y + 3x^2 y$.

14. $f(x, y) = x^3 / y$.

15. $f(x, y) = \sin(x^2 + y^3)$.

16. $f(x, y) = \sqrt{x^2 + y^2 - 1}$.

17. Let f be a function of two variables with continuous third- and fourth-order partial derivatives.

   (a) How many of the third-order partial derivatives can be distinct?

   (b) How many of the fourth order?

18. Let $f(x, y) = e^{xy^2}$. Find f_{xyx}, f_{xxy}, and f_{yxx} and verify their equality.

In Exercises 19–24, find dz/dt using the chain rule.

19. $z = 3x^2 y^3$; $x = t^4$, $y = t^2$.

20. $z = \ln(2x^2 + y)$; $x = \sqrt{t}$, $y = t^{2/3}$.

21. $z = 3\cos x - \sin xy$; $x = 1/t$, $y = 3t$.

22. $z = \sqrt{1 + x - 2xy^4}$; $x = \ln t$, $y = t$.

23. $z = e^{1-xy}$; $x = t^{1/3}$, $y = t^3$.

24. $z = \cosh^2 xy$; $x = t/2$, $y = e^t$.

In Exercises 25–31, find $\partial z/\partial u$ and $\partial z/\partial v$ by the chain rule.

25. $z = 8x^2 y - 2x + 3y$; $x = uv$, $y = u - v$.

26. $z = x^2 - y\tan x$; $x = u/v$, $y = u^2 v^2$.

27. $z = x/y$; $x = 2\cos u$, $y = 3\sin v$.

28. $z = 3x - 2y$; $x = u + v\ln u$, $y = u^2 - v\ln v$.

29. $z = e^{x^2 y}$; $x = \sqrt{uv}$, $y = 1/v$.

30. $z = \cos x \sin y$; $x = u - v$, $y = u^2 + v^2$.

31. $z = \tan^{-1}(x^2 + y^2)$; $x = e^u \sin v$, $y = e^u \cos v$.

32. Let $w = rs/(r^2 + s^2)$; $r = uv$, $s = u - 2v$. Use the chain rule to find $\partial w/\partial u$ and $\partial w/\partial v$.

33. Let $T = x^2 y - xy^3 + 2$; $x = r\cos\theta$, $y = r\sin\theta$. Use the chain rule to find $\partial T/\partial r$ and $\partial T/\partial\theta$.

34. Let $R = e^{2s-t^2}$; $s = 3\phi$, $t = \phi^{1/2}$. Use the chain rule to find $dR/d\phi$.

35. Let $t = u/v$; $u = x^2 - y^2$, $v = 4xy^3$. Use the chain rule to find $\partial t/\partial x$ and $\partial t/\partial y$.

36. Use the chain rule to find $\dfrac{dz}{dt}\bigg|_{t=3}$ if $z = x^2 y$; $x = t^2$, $y = t + 7$.

37. Use the chain rule to find the value of $\dfrac{dw}{ds}\bigg|_{s=1/4}$ if $w = r^2 - r \tan \theta$; $r = \sqrt{s}$, $\theta = \pi s$.

38. Use the chain rule to find the value of
$$\frac{\partial f}{\partial u}\bigg|_{u=1,\,v=-2} \quad \text{and} \quad \frac{\partial f}{\partial v}\bigg|_{u=1,\,v=-2}$$
if $f(x, y) = x^2 y^2 - x + 2y$; $x = \sqrt{u}$, $y = uv^3$.

39. Use the chain rule to find the value of
$$\frac{\partial z}{\partial r}\bigg|_{r=2,\,\theta=\pi/6} \quad \text{and} \quad \frac{\partial z}{\partial \theta}\bigg|_{r=2,\,\theta=\pi/6}$$
if $z = xye^{x/y}$; $x = r\cos\theta$, $y = r\sin\theta$.

In Exercises 40–43, use (17) to find dy/dx and check your result using implicit differentiation.

40. $x^2 y^3 + \cos y = 0$.

41. $x^3 - 3xy^2 + y^3 = 5$.

42. $e^{xy} + ye^y = 1$.

43. $x - \sqrt{xy} + 3y = 4$.

44. The portion of a tree that is usable for lumber may be viewed as a right-circular cylinder. If the height of a tree increases 2 ft per year and the diameter increases 3 in. per year, how fast is the volume of usable lumber increasing when the tree is 20 ft high and the diameter is 30 in.?

45. Two straight roads intersect at right angles. Car A, moving on one of the roads, approaches the intersection at 25 mi/hr and car B, moving on the other road, approaches the intersection at 30 mi/hr. At what rate is the distance between the cars changing when A is 0.3 mile from the intersection and B is 0.4 mile from the intersection?

46. Use the ideal gas law, $P = kT/V$, with $k = 10$ to find the rate at which the temperature of a gas is changing when the volume is 200 in^3 and increasing at the rate of 4 in^3/sec, while the pressure is 5 lb/in^2 and decreasing at the rate of 1 lb/in^2/sec.

47. Two sides of a triangle have lengths $a = 4$ cm and $b = 3$ cm, but are increasing at the rate of 1 cm/sec. If the area of the triangle remains constant, at what rate is the angle θ between a and b changing when $\theta = \pi/6$?

48. Two sides of a triangle have lengths $a = 5$ cm and $b = 10$ cm, and the included angle is $\theta = \pi/3$. If a is increasing at a rate of 2 cm/sec, b is increasing at a rate of 1 cm/sec, and θ remains constant, at what rate is the third side changing? Is it increasing or decreasing? [*Hint:* Use the law of cosines.]

49. Suppose that a particle moving along a metal plate in the xy-plane has velocity $\mathbf{v} = \mathbf{i} - 4\mathbf{j}$ (cm/sec) at the point

$(3, 2)$. If the temperature of the plate at points in the xy-plane is $T(x, y) = y^2 \ln x$, $x \geq 1$, in degrees Celsius, find dT/dt at $(3, 2)$.

In Exercises 50–67, assume that the derivatives satisfy all continuity requirements needed for the problem.

50. Let $z = f(u)$ where $u = g(x, y)$. Express $\partial z/\partial x$ and $\partial z/\partial y$ in terms of dz/du, $\partial u/\partial x$, and $\partial u/\partial y$.

51. Let $z = f(x^2 - y^2)$. Show that $y\partial z/\partial x + x\partial z/\partial y = 0$.

52. Let $z = f(xy)$. Show that $x\,\partial z/\partial x - y\,\partial z/\partial y = 0$.

53. Let $z = f(u)$ where $u = g(x, y)$. Show that

(a) $\dfrac{\partial^2 z}{\partial x^2} = \dfrac{dz}{du}\dfrac{\partial^2 u}{\partial x^2} + \dfrac{d^2 z}{du^2}\left(\dfrac{\partial u}{\partial x}\right)^2$

and $\dfrac{\partial^2 z}{\partial y^2} = \dfrac{dz}{du}\dfrac{\partial^2 u}{\partial y^2} + \dfrac{d^2 z}{du^2}\left(\dfrac{\partial u}{\partial y}\right)^2$

(b) $\dfrac{\partial^2 z}{\partial y\,\partial x} = \dfrac{dz}{du}\dfrac{\partial^2 u}{\partial y\,\partial x} + \dfrac{d^2 z}{du^2}\dfrac{\partial u}{\partial x}\dfrac{\partial u}{\partial y}$.

54. Let $r = \sqrt{x^2 + y^2}$. Show that

(a) $\dfrac{\partial r}{\partial x} = \dfrac{x}{r}$

(b) $\dfrac{\partial r}{\partial y} = \dfrac{y}{r}$

(c) $\dfrac{\partial^2 r}{\partial x^2} = \dfrac{y^2}{r^3}$

(d) $\dfrac{\partial^2 r}{\partial y^2} = \dfrac{x^2}{r^3}$.

55. Let $z = f(r)$ where $r = \sqrt{x^2 + y^2}$, $r \neq 0$.

(a) Use the formulas in Exercises 53 and 54 to show that if z satisfies Laplace's equation, $\dfrac{\partial^2 z}{\partial x^2} + \dfrac{\partial^2 z}{\partial y^2} = 0$, then
$$r\frac{d^2 z}{dr^2} + \frac{dz}{dr} = 0.$$

(b) Use the result of part (a) to solve Laplace's equation for $z = f(r)$. [*Hint:* Write
$$r\frac{d^2 z}{dr^2} + \frac{dz}{dr}$$
as the derivative of a product.]

56. Let $z = f(x - y, y - x)$. Show that $\partial z/\partial x + \partial z/\partial y = 0$.

57. Let $z = f(y + cx) + g(y - cx)$, where $c \neq 0$. Show that
$$\frac{\partial^2 z}{\partial x^2} = c^2 \frac{\partial^2 z}{\partial y^2}$$

58. Let $z = f(x, y)$ where $x = g(t)$ and $y = h(t)$.

(a) Show that
$$\frac{d}{dt}\left(\frac{\partial z}{\partial x}\right) = \frac{\partial^2 z}{\partial x^2}\frac{dx}{dt} + \frac{\partial^2 z}{\partial y\,\partial x}\frac{dy}{dt}$$
and
$$\frac{d}{dt}\left(\frac{\partial z}{\partial y}\right) = \frac{\partial^2 z}{\partial x\,\partial y}\frac{dx}{dt} + \frac{\partial^2 z}{\partial y^2}\frac{dy}{dt}$$

(b) Use the formulas in part (a) to help find a formula for $d^2 z/dt^2$.

59. Let $z = f(x, y)$ where $x = g(u, v)$ and $y = h(u, v)$. Show that

$$\frac{\partial}{\partial u}\left(\frac{\partial z}{\partial x}\right) = \frac{\partial^2 z}{\partial x^2}\frac{\partial x}{\partial u} + \frac{\partial^2 z}{\partial y\,\partial x}\frac{\partial y}{\partial u}$$

60. Let $z = f(x, y)$ where $x = u + v$ and $y = u - v$. Show that

$$\frac{\partial^2 z}{\partial v\,\partial u} = \frac{\partial^2 z}{\partial x^2} - \frac{\partial^2 z}{\partial y^2}$$

61. The equations $x = r\cos\theta$ and $y = r\sin\theta$, which relate Cartesian and polar coordinates, define r and θ implicitly as functions of x and y.

(a) Use implicit differentiation with respect to the variable x on both equations to show that

$$\frac{\partial r}{\partial x} = \cos\theta \quad \text{and} \quad \frac{\partial\theta}{\partial x} = -\frac{\sin\theta}{r}$$

(b) Use implicit differentiation with respect to y on both equations to show that

$$\frac{\partial r}{\partial y} = \sin\theta \quad \text{and} \quad \frac{\partial\theta}{\partial y} = \frac{\cos\theta}{r}$$

62. Let $z = f(r, \theta)$, where r and θ are defined implicitly as functions of x and y by the equations $x = r\cos\theta$ and $y = r\sin\theta$. Use the results in Exercise 61 to show that

(a) $\dfrac{\partial z}{\partial x} = \dfrac{\partial z}{\partial r}\cos\theta - \dfrac{1}{r}\dfrac{\partial z}{\partial\theta}\sin\theta$

(b) $\dfrac{\partial z}{\partial y} = \dfrac{\partial z}{\partial r}\sin\theta + \dfrac{1}{r}\dfrac{\partial z}{\partial\theta}\cos\theta.$

63. Use the formulas in Exercise 62 to show that

$$\left(\frac{\partial z}{\partial x}\right)^2 + \left(\frac{\partial z}{\partial y}\right)^2 = \left(\frac{\partial z}{\partial r}\right)^2 + \frac{1}{r^2}\left(\frac{\partial z}{\partial\theta}\right)^2$$

64. Use the formulas in Exercises 61 and 62 to show that Laplace's equation

$$\frac{\partial^2 z}{\partial x^2} + \frac{\partial^2 z}{\partial y^2} = 0$$

becomes

$$\frac{\partial^2 z}{\partial r^2} + \frac{1}{r^2}\frac{\partial^2 z}{\partial\theta^2} + \frac{1}{r}\frac{\partial z}{\partial r} = 0$$

in polar coordinates.

65. A function $f(x, y)$ is said to be **homogeneous of degree n** if $f(tx, ty) = t^n f(x, y)$ for $t > 0$. Show that the following functions are homogeneous and find the degree of each.

(a) $f(x, y) = 3x^2 + y^2$

(b) $f(x, y) = \sqrt{x^2 + y^2}$

(c) $f(x, y) = x^2 y - 2y^3$

(d) $f(x, y) = \dfrac{5}{(x^2 + 2y^2)^2}, (x, y) \neq (0, 0).$

66. Show that for a homogeneous function $f(x, y)$ of degree n (see Exercise 65)

$$x\frac{\partial f}{\partial x} + y\frac{\partial f}{\partial y} = nf$$

[*Hint:* Let $u = tx$ and $v = ty$ in $f(tx, ty)$, and differentiate both sides of $f(u, v) = t^n f(x, y)$ with respect to t.]

67. Let $w = f(x, y)$ where $y = g(x, z)$. Taking x and z as the independent variables, express each of the following in terms of $\partial f/\partial x$, $\partial f/\partial y$, $\partial y/\partial x$, and $\partial y/\partial z$.

(a) $\dfrac{\partial w}{\partial x}$ (b) $\dfrac{\partial w}{\partial z}.$

68. Show that if $u(x, y)$ and $v(x, y)$ satisfy the Cauchy–Riemann equations (Exercise 54, Section 16.3), and if $x = r\cos\theta$ and $y = r\sin\theta$, then

$$\frac{\partial u}{\partial r} = \frac{1}{r}\frac{\partial v}{\partial\theta} \quad \text{and} \quad \frac{\partial v}{\partial r} = -\frac{1}{r}\frac{\partial u}{\partial\theta}$$

69. Prove: If f, f_x, and f_y are continuous on a circular region containing $A(x_0, y_0)$ and $B(x_1, y_1)$, then there is a point (x^*, y^*) on the line segment joining A and B such that

$$f(x_1, y_1) - f(x_0, y_0)$$
$$= f_x(x^*, y^*)(x_1 - x_0) + f_y(x^*, y^*)(y_1 - y_0)$$

This result is the two-dimensional version of the Mean-Value Theorem. [*Hint:* Express the line segment joining A and B in parametric form and use the Mean-Value Theorem for functions of one variable.]

70. Prove: If $f_x(x, y) = 0$ and $f_y(x, y) = 0$ throughout a circular region, then $f(x, y)$ is constant on that region. [*Hint:* Use the result of Exercise 69.]

■ **16.5** TANGENT PLANES; TOTAL DIFFERENTIALS FOR FUNCTIONS OF TWO VARIABLES

In this section we shall discuss tangent planes to surfaces in three-dimensional space. We are concerned with three main questions: What is a tangent plane? When do tangent planes exist? How do we find equations of tangent planes? We shall use our results on tangent planes to extend the concept of a differential to functions of two variables.

☐ **TANGENT PLANES**

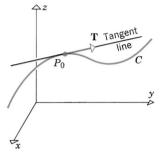

Figure 16.5.1

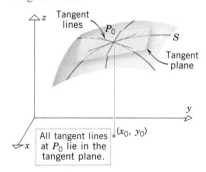

Figure 16.5.2

Recall that if C is a smooth parametric curve in 3-space, then the tangent line to C at a point P_0 is the line through P_0 along the unit tangent vector to C at P_0 (Figure 16.5.1). The concept of a *tangent plane* builds on this definition. If $P_0(x_0, y_0, z_0)$ is a point on a surface S, and if the tangent lines at P_0 to all smooth curves that pass through P_0 and lie on the surface S all lie in a common plane, then we shall regard that plane to be the **tangent plane** to the surface at P_0 (Figure 16.5.2).

The following theorem states conditions that ensure the existence of a tangent plane and tells us how to find its equation.

16.5.1   THEOREM.   *Let $P_0(x_0, y_0, z_0)$ be any point on the surface $z = f(x, y)$. If $f(x, y)$ is differentiable at (x_0, y_0), then the surface has a tangent plane at P_0, and this plane has the equation*

$$f_x(x_0, y_0)(x - x_0) + f_y(x_0, y_0)(y - y_0) - (z - z_0) = 0 \qquad (1)$$

Proof.   To prove the existence of a tangent plane at P_0, we must show that all smooth curves on the surface $z = f(x, y)$ that pass through P_0 have tangent lines that lie in a common plane. We shall do this by showing that these curves all have unit tangent vectors at P_0 that are perpendicular to the vector

$$\mathbf{n} = \langle f_x(x_0, y_0), f_y(x_0, y_0), -1 \rangle \qquad (2)$$

This will force all the tangent lines at P_0 to lie in the plane through P_0 with $\mathbf{n}$ as a normal. Moreover, it will follow from (2) that the point-normal equation of this plane is (1), thereby completing the proof.

Assume that C is any smooth curve that lies on the surface $z = f(x, y)$ and passes through $P_0(x_0, y_0, z_0)$. Moreover, assume that C has parametric equations

$$x = x(s), \quad y = y(s), \quad z = z(s)$$

where s is an arc-length parameter, and assume that $P_0(x_0, y_0, z_0)$ is the point on C that corresponds to the parameter value $s = s_0$. Thus, $x_0 = x(s_0)$, $y_0 = y(s_0)$, and $z_0 = z(s_0)$. Because C lies on the surface $z = f(x, y)$, every point $(x(s), y(s), z(s))$ on C must satisfy this equation for all s, so

$$z(s) = f(x(s), y(s))$$

for all s. If we differentiate both sides of this equation and apply the version of the chain rule in Theorem 16.4.7 with s replacing t, we obtain

$$\frac{dz}{ds} = \frac{\partial f}{\partial x}\frac{dx}{ds} + \frac{\partial f}{\partial y}\frac{dy}{ds}$$

or equivalently,

$$\frac{\partial f}{\partial x}\frac{dx}{ds} + \frac{\partial f}{\partial y}\frac{dy}{ds} - \frac{dz}{ds} = 0$$

The left side of this equation can be rewritten as a dot product of vectors:

$$\left\langle \frac{\partial f}{\partial x}, \frac{\partial f}{\partial y}, -1 \right\rangle \cdot \left\langle \frac{dx}{ds}, \frac{dy}{ds}, \frac{dz}{ds} \right\rangle = 0$$

or in an alternative notation as

$$\langle f_x(x, y), f_y(x, y), -1 \rangle \cdot \langle x'(s), y'(s), z'(s) \rangle = 0$$

In particular, if $s = s_0$, we have

$$\langle f_x(x_0, y_0), f_y(x_0, y_0), -1 \rangle \cdot \langle x'(s_0), y'(s_0), z'(s_0) \rangle = 0 \qquad (3)$$

But the second vector in the dot product is the unit tangent vector to C at the point $P_0(x_0, y_0, z_0)$ [see Formula (2) of Section 15.4], so that from (3) the unit tangent vector to C at P_0 is perpendicular to the vector

$$\mathbf{n} = \langle f_x(x_0, y_0), f_y(x_0, y_0), -1 \rangle$$

which completes the proof. ∎

If $f(x, y)$ is differentiable at (x_0, y_0), then the vector

$$\mathbf{n} = \langle f_x(x_0, y_0), f_y(x_0, y_0), -1 \rangle \tag{4}$$

is called a ***normal vector*** to the surface $z = f(x, y)$ at $P_0(x_0, y_0, z_0)$, and the line through P_0 parallel to $\mathbf{n}$ is called the ***normal line*** to the surface at P_0 (Figure 16.5.3). Parametric equations of the normal line are

$$x = x_0 + f_x(x_0, y_0)t, \quad y = y_0 + f_y(x_0, y_0)t, \quad z = z_0 - t \tag{5}$$

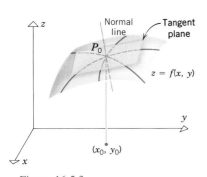

Figure 16.5.3

Example 1 Find equations of the tangent plane and normal line to the surface $z = x^2 y$ at the point $(2, 1, 4)$.

Solution. Since $f(x, y) = x^2 y$, it follows that

$$f_x(x, y) = 2xy \quad \text{and} \quad f_y(x, y) = x^2$$

so that with $x = 2$, $y = 1$,

$$f_x(2, 1) = 4 \quad \text{and} \quad f_y(2, 1) = 4$$

Thus, a vector normal to the surface at $(2, 1, 4)$ is

$$\mathbf{n} = f_x(2, 1)\mathbf{i} + f_y(2, 1)\mathbf{j} - \mathbf{k} = 4\mathbf{i} + 4\mathbf{j} - \mathbf{k}$$

Therefore, the tangent plane has the equation

$$4(x - 2) + 4(y - 1) - (z - 4) = 0 \quad \text{or} \quad 4x + 4y - z = 8$$

and the normal line has equations

$$x = 2 + 4t, \quad y = 1 + 4t, \quad z = 4 - t \quad ◀$$

REMARK. In the preceding section we set two goals for the definition of differentiability of a function $f(x, y)$ of two variables at a point (x_0, y_0)—we wanted f to be continuous at (x_0, y_0) and we wanted the surface $z = f(x, y)$ to have a nonvertical tangent plane at (x_0, y_0). Both of these goals have now been achieved, for we showed in Theorem 16.4.3 that differentiability implies continuity and now Theorem 16.5.1 shows that differentiability implies the existence of a nonvertical tangent plane. [The tangent plane given by (1) is nonvertical because the third component of the normal vector $\mathbf{n}$ in (2) is not zero.]

☐ **DIFFERENTIALS**

Recall that if $y = f(x)$ is a function of one variable, then the differential

$$dy = f'(x_0) \, dx$$

represents the change in y along the *tangent line* at (x_0, y_0) produced by a change dx in x and

$$\Delta y = f(x_0 + \Delta x) - f(x_0)$$

represents the change in y along the *curve* $y = f(x)$ produced by a change Δx in x. Analogously, if $z = f(x, y)$ is a function of two variables, we shall define dz to be the change in z along the *tangent plane* at (x_0, y_0, z_0) to the surface $z = f(x, y)$ produced by changes dx and dy in x and y, respectively. This is in contrast to

$$\Delta z = f(x_0 + \Delta x, y_0 + \Delta y) - f(x_0, y_0)$$

which represents the change in z *along the surface* produced by changes Δx and Δy in x and y. A comparison of dz and Δz is shown in Figure 16.5.4 in the case where $dx = \Delta x$ and $dy = \Delta y$.

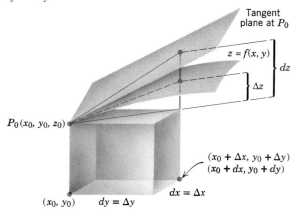

Figure 16.5.4

To derive a formula for dz, let $P_0(x_0, y_0, z_0)$ be a fixed point on the surface $z = f(x, y)$. If we assume f to be differentiable at (x_0, y_0), then the surface has a tangent plane at P_0, given by the equation

$$f_x(x_0, y_0)(x - x_0) + f_y(x_0, y_0)(y - y_0) - (z - z_0) = 0$$

or

$$z = z_0 + f_x(x_0, y_0)(x - x_0) + f_y(x_0, y_0)(y - y_0) \tag{6}$$

From (6) it follows that the tangent plane has height z_0 when $x = x_0$, $y = y_0$, and it has height

$$z_0 + f_x(x_0, y_0)\, dx + f_y(x_0, y_0)\, dy \tag{7}$$

when $x = x_0 + dx$, $y = y_0 + dy$. Thus, the change dz in the height of the tangent plane as (x, y) varies from (x_0, y_0) to $(x_0 + dx, y_0 + dy)$ is obtained by subtracting z_0 from expression (7). This yields

$$dz = f_x(x_0, y_0)\, dx + f_y(x_0, y_0)\, dy$$

This quantity is called the ***total differential*** of z at (x_0, y_0). Often, we omit the subscripts on x_0 and y_0 and write

$$dz = f_x(x, y)\, dx + f_y(x, y)\, dy \tag{8}$$

In this formula, dx and dy are usually viewed as variables and x and y as fixed numbers. Formula (8) can also be written using df in place of dz.

If $z = f(x, y)$ is differentiable at the point (x, y), then the increment Δz can be written as

$$\Delta z = f_x(x, y)\, \Delta x + f_y(x, y)\, \Delta y + \epsilon_1 \Delta x + \epsilon_2 \Delta y \tag{9}$$

where $\epsilon_1 \to 0$, $\epsilon_2 \to 0$ as $(\Delta x, \Delta y) \to (0, 0)$. In the case where $dx = \Delta x$ and $dy = \Delta y$, it follows from (8) and (9) that

$$\Delta z = dz + \epsilon_1 \Delta x + \epsilon_2 \Delta y$$

Thus, when $\Delta x = dx$ and $\Delta y = dy$ are small, we can approximate Δz by

$$\Delta z \approx dz \tag{10}$$

Geometrically, this approximation tells us that the change in z along the surface and the change in z along the tangent plane are approximately equal when $\Delta x = dx$ and $\Delta y = dy$ are small (see Figure 16.5.4).

Example 2 Let $z = 4x^3y^2$. Find dz.

Solution. Since $f(x, y) = 4x^3y^2$,

$$f_x(x, y) = 12x^2y^2 \quad \text{and} \quad f_y(x, y) = 8x^3y$$

so

$$dz = 12x^2y^2\,dx + 8x^3y\,dy \quad \blacktriangleleft$$

Example 3 Let $f(x, y) = \sqrt{x^2 + y^2}$. Use a total differential to approximate the change in $f(x, y)$ as (x, y) varies from the point $(3, 4)$ to the point $(3.04, 3.98)$.

Solution. We shall approximate Δf (the change in f) by

$$df = f_x(x, y)\,dx + f_y(x, y)\,dy = \frac{x}{\sqrt{x^2 + y^2}}\,dx + \frac{y}{\sqrt{x^2 + y^2}}\,dy$$

Since $x = 3$, $y = 4$, $dx = 0.04$, and $dy = -0.02$, we obtain

$$\Delta f \approx df = \frac{3}{\sqrt{3^2 + 4^2}}(0.04) + \frac{4}{\sqrt{3^2 + 4^2}}(-0.02)$$

$$= \tfrac{3}{5}(0.04) - \tfrac{4}{5}(0.02) = 0.008$$

The reader may want to use a calculator to show that the true value of Δf to five decimal places is

$$\Delta f = \sqrt{(3.04)^2 + (3.98)^2} - \sqrt{3^2 + 4^2} \approx 0.00819 \quad \blacktriangleleft$$

Example 4 The radius of a right-circular cylinder is measured with an error of at most 2%, and the height is measured with an error of at most 4%. Approximate the maximum possible percentage error in the volume V calculated from these measurements.

Solution. Let r, h, and V be the true radius, height, and volume of the cylinder, and let Δr, Δh, and ΔV be the errors in these quantities. We are given that

$$\left|\frac{\Delta r}{r}\right| \le 0.02 \quad \text{and} \quad \left|\frac{\Delta h}{h}\right| \le 0.04$$

We want to find the maximum possible value of $|\Delta V / V|$. Since the volume of the cylinder is $V = \pi r^2 h$, it follows from (8) that

$$dV = \frac{\partial V}{\partial r}\,dr + \frac{\partial V}{\partial h}\,dh = 2\pi r h\,dr + \pi r^2\,dh$$

If we choose $dr = \Delta r$ and $dh = \Delta h$, then we can use the approximations

$$\Delta V \approx dV \quad \text{and} \quad \frac{\Delta V}{V} \approx \frac{dV}{V}$$

But

$$\frac{dV}{V} = \frac{2\pi r h\,dr + \pi r^2\,dh}{\pi r^2 h} = 2\frac{dr}{r} + \frac{dh}{h}$$

so by the triangle inequality (Theorem 1.2.5)

$$\left|\frac{dV}{V}\right| = \left|2\frac{dr}{r} + \frac{dh}{h}\right| \le 2\left|\frac{dr}{r}\right| + \left|\frac{dh}{h}\right| \le 2(0.02) + (0.04) = 0.08$$

Thus, the maximum percentage error in V is approximately 8%. $\quad \blacktriangleleft$

▶ Exercise Set 16.5   C 24, 32, 34

In Exercises 1–8, find equations for the tangent plane and normal line to the given surface at the point P.

1. $z = 4x^3y^2 + 2y$; $P(1, -2, 12)$.

2. $z = \frac{1}{2}x^7y^{-2}$; $P(2, 4, 4)$.

3. $z = xe^{-y}$; $P(1, 0, 1)$.

4. $z = \ln \sqrt{x^2 + y^2}$; $P(-1, 0, 0)$.

5. $z = e^{3y} \sin 3x$; $P(\pi/6, 0, 1)$.

6. $z = x^{1/2} + y^{1/2}$; $P(4, 9, 5)$.

7. $x^2 + y^2 + z^2 = 25$; $P(-3, 0, 4)$.

8. $x^2y - 4z^2 = -7$; $P(-3, 1, -2)$.

In Exercises 9–12, find dz.

9. $z = 7x - 2y$.

10. $z = 5x^2y^5 - 2x + 4y + 7$.

11. $z = \tan^{-1} xy$.   12. $z = \sec^2(x - 3y)$.

In Exercises 13–16, use a total differential to approximate the change in $f(x, y)$ as (x, y) varies from P to Q.

13. $f(x, y) = x^2 + 2xy - 4x$; $P(1, 2)$, $Q(1.01, 2.04)$.

14. $f(x, y) = x^{1/3}y^{1/2}$; $P(8, 9)$, $Q(7.78, 9.03)$.

15. $f(x, y) = \dfrac{x + y}{xy}$; $P(-1, -2)$, $Q(-1.02, -2.04)$.

16. $f(x, y) = \ln \sqrt{1 + xy}$; $P(0, 2)$, $Q(-0.09, 1.98)$.

17. Find all points on the surface at which the tangent plane is horizontal.

(a) $z = x^3y^2$

(b) $z = x^2 - xy + y^2 - 2x + 4y$.

18. Find a point on the surface $z = 3x^2 - y^2$ at which the tangent plane is parallel to the plane $6x + 4y - z = 5$.

19. Find a point on the surface $z = 8 - 3x^2 - 2y^2$ at which the tangent plane is perpendicular to the line $x = 2 - 3t$, $y = 7 + 8t$, $z = 5 - t$.

20. Show that the surfaces

$$z = \sqrt{x^2 + y^2} \quad \text{and} \quad z = \tfrac{1}{10}(x^2 + y^2) + \tfrac{5}{2}$$

intersect at $(3, 4, 5)$ and have a common tangent plane at that point.

21. Show that the surfaces $z = \sqrt{16 - x^2 - y^2}$ and $z = \sqrt{x^2 + y^2}$ intersect at $(2, 2, 2\sqrt{2})$ and have tangent planes that are perpendicular at that point.

22. (a) Show that every line normal to the cone $z = \sqrt{x^2 + y^2}$ passes through the z-axis.

(b) Show that every tangent plane intersects the cone in a line passing through the origin.

23. One leg of a right triangle increases from 3 cm to 3.2 cm, while the other leg decreases from 4 cm to 3.96 cm. Use a total differential to approximate the change in the length of the hypotenuse.

24. The volume V of a right-circular cone of radius r and height h is given by $V = \frac{1}{3}\pi r^2 h$. Suppose that the height decreases from 20 in. to 19.95 in., while the radius increases from 4 in. to 4.05 in. Use a total differential to approximate the change in volume.

25. The length and width of a rectangle are measured with errors of at most 3% and 5%, respectively. Use differentials to approximate the maximum percentage error in the calculated area.

26. The radius and height of a right-circular cone are measured with errors of at most 1% and 4%, respectively. Use differentials to approximate the maximum percentage error in the calculated volume.

27. The length and width of a rectangle are measured with errors of at most $r\%$. Use differentials to approximate the maximum percentage error in the calculated length of the diagonal.

28. The legs of a right triangle are measured to be 3 cm and 4 cm, with a maximum error of 0.05 cm in each measurement. Use differentials to approximate the maximum possible error in the calculated value of (a) the hypotenuse and (b) the area of the triangle.

29. The total resistance R of two resistances R_1 and R_2, connected in parallel, is

$$R = \frac{R_1 R_2}{R_1 + R_2}$$

Suppose that R_1 and R_2 are measured to be 200 ohms and 400 ohms, respectively, with a maximum error of 2% in each. Use differentials to approximate the maximum percentage error in the calculated value of R.

30. According to the ideal gas law, the pressure, temperature, and volume of a confined gas are related by $P = kT/V$, where k is a constant. Use differentials to approximate the percentage change in pressure if the temperature of a gas is increased 3% and the volume is increased 5%.

31. An angle θ of a right triangle is calculated by the formula

$$\theta = \sin^{-1} \frac{a}{c}$$

where a is the length of the side opposite to θ and c is the length of the hypotenuse. Suppose that the measurements $a = 3$ in. and $c = 5$ in. each have a maximum possible error of 0.01 in. Use differentials to approximate the maximum possible error in the the calculated value of θ.

32. An open cylindrical can has an inside radius of 2 cm and an inside height of 5 cm. Use differentials to approximate the volume of metal in the can if it is 0.01 cm thick. [*Hint:* The volume of metal is the difference, ΔV, in the volumes of two cylinders.]

33. The period T of a simple pendulum with small oscillations is calculated from the formula $T = 2\pi\sqrt{L/g}$, where L is the length of the pendulum and g is the acceleration due to gravity. Suppose that values of L and g have errors of at most 0.5% and 0.1%, respectively. Use differentials to approximate the maximum percentage error in the calculated value of T.

34. The angle of elevation from a point on the ground to the top of a building is measured as 60° with a maximum possible error of 0.2°. Suppose that the distance from the point to the building is measured as 100 ft with a maximum possible error of 2 in. Use differentials to approximate the maximum possible error in the calculated height of the building.

35. Suppose that x and y have errors of at most $r\%$ and $s\%$, respectively. For each of the following formulas in x and y, use differentials to approximate the maximum possible error in the calculated result.

  (a) xy (b) x/y (c) x^2y^3 (d) $x^3\sqrt{y}$.

36. Show that the volume of the solid bounded by the coordinate planes and the plane tangent to the portion of the surface $xyz = k$, $k > 0$, in the first octant does not depend on the point of tangency.

37. (a) Find all points of intersection of the line $x = -1 + t$, $y = 2 + t$, $z = 2t + 7$ and the surface $z = x^2 + y^2$.

  (b) At each point of intersection, find the cosine of the acute angle between the given line and the line normal to the surface.

38. Show that if f is differentiable and $z = xf(x/y)$, then all tangent planes to the graph of this equation pass through a common point.

39. Show that the equation of the plane that is tangent to the ellipsoid

$$\frac{x^2}{a^2} + \frac{y^2}{b^2} + \frac{z^2}{c^2} = 1$$

at (x_0, y_0, z_0) can be written in the form

$$\frac{x_0 x}{a^2} + \frac{y_0 y}{b^2} + \frac{z_0 z}{c^2} = 1$$

40. Show that the equation of the plane that is tangent to the paraboloid

$$z = \frac{x^2}{a^2} + \frac{y^2}{b^2}$$

at (x_0, y_0, z_0) can be written in the form

$$z + z_0 = \frac{2x_0 x}{a^2} + \frac{2y_0 y}{b^2}$$

41. Prove: If the surfaces $z = f(x, y)$ and $z = g(x, y)$ intersect at $P(x_0, y_0, z_0)$, and if f and g are differentiable at (x_0, y_0), then the normal lines at P are perpendicular if and only if

$$f_x(x_0, y_0)g_x(x_0, y_0) + f_y(x_0, y_0)g_y(x_0, y_0) = -1$$

■ **16.6 DIRECTIONAL DERIVATIVES AND GRADIENTS FOR FUNCTIONS OF TWO VARIABLES**

> *The partial derivatives $f_x(x, y)$ and $f_y(x, y)$ represent the rates of change of $f(x, y)$ in directions parallel to the x- and y-axes. In this section we shall investigate rates of change of $f(x, y)$ in other directions.*

☐ **DIRECTIONAL DERIVATIVES**

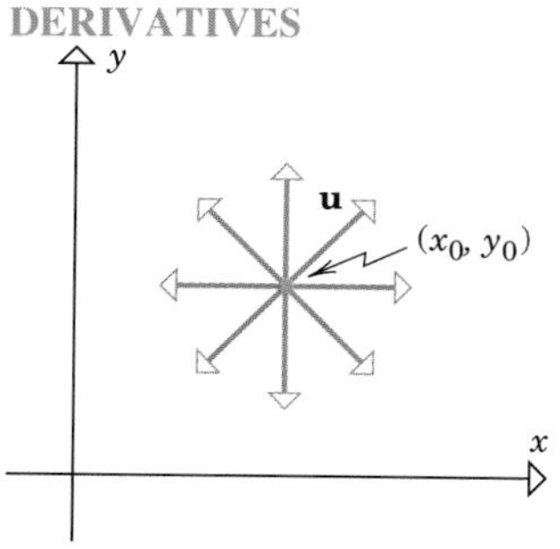

Figure 16.6.1

Just as a person standing on the slope of a hill with mounds and gullies may encounter different grades by walking in different directions, so a surface $z = f(x, y)$ can have different slopes in different directions from a point (x_0, y_0, z_0) on the surface.

To make this idea more precise, we shall use a *unit vector* $\mathbf{u} = \langle u_1, u_2 \rangle$ in the xy-plane to designate the direction in which the slope is to be examined (Figure 16.6.1), and we shall let l be the line in the xy-plane that is parallel to $\mathbf{u}$ and passes through the point $P_0(x_0, y_0)$. As shown in Example 7 of Section 15.3, this line can be represented by the parametric equations

$$x = x_0 + su_1, \quad y = y_0 + su_2 \tag{1}$$

where s is an arc-length parameter with its reference point at $P_0(x_0, y_0)$, and the positive direction is in the direction of $\mathbf{u}$ (Figure 16.6.2). As s increases, the point $P(x, y)$ moves in the direction of $\mathbf{u}$ along l, and a companion point Q with z-coordinate

$$z = f(x, y) = f(x_0 + su_1, y_0 + su_2)$$

moves directly above (or below) along the surface, tracing out a curve C (Figure 16.6.3).

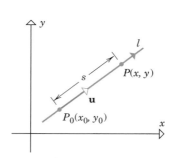

Figure 16.6.2

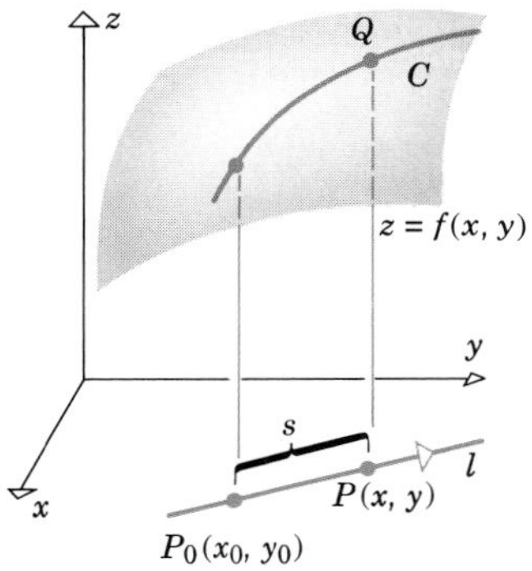

Figure 16.6.3

The rate of change of z with respect to s can be calculated using the chain rule. We obtain

$$\frac{dz}{ds} = f_x(x, y)\frac{dx}{ds} + f_y(x, y)\frac{dy}{ds}$$

or from (1)

$$\frac{dz}{ds} = f_x(x, y)u_1 + f_y(x, y)u_2 \tag{2}$$

where x and y in this formula are expressed in terms of s by the equations in (1). But $P_0(x_0, y_0)$ is the point on l that corresponds to $s = 0$ [see (1)], so setting $s = 0$ in (2) yields the instantaneous rate of change of z with respect to s at the point $P_0(x_0, y_0)$:

$$\left.\frac{dz}{ds}\right|_{s=0} = f_x(x_0, y_0)u_1 + f_y(x_0, y_0)u_2 \tag{3}$$

This quantity is called a *directional derivative of f* and is commonly denoted by $D_{\mathbf{u}}z(x_0, y_0)$ or $D_{\mathbf{u}}f(x_0, y_0)$. More precisely, we make the following definition.

16.6.1   DEFINITION.   If $f(x, y)$ is differentiable at (x_0, y_0), and if $\mathbf{u} = \langle u_1, u_2 \rangle$ is a unit vector, then the ***directional derivative*** of f at (x_0, y_0) in the direction of $\mathbf{u}$ is defined by

$$D_{\mathbf{u}}f(x_0, y_0) = f_x(x_0, y_0)u_1 + f_y(x_0, y_0)u_2 \tag{4}$$

REMARK.   It should be kept in mind that there are infinitely many directional derivatives of $z = f(x, y)$ at a point (x_0, y_0), one for each possible choice of the direction vector $\mathbf{u}$ (Figure 16.6.4). In the special case where $\mathbf{u} = \mathbf{i} = \langle 1, 0 \rangle$, it follows from (4) that $D_{\mathbf{i}}f(x_0, y_0) = f_x(x_0, y_0)$, and in the case where $\mathbf{u} = \mathbf{j} = \langle 0, 1 \rangle$, it follows from (4) that $D_{\mathbf{j}}f(x_0, y_0) = f_y(x_0, y_0)$, so that the partial derivatives with respect to x and with respect to y can be viewed as directional derivatives in the positive x- and positive y-directions, respectively.

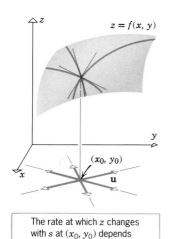

The rate at which z changes with s at (x_0, y_0) depends on the direction of $\mathbf{u}$.

Figure 16.6.4

Example 1   Find the directional derivative of $f(x, y) = 3x^2y$ at the point $(1, 2)$ in the direction of the vector $\mathbf{a} = 3\mathbf{i} + 4\mathbf{j}$.

Solution.   The vector $\mathbf{a}$ is not a unit vector, so we must normalize it to apply Formula (4). This yields

$$\mathbf{u} = \frac{\mathbf{a}}{\|\mathbf{a}\|} = \frac{1}{\sqrt{25}}(3\mathbf{i} + 4\mathbf{j}) = \tfrac{3}{5}\mathbf{i} + \tfrac{4}{5}\mathbf{j}$$

from which we obtain $u_1 = \frac{3}{5}$ and $u_2 = \frac{4}{5}$. Since the partial derivatives of f are

$$f_x(x, y) = 6xy, \quad f_y(x, y) = 3x^2$$

we have $f_x(1, 2) = 12$ and $f_y(1, 2) = 3$, so from (4)

$$D_{\mathbf{u}}f(1, 2) = f_x(1, 2)u_1 + f_y(1, 2)u_2 = 12u_1 + 3u_2$$
$$= 12(\tfrac{3}{5}) + 3(\tfrac{4}{5}) = \tfrac{48}{5} \quad \blacktriangleleft$$

The following alternative version of Formula (4) can be obtained by expressing the unit direction vector as $\mathbf{u} = \cos\theta\,\mathbf{i} + \sin\theta\,\mathbf{j}$, where θ is the angle that $\mathbf{u}$ makes with the positive x-axis [Formula (8) of Section 14.2]:

$$D_{\mathbf{u}}f(x_0, y_0) = f_x(x_0, y_0)\cos\theta + f_y(x_0, y_0)\sin\theta \tag{5}$$

Example 2 Find the directional derivative of $f(x, y) = e^{xy}$ at $(-2, 0)$ in the direction of the unit vector $\mathbf{u}$ that makes an angle of $\pi/3$ with the positive x-axis.

Solution. Let $f(x, y) = e^{xy}$ so that

$$f_x(x, y) = ye^{xy}, \quad f_y(x, y) = xe^{xy}$$
$$f_x(-2, 0) = 0, \quad f_y(-2, 0) = -2$$

From (5)

$$D_{\mathbf{u}}f(-2, 0) = f_x(-2, 0)\cos\frac{\pi}{3} + f_y(-2, 0)\sin\frac{\pi}{3}$$
$$= 0\left(\frac{1}{2}\right) + (-2)\left(\frac{\sqrt{3}}{2}\right) = -\sqrt{3} \quad \blacktriangleleft$$

REMARK. It is worth noting that reversing the direction of $\mathbf{u}$ reverses the sign of the directional derivative $D_{\mathbf{u}}$, since

$$D_{-\mathbf{u}}f(x_0, y_0) = f_x(x_0, y_0)(-u_1) + f_y(x_0, y_0)(-u_2) = -D_{\mathbf{u}}f(x_0, y_0)$$

Thus, we can conclude from the result in Example 2 that the directional derivative of $f(x, y) = e^{xy}$ at $(-2, 0)$ in the direction of the vector making an angle of $4\pi/3$ with the positive x-axis is $\sqrt{3}$.

☐ **THE GRADIENT**

The directional derivative formula

$$D_{\mathbf{u}}f(x, y) = f_x(x, y)u_1 + f_y(x, y)u_2$$

can be expressed in the form of a dot product by writing

$$D_{\mathbf{u}}f(x, y) = (f_x(x, y)\mathbf{i} + f_y(x, y)\mathbf{j}) \cdot (u_1\mathbf{i} + u_2\mathbf{j}) \tag{6}$$

The second vector in the dot product is $\mathbf{u}$. However, the first vector is new; it is called the *gradient of f* and is denoted by the symbol ∇f or $\nabla f(x, y)$.*

16.6.2 DEFINITION. If f is a function of x and y, then the **gradient of f** is defined by

$$\nabla f(x, y) = f_x(x, y)\mathbf{i} + f_y(x, y)\mathbf{j} \tag{7}$$

---

*The symbol ∇ (read, "del") is an inverted delta. In older books this symbol is sometimes called a "nabla" because of its similarity in form to an ancient Hebrew ten-stringed harp of that name.

With the gradient notation, Formula (6) for the directional derivative can be written in the following compact form:

$$D_{\mathbf{u}} f(x, y) = \nabla f(x, y) \cdot \mathbf{u} \tag{8}$$

In words, the dot product of the gradient of f with a unit vector $\mathbf{u}$ produces the directional derivative of f in the direction of $\mathbf{u}$.

Example 3 Find the gradient of $f(x, y) = 3x^2 y$ at the point $(1, 2)$ and use it to calculate the directional derivative of f at $(1, 2)$ in the direction of the vector $\mathbf{a} = 3\mathbf{i} + 4\mathbf{j}$.

Solution. From (7)

$$\nabla f(x, y) = f_x(x, y)\mathbf{i} + f_y(x, y)\mathbf{j} = 6xy\mathbf{i} + 3x^2\mathbf{j}$$

so that the gradient of f at $(1, 2)$ is

$$\nabla f(1, 2) = 12\mathbf{i} + 3\mathbf{j}$$

The unit vector in the direction of $\mathbf{a}$ is

$$\mathbf{u} = \frac{\mathbf{a}}{\|\mathbf{a}\|} = \tfrac{1}{5}(3\mathbf{i} + 4\mathbf{j}) = \tfrac{3}{5}\mathbf{i} + \tfrac{4}{5}\mathbf{j}$$

Thus, from (9)

$$D_{\mathbf{u}} f(1, 2) = \nabla f(1, 2) \cdot \mathbf{u} = (12\mathbf{i} + 3\mathbf{j}) \cdot (\tfrac{3}{5}\mathbf{i} + \tfrac{4}{5}\mathbf{j}) = \tfrac{48}{5}$$

which agrees with the result obtained in Example 1. ◀

☐ **PROPERTIES OF THE GRADIENT**

The gradient is not merely a notational device to simplify the formula for the directional derivative; the length and direction of the gradient ∇f provide important information about the function f.

16.6.3 THEOREM. *Let f be a function of two variables that is differentiable at (x_0, y_0).*

(a) *If $\nabla f(x_0, y_0) = \mathbf{0}$, then all directional derivatives of f at (x_0, y_0) are zero.*

(b) *If $\nabla f(x_0, y_0) \neq \mathbf{0}$, then among all possible directional derivatives of f at (x_0, y_0), the derivative in the direction of $\nabla f(x_0, y_0)$ has the largest value. The value of that directional derivative is $\|\nabla f(x_0, y_0)\|$.*

(c) *If $\nabla f(x_0, y_0) \neq \mathbf{0}$, then among all possible directional derivatives of f at (x_0, y_0), the derivative in the direction opposite to that of $\nabla f(x_0, y_0)$ has the smallest value. The value of that directional derivative is $-\|\nabla f(x_0, y_0)\|$.*

Proof (a). If $\nabla f(x_0, y_0) = \mathbf{0}$, then for all choices of $\mathbf{u}$ we have

$$D_{\mathbf{u}} f(x_0, y_0) = \nabla f(x_0, y_0) \cdot \mathbf{u} = \mathbf{0} \cdot \mathbf{u} = 0$$

Proofs (b) and (c). Assume that $\nabla f(x_0, y_0) \neq \mathbf{0}$ and let θ be the angle between $\nabla f(x_0, y_0)$ and an arbitrary unit vector $\mathbf{u}$. By the definition of dot product,

$$D_{\mathbf{u}} f(x_0, y_0) = \nabla f(x_0, y_0) \cdot \mathbf{u} = \|\nabla f(x_0, y_0)\| \, \|\mathbf{u}\| \cos \theta$$

or, since $\|\mathbf{u}\| = 1$,

$$D_{\mathbf{u}} f(x_0, y_0) = \|\nabla f(x_0, y_0)\| \cos \theta$$

Thus, the maximum value of $D_{\mathbf{u}} f(x_0, y_0)$ is $\|\nabla f(x_0, y_0)\|$, and this occurs when $\cos \theta = 1$, or when $\mathbf{u}$ has the same direction as $\nabla f(x_0, y_0)$ (since $\theta = 0$). The minimum value of

$D_{\mathbf{u}}f(x_0, y_0)$ is $-\|\nabla f(x_0, y_0)\|$, and this occurs when $\cos \theta = -1$ or when $\mathbf{u}$ and $\nabla f(x_0, y_0)$ are oppositely directed (since $\theta = \pi$). ∎

Example 4 For the function $f(x, y) = x^2 e^y$, find the maximum value of a directional derivative at $(-2, 0)$, and give a unit vector in the direction in which the maximum value occurs.

Solution. Since

$$\nabla f(x, y) = f_x(x, y)\mathbf{i} + f_y(x, y)\mathbf{j} = 2xe^y\mathbf{i} + x^2 e^y\mathbf{j}$$

the gradient of f at $(-2, 0)$ is

$$\nabla f(-2, 0) = -4\mathbf{i} + 4\mathbf{j}$$

By Theorem 16.6.3, the maximum value of the directional derivative is

$$\|\nabla f(-2, 0)\| = \sqrt{(-4)^2 + 4^2} = \sqrt{32} = 4\sqrt{2}$$

This maximum occurs in the direction of $\nabla f(-2, 0)$. A unit vector in this direction is

$$\frac{\nabla f(-2, 0)}{\|\nabla f(-2, 0)\|} = \frac{1}{4\sqrt{2}}(-4\mathbf{i} + 4\mathbf{j}) = -\frac{1}{\sqrt{2}}\mathbf{i} + \frac{1}{\sqrt{2}}\mathbf{j} \quad ◀$$

Our next objective is to establish a geometric relationship between the level curves and the gradient of a function f of two variables. For this purpose we note that if (x_0, y_0) is any point in the domain of f, and if $f(x_0, y_0) = c$, then under appropriate conditions the equation

$$f(x, y) = c$$

defines a unique level curve of f that can be smoothly parametrized in terms of arc length and that passes through (x_0, y_0).* It follows that under these conditions the level curve has a unit tangent vector at (x_0, y_0) (Figure 16.6.5).

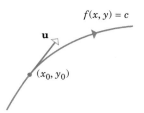

$f(x, y) = c$

$\mathbf{u}$

(x_0, y_0)

Figure 16.6.5

If $f(x, y) = c$ is the level curve for f that passes through the point (x_0, y_0), then the value of $f(x, y)$ remains constant as (x, y) moves away from (x_0, y_0) along the level curve. It seems plausible, therefore, that the directional derivative of f at (x_0, y_0) is zero in the direction of the unit tangent vector $\mathbf{u}$ at (x_0, y_0) to the level curve $f(x, y) = c$. To see that this is so, suppose that $f(x, y) = c$ is smoothly parametrized in terms of an arc-length parameter s by the equations $x = x(s)$, $y = y(s)$, so that

$$f(x(s), y(s)) = c$$

Differentiating this equation with respect to s by the chain rule yields

$$\frac{\partial f}{\partial x}\frac{dx}{ds} + \frac{\partial f}{\partial y}\frac{dy}{ds} = 0$$

which can be rewritten as

$$\left(\frac{\partial f}{\partial x}\mathbf{i} + \frac{\partial f}{\partial y}\mathbf{j}\right) \cdot \left(\frac{dx}{ds}\mathbf{i} + \frac{dy}{ds}\mathbf{j}\right) = 0 \tag{9}$$

At (x_0, y_0) the first vector is $\nabla f(x_0, y_0)$, and from Formula (2) of Section 15.4 the second vector is $\mathbf{u}$. Thus, from (8) and (9)

$$D_{\mathbf{u}}f(x_0, y_0) = \nabla f(x_0, y_0) \cdot \mathbf{u} = 0 \tag{10}$$

which is what we wanted to show.

---

*The conditions under which this occurs are specified in a theorem from advanced calculus called the "implicit function theorem." We shall assume in the remainder of this section that f is differentiable at (x_0, y_0) and that a smooth arc-length parametrization of the level curve through (x_0, y_0) is possible.

The last equality in (10) implies that the gradient of f is perpendicular to the tangent vector $\mathbf{u}$ at the point (x_0, y_0), and hence is normal to the level curve at that point. In summary, we have the following theorem.

16.6.4   THEOREM.   *If f is differentiable at (x_0, y_0), then $\nabla f(x_0, y_0)$ is normal to the level curve of f through (x_0, y_0).*

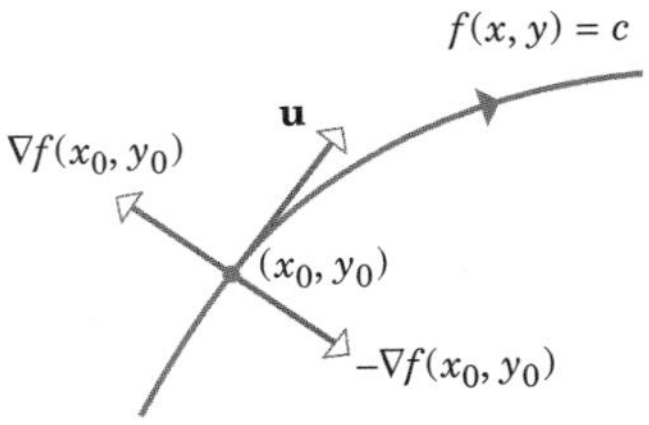

Figure 16.6.6

REMARK.   Figure 16.6.6 shows the relationship between the vectors $\mathbf{u}$, $\nabla f(x_0, y_0)$, and $-\nabla f(x_0, y_0)$ for the level curve of f that passes through the point (x_0, y_0). If the gradient of f is nonzero at that point, then the function f has its maximum rate of increase in the direction of $\nabla f(x_0, y_0)$, its maximum rate of decrease in the direction of $-\nabla f(x_0, y_0)$, and is neither increasing nor decreasing in the direction of $\mathbf{u}$, since $D_{\mathbf{u}} f(x_0, y_0) = 0$. Good skiers use these facts intuitively to control their speed by zigzagging down ski slopes: They ski across the slope with their skis tangential to a level curve to stop their downhill motion and turn their skis down the slope and perpendicular to the level curves to obtain the most rapid downhill motion.

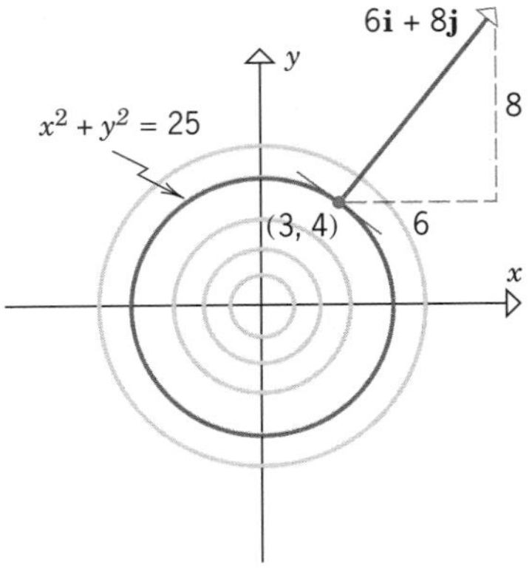

Example 5   For the function $f(x, y) = x^2 + y^2$, sketch the level curve through the point $(3, 4)$, and draw the gradient vector at this point.

Figure 16.6.7

Solution.   Since $f(3, 4) = 25$, the level curve through the point $(3, 4)$ has the equation $f(x, y) = 25$, which is the circle

$$x^2 + y^2 = 25$$

Since

$$\nabla f(x, y) = f_x(x, y)\mathbf{i} + f_y(x, y)\mathbf{j} = 2x\mathbf{i} + 2y\mathbf{j}$$

the gradient vector at $(3, 4)$ is

$$\nabla f(3, 4) = 6\mathbf{i} + 8\mathbf{j}$$

(Figure 16.6.7). Note that the gradient vector is perpendicular to the circle at $(3, 4)$, as guaranteed by Theorem 16.6.4.   ◀

▶ Exercise Set 16.6

In Exercises 1–4, find ∇z.

1. $z = 4x - 8y$.
2. $z = e^{-3y} \cos 4x$.
3. $z = \ln \sqrt{x^2 + y^2}$.
4. $z = e^{-5x} \sec x^2 y$.

In Exercises 5–8, find the gradient of f at the indicated point.

5. $f(x, y) = (x^2 + xy)^3$; $(-1, -1)$.
6. $f(x, y) = (x^2 + y^2)^{-1/2}$; $(3, 4)$.
7. $f(x, y) = y \ln (x + y)$; $(-3, 4)$.
8. $f(x, y) = y^2 \tan^3 x$; $(\pi/4, -3)$.

In Exercises 9–12, find $D_{\mathbf{u}} f$ at P.

9. $f(x, y) = (1 + xy)^{3/2}$; $P(3, 1)$; $\mathbf{u} = \dfrac{1}{\sqrt{2}}\mathbf{i} + \dfrac{1}{\sqrt{2}}\mathbf{j}$.
10. $f(x, y) = e^{2xy}$; $P(4, 0)$; $\mathbf{u} = -\frac{3}{5}\mathbf{i} + \frac{4}{5}\mathbf{j}$.

11. $f(x, y) = \ln (1 + x^2 + y)$; $P(0, 0)$; $\mathbf{u} = -\dfrac{1}{\sqrt{10}}\mathbf{i} - \dfrac{3}{\sqrt{10}}\mathbf{j}$.
12. $f(x, y) = \dfrac{cx + dy}{x - y}$; $P(3, 4)$; $\mathbf{u} = \frac{4}{5}\mathbf{i} + \frac{3}{5}\mathbf{j}$.

In Exercises 13–18, find the directional derivative of f at P in the direction of $\mathbf{a}$.

13. $f(x, y) = 4x^3y^2$; $P(2, 1)$; $\mathbf{a} = 4\mathbf{i} - 3\mathbf{j}$.
14. $f(x, y) = x^2 - 3xy + 4y^3$; $P(-2, 0)$; $\mathbf{a} = \mathbf{i} + 2\mathbf{j}$.
15. $f(x, y) = y^2 \ln x$; $P(1, 4)$; $\mathbf{a} = -3\mathbf{i} + 3\mathbf{j}$.
16. $f(x, y) = e^x \cos y$; $P(0, \pi/4)$; $\mathbf{a} = 5\mathbf{i} - 2\mathbf{j}$.
17. $f(x, y) = \tan^{-1}(y/x)$; $P(-2, 2)$; $\mathbf{a} = -\mathbf{i} - \mathbf{j}$.
18. $f(x, y) = xe^y - ye^x$; $P(0, 0)$; $\mathbf{a} = 5\mathbf{i} - 2\mathbf{j}$.

In Exercises 19–22, find the directional derivative of f at P in the direction of a vector making the angle θ with the positive x-axis.

19. $f(x, y) = \sqrt{xy}$; $P(1, 4)$; $\theta = \pi/3$.

20. $f(x, y) = \dfrac{x - y}{x + y}$; $P(-1, -2)$; $\theta = \pi/2$.

21. $f(x, y) = \tan(2x + y)$; $P(\pi/6, \pi/3)$; $\theta = 7\pi/4$.

22. $f(x, y) = \sinh x \cosh y$; $P(0, 0)$; $\theta = \pi$.

In Exercises 23–26, sketch the level curve of $f(x, y)$ that passes through P and draw the gradient vector at P.

23. $f(x, y) = 4x - 2y + 3$; $P(1, 2)$.

24. $f(x, y) = y/x^2$; $P(-2, 2)$.

25. $f(x, y) = x^2 + 4y^2$; $P(-2, 0)$.

26. $f(x, y) = x^2 - y^2$; $P(2, -1)$.

In Exercises 27–30, find a unit vector in the direction in which f increases most rapidly at P; and find the rate of change of f at P in that direction.

27. $f(x, y) = 4x^3 y^2$; $P(-1, 1)$.

28. $f(x, y) = 3x - \ln y$; $P(2, 4)$.

29. $f(x, y) = \sqrt{x^2 + y^2}$; $P(4, -3)$.

30. $f(x, y) = \dfrac{x}{x + y}$; $P(0, 2)$.

In Exercises 31–34, find a unit vector in the direction in which f decreases most rapidly at P; and find the rate of change of f at P in that direction.

31. $f(x, y) = 20 - x^2 - y^2$; $P(-1, -3)$.

32. $f(x, y) = e^{xy}$; $P(2, 3)$.

33. $f(x, y) = \cos(3x - y)$; $P(\pi/6, \pi/4)$.

34. $f(x, y) = \sqrt{\dfrac{x - y}{x + y}}$; $P(3, 1)$.

35. Find the directional derivative of $f(x, y) = \dfrac{x}{x + y}$ at $P(1, 0)$ in the direction to $Q(-1, -1)$.

36. Find the directional derivative of $f(x, y) = e^{-x} \sec y$ at $P(0, \pi/4)$ in the direction of the origin.

37. Find the directional derivative of $f(x, y) = \sqrt{xy}\, e^y$ at $P(1, 1)$ in the direction of the negative y-axis.

38. Let $f(x, y) = \dfrac{y}{x + y}$. Find a unit vector $\mathbf{u}$ for which $D_{\mathbf{u}} f(2, 3) = 0$.

39. Find a unit vector $\mathbf{u}$ that is perpendicular at $P(1, -2)$ to the level curve of $f(x, y) = 4x^2 y$ through P.

40. Find a unit vector $\mathbf{u}$ that is perpendicular at $P(2, -3)$ to the level curve of $f(x, y) = 3x^2 y - xy$ through P.

41. Given that $D_{\mathbf{u}} f(1, 2) = -5$ if $\mathbf{u} = \frac{3}{5}\mathbf{i} - \frac{4}{5}\mathbf{j}$ and $D_{\mathbf{v}} f(1, 2) = 10$ if $\mathbf{v} = \frac{4}{5}\mathbf{i} + \frac{3}{5}\mathbf{j}$, find

    (a) $f_x(1, 2)$ (b) $f_y(1, 2)$

    (c)   the directional derivative of f at $(1, 2)$ in the direction of the origin.

42. Given that $f_x(-5, 1) = -3$ and $f_y(-5, 1) = 2$, find the directional derivative of f at $P(-5, 1)$ in the direction of the vector from P to $Q(-4, 3)$.

43. Given that $\nabla f(4, -5) = 2\mathbf{i} - \mathbf{j}$, find the directional derivative of the function f at $(4, -5)$ in the direction of $\mathbf{a} = 5\mathbf{i} + 2\mathbf{j}$.

44. Given that $\nabla f(x_0, y_0) = \mathbf{i} - 2\mathbf{j}$ and $D_{\mathbf{u}} f(x_0, y_0) = -2$, find $\mathbf{u}$.

45. Let $z = 3x^2 - y^2$. Find all points where $\|\nabla z\| = 6$.

46. Given that $z = 3x + y^2$, find the maximum value of $d\|\nabla z\|/ds$ at the point $(5, 2)$ and a unit vector in the direction in which the maximum is attained.

47. A particle moves along a path C given by $x = t$ and $y = -t^2$. If $z = x^2 + y^2$, find dz/ds along C at the instant when the particle is at the point $(2, -4)$.

48. The temperature at a point (x, y) on a metal plate in the xy-plane is $T(x, y) = \dfrac{xy}{1 + x^2 + y^2}$ degrees Celsius.

    (a)   Find the rate of change of temperature at $(1, 1)$ in the direction of $\mathbf{a} = 2\mathbf{i} - \mathbf{j}$.

    (b)   An ant at $(1, 1)$ wants to walk in the direction in which the temperature drops most rapidly. Find a unit vector in that direction.

49. If the electric potential at a point (x, y) in the xy-plane is $V(x, y)$, then the **electric intensity vector** at (x, y) is $\mathbf{E} = -\nabla V(x, y)$. Suppose that $V(x, y) = e^{-2x} \cos 2y$.

    (a)   Find the electric intensity vector at $(\pi/4, 0)$.

    (b)   Show that at each point in the plane, the electric potential decreases most rapidly in the direction of the vector $\mathbf{E}$.

50. On a certain mountain, the elevation z above a point (x, y) in a horizontal xy-plane that lies at sea level is $z = 2000 - 2x^2 - 4y^2$ ft. The positive x-axis points east, and the positive y-axis north. A climber is at the point $(-20, 5, 1100)$.

    (a)   If the climber uses a compass reading to walk due west, will he begin to ascend or descend?

    (b)   If the climber uses a compass reading to walk northeast, will he ascend or descend? At what rate?

    (c)   In what compass direction should the climber walk to travel a level path?

51. Let $r = \sqrt{x^2 + y^2}$.

    (a)   Show that $\nabla r = \dfrac{\mathbf{r}}{r}$, where $\mathbf{r} = x\mathbf{i} + y\mathbf{j}$.

    (b)   Show that $\nabla f(r) = f'(r)\nabla r = \dfrac{f'(r)}{r}\mathbf{r}$.

52. Use the formula of part (b) in Exercise 51 to find

(a) $\nabla f(r)$ if $f(r) = re^{-3r}$

(b) $f(r)$ if $\nabla f(r) = 3r^2\mathbf{r}$ and $f(2) = 1$.

53. Let $\mathbf{u}_r$ be a unit vector making an angle θ with the positive x-axis, and let $\mathbf{u}_\theta$ be a unit vector $90°$ counterclockwise from $\mathbf{u}_r$. Show that if $z = f(x, y)$, $x = r\cos\theta$, and $y = r\sin\theta$, then

$$\nabla z = \frac{\partial z}{\partial r}\mathbf{u}_r + \frac{1}{r}\frac{\partial z}{\partial \theta}\mathbf{u}_\theta$$

[*Hint:* Use parts (a) and (b) of Exercise 62, Section 16.4.]

54. Prove: If f and g are differentiable, then

(a) $\nabla(f + g) = \nabla f + \nabla g$

(b) $\nabla(cf) = c\nabla f$ (c constant)

(c) $\nabla(fg) = f\nabla g + g\nabla f$

(d) $\nabla\left(\dfrac{f}{g}\right) = \dfrac{g\nabla f - f\nabla g}{g^2}$

(e) $\nabla(f^p) = pf^{p-1}\nabla f$.

55. Prove: If $x = x(t)$ and $y = y(t)$ are differentiable at t, and if $z = f(x, y)$ is differentiable at the point $(x(t), y(t))$, then

$$\frac{dz}{dt} = \nabla z \cdot \mathbf{r}'(t)$$

where $\mathbf{r}(t) = x(t)\mathbf{i} + y(t)\mathbf{j}$.

56. Prove: If f, f_x, and f_y are continuous on a circular region, and if $\nabla f(x, y) = \mathbf{0}$ throughout the region, then $f(x, y)$ is constant on the region. [*Hint:* See Exercise 70, Section 16.4.]

57. Prove: If $D_{\mathbf{u}}f(x, y) = 0$ in two nonparallel directions, then $D_{\mathbf{u}}f(x, y) = 0$ in all directions.

16.7 DIFFERENTIABILITY, DIRECTIONAL DERIVATIVES, AND GRADIENTS FOR FUNCTIONS OF THREE VARIABLES

In this section we shall extend most of the results obtained in the last two sections to functions of three variables. The main difference between functions of two and three variables is geometric: The graph of $z = f(x, y)$ represents a surface in 3-space, whereas $w = f(x, y, z)$ has no analogous interpretation.

☐ **DIFFERENTIABILITY**

The definition of differentiability for functions of three variables and the basic theorems about differentiability are direct generalizations of the corresponding results for functions of two variables. (Compare the following with Definition 16.4.2, Theorem 16.4.4, and Theorem 16.4.7.)

16.7.1 DEFINITION.   A function f of three variables is defined to be **differentiable** at the point (x_0, y_0, z_0) if the partial derivatives $f_x(x_0, y_0, z_0)$, $f_y(x_0, y_0, z_0)$, and $f_z(x_0, y_0, z_0)$ exist and

$$\Delta f = f(x_0 + \Delta x, y_0 + \Delta y, z_0 + \Delta z) - f(x_0, y_0, z_0)$$

can be written in the form

$$\Delta f = f_x(x_0, y_0, z_0)\,\Delta x + f_y(x_0, y_0, z_0)\,\Delta y + f_z(x_0, y_0, z_0)\,\Delta z + \epsilon_1\,\Delta x + \epsilon_2\,\Delta y + \epsilon_3\,\Delta z$$

where ϵ_1, ϵ_2, and ϵ_3 are functions of Δx, Δy, and Δz such that $\epsilon_1 \to 0$, $\epsilon_2 \to 0$, and $\epsilon_3 \to 0$ as $(\Delta x, \Delta y, \Delta z) \to (0, 0, 0)$.

16.7.2 THEOREM.   *If f has first-order partial derivatives at each point of some spherical region centered at (x_0, y_0, z_0), and if these partial derivatives are continuous at (x_0, y_0, z_0), then f is differentiable at (x_0, y_0, z_0).*

16.7.3 THEOREM (*Chain Rule*). *If* $x = x(t)$, $y = y(t)$, *and* $z = z(t)$, *are differentiable at* t *and* $w = f(x, y, z)$ *is differentiable at the point* $(x(t), y(t), z(t))$, *then* $w = f(x(t), y(t), z(t))$ *is differentiable at* t, *and*

$$\frac{dw}{dt} = \frac{\partial w}{\partial x}\frac{dx}{dt} + \frac{\partial w}{\partial y}\frac{dy}{dt} + \frac{\partial w}{\partial z}\frac{dz}{dt}$$

The meaning of such terms as **differentiable on a region R** and **differentiable** (i.e., **differentiable everywhere**) should be clear, keeping in mind the regions involved are in 3-space. Moreover, as for functions of one and two variables, a function of three variables is continuous at a point if it is differentiable at that point.

Example 1   Suppose that

$$w = x^3 y^2 z, \quad x = t^2, \quad y = t^3, \quad z = t^4$$

Use the chain rule to find dw/dt.

Solution.   By the chain rule,

$$\frac{dw}{dt} = \frac{\partial w}{\partial x}\frac{dx}{dt} + \frac{\partial w}{\partial y}\frac{dy}{dt} + \frac{\partial w}{\partial z}\frac{dz}{dt}$$

$$= (3x^2 y^2 z)(2t) + (2x^3 yz)(3t^2) + (x^3 y^2)(4t^3)$$

$$= (3t^{14})(2t) + (2t^{13})(3t^2) + (t^{12})(4t^3) = 16t^{15} \quad \blacktriangleleft$$

Other variations of the chain rule for functions of three variables will be considered in the next section.

REMARK.   The most significant difference between working with functions of two variables and functions of three variables is geometric. For a function of two variables the equation $z = f(x, y)$ can be graphed as a surface in three-dimensional space. However, for a function of three variables, there is no direct way to graph $w = f(x, y, z)$, since "four dimensions" would be required (one dimension for each variable). This is not devastating, however; it simply means that we must rely more heavily on the analytic formulas than on the geometry.

☐ **DIRECTIONAL DERIVATIVES**

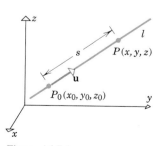

Figure 16.7.1

To define a directional derivative at a point (x_0, y_0, z_0) for a function f of three variables, we shall use a unit vector $\mathbf{u} = \langle u_1, u_2, u_3 \rangle$ to designate the direction, and we shall let l be the line through (x_0, y_0, z_0) that is parallel to $\mathbf{u}$ (Figure 16.7.1). This line can be represented parametrically as

$$x = x_0 + su_1, \quad y = y_0 + su_2, \quad z = z_0 + su_3$$

where s is an arc-length parameter with its reference point at $P_0(x_0, y_0, z_0)$ and the positive direction in the direction of $\mathbf{u}$ [compare with Formula (1) of Section 16.6]. As s increases, the point $P(x, y, z)$ moves in the direction of $\mathbf{u}$ along l, and the value of $w = f(x, y, z)$ changes with s. As for functions of two variables, we define $D_{\mathbf{u}}f(x_0, y_0, z_0)$ to be the instantaneous rate of change of w with respect to s at (x_0, y_0, z_0). Proceeding as in the derivation of Formula (4) in Section 16.6 leads to the following definition.

16.7.4 DEFINITION.   If f is differentiable at (x_0, y_0, z_0), and if $\mathbf{u} = \langle u_1, u_2, u_3 \rangle$ is a unit vector, then the ***directional derivative*** of f at (x_0, y_0, z_0) in the direction of $\mathbf{u}$ is defined by

$$D_{\mathbf{u}}f(x_0, y_0, z_0) = f_x(x_0, y_0, z_0)u_1 + f_y(x_0, y_0, z_0)u_2 + f_z(x_0, y_0, z_0)u_3 \tag{1}$$

☐ **THE GRADIENT**

The definition of the gradient of a function of three variables is

$$\nabla f(x, y, z) = f_x(x, y, z)\mathbf{i} + f_y(x, y, z)\mathbf{j} + f_z(x, y, z)\mathbf{k} \tag{2}$$

which is identical to Definition 16.6.2, except for the additional third component. It follows from (1) and (2) that

$$D_{\mathbf{u}}f(x, y, z) = \nabla f(x, y, z) \cdot \mathbf{u} \tag{3}$$

which parallels Formula (8) of Section 16.6. Using this result, the reader should have no trouble proving the following extension of Theorem 16.6.3.

16.7.5 THEOREM.   *Let f be a function of three variables that is differentiable at (x_0, y_0, z_0).*

(a) *If $\nabla f(x_0, y_0, z_0) = \mathbf{0}$, then all directional derivatives of f at (x_0, y_0, z_0) are zero.*

(b) *If $\nabla f(x_0, y_0, z_0) \neq \mathbf{0}$, then among all possible directional derivatives of f at (x_0, y_0, z_0), the derivative in the direction of $\nabla f(x_0, y_0, z_0)$ has the largest value. The value of that directional derivative is $\|\nabla f(x_0, y_0, z_0)\|$.*

(c) *If $\nabla f(x_0, y_0, z_0) \neq \mathbf{0}$, then among all possible directional derivatives of f at (x_0, y_0, z_0), the derivative in the direction opposite to that of $\nabla f(x_0, y_0, z_0)$ has the smallest value. The value of that directional derivative is $-\|\nabla f(x_0, y_0, z_0)\|$.*

Example 2   Find the directional derivative of $f(x, y, z) = x^2 y - yz^3 + z$ at the point $P(1, -2, 0)$ in the direction of the vector $\mathbf{a} = 2\mathbf{i} + \mathbf{j} - 2\mathbf{k}$, and find the maximum rate of increase of f at P.

Solution.   Since

$$f_x(x, y, z) = 2xy, \quad f_y(x, y, z) = x^2 - z^3, \quad f_z(x, y, z) = -3yz^2 + 1$$

it follows that

$$\nabla f(x, y, z) = 2xy\mathbf{i} + (x^2 - z^3)\mathbf{j} + (-3yz^2 + 1)\mathbf{k}$$

$$\nabla f(1, -2, 0) = -4\mathbf{i} + \mathbf{j} + \mathbf{k}$$

A unit vector in the direction of $\mathbf{a}$ is

$$\mathbf{u} = \frac{\mathbf{a}}{\|\mathbf{a}\|} = \frac{1}{\sqrt{9}}(2\mathbf{i} + \mathbf{j} - 2\mathbf{k}) = \tfrac{2}{3}\mathbf{i} + \tfrac{1}{3}\mathbf{j} - \tfrac{2}{3}\mathbf{k}$$

Therefore

$$D_{\mathbf{u}}f(1, -2, 0) = \nabla f(1, -2, 0) \cdot \mathbf{u} = (-4)(\tfrac{2}{3}) + (1)(\tfrac{1}{3}) + (1)(-\tfrac{2}{3}) = -3$$

The maximum rate of increase of f at P is

$$\|\nabla f(1, -2, 0)\| = \sqrt{(-4)^2 + (1)^2 + (1)^2} = 3\sqrt{2} \qquad \blacktriangleleft$$

Our next objective is to establish a geometric relationship between the level surfaces and the gradient of a function f of three variables. For this purpose we note that if (x_0, y_0, z_0)

is any point in the domain of f, and if $f(x_0, y_0, z_0) = c$, then under appropriate conditions the equation

$$f(x, y, z) = c$$

defines a unique level surface of f through (x_0, y_0, z_0) that has a tangent plane at (x_0, y_0, z_0).*

If $f(x, y, z) = c$ is the level surface of f that passes through the point (x_0, y_0, z_0), then the value of $f(x, y, z)$ remains constant as (x, y, z) moves away from (x_0, y_0, z_0) along the level surface. It seems plausible, therefore, that the directional derivative of f at (x_0, y_0, z_0) is zero in the direction of any unit vector $\mathbf{u}$ at (x_0, y_0, z_0) that lies in the tangent plane to the level surface at (x_0, y_0, z_0). That is, for any such $\mathbf{u}$

$$D_{\mathbf{u}} f(x_0, y_0, z_0) = \nabla f(x_0, y_0, z_0) \cdot \mathbf{u} = 0 \tag{4}$$

which is analogous to Formula (10) in Section 16.6. We shall omit the formal proof.

The last equality in (4) implies that the gradient of f is perpendicular to every unit vector $\mathbf{u}$ at (x_0, y_0, z_0) that lies in the tangent plane and hence is perpendicular to the tangent plane itself. This means that the gradient at (x_0, y_0, z_0) is normal to the level surface at (x_0, y_0, z_0) (Figure 16.7.2). In summary, we have the following theorem.

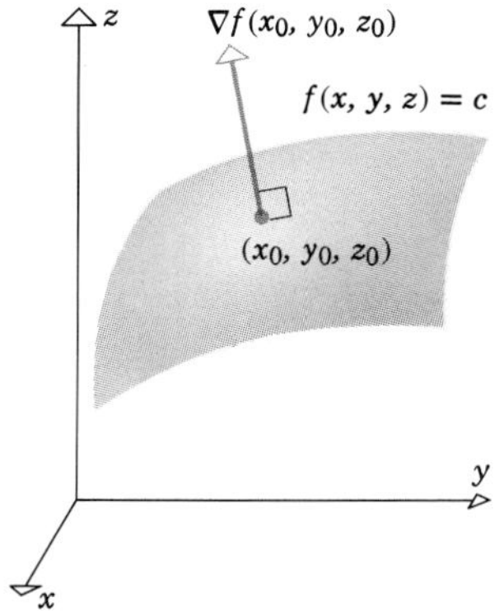

Figure 16.7.2

16.7.6 THEOREM. *If f is differentiable at (x_0, y_0, z_0), then $\nabla f(x_0, y_0, z_0)$ is normal to the level surface of $f(x, y, z)$ through (x_0, y_0, z_0).*

Example 3 Find an equation of the plane that is tangent to the ellipsoid $x^2 + 4y^2 + z^2 = 18$ at the point $(1, 2, -1)$.

Solution. If we let $F(x, y, z) = x^2 + 4y^2 + z^2$, then the given equation has the form $F(x, y, z) = 18$, which may be viewed as the equation of a level surface for F. Thus, the vector $\nabla F(1, 2, -1)$ is normal to the ellipsoid at the point $(1, 2, -1)$. To find this vector we write

$$\nabla F(x, y, z) = \frac{\partial F}{\partial x} \mathbf{i} + \frac{\partial F}{\partial y} \mathbf{j} + \frac{\partial F}{\partial z} \mathbf{k} = 2x\mathbf{i} + 8y\mathbf{j} + 2z\mathbf{k}$$

$$\nabla F(1, 2, -1) = 2\mathbf{i} + 16\mathbf{j} - 2\mathbf{k}$$

Using this normal and the point $(1, 2, -1)$, we obtain as the equation of the tangent plane

$$2(x - 1) + 16(y - 2) - 2(z + 1) = 0$$

or

$$x + 8y - z = 18 \quad \blacktriangleleft$$

☐ **TOTAL DIFFERENTIALS**

If $w = f(x, y, z)$, then we define the **increment** Δw (also written Δf) to be

$$\Delta w = f(x + \Delta x, y + \Delta y, z + \Delta z) - f(x, y, z)$$

and we define the **total differential** dw (also written df) to be

$$dw = f_x(x, y, z)\,dx + f_y(x, y, z)\,dy + f_z(x, y, z)\,dz$$

where Δx, Δy, Δz, dx, dy, and dz are all variables representing changes in the values of x, y, and z.

---

*The conditions under which this occurs are specified in a theorem from advanced calculus called the "implicit function theorem." We shall assume in the remainder of this section that f is differentiable at (x_0, y_0, z_0) and that the level surface through the point (x_0, y_0, z_0) has a tangent plane at that point.

The increment Δw represents the change in the value of $w = f(x, y, z)$ when x, y, and z are changed by amounts Δx, Δy, and Δz, respectively. However, for functions of three variables, the total differential dw has no natural geometric interpretation.

If we let

$$dx = \Delta x, \quad dy = \Delta y, \quad dz = \Delta z$$

and if $w = f(x, y, z)$ is differentiable at (x, y, z), then it follows from Definition 16.7.1 that

$$\Delta w = dw + \epsilon_1 \Delta x + \epsilon_2 \Delta y + \epsilon_3 \Delta z \tag{5}$$

where $\epsilon_1 \rightarrow 0$, $\epsilon_2 \rightarrow 0$, $\epsilon_3 \rightarrow 0$ as $(\Delta x, \Delta y, \Delta z) \rightarrow (0, 0, 0)$. Thus, when Δx, Δy, and Δz are small, it follows from (5) that

$$\Delta w \approx dw$$

Example 4   The length, width, and height of a rectangular box are each measured with an error of at most 5%. Find an upper bound on the maximum possible percentage error that results if these quantities are used to calculate the diagonal of the box.

Solution.   Let x, y, z, and D be the true length, width, height, and diagonal of the box, respectively; and let Δx, Δy, Δz, and ΔD be the errors in these quantities. We are given that

$$|\Delta x / x| \leq 0.05, \quad |\Delta y / y| \leq 0.05, \quad |\Delta z / z| \leq 0.05$$

We want to estimate $|\Delta D / D|$. Since the diagonal D is related to the length, width, and height by

$$D = \sqrt{x^2 + y^2 + z^2}$$

it follows that

$$dD = \frac{\partial D}{\partial x} dx + \frac{\partial D}{\partial y} dy + \frac{\partial D}{\partial z} dz$$

$$= \frac{x}{\sqrt{x^2 + y^2 + z^2}} dx + \frac{y}{\sqrt{x^2 + y^2 + z^2}} dy + \frac{z}{\sqrt{x^2 + y^2 + z^2}} dz$$

If we choose $\Delta x = dx$, $\Delta y = dy$, $\Delta z = dz$, then we can use the approximation $\Delta D / D \approx dD / D$. But,

$$\frac{dD}{D} = \frac{x}{x^2 + y^2 + z^2} dx + \frac{y}{x^2 + y^2 + z^2} dy + \frac{z}{x^2 + y^2 + z^2} dz$$

or

$$\frac{dD}{D} = \frac{x^2}{x^2 + y^2 + z^2} \frac{dx}{x} + \frac{y^2}{x^2 + y^2 + z^2} \frac{dy}{y} + \frac{z^2}{x^2 + y^2 + z^2} \frac{dz}{z}$$

Thus,

$$\left| \frac{dD}{D} \right| = \left| \frac{x^2}{x^2 + y^2 + z^2} \frac{dx}{x} + \frac{y^2}{x^2 + y^2 + z^2} \frac{dy}{y} + \frac{z^2}{x^2 + y^2 + z^2} \frac{dz}{z} \right|$$

$$\leq \left| \frac{x^2}{x^2 + y^2 + z^2} \frac{dx}{x} \right| + \left| \frac{y^2}{x^2 + y^2 + z^2} \frac{dy}{y} \right| + \left| \frac{z^2}{x^2 + y^2 + z^2} \frac{dz}{z} \right|$$

$$\leq \frac{x^2}{x^2 + y^2 + z^2} (0.05) + \frac{y^2}{x^2 + y^2 + z^2} (0.05) + \frac{z^2}{x^2 + y^2 + z^2} (0.05)$$

$$= 0.05$$

Therefore, the maximum percentage error in D is at most 5%.   ◀

▶ Exercise Set 16.7 $\boxed{C}$ *41*

In Exercises 1–4, find dw/dt using the chain rule.

1. $w = 5x^2y^3z^4$; $x = t^2, y = t^3, z = t^5$.

2. $w = \ln(3x^2 - 2y + 4z^3)$; $x = t^{1/2}, y = t^{2/3}, z = t^{-2}$.

3. $w = 5\cos xy - \sin xz$; $x = 1/t, y = t, z = t^3$.

4. $w = \sqrt{1 + x - 2yz^4x}$; $x = \ln t, y = t, z = 4t$.

5. Use the chain rule to find $\dfrac{dw}{dt}\Big|_{t=1}$ if $w = x^3y^2z^4$; $x = t^2$, $y = t + 2$, $z = 2t^4$.

6. Use the chain rule to find $\dfrac{dw}{dt}\Big|_{t=0}$ if $w = x\sin yz^2$; $x = \cos t, y = t^2, z = e^t$.

In Exercises 7–10, find the gradient of f at P, and then use the gradient to calculate $D_{\mathbf{u}}f$ at P.

7. $f(x, y, z) = 4x^5y^2z^3$; $P(2, -1, 1)$; $\mathbf{u} = \frac{1}{3}\mathbf{i} + \frac{2}{3}\mathbf{j} - \frac{2}{3}\mathbf{k}$.

8. $f(x, y, z) = ye^{xz} + z^2$; $P(0, 2, 3)$; $\mathbf{u} = \frac{2}{7}\mathbf{i} - \frac{3}{7}\mathbf{j} + \frac{6}{7}\mathbf{k}$.

9. $f(x, y, z) = \ln(x^2 + 2y^2 + 3z^2)$; $P(-1, 2, 4)$; $\mathbf{u} = -\frac{3}{13}\mathbf{i} - \frac{4}{13}\mathbf{j} - \frac{12}{13}\mathbf{k}$.

10. $f(x, y, z) = \sin xyz$; $P(\frac{1}{2}, \frac{1}{3}, \pi)$; $\mathbf{u} = \dfrac{1}{\sqrt{3}}\mathbf{i} - \dfrac{1}{\sqrt{3}}\mathbf{j} + \dfrac{1}{\sqrt{3}}\mathbf{k}$.

In Exercises 11–14, find the directional derivative of f at P in the direction of $\mathbf{a}$.

11. $f(x, y, z) = x^3z - yx^2 + z^2$; $P(2, -1, 1)$; $\mathbf{a} = 3\mathbf{i} - \mathbf{j} + 2\mathbf{k}$.

12. $f(x, y, z) = y - \sqrt{x^2 + z^2}$; $P(-3, 1, 4)$; $\mathbf{a} = 2\mathbf{i} - 2\mathbf{j} - \mathbf{k}$.

13. $f(x, y, z) = \dfrac{z - x}{z + y}$; $P(1, 0, -3)$; $\mathbf{a} = -6\mathbf{i} + 3\mathbf{j} - 2\mathbf{k}$.

14. $f(x, y, z) = e^{x+y+3z}$; $P(-2, 2, -1)$; $\mathbf{a} = 20\mathbf{i} - 4\mathbf{j} + 5\mathbf{k}$.

In Exercises 15–18, find a unit vector in the direction in which f increases most rapidly at P, and find the rate of increase of f in that direction.

15. $f(x, y, z) = x^3z^2 + y^3z + z - 1$; $P(1, 1, -1)$.

16. $f(x, y, z) = \sqrt{x - 3y + 4z}$; $P(0, -3, 0)$.

17. $f(x, y, z) = \dfrac{x}{z} + \dfrac{z}{y^2}$; $P(1, 2, -2)$.

18. $f(x, y, z) = \tan^{-1}\left(\dfrac{x}{y + z}\right)$; $P(4, 2, 2)$.

In Exercises 19 and 20, find a unit vector in the direction in which f decreases most rapidly at P, and find the rate of change of f in that direction.

19. $f(x, y, z) = \dfrac{x + z}{z - y}$; $P(5, 7, 6)$.

20. $f(x, y, z) = 4e^{xy}\cos z$; $P(0, 1, \pi/4)$.

21. Find the directional derivative of
$$f(x, y, z) = \frac{y}{x + z}$$
at $P(2, 1, -1)$ in the direction from P to $Q(-1, 2, 0)$.

22. Find the directional derivative of the function
$$f(x, y, z) = x^3y^2z^5 - 2xz + yz + 3x$$
at $P(-1, -2, 1)$ in the direction of the negative z-axis.

23. Given that the directional derivative of $f(x, y, z)$ at the point $(3, -2, 1)$ in the direction of $\mathbf{a} = 2\mathbf{i} - \mathbf{j} - 2\mathbf{k}$ is -5 and that $\|\nabla f(3, -2, 1)\| = 5$, find $\nabla f(3, -2, 1)$.

24. The temperature (in degrees Celsius) at a point (x, y, z) in a metal solid is
$$T(x, y, z) = \frac{xyz}{1 + x^2 + y^2 + z^2}$$
(a) Find the rate of change of temperature at $(1, 1, 1)$ in the direction of the origin.

(b) Find the direction in which the temperature rises most rapidly at $(1, 1, 1)$. (Express your answer as a unit vector.)

(c) Find the rate at which the temperature rises moving from $(1, 1, 1)$ in the direction obtained in part (b).

In Exercises 25–28, find equations for the tangent plane and the line that is normal to the given surface at the point P.

25. $x^2 + y^2 + z^2 = 49$; $P(-3, 2, -6)$.

26. $xz - yz^3 + yz^2 = 2$; $P(2, -1, 1)$.

27. $\sqrt{\dfrac{z + x}{y - 1}} = z^2$; $P(3, 5, 1)$.

28. $\sin xz - 4\cos yz = 4$; $P(\pi, \pi, 1)$.

29. Show that every line that is normal to the sphere
$$x^2 + y^2 + z^2 = 1$$
passes through the origin.

30. Find all points on the ellipsoid $2x^2 + 3y^2 + 4z^2 = 9$ at which the tangent plane is parallel to the plane $x - 2y + 3z = 5$.

31. Find all points on the surface $x^2 + y^2 - z^2 = 1$ at which the normal line is parallel to the line through $P(1, -2, 1)$ and $Q(4, 0, -1)$.

32. Show that the ellipsoid $2x^2 + 3y^2 + z^2 = 9$ and the sphere
$$x^2 + y^2 + z^2 - 6x - 8y - 8z + 24 = 0$$
have a common tangent plane at the point $(1, 1, 2)$.

In Exercises 33–36, find dw.

33. $w = 8x - 3y + 4z$.

34. $w = 4x^2y^3z^7 - 3xy + z + 5$.

35. $w = \tan^{-1}(xyz)$.   **36.** $w = \sqrt{x} + \sqrt{y} + \sqrt{z}$.

37. Use a total differential to approximate the change in $f(x, y, z) = 2xy^2z^3$ as (x, y, z) varies from $P(1, -1, 2)$ to $Q(0.99, -1.02, 2.02)$.

38. Use a total differential to approximate the change in $f(x, y, z) = xyz/(x + y + z)$ as (x, y, z) varies from $P(-1, -2, 4)$ to $Q(-1.04, -1.98, 3.97)$.

39. The length, width, and height of a rectangular box are measured to be 3 cm, 4 cm, and 5 cm, respectively, with a maximum error of 0.05 cm in each measurement. Use differentials to approximate the maximum error in the calculated volume.

40. The total resistance R of three resistances R_1, R_2, and R_3, connected in parallel, is given by

$$\frac{1}{R} = \frac{1}{R_1} + \frac{1}{R_2} + \frac{1}{R_3}$$

Suppose that R_1, R_2, and R_3 are measured to be 100 ohms, 200 ohms, and 500 ohms, respectively, with a maximum error of 10% in each. Use differentials to approximate the maximum percentage error in the calculated value of R.

41. The area of a triangle is to be computed from the formula $A = \frac{1}{2}ab \sin \theta$, where a and b are the lengths of two sides and θ is the included angle. Suppose that a, b, and θ are measured to be 40 ft, 50 ft, and 30°, respectively. Use differentials to approximate the maximum error in the calculated value of A if the maximum errors in a, b, and θ are $\frac{1}{2}$ ft, $\frac{1}{4}$ ft, and 2°, respectively.

42. The length, width, and height of a rectangular box are measured with errors of at most $r\%$. Use differentials to approximate the maximum percentage error in the computed value of the volume.

43. Use differentials to approximate the maximum percentage error in $w = xy^2z^3$ if x, y, and z have errors of at most 1%, 2%, and 3%, respectively.

44. Two surfaces are said to be **orthogonal** at a point of intersection if their normal lines are perpendicular at that point. Prove that the surfaces $f(x, y, z) = 0$ and $g(x, y, z) = 0$ are orthogonal at a point of intersection, (x_0, y_0, z_0), if and only if

$$f_x g_x + f_y g_y + f_z g_z = 0$$

at (x_0, y_0, z_0). [Assume that $\nabla f(x_0, y_0, z_0) \neq \mathbf{0}$ and $\nabla g(x_0, y_0, z_0) \neq \mathbf{0}$.]

45. Use the result of Exercise 44 to show that the sphere $x^2 + y^2 + z^2 = a^2$ and the cone $z^2 = x^2 + y^2$ are orthogonal at every point of intersection.

■ **16.8** FUNCTIONS OF n VARIABLES; MORE ON THE CHAIN RULE

In this section we shall discuss functions involving more than three variables. Our main objective is to develop forms of the chain rule for such functions.

Most of the definitions and theorems we have stated for functions of two and three variables can be extended to functions of four or more variables. Recall that if

$$w = f(v_1, v_2, \ldots, v_n)$$

is a function of n variables, then there are n partial derivatives

$$\frac{\partial w}{\partial v_1}, \frac{\partial w}{\partial v_2}, \ldots, \frac{\partial w}{\partial v_n}$$

each of which is calculated by holding $n - 1$ of the variables fixed and differentiating with respect to the remaining variable. In order to define directional derivatives for a function of n variables, it is first necessary to define the notion of a vector in "n-dimensional space." This topic is studied in a branch of mathematics called **linear algebra**, and will not be considered in this text.

☐ **TOTAL DIFFERENTIALS**

If $w = f(v_1, v_2, \ldots, v_n)$, we define the ***increment*** Δw and the ***total differential*** dw to be

$$\Delta w = f(v_1 + \Delta v_1, v_2 + \Delta v_2, \ldots, v_n + \Delta v_n) - f(v_1, v_2, \ldots, v_n) \tag{1}$$

and

$$dw = \frac{\partial w}{\partial v_1} dv_1 + \frac{\partial w}{\partial v_2} dv_2 + \cdots + \frac{\partial w}{\partial v_n} dv_n \tag{2}$$

where $\Delta v_1, \Delta v_2, \ldots, \Delta v_n$ and $dv_1, dv_2, \ldots, dv_n$ are variables representing changes in the values of $v_1, v_2, \ldots, v_n$.

If $v_1, v_2, \ldots, v_n$ are functions of a single variable t, then $w = f(v_1, v_2, \ldots, v_n)$ is a function of t, and a chain-rule formula for dw/dt is

$$\frac{dw}{dt} = \frac{\partial w}{\partial v_1} \frac{dv_1}{dt} + \frac{\partial w}{\partial v_2} \frac{dv_2}{dt} + \cdots + \frac{\partial w}{\partial v_n} \frac{dv_n}{dt} \tag{3}$$

which is a natural extension of the chain-rule formulas in Theorems 16.4.7 and 16.7.3. Observe that (3) results if we formally divide both sides of (2) by dt.

☐ **CHAIN RULES**

Other forms of the chain rule arise, depending on the number of variables involved. For example, in Theorem 16.4.8 we obtained the chain-rule formulas

$$\frac{\partial z}{\partial u} = \frac{\partial z}{\partial x} \frac{\partial x}{\partial u} + \frac{\partial z}{\partial y} \frac{\partial y}{\partial u} \tag{4}$$

$$\frac{\partial z}{\partial v} = \frac{\partial z}{\partial x} \frac{\partial x}{\partial v} + \frac{\partial z}{\partial y} \frac{\partial y}{\partial v} \tag{5}$$

for the case where z is a function of two variables $z = f(x, y)$, and x and y in turn are functions of two other variables, $x = x(u, v)$, $y = y(u, v)$.

Formulas (4) and (5) can be represented schematically by a "tree diagram" constructed as follows (Figure 16.8.1). Starting with z at the top of the diagram and moving downward, join each variable by lines (or branches) to those variables on which it depends *directly*. Thus, z is joined to x and y and these in turn are each joined to u and v. Next, label each branch with a derivative whose "numerator" contains the variable at the top end of that branch, and whose "denominator" contains the variable at the bottom end of that branch. This completes the "tree." To find the formula for $\partial z/\partial u$ trace all paths through the tree that start at z and end at u. Each such path produces one of the terms in the formula for $\partial z/\partial u$ (Figure 16.8.1a). Similarly, each term in the formula for $\partial z/\partial v$ corresponds to a path starting at z and ending at v (Figure 16.8.1b).

The following examples illustrate how tree diagrams can be used to construct other forms of the chain rule.

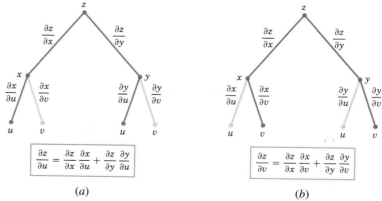

Figure 16.8.1 (a) (b)

Example 1 Suppose that

$$w = e^{xyz}, \quad x = 3r + s, \quad y = 3r - s, \quad z = r^2 s$$

Use appropriate forms of the chain rule to find $\partial w/\partial r$ and $\partial w/\partial s$.

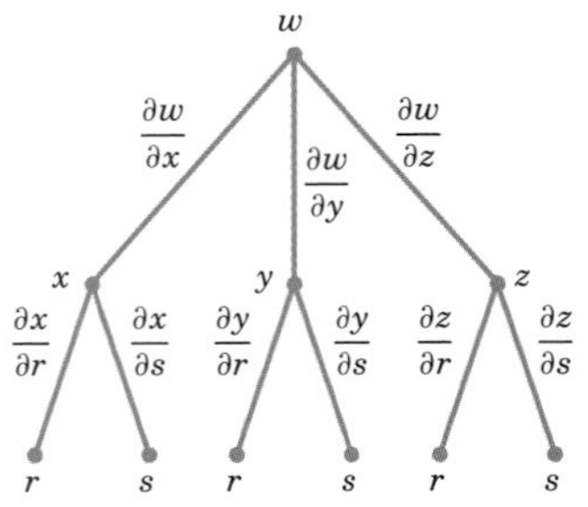

$$\frac{\partial w}{\partial r} = \frac{\partial w}{\partial x}\frac{\partial x}{\partial r} + \frac{\partial w}{\partial y}\frac{\partial y}{\partial r} + \frac{\partial w}{\partial z}\frac{\partial z}{\partial r}$$

$$\frac{\partial w}{\partial s} = \frac{\partial w}{\partial x}\frac{\partial x}{\partial s} + \frac{\partial w}{\partial y}\frac{\partial y}{\partial s} + \frac{\partial w}{\partial z}\frac{\partial z}{\partial s}$$

Figure 16.8.2

Solution. From the tree diagram and corresponding formulas in Figure 16.8.2 we obtain

$$\frac{\partial w}{\partial r} = yze^{xyz}(3) + xze^{xyz}(3) + xye^{xyz}(2rs) = e^{xyz}(3yz + 3xz + 2xyrs)$$

and

$$\frac{\partial w}{\partial s} = yze^{xyz}(1) + xze^{xyz}(-1) + xye^{xyz}(r^2) = e^{xyz}(yz - xz + xyr^2)$$

If desired, we can express $\partial w/\partial r$ and $\partial w/\partial s$ in terms of r and s alone by replacing x, y, and z by their expressions in terms of r and s. ◀

Example 2 Suppose that $w = x^2 + y^2 - z^2$ and

$$x = \rho \sin\phi \cos\theta, \quad y = \rho \sin\phi \sin\theta, \quad z = \rho\cos\phi$$

Use appropriate forms of the chain rule to find $\partial w/\partial \rho$ and $\partial w/\partial \theta$.

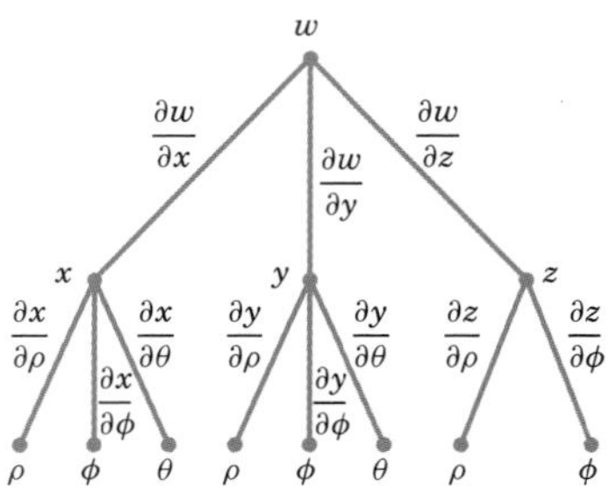

$$\frac{\partial w}{\partial \rho} = \frac{\partial w}{\partial x}\frac{\partial x}{\partial \rho} + \frac{\partial w}{\partial y}\frac{\partial y}{\partial \rho} + \frac{\partial w}{\partial z}\frac{\partial z}{\partial \rho}$$

$$\frac{\partial w}{\partial \theta} = \frac{\partial w}{\partial x}\frac{\partial x}{\partial \theta} + \frac{\partial w}{\partial y}\frac{\partial y}{\partial \theta}$$

Figure 16.8.3

Solution. From the tree diagram and corresponding formulas in Figure 16.8.3 we obtain

$$\begin{aligned}
\frac{\partial w}{\partial \rho} &= (2x)\sin\phi\cos\theta + 2y\sin\phi\sin\theta - 2z\cos\phi \\
&= 2\rho\sin^2\phi\cos^2\theta + 2\rho\sin^2\phi\sin^2\theta - 2\rho\cos^2\phi \\
&= 2\rho\sin^2\phi\,(\cos^2\theta + \sin^2\theta) - 2\rho\cos^2\phi \\
&= 2\rho\,(\sin^2\phi - \cos^2\phi) \\
&= -2\rho\cos 2\phi
\end{aligned}$$

$$\begin{aligned}
\frac{\partial w}{\partial \theta} &= (2x)(-\rho\sin\phi\sin\theta) + (2y)\,\rho\sin\phi\cos\theta \\
&= -2\rho^2\sin^2\phi\sin\theta\cos\theta + 2\rho^2\sin^2\phi\sin\theta\cos\theta \\
&= 0
\end{aligned}$$

This result is explained by the fact that w does not vary with θ. We may see this directly by expressing w in terms of ρ, ϕ, and θ. If this is done, the expressions involving θ will cancel, leaving w as a function of ρ and ϕ alone. (Verify that $w = -\rho^2\cos 2\phi$.) ◀

In many applications of the chain rule, some of the variables in the function $w = f(v_1, v_2, \ldots, v_n)$ are functions of the remaining variables. Tree diagrams are especially helpful in such situations.

Example 3 Suppose that

$$w = xy + yz, \quad y = \sin x, \quad z = e^x$$

Use an appropriate form of the chain rule to find dw/dx.

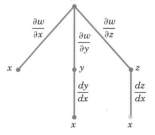

$$\frac{dw}{dx} = \frac{\partial w}{\partial x} + \frac{\partial w}{\partial y}\frac{dy}{dx} + \frac{\partial w}{\partial z}\frac{dz}{dx}$$

Figure 16.8.4

Solution. From the tree diagram and corresponding formulas in Figure 16.8.4 we obtain

$$\begin{aligned}
\frac{dw}{dx} &= y + (x + z)\cos x + ye^x \\
&= \sin x + (x + e^x)\cos x + e^x\sin x
\end{aligned}$$

This same result can be obtained by first expressing w explicitly in terms of x,

$$w = x \sin x + e^x \sin x$$

and then differentiating with respect to x; however, such direct substitution is not always convenient.   ◄

REMARK.   Unlike the differential dz, a partial symbol ∂z has no meaning of its own. For example, if we were to "cancel" partial symbols in the chain-rule formula

$$\frac{\partial z}{\partial u} = \frac{\partial z}{\partial x}\frac{\partial x}{\partial u} + \frac{\partial z}{\partial y}\frac{\partial y}{\partial u}$$

we would obtain

$$\frac{\partial z}{\partial u} = \frac{\partial z}{\partial u} + \frac{\partial z}{\partial u}$$

which is false in cases where $\partial z/\partial u \neq 0$.

In each of the expressions

$$z = \sin xy, \quad z = \frac{xy}{1 + xy}, \quad z = e^{xy}$$

the independent variables occur only in the combination xy, so the substitution $t = xy$ reduces the expression to a function of one variable:

$$z = \sin t, \quad z = \frac{t}{1 + t}, \quad z = e^t$$

Conversely, if we begin with a function of one variable $z = f(t)$ and substitute $t = xy$, we obtain a function $z = f(xy)$ in which the variables appear only in the combination xy. Functions whose variables occur in fixed combinations arise frequently in applications.

Example 4   Show that a function of the form $z = f(xy)$ satisfies the equation

$$x\frac{\partial z}{\partial x} - y\frac{\partial z}{\partial y} = 0$$

(assuming the derivatives exist).

Solution.   Let $t = xy$, so that $z = f(t)$. From the tree diagram in Figure 16.8.5 we obtain the formulas

$$\frac{\partial z}{\partial x} = \frac{dz}{dt}\frac{\partial t}{\partial x} = y\frac{dz}{dt} \quad \text{and} \quad \frac{\partial z}{\partial y} = \frac{dz}{dt}\frac{\partial t}{\partial y} = x\frac{dz}{dt}$$

from which it follows that

$$x\frac{\partial z}{\partial x} - y\frac{\partial z}{\partial y} = xy\frac{dz}{dt} - yx\frac{dz}{dt} = 0 \quad ◄$$

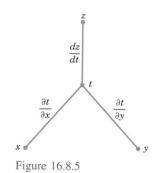

Figure 16.8.5

► Exercise Set 16.8

1. Let $f(v, w, x, y) = 4v^2w^3x^4y^5$. Find $\partial f/\partial v$, $\partial f/\partial w$, $\partial f/\partial x$, and $\partial f/\partial y$.

2. Let $w = r \cos st + e^u \sin ur$. Find $\partial w/\partial r$, $\partial w/\partial s$, $\partial w/\partial t$, and $\partial w/\partial u$.

3. Let $f(v_1, v_2, v_3, v_4) = \dfrac{v_1{}^2 - v_2{}^2}{v_3{}^2 + v_4{}^2}$. Find $\partial f/\partial v_1$, $\partial f/\partial v_2$, $\partial f/\partial v_3$, and $\partial f/\partial v_4$.

4. Let $V = xe^{2x-y} + we^{zw} + yw$. Find V_x, V_y, V_z, and V_w.

5. Let $f(v, w, x, y) = 2v^{1/2}w^4x^{1/2}y^{2/3}$. Find $f_v(1, -2, 4, 8)$, $f_w(1, -2, 4, 8)$, $f_x(1, -2, 4, 8)$, and $f_y(1, -2, 4, 8)$.

6. Let $u(w, x, y, z) = xe^{yw} \sin^2 z$. Find

(a) $\dfrac{\partial u}{\partial x}(0, 0, 1, \pi)$

(b) $\dfrac{\partial u}{\partial y}(0, 0, 1, \pi)$

(c) $\dfrac{\partial u}{\partial w}(0, 0, 1, \pi)$

(d) $\dfrac{\partial u}{\partial z}(0, 0, 1, \pi)$

(e) $\dfrac{\partial^4 u}{\partial x\,\partial y\,\partial w\,\partial z}$

(f) $\dfrac{\partial^4 u}{\partial w\,\partial z\,\partial y^2}$.

In Exercises 7–13, use appropriate forms of the chain rule to find the derivatives.

7. Let $v = 7w^2x^3y^4z^5$, where $w = t^4$, $x = t^3$, $y = t^2$, $z = t$. Find dv/dt.

8. Let $z = u^7$, where $u = 5x^2 - 2y^3$. Find $\partial z/\partial x$ and $\partial z/\partial y$.

9. Let $z = \ln(x^2 + 1)$, where $x = r\cos\theta$. Find $\partial z/\partial r$ and $\partial z/\partial\theta$.

10. Let $u = rs^2 \ln t$, $r = x^2$, $s = 4y + 1$, $t = xy^3$. Find $\partial u/\partial x$ and $\partial u/\partial y$.

11. Let $w = 4x^2 + 4y^2 + z^2$, $x = \rho\sin\phi\cos\theta$, $y = \rho\sin\phi\sin\theta$, $z = \rho\cos\phi$. Find $\partial w/\partial\rho$, $\partial w/\partial\phi$, and $\partial w/\partial\theta$.

12. Let $w = 3xy^2z^3$, $y = 3x^2 + 2$, $z = \sqrt{x - 1}$. Find dw/dx.

13. Let $w = \sqrt{x^2 + y^2 + z^2}$, $x = \cos 2y$, and $z = \sqrt{y}$. Find dw/dy.

14. The length, width, and height of a rectangular box are increasing at rates of 1 in/sec, 2 in/sec, and 3 in/sec, respectively.

(a) At what rate is the volume increasing when the length is 2 in., the width is 3 in., and the height is 6 in.?

(b) At what rate is the length of the diagonal increasing at that instant?

15. Angle A of triangle ABC is increasing at a rate of $\pi/60$ rad/sec, side AB is increasing at a rate of 2 cm/sec, and side AC is increasing at a rate of 4 cm/sec. At what rate is the length of BC changing when angle A is $\pi/3$ rad, $AB = 20$ cm, and $AC = 10$ cm? Is the length of BC increasing or decreasing? [*Hint:* Use the law of cosines.]

16. The area A of a triangle is given by $A = \frac{1}{2}ab\sin\theta$, where a and b are the lengths of two sides and θ is the angle between these sides. Suppose that $a = 5$, $b = 10$, and $\theta = \pi/3$. Find

(a) the rate at which A changes with a if b and θ are held constant

(b) the rate at which A changes with θ if a and b are held constant

(c) the rate at which b changes with a if A and θ are held constant.

17. Suppose that $x^2 + 4xz + z^2 - 3yz + 5 = 0$. Find $\partial z/\partial x$ and $\partial z/\partial y$ by implicit differentiation.

18. Suppose that $e^{xy}\cos yz - e^{yz}\sin xz + 2 = 0$. Find $\partial z/\partial x$ and $\partial z/\partial y$ by implicit differentiation.

19. Let $f(w, x, y, z) = wz\tan^{-1}\dfrac{x}{y} + 5w$. Show that
$$f_{ww} + f_{xx} + f_{yy} + f_{zz} = 0$$

In the remaining exercises, you may assume that all derivatives mentioned exist.

20. Let f be a function of one variable, and let $z = f(x + 2y)$. Show that
$$2\frac{\partial z}{\partial x} - \frac{\partial z}{\partial y} = 0$$

21. Let f be a function of one variable and let $z = f(x^2 + y^2)$. Show that
$$y\frac{\partial z}{\partial x} - x\frac{\partial z}{\partial y} = 0$$

22. Let f be a function of one variable, and let $w = f(\rho)$, where $\rho = (x^2 + y^2 + z^2)^{1/2}$. Show that
$$\left(\frac{\partial w}{\partial x}\right)^2 + \left(\frac{\partial w}{\partial y}\right)^2 + \left(\frac{\partial w}{\partial z}\right)^2 = \left(\frac{dw}{d\rho}\right)^2$$

23. Let f be a function of three variables and suppose that $w = f(x - y, y - z, z - x)$. Show that
$$\frac{\partial w}{\partial x} + \frac{\partial w}{\partial y} + \frac{\partial w}{\partial z} = 0$$

24. Let $w = f(x, y, z)$, $x = \rho\sin\phi\cos\theta$, $y = \rho\sin\phi\sin\theta$, and $z = \rho\cos\phi$. Express $\partial w/\partial\rho$, $\partial w/\partial\phi$, and $\partial w/\partial\theta$ in terms of $\partial w/\partial x$, $\partial w/\partial y$, and $\partial w/\partial z$.

25. Assume that $F(x, y, z) = 0$ defines z implicitly as a function of x and y. Show that
$$\frac{\partial z}{\partial x} = -\frac{\partial F/\partial x}{\partial F/\partial z} \quad\text{and}\quad \frac{\partial z}{\partial y} = -\frac{\partial F/\partial y}{\partial F/\partial z}$$

In Exercises 26–28, use the formulas in Exercise 25 to find $\partial z/\partial x$ and $\partial z/\partial y$.

26. $x^2 - 3yz^2 + xyz - 2 = 0$.

27. $ye^x - 5\sin 3z = 3z$.

28. $\ln(1 + z) + xy^2 + z = 1$.

29. Given that $u = u(x, y, z)$, $v = v(x, y, z)$, and $w = w(x, y, z)$, show that
$$\nabla f(u, v, w) = \frac{\partial f}{\partial u}\nabla u + \frac{\partial f}{\partial v}\nabla v + \frac{\partial f}{\partial w}\nabla w$$

30. Let $w = f(x, y, z)$ where $z = g(x, y)$. Taking x and y as the independent variables, express each of the following in terms of $\partial f/\partial x$, $\partial f/\partial y$, $\partial f/\partial z$, $\partial z/\partial x$, and $\partial z/\partial y$.

(a) $\partial w/\partial x$ (b) $\partial w/\partial y$.

31. Let $w = \ln(e^r + e^s + e^t + e^u)$. Show that
$$w_{rstu} = -6e^{r+s+t+u-4w}$$
[*Hint:* Take advantage of the relationship $e^w = e^r + e^s + e^t + e^u$.]

32. Suppose that w is a function of x_1, x_2, and x_3, and
$$x_1 = a_1y_1 + b_1y_2$$
$$x_2 = a_2y_1 + b_2y_2$$
$$x_3 = a_3y_1 + b_3y_2$$
where the a's and b's are constants. Express $\partial w/\partial y_1$ and $\partial w/\partial y_2$ in terms of $\partial w/\partial x_1$, $\partial w/\partial x_2$, and $\partial w/\partial x_3$.

33. (a) Let w be a function of x_1, x_2, x_3, and x_4, and let each x_i be a function of t. Find a chain-rule formula for dw/dt.

(b) Let w be a function of x_1, x_2, x_3, and x_4, and let each x_i be a function of v_1, v_2, and v_3. Find chain-rule formulas for $\partial w/\partial v_1$, $\partial w/\partial v_2$, and $\partial w/\partial v_3$.

34. Let $w = (x_1^2 + x_2^2 + \cdots + x_n^2)^k$, where $n > 2$. For what values of k does

$$\frac{\partial^2 w}{\partial x_1^2} + \frac{\partial^2 w}{\partial x_2^2} + \cdots + \frac{\partial^2 w}{\partial x_n^2} = 0$$

hold?

35. Show that

$$\frac{d}{dx}\left[\int_{a(x)}^{b(x)} f(t)\, dt\right] = f(b(x))b'(x) - f(a(x))a'(x)$$

This result is called **Leibniz' rule**. [*Hint:* Let $u = a(x)$, $v = b(x)$, and

$$F(u, v) = \int_u^v f(t)\, dt$$

Then use a chain rule and Theorem 5.9.1.]

36. Use the result of Exercise 35 to compute the following derivatives without performing the integrations.

(a) $\dfrac{d}{dx}\displaystyle\int_x^{x^2} e^{t^2}\, dt$

(b) $\dfrac{d}{dx}\displaystyle\int_{\sin x}^{\cos x} (t^3 + 2)^{2/3}\, dt$

(c) $\dfrac{d}{dx}\displaystyle\int_{3x}^{x^3} \sin^5 t\, dt$

(d) $\dfrac{d}{dx}\displaystyle\int_{e^x}^{e^{2x}} (\ln t)^4\, dt.$

■ 16.9 MAXIMA AND MINIMA OF FUNCTIONS OF TWO VARIABLES

Earlier in this text we learned how to find maximum and minimum values of a function of one variable. In this section we shall develop similar techniques for functions of two variables.

□ EXTREMA

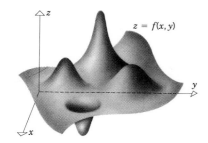

Figure 16.9.1

If we imagine the graph of a function f of two variables to be a portion of the earth's terrain (Figure 16.9.1), then the mountaintops, which are the high points in their immediate vicinity, are called *relative maxima* of f, and the valley bottoms, which are the low points in their immediate vicinity, are called *relative minima* of f.

Just as a geologist might be interested in finding the highest mountain and deepest valley in an entire mountain range, so a mathematician might be interested in finding the largest and smallest values of $f(x, y)$ over the *entire* domain of f. These are called the *absolute maximum* and *absolute minimum* values of f. The following definitions make these informal ideas precise.

16.9.1 DEFINITION. A function f of two variables is said to have a ***relative maximum*** at a point (x_0, y_0) if there is a circle centered at (x_0, y_0) such that $f(x_0, y_0) \geq f(x, y)$ for all points (x, y) in the domain of f that lie inside the circle, and f is said to have an ***absolute maximum*** at (x_0, y_0) if $f(x_0, y_0) \geq f(x, y)$ for all points (x, y) in the domain of f.

16.9.2 DEFINITION. A function f of two variables is said to have a ***relative minimum*** at a point (x_0, y_0) if there is a circle centered at (x_0, y_0) such that $f(x_0, y_0) \leq f(x, y)$ for all points (x, y) in the domain of f that lie inside the circle, and f is said to have an ***absolute minimum*** at (x_0, y_0) if $f(x_0, y_0) \leq f(x, y)$ for all points (x, y) in the domain of f.

If f has a relative maximum or a relative minimum at (x_0, y_0), then we say that f has a ***relative extremum*** at (x_0, y_0), and if f has an absolute maximum or absolute minimum at (x_0, y_0), then we say that f has an ***absolute extremum*** at (x_0, y_0).

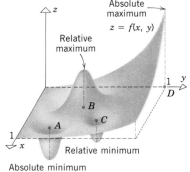

Figure 16.9.2

In Figure 16.9.2 we have sketched the graph of a function f whose domain is the closed square region in the xy-plane whose points satisfy the inequalities $0 \le x \le 1$, $0 \le y \le 1$. The function f has relative minima at the points A and C and a relative maximum at B. There is an absolute minimum at A and an absolute maximum at D.

For functions of two variables we shall be concerned with two important questions:

- Are there any relative or absolute extrema?

- If so, where are they located?

For functions of one variable that are continuous on a closed interval, the Extreme-Value Theorem (4.6.4) answered the existence question for absolute extrema. The following theorem, which we state without proof, is the corresponding result for functions of two variables.

16.9.3   THEOREM (*Extreme-Value Theorem*).   *If $f(x, y)$ is continuous on a closed and bounded set R, then f has both an absolute maximum and an absolute minimum on R.*

Example 1   The square region R whose points satisfy the inequalities

$$0 \le x \le 1 \quad \text{and} \quad 0 \le y \le 1$$

is a closed and bounded set in the xy-plane. The function f whose graph is sketched in Figure 16.9.2 is continuous on R; thus, it is guaranteed to have an absolute maximum and minimum on R by the preceding theorem. These occur at points D and A that are shown in the figure.   ◄

REMARK.   If any of the conditions in the Extreme-Value Theorem fail to hold, then there is no guarantee that an absolute maximum or absolute minimum exists on the region R. Thus, a discontinuous function on a closed and bounded set need not have any absolute extrema, and a continuous function on a set that is not closed and bounded also need not have any absolute extrema. (See Exercise 49.)

☐ **FINDING RELATIVE EXTREMA**

Recall that if a function g of one variable has a relative extremum at a point x_0 where g is differentiable, then $g'(x_0) = 0$. To obtain the analog of this result for functions of two variables, suppose that $f(x, y)$ has a relative maximum at a point (x_0, y_0) and that the partial derivatives of f exist at (x_0, y_0). It seems plausible geometrically that the traces of the surface $z = f(x, y)$ on the planes $x = x_0$ and $y = y_0$ have horizontal tangent lines at (x_0, y_0) (Figure 16.9.3), so

$$f_x(x_0, y_0) = 0 \quad \text{and} \quad f_y(x_0, y_0) = 0$$

The same conclusion holds if f has a relative minimum at (x_0, y_0), all of which suggests the following result.

16.9.4   THEOREM.   *If f has a relative extremum at a point (x_0, y_0), and if the first-order partial derivatives of f exist at this point, then*

$$f_x(x_0, y_0) = 0 \quad \text{and} \quad f_y(x_0, y_0) = 0$$

Proof.   Assume that f has a relative maximum at (x_0, y_0) and that both partial derivatives of f exist at (x_0, y_0). We leave it as an exercise to show that $G(x) = f(x, y_0)$ has a relative maximum at $x = x_0$ and $H(y) = f(x_0, y)$ has a relative maximum at $y = y_0$ (Figure 16.9.3). It follows from these results that

$$G'(x_0) = f_x(x_0, y_0) = 0$$

and

$$H'(y_0) = f_y(x_0, y_0) = 0$$

(Exercise 48). The proof for a relative minimum is similar. ∎

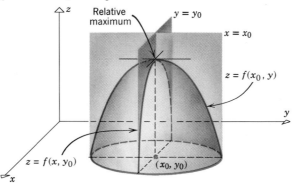

Figure 16.9.3

Recall that for a function f of one variable we defined a critical point to be a point x_0 at which $f'(x_0) = 0$ or at which $f'(x_0)$ does not exist. Analogously, for a function f of two variables we define a ***critical point*** of f to be a point (x_0, y_0) in the interior of the domain of f at which

$$f_x(x_0, y_0) = 0 \quad \text{and} \quad f_y(x_0, y_0) = 0$$

or at which one or both of the partial derivatives does not exist.

For a function of one variable, the condition $f'(x_0) = 0$ is *not* sufficient to guarantee that f has a relative extremum at x_0. (The graph of f may have an inflection point with a horizontal tangent line at x_0.) Similarly, the conditions $f_x(x_0, y_0) = 0$ and $f_y(x_0, y_0) = 0$ are not sufficient to guarantee that a function f of two variables has a relative extremum at (x_0, y_0).

Example 2 The graphs of the functions

$$z = f(x, y) = x^2 + y^2 \quad \boxed{\text{Paraboloid}}$$

$$z = g(x, y) = 1 - x^2 - y^2 \quad \boxed{\text{Paraboloid}}$$

$$z = h(x, y) = y^2 - x^2 \quad \boxed{\text{Hyperbolic paraboloid}}$$

are the quadric surfaces graphed in Figure 16.9.4. We have

$$f_x(x, y) = 2x, \quad f_y(x, y) = 2y$$

$$g_x(x, y) = -2x, \quad g_y(x, y) = -2y$$

$$h_x(x, y) = -2x, \quad h_y(x, y) = 2y$$

so in all three cases the partial derivatives are zero at $(0, 0)$, which means that $(0, 0)$ is a critical point for all three functions. The function f has a relative minimum at $(0, 0)$ and the function g a relative maximum. However, the function h has neither. To see this, observe that inside any circle in the xy-plane centered at $(0, 0)$, there exist points where $h(x, y)$ is positive (points on the y-axis) and there exist points where $h(x, y)$ is negative (points on the x-axis). Thus, $h(0, 0) = 0$ is neither the largest nor the smallest value of $h(x, y)$ in the circle. ◀

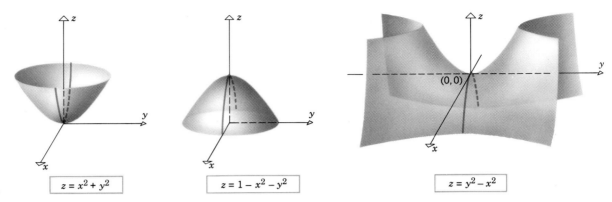

Figure 16.9.4

$$z = x^2 + y^2$$ $$z = 1 - x^2 - y^2$$ $$z = y^2 - x^2$$

A critical point at which a function does not have a relative extremum is called a **saddle point** of the function. Thus, the point $(0, 0)$ is a saddle point of the function $h(x, y) = y^2 - x^2$. (More advanced books distinguish between various types of saddle points. We shall not do this, however.)

Example 3   The graph of the function

$$f(x, y) = \sqrt{x^2 + y^2}$$

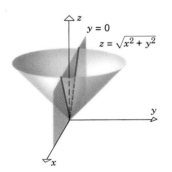

Figure 16.9.5

is the circular cone shown in Figure 16.9.5. The point $(0, 0)$ is a critical point of f because the partial derivatives do not both exist there. It is evident geometrically that $f_x(0, 0)$ does not exist because the trace of the cone in the plane $y = 0$ has a corner at the origin. The fact that $f_x(0, 0)$ does not exist can also be seen algebraically by noting that $f_x(0, 0)$ can be interpreted as the derivative with respect to x of the function

$$f(x, 0) = \sqrt{x^2 + 0} = |x|$$

at $x = 0$. But $|x|$ is not differentiable at $x = 0$, so $f_x(0, 0)$ does not exist. Similarly, $f_y(0, 0)$ does not exist. The function f has a relative minimum at the critical point $(0, 0)$.   ◀

For functions of one variable the second derivative test (Theorem 4.3.7) was used to determine the behavior of a function at a critical point. The following theorem, which is usually proved in advanced calculus, is the analog of that theorem for functions of two variables.

16.9.5   THEOREM (**The Second Partials Test**).   *Let f be a function of two variables with continuous second-order partial derivatives in some circle centered at a critical point (x_0, y_0), and let*

$$D = f_{xx}(x_0, y_0) f_{yy}(x_0, y_0) - f_{xy}^2(x_0, y_0)$$

(a)   *If $D > 0$ and $f_{xx}(x_0, y_0) > 0$, then f has a relative minimum at (x_0, y_0).*
(b)   *If $D > 0$ and $f_{xx}(x_0, y_0) < 0$, then f has a relative maximum at (x_0, y_0).*
(c)   *If $D < 0$, then f has a saddle point at (x_0, y_0).*
(d)   *If $D = 0$, then no conclusion can be drawn.*

Example 4   Locate all relative extrema and saddle points of

$$f(x, y) = 3x^2 - 2xy + y^2 - 8y$$

Solution.   Since $f_x(x, y) = 6x - 2y$ and $f_y(x, y) = -2x + 2y - 8$, the critical points of f satisfy the equations

$$6x - 2y = 0$$
$$-2x + 2y - 8 = 0$$

Solving these for x and y yields $x = 2$, $y = 6$ (verify), so $(2, 6)$ is the only critical point. To apply Theorem 16.9.5 we need the second-order partial derivatives

$$f_{xx}(x, y) = 6, \quad f_{yy}(x, y) = 2, \quad f_{xy}(x, y) = -2$$

At the point $(2, 6)$ we have

$$D = f_{xx}(2, 6)f_{yy}(2, 6) - f_{xy}^2(2, 6) = (6)(2) - (-2)^2 = 8 > 0$$

and

$$f_{xx}(2, 6) = 6 > 0$$

so that f has a relative minimum at $(2, 6)$ by part (a) of the second partials test.   ◀

Example 5   Locate all relative extrema and saddle points of

$$f(x, y) = 4xy - x^4 - y^4$$

Solution.   Since

$$f_x(x, y) = 4y - 4x^3 \tag{1}$$
$$f_y(x, y) = 4x - 4y^3$$

the critical points of f have coordinates satisfying the equations

$$\begin{array}{ccc} 4y - 4x^3 = 0 & & y = x^3 \\ & \text{or} & \tag{2} \\ 4x - 4y^3 = 0 & & x = y^3 \end{array}$$

Substituting the top equation in the bottom yields $x = (x^3)^3$ or $x^9 - x = 0$ or $x(x^8 - 1) = 0$, which has solutions $x = 0$, $x = 1$, $x = -1$. Substituting these values in the top equation of (2) we obtain the corresponding y values $y = 0$, $y = 1$, $y = -1$. Thus, the critical points of f are $(0, 0)$, $(1, 1)$, and $(-1, -1)$.

From (1),

$$f_{xx}(x, y) = -12x^2, \quad f_{yy}(x, y) = -12y^2, \quad f_{xy}(x, y) = 4$$

which yields the following table:

CRITICAL POINT (x_0, y_0)	$f_{xx}(x_0, y_0)$	$f_{yy}(x_0, y_0)$	$f_{xy}(x_0, y_0)$	$D = f_{xx}f_{yy} - f_{xy}^2$
$(0, 0)$	0	0	4	-16
$(1, 1)$	-12	-12	4	128
$(-1, -1)$	-12	-12	4	128

At the points $(1, 1)$ and $(-1, -1)$, we have $D > 0$ and $f_{xx} < 0$, so relative maxima occur at these critical points. At $(0, 0)$ there is a saddle point since $D < 0$. The surface is shown in Figure 16.9.6.   ◀

The following analog of Theorem 4.6.5 is a key result for locating the absolute extreme values of a function of two variables.

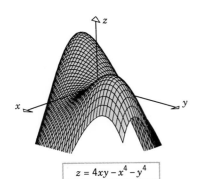

$$z = 4xy - x^4 - y^4$$

Figure 16.9.6

16.9.6 THEOREM. *If a function f of two variables has an absolute extremum (either an absolute maximum or an absolute minimum) at an interior point of its domain, then this extremum occurs at a critical point.*

Proof. We shall prove the result for an absolute maximum. The proof for an absolute minimum is similar.

If f has an absolute maximum at the point (x_0, y_0) in the interior of the domain of f, then f has a relative maximum at (x_0, y_0) (why?). If both partial derivatives exist at (x_0, y_0), then

$$f_x(x_0, y_0) = 0 \quad \text{and} \quad f_y(x_0, y_0) = 0$$

by Theorem 16.9.4, so (x_0, y_0) is a critical point of f. If either partial derivative does not exist, then again (x_0, y_0) is a critical point, so (x_0, y_0) is a critical point in all cases. ∎

☐ **FINDING ABSOLUTE EXTREMA ON CLOSED AND BOUNDED SETS**

If $f(x, y)$ is continuous on a closed and bounded set R, then the Extreme-Value Theorem (16.9.3) guarantees the existence of an absolute maximum and an absolute minimum of f on R. These absolute extrema can occur either on the boundary of R or in the interior of R, but if an absolute extremum occurs in the interior, then it occurs at a critical point by Theorem 16.9.6. Thus, we are led to the following procedure for finding absolute extrema:

How to Find the Absolute Extrema of a Continuous Function f of Two Variables on a Closed and Bounded Set R

Step 1. Find the critical points of f that lie in the interior of R.

Step 2. Find all boundary points at which the absolute extrema can occur.

Step 3. Evaluate $f(x, y)$ at the points obtained in the preceding steps. The largest of these values is the absolute maximum and the smallest the absolute minimum.

Example 6 Find the absolute maximum and minimum values of

$$f(x, y) = 3xy - 6x - 3y + 7 \tag{3}$$

on the closed triangular region R with vertices $(0, 0)$, $(3, 0)$, and $(0, 5)$.

Solution. The region R is shown in Figure 16.9.7. We have

$$\frac{\partial f}{\partial x} = 3y - 6 \quad \text{and} \quad \frac{\partial f}{\partial y} = 3x - 3$$

so all critical points occur where

$$3y - 6 = 0 \quad \text{and} \quad 3x - 3 = 0$$

Solving these equations yields $x = 1$ and $y = 2$, so $(1, 2)$ is the only critical point. As shown in Figure 16.9.7, this critical point is in the interior of R.

Next, we want to determine the location of the points on the boundary of R at which the absolute extrema might occur. The boundary of R consists of three line segments, each of which we shall treat separately:

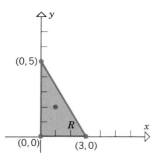

Figure 16.9.7

The line segment between (0, 0) *and* (3, 0):   On this line segment we have $y = 0$, so (3) simplifies to a function of the single variable x,

$$u(x) = f(x, 0) = -6x + 7, \quad 0 \le x \le 3$$

This function has no critical points because $u'(x) = -6$ is nonzero for all x. Thus the extreme values of $u(x)$ occur at the endpoints $x = 0$ and $x = 3$, which correspond to the points (0, 0) and (3, 0) of R.

The line segment between (0, 0) *and* (0, 5):   On this line segment we have $x = 0$, so (3) simplifies to a function of the single variable y,

$$v(y) = f(0, y) = -3y + 7, \quad 0 \le y \le 5$$

This function has no critical points because $v'(y) = -3$ is nonzero for all y. Thus, the extreme values of $v(y)$ occur at the endpoints $y = 0$ and $y = 5$, which correspond to the points (0, 0) and (0, 5) of R.

The line segment between (3, 0) *and* (0, 5):   In the xy-plane, an equation for this line segment is

$$y = -\tfrac{5}{3}x + 5, \quad 0 \le x \le 3 \tag{4}$$

so (3) simplifies to a function of the single variable x,

$$w(x) = f(x, -\tfrac{5}{3}x + 5) = 3x(-\tfrac{5}{3}x + 5) - 6x - 3(-\tfrac{5}{3}x + 5) + 7$$

$$= -5x^2 + 14x - 8, \quad 0 \le x \le 3$$

Since $w'(x) = -10x + 14$, the equation $w'(x) = 0$ yields $x = \tfrac{7}{5}$ as the only critical point of w. Thus, the extreme values of w occur either at the critical point $x = \tfrac{7}{5}$ or at the endpoints $x = 0$ and $x = 3$. The endpoints correspond to the points (0, 5) and (3, 0) of R, and from (4) the critical point corresponds to $(\tfrac{7}{5}, \tfrac{8}{3})$.

Finally, Table 16.9.1 lists the values of $f(x, y)$ at the interior critical point and at the points on the boundary where an absolute extremum can occur. From the table we conclude that the absolute maximum value of f is $f(0, 0) = 7$ and the absolute minimum value is $f(3, 0) = -11$.   ◄

Table 16.9.1

(x, y)	$(0, 0)$	$(3, 0)$	$(0, 5)$	$(\tfrac{7}{5}, \tfrac{8}{3})$	$(1, 2)$
$f(x, y)$	7	-11	-8	$\tfrac{9}{5}$	1

Example 7   Determine the dimensions of a rectangular box, open at the top, having a volume of 32 ft³, and requiring the least amount of material for its construction.

Solution.   Let

$x = $ length of the box (in feet)

$y = $ width of the box (in feet)

$z = $ height of the box (in feet)

$S = $ surface area of the box (in square feet)

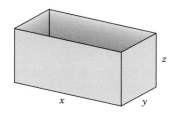

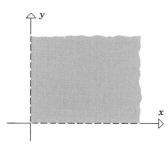

Two sides have area xz.
Two sides have area yz.
The base has area xy.

Figure 16.9.8

Figure 16.9.9

We may reasonably assume that the box with least surface area requires the least amount of material, so our objective is to minimize the surface area

$$S = xy + 2xz + 2yz \tag{5}$$

(see Figure 16.9.8) subject to the volume requirement

$$xyz = 32 \tag{6}$$

From (6) we obtain $z = 32/xy$, so (5) can be rewritten as

$$S = xy + \frac{64}{y} + \frac{64}{x} \tag{7}$$

which expresses S as a function of two variables. The dimensions x and y in this formula must be positive, but otherwise have no limitation, so our problem reduces to finding the absolute minimum value of S over the region for which $x > 0$ and $y > 0$ (Figure 16.9.9). Because this region is not bounded, we have no mathematical guarantee at this stage that an absolute minimum exists. However, if it does, then it occurs at a critical point of S, so we shall begin by finding the critical points. Differentiating (7) we obtain

$$\frac{\partial S}{\partial x} = y - \frac{64}{x^2}, \quad \frac{\partial S}{\partial y} = x - \frac{64}{y^2} \tag{8}$$

so the coordinates of the critical points of S satisfy

$$y - \frac{64}{x^2} = 0, \quad x - \frac{64}{y^2} = 0$$

Solving the first equation for y yields

$$y = \frac{64}{x^2} \tag{9}$$

and substituting this expression in the second equation yields

$$x - \frac{64}{(64/x^2)^2} = 0$$

which can be rewritten as

$$x\left(1 - \frac{x^3}{64}\right) = 0$$

The solutions of this equation are $x = 0$ and $x = 4$. Since we require $x > 0$, the only solution of significance is $x = 4$. Substituting this value in (9) yields $y = 4$. To see that we have located a relative minimum, we use the second partials test. From (8),

$$\frac{\partial^2 S}{\partial x^2} = \frac{128}{x^3}, \quad \frac{\partial^2 S}{\partial y^2} = \frac{128}{y^3}, \quad \frac{\partial^2 S}{\partial y \, \partial x} = 1$$

Thus, when $x = 4$ and $y = 4$, we have

$$\frac{\partial^2 S}{\partial x^2} = 2, \quad \frac{\partial^2 S}{\partial y^2} = 2, \quad \frac{\partial^2 S}{\partial y \, \partial x} = 1$$

and

$$D = \frac{\partial^2 S}{\partial x^2} \frac{\partial^2 S}{\partial y^2} - \left(\frac{\partial^2 S}{\partial y \, \partial x}\right)^2 = (2)(2) - (1)^2 = 3$$

Since $\partial^2 S/\partial x^2 > 0$ and $D > 0$, it follows from the second partials test that a relative minimum occurs when $x = y = 4$. Substituting these values in (6) yields $z = 2$, so the box using least material has a height of 2 ft and a square base whose edges are 4 ft long. ◄

REMARK. Strictly speaking, the solution in the last example is incomplete since we have not shown that an *absolute minimum* for S occurs when $x = y = 4$ and $z = 2$, only a relative minimum. The problem of showing that a relative extremum is also an absolute extremum can be difficult for functions of two or more variables and will not be considered in this text. However, in applied problems such as this it is often obvious from physical or geometric considerations that an absolute extremum has been found.

☐ **RELATIVE EXTREMA FOR FUNCTIONS OF THREE OR MORE VARIABLES**

Definitions of relative extrema can be given for functions of three or more variables. For example, if f is a function of three variables, then f has a *relative maximum* at (x_0, y_0, z_0) if $f(x_0, y_0, z_0) \geq f(x, y, z)$ for all points (x, y, z) in the domain of f that lie within some sphere centered at (x_0, y_0, z_0), and for a *relative minimum* the inequality is reversed. If $f(x, y, z)$ has first partial derivatives, then the relative extrema occur at *critical points*, that is, points where

$$f_x(x_0, y_0, z_0) = f_y(x_0, y_0, z_0) = f_z(x_0, y_0, z_0) = 0$$

The extension of the second partials test (Theorem 16.9.5) to functions of three or more variables is given in advanced calculus texts.

▶ Exercise Set 16.9

In Exercises 1–20, locate all relative maxima, relative minima, and saddle points.

1. $f(x, y) = 3x^2 + 2xy + y^2$.

2. $f(x, y) = x^3 - 3xy - y^3$.

3. $f(x, y) = y^2 + xy + 3y + 2x + 3$.

4. $f(x, y) = x^2 + xy - 2y - 2x + 1$.

5. $f(x, y) = x^2 + xy + y^2 - 3x$.

6. $f(x, y) = xy - x^3 - y^2$.

7. $f(x, y) = x^2 + 2y^2 - x^2y$.

8. $f(x, y) = 2x^2 - 4xy + y^4 + 2$.

9. $f(x, y) = x^2 + y^2 + 2/(xy)$.

10. $f(x, y) = x^3 + y^3 - 3x - 3y$.

11. $f(x, y) = x^2 + y - e^y$.

12. $f(x, y) = xe^y$.

13. $f(x, y) = e^x \sin y$.

14. $f(x, y) = xy + 2/x + 4/y$.

15. $f(x, y) = 2y^2x - yx^2 + 4xy$.

16. $f(x, y) = y \sin x$.

17. $f(x, y) = e^{-(x^2+y^2+2x)}$.

18. $f(x, y) = xy + \dfrac{a^3}{x} + \dfrac{b^3}{y}$ $(a \neq 0, b \neq 0)$.

19. $f(x, y) = \sin x + \sin y$, $0 < x < \pi$, $0 < y < \pi$.

20. $f(x, y) = \sin x + \sin y + \sin (x + y)$, $0 < x < \pi/2$, $0 < y < \pi/2$.

21. (a) Show that the second partials test provides no information about the critical points of $f(x, y) = x^4 + y^4$.

(b) Classify all critical points of f as relative maxima, relative minima, or saddle points.

22. (a) Show that the second partials test provides no information about the critical points of $f(x, y) = x^4 - y^4$.

(b) Classify all critical points of f as relative maxima, relative minima, or saddle points.

23. If f is a function of one variable, and f is continuous on an interval I and has exactly one relative extremum on I, say at x_0, then f has an absolute extremum at x_0 (Theorem 4.6.6). This exercise shows that a similar result does not hold for functions of two variables.

(a) Show that $f(x, y) = 3xe^y - x^3 - e^{3y}$ has only one critical point and that a relative maximum occurs there. (See Figure 16.9.10.)

(b) Show that f does not have an absolute maximum.

[This exercise is based on the article "The Only Critical Point in Town Test" by Ira Rosenholtz and Lowell Smylie, *Mathematics Magazine*, Vol. 58, No. 3, May 1985, pp. 149–150.]

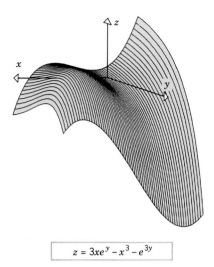

$$z = 3xe^y - x^3 - e^{3y}$$

Figure 16.9.10

24. Let f be a function of two variables that is continuous everywhere. One might think that if f has relative maxima at two points, then f must have another critical point because it is impossible to have two mountains without some sort of valley in between. This exercise shows that this is not true. Let $f(x, y) = 4x^2e^y - 2x^4 - e^{4y}$. Show that f has exactly two critical points and that a relative maximum occurs at each one (Figure 16.9.11).

[This exercise is based on the problem *Two Mountains Without a Valley*, proposed and solved by Ira Rosenholtz, *Mathematics Magazine*, Vol. 60, No. 1, February 1987, p. 48.]

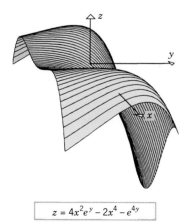

$$z = 4x^2e^y - 2x^4 - e^{4y}$$

Figure 16.9.11

In Exercises 25–30, find the absolute extrema of the given function on the indicated closed and bounded set R.

25. $f(x, y) = xy - x - 3y$; R is the triangular region with vertices $(0, 0)$, $(0, 4)$, and $(5, 0)$.

26. $f(x, y) = xy - 2x$; R is the triangular region with vertices $(0, 0)$, $(0, 4)$, and $(4, 0)$.

27. $f(x, y) = x^2 - 3y^2 - 2x + 6y$; R is the square region with vertices $(0, 0)$, $(0, 2)$, $(2, 2)$, and $(2, 0)$.

28. $f(x, y) = xe^y - x^2 - e^y$; R is the rectangular region with vertices $(0, 0)$, $(0, 1)$, $(2, 1)$, and $(2, 0)$.

29. $f(x, y) = x^2 + 2y^2 - x$; R is the circular region $x^2 + y^2 \leq 4$.

30. $f(x, y) = xy^2$; R is the region that satisfies the inequalities $x \geq 0$, $y \geq 0$, and $x^2 + y^2 \leq 1$.

31. Show that $f(x, y) = y^2 - 2xy + x^2$ has an absolute minimum at each point on the line $y = x$.

32. Find three positive numbers whose sum is 48 and such that their product is as large as possible.

33. Find three positive numbers whose sum is 27 and such that the sum of their squares is as small as possible.

34. Find all points on the plane $x + y + z = 5$ in the first octant at which $f(x, y, z) = xy^2z^2$ has a maximum value.

35. Find the points on the surface $x^2 - yz = 5$ that are closest to the origin.

36. Find the dimensions of the rectangular box of maximum volume that can be inscribed in a sphere of radius a.

37. Find the maximum volume of a rectangular box with three faces in the coordinate planes and a vertex in the first octant on the plane $x + y + z = 1$.

38. A manufacturer makes two models of an item, standard and deluxe. It costs \$40 to manufacture the standard model and \$60 for the deluxe. A market research firm estimates that if the standard model is priced at x dollars and the deluxe at y dollars, then the manufacturer will sell $500(y - x)$ of the standard items and $45,000 + 500(x - 2y)$ of the deluxe each year. How should the items be priced to maximize the profit?

39. A closed rectangular box with a volume of 16 ft^3 is made from two kinds of materials. The top and bottom are made of material costing 10¢ per square foot and the sides from material costing 5¢ per square foot. Find the dimensions of the box so that the cost of materials is minimized.

40. Use the methods of this section to find the distance between the lines

$$\begin{array}{ll} x = 3t & x = 2t \\ y = 2t \quad \text{and} & y = 2t + 3 \\ z = t & z = 2t \end{array}$$

41. Use the methods of this section to find the distance from the point $(-1, 3, 2)$ to the plane $x - 2y + z = 4$.

42. Show that among all parallelograms with perimeter l, a square with sides of length $l/4$ has maximum area. [*Hint:* The area of a parallelogram is given by the formula $A = ab \sin \alpha$, where a and b are the lengths of two adjacent sides and α is the angle between them.]

43. Determine the dimensions of a rectangular box, open at the top, having volume V, and requiring the least amount of material for its construction.

44. A length of sheet metal 27 in. wide is to be made into a water trough by bending up two sides as shown in Figure 16.9.12. Find x and ϕ so that the trapezoid-shaped cross section has a maximum area.

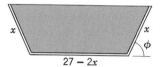

Figure 16.9.12

45. A common problem in experimental work is to obtain a mathematical relationship $y = f(x)$ between two variables x and y by "fitting" a curve to points in the plane corresponding to various experimentally determined values of x and y, say

$$(x_1, y_1), (x_2, y_2), \dots, (x_n, y_n)$$

Based on theoretical considerations, or simply on the pattern of the points, one decides on the general form of the curve $y = f(x)$ to be fitted. Often, the "curve" to be fitted is a straight line, $y = ax + b$. However, because of experimental errors in the data it is often impossible to find a line that passes through all of the points (Figure 16.9.13), so one looks for a line that "best fits" the data.

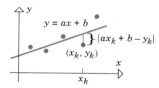

Figure 16.9.13

One criterion for selecting a line of "best fit" is to choose a and b to minimize the function

$$f(a, b) = \sum_{k=1}^{n} (ax_k + b - y_k)^2$$

Geometrically, $|ax_k + b - y_k|$ is the vertical distance between the data point (x_k, y_k) and the line $y = ax + b$, so in effect, minimizing $f(a, b)$ minimizes the sum of the squares of these vertical distances (Figure 16.9.13). This procedure is called the method of **least squares**.

(a) Show that the conditions $\dfrac{\partial f}{\partial a} = 0$ and $\dfrac{\partial f}{\partial b} = 0$ result in the equations

$$(\Sigma x_k^2)a + (\Sigma x_k)b = \Sigma x_k y_k$$
$$(\Sigma x_k)a + nb = \Sigma y_k$$

where $\Sigma = \displaystyle\sum_{k=1}^{n}$.

(b) Solve the equations in part (a) for a and b to show that

$$a = \frac{n(\Sigma x_k y_k) - (\Sigma x_k)(\Sigma y_k)}{n(\Sigma x_k^2) - (\Sigma x_k)^2}$$

and

$$b = \frac{(\Sigma x_k^2)(\Sigma y_k) - (\Sigma x_k)(\Sigma x_k y_k)}{n(\Sigma x_k^2) - (\Sigma x_k)^2}$$

46. This exercise shows that the values of a and b obtained in Exercise 45 produce the absolute minimum value of $f(a, b)$.

(a) Given that the arithmetic average

$$\bar{x} = \frac{1}{n} \sum_{k=1}^{n} x_k$$

minimizes $\Sigma(x_k - \bar{x})^2$ (see Exercise 60, Section 4.7), show that

$$n(\Sigma x_k^2) - (\Sigma x_k)^2 > 0$$

[*Note:* $\Sigma(x_k - \bar{x})^2 > 0$ if the x_k's are not all the same.]

(b) Find $f_{aa}(a, b)$, $f_{bb}(a, b)$, and $f_{ab}(a, b)$.

(c) Use Theorem 16.9.5 and the results of parts (a) and (b) to show that f has a relative minimum at the critical point found in Exercise 45.

(d) Based on the result of part (c) and the nature of the function $f(a, b)$, show that f takes on its absolute minimum value at the critical point.

47. Use the formulas in part (b) of Exercise 45 to find the equation of the least squares line $y = ax + b$ for the following data:

x	1	2	3	4
y	1.5	1.6	2.1	3.0

48. Suppose that f has a relative maximum at (x_0, y_0), and both $f_x(x_0, y_0)$ and $f_y(x_0, y_0)$ exist. Prove that $G(x) = f(x, y_0)$ has a relative maximum at $x = x_0$ and $H(y) = f(x_0, y)$ has a relative maximum at $y = y_0$.

49. Find an example to show that a function that has a discontinuity on a closed and bounded set need not have any absolute extrema and give an example to show that a continuous function on a set that is not closed and bounded also need not have any absolute extrema.

■ 16.10   LAGRANGE MULTIPLIERS

In this section we shall study a powerful method for solving certain types of optimization problems that is sometimes simpler to apply than the methods studied in the last section.

☐ **EXTREMUM PROBLEMS WITH CONSTRAINTS**

In the preceding section, we considered the problem of minimizing

$$S = xy + 2xz + 2yz$$

subject to the constraint

$$xyz - 32 = 0$$

This is a special case of the following general problem, which we shall study in this section:

Three-Variable Extremum Problem with One Constraint
Maximize or minimize the function $f(x, y, z)$ subject to the constraint $g(x, y, z) = 0$.

We shall also be interested in the two-variable version of this problem.

Two-Variable Extremum Problem with One Constraint
Maximize or minimize the function $f(x, y)$ subject to the constraint $g(x, y) = 0$.

☐ **LAGRANGE MULTIPLIERS**

One way to attack these problems is to solve the constraint equation for one of the variables in terms of the rest and substitute the result into f. The resulting function of one or two variables can then be maximized or minimized by finding its critical points. (See the solution of Example 7 in the preceding section.) However, if the constraint equation is too complicated to solve for one of the variables in terms of the rest, then other techniques must be used. We shall discuss one such technique, called *the method of Lagrange multipliers*. (See p. 569 for biography.) Since a rigorous discussion of this topic requires results from advanced calculus, we shall not be too concerned about all the technical details; we shall emphasize computational techniques.

Let us begin with the two-variable problem of maximizing or minimizing $f(x, y)$ subject to the constraint $g(x, y) = 0$. The graph of $g(x, y) = 0$ is usually some curve C in the xy-plane. Geometrically, we are concerned with finding the maximum or minimum value of $f(x, y)$ as (x, y) varies over the constraint curve C. If (x_0, y_0) is a point on the constraint curve C, then we shall say that $f(x, y)$ has a *constrained relative maximum* at (x_0, y_0) if there is a circle centered at (x_0, y_0) such that

$$f(x_0, y_0) \geq f(x, y) \tag{1}$$

for all points (x, y) on C within the circle (Figure 16.10.1). For a *constrained relative minimum* at (x_0, y_0), the inequality in (1) is reversed. We shall say that f has a *constrained relative extremum* at (x_0, y_0) if f has either a constrained relative maximum or a constrained relative minimum at (x_0, y_0). The following result is the key to finding constrained relative extrema.

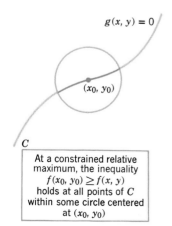

$g(x, y) = 0$

(x_0, y_0)

C

At a constrained relative maximum, the inequality $f(x_0, y_0) \geq f(x, y)$ holds at all points of C within some circle centered at (x_0, y_0)

Figure 16.10.1

> **16.10.1**   THEOREM (*The Constrained-Extremum Principle for Two Variables*).   *Let f and g be functions of two variables with continuous first partial derivatives on some open set containing the constraint curve g(x, y) = 0, and assume that $\nabla g \neq \mathbf{0}$ at any point on this curve. If f has a constrained relative extremum, then this extremum occurs at a point (x_0, y_0) on the constraint curve where the gradient vectors $\nabla f(x_0, y_0)$ and $\nabla g(x_0, y_0)$ are parallel; that is, there is some number λ, called a **Lagrange multiplier**, such that*
>
> $$\nabla f(x_0, y_0) = \lambda \nabla g(x_0, y_0) \tag{2}$$

To make this result plausible, let $z = f(x, y)$, and assume that the constraint curve $g(x, y) = 0$ can be smoothly parametrized in terms of an arc-length parameter s as

$$x = x(s), \quad y = y(s)$$

where the point (x_0, y_0) corresponds to $s = s_0$. Thus, for a point (x, y) on the constraint curve we have

$$z = f(x(s), y(s))$$

Since a relative extremum occurs at (x_0, y_0), we must have $dz/ds = 0$ if $s = s_0$. Thus, from the chain rule we have

$$\frac{dz}{ds} = \frac{\partial f}{\partial x}\frac{dx}{ds} + \frac{\partial f}{\partial y}\frac{dy}{ds} = \left(\frac{\partial f}{\partial x}\mathbf{i} + \frac{\partial f}{\partial y}\mathbf{j}\right) \cdot \left(\frac{dx}{ds}\mathbf{i} + \frac{dy}{ds}\mathbf{j}\right) = 0$$

At $s = s_0$, the first vector is the gradient of f at (x_0, y_0) and the second vector is the unit tangent vector $\mathbf{u}$ to $g(x, y) = 0$ at (x_0, y_0). Thus, at $s = s_0$ we have

$$\nabla f(x_0, y_0) \cdot \mathbf{u} = 0$$

which implies that the gradient of f at (x_0, y_0) is perpendicular to the unit tangent vector to the constraint curve at (x_0, y_0). But the unit tangent vector to the constraint curve $g(x, y) = 0$ is perpendicular to the gradient of g, and hence $\nabla f(x_0, y_0)$ is parallel to $\nabla g(x_0, y_0)$, which is what we wanted to show.

Example 1   At what point on the circle $x^2 + y^2 = 1$ does the product xy have a maximum?

Solution.   We want to maximize $f(x, y) = xy$ subject to the constraint

$$g(x, y) = x^2 + y^2 - 1 = 0 \tag{3}$$

We have

$$\nabla f = y\mathbf{i} + x\mathbf{j} \quad \text{and} \quad \nabla g = 2x\mathbf{i} + 2y\mathbf{j}$$

From the formula for ∇g we see that $\nabla g = \mathbf{0}$ if and only if $x = 0$ and $y = 0$, so $\nabla g \neq \mathbf{0}$ at any point on the circle $x^2 + y^2 = 1$. Thus, it follows from Theorem 16.10.1 that at a constrained relative extremum we must have

$$\nabla f = \lambda \nabla g \quad \text{or} \quad y\mathbf{i} + x\mathbf{j} = \lambda(2x\mathbf{i} + 2y\mathbf{j})$$

which is equivalent to the pair of equations

$$y = 2x\lambda \quad \text{and} \quad x = 2y\lambda \tag{4}$$

Since the maximum value of xy on the circle $x^2 + y^2 = 1$ is obviously greater than zero, we must have $x \neq 0$ and $y \neq 0$ at a constrained maximum. Thus, the equations in (4) can be rewritten as

$$\lambda = \frac{y}{2x} \quad \text{and} \quad \lambda = \frac{x}{2y}$$

from which we obtain

$$\frac{y}{2x} = \frac{x}{2y}$$

or

$$y^2 = x^2 \tag{5}$$

Substituting this in (3) yields

$$2x^2 - 1 = 0$$

or

$$x = 1/\sqrt{2} \quad \text{and} \quad x = -1/\sqrt{2}$$

Substituting $x = 1/\sqrt{2}$ in (5) yields $y = \pm 1/\sqrt{2}$ and substituting $x = -1/\sqrt{2}$ in (5) yields $y = \pm 1/\sqrt{2}$, so there are four candidates for the location of a maximum:

$$(1/\sqrt{2}, 1/\sqrt{2}), \quad (1/\sqrt{2}, -1/\sqrt{2}), \quad (-1/\sqrt{2}, 1/\sqrt{2}), \quad (-1/\sqrt{2}, -1/\sqrt{2})$$

At the first and fourth points the function $f(x, y) = xy$ has value $\frac{1}{2}$, while at the second and third points the value is $-\frac{1}{2}$. Thus, the constrained maximum value of $\frac{1}{2}$ occurs at $(1/\sqrt{2}, 1/\sqrt{2})$ and $(-1/\sqrt{2}, -1/\sqrt{2})$. ◀

REMARK.   If c is a constant, then the functions $g(x, y)$ and $g(x, y) - c$ have the same gradient since the constant c drops out when we differentiate. Consequently, it is *not* essential to rewrite a constraint of the form $g(x, y) = c$ as $g(x, y) - c = 0$ in order to apply the constrained-extremum principle. Thus, in the last example, we could have kept the constraint in the form $x^2 + y^2 = 1$ and then taken $g(x, y) = x^2 + y^2$ rather than $g(x, y) = x^2 + y^2 - 1$.

In Exercise 12 of Section 4.7, it was stated that among all rectangles of perimeter p, a square has maximum area. This result can be obtained using Lagrange multipliers.

Example 2   Find the dimensions of a rectangle having perimeter p and maximum area.

Solution.   Let

$x = $ length of the rectangle

$y = $ width of the rectangle

$A = $ area of the rectangle

We want to maximize $A = xy$ subject to the perimeter constraint

$$2x + 2y = p \tag{6}$$

If we let $f(x, y) = xy$ and $g(x, y) = 2x + 2y$, then we have

$$\nabla f = y\mathbf{i} + x\mathbf{j} \quad \text{and} \quad \nabla g = 2\mathbf{i} + 2\mathbf{j}$$

Noting that $\nabla g \neq \mathbf{0}$, it follows from the constrained-extremum principle (Theorem 16.10.1) that

$$\nabla f = \lambda \nabla g \quad \text{or} \quad y\mathbf{i} + x\mathbf{j} = \lambda(2\mathbf{i} + 2\mathbf{j})$$

at a constrained relative maximum. This is equivalent to the two equations

$$y = 2\lambda \quad \text{and} \quad x = 2\lambda$$

Eliminating λ from these equations we obtain $x = y$, which shows that the rectangle is actually a square. Using this condition and constraint (6), we obtain $x = p/4$, $y = p/4$. ◀

Lagrange multipliers can also be used in the three-variable problem of maximizing or minimizing $f(x, y, z)$ subject to the constraint $g(x, y, z) = 0$. The graph of $g(x, y, z) = 0$ is generally some surface σ in 3-space. Geometrically, we are concerned with finding the maximum or minimum of the function $f(x, y, z)$ as (x, y, z) varies over the surface σ. We shall say that $f(x, y, z)$ has a **constrained relative maximum** at (x_0, y_0, z_0) if there is a sphere centered at (x_0, y_0, z_0) such that

$$f(x_0, y_0, z_0) \geq f(x, y, z)$$

for all points (x, y, z) on σ within the sphere (Figure 16.10.2). The meaning of the terms **constrained relative minimum** and **constrained relative extremum** should be clear. It can be shown that if f and g have continuous first partial derivatives and $\nabla g \neq \mathbf{0}$ on the surface $g(x, y, z) = 0$, then a constrained relative extremum can only occur at a point (x_0, y_0, z_0) where $\nabla f(x_0, y_0, z_0)$ and $\nabla g(x_0, y_0, z_0)$ are parallel, that is,

$$\nabla f(x_0, y_0, z_0) = \lambda \nabla g(x_0, y_0, z_0)$$

for some number λ.

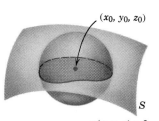

(x_0, y_0, z_0)

S

$g(x, y, z) = 0$

At a constrained relative maximum, the inequality $f(x_0, y_0, z_0) \geq f(x, y, z)$ holds at all points of S inside some sphere centered at (x_0, y_0, z_0).

Figure 16.10.2

Example 3 Find the points on the sphere $x^2 + y^2 + z^2 = 36$ that are closest to and farthest from $(1, 2, 2)$.

Solution. To avoid radicals, we shall find points on the sphere that minimize and maximize the *square* of the distance to $(1, 2, 2)$. Thus, we want to find the extrema of

$$f(x, y, z) = (x - 1)^2 + (y - 2)^2 + (z - 2)^2$$

subject to the constraint

$$x^2 + y^2 + z^2 = 36 \tag{7}$$

Therefore, with $g(x, y, z) = x^2 + y^2 + z^2$, we must have

$$\nabla f(x, y, z) = \lambda \nabla g(x, y, z)$$

at a constrained relative extremum; that is,

$$2(x - 1)\mathbf{i} + 2(y - 2)\mathbf{j} + 2(z - 2)\mathbf{k} = \lambda(2x\mathbf{i} + 2y\mathbf{j} + 2z\mathbf{k})$$

which leads to the equations

$$2(x - 1) = 2x\lambda, \quad 2(y - 2) = 2y\lambda, \quad 2(z - 2) = 2z\lambda \tag{8}$$

We may assume that x, y, and z are nonzero since $x = 0$ does not satisfy the first equation, $y = 0$ does not satisfy the second, and $z = 0$ does not satisfy the third. Thus, we can rewrite (8) as

$$\frac{x - 1}{x} = \lambda, \quad \frac{y - 2}{y} = \lambda, \quad \frac{z - 2}{z} = \lambda$$

The first two equations imply that

$$\frac{x - 1}{x} = \frac{y - 2}{y}$$

from which it follows that

$$y = 2x \tag{9}$$

Similarly, the first and third equations imply that

$$z = 2x \tag{10}$$

Substituting (9) and (10) in the constraint equation (7), we obtain

$$9x^2 = 36, \quad \text{or} \quad x = \pm 2$$

Substituting these values in (9) and (10) yields two points

$$(2, 4, 4) \quad \text{and} \quad (-2, -4, -4)$$

Since $f(2, 4, 4) = 9$ and $f(-2, -4, -4) = 81$, it follows that $(2, 4, 4)$ is the point on the sphere closest to $(1, 2, 2)$, and $(-2, -4, -4)$ is the point that is farthest (Figure 16.10.3).   ◄

Next we shall use Lagrange multipliers to solve the problem of Example 7 in the preceding section.

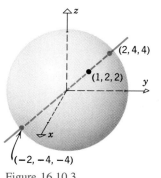

Figure 16.10.3

Example 4   Use Lagrange multipliers to determine the dimensions of a rectangular box, open at the top, having a volume of 32 ft³, and requiring the least amount of material for its construction.

Solution.   With the notation of Example 7 in Section 16.9, the problem is to minimize the surface area

$$S = xy + 2xz + 2yz$$

subject to the volume constraint

$$xyz = 32 \tag{11}$$

If we let $f(x, y, z) = xy + 2xz + 2yz$ and $g(x, y, z) = xyz$, then

$$\nabla f = (y + 2z)\mathbf{i} + (x + 2z)\mathbf{j} + (2x + 2y)\mathbf{k} \quad \text{and} \quad \nabla g = yz\mathbf{i} + xz\mathbf{j} + xy\mathbf{k}$$

It follows that $\nabla g \neq \mathbf{0}$ at any point on the surface $xyz = 32$, since x, y, and z are all nonzero on this surface. Thus, at a constrained relative extremum we must have $\nabla f = \lambda \nabla g$, that is,

$$(y + 2z)\mathbf{i} + (x + 2z)\mathbf{j} + (2x + 2y)\mathbf{k} = \lambda(yz\mathbf{i} + xz\mathbf{j} + xy\mathbf{k})$$

This condition yields the three equations

$$y + 2z = \lambda yz, \quad x + 2z = \lambda xz, \quad 2x + 2y = \lambda xy$$

Because x, y, and z are nonzero these equations can be rewritten as

$$\frac{1}{z} + \frac{2}{y} = \lambda, \quad \frac{1}{z} + \frac{2}{x} = \lambda, \quad \frac{2}{y} + \frac{2}{x} = \lambda$$

From the first two equations,

$$y = x \tag{12}$$

and from the second and third equations, $z = \frac{1}{2}y$. This and (12) imply that

$$z = \frac{1}{2}x \tag{13}$$

Substituting (12) and (13) in the volume constraint (11) yields

$$\tfrac{1}{2}x^3 = 32$$

This equation, together with (12) and (13), yields

$$x = 4, \quad y = 4, \quad z = 2$$

which agrees with the result that was obtained in Example 7 of the preceding section.   ◄

Lagrange multipliers can also be used in problems involving two or more constraints. However, we shall not pursue this topic here.

▶ **Exercise Set 16.10**

In Exercises 1–7, use Lagrange multipliers to find the maximum and minimum values of f subject to the given constraint. Also, find the points at which these extreme values occur.

1. $f(x, y) = xy$; $4x^2 + 8y^2 = 16$.

2. $f(x, y) = x^2 - y$; $x^2 + y^2 = 25$.

3. $f(x, y) = 4x^3 + y^2$; $2x^2 + y^2 = 1$.

4. $f(x, y) = x - 3y - 1$; $x^2 + 3y^2 = 16$.

5. $f(x, y, z) = 2x + y - 2z$; $x^2 + y^2 + z^2 = 4$.

6. $f(x, y, z) = 3x + 6y + 2z$; $2x^2 + 4y^2 + z^2 = 70$.

7. $f(x, y, z) = xyz$; $x^2 + y^2 + z^2 = 1$.

In Exercises 8–16, solve using Lagrange multipliers.

8. Find the point on the line $2x - 4y = 3$ that is closest to the origin.

9. Find the point on the line $y = 2x + 3$ that is closest to $(4, 2)$.

10. Find the point on the plane $x + 2y + z = 1$ that is closest to the origin.

11. Find the point on the plane $4x + 3y + z = 2$ that is closest to $(1, -1, 1)$.

12. Find the points on the surface $xy - z^2 = 1$ that are closest to the origin.

13. Find the maximum value of $\sin x \sin y$, where x and y denote the acute angles of a right triangle.

14. Find a vector in 3-space whose length is 5 and whose components have the largest possible sum.

15. Find the points on the circle $x^2 + y^2 = 45$ that are closest to and farthest from $(1, 2)$.

16. The temperature at a point (x, y) on a metal plate is $T(x, y) = 4x^2 - 4xy + y^2$. An ant, walking on the plate, traverses a circle of radius 5 centered at the origin. What are the highest and lowest temperatures encountered by the ant?

In Exercises 17–24, use Lagrange multipliers to solve the indicated problems from Section 16.9.

17. Exercise 33.   18. Exercise 34.

19. Exercise 35.   20. Exercise 36.

21. Exercise 39.   22. Exercise 41.

23. Exercise 42.   24. Exercise 43.

◆ **TECHNOLOGY EXERCISES**   Chapter 16

Most of these exercises require access to a graphing calculator or a computer algebra system (CAS) such as *Mathematica*, *Maple*, or *Derive*. When you are asked to *find* an answer or to *solve* an equation, you may choose to find an exact result or a numerical approximation, depending on the particular technology you are using and on your own imagination. The form of your answers may differ from those of other students or from those in the answer section of the text, depending on how you solve the problems and the accuracy you use in your numerical approximations. Those exercises that are more appropriate for a CAS than a graphing calculator are labeled with the icon ◆.

◆ 1. **Surfaces defined parametrically—a helicoid:** Graph the portion of the helicoid (spiral ramp)

$$x = u \cos v$$
$$y = u \sin v$$
$$z = v$$

for $0 \le u \le 2$ and $0 \le v \le \pi$.

◆ 2. **Surfaces defined parametrically—a Möbius strip:** Graph the Möbius strip

$$x = \cos v + u \cos (v/2) \cos v$$
$$y = \sin v + u \cos (v/2) \sin v$$
$$z = u \sin (v/2)$$

for $-0.2 \le u \le 0.2$ and $0 \le v \le 2\pi$.

3. Collision of a particle with a surface: Suppose that the equations of motion of a particle are $x = t - 1$, $y = 4e^{-t}$, $z = 2 - \sqrt{t}$, where $t > 0$. Find, to the nearest tenth of a degree, the acute angle between the velocity vector and the normal line to the surface $x^2/4 + y^2 + z^2 = 1$ at the point where the particle collides with the surface.

4. Directional derivative: Let $f(x, y) = x + \sin(xy)$. Find all unit vectors $\mathbf{u}$ such that the directional derivative of f at the point $(2, 3)$ in the direction of $\mathbf{u}$ is 1.

5. Maximum rate of change of temperature at a point: Assume that the temperature T at a point (x, y, z) in 3-space is given by $T = f(x, y, z)$. If a particle starts at the point $P(-1, 3, 2)$, then the instantaneous rate of change of T with respect to distance will depend on the direction in which the particle moves. Suppose that the temperature increases at the rate of 2°/m if the particle moves toward the origin, increases at the rate of 1°/m if it moves toward the point $(-2, 4, 1)$, and decreases at the rate of 3°/m if it moves toward the point $(1, 2, 1)$. Find the unit vector that points in the direction in which T increases most rapidly starting at P. What is the maximum rate of increase of T at the point P?

6. Motion on a surface: Suppose that a particle is moving on the surface $z = 20/(3 + 2x^2 + y^2)$, where x, y, and z are in meters, and that the speed of the particle is 2 m/sec at the instant when the particle is at the point $P(2, 3, 1)$.

(a) Find the two possible velocity vectors at point P if the instantaneous rate of change of the elevation z at the point P is 0.3 m/sec.

(b) Find the velocity vector at P for which the elevation z increases most rapidly. For this velocity vector, what is the instantaneous rate of increase of elevation?

7. Differentiation under the integral sign: Let

$$F(x) = \int_c^d f(x, y)\, dy, \quad a \le x \le b$$

It can be shown that if $f(x, y)$ and $\partial f/\partial x$ are continuous for $a \le x \le b$ and $c \le y \le d$, then

$$F'(x) = \int_c^d \frac{\partial f}{\partial x}\, dy$$

(a) Use this result to find $F'(x)$ if

$$F(x) = \int_0^1 \sin(xe^y)\, dy, \quad 0 \le x \le 2$$

[*Note:* Express $F'(x)$ as an integral and perform the integration.]

(b) Use the result in part (a) to find the maximum value of $F(x)$ for $0 \le x \le 2$.

8. Minimum distance between two curves: Find the minimum distance between a point on the graph of $y = x^2$ and a point on the graph of $y = \ln x$. [*Suggestion:* Let (a, a^2) be a point on the graph of $y = x^2$ and $(b, \ln b)$ a point on the graph of $y = \ln x$.]

9. Manufacturing cost: A metal box with no lid is to be assembled from rectangular pieces of sheet metal and is to have a volume of 1 ft³. Suppose that material costs \$2/ft² for the base, \$3/ft² for the sides, and that it costs \$1/ft to weld each of the eight seams. Find the dimensions and cost of the most economical box.

10. Minimum distance from a point to a surface: Find the shortest distance from the point $(1, 0, 0)$ to a point on the surface $z = 1/(xy)$, where $x > 0$ and $y > 0$.

Pierre-Simon de Laplace (1749–1827)

17 MULTIPLE INTEGRALS

■ 17.1 DOUBLE INTEGRALS

The notion of a definite integral can be extended to functions of two or more variables. In this section we shall discuss the double integral, which is the extension to functions of two variables.

☐ **DEFINITION OF A**
DOUBLE INTEGRAL

Recall that the definite integral of a function of one variable

$$\int_a^b f(x)\, dx = \lim_{n \to +\infty} \sum_{k=1}^{n} f(x_k^*)\, \Delta x_k \tag{1}$$

arose from the problem of finding areas under curves. Integrals of functions of two variables arise from a volume problem, which can be stated as follows:

17.1.1 PROBLEM. *Find the volume of the solid consisting of all points that lie between a region R in the xy-plane and a surface z = f(x, y), where f is continuous on R and f(x, y) ≥ 0 for all (x, y) in R (Figure 17.1.1).*

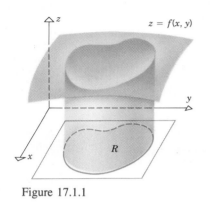

Figure 17.1.1

Later, we shall place more restrictions on the region R, but for now let us just assume that the entire region R can be enclosed within some suitably large rectangle with sides parallel to the coordinate axes. This ensures that R does not extend indefinitely in any direction.

The procedure for finding the volume V of the solid in Figure 17.1.1 will be similar to the limiting process used for finding areas, except that now the approximating elements will be rectangular parallelepipeds rather than rectangles. We proceed as follows:

- Using lines parallel to the coordinate axes, divide the rectangle enclosing the region R into subrectangles, and exclude from consideration all those subrectangles that contain any points outside of R. This leaves only rectangles that are subsets of R (Figure 17.1.2). Assume that there are n such rectangles, and denote the area of the kth such rectangle by ΔA_k.

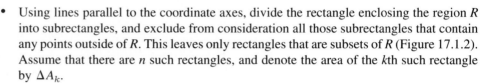

Figure 17.1.2

- Choose any arbitrary point in each subrectangle, and denote the point in the kth subrectangle by (x_k^*, y_k^*). As shown in Figure 17.1.3, the product $f(x_k^*, y_k^*)\,\Delta A_k$ is the volume of a rectangular parallelepiped with base area ΔA_k and height $f(x_k^*, y_k^*)$, so the sum

$$\sum_{k=1}^{n} f(x_k^*, y_k^*)\,\Delta A_k$$

can be viewed as an approximation to the volume V of the entire solid.

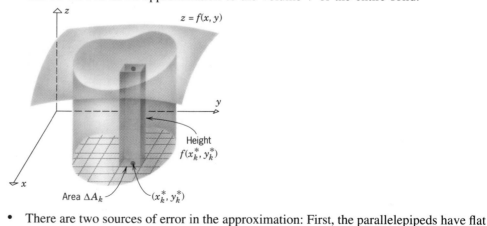

Figure 17.1.3

- There are two sources of error in the approximation: First, the parallelepipeds have flat tops, whereas the surface $z = f(x, y)$ may be curved; second, the rectangles that form the bases of the parallelepipeds do not completely cover the region R. However, if we repeat the above process with more and more subdivisions in such a way that the lengths and widths of the base rectangles approach zero, then it is plausible that the errors of both types approach zero, and the exact volume of the solid is

$$V = \lim_{n \to +\infty} \sum_{k=1}^{n} f(x_k^*, y_k^*)\,\Delta A_k \tag{2}$$

The sums in (2) are called **Riemann sums**, and the limit of the Riemann sums is denoted by

$$\iint\limits_{R} f(x, y)\, dA = \lim_{n \to +\infty} \sum_{k=1}^{n} f(x_k^*, y_k^*)\, \Delta A_k \tag{3}$$

which is called the **double integral** of $f(x, y)$ over R. With this notation, the volume V of the solid can be expressed as

$$V = \iint\limits_{R} f(x, y)\, dA \tag{4}$$

This result was derived under the assumption that f is continuous and nonnegative on the region R. In the case where $f(x, y)$ can have both positive and negative values on R, the double integral over R represents a difference of two volumes: the volume of the solid that is above R but below $z = f(x, y)$ minus the volume of the solid that is below R but above $z = f(x, y)$. We call this difference the **net signed volume** between $z = f(x, y)$ and R; it is analogous to the concept of net signed area defined in Section 5.6. Thus, a positive value for a double integral means that there is more volume above R than below, a negative value means that there is more volume below than above, and a value of zero means that the two volumes are the same.

A precise definition of expression (3) (in terms of ϵ's and δ's) and conditions under which the double integral exists are studied in advanced calculus. However, for our purposes it suffices to say that existence is ensured when f is continuous on R and the region R is not too "complicated."

Observe the similarity between (1) and (3). Because of this it should not be surprising that double integrals have many of the same properties as definite integrals (which we now also call **single integrals**):

$$\iint\limits_{R} c f(x, y)\, dA = c \iint\limits_{R} f(x, y)\, dA \qquad (c \text{ a constant}) \tag{5}$$

$$\iint\limits_{R} [f(x, y) + g(x, y)]\, dA = \iint\limits_{R} f(x, y)\, dA + \iint\limits_{R} g(x, y)\, dA \tag{6}$$

$$\iint\limits_{R} [f(x, y) - g(x, y)]\, dA = \iint\limits_{R} f(x, y)\, dA - \iint\limits_{R} g(x, y)\, dA \tag{7}$$

It is evident intuitively that if $f(x, y)$ is nonnegative on a region R, then subdividing R into two regions R_1 and R_2 has the effect of subdividing the solid between R and $z = f(x, y)$ into two solids, the sum of whose volumes is the volume of the entire solid (Figure 17.1.4). This suggests the following result, which holds even if f has negative values:

$$\iint\limits_{R} f(x, y)\, dA = \iint\limits_{R_1} f(x, y)\, dA + \iint\limits_{R_2} f(x, y)\, dA \tag{8}$$

The proofs of these results will be omitted.

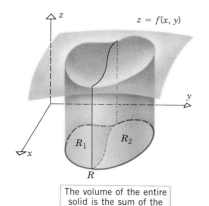

The volume of the entire solid is the sum of the volumes of the solids above R_1 and R_2.

Figure 17.1.4

☐ **EVALUATING DOUBLE INTEGRALS**

Except in the simplest cases, it is impractical to obtain the value of a double integral from the limit in (3). However, we shall now show how to evaluate double integrals by calculating two successive single integrals. For the remainder of this section, we shall limit our discussion to the case where R is a rectangle. In the next section we shall consider double integrals over more complicated regions.

The partial derivatives of a function $f(x, y)$ are calculated by holding one of the variables fixed and differentiating with respect to the other variable. Let us consider the reverse of this process, *partial integration*. The symbols

$$\int_a^b f(x, y)\, dx \quad \text{and} \quad \int_c^d f(x, y)\, dy$$

denote *partial definite integrals*; the first integral, called the *partial definite integral with respect to x*, is evaluated by holding y fixed and integrating with respect to x, and the second integral, called the *partial definite integral with respect to y*, is evaluated by holding x fixed and integrating with respect to y. As the following example shows, the partial definite integral with respect to x is a function of y, and the partial definite integral with respect to y is a function of x.

Example 1

$$\int_0^1 xy^2\, dx = y^2 \int_0^1 x\, dx = \frac{y^2 x^2}{2}\bigg]_{x=0}^1 = \frac{y^2}{2}$$

$$\int_0^1 xy^2\, dy = x \int_0^1 y^2\, dy = \frac{xy^3}{3}\bigg]_{y=0}^1 = \frac{x}{3} \quad \blacktriangleleft$$

A partial definite integral with respect to x is a function of y and hence can be integrated with respect to y; similarly, a partial definite integral with respect to y can be integrated with respect to x. This two-stage integration process is called *iterated* (or *repeated*) *integration*. We introduce the following notation:

$$\int_c^d \int_a^b f(x, y)\, dx\, dy = \int_c^d \left[\int_a^b f(x, y)\, dx \right] dy \tag{9a}$$

$$\int_a^b \int_c^d f(x, y)\, dy\, dx = \int_a^b \left[\int_c^d f(x, y)\, dy \right] dx \tag{9b}$$

These integrals are called *iterated integrals*.

Example 2   Evaluate

(a) $\displaystyle \int_0^3 \int_1^2 (1 + 8xy)\, dy\, dx$ (b) $\displaystyle \int_1^2 \int_0^3 (1 + 8xy)\, dx\, dy$

Solution (a).

$$\int_0^3 \int_1^2 (1 + 8xy)\, dy\, dx = \int_0^3 \left[\int_1^2 (1 + 8xy)\, dy \right] dx$$

$$= \int_0^3 \left[y + 4xy^2 \right]_{y=1}^2 dx$$

$$= \int_0^3 [(2 + 16x) - (1 + 4x)]\, dx$$

$$= \int_0^3 (1 + 12x)\, dx$$

$$= x + 6x^2 \bigg]_0^3 = 57$$

Solution (b).

$$\int_1^2 \int_0^3 (1 + 8xy)\,dx\,dy = \int_1^2 \left[\int_0^3 (1 + 8xy)\,dx \right] dy$$

$$= \int_1^2 \left[x + 4x^2 y \right]_{x=0}^3 dy$$

$$= \int_1^2 (3 + 36y)\,dy$$

$$= 3y + 18y^2 \Big]_1^2 = 57 \qquad \blacktriangleleft$$

It is no accident that the two iterated integrals in the last example have the same value; it is a consequence of the following theorem.

17.1.2 THEOREM. *Let R be the rectangle defined by the inequalities*

$$a \le x \le b, \quad c \le y \le d$$

If $f(x, y)$ is continuous on this rectangle, then

$$\iint_R f(x, y)\,dA = \int_c^d \int_a^b f(x, y)\,dx\,dy = \int_a^b \int_c^d f(x, y)\,dy\,dx$$

This major theorem enables us to evaluate a double integral over a rectangle by calculating an iterated integral. Moreover, the theorem tells us that the order of integration in the iterated integral does not matter. We shall not formally prove this result; however, we will give a geometric argument for the case where $f(x, y)$ is nonnegative on R. In this case the double integral can be interpreted as the volume of the solid S bounded above by the surface $z = f(x, y)$ and below by the region R, so it suffices to show that the two iterated integrals also represent this volume.

For a fixed value of y, the function $f(x, y)$ is a function of x, and hence the integral

$$A(y) = \int_a^b f(x, y)\,dx$$

represents the area under the graph of this function of x. This area, shown in yellow in Figure 17.1.5, is the cross-sectional area at y of the solid S bounded above by $z = f(x, y)$ and below by the region R. Thus, by the method of slicing discussed in Section 6.2, the volume of the solid S is

$$\text{Vol}(S) = \int_c^d A(y)\,dy = \int_c^d \left[\int_a^b f(x, y)\,dx \right] dy = \int_c^d \int_a^b f(x, y)\,dx\,dy \qquad (10)$$

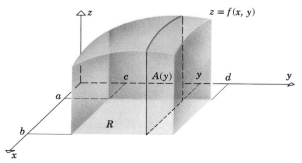

Figure 17.1.5

Similarly, the integral

$$A(x) = \int_c^d f(x, y)\, dy$$

represents the area of the cross section of S at x (Figure 17.1.6), and the method of slicing again yields

$$\text{Vol}(S) = \int_a^b A(x)\, dx = \int_a^b \left[\int_c^d f(x, y)\, dy \right] dx = \int_a^b \int_c^d f(x, y)\, dy\, dx \tag{11}$$

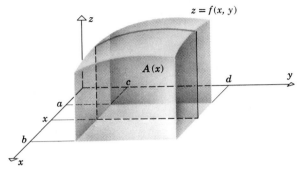

Figure 17.1.6

This establishes the result in Theorem 17.1.2 for the case where $f(x, y)$ is nonnegative on R.

Example 3   Evaluate the double integral

$$\iint_R y^2 x\, dA$$

over the rectangle $R = \{ (x, y) : -3 \leq x \leq 2,\ 0 \leq y \leq 1 \}$.

Solution.   In view of Theorem 17.1.2, the value of the double integral may be obtained from either of the iterated integrals

$$\int_{-3}^2 \int_0^1 y^2 x\, dy\, dx \quad \text{or} \quad \int_0^1 \int_{-3}^2 y^2 x\, dx\, dy \tag{12}$$

Using the first of these, we obtain

$$\iint_R y^2 x\, dA = \int_{-3}^2 \int_0^1 y^2 x\, dy\, dx = \int_{-3}^2 \left[\frac{1}{3} y^3 x \right]_{y=0}^1 dx$$

$$= \int_{-3}^2 \frac{1}{3} x\, dx = \frac{x^2}{6} \Big]_{-3}^2 = -\frac{5}{6}$$

The reader can check this result by evaluating the second integral in (12).   ◀

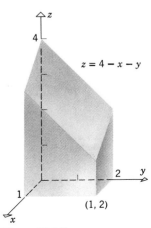

Figure 17.1.7

Example 4   Use a double integral to find the volume of the solid that is bounded above by the plane $z = 4 - x - y$ and below by the rectangle $R = \{ (x, y) : 0 \leq x \leq 1,\ 0 \leq y \leq 2 \}$ (Figure 17.1.7).

Solution.

$$V = \iint\limits_{R} (4 - x - y) \, dA = \int_0^2 \int_0^1 (4 - x - y) \, dx \, dy$$

$$= \int_0^2 \left[4x - \frac{x^2}{2} - xy \right]_{x=0}^1 dy = \int_0^2 \left(\frac{7}{2} - y \right) dy$$

$$= \left[\frac{7}{2} y - \frac{y^2}{2} \right]_0^2 = 5$$

The volume can also be obtained by first integrating with respect to y and then with respect to x. ◀

▶ Exercise Set 17.1

In Exercises 1–14, evaluate the iterated integrals.

1. $\int_0^1 \int_0^2 (x + 3) \, dy \, dx.$ 2. $\int_1^3 \int_{-1}^1 (2x - 4y) \, dy \, dx.$

3. $\int_2^4 \int_0^1 x^2 y \, dx \, dy.$ 4. $\int_{-2}^0 \int_{-1}^2 (x^2 + y^2) \, dx \, dy.$

5. $\int_0^{\ln 3} \int_0^{\ln 2} e^{x+y} \, dy \, dx.$ 6. $\int_0^2 \int_0^1 y \sin x \, dy \, dx.$

7. $\int_0^3 \int_0^1 x(x^2 + y)^{1/2} \, dx \, dy.$

8. $\int_{-1}^2 \int_2^4 (2x^2 y + 3xy^2) \, dx \, dy.$

9. $\int_{-1}^0 \int_2^5 dx \, dy.$ 10. $\int_4^6 \int_{-3}^7 dy \, dx.$

11. $\int_0^1 \int_0^1 \frac{x}{(xy + 1)^2} \, dy \, dx.$ 12. $\int_{\pi/2}^{\pi} \int_1^2 x \cos xy \, dy \, dx.$

13. $\int_0^{\ln 2} \int_0^1 xy \, e^{y^2 x} \, dy \, dx.$ 14. $\int_3^4 \int_1^2 \frac{1}{(x + y)^2} \, dy \, dx.$

In Exercises 15–19, evaluate the double integral over the rectangular region R.

15. $\iint\limits_{R} 4xy^3 \, dA; \ R = \{(x, y): -1 \le x \le 1, -2 \le y \le 2\}.$

16. $\iint\limits_{R} \frac{xy}{\sqrt{x^2 + y^2 + 1}} \, dA;$
     $R = \{(x, y): 0 \le x \le 1, 0 \le y \le 1\}.$

17. $\iint\limits_{R} x\sqrt{1 - x^2} \, dA; \ R = \{(x, y): 0 \le x \le 1, 2 \le y \le 3\}.$

18. $\iint\limits_{R} (x \sin y - y \sin x) \, dA;$
     $R = \{(x, y): 0 \le x \le \pi/2, 0 \le y \le \pi/3\}.$

19. $\iint\limits_{R} \cos(x + y) \, dA;$
     $R = \{(x, y): -\pi/4 \le x \le \pi/4, 0 \le y \le \pi/4\}.$

In Exercises 20–25, the iterated integral represents the volume of a solid. Make an accurate sketch of the solid. (You do *not* have to find the volume.)

20. $\int_0^5 \int_1^2 4 \, dx \, dy.$ 21. $\int_0^1 \int_0^1 (2 - x - y) \, dy \, dx.$

22. $\int_2^3 \int_3^4 y \, dx \, dy.$

23. $\int_0^3 \int_0^4 \sqrt{25 - x^2 - y^2} \, dy \, dx.$

24. $\int_{-2}^2 \int_{-2}^2 (x^2 + y^2) \, dx \, dy.$ 25. $\int_0^1 \int_{-1}^1 \sqrt{4 - x^2} \, dy \, dx.$

In Exercises 26–30, use a double integral to find the volume.

26. The volume under the plane $z = 2x + y$ and over the rectangle $R = \{(x, y): 3 \le x \le 5, 1 \le y \le 2\}$.

27. The volume under the surface $z = 3x^3 + 3x^2 y$ and over the rectangle $R = \{(x, y): 1 \le x \le 3, 0 \le y \le 2\}$.

28. The volume in the first octant bounded by the coordinate planes, the plane $y = 4$, and the plane $x/3 + z/5 = 1$.

29. The volume of the solid in the first octant enclosed by the surface $z = x^2$ and the planes $x = 2$, $y = 3$, $y = 0$, and $z = 0$.

30. The volume of the solid in the first octant that is enclosed by the planes $x = 0$, $z = 0$, $x = 5$, $z - y = 0$, and $z = -2y + 6$. [*Hint:* Break the solid into two parts.]

31. Suppose that $f(x, y) = g(x)h(y)$ and $R = \{(x, y) : a \leq x \leq b, \, c \leq y \leq d\}$. Show that

$$\iint_R f(x, y) \, dA = \left[\int_a^b g(x) \, dx\right]\left[\int_c^d h(y) \, dy\right]$$

32. Evaluate

$$\iint_R x \cos(xy) \cos^2 \pi x \, dA$$

where $R = \{(x, y) : 0 \leq x \leq \frac{1}{2}, \, 0 \leq y \leq \pi\}$. [*Hint:* One order of integration leads to a simpler solution than the other.]

■ 17.2 DOUBLE INTEGRALS OVER NONRECTANGULAR REGIONS

In this section we shall show how to evaluate double integrals over regions other than rectangles.

□ **ITERATED INTEGRALS WITH NONCONSTANT LIMITS OF INTEGRATION**

Later in this section we will see that double integrals over nonrectangular regions can often be evaluated as iterated integrals of the following types:

$$\int_a^b \int_{g_1(x)}^{g_2(x)} f(x, y) \, dy \, dx = \int_a^b \left[\int_{g_1(x)}^{g_2(x)} f(x, y) \, dy\right] dx \tag{1a}$$

$$\int_c^d \int_{h_1(y)}^{h_2(y)} f(x, y) \, dx \, dy = \int_c^d \left[\int_{h_1(y)}^{h_2(y)} f(x, y) \, dx\right] dy \tag{1b}$$

We begin with an example that illustrates how to evaluate such integrals.

Example 1 Evaluate

(a) $\displaystyle\int_0^2 \int_{x^2}^{x} y^2 x \, dy \, dx$ (b) $\displaystyle\int_0^\pi \int_0^{\cos y} x \sin y \, dx \, dy$

Solution (a).

$$\int_0^2 \int_{x^2}^{x} y^2 x \, dy \, dx = \int_0^2 \left[\int_{x^2}^{x} y^2 x \, dy\right] dx = \int_0^2 \left[\frac{y^3 x}{3}\right]_{y=x^2}^{x} dx$$

$$= \int_0^2 \left(\frac{x^4}{3} - \frac{x^7}{3}\right) dx = \left[\frac{x^5}{15} - \frac{x^8}{24}\right]_0^2$$

$$= \frac{32}{15} - \frac{256}{24} = -\frac{128}{15}$$

Solution (b).

$$\int_0^\pi \int_0^{\cos y} x \sin y \, dx \, dy = \int_0^\pi \left[\int_0^{\cos y} x \sin y \, dx\right] dy$$

$$= \int_0^\pi \left[\frac{x^2}{2} \sin y\right]_{x=0}^{\cos y} dy$$

$$= \int_0^\pi \frac{1}{2} \cos^2 y \sin y \, dy$$

$$= \left[-\frac{1}{6} \cos^3 y\right]_0^\pi = \frac{1}{3} \quad \blacktriangleleft$$

DOUBLE INTEGRALS OVER NONRECTANGULAR REGIONS

Plane regions can be extremely complex, and the theory of double integrals over very general regions is a topic for advanced courses in mathematics. We shall limit our study of double integrals to two basic types of regions, which we shall call *type I* and *type II*; they are defined as follows:

17.2.1   DEFINITION.

(a)   A *type I region* is bounded on the left and right by vertical lines $x = a$ and $x = b$ and is bounded below and above by continuous curves $y = g_1(x)$ and $y = g_2(x)$, where $g_1(x) \leq g_2(x)$ for $a \leq x \leq b$ (Figure 17.2.1a).

(b)   A *type II region* is bounded below and above by horizontal lines $y = c$ and $y = d$ and is bounded on the left and right by continuous curves $x = h_1(y)$ and $x = h_2(y)$ satisfying $h_1(y) \leq h_2(y)$ for $c \leq y \leq d$ (Figure 17.2.1b).

The following theorem will enable us to evaluate double integrals over type I and type II regions using iterated integrals.

17.2.2   THEOREM.

(a)   *If R is a type I region on which $f(x, y)$ is continuous, then*

$$\iint\limits_R f(x, y)\, dA = \int_a^b \int_{g_1(x)}^{g_2(x)} f(x, y)\, dy\, dx \qquad (2a)$$

(b)   *If R is a type II region on which $f(x, y)$ is continuous, then*

$$\iint\limits_R f(x, y)\, dA = \int_c^d \int_{h_1(y)}^{h_2(y)} f(x, y)\, dx\, dy \qquad (2b)$$

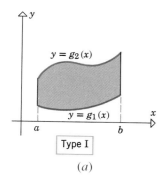

Type I

(a)

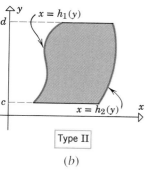

Type II

(b)

Figure 17.2.1

We will not prove this theorem, but for the case where $f(x, y)$ is nonnegative on the region R, it can be made plausible by a geometric argument that is similar to that given for Theorem 17.1.2. Since $f(x, y)$ is nonnegative, the double integral can be interpreted as the volume of the solid S that is bounded above by the surface $z = f(x, y)$ and below by the region R, so it suffices to show that the iterated integrals also represent this volume. Consider the iterated integral in (2a), for example. For a fixed value of x, the function $f(x, y)$ is a function of y, and hence the integral

$$A(x) = \int_{g_1(x)}^{g_2(x)} f(x, y)\, dy$$

represents the area under the graph of this function of y between the points $y = g_1(x)$ and $y = g_2(x)$. This area, shown in yellow in Figure 17.2.2, is the cross-sectional area at x of the solid S, and hence by the method of slicing, the volume of the solid S is

$$\text{Vol}(S) = \int_a^b \int_{g_1(x)}^{g_2(x)} f(x, y)\, dy\, dx$$

which shows that in (2a) the iterated integral is equal to the double integral. Similarly for (2b).

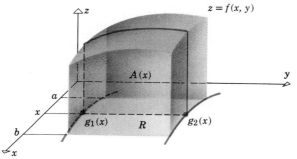

Figure 17.2.2

□ SETTING UP LIMITS OF
INTEGRATION FOR
EVALUATING DOUBLE
INTEGRALS

To apply Theorem 17.2.2, it is usual to start with a two-dimensional sketch of the region R. [It is not necessary to graph $f(x, y)$.] For a type I region, the limits of integration in Formula (2a) can be obtained as follows:

> **Step 1.** Since x is held fixed for the first integration, we draw a vertical line through the region R at an arbitrary fixed point x (Figure 17.2.3). This line crosses the boundary of R twice. The lower point of intersection is on the curve $y = g_1(x)$ and the higher point is on the curve $y = g_2(x)$. These two intersections determine the lower and upper y-limits of integration in Formula (2a).
>
> **Step 2.** Imagine moving the line drawn in Step 1 first to the left and then to the right (Figure 17.2.3). The leftmost position where the line intersects the region R is $x = a$ and the rightmost position where the line intersects the region R is $x = b$. This yields the limits for the x-integration in Formula (2a).

Figure 17.2.3

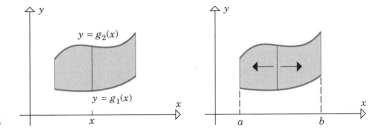

Example 2   Evaluate

$$\iint\limits_R xy \, dA$$

over the region R enclosed between $y = \frac{1}{2}x$, $y = \sqrt{x}$, $x = 2$, and $x = 4$.

Solution.   We view R as a type I region. The region R and a vertical line corresponding to a fixed x are shown in Figure 17.2.4. This line meets the region R at the lower boundary $y = \frac{1}{2}x$ and the upper boundary $y = \sqrt{x}$. These are the y-limits of integration. Moving this line first left and then right yields the x-limits of integration, $x = 2$ and $x = 4$. Thus,

$$\iint\limits_R xy \, dA = \int_2^4 \int_{x/2}^{\sqrt{x}} xy \, dy \, dx = \int_2^4 \left[\frac{xy^2}{2} \right]_{y=x/2}^{\sqrt{x}} dx = \int_2^4 \left(\frac{x^2}{2} - \frac{x^3}{8} \right) dx$$

$$= \left[\frac{x^3}{6} - \frac{x^4}{32} \right]_2^4 = \left(\frac{64}{6} - \frac{256}{32} \right) - \left(\frac{8}{6} - \frac{16}{32} \right) = \frac{11}{6} \qquad \blacktriangleleft$$

Figure 17.2.4

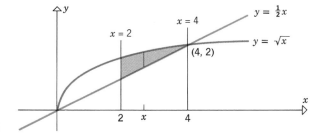

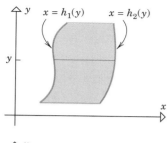

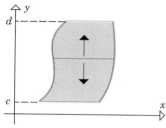

Figure 17.2.5

If R is a type II region, then the limits of integration in Formula (2b) can be obtained as follows:

> **Step 1.** Since y is held fixed for the first integration, we draw a horizontal line through the region R at a fixed point y (Figure 17.2.5). This line crosses the boundary of R twice. The leftmost point of intersection is on the curve $x = h_1(y)$ and the rightmost point is on the curve $x = h_2(y)$. These intersections determine the x-limits of integration in (2b).
>
> **Step 2.** Imagine moving the line drawn in Step 1 first down and then up (Figure 17.2.5). The lowest position where the line intersects the region R is $y = c$, and the highest position where the line intersects the region R is $y = d$. This yields the y-limits of integration in (2b).

Example 3 Evaluate

$$\iint\limits_R (2x - y^2)\, dA$$

over the triangular region R enclosed between the lines $y = -x + 1$, $y = x + 1$, and $y = 3$.

Solution. We view R as a type II region. The region R and a horizontal line corresponding to a fixed y are shown in Figure 17.2.6. This line meets the region R at its left-hand boundary $x = 1 - y$ and its right-hand boundary $x = y - 1$. These are the x-limits of integration. Moving this line first down and then up yields the y-limits, $y = 1$ and $y = 3$. Thus,

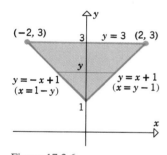

Figure 17.2.6

$$\iint\limits_R (2x - y^2)\, dA = \int_1^3 \int_{1-y}^{y-1} (2x - y^2)\, dx\, dy = \int_1^3 \left[x^2 - y^2 x \right]_{x=1-y}^{y-1} dy$$

$$= \int_1^3 [(1 - 2y + 2y^2 - y^3) - (1 - 2y + y^3)]\, dy$$

$$= \int_1^3 (2y^2 - 2y^3)\, dy = \left[\frac{2y^3}{3} - \frac{y^4}{2} \right]_1^3 = -\frac{68}{3} \quad \blacktriangleleft$$

REMARK. To integrate over a type II region, the left- and right-hand boundaries must be expressed in the form $x = h_1(y)$ and $x = h_2(y)$. This is why we rewrote the boundary equations $y = -x + 1$ and $y = x + 1$ as $x = 1 - y$ and $x = y - 1$ in the last example.

In Example 3 we could have treated R as a type I region, but with an added complication. Viewed as a type I region, the upper boundary of R is the line $y = 3$ (Figure 17.2.7) and the lower boundary consists of two parts, the line $y = -x + 1$ to the left of the origin and the line $y = x + 1$ to the right of the origin. To carry out the integration it is necessary to decompose the region R into two parts, R_1 and R_2, as shown in Figure 17.2.7, and write

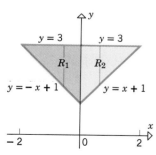

Figure 17.2.7

$$\iint\limits_R (2x - y^2)\, dA = \iint\limits_{R_1} (2x - y^2)\, dA + \iint\limits_{R_2} (2x - y^2)\, dA$$

$$= \int_{-2}^0 \int_{-x+1}^3 (2x - y^2)\, dy\, dx + \int_0^2 \int_{x+1}^3 (2x - y^2)\, dy\, dx$$

This will yield the same result that was obtained in Example 3.

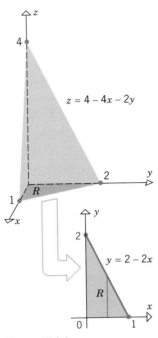

Figure 17.2.8

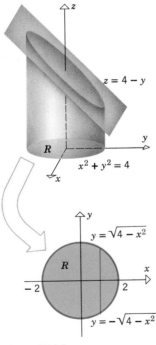

Figure 17.2.9

☐ **REVERSING THE ORDER OF INTEGRATION**

Example 4 Use a double integral to find the volume of the tetrahedron bounded by the coordinate planes and the plane $z = 4 - 4x - 2y$.

Solution. The tetrahedron in question is bounded above by the plane

$$z = 4 - 4x - 2y \tag{3}$$

and below by the triangular region R shown in Figure 17.2.8. Thus, the volume is given by

$$V = \iint_R (4 - 4x - 2y)\, dA$$

The region R is bounded by the x-axis, the y-axis, and the line $y = 2 - 2x$ [set $z = 0$ in (3)], so that treating R as a type I region yields

$$V = \iint_R (4 - 4x - 2y)\, dA = \int_0^1 \int_0^{2-2x} (4 - 4x - 2y)\, dy\, dx$$

$$= \int_0^1 \left[4y - 4xy - y^2 \right]_{y=0}^{2-2x} dx = \int_0^1 (4 - 8x + 4x^2)\, dx = \frac{4}{3} \quad ◀$$

Example 5 Find the volume of the solid bounded by the cylinder $x^2 + y^2 = 4$ and the planes $y + z = 4$ and $z = 0$.

Solution. The solid shown in Figure 17.2.9 is bounded above by the plane $z = 4 - y$ and below by the region R within the circle $x^2 + y^2 = 4$. The volume is given by

$$V = \iint_R (4 - y)\, dA$$

Treating R as a type I region we obtain

$$V = \int_{-2}^2 \int_{-\sqrt{4-x^2}}^{\sqrt{4-x^2}} (4 - y)\, dy\, dx = \int_{-2}^2 \left[4y - \frac{1}{2} y^2 \right]_{y=-\sqrt{4-x^2}}^{\sqrt{4-x^2}} dx$$

$$= \int_{-2}^2 8\sqrt{4 - x^2}\, dx = 8(2\pi) = 16\pi \quad \boxed{\text{See Example 4 of Section 9.5.}} \quad ◀$$

Sometimes the evaluation of an iterated integral can be simplified by reversing the order of integration. The next example illustrates how this is done.

Example 6 Since there is no elementary antiderivative of e^{x^2}, the integral

$$\int_0^2 \int_{y/2}^1 e^{x^2}\, dx\, dy$$

cannot be evaluated by performing the x-integration first. Evaluate this integral by expressing it as an equivalent iterated integral with the order of integration reversed.

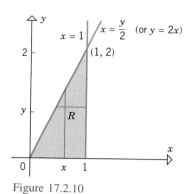

Figure 17.2.10

For the inside integration, y is fixed and x varies from the line $x = y/2$ to the line $x = 1$ (Figure 17.2.10). For the outside integration, y varies from 0 to 2, so the given iterated integral is equal to a double integral over the triangular region R in Figure 17.2.10.

To reverse the order of integration, we treat R as a type I region, which enables us to write the given integral as

$$\int_0^2 \int_{y/2}^1 e^{x^2} \, dx \, dy = \iint_R e^{x^2} \, dA = \int_0^1 \int_0^{2x} e^{x^2} \, dy \, dx = \int_0^1 \left[e^{x^2} y \right]_{y=0}^{2x} dx$$

$$= \int_0^1 2x e^{x^2} \, dx = e^{x^2} \Big]_0^1 = e - 1 \quad \blacktriangleleft$$

□ **AREA CALCULATED AS A DOUBLE INTEGRAL**

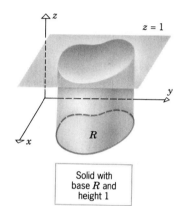

Figure 17.2.11

Although double integrals arose in the context of calculating volumes, they can also be used to calculate areas. For this purpose, we consider the solid consisting of the points between the plane $z = 1$ and a region R in the xy-plane (Figure 17.2.11). The volume V of this solid is

$$V = \iint_R 1 \, dA = \iint_R dA \tag{4}$$

However, the solid has congruent cross sections taken parallel to the xy-plane, so that

$$V = \text{area of base} \cdot \text{height} = \text{area of } R \cdot 1 = \text{area of } R$$

Combining this with (4) yields the area formula

$$\text{area of } R = \iint_R dA \tag{5}$$

REMARK. Formula (5) is sometimes confusing because it equates an area and a volume; the formula is intended to equate only the *numerical values* of the area and volume and not the units, which must, of course, be different.

Example 7 Use a double integral to find the area of the region R enclosed between the parabola $y = \frac{1}{2}x^2$ and the line $y = 2x$.

Solution. The region R may be treated equally well as type I (Figure 17.2.12a) or type II (Figure 17.2.12b). Treating R as type I yields

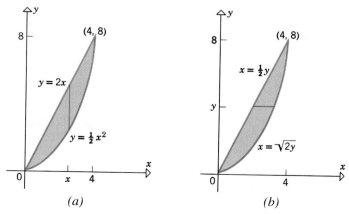

Figure 17.2.12

(a)

(b)

$$\text{area of } R = \iint\limits_{R} dA = \int_0^4 \int_{x^2/2}^{2x} dy\, dx = \int_0^4 \left[y \right]_{y=x^2/2}^{2x} dx$$

$$= \int_0^4 \left(2x - \frac{1}{2}x^2 \right) dx = \left[x^2 - \frac{x^3}{6} \right]_0^4 = \frac{16}{3}$$

Treating R as type II yields

$$\text{area of } R = \iint\limits_{R} dA = \int_0^8 \int_{y/2}^{\sqrt{2y}} dx\, dy = \int_0^8 \left[x \right]_{x=y/2}^{\sqrt{2y}} dy$$

$$= \int_0^8 \left(\sqrt{2y} - \frac{1}{2}y \right) dy = \left[\frac{2\sqrt{2}}{3} y^{3/2} - \frac{y^2}{4} \right]_0^8 = \frac{16}{3} \quad \blacktriangleleft$$

▶ Exercise Set 17.2

In Exercises 1–12, evaluate the iterated integral.

1. $\displaystyle\int_0^1 \int_{x^2}^x xy^2\, dy\, dx.$

2. $\displaystyle\int_1^2 \int_y^{3-y} y\, dx\, dy.$

3. $\displaystyle\int_0^3 \int_0^{\sqrt{9-y^2}} y\, dx\, dy.$

4. $\displaystyle\int_{1/4}^1 \int_{x^2}^x \sqrt{\frac{x}{y}}\, dy\, dx.$

5. $\displaystyle\int_{\sqrt{\pi}}^{\sqrt{2\pi}} \int_0^{x^3} \sin\frac{y}{x}\, dy\, dx.$

6. $\displaystyle\int_{-1}^1 \int_{-x^2}^{x^2} (x^2 - y)\, dy\, dx.$

7. $\displaystyle\int_{\pi/2}^{\pi} \int_0^{x^2} \frac{1}{x}\cos\frac{y}{x}\, dy\, dx.$

8. $\displaystyle\int_0^{\pi/2} \int_0^{\sin y} e^x \cos y\, dx\, dy.$

9. $\displaystyle\int_0^a \int_0^{\sqrt{a^2-x^2}} (x+y)\, dy\, dx \quad (a > 0).$

10. $\displaystyle\int_1^2 \int_0^{y^2} e^{x/y^2}\, dx\, dy.$

11. $\displaystyle\int_0^1 \int_0^x y\sqrt{x^2-y^2}\, dy\, dx.$

12. $\displaystyle\int_0^1 \int_0^x e^{x^2}\, dy\, dx.$

In Exercises 13–28, evaluate the double integral.

13. $\displaystyle\iint\limits_{R} 6xy\, dA;$ R is the region bounded by $y = 0$, $x = 2$, and $y = x^2$.

14. $\displaystyle\iint\limits_{R} xy\, dA;$ R is the region bounded by the trapezoid with vertices $(1, 3)$, $(5, 3)$, $(2, 1)$, and $(4, 1)$.

15. $\displaystyle\iint\limits_{R} x\cos xy\, dA;$ R is the region enclosed by $x = 1$, $x = 2$, $y = \pi/2$, and $y = 2\pi/x$.

16. $\displaystyle\iint\limits_{R} (x + y)\, dA;$ R is the region enclosed between the curves $y = x^2$ and $y = \sqrt{x}$.

17. $\displaystyle\iint\limits_{R} x^2\, dA;$ R is the region bounded by $y = 16/x$, $y = x$, and $x = 8$.

18. $\displaystyle\iint\limits_{R} xy^2\, dA;$ R is the region enclosed by $y = 1$, $y = 2$, $x = 0$, and $y = x$.

19. $\displaystyle\iint\limits_{R} x(1 + y^2)^{-1/2}\, dA;$ R is the region in the first quadrant enclosed by $y = x^2$, $y = 4$, and $x = 0$. [*Hint:* Choose your order of integration carefully.]

20. $\displaystyle\iint\limits_{R} x\cos y\, dA;$ R is the triangular region bounded by $y = x$, $y = 0$, and $x = \pi$.

21. $\displaystyle\iint\limits_{R} (3x - 2y)\, dA;$ R is the region enclosed by the circle $x^2 + y^2 = 1$.

22. $\displaystyle\iint\limits_{R} y\, dA;$ R is the region in the first quadrant enclosed between the circle $x^2 + y^2 = 25$ and the line $x + y = 5$.

23. $\displaystyle\iint\limits_{R} \frac{1}{1 + x^2}\, dA;$ R is the triangular region with vertices $(0, 0)$, $(1, 1)$, and $(0, 1)$.

24. $\displaystyle\iint\limits_{R} (x^2 - xy)\, dA;$ R is the region enclosed by $y = x$ and $y = 3x - x^2$.

25. $\iint\limits_{R} xy\, dA$; R is the region enclosed by $y = \sqrt{x}$, $y = 6 - x$, and $y = 0$.

26. $\iint\limits_{R} x\, dA$; R is the region enclosed by $y = \sin^{-1} x$, $x = 1/\sqrt{2}$, and $y = 0$.

27. $\iint\limits_{R} (x - 1)\, dA$; R is the region enclosed between $y = x$ and $y = x^3$.

28. $\iint\limits_{R} x^2\, dA$; R is the region in the first quadrant enclosed by $xy = 1$, $y = x$, and $y = 2x$.

In Exercises 29–34, use double integration to find the area of the plane region enclosed by the given curves.

29. $x + y = 5$, $x = 0$, and $y = 0$.

30. $y = x^2$ and $y = 4x$.

31. $y = \sin x$ and $y = \cos x$, for $0 \le x \le \pi/4$.

32. $y^2 = -x$ and $3y - x = 4$.

33. $y^2 = 9 - x$ and $y^2 = 9 - 9x$.

34. $y = \cosh x$, $y = \sinh x$, $x = 0$, and $x = 1$.

In Exercises 35–48, use double integration to find the volume of each solid.

35. The tetrahedron that lies in the first octant that is bounded by the three coordinate planes and the plane $z = 5 - 2x - y$.

36. The solid bounded by the cylinder $x^2 + y^2 = 9$ and the planes $z = 0$ and $z = 3 - x$.

37. The solid that is bounded above by the plane $z = x + 2y + 2$, below by the xy-plane, and laterally by $y = 0$ and $y = 1 - x^2$.

38. The solid in the first octant bounded above by the paraboloid $z = x^2 + 3y^2$, below by the plane $z = 0$, and laterally by $y = x^2$ and $y = x$.

39. The solid that is bounded above by the paraboloid $z = 9x^2 + y^2$, below by the plane $z = 0$, and laterally by the planes $x = 0$, $y = 0$, $x = 3$, and $y = 2$.

40. The solid enclosed by $y^2 = x$, $z = 0$, and $x + z = 1$.

41. The wedge cut from the cylinder $4x^2 + y^2 = 9$ by the planes $z = 0$ and $z = y + 3$.

42. The solid in the first octant bounded above by $z = 9 - x^2$, below by $z = 0$, and laterally by $y^2 = 3x$.

43. The solid in the first octant bounded by the three coordinate planes and the planes
$$x + 2y = 4 \quad \text{and} \quad x + 8y - 4z = 0$$

44. The solid in the first octant bounded by the surface $z = e^{y-x}$, the plane $x + y = 1$, and the coordinate planes.

45. The solid bounded above by the paraboloid $z = 1 - x^2 - y^2$ and below by the xy-plane. [*Hint:* Use a trigonometric substitution to evaluate the integral.]

46. The solid in the first octant bounded by the paraboloid $z = x^2 + y^2$ and the cylinder $x^2 + y^2 = 4$. [*Hint:* Use a trigonometric substitution to evaluate the integral.]

47. The solid common to the cylinders $x^2 + y^2 = 25$ and $x^2 + z^2 = 25$.

48. The solid that is bounded above by the paraboloid $z = x^2 + y^2$, bounded laterally by the circular cylinder $x^2 + (y - 1)^2 = 1$, and bounded below by the xy-plane.

In Exercises 49–56, express the integral as an equivalent integral with the order of integration reversed.

49. $\displaystyle\int_0^2 \int_0^{\sqrt{x}} f(x, y)\, dy\, dx.$ **50.** $\displaystyle\int_0^4 \int_{2y}^8 f(x, y)\, dx\, dy.$

51. $\displaystyle\int_0^2 \int_1^{e^y} f(x, y)\, dx\, dy.$ **52.** $\displaystyle\int_1^e \int_0^{\ln x} f(x, y)\, dy\, dx.$

53. $\displaystyle\int_{-2}^2 \int_{-\sqrt{1-(x^2/4)}}^{\sqrt{1-(x^2/4)}} f(x, y)\, dy\, dx.$

54. $\displaystyle\int_0^1 \int_{y^2}^{\sqrt{y}} f(x, y)\, dx\, dy.$ **55.** $\displaystyle\int_0^1 \int_{\sin^{-1}y}^{\pi/2} f(x, y)\, dx\, dy.$

56. $\displaystyle\int_{-3}^1 \int_{x^2+6x}^{4x+3} f(x, y)\, dy\, dx.$

In Exercises 57–62, evaluate the integral by first reversing the order of integration.

57. $\displaystyle\int_0^1 \int_{4x}^4 e^{-y^2}\, dy\, dx.$ **58.** $\displaystyle\int_0^2 \int_{y/2}^1 \cos(x^2)\, dx\, dy.$

59. $\displaystyle\int_0^4 \int_{\sqrt{y}}^2 e^{x^3}\, dx\, dy.$ **60.** $\displaystyle\int_1^3 \int_0^{\ln x} x\, dy\, dx.$

61. $\displaystyle\int_0^1 \int_0^{\cos^{-1} x} x\, dy\, dx.$

62. $\displaystyle\int_0^1 \int_{\sin^{-1}y}^{\pi/2} \sec^2(\cos x)\, dx\, dy.$

63. Evaluate $\displaystyle\iint\limits_{R} \sin(y^3)\, dA$, where R is the region bounded by $y = \sqrt{x}$, $y = 2$, and $x = 0$. [*Hint:* Choose the order of integration carefully.]

64. Evaluate $\displaystyle\iint\limits_{R} x\, dA$, where R is the region bounded by $x = \ln y$, $x = 0$, and $y = e$. [*Hint:* Choose the order of integration carefully.]

65. In each part evaluate $\displaystyle\iint_R xy^2\,dA.$

(a)

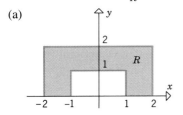

(b)

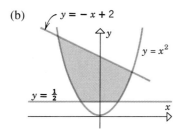

66. Assume that the region R shown in the following figure is symmetric about the y-axis and that the y-axis divides R into the subregions R_1 and R_2 shown. Suppose also that

$$\iint_{R_1} f(x,y)\,dA = 3.$$

(a) Find $\displaystyle\iint_R f(x,y)\,dA$ if $f(-x,y) = f(x,y).$

(b) Find $\displaystyle\iint_R f(x,y)\,dA$ if $f(-x,y) = -f(x,y).$

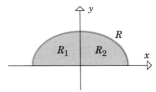

In Exercises 67–69, use symmetry to evaluate the integral without performing an integration. [*Hint:* See Exercise 66.]

67. $\displaystyle\int_{-1}^{1}\int_{0}^{\sqrt{1-x^2}} (2 + x\sqrt{9 - y^2})\,dy\,dx.$

68. $\displaystyle\int_{0}^{2}\int_{x-2}^{2-x} \sin(xy^3)\,dy\,dx.$

69. $\displaystyle\iint_R x^3 y\,dA,$ where R is the semicircular region bounded by $y = \sqrt{4 - x^2}$ and $x = 0.$

■ 17.3 DOUBLE INTEGRALS IN POLAR COORDINATES

In this section we shall study double integrals in which the integrand and the region of integration are expressed in polar coordinates. Such integrals are important for two reasons: First, they arise naturally in many applications, and second, many double integrals in rectangular coordinates are more easily evaluated if they are converted to polar coordinates.

☐ DEFINITION OF DOUBLE
INTEGRALS IN POLAR
COORDINATES

Some problems that lead to double integrals are most easily solved if the integrand and the region of integration are expressed in polar coordinates rather than rectangular coordinates. This is often the case when the region of integration has the form described in the following definition.

> **17.3.1 DEFINITION.** A *simple polar region* is a region (in polar coordinates) enclosed between two rays $\theta = \alpha$ and $\theta = \beta$ and two continuous polar curves $r = r_1(\theta)$ and $r = r_2(\theta)$, where
>
> (i) $\alpha \le \beta$ and $\beta - \alpha \le 2\pi$
> (ii) $0 \le r_1(\theta) \le r_2(\theta)$

Figure 17.3.1*a* shows a typical simple polar region R. Figure 17.3.1*b* shows the special case where $r_1(\theta)$ is identically zero, so the inner boundary reduces to a single point (the

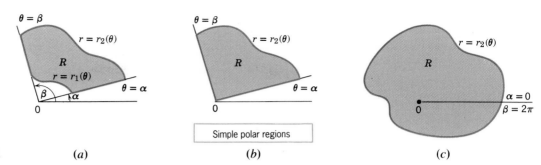

Figure 17.3.1

(a) Simple polar regions (b) (c)

origin); and Figure 17.3.1c shows the special case where $r_1(\theta)$ is identically zero and the rays are $\alpha = 0$ and $\beta = 2\pi$, so the side boundaries coincide.

The concept of a double integral in polar coordinates can be motivated by the following volume problem:

> **17.3.2** PROBLEM. *Find the volume of the solid consisting of all points that lie between a simple polar region R in the xy-plane and a surface whose equation in cylindrical coordinates is $z = f(r, \theta)$, where f is continuous on R and $f(r, \theta) \geq 0$ for all (r, θ) in R (Figure 17.3.2).*

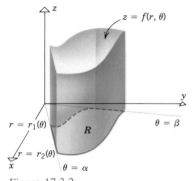

Figure 17.3.2

To find the volume V of the solid in Figure 17.3.2, we shall use a limiting process similar to that used to derive Formula (4) of Section 17.1, except that here we shall use circular arcs and rays to subdivide the region R into blocks, called **polar rectangles**. As shown in Figure 17.3.3, we exclude from consideration all polar rectangles that contain any points outside of R, leaving only polar rectangles that are subsets of R. Assume that there are n such polar rectangles, and denote the area of the kth polar rectangle by ΔA_k. Let (r_k^*, θ_k^*) be any point in this polar rectangle. As shown in Figure 17.3.4, the product $f(r_k^*, \theta_k^*)\,\Delta A_k$ is the volume of a solid with base area ΔA_k and height $f(r_k^*, \theta_k^*)$, so the sum

$$\sum_{k=1}^{n} f(r_k^*, \theta_k^*)\,\Delta A_k$$

can be viewed as an approximation to the volume V of the entire solid.

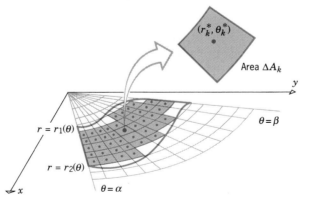

Figure 17.3.3

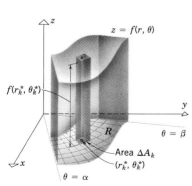

Figure 17.3.4

If we now increase the number of subdivisions in such a way that the dimensions of the polar rectangles approach zero, then it seems plausible that the errors in the approximations approach zero, and the exact volume of the solid is

$$V = \lim_{n \to +\infty} \sum_{k=1}^{n} f(r_k^*, \theta_k^*)\,\Delta A_k \qquad (1)$$

The sums in this equation are called *polar Riemann sums*, and the limit is called a *polar double integral*, which we denote as

$$\iint\limits_{R} f(r, \theta)\, dA = \lim_{n \to +\infty} \sum_{k=1}^{n} f(r_k^*, \theta_k^*)\, \Delta A_k \tag{2}$$

so the volume V of the solid can be expressed as

$$V = \iint\limits_{R} f(r, \theta)\, dA \tag{3}$$

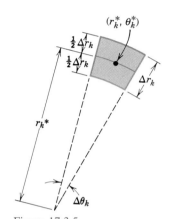

Figure 17.3.5

For computational purposes, it will be helpful to express the double polar integral in (3) as an iterated integral. For this purpose, let us choose the arbitrary point (r_k^*, θ_k^*) in (2) to be at the "center" of the kth polar rectangle as shown in Figure 17.3.5. Suppose also that this polar rectangle has a central angle $\Delta\theta_k$ and a "radial thickness" Δr_k. Thus, the inner radius of this polar rectangle is $r_k^* - \frac{1}{2}\Delta r_k$ and the outer radius is $r_k^* + \frac{1}{2}\Delta r_k$. Treating the area ΔA_k of this polar rectangle as the difference in area of two sectors, we obtain

$$\Delta A_k = \tfrac{1}{2}(r_k^* + \tfrac{1}{2}\Delta r_k)^2\, \Delta\theta_k - \tfrac{1}{2}(r_k^* - \tfrac{1}{2}\Delta r_k)^2\, \Delta\theta_k$$

which simplifies to

$$\Delta A_k = r_k^*\, \Delta r_k\, \Delta\theta_k \tag{4}$$

Thus, from (2) and (3)

$$V = \iint\limits_{R} f(r, \theta)\, dA = \lim_{n \to +\infty} \sum_{k=1}^{n} f(r_k^*, \theta_k^*) r_k^*\, \Delta r_k\, \Delta\theta_k$$

which suggests that the volume V can be expressed as the iterated integral

$$V = \iint\limits_{R} f(r, \theta)\, dA = \int_{\alpha}^{\beta} \int_{r_1(\theta)}^{r_2(\theta)} f(r, \theta) r\, dr\, d\theta \tag{5}$$

in which the limits of integration are chosen to cover the region R; that is, with θ fixed between α and β, the value of r varies from $r_1(\theta)$ to $r_2(\theta)$ (Figure 17.3.6).

If $f(r, \theta)$ can have both positive and negative values on R, then, in keeping with past experience, the polar double integral over R represents the *net signed volume* between $z = f(r, \theta)$ and R, that is, the difference between the volume of the solid above R but below $z = f(r, \theta)$ and the volume of the solid that is below R but above $z = f(r, \theta)$. Moreover, the relationship between the double polar integral and the iterated integral in (5) holds even if $f(r, \theta)$ is not positive on R.

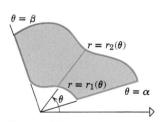

Figure 17.3.6

17.3.3   THEOREM.   *If R is a region of the type shown in* Figure 17.3.1, *and if $f(r, \theta)$ is continuous on R, then*

$$\iint\limits_{R} f(r, \theta)\, dA = \int_{\alpha}^{\beta} \int_{r_1(\theta)}^{r_2(\theta)} f(r, \theta) r\, dr\, d\theta \tag{6}$$

REMARK.   We mention without proof that polar double integrals satisfy the usual properties of integrals such as (5), (6), (7), and (8) in Section 17.1.

☐ **EVALUATION OF POLAR DOUBLE INTEGRALS**

To apply Theorem 17.3.3, we start with a sketch of the region R. From this sketch the limits of integration in (6) can be obtained as follows:

Step 1. Since θ is held fixed for the first integration, draw a radial line from the origin through the region R at a fixed angle θ (Figure 17.3.6). This line crosses the boundary of R at most twice. The innermost point of intersection is on the curve $r = r_1(\theta)$ and the outermost point is on the curve $r = r_2(\theta)$. These intersections determine the r-limits of integration in (6).

Step 2. Imagine rotating a ray along the polar x-axis one revolution counterclockwise about the origin. The smallest angle at which this ray intersects the region R is $\theta = \alpha$ and the largest angle is $\theta = \beta$. This yields the θ-limits of integration.

Example 1 Evaluate

$$\iint_R \sin\theta \, dA$$

where R is the region in the first quadrant that is outside the circle $r = 2$ and inside the cardioid $r = 2(1 + \cos\theta)$.

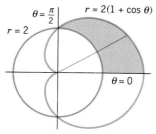

Figure 17.3.7

Solution. The region R is sketched in Figure 17.3.7. Following the two steps outlined above we obtain

$$\iint_R \sin\theta \, dA = \int_0^{\pi/2} \int_2^{2(1+\cos\theta)} (\sin\theta)r \, dr \, d\theta$$

$$= \int_0^{\pi/2} \frac{1}{2} r^2 \sin\theta \bigg]_{r=2}^{2(1+\cos\theta)} d\theta$$

$$= 2 \int_0^{\pi/2} [(1 + \cos\theta)^2 \sin\theta - \sin\theta] \, d\theta$$

$$= 2 \left[-\frac{1}{3}(1 + \cos\theta)^3 + \cos\theta \right]_0^{\pi/2}$$

$$= 2 \left[-\frac{1}{3} - \left(-\frac{5}{3} \right) \right] = \frac{8}{3} \quad \blacktriangleleft$$

Example 2 In cylindrical coordinates, the equation $r^2 + z^2 = a^2$ represents a sphere of radius a centered at the origin. (It is the sphere whose equation in rectangular coordinates is $x^2 + y^2 + z^2 = a^2$.) Use a double polar integral to calculate the volume of this sphere.

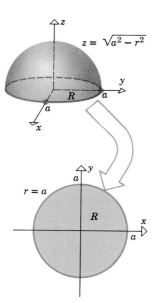

Figure 17.3.8

Solution. In cylindrical coordinates the upper hemisphere is given by the equation

$$z = \sqrt{a^2 - r^2}$$

so the volume enclosed by the entire sphere is

$$V = 2 \iint_R \sqrt{a^2 - r^2} \, dA$$

where R is the circular region shown in Figure 17.3.8. Thus,

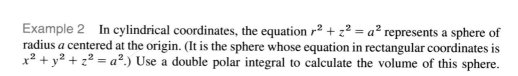

$$V = 2 \iint\limits_{R} \sqrt{a^2 - r^2} \, dA = \int_0^{2\pi} \int_0^a \sqrt{a^2 - r^2} \, (2r) \, dr \, d\theta$$

$$= \int_0^{2\pi} \left[-\frac{2}{3}(a^2 - r^2)^{3/2} \right]_{r=0}^a d\theta = \int_0^{2\pi} \frac{2}{3} a^3 \, d\theta$$

$$= \left[\frac{2}{3} a^3 \theta \right]_0^{2\pi} = \frac{4}{3} \pi a^3 \quad \blacktriangleleft$$

The argument used to derive Formula (5) of Section 17.2 also applies to polar double integrals, so the area of a simple polar region R can be expressed as

$$\text{area of } R = \iint\limits_{R} dA \tag{7}$$

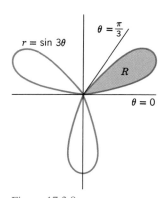

$r = \sin 3\theta$

$\theta = \frac{\pi}{3}$

R

$\theta = 0$

Figure 17.3.9

Example 3   Use a double polar integral to find the area enclosed by the three-petaled rose $r = \sin 3\theta$.

Solution.   The rose is sketched in Figure 17.3.9. We shall use Formula (7) to calculate the area of the petal R in the first quadrant and multiply by three.

$$A = 3 \iint\limits_{R} dA = 3 \int_0^{\pi/3} \int_0^{\sin 3\theta} r \, dr \, d\theta$$

$$= \frac{3}{2} \int_0^{\pi/3} \sin^2 3\theta \, d\theta = \frac{3}{4} \int_0^{\pi/3} (1 - \cos 6\theta) \, d\theta$$

$$= \frac{3}{4} \left[\theta - \frac{\sin 6\theta}{6} \right]_0^{\pi/3} = \frac{1}{4} \pi \quad \blacktriangleleft$$

□ **CONVERTING DOUBLE INTEGRALS FROM RECTANGULAR TO POLAR COORDINATES**

Polar double integrals are sometimes called **double integrals in polar coordinates** in contrast to the double integrals studied in the preceding two sections, which are called **double integrals in rectangular coordinates**. Sometimes a double integral that is difficult to evaluate in rectangular coordinates can be evaluated more easily by making the substitution $x = r \cos \theta$, $y = r \sin \theta$ to convert it to an integral in polar coordinates. Under such a substitution, the rectangular and polar integrals are related by the equation

$$\iint\limits_{R} f(x, y) \, dA = \iint\limits_{\substack{\text{appropriate} \\ \text{limits}}} f(r, \theta) r \, dr \, d\theta \tag{8}$$

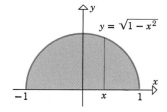

y

$y = \sqrt{1 - x^2}$

-1 x 1 x

Figure 17.3.10

Example 4   Use polar coordinates to evaluate $\int_{-1}^{1} \int_0^{\sqrt{1-x^2}} (x^2 + y^2)^{3/2} \, dy \, dx$.

Solution.   The first step is to express this iterated integral as a double integral in rectangular coordinates. For fixed x, the y-integration runs from the lower boundary $y = 0$ up to the semicircle $y = \sqrt{1 - x^2}$. The fixed x can extend from -1 on the left to $+1$ on the right, so the region of integration R for the double integral is as shown in Figure 17.3.10. In polar coordinates, this is the region swept out as r varies between 0 and 1 and θ varies between 0 and π. Thus,

$$\int_{-1}^{1} \int_{0}^{\sqrt{1-x^2}} (x^2 + y^2)^{3/2} \, dy \, dx = \iint\limits_{R} (x^2 + y^2)^{3/2} \, dA$$

$$= \int_{0}^{\pi} \int_{0}^{1} (r^3)r \, dr \, d\theta = \int_{0}^{\pi} \frac{1}{5} \, d\theta = \frac{\pi}{5} \quad \blacktriangleleft$$

▶ Exercise Set 17.3

In Exercises 1–6, evaluate the iterated integral.

1. $\displaystyle\int_{0}^{\pi/2} \int_{0}^{\sin\theta} r \cos\theta \, dr \, d\theta.$ **2.** $\displaystyle\int_{0}^{\pi} \int_{0}^{1+\cos\theta} r \, dr \, d\theta.$

3. $\displaystyle\int_{-\pi/2}^{\pi/2} \int_{0}^{a \sin\theta} r^2 \, dr \, d\theta.$ **4.** $\displaystyle\int_{0}^{\pi/3} \int_{0}^{\cos 3\theta} r \, dr \, d\theta.$

5. $\displaystyle\int_{0}^{\pi} \int_{0}^{1-\sin\theta} r^2 \cos\theta \, dr \, d\theta.$

6. $\displaystyle\int_{0}^{\pi} \int_{0}^{\cos\theta} r^3 \, dr \, d\theta.$

In Exercises 7–12, use a double integral in polar coordinates to find the area of the region described.

7. The region enclosed by the cardioid $r = 1 - \cos\theta$.

8. The region enclosed by the rose $r = \sin 2\theta$.

9. The region in the first quadrant bounded by $r = 1$ and $r = \sin 2\theta$, with $\pi/4 \le \theta \le \pi/2$.

10. The region inside the circle $x^2 + y^2 = 4$ and to the right of the line $x = 1$.

11. The region inside the circle $r = 4 \sin\theta$ and outside the circle $r = 2$.

12. The region inside the circle $r = 1$ and outside the cardioid $r = 1 + \cos\theta$.

In Exercises 13–18, use a double integral in polar coordinates to find the volume of the solid.

13. The solid enclosed by the sphere $x^2 + y^2 + z^2 = 9$ and the cylinder $x^2 + y^2 = 1$.

14. The solid enclosed by the sphere $r^2 + z^2 = 4$ and the cylinder $r = 2 \cos\theta$.

15. The solid that is bounded above by the cone $z = \sqrt{x^2 + y^2}$, below by the xy-plane, and laterally by the cylinder $x^2 + y^2 = 2y$.

16. The solid that is bounded above by the surface $z = (x^2 + y^2)^{-1/2}$, below by the xy-plane, and enclosed between the cylinders $x^2 + y^2 = 1$ and $x^2 + y^2 = 9$.

17. The solid that is bounded above by the paraboloid $z = 1 - x^2 - y^2$, below by the xy-plane, and laterally by the cylinder $x^2 + y^2 - x = 0$.

18. The solid in the first octant bounded above by the plane $z = r \sin\theta$, below by the xy-plane, and laterally by the plane $x = 0$ and the cylinder $r = 3 \sin\theta$.

In Exercises 19–22, use polar coordinates to evaluate the double integral.

19. $\displaystyle\iint\limits_{R} e^{-(x^2+y^2)} \, dA$, where R is the region enclosed by the circle $x^2 + y^2 = 1$.

20. $\displaystyle\iint\limits_{R} \sqrt{9 - x^2 - y^2} \, dA$, where R is the region in the first quadrant within the circle $x^2 + y^2 = 9$.

21. $\displaystyle\iint\limits_{R} \frac{1}{1 + x^2 + y^2} \, dA$, where R is the sector in the first quadrant bounded by $y = 0$, $y = x$, and $x^2 + y^2 = 4$.

22. $\displaystyle\iint\limits_{R} 2y \, dA$, where R is the region in the first quadrant bounded above by the circle $(x - 1)^2 + y^2 = 1$ and below by the line $y = x$.

In Exercises 23–30, evaluate the iterated integral by converting to polar coordinates.

23. $\displaystyle\int_{0}^{1} \int_{0}^{\sqrt{1-x^2}} (x^2 + y^2) \, dy \, dx.$

24. $\displaystyle\int_{-2}^{2} \int_{-\sqrt{4-y^2}}^{\sqrt{4-y^2}} e^{-(x^2+y^2)} \, dx \, dy.$

25. $\displaystyle\int_{0}^{2} \int_{0}^{\sqrt{2x-x^2}} \sqrt{x^2 + y^2} \, dy \, dx.$

26. $\displaystyle\int_{0}^{1} \int_{0}^{\sqrt{1-y^2}} \cos(x^2 + y^2) \, dx \, dy.$

27. $\displaystyle\int_{0}^{a} \int_{0}^{\sqrt{a^2-x^2}} \frac{dy \, dx}{(1 + x^2 + y^2)^{3/2}} \quad (a > 0).$

28. $\displaystyle\int_{0}^{1} \int_{y}^{\sqrt{y}} \sqrt{x^2 + y^2} \, dx \, dy.$

29. $\int_0^{\sqrt{2}} \int_y^{\sqrt{4-y^2}} \frac{1}{\sqrt{1+x^2+y^2}} \, dx \, dy.$

30. $\int_0^4 \int_3^{\sqrt{25-x^2}} dy \, dx.$

31. Use polar coordinates to find the volume of the solid that is bounded above by the ellipsoid

$$x^2/a^2 + y^2/a^2 + z^2/c^2 = 1$$

below by the *xy*-plane, and laterally by the cylinder $x^2 + y^2 - ay = 0$.

32. Find the area of the region enclosed by the lemniscate $r^2 = 2a^2 \cos 2\theta$.

33. Find the area in the first quadrant inside the circle $r = 4 \sin \theta$ and outside the lemniscate $r^2 = 8 \cos 2\theta$.

34. Show that the shaded area in the following figure is $a^2\phi - \frac{1}{2}a^2 \sin 2\phi$.

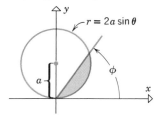

35. The integral $\int_0^{+\infty} e^{-x^2} \, dx$, which arises in probability theory, can be evaluated using the following method. Let the value of the integral be *I*. Thus,

$$I = \int_0^{+\infty} e^{-x^2} \, dx = \int_0^{+\infty} e^{-y^2} \, dy$$

since the letter used for the variable of integration in a definite integral does not matter.

(a) Show that

$$I^2 = \int_0^{+\infty} \int_0^{+\infty} e^{-(x^2+y^2)} \, dx \, dy$$

(b) Evaluate the iterated integral in part (a) by converting to polar coordinates.

(c) Use the result in part (b) to find *I*.

36. In each part evaluate $\iint_R x^2 \, dA.$

(a)

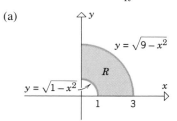

(b)

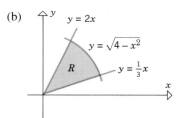

■ 17.4 SURFACE AREA

In Section 6.5 we showed how to find the surface area of a surface of revolution. In this section we shall consider more general surface area problems.

☐ **THE SURFACE AREA FORMULA**

The first surface area problem of concern in this section can be posed as follows:

17.4.1 PROBLEM. *Let f be a function defined on a closed region R of the xy-plane. Find the surface area of that portion of the surface z = f(x, y) whose projection on the xy-plane is the region R (Figure 17.4.1).*

We will attack this problem in stages, first solving it in the special case where the surface is a plane and the region *R* is a rectangle, then using the special case to solve the general problem.

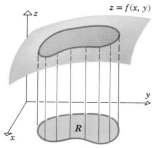

Figure 17.4.1

17.4.2 THEOREM. *Let R be a closed rectangular region in the xy-plane with sides parallel to the coordinate axes. If the lengths of the sides are l and w, then the surface area S of that portion of the plane z = ax + by + c that projects onto the region R is given by*

$$S = \sqrt{a^2 + b^2 + 1} \, lw \qquad (1)$$

Proof. Suppose that the four corners of R are

$$A(x_0, y_0), \quad B(x_0 + l, y_0), \quad C(x_0, y_0 + w), \quad D(x_0 + l, y_0 + w)$$

from which it follows that the points on the plane $z = ax + by + c$ that project onto A, B, and C are

$$A'(x_0, y_0, ax_0 + by_0 + c)$$
$$B'(x_0 + l, y_0, a(x_0 + l) + by_0 + c)$$
$$C'(x_0, y_0 + w, ax_0 + b(y_0 + w) + c)$$

(Figure 17.4.2).

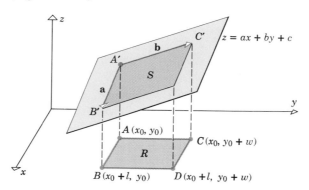

Figure 17.4.2

These points determine a parallelogram with sides

$$\mathbf{a} = \overrightarrow{A'B'} = l\mathbf{i} + 0\mathbf{j} + al\mathbf{k}$$
$$\mathbf{b} = \overrightarrow{A'C'} = 0\mathbf{i} + w\mathbf{j} + bw\mathbf{k}$$

which constitutes the portion of the plane that projects onto the region R. The cross-product of these vectors is

$$\mathbf{a} \times \mathbf{b} = \begin{vmatrix} \mathbf{i} & \mathbf{j} & \mathbf{k} \\ l & 0 & al \\ 0 & w & bw \end{vmatrix} = -alw\mathbf{i} - lbw\mathbf{j} + lw\mathbf{k}$$

and hence from Formula (7) of Section 14.4 the area S of the parallelogram determined by **a** and **b** is

$$S = \|\mathbf{a} \times \mathbf{b}\| = \sqrt{(-alw)^2 + (-lbw)^2 + (lw)^2} = \sqrt{a^2 + b^2 + 1} \, lw \qquad ∎$$

We shall now show how Theorem 17.4.2 can be used to derive the following more general result.

17.4.3 SURFACE AREA FORMULA. *If f has continuous first partial derivatives on a closed region R of the xy-plane, then the area S of that portion of the surface z = f(x, y) that projects onto R is*

$$S = \iint\limits_{R} \sqrt{\left(\frac{\partial z}{\partial x}\right)^2 + \left(\frac{\partial z}{\partial y}\right)^2 + 1} \, dA \qquad (2)$$

To motivate Formula (2) we shall use a now familiar limiting process:

- As shown in Figure 17.4.3, enclose R within a rectangle whose sides are parallel to the x- and y-axes, and subdivide that rectangle into smaller rectangles by lines parallel to these axes. Exclude from consideration all rectangles that contain points outside of R, leaving only rectangles that are subsets of R. Assume that there are n such rectangles, and denote the kth rectangle by R_k. Suppose that the sides of R_k have dimensions Δx_k and Δy_k, and let ΔS_k be the area of the patch on the surface $z = f(x, y)$ determined by projecting R_k onto this surface.

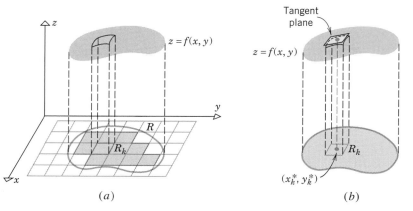

Figure 17.4.3   (a)   (b)

- If we let (x_k^*, y_k^*) be any point in R_k, then the area ΔS_k can be approximated by the area of that portion of the tangent plane to the surface $z = f(x, y)$ at (x_k^*, y_k^*) that projects onto R_k. It follows from Theorem 16.5.1 that this tangent plane has an equation of the form

$$z = f_x(x_k^*, y_k^*)x + f_y(x_k^*, y_k^*)y + c$$

for an appropriate constant c, and hence from Theorem 17.4.2

$$\Delta S_k \approx \sqrt{f_x(x_k^*, y_k^*)^2 + f_y(x_k^*, y_k^*)^2 + 1} \,\Delta x_k \,\Delta y_k \tag{3}$$

Thus, if we let $\Delta A_k = \Delta x_k \,\Delta y_k$ and add the areas of the patches, we obtain the following approximation to the entire area S:

$$S \approx \sum_{k=1}^{n} \sqrt{f_x(x_k^*, y_k^*)^2 + f_y(x_k^*, y_k^*)^2 + 1} \,\Delta A_k$$

- There are two sources of error in this approximation, one resulting from the fact that rectangles $R_1, R_2, \ldots, R_n$ do not fill up the region R completely, and the other resulting from the fact that the patches of area on the surface have been approximated by areas on tangent planes. However, it seems plausible that if we repeat the subdivision process using more and more rectangles with dimensions approaching zero, then the errors will diminish and the exact surface area S will be given by

$$S = \lim_{n \to +\infty} \sum_{k=1}^{n} \sqrt{f_x(x_k^*, y_k^*)^2 + f_y(x_k^*, y_k^*)^2 + 1} \,\Delta A_k$$

$$= \iint_R \sqrt{f_x(x, y)^2 + f_y(x, y)^2 + 1} \,dA$$

which is just Formula (2) with a different notation for the partial derivatives.

Example 1   Find the surface area of the portion of the cylinder $x^2 + z^2 = 4$ above the rectangle $R = \{(x, y) : 0 \le x \le 1, 0 \le y \le 4\}$ in the xy-plane.

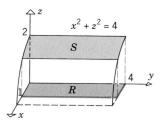

Figure 17.4.4

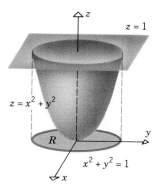

Figure 17.4.5

Solution. The surface is shown in Figure 17.4.4. The portion of the cylinder $x^2 + z^2 = 4$ that lies above the *xy*-plane has the equation $z = \sqrt{4 - x^2}$. Thus, from (2)

$$S = \iint_R \sqrt{\left(\frac{\partial z}{\partial x}\right)^2 + \left(\frac{\partial z}{\partial y}\right)^2 + 1}\; dA$$

$$= \iint_R \sqrt{\left(-\frac{x}{\sqrt{4 - x^2}}\right)^2 + 0 + 1}\; dA = \int_0^4 \int_0^1 \frac{2}{\sqrt{4 - x^2}}\, dx\, dy$$

$$= 2\int_0^4 \left[\sin^{-1}\left(\frac{1}{2}x\right)\right]_{x=0}^1 dy = 2\int_0^4 \frac{\pi}{6}\, dy = \frac{4}{3}\pi \qquad \blacktriangleleft$$

> Formula (21)
> of Section 8.2

Example 2 Find the surface area of the portion of the paraboloid $z = x^2 + y^2$ below the plane $z = 1$.

Solution. The surface is shown in Figure 17.4.5. From the equations $z = 1$ and $z = x^2 + y^2$, we see that the plane and the paraboloid intersect in a circle whose projection on the *xy*-plane has the equation $x^2 + y^2 = 1$. Consequently, the surface whose area we seek projects onto the region R enclosed by this circle. Since the surface has the equation

$$z = x^2 + y^2$$

it follows from (2) that

$$S = \iint_R \sqrt{4x^2 + 4y^2 + 1}\; dA$$

This integral is best evaluated in polar coordinates. Replacing $x^2 + y^2$ by r^2 and substituting $r\, dr\, d\theta$ for dA we obtain

$$S = \int_0^{2\pi} \int_0^1 \sqrt{4r^2 + 1}\; r\, dr\, d\theta = \int_0^{2\pi} \left[\frac{1}{12}(4r^2 + 1)^{3/2}\right]_{r=0}^1 d\theta$$

$$= \int_0^{2\pi} \frac{1}{12}(5\sqrt{5} - 1)\, d\theta = \frac{1}{6}\pi(5\sqrt{5} - 1) \qquad \blacktriangleleft$$

□ **SURFACE AREA OF PARAMETRIC SURFACES**

We now consider the problem of finding the surface area S of a surface σ that is represented parametrically as

$$\mathbf{r}(u, v) = x(u, v)\mathbf{i} + y(u, v)\mathbf{j} + z(u, v)\mathbf{k}$$

where (u, v) varies over some region R of the *uv*-plane. We shall say that $\mathbf{r}$ is a *smooth function* of u and v or that σ is a *smooth parametric surface* if $\partial\mathbf{r}/\partial u$ and $\partial\mathbf{r}/\partial v$ are continuous and $\partial\mathbf{r}/\partial u \times \partial\mathbf{r}/\partial v \neq \mathbf{0}$ on R. Geometrically, this means that σ has a unit normal (and hence a tangent plane) at each point of σ and that the unit normal $\mathbf{n} = \mathbf{n}(u, v)$ is continuous on R [see (6) of Section 16.3]. Thus, on a smooth parametric surface the unit normal varies continuously with no abrupt changes in direction. We shall derive the formula for the surface area of a smooth parametric surface that has no self-intersections.

We begin by subdividing R into rectangular regions by lines parallel to the u and v axes, discarding any nonrectangular portions that contain points of the boundary. Assume that there are n rectangles, and let R_k denote the *k*th rectangle. Let (u_k, v_k) be the lower left corner of R_k, and assume that R_k has area $\Delta A_k = \Delta u_k\, \Delta v_k$, where Δu_k and Δv_k are the dimensions of R_k (Figure 17.4.6a). The image of R_k will be some *curvilinear patch* σ_k on the surface σ that has a corner at $\mathbf{r}(u_k, v_k)$; denote the area of this patch by ΔS_k (Figure 17.4.6b).

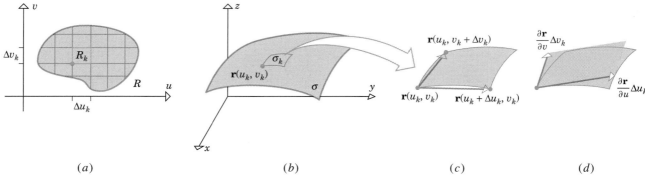

Figure 17.4.6

As suggested by Figure 17.4.6c, the two edges of the patch that meet at $\mathbf{r}(u_k, v_k)$ can be approximated by the "secant" vectors

$$\mathbf{r}(u_k + \Delta u_k, v_k) - \mathbf{r}(u_k, v_k)$$

$$\mathbf{r}(u_k, v_k + \Delta v_k) - \mathbf{r}(u_k, v_k)$$

and hence the area of σ_k can be approximated by the area of the parallelogram determined by these vectors. However, it follows from Formulas (4) and (5) of Section 16.3 that if Δu_k and Δv_k are small, then these secant vectors can in turn be approximated by the tangent vectors

$$\frac{\partial \mathbf{r}}{\partial u} \Delta u_k \quad \text{and} \quad \frac{\partial \mathbf{r}}{\partial v} \Delta v_k$$

where the partial derivatives are evaluated at (u_k, v_k). Thus, the area of the patch σ_k can be approximated by the area of the parallelogram determined by these vectors (Figure 17.4.6d); that is,

$$\Delta S_k \approx \left\| \frac{\partial \mathbf{r}}{\partial u} \Delta u_k \times \frac{\partial \mathbf{r}}{\partial v} \Delta v_k \right\| = \left\| \frac{\partial \mathbf{r}}{\partial u} \times \frac{\partial \mathbf{r}}{\partial v} \right\| \Delta u_k \Delta v_k = \left\| \frac{\partial \mathbf{r}}{\partial u} \times \frac{\partial \mathbf{r}}{\partial v} \right\| \Delta A_k \qquad (4)$$

It follows that the surface area S of the entire surface σ can be approximated as

$$S \approx \sum_{k=1}^{n} \left\| \frac{\partial \mathbf{r}}{\partial u} \times \frac{\partial \mathbf{r}}{\partial v} \right\| \Delta A_k$$

Thus, if we assume that the errors in the approximations approach zero as n increases in such a way that the dimensions of the rectangles approach zero, then it is plausible that the exact value of S is

$$S = \lim_{n \to +\infty} \sum_{k=1}^{n} \left\| \frac{\partial \mathbf{r}}{\partial u} \times \frac{\partial \mathbf{r}}{\partial v} \right\| \Delta A_k$$

or equivalently,

$$S = \iint\limits_{R} \left\| \frac{\partial \mathbf{r}}{\partial u} \times \frac{\partial \mathbf{r}}{\partial v} \right\| dA \qquad (5)$$

REMARK.   If the surface σ is given by $z = f(x, y)$, then with x and y as parameters, the surface can be represented parametrically by

$$\mathbf{r}(x, y) = x\mathbf{i} + y\mathbf{j} + f(x, y)\mathbf{k}$$

We leave it for the reader to use this parametrization to obtain Formula (2) as a special case of Formula (5).

Example 3   The surface represented by

$$\mathbf{r}(u, v) = u \cos v\mathbf{i} + u \sin v\mathbf{j} + 3u\mathbf{k}$$

where $0 \le u \le 2$ and $0 \le v \le 2\pi$ is a portion of a cone (verify). Find its surface area.

Solution.

$$\frac{\partial \mathbf{r}}{\partial u} = \cos v\mathbf{i} + \sin v\mathbf{j} + 3\mathbf{k} \quad \text{and} \quad \frac{\partial \mathbf{r}}{\partial v} = -u \sin v\mathbf{i} + u \cos v\mathbf{j}$$

$$\frac{\partial \mathbf{r}}{\partial u} \times \frac{\partial \mathbf{r}}{\partial v} = \begin{vmatrix} \mathbf{i} & \mathbf{j} & \mathbf{k} \\ \cos v & \sin v & 3 \\ -u \sin v & u \cos v & 0 \end{vmatrix} = -3u \cos v\mathbf{i} - 3u \sin v\mathbf{j} + u\mathbf{k}$$

Thus, from (5)

$$S = \iint_R \left\| \frac{\partial \mathbf{r}}{\partial u} \times \frac{\partial \mathbf{r}}{\partial v} \right\| dA = \int_0^{2\pi} \int_0^2 \sqrt{10}\,u\,du\,dv = 2\sqrt{10} \int_0^{2\pi} dv = 4\sqrt{10}\,\pi \quad \blacktriangleleft$$

▶ Exercise Set 17.4

1. Find the surface area of the portion of the cylinder $y^2 + z^2 = 9$ that is above the rectangle $R = \{(x, y):0 \le x \le 2, -3 \le y \le 3\}$.

2. By integration, find the surface area of the portion of the plane $2x + 2y + z = 8$ in the first octant that is cut off by the three coordinate planes.

3. Find the surface area of the portion of the cone $z^2 = 4x^2 + 4y^2$ that is above the region in the first quadrant bounded by the line $y = x$ and the parabola $y = x^2$.

4. Find the surface area of the portion of the cone $z = \sqrt{x^2 + y^2}$ that lies inside the cylinder $x^2 + y^2 = 2x$.

5. Find the surface area of the portion of the paraboloid $z = 1 - x^2 - y^2$ that is above the xy-plane.

6. Find the area of the portion of the surface $z = 2x + y^2$ that is above the triangular region with vertices $(0, 0)$, $(0, 1)$, and $(1, 1)$.

7. Find the area of the portion of the surface $z = xy$ that is above the sector in the first quadrant bounded by the lines $y = x/\sqrt{3}$, $y = 0$, and the circle $x^2 + y^2 = 9$.

8. Find the surface area of the portion of the paraboloid $2z = x^2 + y^2$ that is inside the cylinder $x^2 + y^2 = 8$.

9. Find the surface area of the portion of the sphere $x^2 + y^2 + z^2 = 16$ between the planes $z = 1$ and $z = 2$.

10. Find the surface area of the portion of the sphere $x^2 + y^2 + z^2 = 8$ that is cut out by the cone $z = \sqrt{x^2 + y^2}$.

11. Find the total surface area of the portion of the sphere $x^2 + y^2 + z^2 = a^2$ inside the cylinder $x^2 + y^2 = ay$.

12. Use a double integral in Cartesian coordinates to derive the formula for the surface area of a sphere of radius a.

13. Find the surface area of the portion of the cylinder $x^2 + z^2 = 16$ that lies inside the circular cylinder $x^2 + y^2 = 16$. [*Hint:* Find the area in the first octant and use symmetry.]

14. Find the surface area of the portion of the cylinder $x^2 + z^2 = 5x$ inside the sphere $x^2 + y^2 + z^2 = 25$.

15. The portion of the surface

$$z = \frac{h}{a}\sqrt{x^2 + y^2} \quad (a, h > 0)$$

between the xy-plane and the plane $z = h$ is a right-circular cone of height h and radius a. Use a double integral to show that the lateral surface area of this cone is $S = \pi a \sqrt{a^2 + h^2}$.

16. Find the surface area of the portion of the cone $z = \sqrt{x^2 + y^2}$ that lies between the planes $z = 0$ and $z = y/2 + 3$.

17. Find the surface area of the portion of the paraboloid

$$\mathbf{r}(u, v) = u \cos v\mathbf{i} + u \sin v\mathbf{j} + u^2\mathbf{k}$$

for which $1 \le u \le 2$, $0 \le v \le 2\pi$.

18. Find the surface area of the portion of the cone

$$\mathbf{r}(u, v) = u \cos v\mathbf{i} + u \sin v\mathbf{j} + u\mathbf{k}$$

for which $0 \le u \le 2v$, $0 \le v \le \pi/2$.

19. Find the surface area of the sphere of radius a

$$\mathbf{r}(u, v) = a \sin u \cos v\mathbf{i} + a \sin u \sin v\mathbf{j} + a \cos u\mathbf{k}$$

for which $0 \le u \le \pi$, $0 \le v \le 2\pi$.

20. Find the surface area of the portion of the hyperbolic paraboloid

$$\mathbf{r}(u, v) = (u + v)\mathbf{i} + (u - v)\mathbf{j} + uv\mathbf{k}$$

for which $u^2 + v^2 \leq 4$.

21. Find the surface area of the portion of the spiral ramp

$$\mathbf{r}(u, v) = u \cos v\mathbf{i} + u \sin v\mathbf{j} + v\mathbf{k}$$

for which $0 \leq u \leq 2$, $0 \leq v \leq 3u$.

22. The torus in Figure 16.1.28 can be represented by

$$\mathbf{r}(u, v) = (a + b \cos v) \cos u\mathbf{i} + (a + b \cos v) \sin u\mathbf{j} + b \sin v\mathbf{k}$$

where

$$0 < b \leq a \quad \text{and} \quad 0 \leq u \leq 2\pi, 0 \leq v \leq 2\pi$$

Find the surface area.

■ **17.5** TRIPLE INTEGRALS

In the preceding sections we defined and discussed properties of double integrals for functions of two variables. In this section we shall define triple integrals for functions of three variables.

☐ **DEFINITION OF A TRIPLE INTEGRAL**

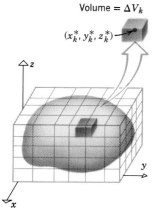

Volume = ΔV_k

(x_k^*, y_k^*, z_k^*)

Figure 17.5.1

Whereas a double integral of a function $f(x, y)$ is defined over a closed region R in the xy-plane, a triple integral of a function $f(x, y, z)$ is defined over a closed three-dimensional solid region G. To ensure that G does not extend indefinitely in any direction, we shall assume that G can be enclosed within some suitably large box (rectangular parallelepiped) with sides parallel to the coordinate planes (Figure 17.5.1).

To define the triple integral of $f(x, y, z)$ over G, we first divide the box into n "subboxes" by planes parallel to the coordinate planes. We then discard those subboxes that contain any points outside of G and choose an arbitrary point in each of the remaining subboxes. As shown in Figure 17.5.1, we denote the volume of the kth remaining subbox by ΔV_k and the point selected in the kth subbox by (x_k^*, y_k^*, z_k^*). Next, we form the product

$$f(x_k^*, y_k^*, z_k^*) \, \Delta V_k$$

for each subbox, then add the products for all of the subboxes to obtain the ***Riemann sum***

$$\sum_{k=1}^{n} f(x_k^*, y_k^*, z_k^*) \, \Delta V_k$$

Finally, we repeat this process with more and more subdivisions in such a way that the length, width, and height of each subbox approaches zero, and n approaches $+\infty$. The limit

$$\iiint_G f(x, y, z) \, dV = \lim_{n \to +\infty} \sum_{k=1}^{n} f(x_k^*, y_k^*, z_k^*) \, \Delta V_k \tag{1}$$

is called the ***triple integral*** of $f(x, y, z)$ over the region G. Conditions under which the triple integral exists are studied in advanced calculus. However, for our purposes it suffices to say that existence is ensured when f is continuous on G and the region G is not too "complicated."

☐ **PROPERTIES OF TRIPLE INTEGRALS**

Triple integrals enjoy many properties of single and double integrals:

$$\iiint_G cf(x, y, z)\, dV = c \iiint_G f(x, y, z)\, dV \quad (c \text{ a constant})$$

$$\iiint_G [f(x, y, z) + g(x, y, z)]\, dV = \iiint_G f(x, y, z)\, dV + \iiint_G g(x, y, z)\, dV$$

$$\iiint_G [f(x, y, z) - g(x, y, z)]\, dV = \iiint_G f(x, y, z)\, dV - \iiint_G g(x, y, z)\, dV$$

Moreover, if the region G is subdivided into two subregions G_1 and G_2 (Figure 17.5.2), then

$$\iiint_G f(x, y, z)\, dV = \iiint_{G_1} f(x, y, z)\, dV + \iiint_{G_2} f(x, y, z)\, dV$$

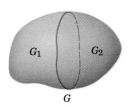

Figure 17.5.2

We omit the proofs.

☐ **EVALUATING TRIPLE INTEGRALS OVER RECTANGULAR BOXES**

Just as a double integral can be evaluated by two successive single integrations, so a triple integral can be evaluated by three successive integrations. The following theorem, which we state without proof, is the analog of Theorem 17.1.2.

17.5.1 THEOREM. *Let G be the rectangular box defined by the inequalities*

$$a \le x \le b, \quad c \le y \le d, \quad k \le z \le l$$

If f is continuous on the region G, then

$$\iiint_G f(x, y, z)\, dV = \int_a^b \int_c^d \int_k^l f(x, y, z)\, dz\, dy\, dx \tag{2}$$

Moreover, the iterated integral on the right can be replaced with any of the five other iterated integrals that result by altering the order of integration.

Example 1   Evaluate the triple integral

$$\iiint_G 12xy^2z^3\, dV$$

over the rectangular box G defined by the inequalities $-1 \le x \le 2,\ 0 \le y \le 3,\ 0 \le z \le 2$.

Solution.   Of the six possible iterated integrals we might use, we shall choose the one in (2). Thus, we shall first integrate with respect to z, holding x and y fixed, then with respect to y, holding x fixed, and finally with respect to x.

$$\iiint_G 12xy^2z^3\, dV = \int_{-1}^2 \int_0^3 \int_0^2 12xy^2z^3\, dz\, dy\, dx$$

$$= \int_{-1}^2 \int_0^3 \left[3xy^2z^4 \right]_{z=0}^2 dy\, dx = \int_{-1}^2 \int_0^3 48xy^2\, dy\, dx$$

$$= \int_{-1}^2 \left[16xy^3 \right]_{y=0}^3 dx = \int_{-1}^2 432x\, dx$$

$$= 216x^2 \Big]_{-1}^2 = 648 \quad \blacktriangleleft$$

☐ **EVALUATING TRIPLE INTEGRALS OVER MORE GENERAL REGIONS**

We shall also be concerned with evaluating triple integrals over solid regions other than rectangular boxes. For simplicity, we shall restrict our discussion to solid regions constructed as follows. Let R be a closed region in the xy-plane and let $g_1(x, y)$ and $g_2(x, y)$ be continuous functions such that $g_1(x, y) \leq g_2(x, y)$ for all (x, y) in R. Geometrically, this condition states that the surface $z = g_2(x, y)$ does not dip below the surface $z = g_1(x, y)$ over R (Figure 17.5.3a). We shall call $z = g_1(x, y)$ the **lower surface** and $z = g_2(x, y)$ the **upper surface**. Let G be the solid consisting of all points above or below the region R that lie between the upper surface and the lower surface (Figure 17.5.3b). A solid G constructed in this way will be called a **simple solid**, and the region R will be called the **projection** of G on the xy-plane.

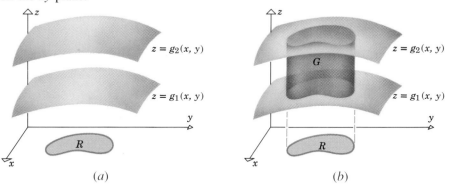

Figure 17.5.3 (a) (b)

The following theorem, which we state without proof, will enable us to evaluate triple integrals over simple solids.

17.5.2   THEOREM.   *Let G be a simple solid with upper surface $z = g_2(x, y)$ and lower surface $z = g_1(x, y)$, and let R be the projection of G on the xy-plane. If $f(x, y, z)$ is continuous on G, then*

$$\iiint_G f(x, y, z)\, dV = \iint_R \left[\int_{g_1(x, y)}^{g_2(x, y)} f(x, y, z)\, dz \right] dA \tag{3}$$

In (3), the first integration is with respect to z, after which a function of x and y remains. This function of x and y is then integrated over the region R in the xy-plane. To apply (3), it is usual to begin with a three-dimensional sketch of the solid G, from which the limits of integration can be obtained as follows:

Step 1. Find an equation $z = g_2(x, y)$ for the upper surface and an equation $z = g_1(x, y)$ for the lower surface of G. The functions $g_1(x, y)$ and $g_2(x, y)$ determine the lower and upper z-limits of integration.

Step 2. Make a two-dimensional sketch of the projection R of the solid on the xy-plane. From this sketch determine the limits of integration for the double integral over R in (3).

Example 2 Let G be the wedge in the first octant cut from the cylindrical solid $y^2 + z^2 \leq 1$ by the planes $y = x$ and $x = 0$. Evaluate

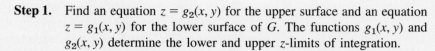

Solution. The solid G and its projection R on the xy-plane are shown in Figure 17.5.4. The upper surface of the solid is formed by the cylinder and the lower surface by the xy-plane.

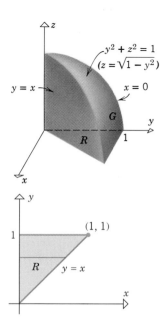

Figure 17.5.4

Since the portion of the cylinder $y^2 + z^2 = 1$ that lies above the xy-plane has the equation $z = \sqrt{1 - y^2}$, and the xy-plane has the equation $z = 0$, it follows from (3) that

$$\iiint_G z \, dV = \iint_R \left[\int_0^{\sqrt{1-y^2}} z \, dz \right] dA \tag{4}$$

For the double integral over R, the x and y integrations can be performed in either order, since R is both a type I and type II region. We shall integrate with respect to x first. With this choice, (4) yields

$$\iiint_G z \, dV = \int_0^1 \int_0^y \int_0^{\sqrt{1-y^2}} z \, dz \, dx \, dy = \int_0^1 \int_0^y \frac{1}{2} z^2 \bigg]_{z=0}^{\sqrt{1-y^2}} dx \, dy$$

$$= \int_0^1 \int_0^y \frac{1}{2}(1 - y^2) \, dx \, dy = \frac{1}{2} \int_0^1 (1 - y^2)x \bigg]_{x=0}^y dy$$

$$= \frac{1}{2} \int_0^1 (y - y^3) \, dy = \frac{1}{2} \left[\frac{1}{2} y^2 - \frac{1}{4} y^4 \right]_0^1 = \frac{1}{8} \quad \blacktriangleleft$$

□ VOLUME CALCULATED
AS A TRIPLE INTEGRAL

Triple integrals have a number of physical interpretations, some of which we shall consider in the next section. However, in the special case where $f(x, y, z) = 1$, Formula (1) yields

$$\iiint_G dV = \lim_{n \to +\infty} \sum_{k=1}^n \Delta V_k$$

which Figure 17.5.1 suggests is the volume of G; that is,

$$\text{volume of } G = \iiint_G dV \tag{5}$$

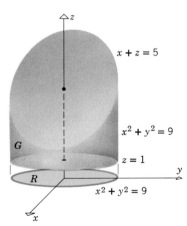

Example 3 Use a triple integral to find the volume of the solid enclosed between the cylinder $x^2 + y^2 = 9$ and the planes $z = 1$ and $x + z = 5$.

Solution. The solid G and its projection R on the xy-plane are shown in Figure 17.5.5. The lower surface of the solid is the plane $z = 1$, and the upper surface is the plane $x + z = 5$, or equivalently, $z = 5 - x$. Thus, from (3) and (5)

$$\text{volume of } G = \iiint_G dV = \iint_R \left[\int_1^{5-x} dz \right] dA \tag{6}$$

For the double integral over R, we shall integrate with respect to y first. Thus, (6) yields

$$\text{volume of } G = \int_{-3}^3 \int_{-\sqrt{9-x^2}}^{\sqrt{9-x^2}} \int_1^{5-x} dz \, dy \, dx = \int_{-3}^3 \int_{-\sqrt{9-x^2}}^{\sqrt{9-x^2}} z \bigg]_{z=1}^{5-x} dy \, dx$$

$$= \int_{-3}^3 \int_{-\sqrt{9-x^2}}^{\sqrt{9-x^2}} (4 - x) \, dy \, dx = \int_{-3}^3 (8 - 2x)\sqrt{9 - x^2} \, dx$$

$$= 8 \int_{-3}^3 \sqrt{9 - x^2} \, dx - \int_{-3}^3 2x\sqrt{9 - x^2} \, dx \quad \boxed{\text{For the first integral, see Example 4 of Section 9.5.}}$$

$$= 8 \left(\frac{9}{2} \pi \right) - \int_{-3}^3 2x\sqrt{9 - x^2} \, dx \quad \boxed{\text{Let } u = 9 - x^2, \text{ or better, apply the result in Exercise 48(a) of Section 5.8.}}$$

$$= 8 \left(\frac{9}{2} \pi \right) - 0 = 36\pi \quad \blacktriangleleft$$

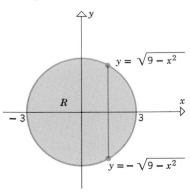

Figure 17.5.5

$z = 5x^2 + 5y^2$

$z = 6 - 7x^2 - y^2$

R $2x^2 + y^2 = 1$

$y = \sqrt{1 - 2x^2}$

R

$-1/\sqrt{2}$ $1/\sqrt{2}$

$y = -\sqrt{1 - 2x^2}$

Figure 17.5.7

Example 4   Find the volume of the solid enclosed by the paraboloids

$$z = 5x^2 + 5y^2 \quad \text{and} \quad z = 6 - 7x^2 - y^2$$

Solution.   The solid G and its projection R on the xy-plane are shown in Figure 17.5.6. The projection R is obtained by solving the given equations simultaneously to determine where the paraboloids intersect. We obtain

$$5x^2 + 5y^2 = 6 - 7x^2 - y^2$$

or

$$2x^2 + y^2 = 1 \tag{7}$$

which tells us that the paraboloids intersect in a curve on the elliptic cylinder given by (7). The projection of this intersection on the xy-plane is an ellipse with this same equation. Therefore,

$$\text{volume of } G = \iiint\limits_{G} dV = \iint\limits_{R} \left[\int_{5x^2+5y^2}^{6-7x^2-y^2} dz \right] dA$$

$$\text{volume of } G = \int_{-1/\sqrt{2}}^{1/\sqrt{2}} \int_{-\sqrt{1-2x^2}}^{\sqrt{1-2x^2}} \int_{5x^2+5y^2}^{6-7x^2-y^2} dz \, dy \, dx$$

$$= \int_{-1/\sqrt{2}}^{1/\sqrt{2}} \int_{-\sqrt{1-2x^2}}^{\sqrt{1-2x^2}} (6 - 12x^2 - 6y^2) \, dy \, dx$$

$$= \int_{-1/\sqrt{2}}^{1/\sqrt{2}} \left[6(1 - 2x^2)y - 2y^3 \right]_{y=-\sqrt{1-2x^2}}^{\sqrt{1-2x^2}} dx$$

$$= 8 \int_{-1/\sqrt{2}}^{1/\sqrt{2}} (1 - 2x^2)^{3/2} \, dx = \frac{8}{\sqrt{2}} \int_{-\pi/2}^{\pi/2} \cos^4 \theta \, d\theta = \frac{3\pi}{\sqrt{2}} \quad \blacktriangleleft$$

Let $x = \dfrac{1}{\sqrt{2}} \sin \theta$.

Exercise 36, Section 9.3

□ **INTEGRATION IN OTHER ORDERS**

For certain regions, triple integrals are best evaluated by integrating first with respect to x or y rather than z. For example, if the solid G is bounded on the left and right by the surfaces $y = g_1(x, z)$ and $y = g_2(x, z)$ and bounded laterally by a cylinder extending in the y-direction (Figure 17.5.7a), then

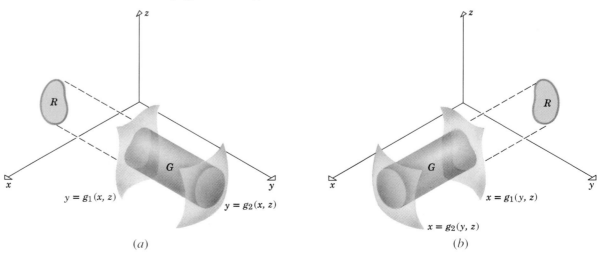

Figure 17.5.7

R

G

$y = g_1(x, z)$

$y = g_2(x, z)$

(a)

R

G

$x = g_1(y, z)$

$x = g_2(y, z)$

(b)

$$\iiint_G f(x, y, z)\, dV = \iint_R \left[\int_{g_1(x,z)}^{g_2(x,z)} f(x, y, z)\, dy \right] dA$$

where R is the projection of the solid G on the xz-plane.

Similarly, if the solid G is bounded in the back and front by the surfaces $x = g_1(y, z)$ and $x = g_2(y, z)$ and bounded laterally by a cylinder extending in the x-direction, then

$$\iiint_G f(x, y, z)\, dV = \iint_R \left[\int_{g_1(y,z)}^{g_2(y,z)} f(x, y, z)\, dx \right] dA$$

where R is the projection of the solid on the yz-plane (Figure 17.5.7b).

Example 5 In Example 2, we evaluated

$$\iiint_G z\, dV$$

over the wedge in Figure 17.5.4 by integrating first with respect to z. Evaluate this integral by integrating first with respect to x.

Solution. The solid is bounded in the back by the plane $x = 0$ and in the front by the plane $x = y$, so

$$\iiint_G z\, dV = \iint_R \left[\int_0^y z\, dx \right] dA$$

where R is the projection of G on the yz-plane (Figure 17.5.8). The integration over R can be performed first with respect to z and then y or vice versa. Performing the z-integration first yields

$$\iiint_G z\, dV = \int_0^1 \int_0^{\sqrt{1-y^2}} \int_0^y z\, dx\, dz\, dy = \int_0^1 \int_0^{\sqrt{1-y^2}} zx \Big]_{x=0}^y dz\, dy$$

$$= \int_0^1 \int_0^{\sqrt{1-y^2}} zy\, dz\, dy = \int_0^1 \frac{1}{2} z^2 y \Big]_{z=0}^{\sqrt{1-y^2}} dy = \int_0^1 \frac{1}{2}(1 - y^2)y\, dy = \frac{1}{8}$$

which agrees with the result in Example 2. ◀

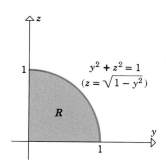

$y^2 + z^2 = 1$
$(z = \sqrt{1 - y^2})$

R

Figure 17.5.8

▶ Exercise Set 17.5

In Exercises 1–8, evaluate the iterated integral.

1. $\displaystyle\int_{-1}^1 \int_0^2 \int_0^1 (x^2 + y^2 + z^2)\, dx\, dy\, dz.$

2. $\displaystyle\int_{1/3}^{1/2} \int_0^\pi \int_0^1 zx \sin xy\, dz\, dy\, dx.$

3. $\displaystyle\int_0^2 \int_{-1}^{y^2} \int_1^z yz\, dx\, dz\, dy.$

4. $\displaystyle\int_0^{\pi/4} \int_0^1 \int_0^{x^2} x \cos y\, dz\, dx\, dy.$

5. $\displaystyle\int_0^3 \int_0^{\sqrt{9-z^2}} \int_0^x xy\, dy\, dx\, dz.$

6. $\displaystyle\int_1^3 \int_x^{x^2} \int_0^{\ln z} x e^y\, dy\, dz\, dx.$

7. $\displaystyle\int_0^2 \int_0^{\sqrt{4-x^2}} \int_{-5+x^2+y^2}^{3-x^2-y^2} x\, dz\, dy\, dx.$

8. $\displaystyle\int_1^2 \int_z^2 \int_0^{\sqrt{3}y} \frac{y}{x^2 + y^2}\, dx\, dy\, dz.$

In Exercises 9–12, evaluate the triple integral.

9. $\iiint\limits_{G} xy \sin yz \, dV$, where G is the rectangular box defined

by the inequalities $0 \le x \le \pi$, $0 \le y \le 1$, $0 \le z \le \pi/6$.

10. $\iiint\limits_{G} y \, dV$, where G is the solid enclosed by the plane

$z = y$, the xy-plane, and the parabolic cylinder $y = 1 - x^2$.

11. $\iiint\limits_{G} xyz \, dV$, where G is the solid in the first octant that is

bounded by the parabolic cylinder $z = 2 - x^2$ and the planes $z = 0$, $y = x$, and $y = 0$.

12. $\iiint\limits_{G} \cos (z/y) \, dV$, where G is the solid defined by the

inequalities, $\pi/6 \le y \le \pi/2$, $y \le x \le \pi/2$, $0 \le z \le xy$.

In Exercises 13–21, use a triple integral to find the volume of the solid.

13. The solid in the first octant bounded by the coordinate planes and the plane $3x + 6y + 4z = 12$.

14. The solid bounded by the surface $z = \sqrt{y}$ and the planes $x + y = 1$, $x = 0$, and $z = 0$.

15. The solid bounded by the surface $y = x^2$ and the planes $y + z = 4$ and $z = 0$.

16. The wedge in the first octant cut from the cylinder $y^2 + z^2 \le 1$ by the planes $y = x$ and $x = 0$.

17. The solid enclosed between the elliptic cylinder $x^2 + 9y^2 = 9$ and the planes $z = 0$ and $z = x + 3$.

18. The solid enclosed by the cylinders $x^2 + y^2 = 1$ and $x^2 + z^2 = 1$.

19. The solid bounded by the paraboloid $z = 4x^2 + y^2$ and the parabolic cylinder $z = 4 - 3y^2$.

20. The solid that is enclosed between the paraboloids $z = 8 - x^2 - y^2$ and $z = 3x^2 + y^2$.

21. The solid that is enclosed between the surfaces $x^2 + y^2 + z^2 = 2a^2$ and $az = x^2 + y^2$ $(a > 0)$.

22. Sketch the solid whose volume is given by the integral.

(a) $\int_0^3 \int_{x^2}^9 \int_0^2 dz \, dy \, dx$

(b) $\int_0^2 \int_0^{2-y} \int_0^{2-x-y} dz \, dx \, dy$.

23. Sketch the solid whose volume is given by the integral.

(a) $\int_{-1}^1 \int_{-\sqrt{1-x^2}}^{\sqrt{1-x^2}} \int_0^{y+1} dz \, dy \, dx$

(b) $\int_0^9 \int_0^{y/3} \int_0^{\sqrt{y^2-9x^2}} dz \, dx \, dy$.

24. Sketch the solid whose volume is given by the integral.

(a) $\int_0^1 \int_0^{\sqrt{1-x^2}} \int_0^2 dy \, dz \, dx$

(b) $\int_{-2}^2 \int_0^{4-y^2} \int_0^2 dx \, dz \, dy$.

25. Let G be the tetrahedron in the first octant bounded by the coordinate planes and the plane

$$\frac{x}{a} + \frac{y}{b} + \frac{z}{c} = 1 \quad (a > 0, b > 0, c > 0)$$

(a) List six different iterated integrals that represent the volume of G.

(b) Evaluate any one of the six to show that the volume of G is $\frac{1}{6}abc$.

26. In parts (a)–(c), express the integral as an equivalent integral in which the z-integration is performed first, the y-integration second, and the x-integration last.

(a) $\int_0^3 \int_0^{\sqrt{9-z^2}} \int_0^{\sqrt{9-y^2-z^2}} f(x, y, z) \, dx \, dy \, dz$

(b) $\int_0^4 \int_0^2 \int_0^{x/2} f(x, y, z) \, dy \, dz \, dx$

(c) $\int_0^4 \int_0^{4-y} \int_0^{\sqrt{z}} f(x, y, z) \, dx \, dz \, dy$.

27. Use a triple integral to find the volume of the tetrahedron with vertices $(0, 0, 0)$, $(a, a, 0)$, $(a, 0, 0)$, and $(a, 0, a)$.

28. Let G be the rectangular box defined by the inequalities $a \le x \le b$, $c \le y \le d$, $k \le z \le l$. Show that

$$\iiint\limits_{G} f(x)g(y)h(z) \, dV$$

$$= \left[\int_a^b f(x) \, dx \right] \left[\int_c^d g(y) \, dy \right] \left[\int_k^l h(z) \, dz \right]$$

29. Use the result of Exercise 28 to evaluate

(a) $\iiint\limits_{G} xy^2 \sin z \, dV$, where G is the set of points satis-

fying $-1 \le x \le 1$, $0 \le y \le 1$, $0 \le z \le \pi/2$

(b) $\iiint\limits_{G} e^{2x+y-z} \, dV$, where G is the set of points

satisfying $0 \le x \le 1$, $0 \le y \le \ln 3$, $0 \le z \le \ln 2$.

30. Use a triple integral to find the volume of the ellipsoid

$$\frac{x^2}{a^2} + \frac{y^2}{b^2} + \frac{z^2}{c^2} = 1$$

31. Let G be a region that is symmetric about the xy-plane, and let G_1 and G_2 be the portions of G above and below the xy-plane, respectively. Suppose that

$$\iiint\limits_{G_1} f(x, y, z) \, dV = 5$$

(a)   Find $\displaystyle\iiint_G f(x, y, z)\, dV$ given that

$f(x, y, -z) = f(x, y, z).$

(b)   Find $\displaystyle\iiint_G f(x, y, z)\, dV$ given that

$f(x, y, -z) = -f(x, y, z).$

> In Exercises 32–34, use symmetry and Exercise 31 to evaluate the integral without performing an integration.

32. $\displaystyle\int_{-2}^{2}\int_{0}^{\sqrt{4-x^2}}\int_{-\sqrt{4-x^2-y^2}}^{\sqrt{4-x^2-y^2}} [z^3 \cos{(xyz)} - 3]\, dz\, dy\, dx.$

33. $\displaystyle\int_{-3}^{3}\int_{-\sqrt{9-y^2}}^{\sqrt{9-y^2}}\int_{x^2+y^2}^{9} x^2 y^3 e^z\, dz\, dx\, dy.$

34. $\displaystyle\iiint_G e^{xy} \sin z\, dV$, where G is the spherical solid enclosed by $x^2 + y^2 + z^2 = R^2$.

■ 17.6   CENTROID, CENTER OF GRAVITY, THEOREM OF PAPPUS

> *Suppose that a physical body is acted on by a gravitational field. Because the body is composed of many particles, each of which is affected by gravity, the action of the gravitational field on the body consists of a large number of forces distributed over the entire body. However, these individual forces can be replaced by a single force acting at a point called the **center of gravity** of the body. In this section we shall show how double and triple integrals can be used to locate centers of gravity.*

□ DENSITY OF A LAMINA

The thickness of a lamina is negligible.

Figure 17.6.1

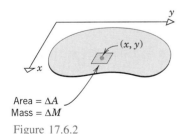

Area = ΔA
Mass = ΔM

Figure 17.6.2

To begin, let us consider an idealized flat object that is sufficiently thin that it can be viewed as two-dimensional (Figure 17.6.1). Such an object is called a **lamina**. A lamina is called **homogeneous** if its composition and structure are uniform throughout and **inhomogeneous** otherwise. The **density** of a *homogeneous* lamina is defined to be its mass per unit area. Thus, the density δ of a homogeneous lamina of mass M and area A is given by $\delta = M/A$.

For an inhomogeneous lamina the composition may vary from point to point, and hence an appropriate definition of ''density'' must reflect this. To motivate such a definition, suppose that the lamina is placed in an xy-plane. The density at a point (x, y) can be specified by a function $\delta(x, y)$, called the **density function**, which can be interpreted as follows. Construct a small rectangle centered at (x, y) and let ΔM and ΔA be the mass and area of the portion of the lamina enclosed by this rectangle (Figure 17.6.2). If the ratio $\Delta M/\Delta A$ approaches a limiting value as the dimensions (and hence the area) of the rectangle approach zero, then this limit is considered to be the density of the lamina at (x, y). Symbolically,

$$\delta(x, y) = \lim_{\Delta A \to 0} \frac{\Delta M}{\Delta A} \tag{1}$$

From this relationship we obtain the approximation

$$\Delta M \approx \delta(x, y)\, \Delta A \tag{2}$$

which relates the mass and area of a small rectangular portion of the lamina centered at (x, y). It is assumed that as the dimensions of the rectangle tend to zero, the error in this approximation also tends to zero.

REMARK.   Note that a lamina with a constant density function is homogeneous.

☐ MASS OF A LAMINA

The mass of a lamina can be found from its density function by the following formula.

17.6.1   MASS OF A LAMINA.   If a lamina with a continuous density function $\delta(x, y)$ occupies a region R in the xy-plane, then its total mass M is given by

$$M = \iint\limits_{R} \delta(x, y)\, dA \tag{3}$$

This formula can be motivated by a familiar limiting process that can be outlined as follows: Imagine the lamina to be subdivided into rectangular pieces using lines parallel to the coordinate axes and excluding from consideration any nonrectangular parts at the boundary (Figure 17.6.3). Assume that there are n such rectangular pieces, and suppose that the kth piece has area ΔA_k. If we let (x_k^*, y_k^*) denote the center of the kth piece, then from Formula (2), the mass ΔM_k of this piece can be approximated by

$$\Delta M_k \approx \delta(x_k^*, y_k^*)\, \Delta A_k \tag{4}$$

and hence the mass M of the entire lamina can be approximated by

$$M \approx \sum_{k=1}^{n} \delta(x_k^*, y_k^*)\, \Delta A_k$$

If we now increase n in such a way that the dimensions of the rectangles tend to zero, then it is plausible that the errors in our approximations will approach zero, so

$$M = \lim_{n \to +\infty} \sum_{k=1}^{n} \delta(x_k^*, y_k^*)\, \Delta A_k = \iint\limits_{R} \delta(x, y)\, dA$$

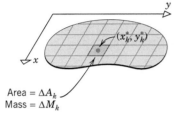

Area = ΔA_k
Mass = ΔM_k

Figure 17.6.3

Example 1   A triangular lamina with vertices $(0, 0)$, $(0, 1)$, and $(1, 0)$ has density function $\delta(x, y) = xy$. Find its total mass.

Solution.   Referring to (3) and Figure 17.6.4, the mass M of the lamina is

$$M = \iint\limits_{R} \delta(x, y)\, dA = \iint\limits_{R} xy\, dA = \int_0^1 \int_0^{-x+1} xy\, dy\, dx$$

$$= \int_0^1 \left[\frac{1}{2} xy^2 \right]_{y=0}^{-x+1} dx = \int_0^1 \left[\frac{1}{2} x^3 - x^2 + \frac{1}{2} x \right] dx$$

$$= \frac{1}{24} \text{ (unit of mass)} \quad \blacktriangleleft$$

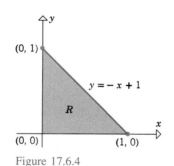

Figure 17.6.4

☐ CENTER OF GRAVITY
OF A LAMINA

Assuming that the force of gravity is constant and acts downward, consider the following problem.

17.6.2   PROBLEM.   *Suppose that a lamina with a continuous density function $\delta(x, y)$ occupies a region R in a horizontal xy-plane. Find the coordinates $(\bar{x}, \bar{y})$ of the center of gravity.*

To motivate the solution, consider what happens if we try to balance the lamina on a knife-edge parallel to the x-axis. Suppose the lamina in Figure 17.6.5 is placed on a knife-

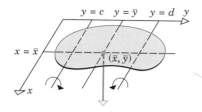

Force of gravity acting on the center of gravity of the lamina

Figure 17.6.5

edge along a line $y = c$ that does not pass through the center of gravity. Because the lamina behaves as if its entire mass is concentrated at the center of gravity $(\bar{x}, \bar{y})$, the lamina will be rotationally unstable and the force of gravity will cause a rotation about $y = c$. Similarly, the lamina will undergo a rotation if placed on a knife-edge along $y = d$. However, if the knife-edge runs along the line $y = \bar{y}$ through the center of gravity, the lamina will be in perfect balance. Similarly, the lamina will be in perfect balance on a knife-edge along the line $x = \bar{x}$ through the center of gravity. This suggests that the center of gravity of a lamina can be determined as the intersection of two lines of balance, one parallel to the x-axis and the other parallel to the y-axis. In order to find these lines of balance, we shall need some preliminary results about rotations.

Children on a seesaw learn by experience that a lighter child can balance a heavier one by sitting farther from the fulcrum or pivot point. This is because the tendency for an object to produce rotation is proportional not only to its mass but also to the distance between the object and the fulcrum. To be precise, if a point-mass m is located on a coordinate axis at a point x, then the tendency for that mass to produce a rotation about a point a on the axis is measured by the following quantity, called the ***moment of m about $x = a$***:

$$\left[\begin{array}{c} \text{moment of } m \\ \text{about } a \end{array} \right] = m(x - a)$$

The number $x - a$ is called the ***lever arm***. Depending on whether the mass is to the right or left of a, the lever arm is either the distance between x and a or the negative of this distance (Figure 17.6.6). Positive lever arms result in positive moments and clockwise rotations, while negative lever arms result in negative moments and counterclockwise rotations.

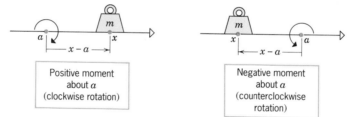

Figure 17.6.6

Suppose that masses $m_1, m_2, \ldots, m_n$ are located at points $x_1, x_2, \ldots, x_n$ on a coordinate axis and a fulcrum is positioned at the point a (Figure 17.6.7). Depending on whether the sum of the moments about a,

$$\sum_{k=1}^{n} m_k(x_k - a) = m_1(x_1 - a) + m_2(x_2 - a) + \cdots + m_n(x_n - a)$$

is positive, negative, or zero, the axis will rotate clockwise about a, rotate counterclockwise about a, or balance perfectly. In the last case, the system of masses is said to be in ***equilibrium***.

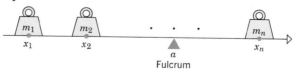

Figure 17.6.7

The preceding ideas can be extended to masses distributed in two-dimensional space: If we imagine the xy-plane to be a weightless sheet supporting a point-mass m located at a point (x, y), then the tendency for the mass to produce a rotation of the sheet about the line $x = a$ is $m(x - a)$, called the ***moment of m about $x = a$***, and the tendency for the mass to produce a rotation about the line $y = c$ is $m(y - c)$, called the ***moment of m about y = c*** (Figure 17.6.8). In summary,

$$
\begin{bmatrix} \text{moment of } m \\ \text{about the} \\ \text{line } x = a \end{bmatrix} = m(x - a) \quad \text{and} \quad \begin{bmatrix} \text{moment of } m \\ \text{about the} \\ \text{line } y = c \end{bmatrix} = m(y - c) \tag{5--6}
$$

If a number of masses are distributed throughout the xy-plane, then the plane (viewed as a weightless sheet) will balance on a knife-edge along the line $x = a$ if the sum of the moments about the line is zero. Similarly for the line $y = c$.

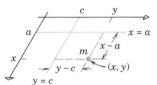

Figure 17.6.8

We are now ready to solve Problem 17.6.2. We imagine the lamina to be subdivided into rectangular pieces using lines parallel to the coordinate axes and excluding from consideration any nonrectangular pieces at the boundary (Figure 17.6.3). We assume that there are n such rectangular pieces and that the kth piece has area ΔA_k and mass ΔM_k. We shall let (x_k^*, y_k^*) be the center of the kth piece, and we shall assume that the entire mass of the kth piece is concentrated at its center. From (4), the mass of the kth piece can be approximated by

$$
\Delta M_k \approx \delta(x_k^*, y_k^*)\, \Delta A_k
$$

Since the lamina balances on the lines $x = \bar{x}$ and $y = \bar{y}$, the sum of the moments of the rectangular pieces about those lines should be close to zero; that is,

$$
\sum_{k=1}^{n} (x_k^* - \bar{x})\, \Delta M_k = \sum_{k=1}^{n} (x_k^* - \bar{x})\, \delta(x_k^*, y_k^*)\, \Delta A_k \approx 0
$$

$$
\sum_{k=1}^{n} (y_k^* - \bar{y})\, \Delta M_k = \sum_{k=1}^{n} (y_k^* - \bar{y})\, \delta(x_k^*, y_k^*)\, \Delta A_k \approx 0
$$

If we now increase n in such a way that the dimensions of the rectangles tend to zero, then it is plausible that the errors in our approximations will approach zero, so that

$$
\lim_{n \to +\infty} \sum_{k=1}^{n} (x_k^* - \bar{x})\, \delta(x_k^*, y_k^*)\, \Delta A_k = 0
$$

$$
\lim_{n \to +\infty} \sum_{k=1}^{n} (y_k^* - \bar{y})\, \delta(x_k^*, y_k^*)\, \Delta A_k = 0
$$

from which we obtain

$$
\iint_R (x - \bar{x})\, \delta(x, y)\, dA = 0 \quad \text{and} \quad \iint_R (y - \bar{y})\, \delta(x, y)\, dA = 0
$$

Since $\bar{x}$ and $\bar{y}$ are constant, these equations can be rewritten as

$$
\iint_R x\delta(x, y)\, dA = \bar{x} \iint_R \delta(x, y)\, dA
$$

$$
\iint_R y\delta(x, y)\, dA = \bar{y} \iint_R \delta(x, y)\, dA
$$

from which we obtain the following formulas for the center of gravity of the lamina:

Center of Gravity $(\bar{x}, \bar{y})$ of a Lamina

$$\bar{x} = \frac{\displaystyle\iint_R x\delta(x, y)\, dA}{\displaystyle\iint_R \delta(x, y)\, dA}, \qquad \bar{y} = \frac{\displaystyle\iint_R y\delta(x, y)\, dA}{\displaystyle\iint_R \delta(x, y)\, dA} \tag{7-8}$$

Observe that in both formulas the denominator is the mass M of the lamina [see (3)]. The numerator in the formula for $\bar{x}$ is denoted by M_y and is called the **first moment of the lamina about the y-axis**; the numerator in the formula for $\bar{y}$ is denoted by M_x and is called the **first moment of the lamina about the x-axis**. Thus, Formulas (7) and (8) can be expressed as

$$\bar{x} = \frac{M_y}{M} = \frac{1}{\text{mass of } R} \iint_R x\delta(x, y)\, dA \tag{9a}$$

$$\bar{y} = \frac{M_x}{M} = \frac{1}{\text{mass of } R} \iint_R y\delta(x, y)\, dA \tag{9b}$$

Example 2 Find the center of gravity of the triangular lamina with vertices $(0, 0)$, $(0, 1)$, and $(1, 0)$ and density function $\delta(x, y) = xy$.

Solution. The lamina is shown in Figure 17.6.4. In Example 1 we found the mass of the lamina to be

$$M = \iint_R \delta(x, y)\, dA = \iint_R xy\, dA = \frac{1}{24}$$

The moment of the lamina about the y-axis is

$$M_y = \iint_R x\delta(x, y)\, dA = \iint_R x^2 y\, dA = \int_0^1 \int_0^{-x+1} x^2 y\, dy\, dx$$

$$= \int_0^1 \left[\frac{1}{2} x^2 y^2 \right]_{y=0}^{-x+1} dx = \int_0^1 \left(\frac{1}{2} x^4 - x^3 + \frac{1}{2} x^2 \right) dx = \frac{1}{60}$$

and the moment about the x-axis is

$$M_x = \iint_R y\delta(x, y)\, dA = \iint_R xy^2\, dA = \int_0^1 \int_0^{-x+1} xy^2\, dy\, dx$$

$$= \int_0^1 \left[\frac{1}{3} xy^3 \right]_{y=0}^{-x+1} dx = \int_0^1 \left(-\frac{1}{3} x^4 + x^3 - x^2 + \frac{1}{3} x \right) dx = \frac{1}{60}$$

From (9a) and (9b),

$$\bar{x} = \frac{M_y}{M} = \frac{1/60}{1/24} = \frac{2}{5}, \qquad \bar{y} = \frac{M_x}{M} = \frac{1/60}{1/24} = \frac{2}{5}$$

so the center of gravity is $\left(\frac{2}{5}, \frac{2}{5} \right)$. ◄

☐ **CENTROIDS**

In the special case of a *homogeneous* lamina, the center of gravity is called the **centroid of the lamina** or sometimes the **centroid of the region R**. Because the density function δ is constant for a homogeneous lamina, the factor δ may be moved through the integral signs in (7) and (8) and canceled. Thus, the centroid $(\bar{x}, \bar{y})$ of a region R is given by the following formulas:

Centroid of a Region R

$$\bar{x} = \frac{\displaystyle\iint_R x \, dA}{\displaystyle\iint_R dA} = \frac{1}{\text{area of } R} \iint_R x \, dA \qquad (10)$$

$$\bar{y} = \frac{\displaystyle\iint_R y \, dA}{\displaystyle\iint_R dA} = \frac{1}{\text{area of } R} \iint_R y \, dA \qquad (11)$$

Example 3   Find the centroid of the semicircular region in Figure 17.6.9.

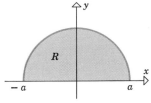

Figure 17.6.9

Solution.   By symmetry, $\bar{x} = 0$ since the y-axis is obviously a line of balance. From (11),

$$\bar{y} = \frac{1}{\text{area of } R} \iint_R y \, dA = \frac{1}{\frac{1}{2}\pi a^2} \iint_R y \, dA$$

$$= \frac{1}{\frac{1}{2}\pi a^2} \int_0^\pi \int_0^a (r \sin\theta) r \, dr \, d\theta = \frac{1}{\frac{1}{2}\pi a^2} \int_0^\pi \left[\frac{1}{3} r^3 \sin\theta \right]_{r=0}^a d\theta$$

Evaluating in polar coordinates

$$= \frac{1}{\frac{1}{2}\pi a^2} \left(\frac{2}{3} a^3 \right) = \frac{4a}{3\pi}$$

so the centroid is $\left(0, \dfrac{4a}{3\pi} \right)$.   ◀

□ **CENTER OF GRAVITY OF A SOLID**

For a three-dimensional solid G, the formulas for moments, center of gravity, and centroid are similar to those for laminas. If G is *homogeneous,* then its **density** is defined to be its mass per unit volume. Thus, if G is a homogeneous solid of mass M and volume V, then its density δ is given by $\delta = M/V$. If G is inhomogeneous and is in an xyz-coordinate system, then its density at a general point (x, y, z) is specified by a **density function** $\delta(x, y, z)$ whose value at a point can be viewed as a limit:

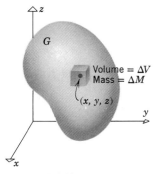

Figure 17.6.10

$$\delta(x, y, z) = \lim_{\Delta V \to 0} \frac{\Delta M}{\Delta V}$$

where ΔM and ΔV represent the mass and volume of a rectangular parallelepiped, centered at (x, y, z), whose dimensions tend to zero (Figure 17.6.10).

Using the discussion of laminas as a model, the reader should be able to show that the mass M of a solid with a continuous density function $\delta(x, y, z)$ is

$$M = \text{mass of } G = \iiint_G \delta(x, y, z) \, dV \qquad (12)$$

The formulas for center of gravity and centroid are

Center of Gravity $(\bar{x}, \bar{y}, \bar{z})$ of a Solid G	**Centroid $(\bar{x}, \bar{y}, \bar{z})$ of a Solid G**
$\bar{x} = \dfrac{1}{M} \iiint\limits_{G} x\delta(x, y, z)\, dV$	$\bar{x} = \dfrac{1}{V} \iiint\limits_{G} x\, dV$
$\bar{y} = \dfrac{1}{M} \iiint\limits_{G} y\delta(x, y, z)\, dV$	$\bar{y} = \dfrac{1}{V} \iiint\limits_{G} y\, dV$
$\bar{z} = \dfrac{1}{M} \iiint\limits_{G} z\delta(x, y, z)\, dV$	$\bar{z} = \dfrac{1}{V} \iiint\limits_{G} z\, dV$

(13–14)

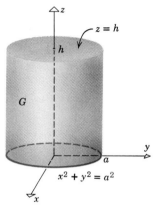

Figure 17.6.11

Example 4   Find the mass and the center of gravity of a cylindrical solid of height h and radius a (Figure 17.6.11), assuming that the density at each point is proportional to the distance between the point and the base of the solid.

Solution.   Since the density is proportional to the distance z from the base, the density function has the form $\delta(x, y, z) = kz$, where k is some (unknown) positive constant of proportionality. From (12) the mass of the solid is

$$M = \iiint\limits_{G} \delta(x, y, z)\, dV = \int_{-a}^{a} \int_{-\sqrt{a^2-x^2}}^{\sqrt{a^2-x^2}} \int_{0}^{h} kz\, dz\, dy\, dx$$

$$= k \int_{-a}^{a} \int_{-\sqrt{a^2-x^2}}^{\sqrt{a^2-x^2}} \tfrac{1}{2}h^2\, dy\, dx$$

$$= kh^2 \int_{-a}^{a} \sqrt{a^2 - x^2}\, dx$$

$$= \tfrac{1}{2}kh^2\pi a^2 \qquad \boxed{\text{See Example 4} \atop \text{of Section 9.5.}}$$

Without additional information, the constant k cannot be determined. However, as we shall now see, the value of k does not affect the center of gravity.

From (14),

$$\bar{z} = \frac{1}{\text{mass }G} \iiint\limits_{G} z\delta(x, y, z)\, dV = \frac{1}{\tfrac{1}{2}kh^2\pi a^2} \iiint\limits_{G} z\delta(x, y, z)\, dV$$

$$= \frac{1}{\tfrac{1}{2}kh^2\pi a^2} \int_{-a}^{a} \int_{-\sqrt{a^2-x^2}}^{\sqrt{a^2-x^2}} \int_{0}^{h} z(kz)\, dz\, dy\, dx$$

$$= \frac{k}{\tfrac{1}{2}kh^2\pi a^2} \int_{-a}^{a} \int_{-\sqrt{a^2-x^2}}^{\sqrt{a^2-x^2}} \tfrac{1}{3}h^3\, dy\, dx = \frac{\tfrac{1}{3}kh^3}{\tfrac{1}{2}kh^2\pi a^2} \int_{-a}^{a} 2\sqrt{a^2 - x^2}\, dx$$

$$= \frac{\tfrac{1}{3}kh^3\pi a^2}{\tfrac{1}{2}kh^2\pi a^2} = \tfrac{2}{3}h$$

Similar calculations using (13) will yield $\bar{x} = \bar{y} = 0$. However, this is evident by inspection, since it follows from the symmetry of the solid and the form of its density function that the center of gravity is on the z-axis. Thus, the center of gravity is $(0, 0, \tfrac{2}{3}h)$.   ◄

☐ **THEOREM OF PAPPUS**

The next theorem, due to the Greek mathematician Pappus,* gives an interesting and useful relationship between the centroid of a plane region R and the volume of the solid generated when the region is revolved about a line.

17.6.3 THEOREM. *If R is a plane region and L is a line that lies in the plane of R, but does not intersect R, then the volume of the solid formed by revolving R about L is given by*

$$\text{volume} = (\text{area of } R) \cdot \begin{pmatrix} \text{distance traveled} \\ \text{by the centroid} \end{pmatrix}$$

Proof. Introduce an xy-coordinate system so that L is along the y-axis and the region R is in the first quadrant (Figure 17.6.12). Let R be partitioned into subregions in the usual way and let R_k be a typical rectangle interior to R. If (x_k^*, y_k^*) is the center of R_k, and if the area of R_k is $\Delta A_k = \Delta x_k \, \Delta y_k$, then the volume generated by R_k as it revolves about L is

$$2\pi x_k^* \, \Delta x_k \, \Delta y_k = 2\pi x_k^* \, \Delta A_k$$

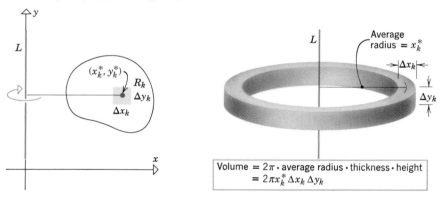

Figure 17.6.12

[See Figure 17.6.12 and Formula (1) of Section 6.3.] Therefore, the total volume of the solid is approximately

$$V \approx \sum_{k=1}^{n} 2\pi x_k^* \, \Delta A_k$$

from which it follows that the exact volume is

$$V = \iint\limits_{R} 2\pi x \, dA = 2\pi \iint\limits_{R} x \, dA$$

Thus, from (10)

$$V = 2\pi \cdot \bar{x} \cdot [\text{area of } R]$$

---

*PAPPUS OF ALEXANDRIA (4th century A.D.). Greek mathematician. Pappus lived during the early Christian era when mathematical activity was in a period of decline. His main contributions to mathematics appeared in a series of eight books called *The Collection* (written about 340 A.D.). This work, which survives only partially, contained some original results, but was devoted mostly to statements, refinements, and proofs of results by earlier mathematicians. Pappus' Theorem, stated without proof in Book VII of *The Collection*, was probably known and proved in earlier times. This result is sometimes called Guldin's Theorem in recognition of the Swiss mathematician, Paul Guldin (1577–1643), who rediscovered it independently.

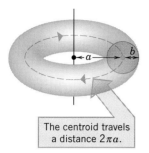

The centroid travels a distance $2\pi a$.

Figure 17.6.13

This completes the proof since $2\pi\bar{x}$ is the distance traveled by the centroid when R is revolved about the y-axis. ∎

Example 5 Use Pappus' Theorem to find the volume V of the torus generated by revolving a circular region of radius b about a line at a distance a (greater than b) from the center of the circle (Figure 17.6.13).

Solution. By symmetry, the centroid of a circular region is its center. Thus, the distance traveled by the centroid is $2\pi a$. Since the area of a circle is πr^2, it follows from Pappus' Theorem that the volume of the torus is

$$V = (2\pi a)(\pi b^2) = 2\pi^2 ab^2 \quad \blacktriangleleft$$

▶ **Exercise Set 17.6**

1. Where should the fulcrum be placed so that the seesaw is in equilibrium?

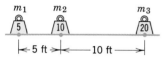

2. A rectangular lamina with density $\delta(x, y) = xy^2$ has vertices $(0, 0), (0, 2), (3, 0), (3, 2)$. Find its mass and center of gravity.

3. A lamina with density $\delta(x, y) = x + y$ is bounded by the x-axis, the line $x = 1$, and the curve $y = \sqrt{x}$. Find its mass and center of gravity.

4. A lamina with density $\delta(x, y) = y$ is bounded by $y = \sin x$, $y = 0$, $x = 0$, and $x = \pi$. Find its mass and center of gravity.

5. A lamina with density $\delta(x, y) = xy$ is in the first quadrant and is bounded by the circle $x^2 + y^2 = a^2$ and the coordinate axes. Find its mass and center of gravity.

6. A lamina with density $\delta(x, y) = x^2 + y^2$ is bounded by the x-axis and the upper half of the circle $x^2 + y^2 = 1$. Find its mass and center of gravity.

In Exercises 7–13, find the centroid of the region.

7. The triangular region bounded by $y = x$, $x = 1$, and the x-axis.

8. The region bounded by $y = x^2$, $x = 1$, and the x-axis.

9. The region that is enclosed between $y = x$ and $y = 2 - x^2$.

10. The region enclosed by the parabolas $x = 3y^2 - 6y$ and $x = 2y - y^2$.

11. The region above the x-axis and between the circles $x^2 + y^2 = a^2$ and $x^2 + y^2 = b^2$ $(a < b)$.

12. The region enclosed between the y-axis and the right half of the circle $x^2 + y^2 = a^2$.

13. The region enclosed between $y = |x|$ and the line $y = 4$.

14. A rectangular box that is defined by the three inequalities $0 \le x \le 1$, $0 \le y \le 1$, and $0 \le z \le 1$ has density function $\delta(x, y, z) = 3xyz$. Find its mass and center of gravity.

15. A cube that is defined by the three inequalities $0 \le x \le a$, $0 \le y \le a$, $0 \le z \le a$ has density $\delta(x, y, z) = a - x$. Find its mass and center of gravity.

16. The cylindrical solid enclosed by $x^2 + y^2 = a^2$, $z = 0$, and $z = h$ has density $\delta(x, y, z) = h - z$. Find its mass and center of gravity.

17. The solid enclosed by the surface $z = 1 - y^2$ (where $y \ge 0$) and the planes $z = 0$, $x = -1$, $x = 1$ has density $\delta(x, y, z) = yz$. Find its mass and center of gravity.

18. The solid enclosed by the surface $y = 9 - x^2$ (where $x \ge 0$) and the planes $y = 0$, $z = 0$, and $z = 1$ has density $\delta(x, y, z) = xz$. Find its mass and center of gravity.

In Exercises 19–23, find the centroid of the solid.

19. The tetrahedron in the first octant enclosed by the coordinate planes and the plane $x + y + z = 1$.

20. The solid enclosed by the xy-plane and the hemisphere $z = \sqrt{a^2 - x^2 - y^2}$.

21. The solid bounded by the surface $z = y^2$, and the planes $x = 0$, $x = 1$, and $z = 1$.

22. The solid in the first octant bounded by the surface $z = xy$ and the planes $z = 0$, $x = 2$, and $y = 2$.

23. The solid in the first octant bounded by the sphere $x^2 + y^2 + z^2 = a^2$ and the coordinate planes.

24. Find the center of gravity of the square lamina with vertices $(0, 0)$, $(1, 0)$, $(0, 1)$, and $(1, 1)$ if

   (a) the density is proportional to the square of the distance from the origin

   (b) the density is proportional to the distance from the y-axis.

25. Find the mass of a circular lamina of radius a whose density at a point is proportional to the distance from the center. (Let k denote the constant of proportionality.)

26. Find the center of gravity of the cube that is determined by the inequalities $0 \le x \le 1$, $0 \le y \le 1$, $0 \le z \le 1$ if

   (a) the density is proportional to the square of the distance to the origin

(b)   the density is proportional to the sum of the distances to the faces that lie in the coordinate planes.

27. Show that in polar coordinates the formulas for the centroid $(\bar{x}, \bar{y})$ of a region R are

$$\bar{x} = \frac{1}{\text{area of } R} \iint_R r^2 \cos\theta \, dr \, d\theta$$

$$\bar{y} = \frac{1}{\text{area of } R} \iint_R r^2 \sin\theta \, dr \, d\theta$$

28. Use the result of Exercise 27 to find the centroid $(\bar{x}, \bar{y})$ of the region enclosed by the cardioid $r = a(1 + \sin\theta)$.

29. Use the result of Exercise 27 to find the centroid $(\bar{x}, \bar{y})$ of the petal of the rose $r = \sin 2\theta$ in the first quadrant.

30. Let R be the rectangle bounded by the lines $x = 0$, $x = 3$, $y = 0$, and $y = 2$. By inspection, find the centroid of R and use it to evaluate

$$\iint_R x \, dA \quad \text{and} \quad \iint_R y \, dA$$

31. Use the Theorem of Pappus and the fact that the volume of a sphere of radius a is $V = \frac{4}{3}\pi a^3$ to show that the centroid of the lamina that is bounded by the x-axis and the semicircle $y = \sqrt{a^2 - x^2}$ is $(0, 4a/(3\pi))$. (This problem was solved directly in Example 3.)

32. Use the Theorem of Pappus and the result of Exercise 31 to find the volume of the solid generated when the region bounded by the x-axis and the semicircle $y = \sqrt{a^2 - x^2}$ is revolved about

(a)   the line $y = -a$ (b)   the line $y = x - a$.

33. Use the Theorem of Pappus and the fact that the area of an ellipse with semiaxes a and b is πab to find the volume of the elliptical torus generated by revolving the ellipse

$$\frac{(x - k)^2}{a^2} + \frac{y^2}{b^2} = 1$$

about the y-axis. Assume that $k > a$.

34. Use the Theorem of Pappus to find the volume of the solid generated when the region enclosed by $y = x^2$ and $y = 8 - x^2$ is revolved about the x-axis.

35. Use the Theorem of Pappus to find the centroid of the triangular region with vertices $(0, 0)$, $(a, 0)$, and $(0, b)$, where $a > 0$ and $b > 0$. [*Hint:* Revolve the region about the x-axis to obtain $\bar{y}$ and about the y-axis to obtain $\bar{x}$.]

36. If a lamina with a continuous density function $\delta(x, y)$ occupies a region R in the xy-plane, then the **second moments** (or **moments of inertia**) of the lamina about the x-axis, y-axis, and z-axis, respectively, are defined by

$$I_x = \iint_R y^2 \delta \, dA, \quad I_y = \iint_R x^2 \delta \, dA,$$

$$I_z = \iint_R (x^2 + y^2)\delta \, dA$$

The corresponding definitions for a solid with density function $\delta(x, y, z)$ that occupies a region G are

$$I_x = \iiint_G (y^2 + z^2)\delta(x, y, z) \, dV$$

$$I_y = \iiint_G (x^2 + z^2)\delta(x, y, z) \, dV$$

$$I_z = \iiint_G (x^2 + y^2)\delta(x, y, z) \, dV$$

Find I_x, I_y, and I_z for the homogeneous (constant δ) rectangular lamina described by the inequalities $0 \le x \le a$ and $0 \le y \le b$. Express your answers in terms of the mass M of the lamina. [*Note:* Moments of inertia are important in problems involving rotational motion.]

In Exercises 37–44, a homogeneous lamina or solid is given. Use the formulas in Exercise 36 to find the indicated moment of inertia. Express your answers in terms of the mass M of the lamina or solid.

---

37. The moment I_z for the rectangular lamina $-a/2 \le x \le a/2$, $-b/2 \le y \le b/2$.

38. The moment I_x for the rectangular lamina $-a/2 \le x \le a/2$, $-b/2 \le y \le b/2$.

39. The circular lamina $x^2 + y^2 \le R^2$; I_y.

40. The circular lamina $x^2 + y^2 \le R^2$; I_z.

41. The circular lamina $x^2 + (y - R)^2 \le R^2$; I_z.

42. The triangular lamina $0 \le y \le 2 - 2x$, $x \ge 0$; I_y.

43. The solid $0 \le x \le a$, $0 \le y \le a$, $0 \le z \le a$; I_z.

44. The rectangular solid $0 \le x \le a$, $-b/2 \le y \le b/2$, $-c/2 \le z \le c/2$; I_x.

■ **17.7   TRIPLE INTEGRALS IN CYLINDRICAL AND SPHERICAL COORDINATES**

In Section 17.3 we saw that some double integrals are easier to evaluate in polar coordinates. Similarly, some triple integrals are easier to evaluate in cylindrical or spherical coordinates. In this section we shall study triple integrals in these coordinate systems.

☐ **TRIPLE INTEGRALS IN CYLINDRICAL COORDINATES**

Recall that in rectangular coordinates the triple integral of a continuous function f over a solid region G is defined as

$$\iiint\limits_{G} f(x, y, z)\, dV = \lim_{n \to +\infty} \sum_{k=1}^{n} f(x_k^*, y_k^*, z_k^*)\, \Delta V_k$$

where ΔV_k denotes the volume of a rectangular parallelepiped interior to G and (x_k^*, y_k^*, z_k^*) is a point in this parallelepiped (Figure 17.5.1). Triple integrals in cylindrical and spherical coordinates are defined similarly, except that the region G is divided not into rectangular parallelepipeds but into regions more appropriate to these coordinate systems.

In cylindrical coordinates, the simplest equations are of the form

$$r = \text{constant}, \quad \theta = \text{constant}, \quad z = \text{constant}$$

As indicated in Figure 14.8.2b, the first equation represents a right-circular cylinder centered on the z-axis, the second a vertical half-plane hinged on the z-axis, and the third a horizontal plane. These surfaces can be paired up to determine solids called ***cylindrical wedges*** or ***cylindrical elements of volume***. To be precise, a cylindrical wedge is a solid enclosed between six surfaces of the following type:

two cylinders $r = r_1, \quad r = r_2 \quad (r_1 < r_2)$

two half-planes $\theta = \theta_1, \quad \theta = \theta_2 \quad (\theta_1 < \theta_2)$

two planes $z = z_1, \quad z = z_2 \quad (z_1 < z_2)$

(Figure 17.7.1). The dimensions $\theta_2 - \theta_1$, $r_2 - r_1$, and $z_2 - z_1$ are called the ***central angle***, ***thickness***, and ***height*** of the wedge.

To define the triple integral over G of a function $f(r, \theta, z)$ in cylindrical coordinates we proceed as follows:

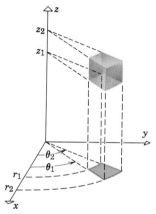

Figure 17.7.1

- Subdivide G into pieces by a three-dimensional grid consisting of concentric circular cylinders centered on the z-axis, half-planes hinged on the z-axis, and horizontal planes. Exclude from consideration all pieces that contain any points outside of G, thereby leaving only cylindrical wedges that are subsets of G.

- Assume that there are n such cylindrical wedges, and denote the volume of the kth cylindrical wedge by ΔV_k. As indicated in Figure 17.7.2, let $(r_k^*, \theta_k^*, z_k^*)$ be any point in the kth cylindrical wedge.

- Repeat this process with more and more subdivisions, so that as n increases the height, thickness, and central angle of the cylindrical wedges approach zero. Define

$$\iiint\limits_{G} f(r, \theta, z)\, dV = \lim_{n \to +\infty} \sum_{k=1}^{n} f(r_k^*, \theta_k^*, z_k^*)\, \Delta V_k \tag{1}$$

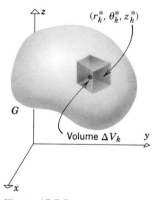

Figure 17.7.2

For computational purposes, it will be helpful to express (1) as an iterated integral. Toward this end we note that the volume ΔV_k of the kth cylindrical wedge can be expressed as

$$\Delta V_k = [\text{area of base}] \cdot [\text{height}] \tag{2}$$

If we denote the thickness, central angle, and height of this wedge by Δr_k, $\Delta \theta_k$, and Δz_k, and if we choose the arbitrary point $(r_k^*, \theta_k^*, z_k^*)$ to lie above the "center" of the base (Figures 17.3.5 and 17.7.3), then it follows from (4) of Section 17.3 that the base has area $r_k^* \, \Delta r_k \, \Delta \theta_k$. Thus, (2) can be written as

$$\Delta V_k = r_k^* \, \Delta r_k \, \Delta \theta_k \, \Delta z_k = r_k^* \, \Delta z_k \, \Delta r_k \, \Delta \theta_k$$

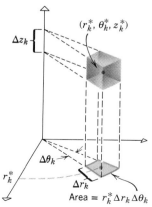

Figure 17.7.3

Substituting this expression in (1) yields

$$\iiint\limits_{G} f(r, \theta, z)\, dV = \lim_{n \to +\infty} \sum_{k=1}^{n} f(r_k^*, \theta_k^*, z_k^*) r_k^*\, \Delta z_k\, \Delta r_k\, \Delta \theta_k$$

which suggests that a triple integral in cylindrical coordinates can be evaluated as an iterated integral of the form

$$\iiint\limits_{G} f(r, \theta, z)\, dV = \iiint\limits_{\substack{\text{appropriate} \\ \text{limits}}} f(r, \theta, z) r\, dz\, dr\, d\theta \tag{3}$$

REMARK.   Note the extra factor of r that appears in the integrand on converting from the triple integral to the iterated integral. In this formula the integration with respect to z is done first, then with respect to r, and then with respect to θ, but any order of integration is allowable.

The following theorem, which we state without proof, makes the preceding ideas more precise.

17.7.1   THEOREM.   *Let G be a simple solid whose upper surface has the equation $z = g_2(r, \theta)$ and whose lower surface has the equation $z = g_1(r, \theta)$ in cylindrical coordinates. If R is the projection of the solid on the xy-plane, and if $f(r, \theta, z)$ is continuous on G, then*

$$\iiint\limits_{G} f(r, \theta, z)\, dV = \iint\limits_{R} \left[\int_{g_1(r,\theta)}^{g_2(r,\theta)} f(r, \theta, z)\, dz \right] dA \tag{4}$$

where the double integral over R is evaluated in polar coordinates. In particular, if the projection R is as shown in Figure 17.7.4, *then* (4) *can be written as*

$$\iiint\limits_{G} f(r, \theta, z)\, dV = \int_{\theta_1}^{\theta_2} \int_{r_1(\theta)}^{r_2(\theta)} \int_{g_1(r,\theta)}^{g_2(r,\theta)} f(r, \theta, z) r\, dz\, dr\, d\theta \tag{5}$$

The type of solid to which Formula (5) *applies is illustrated in* Figure 17.7.4.

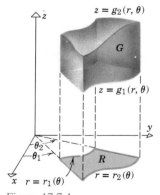

Figure 17.7.4

To apply (4) and (5) it is best to begin with a three-dimensional sketch of the solid G, from which the limits of integration can be obtained as follows:

Step 1.   Identify the upper surface $z = g_2(r, \theta)$ and the lower surface $z = g_1(r, \theta)$ of the solid. The functions $g_1(r, \theta)$ and $g_2(r, \theta)$ determine the z-limits of integration. (If the upper and lower surfaces are given in rectangular coordinates, convert them to cylindrical coordinates.)

Step 2.   Make a two-dimensional sketch of the projection R of the solid on the xy-plane. From this sketch the r- and θ-limits of integration may be obtained exactly as with double integrals in polar coordinates.

Example 1   Use triple integration in cylindrical coordinates to find the volume and the centroid of the solid G that is bounded above by the hemisphere $z = \sqrt{25 - x^2 - y^2}$, below by the xy-plane, and laterally by the cylinder $x^2 + y^2 = 9$.

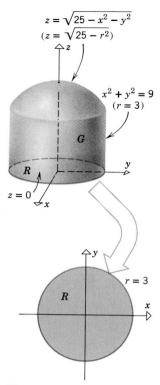

Figure 17.7.5

Solution.   The solid G and its projection R on the xy-plane are shown in Figure 17.7.5. In cylindrical coordinates, the upper surface of G is the hemisphere $z = \sqrt{25 - r^2}$ and the lower surface is the plane $z = 0$. Thus, from (4), the volume of G is

$$V = \iiint\limits_{G} dV = \iint\limits_{R} \left[\int_0^{\sqrt{25-r^2}} dz \right] dA$$

For the double integral over R, we use polar coordinates:

$$V = \int_0^{2\pi} \int_0^3 \int_0^{\sqrt{25-r^2}} r\, dz\, dr\, d\theta = \int_0^{2\pi} \int_0^3 \left[rz \right]_{z=0}^{\sqrt{25-r^2}} dr\, d\theta$$

$$= \int_0^{2\pi} \int_0^3 r\sqrt{25 - r^2}\, dr\, d\theta = \int_0^{2\pi} \left[-\frac{1}{3}(25 - r^2)^{3/2} \right]_{r=0}^{3} d\theta$$

$$= \int_0^{2\pi} \frac{61}{3}\, d\theta = \frac{122}{3}\pi$$

$$\boxed{\begin{array}{l} u = 25 - r^2 \\ du = -2r\, dr \end{array}}$$

From this result and (14) of Section 17.6,

$$\bar{z} = \frac{1}{V} \iiint\limits_{G} z\, dV = \frac{3}{122\pi} \iiint\limits_{G} z\, dV = \frac{3}{122\pi} \iint\limits_{R} \left[\int_0^{\sqrt{25-r^2}} z\, dz \right] dA$$

$$= \frac{3}{122\pi} \int_0^{2\pi} \int_0^3 \int_0^{\sqrt{25-r^2}} zr\, dz\, dr\, d\theta = \frac{3}{122\pi} \int_0^{2\pi} \int_0^3 \left[\frac{1}{2} rz^2 \right]_{z=0}^{\sqrt{25-r^2}} dr\, d\theta$$

$$= \frac{3}{244\pi} \int_0^{2\pi} \int_0^3 (25r - r^3)\, dr\, d\theta = \frac{3}{244\pi} \int_0^{2\pi} \frac{369}{4}\, d\theta = \frac{1107}{488}$$

By symmetry, the centroid $(\bar{x}, \bar{y}, \bar{z})$ of G lies on the z-axis, so $\bar{x} = \bar{y} = 0$. Thus, the centroid is at the point $(0, 0, 1107/488)$. ◄

□ **CONVERTING TRIPLE INTEGRALS FROM RECTANGULAR TO CYLINDRICAL COORDINATES**

Sometimes a triple integral that is difficult to integrate in rectangular coordinates can be evaluated more easily by making the substitution $x = r \cos\theta$, $y = r \sin\theta$, $z = z$ to convert it to an integral in cylindrical coordinates. Under such a substitution, the rectangular and cylindrical triple integrals are related by the equation

$$\iiint\limits_{G} f(x, y, z)\, dV = \iiint\limits_{\substack{\text{appropriate} \\ \text{limits}}} f(r \cos\theta, r \sin\theta, z)\, r\, dz\, dr\, d\theta \qquad (6)$$

REMARK.   In (6), the order of integration is first with respect to z, then r, and then θ. However, when appropriate, the order of integration can be changed, provided the limits of integration are adjusted accordingly.

Example 2   Use cylindrical coordinates to evaluate

$$\int_{-3}^{3} \int_{-\sqrt{9-x^2}}^{\sqrt{9-x^2}} \int_0^{9-x^2-y^2} x^2\, dz\, dy\, dx$$

Solution.   In problems of this type, it is helpful to sketch the region of integration G and its projection R on the xy-plane. From the z-limits of integration, the upper surface of G is the paraboloid $z = 9 - x^2 - y^2$ and the lower surface is the xy-plane $z = 0$. From the x-

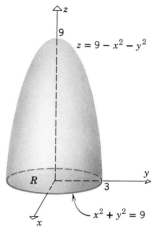

Figure 17.7.6

and y-limits of integration, the projection R is the region in the xy-plane enclosed by the circle $x^2 + y^2 = 9$ (Figure 17.7.6). Thus,

$$\int_{-3}^{3} \int_{-\sqrt{9-x^2}}^{\sqrt{9-x^2}} \int_{0}^{9-x^2-y^2} x^2 \, dz \, dy \, dx = \iiint_{G} x^2 \, dV$$

$$= \iint_{R} \left[\int_{0}^{9-r^2} r^2 \cos^2 \theta \, dz \right] dA = \int_{0}^{2\pi} \int_{0}^{3} \int_{0}^{9-r^2} (r^2 \cos^2 \theta) \, r \, dz \, dr \, d\theta$$

$$= \int_{0}^{2\pi} \int_{0}^{3} \int_{0}^{9-r^2} r^3 \cos^2 \theta \, dz \, dr \, d\theta = \int_{0}^{2\pi} \int_{0}^{3} \left[zr^3 \cos^2 \theta \right]_{z=0}^{9-r^2} dr \, d\theta$$

$$= \int_{0}^{2\pi} \int_{0}^{3} (9r^3 - r^5) \cos^2 \theta \, dr \, d\theta = \int_{0}^{2\pi} \left[\left(\frac{9r^4}{4} - \frac{r^6}{6} \right) \cos^2 \theta \right]_{r=0}^{3} d\theta$$

$$= \frac{243}{4} \int_{0}^{2\pi} \cos^2 \theta \, d\theta = \frac{243}{4} \int_{0}^{2\pi} \frac{1}{2} (1 + \cos 2\theta) \, d\theta = \frac{243\pi}{4} \quad \blacktriangleleft$$

☐ TRIPLE INTEGRALS IN SPHERICAL COORDINATES

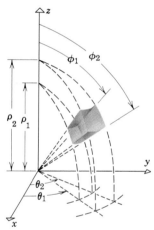

Figure 17.7.7

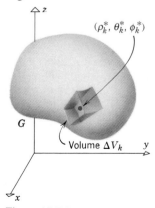

Figure 17.7.8

In spherical coordinates, the simplest equations are of the form

$$\rho = \text{constant}, \quad \theta = \text{constant}, \quad \phi = \text{constant}$$

As indicated in Figure 14.8.2c, the first equation represents a sphere centered at the origin, the second a half-plane hinged on the z-axis, and the third a right-circular cone with its vertex at the origin and its line of symmetry along the z-axis. By a *spherical wedge* or *spherical element of volume* we mean a solid enclosed between six surfaces of the following form:

two spheres	$\rho = \rho_1, \quad \rho = \rho_2$	$(\rho_1 < \rho_2)$
two half-planes	$\theta = \theta_1, \quad \theta = \theta_2$	$(\theta_1 < \theta_2)$
two right-circular cones	$\phi = \phi_1, \quad \phi = \phi_2$	$(\phi_1 < \phi_2)$

(Figure 17.7.7). We shall refer to the numbers $\rho_2 - \rho_1$, $\theta_2 - \theta_1$, and $\phi_2 - \phi_1$ as the *dimensions* of a spherical wedge.

If G is a solid region in three-dimensional space, then the triple integral over G of a continuous function $f(\rho, \theta, \phi)$ in spherical coordinates is similar in definition to the triple integral in cylindrical coordinates, except that the solid G is partitioned into *spherical wedges* by a three-dimensional grid consisting of spheres centered at the origin, half-planes hinged on the z-axis, and right-circular cones with vertices at the origin and lines of symmetry along the z-axis (Figure 17.7.8).

The defining equation of a triple integral in spherical coordinates is

$$\iiint_{G} f(\rho, \theta, \phi) \, dV = \lim_{n \to +\infty} \sum_{k=1}^{n} f(\rho_k^*, \theta_k^*, \phi_k^*) \, \Delta V_k \tag{7}$$

where ΔV_k is the volume of the kth spherical wedge that is interior to G, $(\rho_k^*, \theta_k^*, \phi_k^*)$ is an arbitrary point in this wedge, and n increases in such a way that the dimensions of each interior spherical wedge tend to zero.

For computational purposes, it will be desirable to express (7) as an iterated integral. In the exercises we will help the reader to show that if the point $(\rho_k^*, \theta_k^*, \phi_k^*)$ is suitably chosen, then the volume ΔV_k in (7) can be written as

$$\Delta V_k = \rho_k^{*2} \sin \phi_k^* \, \Delta \rho_k \, \Delta \phi_k \, \Delta \theta_k$$

where $\Delta\rho_k$, $\Delta\phi_k$, and $\Delta\theta_k$ are the dimensions of the wedge. Substituting this in (7) we obtain

$$\iiint\limits_G f(\rho, \theta, \phi)\, dV = \lim_{n \to +\infty} \sum_{k=1}^{n} f(\rho_k^*, \theta_k^*, \phi_k^*)\rho_k^{*2} \sin\phi_k^* \,\Delta\rho_k\,\Delta\phi_k\,\Delta\theta_k$$

which suggests that a triple integral in spherical coordinates can be evaluated as an iterated integral of the form

$$\iiint\limits_G f(\rho, \theta, \phi)\, dV = \iiint\limits_{\substack{\text{appropriate}\\\text{limits}}} f(\rho, \theta, \phi)\rho^2 \sin\phi\, d\rho\, d\phi\, d\theta \tag{8}$$

REMARK.   Note the extra factor of $\rho^2 \sin\phi$ that appears in the integrand of the iterated integral. This is analogous to the extra factor of r that appeared when we integrated in cylindrical coordinates.

The analog of Theorem 17.7.1 for triple integrals in spherical coordinates is tedious to state, so instead we shall give some examples that illustrate techniques for obtaining the limits of integration. In all of our examples we shall use the same order of integration— first with respect to ρ, then ϕ, and then θ. Once the reader has mastered the basic ideas, there should be no trouble using other orders of integration.

Suppose that we want to integrate $f(\rho, \theta, \phi)$ over the spherical solid G enclosed by the sphere $\rho = \rho_0$. The basic idea is to choose the limits of integration so that every point of the solid is accounted for in the integration process. Figure 17.7.9 illustrates one way of doing this:

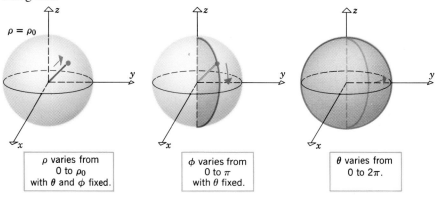

Figure 17.7.9

| ρ varies from 0 to ρ_0 with θ and ϕ fixed. | ϕ varies from 0 to π with θ fixed. | θ varies from 0 to 2π. |

Holding θ and ϕ fixed for the first integration, we let ρ vary from 0 to ρ_0. This covers a radial line from the origin to the surface of the sphere. Next, keeping θ fixed, we let ϕ vary from 0 to π so that the radial line sweeps out a fan-shaped region. Finally, we let θ vary from 0 to 2π so that the fan-shaped region makes a complete revolution, thereby sweeping out the entire sphere. Thus, the triple integral of $f(\rho, \theta, \phi)$ over the spherical solid G may be evaluated by writing

$$\iiint\limits_G f(\rho, \theta, \phi)\, dV = \int_0^{2\pi} \int_0^{\pi} \int_0^{\rho_0} f(\rho, \theta, \phi)\rho^2 \sin\phi\, d\rho\, d\phi\, d\theta$$

Table 17.7.1 (pages 900–901) suggests how the limits of integration in spherical coordinates can be obtained for some other common solids.

Table 17.7.1

DETERMINATION OF LIMITS	INTEGRAL

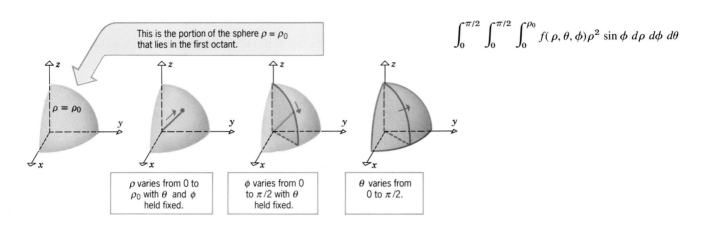

This is the portion of the sphere $\rho = \rho_0$ that lies in the first octant.

ρ varies from 0 to ρ_0 with θ and ϕ held fixed.

ϕ varies from 0 to $\pi/2$ with θ held fixed.

θ varies from 0 to $\pi/2$.

$$\int_0^{\pi/2} \int_0^{\pi/2} \int_0^{\rho_0} f(\rho, \theta, \phi)\rho^2 \sin\phi \, d\rho \, d\phi \, d\theta$$

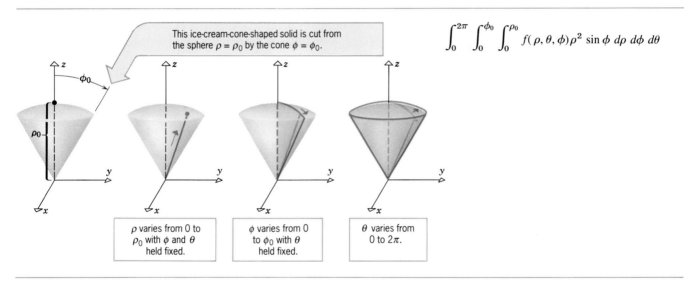

This ice-cream-cone-shaped solid is cut from the sphere $\rho = \rho_0$ by the cone $\phi = \phi_0$.

ρ varies from 0 to ρ_0 with ϕ and θ held fixed.

ϕ varies from 0 to ϕ_0 with θ held fixed.

θ varies from 0 to 2π.

$$\int_0^{2\pi} \int_0^{\phi_0} \int_0^{\rho_0} f(\rho, \theta, \phi)\rho^2 \sin\phi \, d\rho \, d\phi \, d\theta$$

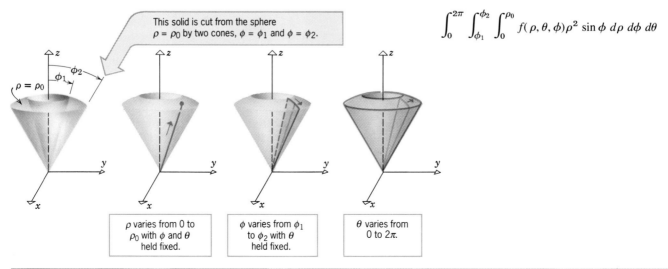

This solid is cut from the sphere $\rho = \rho_0$ by two cones, $\phi = \phi_1$ and $\phi = \phi_2$.

ρ varies from 0 to ρ_0 with ϕ and θ held fixed.

ϕ varies from ϕ_1 to ϕ_2 with θ held fixed.

θ varies from 0 to 2π.

$$\int_0^{2\pi} \int_{\phi_1}^{\phi_2} \int_0^{\rho_0} f(\rho, \theta, \phi)\rho^2 \sin\phi \, d\rho \, d\phi \, d\theta$$

Table 17.7.1

DETERMINATION OF LIMITS	INTEGRAL

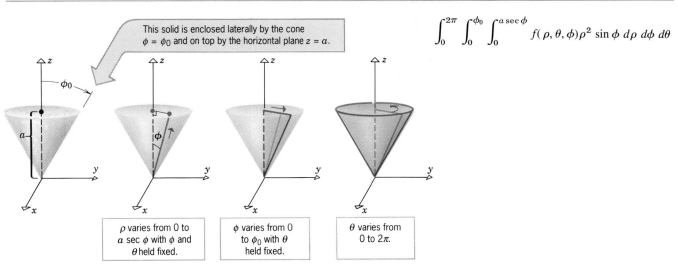

This solid is enclosed laterally by the cone $\phi = \phi_0$ and on top by the horizontal plane $z = a$.

$$\int_0^{2\pi} \int_0^{\phi_0} \int_0^{a\sec\phi} f(\rho, \theta, \phi)\rho^2 \sin\phi \, d\rho \, d\phi \, d\theta$$

ρ varies from 0 to $a \sec \phi$ with ϕ and θ held fixed.	ϕ varies from 0 to ϕ_0 with θ held fixed.	θ varies from 0 to 2π.

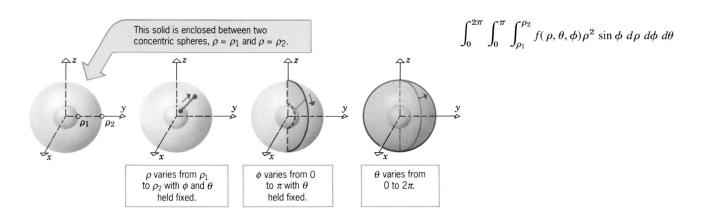

This solid is enclosed between two concentric spheres, $\rho = \rho_1$ and $\rho = \rho_2$.

$$\int_0^{2\pi} \int_0^{\pi} \int_{\rho_1}^{\rho_2} f(\rho, \theta, \phi)\rho^2 \sin\phi \, d\rho \, d\phi \, d\theta$$

ρ varies from ρ_1 to ρ_2 with ϕ and θ held fixed.	ϕ varies from 0 to π with θ held fixed.	θ varies from 0 to 2π.

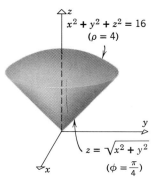

Figure 17.7.10

Example 3   Use spherical coordinates to find the volume and the centroid of the solid G bounded above by the sphere $x^2 + y^2 + z^2 = 16$ and below by the cone $z = \sqrt{x^2 + y^2}$.

Solution.   The solid G is sketched in Figure 17.7.10.

In spherical coordinates, the equation of the sphere $x^2 + y^2 + z^2 = 16$ is $\rho = 4$ and the equation of the cone $z = \sqrt{x^2 + y^2}$ is

$$\rho \cos \phi = \sqrt{\rho^2 \sin^2 \phi \cos^2 \theta + \rho^2 \sin^2 \phi \sin^2 \theta}$$

which simplifies to

$$\rho \cos \phi = \rho \sin \phi$$

or, on dividing both sides by $\rho \cos \phi$,

$$\tan \phi = 1$$

Thus $\phi = \pi/4$, so the volume of G is

$$V = \iiint\limits_{G} dV = \int_0^{2\pi} \int_0^{\pi/4} \int_0^4 \rho^2 \sin\phi \, d\rho \, d\phi \, d\theta$$

$$= \int_0^{2\pi} \int_0^{\pi/4} \left[\frac{\rho^3}{3} \sin\phi \right]_{\rho=0}^4 d\phi \, d\theta$$

$$= \int_0^{2\pi} \int_0^{\pi/4} \frac{64}{3} \sin\phi \, d\phi \, d\theta$$

$$= \frac{64}{3} \int_0^{2\pi} \left[-\cos\phi \right]_{\phi=0}^{\pi/4} d\theta = \frac{64}{3} \int_0^{2\pi} \left(1 - \frac{\sqrt{2}}{2} \right) d\theta$$

$$= \frac{64\pi}{3}(2 - \sqrt{2})$$

By symmetry, the centroid $(\bar{x}, \bar{y}, \bar{z})$ is on the z-axis, so $\bar{x} = \bar{y} = 0$. From (14) of Section 17.6 and the volume calculated above,

$$\bar{z} = \frac{1}{V} \iiint\limits_{G} z \, dV = \frac{1}{V} \int_0^{2\pi} \int_0^{\pi/4} \int_0^4 (\rho \cos\phi)\rho^2 \sin\phi \, d\rho \, d\phi \, d\theta$$

$$= \frac{1}{V} \int_0^{2\pi} \int_0^{\pi/4} \left[\frac{\rho^4}{4} \cos\phi \sin\phi \right]_{\rho=0}^4 d\phi \, d\theta$$

$$= \frac{64}{V} \int_0^{2\pi} \int_0^{\pi/4} \sin\phi \cos\phi \, d\phi \, d\theta = \frac{64}{V} \int_0^{2\pi} \left[\frac{1}{2} \sin^2\phi \right]_{\phi=0}^{\pi/4} d\theta$$

$$= \frac{16}{V} \int_0^{2\pi} d\theta = \frac{32\pi}{V} = \frac{3}{2(2 - \sqrt{2})}$$

With the help of a calculator, $\bar{z} \approx 2.56$ (to two decimal places), so the approximate location of the centroid in the xyz-coordinate system is $(0, 0, 2.56)$.   ◀

☐ CONVERTING TRIPLE INTEGRALS FROM RECTANGULAR TO SPHERICAL COORDINATES

Referring to Table 14.8.1, triple integrals can be converted from rectangular coordinates to spherical coordinates by making the substitution $x = \rho \sin\phi \cos\theta$, $y = \rho \sin\phi \sin\theta$, $z = \rho \cos\phi$. The two integrals are related by the equation

$$\iiint\limits_{G} f(x, y, z) \, dV = \iiint\limits_{\substack{\text{appropriate} \\ \text{limits}}} f(\rho \sin\phi \cos\theta, \rho \sin\phi \sin\theta, \rho \cos\phi) \, \rho^2 \sin\phi \, d\rho \, d\phi \, d\theta$$

(9)

Example 4   Use spherical coordinates to evaluate

$$\int_{-2}^2 \int_{-\sqrt{4-x^2}}^{\sqrt{4-x^2}} \int_0^{\sqrt{4-x^2-y^2}} z^2 \sqrt{x^2 + y^2 + z^2} \, dz \, dy \, dx$$

Solution.   In problems like this, it is helpful to begin (when possible) with a sketch of the region G of integration. From the z-limits of integration, the upper surface of G is the hemisphere $z = \sqrt{4 - x^2 - y^2}$ and the lower surface is the xy-plane $z = 0$. From the x- and y-limits of integration, the projection of the solid G on the xy-plane is the region

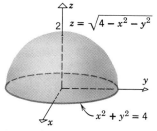

Figure 17.7.11

enclosed by the circle $x^2 + y^2 = 4$. From this information we obtain the sketch of G in Figure 17.7.11. Thus,

$$\int_{-2}^{2} \int_{-\sqrt{4-x^2}}^{\sqrt{4-x^2}} \int_{0}^{\sqrt{4-x^2-y^2}} z^2 \sqrt{x^2 + y^2 + z^2}\, dz\, dy\, dx$$

$$= \iiint_{G} z^2 \sqrt{x^2 + y^2 + z^2}\, dV$$

$$= \int_{0}^{2\pi} \int_{0}^{\pi/2} \int_{0}^{2} \rho^5 \cos^2 \phi \sin \phi \, d\rho\, d\phi\, d\theta$$

$$= \int_{0}^{2\pi} \int_{0}^{\pi/2} \frac{32}{3} \cos^2 \phi \sin \phi \, d\phi\, d\theta$$

$$= \frac{32}{3} \int_{0}^{2\pi} \left[-\frac{1}{3} \cos^3 \phi \right]_{\phi=0}^{\pi/2} d\theta = \frac{32}{9} \int_{0}^{2\pi} d\theta = \frac{64}{9} \pi \quad \blacktriangleleft$$

▶ Exercise Set 17.7

In Exercises 1–4, evaluate the iterated integral.

1. $\int_{0}^{2\pi} \int_{0}^{1} \int_{0}^{\sqrt{1-r^2}} zr\, dz\, dr\, d\theta.$

2. $\int_{0}^{\pi/2} \int_{0}^{\cos \theta} \int_{0}^{r^2} r \sin \theta \, dz\, dr\, d\theta.$

3. $\int_{0}^{\pi/2} \int_{0}^{\pi/2} \int_{0}^{1} \rho^3 \sin \phi \cos \phi \, d\rho\, d\phi\, d\theta.$

4. $\int_{0}^{2\pi} \int_{0}^{\pi/4} \int_{0}^{a \sec \phi} \rho^2 \sin \phi \, d\rho\, d\phi\, d\theta \quad (a > 0).$

In Exercises 5–9, use cylindrical coordinates to find the volume of the solid.

5. The solid bounded by the paraboloid $z = x^2 + y^2$ and the plane $z = 9$.

6. The solid that is bounded above and below by the sphere $x^2 + y^2 + z^2 = 9$ and inside the cylinder $x^2 + y^2 = 4$.

7. The solid that is enclosed between the surface $r^2 + z^2 = 20$ and the surface $z = r^2$.

8. The solid that is enclosed between the cone $z = hr/a$ and the plane $z = h$.

9. The solid in the first octant that is below by the sphere $x^2 + y^2 + z^2 = 16$ and lies inside the sphere $x^2 + y^2 = 4x$.

In Exercises 10–14, use spherical coordinates to find the volume of the solid.

10. The solid bounded above by the sphere $\rho = 4$ and below by the cone $\phi = \pi/3$.

11. The solid in the first octant bounded by the sphere $\rho = 2$, the coordinate planes, and the cones $\phi = \pi/6$ and $\phi = \pi/3$.

12. The solid within the cone $\phi = \pi/4$ and between the spheres $\rho = 1$ and $\rho = 2$.

13. The solid within the sphere $x^2 + y^2 + z^2 = 9$, outside the cone $z = \sqrt{x^2 + y^2}$, and above the xy-plane.

14. The solid bounded by the sphere $x^2 + y^2 + z^2 = 4a^2$ and the planes $z = 0$ and $z = a$.

15. Find the volume of $x^2 + y^2 + z^2 = a^2$ using

(a) cylindrical coordinates

(b) spherical coordinates.

In Exercises 16–18, use cylindrical coordinates.

16. Find the mass of the solid that is bounded by the cone $z = \sqrt{x^2 + y^2}$ and the plane $z = 3$ if the density of the solid is $\delta(x, y, z) = 3 - z$.

17. Find the mass of the solid in the first octant bounded above by the paraboloid $z = 4 - x^2 - y^2$, below by the plane $z = 0$, and laterally by the cylinder $x^2 + y^2 = 2x$ and the planes $x = 0$ and $y = 0$, assuming the density to be $\delta(x, y, z) = z$. [*Hint:* The Wallis formulas in Exercises 34 and 36 of Section 9.3 will help with the integration.]

18. Find the mass of a right-circular cylinder of radius a and height h if the density is proportional to the distance from the base. (Let k be the constant of proportionality.)

In Exercises 19–21, use spherical coordinates.

19. Find the mass of the solid that is enclosed by the sphere $x^2 + y^2 + z^2 = 1$ and lies within the cone $z = \sqrt{x^2 + y^2}$ if the density is $\delta(x, y, z) = \sqrt{x^2 + y^2 + z^2}$.

20. Find the mass of the solid enclosed between the spheres $x^2 + y^2 + z^2 = 1$ and $x^2 + y^2 + z^2 = 4$ if the density is $\delta(x, y, z) = (x^2 + y^2 + z^2)^{-1/2}$.

21. Find the mass of a spherical solid of radius a if the density is proportional to the distance from the center. (Let k be the constant of proportionality.)

In Exercises 22–24, use cylindrical coordinates to find the centroid of the solid.

22. The solid bounded by the cone $z = \sqrt{x^2 + y^2}$ and the plane $z = 2$.

23. The solid that is bounded above by the sphere
$$x^2 + y^2 + z^2 = 2$$
and below by the paraboloid $z = x^2 + y^2$.

24. The solid bounded above by the paraboloid $z = x^2 + y^2$, below by the plane $z = 0$, and laterally by the cylinder $(x - 1)^2 + y^2 = 1$. [*Hint:* Use the Wallis formulas in Exercises 34 and 36 of Section 9.3 for the integration.]

In Exercises 25–27, use spherical coordinates to find the centroid.

25. The solid in the first octant bounded by the coordinate planes and the sphere $x^2 + y^2 + z^2 = a^2$.

26. The solid bounded above by the sphere $\rho = 4$ and below by the cone $\phi = \pi/3$.

27. The solid that is enclosed by the hemispheres $y = \sqrt{9 - x^2 - z^2}$, $y = \sqrt{4 - x^2 - z^2}$, and the plane $y = 0$.

In Exercises 28–31, use cylindrical or spherical coordinates to evaluate the integral.

28. $\displaystyle\int_0^a \int_0^{\sqrt{a^2 - x^2}} \int_0^{a^2 - x^2 - y^2} x^2 \, dz \, dy \, dx \quad (a > 0)$.

29. $\displaystyle\int_{-1}^1 \int_0^{\sqrt{1 - x^2}} \int_0^{\sqrt{1 - x^2 - y^2}} e^{-(x^2 + y^2 + z^2)^{3/2}} \, dz \, dy \, dx$.

30. $\displaystyle\int_0^2 \int_0^{\sqrt{4 - y^2}} \int_{\sqrt{x^2 + y^2}}^{\sqrt{8 - x^2 - y^2}} z^2 \, dz \, dx \, dy$.

31. $\displaystyle\int_{-3}^3 \int_{-\sqrt{9 - y^2}}^{\sqrt{9 - y^2}} \int_{-\sqrt{9 - x^2 - y^2}}^{\sqrt{9 - x^2 - y^2}} \sqrt{x^2 + y^2 + z^2} \, dz \, dx \, dy$.

32. Let G be the solid in the first octant bounded by the sphere $x^2 + y^2 + z^2 = 4$ and the coordinate planes. In each part evaluate
$$\iiint_G xyz \, dV$$

(a) using rectangular coordinates

(b) using cylindrical coordinates

(c) using spherical coordinates.

Solve Exercises 33–38 using either cylindrical or spherical coordinates, whichever seems appropriate.

33. Find the center of gravity of the solid hemisphere bounded by $z = \sqrt{a^2 - x^2 - y^2}$ and $z = 0$ if the density is proportional to the distance from the origin.

34. Find the center of gravity of the solid in the first octant bounded by the cylinder $x^2 + y^2 = a^2$, the coordinate planes, and the plane $z = a$ if the density of the solid is $\delta(x, y, z) = xyz$.

35. Find the center of gravity of the solid bounded by the paraboloid $z = 1 - x^2 - y^2$ and the xy-plane if the density is $\delta(x, y, z) = x^2 + y^2 + z^2$.

36. Find the center of gravity of the solid that is bounded by the cylinder $x^2 + y^2 = 1$, the cone $z = \sqrt{x^2 + y^2}$, and the xy-plane if the density is $\delta(x, y, z) = z$.

37. Find the volume that is enclosed by the spheres
$$x^2 + y^2 + z^2 = 9 \quad \text{and} \quad x^2 + y^2 + (z - 2)^2 = 4$$

38. Suppose that the density at a point on a spherical planet is assumed to be
$$\delta = \delta_0 e^{[(\rho/R)^3 - 1]}$$
where δ_0 is a positive constant, R is the radius of the planet, and ρ is the distance from the point to the planet's center. Calculate the mass of the planet.

39. In this exercise we shall obtain a formula for the volume of the spherical wedge in Figure 17.7.7.

(a) Use a triple integral in cylindrical coordinates to show that the volume of the solid bounded above by a sphere $\rho = \rho_0$, below by a cone $\phi = \phi_0$, and on the sides by $\theta = \theta_1$ and $\theta = \theta_2$ ($\theta_1 < \theta_2$) is
$$V = \tfrac{1}{3}\rho_0^3(1 - \cos\phi_0)(\theta_2 - \theta_1)$$
[*Hint:* In cylindrical coordinates, the sphere has the equation $r^2 + z^2 = \rho_0^2$ and the cone has the equation $z = r \cot\phi_0$. For simplicity, consider only the case $0 < \phi_0 < \pi/2$.]

(b) Subtract appropriate volumes and use the result in part (a) to deduce that the volume ΔV of the spherical wedge is
$$\Delta V = \frac{\rho_2^3 - \rho_1^3}{3}(\cos\phi_1 - \cos\phi_2)(\theta_2 - \theta_1)$$

(c) Apply the Mean-Value Theorem to the functions $\cos\phi$ and ρ^3 to deduce that the formula in part (b) can be written as
$$\Delta V = \rho^{*2}\sin\phi^* \, \Delta\rho \, \Delta\phi \, \Delta\theta$$
where ρ^* is between ρ_1 and ρ_2, ϕ^* is between ϕ_1 and ϕ_2, and $\Delta\rho = \rho_2 - \rho_1$, $\Delta\phi = \phi_2 - \phi_1$, $\Delta\theta = \theta_2 - \theta_1$.

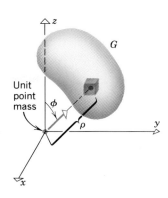

In Exercises 40–44, use the formulas in Exercise 36, Section 17.6, to find the indicated moment of inertia for the given homogeneous solid. Express your answer in terms of the mass M of the solid.

40. I_z for the solid cylinder $x^2 + y^2 \leq R^2$, $0 \leq z \leq h$.

41. I_y for the solid cylinder $x^2 + y^2 \leq R^2$, $0 \leq z \leq h$.

42. I_z for the hollow cylinder $R_1^2 \leq x^2 + y^2 \leq R_2^2$, $0 \leq z \leq h$.

43. I_z for the solid sphere $x^2 + y^2 + z^2 \leq R^2$.

44. I_z for the solid cone $\dfrac{h}{R}\sqrt{x^2 + y^2} \leq z \leq h$.

45. If d is the distance between two point masses m and m', then according to Newton's ***inverse-square law of gravitational attraction*** each point mass attracts the other with a force of magnitude kmm'/d^2, where k is a constant. Because force is a vector, the attraction between a lamina or solid and a point mass can be found by considering the components of the force. For example, assume that a solid G with mass density δ attracts a unit point mass at the origin (see the figure). The component in the z-direction of the force that G exerts on the unit mass can be found by approximating the corresponding components exerted by small subregions, summing, and taking the limit. This leads to the formula, using the spherical coordinates ρ and ϕ,

$$\iiint_G \frac{k\delta \cos \phi}{\rho^2}\, dV$$

for the component of force in the z-direction.

(a) Use cylindrical coordinates to find the component in the z-direction of the force of attraction on a unit mass at the origin by the homogeneous solid cylinder $x^2 + y^2 \leq R^2$, $0 < a \leq z \leq a + h$.

(b) Explain why the force in part (a) is the total force that acts on the unit mass.

In Exercises 46–48, use the formula in Exercise 45 to find the magnitude of the force of attraction on a unit mass at the origin by the given homogeneous lamina or solid. (The regions are all such that the force in the z-direction is the only force of attraction.)

46. The circular lamina $x^2 + y^2 \leq R^2$, $z = a > 0$. [*Hint:* If S is a lamina with area A, and δ is the *surface* density, then the formula is

$$\iint_S \frac{k\delta \cos \phi}{\rho^2}\, dA$$

Use polar coordinates to do the integration.]

47. The solid cone $\dfrac{h}{R}\, r \leq z \leq h$. (Use spherical coordinates.)

48. The solid sphere of radius R with its center on the z-axis at $z = a > R$. (Use spherical coordinates.) Show that the force is the same as if the total mass of the sphere were concentrated at its center.

■ **17.8 CHANGE OF VARIABLES IN MULTIPLE INTEGRALS; JACOBIANS**

> *In this section we shall discuss a general method for evaluating double and triple integrals by substitution. Most of the results in this section are very difficult to prove, so our approach will be informal and motivational. Our goal is to provide a geometric understanding of the basic principles and an exposure to computational techniques.*

☐ **CHANGE OF VARIABLE IN A SINGLE INTEGRAL**

To motivate techniques for evaluating double and triple integrals by substitution, it will be helpful to consider the effect of a substitution $x = g(u)$ on a single definite integral over an interval $[a, b]$, where g is differentiable and either increasing $[g'(u) > 0]$ or decreasing $[g'(u) < 0]$. In either case, g is one-to-one and

$$\int_a^b f(x)\, dx = \int_{g^{-1}(a)}^{g^{-1}(b)} f(g(u))g'(u)\, du$$

In this relationship $f(x)$ and dx are expressed in terms of u, and the u-limits of integration result from solving the equations

$$a = g(u) \quad \text{and} \quad b = g(u)$$

In the case where g is decreasing we have $g^{-1}(b) < g^{-1}(a)$, which is contrary to our usual convention of writing definite integrals with the larger limit of integration at the top. We can remedy this by reversing the limits of integration and writing

$$\int_a^b f(x)\, dx = -\int_{g^{-1}(b)}^{g^{-1}(a)} f(g(u))g'(u)\, du = \int_{g^{-1}(b)}^{g^{-1}(a)} f(g(u))|g'(u)|\, du$$

where the absolute value results from the fact that $g'(u)$ is negative. Thus, regardless of whether g is increasing or decreasing we can write

$$\int_a^b f(x)\, dx = \int_\alpha^\beta f(g(u))|g'(u)|\, du \tag{1}$$

where α and β are the u-limits of integration and $\alpha < \beta$.

The expression $g'(u)$ that appears in (1) is called the **Jacobian** of the change of variable $x = g(u)$ in honor of C. G. J. Jacobi,* who made the first serious study of change of variables in multiple integrals in the mid 1800s. Formula (1) reveals three effects of the change of variable $x = g(u)$:

- The new integrand becomes $f(g(u))$ times the absolute value of the Jacobian.
- dx becomes du.
- The x-interval of integration is transformed into a u-interval of integration.

Our goal in this section is to show that analogous results hold for changing variables in double and triple integrals.

☐ TRANSFORMATIONS OF THE PLANE

Equations of the form

$$x = x(u, v), \quad y = y(u, v) \tag{2}$$

associate a point (x, y) with a point (u, v), and hence the two equations define a function T that associates points in the xy-plane with points in the uv-plane according to the formula

$$T(u, v) = (x(u, v), y(u, v))$$

---

*CARL GUSTAV JACOB JACOBI (1804–1851). German mathematician. Jacobi, the son of a banker, grew up in a background of wealth and culture and showed brilliance in mathematics early. He resisted studying mathematics by rote, preferring instead to learn general principles from the works of the masters, Euler and Lagrange. He entered the University of Berlin at age 16 as a student of mathematics and classical studies. However, he soon realized that he could not do both and turned fully to mathematics with a blazing intensity that he would maintain throughout his life. He received his Ph.D. in 1825 and was able to secure a position as a lecturer at the University of Berlin by giving up Judaism and becoming a Christian. However, his promotion opportunities remained limited and he moved on to the University of Königsberg. Jacobi was born to teach—he had a dynamic personality and delivered his lectures with a clarity and enthusiasm that frequently left his audience spellbound. However, in spite of extensive teaching commitments, he was able to publish volumes of revolutionary mathematical research that eventually made him the leading European mathematician after Gauss. His main body of research was in the area of elliptic functions, a branch of mathematics with important applications in astronomy and physics as well as in other fields of mathematics. Because of his family wealth, Jacobi was not dependent on his teaching salary in his early years. However, his comfortable world eventually collapsed. In 1840 his family went bankrupt and he was personally wiped out financially. In 1842 he had a nervous breakdown from overwork. In 1843 he became seriously ill with diabetes and moved to Berlin with the help of a government grant to defray his medical expenses. In 1848 he made a stupid political remark that caused the government to withdraw the grant, eventually resulting in the loss of his home. His health continued to decline and in 1851 he finally succumbed to successive bouts of influenza and smallpox. In spite of all his problems, Jacobi was a tireless worker to the end. When a friend expressed concern about the effect of the hard work on his health, Jacobi replied, "Certainly, I have sometimes endangered my health by overwork, but what of it? Only cabbages have no nerves, no worries. And what do they get out of their perfect well-being?''

We call T a ***transformation*** from the uv-plane to the xy-plane and (x, y) the ***image*** of (u, v) under the transformation T. We also say that T ***maps*** (x, y) into (u, v). The set R of all images in the xy-plane of a set S in the uv-plane is called the ***image of S under T***. If distinct points in the uv-plane have distinct images in the xy-plane, then T is said to be ***one-to-one***. In this case the equations in (2) define u and v as functions of x and y, say

$$u = u(x, y), \quad v = v(x, y)$$

These equations, which can often be obtained by solving (2) for u and v in terms of x and y, define a transformation from the xy-plane to the uv-plane that maps the image of (u, v) under T back into (u, v). This transformation is denoted by T^{-1} and is called the ***inverse of T*** (Figure 17.8.1).

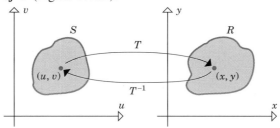

Figure 17.8.1

One way to visualize the geometric effect of a transformation T is to determine the images in the xy-plane of the vertical and horizontal lines in the uv-plane. Sets of points in the xy-plane that are images of horizontal lines (v constant) are called ***u-curves***, and sets of points that are images of vertical lines (u constant) are called ***v-curves*** (Figure 17.8.2).

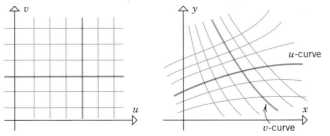

Figure 17.8.2

Example 1 Let T be the transformation from the uv-plane to the xy-plane defined by the equations

$$x = \tfrac{1}{4}(u + v), \quad y = \tfrac{1}{2}(u - v) \tag{3}$$

(a) Find $T(1, 3)$.

(b) Sketch the u-curves corresponding to $v = -2, -1, 0, 1, 2$.

(c) Sketch the v-curves corresponding to $u = -2, -1, 0, 1, 2$.

(d) Sketch the image under T of the square region in the uv-plane bounded by the lines $u = -2$, $u = 2$, $v = -2$, and $v = 2$.

Solution (a). Substituting $u = 1$ and $v = 3$ in (3) yields $T(1, 3) = (1, -1)$.

Solutions (b and c). In these parts it will be convenient to express the transformation equations with u and v as functions of x and y. We leave it for the reader to show that

$$u = 2x + y, \quad v = 2x - y$$

Thus, the u-curves corresponding to $v = -2, -1, 0, 1,$ and 2 are

$$2x - y = -2, \quad 2x - y = -1, \quad 2x - y = 0, \quad 2x - y = 1, \quad 2x - y = 2$$

and the v-curves corresponding to $u = -2, -1, 0, 1$, and 2 are

$$2x + y = -2, \quad 2x + y = -1, \quad 2x + y = 0, \quad 2x + y = 1, \quad 2x + y = 2$$

In Figure 17.8.3 the u-curves are shown in green and the v-curves in blue.

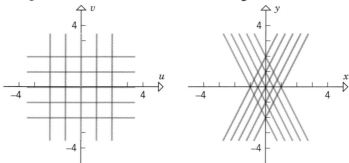

Figure 17.8.3

Solution (d). The image of a region can often be found by finding the image of its boundary. In this case the images of the boundary lines $u = -2, u = 2, v = -2$, and $v = 2$ enclose the diamond-shaped region in the xy-plane shown in Figure 17.8.4.

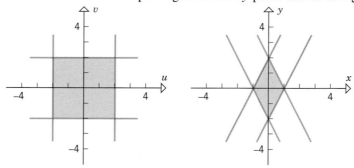

Figure 17.8.4

☐ **JACOBIANS IN TWO VARIABLES**

To derive the change-of-variable formula for double integrals, we will need to understand the relationship between the area of a *small* rectangular region in the uv-plane and the area of its image in the xy-plane under a transformation T given by the equations

$$x = x(u, v), \quad y = y(u, v)$$

For this purpose, suppose that Δu and Δv are positive, and consider a rectangular region S in the uv-plane enclosed by the lines

$$u = u_0, \quad u = u_0 + \Delta u, \quad v = v_0, \quad v = v_0 + \Delta v$$

If the functions $x(u, v)$ and $y(u, v)$ are continuous, and if Δu and Δv are not too large, then the image of S in the xy-plane is a "curvilinear" rectangular region R enclosed by the u-curves and v-curves corresponding to these lines (Figure 17.8.5).

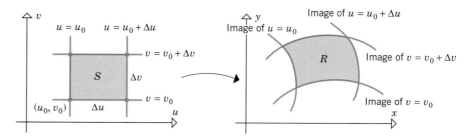

Figure 17.8.5

If we let

$$\mathbf{r} = \mathbf{r}(u, v) = x(u, v)\mathbf{i} + y(u, v)\mathbf{j}$$

be the position vector to the point in the xy-plane that corresponds to the point (u, v) in the uv-plane, then the u-curve corresponding to $v = v_0$ and the v-curve corresponding to $u = u_0$ can be represented in vector form as

$$\mathbf{r}(u, v_0) = x(u, v_0)\mathbf{i} + y(u, v_0)\mathbf{j} \qquad \boxed{u\text{-curve}}$$

$$\mathbf{r}(u_0, v) = x(u_0, v)\mathbf{i} + y(u_0, v)\mathbf{j} \qquad \boxed{v\text{-curve}}$$

Since we are assuming Δu and Δv to be small, the region R can be approximated by a parallelogram determined by the "secant vectors"

$$\mathbf{a} = \mathbf{r}(u_0 + \Delta u, v_0) - \mathbf{r}(u_0, v_0) \tag{4}$$

$$\mathbf{b} = \mathbf{r}(u_0, v_0 + \Delta v) - \mathbf{r}(u_0, v_0) \tag{5}$$

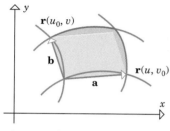

Figure 17.8.6

shown in Figure 17.8.6. A more useful approximation of R can be obtained by using Formulas (4) and (5) of Section 16.3 to approximate these secant vectors by tangent vectors as follows:

$$\mathbf{a} = \frac{\mathbf{r}(u_0 + \Delta u, v_0) - \mathbf{r}(u_0, v_0)}{\Delta u} \Delta u$$

$$\approx \frac{\partial \mathbf{r}}{\partial u} \Delta u = \left(\frac{\partial x}{\partial u} \mathbf{i} + \frac{\partial y}{\partial u} \mathbf{j} \right) \Delta u$$

$$\mathbf{b} = \frac{\mathbf{r}(u_0, v_0 + \Delta v) - \mathbf{r}(u_0, v_0)}{\Delta v} \Delta v$$

$$\approx \frac{\partial \mathbf{r}}{\partial v} \Delta v = \left(\frac{\partial x}{\partial v} \mathbf{i} + \frac{\partial y}{\partial v} \mathbf{j} \right) \Delta v$$

where the partial derivatives are evaluated at (u_0, v_0). But at $T(u_0, v_0)$ the vectors $\partial \mathbf{r}/\partial u$ and $\partial \mathbf{r}/\partial v$ are tangent to the u-curve and v-curve, respectively, and consequently so are $(\partial \mathbf{r}/\partial u) \Delta u$ and $(\partial \mathbf{r}/\partial v) \Delta v$. Thus, the region R can be approximated by the parallelogram determined by the tangent vectors

$$\frac{\partial \mathbf{r}}{\partial u} \Delta u = \left(\frac{\partial x}{\partial u} \mathbf{i} + \frac{\partial y}{\partial u} \mathbf{j} \right) \Delta u$$

$$\frac{\partial \mathbf{r}}{\partial v} \Delta v = \left(\frac{\partial x}{\partial v} \mathbf{i} + \frac{\partial y}{\partial v} \mathbf{j} \right) \Delta v$$

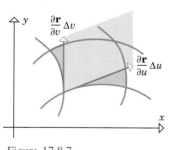

Figure 17.8.7

(Figure 17.8.7). Hence, it follows that the area of the region R, which we shall denote by ΔA, can be approximated by the area of the parallelogram determined by these vectors. Thus, from Formula (7) of Section 14.4 we have

$$\Delta A \approx \left\| \frac{\partial \mathbf{r}}{\partial u} \Delta u \times \frac{\partial \mathbf{r}}{\partial v} \Delta v \right\| = \left\| \frac{\partial \mathbf{r}}{\partial u} \times \frac{\partial \mathbf{r}}{\partial v} \right\| \Delta u \, \Delta v \tag{6}$$

where the derivatives are evaluated at (u_0, v_0). Computing the cross product, we obtain

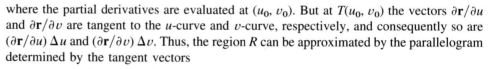

$$\frac{\partial \mathbf{r}}{\partial u} \times \frac{\partial \mathbf{r}}{\partial v} = \begin{vmatrix} \mathbf{i} & \mathbf{j} & \mathbf{k} \\ \dfrac{\partial x}{\partial u} & \dfrac{\partial y}{\partial u} & 0 \\ \dfrac{\partial x}{\partial v} & \dfrac{\partial y}{\partial v} & 0 \end{vmatrix} = \begin{vmatrix} \dfrac{\partial x}{\partial u} & \dfrac{\partial y}{\partial u} \\ \dfrac{\partial x}{\partial v} & \dfrac{\partial y}{\partial v} \end{vmatrix} \mathbf{k} = \begin{vmatrix} \dfrac{\partial x}{\partial u} & \dfrac{\partial x}{\partial v} \\ \dfrac{\partial y}{\partial u} & \dfrac{\partial y}{\partial v} \end{vmatrix} \mathbf{k} \tag{7}$$

The determinant in (7) is sufficiently important that it has its own terminology and notation.

17.8.1 DEFINITION. If T is the transformation from the uv-plane to the xy-plane defined by the equations $x = x(u, v)$, $y = y(u, v)$, then the **Jacobian of T** is denoted by $J(u, v)$ or by $\partial(x, y)/\partial(u, v)$ and is defined by

$$J(u, v) = \frac{\partial(x, y)}{\partial(u, v)} = \begin{vmatrix} \dfrac{\partial x}{\partial u} & \dfrac{\partial x}{\partial v} \\[2mm] \dfrac{\partial y}{\partial u} & \dfrac{\partial y}{\partial v} \end{vmatrix} = \frac{\partial x}{\partial u} \frac{\partial y}{\partial v} - \frac{\partial y}{\partial u} \frac{\partial x}{\partial v}$$

Using the notation in this definition, it follows from (6) and (7) that

$$\Delta A \approx \left\| \frac{\partial(x, y)}{\partial(u, v)} \mathbf{k} \right\| \Delta u \, \Delta v$$

or since $\mathbf{k}$ is a unit vector,

$$\Delta A \approx \left| \frac{\partial(x, y)}{\partial(u, v)} \right| \Delta u \, \Delta v \tag{8}$$

At the point (u_0, v_0) this important formula relates the areas of the regions R and S in Figure 17.8.5: It tells us that *for small values of Δu and Δv, the area of R is approximately the absolute value of the Jacobian times the area of S.* Moreover, it is proved in advanced calculus courses that the error in the approximation approaches zero as $\Delta u \to 0$ and $\Delta v \to 0$.

□ **CHANGE OF VARIABLES IN DOUBLE INTEGRALS**

Our next objective is to provide a geometric motivation for the following result.

17.8.2 CHANGE-OF-VARIABLE FORMULA FOR DOUBLE INTEGRALS. If the transformation $x = x(u, v)$, $y = y(u, v)$ maps the region S in the uv-plane into the region R in the xy-plane, and if the Jacobian $\partial(x, y)/\partial(u, v)$ is nonzero and does not change sign on S, then with appropriate restrictions on the transformation and the regions it follows that

$$\iint\limits_{R} f(x, y) \, dA_{xy} = \iint\limits_{S} f(x(u, v), y(u, v)) \left| \frac{\partial(x, y)}{\partial(u, v)} \right| dA_{uv} \tag{9}$$

where we have attached subscripts to the dA's to help identify the associated variables.

REMARK.   A precise statement of conditions under which Formula (9) holds would take us beyond the scope of this course. Suffice it to say that the formula holds if T is a one-to-one transformation, $f(x, y)$ is continuous on R, the partial derivatives of $x(u, v)$ and $y(u, v)$ exist and are continuous on S, and the regions R and S are not too complicated.

To motivate Formula (9), we proceed as follows:

- Subdivide the region S in the uv-plane into pieces by lines parallel to the coordinate axes, and exclude from consideration any pieces that contain points outside of S. This leaves only rectangular regions that are subsets of S. Assume that there are n such regions and denote the kth such region by S_k. Assume that S_k has dimensions Δu_k by Δv_k and, as shown in Figure 17.8.8a, let (u_k^*, v_k^*) be its "lower left corner."

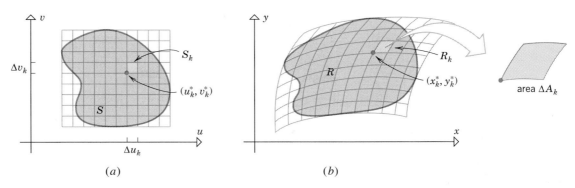

Figure 17.8.8   (a)   (b)

- As shown in Figure 17.8.8b, the transformation T defined by the equations $x = x(u, v)$, $y = y(u, v)$ maps S_k into a curvilinear rectangular region R_k in the xy-plane and maps the point (u_k^*, v_k^*) into the point $(x_k^*, y_k^*) = (x(u_k^*, v_k^*), y(u_k^*, v_k^*))$ in R_k. Denote the area of R_k by ΔA_k.

- In rectangular coordinates the double integral of $f(x, y)$ over a region R is defined as a limit of Riemann sums in which R is subdivided into *rectangular* subregions. It is proved in advanced calculus courses that under appropriate conditions subdivisions into *curvilinear* rectangular subregions can be used instead. Accepting this to be so, we can approximate the double integral of $f(x, y)$ over R as

$$\iint_R f(x, y)\, dA_{xy} \approx \sum_{k=1}^{n} f(x_k^*, y_k^*)\, \Delta A_k$$

$$\approx \sum_{k=1}^{n} f(x(u_k^*, v_k^*), y(u_k^*, v_k^*)) \left| \frac{\partial(x, y)}{\partial(u, v)} \right| \Delta u_k\, \Delta v_k$$

where the Jacobian is evaluated at (u_k^*, v_k^*). But the last expression is a Riemann sum for the integral

$$\iint_S f(x(u, v), y(u, v)) \left| \frac{\partial(x, y)}{\partial(u, v)} \right| dA_{uv}$$

so Formula (9) follows if we assume that the errors in the approximations approach zero as $n \to +\infty$.

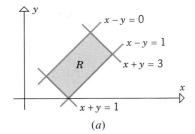

Figure 17.8.9

Example 2 Evaluate

$$\iint_R \frac{x - y}{x + y}\, dA$$

where R is the region enclosed by the lines $x - y = 0$, $x - y = 1$, $x + y = 1$, and $x + y = 3$ (Figure 17.8.9a).

Solution. This integral would be tedious to evaluate directly because the region R is oriented in such a way that we would have to subdivide it and integrate over each part separately. However, the occurrence of the expressions $x - y$ and $x + y$ in the equations of the boundary suggests that the transformation

$$u = x + y, \quad v = x - y \tag{10}$$

would be helpful, since with this transformation the boundary lines

$$x + y = 1, \quad x + y = 3, \quad x - y = 0, \quad x - y = 1$$

are u-curves and v-curves corresponding to the lines

$$u = 1, \quad u = 3, \quad v = 0, \quad v = 1$$

in the uv-plane. These lines enclose the region S shown in Figure 17.8.9b. To find the Jacobian $\partial(x, y)/\partial(u, v)$ of this transformation, we first solve (10) for x and y in terms of u and v. This yields

$$x = \tfrac{1}{2}(u + v), \quad y = \tfrac{1}{2}(u - v)$$

from which we obtain

$$\frac{\partial(x, y)}{\partial(u, v)} = \begin{vmatrix} \dfrac{\partial x}{\partial u} & \dfrac{\partial x}{\partial v} \\ \dfrac{\partial y}{\partial u} & \dfrac{\partial y}{\partial v} \end{vmatrix} = \begin{vmatrix} \tfrac{1}{2} & \tfrac{1}{2} \\ \tfrac{1}{2} & -\tfrac{1}{2} \end{vmatrix} = -\tfrac{1}{4} - \tfrac{1}{4} = -\tfrac{1}{2}$$

Thus, from Formula (9), but with the notation dA rather than dA_{xy},

$$\iint\limits_{R} \frac{x - y}{x + y} \, dA = \iint\limits_{S} \frac{v}{u} \left| \frac{\partial(x, y)}{\partial(u, v)} \right| dA_{uv}$$

$$= \iint\limits_{S} \frac{v}{u} \left| -\frac{1}{2} \right| dA_{uv} = \frac{1}{2} \int_{0}^{1} \int_{1}^{3} \frac{v}{u} \, du \, dv$$

$$= \frac{1}{2} \int_{0}^{1} v \ln |u| \Big]_{u=1}^{3} dv$$

$$= \frac{1}{2} \ln 3 \int_{0}^{1} v \, dv = \frac{1}{4} \ln 3 \quad \blacktriangleleft$$

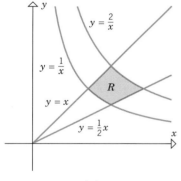

(a)

(b)

Figure 17.8.10

Example 3   Evaluate

$$\iint\limits_{R} e^{xy} \, dA$$

where R is the region enclosed by the lines $y = \tfrac{1}{2}x$ and $y = x$ and the hyperbolas $y = 1/x$ and $y = 2/x$ (Figure 17.8.10a).

Solution.   As in the preceding example, we look for a transformation in which the boundary curves in the xy-plane become u-curves and v-curves. For this purpose we rewrite the four boundary curves as

$$\frac{y}{x} = \tfrac{1}{2}, \quad \frac{y}{x} = 1, \quad xy = 1, \quad xy = 2$$

which suggests the transformation

$$u = \frac{y}{x}, \quad v = xy \tag{11}$$

With this transformation the boundary curves in the xy-plane are u-curves and v-curves corresponding to the lines

$$u = \tfrac{1}{2}, \quad u = 1, \quad v = 1, \quad v = 2$$

in the uv-plane. These lines enclose the region S shown in Figure 17.8.10b. To find the Jacobian $\partial(x, y)/\partial(u, v)$ of this transformation, we first solve (11) for x and y in terms of u and v. This yields

$$x = \sqrt{v/u}, \quad y = \sqrt{uv}$$

from which we obtain

$$\frac{\partial(x, y)}{\partial(u, v)} = \begin{vmatrix} \dfrac{\partial x}{\partial u} & \dfrac{\partial x}{\partial v} \\ \dfrac{\partial y}{\partial u} & \dfrac{\partial y}{\partial v} \end{vmatrix} = \begin{vmatrix} -\dfrac{1}{2u}\sqrt{\dfrac{v}{u}} & \dfrac{1}{2\sqrt{uv}} \\ \dfrac{1}{2}\sqrt{\dfrac{v}{u}} & \dfrac{1}{2}\sqrt{\dfrac{u}{v}} \end{vmatrix} = -\frac{1}{4u} - \frac{1}{4u} = -\frac{1}{2u}$$

Thus, from Formula (9), but with the notation dA rather than dA_{xy},

$$\iint\limits_{R} e^{xy}\, dA = \iint\limits_{S} e^{v}\left| -\frac{1}{2u} \right| dA_{uv} = \frac{1}{2} \iint\limits_{S} \frac{1}{u} e^{v}\, dA_{uv}$$

$$= \frac{1}{2} \int_{1}^{2} \int_{1/2}^{1} \frac{1}{u} e^{v}\, du\, dv = \frac{1}{2} \int_{1}^{2} e^{v} \ln|u| \Big]_{u=1/2}^{1} dv$$

$$= \frac{1}{2} \ln 2 \int_{1}^{2} e^{v}\, dv = \frac{1}{2}(e^{2} - e) \ln 2 \qquad \blacktriangleleft$$

☐ **CHANGE OF VARIABLES IN TRIPLE INTEGRALS**

Equations of the form

$$x = x(u, v, w), \quad y = y(u, v, w), \quad z = z(u, v, w) \tag{12}$$

define a **transformation** T from uvw-space to xyz-space. Just as a transformation $x = x(u, v)$, $y = y(u, v)$ in two variables maps small rectangles in the uv-plane into curvilinear rectangles in the xy-plane, so (12) maps small rectangular parallelepipeds in uvw-space into curvilinear parallelepipeds in xyz-space (Figure 17.8.11). The definition of the Jacobian of (12) is similar to Definition 17.8.1.

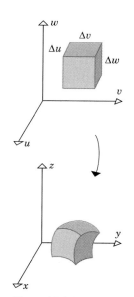

Figure 17.8.11

17.8.3 DEFINITION. If T is the transformation from uvw-space to xyz-space defined by the equations $x = x(u, v, w)$, $y = y(u, v, w)$, $z = z(u, v, w)$, then the **Jacobian of T** is denoted by $J(u, v, w)$ or $\partial(x, y, z)/\partial(u, v, w)$ and is defined by

$$J(u, v, w) = \frac{\partial(x, y, z)}{\partial(u, v, w)} = \begin{vmatrix} \dfrac{\partial x}{\partial u} & \dfrac{\partial x}{\partial v} & \dfrac{\partial x}{\partial w} \\ \dfrac{\partial y}{\partial u} & \dfrac{\partial y}{\partial v} & \dfrac{\partial y}{\partial w} \\ \dfrac{\partial z}{\partial u} & \dfrac{\partial z}{\partial v} & \dfrac{\partial z}{\partial w} \end{vmatrix}$$

For small values of Δu, Δv, and Δw, the volume ΔV of the curvilinear parallelepiped in Figure 17.8.11 is related to the volume $\Delta u\, \Delta v\, \Delta w$ of the rectangular parallelepiped by

$$\Delta V \approx \left| \frac{\partial(x, y, z)}{\partial(u, v, w)} \right| \Delta u\, \Delta v\, \Delta w \tag{13}$$

which is the analog of Formula (8). Using this relationship and an argument similar to the one that led to Formula (9), one can obtain the following result.

17.8.4 CHANGE-OF-VARIABLE FORMULA FOR TRIPLE INTEGRALS. If the transformation $x = x(u, v, w)$, $y = y(u, v, w)$, $z = z(u, v, w)$ maps the region S in uvw-space into the region R in xyz-space, and if the Jacobian $\partial(x, y, z)/\partial(u, v, w)$ is nonzero and does not change sign on S, then with appropriate restrictions on the transformation and the regions it follows that

$$\iiint\limits_{R} f(x, y, z)\, dV_{xyz} = \iiint\limits_{S} f(x(u, v, w), y(u, v, w), z(u, v, w)) \left| \frac{\partial(x, y, z)}{\partial(u, v, w)} \right| dV_{uvw}$$

$$\tag{14}$$

Example 4 Find the volume of the region G enclosed by the ellipsoid

$$\frac{x^2}{a^2} + \frac{y^2}{b^2} + \frac{z^2}{c^2} = 1$$

Solution. The volume V is given by the triple integral

$$V = \iiint\limits_{G} dV$$

To evaluate this integral, we make the change of variables

$$x = au, \quad y = bv, \quad z = cw \tag{15}$$

which maps the region G in xyz-space into the region S in uvw-space enclosed by a sphere of radius 1. This can be seen from (15) by noting that

$$\frac{x^2}{a^2} + \frac{y^2}{b^2} + \frac{z^2}{c^2} = 1 \quad \text{becomes} \quad u^2 + v^2 + w^2 = 1$$

The Jacobian of (15) is

$$\frac{\partial(x, y, z)}{\partial(u, v, w)} = \begin{vmatrix} \dfrac{\partial x}{\partial u} & \dfrac{\partial x}{\partial v} & \dfrac{\partial x}{\partial w} \\[2mm] \dfrac{\partial y}{\partial u} & \dfrac{\partial y}{\partial v} & \dfrac{\partial y}{\partial w} \\[2mm] \dfrac{\partial z}{\partial u} & \dfrac{\partial z}{\partial v} & \dfrac{\partial z}{\partial w} \end{vmatrix} = \begin{vmatrix} a & 0 & 0 \\ 0 & b & 0 \\ 0 & 0 & c \end{vmatrix} = abc$$

Thus, from Formula (14), but with the notation dV rather than dV_{xyz},

$$V = \iiint\limits_{G} dV = \iiint\limits_{S} \left| \frac{\partial(x, y, z)}{\partial(u, v, w)} \right| dV_{uvw} = abc \iiint\limits_{S} dV_{uvw}$$

The last integral is the volume enclosed by a sphere of radius 1, which we know to be $\frac{4}{3}\pi$. Thus, the volume enclosed by the ellipsoid is $V = \frac{4}{3}\pi abc$. ◀

Jacobians also relate to the problems of converting double and triple integrals in rectangular coordinates to iterated integrals in polar, cylindrical, or spherical coordinates. For example, from Formula (14) the relationship between a triple integral in rectangular coordinates and the corresponding iterated integral in cylindrical coordinates is

$$\iiint\limits_{G} f(x, y, z)\, dV_{xyz} = \iiint\limits_{S} f(r\cos\theta, r\sin\theta, z) \left| \frac{\partial(x, y, z)}{\partial(r, \theta, z)} \right| dV_{r\theta z}$$

$$= \iiint\limits_{\substack{\text{appropriate} \\ r\theta z\text{-limits}}} f(r\cos\theta, r\sin\theta, z) r\, dr\, d\theta\, dz \tag{16}$$

The factor r that appears in the final integrand is precisely the absolute value of the Jacobian of the transformation

$$x = r\cos\theta, \quad y = r\sin\theta, \quad z = z \tag{17}$$

from rectangular to cylindrical coordinates, since

$$\frac{\partial(x, y, z)}{\partial(r, \theta, z)} = \begin{vmatrix} \dfrac{\partial x}{\partial r} & \dfrac{\partial x}{\partial \theta} & \dfrac{\partial x}{\partial z} \\ \dfrac{\partial y}{\partial r} & \dfrac{\partial y}{\partial \theta} & \dfrac{\partial y}{\partial z} \\ \dfrac{\partial z}{\partial r} & \dfrac{\partial z}{\partial \theta} & \dfrac{\partial z}{\partial z} \end{vmatrix} = \begin{vmatrix} \cos\theta & -r\sin\theta & 0 \\ \sin\theta & r\cos\theta & 0 \\ 0 & 0 & 1 \end{vmatrix}$$

$$= r\cos^2\theta + r\sin^2\theta = r$$

and $r \geq 0$ in (17).

▶ Exercise Set 17.8

In Exercises 1–4, find the Jacobian $\partial(x, y)/\partial(u, v)$.

1. $x = u + 4v$, $y = 3u - 5v$.

2. $x = u + 2v^2$, $y = 2u^2 - v$.

3. $x = \sin u + \cos v$, $y = -\cos u + \sin v$.

4. $x = \dfrac{2u}{u^2 + v^2}$, $y = -\dfrac{2v}{u^2 + v^2}$.

In Exercises 5–8, solve for x and y in terms of u and v, and then find the Jacobian $\partial(x, y)/\partial(u, v)$.

5. $u = 2x - 5y$, $v = x + 2y$.

6. $u = e^x$, $v = ye^{-x}$.

7. $u = x^2 - y^2$, $v = x^2 + y^2$ $(x > 0, y > 0)$.

8. $u = xy$, $v = xy^3$ $(x > 0, y > 0)$.

In Exercises 9–11, find the Jacobian $\partial(x, y, z)/\partial(u, v, w)$.

9. $x = 3u + v$, $y = u - 2w$, $z = v + w$.

10. $x = u - uv$, $y = uv - uvw$, $z = uvw$.

11. $u = xy$, $v = y$, $w = x + z$.

12. (a) Consider the transformation $x = r\cos\theta$, $y = r\sin\theta$ from rectangular to polar coordinates, where $r \geq 0$. Show that

$$\left| \frac{\partial(x, y)}{\partial(r, \theta)} \right| = r$$

 (b) Consider the transformation $x = \rho\sin\phi\cos\theta$, $y = \rho\sin\phi\sin\theta$, $z = \rho\cos\phi$ from rectangular to spherical coordinates, where $0 \leq \phi \leq \pi$. Show that

$$\left| \frac{\partial(x, y, z)}{\partial(\rho, \theta, \phi)} \right| = \rho^2 \sin\phi$$

13. Use the transformation $u = x - 2y$, $v = 2x + y$ to find

$$\iint_R \frac{x - 2y}{2x + y} \, dA$$

where R is the rectangular region enclosed by the lines $x - 2y = 1$, $x - 2y = 4$, $2x + y = 1$, $2x + y = 3$.

14. Use the transformation $u = x + y$, $v = x - y$ to find

$$\iint_R (x - y)e^{x^2 - y^2} \, dA$$

where R is the rectangular region enclosed by the lines $x + y = 0$, $x + y = 1$, $x - y = 1$, $x - y = 4$.

15. Use the transformation $u = \frac{1}{2}(x + y)$, $v = \frac{1}{2}(x - y)$ to find

$$\iint_R \sin\tfrac{1}{2}(x + y) \cos\tfrac{1}{2}(x - y) \, dA$$

where R is the triangular region with vertices $(0, 0)$, $(2, 0)$, $(1, 1)$.

16. Use the transformation $u = y/x$, $v = xy$ to find

$$\iint_R xy^3 \, dA$$

where R is the region in the first quadrant enclosed by $y = x$, $y = 3x$, $xy = 1$, $xy = 4$.

17. Use the transformation $x = 3u$, $y = 4v$ to find

$$\iint_R \sqrt{16x^2 + 9y^2} \, dA$$

where R is the region enclosed by the ellipse $x^2/9 + y^2/16 = 1$. [*Hint:* Use polar coordinates to evaluate the transformed integral.]

18. Use the transformation $x = 2u$, $y = v$ to find

$$\iint_R e^{-(x^2 + 4y^2)} \, dA$$

where R is the region enclosed by the ellipse $x^2/4 + y^2 = 1$. [*Hint:* Use polar coordinates to evaluate the transformed integral.]

It will be shown in Exercise 38 that the following relationship holds for a one-to-one transformation:

$$\frac{\partial(x, y)}{\partial(u, v)} \cdot \frac{\partial(u, v)}{\partial(x, y)} = 1 \qquad \text{(A)}$$

In Exercises 19–22, verify this result by finding the two factors on the left side of this equation.

19. $x = u - uv$, $y = uv$.

20. $x = uv$, $y = v^2$.

21. $x = v^2/u$, $y = v/u$.

22. $x = \frac{1}{2}(u^2 + v^2)$, $y = \frac{1}{2}(u^2 - v^2)$ $(u > 0, v > 0)$.

23. Evaluate the integral in Exercise 16 by using Formula (A) above to find $\partial(x, y)/\partial(u, v)$ from $\partial(u, v)/\partial(x, y)$ without solving for x and y in terms of u and v.

In Exercises 24–26, use the method of Exercise 23 to find the Jacobian $\partial(x, y)/\partial(u, v)$ required for the integration.

24. Use the transformation $u = x^2 - y^2$, $v = x^2 + y^2$ to find

$$\iint_R xy \, dA$$

where R is the region in the first quadrant enclosed by the hyperbolas $x^2 - y^2 = 1$, $x^2 - y^2 = 4$ and the circles $x^2 + y^2 = 9$, $x^2 + y^2 = 16$.

25. Use the transformation $u = xy$, $v = xy^4$ to find

$$\iint_R \sin(xy) \, dA$$

where R is the region enclosed by the curves $xy = \pi$, $xy = 2\pi$, $xy^4 = 1$, $xy^4 = 2$.

26. Use the transformation $u = xy$, $v = x^2 - y^2$ to find

$$\iint_R (x^4 - y^4)e^{xy} \, dA$$

where R is the region in the first quadrant enclosed by the hyperbolas $xy = 1$, $xy = 3$, $x^2 - y^2 = 3$, $x^2 - y^2 = 4$.

In Exercises 27–30, evaluate the integral by making an appropriate change of variables.

27. $\displaystyle\iint_R \frac{y - 4x}{y + 4x} \, dA$, where R is the region enclosed by the lines $y = 4x$, $y = 4x + 2$, $y = 2 - 4x$, $y = 5 - 4x$.

28. $\displaystyle\iint_R (x^2 - y^2) \, dA$, where R is the rectangular region enclosed by the lines $y = -x$, $y = 1 - x$, $y = x$, $y = x + 2$.

29. $\displaystyle\iint_R \frac{\sin(x - y)}{\cos(x + y)} \, dA$, where R is the triangular region enclosed by the lines $y = 0$, $y = x$, $x + y = \pi/4$.

30. $\displaystyle\iint_R e^{(y-x)/(y+x)} \, dA$, where R is the region in the first quadrant enclosed by the trapezoid with vertices $(0, 1)$, $(1, 0)$, $(0, 4)$, $(4, 0)$.

31. Use an appropriate change of variables to find the area of the region in the first quadrant enclosed by the curves $y = x$, $y = 2x$, $x = y^2$, $x = 4y^2$.

32. Use an appropriate change of variables to find the volume of the solid bounded above by the plane $x + y + z = 9$, below by the xy-plane, and laterally by the elliptic cylinder $4x^2 + 9y^2 = 36$. [*Hint:* Express the volume as a double integral in xy-coordinates, then use polar coordinates to evaluate the transformed integral.]

33. Use the transformation $u = x$, $v = z - y$, $w = xy$ to find

$$\iiint_G (z - y)^2 xy \, dV$$

where G is the region enclosed by the surfaces $x = 1$, $x = 3$, $z = y$, $z = y + 1$, $xy = 2$, $xy = 4$.

34. Use the transformation $u = xy$, $v = yz$, $w = xz$ to find

$$\iiint_G \sqrt{xyz} \, dV$$

where G is the region in the first octant enclosed by the hyperbolic cylinders $xy = 1$, $xy = 2$, $yz = 1$, $yz = 3$, $xz = 1$, $xz = 4$.

35. Use the transformation $x = au$, $y = bv$, $z = cw$ to find

$$\iiint_G x^2 \, dV$$

where G is the region enclosed by the ellipsoid $x^2/a^2 + y^2/b^2 + z^2/c^2 = 1$. [*Hint:* Use spherical coordinates to evaluate the transformed integral.]

36. Find the volume of the region G in Exercise 34.

37. Use the transformation $u = y/z$, $v = 4x - y$, $w = y/z^2$ to find the volume of the region enclosed by the surfaces $y = z$, $y = 2z$, $y = 4x$, $y = 4x - 12$, $y = z^2$, $y = 4z^2$.

38. (a) Verify that

$$\begin{vmatrix} a_1 & b_1 \\ c_1 & d_1 \end{vmatrix} \begin{vmatrix} a_2 & b_2 \\ c_2 & d_2 \end{vmatrix} = \begin{vmatrix} a_1 a_2 + b_1 c_2 & a_1 b_2 + b_1 d_2 \\ c_1 a_2 + d_1 c_2 & c_1 b_2 + d_1 d_2 \end{vmatrix}$$

(b) If $x = x(u, v)$, $y = y(u, v)$ is a one-to-one transformation, then $u = u(x, y)$, $v = v(x, y)$. Assuming differentiability, use the result in part (a) and the chain rule to show that

$$\frac{\partial(x, y)}{\partial(u, v)} \cdot \frac{\partial(u, v)}{\partial(x, y)} = 1$$

◆ TECHNOLOGY EXERCISES   Chapter 17

Most of these exercises require access to a graphing calculator or a computer algebra system (CAS) such as *Mathematica*, *Maple*, or *Derive*. When you are asked to *find* an answer or to *solve* an equation, you may choose to find an exact result or a numerical approximation, depending on the particular technology you are using and on your own imagination. The form of your answers may differ from those of other students or from those in the answer section of the text, depending on how you solve the problems and the accuracy you use in your numerical approximations. Those exercises that are more appropriate for a CAS than a graphing calculator are labeled with the icon ◆.

◆ **1. Volume:** Find the volume of the solid in the first octant beneath the surface $z = \sqrt{1 + x + y}$ and above the region in the xy-plane enclosed by $y = \sin x$ and $y = x/2$.

◆ **2. Volume in polar coordinates:** Use polar coordinates to find the volume of the solid that lies beneath the surface $z = x^2\sqrt{1 + x^2 + y^2}$ and above the region in the xy-plane enclosed by $r = 1 + \cos \theta$.

◆ **3. Surface area:** Find the area of the portion of the surface $z = e^{-x^2-y^2}$ that is inside the cylinder $x^2 + y^2 = 1$.

◆ **4. Spherical coordinates:** Use spherical coordinates to find the exact value of

$$\iiint_G \frac{1}{x^2 + y^2 + (z - 2)^2} \, dV$$

where G is the solid that is enclosed by the sphere $x^2 + y^2 + z^2 = 1$.

◆ **5. The Theorem of Pappus:** Let R be the region in the xy-plane enclosed by the graphs of $y = 3 + x - x^2$ and $y = 1 + \cos x$.

(a) Find the area of R.

(b) Find the y-coordinate of the centroid of R.

(c) Use the results in parts (a) and (b) and the Theorem of Pappus to find the volume of the solid generated by revolving R about the x-axis. Check your result by using the method of washers to find the volume.

◆ **6. Centroid:** Find the centroid of the solid beneath the surface $z = 1/(1 + x^2 + y^2)$ and above the region in the xy-plane enclosed by $y = \sin x$ and the x-axis from $x = 0$ to $x = \pi$.

7. Centroid: Let G be the solid that is enclosed by the surface $z = 1/(1 + x^2 + y^2)$, the xy-plane, and the cylinder $x^2 + y^2 = a^2$. Find the value of a for which the centroid of G is $(0, 0, \frac{1}{4})$.

◆ **8. Centroid of a solid of infinite extent:** Find the centroid of the solid beneath the surface $z = e^{2x+y-x^2-y^2}$ and above the xy-plane.

$\mathcal{C.F.Gauss.}$

Carl Friedrich Gauss (1777–1855)

18 TOPICS IN VECTOR CALCULUS

■ 18.1 VECTOR FIELDS

In this section we consider functions that associate vectors with points in 2-space or 3-space. We shall see that such functions play an important role in the study of fluid flow, gravitational force fields, electromagnetic force fields, and a wide range of other applied problems.

☐ **VECTOR FIELDS**

To motivate the mathematical ideas in this section, consider a unit point mass located at any point in the universe. According to Newton's Universal Law of Gravitation, the earth exerts an attractive force on the mass that is directed toward the earth's center and has a magnitude that is inversely proportional to the square of the distance from the mass to the earth's center (Figure 18.1.1). This association of force vectors with points in space is called the earth's *gravitational field*. A similar idea arises in fluid flow. Imagine a stream in which the water flows horizontally at every level, and consider the layer of water at a

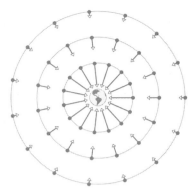

Figure 18.1.1

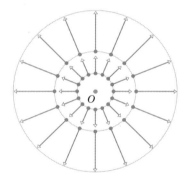

Figure 18.1.2

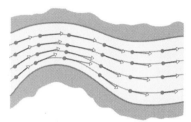

Figure 18.1.3

specific depth. At each point of the layer, the water has a certain velocity, which we can represent by a vector at that point (Figure 18.1.2). This association of velocity vectors with points in the two-dimensional layer is called the *flow field* at that layer. These ideas are captured in the following definition.

> **18.1.1**   DEFINITION.   A ***vector field*** is a function that associates a unique vector $\mathbf{F}(P)$ with each point P in a region of 2-space or 3-space.

Example 1   Let O be a fixed point in 2-space, and for each point P in 2-space define the vector field $\mathbf{F}(P)$ by $\mathbf{F}(P) = \overrightarrow{OP}$. Some typical vectors in this vector field are shown in Figure 18.1.3. In that figure we have followed the standard convention of positioning the vector $\mathbf{F}(P)$ with its initial point at P.   ◄

Observe that the concept of a vector field has been defined without reference to a coordinate system; it is said to be a ***coordinate-free*** definition. However, for computational purposes it is often desirable to work with vector fields in coordinate systems. If $\mathbf{F}(P)$ is a vector field in 2-space with an xy-coordinate system, then the point P has coordinates (x, y), and the components of the vector $\mathbf{F}(P)$ are functions of x and y. Thus, $\mathbf{F}(P)$ can be expressed as

$$\mathbf{F}(x, y) = f(x, y)\mathbf{i} + g(x, y)\mathbf{j}$$

Similarly, in 3-space with an xyz-coordinate system, a vector field $\mathbf{F}(P)$ can be expressed as

$$\mathbf{F}(x, y, z) = f(x, y, z)\mathbf{i} + g(x, y, z)\mathbf{j} + h(x, y, z)\mathbf{k}$$

Just as it is impossible to describe a curve completely by plotting finitely many points, so it is impossible to describe a vector field completely by drawing finitely many vectors. Nevertheless, it is often possible to get a useful picture of a vector field by sketching a finite number of vectors that are well chosen.

Example 2   Figure 18.1.4 shows sketches of three vector fields in 2-space. For simplicity, we have omitted the scales and selected vectors that do not overlap; nevertheless, the sketches still provide some useful geometric insight into the behavior of the fields.   ◄

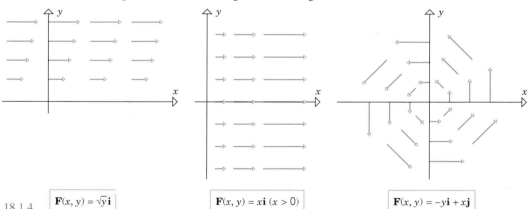

Figure 18.1.4

$\boxed{\mathbf{F}(x, y) = \sqrt{y}\,\mathbf{i}}$ $\boxed{\mathbf{F}(x, y) = x\mathbf{i}\ (x > 0)}$ $\boxed{\mathbf{F}(x, y) = -y\mathbf{i} + x\mathbf{j}}$

REMARK.   Sometimes it is helpful to denote the vector fields $\mathbf{F}(x, y)$ and $\mathbf{F}(x, y, z)$ entirely in vector notation by identifying (x, y) with the radius vector $\mathbf{r} = x\mathbf{i} + y\mathbf{j}$ and (x, y, z) with the radius vector $\mathbf{r} = x\mathbf{i} + y\mathbf{j} + z\mathbf{k}$. With this notation a vector field in either 2-space or 3-space can be written as $\mathbf{F}(\mathbf{r})$. When no confusion is likely to arise, we shall sometimes omit the $\mathbf{r}$ altogether and denote the vector field as $\mathbf{F}$.

□ INVERSE-SQUARE FIELDS

According to Newton's Universal Law of Gravitation, objects with masses m and M attract each other with a force $\mathbf{F}$ of magnitude

$$\|\mathbf{F}\| = \frac{GmM}{r^2} \tag{1}$$

where r is the distance between the objects (treated as point masses) and G is a constant. If we assume that the object of mass M is located at the origin of an xyz-coordinate system and $\mathbf{r}$ is the radius vector to the object of mass m, then $r = \|\mathbf{r}\|$, and the force $\mathbf{F}(\mathbf{r})$ exerted by the object of mass M on the object of mass m is in the direction of the unit vector $-\mathbf{r}/\|\mathbf{r}\|$. Thus, from (1)

$$\mathbf{F}(\mathbf{r}) = -\frac{GmM}{\|\mathbf{r}\|^2} \frac{\mathbf{r}}{\|\mathbf{r}\|} = -\frac{GmM}{\|\mathbf{r}\|^3} \mathbf{r}$$

If m and M are constant, and we let $c = -GmM$, then this formula can be expressed as

$$\mathbf{F}(\mathbf{r}) = \frac{c}{\|\mathbf{r}\|^3} \mathbf{r}$$

Vector fields of this form arise in electromagnetic as well as gravitational problems. Such fields are so important that they have their own terminology.

18.1.2 DEFINITION. If $\mathbf{r}$ is a radius vector in 2-space or 3-space, and if c is a constant, then a vector field of the form

$$\mathbf{F}(\mathbf{r}) = \frac{c}{\|\mathbf{r}\|^3} \mathbf{r} \tag{2}$$

is called an ***inverse-square field***.

Observe that if $c > 0$ in (2), then $\mathbf{F}(\mathbf{r})$ has the same direction as $\mathbf{r}$, so each vector in the field is directed away from the origin; and if $c < 0$, then $\mathbf{F}(\mathbf{r})$ is oppositely directed to $\mathbf{r}$, so each vector in the field is directed toward the origin. In either case the magnitude of $\mathbf{F}(\mathbf{r})$ is inversely proportional to the square of distance from the tip of $\mathbf{r}$ to the origin, since

$$\|\mathbf{F}(\mathbf{r})\| = \frac{|c|}{\|\mathbf{r}\|^3} \|\mathbf{r}\| = \frac{|c|}{\|\mathbf{r}\|^2}$$

Example 3 ***Coulomb's law*** states that *the electrostatic force exerted by one charged particle on another is directly proportional to the product of the charges and inversely proportional to the square of the distance between them.* This has the same form as Newton's Universal Law of Gravitation, so the electrostatic force field exerted by a charged particle is an inverse-square field. Specifically, if a particle of charge Q is at the origin of a coordinate system, and if $\mathbf{r}$ is the radius vector to a particle of charge q, then the force $\mathbf{F}(\mathbf{r})$ that the particle of charge Q exerts on the particle of charge q is of the form

$$\mathbf{F}(\mathbf{r}) = \frac{qQ}{4\pi\epsilon_0 \|\mathbf{r}\|^3} \mathbf{r}$$

where ϵ_0 is a positive constant (called the ***permittivity constant***). This formula is of form (2) with $c = qQ/4\pi\epsilon_0$. ◄

In our later work it will sometimes be convenient to express (2) in terms of (x, y) or (x, y, z) rather than $\mathbf{r}$. In the three-dimensional case $\|\mathbf{r}\| = \sqrt{x^2 + y^2 + z^2}$, so (2) can be written as

$$\mathbf{F}(x, y, z) = \frac{c}{(x^2 + y^2 + z^2)^{3/2}} (x\mathbf{i} + y\mathbf{j} + z\mathbf{k}) \tag{3}$$

Similarly, in the two-dimensional case

$$\mathbf{F}(x, y) = \frac{c}{(x^2 + y^2)^{3/2}} (x\mathbf{i} + y\mathbf{j}) \tag{4}$$

◻ **GRADIENT FIELDS**

An important class of vector fields arises from the process of finding gradients. Recall that if ϕ is a function of three variables, then the gradient of ϕ is defined as

$$\nabla\phi = \frac{\partial\phi}{\partial x}\mathbf{i} + \frac{\partial\phi}{\partial y}\mathbf{j} + \frac{\partial\phi}{\partial z}\mathbf{k}$$

This formula defines a vector field in 3-space called the **gradient field of ϕ**. Similarly, the gradient of a function of two variables defines a gradient field in 2-space. At each point in a gradient field where the gradient is nonzero, the vector points in the direction in which the rate of increase of ϕ is maximum.

Example 4   Sketch the gradient field of $\phi(x, y) = x + y$.

Solution.   The gradient of ϕ is

$$\nabla\phi = \frac{\partial\phi}{\partial x}\mathbf{i} + \frac{\partial\phi}{\partial y}\mathbf{j} = \mathbf{i} + \mathbf{j}$$

which is the same at each point. A portion of the vector field is sketched in Figure 18.1.5.   ◀

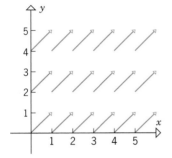

Figure 18.1.5

If $\mathbf{F}$ is an arbitrary vector field, one can ask whether it is the gradient field for some *scalar* function ϕ. We shall see later that not every vector field is a gradient field. However, those vector fields that are gradient fields are sufficiently important that they have a special name.

> **18.1.3**   DEFINITION.   A vector field $\mathbf{F}$ is said to be **conservative** in a region if it is the gradient field for some function ϕ in that region. The function ϕ is called a **potential function** for $\mathbf{F}$ in the region.

Example 5   Inverse-square fields are conservative in any region that does not contain the origin. For example, in the three-dimensional case the function

$$\phi(x, y, z) = -\frac{c}{(x^2 + y^2 + z^2)^{1/2}} \tag{5}$$

is a potential function for (3) in any region not containing the origin, since

$$\nabla\phi(x, y, z) = \frac{\partial\phi}{\partial x}\mathbf{i} + \frac{\partial\phi}{\partial y}\mathbf{j} + \frac{\partial\phi}{\partial z}\mathbf{k}$$

$$= \frac{cx}{(x^2 + y^2 + z^2)^{3/2}}\mathbf{i} + \frac{cy}{(x^2 + y^2 + z^2)^{3/2}}\mathbf{j} + \frac{cz}{(x^2 + y^2 + z^2)^{3/2}}\mathbf{k}$$

$$= \frac{c}{(x^2 + y^2 + z^2)^{3/2}} (x\mathbf{i} + y\mathbf{j} + z\mathbf{k})$$

$$= \mathbf{F}(x, y, z)$$

In a later section we shall discuss methods for finding potential functions for conservative vector fields.   ◀

◻ **DIVERGENCE AND CURL**

We shall now define two important operations on vector fields in 3-space—the *divergence* and the *curl* of the field. These names originate in the study of fluid flow, in which case the divergence relates to the way in which fluid flows toward or away from a point and the curl relates to the rotational properties of the fluid at a point. We will investigate the physical interpretations of these operations in more detail later, but for now we shall focus only on their computation.

18.1.4 DEFINITION. If $\mathbf{F}(x, y, z) = f(x, y, z)\mathbf{i} + g(x, y, z)\mathbf{j} + h(x, y, z)\mathbf{k}$, then we define the **_divergence of_ F**, written div **F**, by

$$\operatorname{div} \mathbf{F} = \frac{\partial f}{\partial x} + \frac{\partial g}{\partial y} + \frac{\partial h}{\partial z} \tag{6}$$

18.1.5 DEFINITION. If $\mathbf{F}(x, y, z) = f(x, y, z)\mathbf{i} + g(x, y, z)\mathbf{j} + h(x, y, z)\mathbf{k}$, then we define the **_curl of_ F**, written curl **F**, by

$$\operatorname{curl} \mathbf{F} = \left(\frac{\partial h}{\partial y} - \frac{\partial g}{\partial z} \right)\mathbf{i} + \left(\frac{\partial f}{\partial z} - \frac{\partial h}{\partial x} \right)\mathbf{j} + \left(\frac{\partial g}{\partial x} - \frac{\partial f}{\partial y} \right)\mathbf{k} \tag{7}$$

REMARK. Observe that div **F** and curl **F** depend on the point at which they are computed, and hence are more properly written as div $\mathbf{F}(x, y, z)$ and curl $\mathbf{F}(x, y, z)$. However, even though these functions are expressed in terms of x, y, and z, it can be proved that their values at a fixed point depend on the point but not on the coordinate system selected. This is important in applications, since it allows physicists and engineers to compute the curl and divergence in any convenient coordinate system.

Before proceeding to some examples, we note that div **F** has scalar values, whereas curl **F** has vector values (i.e., curl **F** is itself a vector field). Moreover, for computational purposes it is useful to note that the formula for the curl can be expressed in the determinant form

$$\operatorname{curl} \mathbf{F} = \begin{vmatrix} \mathbf{i} & \mathbf{j} & \mathbf{k} \\ \dfrac{\partial}{\partial x} & \dfrac{\partial}{\partial y} & \dfrac{\partial}{\partial z} \\ f & g & h \end{vmatrix} \tag{8}$$

The reader should verify that Formula (7) results if we compute this determinant by interpreting a "product" such as $(\partial/\partial x)(g)$ to mean $\partial g/\partial x$. Keep in mind, however, that (8) is just a mnemonic device and not a true determinant, since the entries in a determinant must be numbers, not vectors and partial derivative symbols.

Example 6 Find the divergence and the curl of the vector field

$$\mathbf{F}(x, y, z) = x^2 y\mathbf{i} + 2y^3 z\mathbf{j} + 3z\mathbf{k}$$

Solution. From (6)

$$\operatorname{div} \mathbf{F} = \frac{\partial}{\partial x}(x^2 y) + \frac{\partial}{\partial y}(2y^3 z) + \frac{\partial}{\partial z}(3z)$$

$$= 2xy + 6y^2 z + 3$$

and from (8)

$$\text{curl } \mathbf{F} = \begin{vmatrix} \mathbf{i} & \mathbf{j} & \mathbf{k} \\ \dfrac{\partial}{\partial x} & \dfrac{\partial}{\partial y} & \dfrac{\partial}{\partial z} \\ x^2 y & 2y^3 z & 3z \end{vmatrix}$$

$$= \left[\frac{\partial}{\partial y}(3z) - \frac{\partial}{\partial z}(2y^3 z) \right] \mathbf{i} + \left[\frac{\partial}{\partial z}(x^2 y) - \frac{\partial}{\partial x}(3z) \right] \mathbf{j}$$

$$+ \left[\frac{\partial}{\partial x}(2y^3 z) - \frac{\partial}{\partial y}(x^2 y) \right] \mathbf{k}$$

$$= -2y^3 \mathbf{i} - x^2 \mathbf{k} \quad \blacktriangleleft$$

Example 7 Show that the divergence of the inverse-square field

$$\mathbf{F}(x, y, z) = \frac{c}{(x^2 + y^2 + z^2)^{3/2}}(x\mathbf{i} + y\mathbf{j} + z\mathbf{k})$$

is zero.

Solution.   The computations can be simplified by letting $r = (x^2 + y^2 + z^2)^{1/2}$, in which case $\mathbf{F}$ can be expressed as

$$\mathbf{F}(x, y, z) = \frac{cx\mathbf{i} + cy\mathbf{j} + cz\mathbf{k}}{r^3} = \frac{cx}{r^3}\mathbf{i} + \frac{cy}{r^3}\mathbf{j} + \frac{cz}{r^3}\mathbf{k}$$

We leave it for the reader to show that

$$\frac{\partial r}{\partial x} = \frac{x}{r}, \quad \frac{\partial r}{\partial y} = \frac{y}{r}, \quad \frac{\partial r}{\partial z} = \frac{z}{r}$$

Thus,

$$\text{div } \mathbf{F} = c\left[\frac{\partial}{\partial x}\left(\frac{x}{r^3}\right) + \frac{\partial}{\partial y}\left(\frac{y}{r^3}\right) + \frac{\partial}{\partial z}\left(\frac{z}{r^3}\right) \right] \tag{9}$$

But

$$\frac{\partial}{\partial x}\left(\frac{x}{r^3}\right) = \frac{r^3 - x(3r^2)(x/r)}{(r^3)^2} = \frac{1}{r^3} - \frac{3x^2}{r^5}$$

$$\frac{\partial}{\partial y}\left(\frac{y}{r^3}\right) = \frac{1}{r^3} - \frac{3y^2}{r^5}$$

$$\frac{\partial}{\partial z}\left(\frac{z}{r^3}\right) = \frac{1}{r^3} - \frac{3z^2}{r^5}$$

Substituting these expressions in (9) yields

$$\text{div } \mathbf{F} = c\left[\frac{3}{r^3} - \frac{3x^2 + 3y^2 + 3z^2}{r^5} \right] = c\left[\frac{3}{r^3} - \frac{3r^2}{r^5} \right] = 0 \quad \blacktriangleleft$$

☐ **THE ∇ OPERATOR**

Thus far, the symbol ∇ that appears in the gradient expression $\nabla\phi$ has not been given a meaning of its own. However, it is often convenient to view ∇ as an operator

$$\nabla = \frac{\partial}{\partial x}\mathbf{i} + \frac{\partial}{\partial y}\mathbf{j} + \frac{\partial}{\partial z}\mathbf{k} \tag{10}$$

which when applied to $\phi(x, y, z)$ produces the gradient

$$\nabla\phi = \frac{\partial\phi}{\partial x}\mathbf{i} + \frac{\partial\phi}{\partial y}\mathbf{j} + \frac{\partial\phi}{\partial z}\mathbf{k}$$

We call (10) the ***del operator***. This is analogous to the derivative operator d/dx, which when applied to $f(x)$ produces the derivative $f'(x)$.

The del operator allows us to express the divergence of a vector field

$$\mathbf{F} = f(x, y, z)\mathbf{i} + g(x, y, z)\mathbf{j} + h(x, y, z)\mathbf{k}$$

in dot product notation as

$$\operatorname{div} \mathbf{F} = \nabla \cdot \mathbf{F} = \frac{\partial f}{\partial x} + \frac{\partial g}{\partial y} + \frac{\partial h}{\partial z} \tag{11}$$

and the curl of this field in cross-product notation as

$$\operatorname{curl} \mathbf{F} = \nabla \times \mathbf{F} = \begin{vmatrix} \mathbf{i} & \mathbf{j} & \mathbf{k} \\ \dfrac{\partial}{\partial x} & \dfrac{\partial}{\partial y} & \dfrac{\partial}{\partial z} \\ f & g & h \end{vmatrix} \tag{12}$$

☐ **THE LAPLACIAN** ∇^2

The operator that results by taking the dot product of the del operator with itself is denoted by ∇^2 and is called the ***Laplacian* operator***. This operator has the form

$$\nabla^2 = \nabla \cdot \nabla = \frac{\partial^2}{\partial x^2} + \frac{\partial^2}{\partial y^2} + \frac{\partial^2}{\partial z^2} \tag{13}$$

When applied to $\phi(x, y, z)$ the Laplacian operator produces the function

$$\nabla^2 \phi = \frac{\partial^2 \phi}{\partial x^2} + \frac{\partial^2 \phi}{\partial y^2} + \frac{\partial^2 \phi}{\partial z^2}$$

Note that $\nabla^2 \phi$ can also be expressed as div $(\nabla \phi)$. The equation $\nabla^2 \phi = 0$ or, equivalently,

$$\frac{\partial^2 \phi}{\partial x^2} + \frac{\partial^2 \phi}{\partial y^2} + \frac{\partial^2 \phi}{\partial z^2} = 0$$

---

*PIERRE-SIMON DE LAPLACE (1749–1827). French mathematician and physicist. Laplace is sometimes referred to as the French Isaac Newton because of his work in celestial mechanics. In a five-volume treatise entitled *Traité de Mécanique Céleste*, he solved extremely difficult problems involving gravitational interactions between the planets. In particular, he was able to show that our solar system is stable and not prone to catastrophic collapse as a result of these interactions. This was an issue of major concern at the time because Jupiter's orbit appeared to be shrinking and Saturn's expanding; Laplace showed that these were expected periodic anomalies. In addition to his work in celestial mechanics, he founded modern probability theory, showed with Lavoisier that respiration is a form of combustion, and developed methods that fostered many new branches of pure mathematics.

Laplace was born to moderately successful parents in Normandy, his father being a farmer and cider merchant. He matriculated in the theology program at the University of Caen at age 16 but left for Paris at age 18 with a letter of introduction to the influential mathematician d'Alembert, who eventually helped him undertake a career in mathematics. Laplace was a prolific writer, and after his election to the Academy of Sciences in 1773, the secretary wrote that the Academy had never received so many important research papers by so young a person in such a short time. Laplace had little interest in pure mathematics—he regarded mathematics merely as a tool for solving applied problems. In his impatience with mathematical detail, he frequently omitted complicated arguments with the statement, "It is easy to show that. . . ." He admitted, however, that as time passed he often had trouble reconstructing the omitted details himself!

At the height of his fame, Laplace served on many government committees and held the posts of Minister of the Interior and chancellor of the Senate. He barely escaped imprisonment and execution during the period of the Revolution, probably because he was able to convince each opposing party that he sided with them. Napoleon described him as a great mathematician but a poor administrator who "sought subtleties everywhere, had only doubtful ideas, and . . . carried the spirit of the infinitely small into administration." In spite of his genius, Laplace was both egotistic and insecure, attempting to ensure his place in history by conveniently failing to credit mathematicians whose work he used—an unnecessary pettiness since his own work was so brilliant. However, on the positive side he was supportive of young mathematicians, often treating them as his own children. Laplace ranks as one of the most influential mathematicians in history.

is known as *Laplace's equation*. This equation plays an important role in a wide variety of applications, resulting from the fact that it is satisfied by the potential function (5) for the inverse-square field (3).

▶ Exercise Set 18.1

In Exercises 1–4, sketch the vector field by drawing some typical nonintersecting vectors. The vectors need not be drawn to the same scale as the coordinate axes, but they should be in the correct proportions relative to each other.

1. $F(x, y) = 2i - j$.
2. $F(x, y) = yj$, $y > 0$.
3. $F(x, y) = yi - xj$. [*Note:* Each vector in the field is perpendicular to the position vector $r = xi + yj$.]
4. $F(x, y) = \dfrac{xi + yj}{\sqrt{x^2 + y^2}}$. [*Note:* Each vector in the field is a unit vector in the same direction as the position vector $r = xi + yj$.]

In Exercises 5–10, find div F and curl F.

5. $F(x, y, z) = x^2 i - 2j + yz k$.
6. $F(x, y, z) = xz^3 i + 2y^4 x^2 j + 5z^2 y k$.
7. $F(x, y, z) = 7y^3 z^2 i - 8x^2 z^5 j - 3xy^4 k$.
8. $F(x, y, z) = e^{xy} i - \cos y j + \sin^2 z k$.
9. $F(x, y, z) = \dfrac{1}{\sqrt{x^2 + y^2 + z^2}} (xi + yj + zk)$.
10. $F(x, y, z) = \ln x i + e^{xyz} j + \tan^{-1}(z/x) k$.

In Exercises 11–18, let k be a constant, and let $F = F(x, y, z)$, $G = G(x, y, z)$, and $\phi = \phi(x, y, z)$. Prove the following identities, assuming that all derivatives involved exist and are continuous.

11. div $(kF) = k$ div F.
12. curl $(kF) = k$ curl F.
13. div $(F + G) =$ div $F +$ div G.
14. curl $(F + G) =$ curl $F +$ curl G.
15. div $(\phi F) = \phi$ div $F + \nabla\phi \cdot F$.
16. curl $(\phi F) = \phi$ curl $F + \nabla\phi \times F$.
17. div (curl F) = 0.
18. curl $(\nabla\phi) = 0$.

In Exercises 19 and 20, let $r = xi + yj + zk$, and let $u = ai + bj + ck$, where a, b, and c are constants.

19. Show that div $(u \times r) = 0$.
20. Show that curl $(u \times r) = 2u$.

In Exercises 21 and 22, let $F = f(x)i + g(x)j + h(x)k$, where f, g, and h are differentiable functions.

21. Find div F.
22. Find curl F.

In Exercises 23–29, prove the result, assuming that $r = xi + yj + zk$, $r = \|r\|$, and $F = f(r)r$, where f is a differentiable function of r.

23. div $r = 3$.
24. curl $r = 0$.
25. $\nabla r = \dfrac{1}{r} r$.
26. $\nabla f(r) = \dfrac{f'(r)}{r} r$. [Use the chain rule and Exercise 25.]
27. div $F = 3f(r) + rf'(r)$. [Use Exercises 15, 23, and 26.]
28. curl $F = 0$. [Use Exercises 16, 24, and 26.]
29. $\nabla^2 f(r) = 2\dfrac{f'(r)}{r} + f''(r)$. [Use Exercises 15, 23, and 26.]

30. (a) Use the result in Exercise 27 to show that the divergence of the inverse-square field $F = r/r^3$ is zero.
(b) Use the result of Exercise 27 to show that if F is a vector field of the form $F = f(r)r$ and if div $F = 0$, then F is an inverse-square field. [*Suggestion:* Multiply $3f(r) + rf'(r) = 0$ through by r^2, and write the result as a derivative of a product.]

31. A curve C is called a *flow line* of a vector field F if F is a tangent vector to C at each point along C (Figure 18.1.6).

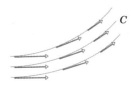

Flow lines of a vector field

Figure 18.1.6

(a) Let C be a flow line for the vector field $F(x, y) = -yi + xj$, and let (x, y) be a point on C for which $y \neq 0$. Show that the flow lines satisfy the differential equation
$$\frac{dy}{dx} = -\frac{x}{y}$$
(b) Solve the differential equation in part (a) by separation of variables, and show that the flow lines are concentric circles centered at the origin. [*Note:* See the third part of Figure 18.1.4.]

In Exercises 32–34, find a differential equation satisfied by the flow lines of **F** (see Exercise 31), and solve it to find equations for the flow lines of **F**. Sketch some typical flow lines and tangent vectors.

32. $\mathbf{F}(x, y) = \mathbf{i} + x\mathbf{j}$.

33. $\mathbf{F}(x, y) = x\mathbf{i} + \mathbf{j}$, $x > 0$.

34. $\mathbf{F}(x, y) = x\mathbf{i} - y\mathbf{j}$, $x > 0$ and $y > 0$.

■ 18.2 LINE INTEGRALS

In previous chapters we considered three kinds of integrals in rectangular coordinates: single integrals over intervals, double integrals over two-dimensional regions, and triple integrals over three-dimensional regions. In this section we shall discuss line integrals, which are integrals over curves in two- or three-dimensional space.

□ **SURFACE AREA AS A MOTIVATION FOR LINE INTEGRALS**

Integrals over curves arise in a variety of problems. One such problem can be stated as follows:

18.2.1 PROBLEM. *Let C be the graph in the xy-plane of a smooth vector-valued function* $\mathbf{r}(t) = x(t)\mathbf{i} + y(t)\mathbf{j}$, *and let* $f(x, y) = f(x(t), y(t))$ *be continuous and nonnegative for* $a \leq t \leq b$. *Find the area of the surface swept out by the vertical line segment from the point* $P(x(t), y(t), 0)$ *to the point* $Q(x(t), y(t), f(x(t), y(t)))$ *as t varies from a to b (Figure 18.2.1).*

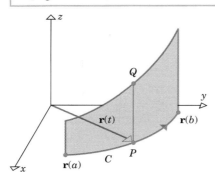

Figure 18.2.1

- Divide the curve C into n arcs by choosing a succession of distinct points $P_1, P_2, \ldots,$ P_{n-1} on C between $\mathbf{r}(a)$ and $\mathbf{r}(b)$ in the direction of increasing t. As shown in Figure 18.2.2, these points divide the surface into n strips. If we denote the area of the kth strip by ΔA_k, then the total area A can be expressed as

$$A = \Delta A_1 + \Delta A_2 + \cdots + \Delta A_n = \sum_{k=1}^{n} \Delta A_k$$

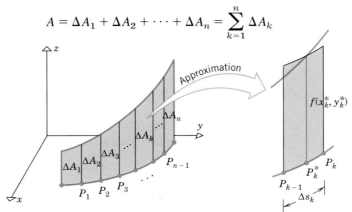

Figure 18.2.2

- Let us now approximate the area ΔA_k as follows: Denote the length of the kth arc by Δs_k, and choose an arbitrary point $P_k^*(x_k^*, y_k^*)$ on this arc. If the kth arc is small, then the value of f cannot change much over the arc, since f is continuous on C. Thus, we can reasonably assume that f has a constant value of $f(x_k^*, y_k^*)$ over the kth arc and that ΔA_k can be approximated by the area of a rectangle of base Δs_k and height $f(x_k^*, y_k^*)$ (Figure 18.2.2); that is,

$$\Delta A_k \approx f(x_k^*, y_k^*)\,\Delta s_k$$

from which it follows that

$$A \approx \sum_{k=1}^{n} f(x_k^*, y_k^*)\,\Delta s_k$$

- If we now increase n in such a way that the length of each arc approaches zero, then it is plausible that the error in this approximation approaches zero, and the exact surface area is

$$A = \lim_{n \to +\infty} \sum_{k=1}^{n} f(x_k^*, y_k^*)\,\Delta s_k \tag{1}$$

The limit in (1) is called the **line integral of f over (or along) C** and is denoted by

$$\int_C f(x, y)\,ds = \lim_{n \to +\infty} \sum_{k=1}^{n} f(x_k^*, y_k^*)\,\Delta s_k \tag{2}$$

With this notation, the area of the surface in Problem 18.2.1 can be expressed as

$$A = \int_C f(x, y)\,ds \tag{3}$$

This result was derived under the assumption that f is nonnegative on C. In the case where $f(x, y)$ can have both positive and negative values on C, the line integral over C represents a difference of two areas, the area above C and below $z = f(x, y)$ minus the area below C and above $z = f(x, y)$.

☐ **EVALUATING LINE INTEGRALS**

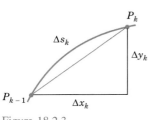

Figure 18.2.3

Except in the simplest cases, it is impractical to evaluate a line integral directly from (2), so we shall now show how to express a line integral as an ordinary definite integral. For this purpose, assume that the points P_{k-1} and P_k in Figure 18.2.3 correspond to parameter values of t_{k-1} and t_k, respectively, and that $P_k^*(x_k^*, y_k^*)$ corresponds to the parameter value t_k^*. If we let $\Delta t_k = t_k - t_{k-1}$, then we can approximate Δs_k as

$$\Delta s_k \approx \sqrt{(\Delta x_k)^2 + (\Delta y_k)^2} = \sqrt{\left(\frac{\Delta x_k}{\Delta t_k}\right)^2 + \left(\frac{\Delta y_k}{\Delta t_k}\right)^2}\,\Delta t_k \tag{4}$$

from which it follows that (2) can be expressed as

$$\int_C f(x, y)\,ds = \lim_{n \to +\infty} \sum_{k=1}^{n} f(x(t_k^*), y(t_k^*)) \sqrt{\left(\frac{\Delta x_k}{\Delta t_k}\right)^2 + \left(\frac{\Delta y_k}{\Delta t_k}\right)^2}\,\Delta t_k$$

which suggests that

$$\int_C f(x, y)\,ds = \int_a^b f(x(t), y(t)) \sqrt{\left(\frac{dx}{dt}\right)^2 + \left(\frac{dy}{dt}\right)^2}\,dt \tag{5}$$

Recall from Formula (16) of Section 15.3 that if s is an arc-length parameter for C, then

$$\frac{ds}{dt} = \sqrt{\left(\frac{dx}{dt}\right)^2 + \left(\frac{dy}{dt}\right)^2}$$

which is commonly expressed as

$$ds = \sqrt{\left(\frac{dx}{dt}\right)^2 + \left(\frac{dy}{dt}\right)^2}\, dt \tag{6}$$

Thus, the definite integral on the right side of (5) can be obtained from the line integral on the left side by expressing $f(x, y)$ in terms of t and using (6) to express ds in terms of dt. Because ds is so closely related to arc length, (5) is sometimes called the **integral of f over C with respect to arc length**. We leave it as an exercise to show that in the special case where C is expressed in terms of an arc-length parameter s (i.e., $t = s$), then (5) simplifies to

$$\int_C f(x, y)\, ds = \int_a^b f(x(s), y(s))\, ds \tag{7}$$

Example 1 Find the area of the surface extending upward from the circle $x^2 + y^2 = 1$ to the parabolic cylinder $z = 1 - x^2$ (Figure 18.2.4).

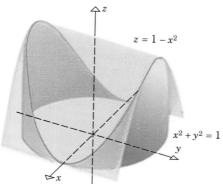

Figure 18.2.4

Solution. Denote the circle by C and represent it as

$$\mathbf{r}(t) = \cos t\mathbf{i} + \sin t\mathbf{j} \quad (0 \le t \le 2\pi)$$

From (3), the area A of the surface is

$$A = \int_C (1 - x^2)\, ds = \int_0^{2\pi} (1 - \cos^2 t)\, \sqrt{(-\sin t)^2 + (\cos t)^2}\, dt$$

$$= \int_0^{2\pi} \sin^2 t\, dt = \frac{1}{2}\int_0^{2\pi} (1 - \cos 2t)\, dt = \pi \quad \blacktriangleleft$$

☐ **LINE INTEGRALS WITH RESPECT TO x AND y**

In addition to line integrals with respect to arc length, there are two other important types of line integrals; these result from replacing Δs_k in (2) by $\Delta x_k = x(t_k) - x(t_{k-1})$ and $\Delta y_k = y(t_k) - y(t_{k-1})$, respectively. We define

$$\int_C f(x, y)\, dx = \lim_{n \to +\infty} \sum_{k=1}^{n} f(x(t_k^*), y(t_k^*))\, \Delta x_k$$

$$= \lim_{n \to +\infty} \sum_{k=1}^{n} f(x(t_k^*), y(t_k^*))\, \frac{\Delta x_k}{\Delta t_k}\, \Delta t_k \tag{8}$$

$$\int_C f(x, y)\, dy = \lim_{n \to +\infty} \sum_{k=1}^{n} f(x(t_k^*), y(t_k^*))\, \Delta y_k$$

$$= \lim_{n \to +\infty} \sum_{k=1}^{n} f(x(t_k^*), y(t_k^*))\, \frac{\Delta y_k}{\Delta t_k}\, \Delta t_k \tag{9}$$

from which it follows that

$$\int_C f(x, y)\, dx = \int_a^b f(x(t), y(t))x'(t)\, dt \tag{10}$$

$$\int_C f(x, y)\, dy = \int_a^b f(x(t), y(t))y'(t)\, dt \tag{11}$$

We call (10) the **line integral of f over C with respect to x** and (11) the **line integral of f over C with respect to y**.

REMARK.   In words, the line integrals with respect to x and y can be written as definite integrals by expressing $f(x, y)$ in terms of t and expressing dx and dy in terms of t as $dx = x'(t)\, dt$ and $dy = y'(t)\, dt$. Note that a line integral with respect to x is zero over a vertical line segment [since $x'(t) = 0$], and a line integral with respect to y is zero over a horizontal line segment [since $y'(t) = 0$].

Frequently, the line integrals in (10) and (11) occur in combination, in which case we dispense with one of the integral signs and write

$$\int_C f(x, y)\, dx + g(x, y)\, dy = \int_C f(x, y)\, dx + \int_C g(x, y)\, dy \tag{12}$$

Example 2   Evaluate

$$\int_C 2xy\, dx + (x^2 + y^2)\, dy$$

over the circular arc C given by $x = \cos t$, $y = \sin t$ $(0 \le t \le \pi/2)$ (Figure 18.2.5).

Solution.   From (10) and (11)

$$\int_C 2xy\, dx = \int_0^{\pi/2} (2\cos t \sin t) \left[\frac{d}{dt} (\cos t) \right] dt$$

$$= -2 \int_0^{\pi/2} \sin^2 t \cos t\, dt = \left. -\frac{2}{3} \sin^3 t \right]_0^{\pi/2} = -\frac{2}{3}$$

$$\int_C (x^2 + y^2)\, dy = \int_0^{\pi/2} (\cos^2 t + \sin^2 t) \left[\frac{d}{dt} (\sin t) \right] dt$$

$$= \int_0^{\pi/2} \cos t\, dt = \left. \sin t \right]_0^{\pi/2} = 1$$

Thus, from (12)

$$\int_C 2xy\, dx + (x^2 + y^2)\, dy = \int_C 2xy\, dx + \int_C (x^2 + y^2)\, dy$$

$$= -\frac{2}{3} + 1 = \frac{1}{3} \quad \blacktriangleleft$$

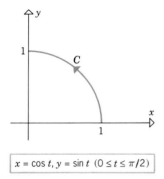

$x = \cos t, y = \sin t$ $(0 \le t \le \pi/2)$

Figure 18.2.5

□ **LINE INTEGRALS IN 3-SPACE**

The notion of a line integral can be extended to 3-space. If C is a curve in 3-space represented by a smooth vector-valued function

$$\mathbf{r}(t) = x(t)\mathbf{i} + y(t)\mathbf{j} + z(t)\mathbf{k} \quad (a \le t \le b)$$

and if $f(x, y, z)$, $g(x, y, z)$, and $h(x, y, z)$ are continuous functions of t on C, then

$$\int_C f(x, y, z)\, ds = \int_a^b f(x(t), y(t), z(t)) \sqrt{\left(\frac{dx}{dt}\right)^2 + \left(\frac{dy}{dt}\right)^2 + \left(\frac{dz}{dt}\right)^2}\, dt \tag{13}$$

$$\int_C f(x, y, z)\, dx = \int_a^b f(x(t), y(t), z(t)) x'(t)\, dt \tag{14a}$$

$$\int_C g(x, y, z)\, dy = \int_a^b g(x(t), y(t), z(t)) y'(t)\, dt \tag{14b}$$

$$\int_C h(x, y, z)\, dz = \int_a^b h(x(t), y(t), z(t)) z'(t)\, dt \tag{14c}$$

$$\int_C f(x, y, z)\, dx + g(x, y, z)\, dy + h(x, y, z)\, dz$$
$$= \int_C f(x, y, z)\, dx + \int_C g(x, y, z)\, dy + \int_C h(x, y, z)\, dz \tag{15}$$

Example 3   Evaluate

$$\int_C (xy + z^3)\, ds$$

where C is the portion of the helix given by the parametric equations

$$x = \cos t, \quad y = \sin t, \quad z = t \quad (0 \le t \le \pi)$$

Solution.   From Formula (13) we can evaluate the given line integral as a definite integral by using the parametric equations to express x, y, and z in terms of t and replacing ds by

$$ds = \sqrt{\left(\frac{dx}{dt}\right)^2 + \left(\frac{dy}{dt}\right)^2 + \left(\frac{dz}{dt}\right)^2}\, dt = \sqrt{(-\sin t)^2 + (\cos t)^2 + 1}\, dt$$
$$= \sqrt{2}\, dt$$

which yields

$$\int_C (xy + z^3)\, ds = \int_0^\pi (\cos t \sin t + t^3) \sqrt{2}\, dt = \sqrt{2} \left[\frac{\sin^2 t}{2} + \frac{t^4}{4} \right]_0^\pi$$
$$= \frac{\sqrt{2}\, \pi^4}{4} \quad \blacktriangleleft$$

☐ **LINE INTEGRALS OVER PIECEWISE SMOOTH CURVES**

Thus far, we have considered only line integrals over smooth curves. However, the notion of a line integral can be extended to curves formed from finitely many smooth curves $C_1, C_2, \ldots, C_n$ joined end to end. Such a curve is called **piecewise smooth** (Figure 18.2.6). We define a line integral over a piecewise smooth curve C to be the sum of the integrals over the pieces:

$$\int_C = \int_{C_1} + \int_{C_2} + \cdots + \int_{C_n}$$

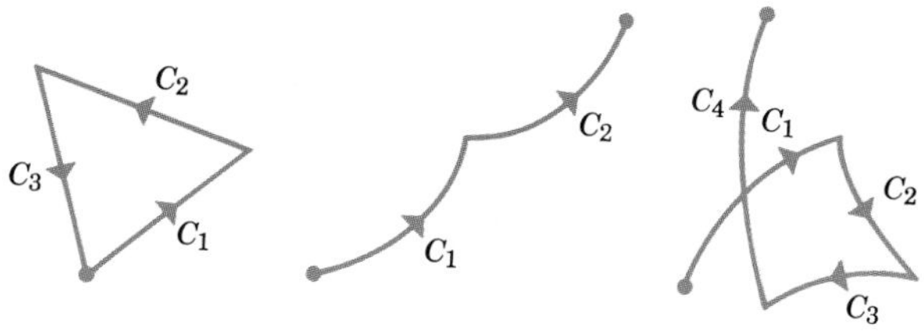

Figure 18.2.6

Example 4   Evaluate

$$\int_C x^2 y\, dx + x\, dy$$

in a counterclockwise direction around the triangular path shown in Figure 18.2.7.

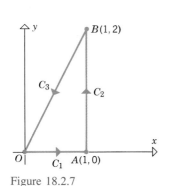

Figure 18.2.7

Solution.   We shall integrate over C_1, C_2, and C_3 separately and add the results. For each of the three integrals we must find parametric equations that trace the path of integration in the correct direction. For this purpose recall from Example 4 of Section 15.1 that the graph of the vector-valued function

$$\mathbf{r}(t) = (1 - t)\mathbf{r}_0 + t\mathbf{r}_1 \quad (0 \le t \le 1)$$

is the line segment joining $\mathbf{r}_0$ and $\mathbf{r}_1$, oriented in the direction from $\mathbf{r}_0$ to $\mathbf{r}_1$. Thus, the line segments C_1, C_2, and C_3 can be represented in vector notation as

$$C_1: \mathbf{r}(t) = (1 - t)\langle 0, 0\rangle + t\langle 1, 0\rangle = \langle t, 0\rangle$$

$$C_2: \mathbf{r}(t) = (1 - t)\langle 1, 0\rangle + t\langle 1, 2\rangle = \langle 1, 2t\rangle$$

$$C_3: \mathbf{r}(t) = (1 - t)\langle 1, 2\rangle + t\langle 0, 0\rangle = \langle 1 - t, 2 - 2t\rangle$$

where t varies from 0 to 1 in each case. From these equations we obtain

$$\int_{C_1} x^2 y\, dx + x\, dy = \int_0^1 (t^2)(0)\frac{d}{dt}[t]\, dt + \int_0^1 (t)\frac{d}{dt}[0]\, dt = 0$$

$$\int_{C_2} x^2 y\, dx + x\, dy = \int_0^1 (1^2)(2t)\frac{d}{dt}[1]\, dt + \int_0^1 (1)\frac{d}{dt}[2t]\, dt = 0 + 2 = 2$$

$$\int_{C_3} x^2 y\, dx + x\, dy = \int_0^1 (1 - t)^2(2 - 2t)\frac{d}{dt}[1 - t]\, dt + \int_0^1 (1 - t)\frac{d}{dt}[2 - 2t]\, dt$$

$$= 2\int_0^1 (t - 1)^3\, dt + 2\int_0^1 (t - 1)\, dt = -\tfrac{1}{2} - 1 = -\tfrac{3}{2}$$

Thus,

$$\int_C x^2 y\, dx + x\, dy = 0 + 2 + (-\tfrac{3}{2}) = \tfrac{1}{2} \quad \blacktriangleleft$$

☐ **CHANGE OF PARAMETER IN LINE INTEGRALS**

Since the parametric equations of a curve are used to evaluate line integrals over that curve, it seems possible that two different parametrizations of a curve C might produce different values for the same line integral over C. The following theorems deal with this question.

18.2.2   THEOREM (*Independence of Parametrization*).   *If C is a smooth parametric curve, then the value of any line integral over C is unchanged by a smooth change of parameter that preserves the orientation of C.*

18.2.3   THEOREM (*Reversal of Orientation*).   *If C is a smooth parametric curve, then a smooth change of parameter that reverses the orientation of C changes the sign of a line integral over C with respect to x, y, or z, but leaves the value of a line integral over C with respect to arc length unchanged.*

REMARK.   We will not prove these results, but some insight into the second theorem can be obtained by noting, for example, that reversing the orientation of C changes the sign of Δx_k in (8) and of Δy_k in (9) and hence changes the sign of the integrals; but reversing the orientation has no effect on Δs_k in (2), since Δs_k is an arc length and hence is positive, regardless of the orientation.

If a curve C is traced in a certain direction, then the same curve traced in the opposite direction is often denoted by the symbol $-C$. Thus, it follows from Theorem 18.2.3 that

$$\int_{-C} f(x, y)\, dx + g(x, y)\, dy = -\int_{C} f(x, y)\, dx + g(x, y)\, dy \tag{16}$$

$$\int_{-C} f(x, y)\, ds = \int_{C} f(x, y)\, ds \tag{17}$$

and similarly for line integrals in 3-space.

☐ **ARC LENGTH AS A LINE INTEGRAL**

If C is the graph in 3-space of a smooth vector-valued function $\mathbf{r}(t) = x(t)\mathbf{i} + y(t)\mathbf{j} + z(t)\mathbf{k}$, where $a \le t \le b$, then it follows from (13) that

$$\int_{C} ds = \int_{a}^{b} \sqrt{\left(\frac{dx}{dt}\right)^2 + \left(\frac{dy}{dt}\right)^2 + \left(\frac{dz}{dt}\right)^2}\, dt$$

which by Formula (10) of Section 15.3 is the arc length L of the curve C. Thus, the arc length L of a smooth parametric curve C can be expressed as

$$L = \int_{C} ds \tag{18}$$

The same result holds in 2-space (verify).

☐ **MASS OF A WIRE AS A LINE INTEGRAL**

We consider an idealized bent wire in 3-space that is sufficiently thin that it can be represented as a curve C. If the composition of the wire is uniform throughout, then the wire is called **homogeneous** and its **linear mass density** is defined to be the total mass divided by the total length. However, if the wire is not homogeneous, then the composition may vary from point to point, and the density is specified by a **density function** $\delta(x, y, z)$, whose value at a point (x, y, z) is viewed as a limit

$$\delta(x, y, z) = \lim_{\Delta s \to 0} \frac{\Delta M}{\Delta s} \tag{19}$$

in which ΔM and Δs represent the mass and length of a small section of wire centered at (x, y, z). Thus, if the wire is subdivided into n small pieces, and if (x_k^*, y_k^*, z_k^*) is the center of the kth piece and Δs_k is the length of the kth piece, then from (19) the mass ΔM_k of that piece can be approximated as

$$\Delta M_k \approx \delta(x_k^*, y_k^*, z_k^*)\, \Delta s_k$$

and hence the mass M of the entire wire can be approximated as

$$M = \sum_{k=1}^{n} \Delta M_k \approx \sum_{k=1}^{n} \delta(x_k^*, y_k^*, z_k^*)\, \Delta s_k \tag{20}$$

If we now increase n in such a way that the lengths of the pieces approach zero, then it is plausible that the error in (20) will approach zero, and the exact value of M will be given by the line integral

$$M = \int_{C} \delta(x, y, z)\, ds \tag{21}$$

The same result holds for wires in 2-space, but with two variables rather than three.

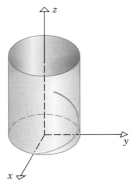

Figure 18.2.8

Example 5 Find the mass of a thin wire shaped in the form of a helix with the parametric equations

$$x = \cos t, \quad y = \sin t, \quad z = 2t \qquad (0 \leq t \leq \pi)$$

if the density function is $\delta(x, y, z) = kz$ $(k > 0)$ (Figure 18.2.8).

Solution. From (21) with C denoting the helix, the mass M is

$$M = \int_C kz \, ds = k \int_0^\pi 2t \sqrt{(-\sin t)^2 + (\cos t)^2 + 2^2} \, dt$$

$$= 2\sqrt{5} \, k \int_0^\pi t \, dt = \sqrt{5} \, k\pi^2 \qquad \blacktriangleleft$$

☐ **WORK AS A LINE INTEGRAL**

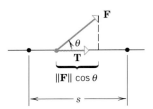

Figure 18.2.9

In everyday language we use the word *work* to denote the exertion of muscular or intellectual effort. However, in physics and engineering the term has a more definite and technical meaning—work is done whenever an object undergoes a displacement while subjected to a force that has a component in the direction of motion. More precisely:

If a constant force $\mathbf{F}$ acts on a particle that moves a distance s along a straight line, and if the force vector $\mathbf{F}$ makes an angle θ with a unit vector $\mathbf{T}$ in the direction of motion, then the work W done by $\mathbf{F}$ on the particle is defined to be

$$W = (\|\mathbf{F}\| \cos \theta)s = (\mathbf{F} \cdot \mathbf{T})s \tag{22}$$

that is, the work W is the scalar component of force in the direction of motion times the distance traveled by the particle (Figure 18.2.9).

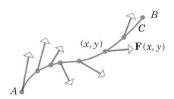

Figure 18.2.10

Although this notion of work is adequate for certain simple applications, it does not apply to problems in which the particle is subjected to a variable force as it moves along a curved path. To deal with such problems we must extend the notion of work to allow for a particle that moves from a point A to a point B along a curve C in 2-space or 3-space, while subjected to a force $\mathbf{F}$ that varies from point to point along the curve (Figure 18.2.10). Focusing on 3-space for the moment, suppose that a particle with position vector $\mathbf{r}(t)$ moves from A to B along a curve C as t increases from a to b. At each point (x, y, z) on C we denote the unit tangent vector to C by $\mathbf{T}$ [or $\mathbf{T}(x, y, z)$] and the force on the particle by $\mathbf{F}$ [or $\mathbf{F}(x, y, z)$] (Figure 18.2.11). We shall assume that $\mathbf{T}$ and $\mathbf{F}$ are continuous on C. To motivate an appropriate definition of work in this case we shall use a limiting process:

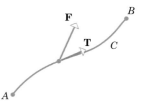

Figure 18.2.11

- As shown in Figure 18.2.12, subdivide C into n arcs by successively inserting distinct points $P_1, P_2, P_3, \ldots, P_{n-1}$ between A and B in the direction of increasing t, and denote the length of the kth arc by Δs_k. Let (x_k^*, y_k^*, z_k^*) be any point on this arc, and let $\mathbf{F}_k^* = \mathbf{F}(x_k^*, y_k^*, z_k^*)$ and $\mathbf{T}_k^* = \mathbf{T}(x_k^*, y_k^*, z_k^*)$ be the force vector and the unit tangent vector at this point.

- If P_{k-1} and P_k are close together, then the kth arc is nearly straight, and the vectors $\mathbf{F}$ and $\mathbf{T}$ cannot vary much from $\mathbf{F}_k^*$ and $\mathbf{T}_k^*$ over the small arc. Thus from (22), the work W_k performed as the particle moves along C from P_{k-1} to P_k can be approximated by

$$W_k \approx (\mathbf{F}_k^* \cdot \mathbf{T}_k^*) \, \Delta s_k$$

and the total work W done as the particle moves along C from A to B can be approximated by

$$W \approx \sum_{k=1}^n (\mathbf{F}_k^* \cdot \mathbf{T}_k^*) \, \Delta s_k$$

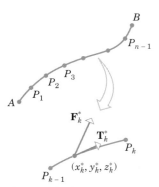

Figure 18.2.12

- There are two sources of error in this approximation, the error in assuming the kth arc to be straight and the error in assuming the force and unit tangent vectors to be constant over the kth arc. However, it seems plausible that if we increase n in such a way that the lengths of the arcs approach zero, then the errors will diminish and the exact value of the work will be given by

$$W = \lim_{n \to +\infty} \sum_{k=1}^{n} (\mathbf{F}_k^* \cdot \mathbf{T}_k^*) \, \Delta s_k$$

• If we now assume that the point (x_k^*, y_k^*, z_k^*) corresponds to the parameter value t_k^*, then

$$\mathbf{F}_k^* = \mathbf{F}(x(t_k^*), y(t_k^*), z(t_k^*)) \quad \text{and} \quad \mathbf{T}_k^* = \mathbf{T}(x(t_k^*), y(t_k^*), z(t_k^*))$$

so from (2), W can be expressed as the line integral

$$W = \int_C \mathbf{F}(x, y, z) \cdot \mathbf{T}(x, y, z) \, ds$$

or more briefly,

$$W = \int_C \mathbf{F} \cdot \mathbf{T} \, ds \tag{23}$$

We take this to be the definition of the **work done by F along C**. In words, *the work done by the force* **F** *along the curve C is obtained by integrating the tangential component of force along the curve.* The definition in 2-space is the same, except for the number of variables.

☐ **A METHOD FOR CALCULATING WORK**

For calculations it is helpful to write (23) in a different form: Recall from Formula (2) of Section 15.4 that the unit tangent vector **T** can be expressed as

$$\mathbf{T} = \frac{d\mathbf{r}}{ds}$$

which suggests that (23) can be expressed as

$$W = \int_C \mathbf{F} \cdot d\mathbf{r} \tag{24}$$

in which $d\mathbf{r}$ is interpreted as

$$d\mathbf{r} = dx\mathbf{i} + dy\mathbf{j} \quad \text{or} \quad d\mathbf{r} = dx\mathbf{i} + dy\mathbf{j} + dz\mathbf{k} \tag{25}$$

depending on whether C is in 2-space or 3-space. Alternatively, if

$$\mathbf{F} = f(x, y)\mathbf{i} + g(x, y)\mathbf{j} \quad \text{or} \quad \mathbf{F} = f(x, y, z)\mathbf{i} + g(x, y, z)\mathbf{j} + h(x, y, z)\mathbf{k}$$

then (24) can be expressed in scalar form as

$$W = \int_C f(x, y) \, dx + g(x, y) \, dy \tag{26}$$

$$W = \int_C f(x, y, z) \, dx + g(x, y, z) \, dy + h(x, y, z) \, dz \tag{27}$$

Example 6   Find the work done by the force field

$$\mathbf{F}(x, y) = x^3 y \mathbf{i} + (x - y)\mathbf{j}$$

on a particle that moves along the parabola $y = x^2$ from $(-2, 4)$ to $(1, 1)$.

Solution.   If we use $x = t$ as the parameter, the path C of the particle is represented by

$$x = t, \quad y = t^2 \qquad (-2 \le t \le 1)$$

or in vector notation as

$$\mathbf{r}(t) = t\mathbf{i} + t^2\mathbf{j} \quad (-2 \le t \le 1)$$

Thus, from (24) the work W done by $\mathbf{F}$ is

$$W = \int_C \mathbf{F} \cdot d\mathbf{r} = \int_C (x^3 y\mathbf{i} + (x - y)\mathbf{j}) \cdot (dx\mathbf{i} + dy\mathbf{j})$$

$$= \int_C x^3 y \, dx + (x - y) \, dy = \int_{-2}^{1} (t^5 + (t - t^2)(2t)) \, dt$$

$$= \frac{1}{6} t^6 + \frac{2}{3} t^3 - \frac{1}{2} t^4 \Bigg]_{-2}^{1} = 3$$

where the units for W depend on the units chosen for force and distance. ◀

REMARK.   Reversing the orientation of C in (23) reverses the direction of the vector $\mathbf{T}$ in the integrand, and hence reverses the sign of W. Thus, for the force field in the preceding example, the work performed on a particle moving from $(1, 1)$ to $(-2, 4)$ along the parabola is -3.

Example 7   Find the work done by the force field $\mathbf{F}(x, y, z) = yz\mathbf{i} + xz\mathbf{j} + xy\mathbf{k}$ on a particle that moves along the twisted cubic $\mathbf{r}(t) = t\mathbf{i} + t^2\mathbf{j} + t^3\mathbf{k}$ $(0 \le t \le 1)$.

Solution.   From (24)

$$W = \int_C \mathbf{F} \cdot d\mathbf{r} = \int_C yz \, dx + xz \, dy + xy \, dz$$

$$= \int_0^1 (t^2 t^3 + (tt^3)(2t) + (tt^2)(3t^2)) \, dt$$

$$= \int_0^1 (t^5 + 2t^5 + 3t^5) \, dt = \int_0^1 6t^5 \, dt = 1 \quad ◀$$

▶ Exercise Set 18.2

1. Find the area of the surface extending upward from the parabola $y = x^2$ ($0 \le x \le 2$) to the plane $z = 3x$.

2. Find the area of the surface extending upward from the semicircle $y = \sqrt{4 - x^2}$ to the surface $z = x^2 y$.

In Exercises 3–6, evaluate the line integral.

3. $\int_C \dfrac{1}{1 + x} \, ds$, where C is the curve $x = t$, $y = \frac{2}{3}t^{3/2}$, $0 \le t \le 3$.

4. $\int_C \dfrac{x}{1 + y^2} \, ds$, where C is the line $x = 1 + 2t$, $y = t$, $0 \le t \le 1$.

5. $\int_C 3x^2 yz \, ds$, where C is the curve $x = t$, $y = t^2$, $z = \frac{2}{3}t^3$, $0 \le t \le 1$.

6. $\int_C \dfrac{e^{-z}}{x^2 + y^2} \, ds$, where C is the helix $x = 2 \cos t$, $y = 2 \sin t$, $z = t$, $0 \le t \le 2\pi$.

7. Let C be the curve $x = t$, $y = t^2$, $0 \le t \le 1$. Find

   (a) $\int_C (2x + y) \, dx$ (b) $\int_C (x^2 - y) \, dy$

   (c) $\int_C (2x + y) \, dx + (x^2 - y) \, dy$.

8. Find $\int_C (3x + 2y) \, dx + (2x - y) \, dy$, where C is

   (a)   the line segment from $(0, 0)$ to $(1, 1)$

   (b)   the parabolic arc $y = x^2$ from $(0, 0)$ to $(1, 1)$

   (c)   the curve $y = \sin (\pi x/2)$ from $(0, 0)$ to $(1, 1)$

   (d)   the curve $x = y^3$ from $(0, 0)$ to $(1, 1)$.

In Exercises 9–16, evaluate the line integral.

9. $\int_C y \, dx - x^2 \, dy$, where C is the curve $x = t$, $y = \frac{1}{2}t^2$, $0 \le t \le 2$.

10. $\int_C e^{y-x} dx + xy\, dy$, where C is the curve $x = 2 + t$, $y = 3 - t$, $0 \le t \le 1$.

11. $\int_C (x + 2y)\, dx + (x - y)\, dy$, where C is the curve $x = 2\cos t$, $y = 4\sin t$, $0 \le t \le \pi/4$.

12. $\int_C (x^2 - y^2)\, dx + x\, dy$, where C is the curve $x = t^{2/3}$, $y = t$, $-1 \le t \le 1$.

13. $\int_C -y\, dx + x\, dy$ along $y^2 = 3x$ from the point $(3, 3)$ to the point $(0, 0)$.

14. $\int_C (y - x)\, dx + x^2 y\, dy$ along $y^2 = x^3$ from the point $(1, -1)$ to the point $(1, 1)$.

15. $\int_C (x^2 + y^2)\, dx - x\, dy$ along the quarter-circle $x^2 + y^2 = 1$ from $(1, 0)$ to $(0, 1)$.

16. $\int_C (y - x)\, dx + xy\, dy$ along the line segment from $(3, 4)$ to $(2, 1)$.

17. Find $\int_C \mathbf{F} \cdot d\mathbf{r}$, where $\mathbf{F}(x, y) = x^2\mathbf{i} + xy\mathbf{j}$ and C is the semicircle given by $\mathbf{r}(t) = 2\cos t\,\mathbf{i} + 2\sin t\,\mathbf{j}$, $0 \le t \le \pi$.

18. Find $\int_C \mathbf{F} \cdot d\mathbf{r}$, where $\mathbf{F}(x, y) = x^2 y\mathbf{i} + 4\mathbf{j}$ and C is the curve $\mathbf{r}(t) = e^t\mathbf{i} + e^{-t}\mathbf{j}$, $0 \le t \le 1$.

19. Find $\int_C \mathbf{F} \cdot d\mathbf{r}$, where $\mathbf{F}(x, y) = (x^2 + y^2)^{-3/2}(x\mathbf{i} + y\mathbf{j})$ and C is the curve $\mathbf{r}(t) = e^t \sin t\,\mathbf{i} + e^t \cos t\,\mathbf{j}$, $0 \le t \le 1$.

20. Evaluate $\int_C y\, dx - x\, dy$ over each of the curves:

(a)
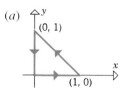

(b)
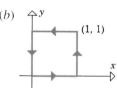

(c)
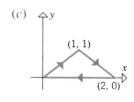

(d)
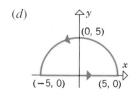

In Exercises 21–24, evaluate the line integral.

21. $\int_C yz\, dx - xz\, dy + xy\, dz$, where C is the curve $x = e^t$, $y = e^{3t}$, $z = e^{-t}$, $0 \le t \le 1$.

22. $\int_C y\, dx + z\, dy - x\, dz$, where C is the line segment from $(1, 1, 1)$ to $(-3, 2, 0)$.

23. $\int_C \mathbf{F} \cdot d\mathbf{r}$, where $\mathbf{F}(x, y, z) = z\mathbf{i} + x\mathbf{j} + y\mathbf{k}$ and C is the curve $\mathbf{r}(t) = \sin t\,\mathbf{i} + 3\sin t\,\mathbf{j} + \sin^2 t\,\mathbf{k}$, $0 \le t \le \pi/2$.

24. $\int_C x^2 z\, dx - yx^2\, dy + 3xz\, dz$, where C is the triangular path from $(0, 0, 0)$ to $(1, 1, 0)$ to $(1, 1, 1)$ to $(0, 0, 0)$.

25. Find the mass of a thin wire shaped in the form of the circular arc $y = \sqrt{9 - x^2}$ $(0 \le x \le 3)$ if the density function is $\delta(x, y) = kx\sqrt{y}$ $(k > 0)$.

26. Find the mass of a thin wire shaped in the form of the curve $x = e^t \cos t$, $y = e^t \sin t$ $(0 \le t \le 1)$ if the density function δ is proportional to the distance from the origin.

27. Find the mass of a thin wire shaped in the form of the helix $x = 3\cos t$, $y = 3\sin t$, $z = 4t$ $(0 \le t \le \pi/2)$ if the density function is $\delta = kx/(1 + y^2)$ $(k > 0)$.

28. Find the mass of a thin wire shaped in the form of the curve $x = 2t$, $y = \ln t$, $z = 4\sqrt{t}$ $(1 \le t \le 4)$ if the density function is proportional to the distance above the xy-plane.

29. Find the work done if a particle moves along the parabolic arc $x = y^2$ from $(0, 0)$ to $(1, 1)$ while subject to the force $\mathbf{F}(x, y) = xy\mathbf{i} + x^2\mathbf{j}$.

30. Find the work done if a particle moves along the curve $x = t$, $y = 1/t$, $1 \le t \le 3$ while subject to the force $\mathbf{F}(x, y) = (x^2 + xy)\mathbf{i} + (y - x^2 y)\mathbf{j}$.

31. Find the work done by the force
$$\mathbf{F}(x, y) = \frac{1}{x^2 + y^2}\mathbf{i} + \frac{4}{x^2 + y^2}\mathbf{j}$$
acting on a particle that moves along the curve C shown.

(a)
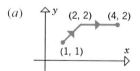

(b)
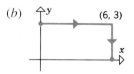

(c)
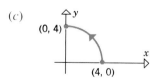

32. Find the work that is done by the force
$$\mathbf{F}(x, y, z) = xy\mathbf{i} + yz\mathbf{j} + xz\mathbf{k}$$
acting on a particle that moves along the curve given by $\mathbf{r}(t) = t\mathbf{i} + t^2\mathbf{j} + t^3\mathbf{k}$, $0 \le t \le 1$.

33. Find the work that is done by a force
$$\mathbf{F}(x, y, z) = (x + y)\mathbf{i} + xy\mathbf{j} - z^2\mathbf{k}$$
acting on a particle that moves along the line segment

from $(0, 0, 0)$ to $(1, 3, 1)$ and then along the line segment from $(1, 3, 1)$ to $(2, -1, 4)$.

34. Evaluate the integral $\displaystyle\int_{-C} \frac{x\,dy - y\,dx}{x^2 + y^2}$, where C is the circle $x^2 + y^2 = a^2$ traversed counterclockwise.

35. A particle moves from the point $(0, 0)$ to the point $(1, 0)$ along the curve $x = t$, $y = \lambda t(1 - t)$ while subject to the force $\mathbf{F}(x, y) = xy\mathbf{i} + (x - y)\mathbf{j}$. For what value of λ is the work done equal to 1?

36. A farmer weighing 150 lb carries a sack of grain weighing 20 lb up a circular helical staircase around a silo of radius 25 ft. As the farmer climbs, grain leaks from the sack at a rate of 1 lb per 10 ft of ascent. How much work is performed by the farmer in climbing through a vertical distance of 60 ft in exactly four revolutions?

■ 18.3 INDEPENDENCE OF PATH; CONSERVATIVE VECTOR FIELDS

In general, the value of a line integral $\int_C \mathbf{F} \cdot d\mathbf{r}$ depends on the curve C. However, we shall show in this section that when the integrand satisfies appropriate conditions, the value of the integral depends only on the location of the endpoints of the curve C and not on the shape of the curve, in which case the evaluation of the line integral is greatly simplified. Vector fields in which line integrals depend only on the endpoints of the curve of integration are of special importance in physics and engineering, and we shall discuss such vector fields in this section.

☐ **INDEPENDENCE OF PATH**

The curve C in a line integral is often called the **path of integration**. The following example illustrates how different paths of integration from one point to another can result in the same value for a line integral.

Example 1   Let $\mathbf{F}(x, y) = y\mathbf{i} + x\mathbf{j}$. Evaluate the line integral

$$\int_C \mathbf{F} \cdot d\mathbf{r}$$

over the following curves (Figure 18.3.1):

(a)   The line segment $y = x$ from $(0, 0)$ to $(1, 1)$.
(b)   The parabola $y = x^2$ from $(0, 0)$ to $(1, 1)$.
(c)   The cubic $y = x^3$ from $(0, 0)$ to $(1, 1)$.

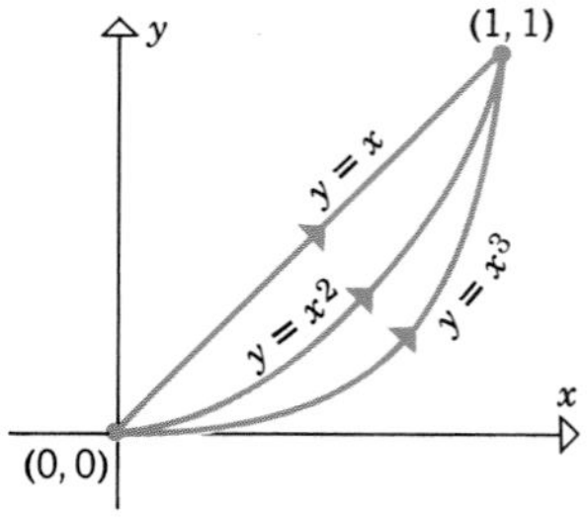

Figure 18.3.1

Solution (a).   With $x = t$ as the parameter, the path of integration is given by

$$x = t, \quad y = t \quad (0 \le t \le 1)$$

Thus,

$$\int_C \mathbf{F} \cdot d\mathbf{r} = \int_C (y\mathbf{i} + x\mathbf{j}) \cdot (dx\mathbf{i} + dy\mathbf{j}) = \int_C y\,dx + x\,dy$$

$$= \int_0^1 2t\,dt = 1$$

Solution (b).   With $x = t$ as the parameter, the path of integration is given by

$$x = t, \quad y = t^2 \quad (0 \le t \le 1)$$

Thus,

$$\int_C \mathbf{F} \cdot d\mathbf{r} = \int_C y\, dx + x\, dy = \int_0^1 3t^2\, dt = 1$$

Solution (c).   With $x = t$ as the parameter, the path of integration is given by

$$x = t, \quad y = t^3 \qquad (0 \le t \le 1)$$

Thus,

$$\int_C \mathbf{F} \cdot d\mathbf{r} = \int_C y\, dx + x\, dy = \int_0^1 4t^3\, dt = 1 \qquad \blacktriangleleft$$

The results in this example are not accidental; we shall soon see that the value of this line integral is the same over *all* piecewise smooth paths from $(0, 0)$ to $(1, 1)$. A line integral whose value is the same over all piecewise smooth paths from one endpoint to the other is said to be *independent of path*.

Recall from the First Fundamental Theorem of Calculus (Theorem 5.7.1) that if F is an antiderivative of f, then

$$\int_a^b f(x)\, dx = F(b) - F(a)$$

The following result is the analog of that theorem for line integrals.

18.3.1   THEOREM (*The Fundamental Theorem of Line Integrals*).   *Suppose that* $\mathbf{F}(x, y) = f(x, y)\mathbf{i} + g(x, y)\mathbf{j}$, *where f and g are continuous in some open region containing the points (x_0, y_0) and (x_1, y_1). If*

$$\mathbf{F}(x, y) = \nabla\phi(x, y)$$

at each point of this region, then for any piecewise smooth curve C that starts at (x_0, y_0), ends at (x_1, y_1), and lies entirely in the region, we have

$$\int_C \mathbf{F}(x, y) \cdot d\mathbf{r} = \phi(x_1, y_1) - \phi(x_0, y_0) \tag{1}$$

We shall give the proof for a smooth curve C. The proof for piecewise smooth curves is obtained by considering each piece separately. The details are left for the exercises.

Proof.   If C is given parametrically by $x = x(t)$, $y = y(t)$ ($a \le t \le b$), then the initial and final points of the curve C are

$$(x_0, y_0) = (x(a), y(a)) \quad \text{and} \quad (x_1, y_1) = (x(b), y(b))$$

Since $\mathbf{F}(x, y) = \nabla\phi$, it follows that

$$\mathbf{F}(x, y) = \frac{\partial\phi}{\partial x}\mathbf{i} + \frac{\partial\phi}{\partial y}\mathbf{j}$$

so

$$\int_C \mathbf{F}(x, y) \cdot d\mathbf{r} = \int_C \frac{\partial\phi}{\partial x}\, dx + \frac{\partial\phi}{\partial y}\, dy = \int_a^b \left[\frac{\partial\phi}{\partial x}\frac{dx}{dt} + \frac{\partial\phi}{\partial y}\frac{dy}{dt}\right] dt$$

$$= \int_a^b \frac{d}{dt}[\phi(x(t), y(t))]\, dt = \phi[x(t), y(t)]\Big]_{t=a}^b$$

$$= \phi(x(b), y(b)) - \phi(x(a), y(a))$$

$$= \phi(x_1, y_1) - \phi(x_0, y_0) \qquad \blacksquare$$

For line integrals that are independent of path, it is common to write (1) as

$$\int_{(x_0,\, y_0)}^{(x_1,\, y_1)} \mathbf{F} \cdot d\mathbf{r} = \phi(x_1, y_1) - \phi(x_0, y_0) \tag{2}$$

Example 2   The Fundamental Theorem of Line Integrals explains the results in Example 1, since $\mathbf{F}(x, y) = y\mathbf{i} + x\mathbf{j}$ is the gradient of the function $\phi(x, y) = xy$ (verify). Thus, for any piecewise smooth curve C from $(0, 0)$ to $(1, 1)$, it follows from (2) that

$$\int_C \mathbf{F} \cdot d\mathbf{r} = \int_{(0,\, 0)}^{(1,\, 1)} \mathbf{F} \cdot d\mathbf{r} = \phi(1, 1) - \phi(0, 0) = 1 - 0 = 1 \quad \blacktriangleleft$$

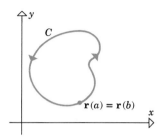

Figure 18.3.2

A curve C represented by a vector-valued function $\mathbf{r}(t)$ ($a \le t \le b$) is said to be **closed** if the initial point $\mathbf{r}(a)$ and the terminal point $\mathbf{r}(b)$ coincide (Figure 18.3.2). If C is a piecewise smooth closed curve in 2-space, and if $\mathbf{F}$ satisfies the conditions of Theorem 18.3.1, then the points (x_0, y_0) and (x_1, y_1) in the theorem are the same and hence

$$\int_C \mathbf{F} \cdot d\mathbf{r} = \phi(x_1, y_1) - \phi(x_0, y_0) = 0 \tag{3}$$

☐ **LINE INTEGRALS OF CONSERVATIVE VECTOR FIELDS**

Recall from Definition 18.1.3 that $\mathbf{F}(x, y)$ is called a *conservative vector field* if $\mathbf{F}$ is the gradient of some potential function ϕ, that is, $\mathbf{F}(x, y) = \nabla\phi(x, y)$. In this case the Fundamental Theorem of Line Integrals implies that the line integral $\int_C \mathbf{F} \cdot d\mathbf{r}$ is independent of the path and has a value of zero if the path is closed. We want to show that the converse results are also true; that is, a vector field in which all line integrals are independent of path or in which all line integrals around closed paths are zero must be conservative. However, to do so we will need to assume that the domain D of $\mathbf{F}(x, y)$ is **connected**; that is, any two points in D can be connected by a piecewise smooth curve that lies entirely in D (Figure 18.3.3).

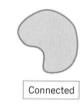

Connected

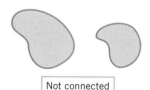

Not connected

Figure 18.3.3

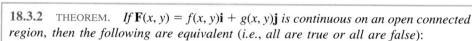

18.3.2   THEOREM.   *If* $\mathbf{F}(x, y) = f(x, y)\mathbf{i} + g(x, y)\mathbf{j}$ *is continuous on an open connected region, then the following are equivalent (i.e., all are true or all are false):*

(*a*) $\mathbf{F}(x, y)$ *is a conservative vector field in the region.*

(*b*) $\displaystyle\int_C \mathbf{F} \cdot d\mathbf{r} = 0$ *for every piecewise smooth closed curve C in the region.*

(*c*) $\displaystyle\int_C \mathbf{F} \cdot d\mathbf{r}$ *is independent of path for every piecewise smooth curve C in the region.*

Proof.   To prove a theorem of this form, it suffices to show that (*a*) implies (*b*), (*b*) implies (*c*), and (*c*) implies (*a*). We have already seen that (*a*) implies (*b*) [see (3)], and we shall leave it as an exercise to show that (*b*) implies (*c*). Thus, it remains to show that if $\int_C \mathbf{F} \cdot d\mathbf{r}$ is independent of path for every piecewise smooth curve C in the region, then $\mathbf{F}(x, y)$ is a conservative vector field in the region. To prove this, assume that

$$\mathbf{F} = f(x, y)\mathbf{i} + g(x, y)\mathbf{j}$$

We must find a function ϕ such that $\nabla\phi = \mathbf{F}$, that is,

$$\frac{\partial\phi}{\partial x} = f(x, y) \quad \text{and} \quad \frac{\partial\phi}{\partial y} = g(x, y)$$

For this purpose, let D denote the domain of $\mathbf{F}$, and let (a, b) be any fixed point in D. For any point (x, y) in D define

$$\phi(x, y) = \int_{(a, b)}^{(x, y)} \mathbf{F} \cdot d\mathbf{r} \tag{4}$$

(which is possible because the integral is assumed to be independent of path in D). We will show that $\nabla\phi = \mathbf{F}$. Since D is open, we can find a circular disk centered at (x, y) whose points lie entirely in D. As shown in Figure 18.3.4, choose any point (x_1, y) in this disk that lies on the same horizontal line as (x, y) but which is different from (x, y). Because the integral in (4) is independent of path, we can evaluate it by first integrating from (a, b) to (x_1, y) along an arbitrary piecewise smooth curve C_1 in D, and then continuing along the horizontal line segment C_2 from (x_1, y) to (x, y). This yields

$$\phi(x, y) = \int_{C_1} \mathbf{F} \cdot d\mathbf{r} + \int_{C_2} \mathbf{F} \cdot d\mathbf{r} = \int_{(a, b)}^{(x_1, y)} \mathbf{F} \cdot d\mathbf{r} + \int_{C_2} \mathbf{F} \cdot d\mathbf{r}$$

Since the first term does not depend on x, its partial derivative with respect to x is zero and hence

$$\frac{\partial\phi}{\partial x} = \frac{\partial}{\partial x} \int_{C_2} \mathbf{F} \cdot d\mathbf{r} = \frac{\partial}{\partial x} \int_{C_2} f(x, y)\, dx + g(x, y)\, dy$$

However, the line integral with respect to y is zero along the horizontal line segment C_2, so this equation simplifies to

$$\frac{\partial\phi}{\partial x} = \frac{\partial}{\partial x} \int_{C_2} f(x, y)\, dx \tag{5}$$

But the line segment C_2 from (x_1, y) to (x, y) can be expressed parametrically as

$$x = t, \quad y = y$$

where t varies from x_1 to x. Thus, (5) can be written as

$$\frac{\partial\phi}{\partial x} = \frac{\partial}{\partial x} \int_{x_1}^{x} f(t, y)\, dt = f(x, y)$$

where the last equality was obtained from the Second Fundamental Theorem of Calculus (Theorem 5.9.1) by treating y as a constant and the integrand as a function of t alone. The proof that $\partial\phi/\partial y = g(x, y)$ can be obtained in a similar manner by joining (x, y) to a point (x, y_1) with a vertical line segment (Exercise 26). ∎

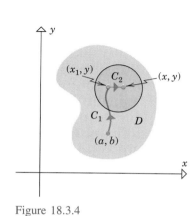

Figure 18.3.4

□ **A TEST FOR CONSERVATIVE VECTOR FIELDS**

Although Theorem 18.3.2 is an important characterization of conservative vector fields, it is not an effective computational tool for determining if a vector field is conservative. To find a useful way of doing this we will need some terminology. A plane curve $\mathbf{r}(t)$, where $a \leq t \leq b$, is said to be *simple* if it does not intersect itself anywhere between its endpoints. As shown in Figure 18.3.5, a simple curve may or may not be closed.

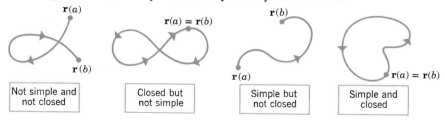

$\mathbf{r}(a)$ $\mathbf{r}(b)$ $\mathbf{r}(a) = \mathbf{r}(b)$ $\mathbf{r}(b)$
$\mathbf{r}(b)$ $\mathbf{r}(a) = \mathbf{r}(b)$ $\mathbf{r}(a)$ $\mathbf{r}(a) = \mathbf{r}(b)$

| Not simple and not closed | Closed but not simple | Simple but not closed | Simple and closed |

Figure 18.3.5

A connected set D in 2-space is called *simply connected* if no simple closed curve in D encloses points not in D. Informally stated, a simply connected set in 2-space is connected and has no holes (Figure 18.3.6).

The following theorem is the primary tool for determining whether a given vector field in 2-space is conservative.

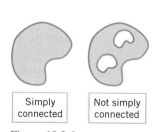

| Simply connected | Not simply connected |

Figure 18.3.6

> **18.3.3**   THEOREM (**The Conservative Field Test**).   *Let* $\mathbf{F}(x, y) = f(x, y)\mathbf{i} + g(x, y)\mathbf{j}$, *where f and g have continuous first partial derivatives in an open simply connected region. Then* $\mathbf{F}$ *is a conservative vector field on that region if and only if*
>
> $$\frac{\partial f}{\partial y} = \frac{\partial g}{\partial x} \qquad (6)$$
>
> *at each point of the region.*

The complete proof of this theorem requires results from advanced calculus and will be omitted. However, it is not hard to see why Formula (6) holds when $\mathbf{F}$ is conservative, that is, when $\mathbf{F}$ is the gradient of a potential function ϕ. In this case we have

$$\frac{\partial \phi}{\partial x} = f \quad \text{and} \quad \frac{\partial \phi}{\partial y} = g \qquad (7)$$

so that

$$\frac{\partial f}{\partial y} = \frac{\partial}{\partial y}\left(\frac{\partial \phi}{\partial x}\right) = \frac{\partial^2 \phi}{\partial y\, \partial x} \quad \text{and} \quad \frac{\partial g}{\partial x} = \frac{\partial}{\partial x}\left(\frac{\partial \phi}{\partial y}\right) = \frac{\partial^2 \phi}{\partial x\, \partial y}$$

But the mixed partial derivatives in these equations are equal (Theorem 16.4.6), so (6) follows.

WARNING.   In (6), the $\mathbf{i}$-component of $\mathbf{F}$ is differentiated with respect to y and the $\mathbf{j}$-component with respect to x. It is easy to get this backwards by mistake.

Once it is established that a vector field is conservative, a potential function for the field can be obtained by first integrating either of the equations in (7). This is illustrated in the following example.

Example 3   Let $\mathbf{F}(x, y) = 2xy^3\mathbf{i} + (1 + 3x^2y^2)\mathbf{j}$.

(a)   Show that $\mathbf{F}$ is a conservative vector field on the entire xy-plane.

(b)   Find ϕ by first integrating $\partial\phi/\partial x$.

(c)   Find ϕ by first integrating $\partial\phi/\partial y$.

Solution (a).   Since $f(x, y) = 2xy^3$ and $g(x, y) = 1 + 3x^2y^2$, we have

$$\frac{\partial f}{\partial y} = 6xy^2 = \frac{\partial g}{\partial x}$$

so (6) holds for all (x, y).

Solution (b).   Since the field $\mathbf{F}$ is conservative, there is a potential function ϕ such that

$$\frac{\partial \phi}{\partial x} = 2xy^3 \quad \text{and} \quad \frac{\partial \phi}{\partial y} = 1 + 3x^2y^2 \qquad (8)$$

Integrating the first of these equations with respect to x (and treating y as a constant) yields

$$\phi = \int 2xy^3\, dx = x^2y^3 + k(y) \qquad (9)$$

where $k(y)$ represents the ''constant'' of integration. We are justified in treating the constant of integration as a function of y, since y is held constant in the integration process. To find $k(y)$ we differentiate (9) with respect to y and use the second equation in (8) to obtain

$$\frac{\partial \phi}{\partial y} = 3x^2y^2 + k'(y) = 1 + 3x^2y^2$$

from which it follows that $k'(y) = 1$. Thus,

$$k(y) = \int k'(y)\,dy = \int 1\,dy = y + K$$

where K is a (numerical) constant of integration. Substituting in (9) we obtain

$$\phi = x^2 y^3 + y + K$$

The appearance of the arbitrary constant K tells us that ϕ is not unique. The reader may want to verify that $\nabla \phi = \mathbf{F}$.

Solution (c).   Integrating the second equation in (8) with respect to y (and treating x as a constant) yields

$$\phi = \int (1 + 3x^2 y^2)\,dy = y + x^2 y^3 + k(x) \tag{10}$$

where $k(x)$ is the "constant" of integration. Differentiating (10) with respect to x and using the first equation in (8) yields

$$\frac{\partial \phi}{\partial x} = 2xy^3 + k'(x) = 2xy^3$$

from which it follows that $k'(x) = 0$ and consequently that $k(x) = K$, where K is a numerical constant of integration. Substituting this in (10) yields

$$\phi = y + x^2 y^3 + K$$

which agrees with the solution in part (b).   ◀

Example 4   Use the potential function obtained in Example 3 to evaluate the integral

$$\int_{(1,4)}^{(3,1)} 2xy^3\,dx + (1 + 3x^2 y^2)\,dy$$

Solution.   The integrand can be expressed as $\mathbf{F} \cdot d\mathbf{r}$, where $\mathbf{F}$ is the vector field in Example 3. Thus, using Formula (2) and the potential function $\phi = y + x^2 y^3 + K$ for $\mathbf{F}$ we obtain

$$\int_{(1,4)}^{(3,1)} 2xy^3\,dx + (1 + 3x^2 y^2)\,dy = \int_{(1,4)}^{(3,1)} \mathbf{F} \cdot d\mathbf{r} = \phi(3,1) - \phi(1,4)$$
$$= (10 + K) - (68 + K) = -58 \qquad ◀$$

REMARK.   Note that the constant K drops out. In future integration problems we shall omit K from the computations.

Example 5   Let $\mathbf{F}(x, y) = e^y \mathbf{i} + xe^y \mathbf{j}$.

(a)   Verify that the vector field $\mathbf{F}$ is conservative on the entire xy-plane.
(b)   Find the work done by the field on a particle that moves from $(1, 0)$ to $(-1, 0)$ over the semicircular path C shown in Figure 18.3.7.

Solution (a).   For the given field we have $f(x, y) = e^y$ and $g(x, y) = xe^y$. Thus,

$$\frac{\partial}{\partial y}(e^y) = e^y = \frac{\partial}{\partial x}(xe^y)$$

so (6) holds for all (x, y) and hence $\mathbf{F}$ is conservative on the entire xy-plane.

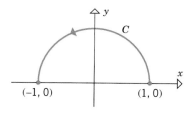

Figure 18.3.7

Solution (b).   From Formula (24) of Section 18.2, the work done by the field is

$$W = \int_C \mathbf{F} \cdot d\mathbf{r} = \int_C e^y \, dx + x e^y \, dy \tag{11}$$

However, the calculations involved in integrating along C are tedious, so it is preferable to apply Theorem 18.3.1, taking advantage of the fact that the field is conservative and the integral is independent of path. Thus, we write (11) as

$$W = \int_{(1,0)}^{(-1,0)} e^y \, dx + x e^y \, dy = \phi(-1,0) - \phi(1,0) \tag{12}$$

As illustrated in Example 3, we can find ϕ by integrating either of the equations

$$\frac{\partial \phi}{\partial x} = e^y \quad \text{and} \quad \frac{\partial \phi}{\partial y} = x e^y \tag{13}$$

We shall integrate the first. We obtain

$$\phi = \int e^y \, dx = x e^y + k(y) \tag{14}$$

Differentiating this equation with respect to y and using the second equation in (13) yields

$$\frac{\partial \phi}{\partial y} = x e^y + k'(y) = x e^y$$

from which it follows that $k'(y) = 0$ or $k(y) = K$. Thus, from (14)

$$\phi = x e^y + K$$

and hence from (12)

$$W = \phi(-1,0) - \phi(1,0) = (-1)e^0 - 1e^0 = -2 \quad \blacktriangleleft$$

CONSERVATIVE VECTOR FIELDS IN 3-SPACE

All of the results in this section have analogs in 3-space: Theorems 18.3.1 and 18.3.2 can be extended to vector fields in 3-space simply by adding a third variable and modifying the hypotheses appropriately. For example, in 3-space, Formula (1) becomes

$$\int_C \mathbf{F}(x, y, z) \cdot d\mathbf{r} = \phi(x_1, y_1, z_1) - \phi(x_0, y_0, z_0) \tag{15}$$

Theorem 18.3.3 can also be extended to vector fields in 3-space. We leave it for the exercises to show that if $\mathbf{F}(x, y, z) = f(x, y, z)\mathbf{i} + g(x, y, z)\mathbf{j} + h(x, y, z)\mathbf{k}$ is a conservative field, then

$$\frac{\partial f}{\partial y} = \frac{\partial g}{\partial x}, \quad \frac{\partial f}{\partial z} = \frac{\partial h}{\partial x}, \quad \frac{\partial g}{\partial z} = \frac{\partial h}{\partial y} \tag{16}$$

that is, curl $\mathbf{F} = \mathbf{0}$. Conversely, a vector field satisfying these conditions on a suitably restricted region is conservative on that region if f, g, and h are continuous and have continuous first partial derivatives in the region. Some problems involving Formulas (15) and (16) are given in the exercises.

POTENTIAL ENERGY

If $\mathbf{F}(x, y, z)$ is a conservative force field with a potential function $\phi(x, y, z)$, then physicists call $V(x, y, z) = -\phi(x, y, z)$ the ***potential energy*** of the field at the point (x, y, z). Thus, it follows from the 3-space version of Theorem 18.3.1 that the work W done by $\mathbf{F}$ on a particle that moves along any path C from a point (x_0, y_0, z_0) to a point (x_1, y_1, z_1) is related to the potential energy by the equation

$$W = \int_C \mathbf{F} \cdot d\mathbf{r} = \phi(x_1, y_1, z_1) - \phi(x_0, y_0, z_0)$$
$$= -[V(x_1, y_1, z_1) - V(x_0, y_0, z_0)]$$

That is, the work done by the force field is the negative of the change in potential energy. In particular, it follows from the 3-space analog of Theorem 18.3.2(*b*) that if a particle traverses a closed path, then the work done by the field is zero, and there is no change in potential energy.

▶ Exercise Set 18.3

In Exercises 1–6, determine whether **F** is conservative. If it is, find a potential function for it.

1. $\mathbf{F}(x, y) = x\mathbf{i} + y\mathbf{j}.$ **2.** $\mathbf{F}(x, y) = 3y^2\mathbf{i} + 6xy\mathbf{j}.$

3. $\mathbf{F}(x, y) = x^2 y\mathbf{i} + 5xy^2\mathbf{j}.$

4. $\mathbf{F}(x, y) = e^x \cos y\mathbf{i} - e^x \sin y\mathbf{j}.$

5. $\mathbf{F}(x, y) = (\cos y + y \cos x)\mathbf{i} + (\sin x - x \sin y)\mathbf{j}.$

6. $\mathbf{F}(x, y) = x \ln y\mathbf{i} + y \ln x\mathbf{j}.$

7. Show that $\displaystyle\int_{(-1, 2)}^{(1, 3)} y^2\, dx + 2xy\, dy$ is independent of the path and evaluate the integral by

  (a)   using Theorem 18.3.1

  (b)   integrating along the line segment from $(-1, 2)$ to $(1, 3)$.

8. Show that $\displaystyle\int_{(0, 1)}^{(\pi, -1)} y \sin x\, dx - \cos x\, dy$ is independent of the path, and evaluate the integral by

  (a)   using Theorem 18.3.1

  (b)   integrating along the line segment from $(0, 1)$ to $(\pi, -1)$.

In Exercises 9–14, show that the integral is independent of the path, and find its value.

9. $\displaystyle\int_{(1, 2)}^{(4, 0)} 3y\, dx + 3x\, dy.$

10. $\displaystyle\int_{(0, 0)}^{(1, \pi/2)} e^x \sin y\, dx + e^x \cos y\, dy.$

11. $\displaystyle\int_{(0, 0)}^{(3, 2)} 2xe^y\, dx + x^2 e^y\, dy.$

12. $\displaystyle\int_{(-1, 2)}^{(0, 1)} (3x - y + 1)\, dx - (x + 4y + 2)\, dy.$

13. $\displaystyle\int_{(2, -2)}^{(-1, 0)} 2xy^3\, dx + 3y^2 x^2\, dy.$

14. $\displaystyle\int_{(1, 1)}^{(3, 3)} \left(e^x \ln y - \frac{e^y}{x} \right) dx + \left(\frac{e^x}{y} - e^y \ln x \right) dy,$ where x and y are positive.

In Exercises 15–18, find the work done by the conservative force **F** as it acts on a particle moving from P to Q.

15. $\mathbf{F}(x, y) = xy^2\mathbf{i} + x^2 y\mathbf{j};\ P(1, 1), Q(0, 0).$

16. $\mathbf{F}(x, y) = ye^{xy}\mathbf{i} + xe^{xy}\mathbf{j};\ P(-1, 1), Q(2, 0).$

17. $\mathbf{F}(x, y) = \dfrac{y}{x^2 + y^2}\mathbf{i} - \dfrac{x}{x^2 + y^2}\mathbf{j};\ P(0, 2), Q(3, 3).$

  (Assume that $y > 0$.)

18. $\mathbf{F}(x, y) = e^{-y} \cos x\mathbf{i} - e^{-y} \sin x\mathbf{j};\ P(\pi/2, 1),$ $Q(-\pi/2, 0).$

19. Find $\displaystyle\int_C \mathbf{F} \cdot d\mathbf{r},$ where

$$\mathbf{F}(x, y) = (e^y + ye^x)\mathbf{i} + (xe^y + e^x)\mathbf{j}$$

  and C is the curve

$$\mathbf{r}(t) = \sin(\pi t/2)\mathbf{i} + \ln t\mathbf{j}, \quad 1 \le t \le 2$$

  [*Hint:* First determine whether **F** is conservative.]

20. Find $\displaystyle\int_C \mathbf{F} \cdot d\mathbf{r},$ where

$$\mathbf{F}(x, y) = 2xy\mathbf{i} + (x^2 + \cos y)\mathbf{j}$$

  and C is the curve

$$\mathbf{r}(t) = t\mathbf{i} + t \cos(t/3)\mathbf{j}, \quad 0 \le t \le \pi$$

21. (a)   Show that the inverse-square field

$$\mathbf{F}(\mathbf{r}) = \frac{1}{\|\mathbf{r}\|^3}\mathbf{r} \quad (\mathbf{r} = x\mathbf{i} + y\mathbf{j})$$

  is conservative everywhere, except at the origin, by finding a potential function for it.

  (b)   Find $\displaystyle\int_C \mathbf{F} \cdot d\mathbf{r},$ where C is the straight line segment from $P(0, -1)$ to $Q(-3, 4)$.

  (c)   Find $\displaystyle\int_C \mathbf{F} \cdot d\mathbf{r},$ where C is any piecewise smooth closed curve that does not go through the origin.

22. Show that the vector field

$$\mathbf{F}(x, y) = \frac{2xy}{(x^2 + y^2)^2}\mathbf{i} + \frac{y^2 - x^2}{(x^2 + y^2)^2}\mathbf{j}$$

  is conservative, except at the origin, by finding a potential function for it.

23. Find a function h for which

$$\mathbf{F}(x, y) = h(x)[x \sin y + y \cos y]\mathbf{i}$$
$$+ h(x)[x \cos y - y \sin y]\mathbf{j}$$

  is conservative.

24. Prove Theorem 18.3.1 if C is a piecewise smooth curve composed of smooth curves $C_1, C_2, \ldots, C_n$.

25. Prove that (*b*) implies (*c*) in Theorem 18.3.2. [*Hint:* Consider any two piecewise smooth oriented curves C_1 and C_2 in the region from a point P to a point Q, and integrate around the closed curve consisting of C_1 and $-C_2$.]

26. Complete the proof of Theorem 18.3.2 by showing that $\partial\phi/\partial y = g(x, y)$, where $\phi(x, y)$ is the function in (4).

27. Prove: If $\mathbf{F}(x, y, z) = f(x, y, z)\mathbf{i} + g(x, y, z)\mathbf{j} + h(x, y, z)\mathbf{k}$ is a conservative field and f, g, and h have continuous first partial derivatives in a region, then

$$\frac{\partial f}{\partial y} = \frac{\partial g}{\partial x}, \quad \frac{\partial f}{\partial z} = \frac{\partial h}{\partial x}, \quad \frac{\partial g}{\partial z} = \frac{\partial h}{\partial y}$$

in the region.

28. Use the result of Exercise 27 to show that curl $\mathbf{F} = \mathbf{0}$ in the region if $\mathbf{F}$ satisfies the stated conditions.

It can be shown that if the components of $\mathbf{F}(x, y, z)$ have continuous first partial derivatives everywhere, except possibly at finitely many points, then $\mathbf{F}$ is conservative (except at those points) if and only if curl $\mathbf{F} = \mathbf{0}$. Use this result in Exercises 29–34 to determine whether $\mathbf{F}$ is conservative; if so, find a potential function for it.

29. $\mathbf{F}(x, y, z) = z^2\mathbf{i} + e^{-y}\mathbf{j} + 2xz\mathbf{k}$.

30. $\mathbf{F}(x, y, z) = xy\mathbf{i} + x^2\mathbf{j} + \sin z\mathbf{k}$.

31. $\mathbf{F}(x, y, z) = yz\mathbf{i} + xz\mathbf{j} + (xy + 3z^2)\mathbf{k}$.

32. $\mathbf{F}(x, y, z) = \sin x\mathbf{i} + z\mathbf{j} + y\mathbf{k}$.

33. $\mathbf{F}(x, y, z) = z\mathbf{i} + 2yz\mathbf{j} + y^2\mathbf{k}$.

34. $\mathbf{F}(x, y, z) = 2xz\mathbf{i} + \cos y\mathbf{j} + (x^2 + 2z)\mathbf{k}$.

35. In Formula (5) of Section 18.1 a potential function was given for the inverse-square field

$$\mathbf{F}(\mathbf{r}) = \frac{1}{\|\mathbf{r}\|^3}\mathbf{r}$$

Use this potential function to evaluate the integrals:

(a) $\displaystyle\int_C \mathbf{F} \cdot d\mathbf{r}$, where C is the line segment from $P(1, 1, 2)$ to $Q(3, 2, 1)$

(b) $\displaystyle\int_C \mathbf{F} \cdot d\mathbf{r}$, where C is the curve
$\mathbf{r}(t) = (2t^2 + 1)\mathbf{i} + (t^3 + 1)\mathbf{j} + (2 - \sqrt{t})\mathbf{k}, 0 \le t \le 1$

(c) $\displaystyle\int_C \mathbf{F} \cdot d\mathbf{r}$, where C is any piecewise smooth closed curve that does not pass through the origin.

36. (a) Show that the vector field

$$\mathbf{F}(\mathbf{r}) = \frac{1}{\|\mathbf{r}\|^2}\mathbf{r} \quad (\mathbf{r} = x\mathbf{i} + y\mathbf{j} + z\mathbf{k})$$

is conservative everywhere, except at the origin, by finding a potential function for it.

(b) Find $\displaystyle\int_C \mathbf{F} \cdot d\mathbf{r}$, where C is any piecewise smooth curve from $P(2, 2, 1)$ to $Q(a, b, c)$ that does not go through the origin.

(c) Find $\displaystyle\int_C \mathbf{F} \cdot d\mathbf{r}$, where C is any piecewise smooth closed curve that does not go through the origin.

■ 18.4 GREEN'S THEOREM

In this section we shall discuss a remarkable and beautiful theorem that expresses the double integral over a plane region in terms of a line integral over its boundary.

☐ **GREEN'S THEOREM**

18.4.1 THEOREM (*Green's* Theorem*). *Let R be a simply connected plane region whose boundary is a simple, closed, piecewise smooth curve C oriented counterclockwise. If $f(x, y)$ and $g(x, y)$ have continuous first partial derivatives on some open set containing R, then*

$$\int_C f(x, y)\, dx + g(x, y)\, dy = \iint\limits_R \left(\frac{\partial g}{\partial x} - \frac{\partial f}{\partial y}\right) dA \tag{1}$$

Proof. For simplicity, we shall prove the theorem only for regions that are simultaneously type I and type II (see Section 17.2). Such a region is shown in Figure 18.4.1. The crux of the proof is to show that

$$\int_C f(x, y)\, dx = -\iint\limits_R \frac{\partial f}{\partial y}\, dA \quad \text{and} \quad \int_C g(x, y)\, dy = \iint\limits_R \frac{\partial g}{\partial x}\, dA \tag{2a–b}$$

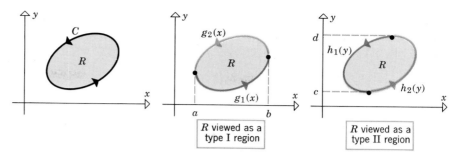

Figure 18.4.1

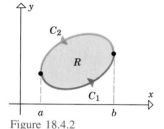

Figure 18.4.2

To prove (2a), view R as a type I region and let C_1 and C_2 be the lower and upper boundary curves, oriented as in Figure 18.4.2. Then

$$\int_C f(x, y)\, dx = \int_{C_1} f(x, y)\, dx + \int_{C_2} f(x, y)\, dx$$

or, equivalently,

$$\int_C f(x, y)\, dx = \int_{C_1} f(x, y)\, dx - \int_{-C_2} f(x, y)\, dx \tag{3}$$

(This step will help simplify our calculations since C_1 and $-C_2$ are then both oriented left to right.) With $x = t$ as the parameter, C_1 and $-C_2$ are represented by

$$C_1 : x = t, y = g_1(t) \quad (a \le t \le b)$$

$$-C_2 : x = t, y = g_2(t) \quad (a \le t \le b)$$

Thus, (3) yields

$$\int_C f(x, y)\, dx = \int_a^b f(t, g_1(t)) \left(\frac{dx}{dt}\right) dt - \int_a^b f(t, g_2(t)) \left(\frac{dx}{dt}\right) dt$$

$$= \int_a^b f(t, g_1(t))\, dt - \int_a^b f(t, g_2(t))\, dt$$

$$= -\int_a^b [f(t, g_2(t)) - f(t, g_1(t))]\, dt$$

$$= -\int_a^b \left[f(t, y) \right]_{y=g_1(t)}^{y=g_2(t)} dt = -\int_a^b \left[\int_{g_1(t)}^{g_2(t)} \frac{\partial f}{\partial y}\, dy \right] dt$$

$$= -\int_a^b \int_{g_1(x)}^{g_2(x)} \frac{\partial f}{\partial y}\, dy\, dx = -\iint_R \frac{\partial f}{\partial y}\, dA$$

Since $x = t$

---

* GEORGE GREEN (1793–1841). English mathematician and physicist. Green left school at an early age to work in his father's bakery and consequently had little early formal education. When his father opened a mill, the boy used the top room as a study in which he taught himself physics and mathematics from library books. In 1828 Green published his most important work, *An Essay on the Application of Mathematical Analysis to the Theories of Electricity and Magnetism*. Although Green's Theorem appeared in that paper, the result went virtually unnoticed because of the small pressrun and local distribution. Following the death of his father in 1829, Green was urged by friends to seek a college education. In 1833, after four years of self-study to close the gaps in his elementary education, Green was admitted to Caius College, Cambridge. He graduated four years later, but with a disappointing performance on his final examinations—possibly because he was more interested in his own research. After a succession of works on light and sound, he was named to be Perse Fellow at Caius College. Two years later he died. In 1845, four years after his death, his paper of 1828 was published and the theories developed therein by this obscure, self-taught baker's son helped pave the way to the modern theories of electricity and magnetism.

The proof of (2b) is obtained similarly by treating R as a type II region. The details are omitted. ▪

Example 1   Use Green's Theorem to evaluate

$$\int_C x^2 y\, dx + x\, dy$$

over the triangular path shown in Figure 18.4.3.

Solution.   Since $f(x, y) = x^2 y$ and $g(x, y) = x$, it follows from (1) that

$$\int_C x^2 y\, dx + x\, dy = \iint_R \left[\frac{\partial}{\partial x}(x) - \frac{\partial}{\partial y}(x^2 y) \right] dA = \int_0^1 \int_0^{2x} (1 - x^2)\, dy\, dx$$

$$= \int_0^1 (2x - 2x^3)\, dx = \left[x^2 - \frac{x^4}{2} \right]_0^1 = \frac{1}{2}$$

This agrees with the result obtained in Example 4 of Section 18.2, where we evaluated the line integral directly. Note how much simpler this solution is. ◀

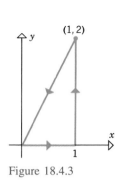

Figure 18.4.3

□ **FINDING WORK USING GREEN'S THEOREM**

Because Formula (1) requires the curve C to be closed, it is desirable to have a notation for indicating this. One way of doing this is to write (1) as

$$\oint_C f(x, y)\, dx + g(x, y)\, dy = \iint_R \left(\frac{\partial g}{\partial x} - \frac{\partial f}{\partial y} \right) dA$$

where the symbol ○ on the integral sign indicates that C is a simple closed curve.

Example 2   Find the work done by the force field $\mathbf{F}(x, y) = (e^x - y^3)\mathbf{i} + (\cos y + x^3)\mathbf{j}$ on a particle that travels once around the unit circle $x^2 + y^2 = 1$ in the counterclockwise direction.

Solution.   From Formula (24) of Section 18.2, the work W done by the field is

$$W = \oint_C \mathbf{F} \cdot d\mathbf{r} = \oint_C (e^x - y^3)\, dx + (\cos y + x^3)\, dy \tag{4}$$

However, the force field $\mathbf{F}$ is not conservative, since Equation (6) in Theorem 18.3.3 does not hold (verify), and consequently the work done by $\mathbf{F}$ need not be zero, even though C is a simple closed curve. Moreover, evaluating (4) directly by parametrizing the circle leads to a complicated line integral in terms of t, so the method of choice for solving this problem is Green's Theorem. This yields

$$W = \oint_C (e^x - y^3)\, dx + (\cos y + x^3)\, dy$$

$$= \iint_R \left[\frac{\partial}{\partial x}(\cos y + x^3) - \frac{\partial}{\partial y}(e^x - y^3) \right] dA$$

$$= \iint_R (3x^2 + 3y^2)\, dA = 3 \iint_R (x^2 + y^2)\, dA$$

$$= 3 \int_0^{2\pi} \int_0^1 (r^2) r\, dr\, d\theta = \frac{3}{4} \int_0^{2\pi} d\theta = \frac{3\pi}{2} \quad ◀$$

We converted to polar coordinates.

REMARK.   Green's Theorem provides another viewpoint about work in conservative vector fields. In a force field $\mathbf{F}(x, y) = f(x, y)\mathbf{i} + g(x, y)\mathbf{j}$, the line integral in (1) represents the work done by the field on a particle moving over the closed curve C. However, if the field is conservative, then from Theorem 18.3.3 the integrand of the double integral in (1) is zero, and hence Green's Theorem implies that the work done by the field is zero, which agrees with results obtained in the preceding section.

☐ FINDING AREAS USING GREEN'S THEOREM

Green's Theorem leads to some new formulas for the area A of a region R that satisfies the conditions of the theorem. Two such formulas can be obtained as follows:

$$A = \iint\limits_{R} dA = \oint_{C} x \, dy \quad \text{and} \quad A = \iint\limits_{R} dA = \oint_{C} (-y) \, dx$$

Set $f(x, y) = 0$ and $g(x, y) = x$ in (1).   Set $f(x, y) = -y$ and $g(x, y) = 0$ in (1).

A third formula can be obtained by adding these two equations together. Thus, we have the following three formulas that express the area A of a region R in terms of line integrals over the boundary:

$$A = \oint_{C} x \, dy = -\oint_{C} y \, dx = \frac{1}{2} \oint_{C} -y \, dx + x \, dy \tag{5}$$

REMARK.   Although the third formula in (5) looks more complicated than the other two, it often leads to simpler integrations; but each has its use.

Example 3   Use the third formula in (5) to find the area enclosed by the ellipse $x^2/a^2 + y^2/b^2 = 1$.

Solution.   The ellipse, with counterclockwise orientation, can be represented parametrically by

$$x = a \cos t, \quad y = b \sin t \quad (0 \le t \le 2\pi)$$

If we denote this curve by C, then the area A enclosed by the ellipse is

$$A = \frac{1}{2} \oint_{C} -y \, dx + x \, dy$$

$$= \frac{1}{2} \int_{0}^{2\pi} [(-b \sin t)(-a \sin t) + (a \cos t)(b \cos t)] \, dt$$

$$= \frac{1}{2} ab \int_{0}^{2\pi} (\sin^2 t + \cos^2 t) \, dt = \frac{1}{2} ab \int_{0}^{2\pi} dt = \pi ab \quad \blacktriangleleft$$

☐ GREEN'S THEOREM FOR MULTIPLY CONNECTED REGIONS

Recall that a region in 2-space is called *simply connected* if it is connected and has no holes. Regions of 2-space that are connected but have finitely many holes are sometimes called *multiply connected*. Although Green's Theorem (18.4.1) was stated for simply connected regions, it can be extended to multiply connected regions with some appropriate modifications. For example, let R be the multiply connected region shown in Figure 18.4.4a. The boundary of R consists of two disjoint curves C_1 and C_2, where C_1 is the "outside boundary" and C_2 is the boundary of the hole in R. As shown in the figure, suppose that the boundary curves are oriented so that the region R is on the left as the curves are traversed in the direction of orientation. Thus, the outside boundary is oriented counterclockwise, and the boundary of the hole is oriented clockwise. As shown in Figure

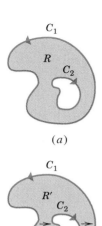

(a)

(b)

Figure 18.4.4

18.4.4*b*, let us divide R into two regions R' and R'' by introducing two "cuts" in R. The cuts are shown as line segments, but any piecewise smooth curves will suffice. If we assume that f and g satisfy the hypotheses of Green's Theorem on R (and hence on R' and R''), then we can apply this theorem to both R' and R'' to obtain

$$\iint_R \left(\frac{\partial g}{\partial x} - \frac{\partial f}{\partial y}\right) dA = \iint_{R'} \left(\frac{\partial g}{\partial x} - \frac{\partial f}{\partial y}\right) dA + \iint_{R''} \left(\frac{\partial g}{\partial x} - \frac{\partial f}{\partial y}\right) dA$$

$$= \underbrace{\oint f(x,y)\,dx + g(x,y)\,dy}_{\substack{\text{Boundary}\\\text{of } R'}} + \underbrace{\oint f(x,y)\,dx + g(x,y)\,dy}_{\substack{\text{Boundary}\\\text{of } R''}}$$

However, the two line integrals are taken in opposite directions along the cuts, and hence cancel there, leaving only the contributions along C_1 and C_2. Thus,

$$\iint_R \left(\frac{\partial g}{\partial x} - \frac{\partial f}{\partial y}\right) dA = \oint_{C_1} f(x,y)\,dx + g(x,y)\,dy + \oint_{C_2} f(x,y)\,dx + g(x,y)\,dy \qquad (6)$$

In general, Green's Theorem applies to a multiply connected region R, provided f and g satisfy the hypotheses of the theorem, and we integrate over the *entire* boundary with the boundary curves oriented so that the region is on the left as we travel in the direction of orientation. For a region with n holes, the analog of Formula (6) would involve $n + 1$ line integrals, one for the outside boundary and one for the boundary of each hole.

Example 4   Evaluate the integral

$$\oint_C \frac{-y\,dx + x\,dy}{x^2 + y^2}$$

if C is a piecewise smooth simple closed curve oriented counterclockwise such that:

(a) C does not enclose the origin;

(b) C encloses the origin.

Solution (a).   Let

$$f(x,y) = -\frac{y}{x^2 + y^2}, \quad g(x,y) = \frac{x}{x^2 + y^2}$$

so that

$$\frac{\partial g}{\partial x} = \frac{y^2 - x^2}{(x^2 + y^2)^2} = \frac{\partial f}{\partial y}$$

if x and y are not both zero. Thus, if C does not enclose the origin, we have

$$\frac{\partial g}{\partial x} - \frac{\partial f}{\partial y} = 0$$

on the simply connected region enclosed by C, and hence the given integral is zero by Green's Theorem.

Solution (b).   Because f and g are undefined at the origin, the hypotheses of Green's Theorem are not satisfied on the region enclosed by C. Moreover, we cannot evaluate the integral directly since we are not given a specific curve C. To circumvent these problems, we construct a circle C_a with *counterclockwise* orientation, centered at the origin, and of sufficiently small radius a that it lies inside the region enclosed by C (Figure 18.4.5). The functions f and g satisfy the hypotheses of Green's Theorem on the *multiply connected* region R enclosed between C and C_a, and hence

$$\oint_C \frac{-y\,dx + x\,dy}{x^2 + y^2} + \oint_{-C_a} \frac{-y\,dx + x\,dy}{x^2 + y^2} = \iint_R 0\,dA = 0 \qquad (7)$$

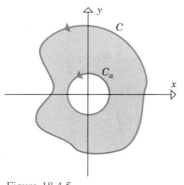

Figure 18.4.5

(where the integration is over $-C_a$ to keep the region R to the left). It follows from (7) that

$$\oint_C \frac{-y\,dx + x\,dy}{x^2 + y^2} = \oint_{C_a} \frac{-y\,dx + x\,dy}{x^2 + y^2}$$

This shows that the integral over the arbitrary curve C is the same as the integral over the circle. Therefore, parametrizing the circle as $x = a\cos t$, $y = a\sin t$ $(0 \le t \le 2\pi)$ yields

$$\oint_C \frac{-y\,dx + x\,dy}{x^2 + y^2} = \int_0^{2\pi} \frac{(-a\sin t)(-a\sin t)\,dt + (a\cos t)(a\cos t)\,dt}{(a\cos t)^2 + (a\sin t)^2}$$

$$= \int_0^{2\pi} 1\,dt = 2\pi \quad \blacktriangleleft$$

▶ Exercise Set 18.4

In Exercises 1 and 2, evaluate the line integral using Green's Theorem and check the answer by evaluating directly.

1. $\oint_C y^2\,dx + x^2\,dy$, where C is the square with vertices $(0, 0)$, $(1, 0)$, $(1, 1)$, and $(0, 1)$ oriented counterclockwise.

2. $\oint_C y\,dx + x\,dy$, where C is the unit circle oriented counterclockwise.

In Exercises 3–13, use Green's Theorem to evaluate the integral. In each exercise, assume that the curve C is oriented counterclockwise.

3. $\oint_C 3xy\,dx + 2xy\,dy$, where C is the rectangle bounded by $x = -2$, $x = 4$, $y = 1$, and $y = 2$.

4. $\oint_C (x^2 - y^2)\,dx + x\,dy$, where C is the circle $x^2 + y^2 = 9$.

5. $\oint_C x\cos y\,dx - y\sin x\,dy$, where C is the square with vertices $(0, 0)$, $(0, \pi/2)$, $(\pi/2, \pi/2)$, and $(\pi/2, 0)$.

6. $\oint_C y\tan^2 x\,dx + \tan x\,dy$, where C is the circle $x^2 + (y + 1)^2 = 1$.

7. $\oint_C (x^2 - y)\,dx + x\,dy$, where C is the circle $x^2 + y^2 = 4$.

8. $\oint_C (e^x + y^2)\,dx + (e^y + x^2)\,dy$, where C is the boundary of the region between $y = x^2$ and $y = x$.

9. $\oint_C \ln(1 + y)\,dx - \frac{xy}{1 + y}\,dy$, where C is the triangle with vertices $(0, 0)$, $(2, 0)$, and $(0, 4)$.

10. $\oint_C x^2 y\,dx - y^2 x\,dy$, where C is the boundary of the region in the first quadrant, enclosed between the coordinate axes and the circle $x^2 + y^2 = 16$.

11. $\oint_C \tan^{-1} y\,dx - \frac{y^2 x}{1 + y^2}\,dy$, where C is the square with vertices $(0, 0)$, $(0, 1)$, $(1, 1)$, and $(1, 0)$.

12. $\oint_C \cos x\sin y\,dx + \sin x\cos y\,dy$, where C is the triangle with vertices $(0, 0)$, $(3, 3)$, and $(0, 3)$.

13. $\oint_C x^2 y\,dx + (y + xy^2)\,dy$, where C is the boundary of the region enclosed by $y = x^2$ and $x = y^2$.

14. Let C be the boundary of the region enclosed between $y = x^2$ and $y = 2x$. Assuming that C is oriented counterclockwise, evaluate the following integrals by Green's Theorem:

   (a) $\oint_C (6xy - y^2)\,dx$ (b) $\oint_C (6xy - y^2)\,dy$.

15. Find the area of the ellipse in Example 3 using
   (a) the first formula in (5)
   (b) the second formula in (5).

16. Use a line integral to find the area of the region enclosed by
   $$x = a\cos^3 t, \quad y = a\sin^3 t \quad (0 \le t \le 2\pi)$$

17. Use a line integral to find the area of the triangle with vertices $(0, 0)$, $(a, 0)$, and $(0, b)$, where $a > 0$ and $b > 0$.

18. Use a line integral to find the area of the region in the first quadrant enclosed by $y = x$, $y = 1/x$, and $y = x/9$.

19. Use the formula
   $$A = \frac{1}{2}\oint_C -y\,dx + x\,dy$$
   to find the area of the region swept out by the line from the origin to the ellipse $x = a\cos t$, $y = b\sin t$ if t varies from $t = 0$ to $t = t_0$ $(0 \le t_0 \le 2\pi)$.

20. Use the formula
   $$A = \frac{1}{2}\oint_C -y\,dx + x\,dy$$
   to find the area of the region swept out by the line from the origin to the hyperbola $x = a\cosh t$, $y = b\sinh t$ if t varies from $t = 0$ to $t = t_0$ $(t_0 \ge 0)$.

21. A particle, starting at $(5, 0)$, traverses the upper semicircle $x^2 + y^2 = 25$ and returns to its starting point along the x-axis. Use Green's Theorem to find the work done on the particle by a force $\mathbf{F}(x, y) = xy\mathbf{i} + (\frac{1}{2}x^2 + yx)\mathbf{j}$.

22. A particle moves counterclockwise one time around the closed curve formed by $y = 0$, $x = 2$, and $y = x^3/4$. Use Green's Theorem to find the work done by a force $\mathbf{F}(x, y) = \sqrt{y}\mathbf{i} + \sqrt{x}\mathbf{j}$.

23. (a) Let R be a plane region with area A whose boundary is a piecewise smooth simple closed curve C. Use Green's Theorem to prove that the centroid $(\bar{x}, \bar{y})$ of R is given by

$$\bar{x} = \frac{1}{2A}\oint_C x^2\, dy, \quad \bar{y} = -\frac{1}{2A}\oint_C y^2\, dx$$

(b) Use the result in part (a) to find the centroid of the region enclosed between the x-axis and the upper half of the circle $x^2 + y^2 = a^2$.

24. Evaluate $\oint_C y\, dx - x\, dy$, where C is the cardioid

$$r = a(1 + \cos\theta), \quad 0 \le \theta \le 2\pi$$

25. (a) Let C be the line segment from a point (a, b) to a point (c, d). Show that

$$\int_C -y\, dx + x\, dy = ad - bc$$

(b) Use the result in part (a) to show that the area A of a triangle with successive vertices (x_1, y_1), (x_2, y_2), and (x_3, y_3) going counterclockwise is

$$A = \frac{1}{2}[(x_1 y_2 - x_2 y_1) + (x_2 y_3 - x_3 y_2) + (x_3 y_1 - x_1 y_3)]$$

(c) Find a formula for the area of a polygon with successive vertices (x_1, y_1), (x_2, y_2), ..., (x_n, y_n) going counterclockwise.

(d) Use the result in part (c) to find the area of a quadrilateral with vertices $(0, 0)$, $(3, 4)$, $(-2, 2)$, $(-1, 0)$.

26. Find a simple closed curve C that maximizes the value of

$$\oint_C \frac{1}{3}y^3\, dx + (x - \frac{1}{3}x^3)\, dy$$

In Exercises 27 and 28, assume that the boundary curves of the regions are oriented so that the region is on the left traveling in the direction of orientation.

27. Find $\oint_C \mathbf{F} \cdot d\mathbf{r}$, where $\mathbf{F}(x, y) = (x^2 + y)\mathbf{i} + (4x - \cos y)\mathbf{j}$ and C is the boundary of the region R that is inside the square with vertices $(0, 0)$, $(5, 0)$, $(5, 5)$, $(0, 5)$ but is outside the rectangle with vertices $(1, 1)$, $(3, 1)$, $(3, 2)$, $(1, 2)$.

28. Find $\oint_C \mathbf{F} \cdot d\mathbf{r}$, where $\mathbf{F}(x, y) = (e^{-x} + 3y)\mathbf{i} + x\mathbf{j}$ and C is the boundary of the region R between the circles $x^2 + y^2 = 16$ and $x^2 - 2x + y^2 = 3$.

18.5 INTRODUCTION TO SURFACE INTEGRALS

In previous sections we considered four kinds of integrals—integrals over intervals, double integrals over two-dimensional regions, triple integrals over three-dimensional solids, and line integrals over curves in two- or three-dimensional space. In this section we shall discuss integrals over surfaces in three-dimensional space. Such integrals occur in problems involving fluid and heat flow, electricity, magnetism, mass, and center of gravity.

☐ DEFINITION OF A
SURFACE INTEGRAL

Recall that if C is a smooth curve in 3-space, and $f(x, y, z)$ is continuous on C, then the line integral of f over C with respect to arc length is defined by subdividing C into n arcs and defining the line integral as the limit

$$\int_C f(x, y, z)\, ds = \lim_{n \to +\infty} \sum_{k=1}^{n} f(x_k^*, y_k^*, z_k^*)\, \Delta s_k$$

where (x_k^*, y_k^*, z_k^*) is a point on the kth arc and Δs_k is the length of the kth arc. We will define *surface integrals* in an analogous manner.

Let σ be a surface in 3-space with finite surface area, and let $f(x, y, z)$ be a continuous function defined on σ. As shown in Figure 18.5.1, subdivide σ into parts, $\sigma_1, \sigma_2, \ldots, \sigma_n$ with areas $\Delta S_1, \Delta S_2, \ldots, \Delta S_n$, and form the sum

$$\sum_{k=1}^{n} f(x_k^*, y_k^*, z_k^*)\, \Delta S_k \tag{1}$$

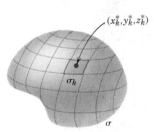

Figure 18.5.1

where (x_k^*, y_k^*, z_k^*) is an arbitrary point on σ_k. Now repeat the subdivision process, dividing σ into more and more parts in such a way that the maximum dimension of each part approaches zero as $n \to +\infty$. If (1) approaches a limit that does not depend on the way the subdivisions are made or how the points (x_k^*, y_k^*, z_k^*) are chosen, then this limit is called the *surface integral* of $f(x, y, z)$ over σ and is denoted by

$$\iint_\sigma f(x, y, z)\, dS = \lim_{n \to +\infty} \sum_{k=1}^{n} f(x_k^*, y_k^*, z_k^*)\, \Delta S_k \tag{2}$$

☐ **EVALUATING SURFACE INTEGRALS**

There are various procedures for evaluating surface integrals, depending on how the surface σ is represented. The following theorem provides a method for evaluating a surface integral when σ is of the form $z = g(x, y)$, or of the form $y = g(x, z)$ or $x = g(y, z)$.

18.5.1   THEOREM.

(a)   Let σ be a surface with equation $z = g(x, y)$ and let R be its projection on the *xy*-plane. If g has continuous first partial derivatives on R and $f(x, y, z)$ is continuous on σ, then

$$\iint_\sigma f(x, y, z)\, dS = \iint_R f[x, y, g(x, y)] \sqrt{\left(\frac{\partial z}{\partial x}\right)^2 + \left(\frac{\partial z}{\partial y}\right)^2 + 1}\, dA \tag{3}$$

(b)   Let σ be a surface with equation $y = g(x, z)$ and let R be its projection on the *xz*-plane. If g has continuous first partial derivatives on R and $f(x, y, z)$ is continuous on σ, then

$$\iint_\sigma f(x, y, z)\, dS = \iint_R f[x, g(x, z), z] \sqrt{\left(\frac{\partial y}{\partial x}\right)^2 + \left(\frac{\partial y}{\partial z}\right)^2 + 1}\, dA \tag{4}$$

(c)   Let σ be a surface with equation $x = g(y, z)$ and let R be its projection on the *yz*-plane. If g has continuous first partial derivatives on R and $f(x, y, z)$ is continuous on σ, then

$$\iint_\sigma f(x, y, z)\, dS = \iint_R f[g(y, z), y, z] \sqrt{\left(\frac{\partial x}{\partial y}\right)^2 + \left(\frac{\partial x}{\partial z}\right)^2 + 1}\, dA \tag{5}$$

To illustrate how the formulas in this theorem can be obtained, consider the case where σ is represented by an equation of the form $z = g(x, y)$ over a region R of the *xy*-plane. In this case it follows from Formula (3) of Section 17.4 (with g instead of f) that the surface area ΔS_k in Formula (2) can be approximated by

$$\Delta S_k \approx \sqrt{g_x(x_k^*, y_k^*)^2 + g_y(x_k^*, y_k^*)^2 + 1}\, \Delta A_k$$

in which case (2) becomes

$$\iint_\sigma f(x, y, z)\, dS = \lim_{n \to +\infty} \sum_{k=1}^{n} f(x_k^*, y_k^*, z_k^*) \sqrt{g_x(x_k^*, y_k^*)^2 + g_y(x_k^*, y_k^*)^2 + 1}\, \Delta A_k$$

$$= \lim_{n \to +\infty} \sum_{k=1}^{n} f(x_k^*, y_k^*, g(x_k^*, y_k^*)) \sqrt{g_x(x_k^*, y_k^*)^2 + g_y(x_k^*, y_k^*)^2 + 1}\, \Delta A_k$$

Noting that the limit in this equation is a double integral over the region R, we can rewrite the equation as

$$\iint_{\sigma} f(x, y, z)\, dS = \iint_{R} f(x, y, g(x, y)) \sqrt{g_x(x, y)^2 + g_y(x, y)^2 + 1}\, dA$$

or, equivalently,

$$\iint_{\sigma} f(x, y, z)\, dS = \iint_{R} f(x, y, g(x, y)) \sqrt{\left(\frac{\partial z}{\partial x}\right)^2 + \left(\frac{\partial z}{\partial y}\right)^2 + 1}\, dA$$

Example 1 Evaluate the surface integral

$$\iint_{\sigma} xz\, dS$$

where σ is the part of the plane $x + y + z = 1$ that lies in the first octant.

Solution. The equation of the plane can be written as

$$z = 1 - x - y$$

which is of the form $z = g(x, y)$. Consequently, we can apply Formula (3) with $z = g(x, y) = 1 - x - y$ and $f(x, y, z) = xz$. Thus,

$$\frac{\partial z}{\partial x} = -1 \quad \text{and} \quad \frac{\partial z}{\partial y} = -1$$

so (3) becomes

$$\iint_{\sigma} xz\, dS = \iint_{R} x(1 - x - y)\sqrt{(-1)^2 + (-1)^2 + 1}\, dA \tag{6}$$

where R is the projection of σ on the xy-plane (Figure 18.5.2). Rewriting the double integral in (6) as an iterated integral yields

$$\iint_{\sigma} xz\, dS = \sqrt{3} \int_0^1 \int_0^{1-x} (x - x^2 - xy)\, dy\, dx$$

$$= \sqrt{3} \int_0^1 \left[xy - x^2 y - \frac{xy^2}{2} \right]_{y=0}^{1-x} dx$$

$$= \sqrt{3} \int_0^1 \left(\frac{x}{2} - x^2 + \frac{x^3}{2} \right) dx$$

$$= \sqrt{3} \left[\frac{x^2}{4} - \frac{x^3}{3} + \frac{x^4}{8} \right]_0^1 = \frac{\sqrt{3}}{24} \quad \blacktriangleleft$$

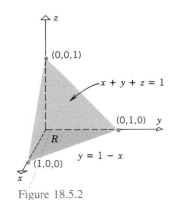

Figure 18.5.2

Example 2 Evaluate the surface integral

$$\iint_{\sigma} y^2 z^2\, dS$$

where σ is the part of the cone $z = \sqrt{x^2 + y^2}$ that lies between the planes $z = 1$ and $z = 2$ (Figure 18.5.3)

Solution. We shall apply Formula (3) with $z = g(x, y) = \sqrt{x^2 + y^2}$ and $f(x, y, z) = y^2 z^2$. Thus,

$$\frac{\partial z}{\partial x} = \frac{x}{\sqrt{x^2 + y^2}} \quad \text{and} \quad \frac{\partial z}{\partial y} = \frac{y}{\sqrt{x^2 + y^2}}$$

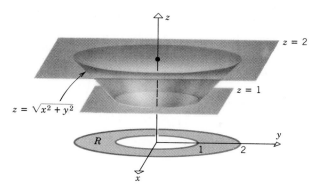

Figure 18.5.3

so

$$\sqrt{\left(\frac{\partial z}{\partial x}\right)^2 + \left(\frac{\partial z}{\partial y}\right)^2 + 1} = \sqrt{2}$$

(verify), and (3) yields

$$\iint_\sigma y^2 z^2 \, dS = \iint_R y^2 (\sqrt{x^2 + y^2})^2 \sqrt{2} \, dA = \sqrt{2} \iint_R y^2 (x^2 + y^2) \, dA$$

where R is the annulus enclosed between $x^2 + y^2 = 1$ and $x^2 + y^2 = 4$ (Figure 18.5.3). Using polar coordinates to evaluate this double integral over the annulus R yields

$$\iint_\sigma y^2 z^2 \, dS = \sqrt{2} \int_0^{2\pi} \int_1^2 (r \sin\theta)^2 (r^2) r \, dr \, d\theta$$

$$= \sqrt{2} \int_0^{2\pi} \int_1^2 r^5 \sin^2\theta \, dr \, d\theta$$

$$= \sqrt{2} \int_0^{2\pi} \frac{r^6}{6} \sin^2\theta \Big]_{r=1}^2 d\theta = \frac{21}{\sqrt{2}} \int_0^{2\pi} \sin^2\theta \, d\theta$$

$$= \frac{21}{\sqrt{2}} \left[\frac{1}{2}\theta - \frac{1}{4}\sin 2\theta \right]_0^{2\pi} = \frac{21\pi}{\sqrt{2}} \quad \boxed{\text{See Example 1, Section 9.3.}} \qquad \blacktriangleleft$$

☐ SURFACE AREA AS A SURFACE INTEGRAL

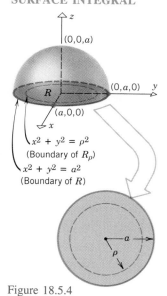

$x^2 + y^2 = \rho^2$
(Boundary of R_ρ)

$x^2 + y^2 = a^2$
(Boundary of R)

Figure 18.5.4

In the special case where $f(x, y, z)$ is 1, Formula (2) becomes

$$\iint_\sigma dS = \lim_{n \to +\infty} \sum_{k=1}^n \Delta S_k$$

But for all values of n the sum is the surface area of the surface σ; and hence so is the limit. Thus, the surface area S of the surface σ can be expressed as

$$S = \iint_\sigma dS \qquad (7)$$

Example 3 Use Formula (3) to show that the surface area of the upper hemisphere of radius a given by the equation $z = \sqrt{a^2 - x^2 - y^2}$ is $2\pi a^2$.

Solution. As shown in Figure 18.5.4, the projection of the hemisphere on the xy-plane is the circular region R of radius a centered at the origin. However, we cannot apply Theorem 18.5.1 directly because the partial derivatives

$$\frac{\partial z}{\partial x} = \frac{-x}{\sqrt{a^2 - x^2 - y^2}} \quad \text{and} \quad \frac{\partial z}{\partial y} = \frac{-y}{\sqrt{a^2 - x^2 - y^2}}$$

do not exist on the boundary of the region R since $x^2 + y^2 = a^2$ there. In order to overcome this problem we evaluate the double integral over a slightly smaller circular region R_ρ of radius ρ, and then let ρ approach a. The computations are as follows:

$$\iint_\sigma dS = \lim_{\rho \to a^-} \iint_{R_\rho} \sqrt{\left(\frac{\partial z}{\partial x}\right)^2 + \left(\frac{\partial z}{\partial y}\right)^2 + 1}\, dA$$

$$= \lim_{\rho \to a^-} \iint_{R_\rho} \sqrt{\frac{x^2}{a^2 - x^2 - y^2} + \frac{y^2}{a^2 - x^2 - y^2} + 1}\, dA$$

$$= \lim_{\rho \to a^-} \iint_{R_\rho} \frac{a}{\sqrt{a^2 - x^2 - y^2}}\, dA = \lim_{\rho \to a^-} \int_0^{2\pi} \int_0^\rho \frac{a}{\sqrt{a^2 - r^2}}\, r\, dr\, d\theta$$

$$= \lim_{\rho \to a^-} \int_0^{2\pi} -a\sqrt{a^2 - r^2} \,\Big]_{r=0}^\rho \, d\theta = \lim_{\rho \to a^-} \int_0^{2\pi} (a^2 - a\sqrt{a^2 - \rho^2})\, d\theta$$

$$= \lim_{\rho \to a^-} 2\pi(a^2 - a\sqrt{a^2 - \rho^2}) = 2\pi a^2 \quad \blacktriangleleft$$

☐ **MASS OF A CURVED LAMINA AS A SURFACE INTEGRAL**

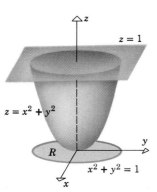

The thickness of a curved lamina is negligible.

Figure 18.5.5

(x, y, z)

Figure 18.5.6

In Section 17.6 we defined a *lamina* to be an idealized thin object in 2-space. There is an analogous concept in 3-space: We define a **curved lamina** to be an idealized object in 3-space that is sufficiently thin that it can be regarded as a surface. A curved lamina may look like a bent plate, as in Figure 18.5.5, or it may enclose a region of 3-space, like the shell of an egg. If the composition of a curved lamina is uniform throughout, then the lamina is called **homogeneous** and its **density** is defined to be the total mass divided by the total surface area. However, if the lamina is not homogeneous, then the density varies from point to point, in which case the density of a curved lamina σ is described by a **density function** $\delta(x, y, z)$, whose value at point (x, y, z) is viewed as a limit

$$\delta(x, y, z) = \lim_{\Delta S \to 0} \frac{\Delta M}{\Delta S} \tag{8}$$

in which ΔM and ΔS represent the mass and surface area of a small section of lamina containing the point (x, y, z) (Figure 18.5.6). Thus, if the lamina is subdivided into n small pieces, and if (x_k^*, y_k^*, z_k^*) is a point in the kth piece and ΔS_k is the surface area of the kth piece, then from (8) the mass ΔM_k of that piece can be approximated as

$$\Delta M_k \approx \delta(x_k^*, y_k^*, z_k^*)\, \Delta S_k$$

and hence the mass M of the entire lamina can be approximated as

$$M = \sum_{k=1}^n \Delta M_k \approx \sum_{k=1}^n \delta(x_k^*, y_k^*, z_k^*)\, \Delta S_k \tag{9}$$

If we now increase n in such a way that the dimensions of the pieces approach zero, then it is plausible that the error in (9) will approach zero, and the exact value of M will be given by the surface integral

$$M = \iint_\sigma \delta(x, y, z)\, dS \tag{10}$$

Example 4   A curved lamina σ is the portion of the paraboloid $z = x^2 + y^2$ below the plane $z = 1$ (Figure 18.5.7) and has constant density $\delta(x, y, z) = \delta_0$. Find the mass of the lamina.

Solution.   Since $z = g(x, y) = x^2 + y^2$, it follows that

$$\frac{\partial z}{\partial x} = 2x \quad \text{and} \quad \frac{\partial z}{\partial y} = 2y$$

$z = 1$

$z = x^2 + y^2$

R

$x^2 + y^2 = 1$

Figure 18.5.7

Substituting these expressions and $\delta(x, y, z) = \delta(x, y, g(x, y)) = \delta_0$ into (10) yields

$$M = \iint_\sigma \delta_0 \, dS = \iint_R \delta_0 \sqrt{(2x)^2 + (2y)^2 + 1} \, dA = \delta_0 \iint_R \sqrt{4x^2 + 4y^2 + 1} \, dA \qquad (11)$$

where R is the circular region enclosed by $x^2 + y^2 = 1$. To evaluate (11) we use polar coordinates:

$$M = \delta_0 \int_0^{2\pi} \int_0^1 \sqrt{4r^2 + 1} \, r \, dr \, d\theta = \frac{\delta_0}{12} \int_0^{2\pi} \left.(4r^2 + 1)^{3/2}\right]_{r=0}^1 d\theta$$

$$= \frac{\delta_0}{12} \int_0^{2\pi} (5^{3/2} - 1) \, d\theta = \frac{\pi \delta_0}{6} (5\sqrt{5} - 1) \qquad \blacktriangleleft$$

☐ **SURFACE INTEGRALS IN PARAMETRIC FORM**

In Section 17.4 we showed that if a surface σ is represented by a vector-valued function

$$\mathbf{r}(u, v) = x(u, v)\mathbf{i} + y(u, v)\mathbf{j} + z(u, v)\mathbf{k}$$

where (u, v) varies over a region R in the uv-plane, then the surface area ΔS_k of a small section σ_k of the surface σ can be approximated by

$$\Delta S_k \approx \left\| \frac{\partial \mathbf{r}}{\partial u} \times \frac{\partial \mathbf{r}}{\partial v} \right\| \Delta A_k$$

[see Formula (4) of Section 17.4]. If we use this formula to approximate ΔS_k in (2), and if we express the x, y, and z coordinates in terms of u and v, then the resulting formula for the surface integral is

$$\iint_\sigma f(x, y, z) \, dS = \iint_R f(x(u, v), y(u, v), z(u, v)) \left\| \frac{\partial \mathbf{r}}{\partial u} \times \frac{\partial \mathbf{r}}{\partial v} \right\| dA \qquad (12)$$

Some problems that use this formula are given in the exercises.

▶ Exercise Set 18.5

In Exercises 1–6, evaluate the surface integrals.

1. $\iint_\sigma z^2 \, dS$, where σ is the portion of the cone
$z = \sqrt{x^2 + y^2}$ between the planes $z = 1$ and $z = 2$.

2. $\iint_\sigma xy \, dS$, where σ is the portion of the plane
$x + y + z = 1$ lying in the first octant.

3. $\iint_\sigma x^2 y \, dS$, where σ is the portion of the cylinder
$x^2 + z^2 = 1$ between the planes $y = 0$, $y = 1$, and above the xy-plane.

4. $\iint_\sigma (x^2 + y^2) z \, dS$, where σ is the portion of the sphere
$x^2 + y^2 + z^2 = 4$ above the plane $z = 1$.

5. $\iint_\sigma (x + y + z) \, dS$, where σ is the portion of the plane
$x + y = 1$ in the first octant between $z = 0$ and $z = 1$.

6. $\iint_\sigma (x + y) \, dS$, where σ is the portion of the plane
$z = 6 - 2x - 3y$ in the first octant.

7. Evaluate

$$\iint_\sigma (x + y + z) \, dS$$

over the surface of the cube defined by the inequalities
$0 \le x \le 1, 0 \le y \le 1, 0 \le z \le 1$. [*Hint:* Integrate over each face separately.]

8. Evaluate

$$\iint_\sigma (z + 1) \, dS$$

where σ is the upper hemisphere

$$z = \sqrt{1 - x^2 - y^2}$$

9. Evaluate

$$\iint_\sigma \sqrt{x^2 + y^2 + z^2} \, dS$$

over the portion of the cone $z = \sqrt{x^2 + y^2}$ below the plane $z = 1$.

10. Evaluate

$$\iint_{\sigma} (x^2 + y^2)\, dS$$

over the surface of the sphere $x^2 + y^2 + z^2 = a^2$. [*Hint:* Divide the sphere into two parts and integrate over each part separately.]

In Exercises 11 and 12, set up, but do not evaluate, an iterated integral equal to the given surface integral by projecting σ on (a) the xy-plane, (b) the yz-plane, and (c) the xz-plane.

11. $\displaystyle\iint_{\sigma} xyz\, dS$, where σ is the portion of the plane

$2x + 3y + 4z = 12$ in the first octant.

12. $\displaystyle\iint_{\sigma} xz\, dS$, where σ is the portion of the sphere

$x^2 + y^2 + z^2 = a^2$ in the first octant.

In Exercises 13 and 14, set up, but do not evaluate, two different iterated integrals equal to the given integral.

13. $\displaystyle\iint_{\sigma} xyz\, dS$, where σ is the portion of the surface $y^2 = x$

between the planes $z = 0$, $z = 4$, $y = 1$, and $y = 2$.

14. $\displaystyle\iint_{\sigma} x^2 y\, dS$, where σ is the portion of the cylinder

$y^2 + z^2 = a^2$ in the first octant between the planes $x = 0$, $x = 9$, $z = y$, and $z = 2y$.

In Exercises 15–18, find the mass of the given lamina assuming the density to be a constant δ_0.

15. The lamina that is the portion of the paraboloid $z = 1 - x^2 - y^2$ above the xy-plane.

16. The lamina that is the portion of the plane $2x + 2y + z = 8$ in the first octant.

17. The lamina that is the portion of the circular cylinder $x^2 + z^2 = 4$ that lies directly above the rectangle $R = \{(x, y) : 0 \le x \le 1,\ 0 \le y \le 4\}$ in the xy-plane.

18. The lamina that is the portion of the paraboloid $2z = x^2 + y^2$ inside the cylinder $x^2 + y^2 = 8$.

19. Find the mass of the lamina that is the portion of the surface $y^2 = 4 - z$ between the planes $x = 0$, $x = 3$, $y = 0$, and $y = 3$ if the density is $\delta(x, y, z) = y$.

20. Find the mass of the lamina that is the portion of the cone $z = \sqrt{x^2 + y^2}$ between $z = 1$ and $z = 4$ if the density is $\delta(x, y, z) = x^2 z$.

21. Use (10) to verify that if a curved lamina has constant density δ_0, then its mass is the density times the surface area.

22. Show that the mass of the spherical lamina given by $x^2 + y^2 + z^2 = a^2$ is $2\pi a^3$ if the density at any point is equal to the distance from the point to the xy-plane.

The centroid of a curved lamina σ is defined by

$$\bar{x} = \frac{\displaystyle\iint_{\sigma} x\, dS}{\text{area of } \sigma}, \quad \bar{y} = \frac{\displaystyle\iint_{\sigma} y\, dS}{\text{area of } \sigma}, \quad \bar{z} = \frac{\displaystyle\iint_{\sigma} z\, dS}{\text{area of } \sigma}$$

In Exercises 23 and 24, find the centroid of the lamina.

23. The lamina that is the portion of the paraboloid $z = \frac{1}{2}(x^2 + y^2)$ below the plane $z = 4$.

24. The lamina that is the portion of the sphere $x^2 + y^2 + z^2 = 4$ above the plane $z = 1$.

In Exercises 25–28, evaluate the surface integrals.

25. $\displaystyle\iint_{\sigma} xyz\, dS$, where σ is the portion of the cone

$$\mathbf{r}(u, v) = u \cos v \mathbf{i} + u \sin v \mathbf{j} + 3u \mathbf{k}$$

for which $1 \le u \le 2$, $0 \le v \le \pi/2$.

26. $\displaystyle\iint_{\sigma} \frac{x^2 + z^2}{y}\, dS$, where σ is the portion of the cylinder

$$\mathbf{r}(u, v) = 2 \cos v \mathbf{i} + u \mathbf{j} + 2 \sin v \mathbf{k}$$

for which $1 \le u \le 3$, $0 \le v \le 2\pi$.

27. $\displaystyle\iint_{\sigma} \frac{1}{\sqrt{1 + 4x^2 + 4y^2}}\, dS$, where σ is the portion of the paraboloid

$$\mathbf{r}(u, v) = u \cos v \mathbf{i} + u \sin v \mathbf{j} + u^2 \mathbf{k}$$

for which $0 \le u \le \sin v$, $0 \le v \le \pi$.

28. $\displaystyle\iint_{\sigma} e^{-z}\, dS$, where σ is the hemisphere

$$\mathbf{r}(u, v) = 2 \sin u \cos v \mathbf{i} + 2 \sin u \sin v \mathbf{j} + 2 \cos u \mathbf{k}$$

for which $0 \le u \le \pi/2$, $0 \le v \le 2\pi$.

■ 18.6 SURFACE INTEGRALS OF VECTOR FIELDS; FLUX

In this section we shall focus on applications of surface integrals to vector fields such as those associated with fluid flow and electrostatic forces. However, the ideas that we shall develop are quite general and are applicable to almost all vector fields.

☐ **FLOW FIELDS**

Throughout this section

$$\mathbf{F}(x, y, z) = f(x, y, z)\mathbf{i} + g(x, y, z)\mathbf{j} + h(x, y, z)\mathbf{k}$$

can be any vector field in 3-space. However, we will motivate most of our ideas by focusing on vector fields arising in fluid flow and electrostatics. In the case of fluid flow, $\mathbf{F}$ will represent the velocity of a fluid particle at the point (x, y, z) (Figure 18.6.1a). In the case of electrostatics, we will assume that charged particles are distributed at various fixed points in 3-space and that $\mathbf{F}(x, y, z)$ is the combined force that these charges exert on a *test particle* of unit positive charge located at the point (x, y, z) (Figures 18.6.1b and 18.6.1c). In both the fluid flow and electrostatic cases it will be convenient to call $\mathbf{F}$ a *flow field*, even though nothing is actually flowing in the electrostatic case.

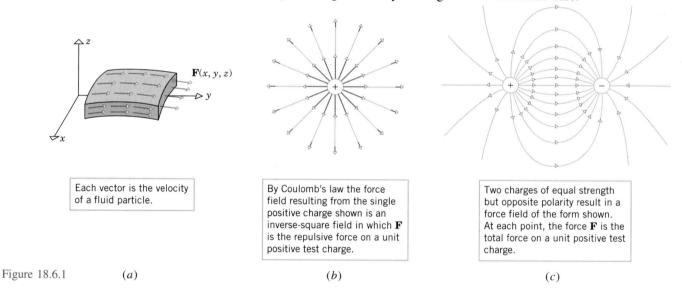

Each vector is the velocity of a fluid particle.

By Coulomb's law the force field resulting from the single positive charge shown is an inverse-square field in which $\mathbf{F}$ is the repulsive force on a unit positive test charge.

Two charges of equal strength but opposite polarity result in a force field of the form shown. At each point, the force $\mathbf{F}$ is the total force on a unit positive test charge.

Figure 18.6.1   (a)   (b)   (c)

☐ **ORIENTED SURFACES**

Suppose that a surface σ is constructed from permeable material through which fluid can flow freely. We shall be concerned with the flow of fluid in directions *normal* to the surface σ. However, because surfaces in 3-space can be extremely complex, it will be necessary to impose certain restrictions on σ to keep the mathematics from getting out of hand.

We will call a surface σ *smooth* if it has a tangent plane (and hence a normal vector) at every point that is not on the boundary of σ. In general, if (x, y, z) is a point on a smooth surface σ, then there are two oppositely directed unit normal vectors at the point, each of which defines a possible direction of fluid flow normal to the surface. For example, in Figure 18.6.2 the fluid can flow roughly "upward" or "downward" through the surface, and in Figure 18.6.3 the fluid can flow "inward" or "outward" through the sphere.

A smooth surface σ is said to be *orientable* if it is possible to construct a unit normal vector at each point of the surface in such a way that the vectors vary continuously (i.e., have no abrupt changes in direction) as we traverse any smooth curve on the surface. The unit normal vectors are then said to form an *orientation* of the surface. It is proved in

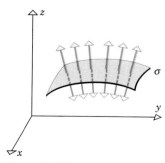

Figure 18.6.2

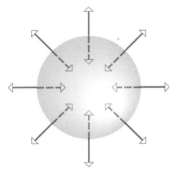

Figure 18.6.3

advanced mathematics courses that an orientable surface has only two possible orientations. For example, the surface in Figure 18.6.4 can be oriented upward, as in part (*a*), or downward, as in part (*b*). However, the vectors in part (*c*) do not define an orientation because the directions of those vectors change abruptly as we cross the curve drawn on the surface. As shown in Figure 18.6.3 a sphere can be oriented by *inward normals* or by *outward normals*.

   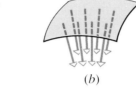   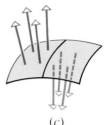

Figure 18.6.4   (*a*)   (*b*)   (*c*)

REMARK.   Observe that the notion of orientability applies only to smooth surfaces. However, many important surfaces are not themselves smooth, but are unions of finitely many smooth orientable surfaces. A box, for example, is not smooth, since no tangent planes exist at the edges. However, each face is a smooth orientable surface. It should also be noted that not every smooth surface is orientable. In the exercises we discuss the ***Möbius strip***, which is a famous nonorientable surface.

☐ **FLUX**

In physics, the term *fluid* is used to describe both liquids and gases. Liquids (such as water) are usually regarded to be ***incompressible***, meaning that the liquid has a uniform density (mass per unit volume) that cannot be altered by compressive forces. Gases (such as air) are regarded to be ***compressible***, meaning that the density may vary from point to point and can be altered by compressive forces. In all of our discussions we shall be concerned only with incompressible fluids. Moreover, we shall assume that the velocity of the fluid at a fixed point does not vary with time. Fluid flows with this property are said to be in a ***steady state***.

Our first goal in this section is to define a fundamental concept of physics known as *flux* (from the Latin word *fluxus*, meaning ''flow''). This concept is applicable in any vector field, but we shall motivate it in the context of steady-state flow of an incompressible fluid. We consider the following problem:

18.6.1   PROBLEM.   *Let* $\mathbf{F}(x, y, z) = f(x, y, z)\mathbf{i} + g(x, y, z)\mathbf{j} + h(x, y, z)\mathbf{k}$ *be a fluid flow field for which f, g, and h are continuous and have continuous first partial derivatives, and let* σ *be an orientable surface in the fluid with an orientation defined by the unit normal vectors* $\mathbf{n} = \mathbf{n}(x, y, z)$. *Find the* net mass Φ *of fluid that passes through the surface per unit time, where the net mass is interpreted to mean the mass passing through the surface in the direction of orientation minus the mass passing through the surface in the direction opposite to the orientation.*

REMARK.   For simplicity, we shall assume that the fluid in this problem has density 1. Since density is mass per unit volume, it follows from this assumption that the net mass of fluid that passes through the surface per unit time is numerically the same as the net volume of fluid that passes through the surface per unit time (although the units differ). Thus, we can replace the word ''mass'' by ''volume'' in this problem.

Before attempting to solve this problem, we note that at each point (x, y, z) on the surface the velocity vector $\mathbf{F}(x, y, z)$ can be resolved into two components, a component $\mathbf{F} \cdot \mathbf{n}$ along the unit normal vector $\mathbf{n}$ in the orientation of σ and a component $\mathbf{F} \cdot \mathbf{T}$ tangent to the surface (Figure 18.6.5). The component $\mathbf{F} \cdot \mathbf{T}$ is the component of flow *along* the surface,

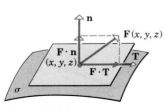

Figure 18.6.5

and $\mathbf{F} \cdot \mathbf{n}$ is the component *across* the surface. Fluid crosses σ in the direction of orientation where $\mathbf{F} \cdot \mathbf{n}$ is positive and opposite to that direction where $\mathbf{F} \cdot \mathbf{n}$ is negative.

To solve Problem 18.6.1, we subdivide σ into n parts $\sigma_1, \sigma_2, \ldots, \sigma_n$ with areas

$$\Delta S_1, \Delta S_2, \ldots, \Delta S_n$$

If the parts are small and the flow is not too erratic, it is reasonable to assume that the velocity does not vary much on each part. Thus, if (x_k^*, y_k^*, z_k^*) is any point in the kth part, we can assume that $\mathbf{F}(x, y, z)$ is constant and equal to $\mathbf{F}(x_k^*, y_k^*, z_k^*)$ throughout the part, and that the component of flow across the surface σ_k is

$$\mathbf{F}(x_k^*, y_k^*, z_k^*) \cdot \mathbf{n}(x_k^*, y_k^*, z_k^*) \tag{1}$$

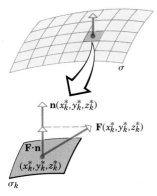

Figure 18.6.6

(Figure 18.6.6). If we observe the flow across σ_k for one unit of time, then the section of fluid initially on the surface will move to some new position, sweeping out a solid whose volume ΔV_k can be approximated by

$$\Delta V_k \approx \begin{bmatrix} \text{distance traveled by} \\ \text{the fluid in one unit} \\ \text{of time} \end{bmatrix} \Delta S_k \tag{2}$$

(Figure 18.6.7). But (1) is the component of fluid velocity across the surface and therefore represents either the distance traveled by the fluid in one unit of time or the negative of that distance, depending on whether $\mathbf{F} \cdot \mathbf{n}$ is positive or negative. Thus, from (2)

$$\mathbf{F}(x_k^*, y_k^*, z_k^*) \cdot \mathbf{n}(x_k^*, y_k^*, z_k^*) \, \Delta S_k$$

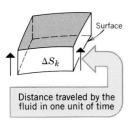

Distance traveled by the fluid in one unit of time

Figure 18.6.7

is either ΔV_k [if the fluid is moving in the direction of $\mathbf{n}(x_k^*, y_k^*, z_k^*)$] or $-\Delta V_k$ [if the fluid is moving in the direction opposite to $\mathbf{n}(x_k^*, y_k^*, z_k^*)$]. Thus, the net volume of fluid passing through σ in one unit of time can be approximated as

$$\Phi \approx \sum_{k=1}^{n} \mathbf{F}(x_k^*, y_k^*, z_k^*) \cdot \mathbf{n}(x_k^*, y_k^*, z_k^*) \, \Delta S_k$$

If we now increase n in such a way that dimensions of the surface parts approach zero, then it is plausible that the errors in the approximations approach zero, and the exact value of Φ is

$$\Phi = \lim_{n \to +\infty} \sum_{k=1}^{n} \mathbf{F}(x_k^*, y_k^*, z_k^*) \cdot \mathbf{n}(x_k^*, y_k^*, z_k^*) \, \Delta S_k$$

which can be expressed as the surface integral

$$\Phi = \iint_{\sigma} \mathbf{F}(x, y, z) \cdot \mathbf{n}(x, y, z) \, dS \tag{3}$$

The quantity Φ defined by this integral is called the ***flux of F across*** $\boldsymbol{\sigma}$. The concept of flux is applicable to any vector field in 3-space. However, in the case of incompressible steady-state fluid flow with a fluid of mass 1, we have seen that Φ represents the net mass of fluid passing through σ per unit time.

□ **CALCULATING FLUX**

To calculate the flux of $\mathbf{F}$ across an oriented surface σ from (3), it is first necessary to find a formula for the normal vector $\mathbf{n}(x, y, z)$ that appears in the integrand. Such formulas depend on the form in which the surface σ is expressed. For example, Table 18.6.1 lists the two possible orientations and the formulas for $\mathbf{n} = \mathbf{n}(x, y, z)$ for surfaces of the form $z = g(x, y)$, $y = g(x, z)$, and $x = g(y, z)$.

The formulas in Table 18.6.1 can be derived by noting that the surfaces $z = g(x, y)$, $y = g(x, z)$, and $x = g(y, z)$ can all be expressed in the form $G(x, y, z) = 0$ by taking the function g to the left side of the equation. In all three cases the surface σ is a level surface

Table 18.6.1

$z = g(x, y)$		$y = g(x, z)$		$x = g(y, z)$	
Positive **k** component	$\mathbf{n} = \dfrac{-\dfrac{\partial z}{\partial x}\mathbf{i} - \dfrac{\partial z}{\partial y}\mathbf{j} + \mathbf{k}}{\sqrt{\left(\dfrac{\partial z}{\partial x}\right)^2 + \left(\dfrac{\partial z}{\partial y}\right)^2 + 1}}$	Positive **j** component	$\mathbf{n} = \dfrac{-\dfrac{\partial y}{\partial x}\mathbf{i} + \mathbf{j} - \dfrac{\partial y}{\partial z}\mathbf{k}}{\sqrt{\left(\dfrac{\partial y}{\partial x}\right)^2 + \left(\dfrac{\partial y}{\partial z}\right)^2 + 1}}$	Positive **i** component	$\mathbf{n} = \dfrac{\mathbf{i} - \dfrac{\partial x}{\partial y}\mathbf{j} - \dfrac{\partial x}{\partial z}\mathbf{k}}{\sqrt{\left(\dfrac{\partial x}{\partial y}\right)^2 + \left(\dfrac{\partial x}{\partial z}\right)^2 + 1}}$
	Positive orientation (up)		Positive orientation (right)		Positive orientation (front)
Negative **k** component	$\mathbf{n} = \dfrac{\dfrac{\partial z}{\partial x}\mathbf{i} + \dfrac{\partial z}{\partial y}\mathbf{j} - \mathbf{k}}{\sqrt{\left(\dfrac{\partial z}{\partial x}\right)^2 + \left(\dfrac{\partial z}{\partial y}\right)^2 + 1}}$	Negative **j** component	$\mathbf{n} = \dfrac{\dfrac{\partial y}{\partial x}\mathbf{i} - \mathbf{j} + \dfrac{\partial y}{\partial z}\mathbf{k}}{\sqrt{\left(\dfrac{\partial y}{\partial x}\right)^2 + \left(\dfrac{\partial y}{\partial z}\right)^2 + 1}}$	Negative **i** component	$\mathbf{n} = \dfrac{-\mathbf{i} + \dfrac{\partial x}{\partial y}\mathbf{j} + \dfrac{\partial x}{\partial z}\mathbf{k}}{\sqrt{\left(\dfrac{\partial x}{\partial y}\right)^2 + \left(\dfrac{\partial x}{\partial z}\right)^2 + 1}}$
	Negative orientation (down)		Negative orientation (left)		Negative orientation (back)

for $G(x, y, z)$, so it follows from Theorem 16.7.6 that ∇G is normal to σ at (x, y, z) and hence that

$$\frac{\nabla G}{\|\nabla G\|} \quad \text{and} \quad -\frac{\nabla G}{\|\nabla G\|} \tag{4}$$

are unit normal vectors to σ at (x, y, z). These are the vectors listed in Table 18.6.1. For example, if $z = g(x, y)$, then

$$G(x, y, z) = z - g(x, y)$$

so

$$\nabla G = -\frac{\partial g}{\partial x}\mathbf{i} - \frac{\partial g}{\partial y}\mathbf{j} + \mathbf{k} = -\frac{\partial z}{\partial x}\mathbf{i} - \frac{\partial z}{\partial y}\mathbf{j} + \mathbf{k} \tag{5}$$

and hence

$$\frac{\nabla G}{\|\nabla G\|} = \frac{-\dfrac{\partial z}{\partial x}\mathbf{i} - \dfrac{\partial z}{\partial y}\mathbf{j} + \mathbf{k}}{\sqrt{\left(\dfrac{\partial z}{\partial x}\right)^2 + \left(\dfrac{\partial z}{\partial y}\right)^2 + 1}}$$

which is the normal for the positive orientation of $z = g(x, y)$ shown in Table 18.6.1. Multiplying by -1 produces the negative orientation.

Example 1 Let σ be the portion of the surface $z = 1 - x^2 - y^2$ that lies above the xy-plane, and suppose that σ is oriented upward (Figure 18.6.8). Find the flux Φ of the flow field $\mathbf{F}(x, y, z) = x\mathbf{i} + y\mathbf{j} + z\mathbf{k}$ across σ.

Solution. Because the surface σ is of the form $z = g(x, y)$, we use Formula (3) of Section 18.5 to express (3) as a double integral. This yields

$$\Phi = \iint_\sigma \mathbf{F} \cdot \mathbf{n}\, dS = \iint_R (\mathbf{F} \cdot \mathbf{n})\sqrt{\left(\frac{\partial z}{\partial x}\right)^2 + \left(\frac{\partial z}{\partial y}\right)^2 + 1}\, dA$$

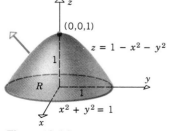

Figure 18.6.8

where R is the projection of the surface σ on the xy-plane (Figure 18.6.8). Since σ is oriented upward, we substitute the formula for $\mathbf{n}$ given in Table 18.6.1 for the positive (up) orientation of $z = g(x, y)$; this yields

$$\Phi = \iint_R \mathbf{F} \cdot \left[\frac{-\frac{\partial z}{\partial x}\mathbf{i} - \frac{\partial z}{\partial y}\mathbf{j} + \mathbf{k}}{\sqrt{\left(\frac{\partial z}{\partial x}\right)^2 + \left(\frac{\partial z}{\partial y}\right)^2 + 1}} \right] \sqrt{\left(\frac{\partial z}{\partial x}\right)^2 + \left(\frac{\partial z}{\partial y}\right)^2 + 1} \, dA$$

$$= \iint_R \mathbf{F} \cdot \left(-\frac{\partial z}{\partial x}\mathbf{i} - \frac{\partial z}{\partial y}\mathbf{j} + \mathbf{k} \right) dA$$

$$= \iint_R (x\mathbf{i} + y\mathbf{j} + z\mathbf{k}) \cdot (2x\mathbf{i} + 2y\mathbf{j} + \mathbf{k}) \, dA$$

$$= \iint_R (x^2 + y^2 + 1) \, dA \quad \boxed{\text{Since } z = 1 - x^2 - y^2 \text{ on the surface.}}$$

$$= \int_0^{2\pi} \int_0^1 (r^2 + 1)r \, dr \, d\theta \quad \boxed{\text{Using polar coordinates to evaluate the integral}}$$

$$= \int_0^{2\pi} \left(\frac{3}{4}\right) d\theta = \frac{3\pi}{2} \quad \blacktriangleleft$$

The cancelation of the radicals in the preceding example was not accidental; it is a consequence of the following theorem.

18.6.2   THEOREM. *Let σ be a smooth surface of the form $z = g(x, y)$, $y = g(x, z)$, or $x = g(y, z)$, and suppose that the equation is rewritten as $G(x, y, z) = 0$ by taking g to the left side. Let R be the projection of σ on the xy-plane if $z = g(x, y)$, on the xz-plane if $y = g(x, z)$, and on the yz-plane if $x = g(y, z)$. If g is continuous, and has continuous first partial derivatives on R, then*

$$\iint_\sigma \mathbf{F} \cdot \mathbf{n} \, dS = \pm \iint_R \mathbf{F} \cdot \nabla G \, dA \tag{6}$$

where the $+$ sign is used if σ has positive orientation and the $-$ sign if it has negative orientation.

Although we omit the formal proof, the basic idea is to apply Theorem 18.5.1. In all three parts of that theorem the radical can be expressed in terms of the function G as $\|\nabla G\|$ (verify). Thus, from Theorem 18.5.1 and the formulas in (4) for the normal it follows that

$$\iint_\sigma \mathbf{F} \cdot \mathbf{n} \, dS = \iint_R \mathbf{F} \cdot \left(\pm \frac{\nabla G}{\|\nabla G\|} \right) \|\nabla G\| \, dA$$

$$= \pm \iint_R \mathbf{F} \cdot \nabla G \, dA$$

In the case where $z = g(x, y)$, we have $G(x, y, z) = z - g(x, y)$, so that

$$\nabla G = -\frac{\partial g}{\partial x}\mathbf{i} - \frac{\partial g}{\partial y}\mathbf{j} + \mathbf{k} = -\frac{\partial z}{\partial x}\mathbf{i} - \frac{\partial z}{\partial y}\mathbf{j} + \mathbf{k}$$

Substituting this expression for ∇G in (6) and taking R to be the projection of the surface $z = g(x, y)$ on the xy-plane yields the following formulas:

$$\iint_{\sigma} \mathbf{F} \cdot \mathbf{n}\, dS = \iint_{R} \mathbf{F} \cdot \left(-\frac{\partial z}{\partial x}\mathbf{i} - \frac{\partial z}{\partial y}\mathbf{j} + \mathbf{k}\right) dA \qquad \boxed{\sigma \text{ oriented up}} \qquad (7)$$

$$\iint_{\sigma} \mathbf{F} \cdot \mathbf{n}\, dS = \iint_{R} \mathbf{F} \cdot \left(\frac{\partial z}{\partial x}\mathbf{i} + \frac{\partial z}{\partial y}\mathbf{j} - \mathbf{k}\right) dA \qquad \boxed{\sigma \text{ oriented down}} \qquad (8)$$

The derivation of the corresponding formulas for the cases where $y = g(x, z)$ and $x = g(y, z)$ are left as exercises.

Example 2   Let σ be the sphere $x^2 + y^2 + z^2 = a^2$ oriented by outward normals (Figure 18.6.9), and let $\mathbf{F}(x, y, z) = z\mathbf{k}$. Evaluate

$$\iint_{\sigma} \mathbf{F} \cdot \mathbf{n}\, dS$$

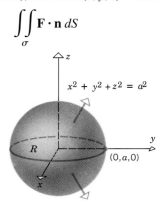

Figure 18.6.9

Solution.   On the upper hemisphere the outward unit normal is upward, while on the lower hemisphere it is downward. Since different formulas for these normals apply on the two hemispheres, it is desirable to write

$$\iint_{\sigma} \mathbf{F} \cdot \mathbf{n}\, dS = \iint_{\sigma_1} \mathbf{F} \cdot \mathbf{n}\, dS + \iint_{\sigma_2} \mathbf{F} \cdot \mathbf{n}\, dS \qquad (9)$$

where σ_1 is the upper hemisphere and σ_2 is the lower hemisphere.

The upper hemisphere σ_1 has the equation

$$z = \sqrt{a^2 - x^2 - y^2} \qquad (10)$$

so that (7) yields

$$\iint_{\sigma_1} \mathbf{F} \cdot \mathbf{n}\, dS = \iint_{R} (z\mathbf{k}) \cdot \left(\frac{x}{\sqrt{a^2 - x^2 - y^2}}\mathbf{i} + \frac{y}{\sqrt{a^2 - x^2 - y^2}}\mathbf{j} + \mathbf{k}\right) dA$$

$$= \iint_{R} z\, dA = \iint_{R} \sqrt{a^2 - x^2 - y^2}\, dA \qquad \boxed{\begin{array}{l} R \text{ is the region shown} \\ \text{in Figure 18.6.9.}\end{array}}$$

$$= \int_0^{2\pi} \int_0^{a} \sqrt{a^2 - r^2}\, r\, dr\, d\theta$$

$$= \int_0^{2\pi} \left. -\frac{1}{3}(a^2 - r^2)^{3/2}\right]_0^{a} d\theta$$

$$= \int_0^{2\pi} \frac{1}{3}a^3\, d\theta = \frac{2\pi a^3}{3}$$

The lower hemisphere σ_2 has the equation

$$z = -\sqrt{a^2 - x^2 - y^2}$$

so that (8) yields

$$\iint\limits_{\sigma_2} \mathbf{F} \cdot \mathbf{n} \, dS = \iint\limits_{R} (z\mathbf{k}) \cdot \left(\frac{x}{\sqrt{a^2 - x^2 - y^2}}\mathbf{i} + \frac{y}{\sqrt{a^2 - x^2 - y^2}}\mathbf{j} - \mathbf{k} \right) dA$$

$$= \iint\limits_{R} -z \, dA = \iint\limits_{R} \sqrt{a^2 - x^2 - y^2} \, dA$$

$$= \frac{2\pi a^3}{3} \quad \boxed{\begin{array}{l} \text{The computations} \\ \text{are identical to} \\ \text{those above.} \end{array}}$$

Thus, from (9)

$$\iint\limits_{\sigma} \mathbf{F} \cdot \mathbf{n} \, dS = \frac{2\pi a^3}{3} + \frac{2\pi a^3}{3} = \frac{4\pi a^3}{3} \qquad \blacktriangleleft$$

▶ Exercise Set 18.6

1. Use three different formulas in Table 18.6.1 to calculate the unit normal to $2x + 3y + 4z = 9$ at $(1, 1, 1)$ that has positive components.

2. Use three different formulas in Table 18.6.1 to calculate the unit normal to $x^2 + y^2 + z^2 = 9$ at $(2, 1, -2)$ that points below the xy-plane.

3. In parts (a)–(c), use any appropriate formula to calculate the indicated unit normal.

   (a) The unit normal to $z = x^2 + y^2$ at $(1, 2, 5)$ that points toward the z-axis

   (b) The unit normal to $z = \sqrt{x^2 + y^2}$ at $(-3, 4, 5)$ that points toward the xz-plane

   (c) The unit normal to the cylinder $x^2 + z^2 = 25$ at $(3, 2, -4)$ that points away from the xy-plane.

4. In each part, use any appropriate formula to calculate the indicated unit normal.

   (a) The unit normal to the surface $y^2 = x$ at $(1, 1, 2)$ that points toward the xz-plane

   (b) The unit normal to the hyperbolic paraboloid $y = z^2 - x^2$ at $(1, 3, 2)$ that points toward the yz-plane

   (c) The unit normal to the cone $x^2 = y^2 + z^2$ at $(\sqrt{2}, -1, -1)$ that points away from the xy-plane.

In Exercises 5–15, evaluate $\displaystyle\iint\limits_{\sigma} \mathbf{F} \cdot \mathbf{n} \, dS$.

5. $\mathbf{F}(x, y, z) = x\mathbf{i} + y\mathbf{j} + 2z\mathbf{k}$; σ is the portion of the surface $z = 1 - x^2 - y^2$ above the xy-plane, oriented by upward normals.

6. $\mathbf{F}(x, y, z) = (x + y)\mathbf{i} + (y + z)\mathbf{j} + (z + x)\mathbf{k}$; σ is the portion of the plane $x + y + z = 1$ in the first octant, oriented by unit normals with positive components.

7. $\mathbf{F}(x, y, z) = z^2\mathbf{k}$; σ is the upper hemisphere given by $z = \sqrt{1 - x^2 - y^2}$, oriented by upward unit normals.

8. $\mathbf{F}(x, y, z) = x^2\mathbf{i} + yx\mathbf{j} + zx\mathbf{k}$; σ is the portion of the plane $6x + 3y + 2z = 6$ in the first octant, oriented by unit normals with positive components.

9. $\mathbf{F}(x, y, z) = x\mathbf{i} + y\mathbf{j} + z\mathbf{k}$; σ is the upper hemisphere $z = \sqrt{9 - x^2 - y^2}$, oriented by upward unit normals.

10. $\mathbf{F}(x, y, z) = \mathbf{i} + \mathbf{j} + \mathbf{k}$; σ is the portion of the cone $z = \sqrt{x^2 + y^2}$ below the plane $z = 1$, oriented by downward unit normals.

11. $\mathbf{F}(x, y, z) = x\mathbf{i} + y\mathbf{j} + 2z\mathbf{k}$; σ is the portion of the cone $z^2 = x^2 + y^2$ between the planes $z = 1$ and $z = 2$, oriented by upward unit normals.

12. $\mathbf{F}(x, y, z) = y\mathbf{j} + \mathbf{k}$; σ is the portion of the paraboloid $z = x^2 + y^2$ below the plane $z = 4$, oriented by downward unit normals.

13. $\mathbf{F}(x, y, z) = x\mathbf{k}$; σ is the portion of the paraboloid $z = x^2 + y^2$ below the plane $z = y$, oriented by downward unit normals.

14. $\mathbf{F}(x, y, z) = x\mathbf{i} + y\mathbf{j} + z\mathbf{k}$; σ is the portion of the cylinder $z^2 = 1 - x^2$ between the planes $y = 1$ and $y = -2$, oriented by outward unit normals.

15. $\mathbf{F}(x, y, z) = x\mathbf{i} + y\mathbf{j} + z\mathbf{k}$, where σ is the sphere $x^2 + y^2 + z^2 = a^2$ oriented by outward unit normals.

16. Let σ be the surface of the cube bounded by the planes $x = \pm 1, y = \pm 1, z = \pm 1$, oriented by outward unit normals. In each part, find the flux of $\mathbf{F}$ over σ.

    (a) $\mathbf{F}(x, y, z) = x\mathbf{i}$

    (b) $\mathbf{F}(x, y, z) = x\mathbf{i} + y\mathbf{j} + z\mathbf{k}$

    (c) $\mathbf{F}(x, y, z) = x^2\mathbf{i} + y^2\mathbf{j} + z^2\mathbf{k}$.

17. Show that reversing the orientation of σ reverses the sign of

$$\iint_{\sigma} \mathbf{F} \cdot \mathbf{n}\, dS$$

18. Let $\mathbf{F} = \|\mathbf{r}\|^k \mathbf{r}$, where $\mathbf{r} = x\mathbf{i} + y\mathbf{j} + z\mathbf{k}$ and k is a constant. (If $k = -3$, this is an inverse-square field.) Let σ be the sphere of radius a centered at the origin and oriented by the outward normal $\mathbf{n} = \mathbf{r}/\|\mathbf{r}\| = \mathbf{r}/a$.

    (a) Evaluate $\displaystyle\iint_{\sigma} \mathbf{F} \cdot \mathbf{n}\, dS$ without performing any integrations. [*Hint:* The surface area of a sphere of radius a is $4\pi a^2$.]

    (b) For what value of k is the integral in part (a) independent of the radius of the sphere?

19. Obtain the analogs of Formulas (7) and (8) for

    (a) surfaces of the form $x = g(y, z)$

    (b) surfaces of the form $y = g(x, z)$.

20. Evaluate $\displaystyle\iint_{\sigma} \mathbf{F} \cdot \mathbf{n}\, dS$ where σ is the portion of the paraboloid $x = y^2 + z^2$ with $x \leq 1$ and $z \geq 0$ oriented by backward unit normals and

    $$\mathbf{F}(x, y, z) = y\mathbf{i} - z\mathbf{j} + 8\mathbf{k}$$

    [*Hint:* Exercise 19(a).]

21. Evaluate $\displaystyle\iint_{\sigma} \mathbf{F} \cdot \mathbf{n}\, dS$ if σ is the hemisphere $y = \sqrt{1 - x^2 - z^2}$ oriented by right unit normals and $\mathbf{F}(x, y, z) = x\mathbf{i} + y\mathbf{j} + z\mathbf{k}$. [*Hint:* Exercise 19(b).]

22. The best-known example of a nonorientable surface is the **Möbius strip**, which can be visualized by taking a band of paper, twisting it once, and gluing the ends together (Figures 18.6.10a and 18.6.10b). In Figure 18.6.10c, we have tried to construct unit normal vectors whose directions vary continuously moving counterclockwise around the dashed curve starting and finishing on the vertical line.

    (a) Explain why these vectors are not part of an orientation of the surface.

    (b) Explain why the surface is not orientable.

If a surface is represented parametrically by

$$\mathbf{r}(u, v) = x(u, v)\mathbf{i} + y(u, v)\mathbf{j} + z(u, v)\mathbf{k}$$

where (u, v) varies over a region R in the uv-plane, then from (6) in Section 16.3 a unit normal vector to the surface is given by

$$\mathbf{n} = \dfrac{\dfrac{\partial \mathbf{r}}{\partial u} \times \dfrac{\partial \mathbf{r}}{\partial v}}{\left\|\dfrac{\partial \mathbf{r}}{\partial u} \times \dfrac{\partial \mathbf{r}}{\partial v}\right\|}$$

If a surface σ has the orientation determined by this $\mathbf{n}$, then from (12) in Section 18.5 with $f(x, y, z) = \mathbf{F} \cdot \mathbf{n}$ we obtain

$$\iint_{\sigma} \mathbf{F} \cdot \mathbf{n}\, dS = \iint_{R} \mathbf{F} \cdot \left(\frac{\partial \mathbf{r}}{\partial u} \times \frac{\partial \mathbf{r}}{\partial v}\right) dA$$

where the integrand on the right is expressed in terms of u and v. In Exercises 23–26, use this to evaluate $\displaystyle\iint_{\sigma} \mathbf{F} \cdot \mathbf{n}\, dS$.

23. $\mathbf{F}(x, y, z) = x\mathbf{i} + y\mathbf{j} + \mathbf{k}$; σ is the portion of the paraboloid

    $$\mathbf{r}(u, v) = u \cos v\mathbf{i} + u \sin v\mathbf{j} + (1 - u^2)\mathbf{k}$$

    for which $1 \leq u \leq 2$, $0 \leq v \leq 2\pi$.

24. $\mathbf{F}(x, y, z) = e^{-y}\mathbf{i} - y\mathbf{j} + x \sin z\mathbf{k}$; σ is the portion of the elliptic cylinder

    $$\mathbf{r}(u, v) = 2 \cos v\mathbf{i} + \sin v\mathbf{j} + u\mathbf{k}$$

    for which $0 \leq u \leq 5$, $0 \leq v \leq 2\pi$.

25. $\mathbf{F}(x, y, z) = \sqrt{x^2 + y^2}\mathbf{k}$; σ is the portion of the cone

    $$\mathbf{r}(u, v) = u \cos v\mathbf{i} + u \sin v\mathbf{j} + 2u\mathbf{k}$$

    for which $0 \leq u \leq \sin v$, $0 \leq v \leq \pi$.

26. $\mathbf{F}(x, y, z) = z^2\mathbf{k}$; σ is the portion of the sphere

    $$\mathbf{r}(u, v) = 2 \sin u \cos v\mathbf{i} + 2 \sin u \sin v\mathbf{j} + 2 \cos u\mathbf{k}$$

    for which $0 \leq u \leq \pi/3$, $0 \leq v \leq 2\pi$.

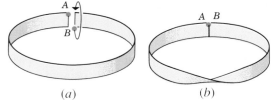

(a)

(b)

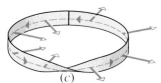

(c)

Figure 18.6.10

■ 18.7   THE DIVERGENCE THEOREM

> *Recall that Green's Theorem establishes a relationship between a double integral over a plane region and a line integral over the boundary of the region. In this section we shall establish a relationship between a triple integral over a solid and a surface integral over the boundary of the solid. This relationship is the cornerstone of many important principles in physics.*

☐ **THE DIVERGENCE THEOREM**

In this section we will be concerned with surfaces that are boundaries of solids in 3-space—for example, the surface of a solid sphere, the surface of a solid box, or the surface of a solid cylinder (the wall and the end caps). Such surfaces are said to be ***closed***. If a closed surface is orientable or piecewise orientable, then the two possible orientations are called ***inward*** (toward the solid) and ***outward*** (away from the solid).

In Section 18.1 we defined the *divergence* of a vector field

$$\mathbf{F}(x, y, z) = f(x, y, z)\mathbf{i} + g(x, y, z)\mathbf{j} + h(x, y, z)\mathbf{k}$$

as

$$\operatorname{div} \mathbf{F} = \frac{\partial f}{\partial x} + \frac{\partial g}{\partial y} + \frac{\partial h}{\partial z}$$

The following result, known as the ***Divergence Theorem*** or ***Gauss'* Theorem***, shows that under appropriate conditions the flux of a vector field across a closed surface with outward

---

* CARL FRIEDRICH GAUSS (1777–1855). German mathematician and scientist. Sometimes called the ''prince of mathematicians,'' Gauss ranks with Newton and Archimedes as one of the three greatest mathematicians who ever lived. His father, a laborer, was an uncouth but honest man who would have liked Gauss to take up a trade such as gardening or bricklaying; but the boy's genius for mathematics was not to be denied. In the entire history of mathematics there may never have been a child so precocious as Gauss—by his own account he worked out the rudiments of arithmetic before he could talk. One day, before he was even three years old, his genius became apparent to his parents in a very dramatic way. His father was preparing the weekly payroll for the laborers under his charge while the boy watched quietly from a corner. At the end of the long and tedious calculation, Gauss informed his father that there was an error in the result and stated the answer, which he had worked out in his head. To the astonishment of his parents, a check of the computations showed Gauss to be correct!

For his elementary education Gauss was enrolled in a squalid school run by a man named Büttner whose main teaching technique was thrashing. Büttner was in the habit of assigning long addition problems which, unknown to his students, were arithmetic progressions that he could sum up using formulas. On the first day that Gauss entered the arithmetic class, the students were asked to sum the numbers from 1 to 100. But no sooner had Büttner stated the problem than Gauss turned over his slate and exclaimed in his peasant dialect, ''Ligget se'.'' (Here it lies.) For nearly an hour Büttner glared at Gauss, who sat with folded hands while his classmates toiled away. When Büttner examined the slates at the end of the period, Gauss' slate contained a single number, 5050—the only correct solution in the class.

To his credit, Büttner recognized the genius of Gauss and with the help of his assistant, John Bartels, had him brought to the attention of Karl Wilhelm Ferdinand, Duke of Brunswick. The shy and awkward boy, who was then fourteen, so captivated the Duke that he subsidized him through preparatory school, college, and the early part of his career.

From 1795 to 1798 Gauss studied mathematics at the University of Göttingen, receiving his degree in absentia from the University of Helmstadt. For his dissertation, he gave the first complete proof of the fundamental theorem of algebra, which states that every polynomial equation has as many solutions as its degree. At age 19 he solved a problem that baffled Euclid, inscribing a regular polygon of seventeen sides in a circle using straightedge and compass; and in 1801, at age 24, he published his first masterpiece, *Disquisitiones Arithmeticae,* considered by many to be one of the most brilliant achievements in mathematics. In that book Gauss systematized the study of number theory (properties of the integers) and formulated the basic concepts that form the foundation of that subject.

(continued on next page)

orientation is equal to the triple integral of the divergence of the field over the solid region enclosed by the surface.

18.7.1 THEOREM (*Divergence Theorem*). *Let G be a solid with surface σ oriented outward. If*

$$\mathbf{F}(x, y, z) = f(x, y, z)\mathbf{i} + g(x, y, z)\mathbf{j} + h(x, y, z)\mathbf{k}$$

where f, g, and h have continuous first partial derivatives on some open set containing G, then

$$\iint_\sigma \mathbf{F} \cdot \mathbf{n}\, dS = \iiint_G \operatorname{div} \mathbf{F}\, dV \tag{1}$$

The proof of this theorem for a general solid G is too difficult to present here. However, we can give a proof for the special case where G is a simple solid (see Figure 17.5.3 and the discussion preceding it).

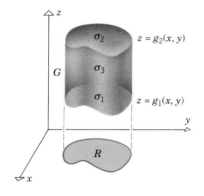

Proof (for simple solids). Let G be a simple solid with upper surface $z = g_2(x, y)$, lower surface $z = g_1(x, y)$, and projection R on the xy-plane. Let σ_1 denote the lower surface, σ_2 the upper surface, and σ_3 the lateral surface (Figure 18.7.1*a*). If the upper surface and lower surface meet as in Figure 18.7.1*b*, then there is no lateral surface σ_3. Our proof will allow for both cases shown in those figures.

We want to show that

$$\iint_\sigma \mathbf{F} \cdot \mathbf{n}\, dS = \iiint_G \operatorname{div} \mathbf{F}\, dV$$

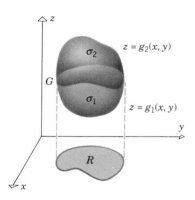

Figure 18.7.1

(*continued*)

In the same year that the *Disquisitiones* was published, Gauss again applied his phenomenal computational skills in a dramatic way. The astronomer Giuseppi Piazzi had observed the asteroid Ceres for $\frac{1}{40}$ of its orbit, but lost it in the sun. Using only three observations and the "method of least squares" that he had developed in 1795, Gauss computed the orbit with such accuracy that astronomers had no trouble relocating it the following year. This achievement brought him instant recognition as the premier mathematician in Europe, and in 1807 he was made Professor of Astronomy and head of the astronomical observatory at Göttingen.

In the years that followed, Gauss revolutionized mathematics by bringing to it standards of precision and rigor undreamed of by his predecessors. He had a passion for perfection that drove him to polish and rework his papers rather than publish less finished work in greater numbers—his favorite saying was "Pauca, sed matura" (Few, but ripe). As a result, many of his important discoveries were squirreled away in diaries that remained unpublished until years after his death.

Among his myriad achievements, Gauss discovered the Gaussian or "bell-shaped" error curve fundamental in probability, gave the first geometric interpretation of complex numbers and established their fundamental role in mathematics, developed methods of characterizing surfaces intrinsically by means of the curves that they contain, developed the theory of conformal (angle-preserving) maps, and discovered non-Euclidean geometry 30 years before the ideas were published by others. In physics he made major contributions to the theory of lenses and capillary action, and with Wilhelm Weber he did fundamental work in electromagnetism. Gauss invented the heliotrope, bifilar magnetometer, and an electrotelegraph.

Gauss was deeply religious and aristocratic in demeanor. He mastered foreign languages with ease, read extensively, and enjoyed minerology and botany as hobbies. He disliked teaching and was usually cool and discouraging to other mathematicians, possibly because he had already anticipated their work. It has been said that if Gauss had published all of his discoveries, the current state of mathematics would be advanced by 50 years. He was without a doubt the greatest mathematician of the modern era.

or equivalently,

$$\iint\limits_{\sigma} [f(x, y, z)\mathbf{i} + g(x, y, z)\mathbf{j} + h(x, y, z)\mathbf{k}] \cdot \mathbf{n} \, dS = \iiint\limits_{G} \left(\frac{\partial f}{\partial x} + \frac{\partial g}{\partial y} + \frac{\partial h}{\partial z} \right) dV$$

To prove this, it suffices to prove the following three equalities:

$$\iint\limits_{\sigma} [f(x, y, z)\mathbf{i} \cdot \mathbf{n}] \, dS = \iiint\limits_{G} \frac{\partial f}{\partial x} \, dV \tag{2a}$$

$$\iint\limits_{\sigma} [g(x, y, z)\mathbf{j} \cdot \mathbf{n}] \, dS = \iiint\limits_{G} \frac{\partial g}{\partial y} \, dV \tag{2b}$$

$$\iint\limits_{\sigma} [h(x, y, z)\mathbf{k} \cdot \mathbf{n}] \, dS = \iiint\limits_{G} \frac{\partial h}{\partial z} \, dV \tag{2c}$$

Since the proofs of all three formulas are similar, we shall prove only the third. It follows from Theorem 17.5.2 that

$$\iiint\limits_{G} \frac{\partial h}{\partial z} \, dV = \iint\limits_{R} \left[\int_{g_1(x, y)}^{g_2(x, y)} \frac{\partial h}{\partial z} \, dz \right] dA = \iint\limits_{R} \left[h(x, y, z) \right]_{z=g_1(x, y)}^{g_2(x, y)} dA$$

so

$$\iiint\limits_{G} \frac{\partial h}{\partial z} \, dV = \iint\limits_{R} [h(x, y, g_2(x, y)) - h(x, y, g_1(x, y))] \, dA \tag{3}$$

We shall evaluate the surface integral in (2c) by integrating over each surface of G separately. If there is a lateral surface σ_3, then at each point of this surface $\mathbf{n} \cdot \mathbf{k} = 0$ since $\mathbf{n}$ is horizontal and $\mathbf{k}$ is vertical. Thus,

$$\iint\limits_{\sigma_3} [h(x, y, z)\mathbf{k} \cdot \mathbf{n}] \, dS = 0$$

Therefore, regardless of whether or not G has a lateral surface, we can write

$$\iint\limits_{\sigma} [h(x, y, z)\mathbf{k} \cdot \mathbf{n}] \, dS = \iint\limits_{\sigma_1} [h(x, y, z)\mathbf{k} \cdot \mathbf{n}] \, dS + \iint\limits_{\sigma_2} [h(x, y, z)\mathbf{k} \cdot \mathbf{n}] \, dS \tag{4}$$

On the upper surface σ_2, the outer normal is an upward normal, and on the lower surface σ_1, the outer normal is a downward normal. Thus, Formulas (7) and (8) of Section 18.6 imply that

$$\iint\limits_{\sigma_2} [h(x, y, z)\mathbf{k} \cdot \mathbf{n}] \, dS = \iint\limits_{R} \left[h(x, y, g_2(x, y))\mathbf{k} \cdot \left(-\frac{\partial z}{\partial x}\mathbf{i} - \frac{\partial z}{\partial y}\mathbf{j} + \mathbf{k} \right) \right] dA$$

$$= \iint\limits_{R} [h(x, y, g_2(x, y))] \, dA \tag{5}$$

and

$$\iint\limits_{\sigma_1} [h(x, y, z)\mathbf{k} \cdot \mathbf{n}] \, dS = \iint\limits_{R} \left[h(x, y, g_1(x, y))\mathbf{k} \cdot \left(\frac{\partial z}{\partial x}\mathbf{i} + \frac{\partial z}{\partial y}\mathbf{j} - \mathbf{k} \right) \right] dA$$

$$= -\iint\limits_{R} [h(x, y, g_1(x, y))] \, dA \tag{6}$$

Substituting (5) and (6) into (4) and combining the terms into a single integral yields

$$\iint\limits_{\sigma} [h(x, y, z)\mathbf{k} \cdot \mathbf{n}] \, dS = \iint\limits_{R} [h(x, y, g_2(x, y)) - h(x, y, g_1(x, y))] \, dA \qquad (7)$$

Equation (2c) now follows from (3) and (7).   ∎

Example 1   Let σ be the sphere $x^2 + y^2 + z^2 = a^2$ oriented outward. Use the Divergence Theorem to find the flux of the vector field $\mathbf{F}(x, y, z) = z\mathbf{k}$ across σ.

Solution.   The divergence of the vector field is

$$\operatorname{div} \mathbf{F} = \frac{\partial z}{\partial z} = 1$$

Thus, if G denotes the spherical solid enclosed by σ, then it follows from (1) that the flux Φ across σ is

$$\Phi = \iint\limits_{\sigma} \mathbf{F} \cdot \mathbf{n} \, dS = \iiint\limits_{G} dV = \text{volume of } G = \frac{4\pi a^3}{3} \qquad \blacktriangleleft$$

REMARK.   We solved this same problem in Example 2 of Section 18.6 by evaluating the integral for Φ directly. Note how much simpler this solution is.

Example 2   Let σ be the surface of the cube shown in Figure 18.7.2 oriented outward. Use the Divergence Theorem to find the flux of the vector field

$$\mathbf{F}(x, y, z) = 2x\mathbf{i} + 3y\mathbf{j} + z^2\mathbf{k}$$

across σ.

Solution.   The divergence of the vector field is

$$\operatorname{div} \mathbf{F} = \frac{\partial}{\partial x}(2x) + \frac{\partial}{\partial y}(3y) + \frac{\partial}{\partial z}(z^2) = 5 + 2z$$

Thus, if G denotes the solid cube enclosed by σ, then it follows from (1) that the flux Φ across σ is

$$\Phi = \iint\limits_{\sigma} \mathbf{F} \cdot \mathbf{n} \, dS = \iiint\limits_{G} (5 + 2z) \, dV = \int_0^1 \int_0^1 \int_0^1 (5 + 2z) \, dz \, dy \, dx$$

$$= \int_0^1 \int_0^1 [5z + z^2]_{z=0}^1 \, dy \, dx = \int_0^1 \int_0^1 6 \, dy \, dx = 6 \qquad \blacktriangleleft$$

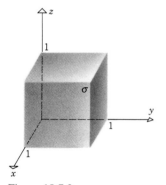

Figure 18.7.2

Example 3   Let σ be the surface of the solid enclosed by the circular cylinder $x^2 + y^2 = 9$ and the planes $z = 0$ and $z = 2$, oriented outward (Figure 18.7.3). Use the Divergence Theorem to find the flux of the vector field $\mathbf{F}(x, y, z) = x^3\mathbf{i} + y^3\mathbf{j} + z^2\mathbf{k}$ across σ.

Solution.   The divergence of the vector field is

$$\operatorname{div} \mathbf{F} = \frac{\partial}{\partial x}(x^3) + \frac{\partial}{\partial y}(y^3) + \frac{\partial}{\partial z}(z^2) = 3x^2 + 3y^2 + 2z$$

Thus, if G denotes the solid circular cylinder enclosed by σ, then it follows from (1) that the flux Φ across σ is

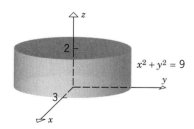

Figure 18.7.3

$$\Phi = \iint_{\sigma} \mathbf{F} \cdot \mathbf{n} \, dS = \iiint_{G} (3x^2 + 3y^2 + 2z) \, dV$$

$$= \int_0^{2\pi} \int_0^3 \int_0^2 (3r^2 + 2z) r \, dz \, dr \, d\theta \qquad \boxed{\text{Using cylindrical coordinates}}$$

$$= \int_0^{2\pi} \int_0^3 \left[3r^3 z + z^2 r \right]_{z=0}^{2} \, dr \, d\theta$$

$$= \int_0^{2\pi} \int_0^3 (6r^3 + 4r) \, dr \, d\theta$$

$$= \int_0^{2\pi} \left[\frac{3r^4}{2} + 2r^2 \right]_0^3 \, d\theta$$

$$= \int_0^{2\pi} \frac{279}{2} \, d\theta = 279\pi \qquad \blacktriangleleft$$

Example 4 Let σ be the surface of the solid enclosed by the hemisphere $z = \sqrt{a^2 - x^2 - y^2}$ and the plane $z = 0$, oriented outward (Figure 18.7.4). Use the Divergence Theorem to find the flux of the vector field $\mathbf{F}(x, y, z) = x^3 \mathbf{i} + y^3 \mathbf{j} + z^3 \mathbf{k}$ across σ.

Solution. The divergence of the vector field is

$$\text{div } \mathbf{F} = \frac{\partial}{\partial x}(x^3) + \frac{\partial}{\partial y}(y^3) + \frac{\partial}{\partial z}(z^3) = 3x^2 + 3y^2 + 3z^2$$

Thus, if G denotes the hemispheric solid enclosed by σ, then it follows from (1) that the flux Φ across σ is

$$\Phi = \iint_{\sigma} \mathbf{F} \cdot \mathbf{n} \, dS = \iiint_{G} (3x^2 + 3y^2 + 3z^2) \, dV$$

$$= \int_0^{2\pi} \int_0^{\pi/2} \int_0^a (3\rho^2) \rho^2 \sin\phi \, d\rho \, d\phi \, d\theta \qquad \boxed{\text{Using spherical coordinates}}$$

$$= 3 \int_0^{2\pi} \int_0^{\pi/2} \int_0^a \rho^4 \sin\phi \, d\rho \, d\phi \, d\theta$$

$$= 3 \int_0^{2\pi} \int_0^{\pi/2} \left[\frac{\rho^5}{5} \sin\phi \right]_{\rho=0}^{a} \, d\phi \, d\theta$$

$$= \frac{3a^5}{5} \int_0^{2\pi} \int_0^{\pi/2} \sin\phi \, d\phi \, d\theta$$

$$= \frac{3a^5}{5} \int_0^{2\pi} \left[-\cos\phi \right]_0^{\pi/2} \, d\theta$$

$$= \frac{3a^5}{5} \int_0^{2\pi} d\theta = \frac{6\pi a^5}{5} \qquad \blacktriangleleft$$

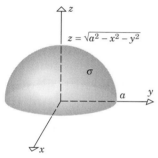

$z = \sqrt{a^2 - x^2 - y^2}$

σ

Figure 18.7.4

☐ **DIVERGENCE VIEWED AS FLUX DENSITY**

The Divergence Theorem provides a useful way of interpreting the divergence of a vector field $\mathbf{F}$ in 3-space. Suppose that G is a *small* spherical region centered at the point P_0 and that its surface $\sigma(G)$ is oriented outward. Denote the volume of the region by $\text{vol}(G)$ and the flux of $\mathbf{F}$ across $\sigma(G)$ by $\Phi(G)$. If div $\mathbf{F}$ is continuous on G, then over the small region G

the value of div $\mathbf{F}$ will not vary much from its value div $\mathbf{F}(P_0)$ at the center, and we can reasonably approximate div $\mathbf{F}$ by the constant div $\mathbf{F}(P_0)$ on G. Thus, the Divergence Theorem implies that the flux $\Phi(G)$ of $\mathbf{F}$ across $\sigma(G)$ can be approximated as

$$\Phi(G) = \iint\limits_{\sigma(G)} \mathbf{F} \cdot \mathbf{n} \, dS = \iiint\limits_{G} \text{div } \mathbf{F} \, dA \approx \text{div } \mathbf{F}(P_0) \iiint\limits_{G} dV = \text{div } \mathbf{F}(P_0) \, \text{vol}(G)$$

from which we obtain the following approximation of div $\mathbf{F}(P_0)$:

$$\text{div } \mathbf{F}(P_0) \approx \frac{\Phi(G)}{\text{vol}(G)} \tag{8}$$

The expression on the right side of (8) is called the *flux density of* $\mathbf{F}$ *over* G. If we now let the radius of the sphere approach zero [so that vol(G) approaches zero], then it is plausible that the error in this approximation approaches zero, and the divergence of $\mathbf{F}$ at the point P_0 is given exactly by

$$\text{div } \mathbf{F}(P_0) = \lim_{\text{vol}(G) \to 0} \frac{\Phi(G)}{\text{vol}(G)}$$

or equivalently,

$$\text{div } \mathbf{F}(P_0) = \lim_{\text{vol}(G) \to 0} \frac{1}{\text{vol}(G)} \iint\limits_{\sigma(G)} \mathbf{F} \cdot \mathbf{n} \, dS \tag{9}$$

This limit, called the *flux density of* $\mathbf{F}$ *at the point* P_0, is sometimes taken as the definition of divergence. This results in a definition of divergence that does not require the introduction of a coordinate system, unlike the definition of divergence in Section 18.1.

☐ **INTERPRETATION OF DIVERGENCE IN FLUID FLOW**

If P_0 is a point in an incompressible fluid at which div $\mathbf{F}(P_0) > 0$, then it follows from (8) that $\Phi(G) > 0$ for a sufficiently small sphere G centered at P_0. Thus, there is a greater volume of fluid going out through the surface of G than coming in. But this can only happen if there is some point *inside* the sphere at which fluid is flowing in; otherwise the net outward flow through the surface would result in a decrease in density within the sphere, contradicting the incompressibility assumption. Similarly, if div $\mathbf{F}(P_0) < 0$, there would have to be a point *inside* the sphere at which fluid is draining out; otherwise the net inward flow through the surface would result in an increase in density within the sphere. In an incompressible fluid, points at which div $\mathbf{F}(P_0) > 0$ are called *sources* and points at which div $\mathbf{F}(P_0) < 0$ are called *sinks*. Fluid enters the flow at a source and drains out at a sink. In an incompressible fluid without sources or sinks we must have

$$\text{div } \mathbf{F}(P) = 0$$

at every point P. In hydrodynamics this is called the *continuity equation for incompressible fluids* and is sometimes taken as the defining characteristic of an incompressible fluid.

☐ **GAUSS' LAW FOR INVERSE-SQUARE FIELDS**

Some of the major principles of physics are consequences of the following result, which we shall obtain by applying the Divergence Theorem to inverse-square fields (see Definition 18.1.2).

18.7.2 GAUSS' LAW FOR INVERSE-SQUARE FIELDS.   If

$$\mathbf{F}(\mathbf{r}) = \frac{c}{\|\mathbf{r}\|^3} \mathbf{r}$$

is an inverse-square field in 3-space, and if σ is a closed orientable surface that surrounds the origin and has outward orientation, then the flux Φ of $\mathbf{F}$ across σ is

$$\Phi = \iint\limits_{\sigma} \mathbf{F} \cdot \mathbf{n} \, dS = 4\pi c \tag{10}$$

Recall from Formula (3) of Section 18.1 that $\mathbf{F}$ can be expressed in component form as

$$\mathbf{F}(x, y, z) = \frac{c}{(x^2 + y^2 + z^2)^{3/2}} (x\mathbf{i} + y\mathbf{j} + z\mathbf{k}) \tag{11}$$

Since the components of $\mathbf{F}$ are not continuous at the origin, we cannot apply the Divergence Theorem over the solid enclosed by σ. However, we can circumvent this difficulty by constructing a sphere of radius a centered at the origin, where the radius is sufficiently small that the sphere lies entirely within the region enclosed by σ (Figure 18.7.5). We shall denote the surface of this sphere by σ_a. The solid G enclosed between σ_a and σ is a three-dimensional *simply connected solid* in which σ is the outer boundary and σ_a is the boundary of a cavity inside the solid. The components of $\mathbf{F}$ satisfy the hypotheses of the Divergence Theorem on G.

Just as we were able to extend Green's Theorem to multiply connected regions in the plane, so it is possible to extend the Divergence Theorem to multiply connected solids in 3-space, provided the surface integral in the theorem is taken over the *entire* boundary with the outside boundary oriented outward (away from G) and the boundaries of the cavities oriented inward (toward the cavities). Thus, if $\mathbf{F}$ is the inverse-square field in (11), and if σ_a is oriented inward, then the Divergence Theorem yields

$$\iiint\limits_{G} \operatorname{div} \mathbf{F} \, dV = \iint\limits_{\sigma} \mathbf{F} \cdot \mathbf{n} \, dS + \iint\limits_{\sigma_a} \mathbf{F} \cdot \mathbf{n} \, dS \tag{12}$$

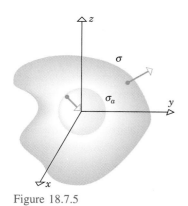

Figure 18.7.5

But we showed in Example 7 of Section 18.1 that $\operatorname{div} \mathbf{F} = 0$, so (12) yields

$$\iint\limits_{\sigma} \mathbf{F} \cdot \mathbf{n} \, dS = - \iint\limits_{\sigma_a} \mathbf{F} \cdot \mathbf{n} \, dS \tag{13}$$

We can evaluate the surface integral over σ_a by expressing the integrand in terms of components; however, it is easier to leave it in vector form. At each point on the sphere the unit normal $\mathbf{n}$ points inward along a radius from the origin, and hence $\mathbf{n} = -\mathbf{r}/\|\mathbf{r}\|$. Thus, (13) yields

$$\iint\limits_{\sigma} \mathbf{F} \cdot \mathbf{n} \, dS = - \iint\limits_{\sigma_a} \frac{c}{\|\mathbf{r}\|^3} \mathbf{r} \cdot \left(-\frac{\mathbf{r}}{\|\mathbf{r}\|}\right) dS$$

$$= \iint\limits_{\sigma_a} \frac{c}{\|\mathbf{r}\|^4} (\mathbf{r} \cdot \mathbf{r}) \, dS$$

$$= \iint\limits_{\sigma_a} \frac{c}{\|\mathbf{r}\|^2} \, dS$$

$$= \frac{c}{a^2} \iint\limits_{\sigma_a} dS \quad \boxed{\|\mathbf{r}\| = a \text{ on } \sigma_a}$$

$$= \frac{c}{a^2} (4\pi a^2) \quad \boxed{\begin{array}{l}\text{The integral is the surface}\\\text{area of the sphere.}\end{array}}$$

$$= 4\pi c$$

which establishes (10). ■

☐ **GAUSS' LAW IN ELECTROSTATICS**

It follows from Example 3 of Section 18.1 with $q = 1$ that a single charged particle of charge Q located at the origin creates an inverse-square field

$$\mathbf{F}(\mathbf{r}) = \frac{Q}{4\pi\epsilon_0 \|\mathbf{r}\|^3} \mathbf{r}$$

in which $\mathbf{F}(\mathbf{r})$ is the electrical force exerted by Q on a unit positive charge ($q = 1$) located at the point with position vector $\mathbf{r}$. In this case Gauss' law (18.7.2) states that the outward flux Φ across any closed orientable surface σ that surrounds Q is

$$\Phi = \iint_{\sigma} \mathbf{F} \cdot \mathbf{n} \, dS = 4\pi \left(\frac{Q}{4\pi\epsilon_0} \right) = \frac{Q}{\epsilon_0}$$

This result can be extended to fields resulting from more than one charge. It is one of the fundamental laws in the study of electricity and magnetism.

► Exercise Set 18.7

In Exercises 1–13, use the Divergence Theorem to evaluate $\iint_{\sigma} \mathbf{F} \cdot \mathbf{n} \, dS$, where $\mathbf{n}$ is the outer unit normal to σ.

1. $\mathbf{F}(x, y, z) = 4x\mathbf{i} - 3y\mathbf{j} + 7z\mathbf{k}$; σ is the surface of the cube bounded by the coordinate planes and the planes $x = 1$, $y = 1$, and $z = 1$.

2. $\mathbf{F}(x, y, z) = (x^2 + y)\mathbf{i} + z^2\mathbf{j} + (e^y - z)\mathbf{k}$; σ is the surface of the rectangular solid bounded by the coordinate planes and the planes $x = 3$, $y = 1$, and $z = 2$.

3. $\mathbf{F}(x, y, z) = 2x\mathbf{i} + 2y\mathbf{j} + 2z\mathbf{k}$, where σ is the sphere $x^2 + y^2 + z^2 = 9$.

4. $\mathbf{F}(x, y, z) = z^3\mathbf{i} - x^3\mathbf{j} + y^3\mathbf{k}$, where σ is the sphere $x^2 + y^2 + z^2 = a^2$.

5. $\mathbf{F}(x, y, z) = (x - z)\mathbf{i} + (y - x)\mathbf{j} + (z - y)\mathbf{k}$; σ is the surface of the cylindrical solid bounded by $x^2 + y^2 = a^2$, $z = 0$, and $z = 1$.

6. $\mathbf{F}(x, y, z) = x\mathbf{i} + y\mathbf{j} + z\mathbf{k}$; σ is the surface of the solid bounded by the paraboloid $z = 1 - x^2 - y^2$ and the xy-plane.

7. $\mathbf{F}(x, y, z) = x^3\mathbf{i} + y^3\mathbf{j} + z^3\mathbf{k}$; σ is the surface of the cylindrical solid bounded by $x^2 + y^2 = 4$, $z = 0$, and $z = 3$.

8. $\mathbf{F}(x, y, z) = (x^3 - e^y)\mathbf{i} + (y^3 + \sin z)\mathbf{j} + (z^3 - xy)\mathbf{k}$, where σ is the surface of the solid bounded by $z = \sqrt{4 - x^2 - y^2}$ and the xy-plane. [*Hint:* Use spherical coordinates.]

9. $\mathbf{F}(x, y, z) = (x^2 + y)\mathbf{i} + xy\mathbf{j} - (2xz + y)\mathbf{k}$; σ is the surface of the tetrahedron in the first octant bounded by $x + y + z = 1$ and the coordinate planes.

10. $\mathbf{F}(x, y, z) = 2xz\mathbf{i} + yz\mathbf{j} + z^2\mathbf{k}$, where σ is the surface of the hemispherical solid bounded above by $z = \sqrt{a^2 - x^2 - y^2}$ and below by the xy-plane.

11. $\mathbf{F}(x, y, z) = x^2\mathbf{i} + y^2\mathbf{j} + z^2\mathbf{k}$; σ is the surface of the conical solid bounded by $z = \sqrt{x^2 + y^2}$ and $z = 1$.

12. $\mathbf{F}(x, y, z) = x^2y\mathbf{i} - xy^2\mathbf{j} + (z + 2)\mathbf{k}$; σ is the surface of the solid bounded above by the plane $z = 2x$ and below by the paraboloid $z = x^2 + y^2$.

13. $\mathbf{F}(x, y, z) = x^3\mathbf{i} + x^2y\mathbf{j} + xy\mathbf{k}$; σ is the surface of the solid bounded by $z = 4 - x^2$, $y + z = 5$, $z = 0$, and $y = 0$.

14. Find $\iint_{\sigma} \mathbf{F} \cdot \mathbf{n} \, dS$, where $\mathbf{F}(x, y, z) = a\mathbf{i} + b\mathbf{j} + c\mathbf{k}$, σ is the surface of a solid G, and $\mathbf{n}$ is an outward unit normal (a, b, and c constants).

15. Prove that if $\mathbf{F}(x, y, z) = x\mathbf{i} + y\mathbf{j} + z\mathbf{k}$ and σ is the surface of a solid G oriented by outward unit normals, then
$$\iint_{\sigma} \mathbf{F} \cdot \mathbf{n} \, dS = 3V$$
where V is the volume of G.

16. Use the result in Exercise 15 to find $\iint_{\sigma} \mathbf{F} \cdot \mathbf{n} \, dS$, where $\mathbf{F}(x, y, z) = x\mathbf{i} + y\mathbf{j} + z\mathbf{k}$ and S is the surface of the cylindrical solid bounded by $x^2 + 4x + y^2 = 5$, $z = -1$, and $z = 4$.

In Exercises 17–20, determine whether the flow field $\mathbf{F}(x, y, z)$ is free of sources and sinks. If it is not, find the location of all sources and sinks.

17. $\mathbf{F}(x, y, z) = (y + z)\mathbf{i} - xz^3\mathbf{j} + (x^2 \sin y)\mathbf{k}$.

18. $\mathbf{F}(x, y, z) = xy\mathbf{i} - xy\mathbf{j} + y^2\mathbf{k}$.

19. $\mathbf{F}(x, y, z) = x^3\mathbf{i} + y^3\mathbf{j} + z^3\mathbf{k}$.

20. $\mathbf{F}(x, y, z) = (x^3 - x)\mathbf{i} + (y^3 - y)\mathbf{j} + (z^3 - z)\mathbf{k}$.

■ 18.8 STOKES' THEOREM

*In this section we shall discuss a generalization of Green's Theorem to three dimensions, called **Stokes'* Theorem**. This theorem has applications in various branches of physics and plays an important role in analyzing the rotational motion of fluids. It also provides a physical interpretation of the curl of a vector field.*

□ **RELATIVE ORIENTATION OF CURVES AND SURFACES**

In this section we shall be concerned with oriented surfaces in 3-space that are bounded by simple closed curves. If an oriented surface σ is bounded by a curve C (Figure 18.8.1a), then there are two possible orientations for C, which can be described as follows: Imagine a person walking along the curve C so the person's head is in the direction of orientation of σ. The person is said to be walking in the ***positive direction*** of C if the surface σ is on the left (Figure 18.8.1b) and is said to be walking in the ***negative direction*** of C if the surface σ is on the right (Figure 18.8.1c). This establishes a right-hand relationship between the orientations of σ and C in the sense that if the fingers of the right hand are cupped in the positive direction of C, then the thumb points (roughly) in the direction of orientation of σ.

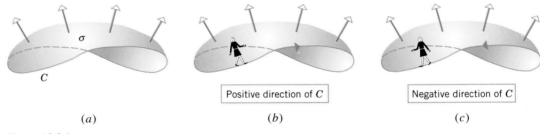

| Positive direction of C | Negative direction of C |

(a)   (b)   (c)

Figure 18.8.1

---

*GEORGE GABRIEL STOKES (1819–1903). Irish mathematician and physicist. Born in Skreen, Ireland, Stokes came from a family deeply rooted in the Church of Ireland. His father was a rector, his mother the daughter of a rector, and three of his brothers took holy orders. He received his early education from his father and a local parish clerk. In 1837, he entered Pembroke College and after graduating with top honors accepted a fellowship at the college. In 1847 he was appointed Lucasian professor of mathematics at Cambridge, a position once held by Isaac Newton, but one that had lost its esteem through the years. By virtue of his accomplishments, Stokes ultimately restored the position to the eminence it once held. Unfortunately, the position paid very little and Stokes was forced to teach at the Government School of Mines during the 1850s to supplement his income.

Stokes was one of several outstanding nineteenth century scientists who helped turn the physical sciences in a more empirical direction. He systematically studied hydrodynamics, elasticity of solids, behavior of waves in elastic solids, and diffraction of light. For Stokes, mathematics was a tool for his physical studies. He wrote classic papers on the motion of viscous fluids that laid the foundation for modern hydrodynamics; he elaborated on the wave theory of light; and he wrote papers on gravitational variation that established him as a founder of the modern science of geodesy.

Stokes was honored in his later years with degrees, medals, and memberships in foreign societies. He was knighted in 1889. Throughout his life, Stokes gave generously of his time to learned societies and readily assisted those who sought his help in solving problems. He was deeply religious and vitally concerned with the relationship between science and religion.

Before proceeding to the main theorem, it will be helpful to recall from Section 18.1 that the curl of a vector field

$$\mathbf{F}(x, y, z) = f(x, y, z)\mathbf{i} + g(x, y, z)\mathbf{j} + h(x, y, z)\mathbf{k}$$

is defined as

$$\operatorname{curl} \mathbf{F} = \left(\frac{\partial h}{\partial y} - \frac{\partial g}{\partial z}\right)\mathbf{i} + \left(\frac{\partial f}{\partial z} - \frac{\partial h}{\partial x}\right)\mathbf{j} + \left(\frac{\partial g}{\partial x} - \frac{\partial f}{\partial y}\right)\mathbf{k} = \begin{vmatrix} \mathbf{i} & \mathbf{j} & \mathbf{k} \\ \dfrac{\partial}{\partial x} & \dfrac{\partial}{\partial y} & \dfrac{\partial}{\partial z} \\ f & g & h \end{vmatrix} \quad (1)$$

☐ **STOKES' THEOREM**

The following result, known as **Stokes' Theorem**, shows that under appropriate conditions the surface integral of the normal component of curl **F** is equal to the line integral over the boundary of the tangential component of **F**. The proof is difficult and is left for advanced courses.

18.8.1 THEOREM (*Stokes' Theorem*). *Let σ be a piecewise smooth orientable surface that is bounded by a simple, closed, piecewise smooth curve C with positive orientation. If the components of the vector field*

$$\mathbf{F}(x, y, z) = f(x, y, z)\mathbf{i} + g(x, y, z)\mathbf{j} + h(x, y, z)\mathbf{k}$$

are continuous and have continuous first partial derivatives on some open set containing σ, and if $\mathbf{T}$ is the unit tangent vector to C, then

$$\oint_C \mathbf{F} \cdot \mathbf{T}\, ds = \iint_\sigma (\operatorname{curl} \mathbf{F}) \cdot \mathbf{n}\, dS \quad (2)$$

For computational purposes, it is desirable to express (2) in a different form. Recall from Formula (2) of Section 15.4 that

$$\frac{d\mathbf{r}}{ds} = \mathbf{T}$$

from which we obtain the following alternative form of (2):

$$\oint_C \mathbf{F} \cdot d\mathbf{r} = \iint_\sigma (\operatorname{curl} \mathbf{F}) \cdot \mathbf{n}\, dS \quad (3)$$

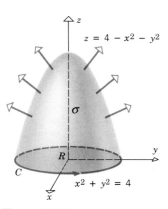

Figure 18.8.2

Example 1 Let σ be the portion of the paraboloid $z = 4 - x^2 - y^2$ for which $z \ge 0$, and let C be the circle $x^2 + y^2 = 4$ that forms the boundary of σ in the xy-plane (Figure 18.8.2). Verify Stokes' Theorem for the vector field $\mathbf{F}(x, y, z) = 2z\mathbf{i} + 3x\mathbf{j} + 5y\mathbf{k}$ if σ is oriented upward.

Solution. We will verify (3). Since σ is oriented upward, the positive orientation of C is counterclockwise looking down the positive z-axis. Thus, C can be represented parametrically (with positive orientation) by

$$x = 2\cos t, \quad y = 2\sin t, \quad z = 0 \qquad (0 \le t \le 2\pi) \quad (4)$$

Therefore

$$\oint_C \mathbf{F} \cdot d\mathbf{r} = \oint_C 2z \, dx + 3x \, dy + 5y \, dz$$

$$= \int_0^{2\pi} [0 + (6 \cos t)(2 \cos t) + 0] \, dt$$

$$= \int_0^{2\pi} 12 \cos^2 t \, dt = 12 \left[\frac{1}{2} t + \frac{1}{4} \sin 2t \right]_0^{2\pi} = 12\pi$$

To evaluate the right side of (3), we note that

$$\text{curl } \mathbf{F} = \begin{vmatrix} \mathbf{i} & \mathbf{j} & \mathbf{k} \\ \dfrac{\partial}{\partial x} & \dfrac{\partial}{\partial y} & \dfrac{\partial}{\partial z} \\ 2z & 3x & 5y \end{vmatrix} = 5\mathbf{i} + 2\mathbf{j} + 3\mathbf{k}$$

Since σ is oriented up and is expressed in the form $z = g(x, y) = 4 - x^2 - y^2$, it follows from Formula (7) of Section 18.6 with curl $\mathbf{F}$ replacing $\mathbf{F}$ that

$$\iint_\sigma (\text{curl } \mathbf{F}) \cdot \mathbf{n} \, dS = \iint_R (\text{curl } \mathbf{F}) \cdot \left(-\frac{\partial z}{\partial x} \mathbf{i} - \frac{\partial z}{\partial y} \mathbf{j} + \mathbf{k} \right) dA$$

$$= \iint_R (5\mathbf{i} + 2\mathbf{j} + 3\mathbf{k}) \cdot (2x\mathbf{i} + 2y\mathbf{j} + \mathbf{k}) \, dA$$

$$= \iint_R (10x + 4y + 3) \, dA$$

$$= \int_0^{2\pi} \int_0^2 (10r \cos \theta + 4r \sin \theta + 3) r \, dr \, d\theta$$

$$= \int_0^{2\pi} \left[\frac{10r^3}{3} \cos \theta + \frac{4r^3}{3} \sin \theta + \frac{3r^2}{2} \right]_{r=0}^2 d\theta$$

$$= \int_0^{2\pi} \left(\frac{80}{3} \cos \theta + \frac{32}{3} \sin \theta + 6 \right) d\theta$$

$$= \left[\frac{80}{3} \sin \theta - \frac{32}{3} \cos \theta + 6\theta \right]_0^{2\pi} = 12\pi$$

which agrees with the value of the line integral obtained above. Note, however, that the line integral was easier to evaluate, and hence would be the computational method of choice in this case.   ◀

REMARK.   Note that if σ_1 and σ_2 are surfaces with the same positively oriented boundary C, then for any vector field $\mathbf{F}$ that satisfies the hypotheses of Stokes' Theorem, it must be the case that

$$\iint_{\sigma_1} \text{curl } \mathbf{F} \cdot \mathbf{n} \, dS = \iint_{\sigma_2} \text{curl } \mathbf{F} \cdot \mathbf{n} \, dS$$

since (2) implies that the two surface integrals are equal to the same line integral over C. For example, the parabolic surface in Example 1 can be replaced by the upper hemisphere of radius 2 (with upward orientation), or even by the circular region R in Figure 18.8.2 (with upward orientation), without altering the value of the surface integral, since the parabolic surface, the upper hemisphere, and the region R have the same oriented boundary C in the xy-plane.

Example 2   Let C be the rectangle in the plane $z = y$ oriented as in Figure 18.8.3, and let $\mathbf{F}(x, y, z) = x^2\mathbf{i} + 4xy^3\mathbf{j} + y^2x\mathbf{k}$. Find

$$\oint_C \mathbf{F} \cdot d\mathbf{r}$$

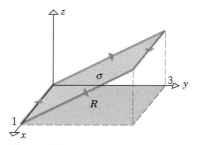

Figure 18.8.3

Solution.   To evaluate the integral directly would require four separate integrations, one over each side of the rectangle. Instead, we shall apply Stokes' Theorem, so we need only evaluate a single surface integral over the rectangular surface σ bounded by C. For the positive direction of C to be as shown in Figure 18.8.3, the surface σ must be oriented by downward normals.

Since the surface σ has equation $z = y$ and

$$\text{curl } \mathbf{F} = \begin{vmatrix} \mathbf{i} & \mathbf{j} & \mathbf{k} \\ \dfrac{\partial}{\partial x} & \dfrac{\partial}{\partial y} & \dfrac{\partial}{\partial z} \\ x^2 & 4xy^3 & y^2x \end{vmatrix} = 2yx\mathbf{i} - y^2\mathbf{j} + 4y^3\mathbf{k}$$

it follows from Formula (8) of Section 18.6 with curl $\mathbf{F}$ replacing $\mathbf{F}$ that

$$\iint_\sigma (\text{curl } \mathbf{F}) \cdot \mathbf{n}\, dS = \iint_R (\text{curl } \mathbf{F}) \cdot \left(\frac{\partial z}{\partial x}\mathbf{i} + \frac{\partial z}{\partial y}\mathbf{j} - \mathbf{k} \right) dA$$

$$= \iint_R (2yx\mathbf{i} - y^2\mathbf{j} + 4y^3\mathbf{k}) \cdot (0\mathbf{i} + \mathbf{j} - \mathbf{k})\, dA$$

$$= \int_0^1 \int_0^3 (-y^2 - 4y^3)\, dy\, dx$$

$$= -\int_0^1 \left[\frac{y^3}{3} + y^4 \right]_{y=0}^3 dx$$

$$= -\int_0^1 90\, dx = -90 \qquad \blacktriangleleft$$

□ **RELATIONSHIP BETWEEN GREEN'S THEOREM AND STOKES' THEOREM**

It is sometimes convenient to regard a vector field

$$\mathbf{F}(x, y) = f(x, y)\mathbf{i} + g(x, y)\mathbf{j}$$

in 2-space as a vector field in 3-space by expressing it as

$$\mathbf{F}(x, y) = f(x, y)\mathbf{i} + g(x, y)\mathbf{j} + 0\mathbf{k} \tag{5}$$

If R is a region in the xy-plane enclosed by a curve C, then we can treat R as a *flat* surface, and we can treat a surface integral over R as an ordinary double integral over R. Thus, if we orient R upward and C counterclockwise looking down the positive z-axis, then Formula (3) applied to (5) yields

$$\oint_C \mathbf{F} \cdot d\mathbf{r} = \iint_R \text{curl } \mathbf{F} \cdot \mathbf{k}\, dA \tag{6}$$

But

$$\text{curl } \mathbf{F} = \begin{vmatrix} \mathbf{i} & \mathbf{j} & \mathbf{k} \\ \dfrac{\partial}{\partial x} & \dfrac{\partial}{\partial y} & \dfrac{\partial}{\partial z} \\ f & g & 0 \end{vmatrix} = -\frac{\partial g}{\partial z}\mathbf{i} + \frac{\partial f}{\partial z}\mathbf{j} + \left(\frac{\partial g}{\partial x} - \frac{\partial f}{\partial y}\right)\mathbf{k} = \left(\frac{\partial g}{\partial x} - \frac{\partial f}{\partial y}\right)\mathbf{k}$$

since $\partial g/\partial z = \partial f/\partial z = 0$ (why?) Substituting this expression in (6) and expressing the integrals in terms of components yields

$$\oint_C f\, dx + g\, dy = \iint\limits_R \left(\frac{\partial g}{\partial x} - \frac{\partial f}{\partial y}\right) dA$$

(verify) which is Green's Theorem (18.4.1). Thus, Green's Theorem is a special case of Stokes' Theorem.

RELATIONSHIP BETWEEN CURL AND CIRCULATION

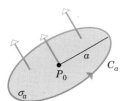

Figure 18.8.4

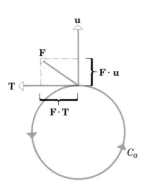

Figure 18.8.5

Stokes' Theorem provides a useful way of interpreting the curl of a vector field $\mathbf{F}$ in 3-space. The ideas that we shall develop here are applicable to most vector fields, but we shall motivate them in the context of incompressible steady-state fluid flow.

If we imagine a small piece of straw floating on the surface of a stream, the straw will be carried downstream by the current, but at the same time it may spin due to whirlpools or eddies in the flow. It is the curl of the flow field that relates to the spinning motion of the straw. To see why this is so, let σ_a be a *small* oriented disk-shaped region of radius a and centered at P_0 (Figure 18.8.4). Denote the circular boundary of the region by C_a and the area of σ_a by $A(\sigma_a)$. At each point of C_a the flow field $\mathbf{F}$ has a component $\mathbf{F} \cdot \mathbf{u}$ along the outward unit normal to C_a and a component $\mathbf{F} \cdot \mathbf{T}$ along the unit tangent to C_a (Figure 18.8.5). Fluid moving in the direction of $\mathbf{u}$ flows *through* the circle, and fluid moving in the direction of $\mathbf{T}$ moves *along* the circle. Physicists call the integral

$$\oint_{C_a} \mathbf{F} \cdot \mathbf{T}\, ds$$

the ***circulation of $\mathbf{F}$ around C_a*** and use it as a measure of the tendency for fluid to flow in the positive direction around the circle C_a. For example, in the extreme case where the flow is normal to the circle at each point, the circulation around C_a is zero, since $\mathbf{F} \cdot \mathbf{T} = 0$ at each point of the circle (Figure 18.8.6a). The more closely that $\mathbf{F}$ aligns with $\mathbf{T}$ along the circle, the larger the value of $\mathbf{F} \cdot \mathbf{T}$, and the larger the value of the circulation. The maximum circulation occurs when $\mathbf{F}$ is aligned with $\mathbf{T}$ at each point of the circle (Figure 18.8.6b).

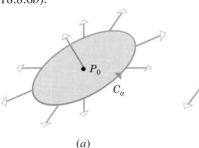

Figure 18.8.6   (a)   (b)

To see the relationship between circulation and curl, suppose that curl $\mathbf{F}$ is continuous on σ_a, so that when σ_a is small the value of curl $\mathbf{F}$ at any point of σ_a will not vary much from the value of curl $\mathbf{F}(P_0)$ at the center. Thus, for a small disk σ_a we can reasonably assume that curl $\mathbf{F}$ has a constant value of curl $\mathbf{F}(P_0)$ on σ_a. Moreover, because the surface σ_a is flat, the unit normal vectors that orient σ_a are all equal; that is, the orienting vector $\mathbf{n}$ is constant on σ_a. Thus, version (2) of Stokes' Theorem implies that

$$\oint_{C_a} \mathbf{F} \cdot \mathbf{T} \, ds = \iint_{\sigma_a} (\text{curl } \mathbf{F}) \cdot \mathbf{n} \, dS \approx \text{curl } \mathbf{F}(P_0) \cdot \mathbf{n} \iint_{\sigma_a} dS$$

where the line integral is taken in the positive direction of C_a. But the double integral in this equation represents the surface area of σ_a, so it follows that

$$\oint_{C_a} \mathbf{F} \cdot \mathbf{T} \, ds \approx [\text{curl } \mathbf{F}(P_0) \cdot \mathbf{n}] \, A(\sigma_a)$$

from which we obtain the following relationship between curl and circulation:

$$\text{curl } \mathbf{F}(P_0) \cdot \mathbf{n} \approx \frac{1}{A(\sigma_a)} \oint_{C_a} \mathbf{F} \cdot \mathbf{T} \, ds \tag{7}$$

The quantity on the right side of (7) is called the *circulation density of* $\mathbf{F}$ *around* C_a. If we now let the radius a of the disk approach zero (with $\mathbf{n}$ fixed), then it is plausible that the error in this approximation will approach zero and the exact value of curl $\mathbf{F}(P_0) \cdot \mathbf{n}$ will be given by

$$\text{curl } \mathbf{F}(P_0) \cdot \mathbf{n} = \lim_{a \to 0} \frac{1}{A(\sigma_a)} \oint_{C_a} \mathbf{F} \cdot \mathbf{T} \, ds \tag{8}$$

Thus, we call curl $\mathbf{F}(P_0) \cdot \mathbf{n}$ the *circulation density of* $\mathbf{F}$ *at* P_0 *in the direction of* $\mathbf{n}$. Since curl $\mathbf{F}(P_0) \cdot \mathbf{n}$ has its maximum value when $\mathbf{n}$ is in the same direction as curl $\mathbf{F}(P_0)$, it follows that in the vicinity of the point P_0 the maximum circulation occurs around small circles in the plane normal to curl $\mathbf{F}(P_0)$. Physically, if a small paddle wheel is immersed in the fluid so that the pivot point is at P_0, then the paddles will turn most rapidly when the spindle is aligned with curl $\mathbf{F}(P_0)$ (Figure 18.8.7). If curl $\mathbf{F} = \mathbf{0}$ at each point of a region, then $\mathbf{F}$ is said to be *irrotational* in that region, since no circulation occurs about any point of the region.

We conclude by noting that if mutually orthogonal unit vectors $\mathbf{i}$, $\mathbf{j}$, and $\mathbf{k}$ are substituted for $\mathbf{n}$ in (8), then the resulting equations produce the components of curl $\mathbf{F}(P_0)$ in the directions of those vectors. Thus (8) is sometimes taken as the definition of curl. This definition of curl does not require the introduction of a coordinate system, unlike the definition in Section 18.1.

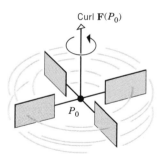

Curl $\mathbf{F}(P_0)$

P_0

Figure 18.8.7

▶ Exercise Set 18.8

In Exercises 1–8, use Stokes' Theorem to evaluate the integral $\oint_C \mathbf{F} \cdot d\mathbf{r}$.

1. $\mathbf{F}(x, y, z) = z^2 \mathbf{i} + 2x \mathbf{j} - y^3 \mathbf{k}$; C is the circle $x^2 + y^2 = 1$ in the xy-plane with counterclockwise orientation looking down the positive z-axis.

2. $\mathbf{F}(x, y, z) = xz \mathbf{i} + 3x^2 y^2 \mathbf{j} + yx \mathbf{k}$; C is the rectangle in the plane $z = y$ shown in Figure 18.8.3.

3. $\mathbf{F}(x, y, z) = 3z \mathbf{i} + 4x \mathbf{j} + 2y \mathbf{k}$; C is the boundary of the paraboloid shown in Figure 18.8.2.

4. $\mathbf{F}(x, y, z) = -3y^2 \mathbf{i} + 4z \mathbf{j} + 6x \mathbf{k}$; C is the triangle in the plane $z = \frac{1}{2} y$ with vertices $(2, 0, 0)$, $(0, 2, 1)$, and $(0, 0, 0)$ with a counterclockwise orientation looking down the positive z-axis.

5. $\mathbf{F}(x, y, z) = xy \mathbf{i} + x^2 \mathbf{j} + z^2 \mathbf{k}$; C is the intersection of the paraboloid $z = x^2 + y^2$ and the plane $z = y$ with a counterclockwise orientation looking down the positive z-axis.

6. $\mathbf{F}(x, y, z) = xy \mathbf{i} + yz \mathbf{j} + zx \mathbf{k}$; C is the triangle in the plane $x + y + z = 1$ with vertices $(1, 0, 0)$, $(0, 1, 0)$, and $(0, 0, 1)$ with a counterclockwise orientation looking from the first octant toward the origin.

7. $\mathbf{F}(x, y, z) = (x - y) \mathbf{i} + (y - z) \mathbf{j} + (z - x) \mathbf{k}$; C is the circle $x^2 + y^2 = a^2$ in the xy-plane with counterclockwise orientation looking down the positive z-axis.

8. $\mathbf{F}(x, y, z) = (z + \sin x) \mathbf{i} + (x + y^2) \mathbf{j} + (y + e^z) \mathbf{k}$; C is the intersection of the sphere $x^2 + y^2 + z^2 = 1$ and the cone $z = \sqrt{x^2 + y^2}$ with counterclockwise orientation looking down the positive z-axis.

In Exercises 9–12, verify Stokes' Theorem by computing the line and surface integrals in (3) and showing that they are equal.

9. $\mathbf{F}(x, y, z) = (x - y)\mathbf{i} + (y - z)\mathbf{j} + (z - x)\mathbf{k}$; σ is the portion of the plane $x + y + z = 1$ in the first octant.

10. $\mathbf{F}(x, y, z) = x^2\mathbf{i} + y^2\mathbf{j} + z^2\mathbf{k}$; σ is the portion of the cone $z = \sqrt{x^2 + y^2}$ below the plane $z = 1$.

11. $\mathbf{F}(x, y, z) = x\mathbf{i} + y\mathbf{j} + z\mathbf{k}$; σ is the upper hemisphere $z = \sqrt{a^2 - x^2 - y^2}$.

12. $\mathbf{F}(x, y, z) = (z - y)\mathbf{i} + (z + x)\mathbf{j} - (x + y)\mathbf{k}$; σ is the portion of the paraboloid $z = 9 - x^2 - y^2$ above the xy-plane.

13. Use the result in Exercise 17 of Section 18.1 to help prove that

$$\iint_\sigma (\operatorname{curl}\mathbf{F}) \cdot \mathbf{n}\, dS = 0$$

where σ is the surface of a solid G, $\mathbf{n}$ is the unit outer normal to σ, and the components of $\mathbf{F}$ have continuous second partial derivatives on and within σ.

14. Consider the flow field given by the formula

$$\mathbf{F}(x, y, z) = (x - z)\mathbf{i} + (y - x)\mathbf{j} + (z - xy)\mathbf{k}$$

(a) Use Stokes' Theorem to find the circulation around the triangle with vertices $A(1, 0, 0)$, $B(0, 2, 0)$, and $C(0, 0, 1)$ oriented counterclockwise looking from the origin toward the first octant.

(b) Find the circulation density of $\mathbf{F}$ at the origin in the direction of $\mathbf{k}$.

(c) Find the unit vector $\mathbf{n}$ such that the circulation density of $\mathbf{F}$ at the origin is maximum in the direction of $\mathbf{n}$.

◆ TECHNOLOGY EXERCISES   Chapter 18

Most of these exercises require access to a graphing calculator or a computer algebra system (CAS) such as *Mathematica*, *Maple*, or *Derive*. When you are asked to *find* an answer or to *solve* an equation, you may choose to find an exact result or a numerical approximation, depending on the particular technology you are using and on your own imagination. The form of your answers may differ from those of other students or from those in the answer section of the text, depending on how you solve the problems and the accuracy you use in your numerical approximations. Those exercises that are more appropriate for a CAS than a graphing calculator are labeled with the icon ◆.

◆ **1. Work:** For any constant c, the curve $y = c(x - x^2)$ passes through the origin and the point $(1, 0)$. Suppose that a particle moves along such a curve from the origin to the point $(1, 0)$ while subject to the force

$$\mathbf{F}(x, y) = (3x^3 + y^2 - y^4)\mathbf{i} + (2y^3 - x^2)\mathbf{j}$$

Find the values of c such that the work done by $\mathbf{F}$ is zero.

◆ **2. Maximum work:** Find the value of c in Exercise 1 so that the work done by $\mathbf{F}$ is maximum.

◆ **3. Work and kinetic energy:** The *kinetic energy* of a particle is defined as $K = \frac{1}{2}mv^2$, where m is the mass of the

particle and v is its speed. If m is measured in kilograms (kg) and v is measured in meters per second (m/sec), then K is in joules (J). The *work–energy principle* of physics states that if $\mathbf{F}$ is the only force acting on a particle, then the change in kinetic energy is equal to the work done on the particle by $\mathbf{F}$, where $\mathbf{F}$ is measured in newtons and distance is measured in meters. Assume that

$$\mathbf{F}(x, y, z) = y^2\mathbf{i} + xz\mathbf{j} + x\mathbf{k}$$

is the only force acting on a particle of mass 0.13 kg that moves along the curve C described parametrically by

$$\mathbf{r}(t) = t\mathbf{i} + \sin t\mathbf{j} + e^t\mathbf{k}$$

where $\mathbf{F}$ is in newtons and $\mathbf{r}$ is in meters. Suppose that the particle is moving with a speed of 1.5 m/sec in the positive direction along C if $t = 0$. Find its speed when it reaches the point at which $t = 2$.

4. **Conservative vector fields:** Suppose that a particle moves along a curve C from the point $(-1, 0)$ to the point $(1, 3)$ while subject to the force

$$\mathbf{F}(x, y) = \frac{2xe^y}{x^2e^y + 1}\mathbf{i} + \frac{x^2e^y}{x^2e^y + 1}\mathbf{j}$$

(a) Find the work done by $\mathbf{F}$ along the line segment C from $(-1, 0)$ to $(1, 3)$ by evaluating $\int_C \mathbf{F} \cdot d\mathbf{r}$.

(b) Show that $\mathbf{F}$ is conservative by finding a potential function $\phi(x, y)$.

(c) Use the function $\phi(x, y)$ obtained in part (b) to check your result in part (a).

5. **Independence of path:** Suppose that a particle moves along a curve C from the origin to the point $(1, 1, 1)$ while subject to the force

$$\mathbf{F}(x, y, z) = 3x^2y^2z^4\mathbf{i} + 2x^3yz^4\mathbf{j} + 4x^3y^2z^3\mathbf{k}$$

(a) The parametric curves

$$C_1 : \mathbf{r}(t) = t\mathbf{i} + t^2\mathbf{j} + t^3\mathbf{k}$$
$$C_2 : \mathbf{r}(t) = \sin\tfrac{1}{2}\pi t\,\mathbf{i} + t^3\mathbf{j} + \sin\tfrac{5}{2}\pi t\,\mathbf{k}$$

are traced from the origin to the point $(1, 1, 1)$ as t varies from 0 to 1. Show that the work done by $\mathbf{F}$ along C_1 is equal to the work done by $\mathbf{F}$ along C_2 by evaluating $\int_{C_1} \mathbf{F} \cdot d\mathbf{r}$ and $\int_{C_2} \mathbf{F} \cdot d\mathbf{r}$.

(b) Show that $\mathbf{F}$ is conservative by finding a potential function $\phi(x, y, z)$.

(c) Use the function $\phi(x, y, z)$ obtained in part (b) to check your result in part (a).

6. **Green's Theorem:** Suppose that C is the ellipse $x^2/3 + y^2/5 = 1$ oriented counterclockwise, and R is the region enclosed by C. Verify Green's Theorem for $f(x, y) = x^3y^2$ and $g(x, y) = xy^2$ by evaluating

$$\oint_C f(x, y)\,dx + g(x, y)\,dy$$

and

$$\iint_R \left(\frac{\partial g}{\partial x} - \frac{\partial f}{\partial y}\right) dA$$

[*Suggestion:* Parametrize C.]

7. **Mass of a lamina:** Find the mass of the curved lamina $z = e^{-x^2-y^2}$ that lies above the region in the xy-plane enclosed by $x^2 + y^2 = 9$ given that the density function is $\delta(x, y, z) = \sqrt{x^2 + y^2}$.

8. **The Divergence Theorem:** Let σ be the surface of the solid G enclosed by the paraboloid $z = 1 - x^2 - y^2$ and the plane $z = 0$. Assuming that σ is oriented outward, verify the Divergence Theorem for the vector field

$$\mathbf{F} = (x^2y - z^2)\mathbf{i} + (y^3 - x)\mathbf{j} + (2x + 3z - 1)\mathbf{k}$$

by evaluating

$$\iint_\sigma \mathbf{F} \cdot \mathbf{n}\,dS \quad \text{and} \quad \iiint_G \text{div}\,\mathbf{F}\,dV$$

9. **Stokes' Theorem:** Let σ be the portion of the paraboloid $z = 1 - x^2 - y^2$ for which $z \geq 0$, and let C be the circle $x^2 + y^2 = 1$ that forms the boundary of σ in the xy-plane. Assuming that σ is oriented upward, verify Stokes' Theorem for the vector field

$$\mathbf{F} = (x^2y - z^2)\mathbf{i} + (y^3 - x)\mathbf{j} + (2x + 3z - 1)\mathbf{k}$$

by evaluating

$$\oint_C \mathbf{F} \cdot d\mathbf{r} \quad \text{and} \quad \iint_\sigma (\text{curl}\,\mathbf{F}) \cdot \mathbf{n}\,dS$$

10. **Flux across a sphere:** Let

$$\mathbf{F}(x, y, z) = a^2x\mathbf{i} + (y/a)\mathbf{j} + az^2\mathbf{k}$$

and let σ be the sphere of radius 1, centered at the origin and oriented outward. Find all values of a such that the flux of $\mathbf{F}$ across σ is 10.

Gabriel Cramer (1704–1752)

19 SECOND-ORDER DIFFERENTIAL EQUATIONS

■ **19.1** SECOND-ORDER LINEAR HOMOGENEOUS DIFFERENTIAL EQUATIONS WITH CONSTANT COEFFICIENTS

In Section 7.7 we showed how to solve first-order linear differential equations. In this section we shall show how to solve certain second-order linear differential equations.

□ **SECOND-ORDER LINEAR DIFFERENTIAL EQUATIONS**

In this section we shall be concerned with the general ***second-order linear differential equation***

$$\frac{d^2y}{dx^2} + p(x)\frac{dy}{dx} + q(x)y = r(x) \tag{1}$$

or in an alternative notation,

$$y'' + p(x)y' + q(x)y = r(x)$$

where $p(x)$, $q(x)$, and $r(x)$ are continuous functions. If $r(x) = 0$ for all x, then (1) reduces to

$$\frac{d^2y}{dx^2} + p(x)\frac{dy}{dx} + q(x)y = 0$$

which is called the general second-order linear **homogeneous** differential equation. If $r(x)$ is not identically zero, then (1) is said to be **nonhomogeneous**. Some examples of second-order linear differential equations are

$$\frac{d^2y}{dx^2} + x^2\frac{dy}{dx} - xy = e^x \qquad \boxed{p(x) = x^2, \quad q(x) = -x, \quad r(x) = e^x}$$

$$y'' + y' - 3y = \sin x \qquad \boxed{p(x) = 1, \quad q(x) = -3, \quad r(x) = \sin x}$$

$$y'' + e^x y = 3 \qquad \boxed{p(x) = 0, \quad q(x) = e^x, \quad r(x) = 3}$$

$$\frac{d^2y}{dx^2} - \frac{dy}{dx} + 2y = 0 \qquad \boxed{p(x) = -1, \quad q(x) = 2, \quad r(x) = 0}$$

The last equation is homogeneous and the first three are nonhomogeneous.

In Section 7.7 we gave a general procedure for solving first-order linear differential equations by integration. For second-order linear differential equations the situation is more complicated and simple general procedures for solving such equations can only be given in special cases. Before we can pursue this matter further, we shall need some preliminary results.

☐ **LINEAR INDEPENDENCE**

Two continuous functions f and g are said to be **linearly dependent** if one is a constant multiple of the other. If neither is a constant multiple of the other, then they are called **linearly independent**. Thus,

$$f(x) = \sin x \quad \text{and} \quad g(x) = 3\sin x$$

are linearly dependent, but

$$f(x) = x \quad \text{and} \quad g(x) = x^2$$

are linearly independent.

The following theorem is central to the study of second-order linear differential equations.

19.1.1 THEOREM. *If $y_1 = y_1(x)$ and $y_2 = y_2(x)$ are linearly independent solutions of the homogeneous equation*

$$\frac{d^2y}{dx^2} + p(x)\frac{dy}{dx} + q(x)y = 0 \tag{2}$$

then

$$y(x) = c_1 y_1(x) + c_2 y_2(x) \tag{3}$$

is the general solution of (2) in the sense that every solution of (2) can be obtained from (3) by choosing appropriate values for the arbitrary constants c_1 and c_2; conversely, (3) is a solution of (2) for all choices of c_1 and c_2.

(Readers interested in a proof of this theorem are referred to *Elementary Differential Equations and Boundary Value Problems*, John Wiley & Sons, New York, 1992, by William E. Boyce and Richard C. DiPrima.)

REMARK.   The expression on the right side of (3) is called a **linear combination** of $y_1(x)$ and $y_2(x)$. Thus, Theorem 19.1.1 tells us that once we find two linearly independent solutions of (2), we know *all* the solutions because every other solution can be expressed as a linear combination of those two.

□ CONSTANT COEFFICIENTS

For the remainder of this section we shall restrict our attention to second-order linear homogeneous equations of the form

$$\frac{d^2y}{dx^2} + p\frac{dy}{dx} + qy = 0 \tag{4}$$

where p and q are *constants*. Our objective is to find two linearly independent solutions of this equation. We shall start by looking for solutions of the form $y = e^{mx}$. This is motivated by the fact that the first and second derivatives of this function are multiples of y, suggesting that a solution of (4) might result by choosing m appropriately. To find such an m, we substitute

$$y = e^{mx}, \quad \frac{dy}{dx} = me^{mx}, \quad \frac{d^2y}{dx^2} = m^2e^{mx} \tag{5}$$

into (4) to obtain

$$(m^2 + pm + q)e^{mx} = 0 \tag{6}$$

which is satisfied if and only if

$$m^2 + pm + q = 0 \tag{7}$$

since $e^{mx} \neq 0$ for every x.

Equation (7), which is called the **auxiliary equation** for (4), can be obtained from (4) by replacing d^2y/dx^2 by m^2, dy/dx by m $(= m^1)$, and y by 1 $(= m^0)$. The solutions, m_1 and m_2, of the auxiliary equation can be obtained by factoring or by the quadratic formula. These solutions are

$$m_1 = \frac{-p + \sqrt{p^2 - 4q}}{2}, \quad m_2 = \frac{-p - \sqrt{p^2 - 4q}}{2} \tag{8}$$

Depending on whether $p^2 - 4q$ is positive, zero, or negative, these roots will be distinct and real, equal and real, or complex conjugates.* We shall consider each of these cases separately.

□ DISTINCT REAL ROOTS

If m_1 and m_2 are distinct real roots, then (4) has the two solutions

$$y_1 = e^{m_1x}, \quad y_2 = e^{m_2x}$$

Neither of the functions e^{m_1x} and e^{m_2x} is a constant multiple of the other (Exercise 29), so the general solution of (4) in this case is

$$y(x) = c_1e^{m_1x} + c_2e^{m_2x} \tag{9}$$

Example 1 Find the general solution of $y'' - y' - 6y = 0$.

Solution. The auxiliary equation is

$$m^2 - m - 6 = 0 \quad \text{or equivalently,} \quad (m + 2)(m - 3) = 0$$

so its roots are $m = -2$, $m = 3$. Thus, from (9) the general solution of the differential equation is

$$y = c_1e^{-2x} + c_2e^{3x}$$

where c_1 and c_2 are arbitrary constants. ◀

---

*Recall that the complex solutions of a polynomial equation, and in particular of a quadratic equation, occur as conjugate pairs $a + bi$ and $a - bi$.

☐ **EQUAL REAL ROOTS**

If m_1 and m_2 are equal real roots, say $m_1 = m_2 \,(= m)$, then the auxiliary equation yields only one solution of (4):

$$y_1(x) = e^{mx}$$

We shall now show that

$$y_2(x) = xe^{mx} \tag{10}$$

is a second linearly independent solution. To see that this is so, note that $p^2 - 4q = 0$ in (8) since the roots are equal. Thus,

$$m = m_1 = m_2 = -p/2$$

and (10) becomes

$$y_2(x) = xe^{(-p/2)x}$$

Differentiating yields

$$y_2'(x) = \left(1 - \frac{p}{2}x\right)e^{(-p/2)x} \quad \text{and} \quad y_2''(x) = \left(\frac{p^2}{4}x - p\right)e^{-(p/2)x}$$

so

$$y_2''(x) + py_2'(x) + qy_2(x) = \left[\left(\frac{p^2}{4}x - p\right) + p\left(1 - \frac{p}{2}x\right) + qx\right]e^{(-p/2)x}$$

$$= \left[-\frac{p^2}{4} + q\right]xe^{(-p/2)x} \tag{11}$$

But $p^2 - 4q = 0$ implies that $(-p^2/4) + q = 0$, so (11) becomes

$$y_2''(x) + py_2'(x) + qy_2(x) = 0$$

which tells us that $y_2(x)$ is a solution of (4). It can be shown that

$$y_1(x) = e^{mx} \quad \text{and} \quad y_2(x) = xe^{mx}$$

are linearly independent (Exercise 29), so the general solution of (4) in this case is

$$y = c_1 e^{mx} + c_2 xe^{mx} \tag{12}$$

Example 2   Find the general solution of $y'' - 8y' + 16y = 0$.

Solution.   The auxiliary equation is

$$m^2 - 8m + 16 = 0 \quad \text{or equivalently,} \quad (m - 4)^2 = 0$$

so $m = 4$ is the only root. Thus, from (12) the general solution of the differential equation is

$$y = c_1 e^{4x} + c_2 xe^{4x} \quad \blacktriangleleft$$

☐ **COMPLEX ROOTS**

If the auxiliary equation has complex roots $m_1 = a + bi$ and $m_2 = a - bi$, then $y_1(x) = e^{ax}\cos bx$ and $y_2(x) = e^{ax}\sin bx$ are linearly independent solutions of (4) and

$$y = e^{ax}(c_1 \cos bx + c_2 \sin bx) \tag{13}$$

is the general solution. The proof is discussed in the exercises (Exercise 30).

Example 3   Find the general solution of $y'' + y' + y = 0$.

Solution. The auxiliary equation $m^2 + m + 1 = 0$ has roots

$$m_1 = \frac{-1 + \sqrt{1 - 4}}{2} = -\frac{1}{2} + \frac{\sqrt{3}}{2}i$$

$$m_2 = \frac{-1 - \sqrt{1 - 4}}{2} = -\frac{1}{2} - \frac{\sqrt{3}}{2}i$$

Thus, from (13) with $a = -1/2$ and $b = \sqrt{3}/2$, the general solution of the differential equation is

$$y = e^{-x/2}\left(c_1 \cos \frac{\sqrt{3}}{2}x + c_2 \sin \frac{\sqrt{3}}{2}x\right) \quad \blacktriangleleft$$

☐ **INITIAL-VALUE PROBLEMS**

When a physical problem leads to a second-order differential equation, there are usually two conditions in the problem that determine specific values for the two arbitrary constants in the general solution of the equation. Conditions that specify the value of the solution $y(x)$ and its derivative $y'(x)$ at some point $x = x_0$ are called ***initial conditions***. A second-order differential equation with initial conditions is called a ***second-order initial-value problem***.

Example 4 Solve the initial-value problem

$$y'' - y = 0, \quad y(0) = 1, \quad y'(0) = 0$$

Solution. We must first solve the differential equation. The auxiliary equation

$$m^2 - 1 = 0$$

has distinct real roots $m_1 = 1$, $m_2 = -1$, so from (9) the general solution is

$$y(x) = c_1 e^x + c_2 e^{-x} \tag{14}$$

and the derivative of this solution is

$$y'(x) = c_1 e^x - c_2 e^{-x} \tag{15}$$

Substituting $x = 0$ in (14) and (15) and using the initial conditions $y(0) = 1$ and $y'(0) = 0$ yields the system of equations

$$c_1 + c_2 = 1$$

$$c_1 - c_2 = 0$$

Solving this system yields $c_1 = \frac{1}{2}$, $c_2 = \frac{1}{2}$, so from (14) the solution of the initial-value problem is

$$y(x) = \tfrac{1}{2}e^x + \tfrac{1}{2}e^{-x} = \cosh x \quad \blacktriangleleft$$

The following summary is included as a ready reference for the solution of second-order homogeneous linear differential equations with constant coefficients.

Summary

EQUATION: $y'' + py' + qy = 0$
AUXILIARY EQUATION: $m^2 + pm + q = 0$

CASE	GENERAL SOLUTION
Distinct real roots m_1, m_2 of the auxiliary equation	$y = c_1 e^{m_1 x} + c_2 e^{m_2 x}$
Equal real roots $m_1 = m_2\ (= m)$ of the auxiliary equation	$y = c_1 e^{mx} + c_2 x e^{mx}$
Complex roots $m_1 = a + bi$, $m_2 = a - bi$ of the auxiliary equation	$y = e^{ax}(c_1 \cos bx + c_2 \sin bx)$

▶ Exercise Set 19.1

1. Verify that the following are solutions of the differential equation $y'' - y' - 2y = 0$ by substituting these functions into the equation:

    (a) e^{2x} and e^{-x}

    (b) $c_1 e^{2x} + c_2 e^{-x}$ (c_1, c_2 constants).

2. Verify that the following are solutions of the differential equation $y'' + 4y' + 4y = 0$ by substituting these functions into the equation:

    (a) e^{-2x} and xe^{-2x}

    (b) $c_1 e^{-2x} + c_2 xe^{-2x}$ (c_1, c_2 constants).

In Exercises 3–16, find the general solution of the differential equation.

3. $y'' + 3y' - 4y = 0$.

4. $y'' + 6y' + 5y = 0$.

5. $y'' - 2y' + y = 0$.

6. $y'' + 6y' + 9y = 0$.

7. $y'' + 5y = 0$.

8. $y'' + y = 0$.

9. $\dfrac{d^2y}{dx^2} - \dfrac{dy}{dx} = 0$.

10. $\dfrac{d^2y}{dx^2} + 3\dfrac{dy}{dx} = 0$.

11. $\dfrac{d^2y}{dt^2} + 4\dfrac{dy}{dt} + 4y = 0$.

12. $\dfrac{d^2y}{dt^2} - 10\dfrac{dy}{dt} + 25y = 0$.

13. $\dfrac{d^2y}{dx^2} - 4\dfrac{dy}{dx} + 13y = 0$.

14. $\dfrac{d^2y}{dx^2} - 6\dfrac{dy}{dx} + 25y = 0$.

15. $8y'' - 2y' - y = 0$.

16. $9y'' - 6y' + y = 0$.

In Exercises 17–22, solve the initial-value problem.

17. $y'' + 2y' - 3y = 0$; $y(0) = 1$, $y'(0) = 5$.

18. $y'' - 6y' - 7y = 0$; $y(0) = 5$, $y'(0) = 3$.

19. $y'' - 6y' + 9y = 0$; $y(0) = 2$, $y'(0) = 1$.

20. $y'' + 4y' + y = 0$; $y(0) = 5$, $y'(0) = 4$.

21. $y'' + 4y' + 5y = 0$; $y(0) = -3$, $y'(0) = 0$.

22. $y'' - 6y' + 13y = 0$; $y(0) = -1$, $y'(0) = 1$.

23. In each part find a second-order linear homogeneous differential equation with constant coefficients that has the given functions as solutions.

    (a) $y_1 = e^{5x}$, $y_2 = e^{-2x}$

    (b) $y_1 = e^{4x}$, $y_2 = xe^{4x}$

    (c) $y_1 = e^{-x}\cos 4x$, $y_2 = e^{-x}\sin 4x$.

24. Show that if e^x and e^{-x} are solutions of a second-order linear homogeneous differential equation, then so are $\cosh x$ and $\sinh x$.

25. Find all values of k for which the differential equation $y'' + ky' + ky = 0$ has a general solution of the given form.

    (a) $y = c_1 e^{ax} + c_2 e^{bx}$ (b) $y = c_1 e^{ax} + c_2 xe^{ax}$

    (c) $y = c_1 e^{ax}\cos bx + c_2 e^{ax}\sin bx$.

26. The equation
$$x^2\frac{d^2y}{dx^2} + px\frac{dy}{dx} + qy = 0 \quad (x > 0)$$
where p and q are constants, is called *Euler's equidimensional equation*. Show that the substitution $x = e^z$ transforms this equation into the equation
$$\frac{d^2y}{dz^2} + (p - 1)\frac{dy}{dz} + qy = 0$$

27. Use the result in Exercise 26 to find the general solution of

    (a) $x^2\dfrac{d^2y}{dx^2} + 3x\dfrac{dy}{dx} + 2y = 0$ ($x > 0$)

    (b) $x^2\dfrac{d^2y}{dx^2} - x\dfrac{dy}{dx} - 2y = 0$ ($x > 0$).

28. Let $y(x)$ be a solution of $y'' + py' + qy = 0$. Prove: If p and q are positive constants, then $\lim\limits_{x \to +\infty} y(x) = 0$.

29. The *Wronskian* of two differentiable functions y_1 and y_2 is denoted by $W(y_1, y_2)$ and is defined to be the function
$$W(y_1, y_2) = y_1 y_2' - y_1' y_2 = \begin{vmatrix} y_1 & y_2 \\ y_1' & y_2' \end{vmatrix}$$
The value of $W(y_1, y_2)$ at a point x is denoted by $W(y_1, y_2)(x)$ or often more simply by $W(x)$. It can be proved that two solutions, $y_1 = y_1(x)$ and $y_2 = y_2(x)$, of Equation (2) are linearly dependent if and only if $W(x) = 0$ for all x. Equivalently, the functions are linearly independent if and only if $W(x) \neq 0$ for at least one value of x. Use this result to prove that the following solutions of Equation (4) are linearly independent:

    (a) $y_1 = e^{m_1 x}$, $y_2 = e^{m_2 x}$ ($m_1 \neq m_2$)

    (b) $y_1 = e^{mx}$, $y_2 = xe^{mx}$.

30. Prove: If the auxiliary equation of
$$y'' + py' + qy = 0$$
has complex roots $a + bi$ and $a - bi$, then the general solution of this differential equation is
$$y(x) = e^{ax}(c_1 \cos bx + c_2 \sin bx)$$
[*Hint:* By substitution, verify that $y_1 = e^{ax}\cos bx$ and $y_2 = e^{ax}\sin bx$ are solutions of the differential equation. Then use Exercise 29 to prove that y_1 and y_2 are linearly independent.]

31. Suppose that the auxiliary equation of the equation $y'' + py' + qy = 0$ has distinct real roots μ and m.

    (a) Show that the function
    $$g_\mu(x) = \frac{e^{\mu x} - e^{mx}}{\mu - m}$$
    is a solution of the differential equation.

(b)   Use L'Hôpital's rule to show that

$$\lim_{\mu \to m} g_\mu(x) = xe^{mx}$$

[*Note:* Can you see how the result in part (b) makes it plausible that the function $y(x) = xe^{mx}$ is a solution of $y'' + py' + qy = 0$ when m is a repeated root of the auxiliary equation?]

32. Consider the problem of solving the differential equation

$$y'' + \lambda y = 0$$

subject to the conditions $y(0) = 0$, $y(\pi) = 0$.

(a)   Show that if $\lambda \leq 0$, then $y = 0$ is the only solution.

(b)   Show that if $\lambda > 0$, then the solution is

$$y = c \sin \sqrt{\lambda} x$$

where c is an arbitrary constant, if

$$\lambda = 1, 2^2, 3^2, 4^2, \ldots$$

and the only solution is $y = 0$ otherwise.

19.2 SECOND-ORDER LINEAR NONHOMOGENEOUS DIFFERENTIAL EQUATIONS WITH CONSTANT COEFFICIENTS; UNDETERMINED COEFFICIENTS

In this section we shall study techniques for solving second-order linear differential equations that are not homogeneous.

☐ THE COMPLEMENTARY EQUATION

The following theorem is the key result for solving a second-order *nonhomogeneous* linear differential equation with constant coefficients

$$y'' + py' + qy = r(x) \tag{1}$$

where p and q are constants and $r(x)$ is a continuous function of x.

19.2.1   THEOREM.   *The general solution of* (1) *is*

$$y(x) = c_1 y_1(x) + c_2 y_2(x) + y_p(x)$$

where $c_1 y_1(x) + c_2 y_2(x)$ is the general solution of the homogeneous equation

$$y'' + py' + qy = 0 \tag{2}$$

and $y_p(x)$ is any solution of (1).

The proof is deferred to the end of the section.

REMARK.   Equation (2) is called the ***complementary equation*** to (1) and $y_p(x)$ is called a ***particular solution*** of (1). Thus, Theorem 19.2.1 states that *the general solution of* (1) *is obtained by adding a particular solution of* (1) *to the general solution of the complementary equation.*

☐ THE METHOD OF UNDETERMINED COEFFICIENTS

Since we already know how to obtain the general solution of (2), we shall focus our attention on the problem of obtaining a particular solution of (1). In this section we shall discuss a procedure for doing this called the method of ***undetermined coefficients***, and in the next section we shall discuss a second procedure called *variation of parameters*.

The first step in the method of undetermined coefficients is to make a reasonable guess about the form of the particular solution. This guess will involve one or more unknown coefficients that we shall then determine from conditions obtained by substituting the proposed solution into the differential equation. We shall begin with some examples that illustrate the basic idea.

Example 1 Find a particular solution of the differential equation

$$y'' + 2y' - 8y = e^{3x} \tag{3}$$

Solution. It should be clear that such functions as $\sin x$, $\cos x$, $\ln x$, or x^3 are not reasonable possibilities for $y_p(x)$ because the derivatives of such functions do not yield expressions involving e^{3x}. Since the obvious choice for $y_p(x)$ is an expression involving e^{3x}, we shall *guess* that $y_p(x)$ has the form

$$y_p(x) = Ae^{3x} \tag{4}$$

where A is an unknown constant to be determined. It follows that

$$y_p'(x) = 3Ae^{3x} \tag{5}$$

$$y_p''(x) = 9Ae^{3x} \tag{6}$$

Substituting expressions (4), (5), and (6) for y, y', and y'' in (3) yields

$$9Ae^{3x} + 2(3Ae^{3x}) - 8(Ae^{3x}) = e^{3x}$$

or

$$7Ae^{3x} = e^{3x}$$

Thus, $A = \frac{1}{7}$. From this result and (4), a particular solution of (3) is

$$y_p(x) = \tfrac{1}{7}e^{3x} \qquad \blacktriangleleft$$

Example 2 Find a particular solution of the differential equation

$$y'' - y' - 6y = e^{3x} \tag{7}$$

Solution. As in Example 1, a reasonable guess is $y_p(x) = Ae^{3x}$. However, if we substitute (4), (5), and (6) in (7), we obtain

$$9Ae^{3x} - 3Ae^{3x} - 6Ae^{3x} = e^{3x} \quad \text{or} \quad 0 = e^{3x}$$

which is a contradiction. The problem here is that $y_p(x) = Ae^{3x}$ is a solution of the complementary equation

$$y'' - y' - 6y = 0$$

which makes it impossible for y_p to satisfy (7), no matter how A is selected. How should we proceed in this case? Experience has shown that multiplying Ae^{3x} by x produces the correct form for the solution. Thus, we shall try

$$y_p(x) = Axe^{3x} \tag{8}$$

It follows that

$$y_p'(x) = 3Axe^{3x} + Ae^{3x} \tag{9}$$

$$y_p''(x) = 9Axe^{3x} + 6Ae^{3x} \tag{10}$$

Substituting expressions (8), (9), and (10) for y, y', and y'' in (7) yields

$$(9Axe^{3x} + 6Ae^{3x}) - (3Axe^{3x} + Ae^{3x}) - 6(Axe^{3x}) = e^{3x}$$

or

$$5Ae^{3x} = e^{3x}$$

Thus, $A = \frac{1}{5}$, so (8) implies that a particular solution of (7) is

$$y_p(x) = \tfrac{1}{5}xe^{3x} \qquad \blacktriangleleft$$

REMARK. Had it turned out that Axe^{3x} was also a solution of the complementary equation, then we would have tried $y_p(x) = Ax^2e^{3x}$ as our candidate for a particular solution.

☐ **SUMMARY**

In summary, a particular solution of an equation of the form

$$y'' + py' + qy = ke^{ax}$$

can be obtained as follows:

> **Step 1.** Start with $y_p = Ae^{ax}$ as an initial guess.
>
> **Step 2.** Determine if the initial guess is a solution of the complementary equation $y'' + py' + qy = 0$.
>
> **Step 3.** If the initial guess is not a solution of the complementary equation, then $y_p = Ae^{ax}$ is the correct form of a particular solution.
>
> **Step 4.** If the initial guess is a solution of the complementary equation, then multiply it by the smallest positive integer power of x required to produce a function that is not a solution of the complementary equation. This will yield either $y_p = Axe^{ax}$ or $y_p = Ax^2e^{ax}$.

Table 19.2.1, which we provide without proof, restates the preceding procedure more compactly and also explains how to find a particular solution of (1) when $r(x)$ is a polynomial or a combination of sine and cosine functions.

Table 19.2.1

EQUATION	INITIAL GUESS FOR y_p
$y'' + py' + qy = ke^{ax}$	$y_p = Ae^{ax}$
$y'' + py' + qy = a_0 + a_1x + \cdots + a_nx^n$	$y_p = A_0 + A_1x + \cdots + A_nx^n$
$y'' + py' + qy = a_1 \cos bx + a_2 \sin bx$	$y_p = A_1 \cos bx + A_2 \sin bx$

MODIFICATION RULE

If any term in the initial guess is a solution of the complementary equation, then the correct form for y_p is obtained by multiplying the initial guess by the smallest positive integer power of x required so that no term is a solution of the complementary equation.

Example 3 Find the general solution of

$$y'' + y' = 4x^2 \tag{11}$$

Solution. We begin by finding the general solution of the complementary equation

$$y'' + y' = 0$$

The auxiliary equation for this homogeneous equation is

$$m^2 + m = 0$$

which has roots $m_1 = 0$, $m_2 = -1$. Thus, the general solution, $y_c(x)$, of the complementary equation is

$$y_c(x) = c_1e^{0x} + c_2e^{-x} = c_1 + c_2e^{-x}$$

Since the right-hand side of (11) is a second-degree polynomial, our initial guess for y_p is

$$y_p = A_0 + A_1x + A_2x^2$$

But A_0 is a solution of the complementary equation (take $c_1 = A_0$, $c_2 = 0$), so our second guess for y_p is

$$y_p = x(A_0 + A_1 x + A_2 x^2) = A_0 x + A_1 x^2 + A_2 x^3 \tag{12}$$

Since no term of this function is a solution of the complementary equation, this is the correct form for y_p. Differentiating (12) we obtain

$$y_p'(x) = A_0 + 2A_1 x + 3A_2 x^2 \tag{13}$$

$$y_p''(x) = 2A_1 + 6A_2 x \tag{14}$$

Substituting (13) and (14) in (11) yields

$$(2A_1 + 6A_2 x) + (A_0 + 2A_1 x + 3A_2 x^2) = 4x^2$$

or

$$(A_0 + 2A_1) + (2A_1 + 6A_2)x + 3A_2 x^2 = 4x^2$$

To solve for A_0, A_1, and A_2, we equate corresponding coefficients on the two sides of this equation; this yields

$$A_0 + 2A_1 = 0$$

$$2A_1 + 6A_2 = 0$$

$$3A_2 = 4$$

from which we obtain

$$A_2 = \tfrac{4}{3}, \quad A_1 = -4, \quad A_0 = 8$$

Substituting these values in (12) yields the particular solution

$$y_p(x) = 8x - 4x^2 + \tfrac{4}{3}x^3$$

Thus, the general solution of (11) is

$$y(x) = y_c(x) + y_p(x) = c_1 + c_2 e^{-x} + 8x - 4x^2 + \tfrac{4}{3}x^3 \quad \blacktriangleleft$$

Example 4   Find the general solution of

$$y'' - 2y' + y = \sin 2x \tag{15}$$

Solution.   We begin by finding the general solution of the complementary equation

$$y'' - 2y' + y = 0$$

The auxiliary equation for this homogeneous equation is

$$m^2 - 2m + 1 = 0$$

which has the repeated root $m_1 = 1$, $m_2 = 1$. Thus, the general solution of the complementary equation is

$$y_c(x) = c_1 e^x + c_2 x e^x \tag{16}$$

Since $\sin 2x$ has the form $a_1 \cos bx + a_2 \sin bx$ ($a_1 = 0$, $a_2 = 1$, $b = 2$), our initial guess for y_p is

$$y_p(x) = A_1 \cos 2x + A_2 \sin 2x \tag{17}$$

Since no term of y_p is a solution of the complementary equation [see (16)], this is the correct form for a particular solution. Differentiating (17) we obtain

$$y_p'(x) = -2A_1 \sin 2x + 2A_2 \cos 2x \tag{18}$$

$$y_p''(x) = -4A_1 \cos 2x - 4A_2 \sin 2x \tag{19}$$

Substituting (17), (18), and (19) in (15) yields

$$(-4A_1 \cos 2x - 4A_2 \sin 2x) - 2(-2A_1 \sin 2x + 2A_2 \cos 2x)$$
$$+ (A_1 \cos 2x + A_2 \sin 2x) = \sin 2x$$

or

$$(-3A_1 - 4A_2) \cos 2x + (4A_1 - 3A_2) \sin 2x = \sin 2x$$

To solve for A_1 and A_2 we equate the coefficients of $\sin 2x$ and $\cos 2x$ on the two sides of this equation; this yields

$$-3A_1 - 4A_2 = 0$$
$$4A_1 - 3A_2 = 1$$

from which we obtain

$$A_1 = \tfrac{4}{25}, \quad A_2 = -\tfrac{3}{25}$$

(Verify.) From this result and (17), a particular solution of (15) is

$$y_p(x) = \tfrac{4}{25} \cos 2x - \tfrac{3}{25} \sin 2x$$

Thus, the general solution of (15) is

$$y(x) = y_c(x) + y_p(x) = c_1 e^x + c_2 x e^x + \tfrac{4}{25} \cos 2x - \tfrac{3}{25} \sin 2x \qquad \blacktriangleleft$$

We conclude this section with a proof of Theorem 19.2.1.

■ PROOF

Proof of Theorem 19.2.1. To prove that

$$y(x) = c_1 y_1(x) + c_2 y_2(x) + y_p(x) \tag{20}$$

is the general solution of (1) we must prove two results: first, that for all choices of c_1 and c_2 the function $y(x)$ defined by (20) satisfies (1); and second, that every solution of (1) can be obtained from (20) by choosing appropriate values for the constants c_1 and c_2.

To prove the first statement, let c_1 and c_2 have any real values. Then differentiating (20) yields

$$y'(x) = c_1 y_1'(x) + c_2 y_2'(x) + y_p'(x)$$
$$y''(x) = c_1 y_1''(x) + c_2 y_2''(x) + y_p''(x)$$

so

$$y''(x) + p y'(x) + q y(x) = c_1 \big(y_1''(x) + p y_1'(x) + q y_1(x) \big)$$
$$+ c_2 \big(y_2''(x) + p y_2'(x) + q y_2(x) \big)$$
$$+ y_p''(x) + p y_p'(x) + q y_p(x) \tag{21}$$

But $y_1(x)$ and $y_2(x)$ satisfy (2) and $y_p(x)$ satisfies (1), so (21) reduces to

$$y''(x) + p y'(x) + q y(x) = 0 + 0 + r(x) = r(x)$$

which proves that $y(x)$ satisfies (1).

To prove the second statement, let $y(x)$ be any solution of (1). We must find values of c_1 and c_2 such that

$$y(x) = c_1 y_1(x) + c_2 y_2(x) + y_p(x) \tag{22}$$

But $y(x) - y_p(x)$ satisfies (2) since

$$(y(x) - y_p(x))'' + p(y(x) - y_p(x))' + q(y(x) - y_p(x))$$
$$= [y''(x) + py'(x) + qy(x)] - [y_p''(x) + py_p'(x) + qy_p(x)]$$
$$= r(x) - r(x) = 0$$

Thus, since $c_1 y_1(x) + c_2 y_2(x)$ is the general solution of (2), there exist values of c_1 and c_2 such that the solution $y(x) - y_p(x)$ can be written as

$$y(x) - y_p(x) = c_1 y_1(x) + c_2 y_2(x)$$

from which (22) follows.   ∎

▶ Exercise Set 19.2

In Exercises 1–24, use the method of undetermined coefficients to find the general solution of the equation.

1. $y'' + 6y' + 5y = 2e^{3x}$.　**2.** $y'' + 3y' - 4y = 5e^{7x}$.

3. $y'' - 9y' + 20y = -3e^{5x}$.

4. $y'' + 7y' - 8y = 7e^x$.　**5.** $y'' + 2y' + y = e^{-x}$.

6. $y'' + 4y' + 4y = 4e^{-2x}$.　**7.** $y'' + y' - 12y = 4x^2$.

8. $y'' - 4y' - 5y = -6x^2$.　**9.** $y'' - 6y' = x - 1$.

10. $y'' + 3y' = 2x + 2$.　**11.** $y'' - x^3 + 1 = 0$.

12. $y'' + 3x^3 + x = 0$.　**13.** $y'' - y' - 2y = 10 \cos x$.

14. $y'' - 3y' - 4y = 2 \sin x$.

15. $y'' - 4y = 2 \sin 2x + 3 \cos 2x$.

16. $y'' - 9y = \cos 3x - \sin 3x$.

17. $y'' + y = \sin x$.　**18.** $y'' + 4y = \cos 2x$.

19. $y'' - 3y' + 2y = x$.　**20.** $y'' + 4y' + 4y = 3x + 3$.

21. $y'' + 4y' + 9y = x^2 + 3x$.

22. $y'' - y = 1 + x + x^2$.　**23.** $y'' + 4y = \sin x \cos x$.

24. $y'' + 4y = \cos^2 x - \sin^2 x$.

25. (a) Prove: If $y_1(x)$ is a solution of
$$y'' + p(x)y' + q(x)y = r_1(x)$$
and $y_2(x)$ is a solution of
$$y'' + p(x)y' + q(x)y = r_2(x)$$
then $y_1(x) + y_2(x)$ is a solution of
$$y'' + p(x)y' + q(x)y = r_1(x) + r_2(x)$$

(b) Use the result in part (a) to find a particular solution of the equation
$$y'' + 3y' - 4y = x + e^x$$

(c) State a generalization of the result in part (a) that is applicable to the equation
$$y'' + p(x)y' + q(x)y = r_1(x) + r_2(x) + \cdots + r_n(x)$$

In Exercises 26–34, use the results in Exercise 25 to find the general solution of the differential equation.

26. $y'' - y' - 2y = x + e^{-x}$.　**27.** $y'' - y = 1 + e^x$.

28. $y'' - 4y' + 3y = 2 \cos x + 4 \sin x$.

29. $y'' + 4y = 1 + x + \sin x$.

30. $y'' + 2y' + y = 2 + 3x + 3e^x + 2 \cos 2x$.

31. $y'' - 2y' + y = \sinh x$. [*Hint:* $\sinh x = \frac{1}{2}(e^x - e^{-x})$.]

32. $y'' + 4y' - 5y = \cosh x$. [*Hint:* $\cosh x = \frac{1}{2}(e^x + e^{-x})$.]

33. $y'' + y = 12 \cos^2 x$. [*Hint:* $\cos^2 x = \frac{1}{2}(1 + \cos 2x)$.]

34. $y'' + 2y' + y = \sin^2 x$. [*Hint:* $\sin^2 x = \frac{1}{2}(1 - \cos 2x)$.]

35. (a) Find the general solution of
$$y'' + \mu^2 y = a \sin bx$$
where a is an arbitrary constant and μ and b are positive constants such that $\mu \neq b$.

(b) Use part (a) and Exercise 25 to find the general solution of
$$y'' + \mu^2 y = \sum_{k=1}^{n} a_k \sin k\pi x$$
where $\mu > 0$ and $\mu \neq k\pi$, $k = 1, 2, \ldots, n$.

36. Find the general solution of
$$y'' + \lambda^2 y = \sum_{k=1}^{n} a_k \cos k\pi x$$
where $\lambda > 0$ and $\lambda \neq k\pi$, $k = 1, 2, \ldots, n$. [*Hint:* See Exercise 35.]

37. Find all solutions of the equation
$$y'' - y' = 4 - 4x$$
(if any) with the property that $y'(x_0) = y''(x_0) = 0$ at some point x_0.

■ **19.3** VARIATION OF PARAMETERS

In this section* we shall consider an alternative method for finding a particular solution of a second-order linear nonhomogeneous equation that often works when the method of undetermined coefficients cannot be applied.

As in the preceding section, we shall be concerned with equations of the form

$$y'' + py' + qy = r(x) \tag{1}$$

where p and q are constants and $r(x)$ is a continuous function of x. The method we shall discuss, called *variation of parameters*, is based on the (not very obvious) fact that if

$$y_c(x) = c_1 y_1(x) + c_2 y_2(x) \tag{2}$$

is the general solution of the complementary equation

$$y'' + py' + qy = 0$$

then it is possible to find functions $u(x)$ and $v(x)$ such that

$$y_p(x) = u(x)y_1(x) + v(x)y_2(x) \tag{3}$$

is a particular solution of (1).

To see how the functions $u(x)$ and $v(x)$ can be found, consider the derivative of (3):

$$y_p' = (uy_1' + vy_2') + (u'y_1 + v'y_2) \tag{4}$$

If we now differentiate y_p', we shall introduce second derivatives of the unknown functions $u(x)$ and $v(x)$. To avoid this complication we shall require that $u(x)$ and $v(x)$ satisfy the condition

$$u'y_1 + v'y_2 = 0 \tag{5}$$

so (4) simplifies to

$$y_p' = uy_1' + vy_2' \tag{6}$$

Thus,

$$y_p'' = uy_1'' + u'y_1' + vy_2'' + v'y_2' \tag{7}$$

On substituting (3), (6), and (7) in (1) and rearranging terms we obtain

$$u(y_1'' + py_1' + qy_1) + v(y_2'' + py_2' + qy_2) + u'y_1' + v'y_2' = r(x) \tag{8}$$

Since y_1 and y_2 are solutions of the complementary equation, we have

$$y_1'' + py_1' + qy_1 = 0 \quad \text{and} \quad y_2'' + py_2' + qy_2 = 0$$

so (8) simplifies to

$$u'y_1' + v'y_2' = r(x)$$

This equation together with (5) yields two equations in the two unknowns $u'(x)$ and $v'(x)$:

$$\begin{aligned} u'y_1 + v'y_2 &= 0 \\ u'y_1' + v'y_2' &= r(x) \end{aligned} \tag{9}$$

---

*It is assumed in this section that the reader is familiar with Cramer's rule, which is discussed in Appendix D.

It can be shown (see Boyce and DiPrima, *Elementary Differential Equations and Boundary Value Problems*, John Wiley & Sons, New York, 1992) that this system has a unique solution for u' and v'. Once this solution is found, u and v can be obtained by integration.

Example 1   Find the general solution of

$$y'' + y = \sec x \qquad (-\pi/2 < x < \pi/2) \tag{10}$$

Solution.   We begin by finding the general solution of the complementary equation

$$y'' + y = 0$$

The auxiliary equation for this homogeneous equation is

$$m^2 + 1 = 0$$

which has roots $m_1 = i$ and $m_2 = -i$. Thus, the general solution of the complementary equation is

$$y_c(x) = c_1 \cos x + c_2 \sin x$$

Comparing this with (2) yields $y_1(x) = \cos x$ and $y_2(x) = \sin x$, so we shall look for a particular solution of the form

$$y_p(x) = u(x) \cos x + v(x) \sin x \tag{11}$$

Substituting $y_1 = \cos x$, $y_2 = \sin x$, and $r(x) = \sec x$ in (9) yields

$$u' \cos x + v' \sin x = 0$$
$$-u' \sin x + v' \cos x = \sec x \tag{12}$$

Solving this system for u' and v' by Cramer's rule (or any suitable method) we obtain

$$u' = \frac{\begin{vmatrix} 0 & \sin x \\ \sec x & \cos x \end{vmatrix}}{\begin{vmatrix} \cos x & \sin x \\ -\sin x & \cos x \end{vmatrix}} = \frac{-\sec x \sin x}{\cos^2 x + \sin^2 x} = \frac{-\tan x}{1} = -\tan x$$

$$v' = \frac{\begin{vmatrix} \cos x & 0 \\ -\sin x & \sec x \end{vmatrix}}{\begin{vmatrix} \cos x & \sin x \\ -\sin x & \cos x \end{vmatrix}} = \frac{\cos x \sec x}{\cos^2 x + \sin^2 x} = \frac{1}{1} = 1$$

Integrating u' and v' yields

$$u(x) = \int -\tan x \, dx = \ln |\cos x| \quad \text{and} \quad v(x) = \int 1 \, dx = x$$

[We set the constants of integration equal to zero because any functions $u(x)$ and $v(x)$ satisfying (12) will suffice.] Substituting these functions in (11) yields

$$y_p(x) = (\ln |\cos x|) \cos x + x \sin x$$

Thus, the general solution of (10) is

$$y(x) = y_c(x) + y_p(x) = c_1 \cos x + c_2 \sin x + (\ln |\cos x|) \cos x + x \sin x \qquad \blacktriangleleft$$

▶ **Exercise Set 19.3**

In Exercises 1–6, use variation of parameters to find the general solution of the differential equation and verify that the same solution results using undetermined coefficients.

1. $y'' + y = x^2$.

2. $y'' + 9y = 3x$.

3. $y'' + y' - 2y = 2e^x$.

4. $y'' + 5y' + 6y = e^{-x}$.

5. $y'' + 4y = \sin 2x$.

6. $y'' + 9y = \cos 3x$.

In Exercises 7–28, use variation of parameters to find the general solution of the differential equation.

7. $y'' + y = \tan x$.

8. $y'' + y = \cot x$.

9. $y'' - 2y' + y = e^x/x$.

10. $y'' - 4y' + 4y = e^{2x}/x$.

11. $y'' + y = 3\sin^2 x$.

12. $y'' + y = 6\cos^2 x$.

13. $y'' + y = \csc x$.

14. $y'' + 9y = 6\sec 3x$.

15. $y'' + y = \sec x \tan x$.

16. $y'' + y = \csc x \cot x$.

17. $y'' + 2y' + y = e^{-x}/x^2$.

18. $y'' - y = x^2 e^x$.

19. $y'' + 4y' + 4y = xe^{-x}$.

20. $y'' + 4y' + 4y = xe^{2x}$.

21. $y'' + y = \sec^2 x$.

22. $y'' + y = \sec^3 x$.

23. $y'' - 2y' + y = e^x/x^2$.

24. $y'' - 2y' + y = x^3 e^x$.

25. $y'' - y = e^x \cos x$.

26. $y'' - 2y' + 2y = e^{2x}\sin x$.

27. $y'' + 2y' + y = e^{-x}\ln|x|$.

28. $y'' - 3y' + 2y = \dfrac{e^x}{1 + e^x}$.

29. Use the method of variation of parameters to show that the general solution of $y'' + y = r(x)$ is

$$y = c_1 \cos x + c_2 \sin x + g(x)\cos x + h(x)\sin x$$

where

$$g(x) = -\int r(x)\sin x \, dx$$

$$h(x) = \int r(x)\cos x \, dx$$

30. Use the method of variation of parameters to show that if y_1 and y_2 are linearly independent solutions of the equation $y'' + py' + qy = 0$, then a particular solution of $y'' + py' + qy = r(x)$ is given by

$$y_p(x) = -y_1(x)\int \frac{y_2(x)r(x)}{W(x)}dx + y_2(x)\int \frac{y_1(x)r(x)}{W(x)}dx$$

where $W(x) = y_1(x)y_2'(x) - y_1'(x)y_2(x)$.

◼ **19.4** VIBRATION OF A SPRING

> *In this section we shall use second-order linear differential equations with constant coefficients to study the vibration of a spring.*

We shall be interested in solving the following problem:

> **19.4.1**   THE VIBRATING SPRING PROBLEM.   *As shown in* Figure 19.4.1, *let a mass attached to a vertical spring be allowed to settle into an equilibrium position. Then, let the spring be stretched (or compressed) by pulling (or pushing) the mass, and finally, let the mass be released, thereby causing it to undergo a vibratory motion. We shall be interested in finding a formula for the position of the mass at any time t.*

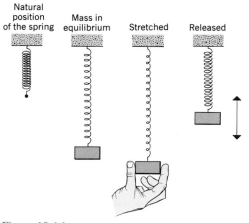

Figure 19.4.1

To solve this problem we shall need three results from physics:

> **19.4.2** HOOKE'S LAW. If a spring is stretched (or compressed) l units beyond its natural position, then it pulls back (or pushes) with a force of magnitude
>
> $$F = kl$$
>
> where k is a positive constant, called the **spring constant**. This constant depends on such factors as the thickness of the spring, the material from which it is made, and the units of force and length; it is measured in units of force per unit of length.

> **19.4.3** NEWTON'S SECOND LAW OF MOTION. If an object with mass M is subjected to a force **F**, then it undergoes an acceleration **a** satisfying
>
> $$\mathbf{F} = M\mathbf{a}$$

> **19.4.4** WEIGHT. The gravitational force exerted by the earth on an object is called the **weight** (more precisely, **earth weight**) of the object. It follows from Newton's Second Law of Motion that an object with mass M has a weight whose magnitude w is given by
>
> $$w = Mg \tag{1}$$
>
> where g is a constant, called the **acceleration due to gravity**. Near the surface of the earth an approximate value of g is $g \approx 32$ ft/sec^2 if length is measured in feet and time in seconds, or $g \approx 9.8$ m/sec^2 if length is measured in meters and time in seconds.

In any problem involving length, mass, time, and force it is important that the units of measurement be consistent. The most important systems of measurement are summarized in Table 19.4.1.

Table 19.4.1

SYSTEM OF MEASUREMENT	LENGTH	TIME	MASS	FORCE
British system	foot (ft)	second (sec)	slug	pound (lb)
SI system	meter (m)	second (sec)	kilogram (kg)	newton (N)
CGS system	centimeter (cm)	second (sec)	gram (g)	dyne

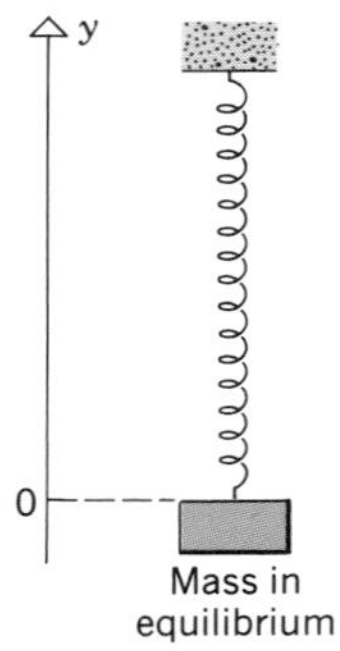

Mass in equilibrium

Figure 19.4.2

To solve the vibrating spring problem posed above, we introduce a coordinate axis (a y-axis) with the positive direction up and the origin at the bottom end of the spring when the mass is in its equilibrium position (Figure 19.4.2). If we take $t = 0$ to be the time at which the mass is released, then at each subsequent time t, the end of the spring has a position $y(t)$, a velocity $y'(t)$, and an acceleration $y''(t)$.* Because the positive direction of the y-axis is up, force, velocity, and acceleration are positive when directed up and negative when directed down.

Before we can solve the spring problem, it will be necessary to derive a preliminary result from the equilibrium conditions for the mass. Let us assume that the spring constant is k, the mass of the object is M, and the spring is stretched l units beyond its natural length when the mass is in equilibrium (Figure 19.4.3). When the mass is in its equilibrium

---

*For convenience of terminology we shall refer to $y(t)$, $y'(t)$, and $y''(t)$ as the position, velocity, and acceleration of the mass, respectively.

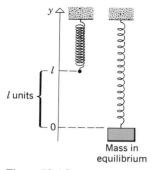

Figure 19.4.3

position its downward weight, $-Mg$, is balanced exactly by the upward force, kl, of the spring, so the sum of these forces must be zero;* that is,

$$kl - Mg = 0$$

or

$$kl = Mg \tag{2}$$

The basic strategy for solving the spring problem is to find the total force $F(t)$ that acts on the mass at time t. Then, since the acceleration of the mass at time t is $y''(t)$, it will follow from Newton's Second Law of Motion that

$$My''(t) = F(t)$$

or

$$y''(t) = F(t)/M \tag{3}$$

which is a second-order linear differential equation that can be solved for the position function $y(t)$.

At each instant, the force $F(t)$ in (3) consists of four possible components:

$F_g(t)$ = the force of gravity (weight of the mass)

$F_s(t)$ = the force of the spring

$F_d(t)$ = the ***damping force*** or frictional force exerted on the mass by the surrounding medium (air, water, oil, etc.)

$F_e(t)$ = external forces due to such factors as movement in the spring support, magnetic forces acting on the mass, and so on

□ UNDAMPED FREE VIBRATIONS

The simplest case occurs when there is no damping ($F_d = 0$) and the mass is *free* of external forces ($F_e = 0$). In this case the only forces acting on the mass are the force of gravity, F_g, and the spring force, F_s. Let us try to calculate these forces when the mass is at an arbitrary point $y(t)$.

When the mass is at the point $y(t)$, the spring is stretched (or compressed) $l - y(t)$ units from its natural length (Figure 19.4.4), so by Hooke's law the spring exerts a force of

$$F_s(t) = k(l - y(t))$$

on the mass. Adding this to the force of gravity,

$$F_g(t) = -Mg$$

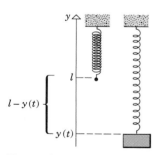

Figure 19.4.4

acting on the mass yields the total force acting on the mass:

$$F_s(t) + F_g(t) = k(l - y(t)) - Mg$$

But $kl = Mg$ from (2), so

$$F_s(t) + F_g(t) = -ky(t) \tag{4}$$

Substituting this expression for $F(t)$ in (3) yields

$$y''(t) = -\frac{k}{M}y(t)$$

or

$$y''(t) + \frac{k}{M}y(t) = 0 \tag{5}$$

which is a second-order linear differential equation with constant coefficients.

---

*We shall assume that the weight of the spring is small relative to the weight of the mass and can be neglected.

Because the mass is *released* (i.e., has zero initial velocity) at time $t = 0$, we have $y'(0) = 0$. If we assume, in addition, that the position of the mass at time $t = 0$ is $y(0) = y_0$, then we have two initial conditions that can be combined with (5) to yield an initial-value problem for $y(t)$:

$$y'' + \frac{k}{M} y = 0$$
$$y(0) = y_0, \quad y'(0) = 0$$

(6)

The auxiliary equation for the differential equation in (6) is

$$m^2 + \frac{k}{M} = 0$$

which has roots $m_1 = \sqrt{k/M}\, i$, $m_2 = -\sqrt{k/M}\, i$ (since k and M are positive), so the general solution of the differential equation is

$$y(t) = c_1 \cos\left(\sqrt{k/M}\, t\right) + c_2 \sin\left(\sqrt{k/M}\, t\right)$$

From the initial conditions $y(0) = y_0$ and $y'(0) = 0$ it follows that $c_1 = y_0$, $c_2 = 0$ (verify), so the solution of (6) is

$$y(t) = y_0 \cos\left(\sqrt{k/M}\, t\right)$$

(7)

This formula describes a periodic vibration with an **amplitude** of $|y_0|$ and a **period T** given by

$$T = \frac{2\pi}{\sqrt{k/M}} = 2\pi\sqrt{M/k}$$

(8)

Physically, the period is the time in seconds required for one complete oscillation (Figure 19.4.5). The reciprocal of the period is called the **frequency** of the oscillation. In the SI system the unit of frequency is the **hertz** (Hz); that is, 1 hertz = 1 Hz = 1 oscillation per second. Thus the frequency f is given by

$$f = \frac{1}{T} = \frac{1}{2\pi}\sqrt{k/M}$$

(9)

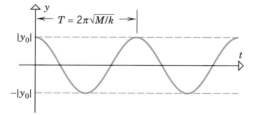

Figure 19.4.5

REMARK.   In the preceding problem, the spring oscillates indefinitely with constant amplitude because there are no frictional forces to dissipate the energy in the mass–spring system.

Example 1   Suppose that the top of a spring is fixed to a ceiling and a mass attached to the bottom end stretches the spring 0.2 m. If the mass is pulled 0.5 m below its equilibrium position and released, find

(a)   a formula for the position of the mass at any time t;

(b)   the amplitude, period, and frequency of the motion.

Solution (a).   The appropriate formula is (7). Although we are not given the mass M or the spring constant k, it does not matter because we are given that the mass stretches the spring $l = 0.2$ m and we know that $g = 9.8$ m/sec^2, so (2) implies that

$$\frac{k}{M} = \frac{g}{l} = \frac{9.8}{0.2} = 49 \text{ sec}^{-2}$$

Also, the mass is initially 0.5 m *below* its equilibrium position, so $y_0 = -0.5$. Therefore, from (7) the formula for the position of the mass at time t is

$$y(t) = -0.5 \cos 7t$$

Solution (b).   From (8) and (9)

$$\text{amplitude} = |y_0| = |-0.5| = 0.5 \text{ m}$$
$$\text{period} = T = 2\pi\sqrt{M/k} = 2\pi\sqrt{1/49} = 2\pi/7 \text{ sec}$$
$$\text{frequency} = f = 1/T = 7/(2\pi) \text{ Hz} \quad \blacktriangleleft$$

☐ **DAMPED FREE VIBRATIONS**

We shall now consider the solution of the spring problem in the case where damping cannot be neglected; that is, $F_d \neq 0$.

Physicists have shown that under appropriate conditions the damping force F_d is opposite to the direction of motion of the mass (i.e., tends to slow the mass down) and has a magnitude that is proportional to the speed of the mass (the greater the speed, the greater the effect of friction). This type of damping force is described by an equation of the form

$$F_d(t) = -cy'(t) \tag{10}$$

where c is a positive constant, called the ***damping constant***. The damping constant depends on the viscosity of the surrounding medium; it is measured in units of force per unit of velocity (lb/ft/sec or, equivalently, lb · sec/ft, for example).

It follows from (4) and (10) that

$$F_s(t) + F_g(t) + F_d(t) = -ky(t) - cy'(t)$$

If we substitute this expression in (3) for $F(t)$, we obtain

$$y''(t) = -\frac{k}{M}y(t) - \frac{c}{M}y'(t)$$

or

$$y''(t) + \frac{c}{M}y'(t) + \frac{k}{M}y(t) = 0$$

Combining this equation with the initial conditions $y(0) = y_0$, $y'(0) = 0$ yields the following initial-value problem whose solution describes the motion of the mass subject to damping:

$$y'' + \frac{c}{M}y' + \frac{k}{M}y = 0$$
$$y(0) = y_0, \quad y'(0) = 0 \tag{11}$$

The form of the solution to (11) will depend on whether the auxiliary equation

$$m^2 + \frac{c}{M}m + \frac{k}{M} = 0 \tag{12}$$

has distinct real roots, equal real roots, or complex roots. We leave it for the reader to show that the roots are distinct and real if $c^2 > 4kM$, equal and real if $c^2 = 4kM$, and complex if $c^2 < 4kM$ (Exercise 22).

The cases $c^2 > 4kM$ and $c^2 = 4kM$ are called ***overdamped*** and ***critically damped***, respectively. In these cases vibration in the usual sense does not occur. We shall leave the analysis of these cases for the exercises and concentrate on the case $c^2 < 4kM$ in which true vibratory motion occurs. This is called the ***underdamped case***.

In the underdamped case the roots of the auxiliary equation in (12) are

$$m_1 = -\frac{c}{2M} + \frac{\sqrt{4Mk - c^2}}{2M}\, i, \quad m_2 = -\frac{c}{2M} - \frac{\sqrt{4Mk - c^2}}{2M}\, i$$

(Verify.) For convenience, let

$$\alpha = \frac{c}{2M}, \quad \beta = \frac{\sqrt{4Mk - c^2}}{2M} \tag{13}$$

so the solution of the differential equation in (11) is

$$y(t) = e^{-\alpha t}(c_1 \cos \beta t + c_2 \sin \beta t) \tag{14}$$

It follows that

$$y'(t) = -\alpha e^{-\alpha t}(c_1 \cos \beta t + c_2 \sin \beta t) + \beta e^{-\alpha t}(-c_1 \sin \beta t + c_2 \cos \beta t)$$

so the initial conditions $y(0) = y_0$, $y'(0) = 0$ yield the equations

$$c_1 = y_0$$
$$-\alpha c_1 + \beta c_2 = 0$$

which can be solved to obtain

$$c_1 = y_0, \quad c_2 = \alpha y_0/\beta$$

Substituting these values in (14) yields the solution of (11):

$$y(t) = \frac{y_0}{\beta} e^{-\alpha t}(\beta \cos \beta t + \alpha \sin \beta t) \tag{15}$$

In the exercises (Exercise 23) we ask the reader to use the cosine addition formula to show that this solution can be rewritten in the alternative form

$$y(t) = \frac{y_0 \sqrt{\alpha^2 + \beta^2}}{\beta} e^{-\alpha t} \cos(\beta t - \omega) \tag{16a}$$

where

$$\omega = \tan^{-1}\left(\frac{\alpha}{\beta}\right), \quad 0 < \omega < \pi/2 \tag{16b}$$

Since $\cos(\beta t - \omega)$ has values between $+1$ and -1, it follows from (16a) and (16b) that the graph of $y(t)$ oscillates between the curves

$$y = \frac{y_0 \sqrt{\alpha^2 + \beta^2}}{\beta} e^{-\alpha t} \quad \text{and} \quad y = -\frac{y_0 \sqrt{\alpha^2 + \beta^2}}{\beta} e^{-\alpha t}$$

Thus, the graph of $y(t)$ resembles a cosine curve, but with decreasing amplitude (Figure 19.4.6). Strictly speaking, the function $y(t)$ is not periodic. However, $\cos(\beta t - \omega)$ has a period of $2\pi/\beta$, so the displacement $y(t)$ reaches a relative maximum at times spaced $2\pi/\beta$ units apart. Thus, we shall *define* the ***period*** T and ***frequency*** f of the motion to be (Figure 19.4.6)

$$T = \frac{2\pi}{\beta} = \frac{4M\pi}{\sqrt{4Mk - c^2}} \quad \text{and} \quad f = \frac{1}{T} = \frac{\sqrt{4Mk - c^2}}{4M\pi} \tag{17-18}$$

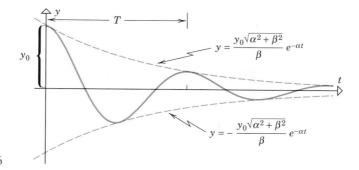

Figure 19.4.6

If we think of a relative maximum as marking the end of one cycle of motion and the start of the next, then the period T is the time required for the completion of one cycle, and the frequency is the number of cycles that occur per unit time. The physically significant fact is that the frequency and period of the mass–spring system do not change as the amplitudes of the vibrations get smaller (why?). Galileo, who first observed this fact, used it to help design clocks.

Example 2 A spring with a spring constant of $k = 4$ lb/ft is attached to a ceiling and a 64-lb weight is attached to the bottom end. The weight is pushed 2 ft above its equilibrium position and released with an initial velocity of 0. Assuming that the damping constant due to air resistance is $c = 4$ lb · sec/ft, find

(a) a formula for the position of the weight at any time t;

(b) the period and frequency of the vibration.

Solution (a).   We shall first use (13) to find α and β, after which we can use either (15) or (16a) and (16b). The attached weight is 64 lb, so from (1) its mass is

$$M = \frac{w}{g} = \frac{64}{32} = 2 \text{ slugs}$$

Thus,

$$\alpha = \frac{c}{2M} = \frac{4}{4} = 1, \quad \beta = \frac{\sqrt{4Mk - c^2}}{2M} = \frac{4}{4} = 1$$

Since the weight is initially 2 ft above its equilibrium position, we have $y_0 = 2$, so from (15) the position function of the weight is

$$y(t) = 2e^{-t}(\cos t + \sin t)$$

Alternatively, we can apply (16a) and (16b), but we must first calculate ω. Since

$$\omega = \tan^{-1}\left(\frac{\alpha}{\beta}\right) = \tan^{-1}(1) = \frac{\pi}{4}$$

(16a) and (16b) yield the alternative formula

$$y(t) = 2\sqrt{2}e^{-t}\cos\left(t - \frac{\pi}{4}\right)$$

Solution (b).   From (17) and (18)

$$T = \frac{2\pi}{\beta} = 2\pi \text{ sec}$$

$$f = \frac{1}{T} = \frac{1}{2\pi} \text{ Hz} \qquad \blacktriangleleft$$

☐ **FORCED VIBRATIONS**

If there are external forces such as movement of the spring support or magnetic forces acting on the mass, then $F_e \neq 0$ and the vibrations are said to be *forced*. The study of forced vibrations leads to important phenomena such as *resonance* and *beats*. These ideas are touched on in the Technology Exercises. Readers interested in this topic are referred to *Elementary Differential Equations and Boundary Value Problems*, John Wiley & Sons, New York, 1992, by William E. Boyce and Richard C. DiPrima.

▶ Exercise Set 19.4

In this exercise set assume that the y-axis is oriented as in Figure 19.4.2.

1. A mass of 1 kg is attached to a vertical spring with spring constant $k = 0.25$ N/m. The mass is pushed 0.3 m above its equilibrium position and released.
   (a) Find an initial-value problem whose solution $y(t)$ is the position function of the mass, assuming that there is no damping.
   (b) Solve the initial-value problem.
   (c) Check the solution in part (b) using Formula (7).

2. A weight of 64 lb is attached to a vertical spring with spring constant $k = 8$ lb/ft. The weight is pushed 1 ft above its equilibrium position and released.
   (a) Find an initial-value problem whose solution $y(t)$ is the position function of the weight, assuming that there is no damping.
   (b) Solve the initial-value problem.
   (c) Check the solution in part (b) using Formula (7).

3. A mass attached to a vertical spring stretches the spring 0.05 m. The mass is pulled 0.12 m below its equilibrium position and released.
   (a) Find an initial-value problem whose solution $y(t)$ is the position function of the mass, assuming that there is no damping.
   (b) Solve the initial-value problem.
   (c) Check the solution in part (b) using Formula (7).

4. A mass attached to a vertical spring stretches the spring 0.2 m. The mass is pulled 0.1 m below its equilibrium position and released.
   (a) Find an initial-value problem whose solution $y(t)$ is the position function of the mass, assuming that there is no damping.
   (b) Solve the initial-value problem.
   (c) Check the solution in part (b) using Formula (7).

5. A mass of 0.5 kg is attached to a vertical spring with spring constant $k = 32$ N/m. The mass is pushed 0.2 m above its equilibrium position and released. Use Formulas (7), (8), and (9) to find

   (a) the position function of the weight
   (b) the amplitude of the vibration
   (c) the period of the vibration
   (d) the frequency of the vibration.

6. A mass of 2 slugs is attached to a vertical spring with spring constant $k = 4$ lb/ft. The mass is pushed 1 ft above its equilibrium position and released. Use Formulas (7), (8), and (9) to find
   (a) the position function of the mass
   (b) the amplitude of the vibration
   (c) the period of the vibration
   (d) the frequency of the vibration.

7. A mass attached to a vertical spring stretches the spring 1 in. The mass is pulled 3 in. below its equilibrium position and released. Use Formulas (7), (8), and (9) to find
   (a) the position function of the mass
   (b) the amplitude of the vibration
   (c) the period of the vibration
   (d) the frequency of the vibration.

8. A mass attached to a vertical spring stretches the spring 0.8 m. The mass is pulled 0.2 m below its equilibrium position and released. Use Formulas (7), (8), and (9) to find
   (a) the position function of the mass
   (b) the amplitude of the vibration
   (c) the period of the vibration
   (d) the frequency of the vibration.

9. A weight of 32 lb is attached to a vertical spring with spring constant $k = 8$ lb/ft. The surrounding medium has a damping constant of $c = 4$ lb · sec/ft. The weight is pulled 3 ft below its equilibrium position and released.
   (a) Find an initial-value problem whose solution $y(t)$ is the position function of the weight.
   (b) Solve the initial-value problem.
   (c) Check the solution in part (b) using Formula (15).
   (d) Express the solution in the form of (16a).
   (e) Find the period of the vibration.
   (f) Find the frequency of the vibration.

10. A mass of 3 slugs is attached to a vertical spring with spring constant $k = 9$ lb/ft. The surrounding medium has a damping constant of $c = 6$ lb · sec/ft. The mass is pulled 1 ft below its equilibrium position and released.

    (a) Find an initial-value problem whose solution $y(t)$ is the position function of the mass.

    (b) Solve the initial-value problem.

    (c) Check the solution in part (b) using Formula (15).

    (d) Express the solution in the form of (16a).

    (e) Find the period of the vibration.

    (f) Find the frequency of the vibration.

11. A mass of 25 g is attached to a vertical spring with spring constant $k = 3$ dynes/cm. The surrounding medium has a damping constant of $c = 10$ dynes · sec/cm. The mass is pushed 5 cm above its equilibrium position and released.

    (a) Use Formula (16a) to find the position function of the mass.

    (b) Find the period of the vibration.

    (c) Find the frequency of the vibration.

12. A mass of 490 g is attached to a vertical spring with spring constant $k = 40$ dynes/cm. The surrounding medium has a damping constant of $c = 140$ dynes · sec/cm. The mass is pushed 20 cm above its equilibrium position and released.

    (a) Use Formula (16a) to find the position function of the mass.

    (b) Find the period of the vibration.

    (c) Find the frequency of the vibration.

> If the object in Figure 19.4.1 is given an initial velocity v_0 rather than being released with initial velocity 0, then in Formulas (6) and (11) the initial conditions become $y(0) = y_0$, $y'(0) = v_0$. Use this fact in Exercises 13–16.

13. A weight of 3 lb attached to a vertical spring stretches the spring $\frac{1}{2}$ ft. While in its equilibrium position, the weight is struck to give it a downward initial velocity of 2 ft/sec. Assuming that there is no damping, find

    (a) the position function of the weight

    (b) the amplitude of the vibration

    (c) the period of the vibration

    (d) the frequency of the vibration.

14. A mass of 64 slugs attached to a vertical spring stretches the spring $1\frac{1}{2}$ in. The mass is pulled 4 in. below its equilibrium position and struck to give it a downward initial velocity of 8 ft/sec. Assuming that there is no damping, find

    (a) the position function of the weight

    (b) the amplitude of the vibration

    (c) the period of the vibration

    (d) the frequency of the vibration.

15. A weight of 4 lb is attached to a vertical spring with spring constant $k = 6\frac{1}{4}$ lb/ft. The weight is pushed $\frac{1}{3}$ ft above its equilibrium position and struck to give it a downward initial velocity of 5 ft/sec. Assuming that the damping constant is $c = \frac{1}{4}$ lb · sec/ft, find the position function of the weight.

16. A weight of 49 dynes is attached to a vertical spring with spring constant $k = \frac{1}{4}$ dyne/cm. The weight is pushed 1 cm above its equilibrium position and struck to give it an upward initial velocity of 2 cm/sec. Assuming that the damping constant is $c = \frac{1}{5}$ dyne · sec/cm, find the position function of the weight.

17. A weight of w pounds is attached to a spring, then pulled below its equilibrium position and released, thereby causing it to vibrate with a period of 3 sec. When 4 additional pounds are added, the period becomes 5 sec. Assuming there is no damping, find

    (a) the spring constant k (b) the weight w.

18. As illustrated in the following figure, let a toy cart of mass M be attached to a wall by a spring with spring constant k, and let an x-axis be introduced as shown with its origin at the point where the cart is in equilibrium. Suppose that the cart is pulled or pushed horizontally, then released. Find a differential equation for the position function of the cart if

    (a) there is no damping

    (b) there is a damping constant of c.

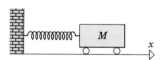

19. A cylindrical buoy of height h and radius r floats in the water with its axis vertical. By ***Archimedes' principle*** the water exerts an upward force on the buoy (the buoyancy force) with magnitude equal to the weight of the water displaced. Neglecting all forces except the buoyancy force and the force of gravity, determine the period with which the buoy will vibrate vertically if it is depressed slightly from its equilibrium position and released. Let h and r be measured in inches and take ρ lb/in^3 to be the density of water and δ lb/in^3 to be the density of the buoy material (density = weight per unit of volume).

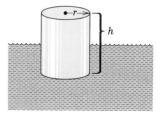

20. Suppose that the mass of the object in Figure 19.4.1 is M and the spring constant is k. Show that if the mass is displaced y_0 units from its equilibrium position and given an initial velocity of v_0 rather than being released with velocity 0, then the solution of (5) is given by

$$y(t) = y_0 \cos\left(\sqrt{\frac{k}{M}}\, t\right) + v_0 \sqrt{\frac{M}{k}} \sin\left(\sqrt{\frac{k}{M}}\, t\right)$$

21. A mass attached to a vertical spring is displaced from its equilibrium position and released, thereby causing it to vibrate with amplitude $|y_0|$ and period T (no damping).

  (a) Show that the velocity of the mass has maximum magnitude $2\pi |y_0|/T$ and that the maximum occurs when the mass is at its equilibrium position.

  (b) Show that the acceleration of the mass has maximum magnitude $4\pi^2 |y_0|/T^2$ and that the maximum occurs when the mass is at a top or bottom point of its motion.

22. Prove that the roots of Equation (12) are distinct and real, equal and real, or complex according to whether $c^2 > 4kM$, $c^2 = 4kM$, or $c^2 < 4kM$.

23. Use the addition formula for cosine to show that Formula (15) can be rewritten as (16a).

24. Overdamped Motion

  (a) Prove that in the case of overdamped motion $(c^2 > 4kM)$ the solution of (11) is

$$y(t) = \frac{y_0}{m_2 - m_1}(m_2 e^{m_1 t} - m_1 e^{m_2 t})$$

  where m_1 and m_2 are the distinct real roots of (12).

  (b) Prove that $\lim\limits_{t \to +\infty} y(t) = 0$. [*Hint:* Show that m_1 and m_2 are negative.]

25. Critically Damped Motion

  (a) Prove that in the case of critically damped motion $(c^2 = 4kM)$ the solution of (11) is

$$y(t) = y_0 e^{-\alpha t}(1 + \alpha t)$$

  (b) Prove that $\lim\limits_{t \to +\infty} y(t) = 0$.

  (c) Prove that $y(t) \neq 0$ for any positive value of t (assuming that the initial displacement y_0 is not zero).

◆ **TECHNOLOGY EXERCISES**   Chapter 19

Most of these exercises require access to a graphing calculator or a computer algebra system (CAS) such as *Mathematica, Maple,* or *Derive*. When you are asked to *find* an answer or to *solve* an equation, you may choose to find an exact result or a numerical approximation, depending on the particular technology you are using and on your own imagination. The form of your answers may differ from those of other students or from those in the answer section of the text, depending on how you solve the problems and the accuracy you use in your numerical approximations. Those exercises that are more appropriate for a CAS than a graphing calculator are labeled with the icon ◆.

The accompanying figure shows a mass–spring system in which an object of mass M is suspended by a spring and linked to a piston that moves in a *dashpot* containing a viscous fluid. If there are no external forces acting on the system, then the object is said to have **free motion** and the motion of the object is completely determined by the displacement and velocity of the object at time $t = 0$, the stiffness of the spring as measured by the spring constant k, and the viscosity of the fluid in the dashpot as measured by the damping constant c. Mathematically, the displacement $y = y(t)$ of the object from its equilibrium position is the solution of an initial-value problem of the

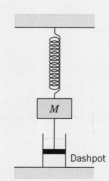

Dashpot

form

$$y'' + Ay' + By = 0, \quad y(0) = y_0, \quad y'(0) = v_0$$

where y_0 is the initial displacement of the object from equilibrium and v_0 is its initial velocity. We showed in Section 19.4 that the coefficient A is determined by the mass and the damping constant, and the coefficient B is determined by the mass and the spring constant. If an external force $F(t)$ is applied to the object, then the object is said to have **forced motion** and its displacement from equilibrium is the solution of an initial-value problem of the form

$$y'' + Ay' + By = f(t), \quad y(0) = y_0, \quad y'(0) = v_0$$

where $f(t) = F(t)/M$. In Section 19.4 we considered only motion in which the object is released from rest, that is, $v_0 = 0$. In the following exercises we shall allow the object to have a nonzero initial velocity, and we shall explore phenomena associated with forced motion. In all of the exercises assume that distance is measured in centimeters and time in seconds.

1. **Critically damped free motion**

   (a) Solve the initial-value problem $y'' + 2.4y' + 1.44y = 0$, $y(0) = 1$, $y'(0) = 2$ and graph $y = y(t)$ on the interval $[0, 5]$.

   (b) Find the maximum distance above the equilibrium position attained by the object.

   (c) The graph of $y(t)$ suggests that the object does not pass through the equilibrium position. Show that this is so.

2. **Overdamped free motion**

   (a) Solve the initial-value problem $y'' + 5y' + 2y = 0$, $y(0) = 1/2$, $y'(0) = -4$ and graph $y = y(t)$ on the interval $[0, 5]$.

   (b) Find the maximum distance below the equilibrium position attained by the object.

   (c) The graph of $y(t)$ suggests that the object passes through the equilibrium position exactly once. With what speed does the object pass through the equilibrium position?

3. **Underdamped free oscillation**

   (a) Solve the initial-value problem $y'' + y' + 5y = 0$, $y(0) = 1$, $y'(0) = -3.5$ and graph $y = y(t)$ on the interval $[0, 8]$.

   (b) Find the maximum distance below the equilibrium position attained by the object.

   (c) Find the velocity of the object when it passes through the equilibrium position the first time.

   (d) Find, by inspection, the acceleration of the object when it passes through the equilibrium position the first time. [*Hint:* Examine the differential equation and use the result in part (c).]

4. **Underdamped free oscillation**

   (a) Solve the initial-value problem $y'' + y' + 3y = 0$, $y(0) = -2$, $y'(0) = v_0$.

   (b) Find the largest positive value of v_0 for which the object will rise no higher than 1 cm above the equilibrium position. [*Hint:* Use a trial-and-error strategy. Estimate v_0 to the nearest hundredth of a cm/sec.]

   (c) Graph the solution of the initial-value problem on the interval $[0, 8]$ using the value of v_0 obtained in part (b).

In the absence of an external force, the frequency of oscillation of an undamped mass–spring system is called the **natural frequency** of the system. This terminology is used in Exercises 5 and 6.

5. **Undamped forced oscillation—beats:** Beats occur when there is a periodic external force with a frequency that is close to the natural frequency of the system. Physically, this results in a periodic variation of the amplitude of the oscillations. To illustrate this idea, consider a mass–spring system in which the motion is governed by the initial-value problem $y'' + 36y = \sin 5t$, $y(0) = 0$, $y'(0) = 0$.

   (a) Find the natural frequency of the system and the frequency of the external force.

   (b) Solve the initial-value problem and graph $y = y(t)$ on the interval $[0, 4\pi]$.

   (c) Solve the initial-value problems

   $$y'' + 36y = \sin 5.5t, \quad y(0) = 0, \quad y'(0) = 0$$
   $$y'' + 36y = \sin 5.8t, \quad y(0) = 0, \quad y'(0) = 0$$

   and graph their solutions. Make a conjecture about the effect on the oscillations as the frequency of the external force is made closer to the natural frequency of the system.

6. **Undamped forced oscillation—resonance:** Resonance occurs when there is a periodic external force with a frequency that is equal to the natural frequency of the system.

In this case the oscillations of the object increase without bound. To illustrate this idea, consider a mass–spring system in which the motion is governed by the initial-value problem $y'' + 36y = \sin 6t$, $y(0) = 0$, $y'(0) = 0$.

(a) Show that the natural frequency of the system is the same as the frequency of the external force.

(b) Solve the initial-value problem and graph $y = y(t)$ on the interval $[0, 4\pi]$.

(c) As a practical matter, the displacement of the object will be limited by factors such as the natural length of the spring, the breaking point of the spring, or the presence of obstructions above or below the object. Suppose that the bottom of the object encounters an obstruction when it is 1.3 cm below the equilibrium position. When and with what speed will the object collide with the obstruction?

7. Effect of initial velocity on overdamped free motion

(a) Solve the initial-value problem $y'' + 3.5y' + 3y = 0$, $y(0) = 1$, $y'(0) = v_0$.

(b) Use the result in part (a) to find the solutions for $v_0 = 2$, $v_0 = -1$, and $v_0 = -4$ and graph all three solutions on the interval $[0, 4]$ in the same coordinate system.

(c) Discuss the effect of the initial velocity on the motion of the object.

8. Effect of spring stiffness on forced motion

(a) Solve the initial-value problem $y'' + 3y' + y = 3 \sin 2t$, $y(0) = 2$, $y'(0) = -1$ and graph the solution on the interval $[0, 20]$.

(b) Suppose that the spring in part (a) is replaced by a stiffer one. Mathematically, this increases the coefficient of y in the differential equation. Make a conjecture about the effect that an increase in spring stiffness has on the motion of the object. Test your conjecture by solving the initial-value problem $y'' + 3y' + 2y = 3 \sin 2t$, $y(0) = 2$, $y'(0) = -1$ and graphing the solution on the interval $[0, 20]$.

APPENDIX A

Review of Sets

A *set* is a collection of objects, called *elements* or *members* of the set. In this text we shall be concerned primarily with sets whose members are either numbers or points that lie on a line, in a plane, or in three-dimensional space. We shall denote sets by capital letters and elements by lowercase letters. To indicate that a is a member of the set A we write $a \in A$ (read, "*a* belongs to *A*"), and to indicate that a is not a member of the set A we write $a \notin A$ (read, "*a* does not belong to *A*"). For example, if A is the set of positive integers, then $5 \in A$, but $-5 \notin A$. Sometimes sets arise that have no members; such sets are said to be *empty*. For example, the set of odd integers that are divisible by 2 is empty. The set with no members is called the *empty set* or the *null set* and is denoted by the symbol $\varnothing$.

If every member of a set A is also a member of a set B, then we say that A is a *subset* of B and write $A \subset B$. For example, if A is the set of positive integers and B is the set of all integers, then $A \subset B$. If two sets A and B have the same members (i.e., $A \subset B$ and $B \subset A$), then we say that A and B are *equal* and write $A = B$. For example, if A is the set of positive even integers and B is the set of positive integers that are divisible by 2, then $A = B$.

Sets can sometimes be described by listing the members, separated by commas, between a pair of braces. The order in which the elements are listed does not matter. For example, the set A of positive integers less than 6 can be expressed as

$$A = \{1, 2, 3, 4, 5\} \quad \text{or} \quad A = \{2, 3, 1, 5, 4\}$$

When listing the elements is inconvenient or impossible, a set can be described in the *set-builder* notation $\{x: \underline{\qquad}\}$ (read, "the set of all x such that") in which the line is replaced by a property of the members that defines the set. For example, the set A of all positive real numbers can be expressed as

$$A = \{x: x \text{ is a real number satisfying } x > 0\}$$

Usually, we would write this more briefly as

$$A = \{x: x > 0\}$$

leaving it understood that x is a real number.

If A and B are sets, then the *union* of A and B (denoted by $A \cup B$) is the set whose members belong to A or B (or both), and the *intersection* of A and B (denoted by $A \cap B$) is the set whose members belong to both A and B. For example,

$$\{x: 0 < x < 5\} \cup \{x: 1 < x < 7\} = \{x: 0 < x < 7\}$$

$$\{x: x < 1\} \cap \{x: x > 0\} = \{x: 0 < x < 1\}$$

$$\{x: x < 1\} \cap \{x: x > 1\} = \varnothing$$

If A and B are the sets of points in the plane enclosed by the circles in Figure A.1, then the union and intersection are the shaded regions shown in the figure. These are called *Venn diagrams* for the union and intersection.

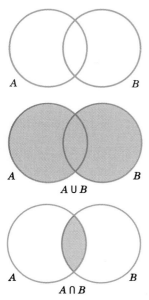

A B

A B
$A \cup B$

A B
$A \cap B$

Figure A.1

APPENDIX B

Trigonometry Review

■ B.I TRIGONOMETRIC FUNCTIONS AND IDENTITIES

☐ ANGLES

Angles in the plane can be generated by rotating a ray about its endpoint. The starting position of the ray is called the *initial side* of the angle, the final position is called the *terminal side* of the angle, and the point at which the initial and terminal sides meet is called the *vertex* of the angle. We allow for the possibility that the ray may make more than one complete revolution. Angles are considered to be *positive* if generated counterclockwise and *negative* if generated clockwise (Figure B.1).

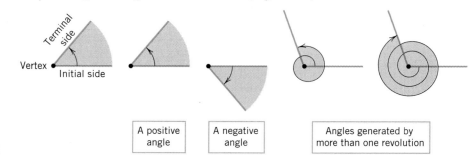

Figure B.1

There are two standard measurement systems for describing the size of an angle, *degree measure* and *radian measure*. In degree measure, one degree (written 1°) is the measure of an angle generated by 1/360 of one revolution. Thus, there are 360° in an angle of one revolution, 180° in an angle of one-half revolution, 90° in an angle of one-quarter revolution (a *right angle*), and so forth. Degrees are divided into sixty equal parts, called *minutes*, and minutes are divided into sixty equal parts, called *seconds*. Thus, one minute (written 1′) is 1/60 of a degree, and one second (written 1″) is 1/60 of a minute. Smaller subdivisions of a degree are expressed as fractions of a second.

In radian measure, angles are measured by the length of arc that the angle subtends on a circle of radius 1 when the vertex is at the center. One unit of arc on a circle of radius 1 is called one *radian* (written 1 radian or 1 rad) (Figure B.2), and hence the entire circumference of a circle of radius 1 is 2π radians. It follows that an angle of 360° subtends an arc of 2π radians, an angle of 180° subtends an arc of π radians, and angle of 90° subtends an arc of $\pi/2$ radians, and so forth. Figure B.3 and Table 1 show the relationship between degree measure and radian measure for some important positive angles.

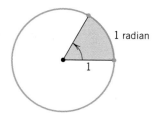

Figure B.2

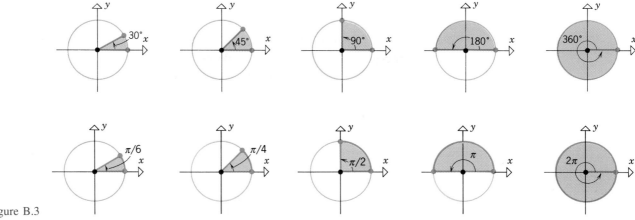

Figure B.3

Table 1

DEGREES	30°	45°	60°	90°	120°	135°	150°	180°	270°	360°
RADIANS	$\dfrac{\pi}{6}$	$\dfrac{\pi}{4}$	$\dfrac{\pi}{3}$	$\dfrac{\pi}{2}$	$\dfrac{2\pi}{3}$	$\dfrac{3\pi}{4}$	$\dfrac{5\pi}{6}$	π	$\dfrac{3\pi}{2}$	2π

REMARK.   Observe that in Table 1, angles in degrees are designated by the degree symbol, but angles in radians have no units specified. This is standard practice—when no units are specified for an angle, it is understood that the units are radians.

From the fact that π radians corresponds to 180°, we obtain the following formulas, which are useful for converting from degrees to radians and conversely.

$$1° = \frac{\pi}{180} \text{ rad} \approx 0.01745 \text{ rad} \tag{1}$$

$$1 \text{ rad} = \left(\frac{180}{\pi}\right)^{\circ} \approx 57° \, 17' \, 44.8'' \tag{2}$$

Example 1

(a)   Express 146° in radians.   (b)   Express 3 radians in degrees.

Solution (a).   From (1), degrees can be converted to radians by multiplying by a conversion factor of $\pi/180$. Thus,

$$146° = \left(\frac{\pi}{180} \cdot 146\right) \text{ rad} = \frac{73\pi}{90} \text{ rad} \approx 2.5482 \text{ rad}$$

Solution (b).   From (2), radians can be converted to degrees by multiplying by a conversion factor of $180/\pi$. Thus,

$$3 \text{ rad} = \left(3 \cdot \frac{180}{\pi}\right)^{\circ} = \left(\frac{540}{\pi}\right)^{\circ} \approx 171.9° \quad \blacktriangleleft$$

☐ **RELATIONSHIPS BETWEEN ARC LENGTH, ANGLE, RADIUS, AND AREA**

There is a theorem from plane geometry which states that for two concentric circles, the ratio of the arc lengths subtended by a central angle is equal to the ratio of the corresponding radii (Figure B.4). In particular, if s is the arc length subtended on a circle of radius r by a central angle of θ radians, then by comparison with the arc length subtended by that angle on a circle of radius 1 we obtain

$$\frac{s}{\theta} = \frac{r}{1}$$

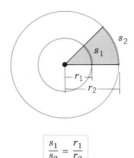

Figure B.4

$$\frac{s_1}{s_2} = \frac{r_1}{r_2}$$

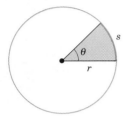

If θ is in radians,
then $\theta = s/r$.

Figure B.5

from which we obtain the following relationships between the central angle θ, the radius r, and the subtended arc length s when θ is in radians (Figure B.5):

$$\theta = \frac{s}{r} \qquad \text{and} \qquad s = r\theta \qquad (3\text{–}4)$$

The shaded region in Figure B.5 is called a **sector**. It is a theorem from plane geometry that the ratio of the area A of this sector to the area of the entire circle is the same as the ratio of the central angle of the sector to the central angle of the entire circle; thus, if the angles are in radians, we have

$$\frac{A}{\pi r^2} = \frac{\theta}{2\pi}$$

Solving for A yields the following formula for the area of a sector in terms of the radius r and the angle θ in radians:

$$A = \tfrac{1}{2} r^2 \theta \qquad (5)$$

☐ **TRIGONOMETRIC FUNCTIONS FOR RIGHT TRIANGLES**

The **sine**, **cosine**, **tangent**, **cosecant**, **secant**, and **cotangent** of a positive acute angle θ can be defined as ratios of the sides of a right triangle. Using the notation from Figure B.6, these definitions take the following form:

$$\sin \theta = \frac{\text{side opposite } \theta}{\text{hypotenuse}} = \frac{y}{r}, \qquad \csc \theta = \frac{\text{hypotenuse}}{\text{side opposite } \theta} = \frac{r}{y}$$

$$\cos \theta = \frac{\text{side adjacent to } \theta}{\text{hypotenuse}} = \frac{x}{r}, \qquad \sec \theta = \frac{\text{hypotenuse}}{\text{side adjacent to } \theta} = \frac{r}{x} \qquad (6)$$

$$\tan \theta = \frac{\text{side opposite } \theta}{\text{side adjacent to } \theta} = \frac{y}{x}, \qquad \cot \theta = \frac{\text{side adjacent to } \theta}{\text{side opposite } \theta} = \frac{x}{y}$$

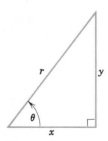

Figure B.6

We shall call sin, cos, tan, csc, sec, and cot the **trigonometric functions**. Because similar triangles have proportional sides, the values of the trigonometric functions depend only on the size of θ and not on the particular right triangle used to compute the ratios. Moreover, in these definitions it does not matter whether θ is measured in degrees or radians.

Example 2   Recall from geometry that the two legs of a 45°–45°–90° triangle are of equal size and that the hypotenuse of a 30°–60°–90° triangle is twice the shorter leg, where the shorter leg is opposite the 30° angle. These facts and the Theorem of Pythagoras yield Figure B.7. From that figure we obtain the results in the accompanying table:

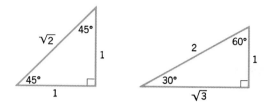

Figure B.7

$$\begin{array}{lll}
\sin 45° = 1/\sqrt{2}, & \cos 45° = 1/\sqrt{2}, & \tan 45° = 1 \\
\csc 45° = \sqrt{2}, & \sec 45° = \sqrt{2}, & \cot 45° = 1 \\
\\
\sin 30° = 1/2, & \cos 30° = \sqrt{3}/2, & \tan 30° = 1/\sqrt{3} \\
\csc 30° = 2, & \sec 30° = 2/\sqrt{3}, & \cot 30° = \sqrt{3} \\
\\
\sin 60° = \sqrt{3}/2, & \cos 60° = 1/2, & \tan 60° = \sqrt{3} \\
\csc 60° = 2/\sqrt{3}, & \sec 60° = 2, & \cot 60° = 1/\sqrt{3}
\end{array}$$

◄

□ **ANGLES IN RECTANGULAR COORDINATE SYSTEMS**

Because the angles of a right triangle are between 0° and 90°, the formulas in (6) are not directly applicable to negative angles or to angles greater than 90°. To extend the trigonometric functions to include these cases, it will be convenient to consider angles in rectangular coordinate systems. An angle is said to be in ***standard position*** in an *xy*-coordinate system if its vertex is at the origin and its initial side is on the positive *x*-axis (Figure B.8).

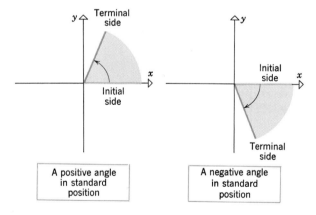

Figure B.8

To define the trigonometric functions of an angle θ in standard position, construct a circle of radius r, centered at the origin, and let $P(x, y)$ be the intersection of the terminal side of θ with this circle (Figure B.9). We make the following definition:

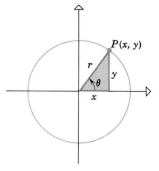

Figure B.9

DEFINITION **1**.

$$\sin \theta = \frac{y}{r}, \quad \cos \theta = \frac{x}{r}, \quad \tan \theta = \frac{y}{x}$$

$$\csc \theta = \frac{r}{y}, \quad \sec \theta = \frac{r}{x}, \quad \cot \theta = \frac{x}{y}$$

Note that the formulas in this definition agree with those in (6), so there is no conflict with the earlier definition of the trigonometric functions for triangles. However, this definition applies to all angles (except for cases where a zero denominator occurs).

In the special case where $r = 1$, we have $\sin \theta = y$ and $\cos \theta = x$, so the terminal side of the angle θ intersects the unit circle at the point $(\cos \theta, \sin \theta)$ (Figure B.10). It follows from Definition 1 that the remaining trigonometric functions of θ are expressible as (verify)

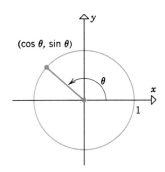

(cos θ, sin θ)

Figure B.10

$$\tan \theta = \frac{\sin \theta}{\cos \theta}, \quad \cot \theta = \frac{\cos \theta}{\sin \theta} = \frac{1}{\tan \theta}, \quad \sec \theta = \frac{1}{\cos \theta}, \quad \csc \theta = \frac{1}{\sin \theta} \quad (7\text{--}10)$$

These observations suggest the following procedure for evaluating the trigonometric functions of common angles:

- Construct the angle θ in standard position in an xy-coordinate system.
- Find the coordinates of the intersection of the terminal side of the angle and the unit circle; the x- and y-coordinates of this intersection are the values of $\cos \theta$ and $\sin \theta$, respectively.
- Use Formulas (7) through (10) to find the values of the remaining trigonometric functions from the values of $\cos \theta$ and $\sin \theta$.

Example 3   Evaluate the trigonometric functions of $\theta = 150°$.

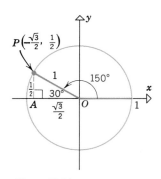

$P\left(-\frac{\sqrt{3}}{2}, \frac{1}{2}\right)$

Figure B.11

Solution.   Construct a unit circle and place the angle $\theta = 150°$ in standard position (Figure B.11). Since $\angle AOP$ is $30°$ and $\triangle OAP$ is a $30°$–$60°$–$90°$ triangle, the leg AP has length $\frac{1}{2}$ (half the hypotenuse) and the leg OA has length $\sqrt{3}/2$ by the Theorem of Pythagoras. Thus, the coordinates of P are $(-\sqrt{3}/2, 1/2)$, from which we obtain

$$\sin 150° = \frac{1}{2}, \quad \cos 150° = -\frac{\sqrt{3}}{2}, \quad \tan 150° = \frac{\sin 150°}{\cos 150°} = \frac{1/2}{-\sqrt{3}/2} = -\frac{1}{\sqrt{3}}$$

$$\csc 150° = \frac{1}{\sin 150°} = 2, \quad \sec 150° = \frac{1}{\cos 150°} = -\frac{2}{\sqrt{3}},$$

$$\cot 150° = \frac{1}{\tan 150°} = -\sqrt{3} \quad \blacktriangleleft$$

Example 4   Evaluate the trigonometric functions of $\theta = 5\pi/6$.

Solution.   Since $5\pi/6 = 150°$, this problem is equivalent to that of Example 3. From that example we obtain

$$\sin \frac{5\pi}{6} = \frac{1}{2}, \quad \cos \frac{5\pi}{6} = -\frac{\sqrt{3}}{2}, \quad \tan \frac{5\pi}{6} = -\frac{1}{\sqrt{3}}$$

$$\csc \frac{5\pi}{6} = 2, \quad \sec \frac{5\pi}{6} = -\frac{2}{\sqrt{3}}, \quad \cot \frac{5\pi}{6} = -\sqrt{3} \quad \blacktriangleleft$$

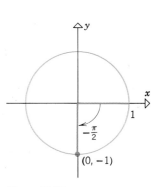

$-\frac{\pi}{2}$

$(0, -1)$

Figure B.12

Example 5   Evaluate the trigonometric functions of $\theta = -\pi/2$.

Solution.   As shown in Figure B.12, the terminal side of $\theta = -\pi/2$ intersects the unit circle at the point $(0, -1)$, so

$$\sin(-\pi/2) = -1, \quad \cos(-\pi/2) = 0$$

and from Formulas (7) to (10)

$$\tan(-\pi/2) = \frac{\sin(-\pi/2)}{\cos(-\pi/2)} = \frac{-1}{0} \quad \text{(undefined)}$$

$$\cot(-\pi/2) = \frac{\cos(-\pi/2)}{\sin(-\pi/2)} = \frac{0}{-1} = 0$$

$$\sec(-\pi/2) = \frac{1}{\cos(-\pi/2)} = \frac{1}{0} \quad \text{(undefined)}$$

$$\csc(-\pi/2) = \frac{1}{\sin(-\pi/2)} = \frac{1}{-1} = -1 \qquad ◄$$

The reader should be able to obtain all of the results in Table 2 by the methods illustrated in the last three examples. The dashes indicate quantities that are undefined.

Table 2

	$\theta = 0$ (0°)	$\pi/6$ (30°)	$\pi/4$ (45°)	$\pi/3$ (60°)	$\pi/2$ (90°)	$2\pi/3$ (120°)	$3\pi/4$ (135°)	$5\pi/6$ (150°)	π (180°)	$3\pi/2$ (270°)	2π (360°)
$\sin\theta$	0	$1/2$	$1/\sqrt{2}$	$\sqrt{3}/2$	1	$\sqrt{3}/2$	$1/\sqrt{2}$	$1/2$	0	-1	0
$\cos\theta$	1	$\sqrt{3}/2$	$1/\sqrt{2}$	$1/2$	0	$-1/2$	$-1/\sqrt{2}$	$-\sqrt{3}/2$	-1	0	1
$\tan\theta$	0	$1/\sqrt{3}$	1	$\sqrt{3}$	—	$-\sqrt{3}$	-1	$-1/\sqrt{3}$	0	—	0
$\csc\theta$	—	2	$\sqrt{2}$	$2/\sqrt{3}$	1	$2/\sqrt{3}$	$\sqrt{2}$	2	—	-1	—
$\sec\theta$	1	$2/\sqrt{3}$	$\sqrt{2}$	2	—	-2	$-\sqrt{2}$	$-2/\sqrt{3}$	-1	—	1
$\cot\theta$	—	$\sqrt{3}$	1	$1/\sqrt{3}$	0	$-1/\sqrt{3}$	-1	$-\sqrt{3}$	—	0	—

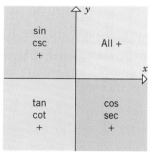

Figure B.13

REMARK.   It is only in special cases that exact values for trigonometric functions can be obtained; usually, a calculator or a computer will be required.

The signs of the trigonometric functions of an angle are determined by the quadrant in which the terminal side of the angle falls. For example, if the terminal side falls in the first quadrant, then x and y are positive in Definition 1, so all of the trigonometric functions have positive values. If the terminal side falls in the second quadrant, then x is negative and y is positive, so sin and csc are positive, but all other trigonometric functions are negative. The diagram in Figure B.13 shows which trigonometric functions are positive in the various quadrants. The reader will find it instructive to check that the results in Table 2 are consistent with Figure B.13.

TRIGONOMETRIC IDENTITIES

A *trigonometric identity* is an equation involving trigonometric functions that is true for all angles for which both sides of the equation are defined. One of the most important identities in trigonometry can be derived by applying the Theorem of Pythagoras to the triangle in Figure B.9 to obtain

$$x^2 + y^2 = r^2$$

Dividing both sides by r^2 and using the definitions of $\sin\theta$ and $\cos\theta$ (Definition 1), we obtain the following fundamental result:

$$\sin^2\theta + \cos^2\theta = 1 \qquad (11)$$

The following identities can be obtained from (11) by dividing through by $\cos^2\theta$ and $\sin^2\theta$, respectively, then applying Formulas (7) through (10):

$$\tan^2\theta + 1 = \sec^2\theta \tag{12}$$

$$1 + \cot^2\theta = \csc^2\theta \tag{13}$$

If (x, y) is a point on the unit circle, then the points $(-x, y)$, $(-x, -y)$, and $(x, -y)$ also lie on the unit circle (why?), and the four points form corners of a rectangle with sides parallel to the coordinate axes (Figure B.14a). The x- and y-coordinates of each corner represent the cosine and sine of an angle in standard position whose terminal side passes through the corner; hence we obtain the identities in parts (b), (c), and (d) of Figure B.14 for sine and cosine. Dividing those identities leads to identities for the tangent. In summary:

$$\sin(\pi - \theta) = \sin\theta, \qquad \sin(\pi + \theta) = -\sin\theta, \qquad \sin(-\theta) = -\sin\theta \tag{14--16}$$

$$\cos(\pi - \theta) = -\cos\theta, \qquad \cos(\pi + \theta) = -\cos\theta, \qquad \cos(-\theta) = \cos\theta \tag{17--19}$$

$$\tan(\pi - \theta) = -\tan\theta, \qquad \tan(\pi + \theta) = \tan\theta, \qquad \tan(-\theta) = -\tan\theta \tag{20--22}$$

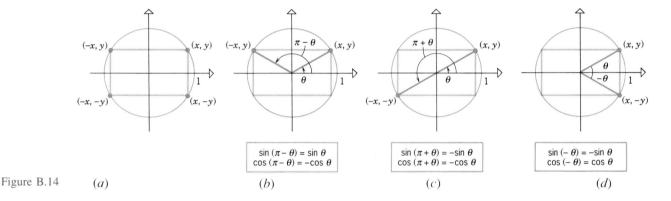

Figure B.14　　(a)　　　　　　　(b)　　　　　　　(c)　　　　　　　(d)

Two angles in standard position that have the same terminal side must have the same values for their trigonometric functions since their terminal sides intersect the unit circle at the same point. In particular, two angles whose radian measures differ by a multiple of 2π have the same terminal side and hence have the same values for their trigonometric functions. This yields the identities

$$\sin\theta = \sin(\theta + 2\pi) = \sin(\theta - 2\pi) \tag{23}$$

$$\cos\theta = \cos(\theta + 2\pi) = \cos(\theta - 2\pi) \tag{24}$$

and more generally,

$$\sin\theta = \sin(\theta \pm 2n\pi), \quad n = 0, 1, 2, \ldots \tag{25}$$

$$\cos\theta = \cos(\theta \pm 2n\pi), \quad n = 0, 1, 2, \ldots \tag{26}$$

Identities (20) to (22) imply that

$$\tan\theta = \tan(\theta + \pi) \qquad \text{and} \qquad \tan\theta = \tan(\theta - \pi) \tag{27--28}$$

Identity (27) is just (21) with the terms in the sum reversed, and identity (28) follows from (20) and (22) (verify). These two identities state that adding or subtracting π from an angle does not affect the value of the tangent of the angle. It follows that the same is true for any multiple of π; thus,

$$\tan \theta = \tan (\theta \pm n\pi), \quad n = 0, 1, 2, \ldots \tag{29}$$

Figure B.15 shows complementary angles θ and $\pi/2 - \theta$ of a right triangle. It follows from (6) that

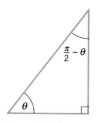

Figure B.15

$$\sin \theta = \frac{\text{side opposite } \theta}{\text{hypotenuse}} = \frac{\text{side adjacent to } (\pi/2 - \theta)}{\text{hypotenuse}} = \cos \left(\frac{\pi}{2} - \theta \right)$$

$$\cos \theta = \frac{\text{side adjacent to } \theta}{\text{hypotenuse}} = \frac{\text{side opposite } (\pi/2 - \theta)}{\text{hypotenuse}} = \sin \left(\frac{\pi}{2} - \theta \right)$$

which yields the identities

$$\sin \left(\frac{\pi}{2} - \theta \right) = \cos \theta, \quad \cos \left(\frac{\pi}{2} - \theta \right) = \sin \theta, \quad \tan \left(\frac{\pi}{2} - \theta \right) = \cot \theta \tag{30–32}$$

where the third identity results from dividing the first two. These identities are also valid for angles that are not acute and for negative angles as well.

☐ **THE LAW OF COSINES**

The next theorem, called the ***law of cosines***, generalizes the Theorem of Pythagoras. This result is important in its own right and is also the starting point for some important trigonometric identities.

> THEOREM **2** (***Law of Cosines***). *If the sides of a triangle have lengths a, b, and c, and if θ is the angle between the sides with lengths a and b, then*
>
> $$c^2 = a^2 + b^2 - 2ab \cos \theta$$

Proof. Introduce a coordinate system so that θ is in standard position and the side of length a falls along the positive x-axis. As shown in Figure B.16, the side of length a extends from the origin to $(a, 0)$ and the side of length b extends from the origin to some point (x, y). From the definition of $\sin \theta$ and $\cos \theta$ we have $\sin \theta = y/b$ and $\cos \theta = x/b$, so

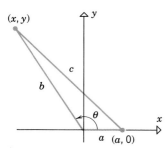

Figure B.16

$$y = b \sin \theta, \quad x = b \cos \theta \tag{33}$$

From the distance formula in Theorem 1.5.1 we obtain

$$c^2 = (x - a)^2 + (y - 0)^2$$

so that, from (33),

$$c^2 = (b \cos \theta - a)^2 + b^2 \sin^2 \theta$$

$$= a^2 + b^2 (\cos^2 \theta + \sin^2 \theta) - 2ab \cos \theta$$

$$= a^2 + b^2 - 2ab \cos \theta$$

which completes the proof. ∎

We will now show how the law of cosines can be used to obtain the following identities, called the ***addition formulas*** for sine and cosine:

$$\sin (\alpha + \beta) = \sin \alpha \cos \beta + \cos \alpha \sin \beta \tag{34}$$

$$\cos (\alpha + \beta) = \cos \alpha \cos \beta - \sin \alpha \sin \beta \tag{35}$$

$$\sin (\alpha - \beta) = \sin \alpha \cos \beta - \cos \alpha \sin \beta \tag{36}$$

$$\cos (\alpha - \beta) = \cos \alpha \cos \beta + \sin \alpha \sin \beta \tag{37}$$

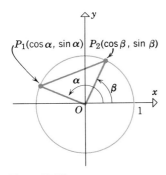

Figure B.17

We will derive (37) first. In our derivation we will assume that $0 \le \beta < \alpha < 2\pi$ (Figure B.17). As shown in the figure, the terminal sides of α and β intersect the unit circle at the points $P_1(\cos \alpha, \sin \alpha)$ and $P_2(\cos \beta, \sin \beta)$. If we denote the lengths of the sides of triangle OP_1P_2 by OP_1, P_1P_2, and OP_2, then $OP_1 = OP_2 = 1$ and, from the distance formula in Theorem 1.5.1,

$$(P_1P_2)^2 = (\cos \beta - \cos \alpha)^2 + (\sin \beta - \sin \alpha)^2$$
$$= (\sin^2 \alpha + \cos^2 \alpha) + (\sin^2 \beta + \cos^2 \beta) - 2(\cos \alpha \cos \beta + \sin \alpha \sin \beta)$$
$$= 2 - 2(\cos \alpha \cos \beta + \sin \alpha \sin \beta)$$

But angle $P_2OP_1 = \alpha - \beta$, so that the law of cosines yields

$$(P_1P_2)^2 = (OP_1)^2 + (OP_2)^2 - 2(OP_1)(OP_2) \cos (\alpha - \beta)$$
$$= 2 - 2 \cos (\alpha - \beta)$$

Equating the two expressions for $(P_1P_2)^2$ and simplifying, we obtain

$$\cos (\alpha - \beta) = \cos \alpha \cos \beta + \sin \alpha \sin \beta$$

which completes the derivation of (37).

We can use (31) and (37) to derive (36) as follows:

$$\sin (\alpha - \beta) = \cos \left[\frac{\pi}{2} - (\alpha - \beta) \right] = \cos \left[\left(\frac{\pi}{2} - \alpha \right) - (-\beta) \right]$$
$$= \cos \left(\frac{\pi}{2} - \alpha \right) \cos (-\beta) + \sin \left(\frac{\pi}{2} - \alpha \right) \sin (-\beta)$$
$$= \cos \left(\frac{\pi}{2} - \alpha \right) \cos \beta - \sin \left(\frac{\pi}{2} - \alpha \right) \sin \beta$$
$$= \sin \alpha \cos \beta - \cos \alpha \sin \beta$$

Identities (34) and (35) can be obtained from (36) and (37) by substituting $-\beta$ for β and using the identities

$$\sin (-\beta) = -\sin \beta, \quad \cos (-\beta) = \cos \beta$$

We leave it for the reader to derive the identities

$$\tan (\alpha + \beta) = \frac{\tan \alpha + \tan \beta}{1 - \tan \alpha \tan \beta} \qquad \tan (\alpha - \beta) = \frac{\tan \alpha - \tan \beta}{1 + \tan \alpha \tan \beta} \qquad (38\text{–}39)$$

Identity (38) can be obtained by dividing (34) by (35) and then simplifying. Identity (39) can be obtained from (38) by substituting $-\beta$ for β and simplifying.

In the special case where $\alpha = \beta$, identities (34), (35), and (38) yield the ***double-angle formulas***

$$\sin 2\alpha = 2 \sin \alpha \cos \alpha \qquad (40)$$

$$\cos 2\alpha = \cos^2 \alpha - \sin^2 \alpha \qquad (41)$$

$$\tan 2\alpha = \frac{2 \tan \alpha}{1 - \tan^2 \alpha} \qquad (42)$$

By using the identity $\sin^2 \alpha + \cos^2 \alpha = 1$, (41) can be rewritten in the alternative forms

$$\cos 2\alpha = 2 \cos^2 \alpha - 1 \qquad \text{and} \qquad \cos 2\alpha = 1 - 2 \sin^2 \alpha \qquad (43\text{–}44)$$

If we replace α by $\alpha/2$ in (43) and (44) and use some algebra, we obtain the **half-angle formulas**

$$\cos^2\frac{\alpha}{2} = \frac{1 + \cos\alpha}{2} \qquad \text{and} \qquad \sin^2\frac{\alpha}{2} = \frac{1 - \cos\alpha}{2} \qquad (45\text{--}46)$$

We leave it for the exercises to derive the following **product-to-sum formulas** from (34) to (37):

$$\sin\alpha\cos\beta = \frac{1}{2}[\sin(\alpha - \beta) + \sin(\alpha + \beta)] \qquad (47)$$

$$\sin\alpha\sin\beta = \frac{1}{2}[\cos(\alpha - \beta) - \cos(\alpha + \beta)] \qquad (48)$$

$$\cos\alpha\cos\beta = \frac{1}{2}[\cos(\alpha - \beta) + \cos(\alpha + \beta)] \qquad (49)$$

We also leave it for the exercises to derive the following **sum-to-product formulas**:

$$\sin\alpha + \sin\beta = 2\sin\frac{\alpha + \beta}{2}\cos\frac{\alpha - \beta}{2} \qquad (50)$$

$$\sin\alpha - \sin\beta = 2\cos\frac{\alpha + \beta}{2}\sin\frac{\alpha - \beta}{2} \qquad (51)$$

$$\cos\alpha + \cos\beta = 2\cos\frac{\alpha + \beta}{2}\cos\frac{\alpha - \beta}{2} \qquad (52)$$

$$\cos\alpha - \cos\beta = -2\sin\frac{\alpha + \beta}{2}\sin\frac{\alpha - \beta}{2} \qquad (53)$$

☐ **FINDING AN ANGLE FROM THE VALUE OF ITS TRIGONOMETRIC FUNCTIONS**

There are numerous situations in which it is necessary to find an unknown angle from a known value of one of its trigonometric functions. The following example illustrates a method for doing this.

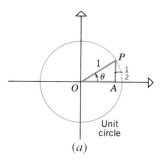

(a)

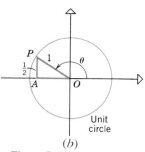

(b)

Figure B.18

Example 6   Find θ if $\sin\theta = \frac{1}{2}$.

Solution.   We begin by looking for positive angles that satisfy the equation. Because $\sin\theta$ is positive, the angle θ must terminate in the first or second quadrant. If it terminates in the first quadrant, then the hypotenuse of $\triangle OAP$ in Figure B.18a is double the leg AP, so

$$\theta = 30° = \frac{\pi}{6} \text{ radians}$$

If θ terminates in the second quadrant (Figure B.18b), then the hypotenuse of $\triangle OAP$ is double the leg AP, so $\angle AOP = 30°$, which implies that

$$\theta = 180° - 30° = 150° = \frac{5\pi}{6} \text{ radians}$$

Now that we have found these two solutions, all other solutions are obtained by adding or subtracting multiples of 360° (2π radians) to them. Thus, the entire set of solutions is given by the formulas

$$\theta = 30° \pm n \cdot 360°, \quad n = 0, 1, 2, \ldots$$

and

$$\theta = 150° \pm n \cdot 360°, \quad n = 0, 1, 2, \ldots$$

or in radian measure,

$$\theta = \frac{\pi}{6} \pm n \cdot 2\pi, \quad n = 0, 1, 2, \dots$$

and

$$\theta = \frac{5\pi}{6} \pm n \cdot 2\pi, \quad n = 0, 1, 2, \dots \quad \blacktriangleleft$$

▶ Exercise Set B.I $\boxed{C}$ *17, 18, 41, 42*

In Exercises 1 and 2, express the angles in radians.

1. (a) 75°   (b) 390°   (c) 20°   (d) 138°.
2. (a) 420°   (b) 15°   (c) 225°   (d) 165°.

In Exercises 3 and 4, express the angles in degrees.

3. (a) $\pi/15$ (b) 1.5   (c) $8\pi/5$ (d) 3π.
4. (a) $\pi/10$ (b) 2   (c) $2\pi/5$ (d) $7\pi/6$.

In Exercises 5 and 6, find the exact values of all six trigonometric functions of θ.

5. (a)   (b)   (c)

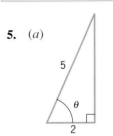

6. (a)   (b)   (c)

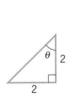 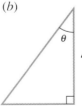 

In Exercises 7–12, the angle θ is an acute angle of a right triangle. Solve the problems by drawing an appropriate right triangle. Do *not* use a calculator.

7. Find $\sin \theta$ and $\cos \theta$ given that $\tan \theta = 3$.
8. Find $\sin \theta$ and $\tan \theta$ given that $\cos \theta = 2/3$.
9. Find $\tan \theta$ and $\csc \theta$ given that $\sec \theta = 5/2$.
10. Find $\cot \theta$ and $\sec \theta$ given that $\csc \theta = 4$.
11. Find the length of the side adjacent to θ given that the hypotenuse has length 6 and $\cos \theta = 0.3$.
12. Find the length of the hypotenuse given that the side opposite θ has length 2.4 and $\sin \theta = 0.8$.

In Exercises 13 and 14, the value of an angle θ is given. Find the values of all six trigonometric functions of θ without using a calculator.

13. (a) 225°   (b) −210°   (c) $5\pi/3$ (d) $-3\pi/2$.
14. (a) 330°   (b) −120°   (c) $9\pi/4$ (d) -3π.

In Exercises 15 and 16, use the information to find the exact values of the remaining five trigonometric functions of θ.

15. (a) $\cos \theta = \frac{3}{5}$, $0 < \theta < \pi/2$
    (b) $\cos \theta = \frac{3}{5}$, $-\pi/2 < \theta < 0$
    (c) $\tan \theta = -1/\sqrt{3}$, $\pi/2 < \theta < \pi$
    (d) $\tan \theta = -1/\sqrt{3}$, $-\pi/2 < \theta < 0$
    (e) $\csc \theta = \sqrt{2}$, $0 < \theta < \pi/2$
    (f) $\csc \theta = \sqrt{2}$, $\pi/2 < \theta < \pi$.
16. (a) $\sin \theta = \frac{1}{4}$, $0 < \theta < \pi/2$
    (b) $\sin \theta = \frac{1}{4}$, $\pi/2 < \theta < \pi$
    (c) $\cot \theta = \frac{1}{3}$, $0 < \theta < \pi/2$
    (d) $\cot \theta = \frac{1}{3}$, $\pi < \theta < 3\pi/2$
    (e) $\sec \theta = -\frac{5}{2}$, $\pi/2 < \theta < \pi$
    (f) $\sec \theta = -\frac{5}{2}$, $\pi < \theta < 3\pi/2$.

In Exercises 17 and 18, use a calculator to find x to four decimal places.

17. (a)   (b)

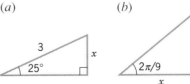

18. (a)   (b)

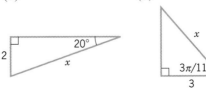

19. In each part, let θ be an acute angle of a right triangle. Express the remaining five trigonometric functions in terms of a.

    (a) $\sin\theta = a/3$ (b) $\tan\theta = a/5$ (c) $\sec\theta = a$.

In Exercises 20–27, find all values of θ (in radians) that satisfy the given equation. Do not use a calculator.

---

20. (a) $\cos\theta = -1/\sqrt{2}$ (b) $\sin\theta = -1/\sqrt{2}$.

21. (a) $\tan\theta = -1$ (b) $\cos\theta = 1/2$.

22. (a) $\sin\theta = -1/2$ (b) $\tan\theta = \sqrt{3}$.

23. (a) $\tan\theta = 1/\sqrt{3}$ (b) $\sin\theta = -\sqrt{3}/2$.

24. (a) $\sin\theta = -1$ (b) $\cos\theta = -1$.

25. (a) $\cot\theta = -1$ (b) $\cot\theta = \sqrt{3}$.

26. (a) $\sec\theta = -2$ (b) $\csc\theta = -2$.

27. (a) $\csc\theta = 2/\sqrt{3}$ (b) $\sec\theta = 2/\sqrt{3}$.

In Exercises 28 and 29, find the values of all six trigonometric functions of θ.

28.

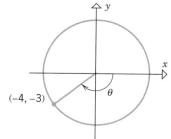

29.

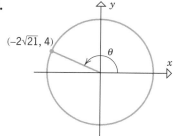

30. Find all values of θ (in radians) such that

    (a) $\sin\theta = 1$ (b) $\cos\theta = 1$ (c) $\tan\theta = 1$

    (d) $\csc\theta = 1$ (e) $\sec\theta = 1$ (f) $\cot\theta = 1$.

31. Find all values of θ (in radians) such that

    (a) $\sin\theta = 0$ (b) $\cos\theta = 0$ (c) $\tan\theta = 0$

    (d) $\csc\theta$ is undefined   (e) $\sec\theta$ is undefined

    (f) $\cot\theta$ is undefined.

32. How could you use a ruler and protractor to approximate $\sin 17°$ and $\cos 17°$?

33. Find the length of the circular arc on a circle of radius 4 cm subtended by an angle of

    (a) $\pi/6$ (b) 150°.

34. Find the radius of a circular sector that has an angle of $\pi/3$ and a circular arc length of 7 units.

35. A point P moving counterclockwise on a circle of radius 5 cm traverses an arc length of 2 cm. What is the angle swept out by a radius from the center to P?

36. Find a formula for the area A of a circular sector in terms of its radius r and arc length s.

37. A right-circular cone is made from a circular piece of paper of radius R by cutting out a sector of angle θ radians and gluing the cut edges of the remaining piece together (figure below). Find

    (a) the radius r of the base of the cone in terms of R and θ

    (b) the height h of the cone in terms of R and θ.

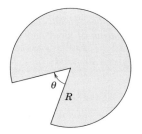
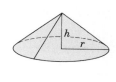

38. Let r and L be the radius of the base and the slant height of a right-circular cone (figure below). Show that the lateral surface area, S, of the cone is $S = \pi r L$. [*Hint:* As shown in the figure in Exercise 37, the lateral surface of the cone becomes a circular sector when cut along a line from the vertex to the base and flattened.]

39. Two sides of a triangle have lengths of 3 cm and 7 cm and meet at an angle of 60°. Find the area of the triangle.

40. Let ABC be a triangle whose angles at A and B are 30° and 45°. If the side opposite the angle B has length 9, find the lengths of the remaining sides and the size of the angle C.

41. A 10-foot ladder leans against a house and makes an angle of 67° with level ground. How far is the top of the ladder above the ground? Express your answer to the nearest tenth of a foot.

42. From a point 120 feet on level ground from a building, the angle of elevation to the top of the building is 76°. Find the height of the building. Express your answer to the nearest foot.

43. An observer on level ground is at a distance d from a building. The angles of elevation to the bottom of the windows on the second and third floors are α and β, respectively. Find the distance h between the bottoms of the windows in terms of α, β, and d.

44. From a point on level ground, the angle of elevation to the top of a tower is α. From a point that is d units closer to the tower, the angle of elevation is β. Find the height h of the tower in terms of α, β, and d.

In Exercises 45 and 46, do *not* use a calculator.

---

45. If $\cos\theta = 2/3$ and $0 < \theta < \pi/2$, find

(a) $\sin 2\theta$ (b) $\cos 2\theta$.

46. If $\tan\alpha = 3/4$ and $\tan\beta = 2$, where $0 < \alpha < \pi/2$ and $0 < \beta < \pi/2$, find

(a) $\sin(\alpha - \beta)$ (b) $\cos(\alpha + \beta)$.

47. Express $\sin 3\theta$ and $\cos 3\theta$ in terms of $\sin\theta$ and $\cos\theta$.

In Exercises 48–58, derive the given identities.

---

48. $\dfrac{\cos\theta \sec\theta}{1 + \tan^2\theta} = \cos^2\theta$.

49. $\dfrac{\cos\theta \tan\theta + \sin\theta}{\tan\theta} = 2\cos\theta$.

50. $2\csc 2\theta = \sec\theta \csc\theta$.   **51.** $\tan\theta + \cot\theta = 2\csc 2\theta$.

52. $\dfrac{\sin 2\theta}{\sin\theta} - \dfrac{\cos 2\theta}{\cos\theta} = \sec\theta$. **53.** $\dfrac{\sin\theta + \cos 2\theta - 1}{\cos\theta - \sin 2\theta} = \tan\theta$.

54. $\sin 3\theta + \sin\theta = 2\sin 2\theta \cos\theta$.

55. $\sin 3\theta - \sin\theta = 2\cos 2\theta \sin\theta$.

56. $\tan\dfrac{\theta}{2} = \dfrac{1 - \cos\theta}{\sin\theta}$.   **57.** $\tan\dfrac{\theta}{2} = \dfrac{\sin\theta}{1 + \cos\theta}$.

58. $\cos\left(\dfrac{\pi}{3} + \theta\right) + \cos\left(\dfrac{\pi}{3} - \theta\right) = \cos\theta$.

Exercises 59 and 60 refer to an arbitrary triangle ABC in which the side of length a is opposite angle A, the side of length b is opposite angle B, and the side of length c is opposite angle C.

---

59. Prove: The area of a triangle ABC can be written as

$$\text{area} = \tfrac{1}{2}bc \sin A$$

Find two other similar formulas for the area.

60. Prove the **law of sines**: In any triangle, the ratios of the sides to the sines of the opposite angles are equal; that is,

$$\frac{a}{\sin A} = \frac{b}{\sin B} = \frac{c}{\sin C}$$

61. Use identities (34) to (37) to express each of the following in terms of $\sin\theta$ or $\cos\theta$.

(a) $\sin\left(\dfrac{\pi}{2} + \theta\right)$ (b) $\cos\left(\dfrac{\pi}{2} + \theta\right)$

(c) $\sin\left(\dfrac{3\pi}{2} - \theta\right)$ (d) $\cos\left(\dfrac{3\pi}{2} + \theta\right)$.

62. Derive identities (38) and (39).

63. Derive identity

(a) (47)   (b) (48)   (c) (49).

64. If $A = \alpha + \beta$ and $B = \alpha - \beta$, then $\alpha = \tfrac{1}{2}(A + B)$ and $\beta = \tfrac{1}{2}(A - B)$ (verify). Use this result and identities (47) to (49) to derive identity

(a) (50)   (b) (52)   (c) (53).

65. Substitute $-\beta$ for β in identity (50) to derive identity (51).

66. (a) Express $3\sin\alpha + 5\cos\alpha$ in the form

$$C\sin(\alpha + \phi)$$

(b) Show that a sum of the form

$$A\sin\alpha + B\cos\alpha$$

can be rewritten in the form $C\sin(\alpha + \phi)$.

67. Show that the length of the diagonal of the parallelogram in the figure below is

$$d = \sqrt{a^2 + b^2 + 2ab\cos\theta}$$

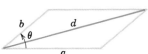

■ **B.II** GRAPHS OF TRIGONOMETRIC FUNCTIONS

☐ **THE GRAPHS OF THE SINE AND COSINE FUNCTIONS**

Recall that if an angle is written without units, then it is understood to be in radians. Thus, when we speak of the graph of $y = \sin x$, the variable x is in radian measure. The graphs of $y = \sin x$ and $y = \cos x$ are shown in Figure B.19.

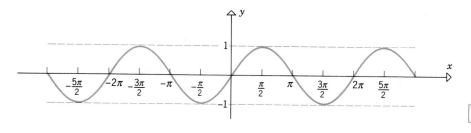

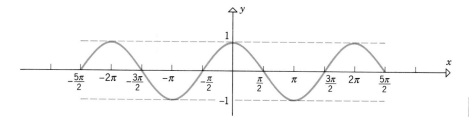

Figure B.19

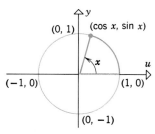

Figure B.20

These graphs can be obtained using a graphing calculator set to the radian mode; however, their general shapes can be deduced by observing the coordinates $(\cos x, \sin x)$ of the intersection with the unit circle of the terminal side of the angle x in standard position in a uy-coordinate system (Figure B.20). (We use u for the horizontal axis, since we want to reserve x for the angle.) Counterclockwise rotations of the terminal side correspond to positive values of x and clockwise rotations to negative values of x. For example, suppose that x increases from 0 to $\pi/2$ (a counterclockwise rotation); then the point $(\cos x, \sin x)$ moves along the unit circle from $(1, 0)$ to $(0, 1)$, so that $\cos x$ decreases from 1 to 0 and $\sin x$ increases from 0 to 1, consistent with the graphs in Figure B.19. As x continues to increase from $\pi/2$ to π, the point $(\cos x, \sin x)$ moves along the unit circle from $(0, 1)$ to $(-1, 0)$, so that $\cos x$ decreases from 0 to -1 and $\sin x$ decreases from 1 to 0, again consistent with the graphs in Figure B.19. We leave it for the reader to continue this analysis as x increases from π to 2π and to try it for a counterclockwise rotation with x decreasing from 0 to -2π.

☐ **THE GRAPH OF THE TANGENT FUNCTION**

The graph of $y = \tan x$, shown in Figure B.21, can be obtained using a graphing calculator set to the radian mode.

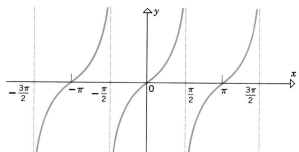

Figure B.21

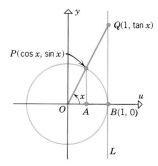

Figure B.22

However, the general shape of the graph for $-\pi/2 < x < \pi/2$ can be deduced by noting that if x is in standard position in a uy-coordinate system, then the terminal side of x intersects the vertical line through $(1, 0)$ at the point $(1, \tan x)$ (Figure B.22). To see that this is so for $0 \le x < \pi/2$, observe that $OB = 1$, so the y-coordinate of Q is

$$y = BQ = \frac{BQ}{OB} = \tan x \qquad (1)$$

With appropriate modifications this argument can be adapted to show that (1) also holds for $-\pi/2 < x \le 0$. It follows that if x increases from 0 toward $\pi/2$ (a counterclockwise rotation), then the point Q moves upward indefinitely along L, starting from $(1, 0)$; hence, $\tan x$ increases indefinitely from an initial value of 0, which is consistent with Figure B.21. Similarly, if x decreases from 0 to $-\pi/2$ (a clockwise rotation), then $\tan x$ decreases indefinitely from an initial value of 0, again consistent with Figure B.21. Moreover, it follows from the identities

$$\tan(x + \pi) = \tan x \quad \text{and} \quad \tan(x - \pi) = \tan x$$

which were obtained in the preceding section [identities (27) and (28)], that the graph of $y = \tan x$ repeats every π units. This produces the complete graph shown in Figure B.21. Observe that unlike the trigonometric functions $\sin x$ and $\cos x$, which are defined for all x, the function $\tan x \, (= \sin x / \cos x)$ is undefined at the points where $\cos x = 0$, namely,

$$x = \pm \frac{\pi}{2}, \pm \frac{3\pi}{2}, \pm \frac{5\pi}{2}, \dots$$

☐ **THE GRAPHS OF THE COTANGENT, SECANT, COSECANT FUNCTIONS**

The graphs of $y = \cot x$, $y = \sec x$, and $y = \csc x$ are shown in Figure B.23 in dark blue. We have also shown the graphs of $y = \tan x$, $y = \cos x$, and $y = \sin x$ in light blue to illustrate the reciprocal relationships $\cot x = 1/\tan x$, $\sec x = 1/\cos x$, and $\csc x = 1/\sin x$ graphically.

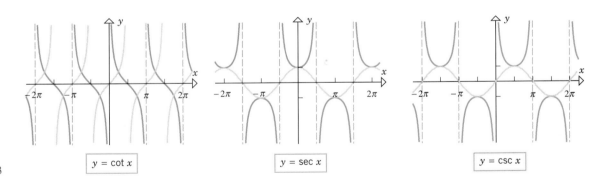

| $y = \cot x$ | $y = \sec x$ | $y = \csc x$ |

Figure B.23

☐ **PERIOD AND AMPLITUDE**

The trigonometric functions are *periodic* in the sense that their values repeat. A trigonometric function that repeats every p units, but no sooner, is said to have *period p*. The sine and cosine functions repeat every 2π units (but no sooner) and the tangent function repeats every π units (but no sooner). Thus, $\sin x$ and $\cos x$ have period 2π and $\tan x$ has period π. It follows by taking reciprocals that $\sec x$ and $\csc x$ have period 2π and that $\cot x$ has period π.

For a trigonometric function, multiplying x by a *positive* factor b has the effect of multiplying the period by $1/b$. For example, $\sin 4x$ has a period of $2\pi/4 = \pi/2$ and $\tan \frac{1}{5}x$ has a period of 5π.

Example 1 Sketch the graph of $y = \sin 4x$.

Solution. The graph of $y = \sin x$ repeats when x changes by 2π. Thus, the graph of $y = \sin 4x$ repeats whenever $4x$ changes by 2π, or equivalently, when x changes by $\frac{1}{4}(2\pi) = \pi/2$. Thus, the factor of 4 in $y = \sin 4x$ has the effect of reducing the fundamental period by a factor of $\frac{1}{4}$ (Figure B.24). ◄

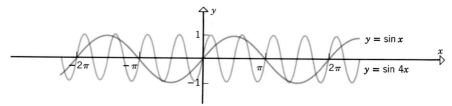

Figure B.24

A trigonometric function whose values oscillate between $-a$ and a (where $a > 0$) is said to have **amplitude** a. Thus, trigonometric functions of the form $\sin bx$ and $\cos bx$ have amplitude 1, since their values oscillate between -1 and 1. Multiplying $\sin bx$ or $\cos bx$ by a *positive* number a has the effect of multiplying the amplitude by a. Thus, $2 \sin 4x$ has amplitude 2 and $\frac{1}{5} \cos 7x$ has amplitude $\frac{1}{5}$.

Example 2 Sketch the graph of $y = 2 \sin 4x$.

Solution. The function has amplitude 2 and period $\pi/2$. The graph is shown in Figure B.25. ◄

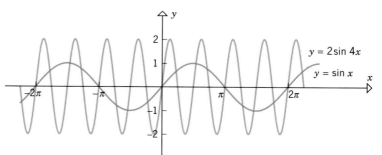

Figure B.25

► Exercise Set B.II

1. Find the period of
    (a) $\sin 5x$ (b) $\cos \frac{1}{3} x$ (c) $\tan \frac{1}{8} x$
    (d) $\tan 7x$ (e) $\sin \pi x$ (f) $\sec \frac{\pi}{5} x$
    (g) $\cot \pi x$ (h) $\csc \frac{x}{k}$.

In Exercises 2 and 3, find the amplitudes of the functions.

2. (a) $4 \sin 6x$ (b) $-8 \cos \pi x$
    (c) $\frac{1}{2} \sin x$ (d) $-\frac{1}{8} \cos 2x$.

3. (a) $5 \cos 2x$ (b) $\frac{1}{3} \sin 4x$
    (c) $-\frac{1}{2} \cos x$ (d) $-\sin \pi x$.

In Exercises 4–11, sketch the graph of the equation.

4. $y = \sin 3x$.   5. $y = 2 \sin 3x$.

6. $y = 2 \sin x$.   7. $y = \frac{1}{2} \cos \pi x$.

8. $y = -4 \cos x$.   9. $y = \cos(-4x)$.

10. $y = \sin(2\pi x)$.   11. $y = -2 \sin(-2x)$.

12. (a) How are the graphs of $y = \cos x$ and $y = \cos(-x)$ related?

     (b) How are the graphs of $y = \sin x$ and $y = \sin(-x)$ related?

13. Classify the following functions as even, odd, or neither. (See Exercise 49 in Section 2.3.)

(a) $\sin x$ (b) $\cos x$ (c) $\tan x$

(d) $|\sec x|$ (e) $\cot (x^2)$ (f) $3 \sec (-2x)$

(g) $-4 \csc x$ (h) $\dfrac{\sin x}{x}$.

In Exercises 14–17, sketch the graph of the equation.

14. $y = 3 \cot 2x$.   **15.** $y = \frac{1}{4} \sec \pi x$.

16. $y = 2 \csc \dfrac{x}{3}$.   **17.** $y = 2 \cot \dfrac{x}{4}$.

In Exercises 18–32, sketch the graph of the equation.

18. $y = \sin (x + \pi)$.   **19.** $y = \cos (x - \pi)$.

20. $y = \cos \left(x - \dfrac{\pi}{4} \right)$.   **21.** $y = \sin \left(x + \dfrac{\pi}{2} \right)$.

22. $y = 2 - \cos x$.   **23.** $y = 1 + \sin x$.

24. $y = 2 \cos \left(4x + \dfrac{\pi}{2} \right)$.   **25.** $y = 3 \sin \left(2x + \dfrac{\pi}{4} \right)$.

26. $y = \tan (\pi x - \frac{1}{2})$.   **27.** $y = 2 \tan (2\pi x + 1)$.

28. $y = |\tan x|$.   **29.** $y = |\cos x|$.

30. $y = \sin x - \cos x$.   **31.** $y = \sin x + \cos x$.

32. $y = 2 \sin x + 3 \cos 2x$.

APPENDIX C

Proofs

◼ C.I LIMITS: A RIGOROUS APPROACH; δ-ϵ PROOFS

In Section 2.4 we discussed the one-sided limits

$$\lim_{x \to a^-} f(x) = L \quad \text{and} \quad \lim_{x \to a^+} f(x) = L$$

from an intuitive viewpoint. We interpreted the first statement to mean that the value of $f(x)$ approaches L as x approaches a from the left side, and the second statement to mean that the value of $f(x)$ approaches L as x approaches a from the right side. We can formalize these ideas mathematically as follows:

DEFINITION 1.   Let f be defined on some open interval extending to the left from a. We shall write

$$\lim_{x \to a^-} f(x) = L$$

if given any number $\epsilon > 0$, we can find a number $\delta > 0$ such that $f(x)$ satisfies $|f(x) - L| < \epsilon$ whenever x satisfies $a - \delta < x < a$.

DEFINITION 2.   Let f be defined on some open interval extending to the right from a. We shall write

$$\lim_{x \to a^+} f(x) = L$$

if given any number $\epsilon > 0$, we can find a number $\delta > 0$ such that $f(x)$ satisfies $|f(x) - L| < \epsilon$ whenever x satisfies $a < x < a + \delta$.

Figure C.1 illustrates the difference between two-sided limits and one-sided limits: For a two-sided limit the condition $|f(x) - L|$ is required to hold on the intervals $(a - \delta, a)$ *and* $(a, a + \delta)$, whereas for a one-sided limit this condition is required to hold only on $(a - \delta, a)$ *or* $(a, a + \delta)$.

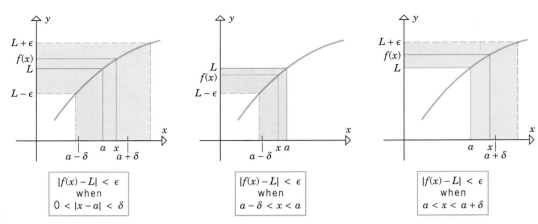

Figure C.1

Example 1   Prove that $\lim\limits_{x \to 0^+} \sqrt{x} = 0$.

Solution.   We shall apply Definition 2 with $f(x) = \sqrt{x}$ and $L = 0$. Thus, given $\epsilon > 0$, we must find a number $\delta > 0$ such that

$$\left|\sqrt{x} - 0\right| < \epsilon \quad \text{whenever } x \text{ satisfies} \quad 0 < x < 0 + \delta$$

or on simplifying and noting that $\left|\sqrt{x}\right| = \sqrt{x}$,

$$\sqrt{x} < \epsilon \quad \text{whenever } x \text{ satisfies} \quad 0 < x < \delta \tag{1}$$

But $\sqrt{x} < \epsilon$ if $0 < x < \epsilon^2$, so a value of δ that satisfies (1) is

$$\delta = \epsilon^2$$

thereby completing the proof.   ◀

REMARK.   In the preceding example we were concerned with only the limit from the right at $x = 0$ because $f(x) = \sqrt{x}$ does not have real values for $x < 0$; the two-sided limit and the limit from the left at $x = 0$ do not apply in this case.

Example 2   Prove that $\lim\limits_{x \to 1^-} f(x) = 2$ if $f(x) = \begin{cases} 2x, & x < 1 \\ x^3, & x \geq 1 \end{cases}$.

Solution.   We shall apply Definition 1 with $L = 2$ and $f(x) = 2x$. [The values of $f(x)$ for $x \geq 1$ are not relevant, since we are concerned with the limit from the left at 1.] Thus, given $\epsilon > 0$, we must find a number $\delta > 0$ such that

$$\left|2x - 2\right| < \epsilon \quad \text{whenever } x \text{ satisfies} \quad 1 - \delta < x < 1$$

or equivalently,

$$\left|x - 1\right| < \frac{\epsilon}{2} \quad \text{whenever } x \text{ satisfies} \quad 1 - \delta < x < 1$$

If $x < 1$, we have $\left|x - 1\right| = -(x - 1) = 1 - x$, so the preceding statement can be rewritten as

$$1 - x < \frac{\epsilon}{2} \quad \text{whenever } x \text{ satisfies} \quad 1 - \delta < x < 1$$

or alternatively,

$$1 - \frac{\epsilon}{2} < x \quad \text{whenever } x \text{ satisfies} \quad 1 - \delta < x < 1$$

A value of δ that meets this requirement is $\delta = \epsilon/2$.   ◀

□ **LIMITS AS x**
APPROACHES $+\infty$ OR $-\infty$

In Section 2.4 we discussed the limits

$$\lim_{x \to -\infty} f(x) = L \quad \text{and} \quad \lim_{x \to +\infty} f(x) = L$$

from an intuitive viewpoint. Informally stated, the first limit means that the values of $f(x)$ can be made as close as we like to L by restricting x to be sufficiently far from the origin on the negative x-axis. Similarly, the second limit means that the values of $f(x)$ can be made as close as we like to L by restricting x to be sufficiently far from the origin on the positive x-axis. The following definitions make these ideas precise.

DEFINITION **3.**   Let f be a function that is defined on some infinite open interval $(-\infty, x_0)$. We shall write

$$\lim_{x \to -\infty} f(x) = L$$

if given any number $\epsilon > 0$, there corresponds a negative number N such that

$$|f(x) - L| < \epsilon \quad \text{whenever x satisfies} \quad x < N$$

DEFINITION **4.**   Let f be a function that is defined on some infinite open interval $(x_0, +\infty)$. We shall write

$$\lim_{x \to +\infty} f(x) = L$$

if given any number $\epsilon > 0$, there corresponds a positive number N such that

$$|f(x) - L| < \epsilon \quad \text{whenever x satisfies} \quad x > N$$

These definitions are illustrated in Figure C.2.

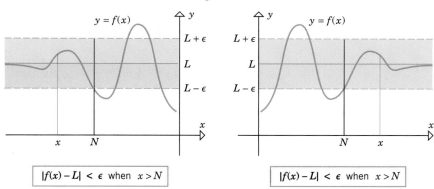

Figure C.2

$$|f(x) - L| < \epsilon \text{ when } x > N \qquad |f(x) - L| < \epsilon \text{ when } x > N$$

Example 3   Prove that $\displaystyle\lim_{x \to +\infty} \frac{1}{x} = 0$.

Solution.   We shall apply Definition 4 with $f(x) = 1/x$ and $L = 0$. Thus, given $\epsilon > 0$, we must find a number $N > 0$ such that

$$\left| \frac{1}{x} - 0 \right| < \epsilon \quad \text{whenever x satisfies} \quad x > N$$

If $x > 0$, we can omit the absolute values and rewrite this statement as

$$\frac{1}{x} < \epsilon \quad \text{whenever x satisfies} \quad x > N$$

or, on taking reciprocals,

$$x > \frac{1}{\epsilon} \quad \text{whenever } x \text{ satisfies} \quad x > N$$

It is self-evident that a value of N that meets this requirement is $N = 1/\epsilon$, which completes the proof.   ◄

□ **INFINITE LIMITS**

In Section 2.4 we discussed limits of the type

$$\lim_{x \to a} f(x) = +\infty, \qquad \lim_{x \to a} f(x) = -\infty \tag{2-3}$$

$$\lim_{x \to a^+} f(x) = +\infty, \qquad \lim_{x \to a^+} f(x) = -\infty \tag{4-5}$$

$$\lim_{x \to a^-} f(x) = +\infty, \qquad \lim_{x \to a^-} f(x) = -\infty \tag{6-7}$$

$$\lim_{x \to +\infty} f(x) = +\infty, \qquad \lim_{x \to +\infty} f(x) = -\infty \tag{8-9}$$

$$\lim_{x \to -\infty} f(x) = +\infty, \qquad \lim_{x \to -\infty} f(x) = -\infty \tag{10-11}$$

from an intuitive viewpoint. Recall that each of these expressions describes a particular way in which the limit fails to exist. The $+\infty$ to the right of the equal sign indicates that the limit fails to exist because $f(x)$ is increasing without bound, and the $-\infty$ indicates that the limit fails to exist because $f(x)$ is decreasing without bound.

It would be too time-consuming to examine each of these limits in detail, so we shall focus only on the two-sided limits

$$\lim_{x \to a} f(x) = +\infty \quad \text{and} \quad \lim_{x \to a} f(x) = -\infty$$

Informally stated, the first limit means that the values of $f(x)$ can be made larger than any specified positive number by restricting x sufficiently close to but different from a; and the second limit means that the values of $f(x)$ can be made smaller than any specified negative number by restricting x sufficiently close to but different from a. The following definitions make these ideas precise.

DEFINITION **5**.   Let f be defined in some open interval containing the number a, except that f need not be defined at a. We shall write

$$\lim_{x \to a} f(x) = +\infty$$

if given any positive number N, we can find a number $\delta > 0$ such that $f(x)$ satisfies

$$f(x) > N \quad \text{whenever } x \text{ satisfies} \quad 0 < |x - a| < \delta$$

DEFINITION **6**.   Let f be defined in some open interval containing the number a, except that f need not be defined at a. We shall write

$$\lim_{x \to a} f(x) = -\infty$$

if given any negative number N, we can find a number $\delta > 0$ such that $f(x)$ satisfies

$$f(x) < N \quad \text{whenever } x \text{ satisfies} \quad 0 < |x - a| < \delta$$

These definitions are illustrated in Figure C.3.

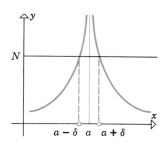

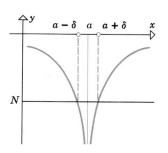

$f(x) > N$ whenever
$0 < |x - a| < \delta$

$f(x) < N$ whenever
$0 < |x - a| < \delta$

Figure C.3

Example 4 Prove that $\lim\limits_{x \to 0} \dfrac{1}{x^2} = +\infty$.

Solution. We shall apply Definition 5 with $f(x) = 1/x^2$ and $a = 0$. Thus, given a number $N > 0$, we must find a number $\delta > 0$ such that

$$\frac{1}{x^2} > N \quad \text{whenever } x \text{ satisfies} \quad 0 < |x - 0| < \delta$$

or, on taking reciprocals and simplifying,

$$x^2 < \frac{1}{N} \quad \text{whenever } x \text{ satisfies} \quad 0 < |x| < \delta \tag{12}$$

But $x^2 < 1/N$ if $|x| < 1/\sqrt{N}$. Thus $\delta = 1/\sqrt{N}$ satisfies (12). ◄

▶ Exercise Set C.I

In Exercises 1–6, use Definitions 1 and 2 to prove that the stated limit is correct.

1. $\lim\limits_{x \to 2^+} (x + 1) = 3$. **2.** $\lim\limits_{x \to 1^-} (3x + 2) = 5$.

3. $\lim\limits_{x \to 4^+} \sqrt{x - 4} = 0$. **4.** $\lim\limits_{x \to 0^-} \sqrt{-x} = 0$.

5. $\lim\limits_{x \to 2^+} f(x) = 2$, where $f(x) = \begin{cases} x, & x > 2 \\ 3x, & x \le 2. \end{cases}$

6. $\lim\limits_{x \to 2^-} f(x) = 6$, where $f(x) = \begin{cases} x, & x > 2 \\ 3x, & x \le 2. \end{cases}$

In Exercises 7–14, use Definitions 3 and 4 to prove that the stated limit is correct.

7. $\lim\limits_{x \to +\infty} \dfrac{1}{x^2} = 0$. **8.** $\lim\limits_{x \to -\infty} \dfrac{1}{x} = 0$.

9. $\lim\limits_{x \to -\infty} \dfrac{1}{x + 2} = 0$. **10.** $\lim\limits_{x \to +\infty} \dfrac{1}{x + 2} = 0$.

11. $\lim\limits_{x \to +\infty} \dfrac{x}{x + 1} = 1$. **12.** $\lim\limits_{x \to -\infty} \dfrac{x}{x + 1} = 1$.

13. $\lim\limits_{x \to -\infty} \dfrac{4x - 1}{2x + 5} = 2$. **14.** $\lim\limits_{x \to +\infty} \dfrac{4x - 1}{2x + 5} = 2$.

In Exercises 15–20, use Definitions 5 and 6 to prove that the stated limit is correct.

15. $\lim\limits_{x \to 3} \dfrac{1}{(x - 3)^2} = +\infty$. **16.** $\lim\limits_{x \to 3} \dfrac{-1}{(x - 3)^2} = -\infty$.

17. $\lim\limits_{x \to 0} \dfrac{1}{|x|} = +\infty$. **18.** $\lim\limits_{x \to 1} \dfrac{1}{|x - 1|} = +\infty$.

19. $\lim\limits_{x \to 0} \left(-\dfrac{1}{x^4}\right) = -\infty$. **20.** $\lim\limits_{x \to 0} \dfrac{1}{x^4} = +\infty$.

21. Define limit statements in (4)–(7).

22. Use the definitions in Exercise 21 to prove

(a) $\lim\limits_{x \to 0^+} \dfrac{1}{x} = +\infty$ (b) $\lim\limits_{x \to 0^-} \dfrac{1}{x} = -\infty$.

23. Use the definitions in Exercise 21 to prove

(a) $\lim\limits_{x \to 1^+} \dfrac{1}{1 - x} = -\infty$ (b) $\lim\limits_{x \to 1^-} \dfrac{1}{1 - x} = +\infty$.

24. Define limit statements in (8)–(11).

25. Use the definitions in Exercise 24 to prove

(a) $\lim\limits_{x \to +\infty} (x + 1) = +\infty$ (b) $\lim\limits_{x \to -\infty} (x + 1) = -\infty$.

26. Use the definitions in Exercise 24 to prove

(a) $\lim\limits_{x \to +\infty} (x^2 - 3) = +\infty$ (b) $\lim\limits_{x \to -\infty} (x^3 + 5) = -\infty$.

■ C.II SELECTED PROOFS

An extensive excursion into proofs of limit theorems is too time-consuming to undertake, so we shall focus on a few proofs that are instructive and illustrate the basic techniques. The following theorem contains a selection of results from Section 2.5.

THEOREM 7. *Let k be a constant, and suppose that $\lim\limits_{x \to a} f(x) = L_1$ and $\lim\limits_{x \to a} g(x) = L_2$. Then*

(a) $\lim\limits_{x \to a} k = k$

(b) $\lim\limits_{x \to a} [f(x) + g(x)] = \lim\limits_{x \to a} f(x) + \lim\limits_{x \to a} g(x) = L_1 + L_2$

(c) $\lim\limits_{x \to a} [f(x)g(x)] = \lim\limits_{x \to a} f(x) \lim\limits_{x \to a} g(x) = L_1 L_2$

Proof (a).   We shall apply Definition 2.6.3 with $f(x) = k$ and $L = k$. Thus, given $\epsilon > 0$, we must find a number $\delta > 0$ such that

$$|k - k| < \epsilon \quad \text{whenever } x \text{ satisfies} \quad 0 < |x - a| < \delta$$

or equivalently,

$$0 < \epsilon \quad \text{whenever } x \text{ satisfies} \quad 0 < |x - a| < \delta$$

But the condition on the left side of this statement is *always* true, no matter how δ is chosen. Thus, any positive value for δ will suffice.

Proof (b).   We must show that given $\epsilon > 0$ we can find a number $\delta > 0$ such that

$$|(f(x) + g(x)) - (L_1 + L_2)| < \epsilon \quad \text{whenever} \quad 0 < |x - a| < \delta \tag{1}$$

However, from the limits of f and g in the hypothesis of the theorem we can find numbers δ_1 and δ_2 such that

$$|f(x) - L_1| < \epsilon/2 \quad \text{whenever} \quad 0 < |x - a| < \delta_1$$
$$|g(x) - L_2| < \epsilon/2 \quad \text{whenever} \quad 0 < |x - a| < \delta_2$$

Moreover, the inequalities on the left sides of these statements *both* hold if we replace δ_1 and δ_2 by any positive number δ that is less than both δ_1 and δ_2. Thus, for any such δ it follows that

$$|f(x) - L_1| + |g(x) - L_2| < \epsilon \quad \text{whenever} \quad 0 < |x - a| < \delta \tag{2}$$

However, it follows from the triangle inequality (Theorem 1.2.5) that

$$|(f(x) + g(x)) - (L_1 + L_2)| = |(f(x) - L_1) + (g(x) - L_2)|$$
$$\leq |f(x) - L_1| + |g(x) - L_2|$$

so that (1) follows from (2).

Proof (c).   We must show that given $\epsilon > 0$ we can find a number $\delta > 0$ such that

$$|f(x)g(x) - L_1 L_2| < \epsilon \quad \text{whenever} \quad 0 < |x - a| < \delta \tag{3}$$

To find δ it will be helpful to express (3) in a different form. If we rewrite $f(x)$ and $g(x)$ as

$$f(x) = L_1 + (f(x) - L_1) \quad \text{and} \quad g(x) = L_2 + (g(x) - L_2)$$

then the inequality on the left side of (3) can be expressed as (verify)

$$|L_1(g(x) - L_2) + L_2(f(x) - L_1) + (f(x) - L_1)(g(x) - L_2)| < \epsilon \tag{4}$$

Since

$$\lim_{x \to a} f(x) = L_1 \quad \text{and} \quad \lim_{x \to a} g(x) = L_2$$

we can find positive numbers δ_1, δ_2, δ_3, and δ_4 such that

$$|f(x) - L_1| < \sqrt{\epsilon/3} \qquad \text{whenever } x \text{ satisfies} \quad 0 < |x - a| < \delta_1$$

$$|f(x) - L_1| < \frac{\epsilon}{3(1 + |L_2|)} \qquad \text{whenever } x \text{ satisfies} \quad 0 < |x - a| < \delta_2$$

$$|g(x) - L_2| < \sqrt{\epsilon/3} \qquad \text{whenever } x \text{ satisfies} \quad 0 < |x - a| < \delta_3 \qquad (5)$$

$$|g(x) - L_2| < \frac{\epsilon}{3(1 + |L_1|)} \qquad \text{whenever } x \text{ satisfies} \quad 0 < |x - a| < \delta_4$$

Moreover, the inequalities on the left sides of these four statements *all* hold if we replace δ_1, δ_2, δ_3, and δ_4 by any number δ that is smaller than δ_1, δ_2, δ_3, and δ_4. Thus, for any such δ it follows with the help of the triangle inequality (Theorem 1.2.5) that

$$|L_1(g(x) - L_2) + L_2(f(x) - L_1) + (f(x) - L_1)(g(x) - L_2)|$$

$$\leq |L_1(g(x) - L_2)| + |L_2(f(x) - L_1)| + |(f(x) - L_1)(g(x) - L_2)|$$

$$= |L_1|\,|g(x) - L_2| + |L_2|\,|f(x) - L_1| + |f(x) - L_1|\,|g(x) - L_2|$$

$$< |L_1|\frac{\epsilon}{3(1 + |L_1|)} + |L_2|\frac{\epsilon}{3(1 + |L_2|)} + \sqrt{\epsilon/3}\sqrt{\epsilon/3} \quad \boxed{\text{From (5)}}$$

$$= \frac{\epsilon}{3}\frac{|L_1|}{1 + |L_1|} + \frac{\epsilon}{3}\frac{|L_2|}{1 + |L_2|} + \frac{\epsilon}{3}$$

$$< \frac{\epsilon}{3} + \frac{\epsilon}{3} + \frac{\epsilon}{3} = \epsilon \quad \boxed{\text{Since } \frac{|L_1|}{1 + |L_1|} < 1 \text{ and } \frac{|L_2|}{1 + |L_2|} < 1}$$

which shows that (4) holds for the δ selected.   ∎

REMARK.   Do not be alarmed if the proof of part (c) seems difficult; it takes some experience with proofs of this type to develop a feel for choosing the right δ. Your initial goal should be to understand the ideas and the computations.

☐ **PROOF OF A BASIC CONTINUITY PROPERTY**

Next, we shall prove Theorem 2.7.5 for two-sided limits.

> THEOREM **8** (*Theorem 2.7.5*).   *If* $\lim_{x \to c} g(x) = L$ *and if the function f is continuous at L,*
> *then* $\lim_{x \to c} f(g(x)) = f(L)$; *that is,* $\lim_{x \to c} f(g(x)) = f(\lim_{x \to c} g(x))$.

Proof.   We must show that given $\epsilon > 0$, we can find a number $\delta > 0$ such that

$$|f(g(x)) - f(L)| < \epsilon \quad \text{whenever } x \text{ satisfies} \quad 0 < |x - c| < \delta \qquad (6)$$

Since f is continuous at L we have

$$\lim_{u \to L} f(u) = f(L)$$

and hence we can find a number $\delta_1 > 0$ such that

$$|f(u) - f(L)| < \epsilon \quad \text{whenever } u \text{ satisfies} \quad |u - L| < \delta_1$$

In particular, if $u = g(x)$, then

$$|f(g(x)) - f(L)| < \epsilon \quad \text{whenever } g(x) \text{ satisfies} \quad |g(x) - L| < \delta_1 \qquad (7)$$

But $\lim\limits_{x \to c} g(x) = L$, and hence there is a number $\delta > 0$ such that

$$|g(x) - L| < \delta_1 \quad \text{whenever } x \text{ satisfies} \quad 0 < |x - c| < \delta \tag{8}$$

Thus, if x satisfies the condition on the right side of statement (8), then it follows that $g(x)$ satisfies the condition on the right side of statement (7), and this implies that the condition on the left side of statement (6) is satisfied, completing the proof. ∎

☐ **PROOF OF THE CHAIN RULE**

Our next objective is to prove the chain rule (Theorem 3.5.2); but first we need a preliminary result.

THEOREM **9**.   *If f is differentiable at x and if $y = f(x)$, then*

$$\Delta y = f'(x)\,\Delta x + \epsilon\,\Delta x$$

where $\epsilon \to 0$ as $\Delta x \to 0$ and $\epsilon = 0$ if $\Delta x = 0$.

Proof.   Define

$$\epsilon = \begin{cases} \dfrac{f(x + \Delta x) - f(x)}{\Delta x} - f'(x) & \text{if } \Delta x \neq 0 \\ \\ 0 & \text{if } \Delta x = 0 \end{cases} \tag{9}$$

If $\Delta x \neq 0$, it follows from (9) that

$$\epsilon\,\Delta x = [\,f(x + \Delta x) - f(x)\,] - f'(x)\,\Delta x \tag{10}$$

But

$$\Delta y = f(x + \Delta x) - f(x) \tag{11}$$

so (10) can be written as

$$\epsilon\,\Delta x = \Delta y - f'(x)\,\Delta x$$

or

$$\Delta y = f'(x)\,\Delta x + \epsilon\,\Delta x \tag{12}$$

If $\Delta x = 0$, then (12) still holds (why?), so (12) is valid for all values of Δx. It remains to show that $\epsilon \to 0$ as $\Delta x \to 0$. But this follows from the assumption that f is differentiable at x, since

$$\lim_{\Delta x \to 0} \epsilon = \lim_{\Delta x \to 0} \left[\frac{f(x + \Delta x) - f(x)}{\Delta x} - f'(x) \right] = f'(x) - f'(x) = 0 \quad ∎$$

We are now ready to prove the chain rule.

THEOREM **10** (*Theorem* **3.5.2**).   *If g is differentiable at the point x and f is differentiable at the point $g(x)$, then the composition $f \circ g$ is differentiable at the point x. Moreover, if $y = f(g(x))$ and $u = g(x)$, then*

$$\frac{dy}{dx} = \frac{dy}{du} \cdot \frac{du}{dx}$$

Proof.   Since g is differentiable at x and $u = g(x)$, it follows from Theorem 9 that

$$\Delta u = g'(x)\,\Delta x + \epsilon_1\,\Delta x \tag{13}$$

where $\epsilon_1 \to 0$ as $\Delta x \to 0$. And since $y = f(u)$ is differentiable at $u = g(x)$, it follows from Theorem 9 that

$$\Delta y = f'(u)\,\Delta u + \epsilon_2\,\Delta u \tag{14}$$

where $\epsilon_2 \to 0$ as $\Delta u \to 0$.

Factoring out the Δu in (14) and then substituting (13) yields

$$\Delta y = [f'(u) + \epsilon_2][g'(x)\,\Delta x + \epsilon_1\,\Delta x]$$

or

$$\Delta y = [f'(u) + \epsilon_2][g'(x) + \epsilon_1]\,\Delta x$$

or if $\Delta x \neq 0$,

$$\frac{\Delta y}{\Delta x} = [f'(u) + \epsilon_2][g'(x) + \epsilon_1] \tag{15}$$

But (13) implies that $\Delta u \to 0$ as $\Delta x \to 0$, and hence $\epsilon_1 \to 0$ and $\epsilon_2 \to 0$ as $\Delta x \to 0$. Thus, from (15)

$$\lim_{\Delta x \to 0} \frac{\Delta y}{\Delta x} = f'(u)g'(x)$$

or

$$\frac{dy}{dx} = f'(u)g'(x) = \frac{dy}{du}\cdot\frac{du}{dx} \qquad \blacksquare$$

☐ **PROOFS OF THE FIRST AND SECOND DERIVATIVE TESTS**

We shall now prove the first and second derivative tests (Theorems 4.3.6 and 4.3.7), which we shall restate here for ease of reference.

THEOREM **11** (*Theorem* **4.3.6**). *Suppose that f is continuous at a critical point x_0.*

(*a*) *If $f'(x) > 0$ on an open interval extending left from x_0 and $f'(x) < 0$ on an open interval extending right from x_0, then f has a relative maximum at x_0.*

(*b*) *If $f'(x) < 0$ on an open interval extending left from x_0 and $f'(x) > 0$ on an open interval extending right from x_0, then f has a relative minimum at x_0.*

(*c*) *If $f'(x)$ has the same sign [either $f'(x) > 0$ or $f'(x) < 0$] on an open interval extending left from x_0 and on an open interval extending right from x_0, then f does not have a relative extremum at x_0.*

Proof (a). In accordance with the hypothesis, assume that $f'(x) > 0$ on the interval (a, x_0) and $f'(x) < 0$ on the interval (x_0, b). To prove that f has a relative maximum at x_0, we shall show that

$$f(x_0) \geq f(x) \tag{16}$$

for all x in (a, b). From Theorem 4.2.2, f is increasing on the interval $(a, x_0]$, and decreasing on the interval $[x_0, b)$, so $f(x_0) \geq f(x)$ for all x in (a, b) with equality only at $x = x_0$, which proves (16). $\blacksquare$

The proofs of parts (*b*) and (*c*) are left as exercises.

THEOREM **12** (*Theorem* **4.3.7**). *Suppose that f is twice differentiable at a stationary point x_0.*

(*a*) *If $f''(x_0) > 0$, then f has a relative minimum at x_0.*

(*b*) *If $f''(x_0) < 0$, then f has a relative maximum at x_0.*

Proof (a).   Using the definition of a derivative we can write

$$f''(x_0) = \lim_{h \to 0} \frac{f'(x_0 + h) - f'(x_0)}{h} \tag{17}$$

By hypothesis, $f''(x_0) > 0$, so we can use $\epsilon = \frac{1}{2}f''(x_0)$ in the definition of a limit and deduce from (17) that there exists a $\delta > 0$ such that

$$\left| \frac{f'(x_0 + h) - f'(x_0)}{h} - f''(x_0) \right| < \frac{1}{2} f''(x_0) \tag{18}$$

whenever h satisfies

$$0 < |h| < \delta \tag{19}$$

To prove that f has a relative minimum at x_0, we shall show that

$$f'(x) > 0 \quad \text{for all } x \text{ in } (x_0, x_0 + \delta) \tag{20}$$

and

$$f'(x) < 0 \quad \text{for all } x \text{ in } (x_0 - \delta, x_0) \tag{21}$$

It will then follow from the first derivative test (Theorem 11) that f has a relative minimum at x_0.

From (18) and (19) it follows that

$$\frac{1}{2}f''(x_0) < \frac{f'(x_0 + h) - f'(x_0)}{h} < \frac{3}{2}f''(x_0) \tag{22}$$

whenever h satisfies

$$0 < |h| < \delta$$

By hypothesis, x_0 is a stationary point for f, so that $f'(x_0) = 0$. From this, the fact that $f''(x_0) > 0$, and the left-hand inequality in (22), it follows that

$$0 < \frac{f'(x_0 + h)}{h} \tag{23}$$

whenever h satisfies (19).

To prove (20), let x be any point in $(x_0, x_0 + \delta)$. If we let

$$h = x - x_0 \tag{24}$$

then $h > 0$ and h satisfies (19), so that (23) holds.

Multiplying (23) by h yields

$$0 < f'(x_0 + h) \tag{25}$$

Then substituting (24) in (25) yields $0 < f'(x)$, which establishes (20). To obtain (21) let x be any point in $(x_0 - \delta, x_0)$. If we let

$$h = x - x_0 \tag{26}$$

then $h < 0$ and h satisfies (19) so that (23) holds.

Multiplying (23) by h yields

$$f'(x_0 + h) < 0 \tag{27}$$

Then substituting (26) in (27) yields $f'(x) < 0$, which establishes (21).   ∎

The proof of part (b) is similar and is left as an exercise.

► Exercise Set C.II

1. Prove: For any constant k, $\lim\limits_{x \to +\infty} k = k$.

2. Prove: For any constant k, $\lim\limits_{x \to -\infty} k = k$.

3. Prove: If $\lim\limits_{x \to -\infty} f(x) = L_1$ and $\lim\limits_{x \to -\infty} g(x) = L_2$, then

$$\lim\limits_{x \to -\infty} [f(x) + g(x)] = L_1 + L_2.$$

4. Prove: If $\lim\limits_{x \to +\infty} f(x) = L_1$ and $\lim\limits_{x \to +\infty} g(x) = L_2$, then

$$\lim\limits_{x \to +\infty} [f(x) + g(x)] = L_1 + L_2.$$

5. Use Theorem 7 of this section to prove: If $\lim\limits_{x \to a} f(x) = L_1$ and $\lim\limits_{x \to a} g(x) = L_2$, then

$$\lim\limits_{x \to a} [f(x) - g(x)] = \lim\limits_{x \to a} f(x) - \lim\limits_{x \to a} g(x)$$
$$= L_1 - L_2$$

6. Suppose that $\lim\limits_{x \to a} f(x) = +\infty$ and $\lim\limits_{x \to a} g(x) = +\infty$.

  (a) Prove: $\lim\limits_{x \to a} [f(x) + g(x)] = +\infty$.

  (b) Is it true that $\lim\limits_{x \to a} [f(x) - g(x)] = 0$?

7. Suppose that $\lim\limits_{x \to a} f(x) = -\infty$ and $\lim\limits_{x \to a} g(x) = +\infty$.

  (a) Prove: $\lim\limits_{x \to a} [f(x) - g(x)] = -\infty$.

  (b) Is it true that $\lim\limits_{x \to a} [f(x) + g(x)] = 0$?

8. Use parts (a) and (c) of Theorem 7 of this section to prove: If $\lim\limits_{x \to a} f(x) = L$ and k is a constant, then

$$\lim\limits_{x \to a} [kf(x)] = kL$$

9. Prove: $\lim\limits_{x \to a} f(x) = L$ if and only if

$$\lim\limits_{x \to a} [f(x) - L] = 0$$

10. Prove: If $\lim\limits_{x \to a} f(x) = L$, then $\lim\limits_{x \to a} |f(x)| = |L|$.

11. Prove:

  (a) part (b) of Theorem 11

  (b) part (c) of Theorem 11.

12. Prove part (b) of Theorem 12.

13. A function f is called **nondecreasing** on a given interval if

$$f(x_2) \geq f(x_1) \quad \text{whenever} \quad x_2 > x_1$$

where x_1 and x_2 are points in the interval. Similarly, f is called **nonincreasing** on the interval if

$$f(x_2) \leq f(x_1) \quad \text{whenever} \quad x_2 > x_1$$

  (a) Sketch the graph of a function that is nondecreasing, yet is not increasing.

  (b) Sketch the graph of a function that is nonincreasing, yet is not decreasing.

[*Remark:* Unfortunately, terminology is not always consistent in the mathematical literature. Some writers use the terms "strictly increasing" and "strictly decreasing" where we have used the terms increasing and decreasing. These writers then use the terms "increasing" and "decreasing" where we have used the terms nondecreasing and nonincreasing.]

14. Use the Mean-Value Theorem to prove each of the following. (See Exercise 13 for the definitions of nondecreasing and nonincreasing.)

  (a) If $f'(x) \geq 0$ on an open interval (a, b), then f is nondecreasing on (a, b).

  (b) If $f'(x) \leq 0$ on an open interval (a, b), then f is nonincreasing on (a, b).

15. · Prove: If $f'(x) < g'(x)$ for all x in (a, b), then for all points x_1, x_2 in (a, b), where $x_2 > x_1$,

$$f(x_2) - f(x_1) < g(x_2) - g(x_1)$$

16. (a) Prove: If $f'(x_0) > 0$ and f' is continuous at x_0, then there is an open interval containing x_0 on which f is increasing.

  (b) Prove: If $f'(x_0) < 0$ and f' is continuous at x_0, then there is an open interval containing x_0 on which f is decreasing.

17. (a) Use Exercise 16(a) to prove: If $f''(x_0) > 0$ and f'' is continuous at x_0, then there is an open interval containing x_0 on which f is concave up.

  (b) Use Exercise 16(b) to prove: If $f''(x_0) < 0$ and f'' is continuous at x_0, then there is an open interval containing x_0 on which f is concave down.

APPENDIX D

Cramer's Rule

We are concerned here with solving systems of equations of the forms

$$a_1x + b_1y = k_1$$
$$a_2x + b_2y = k_2.$$

and

$$a_1x + b_1y + c_1z = k_1$$
$$a_2x + b_2y + c_2z = k_2$$
$$a_3x + b_3y + c_3z = k_3$$

(1–2)

The first is a system of **two linear equations in two unknowns**, x and y; and the second is a system of **three linear equations in three unknowns**, x, y, and z. Values of the unknowns that satisfy *all* of the equations of a system are said to form a **solution** of the system.

The graph of each equation in (1) is a line in 2-space. Thus, if (x, y) is a point that lies on both lines, then the coordinates of this point satisfy both equations in (1), and x and y form a solution of the system. Similarly, the graph of each equation in (2) is a plane in 3-space, and the coordinates of any point (x, y, z) that lies on all three planes form a solution of the system. Figures D.1 and D.2 illustrate that in both cases there are only three possibilities—the system has no solutions, the system has exactly one solution, or the system has infinitely many solutions.

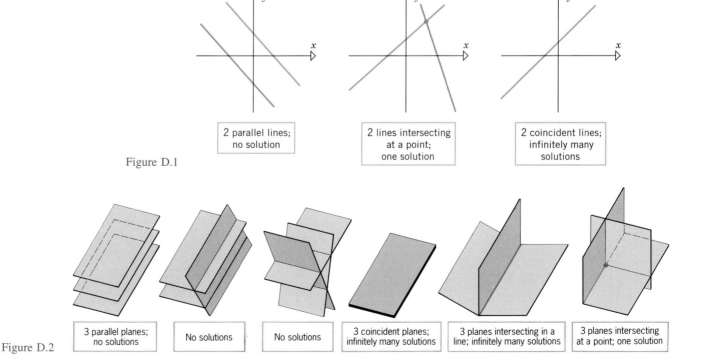

| 2 parallel lines; no solution | 2 lines intersecting at a point; one solution | 2 coincident lines; infinitely many solutions |

Figure D.1

| 3 parallel planes; no solutions | No solutions | No solutions | 3 coincident planes; infinitely many solutions | 3 planes intersecting in a line; infinitely many solutions | 3 planes intersecting at a point; one solution |

Figure D.2

The determinants

$$\begin{vmatrix} a_1 & b_1 \\ a_2 & b_2 \end{vmatrix} \quad \text{and} \quad \begin{vmatrix} a_1 & b_1 & c_1 \\ a_2 & b_2 & c_2 \\ a_3 & b_3 & c_3 \end{vmatrix}$$

are called the ***coefficient determinants*** for the systems in (1) and (2). It is proved in linear algebra courses that in both cases the system has a unique solution if and only if its coefficient determinant is *not* zero. In this event the solution can be expressed as a ratio of determinants, using one of the following theorems, which are special cases of a general result called ***Cramer's rule***.*

THEOREM **1** (***Cramer's Rule for Two Unknowns***).   *The system of equations*

$$a_1 x + b_1 y = k_1$$

$$a_2 x + b_2 y = k_2$$

has a unique solution if and only if its coefficient determinant is not zero, in which case the solution can be expressed as

$$x = \frac{\begin{vmatrix} k_1 & b_1 \\ k_2 & b_2 \end{vmatrix}}{\begin{vmatrix} a_1 & b_1 \\ a_2 & b_2 \end{vmatrix}}, \quad y = \frac{\begin{vmatrix} a_1 & k_1 \\ a_2 & k_2 \end{vmatrix}}{\begin{vmatrix} a_1 & b_1 \\ a_2 & b_2 \end{vmatrix}}$$

THEOREM **2** (***Cramer's Rule for Three Unknowns***).   *The system of equations*

$$a_1 x + b_1 y + c_1 z = k_1$$

$$a_2 x + b_2 y + c_2 z = k_2$$

$$a_3 x + b_3 y + c_3 z = k_3$$

has a unique solution if and only if its coefficient determinant is not zero, in which case the solution can be expressed as

$$x = \frac{\begin{vmatrix} k_1 & b_1 & c_1 \\ k_2 & b_2 & c_2 \\ k_3 & b_3 & c_3 \end{vmatrix}}{\begin{vmatrix} a_1 & b_1 & c_1 \\ a_2 & b_2 & c_2 \\ a_3 & b_3 & c_3 \end{vmatrix}}, \quad y = \frac{\begin{vmatrix} a_1 & k_1 & c_1 \\ a_2 & k_2 & c_2 \\ a_3 & k_3 & c_3 \end{vmatrix}}{\begin{vmatrix} a_1 & b_1 & c_1 \\ a_2 & b_2 & c_2 \\ a_3 & b_3 & c_3 \end{vmatrix}}, \quad z = \frac{\begin{vmatrix} a_1 & b_1 & k_1 \\ a_2 & b_2 & k_2 \\ a_3 & b_3 & k_3 \end{vmatrix}}{\begin{vmatrix} a_1 & b_1 & c_1 \\ a_2 & b_2 & c_2 \\ a_3 & b_3 & c_3 \end{vmatrix}}$$

REMARK.   There is a pattern to the formulas in these theorems. In each formula the determinant in the denominator is formed from the coefficients of the unknowns. The determinant in the numerator differs from the determinant in the denominator in that the coefficients of the unknown being calculated are replaced by the k's. It is assumed in these theorems that the system is written so that like unknowns are aligned vertically and the constants appear by themselves on the right side of each equation.

---

*GABRIEL CRAMER (see p. A32 for biography).

Example 1 Use Cramer's rule to solve

$$\text{(a)} \quad \begin{matrix} 5x - 2y = -1 \\ 2x + 3y = 3 \end{matrix} \qquad \text{(b)} \quad \begin{matrix} x + 2z = 6 \\ -3x + 4y + 6z = 30 \\ -x - 2y + 3z = 8 \end{matrix}$$

Solution (*a*).

$$x = \frac{\begin{vmatrix} -1 & -2 \\ 3 & 3 \end{vmatrix}}{\begin{vmatrix} 5 & -2 \\ 2 & 3 \end{vmatrix}} = \frac{3}{19}; \quad y = \frac{\begin{vmatrix} 5 & -1 \\ 2 & 3 \end{vmatrix}}{\begin{vmatrix} 5 & -2 \\ 2 & 3 \end{vmatrix}} = \frac{17}{19}$$

Solution (*b*).

$$x = \frac{\begin{vmatrix} 6 & 0 & 2 \\ 30 & 4 & 6 \\ 8 & -2 & 3 \end{vmatrix}}{\begin{vmatrix} 1 & 0 & 2 \\ -3 & 4 & 6 \\ -1 & -2 & 3 \end{vmatrix}} = \frac{-10}{11}; \quad y = \frac{\begin{vmatrix} 1 & 6 & 2 \\ -3 & 30 & 6 \\ -1 & 8 & 3 \end{vmatrix}}{\begin{vmatrix} 1 & 0 & 2 \\ -3 & 4 & 6 \\ -1 & -2 & 3 \end{vmatrix}} = \frac{18}{11}; \quad z = \frac{\begin{vmatrix} 1 & 0 & 6 \\ -3 & 4 & 30 \\ -1 & -2 & 8 \end{vmatrix}}{\begin{vmatrix} 1 & 0 & 2 \\ -3 & 4 & 6 \\ -1 & -2 & 3 \end{vmatrix}} = \frac{38}{11}$$

◀

▶ Exercise Set D

In Exercises 1–8, solve the system using Cramer's rule.

1. $\begin{matrix} 3x - 4y = -5 \\ 2x + y = 4. \end{matrix}$

2. $\begin{matrix} -x + 3y = 8 \\ 2x + 5y = 7. \end{matrix}$

3. $\begin{matrix} 2x_1 - 5x_2 = -2 \\ 4x_1 + 6x_2 = 1. \end{matrix}$

4. $\begin{matrix} 3a + 2b = 4 \\ -a + b = 7. \end{matrix}$

5. $\begin{matrix} x + 2y + z = 3 \\ 2x + y - z = 0 \\ x - y + z = 6. \end{matrix}$

6. $\begin{matrix} x - 3y + z = 4 \\ 2x - y = -2 \\ 4x - 3z = 0. \end{matrix}$

7. $\begin{matrix} x_1 + x_2 - 2x_3 = 1 \\ 2x_1 - x_2 + x_3 = 2 \\ x_1 - 2x_2 - 4x_3 = -4. \end{matrix}$

8. $\begin{matrix} r + s + t = 2 \\ r - s - 2t = 0 \\ -r + 2s + t = 4. \end{matrix}$

*GABRIEL CRAMER (1704–1752). Swiss mathematician. Although Cramer does not rank with the great mathematicians of his time, his contributions as a disseminator of mathematical ideas have earned him a well-deserved place in the history of mathematics. The son of a physician, Cramer was born and educated in Geneva, Switzerland. At age 20 he competed for, but failed to secure, the chair of philosophy at the Académie de Calvin at Geneva. However, the awarding magistrates were sufficiently impressed with Cramer and a fellow competitor to create a new chair of mathematics for both men to share. Alternately, each assumed the full responsibility and salary associated with the chair for two or three years while the other traveled. During his travels Cramer met many of the great mathematicians and scientists of his day: the Bernoullis, Euler, Halley, D'Alembert, and others. Many of these contacts and friendships led to extensive correspondence in which information about new mathematical discoveries was transmitted. Eventually, Cramer became sole possessor of the mathematics chair and the chair of philosophy as well.

Cramer's mathematical work was primarily in geometry and probability; he had relatively little knowledge of calculus and did not use it to any great extent in his work. In 1730 he finished second to Johann I Bernoulli in a competition for a prize offered by the Paris Academy to explain properties of planetary orbits.

Cramer's most widely known work, *Introduction à l'analyse des lignes courbes algébriques* (1750), was a study and classification of algebraic curves; Cramer's rule appeared in the appendix. Although the rule bears his name, variations of the basic idea were formulated earlier by Leibniz (and even earlier by Chinese mathematicians). However, Cramer's superior notation helped clarify and popularize the technique.

Perhaps Cramer's most important contributions stemmed from his work as an editor of the mathematical creations of others. He edited and published the works of Jacob I Bernoulli and Leibniz.

Overwork combined with a fall from a carriage eventually led to his death in 1752. Cramer was apparently a good-natured and pleasant person, though he never married. His interests were broad. He wrote on philosophy of law and government and the history of mathematics. He served in public office, participated in artillery and fortifications activities for the government, instructed workers on techniques of cathedral repair, and undertook excavations of cathedral archives. Cramer received numerous honors for his activities.

9. Use Cramer's rule to solve the rotation equations

$$x = x' \cos \theta - y' \sin \theta$$
$$y = x' \sin \theta + y' \cos \theta$$

for x' and y' in terms of x and y.

10. Solve the following system of equations for the unknown angles α, β, and γ, where $-\pi/2 \le \alpha \le \pi/2$, $0 \le \beta \le \pi$, and $-\pi/2 < \gamma < \pi/2$:

$$2 \sin \alpha - \cos \beta + 3 \tan \gamma = 3$$
$$4 \sin \alpha + 2 \cos \beta - 2 \tan \gamma = 2$$
$$6 \sin \alpha - 3 \cos \beta + \tan \gamma = 9$$

[*Hint:* First solve for $\sin \alpha$, $\cos \beta$, and $\tan \gamma$.]

APPENDIX E

Complex Numbers

■ E.I DEFINITIONS; ALGEBRA OF COMPLEX NUMBERS

☐ **DEFINITION OF A COMPLEX NUMBER**

Since $x^2 \geq 0$ for every real number x, the equation $x^2 = -1$ has no real solutions. To deal with this problem, mathematicians of the eighteenth century introduced the *imaginary number* $i = \sqrt{-1}$, which they assumed had the property

$$i^2 = (\sqrt{-1})^2 = -1$$

Expressions of the form $a + bi$ were called *complex numbers*, and these were manipulated according to the standard rules of algebra with the added property that $i^2 = -1$. By the beginning of the nineteenth century it was recognized that a complex number $a + bi$ could be viewed as an ordered pair of real numbers (a, b) or as a vector $\langle a, b \rangle$. This is the approach we will follow.

> DEFINITION 1. A *complex number* is an ordered pair of real numbers, denoted either by (a, b) or $a + bi$. The number a is called the *real part* of the complex number, and the number b is called the *imaginary part*.

Sometimes it is convenient to use a single letter, such as z, to denote a complex number. Thus, we might write

$$z = a + bi$$

in which case the real and imaginary parts of z are denoted by $\mathrm{Re}(z)$ and $\mathrm{Im}(z)$, respectively; that is, $\mathrm{Re}(z) = a$ and $\mathrm{Im}(z) = b$. Real numbers can be regarded as complex numbers that have an imaginary part of zero; that is, $a = a + 0i$. In this sense the complex number system is an extension of the real number system.

Example 1

z (ORDERED PAIR NOTATION)	z (EQUIVALENT NOTATION)	$\mathrm{Re}(z)$	$\mathrm{Im}(z)$
$(3, 4)$	$3 + 4i$	3	4
$(-1, 2)$	$-1 + 2i$	-1	2
$(0, 1)$	$0 + i = i$	0	1
$(2, 0)$	$2 + 0i = 2$	2	0
$(4, -2)$	$4 + (-2)i = 4 - 2i$	4	-2

◀

☐ THE COMPLEX PLANE

A complex number $z = a + bi$ can be represented geometrically as a point or a vector in an xy-coordinate system, in which case we call the x-axis the **real axis**, the y-axis the **imaginary axis**, and the plane itself the **complex plane** (Figure E.1). Figure E.2 shows the complex numbers in Example 1 as points and vectors in the complex plane.

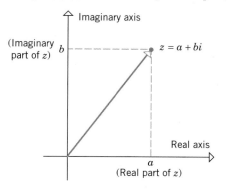

Figure E.1

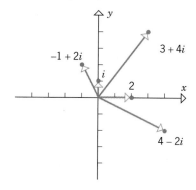

Figure E.2

☐ OPERATIONS ON COMPLEX NUMBERS

The **sum** and **difference** of two complex numbers are defined by adding or subtracting corresponding real and imaginary parts; that is,

$$(a + bi) + (c + di) = (a + c) + (b + d)i \tag{1}$$

$$(a + bi) - (c + di) = (a - c) + (b - d)i \tag{2}$$

Example 2

$$(4 - 5i) + (-1 + 6i) = (4 - 1) + (-5 + 6)i = 3 + i$$

$$(4 - 5i) - (-1 + 6i) = (4 + 1) + (-5 - 6)i = 5 - 11i \qquad \blacktriangleleft$$

The sum and difference of complex numbers z_1 and z_2 can be visualized geometrically by viewing the numbers as vectors (Figure E.3).

Multiplication of two complex numbers can be motivated by expanding the expression $(a + bi)(c + di)$ according to the usual laws of algebra, but treating i^2 as -1; that is,

$$(a + bi)(c + di) = ac + bdi^2 + adi + bci = (ac - bd) + (ad + bc)i$$

Thus, we define the **product** of complex numbers as

$$(a + bi)(c + di) = (ac - bd) + (ad + bc)i \tag{3}$$

For computational purposes, we suggest using the method that led to the formula in this definition, rather than substituting in the formula itself.

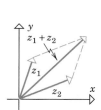

The sum of two complex numbers

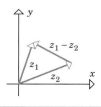

The difference of two complex numbers

Figure E.3

Example 3

$$5(2 + 4i) = 10 + 20i$$

$$(3 + 2i)(4 + i) = 12 + 3i + 8i + 2i^2 = 12 + 11i - 2 = 10 + 11i$$

$$i(1 + i)(1 - 2i) = i(1 - 2i + i - 2i^2) = i(3 - i) = 3i - i^2 = 1 + 3i \qquad \blacktriangleleft$$

Division of complex numbers can be motivated by a process that is similar to "rationalizing" quotients in algebra. To obtain the real and imaginary parts of the quotient

$$\frac{a + bi}{c + di}$$

we multiply the numerator and denominator by $c - di$ and simplify:

$$\frac{a + bi}{c + di} = \frac{a + bi}{c + di} \cdot \frac{c - di}{c - di} = \frac{(ac + bd) + (bc - ad)i}{c^2 - d^2 i^2}$$

$$= \frac{(ac + bd) + (bc - ad)i}{c^2 + d^2}$$

which suggests that we define the **quotient** of complex numbers as

$$\frac{a + bi}{c + di} = \left(\frac{ac + bd}{c^2 + d^2}\right) + \left(\frac{bc - ad}{c^2 + d^2}\right) i \qquad (4)$$

Again, we recommend that computations be performed using the procedure that led to the formula in this definition rather than substituting in the formula itself. Moreover, note that division of complex numbers is undefined if the denominator is zero, just as for real numbers.

Example 4

$$\frac{3 + 4i}{1 - 2i} = \frac{3 + 4i}{1 - 2i} \cdot \frac{1 + 2i}{1 + 2i} = \frac{-5 + 10i}{5} = -1 + 2i \qquad \blacktriangleleft$$

☐ **COMPLEX CONJUGATES**

If $z = a + bi$ is any complex number, then the **complex conjugate** (or **conjugate**) *of z* is denoted by $\bar{z}$ and is defined as

$$\bar{z} = a - bi$$

Example 5

z	$\bar{z}$
$3 + 4i$	$3 - 4i$
$-2 - 3i$	$-2 + 3i$
i	$-i$
5	5

$\blacktriangleleft$

We have already seen that complex conjugates arise in the division of complex numbers. They also arise in solving polynomial equations

$$a_0 x^n + a_1 x^{n-1} + \cdots + a_{n-1} x + a_n = 0$$

with real coefficients. It can be proved that such equations always have solutions in the complex number system and that if z is a solution, then so is $\bar{z}$. Thus, the solutions with nonzero imaginary parts occur in conjugate pairs. For example, from the quadratic formula the solutions of the equation $x^2 + x + 2 = 0$ are

$$x = \frac{-1 \pm \sqrt{-7}}{2} = \frac{-1 \pm \sqrt{7}i}{2}$$

so the solutions $x = -\frac{1}{2} + \frac{1}{2}\sqrt{7}i$ and $x = -\frac{1}{2} - \frac{1}{2}\sqrt{7}i$ are complex conjugates.

Geometrically, complex conjugates are reflections of one another about the real axis

(Figure E.4). Moreover, we leave it for the exercises to show that complex conjugates have the following properties:

$$\bar{\bar{z}} = z, \quad \overline{z_1 + z_2} = \bar{z}_1 + \bar{z}_2, \quad \overline{z_1 - z_2} = \bar{z}_1 - \bar{z}_2$$
$$\overline{z_1 z_2} = \bar{z}_1 \bar{z}_2, \quad \overline{(z_1/z_2)} = \bar{z}_1/\bar{z}_2 \tag{5}$$

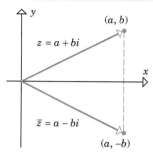

Figure E.4

□ MODULUS

If a complex number z is viewed as a vector, then its length is called the **modulus** of z and is denoted by $|z|$. Thus, if $z = a + bi$, then

$$|z| = \sqrt{a^2 + b^2} \tag{6}$$

Figure E.5

(Figure E.5). For example, if $z = 3 - 4i$, then

$$|z| = |3 - 4i| = \sqrt{3^2 + (-4)^2} = 5$$

If z is a real number, say $z = a + 0i = a$, then

$$|z| = \sqrt{a^2 + 0^2} = \sqrt{a^2} = |a|$$

so the modulus of a real number is the same as the absolute value of that number. Observe that

$$z\bar{z} = (a + bi)(a - bi) = a^2 - abi + bai - b^2 i^2 = a^2 + b^2 = |z|^2$$

so the modulus and complex conjugate of z are related by

$$z\bar{z} = |z|^2 \tag{7}$$

Moreover, it can be shown that

$$|z_1 z_2| = |z_1||z_2| \quad \text{and} \quad |z_1/z_2| = |z_1|/|z_2| \tag{8}$$

▶ Exercise Set E.I

1. In each part plot the point and sketch the vector that corresponds to the given complex number.

(a) $2 + 3i$ (b) -4

(c) $-3 - 2i$ (d) $-5i$.

2. Express each complex number in Exercise 1 as an ordered pair of real numbers.

3. In each part use the given information to find the real numbers x and y.

(a) $x - iy = -2 + 3i$

(b) $(x + y) + (x - y)i = 3 + i$.

4. Given that $z_1 = 1 - 2i$ and $z_2 = 4 + 5i$, find

(a) $z_1 + z_2$ (b) $z_1 - z_2$

(c) $4z_1$ (d) $-z_2$

(e) $3z_1 + 4z_2$ (f) $\frac{1}{2}z_1 - \frac{3}{2}z_2$.

5. In each part solve for z.

(a) $z + (1 - i) = 3 + 2i$ (b) $-5z = 5 + 10i$

(c) $(i - z) + (2z - 3i) = -2 + 7i$.

6. In each part sketch the vectors z_1, z_2, $z_1 + z_2$, and $z_1 - z_2$.

(a) $z_1 = 3 + i$, $z_2 = 1 + 4i$

(b) $z_1 = -2 + 2i$, $z_2 = 4 + 5i$.

7. In each part sketch the vectors z and kz.

(a) $z = 1 + i$, $k = 2$

(b) $z = -3 - 4i$, $k = -2$

(c) $z = 4 + 6i$, $k = \frac{1}{2}$.

8. In each part find real numbers k_1 and k_2 that satisfy the equation.

   (a) $k_1 i + k_2(1 + i) = 3 - 2i$

   (b) $k_1(2 + 3i) + k_2(1 - 4i) = 7 + 5i$.

9. In each part find $z_1 z_2$, z_1^2, and z_2^2.

   (a) $z_1 = 3i$, $z_2 = 1 - i$

   (b) $z_1 = 4 + 6i$, $z_2 = 2 - 3i$

   (c) $z_1 = \frac{1}{3}(2 + 4i)$, $z_2 = \frac{1}{2}(1 - 5i)$.

10. Given that $z_1 = 2 - 5i$ and $z_2 = -1 - i$, find

   (a) $z_1 - z_1 z_2$ (b) $(z_1 + 3z_2)^2$

   (c) $[z_1 + (1 + z_2)]^2$ (d) $iz_2 - z_1^2$.

In Exercises 11–18, perform the calculations and express the result in the form $a + bi$.

---

11. $(1 + 2i)(4 - 6i)^2$.   **12.** $(2 - i)(3 + i)(4 - 2i)$.

13. $(1 - 3i)^3$.   **14.** $i(1 + 7i) - 3i(4 + 2i)$.

15. $[(2 + i)(\frac{1}{2} + \frac{3}{4}i)]^2$.   **16.** $(\sqrt{2} + i) - \sqrt{2}i(1 + \sqrt{2}i)$.

17. $(1 + i + i^2 + i^3)^{100}$.   **18.** $(3 - 2i)^2 - (3 + 2i)^2$.

19. In each part find $\bar{z}$.

   (a) $z = 2 + 7i$ (b) $z = -3 - 5i$

   (c) $z = 5i$ (d) $z = -i$

   (e) $z = -9$ (f) $z = 0$.

20. In each part find $|z|$.

   (a) $z = i$ (b) $z = -7i$

   (c) $z = -3 - 4i$ (d) $z = 1 + i$

   (e) $z = -8$ (f) $z = 0$.

21. Verify that $z\bar{z} = |z|^2$ for

   (a) $z = 2 - 4i$ (b) $z = -3 + 5i$

   (c) $z = \sqrt{2} - \sqrt{2}i$.

22. Given that $z_1 = 1 - 5i$ and $z_2 = 3 + 4i$, find

   (a) z_1/z_2 (b) $\bar{z}_1/z_2$

   (c) $z_1/\bar{z}_2$ (d) $\overline{(z_1/z_2)}$

   (e) $z_1/|z_2|$ (f) $|z_1/z_2|$.

23. In each part find $1/z$.

   (a) $z = i$ (b) $z = 1 - 5i$

   (c) $z = \dfrac{-i}{7}$.

24. Given that $z_1 = 1 + i$ and $z_2 = 1 - 2i$, find

   (a) $z_1 - \dfrac{z_1}{z_2}$ (b) $\dfrac{z_1 - 1}{z_2}$

   (c) $z_1^2 - \dfrac{iz_1}{z_2}$ (d) $\dfrac{z_1}{iz_2}$.

In Exercises 25–32, perform the calculations and express the result in the form $a + bi$.

---

25. $\dfrac{i}{1 + i}$.   **26.** $\dfrac{2}{(1 - i)(3 + i)}$.

27. $\dfrac{1}{(3 + 4i)^2}$.   **28.** $\dfrac{2 + i}{i(-3 + 4i)}$.

29. $\dfrac{\sqrt{3} + i}{(1 - i)(\sqrt{3} - i)}$.   **30.** $\dfrac{1}{i(3 - 2i)(1 + i)}$.

31. $\dfrac{i}{(1 - i)(1 - 2i)(1 + 2i)}$. **32.** $\dfrac{1 - 2i}{3 + 4i} - \dfrac{2 + i}{5i}$.

33. In each part solve for z.

   (a) $iz = 2 - i$ (b) $(4 - 3i)\bar{z} = i$.

34. Use the properties in (5) to show that

   (a) $\overline{\bar{z} + 5i} = z - 5i$ (b) $\overline{iz} = -i\bar{z}$

   (c) $\overline{\dfrac{i + \bar{z}}{i - z}} = -1$.

35. In each part sketch the set of points in the complex plane that satisfies the equation.

   (a) $|z| = 2$ (b) $|z - (1 + i)| = 1$

   (c) $|z - i| = |z + i|$ (b) $\text{Im}(\bar{z} + i) = 3$.

36. In each part sketch the set of points in the complex plane that satisfies the given condition(s).

   (a) $|z + i| \le 1$ (b) $1 < |z| < 2$

   (c) $|2z - 4i| < 1$ (d) $|z| \le |z + i|$.

37. Show that

   (a) $\text{Im}(iz) = \text{Re}(z)$ (b) $\text{Re}(iz) = -\text{Im}(z)$

   (c) $\text{Re}(i\bar{z}) = -\text{Im}(z)$ (d) $\text{Im}(i\bar{z}) = -\text{Re}(z)$

   (e) $\text{Re}(i\bar{z}) = \text{Im}(z)$ (f) $\text{Im}(i\bar{z}) = \text{Re}(z)$.

38. In each part solve the equation by the quadratic formula and check your results by substituting the solutions into the given equation.

   (a) $x^2 + 2x + 2 = 0$ (b) $x^2 - x + 1 = 0$.

39. (a) Show that if n is a positive integer, then the only possible values for i^n are 1, -1, i, and $-i$. [*Hint:* The value of i^n can be determined from the remainder when n is divided by 4.]

   (b) Find i^{2509}.

40. (a) Use the result in part (a) of Exercise 39 to show that if n is a positive integer, then the only possible values for $(1/i)^n$ are 1, -1, i, and $-i$.

   (b) Find $(1/i)^{2509}$.

41. Prove:

   (a) $\dfrac{1}{2}(z + \bar{z}) = \text{Re}(z)$ (b) $\dfrac{1}{2i}(z - \bar{z}) = \text{Im}(z)$.

42. Prove: $z = \bar{z}$ if and only if z is a real number.

43. Given that $z_1 = x_1 + iy_1$ and $z_2 = x_2 + iy_2$, find

   (a) $\text{Re}\left(\dfrac{z_1}{z_2}\right)$ (b) $\text{Im}\left(\dfrac{z_1}{z_2}\right)$.

44. Show: If $(\bar{z})^2 = z^2$, then z is either real or pure imaginary.

45. Show that $|z| = |\bar{z}|$.

46. Show:

(a) $\overline{z_1 - z_2} = \bar{z}_1 - \bar{z}_2$ (b) $\overline{z_1 z_2} = \bar{z}_1 \bar{z}_2$

(c) $\overline{(z_1/z_2)} = \bar{z}_1/\bar{z}_2$ (d) $\bar{\bar{z}} = z$.

47. (a) Use the properties in (5) to show that $\overline{z^2} = (\bar{z})^2$.

(b) Show that if n is a positive integer, then $\overline{z^n} = (\bar{z})^n$.

(c) Is the result in part (b) true if n is a negative integer? Explain.

48. Let $p(x) = a_0 + a_1 x + a_2 x^2 + \cdots + a_n x^n$ be a polynomial for which the coefficients $a_0, a_1, a_2, \ldots, a_n$ are real. Use (5) and the result in part (b) of Exercise 47 to show that if z is a solution of the equation $p(x) = 0$, then so is $\bar{z}$.

49. Show: For any complex number z, $|\text{Re}(z)| \le |z|$ and $|\text{Im}(z)| \le |z|$.

50. Show that
$$\frac{|\text{Re}(z)| + |\text{Im}(z)|}{\sqrt{2}} \le |z|$$

[*Hint:* Let $z = x + iy$ and use the fact that $(|x| - |y|)^2 \ge 0$.]

51. Show that if $z_1 z_2 = 0$, then $z_1 = 0$ or $z_2 = 0$.

■ E.II POLAR FORM; DEMOIVRE'S THEOREM

☐ **POLAR FORM OF A COMPLEX NUMBER**

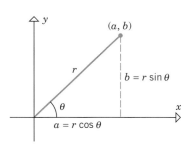

Figure E.6

If we view $z = a + bi$ as a point (a, b) in the complex plane, and if we let r and θ be polar coordinates of that point with $r \ge 0$, then it follows that

$$a = r \cos \theta, \quad b = r \sin \theta$$

(Figure E.6), so z can be expressed as

$$z = a + bi = (r \cos \theta) + (r \sin \theta)i = r(\cos \theta + i \sin \theta)$$

from which it follows that

$$z = r(\cos \theta + i \sin \theta) \tag{1}$$

We call this a **polar form of z**. Observe that the quantity r in this formula is the modulus of z and that r and θ are related to the real and imaginary parts of z by

$$r = |z| = \sqrt{a^2 + b^2} \quad \text{and} \quad \tan \theta = b/a \tag{2-3}$$

The angle θ is called an **argument of z** and is denoted by

$$\theta = \arg z$$

The argument of z is not uniquely determined because we can add or subtract any multiple of 2π from θ to produce another value of the argument. However, if $z \ne 0$, then there is only one value of the argument in radians that satisfies

$$-\pi < \theta \le \pi$$

This is called the **principal argument of z** and is denoted by

$$\theta = \text{Arg } z$$

Example 1 Express the following complex numbers in polar form using their principal arguments:

(a) $z = 1 + \sqrt{3}i$ (b) $z = -1 - i$

Solution (a). To find r and Arg(z), we start with Formulas (2) and (3). We have $a = 1$ and $b = \sqrt{3}$, so

$$r = |z| = \sqrt{1^2 + (\sqrt{3})^2} = \sqrt{4} = 2 \quad \text{and} \quad \tan \theta = \sqrt{3}$$

From these values and the fact that z corresponds to a point in the first quadrant (Figure E.7), we conclude that $\theta = \text{Arg}(z) = \pi/3$, and hence a polar form of z is

$$z = 2\left(\cos\frac{\pi}{3} + i\sin\frac{\pi}{3}\right)$$

Solution (b).   In this case we have $a = -1$ and $b = -1$, so it follows from (2) and (3) that

$$r = |z| = \sqrt{(-1)^2 + (-1)^2} = \sqrt{2} \quad \text{and} \quad \tan\theta = 1$$

From these values and the fact that z corresponds to a point in the third quadrant (Figure E.8), we conclude that $\theta = \text{Arg}(z) = -3\pi/4$, and hence a polar form of z is

$$z = \sqrt{2}\left[\cos\left(\frac{-3\pi}{4}\right) + i\sin\left(\frac{-3\pi}{4}\right)\right] \quad \blacktriangleleft$$

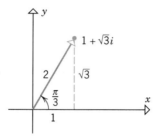

Figure E.7   Figure E.8

☐ **GEOMETRIC INTERPRETATION OF MULTIPLICATION AND DIVISION**

We now show how polar forms can be used to give geometric interpretations of multiplication and division of complex numbers. Let

$$z_1 = r_1(\cos\theta_1 + i\sin\theta_1) \quad \text{and} \quad z_2 = r_2(\cos\theta_2 + i\sin\theta_2)$$

Multiplying, we obtain

$$z_1 z_2 = r_1 r_2[(\cos\theta_1\cos\theta_2 - \sin\theta_1\sin\theta_2) + i(\sin\theta_1\cos\theta_2 + \cos\theta_1\sin\theta_2)]$$

Recall the trigonometric identities

$$\cos(\theta_1 + \theta_2) = \cos\theta_1\cos\theta_2 - \sin\theta_1\sin\theta_2$$

$$\sin(\theta_1 + \theta_2) = \sin\theta_1\cos\theta_2 + \cos\theta_1\sin\theta_2$$

We obtain the following polar form of $z_1 z_2$:

$$z_1 z_2 = r_1 r_2[\cos(\theta_1 + \theta_2) + i\sin(\theta_1 + \theta_2)] \tag{4}$$

In words, *the product of two complex numbers can be obtained by multiplying their moduli and adding their arguments* (Figure E.9).

We leave it as an exercise to show that if $z_2 \neq 0$, then

$$\frac{z_1}{z_2} = \frac{r_1}{r_2}[\cos(\theta_1 - \theta_2) + i\sin(\theta_1 - \theta_2)] \tag{5}$$

In words, *the quotient of two complex numbers can be obtained by dividing their moduli and subtracting their arguments.*

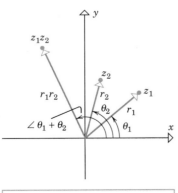

The product of two complex numbers

Figure E.9

Example 2   Use the polar forms obtained in Example 1 to express the following product and quotient in polar form.

(a) $(1 + \sqrt{3}i)(-1 - i)$ (b) $\dfrac{1 + \sqrt{3}i}{-1 - i}$

Solution. From Example 1 we have

$$1 + \sqrt{3}i = 2\left(\cos\frac{\pi}{3} + i\sin\frac{\pi}{3}\right)$$

$$-1 - i = \sqrt{2}\left[\cos\left(-\frac{3\pi}{4}\right) + i\sin\left(-\frac{3\pi}{4}\right)\right]$$

Thus, by multiplying the moduli and adding the arguments we obtain

$$(1 + \sqrt{3}i)(-1 - i) = 2\sqrt{2}\left[\cos\left(-\frac{5\pi}{12}\right) + i\sin\left(-\frac{5\pi}{12}\right)\right]$$

and by dividing the moduli and subtracting the arguments we obtain

$$\frac{1 + \sqrt{3}i}{-1 - i} = \frac{2}{\sqrt{2}}\left[\cos\left(\frac{13\pi}{12}\right) + i\sin\left(\frac{13\pi}{12}\right)\right]$$

$$= \sqrt{2}\left[\cos\left(\frac{13\pi}{12}\right) + i\sin\left(\frac{13\pi}{12}\right)\right] \qquad \blacktriangleleft$$

REMARK. Applying Formulas (4) and (5) may not result in polar forms with principal arguments, even if θ_1 and θ_2 are principal arguments. For example, in the preceding computations the argument that resulted for the product was the principal argument, but the argument for the quotient was not. If a principal argument is desired, then it may be necessary to add or subtract an appropriate multiple of 2π. For example, subtracting 2π from the argument for the quotient in part (b) yields $-11\pi/12$, which is the principal argument for the quotient.

Example 3 The complex number i has a modulus of 1 and an argument of $\pi/2$, so for any complex number z the product iz has the same modulus as z, but its argument is $\pi/2$ ($= 90°$) greater than that of z. Thus, multiplying z by i has the geometric effect of rotating z counterclockwise by $90°$ (Figure E.10). $\qquad \blacktriangleleft$

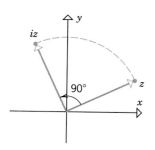

Figure E.10

☐ **DEMOIVRE'S FORMULA**

If n is a positive integer and $z = r(\cos\theta + i\sin\theta)$, then from Formula (4),

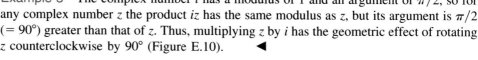

$$z^n = \underbrace{z \cdot z \cdot z \cdots z}_{n \text{ factors}} = r^n[\cos\underbrace{(\theta + \theta + \cdots + \theta)}_{n \text{ terms}} + i\sin\underbrace{(\theta + \theta + \cdots + \theta)}_{n \text{ terms}}]$$

that is,

$$z^n = r^n(\cos n\theta + i\sin n\theta) \qquad (6)$$

In words, *a complex number can be raised to the nth power by raising the modulus to the nth power and multiplying the argument by n.*

Although we derived (6) for positive integer n, this formula is valid for all integer values of n under appropriate conditions (Exercise 17). In the special case where $r = 1$, we have $z = \cos\theta + i\sin\theta$, so (6) becomes

$$(\cos\theta + i\sin\theta)^n = \cos n\theta + i\sin n\theta \qquad (7)$$

which is called **DeMoivre's* Formula.**

---

* ABRAHAM DEMOIVRE (1667–1754) was a French mathematician who made important contributions to probability, statistics, and trigonometry. He developed the concept of statistically independent events, wrote a major influential treatise on probability, and helped transform trigonometry from a branch of geometry into a branch of analysis through his use of complex numbers. In spite of his important work, he barely managed to eke out a living as a tutor and a consultant on gambling and insurance.

Example 4   Find $(1 + \sqrt{3}i)^9$.

Solution.   We could use the binomial formula and simplify, but it is much easier to apply (6). Using the polar form obtained in Example 1 we have

$$(1 + \sqrt{3}i)^9 = \left[2\left(\cos\frac{\pi}{3} + i\sin\frac{\pi}{3} \right) \right]^9 = 2^9(\cos 3\pi + i\sin 3\pi) = -2^9 = -512 \quad \blacktriangleleft$$

☐ **FINDING _n_th ROOTS OF COMPLEX NUMBERS**

We define an ***n*th root** of a nonzero complex number z to be any complex number w that satisfies the equation

$$w^n = z \tag{8}$$

A formula for the nth roots of z can be obtained by expressing w and z in polar form, say

$$w = \rho(\cos\alpha + i\sin\alpha) \quad \text{and} \quad z = r(\cos\theta + i\sin\theta)$$

It now follows from (8) with (6) applied to w that

$$\rho^n(\cos n\alpha + i\sin n\alpha) = r(\cos\theta + i\sin\theta)$$

which implies that

$$\rho^n = r, \quad \cos n\alpha = \cos\theta, \quad \sin n\alpha = \sin\theta$$

Since ρ and r are nonnegative real numbers, it follows from the first of these equations that ρ is the nonnegative real nth root of r, that is, $\rho = \sqrt[n]{r}$; and it follows from the last two equations that

$$n\alpha = \theta + 2k\pi \quad \text{or} \quad \alpha = \frac{\theta}{n} + \frac{2k\pi}{n}, \quad k = 0, \pm 1, \pm 2, \ldots$$

Thus, the values of $w = \rho(\cos\alpha + i\sin\alpha)$ that satisfy (8) are given by

$$w = \sqrt[n]{r}\left[\cos\left(\frac{\theta}{n} + \frac{2k\pi}{n} \right) + i\sin\left(\frac{\theta}{n} + \frac{2k\pi}{n} \right) \right], \quad k = 0, \pm 1, \pm 2, \ldots$$

Although there are infinitely many values of k, it can be shown (Exercise 16) that $k = 0$, 1, 2, ..., $n - 1$ produce distinct values of w satisfying (8), but all other choices of k yield duplicates of these. Therefore, there are exactly n different nth roots of $z = r(\cos\theta + i\sin\theta)$, and these are given by

$$w = \sqrt[n]{r}\left[\cos\left(\frac{\theta}{n} + \frac{2k\pi}{n} \right) + i\sin\left(\frac{\theta}{n} + \frac{2k\pi}{n} \right) \right], \quad k = 0, 1, 2, \ldots, n - 1 \tag{9}$$

REMARK.   Observe that each of these nth roots has the same modulus, namely $\sqrt[n]{r}$ $(= \sqrt[n]{|z|})$, and that the nth roots corresponding to successive values of k have arguments that differ by $2\pi/n$ radians. This implies that the nth roots of z are equally spaced around a circle of radius $\sqrt[n]{|z|}$ and that if one nth root of z is found, then the remaining $n - 1$ roots can be obtained by rotating this root through successive increments of $2\pi/n$ radians.

Example 5   Find all cube roots of -8.

Solution.   Since -8 lies on the negative real axis, it can be expressed in polar form as

$$-8 = 8(\cos\pi + i\sin\pi)$$

Thus, from (9) with $n = 3$ the cube roots of -8 are given by

$$w = \sqrt[3]{8}\left[\cos\left(\frac{\pi}{3} + \frac{2k\pi}{3} \right) + i\sin\left(\frac{\pi}{3} + \frac{2k\pi}{3} \right) \right], \quad k = 0, 1, 2$$

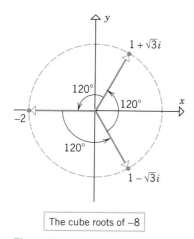

1 + √3i

120°

120°

−2

120°

1 − √3i

The cube roots of −8

Figure E.11

If we denote these cube roots by w_0, w_1, and w_2, respectively, then it follows from this formula that

$$w_0 = 2\left(\cos\frac{\pi}{3} + i\sin\frac{\pi}{3}\right) = 2\left(\frac{1}{2} + \frac{\sqrt{3}}{2}i\right) = 1 + \sqrt{3}i$$

$$w_1 = 2(\cos\pi + i\sin\pi) = 2(-1) = -2$$

$$w_2 = 2\cos\left(\frac{5\pi}{3} + i\sin\frac{5\pi}{3}\right) = 2\left(\frac{1}{2} - \frac{\sqrt{3}}{2}i\right) = 1 - \sqrt{3}i$$

Alternative Solution.   By inspection, -2 is a cube root of -8. If we represent this root as a point or a vector in the complex plane, then the remaining two roots can be obtained by rotating this root through two increments of $2\pi/3$ ($= 120°$), as shown in Figure E.11.   ◄

□ **COMPLEX**
EXPONENTIALS

To define the complex exponential function e^z ($= e^{a+ib}$), we start by defining e^{ib}. A motivation for the definition of this function can be obtained by formally substituting ib for x in the Maclaurin series for e^x and simplifying using the relationships

$$i^2 = -1, \quad i^3 = i \cdot i^2 = -i, \quad i^4 = i \cdot i^3 = 1, \quad i^5 = i \cdot i^4 = i, \quad i^6 = i \cdot i^5 = -1, \ldots$$

This yields

$$e^{ib} = 1 + ib + \frac{(ib)^2}{2!} + \frac{(ib)^3}{3!} + \frac{(ib)^4}{4!} + \frac{(ib)^5}{5!} + \frac{(ib)^6}{6!} + \cdots$$

$$= 1 + ib - \frac{b^2}{2!} - i\frac{b^3}{3!} + \frac{b^4}{4!} + i\frac{b^5}{5!} - \frac{b^6}{6!} - \cdots$$

$$= \left(1 - \frac{b^2}{2!} + \frac{b^4}{4!} - \frac{b^6}{6!} + \cdots\right) + i\left(b - \frac{b^3}{3!} + \frac{b^5}{5!} - \cdots\right)$$

$$= \cos b + i\sin b$$

where the last step follows from the Maclaurin series for $\cos b$ and $\sin b$. This suggests that e^{ib} be defined as

$$e^{ib} = \cos b + i\sin b \tag{10}$$

which is sometimes called the ***Euler formula***. The definition of e^z can now be motivated by writing $e^z = e^{a+ib} = e^a e^{ib}$, which suggests that we define

$$e^z = e^{a+ib} = e^a(\cos b + i\sin b) \tag{11}$$

Example 6

$$e^{i\pi} = \cos\pi + i\sin\pi = -1$$

$$e^{1+2\pi i} = e^1(\cos 2\pi + i\sin 2\pi) = e$$

$$e^{2+(\pi i/2)} = e^2(\cos\pi/2 + i\sin\pi/2) = e^2 i \quad ◄$$

□ **EXPONENTIAL NOTATION**
FOR POLAR FORMS

It follows from (10) that the polar form of a complex number can be expressed in exponential notation as

$$z = r(\cos\theta + i\sin\theta) = re^{i\theta} \tag{12}$$

It can be proved that complex exponents follow the same algebraic laws as real exponents. For example, if $z_1 = r_1 e^{i\theta_1}$ and $z_2 = r_2 e^{i\theta_2}$, then

$$z_1 z_2 = r_1 r_2 e^{i\theta_1 + i\theta_2} = r_1 r_2 e^{i(\theta_1 + \theta_2)}$$

$$\frac{z_1}{z_2} = \frac{r_1}{r_2} e^{i\theta_1 - i\theta_2} = \frac{r_1}{r_2} e^{i(\theta_1 - \theta_2)}$$

which are precisely Formulas (4) and (5), but expressed in exponential notation. Moreover, it follows from (12) that

$$z^n = (re^{i\theta})^n = r^n e^{in\theta} \tag{13}$$

which is precisely Formula (6), but in exponential notation.

Finally, if z has modulus r and argument θ, then $\bar{z}$ has modulus r and argument $-\theta$, since z and $\bar{z}$ are symmetric about the real axis. Thus, if $z = re^{i\theta}$, then $\bar{z} = re^{-i\theta}$, that is,

$$\overline{re^{i\theta}} = re^{-i\theta} \tag{14}$$

▶ Exercise Set E.II

1. In each part find the principal argument of z.
   - (a) $z = 1$
   - (b) $z = i$
   - (c) $z = -i$
   - (d) $z = 1 + i$
   - (e) $z = -1 + \sqrt{3}i$
   - (f) $z = 1 - i$.

2. In each part find the value of $\theta = \arg(1 - \sqrt{3}i)$ that satisfies the given condition.
   - (a) $0 \le \theta < 2\pi$
   - (b) $-\pi < \theta \le \pi$
   - (c) $-\dfrac{\pi}{6} \le \theta < \dfrac{11\pi}{6}$.

3. In each part express the complex number in polar form using its principal argument.
   - (a) $2i$
   - (b) -4
   - (c) $5 + 5i$
   - (d) $-6 + 6\sqrt{3}i$
   - (e) $-3 - 3i$
   - (f) $2\sqrt{3} - 2i$.

4. Given that $z_1 = 2(\cos \pi/4 + i \sin \pi/4)$ and $z_2 = 3(\cos \pi/6 + i \sin \pi/6)$, find a polar form of
   - (a) $z_1 z_2$
   - (b) z_1/z_2
   - (c) z_2/z_1
   - (d) z_1^5/z_2^2.

5. Express $z_1 = i$, $z_2 = 1 - \sqrt{3}i$, and $z_3 = \sqrt{3} + i$ in polar form, and use your results to find $z_1 z_2/z_3$. Check your results by performing the calculations without using polar forms.

6. Use Formula (6) to find
   - (a) $(1 + i)^{12}$
   - (b) $\left(\dfrac{1}{\sqrt{2}} - \dfrac{1}{\sqrt{2}}i\right)^{-6}$
   - (c) $(\sqrt{3} + i)^7$
   - (d) $(1 - \sqrt{3}i)^{-10}$.

7. In each part find all the roots and sketch them as vectors in the complex plane.
   - (a) $(-i)^{1/2}$
   - (b) $(1 + \sqrt{3}i)^{1/2}$
   - (c) $(-27)^{1/3}$
   - (d) $(i)^{1/3}$
   - (e) $(-1)^{1/4}$
   - (f) $(-8 + 8\sqrt{3}i)^{1/4}$.

8. Use the method in the alternative solution of Example 5 to find all cube roots of 1.

9. Use the method in the alternative solution of Example 5 to find all sixth roots of 1.

10. Find all square roots of $1 + i$ and express your results in polar form.

11. In each part find all solutions of the equation.
    - (a) $z^4 - 16 = 0$
    - (b) $z^4 + 64 = 0$.

12. Find the four solutions of the equation $z^4 + 8 = 0$ and use your results to factor $z^4 + 8$ into two quadratic factors with real coefficients.

13. It was shown in Example 3 that multiplying z by i rotates z counterclockwise by 90°. What is the geometric effect of dividing z by i?

14. In each part use (13) to calculate the given power.
    - (a) $(1 + i)^8$
    - (b) $(-2\sqrt{3} + 2i)^{-9}$.

15. In each part find $\text{Re}(z)$ and $\text{Im}(z)$.
    - (a) $z = 3e^{i\pi}$
    - (b) $z = 3e^{-i\pi}$
    - (c) $\bar{z} = \sqrt{2}e^{\pi i/2}$
    - (d) $\bar{z} = -3e^{-2\pi i}$.

16. (a) Show that if $z \ne 0$, then the values of w in Formula (9) are all different.
    (b) Show that integer values of k other than $k = 0, 1, 2, \ldots, n - 1$ produce values of w that are duplicates of those in Formula (9).

17. Assuming that $z \ne 0$ and $z^0 = 1$, show that Formula (6) is valid if $n = 0$ or n is a negative integer.

18. Derive Formula (5).

ANSWERS TO
ODD-NUMBERED EXERCISES

▶ **Exercise Set 1.1 (Page 10)**

1. (a) rational   (b) integer, rational   (c) integer, rational
   (d) rational   (e) integer, rational   (f) irrational
   (g) rational   (h) integer, rational

3. (a) $\frac{41}{333}$ (b) $\frac{115}{9}$ (c) $\frac{20943}{550}$ (d) $\frac{537}{1250}$

5. (a) $\frac{256}{81}$ (b) worse

7.
Line	Blocks
2	3, 4
3	1, 2
4	3, 4
5	2, 4, 5
6	1, 2
7	3, 4

9. (a), (d), (f)

11. (a) all values   (b) none   13. (a) yes   (b) no

15. (a) $\{x : x$ is a positive odd integer$\}$
    (b) $\{x : x$ is an even integer$\}$ (c) $\{x : x$ is irrational$\}$
    (d) $\{x : x$ is an integer and $7 \le x \le 10\}$

17. (a) false   (b) true   (c) true   (d) false   (e) true
    (f) true   (g) true

19. (a) [number line, filled circle at 4, ray right]   (b) [number line, open circle at −3, ray left]

    (c) [number line, filled circles at −1 and 7]   (d) [number line, filled circles at −3 and 3]

    (e) [number line, open circles at −3 and 3]   (f) [number line, filled circles at −3 and 3]

21. (a) $[-2, 2]$ (b) $(-\infty, -2) \cup (2, +\infty)$

23. $(-\infty, \frac{10}{3})$ [number line, open circle at $\frac{10}{3}$, ray left]

25. $(-\infty, -\frac{11}{2}]$ [number line, filled circle at $-\frac{11}{2}$, ray left]

27. $(-\frac{3}{2}, \frac{1}{2}]$ [number line, open circle at $-\frac{3}{2}$, filled at $\frac{1}{2}$]

29. $(-\infty, 3) \cup (4, +\infty)$ [number line, open circles at 3 and 4]

31. $(-\frac{3}{2}, 2)$ [number line, open circles at $-\frac{3}{2}$ and 2]

33. $(-\infty, -2] \cup (2, +\infty)$ [number line, filled at −2, open at 2]

35. $(-\infty, -3) \cup (3, +\infty)$ [number line, open circles at −3 and 3]

37. $(-\infty, -2) \cup (4, +\infty)$ [number line, open circles at −2 and 4]

39. $[4, 5]$ [number line, filled circles at 4 and 5]

41. $(-8, 0) \cup (4, +\infty)$ [number line, open circles at −8, 0, 4]

43. $(2, +\infty)$ [number line, open circle at 2, ray right]

45. $(-\infty, -3] \cup [2, +\infty)$ **47.** $77 \le F \le 104$

55. $(-\infty, -\frac{1}{2})$

▶ **Exercise Set 1.2 (Page 17)**

1. (a) 7   (b) $\sqrt{2}$ (c) k^2 (d) k^2 3. $x \le 3$

5. all real x 7. $x \ge 0$ or $x = -\frac{2}{3}$ 9. $x \ge -5$

13. (a) 2   (b) 1   (c) 14   (d) $3 + \sqrt{2}$ (e) 7   (f) 5

15. (a) -9 (b) 7   (c) 12   17. $-\frac{5}{6}, \frac{3}{2}$ 19. $\frac{1}{2}, \frac{5}{2}$

21. $-\frac{11}{10}, \frac{11}{8}$ 23. $1, \frac{17}{5}$ 25. $(-9, -3)$ 27. $[-\frac{3}{2}, \frac{9}{2}]$

29. $(-\infty, -3) \cup (-1, +\infty)$ 31. $(-\infty, \frac{1}{2}] \cup [\frac{9}{2}, +\infty)$

33. $(-\infty, \frac{1}{2}) \cup (\frac{3}{2}, +\infty)$ 35. $[\frac{1}{8}, \frac{1}{2}) \cup (\frac{1}{2}, \frac{7}{8}]$

37. $x \in (-\infty, 2] \cup [3, +\infty)$ 39. $-3, 9$

▶ **Exercise Set 1.3 (Page 25)**

1. [graph with points $(-2, 5)$, $(3, 4)$, $(4, 0)$, $(-2.5, -3)$, $(1.7, -2)$, $(0, -6)$]

3. (a) [graph]   (b) [graph]

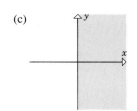

(c) [graph]   (d) [graph]

(e) [graph]   (f) [graph]

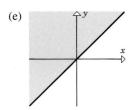
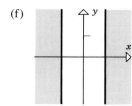
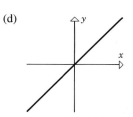

5. (a) vertical   (b) horizontal   (c) vertical
7. (a) yes   (b) no   (c) yes   (d) yes
9. (a) origin   (b) x-axis   (c) y-axis   (d) none
11. (a) $x = y^2$, $x = -y^2$
    (b) $y = x^2$, $y = -x^2$, $y = 1/x^2$, $y = -1/x^2$
    (c) $y = x^3$, $y = \sqrt[3]{x}$, $y = 1/x$, $y = -1/x$
13. (a) y-axis   (b) origin   (c) x-axis, y-axis, origin
15. 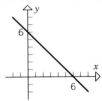   17.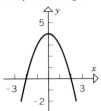

19. 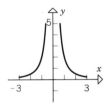   21.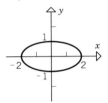

23. 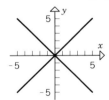   25.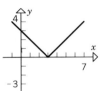

27. the union of the graphs of $x - y = 0$ and $x + y = 0$

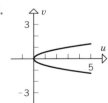

29.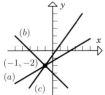

▶ **Exercise Set 1.4 (Page 35)**
1. (a) $\frac{1}{2}$ (b) -1 (c) 0 (d) not defined
3. (a) yes   (b) no
5.

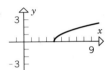

7. (c), (b), (d), (a)   9. (a) $\frac{3}{2}$ (b) $-\frac{3}{4}$
11. (a) 14 (b) $-\frac{1}{3}$ 13. 29 15. $\frac{13}{7}$

17. (a) $1/\sqrt{3}$ (b) -1 (c) $\sqrt{3}$
19. (a) $153°$ (b) $45°$ (c) $117°$ (d) $89°$
21. $(2, 0)$, $(7, 0)$
25. (a)   (b)   (c)   (d)

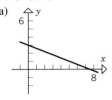

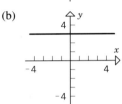

27. (a)   (b)   (c)

29.

	(a)	(b)	(c)	(d)	(e)
Slope	3	$-1/4$	$-3/5$	0	$-b/a$
y-intercept	2	3	$8/5$	1	b

31. (a) $60°$ (b) $117°$ 33. $y = -2x + 4$ 35. $y = 4x + 7$
37. $y = -\frac{1}{5}x + 6$ 39. $y = 11x - 18$ 41. $y = \dfrac{1}{\sqrt{3}}x - 3$
43. $y = \frac{1}{2}x + 2$ 45. $y = 1$ 47. $x = 5$
49. (a) parallel   (b) perpendicular   (c) parallel
    (d) perpendicular   (e) neither
51. (a) $-\frac{3}{2}$ (b) $\frac{4}{5}$ (c) $\frac{5}{2}$ (d) $-\frac{15}{2}$ (e) -4
53. about 240 N/m   55. about 0.6 m/sec/°C
57. (a) $F = \frac{9}{5}C + 32$ (b) $\frac{5}{9}$ (c) $-40°$ (d) $37°$ C
59. (a) $p = 0.098h + 1$ (b) approximately 10.2 m
61. (a) $r = -0.0125t + 0.8$ (b) 64 days   65. $\frac{49}{6}$
67. (a) $(4, -1)$ (b) $(1, -2)$

▶ **Exercise Set 1.5 (Page 45)**
1. in the proof of Theorem 1.5.1   3. (a) 10   (b) $(4, 5)$
5. (a) $\sqrt{29}$ (b) $(-\frac{9}{5}, -5)$ 11. 0   13. $y = -3x + 4$
15. $(-\frac{29}{8}, -\frac{23}{4})$ 17. 3   21. 4
23. (a) $(0, 0)$; 5   (b) $(1, 4)$; 4   (c) $(-1, -3)$; $\sqrt{5}$
    (d) $(0, -2)$; 1
25. $(x - 3)^2 + (y + 2)^2 = 16$ 27. $(x + 4)^2 + (y - 8)^2 = 64$

29. $(x + 3)^2 + (y + 4)^2 = 25$ **31.** $(x - 1)^2 + (y - 1)^2 = 2$

33. circle; center (1, 2), radius 4

35. circle; center $(-1, 1)$, radius $\sqrt{2}$

37. the point $(-1, -1)$ **39.** circle; center (0, 0), radius $\frac{1}{3}$

41. no graph   **43.** circle; center $(-\frac{5}{4}, -\frac{1}{2})$, radius $\frac{3}{2}$

45. (a) $y = -\sqrt{16 - x^2}$ (b) $y = 2 + \sqrt{3 - 2x - x^2}$

47. (a)   (b)

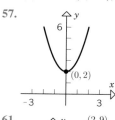

49. $y = -\frac{3}{4}x + \frac{25}{4}$

51. (a) inside   (b) largest $3\sqrt{5}$, smallest $\sqrt{5}$

53. $(1/3, \pm\sqrt{8}/3)$

55. (a) equation: $2x^2 + 2y^2 - 12x + 8y + 1 = 0$

(b) center $(3, -2)$, radius $5/\sqrt{2}$

57.   **59.**

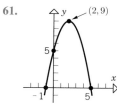   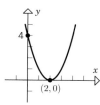

61.   **63.**

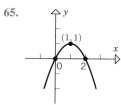   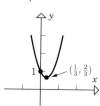

65.   **67.**

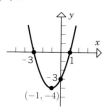

69.

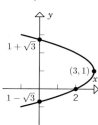

71. (a) $x = \sqrt{3 - y}$ (b) $x = 1 - \sqrt{y + 1}$

73. (a)

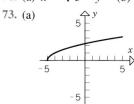   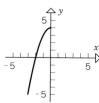

75. (a)

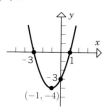

77. (a) $y = 150 - \frac{3}{2}x$ (b) $A = 150x - \frac{3}{2}x^2$ (c) 3750 ft^2

79. (a) $(-5 - \sqrt{33})/4 < x < (-5 + \sqrt{33})/4$

(b) $-\infty < x < +\infty$

81. (a) 30 ft   (b) 2.6 sec   (c) 2.1 sec

▶ **TECHNOLOGY EXERCISES, CHAPTER 1 (Page 48)**

1. 1.115088, 3.934317   **3.** -0.845838, 0, 2, 3.220645

5. $x < -0.871925$, $x > 4.205258$ **7.** $x < 2.478052$, $x > 3$

9. $-0.770917 < x < 1.670244$ **11.** 1.358094

13. 0.48 feet   **15.** 3.46×10^8 meters

17. (a) 9.35 years   (b) 220 sheep

▶ **Exercise Set 2.1 (Page 59)**

1. (a) 14   (b) 50   (c) 2   (d) 11   (e) $3a^2 + 6a + 5$

(f) $27t^2 + 2$

3. (a) -8 (b) $\frac{1}{4}$ (c) 0   (d) 6   (e) 5.8   (f) $\dfrac{1}{t^2 + 5}$

5. $(-\infty, 3) \cup (3, +\infty)$ **7.** $(-\infty, -\sqrt{3}] \cup [\sqrt{3}, +\infty)$

9. $(-\infty, -2) \cup [1, +\infty)$ **11.** $(-\infty, +\infty)$ **13.** [5, 8]

15. $(-\infty, +\infty)$ **17.** $(-\infty, 0) \cup (0, +\infty)$ **19.** $[0, +\infty)$

21. $x \neq \dfrac{\pi}{2} + 2k\pi$, $k = 0, \pm 1, \pm 2, \ldots$

23. domain $(-\infty, 3]$, range $[0, +\infty)$

25. domain $[-2, 2]$, range $[0, 2]$

27. domain $[0, +\infty)$, range $[3, +\infty)$

29. domain $(-\infty, +\infty)$, range $[3, +\infty)$

31. domain $(-\infty, +\infty)$, range $(-\infty, +\infty)$

33. domain $(-\infty, +\infty)$, range $[-3, 3]$

35. domain $(-\infty, +\infty)$, range $[1, 3]$

37. $f(x) = \begin{cases} 2x + 1, & x < 0 \\ 4x + 1, & x \geq 0 \end{cases}$ **39.** $g(x) = \begin{cases} 1 - 2x, & x < 0 \\ 1, & 0 \leq x < 1 \\ 2x - 1, & x \geq 1 \end{cases}$

41. $\frac{38}{3}$ **43.** $\pm\sqrt{2}$ **45.** $2k\pi$, $k = 0, \pm 1, \pm 2, \ldots$

47. $(\frac{1}{6} + 2k)^2\pi^2$ or $(\frac{5}{6} + 2k)^2\pi^2$ for $k = 0, 1, 2, \ldots$

49. $A = \dfrac{C^2}{4\pi}$ **51.** (a) $S = 6x^2$ (b) $S = 6V^{2/3}$

53. $V = 4x^3 - 46x^2 + 120x$ **55.** $h = L(1 - \cos\theta)$

57. (a) $25° \text{F}$ (b) $2° \text{F}$ (c) $-15° \text{F}$ **59.** $5° \text{F}$

61. $\dfrac{1 - (1/x)}{1 + (1/x)} = \dfrac{x - 1}{x + 1}$, $x \neq 0$ **63.** $x - 1$, $x \neq -1$ or -2

65. $\sqrt{x + 1} + 1$, $x \neq -1$ **67.** x, $x \neq -3$ or 1

▶ **Exercise Set 2.2 (Page 66)**

1. (a) $t^2 + 1$ (b) $t^2 + 4t + 5$ (c) $x^2 + 4x + 5$

(d) $\dfrac{1}{x^2} + 1$ (e) $x^2 + 2hx + h^2 + 1$ (f) $x^2 + 1$

(g) $x + 1$, $x \geq 0$ (h) $9x^2 + 1$

3. (a) 1   (b) 12   (c) -5 (d) 4

5. (a) $x^2 + 2x + 1$ (b) $-x^2 + 2x - 1$ (c) $2x(x^2 + 1)$

(d) $\dfrac{2x}{x^2 + 1}$ (e) $2(x^2 + 1)$ (f) $4x^2 + 1$

7. (a) $\sqrt{x+1} + x - 2$ (b) $\sqrt{x+1} - x + 2$
(c) $(x-2)\sqrt{x+1}$ (d) $\dfrac{\sqrt{x+1}}{x-2}$ (e) $\sqrt{x-1}$
(f) $\sqrt{x+1} - 2$

9. (a) $\sqrt{x-2} + \sqrt{x-3}$ (b) $\sqrt{x-2} - \sqrt{x-3}$
(c) $\sqrt{x-2}\sqrt{x-3}$ (d) $\dfrac{\sqrt{x-2}}{\sqrt{x-3}}$ (e) $\sqrt{\sqrt{x-3}-2}$
(f) $\sqrt{\sqrt{x-2}-3}$

11. (a) $\sqrt{1-x^2} + \sin 3x$ (b) $\sqrt{1-x^2} - \sin 3x$
(c) $\sqrt{1-x^2}\sin 3x$ (d) $\sqrt{1-x^2}/\sin 3x$ (e) $|\cos 3x|$
(f) $\sin(3\sqrt{1-x^2})$

13. $6x + 3h,\ h \neq 0$ **15.** $-\dfrac{1}{x(x+h)},\ h \neq 0$

17. $(f \circ g)(x) = \dfrac{4}{x+5},\ x \geq 0;\ (g \circ f)(x) = \dfrac{2}{\sqrt{x^2+5}}$

19. (a) $4x - 15$ (b) $4x^2 - 20x + 25$

25. $g(x) = \sqrt{x},\ h(x) = x + 2$ **27.** $g(x) = x^7,\ h(x) = x - 5$

29. $g(x) = |x|,\ h(x) = x^2 - 3x + 5$

31. $g(x) = x^2,\ h(x) = \sin x$ **33.** $g(x) = \dfrac{3}{5+x},\ h(x) = \cos x$

35. $g(x) = \dfrac{x}{3+x},\ h(x) = \tan x$

37. $f(x) = \sqrt{x},\ g(x) = 3 - x^2,\ h(x) = \sin x$

39. $f(x) = x^2 + x + 3$ **41.** $0, \frac{3}{2}$ **43.** $g(x) = 9x^2 - 5$

47. (a) monomial, polynomial, rational, explicit algebraic
(b) explicit algebraic (c) rational, explicit algebraic
(d) polynomial, rational, explicit algebraic

49. (a) explicit algebraic (b) rational, explicit algebraic
(c) monomial, polynomial, rational, explicit algebraic
(d) explicit algebraic

▶ **Exercise Set 2.3 (Page 75)**

1. (a) $-4, -3, -2, 2, 3$ (b) $0, 4$
(c) $-4 \leq x \leq -3,\ -2 \leq x \leq 2,\ x \geq 3$
(d) $x \leq -4,\ -3 \leq x \leq -2,\ 2 \leq x \leq 3$

3. 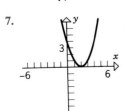 **5.**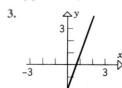

7. 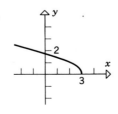 **9.**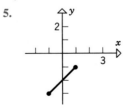

11. 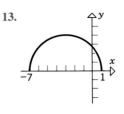 **13.**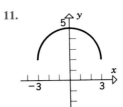

15. 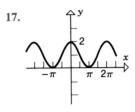 **17.**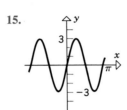

19. 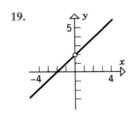 **21.**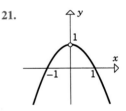

23. 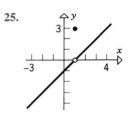 **25.**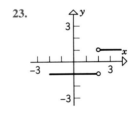

27. 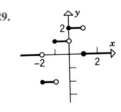 **29.**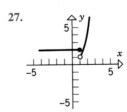

31. $f(x) = \begin{cases} x + 2, & x \leq 2 \\ 3x - 2, & x > 2 \end{cases}$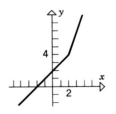

33. $g(x) = \begin{cases} 2, & x < 3 \\ 8 - 2x, & 3 \leq x < 5 \\ -2, & x \geq 5 \end{cases}$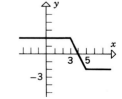

35. (a) 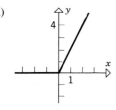 (b)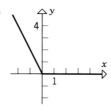

(c)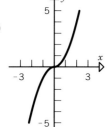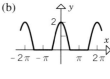

37. $A = \begin{cases} x^2, & 0 \le x \le 1 \\ 2x - 1, & x > 1 \end{cases}$

39. $g(x) = \begin{cases} 2, & x < -1 \\ 1 - x, & -1 \le x < 1 \\ \frac{1}{2}(x - 1), & x \ge 1 \end{cases}$

41. (a) 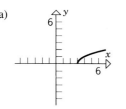 (b)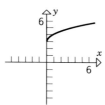

(c) 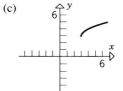 (d)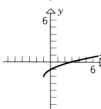

43. (a) 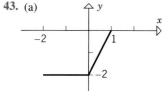 (b)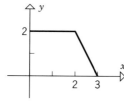

(c) 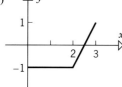 (d)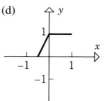

45.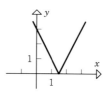

47. (a) 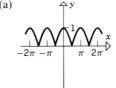 (b)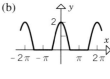

49. (a) even (b) odd (c) even (d) neither (e) odd
(f) even

51. (a) even (b) odd (c) odd (d) neither

55. (a) both (b) y is a function of x. (c) neither (d) both

57. (a) $y = \dfrac{1}{x^2}$ (b) $y = \dfrac{1 - x}{1 + x}$ (c) $y = -x$

61. (a) both (b) a function of x (c) a function of y
(d) neither

▶ **Exercise Set 2.4 (Page 85)**

1. (a) -1 (b) 3 (c) does not exist (d) 1 (e) -1
(f) 3

3. (a) 1 (b) 1 (c) 1 (d) 1 (e) $-\infty$ (f) $+\infty$

5. (a) 0 (b) 0 (c) 0 (d) 3 (e) $+\infty$ (f) $+\infty$

7. (a) $-\infty$ (b) $+\infty$ (c) does not exist (d) not defined
(e) 2 (f) 0

9. (a) $-\infty$ (b) $-\infty$ (c) $-\infty$ (d) 1 (e) 2 (f) 2

11. (a) 0 (b) 0 (c) 0 (d) 0 (e) does not exist
(f) does not exist

13. all values except -4

▶ **Exercise Set 2.5 (Page 96)**

1. 7   **3.** π **5.** 36   **7.** $\sqrt{109}$ **9.** 14   **11.** 0

13. 8   **15.** 4   **17.** $-\frac{4}{5}$ **19.** $\frac{3}{2}$ **21.** 0   **23.** 0

25. $-\sqrt{5}$ **27.** $1/\sqrt{6}$ **29.** $\sqrt{3}$ **31.** $+\infty$

33. does not exist   **35.** $-\infty$ **37.** $+\infty$

39. does not exist   **41.** $+\infty$ **43.** $-\infty$ **45.** $-\frac{1}{7}$

47. -1 **49.** 6   **51.** $+\infty$ **53.** $+\infty$ **55.** $-\infty$

57. $\frac{1}{2}$ if $a \ne 0$, 1 if $a = 0$ **59.** (a) 2 (b) 2 (c) 2

61. 4   **65.** $\frac{1}{4}$ **67.** $\frac{5}{6}$ **69.** 0   **71.** $\frac{5}{2}$ **73.** $a/2$

75. if $r(a)$ is defined

▶ **Exercise Set 2.6 (Page 103)**

Note: There are other possible answers for Exercises 1–23, 29.

1. 0.05   **3.** 1/700   **5.** 0.05   **7.** 1/9000   **9.** 1

11. $\delta = \epsilon/3$ **13.** $\delta = \epsilon/2$ **15.** $\delta = \epsilon$

17. $\delta = \min(\epsilon/6, 1)$ **19.** $\delta = \min(\epsilon/36, \frac{1}{4})$

21. $\delta = \min(2\epsilon, 4)$ **23.** $\delta = \epsilon$ **29.** $\delta = \min(\epsilon/8, 2)$

▶ **Exercise Set 2.7 (Page 110)**

1. continuous on $(1, 2)$, $[2, 3]$, $(2, 3)$;
discontinuous on $[1, 3]$, $(1, 3)$, and $[1, 2]$ at $x = 2$

3. continuous on $(1, 3)$, $(1, 2)$, $(2, 3)$;
discontinuous on $[1, 3]$ at $x = 1$ and $x = 3$; on $[1, 2]$ at
$x = 1$, and on $[2, 3]$ at $x = 3$

5. none   **7.** none   **9.** $x = \pm 4$ **11.** $x = \pm 3$

13. none   **15.** none   **17.** (a) 5   (b) $\frac{4}{3}$

21. $0, \pm 1, \pm 2, \ldots$

23. (a) $x = 0$; not removable   (b) $x = -3$; removable
(c) $x = 2$; removable, $x = -2$; not removable

33. -1.65, 1.35 **35.** (a) 2.25   (b) 2.235

▶ **Exercise Set 2.8 (Page 117)**

1. none   **3.** $x = n\pi$, $n = 0, \pm 1, \pm 2, \ldots$

5. $x = n\pi$, $n = 0, \pm 1, \pm 2, \ldots$ **7.** none

9. $x = \dfrac{\pi}{6} + 2n\pi$ or $\dfrac{5\pi}{6} + 2n\pi$, $n = 0, \pm 1, \pm 2, \ldots$

13. 1   **15.** $-\sqrt{3}/2$ **17.** 3   **19.** -1 **21.** 0

23. $\frac{7}{3}$ **25.** 1   **27.** 2   **29.** 0   **31.** $-\frac{25}{49}$

33. does not exist   **35.** 3   **37.** $\frac{1}{2}$

39. (a) 1   (b) 0   (c) 1   **41.** $-\pi$ **43.** (a) 0   (b) 0

45. 1   **53.** (a) 0.17365   (b) 0.17453

55. (a) 0.08749   (b) 0.08727

▶ **TECHNOLOGY EXERCISES, CHAPTER 2 (Page 118)**

1. 2.718282   **3.** 0.540302   **5.** $\frac{1}{2}$

7. (b) $3° F$, $-11° F$, $-18° F$, $-22° F$
(c) 34 mi/hr, 19 mi/hr, 12 mi/hr, 7 mi/hr

9. domain $[-0.724492, 1.220744]$
range $[-1.055084, 1.490152]$

11. 0.077471   **13.** (b) 24.61 ft/sec   (c) 3.01 sec

15. (d) 1.324702   **17.** 1.267168

▶ **Exercise Set 3.1 (Page 126)**

1. (a) 7/2   (b) 3   (c) x_0 (d)

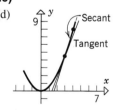

3. (a) $-1/6$ (b) $-1/4$ (c) $-1/x_0^2$ (d)

5. (a) $2x_0$ (b) 4   **7.** (a) $\dfrac{1}{2\sqrt{x_0}}$ (b) $\frac{1}{2}$

9. (a) 4 m/sec   (b)

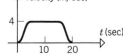

11. (a) t_0 (b) 0   (c) speeding up   (d) slowing down

13. It is a straight line with slope equal to the velocity.

15. (a) $72° F$ at about 4:30 P.M.   (b) $4° F/hr$
(c) $-7° F/hr$ at about 9 P.M.

---

17. (a) first year   (b) 6 cm/yr
(c) 10 cm/yr at about age 14
(d)

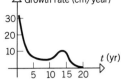

19. (a) 320,000 ft   (b) 8000 ft/sec   (c) 45 ft/sec
(d) 24,000 ft/sec

21. (a) 720 ft/min   (b) 192 ft/min   **23.** (a) 3π (b) 4π

▶ **Exercise Set 3.2 (Page 137)**

1. $6x$ **3.** $3x^2$ **5.** $\dfrac{1}{2\sqrt{x+1}}$ **7.** $-\dfrac{1}{x^2}$ **9.** $2ax$

11. $-\dfrac{1}{2x^{3/2}}$ **13.** 18; $y = 18x - 27$ **15.** 0; $y = 0$

17. $\frac{1}{6}$; $y = \frac{1}{6}x + \frac{5}{3}$ **19.** $y = 5x - 16$ **21.** (a) $8x$ (b) 8

23. $8t + 1$ **25.** $6\lambda - 1$

27. (a) D   (b) F   (c) B   (d) C   (e) A   (f) E

29. 0.08 and 0.018 mol/L/sec

31. (a) $F \approx 200$ lb, $dF/d\theta \approx 60$ lb/rad   (b) $\mu \approx 0.3$

33.   **35.**

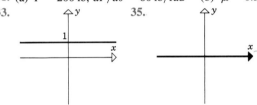

37.   **39.**

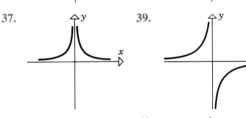

41.   **43.**

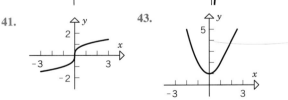

45. $f(1) = 0$, $f'(1) = 5$

▶ **Exercise Set 3.3 (Page 149)**

1. $28x^6$ **3.** $24x^7 + 2$ **5.** 0   **7.** $-\frac{1}{3}(7x^6 + 2)$

9. $3ax^2 + 2bx + c$ **11.** $24x^{-9} + 1/\sqrt{x}$

13. $-3x^{-4} - 7x^{-8}$ **15.** $18x^2 - \frac{3}{2}x + 12$

17. $-15x^{-2} - 14x^{-3} + 48x^{-4} + 32x^{-5}$

19. $12x(3x^2 + 1)$ **21.** $-5/(5x - 3)^2$ **23.** $3/(2x + 1)^2$

25. $7/(x + 3)^2$ **27.** $\left(\dfrac{3x + 2}{x}\right)(-5x^{-6}) + (x^{-5} + 1)\left(-\dfrac{2}{x^2}\right)$

29. (a) $-\frac{37}{4}$ (b) $-\frac{23}{16}$

31. (a) 10   (b) 19   (c) 19   (d) -11 **33.** $32t$

35. $3\pi r^2$ **37.** $\dfrac{7 - 2t^3}{(t^3 + 7)^2}$ **39.** $-\dfrac{2GmM}{r^3}$

41. (a) $42x - 10$ (b) 24 (c) $2/x^3$ (d) $700x^3 - 96x$
43. (a) $-210x^{-8} + 60x^2$ (b) $-6x^{-4}$ (c) $6a$
45. (a) 0 (b) 112 (c) 360 **49.** $F''(x) = xf''(x) + 2f'(x)$
51. $(1, \frac{5}{6})$, $(2, \frac{2}{3})$ **53.** $y = 5x + 17$ **55.** $a = 3$, $b = 2$
57. $y = 3x^2 - x - 2$ **59.** $\frac{1}{2}$ **61.** $2 \pm \sqrt{3}$ **63.** $-2x_0$
69. (a) $2(1 + 1/x)(x^{-3} + 7)$
$\qquad + (2x + 1)(-1/x^2)(x^{-3} + 7)$
$\qquad\qquad + (2x + 1)(1 + 1/x)(-3x^{-4})$
$\quad$ (b) $(-5x^{-6})(x^2 + 2x)(4 - 3x)(2x^9 + 1)$
$\qquad + x^{-5}(2x + 2)(4 - 3x)(2x^9 + 1)$
$\qquad + x^{-5}(x^2 + 2x)(-3)(2x^9 + 1)$
$\qquad\qquad + x^{-5}(x^2 + 2x)(4 - 3x)(18x^8)$
$\quad$ (c) $3(7x^6 + 2)(x^7 + 2x - 3)^2$ (d) $100x(x^2 + 1)^{49}$
71. $2(2x^3 - 5x^2 + 7x - 2)(6x^2 - 10x + 7)$
73. not differentiable at $x = 1$ **75.** $a = 6$, $b = -3$
77. (a) $x = \frac{2}{3}$ (b) $x = \pm 2$
79. (a) $n(n - 1)(n - 2) \cdots 1$ (b) 0
$\quad$ (c) $a_n n(n - 1)(n - 2) \cdots 1$
85. (b) f and all its derivatives up to $f^{(n-1)}(x)$ are continuous
$\quad$ on (a, b).

▶ **Exercise Set 3.4 (Page 154)**

1. $-2 \sin x - 3 \cos x$ **3.** $\dfrac{x \cos x - \sin x}{x^2}$

5. $x^3 \cos x + (3x^2 + 5) \sin x$ **7.** $\sec x \tan x - \sqrt{2} \sec^2 x$
9. $\sec^3 x + \sec x \tan^2 x$ **11.** $1 + 4 \csc x \cot x - 2 \csc^2 x$

13. $-\dfrac{\csc x}{1 + \csc x}$ **15.** 0 **17.** $\dfrac{1}{(1 + x \tan x)^2}$

19. $-x \cos x - 2 \sin x$ **21.** $-x \sin x + 5 \cos x$
23. $-4 \sin x \cos x$
25. (a) $x = n\pi$, $n = 0, \pm 1, \pm 2, \ldots$ (b) none
$\quad$ (c) $x = \dfrac{\pi}{2} + n\pi$, $n = 0, \pm 1, \pm 2, \ldots$

27. (a) $y = x$ (b) $y = 2x - \pi/2 + 1$ (c) $y = 2x + \pi/2 - 1$
29. 0.087 ft/degree   **31.** 1.75 m/degree
33. (a) all x (b) all x
$\quad$ (c) $x \neq \pi/2 + n\pi$, $n = 0, \pm 1, \pm 2, \ldots$
$\quad$ (d) $x \neq n\pi$, $n = 0, \pm 1, \pm 2, \ldots$
$\quad$ (e) $x \neq \pi/2 + n\pi$, $n = 0, \pm 1, \pm 2, \ldots$
$\quad$ (f) $x \neq n\pi$, $n = 0, \pm 1, \pm 2, \ldots$
$\quad$ (g) $x \neq \pi + 2n\pi$, $n = 0, \pm 1, \pm 2, \ldots$
$\quad$ (h) $x \neq n\pi/2$, $n = 0, \pm 1, \pm 2, \ldots$ (i) all x
35. $3, 7, 11, \ldots$ **37.** $\sec^2 y$

▶ **Exercise Set 3.5 (Page 160)**

1. $37(x^3 + 2x)^{36}(3x^2 + 2)$ **3.** $-2\left(x^3 - \dfrac{7}{x}\right)^{-3}\left(3x^2 + \dfrac{7}{x^2}\right)$

5. $\dfrac{24(1 - 3x)}{(3x^2 - 2x + 1)^4}$ **7.** $\dfrac{3}{4\sqrt{x}\sqrt{4 + 3\sqrt{x}}}$ **9.** $3x^2 \cos(x^3)$

11. $8x \sec^2(4x^2)$ **13.** $-20 \cos^4 x \sin x$

15. $-\dfrac{2}{x^3} \cos\left(\dfrac{1}{x^2}\right)$ **17.** $28x^6 \sec^2(x^7) \tan(x^7)$

19. $-\dfrac{5 \sin(5x)}{2\sqrt{\cos(5x)}}$

---

21. $-3[x + \csc(x^3 + 3)]^{-4} \cdot [1 - 3x^2 \csc(x^3 + 3) \cot(x^3 + 3)]$

23. $\dfrac{x(10 - 3x^2)}{\sqrt{5 - x^2}}$ **25.** $10x^3 \sin 5x \cos 5x + 3x^2 \sin^2 5x$

27. $-x^3 \sec\left(\dfrac{1}{x}\right) \tan\left(\dfrac{1}{x}\right) + 5x^4 \sec\left(\dfrac{1}{x}\right)$

29. $\sin x \sin(\cos x)$ **31.** $-6 \cos^2(\sin 2x) \sin(\sin 2x) \cos 2x$
33. $12(5x + 8)^{13}(x^3 + 7x)^{11}(3x^2 + 7)$
$\qquad\qquad\qquad\qquad + 65(x^3 + 7x)^{12}(5x + 8)^{12}$

35. $\dfrac{33(x - 5)^2}{(2x + 1)^4}$ **37.** $-\dfrac{2(2x + 3)^2(52x^2 + 96x + 3)}{(4x^2 - 1)^9}$

39. $5[x \sin 2x + \tan^4(x^7)]^4[2x \cos 2x + \sin 2x$
$\qquad\qquad\qquad\qquad + 28x^6 \tan^3(x^7) \sec^2(x^7)]$
41. $-25x \cos(5x) - 10 \sin(5x) - 2 \cos(2x)$
43. $y = -x$ **45.** $y = -1$ **47.** $3 \cot^2 \theta \csc^2 \theta$
49. $\pi(b - a) \sin 2\pi\omega$
51. (c) $f = 1/T$
$\quad$ (d) amplitude $= 0.6$ cm, $T = 2\pi/15$, $f = 15/(2\pi)$
53. (a) 10 lb/in^2, -2 lb/(in$^3 \cdot$ mi) (b) -0.6 lb/(in$^2 \cdot$ sec)

55. $3x^2 y^2 \dfrac{dy}{dx} + 2xy^3$ **57.** $\left(x\dfrac{dy}{dx} + y\right) \cos(xy)$

59. $2x\dfrac{dx}{dt} + 2y\dfrac{dy}{dt}$ **61.** $\dfrac{x^2}{2\sqrt{y}}\dfrac{dy}{dt} + 2x\sqrt{y}\dfrac{dx}{dt}$

63. $\begin{cases} \cos x, & 0 < x < \pi \\ -\cos x, & -\pi < x < 0 \end{cases}$ **65.** (a) $-\dfrac{1}{x}\cos\dfrac{1}{x} + \sin\dfrac{1}{x}$

67. (a) 21 (b) -36 **69.** 6 **71.** $1/(2x)$ **73.** $\frac{2}{3}x$
75. $f'(g(h(x)))g'(h(x))h'(x)$

▶ **Exercise Set 3.6 (Page 168)**

1. $\frac{2}{3}(2x - 5)^{-2/3}$ **3.** $\dfrac{9}{2(x + 2)^2}\left(\dfrac{x - 1}{x + 2}\right)^{1/2}$

5. $\frac{1}{3}x^2(5x^2 + 1)^{-5/3}(25x^2 + 9)$

7. $-\dfrac{15[\sin(3/x)]^{3/2} \cos(3/x)}{2x^2}$

9. $-\frac{2}{3}(2x - 1)^{-4/3} \sec^2[(2x - 1)^{-1/3}]$ **11.** $-1, \frac{2}{3}$

13. $-\dfrac{x}{y}$ **15.** $\dfrac{1 - 2xy - 3y^3}{x^2 + 9xy^2}$ **17.** $-\dfrac{y^2}{x^2}$

19. $-\dfrac{\sqrt{y}}{\sqrt{x}}$ **21.** $\dfrac{1 - 70x(x^2 + 3y^2)^{34}}{210y(x^2 + 3y^2)^{34}}$

23. $\dfrac{\frac{3}{2}x^2(x^3 + y^2)^{1/2} - y}{x - y(x^3 + y^2)^{1/2}}$ **25.** $\dfrac{1 - 2xy^2 \cos(x^2y^2)}{2x^2y \cos(x^2y^2)}$

27. $\dfrac{1 - 3y^2 \tan^2(xy^2 + y) \sec^2(xy^2 + y)}{3(2xy + 1) \tan^2(xy^2 + y) \sec^2(xy^2 + y)}$

29. $\dfrac{3y^2 \sin^2(xy^2) \cos(xy^2)}{2\sqrt{1 + \sin^3(xy^2)} - 6xy \sin^2(xy^2) \cos(xy^2)}$ **31.** $-\frac{14}{13}$

33. 2 **35.** -2 **37.** 1 **39.** $\frac{6}{5}$ **41.** $-2x/y^5$

43. $-3/(y - x)^3$ **45.** $-\dfrac{\sin 2y + y(\sin^2 y + 1)}{(1 + x \sin y)^3}$

47. $\dfrac{2t^3 + 3a^2}{2a^3 - 6at}$ **49.** $-\dfrac{b^2\lambda}{a^2\omega}$ **51.** $-\dfrac{2y^3 + 3t^2y}{(6ty^2 + t^3) \cos t}$

53. $\dfrac{dy}{dt} = \dfrac{3 \cos 3x - y^2}{2xy}\dfrac{dx}{dt}$ **55.** $a = \frac{1}{4}$, $b = \frac{5}{4}$

57. $y = x/\sqrt{3}$, $y = -x/\sqrt{3}$
59. (a) $(0, 0)$, $(-5/\sqrt{2}, 0)$, $(5/\sqrt{2}, 0)$ (b) $9x + 13y = 40$

▶ **Exercise Set 3.7 (Page 174)**

1. (a) 5   (b) 4   (c)

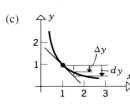

3. (a) $-\frac{1}{3}$ (b) -0.5 (c)

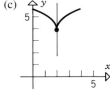

5. $dy = 3x^2\, dx$, $\Delta y = 3x^2\, \Delta x + 3x(\Delta x)^2 + (\Delta x)^3$
7. $dy = (2x - 2)\, dx$, $\Delta y = 2x\, \Delta x + (\Delta x)^2 - 2\, \Delta x$
9. $dy = (12x^2 - 14x + 2)\, dx$ **11.** $dy = (\cos x - x \sin x)\, dx$
13. $2x$ **15.** -1 **17.** 83.16   **19.** 8.0625
21. 8.9944   **23.** 2.005   **25.** 0.8573   **27.** 0.6947
29. 0.0225   **31.** 0.0048
33. (a) $\pm 2\ \text{ft}^2$ (b) side: $\pm 1\%$, area: $\pm 2\%$
35. (a) opposite: $\pm 0.151''$, adjacent: $\pm 0.087''$
    (b) opposite: $\pm 3.0\%$, adjacent: $\pm 1.0\%$
37. $\pm 10\%$ **39.** $\pm 0.017\ \text{cm}^2$ **41.** $\pm 6\%$ **43.** $\pm 0.5\%$
45. $0.236\ \text{cm}^3$ **47.** (a) $\alpha = 1.5 \times 10^{-5}/°C$ (b) 180.1 cm
49. $x = 0.1$: 0.9091, 0.9; $x = -0.02$: 1.0204, 1.02

▶ **TECHNOLOGY EXERCISES, CHAPTER 3 (Page 176)**

1. 3.6   **3.** -0.777778 **5.** 2.772589   **7.** 6.772589
9. 58.75 ft/sec   **11.** ± 0.535428

▶ **Exercise Set 4.1 (Page 182)**

1. (a) $\dfrac{dA}{dt} = 2x\dfrac{dx}{dt}$ (b) $12\ \text{ft}^2/\text{min}$

3. (a) $\dfrac{dV}{dt} = \pi\left(r^2\dfrac{dh}{dt} + 2rh\dfrac{dr}{dt}\right)$
    (b) $-20\pi\ \text{in}^3/\text{sec}$; decreasing

5. (a) $\dfrac{d\theta}{dt} = \dfrac{\cos^2 \theta}{x^2}\left(x\dfrac{dy}{dt} - y\dfrac{dx}{dt}\right)$
    (b) $-\frac{5}{16}$ radian/sec; decreasing
7. $4\pi/15\ \text{in}^2/\text{min}$ **9.** $1/\sqrt{\pi}$ mi/hr
11. $4860\pi\ \text{cm}^3/\text{min}$ **13.** $\frac{5}{6}$ ft/sec   **15.** $\frac{8}{5}\ \text{in}^2/\text{min}$
17. 704 ft/sec
19. (a) 500 mi, 1716 mi   (b) 1354 mi; 27.7 mi/min
21. $9/(20\pi)$ ft/min   **23.** $125\pi\ \text{ft}^3/\text{min}$ **25.** 250 mi/hr
27. $\frac{36}{25}\sqrt{69}$ ft/min   **29.** $8\pi/5$ km/sec   **31.** $600\sqrt{7}$ mi/hr
33. (a) $-60/7$ units/sec   (b) falling   **35.** -4 units/sec
37. $\frac{3}{2}$ **39.** 4.5 cm/sec; away   **43.** $20/(9\pi)$ cm/sec

▶ **Exercise Set 4.2 (Page 190)**

1. (a) (d, f) (b) $(a, d), (f, g)$ (c) $(a, b), (c, e)$
    (d) $(b, c), (e, g)$

3. at A: negative, positive; at B: positive, negative;
    at C: negative, negative
5. (a) $[\frac{5}{2}, +\infty)$ (b) $(-\infty, \frac{5}{2}]$ (c) $(-\infty, +\infty)$ (d) none
    (e) none
7. (a) $(-\infty, +\infty)$ (b) none   (c) $(-2, +\infty)$ (d) $(-\infty, -2)$
    (e) -2
9. (a) $(-\infty, -\frac{2}{3}], [\frac{2}{3}, +\infty)$ (b) $[-\frac{2}{3}, \frac{2}{3}]$ (c) $(0, +\infty)$
    (d) $(-\infty, 0)$ (e) 0
11. (a) $[1, +\infty)$ (b) $(-\infty, 1]$ (c) $(-\infty, 0), (\frac{2}{3}, +\infty)$
    (d) $(0, \frac{2}{3})$ (e) $0, \frac{2}{3}$
13. (a) $[\pi, 2\pi]$ (b) $(0, \pi]$ (c) $(\pi/2, 3\pi/2)$
    (d) $(0, \pi/2), (3\pi/2, 2\pi)$ (e) $\pi/2, 3\pi/2$
15. (a) $(-\pi/2, \pi/2)$ (b) none   (c) $(0, \pi/2)$
    (d) $(-\pi/2, 0)$ (e) 0
17. (a) $(-\infty, +\infty)$ (b) none   (c) $(-\infty, -2)$ (d) $(-2, +\infty)$
    (e) -2
19. (a) $[-1, +\infty)$ (b) $(-\infty, -1]$ (c) $(-\infty, 0), (2, +\infty)$
    (d) $(0, 2)$ (e) $0, 2$
21. $(-\infty, +\infty)$
27. (a)

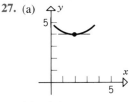

(b)

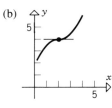

(c)

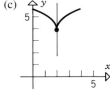

29. (a)

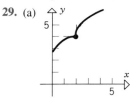

(b)

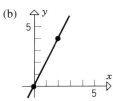

31. none   **39.** $f(x) = x$, $g(x) = 2x$ on $(-\infty, +\infty)$

▶ **Exercise Set 4.3 (Page 196)**

1. $x = \frac{5}{2}$ (stationary)   **3.** $x = -3, 1$ (stationary)
5. $x = 0, \pm\sqrt{3}$ (stationary)   **7.** $x = \pm\sqrt{2}$ (stationary)
9. $x = 0$ (not differentiable)
11. $x = n\pi/3$, $n = 0, \pm 1, \pm 2, \ldots$ (stationary)
13. $x = n\pi/4$, $n = 1, 2, 3, 4, 5, 6, 7$ (stationary)
15. $x = -1$ (stationary); $x = 0$ (not differentiable)
17. (a) $x = 2$ (b) $x = 0$ (c) $x = 1, x = 3$
19. relative maximum at $x = 0$,
    relative minimum at $x = \pm\sqrt{5}$
21. relative maximum at $x = \pm\frac{3}{2}$,
    relative minimum at $x = -1$
23. relative max of 5 at $x = -2$
25. relative min of 0 at $x = \pi$;
    relative max of 1 at $x = \pi/2, 3\pi/2$
27. none

29. relative min of 0 at $x = 1$;
relative max of $\frac{4}{27}$ at $x = \frac{1}{3}$

31. relative min of 0 at $x = 0$;
relative max of 1 at $x = -1, 1$

33. relative min of 0 at $x = 0$

35. relative min of 0 at $x = 0$

37. relative min of 0 at $x = -2, 2$;
relative max of 4 at $x = 0$

39. relative min of 0 at $x = \pm\pi/2, \pm3\pi/2, \pm5\pi/2, \ldots$;
relative max of 1 at $x = 0, \pm\pi, \pm2\pi, \ldots$

41. relative min of $\tan(1)$ at $x = 0$

43. relative min of 0 at $x = \pi/2, \pi, 3\pi/2$;
relative max of 1 at $x = \pi/4, 3\pi/4, 5\pi/4, 7\pi/4$

45. 54

47. $f(x) = -x^4$ has a relative maximum at $x = 0$,
$f(x) = x^4$ has a relative minimum at $x = 0$,
$f(x) = x^3$ has neither at $x = 0$;
$f'(0) = 0$ for all three functions.

49. (a)

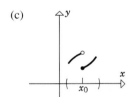

$f(x_0)$ is not an extreme value.

(b)

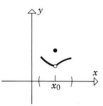

$f(x_0)$ is a relative maximum.

(c)

$f(x_0)$ is a relative minimum.

▶ **Exercise Set 4.4 (Page 203)**

1.

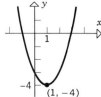

3.

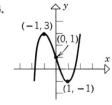

5.

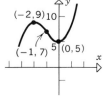

7.

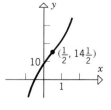

9.

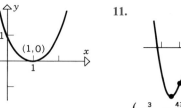

11.

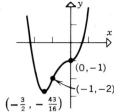

13.

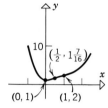

15.

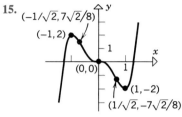

17.

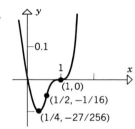

19. vertical: $x = 2$; horizontal: $y = 3$

21. vertical: $x = \pm\sqrt{5}$; horizontal: none

23. vertical: $x = -1$, $x = 3$; horizontal: $y = 1$

25. $x = -\frac{5}{2}$ **27.** $x = -\frac{1}{3}$

29.

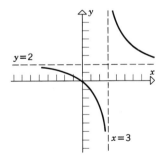

31.

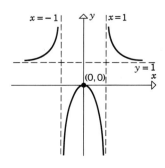

33.

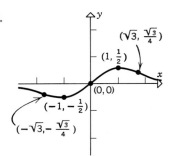

35.

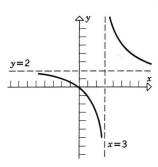

37.

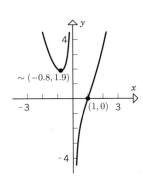

39.

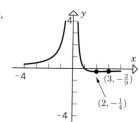

41.

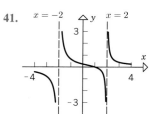

43.

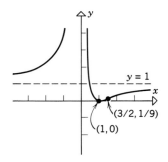

45.

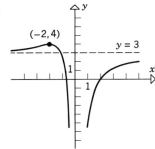

47.

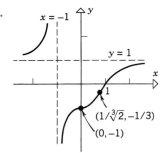

49.

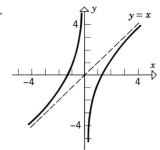

51.

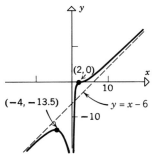

53.

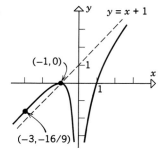

55.

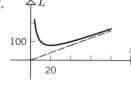

57.

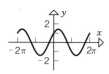

59.

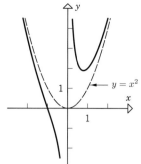

21.

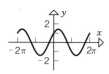

23.

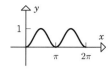

▶ **Exercise Set 4.6 (Page 214)**

1. maximum value 1 when $x = 0, 1$;
   minimum value 0 when $x = \frac{1}{2}$

3. maximum value 27 when $x = 4$;
   minimum value -1 when $x = 0$

5. maximum value $3/\sqrt{5}$ when $x = 1$;
   minimum value $-3/\sqrt{5}$ when $x = -1$

7. maximum value 48 when $x = 8$;
   minimum value 0 when $x = 0, 20$

9. maximum value $1 - (\pi/4)$ when $x = -\pi/4$;
   minimum value $(\pi/4) - 1$ when $x = \pi/4$

11. maximum value 2 when $x = 0$;
    minimum value $\sqrt{3}$ when $x = \pi/6$

13. maximum value 17 when $x = -5$;
    minimum value 1 when $x = -3$

15. minimum $-\frac{13}{4}$, no maximum

17. maximum 1, no minimum

19. minimum 0, no maximum

21. no maximum or minimum

23. no maximum or minimum

25. maximum -4, no minimum

27. maximum value $3\sqrt{3}/2$ when
    $x = (\pi/6) + n\pi$ $(n = 0, \pm1, \pm2, \dots)$
    minimum value $-3\sqrt{3}/2$ when
    $x = (5\pi/6) + n\pi$ $(n = 0, \pm1, \pm2, \dots)$

29. maximum $\sin(1)$, minimum $-\sin(1)$

31. maximum value 2; minimum value $-\frac{1}{4}$

33. (a) relative min of 0 at $x = a$ (b) none   **35.** (b) 125

37. $2/(3\sqrt{3})$ **39.** $a_0 = 9$, $a_1 = -8$, $a_2 = 2$

47. $(\frac{1}{2}, -\frac{1}{4})$ closest, $(-1, -1)$ farthest

▶ **Exercise Set 4.7 (Page 224)**

1. $5 + 5$ **3.** (a) 1   (b) $\frac{1}{2}$ **5.** 500 ft by 750 ft

7. 5 in. by $\frac{12}{5}$ in.   **9.** $10\sqrt{2}$ by $10\sqrt{2}$

11. 80 ft ($1 fencing), 40 ft ($2 fencing)   **15.** $p/(4 + \pi)$

17. 2 in. square   **19.** $\frac{200}{27}$ ft^3

21. base: 10 cm square, height: 20 cm

23. ends: $\sqrt[3]{3V/4}$ units square, length: $\frac{4}{3}\sqrt[3]{3V/4}$ units

25. height: $2\sqrt{(5 - \sqrt{5})/10}\,R$, radius: $\sqrt{(5 + \sqrt{5})/10}\,R$

29. height = radius = $\sqrt[3]{500/\pi}$ **31.** $L/12$ by $L/12$ by $L/12$

33. height: $L/\sqrt{3}$, radius: $\sqrt{2/3}\,L$

35. radius: $\sqrt[6]{\dfrac{450}{\pi^2}}$ cm, height: $\dfrac{30}{\pi}\sqrt[3]{\dfrac{\pi^2}{450}}$ cm

37. height: $4R$, radius: $\sqrt{2}R$

39. $\pi/3$ **41.** $5\sqrt{5}$ ft   **43.** (a) 7000   (b) yes

45. 13,722 lb   **47.** $1/\sqrt{5}$ **51.** $(-\sqrt{2}, 1)$, $(\sqrt{2}, 1)$

53. $(\sqrt{2}, 1/2)$ **55.** (a) no maximum   (b) -3

57. $(-1/\sqrt{3}, 3/4)$ **59.** $4(1 + 2^{2/3})^{3/2}$ ft

61. 30 cm from the weaker source

▶ **Exercise Set 4.5 (Page 207)**

1.

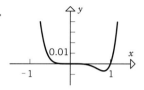

3.

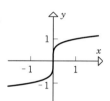

5.

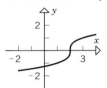

7.

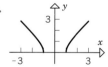

9.

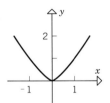

11.

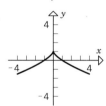

13.

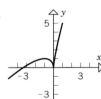

15.

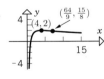

17.

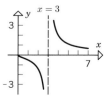

19.

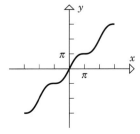

65. (c) $\frac{1}{4}$ mi downstream from the house

▶ **Exercise Set 4.8 (Page 232)**

1. 1.414213562   **3.** 1.817120593   **5.** -1.671699882
7. 1.224439550   **9.** 0.581138830   **11.** -1.452626879
13. 1.895494267   **15.** 4.493409458
17. (b) 3.162277660   **19.** -1.165373043
21. -0.474626618, 1.395336994
23. -1.220744085, 0.724491959   **25.** -4.098859132
27. (0.589754512, 0.347810385)   **29.** 171°

▶ **Exercise Set 4.9 (Page 237)**

1. 3   **3.** π **5.** 1   **7.** 1   **9.** 5/4   **11.** $-\sqrt{5}$
13. (b) $\tan x$ is not continuous on $[0, \pi]$.   **29.** 8
31. $f(x) = x^3 - 4x + 5$

▶ **Exercise Set 4.10 (Page 242)**

1. (a) positive, negative, slowing down
    (b) positive, positive, speeding up
    (c) negative, positive, slowing down
3. (a) left   (b) negative   (c) speeding up
    (d) slowing down
5. (a) 6.7 ft/sec²   (b) $t = 0$ sec

7.

t	s	v	$|v|$	a	Direction; Motion
1	-5	-9	9	-6	Left; speeding up
2	-16	-12	12	0	Left; neither
3	-27	-9	9	6	Left; slowing down
4	-32	0	0	12	Stopped
5	-25	15	15	18	Right; speeding up

9. (a) 12   (b) $t = 2.2$, $s = -24.2$

11.
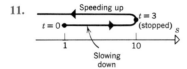

13.
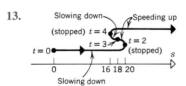

15.
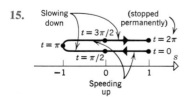

17. (a) $s = 5/16$, $v = 3/2$ (b) $s = 1$, $a = -3$

19. (b) $v = \dfrac{3}{2\sqrt{3t + 7}}$; $a = -9/500$

21. (b) 2/3 unit   (c) $0 \le t < 1$ and $t > 2$
23. (a) -1.25 ft/sec/ft   (b) -2500 ft/sec²

▶ **TECHNOLOGY EXERCISES, CHAPTER 4 (Page 244)**

1. (c) minimum $(-2.111985, -0.355116)$
     maximum $(0.372591, 2.012931)$
3. relative minimum -0.232466 at $x = 0.450184$
5. relative maximum 0.355977 at $x = 1.244155$
    relative minimum -0.876839 at $x = -0.886352$
7. relative maximum at $x = -1.414214$, 1.5
    relative minimum at $x = 1.414214$, 2
    inflection points at $x = -0.657920$, 1.455739, 1.827181
9. $y = x$
11. (a) base: 1.508 ft by 1.508 ft, height: 1.758 ft
    (b) \$130,890   (c) \$319
13. (a) $0 \le t \le 0.64$ sec   (b) 1.10 sec
15. 8.4 years   **17.** 248.938×10^6 km

▶ **Exercise Set 5.2 (Page 256)**

1. $\frac{1}{9}x^9 + C$ **3.** $\frac{7}{12}x^{12/7} + C$ **5.** $8\sqrt{t} + C$
7. $\frac{2}{9}x^{9/2} + C$ **9.** $-\frac{1}{2}x^{-2} + \frac{2}{3}x^{3/2} - \frac{12}{5}x^{5/4} + \frac{1}{3}x^3 + C$
11. $28y^{1/4} - \frac{3}{4}y^{4/3} + \frac{8}{3}y^{3/2} + C$ **13.** $\frac{1}{2}x^2 + \frac{1}{5}x^5 + C$
15. $3x^{4/3} - \frac{12}{7}x^{7/3} + \frac{3}{10}x^{10/3} + C$
17. $\frac{1}{2}x^2 - \frac{2}{x} + \frac{1}{3x^3} + C$ **19.** $-4\cos x + 2\sin x + C$
21. $\tan x + \sec x + C$ **23.** $\sec x + x + C$
25. $\sec x + C$ **27.** $\theta - \cos\theta + C$
29. $\sin\theta - 5\tan\theta + C$ **31.** $F(x) = \frac{3}{4}x^{4/3} + \frac{5}{4}$
33. $f(x) = \frac{4}{15}x^{5/2} + C_1 x + C_2$ **35.** $y = x^2 + x - 6$
37. $y = x^3 - 6x + 7$ **39.** $\displaystyle\int \frac{3x^2}{2\sqrt{x^3 + 5}}\,dx = \sqrt{x^3 + 5} + C$
41. $\displaystyle\int \frac{\cos(2\sqrt{x})}{\sqrt{x}}\,dx = \sin(2\sqrt{x}) + C$
43. (b) $F(x) = G(x) + \frac{8}{3}$ **45.** $f(x) = 15x^2 - 3$
47. $\tan x - x + C$ **49.** $\frac{1}{2}(x - \sin x) + C$

▶ **Exercise Set 5.3 (Page 261)**

1. (a) $\frac{1}{24}(x^2 + 1)^{24} + C$ (b) $-\frac{1}{4}\cos^4 x + C$
    (c) $-2\cos\sqrt{x} + C$ (d) $\frac{3}{4}\sqrt{4x^2 + 5} + C$
3. (a) $-\frac{1}{2}\cot^2 x + C$ (b) $\frac{1}{10}(1 + \sin t)^{10} + C$
    (c) $\frac{2}{7}(1 + x)^{7/2} - \frac{4}{5}(1 + x)^{5/2} + \frac{2}{3}(1 + x)^{3/2} + C$
    (d) $-\cot(\sin x) + C$
5. $-\frac{1}{8}(2 - x^2)^4 + C$ **7.** $\frac{1}{8}\sin 8x + C$ **9.** $\frac{1}{4}\sec 4x + C$
11. $\frac{1}{21}(7t^2 + 12)^{3/2} + C$ **13.** $\frac{2}{3}\sqrt{x^3 + 1} + C$
15. $-\frac{1}{16}(4x^2 + 1)^{-2} + C$ **17.** $\frac{1}{5}\cos(5/x) + C$
19. $\frac{1}{3}\tan(x^3) + C$ **21.** $\frac{1}{18}\sin^6 3t + C$
23. $-\frac{1}{6}(2 - \sin 4\theta)^{3/2} + C$ **25.** $\frac{1}{6}\sec^3 2x + C$
27. $-\frac{1}{3}\tan(\cos 3\theta) + C$ **29.** $\dfrac{1}{b(n + 1)}\sin^{n+1}(a + bx) + C$
31. $\frac{2}{5}(x - 3)^{5/2} + 2(x - 3)^{3/2} + C$
33. $\frac{2}{3}(y + 1)^{3/2} - 2(y + 1)^{1/2} + C$ **35.** $\frac{1}{3}\tan 3\theta - \theta + C$
39. (a) $\frac{25}{3}x^3 - 5x^2 + x + C$; $\frac{1}{15}(5x - 1)^3 + C$
    (b) They differ by a constant.
41. $f(x) = 6x + \frac{5}{2}\cos 2x + \frac{1}{2}$ **43.** $\frac{1}{3}f(3x + 2) + C$
45. $-\frac{1}{2}f(2/x) + C$

▶ **Exercise Set 5.4 (Page 267)**

1. (a) 36   (b) 55   (c) 40   (d) 6

3. $\displaystyle\sum_{k=1}^{10} k$ **5.** $\displaystyle\sum_{k=1}^{49} k(k+1)$ **7.** $\displaystyle\sum_{k=1}^{10} 2k$

9. $\displaystyle\sum_{k=1}^{6} (-1)^{k+1}(2k-1)$ **11.** $\displaystyle\sum_{k=1}^{5} (-1)^k \frac{1}{k}$

13. $\displaystyle\sum_{k=1}^{4} \sin\frac{(2k-1)\pi}{8}$ **15.** $\displaystyle\sum_{k=1}^{5} \frac{k}{k+1}$

17. (a) $\displaystyle\sum_{k=1}^{5} (-1)^{k+1}a_k$ (b) $\displaystyle\sum_{k=0}^{5} (-1)^{k+1}b_k$

(c) $\displaystyle\sum_{k=0}^{n} a_k x^k$ (d) $\displaystyle\sum_{k=0}^{5} a^{5-k}b^k$

19. 5047   **21.** 2870   **23.** 1728   **25.** 214,365
27. $\frac{3}{2}(n+1)$ **29.** $\frac{1}{4}(n-1)^2$ **31.** $\frac{1}{2}$ **33.** $\frac{5}{2}$ **35.** $\frac{17}{2}$
39. $3^{17}-3^4$ **41.** $-\frac{399}{400}$ **43.** a_n-a_0 **45.** (b) $\frac{1}{2}$
47. (a) n^2 (b) -3 (c) $\frac{1}{2}n(n+1)x$ (d) $(n-m+1)c$

49. (a) $\displaystyle\sum_{k=0}^{14} (k+4)(k+1)$ (b) $\displaystyle\sum_{k=5}^{19} (k-1)(k-4)$

51. $\displaystyle\sum_{k=1}^{18} k\sin\frac{\pi}{k}$ **53.** Both are valid.

55. (a) $\frac{3}{2}(3^{20}-1)$ (b) $2^{31}-2^5$ (c) $-\frac{2}{3}\left(1+\frac{1}{2^{101}}\right)$

57. 110

▶ **Exercise Set 5.5 (Page 274)**
1. (a) 46   (b) 58
3. (a) $(1+\sqrt{2})\pi/4 \approx 1.896$ (b) same as in (a)
5. $\frac{15}{4}$ **7.** $\frac{1}{3}$ **9.** 320   **11.** $\frac{15}{4}$ **13.** $\frac{1}{3}$
15. (b) $\frac{1}{4}(b^4-a^4)$ **17.** $\frac{1}{2}(b^2-a^2)$
19. (a) 0.761923639, 0.712712753, 0.684701150
(b) 0.584145862, 0.623823864, 0.649145594
(c) 0.663501867, 0.665867079, 0.666538346
21. (a) 0.919403170, 0.960215997, 0.984209789
(b) 1.076482803, 1.038755813, 1.015625715
(c) 1.001028824, 1.000257067, 1.000041125

▶ **Exercise Set 5.6 (Page 284)**
1. (a) $\frac{71}{6}$ (b) 2   **3.** (a) $-\frac{117}{16}$ (b) 3
5. smallest 9.5, largest 20   **7.** $\displaystyle\int_{-3}^{3} 4x(1-3x)\,dx$

9. $\displaystyle\lim_{\max\Delta x_k\to 0}\sum_{k=1}^{n} 2x_k^*\,\Delta x_k;\ a=1,\ b=2$

11. $\displaystyle\lim_{\max\Delta x_k\to 0}\sum_{k=1}^{n}\frac{x_k^*}{x_k^*+1}\,\Delta x_k;\ a=0,\ b=1$

13. (a) 0.8   (b) -2.6 (c) -1.8 (d) -0.3
15. (a) 1   (b) 2   (c) $-\frac{1}{4}$ (d) $\frac{3}{4}$
17. -4 **19.** $\frac{13}{2}$ **21.** $\pi/2$ **23.** $25\pi/2$ **25.** 0
27. 14   **29.** -1 **31.** 3   **33.** negative   **35.** positive
37. negative   **39.** not integrable   **45.** (b), (c)

▶ **Exercise Set 5.7 (Page 291)**
1. $\frac{65}{4}$ **3.** $\frac{81}{10}$ **5.** $\frac{22}{3}$ **7.** $\frac{2}{3}$ **9.** $-\frac{1}{3}$ **11.** $\frac{52}{3}$
13. $\frac{844}{5}$ **15.** $-\frac{55}{3}$ **17.** 0   **19.** $\sqrt{2}$

---

21. $\pi^2/9+2\sqrt{3}$ **23.** $\frac{5}{2}$ **25.** $2-\sqrt{2}/2$ **27.** $-\frac{11}{6}$
29. $0.665867079;\ \frac{2}{3}$ **31.** $\frac{2}{3}$ **33.** 12   **35.** $\frac{9}{2}$ **37.** $\frac{203}{2}$
41. $m=1.092599583,\ M=1.104669194$ **43.** 6   **45.** $2/\pi$
47. $\frac{4}{3}$ **49.** $\pi/2$
51. (a) $\frac{4}{3}$ (b) $2/\sqrt{3}$ (c)

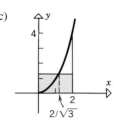

53. $f_{\text{ave}}=2;\ x^*=4$
55. $f_{\text{ave}}=\frac{1}{2}\alpha(x_0+x_1)+\beta;\ x^*=\frac{1}{2}(x_0+x_1)$
59. 1404π lb   **61.** (b) no

▶ **Exercise Set 5.8 (Page 296)**

1. (a) $\displaystyle\int_{1}^{3} u^7\,du$ (b) $-\dfrac{1}{2}\displaystyle\int_{7}^{4} u^{1/2}\,du$ (c) $\dfrac{1}{\pi}\displaystyle\int_{-\pi}^{\pi}\sin u\,du$

(d) $\displaystyle\int_{0}^{1} u^2\,du$ (e) $\dfrac{1}{2}\displaystyle\int_{3}^{4} (u-3)u^{1/2}\,du$

(f) $\displaystyle\int_{-3}^{0} (u+5)u^{20}\,du$

3. $\frac{121}{5}$ **5.** 10   **7.** $\frac{1192}{15}$ **9.** $8-4\sqrt{2}$ **11.** $-\frac{1}{48}$
13. $\frac{2}{3}$ **15.** $\frac{2}{3}(\sqrt{10}-2\sqrt{2})$ **17.** $2(\sqrt{7}-\sqrt{3})$ **19.** 0
21. 0   **23.** -4 **25.** $-\frac{1}{9}$ **27.** $\frac{1}{3}(\sqrt{3}-1)$ **29.** $\frac{106}{405}$
31. $3\sqrt{2}/4$ **33.** $120\sqrt{2}$
35. 0.692835360, 0.693069098, 0.693134682
37. 3.142425985, 3.141800987, 3.141625987
39. 2π **41.** $\pi/8$ **43.** 3   **45.** $\frac{5}{3}$ **49.** $2k$ **51.** 0
53. $2/\pi$ **55.** (b) $\frac{3}{2}$ (c) $\pi/4$

▶ **Exercise Set 5.9 (Page 302)**
1. (a) 3   (b) 3   **3.** (a) x^3+1 **5.** $\sin\sqrt{x}$ **7.** $|x|$

9. $\displaystyle\int_{2}^{x}\frac{1}{t-1}\,dt$ **11.** $\displaystyle\int_{0}^{x}\frac{1}{t-1}\,dt$

13. (a) $(0,+\infty)$ (b) $x=1$ **15.** (a) 0   (b) $\frac{1}{3}$ (c) 0
17. $(-\infty,+\infty)$; zero if $x=1$, positive if $x>1$, negative if $x<1$
19. $[-2,2]$; zero if $x=-1$, positive if $-1<x\le 2$,
negative if $-2\le x<-1$

21. $F(x)=\begin{cases}\frac{1}{2}(1-x^2), & x<0 \\ \frac{1}{2}(1+x^2), & x\ge 0\end{cases}$

23. $F(x)=\begin{cases}\frac{1}{3}(x^3+1), & x\le 0 \\ x^2+\frac{1}{3}, & x>0\end{cases}$ **25.** $3/x$

29. (a) $3x^2\sin^2(x^3)-2x\sin^2(x^2)$ (b) $\dfrac{2}{1-x^2}$

▶ **TECHNOLOGY EXERCISES, CHAPTER 5 (Page 304)**
1. $\frac{1}{3}\sqrt{5+2\sin 3x}+C$ **3.** $-\dfrac{1}{3a(ax^3+b)}+C$

5. $f(x)=-\dfrac{\sqrt{9+x^2}}{9x}+1.138889$

7. $f(x)=\frac{1}{3}x^2\sin 3x-\frac{2}{27}\sin 3x+\frac{2}{9}x\cos 3x-0.251607$
9. 1.137630   **11.** 4.368876   **13.** 1.007514
15. 2.073948   **17.** 1.210563

19. (a)

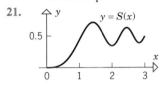

(b) 1.128780
(c) relative minimum at $x = 0$
   inflection point at $x = 1.587401$

21.

y = S(x)

max 0.713972 at $x = \sqrt{2}$

23.

y = E(x)

$E(0.174139) = 1.5$

25. 20

▶ **Exercise Set 6.1 (Page 311)**

1. $\frac{9}{2}$ **3.** 1   **5.** $\frac{32}{3}$ **7.** $\frac{9}{4}$ **9.** $\frac{49}{192}$ **11.** $\frac{1}{2}$
13. $\frac{32}{3}$ **15.** $\pi - 2$ **17.** $\frac{9}{2}$ **19.** $\frac{355}{6}$ **21.** 24
23. $\frac{1}{2}$ **25.** $4\sqrt{2}$ **27.** $\frac{11}{2}$ **29.** $\frac{2}{3}$
31. (a) $2(\sqrt{b} - 1)$ (b) $+\infty$ **33.** $9/\sqrt[3]{4}$
39. 1.180898334

▶ **Exercise Set 6.2 (Page 318)**

1. 8π **3.** $13\pi/6$ **5.** $32\pi/5$ **7.** $373\pi/14$
9. $1296\pi/5$ **11.** $2048\pi/15$ **13.** $\pi/2$ **15.** $\pi/6$
17. $3\pi/5$ **19.** 8π **21.** 2π **23.** $28\pi/3$
25. $58\pi/5$ **27.** $72\pi/5$ **29.** $256\pi/3$
31. (a) $\pi(1 - 1/b)$ (b) π **33.** $2/\sqrt{\pi}$ **35.** $648\pi/5$
37. $\pi/2$ **39.** $\frac{4}{3}\pi ab^2$ **41.** π **43.** $\frac{1}{3}\pi r^2 h$
45. $V = \frac{1}{6}\pi L^3$ **47.** $V = \begin{cases} 3\pi h^2, & 0 \le h < 2 \\ \frac{1}{3}\pi(12h^2 - h^3 - 4), & 2 \le h \le 4 \end{cases}$
49. $40,000\pi$ ft^3 **51.** $\frac{1}{30}$ **53.** $36\sqrt{3}$ **55.** $\frac{1}{4}(\pi + 2)$
57. $\frac{2}{3}r^3 \tan\theta$ **59.** $\frac{16}{3}r^3$ **61.** (b) h_0/k

▶ **Exercise Set 6.3 (Page 325)**

1. $15\pi/2$ **3.** $\pi/3$ **5.** $2\pi/5$ **7.** 4π **9.** $20\pi/3$
11. $3\pi\sqrt[3]{4}(1 + 3\sqrt[3]{3})$ **13.** $\pi/2$ **15.** $\pi/5$
17. (b) $2\pi^2$ **19.** (a) $7\pi/30$ **21.** $9\pi/14$ **23.** $\frac{1}{3}\pi r^2 h$
25. $\frac{4\pi}{3}[r^3 - (r^2 - a^2)^{3/2}]$ **27.** $b = 1$

▶ **Exercise Set 6.4 (Page 329)**

1. $\sqrt{5}$ **3.** $\frac{1}{243}(85\sqrt{85} - 8)$ **5.** $\frac{1}{27}(80\sqrt{10} - 13\sqrt{13})$
7. $\frac{17}{6}$
9. (a)

(8, 4)

(−1, 1)

(b) dy/dx does not exist at $x = 0$.
(c) $\frac{1}{27}(13\sqrt{13} + 80\sqrt{10} - 16)$
13. 4.645975301

▶ **Exercise Set 6.5 (Page 332)**

1. $35\sqrt{2\pi}$ **3.** 8π **5.** $\frac{16\pi}{9}$ **7.** $40\pi\sqrt{82}$ **9.** 24π
11. $\frac{16,911\pi}{1024}$ **15.** $S = 2\pi rh$ **19.** (b) constant functions

▶ **Exercise Set 6.6 (Page 336)**

1. $s(t) = t^2 - 3t + 7$ **3.** $s(t) = \frac{1}{4}t^4 - \frac{2}{3}t^3 + t + 1$
5. $s(t) = 2t^2 + t$ **7.** $s(t) = -\cos 2t - t - 2$
9. (a) $s = 2/\pi, v = 1, |v| = 1, a = 0$
   (b) $s = \frac{1}{2}, v = -\frac{3}{2}, |v| = \frac{3}{2}, a = -3$
11. $-\frac{968}{45}$ ft/sec^2
13. (a) $v(3) = 16$ ft/sec, $v(5) = -48$ ft/sec
   (b) 196 ft   (c) 112 ft/sec
15. (a) 1 sec   (b) $\frac{1}{2}$ sec
17. (a) $\frac{5}{8}(1 + \sqrt{33})$ sec   (b) $20\sqrt{33}$ ft/sec
19. (a) 5 sec   (b) 272.5 m   (c) 10 sec   (d) -49 m/sec
   (e) 12.46 sec   (f) 73.1 m/sec
21. $80\sqrt{10}$ ft/sec   **23.** 256 ft
25. (a) negative   (b) increasing   (c) negative
27. $\frac{2}{3}; 3$ **29.** 0; 2   **31.** $\frac{9}{4}; \frac{11}{4}$ **33.** $-\frac{10}{3}; \frac{17}{3}$
35. $\frac{204}{25}; \frac{204}{25}$

▶ **Exercise Set 6.7 (Page 341)**

1. (a) 210 ft · lb   (b) 5/6 ft · lb   **3.** 160 J   **5.** 20 lb/ft
7. $900\pi\rho$ ft · lb   **9.** 261,600 J
11. (a) 926,640 ft · lb   (b) 0.468 hp   **13.** 75,000 ft · lb
15. (a) 96×10^9 (b) 4,800,000 mi · lb

▶ **Exercise Set 6.8 (Page 345)**

1. (a) 2808 lb   (b) 3600 lb   **3.** 156,960 N
5. 6988.8 lb   **7.** 8.175×10^5 N   **9.** $\rho a^3/\sqrt{2}$
11. $14,976\sqrt{17}$ lb   **13.** (b) $80\rho_0$ lb/min

▶ **TECHNOLOGY EXERCISES, CHAPTER 6 (Page 346)**

1. 2.542696   **3.** 3.268669   **5.** 5.498059   **7.** 2.242872
9. 2.854368   **11.** 6.65 meters
13. (a) 4.21 cm   (b) 2.78 sec   **15.** 4.425822

▶ **Exercise Set 7.1 (Page 357)**

1. (a) -4 (b) 4   (c) $\frac{1}{4}$ **3.** (a) 2.9690   (b) 0.0341
5. (a) 4   (b) -5 (c) 1   (d) $\frac{1}{2}$
7. (a) 1.3655   (b) -0.3011
9. (a) $2r + s/2 + t/2$ (b) $s - 3r - t$
11. (a) $1 + \log x + \frac{1}{2}\log(x - 3)$
   (b) $2\ln|x| + 3\ln\sin x - \frac{1}{2}\ln(x^2 + 1)$
13. $\log\frac{256}{3}$ **15.** $\ln\frac{\sqrt[3]{x}(x + 1)^2}{\cos x}$ **17.** 0.01   **19.** e^2
21. 4   **23.** 10^5 **25.** $\sqrt{3/2}$ **27.** $-\frac{\ln 3}{2\ln 5}$ **29.** $\frac{1}{3}\ln\frac{7}{2}$
31. -2 **33.** $0, -\ln 2$ **35.** $-\ln 2$
37. $\log\frac{1}{2} < 0$, so $3\log\frac{1}{2} < 2\log\frac{1}{2}$ **39.** (b) 2.8777, -0.3174
41. 201 days
43. (a) 7.4, basic   (b) 4.2, acidic   (c) 6.4, acidic
   (d) 5.9, acidic

45. (a) 140 dB, damage   (b) 120 dB, damage
 (c) 80 dB, no damage   (d) 75 dB, no damage
47. ~200   **49.** (a) ~5 × 10^{16} J   (b) ~0.67   **51.** e^{-2}

▶ **Exercise Set 7.2 (Page 365)**

1. $\dfrac{1}{x}$ **3.** $\dfrac{2\ln x}{x}$ **5.** $\dfrac{\sec^2 x}{\tan x}$ **7.** $\dfrac{1-x^2}{x(1+x^2)}$

9. $\dfrac{3x^2-14x}{x^3-7x^2-3}$ **11.** $\dfrac{1}{2x\sqrt{\ln x}}$ **13.** $-\dfrac{5\cos(5/\ln x)}{x(\ln x)^2}$

15. $-\dfrac{2x^3}{3-2x}+3x^2\ln(3-2x)$

17. $4x\ln(x^2+1)+2x[\ln(x^2+1)]^2$ **19.** $\dfrac{x(1+2\ln x)}{(1+\ln x)^2}$

21. $-\dfrac{y}{x(y+1)}$ **23.** $\frac{1}{2}\ln|x|+C$ **25.** $\frac{1}{3}\ln|x^3-4|+C$

27. $\ln|\tan x|+C$ **29.** $-\frac{1}{3}\ln(1+\cos 3\theta)+C$
31. $\frac{1}{2}x^2-\frac{1}{2}\ln(x^2+1)+C$ **33.** $\frac{1}{4}(\ln y)^4+C$

35. $\frac{1}{3}\ln\frac{5}{2}$ **37.** $\frac{1}{2}\ln\frac{5}{6}$ **39.** $-\tan x+\dfrac{3x}{4-3x^2}$

41. $\dfrac{1}{2x}+\dfrac{1}{3(x+3)}+\dfrac{3}{5(3x-2)}$

43. $x\sqrt[3]{1+x^2}\left[\dfrac{1}{x}+\dfrac{2x}{3(1+x^2)}\right]$

45. $\left[\dfrac{(x^2-8)^{1/3}\sqrt{x^3+1}}{x^6-7x+5}\right]$
$\left[\dfrac{2x}{3(x^2-8)}+\dfrac{3x^2}{2(x^3+1)}-\dfrac{6x^5-7}{x^6-7x+5}\right]$

47. (a) $-\dfrac{1}{x(\ln x)^2}$ (b) $-\dfrac{\ln 2}{x(\ln x)^2}$ **49.** $\frac{1}{2}\ln 2$

51. $\frac{1}{2}(1+\ln 2)$
53. relative minimum $(1,0)$, relative maximum $(e^2,4e^{-2})$
59. $\ln 2$ **61.** $\pi\ln 4$ **63.** $0.158594340,\ 3.146193221$
65. $-10xe^{-5x^2}$ **67.** $x^2e^x(x+3)$ **69.** $\dfrac{4}{(e^x+e^{-x})^2}$

71. $(x\sec^2 x+\tan x)e^{x\tan x}$ **73.** $(1-3e^{3x})e^{(x-e^{3x})}$

75. $\dfrac{x-1}{e^x-x}$ **77.** $e^{ax}(a\cos bx-b\sin bx)$ **79.** $3x^2$

81. $-3^{-x}\ln 3$ **83.** $\pi^{x\tan x}(\ln\pi)(x\sec^2 x+\tan x)$

85. $\dfrac{d}{dx}(u^v)=vu^{v-1}\dfrac{du}{dx}+u^v(\ln u)\dfrac{dv}{dx}$

87. $(x^3-2x)^{\ln x}\left[\dfrac{3x^2-2}{x^3-2x}\ln x+\dfrac{1}{x}\ln(x^3-2x)\right]$

89. $(\ln x)^{\tan x}\left[\dfrac{\tan x}{x\ln x}+(\sec^2 x)\ln(\ln x)\right]$

91. $x^{(e^x)}\left[\dfrac{e^x}{x}+e^x\ln x\right]$

95. (a) $k^n e^{kx}$ (b) $(-1)^n k^n e^{-kx}$

97. $-\dfrac{1}{\sqrt{2\pi}\sigma^3}(x-\mu)\exp\left[-\frac{1}{2}\left(\dfrac{x-\mu}{\sigma}\right)^2\right]$

99. $-\frac{1}{5}e^{-5x}+C$ **101.** $e^{\sin x}+C$
103. $-\frac{1}{6}e^{-2x^3}+C$ **105.** $\ln(1+e^x)+C$
107. $\frac{1}{3}(1+e^{2t})^{3/2}+C$ **109.** $-\exp(\cos x)+C$

111. $\tan(2-e^{-x})+C$ **113.** $\dfrac{\pi^{\sin x}}{\ln\pi}+C$

115. $\frac{1}{2}x^2\ln 3-4\pi e^2\sin x+C$ **117.** C **119.** $2e^{\sqrt{y}}+C$

121. -36 **123.** $3+e-e^2$ **125.** $\ln\frac{21}{13}$ **129.** $\dfrac{\ln 3}{\ln(2/3)}$

131. exe^{-1} **133.** $\dfrac{-qk_0}{2T^2}\exp\left[-\dfrac{q(T-T_0)}{2T_0T}\right]$

135. $e^x-3\ln(e^x+3)+C$ **137.** $\frac{27}{8}e^{-3}$ **139.** $3\ln 3-2$
141. 4π **143.** $5\ln 5-4$
145. $-1.315973778<x<0.537274449$
147. (b) 2.821439372 (c) $f=5.87\times 10^{10}T$ **149.** $e^{-1/e}$
151. $A=\frac{1}{2}(1-e^{-2b})=\frac{1}{4}$ when $b=\ln\sqrt{2}$; $A\to\frac{1}{2}$ as $b\to+\infty$

▶ **Exercise Set 7.3 (Page 373)**
1. (a) $+\infty$ (b) 0 **3.** (a) $+\infty$ (b) $+\infty$
5. (a) 1 (b) 1 **7.** (a) $+\infty$ (b) 0

9. 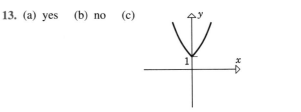   **11.**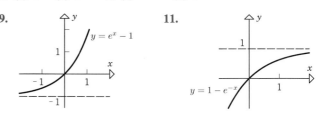

13. (a) yes   (b) no   (c)
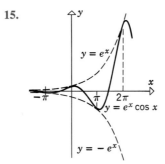

15.
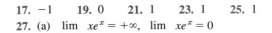

17. -1 **19.** 0 **21.** 1 **23.** 1 **25.** 1
27. (a) $\lim\limits_{x\to+\infty} xe^x=+\infty,\ \lim\limits_{x\to-\infty} xe^x=0$

 (b)
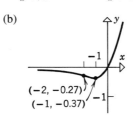

29. (a) $\lim\limits_{x\to+\infty}\dfrac{x^2}{e^{2x}}=0,\ \lim\limits_{x\to-\infty}\dfrac{x^2}{e^{2x}}=+\infty$

(b)

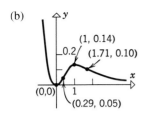

31. (a)

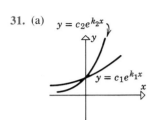

(b)

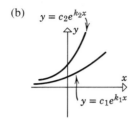

(c)

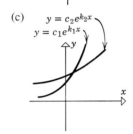

33. $\dfrac{3 - e}{2e}$

41.

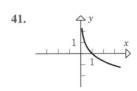

43.

45.

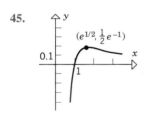

▶ **Exercise Set 7.4 (Page 382)**

1. (a) yes   (b) no   (c) yes   (d) no

3. yes   5. no   7. no   9. yes   11. yes   13. yes

15. $x^{1/5}$ 17. $\frac{1}{7}(x + 6)$ 19. $\sqrt[3]{\dfrac{x + 5}{3}}$ 21. $\frac{1}{2}(x^3 + 1)$

23. $-\sqrt{3}/x$ 25. $\dfrac{1}{\ln x}$ 27. $\frac{1}{3}e^{1-x}$

29. $\begin{cases} 5/2 - x, & x > 1/2 \\ 1/x, & 0 < x \le 1/2 \end{cases}$ 31. $\dfrac{1}{15y^2 + 1}$ 33. $\dfrac{1}{2 \sec^2 2y}$

35. $\dfrac{1}{10y^4 + 3y^2}$

37. (b)

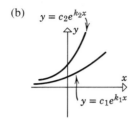

(c) No; $f(g(x)) = x$ for $x > 1$, but the domain of g is $x > 0$.

39. $x^{1/4} - 2$ for $x \ge 16$ 41. $\frac{1}{2}(3 - x^2)$ for $x \le 0$

43. $\frac{1}{10}(1 + \sqrt{1 - 20x})$ for $x \le -4$

45. (b) symmetric about the line $y = x$ 47. (b) $1 - \sqrt{3}/3$

49. 10   51. $\frac{1}{13}$ 53. $1/\sqrt{3}$ 55. (b) $1/\sqrt[3]{2}$

57.

59. $\frac{88}{7}$

▶ **Exercise Set 7.5 (Page 391)**

1. $x > -\frac{2}{3}$ 3. $-2 < x < 2$ 5. $x \ge e^{-1}$

7. (a) all $x \ne 0$ (b) $x > 0$

9. (a) $1/x, x > 0$ (b) $x^2, x \ne 0$ (c) $-x^2, -\infty < x < +\infty$
   (d) $-x, -\infty < x < +\infty$ (e) $x^3, x > 0$ (f) $x + \ln x, x > 0$
   (g) $x - \sqrt[3]{x}, -\infty < x < +\infty$ (h) $e^x/x, x > 0$

11. $\frac{7}{2}$ 13. (a) $e^{-x \ln \pi}$ (b) $e^{2x \ln x}$

▶ **Exercise Set 7.6 (Page 396)**

1.

	$\sinh x_0$	$\cosh x_0$	$\tanh x_0$	$\coth x_0$	$\operatorname{sech} x_0$	$\operatorname{csch} x_0$
(a)	-2	$\sqrt{5}$	$-2/\sqrt{5}$	$-\sqrt{5}/2$	$1/\sqrt{5}$	$-1/2$
(b)	$-3/4$	$5/4$	$-3/5$	$-5/3$	$4/5$	$-4/3$
(c)	$-4/3$	$5/3$	$-4/5$	$-5/4$	$3/5$	$-3/4$
(d)	$1/\sqrt{3}$	$2/\sqrt{3}$	$1/2$	2	$\sqrt{3}/2$	$\sqrt{3}$
(e)	$8/15$	$17/15$	$8/17$	$17/8$	$15/17$	$15/8$
(f)	-1	$\sqrt{2}$	$-1/\sqrt{2}$	$-\sqrt{2}$	$1/\sqrt{2}$	-1

17. $4 \cosh (4x - 8)$ 19. $-\dfrac{\operatorname{csch}^2 (\ln x)}{x}$

21. $\dfrac{\operatorname{csch} (1/x) \coth (1/x)}{x^2}$ 23. $\dfrac{2 + 5 \cosh (5x) \sinh (5x)}{\sqrt{4x + \cosh^2 (5x)}}$

25. $x^{5/2} \tanh (\sqrt{x}) \operatorname{sech}^2 (\sqrt{x}) + 3x^2 \tanh^2 (\sqrt{x})$

27. $\frac{1}{7} \sinh^7 x + C$ 29. $\frac{2}{3}(\tanh x)^{3/2} + C$

31. $\ln (\cosh x) + C$ 33. $-\frac{1}{3} \operatorname{sech}^3 x + C$

35. $2 \sinh (\sqrt{x}) + C$

37. For $x > 0$:
   (a) $\dfrac{x^2 + 1}{2x}$ (b) $\dfrac{x^2 - 1}{2x}$ (c) $\dfrac{x^4 - 1}{x^4 + 1}$ (d) $\dfrac{x^2 + 1}{2x}$

45. (a) $\frac{1}{2}$ (b) $+\infty$ if $a > 1$, $\frac{1}{2}$ if $a = 1$, 0 if $0 < a < 1$

47. $\frac{16}{9}$ 49. 5π 51. $\frac{3}{4}$ 55. 405.9 ft

▶ **Exercise Set 7.7 (Page 406)**

1. $y = Cx$ **3.** $y = Ce^{-\sqrt{1+x^2}} - 1$ **5.** $y = \ln(\sec x + C)$

7. $y = e^{-2x} + Ce^{-3x}$ **9.** $y = e^{-x}\sin(e^x) + Ce^{-x}$

11. $y = -\frac{2}{7}x^4 + Cx^{-3}$ **13.** $y = -1 + 4e^{x^2/2}$

15. $y = 2 - e^{-t}$ **17.** $y = \sqrt[3]{3t - 3\ln t + 24}$

19. $y = -3/(2\sqrt{x} + 1)$ **21.** $y = -\ln(3 - x^2/2)$

23. (a) $h(t) = (2 - 0.003979t)^2$ (b) about 8.4 min

25. $v = 50/(2t + 1)$, $x = 25\ln(2t + 1)$

27. (a) $v(t) = c\ln\dfrac{m_0}{m_0 - kt} - gt$ (b) 3044 m/sec

29. (a) $v^2 = 2gR^2/x + v_0^2 - 2gR$

31. (a) $y = 200 - 175e^{-t/25}$ (b) 136 lb

33. 25 lb   **35.** (a) $y = 10e^{-0.005t}$ (b) 7 mg

37. 196 days   **39.** 6.8 years

41. (a) 14,400   (b) 38 years

▶ **TECHNOLOGY EXERCISES, CHAPTER 7 (Page 409)**

1. 4.319137   **3.** -1.940953

5. (a) 0.45 sec   (b) 4.74 sec   **7.** $\ln 4 - 2$ **9.** $1/e$

11. relative maximum 1.898286 at $x = 2.127616$
     relative minimum 0.697684 at $x = 0.352215$
     relative minimum 0.209277 at $x = 4.842558$

13. (a) 8686   (b) 9629

15. (b) 0.67 sec   (c) 4.48 m/sec

17. (a) -0.914856
     (b) domain: $[-1.54046, +\infty)$, range: $[-0.91485, +\infty)$
     $f^{-1}(5) = 1.425366$

19. (a) 1.335729   (b) $y = e^{\sin x}$; 0.6%

▶ **Exercise Set 8.1 (Page 419)**

1. (a) $-\pi/2$ (b) π (c) $-\pi/4$ (d) $\pi/4$ (e) 0
     (f) $\pi/2$

3. $\frac{1}{2}$, $-\sqrt{3}$, $-1/\sqrt{3}$, 2, $-2/\sqrt{3}$ **5.** $\frac{4}{5}$, $\frac{3}{5}$, $\frac{3}{4}$, $\frac{5}{3}$, $\frac{5}{4}$

7. (a) $\pi/7$ (b) 0   (c) $2\pi/7$ (d) $201\pi - 630$

9. (a) $0 \le x \le \pi$ (b) $-1 \le x \le 1$ (c) $-\pi/2 < x < \pi/2$
     (d) $-\infty < x < +\infty$ (e) $0 < x \le \pi/2$, $-\pi < x \le -\pi/2$
     (f) $|x| \ge 1$

11. $\frac{24}{25}$ **13.** $\pi/2$ **15.** $-4\sqrt{5}$

19. (a) $\dfrac{1}{\sqrt{1+x^2}}$ (b) $1/x$ (c) $\dfrac{\sqrt{x^2-1}}{x}$ (d) $\sqrt{x^2-1}$

21. (a)   (b)

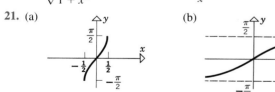

27. (a) 55.0°   (b) 33.6°   (c) 25.8°

29. $x = \pi + \tan^{-1}k$ **31.** 2.7626   **33.** -1.8773

35. $-76.7°$ **37.** (b) 23°   **39.** 32° or 58°; 32°   **41.** 29°

▶ **Exercise Set 8.2 (Page 425)**

1. (a) $\dfrac{1}{\sqrt{9-x^2}}$ (b) $-\dfrac{2}{\sqrt{1-(2x+1)^2}}$

3. (a) $\dfrac{7}{x\sqrt{x^{14}-1}}$ (b) $-\dfrac{1}{\sqrt{e^{2x}-1}}$

5. (a) $-\dfrac{1}{|x|\sqrt{x^2-1}}$ (b) $\begin{cases} 1, & \sin x > 0 \\ -1, & \sin x < 0 \end{cases}$

7. (a) $\dfrac{e^x}{x\sqrt{x^2-1}} + e^x \sec^{-1}x$

     (b) $\dfrac{3x^2(\sin^{-1}x)^2}{\sqrt{1-x^2}} + 2x(\sin^{-1}x)^3$

9. (a) $-\dfrac{1}{x^2+1}$

     (b) $10(1 + x\csc^{-1}x)^9\left(-\dfrac{1}{\sqrt{x^2-1}} + \csc^{-1}x\right)$

11. (a) $-\dfrac{1}{2\sqrt{1-x^2}}$ (b) $\dfrac{x + 2x\ln x}{\sqrt{1-x^4\ln^2 x}}$

13. $\dfrac{y\sqrt{1-(x-y)^2} + \sqrt{1-x^2y^2}}{\sqrt{1-x^2y^2} - x\sqrt{1-(x-y)^2}}$ **15.** $\pi/2$ **17.** $-\pi/12$

19. $\frac{1}{4}\tan^{-1}4x + C$ **21.** $\tan^{-1}(e^x) + C$ **23.** $\pi/6$

25. $\sin^{-1}(\tan x) + C$ **27.** $\sin^{-1}(\ln x) + C$

29. (a) $\sin^{-1}\left(\dfrac{x}{3}\right) + C$ (b) $\dfrac{1}{\sqrt{5}}\tan^{-1}\left(\dfrac{x}{\sqrt{5}}\right) + C$

     (c) $\dfrac{1}{\sqrt{\pi}}\sec^{-1}\left(\dfrac{x}{\sqrt{\pi}}\right) + C$

31. $\pi/(6\sqrt{3})$ **33.** $\pi/18$ **35.** $\pi/18$ **37.** $\pi^2/4$

39. $\pi/2 - 1$ **41.** $1 + 2\sqrt{2}$ **43.** $52\pi/3$ mi/min

45. $2\sqrt{6}$ ft   **49.** (b) 0.93

▶ **Exercise Set 8.3 (Page 430)**

5. (a) $\ln(3 + \sqrt{8})$ (b) $\ln(\sqrt{5} - 2)$

7. (a) $\dfrac{1}{\sqrt{9+x^2}}$ (b) $\dfrac{2}{\sqrt{(2x+1)^2-1}}$

9. (a) $-\dfrac{7}{x\sqrt{1-x^{14}}}$ (b) $-\dfrac{1}{\sqrt{1+e^{2x}}}$

11. (a) $-\dfrac{1}{|x|\sqrt{x^2+1}}$ (b) $\begin{cases} 1, & x > 0 \\ -1, & x < 0 \end{cases}$

13. (a) $-\dfrac{e^x}{x\sqrt{1-x^2}} + e^x\,\text{sech}^{-1}x$

     (b) $\dfrac{3x^2(\sinh^{-1}x)^2}{\sqrt{1+x^2}} + 2x(\sinh^{-1}x)^3$

15. (a) $-1/2x$

     (b) $10(1 + x\,\text{csch}^{-1}x)^9\left(-\dfrac{x}{|x|\sqrt{1+x^2}} + \text{csch}^{-1}x\right)$

17. $\cosh^{-1}\left(\dfrac{x}{\sqrt{2}}\right) + C$ **19.** $-\text{sech}^{-1}(e^x) + C$

21. $-\frac{1}{3}\text{csch}^{-1}|x^3| + C$ **23.** -0.2028

33. (a) $v = \sqrt{16x^2-1}$ (b) $x = \frac{1}{4}\cosh 4t$; 0.7 sec

▶ **TECHNOLOGY EXERCISES, CHAPTER 8 (Page 432)**

1. (a) 0.446593   (b) 0.300283   **3.** 0.793387

5. 0.174192   **7.** 1.345874   **9.** 1.468384

11. (a) 317.26 gal   (c) 11.2 gal/in   (d) 9.8 in

▶ Exercise Set 9.1 (Page 439)

1. $\frac{3}{4}x + \frac{3}{16}\ln|4x - 1| + C$ **3.** $\frac{1}{5}\ln\left|\frac{x}{2x+5}\right| + C$

5. $\frac{1}{5}(x+1)(2x-3)^{3/2} + C$ **7.** $\frac{1}{2}\ln\left|\frac{\sqrt{4-3x}-2}{\sqrt{4-3x}+2}\right| + C$

9. $\frac{1}{2\sqrt{5}}\ln\left|\frac{x+\sqrt{5}}{x-\sqrt{5}}\right| + C$

11. $\frac{1}{2}x\sqrt{x^2-3} - \frac{3}{2}\ln|x+\sqrt{x^2-3}| + C$

13. $\frac{1}{2}x\sqrt{x^2+4} - 2\ln|x+\sqrt{x^2+4}| + C$

15. $\frac{1}{2}x\sqrt{9-x^2} + \frac{9}{2}\sin^{-1}\frac{x}{3} + C$

17. $\sqrt{3-x^2} - \sqrt{3}\ln\left|\frac{\sqrt{3}+\sqrt{3-x^2}}{x}\right| + C$

19. $-\frac{1}{10}\sin 5x + \frac{1}{2}\sin x + C$ **21.** $\frac{1}{16}x^4(4\ln x - 1) + C$

23. $-\frac{1}{13}e^{-2x}(2\sin 3x + 3\cos 3x) + C$

25. $\frac{1}{18}\left[\frac{4}{4-3e^{2x}} + \ln|4 - 3e^{2x}|\right] + C$

27. $\frac{1}{3}\tan^{-1}\frac{3\sqrt{x}}{2} + C$ **29.** $\frac{1}{3}\ln|3x + \sqrt{9x^2-4}| + C$

31. $\frac{1}{54}\left[-\frac{3}{2}x^2\sqrt{5-9x^4} + \frac{5}{2}\sin^{-1}\left(\frac{3x^2}{\sqrt{5}}\right)\right] + C$

33. $\frac{1}{2}\ln x - \frac{1}{4}\sin(2\ln x) + C$ **35.** $-\frac{1}{4}e^{-2x}(2x+1) + C$

37. $-\frac{1}{3}\left[\frac{1}{1+\cos 3x} + \ln\left|\frac{\cos 3x}{1+\cos 3x}\right|\right] + C$

39. $\frac{1}{16}\ln\left|\frac{4x^2-1}{4x^2+1}\right| + C$

41. $\frac{1}{2}\left[e^x\sqrt{3-4e^{2x}} + \frac{3}{2}\sin^{-1}\frac{2e^x}{\sqrt{3}}\right] + C$

43. $\frac{1}{3}\left[\frac{18x-5}{12}\sqrt{5x-9x^2} + \frac{25}{72}\sin^{-1}\frac{18x-5}{5}\right] + C$

45. $\frac{1}{9}(\sin 3x - 3x\cos 3x) + C$ **47.** $-2e^{-\sqrt{x}}(\sqrt{x}+1) + C$

49. $\frac{1}{6}\ln\left|\frac{x-1}{x+5}\right| + C$

51. $-\sqrt{5+4x-x^2} + 2\sin^{-1}\frac{x-2}{3} + C$ **53.** 3.523188312

55. 17.59119022   **57.** 0.054930614   **59.** 3.586419094

61. 5.031899801   **63.** 8.409316783   **65.** 14.42359944

▶ Exercise Set 9.2 (Page 446)

1. $-xe^{-x} - e^{-x} + C$ **3.** $x^2e^x - 2xe^x + 2e^x + C$

5. $-\frac{1}{2}x\cos 2x + \frac{1}{4}\sin 2x + C$

7. $x^2\sin x + 2x\cos x - 2\sin x + C$

9. $\frac{2}{3}x^{3/2}\ln x - \frac{4}{9}x^{3/2} + C$ **11.** $x(\ln x)^2 - 2x\ln x + 2x + C$

13. $x\ln(2x+3) - x + \frac{3}{2}\ln(2x+3) + C$

15. $x\sin^{-1}x + \sqrt{1-x^2} + C$

17. $x\tan^{-1}(2x) - \frac{1}{4}\ln(1+4x^2) + C$

19. $\frac{1}{2}e^x(\sin x - \cos x) + C$

21. $\frac{e^{ax}}{a^2+b^2}(a\sin bx - b\cos bx) + C$

23. $\frac{1}{2}x[\sin(\ln x) - \cos(\ln x)] + C$

25. $x\tan x + \ln|\cos x| + C$ **27.** $\frac{1}{2}x^2e^{x^2} - \frac{1}{2}e^{x^2} + C$

29. $\frac{1}{25}(1 - 6e^{-5})$ **31.** $\frac{1}{9}(2e^3 + 1)$ **33.** $5\ln 5 - 4$

35. $\frac{5\pi}{6} - \sqrt{3} + 1$ **37.** $-\pi/8$

39. $\frac{1}{3}(2\sqrt{3}\pi - \frac{1}{2}\pi - 2 + \ln 2)$ **41.** $\frac{1}{3}(2 - \sqrt{2})$

43. (a) 1   (b) $\pi(e-2)$ **45.** $2\pi^2$

47. (a) $-\frac{1}{3}\sin^2 x\cos x - \frac{2}{3}\cos x + C$ (b) $\frac{1}{32}(3\pi - 8)$

49. (a) $\frac{1}{15}\cos^2 5x\sin 5x + \frac{2}{15}\sin 5x + C$

(b) $\frac{1}{8}\cos^3(x^2)\sin(x^2) + \frac{3}{16}\cos(x^2)\sin(x^2) + \frac{3}{16}x^2 + C$

53. (b) $x^3e^x - 3x^2e^x + 6xe^x - 6e^x + C$

▶ Exercise Set 9.3 (Page 450)

1. $-\frac{1}{6}\cos^6 x + C$ **3.** $\frac{1}{2a}\sin^2 ax + C$

5. $\frac{1}{2}\theta - \frac{1}{20}\sin 10\theta + C$ **7.** $\frac{3}{8}x + \sin\left(\frac{x}{2}\right) + \frac{1}{8}\sin x + C$

9. $\sin\theta - \frac{2}{3}\sin^3\theta + \frac{1}{5}\sin^5\theta + C$

11. $\frac{1}{6}\sin^3 2t - \frac{1}{10}\sin^5 2t + C$ **13.** $-\frac{1}{5}\cos^5 x + \frac{1}{7}\cos^7 x + C$

15. $-\frac{1}{5}\cos^5\theta + \frac{2}{7}\cos^7\theta - \frac{1}{9}\cos^9\theta + C$

17. $\frac{1}{8}x - \frac{1}{32}\sin 4x + C$ **19.** $-\frac{1}{6}\cos 3x + \frac{1}{2}\cos x + C$

21. $-\frac{1}{3}\cos\left(\frac{3x}{2}\right) - \cos\left(\frac{x}{2}\right) + C$ **23.** $\frac{1}{7\cos^7 x} + C$

25. $\frac{5\sqrt{2}}{12}$ **27.** 0   **29.** $\frac{1}{24}$ **33.** $\pi/2$

35. (a) $\frac{2}{3}$ (b) $3\pi/16$ (c) $\frac{8}{15}$ (d) $5\pi/32$

▶ Exercise Set 9.4 (Page 453)

1. $\frac{1}{3}\tan(3x+1) + C$ **3.** $\frac{1}{2}\ln|\cos(e^{-2x})| + C$

5. $\frac{1}{2}\ln|\sec 2x + \tan 2x| + C$ **7.** $\frac{1}{3}\tan^3 x + C$

9. $\frac{1}{16}\tan^4(4x) + \frac{1}{24}\tan^6(4x) + C$

11. $\frac{1}{7}\sec^7 x - \frac{1}{5}\sec^5 x + C$

13. $\frac{1}{4}\sec^3 x\tan x - \frac{5}{8}\sec x\tan x + \frac{3}{8}\ln|\sec x + \tan x| + C$

15. $\frac{1}{6}\sec^3(2t) + C$ **17.** $\tan x + \frac{1}{3}\tan^3 x + C$

19. $\frac{1}{5\pi}\sec^4(\pi x)\tan(\pi x) + \frac{4}{15\pi}\sec^2(\pi x)\tan(\pi x)$

$+ \frac{8}{15\pi}\tan(\pi x) + C$

21. $\frac{1}{3}\tan^3 x - \tan x + x + C$ **23.** $\frac{1}{6}\tan^3(x^2) + C$

25. $-\frac{1}{5}\csc^5 x + \frac{1}{3}\csc^3 x + C$ **27.** $-\frac{1}{2}\csc^2 x - \ln|\sin x| + C$

29. $\frac{2}{3}\tan^{3/2}x + \frac{2}{7}\tan^{7/2}x + C$ **31.** $\sqrt{3}/2 - \pi/6$

33. $-\frac{1}{2} + \ln 2$ **35.** $\ln(\sqrt{2} + 1)$

39. $-\frac{1}{\sqrt{a^2+b^2}}\ln|\csc(x+\theta) + \cot(x+\theta)| + C$, where

θ satisfies $\cos\theta = \frac{a}{\sqrt{a^2+b^2}}$ and $\sin\theta = \frac{b}{\sqrt{a^2+b^2}}$

▶ Exercise Set 9.5 (Page 458)

1. $2\sin^{-1}\left(\frac{x}{2}\right) + \frac{1}{2}x\sqrt{4-x^2} + C$

3. $\frac{9}{2}\sin^{-1}\left(\frac{x}{3}\right) - \frac{1}{2}x\sqrt{9-x^2} + C$

5. $\frac{1}{16}\tan^{-1}\frac{x}{2} + \frac{x}{8(4+x^2)} + C$

7. $\sqrt{x^2-9} - 3\sec^{-1}\frac{x}{3} + C$

9. $-2\sqrt{2-x^2} + \frac{1}{3}(2-x^2)^{3/2} + C$

11. $\dfrac{x}{3\sqrt{3+x^2}}+C$ 13. $\dfrac{\sqrt{4x^2-9}}{9x}+C$

15. $\dfrac{x}{\sqrt{1-x^2}}+C$ 17. $\ln|x+\sqrt{x^2-1}|+C$

19. $-\dfrac{\sqrt{9-4x^2}}{9x}+C$ 21. $-\dfrac{x}{\sqrt{9x^2-1}}+C$

23. $\tfrac12\sin^{-1}(e^x)+\tfrac12 e^x\sqrt{1-e^{2x}}+C$ 25. $\tfrac{2048}{15}$

27. $\tfrac12(\sqrt3-\sqrt2)$ 29. $\tfrac{1}{243}(10\sqrt3+18)$

31. $\tfrac12\ln(x^2+4)+C$ 33. $\sqrt5-\sqrt2+\ln\left(\dfrac{2+2\sqrt2}{1+\sqrt5}\right)$

35. $\dfrac{\pi}{32}[18\sqrt5-\ln(2+\sqrt5)]$

37. (a) $\sinh^{-1}\left(\dfrac{x}{3}\right)+C$ (b) $\ln\left(\dfrac{\sqrt{x^2+9}}{3}+\dfrac{x}{3}\right)+C$

39. $\tfrac13\tan^{-1}\left(\dfrac{x-2}{3}\right)+C$ 41. $\sin^{-1}\left(\dfrac{x-1}{3}\right)+C$

43. $\ln(\sqrt{x^2-6x+10}+x-3)+C$

45. $2\sin^{-1}\left(\dfrac{x+1}{2}\right)+\tfrac12(x+1)\sqrt{3-2x-x^2}+C$

47. $\dfrac{1}{\sqrt{10}}\tan^{-1}\dfrac{\sqrt2(x+1)}{\sqrt5}+C$

49. $\ln(x^2+2x+5)+\dfrac32\tan^{-1}\left(\dfrac{x+1}{2}\right)+C$

51. $\sqrt{x^2+2x+2}+2\ln(\sqrt{x^2+2x+2}+x+1)+C$

53. $\dfrac{2\pi}{3}-\dfrac{\sqrt3}{2}$

▶ **Exercise Set 9.6 (Page 467)**

1. $\dfrac{A}{x-2}+\dfrac{B}{x+5}$ 3. $\dfrac{A}{x}+\dfrac{B}{x^2}+\dfrac{C}{x-1}$

5. $\dfrac{A}{x}+\dfrac{B}{x^2}+\dfrac{C}{x^3}+\dfrac{Dx+E}{x^2+1}$ 7. $\dfrac{Ax+B}{x^2+5}+\dfrac{Cx+D}{(x^2+5)^2}$

9. $\dfrac15\ln\left|\dfrac{x-1}{x+4}\right|+C$ 11. $-2\ln|x-2|+3\ln|x-3|+C$

13. $\tfrac52\ln|2x-1|+3\ln|x+4|+C$

15. $-\tfrac16\ln|x-1|+\tfrac{1}{15}\ln|x+2|+\tfrac{1}{10}\ln|x-3|+C$

17. $\ln\left|\dfrac{x(x+3)^2}{x-3}\right|+C$ 19. $\tfrac12 x^2-2x+6\ln|x+2|+C$

21. $3x+12\ln|x-2|-\dfrac{2}{x-2}+C$

23. $\tfrac12 x^2+3x-\ln|x-1|+8\ln|x-2|+C$

25. $\tfrac13 x^3+x+\ln\left|\dfrac{(x+1)(x-1)^2}{x}\right|+C$

27. $3\ln|x|-\ln|x-1|-\dfrac{5}{x-1}+C$

29. $\ln\dfrac{(x-3)^2}{|x+1|}+\dfrac{1}{x-3}+C$

31. $\ln|x+2|+\dfrac{4}{x+2}-\dfrac{2}{(x+2)^2}+C$

33. $-\tfrac{7}{34}\ln|4x-1|+\tfrac{6}{17}\ln(x^2+1)+\tfrac{3}{17}\tan^{-1}x+C$

35. $\dfrac{1}{32}\ln\left|\dfrac{x-2}{x+2}\right|-\dfrac{1}{16}\tan^{-1}\dfrac{x}{2}+C$

37. $3\tan^{-1}x+\tfrac12\ln(x^2+3)+C$

39. $\tfrac12 x^2-3x+\tfrac12\ln(x^2+1)+C$

41. $\dfrac{1}{\sqrt2}\tan^{-1}\left(\dfrac{x+1}{\sqrt2}\right)+\dfrac{1}{x^2+2x+3}+C$

43. $\tfrac16\ln\left|\dfrac{\sin\theta-1}{\sin\theta+5}\right|+C$ 45. $\ln\dfrac{e^x}{1+e^x}+C$

47. (a) $\sqrt2,\ -\sqrt2$ 49. $\pi\left(\tfrac{19}{5}-\tfrac94\ln5\right)$

51. $y=\dfrac{3-2Ce^x}{1-Ce^x}$ 53. $y=\dfrac{t-1}{Ct-t+1}$ 55. (b) a/b

57. (a) $(x-1)(x-2)(x-3)$ (b) $(x-4)(x^2+x+5)$
    (c) $(x-2)(x-3)(x^2+1)$

59. $\tfrac18\ln|x-1|-\tfrac15\ln|x-2|$
$\qquad\qquad+\tfrac{1}{12}\ln|x-3|-\tfrac{1}{120}\ln|x+3|+C$

▶ **Exercise Set 9.7 (Page 471)**

1. $\tfrac25(x-2)^{5/2}+\tfrac43(x-2)^{3/2}+C$ 3. $4-\pi$

5. $4-6\ln\tfrac53$ 7. $\tfrac{2}{15}(x^3+1)^{5/2}-\tfrac29(x^3+1)^{3/2}+C$

9. $2x^{1/2}-3x^{1/3}+6x^{1/6}-6\ln(x^{1/6}+1)+C$

11. $4\ln\dfrac{v^{1/4}}{|1-v^{1/4}|}+C$

13. $2t^{1/2}+3t^{1/3}+6t^{1/6}+6\ln|t^{1/6}-1|+C$

15. $\tfrac13(1+x^2)^{3/2}-(1+x^2)^{1/2}+C$

17. $-2\sqrt x\cos\sqrt x+2\sin\sqrt x+C$

19. $\ln\dfrac{\sqrt{e^x+1}-1}{\sqrt{e^x+1}+1}+C$ 21. $\ln\left|\tan\left(\dfrac{x}{2}\right)+1\right|+C$

23. 1 25. $\dfrac{4}{\sqrt3}\tan^{-1}\left(\sqrt3\tan\dfrac{x}{2}\right)-x+C$

29. $\dfrac{2}{\sqrt3}\tan^{-1}\left(\dfrac{2\tanh(x/2)+1}{\sqrt3}\right)+C$

31. $-\dfrac{\sqrt{3-x^2}}{3x}+C$ 33. $\dfrac{(x^2-5)^{3/2}}{15x^3}+C$

▶ **Exercise Set 9.8 (Page 481)**

1. exact value $=\tfrac{14}{3}\approx4.666666667$
   (a) 4.667600663, $|E_M|\approx0.000933996$
   (b) 4.664795679, $|E_T|\approx0.001870988$
   (c) 4.666651630, $|E_S|\approx0.000015037$
3. exact value $=2$
   (a) 2.008248408, $|E_M|\approx0.008248408$
   (b) 1.983523538, $|E_T|\approx0.016476462$
   (c) 2.000109517, $|E_S|\approx0.000109517$
5. exact value $=e^{-1}-e^{-3}\approx0.318092373$
   (a) 0.317562837, $|E_M|\approx0.000529536$
   (b) 0.319151975, $|E_T|\approx0.001059602$
   (c) 0.318095187, $|E_S|\approx0.000002814$
7. (a) 0.002812500 (b) 0.005625000 (c) 0.000126563
9. (a) 0.012919282 (b) 0.025838564 (c) 0.000170011
11. (a) 0.001226265 (b) 0.002452530 (c) 0.000006540
13. (a) 24 (b) 34 (c) 8
15. (a) 36 (b) 51 (c) 8
17. (a) 351 (b) 496 (c) 16
19. (a) 0.747130878 (b) 0.746210796 (c) 0.746824948
21. (a) 2.129469966 (b) 2.130644002 (c) 2.129861595
23. (a) 0.809253858 (b) 0.795924733 (c) 0.805376152

25. (a) 3.142425985, $|E_M| \approx 0.000833331$
(b) 3.139925989, $|E_T| \approx 0.001666665$
(c) 3.141592614, $|E_S| \approx 0.000000040$
29. 116   **33.** 3.820   **35.** 1604 ft   **37.** 37.9 mi
39. 9.3 L
41. right endpoint, trapezoidal, midpoint, left endpoint

▶ **TECHNOLOGY EXERCISES, CHAPTER 9 (Page 483)**

1. $\frac{1}{4}\sin^{-1}(x^4) + C$ **3.** $\sqrt{1-x^2} + \ln(1 - \sqrt{1-x^2}) + C$

5. $-\frac{1}{3}\left(\frac{x+1}{x-1}\right)^{3/2} + C$ **7.** $-\frac{1}{10}\ln|3 + 2x^{-5}| + C$

9. $\frac{\sqrt{2}}{3}[(x+2)^{3/2} - (x-2)^{3/2}] + C$

11. (a) $2\sin^{-1}(\sqrt{x/2}) + C$; $-2\sin^{-1}(\sqrt{2-x}/\sqrt{2}) + C$;
$\sin^{-1}(x-1) + C$; answers may vary with CAS used
13. (a) 3.844880 (b) 18 (c) 0.904741
15. (a) 42.551816 (b) 8 (c) 0.904524

17. $\frac{2}{x-2} - \frac{1}{(x-2)^2} + \frac{3x+1}{x^2 - x + 5}$

19. (a) $b = \frac{1}{2}$ (b) $a = 0$ (c) $b = 1$

▶ **Exercise Set 10.1 (Page 489)**

1. 1   **3.** divergent   **5.** $\ln\frac{5}{3}$ **7.** $\frac{1}{2}$ **9.** $\frac{1}{2(a^2 + 1)}$

11. $-\frac{1}{4}$ **13.** $\frac{1}{3}$ **15.** divergent   **17.** 0   **19.** divergent
21. divergent   **23.** $\pi/2$ **25.** 1   **27.** divergent
29. $\frac{9}{2}$ **31.** divergent   **33.** 2   **35.** 2   **37.** $\frac{1}{5}$
39. $\sqrt{\pi}$ **41.** $\sqrt{\pi/2}$ **45.** $\frac{1}{2}$

49. (a) $\frac{1}{24}$ with $\frac{x}{x^5 + 1} \le \frac{1}{x^4}$ (b) $\frac{1}{2}e^{-1}$ with $e^{-x^2} \le xe^{-x^2}$

51. $\frac{1}{3}$ **53.** (a) $V = \pi$ **55.** $\frac{2\pi NI}{kr}\left(1 - \frac{a}{\sqrt{r^2 + a^2}}\right)$

57. (b) 24,000,000   **59.** $1/s$ **61.** $1/(s^2 + 1)$
63. 1.809   **65.** (a) 1.047, $\pi/3 \approx 1.047197551$

▶ **Exercise Set 10.2 (Page 497)**

1. 1   **3.** 1   **5.** 1   **7.** $1/\pi$ **9.** -1 **11.** 0
13. $\frac{1}{2}$ **15.** $\frac{3}{2}$ **17.** $-1/(2\pi)$ **19.** $\frac{1}{4}$ **21.** $-\frac{1}{12}$
23. $\frac{1}{6}$ **25.** $a - b$ **27.** 2   **29.** $+\infty$ **31.** $\frac{1}{9}$
33. (b) 2   **35.** $k = -1$, $l = \pm 2\sqrt{2}$ **37.** $\frac{1}{2}$ **39.** 2
41. does not exist   **43.** 3

▶ **Exercise Set 10.3 (Page 503)**

1. 0   **3.** $-\infty$ **5.** $+\infty$ **7.** 0   **9.** 0   **11.** π
13. 2   **15.** e^{-3} **17.** e^2 **19.** 1   **21.** $+\infty$
23. $e^{2/\pi}$ **25.** 1   **27.** e^3 **29.** 1   **31.** e^2 **33.** $-\frac{1}{2}$
35. 0   **37.** $\frac{1}{2}$ **39.** $+\infty$ **41.** 2
45. (a) 0 (b) $+\infty$ (c) 0 (d) $-\infty$ (e) $+\infty$ (f) $+\infty$
(g) $-\infty$ (h) $-\infty$

47.

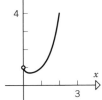

49. -1 **51.** $\frac{1}{9}$ **53.** 2π **55.** $8\sqrt{2}/5$ **57.** 1
59. does not exist
61. (b) $\ln 0.3 \approx -1.203265293$, $\ln 2 \approx 0.693381829$
63. $\frac{1}{2}at^2$ **65.** $1/s^2$

▶ **TECHNOLOGY EXERCISES, CHAPTER 10 (Page 505)**

1. 0.504067   **3.** 2.804364

9. $2\int_0^1 \sqrt{2 - u^2}\, du = 2.570796$ **11.** 3.633168

13. (c) π
15. (a) 4.5×10^{45} (b) Both limits are 0.

▶ **Exercise Set 11.1 (Page 515)**

1. $\frac{1}{3}, \frac{2}{4}, \frac{3}{5}, \frac{4}{6}, \frac{5}{7}$; converges to 1
3. 2, 2, 2, 2, 2; converges to 2
5. $\frac{\ln 1}{1}, \frac{\ln 2}{2}, \frac{\ln 3}{3}, \frac{\ln 4}{4}, \frac{\ln 5}{5}$; converges to 0
7. 0, 2, 0, 2, 0; diverges
9. $-1, \frac{16}{9}, -\frac{54}{28}, \frac{128}{65}, -\frac{250}{126}$; diverges
11. $\frac{6}{2}, \frac{12}{8}, \frac{20}{18}, \frac{30}{32}, \frac{42}{50}$; converges to $\frac{1}{2}$
13. $\cos 3, \cos\frac{3}{2}, \cos 1, \cos\frac{3}{4}, \cos\frac{3}{5}$; converges to 1
15. $e^{-1}, 4e^{-2}, 9e^{-3}, 16e^{-4}, 25e^{-5}$; converges to 0
17. $2, (\frac{5}{3})^2, (\frac{6}{4})^3, (\frac{7}{5})^4, (\frac{8}{6})^5$; converges to e^2

19. $\left\{\frac{2n-1}{2n}\right\}_{n=1}^{+\infty}$; converges to 1

21. $\left\{\frac{1}{3^n}\right\}_{n=1}^{+\infty}$; converges to 0

23. $\left\{\frac{1}{n} - \frac{1}{n+1}\right\}_{n=1}^{+\infty}$; converges to 0

25. $\{\sqrt{n+1} - \sqrt{n+2}\}_{n=1}^{+\infty}$; converges to 0
27. (a) $\sqrt{6}, \sqrt{6 + \sqrt{6}}, \sqrt{6 + \sqrt{6 + \sqrt{6}}}$ (b) 3
29. (a) 1, 1, 2, 3, 5, 8, 13, 21 (b) $(1 + \sqrt{5})/2$
31. (a) $1, \frac{3}{4}, \frac{2}{3}, \frac{5}{8}$ (b) $\frac{1}{2}$ **33.** (a) 3 (b) 11 (c) 1001
39. 3

▶ **Exercise Set 11.2 (Page 522)**

1. decreasing   **3.** increasing   **5.** decreasing
7. increasing   **9.** decreasing   **11.** increasing
13. increasing   **15.** decreasing   **17.** decreasing
19. eventually increasing   **21.** eventually decreasing
23. eventually increasing
27. (a) $\sqrt{2}, \sqrt{2 + \sqrt{2}}, \sqrt{2 + \sqrt{2 + \sqrt{2}}}$ (e) 2
33. (b) converges (decreasing and bounded below by 0)

▶ Exercise Set 11.3 **(Page 529)**

1. (a) converges to $\frac{5}{2}$ (b) converges to $\frac{1}{2}$ (c) diverges

3. $\frac{4}{7}$ 5. 6   7. diverges   9. $\frac{1}{3}$ 11. $\frac{1}{6}$ 13. $\frac{448}{3}$

15. $-\frac{1}{3}$ 17. $\frac{4}{9}$ 19. $\frac{532}{99}$ 21. $\frac{869}{1111}$ 23. diverges

33. $\frac{1}{x^2 - 2x}$, $|x| > 2$ 35. $\frac{2\sin x}{2 + \sin x}$, $-\infty < x < +\infty$

37. 1

41. The series converges to $1/(1 - x)$ only for $-1 < x < 1$.

▶ Exercise Set 11.4 **(Page 536)**

1. $\frac{4}{3}$ 3. $-\frac{1}{36}$

5. (a) converges   (b) diverges   (c) diverges   (d) diverges
   (e) converges   (f) diverges   (g) converges
   (h) converges

9. diverges   11. diverges   13. converges

15. diverges   17. diverges   19. diverges

21. converges   23. diverges   25. converges

27. converges   29. diverges

35. (a) diverges   (b) diverges   (c) diverges   (d) converges

37. (a) 1.1975; $1.2016 < S < 1.2026$ (b) 23

39. (a) $13 < s_{1,000,000} < 15$ (b) 2.69×10^{43}

▶ Exercise Set 11.5 **(Page 541)**

1. converges   3. inconclusive   5. diverges

7. diverges   9. converges   11. diverges

13. converges   15. converges   17. converges

19. converges   21. diverges   23. converges

25. converges   27. diverges   29. converges

31. converges

39. (a) 1.71667; error < 0.00163 (b) 8

41. (a) 0.69226; error < 0.00098 (b) 13

▶ Exercise Set 11.6 **(Page 546)**

1. converges   3. converges   5. diverges

19. converges   21. diverges   23. converges

25. diverges   27. converges   29. diverges

31. converges   33. converges   37. converges

39. $p > 1$ 41. converges

▶ Exercise Set 11.7 **(Page 554)**

1. converges   3. diverges   5. converges

7. absolutely   9. diverges   11. absolutely

13. conditionally   15. divergent   17. conditionally

19. absolutely   21. conditionally   23. divergent

25. conditionally   27. absolutely   29. conditionally

31. 0.125   33. 0.1   35. 9999   37. 39,999

39. $|\text{error}| < 0.00074$; $s_{10} \approx 0.4995$; exact sum $= 0.5$

41. $n = 4$, $s_4 \approx 0.84147$; $\sin(1) \approx 0.841470985$

43. $n = 9$, $s_9 \approx 0.40553$; $\ln\frac{3}{2} \approx 0.405465108$

45. (a) 14   (b) 0.817962176; $|\text{error}| \approx 0.004504858$

▶ Exercise Set 11.8 **(Page 560)**

1. 1, $[-1, 1)$ 3. $+\infty$, $(-\infty, +\infty)$ 5. $\frac{1}{5}$, $[-\frac{1}{5}, \frac{1}{5}]$

7. 1, $[-1, 1]$ 9. 1, $(-1, 1]$ 11. $+\infty$, $(-\infty, +\infty)$

13. $+\infty$, $(-\infty, +\infty)$ 15. 1, $[-1, 1]$ 17. 1, $(-2, 0]$

19. $\frac{4}{3}$, $(-\frac{19}{3}, -\frac{11}{3})$ 21. 1, $[-2, 0]$ 23. $+\infty$, $(-\infty, +\infty)$

25. $x + \frac{1}{2}x^2 + \frac{3}{14}x^3 + \frac{3}{35}x^4 + \cdots$; $R = 3$

27. $x + \frac{3}{2}x^2 + \frac{5}{8}x^3 + \frac{7}{48}x^4 + \cdots$; $R = +\infty$

29. $(a - b, a + b)$ 31. $+\infty$

▶ Exercise Set 11.9 **(Page 567)**

1. $1 - 2x + 2x^2 - \frac{4}{3}x^3 + \frac{2}{3}x^4$ 3. $2x - \frac{4}{3}x^3$ 5. $x + \frac{1}{3}x^3$

7. $x + x^2 + \frac{x^3}{2!} + \frac{x^4}{3!}$ 9. $1 + \frac{1}{2}x^2 + \frac{5}{24}x^4$

11. $\ln 3 + \frac{2}{3}x - \frac{2}{9}x^2 + \frac{8}{81}x^3 - \frac{4}{81}x^4$

13. $e + e(x - 1) + \frac{e}{2!}(x - 1)^2 + \frac{e}{3!}(x - 1)^3$

15. $2 + \frac{1}{4}(x - 4) - \frac{1}{64}(x - 4)^2 + \frac{1}{512}(x - 4)^3$

17. $\frac{\sqrt{2}}{2} - \frac{\sqrt{2}}{2}\left(x - \frac{\pi}{4}\right) - \frac{\sqrt{2}}{4}\left(x - \frac{\pi}{4}\right)^2 + \frac{\sqrt{2}}{12}\left(x - \frac{\pi}{4}\right)^3$

19. $-\frac{\sqrt{3}}{2} + \frac{\pi}{2}\left(x + \frac{1}{3}\right) + \frac{\sqrt{3}\pi^2}{4}\left(x + \frac{1}{3}\right)^2 - \frac{\pi^3}{12}\left(x + \frac{1}{3}\right)^3$

21. $\frac{\pi}{4} + \frac{1}{2}(x - 1) - \frac{1}{4}(x - 1)^2 + \frac{1}{12}(x - 1)^3$

23. $\sum_{k=0}^{\infty} (-1)^k \frac{x^k}{k!}$ 25. $\sum_{k=0}^{\infty} (-1)^k x^k$ 27. $\sum_{k=1}^{\infty} (-1)^{k+1} \frac{x^k}{k}$

29. $\sum_{k=0}^{\infty} (-1)^k \frac{x^{2k}}{4^k (2k)!}$ 31. $\sum_{k=0}^{\infty} \frac{x^{2k}}{(2k)!}$

33. $\sum_{k=0}^{\infty} (-1)(x + 1)^k$ 35. $\sum_{k=1}^{\infty} (-1)^{k+1} \frac{(x - 1)^k}{k}$

37. $\sum_{k=0}^{\infty} (-1)^k \frac{\pi^{2k}}{(2k)!}\left(x - \frac{1}{2}\right)^{2k}$

39. $\sum_{k=0}^{\infty} \left(\frac{16 + (-1)^{k+1}}{8}\right) \frac{(x - \ln 4)^k}{k!}$

▶ Exercise Set 11.10 **(Page 575)**

1. $\frac{2^6 e^{2c}}{6!} x^6$ 3. $-\frac{x^5}{(c + 1)^6}$ 5. $\frac{(4 + c)e^c}{4!} x^4$

7. $-\frac{(1 - 3c^2)}{3(1 + c^2)^3} x^3$ 9. $-\frac{5}{128c^{7/2}}(x - 4)^4$

11. $\frac{\cos c}{5!}\left(x - \frac{\pi}{6}\right)^5$ 13. $\frac{7}{(1 + c)^8}(x + 2)^6$

15. $\frac{x^{n+1}}{(1 - c)^{n+2}}$ 17. $\frac{2^{n+1} e^{2c}}{(n + 1)!} x^{n+1}$

27. $1 - 2x + 2x^2 - \frac{4}{3}x^3 + \cdots$; $(-\infty, +\infty)$

29. $x - x^2 + \frac{1}{2!}x^3 - \frac{1}{3!}x^4 + \cdots$; $(-\infty, +\infty)$

31. $2x - \frac{2^3}{3!}x^3 + \frac{2^5}{5!}x^5 - \frac{2^7}{7!}x^7 + \cdots$; $(-\infty, +\infty)$

33. $x^2 - \frac{1}{2!}x^4 + \frac{1}{4!}x^6 - \frac{1}{6!}x^8 + \cdots$; $(-\infty, +\infty)$

35. $x^2 - \frac{2^3}{4!}x^4 + \frac{2^5}{6!}x^6 - \frac{2^7}{8!}x^8 + \cdots$; $(-\infty, +\infty)$

37. $-x^2 - \frac{1}{2}x^4 - \frac{1}{3}x^6 - \frac{1}{4}x^8 - \cdots$; $(-1, 1)$

39. $1 + 4x^2 + 16x^4 + 64x^6 + \cdots;\ (-\frac{1}{2}, \frac{1}{2})$

41. $x^2 - 3x^3 + 9x^4 - 27x^5 + \cdots;\ (-\frac{1}{3}, \frac{1}{3})$

43. $2x^2 + \frac{2^3}{3!}x^4 + \frac{2^5}{5!}x^6 + \frac{2^7}{7!}x^8 + \cdots;\ (-\infty, +\infty)$

45. $1 + \frac{3}{2}x - \frac{9}{8}x^2 + \frac{27}{16}x^3 - \cdots;\ (-\frac{1}{3}, \frac{1}{3})$

47. $1 + 4x + 12x^2 + 32x^3 + \cdots;\ (-\frac{1}{2}, \frac{1}{2})$

49. $x + \frac{1}{2}x^3 + \frac{3}{8}x^5 + \frac{5}{16}x^7 + \cdots;\ (-1, 1)$

51. $\displaystyle\sum_{k=0}^{\infty} (-1)^k (x-1)^k;\ (0, 2)$ **53.** $\displaystyle\sum_{k=0}^{\infty} \binom{m}{k} x^k$

55. $\sin \pi = 0$ **57.** $e^{-\ln 3} = \frac{1}{3}$

59. (a) $x^2 - 2x^4 + \frac{2}{3}x^6 - \frac{4}{45}x^8 + \cdots$ (b) 0

▶ **Exercise Set 11.11 (Page 581)**

1. (a) 9   (b) 13   **3.** 1.6487   **5.** 0.9877   **7.** 0.223

9. 0.100   **11.** 1.0050   **13.** $|x| < 0.569$ **15.** 9×10^{-8}

17. (a) 1,999,999   (b) 8   **19.** 3.140

▶ **Exercise Set 11.12 (Page 587)**

5. $\displaystyle\sum_{k=1}^{\infty} (-1)^{k+1} k x^{k-1}$ **9.** $\ln \frac{4}{3}$ **11.** $\dfrac{2}{(1-x)^3}$

13. (b) $\displaystyle\sum_{k=0}^{\infty} (-1)^k x^k;\ (-1, 1)$ **15.** 0.764   **17.** 0.494

19. 0.100   **21.** 0.491   **23.** $1 - \frac{3}{2}x^2 + \frac{25}{24}x^4 - \frac{331}{720}x^6 + \cdots$

25. $x - x^2 + \frac{1}{3}x^3 - \frac{1}{30}x^5 + \cdots$

27. $-x^3 - \frac{1}{2}x^5 - \frac{1}{3}x^7 - \frac{1}{4}x^9 - \cdots$

29. $x^2 + \frac{9}{2}x^3 + \frac{79}{8}x^4 + \frac{683}{48}x^5 + \cdots$ **31.** (a) 0   (b) $-\frac{3}{2}$

33. (a) $x - \frac{1}{6}x^3 + \frac{3}{40}x^5 - \frac{5}{112}x^7$

(b) $x + \displaystyle\sum_{k=1}^{\infty} (-1)^k \frac{1 \cdot 3 \cdot 5 \cdots (2k-1)}{2^k k! (2k+1)} x^{2k+1}$ (c) 1

▶ **TECHNOLOGY EXERCISES, CHAPTER 11 (Page 589)**

1. (a) 500   (b) 3.139593; $|\text{error}| < 4 \times 10^{-3}$

(c) $|\text{error}| \approx 2 \times 10^{-3}$

3. (a) 3.141592653589793878   (b) $|\text{error}| \approx 6.4 \times 10^{-16}$

7. (a) 50   (b) 1.106279

9. $B_0 = 1,\ B_1 = -\frac{1}{2},\ B_2 = \frac{1}{6},\ B_3 = 0,\ B_4 = -\frac{1}{30},\ B_5 = 0,\ B_6 = \frac{1}{42}$

11. (a) 0.001615   (b) 0.000333

▶ **Exercise Set 12.2 (Page 598)**

1.

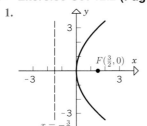

3.

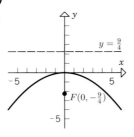

5.

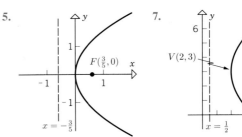

7.

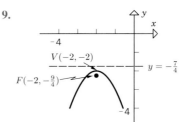

9.

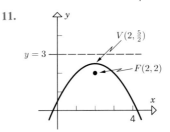

11.

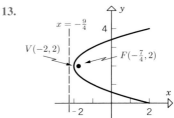

13.

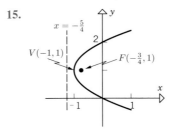

15.

17. $y^2 = 12x$ **19.** $y^2 = -28x$ **21.** $y^2 = 2x$

23. $x^2 = -12y$ **25.** $y^2 = -5(x - \frac{19}{5})$ **27.** $y^2 = 6(x - \frac{3}{2})$

29. $(x-1)^2 = 12(y-1)$ **31.** $(x-5)^2 = 2(y+3)$

33. (a) $y = \frac{7}{6}x^2 - \frac{23}{6}x + 3$ (b) $x = -\frac{7}{6}y^2 + \frac{17}{6}y + 2$

35. vertex: $\left(-\dfrac{B}{2A}, \dfrac{4AC - B^2}{4A} \right)$

focus: $\left(-\dfrac{B}{2A}, \dfrac{4AC - B^2 + 1}{4A} \right)$

directrix: $y = \dfrac{4AC - B^2 - 1}{4A}$

37. 16 ft   **41.** $\frac{1}{16}$ ft

▶ Exercise Set 12.3 **(Page 604)**

1.

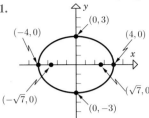

3.

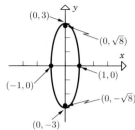

5.

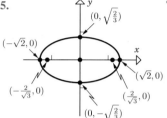

7.

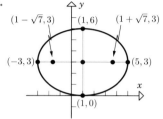

9.

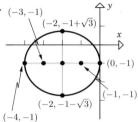

11.

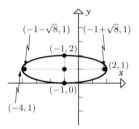

13.

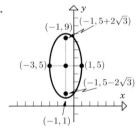

15. $\dfrac{x^2}{9} + \dfrac{y^2}{4} = 1$ **17.** $\dfrac{x^2}{169} + \dfrac{y^2}{144} = 1$ **19.** $\dfrac{x^2}{3} + \dfrac{y^2}{2} = 1$

21. $\dfrac{x^2}{16} + \dfrac{y^2}{4} = 1, \dfrac{x^2}{4} + \dfrac{y^2}{16} = 1$ **23.** $\dfrac{x^2}{36} + \dfrac{y^2}{81/8} = 1$

25. $(x-1)^2 + \dfrac{(y-3)^2}{2} = 1$ **27.** $\dfrac{(x-4)^2}{32} + \dfrac{(y-3)^2}{36} = 1$

29. $\sqrt{3}/2$ **31.** $x^2/36 + y^2/20 = 1$

33. $9x^2 + 25y^2 + 64x - 64 = 0$

35.

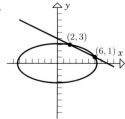

37.

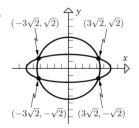

39. $k = -4$ at $(-2, -1)$, $k = 4$ at $(2, 1)$

41. πab **45.** $27\left[4\sin^{-1}\dfrac{h-2}{2} + (h-2)\sqrt{4h-h^2} + 2\pi \right]$

47. $\dfrac{4a^2b^2}{a^2 + b^2}$ **51.** $T = \frac{1}{2}pD, L = D\sqrt{1 + p^2}$

53. major axis $r_1 + r_2$; center at the midpoint of the line segment joining the centers of C_1 and C_2

▶ Exercise Set 12.4 **(Page 613)**

1.

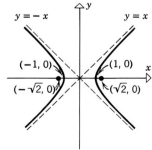

3.

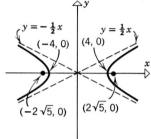

5.

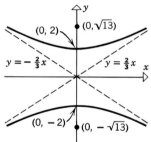

7.

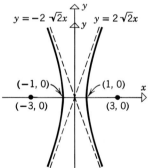

9. $y-4=-\frac{2}{3}(x-2)$ $y-4=\frac{2}{3}(x-2)$

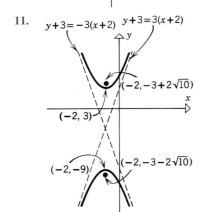

(5, 4)
$(2-\sqrt{13},4)$
$(-1, 4)$ $(2+\sqrt{13},4)$

11. $y+3=-3(x+2)$ $y+3=3(x+2)$

$(-2,-3+2\sqrt{10})$
$(-2, 3)$
$(-2,-9)$ $(-2,-3-2\sqrt{10})$

13. $y-1=\frac{1}{2}(x+1)$
$y-1=-\frac{1}{2}(x+1)$

$(-3, 1)$ $(1, 1)$
$(-1-\sqrt{5}, 1)$ $(-1+\sqrt{5}, 1)$

15. $y+3=-4(x-1)$
$y+3=4(x-1)$

$(-1,-3)$ $(3,-3)$
$(1-2\sqrt{17},-3)$ $(1+2\sqrt{17},-3)$

17. $\dfrac{x^2}{4}-\dfrac{y^2}{5}=1$ **19.** $x^2-\dfrac{y^2}{4}=1$

21. $\dfrac{x^2}{64/9}-\dfrac{y^2}{16}=1,$ $\dfrac{y^2}{36}-\dfrac{x^2}{16}=1$ **23.** $\dfrac{y^2}{56}-\dfrac{x^2}{56}=1$

25. $\dfrac{x^2}{4}-\dfrac{y^2}{4/3}=1$ **27.** $\dfrac{(x-2)^2}{4}-\dfrac{(y+3)^2}{5}=1$

29. $\dfrac{(x-6)^2}{16}-\dfrac{(y-4)^2}{9}=1$ **31.** $8xy-4x-4y+1=0$

33. 5/3   **35.** $x^2/9-y^2/16=1$

37. $8x^2-y^2-18x+9=0$

39.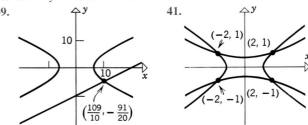

10

10
$\left(\dfrac{109}{10},-\dfrac{91}{20}\right)$

41.

$(-2, 1)$ $(2, 1)$
$(-2,-1)$ $(2, -1)$

43. $(\pm3/\sqrt{5},4/\sqrt{5}),(\pm3/\sqrt{5},-4/\sqrt{5})$

45. $(\frac{3}{2}\sqrt{13},-9),(-\frac{3}{2}\sqrt{13},-9)$

47. The hyperbola flattens and approaches the focal axis, excluding the segment between the vertices.
The hyperbola approaches the lines that are perpendicular to the focal axis at the vertices.

▶ **Exercise Set 12.5 (Page 619)**

1. (a) $(-1+3\sqrt{3}, \sqrt{3}+3)$ (b) $\dfrac{x'^2}{4}-\dfrac{y'^2}{12}=1$

(c)

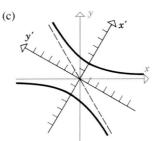

3. $\theta=45°;\dfrac{y'^2}{18}-\dfrac{x'^2}{18}=1,$

hyperbola

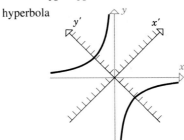

5. $\theta=\tan^{-1}\frac{1}{2};$
$\dfrac{x'^2}{3}-\dfrac{y'^2}{2}=1,$ hyperbola

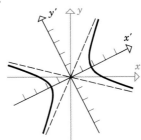

7. $\theta = 60°$; $y' = x'^2$, parabola

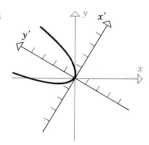

9. $\theta = \tan^{-1}\frac{3}{4}$;
$y'^2 = 4(x' - 1)$, parabola

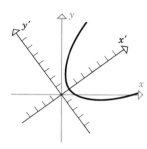

11. $\theta = \tan^{-1}\frac{3}{4}$;
$\frac{(x' + 1)^2}{4} + y'^2 = 1$, ellipse

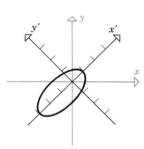

15. $x^2 + xy + y^2 = 3$ **23.** ellipse, point, or no graph
25. parabola, line, pair of parallel lines, or no graph
27. ellipse, point, or no graph

▶ **TECHNOLOGY EXERCISES, CHAPTER 12 (Page 620)**

1. maximum 4.207900, minimum 1.056175
3. 59.3 ft   **5.** (a) 9.36129×10^8 km   (b) 106,864 km/hr
7. (a) 14.3 in, 24 in, 33.7 in   (b) 0.06 in/sec
9. (a) hyperbola   (b) $y = -\frac{9}{2}x - \frac{1}{2} \pm \frac{1}{2}\sqrt{73x^2 + 42x + 17}$

▶ **Exercise Set 13.1 (Page 626)**

1.

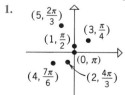

$(5, \frac{2\pi}{3})$ $(3, \frac{\pi}{4})$ $(1, \frac{\pi}{2})$ $(0, \pi)$ $(4, \frac{7\pi}{6})$ $(2, \frac{4\pi}{3})$

3. (a) $(3\sqrt{3}, 3)$ (b) $(-7/2, 7\sqrt{3}/2)$ (c) $(4\sqrt{2}, 4\sqrt{2})$
(d) $(5, 0)$ (e) $(-7\sqrt{3}/2, 7/2)$ (f) $(0, 0)$
5. (a) $(5, \pi)$ (b) $(4, 11\pi/6)$ (c) $(2, 3\pi/2)$
(d) $(8\sqrt{2}, 5\pi/4)$ (e) $(6, 2\pi/3)$ (f) $(\sqrt{2}, \pi/4)$
7. (a) $(-5, 0)$ (b) $(-4, 5\pi/6)$ (c) $(-2, \pi/2)$
(d) $(-8\sqrt{2}, \pi/4)$ (e) $(-6, 5\pi/3)$ (f) $(-\sqrt{2}, 5\pi/4)$

---

9. $x^2 + y^2 = 4$; circle   **11.** $y = 4$; line
13. $x^2 + y^2 - 3x = 0$; circle   **15.** $xy = 4$; hyperbola
17. $x^2 = 4 - 4y$; parabola   **19.** $3x + 2y = 6$; line
21. $r\cos\theta = 7$ **23.** $r = 3$ **25.** $r = 6\sin\theta$
27. $r = 9\sec\theta\tan\theta$ **29.** $r^2\sin 2\theta = \frac{9}{2}$ **31.** $r^2 = \sin 2\theta$
33. The graph consists of the concentric circles $r = 1$ and $r = 2$.
35.

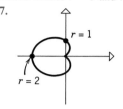

37.

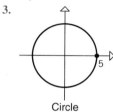

▶ **Exercise Set 13.2 (Page 635)**

1.

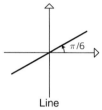

Line

3.

Circle

5.

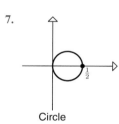

Circle

7.

Circle

9.

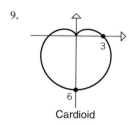

Cardioid

11.

Cardioid

13.

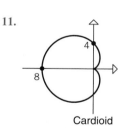

Cardioid

15.

Limaçon

17.

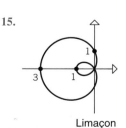

Limaçon

19.

Limaçon

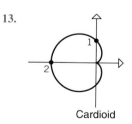

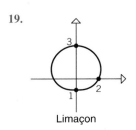

21.

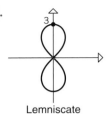

Limaçon

23.

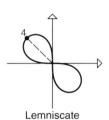

Limaçon

25.

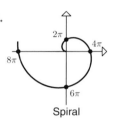

Lemniscate

27.

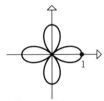

Lemniscate

29.

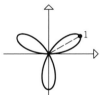

Spiral

31.

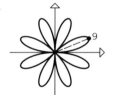

Four-petal rose

33.

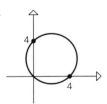

Three-petal rose

35.

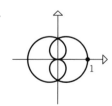

Eight-petal rose

37.

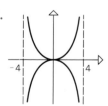

39.

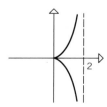

41.

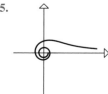

43.

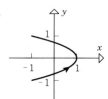

45.

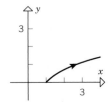

47. $\frac{3}{4}\sqrt{3}$ **51.** $-\frac{1}{4}$

53. $e = 1$

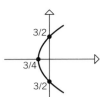

55. $e = \frac{1}{2}$

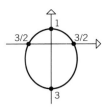

57. $e = \frac{3}{2}$

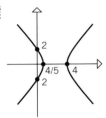

59. The graph of $r = f(\theta + \alpha)$ is the graph of $r = f(\theta)$ rotated α radians clockwise about the pole if $\alpha > 0$, $|\alpha|$ radians counterclockwise if $\alpha < 0$.

▶ **Exercise Set 13.3 (Page 640)**

1. $\dfrac{7\pi^3}{1296}$ **3.** 6π **5.** $\dfrac{8\pi}{3} + \sqrt{3}$ **7.** $\dfrac{9\sqrt{3}}{2} - \pi$ **9.** 1

11. 4π **13.** $\dfrac{5\pi}{4}$ **15.** $100\cos^{-1}\left(\frac{3}{5}\right) - 48$ **17.** $\frac{4}{3}a^2$

19. $\frac{9}{8}(1 - e^{-4\pi})$ **21.** $\ln 6$ **23.** $2\sqrt{3} - \frac{2}{3}\pi$ **25.** (b) $\frac{1}{6}$

▶ **Exercise Set 13.4 (Page 649)**

1.

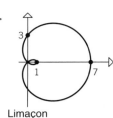

3.

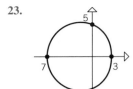

5.

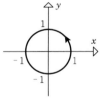

7.

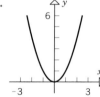

9.

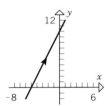

11.

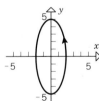

13.

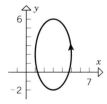

15.

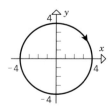

17.
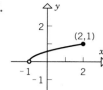

19.
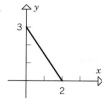

21.
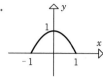

23.
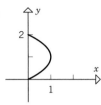

25.
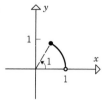

27. $x = -t$, $y = \sqrt{-t}$

29. 1   **31.** 12   **33.** 0   **35.** $\frac{1}{2}$ **37.** 4   **39.** 10

41. $\frac{1}{3}(2\sqrt{2} - 1)$ **43.** π **45.** $\frac{1}{27}(80\sqrt{10} - 13\sqrt{13})$

47. $8a$ **49.** (b) 9.69   (c) 5.16 cm

51. $x = (2 + 3\sin\theta)\cos\theta$, $y = (2 + 3\sin\theta)\sin\theta$

53. $y = -e^{-4}x + 2e^{-2}$ **55.** (a) $-\frac{1}{2}$ (b) 1, 4

57. $y = 5x - 14$, $y = 6x - 17$

59. if $|a| = |b|$, a circle with center at (h, k) and radius $|a|$;
if $|a| \neq |b|$, an ellipse with center at (h, k) and vertices
$(h \pm |a|, k)$ if $|a| > |b|$ and vertices $(h, k \pm |b|)$ if $|a| < |b|$

61. (a)
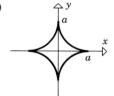

63. $\dfrac{2\sqrt{2}}{5}\pi(2e^{\pi} + 1)$ **65.** $\dfrac{\pi}{24}(17\sqrt{17} - 1)$

▶ **Exercise Set 13.5 (Page 653)**

1. $\dfrac{1}{\sqrt{3}}$ **3.** $\dfrac{\tan 2 - 2}{2\tan 2 + 1}$ **5.** -2 **7.** $\dfrac{\sqrt{10}}{3}(e^6 - 1)$

9. $2\pi a$ **11.** $\dfrac{a}{3}[(\pi^2 + 4)^{3/2} - 8]$ **13.** $8a$

15. (b) 2.42

17. (a) $r = 4\theta + 10$ (b) 38.9 mm

19. $\theta = \dfrac{\pi}{2}, \dfrac{3\pi}{2}$, $\sin^{-1}\frac{1}{4}$, $\pi - \sin^{-1}\frac{1}{4}$

▶ **TECHNOLOGY EXERCISES, CHAPTER 13 (Page 654)**

1. $\alpha = 8\pi$ **3.** 4.454964   **5.** 86,792 km/hr

9. 0.65 m   **11.** (b) 49.64 ft   (c) 69°

▶ **Exercise Set 14.1 (Page 662)**

1. (a) $\sqrt{14}$; $(1, \frac{1}{2}, \frac{3}{2})$ (b) $\sqrt{11}$; $(\frac{9}{2}, \frac{3}{2}, \frac{9}{2})$
    (c) $\sqrt{30}$; $(\frac{1}{2}, -\frac{1}{2}, 4)$ (d) $\sqrt{42}$; $(\frac{3}{2}, 1, -\frac{5}{2})$

3. $(-6, 2, 1)$, $(-6, 2, -2)$, $(-6, 1, -2)$, $(4, 2, 1)$,
    $(4, 1, 1)$, $(4, 1, -2)$, $(4, 2, -2)$, $(-6, 1, 1)$

5. (b) $(2, 1, 6)$ (c) 49

7. distance to x-axis is $\sqrt{y_0^2 + z_0^2}$
    distance to y-axis is $\sqrt{x_0^2 + z_0^2}$

9. $(x + 2)^2 + (y - 4)^2 + (z + 1)^2 = 36$

11. $x^2 + (y - 1)^2 + z^2 = 9$

13. (a) $(x - 2)^2 + (y + 1)^2 + (z + 3)^2 = 9$
    (b) $(x - 2)^2 + (y + 1)^2 + (z + 3)^2 = 1$
    (c) $(x - 2)^2 + (y + 1)^2 + (z + 3)^2 = 4$

15. $(x + \frac{1}{2})^2 + (y - 2)^2 + (z - 2)^2 = \frac{5}{4}$

17. $(x - 3)^2 + (y + 2)^2 + (z - 4)^2 = 41$

19. $x^2 + y^2 + z^2 = 30 \pm 2\sqrt{29}$

21. sphere, center $(-5, -2, -1)$, radius 7

23. sphere, center $(\frac{1}{2}, \frac{3}{4}, -\frac{5}{4})$, radius $\frac{3}{4}\sqrt{6}$ **25.** no graph

27. largest $3 + \sqrt{6}$; smallest $3 - \sqrt{6}$

29. all points outside the circular cylinder
    $(y + 3)^2 + (z - 2)^2 = 16$

31. $(2 - \sqrt{3})R$

33. (a)
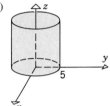
(b)
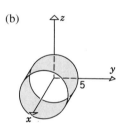

(c)
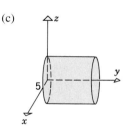

35. (a)
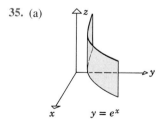
(b)
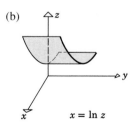

(c)
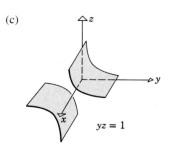

37. (a) 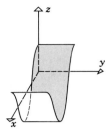 (b)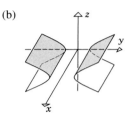

39. (a) 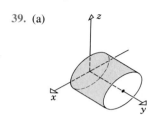 (b)

▶ **Exercise Set 14.2 (Page 670)**

1.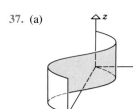

3.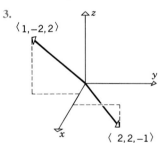

5.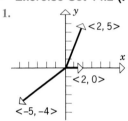

7.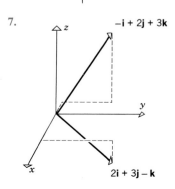

9. (a) $\langle -1, 3 \rangle$ (b) $\langle -7, 2 \rangle$ (c) $\langle 2, 1 \rangle$ (d) $\langle -8, 7 \rangle$

11. (a) $\langle -3, 6, 1 \rangle$ (b) $\langle 1, -3, -5 \rangle$ 13. $(4, -4)$

15. $(4, -4)$ 17. $(8, -1, -3)$

19. (a) $-5\mathbf{i} - 2\mathbf{j}$ (b) $8\mathbf{i} + 10\mathbf{j}$ (c) $-2\mathbf{i} + 4\mathbf{j}$
(d) $40\mathbf{i} + 36\mathbf{j}$ (e) $-20\mathbf{i} + 22\mathbf{j}$ (f) $-8\mathbf{i} - 8\mathbf{j}$

21. (a) $-\mathbf{i} + 4\mathbf{j} - 2\mathbf{k}$ (b) $18\mathbf{i} + 12\mathbf{j} - 6\mathbf{k}$ (c) $-\mathbf{i} - 5\mathbf{j} - 2\mathbf{k}$
(d) $40\mathbf{i} - 4\mathbf{j} - 4\mathbf{k}$ (e) $-2\mathbf{i} - 16\mathbf{j} - 18\mathbf{k}$
(f) $-\mathbf{i} + 13\mathbf{j} - 2\mathbf{k}$

23. (a) $\sqrt{2}$ (b) 2 (c) 3 25. (a) $\sqrt{14}$ (b) 3

27. (a) $\sqrt{41}$ (b) 7 (c) $3\sqrt{29} + 8$ (d) $3\sqrt{10}$
(e) $\langle \frac{3}{5}, \frac{4}{5} \rangle$ (f) 1

29. (a) $2\sqrt{3}$ (b) $\sqrt{14} + \sqrt{2}$ (c) $2\sqrt{14} + 2\sqrt{2}$
(d) $2\sqrt{37}$ (e) $\frac{1}{\sqrt{6}}\mathbf{i} + \frac{1}{\sqrt{6}}\mathbf{j} - \frac{2}{\sqrt{6}}\mathbf{k}$ (f) 1

31. $\langle -\frac{2}{3}, 1 \rangle$ 33. $\mathbf{u} = \frac{5}{7}\mathbf{i} + \frac{2}{7}\mathbf{j} + \frac{1}{7}\mathbf{k}$, $\mathbf{v} = \frac{8}{7}\mathbf{i} - \frac{1}{7}\mathbf{j} - \frac{4}{7}\mathbf{k}$

43. $-\frac{3}{5}\mathbf{i} + \frac{4}{5}\mathbf{j}$ 45. $-\frac{1}{\sqrt{14}}(3\mathbf{i} - 2\mathbf{j} + \mathbf{k})$

47. $\frac{1}{3\sqrt{2}}(4\mathbf{i} + \mathbf{j} - \mathbf{k})$ 49. $\langle -\frac{3}{2}, 2 \rangle$ 51. $6\mathbf{i} - 8\mathbf{j} - 2\mathbf{k}$

53. circle; radius 1, center at (x_0, y_0)

55. (a) sphere; radius 2, center at $(0, 0, 0)$
(b) sphere; radius 3, center at (x_0, y_0, z_0)
(c) all points on and inside the sphere of radius 1, centered at (x_0, y_0, z_0)

57. (a) $\langle 1/\sqrt{2}, -1/\sqrt{2} \rangle$, $\langle -1/\sqrt{2}, 1/\sqrt{2} \rangle$
(b) $\langle 1/\sqrt{2}, 1/\sqrt{2} \rangle$, $\langle -1/\sqrt{2}, -1/\sqrt{2} \rangle$

59. $\left(\frac{13}{4}, \frac{21}{4} \right)$ 61. (a) $\langle 1/2, \sqrt{3}/2 \rangle$ (b) $\langle -2\sqrt{2}, 2\sqrt{2} \rangle$

63. $(1, 1), (-1, -1)$ 77. (a) $\left(\frac{3}{2}, \frac{1}{2} \right)$

▶ **Exercise Set 14.3 (Page 677)**

1. (a) -10 (b) -3 (c) 0 (d) -20

3. (a) obtuse (b) acute (c) obtuse (d) orthogonal

5. (a) $\frac{14}{13}\mathbf{i} + \frac{21}{13}\mathbf{j}$ (b) $\langle 2, 6 \rangle$ (c) $-\frac{11}{13}\mathbf{i} + \mathbf{j} + \frac{55}{13}\mathbf{k}$
(d) $\langle -\frac{32}{89}, -\frac{12}{89}, \frac{73}{89} \rangle$

9. (a) 6 (b) 36 (c) $24\sqrt{5}$ (d) $24\sqrt{5}$

11. $\frac{1}{5\sqrt{2}}, \frac{4}{\sqrt{65}}, \frac{9}{\sqrt{130}}$ 13. The right angle is at vertex B.

15. (a) the line through the origin and perpendicular to $\mathbf{r}_0$
(b) the line through (x_0, y_0) and perpendicular to $\mathbf{r}_0$
(c) circle; center at $(\frac{1}{2}x_0, \frac{1}{2}y_0)$, radius $\frac{1}{2}\|\mathbf{r}_0\|$

17. (a) $-\frac{3}{4}$ (b) $\frac{1}{7}$ (c) $\frac{48 \pm 25\sqrt{3}}{11}$ (d) $\frac{4}{3}$

21. (a) $\frac{9}{5}$ (b) $\frac{13}{5}$ 23. (a) $\frac{4}{3}$ (b) $\frac{1}{3}\sqrt{137}$

25. -12 ft·lb 31. $71°$

▶ **Exercise Set 14.4 (Page 685)**

1. $\langle 7, 10, 9 \rangle$ 3. $\langle -4, -6, -3 \rangle$

5. (a) $\langle -20, -67, -9 \rangle$ (b) $\langle -78, 52, -26 \rangle$
(c) $\langle 24, 0, -16 \rangle$ (d) $\langle -12, -22, -8 \rangle$
(e) $\langle 0, -56, -392 \rangle$ (f) $\langle 0, 56, 392 \rangle$

9. $\pm \frac{1}{\sqrt{5}}(2\mathbf{j} + \mathbf{k})$ 11. $2\mathbf{v} \times \mathbf{u}$

13. (a) $\frac{1}{2}\sqrt{374}$ (b) $9\sqrt{13}$ 15. (a) $\frac{1}{2}\sqrt{26}$ (b) $\frac{1}{3}\sqrt{26}$

17. (a) $2\sqrt{\frac{141}{29}}$ (b) $\frac{1}{3}\sqrt{137}$

19. ambiguous, needs parentheses   **21.** 80   **23.** 1
25. (a) 16   (b) 45
27. (a) 9   (b) $\sqrt{122}$ (c) $\sin^{-1}\left(\frac{9}{14}\right)$ **29.** $\mathbf{u} \times \mathbf{v}$
37. (a) $\frac{2}{3}$ (b) $\frac{1}{2}$

▶ **Exercise Set 14.5 (Page 690)**

1. $x = 3 + 2t, \ y = -2 + 3t$ **3.** $x = 4, \ y = 1 + 2t$
5. $x = 5 - 3t, \ y = -2 + 6t, \ z = 1 + t$
7. $x = -t, \ y = 6t, \ z = t$
9. same as Exercise 1 with $0 \le t \le 1$
11. same as Exercise 3 with $0 \le t \le 1$
13. same as Exercise 5 with $0 \le t \le 1$
15. same as Exercise 7 with $0 \le t \le 1$
17. $x = -5 + 2t, \ y = 2 - 3t$ **19.** $x = 3 + 4t, \ y = -4 + 3t$
21. $x = -1 + 3t, \ y = 2 - 4t, \ z = 4 + t$
23. $x = -2 + 2t, \ y = -t, \ z = 5 + 2t$
25. $x = 3 + t, \ y = 7, \ z = 0$ **27.** $(-1, 1), (3, 9)$
29. (a) $(-2, 10, 0)$ (b) $(-2, 0, -5)$
   (c) does not intersect yz-plane
31. $(0, 4, -2), (4, 0, 6)$
33. $x = x_1 + at, \ y = y_1 + bt, \ z = z_1 + ct$ **35.** $(1, -1, 2)$
39. (a) no   (b) yes   **41.** $(1, \frac{14}{3}, -\frac{5}{3})$
45. (a) $\langle x, y \rangle = \langle 2, -1 \rangle + t\langle -7, 4 \rangle$ (b) $\langle x, y \rangle = \langle 0, 3 \rangle + t\langle 4, 0 \rangle$
47. the line segment joining the points $(1, 0)$ and $(-3, 6)$
49. $2\sqrt{5}$ **51.** $\sqrt{35/6}$
53. (b) 84°   (c) $x = 7 + t, \ y = -1, \ z = -2 + t$
55. $x = t, \ y = 2 + t, \ z = 1 - t$
57. (a) $\sqrt{17}$ cm   (b) $\frac{1}{2}\sqrt{14}$ cm

▶ **Exercise Set 14.6 (Page 696)**

1. $x + 4y + 2z = 28$ **3.** $z = 0$
5. (a) $2y - z = 1$ (b) $x + 9y - 5z = 16$
7. (a) yes   (b) no   **9.** (a) yes   (b) no
11. (a) 35°   (b) 79°
13. (a) $z = 0$ (b) $y = 0$ (c) $x = 0$
15. $4x - 2y + 7z = 0$ **17.** $4x - 13y + 21z = -14$
19. $x = 5 - 2t, \ y = 5t, \ z = -2 + 11t$ **21.** $x + y - 3z = 6$
23. $7x + y + 9z = 25$ **25.** $2x + 4y + 8z = 29$
29. (a) $x = -\frac{11}{7} - 23t, \ y = -\frac{12}{7} + t, \ z = -7t$
   (b) $x = -5t, \ y = -3t, \ z = 0$
31. (b) $\frac{2}{5}$ **33.** $\frac{5}{3}$ **35.** $\frac{137}{21}$ **37.** $\dfrac{5}{3\sqrt{6}}$ **39.** $\dfrac{25}{\sqrt{126}}$
41. $\dfrac{95}{\sqrt{1817}}$ **45.** $(x - 2)^2 + (y - 1)^2 + (z + 3)^2 = \frac{121}{14}$

▶ **Exercise Set 14.7 (Page 706)**

1. (a) $4x^2 + y^2 = 4$; ellipse   (b) $y^2 + z^2 = 3$; circle
   (c) $4x^2 + z^2 = 3$; ellipse
3. (a) $9x^2 - z^2 = 16$; hyperbola   (b) $y^2 + z^2 = 20$; circle
   (c) $9x^2 - y^2 = 20$; hyperbola
5. (a) $z = 4y^2$; parabola   (b) $z = 9x^2 + 16$; parabola
   (c) $9x^2 + 4y^2 = 4$; ellipse

7.
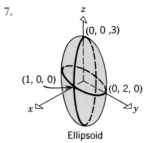
Ellipsoid

9.
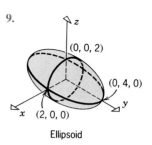
Ellipsoid

11.
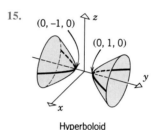
Hyperboloid of one sheet

13.
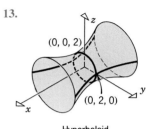
Hyperboloid of one sheet

15.
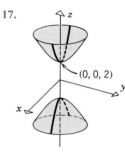
Hyperboloid of two sheets

17.
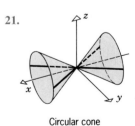
Hyperboloid of two sheets

19.
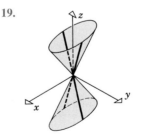
Elliptic cone

21.
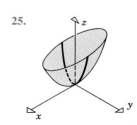
Circular cone

23.
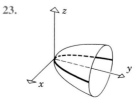
Circular paraboloid

25.

Elliptic paraboloid

27.

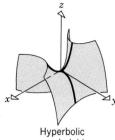

Hyperbolic
paraboloid

29. (a)

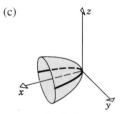

Hyperboloid
of one sheet

(b)

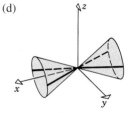

Hyperboloid
of two sheets

(c)

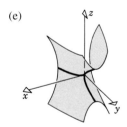

Paraboloid

(d)

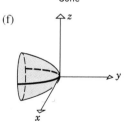

Cone

(e)

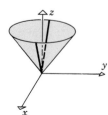

Hyperbolic
paraboloid

(f)

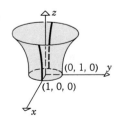

Paraboloid

31.

33.

(0, 1, 0)
(1, 0, 0)

35.

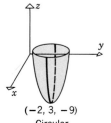

(−2, 3, −9)
Circular
paraboloid

37.

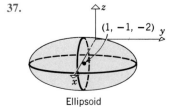

(1, −1, −2)

Ellipsoid

39. (0, −1, 5)

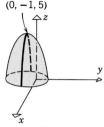

Circular
paraboloid

45. $x^2 + y^2 = 2$; circle   **47.** $x^2 + y^2 - 2x = 0$; circle
49. $y = \frac{1}{2}x^2 - \frac{1}{2}$; parabola
51. $x^2 + y^2 + 4x = 5$, $x \geq 0$; circular arc
53. (a) focus: $\left(k, 0, \frac{k^2}{9} + 1\right)$, vertex: $\left(k, 0, \frac{k^2}{9}\right)$
　　(b) foci: $(\sqrt{5k}, 0, k)$, $(-\sqrt{5k}, 0, k)$;
　　endpoints of major axis: $(3\sqrt{k}, 0, k)$, $(-3\sqrt{k}, 0, k)$;
　　endpoints of minor axis: $(0, 2\sqrt{k}, k)$, $(0, -2\sqrt{k}, k)$
55. $z = \frac{1}{4}(x^2 + y^2)$; paraboloid

▶ **Exercise Set 14.8 (Page 712)**
1. (a) $(8, \pi/6, -4)$ (b) $(5\sqrt{2}, 3\pi/4, 6)$ (c) $(2, \pi/2, 0)$
　　(d) $(8, 5\pi/3, 6)$ (e) $(2, 7\pi/4, 1)$ (f) $(0, 0, 1)$
3. (a) $(2\sqrt{2}, \pi/3, 3\pi/4)$ (b) $(2, 7\pi/4, \pi/4)$
　　(c) $(6, \pi/2, \pi/3)$ (d) $(10, 5\pi/6, \pi/2)$
　　(e) $(8\sqrt{2}, \pi/4, \pi/6)$ (f) $(2\sqrt{2}, 5\pi/3, 3\pi/4)$
5. (a) $(2\sqrt{3}, \pi/6, \pi/6)$ (b) $(\sqrt{2}, \pi/4, 3\pi/4)$
　　(c) $(2, 3\pi/4, \pi/2)$ (d) $(4\sqrt{3}, 1, 2\pi/3)$
　　(e) $(4\sqrt{2}, 5\pi/6, \pi/4)$ (f) $(2\sqrt{2}, 0, 3\pi/4)$

7.

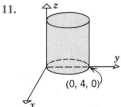

(3, 0, 0)

$x^2 + y^2 = 9$

9.

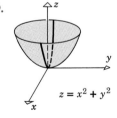

$z = x^2 + y^2$

11.

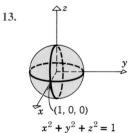

(0, 4, 0)
$x^2 + (y-2)^2 = 4$

13.

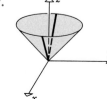

(1, 0, 0)
$x^2 + y^2 + z^2 = 1$

15.

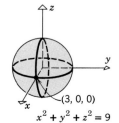

(3, 0, 0)
$x^2 + y^2 + z^2 = 9$

17.

$z = \sqrt{x^2 + y^2}$

19.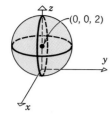

$x^2 + y^2 + (z - 2)^2 = 4$

21.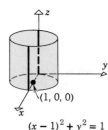

$(x - 1)^2 + y^2 = 1$

23. (a) $z = 3$ (b) $\rho = 3 \sec \phi$
25. (a) $z = 3r^2$ (b) $\rho = \frac{1}{3} \csc \phi \cot \phi$
27. (a) $r = 2$ (b) $\rho = 2 \csc \phi$
29. (a) $r^2 + z^2 = 9$ (b) $\rho = 3$
31. (a) $r(2 \cos \theta + 3 \sin \theta) + 4z = 1$
(b) $\rho(2 \sin \phi \cos \theta + 3 \sin \phi \sin \theta + 4 \cos \phi) = 1$
33. (a) $r^2 \cos^2 \theta = 16 - z^2$ (b) $\rho^2(1 - \sin^2 \phi \sin^2 \theta) = 16$
35. all points on or above the paraboloid $z = x^2 + y^2$ that are also on or below the plane $z = 4$
37. all points on or between the concentric spheres $\rho = 1$ and $\rho = 3$
39. spherical: $(4000, \pi/6, \pi/6)$
rectangular: $(1000\sqrt{3}, 1000, 2000\sqrt{3})$
41.

▶ TECHNOLOGY EXERCISES, CHAPTER 14 (Page 713)
1. (a) $40°$ (b) -0.682328
3. (b) $60°$ (c) $55°, 125°$
   (d) maximum $140°$, minimum $40°$
5. 0.800444
7. $x = 3.932993,\ y = 0.644331,\ z = -0.355669$ **9.** 50.96

▶ Exercise Set 15.1 (Page 719)
1. $(-\infty, +\infty)$; $-\mathbf{i} - 3\pi\mathbf{j}$ **3.** $[2, +\infty)$; $-\mathbf{i} - \ln 3\mathbf{j} + \mathbf{k}$
5. $\mathbf{r} = 3 \cos t\mathbf{i} + (t + \sin t)\mathbf{j}$
7. $\mathbf{r} = 2t\mathbf{i} + 2 \sin 3t\mathbf{j} + 5 \cos 3t\mathbf{k}$
9. $x = 3t^2,\ y = -2,\ z = 0$
11. $x = 2t - 1,\ y = -3\sqrt{t},\ z = \sin 3t$
13. line in the xy-plane through $(2, 0)$, parallel to $-3\mathbf{i} - 4\mathbf{j}$
15. line through $(0, -3, 1)$, parallel to $2\mathbf{i} + 3\mathbf{k}$
17. ellipse in the plane $z = -1$, center at $(0, 0, -1)$, major axis of length 6 parallel to x-axis, minor axis of length 4 parallel to y-axis
19. $-\frac{3}{2}$ **21.** $(\frac{5}{2}, 0, \frac{3}{2})$
23. 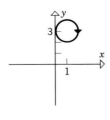   **25.**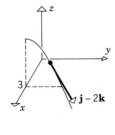

27. 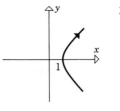   **29.**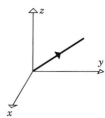

31. 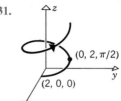   **33.**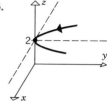

37. lies on the sphere $x^2 + y^2 + z^2 = 4$ and the plane $z = \sqrt{3}x$; center $(0, 0, 0)$, radius 2
39. $3/(2\pi)$ **41.** conical helix
43. $x = t,\ y = \frac{1}{4}t^2 - 1,\ z = \frac{1}{4}t^2 + 1$; parabola
45. $x = 3 \cos t,\ y = 3 \sin t,\ z = 9 \cos^2 t$
47. $x = 1 + \cos t,\ y = \frac{1}{2} \sin t,\ z = 2 + 2 \cos t$

▶ Exercise Set 15.2 (Page 726)
1. $9\mathbf{i} + 6\mathbf{j}$ **3.** $\mathbf{j}$ **5.** $2\mathbf{i} - 3\mathbf{j} + 4\mathbf{k}$ **7.** $\frac{1}{2}\pi\mathbf{i} + \mathbf{k}$
11. $5\mathbf{i} + (1 - 2t)\mathbf{j}$ **13.** $-\dfrac{1}{t^2}\mathbf{i} + \sec^2 t\mathbf{j} + 2e^{2t}\mathbf{k}$

15. 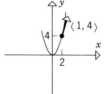   **17.**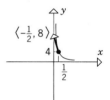

19. 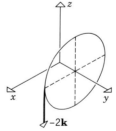   **21.**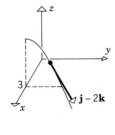

23. $x = 1 + 2t,\ y = 2 - t,\ z = 0$
25. $x = 1 - \sqrt{3}\pi t,\ y = \sqrt{3} + \pi t,\ z = 1 + 3t$
27. $\mathbf{r} = -\mathbf{i} + 2\mathbf{j} + t(2\mathbf{i} + \frac{3}{4}\mathbf{j})$
29. $\mathbf{r} = 4\mathbf{i} + \mathbf{j} + t(-4\mathbf{i} + \mathbf{j} + 4\mathbf{k})$ **31.** $2y + z = \pi/2$
33. (a) $(-2, 4, 6),\ (1, 1, -3)$ (b) $76°, 71°$ **35.** $68°$
37. $3t\mathbf{i} + 2t^2\mathbf{j} + \mathbf{C}$ **39.** $\langle 0, -\frac{2}{3} \rangle$ **41.** $\frac{52}{3}\mathbf{i} + 4\mathbf{j}$
43. $\langle (t - 1)e^t,\ t(\ln t - 1) \rangle + \mathbf{C}$ **45.** $\frac{1}{3}t^3\mathbf{i} - t^2\mathbf{j} + \ln |t|\mathbf{k} + \mathbf{C}$
47. $\frac{1}{2}(e^2 - 1)\mathbf{i} + (1 - e^{-1})\mathbf{j} + \frac{1}{2}\mathbf{k}$
49. (a) $\mathbf{F} = 3\mathbf{i} - 2\mathbf{j} - 3t^2\mathbf{k}$ (b) 30
51. $(1 + \sin t)\mathbf{i} - (\cos t)\mathbf{j}$ **53.** $(t^4 + 2)\mathbf{i} - (t^2 + 4)\mathbf{j}$
55. $2(t - 1)\mathbf{i} + \frac{1}{2} \ln \frac{1}{2}(t^2 + 1)\mathbf{j} + \frac{1}{2}(t^2 - 1)\mathbf{k}$
57. (a) $7t^6$ (b) $12t \sec t \tan t + 12 \sec t - \dfrac{\sin t}{t} - (\cos t) \ln t$

▶ **Exercise Set 15.3 (Page 734)**

3. smooth   5. not smooth   7. $4\mathbf{i} + 8(4\tau + 1)\mathbf{j}$

9. $2\tau e^{\tau^2}\mathbf{i} - 8\tau e^{-\tau^2}\mathbf{j}$ 11. $\sqrt{14}$ 13. 28   15. $e - e^{-1}$

17. $t_0\sqrt{a^2 + c^2}$ 19. $x = \frac{3}{5}s - 2, \ y = \frac{4}{5}s + 3$

21. $x = 3 + \cos s, \ y = 2 + \sin s \ (0 \le s \le 2\pi)$

23. $x = \frac{1}{3}[(3s + 1)^{2/3} - 1]^{3/2}, \ y = \frac{1}{2}[(3s + 1)^{2/3} - 1] \ (s \ge 0)$

25. $x = \left(\dfrac{s}{\sqrt{2}} + 1\right) \cos\left[\ln\left(\dfrac{s}{\sqrt{2}} + 1\right)\right],$

$y = \left(\dfrac{s}{\sqrt{2}} + 1\right) \sin\left[\ln\left(\dfrac{s}{\sqrt{2}} + 1\right)\right]$

where $0 \le s \le \sqrt{2}(e^{\pi/2} - 1)$

27. $x = (\sqrt{2s + 1} - 1) \cos(\sqrt{2s + 1} - 1),$

$y = (\sqrt{2s + 1} - 1) \sin(\sqrt{2s + 1} - 1),$

$z = \frac{2}{3}\sqrt{2}\,(\sqrt{2s + 1} - 1)^{3/2} \ (s \ge 0)$

31. (b) $x = a(\cos\sqrt{2s/a} + \sqrt{2s/a}\,\sin\sqrt{2s/a})$ $(s \ge 0)$
   $y = a(\sin\sqrt{2s/a} - \sqrt{2s/a}\,\cos\sqrt{2s/a})$

33. (a) $\frac{9}{2}$ (b) $9 - 2\sqrt{6}$ 35. (a) $\sqrt{3}(1 - e^{-2})$ (b) $4\sqrt{5}$

37. (a) $2t + \dfrac{1}{t}$ (b) $2t + \dfrac{1}{t}$ (c) $8 + \ln 3$

▶ **Exercise Set 15.4 (Page 739)**

1.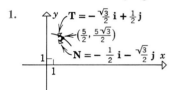

3.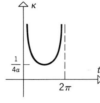

5.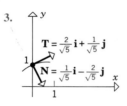

7.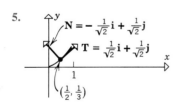

9. $\mathbf{T} = -\dfrac{4}{\sqrt{17}}\mathbf{i} + \dfrac{1}{\sqrt{17}}\mathbf{k}, \ \mathbf{N} = -\mathbf{j}$ 11. $\mathbf{T} = \mathbf{i}, \ \mathbf{N} = \mathbf{j}$

13. $\mathbf{T} = -\dfrac{3}{\sqrt{10}}\mathbf{i} + \dfrac{1}{\sqrt{10}}\mathbf{k}, \ \mathbf{N} = -\mathbf{j}$

15. $\mathbf{T} = \dfrac{1}{\sqrt{5}}\mathbf{j} + \dfrac{2}{\sqrt{5}}\mathbf{k}, \ \mathbf{N} = -\dfrac{2}{\sqrt{5}}\mathbf{j} + \dfrac{1}{\sqrt{5}}\mathbf{k}$

---

17. (a) $\mathbf{T} = \frac{3}{5}\mathbf{i} + \frac{4}{5}\mathbf{j}$
   (b)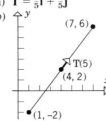

21. See Exercise 1.   23. See Exercise 3.

25. See Exercise 9.   27. See Exercise 11.

29. $(\frac{4}{5}\cos t)\mathbf{i} - (\frac{4}{5}\sin t)\mathbf{j} - \frac{3}{5}\mathbf{k}$

▶ **Exercise Set 15.5 (Page 745)**

1. $\frac{96}{125}$ 3. $\dfrac{6}{5\sqrt{10}}$ 5. $\dfrac{3}{2\sqrt{2}}$ 7. $\frac{4}{17}$ 9. 1   11. $\frac{2}{5}$

13. $\dfrac{2}{5\sqrt{5}}$ 17. 1   19. $\dfrac{1}{\sqrt{2}}$ 21. $\dfrac{4}{5\sqrt{5}}$

23. $\cos x$; maximum for $x = 0$ 25. See Exercise 1.

27. See Exercise 3.   29. See Exercise 5.

31. $\kappa(0) = a/b^2, \ \kappa(\pi/2) = b/a^2$ 33. 1   35. $\dfrac{3}{2\sqrt{2a}}$

37. $\kappa(t) = \dfrac{1}{4a}\csc\dfrac{t}{2}$

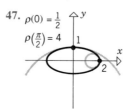

39.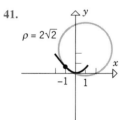

41.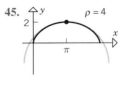

43.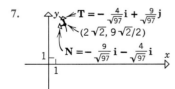

45.

47.

49. $2|p|$ **51.** $(3, 0)$, $(-3, 0)$

53. maximum 2, minimum $1/\sqrt{2}$ **55.** $\kappa = \dfrac{1}{\sqrt{1 + a^2\, r}}$

57. $\sim 2.86°/\text{cm}$ **61.** $\dfrac{1}{2r}$ **69.** $\dfrac{2}{(t^2 + 2)^2}$

71. $-\dfrac{\sqrt{2}}{(e^t + e^{-t})^2}$

▶ **Exercise Set 15.6 (Page 757)**

1.
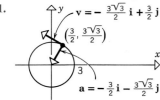

$\mathbf{v} = -\dfrac{3\sqrt{3}}{2}\mathbf{i} + \dfrac{3}{2}\mathbf{j}$

$\left(\dfrac{3}{2}, \dfrac{3\sqrt{3}}{2}\right)$

$\mathbf{a} = -\dfrac{3}{2}\mathbf{i} - \dfrac{3\sqrt{3}}{2}\mathbf{j}$

$\mathbf{v}(t) = -3\sin t\,\mathbf{i} + 3\cos t\,\mathbf{j}$
$\mathbf{a}(t) = -3\cos t\,\mathbf{i} - 3\sin t\,\mathbf{j}$
$\|\mathbf{v}(t)\| = 3$

3.
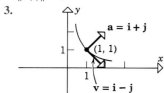

$\mathbf{a} = \mathbf{i} + \mathbf{j}$
$(1, 1)$
$\mathbf{v} = \mathbf{i} - \mathbf{j}$

$\mathbf{v}(t) = e^t\mathbf{i} - e^{-t}\mathbf{j}$
$\mathbf{a}(t) = e^t\mathbf{i} + e^{-t}\mathbf{j}$
$\|\mathbf{v}(t)\| = \sqrt{e^{2t} + e^{-2t}}$

5.
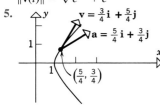

$\mathbf{v} = \dfrac{3}{4}\mathbf{i} + \dfrac{5}{4}\mathbf{j}$
$\mathbf{a} = \dfrac{5}{4}\mathbf{i} + \dfrac{3}{4}\mathbf{j}$
$\left(\dfrac{5}{4}, \dfrac{3}{4}\right)$

$\mathbf{v}(t) = \sinh t\,\mathbf{i} + \cosh t\,\mathbf{j}$
$\mathbf{a}(t) = \cosh t\,\mathbf{i} + \sinh t\,\mathbf{j}$
$\|\mathbf{v}(t)\| = \sqrt{\sinh^2 t + \cosh^2 t}$

7. $\mathbf{v} = \mathbf{i} + \mathbf{j} + \mathbf{k}$, $\|\mathbf{v}\| = \sqrt{3}$, $\mathbf{a} = \mathbf{j} + 2\mathbf{k}$

9. $\mathbf{v} = -\sqrt{2}\mathbf{i} + \sqrt{2}\mathbf{j} + \mathbf{k}$, $\|\mathbf{v}\| = \sqrt{5}$, $\mathbf{a} = -\sqrt{2}\mathbf{i} - \sqrt{2}\mathbf{j}$

11. $\mathbf{v} = e^{\pi/2}\mathbf{i} - e^{\pi/2}\mathbf{j} + \mathbf{k}$, $\|\mathbf{v}\| = \sqrt{2e^\pi + 1}$, $\mathbf{a} = -2e^{\pi/2}\mathbf{j}$

13. $3\sqrt{2}$; $\mathbf{r} = 24\mathbf{i} + 8\mathbf{j}$ **15.** maximum 6, minimum 3

17. $15°$ **19.** $\mathbf{r} = -\dfrac{19}{16}\mathbf{i} + \dfrac{3}{2}\mathbf{j} + \dfrac{3}{16}\mathbf{k}$

21. The particles move counterclockwise around the circle
$x^2 + y^2 = 4$; $\|\mathbf{r}_1'\| = 6$, $\|\mathbf{r}_2'\| = 4t$.

23. $\mathbf{r}(t) = -16t^2\mathbf{j}$, $\mathbf{v}(t) = -32t\mathbf{j}$

25. $\mathbf{r}(t) = (t + \cos t - 1)\mathbf{i} + (\sin t - t + 1)\mathbf{j}$
$\mathbf{v}(t) = (1 - \sin t)\mathbf{i} + (\cos t - 1)\mathbf{j}$

27. $\mathbf{r}(t) = \dfrac{1}{2}t^2\mathbf{i} + \mathbf{j} + \dfrac{1}{6}t^3\mathbf{k}$, $\mathbf{v}(t) = t\mathbf{i} + \dfrac{1}{2}t^2\mathbf{k}$

29. $\mathbf{r}(t) = (t - \sin t - 1)\mathbf{i} + (1 - \cos t)\mathbf{j} + e^t\mathbf{k}$
$\mathbf{v}(t) = (1 - \cos t)\mathbf{i} + \sin t\,\mathbf{j} + e^t\mathbf{k}$

33. $8\mathbf{i} + \dfrac{26}{3}\mathbf{j}$, $\dfrac{1}{3}(13\sqrt{13} - 5\sqrt{5})$ **35.** $\mathbf{0}$, 12π

37. $2\mathbf{i} - \dfrac{2}{3}\mathbf{j} + \sqrt{2}\ln 3\mathbf{k}$, $\dfrac{8}{3}$ **39.** $a_T = 0$, $a_N = 2$

41. $a_T = 0$, $a_N = \sqrt{2}$ **43.** $a_T = 2\sqrt{5}$, $a_N = 2\sqrt{5}$

45. $a_T = 22/\sqrt{14}$, $a_N = \sqrt{38/7}$ **47.** $a_T = 0$, $a_N = 3$

49. $a_T = -3$, $a_N = 2$; $\mathbf{T} = -\mathbf{j}$, $\mathbf{N} = \mathbf{i}$

51. $a_T = \dfrac{4}{3}$, $a_N = \dfrac{1}{3}\sqrt{29}$; $\mathbf{T} = \dfrac{2}{3}\mathbf{i} + \dfrac{2}{3}\mathbf{j} + \dfrac{1}{3}\mathbf{k}$,
$\mathbf{N} = \dfrac{1}{3\sqrt{29}}(\mathbf{i} - 8\mathbf{j} + 14\mathbf{k})$

53. $\dfrac{3}{2}$ **55.** $-\pi/\sqrt{2}$ **57.** $\dfrac{1}{8}$ **59.** $\sqrt{29}/27$

61. $8.41 \times 10^{10}\ \text{km/sec}^2$ **63.** $\dfrac{18}{(1 + 4x^2)^{3/2}}$

65. (a) $x = 160t$, $y = 160\sqrt{3}\,t - 16t^2$, $t \geq 0$
(b) 1200 ft   (c) $1600\sqrt{3}$ ft   (d) 320 ft/sec

67. $40\sqrt{3}$ ft   **69.** 800 ft/sec   **71.** $15°$, $75°$

73. (b) $45°$; v_0^2/g **75.** (a) 2.62 sec   (b) 181.5 ft

▶ **Exercise Set 15.7 (Page 764)**

3. (b) 10.94 km/sec   **7.** (a) 22,278 mi   (b) 6869 mi/hr

9. 248 years

11. (a) 8864.5 km   (b) 0.208   (c) 138 min

13. (a) 17,293 mi/hr
(b) $e = 0.07065$; apogee altitude = 809 mi

15. (b) $r = \dfrac{5.544 \times 10^7}{1 + 0.206\cos\theta}$

▶ **TECHNOLOGY EXERCISES, CHAPTER 15 (Page 766)**

1. 1.338583   **3.** 1.494 m   **5.** 32.63

7. (b) $80°$ (c) $a_T = 0.43\ \text{m/sec}^2$, $a_N = 2.46\ \text{m/sec}^2$

9. (b) $17.8°$ (c) 2.34 hr

▶ **Exercise Set 16.1 (Page 781)**

1. (a) 5   (b) 3   (c) 1   (d) -2 (e) $9a^3 + 1$
(f) $a^3b^2 - a^2b^3 + 1$

3. (a) $x^2 - y^2 + 3$ (b) $3x^3y^4 + 3$ **5.** $x^3e^{x^3(3y+1)}$

7. (a) $t^2 + 3t^{10}$ (b) 0   (c) 3076

9. (a) 19   (b) -9 (c) 3   (d) $a^6 + 3$ (e) $-t^8 + 3$
(f) $(a + b)(a - b)^2b^3 + 3$

11. $(y + 1)e^{x^2(y+1)z^2}$ **13.** (a) t^{14} (b) 0   (c) 16,384

15.
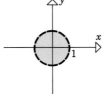

17.
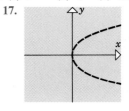

19.
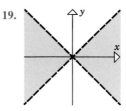

21. all points above or on the line $y = -2$

23. all points above the line $y = 2x$

25. all points not on the plane $x + y + z = 0$

27. all points within the cylinder $x^2 + y^2 = 1$

29.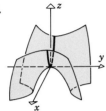

31.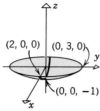

33.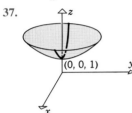

35.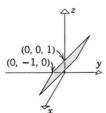

37.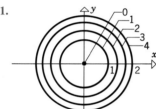

39.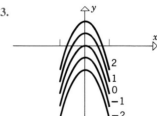

41.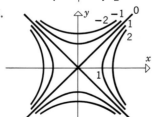

43.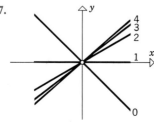

45.

47.

49.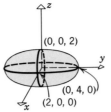

51.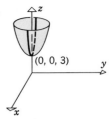

53. concentric spheres, center at $(2, 0, 0)$

55. concentric cylinders, common axis the y-axis

57. (a) $x^2 - 2x^3 + 3xy = 0$
    (b) $x^2 - 2x^3 + 3xy = 0$
    (c) $x^2 - 2x^3 + 3xy = -18$

59. (a) $x^2 + y^2 - z = 5$
    (b) $x^2 + y^2 - z = -2$
    (c) $x^2 + y^2 - z = 0$

61.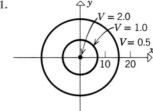

63. (a) A (b) B

65. (a) decrease   (b) increase   (c) increase   (d) decrease

67. (a) open   (b) neither   (c) closed   (d) closed

69. (a) bounded   (b) unbounded   (c) unbounded
    (d) unbounded

71. $x = u$, $y = v$, $z = u^2$, where $-\infty < u < +\infty$, $-\infty < v < +\infty$

73. $x = u$, $y = v$, $z = \sqrt{4 - u^2}$, where $-2 \le u \le 2$, $1 \le v \le 3$

75. $x = r \cos \theta$, $y = r \sin \theta$, $z = 1/(1 + r^2)$, where $r \ge 0$,
    $0 \le \theta < 2\pi$

77. $x = r \cos \theta$, $y = r \sin \theta$, $z = r^2 \sin 2\theta$, where $r \ge 0$,
    $0 \le \theta < 2\pi$

79. $x = r \cos \theta$, $y = r \sin \theta$, $z = \sqrt{9 - r^2}$, where $0 \le r \le \sqrt{5}$,
    $0 \le \theta < 2\pi$

81. $x = r \cos \theta$, $y = r \sin \theta$, $z = 3 - 2r \sin \theta$, where $0 \le r \le 2$,
    $0 \le \theta < 2\pi$

83. $x = \frac{1}{2}\rho \cos \theta$, $y = \frac{1}{2}\rho \sin \theta$, $z = \frac{1}{2}\sqrt{3}\rho$, where $\rho \ge 0$,
    $0 \le \theta < 2\pi$

85. $x = 3 \cos v$, $y = \frac{3}{2} \sin v$, $z = u$, where $0 \le u \le 5$, $0 \le v < 2\pi$

87. the plane $z = x - 2y$

89. the portion of the elliptic cylinder $x^2/9 + y^2/4 = 1$ for
    $2 \le z \le 4$

91. the portion of the elliptic cone $z = \sqrt{x^2/9 + y^2/16}$ for
    $0 \le z \le 1$

▶ **Exercise Set 16.2 (Page 789)**

1.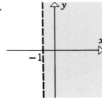

3.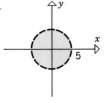

5. 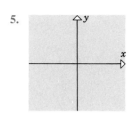   **7.**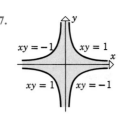

9. all of 3-space

11. all points not on the cylinder $x^2 + z^2 = 1$

13. 35   **15.** -8 **17.** 0   **19.** does not exist   **21.** 1

23. 0   **25.** does not exist   **27.** 0   **29.** 0   **31.** $\frac{8}{3}$

33. 0   **35.** does not exist   **41.** $-\pi/2$ **43.** no

45. choose $\delta = \sqrt{\epsilon}$ **47.** choose $\delta = \sqrt{\epsilon}$

▶ **Exercise Set 16.3 (Page 796)**

1. $\dfrac{\partial z}{\partial x} = 9x^2y^2,\ \dfrac{\partial z}{\partial y} = 6x^3y$

3. $\dfrac{\partial z}{\partial x} = 8xy^3 e^{x^2y^3},\ \dfrac{\partial z}{\partial y} = 12x^2y^2 e^{x^2y^3}$

5. $\dfrac{\partial z}{\partial x} = \dfrac{x^3}{y^{3/5} + x} + 3x^2 \ln\left(1 + xy^{-3/5}\right),\ \dfrac{\partial z}{\partial y} = -\dfrac{3x^4}{5\left(y^{8/5} + xy\right)}$

7. $f_x(x, y) = \frac{3}{2}x^2y(5x^2 - 7)(3x^5y - 7x^3y)^{-1/2}$,
$f_y(x, y) = \frac{1}{2}x^3(3x^2 - 7)(3x^5y - 7x^3y)^{-1/2}$

9. $f_x(x, y) = \dfrac{y^{-1/2}}{y^2 + x^2}$,

$f_y(x, y) = -\dfrac{xy^{-3/2}}{y^2 + x^2} - \dfrac{3}{2}y^{-5/2}\tan^{-1}\left(\dfrac{x}{y}\right)$

11. $f_x(x, y) = -\frac{4}{3}y^2 \sec^2 x (y^2 \tan x)^{-7/3}$,
$f_y(x, y) = -\frac{8}{3}y \tan x (y^2 \tan x)^{-7/3}$

13. (a) -6 (b) -21 **15.** (a) $1/\sqrt{17}$ (b) $8/\sqrt{17}$

17. $\dfrac{\partial z}{\partial x} = -\dfrac{x}{z},\ \dfrac{\partial z}{\partial y} = -\dfrac{y}{z}$

19. $\dfrac{\partial z}{\partial x} = -\dfrac{2x + yz^2 \cos (xyz)}{xyz \cos (xyz) + \sin (xyz)}$

$\dfrac{\partial z}{\partial y} = -\dfrac{xz^2 \cos (xyz)}{xyz \cos (xyz) + \sin (xyz)}$

21. $f_{xx} = 8,\ f_{yy} = -96xy^2 + 140y^3,\ f_{xy} = f_{yx} = -32y^3$

23. $f_{xx} = e^x \cos y,\ f_{yy} = -e^x \cos y,\ f_{xy} = f_{yx} = -e^x \sin y$

25. $f_{xx} = -\dfrac{16}{(4x - 5y)^2},\ f_{yy} = -\dfrac{25}{(4x - 5y)^2}$,

$f_{xy} = f_{yx} = \dfrac{20}{(4x - 5y)^2}$

27. (a) $30xy^4 - 4$ (b) $60x^2y^3$ (c) $60x^3y^2$

29. (a) -30 (b) -125 (c) 150

31. (a) $\dfrac{\partial^3 f}{\partial x^3}$ (b) $\dfrac{\partial^3 f}{\partial y^2 \partial x}$ (c) $\dfrac{\partial^4 f}{\partial x^2 \partial y^2}$ (d) $\dfrac{\partial^4 f}{\partial y^3 \partial x}$

33. $\dfrac{\partial w}{\partial x} = 2xy^4z^3 + y,\ \dfrac{\partial w}{\partial y} = 4x^2y^3z^3 + x,\ \dfrac{\partial w}{\partial z} = 3x^2y^4z^2 + 2z$

35. $\dfrac{\partial w}{\partial x} = \dfrac{2x}{y^2 + z^2},\ \dfrac{\partial w}{\partial y} = -\dfrac{2y(x^2 + z^2)}{(y^2 + z^2)^2},\ \dfrac{\partial w}{\partial z} = \dfrac{2z(y^2 - x^2)}{(y^2 + z^2)^2}$

37. $\dfrac{\partial w}{\partial x} = \dfrac{x}{\sqrt{x^2 + y^2 + z^2}},\ \dfrac{\partial w}{\partial y} = \dfrac{y}{\sqrt{x^2 + y^2 + z^2}}$,

$\dfrac{\partial w}{\partial z} = \dfrac{z}{\sqrt{x^2 + y^2 + z^2}}$

39. $f_x = -\dfrac{y^2z^3}{x^2y^4z^6 + 1},\ f_y = -\dfrac{2xyz^3}{x^2y^4z^6 + 1}$,

$f_z = -\dfrac{3xy^2z^2}{x^2y^4z^6 + 1}$

41. $f_x = 4xyz \cosh \sqrt{z} \sinh (x^2yz) \cosh (x^2yz)$,
$f_y = 2x^2z \cosh \sqrt{z} \sinh (x^2yz) \cosh (x^2yz)$,
$f_z = 2x^2y \cosh \sqrt{z} \sinh (x^2yz) \cosh (x^2yz)$
$\qquad\qquad + \dfrac{\sinh \sqrt{z} \sinh^2 (x^2yz)}{2\sqrt{z}}$

43. (a) -80 (b) 40   (c) -60

45. (a) $2/\sqrt{7}$ (b) $4/\sqrt{7}$ (c) $1/\sqrt{7}$

47. $\dfrac{\partial w}{\partial x} = -\dfrac{x}{w},\ \dfrac{\partial w}{\partial y} = -\dfrac{y}{w},\ \dfrac{\partial w}{\partial z} = -\dfrac{z}{w}$

49. $\dfrac{\partial w}{\partial x} = -\dfrac{yzw \cos (xyz)}{2w + \sin (xyz)},\ \dfrac{\partial w}{\partial y} = -\dfrac{xzw \cos (xyz)}{2w + \sin (xyz)}$,

$\dfrac{\partial w}{\partial z} = -\dfrac{xyw \cos (xyz)}{2w + \sin (xyz)}$

51. (a) $15x^2y^4z^7 + 2y$ (b) $35x^3y^4z^6 + 3y^2$ (c) $21x^2y^5z^6$
     (d) $42x^3y^5z^5$ (e) $140x^3y^3z^6 + 6y$ (f) $30xy^4z^7$
     (g) $105x^2y^4z^6$ (h) $210xy^4z^6$

55. 6   **57.** (a) 8   (b) -2

59. (a) $\dfrac{\partial V}{\partial r} = 2\pi rh$ (b) $\dfrac{\partial V}{\partial h} = \pi r^2$ (c) 48π (d) 64π

61. (a) $\frac{1}{5}$ (b) $-\frac{25}{8}$

63. $\dfrac{\partial z}{\partial x} = -1 - \dfrac{\cos (x - y)}{\cos (x + z)},\ \dfrac{\partial z}{\partial y} = \dfrac{\cos (x - y)}{\cos (x + z)}$,

$\dfrac{\partial^2 z}{\partial x\, \partial y} = -\dfrac{\cos^2 (x + z) \sin (x - y) + \cos^2 (x - y) \sin (x + z)}{\cos^3 (x + z)}$

65. (a) 4 degrees per centimeter   (b) 8 degrees per centimeter

69. (a) Both are positive; $\partial T/\partial x$.   (b) All are negative.

▶ **Exercise Set 16.4 (Page 807)**

1. 0.872   **3.** $\frac{7}{2}$ **5.** $\epsilon_1 = 0,\ \epsilon_2 = \Delta x$

7. $\epsilon_1 = y\Delta x + \Delta x\Delta y,\ \epsilon_2 = 2x\Delta x$

13. $f_{xy} = f_{yx} = 12x^2 + 6x$

15. $f_{xy} = f_{yx} = -6xy^2 \sin (x^2 + y^3)$

17. (a) 4   (b) 5   **19.** $42t^{13}$ **21.** $\dfrac{3 \sin (1/t)}{t^2}$

23. $-\dfrac{10}{3}t^{7/3}e^{1 - t^{10/3}}$

25. $\dfrac{\partial z}{\partial u} = 24u^2v^2 - 16uv^3 - 2v + 3$,

$\dfrac{\partial z}{\partial v} = 16u^3v - 24u^2v^2 - 2u - 3$

27. $\dfrac{\partial z}{\partial u} = -\dfrac{2 \sin u}{3 \sin v},\ \dfrac{\partial z}{\partial v} = -\dfrac{2 \cos u \cos v}{3 \sin^2 v}$

29. $\dfrac{\partial z}{\partial u} = e^u,\ \dfrac{\partial z}{\partial v} = 0$ **31.** $\dfrac{\partial z}{\partial u} = \dfrac{2e^{2u}}{1 + e^{4u}},\ \dfrac{\partial z}{\partial v} = 0$

33. $\dfrac{\partial T}{\partial r} = 3r^2 \sin \theta \cos^2 \theta - 4r^3 \sin^3 \theta \cos \theta$,

$\dfrac{\partial T}{\partial \theta} = -2r^3 \sin^2 \theta \cos \theta + r^4 \sin^4 \theta$
$\qquad\qquad\qquad + r^3 \cos^3 \theta - 3r^4 \cos^2 \theta \sin^2 \theta$

35. $\dfrac{\partial t}{\partial x} = \dfrac{x^2 + y^2}{4x^2y^3},\ \dfrac{\partial t}{\partial y} = \dfrac{y^2 - 3x^2}{4xy^4}$ **37.** $-\pi$

39. $\left.\dfrac{\partial z}{\partial r}\right|_{r=2,\,\theta=\pi/6} = \sqrt{3}e^{\sqrt{3}}, \left.\dfrac{\partial z}{\partial \theta}\right|_{r=2,\,\theta=\pi/6} = (2-4\sqrt{3})e^{\sqrt{3}}$

41. $\dfrac{x^2-y^2}{2xy-y^2}$ **43.** $\dfrac{2\sqrt{xy}-y}{x-6\sqrt{xy}}$ **45.** -39 mi/hr

47. $-\dfrac{7}{36}\sqrt{3}$ rad/sec **49.** $\dfrac{4}{3} - 16\ln 3$ °C/sec

55. (b) $z = C_1 \ln r + C_2$

65. (a) $n=2$ (b) $n=1$ (c) $n=3$ (d) $n=-4$

67. (a) $\dfrac{\partial w}{\partial x} = \dfrac{\partial f}{\partial x} + \dfrac{\partial f}{\partial y}\dfrac{\partial y}{\partial x}$ (b) $\dfrac{\partial w}{\partial z} = \dfrac{\partial f}{\partial y}\dfrac{\partial y}{\partial z}$

▶ **Exercise Set 16.5 (Page 814)**

1. $48x - 14y - z = 64$; $x = 1 + 48t$, $y = -2 - 14t$, $z = 12 - t$

3. $x - y - z = 0$; $x = 1 + t$, $y = -t$, $z = 1 - t$

5. $3y - z = -1$; $x = \pi/6$, $y = 3t$, $z = 1 - t$

7. $3x - 4z = -25$; $x = -3 + \frac{3}{4}t$, $y = 0$, $z = 4 - t$

9. $7\,dx - 2\,dy$ **11.** $\dfrac{y}{1+x^2y^2}\,dx + \dfrac{x}{1+x^2y^2}\,dy$

13. 0.10 **15.** 0.03

17. (a) all points on the x-axis and y-axis (b) $(0, -2, -4)$

19. $(\frac{1}{2}, -2, -\frac{3}{4})$ **23.** 0.088 cm **25.** 8% **27.** $r\%$

29. 2% **31.** 0.004 rad **33.** 0.3%

35. (a) $(r+s)\%$ (b) $(r+s)\%$ (c) $(2r+3s)\%$
(d) $(3r + \frac{1}{2}s)\%$

37. (a) $(-2, 1, 5)$, $(0, 3, 9)$
(b) $4/(3\sqrt{14})$ at $(-2, 1, 5)$, $4/\sqrt{222}$ at $(0, 3, 9)$

▶ **Exercise Set 16.6 (Page 820)**

1. $4\mathbf{i} - 8\mathbf{j}$ **3.** $\dfrac{x}{x^2+y^2}\mathbf{i} + \dfrac{y}{x^2+y^2}\mathbf{j}$ **5.** $-36\mathbf{i} - 12\mathbf{j}$

7. $4\mathbf{i} + 4\mathbf{j}$ **9.** $6\sqrt{2}$ **11.** $-3/\sqrt{10}$ **13.** 0

15. $-8\sqrt{2}$ **17.** $\dfrac{1}{2\sqrt{2}}$ **19.** $\dfrac{1}{2} + \dfrac{\sqrt{3}}{8}$ **21.** $2\sqrt{2}$

23. **25.**

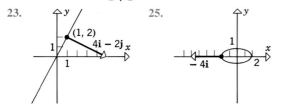

27. $\dfrac{3}{\sqrt{13}}\mathbf{i} - \dfrac{2}{\sqrt{13}}\mathbf{j}$, $4\sqrt{13}$ **29.** $\frac{4}{5}\mathbf{i} - \frac{3}{5}\mathbf{j}$, 1

31. $-\dfrac{1}{\sqrt{10}}\mathbf{i} - \dfrac{3}{\sqrt{10}}\mathbf{j}$, $-2\sqrt{10}$

33. $\dfrac{3}{\sqrt{10}}\mathbf{i} - \dfrac{1}{\sqrt{10}}\mathbf{j}$, $-\sqrt{5}$ **35.** $1/\sqrt{5}$ **37.** $-\frac{3}{2}e$

39. $-\dfrac{4}{\sqrt{17}}\mathbf{i} + \dfrac{1}{\sqrt{17}}\mathbf{j}$, $\dfrac{4}{\sqrt{17}}\mathbf{i} - \dfrac{1}{\sqrt{17}}\mathbf{j}$

41. (a) 5 (b) 10 (c) $-5\sqrt{5}$ **43.** $8/\sqrt{29}$

45. all points on the ellipse $9x^2 + y^2 = 9$

47. $36/\sqrt{17}$ **49.** (a) $2e^{-\pi/2}\mathbf{i}$

▶ **Exercise Set 16.7 (Page 827)**

1. $165t^{32}$ **3.** $-2t\cos(t^2)$ **5.** 3264 **7.** -320
9. $-\frac{314}{741}$ **11.** $72/\sqrt{14}$ **13.** $-\frac{8}{63}$

15. $\dfrac{1}{\sqrt{2}}\mathbf{i} - \dfrac{1}{\sqrt{2}}\mathbf{j}$, $3\sqrt{2}$ **17.** $-\dfrac{1}{\sqrt{2}}\mathbf{i} + \dfrac{1}{\sqrt{2}}\mathbf{j}$, $\dfrac{\sqrt{2}}{2}$

19. $\dfrac{1}{\sqrt{266}}\mathbf{i} - \dfrac{11}{\sqrt{266}}\mathbf{j} + \dfrac{12}{\sqrt{266}}\mathbf{k}$, $-\sqrt{266}$

21. $3/\sqrt{11}$ **23.** $-\frac{10}{3}\mathbf{i} + \frac{5}{3}\mathbf{j} + \frac{10}{3}\mathbf{k}$

25. $3x - 2y + 6z = -49$;
$x = -3 + 3t$, $y = 2 - 2t$, $z = -6 + 6t$

27. $x - y - 15z = -17$; $x = 3 + t$, $y = 5 - t$, $z = 1 - 15t$

31. $(1, \frac{2}{3}, \frac{2}{3})$, $(-1, -\frac{2}{3}, -\frac{2}{3})$ **33.** $8\,dx - 3\,dy + 4\,dz$

35. $\dfrac{yz}{1+(xyz)^2}\,dx + \dfrac{xz}{1+(xyz)^2}\,dy + \dfrac{xy}{1+(xyz)^2}\,dz$

37. 0.96 **39.** 2.35 cm^3 **41.** 39 ft^2 **43.** 14%

▶ **Exercise Set 16.8 (Page 831)**

1. $\dfrac{\partial f}{\partial v} = 8vw^3x^4y^5$, $\dfrac{\partial f}{\partial w} = 12v^2w^2x^4y^5$,

$\dfrac{\partial f}{\partial x} = 16v^2w^3x^3y^5$, $\dfrac{\partial f}{\partial y} = 20v^2w^3x^4y^4$

3. $\dfrac{\partial f}{\partial v_1} = \dfrac{2v_1}{v_3^2 + v_4^2}$, $\dfrac{\partial f}{\partial v_2} = -\dfrac{2v_2}{v_3^2 + v_4^2}$,

$\dfrac{\partial f}{\partial v_3} = -\dfrac{2v_3(v_1^2 - v_2^2)}{(v_3^2 + v_4^2)^2}$, $\dfrac{\partial f}{\partial v_4} = -\dfrac{2v_4(v_1^2 - v_2^2)}{(v_3^2 + v_4^2)^2}$

5. $f_v(1, -2, 4, 8) = 128$, $f_w(1, -2, 4, 8) = -512$,
$f_x(1, -2, 4, 8) = 32$, $f_y(1, -2, 4, 8) = \frac{64}{3}$

7. $210t^{29}$ **9.** $\dfrac{\partial z}{\partial r} = \dfrac{2r\cos^2\theta}{r^2\cos^2\theta + 1}$, $\dfrac{\partial z}{\partial \theta} = -\dfrac{2r^2\sin\theta\cos\theta}{r^2\cos^2\theta + 1}$

11. $\dfrac{\partial w}{\partial \rho} = 2\rho(4\sin^2\phi + \cos^2\phi)$,

$\dfrac{\partial w}{\partial \phi} = 6\rho^2\sin\phi\cos\phi$, $\dfrac{\partial w}{\partial \theta} = 0$

13. $\dfrac{-4\cos 2y\sin 2y + 2y + 1}{2\sqrt{\cos^2 2y + y^2 + y}}$

15. $\sqrt{3} + \pi/6$ cm/sec, increasing

17. $\dfrac{\partial z}{\partial x} = -\dfrac{2x + 4z}{4x + 2z - 3y}$, $\dfrac{\partial z}{\partial y} = \dfrac{3z}{4x + 2z - 3y}$

27. $\dfrac{\partial z}{\partial x} = \dfrac{ye^x}{15\cos 3z + 3}$, $\dfrac{\partial z}{\partial y} = \dfrac{e^x}{15\cos 3z + 3}$

33. (a) $\dfrac{dw}{dt} = \dfrac{\partial w}{\partial x_1}\dfrac{dx_1}{dt} + \dfrac{\partial w}{\partial x_2}\dfrac{dx_2}{dt} + \dfrac{\partial w}{\partial x_3}\dfrac{dx_3}{dt} + \dfrac{\partial w}{\partial x_4}\dfrac{dx_4}{dt}$

(b) $\dfrac{\partial w}{\partial v_1} = \dfrac{\partial w}{\partial x_1}\dfrac{\partial x_1}{\partial v_1} + \dfrac{\partial w}{\partial x_2}\dfrac{\partial x_2}{\partial v_1} + \dfrac{\partial w}{\partial x_3}\dfrac{\partial x_3}{\partial v_1} + \dfrac{\partial w}{\partial x_4}\dfrac{\partial x_4}{\partial v_1}$,

$\dfrac{\partial w}{\partial v_2} = \dfrac{\partial w}{\partial x_1}\dfrac{\partial x_1}{\partial v_2} + \dfrac{\partial w}{\partial x_2}\dfrac{\partial x_2}{\partial v_2} + \dfrac{\partial w}{\partial x_3}\dfrac{\partial x_3}{\partial v_2} + \dfrac{\partial w}{\partial x_4}\dfrac{\partial x_4}{\partial v_2}$,

$\dfrac{\partial w}{\partial v_3} = \dfrac{\partial w}{\partial x_1}\dfrac{\partial x_1}{\partial v_3} + \dfrac{\partial w}{\partial x_2}\dfrac{\partial x_2}{\partial v_3} + \dfrac{\partial w}{\partial x_3}\dfrac{\partial x_3}{\partial v_3} + \dfrac{\partial w}{\partial x_4}\dfrac{\partial x_4}{\partial v_3}$

▶ **Exercise Set 16.9 (Page 841)**

1. $(0, 0)$ relative min **3.** $(1, -2)$ saddle

5. $(2, -1)$ relative min

7. $(2, 1)$, $(-2, 1)$ saddle; $(0, 0)$ relative min

9. $(-1, -1)$, $(1, 1)$ relative min **11.** $(0, 0)$ saddle

13. none

15. $(0, 0)$, $(4, 0)$, $(0, -2)$ saddle; $(\frac{4}{3}, -\frac{2}{3})$ relative min

17. $(-1, 0)$ relative max **19.** $(\pi/2, \pi/2)$ relative max

21. (b) $(0,0)$ relative min   23. critical point $(1,0)$
25. absolute maximum 0 at $(0,0)$;
    absolute minimum -12 at $(0,4)$
27. absolute maximum 3 at $(0,1)$, $(2,1)$;
    absolute minimum -1 at $(1,0)$, $(1,2)$
29. absolute maximum $\frac{33}{4}$ at $(-\frac{1}{2}, \pm\frac{1}{2}\sqrt{15})$;
    absolute minimum $-\frac{1}{4}$ at $(\frac{1}{2},0)$
33. 9, 9, 9   35. $(\sqrt{5},0,0)$, $(-\sqrt{5},0,0)$
37. $\frac{1}{27}$ 39. length and width 2 ft, height 4 ft
41. $\frac{3}{2}\sqrt{6}$ 43. length and width $\sqrt[3]{2V}$, height $\frac{1}{2}\sqrt[3]{2V}$
47. $y = 0.5x + 0.8$
49. For $0 \le x \le 1$ let $f(x,y) = \begin{cases} y, & 0 < y < 1 \\ \frac{1}{2}, & y=0 \text{ or } y=1 \end{cases}$.
    For $-\infty < x < +\infty$ and $y > 0$ let $f(x,y) = y$.

▶ **Exercise Set 16.10 (Page 849)**
1. max $\sqrt{2}$ at $(\sqrt{2},1)$ and $(-\sqrt{2},-1)$,
   min $-\sqrt{2}$ at $(-\sqrt{2},1)$ and $(\sqrt{2},-1)$
3. max $\sqrt{2}$ at $(1/\sqrt{2},0)$, min $-\sqrt{2}$ at $(-1/\sqrt{2},0)$
5. max 6 at $(\frac{4}{3},\frac{2}{3},-\frac{4}{3})$, min -6 at $(-\frac{4}{3},-\frac{2}{3},\frac{4}{3})$
7. max $\frac{1}{3\sqrt{3}}$ at $\left(\frac{1}{\sqrt{3}},\frac{1}{\sqrt{3}},\frac{1}{\sqrt{3}}\right)$, $\left(\frac{1}{\sqrt{3}},-\frac{1}{\sqrt{3}},-\frac{1}{\sqrt{3}}\right)$,
   $\left(-\frac{1}{\sqrt{3}},\frac{1}{\sqrt{3}},-\frac{1}{\sqrt{3}}\right)$, $\left(-\frac{1}{\sqrt{3}},-\frac{1}{\sqrt{3}},\frac{1}{\sqrt{3}}\right)$;
   min $-\frac{1}{3\sqrt{3}}$ at $\left(\frac{1}{\sqrt{3}},\frac{1}{\sqrt{3}},-\frac{1}{\sqrt{3}}\right)$, $\left(\frac{1}{\sqrt{3}},-\frac{1}{\sqrt{3}},\frac{1}{\sqrt{3}}\right)$,
   $\left(-\frac{1}{\sqrt{3}},\frac{1}{\sqrt{3}},\frac{1}{\sqrt{3}}\right)$, $\left(-\frac{1}{\sqrt{3}},-\frac{1}{\sqrt{3}},-\frac{1}{\sqrt{3}}\right)$
9. $(\frac{2}{5},\frac{19}{5})$ 11. $(1,-1,1)$ 13. $\frac{1}{2}$
15. $(3,6)$ closest, $(-3,-6)$ farthest   17. 9, 9, 9
19. $(\sqrt{5},0,0)$, $(-\sqrt{5},0,0)$
21. length and width 2 ft, height 4 ft

▶ **TECHNOLOGY EXERCISES, CHAPTER 16 (Page 849)**
3. $67.7°$ 5. $-0.805428\mathbf{i} - 0.592292\mathbf{j} - 0.021803\mathbf{k}$; $7.37°$/m
7. (a) $\dfrac{\sin(ex) - \sin x}{x}$ (b) 0.909026
9. length and width 1.37 ft, height 0.53 ft; $20.12

▶ **Exercise Set 17.1 (Page 857)**
1. 7   3. 2   5. 2   7. $\frac{2}{15}(31 - 9\sqrt{3})$ 9. 3
11. $1 - \ln 2$ 13. $\frac{1}{2}(1 - \ln 2)$ 15. 0   17. $\frac{1}{3}$ 19. 1
21. 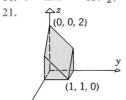   23.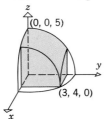

25.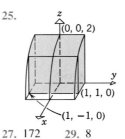
27. 172   29. 8

▶ **Exercise Set 17.2 (Page 864)**
1. $\frac{1}{40}$ 3. 9   5. $\pi/2$ 7. 1   9. $\frac{2}{3}a^3$ 11. $\frac{1}{12}$
13. 32   15. $-2/\pi$ 17. 576   19. $\frac{1}{2}(\sqrt{17} - 1)$
21. 0   23. $\pi/4 - \frac{1}{2}\ln 2$ 25. $\frac{50}{3}$ 27. $-\frac{1}{2}$ 29. $\frac{25}{2}$
31. $\sqrt{2} - 1$ 33. 32   35. $\frac{125}{12}$ 37. $\frac{56}{15}$ 39. 170
41. $\frac{27}{2}\pi$ 43. $\frac{20}{3}$ 45. $\pi/2$ 47. $\frac{2000}{3}$
49. $\int_0^{\sqrt{2}} \int_{y^2}^2 f(x,y)\,dx\,dy$ 51. $\int_1^{e^2} \int_{\ln x}^2 f(x,y)\,dy\,dx$
53. $\int_{-1}^1 \int_{-2\sqrt{1-y^2}}^{2\sqrt{1-y^2}} f(x,y)\,dx\,dy$
55. $\int_0^{\pi/2} \int_0^{\sin x} f(x,y)\,dy\,dx$ 57. $\frac{1}{8}(1 - e^{-16})$
59. $\frac{1}{3}(e^8 - 1)$ 61. $\frac{1}{8}\pi$ 63. $\frac{1}{3}(1 - \cos 8)$
65. (a) 0   (b) $-\frac{603}{40}$ 67. π 69. 0

▶ **Exercise Set 17.3 (Page 871)**
1. $\frac{1}{6}$ 3. 0   5. 0   7. $3\pi/2$ 9. $\pi/16$
11. $4\pi/3 + 2\sqrt{3}$ 13. $\frac{4}{3}(27 - 16\sqrt{2})\pi$ 15. $\frac{32}{9}$
17. $5\pi/32$ 19. $(1 - e^{-1})\pi$ 21. $\frac{\pi}{8}\ln 5$ 23. $\pi/8$
25. $\frac{16}{9}$ 27. $\frac{\pi}{2}\left(1 - \frac{1}{\sqrt{1+a^2}}\right)$ 29. $\frac{1}{4}\pi(\sqrt{5} - 1)$
31. $\frac{1}{9}(3\pi - 4)a^2c$ 33. $\frac{4\pi}{3} + 2\sqrt{3} - 2$
35. (b) $\pi/4$ (c) $\sqrt{\pi}/2$

▶ **Exercise Set 17.4 (Page 877)**
1. 6π 3. $\sqrt{5}/6$ 5. $\frac{\pi}{6}(5\sqrt{5} - 1)$
7. $\frac{\pi}{18}(10\sqrt{10} - 1)$ 9. 8π 11. $2(\pi - 2)a^2$ 13. 128
17. $\frac{1}{6}(17^{3/2} - 5^{3/2})\pi$ 19. $4\pi a^2$ 21. $5\sqrt{5} - 1$

▶ **Exercise Set 17.5 (Page 883)**
1. 8   3. 7   5. $\frac{81}{5}$ 7. $\frac{128}{15}$ 9. $\frac{\pi}{2}(\pi - 3)$ 11. $\frac{1}{6}$
13. 4   15. $\frac{256}{15}$ 17. 9π 19. 2π
21. $\frac{\pi}{6}(8\sqrt{2} - 7)a^3$
23. (a) 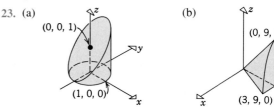   (b)

25. (a) $\displaystyle\int_0^a \int_0^{b(1-x/a)} \int_0^{c(1-x/a-y/b)} dz\,dy\,dx,$

$\displaystyle\int_0^b \int_0^{a(1-y/b)} \int_0^{c(1-x/a-y/b)} dz\,dx\,dy,$

$\displaystyle\int_0^c \int_0^{a(1-z/c)} \int_0^{b(1-x/a-z/c)} dy\,dx\,dz,$

$\displaystyle\int_0^a \int_0^{c(1-x/a)} \int_0^{b(1-x/a-z/c)} dy\,dz\,dx,$

$\displaystyle\int_0^c \int_0^{b(1-z/c)} \int_0^{a(1-y/b-z/c)} dx\,dy\,dz,$

$\displaystyle\int_0^b \int_0^{c(1-y/b)} \int_0^{a(1-y/b-z/c)} dx\,dz\,dy$

27. $\frac{1}{6}a^3$ **29.** (a) 0 (b) $\frac{1}{2}(e^2-1)$ **31.** (a) 10 (b) 0
33. 0

▶ **Exercise Set 17.6 (Page 893)**

1. 10 feet to the right of m_1 **3.** $\frac{13}{20}$, $\left(\frac{190}{273}, \frac{6}{13}\right)$

5. $\frac{a^4}{8}$, $\left(\frac{8a}{15}, \frac{8a}{15}\right)$ **7.** $\left(\frac{2}{3}, \frac{1}{3}\right)$ **9.** $\left(-\frac{1}{2}, \frac{2}{5}\right)$

11. $\left(0, \frac{4(b^3-a^3)}{3\pi(b^2-a^2)}\right)$ **13.** $\left(0, \frac{8}{3}\right)$ **15.** $\frac{a^4}{2}$, $\left(\frac{a}{3}, \frac{a}{2}, \frac{a}{2}\right)$

17. $\frac{1}{6}$, $\left(0, \frac{16}{35}, \frac{1}{2}\right)$ **19.** $\left(\frac{1}{4}, \frac{1}{4}, \frac{1}{4}\right)$ **21.** $\left(\frac{1}{2}, 0, \frac{3}{5}\right)$
23. $\left(\frac{3a}{8}, \frac{3a}{8}, \frac{3a}{8}\right)$ **25.** $\frac{2}{3}\pi k a^3$ **29.** $\left(\frac{128}{105\pi}, \frac{128}{105\pi}\right)$

33. $2\pi^2 kab$ **35.** $\left(\frac{a}{3}, \frac{b}{3}\right)$ **37.** $\frac{1}{12}M(a^2+b^2)$

39. $\frac{1}{4}MR^2$ **41.** $\frac{3}{2}MR^2$ **43.** $\frac{2}{3}Ma^2$

▶ **Exercise Set 17.7 (Page 903)**

1. $\pi/4$ **3.** $\pi/16$ **5.** $81\pi/2$ **7.** $\frac{8}{3}(10\sqrt{5}-19)\pi$
9. $\frac{32}{9}(3\pi-4)$ **11.** $\frac{2}{3}(\sqrt{3}-1)\pi$ **13.** $9\sqrt{2}\pi$
15. $\frac{4}{3}\pi a^3$ **17.** $\frac{11}{6}\pi$ **19.** $\frac{1}{4}(2-\sqrt{2})\pi$ **21.** $\pi k a^4$

23. $\left(0, 0, \dfrac{7}{16\sqrt{2}-14}\right)$ **25.** $\left(\frac{3a}{8}, \frac{3a}{8}, \frac{3a}{8}\right)$

27. $\left(0, \dfrac{195}{152}, 0\right)$ **29.** $\frac{1}{3}(1-e^{-1})\pi$ **31.** 81π

33. $\left(0, 0, \dfrac{2a}{5}\right)$ **35.** $(0, 0, \frac{11}{30})$ **37.** $\frac{63}{8}\pi$

41. $M(\frac{1}{4}R^2 + \frac{1}{3}h^2)$ **43.** $\frac{2}{5}MR^2$
45. (a) $2\pi k\delta(\sqrt{R^2+a^2}-\sqrt{R^2+(a+h)^2}+h)$
   (b) Other components are zero because of symmetry.

47. $2\pi k\delta h\left(1-\dfrac{h}{\sqrt{R^2+h^2}}\right)$

▶ **Exercise Set 17.8 (Page 915)**

1. -17 **3.** $\cos(u-v)$ **5.** $\frac{1}{9}$ **7.** $\dfrac{1}{4\sqrt{v^2-u^2}}$

9. 5 **11.** $1/v$ **13.** $\frac{3}{2}\ln 3$ **15.** $1-\frac{1}{2}\sin 2$
17. 96π **23.** 21 **25.** $-\frac{2}{3}\ln 2$ **27.** $\frac{1}{4}\ln\frac{5}{2}$
29. $\frac{1}{2}[\ln(\sqrt{2}+1)-\pi/4]$ **31.** $\frac{35}{256}$ **33.** $2\ln 3$
35. $\frac{4}{15}\pi a^3 bc$ **37.** $\frac{105}{32}$

▶ **TECHNOLOGY EXERCISES, CHAPTER 17 (Page 917)**

1. 0.676089 **3.** 3.960251
5. (a) 2.859712 (b) 2.153777 (c) 38.699285

7. 1.980291

▶ **Exercise Set 18.1 (Page 926)**

1.

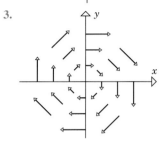

3.

5. $\operatorname{div}\mathbf{F}=2x+y$, $\operatorname{curl}\mathbf{F}=z\mathbf{i}$
7. $\operatorname{div}\mathbf{F}=0$, $\operatorname{curl}\mathbf{F}=(40x^2z^4-12xy^3)\mathbf{i}$
$+ (14y^3z+3y^4)\mathbf{j}-(16xz^5+21y^2z^2)\mathbf{k}$
9. $\operatorname{div}\mathbf{F}=2/\sqrt{x^2+y^2+z^2}$, $\operatorname{curl}\mathbf{F}=\mathbf{0}$ **21.** $f'(x)$
31. (b) $x^2+y^2=K$
33. $y=\ln x + K$

▶ **Exercise Set 18.2 (Page 936)**

1. $(17\sqrt{17}-1)/4$ **3.** 2 **5.** $\frac{13}{20}$
7. (a) $\frac{4}{3}$ (b) 0 (c) $\frac{4}{3}$ **9.** $-\frac{8}{3}$ **11.** $1-\pi$ **13.** 3
15. $-1-\pi/4$ **17.** 0 **19.** $1-e^{-1}$ **21.** $1-e^3$
23. $\frac{23}{6}$ **25.** $6k\sqrt{3}$ **27.** $5k\tan^{-1}3$ **29.** $\frac{3}{5}$
31. (a) $\frac{5}{4}-\pi/8+\frac{1}{2}\tan^{-1}2$ (b) $\frac{1}{3}\tan^{-1}2-\frac{2}{3}\tan^{-1}\frac{1}{2}$ (c) $\frac{3}{4}$
33. $-\frac{37}{2}$ **35.** -12

▶ **Exercise Set 18.3 (Page 945)**

1. $\frac{1}{2}x^2+\frac{1}{2}y^2+K$ **3.** not conservative
5. $x\cos y+y\sin x+K$ **7.** 13 **9.** -6 **11.** $9e^2$
13. 32 **15.** $-\frac{1}{2}$ **17.** $\pi/4$ **19.** $\ln 2-1$
21. (a) $-1/\sqrt{x^2+y^2}$ (b) $\frac{4}{5}$ (c) 0 **23.** Ce^x
29. xz^2-e^{-y} **31.** $xyz+z^3$ **33.** not conservative
35. (a) $1/\sqrt{6}-1/\sqrt{14}$ (b) $1/\sqrt{6}-1/\sqrt{14}$ (c) 0

▶ **Exercise Set 18.4 (Page 951)**

1. 0 **3.** 0 **5.** 0 **7.** 8π **9.** -4 **11.** -1
13. 0 **15.** πab **17.** $\frac{1}{2}ab$ **19.** $\frac{1}{2}abt_0$ **21.** $\frac{250}{3}$
23. (b) $\left(0, \dfrac{4a}{3\pi}\right)$

25. (c) $A = \frac{1}{2}[(x_1 y_2 - x_2 y_1) + (x_2 y_3 - x_3 y_2)$
$+ \cdots + (x_n y_1 - x_1 y_n)]$

(d) 8

27. 69

▶ **Exercise Set 18.5 (Page 957)**

1. $15\pi/\sqrt{2}$ 3. $\pi/4$ 5. $3/\sqrt{2}$ 7. 9 9. $4\pi/3$

11. (a) $\dfrac{\sqrt{29}}{16} \displaystyle\int_0^6 \int_0^{(12-2x)/3} xy(12 - 2x - 3y)\, dy\, dx$

(b) $\dfrac{\sqrt{29}}{4} \displaystyle\int_0^3 \int_0^{(12-4z)/3} yz(12 - 3y - 4z)\, dy\, dz$

(c) $\dfrac{2\sqrt{29}}{9} \displaystyle\int_0^3 \int_0^{6-2z} xz(6 - x - 2z)\, dx\, dz$

13. $\displaystyle\int_0^4 \int_1^2 y^3 z\sqrt{1 + 4y^2}\, dy\, dz$, $\dfrac{1}{2}\displaystyle\int_0^4 \int_1^4 xz\sqrt{1 + 4x}\, dx\, dz$

15. $\dfrac{\pi\delta_0}{6}(5\sqrt{5} - 1)$ 17. $\frac{4}{3}\pi\delta_0$ 19. $\frac{1}{4}[37\sqrt{37} - 1]$

23. $(0, 0, \frac{149}{65})$ 25. $93/\sqrt{10}$ 27. $\pi/4$

▶ **Exercise Set 18.6 (Page 965)**

1. $\dfrac{2}{\sqrt{29}}\mathbf{i} + \dfrac{3}{\sqrt{29}}\mathbf{j} + \dfrac{4}{\sqrt{29}}\mathbf{k}$

3. (a) $-\dfrac{2}{\sqrt{21}}\mathbf{i} - \dfrac{4}{\sqrt{21}}\mathbf{j} + \dfrac{1}{\sqrt{21}}\mathbf{k}$

(b) $\dfrac{3}{5\sqrt{2}}\mathbf{i} - \dfrac{4}{5\sqrt{2}}\mathbf{j} + \dfrac{1}{\sqrt{2}}\mathbf{k}$ (c) $\frac{3}{5}\mathbf{i} - \frac{4}{5}\mathbf{k}$

5. 2π 7. $\pi/2$ 9. 54π 11. $\frac{14}{3}\pi$ 13. 0

15. $4\pi a^3$

19. (a) $\displaystyle\iint_R \mathbf{F} \cdot \left(\mathbf{i} - \frac{\partial x}{\partial y}\mathbf{j} - \frac{\partial x}{\partial z}\mathbf{k}\right) dA$

Front, R = projection on yz-plane

$\displaystyle\iint_R \mathbf{F} \cdot \left(-\mathbf{i} + \frac{\partial x}{\partial y}\mathbf{j} + \frac{\partial x}{\partial z}\mathbf{k}\right) dA$

Back, R = projection on yz-plane

(b) $\displaystyle\iint_R \mathbf{F} \cdot \left(-\frac{\partial y}{\partial x}\mathbf{i} + \mathbf{j} - \frac{\partial y}{\partial z}\mathbf{k}\right) dA$

Right, R = projection on xz-plane

$\displaystyle\iint_R \mathbf{F} \cdot \left(\frac{\partial y}{\partial x}\mathbf{i} - \mathbf{j} + \frac{\partial y}{\partial z}\mathbf{k}\right) dA$

Left, R = projection on xz-plane

21. 2π 23. 18π 25. $\frac{4}{9}$

▶ **Exercise Set 18.7 (Page 974)**

1. 8 3. 216π 5. $3\pi a^2$ 7. 180π 9. $\frac{1}{24}$

11. $\pi/2$ 13. $\frac{4608}{35}$ 17. no sources or sinks

19. sources at all points except the origin; no sinks

▶ **Exercise Set 18.8 (Page 980)**

1. 2π 3. 16π 5. 0 7. πa^2 9. $\frac{3}{2}$ 11. 0

▶ **TECHNOLOGY EXERCISES, CHAPTER 18 (Page 981)**

1. $-4.553124, 6.822229$ 3. 12.4 m/sec

---

5. (a) 1 (b) $x^3 y^2 z^4$ 7. 57.895751 9. $-5\pi/4$

▶ **Exercise Set 19.1 (Page 988)**

3. $y = c_1 e^x + c_2 e^{-4x}$ 5. $y = c_1 e^x + c_2 x e^x$

7. $y = c_1 \cos\sqrt{5}x + c_2 \sin\sqrt{5}x$ 9. $y = c_1 + c_2 e^x$

11. $y = c_1 e^{-2t} + c_2 t e^{-2t}$ 13. $y = e^{2x}(c_1 \cos 3x + c_2 \sin 3x)$

15. $y = c_1 e^{-x/4} + c_2 e^{x/2}$ 17. $y = 2e^x - e^{-3x}$

19. $y = (2 - 5x)e^{3x}$ 21. $y = -e^{-2x}(3\cos x + 6\sin x)$

23. (a) $y'' - 3y' - 10y = 0$ (b) $y'' - 8y' + 16y = 0$
(c) $y'' + 2y' + 17y = 0$

25. (a) $k < 0$ or $k > 4$ (b) 0, 4 (c) $0 < k < 4$

27. (a) $y = (1/x)[c_1 \cos(\ln x) + c_2 \sin(\ln x)]$
(b) $y = c_1 x^{1+\sqrt{3}} + c_2 x^{1-\sqrt{3}}$

▶ **Exercise Set 19.2 (Page 994)**

1. $y = c_1 e^{-x} + c_2 e^{-5x} + \frac{1}{16}e^{3x}$

3. $y = c_1 e^{4x} + c_2 e^{5x} - 3xe^{5x}$

5. $y = (c_1 + c_2 x)e^{-x} + \frac{1}{2}x^2 e^{-x}$

7. $y = c_1 e^{3x} + c_2 e^{-4x} - \frac{13}{216} - \frac{1}{18}x - \frac{1}{3}x^2$

9. $y = c_1 + c_2 e^{6x} + \frac{5}{36}x - \frac{1}{12}x^2$

11. $y = c_1 + c_2 x - \frac{1}{2}x^2 + \frac{1}{20}x^5$

13. $y = c_1 e^{-x} + c_2 e^{2x} - 3\cos x - \sin x$

15. $y = c_1 e^{-2x} + c_2 e^{2x} - \frac{3}{8}\cos 2x - \frac{1}{4}\sin 2x$

17. $y = c_1 \cos x + c_2 \sin x - \frac{1}{2}x\cos x$

19. $y = c_1 e^x + c_2 e^{2x} + \frac{3}{4} + \frac{1}{2}x$

21. $y = e^{-2x}(c_1 \cos\sqrt{5}x + c_2 \sin\sqrt{5}x) - \frac{94}{729} + \frac{19}{81}x + \frac{1}{9}x^2$

23. $y = c_1 \cos 2x + c_2 \sin 2x - \frac{1}{8}x\cos 2x$

25. (b) $-\frac{3}{16} - \frac{1}{4}x + \frac{1}{5}xe^x$ 27. $y = c_1 e^{-x} + c_2 e^x - 1 + \frac{1}{2}xe^x$

29. $y = c_1 \cos 2x + c_2 \sin 2x + \frac{1}{4} + \frac{1}{4}x + \frac{1}{3}\sin x$

31. $y = (c_1 + c_2 x)e^x + \frac{1}{4}x^2 e^x + \frac{1}{8}e^{-x}$

33. $y = c_1 \cos x + c_2 \sin x + 6 - 2\cos 2x$

35. (a) $y = c_1 \cos\mu x + c_2 \sin\mu x + \dfrac{a}{\mu^2 - b^2}\sin bx$

(b) $y = c_1 \cos\mu x + c_2 \sin\mu x + \displaystyle\sum_{k=1}^n \dfrac{a_k}{\mu^2 - k^2\pi^2}\sin k\pi x$

37. $x_0 = 1$; $y = 2x^2 - 4e^{x-1} + c$, c arbitrary

▶ **Exercise Set 19.3 (Page 996)**

1. $y = c_1 \cos x + c_2 \sin x + x^2 - 2$

3. $y = c_1 e^x + c_2 e^{-2x} + \frac{2}{3}xe^x$

5. $y = c_1 \cos 2x + c_2 \sin 2x - \frac{1}{4}x\cos 2x$

7. $y = c_1 \cos x + c_2 \sin x - (\cos x)\ln|\sec x + \tan x|$

9. $y = c_1 e^x + c_2 x e^x + xe^x \ln|x|$

11. $y = c_1 \cos x + c_2 \sin x + \frac{3}{2} + \frac{1}{2}\cos 2x$

13. $y = c_1 \cos x + c_2 \sin x - x\cos x + (\sin x)\ln|\sin x|$

15. $y = c_1 \cos x + c_2 \sin x + x\cos x - (\sin x)\ln|\cos x|$

17. $y = c_1 e^{-x} + c_2 x e^{-x} - e^{-x}\ln|x|$

19. $y = c_1 e^{-2x} + c_2 x e^{-2x} + (x - 2)e^{-x}$

21. $y = c_1 \cos x + c_2 \sin x - 1 + (\sin x)\ln|\sec x + \tan x|$

23. $y = c_1 e^x + c_2 x e^x - e^x \ln|x|$

25. $y = c_1 e^x + c_2 e^{-x} + \frac{1}{5}e^x(2\sin x - \cos x)$

27. $y = c_1 e^{-x} + c_2 x e^{-x} + \frac{1}{2}x^2 e^{-x}\ln|x| - \frac{3}{4}x^2 e^{-x}$

▶ **Exercise Set 19.4 (Page 1004)**

1. (a) $y'' + 0.25y = 0$, $y(0) = 0.3$, $y'(0) = 0$
   (b) $y = 0.3 \cos 0.5t$
3. (a) $y'' + 196y = 0$, $y(0) = -0.12$, $y'(0) = 0$
   (b) $y = -0.12 \cos 14t$
5. (a) $y = 0.2 \cos 8t$ (b) 0.2 m   (c) $\pi/4$ sec   (d) $4/\pi$ Hz
7. (a) $y = -\frac{1}{4} \cos 8\sqrt{6}t$ (b) $\frac{1}{4}$ ft   (c) $\dfrac{\pi}{4\sqrt{6}}$ sec
   (d) $\dfrac{4\sqrt{6}}{\pi}$ Hz
9. (a) $y'' + 4y' + 8y = 0$, $y(0) = -3$, $y'(0) = 0$
   (b) $y = -3e^{-2t}(\cos 2t + \sin 2t)$
   (d) $y = -3\sqrt{2}e^{-2t}\cos(2t - \pi/4)$
   (e) π sec   (f) $1/\pi$ Hz
11. (a) $y = \frac{5}{2}\sqrt{6}e^{-t/5}\cos\left(\dfrac{\sqrt{2}}{5}t - \tan^{-1}\dfrac{1}{\sqrt{2}}\right)$
    (b) $\dfrac{10\pi}{\sqrt{2}}$ sec   (c) $\dfrac{\sqrt{2}}{10\pi}$ Hz
13. (a) $y = -\frac{1}{4}\sin 8t$ (b) $\frac{1}{4}$ ft   (c) $\pi/4$ sec   (d) $4/\pi$ Hz
15. $y = \frac{1}{3}e^{-t}(\cos 7t - 2\sin 7t)$
17. (a) $\frac{1}{32}\pi^2$ lb/ft   (b) $\frac{9}{4}$ lb
19. $T = 2\pi \sqrt{\dfrac{\delta h}{\rho g}}$

▶ **TECHNOLOGY EXERCISES, CHAPTER 19 (Page 1006)**

1. (a) $y = e^{-1.2t} + 3.2te^{-1.2t}$ (b) 1.427364 cm
3. (a) $y = e^{-t/2}\cos(\sqrt{19}t/2) - \frac{6}{19}\sqrt{19}e^{-t/2}\sin(\sqrt{19}t/2)$
   (b) 1.054466 cm   (c) -3.210357 cm/sec
   (d) 3.210357 cm/sec^2
5. (a) $3/\pi$, $5/(2\pi)$ Hz   (b) $y = \frac{1}{11}\sin 5t - \frac{5}{66}\sin 6t$
   (c) $y = \frac{4}{23}\sin 5.5t - \frac{11}{69}\sin 6t$, $y = \frac{25}{59}\sin 5.8t - \frac{145}{354}\sin 6t$
7. (a) $y = (4 + 2v_0)e^{-3t/2} - (3 + 2v_0)e^{-2t}$

▶ **Exercise Set B.I (Page A12)**

1. (a) $\frac{5}{12}\pi$ (b) $\frac{13}{6}\pi$ (c) $\frac{1}{9}\pi$ (d) $\frac{23}{30}\pi$
3. (a) $12°$ (b) $(270/\pi)°$ (c) $288°$ (d) $540°$
5.

	$\sin\theta$	$\cos\theta$	$\tan\theta$	$\csc\theta$	$\sec\theta$	$\cot\theta$
(a)	$\sqrt{21}/5$	$2/5$	$\sqrt{21}/2$	$5/\sqrt{21}$	$5/2$	$2/\sqrt{21}$
(b)	$3/4$	$\sqrt{7}/4$	$3/\sqrt{7}$	$4/3$	$4/\sqrt{7}$	$\sqrt{7}/3$
(c)	$3/\sqrt{10}$	$1/\sqrt{10}$	3	$\sqrt{10}/3$	$\sqrt{10}$	$1/3$

7. $\sin\theta = 3/\sqrt{10}$, $\cos\theta = 1/\sqrt{10}$
9. $\tan\theta = \sqrt{21}/2$, $\csc\theta = 5/\sqrt{21}$ 11. 1.8
13.

	θ	$\sin\theta$	$\cos\theta$	$\tan\theta$	$\csc\theta$	$\sec\theta$	$\cot\theta$
(a)	$225°$	$-1/\sqrt{2}$	$-1/\sqrt{2}$	1	$-\sqrt{2}$	$-\sqrt{2}$	1
(b)	$-210°$	$\frac{1}{2}$	$-\sqrt{3}/2$	$-1/\sqrt{3}$	2	$-2/\sqrt{3}$	$-\sqrt{3}$
(c)	$5\pi/3$	$-\sqrt{3}/2$	$\frac{1}{2}$	$-\sqrt{3}$	$-2/\sqrt{3}$	2	$-1/\sqrt{3}$
(d)	$-3\pi/2$	1	0	$-$	1	$-$	0

15.

	$\sin\theta$	$\cos\theta$	$\tan\theta$	$\csc\theta$	$\sec\theta$	$\cot\theta$
(a)	$4/5$	$3/5$	$4/3$	$5/4$	$5/3$	$3/4$
(b)	$-4/5$	$3/5$	$-4/3$	$-5/4$	$5/3$	$-3/4$
(c)	$1/2$	$-\sqrt{3}/2$	$-1/\sqrt{3}$	2	$-2/\sqrt{3}$	$-\sqrt{3}$
(d)	$-1/2$	$\sqrt{3}/2$	$-1/\sqrt{3}$	-2	$2/\sqrt{3}$	$-\sqrt{3}$
(e)	$1/\sqrt{2}$	$1/\sqrt{2}$	1	$\sqrt{2}$	$\sqrt{2}$	1
(f)	$1/\sqrt{2}$	$-1/\sqrt{2}$	-1	$\sqrt{2}$	$-\sqrt{2}$	-1

17. (a) 1.2679 (b) 3.5753
19.

	$\sin\theta$	$\cos\theta$	$\tan\theta$	$\csc\theta$	$\sec\theta$	$\cot\theta$
(a)	$\dfrac{a}{3}$	$\dfrac{\sqrt{9-a^2}}{3}$	$\dfrac{a}{\sqrt{9-a^2}}$	$\dfrac{3}{a}$	$\dfrac{3}{\sqrt{9-a^2}}$	$\dfrac{\sqrt{9-a^2}}{a}$
(b)	$\dfrac{a}{\sqrt{a^2+25}}$	$\dfrac{5}{\sqrt{a^2+25}}$	$\dfrac{a}{5}$	$\dfrac{\sqrt{a^2+25}}{a}$	$\dfrac{\sqrt{a^2+25}}{5}$	$\dfrac{5}{a}$
(c)	$\dfrac{\sqrt{a^2-1}}{a}$	$\dfrac{1}{a}$	$\sqrt{a^2-1}$	$\dfrac{a}{\sqrt{a^2-1}}$	a	$\dfrac{1}{\sqrt{a^2-1}}$

21. (a) $3\pi/4 \pm n\pi$, $n = 0, 1, 2, \ldots$
    (b) $\pi/3 \pm 2n\pi$ and $5\pi/3 \pm 2n\pi$, $n = 0, 1, 2, \ldots$
23. (a) $\pi/6 \pm n\pi$, $n = 0, 1, 2, \ldots$
    (b) $4\pi/3 \pm 2n\pi$ and $5\pi/3 \pm 2n\pi$, $n = 0, 1, 2, \ldots$
25. (a) $3\pi/4 \pm n\pi$, $n = 0, 1, 2, \ldots$
    (b) $\pi/6 \pm n\pi$, $n = 0, 1, 2, \ldots$
27. (a) $\pi/3 \pm 2n\pi$ and $2\pi/3 \pm 2n\pi$, $n = 0, 1, 2, \ldots$
    (b) $\pi/6 \pm 2n\pi$ and $11\pi/6 \pm 2n\pi$, $n = 0, 1, 2, \ldots$
29. $\sin\theta = 2/5$, $\cos\theta = -\sqrt{21}/5$, $\tan\theta = -2/\sqrt{21}$, $\csc\theta = 5/2$, $\sec\theta = -5/\sqrt{21}$, $\cot\theta = -\sqrt{21}/2$
31. (a) $\theta = \pm n\pi$, $n = 0, 1, 2, \ldots$
    (b) $\theta = \dfrac{\pi}{2} \pm n\pi$, $n = 0, 1, 2, \ldots$
    (c) $\theta = \pm n\pi$, $n = 0, 1, 2, \ldots$
    (d) $\theta = \pm n\pi$, $n = 0, 1, 2, \ldots$
    (e) $\theta = \dfrac{\pi}{2} \pm n\pi$, $n = 0, 1, 2, \ldots$
    (f) $\theta = \pm n\pi$, $n = 0, 1, 2, \ldots$
33. (a) $2\pi/3$ cm   (b) $10\pi/3$ cm   35. $\frac{2}{5}$
37. (a) $\dfrac{2\pi - \theta}{2\pi}R$ (b) $\dfrac{\sqrt{4\pi\theta - \theta^2}}{2\pi}R$
39. $\frac{21}{4}\sqrt{3}$ 41. 9.2 ft   43. $h = (\tan\beta - \tan\alpha)d$
45. (a) $4\sqrt{5}/9$ (b) $-\frac{1}{9}$
47. $\sin 3\theta = 3\sin\theta\cos^2\theta - \sin^3\theta$, $\cos 3\theta = \cos^3\theta - 3\sin^2\theta\cos\theta$
61. (a) $\cos\theta$ (b) $-\sin\theta$ (c) $-\cos\theta$ (d) $\sin\theta$

▶ **Exercise Set B.II (Page A17)**

1. (a) $2\pi/5$ (b) 6π (c) 8π (d) $\pi/7$ (e) 2 (f) 10
   (g) 1 (h) $2\pi k$

3. (a) 5   (b) $\frac{1}{3}$ (c) $\frac{1}{2}$ (d) 1

5. 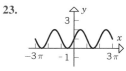   **7.**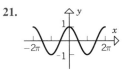

9.    **11.**

13. (a) odd   (b) even   (c) odd   (d) even   (e) even
    (f) even   (g) odd   (h) even

15. 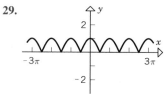   **17.**

19.   **21.**

23.   **25.**

27. 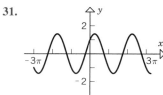   **29.**

31.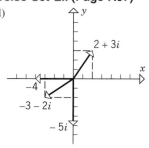

▶ **Exercise Set D (Page A32)**

1. $x = 1, y = 2$ **3.** $x_1 = -\frac{7}{32}, x_2 = \frac{5}{16}$

5. $x = 2, y = -1, z = 3$ **7.** $x_1 = \frac{26}{21}, x_2 = \frac{25}{21}, x_3 = \frac{5}{7}$

9. $x' = x\cos\theta + y\sin\theta, \; y' = -x\sin\theta + y\cos\theta$

▶ **Exercise Set E.I (Page A37)**

1. (a–d)

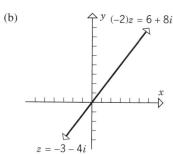

3. (a) $x = -2, y = -3$ (b) $x = 2, y = 1$

5. (a) $2 + 3i$ (b) $-1 - 2i$ (c) $-2 + 9i$

7. (a)

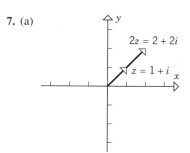

(b)

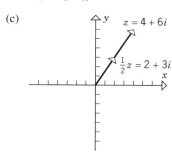

(c)

9. (a) $z_1 z_2 = 3 + 3i, \; z_1^2 = -9, \; z_2^2 = -2i$
   (b) $z_1 z_2 = 26, \; z_1^2 = -20 + 48i, \; z_2^2 = -5 - 12i$
   (c) $z_1 z_2 = \frac{11}{3} - i, \; z_1^2 = -\frac{4}{3} + \frac{16}{9}i, \; z_2^2 = -6 - \frac{5}{2}i$

11. $76 - 88i$ **13.** $-26 + 18i$ **15.** $-\frac{63}{16} + i$ **17.** 0

19. (a) $2 - 7i$ (b) $-3 + 5i$ (c) $-5i$ (d) i (e) -9
   (f) 0

23. (a) $-i$ (b) $\frac{1}{26} + \frac{5}{26}i$ (c) $7i$

25. $\frac{1}{2} + \frac{1}{2}i$ **27.** $-\frac{7}{625} - \frac{24}{625}i$

29. $\frac{1}{4}(1 - \sqrt{3}) + \frac{1}{4}(1 + \sqrt{3})i$ **31.** $-\frac{1}{10} + \frac{1}{10}i$

33. (a) $-1 - 2i$ (b) $-\frac{3}{25} - \frac{4}{25}i$

35. (a)

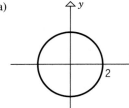

(b)

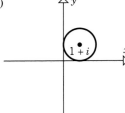

(c)

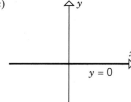

(d)

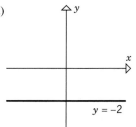

39. (b) i **43.** (a) $\dfrac{x_1x_2 + y_1y_2}{x_2^2 + y_2^2}$ (b) $\dfrac{x_2y_1 - x_1y_2}{x_2^2 + y_2^2}$

47. (c) yes, if $z \neq 0$

▶ **Exercise Set E.II (Page A44)**

1. (a) 0 (b) $\pi/2$ (c) $-\pi/2$ (d) $\pi/4$ (e) $2\pi/3$
(f) $-\pi/4$

3. (a) $2[\cos(\pi/2) + i\sin(\pi/2)]$ (b) $4[\cos\pi + i\sin\pi]$
(c) $5\sqrt{2}[\cos(\pi/4) + i\sin(\pi/4)]$
(d) $12[\cos(2\pi/3) + i\sin(2\pi/3)]$
(e) $3\sqrt{2}[\cos(-3\pi/4) + i\sin(-3\pi/4)]$
(f) $4[\cos(-\pi/6) + i\sin(-\pi/6)]$

5. 1

7. (a)

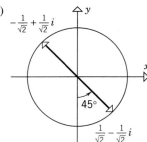

(b)

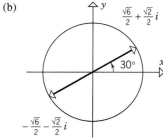

(c)

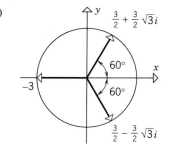

(d)

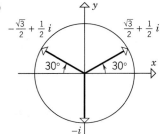

(e)

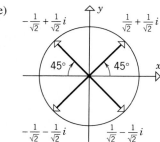

(f)

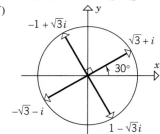

9.

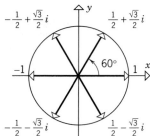

11. (a) $\pm 2, \pm 2i$ (b) $\pm(2 + 2i), \pm(2 - 2i)$

13. rotates z clockwise by $90°$

15. (a) $\text{Re}(z) = -3$, $\text{Im}(z) = 0$ (b) $\text{Re}(z) = -3$, $\text{Im}(z) = 0$
(c) $\text{Re}(z) = 0$, $\text{Im}(z) = -\sqrt{2}$ (d) $\text{Re}(z) = -3$, $\text{Im}(z) = 0$

Index

Photo Credits

Introduction
Culver Pictures, Inc.

Chapter 1
Portrait: H. Josse/Art Resource
Signature: Courtesy Smithsonian Institution

Chapter 2
Portrait: Courtesy of The New York Public Library
Signature: Courtesy Smithsonian Institution

Chapter 3
Portrait: Courtesy of The New York Public Library
Signature: Bettmann Archive

Chapter 4
Portrait: Courtesy of The David Eugene Smith Collection, Columbia University

Chapter 5
Portrait: Courtesy of The David Eugene Smith Collection, Columbia University
Signature: Courtesy Smithsonian Institution

Chapter 6
Portrait: Courtesy of The David Eugene Smith Collection, Columbia University
Signature: Courtesy of The New York Public Library

Chapter 7
Portrait: Culver Pictures, Inc.

Chapter 8
Portrait: Courtesy of The New York Public Library Picture Collection
Signature: Courtesy The British Library

Chapter 9
Portrait: Bettmann Archive

Chapter 10
Portrait: Courtesy of The David Eugene Smith Collection, Columbia University
Signature: Courtesy Smithsonian Institution

Chapter 11
Portrait: Courtesy of The David Eugene Smith Collection, Columbia University
Signature: Courtesy The British Library

Chapter 12
Portrait: Courtesy of The David Eugene Smith Collection, Columbia University
Signature: Courtesy Smithsonian Institution

Chapter 13
Portrait: Courtesy of The David Eugene Smith Collection, Columbia University
Signature: Courtesy Smithsonian Institution

Chapter 14
Portrait: Courtesy of The David Eugene Smith Collection, Columbia University
Signature: Courtesy Universitätsbibliothek Basel

Chapter 15
Portrait: Courtesy AIP Neils Bohr Library
Signature: Courtesy Smithsonian Institution

Chapter 16
Portrait and Signature: Courtesy Smithsonian Institution

Chapter 17
Portrait: Courtesy of The David Eugene Smith Collection, Columbia University
Signature: Courtesy Smithsonian Institution

Chapter 18
Portrait and Signature: Bettmann Archive

Chapter 19
Portrait: Courtesy of The David Eugene Smith Collection, Columbia University
Signature: Courtesy The British Library

☐ **CHAPTER OPENING PORTRAITS**

The chapter opening portraits were created by HRS Electronic Text Management from black and white engravings. They were scanned into and manipulated by computer to create the color illustrations.

Signatures for Chapter Openers 4 and 7 are simulated.

INTEGRALS CONTAINING $2au - u^2$

50. $\int \sqrt{2au - u^2}\, du = \dfrac{u - a}{2} \sqrt{2au - u^2} + \dfrac{a^2}{2} \sin^{-1}\left(\dfrac{u - a}{a}\right) + C$

51. $\int u\sqrt{2au - u^2}\, du = \dfrac{2u^2 - au - 3a^2}{6} \sqrt{2au - u^2} + \dfrac{a^3}{2} \sin^{-1}\left(\dfrac{u - a}{a}\right) + C$

52. $\int \dfrac{\sqrt{2au - u^2}\, du}{u} = \sqrt{2au - u^2} + a \sin^{-1}\left(\dfrac{u - a}{a}\right) + C$

53. $\int \dfrac{\sqrt{2au - u^2}\, du}{u^2} = -\dfrac{2\sqrt{2au - u^2}}{u} - \sin^{-1}\left(\dfrac{u - a}{a}\right) + C$

54. $\int \dfrac{du}{\sqrt{2au - u^2}} = \sin^{-1}\left(\dfrac{u - a}{a}\right) + C$

55. $\int \dfrac{u\, du}{\sqrt{2au - u^2}} = -\sqrt{2au - u^2} + a \sin^{-1}\left(\dfrac{u - a}{a}\right) + C$

56. $\int \dfrac{u^2\, du}{\sqrt{2au - u^2}} = -\dfrac{(u + 3a)}{2} \sqrt{2au - u^2} + \dfrac{3a^2}{2} \sin^{-1}\left(\dfrac{u - a}{a}\right) + C$

57. $\int \dfrac{du}{u\sqrt{2au - u^2}} = -\dfrac{\sqrt{2au - u^2}}{au} + C$

58. $\int \dfrac{du}{(2au - u^2)^{3/2}} = \dfrac{u - a}{a^2\sqrt{2au - u^2}} + C$

59. $\int \dfrac{u\, du}{(2au - u^2)^{3/2}} = \dfrac{u}{a\sqrt{2au - u^2}} + C$

INTEGRALS CONTAINING TRIGONOMETRIC FUNCTIONS

60. $\int \sin u\, du = -\cos u + C$

61. $\int \cos u\, du = \sin u + C$

62. $\int \tan u\, du = \ln |\sec u| + C$

63. $\int \cot u\, du = \ln |\sin u| + C$

64. $\int \sec u\, du = \ln |\sec u + \tan u| + C$
$\qquad = \ln |\tan (\tfrac{1}{4}\pi + \tfrac{1}{2}u)| + C$

65. $\int \csc u\, du = \ln |\csc u - \cot u| + C$
$\qquad = \ln |\tan \tfrac{1}{2}u| + C$

66. $\int \sec^2 u\, du = \tan u + C$

67. $\int \csc^2 u\, du = -\cot u + C$

68. $\int \sec u \tan u\, du = \sec u + C$

69. $\int \csc u \cot u\, du = -\csc u + C$

70. $\int \sin^2 u\, du = \tfrac{1}{2}u - \tfrac{1}{4}\sin 2u + C$

71. $\int \cos^2 u\, du = \tfrac{1}{2}u + \tfrac{1}{4}\sin 2u + C$

72. $\int \tan^2 u\, du = \tan u - u + C$

73. $\int \cot^2 u\, du = -\cot u - u + C$

74. $\int \sin^n u\, du = -\dfrac{1}{n} \sin^{n-1} u \cos u + \dfrac{n - 1}{n} \int \sin^{n-2} u\, du$

75. $\int \cos^n u\, du = \dfrac{1}{n} \cos^{n-1} u \sin u + \dfrac{n - 1}{n} \int \cos^{n-2} u\, du$

76. $\int \tan^n u\, du = \dfrac{1}{n - 1} \tan^{n-1} u - \int \tan^{n-2} u\, du$

77. $\int \cot^n u\, du = -\dfrac{1}{n - 1} \cot^{n-1} u - \int \cot^{n-2} u\, du$

78. $\int \sec^n u\, du = \dfrac{1}{n - 1} \sec^{n-2} u \tan u + \dfrac{n - 2}{n - 1} \int \sec^{n-2} u\, du$

79. $\int \csc^n u\, du = -\dfrac{1}{n - 1} \csc^{n-2} u \cot u + \dfrac{n - 2}{n - 1} \int \csc^{n-2} u\, du$

80. $\int \sin mu \sin nu\, du = -\dfrac{\sin (m + n)u}{2(m + n)} + \dfrac{\sin (m - n)u}{2(m - n)} + C$

81. $\int \cos mu \cos nu\, du = \dfrac{\sin (m + n)u}{2(m + n)} + \dfrac{\sin (m - n)u}{2(m - n)} + C$

82. $\int \sin mu \cos nu\, du = -\dfrac{\cos (m + n)u}{2(m + n)} - \dfrac{\cos (m - n)u}{2(m - n)} + C$

83. $\int u \sin u\, du = \sin u - u \cos u + C$

84. $\int u \cos u\, du = \cos u + u \sin u + C$

85. $\int u^2 \sin u\, du = 2u \sin u + (2 - u^2) \cos u + C$

86. $\int u^2 \cos u\, du = 2u \cos u + (u^2 - 2) \sin u + C$

87. $\int u^n \sin u\, du = -u^n \cos u + n \int u^{n-1} \cos u\, du$

88. $\int u^n \cos u\, du = u^n \sin u - n \int u^{n-1} \sin u\, du$

89. $\int \sin^m u \cos^n u\, du = -\dfrac{\sin^{m-1} u \cos^{n+1} u}{m + n} + \dfrac{m - 1}{m + n} \int \sin^{m-2} u \cos^n u\, du$
$\qquad = \dfrac{\sin^{m+1} u \cos^{n-1} u}{m + n} + \dfrac{n - 1}{m + n} \int \sin^m u \cos^{n-2} u\, du$